Organic Chemistry

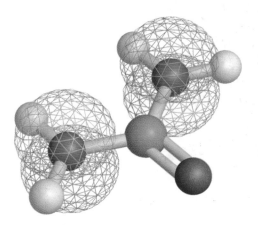

Organic Chemistry

Fifth Edition

John McMurry
Cornell University

Brooks/Cole
Thomson Learning™

Pacific Grove • Albany • Belmont • Boston • Cincinnati • Johannesburg • London • Madrid
Melbourne • Mexico City • New York • Scottsdale • Singapore • Tokyo • Toronto

Sponsoring Editor: *Jennifer Huber*
Developmental Editor: *David Chelton*
Marketing Team: *Heather Woods, Laura Hubrich, Dena Donnelly,* and *Angela Dean*
Editorial Assistant: *Dena Dowsett-Jones*
Senior Assistant Editor: *Melissa Henderson*
Production Service: *Phyllis Niklas*
Production Editor: *Jamie Sue Brooks*
Media Editor: *Claire Masson*
Text Interior Design: *Nancy Benedict*
Cover Design: *Vernon T. Boes*
Cover Photo: © *J. Jamsen, Natural Selection*

Interior Illustration: *Academy Artworks* and *Dovetail Publishing Services*
Chapter Opening Illustrations: *Kenneth Eward, BioGrafx*
Electrostatic Potential Maps and Spartan Models: *Wavefunction, Inc.*
Photo Researcher: *Stuart Kenter Associates*
Manufacturing Buyer: *Barbara Stephan*
Typesetting: *York Graphic Services*
Cover Printing: *R. R. Donnelley & Sons, Willard, Ohio*
Printing and Binding: *R. R. Donnelley & Sons, Willard, Ohio*

For more information about this or any other Brooks/Cole products, contact:
BROOKS/COLE
511 Forest Lodge Road
Pacific Grove, CA 93950 USA
www.brookscole.com
1-800-423-0563 (Thomson Learning Academic Resource Center)

Printed in the United States of America

10   9   8   7   6   5   4   3   2

Library of Congress Cataloging-in-Publication Data
McMurry, John.
    Organic chemistry/John McMurry—fifth ed.
        p.   cm.
    Includes index.
    ISBN 0-534-37366-6 (hardcover)
    1. Chemistry, Organic.   I. Title.
QD251.2.M43   1999
547—dc21   99-16981

Brief Contents

Contents

1 Structure and Bonding   1

2 Polar Bonds and Their Consequences   35

3 Organic Compounds: Alkanes and Cycloalkanes   74

4 Stereochemistry of Alkanes and Cycloalkanes   111

5 An Overview of Organic Reactions   151

6 Alkenes: Structure and Reactivity   188

10 Alkyl Halides   355

11 Reactions of Alkyl Halides: Nucleophilic Substitutions and Eliminations   385

12 Structure Determination: Mass Spectrometry and Infrared Spectroscopy   440

13 Structure Determination: Nuclear Magnetic Resonance Spectroscopy   475

14 Conjugated Dienes and Ultraviolet Spectroscopy   522

15 Benzene and Aromaticity   559

16 Chemistry of Benzene: Electrophilic Aromatic Substitution   592

A Brief Review of Organic Reactions   645

17 Alcohols and Phenols   654

18 Ethers and Epoxides; Thiols and Sulfides   708

A Preview of Carbonyl Compounds   743

19 Aldehydes and Ketones: Nucleophilic Addition Reactions   753

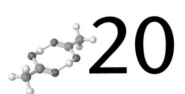

20 Carboxylic Acids   814

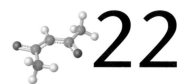

21 Carboxylic Acid Derivatives and Nucleophilic Acyl Substitution Reactions   843

22 Carbonyl Alpha-Substitution Reactions   901

23 Carbonyl Condensation Reactions 937

24 Amines 976

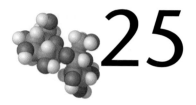

25 Biomolecules: Carbohydrates   1030

26 Biomolecules: Amino Acids, Peptides, and Proteins   1073

27 Biomolecules: Lipids   1118

28 Biomolecules: Heterocycles and Nucleic Acids   1150

29 The Organic Chemistry of Metabolic Pathways   1193

30 Orbitals and Organic Chemistry: Pericyclic Reactions   1235

31 Synthetic Polymers 1264

Appendixes

Index I-1

Preface

I wrote this book for one simple reason: I love writing. I get great pleasure and satisfaction from taking a complicated subject, turning it around until I see it clearly from a new angle, and then explaining it in simple words. I write to explain chemistry to students today the way I wish it had been explained to me years ago.

The enthusiastic response to the four previous editions has been very gratifying and suggests that this book has served students well. I hope you will find that this fifth edition of *Organic Chemistry* builds on the strengths of the first four and serves students even better. I have made every effort to make this edition as effective, clear, and readable as possible, to show the beauty and logic of organic chemistry, and to make it enjoyable to learn.

Organization and Teaching Strategies

This fifth edition, like its predecessors, uses a dual organization that blends the traditional functional-group approach with a mechanistic approach. The primary organization is by functional group, beginning with the simple (alkenes) and progressing to the more complex. Students new to the subject and not yet versed in the subtleties of mechanisms do better with this organization because it is straightforward. In other words, the *what* of chemistry is easier for most students to grasp than the *why*. Within this primary organization, however, I place heavy emphasis on explaining the fundamental mechanistic similarities of reactions. This emphasis is particularly evident in the chapters on carbonyl-group chemistry (Chapters 19–23) where mechanistically related reactions like the aldol and Claisen condensations are covered together. By the time students reach this material, they have seen all the common mechanisms, and the value of mechanisms as an organizing principle has become more evident.

The Lead-Off Reaction: Addition of HBr to Alkenes Students naturally attach great importance to a text's lead-off reaction because it is the first reaction they see and is discussed in such detail. I use the addition of HBr to an alkene as the lead-off to illustrate general principles of organic chemistry for several reasons: It is relatively straightforward; it involves a common but important functional group; no prior knowledge of stereochemistry or kinetics is needed to understand it; and, most importantly, it is a *polar* reaction. As such, I believe that electrophilic addition reactions represent a much more useful and realistic introduction to functional-group chemistry than a lead-off such as radical alkane chlorination.

Reaction Mechanisms In the first edition, I introduced an innovative format for explaining reaction mechanisms in which the reaction steps are printed vertically while the changes taking place in each step are explained next to the reaction arrow. This format allows a reader to see easily what is occurring at each step in a reaction without having to flip back and forth between structures and text. This edition has numerous additional vertical mechanisms, all set off by an orange background.

Organic Synthesis Organic synthesis is treated in this text as a teaching device that helps students organize and deal with a large body of factual information—the same skill so critical in medicine. Two sections, the first in Chapter 8 (Alkynes) and the second in Chapter 16 (Benzene), explain the thought processes involved in working synthesis problems and emphasize the value of starting from what is known and logically working backwards. In addition, new CHEMISTRY@WORK boxes on "The Art of Organic Synthesis," "Combinatorial Chemistry," and "Enantioselective Synthesis" further underscore the importance and timeliness of synthesis.

Modular Presentation Topics are arranged in a roughly modular way. Thus, the chapters on simple hydrocarbons are grouped together (Chapters 3–8), the chapters on spectroscopy are grouped together (Chapters 12–14), and the chapters on carbonyl-group chemistry are grouped together (Chapters 19–23). I believe that this organization brings to these subjects a cohesiveness not found in other texts and allows the instructor the flexibility to teach in an order different from that presented in the book.

Basic Learning Aids Clarity of explanation and smoothness of information flow are crucial requirements for any textbook. In writing and revising this text, I consistently aim for summary sentences at the beginning of paragraphs, lucid explanations, and smooth transitions between paragraphs and between topics. New concepts are introduced only when they are needed, not before, and are immediately illustrated with concrete examples. Frequent cross-references to earlier (but not later) material are given, and numerous summaries are provided to draw information together, both within and at the ends of chapters. In addition, the back of this book contains a wealth of material helpful for learning organic chemistry, including a large glossary, an explanation of how to name polyfunctional organic compounds, and answers to most in-text problems. For still further aid, an accompanying *Study Guide and Solutions Manual* gives a summary of name reactions, a summary of methods for preparing functional groups, a summary of functional-group reactions, and a summary of the uses of important reagents.

Changes and Additions for the Fifth Edition The primary reason for preparing a new edition is to keep the book up-to-date, both in its scientific coverage and in its pedagogy. My overall aim has been to retain and refine the features that made earlier editions so successful, while adding new ones.

- **The writing** has again been revised at the sentence level, streamlining the presentation, improving explanations, and updating a thousand small details. Some reactions have been deleted (the Clemmensen reduction, for instance), and a few new ones have been added (the synthesis of phenols from diazonium salts and the glycal-assembly method of polysaccharide synthesis, for instance).

Particularly noticeable are the changes in Chapter 2. A much expanded coverage of resonance structures has been added (Section 2.6), along with a useful new technique for generating resonance structures. A new section on acid–base strength (Section 2.8), another new section on organic acids and bases (Section 2.10), and a new section introducing the curved-arrow formalism (Section 2.11) have been added. The use of curved arrows for mechanisms is further reinforced several chapters later in a new Section 5.6.

Still other changes include the revised coverage of NMR in Chapter 13 to present ^{13}C before ^{1}H spectroscopy, and a return to the "steering-wheel" method of assigning R,S configuration to chirality centers in Chapter 9.

- **The order of topics**, while remaining basically the same, has been changed to move the coverage of phenols from Chapter 25 to Chapter 17, where it now appears along with the discussion of alcohols. In addition, the coverage of both alkylamines and arylamines has been integrated into a single new Chapter 24, and a new Chapter 31 on polymer chemistry has been added.

- **The problems** within each chapter and at the end of each chapter have been redone, and nearly 25% of them are new. Particularly noticeable are the new problem sections called "Visualizing Chemistry," in which substances are shown as molecular models rather than as typical line structures. These questions are a good deal more challenging than they initially appear, and they provide excellent practice for thinking about chemistry on the atomic level. In addition, all chapters now end with a selection of problems that give students a chance to do their own molecular modeling using the accompanying SpartanView and SpartanBuild software from Wavefunction, Inc.

- **Practice problems** are more numerous in this edition and have been rewritten to make them more useful. They now begin with a "Strategy" discussion that focuses on general approaches to problem solving and on the thought processes used for finding solutions.

- **The artwork** has been completely redone, and a great many new computer-generated models have been added. The use of stereo views to facilitate three-dimensional perception of ball-and-stick molecular models was introduced in the fourth edition, and their number has been nearly doubled in this fifth edition. As before, a stereo viewer is bound into the back of the book.

- **Molecular modeling** receives greatly increased emphasis in this new edition. In addition to the large number of new ball-and-stick models and stereo views, there are approximately 90 images produced using SpartanView molecular modeling software from Wavefunction, Inc. Particularly useful are the many electrostatic potential maps, which show the calculated regions of positive and negative charge within a molecule, thereby emphasizing the nucleophilic/electrophilic character of various functional groups. Many further images and animations are included on the accompanying CD-ROM, and problems at the end of each chapter, developed by Alan Shusterman and Warren Hehre for this edition, provide additional practice in using molecular modeling as a tool for learning organic chemistry.

- **CHEMISTRY@WORK boxes** at the end of each chapter present interesting applications of organic chemistry relevant to the main chapter subject. Including topics from science, industry, and day-to-day life, these applications enliven and reinforce the material presented within the chapter. Topics new to this edition include "The Art of Organic Synthesis," "Combinatorial Chemistry," "Enantioselective Synthesis," "DNA Fingerprinting," and "Biodegradable Polymers."

- **Biomolecules** have received particular attention in this edition to assure that coverage is up-to-date. Chapter 25, for instance, contains new material on the synthesis of polysaccharides (Section 25.10) and other new material on carbohydrate-based vaccines (Section 25.12). Chapter 26 contains a new section on the mechanism of enzyme action, using citrate synthase as the example (Section 26.16), and Chapter 27 contains updated material on cholesterol biosynthesis and on prostaglandins, including a discussion of the recently introduced COX-2 inhibitors. Finally, all the material on nucleic acids in Chapter 28 has been updated, with special treatment of the very latest DNA sequencing technology.

- **Polymer chemistry** is now drawn together in a new Chapter 31. Although much polymer chemistry is still interspersed throughout the text to ensure its full coverage, the new chapter brings a cohesiveness to the subject and makes it possible to relate structure with general physical properties.

Biological Connection ▶

- **Biologically important organic reaction mechanisms** are specially identified by the use of a margin icon. Students often wonder about what topics are "important," and this icon helps biologically inclined students answer that question.

- **The biographies** of the chemists for whom so many organic reactions are named have been expanded. Rather than simply provide dry biographical data, more humanizing, sometimes offbeat, details of the lives of famous chemists have now been added.

A Complete Ancillary Package

Organic Chemistry, Fifth Edition, is supported by a complete set of ancillaries. Each piece has been designed to enhance student understanding. The following resources are available, free of charge, to adopters of the text.

Printed Test Items   Over 1000 multiple-choice and matching questions, with detailed answers, in preprinted test forms corresponding to the main text organization. More than 30% of the questions are new and all structures have been redrawn for clarity and accuracy. ISBN 0-534-37196-5

Transparency Acetates Approximately 200 full-color transparencies, from text illustrations, enlarged for use in the classroom and lecture halls. ISBN 0-534-37193-0

Instructor's Manual: Molecular Modeling in Undergraduate Organic Chemistry Written by Alan Shusterman, Reed College, and Warren Hehre, Wavefunction, Inc., this manual is designed to guide instructors in the effective implementation and use of molecular modeling in undergraduate organic chemistry. ISBN 0-534-37387-9

Brooks/Cole's ChemLink With this cross-platform CD-ROM, creating lectures has never been easier. Using multitiered indexing, search capabilities, and a comprehensive resource bank that includes glossary, graphs, tables, illustrations, and animations, instructors can conduct a quick search to incorporate these materials into presentations and tests. And, any Chem-Link file can be posted to the Web for easy student reference. All tables and figures from the text are included. In addition, annotated animations of reaction mechanisms and three-dimensional Chime™ models of organic compounds are included from *Organic Chemistry Online 2.0.*

Thomson Learning™ Testing Tools An electronic version of the printed test items, with the online testing and classroom management capabilities of the Electronic Learning package. Windows ISBN 0-534-37194-9; Macintosh ISBN 0-534-37195-7

Thomson Learning™ Course Allows instructors to easily post course information, office hours, lesson information, assignments, sample tests, and links to rich Web content, including review and enrichment material from Brooks/Cole. Updates are quick and easy and customer support is available 24 hours a day, 7 days a week. For more information: www.itped.com

Brooks/Cole Chemistry Resource Center
http://www.brookscole.com/chemistry At Brooks/Cole's Web site for chemistry, instructors and students can access a homepage for *Organic Chemistry,* Fifth Edition. All information is arranged according to the *Organic Chemistry* table of contents. Students can access flash cards for all glossary terms, practice quizzes for every chapter, and hyperlinks that relate to each chapter's contents. Instructors can download additional questions by chapter that require the use of either ChemOffice or HyperChem. In addition, students can research the accomplishments of past and present contributors to the field of chemistry in timeline format.

Student Resources
A complete range of student ancillaries is also available, including print, CD-ROM, and online resources.

Study Guide and Solutions Manual   Written by Susan McMurry, this manual provides answers and explanations for all in-text and end-of-chapter exercises. It also includes summaries of name reactions, functional-group synthesis and reactions, lists of reagents and abbreviations, and articles on topics ranging from infrared absorption frequencies to Nobel Prize winners in chemistry. This edition now includes all new artwork, expanded in-text problems, summary quizzes approximately every three chapters, more detailed explanations in solutions, and chapter outlines. ISBN 0-534-37192-2

Organic Chemistry Online 2.0; and Wavefunction's SpartanBuild and SpartanView   This CD-ROM is included with this text and contains rich resources for problem solving, molecular visualization, and model building. *Organic Chemistry Online 2.0*, developed by Paul R. Young, contains a library of over 400 compounds commonly used in a lab, with Web links to a variety of databases such as NIST, NTP, MSDS, and IRIS; over 100 digitized spectra; 30 mechanisms and movies; extensive tutorials including electron pushing tutorials; and sample MCAT questions. In addition, it includes two professional software tools, SpartanView and SpartanBuild, from Wavefunction, for solving end-of-chapter molecular modeling problems in the main text. Spartan material for the text was developed by Alan Shusterman and Warren Hehre. ISBN 0-534-37364-X

Organic Chemistry Online 2.0 Workbook   Written by Paul R. Young, this student workbook focuses on problem solving, and provides additional help, exercises, and practice problems corresponding to *Organic Chemistry Online 2.0* as well as supplemental topic information for a wide variety of organic compounds and reactions. ISBN 0-534-37191-4

Practical Spectroscopy: The Rapid Interpretation of Spectral Data   Also written by Paul R. Young, this workbook is based on the spectroscopy modules in *Organic Chemistry Online 2.0* and contains proton and carbon NMR, infrared and mass spectra for 100 organic molecules, along with expanded tutorial sections to aid undergraduate students in analysis. ISBN 0-534-37230-9

ChemOffice, Ltd. 3.1   This is a suite of software products from CambridgeSoft Corporation that includes student versions of ChemDraw and Chem3D. With ChemOffice, Ltd., students have tools for drawing, modeling, and analyzing organic compounds in two and three dimensions. ISBN 0-534-36529-9

InfoTrac® College Edition   This fully searchable online library is available free with each copy of *Organic Chemistry,* Fifth Edition. Using InfoTrac, students can access full-length articles—not simply abstracts—from more than 700 scholarly and popular periodicals, updated daily, and dating back as much as 4 years. Student subscribers receive a personalized account ID that gives them 4 months of unlimited Internet access—at any hour of the day. (Due to license restrictions, InfoTrac® College Edition is only available free with the purchase of a new book to college students in North America.)

Chemistry Resource Center and Thomson Learning™ Web Tutor
http://www.brookscole.com/chemistry At Brooks/Cole's Web site for chemistry, instructors and students can access a homepage for *Organic Chemistry,* Fifth Edition. All information is arranged according to the *Organic Chemistry* table of contents. Students can access flash cards for all glossary terms, practice quizzes for every chapter, and hyperlinks that relate to each chapter's contents. In addition, students can research the accomplishments of past and present contributors to the field of chemistry in timeline format.

Organic Chemistry Toolbox This electronic study guide provides tutorials for the main concepts of the course—structure, nomenclature, reactions, bioorganic, and spectroscopy. The following visualization tools are included: a molecular modeler, a Lewis-dot structure drawing tool that can check formulas, reaction animation, a way to test knowledge of reactions, and spectral manipulation for infrared and ^{1}H and ^{13}C NMR. ISBN 0-534-35207-3

Beaker™ This sophisticated yet easy-to-use software allows students to explore organic chemistry principles, study and solve problems, and sketch and analyze molecules. Using Beaker™ students can draw a molecule or simply type in an IUPAC name and let the software do the drawing. Beaker™ 2.1 for Macintosh ISBN 0-534-15973-7, Beaker™ 2.2 for Windows ISBN 0-534-13410-6

Acknowledgments

It is a great pleasure to thank the many people whose help and suggestions were so valuable in preparing this fifth edition. Foremost is my wife, Susan, who read, criticized, and improved the manuscript, and who was my constant companion throughout all stages of this book's development.

Fifth Edition Reviewers

Wayne Ayers,
   East Carolina University

Kevin Belfield,
   University of Central Florida

James Canary,
   New York University

Bob Coleman,
   Ohio State University

Patricia DePra,
   Westfield State College

John Michael Ferguson,
   University of Central Oklahoma

Warren Giering,
   Boston University

Dennis Hall,
   University of Alberta

Peter Lillya,
   University of Massachusetts

James Long,
   University of Oregon

Todd Lowary,
   Ohio State University

Fred Matthews,
   Austin Peay State University

Michael Montague-Smith,
   Duke University

Ed Neeland,
   Okanagan University

Mike Oglioruso,
   Virginia Polytechnic Institute and State University

Bob Perkins,
   Kwantlen University—Richmond Campus

Robert Phillips,
   University of Georgia

Carmelo Rizzo,
   Vanderbilt University

Stuart Rosenfeld,
   Smith College

Eric Simanek,
   Texas A&M University

Gary Snyder,
   University of California—San Diego

Douglas Taber,
   University of Delaware

Dennis Taylor,
   University of Adelaide

Barbara Whitlock,
   University of Wisconsin

Previous Edition Reviewers

John Belletire,
   University of Cincinnati

Robert A. Benkeser,
   Purdue University

Donald E. Bergstrom,
   University of North Dakota

Weston J. Borden,
   University of Washington

Larry Bray,
   Miami-Dade Community College

Ronald Caple,
   University of Minnesota—Duluth

John Cawley,
   Villanova University

Clair Cheer,
   San Jose State University

George Clemans,
   Bowling Green State University

William D. Closson,
   State University of New York—Albany

Paul L. Cook,
   Albion College

Dennis Davis,
   New Mexico State University

Otis Dermer,
   Oklahoma State University

Linda Domelsmith,
   McGill University

Dale Drueckhammer,
   Stanford University

Douglas Dyckes,
   University of Colorado—Denver

Kenneth S. Feldman,
   Pennsylvania State University

Martin Feldman,
   Howard University

Kent Gates,
   University of Missouri—Columbia

David Harpp,
   McGill University

David Hart,
   Ohio State University

Dan Harvey,
   University of California—San Diego

John Henderson,
   Jackson Community College

Norbert Hepfinger,
   Rensselaer Polytechnic Institute

Werner Herz,
   Florida State University

Stephen Hixon,
   University of Massachusetts

John Hogg,
   Texas A&M University

Paul Hopkins,
   University of Washington

John Huffman,
   Clemson University

Paul R. Jones,
   North Texas State University

Jack Kampmeier,
   University of Rochester

Thomas Katz,
   Columbia University

Glen Kauffman,
   East Mennonite College

Andrew S. Kende,
   University of Rochester

Paul E. Klinedinst, Jr.,
   California State University—Northridge

Joseph Lambert,
   Northwestern University

John A. Landgrebe,
   University of Kansas

John T. Landrum,
   Florida International University

Thomas Livinghouse,
   University of Minnesota

James Long,
   University of Oregon

James G. Macmillan,
   University of Northern Iowa

Kenneth L. Marsi,
   California State University—Long Beach

Eugene A. Mash,
   University of Arizona

Guy Mattson,
   University of Central Florida

David McKinnon,
   University of Manitoba

Kirk McMichael,
   Washington State University

Monroe Moosnick,
   Transylvania University

Harry Morrison,
   Purdue University

Cary Morrow,
   University of New Mexico

Clarence Murphy,
   East Stroudsburg University

Roger Murray,
University of Delaware

Oliver Muscio,
Murray State University

Wesley A. Pearson,
St. Olaf College

Russell Petter,
University of Pittsburgh

Neil Potter,
Susquehanna University

Naser Pourahmady,
Southwest Missouri State University

Carmelo Rizzo,
Vanderbilt University

Frank P. Robinson,
University of Victoria

William E. Russey,
Juniata College

Neil E. Schore,
University of California—Davis

Gerald Selter,
California State University—San Jose

Jan Simek,
California Polytechnic State University

Ernest Simpson,
California State Polytechnic University—
Pomona

Peter W. Slade,
University College of Fraser Valley

Ronald Starkey,
University of Wisconsin—Green Bay

John R. Stille,
Michigan State University

J. William Suggs,
Brown University

Michelle Sulikowski,
Texas A&M University

Joyce Takahashi,
University of California—Davis

Marcus W. Thomsen,
Franklin & Marshall College

Bruce Toder,
State University of New York—Brockport

Walter Trahanovsky,
Iowa State University

Harry Ungar,
Cabrillo College

Joseph J. Villafranca,
Pennsylvania State University

Daniel Weeks,
Northwestern University

Barbara J. Whitlock,
University of Wisconsin—Madison

David Wiemer,
University of Iowa

Walter Zajac,
Villanova University

Vera Zalkow,
Kennesaw College

A Note for Students

We have the same goals. Yours is to learn organic chemistry; mine is to help you learn. I've done the best I can with my part, and now it's going to take some work from you. The following suggestions should prove helpful.

- **Don't Read the Text Immediately**   As you begin each new chapter, look it over first. Read the introductory paragraphs, find out what topics will be covered, and then read the summary at the end of the chapter. You'll be in a much better position to learn the material if you know where you're going.

- **Work the Problems**   There are no shortcuts; working problems is the only way to learn organic chemistry. The practice problems show you how to approach the material, the in-text problems at the ends of most sections provide immediate practice, and the end-of-chapter problems provide both additional drill and some real challenges. Pay particular attention to the "Visualizing Chemistry" problems, which can help you begin to "see" molecules rather than think of them as vague abstractions. Short answers to in-text problems are given at the back of the book; full answers and explanations for all problems are given in the accompanying *Study Guide and Solutions Manual*.

- **Use the Study Guide**   The *Study Guide and Solutions Manual* that accompanies this text gives complete solutions to all problems as well as a wealth of supplementary material. Included are a summary of how to prepare functional groups, a summary of the reactions that functional groups undergo, a summary of important reagents, a summary of name reactions, and much more. This material can be extremely useful, both as a source of information and as a self-test, particularly when you're studying for an exam. Find out now what's there so you'll know where to go when you need help.

- **Ask Questions**   Faculty members and teaching assistants are there to help you. Most will turn out to be genuinely nice people with a sincere interest in helping you learn.

- **Use Molecular Models**   Organic chemistry is a three-dimensional science. Although this book uses stereo views and many careful drawings to help you visualize molecules, there's no substitute for building a molecular model and turning it around in your own hands.

- **Use the Organic Chemistry Online CD and the Wavefunction Molecular Modeling Software Included with This Book**   Both provide alternative, nontextual ways of approaching chemistry, using reaction animations and other computer-based approaches to learning.

Good luck. I sincerely hope you enjoy learning organic chemistry and come to see the beauty and logic of its structure. I heard from many students who used the first four editions of this book and would be glad to receive more comments and suggestions from those who use this new edition.

John McMurry

How to Use SpartanView and Interpret Molecular Modeling Data

by Alan Shusterman and Warren Hehre

Learning about molecular structure and bonding is a fundamental part of learning organic chemistry. You need to learn how to construct mental pictures of what atoms and molecules look like, and you need to learn how to detect and describe important differences between molecules. This fifth edition of *Organic Chemistry* contains a wealth of accurate, computer-generated figures that will help you learn how to visualize molecular structures. The book also contains a CD with a large number of computer-generated molecular models and SpartanView, a computer program for displaying these models. The models go beyond the book's images in two important ways: first, the computer models are easily manipulated three-dimensional images, and second, many of the computer models provide animations that show how molecular structures change during chemical processes. We strongly encourage you to make full use of these computer models by keeping them handy on your computer as you read this book. (Some of these instructions for using SpartanView and for interpreting molecular modeling data will not make complete sense until you have studied organic chemistry for awhile, so be prepared to refer back to this section as you proceed through the book.)

Installing and Starting SpartanView

The SpartanView program and molecular model library are located on the CD that accompanies this book. You may use it on any Windows or Macintosh-compatible computer, but the program is CD-protected; you must keep the CD in your computer while you use the program. (Copy and sale of the CD are prohibited by the license agreement.)

The following tables contain instructions for using SpartanView. Each table gives instructions for a related group of tasks (install software, change model display, etc.). Computer instructions are listed in the left-hand column, and comments are listed in the right-hand column. You should perform these instructions on your computer as you read along.

Computer Instruction	Comments

Install SpartanView
   Step 1. Insert SpartanView CD.
   Step 2. Double-click on the CD's icon.	

Start SpartanView Step 3. Double-click on the SpartanView icon.	This causes the SpartanView window to open. (The window is blank initially.)
Quit SpartanView Step 4. Select Quit from the File menu.	Restart SpartanView to continue with the tutorial.

Open and Close Models, Select Active Model, Move Model

 The SpartanView models on the CD are keyed to specific figures, discussions, and Molecular Modeling problems located throughout the book, and are marked using the icon shown in the margin. Models for each chapter are grouped together in the same folder on the CD: Chapter1, Chapter2, etc. Models used in figures and Molecular Modeling problems are grouped together by figure and problem number, respectively. For example, model 1.52 contains all of the models used in Problem 1.52. Model names listed under the SpartanView icon identify individual models on the CD.

Computer Instruction	Comments

Open models
   Step 1. Select Open from the File menu.
   Step 2. Double-click on Chapter1, then double-click
         on 1.52.
	Problem 1.52 in Chapter1 contains six models: methanol, sodium methoxide, acetic acid, sodium acetate, hydrochloric acid, and sodium chloride. All the models open at once, but only one model is active.

Make methanol, CH₃OH, the active model
   Step 3. Move the cursor to any part of the methanol
         model and click on it.
	This makes methanol the active model. The name of the active model is displayed at the top of the SpartanView window. Only one model can be active at a time.

Move a model
   Step 4. Rotate, translate, and scale the active model
         using the mouse and keyboard operations listed
         in the table below.
	Rotation and translation affect only the active model, but scaling affects all models on the screen.

Close model
   Step 5. Select Close from the File menu.
   Step 6. Select Close All from the File menu.
	Close affects only the active model. Close All closes all open models.

Computer Instruction

	PC	**Mac**
Select	Click on model with left button depressed.	Click on model.
Rotate	Move mouse with left button depressed.	Move mouse with button depressed.
Translate	Move mouse with right button depressed.	Press option key and move mouse with button depressed.
Scale	Press shift key and move mouse with right button depressed.	Press option key and control key together and move mouse with button depressed.

Change the Model Display

Three types of operations can be applied to models: moving and scaling a model (see above), changing the model display, and measuring molecular geometry. All the model displays show the positions of atomic nuclei in the molecule, but some are more convenient for obtaining certain types of information. In particular, Space-filling displays attempt to show the amount of space each atom occupies as a sphere of fixed radius centered on the atom's nucleus. Sphere radii are taken from a table of experimental atom radii and roughly correspond to the size of the atom's electron cloud.

Computer Instruction	**Comments**
Move into Chapter4 and open butane	Specific colors are used to identify each type of atom (see table below). Make sure you can identify the locations of carbon and hydrogen nuclei in each model.

Change the model display
  Step 1. Select one after another: Wire, Ball and Wire, Tube, Ball and Spoke, and Space-filling from the Model menu.

Atom	**Identifying Color**		**Atom**	**Identifying Color**	
Hydrogen	White		Sulfur	Sky blue	
Carbon	Gray		Chlorine	Tan	
Nitrogen	Blue-gray		Bromine	Orange	
Oxygen	Red		Iodine	Tan	
Fluorine	Green				

Measuring Molecular Geometry

Three types of geometry measurements can be made using SpartanView: distances between pairs of atoms, angles involving any three atoms, and dihedral angles involving any four atoms.

Computer Instruction	Comments
Measure an interatomic distance	
Step 1. Select Distance from the Geometry menu.	The distance (in Å; Å1 = 100 pm) appears at the
Step 2. Click on any two atoms.	bottom of the screen. You can rotate, translate, and
Step 3. Click on Done when finished.	scale a molecule while you select atoms.
Measure the angle made by three atoms	
Step 4. Select Angle from the Geometry menu.	The angle (in degrees) appears at the bottom of the screen.
Step 5. Click on any three atoms.	
Step 6. Click on Done when finished.	
Measure a dihedral angle	
Step 7. Select Dihedral from the Geometry menu.	A dihedral angle is defined as the angle formed by
Step 8. Click on any four atoms.	the intersection of two planes. Atoms #1–3 define
Step 9. Click on Done when finished.	one plane and atoms #2–4 define the other. Atoms #2–3 are common to both planes; that is, they define the line of intersection between the planes.
Close the model	

Quantum Mechanical Models

Nearly all the SpartanView models on the CD have been constructed using quantum mechanical calculations. Calculations offer three important advantages over experiment: (1) the ability to access and display more types of molecular data than one can obtain from experimental measurements (a single calculation not only gives a molecule's geometry, it also gives the molecule's energy and other data); (2) the ability to model unstable molecules; and (3) the ability to create animations showing the structural changes that occur during a conformational change or a chemical reaction.

The models on the CD have been produced using a simplified version of quantum mechanics. This version gives models that closely resemble real molecules, and this book treats models and real molecules as if they were equivalent. You should keep in mind, though, that there are nearly always small differences between models and real molecules.

Measuring and Using Molecular Energies

The most useful information that a model provides is a molecule's structure and energy. Structure tells us what the molecule looks like, while energy tells us which chemical reactions are favorable and how fast they are likely to occur. SpartanView reports energies in atomic units (1 au = 2625.5 kJ/mol). These energies correspond to the energy change that occurs when a molecule is assembled from its component electrons and nuclei. The total energy of the component nuclei and electrons is exactly zero. (This follows from the fact that the nuclei and electrons are taken to be stationary and infinitely far apart.) Combining these components into molecules always releases a great deal of energy, so molecular energies are always very large and negative.

The energies of two molecules (or two groups of molecules) can be compared only if they have the same zero energy reference state, that is, only if they incorporate exactly the same nuclei and contain exactly the same number of electrons. This condition is satisfied by the reactants and products of a balanced chemical reaction. For example, the energy change for the reaction $A + B \rightarrow C + D$ is obtained by subtracting the energies of the reactant molecules from the energies of the product molecules ($= E_C + E_D - E_A - E_B$). This energy change is roughly equivalent to the reaction enthalpy, $\Delta H°$ Reaction energy barriers are obtained in the same way. If a reaction proceeds through a transition state, the barrier for the forward reaction is obtained by subtracting the energies of the reactant molecules from that of the transition state. This energy barrier is roughly equivalent to the reaction's activation enthalpy, $\Delta H^{\ddagger}$.

Computer Instruction	Comments
Move into Chapter12 and open 12.55. Carbon dioxide, CO_2, is the only model.	
Measure the calculated energy	The calculated energy of carbon dioxide (-186.5613 au) is displayed at the bottom of the screen.
Step 1. Select Energy from the Properties menu.	
Step 2. Click on Done when finished.	

Displaying Molecular Vibrations and Measuring Vibrational Frequencies

Molecular vibrations are a useful tool for determining molecular structure. Certain groups of atoms always vibrate at about the same frequency. Consequently, experimental frequency measurements can tell us which atom groups are present in a molecule. You can use SpartanView to display calculated vibrations and frequencies for selected models. Calculated frequencies are listed in units of cm^{-1} and are always a bit larger than observed frequencies. (Observed frequency = 0.9 × Calculated frequency is a good rule-of-thumb.)

Reaction transition states are characterized by a special type of molecular vibration that changes the transition state into the product (and reactant) and its frequency is an imaginary number. You can use SpartanView to display this characteristic vibration and get an idea of how atoms move during a chemical reaction.

Computer Instruction	Comments
Display a list of vibrational frequencies 　Step 1. Select Frequencies from the Properties menu.	Frequencies (in cm^{-1}) are listed in numerical order from smallest (or imaginary) at the top to largest at the bottom. Calculated frequencies are typically about 10% larger than observed frequencies.
Display a vibration 　Step 2. Double-click on a frequency to make it active. 　Step 3. Click on OK to close the window. 　Step 4. Select Ball and Spoke from the Model menu.	A check mark indicates the active vibration (only one vibration can be displayed at a time). Atom motions are exaggerated to make them easier to see. Vibrations appear most clearly when a molecule is displayed as a Ball and Spoke model.
Stop displaying a vibration 　Step 5. Repeat step #1, double-click on the active vibration, and click on OK.	Double-clicking on an active vibration deactivates it.
Close the model	

Electron Density and Electron Density Surfaces

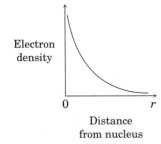

Electron density

0   r

Distance from nucleus

Molecules are made of nuclei and electrons, but the rules of quantum mechanics state that it is not possible to say where the electrons are. The best that one can do is say how probable it is for an electron to be at some particular location. The easiest way to think about this is to give up entirely the idea that an electron is a point-like particle, and picture an electron instead as a fog-like entity or electron cloud that spreads over a region of space. The cloud's electron density (expressed in electrons/volume) varies from one point to the next and indicates the probability of finding the electron at each point. The electron density distribution is especially useful because it can be calculated using quantum mechanics and because it can be converted into a graph showing regions of high and low electron density. The graph shows how electron density in the hydrogen atom varies with distance from the atomic nucleus. The positive nucleus attracts the highest electron density (negative charge), and electron density (negative charge) falls off rapidly with increasing distance from the nucleus.

The data used to plot this graph create a problem for chemists who want to know how big an atom or molecule is. There is no clear edge to the electron cloud; the electron density just gets smaller and smaller with increasing distance from the nucleus. One way to deal with this problem is to consider what happens when two molecules "rub up against each other." In this situation, the electron clouds of the two molecules overlap, and the molecular "boundary" can be defined as the point where each electron cloud has the same electron density. The value of the electron density at this point can then be used to establish the rest of each molecule's "boundary." We inspect each molecule's electron cloud and find all the points where the electron density equals the "boundary" value; then we connect these points together. The result is a skin-like surface that defines the molecule's size and shape. We refer to this surface as a "size density surface" or simply as a "density surface."

Another useful electron density surface is the "bond density surface." This surface is constructed in the same way as the size density surface, except that we look for a much higher value of electron density. Since the points where electron density is high are found near atomic nuclei and in between atoms linked by a covalent bond, a bond density surface is essentially a map of the covalent bonds inside a molecule.

Computer Instruction	**Comments**

Move into Chapter1, open 1.52, and make acetic acid the active model

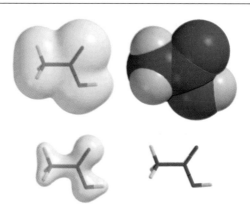

Size density surface (top left), Space-filling model (top right), bond density surface (bottom left), and tube model (bottom right) for acetic acid

Display a size density surface
   Step 1. Select Density from the Surfaces menu, then select Transparent from the sub-menu.

SpartanView uses the word "density" to identify size density surfaces. The size density surface is similar in size and shape to a Space-filling model.

Display a bond density surface
   Step 2. Select Bond Density from the Surfaces menu, then select Solid from the sub-menu.

Multiple surfaces can be displayed at the same time. The (solid) bond density surface can be seen through the (transparent) size density surface.

Stop displaying a surface
   Step 3. Select Density from the Surfaces menu, then select None from the sub-menu.

This removes the size density surface, but does not affect the bond density surface. The bond density surface lies inside the size density surface because electron density increases as one moves to the inside of a surface, and decreases as one moves to the outside of a surface.

Make sodium acetate the active model

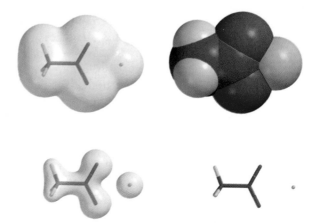

Size density surface (top left), Space-filling model (top right), bond density surface (bottom left), and tube model (bottom right) for sodium acetate

Step 4. Display both the size density (Transparent) and bond density (Solid) surfaces.

Sodium acetate is a salt. The size density surface encloses the entire molecule. The bond density surface, however, splits into two surfaces between oxygen and sodium indicating a low-density ionic oxygen–sodium bond.

Stop displaying all the surfaces when you are done

Electrostatic Potential Maps

The electrostatic potential at any point is defined as the change in energy that occurs when a "probe" particle with +1 charge is brought to this location from another point infinitely far away from the molecule (see the figure below). If the energy goes up (positive potential), then the molecule repels the probe. But, if the energy goes down (negative potential), then the molecule attracts the probe.

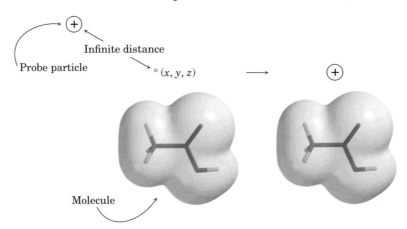

The potential at any point contains contributions from all the nuclei and the entire electron cloud. However, the potential in a given region is usually dominated by the *closest atom*, and variations in potential can normally be assigned to local variations in nuclear charge and electron density. If the potential at a particular point is positive, then it is likely that the atom closest to this point has a net positive charge. Likewise, a negative potential means it is likely that the closest atom has a net negative charge. (These rules only hold for neutral molecules.)

SpartanView uses a two-step procedure to create an electrostatic potential map. First, a size density surface is drawn around the molecule's electron cloud. Second, the electrostatic potential is calculated at all points on the size density surface and the surface is colored according to the value of the potential. Those parts of the surface where the potential is most negative (attractive) are colored red, and parts where the potential is most positive (repulsive) are colored blue. Intermediate potentials are colored according to: red < orange < yellow < green < blue. Electrostatic potential maps of water (left) and methane (right) are shown below. The maps show that the hydrogens carry partial positive charges (the regions closest to these atoms are blue), while oxygen and carbon carry partial negative charges (the regions closest to these atoms are red).

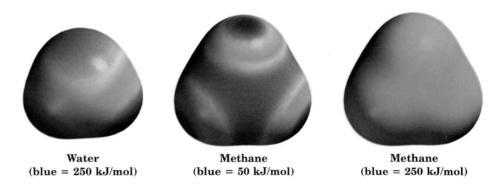

Water
(blue = 250 kJ/mol)   **Methane**
(blue = 50 kJ/mol)   **Methane**
(blue = 250 kJ/mol)

The electrostatic potential maps of water and methane shown above use the same colors, but not the same color–energy scale. The blue regions on water's map correspond to a potential of 250 kJ/mol, while the blue regions on methane's map correspond to a potential of only 50 kJ/mol. Maps with different color–energy scales cannot be used to compare two molecules. Valid comparisons require maps that use a common color–energy scale. The picture on the far right shows a map for methane in which the colors have been assigned using the color-energy scale previously used for water. The new map does not contain any red or blue regions. In other words, no atom in methane is as charged as the atoms in water.

Computer Instruction	Comments

Move into Chapter27 and open detergent.

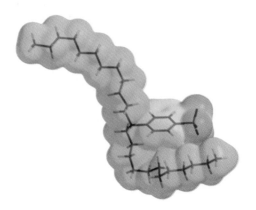

Electrostatic potential map for a typical negatively charged detergent molecule

Display an electrostatic potential map
   Step 1. Select Potential Map from the Surfaces menu; then select Solid from the sub-menu.

SpartanView uses Potential Map to identify electrostatic potential maps. The red part of the map indicates negatively charged atoms, and the blue-green parts indicate neutral atoms.

Close the model

Molecular Orbital Surfaces

A molecular orbital is a mathematical equation used to describe the location of individual electrons. Like the electron density and the electrostatic potential, a molecular orbital has different values at different points in space, and these values can be positive or negative. SpartanView displays orbital surfaces much like it displays electron density surfaces. An orbital surface shows all the points in space where the selected molecular orbital has a particular numerical magnitude, and the surface's color indicates the sign of the orbital (red = negative, blue = positive).

The most important molecular orbitals are the highest-energy occupied molecular orbital (HOMO), and lowest-energy unoccupied molecular orbital (LUMO). The picture in the margin shows a surface for the LUMO of H_2. The surface consists of a red (negative orbital value) region surrounding one hydrogen and a blue (positive orbital value) region surrounding the other. In between the two surfaces is a region where the orbital switches sign. This region appears as an empty space and is called a node. The node indicates that the orbital is antibonding.

Computer Instruction	Comments

Move into Chapter1 and open ethylene

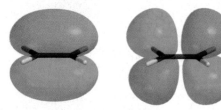

HOMO (left) and LUMO (right) of ethylene

Display an orbital surface
   Step 1. Select LUMO from the Surfaces menu; then select Transparent from the sub-menu.

This displays the LUMO of ethylene. This is an unoccupied antibonding molecular orbital.

Stop displaying an orbital surface
   Step 2. Select LUMO again from the Surfaces menu; then select None from the sub-menu.

The orbital is no longer displayed.

   Step 3. Select HOMO from the Surfaces menu; then select Transparent from the sub-menu.

This displays the HOMO of ethylene. This is an occupied bonding molecular orbital.

Close all of the models

Displaying SpartanView Sequences (Animations)

In addition to geometry changes that occur during a molecular vibration, SpartanView can also be used to display sequences of molecular geometries (animations) that occur during a conformational change or chemical reaction.

Computer Instruction	Comments

Move into Chapter3 and open ethane

Animate a sequence
   Step 1. Click on the arrow button in the lower left-hand corner of the window.

The scroll bar slides back and forth, and the step label is updated during the animation. You can rotate, translate, and scale the model at any point during the animation.

Stop the animation
   Step 2. Click on the double bar button in the lower left-hand corner of the window.

The animation and the scroll bar stop at the current step in the sequence.

Step through a sequence
   Step 3. Click on the bar-arrows at the right end of the scroll bar.

The scroll bar jumps to a new position, and the step label is updated to show the current location in the sequence.

Measure a property for a sequence
   Step 4. Select Energy from the Properties menu.
   Step 5. Repeat step 3 to see other energies.

All properties (energy, dipole moment, atomic charges) and geometry parameters (distance, angle, dihedral angle) can be animated or stepped through.

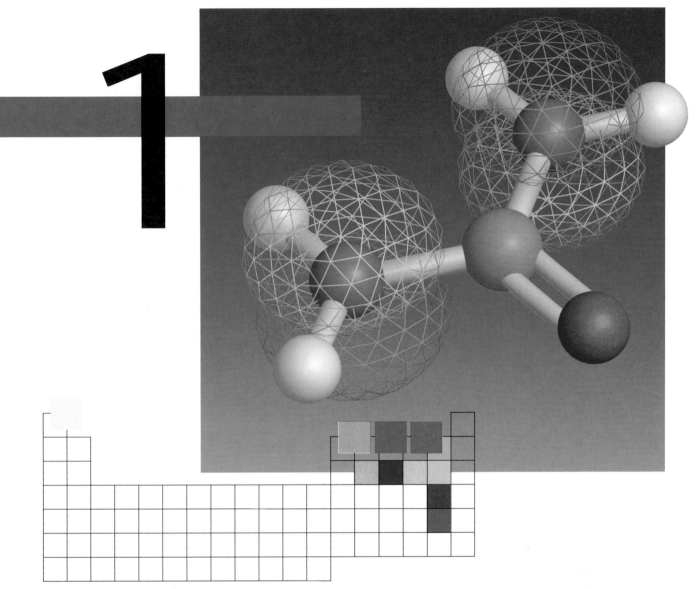

Structure and Bonding

What is organic chemistry? Why have so many millions of people studied it? And why should you study it? The answers to these questions are all around you. Every living organism is made of organic chemicals. The proteins that make up your hair, skin, and muscles; the DNA that controls your genetic heritage; the foods that nourish you; the clothes that keep you warm; and the medicines that heal you are all organic chemicals. Anyone with a curiosity about life and living things must have a basic understanding of organic chemistry.

The foundations of organic chemistry date from the mid-1700s, when chemistry was evolving from an alchemist's art into a modern science. At

Michel-Eugène Chevreul

Michel-Eugène Chevreul (1786–1889) was born in Angers, France. Educated at Paris, he became professor of physics at the Lycée Charlemagne in 1813 and professor of chemistry in 1830. Chevreul's studies of soaps and waxes led him to patent a method for manufacturing candles. He also published work on the psychology of color perception and of aging. All of France celebrated his 100th birthday in 1886.

Friedrich Wöhler

Friedrich Wöhler (1800–1882) was born in Eschersheim, Germany, and studied at Heidelberg under Leopold Gmelin. From 1836 to 1882, he was professor of chemistry at Göttingen. Wöhler developed the first industrial method for preparing aluminum metal, and he discovered several new elements. In addition, he wrote textbooks in both inorganic and organic chemistry.

that time, unexplainable differences were noted between substances obtained from living sources and those obtained from minerals. Compounds obtained from plants and animals were often difficult to isolate and purify. Even when pure, they were often difficult to work with, and they tended to decompose more easily than compounds obtained from minerals. The Swedish chemist Torbern Bergman in 1770 was the first to express this difference between "organic" and "inorganic" substances, and the term *organic chemistry* soon came to mean the chemistry of compounds found in living organisms.

To many chemists of the time, the only explanation for the differences in behavior between organic and inorganic compounds was that organic compounds must contain a peculiar "vital force" as a result of their origin in living sources. One consequence of this vital force, chemists believed, was that organic compounds could not be prepared and manipulated in the laboratory as could inorganic compounds. As early as 1816, however, this vitalistic theory received a heavy blow when Michel Chevreul found that soap, prepared by the reaction of alkali with animal fat, could be separated into several pure organic compounds, which he termed "fatty acids." For the first time, one organic substance (fat) was converted into others (fatty acids plus glycerin) without the intervention of an outside vital force.

$$\text{Animal fat} \xrightarrow[\text{H}_2\text{O}]{\text{NaOH}} \text{Soap} + \text{Glycerin}$$

$$\text{Soap} \xrightarrow{\text{H}_3\text{O}^+} \text{"Fatty acids"}$$

Little more than a decade later, the vitalistic theory suffered still further when Friedrich Wöhler discovered in 1828 that it was possible to convert the "inorganic" salt ammonium cyanate into the "organic" substance urea, which had previously been found in human urine.

$$\text{NH}_4^+ \ {}^-\text{OCN} \xrightarrow{\text{Heat}} \underset{\displaystyle \text{Urea}}{\text{H}_2\text{N}-\overset{\displaystyle \overset{\text{O}}{\|}}{\text{C}}-\text{NH}_2}$$

Ammonium cyanate   **Urea**

By the mid-1800s, the weight of evidence was clearly against the vitalistic theory. As William Brande wrote in 1848: "No definite line can be drawn between organic and inorganic chemistry.... Any distinctions ... must for the present be merely considered as matters of practical convenience calculated to further the progress of students."

Chemistry today is unified. The same principles that explain the simplest inorganic compounds also explain the most complex organic ones. The only distinguishing characteristic of organic chemicals is that *all contain the element carbon*. Nevertheless, the division between organic and inorganic chemistry, which began for historical reasons, maintains its "practical convenience ... to further the progress of students."

Organic chemistry, then, is the study of carbon compounds. Carbon, atomic number 6, is a second-row element whose position in the periodic table is shown in Figure 1.1. Although carbon is the principal element in

FIGURE 1.1 ▼

The position of carbon in the periodic table. Other elements commonly found in organic compounds are shown in color.

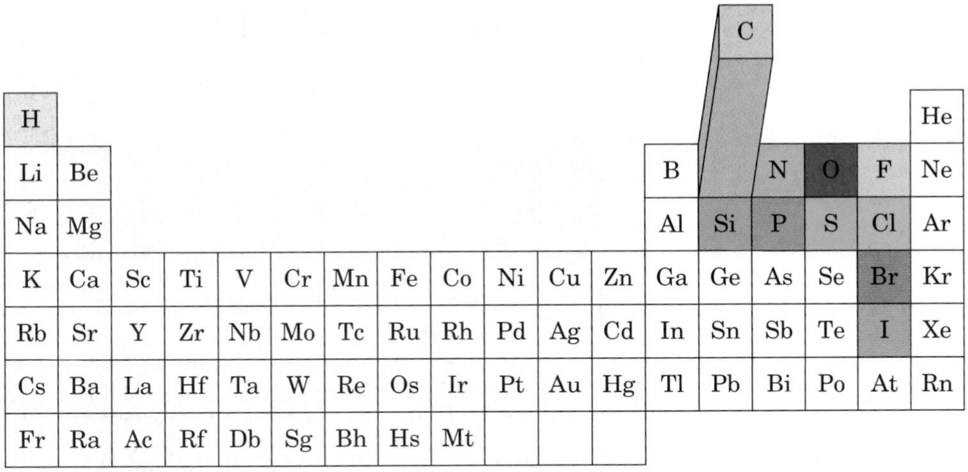

															C		
H																	He
Li	Be											B		N	O	F	Ne
Na	Mg											Al	Si	P	S	Cl	Ar
K	Ca	Sc	Ti	V	Cr	Mn	Fe	Co	Ni	Cu	Zn	Ga	Ge	As	Se	Br	Kr
Rb	Sr	Y	Zr	Nb	Mo	Tc	Ru	Rh	Pd	Ag	Cd	In	Sn	Sb	Te	I	Xe
Cs	Ba	La	Hf	Ta	W	Re	Os	Ir	Pt	Au	Hg	Tl	Pb	Bi	Po	At	Rn
Fr	Ra	Ac	Rf	Db	Sg	Bh	Hs	Mt									

William Thomas Brande

William Thomas Brande (1788–1866) was born in London, England. Trained as an apothecary, he became a lecturer in chemistry at the University of London in 1808 and was a professor at the Royal Institution from 1813 to 1854. His scientific achievements were modest, though he was the first person to discover naphthalene, now used in mothballs.

organic compounds, most also contain hydrogen, and many contain nitrogen, oxygen, phosphorus, sulfur, chlorine, or other elements.

But why is carbon special? What is it that sets carbon apart from all other elements in the periodic table? The answers to these questions come from the unique ability of carbon atoms to bond together, forming long chains and rings. Carbon, alone of all elements, is able to form an immense diversity of compounds, from the simple to the staggeringly complex—from methane, with one carbon atom, to DNA, which can have tens of *billions*.

Not all carbon compounds are derived from living organisms, of course. Modern chemists are extremely sophisticated in their ability to synthesize new organic compounds in the laboratory. Medicines, dyes, polymers, plastics, food additives, pesticides, and a host of other substances are all prepared in the laboratory. Organic chemistry is a science that touches the lives of everyone. Its study is a fascinating undertaking.

1.1 Atomic Structure

Before beginning a study of organic chemistry, let's review some general ideas about atoms and bonds. Atoms consist of a dense, positively charged *nucleus* surrounded at a relatively large distance by negatively charged *electrons* (Figure 1.2). The nucleus consists of subatomic particles called *neutrons*, which are electrically neutral, and *protons*, which are positively charged. Though extremely small—about 10^{-14} to 10^{-15} meter (m) in diameter—the nucleus nevertheless contains essentially all the mass of the atom. Electrons have negligible mass and circulate around the nucleus at a distance of approximately 10^{-10} m. Thus, the diameter of a typical atom is about 2×10^{-10} m, or 200 *picometers* (pm), where 1 pm = 10^{-12} m. To give you an idea of how small this is, a thin pencil line is about 3 *million*

FIGURE 1.2 ▼

A schematic view of an atom. The dense, positively charged nucleus contains most of the atom's mass and is surrounded by negatively charged electrons. The three-dimensional view on the right shows calculated electron-density surfaces (see Preface). Electron density increases steadily toward the nucleus and is 40 times greater at the blue solid surface than at the gray mesh surface.

atom

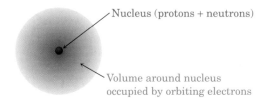

Nucleus (protons + neutrons)

Volume around nucleus occupied by orbiting electrons

carbon atoms across. [Many organic chemists still use the unit *angstrom* (Å) to express atomic distances; 1 Å = 10^{-10} m = 100 pm. Because of the easy decimal conversion, however, the SI unit picometer will be used in this book.]

An atom is described by its **atomic number (Z)**, which gives the number of protons in the atom's nucleus, and its **mass number (A)**, which gives the total of protons plus neutrons. All the atoms of a given element have the same atomic number—1 for hydrogen, 6 for carbon, 17 for chlorine, and so on—but they can have different mass numbers, depending on how many neutrons they contain. Such atoms with the same atomic number but different mass numbers are called **isotopes**. The weighted average mass in *atomic mass units* (amu) of an element's isotopes is called the element's **atomic weight**—1.008 for hydrogen, 12.011 for carbon, 35.453 for chlorine, and so on.

1.2   Atomic Structure: Orbitals

How are the electrons distributed in an atom? According to the *quantum mechanical* model of the atom, the motion of an electron around a nucleus can be described mathematically by what is known as a *wave equation*—the same sort of expression used to describe the motion of waves in a fluid. The solution to a wave equation is called a *wave function,* or **orbital**, and is denoted by the Greek letter psi, ψ.

A good way of viewing an orbital is to think of it as a mathematical expression whose square, ψ^2, predicts the volume of space around a nucleus where an electron can most likely be found. Although we don't know the exact position of an electron at a given moment, the orbital tells us where we would be most likely to find it. You might think of an orbital as looking like a photograph of the atom taken at a slow shutterspeed. Such a photograph would show the orbital as a blurry cloud indicating the region of space around the nucleus where the electron has recently been. This electron cloud doesn't have a sharp boundary, but for practical purposes we can set the limits by saying that an orbital represents the space where an electron spends most (90–95%) of its time.

What shapes do orbitals have? There are four different kinds of orbitals, denoted *s, p, d,* and *f.* Of the four, we'll be concerned primarily with *s* and *p* orbitals because these are the most important in organic chemistry. The *s* orbitals are spherical, with the nucleus at their center; *p* orbitals are dumb-bell-shaped; and four of the five *d* orbitals are cloverleaf-shaped, as shown in Figure 1.3. The fifth *d* orbital is shaped like an elongated dumbbell with a doughnut around its middle.

FIGURE 1.3 ▼

Representations of *s, p,* and *d* orbitals. The *s* orbitals are spherical, the *p* orbitals are dumbbell-shaped, and four of the five *d* orbitals are cloverleaf-shaped. Different lobes of *p* orbitals are often drawn for convenience as "teardrops," but their true shape is more like that of a doorknob, as indicated by the computer-generated representation of a 2*p* orbital on the right.

p-orbital

   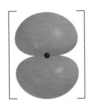

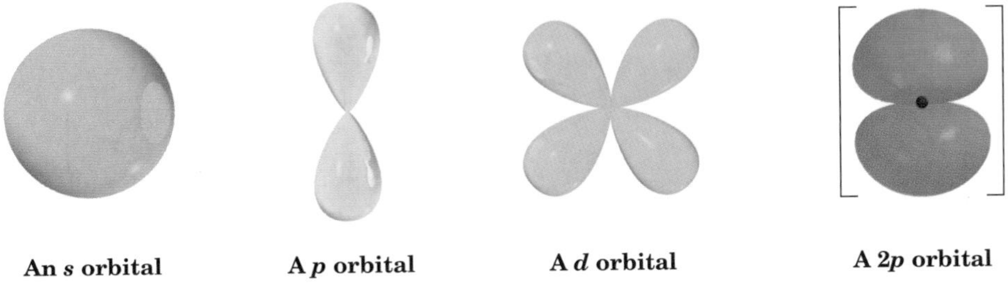

An *s* orbital   A *p* orbital   A *d* orbital   A 2*p* orbital

An atom's electrons can be thought of as being grouped in different layers, or **shells**, around the nucleus according to the amount of energy they have. Different shells have different numbers and kinds of orbitals, each of which can hold a pair of electrons. As indicated in Figure 1.4, the two lowest-energy electrons in an atom are in the first shell, which contains only a single *s* orbital, denoted 1*s*. Next in energy are the two 2*s* electrons, which are farther from the positively charged nucleus on average than 1*s* electrons and thus occupy a somewhat larger orbital.

FIGURE 1.4 ▼

The distribution of electrons in an atom. The first shell holds a maximum of two electrons in one 1*s* orbital; the second shell holds a maximum of eight electrons in one 2*s* and three 2*p* orbitals; the third shell holds a maximum of eighteen electrons in one 3*s*, three 3*p*, and five 3*d* orbitals; and so on. The two electrons in each orbital are represented by up and down arrows ↑↓ .

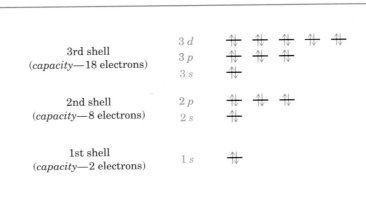

The six 2*p* electrons are next higher in energy. They occupy three orbitals, denoted 2*p_x*, 2*p_y*, and 2*p_z*, which are equal in energy and are oriented in space so that each is perpendicular to the other two (Figure 1.5).

FIGURE 1.5 ▼

Shapes of the 2p orbitals. Each of the three mutually perpendicular dumbbell-shaped orbitals has a node between its two lobes.

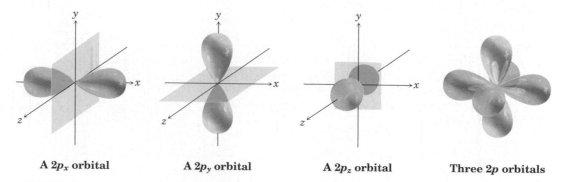

A $2p_x$ orbital   A $2p_y$ orbital   A $2p_z$ orbital   Three $2p$ orbitals

Note that the plane passing between the two lobes of each p orbital in Figure 1.5 is in a region of zero electron density, called a **node**. As we'll see, nodes have important consequences with respect to chemical reactivity.

Still higher in energy are the $3s$ orbital, three $3p$ orbitals, $4s$ orbital, and five $3d$ orbitals. As previously mentioned, we won't be too concerned with d orbitals, but you might note that the $3d$ orbital shown in Figure 1.3 has four lobes and two perpendicular nodal planes.

1.3 Atomic Structure: Electron Configurations

The lowest-energy arrangement, or **ground-state electron configuration**, of an atom is a listing of the orbitals occupied by its electrons. We can predict this arrangement by following three rules:

RULE 1   The lowest-energy orbitals fill up first, a statement called the *aufbau principle*. The ordering is $1s \rightarrow 2s \rightarrow 2p \rightarrow 3s \rightarrow 3p \rightarrow 4s \rightarrow 3d$. Note that the $4s$ orbital lies between the $3p$ and $3d$ orbitals in energy.

RULE 2   Electrons act in some ways as if they were spinning around an axis in much the same way that the earth spins. This spin can have two orientations, denoted as up ↑ and down ↓. Only two electrons can occupy an orbital, and they must be of opposite spin (the *Pauli exclusion principle*).

RULE 3   If two or more orbitals of equal energy are available, one electron occupies each until all orbitals are half-full. Only then does a second electron occupy one of the orbitals (*Hund's rule*). The electrons in the half-filled orbitals all have the same spin.

Some examples of how these rules apply are shown in Table 1.1. Hydrogen, for instance, has only one electron, which must occupy the lowest-energy orbital. Thus, hydrogen has a $1s$ ground-state configuration. Carbon has six electrons and the ground-state configuration $1s^2\ 2s^2\ 2p_x^{\ 1}\ 2p_y^{\ 1}$. Note that a superscript is used to represent the number of electrons in a particular orbital.

TABLE 1.1 Ground-State Electron Configurations of Some Elements

Element	Atomic number	Configuration				Element	Atomic number	Configuration			
Hydrogen	1	$1s$ ↿				Lithium	3	$2s$ ↿ $1s$ ↿⇂			
Carbon	6	$2p$ ↿ ↿ — $2s$ ↿⇂ $1s$ ↿⇂				Neon	10	$2p$ ↿⇂ ↿⇂ ↿⇂ $2s$ ↿⇂ $1s$ ↿⇂			
Sodium	11	$3s$ ↿ $2p$ ↿⇂ ↿⇂ ↿⇂ $2s$ ↿⇂ $1s$ ↿⇂				Argon	18	$3p$ ↿⇂ ↿⇂ ↿⇂ $3s$ ↿⇂ $2p$ ↿⇂ ↿⇂ ↿⇂ $2s$ ↿⇂ $1s$ ↿⇂			

· ·

Problem 1.1 Give the ground-state electron configuration for each of the following elements:
(a) Boron   (b) Phosphorus   (c) Oxygen   (d) Chlorine

Problem 1.2 How many electrons does each of the following elements have in its outermost electron shell?
(a) Potassium   (b) Aluminum   (c) Krypton

· ·

1.4 Development of Chemical Bonding Theory

Friedrich August Kekulé

Friedrich August Kekulé (1829–1896) was born in Darmstadt, Germany. After receiving his doctorate at the University of Giessen, he became a lecturer at Heidelberg in 1855 and a professor of chemistry at Ghent (1858) and Bonn (1867). His realization that carbon can form rings of atoms was said to have come to him in a dream in which he saw a snake biting its tail.

By the mid-1800s, the new science of chemistry was developing rapidly, and chemists had begun to probe the forces holding molecules together. In 1858, August Kekulé and Archibald Couper independently proposed that, in all organic compounds, carbon has four "affinity units." That is, carbon is *tetravalent;* it always forms four bonds when it joins other elements to form stable compounds. Furthermore, said Kekulé, carbon atoms can bond to one another to form extended chains of atoms linked together.

Shortly after the tetravalent nature of carbon was proposed, extensions to the Kekulé–Couper theory were made when the possibility of *multiple* bonding between atoms was suggested. Emil Erlenmeyer proposed a carbon-to-carbon triple bond for acetylene, and Alexander Crum Brown proposed a carbon-to-carbon double bond for ethylene. In 1865, Kekulé provided another major advance when he suggested that carbon chains can double back on themselves to form *rings* of atoms.

Archibald Scott Couper

Archibald Scott Couper (1831–1892) was born in Kirkintilloch, Scotland, and studied at the universities of Glasgow, Edinburgh, and Paris. Couper never received credit for his work, and he suffered a nervous breakdown in 1858. He retired from scientific work and spent his last 30 years in the care of his mother.

Although Kekulé and Couper were correct in describing the tetravalent nature of carbon, chemistry was still viewed in a two-dimensional way until 1874. In that year, Jacobus van't Hoff and Joseph Le Bel added a third dimension to our ideas about molecules. They proposed that the four bonds of carbon are not oriented randomly but have specific spatial directions. Van't Hoff went even further and suggested that the four atoms to which carbon is bonded sit at the corners of a regular tetrahedron, with carbon in the center.

A representation of a tetrahedral carbon atom is shown in Figure 1.6. Note the conventions used to show three-dimensionality: Solid lines represent bonds in the plane of the page, the heavy wedged line represents a bond coming out of the page toward the viewer, and the dashed line represents a bond receding back behind the page, away from the viewer. These representations will be used throughout the text.

FIGURE 1.6 ▼

Van't Hoff's tetrahedral carbon atom. The heavy wedged line comes out of the plane of the paper; the normal lines are in the plane; and the dashed line goes back behind the plane of the page. The three-dimensional stereo view on the right can be seen using the viewer bound inside the back cover of this book.

A tetrahedron   Stereo View

Visualizing Organic Chemistry

The ability to visualize complex organic and biological molecules in three dimensions is a critical skill in organic chemistry. To help you develop this skill, a stereo viewer is bound inside the back cover, and more than 170 three-dimensional stereo views like that in Figure 1.6 are placed throughout this book. Don't overlook this valuable learning tool.

Jacobus Hendricus van't Hoff

Jacobus Hendricus van't Hoff (1852–1911) was born in Rotterdam, Netherlands, and studied at Delft, Leyden, Bonn, Paris, and Utrecht. Widely educated, he served as professor of chemistry, mineralogy, and geology at the University of Amsterdam from 1878 to 1896, and later became professor at Berlin. He received the first Nobel Prize in chemistry in 1901 for his work on chemical equilibrium and osmotic pressure.

Problem 1.3   Draw a molecule of chloroform, $CHCl_3$, using wedged, normal, and dashed lines to show its tetrahedral geometry.

Problem 1.4   Convert the following stereo view of ethane, C_2H_6, into a conventional drawing that uses wedged, normal, and dashed lines to indicate tetrahedral geometry around each carbon (gray = C, ivory = H).

Stereo View

1.5 Covalent Bonds

Why do atoms bond together, and how can bonds be described electronically? The *why* question is relatively easy to answer: Atoms bond together because the compound that results is more stable (has less energy) than the separate atoms. Just as water flows downhill, energy is released and flows *out of* the chemical system when a chemical bond is formed. Conversely, energy is absorbed and must be put *into* the system when a chemical bond is broken. The *how* question is more difficult. To answer it, we need to know more about the properties of atoms.

We know that eight electrons (an electron *octet*) in an atom's outermost shell, or **valence shell**, impart special stability to the noble-gas elements in group 8A of the periodic table—for example, Ne (2 + 8); Ar (2 + 8 + 8); Kr (2 + 8 + 8 + 8). We also know that the chemistry of many main-group elements is governed by their tendency to take on the electron configuration of the nearest noble gas. The alkali metals in group 1A, for example, achieve a noble-gas configuration by losing the single electron from their valence shell to form a cation, while the halogens in group 7A achieve a noble-gas configuration by gaining an electron to form an anion. The resultant ions are held together in compounds like $Na^+ Cl^-$ by an electrostatic attraction that we call an *ionic bond*.

Joseph Achille Le Bel

Joseph Achille Le Bel (1847–1930) was born in Péchelbronn, France, and studied at the École Polytechnique and the Sorbonne in Paris. Freed by his family's wealth from the need to earn a living, he established his own private laboratory.

How, though, do elements in the middle of the periodic table form bonds? Look at methane, CH_4, the main constituent of natural gas, for example. The bonding in methane is not ionic because it would be energetically difficult for carbon ($1s^2\, 2s^2\, 2p^2$) to either gain or lose *four* electrons to achieve a noble-gas configuration. In fact, carbon bonds to other atoms, not by gaining or losing electrons, but by *sharing* them. Such a shared-electron bond, first proposed in 1916 by G. N. Lewis, is called a **covalent bond**. The neutral collection of atoms held together by covalent bonds is called a **molecule**.

A simple shorthand way of indicating the covalent bonds in a molecule is to use what is called a **Lewis structure**, or **electron-dot structure**, in which the valence electrons of an atom are represented as dots. Thus, hydrogen has one dot representing its $1s$ electron, carbon has four dots ($2s^2\, 2p^2$), oxygen has six dots ($2s^2\, 2p^4$), and so on. A stable molecule results whenever a noble-gas configuration is achieved for all the atoms—an octet for main-group atoms or two for hydrogen—as in the following examples:

$$\cdot\overset{\displaystyle\cdot}{\underset{\displaystyle\cdot}{C}}\cdot\; +\; 4\,H\cdot \;\longrightarrow\; H\!:\!\overset{\displaystyle H}{\underset{\displaystyle H}{C}}\!:\!H$$

Methane (CH_4)

$$3\,H\cdot\; +\; \cdot\overset{\displaystyle\cdot}{N}\cdot \;\longrightarrow\; H\!:\!\overset{\displaystyle H}{\underset{\displaystyle\cdot\cdot}{N}}\!:\!H$$

Ammonia (NH_3)

$$2\,H\cdot\; +\; \cdot\overset{\displaystyle\cdot\cdot}{\underset{\displaystyle\cdot}{O}}\!: \;\longrightarrow\; H\!:\!\overset{\displaystyle\cdot\cdot}{\underset{\displaystyle H}{O}}\!:$$

Water (H_2O)

$$3\,H\cdot\; +\; \cdot\overset{\displaystyle\cdot}{\underset{\displaystyle\cdot}{C}}\cdot\; +\; \cdot\overset{\displaystyle\cdot\cdot}{\underset{\displaystyle\cdot}{O}}\!:\; +\; H\cdot \;\longrightarrow\; H\!:\!\overset{\displaystyle H}{\underset{\displaystyle H}{C}}\!:\!\overset{\displaystyle\cdot\cdot}{O}\!:$$

Methanol (CH_3OH)

The number of covalent bonds an atom forms depends both on how many valence electrons it has and on how many additional valence electrons it needs to reach a noble-gas configuration. Atoms with one, two, or three valence electrons form one, two, or three bonds, but atoms with four or more valence electrons form as many bonds as needed to reach an octet. Boron, for instance, has only three valence electrons ($2s^2\, 2p^1$) and can form only three covalent bonds in a neutral molecule, as in BF_3. Carbon has four valence electrons ($2s^2\, 2p^2$) and can fill its valence shell by forming four bonds, as in CH_4. Nitrogen has five valence electrons ($2s^2\, 2p^3$) and needs to form only three bonds, as in NH_3; oxygen has six valence electrons ($2s^2\, 2p^4$) and forms two bonds, as in H_2O.

H— Cl—	—O—	—N— —B—	—C—
Br— F—			
One bond	Two bonds	Three bonds	Four bonds

Valence electrons that are not used for bonding are called **nonbonding electrons**, or **lone-pair electrons**. The nitrogen atom in ammonia, for instance, shares six valence electrons in three covalent bonds and has its remaining two valence electrons in a nonbonding lone pair.

Gilbert Newton Lewis

Gilbert Newton Lewis (1875–1946) was born in Weymouth, Massachusetts, and received his Ph.D. at Harvard in 1899. After a short time as professor of chemistry at the Massachusetts Institute of Technology (1905–1912), he spent the rest of his career at the University of California at Berkeley (1912–1946). In addition to his work on structural theory, Lewis was the first to prepare "heavy water," D_2O, in which the two hydrogens of water are the ^2H isotope deuterium.

Nonbonding, lone-pair electrons →

$:\overset{..}{N}:H$ or $:N-H$

Ammonia

Lewis structures are particularly useful because they make electron bookkeeping possible and act as reminders of the number of valence electrons present. Simpler, however, is the use of **Kekulé structures**, or **line-bond structures**, in which a two-electron covalent bond is indicated as a line drawn between atoms. Lone pairs of nonbonding valence electrons are often not shown when drawing line-bond structures, though it's still necessary to keep track of them mentally. Some examples are shown in Table 1.2.

TABLE 1.2 Lewis and Kekulé Structures of Some Simple Molecules

Name	Lewis structure	Kekulé structure	Name	Lewis structure	Kekulé structure
Water (H_2O)	$H:\overset{..}{\underset{..}{O}}:H$	$H-O-H$	Methane (CH_4)	$H:\overset{H}{\underset{H}{C}}:H$	$H-\overset{H}{\underset{H}{C}}-H$
Ammonia (NH_3)	$H:\overset{..}{N}:H$	$H-\overset{H}{N}-H$	Methanol (CH_3OH)	$H:\overset{H}{\underset{H}{C}}:\overset{..}{\underset{..}{O}}:H$	$H-\overset{H}{\underset{H}{C}}-O-H$

Practice Problem 1.1 How many hydrogen atoms does phosphorus bond to in forming phosphine, $PH_?$?

Strategy Phosphorus is in group 5A of the periodic table and has five valence electrons. It therefore needs to share three more electrons to reach an octet.

Solution Phosphorus bonds to three hydrogen atoms, giving PH_3.

Problem 1.5 What are likely formulas for the following substances?
(a) $GeCl_?$ (b) $AlH_?$ (c) $CH_?Cl_2$ (d) $SiF_?$ (e) $CH_3NH_?$

Problem 1.6   Write both Lewis and line-bond structures for the following substances, showing all nonbonding electrons:
(a) $CHCl_3$, chloroform
(b) H_2S, hydrogen sulfide
(c) CH_3NH_2, methylamine
(d) NaH, sodium hydride
(e) CH_3Li, methyllithium

Problem 1.7   Why can't an organic molecule have the formula C_2H_7?

• •

1.6 Valence Bond Theory and Molecular Orbital Theory

How does electron sharing occur? Two models have been developed to describe covalent bond formation: *valence bond theory* and *molecular orbital theory*. Each model has its strengths and weaknesses, and chemists use them interchangeably depending on the circumstances. Valence bond theory is the more easily visualized of the two, so most of the descriptions we'll be using in this book derive from that approach. We'll take a brief look now at both theories and then return for a second look at molecular orbital theory in Section 1.9.

Valence Bond Theory

According to **valence bond theory**, a covalent bond results when two atoms approach each other closely so that a singly occupied orbital on one atom *overlaps* a singly occupied orbital on the other atom. The electrons are now paired in the overlapping orbitals and are attracted to the nuclei of both atoms, thus bonding the atoms together. In the H_2 molecule, for example, the H–H bond results from the overlap of two singly occupied hydrogen $1s$ orbitals:

| H↑ | + | ↓H | ⟶ | H⇅H |
| 1s | | 1s | | H_2 molecule |

Key Ideas ▶   The key ideas of valence bond theory are as follows:

- Covalent bonds are formed by overlap of two atomic orbitals, each of which contains one electron. The spins of the two electrons are opposite.

- Each of the bonded atoms retains its own atomic orbitals, but the electron pair in the overlapping orbitals is shared by both atoms.

- The greater the amount of orbital overlap, the stronger the bond.

The bond in the H_2 molecule has the elongated egg shape we might get by pressing two spheres together. If a plane were to pass through the middle of the bond, the intersection of the plane and the overlapping orbitals would be a circle. In other words, the H–H bond is *cylindrically symmetrical,* as shown in Figure 1.7. Such bonds, which are formed by the head-on overlap of two atomic orbitals along a line drawn between the nuclei, are called **sigma (σ) bonds.** Although σ bonds are the most common kind, we'll see shortly that there is another type as well.

FIGURE 1.7 ▼

The cylindrical symmetry of the H–H bond. The intersection of a plane cutting through the orbital is a circle.

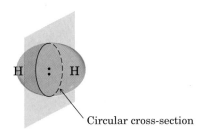

Circular cross-section

During the reaction 2 H· → H$_2$, 436 kJ/mol (104 kcal/mol) of energy is released. Because the product H$_2$ molecule has 436 kJ/mol less energy than the starting 2 H·, we say that the product is more stable than the starting material and that the new H–H bond has a **bond strength** of 436 kJ/mol. In other words, we would have to put 436 kJ/mol of energy *into* the H–H bond to break the H$_2$ molecule apart into H atoms (Figure 1.8.) [Energy values are given in both kilocalories (kcal) and kilojoules (kJ) for convenience: 1 kJ = 0.239 kcal; 1 kcal = 4.184 kJ.]

FIGURE 1.8 ▼

Energy levels of H atoms and the H$_2$ molecule. Because the H$_2$ molecule is lower in energy than the two H atoms by 436 kJ/mol (104 kcal/mol), 436 kJ/mol of energy is released when the H–H bond forms. Conversely, 436 kJ/mol would have to be added to the H$_2$ molecule to break the H–H bond.

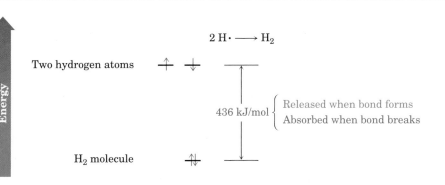

2 H· ⟶ H$_2$

Two hydrogen atoms

Energy

436 kJ/mol { Released when bond forms
Absorbed when bond breaks

H$_2$ molecule

How close are the two nuclei in the H_2 molecule? If they are too close, they will repel each other because both are positively charged, yet if they're too far apart, they won't be able to share the bonding electrons. Thus, there is an optimum distance between nuclei that leads to maximum stability (Figure 1.9). Called the **bond length**, this distance is 74 pm in the H_2 molecule. Every covalent bond has both a characteristic bond strength and bond length.

FIGURE 1.9 ▼

A plot of energy versus internuclear distance for two hydrogen atoms. The distance between nuclei at the lowest-energy point is called the bond length.

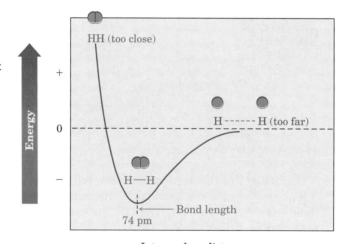

Molecular Orbital Theory

Molecular orbital (MO) theory describes covalent bond formation as arising from a mathematical combination of atomic orbitals (wave functions) to form *molecular orbitals,* so called because they belong to the entire molecule rather than to an individual atom. Just as an *atomic* orbital describes a region of space around an *atom* where an electron is likely to be found, so a *molecular* orbital describes a region of space in a *molecule* where electrons are most likely to be found.

Like an atomic orbital, a molecular orbital has a specific size, shape, and energy. In the H_2 molecule, for example, two singly occupied 1s atomic orbitals combine. There are two ways for the orbital combination to occur—an additive way and a subtractive way. The additive combination leads to formation of a molecular orbital that is roughly egg-shaped, while the subtractive combination leads to formation of a molecular orbital that has a node between nuclei (Figure 1.10). Note that the additive combination is a

FIGURE 1.10 ▼

Computer-generated molecular orbitals of H_2. Combination of two hydrogen 1s atomic orbitals leads to two H_2 molecular orbitals. The lower-energy, bonding MO is filled, and the higher-energy, antibonding MO is unfilled.

hydrogen

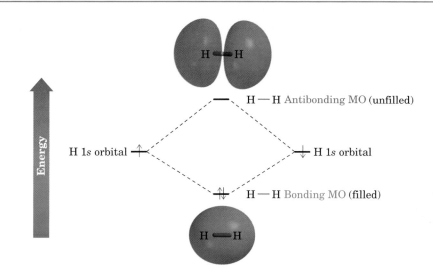

single, egg-shaped, *molecular* orbital; it is not the same as the two overlapping 1s atomic orbitals of the valence bond description.

The additive combination is lower in energy than the two hydrogen 1s atomic orbitals and is called a **bonding MO**. Any electrons in this MO spend most of their time in the region between the two nuclei, thereby bonding the atoms together. The subtractive combination is higher in energy than the two hydrogen 1s orbitals and is called an **antibonding MO**. Any electrons it contains can't occupy the central region between the nuclei where there is a node and can't contribute to bonding. The two nuclei therefore repel each other.

Key Ideas ▶ The key ideas of molecular orbital theory are as follows:

- Molecular orbitals are to molecules what atomic orbitals are to atoms. Molecular orbitals describe regions of space *in a molecule* where electrons are most likely to be found, and they have a specific size, shape, and energy level.

- Molecular orbitals are formed by combining atomic orbitals. The number of MO's formed is the same as the number of atomic orbitals combined.

- Molecular orbitals that are lower in energy than the starting atomic orbitals are bonding; MO's higher in energy than the starting atomic orbitals are antibonding; and MO's with the same energy as the starting atomic orbitals are *nonbonding*.

1.7 Hybridization: *sp³* Orbitals and the Structure of Methane

The bonding in the hydrogen molecule is fairly straightforward, but the situation is more complicated in organic molecules with tetravalent carbon atoms. Let's start with a simple case and consider methane, CH_4. Carbon has four electrons in its valence shell and can form four bonds to hydrogens. In Lewis structures:

$$\cdot \overset{\cdot}{\underset{\cdot}{C}} \cdot \xrightarrow{4\,H\cdot} H : \overset{\overset{\displaystyle H}{\cdot\cdot}}{\underset{\underset{\displaystyle H}{\cdot\cdot}}{C}} : H$$

Because carbon uses two kinds of orbitals (2*s* and 2*p*) to form bonds, we might expect methane to have two kinds of C–H bonds. In fact, though, all four C–H bonds in methane are identical and are spatially oriented toward the corners of a regular tetrahedron (see Figure 1.6). How can we explain this?

An answer was provided in 1931 by Linus Pauling, who showed mathematically how an *s* orbital and three *p* orbitals on an atom can combine, or *hybridize,* to form four equivalent atomic orbitals with tetrahedral orientation. Shown in Figure 1.11, these tetrahedrally oriented orbitals are called ***sp³* hybrids**. (The superscript 3 in the name indicates that three *p* atomic orbitals combine to form the hybrid, not that 3 electrons occupy it.)

FIGURE 1.11 ▼

Four *sp³* hybrid orbitals (green), oriented to the corners of a regular tetrahedron, are formed by combination of an atomic *s* orbital (red) and three atomic *p* orbitals (blue). The *sp³* hybrids are unsymmetrical about the nucleus, giving them a directionality and allowing them to form strong bonds when they overlap an orbital from another atom.

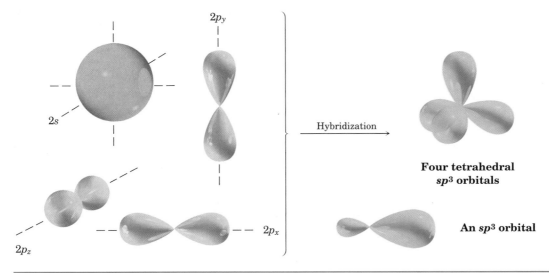

Linus Carl Pauling

Linus Carl Pauling (1901–1994) was born in Portland, Oregon, and obtained a B.S. degree at Oregon State. He received a Ph.D. from the California Institute of Technology in 1925, and remained as professor of chemistry (1925–1967). Pauling was a scientific giant. He made fundamental discoveries in fields ranging from chemical bonding to molecular biology to medicine. A lifelong pacifist, Pauling was the only solo winner of two Nobel Prizes in different fields: one for chemistry (1954) and one for peace (1963).

The concept of hybridization explains *how* carbon forms four equivalent tetrahedral bonds but doesn't explain *why* it does so. Looking at an sp^3 hybrid orbital from the side suggests the answer. When an *s* orbital hybridizes with three *p* orbitals, the resultant sp^3 hybrid orbitals are unsymmetrical about the nucleus. One of the two lobes is much larger than the other and can therefore overlap better with an orbital from another atom when it forms a bond. As a result, sp^3 hybrid orbitals form stronger bonds than do unhybridized *s* or *p* orbitals.

The asymmetry of sp^3 orbitals arises because of a property of orbitals that we have not yet considered. When the wave equation for a *p* orbital is solved, the two lobes have opposite algebraic signs, + and −. Thus, when a *p* orbital hybridizes with an *s* orbital, the positive *p* lobe *adds* to the *s* orbital, but the negative *p* lobe *subtracts* from the *s* orbital. The resultant hybrid orbital is unsymmetrical about the nucleus and is strongly oriented in one direction, as shown in Figure 1.11.

When the four identical orbitals of an sp^3-hybridized carbon atom overlap with the 1*s* orbitals of four hydrogen atoms, four identical C–H bonds are formed and methane results. Each C–H bond in methane has a strength of 438 kJ/mol (105 kcal/mol) and a length of 110 pm. Because the four bonds have a specific geometry, we also can define a property called the **bond angle**. The angle formed by each H–C–H is exactly 109.5°, the so-called *tetrahedral angle*. Methane thus has the structure shown in Figure 1.12.

FIGURE 1.12 ▼

The structure of methane. The drawings are computer-generated.

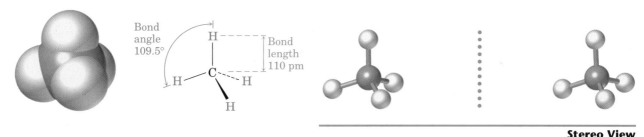

Stereo View

1.8 The Structure of Ethane

The same kind of hybridization that explains the methane structure also explains how carbon atoms can bond together in chains and rings to make possible so many millions of organic compounds. Ethane, C_2H_6, is the simplest molecule containing a carbon–carbon bond:

$$
\begin{array}{ccc}
\text{H H} & \quad \text{H H} & \\
\text{H:}\overset{..}{\text{C}}\text{:}\overset{..}{\text{C}}\text{:H} & \text{H}-\overset{|}{\underset{|}{\text{C}}}-\overset{|}{\underset{|}{\text{C}}}-\text{H} & \quad CH_3CH_3 \\
\overset{..}{\text{H}}\ \overset{..}{\text{H}} & \quad \text{H H} &
\end{array}
$$

Some representations of ethane

We can picture the ethane molecule by imagining that the two carbon atoms bond to each other by σ overlap of an sp^3 hybrid orbital from each. The remaining three sp^3 hybrid orbitals on each carbon overlap with hydrogen $1s$ orbitals to form the six C–H bonds, as shown in Figure 1.13. The C–H bonds in ethane are similar to those in methane, though a bit weaker— 420 kJ/mol (100 kcal/mol) for ethane versus 438 kJ/mol for methane. The C–C bond is 154 pm long and has a strength of 376 kJ/mol (90 kcal/mol). All the bond angles of ethane are near the tetrahedral value of 109.5°.

FIGURE 1.13 ▼

The structure of ethane. The carbon–carbon bond is formed by σ overlap of two carbon sp^3 hybrid orbitals. (For clarity, the smaller lobes of the sp^3 hybrid orbitals are not shown.)

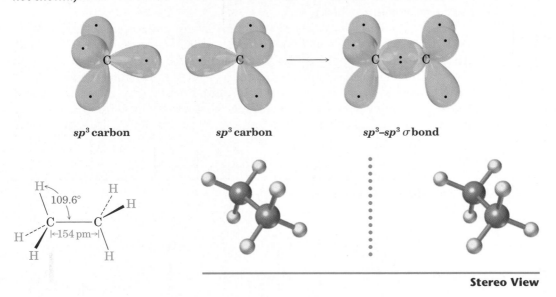

sp^3 **carbon** sp^3 **carbon** sp^3–sp^3 σ **bond**

Stereo View

• •

Problem 1.8   Draw a line-bond structure for propane, $CH_3CH_2CH_3$. Predict the value of each bond angle, and indicate the overall shape of the molecule.

• •

1.9 Hybridization: sp^2 Orbitals and the Structure of Ethylene

Although sp^3 hybridization is the most common electronic state of carbon, it's not the only possibility. Look at ethylene, C_2H_4, for example. It was recognized well over 100 years ago that ethylene carbons can be tetravalent only if the two carbon atoms share *four* electrons and are linked by a *double* bond. Furthermore, ethylene is planar (flat) and has bond angles of approximately 120°.

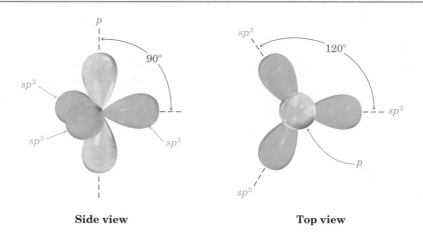

Top view   Side view

Ethylene

When we discussed sp^3 hybrid orbitals in Section 1.7, we said that all four of carbon's valence-shell atomic orbitals combine to form four equivalent sp^3 hybrids. Imagine instead that the 2s orbital combines with only *two* of the three 2p orbitals. Three **sp^2 hybrid orbitals** result, and one 2p orbital remains unchanged. The three sp^2 orbitals lie in a plane at angles of 120° to one another, with the remaining p orbital perpendicular to the sp^2 plane, as shown in Figure 1.14.

FIGURE 1.14 ▼

An *sp²*-hybridized carbon. Three equivalent *sp²* hybrid orbitals (green) lie in a plane at angles of 120° to one another, and a single unhybridized *p* orbital (blue) is perpendicular to the *sp²* plane.

Side view   Top view

When two sp^2-hybridized carbons approach each other, they form a σ bond by sp^2–sp^2 overlap according to valence bond theory. At the same time, the unhybridized p orbitals approach with the correct geometry for *sideways* overlap, leading to the formation of what is called a **pi (π) bond**. Note that the π bond has regions of electron density on either side of a line drawn between nuclei but has no electron density directly between nuclei. The combination of an sp^2–sp^2 σ bond and a 2p–2p π bond results in the sharing of four electrons and the formation of a carbon–carbon double bond (Figure 1.15, p. 20).

To complete the structure of ethylene, four hydrogen atoms form σ bonds with the remaining four sp^2 orbitals. Ethylene has a planar structure with H–C–H and H–C=C bond angles of approximately 120° (the H–C–H bond angles are 116.6°, and the H–C=C bond angles are 121.7°). Each C–H bond has a length of 107.6 pm and a strength of 444 kJ/mol (106 kcal/mol).

FIGURE 1.15 ▼

Orbital overlap of two sp^2-hybridized carbons to form a carbon–carbon double bond. One part of the double bond results from σ (head-on) overlap of sp^2 orbitals (red), and the other part results from π (sideways) overlap of unhybridized p orbitals (blue). The π bond has regions of electron density on either side of a line drawn between nuclei.

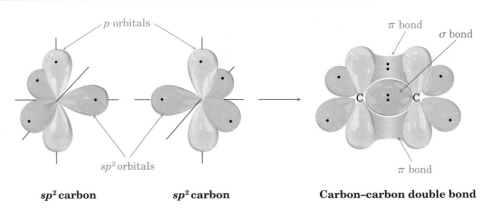

sp^2 **carbon** sp^2 **carbon**   **Carbon–carbon double bond**

As you might expect, the carbon–carbon double bond in ethylene is both shorter and stronger than the single bond in ethane because it results from the sharing of four electrons rather than two. Ethylene has a C=C bond length of 133 pm and a strength of 611 kJ/mol (146 kcal/mol) versus a C–C length of 154 pm and a strength of 376 kJ/mol for ethane. Note, though, that the carbon–carbon double bond is considerably less than twice as strong as a single bond because the overlap in the π part of the double bond is not as effective as the overlap in the σ part. The structure of ethylene is shown in Figure 1.16.

FIGURE 1.16 ▼

The structure of ethylene.

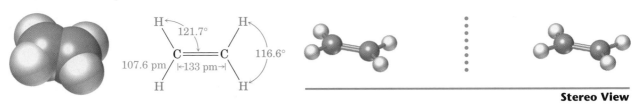

Stereo View

We said in Section 1.6 that chemists use two models for describing covalent bonds: valence bond theory and molecular orbital theory. Having now seen a valence bond description of the double bond in ethylene, let's also look at a molecular orbital description.

Just as bonding and antibonding σ molecular orbitals result from the combination of two s atomic orbitals in H_2 (Section 1.6), so bonding and antibonding π molecular orbitals result from the combination of two p atomic orbitals in ethylene. As shown in Figure 1.17, the π bonding MO has no node between nuclei and results from combination of p orbital lobes with the same algebraic sign. The π antibonding MO has a node between nuclei and results from combination of lobes with opposite algebraic signs. Only the bonding MO is occupied; the higher-energy, antibonding MO is vacant.

FIGURE 1.17 ▼

A molecular orbital description of the C=C π bond. The π bonding MO results from an additive combination of atomic orbitals and is filled. The π antibonding MO results from a subtractive combination of atomic orbitals and is unfilled. The representations on the right are computer-generated for accuracy.

ethylene

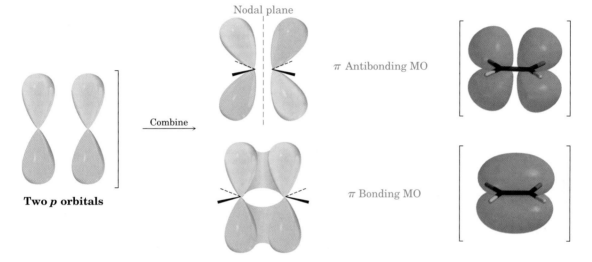

We'll come back to this molecular orbital description of π bonding in future chapters, particularly when we discuss compounds with more than one double bond.

● ●

Practice Problem 1.2   Formaldehyde, CH_2O, contains a carbon–*oxygen* double bond. Draw Lewis and line-bond structures of formaldehyde, and indicate the hybridization of the carbon atom.

Strategy   We know that hydrogen forms one covalent bond, carbon forms four, and oxygen forms two. Trial-and-error, combined with intuition, must be used to fit the atoms together.

Solution   There is only one way that two hydrogens, one carbon, and one oxygen can combine:

Lewis structure   **Line-bond structure**

Like the carbon atoms in ethylene, the doubly bonded carbon atom in formaldehyde is sp^2-hybridized.

Problem 1.9   Draw all the bonds in propene, $CH_3CH=CH_2$. Indicate the hybridization of each carbon, and predict the value of each bond angle.

Problem 1.10   Answer Problem 1.9 for 1,3-butadiene, $H_2C=CH-CH=CH_2$.

Problem 1.11   Draw both a Lewis structure and a line-bond structure for acetaldehyde, CH_3CHO.

Problem 1.12   Shown below is a computer-generated stereo view of aspirin (acetylsalicylic acid). Identify the hybridization of each carbon atom in aspirin, and tell which atoms have lone pairs of electrons (gray = C, red = O, ivory = H).

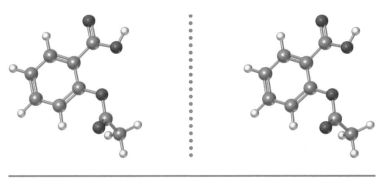

Stereo View

1.10 Hybridization: *sp* Orbitals and the Structure of Acetylene

In addition to forming single and double bonds by sharing two and four electrons, respectively, carbon also can form a *triple* bond by sharing six electrons. To account for the triple bond in a molecule such as acetylene, C_2H_2, we need a third kind of hybrid orbital, an **sp hybrid**.

$$H:C:::C:H \qquad H-C\equiv C-H$$

Acetylene

Imagine that, instead of combining with two or three *p* orbitals, a carbon 2*s* orbital hybridizes with only a single *p* orbital. Two *sp* hybrid orbitals result, and two *p* orbitals remain unchanged. The two *sp* orbitals are linear, or 180° apart on the *x*-axis, while the remaining two *p* orbitals are perpendicular on the *y*-axis and the *z*-axis, as shown in Figure 1.18.

When two *sp*-hybridized carbon atoms approach each other, *sp* hybrid orbitals from each carbon overlap head-on to form a strong *sp–sp* σ bond. In addition, the p_z orbitals from each carbon form a p_z–p_z π bond by sideways overlap, and the p_y orbitals overlap similarly to form a p_y–p_y π bond.

FIGURE 1.18 ▼

An *sp*-hybridized carbon atom. The two *sp* hybrid orbitals (green) are oriented 180° away from each other, perpendicular to the two remaining *p* orbitals (blue).

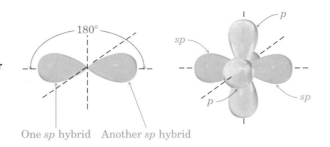

One *sp* hybrid   Another *sp* hybrid

The net effect is the sharing of six electrons and formation of a carbon–carbon triple bond. The remaining *sp* hybrid orbitals each form a σ bond with hydrogen to complete the acetylene molecule (Figure 1.19).

FIGURE 1.19 ▼

The structure of acetylene. The two *sp*-hybridized carbon atoms are joined by one *sp–sp* σ bond and two *p–p* π bonds.

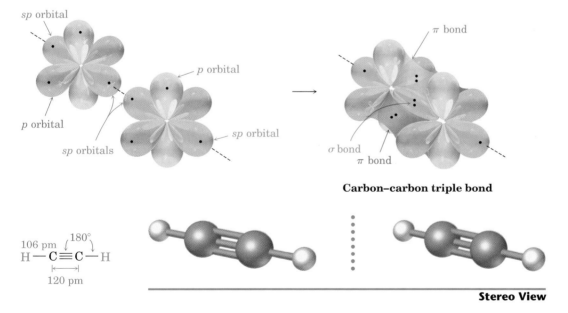

Carbon–carbon triple bond

$$H - C \equiv C - H$$

106 pm   180°

120 pm

Stereo View

As suggested by *sp* hybridization, acetylene is a linear molecule with H–C≡C bond angles of 180°. The C–H bonds have a length of 106 pm and a strength of 552 kJ/mol (132 kcal/mol). The C–C bond length is 120 pm and its strength is about 835 kJ/mol (200 kcal/mol), making the triple bond in acetylene the shortest and strongest of any carbon–carbon bond. A comparison of sp, sp^2, and sp^3 hybridization is given in Table 1.3.

TABLE 1.3 Comparison of C–C and C–H Bonds in Methane, Ethane, Ethylene, and Acetylene

| Molecule | Bond | Bond strength | | Bond length (pm) |
		(kJ/mol)	(kcal/mol)	
Methane, CH_4	C_{sp^3}—H_{1s}	438	105	110
Ethane, CH_3CH_3	C_{sp^3}—C_{sp^3}	376	90	154
	C_{sp^3}—H_{1s}	420	100	110
Ethylene, $H_2C=CH_2$	$C_{sp^2}=C_{sp^2}$	611	146	133
	C_{sp^2}—H_{1s}	444	106	107.6
Acetylene, $HC\equiv CH$	$C_{sp}\equiv C_{sp}$	835	200	120
	C_{sp}—H_{1s}	552	132	106

Problem 1.13   Draw a line-bond structure for propyne, $CH_3C\equiv CH$. Indicate the hybridization of each carbon, and predict a value for each bond angle.

1.11 Hybridization of Other Atoms: Nitrogen and Oxygen

The concept of hybridization described in the previous four sections is not restricted to carbon compounds. Covalent bonds formed by other elements in the periodic table also can be described using hybrid orbitals. Look at the nitrogen atom in ammonia, NH_3, for example. A nitrogen atom has five outer-shell electrons and therefore forms three covalent bonds to complete its valence electron octet.

$$\cdot \ddot{N}\cdot \; + \; 3\,H\cdot \; \longrightarrow \; H\!:\!\overset{\cdot\cdot}{\underset{\overset{\cdot\cdot}{H}}{N}}\!:\!H \quad \text{or} \quad H\!-\!\overset{\cdot\cdot}{\underset{\overset{|}{H}}{N}}\!-\!H$$

The experimentally measured H–N–H bond angle in ammonia is 107.3°, close to the tetrahedral value of 109.5° found in methane. We therefore assume that nitrogen hybridizes to form four sp^3 orbitals, exactly as carbon does. One of the four sp^3 orbitals is occupied by two nonbonding electrons, and the other three hybrid orbitals have one electron each. Sigma overlap of these three half-filled nitrogen sp^3 hybrid orbitals with hydrogen $1s$ orbitals completes the ammonia molecule (Figure 1.20). The N–H bond length is 100.8 pm, and the bond strength is 449 kJ/mol (107 kcal/mol). Note that the unshared lone pair of electrons in the fourth sp^3 hybrid orbital occupies as much or more space as an N–H bond does and is very important to the chemistry of ammonia.

FIGURE 1.20 ▼

Hybridization of nitrogen in ammonia. The nitrogen atom is sp^3-hybridized, resulting in H–N–H bond angles of 107.3°.

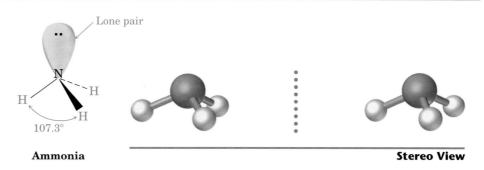

Ammonia   **Stereo View**

Like the carbon atom in methane and the nitrogen atom in ammonia, the oxygen atom in water is also sp^3-hybridized. Because an oxygen atom has six valence-shell electrons, however, it forms only two covalent bonds and has two lone pairs (Figure 1.21). The H–O–H bond angle in water is 104.5°, somewhat less than the 109.5° tetrahedral angle expected for sp^3 hybridization. This diminished bond angle is probably due to a repulsive interaction between the two lone pairs, which forces them apart, thereby compressing the H–O–H angle. The O–H bond length is 95.8 pm, and the bond strength is 498 kJ/mol (119 kcal/mol).

FIGURE 1.21 ▼

The structure of water. The oxygen atom is sp^3-hybridized and has two lone pairs of electrons. The H–O–H bond angle is 104.5°.

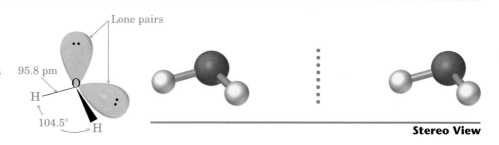

Stereo View

Problem 1.14 Draw Lewis and line-bond structures for formaldimine, CH_2NH. How many electrons are shared in the carbon–nitrogen bond? What is the hybridization of the nitrogen atom?

Problem 1.15 What geometry do you expect for each of the following atoms?

(a) The oxygen atom in methanol, $H_3C — \overset{..}{\underset{..}{O}} — H$

(b) The nitrogen atom in trimethylamine, $H_3C — \overset{..}{N} — CH_3$
    $\hphantom{(b) The nitrogen atom in trimethylamine, H_3C — N —}|$
    $\hphantom{(b) The nitrogen atom in trimethylamine, H_3C — N —}CH_3$

(c) The phosphorus atom in $:PH_3$

Chemical Toxicity and Risk

We hear and read a lot these days about the dangers of "chemicals"—about pesticide residues, toxic wastes, unsafe medicines, and so forth. What's a person to believe?

Life is not risk-free; we all take many risks each day. We decide to ride a bike rather than drive, even though there is a ten times greater likelihood per mile of dying in a bicycling accident than in a car. Some people may decide to smoke cigarettes, even though it increases their chance of getting cancer by 50%. Making judgments that affect our health is something we do every day without thinking about it.

But what about risks from chemicals? Risk evaluation is carried out by exposing test animals (usually rats) to a chemical and then monitoring them for signs of harm. To limit the expense and time needed, the amounts administered are hundreds or thousands of times greater than those a person might normally encounter. Once the animal data are available, the interpretation of those data involves many assumptions. If a substance is harmful to animals, is it necessarily harmful to humans? How can a large dose for a small animal be translated into a small dose for a large human? As pointed out by the sixteenth century Swiss physician Paracelsus, "The dose makes the poison." All substances, including water and table salt, are toxic to some organisms to some extent, and the difference between help and harm is a matter of degree.

The standard method for evaluating acute chemical toxicity, as opposed to long-term toxicity, is to report an *LD$_{50}$ value,* the amount of a substance per kilogram body weight that is lethal to 50% of the test animals. The LD$_{50}$ values of various substances are shown in Table 1.4. The lower the value, the more toxic the substance.

TABLE 1.4 Some LD$_{50}$ Values			
Substance	**LD$_{50}$ (g/kg)**	**Substance**	**LD$_{50}$ (g/kg)**
Aflatoxin B$_1$	4×10^{-4}	Formaldehyde	2.4
Aspirin	1.7	Sodium cyanide	1.5×10^{-2}
Chloroform	3.2	Sodium cyclamate	17
Ethyl alcohol	10.6		

How we respond to risk is strongly influenced by familiarity. The presence of chloroform in municipal water supplies—at a barely detectable

(continued) ▶

level of 0.000 000 01%—has caused an outcry in many cities, yet chloroform has a lower acute toxicity than aspirin. Many foods contain natural ingredients that are far more toxic than synthetic food additives or pesticide residues, but the ingredients are ignored because the foods are familiar. Peanut butter, for example, may contain tiny amounts of aflatoxin, a far more potent cancer threat than sodium cyclamate, an artificial sweetener that has been banned in the United States because of its "risk."

All decisions involve tradeoffs. Does the benefit of a pesticide that will increase the availability of food outweigh the health risk to 1 person in 1 million who are exposed? Do the beneficial effects of a new drug outweigh a potentially dangerous side effect in a small number of users? The answers aren't always obvious, but it's the responsibility of legislators and well-informed citizens to keep their responses on a factual level rather than an emotional one.

We all take many risks each day, some much more dangerous than others.

Summary and Key Words

Organic chemistry is the study of carbon compounds. Although a division into organic and inorganic chemistry occurred historically, there is no scientific reason for the division.

An atom consists of a positively charged nucleus surrounded by one or more negatively charged electrons. The electronic structure of an atom can be described by a quantum mechanical wave equation, in which electrons are considered to occupy **orbitals** around the nucleus. Different orbitals have different energy levels and different shapes. For example, *s* orbitals are spherical and *p* orbitals are dumbbell-shaped. The **electron configuration** of an atom can be found by assigning electrons to the proper orbitals, beginning with the lowest-energy ones.

Covalent bonds are formed when an electron pair is shared between atoms. According to **valence bond theory**, electron sharing occurs by overlap of two atomic orbitals. According to **molecular orbital (MO) theory**, bonds result from the combination of atomic orbitals to give molecular orbitals, which belong to the entire molecule. Bonds that have a circular cross-section and are formed by head-on interaction are called **sigma (σ) bonds**; bonds formed by sideways interaction of *p* orbitals are called **pi (π) bonds**.

Carbon uses hybrid orbitals to form bonds in organic molecules. When forming only single bonds with tetrahedral geometry, carbon has four equivalent **sp^3 hybrid orbitals**. When forming a double bond with planar geometry, carbon has three equivalent **sp^2 hybrid orbitals** and one unhybridized *p* orbital. A carbon–carbon double bond results when two sp^2-hybridized carbon atoms bond together. When forming a triple bond with linear geometry, carbon has two equivalent **sp hybrid orbitals** and two unhybridized *p* orbitals. A carbon–carbon triple bond results when two sp-hybridized carbon atoms bond together.

nonbonding electrons,
  10
orbital, 4
organic chemistry, 2
pi (π) bond, 19
shell (electron), 5
sigma (σ) bond, 13
sp hybrid orbital, 22
sp^2 hybrid orbital, 19
sp^3 hybrid orbital, 16
valence bond theory,
  12
valence shell, 9

Other atoms such as nitrogen and oxygen also hybridize to form strong, oriented bonds. The nitrogen atom in ammonia and the oxygen atom in water are sp^3-hybridized.

Working Problems

There is no surer way to learn organic chemistry than by working problems. Although careful reading and rereading of this text is important, reading alone isn't enough. You must also be able to use the information you've read and be able to apply your knowledge in new situations. Working problems gives you practice at doing this.

Each chapter in this book provides many problems of different sorts. The in-chapter problems are placed for immediate reinforcement of ideas just learned; the end-of-chapter problems provide additional practice and are of several types. They begin with a short section called "Visualizing Chemistry," which helps you "see" the microscopic world of molecules and provides practice for working in three dimensions. After the visualization problems are many "Additional Problems." Early problems in this section are primarily of the drill type, providing an opportunity for you to practice your command of the fundamentals. Later problems tend to be more thought-provoking, and some are real challenges. Finally, each chapter ends with a short section of problems called "A Look Ahead." These are a good test of critical thinking because they must be answered by extending information you've already learned to topics that will be covered more carefully in future chapters.

As you study organic chemistry, take the time to work the problems. Do the ones you can, and ask for help on the ones you can't. If you're stumped by a particular problem, check the accompanying *Study Guide and Solutions Manual* for an explanation that will help clarify the difficulty. Working problems takes effort, but the payoff in knowledge and understanding is immense.

Visualizing Chemistry

• •

(Problems 1.1–1.15 appear within the chapter.)

1.16 Convert each of the following molecular models into a typical line-bond structure, and give the formula of each (gray = C, red = O, blue = N, ivory = H).

(a)   (b)

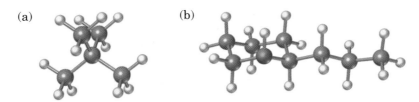

(c)

1.17 Shown below is a model of acetaminophen, a pain-reliever sold in drugstores as Tylenol. Identify the hybridization of each carbon atom in acetaminophen, and tell which atoms have lone pairs of electrons (gray = C, red = O, blue = N, ivory = H).

Stereo View

1.18 Shown below is a model of aspartame, $C_{14}H_{18}N_2O_5$, known commercially as NutraSweet. Only the connections between atoms are shown; multiple bonds are not indicated. Complete the structure by indicating the positions of multiple bonds (gray = C, red = O, blue = N, ivory = H).

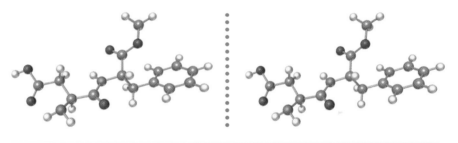

Stereo View

Additional Problems

1.19 How many valence electrons does each of the following atoms have?
(a) Magnesium   (b) Sulfur   (c) Bromine

1.20 Give the ground-state electron configuration for each of the following elements:
(a) Sodium   (b) Aluminum   (c) Silicon   (d) Calcium

1.21 What are likely formulas for the following molecules?
(a) AlCl?   (b) CF_2Cl?   (c) NI?

1.22 Write a Lewis (electron-dot) structure for acetonitrile, C_2H_3N, which contains a carbon–nitrogen triple bond. How many electrons does the nitrogen atom have in its outer shell? How many are bonding, and how many are nonbonding?

1.23 What is the hybridization of each carbon atom in acetonitrile (Problem 1.22)?

1.24 Draw both a Lewis structure and a line-bond structure for vinyl chloride, C_2H_3Cl, the starting material from which PVC [poly(vinyl chloride)] plastic is made.

1.25 Fill in any nonbonding valence electrons that are missing from the following line-bond structures:

(a) H_3C—S—CH_3 (b) 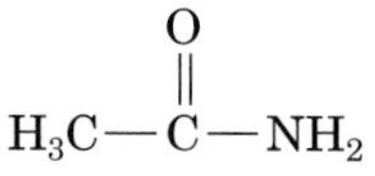 H_3C—C(=O)—NH_2 (c) H_3C—C(=O)—Cl

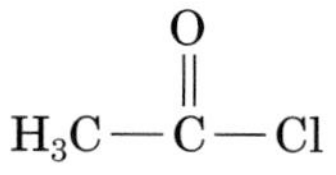

1.26 Convert the following line-bond structures into molecular formulas:

(a)

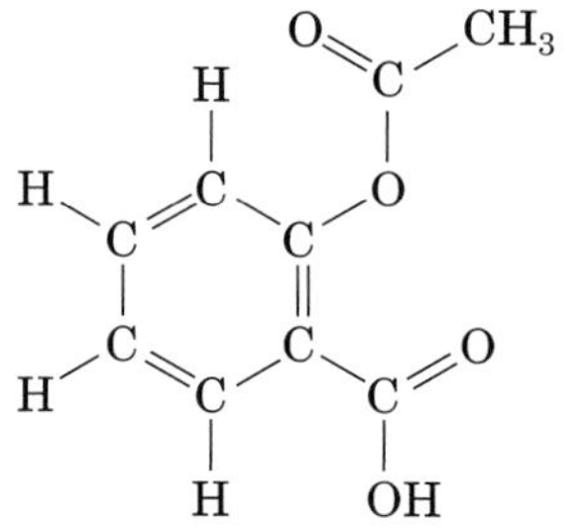

Aspirin

(b)

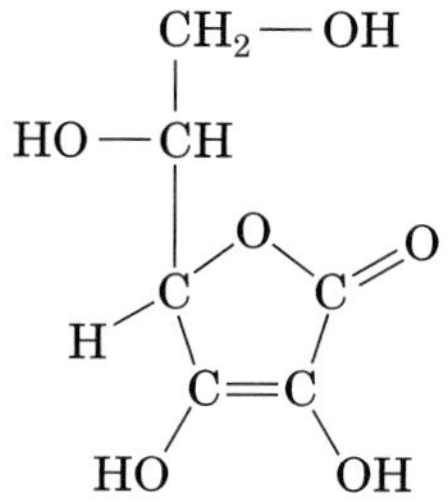

Vitamin C

(c)

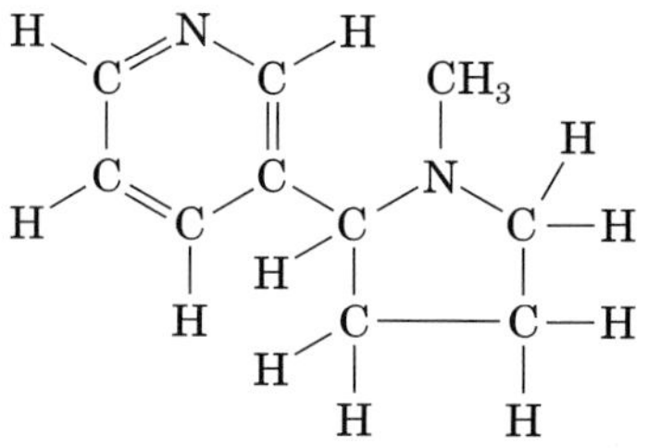

Nicotine

(d)

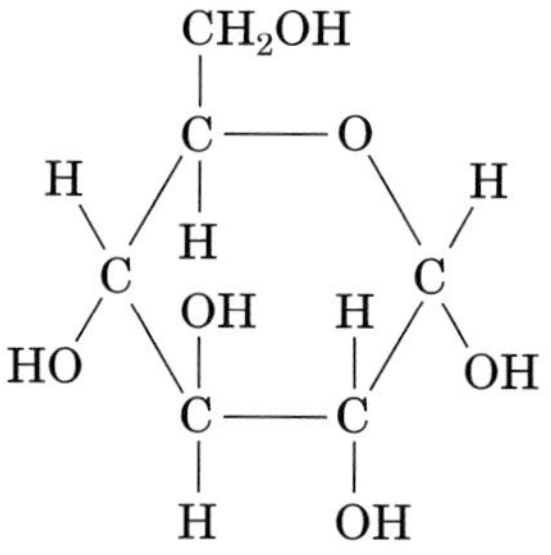

Glucose

1.27 Convert the following molecular formulas into line-bond structures that are consistent with valence rules:
(a) C_3H_8 (b) CH_5N
(c) C_2H_6O (2 possibilities)   (d) C_3H_7Br (2 possibilities)
(e) C_2H_4O (3 possibilities)   (f) C_3H_9N (4 possibilities)

1.28 What kind of hybridization do you expect for each carbon atom in the following molecules?

(a) Propane, $CH_3CH_2CH_3$ (b) 2-Methylpropene, $CH_3\overset{\overset{\textstyle CH_3}{|}}{C}{=}CH_2$

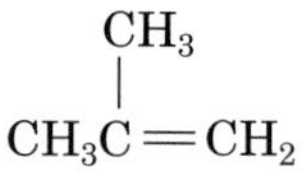

(c) 1-Buten-3-yne, H_2C=CH—C≡CH   (d) Acetic acid, CH_3—$\overset{\overset{\displaystyle O}{\|}}{C}$—OH

1.29 What is the shape of benzene, and what hybridization do you expect for each carbon?

Benzene

1.30 What bond angles do you expect for each of the following, and what kind of hybridization do you expect for the central atom in each?
(a) The C–O–C angle in CH_3–O–CH_3 (b) The C–N–C angle in CH_3–NH–CH_3
(c) The C–N–H angle in CH_3–NH–CH_3 (d) The O=C–O angle in acetic acid
(See Problem 1.28d.)

1.31 Propose structures for molecules that meet the following descriptions:
(a) Contains two sp^2-hybridized carbons and two sp^3-hybridized carbons
(b) Contains only four carbons, all of which are sp^2-hybridized
(c) Contains two sp-hybridized carbons and two sp^2-hybridized carbons

1.32 Why can't molecules with the following formulas exist?
(a) CH_5 (b) C_2H_6N (c) $C_3H_5Br_2$

1.33 Draw a three-dimensional representation of the oxygen-bearing carbon atom in ethanol, CH_3–CH_2–OH, using the standard convention of solid, wedged, and dashed lines.

1.34 Draw line-bond structures for the following molecules:
(a) Acrylonitrile, C_3H_3N, which contains a carbon–carbon double bond and a carbon–nitrogen triple bond
(b) Ethyl methyl ether, C_3H_8O, which contains an oxygen atom bonded to two carbons
(c) Butane, C_4H_{10}, which contains a chain of four carbon atoms
(d) Cyclohexene, C_6H_{10}, which contains a ring of six carbon atoms and one carbon–carbon double bond

1.35 Sodium methoxide, $NaOCH_3$, contains both covalent and ionic bonds. Which do you think is which?

1.36 What kind of hybridization do you expect for each carbon atom in the following molecules?

(a)

Procaine

(b)

Vitamin C

1.37   What bond angles do you expect for the following?
(a) The C–N–H angle in aniline   (b) The C=N–C angle in pyridine

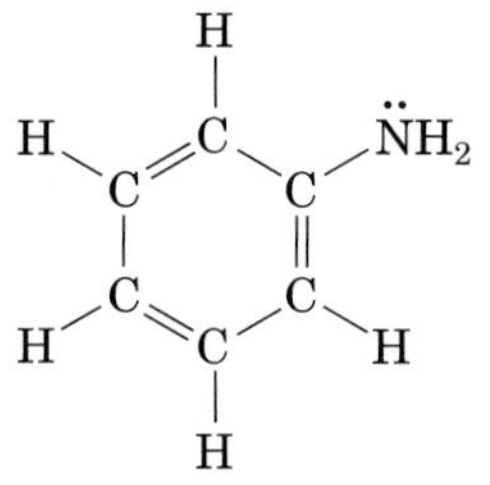

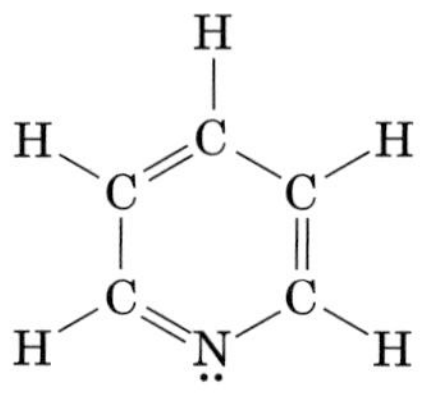

(c) The C–P–C angle in trimethylphosphine, $P(CH_3)_3$

1.38   Identify the bonds in the following compounds as either ionic or covalent:
(a) NaCl   (b) CH_3Cl (c) Cl_2 (d) HOCl

1.39   Why do you suppose no one has ever been able to make cyclopentyne as a stable molecule?

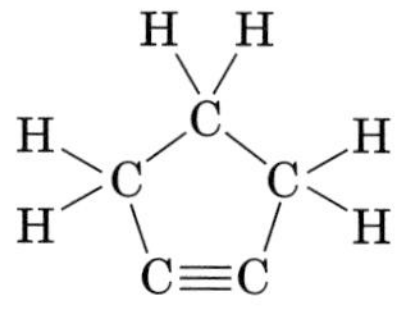

Cyclopentyne

1.40   What is wrong with the following sentence? "The π bonding molecular orbital in ethylene results from sideways overlap of two p atomic orbitals."

1.41   Allene, $H_2C=C=CH_2$, is somewhat unusual in that it has two adjacent double bonds. Draw a picture showing the orbitals involved in the σ and π bonds of allene. Is the central carbon atom sp^2- or sp-hybridized? What about the hybridization of the terminal carbons? What shape do you predict for allene?

1.42   Allene (see Problem 1.41) is related structurally to carbon dioxide, CO_2. Draw a picture showing the orbitals involved in the σ and π bonds of CO_2, and identify the hybridization of carbon.

1.43   Complete the Lewis electron-dot structure of caffeine, showing all lone-pair electrons, and identify the hybridization of the indicated atoms.

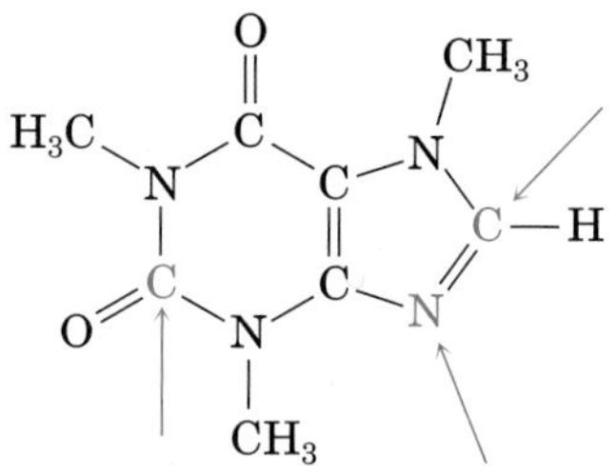

Caffeine

1.44 Although almost all stable organic species have tetravalent carbon atoms, species with trivalent carbon atoms also exist. *Carbocations* are one such class of compounds.

$$H-\overset{\displaystyle H}{\underset{\displaystyle H}{\overset{|}{C^+}}} \quad \textbf{A carbocation}$$

(a) How many valence electrons does the positively charged carbon atom have?
(b) What hybridization do you expect this carbon atom to have?
(c) What geometry is the carbocation likely to have?

1.45 A *carbanion* is a species that contains a negatively charged, trivalent carbon.

$$H-\overset{\displaystyle H}{\underset{\displaystyle H}{\overset{|}{C:^-}}} \quad \textbf{A carbanion}$$

(a) What is the relationship between a carbanion and a trivalent nitrogen compound such as NH_3?
(b) How many valence electrons does the negatively charged carbon atom have?
(c) What hybridization do you expect this carbon atom to have?
(d) What geometry is the carbanion likely to have?

1.46 Divalent carbon species called *carbenes* are capable of fleeting existence. For example, methylene, $:CH_2$, is the simplest carbene. The two unshared electrons in methylene can be either spin-paired in a single orbital or unpaired in different orbitals. Predict the type of hybridization you expect carbon to adopt in singlet (spin-paired) methylene and triplet (spin-unpaired) methylene. Draw a picture of each, and identify the types of carbon orbitals present.

A Look Ahead

• •

1.47 There are two different substances with the formula C_4H_{10}. Draw both, and tell how they differ. (See Section 3.2.)

1.48 There are two different substances with the formula C_3H_6. Draw both, and tell how they differ. (See Section 3.2.)

1.49 There are two different substances with the formula C_2H_6O. Draw both, and tell how they differ. (See Section 3.2.)

1.50 There are three different substances that contain a carbon–carbon double bond and have the formula C_4H_8. Draw them, and tell how they differ. (See Section 6.5.)

Molecular Modeling

1.51   Use SpartanView to compare "low," "medium," and "high" electron-density surfaces of dimethyl ether (CH_3OCH_3). Where is the electron density highest: near the atomic nuclei, in the bonding region between the nuclei, or far from the nuclei? Which electron-density surface most closely resembles the molecule's space-filling model?

1.52   Use SpartanView to compare bond-density surfaces of methanol (CH_3OH) and sodium methoxide (CH_3ONa). Which bond has higher electron density between the nuclei, O–H or O–Na? Repeat for acetic acid (CH_3CO_2H) and sodium acetate (CH_3CO_2Na), and for hydrogen chloride (HCl) and sodium chloride (NaCl). How do your results relate to the covalent versus ionic nature of the bonds?

1.53   Use SpartanView to display the two highest-energy occupied molecular orbitals of acetylene. How do they differ?

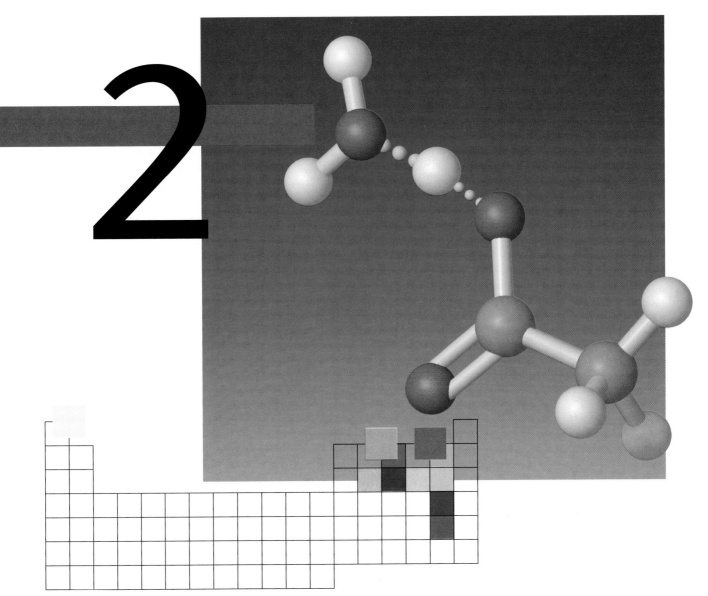

Polar Bonds and
Their Consequences

We saw in the last chapter how covalent bonds between atoms are described, and we looked at the hybrid-orbital model used to depict most organic molecules. Before going on to a systematic study of complex organic substances, however, we still need to review a few fundamental topics. In particular, we need to look more closely at how electrons are distributed in covalent bonds and at some of the consequences that arise when the bonding electrons are not shared equally between atoms.

2.1 Polar Covalent Bonds and Electronegativity

Up to this point, we've treated chemical bonds as though they were either ionic or covalent. In fact, though, chemical bonding is a continuum of possibilities, with ionic and covalent bonds at the two extremes (Figure 2.1).

FIGURE 2.1 ▼

The continuum in bonding from covalent to ionic as a result of unsymmetrical electron distribution. The symbol δ (lowercase Greek delta) means *partial* charge, either partial positive (δ+) for the electron-poor atom or partial negative (δ−) for the electron-rich atom.

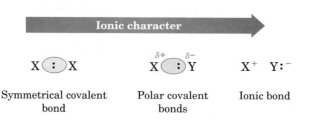

The bond in sodium chloride, for instance, is largely ionic. Sodium has transferred an electron to chlorine to give Na^+ and Cl^- ions, which are held together in the solid by electrostatic attraction. The C–C bond in ethane, however, is fully covalent. The two bonding electrons are shared equally by the two equivalent carbon atoms, resulting in a symmetrical electron distribution in the bond. Between these two extremes lie the great majority of chemical bonds, in which the electrons are attracted *somewhat* more strongly by one atom than by the other. We call such bonds, in which the electron distribution is unsymmetrical, **polar covalent bonds**.

Bond polarity is due to differences in **electronegativity (EN)**, the intrinsic ability of an atom to attract the shared electrons in a covalent bond. Metals on the left side of the periodic table attract electrons weakly, whereas the halogens and other reactive nonmetals on the right side of the periodic table attract electrons strongly. As shown in Figure 2.2, electronegativities are based on an arbitrary scale, with fluorine the most electronegative (EN = 4.0) and cesium the least (EN = 0.7).

Carbon, the most important element for our purposes, has an electronegativity value of 2.5. Any element more electronegative than carbon has a value greater than 2.5, and any element less electronegative than carbon has a value less than 2.5.

As a general rule, bonds between atoms with similar electronegativities are nonpolar covalent, bonds between atoms whose electronegativities differ by 0.3–2.0 units are polar covalent, and bonds between atoms whose electronegativities differ by more than 2 units are largely ionic. Carbon–hydrogen bonds, for example, are relatively nonpolar because carbon and hydrogen have similar electronegativities. Bonds between carbon and more

FIGURE 2.2 ▼

Electronegativity values and trends. Electronegativity generally increases from left to right across the periodic table and decreases from top to bottom, as indicated by the heights of the various columns. The values are on an arbitrary scale, with F = 4.0 and Cs = 0.7. Carbon has an electronegativity value of 2.5. Elements in violet are the most electronegative, those in green are medium, and those in yellow are the least electronegative.

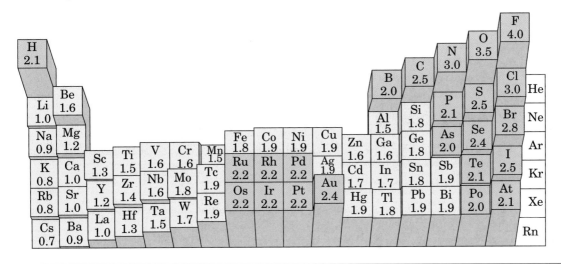

electronegative elements such as oxygen, fluorine, and chlorine, by contrast, are polarized so that the bonding electrons are drawn away from carbon toward the electronegative atom. This leaves carbon with a partial positive charge, denoted by $\delta+$, and the electronegative atom with a partial negative charge, $\delta-$ (δ is the lowercase Greek letter delta). For example, the C–Cl bond in chloromethane is polar covalent:

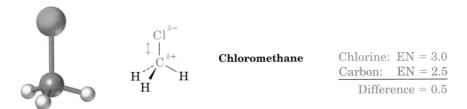

Chloromethane

Chlorine: EN = 3.0
Carbon: EN = 2.5
Difference = 0.5

A crossed arrow ↦ is often used to indicate the direction of bond polarity. By convention, *electrons are displaced in the direction of the arrow*. The tail of the arrow (which looks like a plus sign) is electron-poor ($\delta+$), and the head of the arrow is electron-rich ($\delta-$).

Bonds between carbon and less electronegative elements are polarized so that carbon bears a partial negative charge and the other atom bears a partial positive charge. So-called *organometallic* compounds, such as methylmagnesium bromide (a valuable substance whose use we'll explore in later chapters), are good examples.

$$\overset{\delta+}{\text{MgBr}}$$
$$\updownarrow \quad |$$
$$\overset{\delta-}{\text{C}}$$
$$H \diagup \diagdown H$$
$$H$$

Methylmagnesium bromide

Carbon: EN = 2.5
Magnesium: EN = 1.2
Difference = 1.3

When speaking of an atom's ability to polarize a bond, we often use the term *inductive effect*. An **inductive effect** is the shifting of electrons in a σ bond in response to the electronegativity of nearby atoms. Metals, such as lithium and magnesium, inductively donate electrons, whereas electronegative nonmetals, such as oxygen and chlorine, inductively withdraw electrons. Inductive effects play a major role in understanding chemical reactivity, and we'll use them many times throughout this text to explain a variety of chemical phenomena.

Problem 2.1   Without looking at Figure 2.2, tell which element in each of the following pairs is more electronegative:
(a) Li or H   (b) B or Br   (c) Cl or I   (d) C or H

Problem 2.2   Use the $\delta+/\delta-$ convention to indicate the direction of expected polarity for each of the bonds indicated.
(a) $H_3C–Br$ (b) $H_3C–NH_2$ (c) $H_3C–Li$ (d) $H_2N–H$
(e) $H_3C–OH$ (f) $H_3C–MgBr$ (g) $H_3C–F$

Problem 2.3   Use the electronegativity values shown in Figure 2.2 to rank the following bonds from least polar to most polar: $H_3C–Li$, $H_3C–K$, $H_3C–F$, $H_3C–MgBr$, $H_3C–OH$

2.2 Polar Covalent Bonds and Dipole Moment

Because individual bonds are often polar, molecules as a whole are often polar also. Overall molecular polarity results from the summation of all individual bond polarities and lone-pair contributions in the molecule. The measure of this net molecular polarity is a quantity called the *dipole moment*. As a practical matter, strongly polar substances are often soluble in polar solvents like water, whereas nonpolar substances are insoluble in water.

Dipole moments can be thought of in the following way: Assume that there is a center of mass of all positive charges (nuclei) in a molecule and a center of mass of all negative charges (electrons) in the molecule. If these two centers don't coincide, then the molecule has a net polarity. The **dipole moment, μ** (Greek mu), is defined as the magnitude of the charge Q at either end of the molecular dipole times the distance r between the charges,

$\mu = Q \times r$. Dipole moments are expressed in *debyes* (D), where $1 D = 3.336 \times 10^{-30}$ coulomb meter (C · m) in SI units. For example, the unit charge on an electron is 1.60×10^{-19} C. Thus, if one positive charge and one negative charge were separated by 100 pm (a bit less than the length of an average covalent bond), the dipole moment would be 1.60×10^{-29} C · m, or 4.80 D.

$$\mu = Q \times r$$

$$\mu = (1.60 \times 10^{-19} \text{ C})(100 \times 10^{-12} \text{ m})\left(\frac{1 \text{ D}}{3.336 \times 10^{-30} \text{ C} \cdot \text{m}}\right) = 4.80 \text{ D}$$

It's relatively easy to measure dipole moments, and values for some common substances are given in Table 2.1. Once the dipole moment is known, it's then possible to calculate the amount of charge separation in a molecule. In chloromethane, for example, the measured dipole moment is $\mu = 1.87$ D. If we assume that the contributions of the nonpolar C–H bonds are small, then most of the chloromethane dipole moment is due to the C–Cl bond. Since the C–Cl bond length is 178 pm, the dipole moment of chloromethane would be 1.78×4.8 D $= 8.5$ D if a full negative charge on chlorine were separated from a full positive charge on carbon by a distance of 178 pm (that is, if the C–Cl bond were ionic, $C^+ \text{ Cl}^-$). But because the actual dipole moment of chloromethane is only 1.87 D, the C–Cl bond is only about $(\frac{1.87}{8.54})(100\%) = 22\%$ ionic. Thus, the chlorine atom in chloromethane has an excess of about 0.2 electron, and the carbon atom has a deficiency of about 0.2 electron (Figure 2.3, p. 40).

TABLE 2.1 Dipole Moments of Some Compounds

Compound	Dipole moment (D)	Compound	Dipole moment (D)
NaCl	9.0	NH_3	1.47
		CH_4	0
$H_3C-N\overset{+}{\underset{O^-}{\overset{O}{\diagdown}}}$ Nitromethane	3.46	CCl_4	0
		CH_3CH_3	0
CH_3Cl	1.87	Benzene	0
H_2O	1.85		
CH_3OH	1.70	BF_3	0
$H_2C=\overset{+}{N}=N^-$ Diazomethane	1.50		

FIGURE 2.3 ▼

Chloromethane contains a polar C–Cl bond. In the electrostatic potential map on the right, the charge distribution is shown using colors ranging from red (negative) to blue (positive). Thus, the chlorine atom is electron-rich, and the carbon and hydrogen atoms are electron-poor.

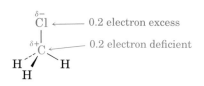

Chloromethane ($\mu = 1.87$ D)

chloromethane

Note that Figure 2.3 shows a representation of chloromethane with what is called an *electrostatic potential map* (see Preface), which uses color to indicate the calculated charge distribution in the molecule. Chlorine carries a partial negative charge (red), and the carbon and hydrogen atoms carry partial positive charges (blue). We'll make use of these maps in numerous places throughout the text to draw correlations between electronic structure and chemical reactivity.

Water and ammonia have relatively large dipole moments (Table 2.1), both because oxygen and nitrogen are more electronegative than hydrogen and because they have lone-pair electrons. The lone-pair electrons on the oxygen atom of water and the nitrogen atom of ammonia stick out into space away from the positively charged nuclei, which gives rise to a considerable charge separation and large contribution to the dipole moment.

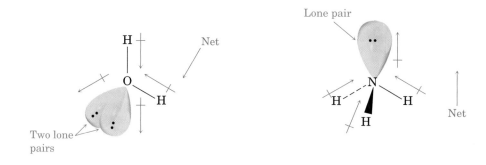

Water, H₂O ($\mu = 1.85$ D)   Ammonia, NH₃ ($\mu = 1.47$ D)

By contrast with water and ammonia, methane, tetrachloromethane, and ethane have zero dipole moments. Because of the symmetrical structures of these molecules, the individual bond polarities exactly cancel.

Methane
($\mu = 0$ D)

Tetrachloromethane
($\mu = 0$ D)

Ethane
($\mu = 0$ D)

Practice Problem 2.1 Make a three-dimensional drawing of methylamine, CH_3NH_2, a substance responsible for the odor of rotting fish, and predict whether it has a dipole moment. If you expect a dipole moment, show its direction.

Strategy Look for any lone-pair electrons, and identify any atom with an electronegativity substantially different from that of carbon. (Usually, this means O, N, F, Cl, or Br.) Electron density will be displaced in the general direction of the electronegative atoms and the lone pairs.

Solution Methylamine contains an electronegative nitrogen atom with two lone-pair electrons. The dipole moment thus points generally from $-CH_3$ toward $-NH_2$.

Problem 2.4 Account for the observed dipole moment of methanol (CH_3OH, 1.70 D) by using a crossed arrow to indicate the direction in which electron density is displaced.

Problem 2.5 Carbon dioxide, CO_2, has zero dipole moment even though carbon–oxygen bonds are strongly polarized. Explain.

Problem 2.6 Make three-dimensional drawings of the following molecules, and predict whether each has a dipole moment. If you expect a dipole moment, show its direction.
(a) $H_2C=CH_2$ (b) $CHCl_3$ (c) CH_2Cl_2 (d) $H_2C=CCl_2$

2.3 Formal Charges

Closely related to the ideas of bond polarity and dipole moment is the occasional need to assign *formal charges* to specific atoms within a molecule. This is particularly common for atoms that have an apparently "abnormal" number of bonds. In nitromethane (CH_3NO_2), for example, the nitrogen atom has four bonds rather than the usual three and has a formal positive charge.

The singly bonded oxygen atom, by contrast, has one bond rather than the usual two and has a formal negative charge. Note that an electrostatic potential map of nitromethane shows the oxygens as negative (red) and the nitrogen as relatively positive (blue-green).

nitromethane

$$H - \overset{\overset{\displaystyle H}{|}}{\underset{\underset{\displaystyle H}{|}}{C}} - \overset{:\ddot{O}:}{\underset{:\ddot{O}:^-}{N^+}}$$

Formal positive charge
Formal negative charge

Nitromethane

Formal charges result from a kind of electron "bookkeeping" and can be thought of in the following way: A typical covalent bond is formed when each atom donates one electron. Although the bonding electrons are shared by both atoms, each atom can still be considered to "own" one electron for bookkeeping purposes. In methane, for example, the carbon atom owns one electron in each of the four bonds, for a total of four. Since a neutral, isolated carbon atom has four valence electrons, and since the carbon atom in methane still owns four, the methane carbon atom is neutral and has no formal charge.

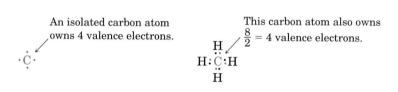

An isolated carbon atom owns 4 valence electrons.

This carbon atom also owns $\frac{8}{2} = 4$ valence electrons.

The same is true for the nitrogen atom in ammonia, which has three covalent N–H bonds and one lone pair. Atomic nitrogen has five valence electrons, and the ammonia nitrogen also has five—one in each of three shared N–H bonds plus two in the lone pair. Thus, the nitrogen atom in ammonia is neutral and has no formal charge.

An isolated nitrogen atom owns 5 valence electrons.

This nitrogen atom also owns $\frac{6}{2} + 2 = 5$ valence electrons.

$$H:\ddot{N}:H$$
$$H$$

The situation is different in nitromethane. Atomic nitrogen has five valence electrons, but the nitromethane nitrogen owns only *four*—one in the C–N bond, one in the N–O single bond, and two in the N=O double bond. Thus, the nitrogen has formally lost an electron and therefore has a positive charge. A similar calculation for the singly bonded oxygen atom shows that it has formally gained an electron and has a negative charge. (Atomic oxygen has six valence electrons, but the singly bonded oxygen in nitromethane has seven—one in the O–N bond and two in each of three lone pairs.)

To express the calculations in a general way, the **formal charge** on an atom is equal to the number of valence electrons in a neutral, isolated atom minus the number of electrons owned by that atom in a molecule:

$$\textbf{Formal charge} = \left(\begin{array}{c}\text{Number of}\\\text{valence electrons}\\\text{in free atom}\end{array}\right) - \left(\begin{array}{c}\text{Number of}\\\text{valence electrons}\\\text{in bound atom}\end{array}\right)$$

$$= \left(\begin{array}{c}\text{Number of}\\\text{valence}\\\text{electrons}\end{array}\right) - \left(\begin{array}{c}\text{Half of}\\\text{bonding}\\\text{electrons}\end{array}\right) - \left(\begin{array}{c}\text{Number of}\\\text{nonbonding}\\\text{electrons}\end{array}\right)$$

For the nitromethane nitrogen:

$$CH_3NO_2 = H\!:\!\overset{\displaystyle H}{\underset{\displaystyle H}{\overset{..}{\underset{..}{C}}}}\!:\!N\overset{\displaystyle :\ddot{O}.}{\underset{\displaystyle :\ddot{O}:}{}}$$

Nitrogen valence electrons   = 5
Nitrogen bonding electrons   = 8
Nitrogen nonbonding electrons = 0

$$\text{Formal charge} = 5 - \tfrac{8}{2} - 0 \quad = +1$$

For the singly bonded nitromethane oxygen:

Oxygen valence electrons   = 6
Oxygen bonding electrons   = 2
Oxygen nonbonding electrons = 6

$$\text{Formal charge} = 6 - \tfrac{2}{2} - 6 \quad = -1$$

A summary of commonly encountered formal charges and the bonding situations in which they occur is given in Table 2.2.

TABLE 2.2 A Summary of Formal Charges on Atoms

Atom	C			N			O		
Structure	$-\overset{+}{C}-$	$-\overset{\|}{\underset{\|}{C}}-$	$-\overset{\|}{\underset{\|}{\ddot{C}}}{}^{-}$	$-\overset{\|}{\underset{\|}{\overset{+}{N}}}-$	$-\overset{..}{\ddot{N}}-$	$-\underset{..}{\overset{..}{\ddot{N}}}{}^{-}$	$-\overset{\|}{\overset{+}{\ddot{O}}}-$	$-\underset{..}{\ddot{O}}-$	$-\underset{..}{\overset{..}{\ddot{O}}}{:}^{-}$
Number of bonds	3	4	3	4	3	2	3	2	1
Lone pairs	0	0	1	0	1	2	1	2	3
Formal charge	+1	0	−1	+1	0	−1	+1	0	−1

Molecules such as nitromethane, which are neutral overall but have plus and minus charges on individual atoms, are said to be **dipolar**. Dipolar character in molecules often has important consequences for chemical reactivity, and it's helpful to be able to identify and calculate the charges correctly.

● ●

Problem 2.7   Dimethyl sulfoxide, a common solvent, has the structure indicated. Show why dimethyl sulfoxide must have formal charges on S and O.

$$:\overset{..}{\underset{..}{O}}:^-$$

$$\underset{H_3C}{\quad}\overset{|}{\underset{..}{\overset{+}{S}}}\underset{\quad CH_3}{}$$

Dimethyl sulfoxide

Problem 2.8   Calculate formal charges for the atoms in the following molecules:

(a) Diazomethane, $H_2C = N = \overset{..}{N}:$ (b) Acetonitrile oxide, $H_3C - C \equiv N - \overset{..}{\underset{..}{O}}:$

(c) Methyl isocyanide, $H_3C - N \equiv C:$

● ●

2.4 Resonance

Most substances can be represented without difficulty by the Lewis structures or Kekulé line-bond structures we've been using up to this point, but an interesting problem sometimes arises. For instance, look again at nitromethane, whose structure we discussed in the previous section. When we draw a Lewis structure for nitromethane, we need to show a double bond to one oxygen and a single bond to the other. But which oxygen is which? Should we draw a double bond to the "top" oxygen and a single bond to the "bottom" oxygen, or vice versa?

Double bond to this oxygen?

$$\underset{\textbf{Nitromethane}}{H - \overset{\overset{\textstyle H}{|}}{\underset{\underset{\textstyle H}{|}}{C}} - \overset{\overset{\textstyle :O:}{\parallel}}{\underset{\underset{\textstyle :O:^-}{}}{N^+}} \quad\longleftrightarrow\quad H - \overset{\overset{\textstyle H}{|}}{\underset{\underset{\textstyle H}{|}}{C}} - \overset{\overset{\textstyle :\overset{..}{O}:^-}{|}}{\underset{\underset{\textstyle :O:}{}}{N^+}}}$$

Or to this oxygen?

Although the two oxygen atoms in nitromethane appear different in Lewis structures, experiments show that they are equivalent. Both nitrogen–oxygen bonds, for example, are 122 pm in length, midway between the length of a typical N–O single bond (130 pm) and a typical N=O double bond (116 pm). In other words, *neither* of the two Lewis structures for nitromethane is correct by itself; the true structure is intermediate between the two.

The two individual Lewis structures for nitromethane are called **resonance forms**, and their relationship is indicated by the double-headed arrow between them. *The only difference between resonance forms is in the*

*placement of their π and nonbonding valence electrons. The atoms them-
selves occupy exactly the same place in both resonance forms, and the connections between atoms are the same.*

The best way to think about resonance forms is to realize that a molecule like nitromethane is no different from any other. Nitromethane doesn't
jump back and forth between two resonance forms, spending part of its time
looking like one and the rest of its time looking like the other. Rather,
nitromethane has a single unchanging structure that is a **resonance
hybrid** of the two individual forms and has characteristics of both. The only
"problem" with nitromethane is that we can't draw it accurately using a
familiar Kekulé line-bond structure. Line-bond structures just don't work
well for resonance hybrids. The difficulty, however, lies with the *representation* of nitromethane on paper, not with nitromethane itself.

Resonance is an extremely useful concept, which we'll return to on
numerous occasions throughout the rest of this book. We'll see in Chapter
15, for example, that the six carbon–carbon bonds in so-called *aromatic* compounds such as benzene are equivalent and that benzene is best represented
as a hybrid of two resonance forms. Although each individual resonance form
seems to imply that benzene has alternating single and double bonds, neither form is correct by itself. The true benzene structure is a hybrid of the
two individual forms, and all six carbon–carbon bonds are equivalent. The
bond-density surface shown on the left below shows that electrons are distributed symmetrically around the molecule.

benzene

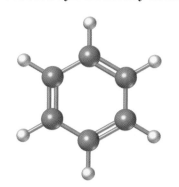

Benzene (two resonance forms)

Similarly, we'll see in Chapter 20 that resonance can account for the equivalency of the two oxygen atoms in carboxylic acid anions such as acetate ion,
CH_3COO^-. Both carbon–oxygen bonds are 127 pm in length, midway between
typical C–O single bonds (135 pm) and C=O double bonds (120 pm). An electrostatic potential map indicates this equivalency of the oxygen atoms by showing that they share the negative charge and have equal electron density (red).

acetate ion

Acetate ion (two resonance forms)

2.5 Rules for Resonance Forms

When first dealing with resonance theory, it's useful to have a set of guidelines that describe how to draw and interpret resonance forms. The following rules should prove helpful:

RULE 1 **Individual resonance forms are imaginary, not real.** The real structure is a composite, or resonance hybrid, of the different forms. Species such as nitromethane, benzene, and the acetate ion are no different from any other. They have single, unchanging structures, and they do not switch back and forth between resonance forms. The only difference between these and other substances is in the way they must be represented on paper.

RULE 2 **Resonance forms differ only in the placement of their π or nonbonding electrons.** Neither the position nor the hybridization of any atom changes from one resonance form to another. In nitromethane, for example, the nitrogen atom is sp^2-hybridized and the oxygen atoms remain in exactly the same place in both resonance forms. Only the positions of the π electrons in the N=O double bond and the lone-pair electrons on oxygen differ from one form to another. This movement of electrons on going from one resonance structure to another is sometimes indicated by using curved arrows. *A curved arrow always indicates the movement of electrons, not the movement of atoms.* An arrow shows that a pair of electrons moves *from* the atom or bond at the tail of the arrow *to* the atom or bond at the head of the arrow.

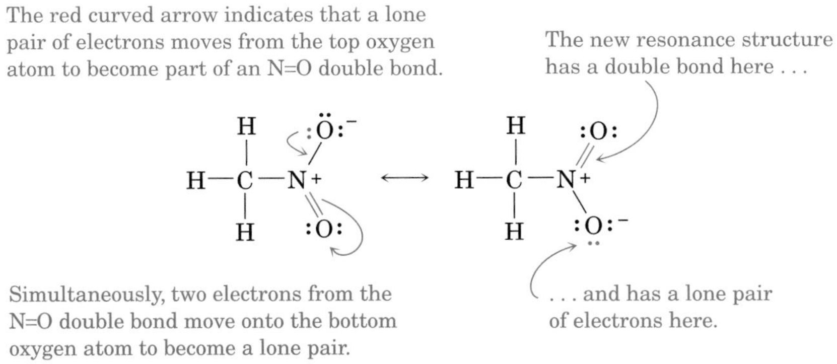

The red curved arrow indicates that a lone pair of electrons moves from the top oxygen atom to become part of an N=O double bond.

The new resonance structure has a double bond here . . .

Simultaneously, two electrons from the N=O double bond move onto the bottom oxygen atom to become a lone pair.

. . . and has a lone pair of electrons here.

The situation with benzene is similar to that with nitromethane: The π electrons in the double bonds move, as shown with curved arrows, but the carbon and hydrogen atoms remain in place.

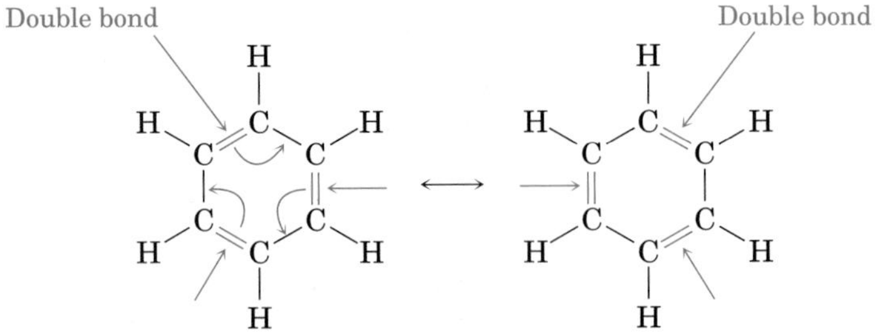

Double bond

Double bond

RULE 3 **Different resonance forms of a substance don't have to be equivalent.** For example, we'll see in Chapter 22 that compounds like acetone (a common industrial solvent) are converted into anions by reaction with a strong base. The resultant acetone anion has two resonance forms. One form contains a carbon–*oxygen* double bond and has a negative charge on *carbon;* the other contains a carbon–*carbon* double bond and has a negative charge on *oxygen*. Even though the two resonance forms aren't equivalent, they both contribute to the overall resonance hybrid.

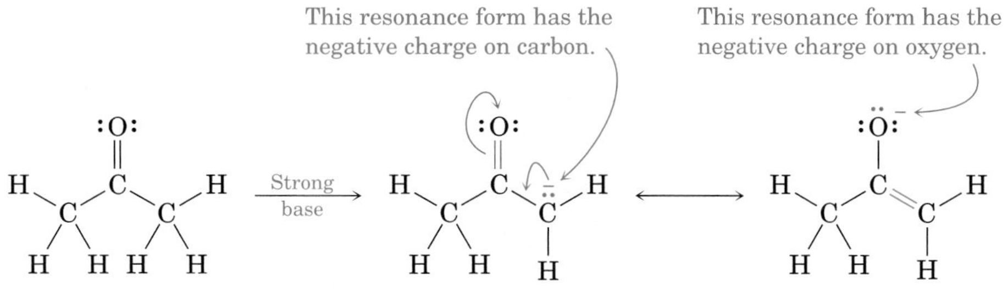

Acetone   Acetone anion (two resonance forms)

When two resonance forms are nonequivalent, the actual structure of the resonance hybrid is closer to the more stable form than the less stable form. Thus, we might expect the true structure of the acetone anion to be more like the resonance form that places the negative charge on an electronegative oxygen atom rather than on a carbon atom.

RULE 4 **Resonance forms must be valid Lewis structures and obey normal rules of valency.** A resonance form is like any other structure: The octet rule still applies. For example, one of the following structures for the acetate ion is not a valid resonance form because the carbon atom has five bonds and ten valence electrons:

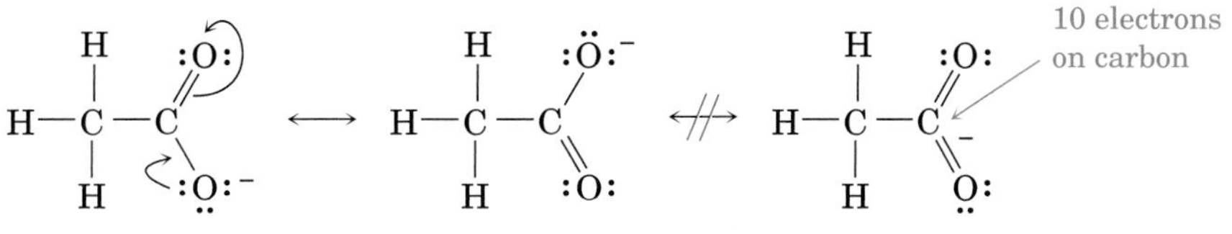

Acetate ion   *NOT* a valid resonance form

RULE 5 **The resonance hybrid is more stable than any individual resonance form.** In other words, resonance leads to stability. Generally speaking, the larger the number of resonance forms, the more stable a substance is. We'll see in Chapter 15, for instance, that a benzene ring is more stable because of resonance than might otherwise be expected.

2.6 A Useful Technique for Drawing Resonance Forms

Look back at the resonance forms of nitromethane and the acetate ion shown in the previous section. The pattern seen in both is a common one that leads to a useful technique for drawing resonance forms. In general, *any three-atom grouping with a multiple bond has two resonance forms:*

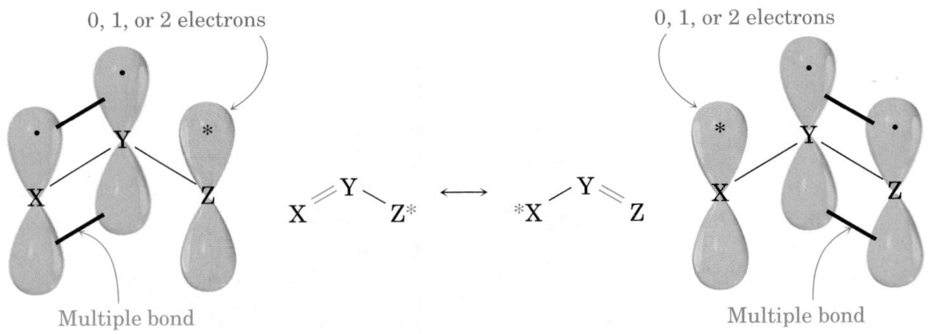

The atoms X, Y, and Z might be C, N, O, P, or S, and the asterisk (*) might mean that the *p* orbital on atom Z is vacant, that it contains a single electron, or that it contains a lone pair of electrons. The two resonance forms differ simply by an exchange of the positions of the multiple bond and the asterisk from one end to the other.

By recognizing such three-atom pieces within larger structures, resonance forms can be systematically generated. Look, for instance, at the anion produced when H^+ is removed from 2,4-pentanedione by reaction with a strong base. How many resonance structures does the resultant anion have?

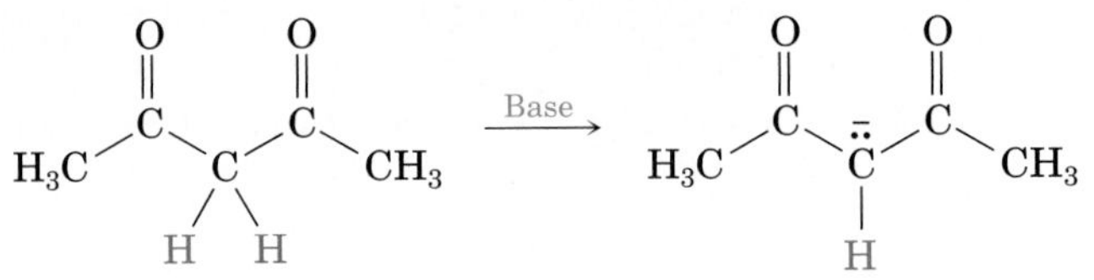

2,4-Pentanedione

The 2,4-pentanedione anion has a lone pair of electrons and a formal negative charge on the central carbon atom, next to a C=O bond on the left. The O=C–C: $^-$ grouping is a typical one for which two resonance structures can be drawn:

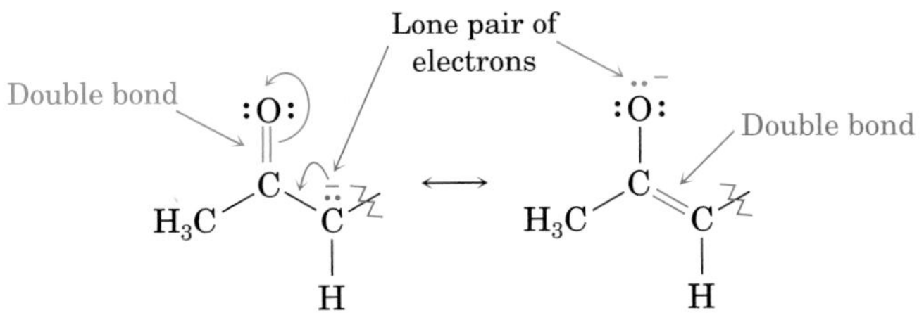

Just as there is a C=O bond to the left of the lone pair, there is a second C=O bond to the right. Thus, we can draw a total of three resonance structures for the 2,4-pentanedione anion:

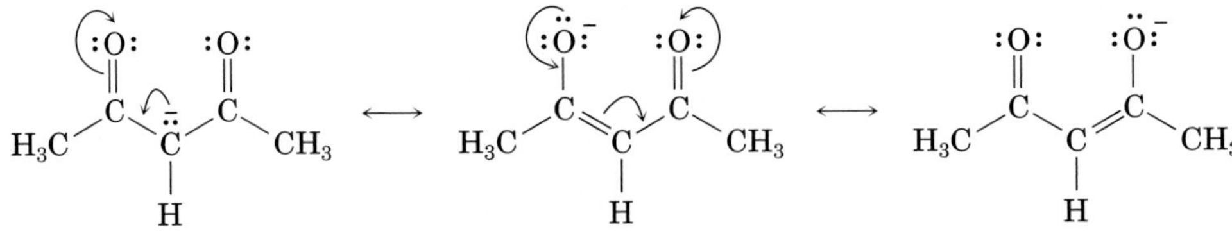

Practice Problem 2.2 Draw three resonance forms for the carbonate ion, CO_3^{2-}.

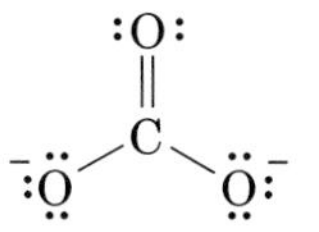

$$:O:$$
$$\|$$
$$\overset{-}{:\ddot{O}}\diagup \overset{C}{}\diagdown \overset{}{\ddot{O}:}^{-}$$
$$:\ddot{O}:$$

Carbonate ion

Strategy Look for three-atom groupings that contain a multiple bond next to an atom with a p orbital. Then exchange the positions of the multiple bond and the electrons in the p orbital. In the carbonate ion, each of the singly bonded oxygen atoms with its lone pairs and negative charge is next to the C=O double bond, giving the grouping $O=C-O:^{-}$.

Solution Exchanging the position of the double bond and an electron lone pair in each grouping generates three resonance structures:

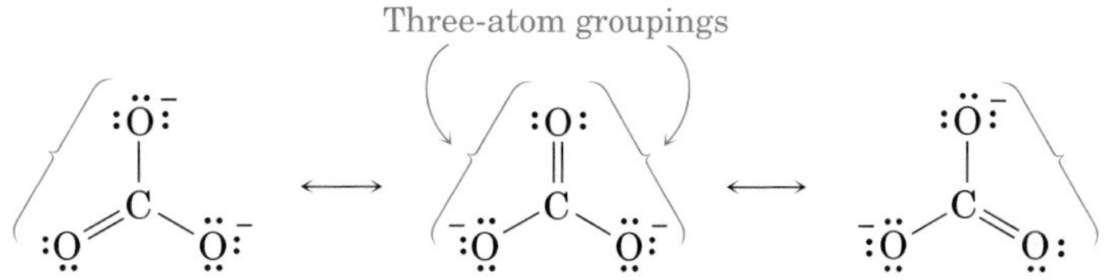

Three-atom groupings

Practice Problem 2.3 Draw three resonance forms for the pentadienyl radical. A *radical* is a substance that contains a single, unpaired electron in one of its orbitals, denoted by a dot (·).

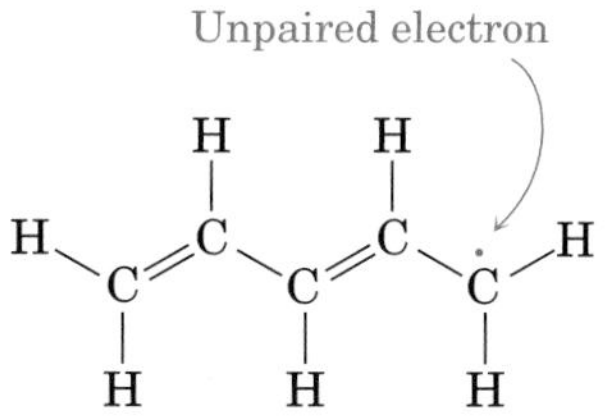

Unpaired electron

Pentadienyl radical

Strategy Find the three-atom groupings that contain a multiple bond next to a p orbital.

Solution The unpaired electron is on a carbon atom next to a C=C bond, giving a typical three-atom grouping that has two resonance forms:

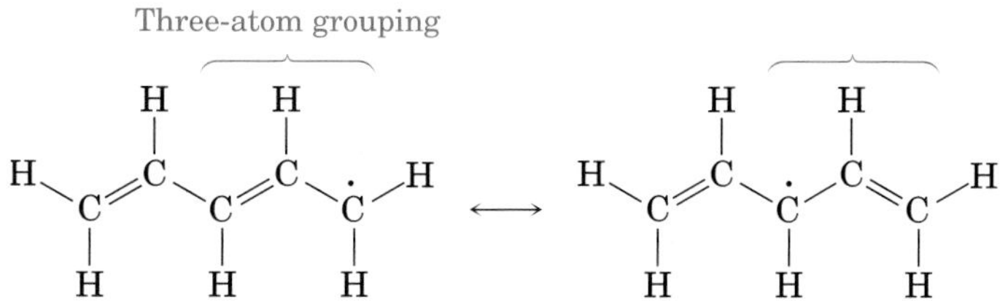

Three-atom grouping

In the second resonance form, the unpaired electron is now next to another double bond, giving another three-atom grouping and leading to another resonance form:

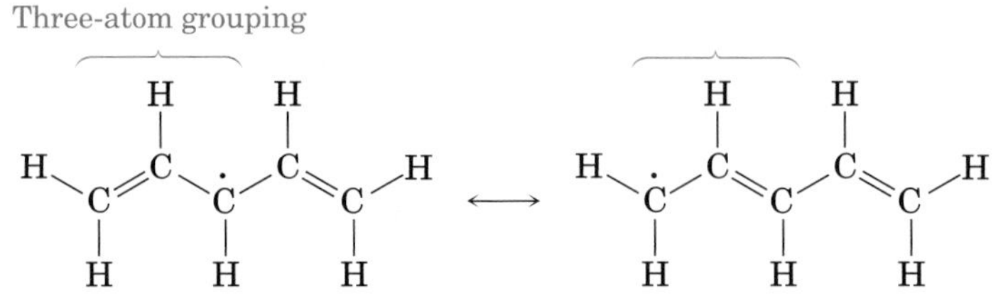

Three-atom grouping

Thus, the three resonance forms for the pentadienyl radical are:

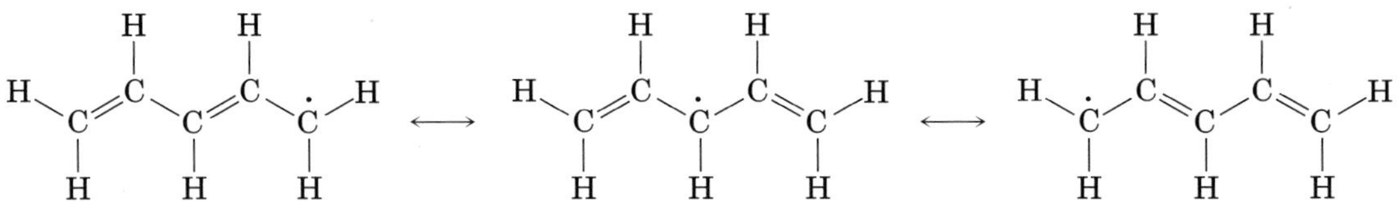

. . . .

Problem 2.9   Draw the indicated number of resonance structures for each of the following species:

(a) The nitrate ion, NO_3^- (3)

(b) The allyl cation, $H_2C\!\!=\!\!CH\!-\!CH_2^+$ (2)

(c) Hydrazoic acid, $:N\!\!\equiv\!\!\overset{+}{N}\!-\!\overset{..}{\underset{..}{N}}\!-\!H$ (2)

(d) (2)

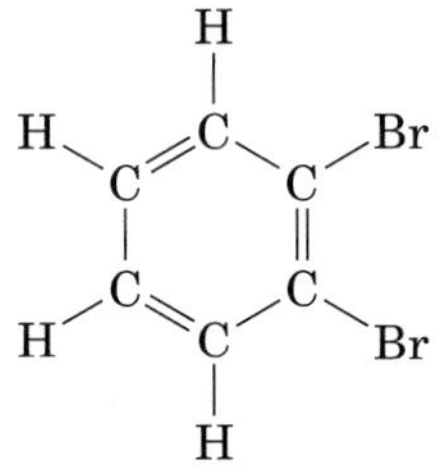

ortho-**Dibromobenzene**

. . . .

2.7 Acids and Bases: The Brønsted–Lowry Definition

Still another important concept related to electronegativity and polarity is that of *acidity* and *basicity*. We'll soon see that the acid–base behavior of organic molecules helps explain much of their chemistry. You may recall from a course in general chemistry that there are two frequently used definitions of acidity, the *Brønsted–Lowry definition* and the *Lewis definition*.

We'll look at the Brønsted–Lowry definition in this and the next three sections, and then discuss the Lewis definition in Section 2.11.

A **Brønsted–Lowry acid** is a substance that donates a hydrogen ion (H^+), and a **Brønsted–Lowry base** is a substance that accepts H^+. (The name *proton* is often used as a synonym for H^+, because loss of the valence electron from a neutral hydrogen atom leaves only the hydrogen nucleus—a proton.) When gaseous hydrogen chloride dissolves in water, for example, an acid–base reaction occurs. A polar HCl molecule donates a proton, and a water molecule accepts the proton, yielding hydronium ion (H_3O^+) and chloride ion (Cl^-).

$HCl(g)$	+	$H_2O(l)$	$\longrightarrow$	$H_3O^+(aq)$	+	$Cl^-(aq)$
Acid		Base		**Conjugate acid**		Conjugate base

Hydronium ion, the product that results when the base H_2O gains a proton, is called the **conjugate acid** of the base; chloride ion, the product that results when the acid HCl loses a proton, is called the **conjugate base** of the acid. Other common mineral acids such as H_2SO_4 and HNO_3 behave similarly, as do organic acids such as acetic acid, CH_3COOH.

In a general sense,

$$\text{H--A} \quad + \quad \text{:B} \quad \rightleftharpoons \quad \text{A:}^- \quad + \quad \text{H--B}^+$$

Acid	Base	Conjugate base	**Conjugate acid**

For example:

Acid	Base	Conjugate base	**Conjugate acid**

Acid	Base	Conjugate base	**Conjugate acid**

Note that water can act *either* as an acid or as a base, depending on the circumstances. In its reaction with HCl, water is a base that accepts a proton to give the hydronium ion, H_3O^+. In its reaction with amide ion, $^-NH_2$, however, water is an acid that donates a proton to give ammonia, NH_3, and hydroxide ion, HO^-.

• •

Problem 2.10   Nitric acid (HNO_3) reacts with ammonia (NH_3) to yield ammonium nitrate. Write the reaction, and identify the acid, the base, the conjugate acid product and the conjugate base product.

• •

2.8 Acid and Base Strength

Acids differ in their ability to donate H^+. Stronger acids such as HCl react almost completely with water, whereas weaker acids such as acetic acid (CH_3COOH) react only slightly. The exact strength of an acid, HA, in water solution is described using the equilibrium constant K_{eq} for the acid-dissociation equilibrium. (Remember from general chemistry that brackets [ ] around a substance mean that the concentration of the enclosed species is given in moles per liter, M.)

$$HA + H_2O \rightleftharpoons A^- + H_3O^+$$

$$K_{eq} = \frac{[H_3O^+][A^-]}{[HA][H_2O]}$$

In the dilute aqueous solution normally used for measuring acidity, the concentration of water, $[H_2O]$, remains nearly constant at approximately 55.6 M. We can therefore rewrite the equilibrium expression using a new quantity called the **acidity constant, K_a**. The acidity constant for any generalized acid HA is simply the equilibrium constant for the acid dissociation multiplied by the molar concentration of pure water, 55.6 M:

$$HA + H_2O \rightleftharpoons A^- + H_3O^+$$

$$K_a = K_{eq}[H_2O] = \frac{[H_3O^+][A^-]}{[HA]}$$

Stronger acids have their equilibria toward the right and thus have larger acidity constants, whereas weaker acids have their equilibria toward the left and have smaller acidity constants. The range of K_a values for different acids is enormous, running from about 10^{15} for the strongest acids to about 10^{-60} for the weakest. The common inorganic acids such as H_2SO_4, HNO_3, and HCl have K_a's in the range 10^2–10^9, while organic acids generally have K_a's in the range 10^{-5}–10^{-15}. As you gain more experience in later chapters, you'll develop a rough feeling for which acids are "strong" and which are "weak" (remembering that the terms are always relative).

Acid strengths are normally expressed using pK_a values rather than K_a values, where the pK_a is the negative common logarithm of the K_a:

$$pK_a = -\log K_a$$

A *stronger* acid (larger K_a) has a *smaller* pK_a, and a *weaker* acid (smaller K_a) has a *larger* pK_a. Table 2.3 lists the pK_a's of some common acids in order of their strength. A more comprehensive table is given in Appendix B.

TABLE 2.3 Relative Strengths of Some Common Acids and Their Conjugate Bases

	Acid	Name	pK_a	Conjugate base	Name	
Weaker acid	CH_3CH_2OH	Ethanol	16.00	$CH_3CH_2O^-$	Ethoxide ion	Stronger base
	H_2O	Water	15.74	HO^-	Hydroxide ion	
	HCN	Hydrocyanic acid	9.31	CN^-	Cyanide ion	
	CH_3COOH	Acetic acid	4.76	CH_3COO^-	Acetate ion	
	HF	Hydrofluoric acid	3.45	F^-	Fluoride ion	
Stronger acid	HNO_3	Nitric acid	−1.3	NO_3^-	Nitrate ion	Weaker base
	HCl	Hydrochloric acid	−7.0	Cl^-	Chloride ion	

Notice that the pK_a value shown in Table 2.3 for water is 15.74, a value that results from the following calculation: The K_a for any acid in water is the equilibrium constant K_{eq} for the acid dissociation multiplied by the molar concentration of pure water. For the acid dissociation of water, we have

$$H_2O + H_2O \rightleftharpoons OH^- + H_3O^+$$

$$K_{eq} = \frac{[H_3O^+][OH^-]}{[H_2O]^2} \quad \text{and} \quad K_a = K_{eq} \times [H_2O] = \frac{[H_3O^+][OH^-]}{[H_2O]}$$

The numerator in this expression, $[H_3O^+][OH^-]$, is the so-called ion-product constant for water, $K_w = 1.0 \times 10^{-14}$, and the denominator is $[H_2O] = 55.6$ M. Thus, we have

$$K_a = \frac{1.0 \times 10^{-14}}{55.6} = 1.80 \times 10^{-16} \quad \text{and} \quad pK_a = 15.74$$

Notice also in Table 2.3 that there is an inverse relationship between the acid strength of an acid and the base strength of its conjugate base. To understand this relationship, think about what happens to the acidic hydrogen in a reaction: A *strong acid* is one that loses an H^+ easily, meaning that its conjugate base has little affinity for the H^+ and is therefore a *weak base*. A *weak acid* is one that loses an H^+ with difficulty, meaning that its conjugate base has a high affinity for the H^+ and is therefore a *strong base*. The fact that HCl is a strong acid, for example, means that Cl^- does not hold the H^+ tightly and is thus a weak base. Water, however, is a weak acid, meaning that OH^- *does* hold the H^+ tightly and is a strong base.

· ·

Problem 2.11   Formic acid, HCOOH, has $pK_a = 3.75$, and picric acid, $C_6H_3N_3O_7$, has $pK_a = 0.38$. Which is the stronger acid?

Problem 2.12   Amide ion, H_2N^-, is a much stronger base than hydroxide ion, HO^-. Which would you expect to be a stronger acid, NH_3 or H_2O? Explain.

· ·

2.9 Predicting Acid–Base Reactions from pK_a Values

Compilations of pK_a values like those in Table 2.3 and Appendix B are very useful for predicting whether a given acid–base reaction will take place, because H^+ will always go from the stronger acid to the stronger base. For example, the data in Table 2.3 indicate that acetic acid ($pK_a = 4.76$) is a stronger acid than water ($pK_a = 15.74$). This means that hydroxide ion has a greater affinity for H^+ than acetate ion has, and that OH^- will accept H^+ from CH_3COOH to yield CH_3COO^- and H_2O. In general, *an acid with a lower pK_a will react with the conjugate base of an acid with a higher pK_a*.

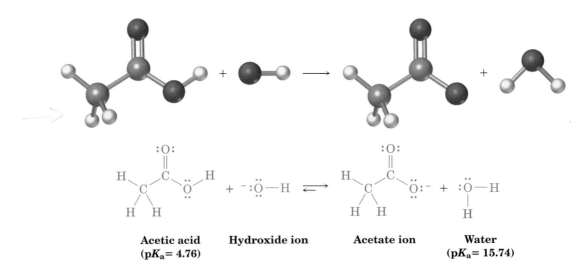

| Acetic acid | Hydroxide ion | Acetate ion | Water |
| (pK_a= 4.76) | | | (pK_a= 15.74) |

Another way to predict acid–base reactivity is to remember that the products of an acid–base reaction must be more stable than the reactants. In other words, the product acid must be weaker and less reactive than the starting acid, and the product base must be weaker and less reactive than the starting base. In the reaction of acetic acid with hydroxide ion, for example, the product conjugate acid (H_2O) is weaker than the starting acid (CH_3COOH), and the product conjugate base (CH_3COO^-) is weaker than the starting base (OH^-).

$$CH_3\overset{\overset{O}{\|}}{C}OH \ + \ HO^- \ \rightleftharpoons \ H_2O \ + \ CH_3\overset{\overset{O}{\|}}{C}O^-$$

| **Stronger acid** | **Stronger base** | **Weaker acid** | **Weaker base** |

• •

Practice Problem 2.4   Water has pK_a = 15.74, and acetylene has pK_a = 25. Which is the stronger acid? Does hydroxide ion react with acetylene?

$$H-C{\equiv}C-H + H-O^- \overset{?}{\longrightarrow} H-C{\equiv}C:^- + H-O-H$$

Strategy   In comparing two acids, the one with the lower pK_a is stronger. Thus, water is a stronger acid than acetylene and gives up H^+ more easily.

Solution   Since water is a stronger acid and gives up H^+ more easily than acetylene does, the HO^- ion must have less affinity for H^+ than the $HC{\equiv}C:^-$ ion has. In other words, the anion of acetylene is a stronger base than hydroxide ion, and the reaction will not proceed as written.

• •

Practice Problem 2.5   According to the data in Table 2.3, acetic acid has pK_a = 4.76. What is its K_a?

Strategy   Since pK_a is the negative logarithm of K_a, it's necessary to use a calculator capable of finding antilogarithms. Enter the value of the pK_a (4.76), change the sign (−4.76), and then find the antilog (1.74 × 10^{-5}).

Solution $K_a = 1.74 \times 10^{-5}$

• •

Problem 2.13   Will either of the following reactions take place as written, according to the pK_a data in Table 2.3?

(a) $HCN + CH_3COO^- \ Na^+ \overset{?}{\longrightarrow} Na^+ \ {}^-CN + CH_3COOH$

(b) $CH_3CH_2OH + Na^+ \ {}^-CN \overset{?}{\longrightarrow} CH_3CH_2O^- \ Na^+ + HCN$

Problem 2.14   Ammonia, NH_3, has $pK_a \approx 36$ and acetone has $pK_a \approx 19$. Will the following reaction take place?

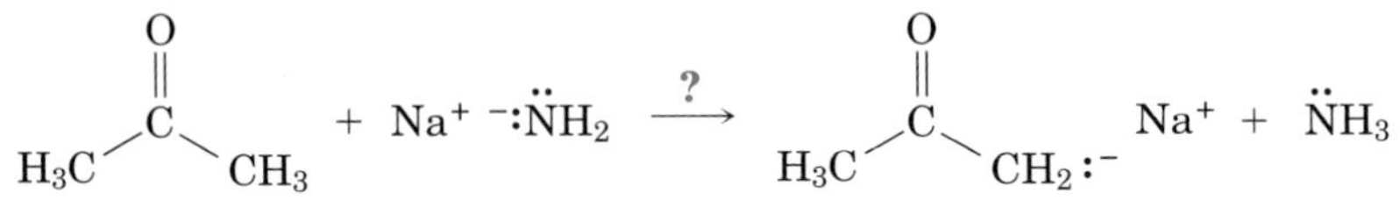

Acetone

Problem 2.15   What is the K_a of HCN if its $pK_a = 9.31$?

· ·

2.10 Organic Acids and Organic Bases

Many of the reactions we'll be seeing in future chapters involve organic acids and organic bases. Although it's much too early to go into the details of these processes now, you might keep the following generalities in mind as your study of organic chemistry progresses.

Organic acids are of two main kinds: those such as methyl alcohol and acetic acid, which contain a hydrogen atom bonded to an oxygen atom (O–H), and those such as acetone, which contain a hydrogen atom bonded to a carbon atom next to a C=O double bond (O=C–C–H).

Some organic acids

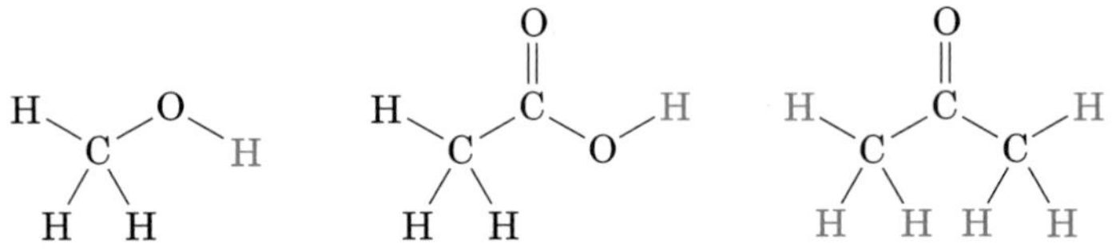

Methyl alcohol   **Acetic acid**   **Acetone**
($pK_a = 15.54$)   **($pK_a = 4.76$)**   **($pK_a = 19.3$)**

Methyl alcohol contains an O–H bond and is a weak acid; acetic acid also contains an O–H bond and is a somewhat stronger acid. In both cases, acidity is due to the fact that the conjugate base resulting from loss of H^+ is stabilized by having its negative charge on a highly electronegative oxygen atom. In addition, the conjugate base of acetic acid is stabilized by resonance (Section 2.4).

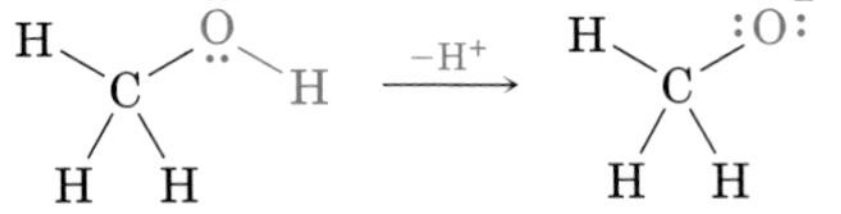

Anion is stabilized by having negative charge on a highly electronegative atom.

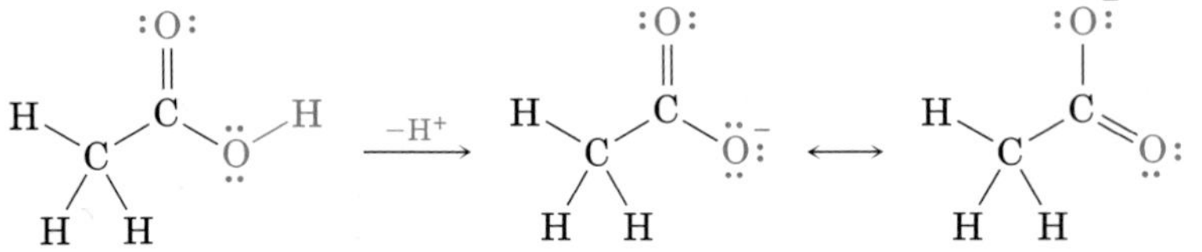

Anion is stabilized by having negative charge on a highly electronegative atom and by resonance.

The acidity of acetone is due to the fact that the conjugate base resulting from loss of H^+ is stabilized by resonance (Section 2.5). In addition, one of the resonance forms stabilizes the negative charge by placing it on an electronegative oxygen atom.

  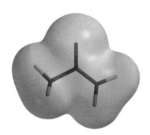

Anion is stabilized by resonance and by having negative charge on a highly electronegative atom.

Electrostatic potential maps of the conjugate bases from methyl alcohol, acetic acid, and acetone are shown in Figure 2.4. As you might expect, all three substances show a substantial amount of negative charge on oxygen.

FIGURE 2.4 ▼

Electrostatic potential maps of the conjugate bases of (a) methyl alcohol, (b) acetic acid, and (c) acetone. The oxygen atoms carry much of the negative charge in all three.

methanol conjugate base, acetic acid conjugate base, acetone conjugate base

(a) CH_3O^- (b) $CH_3CO_2^-$ (c) $CH_3COCH_2^-$

In contrast to organic acids, organic bases are of only one main kind. They usually contain a nitrogen atom with a lone pair of electrons and thus behave in the same way as ammonia, NH_3. Methylamine, for example, reacts with HCl just as ammonia does:

**Methylamine
(an organic base)**

2.11 Acids and Bases: The Lewis Definition

The Brønsted–Lowry definition of acidity discussed in the previous four sections encompasses all compounds containing hydrogen. Of even more use, however, is the *Lewis definition* of acids and bases, which is not limited to compounds that gain or lose protons. A **Lewis acid** is a substance that *accepts an electron pair*, and a **Lewis base** is a substance that *donates an* *electron pair*. The donated electron pair is then shared between acid and base in a covalent bond.

Filled
orbital

Vacant
orbital

B : + A ⟶ B—A

Lewis base   **Lewis acid**

Lewis Acids and the Curved Arrow Formalism

The fact that a Lewis acid must be able to accept an electron pair means that it must have either a vacant, low-energy orbital or a polar bond to hydrogen so it can donate H^+ (which has an empty $1s$ orbital). Thus, the Lewis definition of acidity is much broader than the Brønsted–Lowry definition and includes many other species in addition to H^+. For example, various metal cations such as Mg^{2+} are Lewis acids because they accept a pair of electrons when they form a bond to a base. In the same way, compounds of group 3A elements such as BF_3 and $AlCl_3$ are Lewis acids because they have unfilled valence orbitals and can accept electron pairs from Lewis bases, as shown in Figure 2.5. Similarly, many transition-metal compounds, such as $TiCl_4$, $FeCl_3$, $ZnCl_2$, and $SnCl_4$, are Lewis acids.

FIGURE 2.5 ▼

The reactions of some Lewis acids with some Lewis bases. The Lewis acids accept an electron pair; the Lewis bases donate a pair of nonbonding electrons. Note how the movement of electrons *from* the Lewis base *to* the Lewis acid is indicated by curved arrows.

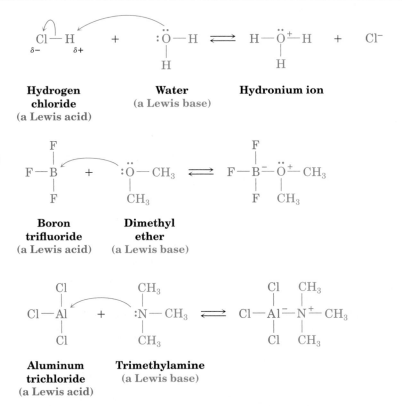

Look closely at the acid–base reactions in Figure 2.5, and note how they are shown. In the first reaction, the Lewis base water uses an electron pair to abstract H^+ from the polar HCl molecule. In the remaining two reactions, a Lewis base donates an electron pair to a vacant valence orbital of a boron or aluminum atom. In all three reactions, the direction of electron-pair flow from the electron-rich Lewis base to the electron-poor Lewis acid is shown using curved arrows, just as the direction of electron flow in going from one resonance structure to another was shown using curved arrows in Section 2.5. A curved arrow always means that a pair of electrons moves *from* the atom at the tail of the arrow *to* the atom at the head of the arrow.

The movement of electrons in Lewis acid–base reactions can be seen clearly with electrostatic potential maps. In the reaction of boron trifluoride with dimethyl ether, for instance, the ether oxygen atom becomes more positive and the boron becomes more negative as electron density is transferred and the B–O bond forms (Figure 2.6).

FIGURE 2.6 ▼

Electrostatic potential maps of boron trifluoride, dimethyl ether, and their Lewis acid–base reaction product. The oxygen atom becomes more positive and the boron becomes more negative as electron density is transferred and the B–O bond forms.

 boron trifluoride, dimethyl ether, Lewis acid–base product

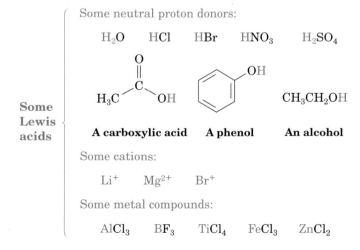

Boron trifluoride + **Dimethyl ether** ⇌ **Lewis acid–base product**

Some further examples of Lewis acids are shown below:

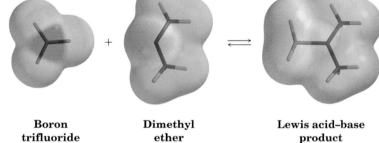

Some Lewis acids

Some neutral proton donors:

H_2O HCl   HBr HNO_3 H_2SO_4

H_3C—C(=O)—OH   (phenol)—OH CH_3CH_2OH

A carboxylic acid   **A phenol**   **An alcohol**

Some cations:

Li^+ Mg^{2+} Br^+

Some metal compounds:

$AlCl_3$ BF_3 $TiCl_4$ $FeCl_3$ $ZnCl_2$

Lewis Bases

The Lewis definition of a base as a compound with a pair of nonbonding electrons that it can use to bond to a Lewis acid is similar to the Brønsted–Lowry definition. Thus, H_2O, with its two pairs of nonbonding electrons on oxygen, acts as a Lewis base by donating an electron pair to an H^+ in forming the hydronium ion, H_3O^+.

$$:\ddot{Cl}-H \; + \; :\underset{H}{\overset{|}{O}}-H \;\; \rightleftharpoons \;\; H-\underset{H}{\overset{|}{\overset{+}{O}}}-H \; + \; :\ddot{Cl}:^-$$

      Acid   Lewis base   **Hydronium ion**

In a more general sense, most oxygen- and nitrogen-containing organic compounds are Lewis bases because they too have lone pairs of electrons. A divalent oxygen compound has two lone pairs of electrons, and a trivalent nitrogen compound has one lone pair. Note in the following examples that some compounds can act as both acids and bases, just as water can. Alcohols and carboxylic acids, for instance, act as acids when they donate an H^+ but as bases when their oxygen atom accepts an H^+.

Some Lewis bases

$CH_3CH_2\ddot{O}H$	$CH_3\ddot{O}CH_3$	$CH_3\overset{:O:}{\overset{\|}{C}}H$	$CH_3\overset{:O:}{\overset{\|}{C}}CH_3$
An alcohol	**An ether**	**An aldehyde**	**A ketone**
$CH_3\overset{:O:}{\overset{\|}{C}}Cl$	$CH_3\overset{:O:}{\overset{\|}{C}}\ddot{O}H$	$CH_3\overset{:O:}{\overset{\|}{C}}\ddot{O}CH_3$	$CH_3\overset{:O:}{\overset{\|}{C}}\ddot{N}H_2$
An acid chloride	**A carboxylic acid**	**An ester**	**An amide**
$CH_3\underset{CH_3}{\overset{|}{\ddot{N}}}CH_3$	$CH_3\ddot{S}CH_3$		
An amine	**A sulfide**		

For example:

$$CH_3-\ddot{O}-H \; + \; HBr \;\; \rightleftharpoons \;\; CH_3-\underset{H}{\overset{+}{\overset{|}{\ddot{O}}}}-H \; + \; Br^-$$

     Methyl   **Hydrogen**   **Methyloxonium bromide**
     alcohol   **bromide**
     (base)   (acid)

$$\underset{H_3C}{}\overset{:O:}{\overset{\|}{C}}\underset{CH_3}{} \; + \; H_2SO_4 \;\; \rightleftharpoons \;\; \underset{H_3C}{}\overset{\overset{+}{:O}-H}{\overset{\|}{C}}\underset{CH_3}{} \; + \; HSO_4^-$$

     Acetone   **Sulfuric**
     (base)   acid

Notice in the list of Lewis bases given above that some compounds, such as carboxylic acids, esters, and amides, have more than one atom with a lone pair of electrons and can therefore react at more than one site. Acetic acid, for example, can be protonated either on the doubly bonded oxygen atom or on the singly bonded oxygen atom:

Acetic acid   **Sulfuric**   **Protonated acetic acid**
(base)   acid

Reaction normally occurs only once in such instances, and the more stable of the two possible protonation products is formed. For acetic acid, protonation occurs on the doubly bonded oxygen.

Practice Problem 2.6   Using curved arrows, show how acetaldehyde can act as a Lewis base.

Strategy   A Lewis base donates an electron pair to a Lewis acid. We therefore need to locate the electron lone pairs on acetaldehyde and use a curved arrow to show their movement toward the H atom of the acid.

Solution   The oxygen atom of acetaldehyde has two lone pairs of electrons that it can donate to a Lewis acid such as H^+.

Acetaldehyde

Problem 2.16   Using curved arrows, show how the species in part (a) can act as Lewis bases in their reactions with HCl, and show how the species in part (b) can act as Lewis acids in their reaction with OH^-.
(a) CH_3CH_2OH, $HN(CH_3)_2$, $P(CH_3)_3$ (b) H_3C^+, $B(CH_3)_3$, $MgBr_2$

Problem 2.17   Explain by calculating formal charges why the following acid–base reaction products have the charges indicated:

(a) $F_3\bar{B}-\overset{+}{O}-CH_3$ (b) $Cl_3\bar{Al}-\overset{+}{N}-CH_3$
              | |
             CH_3 CH_3

with CH_3 above N in (b)

2.12 Drawing Chemical Structures

In the Kekulé structures we've been drawing up to this point, a line between atoms represents the two electrons in a covalent bond. Such structures have been used for many years and comprise a universal chemical language. Two chemists from different countries may not understand each other's words, but a chemical structure means the same to both.

Most organic chemists find themselves drawing many structures each day, and it would soon become awkward if every bond and atom had to be indicated. For example, vitamin A, $C_{20}H_{30}O$, has 51 different chemical bonds uniting the 51 atoms. Vitamin A can be drawn showing each bond and atom, but doing so is a time-consuming process, and the resultant drawing is difficult to read. Chemists have therefore devised several shorthand ways for writing structures. In **condensed structures**, carbon–hydrogen and carbon–carbon single bonds aren't shown; instead, they're understood. If a carbon has three hydrogens bonded to it, we write CH_3; if a carbon has two hydrogens bonded to it, we write CH_2; and so on. The compound called 2-methylbutane, for example, is written as follows:

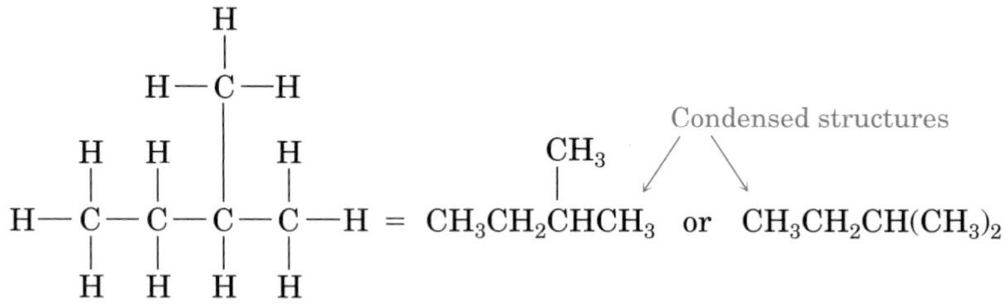

Condensed structures

$$CH_3CH_2CHCH_3 \text{ or } CH_3CH_2CH(CH_3)_2$$

2-Methylbutane

Notice that the horizontal bonds between carbons aren't shown in condensed structures—the CH_3, CH_2, and CH units are simply placed next to each other—but the vertical carbon–carbon bond in the first condensed structure above is shown for clarity. Notice also that in the second condensed structure, the two CH_3 units attached to the CH carbon are grouped together as $(CH_3)_2$.

Simpler still is the use of **skeletal structures** such as those shown in Table 2.4. The rules for drawing skeletal structures are straightforward:

RULE 1   Carbon atoms aren't usually shown. Instead, a carbon atom is assumed to be at each intersection of two lines (bonds) and at the end of each line. Occasionally, a carbon atom might be indicated for emphasis or clarity.

RULE 2   Hydrogen atoms bonded to carbons aren't shown. Since carbon always has a valence of 4, we mentally supply the correct number of hydrogen atoms for each carbon.

RULE 3   Atoms other than carbon and hydrogen *are* shown.

Table 2.4 gives some examples of how these rules are applied.

TABLE 2.4 Kekulé and Skeletal Structures for Some Compounds

Compound	Kekulé structure	Skeletal structure
Isoprene, C_5H_8	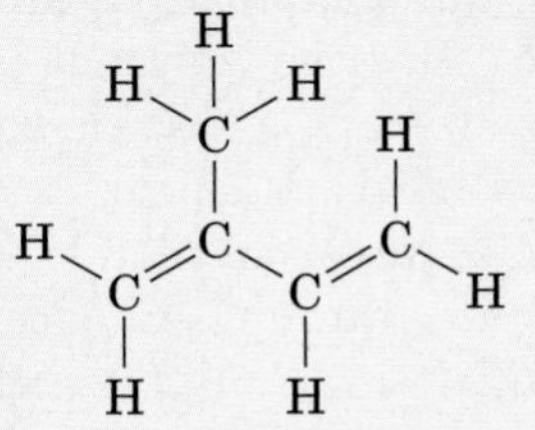	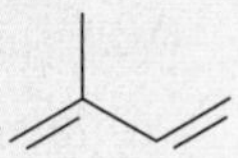
Methylcyclohexane, C_7H_{14}	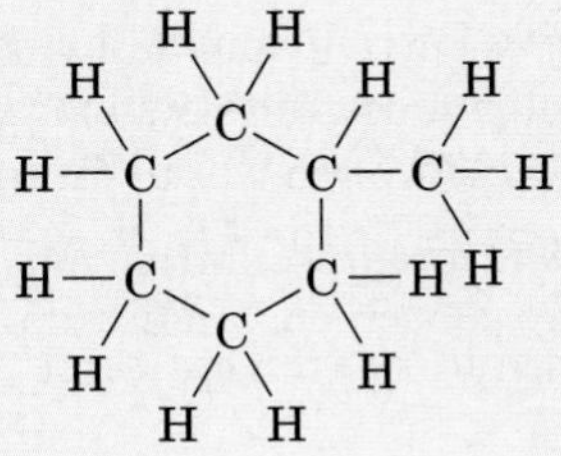	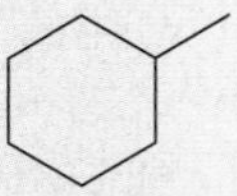
Phenol, C_6H_6O	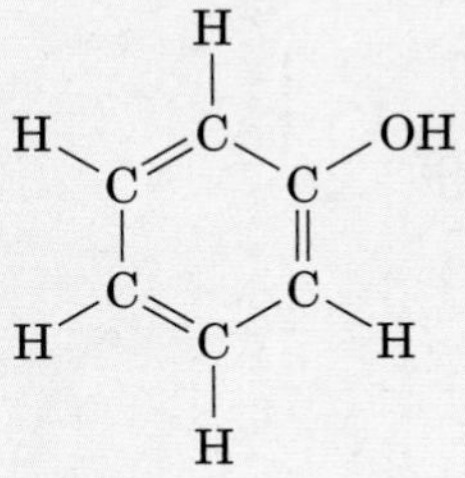	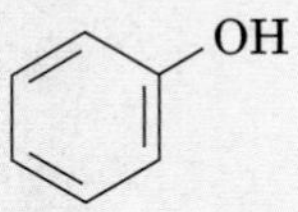

• •

Practice Problem 2.7 Carvone, a substance responsible for the odor of spearmint, has the following structure. Tell how many hydrogens are bonded to each carbon, and give the molecular formula of carvone.

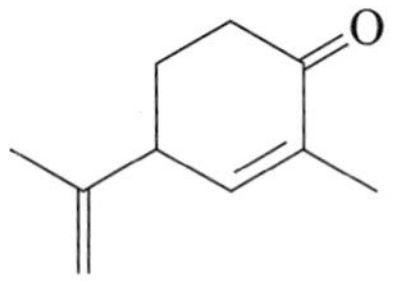

Carvone

Strategy The end of a line represents a carbon atom with 3 hydrogens, CH_3; a two-way intersection is a carbon atom with 2 hydrogens, CH_2; a three-way intersection is a carbon atom with 1 hydrogen, CH; and a four-way intersection is a carbon atom with no attached hydrogens.

Solution

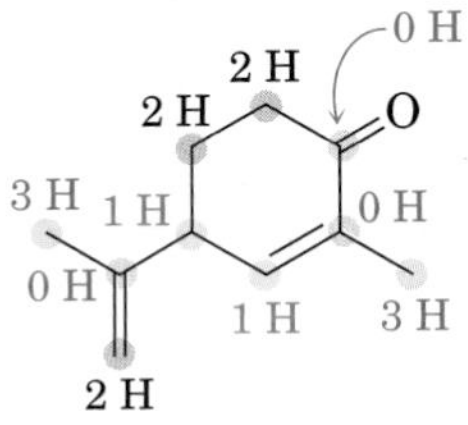

Carvone, $C_{10}H_{14}O$

● ●

Problem 2.18   Tell how many hydrogens are bonded to each carbon in the following compounds, and give the molecular formula of each substance.

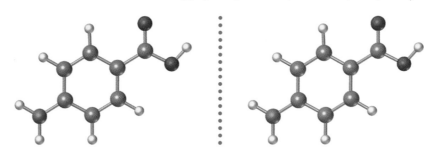

Adrenaline

Estrone (a hormone)

Problem 2.19   Propose skeletal structures for compounds that satisfy the following molecular formulas (there is more than one possibility in each case):
(a) C_5H_{12} (b) C_2H_7N (c) C_3H_6O (d) C_4H_9Cl

Problem 2.20   The following stereo view is a representation of *para*-aminobenzoic acid (PABA), the active ingredient in many sunscreens. Indicate the positions of the multiple bonds, and draw a skeletal structure (gray = C, red = O, blue = N, ivory = H).

Stereo View

● ●

2.13 Molecular Models

Organic chemistry is a three-dimensional science, and molecular shape is often critical in determining the chemistry a compound undergoes. Many computer programs are available that can help you visualize molecules by rotating and manipulating them on the screen. Another helpful technique is to use molecular models. With practice, you can learn to see many spatial relationships even when viewing two-dimensional drawings, but there's no substitute for building a molecular model and turning it in your hands to get different perspectives.

Many kinds of models are available, some at relatively modest cost, and everyone should have access to a set of models while studying this book. So-called *space-filling models* are better for examining the crowding within a molecule, but *ball-and-stick models* are generally the least expensive and most durable for student use. Figure 2.7 shows two kinds of models of acetic acid, CH_3COOH.

FIGURE 2.7 ▼

Molecular models of
acetic acid, CH_3COOH.
(a) Space-filling; (b) ball-
and-stick.

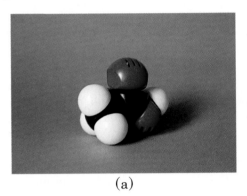

(a)

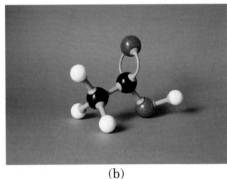

(b)

● ●

Problem 2.21 Build a molecular model of ethane, $H_3C–CH_3$. Sight along the C–C bond to see the relationships between hydrogens on the different carbons, and draw a three-dimensional representation.

● ●

CHEMISTRY @ WORK

Alkaloids: Naturally Occurring Bases

Don't eat this one! The deadly *Amanita muscaria*
contains muscarine and other toxic alkaloids.

Just as ammonia, NH_3, is a weak base, there are a large number of nitrogen-containing organic compounds called *amines* that are also weak bases. In the early days of organic chemistry, basic amines derived from natural sources were known as "vegetable alkali," but they are now referred to as *alkaloids*. The study of alkaloids provided much of the impetus for the growth of organic chemistry in the nineteenth century, and it remains today a fascinating area of research.

Alkaloids vary widely in structure, from the simple to the enormously complex. The odor of rotting fish, for example, is caused by methylamine, a simple relative of ammonia in which one of the NH_3 hydrogens has been replaced by an organic CH_3 group. (In fact, the use of acidic lemon juice to mask fish odors is simply an acid–base reaction.)

(continued) ▶

H—C—N **Methylamine**
(found in rotting fish)

Many alkaloids have pronounced biological properties, and many of the pharmaceutical agents used today are derived from naturally occurring amines. Morphine and related alkaloids from the opium poppy, for instance, are used for pain relief; atropine from the flowering plant *Atropa belladonna,* commonly called the deadly nightshade, is used as an anti-spasmodic agent for the treatment of colitis; and ephedrine from the Chinese plant *Ephedra sinica* is used as a bronchodilator and decongestant.

Morphine   **Atropine**   **Ephedrine**

Summary and Key Words

KEY WORDS	
acidity constant (K_a), 52	
Brønsted–Lowry acid, 51	
Brønsted–Lowry base, 51	
condensed structure, 62	
conjugate acid, 51	
conjugate base, 51	
dipolar, 44	
dipole moment (μ), 38	

Organic molecules often have **polar covalent bonds** as a result of unsymmetrical electron sharing caused by differences in the **electronegativity** of atoms. For example, a carbon–chlorine bond is polar because chlorine attracts the shared electrons more strongly than carbon does. Carbon–hydrogen bonds are relatively nonpolar. Many molecules as a whole are also polar owing to the cumulative effects of individual polar bonds and electron lone pairs. The polarity of a molecule is measured by its **dipole moment, μ.**

Plus ($+$) and minus ($-$) signs are used to indicate the presence of **formal charges** on atoms in molecules. Assigning formal charges to specific atoms is a bookkeeping technique that makes it possible to keep track of the valence electrons around an atom.

Some substances, such as nitromethane, benzene, and acetate ion, can't be represented by a single Lewis or line-bond structure and must be considered as a **resonance hybrid** of two or more structures, neither of which is correct by itself. The only difference between two resonance forms is in the location of their π or nonbonding electrons. The nuclei remain in the same places in both structures.

Acidity and basicity are closely related to polarity and electronegativity. A **Brønsted–Lowry acid** is a compound that can donate a proton (hydrogen ion, H^+), and a **Brønsted–Lowry base** is a compound that can accept a proton. The strength of a Brønsted–Lowry acid or base is expressed by its **acidity constant, K_a,** or by the negative logarithm of the acidity constant, pK_a. The higher the pK_a, the weaker the acid. More useful is the Lewis definition of acids and bases. A **Lewis acid** is a compound that has a low-energy empty orbital that can accept an electron pair; BF_3, $AlCl_3$, and H^+ are examples. A **Lewis base** is a compound that can donate an unshared electron pair; NH_3 and H_2O are examples. Most organic molecules that contain oxygen and nitrogen are Lewis bases.

Organic molecules are usually drawn using either condensed structures or skeletal structures. In **condensed structures**, carbon–carbon and carbon–hydrogen bonds aren't shown. In **skeletal structures**, only the bonds and not the atoms are shown. A carbon atom is assumed to be at the ends and at the junctions of lines (bonds), and the correct number of hydrogens is mentally supplied.

Visualizing Chemistry

(Problems 2.1–2.21 appear within the chapter.)

2.22 Convert each of the following models into a skeletal structure. Only the connections between atoms are shown; multiple bonds are not indicated (gray = C, red = O, blue = N, ivory = H).

(a) (b)

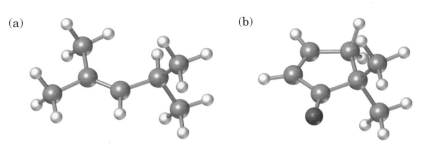

(c)

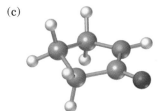

2.23 Fill in the multiple bonds in the following model of naphthalene, $C_{10}H_8$ (gray = C, ivory = H). How many resonance structures does naphthalene have?

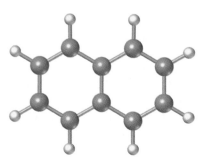

2.24 The following stereo view is a representation of ibuprofen, a common over-the-counter pain reliever. Indicate the positions of the multiple bonds, and draw a skeletal structure (gray = C, red = O, ivory = H).

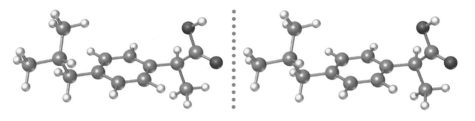

Stereo View

Additional Problems

2.25 Convert the following structures into skeletal drawings:

(a)

Indole

(b)

1,3-Pentadiene

(c)

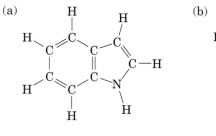

1,2-Dichlorocyclopentane

(d)

Quinone

2.26 Tell the number of hydrogens bonded to each carbon atom in the following substances, and give the molecular formula of each:

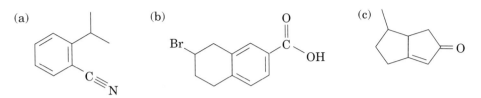

(a)   (b)   (c)

2.27 Identify the most electronegative element in each of the following molecules:
(a) CH_2FCl (b) $FCH_2CH_2CH_2Br$ (c) $HOCH_2CH_2NH_2$ (d) CH_3OCH_2Li

2.28 Use the electronegativity table (Figure 2.2) to predict which bond in each of the following sets is more polar:
(a) $H_3C–Cl$ or $Cl–Cl$ (b) $H_3C–H$ or $H–Cl$
(c) $HO–CH_3$ or $(CH_3)_3Si–CH_3$ (d) $H_3C–Li$ or $Li–OH$

2.29 Indicate the direction of bond polarity for each compound in Problem 2.28.

2.30 Which of the following molecules has a dipole moment? Indicate the expected direction of each.

(a) Cl, Cl / C=C / H, H (b) Cl, H / C=C / H, Cl (c) Cl (d) Cl, Cl

2.31 Phosgene, $Cl_2C=O$, has a smaller dipole moment than formaldehyde, $H_2C=O$. Explain.

2.32 The dipole moment of HCl is 1.08 D, and the H–Cl bond length is 136 pm. What is the percent ionic character of the H–Cl bond?

2.33 Fluoromethane (CH_3F, $\mu = 1.81$ D) has a smaller dipole moment than chloromethane (CH_3Cl, $\mu = 1.87$ D) even though fluorine is more electronegative than chlorine. Explain.

2.34 Calculate the formal charges on the atoms shown in red.

(a) $(CH_3)_2\overset{..}{O}BF_3$ (b) $H_2\overset{..}{C}-N\equiv N:$ (c) $H_2C=N=\overset{..}{N}:$

(d) $:\overset{..}{O}=\overset{..}{O}-\overset{..}{O}:$ (e) $H_2\overset{..}{C}-\underset{\underset{CH_3}{|}}{\overset{\overset{CH_3}{|}}{P}}-CH_3$ (f) [pyridine ring with N and :O:]

2.35 Which of the following pairs of structures represent resonance forms?

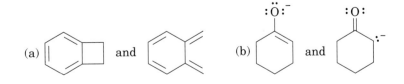

(a)   and   (b)   and

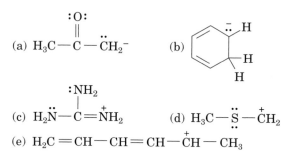

2.36 Draw as many resonance structures as you can for the following species:

(a) $H_3C-\overset{\overset{\displaystyle :O:}{\|}}{C}-\overset{..}{C}H_2^-$ (b)

(c) $H_2\overset{..}{N}-\overset{\overset{\displaystyle :NH_2}{|}}{C}=\overset{+}{N}H_2$ (d) $H_3C-\overset{..}{\underset{..}{S}}-\overset{+}{C}H_2$

(e) $H_2C=CH-CH=CH-\overset{+}{C}H-CH_3$

2.37 Cyclobutadiene is a rectangular molecule with two shorter double bonds and two longer single bonds. Why do the following structures *not* represent resonance forms?

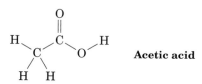

2.38 Alcohols can act either as weak acids or as weak bases, just as water can. Show the reaction of methyl alcohol, CH_3OH, with a strong acid such as HCl and with a strong base such as Na^+ $^-NH_2$.

2.39 The O–H hydrogen in acetic acid is much more acidic than any of the C–H hydrogens. Explain.

$$H-\underset{\underset{\displaystyle H\quad H}{|\quad |}}{C}-\overset{\overset{\displaystyle O}{\|}}{C}-O-H$$ **Acetic acid**

2.40 Which of the following are likely to act as Lewis acids and which as Lewis bases?
(a) $AlBr_3$ (b) $CH_3CH_2NH_2$ (c) BH_3
(d) HF   (e) CH_3SCH_3 (f) $TiCl_4$

2.41 Draw a Lewis electron-dot structure for each of the molecules in Problem 2.40, indicating any unshared electron pairs.

2.42 Write the products of the following acid–base reactions:
(a) $CH_3OH + H_2SO_4 \rightleftharpoons$?
(b) $CH_3OH + NaNH_2 \rightleftharpoons$?
(c) $CH_3NH_3^+Cl^- + NaOH \rightleftharpoons$?

2.43 Assign formal charges to the atoms in each of the following molecules:

(a) $H_3C-\overset{\overset{\displaystyle CH_3}{|}}{\underset{\underset{\displaystyle CH_3}{|}}{N}}-\overset{..}{\underset{..}{O}}:$ (b) $H_3C-\overset{..}{N}-N\equiv N:$ (c) $H_3C-\overset{..}{N}=N=\overset{..}{N}:$

2.44 Rank the following substances in order of increasing acidity:

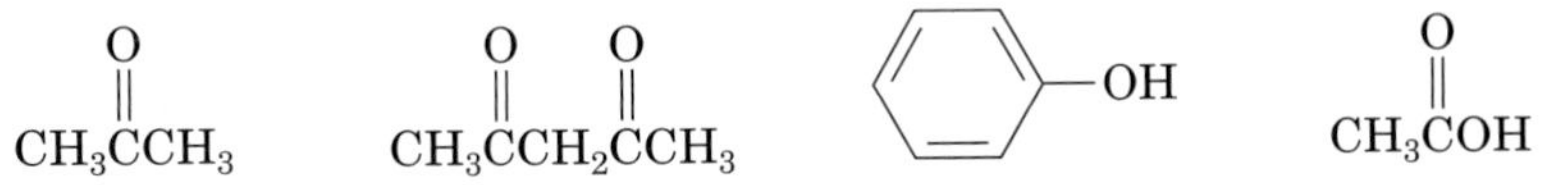

$$
\underset{\substack{\textbf{Acetone}\\(\textbf{p}K_a = 19.3)}}{CH_3\overset{\displaystyle O}{\overset{\|}{C}}CH_3}
\qquad
\underset{\substack{\textbf{2,4-Pentanedione}\\(\textbf{p}K_a = 9)}}{CH_3\overset{\displaystyle O}{\overset{\|}{C}}CH_2\overset{\displaystyle O}{\overset{\|}{C}}CH_3}
\qquad
\underset{\substack{\textbf{Phenol}\\(\textbf{p}K_a = 9.9)}}{\text{⬡—OH}}
\qquad
\underset{\substack{\textbf{Acetic acid}\\(\textbf{p}K_a = 4.76)}}{CH_3\overset{\displaystyle O}{\overset{\|}{C}}OH}
$$

2.45 Which, if any, of the four substances in Problem 2.44 is a strong enough acid to react almost completely with NaOH? (The pK_a of H_2O is 15.74.)

2.46 The ammonium ion ($NH_4{}^+$, $pK_a = 9.25$) has a lower pK_a than the methylammonium ion ($CH_3NH_3{}^+$, $pK_a = 10.66$). Which is the stronger base, ammonia (NH_3) or methylamine (CH_3NH_2)? Explain.

2.47 Is *tert*-butoxide anion a strong enough base to react with water? In other words, can a solution of potassium *tert*-butoxide be prepared in water? The pK_a of *tert*-butyl alcohol is approximately 18.

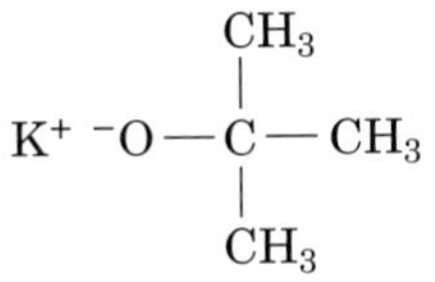

$$
K^+ \ {}^-O-\overset{\displaystyle CH_3}{\underset{\displaystyle CH_3}{\overset{|}{\underset{|}{C}}}}-CH_3 \qquad \textbf{Potassium \textit{tert}-butoxide}
$$

2.48 Predict the structure of the product formed in the reaction of the organic base pyridine with the organic acid acetic acid, and use curved arrows to indicate the direction of electron flow.

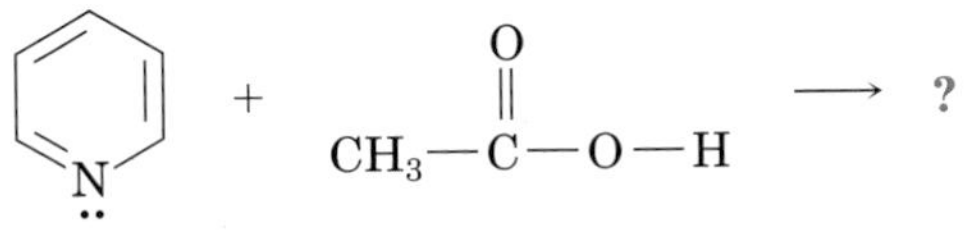

$$
\underset{\textbf{Pyridine}}{\overset{\displaystyle \text{⬡}}{\underset{\displaystyle N}{\vcenter{}}}} \quad + \quad \underset{\textbf{Acetic acid}}{CH_3-\overset{\displaystyle O}{\overset{\|}{C}}-O-H} \quad \longrightarrow \quad ?
$$

2.49 Calculate K_a values from the following pK_a's:
(a) Acetone, $pK_a = 19.3$ (b) Formic acid, $pK_a = 3.75$

2.50 Calculate pK_a values from the following K_a's:
(a) Nitromethane, $K_a = 5.0 \times 10^{-11}$ (b) Acrylic acid, $K_a = 5.6 \times 10^{-5}$

2.51 What is the pH of a 0.050 M solution of formic acid (see Problem 2.49)?

2.52 Sodium bicarbonate, $NaHCO_3$, is the sodium salt of carbonic acid (H_2CO_3), $pK_a = 6.37$. Which of the substances shown in Problem 2.44 will react with sodium bicarbonate?

2.53 Assume that you have two unlabeled bottles, one of which contains phenol ($pK_a = 9.9$) and one of which contains acetic acid ($pK_a = 4.76$). In light of your answer to Problem 2.52, propose a simple way to determine what is in each bottle.

2.54 Identify the acids and bases in the following reactions:

(a) $CH_3OH + H^+ \longrightarrow CH_3\overset{+}{O}H_2$

(b) $CH_3\overset{\displaystyle O}{\overset{\|}{C}}CH_3 + TiCl_4 \longrightarrow H_3C-\overset{\displaystyle \overset{+}{O}-\overset{-}{T}iCl_4}{\overset{\|}{C}}-CH_3$

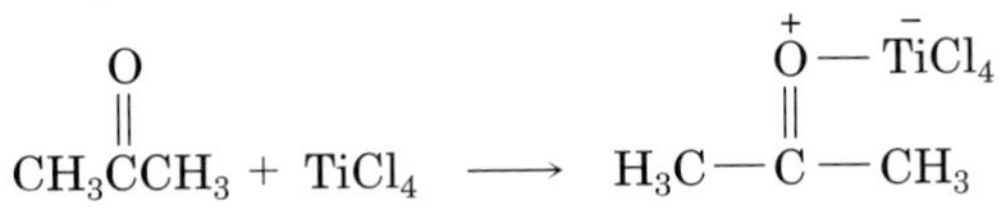

(c)

(d)

2.55 Which of the following pairs represent resonance structures?

(a) $CH_3C \equiv \overset{+}{N} - \overset{..}{\underset{..}{O}} :^-$ and $CH_3\overset{+}{C} = \overset{..}{N} - \overset{..}{\underset{..}{O}} :^-$

(b) $CH_3\overset{\overset{\displaystyle :O:}{\|}}{C} - \overset{..}{\underset{..}{O}} :^-$ and $:\overset{-}{C}H_2\overset{\overset{\displaystyle :O:}{\|}}{C} - \overset{..}{O} - H$

(c)   and

(d) $CH_2 = \overset{+}{N} \begin{smallmatrix} \overset{..}{O}:^- \\ \\ \overset{..}{\underset{..}{O}}:^- \end{smallmatrix}$ and $:\overset{-}{C}H_2 - \overset{+}{N} \begin{smallmatrix} \overset{..}{O}. \\ \\ \overset{..}{\underset{..}{O}}:^- \end{smallmatrix}$

2.56 Draw as many resonance structures as you can for the following species, adding appropriate formal charges to each:

(a) Nitromethane, $H_3C - \overset{+}{N} \begin{smallmatrix} \overset{\displaystyle :O:}{\diagup\diagup} \\ \\ \diagdown \\ :\overset{..}{\underset{..}{O}}:^- \end{smallmatrix}$ (b) Ozone, $:\overset{..}{O} = \overset{+}{\overset{..}{O}} - \overset{..}{\underset{..}{O}} :^-$

(c) Diazomethane, $H_2C = \overset{+}{N} = \overset{..}{\underset{..}{N}} :$

2.57 Dimethyl sulfone has dipole moment $\mu = 4.4$ D. Calculate the formal charges present on oxygen and sulfur, and suggest a geometry for the molecule that is consistent with the observed dipole moment.

$$H_3C - \overset{\overset{\displaystyle :\overset{..}{O}:}{|}}{\underset{\underset{\displaystyle :\overset{..}{\underset{..}{O}}:}{|}}{S}} - CH_3 \qquad \textbf{Dimethyl sulfone}$$

2.58 We said in Section 2.11 that acetic acid can be protonated by H_2SO_4 either on its double-bond oxygen or on its single-bond oxygen. Draw resonance structures of the possible products to explain why the product of protonation on the double-bond oxygen is more stable.

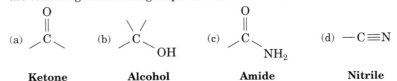

Acetic acid

A Look Ahead

. . . .

2.59 Organic molecules can be classified according to the *functional groups* they contain, where a functional group is a collection of atoms with a characteristic chemical reactivity. Use the electronegativity values given in Figure 2.2 to predict the polarity of the following functional groups. (See Section 3.1.)

(a) Ketone (b) Alcohol (c) Amide (d) —C≡N Nitrile

2.60 Phenol, C_6H_5OH, is a stronger acid than methyl alcohol, CH_3OH, even though both contain an O–H bond. Draw the structures of the anions resulting from loss of H^+ from phenol and methyl alcohol, and use resonance structures to explain the difference in acidity. (See Section 17.3.)

Phenol ($pK_a = 9.89$)　　**Methyl alcohol ($pK_a = 15.54$)**

Molecular Modeling

. . . .

2.61 1,2-Difluoroethane (FCH_2CH_2F) exists in two geometries, called *anti* and *gauche*. Use SpartanView to look at the electrostatic potential map and dipole moment of each. Which geometry has a dipole moment of zero? Why?

2.62 Which atoms in protonated methylamine ($CH_3NH_3^+$) and protonated methyl alcohol ($CH_3OH_2^+$) carry a formal charge? Use SpartanView to examine the electrostatic potential map for each ion, and identify the most positive atom in each. Do the formal charge assignments agree with the electrostatic potential maps?

2.63 Protonation of 4-aminopyridine might give either cation A or cation B. Use SpartanView to look at the electrostatic potential map for each ion, and draw resonance structures for each. Which ion uses resonance more effectively to spread out and stabilize the positive charge?

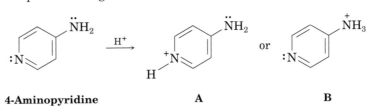

4-Aminopyridine　　　　　　**A**　　　　　**B**

2.64 Use SpartanView to compare electrostatic potential maps of formic acid (HCO_2H), acetic acid (CH_3CO_2H), and pivalic acid [$(CH_3)_3CCO_2H$]. How is the –OH hydrogen different from the others?

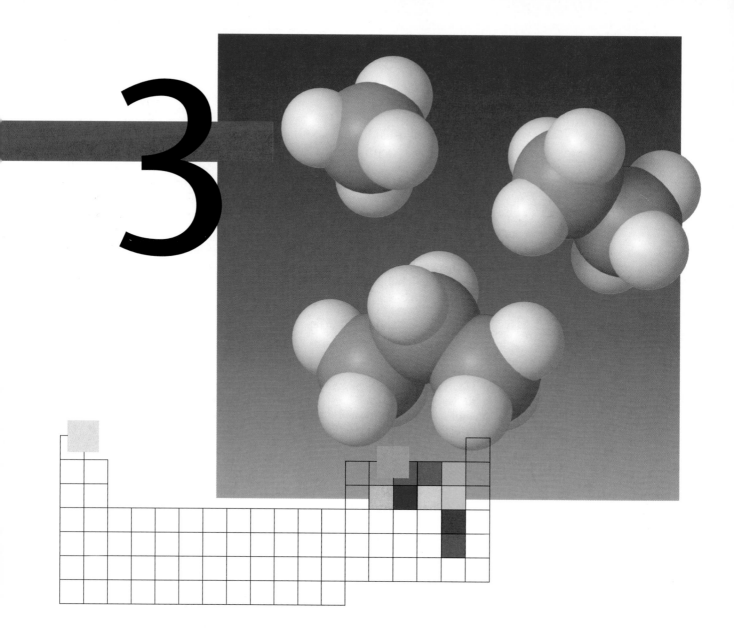

Organic Compounds:
Alkanes and Cycloalkanes

According to *Chemical Abstracts,* the publication that abstracts and indexes the chemical literature, there are more than 18 million known organic compounds. Each of these compounds has its own physical properties, such as melting point and boiling point, and each has its own chemical reactivity.

Chemists have learned through many years of experience that organic compounds can be classified into families according to their structural features and that the members of a given family often have similar chemical behavior. Instead of 18 million compounds with random reactivity, there are a few dozen families of organic compounds whose chemistry is reasonably predictable. We'll study the chemistry of specific families throughout the rest of this book, beginning in the present chapter with a look at the simplest family, the *alkanes*.

3.1 Functional Groups

The structural features that make it possible to classify compounds by reactivity are called *functional groups*. A **functional group** is a group of atoms within a molecule that has a characteristic chemical behavior. Chemically, a given functional group behaves in nearly the same way in every molecule it's a part of. For example, one of the simplest functional groups is the carbon–carbon double bond. Ethylene, the simplest compound with a double bond, undergoes reactions that are remarkably similar to those of cholesterol, a much more complicated molecule that also contains a double bond. Both, for example, react with Br_2 to give products in which a Br atom has added to each of the double-bond carbons (Figure 3.1). This example is typical: *The chemistry of every organic molecule, regardless of size and complexity, is determined by the functional groups it contains.*

FIGURE 3.1 ▼

The reactions of ethylene and cholesterol with bromine. In both molecules, electrostatic potential maps show similar charge patterns for the carbon–carbon double-bond functional group. Bromine therefore reacts with both in exactly the same way. The size and complexity of the remainders of the molecules are not important.

ethylene, cholesterol

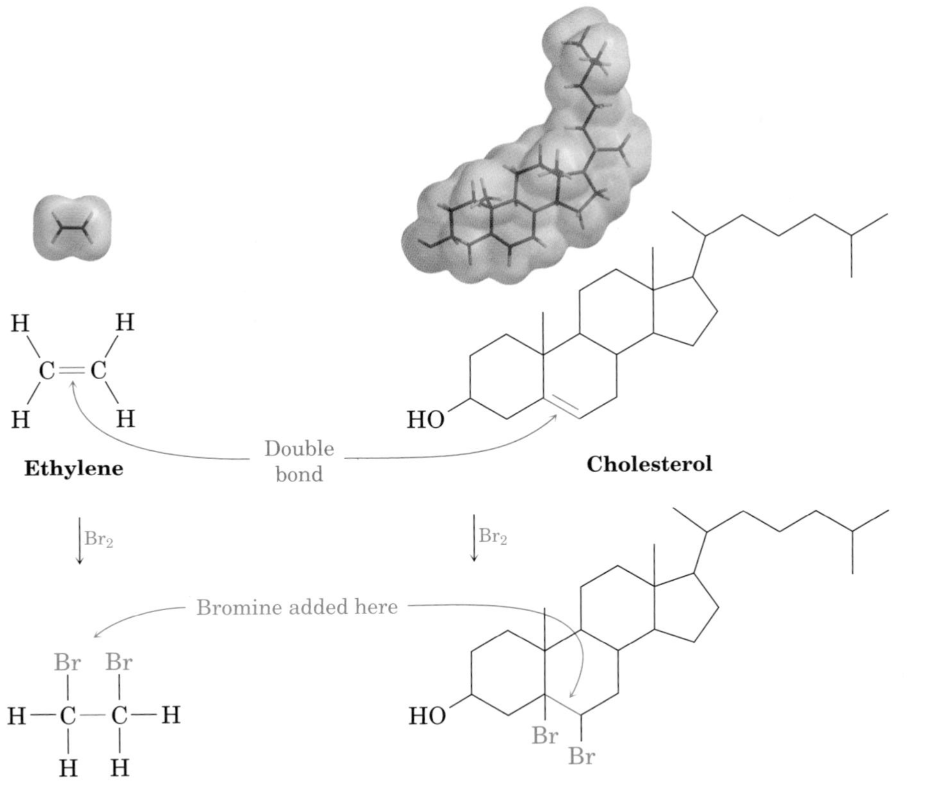

Ethylene

Double bond

Cholesterol

Br₂

Br₂

Bromine added here

Look carefully at Table 3.1, which lists many of the common functional groups and gives simple examples of their occurrence. Some functional groups have only carbon–carbon double or triple bonds; others have halogen atoms; and still others contain oxygen, nitrogen, or sulfur. It's a good idea at this point to familiarize yourself with the structures of the functional groups shown in Table 3.1 so that you'll recognize them when you see them again. Much of the chemistry you'll be studying is the chemistry of these functional groups.

TABLE 3.1 Structures of Some Common Functional Groups

Family name	Functional group structure[a]	Simple example	Name ending
Alkane	(Contains only C—H and C—C single bonds)	CH_3CH_3	*-ane* Ethane
Alkene	C=C	$H_2C=CH_2$	*-ene* Ethene (Ethylene)
Alkyne	—C≡C—	H—C≡C—H	*-yne* Ethyne (Acetylene)
Arene	(benzene ring structure)	(benzene ring structure)	None Benzene
Halide	—C—Ẍ: (X = F, Cl, Br, I)	H_3C—Cl	None Chloromethane
Alcohol	—C—Ö—H	H_3C—O—H	*-ol* Methanol
Ether	—C—Ö—C—	H_3C—O—CH_3	*ether* Dimethyl ether
Amine	—C—N̈—H, —C—N̈—H, —C—N̈—	H_3C—NH_2	*-amine* Methylamine
Nitrile	—C—C≡N:	H_3C—C≡N	*-nitrile* Ethanenitrile (Acetonitrile)
Nitro	—C—N⁺(=Ö:)(Ö:⁻)	H_3C—N⁺(=O)(O⁻)	None Nitromethane
Sulfide	—C—S̈—C—	H_3C—S—CH_3	*sulfide* Dimethyl sulfide
Sulfoxide	—C—S⁺(Ö:⁻)—C—	H_3C—S⁺(O⁻)—CH_3	*sulfoxide* Dimethyl sulfoxide

TABLE 3.1 (Continued)

Family name	Functional group structure[a]	Simple example	Name ending
Sulfone	$-\overset{\vert}{\underset{\vert}{C}}-\overset{\ddot{O}^-}{\underset{\underset{\ddot{}}{\ddot{O}^-}}{\overset{\vert}{\underset{\vert}{S^{2+}}}}}-\overset{\vert}{\underset{\vert}{C}}-$	$H_3C-\overset{O^-}{\underset{O^-}{\overset{\vert}{\underset{\vert}{S^{2+}}}}}-CH_3$	*sulfone* Dimethyl sulfone
Thiol	$-\overset{\vert}{\underset{\vert}{C}}-\ddot{S}-H$	H_3C-SH	*-thiol* Methanethiol
Carbonyl, $-\overset{\overset{\displaystyle :O:}{\|}}{C}-$			
Aldehyde	$-\overset{\overset{\displaystyle :O:}{\|}}{\underset{\vert}{C}}-\overset{\vert}{\underset{\vert}{C}}-H$	$H_3C-\overset{\overset{\displaystyle O}{\|}}{C}-H$	*-al* Ethanal (Acetaldehyde)
Ketone	$-\overset{\vert}{\underset{\vert}{C}}-\overset{\overset{\displaystyle :O:}{\|}}{C}-\overset{\vert}{\underset{\vert}{C}}-$	$H_3C-\overset{\overset{\displaystyle O}{\|}}{C}-CH_3$	*-one* Propanone (Acetone)
Carboxylic acid	$-\overset{\vert}{\underset{\vert}{C}}-\overset{\overset{\displaystyle :O:}{\|}}{C}-\ddot{O}H$	$H_3C-\overset{\overset{\displaystyle O}{\|}}{C}-OH$	*-oic acid* Ethanoic acid (Acetic acid)
Ester	$-\overset{\vert}{\underset{\vert}{C}}-\overset{\overset{\displaystyle :O:}{\|}}{C}-\ddot{O}-\overset{\vert}{\underset{\vert}{C}}-$	$H_3C-\overset{\overset{\displaystyle O}{\|}}{C}-O-CH_3$	*-oate* Methyl ethanoate (Methyl acetate)
Amide	$-\overset{\vert}{\underset{\vert}{C}}-\overset{\overset{\displaystyle :O:}{\|}}{C}-\ddot{N}H_2$	$H_3C-\overset{\overset{\displaystyle O}{\|}}{C}-NH_2$	*-amide* Ethanamide (Acetamide)
	$-\overset{\vert}{\underset{\vert}{C}}-\overset{\overset{\displaystyle :O:}{\|}}{C}-\overset{\vert}{\ddot{N}}-H$		
	$-\overset{\vert}{\underset{\vert}{C}}-\overset{\overset{\displaystyle :O:}{\|}}{C}-\overset{\vert}{\underset{\vert}{\ddot{N}}}-$		
Carboxylic acid chloride	$-\overset{\vert}{\underset{\vert}{C}}-\overset{\overset{\displaystyle :O:}{\|}}{C}-Cl$	$H_3C-\overset{\overset{\displaystyle O}{\|}}{C}-Cl$	*-oyl chloride* Ethanoyl chloride (Acetyl chloride)
Carboxylic acid anhydride	$-\overset{\vert}{\underset{\vert}{C}}-\overset{\overset{\displaystyle :O:}{\|}}{C}-\ddot{O}-\overset{\overset{\displaystyle :O:}{\|}}{C}-\overset{\vert}{\underset{\vert}{C}}-$	$H_3C-\overset{\overset{\displaystyle O}{\|}}{C}-O-\overset{\overset{\displaystyle O}{\|}}{C}-CH_3$	*-oic anhydride* Ethanoic anhydride (Acetic anhydride)

[a]The bonds whose connections aren't specified are assumed to be attached to carbon or hydrogen atoms in the rest of the molecule.

Functional Groups with Carbon–Carbon Multiple Bonds

Alkenes, alkynes, and arenes (aromatic compounds) all contain carbon–carbon multiple bonds. *Alkenes* have a double bond, *alkynes* have a triple bond, and *arenes* have alternating double and single bonds in a six-membered ring of carbon atoms. Because of their structural similarities, these compounds also have chemical similarities.

Alkene	Alkyne	Arene (aromatic ring)

Functional Groups with Carbon Singly Bonded to an Electronegative Atom

Alkyl halides, alcohols, ethers, amines, thiols, and sulfides all have a carbon atom singly bonded to an electronegative atom. *Alkyl halides* have a carbon atom bonded to halogen, *alcohols* have a carbon atom bonded to the oxygen of a hydroxyl group (–OH), *ethers* have two carbon atoms bonded to the same oxygen, *amines* have a carbon atom bonded to a nitrogen, *thiols* have a carbon atom bonded to an –SH group, and *sulfides* have two carbon atoms bonded to the same sulfur. In all cases, the bonds are polar, with the carbon atom bearing a partial positive charge ($\delta+$) and the electronegative atom bearing a partial negative charge ($\delta-$).

Alkyl halide	Alcohol	Ether	Amine	Thiol	Sulfide

Functional Groups with a Carbon–Oxygen Double Bond (Carbonyl Groups)

Note particularly in Table 3.1 the different families of compounds that contain the *carbonyl group*, C=O (pronounced car-bo-**neel**). Carbon–oxygen double bonds are present in some of the most important compounds in organic chemistry. These compounds behave similarly in many respects but differ depending on the identity of the atoms bonded to the carbonyl-group carbon. *Aldehydes* have at least one hydrogen bonded to the C=O, *ketones* have two carbons bonded to the C=O, *carboxylic acids* have an –OH group bonded to the C=O, *esters* have an ether-like oxygen bonded to the C=O, *amides* have an amine-like nitrogen bonded to the C=O, *acid chlorides* have a chlorine bonded to the C=O, and so on.

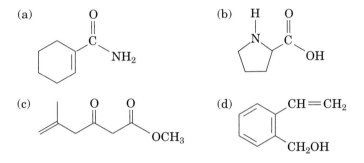

| Aldehyde | Ketone | Carboxylic acid | Ester | Amide | Acid chloride |

• •

Problem 3.1   Identify the functional groups in each of the following molecules:

(a)

(b)

(c)

(d)

Problem 3.2   Propose structures for simple molecules that contain the following functional groups:
(a) Alcohol   (b) Aromatic ring   (c) Carboxylic acid
(d) Amine   (e) Both ketone and amine (f) Two double bonds

Problem 3.3   Identify the functional groups in the following stereo view of arecoline, a veterinary drug used to control worms in animals. Convert the drawing into a line-bond structure and a molecular formula (gray = C, red = O, blue = N, ivory = H).

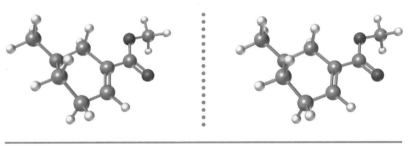

Stereo View

• •

3.2 Alkanes and Alkane Isomers

We saw in Section 1.8 that the carbon–carbon single bond in ethane results from σ (head-on) overlap of carbon sp^3 orbitals. If we imagine joining three, four, five, or even more carbon atoms by C–C single bonds, we can generate the large family of molecules called *alkanes*.

$$H-\underset{\underset{H}{|}}{\overset{\overset{H}{|}}{C}}-H \qquad H-\underset{\underset{H}{|}}{\overset{\overset{H}{|}}{C}}-\underset{\underset{H}{|}}{\overset{\overset{H}{|}}{C}}-H \qquad H-\underset{\underset{H}{|}}{\overset{\overset{H}{|}}{C}}-\underset{\underset{H}{|}}{\overset{\overset{H}{|}}{C}}-\underset{\underset{H}{|}}{\overset{\overset{H}{|}}{C}}-H \qquad H-\underset{\underset{H}{|}}{\overset{\overset{H}{|}}{C}}-\underset{\underset{H}{|}}{\overset{\overset{H}{|}}{C}}-\underset{\underset{H}{|}}{\overset{\overset{H}{|}}{C}}-\underset{\underset{H}{|}}{\overset{\overset{H}{|}}{C}}-H \quad \dots \text{and so on}$$

    Methane   **Ethane**   **Propane**   **Butane**

Alkanes are often described as *saturated hydrocarbons*—**hydrocarbons** because they contain only carbon and hydrogen; **saturated** because they have only C–C and C–H single bonds and thus contain the maximum possible number of hydrogens per carbon. They have the general formula C_nH_{2n+2}, where n is an integer. Alkanes are also occasionally referred to as **aliphatic** compounds, a name derived from the Greek *aleiphas,* meaning "fat." We'll see later that animal fats contain long carbon chains similar to alkanes.

Think about the ways that carbon and hydrogen can combine to make alkanes. With one carbon and four hydrogens, only one structure is possible: methane, CH_4. Similarly, there is only one possible combination of two carbons with six hydrogens (ethane, CH_3CH_3) and only one possible combination of three carbons with eight hydrogens (propane, $CH_3CH_2CH_3$). If larger numbers of carbons and hydrogens combine, however, more than one kind of molecule can result. For example, there are *two* substances with the formula C_4H_{10}: The four carbons can be in a row (butane), or they can branch (isobutane). Similarly, there are three C_5H_{12} molecules, and so on for larger alkanes.

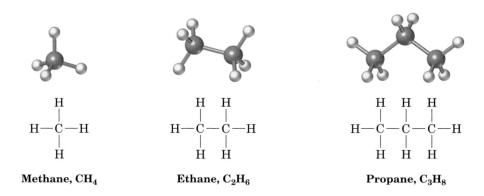

$$H-\underset{\underset{H}{|}}{\overset{\overset{H}{|}}{C}}-H \qquad\qquad H-\underset{\underset{H}{|}}{\overset{\overset{H}{|}}{C}}-\underset{\underset{H}{|}}{\overset{\overset{H}{|}}{C}}-H \qquad\qquad H-\underset{\underset{H}{|}}{\overset{\overset{H}{|}}{C}}-\underset{\underset{H}{|}}{\overset{\overset{H}{|}}{C}}-\underset{\underset{H}{|}}{\overset{\overset{H}{|}}{C}}-H$$

    Methane, CH_4   **Ethane, C_2H_6**   **Propane, C_3H_8**

Compounds like butane and pentane, whose carbons are connected in a row, are called **straight-chain alkanes**, or **normal alkanes**. Compounds like 2-methylpropane (isobutane), 2-methylbutane, and 2,2-dimethylpropane, whose carbon chains branch, are called **branched-chain alkanes**. The difference between the two is that you can draw a line connecting all the carbons of a straight-chain alkane without retracing your path or lifting your pencil from the paper. For a branched-chain alkane, however, you either have to retrace your path or lift your pencil from the paper to draw a line connecting all the carbons.

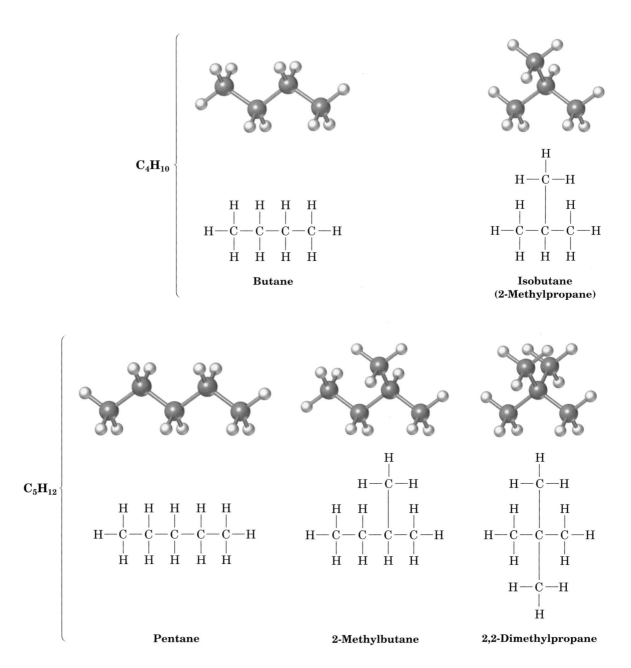

C_4H_{10}

Butane

Isobutane (2-Methylpropane)

C_5H_{12}

Pentane

2-Methylbutane

2,2-Dimethylpropane

Compounds like the two C_4H_{10} molecules and the three C_5H_{12} molecules, which have the same formula but different structures, are called *isomers,* from the Greek *isos + meros,* meaning "made of the same parts." **Isomers** are compounds that have the same numbers and kinds of atoms but differ in the way the atoms are arranged. Compounds like butane and isobutane, whose atoms are connected differently, are called **constitutional isomers**. We'll see shortly that other kinds of isomers are also possible, even among compounds whose atoms are connected in the same order. As Table 3.2 shows, the number of possible alkane isomers increases dramatically as the number of carbon atoms increases.

TABLE 3.2 Number of Alkane Isomers

Formula	Number of isomers	Formula	Number of isomers
C_6H_{14}	5	$C_{10}H_{22}$	75
C_7H_{16}	9	$C_{15}H_{32}$	4,347
C_8H_{18}	18	$C_{20}H_{42}$	366,319
C_9H_{20}	35	$C_{30}H_{62}$	4,111,846,763

Constitutional isomerism is not limited to alkanes—it occurs widely throughout organic chemistry. Constitutional isomers may have different carbon skeletons (as in isobutane and butane), different functional groups (as in ethyl alcohol and dimethyl ether), or different locations of a functional group along the chain (as in isopropylamine and propylamine). Regardless of the reason for the isomerism, constitutional isomers are always different compounds with different properties, but with the same formula.

Different carbon skeletons C_4H_{10}

$CH_3CH(CH_3)CH_3$ (**2-Methylpropane (Isobutane)**) and $CH_3CH_2CH_2CH_3$ (**Butane**)

Different functional groups C_2H_6O

CH_3CH_2OH (**Ethyl alcohol**) and CH_3OCH_3 (**Dimethyl ether**)

Different position of functional groups C_3H_9N

$CH_3CH(NH_2)CH_3$ (**Isopropylamine**) and $CH_3CH_2CH_2NH_2$ (**Propylamine**)

A given alkane can be drawn in many ways. For example, the straight-chain, four-carbon alkane called butane can be represented by any of the structures shown in Figure 3.2. These structures don't imply any particular three-dimensional geometry for butane; they only indicate the connections among atoms. In practice, we usually refer to butane by the condensed

FIGURE 3.2 ▼

Some representations of butane, C_4H_{10}. The molecule is the same regardless of how it's drawn. These structures imply only that butane has a continuous chain of four carbon atoms; they do not imply any specific geometry.

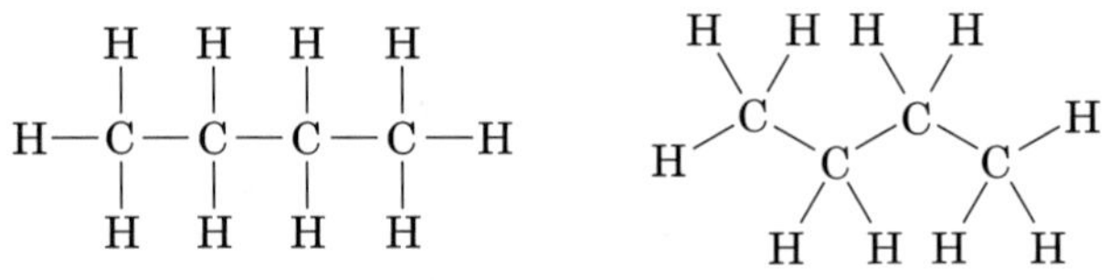

$CH_3—CH_2—CH_2—CH_3$ $CH_3CH_2CH_2CH_3$ $CH_3(CH_2)_2CH_3$

structure $CH_3CH_2CH_2CH_3$, or even more simply as $n\text{-}C_4H_{10}$, where n denotes *normal,* straight-chain butane.

Straight-chain alkanes are named according to the number of carbon atoms in their chain, as shown in Table 3.3. With the exception of the first four compounds—methane, ethane, propane, and butane—whose names have historical roots, the alkanes are named based on Greek numbers. The suffix *-ane* is added to the end of each name to indicate that the molecule identified is an alkane. Thus, pent*ane* is the five-carbon alkane, hex*ane* is the six-carbon alkane, and so on. We'll soon see that these alkane names form the basis for naming all other organic compounds, so at least the first ten should be memorized.

TABLE 3.3 Names of Straight-Chain Alkanes

Number of carbons (n)	Name	Formula (C_nH_{2n+2})	Number of carbons (n)	Name	Formula (C_nH_{2n+2})
1	Methane	CH_4	9	Nonane	C_9H_{20}
2	Ethane	C_2H_6	10	Decane	$C_{10}H_{22}$
3	Propane	C_3H_8	11	Undecane	$C_{11}H_{24}$
4	Butane	C_4H_{10}	12	Dodecane	$C_{12}H_{26}$
5	Pentane	C_5H_{12}	13	Tridecane	$C_{13}H_{28}$
6	Hexane	C_6H_{14}	20	Icosane	$C_{20}H_{42}$
7	Heptane	C_7H_{16}	21	Henicosane	$C_{21}H_{44}$
8	Octane	C_8H_{18}	30	Triacontane	$C_{30}H_{62}$

• •

Practice Problem 3.1 Propose structures for two isomers with the formula C_2H_7N.

Strategy We know that carbon forms four bonds, nitrogen forms three, and hydrogen forms one. Write down the carbon atoms first, and then use a combination of trial and error plus intuition to put the pieces together.

Solution There are two isomeric structures. One has the connection C–C–N, and the other has the connection C–N–C.

These pieces . . . give . . . these structures.

• •

Problem 3.4 Draw structures of the five isomers of C_6H_{14}.

Problem 3.5 There are seven constitutional isomers with the formula $C_4H_{10}O$. Draw as many as you can.

Problem 3.6   Propose structures that meet the following descriptions:
(a) Two isomeric esters with the formula $C_5H_{10}O_2$
(b) Two isomeric nitriles with the formula C_4H_7N

Problem 3.7   How many isomers are there with the following structures?
(a) Alcohols with the formula C_3H_8O (b) Bromoalkanes with the formula C_4H_9Br

3.3 Alkyl Groups

If a hydrogen atom is removed from an alkane, the partial structure that remains is called an **alkyl group**. Note that alkyl groups are not stable compounds themselves; they are simply *parts* of larger compounds. Alkyl groups are named by replacing the *-ane* ending of the parent alkane with an *-yl* ending. For example, removal of a hydrogen from methane, CH_4, generates a *methyl* group, $–CH_3$, and removal of a hydrogen from ethane, CH_3CH_3, generates an *ethyl* group, $–CH_2CH_3$. Similarly, removal of a hydrogen atom from the end carbon of any *n*-alkane gives the series of straight-chain alkyl groups shown in Table 3.4. Combining an alkyl group with any of the functional groups listed earlier makes it possible to generate and name many thousands of compounds. For example:

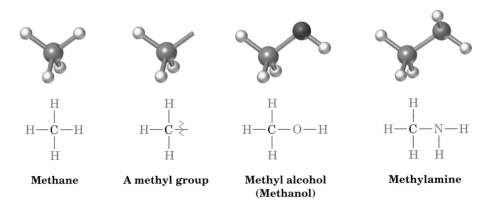

Methane   **A methyl group**   **Methyl alcohol (Methanol)**   **Methylamine**

TABLE 3.4   Some Straight-Chain Alkyl Groups

Alkane	Name	Alkyl group	Name (abbreviation)
CH_4	Methane	$–CH_3$	Methyl (Me)
CH_3CH_3	Ethane	$–CH_2CH_3$	Ethyl (Et)
$CH_3CH_2CH_3$	Propane	$–CH_2CH_2CH_3$	Propyl (Pr)
$CH_3CH_2CH_2CH_3$	Butane	$–CH_2CH_2CH_2CH_3$	Butyl (Bu)
$CH_3CH_2CH_2CH_2CH_3$	Pentane	$–CH_2CH_2CH_2CH_2CH_3$	Pentyl, or Amyl

Just as straight-chain alkyl groups are generated by removing a hydrogen from an *end* carbon, branched alkyl groups are generated by removing a hydrogen atom from an *internal* carbon. Two 3-carbon alkyl groups and four 4-carbon alkyl groups are possible (Figure 3.3).

FIGURE 3.3 ▼

Generation of straight-chain and branched-chain alkyl groups from *n*-alkanes.

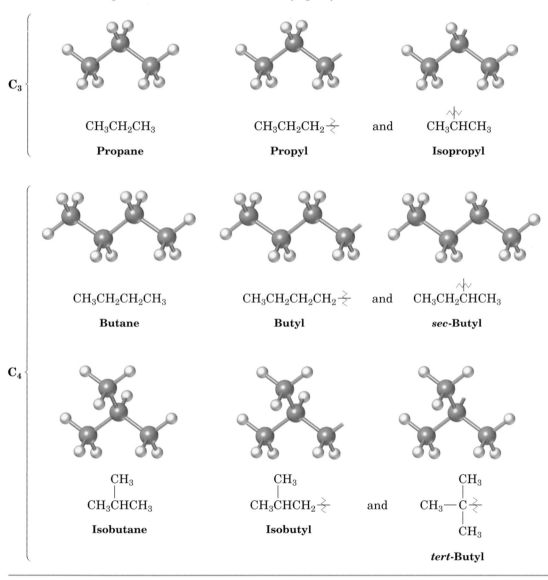

C₃

CH₃CH₂CH₃
Propane

CH₃CH₂CH₂⤙ and CH₃ĊHCH₃
Propyl **Isopropyl**

C₄

CH₃CH₂CH₂CH₃
Butane

CH₃CH₂CH₂CH₂⤙ and CH₃CH₂ĊHCH₃
Butyl ***sec*-Butyl**

CH₃
|
CH₃CHCH₃
Isobutane

CH₃
|
CH₃CHCH₂⤙
Isobutyl

and

CH₃
|
CH₃—C⤙
|
CH₃
***tert*-Butyl**

One further word about naming alkyl groups: The prefixes *sec* (for secondary) and *tert* (for tertiary) used for the C₄ alkyl groups in Figure 3.3 refer to the degree of alkyl substitution at the branching carbon atom. There are four possible degrees of alkyl substitution for carbon, denoted 1° (primary), 2° (secondary), 3° (tertiary), and 4° (quaternary):

$$R-\overset{\overset{\displaystyle H}{|}}{\underset{\underset{\displaystyle H}{|}}{C}}-H \qquad R-\overset{\overset{\displaystyle H}{|}}{\underset{\underset{\displaystyle R}{|}}{C}}-H \qquad R-\overset{\overset{\displaystyle R}{|}}{\underset{\underset{\displaystyle R}{|}}{C}}-H \qquad R-\overset{\overset{\displaystyle R}{|}}{\underset{\underset{\displaystyle R}{|}}{C}}-R$$

Primary carbon (1°) is bonded to one other carbon

Secondary carbon (2°) is bonded to two other carbons

Tertiary carbon (3°) is bonded to three other carbons

Quaternary carbon (4°) is bonded to four other carbons

The symbol **R** is used here and throughout the text to represent a *generalized* organic group. The R group can be methyl, ethyl, propyl, or any of a multitude of others. You might think of R as representing the **R**est of the molecule, which we aren't bothering to specify because it's not important. The terms *primary, secondary, tertiary,* and *quaternary* are routinely used in organic chemistry, and their meanings should become second nature. For example, if we were to say "The product of the reaction is a primary alcohol," we would be talking about the general class of compounds that has an alcohol functional group (–OH) bonded to a carbon atom, which itself is bonded to one R group: $R-CH_2-OH$.

$$R-\overset{\overset{\displaystyle H}{|}}{\underset{\underset{\displaystyle H}{|}}{C}}-OH \qquad\qquad \overset{\overset{\displaystyle CH_3}{|}}{CH_3CHCH_2CH_2OH} \qquad CH_2OH$$

General class of primary alcohols, RCH₂OH

Specific examples of primary alcohols, RCH₂OH

In addition, we also speak about hydrogen atoms as being primary, secondary, or tertiary. Primary hydrogen atoms are attached to primary carbons (RCH_3), secondary hydrogens are attached to secondary carbons (R_2CH_2), and tertiary hydrogens are attached to tertiary carbons (R_3CH). There is, of course, no such thing as a quaternary hydrogen. (Why?)

Primary hydrogens (CH₃)

$$CH_3CH_2\overset{\overset{\displaystyle CH_3}{|}}{C}HCH_3 \;=\; \left[H-\overset{\overset{\displaystyle H}{|}}{\underset{\underset{\displaystyle H}{|}}{C}}-\overset{\overset{\displaystyle H}{|}}{\underset{\underset{\displaystyle H}{|}}{C}}-\overset{\overset{\displaystyle H}{|}}{\underset{\underset{\displaystyle H}{|}}{C}}-\overset{\overset{\displaystyle H}{|}}{\underset{\underset{\displaystyle H}{|}}{C}}-H \right] $$

Secondary hydrogens (CH₂)

A tertiary hydrogen (CH)

• •

Problem 3.8   Draw the eight five-carbon alkyl groups (pentyl isomers).

Problem 3.9   Identify the carbon atoms in the following molecules as primary, secondary, tertiary, or quaternary:

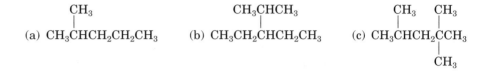

Problem 3.10   Identify the hydrogen atoms on the compounds shown in Problem 3.9 as primary, secondary, or tertiary.

Problem 3.11   Draw structures of alkanes that meet the following descriptions:
(a) An alkane with two tertiary carbons
(b) An alkane that contains an isopropyl group
(c) An alkane that has one quaternary and one secondary carbon

• •

3.4 Naming Alkanes

In earlier times, when relatively few pure organic chemicals were known, new compounds were named at the whim of their discoverer. Thus, urea (CH_4N_2O) is a crystalline substance isolated from urine; morphine ($C_{17}H_{19}NO_3$) is an analgesic (painkiller) named after Morpheus, the Greek god of dreams; and barbituric acid is a tranquilizing agent named by its discoverer in honor of his friend Barbara.

As the science of organic chemistry slowly grew in the nineteenth century, so too did the number of known compounds and the need for a systematic method of naming them. The system of nomenclature we'll use in this book is that devised by the International Union of Pure and Applied Chemistry (IUPAC, usually spoken as **eye**-you-pac).

A chemical name has three parts in the IUPAC system: prefix, parent, and suffix. The parent selects a main part of the molecule and tells how many carbon atoms are in that part; the suffix identifies the functional-group family the molecule belongs to; and the prefix gives the locations of the functional groups and other substituents on the parent.

As we cover new functional groups in later chapters, the applicable IUPAC rules of nomenclature will be given. In addition, Appendix A at the back of this book gives an overall view of organic nomenclature and shows how compounds that contain more than one functional group are named. For the present, let's see how to name branched-chain alkanes.

All but the most complex branched-chain alkanes can be named by following four steps. For a very few compounds, a fifth step is needed.

STEP 1   **Find the parent hydrocarbon.**

(a) Find the *longest continuous chain of carbon atoms* present in the molecule, and use the name of that chain as the parent name. The longest chain may not always be apparent from the manner of writing; you may have to "turn corners."

$$CH_3CH_2CH_2\underset{\displaystyle \overset{|}{CH_2CH_3}}{CH}-CH_3 \qquad \text{Named as a substituted hexane}$$

$$CH_3-\underset{\displaystyle \overset{\displaystyle \overset{\displaystyle CH_3}{\overset{|}{CH_2}}}{|}}{CH}\underset{\displaystyle \overset{|}{CH_2CH_2CH_3}}{CH}-CH_2CH_3 \qquad \text{Named as a substituted heptane}$$

(b) If two different chains of equal length are present, choose the one with the larger number of branch points as the parent:

$$\underset{\displaystyle \overset{|}{CH_2CH_3}}{\underset{\displaystyle \overset{\displaystyle CH_3}{|}}{CH_3CH}CHCH_2CH_2CH_3} \qquad\qquad \textit{NOT} \qquad\qquad CH_3\underset{\displaystyle \overset{\displaystyle CH_3}{|}}{CH}-\underset{\displaystyle \overset{|}{CH_2CH_3}}{CH}CH_2CH_2CH_3$$

Named as a hexane with   as a hexane with
two substituents   *one* substituent

STEP 2   **Number the atoms in the main chain.**

(a) Beginning at the end *nearer the first branch point,* number each carbon atom in the parent chain:

$$CH_3-\underset{\displaystyle \underset{5}{CH_2}\underset{6}{CH_2}\underset{7}{CH_3}}{\underset{3}{CH}}\underset{4}{CH}-CH_2CH_3 \qquad \textit{NOT} \qquad CH_3-\underset{\displaystyle \underset{3}{CH_2}\underset{2}{CH_2}\underset{1}{CH_3}}{\underset{5}{CH}}\underset{4}{CH}-CH_2CH_3$$

The first branch occurs at C3 in the proper system of numbering, not at C4.

(b) If there is branching an equal distance away from both ends of the parent chain, begin numbering at the end nearer the *second* branch point:

CH₃CH₂ — CH₃ CH₂CH₃ ... NOT ... CH₃CH₂ — CH₃ CH₂CH₃

STEP 3 **Identify and number the substituents.**

(a) Assign a number to each substituent according to its point of attachment to the main chain:

Named as a nonane

Substituents: On C3, CH_2CH_3 (3-ethyl)
On C4, CH_3 (4-methyl)
On C7, CH_3 (7-methyl)

(b) If there are two substituents on the same carbon, give them both the same number. There must be as many numbers in the name as there are substituents.

Named as a hexane

Substituents: On C2, CH_3 (2-methyl)
On C4, CH_3 (4-methyl)
On C4, CH_2CH_3 (4-ethyl)

STEP 4 **Write the name as a single word.** Use hyphens to separate the different prefixes, and use commas to separate numbers. If two or more different substituents are present, cite them in alphabetical order. If two or more identical substituents are present, use one of the multiplier prefixes *di-*, *tri-*, *tetra-*, and so forth. Don't use these prefixes for alphabetizing purposes, however. Full names for some of the examples we have been using follow:

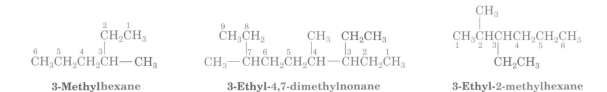

3-Methylhexane **3-Ethyl-4,7-dimethylnonane** **3-Ethyl-2-methylhexane**

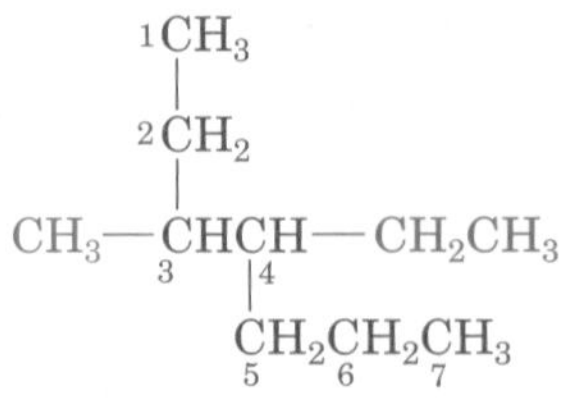

$^1CH_3$

$^2CH_2$

$CH_3-\underset{3}{C}H\underset{|}{C}H-CH_2CH_3$
            ^{4}
$\underset{5}{C}H_2\underset{6}{C}H_2\underset{7}{C}H_3$

4-Ethyl-3-methylheptane

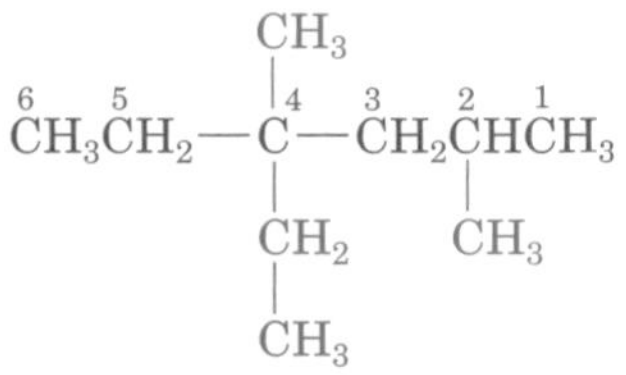

CH_3

$\underset{6}{C}H_3\underset{5}{C}H_2-\overset{4}{C}-\underset{3}{C}H_2\underset{2}{C}H\underset{1}{C}H_3$

CH_2 CH_3

CH_3

4-Ethyl-2,4-dimethylhexane

In some particularly complex cases, a fifth step is necessary. It occasionally happens that a substituent of the main chain has sub-branching:

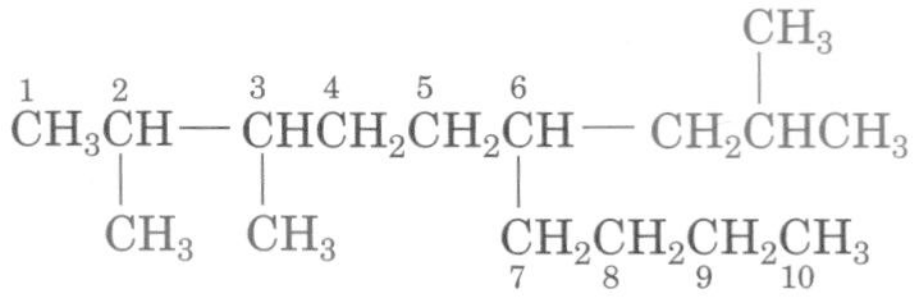

CH_3

$\underset{1}{C}H_3\underset{2}{C}H-\underset{3}{C}H\underset{4}{C}H_2\underset{5}{C}H_2\underset{6}{C}H-\underset{}{C}H_2CHCH_3$

CH_3 CH_3 $\underset{7}{C}H_2\underset{8}{C}H_2\underset{9}{C}H_2\underset{10}{C}H_3$

Named as a 2,3,6-
trisubstituted decane

In this case, the substituent at C6 is a four-carbon unit with a sub-branch. To name the compound fully, the sub-branched substituent must first be named.

STEP 5   **Name a complex substituent just as though it were itself a compound.** For the compound shown above, the complex substituent is a substituted propyl group:

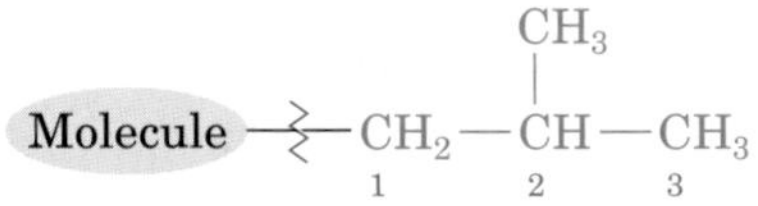

CH_3

Molecule $-\!\!\!\!\!\!\!\!\succ\!\!-\underset{1}{C}H_2-\underset{2}{C}H-\underset{3}{C}H_3$

We begin numbering *at the point of attachment* to the main chain and find that the complex substituent is a 2-methylpropyl group. The substituent is alphabetized according to the first letter of its complete name (including any numerical prefix) and is set off in parentheses when naming the complete molecule:

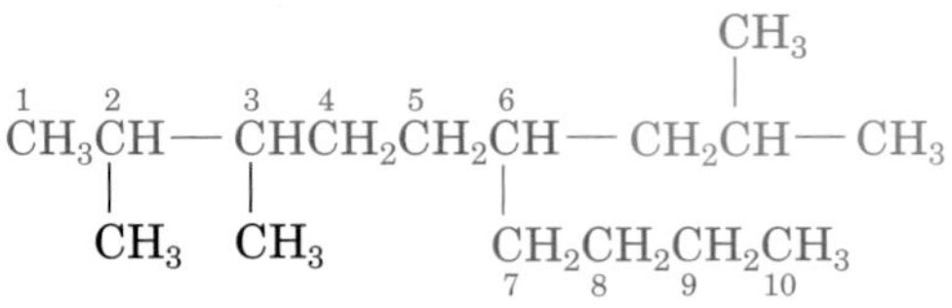

CH_3

$\underset{1}{C}H_3\underset{2}{C}H-\underset{3}{C}H\underset{4}{C}H_2\underset{5}{C}H_2\underset{6}{C}H-\underset{}{C}H_2CH-CH_3$

CH_3 CH_3 $\underset{7}{C}H_2\underset{8}{C}H_2\underset{9}{C}H_2\underset{10}{C}H_3$

2,3-Dimethyl-6-(2-methylpropyl)decane

As a further example:

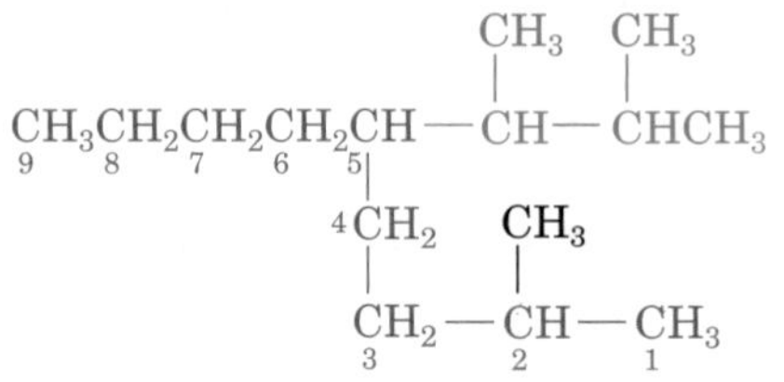

CH_3 CH_3

$\underset{9}{C}H_3\underset{8}{C}H_2\underset{7}{C}H_2\underset{6}{C}H_2\underset{5}{C}H-\underset{}{C}H-CHCH_3$

$^4CH_2$ CH_3

$CH_2-\underset{2}{C}H-\underset{1}{C}H_3$
  $_3$

5-(1,2-Dimethylpropyl)-2-methylnonane

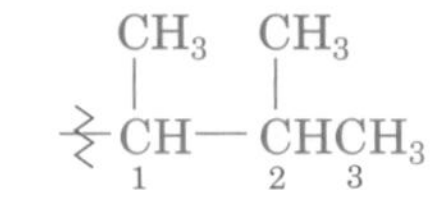

CH_3 CH_3

$-\!\!\!\!\!\!\succ\!\!\underset{1}{C}H-\underset{2}{C}HCH_3$
              $_3$

5-(1,2-Dimethylpropyl)-

For historical reasons, some of the simpler branched-chain alkyl groups also have nonsystematic, or *common,* names, as noted earlier.

1. Three-carbon alkyl group:

$$CH_3\overset{\wedge\!\!\vee}{C}HCH_3$$

Isopropyl (*i*-Pr)

2. Four-carbon alkyl groups:

$$CH_3CH_2\overset{\wedge\!\!\vee}{C}HCH_3 \qquad \underset{\overset{|}{CH_3}}{CH_3}CHCH_2\!\!\Rightarrow \qquad CH_3\!-\!\underset{\overset{|}{CH_3}}{\overset{\overset{CH_3}{|}}{C}}\!\!\Rightarrow$$

***sec*-Butyl**	**Isobutyl**	***tert*-Butyl**
(*sec*-Bu)		**(*t*-Butyl or *t*-Bu)**

3. Five-carbon alkyl groups:

$$\underset{\overset{|}{CH_3}}{CH_3}CHCH_2CH_2\!\!\Rightarrow \qquad CH_3\!-\!\underset{\overset{|}{CH_3}}{\overset{\overset{CH_3}{|}}{C}}\!-\!CH_2\!\!\Rightarrow \qquad CH_3CH_2\!-\!\underset{\overset{|}{CH_3}}{\overset{\overset{CH_3}{|}}{C}}\!\!\Rightarrow$$

Isopentyl, also called	**Neopentyl**	***tert*-Pentyl, also called**
isoamyl (*i*-amyl)		***tert*-amyl (*t*-amyl)**

The common names of these simple alkyl groups are so well entrenched in the chemical literature that IUPAC rules make allowance for them. Thus, the following compound is properly named *either* 4-(1-methylethyl)heptane or 4-isopropylheptane. There is no choice but to memorize these common names; fortunately, there aren't many of them.

$$\underset{CH_3CH_2CH_2CHCH_2CH_2CH_3}{\overset{CH_3CHCH_3}{\overset{|}{}}}$$

4-(1-Methylethyl)heptane   or   4-Isopropylheptane

When writing an alkane name, the nonhyphenated prefix *iso* is considered part of the alkyl-group name for alphabetizing purposes, but the hyphenated prefixes *sec*- and *tert*- are not. Thus, isopropyl and isobutyl are listed alphabetically under *i*, but *sec*-butyl and *tert*-butyl are listed under *b*.

• •

Practice Problem 3.2   What is the IUPAC name of the following alkane?

$$\underset{CH_3CHCH_2CH_2CH_2CHCH_3}{\overset{\overset{CH_2CH_3}{|}\qquad\qquad\overset{CH_3}{|}}{}}$$

Strategy   Find the longest continuous carbon chain in the molecule and use that as the parent name. Then name and number the substituents.

Solution   The molecule has a chain of eight carbons (octane) with two methyl substituents. (You have to turn corners to see it.) Numbering from the end nearer the first methyl substituent indicates that the methyls are at C2 and C6, giving the name 2,6-dimethyloctane.

$$
\begin{array}{c}
\overset{7}{C}H_2\overset{8}{C}H_3 \qquad\qquad CH_3 \\
| \qquad\qquad\qquad\quad | \\
CH_3\underset{6}{C}H\underset{5}{C}H_2\underset{4}{C}H_2\underset{3}{C}H_2\underset{2}{C}H\underset{1}{C}H_3 \qquad \textbf{2,6-Dimethyloctane}
\end{array}
$$

• •

Practice Problem 3.3   Draw the structure of 3-isopropyl-2-methylhexane.

Strategy   This is the reverse of Practice Problem 3.2 and uses a reverse strategy. Look at the parent name (hexane) and draw its carbon structure. Then identify the substituents and attach them.

Solution   Draw the parent compound (hexane):

$$ C{-}C{-}C{-}C{-}C{-}C \qquad \textbf{Hexane} $$

Then place the substituents (3-isopropyl and 2-methyl) on the proper carbons:

$$
\begin{array}{c}
CH_3CHCH_3 \quad\longleftarrow\quad \text{An isopropyl group at C3} \\
| \\
\underset{1}{C}{-}\underset{2}{C}{-}\underset{3}{C}{-}\underset{4}{C}{-}\underset{5}{C}{-}\underset{6}{C} \\
| \\
CH_3 \quad\longleftarrow\quad \text{A methyl group at C2}
\end{array}
$$

Finally, add hydrogens to complete the structure:

$$
\begin{array}{c}
CH_3CHCH_3 \\
| \\
CH_3CHCHCH_2CH_2CH_3 \qquad \textbf{3-Isopropyl-2-methylhexane} \\
| \\
CH_3
\end{array}
$$

• •

Problem 3.12   Give IUPAC names for the following compounds:

(a) The three isomers of C_5H_{12}

(b)
$$
\begin{array}{c}
CH_3 \\
| \\
CH_3CH_2CHCHCH_3 \\
| \\
CH_2CH_3
\end{array}
$$

(c)
$$
\begin{array}{c}
\qquad\quad CH_3 \\
\qquad\quad | \\
(CH_3)_2CHCH_2CHCH_3
\end{array}
$$

(d)
$$
\begin{array}{c}
CH_3 \\
| \\
(CH_3)_3CCH_2CH_2CH \\
| \\
CH_2CH_3
\end{array}
$$

Problem 3.13 Draw structures corresponding to the following IUPAC names:
(a) 3,4-Dimethylnonane (b) 3-Ethyl-4,4-dimethylheptane
(c) 2,2-Dimethyl-4-propyloctane (d) 2,2,4-Trimethylpentane

Problem 3.14 Name the eight five-carbon alkyl groups you drew in Problem 3.8.

Problem 3.15 Give the IUPAC name for the following hydrocarbon, and convert the drawing into a skeletal structure (gray = C, ivory = H):

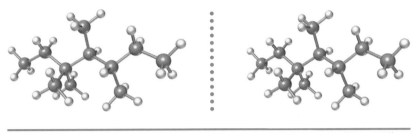

Stereo View

3.5 Properties of Alkanes

Alkanes are sometimes referred to as **paraffins**, a word derived from the Latin *parum affinis,* meaning "slight affinity." This term aptly describes their behavior, for alkanes show little chemical affinity for other substances and are chemically inert to most laboratory reagents. Alkanes do, however, react with oxygen, chlorine, and a few other substances under appropriate conditions.

Reaction with oxygen occurs during combustion in an engine or furnace when the alkane is used as a fuel. Carbon dioxide and water are formed as products, and a large amount of heat is released. For example, methane (natural gas) reacts with oxygen according to the equation

$$CH_4 + 2\,O_2 \longrightarrow CO_2 + 2\,H_2O + 890 \text{ kJ/mol (213 kcal/mol)}$$

The reaction of an alkane with Cl_2 occurs when a mixture of the two is irradiated with ultraviolet light (denoted hν, where ν is the Greek letter nu). Depending on the relative amounts of the two reactants and on the time allowed, a sequential substitution of the alkane hydrogen atoms by chlorine occurs, leading to a mixture of chlorinated products. Methane, for example, reacts with Cl_2 to yield a mixture of CH_3Cl, CH_2Cl_2, $CHCl_3$, and CCl_4. We'll explore the details of this reaction in Section 5.3.

$$CH_4 + Cl_2 \xrightarrow{\ h\nu\ } CH_3Cl + HCl$$
$$\xrightarrow{\ Cl_2\ } CH_2Cl_2 + HCl$$
$$\xrightarrow{\ Cl_2\ } CHCl_3 + HCl$$
$$\xrightarrow{\ Cl_2\ } CCl_4 + HCl$$

Alkanes show regular increases in both boiling point and melting point as molecular weight increases (Figure 3.4), an effect that is due to the presence of weak *van der Waals forces* between molecules. These intermolecular forces, which operate only over very small distances, arise because the electron distribution in an alkane molecule, although uniform over time, is likely to be *nonuniform* at any given instant. One side of a molecule may, by chance, have a slight excess of electrons relative to the opposite side, giving the molecule a temporary dipole moment. This temporary dipole in one molecule causes a nearby molecule to adopt a temporarily opposite dipole, with the result that a tiny electrical attraction is induced between the two (Figure 3.5).

FIGURE 3.4 ▼

A plot of melting and boiling points versus number of carbon atoms for the C_1–C_{14} alkanes. There is a regular increase with molecular size.

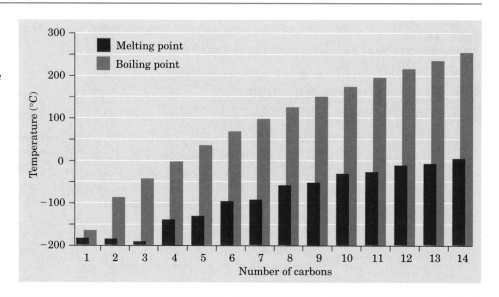

Temporary molecular dipoles have a fleeting existence and are constantly changing, but the cumulative effect of an enormous number of them produces attractive forces sufficient to cause a substance to remain in the liquid or solid state. Only when sufficient energy is applied to overcome these forces does the solid melt or liquid boil. As you might expect, van der Waals forces increase as molecule size increases, accounting for the higher melting and boiling points of larger alkanes.

Another interesting effect seen in alkanes is that increased branching lowers an alkane's boiling point. Thus, pentane has no branches and boils at 36.1°C, isopentane (2-methylbutane) has one branch and boils at 27.85°C, and neopentane (2,2-dimethylpropane) has two branches and boils at 9.5°C. Similarly, octane boils at 125.7°C, whereas isooctane (2,2,4-trimethylpentane) boils at 99.3°C. Branched-chain alkanes are lower-boiling because they are more nearly spherical than straight-chain alkanes, have smaller surface areas, and consequently have smaller van der Waals forces.

FIGURE 3.5 ▼

Attractive van der Waals forces are caused by temporary dipoles in molecules, as shown in these space-filling models of pentane.

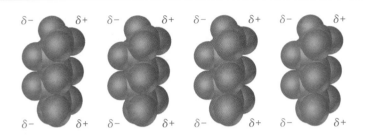

3.6 Cycloalkanes

We've discussed only open-chain alkanes up to this point, but chemists have known for over a century that compounds with *rings* of carbon atoms also exist. Such compounds are called **cycloalkanes**, or **alicyclic compounds** (**ali**phatic **cyclic**). Since cycloalkanes consist of rings of $-CH_2-$ units, they have the general formula $(CH_2)_n$, or C_nH_{2n}, and are represented by polygons in skeletal drawings:

Cyclopropane   **Cyclobutane**   **Cyclopentane**   **Cyclohexane**

Alicyclic compounds with many different ring sizes abound in nature. For example, chrysanthemic acid contains a three-membered (cyclopropane) ring. Various esters of chrysanthemic acid occur naturally as the active insecticidal constituents of chrysanthemum flowers.

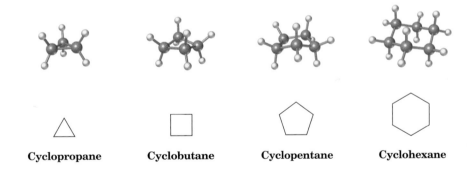

Chrysanthemic acid

Prostaglandins, such as PGE_1, contain a five-membered (cyclopentane) ring. Prostaglandins are potent hormones that control a wide variety of physiological functions in humans, including blood platelet aggregation, bronchial dilation, and inhibition of gastric secretions.

Prostaglandin E$_1$ (PGE$_1$)

Steroids, such as cortisone, contain four rings joined together—three of them six-membered (cyclohexane) and one five-membered (cyclopentane). We'll discuss steroids in more detail in Sections 27.7 and 27.8.

Cortisone

The melting points and boiling points of some simple unsubstituted cycloalkanes are shown in Figure 3.6. Melting points are affected irregularly by increasing molecular weight because the different shapes of the various cycloalkanes cause differences in the efficiency with which molecules pack together in crystals. Boiling points, however, show a regular increase with molecular weight.

FIGURE 3.6 ▼

Melting points and boiling points for cycloalkanes, cyclo-$(CH_2)_n$.

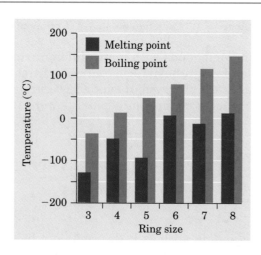

3.7 Naming Cycloalkanes

Substituted cycloalkanes are named by rules similar to those used for open-chain alkanes. For most compounds, there are only two rules:

RULE 1 **Find the parent.** Count the number of carbon atoms in the ring and the number in the largest substituent chain. If the number of carbon atoms in the ring is equal to or greater than the number in the substituent, the compound is named as an alkyl-substituted cycloalkane. If the number of carbon atoms in the largest substituent is greater than the number in the ring, the compound is named as a cycloalkyl-substituted alkane. For example:

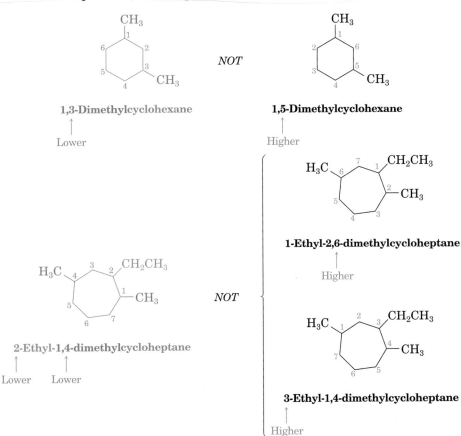

RULE 2 **Number the substituents.** For alkyl- and halo-substituted cycloalkanes, choose a point of attachment as C1 and number the substituents on the ring so that the *second* substituent has as low a number as possible. If ambiguity still exists, number so that the *third* or *fourth* substituent has as low a number as possible, until a point of difference is found:

(a) When two or more different alkyl groups that could potentially receive the same numbers are present, number them by alphabetical priority:

1-Ethyl-2-methylcyclopentane   **2-Ethyl-1-methylcyclopentane**

(b) If halogens are present, treat them exactly like alkyl groups:

1-Bromo-2-methylcyclobutane   **2-Bromo-1-methylcyclobutane**

Some additional examples follow:

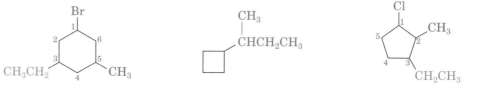

1-Bromo-3-ethyl-5-methyl-   (1-Methylpropyl)cyclobutane   1-Chloro-3-ethyl-2-methyl-
cyclohexane   (or *sec*-Butylcyclobutane)   **cyclopentane**

• •

Problem 3.16   Give IUPAC names for the following cycloalkanes:

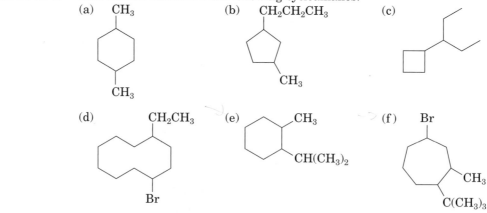

Problem 3.17   Draw structures corresponding to the following IUPAC names:
(a) 1,1-Dimethylcyclooctane   (b) 3-Cyclobutylhexane
(c) 1,2-Dichlorocyclopentane   (d) 1,3-Dibromo-5-methylcyclohexane

• •

3.8 Cis–Trans Isomerism in Cycloalkanes

In many respects, the chemistry of cycloalkanes is like that of open-chain, acyclic alkanes: Both classes of compounds are nonpolar and fairly inert. There are, however, some important differences.

One difference is that cycloalkanes are less flexible than their open-chain counterparts. To see what this means, think about the nature of a carbon–carbon single bond. We know from Section 1.7 that σ bonds are cylindrically symmetrical. In other words, the intersection of a plane cutting through a carbon–carbon single-bond orbital looks like a circle. Because of this cylindrical symmetry, *rotation* is possible around carbon–carbon bonds in open-chain molecules. In ethane, for example, rotation around the C–C bond occurs freely, constantly changing the geometric relationships of the hydrogens on one carbon with those on the other (Figure 3.7).

FIGURE 3.7 ▼

Rotation occurs around the carbon–carbon single bond in ethane because of σ bond cylindrical symmetry.

ethane
(see computer animation on CD-ROM)

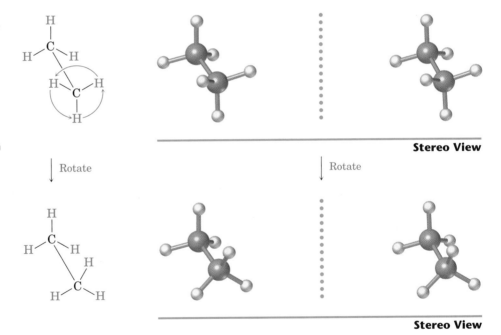

Stereo View

↓ Rotate   ↓ Rotate

Stereo View

In contrast to the rotational freedom around single bonds in open-chain alkanes, there is much less freedom in cycloalkanes. Cyclopropane, for example, must be a rigid, planar molecule (three points define a plane). No bond rotation can take place around a cyclopropane carbon–carbon bond without breaking open the ring (Figure 3.8, p. 100).

Larger cycloalkanes have increasingly more rotational freedom, and the very large rings (C_{25} and up) are so floppy that they are nearly indistinguishable from open-chain alkanes. The common ring sizes (C_3, C_4, C_5, C_6, C_7), however, are severely restricted in their molecular motions.

FIGURE 3.8 ▼

The structure of cyclopropane. No rotation is possible around the carbon–carbon bonds without breaking open the ring.

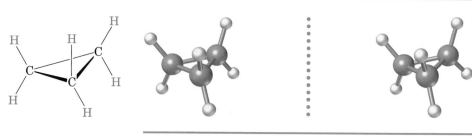

Stereo View

Because of their cyclic structures, cycloalkanes have two sides, a "top" side and a "bottom" side, leading to the possibility of isomerism in substituted cycloalkanes. For example, there are two different 1,2-dimethylcyclopropane isomers, one with the two methyls on the same side of the ring, and one with the methyls on opposite sides (Figure 3.9). Both isomers are stable compounds; neither can be converted into the other without breaking and reforming chemical bonds. Make molecular models to prove this to yourself.

FIGURE 3.9 ▼

There are two different 1,2-dimethylcyclopropane isomers, one with the methyl groups on the same side of the ring and the other with the methyl groups on opposite sides of the ring.

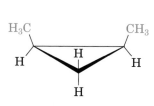  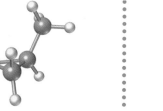 

cis-**1,2-Dimethylcyclopropane**   **Stereo View**

Do *NOT* interconvert

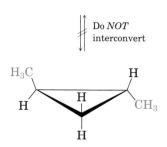  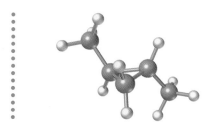

trans-**1,2-Dimethylcyclopropane**   **Stereo View**

Unlike the constitutional isomers butane and isobutane (Figure 3.3), which have their atoms connected in a different order, the two 1,2-dimethylcyclopropanes have the *same* order of connection but differ in the spatial orientation of their atoms. Such compounds, which have their atoms connected in the same order but differ in three-dimensional orientation, are called **stereoisomers**.

Constitutional isomers (different connections between atoms)

$$CH_3-CH-CH_3 \quad \text{and} \quad CH_3-CH_2-CH_2-CH_3$$
$$\overset{|}{CH_3}$$

Stereoisomers (same connections but different three-dimensional geometry)

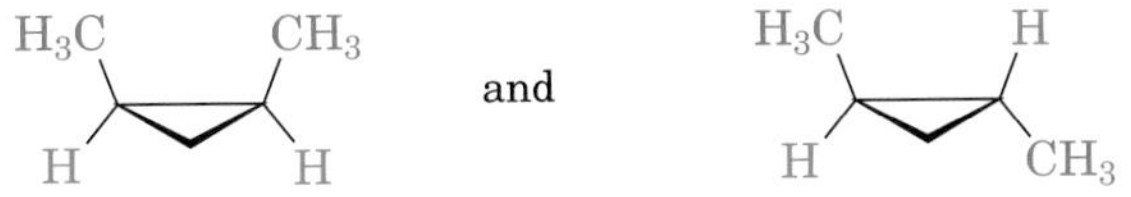

The 1,2-dimethylcyclopropanes are special kinds of stereoisomers called **cis–trans isomers**. The prefixes *cis*- (Latin, "on the same side") and *trans*- (Latin, "across") are used to distinguish between them. Cis–trans isomerism is a common occurrence in substituted cycloalkanes.

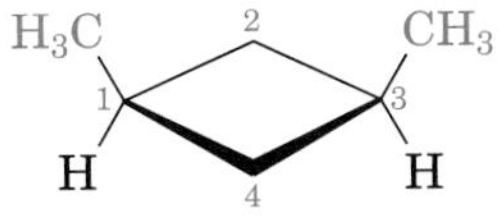

cis-**1,3-Dimethylcyclobutane** 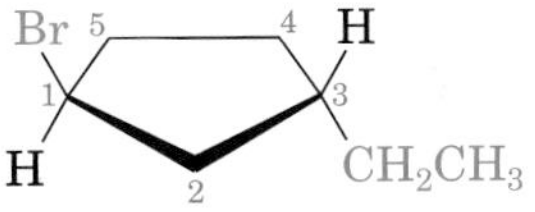   *trans*-**1-Bromo-3-ethylcyclopentane**

· ·

Practice Problem 3.4   Name the following substances, specifying each as cis or trans:

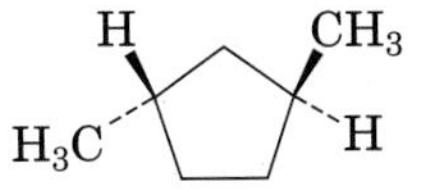

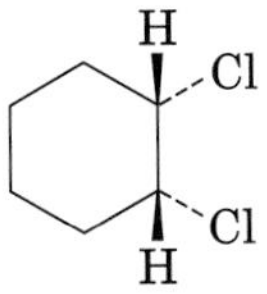

Strategy   In these views, the ring is roughly in the plane of the page, a wedged bond is above the page, and a dashed bond recedes below the page. Two substituents are cis if they are both above or below the page; they are trans if one is above and one is below.

Solution   (a) *trans*-1,3-Dimethylcyclopentane   (b) *cis*-1,2-Dichlorocyclohexane

· ·

Problem 3.18   Name the following substances, specifying each as cis or trans:

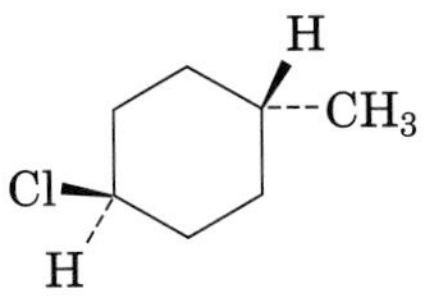

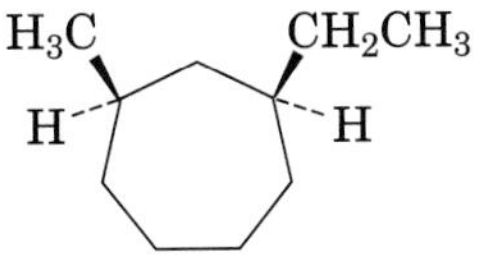

Problem 3.19   Draw the structures of the following molecules:

(a) *trans*-1-Bromo-3-methylcyclohexane   (b) *cis*-1,2-Dimethylcyclopentane

(c) *trans*-1-*tert*-Butyl-2-ethylcyclohexane

 **CHEMISTRY @ WORK**

Gasoline from Petroleum

Natural gas and petroleum deposits make up the world's largest source of alkanes. Laid down eons ago, these deposits are derived from the decomposition of plant and animal matter, primarily of marine origin. *Natural gas* consists chiefly of methane, but also contains ethane, propane, butane, and isobutane. *Petroleum* is a complex mixture of hydrocarbons that must be refined into fractions before it can be used.

Refining begins by distillation of crude oil into three principal cuts: straight-run gasoline (bp 30–200°C), kerosene (bp 175–300°C), and gas oil (bp 275–400°C). Finally, distillation under reduced pressure gives lubricating oils and waxes, and leaves an undistillable tarry residue of asphalt.

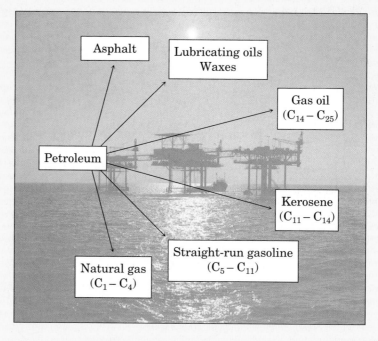

(continued) ▶

The distillation of crude oil is only the first step in gasoline production. Straight-run gasoline turns out to be a poor fuel because of *engine knock*. In the typical four-stroke automobile engine, a piston draws a mixture of fuel and air into a cylinder on its downward stroke and compresses the mixture on its upward stroke. Just before the end of the compression, a spark plug ignites the mixture and combustion occurs, driving the piston downward and turning the crankshaft.

Not all fuels burn equally well, though. When poor fuels are used, uncontrolled combustion can be initiated by a hot surface in the cylinder before the spark plug fires. This preignition, detected as an engine knock, can destroy the engine by putting irregular forces on the crankshaft and raising engine temperature.

The *octane number* of a fuel is the measure by which its antiknock properties are judged. It was recognized long ago that straight-chain hydrocarbons are far more prone to induce engine knock than are highly branched compounds. Heptane, a particularly bad fuel, is assigned a base value of 0 octane number; 2,2,4-trimethylpentane (commonly known as isooctane) has a rating of 100.

$$CH_3CH_2CH_2CH_2CH_2CH_2CH_3$$

$$CH_3\overset{\overset{\displaystyle CH_3}{|}}{C}CH_2\overset{\overset{\displaystyle CH_3}{|}}{C}HCH_3$$
$$\underset{\underset{\displaystyle CH_3}{|}}{}$$

Heptane
(octane number = 0)

2,2,4-Trimethylpentane
(octane number = 100)

Because straight-run gasoline has a high percentage of unbranched alkanes and is therefore a poor fuel, petroleum chemists have devised several methods for producing higher-quality fuels. One of these methods, *catalytic cracking,* involves taking the high-boiling kerosene cut (C_{11}–C_{14}) and "cracking" it into smaller molecules suitable for use in gasoline. The process takes place on a silica–alumina catalyst at temperatures of 400–500°C, and the major products are light hydrocarbons in the C_3–C_5 range. These small hydrocarbons are then catalytically recombined to yield useful C_7–C_{10} alkanes.

The petroleum flowing from the north slope of Alaska through this pipeline is a complex mixture of alkanes and other organic substances.

Summary and Key Words

A **functional group** is a group of atoms within a larger molecule that has a characteristic chemical reactivity. Because functional groups behave approximately the same way in all molecules where they occur, the chemical reactions of an organic molecule are largely determined by its functional groups.

Alkanes are a class of **hydrocarbons** with the general formula C_nH_{2n+2}. They contain no functional groups, are relatively inert, and can be either straight-chain (***normal* alkanes**) or branched. Alkanes are named by a series of IUPAC rules of nomenclature. Compounds that have the same chemical formula but different structures are called **isomers**. More specifically, compounds such as butane and isobutane, which differ in their connections between atoms, are called **constitutional isomers**.

Cycloalkanes contain rings of carbon atoms and have the general formula C_nH_{2n}. Although free rotation is possible around C–C single bonds in open-chain alkanes, rotation is greatly reduced in cycloalkanes. Disubstituted cycloalkanes can therefore exist as **cis–trans isomers**. The cis isomer has both substituents on the same side of the ring; the trans isomer has substituents on opposite sides of the ring. Cis–trans isomers are just one kind of **stereoisomers**—isomers that have the same connections between atoms but differ in their three-dimensional arrangements.

Visualizing Chemistry

• •

(*Problems 3.1–3.19 appear within the chapter.*)

3.20 Identify the functional groups in the following substances, and convert each drawing into a molecular formula (gray = C, red = O, blue = N, ivory = H).

(a)

(b)

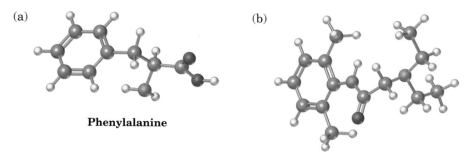

Phenylalanine

Lidocaine

3.21 Give IUPAC names for the following hydrocarbons, and convert each drawing into a skeletal structure (gray = C, yellow-green = Cl, ivory = H).

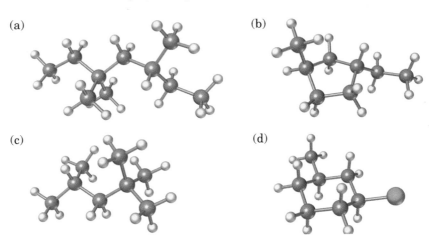

(a)   (b)

(c)   (d)

3.22 The following cyclohexane derivative has three substituents—red, green, and blue. Identify each pair of relationships (red–blue, red–green, and blue–green) as cis or trans.

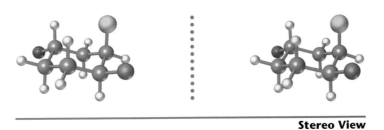

Stereo View

Additional Problems

3.23 Locate and identify the functional groups in the following molecules:

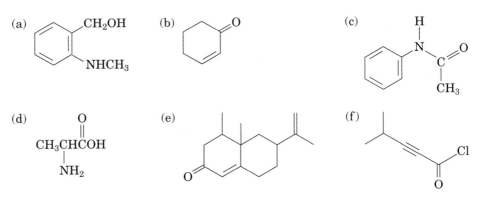

(a)   CH$_2$OH
      NHCH$_3$

(b)   O

(c)   H
      N   O
               C
               CH$_3$

(d)   O
           ‖
      CH$_3$CHCOH
           |
           NH$_2$

(e)   O

(f)   Cl
           O

3.24 Draw structures that meet the following descriptions (there are many possibilities):
(a) Three isomers with the formula C_8H_{18}
(b) Two isomers with the formula $C_4H_8O_2$

3.25 Draw structures of the nine isomers of C_7H_{16}.

3.26 In each of the following sets, which structures represent the same compound, and which represent different compounds?

(a) $CH_3CH(Br)CHCH_3$ $(CH_3)_2CHCH(Br)CH_3$ $CH_3CHCHCH_3$
with CH_3 on the third carbon;   with Br and CH_3 substituents

(b) [structure: benzene with OH, OH (ortho)]   [structure: benzene with HO, HO]   [structure: benzene with HO, OH]

(c) $CH_3CH_2CHCH_2CHCH_3$ with CH_3 and CH_2OH substituents $HOCH_2CHCH_2CHCH_3$ with CH_2CH_3 and CH_3 substituents $CH_3CH_2CHCH_2CHCH_2OH$ with CH_3 and CH_3 substituents

3.27 Propose structures that meet the following descriptions:
(a) A ketone with five carbons   (b) A four-carbon amide
(c) A five-carbon ester   (d) An aromatic aldehyde
(e) A keto ester   (f) An amino alcohol

3.28 Propose structures for the following:
(a) A ketone, C_4H_8O (b) A nitrile, C_5H_9N
(c) A dialdehyde, $C_4H_6O_2$ (d) A bromoalkene, $C_6H_{11}Br$
(e) An alkane, C_6H_{14} (f) A cycloalkane, C_6H_{12}
(g) A diene (dialkene), C_5H_8 (h) A keto alkene, C_5H_8O

3.29 Draw as many compounds as you can that fit the following descriptions:
(a) Alcohols with formula $C_4H_{10}O$ (b) Amines with formula $C_5H_{13}N$
(c) Ketones with formula $C_5H_{10}O$ (d) Aldehydes with formula $C_5H_{10}O$
(e) Esters with formula $C_4H_8O_2$ (f) Ethers with formula $C_4H_{10}O$

3.30 Draw compounds that contain the following:
(a) A primary alcohol   (b) A tertiary nitrile
(c) A secondary bromide   (d) Both primary and secondary alcohols
(e) An isopropyl group   (f) A quaternary carbon

3.31 Draw and name all monobromo derivatives of pentane, $C_5H_{11}Br$.

3.32 Draw and name all monochloro derivatives of 2,5-dimethylhexane, $C_8H_{17}Cl$.

3.33 Predict the hybridization of the carbon atom in each of the following functional groups:
(a) Ketone   (b) Nitrile   (c) Carboxylic acid

3.34 Draw structures for the following:
(a) 2-Methylheptane   (b) 4-Ethyl-2,2-dimethylhexane
(c) 4-Ethyl-3,4-dimethyloctane   (d) 2,4,4-Trimethylheptane
(e) 3,3-Diethyl-2,5-dimethylnonane (f) 4-Isopropyl-3-methylheptane

3.35 Draw a compound that:
(a) Has only primary and tertiary carbons   (b) Has no primary carbons
(c) Has four secondary carbons

3.36 Draw a compound that:
(a) Has no primary hydrogens   (b) Has only primary and tertiary hydrogens

3.37 For each of the following compounds, draw an isomer with the same functional groups:

(a) $CH_3CHCH_2CH_2Br$ (with CH_3 substituent on second carbon)

(b) 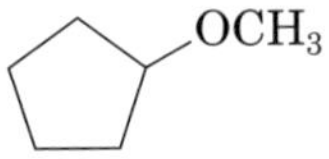 cyclopentane with OCH_3

(c) $CH_3CH_2CH_2C{\equiv}N$

(d) 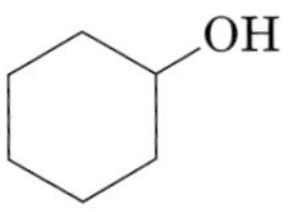 cyclohexane with OH

(e) CH_3CH_2CHO

(f) 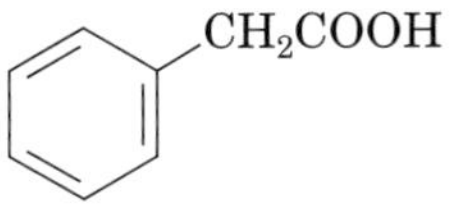 benzene ring with CH_2COOH

3.38 Draw structures for the following compounds:
(a) *trans*-1,3-Dibromocyclopentane   (b) *cis*-1,4-Diethylcyclohexane
(c) *trans*-1-Isopropyl-3-methylcycloheptane   (d) Dicyclohexylmethane

3.39 Identify the kinds of carbons (1°, 2°, 3°, or 4°) in the following molecules:

(a) $CH_3CHCH_2CH_3$ (with CH_3 substituent)

(b) $(CH_3)_2CHCH(CH_2CH_3)_2$

(c) $(CH_3)_3CCH_2CH_2CH$ (with CH_3 substituents)

(d) 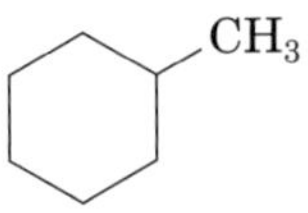 cyclohexane with CH_3

(e) 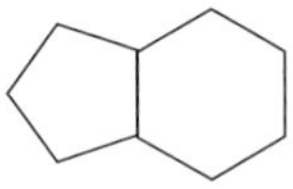 bicyclic structure

(f) 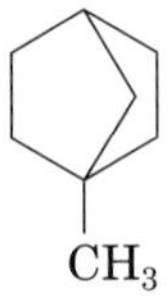 bicyclic structure with CH_3

3.40 Give IUPAC names for the following compounds:

(a) $CH_3CHCH_2CH_2CH_3$ (with CH_3 substituent)

(b) $CH_3CH_2C(CH_3)_2CH_3$

(c) $(CH_3)_2CHC(CH_3)_2CH_2CH_2CH_3$

(d) $CH_3CH_2CHCH_2CH_2CHCH_3$ (with CH_2CH_3 and CH_3 substituents)

(e) $CH_3CH_2CH_2CHCH_2CCH_3$ (with CH_3, CH_2CH_3, and CH_3 substituents)

(f) $(CH_3)_3CC(CH_3)_2CH_2CH_2CH_3$

(g) $CH_3CHCH_2CCH_2CH_3$ (with $CH_2CH_2CH_3$, CH_3CH_2, and CH_3 substituents)

3.41 Name the five isomers of C_6H_{14}.

3.42 Explain why each of the following names is incorrect:
(a) 2,2-Dimethyl-6-ethylheptane   (b) 4-Ethyl-5,5-dimethylpentane
(c) 3-Ethyl-4,4-dimethylhexane   (d) 5,5,6-Trimethyloctane
(e) 2-Isopropyl-4-methylheptane   (f) *cis*-1,5-Dimethylcyclohexane

3.43 Propose structures and give IUPAC names for the following:
(a) A dimethylcyclooctane   (b) A diethyldimethylhexane
(c) A cyclic alkane with three methyl groups   (d) A (3-methylbutyl)-substituted alkane

3.44 Give IUPAC names for the following compounds:

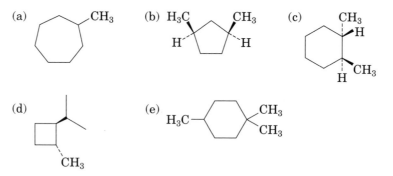

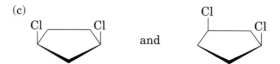

3.45 Draw 1,3,5-trimethylcyclohexane using a hexagon to represent the ring. How many cis–trans stereoisomers are possible?

3.46 Tell whether the following pairs of compounds are identical, constitutional isomers, or stereoisomers:
(a) *cis*-1,3-Dibromocyclohexane and *trans*-1,4-dibromocyclohexane
(b) 2,3-Dimethylhexane and 2,5,5-trimethylpentane

(c)

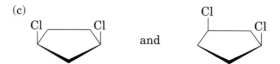

3.47 Draw two constitutional isomers of *cis*-1,2-dibromocyclopentane.

3.48 Draw a stereoisomer of *trans*-1,3-dimethylcyclobutane.

3.49 Malic acid, $C_4H_6O_5$, has been isolated from apples. Since this compound reacts with 2 molar equivalents of base, it is a dicarboxylic acid.
(a) Draw at least five possible structures.
(b) If malic acid is a secondary alcohol, what is its structure?

3.50 Cyclopropane was first prepared by reaction of 1,3-dibromopropane with sodium metal. Formulate the cyclopropane-forming reaction and then predict the product of the following reaction. What geometry do you expect for the product? (Try building a molecular model.)

$$\text{BrCH}_2\!-\!\underset{\underset{\text{CH}_2\text{Br}}{|}}{\overset{\overset{\text{CH}_2\text{Br}}{|}}{\text{C}}}\!-\!\text{CH}_2\text{Br} \xrightarrow{\text{4 Na}} ?$$

3.51 Formaldehyde, $H_2C=O$, is known to all biologists because of its usefulness as a tissue preservative. When pure, formaldehyde *trimerizes* to give trioxane, $C_3H_6O_3$, which, surprisingly enough, has no carbonyl groups. Only one monobromo derivative ($C_3H_5BrO_3$) of trioxane is possible. Propose a structure for trioxane.

3.52 There are four cis–trans isomers of menthol, including the one shown. Draw the other three.

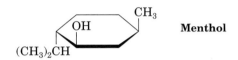

Menthol

3.53 Draw the five cycloalkanes with the formula C_5H_{10}.

A Look Ahead

3.54 There are two different substances named *trans*-1,2-dimethylcyclopentane. Make molecular models and see if you can find the relationship between them. (See Section 9.7.)

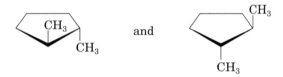

and

3.55 Cyclohexane has a puckered shape like a lounge chair rather than a flat shape. Why? (See Sections 4.6 and 4.9.)

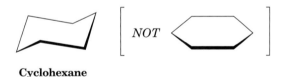

Cyclohexane

Molecular Modeling

3.56 Skeletal structures of cycloalkanes imply that there is empty space in the middle of the rings. Use SpartanView to look at space-filling models of cyclohexane, cyclodecane, and cyclooctadecane. Which, if any, rings have empty space?

3.57 Use SpartanBuild to construct models of the following molecules. Use atom and group fragments for building, and remember to minimize the energy of each.

(a) $CH_3CH_2CH_2CH_3$

(b) $CH_3\overset{\overset{\displaystyle CH_3}{|}}{C}HCH_3$

(c) $CH_3\overset{\overset{\displaystyle Cl}{|}}{\underset{\underset{\displaystyle CH_3}{|}}{C}}CH_3$

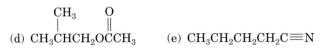

(d) $CH_3CHCH_2OCCH_3$ (e) $CH_3CH_2CH_2CH_2C{\equiv}N$

3.58 Use SpartanBuild to construct models of cubane and adamantane. Use ring fragments for building (two cyclobutane fragments for cubane; one cyclohexane fragment plus four sp^3 carbon atom fragments for adamantane), and remember to minimize the energies.

Cubane   **Adamantane**

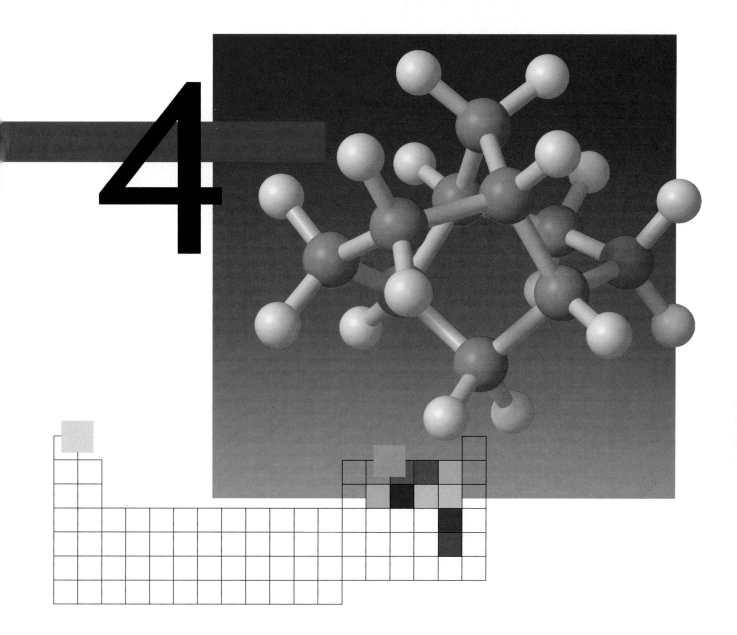

4

Stereochemistry of Alkanes and Cycloalkanes

Up to this point, we've viewed molecules primarily in a two-dimensional way and have given little thought to any consequences that might arise from the spatial arrangement of atoms in molecules. Now it's time to add a third dimension to our study. **Stereochemistry** is the branch of chemistry concerned with the three-dimensional aspects of molecules.

4.1 Conformations of Ethane

We know from Sections 1.7 and 1.8 that an sp^3-hybridized carbon atom has tetrahedral geometry and that the carbon–carbon bonds in alkanes result from σ overlap of carbon sp^3 orbitals. Let's now look into the three-dimensional consequences of such bonding. What are the spatial relationships between the hydrogens on one carbon and the hydrogens on a neighboring carbon? We'll see in later chapters that an understanding of these spatial relationships is often crucial for understanding chemical behavior.

Because of the cylindrical symmetry of σ bonds (Section 3.8), orbital overlap in the C–C single bond of ethane is exactly the same regardless of the geometric relationships among other atoms attached to the carbons (Figure 4.1). The different arrangements of atoms that result from rotation about a single bond are called **conformations**, and a specific conformation is called a **conformer** (**confor**mational iso**mer**). Unlike constitutional isomers, which have different connections of atoms, different conformers have

FIGURE 4.1 ▼

Some conformations of ethane. Rapid rotation around the carbon–carbon single bond interconverts the different conformers.

ethane
(see computer
animation on
CD-Rom)

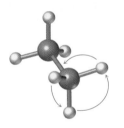      

Stereo View

Rotate

   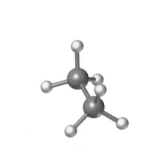   

Stereo View

the same connections of atoms and can't usually be isolated because they interconvert too rapidly.

Chemists represent conformational isomers in two ways, as shown in Figure 4.2. **Sawhorse representations** view the carbon–carbon bond from an oblique angle and indicate spatial orientation by showing all the C–H bonds. **Newman projections** view the carbon–carbon bond directly end-on and represent the two carbon atoms by a circle. Bonds attached to the front carbon are represented by lines going to the center of the circle, and bonds attached to the rear carbon are represented by lines going to the edge of the circle. The advantage of Newman projections is that they're easy to draw and the relationships among substituents on the different carbon atoms are easy to see.

Melvin S. Newman

Melvin S. Newman (1908–1993) was born in New York and received his Ph.D. in 1932 from Yale University. He was professor of chemistry at The Ohio State University (1936–1973), where he was active both in research and in chemical education.

FIGURE 4.2 ▾

A sawhorse representation and a Newman projection of ethane. The sawhorse projection views the molecule from an oblique angle, while the Newman projection views the molecule end-on.

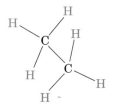

Sawhorse representation

Stereo View

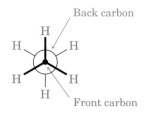

Newman projection

Stereo View

In spite of what we've just said about σ bond symmetry, we don't actually observe *perfectly* free rotation in ethane. Experiments show that there is a small 12 kJ/mol (2.9 kcal/mol) barrier to rotation and that some conformations are more stable than others. The lowest-energy, most stable conformation is the one in which all six C–H bonds are as far away from one another as possible—**staggered** when viewed end-on in a Newman projection. The highest-energy, least stable conformation is the one in which the six C–H bonds are as close as possible—**eclipsed** in a Newman projection. Between these two extremes are an infinite number of other possibilities.

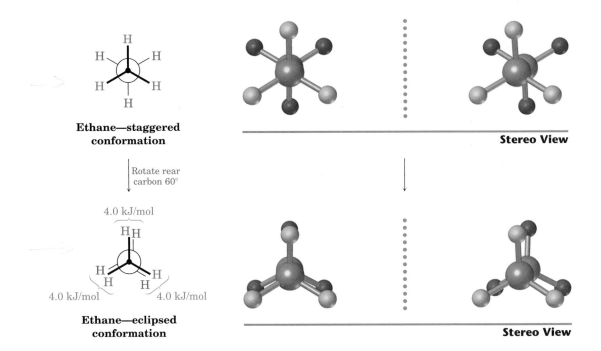

Ethane—staggered
conformation

Rotate rear
carbon 60°

Stereo View

Ethane—eclipsed
conformation

Stereo View

The 12 kJ/mol of extra energy present in the eclipsed conformation of ethane is called **torsional strain**. Its cause was the subject of controversy for some years, but most chemists now believe that torsional strain is due to the slight repulsion between electron clouds in the C–H bonds as they pass close by each other in the eclipsed conformer. Calculations indicate that the hydrogen–hydrogen distance is 255 pm in the staggered conformer but only about 229 pm in the eclipsed conformer.

Since the total strain is 12 kJ/mol, and since the strain is caused by three equal hydrogen–hydrogen eclipsing interactions, we can assign a value of approximately 4.0 kJ/mol (1.0 kcal/mol) to each single interaction. The barrier to rotation that results can be represented on a graph of potential energy versus degree of rotation in which the angle between C–H bonds on front and back carbons as viewed end-on (the *dihedral angle*) goes full circle from 0° to 360°. Energy minima occur at staggered conformations, and energy maxima occur at eclipsed conformations, as shown in Figure 4.3.

FIGURE 4.3 ▼

A graph of potential energy versus bond rotation in ethane. The staggered conformers are 12 kJ/mol lower in energy than the eclipsed conformers.

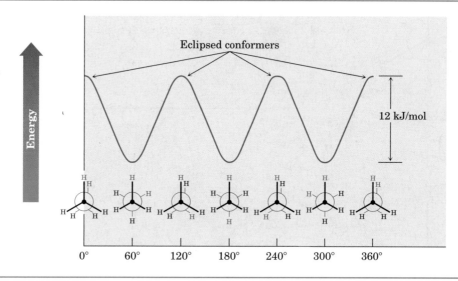

- -

Problem 4.1   Build a molecular model of ethane, and look at the interconversion of staggered and eclipsed forms. Measure the H–H distances in each case, and see if you can detect a difference.

- -

4.2 Conformations of Propane

Propane, the next higher member in the alkane series, also has a torsional barrier that results in hindered rotation around the carbon–carbon bonds. The barrier is slightly higher in propane than in ethane—14 kJ/mol (3.4 kcal/mol) versus 12 kJ/mol. In the eclipsed conformer of propane, there are two ethane-type hydrogen–hydrogen interactions and one additional interaction between a C–H bond and a C–C bond. Since each eclipsing hydrogen–hydrogen interaction has an energy "cost" of 4.0 kJ/mol, we can assign a value of $14 - (2 \times 4.0) = 6.0$ kJ/mol (1.4 kcal/mol) to the eclipsing interaction between the C–C bond and the C–H bond (Figure 4.4).

FIGURE 4.4 ▼

Newman projections of propane showing staggered and eclipsed conformations. The staggered conformer is lower in energy by 14 kJ/mol.

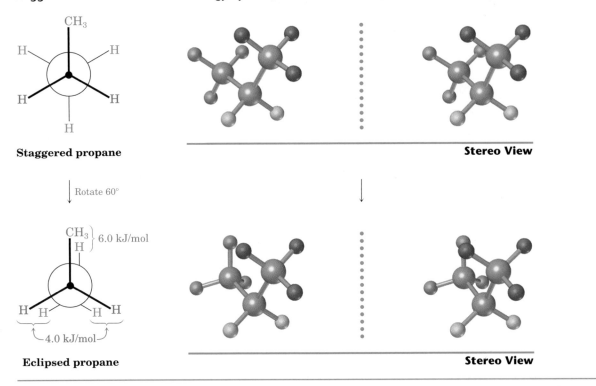

Staggered propane

Rotate 60°

Eclipsed propane

Stereo View

Stereo View

Problem 4.2   Make a graph of potential energy versus angle of bond rotation for propane, and assign values to the energy maxima.

4.3 Conformations of Butane

The conformational situation becomes more complex as the alkane becomes larger. In butane, for instance, a plot of potential energy versus rotation about the C2–C3 bond is shown in Figure 4.5.

Not all staggered conformations of butane have the same energy, and not all eclipsed conformations have the same energy. The lowest-energy arrangement, called the **anti conformation**, is the one in which the two large methyl groups are as far apart as possible—180° away from each other. As rotation around the C2–C3 bond occurs, an eclipsed conformation is reached in which there are two methyl–hydrogen interactions and one hydrogen–hydrogen interaction. If we assign the energy values for eclipsing interactions that were previously derived from ethane and propane, we

FIGURE 4.5 ▼

A plot of potential energy versus rotation for the C2–C3 bond in butane.
The energy maximum occurs when the two methyl groups eclipse each other, and the
energy minimum occurs when the two methyl groups are 180° apart (anti).

**butane
(see computer
animation on
CD-Rom)**

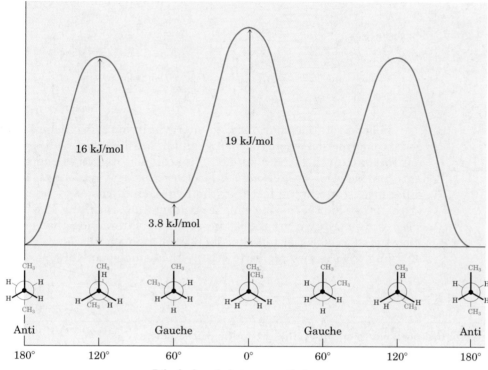

Dihedral angle between methyl groups

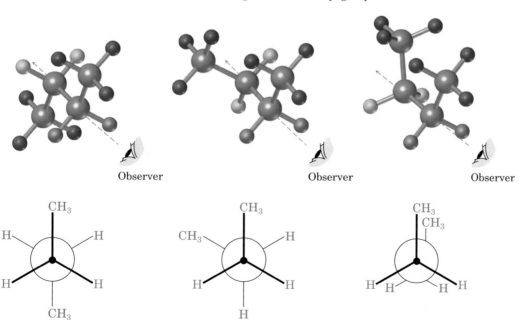

might predict that this eclipsed conformation should be more strained than the anti conformation by 2×6.0 kJ/mol (two methyl–hydrogen interactions) plus 4.0 kJ/mol (one hydrogen–hydrogen interaction), or a total of 16 kJ/mol (3.8 kcal/mol). This is exactly what is found.

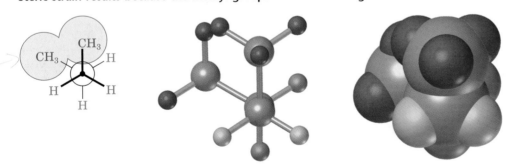

As bond rotation continues, an energy minimum is reached at the staggered conformation where the methyl groups are 60° apart. Called the **gauche conformation**, it lies 3.8 kJ/mol (0.9 kcal/mol) higher in energy than the anti conformation *even though it has no eclipsing interactions*. This energy difference is due to the fact that the hydrogen atoms of the methyl groups are near one another in the gauche conformation, resulting in what is called *steric strain*. **Steric strain** is the repulsive interaction that occurs when atoms are forced closer together than their atomic radii allow. It's the result of trying to force two atoms to occupy the same space (Figure 4.6).

FIGURE 4.6 ▼

The interaction between hydrogen atoms on the methyl groups in gauche butane. Steric strain results because the methyl groups are too close together.

As the dihedral angle between the methyl groups approaches 0°, an energy maximum is reached. Because the methyl groups are forced even closer together than in the gauche conformation, substantial amounts of both torsional strain and steric strain are present. A total strain energy of 19 kJ/mol (4.5 kcal/mol) has been estimated for this conformation, allowing us to calculate a value of 11 kJ/mol (2.6 kcal/mol) for the methyl–methyl eclipsing interaction: total strain (19 kJ/mol), less the strain of two hydrogen–hydrogen eclipsing interactions (2×4.0 kcal/mol), equals 11 kJ/mol.

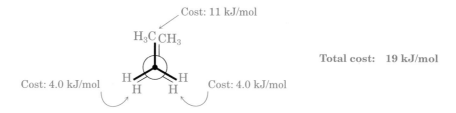

Cost: 11 kJ/mol

Total cost: 19 kJ/mol

Cost: 4.0 kJ/mol   Cost: 4.0 kJ/mol

After 0°, the rotation becomes a mirror image of what we've already seen. Another gauche conformation is reached, another eclipsed conformation, and finally a return to the anti conformation (Figure 4.5).

The notion of assigning definite energy values to specific interactions within a molecule is a very useful one that we'll return to later in this chapter. A summary of what we've seen thus far is given in Table 4.1.

TABLE 4.1   Energy Costs for Interactions in Alkane Conformers

Interaction	Cause	Energy cost	
		(kJ/mol)	**(kcal/mol)**
H ↔ H eclipsed	Torsional strain	4.0	1.0
H ↔ CH$_3$ eclipsed	Mostly torsional strain	6.0	1.4
CH$_3$ ↔ CH$_3$ eclipsed	Torsional plus steric strain	11	2.6
CH$_3$ ↔ CH$_3$ gauche	Steric strain	3.8	0.9

The same principles just developed for butane apply to pentane, hexane, and all higher alkanes. The most favorable conformation for any alkane has the carbon–carbon bonds in staggered arrangements and large substituents arranged anti to one another. A generalized alkane structure is shown in Figure 4.7.

FIGURE 4.7 ▼

The most stable alkane conformation is the one in which all substituents are staggered and the carbon–carbon bonds are arranged anti, as shown in this model of decane.

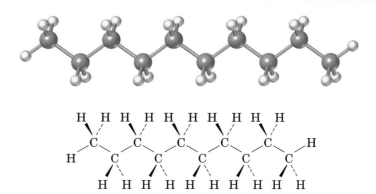

A final point: When we say that one particular conformer is "more stable" than another, we don't mean the molecule adopts and maintains only the more stable conformation. At room temperature, enough thermal energy is present to cause rotation around σ bonds to occur rapidly so that all conformers are in equilibrium. At any given instant, however, a larger percentage of molecules will be found in a more stable conformation than in a less stable one.

* * *

Practice Problem 4.1   Sighting along the C1–C2 bond of 1-chloropropane, draw Newman projections of the most stable conformation and the least stable conformation.

Strategy   The most stable conformation of a substituted alkane is generally a staggered one in which large groups have an anti relationship. The least stable conformation is generally an eclipsed one in which large groups are as close as possible.

Solution

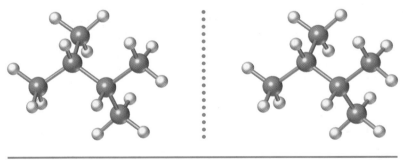

Most stable (staggered)   **Least stable (eclipsed)**

* * *

Problem 4.3   Consider 2-methylpropane (isobutane). Sighting along the C2–C1 bond:
(a) Draw a Newman projection of the most stable conformation.
(b) Draw a Newman projection of the least stable conformation.
(c) Make a graph of energy versus angle of rotation around the C2–C1 bond.
(d) Since a hydrogen–hydrogen eclipsing interaction costs 4.0 kJ/mol and a hydrogen–methyl eclipsing interaction costs 6.0 kJ/mol, assign relative values to the maxima and minima in your graph.

Problem 4.4   Sight along the C2–C3 bond of 2,3-dimethylbutane, and draw a Newman projection of the most stable conformation.

Problem 4.5   Draw a Newman projection along the C2–C3 bond of the following conformation of 2,3-dimethylbutane, and calculate a total strain energy (gray = C, ivory = H):

Stereo View

* * *

4.4 Conformation and Stability of Cycloalkanes: The Baeyer Strain Theory

Chemists in the late 1800s knew that cyclic molecules existed, but the limitations on ring sizes were unclear. Numerous compounds containing five-membered and six-membered rings were known, but smaller and larger ring sizes had not been prepared. For example, no cyclopropanes or cyclobutanes were known, despite many efforts to prepare them.

A theoretical interpretation of this observation was proposed in 1885 by Adolf von Baeyer. Baeyer suggested that, since carbon prefers to have tetrahedral geometry with bond angles of approximately 109°, ring sizes other than five and six may be too *strained* to exist. Baeyer based his hypothesis on the simple geometric notion that a three-membered ring (cyclopropane) should be an equilateral triangle with bond angles of 60°, a four-membered ring (cyclobutane) should be a square with bond angles of 90°, a five-membered ring (cyclopentane) should be a regular pentagon with bond angles of 108°, and so on.

According to Baeyer's analysis, cyclopropane, with a bond-angle compression of $109° - 60° = 49°$, should have a large amount of *angle strain* and must therefore be highly reactive. Cyclobutane ($109° - 90° = 19°$ angle strain) also must be reactive, but cyclopentane ($109° - 108° = 1°$ angle strain) must be nearly strain-free. Cyclohexane ($109° - 120° = -11°$ angle strain) must be somewhat strained, but cycloheptane ($109° - 128° = -19°$ angle strain) and higher cycloalkanes must have bond angles that are forced to be too large. Carrying this line of reasoning further, Baeyer suggested that very large rings should be impossibly strained and incapable of existence.

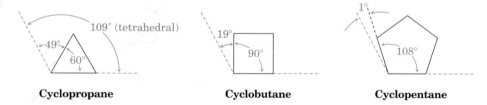

Cyclopropane	**Cyclobutane**	**Cyclopentane**

Although there is some truth to Baeyer's suggestion about angle strain in small rings, he was wrong in believing that small and large rings can't exist. Rings of all sizes from 3 through 30 and beyond can now be prepared easily. Nevertheless, the concept of **angle strain**—the strain induced in a molecule when a bond angle deviates from the ideal tetrahedral value—is a very useful one. Let's look at the facts.

4.5 Heats of Combustion of Cycloalkanes

To measure the amount of strain in a compound, we have to measure the total energy of the compound and then subtract the energy of a strain-free reference compound. The difference between the two values should represent the amount of extra energy in the molecule due to strain.

The simplest way to determine cycloalkane strain energies is to measure their **heats of combustion**, the amount of heat released when a compound burns completely with oxygen. The more energy (strain) a compound contains, the more energy (heat) is released on combustion.

$$(CH_2)_n + \frac{3n}{2} O_2 \longrightarrow n\,CO_2 + n\,H_2O + \text{Heat}$$

Because the heat of combustion of a hydrocarbon depends on its size, it's necessary to look at heats of combustion per CH_2 unit. Subtracting a reference value derived from a strain-free acyclic alkane and then multiplying by the number of CH_2 units in the ring gives overall strain energies. Figure 4.8 shows the results of these calculations.

FIGURE 4.8 ▼

Cycloalkane strain energies, calculated by taking the difference between cycloalkane heat of combustion per CH_2 and acyclic alkane heat of combustion per CH_2, and multiplying by the number of CH_2 units in a ring. Small and medium rings are strained, but cyclohexane rings are strain-free.

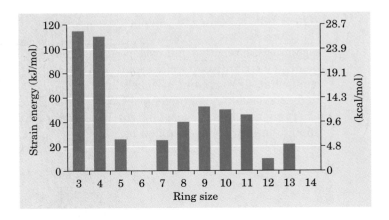

The data in Figure 4.8 show that Baeyer's theory is incorrect. Cyclopropane and cyclobutane are indeed quite strained, just as predicted, but cyclopentane is more strained than predicted, and cyclohexane is strain-free. For cycloalkanes of larger size, there is no regular increase in strain, and rings of more than 14 carbons are strain-free. Why is Baeyer's theory wrong?

Problem 4.6   Figure 4.8 shows that cyclopropane is more strained than cyclohexane by 115 kJ/mol. Which has the higher heat of combustion on a per-gram basis, cyclopropane or cyclohexane?

4.6 The Nature of Ring Strain

Baeyer's theory was wrong for a very simple reason: He assumed that rings are flat. In fact, though, most cycloalkanes are *not* flat; they adopt puckered three-dimensional conformations that allow bond angles to be nearly tetrahedral. Only for three- and four-membered rings is his concept of angle strain important.

Several factors in addition to angle strain are involved in determining the shapes and total strain energies of cycloalkanes. One such factor is the barrier to bond rotation (torsional strain) encountered in Section 4.1 during the discussion of alkane conformations. We said at that time that open-chain alkanes are most stable in a staggered conformation and least stable in an eclipsed conformation. A similar conclusion holds for cycloalkanes: Torsional strain is present in cycloalkanes if any neighboring C–H bonds eclipse each other. For example, cyclopropane must have considerable torsional strain (in addition to angle strain), because C–H bonds on neighboring carbon atoms are eclipsed (Figure 4.9). Larger cycloalkanes minimize torsional strain by adopting puckered, nonplanar conformations.

FIGURE 4.9 ▼

The conformation of cyclopropane, showing the eclipsing of neighboring C–H bonds that gives rise to torsional strain. Part (b) is a Newman projection along a C–C bond.

(a)

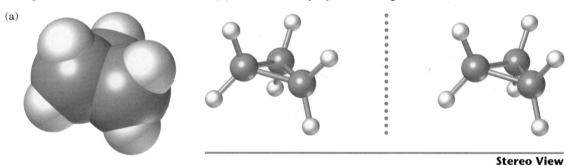

Stereo View

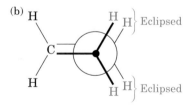

In addition to angle strain and torsional strain, *steric strain* is yet a third factor that contributes to the overall strain energy of cycloalkanes. As in gauche butane (Section 4.3), two nonbonded atoms in a molecule repel each other if they approach too closely and attempt to occupy the same space. Such nonbonding steric interactions are particularly important in determining the minimum-energy conformations of cycloalkanes with medium-size rings (C_7–C_{11}).

Key Ideas ▶ In summary, cycloalkanes adopt their minimum-energy conformations for a combination of three reasons:

- **Angle strain**—the strain due to expansion or compression of bond angles

- **Torsional strain**—the strain due to eclipsing of bonds on neighboring atoms

- **Steric strain**—the strain due to repulsive interactions when atoms approach each other too closely

· ·

Problem 4.7   Each hydrogen–hydrogen eclipsing interaction in ethane costs about 4.0 kJ/mol. How many such interactions are present in cyclopropane? What fraction of the overall 115 kJ/mol (27.5 kcal/mol) strain energy of cyclopropane is due to torsional strain?

Problem 4.8   *cis*-1,2-Dimethylcyclopropane has a larger heat of combustion than *trans*-1,2-dimethylcyclopropane. How can you account for this difference? Which of the two compounds is more stable?

· ·

4.7 Cyclopropane: An Orbital View

Cyclopropane, a colorless gas (bp = −33°C), was first prepared by reaction of sodium with 1,3-dibromopropane:

$$\underset{\textbf{1,3-Dibromopropane}}{BrH_2C \overset{\displaystyle CH_2}{\diagup \quad \diagdown} CH_2Br} \xrightarrow{\text{2 Na}} \underset{\textbf{Cyclopropane}}{H_2C \overset{\displaystyle CH_2}{\diagup \quad \diagdown} CH_2} + 2\ NaBr$$

Because three points (the carbon atoms) define a plane, cyclopropane must be flat. Assuming it's symmetrical, cyclopropane must also have C–C–C bond angles of 60°. How can the hybrid-orbital model of bonding account for this large distortion of bond angles from the normal 109° tetrahedral value? The answer is that cyclopropane has *bent bonds*. In an unstrained alkane, maximum bonding is achieved when two atoms have their overlapping orbitals pointing directly toward each other. In cyclopropane, though, the orbitals can't point directly toward each other; rather, they overlap at a slight angle. The result is that cyclopropane bonds are weaker and more reactive than typical alkane bonds.

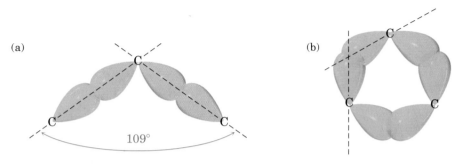

(a)

109°

A typical alkane C–C bond

(b)

A bent cyclopropane C–C bond

Spectroscopic evidence for bent bonds in cyclopropanes has been provided by careful, low-temperature X-ray studies, which are able to map the electron density in molecules. As shown in Figure 4.10, the electron density in a cyclopropane bond is strongly displaced outward from the internuclear axis.

FIGURE 4.10 ▼

An electron-density map provided by low-temperature X-ray studies. A top view looking down at the sample molecule shows how the electron densities in the cyclopropane bonds of the central ring are bent away from the internuclear axis.

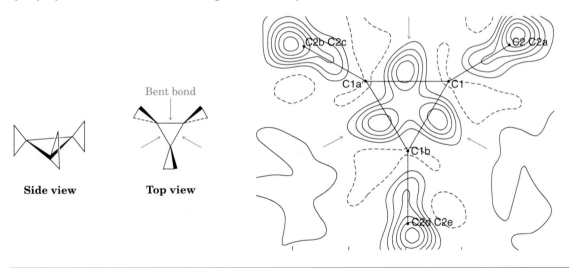

Side view   **Top view**

Bent bond

4.8 Conformations of Cyclobutane and Cyclopentane

Cyclobutane

Cyclobutane has less angle strain than cyclopropane but has more torsional strain because of its larger number of ring hydrogens. As a result, the total strain for the two compounds is nearly the same—110.4 kJ/mol (26.4 kcal/mol) for cyclobutane versus 115 kJ/mol (27.5 kcal/mol) for cyclopropane. Experiments show that cyclobutane is not quite flat but is slightly bent so that one carbon atom lies about 25° above the plane of the other three (Figure 4.11). The effect of this slight bend is to *increase* angle strain but to *decrease* torsional strain, until a minimum-energy balance between the two opposing effects is achieved.

FIGURE 4.11 ▼

The conformation of cyclobutane. Part (a) shows computer-generated molecular models. Part (c) is a Newman projection along the C1–C2 bond, showing that neighboring C–H bonds are not quite eclipsed.

(a)

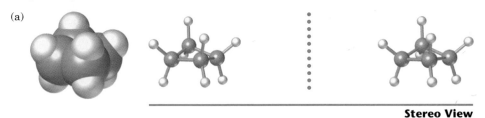

Stereo View

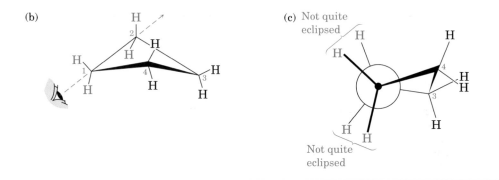

Cyclopentane

Cyclopentane was predicted by Baeyer to be nearly strain-free, but combustion data indicate a total strain energy of 26.0 kJ/mol (6.2 kcal/mol). Although planar cyclopentane has practically no angle strain, it has a large amount of torsional strain. Cyclopentane therefore twists to adopt a puckered, nonplanar conformation that strikes a balance between increased angle strain and decreased torsional strain. Four of the cyclopentane carbon atoms are in approximately the same plane, with the fifth carbon atom bent out of the plane. Most of the hydrogens are nearly staggered with respect to their neighbors (Figure 4.12).

FIGURE 4.12 ▼

The conformation of cyclopentane. Carbons 1, 2, 3, and 4 are nearly planar, but carbon 5 is out of the plane. Part (c) is a Newman projection along the C1–C2 bond, showing that neighboring C–H bonds are nearly staggered.

(a)

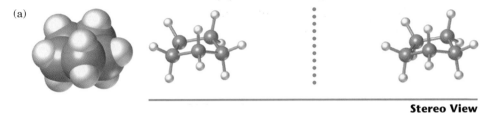

Stereo View

(b)   (c)

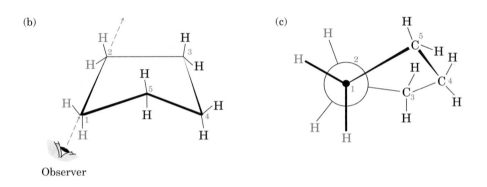

Observer

• •

Problem 4.9   How many hydrogen–hydrogen eclipsing interactions would be present if cyclopentane were planar? Assuming an energy cost of 4.0 kJ/mol for each eclipsing interaction, how much torsional strain would planar cyclopentane have? How much of this strain is relieved by puckering if the measured total strain of cyclopentane is 26.0 kJ/mol?

Problem 4.10   Draw the most stable conformation of *cis*-1,3-dimethylcyclobutane. Draw the least stable conformation.

• •

4.9 Conformations of Cyclohexane

Substituted cyclohexanes are the most common cycloalkanes because of their wide occurrence in nature. A vast number of compounds, including many important pharmaceutical agents, contain cyclohexane rings.

Hermann Sachse

Hermann Sachse (1862–1893) was born in Berlin, Germany, where he also received his Ph.D. (1889) and taught at the Technische Hochschule Charlottenburg-Berlin.

Combustion data show that cyclohexane is strain-free, with neither angle strain nor torsional strain. How can this be? The answer was first suggested in 1890 by Hermann Sachse and later expanded on by Ernst Mohr. Cyclohexane is not flat as Baeyer assumed; instead, it is puckered into a three-dimensional conformation that relieves all strain. The C–C–C angles of cyclohexane can reach the strain-free tetrahedral value if the ring adopts a **chair conformation**, so-called because of its similarity to a lounge chair—a back, a seat, and a footrest (Figure 4.13). Furthermore, sighting along any one of the carbon–carbon bonds in a Newman projection shows that chair cyclohexane has no torsional strain; all neighboring C–H bonds are staggered.

FIGURE 4.13 ▼

The strain-free chair conformation of cyclohexane. All C–C–C bond angles are 111.5° (close to the ideal 109.5° tetrahedral angle), and all neighboring C–H bonds are staggered.

(a)

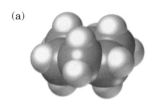

Stereo View

(b)

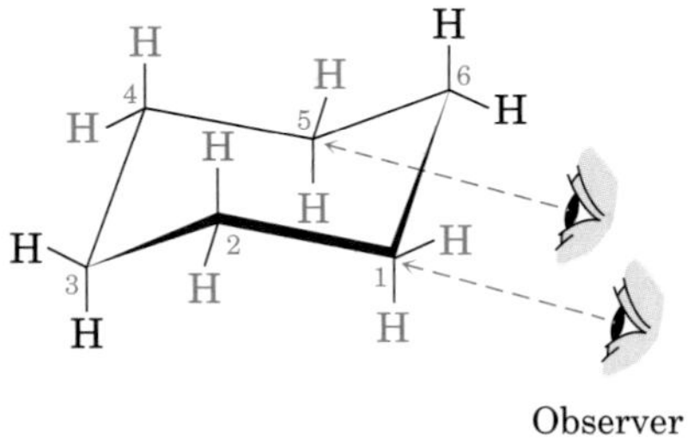

Observer

(c)

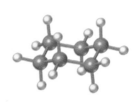

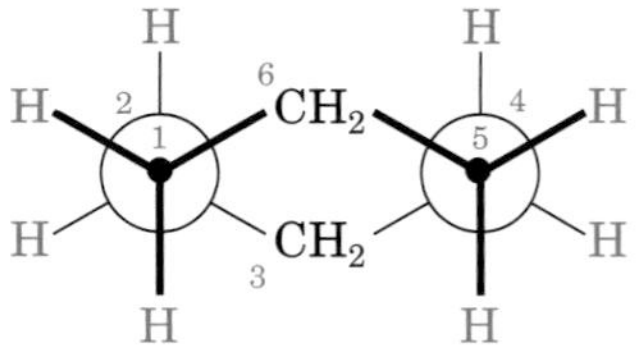

Ernst Mohr

Ernst Mohr (1873–1926) was born in Dresden, Germany, and received his Ph.D. at the University of Kiel (1897). He was then professor of chemistry at the University of Heidelberg.

The easiest way to visualize chair cyclohexane is to build a molecular model. (In fact, do it now.) Two-dimensional drawings such as Figure 4.13 are useful, but there is no substitute for holding, twisting, and turning a three-dimensional model in your own hands. The chair conformation of cyclohexane can be drawn by following the three steps shown in Figure 4.14.

FIGURE 4.14 ▼

How to draw the
cyclohexane chair
conformation.

STEP 1 Draw two parallel lines, slanted
downward and slightly offset from
each other. This means that four of
the cyclohexane carbon atoms lie
in a plane.

STEP 2 Locate the topmost carbon atom
above and to the right of the plane
of the other four and connect the
bonds.

STEP 3 Locate the bottommost carbon
atom below and to the left of the
plane of the middle four and
connect the bonds. Note that the
bonds to the bottommost carbon
atom are parallel to the bonds to
the topmost carbon.

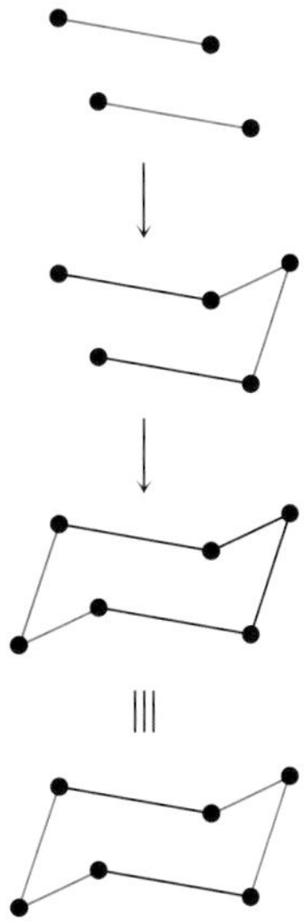

When viewing chair cyclohexane, the lower bond is in front and the
upper bond is in back. If this convention is not defined, an optical illusion
can make the reverse appear true. For clarity, all the cyclohexane rings
drawn in this book will have the front (lower) bond heavily shaded to indicate its nearness to the viewer.

This bond is in back.

This bond is in front.

4.10 Axial and Equatorial Bonds
in Cyclohexane

The chair conformation of cyclohexane has many chemical consequences.
For example, we'll see in Section 11.12 that the chemical behavior of many
substituted cyclohexanes is directly controlled by their conformation.
Another consequence of the chair conformation is that there are two kinds
of positions for substituents on the ring: *axial* positions and *equatorial*

positions (Figure 4.15). Chair cyclohexane has six **axial** hydrogens that are perpendicular to the ring (parallel to the ring axis) and six **equatorial** hydrogens that are in the rough plane of the ring (around the ring equator).

FIGURE 4.15 ▼

Axial and equatorial hydrogen atoms in chair cyclohexane. The six axial hydrogens (red) are parallel to the ring axis, and the six equatorial hydrogens (blue) are in a band around the ring equator.

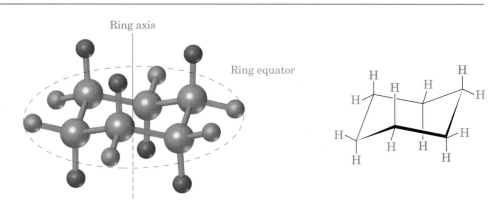

Look carefully at the disposition of the axial and equatorial hydrogens in Figure 4.15. Each carbon atom in cyclohexane has one axial and one equatorial hydrogen, and each side of the ring has three axial and three equatorial hydrogens in an alternating arrangement. For example, if the top side of the ring has axial hydrogens on carbons 1, 3, and 5, then it has equatorial hydrogens on carbons 2, 4, and 6. Exactly the reverse is true for the bottom side: Carbons 1, 3, and 5 have equatorial hydrogens, but carbons 2, 4, and 6 have axial hydrogens (Figure 4.16).

FIGURE 4.16 ▼

Alternating axial and equatorial positions in chair cyclohexane, as shown in a view looking directly down the ring axis. Each carbon atom has one axial and one equatorial position, and each side has alternating axial and equatorial positions.

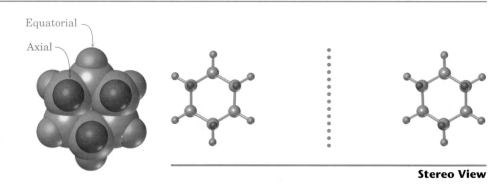

Stereo View

Note that we haven't used the words *cis* and *trans* in this discussion of cyclohexane geometry. Two hydrogens on the same side of the ring are always cis, regardless of whether they're axial or equatorial and regardless of whether they're adjacent. Similarly, two hydrogens on opposite sides of the ring are always trans, regardless of whether they're axial or equatorial.

Axial and equatorial bonds can be drawn following the procedure outlined in Figure 4.17. (Look at a molecular model as you practice.)

FIGURE 4.17 ▼

A procedure for drawing axial and equatorial bonds in chair cyclohexane.

Axial bonds: The six axial bonds, one on each carbon, are parallel and alternate up–down.

Equatorial bonds: The six equatorial bonds, one on each carbon, come in three sets of two parallel lines. Each set is also parallel to two ring bonds. Equatorial bonds alternate between sides around the ring.

Completed cyclohexane

4.11 Conformational Mobility of Cyclohexane

Because chair cyclohexane has two kinds of positions, axial and equatorial, we might expect to find two isomeric forms of a monosubstituted cyclohexane. In fact, though, there is only *one* methylcyclohexane, *one* bromocyclohexane, *one* cyclohexanol, and so on, because cyclohexane rings are *conformationally mobile* at room temperature. Different chair conformations readily interconvert, resulting in the exchange of axial and equatorial positions. This interconversion of chair conformations, usually referred to as a **ring-flip**, is shown in Figure 4.18. Molecular models show the process more clearly, so you should practice ring-flipping with models.

FIGURE 4.18 ▼

A ring-flip in chair cyclohexane interconverts axial and equatorial positions.

cyclohexane
(see computer animation
on CD-Rom)

Stereo View

↕ Ring-flip

↕ Ring-flip

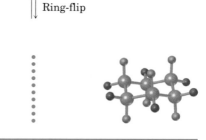

Stereo View

A chair cyclohexane can be ring-flipped by keeping the middle four carbon atoms in place while folding the two ends in opposite directions. An axial substituent in one chair form becomes an equatorial substituent in the ring-flipped chair form, and vice versa. For example, axial bromocyclohexane becomes equatorial bromocyclohexane after ring-flip. Since the energy barrier to chair–chair interconversion is only about 45 kJ/mol (10.8 kcal/mol), the process is extremely rapid at room temperature. We therefore see only what appears to be a single structure, rather than distinct axial and equatorial isomers.

Ring-flip
⇌

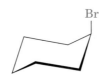

Axial bromocyclohexane   **Equatorial bromocyclohexane**

Practice Problem 4.2   Draw 1,1-dimethylcyclohexane, indicating which methyl group is axial and which is equatorial.

Strategy   Draw a chair cyclohexane ring, and then put two methyl groups on the same carbon. The methyl group in the rough plane of the ring is equatorial, and the other (directly above or below the ring) is axial.

Solution

Problem 4.11   Draw two different chair conformations of cyclohexanol (hydroxycyclohexane), showing all hydrogen atoms. Identify each position as axial or equatorial.

Problem 4.12   A *cis*-1,2-disubstituted cyclohexane, such as *cis*-1,2-dichlorocyclohexane, must have one group axial and one group equatorial. Explain.

Problem 4.13   A *trans*-1,2-disubstituted cyclohexane must either have both groups axial or both groups equatorial. Explain.

Problem 4.14   Draw two different chair conformations of *trans*-1,4-dimethylcyclohexane, and label all positions as axial or equatorial.

4.12 Conformations of Monosubstituted Cyclohexanes

Although cyclohexane rings rapidly flip between conformations at room temperature, the two conformers of a monosubstituted cyclohexane aren't equally stable. In methylcyclohexane, for example, the equatorial conformer is more stable than the axial conformer by 7.6 kJ/mol (1.8 kcal/mol). Similarly for other monosubstituted cyclohexanes: A substituent is more stable in an equatorial position than in an axial position.

You might recall from your general chemistry course that it's possible to calculate the percentages of two isomers at equilibrium using the equation $\Delta E = -RT \ln K$, where ΔE is the energy difference between isomers, R is the gas constant [8.315 J/(K · mol)], T is the Kelvin temperature, and K is the equilibrium constant between isomers. For example, an energy difference of 7.6 kJ/mol means that about 95% of methylcyclohexane molecules have the methyl group equatorial at any given instant, and only 5% have the methyl group axial. Figure 4.19 plots the relationship between energy and isomer percentages.

FIGURE 4.19 ▼

A plot of the percentages of two isomers at equilibrium versus the energy difference between them. The curves are calculated using the equation $\Delta E = -RT \ln K$.

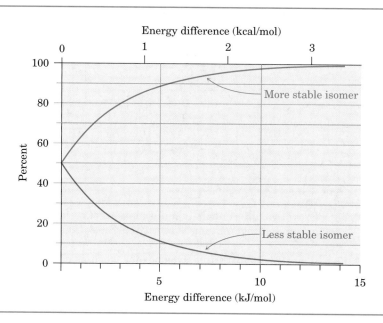

The energy difference between axial and equatorial conformers is due to steric strain caused by so-called **1,3-diaxial interactions**. That is, the axial methyl group on C1 is too close to the axial hydrogens three carbons away on C3 and C5, resulting in 7.6 kJ/mol of steric strain (Figure 4.20).

1,3-Diaxial steric strain is already familiar—we've seen it before as the steric strain between methyl groups in gauche butane (Section 4.3). Recall that gauche butane is less stable than anti butane by 3.8 kJ/mol (0.9 kcal/mol) because of steric interference between hydrogen atoms on the two methyl groups. Comparing a four-carbon fragment of axial methylcyclohexane with gauche butane shows that the steric interaction is the same in both cases (Figure 4.21). Because methylcyclohexane has two such interactions, though, it has $2 \times 3.8 = 7.6$ kJ/mol of steric strain.

Sighting along the C1–C2 bond of axial methylcyclohexane shows that the axial hydrogen at C3 has a gauche butane interaction with the axial methyl group at C1. Sighting similarly along the C1–C6 bond shows that the axial hydrogen at C5 also has a gauche butane interaction with the axial methyl group at C1. Both interactions are absent in equatorial methylcyclohexane, and we therefore find an energy difference of 7.6 kJ/mol between the two forms.

What is true for methylcyclohexane is also true for other monosubstituted cyclohexanes: A substituent is more stable in an equatorial position than in an axial position. The exact amount of 1,3-diaxial steric strain in a specific compound depends on the nature and size of the substituent, as

FIGURE 4.20 ▼

Interconversion of axial and equatorial methylcyclohexane, as represented in several formats. The equatorial conformer is more stable than the axial conformer by 7.6 kJ/mol.

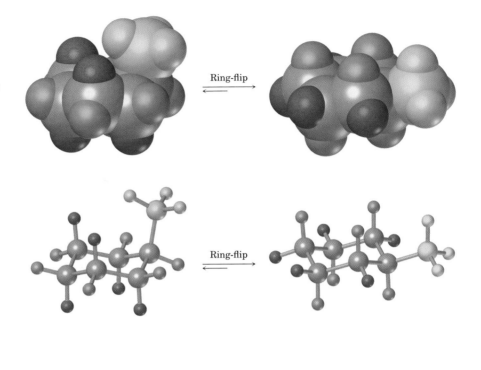

FIGURE 4.21 ▼

Origin of 1,3-diaxial cyclohexane interactions in methylcyclohexane. The steric strain between an axial methyl group and an axial hydrogen atom three carbons away is identical to the steric strain in gauche butane. (Note that the −CH₃ group in methylcyclohexane is displaced slightly away from a true axial position to minimize strain.)

butane–methylcyclohexane
(see computer animation
on CD-ROM)

Gauche butane
(3.8 kJ/mol strain)

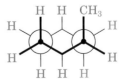

**Axial
methylcyclohexane
(7.6 kJ/mol strain)**

indicated in Table 4.2. Not surprisingly, the amount of steric strain increases through the series $H_3C- < CH_3CH_2- < (CH_3)_2CH- \ll (CH_3)_3C-$, paralleling the increasing bulk of the alkyl groups. Note that the values in Table 4.2 refer to 1,3-diaxial interactions of the substituent with a *single* hydrogen atom. These values must be doubled to arrive at the amount of strain in a monosubstituted cyclohexane.

TABLE 4.2 Steric Strain in Monosubstituted Cyclohexanes		
	Strain of one H–Y 1,3-diaxial interaction	
Y	(kJ/mol)	(kcal/mol)
–F	0.5	0.12
–Cl	1.0	0.25
–Br	1.0	0.25
–OH	2.1	0.5
–CH_3	3.8	0.9
–CH_2CH_3	4.0	0.95
–$CH(CH_3)_2$	4.6	1.1
–$C(CH_3)_3$	11.4	2.7
–C_6H_5	6.3	1.5
–COOH	2.9	0.7
–CN	0.4	0.1

Problem 4.15   How can you account for the fact (Table 4.2) that an axial *tert*-butyl substituent has much larger 1,3-diaxial interactions than isopropyl, but isopropyl is fairly similar to ethyl and methyl? Use molecular models to help with your answer.

Problem 4.16   Why do you suppose an axial cyano substituent causes practically no 1,3-diaxial steric strain (0.4 kJ/mol)? Use molecular models to help with your answer.

Problem 4.17   Look at Figure 4.19 and estimate the percentages of axial and equatorial conformers present at equilibrium in bromocyclohexane.

4.13 Conformational Analysis of Disubstituted Cyclohexanes

Monosubstituted cyclohexanes always have the substituent in an equatorial position. In disubstituted cyclohexanes, however, the situation is more complex because the steric effects of both substituents must be taken into account. All steric interactions in both possible chair conformations must be analyzed before deciding which conformation is favored.

Let's look at 1,2-dimethylcyclohexane as an example. There are two isomers, *cis*-1,2-dimethylcyclohexane and *trans*-1,2-dimethylcyclohexane, which must be considered separately. In the cis isomer, both methyl groups are on the same side of the ring, and the compound can exist in either of the two chair conformations shown at the top of Figure 4.22. (It's often easier to see

FIGURE 4.22 ▼

Conformations of *cis*- and *trans*-1,2-dimethylcyclohexane. In the cis isomer (top), the two chair conformations are equal in energy because each has one axial methyl group and one equatorial methyl group. In the trans isomer (bottom), the conformation with both methyl groups equatorial is favored by 11.4 kJ/mol (2.7 kcal/mol) over the conformation with both methyl groups axial.

cis-**1,2-Dimethylcyclohexane**

One gauche
interaction (3.8 kJ/mol)
Two CH₃–H diaxial
interactions (7.6 kJ/mol)
Total strain: 3.8 + 7.6 = 11.4 kJ/mol

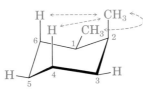

Ring-flip

One gauche
interaction (3.8 kJ/mol)
Two CH₃–H diaxial
interactions (7.6 kJ/mol)
Total strain: 3.8 + 7.6 = 11.4 kJ/mol

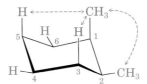

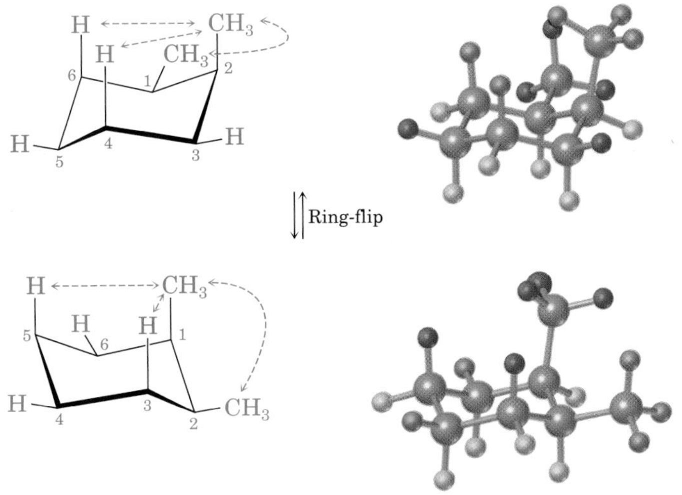

trans-**1,2-Dimethylcyclohexane**

One gauche
interaction (3.8 kJ/mol)

Ring-flip

Four CH₃–H diaxial
interactions (15.2 kJ/mol)

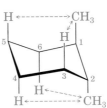

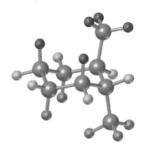

whether a compound is cis- or trans-disubstituted by first drawing the ring as a flat representation and then converting to a chair conformation.)

Both chair conformations of *cis*-1,2-dimethylcyclohexane shown in Figure 4.22 have one axial methyl group and one equatorial methyl group. The top conformation has an axial methyl group at C2, which has 1,3-diaxial interactions with hydrogens on C4 and C6. The ring-flipped conformation has an axial methyl group at C1, which has 1,3-diaxial interactions with hydrogens on C3 and C5. In addition, both conformations have gauche butane interactions between the two methyl groups. *The two conformations are exactly equal in energy,* with a total steric strain of 3 × 3.8 kJ/mol = 11.4 kJ/mol (2.7 kcal/mol).

In *trans*-1,2-dimethylcyclohexane, the two methyl groups are on opposite sides of the ring, and the compound can exist in either of the two chair conformations shown in Figure 4.22. The situation here is quite different from that of the cis isomer. The top trans conformation in Figure 4.22 has both methyl groups equatorial and therefore has only a gauche butane interaction between methyls (3.8 kJ/mol) but no 1,3-diaxial interactions. The ring-flipped conformation, however, has both methyl groups axial. The axial methyl group at C1 interacts with axial hydrogens at C3 and C5, and the axial methyl group at C2 interacts with axial hydrogens at C4 and C6. These four 1,3-diaxial interactions produce a steric strain of 4 × 3.8 kJ/mol = 15.2 kJ/mol and make the diaxial conformation 15.2 − 3.8 = 11.4 kJ/mol less favorable than the diequatorial conformation. We therefore predict that *trans*-1,2-dimethylcyclohexane will exist almost exclusively (>99%) in the diequatorial conformation.

The same kind of **conformational analysis** just carried out for *cis*- and *trans*-1,2-dimethylcyclohexane can be done for any substituted cyclohexane, such as *cis*-1-*tert*-butyl-4-chlorocyclohexane in Practice Problem 4.3. It turns out that the large amount of steric strain caused by an axial *tert*-butyl group effectively holds the cyclohexane ring in a single conformation. Chemists sometimes take advantage of this steric locking to study the chemical reactivity of immobile cyclohexane rings. If, for example, you wanted to study the difference in reactivity between an axial alkyl bromide and an equatorial alkyl bromide, you might compare the behaviors of *cis*- and *trans*-1-bromo-4-*tert*-butylcyclohexane. (We'll see in Section 11.12 that there *is*, in fact, a difference.)

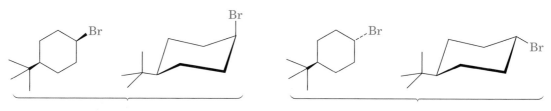

cis-1-Bromo-4-**tert**-butylcyclohexane
(axial bromine)

trans-1-Bromo-4-**tert**-butylcyclohexane
(equatorial bromine)

• •

Practice Problem 4.3   Draw the most stable conformation of *cis*-1-*tert*-butyl-4-chlorocyclohexane. By how much is it favored?

Strategy   Draw the possible conformations, and calculate the strain energy in each. Remember that equatorial substituents cause less strain than axial substituents.

Solution   First draw the two chair conformations of the molecule:

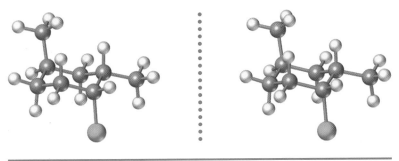

2 × 1.0 = 2.0 kJ/mol steric strain   2 × 11.4 = 22.8 kJ/mol steric strain

In the left-hand conformation, the *tert*-butyl group is equatorial and the chlorine is axial. In the right-hand conformation, the *tert*-butyl group is axial and the chlorine is equatorial. These conformations aren't of equal energy because an axial *tert*-butyl substituent and an axial chloro substituent produce different amounts of steric strain. Table 4.2 shows that the 1,3-diaxial interaction between a hydrogen and a *tert*-butyl group costs 11.4 kJ/mol (2.7 kcal/mol), whereas the interaction between a hydrogen and a chlorine costs only 1.0 kJ/mol (0.25 kcal/mol). An axial *tert*-butyl group therefore produces (2 × 11.4 kJ/mol) − (2 × 1.0 kJ/mol) = 20.8 kJ/mol (4.9 kcal/mol) more steric strain than does an axial chlorine, and the compound preferentially adopts the conformation with the chlorine axial and the *tert*-butyl equatorial.

· ·

Problem 4.18   Draw the most stable chair conformation of the following molecules, and estimate the amount of strain in each.
(a) *trans*-1-Chloro-3-methylcyclohexane   (b) *cis*-1-Ethyl-2-methylcyclohexane
(c) *cis*-1-Bromo-4-ethylcyclohexane   (d) *cis*-1-*tert*-Butyl-4-ethylcyclohexane

Problem 4.19   Name the following compound, identify each substituent as axial or equatorial, and tell whether the conformation shown is the more stable or less stable chair form (gray = C, yellow-green = Cl, ivory = H).

Stereo View

· ·

4.14 Boat Cyclohexane

In addition to the chair conformation of cyclohexane, a second possibility called the **boat conformation** is also free of angle strain. We haven't paid it any attention thus far, however, because boat cyclohexane is less stable than chair cyclohexane (Figure 4.23).

FIGURE 4.23 ▼

The boat conformation of cyclohexane. There is steric strain and torsional strain in this conformation but no angle strain.

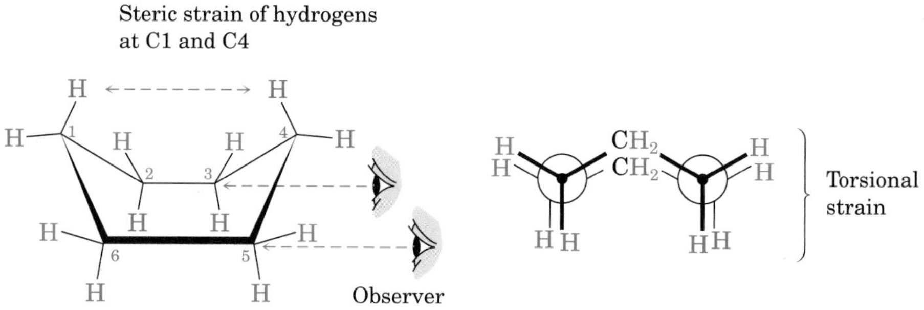

Steric strain of hydrogens at C1 and C4

Observer

Torsional strain

Carbons 2, 3, 5, and 6 in boat cyclohexane lie in a plane, with carbons 1 and 4 above the plane. The inside hydrogen atoms on carbons 1 and 4 approach each other closely enough to produce considerable steric strain, and the four eclipsed pairs of hydrogens on carbons 2, 3, 5, and 6 produce torsional strain. The Newman projection in Figure 4.23, obtained by sighting along the C2–C3 and C5–C6 bonds, shows this eclipsing clearly.

Boat cyclohexane is approximately 29 kJ/mol (7.0 kcal/mol) less stable than chair cyclohexane, although this value is reduced to about 23 kJ/mol (5.5 kcal/mol) by twisting slightly, thereby relieving some torsional strain (Figure 4.24). Even this **twist-boat conformation** is still much more strained than the chair conformation, though, and molecules adopt this geometry only under special circumstances.

Problem 4.20   *trans*-1,3-Di-*tert*-butylcyclohexane is one of the few molecules that exists largely in a twist-boat conformation. Draw both a chair conformation and the likely twist-boat conformation, and then explain why the twist-boat form is favored.

FIGURE 4.24 ▼

Boat and twist-boat conformations of cyclohexane. The twist-boat conformation is lower in energy than the boat conformation by 6 kJ/mol. Both conformations are much more strained than chair cyclohexane.

(a)

Boat cyclohexane
(29 kJ/mol strain)

Stereo View

(b)

Twist-boat cyclohexane
(23 kJ/mol strain)

Stereo View

4.15 Conformations of Polycyclic Molecules

The last point we'll consider about cycloalkane stereochemistry is to see what happens when two or more cycloalkane rings are fused together along a common bond to construct a **polycyclic** molecule—for example, decalin.

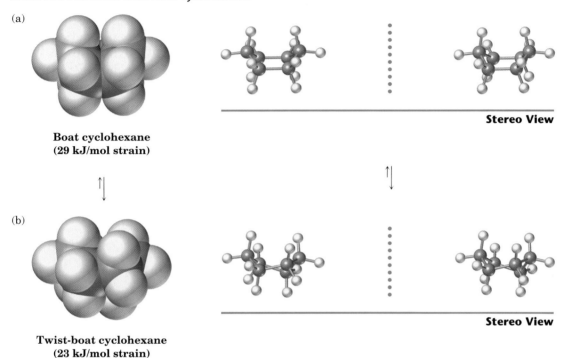

Decalin (two fused cyclohexane rings)

Decalin consists of two cyclohexane rings joined to share two carbon atoms (the *bridgehead* carbons, C1 and C6) and a common bond. Decalin can exist in either of two isomeric forms, depending on whether the rings are trans fused or cis fused. In *trans*-decalin, the hydrogen atoms at the bridgehead carbons are on opposite sides of the rings; in *cis*-decalin, the bridgehead hydrogens are on the same side. Figure 4.25 shows how both compounds can be represented using chair cyclohexane conformations. Note that *trans*- and *cis*-decalin are not interconvertible by ring-flips or other rotations. They are cis–trans stereoisomers (Section 3.8) and have the same relationship to each other that *cis*- and *trans*-1,2-dimethylcyclohexane have (Figure 4.22).

FIGURE 4.25 ▼

Representations of *trans*- and *cis*-decalin. The hydrogen atoms (red) at the bridgehead carbons are on the same side of the rings in the cis isomer but on opposite sides in the trans isomer.

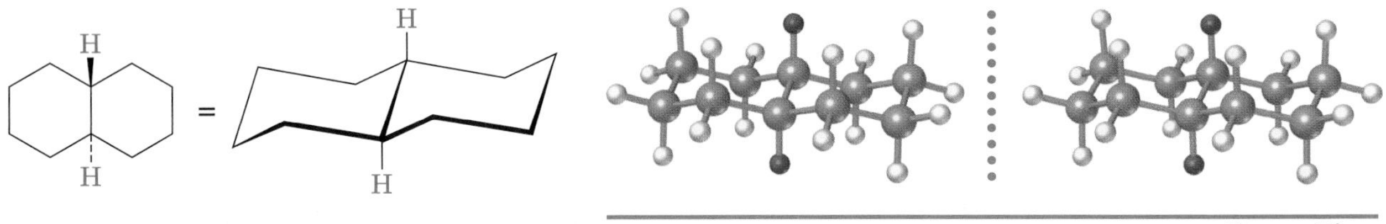

trans-Decalin

Stereo View

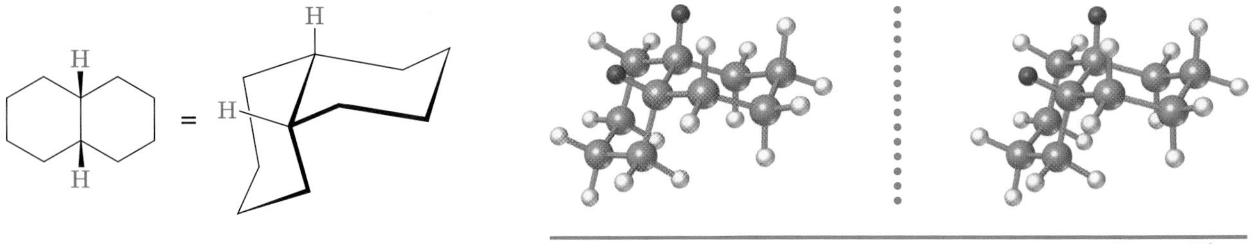

cis-Decalin

Stereo View

Polycyclic compounds are common, and many valuable substances have fused-ring structures. For example, steroids, such as cholesterol, have four rings fused together—three six-membered and one five-membered. Though steroids look complicated compared with cyclohexane or decalin, the same principles that apply to the conformational analysis of simple cyclohexane rings apply equally well (and often better) to steroids.

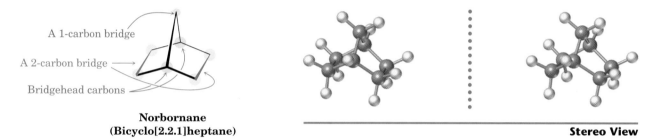

Cholesterol (a steroid)

Another common ring system is the norbornane, or bicyclo[2.2.1]heptane, structure. Like decalin, norbornane is a *bicycloalkane,* so-called because *two* rings would have to be broken open to generate an acyclic structure. Its systematic name, bicyclo[2.2.1]heptane, reflects the fact that the molecule has seven carbons, is bicyclic, and has three "bridges" of 2, 2, and 1 carbon atoms connecting the two bridgehead carbons.

A 1-carbon bridge

A 2-carbon bridge

Bridgehead carbons

Norbornane
(Bicyclo[2.2.1]heptane)

Stereo View

Norbornane has a conformationally locked boat cyclohexane ring in which carbons 1 and 4 are joined by an extra CH_2 group. Note how, in drawing this structure, a break in the rear bond indicates that the vertical bond crosses in front of it. Making a molecular model is particularly helpful when trying to see the three-dimensionality of norbornane.

Substituted norbornanes, such as camphor, are found widely in nature, and many have been important historically in developing organic structural theories.

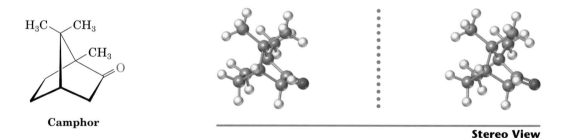

Camphor

Stereo View

Problem 4.21 Which isomer is more stable, *trans*-decalin or *cis*-decalin? Explain.

Molecular Mechanics

All the structural models in this book are computer-drawn. To make sure they accurately portray bond angles, bond lengths, torsional interactions, and steric interactions, the optimum geometry of each molecule has been calculated on a desktop computer using a commercially available *molecular mechanics* program developed by N. L. Allinger of the University of Georgia.

The idea behind molecular mechanics is to begin with a rough geometry for a molecule and then calculate a total strain energy for that starting geometry, using mathematical equations that assign values to specific kinds of molecular interactions. Bond angles that are too large or too small cause angle strain; bond lengths that are too short or too long cause stretching or compressing strain; unfavorable eclipsing interactions around single bonds cause torsional strain; and nonbonded atoms that approach each other too closely cause steric, or *van der Waals,* strain.

$$E_{\text{total}} = E_{\text{bond stretching}} + E_{\text{angle strain}} + E_{\text{torsional strain}} + E_{\text{van der Waals}}$$

After calculating a total strain energy for the starting geometry, the program automatically changes the geometry slightly in an attempt to lower strain—perhaps by lengthening a bond that is too short or decreasing an angle that is too large. Strain is recalculated for the new geometry, more changes are made, and more calculations are done. After dozens or hundreds of iterations, the calculation ultimately converges on a minimum energy that corresponds to the most favorable, least strained conformation of the molecule.

Molecular mechanics calculations have proven to be enormously useful in organic chemistry, particularly in pharmaceutical research where the complementary fit between a drug molecule and a receptor molecule in the body is often a key to designing new pharmaceutical agents. Morphine and other opium alkaloids, for instance, have a specific three-dimensional shape (Figure 4.26) that allows them to nestle into complementary-shaped cavities on opiate receptor proteins in the brain. Once this shape is known, other molecules calculated to have similar shapes can be designed, leading to the possibility of enhanced biological activity.

Computer programs make it possible to portray accurate representations of molecular geometry.

(continued) ▶

FIGURE 4.26 ▼

The structure of morphine and a stereoview of its minimum-energy conformation, as calculated by molecular mechanics.

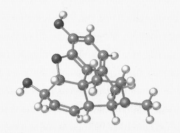

Morphine

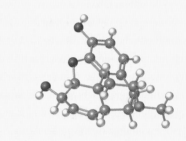

Stereo View

Summary and Key Words

Carbon–carbon single bonds in alkanes are formed by σ overlap of carbon sp^3 hybrid orbitals. Rotation is possible around σ bonds because of their cylindrical symmetry, and alkanes therefore have a large number of rapidly interconverting **conformations**. **Newman projections** make it possible to visualize the spatial consequences of bond rotation by sighting directly along a carbon–carbon bond axis. The **staggered** conformation of ethane is 12 kJ/mol (2.9 kcal/mol) more stable than the **eclipsed** conformation. In general, any alkane is most stable when all its bonds are staggered.

Not all cycloalkanes are equally stable. Three kinds of strain contribute to the overall energy of a cycloalkane: (1) **angle strain**, the resistance of a bond angle to compression or expansion from the normal 109° tetrahedral value; (2) **torsional strain**, the energy cost of having neighboring C–H bonds eclipsed rather than staggered; and (3) **steric strain**, the result of the repulsive interaction that arises when two groups try to occupy the same space.

Cyclopropane (115 kJ/mol strain) and cyclobutane (110.4 kJ/mol strain) have both angle strain and torsional strain. Cyclopentane is free of angle strain but has a substantial torsional strain due to its large number of eclipsing interactions. Both cyclobutane and cyclopentane pucker slightly away from planarity to relieve torsional strain.

Cyclohexane is strain-free because of its puckered **chair conformation**, in which all bond angles are near 109° and all neighboring C–H bonds are staggered. Chair cyclohexane has two kinds of positions: **axial** and **equatorial**. Axial positions are oriented up and down, parallel to the ring axis, whereas equatorial positions lie in a belt around the equator of the ring. Each carbon atom has one axial and one equatorial position.

Chair cyclohexanes are conformationally mobile and can undergo a **ring-flip**, which interconverts axial and equatorial positions:

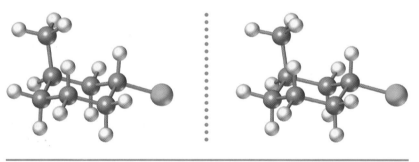

Substituents on the ring are more stable in the equatorial position, because axial substituents cause **1,3-diaxial interactions**. The amount of 1,3-diaxial steric strain caused by an axial substituent depends on its bulk.

Visualizing Chemistry

(Problems 4.1–4.21 appear within the chapter.)

4.22   Name the following compound, identify each substituent as axial or equatorial, and tell whether the conformation shown is the more stable or less stable chair form (gray = C, yellow-green = Cl, ivory = H):

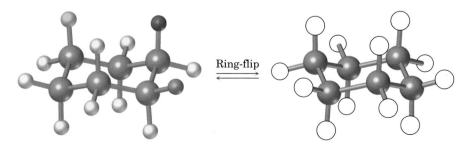

Stereo View

4.23   A trisubstituted cyclohexane with three substituents—red, yellow, and blue—undergoes a ring-flip to its alternative chair conformation. Identify each substituent as axial or equatorial, and show the positions occupied by the three substituents in the ring-flipped form.

4.24 Glucose exists in two forms having a 36:64 ratio at equilibrium. Draw a skeletal structure of each, describe the difference between them, and tell which of the two you think is more stable (gray = C, red = O, ivory = H).

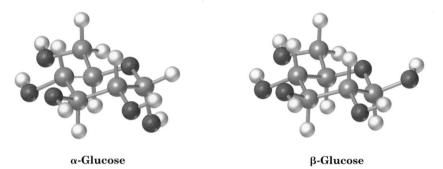

α-Glucose β-Glucose

Additional Problems

4.25 Consider 2-methylbutane (isopentane). Sighting along the C2–C3 bond:
(a) Draw a Newman projection of the most stable conformation.
(b) Draw a Newman projection of the least stable conformation.
(c) Since a CH_3–CH_3 eclipsing interaction costs 11 kJ/mol (2.5 kcal/mol) and a CH_3–CH_3 gauche interaction costs 3.8 kJ/mol (0.9 kcal/mol), make a quantitative plot of energy versus rotation about the C2–C3 bond.

4.26 What are the relative energies of the three possible staggered conformations around the C2–C3 bond in 2,3-dimethylbutane? (See Problem 4.25.)

4.27 Construct a qualitative potential-energy diagram for rotation about the C–C bond of 1,2-dibromoethane. Which conformation would you expect to be more stable? Label the anti and gauche conformations of 1,2-dibromoethane.

4.28 Which conformation of 1,2-dibromoethane (Problem 4.27) would you expect to have the larger dipole moment? The observed dipole moment is $\mu = 1.0$ D. What does this tell you about the actual structure of the molecule?

4.29 The barrier to rotation about the C–C bond in bromoethane is 15 kJ/mol (3.6 kcal/mol).
(a) What energy value can you assign to an H–Br eclipsing interaction?
(b) Construct a quantitative diagram of potential energy versus amount of bond rotation for bromoethane.

4.30 Draw the most stable conformation of pentane, using wedges and dashes to represent bonds coming out of the paper and going behind the paper, respectively.

4.31 Draw the most stable conformation of 1,4-dichlorobutane, using wedges and dashes to represent bonds coming out of the paper and going behind the paper, respectively.

4.32 Draw a chair cyclohexane ring, and label all positions as axial or equatorial.

4.33 Why is a 1,3-cis disubstituted cyclohexane more stable than its trans isomer?

4.34 Why is a 1,2-trans disubstituted cyclohexane more stable than its cis isomer?

4.35 Which is more stable, a 1,4-trans disubstituted cyclohexane or its cis isomer?

4.36 *cis*-1,2-Dimethylcyclobutane is less stable than its trans isomer, but *cis*-1,3-dimethyl-cyclobutane is more stable than its trans isomer. Draw the most stable conformations of both, and explain.

4.37 *N*-Methylpiperidine has the conformation shown below. What does this tell you about the relative steric requirements of a methyl group versus an electron lone pair?

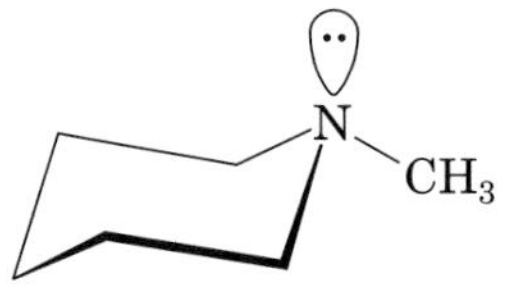

N-Methylpiperidine

4.38 Draw the two chair conformations of *cis*-1-chloro-2-methylcyclohexane. Which is more stable, and by how much?

4.39 Draw the two chair conformations of *trans*-1-chloro-2-methylcyclohexane. Which is more stable, and by how much?

4.40 β-Galactose, a sugar related to glucose, contains a six-membered ring in which all the substituents except the –OH group indicated below in red are equatorial. Draw β-galactose in its more stable chair conformation.

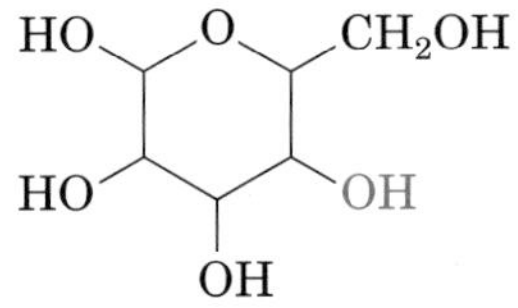

Galactose

4.41 Draw the two chair conformations of menthol, and tell which is more stable.

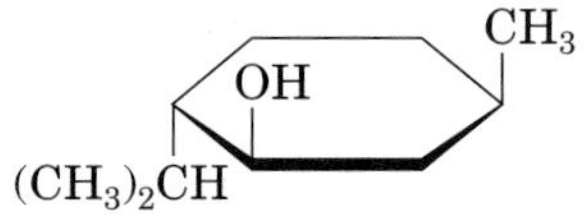

Menthol

4.42 From the data in Figure 4.19 and Table 4.2, estimate the percentages of molecules that have their substituents in an axial orientation for the following compounds:
(a) Isopropylcyclohexane (b) Fluorocyclohexane
(c) Cyclohexanecarbonitrile, $C_6H_{11}CN$ (d) Cyclohexanol, $C_6H_{11}OH$

4.43 Assume that you have a variety of cyclohexanes substituted in the positions indicated. Identify the substituents as either axial or equatorial. For example, a 1,2-cis relationship means that one substituent must be axial and one equatorial, whereas a 1,2-trans relationship means that both substituents are axial or both are equatorial.
(a) 1,3-Trans disubstituted (b) 1,4-Cis disubstituted
(c) 1,3-Cis disubstituted (d) 1,5-Trans disubstituted
(e) 1,5-Cis disubstituted (f) 1,6-Trans disubstituted

4.44 The diaxial conformation of *cis*-1,3-dimethylcyclohexane is approximately 23 kJ/mol (5.4 kcal/mol) less stable than the diequatorial conformation. Draw the two possible chair conformations, and suggest a reason for the large energy difference.

4.45 Approximately how much steric strain does the 1,3-diaxial interaction between the two methyl groups introduce into the diaxial conformation of *cis*-1,3-dimethylcyclohexane? (See Problem 4.44.)

4.46 In light of your answer to Problem 4.45, draw the two chair conformations of 1,1,3-trimethylcyclohexane, and estimate the amount of strain energy in each. Which conformation is favored?

4.47 Draw 1,3,5-trimethylcyclohexane using a regular hexagon to represent the ring. How many cis–trans stereoisomers are there? Which stereoisomer is the most stable?

4.48 We saw in Problem 4.21 that *cis*-decalin is less stable than *trans*-decalin. Assume that the 1,3-diaxial interactions in *trans*-decalin are similar to those in axial methylcyclohexane [that is, one CH_2–H interaction costs 3.8 kJ/mol (0.9 kcal/mol)], and calculate the magnitude of the energy difference between *cis*- and *trans*-decalin.

4.49 Using molecular models as well as structural drawings, explain why *trans*-decalin is rigid and cannot ring-flip, whereas *cis*-decalin can easily ring-flip.

4.50 How many cis–trans stereoisomers of 1,2,3,4,5,6-hexachlorocyclohexane are there? Draw the structure of the most stable isomer.

4.51 Increased substitution around a bond leads to increased strain. Take the four substituted butanes listed below, for example. For each compound, sight along the C2–C3 bond and draw Newman projections of the most stable and least stable conformations. Use the data in Table 4.1 to assign strain energy values to each conformation. Which of the eight conformations is most strained? Which is least strained?
(a) 2-Methylbutane   (b) 2,2-Dimethylbutane
(c) 2,3-Dimethylbutane (d) 2,2,3-Trimethylbutane

4.52 One of the two chair structures of *cis*-1-chloro-3-methylcyclohexane is more stable than the other by 15.5 kJ/mol (3.7 kcal/mol). Which is it? What is the energy cost of a 1,3-diaxial interaction between a chlorine and a methyl group?

4.53 The German chemist J. Bredt proposed in 1935 that bicycloalkenes such as 1-norbornene, which have a double bond to the bridgehead carbon, are too strained to exist. Make a molecular model of 1-norbornene, and explain Bredt's proposal.

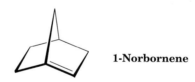

1-Norbornene

4.54 Tell whether each of the following substituents on a steroid is axial or equatorial. (A substituent that is "up" is on the top side of the molecule as drawn, and a substituent that is "down" is on the bottom side.)

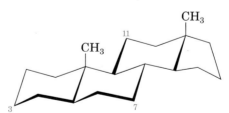

(a) Substituent up at C3   (b) Substituent down at C7
(c) Substituent down at C11

4.55 Amantadine is an antiviral agent that is active against influenza A infection. Draw a three-dimensional representation of amantadine showing the chair cyclohexane rings.

—NH₂   **Amantadine**

A Look Ahead

• •

4.56 Alkyl halides undergo an *elimination* reaction to yield alkenes on treatment with strong base. For example, chlorocyclohexane gives cyclohexene on reaction with NaNH₂:

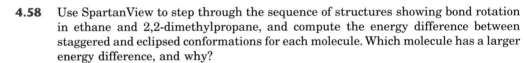

If axial chlorocyclohexanes are generally more reactive than their equatorial isomers, which do you think would react faster, *cis*-1-*tert*-butyl-2-chlorocyclohexane or *trans*-1-*tert*-butyl-2-chlorocyclohexane? Explain. (See Section 11.11.)

4.57 Ketones react with alcohols to yield products called *acetals*. Why does the all-cis isomer of 4-*tert*-butylcyclohexane-1,3-diol react readily with acetone and an acid catalyst to form an acetal, but other stereoisomers do not react? In formulating your answer, draw the more stable chair conformations of all four stereoisomers and the product acetal. Use molecular models for help. (See Section 19.11.)

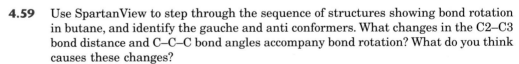

An acetal

Molecular Modeling

• •

4.58 Use SpartanView to step through the sequence of structures showing bond rotation in ethane and 2,2-dimethylpropane, and compute the energy difference between staggered and eclipsed conformations for each molecule. Which molecule has a larger energy difference, and why?

4.59 Use SpartanView to step through the sequence of structures showing bond rotation in butane, and identify the gauche and anti conformers. What changes in the C2–C3 bond distance and C–C–C bond angles accompany bond rotation? What do you think causes these changes?

4.60 Use SpartanBuild to construct a model of isooctane (2,2,4-trimethylpentane). How many different staggered conformations can you generate by bond rotation about the C–C bonds in this molecule? Build each of the staggered conformers and minimize their energies. What are their relative strain energies? Explain.

4.61 Use SpartanBuild to construct a model of *trans*-decalin (Figure 4.25), and then replace one of the CH₂ hydrogens on a ring by a methyl group. How many different methyl-substituted decalins can you make? Minimize the energy of each, and identify the factors that are responsible for the energy differences.

4.62 Use SpartanBuild to construct models of axial and equatorial conformations of methylcyclohexane and *tert*-butylcyclohexane. Minimize each structure, and use the energy differences to predict the relative conformational preferences of methyl and *tert*-butyl groups.

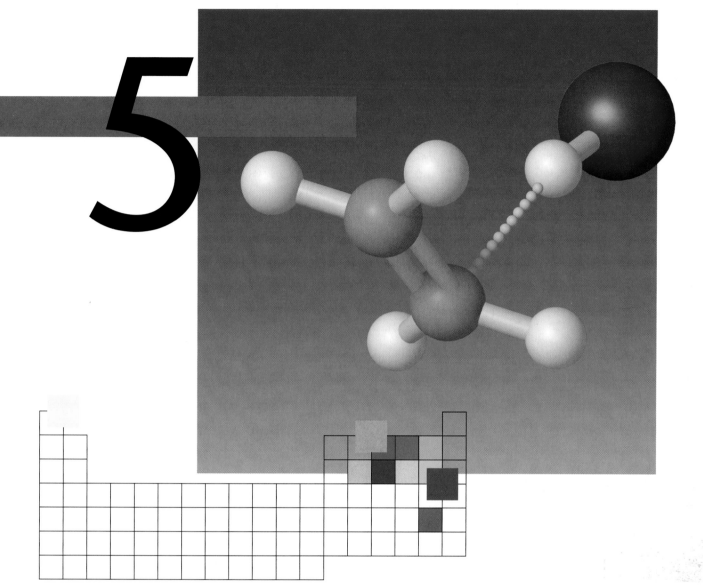

5

An Overview of Organic Reactions

When first approached, organic chemistry can seem like a bewildering collection of millions of compounds, dozens of functional groups, and an endless number of reactions. With study, though, it becomes evident that there are only a few fundamental ideas that underlie all organic reactions.

Far from being a collection of isolated facts, organic chemistry is a beautifully logical subject that is unified by a few broad themes. When these themes are understood, learning organic chemistry becomes much easier and rote memorization can be minimized. The aim of this book is to describe the themes and clarify the patterns that unify organic chemistry. We'll begin by taking an overview of the fundamental kinds of organic reactions that take place and seeing how reactions can be described.

5.1 Kinds of Organic Reactions

Organic chemical reactions can be organized broadly in two ways—by *what kinds* of reactions occur and by *how* reactions occur. Let's look first at the kinds of reactions that take place. There are four general types of organic reactions: *additions*, *eliminations*, *substitutions*, and *rearrangements*.

Addition reactions occur when two reactants add together to form a single new product with no atoms "left over." We can generalize the process as

These reactants add together ... $A + B \longrightarrow C$ ... to give this single product.

An example of an addition reaction that we'll be studying soon is the reaction of an alkene, such as ethylene, with HBr to yield an alkyl bromide:

These two reactants ... 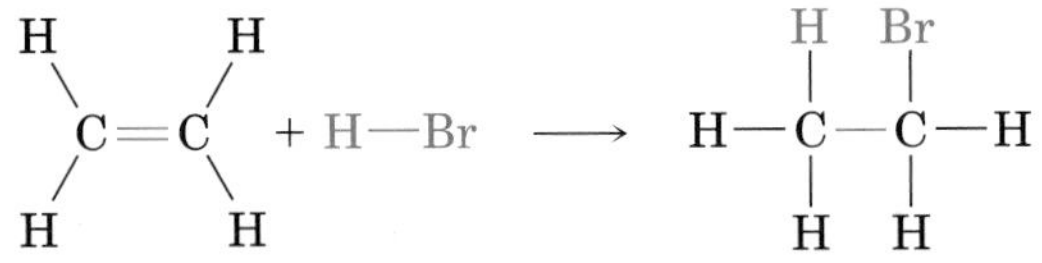
$$\text{H}_2\text{C}=\text{CH}_2 + \text{H—Br} \longrightarrow \text{H}_3\text{C—CH}_2\text{—H}$$
... add to give this product.

Ethylene
(an alkene)
         Bromoethane
         (an alkyl halide)

Elimination reactions are, in a sense, the opposite of addition reactions. Eliminations occur when a single reactant splits into two products:

This one reactant ... $A \longrightarrow B + C$ ... splits apart to give these two products.

An example of an elimination reaction is the reaction of an alkyl halide with base to yield an acid and an alkene:

This one reactant ... 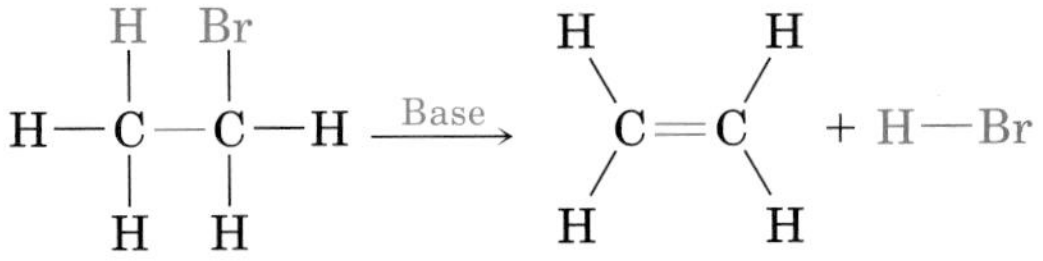
$$\text{H}_3\text{C—CH}_2\text{—H} \xrightarrow{\text{Base}} \text{H}_2\text{C}=\text{CH}_2 + \text{H—Br}$$
... gives these two products.

Bromoethane
(an alkyl halide)
         Ethylene
         (an alkene)

Substitution reactions occur when two reactants exchange parts to give two new products:

These two reactants exchange parts ... $A—B + C—D \longrightarrow A—C + B—D$ ... to give these two new products.

An example of a substitution reaction is the reaction of an alkane with Cl_2 in the presence of ultraviolet light to yield an alkyl chloride. A Cl atom from Cl_2 substitutes for an H atom of the alkane, and two new products result:

These two
reactants . . .

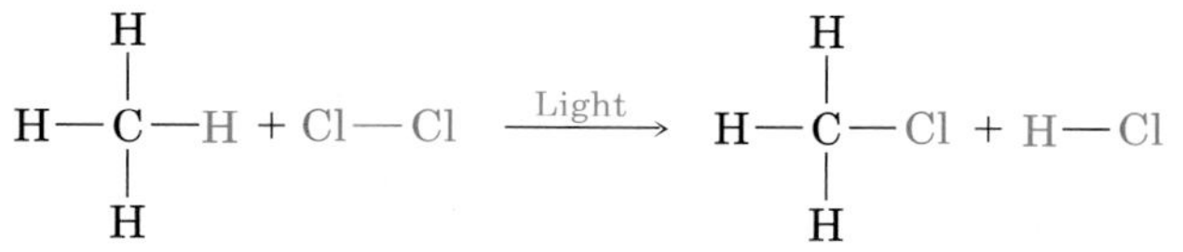

$$H-\underset{\underset{H}{|}}{\overset{\overset{H}{|}}{C}}-H + Cl-Cl \xrightarrow{\text{Light}} H-\underset{\underset{H}{|}}{\overset{\overset{H}{|}}{C}}-Cl + H-Cl$$

. . . give these
two products.

Methane
(an alkane)

Chloromethane
(an alkyl halide)

Rearrangement reactions occur when a single reactant undergoes a reorganization of bonds and atoms to yield an isomeric product:

This single reactant . . .   A ⟶ B   . . . gives this isomeric product.

An example of a rearrangement reaction is the conversion of the alkene 1-butene into its constitutional isomer 2-butene by treatment with an acid catalyst:

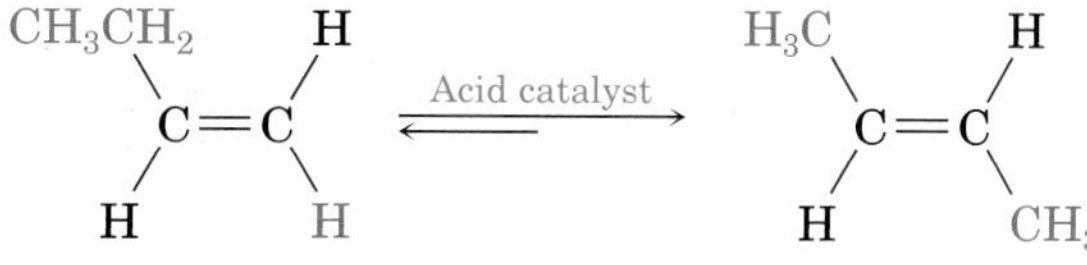

$$\underset{\underset{H}{}}{\overset{CH_3CH_2}{}}C=C\underset{\underset{H}{}}{\overset{H}{}} \underset{\text{Acid catalyst}}{\rightleftharpoons} \underset{\underset{H}{}}{\overset{H_3C}{}}C=C\underset{\underset{CH_3}{}}{\overset{H}{}}$$

1-Butene   **2-Butene**

• •

Problem 5.1   Classify each of the following reactions as an addition, elimination, substitution, or rearrangement:

(a) $CH_3Br + KOH \longrightarrow CH_3OH + KBr$
(b) $CH_3CH_2OH \longrightarrow H_2C{=}CH_2 + H_2O$
(c) $H_2C{=}CH_2 + H_2 \longrightarrow CH_3CH_3$

• •

5.2 How Organic Reactions Occur: Mechanisms

Having looked at the kinds of reactions that take place, let's now see how reactions occur. An overall description of how a reaction occurs is called a **reaction mechanism**. A mechanism describes in detail exactly what takes place at each stage of a chemical transformation. It describes which bonds are broken and in what order, which bonds are formed and in what order, and what the relative rates of the steps are. A complete mechanism must also account for all reactants used, all products formed, and the amount of each.

All chemical reactions involve bond breaking and bond making. When two molecules come together, react, and yield products, specific bonds in the reactant molecules are broken, and specific bonds in the product molecules are formed. Fundamentally, there are two ways in which a covalent two-electron bond can break: A bond can break in an electronically *symmetrical* way so that one electron remains with each product fragment, or a bond can break in an electronically *unsymmetrical* way so that both bonding electrons remain with one product fragment, leaving the other fragment with a vacant orbital. The symmetrical cleavage is said to be **homolytic**, and the unsymmetrical cleavage is said to be **heterolytic**. We'll develop this point in more detail later, but you might notice now that the movement of *one* electron in a homolytic process is indicated using a half-headed, or "fishhook," arrow ($\curvearrowright$), whereas the movement of *two* electrons in a heterolytic process is indicated using a full-headed curved arrow ($\curvearrowright$).

$$A : B \longrightarrow A\cdot + \cdot B$$

Homolytic bond breaking (radical)
(one electron stays with each fragment)

$$A : B \longrightarrow A^+ + :B^-$$

Heterolytic bond breaking (polar)
(two electrons stay with one fragment)

Just as there are two ways in which a bond can break, there are two ways in which a covalent two-electron bond can form: A bond can form in an electronically symmetrical **homogenic** way when one electron is donated to the new bond by each reactant, or a bond can form in an electronically unsymmetrical **heterogenic** way when both bonding electrons are donated to the new bond by one reactant.

$$A\cdot + \cdot B \longrightarrow A : B$$

Homogenic bond making (radical)
(one electron donated by each fragment)

$$A^+ + :B^- \longrightarrow A : B$$

Heterogenic bond making (polar)
(two electrons donated by one fragment)

Processes that involve symmetrical bond breaking and bond making are called **radical reactions**. A **radical** (sometimes called a "free radical") is a neutral chemical species that contains an odd number of electrons and thus has a single, unpaired electron in one of its orbitals. Processes that involve unsymmetrical bond breaking and bond making are called **polar reactions**. Polar reactions involve species that have an even number of electrons and thus have only electron pairs in their orbitals. Polar processes are the more common reaction type in organic chemistry, and a large part of this book is devoted to their description.

In addition to polar and radical reactions, there is a third, less commonly encountered process called a *pericyclic reaction*. Rather than explain pericyclic reactions now, though, we'll study them in more detail in Chapter 30.

5.3 Radical Reactions and How They Occur

Radical reactions are not as common as polar reactions, but they're nevertheless important in organic chemistry, particularly in some industrial processes. Let's see how they occur.

Radicals are highly reactive because they contain an atom with an odd number of electrons (usually seven) in its valence shell, rather than a stable noble-gas octet. A radical can achieve a valence-shell octet in several ways. For example, a radical might abstract an atom from another molecule, leaving behind a new radical. The net result is a radical *substitution* reaction:

$$\text{Rad}\cdot + \text{A}\!:\!\text{B} \longrightarrow \text{Rad}\!:\!\text{A} + \cdot\text{B}$$

Unpaired electron   Unpaired electron

Reactant   **Substitution**   **Product**
radical   **product**   **radical**

Alternatively, a reactant radical might add to an alkene, taking one electron from the alkene double bond and yielding a new radical. The net result is a radical *addition* reaction:

Unpaired electron   Unpaired electron

$$\text{Rad}\cdot + \quad \text{C}=\text{C} \longrightarrow \quad -\overset{\text{Rad}}{\underset{|}{\text{C}}}-\text{C}\cdot$$

Reactant   **Alkene**   **Addition product**
radical   **radical**

Let's look at a specific example of a radical reaction—the chlorination of methane—to see its characteristics. A more detailed discussion of this radical substitution reaction is given in Chapter 10. For the present, it's only necessary to know that methane chlorination is a multistep process.

$$\underset{\text{Methane}}{\text{H}-\overset{\overset{\text{H}}{|}}{\underset{\underset{\text{H}}{|}}{\text{C}}}-\text{H}} + \underset{\text{Chlorine}}{\text{Cl}-\text{Cl}} \xrightarrow{\text{Light}} \underset{\text{Chloromethane}}{\text{H}-\overset{\overset{\text{H}}{|}}{\underset{\underset{\text{H}}{|}}{\text{C}}}-\text{Cl}} + \text{H}-\text{Cl}$$

Radical substitution reactions normally require three kinds of steps: *initiation*, *propagation*, and *termination*.

STEP 1   **Initiation** The initiation step starts off the reaction by producing a small number of reactive radicals. In the present case, the relatively weak Cl–Cl bond is homolytically broken by irradiation with ultraviolet light. Two reactive chlorine radicals are produced:

$$:\!\overset{..}{\underset{..}{\text{Cl}}}\!:\!\overset{..}{\underset{..}{\text{Cl}}}\!: \xrightarrow{\text{Light}} 2\;:\!\overset{..}{\underset{..}{\text{Cl}}}\!\cdot$$

STEP 2   **Propagation** Once a few chlorine radicals have been produced, propagation steps take place. When a reactive chlorine radical collides with a methane molecule, it abstracts a hydrogen atom to produce HCl and a methyl radical ($\cdot$CH$_3$). This methyl radical reacts further with Cl$_2$ in a second propagation step to give the product chloromethane and a new chlorine radical (Cl$\cdot$), which cycles back into the first propagation step. Once the sequence has been initiated, it becomes a self-sustaining cycle of repeating steps (a) and (b), making the overall process a **chain reaction**.

(a) $:\ddot{\underset{..}{Cl}}\cdot + H:CH_3 \longrightarrow H:\ddot{\underset{..}{Cl}}: + \cdot CH_3$

(b) $\cdot CH_3 + :\ddot{\underset{..}{Cl}}:\ddot{\underset{..}{Cl}}: \longrightarrow :\ddot{\underset{..}{Cl}}:CH_3 + :\ddot{\underset{..}{Cl}}\cdot$

(c) Repeat steps (a) and (b) over and over.

STEP 3 Termination Occasionally, two radicals might collide and combine to form a stable product. When this happens, the reaction cycle is broken and the chain is ended. Such termination steps occur infrequently, however, because the concentration of radicals in the reaction at any given moment is very small. Thus, the likelihood that two radicals will collide is also small.

$:\ddot{\underset{..}{Cl}}\cdot + \cdot\ddot{\underset{..}{Cl}}: \longrightarrow :\ddot{\underset{..}{Cl}}:\ddot{\underset{..}{Cl}}:$

$:\ddot{\underset{..}{Cl}}\cdot + \cdot CH_3 \longrightarrow :\ddot{\underset{..}{Cl}}:CH_3$ ⎫ Possible termination steps

$H_3C\cdot + \cdot CH_3 \longrightarrow H_3C:CH_3$ ⎭

The radical substitution reaction just discussed is only one of several different processes that radicals can undergo. The fundamental principle behind all radical reactions is the same, however: *All bonds are broken and formed by reaction of species that have an odd number of electrons.*

· ·

Problem 5.2 Alkane chlorination is not a generally useful reaction because most alkanes have several different kinds of hydrogens, causing mixtures of chlorinated products to result. Draw and name all monochloro substitution products you might obtain by reaction of 2-methylpentane with Cl$_2$.

Problem 5.3 Radical chlorination of pentane is a poor way to prepare 1-chloropentane, $CH_3CH_2CH_2CH_2CH_2Cl$, but radical chlorination of neopentane, $(CH_3)_4C$, is a good way to prepare neopentyl chloride, $(CH_3)_3CCH_2Cl$. Explain.

· ·

5.4 Polar Reactions and How They Occur

Polar reactions occur because of the attraction between positive and negative charges on different functional groups in molecules. To see how these reactions take place, we first need to recall the discussion of polar covalent bonds in Section 2.1 and then we need to look more deeply into the effects of bond polarity on organic molecules.

Most organic molecules are electrically neutral; they have no net charge, either positive or negative. We saw in Section 2.1, however, that certain bonds within a molecule, particularly the bonds in functional groups, are polar. Bond polarity is a consequence of an unsymmetrical electron distribution in a bond and is due to the difference in electronegativity of the bonded atoms. Figure 5.1, which repeats some of the information in Figure 2.2 for convenience, gives the electronegativities of some commonly encountered elements.

FIGURE 5.1 ▼

Electronegativity of some common elements.

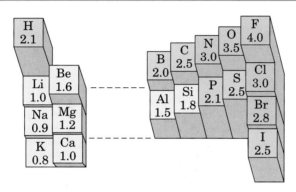

Elements such as oxygen, nitrogen, fluorine, chlorine, and bromine are more electronegative than carbon. Thus, a carbon atom bonded to one of these electronegative atoms has a partial positive charge ($\delta+$). Conversely, metals are less electronegative than carbon, so a carbon atom bonded to a metal has a partial negative charge ($\delta-$). Electrostatic potential maps of chloromethane and methyllithium illustrate these charge distributions, showing that the carbon atom in chloromethane is electron-poor (blue) while the carbon in methyllithium is electron-rich (red).

$Y^{\delta-}$

$C^{\delta+}$

Y = O, N, Cl, Br

$M^{\delta+}$

$C^{\delta-}$

M = A metal

chloromethane, methyllithium

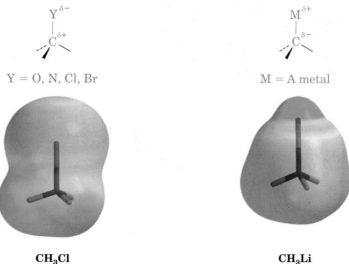

CH_3Cl CH_3Li

The polarity patterns of some common functional groups are shown in Table 5.1. Notice that carbon is always positively polarized except in Grignard reagents and alkyllithiums.

TABLE 5.1   Polarity Patterns in Some Common Functional Groups

Compound type	Functional group structure	Compound type	Functional group structure
Alcohol	$\overset{\delta+}{-\text{C}}-\overset{\delta-}{\text{OH}}$	Carbonyl	$\overset{\delta+}{\text{C}}=\overset{\delta-}{\text{O}}$
Alkene	$\text{C}=\text{C}$ Symmetrical, nonpolar	Carboxylic acid	$\overset{\delta+}{-\text{C}}\overset{\text{O}}{\underset{\overset{\delta-}{\text{OH}}}{\diagup\diagdown}}^{\delta-}$
Alkyl halide	$\overset{\delta+}{-\text{C}}-\overset{\delta-}{\text{X}}$	Carboxylic acid chloride	$\overset{\delta+}{-\text{C}}\overset{\text{O}}{\underset{\overset{\delta-}{\text{Cl}}}{\diagup\diagdown}}^{\delta-}$
Amine	$\overset{\delta+}{-\text{C}}-\overset{\delta-}{\text{NH}_2}$	Aldehyde	$\overset{\delta+}{-\text{C}}\overset{\text{O}}{\underset{\text{H}}{\diagup\diagdown}}^{\delta-}$
Ether	$\overset{\delta+}{-\text{C}}-\overset{\delta-}{\text{O}}-\overset{\delta+}{\text{C}}-$	Ester	$\overset{\delta+}{-\text{C}}\overset{\text{O}}{\underset{\overset{\delta-}{\text{O}}-\text{C}}{\diagup\diagdown}}^{\delta-}$
Nitrile	$\overset{\delta+}{-\text{C}}\equiv\overset{\delta-}{\text{N}}$	Ketone	$\overset{\delta+}{-\text{C}}\overset{\text{O}}{\underset{\text{C}}{\diagup\diagdown}}^{\delta-}$
Grignard reagent	$\overset{\delta-}{-\text{C}}-\overset{\delta+}{\text{MgBr}}$		
Alkyllithium	$\overset{\delta-}{-\text{C}}-\overset{\delta+}{\text{Li}}$		

This discussion of bond polarity is oversimplified in that we've considered only bonds that are inherently polar due to differences in electronegativity. Polar bonds can also result from the interaction of functional groups with solvents and with Lewis acids or bases. For example, the polarity of the carbon–oxygen bond in methanol is greatly enhanced by protonation of the oxygen atom with an acid. In neutral methanol, the carbon atom is somewhat electron-poor because the electronegative oxygen attracts the electrons in the carbon–oxygen bond. In the protonated methanol cation, however, a full positive charge on oxygen *strongly* attracts the electrons in the carbon–oxygen bond and makes the carbon much more electron-poor and much more reactive. We'll thus see numerous examples throughout this book of organic reactions that are catalyzed by acids.

Methanol
(weakly polar C–O bond)

Protonated methanol
(strongly polar C–O bond)

Yet a further consideration is the *polarizability* (as opposed to polarity) of an atom. As the electric field around a given atom changes because of changing interactions with solvent or with other polar molecules, the electron distribution around that atom also changes. The measure of this response to an external influence is called the **polarizability** of the atom. Larger atoms with more loosely held electrons are more polarizable than smaller atoms with tightly held electrons. Thus, iodine is much more polarizable than fluorine. The effect of iodine's high polarizability is that the carbon–iodine bond, although nonpolar according to electronegativity values (Figure 5.1), can nevertheless react as if it were polar.

$I^{\delta-}$
$C^{\delta+}$

Because of iodine's high polarizability, the carbon–iodine bond behaves as if it were polar.

What does functional-group polarity mean with respect to chemical reactivity? *Because unlike charges attract, the fundamental characteristic of all polar organic reactions is that electron-rich sites in one molecule react with electron-poor sites in another molecule.* Bonds are made when an electron-rich atom donates a pair of electrons to an electron-poor atom, and bonds are broken when one atom leaves with both electrons from the former bond.

As we saw in Section 2.11, chemists normally indicate the movement of an electron pair during a polar reaction by using a curved arrow. *A curved arrow shows where electrons move when reactant bonds are broken and product bonds are formed.* It means that an electron pair moves *from* the atom (or bond) at the tail of the arrow *to* the atom at the head of the arrow during the reaction.

A generalized polar reaction

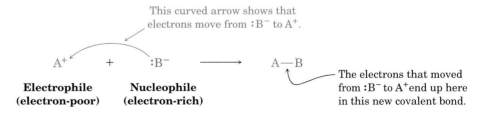

This curved arrow shows that electrons move from $:B^-$ to A^+.

A^+ + $:B^-$ ⟶ A—B

Electrophile
(electron-poor)

Nucleophile
(electron-rich)

The electrons that moved from $:B^-$ to A^+ end up here in this new covalent bond.

In referring to the species involved in a polar reaction, chemists use the words *nucleophile* and *electrophile*. A **nucleophile** is a substance that is "nucleus-loving." (Remember that a nucleus is positively charged.) A nucleophile has an electron-rich atom and can form a bond by donating a pair of electrons to an electron-poor atom. Nucleophiles may be either neutral or negatively charged. Ammonia, water, hydroxide ion, and bromide ion are examples. An **electrophile**, by contrast, is "electron-loving." An electrophile has an electron-poor atom and can form a bond by accepting a pair of electrons from a nucleophile. Electrophiles can be either neutral or positively charged. Acids (H^+ donors), alkyl halides, and carbonyl compounds are examples (Figure 5.2).

FIGURE 5.2 ▼

Some nucleophiles and electrophiles. Electrostatic potential maps identify the nucleophilic (negative) atoms in NH_3 and OH^-, and the electrophilic (positive) atoms in H_3O^+ and CH_3COCH_3.

ammonia, hydroxide ion, hydronium ion, acetone

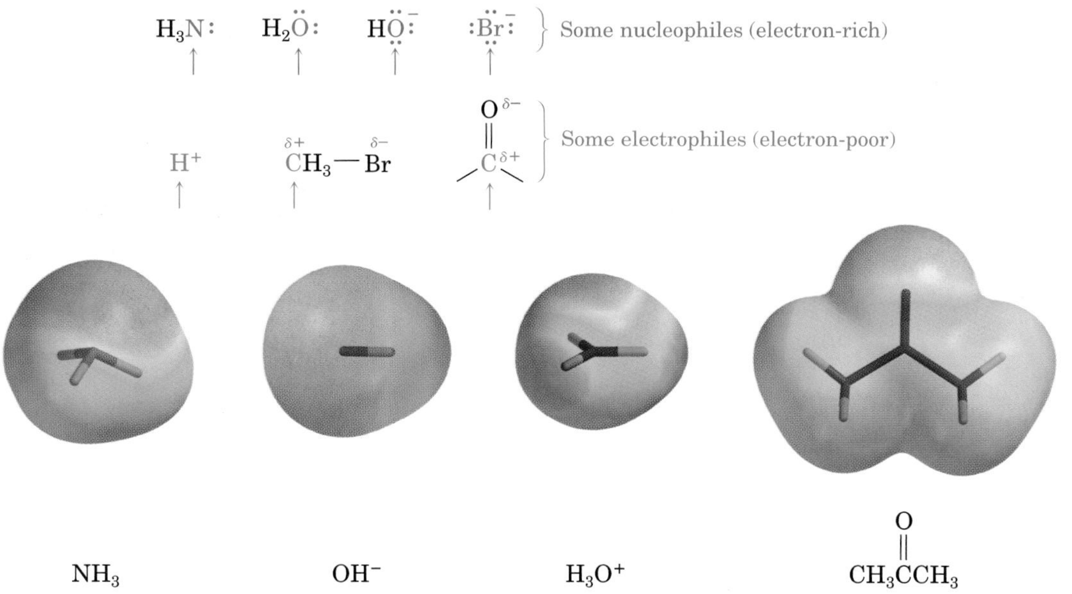

$H_3N:$ $H_2\ddot{O}:$ $H\ddot{O}:^-$ $:\ddot{B}r:^-$ } Some nucleophiles (electron-rich)

H^+ $\overset{\delta+}{C}H_3\overset{\delta-}{-}Br$ $\begin{matrix} O^{\delta-} \\ \| \\ C^{\delta+} \end{matrix}$ } Some electrophiles (electron-poor)

NH_3 OH^- H_3O^+ $\begin{matrix} O \\ \| \\ CH_3CCH_3 \end{matrix}$

Note that it's sometimes possible for a species to be *either* a nucleophile or an electrophile, depending on the circumstances. Water, for instance, can act as a nucleophile if it donates a pair of electrons, yet can act as an electrophile if it donates H^+

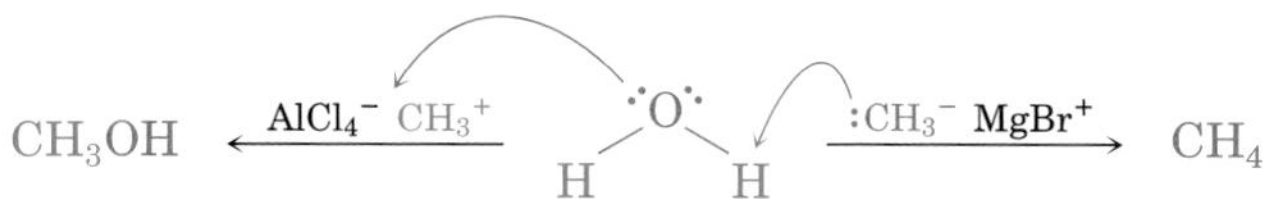

CH_3OH $\xleftarrow{AlCl_4^-\ CH_3^+}$ $\begin{matrix} :\ddot{O}: \\ H \quad H \end{matrix}$ $\xrightarrow{:CH_3^-\ MgBr^+}$ CH_4

Water as a   **Water as an**
nucleophile   **electrophile**

If the definitions of nucleophiles and electrophiles sound similar to those given in Section 2.11 for Lewis acids and Lewis bases, that's because there is indeed a correlation between electrophilicity/nucleophilicity and Lewis acidity/basicity. Lewis bases are electron donors and behave as nucleophiles, whereas Lewis acids are electron acceptors and behave as electrophiles. Therefore, much of organic chemistry is explainable in terms of acid–base reactions. The main difference is that the terms *nucleophile* and *electrophile* are used when bonds to *carbon* are involved. We'll explore these ideas in more detail in Chapter 10.

Practice Problem 5.1   Which of the following species is likely to be an electrophile, and which a nucleophile?
(a) $NO_2{}^+$ (b) CN^- (c) CH_3OH

Strategy   Electrophiles have an electron-poor site, either because they are positively charged or because they have a functional group containing an atom that is positively polarized. Nucleophiles have an electron-rich site, either because they are negatively charged or because they have a functional group containing an atom that has a lone pair of electrons.

Solution   (a) $NO_2{}^+$ (nitronium ion) is likely to be an electrophile because it is positively charged.
(b) $:C{\equiv}N^-$ (cyanide ion) is likely to be a nucleophile because it is negatively charged.
(c) CH_3OH (methyl alcohol) can be either a nucleophile, because it has two lone pairs of electrons on oxygen, or an electrophile, because it has polar C–O and O–H bonds.

$$\overset{\delta+}{CH_3} - \overset{\overset{\delta-}{\underset{..}{\overset{..}{O}}}}{} - \overset{\delta+}{H}$$

Electrophilic   Nucleophilic   Electrophilic

Problem 5.4   Which of the following species is likely to be an electrophile, and which a nucleophile?
(a) HCl   (b) CH_3NH_2 (c) CH_3SH (d) CH_3CHO

5.5 An Example of a Polar Reaction: Addition of HBr to Ethylene

Let's look at a typical polar process—the addition reaction of an alkene such as ethylene with hydrogen bromide. When ethylene is treated with HBr at room temperature, bromoethane is produced. Overall, the reaction can be formulated as follows:

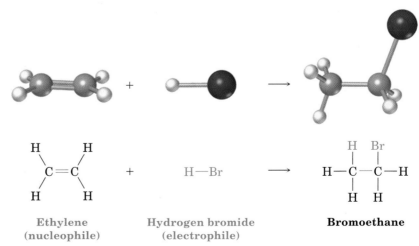

$$\underset{\substack{\text{Ethylene} \\ \text{(nucleophile)}}}{\overset{H}{\underset{H}{\diagdown}} C = C \overset{H}{\underset{H}{\diagup}}} \quad + \quad \underset{\substack{\text{Hydrogen bromide} \\ \text{(electrophile)}}}{H-Br} \quad \longrightarrow \quad \underset{\textbf{Bromoethane}}{\overset{\overset{H \quad Br}{|\quad\ |}}{H - C - C - H}\underset{|\quad |}{\underset{H \quad H}{}}}$$

This reaction, an example of a polar reaction type known as an *electrophilic addition*, can be understood using the general concepts discussed in the previous section. Let's begin by looking at the nature of the two reactants.

What do we know about ethylene? We know from Section 1.9 that a carbon–carbon double bond results from orbital overlap of two sp^2-hybridized carbon atoms. The σ part of the double bond results from sp^2–sp^2 overlap, and the π part results from p–p overlap.

What kind of chemical reactivity might we expect of a carbon–carbon double bond? We know that *alkanes*, such as ethane, are relatively inert because all valence electrons are tied up in strong, nonpolar, C–C and C–H bonds. Furthermore, the bonding electrons in alkanes are relatively inaccessible to approaching reactants because they are sheltered in σ bonds between nuclei. The electronic situation in *alkenes* is quite different, however. For one thing, double bonds have a greater electron density than single bonds—four electrons in a double bond versus only two in a single bond. Equally important, the electrons in the π bond are accessible to approaching reactants because they are located above and below the plane of the double bond rather than being sheltered between the nuclei (Figure 5.3).

FIGURE 5.3 ▼

A comparison of carbon–carbon single and double bonds. A double bond is both more accessible to attack by approaching reactants than a single bond and more electron-rich (more nucleophilic). An electrostatic potential map of 2,3-dimethyl-2-butene indicates that the double bond is the region of highest negative charge (red).

2.3-dimethyl-2-butene

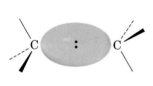

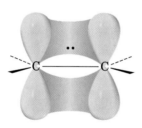

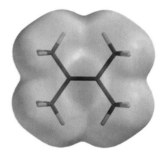

Carbon–carbon σ bond:
stronger; less accessible bonding electrons

Carbon–carbon π bond:
weaker; more accessible electrons

Both electron richness and electron accessibility lead to the prediction that a carbon–carbon double bond should be *nucleophilic*. That is, the chemistry of alkenes should involve reactions of the electron-rich double bond with electron-poor reactants. This is exactly what we find: The most important reaction of alkenes is their reaction with electrophiles.

Now, what about the second reactant, HBr? As a strong acid, HBr is a powerful proton (H^+) donor. Since a proton is positively charged and electron-poor, it is a good electrophile. Thus, the reaction between HBr and ethylene is a typical electrophile–nucleophile combination, characteristic of all polar reactions.

We'll see more details about alkene electrophilic addition reactions shortly, but for the present we can imagine the reaction as taking place by the pathway shown in Figure 5.4. The reaction begins when the alkene donates a pair of electrons from its C=C bond to HBr to form a new C–H bond and Br⁻, as indicated by the path of the curved arrows in the first step of Figure 5.4. One curved arrow begins at the middle of the double bond (the source of the electron pair) and points to the hydrogen atom in HBr (the atom to which a bond will form). This arrow indicates that a new C–H bond forms using electrons from the former C=C bond. A second curved arrow begins in the middle of the H–Br bond and points to the Br, indicating that the H–Br bond breaks and the electrons remain with the Br atom, giving Br⁻.

FIGURE 5.4 ▼

The electrophilic addition reaction of ethylene and HBr. The reaction takes place in two steps, both of which involve electrophile–nucleophile interactions. An electrostatic potential map shows the charge on the carbocation intermediate.

HBr, ethylene, Br⁻, carbocation intermediate, addition product

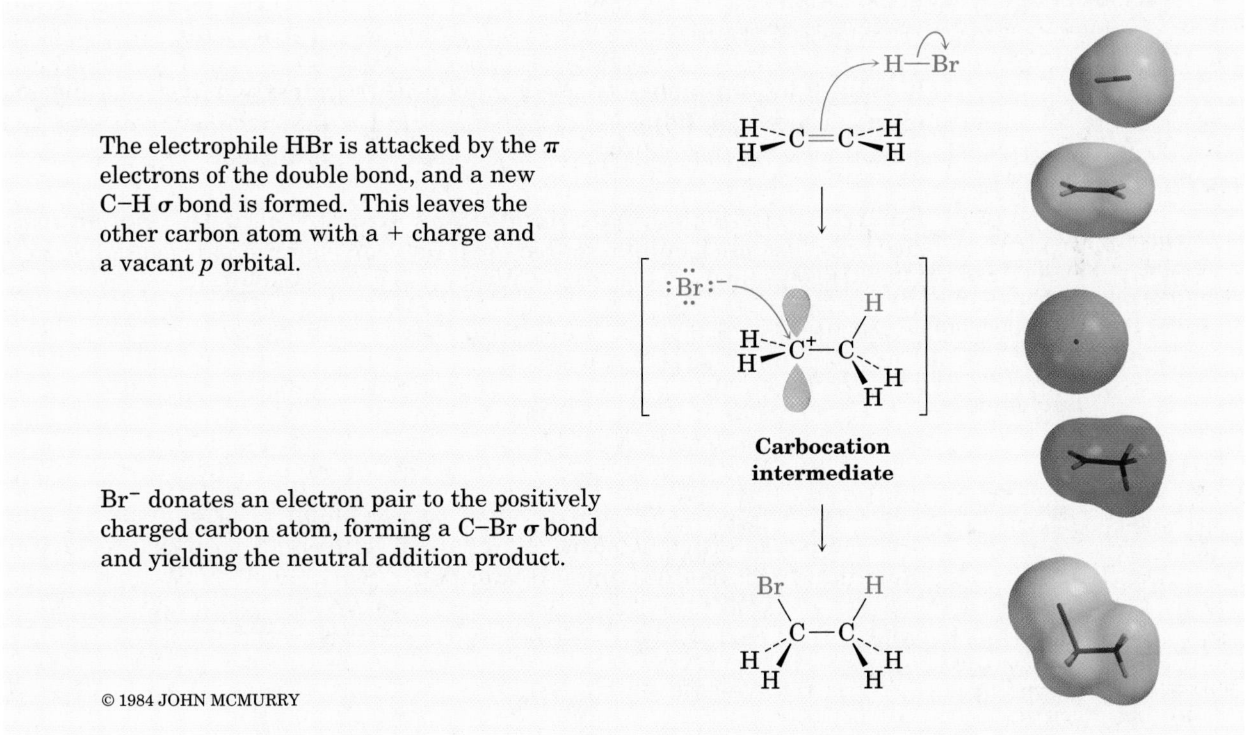

The electrophile HBr is attacked by the π electrons of the double bond, and a new C–H σ bond is formed. This leaves the other carbon atom with a + charge and a vacant *p* orbital.

Br⁻ donates an electron pair to the positively charged carbon atom, forming a C–Br σ bond and yielding the neutral addition product.

Carbocation intermediate

© 1984 JOHN MCMURRY

refer to Mechanisms & Movies

When one of the alkene carbon atoms bonds to the incoming hydrogen, the other carbon atom, having lost its share of the double-bond electrons, now has only six valence electrons and is left with a positive charge. This positively charged species—a carbon cation, or **carbocation**—is itself an electrophile that can accept an electron pair from nucleophilic Br⁻ anion in a second step, forming a C–Br bond and yielding the observed addition product. Once again, a curved arrow in Figure 5.4 shows the electron-pair movement from Br⁻ to the positively charged carbon.

The electrophilic addition of HBr to ethylene is only one example of a polar process; there are many others that we'll study in detail in later chapters. Regardless of the details of individual reactions, *all polar reactions take place between an electron-poor site and an electron-rich site and involve the donation of an electron pair from a nucleophile to an electrophile.*

· ·

Problem 5.5   What product would you expect from reaction of cyclohexene with HBr? With HCl?

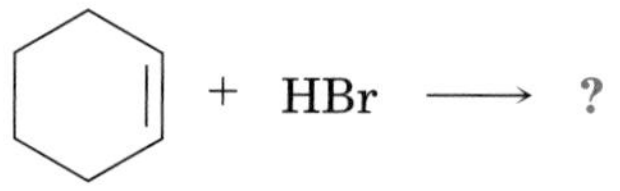

$$\bigcirc + \text{ HBr } \longrightarrow \text{ ?}$$

· ·

5.6 Using Curved Arrows in Polar Reaction Mechanisms

It takes a lot of practice to use curved arrows properly in reaction mechanisms. There are, however, a few rules and a few common patterns you should look for that will help you become more proficient:

RULE 1   **Electrons move *from* a nucleophilic source (Nu:) *to* an electrophilic sink (E).** The nucleophilic source must have an electron pair available, usually either in a lone pair or a multiple bond. For example:

Electrons usually flow *from* one of these nucleophiles:

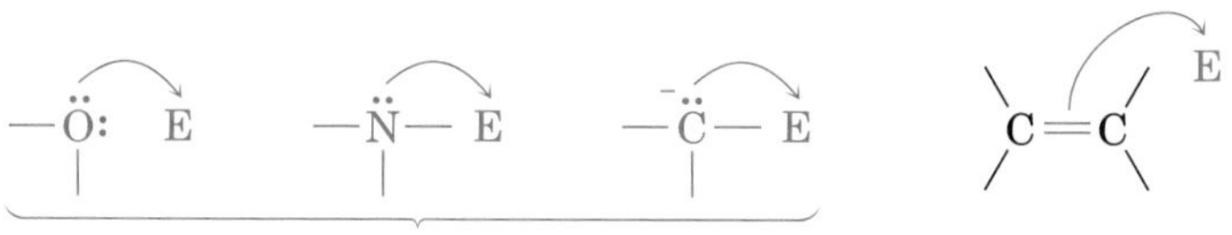

$$-\ddot{\text{O}}: \quad \text{E} \qquad -\ddot{\text{N}}- \text{E} \qquad -\ddot{\text{C}}- \text{E} \qquad \overset{\diagdown}{\underset{\diagup}{\text{C}}}=\overset{\diagdown}{\underset{\diagup}{\text{C}}} \quad \text{E}$$

The electrophilic sink must be able to accept an electron pair, usually because it has either a positively charged atom or a positively polarized atom in a functional group. For example:

Electrons usually flow *to* one of these electrophiles:

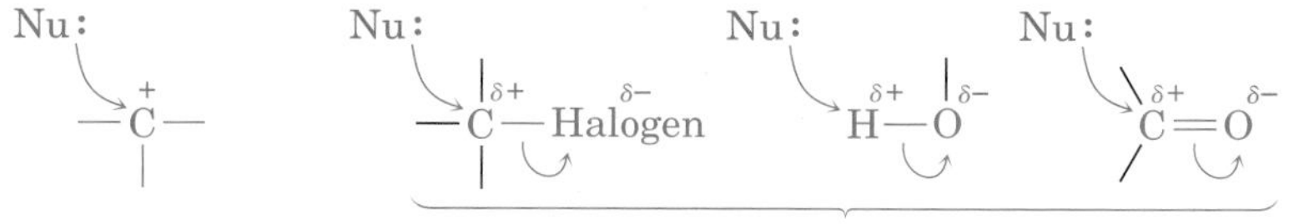

$$\text{Nu:} \qquad \text{Nu:} \qquad \text{Nu:} \qquad \text{Nu:}$$

$$-\overset{+}{\underset{|}{\text{C}}}- \qquad -\overset{\delta+}{\underset{|}{\text{C}}}-\overset{\delta-}{\text{Halogen}} \qquad \overset{\delta+}{\text{H}}-\overset{\delta-}{\text{O}} \qquad \overset{\delta+}{\underset{\diagup}{\text{C}}}=\overset{\delta-}{\text{O}}$$

RULE 2   **The nucleophile can be either negatively charged or neutral.** If the nucleophile is negatively charged, the atom that gives away an electron pair becomes neutral. For example:

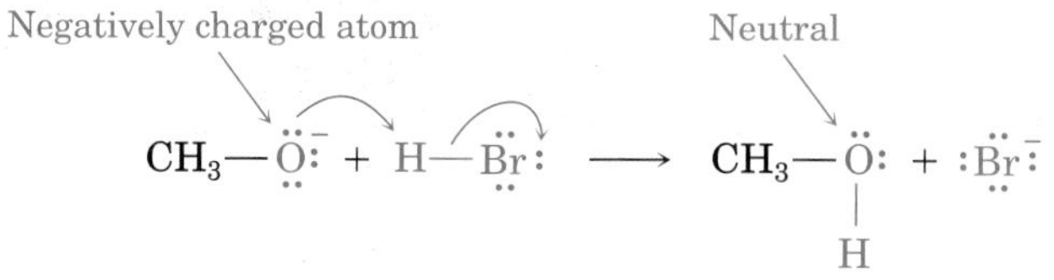

Negatively charged atom   Neutral

$$\text{CH}_3-\ddot{\underset{..}{\text{O}}}:^- + \text{ H}-\ddot{\underset{..}{\text{Br}}}: \longrightarrow \text{ CH}_3-\underset{\underset{\text{H}}{|}}{\ddot{\text{O}}}: + :\ddot{\underset{..}{\text{Br}}}:^-$$

If the nucleophile is neutral, the atom that gives away an electron pair acquires a positive charge. For example:

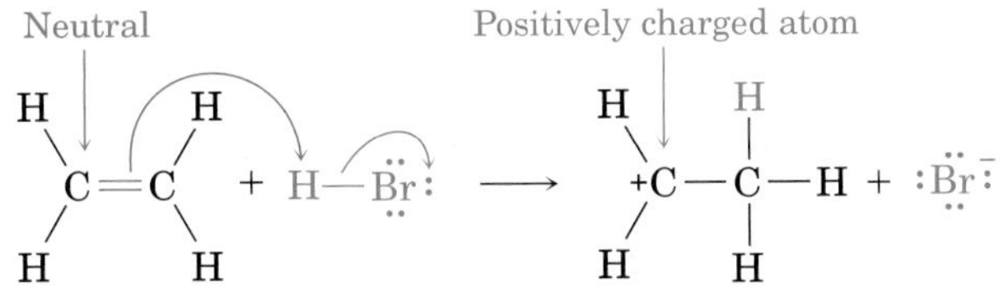

Neutral   Positively charged atom

$$H_2C{=}CH_2 \;+\; H{-}\ddot{B}r\mathbin{:} \;\longrightarrow\; {}^{+}C H_2{-}CH_2{-}H \;+\; \mathbin{:}\ddot{B}\ddot{r}\mathbin{:}^{-}$$

RULE 3 **The electrophile can be either positively charged or neutral.** If the electrophile is positively charged, the atom bearing that charge becomes neutral after accepting an electron pair. For example:

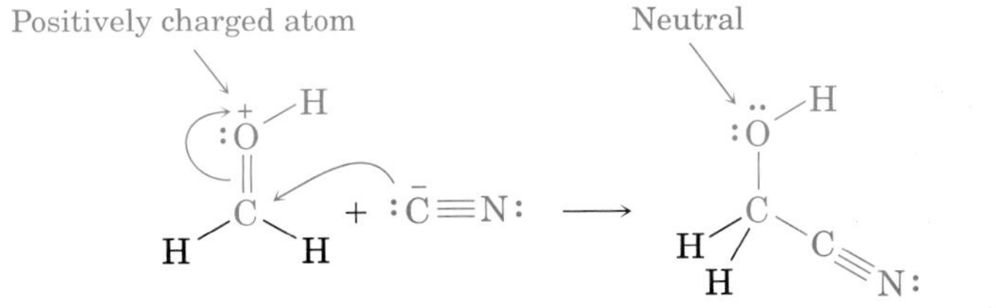

Positively charged atom   Neutral

$$\overset{+}{:}O{=}CH_2 \;+\; \mathbin{:}\bar{C}{\equiv}N\mathbin{:} \;\longrightarrow\; :\ddot{O}{-}CH_2{-}C{\equiv}N\mathbin{:}$$

If the electrophile is neutral, the atom that accepts an electron pair acquires a negative charge. For this to happen, however, the negative charge must be stabilized by being on an electronegative atom such as oxygen or a halogen. For example:

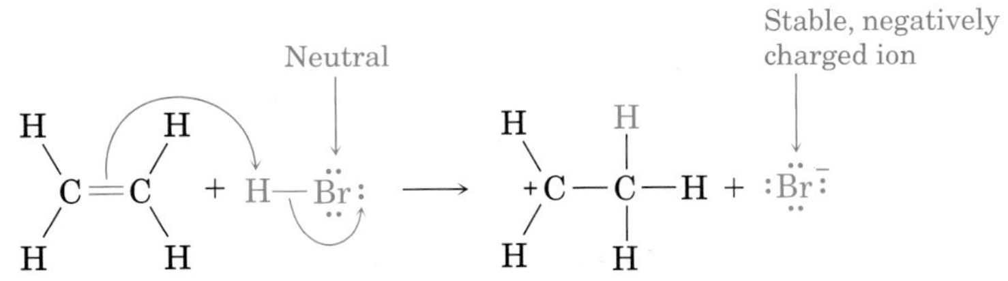

Neutral   Stable, negatively charged ion

$$H_2C{=}CH_2 \;+\; H{-}\ddot{B}r\mathbin{:} \;\longrightarrow\; {}^{+}CH_2{-}CH_2{-}H \;+\; \mathbin{:}\ddot{B}\ddot{r}\mathbin{:}^{-}$$

RULE 4 **The octet rule must be followed.** That is, no second-row atom can be left with ten electrons (or four for hydrogen). If an electron pair moves *to* an atom that already has an octet (or two for hydrogen), another electron pair must concurrently move *from* that atom. When two electrons move from the C=C bond of ethylene to the hydrogen atom of HBr, for example, two electrons must leave that hydrogen. This means that the H–Br bond must break and the electrons must stay with the bromine, giving the stable bromide ion:

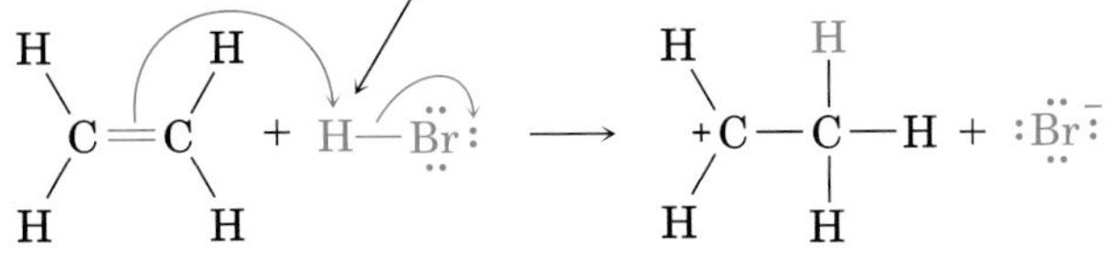

This hydrogen already has two electrons. When another electron pair moves to the hydrogen from the double bond, the electron pair in the H–Br bond must leave.

$$H_2C{=}CH_2 \;+\; H{-}\ddot{B}r\mathbin{:} \;\longrightarrow\; {}^{+}CH_2{-}CH_2{-}H \;+\; \mathbin{:}\ddot{B}\ddot{r}\mathbin{:}^{-}$$

Similarly, when electrons move from cyanide ion (CN^-) to the carbon atom of protonated formaldehyde ($H_2C{=}OH^+$), two electrons must leave that carbon. This means that the C=O double bond must become a single bond, and the two electrons must stay with the oxygen, neutralizing the positive charge.

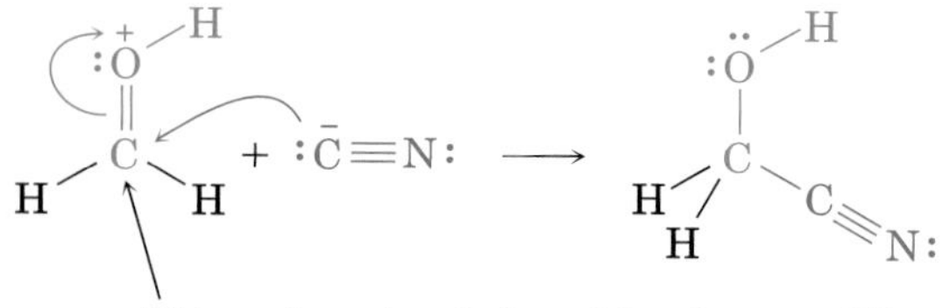

This carbon already has eight electrons. When another electron pair moves to the carbon from CN⁻, an electron pair in the C=O bond must leave.

Practice Problem 5.2 gives another example of drawing curved arrows.

• •

Practice Problem 5.2   Add curved arrows to the following polar reaction to indicate the flow of electrons:

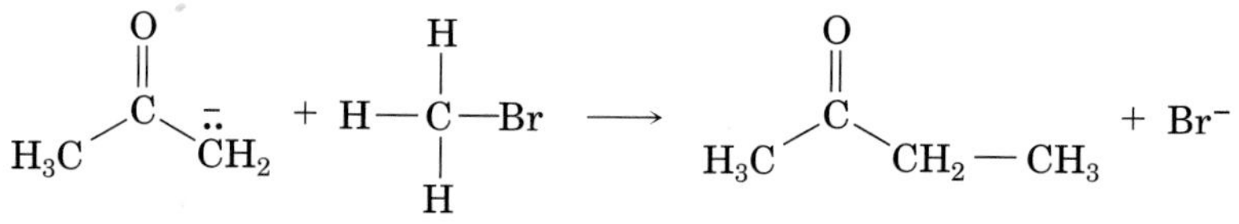

Strategy   First, look at the reaction and identify the bonding changes that have occurred. In this case, a C–Br bond has broken and a C–C bond has formed. The formation of the C–C bond involves donation of an electron pair from the nucleophilic carbon atom of the reactant on the left to the electrophilic carbon atom of CH_3Br, so we draw a curved arrow originating from the lone pair on the negatively charged C atom and pointing to the C atom of CH_3Br. At the same time the C–C bond forms, the C–Br bond must break so that the octet rule is not violated. We therefore draw a second curved arrow from the C–Br bond to Br. The bromine, having gained an electron, is now a stable Br^- ion.

Solution

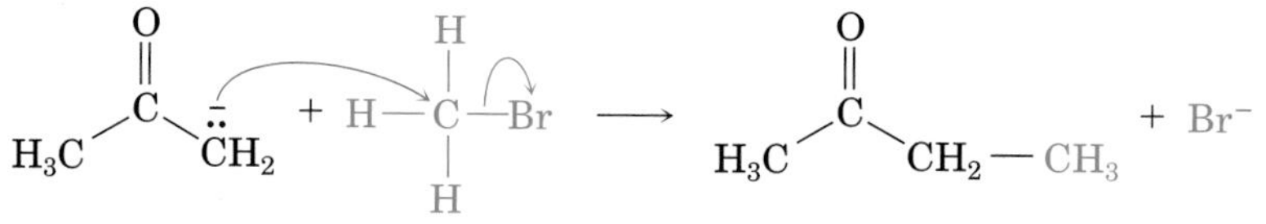

• •

Problem 5.6   Add curved arrows to the following polar reactions to indicate the flow of electrons in each:

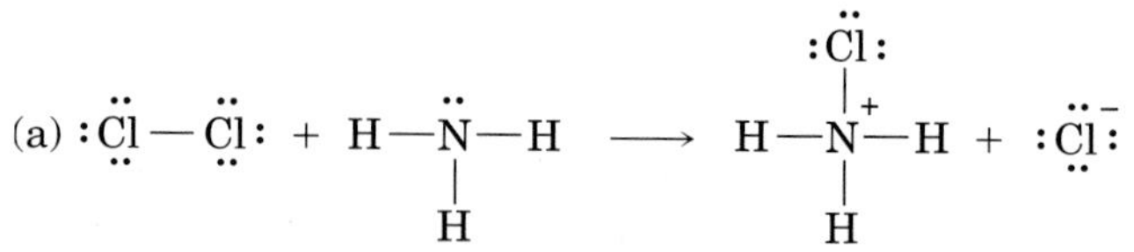

(a)

(b) $CH_3-\overset{..}{\underset{..}{O}}:^- + H-\overset{\overset{\displaystyle H}{|}}{\underset{\underset{\displaystyle H}{|}}{C}}-\overset{..}{\underset{..}{Br}}: \longrightarrow CH_3-\overset{..}{\underset{..}{O}}-CH_3 + :\overset{..}{\underset{..}{Br}}:^-$

(c) $H_3C-\overset{:\overset{..}{\underset{..}{O}}:^-}{\underset{Cl}{\overset{|}{\underset{|}{C}}}}-OCH_3 \longrightarrow \overset{:O:}{\underset{}{\overset{||}{\underset{}{C}}}} \begin{matrix} \\ H_3C \qquad OCH_3 \end{matrix} + :\overset{..}{\underset{..}{Cl}}:^-$

Problem 5.7   Predict the products of the following polar reaction by interpreting the flow of electrons as indicated by the curved arrows.

5.7 Describing a Reaction: Equilibria, Rates, and Energy Changes

Every chemical reaction can go in either forward or reverse direction. Reactants can go forward to products, and products can revert to reactants. The position of the resulting chemical equilibrium is expressed by an equation in which K_{eq}, the equilibrium constant, is equal to the product concentrations multiplied together, divided by the reactant concentrations multiplied together, with each concentration raised to the power of its coefficient in the balanced equation. For the generalized reaction

$$a\text{A} + b\text{B} \rightleftharpoons c\text{C} + d\text{D}$$

we have

$$K_{eq} = \frac{[\text{Products}]}{[\text{Reactants}]} = \frac{[\text{C}]^c[\text{D}]^d}{[\text{A}]^a[\text{B}]^b}$$

The value of the equilibrium constant tells which side of the reaction arrow is energetically favored. If K_{eq} is much larger than 1, then the product concentration term $[\text{C}]^c[\text{D}]^d$ is much larger than the reactant concentration term $[\text{A}]^a[\text{B}]^b$, and the reaction proceeds as written from left to right. If K_{eq} is much smaller than 1, the reaction does not take place as written but instead goes from right to left.

In the reaction of ethylene with HBr, for example, we can write the following equilibrium expression, and we can determine experimentally that the equilibrium constant at room temperature is approximately 7.5×10^7:

$$H_2C{=}CH_2 + HBr \rightleftharpoons CH_3CH_2Br$$

$$K_{eq} = \frac{[CH_3CH_2Br]}{[HBr][H_2C{=}CH_2]} = 7.5 \times 10^7$$

Since K_{eq} is relatively large, the reaction proceeds as written, and greater than 99.999 99% of the ethylene is converted into bromoethane. For practical purposes, an equilibrium constant greater than about 10^3 means that the amount of reactant left over will be barely detectable (less than 0.1%).

What determines the magnitude of the equilibrium constant? *For a reaction to have a favorable equilibrium constant and proceed as written, the energy of the products must be lower than the energy of the reactants.* In other words, energy must be *released*. The situation is analogous to that of a rock poised precariously in a high-energy position near the top of a hill. When it rolls downhill, the rock releases energy until it reaches a more stable low-energy position at the bottom.

The energy change that occurs during a chemical reaction is called the **Gibbs free-energy change, ΔG**. For a favorable reaction, ΔG has a negative value, meaning that energy is released *to* the surroundings. Such reactions are said to be **exergonic**. For an unfavorable reaction, ΔG has a positive value, meaning that energy is absorbed *from* the surroundings. Such reactions are said to be **endergonic**. (Recall from general chemistry that the *standard* free-energy change for a reaction is denoted $\Delta G°$, where the superscript ° means that the reaction is carried out with pure substances in their most stable form at 1 atm pressure and a specified temperature, usually 298 K. For reactions in solution, all reactant concentrations are 1 M. The superscript is dropped and the free-energy change is ΔG if a reaction is carried out under nonstandard conditions.)

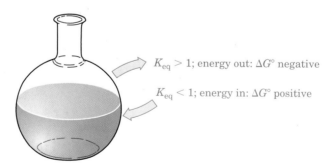

$K_{eq} > 1$; energy out: $\Delta G°$ negative

$K_{eq} < 1$; energy in: $\Delta G°$ positive

Because the equilibrium constant, K_{eq}, and the standard free-energy change, $\Delta G°$, both measure whether a reaction is favored, they are mathematically related:

$$\Delta G° = -RT \ln K_{eq} \qquad \text{or} \qquad K_{eq} = e^{-\Delta G°/RT}$$

where

$$R = 8.315 \text{ J/(K} \cdot \text{mol)} = 1.987 \text{ cal/(K} \cdot \text{mol)}$$
$$T = \text{Kelvin temperature}$$
$$e = 2.718$$
$$\ln K_{eq} = \text{Natural logarithm of } K_{eq}$$

As an example of how this relationship can be used, the reaction of ethylene with HBr has $K_{eq} = 7.5 \times 10^7$. We can therefore calculate that $\Delta G° = -44.8$ kJ/mol (-10.7 kcal/mol) at 298 K:

$$K_{eq} = 7.5 \times 10^7 \quad \text{and} \quad \ln K_{eq} = 18.1$$

$$\Delta G° = -RT \ln K_{eq} = -[8.315 \text{ J/(K} \cdot \text{mol)}] (298 \text{ K})(18.1)$$

$$= -44{,}800 \text{ J/mol} = -44.8 \text{ kJ/mol}$$

The free-energy change ΔG is made up of two terms, an **enthalpy** term, ΔH, and a temperature-dependent **entropy** term, $T\Delta S$, where T is the temperature in kelvins. Of the two terms, the enthalpy term is frequently larger and more dominant.

$$\Delta G° = \Delta H° - T\Delta S° \quad \text{Standard-state conditions}$$

$$\Delta G = \Delta H - T\Delta S \quad \text{Nonstandard-state conditions}$$

For the reaction of ethylene with HBr at room temperature (298 K), the values are $\Delta G° = -44.8$ kJ/mol, $\Delta H° = -84.1$ kJ/mol, and $\Delta S° = -132$ J/(K · mol).

$$\text{H}_2\text{C}{=}\text{CH}_2 + \text{HBr} \rightleftharpoons \text{CH}_3\text{CH}_2\text{Br} \quad \begin{cases} \Delta G° = -44.8 \text{ kJ/mol} \\ \Delta H° = -84.1 \text{ kJ/mol} \\ \Delta S° = -0.132 \text{ kJ/(K} \cdot \text{mol)} \\ T = 298 \text{ K} \end{cases}$$

The enthalpy change, ΔH, is called the **heat of reaction** and is a measure of the change in total bonding energy during a reaction. If ΔH is negative, as in the reaction of HBr with ethylene, the bonds in the products are stronger (more stable) than the bonds in the reactants, heat is released, and the reaction is said to be **exothermic**. If ΔH is positive, the bonds in the products are weaker (less stable) than the bonds in the reactants, heat is absorbed, and the reaction is said to be **endothermic**. For example, if a certain reaction breaks reactant bonds with a total strength of 380 kJ/mol and forms product bonds with a total strength of 400 kJ/mol, then ΔH for the reaction is -20 kJ/mol and the reaction is exothermic. (*Remember*: Breaking bonds absorbs energy, and making bonds releases energy.)

Energy absorbed in breaking reactant bonds:	$\Delta H =$	380 kJ/mol
Energy released in making product bonds:	$\Delta H =$	-400 kJ/mol
Net change:	$\Delta H =$	-20 kJ/mol

The entropy change, ΔS, is a measure of the change in the amount of molecular disorder, or freedom of motion, that accompanies a reaction. For example, in an elimination reaction of the type

$$A \longrightarrow B + C$$

there is more freedom of movement (disorder) in the products than in the reactant because one molecule has split into two. Thus, there is a net increase in entropy during the reaction, and ΔS has a positive value.

On the other hand, for an addition reaction of the type

$$A + B \longrightarrow C$$

the opposite is true. Because such reactions restrict the freedom of movement of two molecules by joining them together, the product has *less* disorder than the reactants, and ΔS has a negative value. The reaction of ethylene and HBr to yield bromoethane is an example [$\Delta S° = -132$ J/(K · mol)].

Table 5.2 describes the thermodynamic terms more fully.

TABLE 5.2   Explanation of Thermodynamic Quantities: $\Delta G° = \Delta H° - T\Delta S°$

Term	Name	Explanation
$\Delta G°$	Gibbs free-energy change	The energy difference between reactants and products. When $\Delta G°$ is negative, the reaction is **exergonic**, has a favorable equilibrium constant, and can occur spontaneously. When $\Delta G°$ is positive, the reaction is **endergonic**, has an unfavorable equilibrium constant, and cannot occur spontaneously.
$\Delta H°$	Enthalpy change	The heat of reaction, or difference in strength between the bonds broken in a reaction and the bonds formed. When $\Delta H°$ is negative, the reaction releases heat and is **exothermic**. When $\Delta H°$ is positive, the reaction absorbs heat and is **endothermic**.
$\Delta S°$	Entropy change	The change in molecular disorder during a reaction. When $\Delta S°$ is negative, disorder decreases; when $\Delta S°$ is positive, disorder increases.

Knowing the value of K_{eq} for a reaction is extremely useful, but it's important to realize the limitations. An equilibrium constant tells only the *position* of the equilibrium, or how much product is theoretically possible. It doesn't tell the rate of reaction, or how fast the equilibrium is established. Some reactions are extremely slow even though they have favorable equilibrium constants. Gasoline is stable at room temperature, for example, because the rate of its reaction with oxygen is slow at 298 K. At higher temperatures, however, such as occur in contact with a lighted match, gasoline reacts rapidly with oxygen and undergoes complete conversion to the equilibrium products water and carbon dioxide. Rates (*how fast* a reaction occurs) and equilibria (*how much* a reaction occurs) are entirely different.

Rate → **Is the reaction fast or slow?**

Equilibrium → **In what direction does the reaction proceed?**

- -

Problem 5.8   Which reaction is more favored, one with $\Delta G° = -44$ kJ/mol or one with $\Delta G° = +44$ kJ/mol?

Problem 5.9   Which reaction is likely to be more exergonic, one with $K_{eq} = 1000$ or one with $K_{eq} = 0.001$?

Problem 5.10   What is the value of $\Delta G°$ at 298 K for reactions where $K_{eq} = 1000$, $K_{eq} = 1$, and $K_{eq} = 0.001$? What is the value of K_{eq} for reactions where $\Delta G° = -40$ kJ/mol, $\Delta G° = 0$ kJ/mol, and $\Delta G° = +40$ kJ/mol?

- -

5.8 Describing a Reaction: Bond Dissociation Energies

We've just seen that heat is released (negative ΔH) when a bond is formed and absorbed (positive ΔH) when a bond is broken. The measure of the heat change that occurs on bond breaking is a quantity called the **bond dissociation energy (D)**, defined as the amount of energy required to break a given bond to produce two radical fragments when the molecule is in the gas phase at 25°C.

$$A : B \xrightarrow[\text{energy}]{\text{Bond dissociation}} A· + ·B$$

Each specific bond has its own characteristic strength, and extensive tables of data are available. For example, a C–H bond in methane has a bond dissociation energy $D = 438.4$ kJ/mol (104.8 kcal/mol), meaning that 438.4 kJ/mol must be added to break a C–H bond of methane to give the two radical fragments $\cdot CH_3$ and $\cdot H$. Conversely, 438.4 kJ/mol of energy is released when a methyl radical and a hydrogen atom combine to form methane. Table 5.3 lists some other bond-strength data.

TABLE 5.3   Some Bond Dissociation Energies

Bond	D (kJ/mol)	Bond	D (kJ/mol)	Bond	D (kJ/mol)
H–H	436	$(CH_3)_3C$–Br	263	CH_3–CH_3	376
H–F	570	$(CH_3)_3C$–I	209	C_2H_5–CH_3	355
H–Cl	432	H_2C=CH–H	444	$(CH_3)_2CH$–CH_3	351
H–Br	366	H_2C=CH–Cl	368	$(CH_3)_3C$–CH_3	339
H–I	298	H_2C=CHCH$_2$–H	361	H_2C=CH–CH_3	406
Cl–Cl	243	H_2C=CHCH$_2$–Cl	289	H_2C=CHCH$_2$–CH_3	310
Br–Br	193	(phenyl)–H	464	H_2C=CH_2	611
I–I	151	(phenyl)–Cl	405	(phenyl)–CH_3	427
CH_3–H	438	(phenyl)CH_2–H	368	(phenyl)CH_2—CH_3	332
CH_3–Cl	351	(phenyl)CH_2–Cl	293	$CH_3\overset{O}{\overset{\|}{C}}$—H	368
CH_3–Br	293	(phenyl)–Br	337	HO–H	498
CH_3–I	234	(phenyl)–OH	469	HO–OH	213
CH_3–OH	380	HC≡C–H	552	CH_3O–H	437
CH_3–NH_2	335			CH_3S–H	371
C_2H_5–H	420			C_2H_5O–H	436
C_2H_5–Cl	338			$CH_3\overset{O}{\overset{\|}{C}}$—$CH_3$	322
C_2H_5–Br	285			CH_3CH_2O–CH_3	339
C_2H_5–I	222			NH_2–H	449
C_2H_5–OH	380			H–CN	518
$(CH_3)_2CH$–H	401				
$(CH_3)_2CH$–Cl	339				
$(CH_3)_2CH$–Br	274				
$(CH_3)_3C$–H	390				
$(CH_3)_3C$–Cl	330				

If enough bond dissociation energies were known, it would seem possible to calculate $\Delta H°$ for any reaction of interest and thus be able to get a rough idea about whether the reaction is favorable. To take the radical substitution reaction of chlorine with methane (Section 5.3) as an example, the bonds formed in this gas-phase reaction (783 kJ/mol) are stronger than the bonds broken (681 kJ/mol), so a net release of heat occurs and we calculate that the reaction is exothermic by about -102 kJ/mol (-24 kcal/mol).

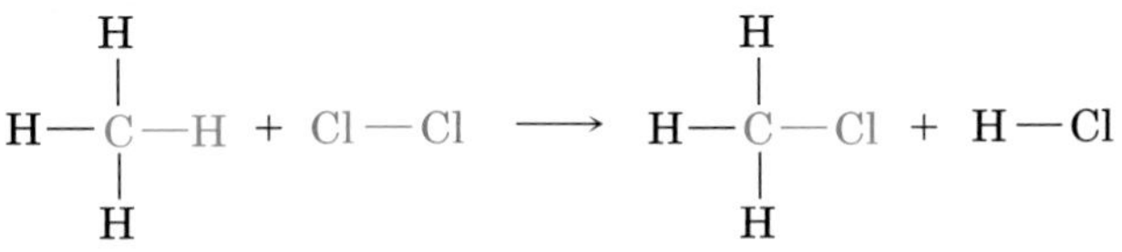

$$
\begin{array}{c}
\quad\text{H} \\
\quad | \\
\text{H} - \overset{\displaystyle |}{\underset{\displaystyle |}{\text{C}}} - \text{H} + \text{Cl} - \text{Cl} \longrightarrow \text{H} - \overset{\displaystyle |}{\underset{\displaystyle |}{\text{C}}} - \text{Cl} + \text{H} - \text{Cl} \\
\quad \text{H} \qquad\qquad\qquad\qquad\qquad \text{H}
\end{array}
$$

Product bonds formed		Reactant bonds broken	
C–Cl	$D = 351$ kJ/mol	C–H	$D = 438$ kJ/mol
H–Cl	$D = 432$ kJ/mol	Cl–Cl	$D = 243$ kJ/mol
Total	$D = 783$ kJ/mol	Total	$D = 681$ kJ/mol

$$\Delta H° = 681 \text{ kJ/mol} - 783 \text{ kJ/mol} = -102 \text{ kJ/mol}$$

Unfortunately, there are several problems with this calculation that limit its value. First, the calculation says nothing about the entropy change $\Delta S°$ for the reaction and thus nothing about the free-energy change $\Delta G°$. Furthermore, the calculation gives no information about the *rate* of reaction even if $\Delta G°$ is favorable. And finally, bond dissociation energies refer to molecules in the gas phase and aren't directly relevant to chemistry in solutions.

In practice, most organic reactions are carried out in solution, where solvent molecules can surround and interact with dissolved reactants, a phenomenon called *solvation*. Solvation can weaken bonds and cause large deviations from the gas-phase value of $\Delta H°$ for a reaction. In addition, the entropy term, $\Delta S°$, also can be different in solution because the solvation of a polar reactant by a polar solvent causes a certain amount of orientation in the solvent and thereby reduces the amount of disorder. Although we can often use bond-strength data to get a rough idea of how thermodynamically favorable a given reaction might be, we have to keep in mind that the answer is only approximate.

● ●

Problem 5.11   Use the data in Table 5.3 to calculate $\Delta H°$ for the gas-phase radical substitution reaction of Br_2 with methane. Is this reaction more exothermic or less exothermic than the corresponding reaction with Cl_2?

$$CH_4 + Br_2 \longrightarrow CH_3Br + HBr$$

Problem 5.12   Calculate $\Delta H°$ for the following reactions:
(a) $CH_3CH_2OCH_3 + HI \longrightarrow CH_3CH_2OH + CH_3I$
(b) $CH_3Cl + NH_3 \longrightarrow CH_3NH_2 + HCl$

● ●

5.9 Describing a Reaction: Energy Diagrams and Transition States

For a reaction to take place, reactant molecules must collide, and reorganization of atoms and bonds must occur. Let's again look at the addition reaction of HBr with ethylene:

Carbocation

As the reaction proceeds, ethylene and HBr must approach each other, the ethylene π bond and H–Br bond must break, a new C–H bond must form in the first step, and a new C–Br bond must form in the second step.

To depict graphically the energy changes that occur during a reaction, chemists use **reaction energy diagrams**, such as that shown in Figure 5.5. The vertical axis of the diagram represents the total energy of all reactants, and the horizontal axis, called the *reaction coordinate,* represents the progress of the reaction from beginning (left) to end (right). Let's see how the addition of HBr to ethylene can be described in a reaction energy diagram.

FIGURE 5.5 ▼

A reaction energy diagram for the first step in the reaction of ethylene with HBr. The energy difference between reactants and transition state, $\Delta G^{\ddagger}$, controls the reaction rate. The energy difference between reactants and carbocation product, $\Delta G°$, controls the position of the equilibrium.

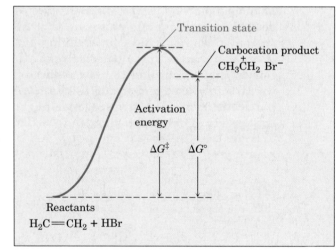

At the beginning of the reaction, ethylene and HBr have the total amount of energy indicated by the reactant level on the left side of the diagram in Figure 5.5. As the two molecules collide and reaction commences, their electron clouds repel each other, causing the energy level to rise. If the collision has occurred with sufficient force and proper orientation, the reactants continue to approach each other despite the rising repulsion until the new C–H bond starts to form. At some point, a structure of maximum energy is reached, a structure we call the *transition state.*

The **transition state** represents the highest-energy structure involved in this step of the reaction. It is unstable and can't be isolated, but we can nevertheless imagine it to be an activated complex of the two reactants in which the carbon–carbon π bond is partially broken and the new carbon–hydrogen bond is partially formed (Figure 5.6).

FIGURE 5.6 ▼

A hypothetical transition-state structure for the first step of the reaction of ethylene with HBr. The carbon–carbon π bond is just beginning to break, the C–H bond is just beginning to form, and the H–Br bond is just beginning to break.

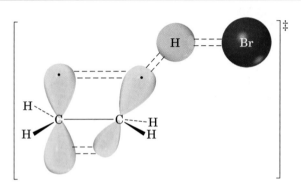

The energy difference between reactants and transition state, called the **activation energy, $\Delta G^{\ddagger}$**, determines how rapidly the reaction occurs at a given temperature. (The double-dagger superscript, ‡, is always used to refer to the transition state.) A large activation energy results in a slow reaction because few collisions occur with enough energy for the reacting molecules to reach the transition state. A small activation energy results in a rapid reaction because almost all collisions occur with enough energy for the reacting molecules to reach the transition state.

The situation of reactants needing enough energy to climb the activation barrier from reactant to transition state is similar to the situation of hikers who need enough energy to climb over a mountain pass. If the pass is a high one, the hikers need a lot of energy and surmount the barrier slowly. If the pass is low, however, the hikers need less energy and reach the top quickly.

As a crude generalization, many organic reactions have activation energies in the range 40–150 kJ/mol (10–35 kcal/mol). The reaction of ethylene with HBr, for example, has an activation energy of approximately 140 kJ/mol (34 kcal/mol). Reactions with activation energies less than 80 kJ/mol take place at or below room temperature, whereas reactions with higher activation energies normally require a higher temperature. Heat provides the energy necessary for the reactants to climb the activation barrier.

Once the transition state is reached, the reaction can either continue on to give the carbocation product or revert back to reactants. When reversion to reactants occurs, the transition-state structure comes apart and an amount of energy corresponding to $-\Delta G^{\ddagger}$ is released. When the reaction continues on to give the carbocation, the new C–H bond forms fully and an amount of energy corresponding to the difference between transition state and carbocation product is released. The net change in energy for the step, $\Delta G°$, is represented in the energy diagram as the difference in level between reactant and product. Since the carbocation is higher in energy than the starting alkene, the step is endergonic, $\Delta G°$ has a positive value, and energy is absorbed.

Not all reaction energy diagrams are like the one shown for the reaction of ethylene and HBr. Each reaction has its own energy profile. Some reactions are fast (small $\Delta G^{\ddagger}$) and some are slow (large $\Delta G^{\ddagger}$); some have a negative $\Delta G°$ and some have a positive $\Delta G°$. Figure 5.7 illustrates some different possibilities for energy profiles.

FIGURE 5.7 ▼

Some hypothetical reaction energy diagrams: (a) a fast exergonic reaction (small $\Delta G^{\ddagger}$, negative $\Delta G°$); (b) a slow exergonic reaction (large $\Delta G^{\ddagger}$, negative $\Delta G°$); (c) a fast endergonic reaction (small $\Delta G^{\ddagger}$, small positive $\Delta G°$); (d) a slow endergonic reaction (large $\Delta G^{\ddagger}$, positive $\Delta G°$).

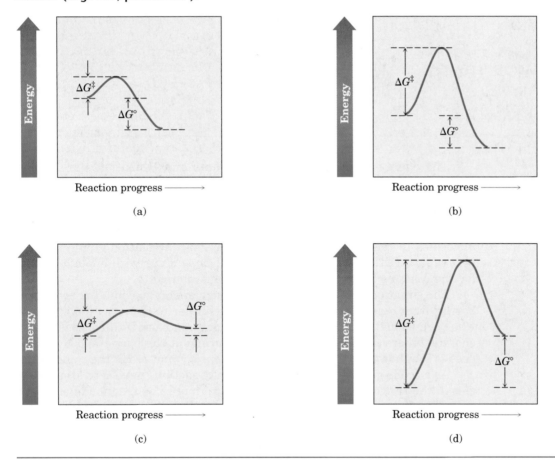

Problem 5.13   Which reaction is faster, one with $\Delta G^{\ddagger} = +45$ kJ/mol or one with $\Delta G^{\ddagger} = +70$ kJ/mol? Which of the two has the larger K_{eq}?

5.10 Describing a Reaction: Intermediates

How can we describe the carbocation formed in the first step of the reaction of ethylene with HBr? The carbocation is clearly different from the reactants, yet it isn't a transition state and it isn't a final product.

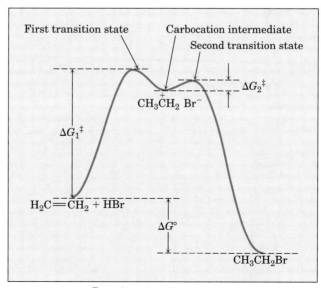

Reaction intermediate

We call the carbocation, which exists momentarily during the course of the multistep reaction, a **reaction intermediate**. As soon as the intermediate is formed in the first step by reaction of ethylene with H^+, it reacts further with Br^- in a second step to give the final product, bromoethane. This second step has its own activation energy ($\Delta G^{\ddagger}$), its own transition state, and its own energy change (ΔG°). We can picture the second transition state as an activated complex between the electrophilic carbocation intermediate and the nucleophilic bromide anion, in which Br^- is donating a pair of electrons to the positively charged carbon atom and the new C–Br bond is just starting to form.

A complete energy diagram for the overall reaction of ethylene with HBr is shown in Figure 5.8. In essence, we draw a diagram for each of the individual steps and then join them in the middle so that the carbocation

FIGURE 5.8 ▼

A reaction energy diagram for the overall reaction of ethylene with HBr. Two separate steps are involved, each with its own transition state. The energy minimum between the two steps represents the carbocation reaction intermediate.

product of step 1 serves as the *reactant* for step 2. As indicated in Figure 5.8, the reaction intermediate lies at an energy minimum between steps 1 and 2. Since the energy level of this intermediate is higher than the level of either the initial reactants (ethylene + HBr) or the final product (bromo-ethane), the intermediate can't be isolated. It is, however, more stable than either of the two transition states that neighbor it.

Each step in a multistep process can always be considered separately. Each step has its own $\Delta G^{\ddagger}$ and its own $\Delta G°$. The *overall* $\Delta G°$ of the reaction, however, is the energy difference between initial reactants (far left) and final products (far right). Figure 5.9 illustrates some different possible cases.

FIGURE 5.9 ▼

Hypothetical reaction energy diagrams for some two-step reactions. The overall $\Delta G°$ for any reaction, regardless of complexity, is the energy difference between initial reactants and final products. Note that reaction (a) is exergonic, whereas reaction (b) is endergonic.

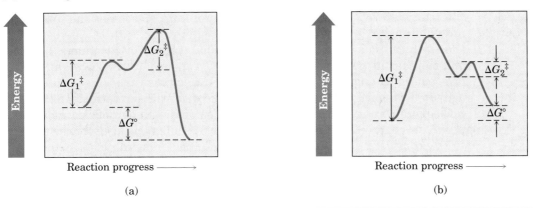

(a)   (b)

Practice Problem 5.3   Sketch a reaction energy diagram for a one-step reaction that is fast and highly exergonic.

Strategy   A fast reaction has a small $\Delta G^{\ddagger}$, and a highly exergonic reaction has a large negative $\Delta G°$.

Solution

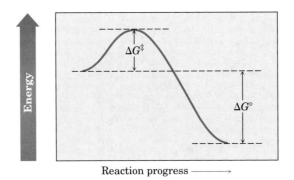

● ●

Problem 5.14 Sketch a reaction energy diagram for a two-step reaction with an endergonic first step and an exergonic second step. Label the parts of the diagram corresponding to reactant, product, and intermediate.

Problem 5.15 Sketch a reaction energy diagram that shows both propagation steps in the radical reaction of chlorine with methane. Is the overall $\Delta G°$ for this reaction positive or negative? Label the parts of your diagram corresponding to $\Delta G°$ and $\Delta G^{\ddagger}$.

$$CH_4 + Cl_2 \xrightarrow{\text{Light}} CH_3Cl + HCl$$

● ●

CHEMISTRY @ WORK

Explosives

Most chemical reactions take place in one or more discrete steps, each of which has a rate, an equilibrium constant, and a well-defined mechanism. The steps can usually be identified, the rates and equilibrium constants can be measured, and the mechanisms can be studied until the reaction is well understood. *Explosions*, however, are different. Their rates are so fast, and their mechanisms are so complex, that the details by which explosions occur defy a complete understanding.

Chemical explosions are characterized by the spontaneous breakdown of molecules into fragments, which then recombine to give the final products—usually stable gases such as N_2, H_2O, and CO_2. The result is a nearly instantaneous release of large quantities of hot gases, which set up a devastating shock wave as they expand. The shock wave can travel at speeds of up to 9000 m/s (approximately 20,000 mi/h) and generate a pressure of up to 700,000 atm, causing enormous physical devastation to the surroundings.

Explosives are categorized as either *primary* or *secondary*, depending on their sensitivity to shock. Primary explosives, such as lead azide, $Pb(N_3)_2$, are the most sensitive. They are used in detonators, blasting caps, and military fuses to initiate the explosion of a less sensitive, secondary explosive. Secondary explosives, or *high explosives*, are less sensitive to heat and shock than primary explosives and are therefore safer to manufacture and transport. Most secondary explosives simply burn rather than explode when ignited in air, and most can be detonated only by the nearby explosion of a primary initiator.

Chemical explosions can generate pressures up to 700,000 atm, devastating the surrounding area.

(continued) ▶

The first commercially important high explosive was nitroglycerin, prepared in 1847 by reaction of glycerin with nitric acid in the presence of sulfuric acid:

$$
\begin{array}{c}
\text{H} \\
| \\
\text{H—C—OH} \\
| \\
\text{H—C—OH} \\
| \\
\text{H—C—OH} \\
| \\
\text{H}
\end{array}
+ 3\ \text{HNO}_3 \xrightarrow{\text{H}_2\text{SO}_4}
\begin{array}{c}
\text{H} \\
| \\
\text{H—C—ONO}_2 \\
| \\
\text{H—C—ONO}_2 \\
| \\
\text{H—C—ONO}_2 \\
| \\
\text{H}
\end{array}
+ 3\ \text{H}_2\text{O}
$$

Glycerin   **Nitroglycerin**

As you might expect, the reaction is extremely hazardous to carry out, and it was not until 1865 that the Swedish chemist Alfred Nobel succeeded in finding a reliable method of producing nitroglycerin and incorporating it into the commercial blasting product called *dynamite*. (The fortune Nobel accumulated from his discovery was subsequently used to fund the Nobel Prizes.) Modern industrial dynamite used for quarrying stone and blasting roadbeds is a mixture of ammonium nitrate and nitroglycerin absorbed onto diatomaceous earth.

The military explosives used as fillings for bombs or shells must have a low sensitivity to impact shock on firing, and must have good stability for long-term storage. TNT (trinitrotoluene), PETN (pentaerythritol tetranitrate), and RDX (research department explosive) are the most commonly used military high explosives. PETN and RDX are also compounded with waxes or synthetic polymers to make so-called *plastic explosives*.

Trinitrotoluene (TNT)

Pentaerythritol tetranitrate (PETN)

$$\text{O}_2\text{NOCH}_2 - \underset{\underset{\text{CH}_2\text{ONO}_2}{|}}{\overset{\overset{\text{CH}_2\text{ONO}_2}{|}}{\text{C}}} - \text{CH}_2\text{ONO}_2$$

RDX

Summary and Key Words

There are four common kinds of reactions: **Addition reactions** take place when two reactants add together to give a single product; **elimination reactions** take place when one reactant splits apart to give two products; **substitution reactions** take place when two reactants exchange parts to give two new products; and **rearrangement reactions** take place when one reactant undergoes a reorganization of bonds and atoms to give an isomeric product.

Addition	$A + B \longrightarrow C$
Elimination	$A \longrightarrow B + C$
Substitution	$A–B + C–D \longrightarrow A–C + B–D$
Rearrangement	$A \longrightarrow B$

A full description of how a reaction occurs is called its **mechanism**. There are two general kinds of mechanisms by which reactions take place: **radical** mechanisms and **polar** mechanisms. Polar reactions, the most common type, occur because of an attractive interaction between a **nucleophilic** (electron-rich) site in one molecule and an **electrophilic** (electron-poor) site in another molecule. A bond is formed in a polar reaction when the nucleophile donates an electron pair to the electrophile. This movement of electrons is indicated by a curved arrow showing the direction of electron travel from the nucleophile to the electrophile. Radical reactions involve species that have an odd number of electrons. A bond is formed when each reactant donates one electron.

Polar $B:^- \quad + \quad A^+ \longrightarrow A:B$

          Nucleophile   Electrophile

Radical $B\cdot + \cdot A \longrightarrow A:B$

The energy changes that take place during reactions can be described by considering both rates (how fast the reactions occur) and equilibria (how much the reactions occur). The position of a chemical equilibrium is determined by the value of the **free-energy change** (ΔG) for the reaction, where $\Delta G = \Delta H - T\Delta S$. The **enthalpy** term (ΔH) corresponds to the net change in strength of chemical bonds broken and formed during reaction; the **entropy** term (ΔS) corresponds to the change in the amount of disorder during reaction. Reactions that have negative values of ΔG release energy, are said to be **exergonic**, and have favorable equilibria. Reactions that have positive values of ΔG absorb energy, are said to be **endergonic**, and have unfavorable equilibria.

A reaction can be described pictorially using a **reaction energy diagram**, which follows the reaction course from reactant through transition state to product. The **transition state** is an activated complex occurring at the highest-energy point of a reaction. The amount of energy needed by reactants to reach this high point is the **activation energy, $\Delta G^{\ddagger}$**. The higher the activation energy, the slower the reaction.

Many reactions take place in more than one step and involve the formation of a **reaction intermediate**. An intermediate is a species that lies at an energy minimum between steps on the reaction curve and is formed briefly during the course of a reaction.

Visualizing Chemistry

(Problems 5.1–5.15 appear within the chapter.)

5.16 The following alkyl halide can be prepared by addition of HBr to two different alkenes. Draw the structures of both (gray = C, reddish brown = Br, ivory = H).

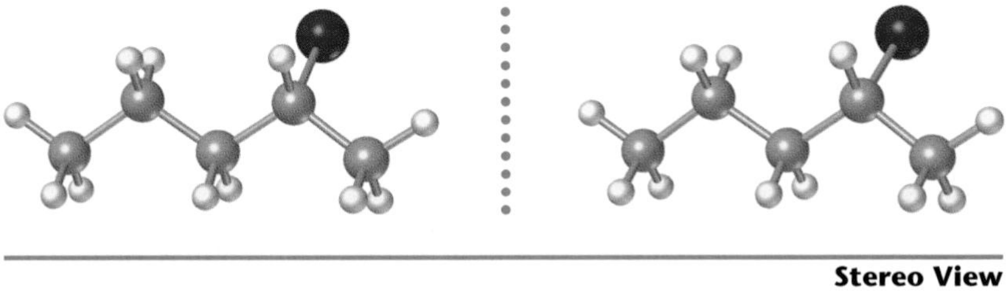
Stereo View

5.17 The following structure represents the carbocation intermediate formed in the addition reaction of HCl to an alkene. Draw the structure of the alkene (gray = C, ivory = H).

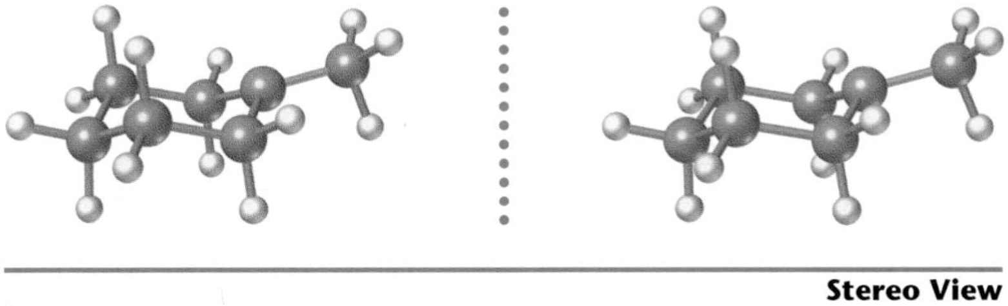
Stereo View

Additional Problems

5.18 Identify the functional groups in the following molecules:

(a) $CH_3CH_2C{\equiv}N$

(b)
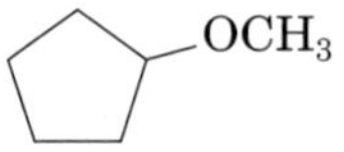
OCH3

(c) $CH_3\overset{O}{\overset{\|}{C}}CH_2\overset{O}{\overset{\|}{C}}OCH_3$

(d)
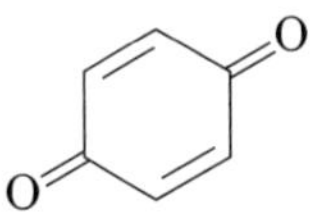
O
O

(e)
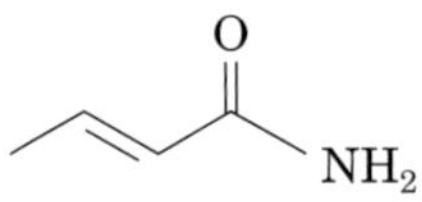
O
NH2

(f)
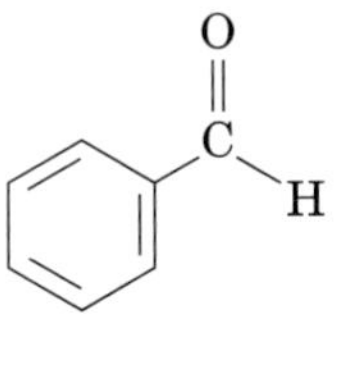
O
C
H

5.19 Show the polarity of the functional groups you identified in Problem 5.18.

5.20 Identify the following reactions as additions, eliminations, substitutions, or rearrangements:

(a) $CH_3CH_2Br + NaCN \longrightarrow CH_3CH_2CN (+ NaBr)$

(b) [cyclohexyl]—OH $\xrightarrow[\text{catalyst}]{\text{Acid}}$ [cyclohexene] $(+ H_2O)$

(c) [cyclopentadiene] + [methyl vinyl ketone] $\xrightarrow{\text{Heat}}$ [bicyclic ketone product]

(d) [benzene] $+ O_2N—NO_2 \xrightarrow{\text{Light}}$ [nitrocyclohexane with NO$_2$] $(+ HNO_2)$

5.21 Give an example of each of the following:
(a) A nucleophile   (b) An electrophile
(c) A polar reaction   (d) A substitution reaction
(e) A heterolytic bond breakage   (f) A homolytic bond breakage

5.22 Which of the following is likely to be a nucleophile and which an electrophile?

(a) Cl^- (b) BF_3 (c) [pyrrolidine] N—H

5.23 What is the difference between a transition state and an intermediate?

5.24 Draw a reaction energy diagram for a one-step endergonic reaction. Label the parts of the diagram corresponding to reactants, products, transition state, $\Delta G°$, and $\Delta G^{\ddagger}$. Is $\Delta G°$ positive or negative?

5.25 Draw a reaction energy diagram for a two-step exergonic reaction. Label the overall $\Delta G°$, transition states, and intermediate. Is $\Delta G°$ positive or negative?

5.26 Draw a reaction energy diagram for a two-step exergonic reaction whose second step is faster than its first step.

5.27 Draw a reaction energy diagram for a reaction with $K_{eq} = 1$. What is the value of $\Delta G°$ in this reaction?

5.28 Look at the reaction energy diagram shown here, and answer the following questions:

 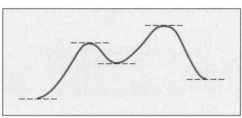

Reaction progress ⟶

(a) Is $\Delta G°$ for the reaction positive or negative? Label it on the diagram.
(b) How many steps are involved in the reaction?
(c) Which step is faster?
(d) How many transition states are there? Label them on the diagram.

5.29 Use the data in Table 5.3 to calculate $\Delta H°$ for the following reactions:
(a) $CH_3OH + HBr \longrightarrow CH_3Br + H_2O$
(b) $CH_3CH_2OH + CH_3Cl \longrightarrow CH_3CH_2OCH_3 + HCl$

5.30 Use the data in Table 5.3 to calculate $\Delta H°$ for the reaction of ethane with chlorine, bromine, and iodine:
(a) $CH_3CH_3 + Cl_2 \longrightarrow CH_3CH_2Cl + HCl$
(b) $CH_3CH_3 + Br_2 \longrightarrow CH_3CH_2Br + HBr$
(c) $CH_3CH_3 + I_2 \longrightarrow CH_3CH_2I + HI$
What can you conclude about the relative energetics of chlorination, bromination, and iodination?

5.31 An alternative course for the reaction of bromine with ethane could result in the formation of bromomethane:

$$H_3C\text{–}CH_3 + Br_2 \longrightarrow 2\ CH_3Br$$

Calculate $\Delta H°$ for this reaction, and compare it with the value you calculated in Problem 5.30 for the formation of bromoethane.

5.32 When a mixture of methane and chlorine is irradiated, reaction commences immediately. When irradiation is stopped, the reaction gradually slows down but does not stop immediately. Explain.

5.33 Radical chlorination of alkanes is not generally useful because mixtures of products often result when more than one kind of C–H bond is present in the substrate. Calculate approximate $\Delta H°$ values for the possible monochlorination reactions of 2-methylbutane. Use the bond dissociation energies measured for $CH_3CH_2\text{–}H$, $H\text{–}CH(CH_3)_2$, and $H\text{–}C(CH_3)_3$ as representative of typical primary, secondary, and tertiary C–H bonds.

5.34 Name each of the products formed in Problem 5.33.

5.35 Despite the limitations of radical chlorination of alkanes, the reaction is still useful for synthesizing certain halogenated compounds. For which of the following compounds does radical chlorination give a single monochloro product?

(a) C_2H_6 (b) $CH_3CH_2CH_3$ (c)

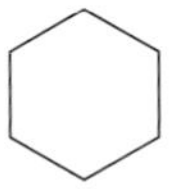

(d) $(CH_3)_3CCH_2CH_3$ (e) 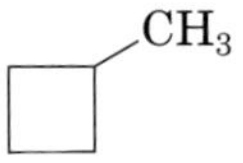   (f) $CH_3C≡CCH_3$

5.36 We've said that the chlorination of methane proceeds by the following steps:

(a) $Cl_2 \xrightarrow{\text{Light}} 2\ Cl\cdot$
(b) $Cl\cdot + CH_4 \longrightarrow HCl + \cdot CH_3$
(c) $\cdot CH_3 + Cl_2 \longrightarrow CH_3Cl + Cl\cdot$

Alternatively, one might propose a different series of steps:

(d) $Cl_2 \longrightarrow 2\ Cl\cdot$
(e) $Cl\cdot + CH_4 \longrightarrow CH_3Cl + H\cdot$
(f) $H\cdot + Cl_2 \longrightarrow HCl + Cl\cdot$

Calculate $\Delta H°$ for each step in both routes. What insight does this provide into the relative merits of each route?

5.37 Add curved arrows to the following reactions to indicate the flow of electrons in each:

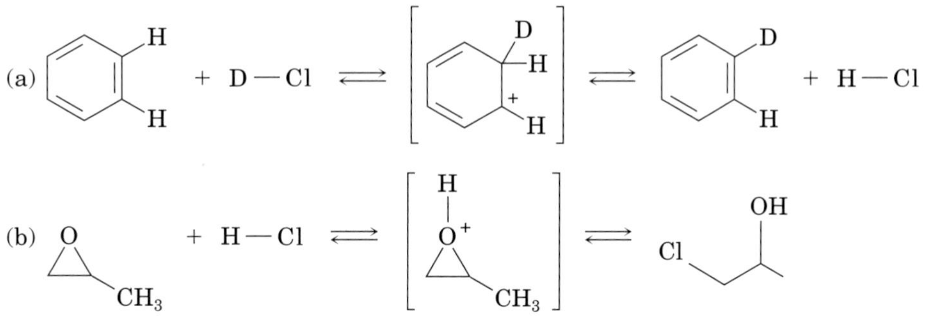

(a) [structure] + D—Cl $\rightleftharpoons$ [structure] $\rightleftharpoons$ [structure] + H—Cl

(b) [structure] + H—Cl $\rightleftharpoons$ [structure] $\rightleftharpoons$ [structure]

5.38 Follow the flow of electrons indicated by the curved arrows in each of the following reactions, and predict the products that result:

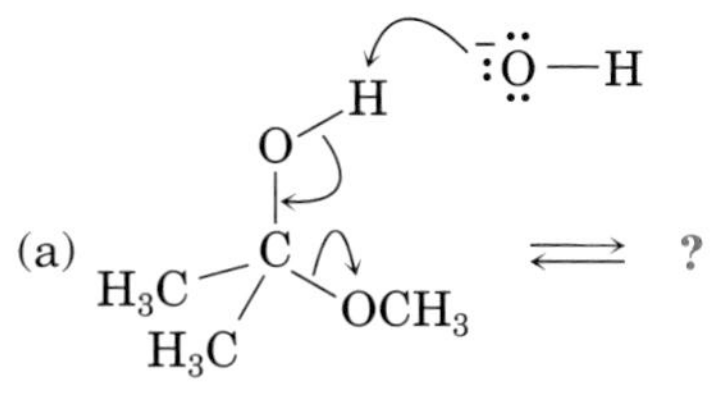

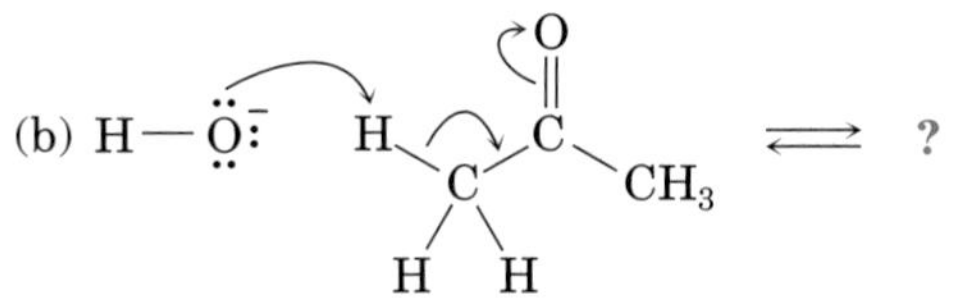

(a) [structure] $\rightleftharpoons$?

(b) [structure] $\rightleftharpoons$?

5.39 When isopropylidenecyclohexane is treated with strong acid at room temperature, isomerization occurs by the mechanism shown below to yield 1-isopropylcyclohexene:

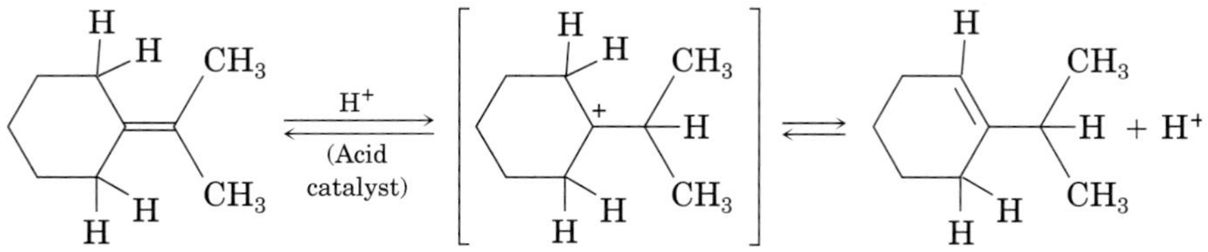

[structure] $\xrightarrow[\text{(Acid catalyst)}]{\text{H}^+}$ [structure] $\rightleftharpoons$ [structure] + H$^+$

Isopropylidenecyclohexane   **1-Isopropylcyclohexene**

At equilibrium, the product mixture contains about 30% isopropylidenecyclohexane and about 70% 1-isopropylcyclohexene.

(a) Calculate K_{eq} for the reaction.
(b) Since the reaction occurs slowly at room temperature, what is its approximate $\Delta G^{\ddagger}$?
(c) Draw a quantitative reaction energy diagram for the reaction.

5.40 Add curved arrows to the mechanism shown in Problem 5.39 to indicate the electron movement in each step.

5.41 2-Chloro-2-methylpropane reacts with water in three steps to yield 2-methyl-2-propanol. The first step is slower than the second, which in turn is much slower than the third. The reaction takes place slowly at room temperature, and the equilibrium constant is near 1.

$$
H_3C-\underset{\underset{CH_3}{|}}{\overset{\overset{CH_3}{|}}{C}}-Cl \;\rightleftharpoons\; \left[\; H_3C-\underset{\underset{CH_3}{|}}{\overset{\overset{CH_3}{|}}{C^+}} \;\xrightarrow{H_2O}\; H_3C-\underset{\underset{CH_3}{|}}{\overset{\overset{CH_3}{|}}{C}}-\overset{+}{O}\!\!\begin{smallmatrix}H\\ \\H\end{smallmatrix} \;\right] \;\xrightarrow{H_2O}\; H_3C-\underset{\underset{CH_3}{|}}{\overset{\overset{CH_3}{|}}{C}}-O-H + H_3O^+ + Cl^-
$$

| 2-Chloro-2-methylpropane | | | 2-Methyl-2-propanol |

(a) Give approximate values for $\Delta G^{\ddagger}$ and ΔG° that are consistent with the above information.

(b) Draw a reaction energy diagram, labeling all points of interest and making sure that the relative energy levels on the diagram are consistent with the information given.

5.42 Add curved arrows to the mechanism shown in Problem 5.41 to indicate the electron movement in each step.

5.43 The reaction of hydroxide ion with chloromethane to yield methanol and chloride ion is an example of a general reaction type called a *nucleophilic substitution reaction*:

$$HO^- + CH_3Cl \;\rightleftharpoons\; CH_3OH + Cl^-$$

The value of ΔH° for the reaction is $-75\,kJ/mol$, and the value of ΔS° is $+54\,J/(K \cdot mol)$. What is the value of ΔG° (in kJ/mol) at 298 K? Is the reaction exothermic or endothermic? Is it exergonic or endergonic?

5.44 Use the value of ΔG° you calculated in Problem 5.43 to find the equilibrium constant K_{eq} for the reaction of hydroxide ion with chloromethane.

A Look Ahead

● ●

5.45 Reaction of 2-methylpropene with HBr might, in principle, lead to a mixture of two bromoalkane addition products. Name them, and draw their structures. (See Section 6.9.)

5.46 Draw the structures of the two carbocation intermediates that might form during the reaction of 2-methylpropene with HBr (Problem 5.45). We'll see in the next chapter that the stability of carbocations depends on the number of alkyl substituents attached to the positively charged carbon—the more alkyl substituents there are, the more stable the cation. Which of the two carbocation intermediates you drew is more stable? (See Section 6.10.)

5.47 Alkenes can be converted into alcohols by acid-catalyzed addition of water. Review the mechanism of the addition of HBr to ethylene (Figure 5.4), and propose a mechanism for the analogous addition of H_2O, using curved arrows to show the electron flow in each step. (See Section 7.4.)

$$H_2C=CH_2 + H_2O \;\xrightarrow{H^+ \text{ catalyst}}\; CH_3CH_2OH$$

Molecular Modeling

• •

5.48 Use SpartanView to examine electrostatic potential maps of trifluoroacetic acid, 3-chloropropene, *tert*-butyl cation, and protonated dimethyl ether. Assuming that the most positive atom is also the most electrophilic, identify the most electrophilic atom in each molecule. (Identify the most electrophilic carbon atom in 3-chloropropene.)

$$\overset{\overset{\textstyle O}{\|}}{CF_3COH} \qquad H_2C{=}CHCH_2Cl \qquad (CH_3)_3C^+ \qquad \overset{\overset{\textstyle H}{\overset{\textstyle |}{\textstyle +}}}{CH_3OCH_3}$$

| **Trifluoroacetic acid** | **3-Chloropropene** | ***tert*-Butyl cation** | **Protonated dimethyl ether** |

5.49 Use SpartanView to examine electrostatic potential maps of deprotonated acetonitrile, *N*-methylimidazole, and 1-methylcyclohexene. Assuming that the most negative atom is also the most nucleophilic, identify the two most nucleophilic atoms in each molecule.

$$^-{:}CH_2C{\equiv}N$$

Deprotonated acetonitrile   ***N*-Methylimidazole**   **1-Methylcyclohexene**

5.50 Use SpartanView to obtain energies for the molecules in the following reaction, and use these energies to estimate $\Delta H°$ for the reaction. Is the reaction exothermic or endothermic?

$$\overset{\overset{\textstyle O}{\|}}{CH_3COH} + CH_2N_2 \longrightarrow \overset{\overset{\textstyle O}{\|}}{CH_3COCH_3} + N_2$$

5.51 Use SpartanView to obtain energies for the reactants and the transition state for each of the following radical reactions, and use these energies to estimate $\Delta H^{\ddagger}$. Which reaction will be faster if they are both carried out under identical conditions?
(a) $CH_4 + Cl{\cdot} \longrightarrow$ [Transition state A] $\longrightarrow CH_3{\cdot} + HCl$
(b) $(CH_3)_4C + Cl{\cdot} \longrightarrow$ [Transition state B] $\longrightarrow (CH_3)_3CCH_2{\cdot} + HCl$

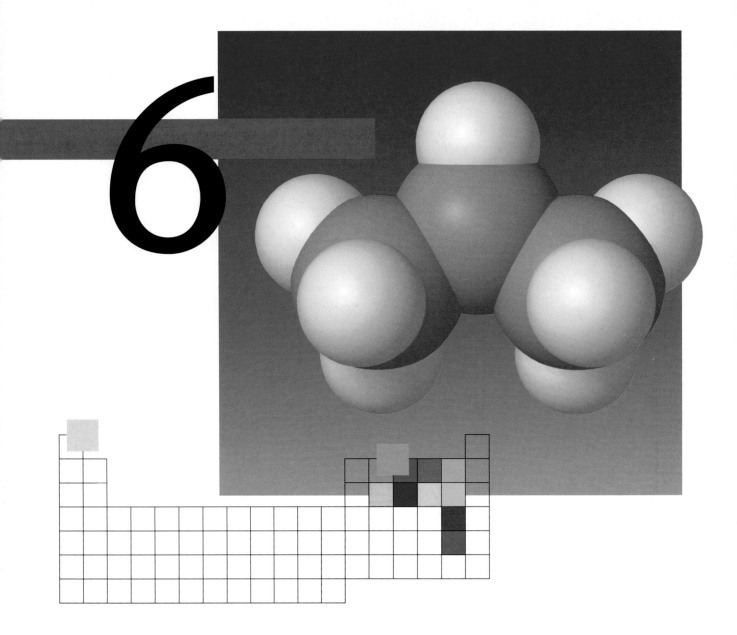

Alkenes: Structure and Reactivity

Alkenes are hydrocarbons that contain a carbon–carbon double bond. The word *olefin* is often used as a synonym, but alkene is the generally preferred term. Alkenes occur abundantly in nature. Ethylene, for example, is a plant hormone that induces ripening in fruit, and α-pinene is the major component of turpentine. Life itself would be impossible without such alkenes as β-carotene, a compound that contains 11 double bonds. An orange pigment responsible for the color of carrots, β-carotene serves as a valuable dietary source of vitamin A and is thought to offer some protection against certain types of cancer.

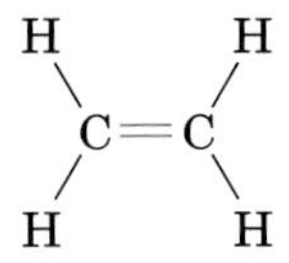

Ethylene   **α-Pinene**

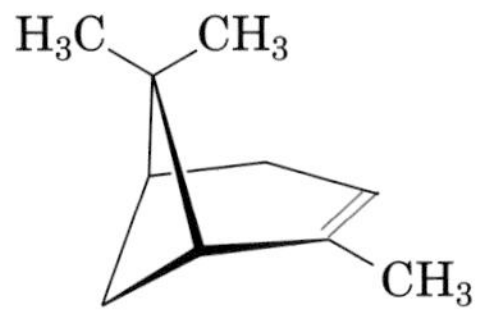

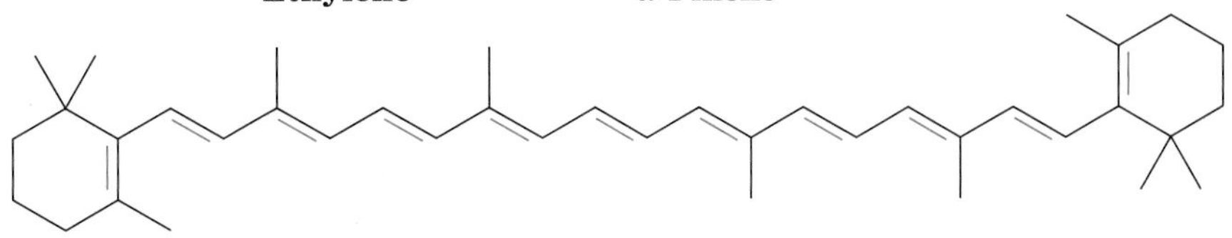

β-Carotene
(orange pigment and vitamin A precursor)

6.1 Industrial Preparation and Use of Alkenes

Ethylene and propylene, the simplest alkenes, are the two most important organic chemicals produced industrially. Approximately 26 million tons of ethylene and 14 million tons of propylene are produced each year in the United States for use in the synthesis of polyethylene, polypropylene, ethylene glycol, acetic acid, acetaldehyde, and a host of other substances (Figure 6.1).

FIGURE 6.1 ▼

Compounds derived industrially from ethylene and propylene.

CH_3CH_2OH	Ethanol
CH_3CHO	Acetaldehyde
CH_3COOH	Acetic acid
CH_2CH_2 (O)	Ethylene oxide
$HOCH_2CH_2OH$	Ethylene glycol
$ClCH_2CH_2Cl$	Ethylene dichloride
$H_2C=CHCl$	Vinyl chloride
$H_2C=CH-O-CCH_3$ (O)	Vinyl acetate
$CH_2CH_2CH_2CH_2$	Polyethylene

$H_2C=CH_2$ →
Ethylene
(26 million tons/yr)

$CH_3CH(OH)CH_3$	Isopropyl alcohol
CH_3CH-CH_2 (O)	Propylene oxide
$CH(CH_3)_2$	Cumene
$CHCH_2CHCH_2$ (CH_3, CH_3)	Polypropylene

$CH_3CH=CH_2$ →
Propylene
(14 million tons/yr)

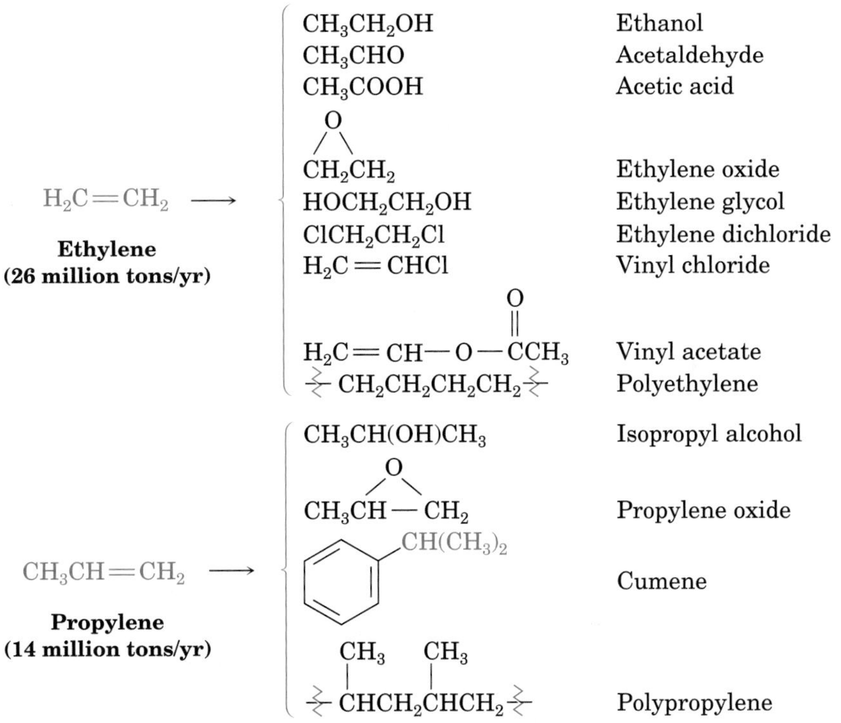

Ethylene, propylene, and butene are synthesized industrially by thermal cracking of natural gas (C_1–C_4 alkanes) and straight-run gasoline (C_4–C_8 alkanes):

$$CH_3(CH_2)_nCH_3 \xrightarrow[\text{Steam}]{850-900°C} H_2 + CH_4 + H_2C=CH_2 + CH_3CH=CH_2 + CH_3CH_2CH=CH_2$$

$$n = 0-6$$

Introduced in 1912, thermal cracking takes place in the absence of catalysts at temperatures up to 900°C. The exact processes are complex, although they undoubtedly involve radical reactions. The high-temperature reaction conditions cause spontaneous homolytic breaking of C–C and C–H bonds, with resultant formation of smaller fragments. We might imagine, for instance, that a molecule of butane splits into two ethyl radicals, each of which then loses a hydrogen atom to generate two molecules of ethylene and H_2:

$$CH_3CH_2 {\large\frown\frown} CH_2CH_3 \xrightarrow{900°C} 2\ \overset{\overset{\displaystyle H}{|}}{CH_2}-CH· \longrightarrow 2\ H_2C=CH_2 + H_2$$

Thermal cracking is an example of a reaction whose energetics are dominated by entropy ($\Delta S°$) rather than by enthalpy ($\Delta H°$) in the free-energy equation $\Delta G° = \Delta H° - T\Delta S°$. Although the bond dissociation energy D for a carbon–carbon single bond is relatively high (about 375 kJ/mol), the large positive entropy change resulting from the fragmentation of one large molecule into several smaller pieces, together with the extremely high temperature, makes the $T\Delta S°$ term larger than the $\Delta H°$ term, thereby favoring the cracking reaction.

6.2 Calculating a Molecule's Degree of Unsaturation

Because of its double bond, an alkene has fewer hydrogens than an alkane with the same number of carbons—C_nH_{2n} for an alkene versus C_nH_{2n+2} for an alkane—and is therefore referred to as **unsaturated**. Ethylene, for example, has the formula C_2H_4, whereas ethane has the formula C_2H_6.

Ethylene: C_2H_4
(fewer hydrogens—*unsaturated*)

Ethane: C_2H_6
(more hydrogens—*saturated*)

In general, each ring or double bond in a molecule corresponds to a loss of two hydrogens from the alkane formula C_nH_{2n+2}. Knowing this relationship, it's possible to work backwards from a molecular formula to calculate a molecule's **degree of unsaturation**—the number of rings and/or multiple bonds present in the molecule.

Let's assume that we want to find the structure of an unknown hydrocarbon. A molecular weight determination on the unknown yields a value of 82, which corresponds to a molecular formula of C_6H_{10}. Since the saturated C_6 alkane (hexane) has the formula C_6H_{14}, the unknown compound has two fewer pairs of hydrogens ($H_{14} - H_{10} = H_4 = 2 H_2$), and its degree of unsaturation is two. The unknown therefore contains two double bonds, one ring and one double bond, two rings, or one triple bond. There's still a long way to go to establish structure, but the simple calculation has told us a lot about the molecule.

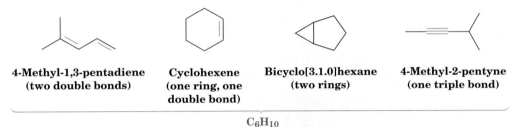

4-Methyl-1,3-pentadiene | **Cyclohexene** | **Bicyclo[3.1.0]hexane** | **4-Methyl-2-pentyne**
(two double bonds) | **(one ring, one double bond)** | **(two rings)** | **(one triple bond)**

$$C_6H_{10}$$

Similar calculations can be carried out for compounds containing elements other than just carbon and hydrogen.

1. **Organohalogen compounds, containing C, H, X (where X = F, Cl, Br, or I).** Because a halogen substituent is simply a replacement for hydrogen in an organic molecule, we can *add* the number of halogens and hydrogens to arrive at an equivalent hydrocarbon formula from which the degree of unsaturation can be found. For example, the alkyl halide formula $C_4H_6Br_2$ is equivalent to the hydrocarbon formula C_4H_8 and thus has one degree of unsaturation:

Replace 2 Br by 2 H

$$BrCH_2CH=CHCH_2Br = HCH_2CH=CHCH_2H$$

$$\underbrace{C_4H_6Br_2}_{Add} = \text{"}C_4H_8\text{"} \quad \text{One unsaturation: one double bond}$$

2. **Organooxygen compounds, containing C, H, O.** Because oxygen forms two bonds, it doesn't affect the formula of an equivalent hydrocarbon and can be ignored when calculating the degree of unsaturation. You can convince yourself of this by seeing what happens when an oxygen atom is inserted into an alkane bond: C–C becomes C–O–C or C–H becomes C–O–H. There's no change in the number of hydrogen atoms. For example, the formula C_5H_8O is equivalent to the hydrocarbon formula C_5H_8 and thus has two degrees of unsaturation:

O removed from here

$$H_2C=CHCH=CHCH_2OH = H_2C=CHCH=CHCH_2-H$$

$$C_5H_8O = \text{"}C_5H_8\text{"} \quad \text{Two unsaturations: two double bonds}$$

3. **Organonitrogen compounds, containing C, H, N.** Because nitrogen forms three bonds, an organonitrogen compound has one more hydrogen than a related hydrocarbon has, and we therefore *subtract* the number of nitrogens from the number of hydrogens to arrive at the equivalent hydrocarbon formula. Again, you can convince yourself of this by seeing what happens when a nitrogen atom is inserted into an alkane bond: C–C becomes C–NH–C or C–H becomes C–NH$_2$. One additional hydrogen atom is required, and we must therefore subtract this extra hydrogen atom to arrive at the equivalent hydrocarbon formula. For example, the formula C$_5$H$_9$N is equivalent to C$_5$H$_8$ and thus has two degrees of unsaturation:

C$_5$H$_9$N   =   "C$_5$H$_8$"   Two unsaturations: one ring and one double bond

Key Ideas ▶   To summarize:

- **Add** the number of halogens to the number of hydrogens.

- **Ignore** the number of oxygens

- **Subtract** the number of nitrogens from the number of hydrogens.

• •

Problem 6.1   Calculate the degree of unsaturation in the following hydrocarbons:
(a) C$_8$H$_{14}$ (b) C$_5$H$_6$ (c) C$_{12}$H$_{20}$
(d) C$_{20}$H$_{32}$ (e) C$_{40}$H$_{56}$ (β-carotene)

Problem 6.2   Calculate the degree of unsaturation in the following formulas, and then draw as many structures as you can for each:
(a) C$_4$H$_8$ (b) C$_4$H$_6$ (c) C$_3$H$_4$

Problem 6.3   Calculate the degree of unsaturation in the following formulas:
(a) C$_6$H$_5$N   (b) C$_6$H$_5$NO$_2$ (c) C$_8$H$_9$Cl$_3$
(d) C$_9$H$_{16}$Br$_2$ (e) C$_{10}$H$_{12}$N$_2$O$_3$ (f) C$_{20}$H$_{32}$ClN

• •

6.3 Naming Alkenes

Alkenes are named using a series of rules similar to those for alkanes (Section 3.4), with the suffix -*ene* used instead of -*ane* to identify the family. There are three steps:

STEP 1   **Name the parent hydrocarbon.** Find the longest carbon chain containing the double bond, and name the compound accordingly, using the suffix *-ene*:

$$CH_3CH_2 \quad\quad H$$
$$\diagdown \quad\quad \diagup$$
$$C=C$$
$$\diagup \quad\quad \diagdown$$
$$CH_3CH_2CH_2 \quad\quad H$$

$$CH_3CH_2 \quad\quad H$$
$$\diagdown \quad\quad \diagup$$
$$C=C$$
$$\diagup \quad\quad \diagdown$$
$$CH_3CH_2CH_2 \quad\quad H$$

Named as a *pentene*   *NOT*   as a hexene, since the double bond is not contained in the six-carbon chain

STEP 2   **Number the carbon atoms in the chain.** Begin at the end nearer the double bond or, if the double bond is equidistant from the two ends, begin at the end nearer the first branch point. This rule ensures that the double-bond carbons receive the lowest possible numbers:

$$\underset{6}{CH_3}\underset{5}{CH_2}\underset{4}{CH_2}\underset{3}{CH}=\underset{2}{CH}\underset{1}{CH_3}$$

$$\begin{array}{c} CH_3 \\ | \end{array}$$
$$\underset{1}{CH_3}\underset{2}{CH}\underset{3}{CH}=\underset{4}{CH}\underset{5}{CH_2}\underset{6}{CH_3}$$

STEP 3   **Write the full name.** Number the substituents according to their positions in the chain, and list them alphabetically. Indicate the position of the double bond by giving the number of the first alkene carbon and placing that number immediately before the parent name. If more than one double bond is present, indicate the position of each, and use one of the suffixes *-diene*, *-triene*, and so on.

$$\underset{6}{CH_3}\underset{5}{CH_2}\underset{4}{CH_2}\underset{3}{CH}=\underset{2}{CH}\underset{1}{CH_3}$$

2-Hexene

$$\begin{array}{c} CH_3 \\ | \end{array}$$
$$\underset{1}{CH_3}\underset{2}{CH}\underset{3}{CH}=\underset{4}{CH}\underset{5}{CH_2}\underset{6}{CH_3}$$

2-Methyl-3-hexene

$$CH_3CH_2 \quad\quad H$$
$$\diagdown \quad \diagup$$
$$\overset{2}{C}=\overset{1}{C}$$
$$\diagup \quad \diagdown$$
$$\underset{5}{CH_3}\underset{4}{CH_2}\underset{3}{CH_2} \quad\quad H$$

2-Ethyl-1-pentene

$$\begin{array}{c} CH_3 \\ | \end{array}$$
$$\underset{1}{H_2C}=\underset{2}{C}-\underset{3}{CH}=\underset{4}{CH_2}$$

2-Methyl-1,3-butadiene

Cycloalkenes are named similarly, but because there is no chain end to begin from, we number the cycloalkene so that the double bond is between C1 and C2 and the first substituent has as low a number as possible. Note that it's not necessary to indicate the position of the double bond in the name because it is always between C1 and C2.

1-Methylcyclohexene   **1,4-Cyclohexadiene**   **1,5-Dimethylcyclopentene**

For historical reasons, there are a few alkenes whose names are firmly entrenched in common usage but don't conform to the rules. For example, the alkene derived from ethane should be called *ethene,* but the name *ethylene* has been used so long that it is accepted by IUPAC. Table 6.1 lists several other common names that are often used and are recognized by IUPAC. Note also that a $=CH_2$ substituent is called a **methylene group**, a $H_2C=CH-$ substituent is called a **vinyl group**, and a $H_2C=CHCH_2-$ substituent is called an **allyl group**:

$$H_2C \underset{\displaystyle\smallfrown}{=}\qquad\qquad H_2C=CH \underset{\displaystyle\smallfrown}{} \qquad\qquad H_2C=CH-CH_2 \underset{\displaystyle\smallfrown}{}$$

A *methylene* group   A *vinyl* group   An *allyl* group

TABLE 6.1   Common Names of Some Alkanes[a]

Compound	Systematic name	Common name
$H_2C=CH_2$	Ethene	Ethylene
$CH_3CH=CH_2$	Propene	Propylene
$\begin{matrix} CH_3 \\ | \\ CH_3C=CH_2 \end{matrix}$	2-Methylpropene	Isobutylene
$\begin{matrix} CH_3 \\ | \\ H_2C=C-CH=CH_2 \end{matrix}$	2-Methyl-1,3-butadiene	Isoprene
$CH_3CH=CHCH=CH_2$	1,3-Pentadiene	Piperylene

[a]Both common and systematic names are recognized by IUPAC for these compounds.

●●

Problem 6.4   Give IUPAC names for the following compounds:

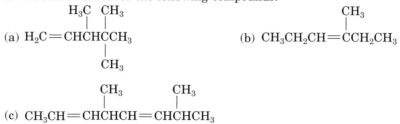

(a) $\begin{matrix} & H_3C\ \ CH_3 \\ & |\ \ \ \ | \\ H_2C=CHCHCCH_3 \\ & | \\ & CH_3 \end{matrix}$ (b) $\begin{matrix} & & CH_3 \\ & & | \\ CH_3CH_2CH=CCH_2CH_3 \end{matrix}$

(c) $\begin{matrix} & CH_3 & & CH_3 \\ & | & & | \\ CH_3CH=CHCHCH=CHCHCH_3 \end{matrix}$

Problem 6.5   Draw structures corresponding to the following IUPAC names:
(a) 2-Methyl-1,5-hexadiene   (b) 3-Ethyl-2,2-dimethyl-3-heptene
(c) 2,3,3-Trimethyl-1,4,6-octatriene   (d) 3,4-Diisopropyl-2,5-dimethyl-3-
(e) 4-*tert*-Butyl-2-methylheptane   hexene

Problem 6.6 Name the following cycloalkenes:

(a) 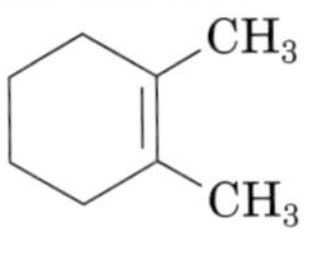 [structure with CH₃ groups]

(b) 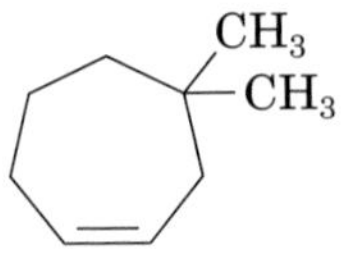 [structure with CH₃, CH₃ groups]

(c) 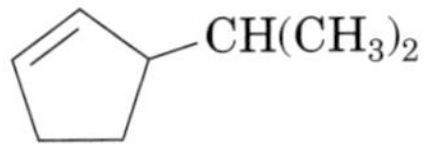 [structure with CH(CH₃)₂]

· ·

6.4 Electronic Structure of Alkenes

We saw in Section 1.9 that the carbon atoms in a double bond are sp^2-hybridized and have three equivalent orbitals that lie in a plane at angles of 120° to one another. The fourth carbon orbital is an unhybridized p orbital perpendicular to the sp^2 plane. When two such carbon atoms approach each other, they form a σ bond by head-on overlap of sp^2 orbitals and a π bond by sideways overlap of p orbitals.

In molecular orbital language, interaction of the p orbitals leads to one bonding and one antibonding π molecular orbital. The π bonding MO has no node between nuclei and results from an additive combination of p orbital lobes with the same algebraic sign. The π antibonding MO has a node between nuclei and results from a subtractive combination of lobes with different algebraic signs (Figure 1.17).

Although free rotation is possible around σ bonds (Section 4.1), the same is not true for double bonds. For rotation to occur around a double bond, the π bond must break temporarily (Figure 6.2). Thus, the barrier to double-bond rotation must be at least as great as the strength of the π bond itself.

FIGURE 6.2 ▼

The π bond must break for rotation to take place around a carbon–carbon double bond.

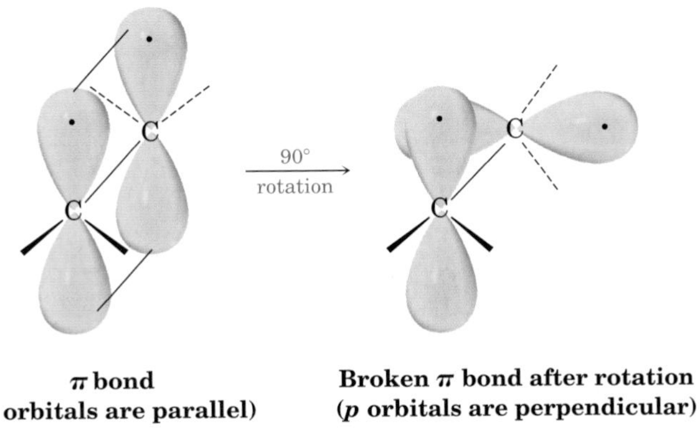

$$\xrightarrow[\text{rotation}]{90°}$$

π bond
(*p* orbitals are parallel)

Broken π bond after rotation
(*p* orbitals are perpendicular)

An experimental determination of how much energy is required to break the π bond of ethylene gives an approximate value of 268 kJ/mol (64 kcal/mol), so it's clear why rotation does not occur. Recall that the barrier to bond rotation in ethane is only 12 kJ/mol.

6.5 Cis–Trans Isomerism in Alkenes

The lack of rotation around the carbon–carbon double bond is of more than just theoretical interest; it also has chemical consequences. Imagine the situation for a disubstituted alkene such as 2-butene. (*Disubstituted* means that two substituents other than hydrogen are bonded to the double-bond carbons.) The two methyl groups in 2-butene can be either on the same side of the double bond or on opposite sides, a situation reminiscent of disubstituted cycloalkanes (Section 3.8). Figure 6.3 shows the two 2-butene isomers.

FIGURE 6.3 ▼

Cis and trans isomers of 2-butene. The cis isomer has the two methyl groups on the same side of the double bond, and the trans isomer has the methyl groups on opposite sides.

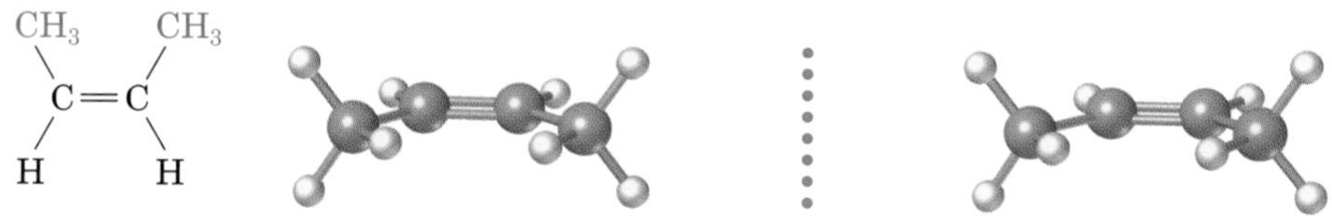

cis-2-Butene

Stereo View

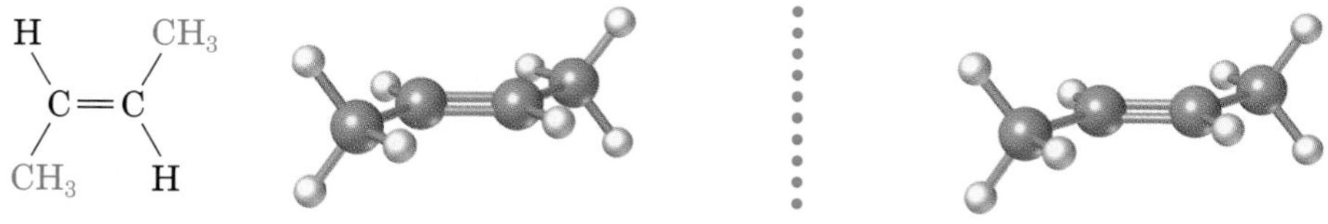

trans-2-Butene

Stereo View

Since bond rotation can't occur, the two 2-butenes can't spontaneously interconvert; they are different, isolable compounds. As with disubstituted cycloalkanes (Section 3.8), we call such compounds *cis–trans stereoisomers*. The compound with substituents on the same side of the double bond is called *cis*-2-butene, and the isomer with substituents on opposite sides is *trans*-2-butene.

Cis–trans isomerism is not limited to *di*substituted alkenes. It can occur in any alkene that has both of its double-bond carbons attached to two different groups. If one of the double-bond carbons is attached to two identical groups, however, then cis–trans isomerism is not possible (Figure 6.4).

FIGURE 6.4 ▼

The requirement for cis–trans isomerism in alkenes. Compounds that have one of their carbons bonded to two identical groups can't exist as cis–trans isomers. Only when both carbons are bonded to two different groups are cis–trans isomers possible.

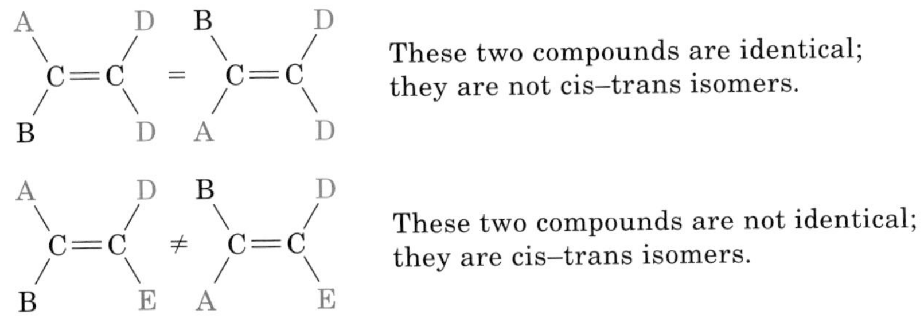

These two compounds are identical; they are not cis–trans isomers.

These two compounds are not identical; they are cis–trans isomers.

· ·

Problem 6.7   Which of the following compounds can exist as pairs of cis–trans isomers? Draw each cis–trans pair, and indicate the geometry of each isomer.
(a) $CH_3CH=CH_2$ (b) $(CH_3)_2C=CHCH_3$
(c) $CH_3CH_2CH=CHCH_3$ (d) $(CH_3)_2C=C(CH_3)CH_2CH_3$
(e) $ClCH=CHCl$ (f) $BrCH=CHCl$

Problem 6.8   Cyclodecene can exist in both cis and trans forms, but cyclohexene cannot. Explain. (Making molecular models will be helpful.)

· ·

6.6 Sequence Rules: The *E,Z* Designation

Which of the following compounds has cis geometry, and which has trans geometry?

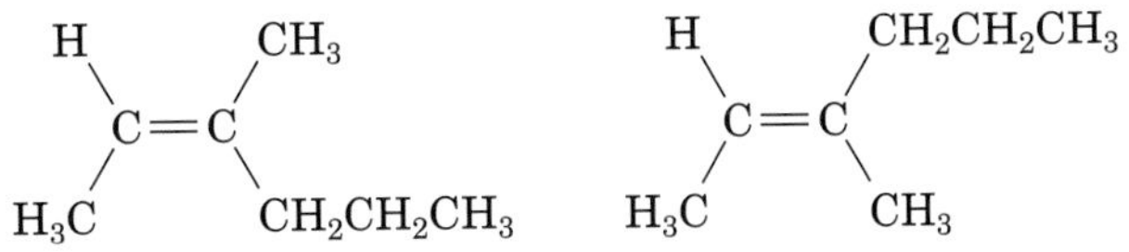

Cis or trans?   Cis or trans?

The question can't be answered because the prefixes *cis* and *trans* describe only the geometry of *disubstituted* double bonds. The cis–trans naming system fails with trisubstituted double bonds like those shown above and with tetrasubstituted double bonds. (*Trisubstituted* means three substituents other than hydrogen are attached to the double-bond carbons; *tetrasubstituted* means four substituents other than hydrogen.)

A more general method for describing double-bond geometry is provided by the **E,Z system** of nomenclature, which uses a series of **sequence rules** to assign priorities to the substituent groups on the double-bond carbons. Considering each carbon atom of the double bond separately, the sequence rules are used to decide which of the groups attached to each carbon is higher in priority. If the higher-priority groups on each carbon are on the same side of the double bond, the alkene is designated Z (for the German *zusammen,* "together"). If the higher-priority groups are on opposite sides, the alkene is designated E (for the German *entgegen,* "opposite"). A simple way to remember which is which is to think with an accent: In the Z isomer, the groups are on "ze zame zide." The assignments are shown in Figure 6.5.

FIGURE 6.5 ▼

The *E,Z* system of nomenclature for substituted alkenes. The higher-priority groups on each carbon are on the same side of the double bond in the *Z* isomer, but are on opposite sides in the *E* isomer.

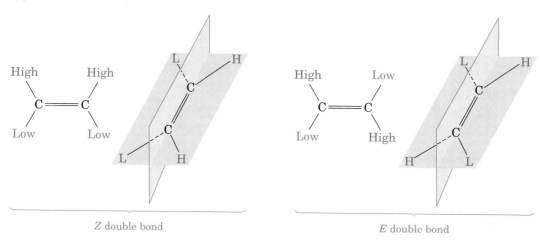

Called the Cahn–Ingold–Prelog rules after the chemists who proposed them, the sequence rules are as follows:

RULE 1

Considering each of the double-bond carbons separately, identify the two atoms directly attached and rank them according to atomic number. An atom with higher atomic number receives higher priority than an atom with lower atomic number. Thus, the atoms commonly found attached to a double bond are assigned the following order:

$$\overset{35}{Br} > \overset{17}{Cl} > \overset{8}{O} > \overset{7}{N} > \overset{6}{C} > \overset{1}{H}$$

For example:

Sir Christopher Kelk Ingold

Sir Christopher Kelk Ingold (1893–1970) was born in Ilford, England, and received his D.Sc. at the University of London. He spent most of his career at University College, London (1930–1961), where he published over 400 scientific papers. Along with Linus Pauling, he was instrumental in developing the theory of resonance.

Low priority H | Cl High priority

C≡C

High priority CH₃ | CH₃ Low priority

(a) (*E*)-2-Chloro-2-butene

Low priority H | CH₃ Low priority

C≡C

High priority CH₃ | Cl High priority

(b) (*Z*)-2-Chloro-2-butene

Because chlorine has a higher atomic number than carbon, a –Cl substituent receives higher priority than a –CH₃ group. Methyl receives higher priority than hydrogen, however, and isomer (a) is assigned *E* geometry because its high-priority groups are on opposite sides of the double bond. Isomer (b) has *Z* geometry because its high-priority groups are on "ze zame zide" of the double bond.

RULE 2 **If a decision can't be reached by ranking the first atoms in the substituent, look at the second, third, or fourth atoms away from the double-bond carbons until the first difference is found.** A –CH₂CH₃ substituent and a –CH₃ substituent are equivalent by rule 1 because both have carbon as the first atom. By rule 2, however, ethyl receives higher priority than methyl because ethyl has a carbon as its highest *second* atom, while methyl has only hydrogen as its second atoms. Look at the following examples to see how the rule works:

Vladimir Prelog

Vladimir Prelog (1906–1998) was born in Sarajevo, Bosnia, where, as a young boy, he heard the shots that killed Archduke Ferdinand and ignited World War I. After receiving a Dr.Ing. degree in 1929 at the Institute of Technology in Prague, Czechoslovakia, he became professor of chemistry at the Swiss Federal Institute of Technology (ETH) in Zürich (1941–1976). He received the 1975 Nobel Prize in chemistry for his lifetime achievements on the stereochemistry of antibiotics, alkaloids, and enzymes.

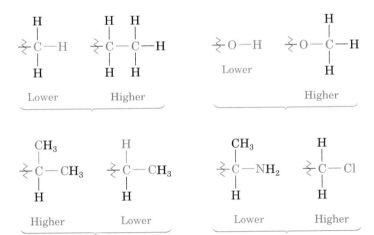

RULE 3 **Multiple-bonded atoms are equivalent to the same number of single-bonded atoms.** For example, an aldehyde substituent (–CH=O), which has a carbon atom *doubly* bonded to *one* oxygen, is equivalent to a substituent having a carbon atom *singly* bonded to *two* oxygen atoms:

H \ C=O is equivalent to H \ C—O / O \ C

This carbon is bonded to H, O, O

This oxygen is bonded to C, C

This carbon is bonded to H, O, O

This oxygen is bonded to C, C

As further examples, the following pairs are equivalent:

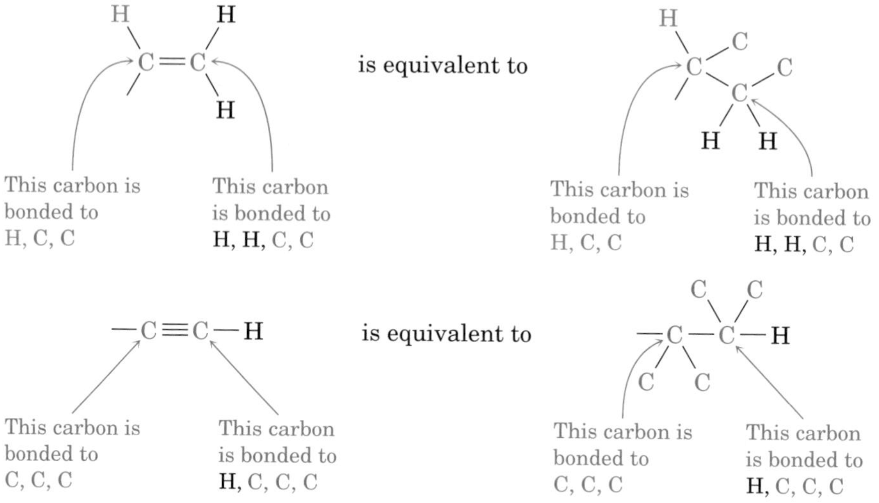

$$H_2C{=}CH_2$$

This carbon is bonded to H, C, C

This carbon is bonded to H, H, C, C

is equivalent to

This carbon is bonded to H, C, C

This carbon is bonded to H, H, C, C

$$-C{\equiv}C-H$$

This carbon is bonded to C, C, C

This carbon is bonded to H, C, C, C

is equivalent to

This carbon is bonded to C, C, C

This carbon is bonded to H, C, C, C

Taking all the sequence rules into account, we can assign the configurations shown in the following examples. Work through each one to convince yourself that the assignments are correct.

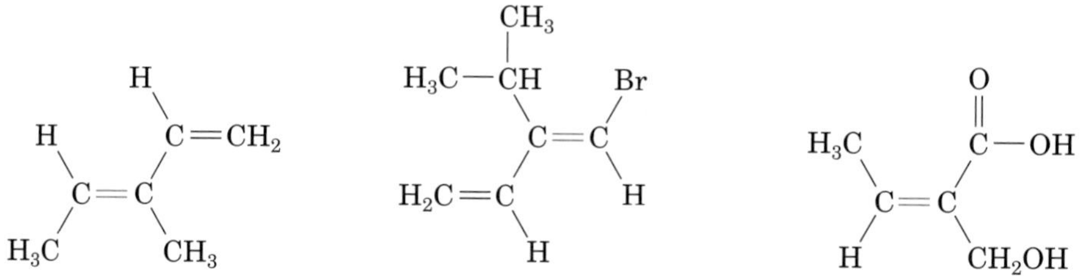

(E)-3-Methyl-1,3-pentadiene

(E)-1-Bromo-2-isopropyl-1,3-butadiene

(Z)-2-Hydroxymethyl-2-butenoic acid

• •

Practice Problem 6.1   Assign *E* or *Z* configuration to the double bond in the following compound:

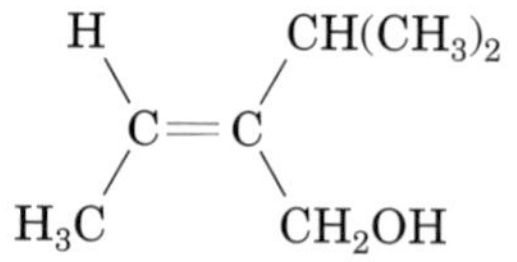

$$H_3C, H \quad C{=}C \quad CH(CH_3)_2, CH_2OH$$

Strategy   Look at the two groups connected to each double-bond carbon, determine the priorities of the groups using the Cahn–Ingold–Prelog rules, and assign configuration.

Solution   The left-hand carbon has –H and –CH₃ substituents, of which –CH₃ receives higher priority by sequence rule 1. The right-hand carbon has –CH(CH₃)₂ and –CH₂OH substituents, which are equivalent by rule 1. By rule 2, however, –CH₂OH receives higher priority than –CH(CH₃)₂. The substituent –CH₂OH has an *oxygen* as its highest second atom, but –CH(CH₃)₂ has a *carbon* as its highest second atom. The two high-priority groups are on the same side of the double bond, so we assign *Z* configuration.

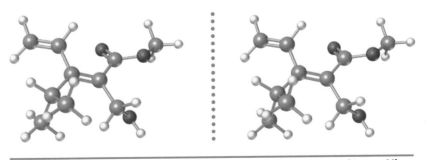

Z configuration

Problem 6.9 Which member in each of the following sets has higher priority?
(a) –H or –Br   (b) –Cl or –Br   (c) –CH₃ or –CH₂CH₃
(d) –NH₂ or –OH   (e) –CH₂OH or –CH₃   (f) –CH₂OH or –CH=O

Problem 6.10 Rank the following sets of substituents in order of Cahn–Ingold–Prelog priorities:
(a) –CH₃, –OH, –H, –Cl
(b) –CH₃, –CH₂CH₃, –CH=CH₂, –CH₂OH
(c) –COOH, –CH₂OH, –C≡N, –CH₂NH₂
(d) –CH₂CH₃, –C≡CH, –C≡N, –CH₂OCH₃

Problem 6.11 Assign E or Z configuration to the following alkenes:

(a)

$$H_3C \quad CH_2OH$$
$$C=C$$
$$CH_3CH_2 \quad Cl$$

(b)

$$Cl \quad CH_2CH_3$$
$$C=C$$
$$CH_3O \quad CH_2CH_2CH_3$$

(c)

$$CH_3 \qquad COOH$$
$$C=C$$
$$CH_2OH$$

(d)

$$H \quad CN$$
$$C=C$$
$$H_3C \quad CH_2NH_2$$

Problem 6.12 Assign stereochemistry (*E* or *Z*) to the following alkene, and convert the drawing into a skeletal structure (red = O):

Stereo View

6.7 Alkene Stability

Although the cis–trans interconversion of alkene isomers does not occur spontaneously, it can be made to happen by treating the alkene with a strong acid catalyst. If we interconvert *cis*-2-butene with *trans*-2-butene and allow them to reach equilibrium, we find that they aren't of equal stability. The trans isomer is more favored than the cis isomer by a ratio of 76 to 24:

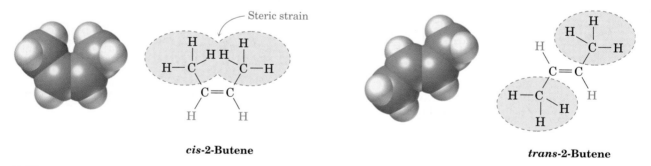

Trans (76%)   **Cis (24%)**

Using the relationship between equilibrium constant and free energy shown previously in Figure 4.19, we can calculate that *cis*-2-butene is less stable than *trans*-2-butene by 2.8 kJ/mol (0.66 kcal/mol) at room temperature.

Cis alkenes are less stable than their trans isomers because of steric (spatial) strain between the two bulky substituents on the same side of the double bond. As shown in Figure 6.6, this is the same kind of steric interference that we saw previously in the axial conformation of methylcyclohexane (Section 4.12).

FIGURE 6.6 ▼

cis-2-Butene is less stable than its trans isomer because of the steric strain that occurs when the two methyl groups try to occupy the same space.

cis-2-Butene

trans-2-Butene

Although it's sometimes possible to find relative stabilities of alkene isomers by establishing a cis–trans equilibrium through treatment with strong acid, there are easier ways to gain the same information. One way is simply to measure the heats of combustion for the two isomers, as we did in determining cycloalkane strain energies (Section 4.5). *cis*-2-Butene is found to be more strained than *trans*-2-butene by 3.3 kJ/mol, a value that is in good agreement with the 2.8 kJ/mol difference found by establishing the cis–trans equilibrium.

H₃C   CH₃
  \   /
   C ═══ C
  /   \
 H   H

cis-2-Butene
$\Delta H°_{combustion} = -2685.5$ kJ/mol

H   CH₃
  \   /
   C ═══ C
  /   \
 H₃C   H

trans-2-Butene
$\Delta H°_{combustion} = -2682.2$ kJ/mol

Another, more general, way to determine the relative stabilities of alkenes is to take advantage of the fact that alkenes undergo a *hydrogenation* reaction on treatment with H₂ gas in the presence of a catalyst such as palladium or platinum:

H   CH₃   H₃C   CH₃
  \   /   \   /
   C ═══ C $\xrightarrow[Pd]{H_2}$ CH₃CH₂CH₂CH₃ $\xleftarrow[Pd]{H_2}$ C ═══ C
  /   \   /   \
 H₃C   H   H   H

trans-2-Butene   **Butane**   **cis-2-Butene**

Energy profiles for the hydrogenation reactions of *cis*- and *trans*-2-butene are shown in Figure 6.7. Since *cis*-2-butene is less stable than *trans*-2-butene by 2.8 kJ/mol, the energy diagram shows the cis alkene at a higher energy level. After reaction, however, both curves are at the same energy level (butane). It therefore follows that $\Delta G°$ for reaction of the cis isomer must be larger than $\Delta G°$ for reaction of the trans isomer by 2.8 kJ/mol. In other words, more energy is released in the hydrogenation of the cis isomer than the trans isomer because the cis isomer has more energy to begin with.

FIGURE 6.7 ▼

Reaction energy diagrams for hydrogenation of *cis*- and *trans*-2-butene. The cis isomer is higher in energy than the trans isomer by about 2.8 kJ/mol and therefore releases more energy in the reaction.

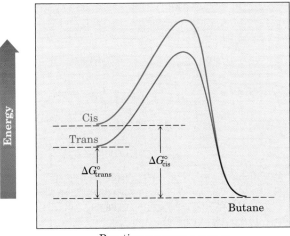

If we were to measure the heats of reaction for the two hydrogenations and find their difference, we could determine the relative stabilities of cis and trans isomers without having to measure an equilibrium position. A large number of such **heats of hydrogenation** ($\Delta H^{\circ}_{hydrog}$) have been measured, and the results bear out our expectation. For *cis*-2-butene, $\Delta H^{\circ}_{hydrog} = -120$ kJ/mol (-28.6 kcal/mol); for the trans isomer, $\Delta H^{\circ}_{hydrog} = -115$ kJ/mol (-27.6 kcal/mol).

Cis isomer
$\Delta H^{\circ}_{hydrog} = -120$ **kJ/mol**

Trans isomer
$\Delta H^{\circ}_{hydrog} = -115$ **kJ/mol**

The energy difference between the 2-butene isomers as calculated from heats of hydrogenation (4 kJ/mol) agrees reasonably well with the energy difference calculated from equilibrium data (2.8 kJ/mol) and from heats of combustion (3.3 kJ/mol), but the numbers aren't exactly the same for two reasons. First, there is probably some experimental error, since heats of hydrogenation require skill and specialized equipment to measure accurately. Second, heats of reaction and equilibrium constants don't measure exactly the same thing. Heats of reaction measure enthalpy changes, ΔH°, whereas equilibrium constants measure free-energy changes, ΔG°. We therefore expect a slight difference between the two.

Table 6.2 lists some representative data for the hydrogenation of different alkenes, and Figure 6.8 plots the results graphically. The data show that alkenes become more stable with increasing substitution. For example, ethylene has $\Delta H^{\circ}_{hydrog} = -137$ kJ/mol (-32.8 kcal/mol), but when one

TABLE 6.2 Heats of Hydrogenation of Some Alkenes

Substitution	Alkene	$\Delta H^{\circ}_{hydrog}$ (kJ/mol)	(kcal/mol)
	$H_2C=CH_2$	-137	-32.8
Monosubstituted	$CH_3CH=CH_2$	-126	-30.1
	$CH_3CH_2CH=CH_2$	-126	-30.1
	$(CH_3)_2CHCH=CH_2$	-127	-30.3
Disubstituted	$CH_3CH=CHCH_3$ (cis)	-120	-28.6
	$CH_3CH=CHCH_3$ (trans)	-115	-27.6
	$(CH_3)_2C=CH_2$	-119	-28.4
Trisubstituted	$(CH_3)_2C=CHCH_3$	-113	-26.9
Tetrasubstituted	$(CH_3)_2C=C(CH_3)_2$	-111	-26.6

alkyl substituent is attached to the double bond, as in 1-butene, the alkene becomes approximately 10 kJ/mol more stable ($\Delta H^\circ_{\text{hydrog}} = -126$ kJ/mol). Further increasing the degree of substitution leads to still further stability. As a general rule, alkenes follow the stability order:

Tetrasubstituted > Trisubstituted > Disubstituted > Monosubstituted

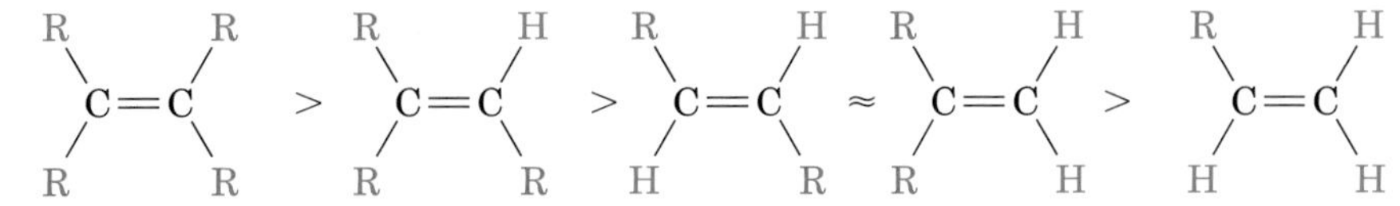

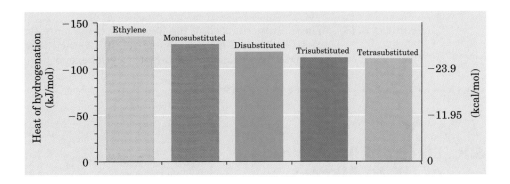

FIGURE 6.8 ▼

A plot of $\Delta H^\circ_{\text{hydrog}}$ versus substitution pattern for alkenes. Alkene stability increases with increasing substitution.

The observed stability order of alkenes is due to a combination of two factors. The first is **hyperconjugation**—a stabilizing interaction between the unfilled antibonding C=C π bond orbital and a filled C–H σ bond orbital on a neighboring substituent (Figure 6.9). The more substituents that are present, the more opportunities exist for hyperconjugation, and the more stable the alkene.

FIGURE 6.9 ▼

Hyperconjugation is a stabilizing interaction between an unfilled π orbital and a neighboring filled C–H σ bond orbital.

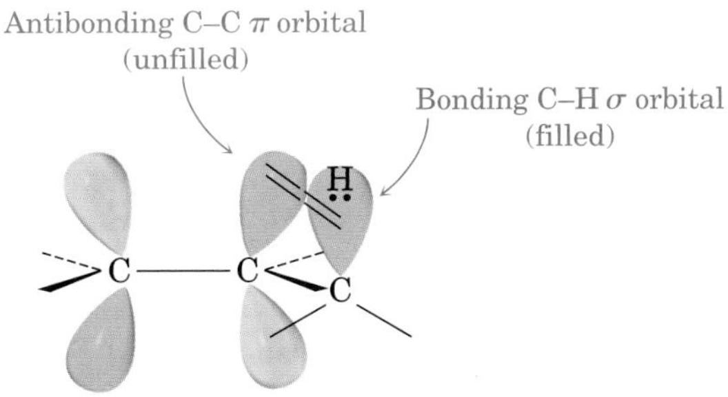

Antibonding C–C π orbital (unfilled)

Bonding C–H σ orbital (filled)

In addition to the effect of hyperconjugation, bond strengths are also important in determining alkene stability. A bond between an sp^2 carbon and an sp^3 carbon is somewhat stronger than a bond between two sp^3 carbons. Thus, in comparing 1-butene and 2-butene, we find that the monosubstituted isomer has one sp^3–sp^3 bond and one sp^3–sp^2 bond, while the disubstituted isomer has two sp^3–sp^2 bonds. More highly substituted alkenes always have a higher ratio of sp^3–sp^2 bonds to sp^3–sp^3 bonds than less highly substituted alkenes and are therefore more stable.

$$sp^3\text{–}sp^2 \qquad sp^3\text{–}sp^2 \qquad\qquad sp^3\text{–}sp^3 \ \ sp^3\text{–}sp^2$$

$$\downarrow \qquad\qquad \downarrow \qquad\qquad\qquad \downarrow \quad\quad \downarrow$$

$$\mathrm{CH_3\!-\!CH\!=\!CH\!-\!CH_3} \qquad\qquad \mathrm{CH_3\!-\!CH_2\!-\!CH\!=\!CH_2}$$

2-Butene   **1-Butene**
(more stable)   **(less stable)**

· ·

Problem 6.13   Which alkene in each of the following sets is more stable?
(a) 1-Butene or 2-methylpropene   (b) (*Z*)-2-Hexene or (*E*)-2-hexene
(c) 1-Methylcyclohexene or 3-methylcyclohexene

· ·

6.8   Electrophilic Addition of HX to Alkenes

Before beginning a detailed discussion of alkene reactions, let's review briefly some conclusions from the previous chapter. We said in Section 5.5 that alkenes behave as nucleophiles (Lewis bases) in polar reactions. The carbon–carbon double bond is electron-rich and can donate a pair of electrons to an electrophile (Lewis acid). For example, reaction of 2-methylpropene with HBr yields 2-bromo-2-methylpropane. A careful study of this and similar reactions by Christopher Ingold and others in the 1930s led to the generally accepted mechanism shown in Figure 6.10 for **electrophilic addition reactions**.

The reaction begins with an attack on the electrophile, HBr, by the electrons of the nucleophilic π bond. Two electrons from the π bond form a new σ bond between the entering hydrogen and an alkene carbon, as shown by the curved arrow at the top of Figure 6.10. The carbocation intermediate that results is itself an electrophile, which can accept an electron pair from nucleophilic Br⁻ ion to form a C–Br bond and yield a neutral addition product.

The energy diagram for the overall electrophilic addition reaction (Figure 6.11) has two peaks (transition states) separated by a valley (carbocation intermediate). The energy level of the intermediate is higher than that of the starting alkene, but the reaction as a whole is exergonic (negative $\Delta G°$). The first step, protonation of the alkene to yield the intermediate cation, is relatively slow, but once formed, the cation intermediate rapidly reacts further to yield the final bromoalkane product. The relative rates of the two steps are indicated in Figure 6.11 by the fact that $\Delta G_1^{\ddagger}$ is larger than $\Delta G_2^{\ddagger}$.

FIGURE 6.10 ▼

Mechanism of the electrophilic addition of HBr to 2-methylpropene. The reaction occurs in two steps and involves a carbocation intermediate.

refer to
**Mechanisms
& Movies**

The electrophile HBr is attacked by the π electrons of the double bond, and a new C–H σ bond is formed. This leaves the other carbon atom with a + charge and a vacant p orbital.

Br⁻ donates an electron pair to the positively charged carbon atom, forming a C–Br σ bond and yielding the neutral addition product.

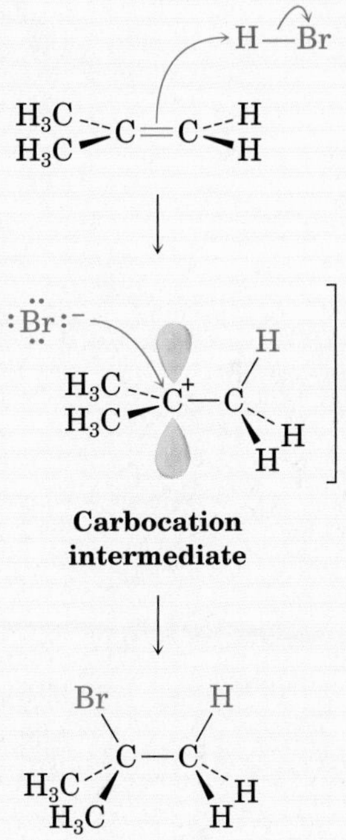

© 1984 JOHN MCMURRY

FIGURE 6.11 ▼

Reaction energy diagram for the two-step electrophilic addition of HBr to 2-methylpropene. The first step is slower than the second step.

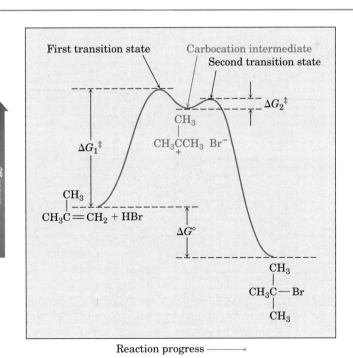

Electrophilic addition of HX to alkenes is successful not only with HBr but with HCl and HI as well. Note that HI is usually generated in the reaction mixture by treating potassium iodide with phosphoric acid.

$$\underset{\textbf{2-Methylpropene}}{\underset{CH_3}{\overset{CH_3}{>}}C=CH_2} + HCl \xrightarrow{\text{Ether}} \underset{\substack{\textbf{2-Chloro-2-methylpropane} \\ \textbf{(94\%)}}}{CH_3-\overset{\overset{\displaystyle Cl}{|}}{\underset{\underset{\displaystyle CH_3}{|}}{C}}-CH_3}$$

$$\underset{\textbf{1-Pentene}}{CH_3CH_2CH_2CH=CH_2} \xrightarrow[\substack{\text{H}_3\text{PO}_4 \\ \textbf{(HI)}}]{\text{KI}} \underset{\textbf{2-Iodopentane}}{CH_3CH_2CH_2\overset{\overset{\displaystyle I}{|}}{C}HCH_3}$$

Writing Organic Reactions

This is a good time to mention that organic reaction equations are sometimes written in different ways to emphasize different points. For example, the reaction of ethylene with HBr might be written in the format A + B ⟶ C to emphasize that both reactants are equally important for the purposes of the discussion. The solvent and notes about other reaction conditions such as temperature are written either above or below the reaction arrow.

$$H_2C=CH_2 + HBr \xrightarrow[25°C]{\text{Ether}} CH_3CH_2Br$$

Solvent

Alternatively, we might write the same reaction in the format

$$A \xrightarrow{B} C$$

to emphasize that A is the reactant whose chemistry is of greater interest. Reactant B is placed above the reaction arrow together with notes about solvent and reaction conditions.

$$H_2C=CH_2 \xrightarrow[\text{Ether, 25°C}]{\text{HBr}} CH_3CH_2Br$$

Reactant

Solvent

Both reaction formats are frequently used in chemistry, and you sometimes have to look carefully at the overall transformation to see the roles of the substances above and below the reaction arrow.

6.9 Orientation of Electrophilic Addition: Markovnikov's Rule

Look carefully at the reactions shown in the previous section. In each case, an unsymmetrically substituted alkene has given a single addition product, rather than the mixture that might have been expected. For example, 2-methylpropene *might* have reacted with HCl to give 1-chloro-2-methylpropane (isobutyl chloride) in addition to 2-chloro-2-methylpropane, but it didn't. We say that such reactions are **regiospecific** (**ree**-jee-oh-specific) when only one of two possible orientations of addition occurs.

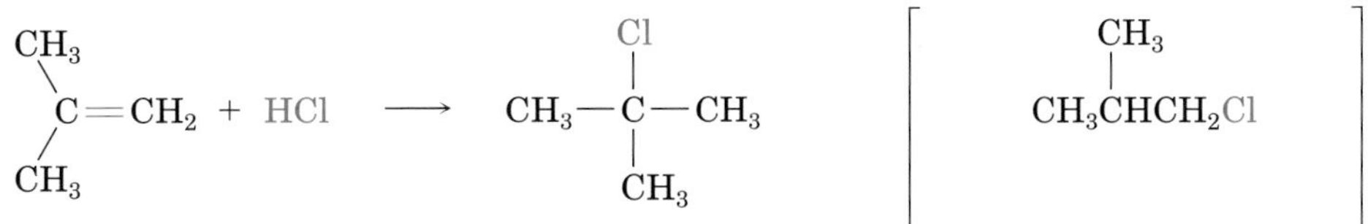

2-Methylpropene   **2-Chloro-2-methylpropane**   **1-Chloro-2-methylpropane**
                                                     (sole product)   *(NOT formed)*

After looking at the results of many such reactions, the Russian chemist Vladimir Markovnikov proposed in 1869 what has become known as **Markovnikov's rule:**

Markovnikov's rule   In the addition of HX to an alkene, the H attaches to the carbon with fewer alkyl substituents and the X attaches to the carbon with more alkyl substituents.

Vladimir Vassilyevich Markovnikov

Vladimir Vassilyevich Markovnikov (1838–1904) was born in Nijni-Novgorod, Russia, and received his Ph.D. working with A. M. Butlerov at the university in Kazan. He was a professor in Kazan (1870), Odessa (1871), and Moscow (1873–1898). In addition to his work on the orientation of addition reactions, he was the first to synthesize a four-membered ring.

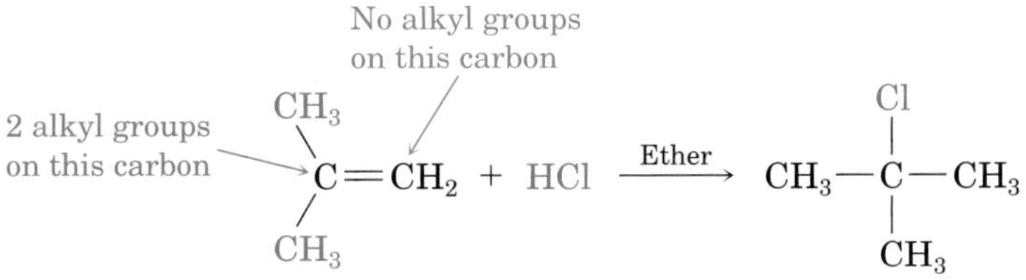

2-Methylpropene   **2-Chloro-2-methylpropane**

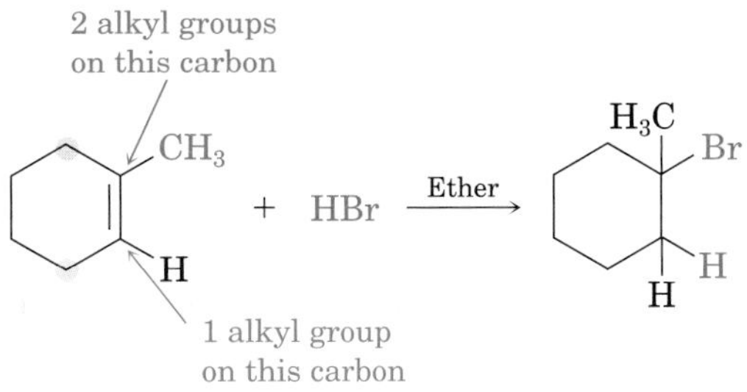

1-Methylcyclohexene   **1-Bromo-1-methylcyclohexane**

When both ends of the double bond have the same degree of substitution, a mixture of products results:

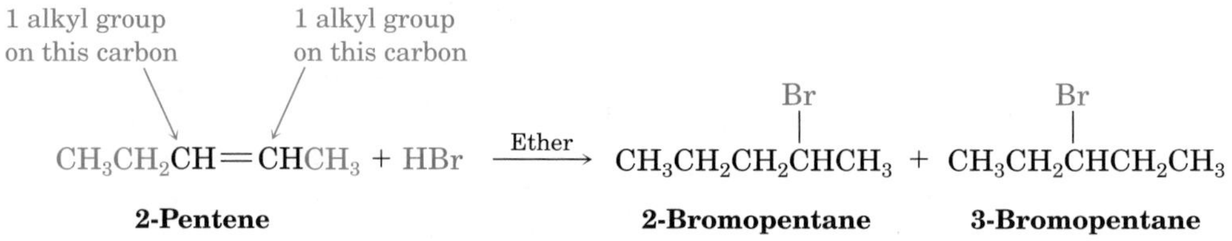

1 alkyl group on this carbon   1 alkyl group on this carbon

$$CH_3CH_2CH{=}CHCH_3 + HBr \xrightarrow{\text{Ether}} CH_3CH_2CH_2\overset{\overset{\displaystyle Br}{|}}{C}HCH_3 + CH_3CH_2\overset{\overset{\displaystyle Br}{|}}{C}HCH_2CH_3$$

2-Pentene   **2-Bromopentane**   **3-Bromopentane**

Since carbocations are involved as intermediates in these reactions, Markovnikov's rule can be restated:

Markovnikov's rule (restated) In the addition of HX to an alkene, the more highly substituted carbocation is formed as the intermediate rather than the less highly substituted one.

For example, addition of H⁺ to 2-methylpropene yields the intermediate *tertiary* carbocation rather than the primary carbocation, and addition to 1-methylcyclohexene yields a tertiary cation rather than a secondary one. Why should this be?

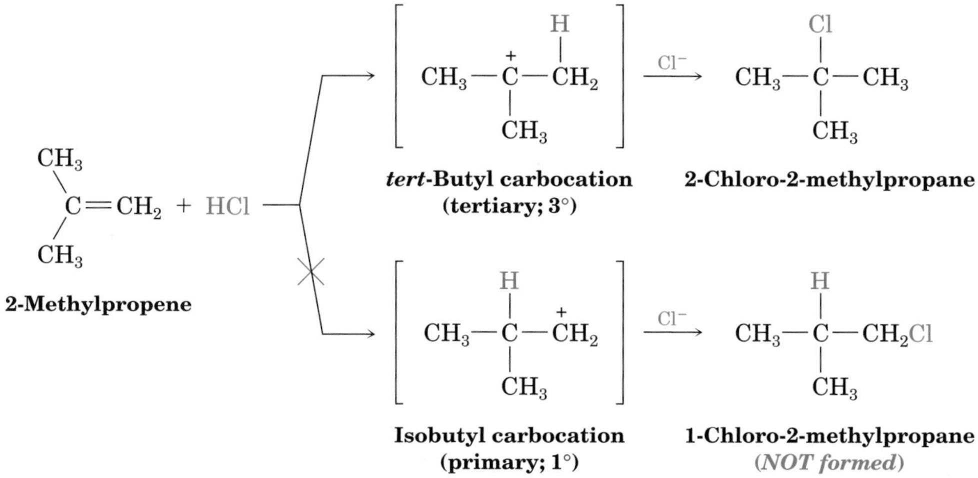

tert-Butyl carbocation (tertiary; 3°)   2-Chloro-2-methylpropane

Isobutyl carbocation (primary; 1°)   1-Chloro-2-methylpropane (*NOT formed*)

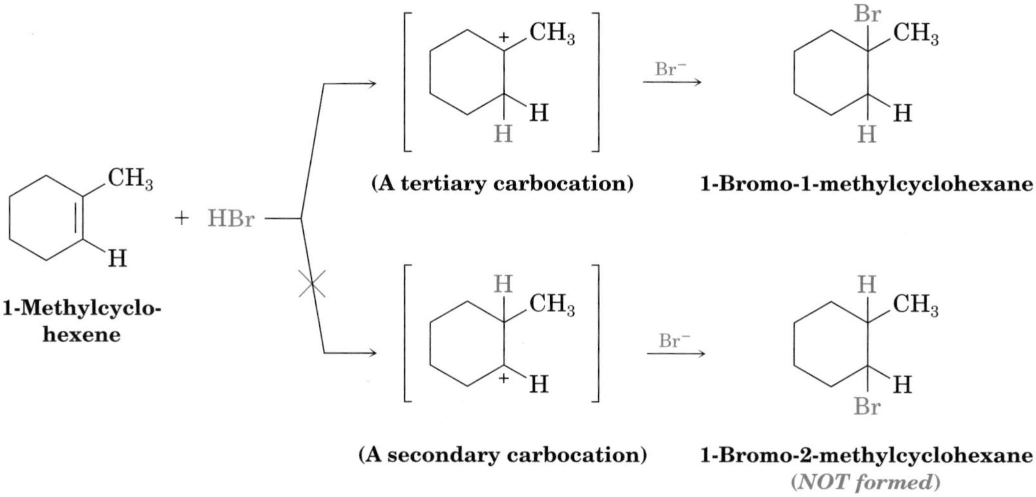

2-Methylpropene

(A tertiary carbocation)   1-Bromo-1-methylcyclohexane

1-Methylcyclohexene

(A secondary carbocation)   1-Bromo-2-methylcyclohexane (*NOT formed*)

● ●

Practice Problem 6.2   What product would you expect from reaction of HCl with 1-ethylcyclopentene?

Strategy   When solving a problem that asks you to predict a reaction product, begin by looking at the functional group(s) in the reactants and deciding what kind of reaction is likely to occur. In the present instance, the reactant is an alkene that will probably undergo an electrophilic addition reaction with HCl. Next, recall what you know about electrophilic addition reactions, and use your knowledge to predict the product. You know that electrophilic addition reactions follow Markovnikov's rule, so the –Cl will add to the more highly substituted carbon.

Solution   Markovnikov's rule predicts that H will add to the double-bond carbon that has one alkyl group (C2 on the ring) and Cl will add to the double-bond carbon that has two alkyl groups (C1 on the ring). The expected product is 1-chloro-1-ethylcyclopentane.

● ●

Practice Problem 6.3   What alkene would you start with to prepare the following alkyl halide? There may be more than one possibility.

$$? \longrightarrow CH_3CH_2\overset{\overset{\displaystyle Cl}{|}}{\underset{\underset{\displaystyle CH_3}{|}}{C}}CH_2CH_2CH_3$$

Strategy   When solving a problem that asks how to prepare a given product, *always work backward.* Look at the product, identify the functional group(s) it contains, and ask yourself, "How can I prepare that functional group?" In the present instance, the product is a tertiary alkyl chloride, which can be prepared by reaction of an alkene with HCl. The carbon atom bearing the –Cl atom in the product must be one of the double-bond carbons in the reactant. Draw and evaluate all possibilities.

Solution   There are three possibilities, any one of which could give the desired product.

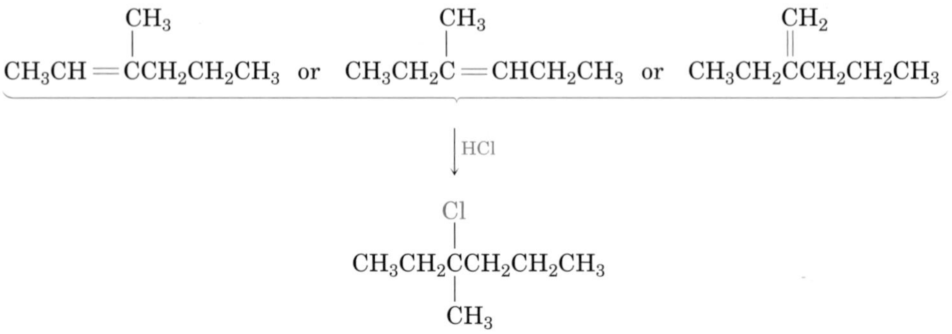

$$
\begin{array}{ccc}
\overset{\displaystyle CH_3}{\underset{\displaystyle |}{}} & \overset{\displaystyle CH_3}{\underset{\displaystyle |}{}} & \overset{\displaystyle CH_2}{\underset{\displaystyle ||}{}} \\
CH_3CH\!=\!CCH_2CH_2CH_3 \ \text{ or } & CH_3CH_2C\!=\!CHCH_2CH_3 \ \text{ or } & CH_3CH_2CCH_2CH_2CH_3
\end{array}
$$

$$\Big\downarrow HCl$$

$$
\underset{\displaystyle \underset{|}{CH_3}}{\overset{\displaystyle \overset{Cl}{|}}{CH_3CH_2CCH_2CH_2CH_3}}
$$

• •

Problem 6.14   Predict the products of the following reactions:

(a) 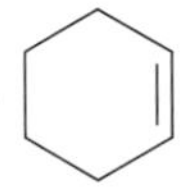 ⬡‖ + HCl ⟶ ?   (b) $(CH_3)_2C\!=\!CHCH_2CH_3 \xrightarrow{\text{HBr}}$?

(c) $CH_3CH_2CH_2CH\!=\!CH_2 \xrightarrow[\text{KI}]{\text{H}_3\text{PO}_4}$?   (d) 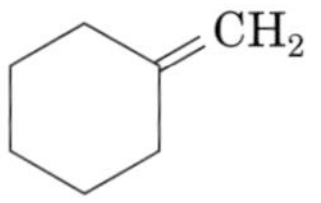 ⬡=CH$_2$ + HBr ⟶ ?

Problem 6.15   What alkenes would you start with to prepare the following alkyl halides?

     (a) Bromocyclopentane   (b) 1-Ethyl-1-iodocyclohexane

(c) 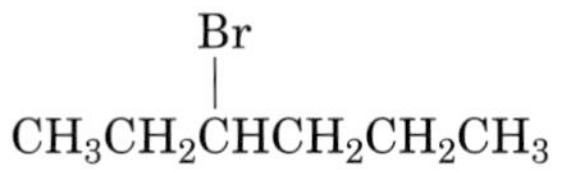 $\underset{\displaystyle \overset{|}{CH_3CH_2CHCH_2CH_2CH_3}}{\overset{\displaystyle \overset{Br}{|}}{}}$ (d) 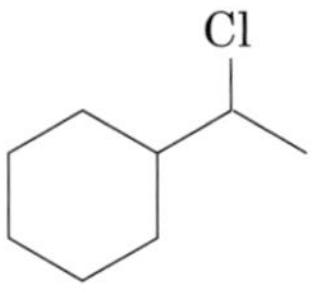 ⬡–CHCl–CH₃

• •

6.10 Carbocation Structure and Stability

To understand the reasons for the Markovnikov orientation of electrophilic addition reactions, we need to learn more about the structure and stability of carbocations and about the general nature of reactions and transition states. The first point to explore involves structure.

A great deal of evidence has shown that carbocations are *planar*. The trivalent carbon is sp^2-hybridized, and the three substituents are oriented to the corners of an equilateral triangle, as indicated in Figure 6.12. Since there are only six valence electrons on carbon, and since all six are used in the three σ bonds, the p orbital extending above and below the plane is unoccupied.

The second point to explore involves carbocation stability. 2-Methylpropene might react with H^+ to form a carbocation having three alkyl substituents (a tertiary ion, 3°), or it might react to form a carbocation having one alkyl substituent (a primary ion, 1°). Since the tertiary chloride,

FIGURE 6.12 ▼

The electronic structure of a carbocation. The trivalent carbon is *sp²*-hybridized and has a vacant *p* orbital extending perpendicular to the plane of the carbon and three attached groups.

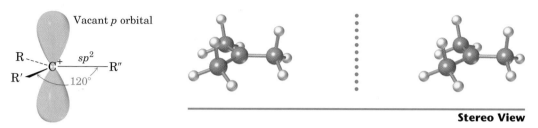

Stereo View

2-chloro-2-methylpropane, is the only product observed, formation of the tertiary cation is evidently favored over formation of the primary cation. Thermodynamic measurements show that, indeed, the stability of carbocations increases with increasing substitution: More highly substituted carbocations are more stable than less highly substituted ones.

One way of determining carbocation stabilities is to measure the amount of energy required to form the carbocation from its corresponding alkyl halide, R–X $\longrightarrow$ R⁺ + :X⁻. As shown in Figure 6.13 (p. 214), tertiary halides dissociate to give carbocations much more readily than secondary or primary halides. As a result, trisubstituted carbocations are more stable than disubstituted ones, which are more stable than monosubstituted ones.

The data in Figure 6.13 are taken from measurements made in the gas phase, but a similar stability order is found for carbocations in solution. The dissociation enthalpies are much lower in solution because polar solvents can stabilize the ions, but the order of carbocation stability remains the same.

Why are more highly substituted carbocations more stable than less highly substituted ones? There are at least two reasons. Part of the answer has to do with inductive effects, and part has to do with hyperconjugation. Inductive effects, discussed in Section 2.1 in connection with polar covalent bonds, result from the shifting of electrons in a σ bond in response to the electronegativity of a nearby atom. In the present instance, electrons from a relatively large and polarizable alkyl group can shift toward a neighboring positive charge more easily than the electron from a hydrogen. Thus, the more alkyl groups there are attached to the positively charged carbon, the more electron density shifts toward the charge and the more inductive stabilization of the cation occurs.

3°: Three alkyl groups donating electrons

2°: Two alkyl groups donating electrons

1°: One alkyl group donating electrons

Methyl: No alkyl groups donating electrons

FIGURE 6.13 ▼

A plot of dissociation enthalpy versus substitution pattern for the gas-phase dissociation of alkyl chlorides to yield carbocations. More highly substituted alkyl halides dissociate more readily than less highly substituted ones. These enthalpies are calculated in the following way:

$$CH_3Cl \longrightarrow CH_3\cdot + Cl\cdot \qquad D = \quad 351 \text{ kJ/mol} \quad \text{(Bond dissociation energy)}$$

$$CH_3\cdot \longrightarrow CH_3^+ + e^- \qquad E_i = \quad 948 \text{ kJ/mol} \quad \text{(Ionization energy)}$$

$$Cl\cdot + e^- \longrightarrow Cl^- \qquad E_{ea} = -348 \text{ kJ/mol} \quad \text{(Electron affinity)}$$

Net: $\quad CH_3{-}Cl \longrightarrow CH_3^+ + Cl^- \qquad\qquad 951 \text{ kJ/mol} \quad \text{(Dissociation enthalpy)}$

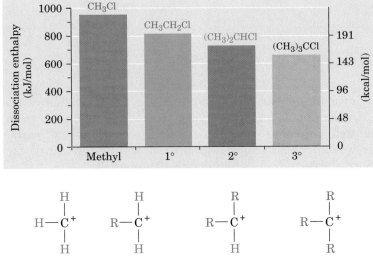

Hyperconjugation, discussed in Section 6.7 in connection with the stabilities of substituted alkenes, is the stabilizing interaction of a vacant p orbital and a properly oriented C–H σ orbital nearby (Figure 6.14). The more alkyl groups there are on the carbocation, the more possibilities there are for hyperconjugation, and the more stable the carbocation. Note in Figure 6.14 that an electrostatic potential map for the *tert*-butyl carbocation, $(CH_3)_3C^+$, shows a difference between the three hydrogens in the plane of the carbons and the six hydrogens above and below the plane. The three in-plane hydrogens have their C–H σ orbital perpendicular to the cation p orbital, while the six out-of-plane hydrogens have their C–H σ orbital more nearly parallel. As a result, only the out-of-plane hydrogens can take part in hyperconjugation, making them more electron-poor (green) than the electron-rich in-plane hydrogens.

FIGURE 6.14 ▼

Stabilization of a carbocation through hyperconjugation. Interaction of a nearby C–H
σ orbital with the vacant carbocation p orbital stabilizes the cation and lowers its
energy. An electrostatic potential map of the *tert*-butyl carbocation, $(CH_3)_3C^+$, shows
that the six hydrogens whose C–H σ orbital is roughly parallel to the cation p orbital
are more electron-poor (green) because of hyperconjugation than the three
hydrogens whose C–H σ orbital is perpendicular to the p orbital.

tert-butyl carbocation

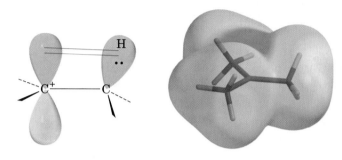

. . . .

Problem 6.16   Show the structures of the carbocation intermediates you would expect in the following reactions:

(a) $CH_3CH_2\overset{\overset{\displaystyle CH_3}{|}}{C}{=}CH\overset{\overset{\displaystyle CH_3}{|}}{C}HCH_3 + HBr \longrightarrow$?

(b)   ⬠=$CHCH_3$ + HI $\longrightarrow$?

Problem 6.17   Draw a skeletal structure of the following carbocation. Identify it as primary, secondary, or tertiary, and identify the hydrogen atoms that are involved in hyperconjugation in the conformation shown.

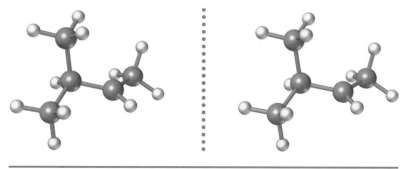

Stereo View

. . . .

6.11 The Hammond Postulate

Key Ideas ▶ To summarize our knowledge of electrophilic addition reactions up to this point, we know that:

- **Electrophilic addition to an unsymmetrically substituted alkene gives the more highly substituted carbocation.** A more highly substituted carbocation forms faster than a less highly substituted one and, once formed, rapidly goes on to give the final product.

- **A more highly substituted carbocation is more stable than a less highly substituted one.** That is, the stability order of carbocations is tertiary > secondary > primary > methyl.

What we have not yet seen is how these two points are related. Why does the *stability* of the carbocation intermediate affect the *rate* at which it's formed and thereby determine the structure of the final product? After all, carbocation stability is determined by $\Delta G°$, but reaction rate is determined by $\Delta G^{\ddagger}$ (activation energy). The two quantities aren't directly related.

Although there is no quantitative relationship between the stability of a high-energy carbocation intermediate and the rate of its formation, there *is* an intuitive relationship. It's generally true when comparing two similar reactions that the more stable intermediate forms faster than the less stable one. The situation is shown graphically in Figure 6.15, where the reaction energy profile in part (a) represents the typical situation. The profile in part (b) is atypical—that is, the curves for two similar reactions don't cross one another.

George Simms Hammond

George Simms Hammond (1921–) was born in Auburn, Maine, the son of a farmer. He received his Ph.D. at Harvard University in 1947 and served as professor of chemistry at Iowa State University, California Institute of Technology (1958–1972), and the University of California at Santa Cruz (1972–1978). He is known for his exploratory work on organic photochemistry—the use of light to bring about organic reactions.

FIGURE 6.15 ▼

Reaction energy diagrams for two similar competing reactions. In (a), the faster reaction yields the more stable intermediate. In (b), the slower reaction yields the more stable intermediate. The curve shown in (a) represents the typical situation.

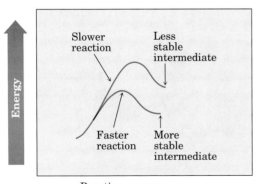

(a)

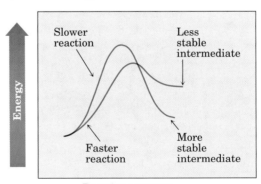

(b)

An explanation of the relationship between reaction rate and intermediate stability was first advanced in 1955. Known as the **Hammond postulate**, this explanation intuitively links reaction rate and intermediate stability by looking at the energy level and structure of the transition state.

Transition states represent energy maxima. They are high-energy activated complexes that occur transiently during the course of a reaction and immediately go on to a more stable species. Although we can't actually *observe* transition states, because they have no finite lifetime, the Hammond postulate says that we can get an *idea* of a particular transition state's structure by looking at the structure of the nearest stable species. Imagine the two cases shown in Figure 6.16, for example. The reaction profile in part (a) shows the energy curve for an endergonic reaction step, and the profile in part (b) shows the curve for an exergonic step.

FIGURE 6.16 ▼

Reaction energy diagrams for endergonic and exergonic steps. (a) In an endergonic step, the energy levels of transition state and *product* are similar. (b) In an exergonic step, the energy levels of transition state and *reactant* are similar.

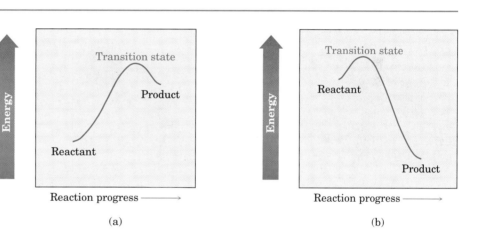

(a)

(b)

In an endergonic reaction (Figure 6.16a), the energy level of the transition state is closer to that of the product than to that of the reactant. Since the transition state is closer *energetically* to the product, we make the natural assumption that it's also closer *structurally*. In other words, *the transition state for an endergonic reaction step structurally resembles the product of that step.* Conversely, the transition state for an exergonic reaction (Figure 6.16b) is closer energetically, and thus structurally, to the reactant than to the product. We therefore say that *the transition state for an exergonic reaction step structurally resembles the reactant for that step.*

Hammond postulate   The structure of a transition state resembles the structure of the nearest stable species. Transition states for endergonic steps structurally resemble products, and transition states for exergonic steps structurally resemble reactants.

How does the Hammond postulate apply to electrophilic addition reactions? We know that the formation of a carbocation by protonation of an alkene is an endergonic step. Therefore, the transition state for alkene protonation should structurally resemble the carbocation intermediate, and any factor that makes the carbocation product more stable should also make the nearby transition state more stable. Since increasing alkyl substitution stabilizes carbocations, it also stabilizes the transition states leading to those ions, thus resulting in faster reaction. *More stable carbocations form faster because their stability is reflected in the transition state leading to them.* A hypothetical transition state for alkene protonation might be expected to look like that shown in Figure 6.17.

FIGURE 6.17 ▼

The hypothetical structure of a transition state for alkene protonation. The transition state is closer in both energy and structure to the carbocation than to the alkene. Thus, an increase in carbocation stability (lower $\Delta G°$) also causes an increase in transition-state stability (lower $\Delta G^{\ddagger}$).

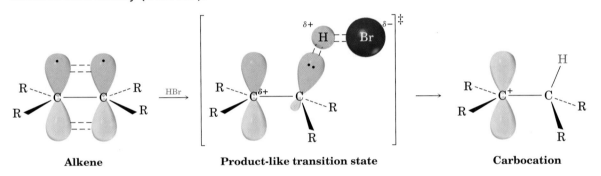

Alkene   Product-like transition state   Carbocation

Because the transition state for alkene protonation resembles the carbocation product, we can imagine it to be a structure in which one of the alkene carbon atoms has almost completely rehybridized from sp^2 to sp^3 and in which the remaining alkene carbon bears much of the positive charge. This transition state is stabilized by hyperconjugation and inductive effects in the same way as the product carbocation. The more alkyl groups that are present, the greater the extent of stabilization in the transition state and the faster the transition state forms. Figure 6.18 summarizes the situation by showing competing reaction energy profiles for the reaction of 2-methylpropene with HCl.

Problem 6.18   What about the second step in the electrophilic addition of HCl to an alkene—the reaction of chloride ion with the carbocation intermediate? Is this step exergonic or endergonic? Does the transition state for this second step resemble the reactant (carbocation) or product (chloroalkane)? Make a rough drawing of what the transition-state structure might look like.

FIGURE 6.18 ▼

A reaction energy diagram for the electrophilic addition of HCl to 2-methylpropene. The tertiary cation intermediate forms faster than the primary cation because it is more stable. The same factors that make the tertiary cation more stable also make the transition state leading to it more stable.

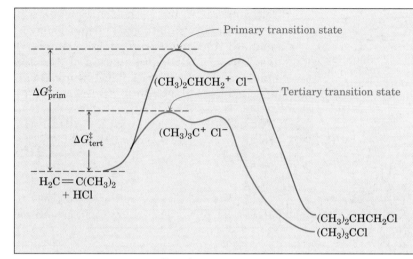

6.12 Evidence for the Mechanism of Electrophilic Addition: Carbocation Rearrangements

Frank C. Whitmore

Frank C. Whitmore (1887–1947) was born in North Attleboro, Massachusetts, and received his Ph.D. at Harvard working with E. L. Jackson. He was professor of chemistry at Minnesota, Northwestern, and The Pennsylvania State University. Nicknamed "Rocky," he wrote an influential advanced textbook in organic chemistry.

How do we know that the carbocation mechanism for addition of HX to alkenes is correct? The answer is that we *don't* know it's correct, or at least we don't know with complete certainty. Although an *incorrect* reaction mechanism can be *disproved* by demonstrating that it doesn't satisfactorily account for observed data, a correct reaction mechanism can never be entirely proven. The best we can do is to show that a proposed mechanism is consistent with all known facts. If enough facts are satisfactorily accounted for, then the mechanism is probably correct.

What evidence is there to support the two-step, carbocation mechanism we've proposed for the reaction of HX with alkenes? How do we know that the two reactants, HX and alkene, don't simply come together in a single step to give the final product without going through a carbocation intermediate? One of the best pieces of evidence for a carbocation mechanism was discovered during the 1930s by F. C. Whitmore, who found that structural *rearrangements* often occur during the reaction of HX with an alkene. For example, reaction of HCl with 3-methyl-1-butene yields a substantial amount of 2-chloro-2-methylbutane in addition to the "expected" product, 2-chloro-3-methylbutane:

3-Methyl-1-butene   **2-Chloro-3-methylbutane**   **2-Chloro-2-methylbutane**
                                (approx. 50%)   (approx. 50%)

How can the formation of 2-chloro-2-methylbutane be explained? If the reaction takes place in a single step, it would be difficult to account for rearrangement, but if the reaction takes place in two steps, rearrangement is more easily explained. Whitmore suggested that it is a carbocation intermediate that undergoes rearrangement. The secondary carbocation intermediate formed by protonation of 3-methyl-1-butene rearranges to a more stable tertiary carbocation by a **hydride shift**—the shift of a hydrogen atom and its electron pair (a hydride ion, :H⁻) between neighboring carbons:

Carbocation rearrangements can also occur by the shift of an *alkyl group* with its electron pair. For example, reaction of 3,3-dimethyl-1-butene with HCl leads to an equal mixture of unrearranged 2-chloro-3,3-dimethylbutane and rearranged 2-chloro-2,3-dimethylbutane. In this instance, a secondary carbocation rearranges to a more stable tertiary carbocation by the shift of a methyl group:

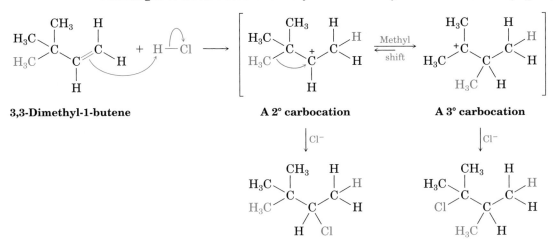

Note the similarities between these two carbocation rearrangements: In both cases, a group (:H⁻ or :CH₃⁻) moves to an adjacent positively charged carbon, *taking its bonding electron pair with it*. Also in both cases, *a less stable carbocation rearranges to a more stable ion.* Rearrangements of this kind are a common feature of carbocation chemistry. We'll see at numerous places in future chapters that their occurrence in a reaction provides strong mechanistic evidence for the presence of carbocation intermediates.

Problem 6.19 On treatment with HBr, vinylcyclohexane undergoes addition and rearrangement to yield 1-bromo-1-ethylcyclohexane. Propose a mechanism to account for this result.

Vinylcyclohexane   **1-Bromo-1-ethylcyclohexane**

CHEMISTRY @ WORK

Carrots, Alkenes, and the Chemistry of Vision

Folk medicine has long held that eating carrots is good for your eyes. Although that's probably not true for healthy adults on a proper diet, there's no question that the chemistry of carrots and the chemistry of vision are related. Carrots are rich in β-carotene, a purple-orange alkene that is an excellent dietary source of vitamin A. β-Carotene is converted to vitamin A by enzymes in the liver, oxidized to an aldehyde called *all-trans-retinal,* and then isomerized by a change in geometry of the C11–C12 double bond to produce 11-*cis*-retinal, the light-sensitive pigment on which the visual systems of all living things are based.

β-Carotene

Vitamin A

11-*cis*-Retinal

(continued) ▶

There are two types of light-sensitive receptor cells in the retina of the human eye, *rod* cells and *cone* cells. The three million or so rod cells are primarily responsible for seeing in dim light, whereas the hundred million cone cells are responsible for seeing in bright light and for the perception of bright colors. In the rod cells of the eye, 11-*cis*-retinal is converted into *rhodopsin,* a light-sensitive substance formed from the protein *opsin* and 11-*cis*-retinal. When light strikes the rod cells, isomerization of the C11–C12 double bond occurs and *trans*-rhodopsin, called *metarhodopsin II,* is produced. This cis–trans isomerization of rhodopsin is accompanied by a change in molecular geometry, which in turn causes a nerve impulse to be sent to the brain where it is perceived as vision. (In the absence of light, the cis–trans isomerization takes approximately 1100 years; in the presence of light, it occurs within 2×10^{-11} seconds!)

Metarhodopsin II is then recycled back into rhodopsin by a multi-step sequence involving cleavage to all-*trans*-retinal and cis–trans isomerization back to 11-*cis*-retinal.

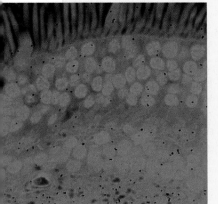

Rod cells in the eye are responsible for seeing in dim light.

Summary and Key Words

Alkenes are hydrocarbons that contain one or more carbon–carbon double bonds. Because they contain fewer hydrogens than alkanes with the same number of carbons, alkenes are often referred to as **unsaturated**.

Rotation around the double bond is restricted, and substituted alkenes can therefore exist as cis–trans stereoisomers. The geometry of a double bond can be specified by application of the Cahn–Ingold–Prelog **sequence**

rules, which assign priorities to double-bond substituents. If the high-priority groups on each carbon are on the same side of the double bond, the geometry is ***Z*** (*zusammen*, "together"); if the high-priority groups on each carbon are on opposite sides of the double bond, the geometry is ***E*** (*entgegen*, "apart"). The stability order of substituted alkenes is

Tetrasubstituted > Trisubstituted > Disubstituted > Monosubstituted

$$R_2C=CR_2 \quad > \quad R_2C=CHR \quad > RCH=CHR \approx R_2C=CH_2 > \quad RCH=CH_2$$

Alkene chemistry is dominated by **electrophilic addition reactions**. When HX reacts with an unsymmetrically substituted alkene, **Markovnikov's rule** predicts that the H will add to the carbon having fewer alkyl substituents and the X group will add to the carbon having more alkyl substituents. Electrophilic additions to alkenes take place through *carbocation* intermediates formed by reaction of the nucleophilic alkene π bond with electrophilic H^+. Carbocation stability follows the order

Tertiary (3°) > Secondary (2°) > Primary (1°) > Methyl

$$R_3C^+ \quad > \quad R_2CH^+ \quad > \quad RCH_2^+ \quad > \quad CH_3^+$$

Markovnikov's rule can be restated by saying that, in the addition of HX to an alkene, the more stable carbocation intermediate is formed. This result is explained by the **Hammond postulate**, which says that the transition state of an exergonic reaction step structurally resembles the reactant, whereas the transition state of an endergonic reaction step structurally resembles the product. Since an alkene protonation step is endergonic, the stability of the more highly substituted carbocation is reflected in the stability of the transition state leading to its formation.

Evidence in support of a carbocation mechanism for electrophilic additions comes from the observation that structural *rearrangements* often take place during reaction. Rearrangements occur by shift of either a hydride ion, $:H^-$ (a **hydride shift**), or an alkyl group anion, $:R^-$, from a carbon atom to the adjacent positively charged carbon. The result is isomerization of a less stable carbocation to a more stable one.

Visualizing Chemistry

● ●

(Problems 6.1–6.19 appear within the chapter.)

6.20 Give IUPAC names for the following alkenes, and convert each drawing into a skeletal structure:

(a)   (b)

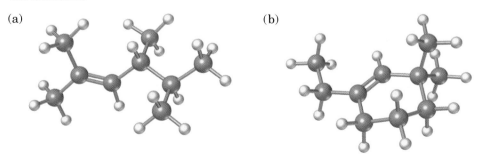

6.21   Assign stereochemistry (*E* or *Z*) to each of the following alkenes, and convert each drawing into a skeletal structure (red = O, yellow-green = Cl).

(a)   (b)

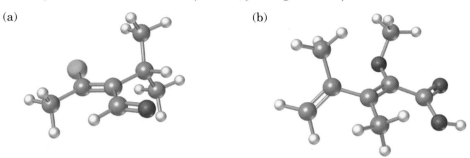

6.22   The following drawing does *not* represent a stable molecule. Why not?

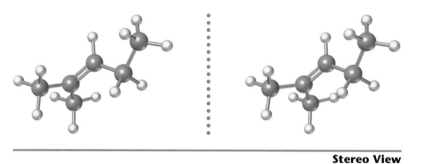

Stereo View

6.23   The following carbocation is an intermediate in the electrophilic addition reaction of HCl with two different alkenes. Identify both, and tell which C–H bonds in the carbocation are aligned for maximum hyperconjugation with the vacant *p* orbital on the positively charged carbon.

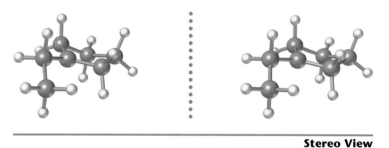

Stereo View

Additional Problems

6.24   Calculate the degree of unsaturation in the following formulas, and draw five possible structures for each:
(a) $C_{10}H_{16}$ (b) C_8H_8O (c) $C_7H_{10}Cl_2$
(d) $C_{10}H_{16}O_2$ (e) $C_5H_9NO_2$ (f) $C_8H_{10}ClNO$

6.25 A compound of formula $C_{10}H_{14}$ undergoes catalytic hydrogenation but absorbs only 2 molar equivalents of hydrogen. How many rings does the compound have?

6.26 A compound of formula $C_{12}H_{13}N$ contains two rings. How many molar equivalents of hydrogen does it absorb if all the remaining unsaturations are double bonds?

6.27 Give IUPAC names for the following alkenes:

(a)

$$CH_3$$
$$\underset{H_3C}{\overset{H}{\diagdown}}C=C\underset{H}{\overset{CHCH_2CH_3}{\diagup}}$$

(b)

$$\underset{CH_3CHCH_2CH_2CH}{\overset{CH_3 \quad CH_2CH_3}{}}$$
$$C=C\underset{H}{\overset{CH_3}{\diagup}}$$
$$\underset{H}{}$$

(c)

$$CH_2CH_3$$
$$H_2C=CCH_2CH_3$$

(d)

$$\underset{H_2C=CHCHCH}{\overset{H \quad CH_3}{\diagdown C=C \diagup}}$$
$$\underset{CH_3}{}$$

(e)

$$\underset{CH_3CH_2CH_2}{\overset{H_3C}{\diagdown}}C=C\underset{CH_3}{\overset{H \quad H}{\diagup}}C=C\underset{CH_3}{\overset{}{\diagdown}}$$

(f) $H_2C=C=CHCH_3$

6.28 Ocimene is a triene found in the essential oils of many plants. What is its IUPAC name, including stereochemistry?

Ocimene

6.29 α-Farnesene is a constituent of the natural wax found on apples. What is its IUPAC name, including stereochemistry?

α-Farnesene

6.30 Draw structures corresponding to the following systematic names:
(a) (4*E*)-2,4-Dimethyl-1,4-hexadiene
(b) *cis*-3,3-Dimethyl-4-propyl-1,5-octadiene
(c) 4-Methyl-1,2-pentadiene
(d) (3*E*,5*Z*)-2,6-Dimethyl-1,3,5,7-octatetraene
(e) 3-Butyl-2-heptene
(f) *trans*-2,2,5,5-Tetramethyl-3-hexene

6.31 Menthene, a hydrocarbon found in mint plants, has the systematic name 1-isopropyl-4-methylcyclohexene. Draw its structure.

6.32 Draw and name the 6 pentene isomers, C_5H_{10}, including *E,Z* isomers.

6.33 Draw and name the 17 hexene isomers, C_6H_{12}, including *E,Z* isomers.

6.34   *trans*-2-Butene is more stable than *cis*-2-butene by only 4 kJ/mol, but *trans*-2,2,5,5-tetramethyl-3-hexene is more stable than *cis*-2,2,5,5-tetramethyl-3-hexene by 39 kJ/mol. Explain.

Alkene	$\Delta H^\circ_{\text{hydrog}}$	
	(kJ/mol)	**(kcal/mol)**
cis-2-Butene	−119.7	−28.6
trans-2-Butene	−115.5	−27.6
cis-2,2,5,5-Tetramethyl-3-hexene	−151.5	−36.2
trans-2,2,5,5-Tetramethyl-3-hexene	−112.6	−26.9

6.35   Normally, a trans alkene is *more* stable than its cis isomer. *trans*-Cyclooctene, however, is *less* stable than *cis*-cyclooctene by 38.5 kJ/mol. Explain.

6.36   *trans*-Cyclooctene is less stable than *cis*-cyclooctene by 38.5 kJ/mol, but *trans*-cyclononene is less stable than *cis*-cyclononene by only 12.2 kJ/mol. Explain.

6.37   Allene (1,2-propadiene), $H_2C=C=CH_2$, has two adjacent double bonds. What kind of hybridization must the central carbon have? Sketch the bonding π orbitals in allene. What shape do you predict for allene?

6.38   The heat of hydrogenation for allene (Problem 6.37) to yield propane is −295 kJ/mol, and the heat of hydrogenation for a typical monosubstituted alkene such as propene is −126 kJ/mol. Is allene more stable or less stable than you might expect for a diene? Explain.

6.39   Predict the major product in each of the following reactions:

(a) $CH_3CH_2CH=\overset{\overset{\displaystyle CH_3}{\displaystyle |}}{C}CH_2CH_3$ + HCl $\longrightarrow$?

(b) 1-Ethylcyclopentene + HBr $\longrightarrow$?

(c) 2,2,4-Trimethyl-3-hexene + HI $\longrightarrow$?

(d) 1,6-Heptadiene + 2 HCl $\longrightarrow$?

(e) 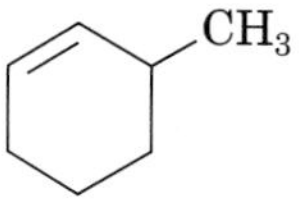 + HBr $\longrightarrow$?

6.40   Predict the major product from addition of HBr to each of the following alkenes:

(a) 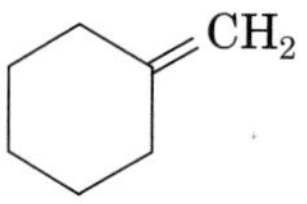   (b) 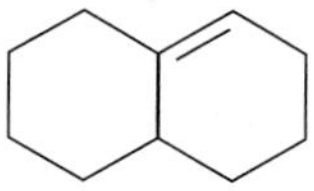   (c) $CH_3CH=\overset{\overset{\displaystyle CH_3}{\displaystyle |}}{C}HCHCH_3$

6.41   Rank the following sets of substituents in order of priority according to the Cahn–Ingold–Prelog sequence rules:

(a) $-CH_3$, $-Br$, $-H$, $-I$

(b) $-OH$, $-OCH_3$, $-H$, $-COOH$

(c) $-COOH$, $-COOCH_3$, $-CH_2OH$, $-CH_3$

$$\overset{\displaystyle O}{\overset{\displaystyle \|}{}}$$

(d) $-CH_3$, $-CH_2CH_3$, $-CH_2CH_2OH$, $-CCH_3$

(e) $-CH=CH_2$, $-CN$, $-CH_2NH_2$, $-CH_2Br$

(f) $-CH=CH_2$, $-CH_2CH_3$, $-CH_2OCH_3$, $-CH_2OH$

6.42 Assign *E* or *Z* configuration to each of the following alkenes:

(a) 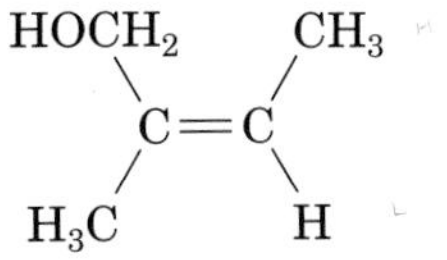
```
HOCH2      CH3
     \    /
      C = C
     /    \
   H3C      H
```

(b) 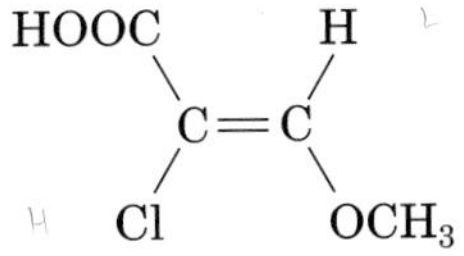
```
HOOC       H
    \     /
     C = C
    /     \
  Cl       OCH3
```

(c)
```
  NC        CH3
    \      /
     C =  C
    /      \
CH3CH2     CH2OH
```

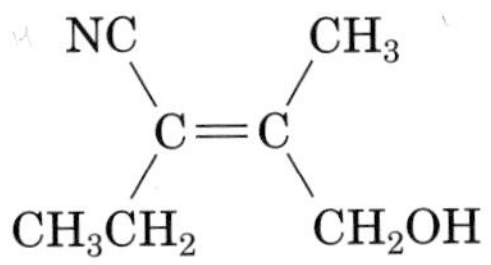

(d) 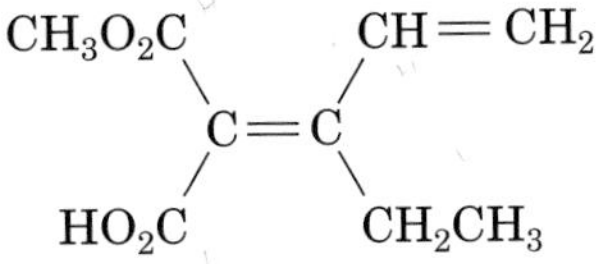
```
CH3O2C        CH=CH2
      \      /
       C =  C
      /      \
  HO2C        CH2CH3
```

6.43 Give IUPAC names for the following cycloalkenes:

(a) 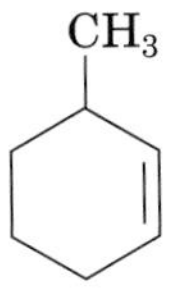 CH₃

(b)

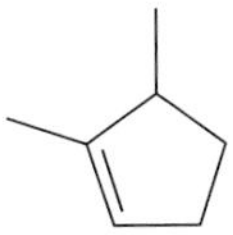

(c)

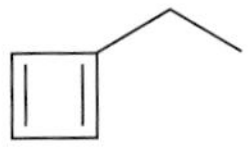

(d)

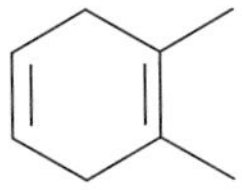

(e)

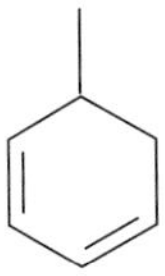

(f)

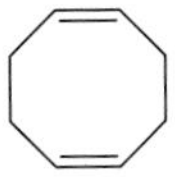

6.44 Which of the following *E,Z* designations are correct, and which are incorrect?

(a) 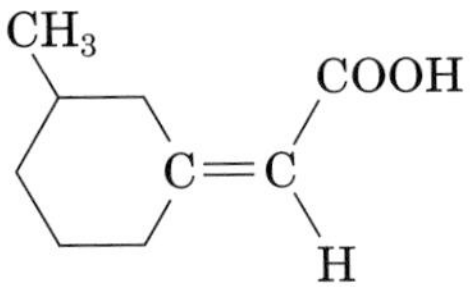
```
CH3
              COOH
          C = C
              H
```
Z

(b) 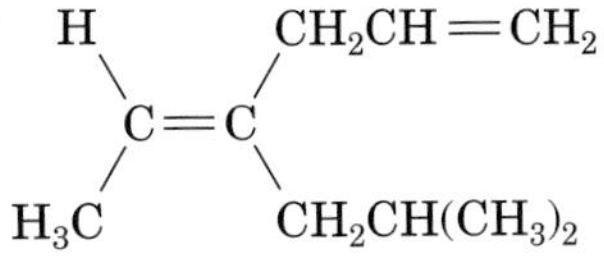
```
 H         CH2CH=CH2
   \      /
    C =  C
   /      \
 H3C       CH2CH(CH3)2
```
E

(c) 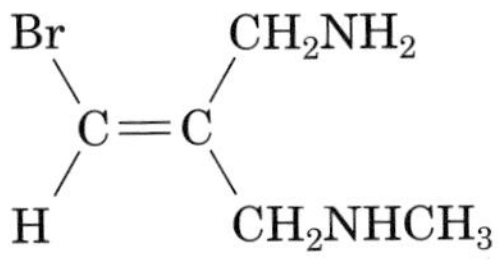
```
 Br        CH2NH2
   \      /
    C =  C
   /      \
  H        CH2NHCH3
```
Z

(d) 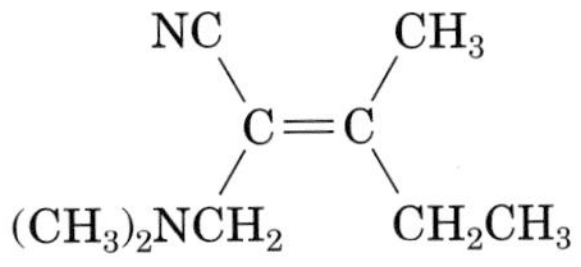
```
   NC          CH3
     \        /
      C  =   C
     /        \
(CH3)2NCH2     CH2CH3
```
E

(e) 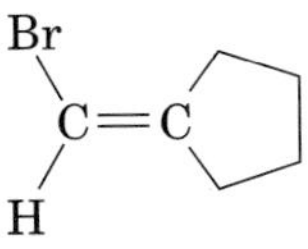
```
 Br
   \
    C = C
   /      
  H
```
Z

(f) 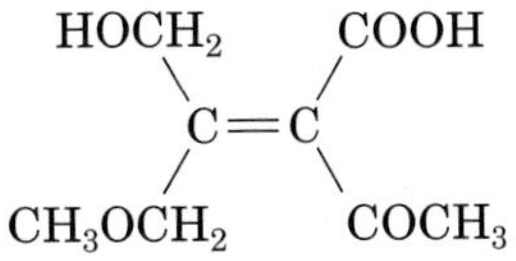
```
 HOCH2       COOH
      \     /
       C = C
      /     \
 CH3OCH2     COCH3
```
E

6.45   Use the bond dissociation energies in Table 5.3 to calculate $\Delta H°$ for the reaction of ethylene with HCl, HBr, and HI. Which reaction is most favorable?

6.46   Addition of HCl to 1-isopropylcyclohexene yields a rearranged product. Propose a mechanism, showing the structures of the intermediates and using curved arrows to indicate electron flow in each step.

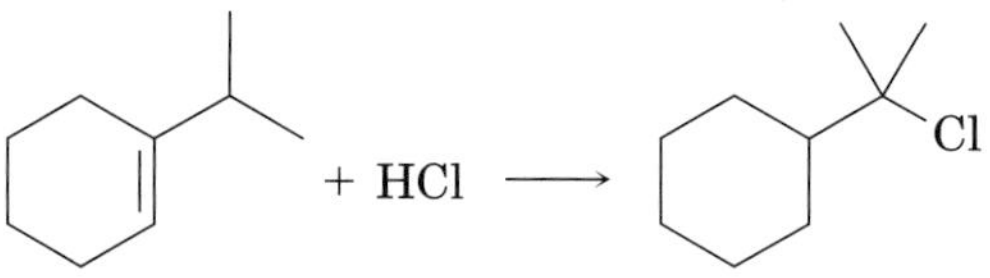

6.47   Addition of HCl to 1-isopropenyl-1-methylcyclopentane yields 1-chloro-1,2,2-trimethylcyclohexane. Propose a mechanism, showing the structures of the intermediates and using curved arrows to indicate electron flow in each step.

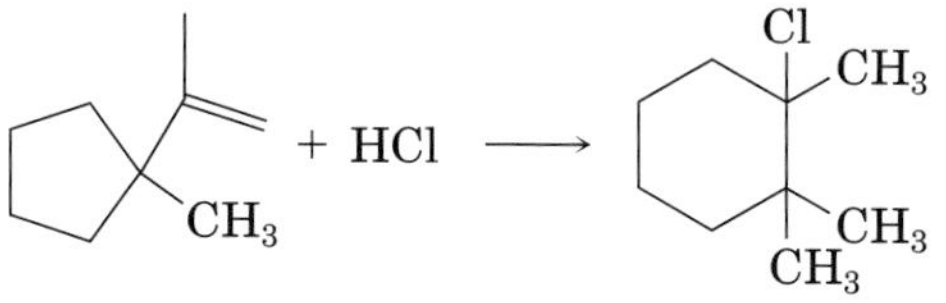

6.48   Vinylcyclopropane reacts with HBr to yield a rearranged alkyl bromide. Follow the flow of electrons as represented by the curved arrows, show the structure of the intermediate in brackets, and show the structure of the final product.

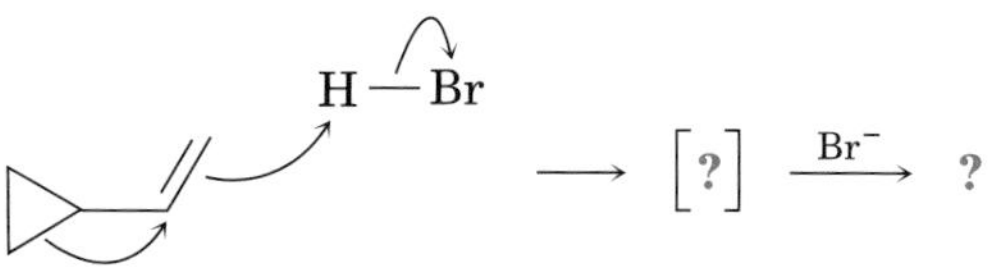

Vinylcyclopropane

6.49   Calculate the degree of unsaturation in each of the following formulas:
(a) Cholesterol, $C_{27}H_{46}O$ (b) DDT, $C_{14}H_9Cl_5$
(c) Prostaglandin E_1, $C_{20}H_{34}O_5$ (d) Caffeine, $C_8H_{10}N_4O_2$
(e) Cortisone, $C_{21}H_{28}O_5$ (f) Atropine, $C_{17}H_{23}NO_3$

6.50   The isobutyl cation spontaneously rearranges to the *tert*-butyl cation by a hydride shift:

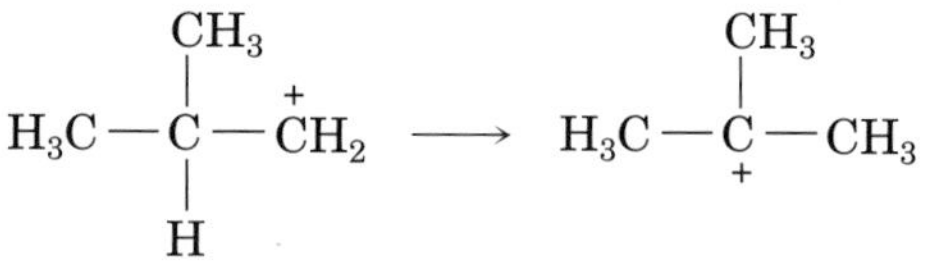

Isobutyl cation   *tert*-**Butyl cation**

Is this rearrangement exergonic or endergonic? Draw what you think the transition state for the hydride shift might look like, according to the Hammond postulate.

6.51   Draw a reaction energy diagram for the addition of HBr to 1-pentene. Let one curve on your diagram show the formation of 1-bromopentane product and another curve on the same diagram show the formation of 2-bromopentane product. Label the positions for all reactants, intermediates, and products. Which curve has the higher-energy carbocation intermediate? Which curve has the higher-energy first transition state?

6.52 Make sketches of the transition-state structures involved in the reaction of HBr with 1-pentene (Problem 6.51). Tell whether each structure resembles reactant or product.

A Look Ahead

. . . .

6.53 Aromatic compounds such as benzene react with alkyl chlorides in the presence of $AlCl_3$ catalyst to yield alkyl benzenes. The reaction occurs through a carbocation intermediate, formed by reaction of the alkyl chloride with $AlCl_3$ ($R-Cl + AlCl_3 \longrightarrow R^+ + AlCl_4^-$). How can you explain the observation that reaction of benzene with 1-chloropropane yields isopropylbenzene as the major product? (See Section 16.3.)

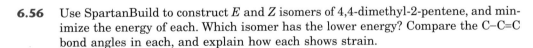

6.54 Alkenes can be converted into alcohols by acid-catalyzed addition of water. Assuming that Markovnikov's rule is valid, predict the major alcohol product from each of the following alkenes. (See Section 7.4.)

(a)

$$CH_3$$
$$CH_3CH_2C = CHCH_3$$

(b)

CH_2 (cyclohexane ring with =CH_2)

(c)

$$CH_3$$
$$CH_3CHCH_2CH = CH_2$$

6.55 Reaction of 2,3-dimethyl-1-butene with HBr leads to a bromoalkane, $C_6H_{13}Br$. On treatment of this bromoalkane with KOH in methanol, elimination of HBr occurs and a hydrocarbon that is isomeric with the starting alkene is formed. What is the structure of this hydrocarbon, and how do you think it is formed from the bromoalkane? (See Sections 11.10–11.11.)

Molecular Modeling

. . . .

6.56 Use SpartanBuild to construct E and Z isomers of 4,4-dimethyl-2-pentene, and minimize the energy of each. Which isomer has the lower energy? Compare the C–C=C bond angles in each, and explain how each shows strain.

6.57 Use SpartanView to compare electrostatic potential maps of the ethyl cation, the isopropyl cation, and the *tert*-butyl cation. How does the number of alkyl groups attached to the positive carbon change the potential at this atom? What does this imply about the direction of electron transfer between an alkyl group and the positive carbon?

6.58 Hyperconjugation in a carbocation is believed to strengthen and shorten C–C$^+$ bonds while making participating C–H bonds weaker and longer. Use SpartanView to compare bond distances in *tert*-butyl cation and 2-methylpropane. Which geometry changes support the hyperconjugation argument?

6.59 Use SpartanView to obtain the energies of the carbocations and the transition state for the following hydride shift. Use these energies to draw a reaction energy diagram for the rearrangement.

$$CH_3\overset{\overset{\displaystyle H}{|}}{\underset{\underset{\displaystyle CH_3}{|}}{C}}-\overset{+}{C}HCH_3 \longrightarrow \left[\begin{array}{c} \text{Transition state for} \\ \text{hydride shift} \end{array} \right] \longrightarrow CH_3\overset{+}{C}-\overset{\overset{\displaystyle H}{|}}{\underset{\underset{\displaystyle CH_3}{|}}{C}}HCH_3$$

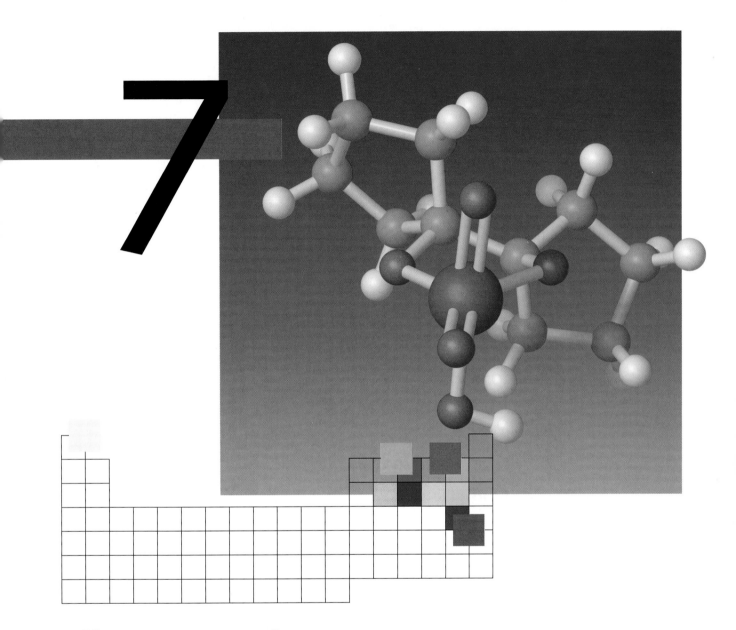

Alkenes: Reactions and Synthesis

The addition of electrophiles to alkenes is a useful and general reaction that makes possible the synthesis of many different kinds of compounds. Although we've studied only the addition of HX thus far, many other electrophiles also add to alkenes. In this chapter, we'll see how alkenes are prepared, we'll discuss many further examples of alkene addition reactions, and we'll review the wide variety of compounds that can be made from alkenes.

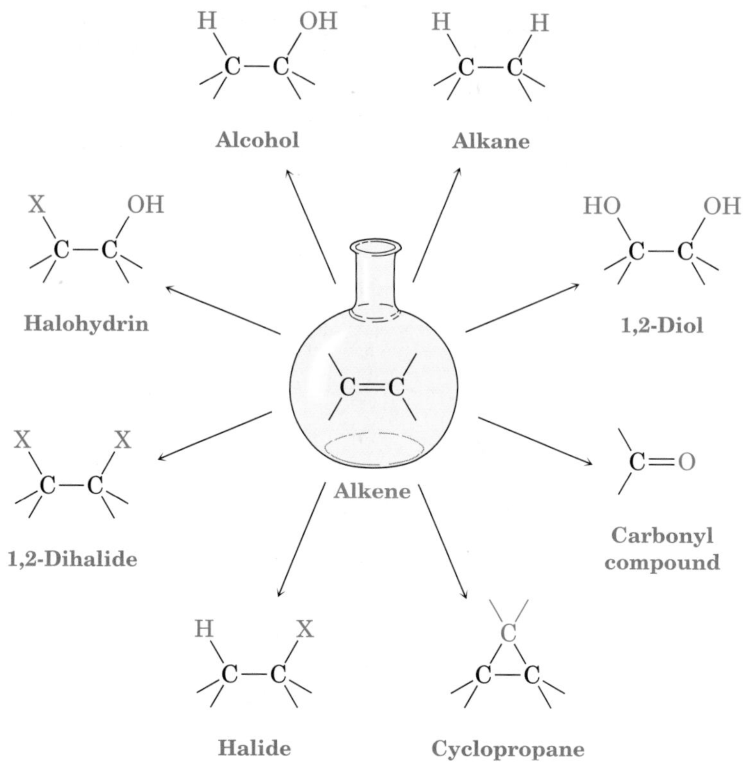

Alcohol

Alkane

Halohydrin

1,2-Diol

1,2-Dihalide

Carbonyl compound

Halide

Cyclopropane

Alkene

7.1 Preparation of Alkenes: A Preview of Elimination Reactions

Before getting to the main subject of this chapter—the reactions of alkenes— let's take a brief look at how alkenes are prepared. The subject is a bit complex, though, so we'll return to this topic in Chapter 11 for a more detailed study.

Just as the chemistry of alkenes is dominated by addition reactions, the preparation of alkenes is dominated by elimination reactions. Additions and eliminations are, in many respects, two sides of the same coin. That is, an addition reaction might involve the *addition* of HBr or H_2O to an alkene to form an alkyl halide or alcohol, whereas an elimination reaction might involve the *loss* of HBr or H_2O from an alkyl halide or alcohol to form an alkene.

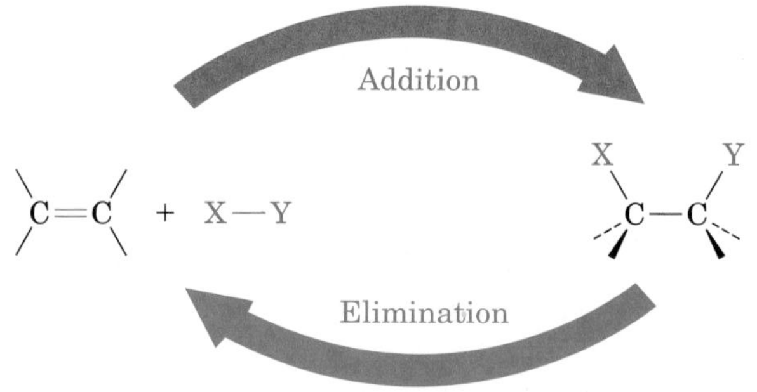

Addition

Elimination

The two most common alkene-forming elimination reactions are **dehydrohalogenation**—the loss of HX from an alkyl halide—and **dehydration**—the loss of water from an alcohol. Dehydrohalogenation usually occurs by reaction of an alkyl halide with strong base, such as potassium hydroxide. For example, bromocyclohexane yields cyclohexene when treated with KOH in ethanol solution:

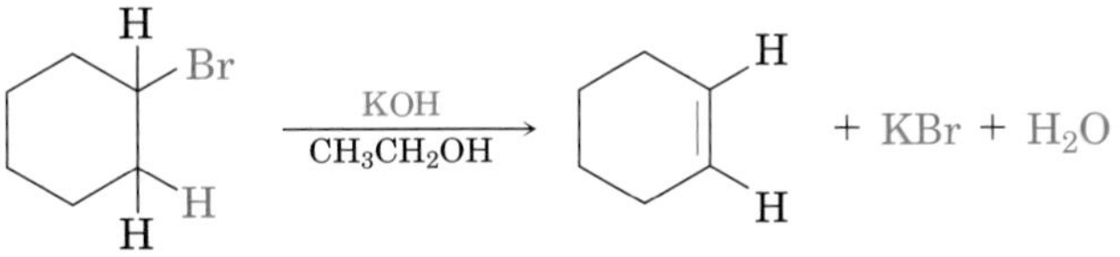
H
Br
KOH
CH3CH2OH
H
H
H
H
+ KBr + H2O

Bromocyclohexane **Cyclohexene (81%)**

Dehydration is often carried out by treatment of an alcohol with a strong acid. For example, loss of water occurs and 1-methylcyclohexene is formed when 1-methylcyclohexanol is warmed with aqueous sulfuric acid in tetrahydrofuran (THF) solvent:

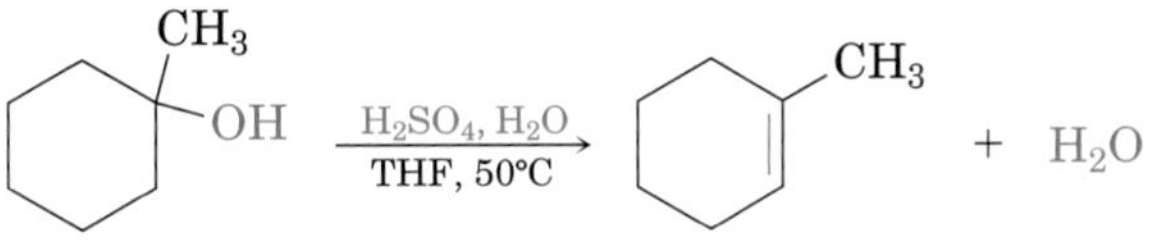
CH3
OH
H2SO4, H2O
THF, 50°C
CH3
+ H2O

1-Methylcyclohexanol **1-Methylcyclohexene (91%)**

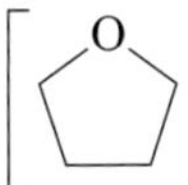
O

Tetrahydrofuran (THF)—a common solvent

Elimination reactions are sufficiently complex that it's best to defer a detailed discussion until Chapter 11. For the present, it's enough to realize that alkenes are readily available from simple precursors.

Problem 7.1 One problem with elimination reactions is that mixtures of products are often formed. For example, treatment of 2-bromo-2-methylbutane with KOH in ethanol yields a mixture of two alkene products. What are their likely structures?

Problem 7.2 How many alkene products, including *E,Z* isomers, might be obtained by dehydration of 3-methyl-3-hexanol with aqueous sulfuric acid?

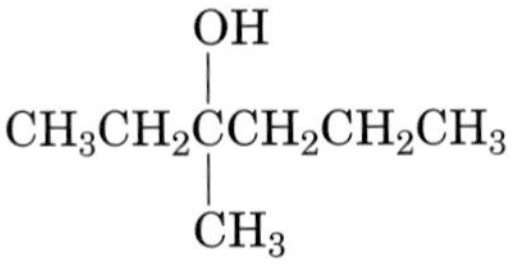
OH
CH3CH2CCH2CH2CH3
CH3

3-Methyl-3-hexanol

7.2 Addition of Halogens to Alkenes

Bromine and chlorine both add readily to alkenes to yield 1,2-dihaloalkanes. For example, approximately 6 million tons per year of 1,2-dichloroethane (ethylene dichloride) are synthesized industrially by the addition of Cl_2 to ethylene. The product is used both as a solvent and as starting material for use in the manufacture of poly(vinyl chloride), PVC.

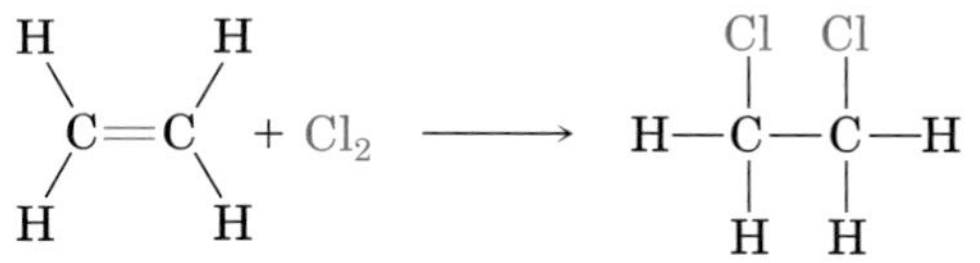

Ethylene

1,2-Dichloroethane (Ethylene dichloride)

Fluorine is too reactive and difficult to control for most laboratory applications, and iodine does not react with most alkenes.

Based on what we've seen thus far, a possible mechanism for the reaction of bromine with alkenes might involve attack by the π electron pair of the alkene on Br_2, breaking the Br–Br bond and displacing Br^- ion. The net result would be electrophilic addition of Br^+ to the alkene, giving a carbocation that could undergo further reaction with Br^- to yield the dibromo addition product:

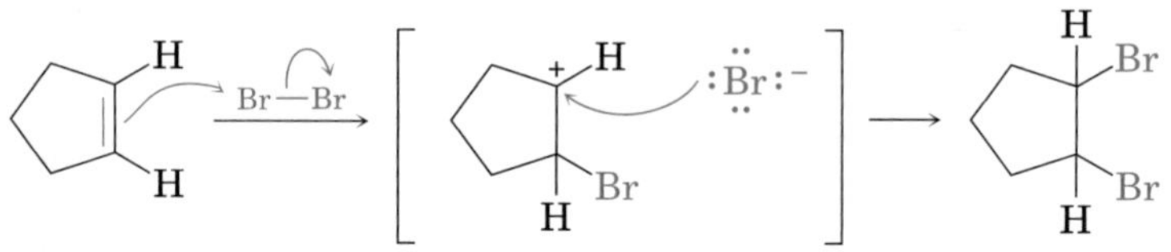

Although this mechanism looks reasonable, it's not completely consistent with known facts. In particular, the proposed mechanism doesn't explain the *stereochemistry* of the addition reaction. That is, the mechanism doesn't tell us which stereoisomer is formed.

Let's look again at the reaction of Br_2 with cyclopentene and assume that Br^+ adds to cyclopentene from the bottom side of the ring to form the carbocation intermediate shown in Figure 7.1. (The addition could equally well occur from the top side, but we'll consider only one possibility to keep things simple.) Since the positively charged carbon in the intermediate is planar and sp^2-hybridized, it might be attacked by Br^- ion in the second step of the reaction from either the top or the bottom to give a *mixture* of products. One product has the two Br atoms on the same side of the ring (cis), and the other has them on opposite sides (trans). We find, however, that only *trans*-1,2-dibromocyclopentane is produced; none of the cis product is formed. We therefore say that the reaction occurs with **anti stereochemistry**, meaning that the two bromines have come from opposite sides of the molecule—one from the top face of the ring and one from the bottom face.

An explanation for the observed anti stereochemistry of addition was suggested in 1937 by George Kimball and Irving Roberts, who proposed that the true reaction intermediate is not a carbocation but is instead a **bromonium**

FIGURE 7.1 ▼

The stereochemistry of the addition reaction of Br_2 with cyclopentene. Only the trans product is formed.

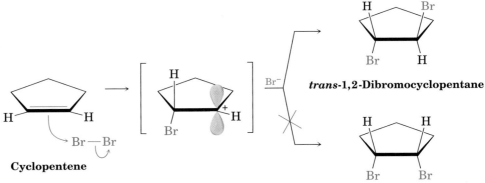

trans-1,2-Dibromocyclopentane

cis-1,2-Dibromocyclopentane
(*NOT formed*)

ion, R_2Br^+. (Similarly, a **chloronium ion** contains a positively charged, divalent chlorine, R_2Cl^+.) In the present instance, the bromonium ion is in a three-membered ring and is formed by donation of bromine lone-pair electrons to the vacant p orbital of the neighboring carbocation (Figure 7.2). Although Figure 7.2 depicts bromonium ion formation as stepwise, this is done only for clarity. The bromonium ion is formed in a single step by interaction of the alkene with Br^+.

FIGURE 7.2 ▼

Formation of a bromonium ion intermediate by electrophilic addition of Br^+ to an alkene.

refer to
**Mechanisms
& Movies**

Alkene π electrons attack bromine, pushing out bromide ion and leaving a bromo carbocation.

$+$ Br^-

The neighboring bromo substituent stabilizes the positive charge by using two of its electrons to overlap the vacant carbon p orbital, giving a three-membered-ring bromonium ion.

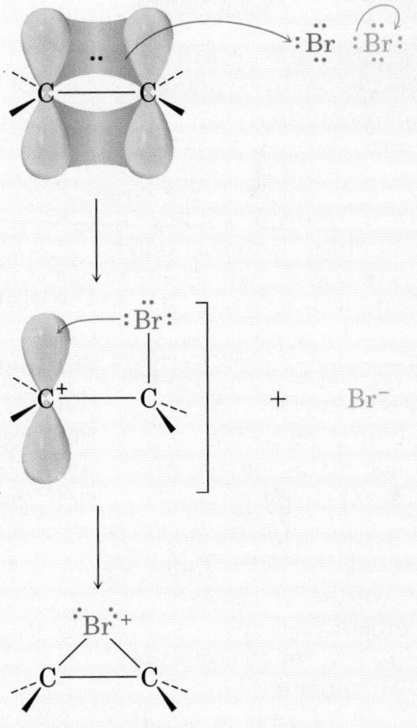

Bromonium ion

How does the formation of a bromonium ion account for anti stereochemistry of addition to cyclopentene? If a bromonium ion is formed as an intermediate, we can imagine that the large bromine atom might "shield" one side of the molecule. Attack by Br⁻ ion in the second step could then occur only from the opposite, unshielded side to give trans product.

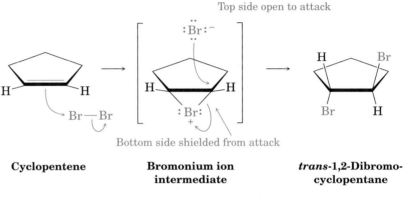

| Cyclopentene | Bromonium ion intermediate | *trans*-1,2-Dibromo-cyclopentane |

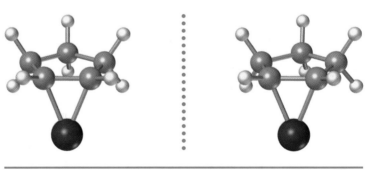

Stereo View

George Andrew Olah

George Andrew Olah (1927–) was born in Budapest, Hungary, and received a doctorate in 1949 at the Technical University of Budapest. During the Hungarian revolution in 1956, he emigrated to Canada and joined the Dow Chemical Company. After moving to the United States, he was professor of chemistry at Case-Western Reserve University (1965–1977) and then at the University of Southern California (1977–). He received the 1994 Nobel Prize in chemistry for his work on carbocations.

The bromonium ion postulate, made more than 60 years ago to explain the stereochemistry of halogen addition to alkenes, is a remarkable example of deductive logic in chemistry. Arguing from experimental results, chemists were able to make a hypothesis about the intimate mechanistic details of alkene electrophilic reactions. More recently, strong evidence supporting the mechanism has come from the work of George Olah, who has prepared and studied *stable* solutions of cyclic bromonium ions in liquid SO_2. There's no question that bromonium ions exist.

**Bromonium ion
(stable in SO_2 solution)**

● ●

Problem 7.3 What product would you expect to obtain from addition of Cl_2 to 1,2-dimethylcyclohexene? Show the stereochemistry of the product.

Problem 7.4 Unlike the reaction in Problem 7.3, addition of HCl to 1,2-dimethylcyclohexene yields a mixture of two products. Show the stereochemistry of each, and explain why a mixture is formed.

● ●

7.3 Halohydrin Formation

Many different kinds of electrophilic additions to alkenes take place. For example, alkenes add HO–Cl or HO–Br under suitable conditions to yield 1,2-halo alcohols, called **halohydrins**. Halohydrin formation doesn't take place by direct reaction of an alkene with HOBr or HOCl, however. Rather, the addition is done indirectly by reaction of the alkene with either Br_2 or Cl_2 in the presence of water.

An alkene   **A halohydrin**

We've seen that, when Br_2 reacts with an alkene, the cyclic bromonium ion intermediate reacts with the only nucleophile present, Br^- ion. If the reaction is carried out *in the presence of an additional nucleophile,* however, the intermediate bromonium ion can be intercepted by the added nucleophile and diverted to a different product. In the presence of water, for example, water competes with Br^- ion as nucleophile and reacts with the bromonium ion intermediate to yield a **bromohydrin**. The net effect is addition of HO–Br to the alkene. The reaction takes place by the pathway shown in Figure 7.3 (p. 238).

In practice, few alkenes are soluble in water, and bromohydrin formation is often carried out in a solvent such as aqueous dimethyl sulfoxide, CH_3SOCH_3 (DMSO), using a reagent called *N*-bromosuccinimide (NBS) as a source of Br_2. NBS is a stable, easily handled compound that slowly decomposes in water to yield Br_2 at a controlled rate. Bromine itself can also be used in the addition reaction, but it is more dangerous and more difficult to handle than NBS.

Styrene   **2-Bromo-1-phenylethanol (76%)**

FIGURE 7.3 ▼

Mechanism of
bromohydrin formation
by reaction of an alkene
with Br_2 in the presence
of water. Water acts as a
nucleophile to react with
the intermediate
bromonium ion.

refer to
**Mechanisms
& Movies**

Reaction of the alkene with Br_2
yields a bromonium ion intermediate.

Water acts as a nucleophile, using a
lone pair of electrons to open the
bromonium ion ring and form a bond
to carbon. Since oxygen donates its
electrons in this step, it now has the
positive charge.

Loss of a proton (H^+) from oxygen
then gives H_3O^+ and the neutral
bromohydrin addition product.

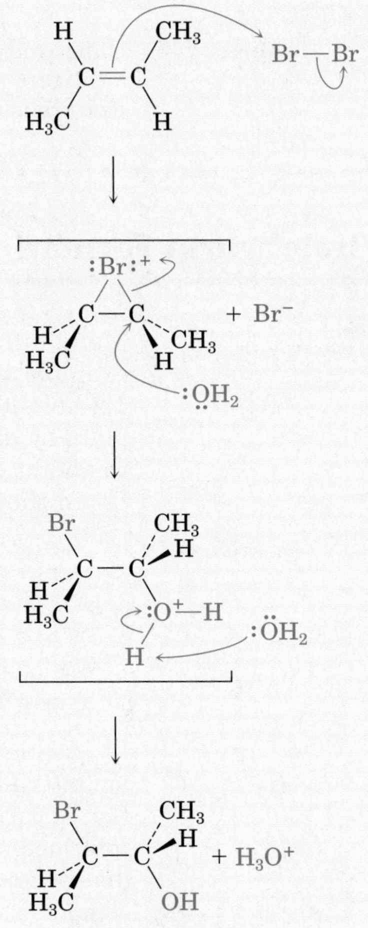

**3-Bromo-2-butanol
(A bromohydrin)**

© 1984 JOHN MCMURRY

Note that the aromatic ring in the example on page 237 is inert to Br_2
under the conditions used, even though it contains three carbon–carbon
double bonds. Aromatic rings are a good deal more stable than might be
expected, a property that will be examined in Chapter 15.

• •

Problem 7.5   What product would you expect from the reaction of cyclopentene with NBS and
water? Show the stereochemistry.

Problem 7.6   When an unsymmetrical alkene, such as propene, is treated with *N*-bromosuccinimide in aqueous dimethyl sulfoxide, the major product has the bromine atom bonded to the less highly substituted carbon atom:

$$CH_3CH\!=\!CH_2 \quad \xrightarrow[\text{DMSO}]{\text{NBS, H}_2\text{O}} \quad CH_3\overset{\displaystyle OH}{\underset{\displaystyle |}{C}}HCH_2Br$$

Is this Markovnikov or non-Markovnikov orientation? Explain.

• •

7.4 Addition of Water to Alkenes: Oxymercuration

Water adds to alkenes to yield alcohols, a process called **hydration**. The reaction takes place on treatment of the alkene with water and a strong acid catalyst (HA) by a mechanism similar to that of HX addition. Thus, protonation of an alkene double bond yields a carbocation intermediate, which reacts with water to yield a protonated alcohol product (ROH_2^+). Loss of H^+ from this protonated alcohol gives the neutral alcohol and regenerates the acid catalyst (Figure 7.4, p. 240).

Acid-catalyzed alkene hydration is suitable for large-scale industrial procedures, and approximately 300,000 tons of ethanol are manufactured each year in the United States by hydration of ethylene. The reaction is of little value in the typical laboratory, however, because it requires high temperatures and strongly acidic conditions. The hydration of ethylene, for example, takes place at 250°C with phosphoric acid as catalyst.

$$\underset{\textbf{Ethylene}}{\overset{\displaystyle H}{\underset{\displaystyle H}{}}C\!=\!C\overset{\displaystyle H}{\underset{\displaystyle H}{}}} \; + \; H_2O \quad \xrightarrow[\text{250°C}]{\text{H}_3\text{PO}_4 \text{ catalyst}} \quad \underset{\textbf{Ethanol}}{CH_3CH_2OH}$$

In the laboratory, alkenes are often hydrated by the **oxymercuration** procedure. When an alkene is treated with mercury(II) acetate [$Hg(O_2CCH_3)_2$, usually abbreviated $Hg(OAc)_2$] in aqueous tetrahydrofuran (THF) solvent, electrophilic addition to the double bond rapidly occurs. The intermediate *organomercury* compound is then treated with sodium borohydride, $NaBH_4$, and an alcohol is produced. For example:

$$\underset{\textbf{1-Methylcyclopentene}}{\text{[structure]}\!-\!CH_3} \quad \xrightarrow[\text{2. NaBH}_4]{\text{1. Hg(OAc)}_2, \text{ H}_2\text{O/THF}} \quad \underset{\substack{\textbf{1-Methylcyclopentanol}\\ \textbf{(92\%)}}}{\text{[structure]}\overset{CH_3}{\underset{OH}{}}}$$

FIGURE 7.4 ▼

Mechanism of the acid-catalyzed hydration of an alkene to yield an alcohol. Protonation of the alkene gives a carbocation intermediate that reacts with water.

refer to
**Mechanisms
& Movies**

Reaction of an alkene with acid (HA) yields a carbocation intermediate.

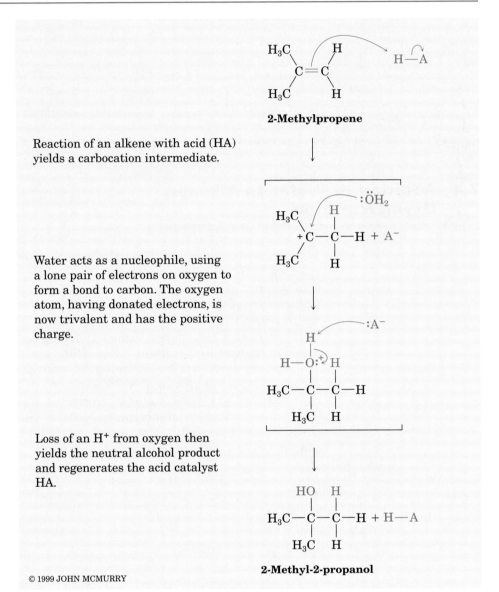

Water acts as a nucleophile, using a lone pair of electrons on oxygen to form a bond to carbon. The oxygen atom, having donated electrons, is now trivalent and has the positive charge.

Loss of an H⁺ from oxygen then yields the neutral alcohol product and regenerates the acid catalyst HA.

© 1999 JOHN MCMURRY

Alkene oxymercuration is closely analogous to halohydrin formation. The reaction is initiated by electrophilic addition of Hg^{2+} (mercuric) ion to the alkene to give an intermediate *mercurinium ion,* whose structure resembles that of a bromonium ion (Figure 7.5). Nucleophilic attack of water, followed by loss of a proton, then yields a stable organomercury addition product. The final step, reaction of the organomercury compound with sodium borohydride, is not fully understood but appears to involve radicals. Note that

FIGURE 7.5 ▼

Mechanism of the oxymercuration of an alkene to yield an alcohol. This electrophilic addition reaction involves a mercurinium ion intermediate, and its mechanism is similar to that of halohydrin formation. The product of the reaction is the more highly substituted alcohol, corresponding to Markovnikov regiochemistry.

Electrophilic addition of mercuric acetate to an alkene produces an intermediate, three-membered mercurinium ion.

A mercurinium ion

Water as nucleophile then displaces mercury by back-side attack at the more highly substituted carbon, breaking the C–Hg bond.

Loss of H⁺ yields a neutral organo-mercury addition product.

Treatment with sodium borohydride replaces the –Hg by –H and reduces the mercury, yielding an alcohol product.

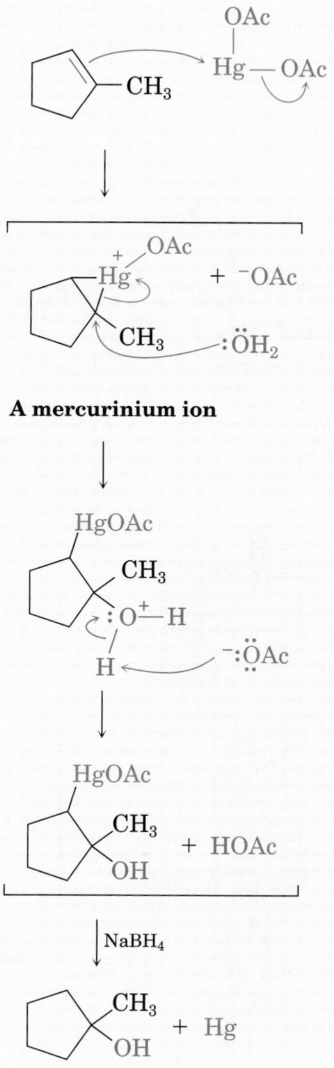

© 1999 JOHN MCMURRY

the regiochemistry of the reaction corresponds to Markovnikov addition of water; that is, the –OH group attaches to the more highly substituted carbon atom, and the –H attaches to the less highly substituted carbon.

• •

Problem 7.7 What products would you expect from oxymercuration of the following alkenes?
(a) $CH_3CH_2CH_2CH=CH_2$ (b) 2-Methyl-2-pentene

Problem 7.8   What alkenes might the following alcohols have been prepared from?

(a)

$$CH_3\underset{\underset{\displaystyle CH_3}{|}}{\overset{\overset{\displaystyle OH}{|}}{C}}CH_2CH_2CH_2CH_3$$

(b)

· ·

7.5 Addition of Water to Alkenes: Hydroboration

One of the most useful methods for preparing alcohols from alkenes is the *hydroboration* reaction reported in 1959 by H. C. Brown. **Hydroboration** involves addition of a B–H bond of borane, BH_3, to an alkene to yield an organoborane intermediate, RBH_2:

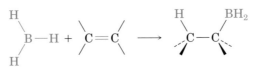

Borane   **An organoborane**

Borane is highly reactive because the boron atom has only six electrons in its valence shell. In tetrahydrofuran (THF) solution, BH_3 accepts an electron pair from a solvent molecule in a Lewis acid–base reaction to complete its octet and form a stable BH_3–THF complex.

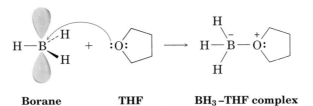

Borane   **THF**   **BH_3–THF complex**

When an alkene reacts with BH_3 in THF solution, rapid addition to the double bond occurs. Since BH_3 has three hydrogens, addition occurs three times, and a *trialkylborane, R_3B,* is formed. For example, 1 molar equivalent of BH_3 adds to 3 molar equivalents of cyclohexene to yield tricyclohexylborane. When tricyclohexylborane is then treated with aqueous hydrogen peroxide (H_2O_2) in basic solution, an oxidation takes place. The three C–B bonds are broken, –OH groups bond to the three carbons, and 3 equivalents of cyclohexanol are produced. The net effect of the two-step hydroboration/oxidation sequence is hydration of the alkene double bond:

Herbert Charles Brown

Herbert Charles Brown (1912–) was born in London to Ukrainian parents. Originally named Brovarnik, he was brought to the United States in 1914. Brown received his Ph.D. in 1938 from the University of Chicago, taught at Chicago and at Wayne State University, and then became professor of chemistry at Purdue University (1947–). The author of more than 1000 scientific papers, he received the Nobel Prize in chemistry in 1979 for his work on organoboranes.

Cyclohexene

Tricyclohexylborane

Cyclohexanol
(87%)

One of the features that makes the hydroboration reaction so useful is the regiochemistry that results when an unsymmetrical alkene is hydroborated. For example, hydroboration/oxidation of 1-methylcyclopentene yields *trans*-2-methylcyclopentanol. Boron and hydrogen both add to the alkene from the *same* face of the double bond—that is, with **syn stereochemistry** (the opposite of anti)—with boron attaching to the less highly substituted carbon. During the oxidation step, the boron is replaced by an —OH with the same stereochemistry, resulting in an overall syn non-Markovnikov addition of water. This stereochemical result is particularly useful because it is *complementary* to the Markovnikov regiochemistry observed for oxymercuration.

1-Methylcyclopentene

Alkylborane
intermediate

***trans*-2-Methylcyclopentanol**
(85%)

Why does alkene hydroboration take place with non-Markovnikov regiochemistry, yielding the less highly substituted alcohol? Hydroboration differs from many other alkene addition reactions in that it occurs in a single step without a carbocation intermediate. We can view the reaction as taking place through a four-center, cyclic transition state, as shown in Figure 7.6 (p. 244). Since both C–H and C–B bonds form at the same time and from the same face of the alkene, syn stereochemistry is observed.

The mechanism shown in Figure 7.6 accounts for not only the reaction's stereochemistry but also its regiochemistry. Although hydroboration does not involve a carbocation intermediate as other alkene addition reactions do, the interaction of borane with an alkene nevertheless has a large amount of polar character to it. Borane, with only six valence electrons on boron, is a Lewis acid and electrophile because of its vacant *p* orbital. Thus, the interaction of BH_3 with an alkene involves a partial transfer of electrons from the alkene to boron, with consequent buildup of polar character and a somewhat unsymmetrical transition state. Boron carries a partial negative charge ($\delta-$) because it has gained electrons, and one of the alkene carbons carries a partial positive charge ($\delta+$) because it has lost electrons.

FIGURE 7.6 ▼

Mechanism of alkene hydroboration. The reaction occurs in a single step, in which both C–H and C–B bonds form at the same time and on the same face of the double bond. Electrostatic potential maps show that boron becomes negative in the transition state, as electrons shift from the alkene to boron, but is positive in the product.

 refer to Mechanisms & Movies

hydroboration reactants, hydroboration transition state, hydroboration product, hydroboration animation

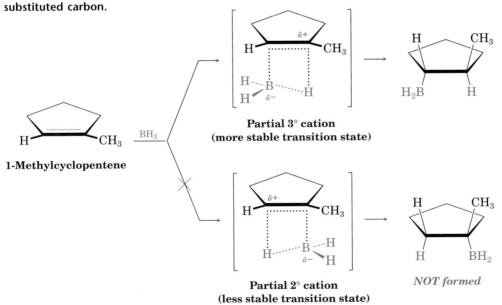

Addition of borane to the alkene π bond occurs in a single step through a cyclic four-membered-ring transition state. The dotted lines indicate partial bonds that are breaking or forming.

A neutral alkylborane addition product is then formed when reaction is complete.

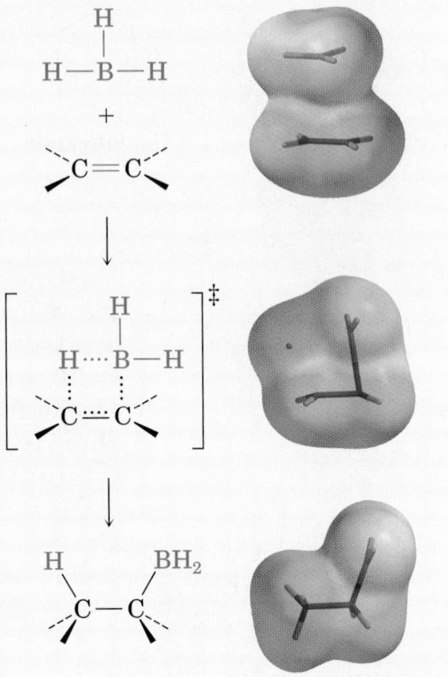

© 1984 JOHN MCMURRY

In the addition of BH_3 to an unsymmetrically substituted alkene such as 1-methylcyclopentene, there are two possible transition states (Figure 7.7). In one transition state, boron adds to the *less* highly substituted car-

FIGURE 7.7 ▼

Mechanism of the hydroboration of 1-methylcyclopentene. The favored transition state is the one that places the partial positive charge on the more highly substituted carbon.

1-Methylcyclopentene

**Partial 3° cation
(more stable transition state)**

**Partial 2° cation
(less stable transition state)**

NOT formed

bon, thereby placing a partial positive charge on the *more* highly substituted carbon. In the other transition state, boron adds to the *more* highly substituted carbon and places a partial positive charge on the *less* highly substituted carbon. The first alternative is favored because it resembles a tertiary carbocation, thus accounting for the observed result.

In addition to electronic factors, a steric factor is probably also involved in determining the regiochemistry of hydroboration. Attachment of boron is favored at the less sterically hindered carbon atom of the alkene, rather than at the more hindered carbon, because there is less steric crowding in the resultant transition state:

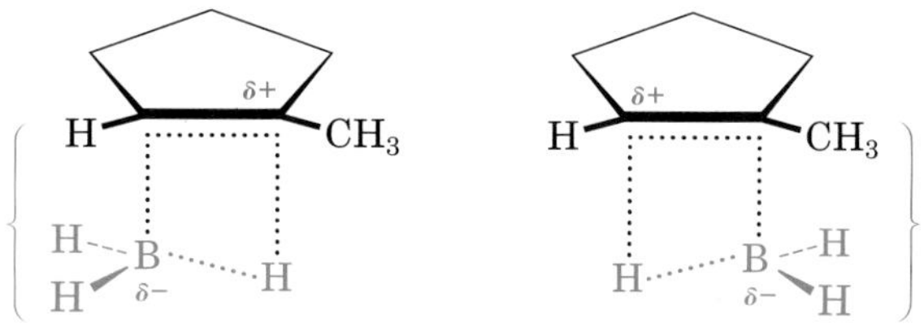

Practice Problem 7.1   What products would you obtain from reaction of 1-ethylcyclopentene with:
(a) BH_3, followed by H_2O_2, OH^- (b) $Hg(OAc)_2$, followed by $NaBH_4$

Strategy   When predicting the product of a reaction, you have to recall what you know about the kind of reaction being carried out and then apply that knowledge to the specific case you're dealing with. In the present instance, recall that the two methods of hydration—hydroboration/oxidation and oxymercuration—give complementary products. Hydroboration/oxidation occurs with syn stereochemistry and gives the non-Markovnikov addition product; oxymercuration gives the Markovnikov product.

Solution

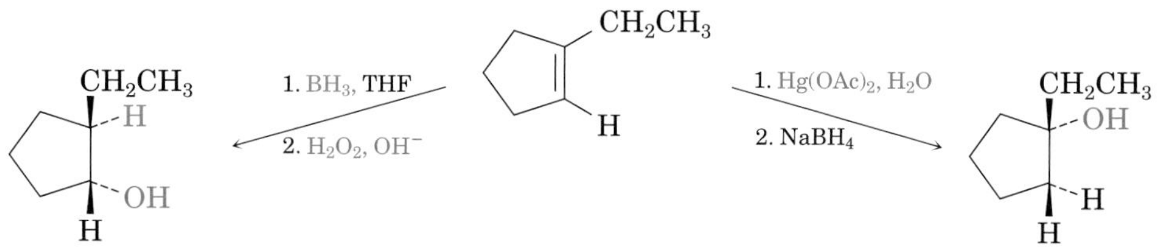

Practice Problem 7.2   How might you prepare the following alcohol?

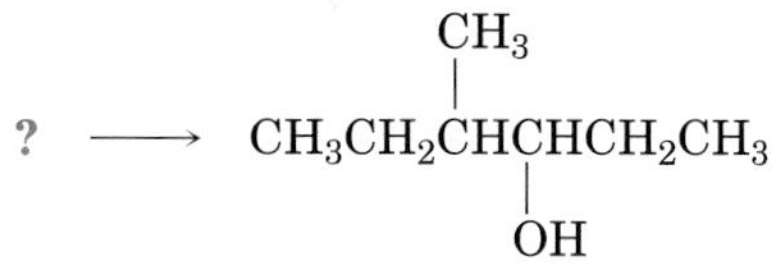

$$? \longrightarrow \underset{\underset{OH}{\overset{\overset{CH_3}{|}}{\underset{}{|}}}{CH_3CH_2CHCHCH_2CH_3}}$$

Strategy   Problems that require the synthesis of a specific target molecule should always be worked backward. Look at the target, identify its functional group(s), and ask yourself, "What are the methods for preparing this functional group?" In the present instance, the target molecule is a secondary alcohol (R_2CHOH), and we've seen that alcohols can be prepared from alkenes by either hydroboration/oxidation or oxymercuration. The –OH bearing carbon in the product must have been a double-bond carbon in the alkene reactant, so there are two possibilities: 4-methyl-2-hexene and 3-methyl-3-hexene:

$$\text{Add} \longrightarrow \text{OH here}$$
$$\underset{\text{CH}_3}{\overset{|}{\underset{}{}}}$$
$$CH_3CH_2CHCH{=}CHCH_3$$

4-Methyl-2-hexene

$$\text{Add} \longrightarrow \text{OH here}$$
$$\underset{\text{CH}_3}{\overset{|}{\underset{}{}}}$$
$$CH_3CH_2C{=}CHCH_2CH_3$$

3-Methyl-3-hexene

4-Methyl-2-hexene has a disubstituted double bond, RCH=CHR′, and would probably give a mixture of two alcohols with either hydration method since Markovnikov's rule does not apply to symmetrically substituted alkenes. 3-Methyl-3-hexene, however, has a trisubstituted double bond, and would give only the desired product on non-Markovnikov hydration using the hydroboration/oxidation method.

Solution

$$\underset{\text{CH}_3}{\overset{|}{\underset{}{}}}$$
$$CH_3CH_2C{=}CHCH_2CH_3 \xrightarrow[\text{2. H}_2\text{O}_2,\ \text{OH}^-]{\text{1. BH}_3,\ \text{THF}} CH_3CH_2\underset{\underset{\text{OH}}{|}}{\overset{\overset{\text{CH}_3}{|}}{C}}HCHCH_2CH_3$$

3-Methyl-3-hexene

⋯⋯⋯⋯⋯⋯⋯⋯⋯⋯⋯⋯⋯⋯⋯⋯⋯⋯⋯⋯⋯⋯⋯

Problem 7.9   What product will result from hydroboration/oxidation of 1-methylcyclopentene with deuterated borane, BD_3? Show both the stereochemistry (spatial arrangement) and the regiochemistry (orientation) of the product.

Problem 7.10   What alkenes might be used to prepare the following alcohols by hydroboration/oxidation?

(a) $(CH_3)_2CHCH_2CH_2OH$ (b) $(CH_3)_2CHCHCH_3$ (c) ⬡—CH_2OH
                                              $\underset{\text{OH}}{\overset{|}{}}$

Problem 7.11   The following cycloalkene gives a mixture of two alcohols on hydroboration followed by oxidation. Draw the structures of both, and explain the result.

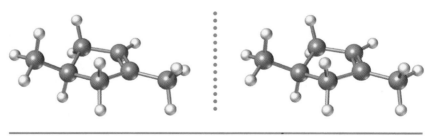

Stereo View

⋯⋯⋯⋯⋯⋯⋯⋯⋯⋯⋯⋯⋯⋯⋯⋯⋯⋯⋯⋯⋯⋯

7.6 Addition of Carbenes to Alkenes: Cyclopropane Synthesis

Yet another kind of alkene addition is the reaction of a *carbene* with an alkene to yield a cyclopropane. A **carbene, R_2C:**, is a neutral molecule containing a divalent carbon with only six electrons in its valence shell. It is therefore highly reactive and can be generated only as a reaction intermediate, rather than as an isolable molecule. Because a carbene has only six valence electrons on carbon, it is electron-deficient and behaves as an electrophile. Thus, carbenes react with nucleophilic C=C bonds much as other electrophiles do. The reaction occurs in a single step without intermediates.

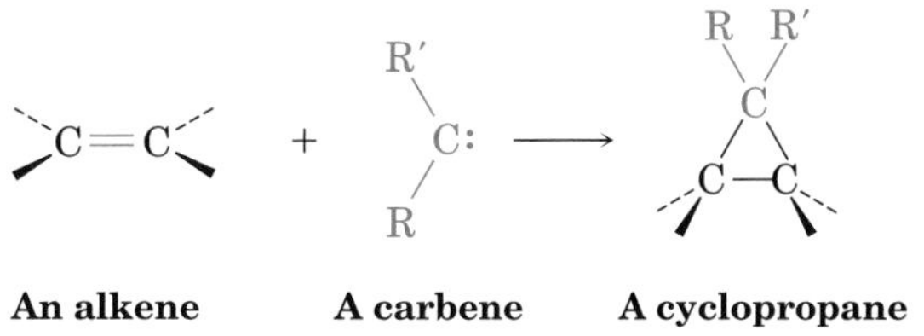

$$>C=C< \quad + \quad \overset{R'}{\underset{R}{>}}C: \quad \longrightarrow \quad \overset{R \quad R'}{\underset{C-C}{C}}$$

An alkene   **A carbene**   **A cyclopropane**

One of the simplest methods for generating a substituted carbene is by treatment of chloroform, $CHCl_3$, with a strong base such as KOH. Loss of a proton from $CHCl_3$ gives the trichloromethanide anion, $^-$:CCl_3, which expels a Cl^- ion to yield dichlorocarbene, :CCl_2 (Figure 7.8).

FIGURE 7.8 ▼

Mechanism of the formation of dichlorocarbene by reaction of chloroform with strong base.

refer to **Mechanisms & Movies**

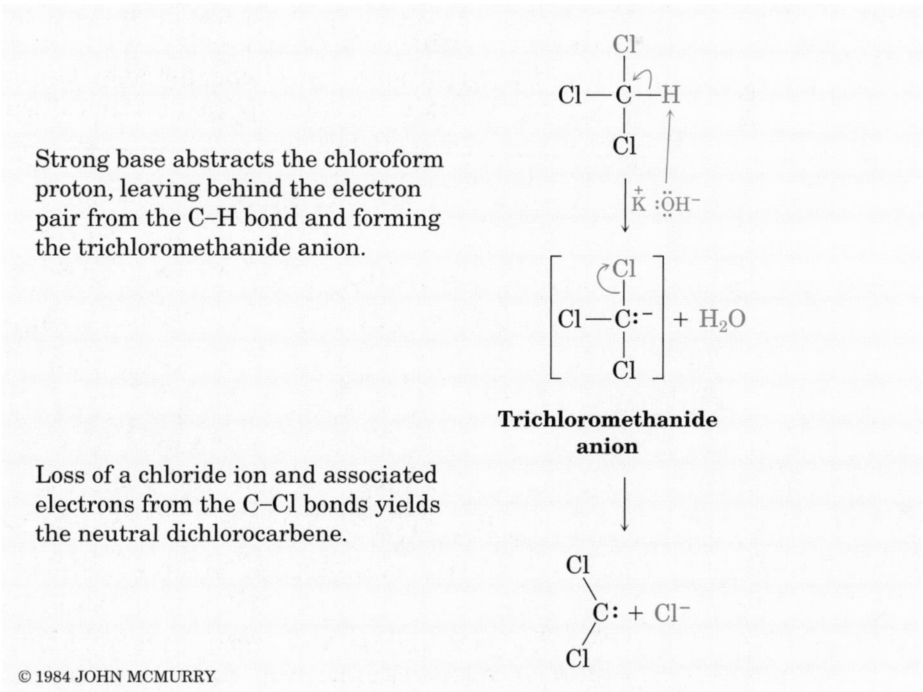

Strong base abstracts the chloroform proton, leaving behind the electron pair from the C–H bond and forming the trichloromethanide anion.

$$Cl-\overset{\overset{\displaystyle Cl}{|}}{\underset{\underset{\displaystyle Cl}{|}}{C}}-H$$

$$\overset{+}{K} : \overset{..}{O}H^-$$

$$\left[Cl-\overset{\overset{\displaystyle Cl}{|}}{\underset{\underset{\displaystyle Cl}{|}}{C}}:^- \right] + H_2O$$

Trichloromethanide anion

Loss of a chloride ion and associated electrons from the C–Cl bonds yields the neutral dichlorocarbene.

$$\overset{Cl}{\underset{Cl}{>}}C: + Cl^-$$

© 1984 JOHN MCMURRY

The dichlorocarbene carbon atom is sp^2-hybridized, with a vacant p orbital extending above and below the plane of the three atoms and with an unshared pair of electrons occupying the third sp^2 lobe. Note that this electronic description of dichlorocarbene is similar to that for a carbocation (Section 6.10) with respect to both the sp^2 hybridization of carbon and the vacant p orbital. Electrostatic potential maps further show this similarity (Figure 7.9)

FIGURE 7.9 ▼

The structure of dichlorocarbene. Electrostatic potential maps show how the positive region coincides with the empty p orbital in both dichlorocarbene and a carbocation (CH_3^+). The negative region in the dichlorocarbene map coincides with the lone-pair electrons.

dichlorocarbene, carbocation

 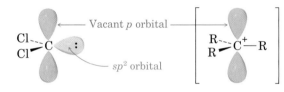

Dichlorocarbene   **A carbocation**
                         (**sp^2-hybridized**)

If dichlorocarbene is generated in the presence of an alkene, addition to the double bond occurs, and a dichlorocyclopropane is formed. As the reaction of dichlorocarbene with *cis*-2-pentene demonstrates, the addition is **stereospecific**, meaning that only a single stereoisomer is formed as product. Starting from a cis alkene, for instance, only cis-disubstituted cyclopropane is produced; starting from a trans alkene, only trans-disubstituted cyclopropane is produced.

cis-**2-Pentene**

Cyclohexene

The best method for preparing nonhalogenated cyclopropanes is by a process called the **Simmons–Smith reaction**. First investigated at the Du Pont company, this reaction does not involve a free carbene. Rather, it utilizes a *carbenoid*—a metal-complexed reagent with carbene-like reactivity. When diiodomethane is treated with a specially prepared zinc–copper alloy, (iodomethyl)zinc iodide, ICH_2ZnI, is formed. In the presence of an alkene, (iodomethyl)zinc iodide transfers a CH_2 group to the double bond and yields the cyclopropane. For example, cyclohexene reacts cleanly and in good yield to give the corresponding cyclopropane. Although we won't discuss the mechanistic details, carbene addition to an alkene is an example of a general class of reactions called *cycloadditions,* which we'll study more carefully in Chapter 30.

$$CH_2I_2 + Zn(Cu) \xrightarrow{\text{Ether}} I—CH_2—Zn—I = \text{“}:CH_2\text{”}$$

Diiodomethane   **(Iodomethyl)zinc iodide**
(a carbenoid)

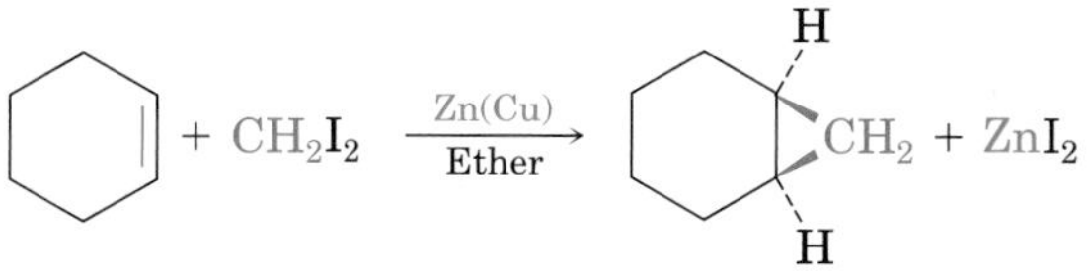

Cyclohexene   **Bicyclo[4.1.0]heptane (92%)**

Problem 7.12   What products would you expect from the following reactions?

(a) 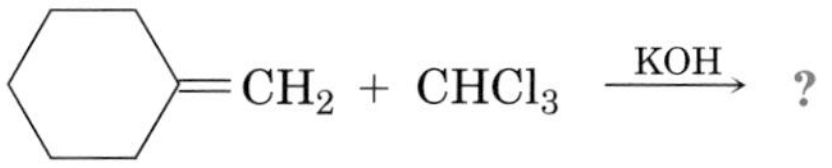 $=CH_2 + CHCl_3 \xrightarrow{\text{KOH}}$?

(b) $(CH_3)_2CHCH_2CH=CHCH_3 + CH_2I_2 \xrightarrow{\text{Zn(Cu)}}$?

7.7 Reduction of Alkenes: Hydrogenation

Alkenes react with H_2 in the presence of a catalyst to yield the corresponding saturated alkane addition products. We describe the result by saying that the double bond has been **hydrogenated**, or *reduced.* Note that the words *oxidation* and *reduction* are used somewhat differently in organic chemistry than in inorganic chemistry. We'll explore oxidation and reduction in more detail in Section 10.10 but will note for the present that an organic oxidation often forms carbon–oxygen bonds, while a reduction often forms carbon–hydrogen bonds.

An alkene   **An alkane**

Roger Adams

Roger Adams (1889–1971) was born in Boston, Massachusetts, and received his Ph.D. in 1912 at Harvard. He taught at the University of Illinois from 1916 until his retirement in 1957, during which time he had an enormous influence on the development of organic chemistry in the United States. Among many other accomplishments, he established the structure of tetrahydrocannabinol, the active ingredient in marijuana.

Platinum and palladium are the most common catalysts for alkene hydrogenations. Palladium is normally used as a very fine powder "supported" on an inert material such as charcoal (Pd/C) to maximize surface area. Platinum is normally used as PtO_2, a reagent known as *Adams' catalyst* after its discoverer, Roger Adams.

Catalytic hydrogenation, unlike most other organic reactions, is a *heterogeneous* process rather than a homogeneous one. That is, the hydrogenation reaction does not occur in a homogeneous solution but instead takes place on the surface of insoluble catalyst particles. Hydrogenation usually occurs with syn stereochemistry—both hydrogens add to the double bond from the same face.

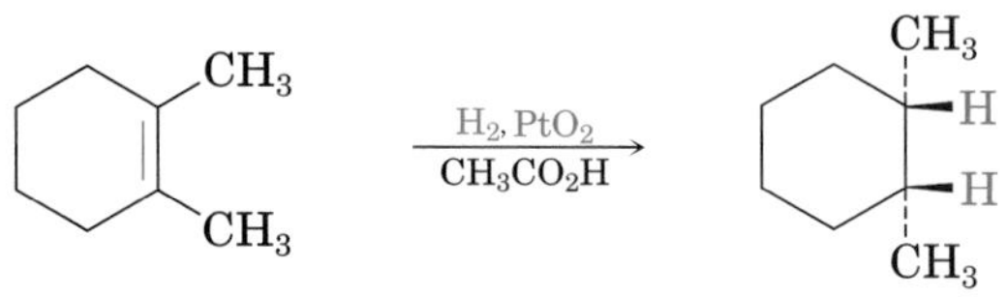

1,2-Dimethylcyclohexene **cis-1,2-Dimethylcyclohexane (82%)**

The first step in the reaction is adsorption of H_2 onto the catalyst surface. Complexation between catalyst and alkene then occurs as a vacant orbital on the metal interacts with the filled alkene π orbital. In the final steps, hydrogen is inserted into the double bond, and the saturated product diffuses away from the catalyst (Figure 7.10). The stereochemistry of hydrogenation is syn because both hydrogens add to the double bond from the same catalyst surface.

An interesting stereochemical feature of catalytic hydrogenation is that the reaction is extremely sensitive to the steric environment around the double bond. As a result, the catalyst often approaches only one face of an

FIGURE 7.10

Mechanism of alkene hydrogenation. The reaction takes place with syn stereochemistry on the surface of insoluble catalyst particles.

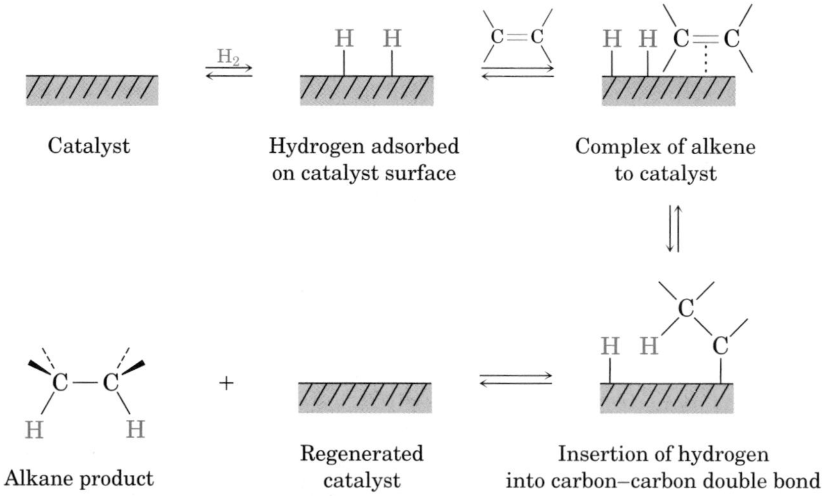

alkene, giving rise to a single product. In α-pinene, for example, one of the methyl groups attached to the four-membered ring hangs over the top face of the double bond and blocks approach of the hydrogenation catalyst from that side. Reduction therefore occurs exclusively from the bottom face to yield the product shown.

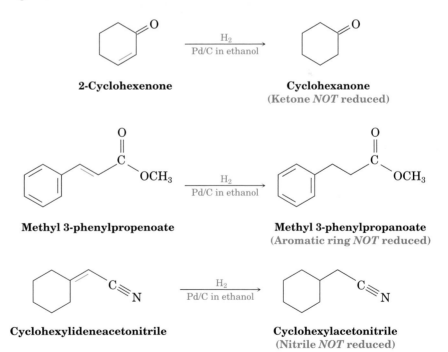

Alkenes are much more reactive than most other functional groups toward catalytic hydrogenation, and the reaction is therefore quite selective. Such other functional groups as aldehydes, ketones, esters, and nitriles survive normal alkene hydrogenation conditions unchanged, although reaction with these groups does occur under more vigorous conditions. Note particularly in the hydrogenation of methyl 3-phenylpropenoate shown below that the aromatic ring is not reduced by hydrogen and palladium even though it contains double bonds.

In addition to its usefulness in the laboratory, catalytic hydrogenation is of great commercial value in the food industry. Unsaturated vegetable oils, which usually contain numerous double bonds, are catalytically hydrogenated on a vast scale to produce the saturated fats used in margarine and solid cooking fats.

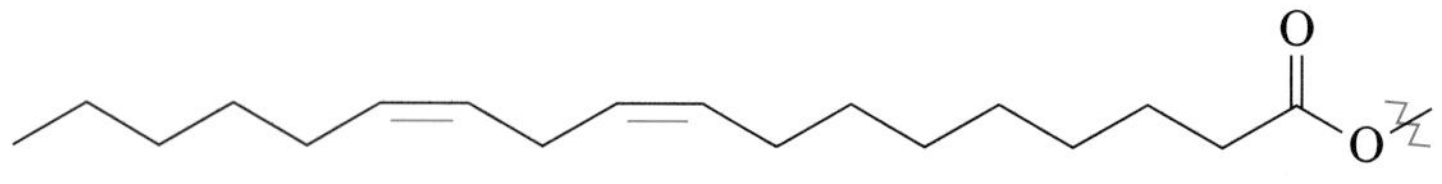

Ester of linoleic acid (a constituent of vegetable oil)

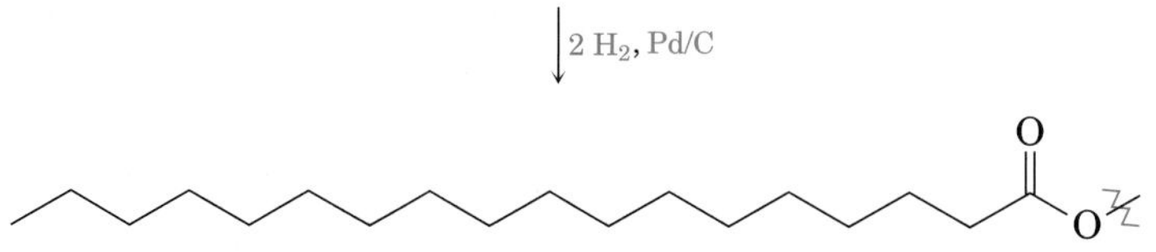

$\downarrow$ 2 H$_2$, Pd/C

Ester of stearic acid

Problem 7.13 What product would you obtain from catalytic hydrogenation of the following alkenes?
(a) (CH$_3$)$_2$C=CHCH$_2$CH$_3$ (b) 3,3-Dimethylcyclopentene

7.8 Oxidation of Alkenes: Hydroxylation and Cleavage

Alkene Hydroxylation

Hydroxylation of an alkene—the addition of an –OH group to each of the two alkene carbons—can be carried out by reaction of the alkene with osmium tetraoxide (OsO$_4$). The reaction occurs with syn stereochemistry and yields a 1,2-dialcohol, or **diol** product (also called a **glycol**).

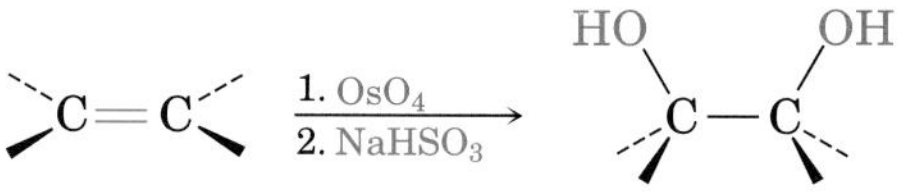

An alkene   **A 1,2-diol**

Alkene hydroxylation does not involve a carbocation intermediate but instead occurs through an intermediate cyclic *osmate,* which is thought to be formed in a single step by addition of OsO$_4$ to the alkene. This cyclic osmate is then cleaved in a second, separate step using aqueous sodium bisulfite, NaHSO$_3$.

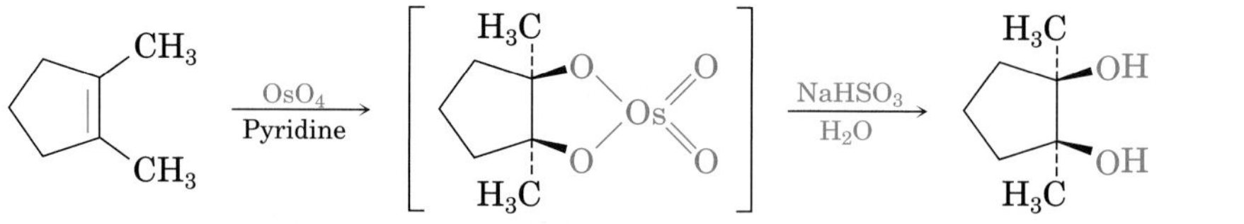

1,2-Dimethylcyclopentene   **A cyclic osmate intermediate**   ***cis*-1,2-Dimethyl-1,2-cyclopentanediol (87%)**

Alkene Cleavage

In all the alkene addition reactions we've seen thus far, the carbon–carbon double bond has been converted into a single bond but the carbon skeleton of the starting material has been left intact. There are, however, powerful oxidizing reagents that will *cleave* C=C bonds and produce two fragments.

Ozone (O_3) is perhaps the most useful double-bond cleavage reagent. Prepared by passing a stream of oxygen through a high-voltage electrical discharge, ozone adds rapidly to an alkene at low temperature to give a cyclic intermediate called a **molozonide**. Once formed, the molozonide then rapidly rearranges to form an **ozonide**. Though we won't study the mechanism of this rearrangement in detail, it involves the molozonide coming apart into two fragments, which then recombine in a different way.

$$3\ O_2 \xrightarrow[\text{discharge}]{\text{Electric}} 2\ O_3$$

A molozonide   **An ozonide**

Low-molecular-weight ozonides are explosive and are therefore never isolated. Instead, ozonides are further treated with a reducing agent such as zinc metal in acetic acid to convert them to carbonyl compounds. The net result of the ozonolysis/reduction sequence is that the C=C bond is cleaved, and oxygen becomes doubly bonded to each of the original alkene carbons. If an alkene with a tetrasubstituted double bond is ozonized, two ketone fragments result; if an alkene with a trisubstituted double bond is ozonized, one ketone and one aldehyde result; and so on.

Isopropylidenecyclohexane   **Cyclohexanone   Acetone**
(tetrasubstituted)
                                   84%; two ketones

$$CH_3(CH_2)_7CH{=}CH(CH_2)_7\overset{\displaystyle O}{\overset{\|}{C}}OCH_3 \xrightarrow[\text{2. Zn, H}_3O^+]{\text{1. O}_3} CH_3(CH_2)_7\overset{\displaystyle O}{\overset{\|}{C}}H + H\overset{\displaystyle O}{\overset{\|}{C}}(CH_2)_7\overset{\displaystyle O}{\overset{\|}{C}}OCH_3$$

Methyl 9-octadecenoate   **Nonanal   Methyl 9-oxononanoate**
(disubstituted)
                                      78%; two aldehydes

Several oxidizing reagents other than ozone also cause double-bond cleavage. For example, potassium permanganate ($KMnO_4$) in neutral or acidic solution cleaves alkenes, giving carbonyl-containing products in low to moderate yield. If hydrogens are present on the double bond, carboxylic acids are produced; if two hydrogens are present on one carbon, CO_2 is formed.

3,7-Dimethyl-1-octene   **2,6-Dimethylheptanoic acid (45%)**

1,2-Diol Cleavage

1,2-Diols are oxidatively cleaved by reaction with periodic acid (HIO_4) to yield carbonyl compounds, a reaction similar to the $KMnO_4$ cleavage of alkenes just discussed. The sequence of (1) alkene hydroxylation with OsO_4 followed by (2) diol cleavage with HIO_4 is often an excellent alternative to direct alkene cleavage with ozone or potassium permanganate.

A 1,2-diol   **Two carbonyl compounds**

If the two –OH groups are on an open chain, two carbonyl compounds result. If the two –OH groups are on a ring, a single, open-chain dicarbonyl compound is formed. As indicated in the following examples, the cleavage reaction is believed to take place through a cyclic periodate intermediate.

A 1,2-diol   **Cyclic periodate intermediate**   **6-Oxoheptanal (86%)**

A 1,2-diol   **Cyclic periodate intermediate**   **Cyclopentanone (81%)**

Practice Problem 7.3   What alkene would yield a mixture of cyclopentanone and propanal on treatment with ozone followed by reduction with zinc?

$$? \xrightarrow[\text{2. Zn, acetic acid}]{\text{1. } O_3} \quad \begin{array}{c} \end{array} = O + CH_3CH_2\overset{\displaystyle O}{\overset{\|}{C}}H$$

Strategy   Reaction of an alkene with ozone, followed by reduction with zinc, cleaves the carbon–carbon double bond and gives two carbonyl-containing fragments. That is, the C=C bond becomes two C=O bonds. Working backward from the carbonyl-containing products, the alkene precursor can be found by removing the oxygen from each product and joining the two carbon atoms to form a double bond.

Solution

$$\begin{array}{c} \end{array} = O + O = CHCH_2CH_3 \quad \longleftarrow \quad \begin{array}{c} \end{array} = CHCH_2CH_3$$

Remove oxygens and join carbons.

• •

Problem 7.14   What alkene would you start with to prepare each of the following compounds?

(a)

H
⋮OH

⋮OH
CH₃

(b) $CH_3CH_2\overset{\displaystyle |}{\underset{\displaystyle \underset{CH_3}{|}}{C}H}\overset{\displaystyle HO \quad OH}{\underset{}{}}\overset{|}{C}CH_3$

(b) $CH_3CH_2CHCCH_3$ with HO OH above and CH₃ below

(c) $HOCH_2CHCHCH_2OH$ with HO OH above

Problem 7.15   What products would you expect from reaction of 1-methylcyclohexene with the following reagents?
(a) Aqueous acidic $KMnO_4$ (b) O_3, followed by Zn, CH_3COOH

Problem 7.16   Propose structures for alkenes that yield the following products on reaction with ozone followed by treatment with Zn.
(a) $(CH_3)_2C=O + H_2C=O$ (b) 2 equiv $CH_3CH_2CH=O$

• •

7.9 Biological Alkene Addition Reactions

Biological Connection ▶

The chemistry of living organisms is a fascinating field of study. The simplest one-celled organism is capable of more complex organic synthesis than any human chemist, yet the same principles that apply to laboratory chemistry also apply to biological chemistry.

Biological organic chemistry takes place in the aqueous medium inside cells rather than in organic solvents, and it involves complex catalysts called *enzymes*. Nevertheless, the *kinds* of biological reactions that occur are remarkably similar to laboratory reactions. Thus, there are many cases of biological addition reactions to alkenes. For example, the enzyme fumarase catalyzes the addition of water to fumaric acid much as sulfuric acid might catalyze the addition of water to ethylene:

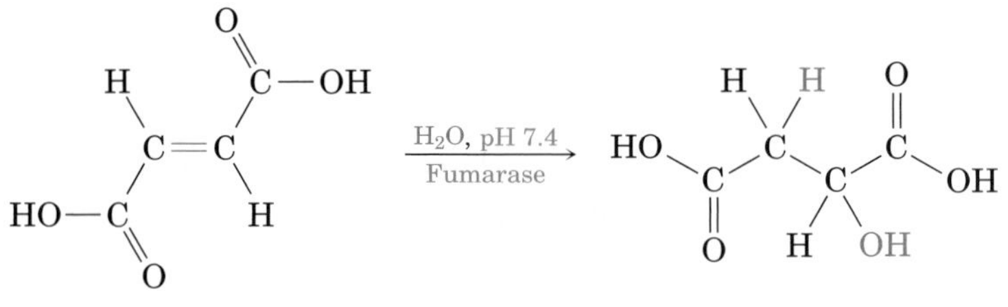

Fumaric acid   **Malic acid**

This reaction is one step in the so-called *citric acid cycle,* which our bodies use to metabolize food. A more complete discussion of the cycle is given in Chapter 29.

Enzyme-catalyzed reactions are usually much more chemically selective than their laboratory counterparts. Fumarase, for example, is completely inert toward maleic acid, the cis isomer of fumaric acid. Nevertheless, the fundamental processes of organic chemistry are the same in the living cell and in the laboratory.

7.10 Addition of Radicals to Alkenes: Polymers

No other group of chemicals has had as great an impact on our day-to-day lives as have the synthetic *polymers.* From carpets to clothes to foam coffee cups, it sometimes seems that we are surrounded by polymers.

A **polymer** is simply a large—sometimes *very* large—molecule built up by repetitive bonding together of many smaller molecules, called **monomers.** Polyethylene, for example, consists of enormous, long-chain alkane molecules prepared by bonding together of several thousand ethylene units. More than 10 million tons per year of polyethylene are manufactured in the United States alone.

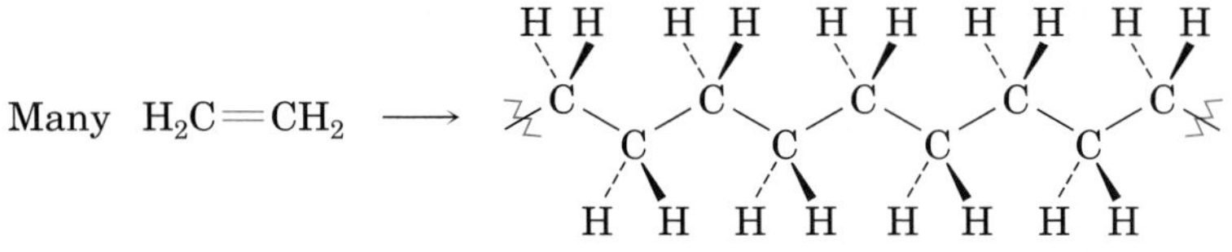

Ethylene   **A section of polyethylene**

Ethylene polymerization is usually carried out at high pressure (1000–3000 atm) and high temperature (100–250°C) in the presence of a catalyst such as benzoyl peroxide. The key step is the addition of a *radical* to the ethylene double bond, a reaction similar in many respects to what takes place in the addition of an *electrophile* to an alkene. As with the radical chain process we saw earlier for the light-induced chlorination of methane (Section 5.3), three kinds of steps are involved in the overall polymerization process: *initiation, propagation,* and *termination.* In writing the mechanism of this radical reaction, recall that a curved half-arrow, or "fishhook" ⌒, is used to show the movement of a single electron, as opposed to the full curved arrow used to show the movement of an electron pair in a polar reaction.

STEP 1 **Initiation** The reaction is initiated in two steps. In the first step, heat-induced homolytic cleavage of the weak O–O bond of benzoyl peroxide generates two benzoyloxy radicals, BzO·.

Benzoyl peroxide   **Benzoyloxy radical**

In the second initiation step, a benzoyloxy radical adds to ethylene to generate an alkyl radical. One electron from the carbon–carbon double bond pairs up with the odd electron on the benzoyloxy initiator to form an O–C bond, and the other electron remains on carbon:

$$BzO\cdot \quad H_2C{=}CH_2 \quad \longrightarrow \quad BzO{-}CH_2CH_2\cdot$$

STEP 2 **Propagation** The alkyl radical produced in the second initiation step adds to another ethylene molecule to yield another radical, and repetition of this radical addition step for hundreds or thousands of times builds the polymer chain.

$$BzOCH_2CH_2\cdot \quad H_2C{=}CH_2 \quad \longrightarrow \quad BzOCH_2CH_2CH_2CH_2\cdot \quad \xrightarrow[\text{many times}]{\text{Repeat}} \quad BzO(CH_2CH_2)_nCH_2CH_2\cdot$$

STEP 3 **Termination** The chain process is eventually ended by a reaction that consumes the radical. Combination of two growing chains is one possible chain-terminating reaction:

$$2\ R\cdot \longrightarrow R{-}R$$

Many substituted ethylenes also undergo radical chain polymerization, yielding polymers with substituent groups regularly spaced at alternating carbon atoms along the chain. Propylene, for example, yields polypropylene (although a different method of polymerization is used in practice), and styrene yields polystyrene (p. 258).

$$H_2C=CHCH_3 \longrightarrow \left(\begin{array}{c} CH_3 \quad CH_3 \quad CH_3 \quad CH_3 \\ | \qquad | \qquad | \qquad | \\ CH_2CHCH_2CHCH_2CHCH_2CH \end{array} \right)$$

Propylene   **Polypropylene**

$$H_2C=CH \longrightarrow \left(CH_2CHCH_2CHCH_2CHCH_2CH \right)$$

Styrene

Polystyrene

When an unsymmetrically substituted alkene monomer such as propylene or styrene is polymerized, the radical addition steps can take place at either end of the double bond to yield either a primary radical intermediate ($RCH_2\cdot$) or a secondary radical ($R_2CH\cdot$). Just as in electrophilic addition reactions, however, we find that only the more highly substituted, secondary radical is formed.

$$BzO\cdot \quad H_2C=CHCH_3 \longrightarrow \begin{array}{c} CH_3 \\ | \\ BzO-CH_2-CH\cdot \end{array} \qquad \left[\begin{array}{c} CH_3 \\ | \\ BzO-CH-CH_2\cdot \end{array} \right]$$

Secondary radical   **Primary radical**
                              (***NOT formed***)

Table 7.1 shows some of the more important alkene polymers, their uses, and the monomer units from which they are made.

Problem 7.17   Show the monomer units you would use to prepare the following polymers:

$$\text{(a)} \quad \left(\begin{array}{c} OCH_3 \qquad OCH_3 \qquad OCH_3 \\ | \qquad\quad | \qquad\quad | \\ CH_2-CH-CH_2-CH-CH_2-CH \end{array} \right)$$

$$\text{(b)} \quad \left(\begin{array}{c} Cl \quad Cl \quad Cl \quad Cl \quad Cl \quad Cl \\ | \quad\, | \quad\, | \quad\, | \quad\, | \quad\, | \\ CH-CH-CH-CH-CH-CH \end{array} \right)$$

Chain Branching During Polymerization

The polymerization of an alkene monomer is complicated in practice by several problems that greatly affect the properties of the product. One such problem is that radical polymerization yields a product that is not linear but has numerous *branches* in it. Branches arise when the radical end of a growing

TABLE 7.1 Some Alkene Polymers and Their Uses

Monomer name	Formula	Trade or common name of polymer	Uses
Ethylene	$H_2C{=}CH_2$	Polyethylene	Packaging, bottles, cable insulation, films and sheets
Propene (propylene)	$H_2C{=}CHCH_3$	Polypropylene	Automotive moldings, rope, carpet fibers
Chloroethylene (vinyl chloride)	$H_2C{=}CHCl$	Poly(vinyl chloride), Tedlar	Insulation, films, pipes
Styrene	$H_2C{=}CHC_6H_5$	Polystyrene, Styron	Foam and molded articles
Tetrafluoroethylene	$F_2C{=}CF_2$	Teflon	Valves and gaskets, coatings
Acrylonitrile	$H_2C{=}CHCN$	Orlon, Acrilan	Fibers
Methyl methacrylate	$H_2C{=}\overset{\displaystyle CH_3}{\underset{\displaystyle }{C}}CO_2CH_3$	Plexiglas, Lucite	Molded articles, paints
Vinyl acetate	$H_2C{=}CHOCOCH_3$	Poly(vinyl acetate)	Paints, adhesives

chain abstracts a hydrogen atom from the middle of the chain to yield an internal radical site that continues the polymerization. The most common kind of branching, termed *short-chain branching,* arises from intramolecular hydrogen atom abstraction from a position four carbon atoms away from the chain end.

Alternatively, intermolecular hydrogen atom abstraction can take place by reaction of the radical end of one chain with the middle of another chain. *Long-chain branching* results from this kind of reaction.

Chain branching is a common occurrence during radical polymerizations and is not restricted to polyethylene. Polypropylene, polystyrene, and poly(methyl methacrylate) all contain branched chains. Studies have shown that short-chain branching occurs about 50 times as often as long-chain branching.

● ●

Problem 7.18   One of the chain-termination steps that sometimes occurs to interrupt polymerization is the following reaction between two radicals:

$$2 \ \xi\text{CH}_2\dot{\text{C}}\text{H}_2 \ \longrightarrow \ \xi\text{CH}_2\text{CH}_3 + \xi\text{CH}{=}\text{CH}_2$$

Propose a mechanism for this reaction, using fishhook arrows to indicate electron flow.

● ●

Cationic Polymerization

Some alkene monomers can be polymerized by a *cationic* initiator, as well as by a radical initiator. Cationic polymerization occurs by a chain-reaction pathway and requires the use of a strong protic or Lewis acid catalyst. The chain-carrying step is the electrophilic addition of a carbocation intermediate to the carbon–carbon double bond of another monomer unit. Not surprisingly, cationic polymerization is most effective when a stable, tertiary carbocation intermediate is involved. Thus, the most common commercial use of cationic polymerization is for the preparation of polyisobutylene by treatment of isobutylene (2-methylpropene) with BF_3 catalyst at $-80°C$. The product is used in the manufacture of inner tubes for truck and bicycle tires.

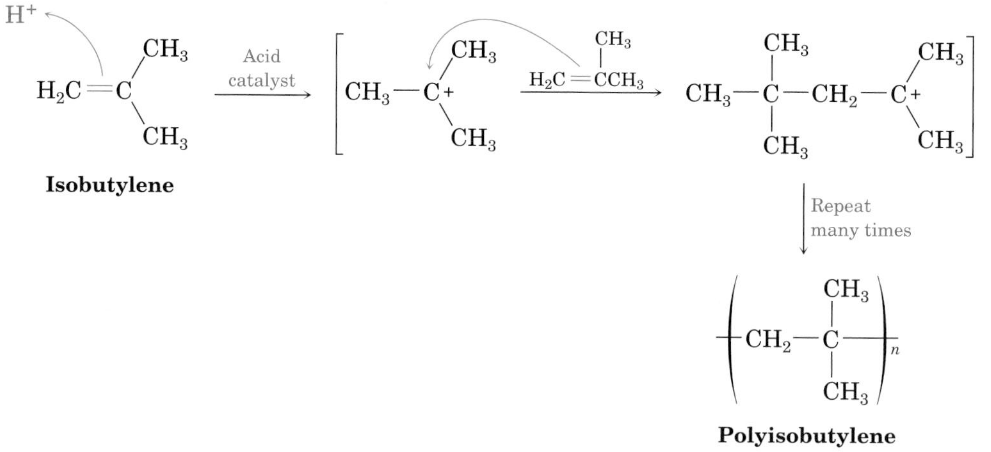

Isobutylene

Polyisobutylene

● ●

Problem 7.19   *tert*-Butyl vinyl ether is polymerized commercially for use in adhesives by a cationic process. Draw a segment of poly(*tert*-butyl vinyl ether), and show the mechanism of the chain-carrying step.

$$tert\text{-Butyl vinyl ether}$$

tert-Butyl vinyl ether

Natural Rubber

Rubber—an unusual name for a most unusual substance—is a naturally occurring alkene polymer produced by more than 400 different plants. The major source is the so-called rubber tree, *Hevea brasiliensis*, from which the crude material is harvested as it drips from a slice made through the bark. The name *rubber* was coined by Joseph Priestley, the discoverer of oxygen and early researcher of rubber chemistry, for the simple reason that one of rubber's early uses was to rub out pencil marks on paper.

Unlike polyethylene and other simple alkene polymers, natural rubber is a polymer of a *diene,* isoprene (2-methyl-1,3-butadiene). The polymerization takes place by addition of each isoprene monomer unit to the growing chain, leading to formation of a polymer that still contains double bonds spaced regularly at four-carbon intervals. As the following structure shows, these double bonds have *Z* stereochemistry:

Many isoprene units   **A segment of natural rubber**

Crude rubber, called latex, is collected from the tree as an aqueous dispersion that is washed, dried, and coagulated by warming in air. The resultant polymer has chains that average about 5000 monomer units in length and have molecular weights of 200,000–500,000. This crude coagulate is too soft and tacky to be useful until it is hardened by heating with elemental sulfur, a process called *vulcanization.* By mechanisms

(continued) ▶

that are still not fully understood, vulcanization introduces cross-links between the rubber chains by forming carbon–sulfur bonds between them, thereby hardening and stiffening the polymer. The exact degree of hardening can be varied, yielding material soft enough for automobile tires or hard enough for bowling balls (*ebonite*).

The remarkable ability of rubber to stretch and then contract to its original shape is due to the irregular shapes of the polymer chains caused by the double bonds. These double bonds introduce bends and kinks into the polymer chains, thereby preventing neighboring chains from nestling together. When stretched, the randomly coiled chains straighten out and orient along the direction of the pull but are kept from sliding over one another by the cross-links. When the stretch is released, the polymer reverts to its original random state.

Natural rubber is obtained from the bark of the rubber tree, *Hevea brasiliensis,* **grown on enormous plantations in Southeast Asia.**

Summary and Key Words

KEY WORDS

anti stereochemistry, 234
bromohydrin, 237
bromonium ion (R_2Br^+), 234
carbene (R_2C), 247
chloronium ion (R_2Cl^+), 235
dehydration, 233
dehydrohalogenation, 233
diol, 252
glycol, 252
halohydrin, 237
hydration, 239
hydroboration, 242
hydrogenation, 249
hydroxylation, 252
initiation steps, 257
molozonide, 253
monomer, 256
oxymercuration, 239

Methods for the preparation of alkenes generally involve *elimination reactions,* such as **dehydrohalogenation**, the elimination of HX from an alkyl halide, and **dehydration**, the elimination of water from an alcohol.

HCl, HBr, and HI add to alkenes by a two-step electrophilic addition mechanism. Initial reaction of the nucleophilic double bond with H^+ gives a carbocation intermediate, which then reacts with halide ion. Bromine and chlorine add to alkenes via three-membered-ring **bromonium ion** or **chloronium ion** intermediates to give addition products having **anti stereochemistry**. If water is present during halogen addition reactions, a **halohydrin** is formed.

Hydration of an alkene—the addition of water—is carried out by either of two procedures, depending on the product desired. **Oxymercuration** involves electrophilic addition of Hg^{2+} to an alkene, followed by trapping of the cation intermediate with water and subsequent treatment with $NaBH_4$. **Hydroboration** involves addition of borane (BH_3) followed by oxidation of the intermediate organoborane with alkaline H_2O_2. The two hydration methods are complementary: Oxymercuration gives the product of Markovnikov addition, whereas hydroboration/oxidation gives the product of non-Markovnikov **syn addition**.

A **carbene, R_2C:**, is a neutral molecule containing a divalent carbon with only six valence electrons. Carbenes are highly reactive toward alkenes, adding to give cyclopropanes. Dichlorocarbene adds to alkenes to give 1,1-dichlorocyclopropanes. Nonhalogenated cyclopropanes are best prepared by treatment of the alkene with CH_2I_2 and zinc–copper alloy—**the Simmons–Smith reaction**.

Alkenes are reduced by addition of H_2 in the presence of a catalyst such as platinum or palladium to yield alkanes, a process called **catalytic hydrogenation**. Cis-1,2-diols can be made directly from alkenes by **hydroxylation** with OsO_4. Alkenes can also be cleaved to produce carbonyl compounds by reaction with ozone, followed by reduction with zinc metal.

Alkene **polymers**—large molecules resulting from repetitive bonding together of many hundreds or thousands of small **monomer** units—are formed by reaction of simple alkenes with a radical initiator at high temperature and pressure. Polyethylene, polypropylene, and polystyrene are common examples.

Learning Reactions

What's seven times nine? Sixty-three, of course. You didn't have to stop and figure it out; you knew the answer immediately because you long ago learned the multiplication tables. Learning the reactions of organic chemistry requires the same approach: Reactions have to be learned for immediate recall if they are to be useful.

Different people take different approaches to learning reactions. Some people make flashcards; others find studying with friends to be helpful. To help guide your study, most chapters in this book end with a summary of the reactions just presented. In addition, the accompanying *Study Guide and Solutions Manual* has several appendixes that organize organic reactions from other viewpoints. Fundamentally, though, there are no shortcuts. Learning organic chemistry takes effort.

Summary of Reactions

Note: No stereochemistry is implied unless specifically indicated with wedged, solid, and dashed lines.

1. Synthesis of alkenes
   (a) Dehydrohalogenation of alkyl halides (Section 7.1)

   (b) Dehydration of alcohols (Section 7.1)

(continued) ▶

2. Addition reactions of alkenes

(a) Addition of HX, where X = Cl, Br, or I (Sections 6.8 and 6.9)

Markovnikov regiochemistry is observed: H adds to the less highly substituted carbon, and X adds to the more highly substituted carbon.

(b) Addition of halogens, where $X_2 = Cl_2$ or Br_2 (Section 7.2)

Anti addition is observed.

(c) Halohydrin formation (Section 7.3)

Markovnikov regiochemistry and anti stereochemistry are observed.

(d) Addition of water by oxymercuration (Section 7.4)

Markovnikov regiochemistry is observed, with the –OH attaching to the more highly substituted carbon.

(e) Addition of water by hydroboration/oxidation (Section 7.5)

Non-Markovnikov syn addition is observed.

(f) Hydrogenation of alkenes (Section 7.7)

Syn addition is observed.

(continued) ▶

(g) Hydroxylation of alkenes (Section 7.8)

$$>C=C< \xrightarrow[\text{2. NaHSO}_3, \text{H}_2\text{O}]{\text{1. OsO}_4} \quad \begin{array}{c} \text{HO} \quad \text{OH} \\ -\text{C}-\text{C}- \end{array}$$

Syn addition is observed.

(h) Addition of carbenes to alkenes to yield cyclopropanes (Section 7.6)

(1) Dichlorocarbene addition

$$>C=C< + \text{CHCl}_3 \xrightarrow{\text{KOH}} \begin{array}{c} \text{Cl} \quad \text{Cl} \\ \text{C} \\ -\text{C}-\text{C}- \end{array}$$

(2) Simmons–Smith reaction

$$>C=C< + \text{CH}_2\text{I}_2 \xrightarrow[\text{Ether}]{\text{Zn(Cu)}} \begin{array}{c} \text{H} \quad \text{H} \\ \text{C} \\ -\text{C}-\text{C}- \end{array}$$

3. Oxidative cleavage of alkenes (Section 7.8)

(a) Treatment with ozone, followed by zinc in acetic acid

$$>C=C< \xrightarrow[\text{2. Zn/H}_3\text{O}^+]{\text{1. O}_3} \quad >C=O + O=C<$$

(b) Reaction with $KMnO_4$ in acidic solution

$$\begin{array}{c} \text{R} \quad\quad \text{R} \\ C=C \\ \text{R} \quad\quad \text{R} \end{array} \xrightarrow{\text{KMnO}_4, \text{H}_3\text{O}^+} \begin{array}{c} \text{R} \quad\quad\quad \text{R} \\ C=O + O=C \\ \text{R} \quad\quad\quad \text{R} \end{array}$$

$$\begin{array}{c} \text{H} \quad\quad \text{H} \\ C=C \\ \text{R} \quad\quad \text{H} \end{array} \xrightarrow{\text{KMnO}_4, \text{H}_3\text{O}^+} \begin{array}{c} \text{O} \\ \parallel \\ \text{R}-\text{C}-\text{OH} \end{array} + \text{CO}_2$$

4. Oxidative cleavage of 1,2-diols (Section 7.8)

$$\begin{array}{c} \text{HO} \quad\quad \text{OH} \\ C-C \end{array} \xrightarrow[\text{H}_2\text{O}]{\text{HIO}_4} \quad >C=O + O=C<$$

Visualizing Chemistry

• •

(Problems 7.1–7.19 appear within the chapter.)

7.20 Name the following alkenes, and predict the products of their reaction with (i) $KMnO_4$ in aqueous acid and (ii) O_3, followed by Zn in acetic acid:

(a)   (b)

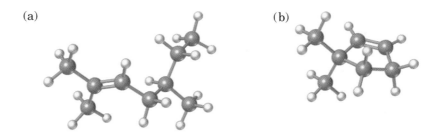

7.21 Draw the structures of alkenes that would yield the following alcohols on hydration (red = O). Tell in each case whether you would use hydroboration/oxidation or oxymercuration.

(a)   (b)

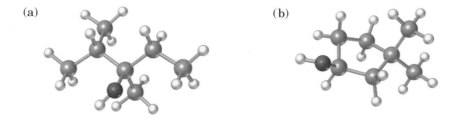

7.22 The following alkene undergoes hydroboration/oxidation to yield a single product rather than a mixture. Explain the result, and draw the product showing its stereochemistry.

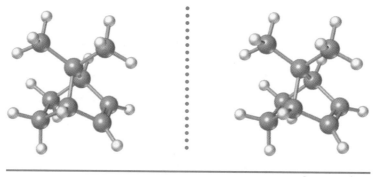

Stereo View

Additional Problems

• •

7.23 Predict the products of the following reactions (the aromatic ring is unreactive in all cases). Indicate regiochemistry when relevant.

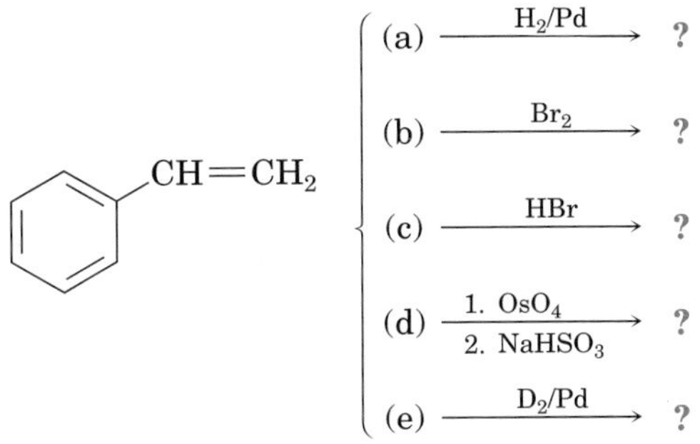

(a) $\xrightarrow{\text{H}_2/\text{Pd}}$?

(b) $\xrightarrow{\text{Br}_2}$?

(c) $\xrightarrow{\text{HBr}}$?

(d) $\xrightarrow[\text{2. NaHSO}_3]{\text{1. OsO}_4}$?

(e) $\xrightarrow{\text{D}_2/\text{Pd}}$?

7.24 Suggest structures for alkenes that give the following reaction products. There may be more than one answer for some cases.

(a) ? $\xrightarrow{\text{H}_2/\text{Pd}}$ 2-Methylhexane

(b) ? $\xrightarrow{\text{H}_2/\text{Pd}}$ 1,1-Dimethylcyclohexane

(c) ? $\xrightarrow{\text{Br}_2/\text{CH}_2\text{Cl}_2}$ 2,3-Dibromo-5-methylhexane

(d) ? $\xrightarrow[\text{2. NaBH}_4]{\text{1. Hg(OAc)}_2,\ \text{H}_2\text{O}}$ $\text{CH}_3\text{CH}_2\text{CH}_2\text{CH(OH)CH}_3$

(e) ? $\xrightarrow{\text{HCl, ether}}$ 2-Chloro-3-methylheptane

7.25 Predict the products of the following reactions, indicating both regiochemistry and stereochemistry where appropriate:

(a) 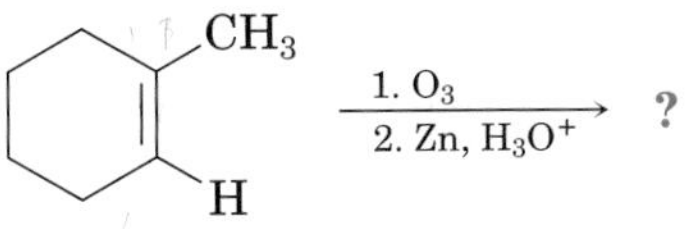 $\xrightarrow[\text{2. Zn, H}_3\text{O}^+]{\text{1. O}_3}$?   (b) 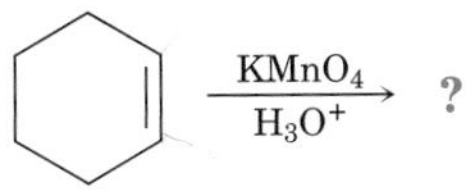 $\xrightarrow[\text{H}_3\text{O}^+]{\text{KMnO}_4}$?

(c) 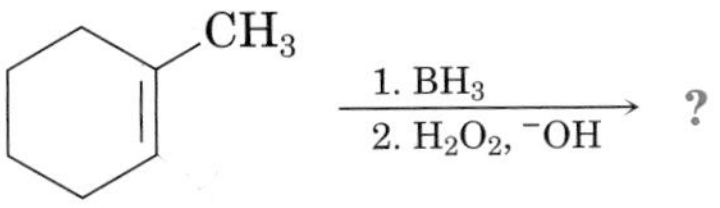 $\xrightarrow[\text{2. H}_2\text{O}_2,\ ^-\text{OH}]{\text{1. BH}_3}$?   (d) 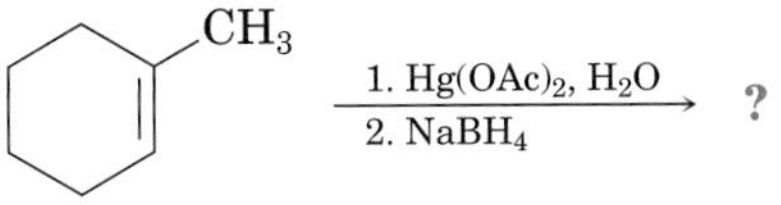 $\xrightarrow[\text{2. NaBH}_4]{\text{1. Hg(OAc)}_2,\ \text{H}_2\text{O}}$?

7.26 How would you carry out the following transformations? Indicate the reagents you would use in each case.

(a) 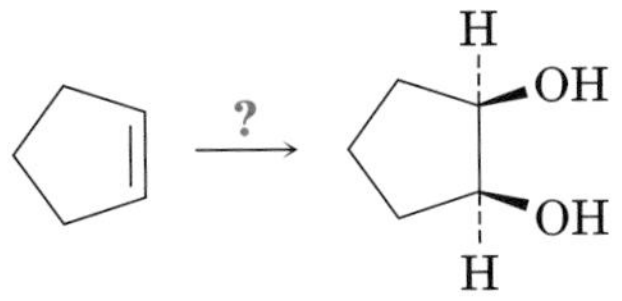 $\xrightarrow{?}$ (b) 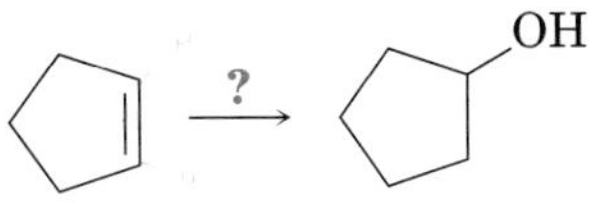 $\xrightarrow{?}$

(c) (d)

CH₃
(e) $CH_3CH=CHCHCH_3$ $\xrightarrow{?}$ $CH_3CH + CH_3CHCH$

CH₃
(f) $CH_3C=CH_2$ $\xrightarrow{?}$ CH_3CHCH_2OH

7.27 Draw the structure of an alkene that yields only acetone, $(CH_3)_2C=O$, on ozonolysis followed by treatment with Zn.

7.28 Draw the structure of a hydrocarbon that reacts with 1 molar equivalent of H_2 on catalytic hydrogenation and gives only pentanal, $CH_3CH_2CH_2CH_2CHO$, on ozonolysis followed by treatment with Zn. Write the reactions involved.

7.29 Show the structures of alkenes that give the following products on oxidative cleavage with $KMnO_4$ in acidic solution:

(a) $CH_3CH_2COOH + CO_2$ (b) $(CH_3)_2C=O + CH_3CH_2CH_2COOH$

(c) $=O + (CH_3)_2C=O$

7.30 Compound A has the formula $C_{10}H_{16}$. On catalytic hydrogenation over palladium, it reacts with only 1 molar equivalent of H_2. Compound A also undergoes reaction with ozone, followed by zinc treatment, to yield a symmetrical diketone, B ($C_{10}H_{16}O_2$).
(a) How many rings does A have?   (b) What are the structures of A and B?
(c) Write the reactions.

7.31 An unknown hydrocarbon A, with formula C_6H_{12}, reacts with 1 molar equivalent of H_2 over a palladium catalyst. Hydrocarbon A also reacts with OsO_4 to give a diol, B. When oxidized with $KMnO_4$ in acidic solution, A gives two fragments. One fragment is propanoic acid, CH_3CH_2COOH, and the other fragment is a ketone, C. What are the structures of A, B, and C? Write all reactions, and show your reasoning.

7.32 Using an oxidative cleavage reaction, explain how you would distinguish between the following two isomeric dienes:

and

7.33 Compound A, $C_{10}H_{18}O$, undergoes reaction with dilute H_2SO_4 at 250°C to yield a mixture of two alkenes, $C_{10}H_{16}$. The major alkene product, B, gives only cyclopentanone after ozone treatment followed by reduction with zinc in acetic acid. Identify A and B, and write the reactions.

7.34 Which reaction would you expect to be faster, addition of HBr to cyclohexene or to 1-methylcyclohexene? Explain.

7.35 Predict the products of the following reactions, and indicate regiochemistry if relevant:

(a) $CH_3CH{=}CHCH_3 \xrightarrow{\text{HBr}}$?

(b) $CH_3CH{=}CHCH_3 \xrightarrow{\text{BH}_3}$ A? $\xrightarrow[^-\text{OH}]{\text{H}_2\text{O}_2}$ B?

(c) $cis\text{-}CH_3CH{=}CHCH_3 \xrightarrow{\text{CH}_2\text{I}_2,\ \text{Zn–Cu}}$?

(d) $trans\text{-}CH_3CH{=}CHCH_3 \xrightarrow{\text{CH}_2\text{I}_2,\ \text{Zn–Cu}}$?

7.36 Iodine azide, IN_3, adds to alkenes by an electrophilic mechanism similar to that of bromine. If a monosubstituted alkene such as 1-butene is used, only one product results:

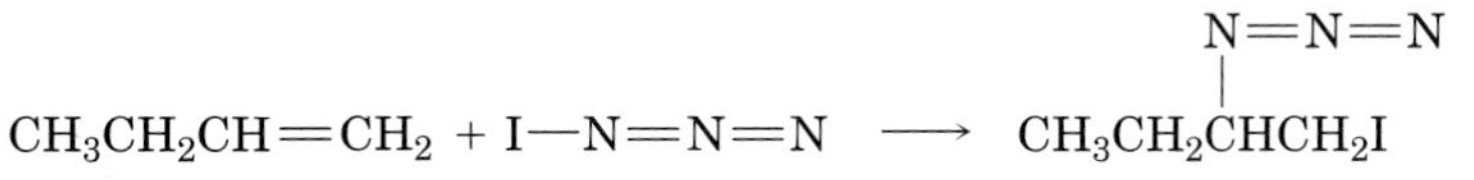

$$CH_3CH_2CH{=}CH_2 + I{-}N{=}N{=}N \longrightarrow CH_3CH_2\overset{\overset{\displaystyle N=N=N}{\displaystyle |}}{C}HCH_2I$$

(a) Add lone-pair electrons to the structure shown for IN_3, and draw a second resonance form for the molecule.
(b) Calculate formal charges for the atoms in both resonance structures you drew for IN_3 in part (a).
(c) In light of the result observed when IN_3 adds to 1-butene, what is the polarity of the $I{-}N_3$ bond? Propose a mechanism for the reaction using curved arrows to show the electron flow in each step.

7.37 Draw the structure of a hydrocarbon that absorbs 2 molar equivalents of H_2 on catalytic hydrogenation and gives only butanedial on ozonolysis.

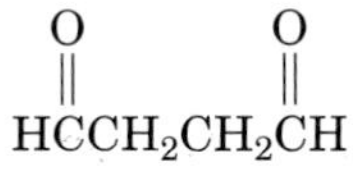

$$\overset{\overset{\displaystyle O}{\displaystyle ||}}{H}CCH_2CH_2\overset{\overset{\displaystyle O}{\displaystyle ||}}{C}H \qquad \textbf{Butanedial}$$

7.38 Simmons–Smith reaction of cyclohexene with diiodomethane gives a single cyclopropane product, but the analogous reaction of cyclohexene with 1,1-diiodoethane gives (in low yield) a mixture of two isomeric methylcyclopropane products. What are the two products, and how do they differ?

7.39 In planning the synthesis of one compound from another, it's just as important to know what *not* to do as to know what to do. The following reactions all have serious drawbacks to them. Explain the potential problems of each.

(a) 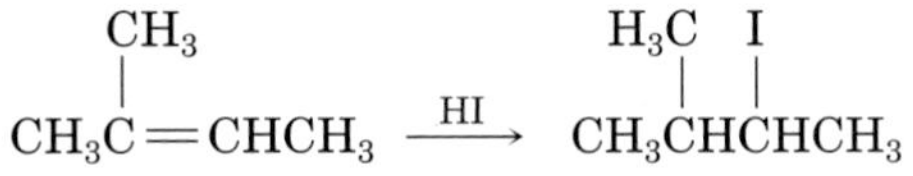 $CH_3\overset{\overset{\displaystyle CH_3}{\displaystyle |}}{C}{=}CHCH_3 \xrightarrow{\text{HI}} CH_3\overset{\overset{\displaystyle H_3C}{\displaystyle |}}{C}H\overset{\overset{\displaystyle I}{\displaystyle |}}{C}HCH_3$

(b) $\xrightarrow[\text{2. NaHSO}_3]{\text{1. OsO}_4}$

(c) $\xrightarrow[\text{2. Zn}]{\text{1. O}_3}$

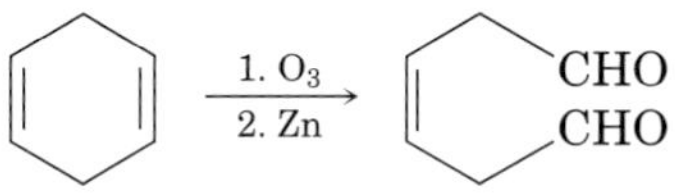

(d) 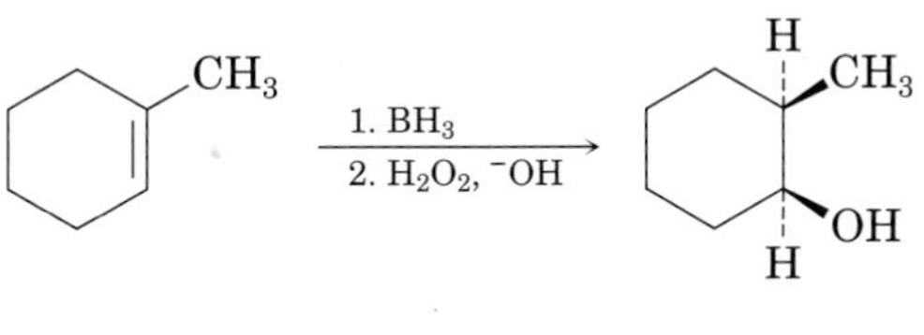 $\xrightarrow[\text{2. H}_2\text{O}_2,\ ^-\text{OH}]{\text{1. BH}_3}$

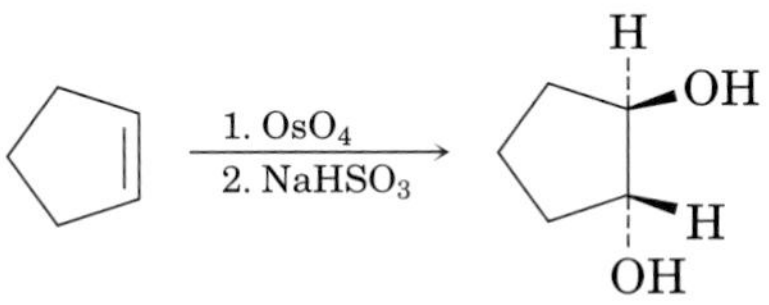

7.40 Which of the following alcohols could *not* be made selectively by hydroboration/oxidation of an alkene? Explain.

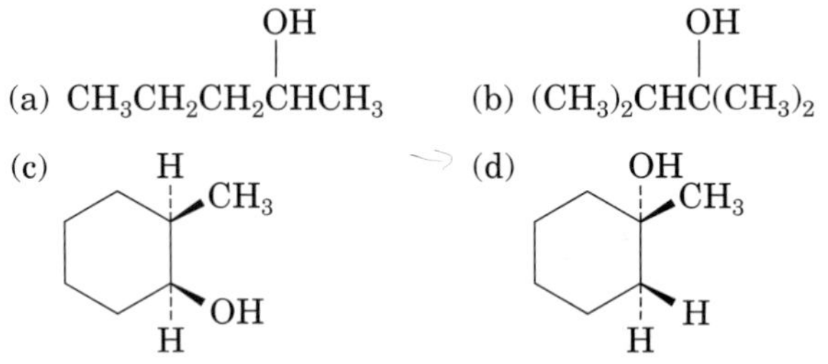

(a) $CH_3CH_2CH_2\overset{\overset{\displaystyle OH}{|}}{C}HCH_3$ (b) $(CH_3)_2CHC(CH_3)_2$ with OH

(c)   (d)

7.41 What alkenes might be used to prepare the following cyclopropanes?

(a) ▷—$CH(CH_3)_2$ (b)

7.42 Predict the products of the following reactions. Don't worry about the size of the molecule; concentrate on the functional groups.

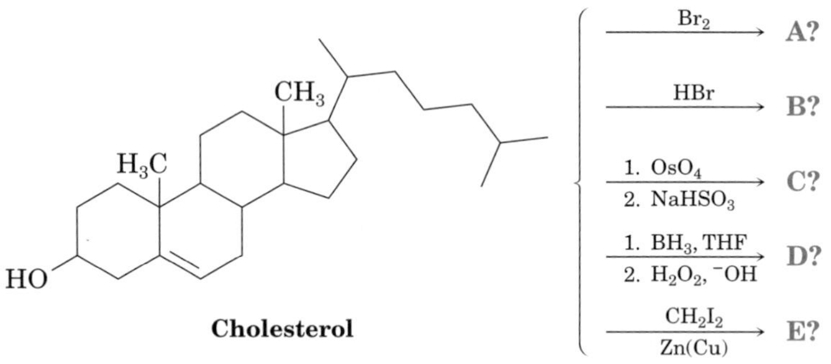

$$\xrightarrow{Br_2} \text{ A?}$$

$$\xrightarrow{HBr} \text{ B?}$$

$$\xrightarrow[\text{2. NaHSO}_3]{\text{1. OsO}_4} \text{ C?}$$

$$\xrightarrow[\text{2. H}_2O_2, \ ^-OH]{\text{1. BH}_3, \text{THF}} \text{ D?}$$

$$\xrightarrow[\text{Zn(Cu)}]{CH_2I_2} \text{ E?}$$

Cholesterol

7.43 The sex attractant of the common housefly is a hydrocarbon with the formula $C_{23}H_{46}$. On treatment with aqueous acidic $KMnO_4$, two products are obtained, $CH_3(CH_2)_{12}COOH$ and $CH_3(CH_2)_7COOH$. Propose a structure.

7.44 Compound A has the formula C_8H_8. It reacts rapidly with $KMnO_4$ to give CO_2 and a carboxylic acid, B ($C_7H_6O_2$), but reacts with only 1 molar equivalent of H_2 on catalytic hydrogenation over a palladium catalyst. On hydrogenation under conditions that reduce aromatic rings, 4 equivalents of H_2 are taken up, and hydrocarbon C (C_8H_{16}) is produced. What are the structures of A, B, and C? Write the reactions.

7.45 Plexiglas, a clear plastic used to make many molded articles, is made by polymerization of methyl methacrylate. Draw a representative segment of Plexiglas.

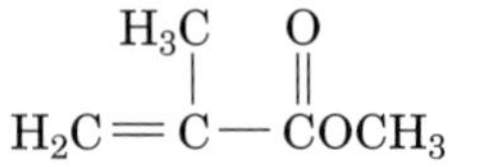

$$H_2C \!=\! \overset{\overset{\displaystyle H_3C}{|}}{C} \!-\! \overset{\overset{\displaystyle O}{||}}{C}OCH_3 \quad \textbf{Methyl methacrylate}$$

7.46 Draw representative segments of polymers made from the following monomers:
(a) Teflon, from $F_2C=CF_2$ (b) Poly(vinyl chloride), from $H_2C=CHCl$

7.47 Reaction of 2-methylpropene with CH_3OH in the presence of H_2SO_4 catalyst yields methyl *tert*-butyl ether, $CH_3OC(CH_3)_3$, by a mechanism analogous to that of acid-catalyzed alkene hydration. Write the mechanism, using curved arrows for each step.

7.48 When 4-penten-1-ol is treated with aqueous Br_2, a cyclic bromo ether is formed, rather than the expected bromohydrin. Propose a mechanism, using curved arrows to show electron movement.

$$H_2C=CHCH_2CH_2CH_2OH \xrightarrow{\ Br_2,\ H_2O\ }$$

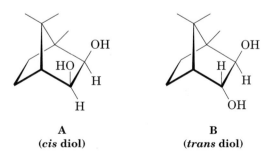

4-Penten-1-ol   **2-(Bromomethyl)tetrahydrofuran**

7.49 How would you distinguish between the following pairs of compounds using simple chemical tests? Tell what you would do and what you would see.
(a) Cyclopentene and cyclopentane   (b) 2-Hexene and benzene

7.50 Dichlorocarbene can be generated by heating sodium trichloroacetate:

$$Cl-\underset{\underset{Cl}{|}}{\overset{\overset{Cl}{|}}{C}}-\overset{\overset{O}{\|}}{C}-O^- Na^+ \xrightarrow{70°C} \underset{Cl}{\overset{Cl}{}}C: + CO_2 + NaCl$$

Propose a mechanism for the reaction, and use curved arrows to indicate the movement of electrons in each step. What relationship does your mechanism bear to the base-induced elimination of HCl from chloroform?

7.51 α-Terpinene, $C_{10}H_{16}$, is a pleasant-smelling hydrocarbon that has been isolated from oil of marjoram. On hydrogenation over a palladium catalyst, α-terpinene reacts with 2 molar equivalents of H_2 to yield a hydrocarbon, $C_{10}H_{20}$. On ozonolysis, followed by reduction with zinc and acetic acid, α-terpinene yields two products, glyoxal and 6-methyl-2,5-heptanedione.

$$\underset{\underset{H}{}\quad\underset{H}{}}{\overset{\overset{O}{\|}\quad\overset{O}{\|}}{C-C}}$$

Glyoxal

$$\underset{\underset{CH_3}{|}}{CH_3\overset{\overset{O}{\|}}{C}CH_2CH_2\overset{\overset{O}{\|}}{C}CHCH_3}$$

6-Methyl-2,5-heptanedione

(a) How many degrees of unsaturation does α-terpinene have?
(b) How many double bonds and how many rings does it have?
(c) Propose a structure for α-terpinene.

7.52 Evidence that cleavage of 1,2-diols by HIO_4 occurs through a five-membered cyclic periodate intermediate is based on *kinetic data*—the measurement of reaction rates. When diols A and B were prepared and the rates of their reaction with HIO_4 were measured, it was found that diol A cleaved approximately 1 million times faster than diol B. Make molecular models of A and B and of potential cyclic periodate intermediates, and then explain the kinetic results.

A
(*cis* diol)   **B**
(*trans* diol)

7.53 Reaction of HBr with 3-methylcyclohexene yields a mixture of four products: *cis-* and *trans*-1-bromo-3-methylcyclohexane and *cis-* and *trans*-1-bromo-2-methylcyclohexane. The analogous reaction of HBr with 3-bromocyclohexene yields only *trans*-1,2-dibromocyclohexane as the sole product. Draw structures of the possible intermediates, and then explain why only a single product is formed in the reaction of HBr with 3-bromocyclohexene.

cis, trans   **cis, trans**

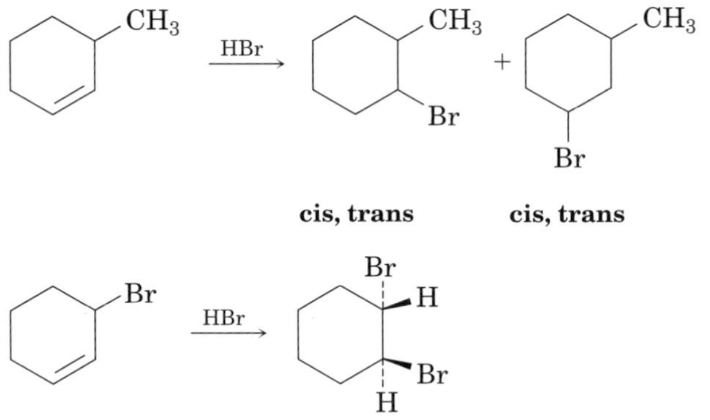

7.54 The following reaction takes place in high yield:

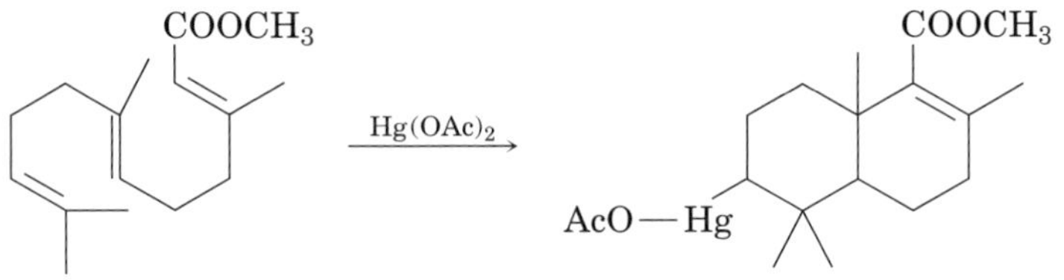

Use your general knowledge of alkene chemistry to propose a mechanism, even though you've never seen this reaction before.

7.55 Hydroboration of 2-methyl-2-pentene at 25°C followed by oxidation with alkaline H_2O_2 yields 2-methyl-3-pentanol, but hydroboration at 160°C followed by oxidation yields 4-methyl-1-pentanol. Explain.

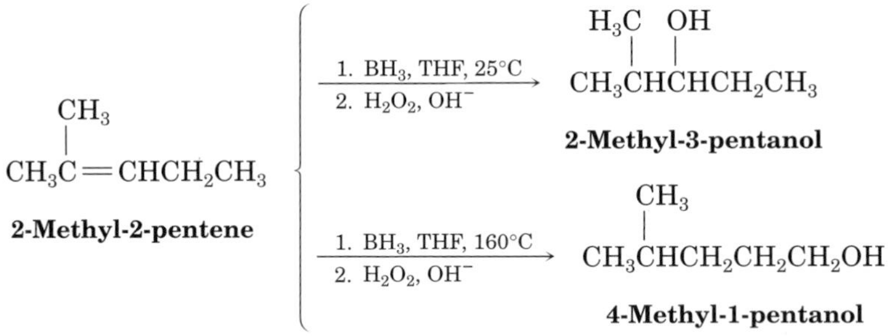

A Look Ahead

7.56 Alkynes undergo many of the same reactions that alkenes do. What product would you expect from each of the following reactions? (See Sections 8.4–8.6.)

$$CH_3$$
$$|$$
$$CH_3CHCH_2CH_2C\equiv CH$$

(a) $\xrightarrow{\text{1 equiv Br}_2}$?

(b) $\xrightarrow{\text{2 equiv H}_2\text{, Pd/C}}$?

(c) $\xrightarrow{\text{1 equiv HBr}}$?

7.57 Explain the observation that hydroxylation of *cis*-2-butene with OsO_4 yields a different product than hydroxylation of *trans*-2-butene. First draw the structure and show the stereochemistry of each product, and then make molecular models. (See Sections 9.15 and 9.16.)

7.58 Reaction of cyclohexene with mercury(II) acetate in CH_3OH rather than H_2O, followed by treatment with $NaBH_4$, yields cyclohexyl methyl ether rather than cyclohexanol. Suggest a mechanism. (See Section 18.4.)

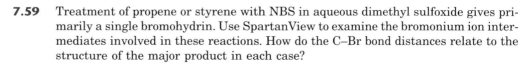

$\xrightarrow[\text{2. NaBH}_4]{\text{1. Hg(OAc)}_2\text{, CH}_3\text{OH}}$

Cyclohexene   **Cyclohexyl methyl ether**

Molecular Modeling

● ●

7.59 Treatment of propene or styrene with NBS in aqueous dimethyl sulfoxide gives primarily a single bromohydrin. Use SpartanView to examine the bromonium ion intermediates involved in these reactions. How do the C–Br bond distances relate to the structure of the major product in each case?

7.60 Use SpartanView to examine the electrostatic potential map of dichlorocarbene, $:CCl_2$. Identify the electrophilic sites on carbon, and go through the sequence of structures showing addition of dichlorocarbene to propene. Is the initial approach of the carbene consistent with the carbene acting as a nucleophile or electrophile? Explain.

7.61 Addition of BH_3 to 1-methylcyclopentene can occur in two orientations. Use SpartanView to examine the two transition states. Does the lower-energy transition state lead to the observed product? Compare distances of the partial B–H and C–H bonds in the lower-energy transition state with other B–H and C–H bonds, and then tell whether the lower-energy transition state is more like the reactant or the product.

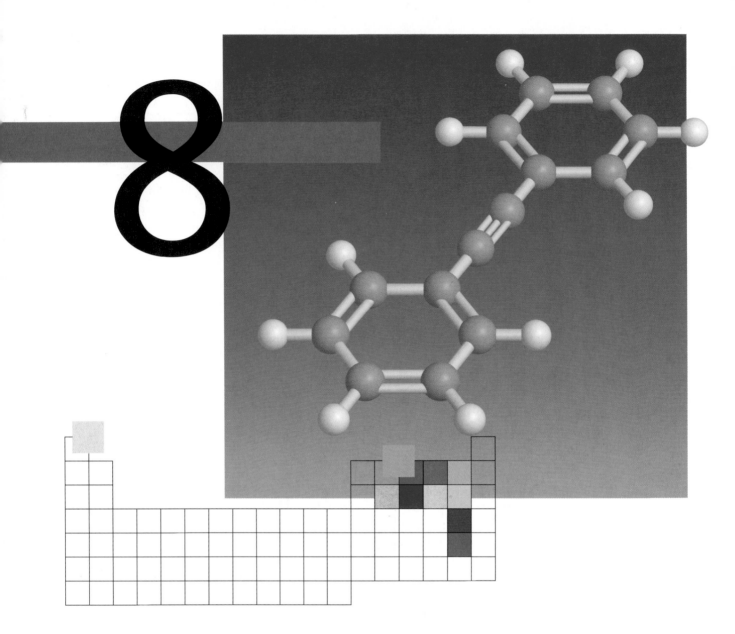

Alkynes:
An Introduction to
Organic Synthesis

Alkynes are hydrocarbons that contain a carbon–carbon triple bond. Acetylene, H–C≡C–H, the simplest alkyne, was once widely used in industry as the starting material for the preparation of acetaldehyde, acetic acid, vinyl chloride, and other high-volume chemicals, but more efficient routes to these substances using ethylene as starting material are now available. Acetylene is still used in the preparation of acrylic polymers, however, and is prepared industrially by high-temperature decomposition (*pyrolysis*) of methane. This method is of no use in the laboratory, however.

$$2\ CH_4 \xrightarrow[1200°C]{Steam} HC\equiv CH + 3\ H_2$$

Methane   **Acetylene**

8.1 Electronic Structure of Alkynes

A carbon–carbon triple bond results from the interaction of two *sp*-hybridized carbon atoms (Section 1.10). Recall that the two *sp* hybrid orbitals of carbon lie at an angle of 180° to each other along an axis perpendicular to the axes of the two unhybridized $2p_y$ and $2p_z$ orbitals. When two *sp*-hybridized carbons approach each other, one *sp–sp* σ bond and two *p–p* π bonds are formed. The two remaining *sp* orbitals form bonds to other atoms at an angle of 180° from the carbon–carbon bond. Thus, acetylene, C_2H_2, is a linear molecule with H–C≡C bond angles of 180° (Figure 8.1).

FIGURE 8.1 ▼

The structure of acetylene, H–C≡C–H. The H–C≡C bond angles are 180°, and the C≡C bond length is 120 pm. The electrostatic potential map shows that the π bonds create a negative belt around the molecule.

acetylene

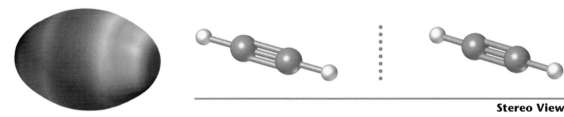

Stereo View

The length of the carbon–carbon triple bond in acetylene is 120 pm, and its strength is approximately 835 kJ/mol (200 kcal/mol), making it the shortest and strongest known carbon–carbon bond. Experiments show that approximately 318 kJ/mol (76 kcal/mol) is needed to break a π bond in acetylene, a value some 50 kJ/mol larger than the amount of energy needed to break an *alkene* π bond (268 kJ/mol; Section 6.4).

8.2 Naming Alkynes

Alkynes follow the general rules of hydrocarbon nomenclature discussed in Sections 3.4 and 6.3. The suffix *-yne* is used, and the position of the triple bond is indicated by giving the number of the first alkyne carbon in the chain. Numbering the main chain begins at the end nearer the triple bond so that the triple bond receives as low a number as possible.

$$\overset{8}{C}H_3\overset{7}{C}H_2\overset{6}{C}H\overset{5}{C}H_2\overset{4}{C}\equiv\overset{3}{C}\overset{2}{C}H_2\overset{1}{C}H_3$$
$$\underset{CH_3}{|}$$

Begin numbering at the end nearer the triple bond.

6-Methyl-3-octyne

Compounds with more than one triple bond are called *diynes, triynes,* and so forth; compounds containing both double and triple bonds are called *enynes* (not *ynenes*). Numbering of an enyne chain starts from the end nearer the first multiple bond, whether double or triple. When there is a choice in numbering, double bonds receive lower numbers than triple bonds. For example:

$$HC\equiv CCH_2CH_2CH_2CH=CH_2$$
$$\;\;7\;\;\;6\;5\;\;\;4\;\;\;3\;\;\;2\;\;\;1$$

1-Hepten-6-yne

$$\underset{CH_3}{\overset{CH_3}{|}}$$
$$HC\equiv CCH_2CHCH_2CH_2CH=CHCH_3$$
$$\;1\;\;\;2\;3\;\;\;4\;\;\;5\;\;\;6\;\;\;7\;\;\;8\;9$$

4-Methyl-7-nonen-1-yne

As with alkyl and alkenyl substituents derived from alkanes and alkenes, respectively, *alkynyl* groups are also possible:

$$CH_3CH_2CH_2CH_2\!\!-\!\!\xi$$

Butyl
(an alkyl group)

$$CH_3CH_2CH=CH\!\!-\!\!\xi$$

1-Butenyl
(a vinylic group)

$$CH_3CH_2C\equiv C\!\!-\!\!\xi$$

1-Butynyl
(an alkynyl group)

● ●

Problem 8.1   Give IUPAC names for the following compounds:

(a) $CH_3\overset{|}{\underset{CH_3}{C}}HC\equiv C\overset{|}{\underset{CH_3}{C}}HCH_3$

(b) $HC\equiv C\overset{|}{\underset{CH_3}{\overset{CH_3}{C}}}CH_3$

(c) $CH_3CH=CHCH=CHC\equiv CCH_3$

(d) $CH_3CH_2\overset{|}{\underset{CH_3}{\overset{CH_3}{C}}}C\equiv CCH_2CH_2CH_3$

(e) $CH_3CH_2\overset{|}{\underset{CH_3}{C}}C\equiv C\overset{|}{\underset{}{C}}HCH_3$ with CH_3 groups

(f)

Problem 8.2   There are seven isomeric alkynes with the formula C_6H_{10}. Draw them, and name them according to IUPAC rules.

● ●

8.3   Preparation of Alkynes: Elimination Reactions of Dihalides

Alkynes can be prepared by elimination of HX from alkyl halides in much the same manner as alkenes (Section 7.1). Treatment of a 1,2-dihalide (a **vicinal** dihalide) with excess strong base such as KOH or $NaNH_2$ results in a twofold elimination of HX and formation of an alkyne. As with the elimination of HX to form an alkene, we'll defer a discussion of the mechanism until Chapter 11.

The necessary vicinal dihalides are themselves readily available by addition of Br_2 or Cl_2 to alkenes. Thus, the overall halogenation/dehydrohalogenation sequence provides a method for going from an alkene to an alkyne. For example, diphenylethylene is converted into diphenylacetylene by reaction with Br_2 and subsequent base treatment.

1,2-Diphenylethylene
(Stilbene)

1,2-Dibromo-1,2-diphenylethane
(a vicinal dibromide)

$$\downarrow \; 2 \text{ KOH, ethanol}$$

Diphenylacetylene (85%)

The twofold dehydrohalogenation takes place through a vinylic halide intermediate, which suggests that vinylic halides themselves should give alkynes when treated with strong base. (*Recall*: A *vinylic* substituent is one that is attached to a double-bond carbon.) This is indeed the case. For example,

3-Chloro-2-buten-1-ol
(a vinylic chloride)

$$\xrightarrow[\text{2. H}_3\text{O}^+]{\text{1. 2 NaNH}_2}$$

$$CH_3C\equiv CCH_2OH$$

2-Butyn-1-ol (85%)

8.4 Reactions of Alkynes: Addition of HX and X_2

Based on the electronic similarity between alkenes and alkynes, you might expect that the chemical reactivity of the two functional groups should also be similar. Alkynes do indeed exhibit much chemistry similar to that of alkenes, but there are also significant differences.

As a general rule, electrophiles undergo electrophilic addition reactions with alkynes much as they do with alkenes. Take the reaction of alkynes with HX, for example. The reaction often can be stopped after addition of 1 equivalent of HX, but reaction with an excess of HX leads to a dihalide product. For example, reaction of 1-hexyne with 2 equivalents of HBr yields 2,2-dibromohexane. As the following examples indicate, the regiochemistry of addition follows Markovnikov's rule: Halogen adds to the more highly substituted side of the alkyne bond, and hydrogen adds to the less highly substituted side. Trans stereochemistry of H and X is normally (though not always) found in the product.

$$CH_3CH_2CH_2CH_2C{\equiv}CH \xrightarrow[CH_3COOH]{HBr} CH_3CH_2CH_2CH_2\overset{\overset{\displaystyle Br}{|}}{C}{=}\overset{\overset{\displaystyle H}{|}}{CH} \xrightarrow{HBr} CH_3CH_2CH_2CH_2\overset{\overset{\displaystyle Br}{|}}{\underset{\underset{\displaystyle Br}{|}}{C}}{-}\overset{\overset{\displaystyle H}{|}}{\underset{\underset{\displaystyle H}{|}}{C}}{-}H$$

1-Hexyne   **2-Bromo-1-hexene**   **2,2-Dibromohexane**

$$CH_3CH_2C{\equiv}CCH_2CH_3 \xrightarrow[CH_3COOH]{HCl,\ NH_4Cl} \underset{CH_3CH_2}{\overset{Cl}{\diagdown}}C{=}C\underset{H}{\overset{CH_2CH_3}{\diagup}}$$

3-Hexyne

(Z)-3-Chloro-3-hexene
(95%)

Bromine and chlorine also add to alkynes to give addition products, and trans stereochemistry again results:

$$CH_3CH_2C{\equiv}CH \xrightarrow[CH_2Cl_2]{Br_2} \underset{Br}{\overset{CH_3CH_2}{\diagdown}}C{=}C\underset{H}{\overset{Br}{\diagup}} \xrightarrow[CH_2Cl_2]{Br_2} CH_3CH_2CBr_2CHBr_2$$

1-Butyne   **1,1,2,2-Tetrabromobutane**

(E)-1,2-Dibromo-1-butene

The mechanism of electrophilic alkyne addition is similar but not identical to that of alkene addition. When an electrophile such as HBr adds to an *alkene* (Sections 6.8 and 6.9), the reaction takes place in two steps and involves an *alkyl* carbocation intermediate. If HBr were to add by the same mechanism to an *alkyne,* an analogous *vinylic carbocation* would be formed as the intermediate.

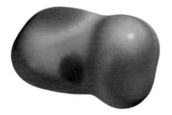

An alkene   An alkyl carbocation   An alkyl bromide

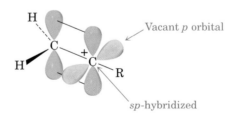

An alkyne   A vinylic carbocation   A vinylic bromide

A vinylic carbocation has an *sp*-hybridized carbon and generally forms less readily than an alkyl carbocation (Figure 8.2). As a rule, a *secondary* vinylic carbocation forms about as readily as a *primary* alkyl carbocation, but a *primary* vinylic carbocation is so difficult to form that there is no clear evidence it even exists. Thus, many alkyne additions occur through more complex mechanistic pathways.

FIGURE 8.2 ▼

The structure of a secondary vinylic carbocation. The cationic carbon atom is *sp*-hybridized and has a vacant *p* orbital perpendicular to the plane of the π bond orbitals. Only one R group is attached to the positively charged carbon rather than two, as in a secondary alkyl carbocation. The electrostatic potential map shows that the most positive (blue) regions coincide with lobes of the vacant *p* orbital and are perpendicular to the most negative (red) regions associated with the π bond.

vinylic carbocation

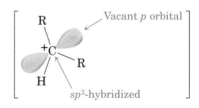

A 2° vinylic carbocation   A 2° alkyl carbocation

· ·

Problem 8.3   What products would you expect from the following reactions?

(a) $CH_3CH_2CH_2C{\equiv}CH + 2\ Cl_2 \longrightarrow$?   (b) ⬠—$C{\equiv}CH + 1\ HBr \longrightarrow$?

(c) $CH_3CH_2CH_2CH_2C{\equiv}CCH_3 + 1\ HBr \longrightarrow$?

· ·

8.5 Hydration of Alkynes

Like alkenes (Sections 7.4 and 7.5), alkynes can be hydrated by either of two methods. Direct addition of water catalyzed by mercury(II) ion yields the Markovnikov product, and indirect addition of water by a hydroboration/oxidation sequence yields the non-Markovnikov product.

Mercury(II)-Catalyzed Hydration of Alkynes

Alkynes don't react directly with aqueous acid but will undergo hydration readily in the presence of mercury(II) sulfate catalyst. The reaction occurs with Markovnikov regiochemistry: The –OH group adds to the more highly substituted carbon, and the –H attaches to the less highly substituted one.

$$CH_3CH_2CH_2CH_2C\equiv CH \xrightarrow[\text{HgSO}_4]{\text{H}_2\text{O, H}_2\text{SO}_4} \left[\underset{\underset{\text{H}}{|}}{\overset{\overset{\text{OH}}{|}}{CH_3CH_2CH_2CH_2C}}{=}CH \right] \longrightarrow CH_3CH_2CH_2CH_2\overset{\overset{\text{O}}{\|}}{C}-CH_3$$

1-Hexyne   An enol   **2-Hexanone**
                                                                              (78%)

Interestingly, the product actually isolated from alkyne hydration is not the vinylic alcohol, or **enol** (*ene* + *ol*), but is instead a *ketone*. Although the enol is an intermediate in the reaction, it immediately rearranges to a ketone by a process called *keto–enol tautomerism*. The individual keto and enol forms are said to be **tautomers**, a word used to describe constitutional isomers that interconvert rapidly. With few exceptions, the keto–enol tautomeric equilibrium lies on the side of the ketone; enols are almost never isolated. We'll look more closely at this equilibrium in Section 22.1.

Enol tautomer   **Keto tautomer**
(less favored)   **(more favored)**

The mechanism of the mercury(II)-catalyzed alkyne hydration reaction is analogous to the oxymercuration reaction of alkenes (Section 7.4). Electrophilic addition of mercury(II) ion to the alkyne gives a vinylic cation, which reacts with water and loses a proton to yield a mercury-containing enol intermediate. In contrast to alkene oxymercuration, no treatment with $NaBH_4$ is necessary to remove the mercury; the acidic reaction conditions alone are sufficient to effect replacement of mercury by hydrogen (Figure 8.3).

A mixture of both possible ketones results when an unsymmetrically substituted internal alkyne (RC≡CR′) is hydrated. The reaction is therefore

FIGURE 8.3 ▼

Mechanism of the mercury(II)-catalyzed hydration of an alkyne to yield a ketone. The reaction yields an intermediate enol, which rapidly tautomerizes to give a ketone.

The alkyne uses a pair of electrons to attack the electrophilic mercury(II) ion, yielding a mercury-containing vinylic carbocation intermediate.

Nucleophilic attack of water on the carbocation forms a C–O bond and yields a protonated mercury-containing enol.

Abstraction of H⁺ from the protonated enol by water gives an organomercury compound.

Replacement of Hg⁺ by H⁺ occurs to give a neutral enol.

The enol undergoes tautomerization to give the final ketone product.

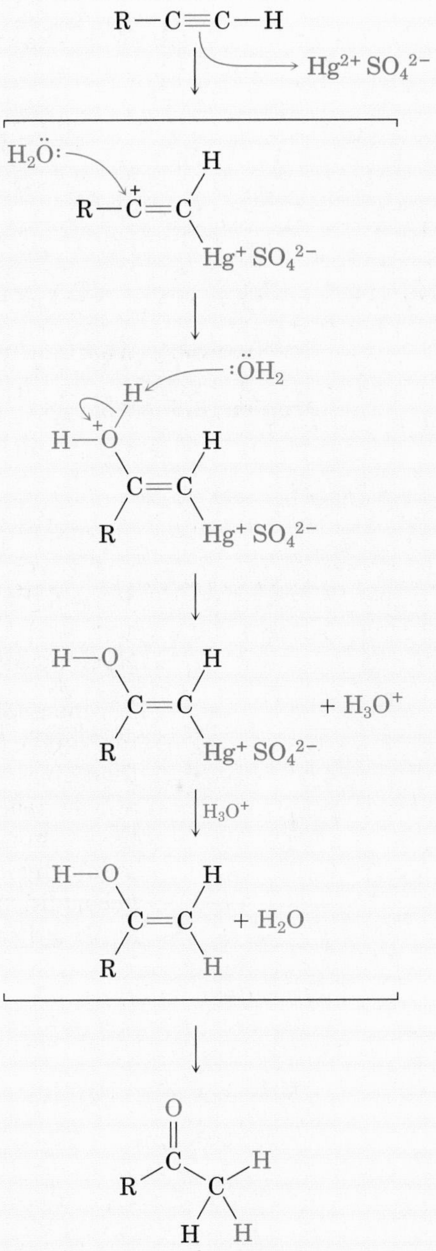

most useful when applied to a terminal alkyne (RC≡CH) because only a methyl ketone is formed.

$$R—C≡C—R' \xrightarrow[\text{HgSO}_4]{\text{H}_3\text{O}^+} \underbrace{\begin{matrix} \overset{O}{\underset{\|}{C}} \\ R \quad \text{CH}_2\text{R}' \end{matrix} + \begin{matrix} \overset{O}{\underset{\|}{C}} \\ \text{RCH}_2 \quad R' \end{matrix}}$$

An internal alkyne   Mixture

$$R—C≡CH \xrightarrow[\text{HgSO}_4]{\text{H}_3\text{O}^+} \begin{matrix} \overset{O}{\underset{\|}{C}} \\ R \quad \text{CH}_3 \end{matrix}$$

A terminal alkyne   **A methyl ketone**

Problem 8.4   What product would you obtain by hydration of 4-octyne? Of 2-methyl-4-octyne?

Problem 8.5   What alkynes would you start with to prepare the following ketones?

(a) $CH_3CH_2CH_2\overset{O}{\underset{\|}{C}}CH_3$ (b) $CH_3CH_2\overset{O}{\underset{\|}{C}}CH_2CH_3$

Hydroboration/Oxidation of Alkynes

Borane adds rapidly to an alkyne just as it does to an alkene, and the resulting vinylic borane can be oxidized by H_2O_2 to yield an enol. Tautomerization then gives either a ketone or an aldehyde, depending on the structure of the alkyne reactant. Hydroboration/oxidation of an internal alkyne such as 3-hexyne gives a ketone, and hydroboration/oxidation of a terminal alkyne gives an aldehyde. Note that the relatively unhindered terminal alkyne undergoes *two* additions, giving a doubly hydroborated intermediate. Oxidation with H_2O_2 at pH 8 then replaces both boron atoms by oxygen and generates the aldehyde.

An internal alkyne

$$3\ CH_3CH_2C≡CCH_2CH_3 \xrightarrow[\text{THF}]{\text{BH}_3} \begin{bmatrix} \overset{\displaystyle H \qquad\quad BR_2}{\diagdown\quad\diagup} \\ C=C \\ \diagup\qquad\quad\diagdown \\ CH_3CH_2 \qquad CH_2CH_3 \end{bmatrix} \xrightarrow[\text{H}_2\text{O, NaOH}]{\text{H}_2\text{O}_2} \begin{bmatrix} \overset{\displaystyle H \qquad\quad OH}{\diagdown\quad\diagup} \\ C=C \\ \diagup\qquad\quad\diagdown \\ CH_3CH_2 \qquad CH_2CH_3 \end{bmatrix}$$

3-Hexyne   **A vinylic borane**   **An enol**

$$3\ CH_3CH_2CH_2\overset{O}{\underset{\|}{C}}CH_2CH_3$$

3-Hexanone

A terminal alkyne

$$CH_3CH_2CH_2CH_2C \equiv CH \xrightarrow{BH_3} \left[\begin{array}{c} BR_2 \\ | \\ CH_3CH_2CH_2CH_2CH_2CH \\ | \\ BR_2 \end{array} \right] \xrightarrow[H_2O]{H_2O_2} CH_3CH_2CH_2CH_2CH_2\overset{\overset{\displaystyle O}{\|}}{C}H$$

1-Hexyne   **Hexanal (70%)**

The hydroboration/oxidation sequence is *complementary* to the direct, mercury(II)-catalyzed hydration reaction of a terminal alkyne because different products result. Direct hydration with aqueous acid and mercury(II) sulfate leads to a methyl ketone, whereas hydroboration/oxidation of the same terminal alkyne leads to an aldehyde:

$$R-C\equiv CH \begin{array}{c} \xrightarrow[HgSO_4]{H_2O, H_2SO_4} R-\overset{\overset{\displaystyle O}{\|}}{C}-CH_3 \\ \\ \xrightarrow[2.\,H_2O_2]{1.\,BH_3,\,THF} R-CH_2-\overset{\overset{\displaystyle O}{\|}}{C}-H \end{array}$$

A terminal alkyne   **A methyl ketone**   **An aldehyde**

. . . .

Problem 8.6   What alkyne would you start with to prepare each of the following compounds by a hydroboration/oxidation reaction?

(a) ⬡—CH₂CHO   (b) $(CH_3)_2CHCH_2\overset{\overset{\displaystyle O}{\|}}{C}CH(CH_3)_2$

. . . .

8.6 Reduction of Alkynes

Alkynes are easily reduced to alkanes by addition of H_2 over a metal catalyst. The reaction occurs in steps through an alkene intermediate, and measurements indicate that the first step in the reaction has a larger $\Delta H°_{hydrog}$ than the second step.

$$HC \equiv CH \xrightarrow[Catalyst]{H_2} H_2C = CH_2 \qquad \Delta H°_{hydrog} = -176 \text{ kJ/mol } (-42 \text{ kcal/mol})$$

$$H_2C = CH_2 \xrightarrow[Catalyst]{H_2} CH_3-CH_3 \qquad \Delta H°_{hydrog} = -137 \text{ kJ/mol } (-33 \text{ kcal/mol})$$

Complete reduction to the alkane occurs when palladium on carbon (Pd/C) is used as catalyst, but hydrogenation can be stopped at the alkene if the less active *Lindlar catalyst* is used. (The Lindlar catalyst is a finely divided palladium metal that has been precipitated onto a calcium carbonate support and then deactivated by treatment with lead acetate and quinoline, an aromatic amine.) The hydrogenation occurs with syn stereochemistry (Section 7.5), giving a cis alkene product.

$$CH_3(CH_2)_3C \equiv C(CH_2)_3CH_3$$

5-Decyne

$$\xrightarrow[\text{Pd/C}]{2\ H_2} CH_3(CH_2)_8CH_3$$

Decane (96%)

$$\xrightarrow[\substack{\text{Lindlar} \\ \text{catalyst}}]{H_2}$$

***cis*-5-Decene (96%)**

The alkyne hydrogenation reaction has been explored extensively by the Hoffmann–LaRoche pharmaceutical company, where it is used in the commercial synthesis of vitamin A. The cis isomer of vitamin A produced on hydrogenation is converted to the trans isomer by heating.

$$\xrightarrow[\substack{\text{Lindlar} \\ \text{catalyst}}]{H_2}$$

7-*cis*-Retinol
(7-*cis*-vitamin A; vitamin A has
a trans double bond at C7)

Another method for the conversion of an alkyne to an alkene uses sodium or lithium metal as the reducing agent in liquid ammonia as solvent. This method is complementary to the Lindlar reduction because it produces trans rather than cis alkenes. For example, 5-decyne gives *trans*-5-decene on treatment with lithium in liquid ammonia.

$$CH_3CH_2CH_2CH_2C \equiv CCH_2CH_2CH_2CH_3 \xrightarrow[\text{NH}_3]{\text{Li}}$$

5-Decyne

***trans*-5-Decene (78%)**

Alkali metals dissolve in liquid ammonia at −33°C to produce a deep blue solution containing the metal cation and ammonia-solvated electrons.

When an alkyne is then added to the solution, an electron adds to the triple bond to yield an intermediate *anion radical*—a species that is both an anion (has a negative charge) and a radical (has an odd number of electrons). This anion radical is a strong base, which removes H^+ from ammonia to give a vinylic radical. Addition of a second electron to the vinylic radical gives a vinylic anion, which abstracts a second H^+ from ammonia to give trans alkene product. The mechanism is shown in Figure 8.4.

FIGURE 8.4 ▼

Mechanism of the lithium/ammonia reduction of an alkyne to produce a trans alkene.

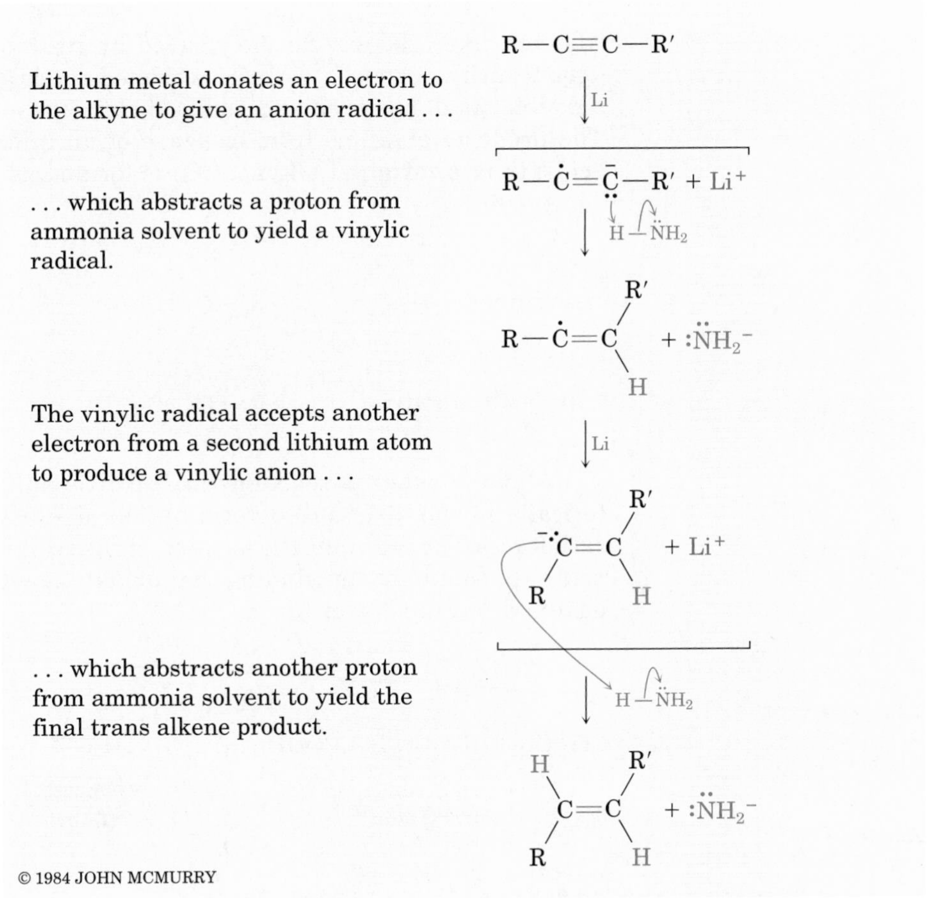

Lithium metal donates an electron to the alkyne to give an anion radical . . .

. . . which abstracts a proton from ammonia solvent to yield a vinylic radical.

The vinylic radical accepts another electron from a second lithium atom to produce a vinylic anion . . .

. . . which abstracts another proton from ammonia solvent to yield the final trans alkene product.

© 1984 JOHN MCMURRY

The trans stereochemistry of the alkene product is established during the second reduction step when the less hindered, trans vinylic anion is formed from the vinylic radical. Vinylic radicals undergo rapid cis–trans equilibration, but vinylic anions equilibrate much less rapidly. Thus, the more stable trans vinylic anion is formed rather than the less stable cis anion and is then protonated without equilibration.

• •

Problem 8.7   Using any alkyne needed, how would you prepare the following alkenes?
(a) *trans*-2-Octene   (b) *cis*-3-Heptene   (c) 3-Methyl-1-pentene

• •

8.7   Oxidative Cleavage of Alkynes

Alkynes, like alkenes, can be cleaved by reaction with powerful oxidizing agents such as ozone or $KMnO_4$. A triple bond is generally less reactive than a double bond, however, and yields of cleavage products are sometimes low. The products obtained from cleavage of an internal alkyne are carboxylic acids; from a terminal alkyne, CO_2 is formed as one product.

$$\text{An internal alkyne} \quad R-C\equiv C-R' \xrightarrow{\text{KMnO}_4 \text{ or } O_3} RCOOH + R'COOH$$

$$\text{A terminal alkyne} \quad R-C\equiv C-H \xrightarrow{\text{KMnO}_4 \text{ or } O_3} RCOOH + CO_2$$

Alkyne oxidation reactions are of little value now but were used historically in the structure determination of substances isolated from natural sources. For example, the location of the triple bond in the chain of tariric acid was established by finding that oxidation with $KMnO_4$ gave dodecanoic acid and hexanedioic acid:

$$CH_3(CH_2)_{10}C\equiv C(CH_2)_4COOH \xrightarrow[H_3O^+]{\text{KMnO}_4} CH_3(CH_2)_{10}COOH + HOC(CH_2)_4COH$$

6-Octadecynoic acid   **Dodecanoic acid**   **1,6-Hexanedioic acid**
(Tariric acid)   **(Lauric acid)**   **(Adipic acid)**

• •

Problem 8.8   Propose structures for alkynes that give the following products on oxidative cleavage by $KMnO_4$:

(a)   (image of benzene ring with COOH) $+ CO_2$ (b) 2 $CH_3(CH_2)_7COOH + HO_2C(CH_2)_7COOH$

• •

8.8 Alkyne Acidity: Formation of Acetylide Anions

The most striking difference between alkenes and alkynes is that terminal alkynes are weakly acidic. When a terminal alkyne is treated with a strong base, such as sodium amide, $Na^+\ ^-NH_2$ the terminal hydrogen is removed and an **acetylide anion** is formed:

$$R-C\equiv C-H + :\ddot{N}H_2\ Na^+ \longrightarrow R-C\equiv C:^-\ Na^+ + :NH_3$$

Acetylide anion

According to the Brønsted–Lowry definition (Section 2.7), an acid is any substance that donates H^+. Although we usually think of oxyacids (H_2SO_4, HNO_3) or halogen acids (HCl, HBr) in this context, *any* compound containing a hydrogen atom can be an acid under the right circumstances. By measuring dissociation constants of different acids and expressing the results as pK_a values, an acidity order can be established. Recall from Section 2.8 that a low pK_a corresponds to a strong acid, and a high pK_a corresponds to a weak acid.

Since a stronger acid donates its proton to the anion of a weaker acid in an acid–base reaction, a rank-ordered list tells which bases are needed to deprotonate which acids. For example, since acetic acid ($pK_a = 4.75$) is a stronger acid than ethanol ($pK_a = 16$), we know that the anion of ethanol (ethoxide ion, $CH_3CH_2O^-$) will remove a proton from acetic acid. Similarly, amide ion ($^-NH_2$), the anion of ammonia ($pK_a = 35$), will remove a proton from ethanol ($pK_a = 16$).

$$CH_3CH_2O^- + CH_3\overset{\overset{\displaystyle O}{\|}}{C}OH \rightleftarrows CH_3CH_2OH + CH_3\overset{\overset{\displaystyle O}{\|}}{C}O^-$$

Ethoxide ion   **Acetic acid**   **Ethanol**   **Acetate ion**

$$H_2N^- + CH_3CH_2OH \rightleftarrows H_2NH + CH_3CH_2O^-$$

Amide ion   **Ethanol**   **Ammonia**   **Ethoxide ion**

Where do hydrocarbons lie on the acidity scale? As the data in Table 8.1 indicate, both methane ($pK_a \approx 60$) and ethylene ($pK_a = 44$) are very

TABLE 8.1 Acidity of Simple Hydrocarbons				
Type	**Example**	K_a	pK_a	
Alkyne	HC≡CH	10^{-25}	25	Stronger acid
Alkene	$H_2C=CH_2$	10^{-44}	44	↑
Alkane	CH_4	$\sim 10^{-60}$	60	Weaker acid

weak acids, and thus do not react with common bases. Acetylene, however, has $pK_a = 25$ and can be deprotonated by the conjugate base of any acid whose pK_a is greater than 25. Amide ion, NH_2^-, for example, can abstract a proton from a terminal alkyne.

Why are terminal alkynes more acidic than alkenes or alkanes? In other words, why are acetylide anions more stable than vinylic or alkyl anions? The simplest explanation involves the hybridization of the negatively charged carbon atom. An acetylide anion has an sp-hybridized carbon, so the negative charge resides in an orbital that has 50% "s character"; a vinylic anion has an sp^2-hybridized carbon and therefore has 33% s character; and an alkyl anion (sp^3) has only 25% s character (Figure 8.5). Since s orbitals are nearer the positive nucleus and lower in energy than p orbitals, the negative charge is stabilized to a greater extent in an orbital with higher s character. As a result, acetylide anions are more stable than vinylic anions, which are more stable than alkyl anions.

FIGURE 8.5 ▼

A comparison of methyl, vinylic, and acetylide anions. The acetylide anion, with sp hybridization, has more s character and is more stable. Electrostatic potential maps show that placing the negative charge closer to a carbon nucleus makes carbon appear less negative (red).

**methyl anion,
vinylic anion,
acetylide anion**

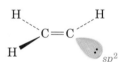

CH$_3$ anion; 25% s Vinylic anion; 33% s Acetylide anion; 50% s

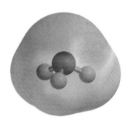

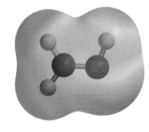

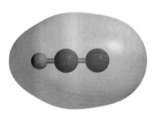

Less   **Stability** →   More
stable   stable

• •

Problem 8.9 The pK_a of acetone, CH_3COCH_3, is 19.3. Which of the following bases is strong enough to deprotonate acetone?
(a) KOH (pK_a of H_2O = 15.7)   (b) Na^+ $^-C{\equiv}CH$ (pK_a of C_2H_2 = 25)
(c) $NaHCO_3$ (pK_a of H_2CO_3 = 6.4)   (d) $NaOCH_3$ (pK_a of CH_3OH = 15.6)

• •

8.9 Alkylation of Acetylide Anions

The presence of a negative charge and an unshared electron pair on carbon makes an acetylide anion strongly nucleophilic. As a result, an acetylide anion can react with an alkyl halide such as bromomethane to substitute for the halogen and yield a new alkyne product:

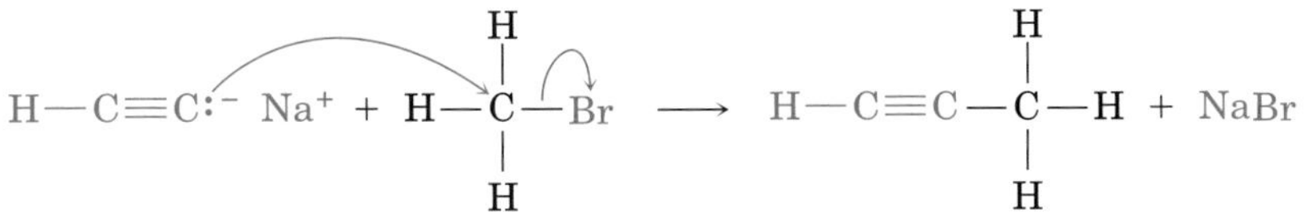

$$H-C{\equiv}C{:}^- \; Na^+ + H-\overset{\displaystyle H}{\underset{\displaystyle H}{C}}-Br \longrightarrow H-C{\equiv}C-\overset{\displaystyle H}{\underset{\displaystyle H}{C}}-H + NaBr$$

We won't study the details of this substitution reaction until Chapter 11, but we can picture it as happening by the pathway shown in Figure 8.6.

FIGURE 8.6 ▼

A mechanism for the alkylation reaction of acetylide anion with bromomethane to give propyne.

 OCOL refer to Mechanisms & Movies

The nucleophilic acetylide anion uses its electron lone pair to form a bond to the positively polarized, electrophilic carbon atom of bromomethane. As the new C–C bond begins to form, the C–Br bond begins to break in the transition state.

$$H-C{\equiv}C{:}^- \; Na^+ \quad \overset{\displaystyle H}{\underset{\displaystyle H}{}}C-Br$$

$$\left[H-C{\equiv}\overset{\delta-}{C}\cdots\underset{\displaystyle H\;H}{C}\cdots\overset{\delta-}{Br} + Na^+ \right]^{\ddagger}$$

Transition state

The new C–C bond is fully formed and the old C–Br bond is fully broken at the end of the reaction.

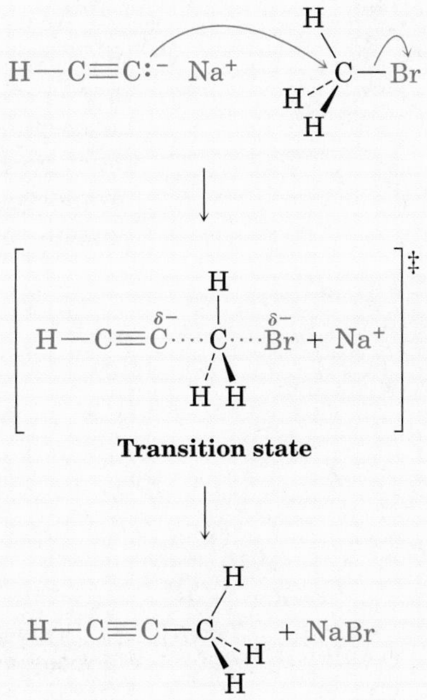

$$H-C{\equiv}C-C\overset{\displaystyle H}{\underset{\displaystyle H}{\diagup}}_H + NaBr$$

The nucleophilic acetylide ion uses an electron pair to attack the positively polarized, electrophilic carbon atom of bromomethane. As the new C–C bond forms, Br⁻ departs, taking with it the electron pair from the former C–Br bond and yielding propyne as product. We call such a reaction an **alkylation** because a new alkyl group has become attached to the starting alkyne.

Alkyne alkylation is not limited to acetylene itself. *Any* terminal alkyne can be converted into its corresponding anion and then alkylated by treatment with an alkyl halide, yielding an internal alkyne. For example, conversion of 1-hexyne into its anion, followed by reaction with 1-bromobutane, yields 5-decyne:

$$CH_3CH_2CH_2CH_2C \equiv CH \quad \xrightarrow[\text{2. } CH_3CH_2CH_2CH_2Br,\, THF]{\text{1. } NaNH_2,\, NH_3} \quad CH_3CH_2CH_2CH_2C \equiv CCH_2CH_2CH_2CH_3$$

1-Hexyne   **5-Decyne (76%)**

Because of its generality, acetylide alkylation is the best method for preparing a substituted alkyne from a simpler precursor. A terminal alkyne can be prepared by alkylation of acetylene itself, and an internal alkyne can be prepared by further alkylation of a terminal alkyne.

$$HC \equiv CH \quad \xrightarrow{NaNH_2} \quad HC \equiv C^- Na^+ \quad \xrightarrow{RCH_2Br} \quad HC \equiv CCH_2R$$

Acetylene   **A terminal alkyne**

$$RC \equiv CH \quad \xrightarrow{NaNH_2} \quad RC \equiv C^- Na^+ \quad \xrightarrow{R'CH_2Br} \quad RC \equiv CCH_2R'$$

A terminal alkyne   **An internal alkyne**

Acetylide ion alkylation is limited to primary alkyl bromides and iodides, RCH_2X, for reasons that will be discussed in detail in Chapter 11. In addition to their reactivity as nucleophiles, acetylide ions are sufficiently strong bases that they cause dehydrohalogenation instead of substitution when they react with secondary and tertiary alkyl halides. For example, reaction of bromocyclohexane with propyne anion yields the elimination product cyclohexene rather than the substitution product cyclohexylpropyne.

Bromocyclohexane
(a secondary alkyl halide)

Cyclohexene

NOT formed

● ●

Problem 8.10 Show the terminal alkyne and alkyl halide from which each of the following products can be obtained. If two routes look feasible, list both.

(a) $CH_3CH_2CH_2C{\equiv}CCH_3$ (b) $(CH_3)_2CHC{\equiv}CCH_2CH_3$ (c) ⬡—$C{\equiv}CCH_3$

(d) 5-Methyl-2-hexyne   (e) 2,2-Dimethyl-3-hexyne

Problem 8.11 How would you prepare *cis*-2-butene starting from propyne, an alkyl halide, and any other reagents needed? This problem can't be worked in a single step. You'll have to carry out more than one reaction.

● ●

8.10 An Introduction to Organic Synthesis

There are many reasons for carrying out the laboratory synthesis of an organic molecule from simpler precursors. In the pharmaceutical industry, new organic molecules are designed and synthesized in the hope that some might be useful new drugs. In the chemical industry, syntheses are done to devise more economical routes to known compounds. In academic laboratories, the synthesis of complex molecules is sometimes done purely for the intellectual challenge involved in mastering so difficult a subject. The successful synthesis route is a highly creative work that is sometimes described by such subjective terms as *elegant* or *beautiful*.

In this book, too, we will often devise syntheses of molecules from simpler precursors. Our purpose, however, is pedagogical. The ability to plan a workable synthetic sequence demands knowledge of a wide variety of organic reactions. Furthermore, it requires the practical ability to fit together the steps in a sequence such that each reaction does only what is desired without causing changes elsewhere in the molecule. *Working synthesis problems is an excellent way to learn organic chemistry.*

Some of the syntheses we plan may appear trivial. Here's an example:

● ●

Practice Problem 8.1 Prepare octane from 1-pentyne.

$$CH_3CH_2CH_2C{\equiv}CH \quad \Longrightarrow \quad CH_3CH_2CH_2CH_2CH_2CH_2CH_2CH_3$$

1-Pentyne   **Octane**

Strategy Compare the product with the starting material, and catalog the differences. In this case, we need to add three carbons and reduce the triple bond.

Solution   First alkylate the acetylide anion of 1-pentyne with 1-bromopropane to add three carbons, and then reduce the product using catalytic hydrogenation:

$$CH_3CH_2CH_2C \equiv CH \xrightarrow[\text{2. BrCH}_2\text{CH}_2\text{CH}_3, \text{ THF}]{\text{1. NaNH}_2, \text{ NH}_3} CH_3CH_2CH_2C \equiv CCH_2CH_2CH_3$$

1-Pentyne   **4-Octyne**

$$\downarrow \text{H}_2/\text{Pd in ethanol}$$

$$CH_3CH_2CH_2\overset{\displaystyle\overset{H}{|}\ \overset{H}{|}}{C} - \underset{\displaystyle\underset{H}{|}\ \underset{H}{|}}{C}CH_2CH_2CH_3$$

Octane

Although the synthesis route just presented will work perfectly well, it has little practical value because a chemist can simply *buy* octane from any of several dozen chemical supply companies. The value of working the problem is that it makes us approach a chemical problem in a logical way, draw on our knowledge of chemical reactions, and organize that knowledge into a workable plan—it helps us *learn* organic chemistry.

There's no secret to planning an organic synthesis. All it takes is a knowledge of the different reactions, some discipline, and a lot of practice. The only real trick is to *always work backward* in what's often referred to as a *retrosynthetic* direction. Don't look at the starting material and ask yourself what reactions it might undergo. Instead, look at the final product and ask, "What was the immediate precursor of that product?" For example, if the final product is an alkyl halide, the immediate precursor might be an alkene (to which you could add HX). Having found an immediate precursor, work backward again, one step at a time, until you get back to the starting material. (You have to keep the starting material in mind, of course, so that you can work back to it, but you don't want that starting material to be your main focus.)

Let's work some examples of increasing complexity.

Practice Problem 8.2   Synthesize *cis*-2-hexene from 1-pentyne and any alkyl halide needed. More than one step is required.

$$CH_3CH_2CH_2C \equiv CH + RX \longrightarrow$$

1-Pentyne   **Alkyl halide**

$$\underset{H}{\overset{CH_3CH_2CH_2}{\diagdown}}C = C\underset{H}{\overset{CH_3}{\diagup}}$$

cis-**2-Hexene**

Strategy   When undertaking any synthesis problem, the idea is to look at the product, identify the functional groups it contains, and then ask yourself how those functional groups can be prepared. Always work in a retrosynthetic sense, one step at a time.

The product in this case is a cis-disubstituted alkene, so the first question is, "What is an immediate precursor of a cis-disubstituted alkene?" We know that an alkene can be prepared from an alkyne by reduction and that the right choice of experimental conditions will allow us to prepare either a trans-disubstituted alkene (using lithium in liquid ammonia) or a cis-disubstituted alkene (using catalytic hydrogenation over the Lindlar catalyst). Thus, reduction of 2-hexyne by catalytic hydrogenation using the Lindlar catalyst should yield *cis*-2-hexene:

$$CH_3CH_2CH_2C{\equiv}CCH_3 \xrightarrow[\text{Lindlar catalyst}]{H_2}$$

2-Hexyne

$$\underset{H}{\overset{CH_3CH_2CH_2}{\diagdown}}C{=}C\underset{H}{\overset{CH_3}{\diagup}}$$

***cis*-2-Hexene**

Next ask, "What is an immediate precursor of 2-hexyne?" We've seen that an internal alkyne can be prepared by alkylation of a terminal alkyne anion. In the present instance, we're told to start with 1-pentyne and an alkyl halide. Thus, alkylation of the anion of 1-pentyne with iodomethane should yield 2-hexyne:

$$CH_3CH_2CH_2C{\equiv}CH + NaNH_2 \xrightarrow{\text{In } NH_3} CH_3CH_2CH_2C{\equiv}C{:}^- \ Na^+$$

1-Pentyne

$$CH_3CH_2CH_2C{\equiv}C{:}^- \ Na^+ + CH_3I \xrightarrow{\text{In THF}} CH_3CH_2CH_2C{\equiv}CCH_3$$

2-Hexyne

Solution   *cis*-2-Hexene can be synthesized from the given starting materials in three steps:

$$CH_3CH_2CH_2C{\equiv}CH \xrightarrow[\text{2. } CH_3I, \text{ THF}]{\text{1. } NaNH_2, \ NH_3} CH_3CH_2CH_2C{\equiv}CCH_3 \xrightarrow[\text{Lindlar catalyst}]{H_2} \underset{H}{\overset{CH_3CH_2CH_2}{\diagdown}}C{=}C\underset{H}{\overset{CH_3}{\diagup}}$$

1-Pentyne   **2-Hexyne**   ***cis*-2-Hexene**

• •

Practice Problem 8.3   Synthesize 2-bromopentane from acetylene and any alkyl halide needed. More than one step is required.

$$HC{\equiv}CH + RX \longrightarrow \underset{\text{2-Bromopentane}}{CH_3CH_2CH_2\overset{\overset{\displaystyle Br}{\displaystyle |}}{C}HCH_3}$$

Acetylene   Alkyl halide

Strategy   Identify the functional group in the product (an alkyl bromide) and work the problem retrosynthetically. "What is an immediate precursor of an alkyl bromide?" Perhaps an alkene plus HBr:

$$CH_3CH_2CH_2CH{=}CH_2$$

$$\text{or} \qquad \xrightarrow[\text{Ether}]{\text{HBr}} \quad CH_3CH_2CH_2\overset{\displaystyle Br}{\underset{\displaystyle |}{C}}HCH_3$$

$$CH_3CH_2CH{=}CHCH_3$$

Of the two possibilities, addition of HBr to 1-pentene looks like a better choice than addition to 2-pentene, because the latter reaction would give a mixture of isomers.

"What is an immediate precursor of an alkene?" Perhaps an alkyne, which could be reduced:

$$CH_3CH_2CH_2C{\equiv}CH \quad \xrightarrow[\text{Lindlar catalyst}]{H_2} \quad CH_3CH_2CH_2CH{=}CH_2$$

"What is an immediate precursor of a terminal alkyne?" Perhaps sodium acetylide and an alkyl halide:

$$Na^+ \; :\bar{C}{\equiv}CH + BrCH_2CH_2CH_3 \quad \longrightarrow \quad CH_3CH_2CH_2C{\equiv}CH$$

Solution The desired product can be synthesized in four steps from acetylene and 1-bromopropane.

$$HC{\equiv}CH \quad \xrightarrow[\text{2. } CH_3CH_2CH_2Br, \text{ THF}]{\text{1. } NaNH_2, \, NH_3} \quad CH_3CH_2CH_2C{\equiv}CH \quad \xrightarrow[\substack{\text{Lindlar} \\ \text{catalyst}}]{H_2} \quad CH_3CH_2CH_2CH{=}CH_2$$

Acetylene   **1-Pentyne**   **1-Pentene**

$$\downarrow \text{HBr, ether}$$

$$CH_3CH_2CH_2\overset{\displaystyle }{\underset{\displaystyle }{C}}HCH_3$$
$$\underset{\displaystyle Br}{\underset{\displaystyle |}{}}$$

2-Bromopentane

• •

Practice Problem 8.4 Synthesize 1-hexanol from acetylene and an alkyl halide.

$$HC{\equiv}CH \; + \; RX \quad \rightleftharpoons \quad CH_3CH_2CH_2CH_2CH_2CH_2OH$$

Acetylene   **Alkyl halide**   **1-Hexanol**

Strategy "What is an immediate precursor of a primary alcohol?" Perhaps an alkene, which could be hydrated with non-Markovnikov regiochemistry by reaction with borane followed by oxidation with H_2O_2:

$$CH_3CH_2CH_2CH_2CH{=}CH_2 \quad \xrightarrow[\text{2. } H_2O_2, \, NaOH]{\text{1. } BH_3} \quad CH_3CH_2CH_2CH_2CH_2CH_2OH$$

"What is an immediate precursor of a terminal alkene?" Perhaps a terminal alkyne, which could be reduced:

$$CH_3CH_2CH_2CH_2C{\equiv}CH \xrightarrow[\text{Lindlar catalyst}]{H_2} CH_3CH_2CH_2CH_2CH{=}CH_2$$

"What is an immediate precursor of 1-hexyne?" Perhaps acetylene and 1-bromobutane:

$$HC{\equiv}CH \xrightarrow{NaNH_2} Na^+ \ {}^-C{\equiv}CH \xrightarrow{CH_3CH_2CH_2CH_2Br} CH_3CH_2CH_2CH_2C{\equiv}CH$$

Solution   The synthesis can be completed in four steps by working backward: (1) formation of sodium acetylide; (2) alkylation with 1-bromobutane to yield 1-hexyne; (3) reduction of 1-hexyne using the Lindlar catalyst to give 1-hexene; (4) hydroboration/oxidation of 1-hexene to give 1-hexanol.

· ·

Problem 8.12   Beginning with 4-octyne as your only source of carbon, and using any inorganic reagents necessary, how would you synthesize the following compounds?
(a) Butanoic acid   (b) *cis*-4-Octene   (c) 4-Bromooctane
(d) 4-Octanol (4-hydroxyoctane)   (e) 4,5-Dichlorooctane

Problem 8.13   Beginning with acetylene and any alkyl halides needed, how would you synthesize the following compounds?
(a) Decane   (b) 2,2-Dimethylhexane   (c) Hexanal   (d) 2-Heptanone

· ·

CHEMISTRY @ WORK

The Art of Organic Synthesis

If you think some of the synthesis problems at the end of this chapter are hard, try planning (and executing) a synthesis of vitamin B_{12}, starting only from simple substances you can buy in a chemical catalog. This extraordinary achievement was reported in 1973 as the culmination of a collaborative effort headed by Robert B. Woodward of Harvard University and Albert Eschenmoser of the Swiss Federal Institute of Technology in Zürich. More than 100 graduate students and postdoctorals contributed to the work, which took over a decade.

(continued) ▶

The chemical structure of Vitamin B_{12} is shown, with the label **Vitamin B$_{12}$**.

Why put such extraordinary effort into the synthesis of a molecule so easily obtained from natural sources? There are many reasons. On a basic human level, a chemist might be motivated primarily by the challenge, much as a climber might be challenged by the ascent of a difficult peak. Beyond the pure challenge, the completion of a difficult synthesis is also valuable for the way in which it establishes new standards and brings the field to a new level of complexity. If vitamin B_{12} can be made, then why can't any molecule found in nature be made? Indeed, the quarter century that has passed since the work of Woodward and Eschenmoser has seen the laboratory synthesis of many enormously complex and valuable substances. Often, these substances—the anticancer compound Taxol, for instance—are not easily available in nature, so laboratory synthesis is the only method for obtaining larger quantities.

But perhaps the most important reason for undertaking a complex synthesis is that in so doing, new reactions and new chemistry are discovered. It invariably happens in synthesis that a point is reached at which the planned route fails. At such a time, the only alternatives are to quit or to devise a way around the unexpected difficulty. New reactions and new principles come from such situations, and it is in this way that the science of organic chemistry grows richer. In the synthesis of vitamin B_{12}, for example, unexpected findings emerged that led to the understanding of an entire new class of reactions—the *pericyclic* reactions that are the subject of Chapter 30 in this book. From synthesizing vitamin B_{12} to understanding pericyclic reactions—no one could have possibly predicted such a link at the beginning of the synthesis, but that is the way of science.

Vitamin B$_{12}$ has been synthesized in the laboratory, but bacteria growing on sludge from municipal sewage plants provide the richest supply.

Summary and Key Words

Alkynes are hydrocarbons that contain one or more carbon–carbon triple bonds. Alkyne carbon atoms are *sp*-hybridized, and the triple bond consists of one *sp–sp* σ bond and two *p–p* π bonds. There are relatively few general methods of alkyne synthesis. The two best are the alkylation of an acetylide anion with a primary alkyl halide and the twofold elimination of HX from a vicinal dihalide.

The chemistry of alkynes is dominated by electrophilic addition reactions, similar to those of alkenes. Alkynes react with HBr and HCl to yield *vinylic* halides, and with Br_2 and Cl_2 to yield 1,2-dihalides (**vicinal** dihalides). Alkynes can be hydrated by reaction with aqueous sulfuric acid in the presence of mercury(II) catalyst. The reaction leads to an intermediate **enol** that immediately **tautomerizes** to yield a ketone. Since the addition reaction occurs with Markovnikov regiochemistry, a methyl ketone is produced from a terminal alkyne. Alternatively, hydroboration/oxidation of a terminal alkyne yields an aldehyde.

Alkynes can be reduced to yield alkenes and alkanes. Complete reduction of the triple bond over a palladium hydrogenation catalyst yields an alkane; partial reduction by catalytic hydrogenation over a *Lindlar catalyst* yields a cis alkene. Reduction of the alkyne with lithium in ammonia yields a trans alkene.

Terminal alkynes are weakly acidic. The alkyne hydrogen can be removed by a strong base such as Na^+ $^-NH_2$ to yield an **acetylide anion**. An acetylide anion acts as a nucleophile and can displace a halide ion from a primary alkyl halide in an **alkylation** reaction. Acetylide anions are more stable than either alkyl anions or vinylic anions because their negative charge is in a hybrid orbital with 50% *s* character, allowing the charge to be closer to the nucleus.

Summary of Reactions

1. Preparation of alkynes
   (a) Dehydrohalogenation of vicinal dihalides (Section 8.3)

$$R-\underset{\underset{Br}{|}}{\overset{\overset{H}{|}}{C}}-\underset{\underset{Br}{|}}{\overset{\overset{H}{|}}{C}}-R' \xrightarrow[\text{or 2 NaNH}_2,\text{ NH}_3]{\text{2 KOH, ethanol}} R-C\equiv C-R' + 2\ H_2O + 2\ KBr$$

$$R-\underset{\underset{}{|}}{\overset{\overset{H}{|}}{C}}=\underset{\underset{Br}{|}}{\overset{\overset{Br}{}}{C}}-R' \xrightarrow[\text{or NaNH}_2,\text{ NH}_3]{\text{KOH, ethanol}} R-C\equiv C-R' + H_2O + KBr$$

   (b) Acetylide ion alkylation (Section 8.9)

$$HC\equiv CH \xrightarrow{\text{NaNH}_2} HC\equiv C^-\ Na^+ \xrightarrow{\text{RCH}_2\text{Br}} HC\equiv CCH_2R$$

   Acetylene   **A terminal alkyne**

(continued) ▶

$$RC \equiv CH \xrightarrow{\text{NaNH}_2} RC \equiv C^- \, Na^+ \xrightarrow{\text{R'CH}_2\text{Br}} RC \equiv CCH_2R'$$

A terminal alkyne   **An internal alkyne**

2. Reactions of alkynes
   (a) Addition of HX, where X = Br or Cl (Section 8.4)

$$R - C \equiv C - H \xrightarrow[\text{Ether}]{\text{HX}} R - \overset{X}{\underset{}{C}} = \overset{H}{\underset{}{C}} - H \xrightarrow[\text{Ether}]{\text{HX}} R - \overset{X}{\underset{X}{C}} - \overset{H}{\underset{H}{C}} - H$$

   (b) Addition of X_2, where X = Br or Cl (Section 8.4)

$$R - C \equiv C - R' \xrightarrow[\text{CH}_2\text{Cl}_2]{\text{X}_2} \overset{X \qquad R'}{\underset{R \qquad X}{C = C}} \xrightarrow[\text{CH}_2\text{Cl}_2]{\text{X}_2} R - \overset{X}{\underset{X}{C}} - \overset{X}{\underset{X}{C}} - R'$$

   (c) Mercuric sulfate-catalyzed hydration (Section 8.5)

$$R - C \equiv CH \xrightarrow[\text{HgSO}_4]{\text{H}_2\text{SO}_4, \text{ H}_2\text{O}} \left[R - \overset{OH}{\underset{}{C}} = CH_2 \right] \longrightarrow R - \overset{O}{\underset{}{C}} - CH_3$$

                                                                          A methyl ketone

   (d) Hydroboration/oxidation (Section 8.5)

$$R - C \equiv C - H \xrightarrow[\text{2. H}_2\text{O}_2]{\text{1. BH}_3} R - CH_2 - \overset{O}{\underset{}{C}} - H$$

   9-BBN (handwritten annotation above BH₃)

   (e) Reduction (Section 8.6)
      (1) Catalytic hydrogenation

$$R - C \equiv C - R' \xrightarrow[\text{Pd/C}]{2 \text{ H}_2} R - \overset{H}{\underset{H}{C}} - \overset{H}{\underset{H}{C}} - R'$$

$$R - C \equiv C - R' \xrightarrow[\text{Lindlar catalyst}]{\text{H}_2} \overset{H \qquad H}{\underset{R \qquad R'}{C = C}}$$

                                                                          A cis alkene

      (2) Lithium/ammonia

$$R - C \equiv C - R \xrightarrow{\text{Li, NH}_3} \overset{H \qquad R}{\underset{R \qquad H}{C = C}}$$

                                                                          A trans alkene   *(continued)* ▶

(f) Acidity: conversion into acetylide anions (Section 8.8)

$$R-C\equiv C-H \xrightarrow[\text{NH}_3]{\text{NaNH}_2} R-C\equiv C\!:^- \text{ Na}^+ + \text{NH}_3$$

(g) Acetylide ion alkylation (Section 8.9)

$$HC\equiv CH \xrightarrow{\text{NaNH}_2} HC\equiv C^- \text{Na}^+ \xrightarrow{\text{RCH}_2\text{Br}} HC\equiv C\,CH_2R$$

Acetylene   **A terminal alkyne**

$$RC\equiv CH \xrightarrow{\text{NaNH}_2} RC\equiv C^- \text{Na}^+ \xrightarrow{\text{R}'\text{CH}_2\text{Br}} RC\equiv C\,CH_2R'$$

A terminal alkyne   **An internal alkyne**

(h) Oxidative cleavage (Section 8.7)

$$R-C\equiv C-R' \xrightarrow[\text{H}_3\text{O}^+]{\text{KMnO}_4} \overset{\overset{\displaystyle O}{\|}}{\text{RCOH}} + \overset{\overset{\displaystyle O}{\|}}{\text{R}'\text{COH}}$$

Visualizing Chemistry

• •

(Problems 8.1–8.13 appear within the chapter.)

8.14 Name the following alkynes, and predict the products of their reaction with (i) H_2 in the presence of a Lindlar catalyst and (ii) H_3O^+ in the presence of $HgSO_4$:

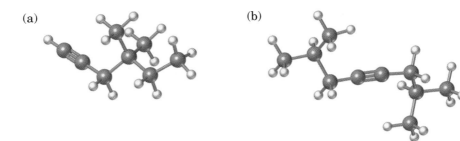

(a)   (b)

8.15 From what alkyne might each of the following substances have been made? (Red = O, yellow-green = Cl.)

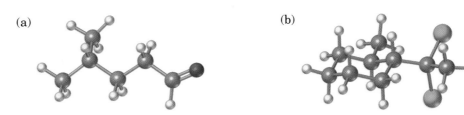

(a)   (b)

8.16 How would you prepare the following substances, starting from any compounds having four carbons or fewer? (Red = O.)

(a)   (b)

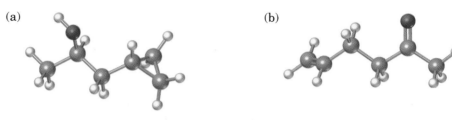

8.17 The following cycloalkyne is too unstable to exist. Explain.

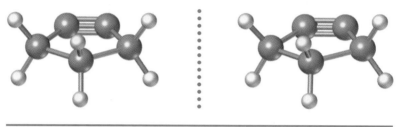

Stereo View

Additional Problems

• •

8.18 Give IUPAC names for the following compounds:

(a) $CH_3CH_2C \equiv CCCH_3$ with CH_3 above and CH_3 below the second-to-last carbon

(b) $CH_3C \equiv CCH_2C \equiv CCH_2CH_3$

(c) $CH_3CH = CC \equiv CCHCH_3$ with CH_3 and CH_3 substituents

(d) $HC \equiv CCCH_2C \equiv CH$ with CH_3 above and CH_3 below

(e) $H_2C = CHCH = CHC \equiv CH$

(f) $CH_3CH_2CHC \equiv CCHCHCH_3$ with CH_2CH_3 above, CH_2CH_3 and CH_3 below

8.19 Draw structures corresponding to the following names:
   (a) 3,3-Dimethyl-4-octyne
   (b) 3-Ethyl-5-methyl-1,6,8-decatriyne
   (c) 2,2,5,5-Tetramethyl-3-hexyne
   (d) 3,4-Dimethylcyclodecyne
   (e) 3,5-Heptadien-1-yne
   (f) 3-Chloro-4,4-dimethyl-1-nonen-6-yne
   (g) 3-*sec*-Butyl-1-heptyne
   (h) 5-*tert*-Butyl-2-methyl-3-octyne

8.20 The following two hydrocarbons have been isolated from various plants in the sunflower family. Name them according to IUPAC rules.
   (a) $CH_3CH=CHC \equiv CC \equiv CCH=CHCH=CHCH=CH_2$ (all trans)
   (b) $CH_3C \equiv CC \equiv CC \equiv CC \equiv CC \equiv CCH=CH_2$

8.21 Predict the products of the following reactions:

A?

B?

8.22 A hydrocarbon of unknown structure has the formula C_8H_{10}. On catalytic hydrogenation over the Lindlar catalyst, 1 equivalent of H_2 is absorbed. On hydrogenation over a palladium catalyst, 3 equivalents of H_2 are absorbed.
(a) How many degrees of unsaturation are present in the unknown?
(b) How many triple bonds are present?
(c) How many double bonds are present?
(d) How many rings are present?
(e) Draw a structure that fits the data.

8.23 Predict the products from reaction of 1-hexyne with the following reagents:
(a) 1 equiv HBr   (b) 1 equiv Cl_2
(c) H_2, Lindlar catalyst   (d) $NaNH_2$ in NH_3, then CH_3Br
(e) H_2O, H_2SO_4, $HgSO_4$ (f) 2 equiv HCl

8.24 Predict the products from reaction of 5-decyne with the following reagents:
(a) H_2, Lindlar catalyst   (b) Li in NH_3
(c) 1 equiv Br_2 (d) BH_3 in THF, then H_2O_2, OH^-
(e) H_2O, H_2SO_4, $HgSO_4$ (f) Excess H_2, Pd/C catalyst

8.25 Predict the products from reaction of 2-hexyne with the following reagents:
(a) 2 equiv Br_2 (b) 1 equiv HBr   (c) Excess HBr
(d) Li in NH_3 (e) H_2O, H_2SO_4, $HgSO_4$

8.26 Predict the products of the following reactions:

(a) $CH_3CH_2CH_2CH_2CH_2C\equiv CH$ $\xrightarrow[\text{2. H}_2\text{O}_2]{\text{1. BH}_3,\ \text{THF}}$?

(b) Br

$\xrightarrow{\text{NaNH}_2,\ \text{NH}_3}$?

(c) Br Br

$\xrightarrow{\text{NaNH}_2,\ \text{NH}_3}$?

8.27 Hydrocarbon A has the formula C_9H_{12} and absorbs 3 equivalents of H_2 to yield B, C_9H_{18}, when hydrogenated over a Pd/C catalyst. On treatment of A with aqueous H_2SO_4 in the presence of mercury(II), two isomeric ketones, C and D, are produced. Oxidation of A with $KMnO_4$ gives a mixture of acetic acid (CH_3COOH) and the tricarboxylic acid E. Propose structures for compounds A–D, and write the reactions.

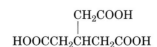

CH_2COOH

$HOOCCH_2CHCH_2COOH$

E

8.28   How would you carry out the following reactions?

(a) $CH_3CH_2C{\equiv}CH \xrightarrow{\ ?\ } CH_3CH_2\overset{\displaystyle O}{\overset{\|}{C}}CH_3$

(b) $CH_3CH_2C{\equiv}CH \xrightarrow{\ ?\ } CH_3CH_2CH_2CHO$

(c) $\underset{}{\text{Ph}}C{\equiv}CH \xrightarrow{\ ?\ } \underset{}{\text{Ph}}C{\equiv}C{-}CH_3$

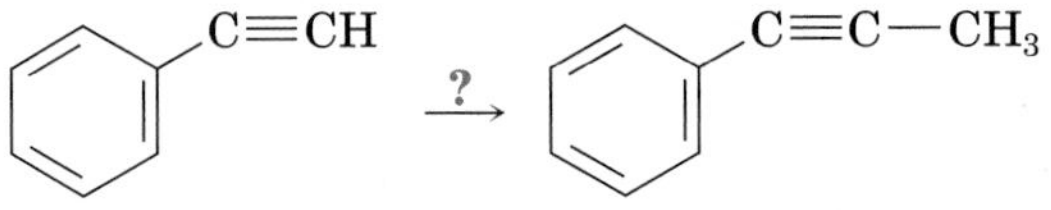

(d) $\underset{}{\text{Ph}}C{\equiv}CCH_3 \xrightarrow{\ ?\ }$ (cis styrene with CH₃)

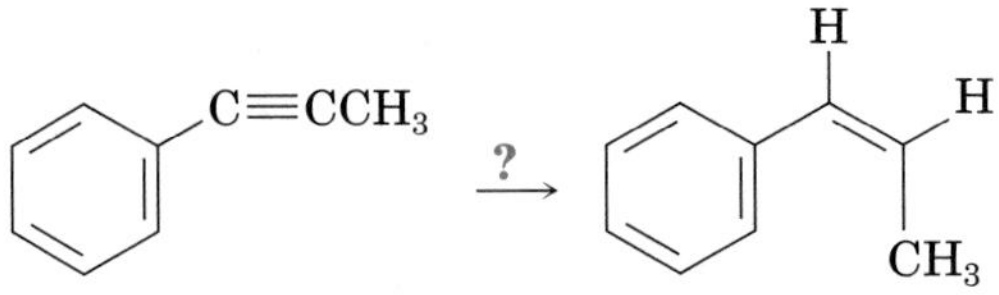

(e) $CH_3CH_2C{\equiv}CH \xrightarrow{\ ?\ } CH_3CH_2COOH$

(f) $CH_3CH_2CH_2CH_2CH{=}CH_2 \xrightarrow[\text{(2 steps)}]{\ ?\ } CH_3CH_2CH_2CH_2C{\equiv}CH$

8.29   Occasionally, chemists need to *invert* the stereochemistry of an alkene—that is, to convert a cis alkene to a trans alkene, or vice versa. There is no one-step method for doing an alkene inversion, but the transformation can be carried out by combining several reactions in the proper sequence. How would you carry out the following reactions?
(a) *trans*-5-Decene $\xrightarrow{\ ?\ }$ *cis*-5-Decene
(b) *cis*-5-Decene $\xrightarrow{\ ?\ }$ *trans*-5-Decene

8.30   Propose structures for hydrocarbons that give the following products on oxidative cleavage by $KMnO_4$ or O_3:

(a) $CO_2 + CH_3(CH_2)_5COOH$ (b) $CH_3COOH +$ (benzene ring with COOH)

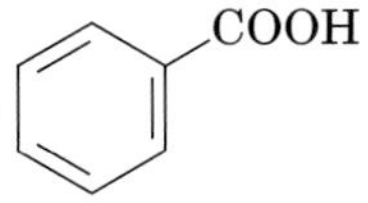

(c) $HOOC(CH_2)_8COOH$ (d) $CH_3CHO + CH_3\overset{\displaystyle O}{\overset{\|}{C}}CH_2CH_2COOH + CO_2$

(e) $HCCH_2CH_2CH_2CH_2CCOOH + CO_2$ (with two C=O groups shown)

8.31   Each of the following syntheses requires more than one step. How would you carry them out?

(a) $CH_3CH_2CH_2C{\equiv}CH \xrightarrow{\ ?\ } CH_3CH_2CH_2CHO$

(b) $(CH_3)_2CHCH_2C{\equiv}CH \xrightarrow{\ ?\ }$ (cis alkene: $(CH_3)_2CHCH_2$ and H on one carbon, H and CH_2CH_3 on the other)

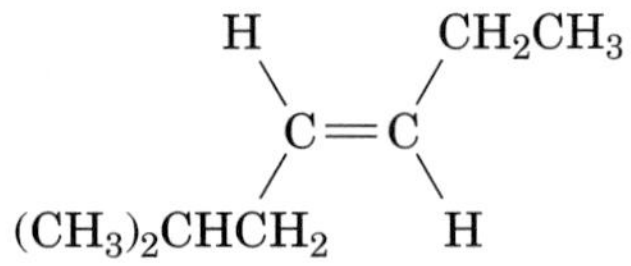

8.32 How would you carry out the following transformation? More than one step is needed.

$$CH_3CH_2CH_2CH_2C\equiv CH \xrightarrow{?}$$

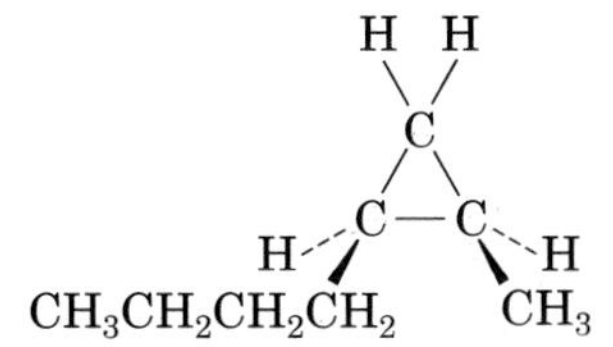

8.33 How would you carry out the following conversion? More than one step is needed.

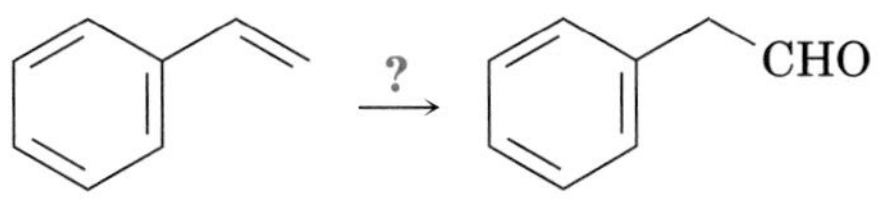

8.34 How would you carry out the following transformation? More than one step is needed.

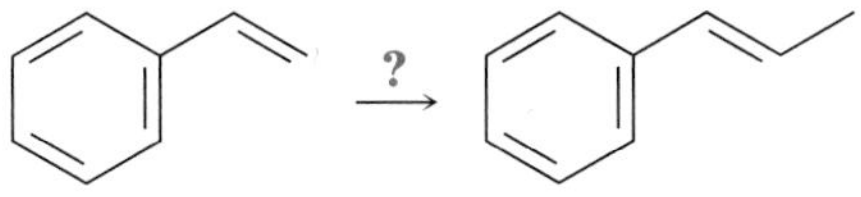

8.35 Synthesize the following compounds using 1-butyne as the only source of carbon, along with any inorganic reagents you need. More than one step may be needed.
(a) 1,1,2,2-Tetrachlorobutane   (b) 1,1-Dichloro-2-ethylcyclopropane
(c) Butanal

8.36 How would you synthesize the following compounds from acetylene and any alkyl halides with four or fewer carbons? More than one step may be required.
(a) $CH_3CH_2CH_2C\equiv CH$ (b) $CH_3CH_2C\equiv CCH_2CH_3$

(c) $(CH_3)_2CHCH_2CH=CH_2$ (d) $CH_3CH_2CH_2\overset{\overset{\displaystyle O}{\|}}{C}CH_2CH_2CH_2CH_3$

(e) $CH_3CH_2CH_2CH_2CH_2CHO$

8.37 How would you carry out the following reactions to introduce deuterium into organic molecules?

(a) $CH_3CH_2C\equiv CCH_2CH_3 \xrightarrow{?}$

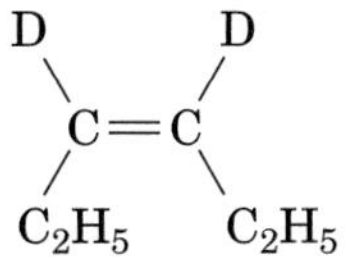

(b) $CH_3CH_2C\equiv CCH_2CH_3 \xrightarrow{?}$

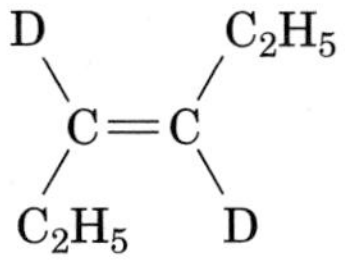

(c) $CH_3CH_2CH_2C\equiv CH \xrightarrow{?} CH_3CH_2CH_2C\equiv CD$

(d)

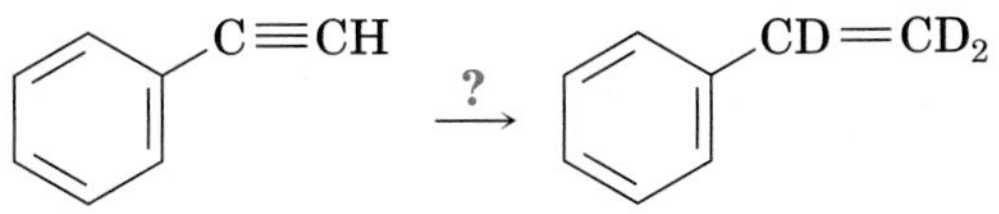

8.38 How would you prepare cyclodecyne starting from acetylene and any alkyl halide needed?

8.39 The sex attractant given off by the common housefly is an alkene named *muscalure*. Propose a synthesis of muscalure starting from acetylene and any alkyl halides needed. What is the IUPAC name for muscalure?

$$cis\text{-}CH_3(CH_2)_7CH{=}CH(CH_2)_{12}CH_3$$

Muscalure

8.40 Compound A (C_9H_{12}) absorbed 3 equivalents of H_2 on catalytic reduction over a palladium catalyst to give B (C_9H_{18}). On ozonolysis, compound A gave, among other things, a ketone that was identified as cyclohexanone. On treatment with $NaNH_2$ in NH_3, followed by addition of iodomethane, compound A gave a new hydrocarbon, C ($C_{10}H_{14}$). What are the structures of A, B, and C?

8.41 Hydrocarbon A has the formula $C_{12}H_8$. It absorbs 8 equivalents of H_2 on catalytic reduction over a palladium catalyst. On ozonolysis, only two products are formed: oxalic acid (HOOCCOOH) and succinic acid ($HOOCCH_2CH_2COOH$). Write the reactions, and propose a structure for A.

8.42 Organometallic reagents such as sodium acetylide undergo an addition reaction with ketones, giving alcohols:

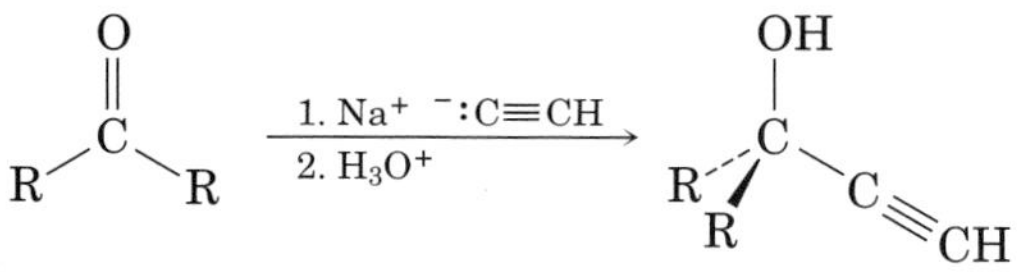

How might you use this reaction to prepare 2-methyl-1,3-butadiene, the starting material used in the manufacture of synthetic rubber?

8.43 Erythrogenic acid, $C_{18}H_{26}O_2$, is an interesting acetylenic fatty acid that turns a vivid red on exposure to light. On catalytic hydrogenation over a palladium catalyst, 5 equivalents of H_2 are absorbed, and stearic acid, $CH_3(CH_2)_{16}COOH$, is produced. Ozonolysis of erythrogenic acid gives four products: formaldehyde, CH_2O; oxalic acid, HOOCCOOH; azelaic acid, $HOOC(CH_2)_7COOH$; and the aldehyde acid $OHC(CH_2)_4COOH$. Draw two possible structures for erythrogenic acid, and suggest a way to tell them apart by carrying out some simple reactions.

8.44 Terminal alkynes react with Br_2 and water to yield bromo ketones. For example:

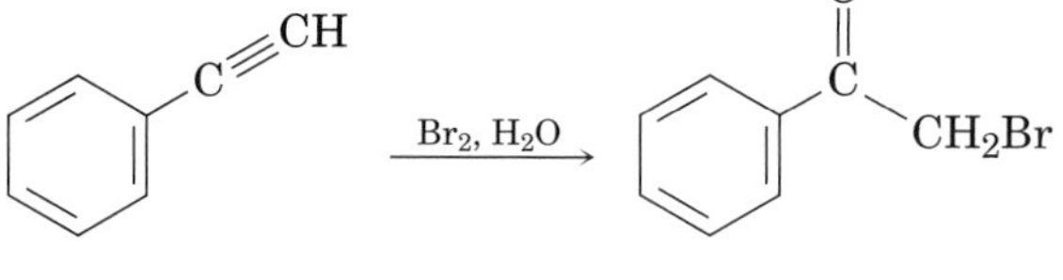

Propose a mechanism for the reaction. To what reaction of alkenes is the process analogous?

8.45 A *cumulene* is a compound with three adjacent double bonds. Draw an orbital picture of a cumulene. What kind of hybridization do the two central carbon atoms have? What is the geometric relationship of the substituents on one end to the substituents on the other end? What kind of isomerism is possible? Make a model to help see the answer.

$$R_2C{=}C{=}C{=}CR_2$$

A cumulene

A Look Ahead

- -

8.46 Reaction of acetone with D_3O^+ yields hexadeuterioacetone. That is, all the hydrogens in acetone are exchanged for deuterium. Review the mechanism of alkyne hydration, and then propose a mechanism for this deuterium incorporation. (See Section 22.2.)

Acetone   **Hexadeuterioacetone**

Molecular Modeling

- -

8.47 Cycloheptene is a stable molecule, but cycloheptyne is not. Use SpartanBuild to build structures, and then minimize the energies of cycloheptene and cycloheptyne. Why is cycloheptyne so reactive?

8.48 Use SpartanView to examine the electrostatic potential map of the acetylenic ether shown below. Assuming that the most positive hydrogen is also the most acidic, what product would be obtained from treatment of the compound with $NaNH_2$ to form an anion, followed by alkylation with 1-bromopropane?

$$\underset{}{}O{\diagdown}O{\diagup}CH_2C{\equiv}CH$$

8.49 Treatment of 3-chloro-2-buten-1-ol with $NaNH_2$ gives primarily 2-butyn-1-ol rather than 2,3-butadien-1-ol. Using SpartanView to obtain the energies of the two products, tell whether the lower-energy molecule corresponds to the observed product.

$$\overset{\overset{\textstyle Cl}{|}}{CH_3C}{=}CHCH_2OH \xrightarrow{\text{2 NaNH}_2} CH_3C{\equiv}CCH_2OH \qquad [H_2C{=}C{=}CHCH_2OH]$$

3-Chloro-2-buten-1-ol   **2-Butyn-1-ol**   **2,3-Butadien-1-ol**
                                                              Not formed

8.50 Use SpartanView to compare energies of cis and trans isomers for a vinylic radical and a vinylic anion. Which intermediate, radical or anion, shows a stronger preference for the trans structure?

Stereochemistry

Are you right-handed or left-handed? Though most of us don't often think about it, handedness plays a surprisingly large role in our daily activities. Many musical instruments, such as oboes and clarinets, have a handedness to them; the last available softball glove always fits the wrong hand; left-handed people write in a "funny" way. The fundamental reason for these difficulties is that our hands aren't identical; rather, they're *mirror images*. When you hold a *right* hand up to a mirror, the image you see looks like a *left* hand. Try it.

Handedness also plays a large role in organic chemistry as a direct consequence of the tetrahedral stereochemistry of sp^3-hybridized carbon. Most drugs and most of the molecules in our bodies, for instance, are handed.

Furthermore, it is molecular handedness that makes possible many of the specific interactions between molecules that are so crucial to biochemistry. Let's see how handedness in organic molecules arises.

9.1 Enantiomers and the Tetrahedral Carbon

Look at the generalized molecules of the type CH_3X, CH_2XY, and CHXYZ shown in Figure 9.1. On the left are three molecules, and on the right are their images reflected in a mirror. The CH_3X and CH_2XY molecules are identical to their mirror images and thus are not handed. If you make a molecular model of each molecule and of its mirror image, you can superimpose one on the other.

FIGURE 9.1 ▼

Tetrahedral carbon atoms and their mirror images. Molecules of the type CH_3X and CH_2XY are identical to their mirror images, but a molecule of the type CHXYZ is not. A CHXYZ molecule is related to its mirror image in the same way that a right hand is related to a left hand.

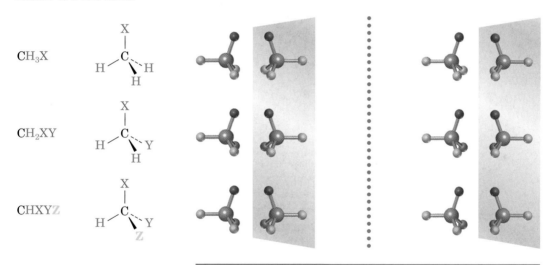

Stereo View

Left hand

Right hand

Unlike the CH_3X and CH_2XY molecules, the CHXYZ molecule is *not* identical to its mirror image. You can't superimpose a model of the molecule on a model of its mirror image for the same reason that you can't superimpose a left hand on a right hand. You might get *two* of the substituents superimposed, X and Y for example, but H and Z would be reversed. If the H and Z substituents were superimposed, X and Y would be reversed.

Mirror-image molecules that are not superimposable are called **enantiomers** (Greek *enantio,* "opposite"). Enantiomers are related to each other as a right hand is related to a left hand and result whenever a tetrahedral carbon is bonded to four different substituents (one need not be H). For example, lactic acid (2-hydroxypropanoic acid) exists as a pair of enantiomers because there are four different groups (–H, –OH, –CH₃, –COOH) bonded to the central carbon atom. The enantiomers are called (+)-lactic acid and (−)-lactic acid.

Lactic acid: a molecule of general formula CHXYZ

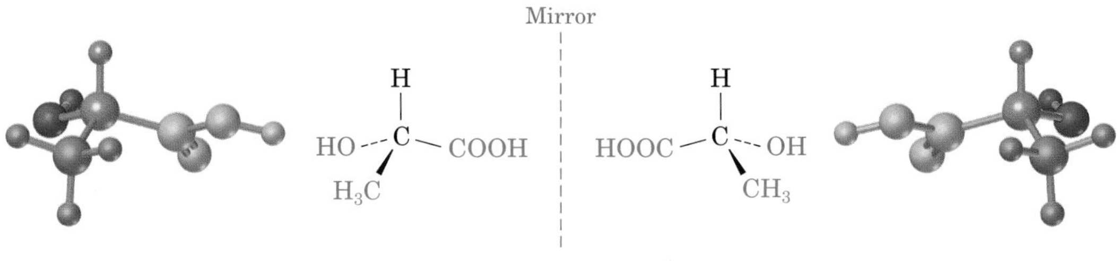

(+)-Lactic acid   **(−)-Lactic acid**

No matter how hard you try, you can't superimpose a molecule of (+)-lactic acid on a molecule of (−)-lactic acid; the two simply aren't identical, as Figure 9.2 shows. If any two groups match up, say –H and –COOH, the remaining two groups don't match.

FIGURE 9.2 ▼

Attempts at superimposing the mirror-image forms of lactic acid: (a) When the –H and –OH substituents match up, the –COOH and –CH₃ substituents don't; (b) when –COOH and –CH₃ match up, –H and –OH don't. Regardless of how the molecules are oriented, they aren't identical.

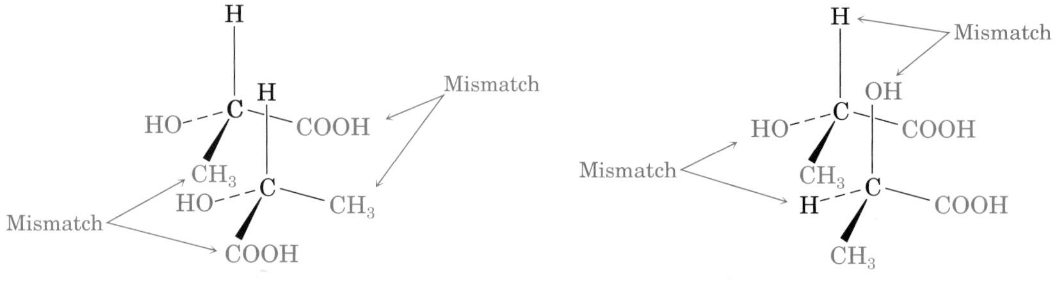

(a)   (b)

9.2 The Reason for Handedness in Molecules: Chirality

Molecules that are not superimposable with their mirror images and thus exist in two enantiomeric forms are said to be **chiral** (ky-ral, from the Greek *cheir,* "hand"). You can't take a chiral molecule and its enantiomer and place one on the other so that all atoms coincide.

How can you predict whether a given molecule is or is not chiral? *A molecule is not chiral if it contains a plane of symmetry.* A **plane of symmetry** is a plane that cuts through the middle of an object (or molecule) in such a way that one half of the object is a mirror image of the other half. For example, a laboratory flask has a plane of symmetry. If you were to cut the flask in half, one half would be a mirror image of the other half. A hand, however, has no plane of symmetry. One "half" of a hand is not a mirror image of the other "half" (Figure 9.3).

FIGURE 9.3 ▼

The meaning of *symmetry plane.* An object like the flask (a) has a symmetry plane cutting through it, making right and left halves mirror images. An object like a hand (b) has no symmetry plane; the right "half" of a hand is not a mirror image of the left "half."

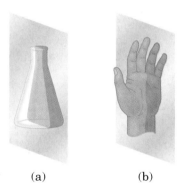

(a)   (b)

A molecule that has a plane of symmetry in any of its possible conformations must be identical to its mirror image and hence must be nonchiral, or **achiral** (a-**ky**-ral). Thus, propanoic acid has a plane of symmetry when it is lined up as shown in Figure 9.4 (p. 310), and it is therefore achiral. Lactic acid (2-hydroxypropanoic acid), however, has no plane of symmetry and is thus chiral.

The most common, although not the only, cause of chirality in an organic molecule is the presence of a carbon atom bonded to four different groups—for example, the central carbon atom in lactic acid. Such carbons are currently referred to as **chirality centers**, although numerous other terms such as *asymmetric center* and *stereogenic center* have been also used. Note that *chirality* is a property of the entire molecule, whereas a *chirality center* is a structural feature within the molecule that gives rise to chirality.

FIGURE 9.4 ▼

The achiral propanoic acid molecule versus the chiral lactic acid molecule. Propanoic acid has a plane of symmetry that makes one side of the molecule a mirror image of the other side. Lactic acid, however, has no such symmetry plane.

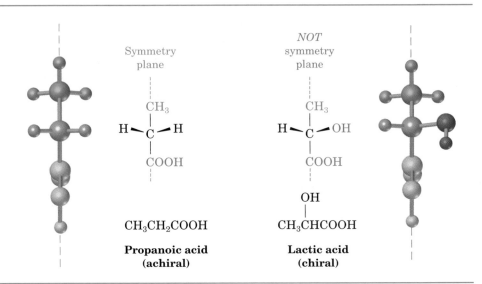

Symmetry plane

CH_3

H —C— H

COOH

CH_3CH_2COOH

Propanoic acid (achiral)

NOT symmetry plane

CH_3

H —C— OH

COOH

OH

$CH_3CHCOOH$

Lactic acid (chiral)

Detecting chirality centers in a complex molecule takes practice because it's not always immediately apparent that four different groups are bonded to a given carbon. The differences don't necessarily appear right next to the chirality center. For example, 5-bromodecane is a chiral molecule because four different groups are bonded to C5, the chirality center (marked by an asterisk):

$$\underset{\overset{|}{H}}{\overset{\overset{Br}{|}}{CH_3CH_2CH_2CH_2CH_2\underset{*}{C}CH_2CH_2CH_2CH_3}}$$

5-Bromodecane (chiral)

Substituents on carbon 5

—H

—Br

— $CH_2CH_2CH_2CH_3$ (butyl)

— $CH_2CH_2CH_2CH_2CH_3$ (pentyl)

A butyl substituent is *similar* to a pentyl substituent but is not identical. The difference isn't apparent until four carbon atoms away from the chirality center, but there's still a difference.

As other examples, look at methylcyclohexane and 2-methylcyclohexanone. Are either of these molecules chiral?

Methylcyclohexane (achiral)

2-Methylcyclohexanone (chiral)

Methylcyclohexane is achiral because no carbon atom in the molecule is bonded to four different groups. You can immediately eliminate all –CH_2–

carbons and the –CH$_3$ carbon from consideration, but what about C1 on the ring? The C1 carbon atom is bonded to a –CH$_3$ group, to an –H atom, and to C2 and C6 of the ring. Carbons 2 and 6 are equivalent, however, as are carbons 3 and 5. Thus, the C6–C5–C4 "substituent" is equivalent to the C2–C3–C4 substituent, and methylcyclohexane is therefore achiral. Another way of reaching the same conclusion is to realize that methylcyclohexane has a symmetry plane passing through the methyl group and through C1 and C4 of the ring. Make a molecular model to see this symmetry plane more clearly.

Methylcyclohexane is achiral because of a symmetry plane.

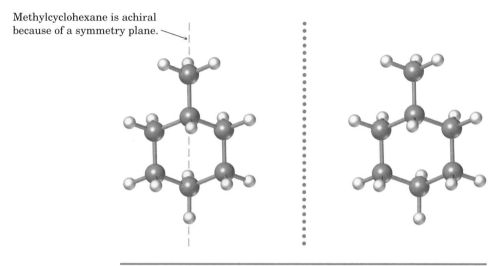

Stereo View

The situation is different for 2-methylcyclohexanone. 2-Methylcyclohexanone has no symmetry plane and is chiral because C2 is bonded to four different groups: a –CH$_3$ group, an –H atom, a –COCH$_2$– ring bond (C1), and a –CH$_2$CH$_2$– ring bond (C3).

Several more examples of chiral molecules are shown below. Check for yourself that the labeled carbons are chirality centers. (Carbons in –CH$_2$–, –CH$_3$, C=C, C=O, and C≡C groups *can't* be chirality centers.)

Carvone (spearmint oil)   Nootkatone (grapefruit oil)

Problem 9.1 Which of the following objects are chiral?
(a) A screwdriver   (b) A screw   (c) A bean stalk
(d) A shoe   (e) A hammer

Problem 9.2 Which of the following compounds are chiral? Build molecular models if you need help.

(a)
Toluene

(b)
CH$_2$CH$_2$CH$_3$
Coniine
(from poison hemlock)

(c)
Phenobarbital
(tranquilizer)

Problem 9.3 Place asterisks at all chirality centers in the following molecules:

(a)
Menthol

(b)
Camphor

(c)
Dextromethorphan
(a cough suppressant)

Problem 9.4 Alanine, an amino acid found in proteins, is chiral. Draw the two enantiomers of alanine using the standard convention of solid, wedged, and dashed lines.

$$NH_2$$
$$CH_3CHCOOH \quad \textbf{Alanine}$$

9.3 Optical Activity

Jean Baptiste Biot

Jean Baptiste Biot (1774–1862) was born in Paris, France, and was educated there at the École Polytechnique. His work on determining the optical rotation of naturally occurring molecules included an experiment on turpentine, which caught fire and nearly burned down the church he was using for his experiments.

The study of stereochemistry has its origins in the work of the nineteenth century French scientist Jean Baptiste Biot, who was investigating the nature of *plane-polarized light*. A beam of ordinary light consists of electromagnetic waves that oscillate in an infinite number of planes at right angles to the direction of light travel. When a beam of ordinary light is passed through a device called a *polarizer*, however, only the light waves oscillating in a *single* plane pass through—hence the name **plane-polarized light**. Light waves in all other planes are blocked out.

Biot made the remarkable observation that, when a beam of plane-polarized light passes through a solution of certain organic molecules such as sugar or camphor, the plane of polarization is *rotated*. Not all organic substances exhibit this property, but those that do are said to be **optically active**.

The amount of rotation can be measured with an instrument known as a *polarimeter*, represented schematically in Figure 9.5. A solution of optically active organic molecules is placed in a sample tube, plane-polarized light is passed through the tube, and rotation of the polarization plane occurs. The light then goes through a second polarizer called the *analyzer*. By rotating the analyzer until the light passes through it, we can find the new plane of polarization and can tell to what extent rotation has occurred. The amount of rotation is denoted α (Greek alpha) and is expressed in degrees.

FIGURE 9.5 ▼

Schematic representation of a polarimeter. Plane-polarized light passes through a solution of optically active molecules, which rotate the plane of polarization.

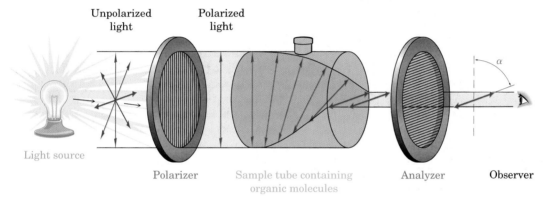

Unpolarized light   Polarized light

Light source

Polarizer   Sample tube containing organic molecules   Analyzer   Observer

In addition to determining the extent of rotation, we can also find the direction. From the vantage point of an observer looking directly end-on at the analyzer, some optically active molecules rotate polarized light to the left (counterclockwise) and are said to be **levorotatory**, whereas others rotate polarized light to the right (clockwise) and are said to be **dextrorotatory**. By convention, rotation to the left is given a minus sign ($-$), and rotation to the right is given a plus sign ($+$). For example, ($-$)-morphine is levorotatory, and ($+$)-sucrose is dextrorotatory.

9.4 Specific Rotation

The amount of rotation observed in a polarimetry experiment depends on the number of optically active molecules that the light beam encounters. The more molecules the light encounters, the greater the observed rotation. Thus, the amount of rotation depends on both sample concentration and sample pathlength. If we double the concentration of sample, the observed rotation doubles. Similarly, if we keep the concentration constant but double the length of the sample tube, the observed rotation doubles. It also turns out that the amount of rotation depends on the wavelength of the light used.

To express optical rotation data in a meaningful way so that comparisons can be made, we have to choose standard conditions. The **specific rotation**, $[\alpha]_D$, of a compound is defined as the observed rotation when the sample pathlength l is 1 decimeter (1 dm = 10 cm), the sample concentration C is 1 g/mL, and light of 589 nanometer (nm) wavelength is used. (Light of 589 nm, the so-called *sodium D line,* is the yellow light emitted from common sodium street lamps; 1 nm = 10^{-9} m.)

$$[\alpha]_D = \frac{\text{Observed rotation (degrees)}}{\text{Pathlength, } l \text{ (dm)} \times \text{Concentration, } C \text{ (g/mL)}} = \frac{\alpha}{l \times C}$$

When optical rotation data are expressed in this standard way, the specific rotation, $[\alpha]_D$, is a physical constant characteristic of a given optically active compound. For example, the (+)-lactic acid that we saw in Section 9.1 has $[\alpha]_D = +3.82°$, and (−)-lactic acid has $[\alpha]_D = -3.82°$. Some additional examples are listed in Table 9.1.

TABLE 9.1   Specific Rotation of Some Organic Molecules

Compound	$[\alpha]_D$ (degrees)	Compound	$[\alpha]_D$ (degrees)
Penicillin V	+233	Cholesterol	−31.5
Sucrose	+66.47	Morphine	−132
Camphor	+44.26	Acetic acid	0
Monosodium glutamate	+25.5	Benzene	0

. . . .

Problem 9.5   A 1.50 g sample of coniine, the toxic extract of poison hemlock, was dissolved in 10.0 mL of ethanol and placed in a sample cell with a 5.00 cm pathlength. The observed rotation at the sodium D line was +1.21°. Calculate $[\alpha]_D$ for coniine.

. . . .

9.5 Pasteur's Discovery of Enantiomers

Little was done after Biot's discovery of optical activity until Louis Pasteur began work in 1849. Pasteur had received his formal training in chemistry but had become interested in the subject of crystallography. He began work on crystalline salts of tartaric acid derived from wine and was repeating some measurements published a few years earlier when he made a surprising observation. On recrystallizing a concentrated solution of sodium ammonium tartrate below 28°C, two distinct kinds of crystals precipitated. Furthermore, the two kinds of crystals were mirror images and were related to each other in the same way that a right hand is related to a left hand.

Working carefully with tweezers, Pasteur was able to separate the crystals into two piles, one of "right-handed" crystals and one of "left-handed" crystals like those shown in Figure 9.6. Although the original sample (a 50:50 mixture of right and left) was optically inactive, *solutions of the crystals from each of the sorted piles were optically active,* and their specific rotations were equal in amount but opposite in sign.

FIGURE 9.6 ▼

Drawings of sodium ammonium tartrate crystals taken from Pasteur's original sketches. One of the crystals is "right-handed" and one is "left-handed."

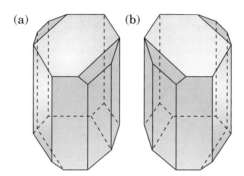

(a)   (b)

$$COO^- \ Na^+$$
$$H-C-OH$$
$$HO-C-H$$
$$COO^- \ NH_4^+$$

Sodium ammonium tartrate

Louis Pasteur

Louis Pasteur (1822–1895) was born at Dôle, the son of leather tanners. After receiving his doctorate from the École Normale Supérieure at age 25, his landmark discovery of tartaric acid enantiomers was made only one year later. Pasteur is best known for his studies in bacteriology and for his discovery of vaccines for anthrax and rabies.

Pasteur was far ahead of his time. Although the structural theory of Kekulé had not yet been proposed, Pasteur explained his results by speaking of the *molecules themselves,* saying, "There is no doubt that [in the *dextro* tartaric acid] there exists an asymmetric arrangement having a non-superimposable image. It is no less certain that the atoms of the *levo* acid possess precisely the inverse asymmetric arrangement." Pasteur's vision was extraordinary, for it was not until 25 years later that the theories of van't Hoff and Le Bel confirmed his ideas regarding the asymmetric carbon atom.

Today, we would describe Pasteur's work by saying that he had discovered the phenomenon of enantiomerism. The enantiomeric tartaric acid salts that Pasteur separated are physically identical in all respects except for their interaction with plane-polarized light. They have the same melting point, the same boiling point, the same solubilities, and the same spectroscopic properties.

9.6 Sequence Rules for Specification of Configuration

Although drawings provide a pictorial representation of stereochemistry, they are difficult to translate into words. Thus, a verbal method for indicating the three-dimensional arrangement of atoms, or **configuration**, at a chirality center is also necessary. The standard method employs the same Cahn–Ingold–Prelog sequence rules used for the specification of E and Z alkene geometry in Section 6.6. Let's briefly review the sequence rules and see how they're used to specify the configuration of a chirality center. Refer to Section 6.6 for an explanation of each rule.

RULE 1   Look at the four atoms directly attached to the chirality center, and assign priorities in order of decreasing atomic number. The atom with highest atomic number is ranked first; the atom with lowest atomic number is ranked fourth.

RULE 2   If a decision about priority can't be reached by applying rule 1, compare atomic numbers of the second atoms in each substituent, continuing on as necessary through the third or fourth atoms until the first point of difference is reached.

RULE 3   Multiple-bonded atoms are equivalent to the same number of single-bonded atoms. For example:

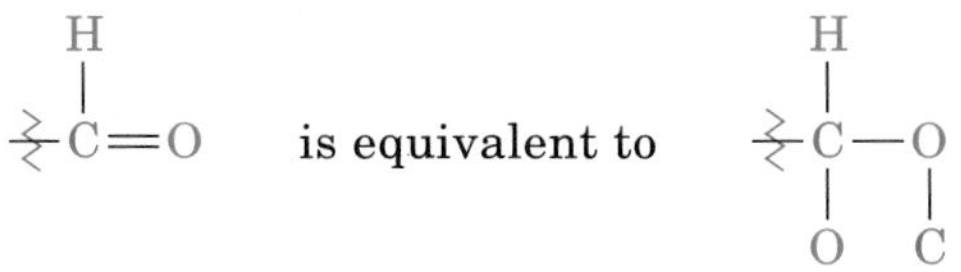

Having assigned priorities to the four groups attached to a chiral carbon, we describe the stereochemical configuration around the carbon by orienting the molecule so that the group of lowest priority (4) is pointing directly back, away from us. We then look at the three remaining substituents, which now appear to radiate toward us like the spokes on a steering wheel (Figure 9.7). If a curved arrow drawn from the highest to second-highest to third-highest-priority substituent ($1 \rightarrow 2 \rightarrow 3$) is clockwise, we say that the chirality center has the R configuration (Latin *rectus,* "right"). If an arrow from $1 \rightarrow 2 \rightarrow 3$ is counterclockwise, the chirality center has the S configuration (Latin *sinister,* "left"). To remember these assignments, think of a car's steering wheel when making a right (clockwise) or left (counterclockwise) turn.

Look at (−)-lactic acid in Figure 9.8 for an example of how configuration is assigned. Sequence rule 1 says that –OH has priority 1 and –H has priority 4, but it doesn't allow us to distinguish between –CH$_3$ and –COOH because both groups have carbon as their first atom. Sequence rule 2, however, says that –COOH is higher priority than –CH$_3$ because O outranks H (the second atom in each group). Now, turn the molecule so that the fourth-priority group (–H) is oriented toward the rear, away from the observer. Since a curved arrow from 1 (–OH) to 2 (–COOH) to 3 (–CH$_3$) is clockwise (right turn of the steering wheel), (−)-lactic acid has the R configuration. Applying the same procedure to (+)-lactic acid leads to the opposite assignment.

Further examples are provided by naturally occurring (−)-glyceraldehyde and (+)-alanine, which both have the S configuration, as shown in Figure 9.9 (p. 318). *Note that the sign of optical rotation,* (+) *or* (−), *is not related to the R,S designation.* (S)-Glyceraldehyde happens to be levorotatory (−) and (S)-alanine happens to be dextrorotatory (+). There is no simple correlation between R,S configuration and direction or magnitude of optical rotation.

FIGURE 9.7 ▼

Assignment of
configuration to a
chirality center. When the
molecule is oriented so
that the group of lowest
priority (4) is toward the
rear, the remaining three
groups radiate toward the
viewer like the spokes of a
steering wheel. If the
direction of travel
1 → 2 → 3 is clockwise
(right turn), the center
has the *R* configuration. If
the direction of travel
1 → 2 → 3 is
counterclockwise (left
turn), the center is *S.*

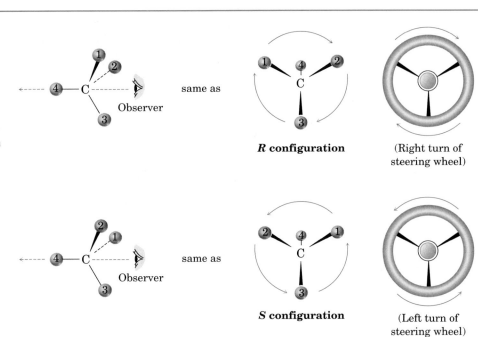

R configuration   (Right turn of steering wheel)

S configuration   (Left turn of steering wheel)

FIGURE 9.8 ▼

Assignment of configuration to (a) (*R*)-(−)-lactic acid and (b) (*S*)-(+)-lactic acid.

(a) (−)-Lactic acid

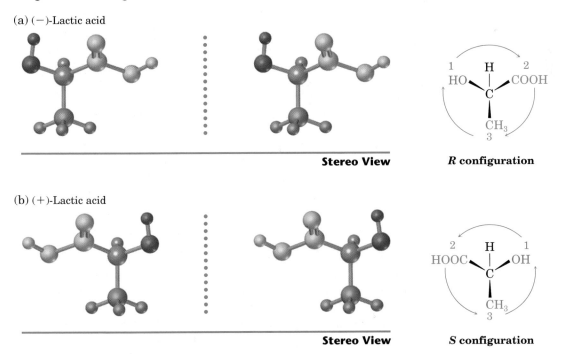

Stereo View   *R* configuration

(b) (+)-Lactic acid

Stereo View   *S* configuration

FIGURE 9.9 ▼

Assignment of configuration to (a) (−)-glyceraldehyde and (b) (+)-alanine. Both happen to have the *S* configuration, although one is levorotatory and the other is dextrorotatory.

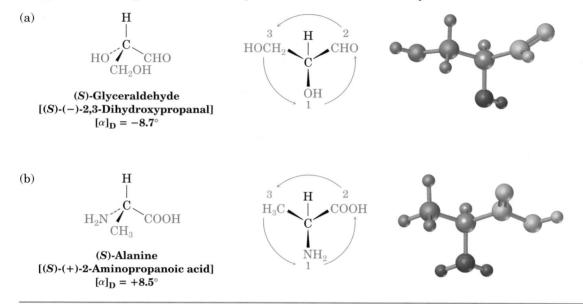

(a)

(S)-Glyceraldehyde
[(S)-(−)-2,3-Dihydroxypropanal]
$[\alpha]_D = -8.7°$

(b)

(S)-Alanine
[(S)-(+)-2-Aminopropanoic acid]
$[\alpha]_D = +8.5°$

One further point needs mentioning: the matter of **absolute configuration**. How do we know that our assignments of *R,S* configuration are correct in an *absolute*, rather than a relative, sense? Since we can't see the molecules themselves, how do we know that the *R* configuration belongs to the dextrorotatory enantiomer of lactic acid? This difficult question was not solved until 1951 when J. M. Bijvoet of the University of Utrecht reported an X-ray spectroscopic method for determining the absolute spatial arrangement of atoms in a molecule. Based on his results, we can say with certainty that the *R,S* conventions are correct.

• •

Practice Problem 9.1   Orient each of the following drawings so that the lowest-priority group is toward the rear, and then assign *R* or *S* configuration:

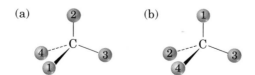

(a)   (b)

Strategy   It takes practice to be able to visualize and orient a chirality center in three dimensions. You might start by indicating where the observer must be located—180° opposite the lowest-priority group. Then imagine yourself in the position of the observer, and redraw what you would see.

Solution In (a), you would be located in front of the page toward the top right of the molecule, and you would see group 2 to your left, group 3 to your right, and group 1 below you. This corresponds to an *R* configuration.

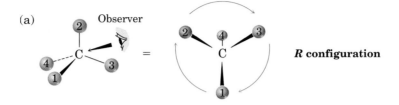

  In (b), you would be located behind the page toward the top *left* of the molecule from your point of view, and you would see group 3 to your left, group 1 to your right, and group 2 below you. This also corresponds to an *R* configuration.

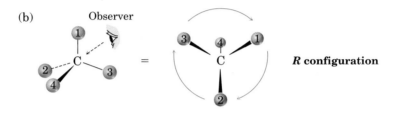

Practice Problem 9.2 Draw a tetrahedral representation of (*R*)-2-chlorobutane.

Strategy Begin by assigning priorities to the four substituents bonded to the chirality center. To draw a tetrahedral representation of the molecule, orient the low-priority –H group away from you and imagine that the other three groups are coming out of the page toward you. Then place the remaining three substituents such that the direction of travel 1 → 2 → 3 is clockwise (right turn), and tilt the molecule toward you by 90° to bring the rear hydrogen into view.

Solution The four substituents bonded to the chiral carbon of (*R*)-2-chlorobutane can be assigned the following priorities: (1) –Cl, (2) –CH$_2$CH$_3$, (3) –CH$_3$, (4) –H. Orienting the low-priority –H group away from you, placing the remaining three substituents such that the direction of travel 1 → 2 → 3 is clockwise, and then tilting the molecule forward, gives the following tetrahedral representation:

$$
\underset{3}{\text{CH}_3}\ \overset{\displaystyle \overset{1}{\text{Cl}}\diagdown \underset{\text{C}}{\overset{\text{H}}{|}} \diagup \overset{2}{\text{CH}_2\text{CH}_3}}{} \quad\longrightarrow\quad \underset{\underset{\text{Cl}}{}}{\text{H}_3\text{C}}\diagup\overset{\text{H}}{\underset{|}{\text{C}}}\diagdown\text{CH}_2\text{CH}_3 \qquad \textbf{(\textit{R})-2-Chlorobutane}
$$

Using molecular models is a great help in working problems of this sort.

Problem 9.6   Assign priorities to the following sets of substituents:
(a) –H, –Br, –CH$_2$CH$_3$, –CH$_2$CH$_2$OH   (b) –CO$_2$H, –CO$_2$CH$_3$, –CH$_2$OH, –OH
(c) –CN, –CH$_2$NH$_2$, –CH$_2$NHCH$_3$, –NH$_2$ (d) –Br, –CH$_2$Br, –Cl, –CH$_2$Cl

Problem 9.7   Orient each of the following drawings so that the lowest-priority group is toward the rear, and then assign R or S configuration:

Problem 9.8   Assign R or S configuration to the chirality center in each of the following molecules:

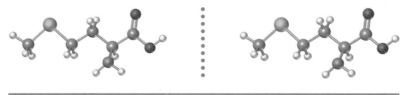

Problem 9.9   Draw a tetrahedral representation of (S)-2-pentanol (2-hydroxypentane).

Problem 9.10   Assign R or S configuration to the chirality center in the following molecular model of the amino acid methionine (red = O, blue = N, yellow = S).

Stereo View

9.7 Diastereomers

Molecules like lactic acid, alanine, and glyceraldehyde are relatively simple because each has only one chirality center and only two stereoisomers. The situation becomes more complex, however, with molecules that have more than one chirality center.

Look at the amino acid threonine (2-amino-3-hydroxybutanoic acid), for example. Since threonine has two chirality centers (C2 and C3), there are four possible stereoisomers, as shown in Figure 9.10. Check for yourself that the R,S configurations are correct as indicated.

The four threonine stereoisomers can be grouped into two pairs of enantiomers. The $2R,3R$ stereoisomer is the mirror image of $2S,3S$, and the $2R,3S$ stereoisomer is the mirror image of $2S,3R$. But what is the relationship

FIGURE 9.10 ▼

The four stereoisomers of 2-amino-3-hydroxybutanoic acid (threonine).

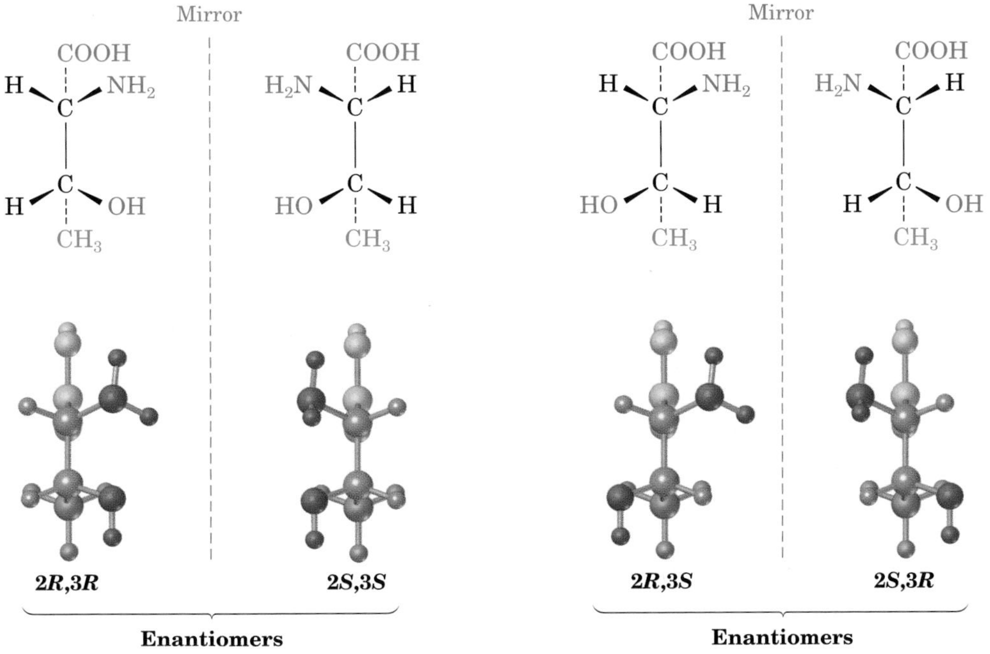

2R,3R   2S,3S   2R,3S   2S,3R

Enantiomers   Enantiomers

between any two molecules that are not mirror images? What, for example, is the relationship between the 2R,3R isomer and the 2R,3S isomer? They are stereoisomers, yet they aren't enantiomers. To describe such a relationship, we need a new term—*diastereomer.*

Diastereomers are stereoisomers that are not mirror images of each other. Chiral diastereomers have opposite configurations at *some* (one or more) chirality centers, but have the same configuration at others. Enantiomers, by contrast, have opposite configurations at *all* chirality centers. A full description of the four threonine stereoisomers is given in Table 9.2.

TABLE 9.2 Relationships Among Four Stereoisomers of Threonine

Stereoisomer	Enantiomeric with	Diastereomeric with
2R,3R	2S,3S	2R,3S and 2S,3R
2S,3S	2R,3R	2R,3S and 2S,3R
2R,3S	2S,3R	2R,3R and 2S,3S
2S,3R	2R,3S	2R,3R and 2S,3S

Of the four stereoisomers of threonine, only the $2S,3R$ isomer, $[\alpha]_D = -29.3°$, occurs naturally in plants and animals. This result is typical: Most biologically important molecules are chiral, and usually only a single stereoisomer is found in nature.

• •

Problem 9.11   Assign R,S configurations to each chirality center in the following molecules. Which are enantiomers, and which are diastereomers?

(a)

$$
\begin{array}{c}
\text{Br} \\
\text{H} \diagdown \overset{|}{\text{C}} \diagup \text{CH}_3 \\
\text{H} \diagup \overset{|}{\text{C}} \diagdown \text{OH} \\
\text{CH}_3
\end{array}
$$

(b)

$$
\begin{array}{c}
\text{CH}_3 \\
\text{H} \diagdown \overset{|}{\text{C}} \diagup \text{Br} \\
\text{H}_3\text{C} \diagup \overset{|}{\text{C}} \diagdown \text{H} \\
\text{OH}
\end{array}
$$

(c)

$$
\begin{array}{c}
\text{CH}_3 \\
\text{Br} \diagdown \overset{|}{\text{C}} \diagup \text{H} \\
\text{H} \diagup \overset{|}{\text{C}} \diagdown \text{CH}_3 \\
\text{OH}
\end{array}
$$

(d)

$$
\begin{array}{c}
\text{H} \\
\text{Br} \diagdown \overset{|}{\text{C}} \diagup \text{CH}_3 \\
\text{H}_3\text{C} \diagup \overset{|}{\text{C}} \diagdown \text{OH} \\
\text{H}
\end{array}
$$

Problem 9.12   Chloramphenicol, a powerful antibiotic isolated in 1949 from the *Streptomyces venezuelae* bacterium, is active against a broad spectrum of bacterial infections and is particularly valuable against typhoid fever. Assign R,S configurations to the chirality centers in chloramphenicol.

$$
\begin{array}{c}
\text{NO}_2 \\
\\
\\
\text{HO} \diagdown \overset{|}{\underset{|}{\text{C}}} \diagup \text{H} \\
\text{H} \diagup \overset{|}{\underset{|}{\text{C}}} \diagdown \text{NHCOCHCl}_2 \\
\text{CH}_2\text{OH}
\end{array}
$$

Chloramphenicol
$[\alpha]_D = +18.6°$

Problem 9.13   Assign R,S configuration to each chirality center in the following molecular model of the amino acid isoleucine (red = O, blue = N):

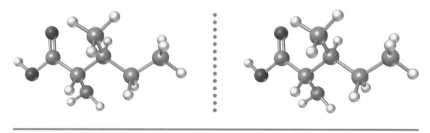

Stereo View

• •

9.8 Meso Compounds

Let's look at one more example of a compound with two chirality centers: tartaric acid. We're already acquainted with tartaric acid because of its role in Pasteur's discovery of optical activity, and we can now draw the four stereoisomers:

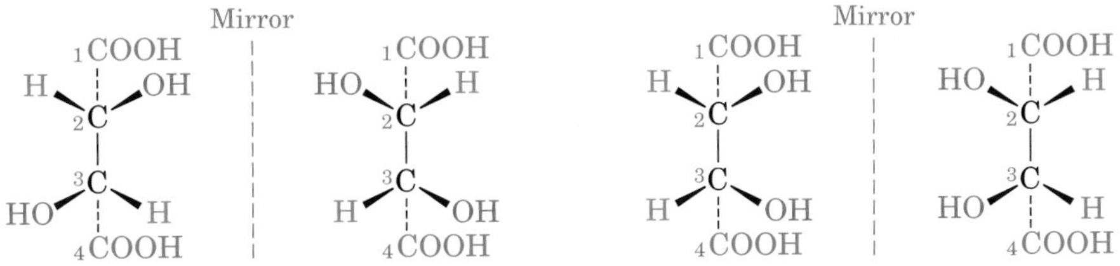

2R,3R   **2S,3S**   **2R,3S**   **2S,3R**

The mirror-image 2R,3R and 2S,3S structures are not identical and are therefore a pair of enantiomers. A careful look, however, shows that the 2R,3S and 2S,3R structures *are* identical, as can be seen by rotating one structure 180°:

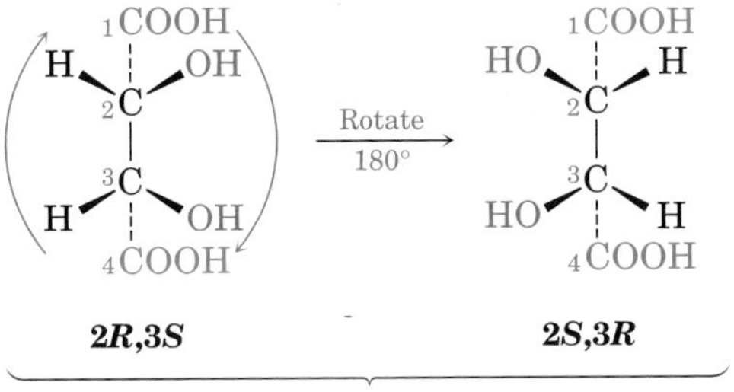

2R,3S   **2S,3R**

Identical

The 2R,3S and 2S,3R structures are identical because the molecule has a plane of symmetry and is therefore achiral. The symmetry plane cuts through the C2–C3 bond, making one half of the molecule a mirror image of the other half (Figure 9.11).

FIGURE 9.11 ▼

A symmetry plane through the C2–C3 bond of *meso*-tartaric acid makes the molecule achiral.

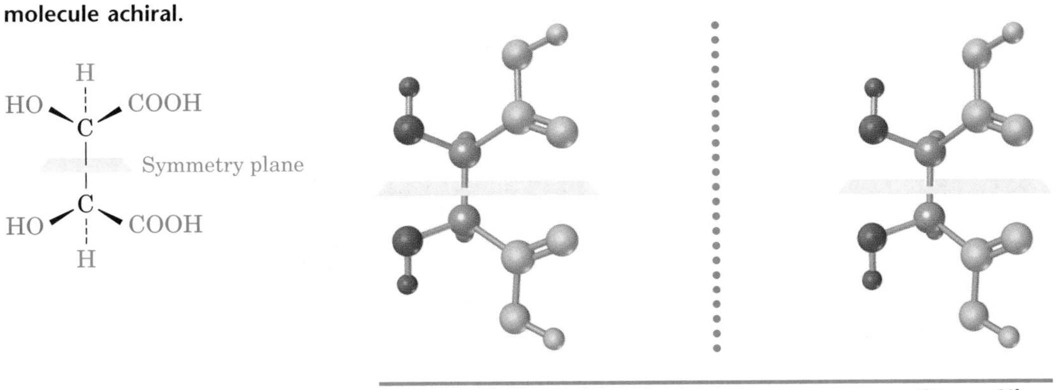

Stereo View

Because of the plane of symmetry, the tartaric acid stereoisomer shown in Figure 9.11 must be achiral, despite the fact that it has two chirality centers. Compounds that are achiral, yet contain chirality centers, are called **meso compounds** (**me**-zo). Thus, tartaric acid exists in three stereoisomeric forms: two enantiomers and one meso form.

- -

Practice Problem 9.3   Does *cis*-1,2-dimethylcyclobutane have any chirality centers? Is it chiral?

Strategy   To see whether a chirality center is present, look for a carbon atom bonded to four different groups. To see whether the molecule is chiral, look for the absence of a symmetry plane. Not all molecules with chirality centers are chiral—meso compounds are an exception.

Solution   A look at the structure of *cis*-1,2-dimethylcyclobutane shows that both methyl-bearing ring carbons (C1 and C2) are chirality centers. Overall, though, the compound is achiral because there is a symmetry plane bisecting the ring between C1 and C2. Thus, the molecule is a meso compound.

Symmetry plane

H_3C CH_3

1   2

H   H

- -

Problem 9.14   Which of the following structures represent meso compounds?

(a) OH
H.
⟍OH
⟍H

(b) OH
H.
⟍OH
⟍H

(c) CH₃
H

(d) H
Br⟍ | ⟍CH₃
    C
    |
    C
H₃C | ⟍H
    Br

Problem 9.15   Which of the following have a meso form?
(a) 2,3-Dibromobutane   (b) 2,3-Dibromopentane   (c) 2,4-Dibromopentane

Problem 9.16   Does the following structure represent a meso compound? If so, indicate the symmetry plane (red = O).

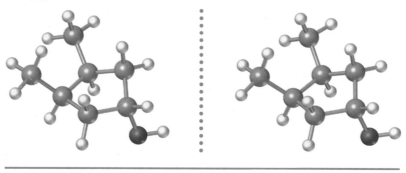

Stereo View

- -

9.9 Molecules with More Than Two Chirality Centers

We've seen now that a single chirality center in a molecule gives rise to two stereoisomers (one pair of enantiomers) and that two chirality centers in a molecule give rise to a maximum of four stereoisomers, or two pairs of enantiomers. In general, a molecule with n chirality centers has a maximum of 2^n stereoisomers, or 2^{n-1} pairs of enantiomers, though it may have less if any stereoisomers are meso compounds. Cholesterol, for example, contains eight chirality centers, making possible $2^8 = 256$ stereoisomers, although many are too strained to exist. Only one is produced in nature.

**Cholesterol
(eight chirality centers)**

· ·

Problem 9.17   How many chirality centers does morphine have? How many stereoisomers of morphine are possible in principle?

Morphine

· ·

9.10 Racemic Mixtures and Their Resolution

To conclude this discussion of stereoisomerism, let's return for a final look at Pasteur's pioneering work. Pasteur took an optically inactive tartaric acid salt and found that he could crystallize from it two optically active forms having the 2R,3R and 2S,3S configurations. But what was the optically inactive form he started with? It couldn't have been *meso*-tartaric acid, because *meso*-tartaric acid is a different chemical compound and can't interconvert with the two chiral enantiomers without breaking and re-forming chemical bonds.

The answer is that Pasteur started with a 50:50 *mixture* of the two chiral tartaric acid enantiomers. Such a mixture is called a **racemic** (ray-**see**-mic) **mixture**, or **racemate**, and is denoted either by the symbol ($\pm$) or by the prefix *d,l* to indicate a mixture of dextrorotatory and levorotatory forms. Racemic mixtures show zero optical rotation because they contain equal amounts of (+) and (−) enantiomers. The (+) rotation from one enantiomer exactly cancels the (−) rotation from the other. Through luck, Pasteur was able to separate, or **resolve**, racemic tartaric acid into its (+) and (−) enantiomers by fractional crystallization. Unfortunately, this method doesn't work for most racemic mixtures, so other techniques are required.

The most common method of resolution uses an acid–base reaction between a racemic mixture of chiral carboxylic acids (RCOOH) and an amine (RNH₂) to yield an ammonium salt:

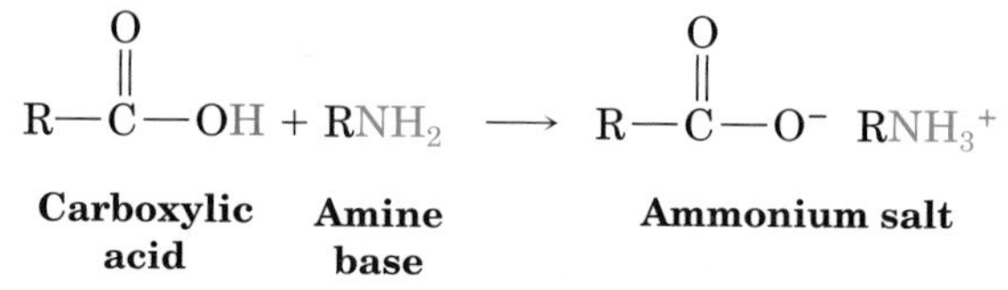

$$R-\overset{\overset{\displaystyle O}{\|}}{C}-OH + RNH_2 \longrightarrow R-\overset{\overset{\displaystyle O}{\|}}{C}-O^- \ RNH_3^+$$

Carboxylic   Amine   Ammonium salt
acid   base

To understand how this method of resolution works, let's see what happens when a racemic mixture of chiral acids, such as (+)- and (−)-lactic acids, reacts with an achiral amine base, such as methylamine, CH_3NH_2, to yield the ammonium salt. Stereochemically, the situation is analogous to what happens when left and right hands (chiral) pick up a tennis ball (achiral). Both left and right hands pick up the ball equally well, and the products— ball in right hand versus ball in left hand—are mirror images. In the same way, both (+)- and (−)-lactic acid react with methylamine equally well, and the product is a racemic mixture of two mirror-image salts, methylammonium (+)-lactate and methylammonium (−)-lactate (Figure 9.12).

FIGURE 9.12 ▼

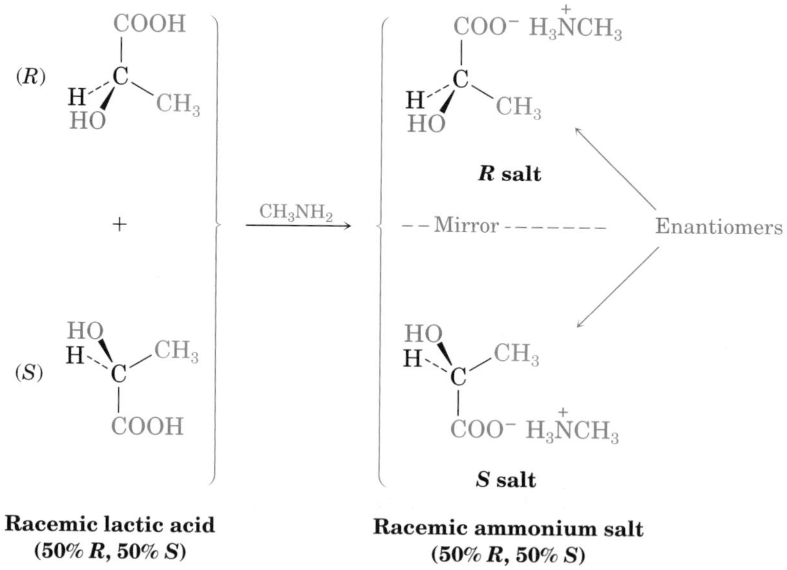

Reaction of racemic lactic acid with methylamine leads to a racemic mixture of ammonium salts.

Racemic lactic acid
(50% *R*, 50% *S*)

Racemic ammonium salt
(50% *R*, 50% *S*)

Now let's see what happens when the racemic mixture of (+)- and (−)-lactic acids reacts with a single enantiomer of a chiral amine base, such as (R)-1-phenylethylamine. Stereochemically, the situation is analogous to what happens when a hand (chiral) puts on a right-handed glove (*also chiral*). *Left and right hands don't put on the same glove in the same way.* The products—right hand in right glove versus left hand in right glove—are not mirror images, they're altogether different.

In the same way, (+)- and (−)-lactic acids react with (R)-1-phenylethylamine to give two different products (Figure 9.13). (R)-Lactic acid reacts with (R)-1-phenylethylamine to give the *R,R* salt, and (S)-lactic acid reacts with the *R* amine to give the *S,R* salt. *These two salts are diastereomers;* they are different compounds, with different chemical and physical properties. It may therefore be possible to separate them by crystallization or some other means. Once separated, acidification of the two diastereomeric salts with strong acid then allows us to isolate the two pure enantiomers of lactic acid and to recover the chiral amine for further use.

FIGURE 9.13 ▼

Reaction of racemic lactic acid with (R)-1-phenylethylamine yields a mixture of diastereomeric ammonium salts.

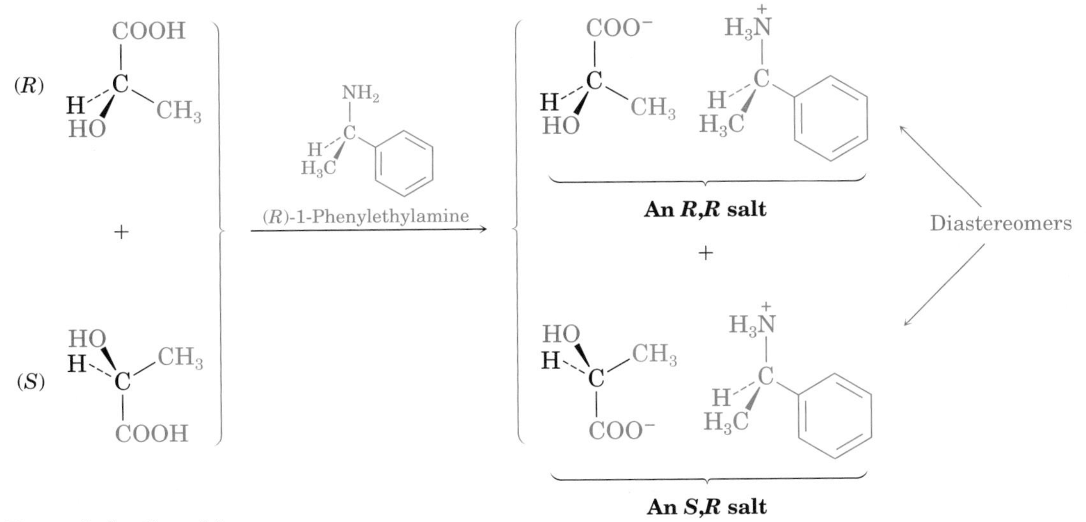

Racemic lactic acid
(50% *R*, 50% *S*)

· ·

Problem 9.18   What stereoisomers would result from reaction of (±)-lactic acid with (S)-1-phenylethylamine, and what is the relationship between them?

· ·

9.11 Physical Properties of Stereoisomers

Some physical properties of the three stereoisomers of tartaric acid and of the racemic mixture are listed in Table 9.3. As indicated, the (+)- and (−)-tartaric acids have identical melting points, solubilities, and densities. They differ only in the sign of their rotation of plane-polarized light. The meso isomer, by contrast, is diastereomeric with the (+) and (−) forms. As such, it has no mirror-image relationship to (+)- and (−)-tartaric acids, is a different compound altogether, and has different physical properties.

The racemic mixture is different still. Though a mixture of enantiomers, racemates usually act as though they were pure compounds, different from either enantiomer. Thus, the physical properties of racemic tartaric acid differ from those of the two enantiomers and from those of the meso form.

TABLE 9.3   Some Properties of the Stereoisomers of Tartaric Acid

Stereoisomer	Melting point (°C)	$[\alpha]_D$ (degrees)	Density (g/cm³)	Solubility at 20°C (g/100 mL H₂O)
(+)	168–170	+12	1.7598	139.0
(−)	168–170	−12	1.7598	139.0
Meso	146–148	0	1.6660	125.0
(±)	206	0	1.7880	20.6

9.12 A Brief Review of Isomerism

As noted on several previous occasions, isomers are compounds that have the same chemical formula but different structures. We've seen several kinds of isomers in the past few chapters, and it's a good idea at this point to see how they relate to one another by looking at the flowchart in Figure 9.14.

Key Ideas ▶ There are two fundamental types of isomers, both of which we've now encountered: constitutional isomers and stereoisomers.

- **Constitutional isomers** (Section 3.2) are compounds whose atoms are connected differently. Among the kinds of constitutional isomers we've seen are skeletal, functional, and positional isomers.

Different carbon skeletons

$$CH_3CHCH_3 \quad \text{and} \quad CH_3CH_2CH_2CH_3$$
|
$$CH_3$$

Isobutane   **Butane**

FIGURE 9.14 ▼

A flow diagram summarizing the different kinds of isomers.

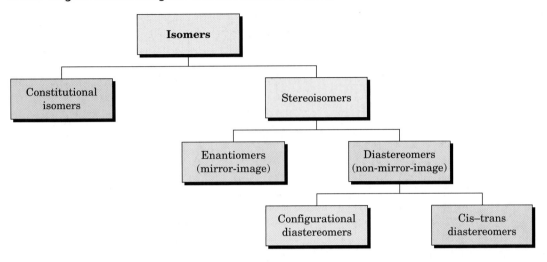

Different functional groups	CH_3CH_2OH and CH_3OCH_3
	Ethyl alcohol **Dimethyl ether**

Different position of functional groups

$$\underset{\textbf{Isopropylamine}}{CH_3\overset{\overset{\displaystyle NH_2}{|}}{C}HCH_3} \quad \text{and} \quad \underset{\textbf{Propylamine}}{CH_3CH_2CH_2NH_2}$$

• **Stereoisomers** (Section 3.8) are compounds whose atoms are connected in the same order but with a different geometry. Among the kinds of stereoisomers we've seen are enantiomers, diastereomers, and cis–trans isomers (both in alkenes and in cycloalkanes). In fact, though, cis–trans isomers are really just another kind of diastereomers, because they are non-mirror-image stereoisomers:

Enantiomers
(nonsuperimposable mirror-image stereoisomers)

(R)-Lactic acid   (S)-Lactic acid

Diastereomers
(nonsuperimposable, non-mirror-image stereoisomers)

Configurational diastereomers

2R,3R-2-Amino-3-hydroxybutanoic acid   **2R,3S-2-Amino-3-hydroxybutanoic acid**

Cis–trans diastereomers (substituents on same side or opposite side of double bond or ring)

H_3C H
   $C=C$ and
 H CH_3

trans-2-Butene

H_3C CH_3
   $C=C$
 H   H

cis-2-Butene

H_3C H
      CH_3 and
 H

trans-1,3-Dimethyl-cyclopentane

H_3C CH_3
 H   H

cis-1,3-Dimethyl-cyclopentane

• •

Problem 9.19   What kinds of isomers are the following pairs?
(a) (S)-5-Chloro-2-hexene and chlorocyclohexane
(b) (2R,3R)-Dibromopentane and (2S,3R)-dibromopentane

• •

9.13   Fischer Projections

When learning to visualize chiral molecules, it's best to begin by building molecular models. As more experience is gained, it becomes easier to draw pictures and work with mental images. To do this successfully, though, a standard method of representation is needed for depicting the three-dimensional arrangement of atoms on a page. In 1891, Emil Fischer suggested a method based on the projection of a tetrahedral carbon atom onto a flat surface. These **Fischer projections** were soon adopted and are now a standard means of depicting stereochemistry at chirality centers, particularly in carbohydrate chemistry.

A tetrahedral carbon atom is represented in a Fischer projection by two crossed lines. The horizontal lines represent bonds coming out of the page, and the vertical lines represent bonds going into the page:

Press flat
↓

W   X
   C
   Y   Z

⟶

Z
W — C — X
Y

⟶

Z
W ┼ X
Y

Fischer projection

For example, (*R*)-lactic acid can be drawn as follows:

$$
\begin{array}{ccc}
& \text{COOH} & \\
\text{H} \cdots & \text{C} & \text{CH}_3 \\
& \text{HO} &
\end{array}
\quad = \quad
\begin{array}{c}
\text{COOH} \\
\text{H}\!-\!\text{C}\!-\!\text{OH} \\
\text{CH}_3
\end{array}
$$

Bonds out of page ⟶

$$
\begin{array}{c}
\text{COOH} \\
\text{H}-\!\!\!+\!\!\!-\text{OH} \\
\text{CH}_3
\end{array}
$$

⟵ Bonds into page

Fischer projection
(*R*)-Lactic acid

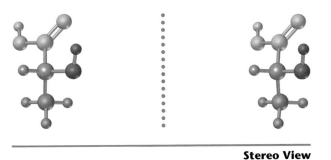

Stereo View

Because a given chiral molecule can be drawn in many different ways, it's often necessary to compare two projections to see if they represent the same or different enantiomers. To test for identity, Fischer projections can be moved around on the paper, but care must be taken not to change the meaning of the projection inadvertently. Only two kinds of motions are allowed:

1. A Fischer projection can be rotated on the page by 180°, but *not by 90° or 270°*. A 180° rotation maintains the Fischer convention by keeping the same substituent groups going into and coming out of the plane. In the following Fischer projection of (*R*)-lactic acid, for example, the –H and –OH groups come out of the plane both before and after a 180° rotation:

$$
\begin{array}{c}
\text{COOH} \\
\text{H} \diagdown\! \diagup \text{OH} \\
\text{C} \\
\text{CH}_3
\end{array}
=
\begin{array}{c}
\text{COOH} \\
\text{H}-\!\!\!+\!\!\!-\text{OH} \\
\text{CH}_3
\end{array}
\xrightarrow[\text{rotation}]{180°}
\begin{array}{c}
\text{CH}_3 \\
\text{HO}-\!\!\!+\!\!\!-\text{H} \\
\text{COOH}
\end{array}
=
\begin{array}{c}
\text{CH}_3 \\
\text{HO} \diagdown\! \diagup \text{H} \\
\text{C} \\
\text{COOH}
\end{array}
$$

(*R*)-Lactic acid　　　　　　　　　　　　　**(*R*)-Lactic acid**

A 90° rotation, however, breaks the Fischer convention by exchanging the groups that go into the plane and those that come out. In the following Fischer projection of (*R*)-lactic acid, the –H and –OH groups come out of the plane before rotation but go into the plane after a 90° rotation. As a result, the rotated projection represents (*S*)-lactic acid:

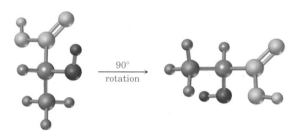

(*R*)-Lactic acid   **(*S*)-Lactic acid**

2. A Fischer projection can have one group held steady while the other three rotate in either a clockwise or a counterclockwise direction. For example:

These are the only kinds of motion allowed. Moving a Fischer projection in any other way inverts its meaning.

Knowing the two rules provides a way to see if two projections represent the same or different enantiomers. For example, three different Fischer projections of 2-butanol follow. Do all represent the same enantiomer, or is one different?

A   **B**   **C**

The simplest way to see if two Fischer projections represent the same enantiomer is to carry out allowed rotations until *two* groups are superimposed. If the other two groups are also superimposed, the Fischer projections are the same; if the other two groups are not superimposed, the Fischer projections are different.

Let's keep projection A unchanged and move B so that the $-CH_3$ and $-H$ substituents match up with those in A:

$$
HO-\overset{CH_2CH_3}{\underset{CH_3}{|}}-H \quad \xrightarrow[\substack{\text{Rotate other} \\ \text{three groups} \\ \text{clockwise}}]{\text{Hold } CH_3} \quad H-\overset{OH}{\underset{CH_3}{|}}-CH_2CH_3 \quad \xrightarrow[\substack{\text{Rotate other} \\ \text{three groups} \\ \text{clockwise}}]{\text{Hold } CH_2CH_3} \quad H_3C-\overset{H}{\underset{OH}{|}}-CH_2CH_3
$$

B   **A**

By performing two allowed movements on B, we find that it is identical to A. Now let's do the same thing to C:

$$
H-\overset{OH}{\underset{CH_2CH_3}{|}}-CH_3 \quad \xrightarrow[]{\substack{\text{Rotate} \\ 180°}} \quad H_3C-\overset{CH_2CH_3}{\underset{OH}{|}}-H \quad \xrightarrow[\substack{\text{Rotate other} \\ \text{three groups} \\ \text{counterclock-} \\ \text{wise}}]{\text{Hold } CH_3} \quad H_3C-\overset{H}{\underset{CH_2CH_3}{|}}-OH
$$

C   ***NOT* A**

By performing two allowed movements on C, we can match up the $-H$ and $-CH_3$ substituents with those in A, but we then find that the $-OH$ and $-CH_2CH_3$ substituents *don't* match up. Thus, C is enantiomeric with A and B.

• •

Practice Problem 9.4   Convert the following tetrahedral representation of (*R*)-2-butanol into a Fischer projection:

$$
H\cdots\overset{\overset{CH_2CH_3}{|}}{C}\diagdown_{CH_3} \\ HO
$$

(*R*)-2-Butanol

Strategy   Rotate the molecule so that two horizontal bonds are facing you and two vertical bonds are receding from you. Then press the molecule flat into the paper, indicating the chiral carbon as the intersection of two crossed lines. Remember that there is no single right answer: A given molecule can be represented by many different Fischer projections.

Solution   Applying the above strategy to (*R*)-2-butanol gives the following Fischer projection:

$$
H\cdots\overset{\overset{CH_2CH_3}{|}}{C}\diagdown_{CH_3} \quad = \quad H\diagup\overset{\overset{CH_2CH_3}{|}}{C}\diagdown_{\substack{OH}} \quad = \quad H-\overset{CH_2CH_3}{\underset{CH_3}{|}}-OH \\ HO \qquad\qquad\qquad\qquad CH_3
$$

(*R*)-2-Butanol

Problem 9.20   Which of the following Fischer projections represent the same enantiomer?

$$
\begin{array}{cccc}
\text{COOH} & \text{COOH} & \text{H} & \text{CH}_3 \\
\text{H}\!-\!\!|\!-\!\text{OH} & \text{H}_3\text{C}\!-\!\!|\!-\!\text{H} & \text{HO}\!-\!\!|\!-\!\text{CH}_3 & \text{HOOC}\!-\!\!|\!-\!\text{H} \\
\text{CH}_3 & \text{OH} & \text{COOH} & \text{OH} \\
\textbf{A} & \textbf{B} & \textbf{C} & \textbf{D}
\end{array}
$$

Problem 9.21   Are the following pairs of Fischer projections the same, or are they enantiomers?

$$
\text{(a)}\quad
\begin{array}{cc}
\text{CH}_3 & \text{H} \\
\text{Cl}\!-\!\!|\!-\!\text{H} \;\text{ and }\; \text{OHC}\!-\!\!|\!-\!\text{Cl} \\
\text{CHO} & \text{CH}_3
\end{array}
\qquad
\text{(b)}\quad
\begin{array}{cc}
\text{CH}_2\text{OH} & \text{CHO} \\
\text{H}\!-\!\!|\!-\!\text{OH} \;\text{ and }\; \text{HO}\!-\!\!|\!-\!\text{CH}_2\text{OH} \\
\text{CHO} & \text{H}
\end{array}
$$

Problem 9.22   Convert the following tetrahedral representation of (*S*)-2-chlorobutane into a Fischer projection:

$$
\underset{\text{CH}_3}{\overset{\text{H}}{\text{Cl}\cdots\text{C}\diagdown\text{CH}_2\text{CH}_3}}
\qquad \textbf{(S)-2-Chlorobutane}
$$

9.14 Assigning *R,S* Configurations to Fischer Projections

The *R,S* stereochemical designations can be assigned to Fischer projections by following three steps:

STEP 1   Assign priorities to the four substituents in the usual way.

STEP 2   Perform one of the two allowed motions to place the group of lowest (fourth) priority at the top of the Fischer projection. This means that the lowest-priority group is oriented back, away from the viewer, as required for assigning configuration.

STEP 3   Determine the direction of rotation $1 \rightarrow 2 \rightarrow 3$ of the remaining three groups, and assign *R* or *S* configuration. Practice Problem 9.5 gives an example.

Fischer projections can also be used to specify more than one chirality center in a molecule simply by "stacking" the centers on top of one another. For example, threose, a simple four-carbon sugar, has the 2*S*,3*R* configuration:

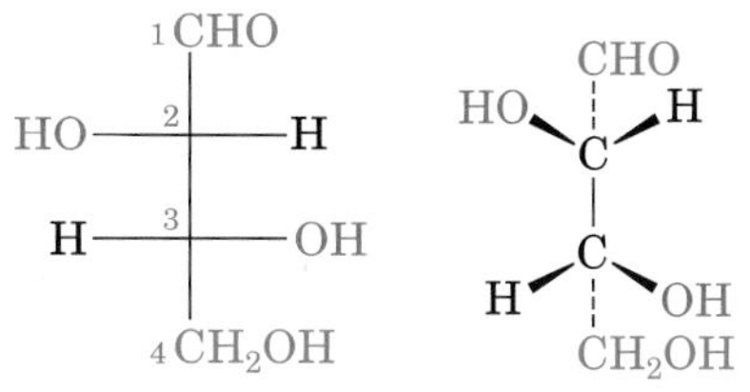

$$_1\text{CHO}$$

HO——²———H

H——³———OH

$$_4\text{CH}_2\text{OH}$$

Threose [(2*S*,3*R*)-2,3,4-Trihydroxybutanal]

Molecular models are particularly helpful in visualizing these structures.

• •

Practice Problem 9.5   Assign *R* or *S* configuration to the following Fischer projection of alanine:

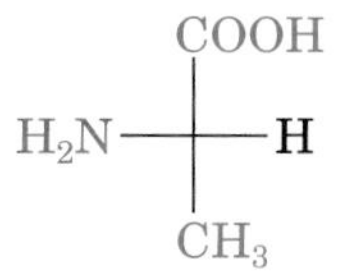

$$\text{COOH}$$

$$\text{H}_2\text{N}——\text{H}$$ **Alanine**

$$\text{CH}_3$$

Strategy   Follow the steps discussed in the text: (1) Assign priorities to the four substituents on the chiral carbon. (2) Manipulate the Fischer projection to place the group of lowest priority at the top by carrying out one of the allowed motions. (3) Determine the direction $1 \rightarrow 2 \rightarrow 3$ of the remaining three groups.

Solution   The priorities of the groups are : (1) –NH$_2$, (2) –COOH, (3) –CH$_3$, and (4) –H. To bring the group of lowest priority (–H) to the top, we might want to hold the –CH$_3$ group steady while rotating the other three groups counterclockwise:

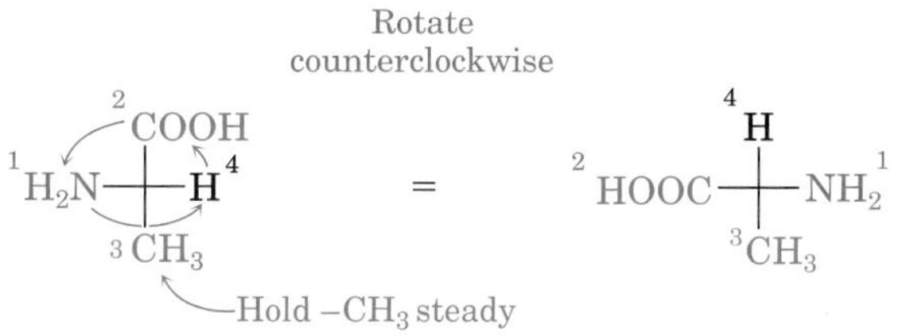

Rotate
counterclockwise

$$^1\text{H}_2\text{N}——\text{H}^4 \quad = \quad ^2\text{HOOC}——\text{NH}_2^1$$

—Hold –CH$_3$ steady

Going now from first to second to third highest priority requires a counterclockwise turn, corresponding to *S* stereochemistry.

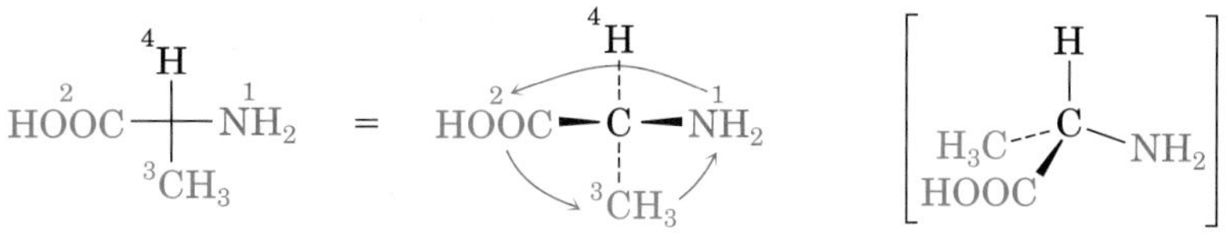

$$^2\text{HOOC}——\text{NH}_2^1 \quad = \quad ^2\text{HOOC}—\text{C}—\text{NH}_2^1$$

S configuration

Problem 9.23   Assign R or S configuration to the chirality centers in the following molecules:

(a)
$$
\begin{array}{c}
\text{COOH} \\
| \\
\text{H} \longrightarrow\!\!\!\!\!\!\!\!\!\!\!\longleftarrow \text{CH}_3 \\
| \\
\text{Br}
\end{array}
$$

(b)
$$
\begin{array}{c}
\text{CH}_3 \\
| \\
\text{HO} \longrightarrow\!\!\!\!\!\!\!\!\!\!\!\longleftarrow \text{CH}_2\text{CH}_3 \\
| \\
\text{H}
\end{array}
$$

(c)
$$
\begin{array}{c}
\text{H} \\
| \\
\text{HO} \longrightarrow\!\!\!\!\!\!\!\!\!\!\!\longleftarrow \text{CHO} \\
| \\
\text{CH}_3
\end{array}
$$

Problem 9.24   Redraw the following molecule as a Fischer projection and assign R or S configuration to the chirality center (red = O, yellow-green = Cl):

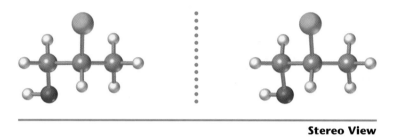

Stereo View

9.15 Stereochemistry of Reactions: Addition of HBr to Alkenes

Most of the biochemical reactions that take place in the body and many organic reactions in the laboratory yield products with chirality centers. For example, addition of HBr to 1-butene yields 2-bromobutane, a chiral molecule. What predictions can we make about the stereochemistry of this chiral product? If a single enantiomer is formed, is it R or S? If a mixture of enantiomers is formed, how much of each? In fact, the 2-bromobutane produced is a racemic mixture of R and S enantiomers. Let's see why.

$$
\text{CH}_3\text{CH}_2\text{CH}\!=\!\text{CH}_2 \quad \xrightarrow[\text{Ether}]{\text{HBr}} \quad \text{CH}_3\text{CH}_2\overset{\overset{\displaystyle \text{Br}}{|}}{\underset{*}{\text{CH}}}\text{CH}_3
$$

1-Butene   **(±)-2-Bromobutane**
(achiral)   **(chiral)**

To understand why a racemic product results from the reaction of HBr with 1-butene, think about how the reaction occurs. 1-Butene is first protonated to yield an intermediate secondary (2°) carbocation. Since the trivalent carbon is sp^2-hybridized and planar, the cation has no chirality centers, has a plane of symmetry, and is achiral. As a result, it can react with Br⁻ ion equally well from either the top or the bottom. Attack from the top leads to (S)-2-bromobutane, and attack from the bottom leads to (R)-2-bromobutane. Since both pathways occur with equal probability, a racemic product mixture results (Figure 9.15).

FIGURE 9.15 ▼

Stereochemistry of the addition of HBr to 1-butene. The achiral intermediate carbocation reacts equally well from both top and bottom, giving a racemic product mixture.

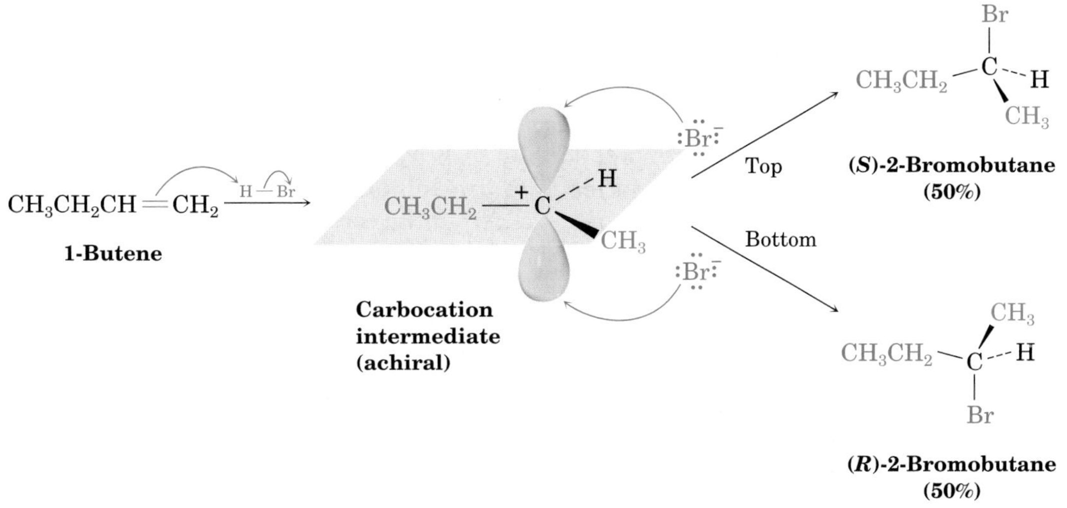
CH3CH2CH=CH2
H—Br
1-Butene
CH3CH2—C—H
CH3
:Br:−
Carbocation intermediate (achiral)
Top
Bottom
Br
CH3CH2—C---H
CH3
(S)-2-Bromobutane (50%)
CH3
CH3CH2—C---H
Br
(R)-2-Bromobutane (50%)

Another way to think about the reaction is in terms of transition states. If the intermediate carbocation is attacked from the top, *S* product is formed through transition state 1 (TS 1) in Figure 9.16. If the cation is attacked from the bottom, *R* product is formed through TS 2. *The two transition states are mirror images.* They therefore have identical energies, form at identical rates, and are equally likely to occur.

FIGURE 9.16 ▼

Attack of Br^- ion on the *sec*-butyl carbocation. Attack from the top leads to *S* product and is the mirror image of attack from the bottom, which leads to *R* product. Since both are equally likely, racemic product is formed. The dotted C· · ·Br bond in the transition state indicates partial bond formation.

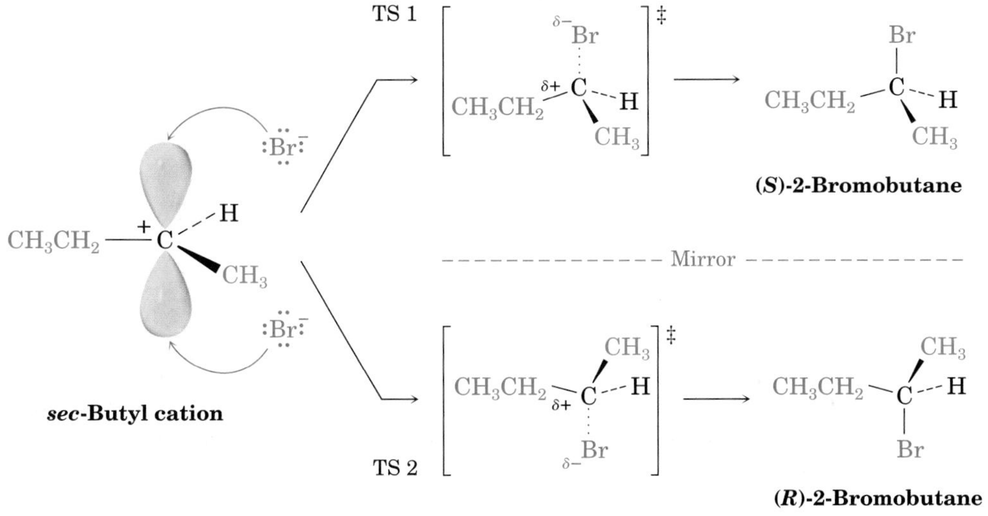
:Br:−
CH3CH2—C---H
CH3
sec-Butyl cation
TS 1
δ−
Br
δ+
CH3CH2—C---H
CH3
‡
Br
CH3CH2—C---H
CH3
(S)-2-Bromobutane
Mirror
CH3
CH3CH2—C---H
δ+
Br
δ−
‡
TS 2
CH3
CH3CH2—C---H
Br
(R)-2-Bromobutane

9.16 Stereochemistry of Reactions: Addition of Br₂ to Alkenes

Addition of Br$_2$ to 2-butene leads to the formation of 2,3-dibromobutane and to the generation of two chirality centers. What stereochemistry should we predict for such a reaction? Starting with planar, achiral *cis*-2-butene, Br$_2$ can add to the double bond equally well from either the top or the bottom face to generate two intermediate bromonium ions. For the sake of simplicity, let's consider only the attack from the top face, keeping in mind that every structure we consider also has a mirror image.

The bromonium ion formed by addition to the top face of *cis*-2-butene can be attacked by Br$^-$ ion from either the right or the left side of the bottom face, as shown in Figure 9.17. Attack from the left (path a) leads to (2*S*,3*S*)-dibromobutane, and attack from the right (path b) leads to (2*R*,3*R*)-dibromobutane. Since both modes of attack on the achiral bromonium ion are equally likely, a 50:50 (racemic) mixture of the two enantiomeric products is formed. Thus, we obtain (±)-2,3-dibromobutane.

FIGURE 9.17 ▼

Stereochemistry of the addition of Br$_2$ to *cis*-2-butene. A racemic mixture of 2*S*,3*S* and 2*R*,3*R* products is formed because attack of Br$^-$ on both carbons of the bromonium ion intermediate is equally likely.

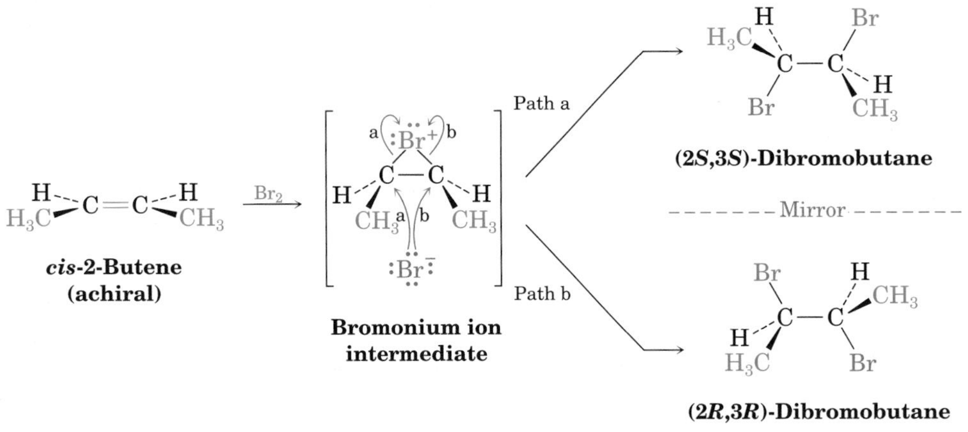

What about the addition of Br$_2$ to *trans*-2-butene? Is the same racemic product mixture formed? Perhaps surprisingly at first glance, the answer is no. *trans*-2-Butene reacts with Br$_2$ to form a bromonium ion, and again we'll consider only top-face attack for simplicity. Attack of Br$^-$ ion on the bromonium ion intermediate takes place equally well from both right and left sides of the bottom face, leading to the formation of 2*R*,3*S* and 2*S*,3*R* products in equal amounts (Figure 9.18). A close look at the two products,

FIGURE 9.18 ▼

Stereochemistry of the addition of Br$_2$ to *trans*-2-butene. A meso product is formed.

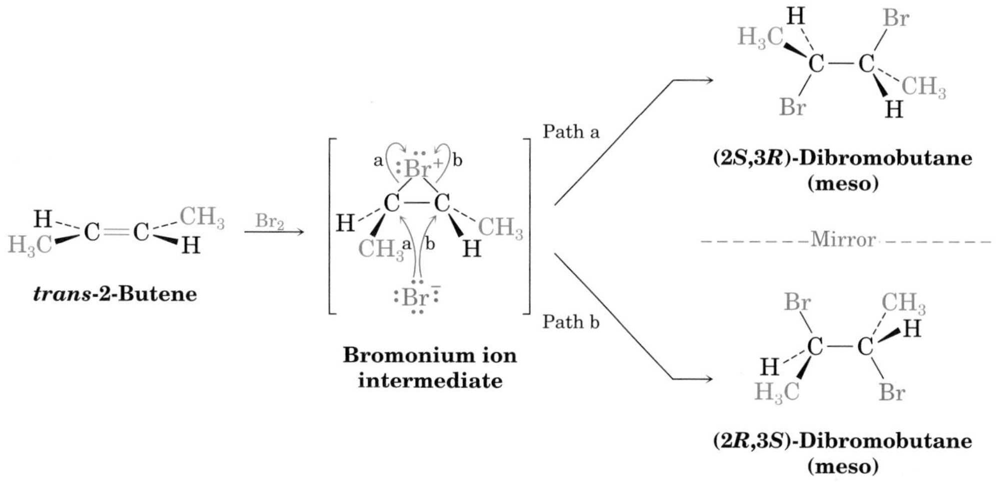

however, shows that they are *identical*. Both structures represent *meso*-2,3-dibromobutane.

The key conclusion from all three addition reactions just discussed in this and the previous section is that an optically inactive product has been formed in each case. *Reaction between two optically inactive (achiral) partners always leads to an optically inactive product—either racemic or meso.* Put another way, optical activity can't come from nowhere; optically active products can't be produced from optically inactive reactants.

- -

Practice Problem 9.6   What is the stereochemistry of the product that results from addition of Br$_2$ to 1-methylcyclohexene? Is the product optically active? Explain.

Strategy   Problems of this sort require careful reasoning, a knowledge of the mechanism by which the reaction occurs, and a good grasp of stereochemical principles. Write the steps of the reaction mechanism, identify the step or steps in which product stereochemistry is determined, and decide what products will be formed.

Solution   Addition of Br$_2$ to an alkene involves two steps: (1) reaction with Br$^+$ to form a bromonium-ion intermediate, and (2) reaction of the bromonium ion with Br$^-$. The first step determines the stereochemistry of one center, and the second step determines the stereochemistry of the other. Because of a symmetry plane in the reactant, the first step can occur equally well from either face of the double bond, to give a 50:50 mixture of two enantiomeric bromonium ions. Each ion can then react with Br$^-$ from either the right or the left side to give two products. But because the bromonium ions do not have symmetry planes, attack from right and left is not equally likely, so an unequal mix of *R,R* and *S,S* products will be formed from each. The *minor* product from one bromonium ion, however, will be the *major* product from the other, so overall a 50:50 (racemic) mixture of *R,R* and *S,S* dibromides will result.

• •

Problem 9.25   Addition of Br_2 to an unsymmetrical alkene such as *cis*-2-hexene leads to racemic product, even though attack of Br^- ion on the unsymmetrical bromonium ion intermediate is not equally likely at both ends. Make drawings of the intermediate and the products, and explain the observed stereochemical result.

Problem 9.26   Predict the stereochemical outcome of the reaction of Br_2 with *trans*-2-hexene, and explain your reasoning.

• •

9.17   Stereochemistry of Reactions: Addition of HBr to a Chiral Alkene

The reactions considered in the previous two sections involve additions to achiral alkenes, and optically inactive products are formed in all cases. What would happen, though, if we were to carry out a reaction on a *single* enantiomer of a *chiral* reactant? For example, what stereochemical result would be obtained from addition of HBr to a chiral alkene, such as (*R*)-4-methyl-1-hexene? The product of the reaction, 2-bromo-4-methylhexane, has two chirality centers and four possible stereoisomers.

(*R*)-4-Methyl-1-hexene   **2-Bromo-4-methylhexane**

Let's think about the two chirality centers separately. What about the configuration at C4, the methyl-bearing carbon atom? Since C4 has the *R* configuration in the starting material, and since this chirality center is unaffected by the reaction, its configuration remains unchanged. Thus, the configuration of C4 in the product remains *R* (assuming that the relative priorities of the four attached groups are not changed by the reaction).

What about the configuration at C2, the newly formed chirality center? As illustrated in Figure 9.19, the stereochemistry at C2 is established by attack of Br⁻ ion on a carbocation intermediate in the usual manner. *But this carbocation does not have a plane of symmetry; it is chiral because of the chirality center at C4.* Since the carbocation has no plane of symmetry, it is not attacked equally well from top and bottom faces. One of the two faces is likely, for steric reasons, to be a bit more accessible than the other face, leading to a mixture of *R* and *S* products in some ratio other than 50:50. Thus, two diastereomeric products, (2*R*,4*R*)-2-bromo-4-methylhexane and (2*S*,4*R*)-2-bromo-4-methylhexane, are formed in unequal amounts, and the mixture is optically active.

FIGURE 9.19 ▼

Stereochemistry of the addition of HBr to the chiral alkene, (*R*)-4-methyl-1-hexene. A mixture of diastereomeric 2*R*,4*R* and 2*S*,4*R* products is formed in unequal amounts because attack on the chiral carbocation intermediate is not equally likely from top and bottom. The product mixture is optically active.

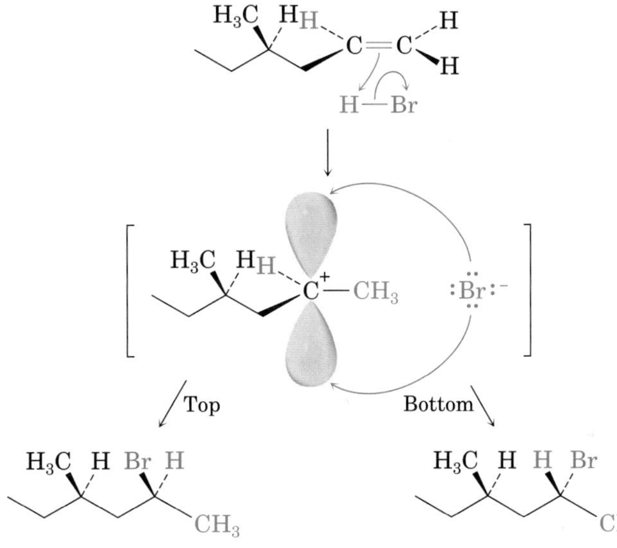

(2*S*,4*R*)-2-Bromo-4-methylhexane   (2*R*,4*R*)-2-Bromo-4-methylhexane

As a general rule, *reaction of a chiral reactant with an achiral reactant leads to unequal amounts of diastereomeric products.* If the chiral reactant is optically active because only one enantiomer is used, then the products are also optically active.

Problem 9.27 What products are formed from reaction of HBr with racemic (±)-4-methyl-1-hexene? What can you say about the relative amounts of the products? Is the product mixture optically active?

Problem 9.28   What products are formed from reaction of HBr with 4-methylcyclopentene? What can you say about the relative amounts of the products?

• •

9.18 Chirality at Atoms Other Than Carbon

Since the most common cause of chirality is the presence of four different substituents bonded to a tetrahedral atom, tetrahedral atoms other than carbon can also be chirality centers. Silicon, nitrogen, phosphorus, and sulfur are all commonly encountered in organic molecules, and all can be chirality centers under the proper circumstances. We know, for example, that trivalent nitrogen is tetrahedral, with its lone pair of electrons acting as the fourth "substituent" (Section 1.11). Is trivalent nitrogen chiral? Does a compound such as ethylmethylamine exist as a pair of enantiomers?

Ethylmethylamine

The answer is both yes and no. Yes in principle, but no in practice. Trivalent nitrogen compounds undergo a rapid umbrella-like inversion that interconverts enantiomers. We therefore can't isolate individual enantiomers except in special cases.

9.19 Chirality in Nature

Biological Connection ▶

Although the different enantiomers of a chiral molecule have the same physical properties, they usually have different biological properties. For example, the dextrorotatory enantiomer of limonene has the odor of oranges, but the levorotatory enantiomer has the odor of lemons.

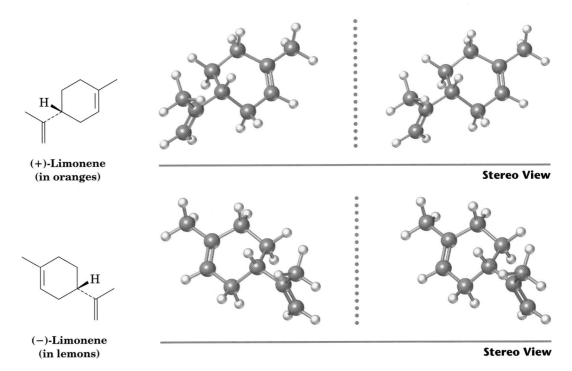

(+)-Limonene
(in oranges)

Stereo View

(−)-Limonene
(in lemons)

Stereo View

More dramatic examples of how a change in chirality can affect the biological properties of a molecule are found in many drugs, such as fluoxetine, a heavily prescribed medication sold under the trade name Prozac. Racemic fluoxetine is an extraordinarily effective antidepressant, but has no activity against migraine. The pure *S* enantiomer, however, works remarkably well in preventing migraine and is now undergoing clinical evaluation. "Chiral Drugs" (p. 344) gives other examples.

(*S*)-Fluoxetine
(prevents migraine)

Why do different stereoisomers have different biological properties? To exert its biological action, a chiral molecule must fit into a chiral receptor at some target site, much as a hand fits into a glove. But just as a right hand can fit only into a right-hand glove, so a particular stereoisomer can fit only into a receptor having the proper complementary shape. Any other stereoisomer will be a misfit, like a right hand in a left-hand glove. A schematic representation of the interaction between a chiral molecule and a chiral biological receptor is shown in Figure 9.20. One enantiomer fits the receptor perfectly, but the other does not.

FIGURE 9.20 ▼

(a) One enantiomer fits easily into a chiral receptor site to exert its biological effect, but (b) the other enantiomer can't fit into the same receptor.

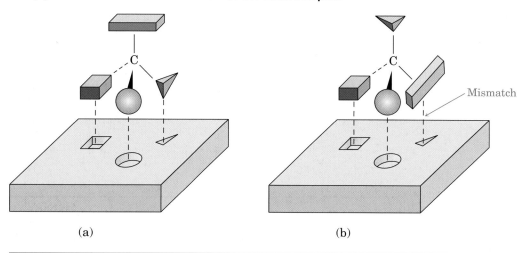

(a)   (b)

CHEMISTRY @ WORK

Chiral Drugs

The hundreds of different pharmaceutical agents approved for use by the U.S. Food and Drug Administration come from many sources. Many drugs are isolated directly from plants or bacteria, others are made by chemical modification of naturally occurring compounds, and still others are made entirely in the laboratory and have no relatives in nature.

Those drugs that come from natural sources, either directly or after chemical modification, are usually chiral and are generally found only as a single enantiomer rather than as a racemic mixture. Penicillin V, for example, an antibiotic isolated from the *Penicillium* mold, has the 2S,5R,6R configuration. Its enantiomer, which does not occur naturally but can be made in the laboratory, has essentially no biological activity.

(continued) ▶

Penicillin V (2S,5R,6R configuration)

In contrast to drugs from natural sources, those drugs that are made entirely in the laboratory are either achiral or, if chiral, are often produced and sold as racemic mixtures. Ibuprofen, for example, contains one chirality center, and only the S enantiomer is active as an analgesic and anti-inflammatory agent. The R enantiomer of ibuprofen is inactive, although it is slowly converted in the body to the active S form. Nevertheless, the substance marketed under such trade names as Advil, Nuprin, and Motrin is a racemic mixture of R and S.

(S)-Ibuprofen
(an active analgesic agent)

Stereo View

Not only is it chemically wasteful to synthesize and administer an enantiomer that does not serve the intended purpose, many examples are now known where the presence of the "wrong" enantiomer in a racemic mixture either affects the body's ability to utilize the "right" enantiomer or has unintended pharmacological effects of its own. The presence of (R)-ibuprofen in the racemic mixture, for instance, slows substantially the rate at which the S enantiomer takes effect in the body, from 12 minutes to 38 minutes.

To get around this problem, pharmaceutical companies are now devising methods of so-called *enantioselective synthesis,* which allows them to prepare only a single enantiomer rather than a racemic mixture. Viable methods have already been developed for the preparation of (S)-ibuprofen, which is now being marketed in Europe. The time may not be far off when television commercials show famous athletes talking about the advantages of chiral drugs.

The *S* enantiomer of ibuprofen soothes the aches and pains of athletic injuries much more effectively than the *R* enantiomer.

Summary and Key Words

KEY WORDS

absolute
    configuration, 318
achiral, 309
chiral, 309
chirality center, 309
configuration, 315
dextrorotatory, 313
diastereomers, 321
enantiomers, 308
Fischer projection,
    330
levorotatory, 313
meso compound, 324
optically active, 312
plane of symmetry,
    309
plane-polarized light,
    312
racemate, 325
racemic mixture, 325
resolve, 326
specific rotation, 314

When a beam of **plane-polarized light** passes through a solution of certain organic molecules, the plane of polarization is rotated. Compounds that exhibit this behavior are called **optically active**. Optical activity is due to the asymmetric structure of the molecules themselves.

An object or molecule that is not superimposable on its mirror image is said to be **chiral**, meaning "handed." For example, a glove is chiral but a coffee cup is nonchiral, or **achiral**. A chiral molecule is one that does not contain a **plane of symmetry** cutting through the molecule so that one half is a mirror image of the other half. The most common cause of chirality in organic molecules is the presence of a tetrahedral, sp^3-hybridized carbon atom bonded to four different groups. Compounds that contain such **chirality centers** exist as a pair of nonsuperimposable, mirror-image stereoisomers called **enantiomers**. Enantiomers are identical in all physical properties except for the direction in which they rotate plane-polarized light.

The stereochemical **configuration** of a carbon atom can be depicted using **Fischer projections**, in which horizontal lines (bonds) are understood to come out of the plane of the paper and vertical bonds are understood to go back into the plane of the paper. The configuration can be specified as either R (*rectus*) or S (*sinister*) by using the Cahn–Ingold–Prelog sequence rules. This is done by first assigning priorities to the four substituents on the chiral carbon atom and then orienting the molecule so that the lowest-priority group points directly back away from the viewer. If a curved arrow drawn in the direction of decreasing priority ($1 \rightarrow 2 \rightarrow 3$) for the remaining three groups is clockwise, the chirality center has the R configuration. If the direction is counterclockwise, the chirality center has the S configuration.

Some molecules have more than one chirality center. Enantiomers have opposite configuration at all chirality centers, whereas **diastereomers** have the same configuration in at least one center but opposite configurations at the others. A compound with n chirality centers can have a maximum of 2^n stereoisomers.

Meso compounds contain chirality centers, but are achiral overall because they have a plane of symmetry. **Racemic mixtures**, or **racemates**, are 50:50 mixtures of (+) and (−) enantiomers. Racemic mixtures and individual diastereomers differ in their physical properties, such as solubility, melting point, and boiling point.

Many reactions give chiral products. If the reactants are optically inactive, the products are also optically inactive—either meso or racemic. If one or both of the reactants is optically active, the product can also be optically active.

Visualizing Chemistry

(Problems 9.1–9.28 appear within the chapter.)

9.29   Which of the following structures are identical? (Red = O, yellow-green = Cl.)

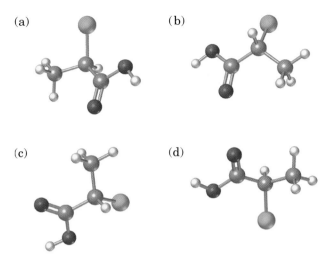

9.30 Assign *R* or *S* configuration to the following molecules (red = O, blue = N):

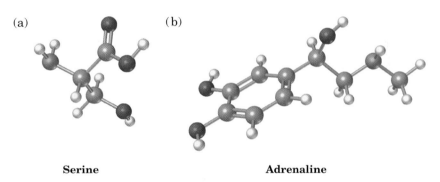

Serine   **Adrenaline**

9.31 Which, if any, of the following structures represent meso compounds? (Red = O, blue = N, yellow-green = Cl.)

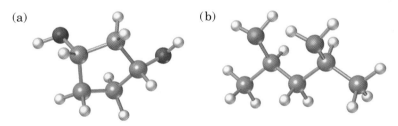

9.32   Assign R or S configuration to each chirality center in pseudoephedrine, an over-the-counter decongestant found in cold remedies (red = O, blue = N).

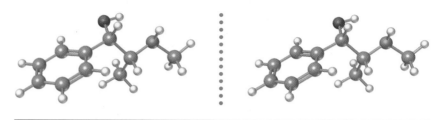

Stereo View

Additional Problems

9.33   Cholic acid, the major steroid in bile, was found to have a rotation of $+2.22°$ when a 3.00 g sample was dissolved in 5.00 mL alcohol and the solution was placed in a sample tube with a 1.00 cm pathlength. Calculate $[\alpha]_D$ for cholic acid.

9.34   Polarimeters for measuring optical rotation are so sensitive that they can measure rotations to $0.001°$, an important fact when only small amounts of sample are available. Ecdysone, for example, is an insect hormone that controls molting in the silkworm moth. When 7.00 mg ecdysone was dissolved in 1.00 mL chloroform and the solution was placed in a cell with a 2.00 cm pathlength, an observed rotation of $+0.087°$ was found. Calculate $[\alpha]_D$ for ecdysone.

9.35   Which of the following compounds are chiral? Draw them, and label the chirality centers.
(a) 2,4-Dimethylheptane   (b) 3-Ethyl-5,5-dimethylheptane
(c) *cis*-1,4-Dichlorocyclohexane   (d) 4,5-Dimethyl-2,6-octadiyne

9.36   Draw chiral molecules that meet the following descriptions:
(a) A chloroalkane, $C_5H_{11}Cl$ (b) An alcohol, $C_6H_{14}O$
(c) An alkene, C_6H_{12} (d) An alkane, C_8H_{18}

9.37   Eight alcohols have the formula $C_5H_{12}O$. Draw them. Which are chiral?

9.38   Draw the nine chiral molecules that have the formula $C_6H_{13}Br$.

9.39   Draw compounds that fit the following descriptions:
(a) A chiral alcohol with four carbons
(b) A chiral carboxylic acid with the formula $C_5H_{10}O_2$
(c) A compound with two chirality centers
(d) A chiral aldehyde with the formula C_3H_5BrO

9.40   Which of the following objects are chiral?
(a) A basketball   (b) A fork   (c) A wine glass
(d) A golf club   (e) A monkey wrench   (f) A snowflake

9.41   Penicillin V (shown at the top of the next page) is an important broad-spectrum antibiotic that contains three chirality centers. Identify them.

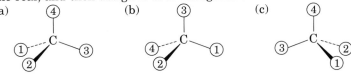

Penicillin V

9.42 Draw examples of the following:
(a) A meso compound with the formula C_8H_{18}
(b) A meso compound with the formula C_9H_{20}
(c) A compound with two chirality centers, one R and the other S

9.43 What is the relationship between the specific rotations of $(2R,3R)$-dichloropentane and $(2S,3S)$-dichloropentane? Between $(2R,3S)$-dichloropentane and $(2R,3R)$-dichloropentane?

9.44 What is the stereochemical configuration of the enantiomer of $(2S,4R)$-dibromo-octane?

9.45 What are the stereochemical configurations of the two diastereomers of $(2S,4R)$-dibromooctane?

9.46 Orient each of the following drawings so that the lowest-priority group is toward the rear, and then assign R or S configuration:

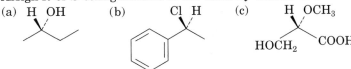

9.47 Assign Cahn–Ingold–Prelog priorities to the following sets of substituents:
(a) $—CH{=}CH_2$, $—CH(CH_3)_2$, $—C(CH_3)_3$, $—CH_2CH_3$

(b) $—C{\equiv}CH$, $—CH{=}CH_2$, $—C(CH_3)_3$, ⬡

(c) $—CO_2CH_3$, $—COCH_3$, $—CH_2OCH_3$, $—CH_2CH_3$

(d) $—C{\equiv}N$, $—CH_2Br$, $—CH_2CH_2Br$, $—Br$

9.48 Assign R or S configurations to the chirality centers in the following molecules:

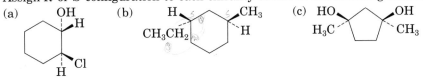

9.49 Assign R or S configuration to each chirality center in the following molecules:
(a) OH H ... Cl
(b) H CH$_3$ CH$_3$CH$_2$ H
(c) HO OH H$_3$C CH$_3$

9.50　Draw tetrahedral representations of the following molecules:
(a) (S)-2-Butanol, $CH_3CH_2CH(OH)CH_3$　　(b) (R)-3-Chloro-1-pentene

9.51　Draw tetrahedral representations of the two enantiomers of the amino acid cysteine, $HSCH_2CH(NH_2)COOH$, and identify each as R or S.

9.52　Which of the following pairs of Fischer projections represent the same enantiomer, and which represent different enantiomers?

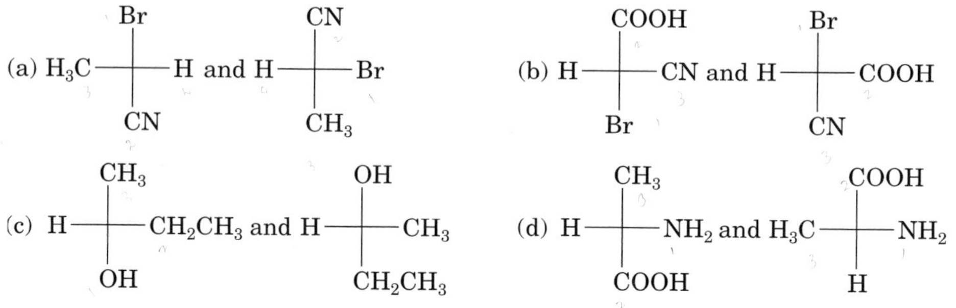

9.53　Assign R or S configurations to the following Fischer projections:

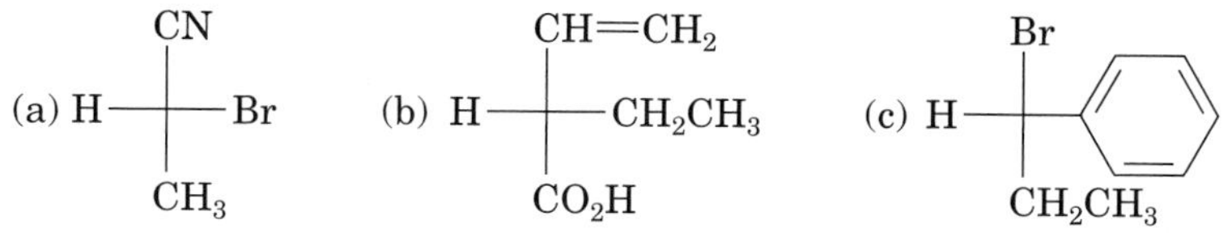

9.54　Assign R or S configuration to each chirality center in the following molecules:

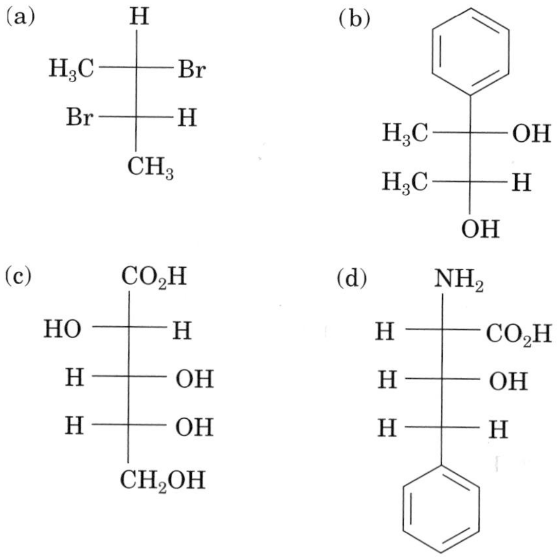

9.55　Draw Fischer projections that fit the following descriptions:
(a) The S enantiomer of 2-bromobutane
(b) The R enantiomer of alanine, $CH_3CH(NH_2)COOH$
(c) The R enantiomer of 2-hydroxypropanoic acid
(d) The S enantiomer of 3-methylhexane

9.56 Assign R or S configurations to the chirality centers in ascorbic acid (vitamin C).

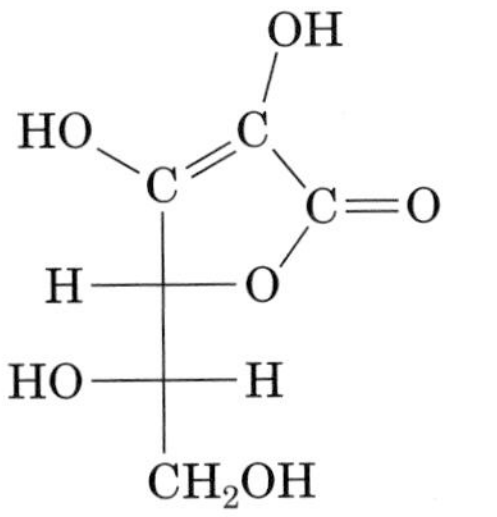

OH
|
HO   C
 \   ‖
  C   C=O
 /   /
H ——— O

H ——|—— O

HO ——|—— H

CH$_2$OH

Ascorbic acid

9.57 Xylose is a common sugar found in many types of wood, including maple and cherry. Because it is much less prone to cause tooth decay than sucrose, xylose has been used in candy and chewing gum. Assign R or S configurations to the chirality centers in xylose.

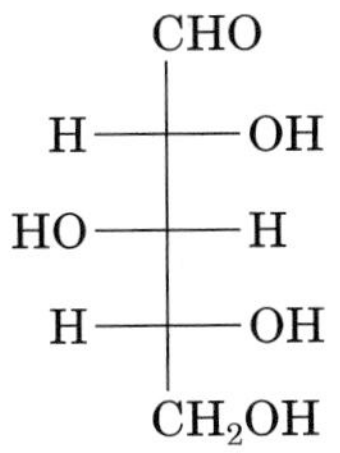

CHO

H ——|—— OH

HO ——|—— H   **(+)-Xylose, $[\alpha]_D = +92°$**

H ——|—— OH

CH$_2$OH

9.58 Hydroxylation of *cis*-2-butene with OsO$_4$ yields butane-2,3-diol. What stereochemistry do you expect for the product? (Review Section 7.8 if necessary.)

9.59 Hydroxylation of *trans*-2-butene with OsO$_4$ also yields butane-2,3-diol. What stereochemistry do you expect for the product?

9.60 Alkenes undergo reaction with peroxycarboxylic acids (RCO$_3$H) to give three-membered-ring cyclic ethers called *epoxides*. For example, 4-octene reacts with a peroxyacid to yield 4,5-epoxyoctane:

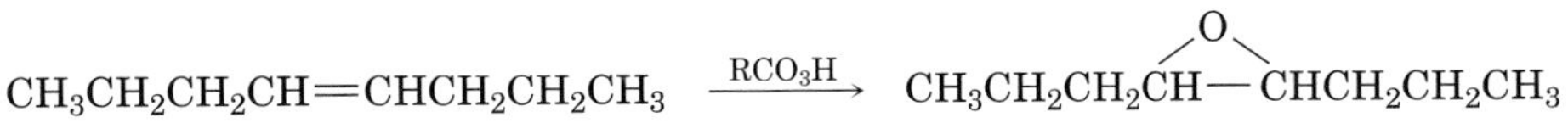

$$CH_3CH_2CH_2CH=CHCH_2CH_2CH_3 \xrightarrow{RCO_3H} CH_3CH_2CH_2CH-CHCH_2CH_2CH_3$$

4-Octene   **4,5-Epoxyoctane**

Assuming that this epoxidation reaction occurs with syn stereochemistry, draw the structure obtained from epoxidation of *cis*-4-octene. Is the product chiral? How many chirality centers does it have? How would you describe it stereochemically?

9.61 Answer Problem 9.60, assuming that the epoxidation reaction is carried out on *trans*-4-octene.

9.62 Write the products of the following reactions, and indicate the stereochemistry obtained in each instance:

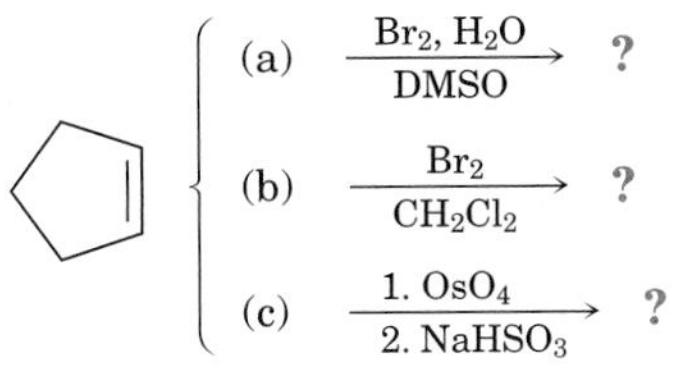

(a) $\xrightarrow[\text{DMSO}]{\text{Br}_2, \text{H}_2\text{O}}$?

(b) $\xrightarrow[\text{CH}_2\text{Cl}_2]{\text{Br}_2}$?

(c) $\xrightarrow[\text{2. NaHSO}_3]{\text{1. OsO}_4}$?

9.63 Draw all possible stereoisomers of cyclobutane-1,2-dicarboxylic acid, and indicate the interrelationships. Which, if any, are optically active? Do the same for cyclobutane-1,3-dicarboxylic acid.

9.64 Compound A, C_7H_{12}, was found to be optically active. On catalytic reduction over a palladium catalyst, 2 equivalents of hydrogen were absorbed, yielding compound B, C_7H_{16}. On ozonolysis of A, two fragments were obtained. One fragment was identified as acetic acid. The other fragment, compound C, was an optically active carboxylic acid, $C_5H_{10}O_2$. Write the reactions, and draw structures for A, B, and C.

9.65 Compound A, $C_{11}H_{16}O$, was found to be an optically active alcohol. Despite its apparent unsaturation, no hydrogen was absorbed on catalytic reduction over a palladium catalyst. On treatment of A with dilute sulfuric acid, dehydration occurred, and an optically inactive alkene B, $C_{11}H_{14}$, was produced as the major product. Alkene B, on ozonolysis, gave two products. One product was identified as propanal, CH_3CH_2CHO. Compound C, the other product, was shown to be a ketone, C_8H_8O. How many degrees of unsaturation does A have? Write the reactions, and identify A, B, and C.

9.66 Draw the structure of (*R*)-2-methylcyclohexanone.

9.67 The so-called tetrahedranes are an interesting class of compounds, the first example of which was synthesized in 1979. Make a model of a substituted tetrahedrane with four different substituents. Is it chiral? Explain.

9.68 *Allenes* are compounds with adjacent carbon–carbon double bonds. Many allenes are chiral, even though they don't contain chirality centers. Mycomycin, for example, a naturally occurring antibiotic isolated from the bacterium *Nocardia acidophilus,* is chiral and has $[\alpha]_D = -130°$. Explain why mycomycin is chiral. Making a molecular model should be helpful.

$$HC \equiv C–C \equiv C–CH = C = CH–CH = CH–CH = CH–CH_2COOH$$
Mycomycin (an allene)

9.69 Long before chiral allenes were known (Problem 9.68), the resolution of 4-methylcyclohexylideneacetic acid into two enantiomers had been carried out. Why is it chiral? What geometric similarity does it have to allenes?

Methylcyclohexylideneacetic acid

9.70 Carboxylic acids react with alcohols to yield esters:

Suppose that racemic lactic acid reacts with methanol, CH_3OH, to yield the ester, methyl lactate. What stereochemistry would you expect the products to have? What is the relationship of one product to another?

9.71 Suppose that (S)-lactic acid reacts with (R)-2-butanol to form an ester (Problem 9.70). What stereochemistry would you expect the product(s) to have? Draw the reactants and product(s).

9.72 Suppose that racemic lactic acid reacts with (S)-2-butanol to form an ester (Problem 9.71). What stereochemistry does the product(s) have? What is the relationship of one product to another? Assuming that esters can be converted back into carboxylic acids, how might you use this reaction to resolve (±)-lactic acid?

9.73 (S)-1-Chloro-2-methylbutane undergoes light-induced reaction with Cl_2 by a radical mechanism to yield a mixture of products. Among the products are 1,4-dichloro-2-methylbutane and 1,2-dichloro-2-methylbutane.
(a) Write the reaction, showing the correct stereochemistry of the reactant.
(b) One of the two products is optically active, but the other is optically inactive. Which is which?
(c) What can you conclude about the stereochemistry of radical chlorination reactions?

9.74 Draw a meso compound that has five carbons and three chirality centers.

9.75 How many stereoisomers of 2,4-dibromo-3-chloropentane are there? Draw them, and indicate which are optically active.

9.76 Draw both *cis*- and *trans*-1,4-dimethylcyclohexane in their most stable chair conformations.
(a) How many stereoisomers are there of *cis*-1,4-dimethylcyclohexane, and how many of *trans*-1,4-dimethylcyclohexane?
(b) Are any of the structures chiral?
(c) What are the stereochemical relationships among the various stereoisomers of 1,4-dimethylcyclohexane?

9.77 Draw both *cis*- and *trans*-1,3-dimethylcyclohexane in their most stable chair conformations.
(a) How many stereoisomers are there of *cis*-1,3-dimethylcyclohexane, and how many of *trans*-1,3-dimethylcyclohexane?
(b) Are any of the structures chiral?
(c) What are the stereochemical relationships among the various stereoisomers of 1,3-dimethylcyclohexane?

9.78 How can you explain the observation that *cis*-1,2-dimethylcyclohexane is optically inactive even though it has two chirality centers?

A Look Ahead

• •

9.79 An alkyl halide reacts with a nucleophile to give a substitution product by a mechanism that involves *inversion* of stereochemistry at carbon:

$$\diagdown\text{C}-\text{X} \xrightarrow{\ :\text{Nu}^-\ } \text{Nu}-\text{C}\diagup + \text{X}^-$$

Formulate the reaction of (S)-2-bromobutane with HS^- ion to yield butane-2-thiol, $CH_3CH_2CH(SH)CH_3$. What is the stereochemistry of the product? (See Section 11.2.)

9.80 *Grignard reagents,* RMgX, react with aldehydes to yield alcohols. For example, the reaction of methylmagnesium bromide with propanal yields 2-butanol:

$$CH_3CH_2 - \overset{\overset{\displaystyle O}{\|}}{C} - H \quad \xrightarrow[\text{2. } H_3O^+]{\text{1. } CH_3MgBr} \quad CH_3CH_2 - \overset{\overset{\displaystyle OH}{|}}{\underset{\underset{\displaystyle H}{|}}{C}} - CH_3$$

Propanal

2-Butanol

(a) Is the product chiral? Is it optically active?

(b) How many stereoisomers of butanol are formed, what are their stereochemical relationships, and what are their relative amounts? (See Section 17.6.)

9.81 Imagine that another Grignard reaction similar to that in Problem 9.80 is carried out between methylmagnesium bromide and (*R*)-2-phenylpropanal to yield 3-phenyl-2-butanol:

$$\xrightarrow[\text{2. } H_3O^+]{\text{1. } CH_3MgBr}$$

(R)-2-Phenylpropanal

$$CH_3 - \underset{\underset{\displaystyle |}{|}}{CH} - \overset{\overset{\displaystyle OH}{|}}{CH} - CH_3$$

3-Phenyl-2-butanol

(a) Is the product chiral? Is it optically active?

(b) How many stereoisomers of 3-phenyl-2-butanol are formed, what are their stereochemical relationships, and what are their relative amounts? (See Section 17.6.)

Molecular Modeling

· ·

9.82 Use SpartanBuild to build all possible stereoisomers of 2-bromo-3-chlorobutane (consider only the conformations in which bromine and chlorine are anti). Minimize the energy of each molecule, and identify stereoisomers with identical energies. Are they enantiomers or diastereomers?

9.83 Use SpartanView to compare energies of the pyramidal and planar forms of *N*-ethyl-*N*-methylpropylamine, *P*-ethyl-*P*-methylpropylphosphine, and ethyl methyl sulfoxide. Which geometry of each is lower in energy? Which molecule will racemize most rapidly?

$$CH_3CH_2 - \overset{\overset{\displaystyle CH_3}{|}}{\underset{\underset{\displaystyle \cdot\cdot}{}}{N}} - CH_2CH_2CH_3$$

***N*-Ethyl-*N*-methylpropylamine**

$$CH_3CH_2 - \overset{\overset{\displaystyle CH_3}{|}}{\underset{\underset{\displaystyle \cdot\cdot}{}}{P}} - CH_2CH_2CH_3$$

***P*-Ethyl-*P*-methylpropylphosphine**

$$CH_3CH_2 - \overset{\overset{\displaystyle O}{\|}}{\underset{\underset{\displaystyle \cdot\cdot}{}}{S}} - CH_3$$

Ethyl methyl sulfoxide

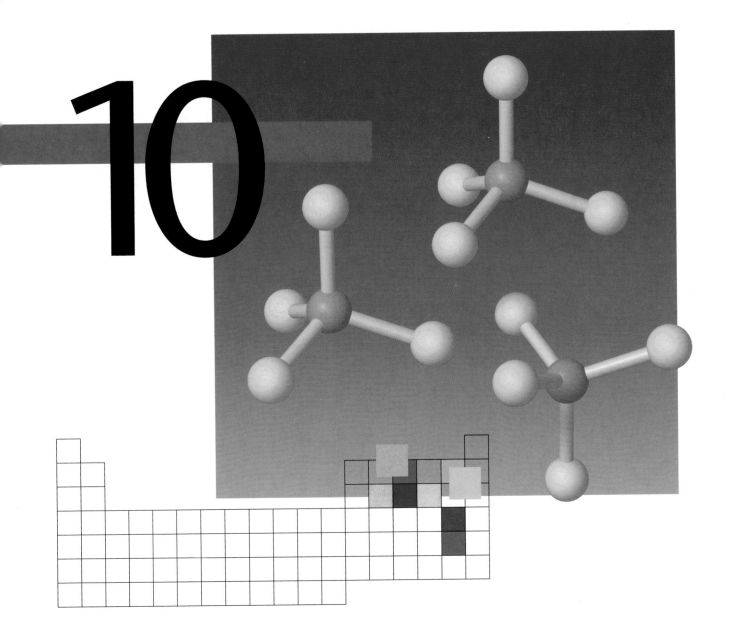

Alkyl Halides

Halogen-substituted organic compounds are widespread throughout nature and have a vast array of uses in modern industrial processes. Several thousand organohalides have been found in algae and various other marine organisms. Chloromethane, for example, is released in large amounts by oceanic kelp, as well as by forest fires and volcanoes. Among their many uses, organohalides are valuable as industrial solvents, inhaled anesthetics in medicine, refrigerants, and pesticides. The modern electronics industry, for example, relies on halogenated solvents such as trichloroethylene for cleaning semiconductor chips and other components.

Trichloroethylene
(a solvent)

Halothane
(an inhaled anesthetic)

Dichlorodifluoromethane
(a refrigerant)

Bromomethane
(a fumigant)

Still other halo-substituted compounds are providing important leads to new pharmaceuticals. The compound *epibatidine,* for instance, has been isolated from the skin of Ecuadorian frogs and found to be more than 200 times as potent as morphine at blocking pain in animals.

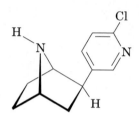

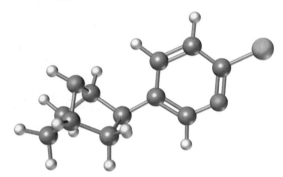

Epibatidine
(from the Ecuadorian frog
*Epipedobates tricolor***)**

In this and the next chapter, we'll be discussing the chemistry of **alkyl halides**—compounds that have a halogen atom bonded to a saturated, sp^3-hybridized carbon atom. We'll begin in this chapter with a look at how to name and prepare alkyl halides, and we'll see several of their reactions. Then in the following chapter, we'll make a detailed study of the substitution and elimination reactions of alkyl halides—two of the most important and well-studied reaction types in organic chemistry.

10.1 Naming Alkyl Halides

Alkyl halides are named in the same way as alkanes (Section 3.4), by treating the halogen as a substituent on a parent alkane chain. There are three rules:

RULE 1   Find the longest carbon chain and name it as the parent. If a double or triple bond is present, the parent chain must contain it.

RULE 2   Number the carbon atoms of the parent chain, beginning at the end nearer the first substituent, regardless of whether it is alkyl or halo. Assign each substituent a number according to its position on the chain. For example:

$$\underset{1}{CH_3}\underset{2}{CH}\underset{3}{CH_2}\underset{4}{CH}\underset{5}{CH}\underset{6}{CH_2}\underset{7}{CH_3}$$

with CH₃ at position 2 and Br at position 5, CH₃ at position 4

5-Bromo-2,4-dimethylheptane

$$\overset{\displaystyle \underset{|}{Br} \qquad \underset{|}{CH_3}}{\underset{\underset{1 \quad 2 \quad 3 \quad \overset{|}{CH_3}}{}}{CH_3CHCH_2CHCHCH_2CH_3}}$$

$$\underset{1 \ \ 2 \ \ 3 \ \ 4 \ 5 \ \ 6 \ \ 7}{}$$

$$\underset{CH_3}{|}$$

2-Bromo-4,5-dimethylheptane

(a) If more than one of the same kind of halogen is present, number each and use one of the prefixes *di-*, *tri-*, *tetra-*, and so on. For example:

$$\overset{\underset{|}{Cl} \ \underset{|}{Cl}}{CH_3CHCHCHCH_2CH_3}$$

$$\underset{1 \ \ 2 \ \ 3 \ \ 4 \ 5 \ \ 6}{}$$

$$\underset{CH_3}{|}$$

2,3-Dichloro-4-methylhexane

(b) If different halogens are present, number all and list them in alphabetical order when writing the name. For example:

$$\overset{\underset{|}{Cl}}{BrCH_2CH_2CHCHCH_3}$$

$$\underset{1 \ \ \ 2 \ \ \ 3 \ \ 4 \ \ 5}{}$$

$$\underset{CH_3}{|}$$

1-Bromo-3-chloro-4-methylpentane

RULE 3 If the parent chain can be properly numbered from either end by rule 2, begin at the end nearer the substituent (either alkyl or halo) that has alphabetical precedence. For example:

$$\overset{\underset{|}{CH_3} \qquad \underset{|}{Br}}{CH_3CHCH_2CH_2CHCH_3}$$

$$\underset{6 \ \ 5 \ \ \ 4 \ \ \ 3 \ \ 2 \ \ 1}{}$$

2-Bromo-5-methylhexane
(***NOT** 5-bromo-2-methylhexane*)

In addition to their systematic names, many simple alkyl halides are also named by identifying first the alkyl group and then the halogen. For example, CH_3I can be called methyl iodide. Such names are well entrenched in the chemical literature and in daily usage, but they won't be used in this book.

$$CH_3I$$

$$\overset{\underset{|}{Cl}}{CH_3CHCH_3}$$

Iodomethane
(or methyl iodide)

2-Chloropropane
(or isopropyl chloride)

Bromocyclohexane
(or cyclohexyl bromide)

· ·

Problem 10.1   Give the IUPAC names of the following alkyl halides:

(a) $CH_3CH_2CH_2CH_2I$

(b) $CH_3CHCH_2CH_2Cl$
with CH_3 on the second carbon

(c) $BrCH_2CH_2CH_2CCH_2Br$
with CH_3 above and CH_3 below the C

(d) $CH_3CCH_2CH_2Cl$
with CH_3 above and Cl below the C

(e) $CH_3CHCHCH_2CH_3$
with I above the first CH and CH_2CH_2Cl above the second CH

(f) $CH_3CHCH_2CH_2CHCH_3$
with Br above the first CH and Cl above the second CH

Problem 10.2   Draw structures corresponding to the following IUPAC names:
(a) 2-Chloro-3,3-dimethylhexane   (b) 3,3-Dichloro-2-methylhexane
(c) 3-Bromo-3-ethylpentane   (d) 1,1-Dibromo-4-isopropylcyclohexane
(e) 4-*sec*-Butyl-2-chlorononane   (f) 1,1-Dibromo-4-*tert*-butylcyclohexane

· ·

10.2 Structure of Alkyl Halides

The carbon–halogen bond in an alkyl halide results from the overlap of a carbon sp^3 hybrid orbital with a halogen orbital. Thus, alkyl halide carbon atoms have an approximately tetrahedral geometry, with H–C–X bond angles near 109°. Halogens increase in size going down the periodic table so the bond lengths of the halomethanes increase accordingly (Table 10.1). Table 10.1 also indicates that C–X bond strengths decrease going down the periodic table. (As we've been doing consistently thus far, we'll continue to use the abbreviation X to represent any of the halogens F, Cl, Br, or I.)

TABLE 10.1   A Comparison of the Halomethanes

Halomethane	Bond length (pm)	Bond strength		Dipole moment (*D*)
		(kJ/mol)	(kcal/mol)	
CH_3F	139	452	108	1.85
CH_3Cl	178	351	84	1.87
CH_3Br	193	293	70	1.81
CH_3I	214	234	56	1.62

In an earlier discussion of bond polarity in functional groups (Section 5.4), we noted that halogens are more electronegative than carbon. The C–X bond is therefore polar, with the carbon atom bearing a slight positive charge ($\delta+$) and the halogen a slight negative charge ($\delta-$). This polarity results in a substantial dipole moment for all the halomethanes (Table 10.1), and implies that the alkyl halide C–X carbon atom should behave as an electrophile in polar reactions. We'll see in the next chapter that much of the chemistry of alkyl halides is indeed dominated by their electrophilic behavior.

10.3 Preparation of Alkyl Halides

We've already seen several methods for preparing alkyl halides, including the reactions of HX and X_2 with alkenes in electrophilic addition reactions (Sections 6.8 and 7.2). The hydrogen halides HCl, HBr, and HI react with alkenes by a polar mechanism to give the product of Markovnikov addition. Bromine and chlorine yield trans 1,2-dihalogenated addition products.

Another method of alkyl halide synthesis is the reaction of an alkane with Cl_2 or Br_2 by a radical chain-reaction pathway (Section 5.3). Although inert to most reagents, alkanes react readily with Cl_2 or Br_2 in the presence of light to give haloalkane substitution products. The reaction occurs by the radical mechanism shown in Figure 10.1 (p. 360) for chlorination.

Recall from Section 5.3 that radical substitution reactions require three kinds of steps: *initiation, propagation,* and *termination.* Once an initiation step has started the process by producing radicals, the reaction continues in a self-sustaining cycle. The cycle requires two repeating propagation steps in which a radical, the halogen, and the alkane yield alkyl halide product plus more radical to carry on the chain. The chain is occasionally terminated by the combination of two radicals.

Though interesting from a mechanistic point of view, alkane halogenation is a poor synthetic method for preparing different haloalkanes. Let's see why.

FIGURE 10.1 ▼

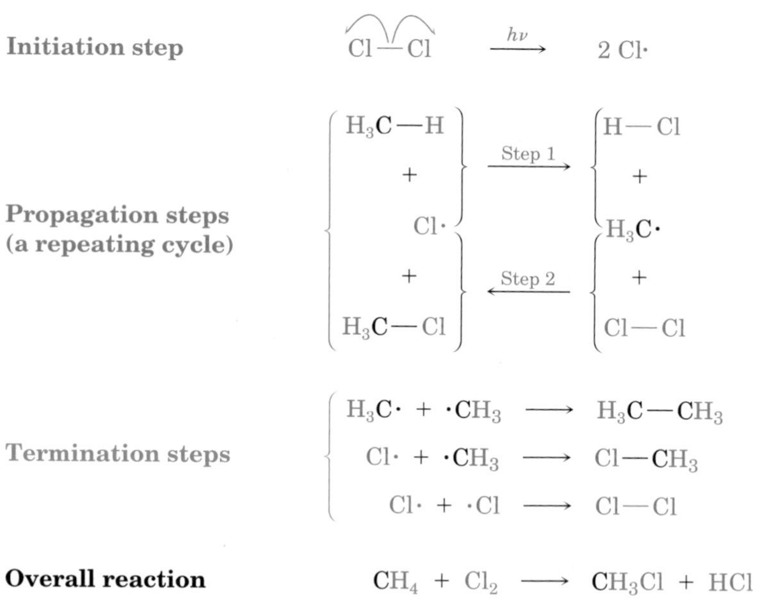

Mechanism of the radical chlorination of methane. Three kinds of steps are required: initiation, propagation, and termination. The propagation steps are a repeating cycle, with Cl· a reactant in step 1 and a product in step 2, and with ·CH$_3$ a product in step 1 and a reactant in step 2. (The symbol $h\nu$ shown in the initiation step is the standard way of indicating irradiation with light.)

Initiation step

$$\text{Cl} - \text{Cl} \xrightarrow{h\nu} 2\text{ Cl·}$$

Propagation steps (a repeating cycle)

$$\left.\begin{array}{c} \text{H}_3\text{C} - \text{H} \\ + \\ \text{Cl·} \\ + \\ \text{H}_3\text{C} - \text{Cl} \end{array}\right\} \begin{array}{c} \xrightarrow{\text{Step 1}} \\ \\ \xleftarrow{\text{Step 2}} \end{array} \left\{\begin{array}{c} \text{H} - \text{Cl} \\ + \\ \text{H}_3\text{C·} \\ + \\ \text{Cl} - \text{Cl} \end{array}\right.$$

Termination steps

$$\left\{\begin{array}{l} \text{H}_3\text{C·} + \text{·CH}_3 \longrightarrow \text{H}_3\text{C} - \text{CH}_3 \\ \text{Cl·} + \text{·CH}_3 \longrightarrow \text{Cl} - \text{CH}_3 \\ \text{Cl·} + \text{·Cl} \longrightarrow \text{Cl} - \text{Cl} \end{array}\right.$$

Overall reaction

$$\text{CH}_4 + \text{Cl}_2 \longrightarrow \text{CH}_3\text{Cl} + \text{HCl}$$

10.4 Radical Halogenation of Alkanes

Alkane halogenation is a poor method of alkyl halide synthesis because mixtures of products invariably result. For example, chlorination of methane does not stop cleanly at the monochlorinated stage. Rather, the reaction continues on to give a mixture of dichloro, trichloro, and even tetrachloro products:

$$\text{CH}_4 + \text{Cl}_2 \xrightarrow{h\nu} \text{CH}_3\text{Cl} + \text{HCl}$$
$$\xrightarrow{\text{Cl}_2} \text{CH}_2\text{Cl}_2 + \text{HCl}$$
$$\xrightarrow{\text{Cl}_2} \text{CHCl}_3 + \text{HCl}$$
$$\xrightarrow{\text{Cl}_2} \text{CCl}_4 + \text{HCl}$$

The situation is even worse for chlorination of alkanes that have more than one type of hydrogen. For example, chlorination of butane gives two monochlorinated products in addition to dichlorobutane, trichlorobutane, and so on. Thirty percent of the monochloro product is 1-chlorobutane, and 70% is 2-chlorobutane:

$$CH_3CH_2CH_2CH_3 + Cl_2 \xrightarrow{h\nu} CH_3CH_2CH_2CH_2Cl + CH_3CH_2\overset{\overset{\displaystyle Cl}{|}}{C}HCH_3 +$$

Butane   **1-Chlorobutane**   **2-Chlorobutane**

Dichloro-, trichloro-, tetrachloro-, and so on

30:70

As another example, 2-methylpropane yields 2-chloro-2-methylpropane and 1-chloro-2-methylpropane in the ratio 35:65, along with more highly chlorinated products:

$$\overset{\overset{\displaystyle CH_3}{|}}{CH_3CHCH_3} + Cl_2 \xrightarrow{h\nu} \overset{\overset{\displaystyle CH_3}{|}}{\underset{\underset{\displaystyle Cl}{|}}{CH_3CCH_3}} + \overset{\overset{\displaystyle CH_3}{|}}{CH_3CHCH_2Cl} +$$

2-Methylpropane

Dichloro-, trichloro-, tetrachloro-, and so on

2-Chloro-2-methylpropane   **1-Chloro-2-methylpropane**

35:65

From these and similar reactions, it's possible to calculate a reactivity order toward chlorination for different types of hydrogen atoms in a molecule. Take the butane chlorination, for instance. Butane has six equivalent primary hydrogens ($-CH_3$) and four equivalent secondary hydrogens ($-CH_2-$). The fact that butane yields 30% of 1-chlorobutane product means that *each one* of the six primary hydrogens is responsible for 30% ÷ 6 = 5% of the product. Similarly, the fact that 70% of 2-chlorobutane is formed means that each of the four secondary hydrogens is responsible for 70% ÷ 4 = 17.5% of the product. Thus, reaction of a secondary hydrogen happens 17.5% ÷ 5% = 3.5 times as often as reaction of a primary hydrogen.

A similar calculation for the chlorination of 2-methylpropane indicates that each of the nine primary hydrogens accounts for 65% ÷ 9 = 7.2% of the product, while the single tertiary hydrogen (R_3CH) accounts for 35% of the product. Thus, a tertiary hydrogen is 35% ÷ 7.2% = 5 times as reactive as a primary hydrogen toward chlorination.

Primary	Secondary	Tertiary
H	H	R
|	|	|
R—C—H	R—C—H	R—C—H
|	|	|
H	R	R
Primary <	Secondary <	Tertiary
1.0	3.5	5.0

Reactivity →

What are the reasons for the observed reactivity order of alkane hydrogens toward radical chlorination? A look at the bond dissociation energies given previously in Table 5.3 hints at the answer. The data in Table 5.3 indicate that a tertiary C–H bond [390 kJ/mol (93 kcal/mol)] is weaker than a secondary C–H bond [401 kJ/mol (96 kcal/mol)], which is in turn weaker than a primary C–H bond [420 kJ/mol (100 kcal/mol)]. Since less energy is needed to break a tertiary C–H bond than to break a primary or secondary C–H bond, the resultant tertiary radical is more stable than a primary or secondary radical.

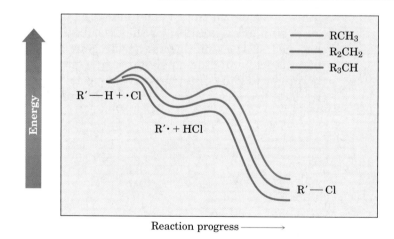

An explanation of the relationship between reactivity and bond strength in radical chlorination reactions relies on the Hammond postulate, developed in Section 6.11 to explain why more stable carbocations form faster than less stable ones in alkene electrophilic addition reactions. A reaction energy diagram for the formation of an alkyl radical during alkane chlorination is shown in Figure 10.2. Although the hydrogen abstraction step is slightly exergonic, there is nevertheless a certain amount of developing radical character in the transition state. Since the increasing alkyl substitution that stabilizes the radical intermediate also stabilizes the transition state leading to that intermediate, the more stable radical forms faster than the less stable one.

FIGURE 10.2 ▼

Reaction energy diagram for alkane chlorination. The relative rate of formation of tertiary, secondary, and primary radicals is the same as their stability order.

In contrast to alkane chlorination, alkane bromination is usually much more selective. In its reaction with 2-methylpropane, for example, bromine abstracts the tertiary hydrogen with greater than 99% selectivity, as opposed to the 35:65 mixture observed in the corresponding chlorination.

$$CH_3CHCH_3 + Br_2 \xrightarrow{h\nu} CH_3CCH_3$$

with CH_3 groups and Br substituents as drawn.

CH₃CHCH₃ + Br₂ —(hν)→ CH₃CCH₃ [CH₃CHCH₂Br]

2-Methylpropane   2-Bromo-2-methylpropane (>99%)   1-Bromo-2-methylpropane (<1%)

The enhanced selectivity of alkane bromination over chlorination can be explained by turning once again to the Hammond postulate. In comparing the abstractions of an alkane hydrogen by Cl· and Br· radicals, reaction with Br· is much less exergonic. As a result, the transition state for bromination resembles the alkyl radical more closely than does the transition state for chlorination, and the stability of that radical is therefore more important for bromination than for chlorination.

$$H_3C-\underset{\underset{CH_3}{|}}{\overset{\overset{CH_3}{|}}{C}}-H + X\cdot \longrightarrow H_3C-\underset{\underset{CH_3}{|}}{\overset{\overset{CH_3}{|}}{C}}\cdot + HX$$

$\Delta H° = -50$ kJ for X = Cl
$\Delta H° = +13$ kJ for X = Br

2-Methylpropane

Problem 10.3 Draw and name all monochloro products you would expect to obtain from radical chlorination of 2-methylpentane. Which, if any, are chiral?

Problem 10.4 Taking the relative reactivities of 1°, 2°, and 3° hydrogen atoms into account, what product(s) would you expect to obtain from monochlorination of 2-methylbutane? What would the approximate percentage of each product be? (Don't forget to take into account the number of each type of hydrogen.)

Problem 10.5 Use the bond dissociation energies listed in Table 5.3 to calculate $\Delta H°$ for the reactions of Cl· and Br· with a secondary hydrogen atom of propane. Which reaction would you expect to be more selective?

10.5 Allylic Bromination of Alkenes

While repeating work done earlier by others, the German chemist Karl Ziegler reported in 1942 that alkenes react with *N*-bromosuccinimide (abbreviated NBS) in the presence of light to give products resulting from substitution of hydrogen by bromine at the **allylic** position—the position *next to* the double bond. Cyclohexene, for example, gives 3-bromocyclohexene in 85% yield.

Cyclohexene   **3-Bromocyclohexene**
                              (85%)

This allylic bromination with NBS looks analogous to the alkane halogenation reaction discussed in the previous section. In both cases, a C–H bond on a saturated carbon is broken and the hydrogen atom is replaced by halogen. The analogy is a good one, for studies have shown that allylic NBS brominations do in fact occur by a two-step, radical chain pathway. As in alkane halogenation, Br· radical abstracts an allylic hydrogen atom of the alkene, thereby forming an allylic radical plus HBr. This allylic radical then reacts with Br_2 to yield the product and a Br· radical, which cycles back into the first step to carry on the chain. The Br_2 results from the reaction of NBS with the HBr formed in the first step.

Allylic radical

Why does bromination with NBS occur exclusively at an allylic position rather than elsewhere in the molecule? The answer, once again, is found by looking at bond dissociation energies to see the relative stabilities of various kinds of radicals.

There are three types of C–H bonds in cyclohexene, and Table 5.3 gives an idea of their relative strengths. Although a typical secondary alkyl C–H bond has a strength of about 400 kJ/mol (96 kcal/mol), and a typical vinylic C–H bond has a strength of 445 kJ/mol (106 kcal/mol), an *allylic* C–H bond has a strength of only about 360 kJ/mol (87 kcal/mol). An allylic radical is therefore more stable than a typical alkyl radical by about 40 kJ/mol (9 kcal/mol).

Allylic
360 kJ/mol (87 kcal/mol)

Alkyl
400 kJ/mol (96 kcal/mol)

Vinylic
445 kJ/mol (106 kcal/mol)

We can thus expand the stability ordering to include vinylic and allylic radicals:

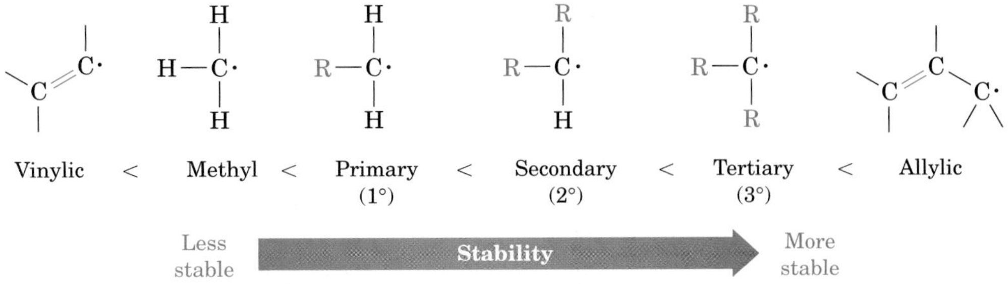

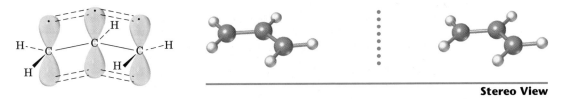

Vinylic < Methyl < Primary < Secondary < Tertiary < Allylic
                          (1°)   (2°)   (3°)

Less   More
stable   **Stability**   stable

10.6 Stability of the Allyl Radical: Resonance Revisited

To see why allylic radicals are so stable, look at the orbital picture in Figure 10.3. The radical carbon atom with an unpaired electron can adopt sp^2 hybridization, placing the unpaired electron in a p orbital and giving a structure that is electronically symmetrical. The p orbital on the central carbon can therefore overlap equally well with a p orbital on *either* of the two neighboring carbons.

FIGURE 10.3 ▼

An orbital view of the allyl radical. The p orbital on the central carbon can overlap equally well with a p orbital on either neighboring carbon because the structure is electronically symmetrical.

Stereo View

Since the allyl radical is electronically symmetrical, it can be drawn in either of two resonance forms—with the unpaired electron on the left and the double bond on the right, or with the unpaired electron on the right and the double bond on the left. Neither structure is correct by itself; the true structure of the allyl radical is a resonance hybrid of the two. (You might want to review Sections 2.4–2.6 if you need to brush up on resonance.) As noted in Section 2.5, the greater the number of resonance forms, the greater the stability of a compound. An allyl radical, with two resonance forms, is therefore more stable than a typical alkyl radical, which has only a single structure.

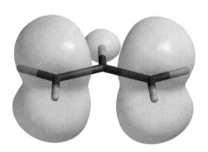

In molecular orbital terms, the stability of the allyl radical is due to the fact that the unpaired electron is **delocalized**, or spread out, over an extended π orbital network rather than localized at only one site. This delocalization is particularly apparent in the so-called *spin surface* in Figure 10.4, which shows the calculated location of the unpaired electron. The two terminal carbons share the unpaired electron equally.

In addition to its effect on stability, delocalization of the unpaired electron in the allyl radical has other chemical consequences. Since the unpaired electron is delocalized over both ends of the π orbital system, reaction with Br_2 can occur at either end. As a result, allylic bromination of an unsymmetrical alkene often leads to a mixture of products. For example, bromination of 1-octene gives a mixture of 3-bromo-1-octene and 1-bromo-2-octene. The two products are not formed in equal amounts, however, because the intermediate allylic radical is not symmetrical and reaction at the two ends is not equally likely. Reaction at the less hindered, primary end is favored.

$$CH_3CH_2CH_2CH_2CH_2CH_2CH{=}CH_2$$

1-Octene

NBS, CCl₄

$$[CH_3CH_2CH_2CH_2CH_2\overset{\cdot}{C}HCH{=}CH_2 \longleftrightarrow CH_3CH_2CH_2CH_2CH_2CH{=}CH\overset{\cdot}{C}H_2]$$

Br
|
$$CH_3CH_2CH_2CH_2CH_2CHCH{=}CH_2 \;+\; CH_3CH_2CH_2CH_2CH_2CH{=}CHCH_2Br$$

3-Bromo-1-octene (17%)   **1-Bromo-2-octene (83%)**
                                   (53:47 trans:cis)

The products of allylic bromination reactions are particularly useful for conversion into dienes by dehydrohalogenation with base. Cyclohexene can be converted into 1,3-cyclohexadiene, for example.

Cyclohexene   3-Bromocyclohexene   1,3-Cyclohexadiene

• •

Practice Problem 10.1 What products would you expect from reaction of 4,4-dimethylcyclohexene with NBS?

Strategy Draw the alkene reactant, and identify the allylic positions. In this case, there are two different allylic positions; we'll label them **A** and **B**. Now abstract an allylic hydrogen from each position to generate the two corresponding allylic radicals. Each of the two allylic radicals can add a Br atom at either end (**A** or **a**; **B** or **b**), to give a mixture of up to four products. Draw and name the products. In the present instance, the "two" products from reaction at position **B** are identical, so a total of only three products are formed in this reaction.

Solution

3-Bromo-4,4-dimethyl-cyclohexene   **3-Bromo-6,6-dimethyl-cyclohexene**

3-Bromo-5,5-dimethyl-cyclohexene

• •

Problem 10.6 Draw as many resonance forms as you can for the cyclohexadienyl radical.

Cyclohexadienyl radical

Problem 10.7   The major product of the reaction of methylenecyclohexane with *N*-bromosuccin-imide is 1-(bromomethyl)cyclohexene. Explain.

$$\text{(cyclohexane with } CH_2) \quad \xrightarrow[CCl_4]{NBS} \quad \text{(cyclohexene with } CH_2Br)$$

Major product

Problem 10.8   What products would you expect from reaction of the following alkenes with NBS? If more than one product is formed, show the structures of all.

(a) 5-Methylcycloheptene

(b) $CH_3CHCH=CHCH_2CH_3$ with CH_3 substituent

- -

10.7 Preparing Alkyl Halides from Alcohols

The most general method for preparing alkyl halides is to make them from alcohols. A great many alcohols are commercially available, and we'll see later that a great many more can be obtained from carbonyl compounds. Because of the importance of the reaction, many different reagents have been used for transforming alcohols into alkyl halides.

The simplest method for converting an alcohol to an alkyl halide involves treating the alcohol with HCl, HBr, or HI:

$$ROH + HX \longrightarrow RX + H_2O \qquad (X = Cl, Br, or I)$$

For reasons that will be discussed in the next chapter (Section 11.16), the reaction works best when applied to tertiary alcohols, R_3COH. Primary and secondary alcohols also react, but at slower rates and at higher reaction temperatures. Although this is not a problem in simple cases, more complicated molecules are sometimes acid-sensitive and are destroyed by the reaction conditions.

Methyl	<	Primary	<	Secondary	<	Tertiary
Less reactive			**Reactivity** →			More reactive

The reaction of HX with a tertiary alcohol is so rapid that it's often carried out simply by bubbling the pure HCl or HBr gas into a cold ether solution of the alcohol. Reaction is usually complete within a few minutes.

1-Methylcyclohexanol   **1-Chloro-1-methylcyclohexane**
                                              (90%)

Primary and secondary alcohols are best converted into alkyl halides by treatment with such reagents as thionyl chloride ($SOCl_2$) or phosphorus tribromide (PBr_3). These reactions, which normally take place readily under mild conditions, are less acidic and less likely to cause acid-catalyzed rearrangements than the HX method.

Benzoin   **Desyl chloride (86%)**

2-Butanol   **2-Bromobutane**
                                   (86%)

As the preceding examples indicate, the yields of these $SOCl_2$ and PBr_3 reactions are generally high, and other functional groups such as ethers, carbonyls, and aromatic rings don't usually interfere. We'll look at the mechanisms of these substitution reactions in the next chapter.

• •

Problem 10.9   How would you prepare the following alkyl halides from the appropriate alcohols?
(a) 2-Chloro-2-methylpropane   (b) 2-Bromo-4-methylpentane

$$\begin{array}{cc}
 & CH_3 \\
 & | \\
\text{(c) } BrCH_2CH_2CH_2CH_2CHCH_3 &
\end{array}$$

$$\begin{array}{c}
CH_3 \quad Cl \\
| \quad\quad | \\
\text{(d) } CH_3CH_2CHCH_2CCH_3 \\
| \\
CH_3
\end{array}$$

• •

10.8 Reactions of Alkyl Halides: Grignard Reagents

Organohalides, RX, react with magnesium metal in ether or tetrahydrofuran (THF) solvent to yield organomagnesium halides, RMgX. The products, called **Grignard reagents** after their discoverer, Victor Grignard, are

François Auguste Victor Grignard

François Auguste Victor Grignard (1871–1935) was born in Cherbourg, France, and received his Ph.D. at the University of Lyon in 1901. During his doctoral work under Philippe Barbier, Grignard discovered the preparation and usefulness of organomagnesium reagents. He became professor of chemistry at Nancy and at Lyon, and won the Nobel Prize in chemistry in 1912. During World War I, he was drafted into the French army as a Corporal (a Nobel-Prize-winning Corporal!), where he developed a method for detecting German war gases.

Grignard reagent

examples of *organometallic* compounds because they contain a carbon–metal bond.

$$\text{R}—\text{X} + \text{Mg} \xrightarrow[\text{or THF}]{\text{Ether}} \text{R}—\text{Mg}—\text{X}$$

where R = 1°, 2°, or 3° alkyl, aryl, or alkenyl
        X = Cl, Br, or I

For example:

Bromobenzene   **Phenylmagnesium bromide**

2-Chlorobutane   ***sec*-Butylmagnesium chloride**

Many different kinds of organohalides form Grignard reagents. Steric hindrance in the halide is not a problem in the formation of Grignard reagents, and 1°, 2°, and 3° alkyl halides all react with similar ease. Aryl and alkenyl halides also react with magnesium, although it's best to use THF as solvent for these cases. The halogen may be Cl, Br, or I, although chlorides are less reactive than bromides and iodides. Organofluorides rarely react with magnesium.

As you might expect from the discussion of electronegativity and bond polarity in Section 5.4, the carbon–magnesium bond is polarized, making the carbon atom both nucleophilic and basic. An electrostatic potential map clearly shows the electron-rich (red) character of the carbon bonded to magnesium:

Basic and nucleophilic site

In a formal sense, a Grignard reagent can be thought of as the magnesium salt, $R_3C^-\ {}^+MgX$, of a hydrocarbon acid, $R_3C–H$. But because hydrocarbons are such weak acids, with pK_a's in the range 44–60 (Section 8.8), carbon anions are very strong bases. Grignard reagents therefore react with such weak acids as H_2O, ROH, RCOOH, and RNH_2 to become protonated and yield hydrocarbons. The overall sequence of Grignard formation followed by acid treatment is a useful method for converting an organohalide into a hydrocarbon, R–X $\longrightarrow$ R–H. For example,

$$CH_3(CH_2)_8CH_2Br \xrightarrow[\text{2. } H_3O^+]{\text{1. Mg}} CH_3(CH_2)_8CH_3$$

1-Bromodecane   **Decane (85%)**

We'll see many more uses of Grignard reagents as nucleophiles in later chapters.

Problem 10.10   Just how strong a base would you expect a Grignard reagent to be? Look at Table 8.1, and then predict whether the following reactions will occur as written. (The pK_a of NH_3 is 35.)
(a) $CH_3MgBr + H–C\equiv C–H \longrightarrow CH_4 + H–C\equiv C–MgBr$
(b) $CH_3MgBr + NH_3 \longrightarrow CH_4 + H_2N–MgBr$

Problem 10.11   How might you replace a halogen substituent by a deuterium atom if you wanted to prepare a deuterated compound?

$$\underset{\displaystyle CH_3CHCH_2CH_3}{\overset{\displaystyle \overset{\textstyle Br}{|}}{}} \xrightarrow{\;?\;} \underset{\displaystyle CH_3CHCH_2CH_3}{\overset{\displaystyle \overset{\textstyle D}{|}}{}}$$

10.9 Organometallic Coupling Reactions

Many other kinds of organometallic compounds can be prepared in a manner similar to that of Grignard reagents. For example, alkyllithium reagents, RLi, can be prepared by the reaction of an alkyl halide with lithium metal. Alkyllithiums are both nucleophiles and bases, and their chemistry is similar in many respects to that of the alkylmagnesium halides.

$$CH_3CH_2CH_2CH_2Br \xrightarrow[\text{Pentane}]{\text{2 Li}} CH_3CH_2CH_2\overset{\delta-}{C}H_2\overset{\delta+}{Li} + LiBr$$

1-Bromobutane   **Butyllithium**

One of the most valuable reactions of alkyllithiums is their use in making lithium diorganocopper compounds, R_2CuLi, called **Gilman reagents**. These reagents are easily prepared by reaction of an alkyllithium with copper(I) iodide, CuI, in ether solvent.

$$2\ CH_3Li\ +\ CuI \xrightarrow{\text{Ether}} (CH_3)_2Cu^-\ Li^+\ +\ LiI$$

Methyllithium   **Lithium dimethylcopper**
(a Gilman reagent)

Gilman reagents are useful because they undergo organometallic *coupling* reactions with alkyl chlorides, bromides and iodides (but not fluorides). One of the alkyl groups from the Gilman reagent replaces the halogen of the alkyl halide, forming a new carbon–carbon bond and yielding a hydrocarbon product. Lithium dimethylcopper, for example, reacts with 1-iododecane to give undecane in 90% yield.

$$(CH_3)_2CuLi + CH_3(CH_2)_8CH_2I \xrightarrow[0°C]{Ether} CH_3(CH_2)_8CH_2CH_3 + LiI + CH_3Cu$$

Lithium   1-Iododecane   Undecane (90%)
dimethylcopper

This organometallic coupling reaction is extremely versatile and very useful in organic synthesis because it makes possible the preparation of larger molecules from smaller pieces. As the following examples indicate, the coupling reaction can be carried out on aryl and vinylic halides as well as on alkyl halides:

$$n\text{-}C_7H_{15}, H \quad C=C \quad H, I + (n\text{-}C_4H_9)_2CuLi \longrightarrow n\text{-}C_7H_{15}, H \quad C=C \quad H, C_4H_9\text{-}n + n\text{-}C_4H_9Cu + LiI$$

***trans*-1-Iodo-1-nonene**   ***trans*-5-Tridecene (71%)**

Iodobenzene $+ (CH_3)_2CuLi \longrightarrow$ Toluene with CH_3 group $+ CH_3Cu + LiI$

Iodobenzene   **Toluene (91%)**

The mechanism of the reaction appears to involve initial formation of a triorganocopper intermediate, followed by coupling and loss of RCu. The coupling is not a typical polar nucleophilic substitution reaction of the sort considered in the next chapter.

$$R\text{—}X + [R'\text{—}Cu\text{—}R']^- \; Li^+ \longrightarrow \left[\begin{array}{c} R \\ | \\ R'\text{—}Cu\text{—}R' \end{array} \right] \longrightarrow R\text{—}R' + R'\text{—}Cu$$

Problem 10.12   How would you prepare the following compounds using an organocopper coupling reaction? More than one step is required in each case.
(a) 3-Methylcyclohexene from cyclohexene   (b) Octane from 1-bromobutane
(c) Decane from 1-pentene

10.10 Oxidation and Reduction in Organic Chemistry

Although we often haven't pointed it out, quite a few of the reactions discussed in this and earlier chapters are either *oxidations* or *reductions*. In inorganic chemistry, where ionic bonds are common, an oxidation is defined as the loss of one or more electrons by an atom. In organic chemistry, however, where polar covalent bonds are common, an **oxidation** is a reaction that results in a *loss of electron density by carbon*. This loss is usually caused either by bond formation between carbon and a more electronegative atom (usually oxygen, nitrogen, or a halogen) or by bond breaking between carbon and a less electronegative atom (usually hydrogen).

Conversely, a reduction in inorganic chemistry is defined as the gain of one or more electrons by an atom, whereas an organic **reduction** is a reaction that results in a *gain of electron density by carbon*. This gain is usually caused either by bond formation between carbon and a less electronegative atom or by bond breaking between carbon and a more electronegative atom. Note that an *oxidation* often adds *oxygen,* while a reduction usually adds hydrogen.

>**Oxidation** Decreases electron density on carbon by:
> forming one of these: C–O, C–N, C–X
> or breaking this: C–H
>
>**Reduction** Increases electron density on carbon by:
> forming this: C–H
> or breaking one of these: C–O, C–N, C–X

Based on these definitions, the chlorination reaction of methane to yield chloromethane is an oxidation because a C–H bond is broken and a C–Cl bond is formed. The conversion of an alkyl chloride to an alkane via a Grignard reagent followed by protonation is a reduction, however, because a C–Cl bond is broken and a C–H bond is formed.

Oxidation: C–H bond broken and C–Cl bond formed

Methane   **Chloromethane**

Reduction: C–Cl bond broken and C–H bond formed

Chloromethane   **Methane**

As still other examples, the reaction of an alkene with Br_2 to yield a 1,2-dibromide is an oxidation because two C–Br bonds are formed, but the reaction of an alkene with H_2 to yield an alkane is a reduction because two C–H bonds are formed. The reaction of an alkene with HBr to yield an alkyl bromide is *neither* an oxidation nor a reduction, because both a C–H and a C–Br bond are formed.

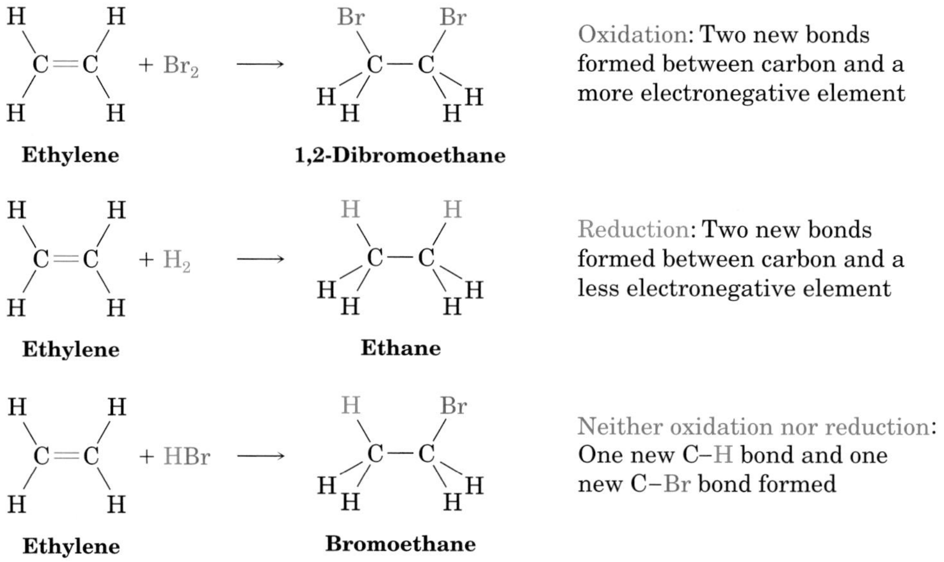

Ethylene   **1,2-Dibromoethane**   Oxidation: Two new bonds formed between carbon and a more electronegative element

Ethylene   **Ethane**   Reduction: Two new bonds formed between carbon and a less electronegative element

Ethylene   **Bromoethane**   Neither oxidation nor reduction: One new C–H bond and one new C–Br bond formed

A list of compounds of increasing oxidation level is shown in Figure 10.5. Alkanes are at the lowest oxidation level because they have the maximum possible number of C–H bonds, and CO_2 is at the highest level because it has the maximum possible number of C–O bonds. Any reaction that converts a compound from a lower level to a higher level is an oxidation, any reaction that converts a compound from a higher level to a lower level is a reduction, and any reaction that doesn't change the level is neither an oxidation nor a reduction.

FIGURE 10.5 ▼

Oxidation levels of some common types of compounds.

CH_3CH_3	$H_2C{=}CH_2$	$HC{\equiv}CH$		
	CH_3OH	$H_2C{=}O$	HCO_2H	CO_2
	CH_3Cl	CH_2Cl_2	$CHCl_3$	CCl_4
	CH_3NH_2	$H_2C{=}NH$	$HC{\equiv}N$	

Low oxidation level   ⟶   High oxidation level

Practice Problem 10.2 shows how to compare the oxidation levels of different compounds.

• •

Practice Problem 10.2   Rank the following compounds in order of increasing oxidation level:

(a) $CH_3CH{=}CH_2$

(b) $\overset{\overset{\displaystyle OH}{|}}{CH_3CHCH_3}$

(c) $CH_3\overset{\overset{\displaystyle O}{\|}}{C}CH_3$

(d) $CH_3CH_2CH_3$

Strategy   Compounds that have the same number of carbon atoms can be compared by adding the number of C–O, C–N, and C–X bonds in each and then subtracting the number of C–H bonds. The larger the resultant value, the higher the oxidation level.

Solution   Compound (a) has six C–H bonds, giving an oxidation level of -6; (b) has one C–O bond and seven C–H bonds, giving an oxidation level of -6; (c) has two C–O bonds and six C–H bonds, giving an oxidation level of -4; and (d) has eight C–H bonds, giving an oxidation level of -8. Thus, the order of increasing oxidation level is (d) < (a) = (b) < (c).

• •

Problem 10.13   Rank each of the following series of compounds in order of increasing oxidation level:

(a)

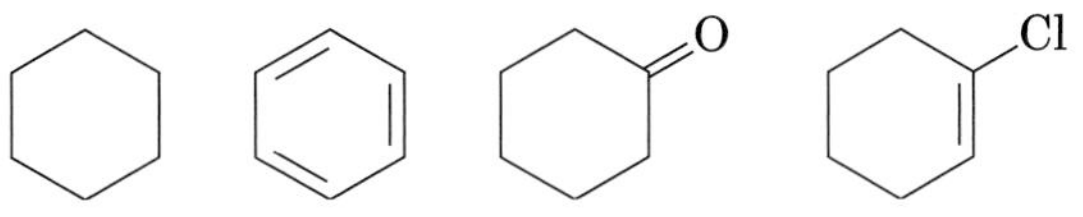

(b) $CH_3CN,$ $CH_3CH_2NH_2,$ $H_2NCH_2CH_2NH_2$

Problem 10.14   Tell whether each of the following reactions is an oxidation, a reduction, or neither. Explain your answers.

(a) $CH_3CH_2\overset{\overset{\displaystyle O}{\|}}{CH} \xrightarrow[\text{H}_2\text{O}]{\text{NaBH}_4} CH_3CH_2CH_2OH$

(b)

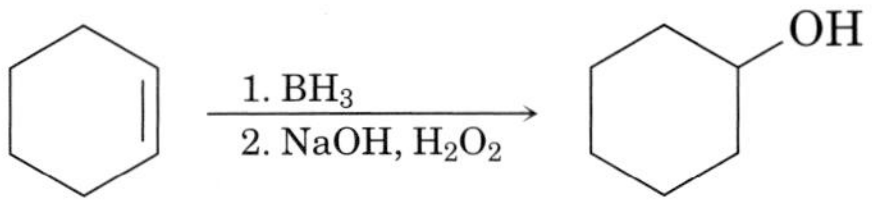

$\xrightarrow[\text{2. NaOH, H}_2\text{O}_2]{\text{1. BH}_3}$

• •

Naturally Occurring Organohalides

As recently as 1968, only about 30 naturally occurring organohalogen compounds were known. It was simply assumed that chloroform, halogenated phenols, chlorinated aromatic compounds called PCB's, and other such substances found in the environment were industrial "pollutants." Now, only a third of a century later, the situation is quite different. More than 3000 organohalogen compounds have been found to occur naturally, and many thousands more surely exist. From a simple compound like chloromethane to extremely complex ones, a remarkably diverse range of organohalogen compounds exists in plants, bacteria, and animals. Many even have unusual physiological activity. For example, the bromine-containing substance called jasplakinolide, discovered by Phillip Crews at the University of California, Santa Cruz, disrupts formation of the actin microtubules that make up the skeleton of cellular organelles.

Jasplakinolide

Some naturally occurring organohalogen compounds are produced in massive quantities. Forest fires, volcanoes, and marine kelp release up to 5 *million* tons of CH_3Cl per year, for example, while annual industrial emissions total only about 26,000 tons. A detailed examination of one species of Okinawan acorn worm in a 1 km^2 study area showed that they released nearly 100 pounds per day of halogenated phenols, compounds previously thought to be nonnatural pollutants.

Why do organisms produce organohalogen compounds, many of which are undoubtedly toxic? The answer seems to be that many organisms use organohalogen compounds for self-defense, either as feeding deterrents, as irritants to predators, or as natural pesticides. Marine sponges, coral, and sea hares, for example, release foul-tasting organohalogen compounds

(continued) ▶

that deter fish, starfish, and other predators from eating them. More remarkably, even humans appear to produce halogenated compounds as part of their defense against infection. The human immune system contains a peroxidase enzyme capable of carrying out halogenation reactions on fungi and bacteria, thereby killing the pathogen.

Much remains to be learned—only a few hundred of the more than 500,000 known species of marine organisms have been examined—but it is already clear that organohalogen compounds are an integral part of the world around us.

Marine corals secrete organohalogen compounds that act as a feeding deterrent to starfish.

Summary and Key Words

KEY WORDS

alkyl halide, 356
allylic position, 363
delocalized, 366
Gilman reagents, 371
Grignard reagents, 369
oxidation, 373
reduction, 373

Alkyl halides are compounds containing halogen bonded to a saturated, sp^3-hybridized carbon atom. The C–X bond is polar, and alkyl halides can therefore behave as electrophiles.

Alkyl halides can be prepared by *radical halogenation* of alkanes, but this method is of little general value since mixtures of products usually result. The reactivity order of alkanes toward halogenation is identical to the stability order of radicals: $R_3C\cdot > R_2CH\cdot > RCH_2\cdot$. Alkyl halides can also be prepared from alkenes by reaction with *N*-bromosuccinimide (NBS) to give the product of **allylic** bromination. The NBS bromination of alkenes takes place through an intermediate allyl radical, which is stabilized by resonance.

Alcohols react with HX to form alkyl halides, but the reaction works well only for tertiary alcohols, R_3COH. Primary and secondary alkyl halides are normally prepared from alcohols using either $SOCl_2$ or PBr_3. Alkyl halides react with magnesium in ether solution to form organomagnesium halides, or **Grignard reagents (RMgX)**. Since Grignard reagents are both nucleophilic and basic, they react with acids to yield hydrocarbons. The overall result of Grignard formation and protonation is the conversion of an alkyl halide into an alkane ($RX \rightarrow RMgX \rightarrow RH$).

Alkyl halides also react with lithium metal to form organolithium reagents, RLi. In the presence of CuI, these form diorganocoppers, or **Gilman reagents (R_2CuLi)**. Gilman reagents react with alkyl halides to yield coupled hydrocarbon products.

In organic chemistry, an **oxidation** is a reaction that causes a decrease in electron density on carbon, either by bond *formation* between carbon and a more electronegative atom (usually oxygen, nitrogen, or a halogen) or by bond *breaking* between carbon and a less electronegative atom (usually hydrogen). Conversely, a **reduction** causes an increase of electron density on carbon, either by bond breaking between carbon and a more electronegative atom or by bond formation between carbon and a less electronegative atom. Thus, the halogenation of an alkane to yield an alkyl halide is an oxidation, while the conversion of an alkyl halide to an alkane by protonation of a Grignard reagent is a reduction.

Summary of Reactions

1. Preparation of alkyl halides
   (a) From alkenes by allylic bromination (Section 10.5)

   (b) From alkenes by addition of HBr and HCl (Sections 6.8 and 6.9)

   (c) From alcohols
   (1) Reaction with HX, where X = Cl, Br, or I (Section 10.7)

Reactivity order: 3° > 2° > 1°

   (2) Reaction of 1° and 2° alcohols with SOCl$_2$ (Section 10.7)

   (3) Reaction of 1° and 2° alcohols with PBr$_3$ (Section 10.7)

(continued) ▶

2. Reaction of alkyl halides
   (a) Grignard reagent formation (Section 10.8)

$$R-X \xrightarrow[\text{Ether}]{\text{Mg}} R-Mg-X$$

where X = Br, Cl, or I
        R = 1°, 2°, or 3° alkyl, aryl, or vinylic

   (b) Diorganocopper (Gilman reagent) formation (Section 10.9)

$$R-X \xrightarrow[\text{Pentane}]{\text{2 Li}} R-Li + LiX$$

where R = 1°, 2°, or 3° alkyl, aryl, or vinylic

$$2\ R-Li + CuI \xrightarrow{\text{In ether}} [R-Cu-R]^- \ Li^+ + LiI$$

   (c) Organometallic coupling (Section 10.9)

$$R_2CuLi + R'-X \xrightarrow{\text{In ether}} R-R' + RCu + LiX$$

   (d) Conversion of alkyl halides to alkanes (Section 10.8)

$$R-X \xrightarrow[\text{Ether}]{\text{Mg}} R-Mg-X \xrightarrow{\text{H}_3\text{O}^+} R-H + HOMgX$$

Visualizing Chemistry

• •

(Problems 10.1–10.14 appear within the chapter.)

10.15 Give a IUPAC name for each of the following alkyl halides (yellow-green = Cl):

(a)

(b)

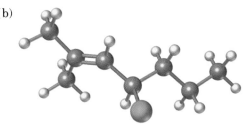

10.16 Show the product(s) of reaction of the following alkenes with NBS:

(a)

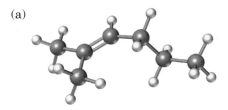

(b)

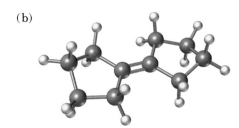

10.17   The following alkyl bromide can be prepared by reaction of the alcohol (*S*)-2-pentanol with PBr_3. Name the compound, assign (*R*) or (*S*) stereochemistry, and tell whether the reaction of the alcohol occurs with retention of the same stereochemistry or with a change in stereochemistry (reddish-brown = Br).

Stereo View

Additional Problems

10.18   Name the following alkyl halides according to IUPAC rules:

(a) $CH_3CHCHCHCH_2CHCH_3$ with substituents H_3C, Br, Br, CH_3

(b) $CH_3CH{=}CHCH_2CHCH_3$ with substituent I

(c) $CH_3CCH_2CHCHCH_3$ with substituents Br, Cl, CH_3, and CH_3

(d) $CH_3CH_2CHCH_2CH_2CH_3$ with substituent CH_2Br

(e) $ClCH_2CH_2CH_2C{\equiv}CCH_2Br$

10.19   Draw structures corresponding to the following IUPAC names:
(a) 2,3-Dichloro-4-methylhexane   (b) 4-Bromo-4-ethyl-2-methylhexane
(c) 3-Iodo-2,2,4,4-tetramethylpentane   (d) *cis*-1-Bromo-2-ethylcyclopentane

10.20   Draw and name the monochlorination products you might obtain by radical chlorination of 2-methylpentane. Which of the products are chiral? Are any of the products optically active?

10.21   A chemist requires a large amount of 1-bromo-2-pentene as starting material for a synthesis and decides to carry out an NBS allylic bromination reaction:

$$CH_3CH_2CH{=}CHCH_3 \xrightarrow[CCl_4]{NBS} CH_3CH_2CH{=}CHCH_2Br$$

What is wrong with this synthesis plan? What side products would form in addition to the desired product?

10.22   What product(s) would you expect from the reaction of 1-methylcyclohexene with NBS? Would you use this reaction as part of a synthesis?

$$\text{(1-methylcyclohexene with } CH_3\text{)} \xrightarrow[CCl_4]{NBS} ?$$

10.23   How would you prepare the following compounds, starting with cyclopentene and any other reagents needed?
(a) Chlorocyclopentane   (b) Methylcyclopentane
(c) 3-Bromocyclopentene   (d) Cyclopentanol
(e) Cyclopentylcyclopentane   (f) 1,3-Cyclopentadiene

10.24 Predict the product(s) of the following reactions:

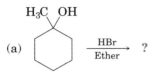

(a) $\xrightarrow[\text{Ether}]{\text{HBr}}$?

(b) $CH_3CH_2CH_2CH_2OH$ $\xrightarrow{\text{SOCl}_2}$?

(c) $\xrightarrow[\text{CCl}_4]{\text{NBS}}$?

(d) $\xrightarrow[\text{Ether}]{\text{PBr}_3}$?

(e) $CH_3CH_2CHBrCH_3$ $\xrightarrow[\text{Ether}]{\text{Mg}}$ A? $\xrightarrow{\text{H}_2\text{O}}$ B?

(f) $CH_3CH_2CH_2CH_2Br$ $\xrightarrow[\text{Pentane}]{\text{Li}}$ A? $\xrightarrow{\text{CuI}}$ B?

(g) $CH_3CH_2CH_2CH_2Br$ + $(CH_3)_2CuLi$ $\xrightarrow{\text{Ether}}$?

10.25 (*S*)-3-Methylhexane undergoes radical bromination to yield optically inactive 3-bromo-3-methylhexane as the major product. Is the product chiral? What conclusions can you draw about the radical intermediate?

10.26 Assume that you have carried out a radical chlorination reaction on (*R*)-2-chloropentane and have isolated (in low yield) 2,4-dichloropentane. How many stereoisomers of the product are formed and in what ratio? Are any of the isomers optically active? (See Problem 10.25.)

10.27 Calculate $\Delta H°$ for the reactions of Cl· and Br· with CH_4, and then draw a reaction energy diagram showing both processes. Which reaction is likely to be faster?

10.28 What product(s) would you expect from the reaction of 1,4-hexadiene with NBS? What is the structure of the most stable radical intermediate?

10.29 Alkylbenzenes such as toluene (methylbenzene) react with NBS to give products in which bromine substitution has occurred at the position next to the aromatic ring (the *benzylic* position). Explain, based on the bond dissociation energies in Table 5.3.

$\xrightarrow[\text{CCl}_4]{\text{NBS}}$

10.30 Draw as many resonance structures as you can for the benzyl radical, $C_6H_5CH_2\cdot$, the intermediate produced in the NBS bromination reaction of toluene (Problem 10.29).

10.31 What product would you expect from the reaction of 1-phenyl-2-butene with NBS? Explain.

1-Phenyl-2-butene

10.32 Draw as many resonance structures as you can for the following species:

(a) $CH_3CH{=}CHCH{=}CHCH{=}CH\overset{+}{C}H_2$ (b) :⁻   (c) $CH_3C{\equiv}\overset{+}{N}-\overset{..}{\underset{..}{O}}{:}^-$

10.33 Rank the compounds in each of the following series in order of increasing oxidation level:

(a) $CH_3CH=CHCH_3$ $CH_3CH_2CH=CH_2$ $\underset{\displaystyle \overset{O}{\|}}{CH_3CH_2CH_2CH}$ $\underset{\displaystyle \overset{O}{\|}}{CH_3CH_2CH_2COH}$

(b) $CH_3CH_2CH_2NH_2$ $CH_3CH_2CH_2Br$ $\underset{\displaystyle \overset{O}{\|}}{CH_3CCH_2Cl}$ $BrCH_2CH_2CH_2Cl$

10.34 Which of the following compounds have the same oxidation level, and which have different levels?

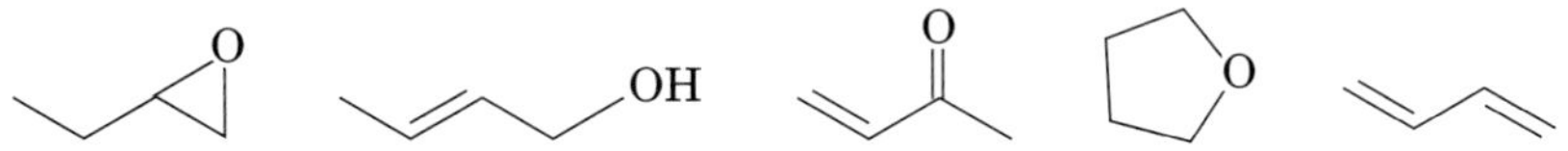

1   **2**   **3**   **4**   **5**

10.35 Tell whether each of the following reactions is an oxidation or a reduction:

(a) $CH_3CH_2OH \xrightarrow{CrO_3} \underset{\displaystyle \overset{O}{\|}}{CH_3CH}$

(b) $H_2C=\underset{\displaystyle \overset{O}{\|}}{CHCCH_3} + NH_3 \longrightarrow H_2N\underset{\displaystyle \overset{O}{\|}}{CH_2CH_2CCH_3}$

(c) $CH_3CH_2\overset{\displaystyle \overset{Br}{|}}{CH}CH_3 \xrightarrow[\text{2. H}_2\text{O}]{\text{1. Mg}} CH_3CH_2CH_2CH_3$

10.36 How would you carry out the following syntheses?
(a) Butylcyclohexane from cyclohexene   (b) Butylcyclohexane from cyclohexanol
(c) Butylcyclohexane from cyclohexane

10.37 The syntheses shown here are unlikely to occur as written. What is wrong with each?

(a) $CH_3CH_2CH_2F \xrightarrow[\text{2. H}_3\text{O}^+]{\text{1. Mg}} CH_3CH_2CH_3$

(b) 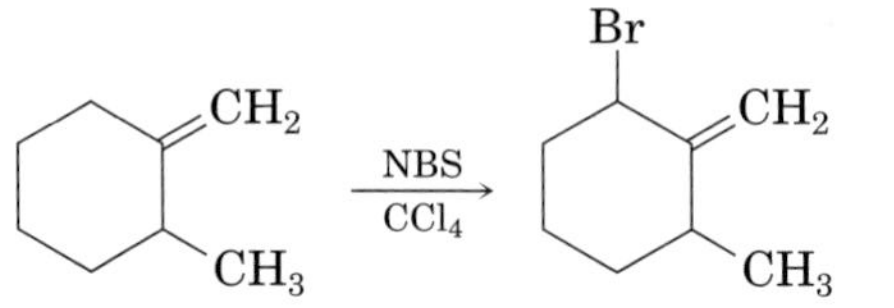 $\xrightarrow[\text{CCl}_4]{\text{NBS}}$

(c) 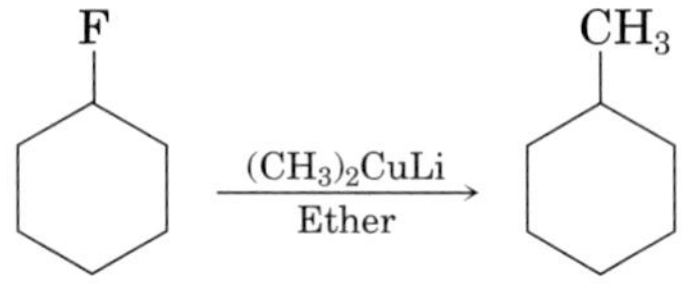 $\xrightarrow[\text{Ether}]{(CH_3)_2CuLi}$

10.38 Why do you suppose it's not possible to prepare a Grignard reagent from a bromo alcohol such as 4-bromo-1-pentanol?

$$CH_3\overset{\displaystyle \overset{Br}{|}}{CH}CH_2CH_2CH_2OH \xrightarrow[]{Mg} \!\!\!\!\not\!\!\longrightarrow CH_3\overset{\displaystyle \overset{MgBr}{|}}{CH}CH_2CH_2CH_2OH$$

Give another example of a molecule that is unlikely to form a Grignard reagent.

10.39 Addition of HBr to a double bond with an ether (–OR) substituent occurs regiospecifically to give a product in which the –Br and –OR are bonded to the same carbon:

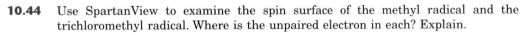

Draw the two possible carbocation intermediates in this electrophilic addition reaction, and explain using resonance why the observed product is formed.

10.40 *Phenols,* compounds that have an –OH group bonded to a benzene ring, are relatively acidic because their anions are stabilized by resonance. Draw as many resonance structures as you can for the phenoxide ion.

Phenoxide ion

10.41 Alkyl halides can be reduced to alkanes by a radical reaction with tributyltin hydride, $(C_4H_9)_3SnH$, in the presence of light $(h\nu)$:

$$R{-}X + (C_4H_9)_3SnH \xrightarrow{h\nu} R{-}H + (C_4H_9)_3SnX$$

Propose a radical chain mechanism by which the reaction might occur. The initiation step is the light-induced homolytic cleavage of the Sn–H bond to yield a tributyltin radical.

A Look Ahead

• •

10.42 Tertiary alkyl halides, R_3CX, undergo spontaneous dissociation to yield a carbocation, R_3C^+. Which do you think reacts faster, $(CH_3)_3CBr$ or $H_2C{=}CHC(CH_3)_2Br$? Explain. (See Section 11.9.)

10.43 Carboxylic acids (RCOOH) are approximately 10^{11} times more acidic than alcohols (ROH). In other words, a carboxylate ion $(RCO_2{^-})$ is more stable than an alkoxide ion (RO^-). Explain, using resonance. (See Section 20.3.)

Molecular Modeling

• •

10.44 Use SpartanView to examine the spin surface of the methyl radical and the trichloromethyl radical. Where is the unpaired electron in each? Explain.

10.45 Grignard reagents dissolve in diethyl ether. Use SpartanView to examine the structures and electrostatic potential maps of methylmagnesium chloride, diethyl ether, and methylmagnesium chloride—diethyl ether complex. What is the nature of the interaction between Grignard reagent and solvent, and what effect does the solvent have on the –CH₃ group of the Grignard reagent?

10.46 Use SpartanView to examine transition states for the reactions of Br· with methane, 2-methylpropane, and propene. How do the breaking C–H bond distances vary? Account for this variation using the Hammond postulate.

10.47 Use SpartanView to examine spin surfaces for the allyl radical and the benzyl radical ($C_6H_5CH_2\cdot$). Draw resonance structures that describe how the unpaired electron is delocalized in each.

10.48 2-Heptene reacts with NBS to give primarily a mixture of 4-bromo-2-heptene and 2-bromo-3-heptene. Use SpartanView to examine the two radicals A and B that might form in the reaction, identify the radical that leads to the observed products, and explain why this radical forms preferentially.

$$CH_3CH_2CH_2CH_2CH\!=\!CHCH_3 \xrightarrow[CCl_4]{NBS} CH_3CH_2CH_2\overset{\overset{\displaystyle Br}{|}}{C}HCH\!=\!CHCH_3 + CH_3CH_2CH_2CH\!=\!CH\overset{\overset{\displaystyle Br}{|}}{C}HCH_3$$

| 2-Heptene | 4-Bromo-2-heptene | 2-Bromo-3-heptene |

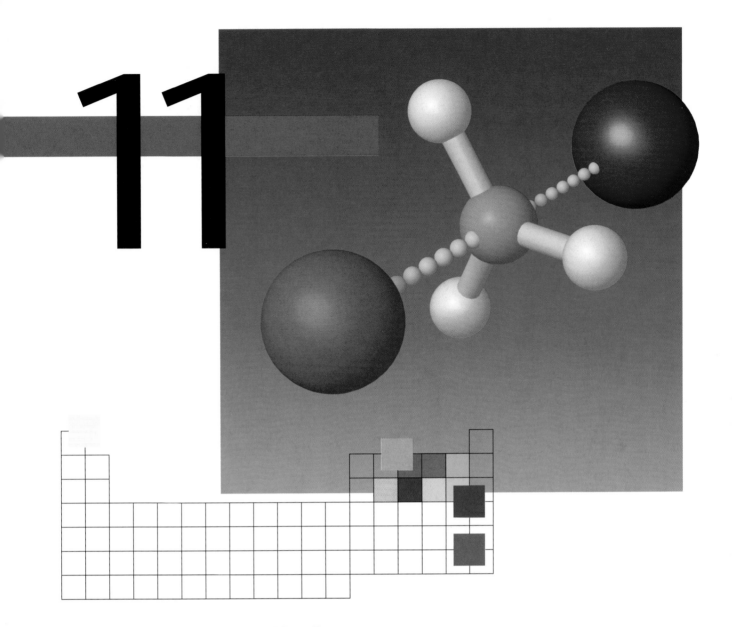

Reactions of Alkyl Halides: Nucleophilic Substitutions and Eliminations

We saw in the preceding chapter that the carbon–halogen bond in alkyl halides is polar and that the carbon atom is electron-poor. Thus, alkyl halides are electrophiles, and much of their chemistry involves polar reactions with nucleophiles and bases.

$$\overset{\delta-}{X}$$

Electrophilic carbon atom

Alkyl halides do one of two things when they react with a nucleophile/base: Either they undergo *substitution* of the X group by the nucleophile (Nu), or they undergo *elimination* of HX to yield an alkene:

Substitution $Nu: + -\overset{|}{\underset{|}{C}}-X \longrightarrow -\overset{|}{\underset{|}{C}}-Nu + X:^-$

Elimination

$$\overset{H}{\underset{X}{C-C}} \longrightarrow \ \ C=C\ + Nu-H + X:^-$$

These two reactions—nucleophilic substitution and base-induced elimination—are two of the most widely occurring and versatile reactions in organic chemistry. We'll take a close look at both in this chapter to see how they occur, what their characteristics are, and how they can be used to synthesize new molecules.

11.1 The Discovery of the Walden Inversion

Paul Walden

Paul Walden (1863–1957) was born in Cesis, Latvia, to German parents who died while he was still a child. He received his Ph.D. in Leipzig, Germany, and returned to Russia as professor of chemistry at Riga Polytechnic (1882–1919). Following the Russian Revolution, he went back to Germany as professor at the University of Rostock (1919–1934) and later at the University of Tübingen.

In 1896, the German chemist Paul Walden made a remarkable discovery. He found that the pure enantiomeric (+)- and (−)-malic acids could be interconverted by a series of simple substitution reactions. When Walden treated (−)-malic acid with PCl$_5$, he isolated (+)-chlorosuccinic acid. This, on treatment with wet Ag$_2$O, gave (+)-malic acid. Similarly, reaction of (+)-malic acid with PCl$_5$ gave (−)-chlorosuccinic acid, which was converted into (−)-malic acid when treated with wet Ag$_2$O. The full cycle of reactions reported by Walden is shown in Figure 11.1.

At the time, the results were astonishing. The eminent chemist Emil Fischer called Walden's discovery "the most remarkable observation made in the field of optical activity since the fundamental observations of Pasteur." Because (−)-malic acid was converted into (+)-malic acid, *some reactions in the cycle must have occurred with an inversion, or change, in configuration at the chirality center.* But which ones, and how? (Recall that the direction of light rotation and the absolute configuration of a molecule aren't directly related. You can't tell by looking at the sign of rotation whether a change in configuration has occurred during a reaction.)

Today, we refer to the transformations taking place in Walden's cycle as **nucleophilic substitution reactions** because each step involves the substitution of one nucleophile (chloride ion, Cl$^-$, or hydroxide ion, HO$^-$) by another. Nucleophilic substitution reactions are one of the most common and versatile reaction types in organic chemistry.

FIGURE 11.1 ▼

Walden's cycle of reactions interconverting (+)- and (−)-malic acids.

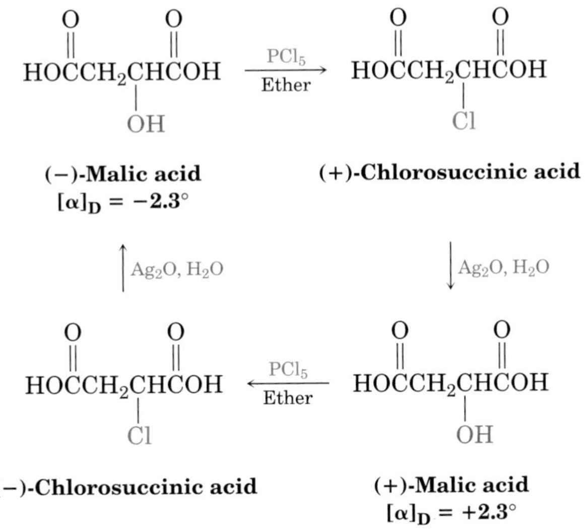

$$\underset{\substack{\text{OH} \\ (-)\text{-Malic acid} \\ [\alpha]_D = -2.3°}}{\text{HOCCH}_2\text{CHCOH}} \xrightarrow[\text{Ether}]{\text{PCl}_5} \underset{\substack{\text{Cl} \\ (+)\text{-Chlorosuccinic acid}}}{\text{HOCCH}_2\text{CHCOH}}$$

↑ Ag$_2$O, H$_2$O   ↓ Ag$_2$O, H$_2$O

$$\underset{\substack{\text{Cl} \\ (-)\text{-Chlorosuccinic acid}}}{\text{HOCCH}_2\text{CHCOH}} \xleftarrow[\text{Ether}]{\text{PCl}_5} \underset{\substack{\text{OH} \\ (+)\text{-Malic acid} \\ [\alpha]_D = +2.3°}}{\text{HOCCH}_2\text{CHCOH}}$$

11.2 Stereochemistry of Nucleophilic Substitution

Joseph Kenyon

Joseph Kenyon (1885–1961) was born in Blackburn, England, and received his D.Sc. at the University of London in 1914. After several years at the British Dyestuffs Corp. (1916–1920), he became professor at Battersea Polytechnic in London (1920–1950).

Although Walden realized that changes in configuration had taken place during his reaction cycle, he didn't know at which steps the changes occurred. In the 1920s, Joseph Kenyon and Henry Phillips began a series of investigations to elucidate the mechanism of nucleophilic substitution reactions and to find out how inversions of configuration occur. They recognized that the presence of the carboxylic acid group in Walden's work with malic acid may have led to complications, and they therefore carried out their own work on simpler cases. (In fact, the particular sequence of reactions studied by Walden *is* unusually complex for reasons we won't go into.)

Among the reaction series studied by Kenyon and Phillips was one that interconverted the two enantiomers of 1-phenyl-2-propanol (Figure 11.2, p. 388). Although this particular series of reactions involves nucleophilic substitution of an alkyl toluenesulfonate (called a *tosylate*) rather than an alkyl halide, exactly the same type of reaction is involved as that studied by Walden. For all practical purposes, the *entire* tosylate group acts as if it were simply a halogen substituent:

$$\text{R—Y} + \text{Nu:}^- \longrightarrow \text{R—Nu} + \text{Y:}^-$$

where   Y = Cl, Br, I, OTos
        Nu = A nucleophile

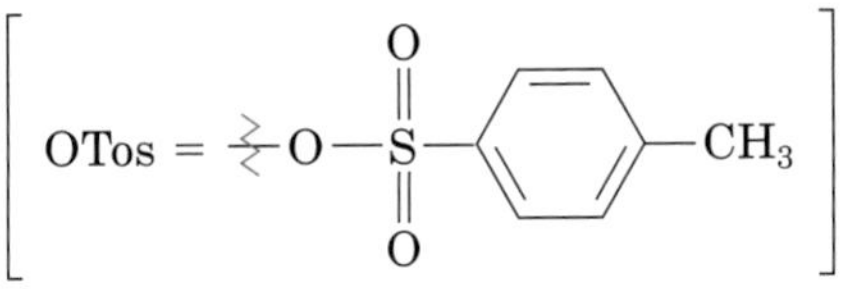

$$\left[\text{OTos} = \; \overset{\displaystyle O}{\underset{\displaystyle O}{\overset{\|}{\underset{\|}{\text{O—S}}}}}\!\!-\!\!\left\langle \text{O} \right\rangle\!\!-\!\!\text{CH}_3 \right]$$

FIGURE 11.2 ▼

A Walden cycle
interconverting (+) and
(−) enantiomers of
1-phenyl-2-propanol.
Chirality centers are
marked by asterisks, and
the bonds broken in each
reaction are indicated by
red wavy lines.

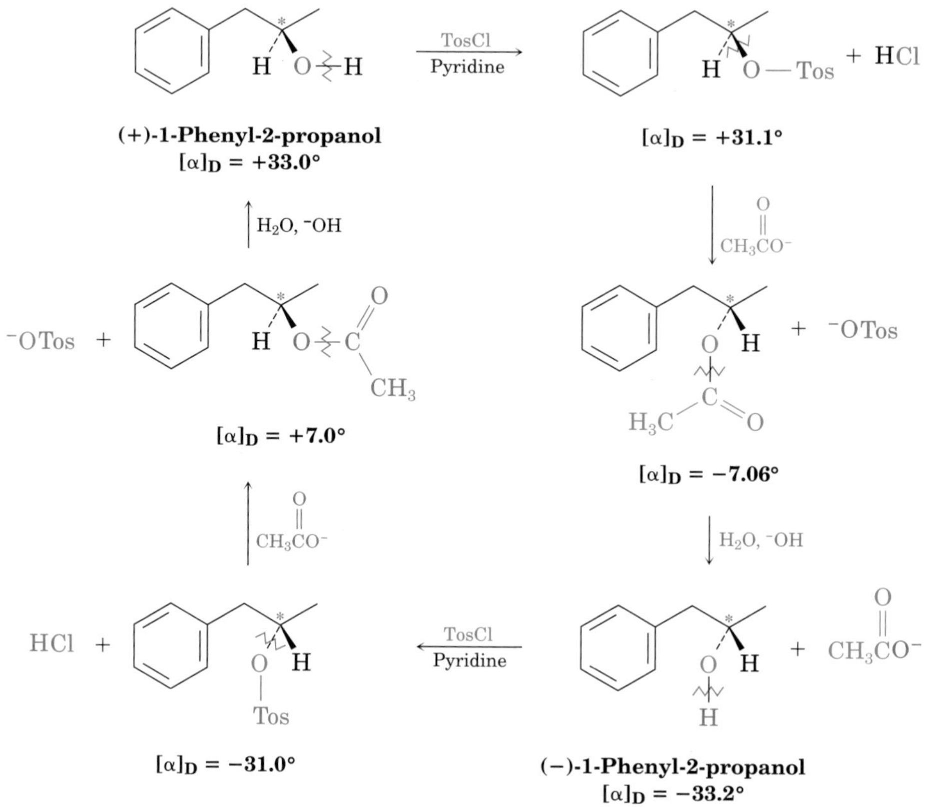

(+)-1-Phenyl-2-propanol
$[\alpha]_D = +33.0°$

$[\alpha]_D = +31.1°$

$H_2O, \ ^-OH$

$[\alpha]_D = +7.0°$

$[\alpha]_D = -7.06°$

$[\alpha]_D = -31.0°$

(−)-1-Phenyl-2-propanol
$[\alpha]_D = -33.2°$

In the three-step reaction sequence shown in Figure 11.2, (+)-1-phenyl-2-propanol is interconverted with its (−) enantiomer, so at least one of the three steps must involve an inversion of configuration at the chirality center. The first step, formation of a toluenesulfonate, occurs by breaking the O–H bond of the alcohol rather than the C–O bond to the chiral carbon, so the configuration around carbon is unchanged. Similarly, the third step, hydroxide ion cleavage of the acetate, also takes place without breaking the C–O bond at the chirality center. *The inversion of stereochemical configuration must therefore take place in the second step, the nucleophilic substitution of tosylate ion by acetate ion.*

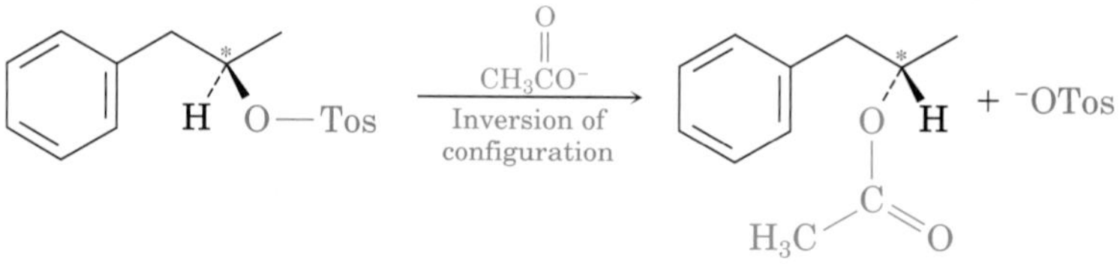

From this and nearly a dozen other series of similar reactions, Kenyon and Phillips concluded that the nucleophilic substitution reactions of primary and secondary alkyl halides and tosylates always proceed with inversion of configuration.

. . . .

Practice Problem 11.1   What product would you expect from a nucleophilic substitution reaction of (R)-1-bromo-1-phenylethane with cyanide ion, $^-C\equiv N$? Show the stereochemistry of both reactant and product, assuming that inversion of configuration occurs.

Strategy   Identify the outgoing group (Br$^-$) and replace it with the incoming nucleophile ($^-$CN), while changing the stereochemistry at the reacting center.

Solution   Draw the R enantiomer of the reactant, and then change the configuration of the chirality center while replacing the –Br with a –CN.

(R)-1-Bromo-1-phenylethane   **(S)-2-Phenylpropanenitrile**

. . . .

Problem 11.1   What product would you expect to obtain from a nucleophilic substitution reaction of (S)-2-bromohexane with acetate ion, CH$_3$COO$^-$? Assume that inversion of configuration occurs, and show the stereochemistry of both reactant and product.

. . . .

11.3 Kinetics of Nucleophilic Substitution

Chemists often speak of a reaction as being "fast" or "slow." The exact rate at which a reactant is converted into product is called the **reaction rate** and can often be measured. The determination of reaction rates and of how those rates depend on reactant concentrations is a powerful tool for probing reaction mechanisms. Let's see what can be learned about the nucleophilic substitution reaction from a study of reaction rates.

In every chemical reaction, there is a direct relationship between reaction rate and reactant concentrations. When we measure this relationship, we measure the **kinetics** of the reaction. For example, let's look at the kinetics of a simple nucleophilic substitution—the reaction of CH_3Br with OH^- to yield CH_3OH plus Br^-.

$$HO:^- + CH_3-Br: \longrightarrow HO-CH_3 + :Br:^-$$

At a given temperature and concentration of reactants, the reaction occurs at a certain rate. If we double the concentration of OH^-, the frequency of encounter between the reaction partners is also doubled, and we might therefore predict that the reaction rate will double. Similarly, if we double the concentration of bromomethane, we might expect that the reaction rate will again double. This behavior is exactly what is found. We call such a reaction, in which the rate is linearly dependent on the concentrations of two species, a **second-order reaction**. Mathematically, we can express this second-order dependence of the nucleophilic substitution reaction by setting up a *rate equation*:

$$\text{Reaction rate} = \text{Rate of disappearance of reactant}$$

$$= k \times [RX] \times [^-OH]$$

where $[RX] = CH_3Br$ concentration
$[^-OH] = ^-OH$ concentration
$k = $ A constant value

This equation says that the rate of disappearance of reactant is equal to a constant k times the alkyl halide concentration times the hydroxide ion concentration. The constant k is called the **rate constant** for the reaction and has units of liters per mole second $(L/mol \cdot s)$. The rate equation says that as either $[RX]$ or $[^-OH]$ changes, the rate of the reaction changes proportionately. If the alkyl halide concentration is doubled, the reaction rate doubles; if the alkyl halide concentration is halved, the reaction rate is halved.

11.4 The S$_N$2 Reaction

Key Ideas▶   At this point, we have two important pieces of information about the nature of nucleophilic substitution reactions on primary and secondary alkyl halides and tosylates:

• The reactions occur with inversion of stereochemistry at the carbon atom.

• The reactions show second-order kinetics, with the rate law:

$$\text{Rate} = k \times [RX] \times [Nu:^-]$$

A mechanism that accounts for both the stereochemistry and the kinetics of nucleophilic substitution reactions was suggested in 1937 by E. D.

Edward Davies Hughes

Edward Davies Hughes
(1906–1963) was born in
Criccieth, North Wales,
and earned two doctoral
degrees: a Ph.D. from
Wales and a D.Sc. from
the University of London,
working with Christopher
Ingold. From 1930–1963,
he was professor of chem-
istry at University College,
London.

Hughes and Christopher Ingold, who formulated what they called the **S$_N$2 reaction**—short for *substitution, nucleophilic, bimolecular*. (**Bimolecular** means that two molecules, nucleophile and alkyl halide, take part in the step whose kinetics are measured.)

The essential feature of the S$_N$2 mechanism is that the reaction takes place in a single step without intermediates when the incoming nucleophile attacks the alkyl halide or tosylate (the *substrate*) from a direction directly opposite the group that leaves. As the nucleophile comes in on one side of the substrate and bonds to the carbon, the halide or tosylate departs from the other side, thereby inverting the stereochemical configuration. The process is shown in Figure 11.3 for the reaction of (S)-2-bromobutane with HO$^-$, leading to (R)-2-butanol.

FIGURE 11.3 ▼

The mechanism of the S$_N$2 reaction. The reaction takes place in a single step when the incoming nucleophile approaches from a direction 180° away from the leaving halide ion, thereby inverting the stereo-chemistry at carbon.

refer to
Mechanisms
& Movies

The nucleophile $^-$OH uses its lone-pair electrons to attack the alkyl halide carbon 180° away from the departing halogen. This leads to a transition state with a partially formed C−OH bond and a partially broken C−Br bond.

The stereochemistry at carbon is inverted as the C−OH bond forms fully and the bromide ion departs with the electron pair from the former C−Br bond.

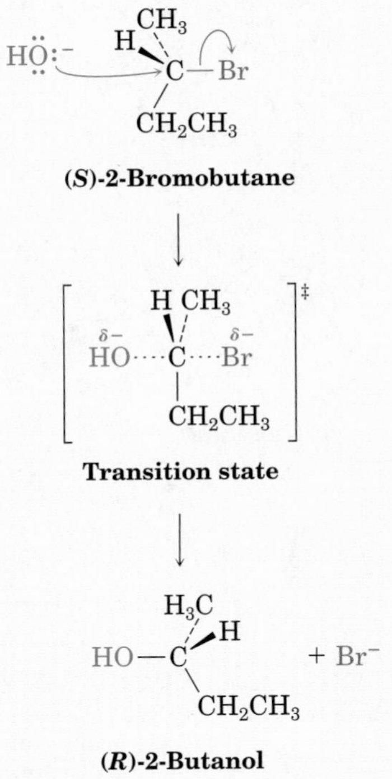

(S)-2-Bromobutane

Transition state

(R)-2-Butanol

© 1984 JOHN MCMURRY

We can picture an S$_N$2 reaction as occurring when an electron pair on the nucleophile Nu:$^-$ forces out the group Y:$^-$, which takes the electron pair from the former C–Y bond. This occurs through a transition state in which the new Nu–C bond is partially forming at the same time that the old C–Y bond is partially breaking, and in which the negative charge is shared by both the incoming nucleophile and the outgoing halide ion. The transition state for this inversion has the remaining three bonds to carbon in a planar arrangement, as shown in Figure 11.4.

FIGURE 11.4 ▼

The transition state of an S_N2 reaction has a planar arrangement of the carbon atom and the remaining three groups. Electrostatic potential maps show that negative charge (red) is delocalized in the transition state.

S_N2 reactants,
S_N2 transition state,
S_N2 products,
S_N2 animation

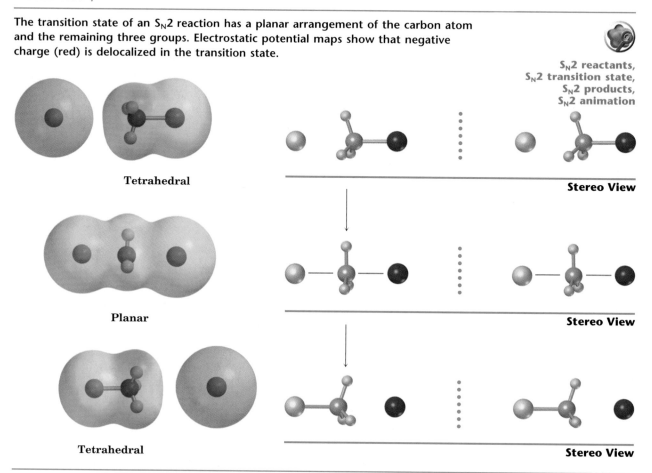

Tetrahedral

Stereo View

Planar

Stereo View

Tetrahedral

Stereo View

The mechanism proposed by Hughes and Ingold is fully consistent with experimental results, explaining both stereochemical and kinetic data. Thus, the requirement for back-side attack of the entering nucleophile from a direction 180° away from the departing Y group causes the stereochemistry of the substrate to invert, much like an umbrella turning inside out in the wind. The Hughes–Ingold mechanism also explains why second-order kinetics are found: The S_N2 reaction occurs in a single step that involves *both* alkyl halide and nucleophile. Two molecules are involved in the step whose rate is measured.

• •

Problem 11.2 What product would you expect to obtain from S_N2 reaction of OH^- with (*R*)-2-bromobutane? Show the stereochemistry of both reactant and product.

Problem 11.3 Assign configuration to the following substance, and draw the structure of the product that would result on nucleophilic substitution reaction with HS^- (reddish-brown = Br):

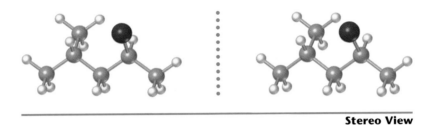

Stereo View

Problem 11.4 A further piece of evidence in support of the requirement for back-side S$_N$2 displacement is the finding that the following alkyl bromide does not undergo a substitution reaction with hydroxide ion. Make a molecular model, and suggest a reason for the lack of reactivity.

Br

11.5 Characteristics of the S$_N$2 Reaction

We now have a good picture of how S$_N$2 reactions occur, but we also need to see how these substitutions can be used and what variables affect them. Some S$_N$2 reactions are fast and some are slow; some take place in high yield and others in low yield. Understanding the factors involved can be of tremendous value to chemists. Let's begin by reviewing what we know about reaction rates in general.

The rate of a chemical reaction is determined by $\Delta G^{\ddagger}$, the energy difference between reactant (ground state) and transition state. A change in reaction conditions can affect $\Delta G^{\ddagger}$ either by changing the reactant energy level or by changing the transition-state energy level. Lowering the reactant energy or raising the transition-state energy increases $\Delta G^{\ddagger}$ and decreases the reaction rate. Conversely, raising the reactant energy or decreasing the transition-state energy decreases $\Delta G^{\ddagger}$ and increases the reaction rate (Figure 11.5, p. 394). We'll see examples of all these effects as we look at S$_N$2 reaction variables.

The Substrate: Steric Effects in the S$_N$2 Reaction

The first S$_N$2 reaction variable we'll look at is the substitution pattern of the alkyl halide substrate. Since the S$_N$2 transition state involves partial bond formation between the incoming nucleophile and the alkyl halide carbon atom, it seems reasonable that a hindered, bulky substrate should prevent easy approach of the nucleophile, making bond formation difficult. In

FIGURE 11.5 ▼

The effects of changes in reactant and transition-state energy levels on reaction rate. (a) A higher reactant energy level (red curve) corresponds to a faster reaction (smaller $\Delta G^{\ddagger}$). (b) A higher transition-state energy level (red curve) corresponds to a slower reaction (larger $\Delta G^{\ddagger}$).

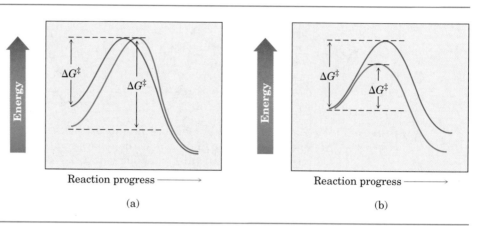

(a)   (b)

other words, the transition state for reaction of a sterically hindered alkyl halide, whose carbon atom is "shielded" from attack of the incoming nucleophile, is higher in energy and forms more slowly than the corresponding transition state for a less hindered alkyl halide (Figure 11.6).

FIGURE 11.6 ▼

Steric hindrance to the S_N2 reaction. As the computer-generated models indicate, the carbon atom in (a) bromomethane is readily accessible, resulting in a fast S_N2 reaction. The carbon atoms in (b) bromoethane (primary), (c) 2-bromopropane (secondary), and (d) 2-bromo-2-methylpropane (tertiary) are successively more hindered, resulting in successively slower S_N2 reactions.

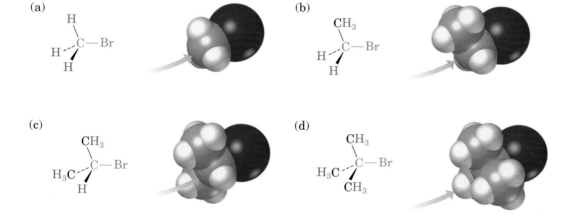

As Figure 11.6 shows, the difficulty of nucleophilic attack increases as the three substituents bonded to the halo-substituted carbon atom increase in size. The relative reactivities for some different substrates are indicated on the next page.

$$R—Br + Cl^- \longrightarrow R—Cl + Br^-$$

CH_3	CH_3	CH_3	H	H
$H_3C—\underset{\underset{CH_3}{\mid}}{\overset{\overset{CH_3}{\mid}}{C}}—Br$	$H_3C—\underset{\underset{CH_3}{\mid}}{\overset{\overset{CH_3}{\mid}}{C}}—CH_2—Br$	$H_3C—\underset{\underset{H}{\mid}}{\overset{\overset{CH_3}{\mid}}{C}}—Br$	$H_3C—\underset{\underset{H}{\mid}}{\overset{\overset{H}{\mid}}{C}}—Br$	$H—\underset{\underset{H}{\mid}}{\overset{\overset{H}{\mid}}{C}}—Br$
Tertiary	Neopentyl	Secondary	Primary	Methyl

Relative reactivity <1 1   500   40,000   2,000,000

Less reactive ⟶ **S$_N$2 reactivity** ⟶ More reactive

Methyl halides are by far the most reactive substrates in S$_N$2 reactions, followed by primary alkyl halides such as ethyl and propyl. Alkyl branching next to the leaving group, as in isopropyl halides (2°), slows the reaction greatly, and further branching, as in *tert*-butyl halides (3°), effectively halts the reaction. Even branching one carbon removed from the leaving group, as in 2,2-dimethylpropyl (*neopentyl*) halides, greatly slows nucleophilic displacement. *S$_N$2 reactions can occur only at relatively unhindered sites,* and are normally useful only with methyl halides, primary halides, and a few simple secondary halides.

Although not shown in the preceding reactivity order, vinylic halides ($R_2C{=}CRX$) and aryl halides are unreactive toward S$_N$2 reaction. This lack of reactivity is probably due to steric factors, because the incoming nucleophile would have to approach in the plane of the carbon–carbon double bond to carry out a back-side displacement.

$$\underset{R}{\overset{R}{}}C{=}C\overset{Cl}{\underset{R}{}} \;\;\xcancel{\longrightarrow}\;\; \text{No reaction}$$
Nu:⁻

Vinylic halide

:Nu⁻ Cl $\xcancel{\longrightarrow}$ No reaction

Aryl halide

The Attacking Nucleophile

The nature of the attacking nucleophile is another variable that has a major effect on the S$_N$2 reaction. Any species, either neutral or negatively charged, can act as a nucleophile as long as it has an unshared pair of electrons (that is, as long as it is a Lewis base). If the nucleophile is negatively charged, the product is neutral; if the nucleophile is neutral, the product is positively charged.

Negatively charged Nu:⁻ $Nu:^- + R{-}Y \longrightarrow R{-}Nu + Y:^-$ Neutral

Neutral Nu: $Nu: + R{-}Y \longrightarrow R{-}Nu^+ + Y:^-$ Positively charged

Because of the great versatility of nucleophilic substitution reactions, many kinds of products can be prepared from alkyl halides. Table 11.1 lists some common nucleophiles in the approximate order of their reactivity and shows the products of their reactions with bromomethane.

TABLE 11.1   Some S_N2 Reactions with Bromomethane: Nu:$^-$ + CH$_3$Br ⟶ NuCH$_3$ + Br$^-$

Nucleophile		Product	
Formula	**Name**	**Formula**	**Name**
$CH_3\ddot{S}:^-$	Methanethiolate	CH_3SCH_3	Dimethyl sulfide
$H\ddot{S}:^-$	Hydrosulfide	$HSCH_3$	Methanethiol
$N{\equiv}C:^-$	Cyanide	$N{\equiv}CCH_3$	Acetonitrile
$N{=}N{=}\ddot{N}:^-$	Azide	N_3CH_3	Azidomethane
$:\ddot{I}:^-$	Iodide	ICH_3	Iodomethane
$CH_3\ddot{O}:^-$	Methoxide	CH_3OCH_3	Dimethyl ether
$H\ddot{O}:^-$	Hydroxide	$HOCH_3$	Methanol
$:\ddot{C}l:^-$	Chloride	$ClCH_3$	Chloromethane
$H_3N:$	Ammonia	$\overset{+}{H_3N}CH_3\ Br^-$	Methylammonium bromide
$CH_3CO_2:^-$	Acetate	$CH_3CO_2CH_3$	Methyl acetate
$(CH_3)_3N:$	Trimethylamine	$(CH_3)_3\overset{+}{N}CH_3\ Br^-$	Tetramethylammonium bromide
$H:^-$	Hydride	CH_4	Methane

Although all the S_N2 reactions shown in Table 11.1 take place, some are much faster than others. What are the reasons for the reactivity differences? Why do some reactants appear to be much more "nucleophilic" than others?

The answers to these questions aren't straightforward. Part of the problem is that the term *nucleophilicity* is imprecise. The term is usually taken to be a measure of the affinity of a nucleophile for a carbon atom in the S_N2 reaction, but the reactivity of a given nucleophile can change from one reaction to the next. The exact nucleophilicity of a species in a given reaction depends on the substrate, the solvent, and even the reactant concentrations. It's therefore best to define a set of standard conditions and study the relative reactivity of various nucleophiles on a single substrate in a single solvent system. Much work has been carried out on the S_N2 reactions of bromomethane in aqueous ethanol, with the following results. Note that hydrosulfide ion (HS^-) is approximately 125,000 times as reactive as water.

$$CH_3—Br + Nu:^- \longrightarrow CH_3—Nu + Br^-$$

	Nu = H$_2$O	CH$_3$CO$_2^-$	NH$_3$	Cl$^-$	OH$^-$	CH$_3$O$^-$	I$^-$	CN$^-$	HS$^-$
Relative reactivity	1	500	700	1,000	16,000	25,000	100,000	125,000	125,000

Less reactive ⟶ **Nucleophile reactivity** ⟶ More reactive

Complete explanations for the observed nucleophilicities aren't known, but some trends can be detected in the data:

- **Nucleophilicity roughly parallels basicity** when comparing nucleophiles that have the same attacking atom (Table 11.2). For example, OH$^-$ is both more basic and more nucleophilic than acetate ion, CH$_3$CO$_2^-$, which in turn is more basic and more nucleophilic than H$_2$O. Since "nucleophilicity" measures the affinity of a Lewis base for a carbon atom in the S$_N$2 reaction, and "basicity" measures the affinity of a base for a proton, it's easy to see why there might be a correlation between the two kinds of behavior.

- **Nucleophilicity usually increases going down a column of the periodic table.** Thus, HS$^-$ is more nucleophilic than HO$^-$, and the halide reactivity order is I$^-$ > Br$^-$ > Cl$^-$. The matter is complex, though, and the nucleophilicity order can change depending on the solvent.

- **Negatively charged nucleophiles are usually more reactive than neutral ones.** As a result, S$_N$2 reactions are often carried out under basic conditions rather than neutral or acidic conditions.

TABLE 11.2 Correlation of Basicity and Nucleophilicity

Nucleophile	CH$_3$O$^-$	HO$^-$	CH$_3$CO$_2^-$	H$_2$O
Rate of S$_N$2 reaction with CH$_3$Br	25	16	0.5	0.001
pK_a of conjugate acid	15.5	15.7	4.7	−1.7

Problem 11.5 What product would you expect from S$_N$2 reaction of 1-bromobutane with each of the following?
(a) NaI   (b) KOH   (c) H–C≡C–Li   (d) NH$_3$

Problem 11.6 Which substance in each of the following pairs is more reactive as a nucleophile? Explain.
(a) (CH$_3$)$_2$N$^-$ or (CH$_3$)$_2$NH   (b) (CH$_3$)$_3$B or (CH$_3$)$_3$N   (c) H$_2$O or H$_2$S

The Leaving Group

Still another variable that can affect the S_N2 reaction is the nature of the group displaced by the attacking nucleophile—the **leaving group**. Because the leaving group is expelled with a negative charge in most S_N2 reactions, we might expect the best leaving groups to be those that best stabilize the negative charge. Furthermore, because the stability of an anion is inversely related to its basicity (Section 2.8), the best leaving groups should be the weakest bases.

As indicated below, the weakest bases (that is, the anions derived from the strongest acids) are indeed the best leaving groups. The *p*-toluene-sulfonate (tosylate) leaving group is very easily displaced, as are I^- and Br^-, but Cl^- and F^- are much less effective as leaving groups.

Relative reactivity	OH^-, NH_2^-, OR^-	F^-	Cl^-	Br^-	I^-	$TosO^-$
	$\ll 1$	1	200	10,000	30,000	60,000

Less reactive ————→ **Leaving group reactivity** ————→ More reactive

The reason that stable anions make good leaving groups can be understood by looking at the transition state. In the transition state for an S_N2 reaction, the charge is distributed over both the incoming nucleophile and the leaving group. The greater the extent of charge stabilization by the leaving group, the lower the energy of the transition state and the more rapid the reaction.

Transition state
(Negative charge is delocalized over both Nu: and Y)

It's just as important to know which are *poor* leaving groups as to know which are good, and the preceding data clearly indicate that F^-, HO^-, RO^-, and H_2N^- are not displaced by nucleophiles. In other words, alkyl fluorides, alcohols, ethers, and amines do not normally undergo S_N2 reactions.

$$R-F \qquad R-OH \qquad R-OR' \qquad R-NH_2$$

These compounds do not undergo S_N2 reactions.

Problem 11.7   Rank the following compounds in order of their expected reactivity toward S_N2 reaction:

$$CH_3Br, \quad CH_3OTos, \quad (CH_3)_3CCl, \quad (CH_3)_2CHCl$$

The Solvent

The rates of many S$_N$2 reactions are affected by the solvent. *Protic solvents*—those that contain –OH or –NH groups—are generally the worst solvents for S$_N$2 reactions. *Polar aprotic solvents,* which have strong dipoles but don't have –OH or –NH groups, are the best.

Protic solvents, such as methanol and ethanol, slow down S$_N$2 reactions by clustering around the reactant nucleophile, a process called **solvation**. Solvent molecules hydrogen bond to the nucleophile, orienting themselves into a "cage" around it and thereby lowering its reactivity.

$$
\begin{array}{c}
\text{OR} \\
| \\
\text{H} \\
\vdots \\
\text{RO—H}\cdots\text{X:}^-\cdots\text{H—OR} \\
\vdots \\
\text{H} \\
| \\
\text{OR}
\end{array}
$$

A solvated anion (reduced nucleophilicity due to enhanced ground-state stability)

In contrast to protic solvents, which decrease the rates of S$_N$2 reactions by *lowering* the ground-state energy of the nucleophile, polar aprotic solvents increase the rates of S$_N$2 reactions by *raising* the ground-state energy of the nucleophile. Particularly valuable are acetonitrile (CH$_3$CN), dimethylformamide [(CH$_3$)$_2$NCHO, abbreviated DMF], dimethyl sulfoxide [(CH$_3$)$_2$SO, abbreviated DMSO], and hexamethylphosphoramide {[(CH$_3$)$_2$N]$_3$PO, abbreviated HMPA}. These solvents can dissolve many salts because of their high polarity, but they tend to solvate metal *cations* rather than nucleophilic anions. As a result, the bare unsolvated anions have a greater nucleophilicity, and S$_N$2 reactions take place at correspondingly faster rates. For instance, a rate increase of 200,000 has been observed on changing from methanol to HMPA for the reaction of azide ion with 1-bromobutane.

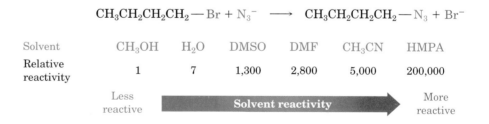

$$\text{CH}_3\text{CH}_2\text{CH}_2\text{CH}_2\text{—Br} + \text{N}_3^- \longrightarrow \text{CH}_3\text{CH}_2\text{CH}_2\text{CH}_2\text{—N}_3 + \text{Br}^-$$

Solvent	CH$_3$OH	H$_2$O	DMSO	DMF	CH$_3$CN	HMPA
Relative reactivity	1	7	1,300	2,800	5,000	200,000

Less reactive   **Solvent reactivity** ⟶   More reactive

Problem 11.8   Organic solvents such as benzene, ether, and chloroform are neither protic nor strongly polar. What effect would you expect these solvents to have on the reactivity of a nucleophile in S$_N$2 reactions?

S$_N$2 Reaction Characteristics: A Summary

Key Ideas ▷   The effects on S$_N$2 reactions of the four variables—substrate structure, nucleophile, leaving group, and solvent—are summarized in the following statements and in the reaction energy diagrams of Figure 11.7:

- **Substrate**   Steric hindrance raises the energy of the transition state, thus increasing $\Delta G^{\ddagger}$ and decreasing the reaction rate. As a result, S$_N$2 reactions are best for methyl and primary substrates.

- **Nucleophile**   More reactive nucleophiles are less stable and have a higher ground-state energy, thereby decreasing $\Delta G^{\ddagger}$ and increasing the reaction rate. Basic, negatively charged nucleophiles are more effective than neutral ones.

FIGURE 11.7 ▼

Reaction energy diagrams showing the effects of (a) substrate, (b) nucleophile, (c) leaving group, and (d) solvent on S$_N$2 reaction rates. Substrate and leaving group effects are felt primarily in the transition state. Nucleophile and solvent effects are felt primarily in the reactant ground state.

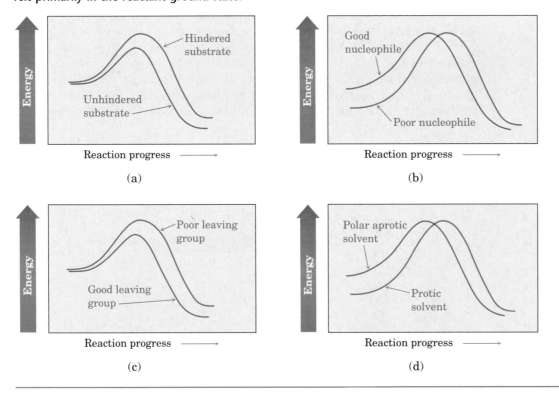

- **Leaving group** — Good leaving groups (more stable anions) lower the energy of the transition state, thus decreasing $\Delta G^{\ddagger}$ and increasing the reaction rate.

- **Solvent** — Protic solvents solvate the nucleophile, thereby lowering its ground-state energy, increasing $\Delta G^{\ddagger}$, and decreasing the reaction rate. Polar aprotic solvents surround the accompanying cation but not the nucleophilic anion, thereby raising the ground-state energy of the nucleophile, decreasing $\Delta G^{\ddagger}$, and increasing the reaction rate.

11.6   The S_N1 Reaction

We've now seen that the S_N2 reaction is worst when carried out with a hindered substrate, a neutral nucleophile, and a protic solvent. You might therefore expect the reaction of a tertiary substrate (hindered) with water (neutral, protic) to be among the slowest of substitution reactions. Remarkably, however, the opposite is true. The reaction of the tertiary halide $(CH_3)_3CBr$ with H_2O to give the alcohol 2-methyl-2-propanol is more than *1 million times* as fast as the corresponding reaction of the methyl halide CH_3Br to give methanol.

$$R-Br + H_2O \longrightarrow R-OH + HBr$$

Relative reactivity	<1	1	12	1,200,000

What's going on here? Clearly, a nucleophilic substitution reaction is occurring, yet the reactivity order seems backward. These reactions can't be taking place by the S_N2 mechanism we've been discussing, and we must therefore conclude that they are occurring by *an alternative substitution mechanism*. This alternative mechanism is called the **S_N1 reaction** (for *substitution, nucleophilic, unimolecular*). Let's see what evidence is available concerning the S_N1 reaction.

11.7   Kinetics of the S_N1 Reaction

The reaction of $(CH_3)_3CBr$ with H_2O looks analogous to the reaction of CH_3Br with OH^-, and we might therefore expect to observe second-order kinetics. In fact, we do not. We find instead that the reaction rate is dependent only on the alkyl halide concentration and is independent of the H_2O concentration. In other words, the reaction is a **first-order process**: Only *one* molecule is involved in the step whose kinetics are measured. *The concentration of the nucleophile does not appear in the rate expression.*

$$\text{Reaction rate} = \text{Rate of disappearance of alkyl halide}$$

$$= k \times [\text{RX}]$$

How can this result be explained? To answer this question, we must first learn more about kinetics measurements.

Many organic reactions are relatively complicated and occur in successive steps. One of these steps is usually slower than the others, and we call this the **rate-limiting step**, or *rate-determining step*. No reaction can proceed faster than its rate-limiting step, which acts as a kind of traffic jam, or bottleneck. The overall reaction rate that we actually measure in a kinetics experiment is determined by the height of the highest energy barrier between a low point and a subsequent high point in the energy diagram of the reaction. The reaction energy diagrams in Figure 11.8 illustrate the idea of the rate-limiting step.

FIGURE 11.8 ▼

Reaction energy diagrams for two hypothetical reactions. The rate-limiting step in each is determined by the difference in height between a low point and a subsequent high point. In (a), the first step is rate-limiting; in (b), the second step is rate-limiting.

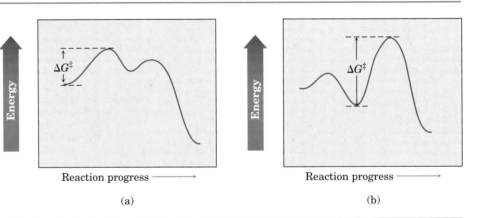

(a)   (b)

The observation of first-order kinetics for the S_N1 reaction of $(CH_3)_3CBr$ with H_2O tells us that the alkyl halide is involved in a unimolecular rate-limiting step. In other words, 2-bromo-2-methylpropane undergoes a spontaneous, rate-limiting reaction without involvement of the nucleophile. The nucleophile must be involved at some other step. The mechanism shown in Figure 11.9 accounts for the kinetic observations.

Unlike what happens in an S_N2 reaction, where the leaving group is displaced *at the same time* the incoming nucleophile is approaching, an S_N1 reaction takes place by loss of the leaving group *before* the incoming nucleophile approaches. 2-Bromo-2-methylpropane spontaneously dissociates to the *tert*-butyl carbocation plus Br^- in a slow, rate-limiting step, and the intermediate carbocation is then immediately trapped by the nucleophile water in a fast step. *Water is not a reactant in the step whose rate is measured by kinetics.* The reaction energy diagram is shown in Figure 11.10.

FIGURE 11.9 ▼

The mechanism of the S_N1 reaction of 2-bromo-2-methylpropane with H_2O involves three steps. The first step—spontaneous, unimolecular dissociation of the alkyl bromide to yield a carbocation—is rate-limiting.

refer to
Mechanisms & Movies

Spontaneous dissociation of the alkyl bromide occurs in a slow, rate-limiting step to generate a carbocation intermediate plus bromide ion.

The carbocation intermediate reacts with water as nucleophile in a fast step to yield protonated alcohol as product.

Loss of a proton from the protonated alcohol intermediate then gives the neutral alcohol product.

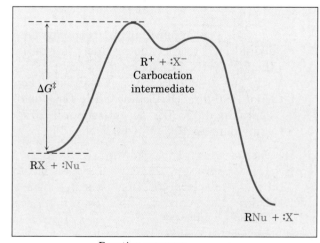

© 1984 JOHN MCMURRY

FIGURE 11.10 ▼

A reaction energy diagram for an S_N1 reaction. The rate-limiting step is spontaneous dissociation of the alkyl halide to give a carbocation intermediate.

11.8 Stereochemistry of the S$_N$1 Reaction

Since an S$_N$1 reaction occurs through a carbocation intermediate, its stereo-chemical outcome should be different from that for an S$_N$2 reaction. Since carbocations are planar and sp^2-hybridized, they are achiral. Thus, if we carry out an S$_N$1 reaction on one enantiomer of a chiral reactant and go through an achiral carbocation intermediate, then the product must be optically inactive. The symmetrical intermediate carbocation can be attacked by a nucleophile equally well from either side, leading to a 50:50 mixture of enantiomers—a racemic mixture (Figure 11.11).

FIGURE 11.11 ▼

Stereochemistry of the S$_N$1 reaction. An enantiomerically pure reactant must give a racemic product.

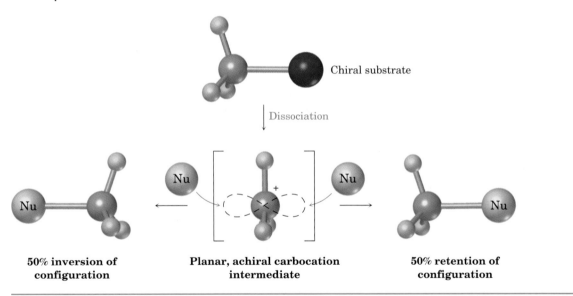

| 50% inversion of configuration | Planar, achiral carbocation intermediate | 50% retention of configuration |

The prediction that S$_N$1 reactions on enantiomerically pure substrates should lead to racemic products is exactly what is observed. Surprisingly, though, few S$_N$1 displacements occur with complete racemization. Most give a minor (0–20%) excess of inversion. For example, the reaction of (R)-6-chloro-2,6-dimethyloctane with H$_2$O leads to an alcohol product that is approximately 80% racemized and 20% inverted (80% R,S + 20% S is equivalent to 40% R + 60% S):

(R)-6-Chloro-2,6-dimethyloctane   **40% R**   **60% S**
                                              (retention)   **(inversion)**

Saul Winstein (1912–1969) was born in Montreal, Canada, and received his Ph.D. in 1938 at Cal Tech. He was professor at the University of California, Los Angeles, where he studied reaction mechanisms, particularly those involving carbocations.

The probable reason for the lack of complete racemization in most S$_N$1 reactions is that **ion pairs** are involved. According to this explanation, first proposed by Saul Winstein, dissociation of the substrate occurs to give a structure in which the two ions are still loosely associated and in which the carbocation is effectively shielded from nucleophilic attack on one side by the departing anion. If a certain amount of substitution occurs before the two ions fully diffuse apart, then a net inversion of configuration will be observed (Figure 11.12).

FIGURE 11.12 ▼

The ion-pair hypothesis in S$_N$1 reactions. The leaving group shields one side of the carbocation intermediate from attack by the nucleophile, thereby leading to some inversion of configuration rather than complete racemization.

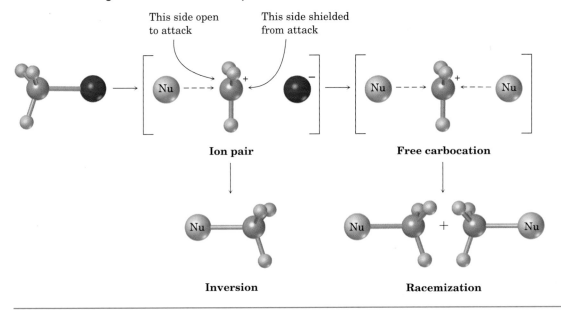

•••••••••••••••••••••••••••••••••••••••

Problem 11.9 What product(s) would you expect from reaction of (S)-3-chloro-3-methyloctane with acetic acid? Show the stereochemistry of both reactant and product.

Problem 11.10 Assign configuration to the following substrate, and show the stereochemistry and identity of the product you would obtain by S$_N$1 reaction with water (reddish brown = Br):

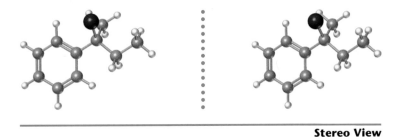

Stereo View

Problem 11.11　Among the numerous examples of S_N1 reactions that occur with incomplete racemization is one reported by Winstein in 1952. The optically pure tosylate of 2,2-dimethyl-1-phenyl-1-propanol ($[\alpha]_D = -30.3°$) was heated in acetic acid to yield the corresponding acetate ($[\alpha]_D = +5.3°$). If complete inversion had occurred, the optically pure acetate would have had $[\alpha]_D = +53.6°$. What percentage racemization and what percentage inversion occurred in this reaction?

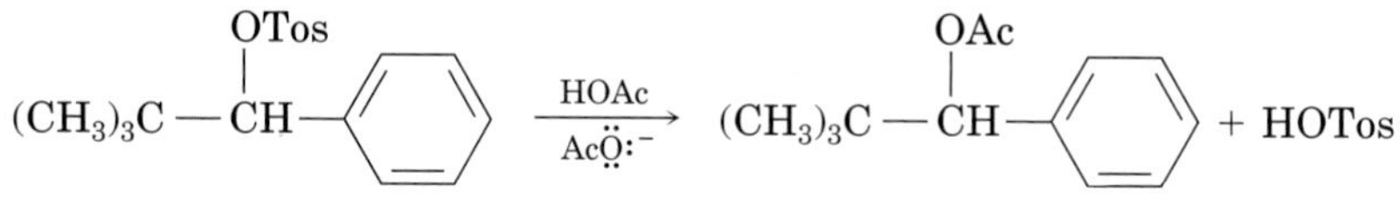

$[\alpha]_D = -30.3°$　　　　Observed $[\alpha]_D = +5.3°$
(optically pure $[\alpha]_D = +53.6°$)

11.9 Characteristics of the S_N1 Reaction

Just as the S_N2 reaction is strongly influenced by such variables as solvent, leaving group, substrate structure, and nature of the attacking nucleophile, the S_N1 reaction is similarly influenced. Factors that lower $\Delta G^{\ddagger}$, either by lowering the energy level of the transition state or by raising the energy level of the ground state, favor faster S_N1 reactions. Conversely, factors that raise $\Delta G^{\ddagger}$, either by raising the energy level of the transition state or by lowering the energy level of the reactant, slow down the S_N1 reaction.

The Substrate

According to the Hammond postulate (Section 6.11), any factor that stabilizes a high-energy intermediate should also stabilize the transition state leading to that intermediate. Since the rate-limiting step in the S_N1 reaction is the spontaneous, unimolecular dissociation of the substrate to yield a carbocation, we would expect the reaction to be favored whenever a stabilized carbocation intermediate is formed. This is exactly what is found: *The more stable the carbocation intermediate, the faster the S_N1 reaction.*

We've already seen (Section 6.10) that the stability order of alkyl carbocations is $3° > 2° > 1° > -CH_3$. To this list we must also add the resonance-stabilized allyl and benzyl cations:

Allyl carbocation

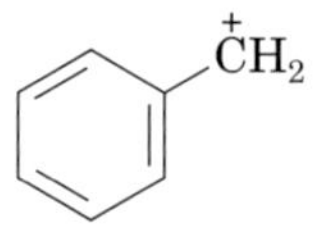

Benzyl carbocation

Just as allylic *radicals* are unusually stable because the unpaired electron can be delocalized over an extended π orbital system (Section 10.6), so allylic and benzylic *carbocations* are unusually stable. (The word **benzylic** means "next to an aromatic ring.") As Figure 11.13 indicates, an allylic cation has two resonance forms. In one form the double bond is on the "left," and in the other form the double bond is on the "right." A benzylic cation, however, has *four* resonance forms, all of which make substantial contributions to the overall resonance hybrid.

FIGURE 11.13 ▼

Resonance forms of the allyl and benzyl carbocations. Electrostatic potential maps show that the positive charge is delocalized over the π system in both.

allyl carbocation, benzyl carbocation

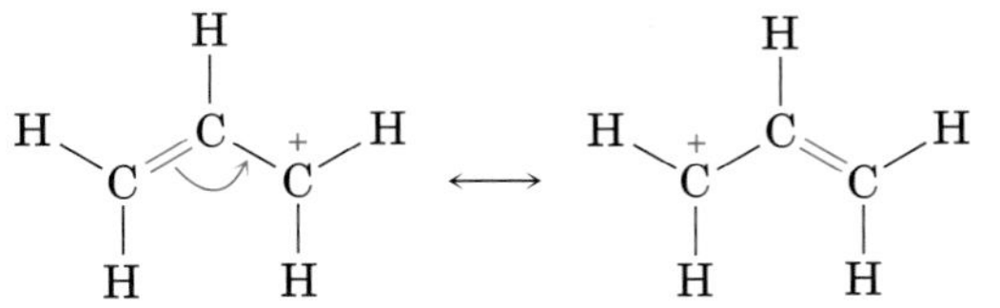

Allyl carbocation

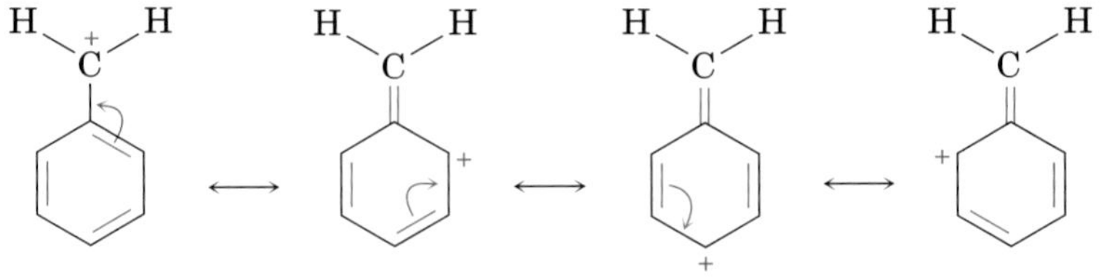

Benzyl carbocation

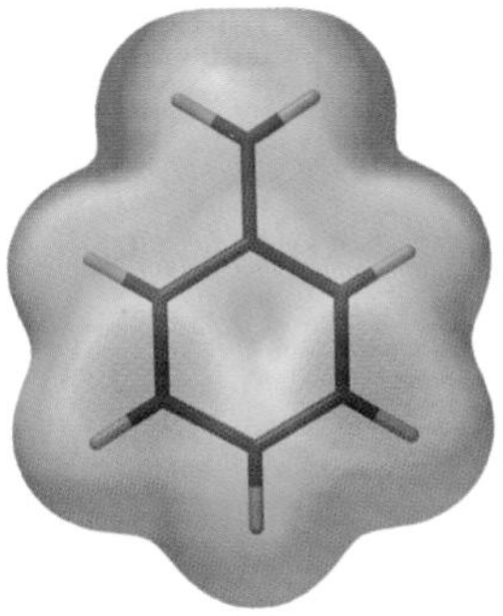

Because of resonance stabilization, a *primary* allylic or benzylic carbocation is about as stable as a *secondary* alkyl carbocation. Similarly, a *secondary* allylic or benzylic carbocation is about as stable as a *tertiary* alkyl carbocation:

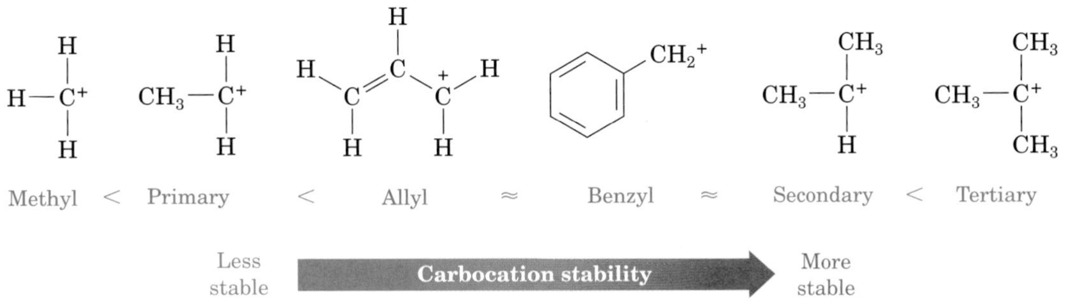

Methyl < Primary < Allyl ≈ Benzyl ≈ Secondary < Tertiary

Less stable ⟶ **Carbocation stability** ⟶ More stable

This stability order of carbocations is exactly the same as the order of S_N1 reactivity for alkyl halides and tosylates.

Parenthetically, it should also be noted that allylic and benzylic substrates are particularly reactive in S_N2 reactions as well as in S_N1 reactions. Allylic and benzylic C–X bonds are about 50 kJ/mol (12 kcal/mol) weaker than the corresponding saturated bonds and are therefore more easily broken.

$$CH_3CH_2 - Cl \qquad H_2C{=}CHCH_2 - Cl$$

338 kJ/mol   289 kJ/mol   293 kJ/mol
(81 kcal/mol)   (69 kcal/mol)   (70 kcal/mol)

• •

Problem 11.12   Rank the following substances in order of their expected S_N1 reactivity:

$$CH_3CH_2Br, \quad H_2C{=}CHCH(Br)CH_3, \quad H_2C{=}CHBr, \quad CH_3CH(Br)CH_3$$

Problem 11.13   3-Bromo-1-butene and 1-bromo-2-butene undergo S_N1 reaction at nearly the same rate even though one is a secondary halide and the other is primary. Explain.

• •

The Leaving Group

We reasoned during the discussion of S_N2 reactivity that the best leaving groups should be those that are most stable—that is, the conjugate bases of strong acids. An identical reactivity order is found for the S_N1 reaction because the leaving group is directly involved in the rate-limiting step. Thus, we find the S_N1 reactivity order to be

$$H_2O \quad < \quad Cl^- \quad < \quad Br^- \quad < \quad I^- \quad \approx \quad TosO^-$$

Less   **Leaving group reactivity** ⟶   More
reactive   reactive

Note that in the S_N1 reaction, which is often carried out under acidic conditions, neutral water can act as a leaving group. This occurs, for example, when an alkyl halide is prepared from a tertiary alcohol by reaction with HBr or HCl (Section 10.7). The alcohol is first protonated and then spontaneously loses H_2O to generate a carbocation. Reaction of the carbocation with halide ion yields the alkyl halide (Figure 11.14). Knowing that an S_N1 reaction is involved in the conversion of alcohols to alkyl halides makes it clear why the reaction works well only for tertiary alcohols: Tertiary alcohols react fastest because they give the most stable carbocation intermediates.

The Nucleophile

The nature of the attacking nucleophile plays a major role in the S_N2 reaction. Is the nucleophile also important in determining the rate of an S_N1 reaction? The answer is no. The S_N1 reaction, by its very nature, occurs

FIGURE 11.14 ▼

The mechanism of the S$_N$1 reaction of a tertiary alcohol with HBr to yield an alkyl halide. Neutral water is the leaving group.

refer to
Mechanisms
& Movies

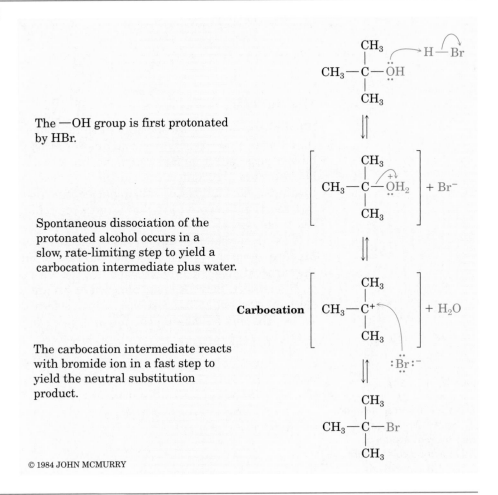

The —OH group is first protonated by HBr.

Spontaneous dissociation of the protonated alcohol occurs in a slow, rate-limiting step to yield a carbocation intermediate plus water.

The carbocation intermediate reacts with bromide ion in a fast step to yield the neutral substitution product.

© 1984 JOHN MCMURRY

through a rate-limiting step in which the added nucleophile has no kinetic role. The nucleophile does not enter into the reaction until after rate-limiting dissociation has occurred and thus cannot affect the reaction rate. The reaction of 2-methyl-2-propanol with HX, for example, occurs at the same rate regardless of whether X is Cl, Br, or I:

$$CH_3-\underset{\underset{CH_3}{|}}{\overset{\overset{CH_3}{|}}{C}}-OH \ + \ HX \longrightarrow CH_3-\underset{\underset{CH_3}{|}}{\overset{\overset{CH_3}{|}}{C}}-X \ + \ H_2O$$

2-Methyl-2-propanol   (Same rate for X = Cl, Br, I)

Furthermore, neutral nucleophiles are just as effective as negatively charged ones, so S$_N$1 reactions frequently occur under neutral or acidic conditions.

● ●

Problem 11.14   1-Chloro-1,2-diphenylethane reacts with the nucleophiles fluoride ion and triethyl-amine at the same rate, even though one is charged and one is neutral. Explain.

● ●

The Solvent

What about solvent? Do solvents have the same effect in S_N1 reactions that they have in S_N2 reactions? The answer is both yes and no. Yes, solvents have a large effect on S_N1 reactions, but no, the reasons for the effects are not the same. Solvent effects in the S_N2 reaction are due largely to stabilization or destabilization of the nucleophile *reactant*. Solvent effects in the S_N1 reaction, however, are due largely to stabilization or destabilization of the *transition state*.

The Hammond postulate says that any factor stabilizing the intermediate carbocation should increase the rate of an S_N1 reaction. Solvation of the carbocation—the interaction of the ion with solvent molecules—has just such an effect. Solvent molecules orient around the carbocation so that the electron-rich ends of the solvent dipoles face the positive charge (Figure 11.15), thereby stabilizing the ion.

FIGURE 11.15 ▼

Solvation of a carbocation by water. The electron-rich oxygen atoms of solvent molecules orient around the positively charged carbocation and thereby stabilize it.

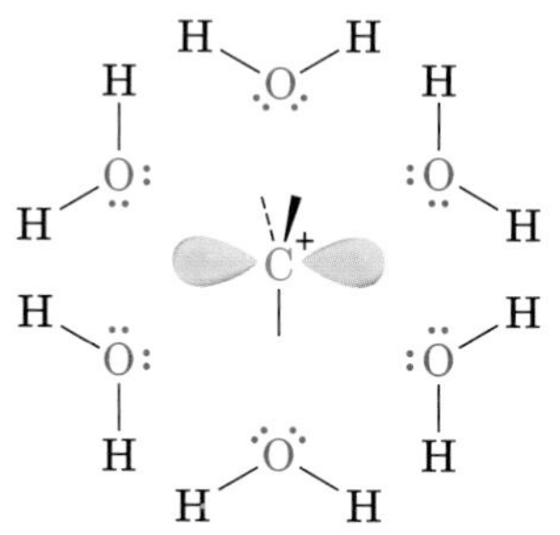

The properties of a solvent that contribute to its ability to stabilize ions by solvation are related to the solvent's polarity. Polar solvents, such as water, methanol, and dimethyl sulfoxide, are good at solvating ions, but most nonpolar ether and hydrocarbon solvents are very poor at solvating ions.

Solvent polarity is expressed in terms of the **dielectric polarization** (**P**) which measures the ability of a solvent to act as an insulator of electric charges. Solvents of low dielectric polarization, such as hydrocarbons, are nonpolar, whereas solvents of high dielectric polarization, such as water, are polar. Table 11.3 lists the dielectric polarizations of some common solvents.

TABLE 11.3 Dielectric Polarizations of Some Common Solvents

Name	Dielectric polarization	Name	Dielectric polarization
Aprotic solvents		Protic solvents	
Hexane	1.9	Acetic acid	6.2
Benzene	2.3	Ethanol	24.3
Diethyl ether	4.3	Methanol	33.6
Chloroform	4.8	Formic acid	58.0
Hexamethylphosphoramide (HMPA)	30	Water	80.4
Dimethylformamide (DMF)	38		
Dimethyl sulfoxide (DMSO)	48		

S$_N$1 reactions take place much more rapidly in polar solvents than in nonpolar solvents. In the reaction of 2-chloro-2-methylpropane, for example, a rate increase of 100,000 is observed on going from ethanol to water. The rate increases on going from hydrocarbon solvents to water are so large that they can't be measured accurately.

$$CH_3-\underset{\underset{CH_3}{|}}{\overset{\overset{CH_3}{|}}{C}}-Cl + ROH \longrightarrow CH_3-\underset{\underset{CH_3}{|}}{\overset{\overset{CH_3}{|}}{C}}-OR + HCl$$

	Ethanol	40% Water/ 60% Ethanol	80% Water/ 20% Ethanol	Water
Relative reactivity	1	100	14,000	100,000

Less reactive ⟶ **Solvent reactivity** ⟶ More reactive

It should be emphasized again that both S$_N$1 and S$_N$2 reactions show large solvent effects, but that they do so for different reasons. S$_N$2 reactions are *disfavored* in protic solvents because the *ground-state energy* of the attacking nucleophile is lowered by solvation. S$_N$1 reactions are *favored* in protic solvents because the *transition-state energy* leading to carbocation intermediate is lowered by solvation. To see the difference, compare the S$_N$1 reaction energy diagram in Figure 11.16 (p. 412) to that in Figure 11.7d, where the effect of solvent on the S$_N$2 reaction was illustrated.

FIGURE 11.16 ▼

The effect of solvent on an S_N1 reaction. The energy level of the transition-state energy is lowered dramatically by solvation in a polar solvent.

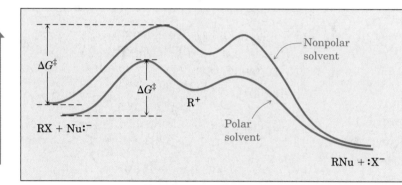

S_N1 Reaction Characteristics: A Summary

Key Ideas ▶

The effects on S_N1 reactions of the four variables—substrate structure, leaving group, nucleophile, and solvent—are summarized as follows:

- **Substrate**
  The best substrates yield the most stable carbocations. As a result, S_N1 reactions are best for tertiary, allylic, and benzylic halides.

- **Leaving group**
  Good leaving groups (more stable anions) increase the reaction rate by lowering the energy level of the transition state leading to carbocation formation.

- **Nucleophile**
  The nucleophile must be nonbasic to prevent a competitive elimination of HX (Section 11.11), but otherwise does not affect the reaction rate. Neutral nucleophiles work well.

- **Solvent**
  Polar solvents stabilize the carbocation intermediate by solvation, thereby increasing the reaction rate.

Practice Problem 11.2   Predict whether each of the following substitution reactions is likely to be S_N1 or S_N2:

Strategy   Look carefully in each reaction at the structure of the substrate, the leaving group, the nucleophile, and the solvent. Then decide from the summaries at the ends of Section 11.5 and this section whether an S_N1 or an S_N2 reaction is likely to be favored. S_N1 reactions are favored by tertiary, allylic, or benzylic substrates, by halide leaving groups, by nonbasic nucleophiles, and by neutral or acidic solvents. S_N2 reactions are favored by primary substrates, by halide leaving groups, by good nucleophiles, and by polar aprotic solvents.

Solution   (a) This is likely to be an S_N1 reaction because the substrate is secondary and benzylic, the nucleophile is weakly basic, and the solvent is acidic.

(b) This is likely to be an S_N2 reaction because the substrate is primary, the nucleophile is reasonably good, and the solvent is polar and aprotic.

· ·

Problem 11.15   Predict whether each of the following substitution reactions is likely to be S_N1 or S_N2:

(a)

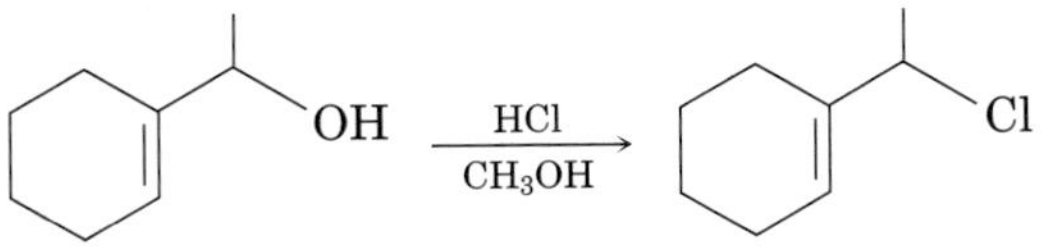

$$\xrightarrow[\text{CH}_3\text{OH}]{\text{HCl}}$$

(b) $\text{H}_2\text{C}=\overset{\overset{\displaystyle \text{CH}_3}{|}}{\text{C}}\text{CH}_2\text{Br} \xrightarrow[\text{CH}_3\text{CN}]{\text{Na}^+ \ ^-\text{SCH}_3} \text{H}_2\text{C}=\overset{\overset{\displaystyle \text{CH}_3}{|}}{\text{C}}\text{CH}_2\text{SCH}_3$

· ·

11.10 Elimination Reactions of Alkyl Halides: Zaitsev's Rule

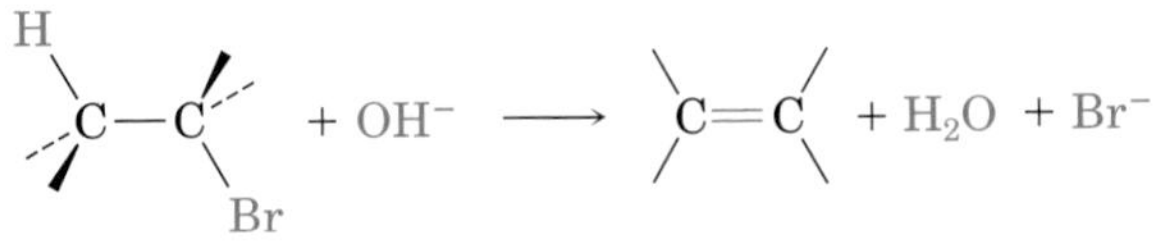

Alexander M. Zaitsev

Alexander M. Zaitsev (1841–1910) was born in Kazan, Russia, and received his Ph.D. from the University of Leipzig in 1866. He was professor at the University of Kazan (1870–1903) and at Kiev University, and many of his students went on to assume faculty positions throughout Russia.

We began this chapter by saying that two kinds of reactions are possible when a nucleophile/Lewis base reacts with an alkyl halide. The nucleophile can either attack at carbon and substitute for the halide or it can attack at a neighboring hydrogen and cause elimination of HX to form an alkene:

Substitution

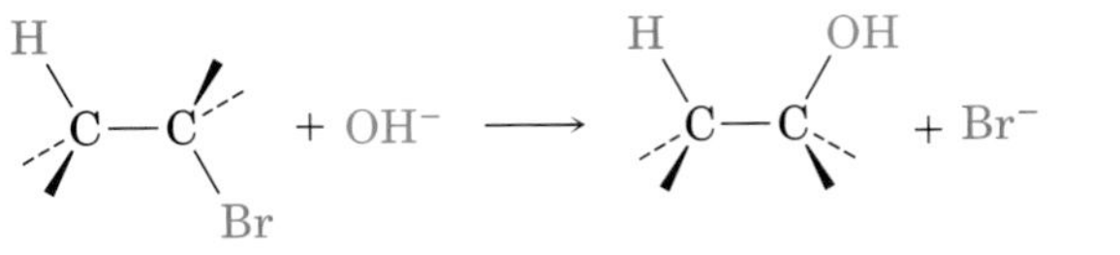

Elimination

Elimination reactions are more complex than substitution reactions for several reasons. There is, for example, the problem of regiochemistry: What products result from loss of HX from an unsymmetrical halide? In fact, elimination reactions almost always give *mixtures* of alkene products, and the best we can usually do is to predict which will be the major product.

According to a rule formulated in 1875 by the Russian chemist Alexander Zaitsev, base-induced elimination reactions generally give the more highly substituted (more stable) alkene product—that is, the alkene with more alkyl substituents on the double-bond carbons. In the following two cases, for example, **Zaitsev's rule** is clearly applicable. The more highly substituted alkene product predominates in both cases when sodium ethoxide in ethanol is used as the base.

Zaitsev's rule   In the elimination of HX from an alkyl halide, the more highly substituted alkene product predominates.

$$\underset{\textbf{2-Bromobutane}}{CH_3CH_2\overset{\overset{\displaystyle Br}{|}}{C}HCH_3} \xrightarrow[CH_3CH_2OH]{CH_3CH_2O^-\ Na^+} \underset{\substack{\textbf{2-Butene}\\\textbf{(81\%)}}}{CH_3CH=CHCH_3} + \underset{\substack{\textbf{1-Butene}\\\textbf{(19\%)}}}{CH_3CH_2CH=CH_2}$$

$$\underset{\textbf{2-Bromo-2-methylbutane}}{CH_3CH_2\overset{\overset{\displaystyle Br}{|}}{\underset{\underset{\displaystyle CH_3}{|}}{C}}CH_3} \xrightarrow[CH_3CH_2OH]{CH_3CH_2O^-\ Na^+} \underset{\substack{\textbf{2-Methyl-2-butene}\\\textbf{(70\%)}}}{CH_3CH=\overset{\overset{\displaystyle CH_3}{|}}{C}CH_3} + \underset{\substack{\textbf{2-Methyl-1-butene}\\\textbf{(30\%)}}}{CH_3CH_2\overset{\overset{\displaystyle CH_3}{|}}{C}=CH_2}$$

The elimination of HX from an alkyl halide is an excellent method for preparing an alkene, but the subject is complex because elimination reactions can take place through different mechanistic pathways, just as substitutions can. We'll consider two of the most common pathways: the *E1* and *E2 reactions*.

● ●

Problem 11.16   What products would you expect from elimination reactions of the following alkyl halides? Which product will be major in each case?

(a) $CH_3CH_2\overset{\overset{\displaystyle Br}{|}}{C}H\overset{\overset{\displaystyle CH_3}{|}}{C}HCH_3$ (b) $CH_3\overset{\overset{\displaystyle CH_3}{|}}{C}HCH_2-\overset{\overset{\displaystyle Cl}{|}}{\underset{\underset{\displaystyle CH_3}{|}}{C}}-\overset{\overset{\displaystyle CH_3}{|}}{C}HCH_3$ (c) ⬡$-\overset{\overset{\displaystyle Br}{|}}{C}HCH_3$

● ●

11.11 The E2 Reaction

The **E2 reaction** (for *elimination, bimolecular*) occurs when an alkyl halide is treated with a strong base, such as hydroxide ion or alkoxide ion (RO^-). It is the most commonly occurring pathway for elimination and can be formulated as shown in Figure 11.17.

FIGURE 11.17 ▼

Mechanism of the E2 reaction of an alkyl halide. The reaction takes place in a single step through a transition state in which the double bond begins to form at the same time the H and X groups are leaving.

refer to
**Mechanisms
& Movies**

Base (B:) attacks a neighboring hydrogen and begins to remove the H at the same time as the alkene double bond starts to form and the X group starts to leave.

Neutral alkene is produced when the C–H bond is fully broken and the X group has departed with the C–X bond electron pair.

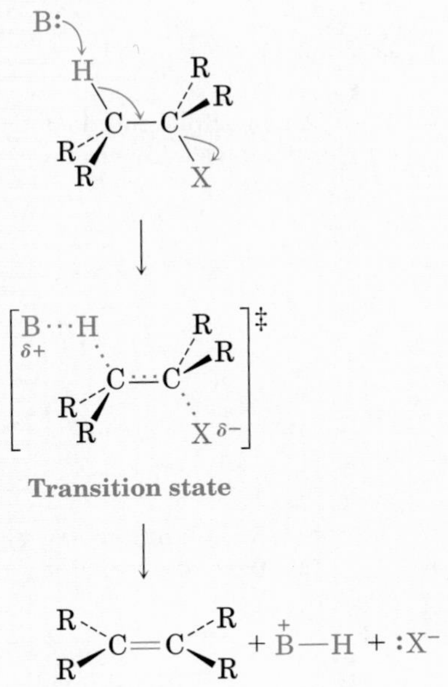

Transition state

© 1984 JOHN MCMURRY

Like the S_N2 reaction, the E2 reaction takes place in one step without intermediates. As the attacking base begins to abstract H^+ from a carbon next to the leaving group, the C–H bond begins to break, a C=C bond begins to form, and the leaving group begins to depart, taking with it the electron pair from the C–X bond.

Among the pieces of evidence supporting this mechanism is the measurement of reaction kinetics. Since both base and alkyl halide enter into the single, rate-limiting step, E2 reactions show second-order kinetics. In other words, E2 reactions follow the rate law

$$\text{Rate} = k \times [\text{RX}] \times [\text{Base}]$$

A second and more compelling piece of evidence involves the stereochemistry of E2 eliminations. As shown by a large number of experiments, E2 reactions always occur with a **periplanar** geometry, meaning that all four reacting atoms—the hydrogen, the two carbons, and the leaving group—lie in the same plane. Two such geometries are possible: **syn periplanar** geometry, in which the H and the X are on the same side of the molecule, and **anti periplanar** geometry, in which the H and the X are on opposite sides of the molecule. Of the two choices, anti periplanar geometry is energetically preferred because it allows the substituents on the two carbons to adopt a staggered relationship, whereas syn geometry requires that the substituents on carbon be eclipsed.

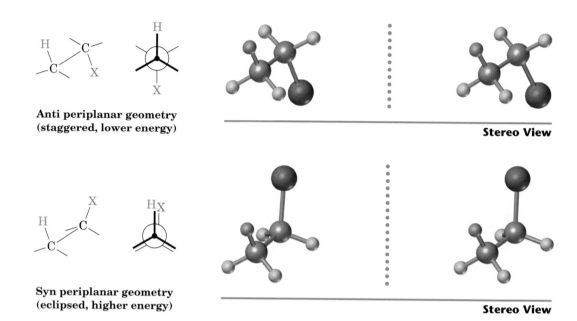

Anti periplanar geometry (staggered, lower energy)

Stereo View

Syn periplanar geometry (eclipsed, higher energy)

Stereo View

What's so special about periplanar geometry? Because the sp^3 σ orbitals in the original C–H and C–X bonds must overlap and become p π orbitals in the alkene product, there must also be some overlap in the transition state. This can occur most easily if all the orbitals are in the same plane to begin with—that is, if they're periplanar (Figure 11.18).

FIGURE 11.18 ▼

The transition state for the E2 reaction of an alkyl halide with base. Overlap of the developing p orbitals in the transition state requires periplanar geometry of the reactant.

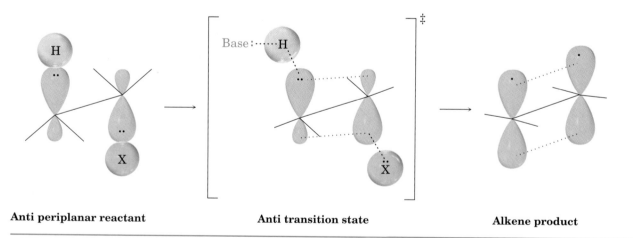

Anti periplanar reactant   **Anti transition state**   **Alkene product**

It might help to think of E2 elimination reactions with periplanar geometry as being similar to S_N2 reactions with 180° geometry. In an S_N2 reaction, an electron pair from the incoming nucleophile pushes out the leaving group on the opposite side of the molecule (back-side attack). In an E2 reaction, an electron pair from a neighboring C–H bond pushes out the leaving group on the opposite side of the molecule (anti periplanar).

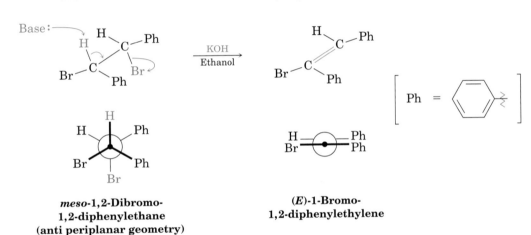

**S$_N$2 reaction
(back-side attack)**   **E2 reaction
(anti periplanar)**

Anti periplanar geometry for E2 eliminations has specific stereochemical consequences that provide strong evidence for the proposed mechanism. To take just one example, *meso*-1,2-dibromo-1,2-diphenylethane undergoes E2 elimination on treatment with base to give only the pure *E* alkene. None of the isomeric *Z* alkene is formed because the transition state leading to the *Z* alkene would have to have syn periplanar geometry.

meso-**1,2-Dibromo-
1,2-diphenylethane
(anti periplanar geometry)**

(*E*)-**1-Bromo-
1,2-diphenylethylene**

· ·

Practice Problem 11.3   What stereochemistry do you expect for the alkene obtained by E2 elimination of (1*S*,2*S*)-1,2-dibromo-1,2-diphenylethane?

Strategy   Draw the reactant with the –H and the leaving group in an anti periplanar arrangement. Then carry out the elimination while keeping all substituents in approximately their same positions, and see what alkene results.

Solution   Draw (1*S*,2*S*)-1,2-dibromo-1,2-diphenylethane so that you can see its stereochemistry and so that the –H and –Br groups to be eliminated are anti periplanar (molecular models are extremely helpful here). Elimination of HBr from this conformation gives (*Z*)-1-bromo-1,2-diphenylethylene.

Problem 11.17   What stereochemistry do you expect for the alkene obtained by E2 elimination of (1R,2R)-1,2-dibromo-1,2-diphenylethane? Draw a Newman projection of the reacting conformation.

Problem 11.18   What stereochemistry do you expect for the trisubstituted alkene obtained by E2 elimination of the following alkyl halide on treatment with KOH? (Reddish brown = Br.)

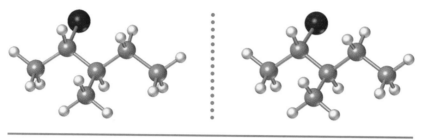

Stereo View

11.12 Elimination Reactions and Cyclohexane Conformation

Derek H. R. Barton

Derek H. R. Barton (1918–1998) was born in Gravesend, England, and received both Ph.D. and D.Sc. degrees from Imperial College, London. Among his numerous positions were those as professor at Imperial College, the University of London, Glasgow, Institut de Chimie des Substances Naturelles, and finally at Texas A and M University. Barton received the Nobel Prize in chemistry in 1969, and was knighted by Queen Elizabeth in 1972.

Anti periplanar geometry for E2 reactions is particularly important in cyclohexane rings, where chair geometry forces a rigid relationship between the substituents on neighboring carbon atoms (Section 4.9). As pointed out by Derek Barton in a landmark 1950 paper, much of the chemical reactivity of substituted cyclohexanes is controlled by their conformation. Let's look at the E2 dehydrohalogenation of chlorocyclohexanes to see an example of such conformational control.

The anti periplanar requirement for E2 reactions can be met in cyclohexanes only if the hydrogen and the leaving group are trans diaxial (Figure 11.19). If either the leaving group or the hydrogen is equatorial, E2 elimination can't occur.

The elimination of HCl from the isomeric menthyl and neomenthyl chlorides shown in Figure 11.20 provides a good illustration of this trans-diaxial requirement. Neomenthyl chloride undergoes elimination of HCl on reaction with ethoxide ion 200 times as fast as menthyl chloride. Furthermore, neomenthyl chloride yields 3-menthene as the major alkene product, whereas menthyl chloride yields 2-menthene.

We can understand the difference in reactivity between the isomeric menthyl chlorides by looking at the more favorable chair conformations of the reactant molecules. Neomenthyl chloride has the conformation shown in Figure 11.20a, with the methyl and isopropyl groups equatorial and the chlorine axial—a perfect geometry for E2 elimination. Loss of the hydrogen atom at C4 occurs easily to yield the more substituted alkene product, 3-menthene, as predicted by Zaitsev's rule.

FIGURE 11.19 ▼

The geometric requirement for E2 reaction in a cyclohexane. The leaving group and the hydrogen must both be axial for anti periplanar elimination to be possible.

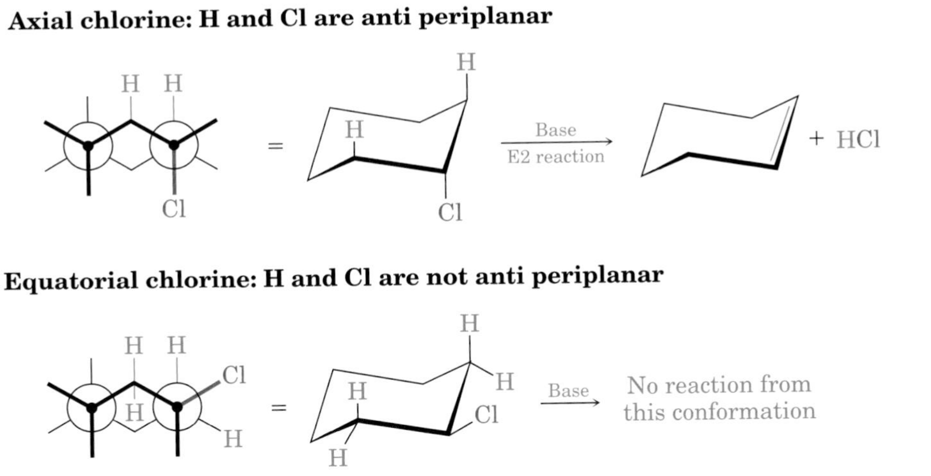

Axial chlorine: H and Cl are anti periplanar

Equatorial chlorine: H and Cl are not anti periplanar

FIGURE 11.20 ▼

Dehydrochlorination of menthyl and neomenthyl chlorides. (a) Neomenthyl chloride loses HCl from its more stable conformation, but (b) menthyl chloride must first ring-flip before HCl loss can occur.

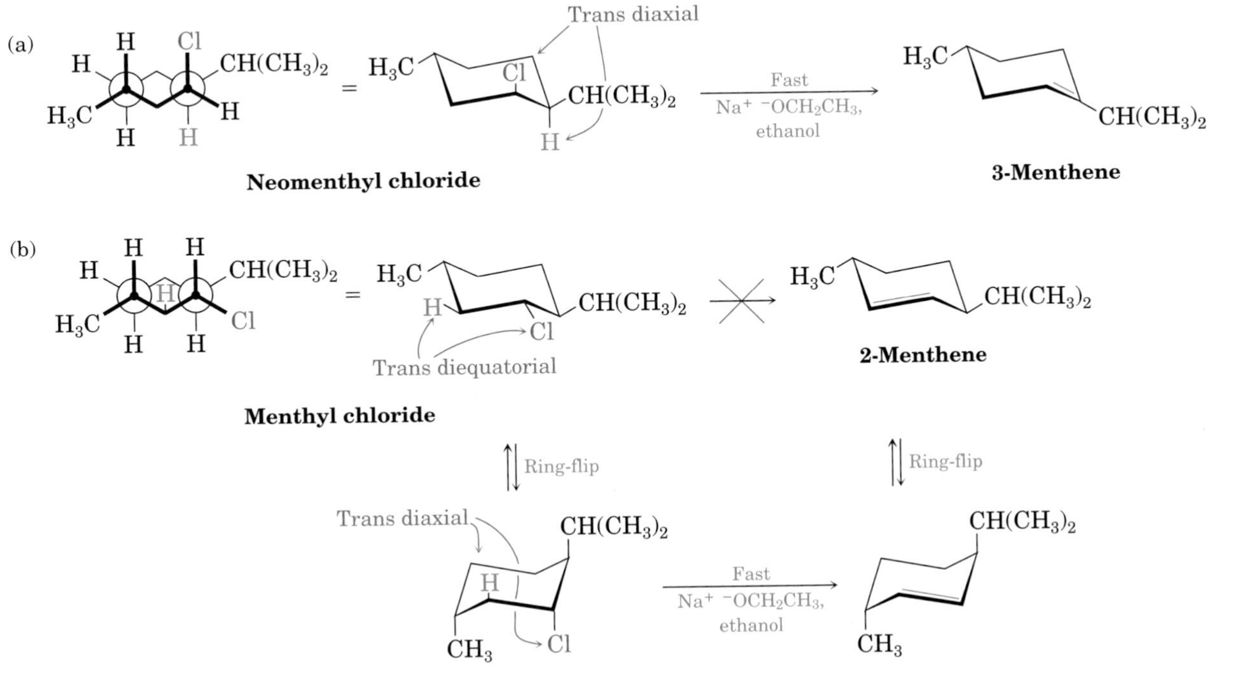

(a)

Neomenthyl chloride

3-Menthene

(b)

Menthyl chloride

2-Menthene

Menthyl chloride, by contrast, has a conformation in which all three substituents are equatorial (Figure 11.20b). To achieve the necessary geometry for elimination, menthyl chloride must first ring-flip to a higher-energy chair conformation, in which all three substituents are axial. E2 elimination then occurs with loss of the only trans-diaxial hydrogen, leading to 2-menthene. The net effect of the simple change in chlorine stereochemistry is a 200-fold change in reaction rate and a complete change of product. The chemistry of the molecule is truly controlled by its conformation.

Problem 11.19   Which isomer would you expect to undergo E2 elimination faster, *trans*-1-bromo-4-*tert*-butylcyclohexane or *cis*-1-bromo-4-*tert*-butylcyclohexane? Draw each molecule in its more stable chair conformation, and explain your answer.

11.13   The Deuterium Isotope Effect

One final piece of evidence in support of the E2 mechanism is provided by a phenomenon known as the **deuterium isotope effect**. For reasons that we won't go into, a carbon–*hydrogen* bond is weaker by a small amount [about 5 kJ/mol (1.2 kcal/mol)] than a corresponding carbon–*deuterium* bond. Thus, a C–H bond is more easily broken than an equivalent C–D bond, and the rate of C–H bond cleavage is faster. As an example of how this effect can be used to obtain mechanistic information, the base-induced elimination of HBr from 1-bromo-2-phenylethane proceeds 7.11 times as fast as the corresponding elimination of DBr from 1-bromo-2,2-dideuterio-2-phenylethane:

Faster reaction

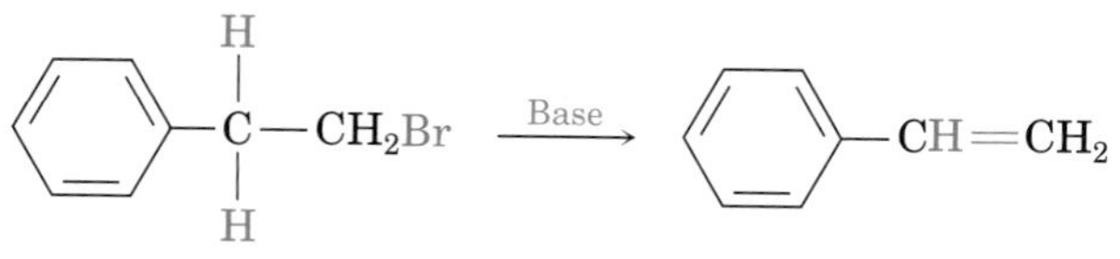

1-Bromo-2-phenylethane

Slower reaction

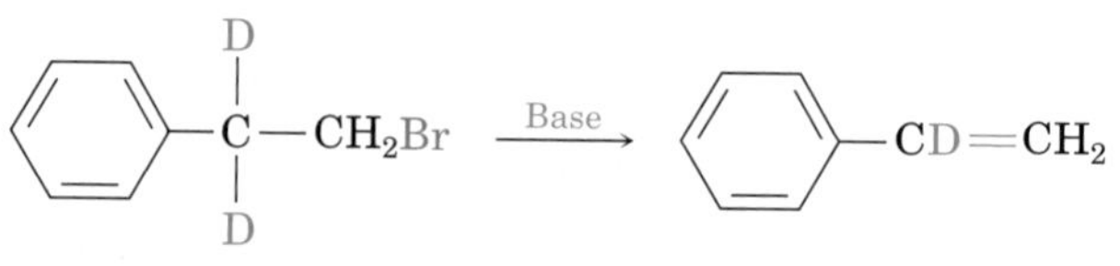

1-Bromo-2,2-dideuterio-2-phenylethane

This result tells us that the C–H (or C–D) bond is broken *in the rate-limiting step*, consistent with our picture of the E2 reaction as a one-step process. If it were otherwise, we couldn't measure a rate difference.

11.14   The E1 Reaction

Just as the E2 reaction is analogous to the S_N2 reaction, there is a close analog to the S_N1 reaction called the **E1 reaction** (for *elimination, unimolecular*). The E1 reaction can be formulated as shown in Figure 11.21 for the elimination of HCl from 2-chloro-2-methylpropane.

Spontaneous dissociation of the tertiary alkyl chloride yields an intermediate carbocation in a slow, rate-limiting step.

$$CH_3-\overset{\overset{\displaystyle Cl}{|}}{\underset{\underset{\displaystyle CH_3}{|}}{C}}-CH_3$$

Rate-limiting

Carbocation

$$\left[\begin{array}{c} H_3C \\ \\ H_3C \end{array} \!\!\!\! \overset{+}{C}-\overset{\overset{\displaystyle H}{|}}{\underset{\underset{\displaystyle H}{|}}{C}}-H + Cl^- \right] \quad :Base$$

Fast

Loss of a neighboring H^+ in a fast step yields the neutral alkene product. The electron pair from the C—H bond goes to form the alkene π bond.

$$\overset{\displaystyle CH_3}{\underset{\displaystyle CH_3}{}}C=C\overset{\displaystyle H}{\underset{\displaystyle H}{}}$$

© 1984 JOHN MCMURRY

E1 eliminations begin with the same unimolecular dissociation we saw in the S_N1 reaction, but the dissociation is followed by loss of H^+ from the intermediate carbocation rather than by substitution. In fact, the E1 and S_N1 reactions normally occur in competition whenever an alkyl halide is treated in a protic solvent with a nonbasic nucleophile. Thus, the best E1 substrates are also the best S_N1 substrates, and mixtures of substitution and elimination products are usually obtained. For example, when 2-chloro-2-methylpropane is warmed to 65°C in 80% aqueous ethanol, a 64:36 mixture of 2-methyl-2-propanol (S_N1) and 2-methylpropene (E1) results:

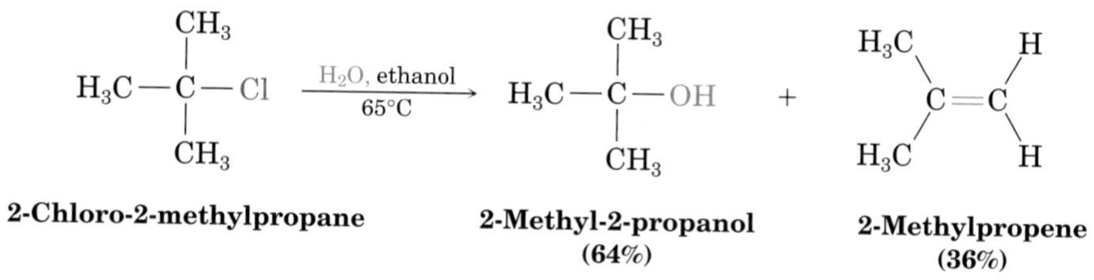

2-Chloro-2-methylpropane → **2-Methyl-2-propanol** **(64%)** + **2-Methylpropene** **(36%)**

Much evidence has been obtained in support of the E1 mechanism. For example, E1 reactions show first-order kinetics, consistent with a rate-limiting spontaneous dissociation process:

$$\text{Rate} = k \times [\text{RX}]$$

Another piece of evidence involves the stereochemistry of elimination. Unlike the E2 reaction, where periplanar geometry is required, there is no geometric requirement on the E1 reaction because the halide and the hydrogen are lost in separate steps. We might therefore expect to obtain the more stable (Zaitsev's rule) product from E1 reaction, which is just what we find. To return to a familiar example, menthyl chloride loses HCl under E1 conditions in a polar solvent to give a mixture of alkenes in which the Zaitsev product, 3-menthene, predominates (Figure 11.22).

FIGURE 11.22 ▼

Elimination reactions of menthyl chloride. E2 conditions (strong base in pure ethanol) lead to 2-menthene, whereas E1 conditions (very dilute base in aqueous ethanol) lead to a mixture of 2-menthene and 3-menthene.

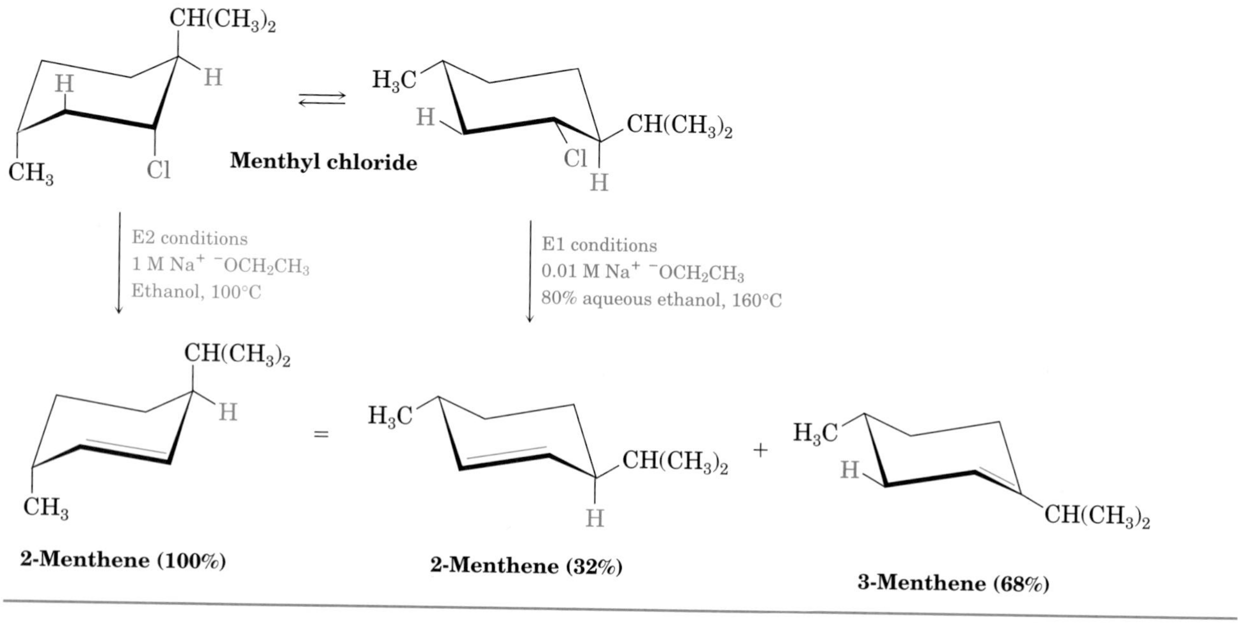

Menthyl chloride

E2 conditions
1 M Na⁺ ⁻OCH₂CH₃
Ethanol, 100°C

E1 conditions
0.01 M Na⁺ ⁻OCH₂CH₃
80% aqueous ethanol, 160°C

2-Menthene (100%)   **2-Menthene (32%)**   **3-Menthene (68%)**

A final piece of evidence about the mechanism of E1 reactions is that they show no deuterium isotope effect. Because rupture of the C–H (or C–D) bond occurs *after* the rate-limiting step rather than during it, we can't measure a rate difference between a deuterated and nondeuterated substrate.

11.15 Summary of Reactivity: S$_N$1, S$_N$2, E1, E2

S$_N$1, S$_N$2, E1, E2: How can you keep it all straight? How can you predict what will happen in any given case? Will substitution or elimination occur? Will the reaction be bimolecular or unimolecular? There are no rigid answers to these questions, but it's possible to recognize some trends and make some generalizations (Table 11.4).

TABLE 11.4 A Summary of Substitution and Elimination Reactions

Halide type	S$_N$1	S$_N$2	E1	E2
RCH$_2$X (primary)	Does not occur	Highly favored	Does not occur	Occurs when strong bases are used
R$_2$CHX (secondary)	Can occur with benzylic and allylic halides	Occurs in competition with E2 reaction	Can occur with benzylic and allylic halides	Favored when strong bases are used
R$_3$CX (tertiary)	Favored in hydroxylic solvents	Does not occur	Occurs in competition with S$_N$1 reaction	Favored when bases are used

- **Primary alkyl halides:** S$_N$2 substitution occurs if a good nucleophile such as RS$^-$, I$^-$, CN$^-$, NH$_3$, or Br$^-$ is used. E2 elimination takes place if a strong, sterically hindered base such as *tert*-butoxide is used.

$$\text{CH}_3\text{CH}_2\text{CH}_2\text{CH}_2\text{Br} \xrightarrow[\text{THF–HMPA}]{\text{Na}^+ \ ^-\text{CN}} \text{CH}_3\text{CH}_2\text{CH}_2\text{CH}_2\text{CN}$$

1-Bromobutane   **Pentanenitrile (90%)**

$$\text{CH}_3\text{CH}_2\text{CH}_2\text{CH}_2\text{Br} \xrightarrow{(\text{CH}_3)_3\text{CO}^- \ \text{K}^+} \text{CH}_3\text{CH}_2\text{CH}=\text{CH}_2$$

1-Bromobutane   **1-Butene (85%)**

- **Secondary alkyl halides:** S$_N$2 substitution and E2 elimination occur in competition, often leading to a mixture of products. If a weakly basic nucleophile is used in a polar aprotic solvent, S_N2 substitution predominates. If a strong base such as $CH_3CH_2O^-$, OH^-, or NH_2^- is used, E2 elimination predominates. For example, 2-bromo-propane undergoes different reactions when treated with ethoxide ion (strong base; E2) and with acetate ion (weak base; S_N2):

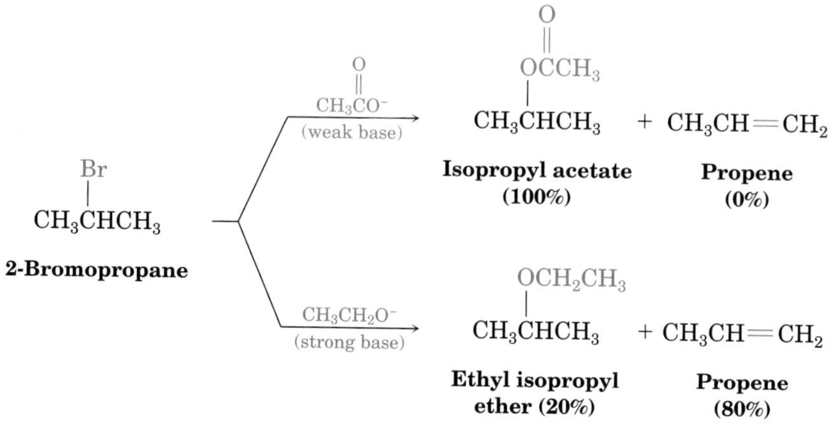

Secondary alkyl halides, particularly allylic and benzylic ones, can also undergo S_N1 and E1 reactions if weakly basic nucleophiles are used in protic solvents such as ethanol or acetic acid.

- **Tertiary alkyl halides:** E2 elimination occurs when a base such as OH^- or RO^- is used. For example, 2-bromo-2-methylpropane gives 97% elimination product when treated with ethoxide ion in ethanol. By contrast, reaction under neutral conditions (heating in pure ethanol) leads to a mixture of products resulting from both S_N1 substitution and E1 elimination.

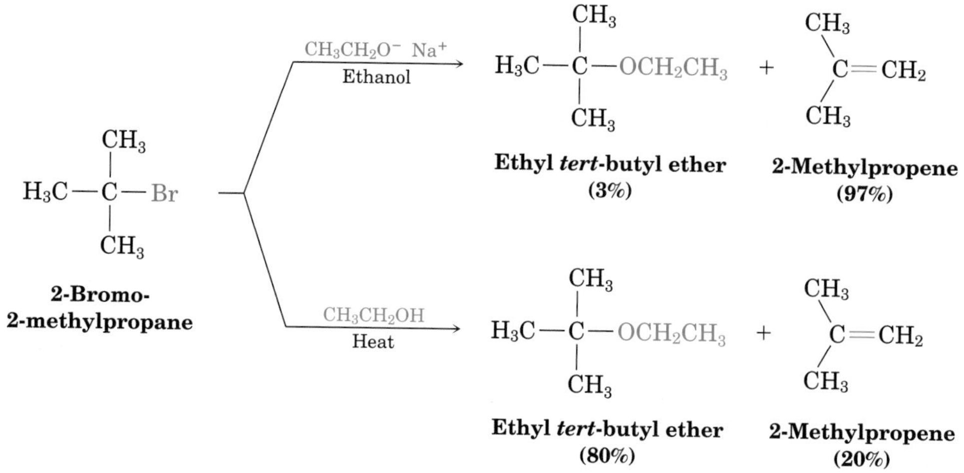

Practice Problem 11.4   Tell whether each of the following reactions is likely to be S_N1, S_N2, E1, or E2, and predict the product of each:

(a) 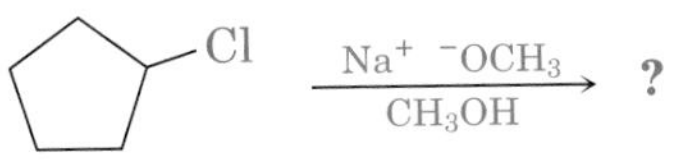 $\xrightarrow[\text{CH}_3\text{OH}]{\text{Na}^+ \ ^-\text{OCH}_3}$?   (b) 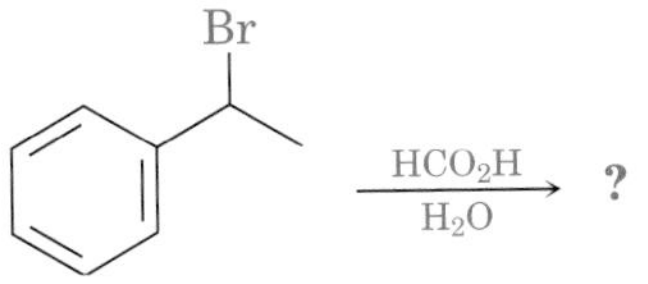 $\xrightarrow[\text{H}_2\text{O}]{\text{HCO}_2\text{H}}$?

Strategy   Look carefully in each reaction at the structure of the substrate, the leaving group, the nucleophile, and the solvent. Then decide from Table 11.4 which kind of reaction is likely to be favored.

Solution   (a) A secondary, nonallylic substrate can undergo an S_N2 reaction with a good nucleophile in a polar aprotic solvent, but will undergo an E2 reaction on treatment with a strong base in a protic solvent. In this case, E2 reaction is likely to predominate.

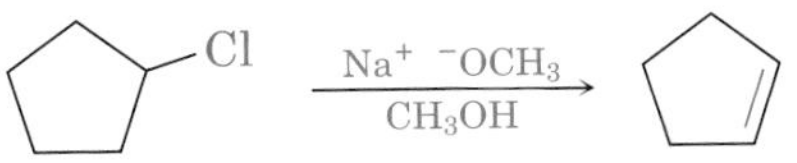

$\xrightarrow[\text{CH}_3\text{OH}]{\text{Na}^+ \ ^-\text{OCH}_3}$ **E2 reaction**

(b) A secondary benzylic substrate can undergo an S_N2 reaction on treatment with a nonbasic nucleophile in a polar aprotic solvent, and will undergo an E2 reaction on treatment with a strong base. Under protic acidic conditions, such as aqueous formic acid (HCO_2H), an S_N1 reaction is likely, along with some E1 reaction.

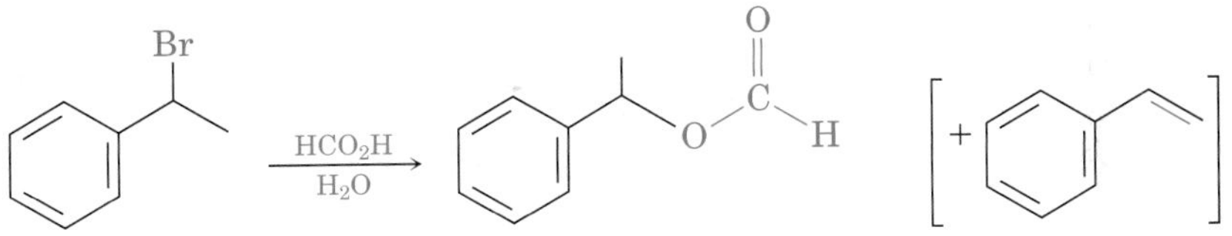

$\xrightarrow[\text{H}_2\text{O}]{\text{HCO}_2\text{H}}$

S_N1   **E1**

Problem 11.20   Tell whether each of the following reactions is likely to be S_N1, S_N2, E1, or E2:
(a) 1-Bromobutane + NaN_3 $\longrightarrow$ 1-Azidobutane

(b) $CH_3CH_2\overset{\overset{\text{Cl}}{|}}{C}HCH_2CH_3$ + KOH $\longrightarrow$ $CH_3CH_2CH{=}CHCH_3$

(c) 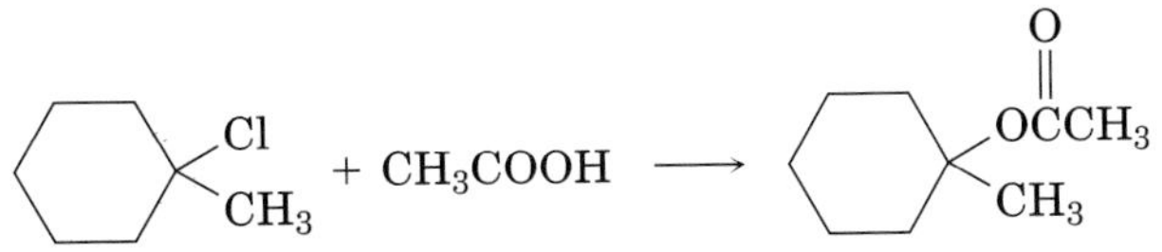 + CH_3COOH $\longrightarrow$

11.16 Substitution Reactions in Synthesis

The reason we've discussed nucleophilic substitution reactions in such detail is that they're so important in organic chemistry. In fact, we've already seen a number of substitution reactions in previous chapters, although they weren't identified as such at the time. For example, we said in Section 8.9 that acetylide anions react well with primary alkyl halides to provide the alkyne product.

$$RC\equiv C:^- Na^+ + R'CH_2X \longrightarrow RC\equiv CCH_2R' + NaX$$

where   X = Br, I, or OTos

Acetylide ion alkylation is an S_N2 reaction, and it's therefore understandable that only primary alkyl halides and tosylates react well. Since acetylide anion is a strong base as well as a good nucleophile, E2 elimination competes with S_N2 alkylation when a secondary or tertiary substrate is used. For example, reaction of sodio 1-hexyne with 2-bromopropane gives primarily the elimination product rather than the substitution product:

$$CH_3(CH_2)_3C\equiv C:^- Na^+ + \underset{\displaystyle CH_3CHCH_3}{\overset{\displaystyle Br}{|}} \longrightarrow$$

Sodio 1-hexyne

$$CH_3(CH_2)_3C\equiv CCH(CH_3)_2$$

7% S_N2

+

$$CH_3(CH_2)_3C\equiv CH + CH_3CH=CH_2$$

93% E2

Other substitution reactions we've seen include some of the reactions used for preparing alkyl halides from alcohols. We said in Section 10.7, for example, that alkyl halides can be prepared by treating alcohols with HX—reactions now recognizable as nucleophilic substitutions of halide on the protonated alcohols. Tertiary alcohols react by an S_N1 pathway involving unimolecular dissociation of the protonated alcohol to yield a carbocation, whereas primary alcohols react by an S_N2 pathway involving direct bimolecular displacement of H_2O from the protonated alcohol (Figure 11.23).

Yet another substitution reaction we've seen is the conversion of primary and secondary alcohols into alkyl bromides by treatment with PBr_3 (Section 10.7). Although OH^- is a poor leaving group and can't be displaced directly by nucleophiles, reaction with PBr_3 transforms the hydroxyl into a better leaving group, thereby activating it for nucleophilic displacement. Alcohols react with PBr_3 to give dibromophosphites ($ROPBr_2$), which are highly reactive substrates in S_N2 reactions. Displacement by Br^- then occurs rapidly on the primary carbon, and alkyl bromides are produced in good yield.

Good leaving group

Poor leaving group

$$CH_3(CH_2)_4CH_2OH \xrightarrow[\text{Ether}]{PBr_3} \left[CH_3(CH_2)_4CH_2-\overset{H}{\underset{+}{O}}-PBr_2 \right] \xrightarrow{S_N2} CH_3(CH_2)_4CH_2Br$$

$:\ddot{Br}:^-$

1-Hexanol   **1-Bromohexane**

FIGURE 11.23 ▼

Mechanisms of reactions of HCl with a tertiary and a primary alcohol. Both reactions involve initial protonation of the alcohol –OH group. A tertiary alcohol reacts by an S_N1 mechanism because it can form a stable tertiary carbocation intermediate by loss of H_2O from the protonated reactant. A primary alcohol reacts by an S_N2 pathway because unhindered back-side attack of a nucleophile on the protonated reactant can occur easily.

Tertiary alcohol—S_N1

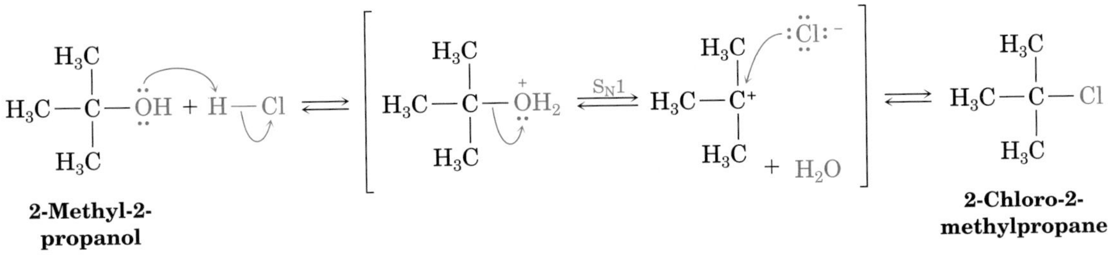

H$_3$C
|
H$_3$C—C—ÖH + H—Cl ⇌ [H$_3$C—C—ÖH$_2$ $\xrightarrow{S_N1}$ H$_3$C—C$^+$ + H$_2$O] ⇌ H$_3$C—C—Cl
|
H$_3$C

2-Methyl-2-propanol   **2-Chloro-2-methylpropane**

Primary alcohol—S_N2

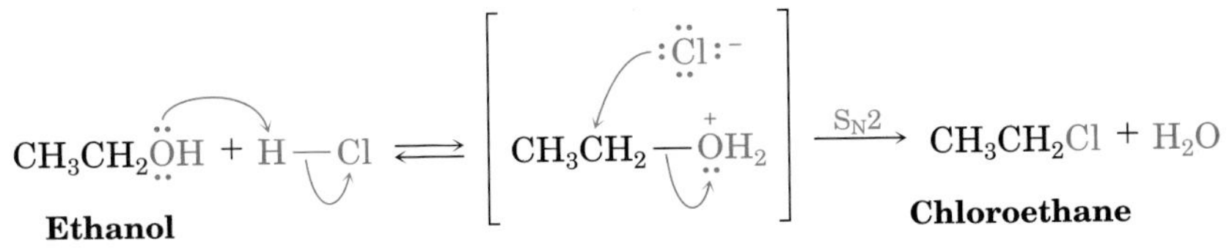

CH$_3$CH$_2$ÖH + H—Cl ⇌ [CH$_3$CH$_2$—ÖH$_2$ $\xrightarrow{S_N2}$ CH$_3$CH$_2$Cl + H$_2$O]

Ethanol   **Chloroethane**

CHEMISTRY @ WORK

Biological Substitution Reactions

All chemistry—whether carried out in flasks by chemists or in cells by living organisms—follows the same rules. Most biological reactions therefore occur by the same addition, substitution, elimination, and rearrangement mechanisms encountered in laboratory reactions.

Perhaps the most common biological substitution reaction is *methylation*, the transfer of a –CH$_3$ group from an electrophilic donor to a nucleophile. A laboratory chemist might choose CH$_3$I for such a reaction, but

(continued) ▶

living organisms use the complex molecule S-adenosylmethionine as the biological methyl-group donor. Since the sulfur atom in S-adenosylmethionine has a positive charge (a *sulfonium* ion), it is an excellent leaving group for S_N2 displacements on the methyl carbon. An example of such a biological methylation takes place in the adrenal medulla during the biological synthesis of adrenaline from norepinephrine.

Norepinephrine + **S-Adenosylmethionine**

S_N2 reaction

Adrenaline +

After dealing only with simple halides such as iodomethane used for laboratory alkylations, it's something of a shock to encounter a molecule as complex as S-adenosylmethionine. From a chemical standpoint, however, CH_3I and S-adenosylmethionine do exactly the same thing: Both transfer a methyl group by an S_N2 reaction. The same principles of reactivity apply to both.

Another example of a biological S_N2 reaction is involved in the response of organisms to certain toxic chemicals. Many reactive S_N2 substrates with deceptively simple structures are quite toxic to living organisms. Methyl bromide, for example, has been widely used as a fumigant to kill termites and as a soil sterilant. The toxicity of these compounds derives from their ability to transfer an alkyl group to a nucleophilic amino group (–NH_2) or mercapto group (–SH) in enzymes, thus altering the enzyme's normal biological activity.

Methyl bromide, the world's most widely used pesticide, is used as a fumigant because it is toxic to termites and other organisms.

(continued) ▶

One of the best-known toxic alkylating agents is *mustard gas,* an early chemical warfare agent that caused an estimated 400,000 casualties in World War I. A primary halide, mustard gas is highly reactive toward S_N2 displacements by nucleophilic amino groups in proteins. It is thought to act through an intermediate sulfonium ion in much the same manner as *S*-adenosylmethionine.

$$ClCH_2CH_2\overset{..}{\underset{..}{S}}CH_2CH_2 - Cl \underset{\text{Internal } S_N2 \text{ reaction}}{\rightleftharpoons} \left[ClCH_2CH_2\overset{+}{\underset{}{S}} \overset{CH_2}{\underset{CH_2}{|}} \quad Cl^- \right]$$

Mustard gas

$$S_N2 \text{ reaction} \downarrow \quad :NH_2 \text{—Protein}$$

$$ClCH_2CH_2SCH_2CH_2NH \sim\sim\sim \boxed{\text{Protein}}$$

Alkylated protein

Summary and Key Words

Reaction of an alkyl halide or tosylate with a nucleophile/base results either in *substitution* or in *elimination*. **Nucleophilic substitutions** are of two types: **S_N2 reactions** and **S_N1 reactions**. In the S_N2 reaction, the entering nucleophile attacks the halide from a direction 180° away from the **leaving group**, resulting in an umbrella-like *Walden inversion of configuration* at the carbon atom. The reaction shows **second-order kinetics** and is strongly inhibited by increasing steric bulk of the reactants. Thus, S_N2 reactions are favored for primary and secondary substrates.

The S_N1 reaction occurs when the substrate spontaneously dissociates to a carbocation in a slow **rate-limiting step**, followed by a rapid attack of nucleophile. As a result, S_N1 reactions show **first-order kinetics** and take place with racemization of configuration at the carbon atom. They are most favored for tertiary substrates.

Eliminations of alkyl halides to yield alkenes also occur by two different mechanisms: **E2 reaction** and **E1 reaction**. In the E2 reaction, a base abstracts H^+ from one carbon at the same time the leaving group departs from the neighboring carbon. The reaction takes place preferentially through an **anti periplanar** transition state in which the four reacting atoms—hydrogen, two carbons, and leaving group—are in the same plane. The reaction shows second-order kinetics and a **deuterium isotope effect**, and occurs when a secondary or tertiary substrate is treated with a strong

periplanar, 415
rate constant, 390
rate-limiting step, 402
reaction rate, 389
second-order reaction, 390
S_N1 reaction, 401
S_N2 reaction, 391
solvation, 399
syn periplanar, 415
Zaitsev's rule, 414

base. These elimination reactions usually give a mixture of alkene products in which the more highly substituted alkene predominates (**Zaitsev's rule**).

The E1 reaction takes place when the substrate spontaneously dissociates to yield a carbocation in the slow rate-limiting step before losing H^+ from a neighboring carbon in a second step. The reaction shows first-order kinetics and no deuterium isotope effect, and occurs when a tertiary substrate reacts in polar, nonbasic solution.

In general, substrates react in the following way:

RCH_2X (primary)	⟶	Mostly S_N2 substitution
R_2CHX (secondary)	⟶	S_N2 substitution with nonbasic nucleophiles E2 elimination with strong bases
R_3CX (tertiary)	⟶	Mostly E2 elimination (S_N1 substitution and E1 elimination in nonbasic solvents)

Summary of Reactions

1. Nucleophilic substitutions
   (a) S_N1 reaction; carbocation intermediate is involved (Sections 11.6–11.9)

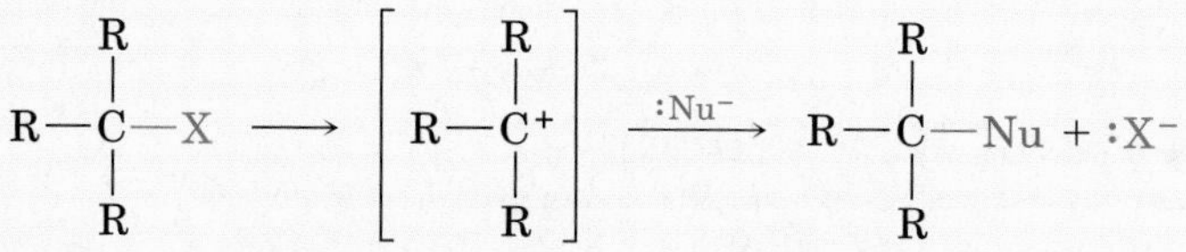

Best for 3°, allylic, and benzylic halides and tosylates

   (b) S_N2 reaction; back-side attack of nucleophile occurs (Sections 11.4–11.5)

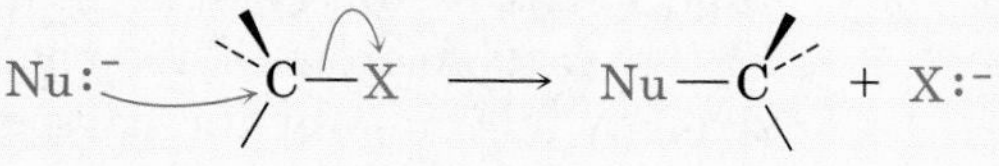

Best for 1° or 2° halides

$Nu:^- = H^-, {}^-CN, I^-, Br^-, Cl^-, {}^-OH, {}^-NH_2, CH_3O^-,$
$CH_3CO_2^-, HS^-, H_2O, NH_3$, and so forth

2. Eliminations
   (a) E1 reaction; more highly substituted alkene is formed (Section 11.14)

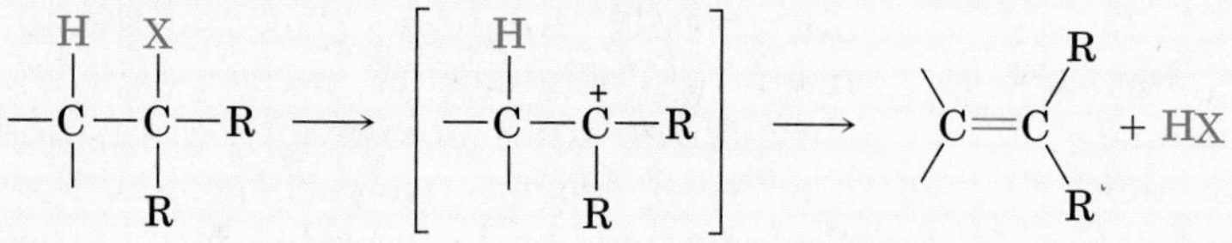

Best for 3° halides

(b) E2 reaction; anti periplanar geometry is required (Section 11.11)

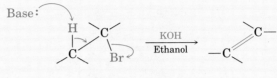

Best for 2° and 3° halides

Visualizing Chemistry

● ●

(Problems 11.1–11.20 appear within the chapter.)

11.21 Write the product you would expect from reaction of each of the following molecules with (i) Na^+ $^-SCH_3$ and (ii) Na^+ ^-OH (yellow-green = Cl):

(a)   (b)   (c)

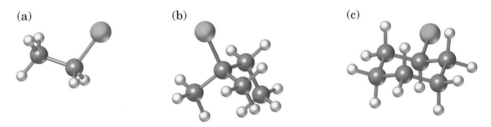

11.22 From what alkyl bromide was the following alkyl acetate made by S_N2 reaction? Write the reaction, showing all stereochemistry.

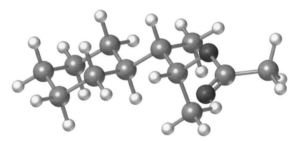

11.23 Assign R or S configuration to the following molecule, write the product you would expect from S_N2 reaction with NaCN, and assign R or S configuration to the product (yellow-green = Cl):

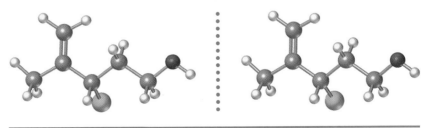

Stereo View

11.24 Draw the structure and assign Z or E stereochemistry to the product you expect from E2 reaction of the following molecule with NaOH (yellow-green = Cl):

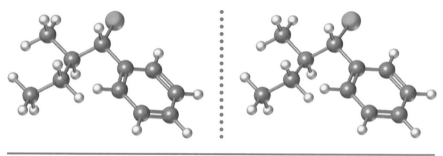

Stereo View

Additional Problems

11.25 Describe the effects of each of the following variables on both S_N2 and S_N1 reactions:
(a) Solvent   (b) Leaving group
(c) Attacking nucleophile   (d) Substrate structure

11.26 Which choice in each of the following pairs will react faster in an S_N2 reaction with OH^-?
(a) CH_3Br or CH_3I (b) CH_3CH_2I in ethanol or in dimethyl sulfoxide
(c) $(CH_3)_3CCl$ or CH_3Cl (d) $H_2C=CHBr$ or $H_2C=CHCH_2Br$

11.27 What effect would you expect the following changes to have on the rate of the reaction of 1-iodo-2-methylbutane with cyanide ion?
(a) The CN^- concentration is halved and the 1-iodo-2-methylbutane concentration is doubled.
(b) Both the CN^- and the 1-iodo-2-methylbutane concentrations are tripled.

11.28 What effect would you expect the following changes to have on the rate of the reaction of ethanol with 2-iodo-2-methylbutane?
(a) The concentration of the halide is tripled.
(b) The concentration of the ethanol is halved by adding diethyl ether as an inert solvent.

11.29 How might you prepare each of the following molecules using a nucleophilic substitution reaction at some step?
(a) $CH_3C{\equiv}CCH(CH_3)_2$ (b) $CH_3CH_2CH_2CH_2CN$
(c) $H_3C{-}O{-}C(CH_3)_3$ (d) $CH_3CH_2CH_2NH_2$
(e)   (f)

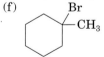

11.30 Which reaction in each of the following pairs would you expect to be faster?
(a) The S_N2 displacement by I^- on CH_3Cl or on CH_3OTos
(b) The S_N2 displacement by $CH_3CO_2^-$ on bromoethane or on bromocyclohexane
(c) The S_N2 displacement on 2-bromopropane by $CH_3CH_2O^-$ or by CN^-
(d) The S_N2 displacement by $HC{\equiv}C^-$ on bromomethane in benzene or in hexamethylphosphoramide

11.31 What products would you expect from the reaction of 1-bromopropane with each of the following?

(a) $NaNH_2$ (b) $KOC(CH_3)_3$ (c) NaI
(d) NaCN (e) $NaC{\equiv}CH$ (f) Mg, then H_2O

11.32 Which reactant in each of the following pairs is more nucleophilic? Explain.
(a) $^-NH_2$ or NH_3 (b) H_2O or CH_3COO^- (c) BF_3 or F^-
(d) $(CH_3)_3P$ or $(CH_3)_3N$ (e) I^- or Cl^- (f) $^-C{\equiv}N$ or $^-OCH_3$

11.33 Among the Walden cycles carried out by Kenyon and Phillips is the following series of reactions reported in 1923. Explain the results, and indicate where Walden inversion is occurring.

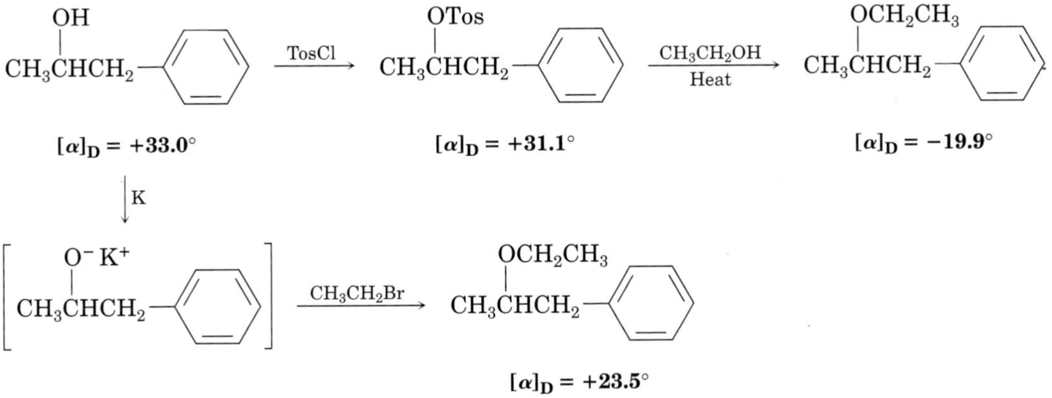

11.34 The synthetic sequences shown below are unlikely to occur as written. Tell what is wrong with each, and predict the true product.

(a) 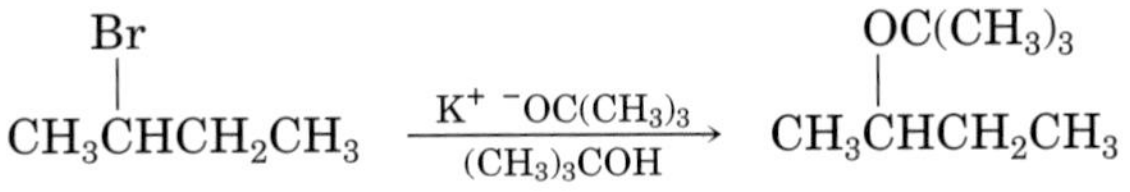

(b) 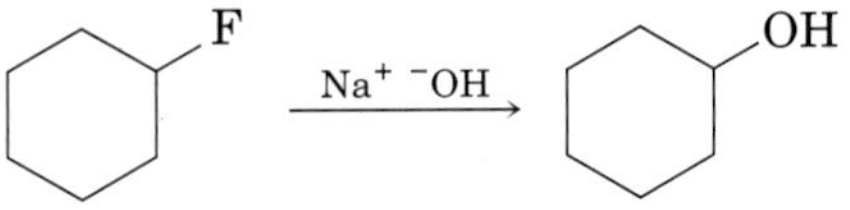

(c) 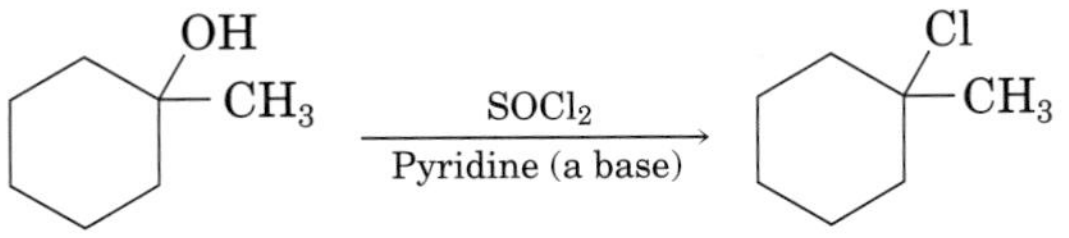

11.35 Order each of the following sets of compounds with respect to S_N1 reactivity:

(a) 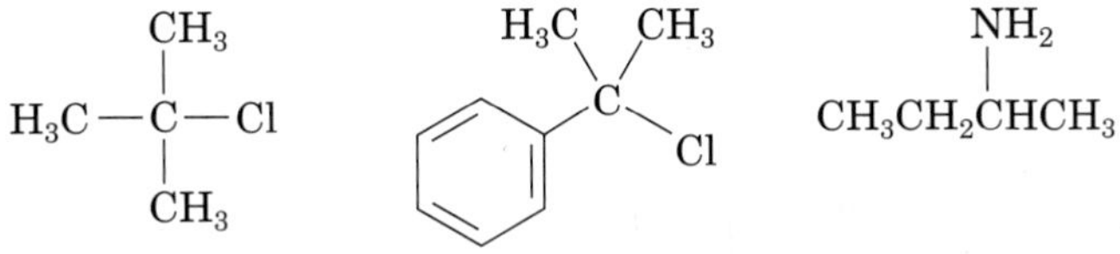

all are tertiary alkyl halide
more stable carbocation

(b) $(CH_3)_3CF$ $(CH_3)_3CBr$ $(CH_3)_3COH$

the larger the leaving grp, the better

(c) 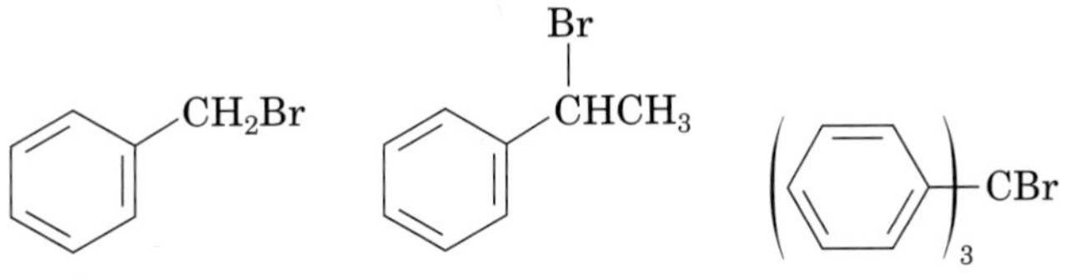

∵ of carbocation stability

11.36   Order each of the following sets of compounds with respect to S_N2 reactivity:

(a) 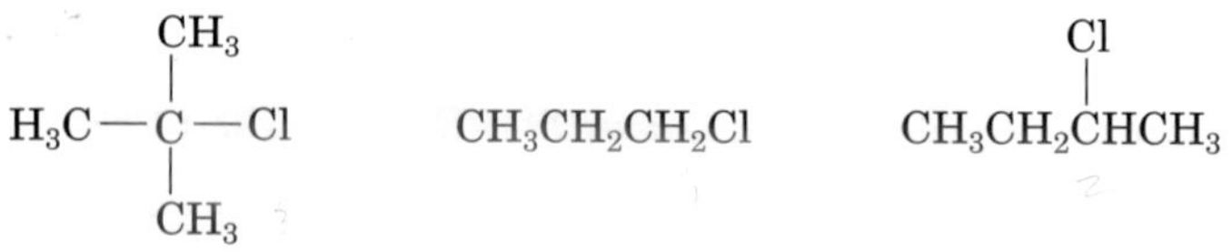
$$\begin{array}{c} CH_3 \\ | \\ H_3C-C-Cl \\ | \\ CH_3 \end{array}$$
$CH_3CH_2CH_2Cl$
$$\begin{array}{c} Cl \\ | \\ CH_3CH_2CHCH_3 \end{array}$$

(b) 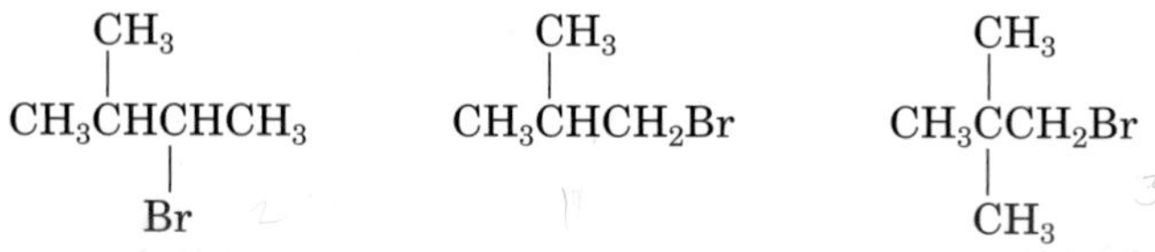
$$\begin{array}{c} CH_3 \\ | \\ CH_3CHCHCH_3 \\ | \\ Br \end{array}$$
$$\begin{array}{c} CH_3 \\ | \\ CH_3CHCH_2Br \end{array}$$
$$\begin{array}{c} CH_3 \\ | \\ CH_3CCH_2Br \\ | \\ CH_3 \end{array}$$

(c) $CH_3CH_2CH_2OCH_3$ $CH_3CH_2CH_2OTos$ $CH_3CH_2CH_2Br$

11.37   Predict the product and give the stereochemistry resulting from reaction of each of the following nucleophiles with (R)-2-bromooctane:
(a) ⁻CN   (b) $CH_3CO_2^-$ (c) CH_3S^-

11.38   (R)-2-Bromooctane undergoes racemization to give (±)-2-bromooctane when treated with NaBr in dimethyl sulfoxide. Explain.

11.39   Ethers can often be prepared by S_N2 reaction of alkoxide ions, RO⁻, with alkyl halides. Suppose you wanted to prepare cyclohexyl methyl ether. Which of the two possible routes shown below would you choose? Explain.

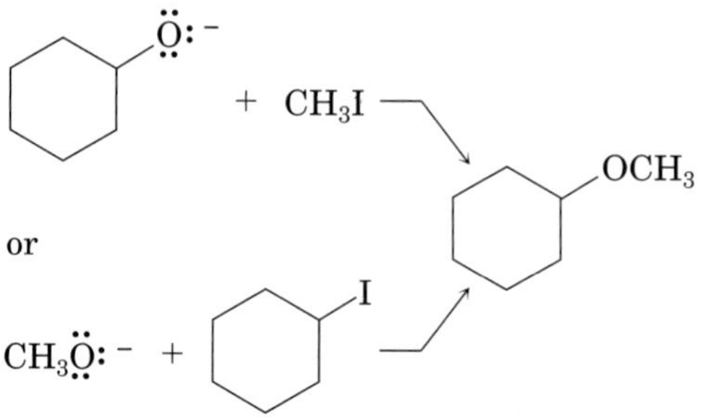

11.40   The S_N2 reaction can occur *intramolecularly* (within the same molecule). What product would you expect from treatment of 4-bromo-1-butanol with base?

$$BrCH_2CH_2CH_2CH_2OH \xrightarrow{Na^+ \ ^-OCH_3} CH_3OH + [BrCH_2CH_2CH_2CH_2O^- \ Na^+] \longrightarrow ?$$

11.41   In light of your answer to Problem 11.40, propose a synthesis of 1,4-dioxane starting only with 1,2-dibromoethane.

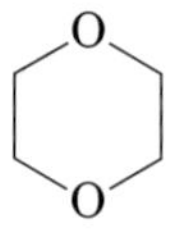

1,4-Dioxane

11.42   As indicated in Problem 11.4, the alkyl halide shown below is inert to S_N2 displacement. Perhaps more surprisingly, it is also unreactive to S_N1 substitution even though it is tertiary. Explain.

11.43   1-Chloro-1,2-diphenylethane can undergo E2 elimination to give either *cis-* or *trans-*1,2-diphenylethylene (stilbene). Draw Newman projections of the reactive conformations leading to both possible products, and suggest a reason why the trans alkene is the major product.

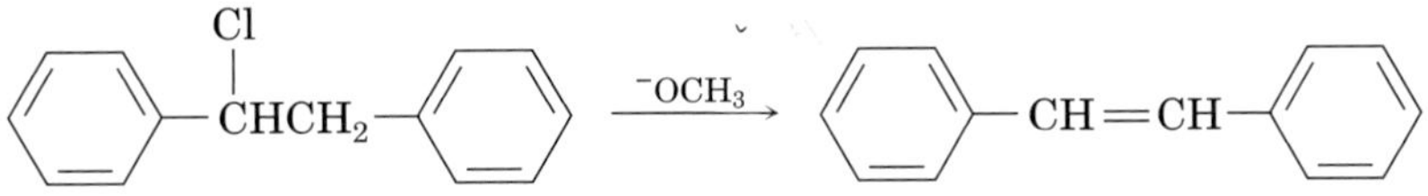

1-Chloro-1,2-diphenylethane   **trans-1,2-Diphenylethylene**

11.44   Predict the major alkene product of the following E1 reaction:

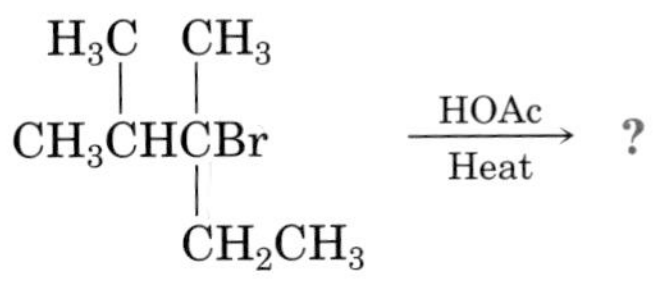

11.45   The tosylate of (2R,3S)-3-phenyl-2-butanol undergoes E2 elimination on treatment with sodium ethoxide to yield (Z)-2-phenyl-2-butene. Explain, using Newman projections.

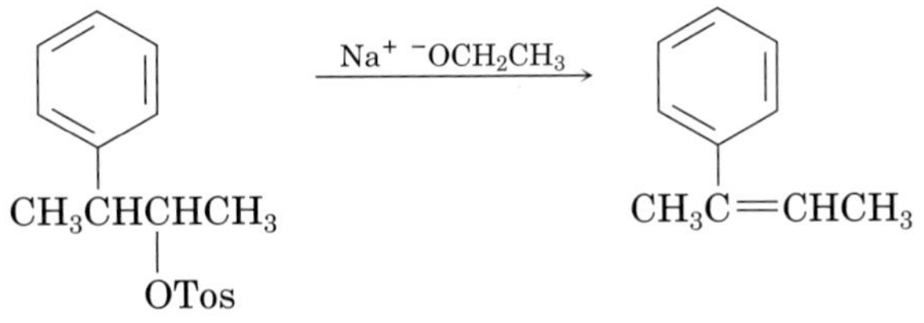

11.46   In light of your answer to Problem 11.45, which alkene, *E* or *Z,* would you expect from an elimination reaction on the tosylate of (2R,3R)-3-phenyl-2-butanol? Which alkene would result from E2 reaction on the (2S,3R) and (2S,3S) tosylates? Explain.

11.47   How can you explain the fact that *trans*-1-bromo-2-methylcyclohexane yields the non-Zaitsev elimination product 3-methylcyclohexene on treatment with base?

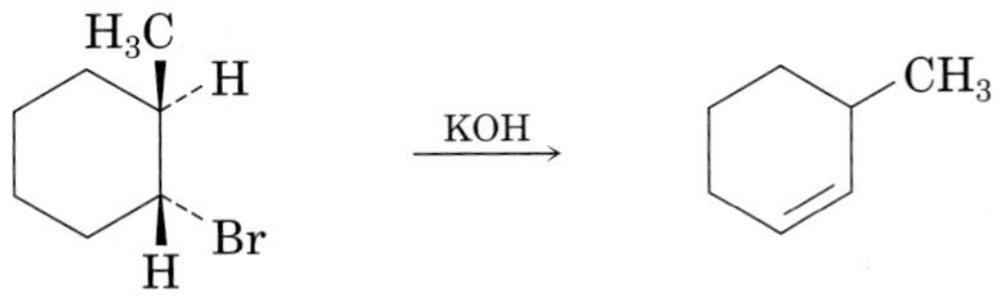

trans-1-Bromo-2-methylcyclohexane   **3-Methylcyclohexene**

11.48   Predict the product(s) of the following reaction, indicating stereochemistry where necessary:

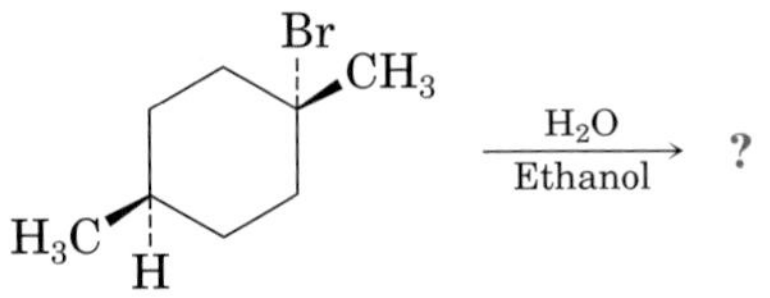

11.49   Draw all isomers of C_4H_9Br, name them, and arrange them in order of decreasing reactivity in the S_N2 reaction.

11.50 Reaction of iodoethane with CN⁻ yields a small amount of *isonitrile,* $CH_3CH_2N≡C$, along with the nitrile $CH_3CH_2C≡N$ as the major product. Write Lewis structures for both products, assign formal charges as necessary, and propose mechanisms to account for their formation.

11.51 Alkynes can be made by dehydrohalogenation of vinylic halides in a reaction that is essentially an E2 process. In studying the stereochemistry of this elimination, it was found that (*Z*)-2-chloro-2-butenedioic acid reacts 50 times as fast as the corresponding *E* isomer. What conclusion can you draw about the stereochemistry of eliminations in vinylic halides? How does this result compare with eliminations of alkyl halides?

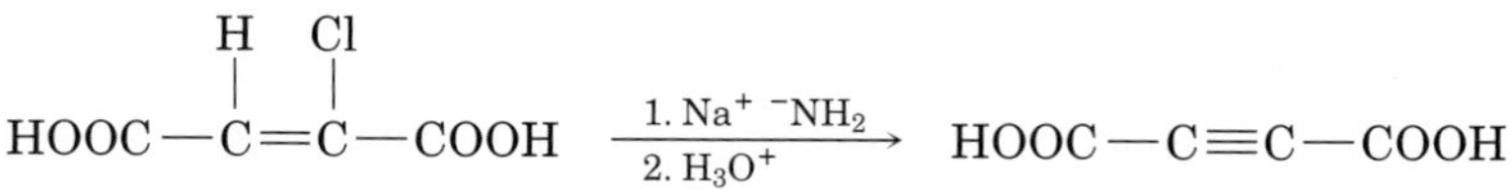

11.52 (*S*)-2-Butanol slowly racemizes on standing in dilute sulfuric acid. Explain.

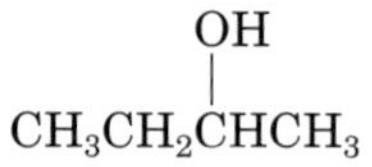

2-Butanol

11.53 Reaction of HBr with (*R*)-3-methyl-3-hexanol leads to (±)-3-bromo-3-methylhexane. Explain.

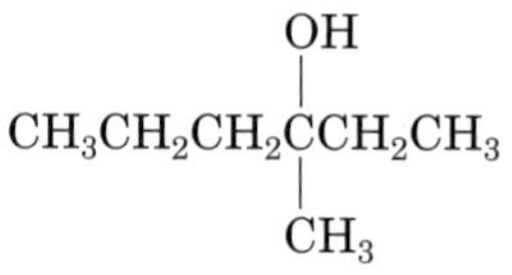

3-Methyl-3-hexanol

11.54 Treatment of 1-bromo-2-deuterio-2-phenylethane with strong base leads to a mixture of deuterated and nondeuterated phenylethylenes in an approximately 7:1 ratio. Explain.

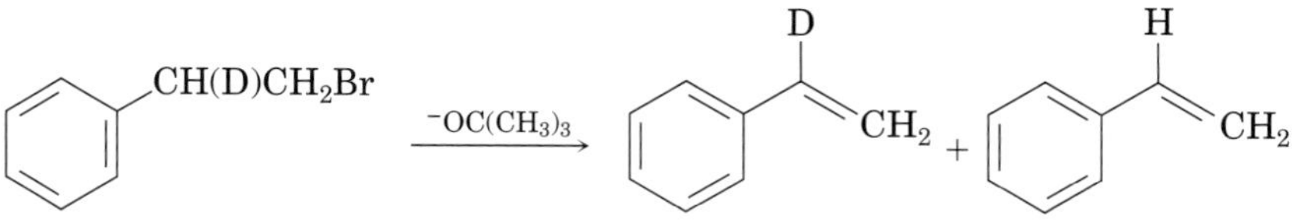

7:1 ratio

11.55 Although anti periplanar geometry is preferred for E2 reactions, it isn't absolutely necessary. The deuterated bromo compound shown here reacts with strong base to yield an undeuterated alkene. Clearly, a syn elimination has occurred. Make a molecular model of the reactant, and explain the result.

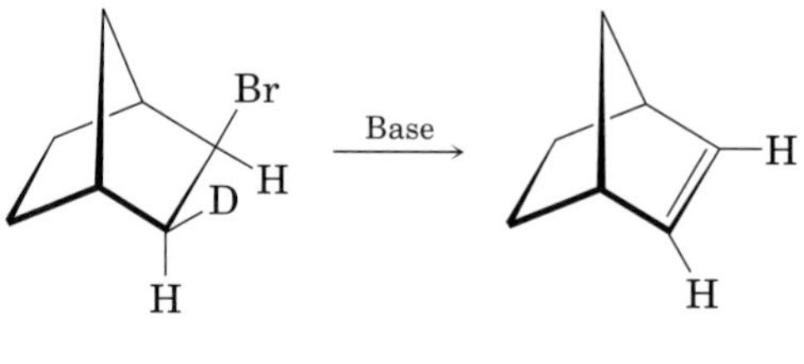

11.56   In light of your answer to Problem 11.55, explain why one of the following isomers undergoes E2 reaction approximately 100 times as fast as the other. Which isomer is more reactive, and why?

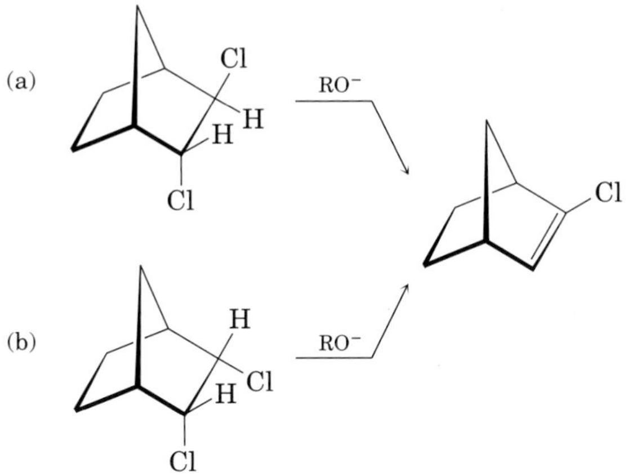

11.57   Propose structures for compounds that fit the following descriptions:
(a) An alkyl halide that gives a mixture of three alkenes on E2 reaction
(b) An organohalide that will not undergo nucleophilic substitution
(c) An alkyl halide that gives the non-Zaitsev product on E2 reaction
(d) An alcohol that reacts rapidly with HCl at 0°C

11.58   There are eight diastereomers of 1,2,3,4,5,6-hexachlorocyclohexane. Draw each in its more stable chair conformation. One isomer loses HCl in an E2 reaction nearly 1000 times more slowly than the others. Which isomer reacts so slowly, and why?

11.59   The tertiary amine quinuclidine reacts with CH_3I 50 times as fast as triethylamine, $(CH_3CH_2)_3N$. Explain.

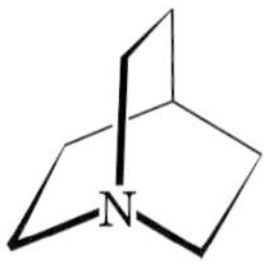

Quinuclidine

11.60   Methyl esters (RCO_2CH_3) undergo a cleavage reaction to yield carboxylate ions plus iodomethane on heating with LiI in dimethylformamide:

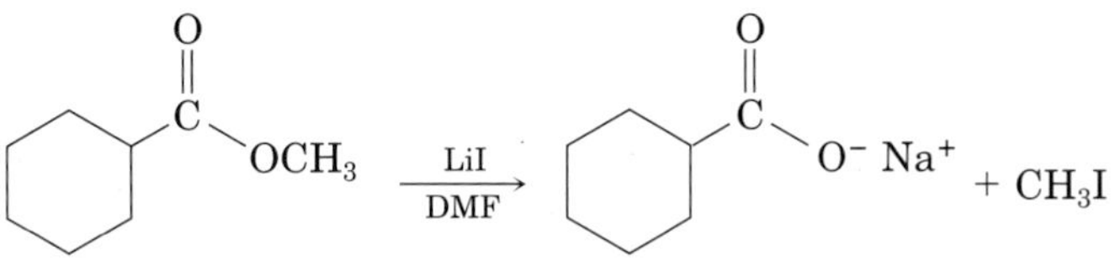

The following evidence has been obtained: (1) The reaction occurs much faster in DMF than in ethanol. (2) The corresponding ethyl ester ($RCO_2CH_2CH_3$) cleaves approximately 10 times more slowly than the methyl ester. Propose a mechanism for the reaction. What other kinds of experimental evidence could you gather to support your hypothesis?

11.61   The reaction of 1-chlorooctane with $CH_3CO_2^-$ to give octyl acetate is greatly accelerated by adding a small quantity of iodide ion. Explain.

11.62 Compound X is optically inactive and has the formula $C_{16}H_{16}Br_2$. On treatment with strong base, X gives hydrocarbon Y, $C_{16}H_{14}$. Compound Y absorbs 2 equivalents of hydrogen when reduced over a palladium catalyst and reacts with ozone to give two fragments. One fragment, Z, is an aldehyde with formula C_7H_6O. The other fragment is glyoxal, $(CHO)_2$. Write the reactions involved, and suggest structures for X, Y, and Z. What is the stereochemistry of X?

11.63 Propose a structure for an alkyl halide that gives only (*E*)-3-methyl-2-phenyl-2-pentene on E2 elimination. Make sure you indicate the stereochemistry.

11.64 When primary alcohols are treated with *p*-toluenesulfonyl chloride at room temperature in the presence of an organic base such as pyridine, a tosylate is formed. When the same reaction is carried out at higher temperature, an alkyl chloride is often formed. Propose a mechanism.

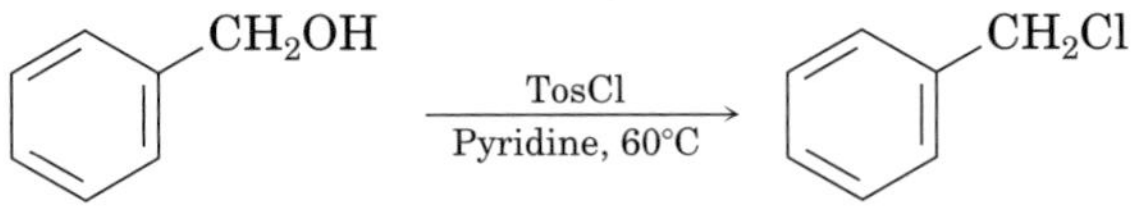

11.65 S_N2 reactions take place with inversion of configuration, and S_N1 reactions take place with racemization. The following substitution reaction, however, occurs with complete *retention* of configuration. Propose a mechanism.

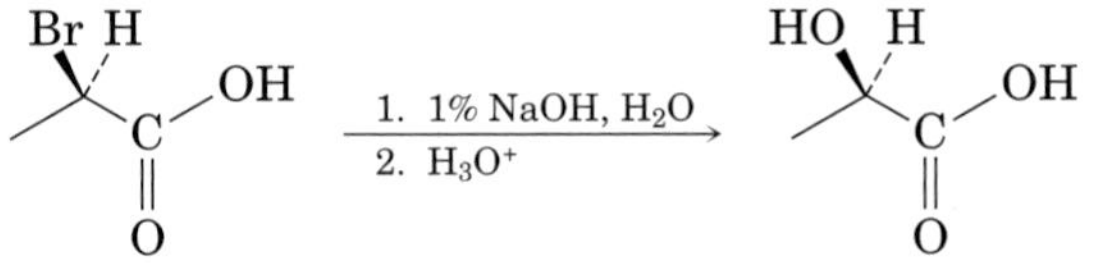

11.66 Propose a mechanism for the following reaction, an important step in the laboratory synthesis of proteins:

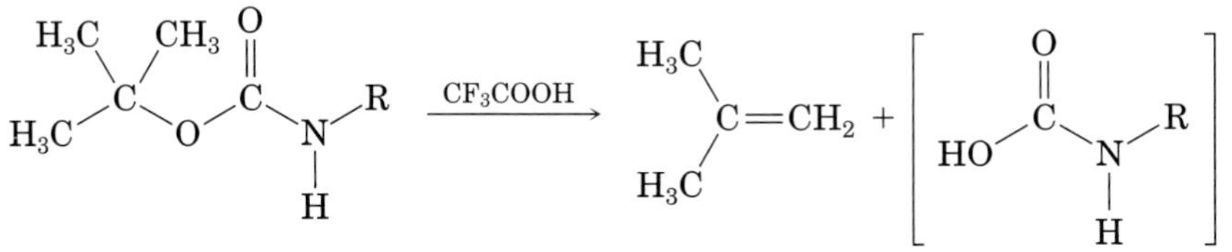

A Look Ahead

11.67 Bromohydrins (Section 7.3) are converted into cyclic ethers called *epoxides* when treated with base. Propose a mechanism, using curved arrows to show the electron flow. (See Section 18.7.)

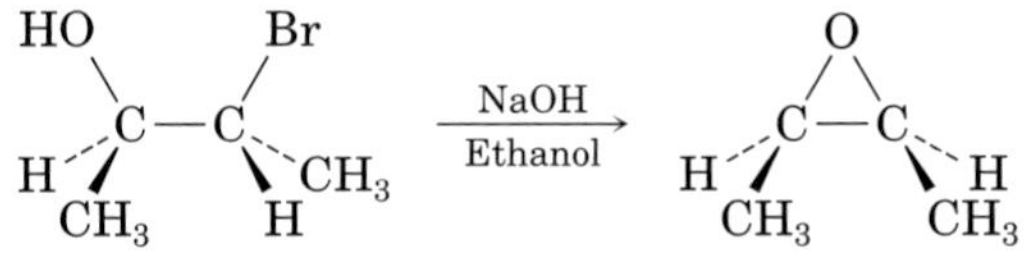

11.68 Show the stereochemistry of the epoxide (Problem 11.67) you would obtain by formation of a bromohydrin from *trans*-2-butene, followed by treatment with base. (See Section 18.8.)

11.69 Amines are converted into alkenes by a two-step process called the *Hofmann elimination*. Reaction of the amine with excess CH_3I in the first step yields an intermediate that undergoes E2 reaction when treated with basic silver oxide. Pentylamine, for example, yields 1-pentene. Propose a structure for the intermediate, and explain why it undergoes ready elimination. (See Section 24.7.)

$$CH_3CH_2CH_2CH_2CH_2NH_2 \xrightarrow[\text{2. Ag}_2\text{O, H}_2\text{O}]{\text{1. Excess CH}_3\text{I}} CH_3CH_2CH_2CH{=}CH_2$$

Molecular Modeling

• •

11.70 The idealized S_N2 transition state has a linear arrangement of Nu···C···X. Use SpartanView to examine S_N2 transition states for the reaction of Cl^- with CH_3Br, CH_3CH_2Br, $(CH_3)_2CHBr$, $(CH_3)_3CBr$, and $(CH_3)_3CCH_2Br$. Which transition state is most ideal? Identify significantly distorted bond angles in the other transition states, and account for the distortions.

11.71 Suppose a chemist wants to use KF as a nucleophilic reagent in an S_N2 reaction. Use SpartanView to compare electrostatic potential maps of water, acetonitrile, and DMSO, and tell which solvent(s) you expect to bind strongly to F^- and to K^+. Which solvent(s) should promote a rapid S_N2 reaction?

11.72 The *cis* and *trans* isomers of 2-bromo-1-methylcyclohexane react at different rates and give different E2 elimination products. Use SpartanBuild to build the two chair conformations for both isomers, and minimize the structures of each. Identify the reactive conformation of each molecule and the expected reaction product from each. Is the reactive conformation the one with lower strain energy? Which isomer, cis or trans, will undergo elimination more readily?

Cis isomer   **Trans isomer**

Structure Determination: Mass Spectrometry and Infrared Spectroscopy

Many of the assertions made in previous chapters have been stated without proof. We said in Section 6.9, for instance, that Markovnikov's rule is followed in alkene electrophilic addition reactions and that treatment of 1-butene with HCl yields 2-chlorobutane rather than 1-chlorobutane. Similarly, we said in Section 11.10 that Zaitsev's rule is followed in elimination reactions and that treatment of 2-chlorobutane with NaOH yields 2-butene rather than 1-butene. But how do we know with certainty that these assertions are correct? The answer to these and many thousands of similar questions is that the *structures* of the reaction products have been elucidated.

$$CH_3CH_2CH{=}CH_2 + HCl \longrightarrow \underset{\textbf{2-Chlorobutane}}{CH_3CH_2\overset{\displaystyle Cl}{\underset{\displaystyle |}{C}}HCH_3} \qquad \left[\textit{NOT} \quad \underset{\textbf{1-Chlorobutane}}{CH_3CH_2CH_2CH_2Cl} \right]$$

1-Butene

$$\underset{\textbf{2-Chlorobutane}}{CH_3CH_2\overset{\displaystyle Cl}{\underset{\displaystyle |}{C}}HCH_3} \xrightarrow{\text{NaOH}} \underset{\textbf{2-Butene}}{CH_3CH{=}CHCH_3} \qquad \left[\textit{NOT} \quad \underset{\textbf{1-Butene}}{CH_3CH_2CH{=}CH_2} \right]$$

Determining the structure of an organic compound was a difficult and time-consuming process in the nineteenth and early twentieth centuries, but extraordinary advances have been made in the past few decades. Powerful techniques are now available that greatly simplify the problem of structure determination. In this and the next two chapters we'll look at four of the most useful techniques—mass spectrometry (MS), infrared spectroscopy (IR), nuclear magnetic resonance spectroscopy (NMR), and ultraviolet spectroscopy (UV)—and we'll see the kind of information that can be obtained from each.

Mass spectrometry	What size and formula?
Infrared spectroscopy	What functional groups are present?
Ultraviolet spectroscopy	Is a conjugated π electron system present?
Nuclear magnetic resonance spectroscopy	What carbon–hydrogen framework is present?

12.1 Mass Spectrometry

At its simplest, **mass spectrometry (MS)** is a technique for measuring the mass, and therefore the molecular weight (MW), of a molecule. In addition, it's often possible to gain structural information about a molecule by measuring the masses of the fragments produced when molecules are broken apart. There are several different kinds of mass spectrometers available, but one of the most common is the electron-ionization, magnetic-sector instrument shown schematically in Figure 12.1 (p. 442).

A small amount of sample is vaporized into the mass spectrometer, where it is bombarded by a stream of high-energy electrons. The energy of the electron beam can be varied but is commonly around 70 electron volts

FIGURE 12.1 ▼

A schematic representation of an electron-ionization, magnetic-sector mass spectrometer. Molecules are ionized by collision with high-energy electrons, causing some of the molecules to fragment. Passage of the charged fragments through a magnetic field then sorts them according to their mass.

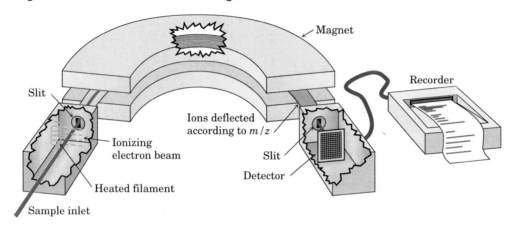

(eV), or 6700 kJ/mol (1600 kcal/mol). When a high-energy electron strikes an organic molecule, it dislodges a valence electron from the molecule, producing a *cation radical—cation* because the molecule has lost an electron and now has a positive charge; *radical* because the molecule now has an odd number of electrons.

$$RH \xrightarrow{\;\;e^-\;\;} RH^{+\cdot} \;+\; e^-$$

Organic   **Cation**
molecule   **radical**

Electron bombardment transfers so much energy to the molecules that most of the cation radicals *fragment* after formation. They fly apart into smaller pieces, some of which retain the positive charge, and some of which are neutral. The fragments then flow through a curved pipe in a strong magnetic field, which deflects them by slightly different amounts according to their mass-to-charge ratio (*m/z*). Neutral fragments are not deflected by the magnetic field and are lost on the walls of the pipe, but positively charged fragments are sorted by the mass spectrometer onto a detector, which records them as peaks at the various *m/z* ratios. Since the number of charges *z* on each ion is usually 1, the value of *m/z* for each ion is simply its mass *m*.

The **mass spectrum** of a compound is usually presented as a bar graph with masses (*m/z* values) on the *x* axis and intensity (number of ions of a given *m/z* striking the detector) on the *y* axis. The tallest peak, called the **base peak**, is arbitrarily assigned an intensity of 100%. Figure 12.2 shows mass spectra of methane and propane.

FIGURE 12.2 ▼

Mass spectra of (a) methane (CH_4; MW = 16) and (b) propane (C_3H_8; MW = 44).

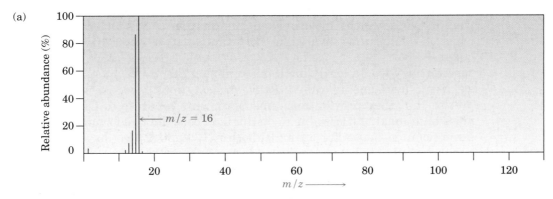

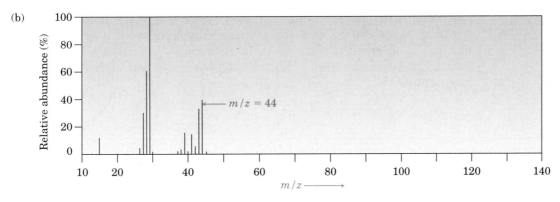

The mass spectrum of methane is relatively simple because few fragmentations are possible. As Figure 12.2a shows, the base peak has $m/z = 16$, which corresponds to the unfragmented methane cation radical, $CH_4^{+\cdot}$, called the **parent peak** or the **molecular ion (M^+)**. The mass spectrum also shows peaks at $m/z = 15$ and 14, corresponding to cleavage of the molecular ion into CH_3^+ and $CH_2^{+\cdot}$ fragments.

$$CH_4 \xrightarrow{-e^-} \left[\begin{array}{c} H \\ H:\overset{\cdot\cdot}{\underset{\cdot\cdot}{C}}\cdot H \\ H \end{array} \right]^{+\cdot}$$

$m/z = 16$
(Molecular ion, M^+)

$\longrightarrow [CH_3]^+ + H\cdot$
  $m/z = 15$

$\longrightarrow [CH_2]^{+\cdot} + 2\,H\cdot$
  $m/z = 14$

The mass spectral fragmentation patterns of larger molecules are usually complex, and the molecular ion is often not the base peak. For example, the mass spectrum of propane shown in Figure 12.2b has a molecular ion at $m/z = 44$ that is only about 30% as high as the base peak at $m/z = 29$. In addition, many other fragment ions are observed.

12.2 Interpreting Mass Spectra

What kinds of information can we get from the mass spectrum of a compound? Certainly the most obvious information is the molecular weight, which in itself can be invaluable. For example, if we were given samples of hexane (MW = 86), 1-hexene (MW = 84), and 1-hexyne (MW = 82), mass spectrometry would easily distinguish among them.

Some instruments, called *double-focusing* mass spectrometers, are so precise that they provide mass measurements accurate to 0.0001 atomic mass unit, making it possible to distinguish between two formulas with the same nominal mass. For example, both C_5H_{12} and C_4H_8O have MW = 72, but they differ slightly beyond the decimal point: C_5H_{12} has an exact mass of 72.0939 amu, whereas C_4H_8O has an exact mass of 72.0575 amu. A high-resolution instrument can easily distinguish between them.

Unfortunately, not every compound shows a molecular ion in its mass spectrum. Although M^+ is usually easy to identify if it's abundant, some compounds, such as 2,2-dimethylpropane, fragment so easily that no molecular ion is observed (Figure 12.3). In such cases, alternative "soft" ionization methods that do not use electron bombardment can sometimes prevent fragmentation.

FIGURE 12.3 ▼

Mass spectrum of 2,2-dimethylpropane (C_5H_{12}; MW = 72). No molecular ion is observed when electron-impact ionization is used. (What do you think is the structure of the M^+ peak at m/z = 57?)

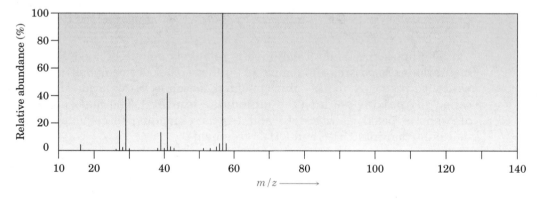

Knowing the molecular weight makes it possible to narrow greatly the choices of molecular formula. For example, if the mass spectrum of an unknown compound shows a molecular ion at m/z = 110, the molecular formula is likely to be C_8H_{14}, $C_7H_{10}O$, $C_6H_6O_2$, or $C_6H_{10}N_2$. There are always a number of molecular formulas possible for all but the lowest molecular weights, and computer programs can easily generate a list of choices.

A further point about mass spectrometry is noticeable in the mass spectra of methane and propane in Figure 12.2. Perhaps surprisingly, the peaks for the molecular ions are not at the highest m/z values in the two spectra. There is also a small peak in each spectrum at M + 1 because of the presence in the samples of small amounts of isotopically substituted molecules. Although ^{12}C is the most abundant carbon isotope, a small amount (1.10% natural abundance) of ^{13}C is also present. Thus, a certain percentage of the molecules analyzed in the mass spectrometer are likely to contain a ^{13}C atom, giving rise to the observed M + 1 peak. In addition, a small amount of ^2H (deuterium; 0.015% natural abundance) is present, making a further contribution to the M + 1 peak.

Practice Problem 12.1 List the possible formulas of molecules with $M^+ = 100$. Assume that C, H, and O may be present.

Strategy A good approach to this kind of problem is to begin by calculating the possible hydrocarbon formulas. First divide the molecular weight by 12 to find the maximum number of carbons possible. Each carbon is equal in mass to 12 hydrogens, so the next step is to replace 1 C by 12 H, giving another possible formula.

Oxygen-containing formulas can be calculated by realizing that one oxygen is equal in mass to CH_4.

Solution Dividing M^+ by 12 gives 100/12 = 8 (remainder 4), so a possible hydrocarbon formula is C_8H_4. Replacing 1 C by 12 H gives the second possible hydrocarbon formula C_7H_{16}.

Starting with the hydrocarbon formula C_8H_4 and replacing CH_4 by O gives C_7O as a possible (but unlikely) formula. Doing the same with C_7H_{16} gives $C_6H_{12}O$. Again replacing CH_4 by O gives $C_5H_8O_2$, and repeating the process a third time gives $C_4H_4O_3$. Thus, there are five likely formulas for a substance with MW = 100. A double-focusing instrument could distinguish among the five.

Problem 12.1 Write as many molecular formulas as you can for compounds that have the following molecular ions in their mass spectra. Assume that all the compounds contain C and H, and that O may or may not be present.
(a) $M^+ = 86$ (b) $M^+ = 128$ (c) $M^+ = 156$

Problem 12.2 Nootkatone, one of the chemicals responsible for the odor and taste of grapefruit, shows a molecular ion at $m/z = 218$ in its mass spectrum and contains C, H, and O. Suggest several possible molecular formulas for nootkatone.

12.3 Interpreting Mass-Spectral Fragmentation Patterns

Mass spectrometry would be useful even if molecular weight and formula were the only information that could be obtained. In fact, though, we can get much more. For example, the mass spectrum of a compound serves as a kind of "molecular fingerprint." Each organic molecule fragments in a unique way depending on its structure, and the likelihood of two compounds having identical mass spectra is small. Thus, it's sometimes possible to identify an unknown by computer-based matching of its mass spectrum to one of the more than 220,000 mass spectra recorded in a computerized data base called the *Registry of Mass Spectral Data*.

It's also possible to derive structural information about a molecule by interpreting the observed fragmentation pattern. Fragmentation occurs when the high-energy cation radical flies apart by spontaneous cleavage of a chemical bond. One of the two fragments retains the positive charge and is a carbocation, while the other fragment is a neutral radical.

Not surprisingly, the positive charge often remains with the fragment that is best able to stabilize it. In other words, a relatively stable carbocation is often formed during fragmentation. For example, 2,2-dimethylpropane tends to fragment in such a way that the positive charge remains with the *tert*-butyl group. 2,2-Dimethylpropane therefore has a base peak at $m/z = 57$, corresponding to $C_4H_9^+$ (Figure 12.3).

$$\left[\begin{array}{c} CH_3 \\ | \\ H_3C-C-CH_3 \\ | \\ CH_3 \end{array} \right]^{+\cdot} \longrightarrow \begin{array}{c} CH_3 \\ | \\ H_3C-C^+ \\ | \\ CH_3 \end{array} + \cdot CH_3$$

$$\textbf{m/z = 57}$$

Because mass-spectral fragmentation patterns are usually complex, it's often difficult to assign definite structures to fragment ions. Most hydrocarbons fragment in many ways, as the mass spectrum of hexane shown in Figure 12.4 demonstrates. The hexane spectrum shows a moderately abundant molecular ion at $m/z = 86$ and fragment ions at $m/z = 71, 57, 43$, and 29. Since all the carbon–carbon bonds of hexane are electronically similar, all break to a similar extent, giving rise to the observed ions.

FIGURE 12.4 ▼

Mass spectrum of hexane (C_6H_{14}; MW = 86). The base peak is at $m/z = 57$, and numerous other ions are present.

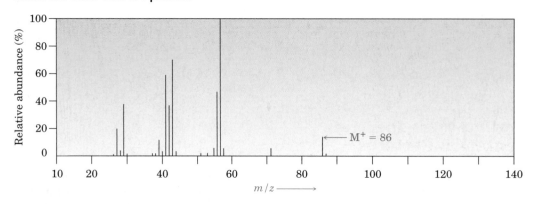

Figure 12.5 shows how the hexane fragments might arise. The loss of a methyl radical from the hexane cation radical ($M^+ = 86$) gives rise to a fragment of mass 71; the loss of an ethyl radical accounts for a fragment of mass 57; the loss of a propyl radical accounts for a fragment of mass 43; and the loss of a butyl radical accounts for a fragment of mass 29. With skill and practice, chemists can learn to analyze the fragmentation patterns of unknown compounds and work backward to a structure that is compatible with the data.

FIGURE 12.5 ▼

Fragmentation of hexane
in a mass spectrometer.

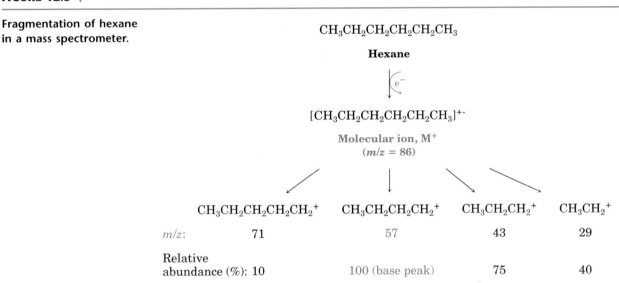

An example of how information from fragmentation patterns can be used to solve structural problems is given in Practice Problem 12.2. This example is a simple one, but the principles used are broadly applicable for organic structure determination by mass spectrometry. We'll see in later chapters that specific functional groups, such as alcohols, ketones, aldehydes, and amines, show specific kinds of mass-spectral fragmentations that can be interpreted to provide structural information.

• •

Practice Problem 12.2 Assume that you have two unlabeled samples, one of methylcyclohexane and the other of ethylcyclopentane. How could you use mass spectrometry to tell them apart? The mass spectra of both are shown in Figure 12.6.

FIGURE 12.6 ▼

Mass spectra of unlabeled samples A and B for Practice Problem 12.2. (Sample A is ethylcyclopentane; sample B is methylcyclohexane.)

(a)

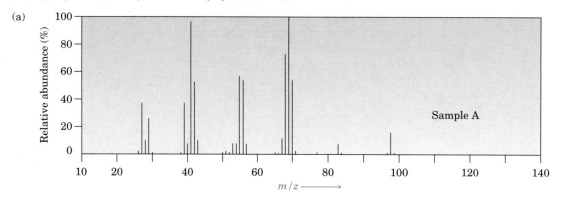

Sample A

(b)

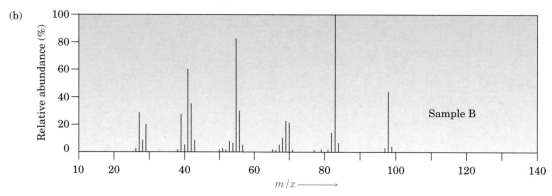

Sample B

Strategy   Look at the two possible structures and decide on how they differ. Then think about how any of these differences might give rise to differences in mass spectra. Methylcyclohexane, for instance, has a –CH_3 group, and ethylcyclopentane has a –CH_2CH_3 group, which should affect the fragmentation patterns.

Solution   The mass spectra of both samples show molecular ions at $M^+ = 98$, corresponding to C_7H_{14}, but the two spectra differ considerably in their fragmentation patterns. Sample B shows a base peak at $m/z = 83$, corresponding to the loss of a CH_3 group (15 mass units) from the molecular ion, but sample A has only a small peak at $m/z = 83$. Conversely, A has its base peak at $m/z = 69$, corresponding to the loss of a CH_2CH_3 group (29 mass units), but B has a rather small peak at $m/z = 69$. We can therefore be reasonably certain that B is methylcyclohexane and A is ethylcyclopentane.

• •

Problem 12.3   Two mass spectra are shown in Figure 12.7. One spectrum corresponds to 2-methyl-2-pentene; the other, to 2-hexene. Which is which? Explain.

FIGURE 12.7 ▼

Mass spectra for Problem 12.3.

(a)

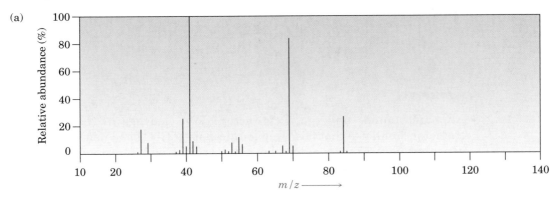

(b)

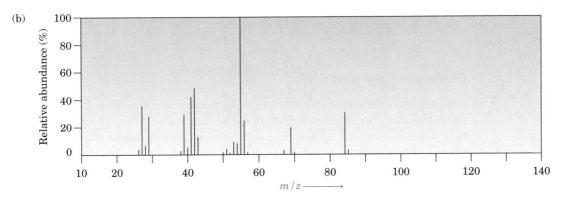

12.4 Mass-Spectral Behavior of Some Common Functional Groups

As each functional group is discussed in future chapters, any mass-spectral fragmentation characteristic of that group will be described. For the present, though, we'll point out some distinguishing features of several common functional groups.

Alcohols

Alcohols undergo fragmentation in the mass spectrometer by two pathways: *alpha (α) cleavage* and *dehydration*. In the α-cleavage pathway, a C–C bond nearest the hydroxyl group is broken, yielding a neutral radical plus a charged oxygen-containing fragment:

$$\left[RCH_2 \overset{|}{\underset{|}{C}} - OH \right]^{+\cdot} \xrightarrow[\text{cleavage}]{\text{Alpha}} \left[\overset{\diagdown}{\underset{\diagup}{C}} - OH \right]^{+} + RCH_2 \cdot$$

In the dehydration pathway, water is eliminated, yielding an alkene radical cation with a mass 18 units less than M^+:

Amines

Aliphatic amines undergo a characteristic α cleavage in the mass spectrometer, similar to that observed for alcohols. A C–C bond nearest the nitrogen atom is broken, yielding an alkyl radical and a nitrogen-containing cation:

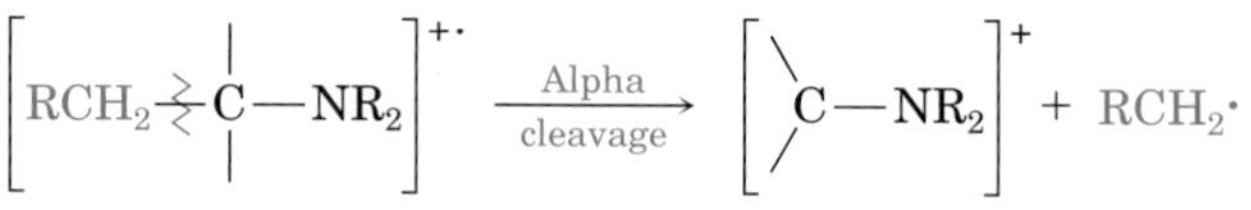

Carbonyl Compounds

Ketones and aldehydes that have a hydrogen on a carbon 3 atoms away from the carbonyl group undergo a characteristic mass-spectral cleavage called the *McLafferty rearrangement*. The hydrogen atom is transferred to the carbonyl oxygen, a C–C bond is broken, and a neutral alkene fragment is produced. The charge remains with the oxygen-containing fragment.

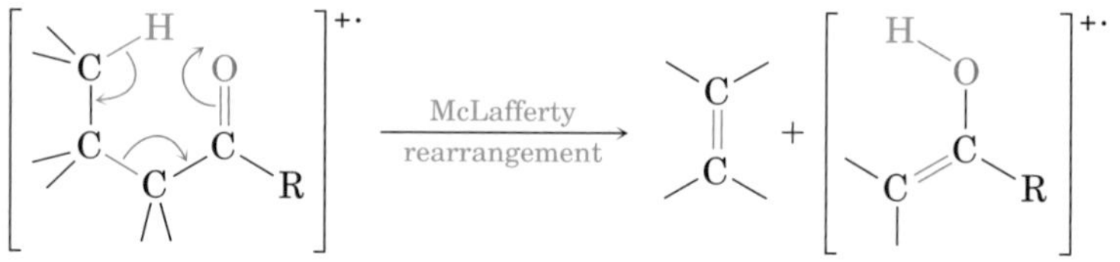

In addition, ketones and aldehydes also undergo α cleavage of the bond between the carbonyl group and the neighboring carbon. Alpha cleavage yields a neutral radical and an oxygen-containing cation.

· ·

Practice Problem 12.3   The mass spectrum of 2-methyl-3-pentanol is shown in Figure 12.8. What fragments can you identify?

Strategy   Calculate the mass of the molecular ion, and identify the functional groups in the molecule. Then write the fragmentation processes you might expect, and compare the masses of the resultant fragments with those peaks present in the spectrum.

FIGURE 12.8 ▼

Mass spectrum of 2-methyl-3-pentanol, Practice Problem 12.3.

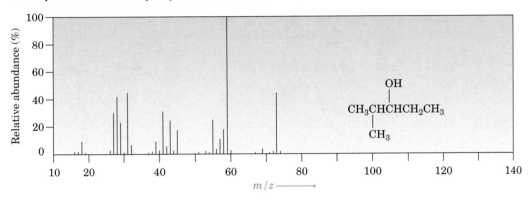

Solution   2-Methyl-3-pentanol, an open-chain alcohol, has $M^+ = 102$ and might be expected to fragment by α cleavage and by dehydration. These processes would lead to fragment ions of $m/z = 84, 73$, and 59. Of the three expected fragments, dehydration is not observed (no $m/z = 84$ peak), but both possible α cleavages do take place ($m/z = 73, 59$).

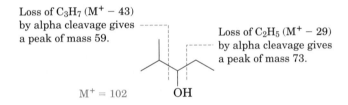

Loss of C_3H_7 ($M^+ - 43$) by alpha cleavage gives a peak of mass 59.

Loss of C_2H_5 ($M^+ - 29$) by alpha cleavage gives a peak of mass 73.

$M^+ = 102$ OH

Problem 12.4   What are the masses of the charged fragments produced in the following cleavage pathways?
(a) Alpha cleavage of 2-pentanone ($CH_3COCH_2CH_2CH_3$)
(b) Dehydration of cyclohexanol (hydroxycyclohexane)
(c) McLafferty rearrangement of 4-methyl-2-pentanone [$CH_3COCH_2CH(CH_3)_2$]
(d) Alpha cleavage of triethylamine [$(CH_3CH_2)_3N$]

12.5 Spectroscopy and the Electromagnetic Spectrum

Infrared, ultraviolet, and nuclear magnetic resonance spectroscopies differ from mass spectrometry in that they involve the interaction of molecules with electromagnetic energy rather than with a high-energy electron beam. Before beginning a study of these techniques, we need to look into the nature of radiant energy and the electromagnetic spectrum.

Visible light, X rays, microwaves, radio waves, and so forth are all different kinds of **electromagnetic radiation**. Collectively, they make up the **electromagnetic spectrum**, shown in Figure 12.9. As indicated, the electromagnetic spectrum is arbitrarily divided into various regions, with the familiar visible region accounting for only a small portion of the overall spectrum, from 3.8×10^{-7} m to 7.8×10^{-7} m in wavelength. The visible region is flanked by the infrared and ultraviolet regions.

FIGURE 12.9 ▼

The electromagnetic spectrum consists of a continuous range of wavelengths and frequencies, from radio waves at the low-frequency end to gamma (γ) rays at the high-frequency end. The familiar visible region accounts for only a small portion near the middle of the spectrum.

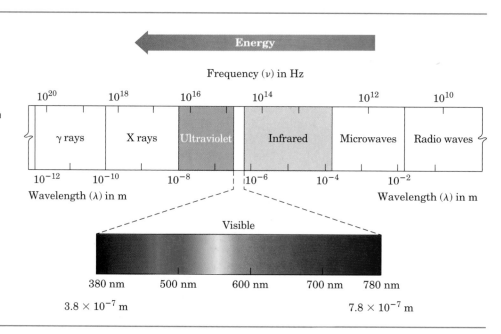

Electromagnetic radiation has dual behavior. In some respects, it has the properties of a particle (called a *photon*), yet in other respects it behaves as an energy wave traveling at the speed of light. Like all waves, electromagnetic radiation is characterized by a *wavelength*, a *frequency*, and an *amplitude* (Figure 12.10). The **wavelength, λ** (Greek lambda), is simply the distance from one wave maximum to the next. The **frequency, ν** (Greek nu), is the number of wave maxima that pass by a fixed point per unit time, usually given in reciprocal seconds (s^{-1}), or **hertz, Hz** (1 Hz $= 1$ s^{-1}). The **amplitude** is the height of a wave, measured from the midpoint to the maximum. The intensity of radiant energy, whether a feeble beam or a blinding glare, is proportional to the square of the wave's amplitude.

Multiplying the wavelength of a wave in meters (m) by its frequency in reciprocal seconds (s^{-1}) gives the speed of the wave in meters per second (m/s). The rate of travel of all electromagnetic radiation in a vacuum is a constant value, commonly called the "speed of light" and abbreviated c. It

FIGURE 12.10 ▼

Electromagnetic waves are characterized by a wavelength, a frequency, and an amplitude. (a) Wavelength (λ) is the distance between two successive wave maxima. Amplitude is the height of the wave measured from the center. (b) What we perceive as different kinds of electromagnetic radiation are simply waves with different wavelengths and frequencies.

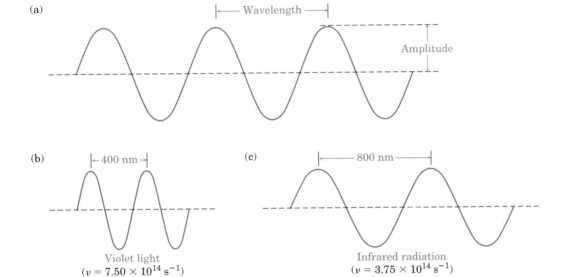

(a) Wavelength ⟶ / Amplitude

(b) 400 nm
Violet light
($\nu = 7.50 \times 10^{14} \text{ s}^{-1}$)

(c) 800 nm
Infrared radiation
($\nu = 3.75 \times 10^{14} \text{ s}^{-1}$)

is one of the most accurately known of all physical constants, with a numerical value of 2.997 924 58 $\times$ 10^8 m/s, usually rounded off to 3.00 $\times$ 10^8 m/s.

$$\text{Wavelength} \times \text{Frequency} = \text{Speed}$$

$$\lambda \text{ (m)} \times \nu \text{ (s}^{-1}) = c \text{ (m/s)}$$

which can be rewritten as:

$$\lambda = \frac{c}{\nu} \quad \text{or} \quad \nu = \frac{c}{\lambda}$$

Electromagnetic energy is transmitted only in discrete amounts, called *quanta*. The amount of energy ε corresponding to 1 quantum of energy (or 1 photon) of a given frequency ν is expressed by the equation

$$\varepsilon = h\nu = \frac{hc}{\lambda}$$

where ε = Energy of 1 photon (1 quantum)
h = Planck's constant (6.62 $\times$ 10^{-34} J·s = 1.58 $\times$ 10^{-34} cal·s)
ν = Frequency (s^{-1})
λ = Wavelength (m)
c = Speed of light (3.00 $\times$ 10^8 m/s)

This equation says that the energy of a given photon varies *directly* with its frequency ν but *inversely* with its wavelength λ. High frequencies and short wavelengths correspond to high-energy radiation such as gamma rays; low frequencies and long wavelengths correspond to low-energy radiation such as radio waves. If we multiply ε by Avogadro's number N_A, we arrive at the same equation expressed in units familiar to organic chemists:

$$ E = \frac{N_A hc}{\lambda} = \frac{1.20 \times 10^{-4}\ \text{kJ/mol}}{\lambda\ (\text{m})} \quad \left[\text{or} \quad \frac{2.86 \times 10^{-5}\ \text{kcal/mol}}{\lambda\ (\text{m})} \right] $$

where E represents the energy of Avogadro's number (a "mole") of photons of wavelength λ.

When an organic compound is struck by a beam of electromagnetic radiation, it absorbs energy of certain wavelengths but transmits energy of other wavelengths. If we irradiate the sample with energy of many different wavelengths and determine which are absorbed and which are transmitted, we can determine the **absorption spectrum** of the compound. The results are displayed on a graph that plots wavelength versus the amount of radiation transmitted.

An example of an absorption spectrum—that of ethyl alcohol exposed to infrared radiation—is shown in Figure 12.11. The horizontal axis records the wavelength, and the vertical axis records the intensity of the various energy absorptions in percent transmittance. The baseline corresponding to 0% absorption (or 100% transmittance) runs along the top of the chart, and a downward spike means that energy absorption has occurred at that wavelength.

FIGURE 12.11 ▼

An infrared absorption spectrum of ethyl alcohol, CH_3CH_2OH. A transmittance of 100% means that all the energy is passing through the sample, whereas a lower transmittance means that some energy is being absorbed. Thus, each downward spike corresponds to an energy absorption.

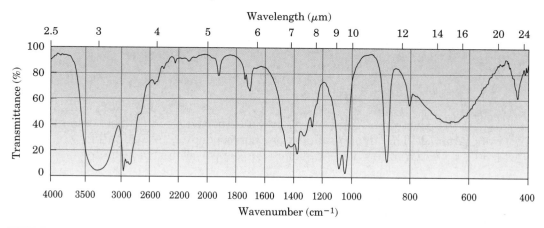

The energy that a molecule gains when it absorbs radiation must be distributed over the molecule in some way. For example, absorption of radiation might increase a molecule's energy by causing bonds to stretch or bend more vigorously. Alternatively, absorption of radiation might cause an electron to jump from a lower-energy orbital to a higher one. Different radiation frequencies affect molecules in different ways, but each can provide structural information if the results are interpreted properly.

There are many kinds of spectroscopies, which differ according to the region of the electromagnetic spectrum that is used. We'll look closely at two types—infrared spectroscopy and nuclear magnetic resonance spectroscopy—and have a brief introduction to a third—ultraviolet spectroscopy. Let's begin by seeing what happens when an organic sample absorbs infrared energy.

Practice Problem 12.4   Which is higher in energy, FM radio waves with a frequency of 1.015×10^8 Hz (101.5 MHz) or visible green light with a frequency of 5×10^{14} Hz?

Strategy   Remember the equations $\varepsilon = h\nu$ and $\varepsilon = hc/\lambda$, which say that energy increases as frequency increases and as wavelength decreases.

Solution   Since visible light has a higher frequency than a radio wave, it is higher in energy.

Problem 12.5   Which has higher energy, infrared radiation with $\lambda = 1.0 \times 10^{-6}$ m or an X ray with $\lambda = 3.0 \times 10^{-9}$ m?

Problem 12.6   Which has the higher energy, radiation with $\nu = 4.0 \times 10^9$ Hz or radiation with $\lambda = 9.0 \times 10^{-6}$ m?

Problem 12.7   It's useful to develop a feeling for the amounts of energy that correspond to different parts of the electromagnetic spectrum. Use the relationships

$$E = \frac{1.20 \times 10^{-4} \text{ kJ/mol}}{\lambda \text{ (m)}} \quad \text{and} \quad \nu = \frac{c}{\lambda}$$

to calculate the energies of each of the following kinds of radiation:
(a) A gamma ray with $\lambda = 5.0 \times 10^{-11}$ m
(b) An X ray with $\lambda = 3.0 \times 10^{-9}$ m
(c) Ultraviolet light with $\nu = 6.0 \times 10^{15}$ Hz
(d) Visible light with $\nu = 7.0 \times 10^{14}$ Hz
(e) Infrared radiation with $\lambda = 2.0 \times 10^{-5}$ m
(f) Microwave radiation with $\nu = 1.0 \times 10^{11}$ Hz

12.6 Infrared Spectroscopy of Organic Molecules

The **infrared (IR)** region of the electromagnetic spectrum covers the range from just above the visible (7.8×10^{-7} m) to approximately 10^{-4} m, but only the midportion from 2.5×10^{-6} m to 2.5×10^{-5} m is used by organic chemists (Figure 12.12). Wavelengths within the IR region are usually given in micrometers (1 μm = 10^{-6} m), and frequencies are expressed in **wavenumbers** ($\tilde{\nu}$) rather than in hertz. The wavenumber, expressed in units of cm^{-1}, is simply the reciprocal of the wavelength in centimeters:

$$\text{Wavenumber } \tilde{\nu} \; = \frac{1}{\lambda \text{ (cm)}}$$

Thus, the useful IR region is from 4000 cm^{-1} to 400 cm^{-1}. Using the equation $E = (1.20 \times 10^{-4} \text{ kJ/mol})/\lambda$, we can calculate that the energy levels of IR radiation range from 48.0 kJ/mol to 4.80 kJ/mol (11.5–1.15 kcal/mol).

FIGURE 12.12 ▼

The infrared region of the electromagnetic spectrum.

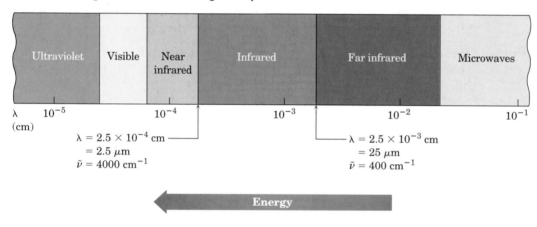

Why does an organic molecule absorb some wavelengths of IR radiation but not others? All molecules have a certain amount of energy distributed throughout their structure, causing bonds to stretch and contract, atoms to wag back and forth, and other molecular vibrations to occur. Some of the kinds of allowed vibrations are shown below:

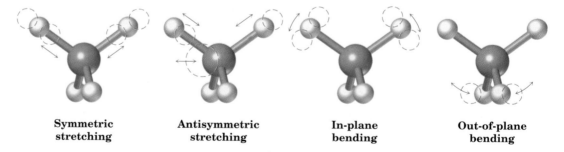

| Symmetric stretching | Antisymmetric stretching | In-plane bending | Out-of-plane bending |

The amount of energy a molecule contains is not continuously variable but is *quantized*. That is, a molecule can stretch or bend only at specific frequencies. Take bond stretching, for example. Although we usually speak of bond lengths as if they were fixed, the numbers given are actually averages. In reality, bonds are constantly changing in length. Thus, a typical C–H bond with an average bond length of 110 pm is actually vibrating at a specific frequency, alternately stretching and contracting as if there were a spring connecting the two atoms. When the molecule is irradiated with electromagnetic radiation, *energy is absorbed when the frequency of the radiation matches the frequency of the vibrational motion.*

When a molecule absorbs IR radiation, the molecular vibration with a frequency matching that of the radiation increases in amplitude. In other words, the "spring" connecting the two atoms stretches and compresses a bit further. Since each frequency absorbed by a molecule corresponds to a specific molecular motion, we can see what kinds of motions a molecule has by measuring its IR spectrum. By then interpreting those motions, we can find out what kinds of bonds (functional groups) are present in the molecule.

IR spectrum⟶ What molecular motions?⟶ What functional groups?

• •

Problem 12.8 Because IR absorptions can be expressed either in micrometers or in wavenumbers, it's useful to be able to interconvert between units. Do the following conversions:
(a) 3.10 μm to cm^{-1} (b) 5.85 μm to cm^{-1} (c) 2250 cm^{-1} to μm
(d) 970 cm^{-1} to μm

• •

12.7 Interpreting Infrared Spectra

The full interpretation of an IR spectrum is difficult because most organic molecules are so large that they have dozens of different bond stretching and bending motions. Thus, an IR spectrum contains dozens of absorption bands. In one sense, this complexity is valuable because an IR spectrum serves as a unique fingerprint of a specific compound. In fact, the complex region of the IR spectrum from 1500 cm^{-1} to around 400 cm^{-1} is called the **fingerprint region**. If two compounds have identical IR spectra, they are almost certainly identical.

Fortunately, we don't need to interpret an IR spectrum fully to get useful structural information. *Most functional groups have characteristic IR absorption bands that don't change from one compound to another.* The C=O absorption of a ketone is almost always in the range 1680–1750 cm^{-1}; the O–H absorption of an alcohol is almost always in the range 3400–3650 cm^{-1}; the C=C absorption of an alkene is almost always in the range 1640–1680 cm^{-1}; and so forth. By learning where characteristic functional-group absorptions occur, it's possible to get structural information from IR spectra. Table 12.1 lists the characteristic IR bands of some common functional groups.

Look at the IR spectra of hexane, 1-hexene, and 1-hexyne in Figure 12.13 (p. 459) to see an example of how infrared spectroscopy can be used. Although all three IR spectra contain many peaks, there are characteristic absorptions of the C=C and C≡C functional groups that allow the three

TABLE 12.1 Characteristic IR Absorptions of Some Functional Groups

Functional group class	Band position (cm^{-1})	Intensity of absorption
Alkanes, alkyl groups		
C—H	2850–2960	Medium to strong
Alkenes		
=C—H	3020–3100	Medium
C=C	1640–1680	Medium
Alkynes		
≡C—H	3300	Strong
—C≡C—	2100–2260	Medium
Alkyl halides		
C—Cl	600–800	Strong
C—Br	500–600	Strong
C—I	500	Strong
Alcohols		
O—H	3400–3650	Strong, broad
C—O	1050–1150	Strong
Aromatics		
C—H	3030	Weak
(ring)	1660–2000	Weak
	1450–1600	Medium
Amines		
N—H	3300–3500	Medium
C—N	1030–1230	Medium
Carbonyl compounds[a]		
C=O	1670–1780	Strong
Carboxylic acids		
O—H	2500–3100	Strong, very broad
Nitriles		
C≡N	2210–2260	Medium
Nitro compounds		
NO$_2$	1540	Strong

[a]Carboxylic acids, esters, aldehydes, and ketones.

compounds to be distinguished. Thus, 1-hexene shows a characteristic C=C absorption at 1660 cm^{-1} and a vinylic =C–H absorption at 3100 cm^{-1}, whereas 1-hexyne has a C≡C absorption at 2100 cm^{-1} and a terminal alkyne ≡C–H absorption at 3300 cm^{-1}.

FIGURE 12.13 ▼

Infrared spectra of (a) hexane, (b) 1-hexene, and (c) 1-hexyne. Spectra like these are easily obtained on milligram amounts of material in a few minutes using commercially available instruments.

hexane, 1-hexene, 1-hexyne (see vibration on CD-Rom)

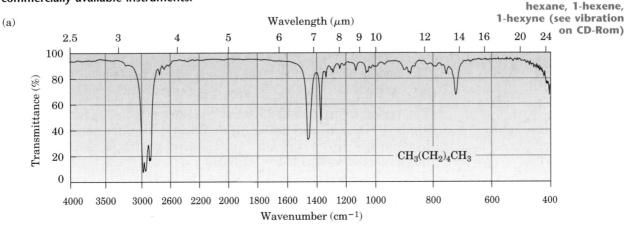

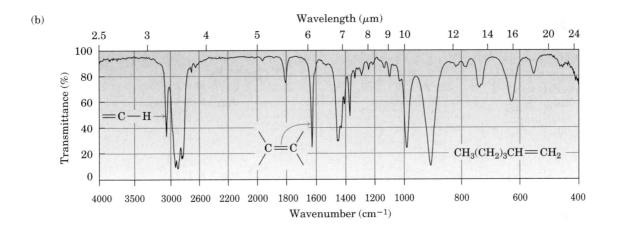

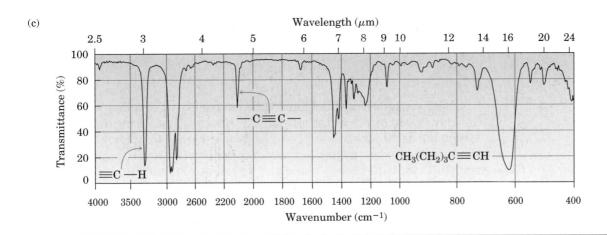

It helps in remembering the position of specific IR absorptions to divide the infrared region from 4000 cm⁻¹ to 400 cm⁻¹ into four parts, as shown in Figure 12.14:

- The region from 4000 to 2500 cm⁻¹ corresponds to absorptions caused by N–H, C–H, and O–H single-bond stretching motions. N–H and O–H bonds absorb in the 3300–3600 cm⁻¹ range; C–H bond stretching occurs near 3000 cm⁻¹.

- The region from 2500 to 2000 cm⁻¹ is where triple-bond stretching occurs. Both nitriles (RC≡N) and alkynes show absorptions here.

- The region from 2000 to 1500 cm⁻¹ is where double bonds of all kinds (C=O, C=N, and C=C) absorb. Carbonyl groups generally absorb in the range 1670–1780 cm⁻¹, and alkene stretching normally occurs in the narrow range 1640–1680 cm⁻¹.

- The region below 1500 cm⁻¹ is the fingerprint portion of the IR spectrum. A large number of absorptions due to a variety of C–C, C–O, C–N, and C–X single-bond vibrations occur here.

FIGURE 12.14 ▼

Regions in the infrared spectrum. The IR spectrum is divided into four regions: single bonds to hydrogen, triple bonds, double bonds, and fingerprint.

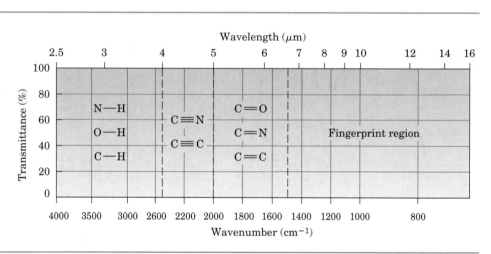

Why do different functional groups absorb where they do? The best analogy is that of two weights (atoms) connected by a spring (a bond). Short, strong bonds vibrate at a higher energy and higher frequency than do long, weak bonds, just as a short, strong spring vibrates faster than a long, weak spring. Thus, triple bonds absorb at a higher frequency than double bonds, which in turn absorb higher than single bonds. In addition, springs connecting small weights vibrate faster than springs connecting large weights. Thus, C–H, O–H, and N–H bonds vibrate at a higher frequency than bonds between heavier C, O, and N atoms.

• •

Problem 12.9 Refer to Table 12.1, and make educated guesses about what functional groups the following molecules might contain.
(a) A compound with a strong absorption at 1710 cm^{-1}
(b) A compound with a strong absorption at 1540 cm^{-1}
(c) A compound with strong absorptions at 1720 cm^{-1} and at 2500–3100 cm^{-1}

Problem 12.10 How might you use IR spectroscopy to distinguish between the following pairs of isomers?
(a) CH_3CH_2OH and CH_3OCH_3 (b) Cyclohexane and 1-hexene
(c) CH_3CH_2COOH and $HOCH_2CH_2CHO$

• •

12.8 Infrared Spectra of Hydrocarbons

Alkanes

The infrared spectrum of an alkane is fairly uninformative because no functional groups are present and all absorptions are due to C–H and C–C bonds. Alkane C–H bonds always show a strong absorption from 2850 to 2960 cm^{-1}, and saturated C–C bonds show a number of bands in the 800–1300 cm^{-1} range. Since most organic compounds contain saturated alkane-like portions, most organic compounds have these characteristic IR absorptions. The C–H and C–C bands are clearly visible in the three spectra shown in Figure 12.13.

Alkanes $-\overset{|}{\underset{|}{C}}-H$ 2850–2960 cm^{-1}

$-\overset{|}{\underset{|}{C}}-\overset{|}{\underset{|}{C}}-$ 800–1300 cm^{-1}

Alkenes

Alkenes show several characteristic stretching absorptions. Vinylic =C–H bonds absorb from 3020 to 3100 cm^{-1}, and alkene C=C bonds usually absorb near 1650 cm^{-1}, although in some cases the peaks can be rather small and difficult to see clearly. Both absorptions are visible in the 1-hexene spectrum in Figure 12.13b.

Mono- and disubstituted alkenes have characteristic =C–H out-of-plane bending absorptions in the 700–1000 cm^{-1} range, thereby allowing the substitution pattern on a double bond to be determined. Monosubstituted alkenes such as 1-hexene show strong characteristic bands at 910 and 990 cm^{-1}, and 2,2-disubstituted alkenes ($R_2C=CH_2$) have an intense band at 890 cm^{-1}.

Alkenes	$=C-H$	3020–3100 cm^{-1}
	$\underset{/}{\overset{\backslash}{C}} = \underset{\backslash}{\overset{/}{C}}$	1640–1680 cm^{-1}
	$RCH=CH_2$	910 and 990 cm^{-1}
	$R_2C=CH_2$	890 cm^{-1}

Alkynes

Alkynes show a C≡C stretching absorption at 2100–2260 cm^{-1}, an absorption that is much more intense for terminal alkynes than for internal alkynes. In fact, symmetrically substituted triple bonds like that in 3-hexyne show no absorption at all, for reasons we won't go into. Terminal alkynes such as 1-hexyne also have a characteristic ≡C–H stretch at 3300 cm^{-1} (Figure 12.13c) This band is diagnostic for terminal alkynes because it is fairly intense and quite sharp.

Alkynes	$-C≡C-$	2100–2260 cm^{-1}
	$≡C-H$	3300 cm^{-1}

One other important point about IR spectroscopy: *It's also possible to get structural information from an IR spectrum by noticing which absorptions are not present.* If the spectrum of a compound has no absorptions at 3300 and 2150 cm^{-1}, the compound is not a terminal alkyne; if the spectrum has no absorption near 3400 cm^{-1}, the compound is not an alcohol; and so on.

●●

Problem 12.11   The infrared spectrum of phenylacetylene is shown in Figure 12.15. What absorption bands can you identify?

FIGURE 12.15 ▼

The IR spectrum of phenylacetylene, Problem 12.11.

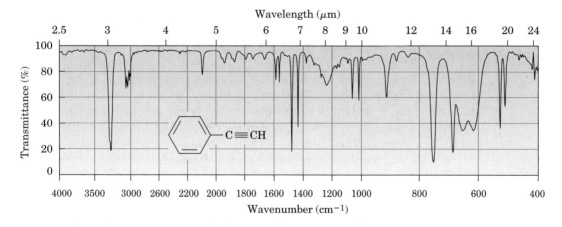

●●●●●●●●●●●●●●●●●●●●●●●●●●●●●●●●●●●●●●●

12.9 Infrared Spectra of Some Common Functional Groups

As each functional group is discussed in future chapters, the spectroscopic behavior of that group will be described. For the present, though, we'll simply point out some distinguishing features of the more common functional groups.

Alcohols

The O–H functional group of alcohols is easy to spot in the IR. Alcohols have a characteristic band in the range 3400–3650 cm^{-1} that is usually broad and intense. If present, it's hard to miss this band or to confuse it with anything else.

> Alcohols   —O—H   3400–3650 cm^{-1} (broad, intense)

Amines

The N–H functional group of amines is also easy to spot in the IR, with a characteristic absorption in the 3300–3500 cm^{-1} range. Although alcohols absorb in the same range, an N–H absorption is much sharper and less intense than an O–H band.

> Amines   —N—H   3300–3500 cm^{-1} (sharp, medium intensity)

Aromatic Compounds

Aromatic compounds such as benzene have a weak C–H stretching absorption at 3030 cm^{-1}, a series of weak absorptions in the 1660–2000 cm^{-1} range, and a second series of medium-intensity absorptions in the 1450–1600 cm^{-1} region. These latter absorptions are due to complex molecular motions of the entire ring. The IR spectrum of phenylacetylene in Figure 12.15 gives an example.

> Aromatic compounds   C–H   3030 cm^{-1} (weak)
>
> Ring   1660–2000 cm^{-1} (weak)
> 1450–1600 cm^{-1} (medium)

Carbonyl Compounds

Carbonyl functional groups are the easiest to identify of all IR absorptions because of their sharp, intense peak in the range 1670–1780 cm^{-1}. Most important, the exact position of absorption within the range can often be used to identify the exact kind of carbonyl functional group—aldehyde, ketone, ester, and so forth.

Aldehydes   Saturated aldehydes absorb at 1730 cm^{-1}; aldehydes next to either a double bond or an aromatic ring absorb at 1705 cm^{-1}.

Aldehydes CH_3CH_2CH $CH_3CH=CHCH$

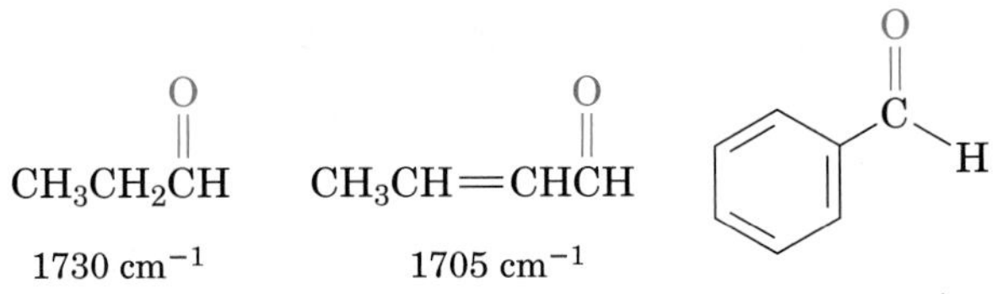

1730 cm^{-1} 1705 cm^{-1}

1705 cm^{-1}

Ketones Saturated open-chain ketones and six-membered-ring cyclic ketones absorb at 1715 cm^{-1}, five-membered-ring ketones absorb at 1750 cm^{-1}, and ketones next to a double bond or an aromatic ring absorb at 1690 cm^{-1}.

Ketones CH_3CCH_3 $CH_3CH=CHCCH_3$

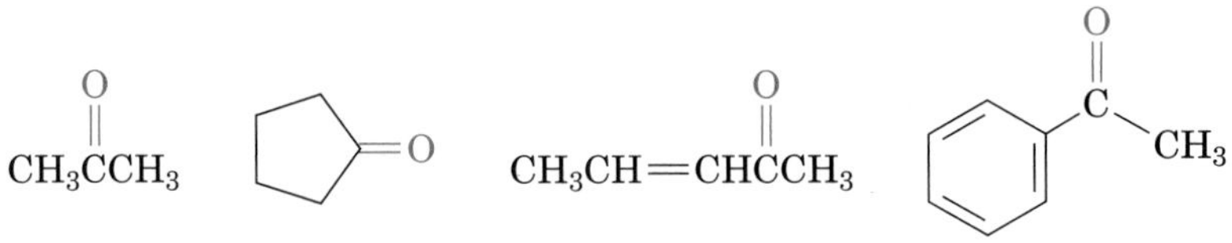

1715 cm^{-1} 1750 cm^{-1} 1690 cm^{-1}

1690 cm^{-1}

Esters Saturated esters absorb at 1735 cm^{-1}; esters next to either an aromatic ring or a double bond absorb at 1715 cm^{-1}.

Esters CH_3COCH_3 $CH_3CH=CHCOCH_3$

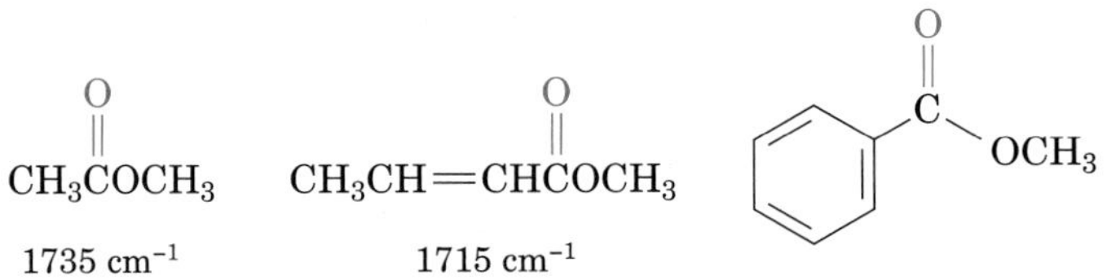

1735 cm^{-1} 1715 cm^{-1}

1715 cm^{-1}

Practice Problem 12.5 Where might the following compounds have IR absorptions?

(a) 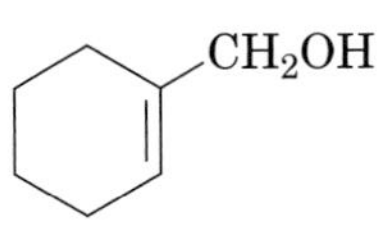 [cyclohexene with] CH_2OH

(b) 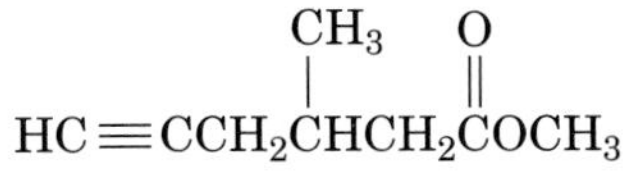 $HC≡CCH_2CHCH_2COCH_3$ with CH_3 substituent

Strategy Identify all the functional groups in each molecule, and then check Table 12.1 to see where those groups absorb.

Solution (a) *Absorptions*: 3400–3650 cm^{-1} (O–H), 3020–3100 cm^{-1} (=C–H), 1640–1680 cm^{-1} (C=C). This molecule has an alcohol O–H group and an alkene double bond.

(b) *Absorptions*: 3300 cm^{-1} (≡C–H), 2100–2260 cm^{-1} (C≡C), 1735 cm^{-1} (C=O). This molecule has a terminal alkyne triple bond and a saturated ester carbonyl group.

• •

Practice Problem 12.6   The IR spectrum of an unknown compound is shown in Figure 12.16. What functional groups does the compound contain?

FIGURE 12.16 ▼

The infrared spectrum for Practice Problem 12.6.

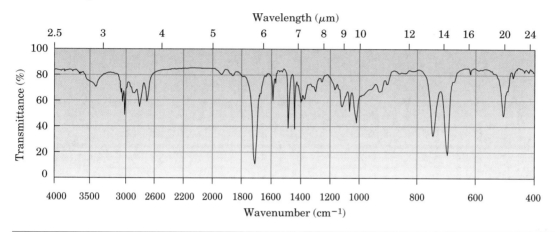

Strategy   All infrared spectra have many absorptions, but those useful for identifying specific functional groups are usually found in the region from 1500 cm^{-1} to 3300 cm^{-1}. Pay particular attention to the carbonyl region (1670–1780 cm^{-1}), the aromatic region (1660–2000 cm^{-1}), the triple-bond region (2000–2500 cm^{-1}), and the C–H region (2500–3500 cm^{-1}).

Solution   The spectrum shows an intense absorption at 1725 cm^{-1} due to a carbonyl group (perhaps an aldehyde, –CHO), a series of weak absorptions from 1800–2000 cm^{-1} characteristic of aromatic compounds, and a C–H absorption near 3030 cm^{-1}, also characteristic of aromatic compounds. In fact, the spectrum is that of phenylacetaldehyde.

$$\text{CH}_2\text{CH}=\text{O}$$

Phenylacetaldehyde

• •

Problem 12.12   Where might the following compounds have IR absorptions?

(a) COCH_3 (with O double bond)   (b) $\text{HC}\equiv\text{CCH}_2\text{CH}_2\text{CH}=\text{O}$ (c) CO_2H and CH_2OH

Problem 12.13 Where might the following compound have IR absorptions? (Red = O, blue = N.)

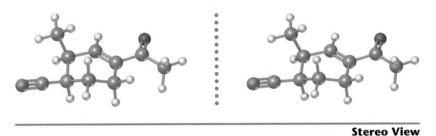

Stereo View

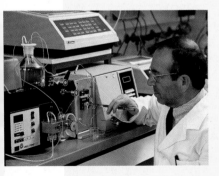

CHEMISTRY @ WORK

Chromatography: Purifying Organic Compounds

Every time a new organic substance is isolated from a plant or animal, and every time a reaction is run, the target compound must be purified by separating it from all solvents and contaminants. Purification was an enormously time-consuming, hit-or-miss proposition in the nineteenth and early twentieth centuries, but the development of powerful instruments in the last few decades now simplifies the problem greatly.

Most organic purification is done by *chromatography* (literally, "color writing"), a separation technique that dates from the work of the Russian chemist Mikhail Tswett in 1903. Tswett accomplished the separation of the pigments in green leaves by dissolving the leaf extract in an organic solvent and allowing the solution to run down through a vertical glass tube packed with chalk powder. Different pigments passed down the column at different rates, leaving a series of colored bands on the white chalk column.

There are a variety of chromatographic techniques in common use, all of which work on a similar principle: The mixture to be separated is dissolved in a solvent, called the *mobile phase*, and passed over an adsorbent material, called the *stationary phase*. Because different compounds adsorb to the stationary phase to different extents, they migrate along the phase at different rates and are separated as they emerge (*elute*) from the end of the chromatography column.

High-pressure liquid chromatography (HPLC) is used to separate and purify the products of laboratory reactions.

(continued) ▶

Liquid chromatography, or *column chromatography*, is perhaps the most often used chromatographic method. As in Tswett's original experiments, a mixture of organic compounds is dissolved in a suitable solvent and adsorbed onto a stationary phase such as alumina (Al_2O_3) or silica gel (hydrated SiO_2) packed into a glass column. More solvent is then passed down the column, and different compounds are eluted at different times.

The time at which a compound is eluted is strongly influenced by its polarity. Molecules with polar functional groups are generally adsorbed more strongly and therefore migrate through the stationary phase more slowly than nonpolar molecules. A mixture of an alcohol and an alkene, for example, can be easily separated by liquid chromatography because the nonpolar alkene passes through the column much faster than the more polar alcohol.

High-performance liquid chromatography (HPLC) is a variant of the simple column technique, based on the discovery that chromatographic separations are vastly improved if the stationary phase is made up of very small, uniformly sized spherical particles. Small particle size ensures a large surface area for better adsorption, and a uniform spherical shape allows a tight, uniform packing. In practice, specially prepared and coated silica microspheres of 10–25 μm size are often used. Only 15 g of these microspheres have a surface area the size of a football field!

High-pressure pumps are required to force solvent through a tightly packed HPLC column, and electronic detectors are used for monitoring the appearance of material eluting from the column. Figure 12.17 shows the results of HPLC analysis of a mixture of 14 common pesticides, using coated silica microspheres as the stationary phase and acetonitrile/water as the mobile phase.

FIGURE 12.17 ▼

The HPLC analysis of a mixture of 14 agricultural pesticides. The structures of the pesticides can be found in the *Merck Index*.

1. Oxamyl
2. Methomyl
3. ANTU
4. Aldicarb
5. Monuron
6. Carbofuran
7. Propoxur
8. Fluometuron
9. Diuron
10. Warfarin
11. Siduron
12. Methiocarb
13. Linuron
14. Mexacarbate

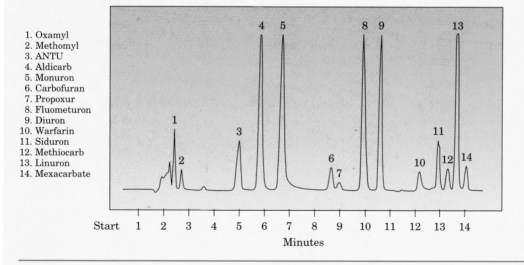

Summary and Key Words

The structure of an organic molecule is usually determined using spectroscopic methods such as mass spectrometry and infrared spectroscopy. **Mass spectrometry (MS)** tells the molecular weight and formula of a molecule; **infrared (IR) spectroscopy** identifies the functional groups present in the molecule.

In mass spectrometry, molecules are first ionized by collision with a high-energy electron beam. The ions then fragment into smaller pieces, which are magnetically sorted according to their mass-to-charge ratio (m/z). The ionized sample molecule is called the **molecular ion, M^+,** and measurement of its mass gives the molecular weight of the sample. Structural clues about unknown samples can be obtained by interpreting the *fragmentation pattern* of the molecular ion. Mass-spectral fragmentations are usually complex, however, and interpretation is often difficult.

Infrared spectroscopy involves the interaction of a molecule with **electromagnetic radiation**. When an organic molecule is irradiated with infrared energy, certain **frequencies** are absorbed by the molecule. The frequencies absorbed correspond to the amounts of energy needed to increase the amplitude of specific molecular vibrations such as bond stretchings and bendings. Since every functional group has a characteristic combination of bonds, every functional group has a characteristic set of infrared absorptions. For example, the terminal alkyne ≡C–H bond absorbs IR radiation of 3300 cm^{-1} frequency, and the alkene C=C bond absorbs in the range $1640–1680 \text{ cm}^{-1}$. By observing which frequencies of infrared radiation are absorbed by a molecule and which are not, it's possible to determine the functional groups a molecule contains.

Visualizing Chemistry

• •

(Problems 12.1–12.13 appear within the chapter.)

12.14   Where in the infrared spectrum would you expect each of the following molecules to absorb? (Red = O, blue = N.)

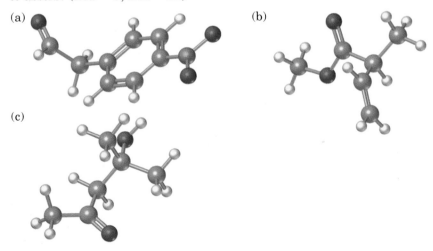

(a)

(b)

(c)

12.15 Show the structures of the likely fragments you would expect in the mass spectra of the following molecules (red = O, blue = N):

(a)   (b)

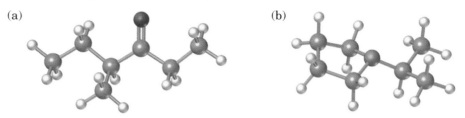

Additional Problems

12.16 Write the molecular formulas of all hydrocarbons corresponding to the following molecular ions. How many degrees of unsaturation (double bonds and/or rings) are indicated by each formula?
(a) $M^+ = 86$ (b) $M^+ = 110$ (c) $M^+ = 146$ (d) $M^+ = 190$

12.17 Draw the structure of a molecule that is consistent with the mass-spectral data in each of the following examples:
(a) A hydrocarbon with $M^+ = 132$ (b) A hydrocarbon with $M^+ = 166$
(c) A hydrocarbon with $M^+ = 84$

12.18 Write as many molecular formulas as you can for compounds that show the following molecular ions in their mass spectra. Assume that C, H, N, and O might be present.
(a) $M^+ = 74$ (b) $M^+ = 131$

12.19 Camphor, a saturated monoketone from the Asian camphor tree, is used as a moth repellent and as a constituent of embalming fluid, among other things. If camphor has $M^+ = 152$, what is a likely molecular formula? How many rings does camphor have?

12.20 The *nitrogen rule* of mass spectrometry says that a compound containing an odd number of nitrogens has an odd-numbered molecular ion. Conversely, a compound containing an even number of nitrogens has an even-numbered M^+ peak. Explain.

12.21 In light of the nitrogen rule mentioned in Problem 12.20, what is the molecular formula of pyridine, $M^+ = 79$?

12.22 Nicotine is a diamino compound that can be isolated from dried tobacco leaves. Nicotine has two rings and $M^+ = 162$ in its mass spectrum. Propose a molecular formula for nicotine, and calculate the number of double bonds. (There is no oxygen.)

12.23 Halogenated compounds are particularly easy to identify by their mass spectra because both chlorine and bromine occur naturally as mixtures of two abundant isotopes. Chlorine occurs as ^{35}Cl (75.8%) and ^{37}Cl (24.2%); bromine occurs as ^{79}Br (50.7%) and ^{81}Br (49.3%). At what masses do the molecular ions occur for the following formulas? What are the relative percentages of each molecular ion?
(a) Bromomethane, CH_3Br (b) 1-Chlorohexane, $C_6H_{13}Cl$

12.24 Molecular ions can be particularly complex for polyhalogenated compounds. Taking the natural abundance of Cl into account (see Problem 12.23), calculate the masses of the molecular ions of the following formulas. What are the relative percentages of each ion?
(a) Chloroform, $CHCl_3$ (b) Freon 12, CF_2Cl_2 (Fluorine occurs only as ^{19}F.)

12.25   By knowing the natural abundances of minor isotopes, it's possible to calculate the relative heights of M^+ and $M + 1$ peaks. If ^{13}C has a natural abundance of 1.10%, what are the relative heights of the M^+ and $M + 1$ peaks in the mass spectrum of benzene, C_6H_6?

12.26   Propose structures for compounds that fit the following data:
(a) A ketone with $M^+ = 86$ and fragments at $m/z = 71$ and $m/z = 43$
(b) An alcohol with $M^+ = 88$ and fragments at $m/z = 73$, $m/z = 70$, and $m/z = 59$

12.27   2-Methylpentane (C_6H_{14}) has the mass spectrum shown. Which peak represents M^+? Which is the base peak? Propose structures for fragment ions of $m/z = 71, 57, 43,$ and 29. Why does the base peak have the mass it does?

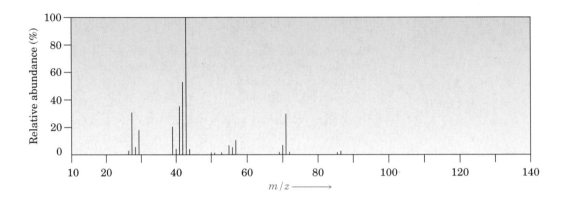

12.28   Assume that you are in a laboratory carrying out the catalytic hydrogenation of cyclohexene to cyclohexane. How could you use a mass spectrometer to determine when the reaction is finished?

12.29   Convert the following infrared absorption values from micrometers to wavenumbers:
(a) An alcohol, 2.98 μm   (b) An ester, 5.81 μm   (c) A nitrile, 4.93 μm

12.30   Convert the following infrared absorption values from wavenumbers to micrometers:
(a) A cyclopentanone, 1755 cm^{-1} (b) An amine, 3250 cm^{-1}
(c) An aldehyde, 1725 cm^{-1} (d) An acid chloride, 1780 cm^{-1}

12.31   How might you use IR spectroscopy to distinguish among the three isomers 1-butyne, 1,3-butadiene, and 2-butyne?

12.32   Would you expect two enantiomers such as (R)-2-bromobutane and (S)-2-bromobutane to have identical or different IR spectra? Explain.

12.33   Would you expect two diastereomers such as *meso*-2,3-dibromobutane and (2R,3R)-dibromobutane to have identical or different IR spectra? Explain.

12.34   Propose structures for compounds that meet the following descriptions:
(a) C_5H_8, with IR absorptions at 3300 and 2150 cm^{-1}
(b) C_4H_8O, with a strong IR absorption at 3400 cm^{-1}
(c) C_4H_8O, with a strong IR absorption at 1715 cm^{-1}
(d) C_8H_{10}, with IR absorptions at 1600 and 1500 cm^{-1}

12.35   How could you use infrared spectroscopy to distinguish between the following pairs of isomers?
(a) HC≡CCH$_2$NH$_2$ and CH$_3$CH$_2$C≡N   (b) CH$_3$COCH$_3$ and CH$_3$CH$_2$CHO

12.36 Two infrared spectra are shown. One is the spectrum of cyclohexane, and the other is the spectrum of cyclohexene. Identify them, and explain your answer.

(a)

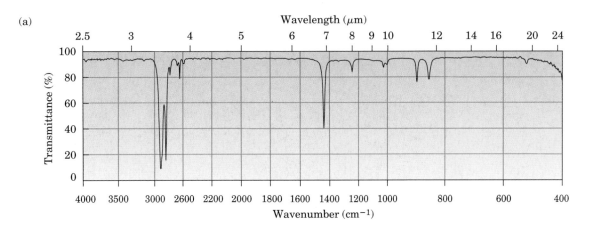

(b)

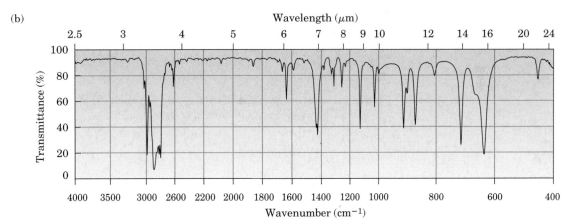

12.37 How would you use infrared spectroscopy to distinguish between the following pairs of constitutional isomers?
(a) $CH_3C\equiv CCH_3$ and $CH_3CH_2C\equiv CH$

(b) $CH_3\overset{\overset{\displaystyle O}{\|}}{C}CH=CHCH_3$ and $CH_3\overset{\overset{\displaystyle O}{\|}}{C}CH_2CH=CH_2$

(c) $H_2C=CHOCH_3$ and CH_3CH_2CHO

12.38 The hormone cortisone contains C, H, and O, and shows a molecular ion at $M^+ = 360.1937$ when analyzed by double-focusing mass spectrometry. What is the molecular formula of cortisone? (Isotopic masses are: ^{12}C, 12.0000 amu; ^1H, 1.007 83 amu; ^{16}O, 15.9949 amu. The degree of unsaturation of cortisone is 8.)

12.39 Assume you are carrying out the dehydration of 1-methylcyclohexanol to yield 1-methylcyclohexene. How could you use infrared spectroscopy to determine when the reaction is complete?

12.40 Assume that you are carrying out the base-induced dehydrobromination of 3-bromo-3-methylpentane (Section 11.10). How could you use IR spectroscopy to tell which of two possible elimination products is formed?

12.41 At what approximate positions might the following compounds show IR absorptions?

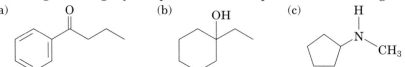

(a) CH₃CH₂CCH₃   (b) (CH₃)₂CHCH₂C≡CH   (c) (CH₃)₂CHCH₂CH=CH₂

(d) CH₃CH₂CH₂COOCH₃   (e)

12.42 What fragments might you expect in the mass spectra of the following compounds?

(a)   (b)   (c)

12.43 Which is stronger, the C=O bond in an ester (1735 cm⁻¹) or the C=O bond in a saturated ketone (1715 cm⁻¹)? Explain.

12.44 Carvone is an unsaturated ketone responsible for the odor of spearmint. If carvone has M⁺ = 150 in its mass spectrum, what molecular formulas are likely? If carvone has three double bonds and one ring, what molecular formula is correct?

12.45 Carvone (Problem 12.44) has an intense infrared absorption at 1690 cm⁻¹. What kind of ketone does carvone contain?

12.46 The mass spectrum (a) and the infrared spectrum (b) of an unknown hydrocarbon are shown. Propose as many structures as you can.

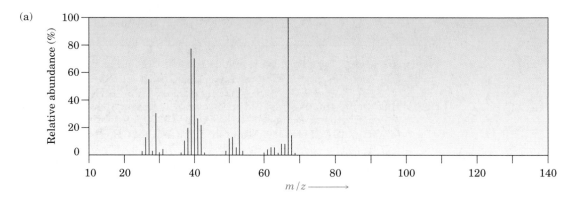

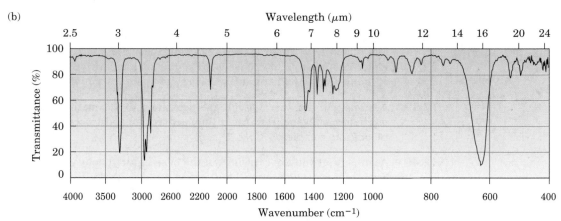

12.47 The mass spectrum (a) and the infrared spectrum (b) of another unknown hydrocarbon are shown. Propose as many structures as you can.

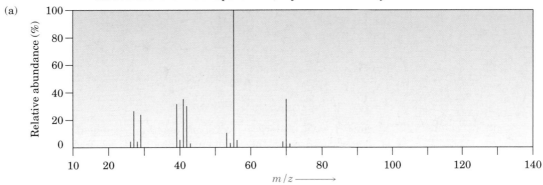

(a)

(b)

12.48 Propose structures for compounds that meet the following descriptions:
(a) An optically active compound $C_5H_{10}O$ with an IR absorption at 1730 cm^{-1}
(b) A non-optically active compound C_5H_9N with an IR absorption at 2215 cm^{-1}

12.49 4-Methyl-2-pentanone and 3-methylpentanal are isomers. Explain how you could tell them apart, both by mass spectrometry and by infrared spectroscopy.

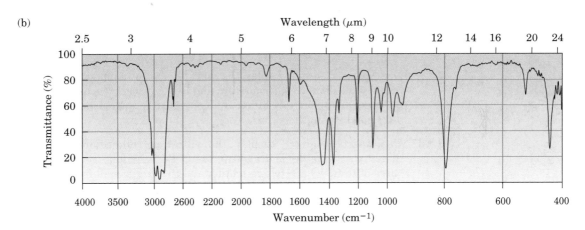

4-Methyl-2-pentanone   **3-Methylpentanal**

A Look Ahead

• •

12.50 Grignard reagents undergo a general and very useful reaction with ketones. Methylmagnesium bromide, for example, reacts with cyclohexanone to yield a product with the formula $C_7H_{14}O$. What is the structure of this product if it has an IR absorption at 3400 cm^{-1}? (See Section 17.6.)

Cyclohexanone

12.51   Ketones undergo a reduction when treated with sodium borohydride, $NaBH_4$. What is the structure of the compound produced by reaction of 2-butanone with $NaBH_4$ if it has an IR absorption at 3400 cm^{-1} and $M^+ = 74$ in the mass spectrum? (See Section 17.5.)

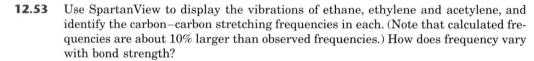

2-Butanone

12.52   Nitriles, R–C≡N, undergo a hydrolysis reaction when heated with aqueous acid. What is the structure of the compound produced by hydrolysis of propanenitrile, $CH_3CH_2C≡N$, if it has IR absorptions at 2500–3100 cm^{-1} and 1710 cm^{-1} and has $M^+ = 74$? (See Section 21.8.)

Molecular Modeling

12.53   Use SpartanView to display the vibrations of ethane, ethylene and acetylene, and identify the carbon–carbon stretching frequencies in each. (Note that calculated frequencies are about 10% larger than observed frequencies.) How does frequency vary with bond strength?

12.54   Use SpartanView to display the vibrations of acetone, methyl benzoate, and dimethylformamide, and identify the C=O stretching frequency in each. What features of the C=O stretching motion and the vibrational frequency make this a good diagnostic tool for identifying the carbonyl group?

12.55   Stretching vibrations can involve individual bonds or groups of bonds. Use SpartanView to display the two C=O stretching vibrations of carbon dioxide, and describe which bond(s) are involved in each vibration.

12.56   Hydrogen bonding can affect O–H stretching frequencies. Use SpartanView to identify the O–H stretching frequencies for the –CO_2H group in acetic acid, acetic acid plus water, and acetic acid dimer. Does hydrogen bonding raise or lower the O–H stretching frequency? (Consider only the highest O–H stretching frequency in acetic acid dimer.)

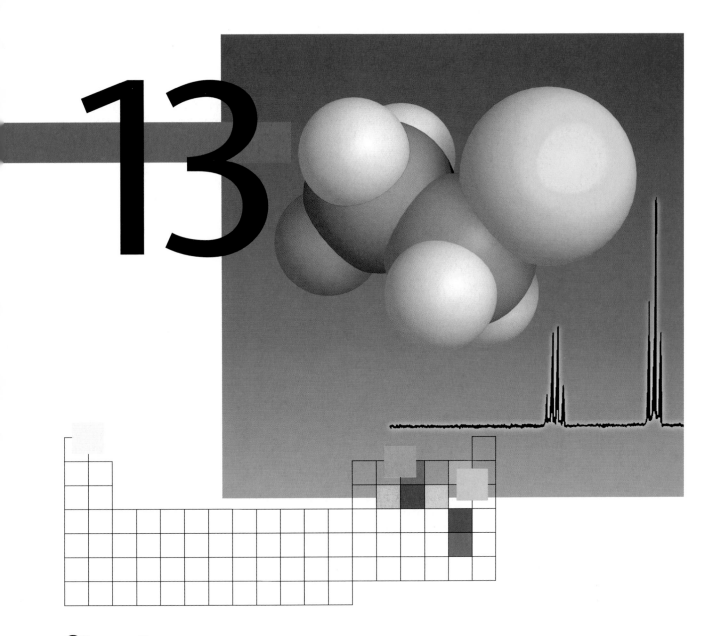

13

Structure Determination: Nuclear Magnetic Resonance Spectroscopy

Nuclear magnetic resonance spectroscopy (NMR) is the most valuable spectroscopic technique available to organic chemists. It's the method of structure determination that organic chemists first turn to for information.

We saw in Chapter 12 that mass spectrometry provides information about a molecule's formula and that infrared spectroscopy provides information about a molecule's functional groups. Nuclear magnetic resonance spectroscopy does not replace either of these techniques; rather, it complements them by providing a "map" of the carbon–hydrogen framework of an organic molecule. Taken together, NMR, IR, and mass spectrometry often make it possible to determine the complete structures of even very complex molecules.

Mass spectrometry   Molecular size and formula
Infrared spectroscopy   Functional groups
NMR spectroscopy   Map of carbon–hydrogen framework

13.1 Nuclear Magnetic Resonance Spectroscopy

Many kinds of atomic nuclei behave as if they were spinning about an axis, much as the earth spins daily. Since they're positively charged, these spinning nuclei act like tiny bar magnets and therefore interact with an external magnetic field, denoted B_0. Not all nuclei act this way, but fortunately for organic chemists, both the proton (^1H) and the ^{13}C nucleus do have spins. (In speaking about NMR, the words *proton* and *hydrogen* are often used interchangeably.) Let's see what the consequences of nuclear spin are and how we can use the results.

In the absence of an external magnetic field, the spins of magnetic nuclei are oriented randomly. When a sample containing these nuclei is placed between the poles of a strong magnet, however, the nuclei adopt specific orientations, much as a compass needle orients in the earth's magnetic field. A spinning ^1H or ^{13}C nucleus can orient so that its own tiny magnetic field is aligned either with (parallel to) or against (antiparallel to) the external field. The two orientations don't have the same energy and therefore aren't equally likely. The parallel orientation is slightly lower in energy by an amount that depends on the strength of the external field, making this spin state very slightly favored over the antiparallel orientation (Figure 13.1).

FIGURE 13.1 ▼

(a) Nuclear spins are oriented randomly in the absence of an external magnetic field but (b) have a specific orientation in the presence of an external field, B_0. Note that some of the spins (red) are aligned parallel to the external field while others (blue) are antiparallel. The parallel spin state is slightly lower in energy.

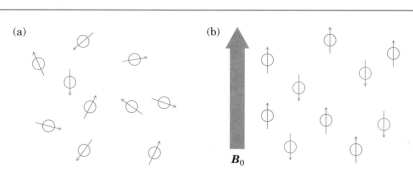

If the oriented nuclei are now irradiated with electromagnetic radiation of the proper frequency, energy absorption occurs and the lower-energy state "spin-flips" to the higher-energy state. When this spin-flip occurs, the magnetic nuclei are said to be in resonance with the applied radiation—hence the name *nuclear magnetic resonance*.

The exact frequency necessary for resonance depends both on the strength of the external magnetic field and on the identity of the nuclei. If a very strong magnetic field is applied, the energy difference between the two spin states is larger, and higher-frequency (higher-energy) radiation is required for a spin-flip. If a weaker magnetic field is applied, less energy is required to effect the transition between nuclear spin states (Figure 13.2).

FIGURE 13.2 ▼

The energy difference ΔE between nuclear spin states depends on the strength of the applied magnetic field. Absorption of energy of frequency ν converts a nucleus from a lower spin state to a higher spin state. (a) Spin states have equal energies in the absence of an applied magnetic field, but (b) have unequal energies in the presence of a magnetic field. At $\nu = 60$ MHz, $\Delta E = 2.4 \times 10^{-5}$ kJ/mol (5.7×10^{-6} kcal/mol). (c) The energy difference between spin states is greater at larger applied fields. At $\nu = 500$ MHz, $\Delta E = 2.0 \times 10^{-4}$ kJ/mol.

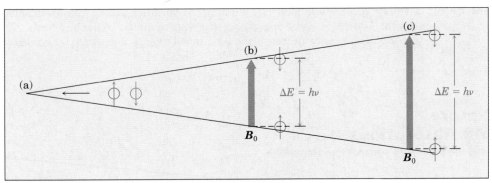

Strength of applied field, B_0 ⟶

In practice, superconducting magnets that produce enormously powerful fields up to 14.1 tesla (T) are sometimes used, but field strengths in the range 1.41–4.7 T are more common. At a magnetic field strength of 1.41 T, so-called *radiofrequency* (rf) energy in the 60 MHz range (1 MHz = 10^6 Hz) is required to bring a ^{1}H nucleus into resonance, and rf energy of 15 MHz is required to bring a ^{13}C nucleus into resonance. These energies needed for NMR are much smaller than those required for infrared spectroscopy; 60 MHz rf energy corresponds to only 2.4×10^{-5} kJ/mol versus 4.8–48 kJ/mol needed for IR spectroscopy.

^{1}H and ^{13}C nuclei are not unique in their ability to exhibit the NMR phenomenon. All nuclei with an odd number of protons (^{1}H, ^{2}H, ^{14}N, ^{19}F, ^{31}P, for example) and all nuclei with an odd number of neutrons (^{13}C, for example) show magnetic properties. Only nuclei with even numbers of both protons and neutrons (^{12}C, ^{16}O) do not give rise to magnetic phenomena (Table 13.1).

TABLE 13.1 The NMR Behavior of Some Common Nuclei		
Magnetic nuclei		**Nonmagnetic nuclei**

^1H
^{13}C
^2H NMR observed
^{14}N
^{19}F
^{31}P

^{12}C
^{16}O No NMR observed
^{32}S

Problem 13.1   The amount of energy required to spin-flip a nucleus depends both on the strength of the external magnetic field and on the nucleus. At a field strength of 1.41 T, rf energy of 60 MHz is required to bring a ^1H nucleus into resonance, but energy of only 56 MHz will bring a ^{19}F nucleus into resonance. Use the equation given in Problem 12.7 (p. 455) to calculate the amount of energy required to spin-flip a ^{19}F nucleus. Is this amount greater or less than that required to spin-flip a ^1H nucleus?

Problem 13.2   Calculate the amount of energy required to spin-flip a proton in a spectrometer operating at 100 MHz. Does increasing the spectrometer frequency from 60 MHz to 100 MHz increase or decrease the amount of energy necessary for resonance?

13.2 The Nature of NMR Absorptions

From the description given thus far, you might expect all ^1H nuclei in a molecule to absorb rf energy at the same frequency and all ^{13}C nuclei to absorb at the same frequency. If this were true, we would observe only a single NMR absorption band in the ^1H or ^{13}C spectrum of a molecule, a situation that would be of little use for structure determination. In fact, the absorption frequency is *not* the same for all ^1H or all ^{13}C nuclei.

All nuclei in molecules are surrounded by electrons. When an external magnetic field is applied to a molecule, the moving electrons set up tiny local magnetic fields of their own. These local magnetic fields act in opposition to the applied field so that the *effective* field actually felt by the nucleus is a bit smaller than the applied field.

$$B_{effective} = B_{applied} - B_{local}$$

In describing this effect, we say that nuclei are **shielded** from the full effect of the applied field by the circulating electrons that surround them. Since each specific nucleus in a molecule is in a slightly different electronic environment, each nucleus is shielded to a slightly different extent, and the effective magnetic field is not the same for each nucleus. If the NMR instrument is sensitive enough, the tiny differences in the effective magnetic fields experienced by different nuclei can be detected, and we can see a distinct NMR signal for each chemically distinct carbon or hydrogen nucleus in a molecule. Thus, the NMR spectrum of an organic compound effectively maps the carbon–hydrogen framework. With practice, it's possible to read the map and thereby derive structural information about an unknown molecule.

Figure 13.3 shows both the ^1H and the ^{13}C NMR spectra of methyl acetate, $CH_3CO_2CH_3$. The horizontal axis shows the effective field strength felt by the nuclei, and the vertical axis indicates intensity of absorption of rf energy. Each peak in the NMR spectrum corresponds to a chemically distinct nucleus in the molecule. [Note that NMR spectra are formatted with the zero absorption line at the *bottom,* whereas IR spectra are formatted with the zero absorption line (100% transmittance) at the *top;* Section 12.5.] Note also that ^1H and ^{13}C spectra can't both be observed at the same time on the same spectrometer because different amounts of energy are required to spin-flip the different kinds of nuclei. The two spectra must be recorded separately.

FIGURE 13.3 ▼

(a) The ^1H NMR spectrum and (b) the ^{13}C NMR spectrum of methyl acetate, $CH_3CO_2CH_3$. The small peaks labeled "TMS" at the far right of each spectrum are calibration peaks, as explained in Section 13.3.

(a)

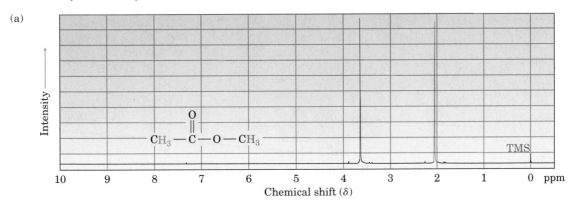

(b)

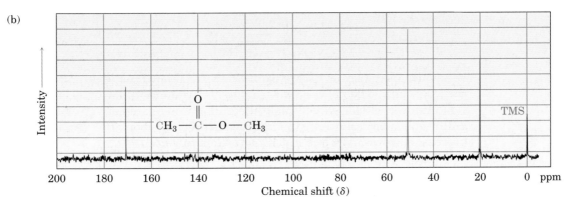

The ^{13}C spectrum of methyl acetate in Figure 13.3b shows three peaks, one for each of the three carbon atoms in the molecule. The ^{1}H NMR spectrum in Figure 13.3a shows only *two* peaks, however, even though methyl acetate has *six* hydrogens. One peak is due to the CH_3CO hydrogens, and the other to the OCH_3 hydrogens. Because the three hydrogens of each methyl group have the same electronic environment, they are shielded to the same extent and are said to be *equivalent*. Chemically equivalent nuclei always show a single absorption. The two methyl groups themselves, however, are nonequivalent and absorb at different positions.

The operation of a typical NMR spectrometer is illustrated schematically in Figure 13.4. An organic sample is dissolved in a suitable solvent (usually deuteriochloroform, $CDCl_3$) and placed in a thin glass tube between the poles of a magnet. The strong magnetic field causes the ^{1}H and ^{13}C nuclei in the molecule to align in one of the two possible orientations, and the sample is irradiated with rf energy. If the frequency of the rf irradiation is held constant and the strength of the applied magnetic field is changed, each nucleus comes into resonance at a slightly different field strength. A sensitive detector monitors the absorption of rf energy, and the electronic signal is then amplified and displayed as a peak on a recorder chart.

FIGURE 13.4 ▼

Schematic operation of an NMR spectrometer. A thin glass tube containing the sample solution is placed between the poles of a strong magnet and irradiated with rf energy.

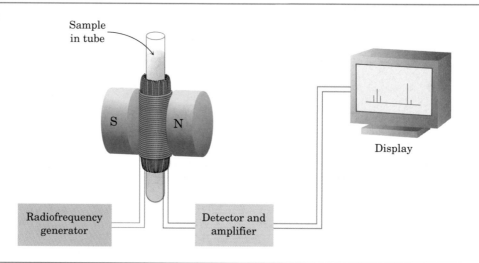

NMR spectroscopy differs from IR spectroscopy (Sections 12.6–12.9) in that the time scales of the two techniques are quite different. The absorption of infrared energy by a molecule giving rise to a change in vibrational amplitude is an essentially instantaneous process (about 10^{-13} s). The NMR process, however, requires much more time (about 10^{-3} s).

The difference in time scales between IR and NMR spectroscopy is comparable to the difference between a camera operating at a very fast shutter speed and a camera operating at a very slow shutter speed. The fast camera (IR) takes an instantaneous picture and "freezes" the action. If two rapidly interconverting species are present, IR spectroscopy records the spectrum of each. The slow camera (NMR), however, takes a blurred, "time-averaged" picture. If two species interconverting faster than 10^3 times per second are present in a sample, NMR records only a single, averaged spectrum, rather than separate spectra of the two discrete species.

Because of this "blurring" effect, NMR spectroscopy can be used to measure the rates and activation energies of very fast processes. In cyclohexane, for example, a ring-flip (Section 4.11) occurs so rapidly at room temperature that axial and equatorial hydrogens can't be distinguished by NMR; only a single ^{1}H NMR absorption is seen for cyclohexane at 25°C. At −90°C, however, the ring-flip is slowed down enough so that two absorption peaks are seen, one for the six axial hydrogens and one for the six equatorial hydrogens. Knowing the temperature and the rate at which signal blurring begins to occur, it's possible to calculate that the activation energy for the cyclohexane ring-flip is 45 kJ/mol (10.8 kcal/mol).

^{1}H NMR: 1 peak at 25°C
2 peaks at −90°C

Problem 13.3 2-Chloropropene shows signals for three kinds of protons in its ^{1}H NMR spectrum. Explain.

13.3 Chemical Shifts

NMR spectra are displayed on charts that show the applied field strength increasing from left to right (Figure 13.5, p. 482). Thus, the left part of the chart is the low-field, or **downfield**, side, and the right part is the high-field, or **upfield**, side. Nuclei that absorb on the downfield side of the chart require a lower field strength for resonance, implying that they have relatively little shielding. Nuclei that absorb on the upfield side require a higher field strength for resonance, implying that they are strongly shielded.

To define the position of an absorption, the NMR chart is calibrated and a reference point is used. In practice, a small amount of tetramethylsilane [TMS, $(CH_3)_4Si$] is added to the sample so that a reference absorption is produced when the spectrum is run. TMS is used as reference for both ^{1}H and ^{13}C measurements because it produces in both kinds of spectra a single peak that occurs upfield of other absorptions normally found in organic compounds. The ^{1}H and ^{13}C spectra of methyl acetate in Figure 13.3 have the TMS reference peak indicated.

FIGURE 13.5 ▼

The NMR chart. The downfield, deshielded side is on the left, and the upfield, shielded side is on the right. The tetramethylsilane (TMS) absorption is used as reference point.

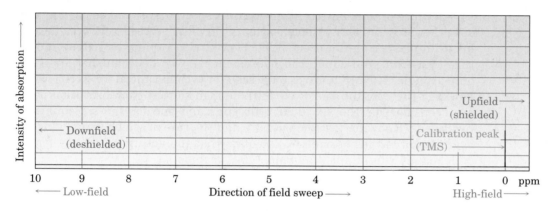

The position on the chart at which a nucleus absorbs is called its **chemical shift**. By convention, the chemical shift of TMS is set as the zero point, and other absorptions normally occur downfield, to the left on the chart. NMR charts are calibrated using an arbitrary scale called the **delta scale**. One delta unit (δ) is equal to 1 part per million (ppm; one-millionth) of the spectrometer operating frequency. For example, if we were measuring the ^{1}H NMR spectrum of a sample using an instrument operating at 60 MHz, 1 δ would be 1 ppm of 60,000,000 Hz, or 60 Hz. Similarly, if we were measuring the spectrum using a 300 MHz instrument, then 1 δ = 300 Hz. The following equation can be used for any absorption:

$$\delta = \frac{\text{Observed chemical shift (number of Hz away from TMS)}}{\text{Spectrometer frequency in MHz}}$$

Although this method of calibrating NMR charts may seem needlessly complex, there's a good reason for it. As we saw earlier, the rf frequency required to bring a given nucleus into resonance depends on the spectrometer's magnetic field strength. But because there are many different kinds of spectrometers with many different magnetic field strengths available, chemical shifts given in frequency units (Hz) vary greatly from one instrument to another. Thus, a resonance that occurs at 120 Hz downfield from TMS on one spectrometer might occur at 600 Hz downfield from TMS on another spectrometer with a more powerful magnet.

By using a system of measurement in which NMR absorptions are expressed in *relative* terms (ppm) rather than absolute terms (Hz), comparisons of spectra obtained on different instruments are possible. *The chemical shift of an NMR absorption given in δ units is constant, regardless of the operating frequency of the spectrometer.* A ^{1}H nucleus that absorbs at 2.0 δ on a 60 MHz instrument also absorbs at 2.0 δ on a 300 MHz instrument.

The range in which most NMR absorptions occur is quite narrow. Almost all ^{1}H NMR absorptions occur 0–10 δ downfield from the proton absorption of TMS, and almost all ^{13}C absorptions occur 1–220 δ downfield from the carbon absorption of TMS. Thus, there is a considerable likelihood that accidental overlap of nonequivalent signals will occur. The advantage of using an instrument with higher field strength (say, 300 MHz NMR) rather than lower field strength (60 MHz NMR) is that different NMR absorptions are more widely separated at the higher field strength. The chances that two signals will accidentally overlap are also lessened, and interpretation of spectra becomes easier. For example, two signals that are only 6 Hz apart at 60 MHz (0.1 ppm) are 30 Hz apart at 300 MHz (still 0.1 ppm).

Problem 13.4   When the ^{1}H NMR spectrum of acetone, CH_3COCH_3, is recorded on an instrument operating at 60 MHz, a single sharp resonance at 2.1 δ is seen.
(a) How many hertz downfield from TMS does the acetone resonance correspond to?
(b) If the ^{1}H NMR spectrum of acetone were recorded at 100 MHz, what would be the position of the absorption in δ units?
(c) How many hertz downfield from TMS does this 100 MHz resonance correspond to?

Problem 13.5   The following ^{1}H NMR peaks were recorded on a spectrometer operating at 60 MHz. Convert each into δ units.
(a) $CHCl_3$; 436 Hz   (b) CH_3Cl; 183 Hz
(c) CH_3OH; 208 Hz   (d) CH_2Cl_2; 318 Hz

13.4 ^{13}C NMR Spectroscopy: Signal Averaging and FT-NMR

Everything we've said thus far about NMR spectroscopy applies to both ^{1}H and ^{13}C spectra, but let's now focus only on ^{13}C spectra because they're much easier to interpret. What we learn now about interpreting ^{13}C spectra will simplify the subsequent discussion of ^{1}H spectra.

In some ways, it's surprising that carbon NMR is even possible. After all, ^{12}C, the most abundant carbon isotope, has no nuclear spin and can't be seen by NMR. Carbon-13 is the only naturally occurring carbon isotope with a nuclear spin, but its natural abundance is only 1.1%. Thus, only about 1 of every 100 carbons in an organic sample is observable by NMR. The problem of low abundance has been overcome, however, by the development of two techniques: *signal averaging* and **Fourier-transform NMR (FT-NMR)**. Signal averaging increases instrument sensitivity, and FT-NMR increases instrument speed.

The low natural abundance of ^{13}C means that any individual NMR spectrum is extremely "noisy." That is, the signals are so weak that they are cluttered with random background electronic noise, as shown in Figure 13.6a. If, however, hundreds (or thousands) of individual runs are added together by computer and then averaged, a greatly improved spectrum results (Figure 13.6b). Background noise, because of its random nature, averages to zero, so the nonzero signals stand out clearly. Unfortunately, the value of signal averaging is limited when using the method of NMR spectrometer operation described in Section 13.2, because it takes about 5–10 minutes to obtain a single spectrum. Thus, a faster way to obtain spectra is needed if signal averaging is to be used.

FIGURE 13.6 ▼

Carbon-13 NMR spectra of 1-pentanol, $CH_3CH_2CH_2CH_2CH_2OH$.
Spectrum (a) is a single run, showing the large amount of background noise.
Spectrum (b) is an average of 200 runs.

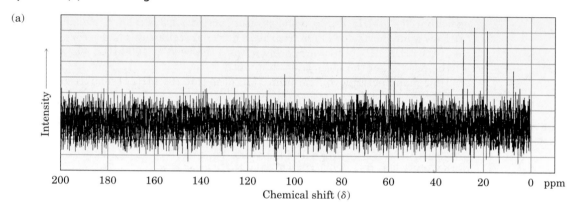

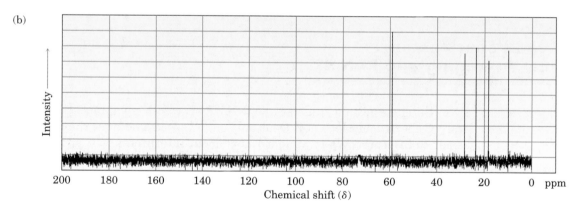

In the method of NMR spectrometer operation described in Section 13.2, either the rf frequency is held constant while the strength of the magnetic field is varied or the strength of the magnetic field is held constant while the rf frequency is varied. In either case, all signals in the spectrum are recorded sequentially. In the FT-NMR technique used by modern spectrometers, however, all the signals are recorded simultaneously. A sample is placed in a magnetic field of constant strength and is irradiated with a short burst, or "pulse," of rf energy that covers the entire range of useful frequencies. All ^{1}H or ^{13}C nuclei in the sample resonate at once, giving a complex, composite signal that must be mathematically manipulated using so-called *Fourier transforms* before it can be displayed in the usual way. Since all resonance signals are collected at once, it takes only a few seconds rather than a few minutes to record an entire spectrum.

Combining the speed of FT-NMR with the sensitivity enhancement of signal averaging is what gives modern NMR spectrometers their power. Literally thousands of spectra can be taken and averaged in a few hours, resulting in sensitivity so high that ^{13}C NMR spectra can be obtained with only a few milligrams of sample, and ^{1}H spectra can be recorded with only a few *micrograms*.

13.5 Characteristics of ^{13}C NMR Spectroscopy

At its simplest, ^{13}C NMR makes it possible to count the number of different carbon atoms in a molecule of unknown structure. Look at the ^{13}C NMR spectra of methyl acetate and 1-pentanol shown previously in Figures 13.3b and 13.6b, for instance. In each case, a single sharp resonance line is observed for each different carbon atom.

Most ^{13}C resonances are between 0 and 220 ppm downfield from the TMS reference line, with the exact chemical shift of each ^{13}C resonance dependent on that carbon's electronic environment within the molecule. Figure 13.7 shows the correlation of environment with chemical shift.

FIGURE 13.7 ▼

Chemical shift correlations for ^{13}C NMR (Hal = Cl, Br, I).

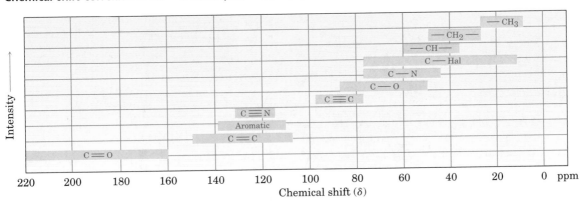

The factors that determine chemical shifts are complex, but it's possible to make some generalizations from the data in Figure 13.7. One trend is that a carbon's chemical shift is affected by the electronegativity of nearby atoms: Carbons bonded to oxygen, nitrogen, or halogen absorb downfield (to the left) of typical alkane carbons. Since electronegative atoms attract electrons, they pull electrons away from neighboring carbon atoms, causing those carbons to be deshielded and to come into resonance at a lower field.

Another trend is that sp^3-hybridized carbons generally absorb in the range 0–90 δ, while sp^2 carbons absorb in the range 110–220 δ. Carbonyl carbons (C=O) are particularly distinct in ^{13}C NMR and are always found at the low-field end of the spectrum, in the range 160–220 δ. Figure 13.8 shows the ^{13}C NMR spectra of 2-butanone and *para*-bromoacetophenone, and indicates the peak assignments. Note that the C=O carbons are at the left edge of the spectrum in each case.

FIGURE 13.8 ▼

Carbon-13 NMR spectra of (a) 2-butanone and (b) *para*-bromoacetophenone.

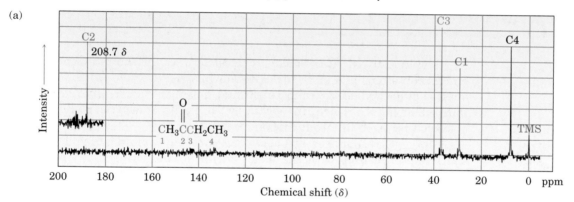

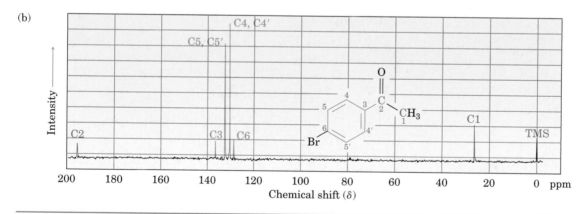

The ^{13}C NMR spectrum of *para*-bromoacetophenone is interesting in several ways. Note particularly that only six carbon absorptions are observed even though the molecule contains eight carbons. *para*-Bromoacetophenone has a symmetry plane that makes ring carbons 4 and 4′, and ring carbons 5 and 5′ equivalent (Figure 13.9). Thus, the six ring carbons show only four absorptions in the range 128–137 δ.

FIGURE 13.9 ▼

As evident in this stereo view, *para*-bromoacetophenone has a plane of symmetry roughly coincident with the plane of the page. As a result, carbons 4 and 4′, and carbons 5 and 5′ are equivalent.

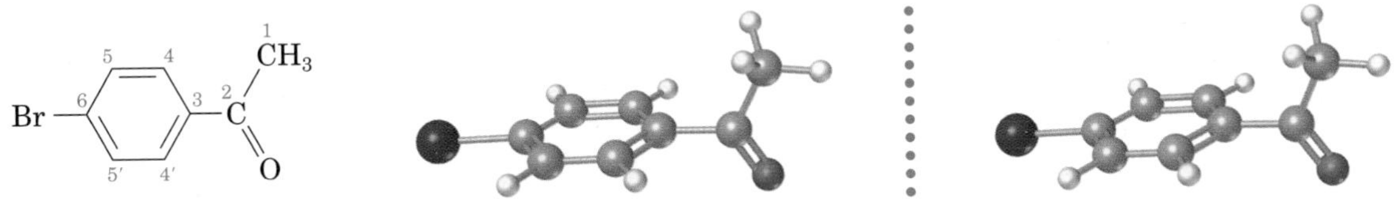

para-Bromoacetophenone

Stereo View

A second interesting point about both spectra in Figure 13.8 is that the peaks aren't uniform in size. Some peaks are larger than others even though they are 1-carbon resonances (except for the two 2-carbon peaks of *para*-bromoacetophenone). This difference in peak size is caused by several factors that we won't go into, but it is a general feature of ^{13}C NMR spectra.

Practice Problem 13.1   At what approximate positions would you expect ethyl acrylate, $H_2C=CHCO_2CH_2CH_3$, to show ^{13}C NMR absorptions?

Strategy   Identify the distinct carbons in the molecule, and note whether each is alkyl, vinylic, aromatic, or in a carbonyl group. Then predict where each absorbs, using Figure 13.7 as necessary.

Solution   Ethyl acrylate has five distinct carbons: two different C=C, one C=O, one O–C, and one alkyl C. From Figure 13.7, the likely absorptions are

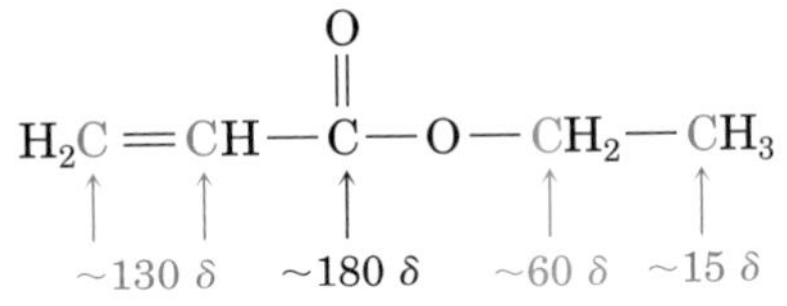

$$H_2C=CH-\overset{\overset{\displaystyle O}{\|}}{C}-O-CH_2-CH_3$$

~130 δ   ~180 δ   ~60 δ   ~15 δ

The actual absorptions are at 14.1, 60.5, 128.5, 130.3, and 166.0 δ.

Problem 13.6   Assign the resonances in the ^{13}C NMR spectrum of methyl propanoate, $CH_3CH_2CO_2CH_3$ (Figure 13.10).

FIGURE 13.10 ▼

¹³C NMR spectrum of methyl propanoate, Problem 13.6.

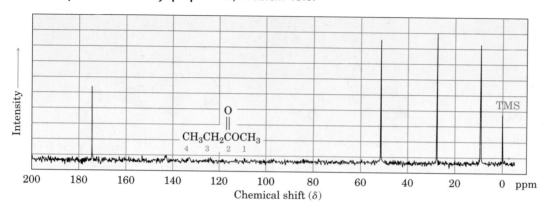

Problem 13.7   Predict the number of carbon resonance lines you would expect in the ¹³C NMR spectra of the following compounds:
(a) Methylcyclopentane   (b) 1-Methylcyclohexene
(c) 1,2-Dimethylbenzene   (d) 2-Methyl-2-butene

Problem 13.8   Propose structures for compounds that fit the following descriptions:
(a) A hydrocarbon with seven lines in its ¹³C NMR spectrum
(b) A 6-carbon compound with only five lines in its ¹³C NMR spectrum
(c) A 4-carbon compound with three lines in its ¹³C NMR spectrum

13.6   DEPT ¹³C NMR Spectroscopy

New techniques developed in recent years have made it possible to obtain enormous amounts of information from ¹³C NMR spectra. Among the most useful of these new techniques is one called **DEPT-NMR**, for *distortionless enhancement by polarization transfer*, which makes it possible to distinguish among signals due to CH₃, CH₂, CH, and quaternary carbons. That is, the number of hydrogens attached to each carbon in a molecule can be determined.

A DEPT experiment is usually done in three stages, as shown in Figure 13.11 for 6-methyl-5-hepten-2-ol. The first stage is to run an ordinary spectrum (called a *broadband-decoupled spectrum*) to locate the chemical shifts of all carbons. Next, a second spectrum called a DEPT-90 is run, using special conditions under which *only signals due to CH carbons appear*. Signals due to CH₃, CH₂, and quaternary carbons are absent. Finally, a third spectrum called a DEPT-135 is run, using conditions under which CH₃ and CH resonances appear as positive signals, CH₂ resonances appear as *negative* signals—that is, as peaks *below* the baseline—and quaternary carbons are again absent.

FIGURE 13.11 ▼

DEPT-NMR spectra for 6-methyl-5-hepten-2-ol. Part (a) is an ordinary broadband-decoupled spectrum, which shows signals for all eight carbons. Part (b) is a DEPT-90 spectrum, which shows only signals for the two CH carbons. Part (c) is a DEPT-135 spectrum, which shows positive signals for the two CH and three CH_3 carbons and negative signals for the two CH_2 carbons.

(a)

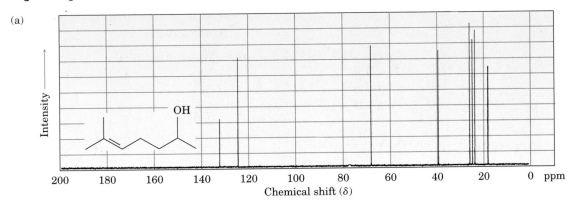

(b)

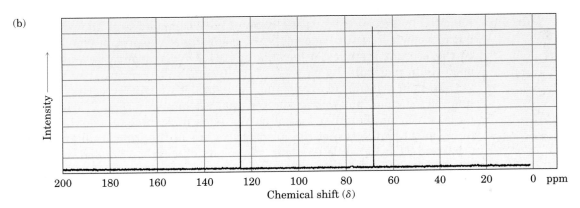

(c)

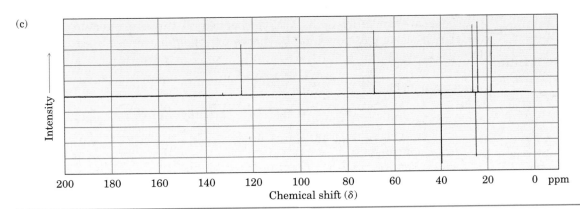

Putting together the information from all three spectra makes it possible to tell the number of hydrogens attached to each carbon. The CH carbons are identified in the DEPT-90 spectrum; the CH_2 carbons are identified as the negative peaks in the DEPT-135 spectrum; the CH_3 carbons are identified by subtracting the CH peaks from the positive peaks in the DEPT-135 spectrum; and quaternary carbons are identified by subtracting all peaks in the DEPT-135 spectrum from the peaks in the broadband-decoupled spectrum.

Broadband decoupled	DEPT-90	DEPT-135
C, CH, CH$_2$, CH$_3$	CH	CH$_3$, CH are positive
		CH$_2$ is negative

C	Subtract DEPT-135 from broadband decoupled
CH	DEPT-90
CH$_2$	Negative DEPT-135
CH$_3$	Subtract DEPT-90 from positive DEPT-135

• •

Practice Problem 13.2   Propose a structure for an alcohol, C$_4$H$_{10}$O, that has the following ^{13}C NMR spectral data. Broadband decoupled ^{13}C NMR: 19.0, 31.7, 69.5 δ; DEPT-90: 31.7 δ; DEPT-135: positive peak at 19.0 δ, negative peak at 69.5 δ.

Strategy   As noted in Section 6.2, it usually helps with compounds of known formula but unknown structure to calculate the substance's degree of unsaturation. In the present instance, a formula of C$_4$H$_{10}$O corresponds to a saturated, open-chain molecule.

To gain information from the ^{13}C data, let's begin by noting that the unknown alcohol has *four* carbon atoms, yet has only *three* NMR absorptions, which implies that two of the carbons must be equivalent. Looking at chemical shifts, two of the absorptions are in the typical alkane region (19.0 and 31.7 δ;), while one is in the region of a carbon bonded to an electronegative atom (69.5 δ)—oxygen in this instance. The DEPT-90 spectrum tells us that the alkyl carbon at 31.7 δ is tertiary (CH); the DEPT-135 spectrum tells us that the alkyl carbon at 19.0 δ is a methyl (CH$_3$) and that the carbon bonded to oxygen (69.5 δ) is secondary (CH$_2$). The two equivalent carbons are probably both methyls bonded to the same tertiary carbon, (CH$_3$)$_2$CH–. We can now put the pieces together to propose a structure.

Solution

2-Methyl-1-propanol

• •

Problem 13.9   Assign a chemical shift to each carbon in 6-methyl-5-hepten-2-ol (Figure 13.11).

Problem 13.10   Estimate the chemical shift of each carbon in the following molecule. Predict which carbons will appear in the DEPT-90 spectrum, which will give positive peaks in the DEPT-135 spectrum, and which will give negative peaks in the DEPT-135 spectrum.

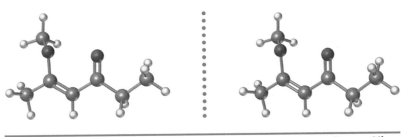

Stereo View

Problem 13.11   Propose a structure for an aromatic hydrocarbon, $C_{11}H_{16}$, that has the following ¹³C NMR spectral data. Broadband decoupled ¹³C NMR: 29.5, 31.8, 50.2, 125.5, 127.5, 130.3, 139.8 δ; DEPT-90: 125.5, 127.5, 130.3 δ; DEPT-135: positive peaks at 29.5, 125.5, 127.5, 130.3 δ, negative peak at 50.2 δ.

13.7   Uses of ¹³C NMR Spectroscopy

The information derived from ¹³C NMR spectroscopy is extraordinarily useful for structure determination. Not only can we count the number of nonequivalent carbon atoms in a molecule, we can also get information about the electronic environment of each and can even find how many protons each is attached to. As a result, we are able to answer many structural questions that go unanswered by infrared spectroscopy or mass spectrometry.

Let's take an example. How might we prove that E2 elimination of an alkyl halide gives the more highly substituted alkene (Zaitsev's rule, Section 11.10)? Does reaction of 1-chloro-1-methylcyclohexane with strong base lead predominantly to 1-methylcyclohexene or to methylenecyclohexane?

H₃C Cl

KOH
Ethanol

CH₃

or

CH₂

?

1-Chloro-1-
methylcyclohexane

1-Methylcyclohexene

Methylenecyclohexane

1-Methylcyclohexene should have five sp^3-carbon resonances in the range 20–50 δ and two sp^2-carbon resonances in the range 100–150 δ. Methylenecyclohexane, however, because of its symmetry, should have only three sp^3-carbon resonance peaks and two sp^2-carbon peaks. The spectrum of the actual reaction product, shown in Figure 13.12, clearly identifies 1-methylcyclohexene as the substance formed in this E2 reaction.

FIGURE 13.12 ▼

The ^{13}C NMR spectrum of 1-methylcyclohexene, the E2 reaction product from
1-chloro-1-methylcyclohexane.

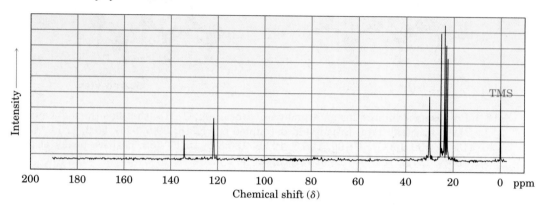

· ·

Problem 13.12   We saw in Section 8.4 that addition of HBr to terminal alkynes leads to the
Markovnikov addition product, with the Br bonding to the more highly substituted
carbon. How could you use ^{13}C NMR to identify the product of the addition of 1
equivalent of HBr to 1-hexyne?

· ·

13.8 ^{1}H NMR Spectroscopy and Proton Equivalence

Having looked at ^{13}C spectra, let's now focus on ^{1}H NMR spectroscopy. Since
each chemically distinct hydrogen in a molecule normally has its own unique
absorption, one use of ^{1}H NMR is to find out how many kinds of nonequivalent hydrogens are present. In the ^{1}H NMR spectrum of methyl
acetate shown previously in Figure 13.3a, for example, there are two signals, corresponding to the two nonequivalent kinds of protons present,
CH_3CO- protons and $-OCH_3$ protons.

A quick look at a structure is usually enough to decide how many kinds
of nonequivalent protons are present in a molecule. If in doubt, though, the
equivalence or nonequivalence of two protons can be determined by seeing
whether the same or different structures would result if some group X were
substituted for one of the protons. If the protons are chemically equivalent,
the same product will be formed regardless of which proton is replaced. If
the protons are not chemically equivalent, different products will be formed
on substitution. In 2,3-dimethyl-2-butene, for example, all 12 protons are
equivalent. No matter which proton we replace by an X group, we get the
same structure. The 12 protons thus give rise to a single, sharp ^{1}H NMR
peak (Figure 13.13).

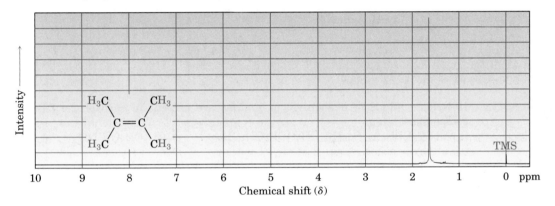

All four methyl groups
are equivalent.

Only one substitution
product is possible.

FIGURE 13.13 ▼

The ¹H NMR spectrum of 2,3-dimethyl-2-butene. Since all 12 protons in the molecule
are chemically equivalent, there is only one peak in the spectrum.

By contrast, the 10 protons of 2-methyl-2-butene are *not* all equivalent.
There are three different kinds of methyl-group protons and one vinylic proton, leading to four different possible substitution products and four different signals in the ¹H NMR spectrum (Figure 13.14)

Four different substitution products

FIGURE 13.14 ▼

The ^{1}H NMR spectrum of 2-methyl-2-butene. There are four kinds of protons and four different signals.

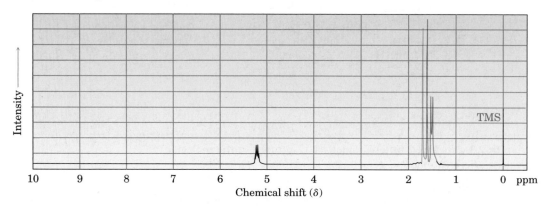

● ●

Problem 13.13   How many kinds of nonequivalent protons are present in each of the following compounds?
(a) CH_3CH_2Br (b) $CH_3OCH_2CH(CH_3)_2$ (c) $CH_3CH_2CH_2NO_2$
(d) Methylbenzene   (e) 2-Methyl-1-butene   (f) *cis*-3-Hexene

Problem 13.14   How many signals would you expect the following compound to have in its ^{1}H NMR spectrum?

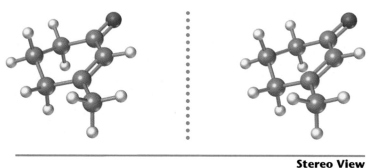

Stereo View

● ●

13.9 Chemical Shifts in ^{1}H NMR Spectroscopy

We said previously that differences in chemical shifts are caused by the small local magnetic fields of electrons surrounding the different nuclei. Nuclei that are more strongly shielded by electrons require a higher applied field to bring them into resonance and therefore absorb on the right side of the NMR chart. Nuclei that are less strongly shielded need a lower applied field for resonance to occur and therefore absorb on the left of the NMR chart.

Most ¹H chemical shifts fall within the range 0–10 δ, which can be conveniently divided into the five regions shown in Table 13.2. By remembering the positions of these regions, it's often possible to tell at a glance what kinds of protons a molecule contains.

TABLE 13.2 Regions of the ¹H NMR Spectrum

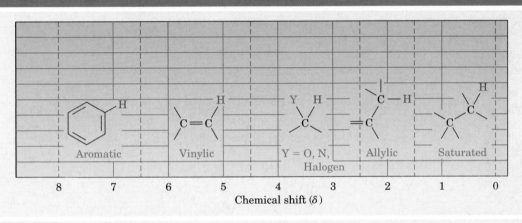

Region (δ)	Proton type	Comments
0–1.5	—C—C—H	Protons on carbon next to saturated centers absorb in this region. Thus, the alkane portions of most organic molecules show complex absorption here.
1.5–2.5	=C—C—H	Protons on carbon next to unsaturated centers (allylic, benzylic, next to carbonyl) show characteristic absorptions in this region, just downfield from other alkane resonances.
2.5–4.5	Y—C—H	Protons on carbon next to electronegative atoms (halogen, O, N) are deshielded because of the electron-withdrawing ability of these atoms. Thus, the protons absorb in this midfield region.
4.5–6.5	C=C with H	Protons on double-bond carbons (vinylic protons) are strongly deshielded by the neighboring π bond and therefore absorb in this characteristic downfield region.
6.5–8.0	aromatic ring with H	Protons on aromatic rings (aryl protons) are strongly deshielded by the π orbitals of the ring and absorb in this characteristic low-field range.

Table 13.3 shows the correlation of ^1H chemical shift with electronic environment in more detail. In general, protons bonded to saturated, sp^3-hybridized carbons absorb at higher fields, whereas protons bonded to sp^2-hybridized carbons absorb at lower fields. Protons on carbons that are bonded to electronegative atoms, such as N, O, or halogen, also absorb at lower fields.

TABLE 13.3   Correlation of ^1H Chemical Shift with Environment

Type of hydrogen		Chemical shift (δ)	Type of hydrogen		Chemical shift (δ)
Reference	$(CH_3)_4Si$	0	Alkyl halide X = Cl, Br, I	X—C—H	2.5–4.0
Saturated primary	—CH_3	0.7–1.3	Alcohol	C—O—H	2.5–5.0 (Variable)
Saturated secondary	—CH_2—	1.2–1.6	Alcohol, ether	O—C—H	3.3–4.5
Saturated tertiary	C—H	1.4–1.8	Vinylic	C=C—H	4.5–6.5
Allylic	C=C—C—H	1.6–2.2	Aromatic	Ar—H	6.5–8.0
Methyl ketone	C(=O)—CH_3	2.0–2.4	Aldehyde	C(=O)—H	9.7–10.0
Aromatic methyl	Ar—CH_3	2.4–2.7	Carboxylic acid	C(=O)—O—H	11.0–12.0
Alkynyl	—C≡C—H	2.5–3.0			

Practice Problem 13.3   Methyl 2,2-dimethylpropanoate $(CH_3)_3COOCH_3$ has two peaks in its ^1H NMR spectrum. What are their approximate chemical shifts?

Strategy   Identify the types of hydrogens in the molecule, and note whether each is alkyl, vinylic, or next to an electronegative atom. Then predict where each absorbs, using Table 13.3 if necessary.

Solution   The –OCH_3 protons absorb around 3.5–4.0 δ because they are on carbon bonded to oxygen. The $(CH_3)_3C$– protons absorb near 1.0 δ because they are typical alkane-like protons.

Problem 13.15   Each of the following compounds has a single ¹H NMR peak. Approximately where would you expect each compound to absorb?
(a) Cyclohexane   (b) CH_3COCH_3 (c) Benzene

$$\text{(d) Glyoxal, } H-\overset{\overset{\displaystyle O}{\|}}{C}-\overset{\overset{\displaystyle O}{\|}}{C}-H \qquad \text{(e) } CH_2Cl_2 \qquad \text{(f) } (CH_3)_3N$$

Problem 13.16   Identify the different kinds of protons in the following molecule, and tell where you would expect each to absorb:

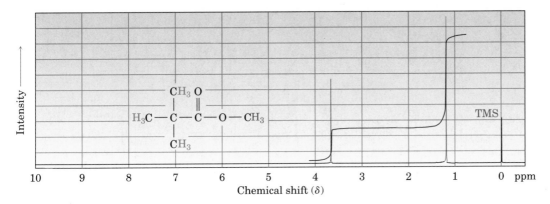

● ●

13.10   Integration of ¹H NMR Absorptions: Proton Counting

Look at the ¹H NMR spectrum of methyl 2,2-dimethylpropanoate in Figure 13.15. There are two peaks, corresponding to the two kinds of protons, but the peaks aren't the same size. The peak at 1.2 δ, due to the $(CH_3)_3C-$ protons, is larger than the peak at 3.7 δ, due to the $-OCH_3$ protons.

FIGURE 13.15 ▼

The ¹H NMR spectrum of methyl 2,2-dimethylpropanoate. Integrating the peaks in a "stair-step" manner shows that they have a 1:3 ratio, corresponding to the ratio of the numbers of protons (3:9) responsible for each peak.

The area under each peak is proportional to the number of protons causing that peak. By electronically measuring, or **integrating**, the area under each peak, it's possible to measure the relative number of each kind of proton in a molecule. Integrated peak areas are superimposed over the spectrum as a "stair-step" line, with the height of each step proportional to the area under the peak, and therefore proportional to the relative number of protons causing the peak. To compare the size of one peak against another, simply take a ruler and measure the heights of the various steps. For example, the two peaks in methyl 2,2-dimethylpropanoate are found to have a 1:3 (or 3:9) ratio when integrated—exactly what we expect since the three $-OCH_3$ protons are equivalent and the nine $(CH_3)_3C-$ protons are equivalent.

● ●

Problem 13.17   How many peaks would you expect in the ^1H NMR spectrum of 1,4-dimethylbenzene (*p*-xylene)? What ratio of peak areas would you expect on integration of the spectrum? Refer to Table 13.3 for approximate chemical shifts, and sketch what the spectrum would look like. (Remember from Section 2.4 that aromatic rings have two resonance forms.)

p-Xylene

● ●

13.11 Spin–Spin Splitting in ^1H NMR Spectra

In the ^1H NMR spectra we've seen thus far, each different kind of proton in a molecule has given rise to a single peak. It often happens, though, that the absorption of a proton splits into *multiple* peaks (a **multiplet**). For example, in the ^1H NMR spectrum of bromoethane shown in Figure 13.16,

FIGURE 13.16 ▼

The ^1H NMR spectrum of bromoethane, CH_3CH_2Br. The $-CH_2Br$ protons appear as a quartet at 3.42 δ, and the $-CH_3$ protons appear as a triplet at 1.68 δ.

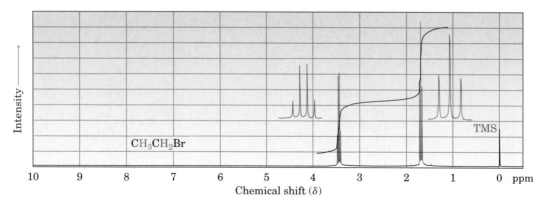

the –CH₂Br protons appear as four peaks (a *quartet*) at 3.42 δ and the –CH₃ protons appear as three peaks (a *triplet*) at 1.68 δ.

Called **spin–spin splitting**, the phenomenon of multiple absorptions is caused by the interaction, or **coupling**, of the spins of nearby nuclei. In other words, the tiny magnetic field produced by one nucleus affects the magnetic field felt by neighboring nuclei. Look at the –CH₃ protons in bromoethane, for example. The three equivalent –CH₃ protons are neighbored by two other magnetic nuclei—the protons on the adjacent –CH₂Br group. Each of the –CH₂Br protons has its own nuclear spin, which can align either with or against the applied field, producing a tiny effect that is felt by the neighboring –CH₃ protons.

There are three ways in which the spins of the two –CH₂Br protons can align, as shown schematically in Figure 13.17. If both proton spins align *with* the applied field, the total effective field felt by the neighboring –CH₃ protons is slightly larger than it would otherwise be. Consequently, the applied field necessary to cause resonance is slightly reduced. Alternatively, if one of the –CH₂Br proton spins aligns *with* the field and one aligns *against* the field, there is no effect on the neighboring –CH₃ protons. (There are two ways this arrangement can occur, depending on which of the two proton spins aligns which way.) Finally, if both –CH₂Br proton spins align *against* the applied field, the effective field felt by the –CH₃ protons is slightly smaller than it would otherwise be, and the applied field needed for resonance is slightly increased.

FIGURE 13.17 ▼

The origin of spin–spin splitting in bromoethane. The nuclear spins of neighboring protons, indicated by horizontal arrows, align either with or against the applied field, causing the splitting of absorptions into multiplets.

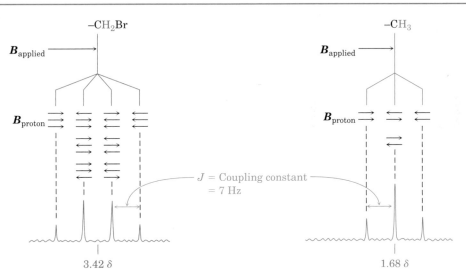

Quartet due to coupling with –CH₃   Triplet due to coupling with –CH₂Br

Any given molecule can adopt only one of the three possible alignments of –CH₂Br spins, but in a large collection of molecules, all three spin states will be represented in a 1:2:1 statistical ratio. We therefore find that the neighboring –CH₃ protons come into resonance at three slightly different values of the applied field, and we see a 1:2:1 triplet in the NMR spectrum.

One resonance is a little above where it would be without coupling, one is at the same place it would be without coupling, and the third resonance is a little below where it would be without coupling.

In the same way that the –CH$_3$ absorption of bromoethane is split into a triplet, the –CH$_2$Br absorption is split into a quartet. The three spins of the neighboring –CH$_3$ protons can align in four possible combinations: all three with the applied field, two with and one against (three ways), one with and two against (three ways), or all three against. Thus, four peaks are produced for the –CH$_2$Br protons in a 1:3:3:1 ratio.

As a general rule, called the **$n + 1$ rule**, protons that have n equivalent neighboring protons show $n + 1$ peaks in their NMR spectrum. For example, the spectrum of 2-bromopropane in Figure 13.18 shows a doublet at 1.71 δ and a seven-line multiplet, or *septet,* at 4.28 δ. The septet is caused by splitting of the –CHBr– proton signal by six equivalent neighboring protons on the two methyl groups ($n = 6$ leads to $6 + 1 = 7$ peaks). The doublet is due to signal splitting of the six equivalent methyl protons by the single –CHBr– proton ($n = 1$ leads to 2 peaks). Integration confirms the expected 6:1 ratio.

FIGURE 13.18 ▼

The ^{1}H NMR spectrum of 2-bromopropane. The –CH$_3$ proton signal at 1.71 δ is split into a doublet, and the –CHBr– proton signal at 4.28 δ is split into a septet.

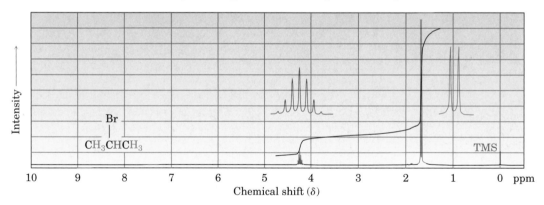

The distance between peaks in a multiplet is called the **coupling constant**, denoted *J*. Coupling constants are measured in hertz and generally fall in the range 0–18 Hz. The exact value of the coupling constant between two neighboring protons depends on the geometry of the molecule, but a typical value for an open-chain alkane is $J = 6$–8 Hz. Note that the same coupling constant is shared by both groups of hydrogens whose spins are coupled and is independent of spectrometer field strength. In bromoethane, for instance, the –CH$_2$Br protons are coupled to the –CH$_3$ protons and appear as a quartet with $J = 7$ Hz. The –CH$_3$ protons appear as a triplet with the same $J = 7$ Hz coupling constant.

Since coupling is a reciprocal interaction between two adjacent groups of protons, it's sometimes possible to tell which multiplets in a complex NMR spectrum are related to each other. If two multiplets have the same coupling constant, they are probably related, and the protons causing those multiplets are therefore adjacent in the molecule.

Spin–spin splitting in ¹H NMR can be summarized by three rules:

RULE 1 **Chemically equivalent protons do not show spin–spin splitting.** The equivalent protons may be on the same carbon or on different carbons, but their signals don't split.

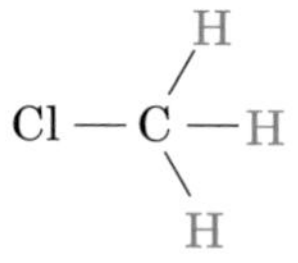

$$Cl - \overset{\displaystyle H}{\underset{\displaystyle H}{C}} - H \qquad\qquad Cl - \overset{\displaystyle H}{\underset{\displaystyle H}{C}} - \overset{\displaystyle H}{\underset{\displaystyle H}{C}} - Cl$$

Three C–H protons are chemically equivalent; no splitting occurs.   Four C–H protons are chemically equivalent; no splitting occurs.

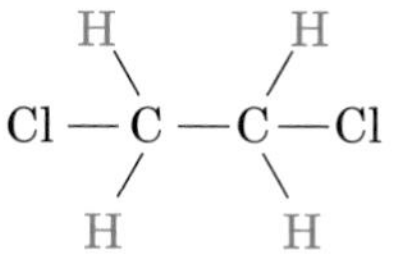

RULE 2 **The signal of a proton that has *n* equivalent neighboring protons is split into a multiplet of *n* + 1 peaks with coupling constant *J*.** Protons that are farther than two carbon atoms apart don't usually couple, although they sometimes show small coupling when they are separated by a π bond.

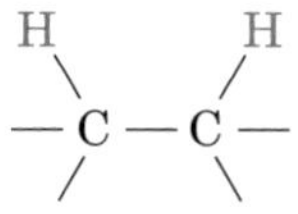

$$-\overset{\displaystyle H}{\underset{\displaystyle}{C}} - \overset{\displaystyle H}{\underset{\displaystyle}{C}} - \qquad\qquad -\overset{\displaystyle H}{\underset{\displaystyle}{C}} - \overset{\displaystyle H}{\underset{\displaystyle}{C}} - \overset{\displaystyle H}{\underset{\displaystyle}{C}} -$$

Splitting observed 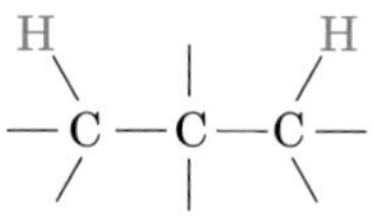   Splitting not usually observed

RULE 3 **Two groups of protons coupled to each other have the same coupling constant, *J*.**

The most commonly observed coupling patterns and the relative intensities of lines in their multiplets are listed in Table 13.4. Note that it's not

TABLE 13.4   Some Common Spin Multiplicities

Number of equivalent adjacent protons	Type of multiplet observed	Ratio of intensities
0	Singlet	1
1	Doublet	1:1
2	Triplet	1:2:1
3	Quartet	1:3:3:1
4	Quintet	1:4:6:4:1
6	Septet	1:6:15:20:15:6:1

possible for a given proton to have *five* equivalent neighboring protons. (Why not?) A six-line multiplet, or sextet, is therefore found only when a proton has five *nonequivalent* neighboring protons that happen to be coupled with an identical coupling constant *J*.

The spectrum of *para*-methoxypropiophenone in Figure 13.19 further illustrates the preceding three rules. The downfield absorptions at 6.91 and 7.93 δ are due to the four aromatic ring protons. There are two kinds of aromatic protons, each of which gives a signal that is split into a doublet by its neighbor. The –OCH$_3$ signal is unsplit and appears as a sharp singlet at 3.84 δ. The –CH$_2$– protons next to the carbonyl group appear at 2.93 δ in the region expected for protons on carbon next to an unsaturated center, and their signal is split into a quartet by coupling with the protons of the neighboring methyl group. The methyl protons appear as a triplet at 1.20 δ in the usual upfield region.

FIGURE 13.19 ▼

The ^{1}H NMR spectrum of *para*-methoxypropiophenone.

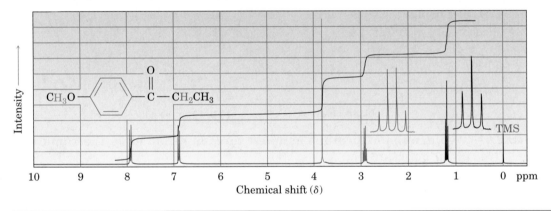

One further question needs to be answered before leaving the topic of spin–spin splitting: Why is spin–spin splitting seen only for ^{1}H NMR? Why is there no splitting of *carbon* signals into multiplets in ^{13}C NMR? After all, you might expect that the spin of a given ^{13}C nucleus would couple with the spin of an adjacent magnetic nucleus, either ^{13}C or ^{1}H. In fact, the spins of ^{13}C nuclei *do* couple with the spins of nearby magnetic nuclei, but the spectrometer operating conditions typically used in ^{13}C NMR are such that the splitting of signals is suppressed.

No coupling of a ^{13}C nucleus with nearby *carbons* is seen because the low natural abundance makes it improbable that two ^{13}C nuclei will be adjacent. No coupling of a ^{13}C nucleus with nearby *hydrogens* is seen because ^{13}C spectra, as previously noted (Section 13.6), are normally recorded using what is called *broadband decoupling*. At the same time that the sample is irradiated with a pulse of rf energy to cover the *carbon* resonance frequencies, it is also irradiated by a second band of rf energy covering all the *hydrogen* resonance frequencies. This second irradiation makes the hydrogens spin-flip so rapidly that their local magnetic fields average to zero, and no coupling with carbon spins occurs.

• •

Practice Problem 13.4   Propose a structure for a compound, $C_5H_{12}O$, that fits the following ¹H NMR data: 0.92 δ (3 H, triplet, $J = 7$ Hz), 1.20 δ (6 H, singlet), 1.50 δ (2 H, quartet, $J = 7$ Hz), 1.64 δ (1 H, broad singlet).

Strategy   As noted in Practice Problem 13.2, it's best to begin solving structural problems by calculating a molecule's degree of unsaturation. In the present instance, a formula of $C_5H_{12}O$ corresponds to a saturated, open-chain molecule, either an alcohol or an ether.

   To interpret the NMR information, let's look at each absorption individually. The 3-proton absorption at 0.92 δ is due to a methyl group in an alkane-like environment, and the triplet splitting pattern implies that the CH_3 is next to a CH_2. Thus, our molecule contains an ethyl group, CH_3CH_2. The 6-proton singlet at 1.20 δ is due to two equivalent alkane-like methyl groups attached to a carbon with no hydrogens, $(CH_3)_2C$, and the 2-proton quartet at 1.50 δ is due to the CH_2 of the ethyl group. All 5 carbons and 11 of the 12 hydrogens in the molecule are now accounted for. The remaining hydrogen, which appears as a broad 1-proton singlet at 1.64 δ, is probably due to an OH group, since there is no other way to account for it. Putting the pieces together gives the structure.

Solution

• •

Problem 13.18   Predict the splitting patterns you would expect for each proton in the following molecules:

(a) $CHBr_2CH_3$ (b) $CH_3OCH_2CH_2Br$ (c) $ClCH_2CH_2CH_2Cl$

(d) $CH_3CHCOCH_2CH_3$ (e) $CH_3CH_2COCHCH_3$ (f)
      | |
      CH_3 (with C=O above COCH₂CH₃) CH_3 (with C=O above COCHCH₃)

Problem 13.19   Draw structures for compounds that meet the following descriptions:

(a) C_2H_6O; one singlet
(b) C_3H_7Cl; one doublet and one septet
(c) $C_4H_8Cl_2O$; two triplets
(d) $C_4H_8O_2$; one singlet, one triplet, and one quartet

Problem 13.20   The integrated ^1H NMR spectrum of a compound of formula $C_4H_{10}O$ is shown in Figure 13.20. Propose a structure consistent with the data.

FIGURE 13.20 ▼

An integrated ^1H NMR spectrum for Problem 13.20.

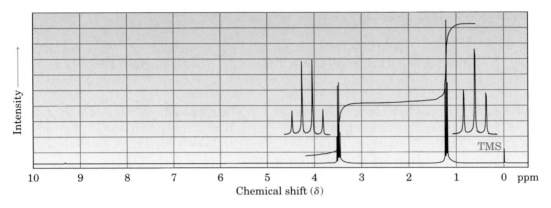

• •

13.12 More Complex Spin–Spin Splitting Patterns

In all the ^1H NMR spectra we've seen so far, the chemical shifts of different protons have been distinct, and the spin–spin splitting patterns have been straightforward. It often happens, however, that different kinds of hydrogens in a molecule have accidentally *overlapping* signals. The spectrum of toluene (methylbenzene) in Figure 13.21, for example, shows that the five aromatic ring protons give a complex, overlapping pattern, even though they aren't all equivalent.

FIGURE 13.21 ▼

The ^1H NMR spectrum of toluene, showing the accidental overlap of the five nonequivalent aromatic ring protons.

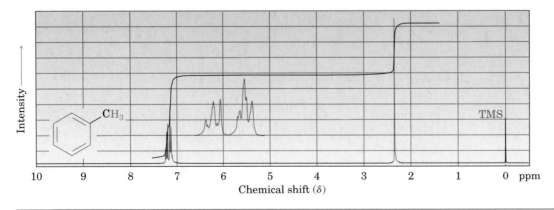

Yet another complication in ^{1}H NMR spectroscopy arises when a signal is split by two or more *nonequivalent* kinds of protons, as is the case with *trans*-cinnamaldehyde, isolated from oil of cinnamon (Figure 13.22). Although the $n + 1$ rule correctly predicts splitting caused by *equivalent* protons, splittings caused by nonequivalent protons are more complex.

FIGURE 13.22 ▼

The ^{1}H NMR spectrum of *trans*-cinnamaldehyde. The signal of the proton at C2 (blue) is split into four peaks—a doublet of doublets—by the two nonequivalent neighboring protons.

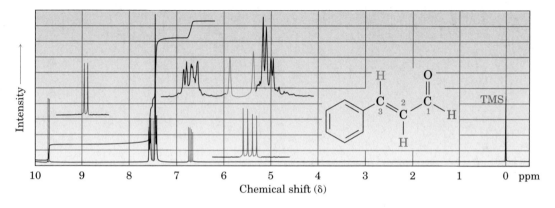

To understand the ^{1}H NMR spectrum of *trans*-cinnamaldehyde, we have to isolate the different parts and look at the signal of each proton individually:

- The five aromatic proton signals (black in Figure 13.22) overlap into a complex pattern with a large peak at 7.42 δ and a broad absorption at 7.57 δ.

- The aldehyde proton signal at C1 (red) appears in the normal downfield position at 9.69 δ and is split into a doublet with $J = 6$ Hz by the adjacent proton at C2.

- The vinylic proton at C3 (green) is next to the aromatic ring and is therefore shifted downfield from the normal vinylic region. This C3 proton signal appears as a doublet centered at 7.49 δ. Because it has one neighbor proton at C2, its signal is split into a doublet, with $J = 12$ Hz.

- The C2 vinylic proton signal (blue) appears at 6.73 δ and shows an interesting, four-line absorption pattern. It is coupled to the two nonequivalent protons at C1 and C3 with two different coupling constants: $J_{1\text{-}2} = 6$ Hz and $J_{2\text{-}3} = 12$ Hz.

The best way to understand the effect of multiple coupling such as occurs for the C2 proton of *trans*-cinnamaldehyde is to draw a *tree diagram,* like that in Figure 13.23. The diagram shows the individual effect of each coupling constant on the overall pattern. Coupling with the C3 proton splits the signal of the C2 proton in *trans*-cinnamaldehyde into a doublet with $J = 12$ Hz. Further coupling with the aldehyde proton then splits each peak of the doublet into new doublets, and we therefore observe a four-line spectrum for the C2 proton.

FIGURE 13.23 ▼

A tree diagram for the C2 proton of *trans*-cinnamaldehyde shows how it is coupled to the C1 and C3 protons with different coupling constants.

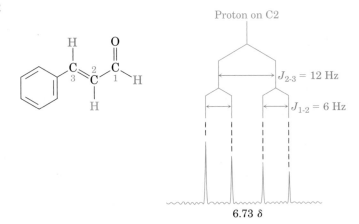

Proton on C2

$J_{2-3} = 12$ Hz

$J_{1-2} = 6$ Hz

6.73 δ

One further point evident in the cinnamaldehyde spectrum is that the four peaks of the C2 proton signal are not all the same size: The two left-hand peaks are somewhat larger than the two right-hand peaks. Such a size difference occurs whenever coupled nuclei have similar chemical shifts—in this case, 7.49 δ for the C3 proton and 6.73 δ for the C2 proton. The peaks nearer the signal of the coupled partner are always larger, and the peaks farther from the signal of the coupled partner are always smaller. Thus, the left-hand peaks of the C2 proton multiplet at 6.73 δ are closer to the C3 proton absorption at 7.49 δ and are larger than the right-hand peaks. At the same time, the *right-hand* peak of the C3 proton doublet at 7.49 δ is larger than the left-hand peak because it is closer to the C2 proton multiplet at 6.73 δ. This skewing effect on multiplets can often be useful because it tells where to look in the spectrum to find the coupled partner: Look toward the direction of the larger peaks.

Problem 13.21   3-Bromo-1-phenyl-1-propene shows a complex NMR spectrum in which the vinylic proton at C2 is coupled with both the C1 vinylic proton ($J = 16$ Hz) and the C3 methylene protons ($J = 8$ Hz). Draw a tree diagram for the C2 proton signal and account for the fact that a five-line multiplet is observed.

3-Bromo-1-phenyl-1-propene

. . . .

13.13 Uses of ¹H NMR Spectroscopy

NMR can be used to help identify the product of nearly every reaction run in the laboratory. For example, we said in Section 7.5 that hydroboration/oxidation of alkenes occurs with non-Markovnikov regiochemistry; that is, the less highly substituted alcohol is formed. With the help of NMR, we can now prove this statement.

Does hydroboration/oxidation of methylenecyclohexane yield cyclohexylmethanol or 1-methylcyclohexanol?

Methylenecyclohexane   Cyclohexylmethanol   1-Methylcyclohexanol

The ¹H NMR spectrum of the reaction product is shown in Figure 13.24a (p. 508). The spectrum shows a 2-proton triplet at 3.40 δ, indicating that the product has a –CH₂– group bonded to an electronegative oxygen atom (–CH₂OH). Furthermore, the spectrum shows *no* large 3-proton singlet absorption near 1 δ, where we would expect the signal of a quaternary –CH₃ group to occur. (Figure 13.24b gives the spectrum of 1-methylcyclohexanol, the alternative product.) Thus, it's clear that cyclohexylmethanol is the reaction product.

. . . .

Problem 13.22   How could you use ¹H NMR to determine the regiochemistry of electrophilic addition to alkenes? For example, does addition of HCl to 1-methylcyclohexene yield 1-chloro-1-methylcyclohexane or 1-chloro-2-methylcyclohexane?

. . . .

FIGURE 13.24 ▼

(a) The ^{1}H NMR spectrum of cyclohexylmethanol, the product from hydroboration/oxidation of methylenecyclohexane, and (b) the ^{1}H NMR spectrum of 1-methylcyclohexanol, the possible alternative reaction product.

(a)

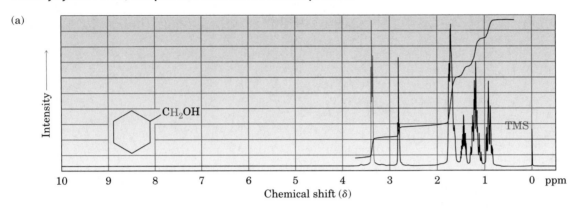

(b)

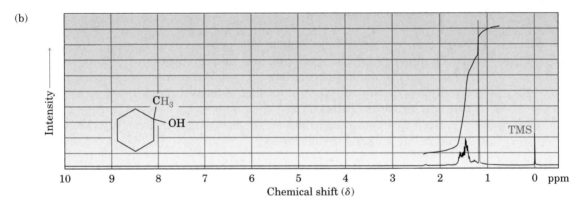

CHEMISTRY @ WORK

Magnetic Resonance Imaging (MRI)

As practiced by organic chemists, NMR spectroscopy is a powerful method of structure determination. A small amount of sample, typically a few milligrams or less, is dissolved in 1 mL or so of solvent, the solution is placed in a thin glass tube, and the tube is placed into the narrow (1–2 cm) gap between the poles of a strong magnet. Imagine, though, that a much larger NMR instrument were available. Instead of a few milligrams, the sample

(continued) ▶

size could be tens of kilograms; instead of a narrow gap between magnet poles, the gap could be large enough for a whole person to climb into so that an NMR spectrum of body parts could be obtained. What you've just imagined is an instrument for *magnetic resonance imaging (MRI)*, a diagnostic technique of enormous value to the medical community because of its advantages over X-ray or radioactive imaging methods.

Like NMR spectroscopy, MRI takes advantage of the magnetic properties of certain nuclei, typically hydrogen, and of the signals emitted when those nuclei are stimulated by radiofrequency energy. Unlike what happens in NMR spectroscopy, though, MRI instruments use powerful computers and data manipulation techniques to look at the three-dimensional *location* of magnetic nuclei in the body rather than at the chemical nature of the nuclei. As noted, most MRI instruments currently look at hydrogen, present in abundance wherever there is water or fat in the body.

The signals produced vary with the density of hydrogen atoms and with the nature of their surroundings, allowing identification of different types of tissue and even allowing the visualization of motion. For example, the volume of blood leaving the heart in a single stroke can be measured, and heart motion can be observed. Soft tissues that do not show up well on X rays can be seen clearly, allowing diagnosis of brain tumors, strokes, and other conditions. The technique is also valuable in diagnosing damage to knees or other joints and is a painless alternative to arthroscopy, in which an endoscope is physically introduced into the knee joint.

Several types of atoms in addition to hydrogen can be detected by MRI, and the applications of images based on ^{31}P atoms are being explored. The technique holds great promise for studies of metabolism.

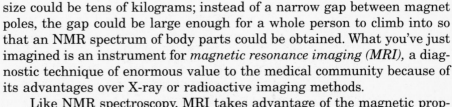

Magnetic resonance imaging is a noninvasive NMR technique used for the diagnosis of brain cancer and many other conditions.

Summary and Key Words

When magnetic nuclei such as ^1H and ^{13}C are placed in a strong magnetic field, their spins orient either with or against the field. On irradiation with radiofrequency (rf) waves, energy is absorbed and the nuclei "spin-flip" from the lower energy state to the higher energy state. This absorption of rf energy is detected, amplified, and displayed as a **nuclear magnetic resonance (NMR) spectrum**.

An NMR spectrum can be obtained by irradiating a sample with rf energy of constant frequency while slowly changing the strength of the applied magnetic field. Each chemically distinct ^1H or ^{13}C nucleus in a molecule comes into resonance at a slightly different value of the applied field, thereby producing a unique absorption signal. The exact position of each peak is called the **chemical shift**. Chemical shifts are caused by electrons setting up tiny local magnetic fields that **shield** a nearby nucleus from the applied field.

The NMR chart is calibrated in **delta units (δ)**, where $1\ \delta = 1$ ppm of spectrometer frequency. Tetramethylsilane (TMS) is used as a reference point to which other peaks are compared because it shows both ^{1}H and ^{13}C absorptions at unusually high values of the applied magnetic field. The TMS absorption occurs at the right-hand (**upfield**) side of the chart and is arbitrarily assigned a value of $0\ \delta$.

Most ^{13}C spectra are run on **Fourier-transform NMR (FT-NMR)** spectrometers using broadband decoupling of proton spins so that each chemically distinct carbon shows a single unsplit resonance line. As with ^{1}H NMR, the chemical shift of each ^{13}C signal provides information about a carbon's chemical environment in the sample. In addition, the number of protons attached to each carbon can be determined using the **DEPT-NMR** technique.

In ^{1}H NMR spectra, the area under each absorption peak can be electronically **integrated** to determine the relative number of hydrogens responsible for each peak. In addition, neighboring nuclear spins can **couple**, causing the **spin–spin splitting** of NMR peaks into **multiplets**. The NMR signal of a hydrogen neighbored by n equivalent adjacent hydrogens splits into $n + 1$ peaks (the **$n + 1$ rule**) with **coupling constant J**.

Visualizing Chemistry

• •

(Problems 13.1–13.22 appear within the chapter.)

13.23   Into how many peaks would you expect the ^{1}H NMR signals of the indicated protons to be split? (Yellow-green = Cl.)

(a)   (b)

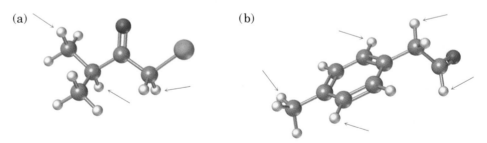

13.24   How many absorptions would you expect the following compound to have in its ^{13}C NMR spectrum?

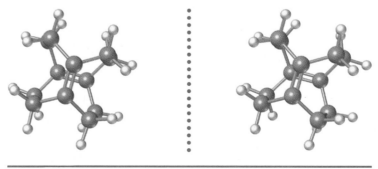

Stereo View

13.25 Sketch what you might expect the ^1H and ^{13}C NMR spectra of the following compound to look like (yellow-green = Cl):

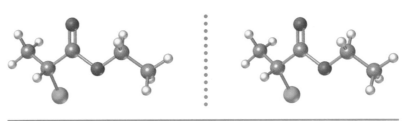

Stereo View

13.26 How many unique kinds of protons and how many unique kinds of carbons are there in the following compound? Don't forget that cyclohexane rings can ring-flip.

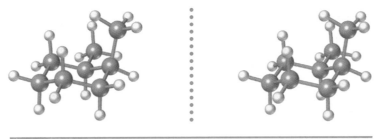

Stereo View

Additional Problems

13.27 The following ^1H NMR absorptions were obtained on a spectrometer operating at 100 MHz and are given in hertz downfield from the TMS standard. Convert the absorptions to δ units.
(a) 218 Hz   (b) 478 Hz   (c) 752 Hz

13.28 The following ^1H NMR absorptions were obtained on a spectrometer operating at 300 MHz. Convert the chemical shifts from δ units to hertz downfield from TMS.
(a) 2.1 δ (b) 3.45 δ (c) 6.30 δ (d) 7.70 δ

13.29 When measured on a spectrometer operating at 60 MHz, chloroform ($CHCl_3$) shows a single sharp absorption at 7.3 δ.
(a) How many parts per million downfield from TMS does chloroform absorb?
(b) How many hertz downfield from TMS would chloroform absorb if the measurement were carried out on a spectrometer operating at 360 MHz?
(c) What would be the position of the chloroform absorption in δ units when measured on a 360 MHz spectrometer?

13.30 How many signals would you expect each of the following molecules to have in its ^1H and ^{13}C spectra?
(a) $(CH_3)_2C \!=\! C(CH_3)_2$ (b) 1,1-Dimethylcyclohexane

(c) $CH_3\overset{\overset{\displaystyle O}{\displaystyle \|}}{C}CH_3$ (d) $(CH_3)_3C\overset{\overset{\displaystyle O}{\displaystyle \|}}{C}OCH_3$

(e)   (f) 1,1-Dimethylcyclopropane

13.31   How many absorptions would you expect to observe in the ^{13}C NMR spectra of the following compounds?
(a) 1,1-Dimethylcyclohexane   (b) $CH_3CH_2OCH_3$
(c) *tert*-Butylcyclohexane   (d) 3-Methyl-1-pentyne
(e) *cis*-1,2-Dimethylcyclohexane   (f)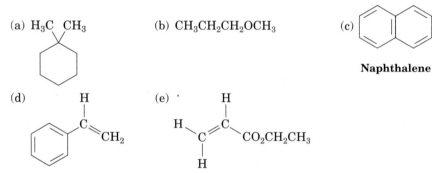

13.32   Suppose you ran a DEPT-135 spectrum for each substance in Problem 13.31. Indicate which carbon atoms in each molecule would show positive peaks and which would show negative peaks.

13.33   Why do you suppose accidental overlap of signals is much more common in ^1H NMR than in ^{13}C NMR?

13.34   Is a nucleus that absorbs at 6.50 δ more shielded or less shielded than a nucleus that absorbs at 3.20 δ? Does the nucleus that absorbs at 6.50 δ require a stronger applied field or a weaker applied field to come into resonance than the nucleus that absorbs at 3.20 δ?

13.35   How many types of nonequivalent protons are in each of the following molecules?

(a) H_3C CH_3 (b) $CH_3CH_2CH_2OCH_3$ (c)

Naphthalene

(d)   H   (e)   H

C═CH₂   C═C $CO_2CH_2CH_3$

H

Styrene   **Ethyl acrylate**

13.36   The following compounds all show a single line in their ^1H NMR spectra. List them in expected order of increasing chemical shift:

CH_4, CH_2Cl_2,   Cyclohexane, CH_3COCH_3, $H_2C{=}CH_2$,   Benzene

13.37   Predict the splitting pattern for each kind of hydrogen in the following molecules:
(a) $(CH_3)_3CH$ (b) $CH_3CH_2COOCH_3$ (c) *trans*-2-Butene

13.38   Predict the splitting pattern for each kind of hydrogen in isopropyl propanoate, $CH_3CH_2COOCH(CH_3)_2$.

13.39   The acid-catalyzed dehydration of 1-methylcyclohexanol yields a mixture of two alkenes. How would you use ^1H NMR to help you decide which was which?

CH_3 CH_2 CH_3

OH $\xrightarrow{H_3O^+}$ +

13.40 How would you use ^1H NMR to distinguish between the following pairs of isomers?

(a) $CH_3CH=CHCH_2CH_3$ and $\overset{CH_2}{\underset{H_2C-CHCH_2CH_3}{\diagdown}}$

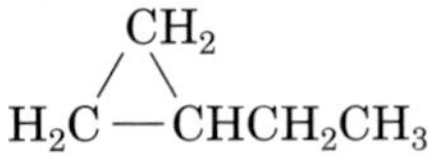

(b) $CH_3CH_2OCH_2CH_3$ and $CH_3OCH_2CH_2CH_3$

(c) $CH_3\overset{O}{\overset{\|}{C}}OCH_2CH_3$ and $CH_3CH_2\overset{O}{\overset{\|}{C}}CH_3$

(d) $H_2C=C(CH_3)\overset{O}{\overset{\|}{C}}CH_3$ and $CH_3CH=CH\overset{O}{\overset{\|}{C}}CH_3$

13.41 Propose structures for compounds with the following formulas that show only one peak in their ^1H NMR spectra:
(a) C_5H_{12} (b) C_5H_{10} (c) $C_4H_8O_2$

13.42 How many ^{13}C NMR absorptions would you expect for *cis*-1,3-dimethylcyclohexane? For *trans*-1,3-dimethylcyclohexane? Explain.

13.43 Assume that you have a compound with formula C_3H_6O.
(a) How many double bonds and/or rings does your compound contain?
(b) Propose as many structures as you can that fit the molecular formula.
(c) If your compound shows an infrared absorption peak at 1715 cm^{-1}, what functional group does it have?
(d) If your compound shows a single ^1H NMR absorption peak at 2.1 δ, what is its structure?

13.44 How would you use ^1H and ^{13}C NMR to help you distinguish among the following isomeric compounds of formula C_4H_8?

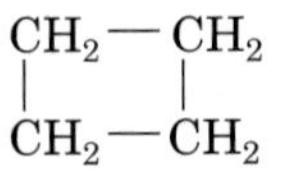

$$\begin{matrix} CH_2-CH_2 \\ | \qquad | \\ CH_2-CH_2 \end{matrix} \qquad H_2C=CHCH_2CH_3$$

$$CH_3CH=CHCH_3 \qquad \overset{CH_3}{\underset{CH_3C=CH_2}{|}}$$

13.45 How could you use ^1H NMR, ^{13}C NMR, and IR spectroscopy to help you distinguish between the following structures?

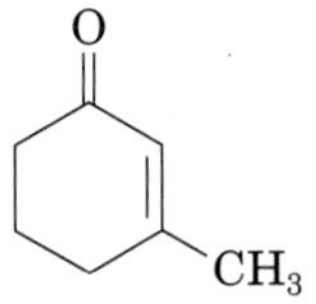

3-Methyl-2-cyclohexenone

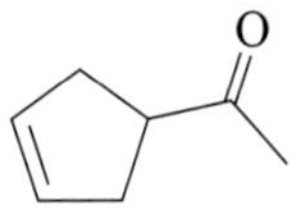

3-Cyclopentenyl methyl ketone

13.46 The compound whose ^{1}H NMR spectrum is shown has the molecular formula $C_3H_6Br_2$. Propose a structure.

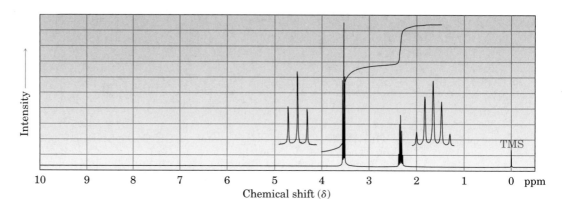

13.47 Propose structures for compounds that fit the following ^{1}H NMR data:

(a) $C_5H_{10}O$
   0.95 δ (6 H, doublet, $J = 7$ Hz)
   2.10 δ (3 H, singlet)
   2.43 δ (1 H, multiplet)

(b) C_3H_5Br
   2.32 δ (3 H, singlet)
   5.35 δ (1 H, broad singlet)
   5.54 δ (1 H, broad singlet)

13.48 The compound whose ^{1}H NMR spectrum is shown has the molecular formula $C_4H_7O_2Cl$ and has an infrared absorption peak at 1740 cm^{-1}. Propose a structure.

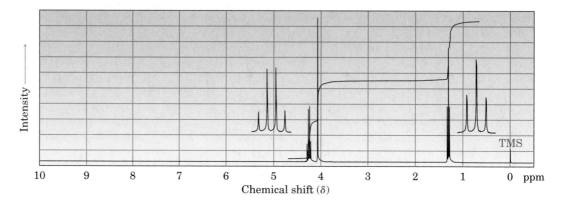

13.49 Propose structures for compounds that fit the following ^{1}H NMR data:

(a) $C_4H_6Cl_2$
   2.18 δ (3 H, singlet)
   4.16 δ (2 H, doublet, $J = 7$ Hz)
   5.71 δ (1 H, triplet, $J = 7$ Hz)

(b) $C_{10}H_{14}$
   1.30 δ (9 H, singlet)
   7.30 δ (5 H, singlet)

(c) C_4H_7BrO
   2.11 δ (3 H, singlet)
   3.52 δ (2 H, triplet, $J = 6$ Hz)
   4.40 δ (2 H, triplet, $J = 6$ Hz)

(d) $C_9H_{11}Br$
   2.15 δ (2 H, quintet, $J = 7$ Hz)
   2.75 δ (2 H, triplet, $J = 7$ Hz)
   3.38 δ (2 H, triplet, $J = 7$ Hz)
   7.22 δ (5 H, singlet)

13.50 Propose structures for the two compounds whose ^1H NMR spectra are shown.

(a) C_4H_9Br

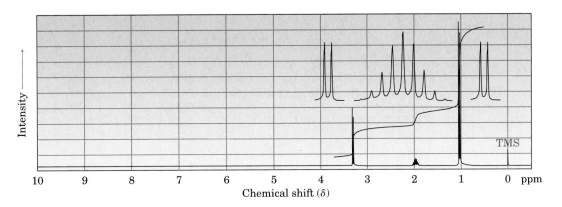

(b) $C_4H_8Cl_2$

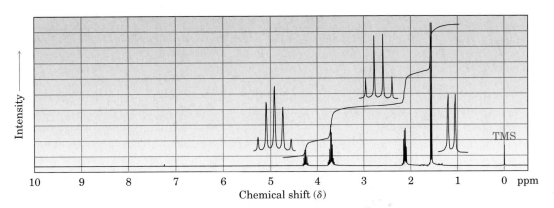

13.51 Long-range coupling between protons more than two carbon atoms apart is some-times observed when π bonds intervene. One example is found in 1-methoxy-1-buten-3-yne, whose ^1H NMR spectrum is shown below:

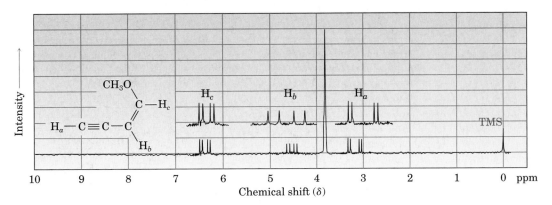

Not only does the acetylenic proton, H_a, couple with the vinylic proton H_b, it also couples with the vinylic proton H_c (*four* carbon atoms away). The coupling constants are

$$J_{a\text{-}b} = 3\ \text{Hz} \qquad J_{a\text{-}c} = 1\ \text{Hz} \qquad J_{b\text{-}c} = 7\ \text{Hz}$$

Construct tree diagrams that account for the observed splitting patterns of H_a, H_b, and H_c.

13.52 Assign as many of the resonances as you can to specific carbon atoms in the ^{13}C NMR spectrum of ethyl benzoate.

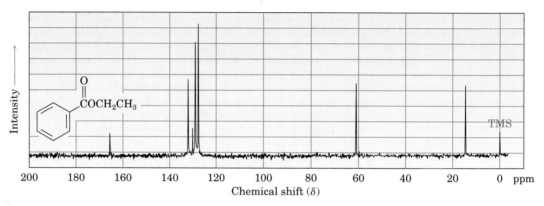

13.53 The ^1H and ^{13}C NMR spectra of compound A, C_8H_9Br, are shown. Propose a structure for A, and assign peaks in the spectra to your structure.

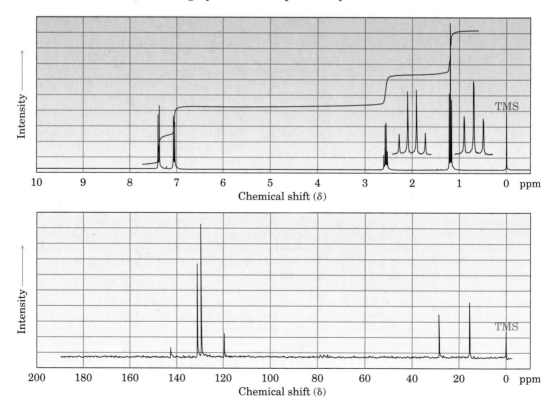

13.54 Propose structures for the three compounds whose ^1H NMR spectra are shown.

(a) $C_5H_{10}O$

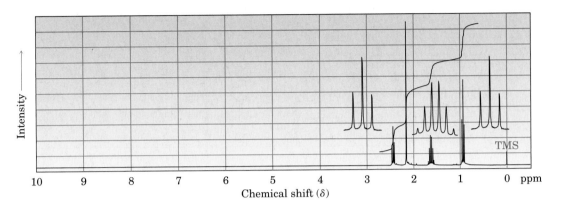

(b) C_7H_7Br

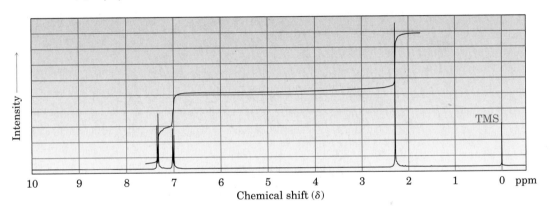

(c) C_8H_9Br

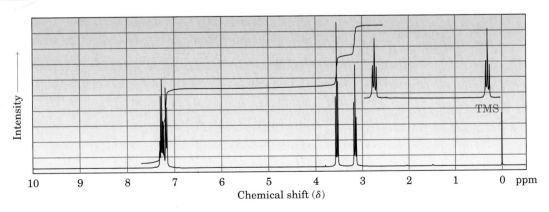

13.55   The mass spectrum and ^{13}C NMR spectrum of a hydrocarbon are shown. Propose a structure for this hydrocarbon, and explain the spectral data.

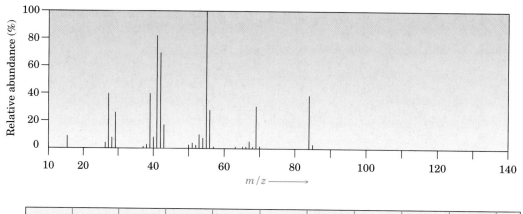

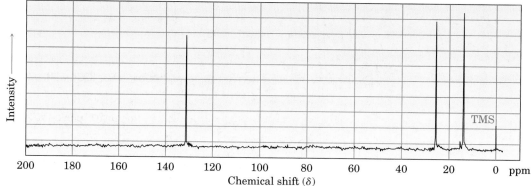

13.56   Compound A, a hydrocarbon with $M^+ = 96$ in its mass spectrum, has the ^{13}C spectral data given below. On reaction with BH_3 followed by treatment with basic H_2O_2, A is converted into B, whose ^{13}C spectral data are also given below. Propose structures for A and B.

Compound A

Broadband-decoupled ^{13}C NMR: 26.8, 28.7, 35.7, 106.9, 149.7 δ

DEPT-90: no peaks

DEPT-135: no positive peaks; negative peaks at 26.8, 28.7, 35.7, 106.9 δ

Compound B

Broadband-decoupled ^{13}C NMR: 26.1, 26.9, 29.9, 40.5, 68.2 δ

DEPT-90: 40.5 δ

DEPT-135: positive peak at 40.5 δ; negative peaks at 26.1, 26.9, 29.9, 68.2 δ

13.57   Propose a structure for compound C, which has $M^+ = 86$ in its mass spectrum, an IR absorption at 3400 cm^{-1}, and the following ^{13}C NMR spectral data:

Compound C

Broadband-decoupled ^{13}C NMR: 30.2, 31.9, 61.8, 114.7, 138.4 δ

DEPT-90: 138.4 δ

DEPT-135: positive peak at 138.4 δ; negative peaks at 30.2, 31.9, 61.8, 114.7 δ

13.58 Compound D is isomeric with compound C (Problem 13.57) and has the following ^{13}C NMR spectral data. Propose a structure.

> **Compound D**
>
> Broadband-decoupled ^{13}C NMR: 9.7, 29.9, 74.4, 114.4, 141.4 δ
>
> DEPT-90: 74.4, 141.4 δ
>
> DEPT-135: positive peaks at 9.7, 74.4, 141.4 δ; negative peaks at 29.9, 114.4 δ

13.59 Propose a structure for compound E, $C_7H_{12}O_2$, which has the following ^{13}C NMR spectral data:

> **Compound E**
>
> Broadband-decoupled ^{13}C NMR: 19.1, 28.0, 70.5, 129.0, 129.8, 165.8 δ
>
> DEPT-90: 28.0, 129.8 δ
>
> DEPT-135: positive peaks at 19.1, 28.0, 129.8 δ; negative peaks at 70.5, 129.0 δ

13.60 Compound F, a hydrocarbon with $M^+ = 96$ in its mass spectrum, undergoes reaction with HBr to yield compound G. Propose structures for F and G, whose ^{13}C NMR spectral data are given below.

> **Compound F**
>
> Broadband-decoupled ^{13}C NMR: 27.6, 29.3, 32.2, 132.4 δ
>
> DEPT-90: 132.4 δ
>
> DEPT-135: positive peak at 132.4 δ; negative peaks at 27.6, 29.3, 32.2 δ
>
> **Compound G**
>
> Broadband-decoupled ^{13}C NMR: 25.1, 27.7, 39.9, 56.0 δ
>
> DEPT-90: 56.0 δ
>
> DEPT-135: positive peak at 56.0 δ; negative peaks at 25.1, 27.7, 39.9 δ

13.61 3-Methyl-2-butanol has five signals in its ^{13}C NMR spectrum at 17.90, 18.15, 20.00, 35.05, and 72.75 δ. Why are the two methyl groups attached to C3 nonequivalent? Making a molecular model should be helpful.

$$\underset{4\quad3\quad2\quad1}{CH_3CHCHCH_3}$$

H₃C OH attached above, **3-Methyl-2-butanol**

13.62 A ^{13}C NMR spectrum of commercially available 2,4-pentanediol, shows *five* peaks at 23.3, 23.9, 46.5, 64.8, and 68.1 δ. Explain.

$$CH_3CHCH_2CHCH_3$$

OH OH attached above, **2,4-Pentanediol**

A Look Ahead

• •

13.63 Carboxylic acids (RCOOH) react with alcohols (R′OH) in the presence of an acid catalyst. The reaction product of propanoic acid with methanol has the following spectroscopic properties. Propose a structure. (See Section 21.3.)

$$CH_3CH_2\overset{\overset{\displaystyle O}{\|}}{C}OH \quad \xrightarrow[\text{H}^+ \text{ catalyst}]{\text{CH}_3\text{OH}} \quad ?$$

Propanoic acid

MS: M$^+$ = 88

IR: 1735 cm^{-1}

^{1}H NMR: 1.11 δ (3 H, triplet, J = 7 Hz); 2.32 δ (2 H, quartet, J = 7 Hz);
3.65 δ (1 H, singlet)

^{13}C NMR: 9.3, 27.6, 51.4, 174.6 δ

13.64 Nitriles (RC≡N) react with Grignard reagents (R′MgBr). The reaction product from 2-methylpropanenitrile with methylmagnesium bromide has the following spectroscopic properties. Propose a structure. (See Section 21.8.)

$$CH_3\overset{\overset{\displaystyle CH_3}{|}}{C}HC≡N \quad \xrightarrow[\text{2. H}_3\text{O}^+]{\text{1. CH}_3\text{MgBr}} \quad ?$$

2-Methylpropanenitrile

MS: M$^+$ = 86

IR: 1715 cm^{-1}

^{1}H NMR: 1.05 δ (6 H, doublet, J = 7 Hz); 2.12 δ (3 H, singlet);
2.67 δ (1 H, septet, J = 7 Hz)

^{13}C NMR: 18.2, 27.2, 41.6, 211.2 δ

Molecular Modeling

13.65 The ^{13}C spectrum of cyclooctatetraene (COT) contains a single peak, but different ^{13}C spectra are observed for 1,2-dimethylcyclooctatetraene and 2,3-dimethylcyclooctatetraene, indicating that these molecules are isomers rather than resonance structures. Use SpartanBuild to build and minimize COT and the two dimethyl derivatives. Examine their structures, and explain the ^{13}C data. How many peaks does the ^{13}C spectrum of each dimethylcyclooctatetraene have?

Cyclooctatetraene

**1,2-Dimethyl-
cyclooctatetraene**

**2,3-Dimethyl-
cyclooctatetraene**

13.66 Aromatic protons are strongly deshielded by the ring's π orbitals, but protons that lie over the face of a ring are shielded. Use SpartanView to examine the structure of *m*-cyclophane, and assign the aromatic proton peaks in the ^{1}H NMR spectrum: δ 4.27, 6.97, and 7.24 in a 1:2:1 ratio.

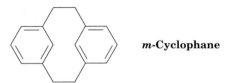

m-Cyclophane

13.67 Coupling is not usually observed for hydrogens separated by two carbon–carbon bonds (H–C–C–C–H), but an exception occurs when these atoms form a "W," or planar zig-zag pattern. Use SpartanBuild to build models of 2-chloronorcamphor and 2-chlorocamphor, and explain why one molecule shows W coupling with H_a ($J \approx 4$ Hz) and the other does not.

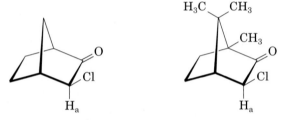

2-Chloronorcamphor   **2-Chlorocamphor**

Conjugated Dienes and Ultraviolet Spectroscopy

Multiple bonds that alternate with single bonds are said to be **conjugated**. Thus, 1,3-butadiene is a *conjugated diene,* whereas 1,4-pentadiene is a non-conjugated diene.

$$H_2C=CH-CH=CH_2$$

1,3-Butadiene
(conjugated; alternating
double and single bonds)

$$H_2C=CH-CH_2-CH=CH_2$$

1,4-Pentadiene
(nonconjugated; nonalternating
double and single bonds)

Many of the pigments responsible for the brilliant colors of fruits and flowers have numerous alternating single and double bonds and are said to be conjugated *polyenes*. Lycopene, the red pigment in tomatoes, is one such molecule. Conjugated *enones* (alkene + ketone) are common structural features of biologically important molecules such as progesterone, the hormone that prepares the uterus for implantation of a fertilized ovum. Cyclic conjugated molecules such as benzene are a major field of study in themselves and will be considered in detail in the next chapter.

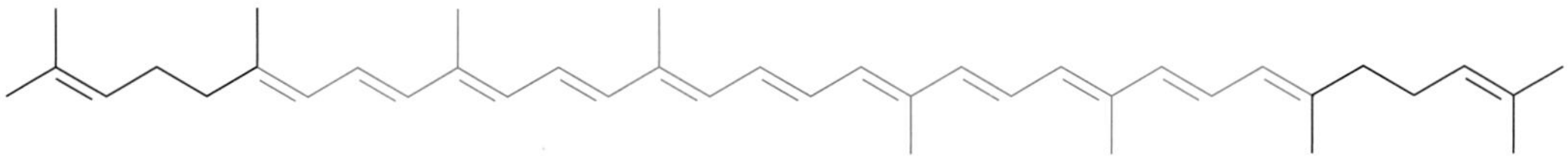

Lycopene, a conjugated polyene

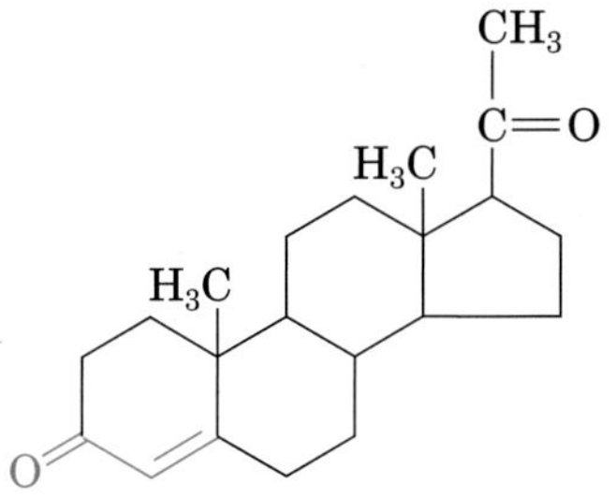

Progesterone, a conjugated enone

Benzene, a cyclic conjugated molecule

Problem 14.1 Which of the following molecules contains conjugation? Circle the conjugated part in each.

(a)

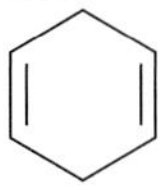

(b)

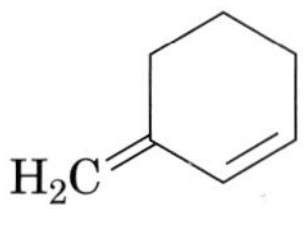

(c) $H_2C=CH-C\equiv N$

(d)

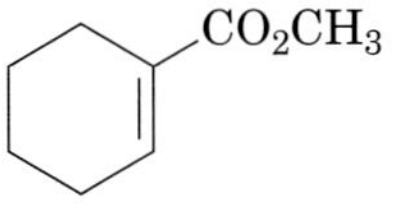

(e)

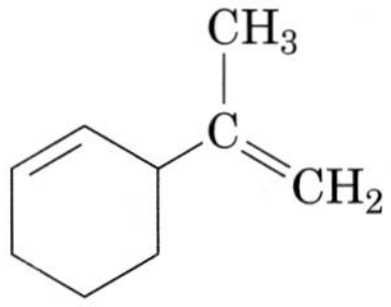

(f)

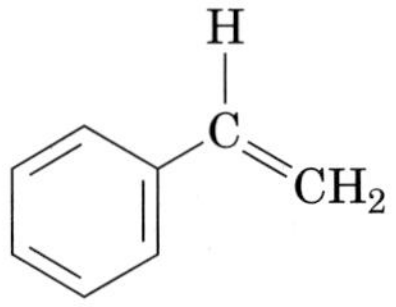

14.1 Preparation of Conjugated Dienes

Conjugated dienes are generally prepared by the methods previously discussed for alkene synthesis. The base-induced elimination of HX from an allylic halide is one such reaction.

Cyclohexene **3-Bromocyclohexene** **1,3-Cyclohexadiene (76%)**

1,3-Butadiene, a substance used industrially to make polymers, is prepared by thermal cracking of butane over a chromium oxide/aluminum oxide catalyst, but this procedure is of little use in the laboratory.

$$CH_3CH_2CH_2CH_3 \xrightarrow[\text{Catalyst}]{600°C} H_2C{=}CHCH{=}CH_2 + 2\,H_2$$

Butane **1,3-Butadiene**

Other simple conjugated dienes used in polymer synthesis include chloroprene (2-chloro-1,3-butadiene) and isoprene (2-methyl-1,3-butadiene). Isoprene has been prepared industrially by several methods, including the acid-catalyzed double dehydration of 3-methyl-1,3-butanediol.

3-Methyl-1,3-butanediol **Isoprene**
(2-Methyl-1,3-butadiene)

14.2 Stability of Conjugated Dienes

Conjugated dienes are similar to other alkenes in much of their chemistry, but there are also important differences. One such difference is *stability*: Conjugated dienes are somewhat more stable than nonconjugated dienes.

Evidence for the extra stability of conjugated dienes comes from measurements of heats of hydrogenation (Table 14.1). We saw earlier in the discussion of alkene stabilities (Section 6.7) that alkenes of similar substitution pattern have remarkably similar $\Delta H°_{hydrog}$ values. Monosubstituted alkenes such as 1-butene have values for $\Delta H°_{hydrog}$ near -126 kJ/mol (-30.1 kcal/mol), whereas disubstituted alkenes such as 2-methylpropene have $\Delta H°_{hydrog}$ values near -119 kJ/mol (-28.4 kcal/mol), approximately 7 kJ/mol less negative. We concluded from these data that more highly substituted alkenes are more stable than less substituted ones. That is, more highly substituted alkenes release less heat on hydrogenation because they contain less energy to start with. A similar conclusion can be drawn for conjugated dienes.

Since a monosubstituted alkene such as 1-butene has $\Delta H°_{hydrog} = -126$ kJ/mol, we might expect that a compound with two monosubstituted

TABLE 14.1 Heats of Hydrogenation for Some Alkenes and Dienes

Alkene or diene	Product	$\Delta H°_{hydrog}$ (kJ/mol)	(kcal/mol)
$CH_3CH_2CH{=}CH_2$	$CH_3CH_2CH_2CH_3$	−126	−30.1
$\underset{\underset{CH_3}{\mid}}{CH_3C}{=}CH_2$	$\underset{\underset{CH_3}{\mid}}{CH_3CHCH_3}$	−119	−28.4
$H_2C{=}CHCH{=}CH_2$	$CH_3CH_2CH_2CH_3$	−236	−56.4
$\underset{\underset{CH_3}{\mid}}{H_2C{=}CHC}{=}CH_2$	$\underset{\underset{CH_3}{\mid}}{CH_3CH_2CHCH_3}$	−229	−54.7
$H_2C{=}CHCH_2CH{=}CH_2$	$CH_3CH_2CH_2CH_2CH_3$	−253	−60.5

double bonds would have a $\Delta H°_{hydrog}$ approximately twice this value, or −252 kJ/mol. Nonconjugated dienes, such as 1,4-pentadiene ($\Delta H°_{hydrog}$ = −253 kJ/mol), meet this expectation, but the conjugated diene 1,3-butadiene ($\Delta H°_{hydrog}$ = −236 kJ/mol) does not. 1,3-Butadiene is approximately 16 kJ/mol (3.8 kcal/mol) more stable than expected.

Confirmation of this unexpected stability comes from data on the *partial* hydrogenation of 1,3-butadiene to yield 1-butene. The amount of energy released is −110 kJ/mol, some 16 kJ/mol less than that for the isolated monosubstituted double bond in 1-butene.

$$\Delta H°_{hydrog} \text{ (kJ/mol)}$$

$$H_2C{=}CHCH_2CH{=}CH_2$$
1,4-Pentadiene

−126 + (−126) = −252	Expected
−253	Observed
1	Difference

$$H_2C{=}CHCH{=}CH_2$$
1,3-Butadiene

−126 + (−126) = −252	Expected
−236	Observed
−16	Difference

$$\underset{\underset{CH_3}{\mid}}{H_2C{=}CHC}{=}CH_2$$
2-Methyl-1,3-butadiene

−126 + (−118) = −244	Expected
−229	Observed
−15	Difference

● ●

Problem 14.2 Use the data in Table 14.1 to calculate an expected heat of hydrogenation for allene, $H_2C{=}C{=}CH_2$. The measured value is −298 kJ/mol (−71.3 kcal/mol). Rank a conjugated diene, a nonconjugated diene, and an allene in order of stability.

● ●

14.3 Molecular Orbital Description of 1,3-Butadiene

Why are conjugated dienes so stable? Two explanations have been advanced. One explanation says that the difference in stability between conjugated and nonconjugated dienes results primarily from differences in orbital hybridization. In a nonconjugated diene, such as 1,4-pentadiene, the C–C single bonds result from σ overlap of an sp^2 orbital on one carbon with an sp^3 orbital on the neighboring carbon. In a conjugated diene, however, the C–C single bond results from σ overlap of sp^2 orbitals on both carbons. Since sp^2 orbitals have more s character than sp^3 orbitals, the electrons in sp^2 orbitals are closer to the nucleus, and the bonds they form are somewhat shorter and stronger. Thus, the "extra" stability of a conjugated diene results from the greater amount of s character in the orbitals forming the C–C single bond.

$$H_2C=CH-CH_2-CH=CH_2 \qquad\qquad H_2C=CH-CH=CH_2$$

Bonds formed by overlap of sp^2 and sp^3 orbitals   Bond formed by overlap of sp^2 and sp^2 orbitals

The second explanation for the stability of conjugated dienes focuses on the interaction between the π orbitals of the two double bonds. To see how this interaction arises, let's briefly review molecular orbital theory (Sections 1.6 and 1.9). When two p atomic orbitals combine to form a π bond, two π molecular orbitals result. One is lower in energy than the starting p orbitals and is therefore bonding; the other is higher in energy, has a node between nuclei, and is antibonding. Both electrons occupy the low-energy, bonding orbital, resulting in formation of a stable bond between atoms (Figure 14.1).

FIGURE 14.1 ▼

Two p orbitals combine to form two π molecular orbitals. When these orbitals are occupied by two electrons, both electrons occupy the low-energy, bonding orbital, leading to a net lowering of energy and formation of a stable bond. The asterisk on $\psi_2{}^*$ indicates an antibonding orbital.

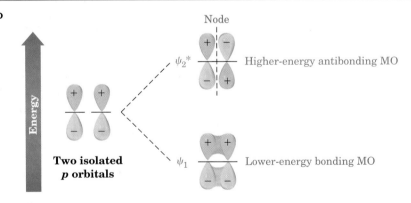

Now let's combine *four* adjacent p atomic orbitals, as occurs in a conjugated diene. In so doing, we generate a set of four π molecular orbitals, two of which are bonding and two of which are antibonding (Figure 14.2). The four π electrons occupy the two bonding orbitals, leaving the antibonding orbitals vacant.

FIGURE 14.2 ▼

Four π molecular orbitals in 1,3-butadiene. Note that the number of nodes between nuclei increases as the energy level of the orbital increases.

1.3-butadiene (see MO's on **CD-Rom**)

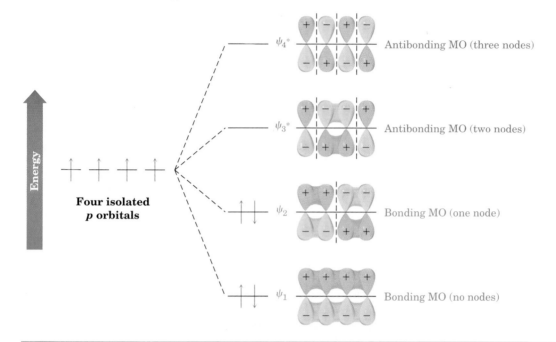

The lowest-energy π molecular orbital (denoted ψ_1, Greek psi) is a fully additive combination that has no nodes between the nuclei and is therefore bonding. The π MO of the next lowest energy, ψ_2, has one node between nuclei and is also bonding. Above ψ_1 and ψ_2 in energy are the two antibonding π MO's, ψ_3^* and ψ_4^*. (The asterisks indicate antibonding orbitals.) Note that the number of nodes between nuclei increases as the energy level of the orbital increases. The ψ_3^* orbital has two nodes between nuclei, and ψ_4^*, the highest-energy MO, has three nodes between nuclei.

Comparing the π molecular orbitals of 1,3-butadiene (two conjugated double bonds) with those of 1,4-pentadiene (two isolated double bonds) shows why the conjugated diene is more stable. In a conjugated diene, the lowest-energy π MO (ψ_1) has a favorable bonding interaction between C2 and C3 that is absent in a nonconjugated diene. As a result, there is a certain amount of double-bond character to the C2–C3 bond, making that bond stronger and stabilizing the molecule.

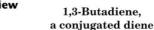

Stereo View

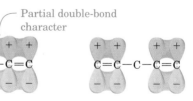

Partial double-bond character

1,3-Butadiene, a conjugated diene

1,4-Pentadiene, a nonconjugated diene

In describing the 1,3-butadiene molecular orbitals, we say that the π electrons are spread out, or *delocalized,* over the entire π framework rather than localized between two specific nuclei. Electron delocalization always leads to lower energy and greater stability of the molecule.

14.4 Bond Lengths in 1,3-Butadiene

Further evidence for the special nature of conjugated dienes comes from data on bond lengths (Table 14.2). If we compare the length of the carbon–carbon single bond in 1,3-butadiene (148 pm) to that in ethane (154 pm), we find that the 1,3-butadiene single bond is shorter by 6 pm.

TABLE 14.2   Some Carbon–Carbon Bond Lengths

Bond	Bond length (pm)	Bond hybridization
$CH_3\!-\!CH_3$	154	$C_{sp^3}\!-\!C_{sp^3}$
$H_2C\!=\!CH_2$	133	$C_{sp^2}\!-\!C_{sp^2}$
$H_2C\!=\!CH\!-\!CH_3$	149	$C_{sp^2}\!-\!C_{sp^3}$
$H_2C\!=\!CH\!-\!CH\!=\!CH_2$	148	$C_{sp^2}\!-\!C_{sp^2}$
$H_2C\!=\!CHCH\!=\!CH_2$	134	$C_{sp^2}\!-\!C_{sp^2}$

Both explanations advanced in the previous section to account for the stability of conjugated dienes also explain the bond shortening. According to the molecular orbital argument, the partial double-bond character of the C2–C3 bond in 1,3-butadiene gives the bond a length midway between a pure single bond and a pure double bond. Alternatively, it can be argued that the shortened single bond is a consequence of orbital hybridization. The C2–C3 bond results from σ overlap of two carbon sp^2 orbitals, whereas a typical alkane C–C bond results from overlap of two carbon sp^3 orbitals. The greater amount of s character in the 1,3-butadiene single bond makes it a bit shorter and stronger than usual. Both explanations are valid, and both contribute to the bond shortening observed for 1,3-butadiene.

14.5 Electrophilic Additions to Conjugated Dienes: Allylic Carbocations

One of the most striking differences between conjugated dienes and typical alkenes is in their electrophilic addition reactions. To review briefly, the addition of an electrophile to a carbon–carbon double bond is a general reaction of alkenes (Section 6.8). Markovnikov regiochemistry is found because the more stable carbocation is involved as an intermediate. Thus, addition of HCl to 2-methylpropene yields 2-chloro-2-methylpropane rather than 1-chloro-2-methylpropane, and addition of 2 mol equiv of HCl to the nonconjugated diene 1,4-pentadiene yields 2,4-dichloropentane.

$$CH_3C{=}CH_2 \xrightarrow[\text{Ether}]{\text{HCl}} \left[CH_3\overset{+}{C}CH_3 \right] \longrightarrow CH_3\overset{Cl}{\underset{CH_3}{C}}CH_3$$

2-Methylpropene   **Tertiary carbocation**   **2-Chloro-2-methylpropane**

$$H_2C{=}CHCH_2CH{=}CH_2 \xrightarrow[\text{Ether}]{\text{HCl}} CH_3\overset{Cl}{C}HCH_2\overset{Cl}{C}HCH_3$$

1,4-Pentadiene
(nonconjugated)   **2,4-Dichloropentane**

Conjugated dienes also undergo electrophilic addition reactions readily, but mixtures of products are invariably obtained. For example, addition of HBr to 1,3-butadiene yields a mixture of two products (not counting cis–trans isomers). 3-Bromo-1-butene is the typical Markovnikov product of **1,2 addition**, but 1-bromo-2-butene appears unusual. The double bond in this product has moved to a position between carbons 2 and 3, and HBr has added to carbons 1 and 4, a result described as **1,4 addition**.

1,3-Butadiene
(a conjugated diene)   **3-Bromo-1-butene**
(71%; 1,2 addition)   **1-Bromo-2-butene**
(29%; 1,4 addition)

Many other electrophiles besides HBr add to conjugated dienes, and mixtures of products are usually formed. For example, Br_2 adds to 1,3-butadiene to give a mixture of 1,4-dibromo-2-butene and 3,4-dibromo-1-butene.

$$H_2C{=}CHCH{=}CH_2 \xrightarrow[25°C]{Br_2} BrCH_2CH{=}CHCH_2Br \ + \ BrCH_2\overset{Br}{C}HCH{=}CH_2$$

1,3-Butadiene   **1,4-Dibromo-2-butene**
(45%; 1,4 addition)   **3,4-Dibromo-1-butene**
(55%; 1,2 addition)

How can we account for the formation of 1,4-addition products? The answer is that *allylic carbocations* are involved as intermediates. When 1,3-butadiene is protonated, two carbocation intermediates are possible: a primary carbocation and a secondary allylic cation (recall that *allylic* means "next to a double bond"). Since an allylic cation is stabilized by resonance between two forms (Section 11.9), it is more stable and forms faster than a nonallylic carbocation.

1,3-Butadiene

Secondary, allylic

Primary, nonallylic
(*NOT formed*)

When the allylic cation reacts with Br⁻ to complete the electrophilic addition reaction, attack can occur either at C1 or at C3 because both carbons share the positive charge (Figure 14.3). Thus, a mixture of 1,2- and

---

FIGURE 14.3 ▼

An electrostatic potential map of the carbocation produced by protonation of 1,3-butadiene shows that the positive charge is shared by carbons 1 and 3. Reaction of Br⁻ with the more positive carbon (purple) gives the 1,2-addition product.

**protonated
1.3-butadiene**

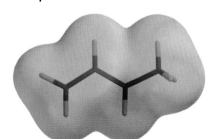

1,4 addition
(29%)

1,2 addition
(71%)

1,4-addition products results. (Recall that a similar product mixture was found in Section 10.5 for NBS bromination of alkenes, a reaction that proceeds through an allylic *radical.*)

• •

Practice Problem 14.1 Give the structures of the likely products from reaction of 1 equivalent of HCl with 2-methyl-1,3-cyclohexadiene. Show both 1,2 and 1,4 adducts.

Strategy Electrophilic addition of HCl to a conjugated diene involves the formation of allylic carbocation intermediates. Thus, the first step is to protonate the two ends of the diene and draw the resonance forms of the two allylic carbocations that result. Then allow each resonance form to react with Cl⁻, generating a maximum of four possible products.

In the present instance, protonation of the C1–C2 double bond gives a carbocation that can react further to give the 1,2 adduct 3-chloro-3-methylcyclohexene and the 1,4 adduct 3-chloro-1-methylcyclohexene. Protonation of the C3–C4 double bond gives a symmetrical carbocation whose two resonance forms are equivalent. Thus, the 1,2 adduct and the 1,4 adduct have the same structure: 6-chloro-1-methylcyclohexene. Of the two possible modes of protonation, the first is more likely because it yields a tertiary allylic cation rather than a secondary allylic cation.

Solution

2-Methyl-1,3-cyclohexadiene

3-Chloro-3-methyl-cyclohexene   **3-Chloro-1-methyl-cyclohexene**   **6-Chloro-1-methyl-cyclohexene**

• •

Problem 14.3 Give the structures of the possible products from reaction of 1 equivalent of HCl with 1,3-pentadiene. Show both 1,2 and 1,4 adducts.

Problem 14.4 Look at the possible carbocation intermediates produced during addition of HCl to 1,3-pentadiene (Problem 14.3), and predict which 1,2 adduct predominates. Which 1,4 adduct predominates?

• •

14.6 Kinetic versus Thermodynamic Control of Reactions

Electrophilic addition to a conjugated diene at or below room temperature normally leads to a mixture of products in which the 1,2 adduct predominates over the 1,4 adduct. When the same reaction is carried out at higher temperatures, though, the product ratio often changes and the 1,4 adduct predominates. For example, addition of HBr to 1,3-butadiene at 0°C yields a 71:29 mixture of 1,2 and 1,4 adducts, but the same reaction carried out at 40°C yields a 15:85 mixture. Furthermore, when the product mixture formed at 0°C is heated to 40°C in the presence of HBr, the ratio of adducts slowly changes from 71:29 to 15:85. Why?

$$H_2C=CHCH=CH_2 + HBr \longrightarrow \overset{\overset{\textstyle Br}{|}}{H_2C=CHCHCH_3} + CH_3CH=CHCH_2Br$$

	1,2 adduct	**1,4 adduct**
At 0°C:	71%	29%
At 40°C:	15%	85%

To understand the effect of temperature on product distribution, let's briefly review what we said in Section 5.7 about rates and equilibria. Imagine a reaction that can give either or both of two products, B and C:

$$A \longrightarrow B + C$$

Let's assume that B forms faster than C (in other words, $\Delta G_B^{\ddagger} < \Delta G_C^{\ddagger}$) but that C is more stable than B (in other words, $\Delta G_C° > \Delta G_B°$). A reaction energy diagram for the two processes might look like that shown in Figure 14.4.

FIGURE 14.4 ▼

A reaction energy diagram for two competing reactions in which the less stable product (B) forms faster than the more stable product (C).

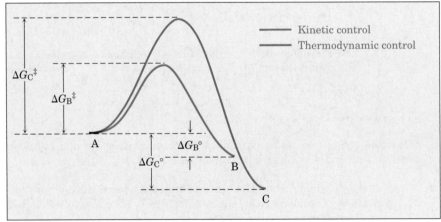

Let's first carry out the reaction at some higher temperature so that both processes are readily reversible and an equilibrium is reached. That is, enough energy is supplied for reactant molecules to surmount the barriers to both products, and for both product molecules to climb the higher barriers back to reactant. Since C is more stable than B, C is the major product obtained. It doesn't matter that C forms more slowly than B, because the two are in equilibrium. *The product of a readily reversible reaction depends only on thermodynamic stability.* Such reactions are said to be under equilibrium control, or **thermodynamic control**.

$$B \rightleftharpoons A \rightleftharpoons C \qquad \text{Thermodynamic control}$$
$$\text{(vigorous conditions; reversible)}$$

Now let's carry out the same reaction at a lower temperature so that both processes are *irreversible* and no equilibrium is reached. That is, only enough energy is supplied for the reactant molecules to surmount the barriers to products, *but not for the product molecules to climb the higher barriers back to reactant.* Since B forms faster than C, B is the major product. It doesn't matter that C is more stable than B, because the two are not in equilibrium. *The product of an irreversible reaction depends only on relative rates.* Such reactions are said to be under **kinetic control**.

$$B \longleftarrow A \longrightarrow C \qquad \text{Kinetic control}$$
$$\text{(mild conditions; irreversible)}$$

We can now explain the effect of temperature on electrophilic addition reactions of conjugated dienes. At low temperature (0°C), HBr adds to 1,3-butadiene under kinetic control to give a 71:29 mixture of products, with the more rapidly formed 1,2 adduct predominating. Since these mild conditions don't allow the reaction to reach equilibrium, the product that forms faster predominates. At higher temperature (40°C), however, the reaction occurs under thermodynamic control to give a 15:85 mixture of products, with the more stable 1,4 adduct predominating. The higher temperature makes the addition process reversible, and an equilibrium mixture of products results. Figure 14.5 shows the situation in a reaction energy diagram.

FIGURE 14.5 ▼

Reaction energy diagram for the electrophilic addition of HBr to 1,3-butadiene. The 1,2 adduct is the kinetic product because it forms faster, but the 1,4 adduct is the thermodynamic product because it is more stable.

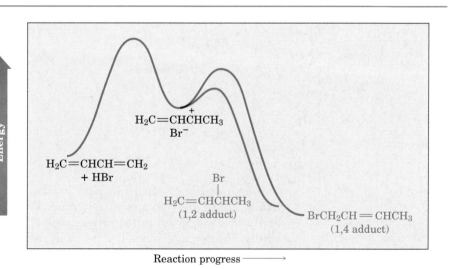

The electrophilic addition of HBr to 1,3-butadiene is a good example of how a change in experimental conditions can change the product of a reaction. The concept of thermodynamic control versus kinetic control is a valuable one that we can often take advantage of in the laboratory.

• •

Problem 14.5   The 1,2 adduct and the 1,4 adduct formed by reaction of HBr with 1,3-butadiene are in equilibrium at 40°C. Propose a mechanism by which the interconversion of products takes place. (See Section 11.6.)

Problem 14.6   Why do you suppose 1,4 adducts of 1,3-butadiene are generally more stable than 1,2 adducts?

• •

14.7 Diene Polymers: Natural and Synthetic Rubbers

Conjugated dienes can be polymerized just as simple alkenes can (Section 7.10). Diene polymers are structurally more complex than simple alkene polymers, though, because double bonds remain every four carbon atoms along the chain, leading to the possibility of cis–trans isomers. The initiator (**In**) for the reaction can be either a radical, as occurs in ethylene polymerization, or an acid. Note that the polymerization is a 1,4 addition of the growing chain to a conjugated diene monomer.

1,3-Butadiene

cis-**Polybutadiene**

trans-**Polybutadiene**

As noted in "Natural Rubber" at the end of Chapter 7, rubber is a naturally occurring polymer of isoprene. The double bonds of rubber have *Z* stereochemistry, but *gutta-percha,* the *E* isomer of rubber, also occurs naturally. Harder and more brittle than rubber, gutta-percha has a variety of minor applications, including occasional use as the covering on golf balls.

natural rubber, Gutta-percha

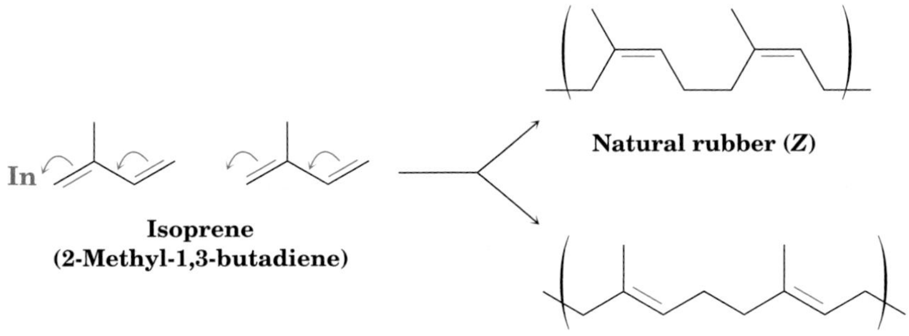

**Isoprene
(2-Methyl-1,3-butadiene)**

Natural rubber (Z)

Gutta-percha (E)

A number of different synthetic rubbers are produced commercially by diene polymerization. Both *cis*- and *trans*-polyisoprene can be made, and the synthetic rubber thus produced is similar to the natural material. Chloroprene (2-chloro-1,3-butadiene) is polymerized to yield neoprene, an excellent, though expensive, synthetic rubber with good weather resistance. Neoprene is used in the production of industrial hoses and gloves, among other things.

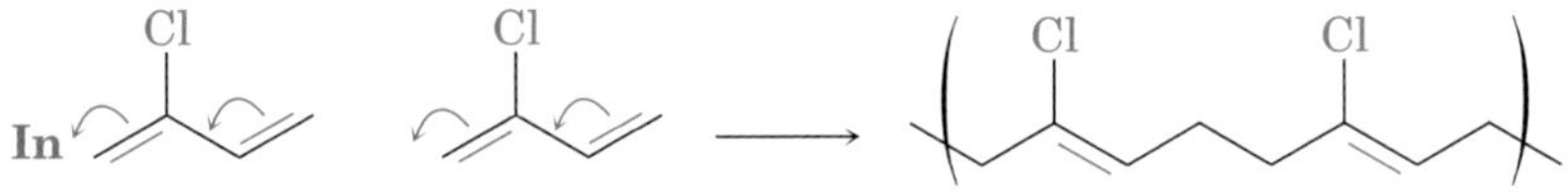

**Chloroprene
(2-Chloro-1,3-butadiene)**

Neoprene (Z)

Both natural and synthetic rubbers are soft and tacky unless hardened by **vulcanization**. Discovered in 1839 by Charles Goodyear, vulcanization involves heating the crude polymer with a few percent by weight of sulfur. Sulfur forms bridges, or cross-links, between polymer chains, locking the chains together into immense molecules that can no longer slip over one another (Figure 14.6). The result is a much harder rubber with greatly improved resistance to wear and abrasion.

FIGURE 14.6 ▼

Sulfur cross-linked chains resulting from vulcanization of poly-1,3-butadiene.

vulcanized rubber

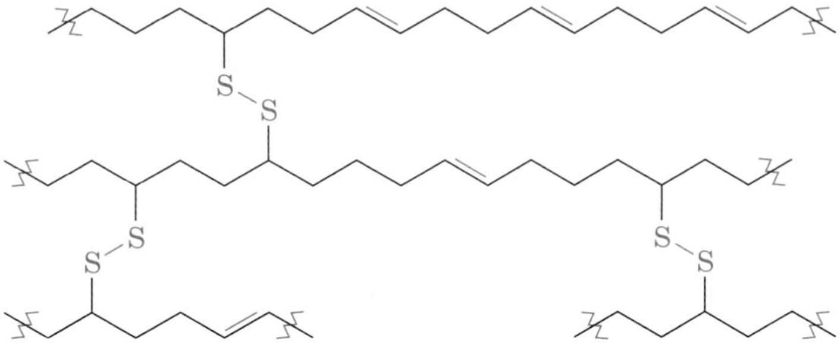

• •

Problem 14.7 Draw a segment of the polymer that might be prepared from 2-phenyl-1,3-butadiene.

Problem 14.8 Show the mechanism of the acid-catalyzed polymerization of 1,3-butadiene.

• •

14.8 The Diels–Alder Cycloaddition Reaction

Another striking difference between conjugated and nonconjugated dienes is that conjugated dienes undergo an addition reaction with alkenes to yield substituted cyclohexene products. For example, 1,3-butadiene and 3-buten-2-one give 3-cyclohexenyl methyl ketone.

1,3-Butadiene   **3-Buten-2-one**   **3-Cyclohexenyl methyl ketone**
                                                                      (96%)

Otto Paul Hermann Diels

Otto Paul Hermann Diels (1876–1954) was born in Hamburg, Germany, and received his Ph.D. at the University of Berlin working with Emil Fischer. He was professor of chemistry both at the University of Berlin (1906–1916) and at Kiel (1916–1948). His most important discovery was the so-called Diels–Alder reaction, which he developed with one of his research students and for which he received the 1950 Nobel Prize in chemistry.

This process, named the **Diels–Alder cycloaddition reaction** after its discoverers, is extremely useful in organic synthesis because it forms two carbon–carbon bonds in a single step and is one of the few methods available for making cyclic molecules. (As you might expect, a *cycloaddition* reaction is one in which two reactants *add* together to give a *cyclic* product.) The 1950 Nobel Prize in chemistry was awarded to Diels and Alder in recognition of the importance of their discovery.

The mechanism of the Diels–Alder cycloaddition is different from that of other reactions we've studied because it is neither polar nor radical. Rather, the Diels–Alder reaction is a *pericyclic* process. Pericyclic reactions, which we'll discuss in more detail in Chapter 30, take place in a single step by a cyclic redistribution of bonding electrons. The two reactants simply join together through a cyclic transition state in which two new carbon–carbon bonds form at the same time.

We can picture a Diels–Alder addition as occurring by head-on (σ) overlap of the two alkene *p* orbitals with the two *p* orbitals on carbons 1 and 4 of the diene (Figure 14.7). This is, of course, a *cyclic* orientation of the reactants.

FIGURE 14.7 ▼

Mechanism of the Diels–Alder cycloaddition reaction. The reaction occurs in a single step through a cyclic transition state in which the two new carbon–carbon bonds form simultaneously.

Diels–Alder reaction (see computer animation on CD-Rom)

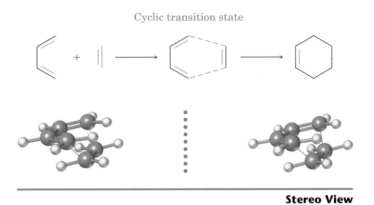

Cyclic transition state

Stereo View

Kurt Alder

Kurt Alder (1902–1958) was born in Königshütte, Prussia, and moved to Germany after World War I. He received his Ph.D. in 1926 at Kiel working with Otto Diels. He worked first at I. G. Farben on the manufacture of plastics but then became professor at the University of Cologne (1940–1958). He shared the 1950 Nobel Prize in chemistry with his mentor, Otto Diels.

In the Diels–Alder transition state, the two alkene carbons and carbons 1 and 4 of the diene rehybridize from sp^2 to sp^3 to form two new single bonds. Carbons 2 and 3 of the diene remain sp^2-hybridized to form the new double bond in the cyclohexene product. We'll study this mechanism at greater length in Chapter 30 and will concentrate for the present on learning more about the chemistry of the Diels–Alder reaction.

14.9 Characteristics of the Diels–Alder Reaction

The Dienophile

The Diels–Alder cycloaddition reaction occurs most rapidly and in highest yield if the alkene component, or **dienophile** ("diene lover"), has an electron-withdrawing substituent group. Thus, ethylene itself reacts sluggishly, but propenal, ethyl propenoate, maleic anhydride, benzoquinone, propenenitrile, and others are highly reactive. Note also that alkynes, such as methyl propynoate, can act as Diels–Alder dienophiles.

Some Diels–Alder dienophiles

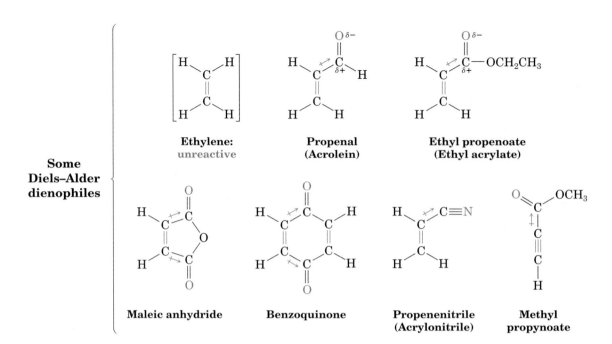

Ethylene: unreactive

Propenal (Acrolein)

Ethyl propenoate (Ethyl acrylate)

Maleic anhydride

Benzoquinone

Propenenitrile (Acrylonitrile)

Methyl propynoate

In all the above cases, the dienophile double or triple bond is next to the positively polarized carbon of a substituent that withdraws electrons. Electrostatic potential maps of propenal and propenenitrile, for instance, show that the double-bond carbons are less negative in these substances than in ethylene (Figure 14.8).

One of the most useful features of the Diels–Alder reaction is that it is **stereospecific**: The stereochemistry of the starting dienophile is maintained during the reaction, and a single product stereoisomer results. If we carry out the cycloaddition with a cis dienophile, such as methyl *cis*-2-

FIGURE 14.8 ▼

Electrostatic potential maps of ethylene, propenal, and propenenitrile show that electron-withdrawing groups make the double-bond carbons less negative.

ethylene, propenal, propenenitrile

Ethylene

Propenal

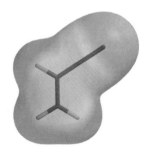

Propenenitrile

butenoate, only the cis-substituted cyclohexene product is formed. Conversely, Diels–Alder reaction with methyl *trans*-2-butenoate yields only the trans-substituted cyclohexene product.

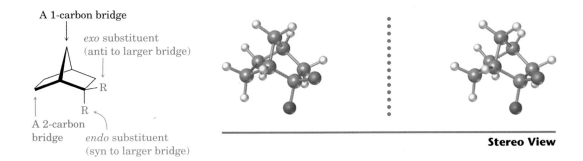

1,3-Butadiene   **Methyl (Z)-2-butenoate**   **Cis product**

1,3-Butadiene   **Methyl (E)-2-butenoate**   **Trans product**

Another stereochemical feature of the Diels–Alder reaction is that the diene and dienophile partners line up so that the *endo* product, rather than the alternative *exo* product, is formed. The words *endo* and *exo* are used to indicate relative stereochemistry when referring to bicyclic structures like substituted norbornanes (Section 4.15). A substituent on one bridge is said to be *exo* if it is anti (trans) to the larger of the other two bridges and is said to be *endo* if it is syn (cis) to the larger of the other two bridges.

A 1-carbon bridge

exo substituent
(anti to larger bridge)

R

A 2-carbon
bridge

R

endo substituent
(syn to larger bridge)

Stereo View

Endo products result from Diels–Alder reactions because the amount of orbital overlap between diene and dienophile is higher when the reactants lie directly on top of one another so that the electron-withdrawing substituent on the dienophile is underneath the diene. In the reaction of 1,3-cyclopentadiene with maleic anhydride, for example, the following result is obtained:

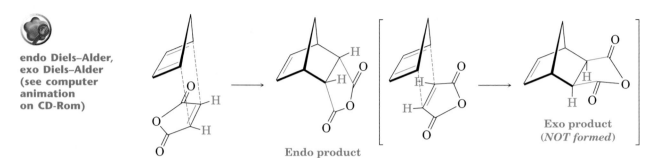

endo Diels–Alder,
exo Diels–Alder
(see computer
animation
on CD-Rom)

Maleic anhydride

Endo product

Exo product
(*NOT formed*)

• •

Practice Problem 14.2   Predict the product of the following Diels–Alder reaction:

Strategy   Draw the diene so that the ends of the two double bonds are near the dienophile double bond. Then form two single bonds between the partners, convert the three double bonds into single bonds, and convert the former single bond of the diene into a new double bond. Note that, because the dienophile double bond is cis to begin with, the two attached hydrogens must remain cis in the product.

Solution

Cis hydrogens

New double bond

• •

Problem 14.9   Predict the product of the following Diels–Alder reaction:

• •

The Diene

A diene must adopt what is called an *s-cis conformation* ("cis-like" about the *s*ingle bond) to undergo the Diels–Alder reaction. Only in the *s*-cis conformation are carbons 1 and 4 of the diene close enough to react through a cyclic transition state. In the alternative *s*-trans conformation, the ends of the diene partner are too far apart to overlap with the dienophile *p* orbitals.

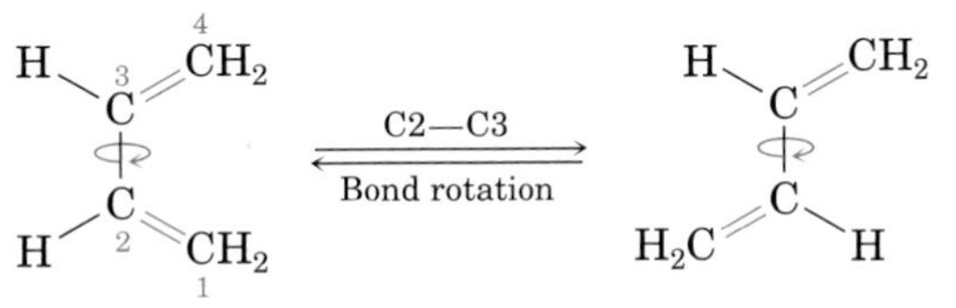

s-Cis conformation   **s-Trans conformation**

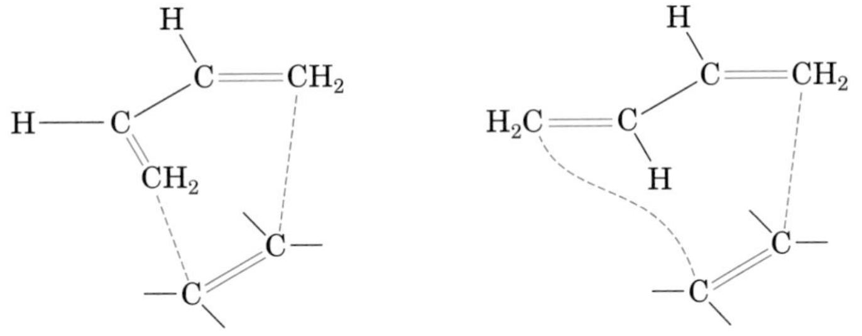

Successful reaction   **No reaction (ends too far apart)**

Two examples of dienes that can't adopt an *s*-cis conformation, and therefore don't undergo Diels–Alder reactions, are shown in Figure 14.9. In the bicyclic diene, the double bonds are rigidly fixed in an *s*-trans arrangement by geometric constraints of the rings. In (2Z,4Z)-hexadiene, steric strain between the two methyl groups prevents the molecule from adopting *s*-cis geometry.

FIGURE 14.9 ▼

Two dienes that can't achieve an *s*-cis conformation and can't undergo Diels–Alder reactions.

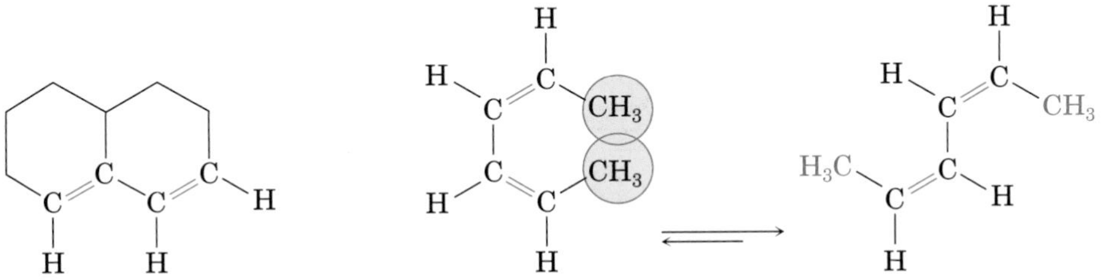

A bicyclic diene
(rigid s-trans diene)   **Severe steric strain**
in s-cis form   **(2Z,4Z)-Hexadiene**
(s-trans, more stable)

In contrast to the unreactive dienes that can't achieve an *s*-cis conformation, other dienes are fixed *only* in the correct *s*-cis geometry and are therefore highly reactive in the Diels–Alder cycloaddition reaction. Cyclopentadiene, for example, is so reactive that it reacts with itself. At room temperature, cyclopentadiene *dimerizes:* One molecule acts as diene and another acts as dienophile in a self Diels–Alder reaction.

1,3-Cyclopentadiene   **Bicyclopentadiene**
(*s*-cis)

Problem 14.10   Which of the following alkenes would you expect to be good Diels–Alder dienophiles?

(a) $H_2C{=}CHCCl$ with O above

(b) $H_2C{=}CHCH_2CH_2COCH_3$ with O above

(c)

(d)

(e)

Problem 14.11   Which of the following dienes have an *s*-cis conformation, and which have an *s*-trans conformation? Of the *s*-trans dienes, which can readily rotate to *s*-cis?

(a)

(b)

(c)

Problem 14.12   Predict the product of the following Diels–Alder reaction:

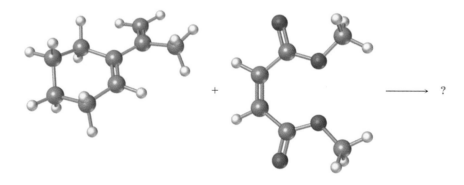

+   ⟶ ?

14.10 Structure Determination in Conjugated Systems: Ultraviolet Spectroscopy

Mass spectrometry, infrared spectroscopy, and nuclear magnetic resonance spectroscopy are techniques of structure determination applicable to all organic molecules. In addition to these three generally useful methods, there's a fourth—**ultraviolet (UV) spectroscopy**—that is applicable only to conjugated systems.

Mass spectrometry	Molecular size and formula
Infrared spectroscopy	Functional groups present
NMR spectroscopy	Carbon–hydrogen framework
Ultraviolet spectroscopy	Nature of conjugated π electron system

Ultraviolet spectroscopy is less commonly used than the other three spectroscopic techniques because of the specialized information it gives. We'll therefore study it only briefly.

The ultraviolet region of the electromagnetic spectrum extends from the low-wavelength end of the visible region (4×10^{-7} m) to 10^{-8} m, but the narrow range from 2×10^{-7} m to 4×10^{-7} m is the portion of greatest interest to organic chemists. Absorptions in this region are usually measured in nanometers (nm), where 1 nm = 10^{-9} m. Thus, the ultraviolet range of interest is from 200 to 400 nm (Figure 14.10).

FIGURE 14.10 ▼

The ultraviolet (UV) region of the electromagnetic spectrum.

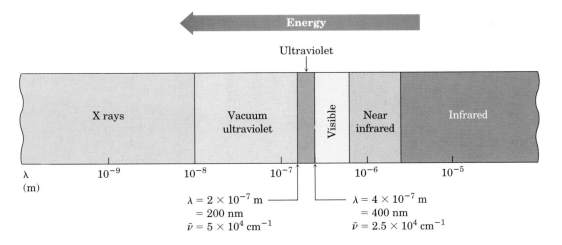

We saw in Section 12.5 that when an organic molecule is irradiated with electromagnetic energy, the radiation either passes through the sample or is absorbed, depending on its energy. With IR irradiation, the energy absorbed corresponds to the amount necessary to increase molecular bending and stretching vibrations. With UV radiation, the energy absorbed corresponds to the amount necessary to promote an electron from one orbital to another. We'll see what this means by looking first at 1,3-butadiene.

• •

Problem 14.13   Calculate the energy range of electromagnetic radiation in the UV region of the spectrum from 200 to 400 nm. Recall the equation

$$E = \frac{N_A hc}{\lambda} = \frac{1.20 \times 10^{-4} \text{ kJ/mol}}{\lambda \text{ (m)}}$$

Problem 14.14   How does the energy you calculated in Problem 14.13 for UV radiation compare with the values calculated previously for IR and NMR spectroscopy?

• •

14.11 Ultraviolet Spectrum of 1,3-Butadiene

1,3-Butadiene has four π molecular orbitals (Section 14.3). The two lower-energy, bonding MO's are occupied in the ground state, and the two higher-energy, antibonding MO's are unoccupied, as illustrated in Figure 14.11.

FIGURE 14.11 ▼

Ultraviolet excitation of 1,3-butadiene results in the promotion of an electron from ψ_2, the highest occupied molecular orbital (HOMO), to ψ_3^*, the lowest unoccupied molecular orbital (LUMO).

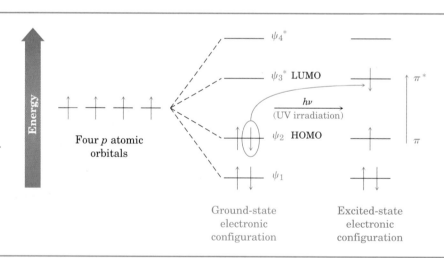

On irradiation with ultraviolet light ($h\nu$), 1,3-butadiene absorbs energy and a π electron is promoted from the **highest occupied molecular orbital**, or **HOMO**, to the **lowest unoccupied molecular orbital**, or **LUMO**. Since the electron is promoted from a bonding π molecular orbital to an antibonding π^* molecular orbital, we call this a $\pi \longrightarrow \pi^*$ excitation (read as "pi to pi star"). The energy gap between the HOMO and the LUMO of 1,3-butadiene is such that UV light of 217 nm wavelength is required to accomplish the $\pi \longrightarrow \pi^*$ electronic transition.

In practice, an ultraviolet spectrum is recorded by irradiating the sample with UV light of continuously changing wavelength. When the wavelength corresponds to the energy level required to excite an electron to a higher level, energy is absorbed. This absorption is detected and displayed on a chart that plots wavelength versus *absorbance* (A), defined as

$$A = \log \frac{I_0}{I}$$

where I_0 is the intensity of the incident light and I is the intensity of the light transmitted through the sample. Note that UV spectra differ from IR spectra in the way they are presented. IR spectra are usually displayed so that the baseline corresponding to zero absorption runs across the top of the chart and a valley indicates an absorption. UV spectra are displayed with the baseline at the bottom of the chart so that a peak indicates an absorption (Figure 14.12).

FIGURE 14.12 ▼

The ultraviolet spectrum of 1,3-butadiene, λ_{max} = 217 nm.

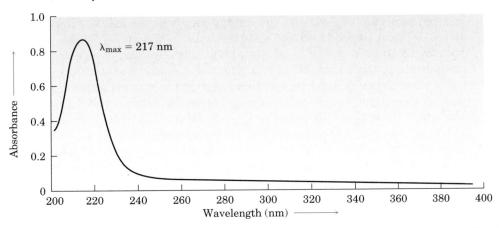

The exact amount of UV light absorbed is expressed as the sample's **molar absorptivity (ε)**, defined by the equation

$$\text{Molar absorptivity } \varepsilon = \frac{A}{C \times l}$$

where   A = Absorbance
        C = Concentration in mol/L
        I = Sample pathlength in cm

Molar absorptivity is a physical constant, characteristic of the particular substance being observed and thus characteristic of the particular π electron system in the molecule. Typical values for conjugated dienes are in the range $\varepsilon = 10,000-25,000$.

Unlike IR and NMR spectra, which show many absorptions for a given molecule, UV spectra are usually quite simple—often only a single peak. The peak is usually broad, however, and we identify its position by noting the wavelength at the very top of the peak (λ_{max}, read as "lambda max").

· ·

Problem 14.15   A knowledge of molar absorptivities is particularly important in biochemistry where UV spectroscopy can provide an extremely sensitive method of analysis. For example, imagine that you wanted to determine the concentration of vitamin A in a sample. If pure vitamin A has $\lambda_{max} = 325$ ($\varepsilon = 50,100$), what is the vitamin A concentration in a sample whose absorbance at 325 nm is $A = 0.735$ in a cell with a pathlength of 1.00 cm?

· ·

14.12 Interpreting Ultraviolet Spectra: The Effect of Conjugation

The exact wavelength necessary to effect the $\pi \longrightarrow \pi^*$ transition in a conjugated molecule depends on the energy gap between HOMO and LUMO, which in turn depends on the nature of the conjugated system. Thus, by measuring the UV spectrum of an unknown, we can derive structural information about the nature of any conjugated π electron system present.

One of the most important factors affecting the wavelength of UV absorption by a molecule is the extent of conjugation. Molecular orbital calculations show that the energy difference between HOMO and LUMO decreases as the extent of conjugation increases. Thus, 1,3-butadiene absorbs at $\lambda_{max} = 217$ nm, 1,3,5-hexatriene absorbs at $\lambda_{max} = 258$ nm, and 1,3,5,7-octatetraene absorbs at $\lambda_{max} = 290$ nm. (*Remember:* Longer wavelength means lower energy.)

Other kinds of conjugated systems, such as conjugated enones and aromatic rings, also have characteristic UV absorptions that are useful in structure determination. The UV absorption maxima of some representative conjugated molecules are given in Table 14.3.

· ·

Problem 14.16   Which of the following compounds would you expect to show ultraviolet absorptions in the 200–400 nm range?
(a) 1,4-Cyclohexadiene   (b) 1,3-Cyclohexadiene   (c) $H_2C=CH-C\equiv N$

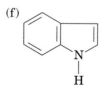

(d) Aspirin   (e)   (f) Indole

· ·

TABLE 14.3 Ultraviolet Absorptions of Some Conjugated Molecules

Name	Structure	λ_{max} (nm)
2-Methyl-1,3-butadiene	$\begin{matrix}CH_3\\ H_2C=C-CH=CH_2\end{matrix}$	220
1,3-Cyclohexadiene		256
1,3,5-Hexatriene	$H_2C=CH-CH=CH-CH=CH_2$	258
1,3,5,7-Octatetraene	$H_2C=CH-CH=CH-CH=CH-CH=CH_2$	290
2,4-Cholestadiene		275
3-Buten-2-one	$\begin{matrix}CH_3\\ H_2C=CH-C=O\end{matrix}$	219
Benzene		254
Naphthalene		275

14.13 Colored Organic Compounds

Why are some organic compounds colored while others aren't? Why is β-carotene orange (the pigment in carrots), while benzene is colorless? The answers involve both the structures of colored molecules and the way we perceive light.

The visible region of the electromagnetic spectrum is adjacent to the ultraviolet region, extending from approximately 400 to 800 nm. Colored compounds have such extended systems of conjugation that their "UV" absorptions extend into the visible region. β-Carotene, for example, has 11 double bonds in conjugation, and its absorption occurs at $\lambda_{max} = 455$ nm (Figure 14.13).

FIGURE 14.13 ▼

Ultraviolet spectrum of β-carotene, a conjugated molecule with 11 double bonds. The absorption occurs in the visible region.

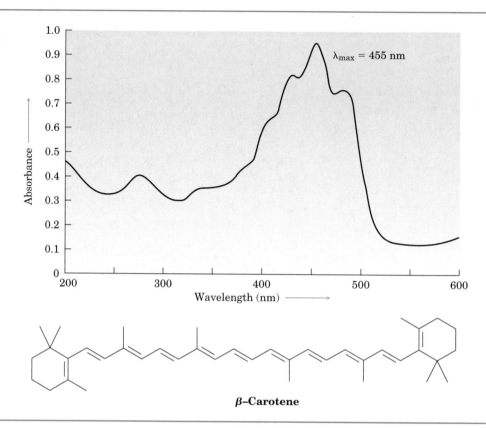

β–Carotene

"White" light from the sun or from a lamp consists of all wavelengths in the visible region. When white light strikes β-carotene, the wavelengths from 400 to 500 nm (blue) are absorbed, while all other wavelengths are transmitted and can reach our eyes. We therefore see the white light with the blue removed, and we perceive a yellow-orange color for β-carotene.

What is true for β-carotene is also true for all other colored organic compounds: All have an extended system of π electron conjugation that gives rise to an absorption in the visible region of the electromagnetic spectrum.

CHEMISTRY @ WORK

Resists for Integrated Circuits

Twenty-five years ago, someone interested in owning a computer would have paid approximately $150,000 for 16 megabytes of random-access memory that would have occupied a volume the size of a small desk.

(continued) ▶

Today, anyone can buy 16 MB of computer memory for under $50 and can fit the chips into their shirt pocket. The difference between then and now is due to improvements in *photolithography*, the process by which integrated circuit chips are made.

Photolithography begins by coating a layer of SiO_2 onto a silicon wafer and further coating with a thin (0.5–1.0 μm) film of a light-sensitive organic polymer called a *resist*. A *mask* is then used to cover those parts of the chip that will become a circuit, and the wafer is irradiated with UV light. The nonmasked sections of the polymer undergo a chemical change when irradiated that makes them more soluble than the masked, unirradiated sections. On washing the irradiated chip with solvent, solubilized polymer is selectively removed from the irradiated areas, exposing the SiO_2 underneath. This SiO_2 is then chemically etched away by reaction with hydrofluoric acid, leaving behind a pattern of polymer-coated SiO_2. Further washing removes the remaining polymer, leaving a positive image of the mask in the form of exposed ridges of SiO_2 (Figure 14.14). Additional cycles of coating, masking, and etching then produce the completed chips.

FIGURE 14.14 ▼

Outline of the photolithography process for producing integrated circuit chips.

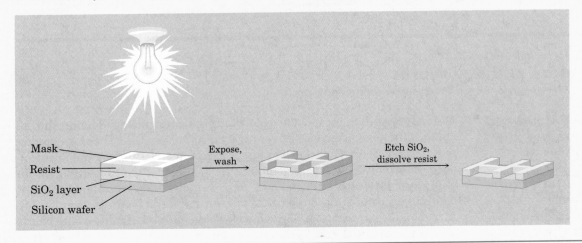

The polymer resist currently used in chip manufacturing is based on the two-component *diazoquinone–novolac system*. Novolac resin is a soft, relatively low-molecular-weight polymer made from methylphenol and formaldehyde, while the diazoquinone is a bicyclic (two-ring) molecule containing a diazo group (=N=N) adjacent to a ketone carbonyl (C=O). The diazoquinone–novolac mix is relatively insoluble when fresh, but on

(continued) ▶

exposure to ultraviolet light and water vapor, the diazoquinone component undergoes reaction to yield N_2 and a carboxylic acid, which can be washed away with dilute base. Novolac–diazoquinone technology is capable of producing features as small as 0.5 μm (5×10^{-7} m), but further improvements in miniaturization will have to come from newer resist materials currently being developed.

Diazonaphthoquinone

Novolac resin

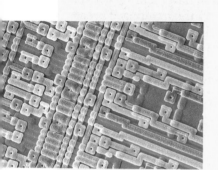

Manufacturing the ultrathin circuitry on this computer chip depends on the organic chemical reactions of special polymers.

Summary and Key Words

OCOL

A **conjugated** diene is one that contains alternating double and single bonds. One characteristic of conjugated dienes is that they are somewhat more stable than their nonconjugated counterparts. This unexpected stability can be explained by a molecular orbital description in which four p atomic orbitals combine to form four π molecular orbitals. Only the two bonding orbitals are occupied; the two antibonding orbitals are unoccupied. A π bonding interaction introduces some partial double-bond character between carbons 2 and 3, thereby strengthening the C2–C3 bond and stabilizing the molecule.

Conjugated dienes undergo two reactions not observed for nonconjugated dienes. The first is **1,4 addition** of electrophiles. When a conjugated diene is treated with an electrophile such as HCl, **1,2** and **1,4 adducts** are formed. Both products are formed from the same resonance-stabilized allylic carbocation intermediate and are produced in varying amounts depending on the reaction conditions. The 1,2 adduct is usually formed faster and is said to be the product of **kinetic control**. The 1,4 adduct is usually more stable and is said to be the product of **thermodynamic control**.

The second reaction unique to conjugated dienes is **Diels–Alder cycloaddition**. Conjugated dienes react with electron-poor alkenes (**dienophiles**) in a single step through a cyclic transition state to yield a cyclohexene product. The reaction can occur only if the diene is able to adopt an *s-cis conformation*.

Ultraviolet (UV) spectroscopy is a method of structure determination applicable specifically to conjugated systems. When a conjugated molecule is irradiated with ultraviolet light, energy absorption occurs and a π electron is promoted from the **highest occupied molecular orbital (HOMO)** to the **lowest unoccupied molecular orbital (LUMO)**. For 1,3-butadiene, radiation of $\lambda_{max} = 217$ nm is required. As a general rule, the greater the extent of conjugation, the less the energy needed (that is, the longer the wavelength of radiation required).

Visualizing Chemistry

• •

(Problems 14.1–14.16 appear within the chapter.)

14.17   Write the structures of all possible adducts of the following diene with 1 equivalent of HCl:

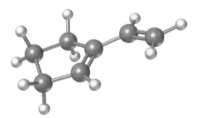

14.18   Write the product of the Diels–Alder reaction of the following diene with 3-buten-2-one, $H_2C=CHCOCH_3$. Make sure you show the full stereochemistry of the reaction product.

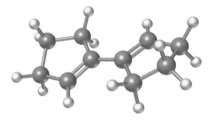

14.19   The following drawing of 4-methyl-1,3-pentadiene represents a high-energy conformation rather than a low-energy conformation. Explain.

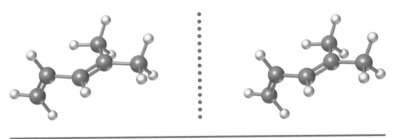

Stereo View

Additional Problems

●●

14.20 Give IUPAC names for the following alkenes:

$$\overset{\displaystyle CH_3}{\overset{\displaystyle |}{}}$$

(a) $CH_3CH{=}CCH{=}CHCH_3$

(b) $H_2C{=}CHCH{=}CHCH{=}CHCH_3$

$$\overset{\displaystyle CH_2CH_2CH_3}{\overset{\displaystyle |}{}}$$

(c) $CH_3CH{=}C{=}CHCH{=}CHCH_3$

(d) $CH_3CH{=}CCH{=}CH_2$

14.21 What product(s) would you expect to obtain from reaction of 1,3-cyclohexadiene with each of the following?
(a) 1 mol Br_2 in CH_2Cl_2 (b) O_3 followed by Zn
(c) 1 mol HCl in ether   (d) 1 mol DCl in ether
(e) 3-Buten-2-one ($H_2C{=}CHCOCH_3$)   (f) Excess OsO_4, followed by $NaHSO_3$

14.22 Draw and name the six possible diene isomers of formula C_5H_8. Which of the six are conjugated dienes?

14.23 Treatment of 3,4-dibromohexane with strong base leads to loss of 2 equivalents of HBr and formation of a product with formula C_6H_{10}. Three products are possible. Name each of the three, and tell how you would use ^1H and ^{13}C NMR spectroscopy to help identify them. How would you use UV spectroscopy?

14.24 Electrophilic addition of Br_2 to isoprene yields the following product mixture:

$$\overset{\displaystyle CH_3}{\overset{\displaystyle |}{}}$$

$H_2C{=}CCH{=}CH_2$ $\xrightarrow{Br_2}$ $H_2C{=}CCHBrCH_2Br$ + $BrCH_2CBrCH{=}CH_2$ + $BrCH_2C{=}CHCH_2Br$

with CH_3 groups above the second, third, and fourth structures respectively

(3%)   **(21%)**   **(76%)**

Of the 1,2-addition products, explain why 3,4-dibromo-3-methyl-1-butene (21%) predominates over 3,4-dibromo-2-methyl-1-butene (3%).

14.25 Propose a structure for a conjugated diene that gives the same product from both 1,2 and 1,4 addition of HBr.

14.26 Draw the possible products resulting from addition of 1 equivalent of HCl to 1-phenyl-1,3-butadiene. Which would you expect to predominate, and why?

$CH{=}CH{-}CH{=}CH_2$

1-Phenyl-1,3-butadiene

14.27 Diene polymers contain occasional vinyl branches along the chain. How do you think these branches might arise?

$CH_2{=}CH{-}CH{=}CH_2$ $\longrightarrow$ $CH_2CH{=}CHCH_2CH_2CHCH_2CH{=}CHCH_2$

A vinyl branch $CH{=}CH_2$

14.28 Tires whose sidewalls are made of natural rubber tend to crack and weather rapidly in areas around cities where high levels of ozone and other industrial pollutants are found. Explain.

14.29 Would you expect allene, $H_2C=C=CH_2$, to show a UV absorption in the 200–400 nm range? Explain.

14.30 Which of the following compounds would you expect to have a $\pi \longrightarrow \pi^*$ UV absorption in the 200–400 nm range?

(a)

=CH$_2$

(b)

Pyridine

(c) $(CH_3)_2C=C=O$

A ketene

14.31 Predict the products of the following Diels–Alder reactions:

(a) + CHO ⟶ ?

(b) + (O,O,O) ⟶ ?

(c) + (O...O) ⟶ ?

14.32 How can you account for the fact that *cis*-1,3-pentadiene is much less reactive than *trans*-1,3-pentadiene in the Diels–Alder reaction?

14.33 Would you expect a conjugated diyne such as 1,3-butadiyne to undergo Diels–Alder reaction with a dienophile? Explain.

14.34 Reaction of isoprene (2-methyl-1,3-butadiene) with ethyl propenoate gives a mixture of two Diels–Alder adducts. Show the structure of each, and explain why a mixture is formed.

$$H_2C=\overset{\overset{\displaystyle CH_3}{|}}{C}-CH=CH_2 + H_2C=CH\overset{\overset{\displaystyle O}{\|}}{C}OCH_2CH_3 \longrightarrow \ ?$$

14.35 Rank the following dienophiles in order of their expected reactivity in the Diels–Alder reaction. Explain.

$H_2C=CHCH_3$ $H_2C=CHCHO$ $(N\equiv C)_2C=C(C\equiv N)_2$ $(CH_3)_2C=C)CH_3)_2$

14.36 Cyclopentadiene is very reactive in Diels–Alder cycloaddition reactions, but 1,3-cyclohexadiene is less reactive, and 1,3-cycloheptadiene is nearly inert. Explain. (Molecular models are helpful.)

14.37   How would you use Diels–Alder reactions to prepare the following products? Show the starting diene and dienophile in each case.

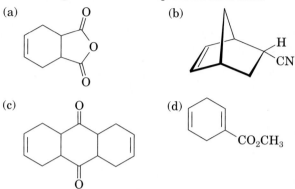

(a)   (b)

(c)   (d)

14.38   Aldrin, a chlorinated insecticide now banned for use in the United States, can be made by Diels–Alder reaction of hexachloro-1,3-cyclopentadiene with norbornadiene. What is the structure of aldrin?

Norbornadiene

14.39   Norbornadiene (Problem 14.38) can be prepared by reaction of chloroethylene with cyclopentadiene, followed by treatment of the product with sodium ethoxide. Write out the overall scheme, and identify the two kinds of reactions.

14.40   We've seen that the Diels–Alder cycloaddition reaction is a one-step, pericyclic process that occurs through a cyclic transition state. Propose a mechanism for the following reaction:

$$\xrightarrow{\text{Heat}} \quad + \text{H}_2\text{C}{=}\text{CH}_2$$

14.41   Propose a mechanism to explain the following reaction (see Problem 14.40):

$$\text{α-Pyrone} \quad + \quad \overset{\text{CO}_2\text{CH}_3}{\underset{\text{CO}_2\text{CH}_3}{\text{C}{\equiv}\text{C}}} \quad \xrightarrow{\text{Heat}} \quad \overset{\text{CO}_2\text{CH}_3}{\underset{\text{CO}_2\text{CH}_3}{}} \quad + \text{CO}_2$$

14.42   The triene shown below reacts with *two* equivalents of maleic anhydride to yield $C_{17}H_{16}O_6$ as product. Predict a structure for the product.

$$+ 2 \quad \longrightarrow \quad C_{17}H_{16}O_6$$

Maleic anhydride

14.43 The following ultraviolet absorption maxima have been measured:

	λ_{max} (nm)
1,3-Butadiene	217
2-Methyl-1,3-butadiene	220
1,3-Pentadiene	223
2,3-Dimethyl-1,3-butadiene	226
2,4-Hexadiene	227
2,4-Dimethyl-1,3-pentadiene	232
2,5-Dimethyl-2,4-hexadiene	240

What conclusion can you draw about the effect of alkyl substitution on UV absorption maxima? Approximately what effect does each added alkyl group have?

14.44 1,3,5-Hexatriene has $\lambda_{max} = 258$ nm. In light of your answer to Problem 14.43, approximately where would you expect 2,3-dimethyl-1,3,5-hexatriene to absorb? Explain.

14.45 β-Ocimene is a pleasant-smelling hydrocarbon found in the leaves of certain herbs. It has the molecular formula $C_{10}H_{16}$ and exhibits a UV absorption maximum at 232 nm. On hydrogenation with a palladium catalyst, 2,6-dimethyloctane is obtained. Ozonolysis of β-ocimene, followed by treatment with zinc and acetic acid, produces four fragments: acetone, formaldehyde, pyruvaldehyde, and malonaldehyde:

$$CH_3\overset{\overset{\displaystyle O}{\|}}{C}CH_3 \qquad H\overset{\overset{\displaystyle O}{\|}}{C}H \qquad CH_3\overset{\overset{\displaystyle O}{\|}}{C}-\overset{\overset{\displaystyle O}{\|}}{C}H \qquad H\overset{\overset{\displaystyle O}{\|}}{C}CH_2\overset{\overset{\displaystyle O}{\|}}{C}H$$

Acetone   **Formaldehyde**   **Pyruvaldehyde**   **Malonaldehyde**

(a) How many double bonds does β-ocimene have?
(b) Is β-ocimene conjugated or nonconjugated?
(c) Propose a structure for β-ocimene.
(d) Formulate the reactions, showing starting material and products.

14.46 Myrcene, $C_{10}H_{16}$, is found in oil of bay leaves and is isomeric with β-ocimene (see Problem 14.45). It shows an ultraviolet absorption at 226 nm and can be catalytically hydrogenated to yield 2,6-dimethyloctane. On ozonolysis followed by zinc/acetic acid treatment, myrcene yields formaldehyde, acetone, and 2-oxopentanedial:

$$H\overset{\overset{\displaystyle O}{\|}}{C}CH_2CH_2\overset{\overset{\displaystyle O}{\|}}{C}-\overset{\overset{\displaystyle O}{\|}}{C}H \qquad \textbf{2-Oxopentanedial}$$

Propose a structure for myrcene, and formulate the reactions, showing starting material and products.

14.47 Addition of HCl to 1-methoxycyclohexene yields 1-chloro-1-methoxycyclohexane as the sole product. Why is none of the other regioisomer formed?

14.48 Hydrocarbon A, $C_{10}H_{14}$, has a UV absorption at $\lambda_{max} = 236$ nm and gives hydrocarbon B, $C_{10}H_{18}$, on catalytic hydrogenation. Ozonolysis of A followed by zinc/acetic acid treatment yields the following diketo dialdehyde:

$$\underset{\overset{\|}{O}}{HC}CH_2CH_2\underset{\overset{\|}{O}}{C}-\underset{\overset{\|}{O}}{C}CH_2CH_2CH_2\underset{\overset{\|}{O}}{C}H$$

(a) Propose two possible structures for A.
(b) Hydrocarbon A reacts with maleic anhydride to yield a Diels–Alder adduct. Which of your structures for A is correct?
(c) Formulate the reactions showing starting material and products.

14.49 Adiponitrile, a starting material used in the manufacture of nylon, can be prepared in three steps from 1,3-butadiene. How would you carry out this synthesis?

$$H_2C{=}CHCH{=}CH_2 \quad \xrightarrow{3\ steps} \quad N{\equiv}CCH_2CH_2CH_2CH_2C{\equiv}N$$

Adiponitrile

14.50 Ergosterol, a precursor of vitamin D, has $\lambda_{max} = 282$ nm and molar absorptivity $\varepsilon = 11,900$. What is the concentration of ergosterol in a solution whose absorbance $A = 0.065$ with a sample pathlength $l = 1.00$ cm?

Ergosterol ($C_{28}H_{44}O$)

14.51 Cyclopentadiene polymerizes slowly at room temperature to yield a polymer that has no double bonds. On heating, the polymer breaks down to regenerate cyclopentadiene. Propose a structure for the product.

14.52 Dimethyl butynedioate undergoes a Diels–Alder reaction with (2E,4E)-hexadiene. Show the structure and stereochemistry of the product.

$$CH_3O\underset{\overset{\|}{O}}{C}-C{\equiv}C-\underset{\overset{\|}{O}}{C}OCH_3 \quad \textbf{Dimethyl butynedioate}$$

14.53 Dimethyl butynedioate also undergoes a Diels–Alder reaction with (2E,4Z)-hexadiene, but the stereochemistry of the product is different from that of the (2E,4E) isomer (Problem 14.52). Explain.

14.54 How would you carry out the following synthesis (more than one step is required)? What stereochemical relationship between the –CO_2CH_3 group attached to the cyclohexane ring and the –CHO groups would your synthesis produce?

A Look Ahead

• •

14.55 The double bond of an *enamine* (alk*ene* + *amine*) is much more nucleophilic than a typical alkene double bond. Assuming that the nitrogen atom in an enamine is *sp²*-hybridized, draw an orbital picture of an enamine, and explain why the double bond is electron-rich. (See Section 23.12.)

An enamine

14.56 Benzene has an ultraviolet absorption at λ_{max} = 204 nm, and *para*-toluidine has λ_{max} = 235 nm. How do you account for this difference? (See Sections 16.5 and 16.6.)

Benzene
(λ_{max} = 204 nm)

H₃C——**NH₂**

***para*-Toluidine**
(λ_{max} = 235 nm)

14.57 Phenol, a weak acid with pK_a = 10.0, has a UV absorption at λ_{max} = 210 nm in ethanol solution. When dilute NaOH is added, the absorption increases to λ_{max} = 235 nm. Explain. (See Section 17.3.)

——OH   **Phenol**

Molecular Modeling

• •

14.58 UV excitation of 1,3-butadiene promotes one electron from ψ_2 (the HOMO) to $\psi_3{}^*$ (the LUMO). Use SpartanView to examine these orbitals in 1,3-butadiene. Compare the structures of 1,3-butadiene and excited 1,3-butadiene, and tell which bonds are weakened and which are strengthened by the UV excitation.

14.59 Use SpartanView to examine bond-rotation sequences about the C2–C3 bond in both 1-butene and 1,3-butadiene. Compare the energies of the lowest-energy and transition-state conformations, and tell in which molecule rotation is more difficult. Identify the two minimum-energy conformations of 1,3-butadiene, and tell which geometry permits Diels–Alder cycloaddition. Is this the preferred geometry?

14.60 Use SpartanView to examine electrostatic potential maps of ethylene, benzoquinone, and 3,3,3-trifluoropropene. Which of these compounds are reactive as Diels–Alder dienophiles?

14.61 Maleic anhydride and 1,3-cyclopentadiene undergo a Diels–Alder cycloaddition reaction to give endo product A rather that exo product B. Use SpartanView to compare the energies of transition states A and B. Which transition state is lower in energy (thus giving the kinetic product)? Next, compare the energies of products A and B. Which is lower in energy (thus being the thermodynamic product)? Is the kinetic product the same as the thermodynamic product?

1,3-Cyclo-   **Maleic**   **Product A (endo)**   **Product B (exo)**
pentadiene   **anhydride**   (***NOT formed***)

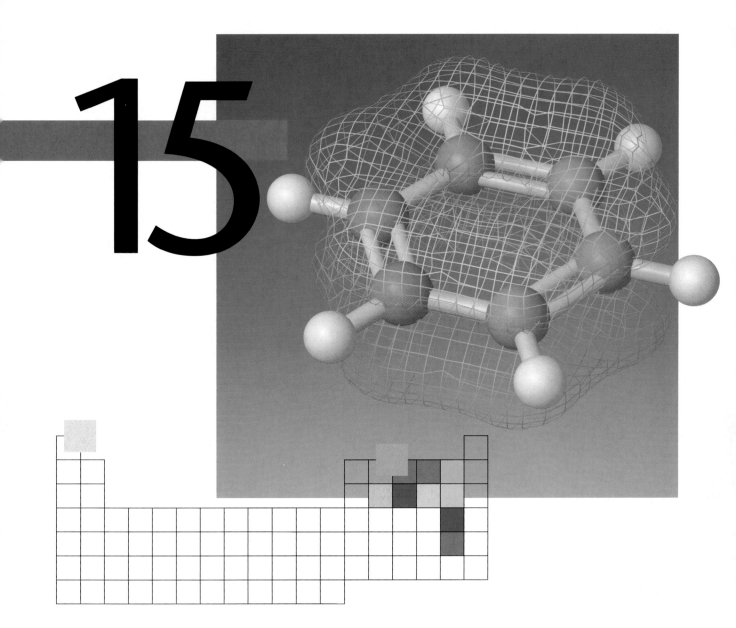

Benzene and Aromaticity

In the early days of organic chemistry, the word *aromatic* was used to describe such fragrant substances as benzaldehyde (from cherries, peaches, and almonds), toluene (from Tolu balsam), and benzene (from coal distillate). It was soon realized, however, that substances grouped as aromatic differed from most other organic compounds in their chemical behavior.

Benzene   **Benzaldehyde**   **Toluene**

Today, we use the word **aromatic** to refer to benzene and its structural relatives. We'll see in this and the next chapter that aromatic compounds show chemical behavior quite different from that of the aliphatic compounds we've studied to this point. Thus, chemists of the early nineteenth century were correct about there being a chemical difference between aromatic compounds and others, but the association of aromaticity with fragrance has long been lost.

Many compounds isolated from natural sources are aromatic in part. In addition to benzene, benzaldehyde, and toluene, such compounds as the steroidal hormone estrone and the well-known analgesic morphine have aromatic rings. Many synthetic drugs are also aromatic in part; the tranquilizer diazepam (Valium) is an example.

Estrone   **Morphine**   **Diazepam (Valium)**

Benzene itself has been found to cause bone-marrow depression and consequent leukopenia (lowered white blood cell count) on prolonged exposure. Benzene should therefore be handled cautiously if used as a laboratory solvent.

15.1 Sources of Aromatic Hydrocarbons

Simple aromatic hydrocarbons come from two main sources: coal and petroleum. Coal is an enormously complex mixture made up primarily of large arrays of benzene-like rings joined together. Thermal breakdown of coal occurs when it is heated to 1000°C in the absence of air, and a mixture of volatile products called *coal tar* boils off. Fractional distillation of coal tar yields benzene, toluene, xylene (dimethylbenzene), naphthalene, and a host of other aromatic compounds (Figure 15.1).

FIGURE 15.1 ▼

Some aromatic hydrocarbons found in coal tar.

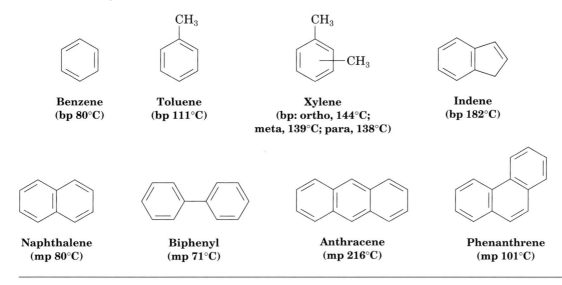

Benzene
(bp 80°C)

Toluene
(bp 111°C)

Xylene
(bp: ortho, 144°C;
meta, 139°C; para, 138°C)

Indene
(bp 182°C)

Naphthalene
(mp 80°C)

Biphenyl
(mp 71°C)

Anthracene
(mp 216°C)

Phenanthrene
(mp 101°C)

Petroleum, unlike coal, contains few aromatic compounds and consists largely of alkanes (See "Gasoline from Petroleum" at the end of Chapter 3). During petroleum refining, however, aromatic molecules are formed when alkanes are passed over a catalyst at about 500°C under high pressure. Heptane (C_7H_{16}), for example, is converted into toluene (C_7H_8) by dehydrogenation and cyclization.

15.2 Naming Aromatic Compounds

Aromatic substances, more than any other class of organic compounds, have acquired a large number of nonsystematic names. Although the use of such names is discouraged, IUPAC rules allow for some of the more widely used names to be retained (Table 15.1). Thus, methylbenzene is known commonly as *toluene*, hydroxybenzene as *phenol*, aminobenzene as *aniline*, and so on.

Monosubstituted benzene derivatives are systematically named in the same manner as other hydrocarbons, with *-benzene* as the parent name. Thus, C_6H_5Br is bromobenzene, $C_6H_5NO_2$ is nitrobenzene, and $C_6H_5CH_2CH_2CH_3$ is propylbenzene.

Bromo**benzene**

Nitro**benzene**

Propyl**benzene**

TABLE 15.1   Common Names of Some Aromatic Compounds

Formula	Name	Formula	Name
⟨benzene ring⟩–CH₃	Toluene (bp 111°C)	⟨benzene ring⟩–CHO	Benzaldehyde (bp 178°C)
⟨benzene ring⟩–OH	Phenol (mp 43°C)	⟨benzene ring⟩–COOH	Benzoic acid (mp 122°C)
⟨benzene ring⟩–NH₂	Aniline (bp 184°C)	⟨benzene ring⟩–CN	Benzonitrile (bp 191°C)
⟨benzene ring⟩–C(=O)CH₃	Acetophenone (mp 21°C)	⟨benzene ring⟩ with two CH₃ (ortho)	*ortho*-Xylene (bp 144°C)
⟨benzene ring⟩–CHCH₃ with CH₃	Cumene (bp 152°C)	⟨benzene ring⟩–CH=CH₂	Styrene (bp 145°C)

Alkyl-substituted benzenes, sometimes referred to as **arenes**, are named in different ways depending on the size of the alkyl group. If the alkyl substituent has six or fewer carbons, the arene is named as an alkyl-substituted benzene. If the alkyl substituent has more than six carbons, the compound is named as a phenyl-substituted alkane. The name **phenyl**, pronounced **fen**-nil and often abbreviated as Ph or Φ (Greek phi), is used for the $-C_6H_5$ unit when the benzene ring is considered as a substituent. The word is derived from the Greek *pheno* ("I bear light"), commemorating the fact that benzene was discovered by Michael Faraday in 1825 from the oily residue left by the illuminating gas used in London street lamps. As mentioned previously, the $C_6H_5CH_2-$ group is called **benzyl**.

A phenyl group   **2-Phenylheptane**   **A benzyl group**

Michael Faraday

Michael Faraday (1791–1867) was born in Newington Butts, Surrey, England, the son of a blacksmith. Though he received little formal schooling, he was one of the greatest scientists of the nineteenth century. As a young man in 1812, he became a laboratory assistant to Sir Humphry Davy at the Royal Institution and learned chemistry through this apprenticeship. By 1820, he was said to know as much chemistry as any living person; by 1825, he was director of a laboratory at the Royal Institution; and by 1833, he was Fullerian Professor of Chemistry. He is best remembered for his work on electricity and magnetism.

Disubstituted benzenes are named using one of the prefixes ***ortho (o)***, ***meta (m)***, or ***para (p)***. An ortho-disubstituted benzene has its two substituents in a 1,2 relationship on the ring; a meta-disubstituted benzene has its two substituents in a 1,3 relationship; and a para-disubstituted benzene has its substituents in a 1,4 relationship.

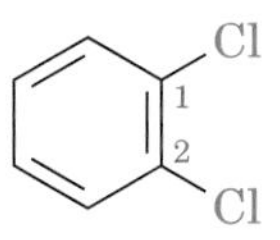

ortho-**Dichlorobenzene**
1,2 disubstituted

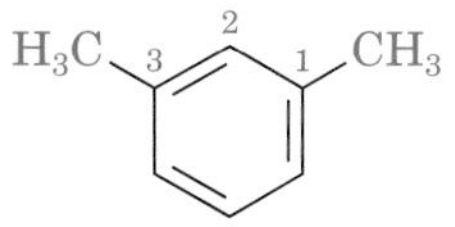

meta-**Xylene**
1,3 disubstituted

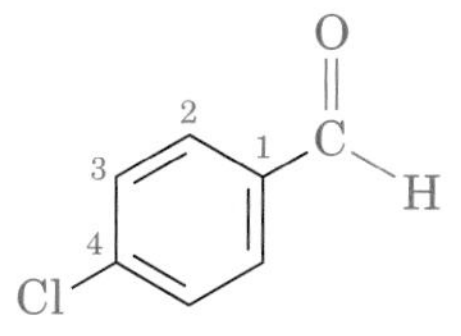

para-**Chlorobenzaldehyde**
1,4 disubstituted

The ortho, meta, para system of nomenclature is also useful when discussing reactions. For example, we might describe the reaction of bromine with toluene by saying, "Reaction occurs at the para position"—in other words, at the position para to the methyl group already present on the ring.

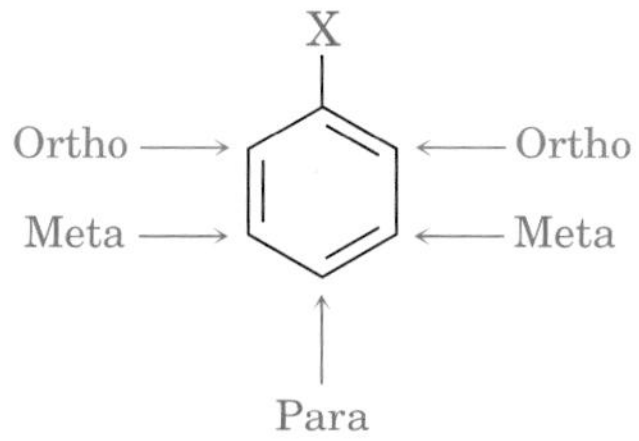

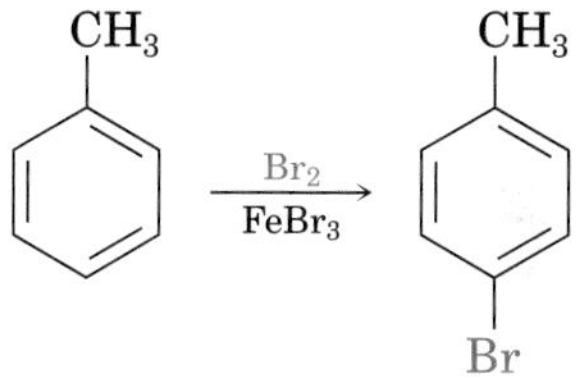

Toluene ***p*-Bromotoluene**

Benzenes with more than two substituents are named by numbering the position of each substituent so that the lowest possible numbers are used. The substituents are listed alphabetically when writing the name.

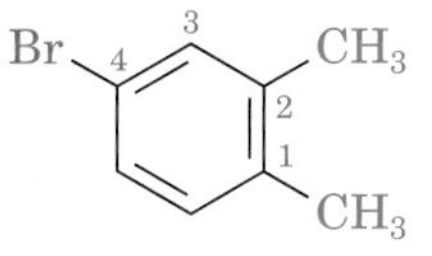

4-Bromo-1,2-dimethyl**benzene**

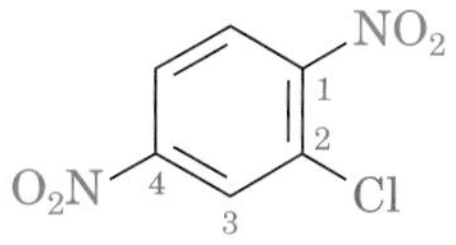

2-Chloro-1,4-dinitro**benzene**

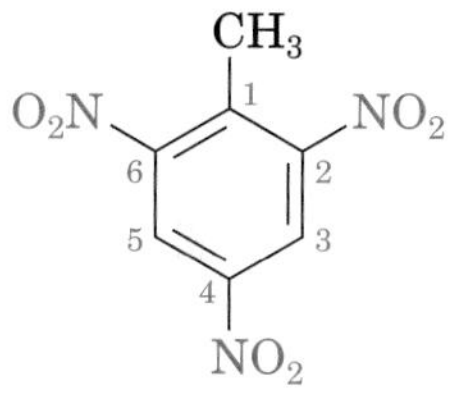

2,4,6-Trinitro**toluene (TNT)**

Note in the third example shown that *-toluene* is used as the parent name rather than *-benzene*. Any of the monosubstituted aromatic compounds shown in Table 15.1 can serve as a parent name, with the principal substituent (–CH_3 in toluene) assumed to be on C1. The following two examples further illustrate this practice:

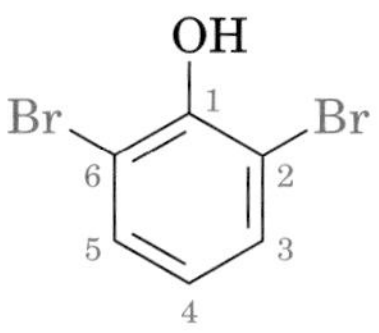

2,6-Dibromo***phenol***

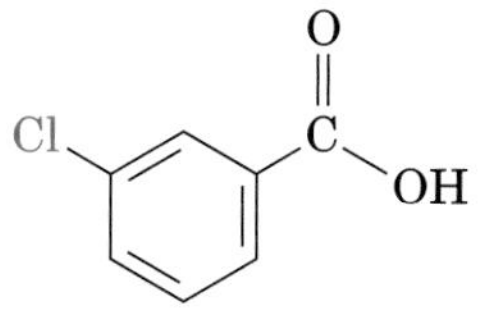

m-Chloro***benzoic acid***

• •

Problem 15.1   Tell whether the following compounds are ortho, meta, or para disubstituted:

(a) Cl $\diagdown$ $\diagup$ CH$_3$ (b) $\diagup$ NO$_2$ Br   (c) $\diagup$ SO$_3$H   OH

Problem 15.2   Give IUPAC names for the following compounds:

(a) Cl $\diagdown$ $\diagup$ Br   (b) CH$_2$CH$_2$CHCH$_3$ CH$_3$ (c) $\diagup$ NH$_2$ Br

(d) Cl $\diagdown$ $\diagup$ CH$_3$ Cl   (e) O$_2$N $\diagdown$ $\diagup$ CH$_2$CH$_3$ NO$_2$ (f) H$_3$C $\diagdown$ CH$_3$ CH$_3$ CH$_3$

Problem 15.3   Draw structures corresponding to the following IUPAC names:
(a) *p*-Bromochlorobenzene   (b) *p*-Bromotoluene
(c) *m*-Chloroaniline   (d) 1-Chloro-3,5-dimethylbenzene

• •

15.3 Structure and Stability of Benzene

Although benzene is clearly unsaturated, it is much more stable than other alkenes, and it fails to undergo typical alkene reactions. Cyclohexene, for instance, reacts rapidly with Br$_2$ and gives the addition product 1,2-dibromocyclohexane, but benzene reacts only slowly with Br$_2$ and gives the *substitution* product C$_6$H$_5$Br. As a result of this substitution, the cyclic conjugation of the benzene ring is retained.

Benzene   **Bromobenzene**   **(Addition product)**
                          (substitution product)   *NOT* formed

We can get a quantitative idea of benzene's stability from the heats of hydrogenation. Cyclohexene, an isolated alkene, has $\Delta H°_{\text{hydrog}} = -118$ kJ/mol (-28.2 kcal/mol), and 1,3-cyclohexadiene, a conjugated diene, has $\Delta H°_{\text{hydrog}} = -230$ kJ/mol (-55.0 kcal/mol). As expected, this value for 1,3-cyclohexadiene is a bit less than twice that for cyclohexene because conjugated dienes are more stable than isolated dienes (Section 14.2).

Carrying the process one step further, we might expect $\Delta H^{\circ}_{\text{hydrog}}$ for "cyclohexatriene" (benzene) to be a bit less than −356 kJ/mol, or three times the cyclohexene value. The actual value, however, is −206 kJ/mol, some 150 kJ/mol (36 kcal/mol) less than expected. Since 150 kJ/mol less heat than expected is released during hydrogenation of benzene, benzene must have 150 kJ/mol less energy than expected to begin with. In other words, benzene has 150 kJ/mol "extra" stability (Figure 15.2).

FIGURE 15.2 ▼

A comparison of the heats of hydrogenation of cyclohexene, 1,3-cyclohexadiene, and benzene. Benzene is 150 kJ/mol (36 kcal/mol) more stable than might be expected for "cyclohexatriene."

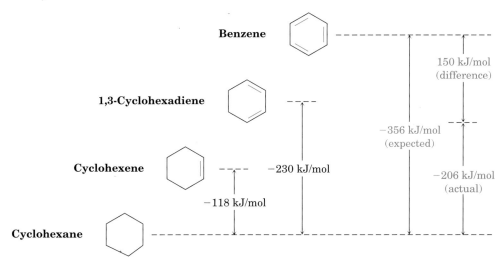

Further evidence for the unusual nature of benzene is that all its carbon–carbon bonds have the same length—139 pm—intermediate between typical single (154 pm) and double (134 pm) bonds. In addition, the electrostatic potential map below shows that the electron density in all six carbon–carbon bonds is identical.

benzene

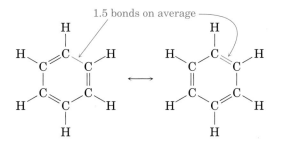

Resonance theory (Sections 2.4–2.5) accounts for the stability and properties of benzene by describing it as a resonance hybrid of two equivalent forms. Neither form is correct by itself; the true structure of benzene is somewhere in between the two resonance forms but is impossible to draw with our usual conventions. Many chemists therefore represent benzene by drawing it with a circle inside to indicate the equivalence of the carbon–carbon bonds. This kind of representation has to be used carefully, however, because it doesn't indicate the number of π electrons in the ring. (How many electrons does a circle represent?) In this book, benzene and other aromatic compounds will be represented by a single line-bond structure. We'll be able to keep count of π electrons this way, but we must be aware of the limitations of the drawings.

Alternative representations of benzene. The "circle" representation must be used carefully since it doesn't indicate the number of π electrons in the ring.

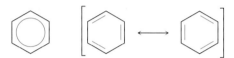

15.4 Molecular Orbital Description of Benzene

Having just seen a resonance description of benzene, let's now look at the alternative molecular orbital description. An orbital view of benzene makes clear the cyclic conjugation of the benzene molecule and the equivalence of the six carbon–carbon bonds. Benzene is a planar molecule with the shape of a regular hexagon. All C–C–C bond angles are 120°, all six carbon atoms are sp^2-hybridized, and each carbon has a p orbital perpendicular to the plane of the six-membered ring.

Since all six carbon atoms and all six p orbitals in benzene are equivalent, it's impossible to define three localized π bonds in which a given p orbital overlaps only one neighboring p orbital. Rather, each p orbital overlaps equally well with *both* neighboring p orbitals, leading to a picture of benzene in which the six π electrons are completely delocalized around the ring.

We can construct π molecular orbitals for benzene just as we did for 1,3-butadiene in Section 14.3. If six p atomic orbitals combine in a cyclic manner, six benzene molecular orbitals result, as shown in Figure 15.3. The three low-energy molecular orbitals, denoted ψ_1, ψ_2, and ψ_3, are bonding combinations, and the three high-energy orbitals are antibonding. Note that two of the bonding orbitals, ψ_2 and ψ_3, have the same energy, as do the antibonding orbitals $\psi_4{}^*$ and $\psi_5{}^*$. Such orbitals are said to be **degenerate**. Note also that two of the orbitals, ψ_3 and $\psi_4{}^*$, have nodes passing through two of the ring carbon atoms, thereby leaving no π electron density on these carbons. The six p electrons of benzene occupy the three bonding molecular orbitals and are delocalized over the entire conjugated system, leading to the observed 150 kJ/mol stabilization of benzene.

FIGURE 15.3 ▼

The six benzene π molecular orbitals. The bonding orbitals ψ_2 and ψ_3 have the same energy and are said to be degenerate, as are the antibonding orbitals ψ_4* and ψ_5*. The orbitals ψ_3 and ψ_4* have no π electron density on two carbons because of a node passing through these atoms.

benzene
(see MO's on
CD-Rom)

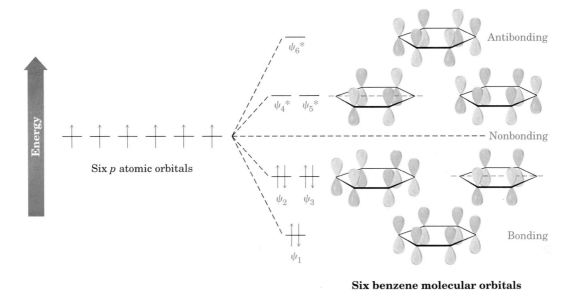

Six benzene molecular orbitals

· ·

Problem 15.4   Pyridine is a flat, hexagonal molecule with bond angles of 120°. It undergoes electrophilic substitution rather than addition and generally behaves like benzene. Draw an orbital picture of pyridine to explain its properties. Check your answer by looking ahead to Section 15.7.

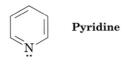   **Pyridine**

· ·

15.5 Aromaticity and the Hückel 4*n* + 2 Rule

Key Ideas ▶   Let's review what we've learned thus far about benzene and, by extension, about other benzene-like aromatic molecules:

- Benzene is a cyclic conjugated molecule.

- Benzene is unusually stable, having a heat of hydrogenation 150 kJ/mol less negative than we might expect for a cyclic triene.

- Benzene is planar and has the shape of a regular hexagon. All bond angles are 120°, and all carbon–carbon bond lengths are 139 pm.

- Benzene undergoes substitution reactions that retain the cyclic conjugation rather than electrophilic addition reactions that would destroy the conjugation.

- Benzene is a resonance hybrid whose structure is intermediate between two line-bond structures.

Erich Hückel

Erich Hückel (1896–1980) was born in Stuttgart, Germany, and received his Ph.D. at the University of Göttingen with Peter Debye. He was professor of physics, first at Stuttgart and later at Marburg (1937–1961).

Although these facts would seem to provide a good description of benzene and other aromatic molecules, they aren't enough. Something else called the **Hückel 4n + 2 rule** is also needed to complete a description of aromaticity. According to a theory devised by the German physicist Erich Hückel in 1931, a molecule is aromatic only if it has a planar, monocyclic system of conjugation with *a total of 4n + 2 π electrons,* where n is an integer (n = 0, 1, 2, 3, . . .). In other words, only molecules with 2, 6, 10, 14, 18 . . . π electrons can be aromatic. Molecules with 4n π electrons (4, 8, 12, 16, . . .) *can't* be aromatic, even though they may be cyclic and apparently conjugated. In fact, planar, conjugated molecules with 4n π electrons are even said to be **antiaromatic**, because they are *destabilized* by delocalization of their π electrons.

Let's look at some examples to see how the Hückel 4n + 2 rule works.

- *Cyclobutadiene* has four π electrons and is antiaromatic:

Two double bonds; four π electrons

Cyclobutadiene

Stereo View

Cyclobutadiene is highly reactive and shows none of the properties associated with aromaticity. In fact, it was not even prepared until 1965, when Rowland Pettit of the University of Texas was able to make it at low temperature. Even at −78°C, however, cyclobutadiene is so reactive that it dimerizes by a self-Diels–Alder reaction. One molecule behaves as a diene and the other as a dienophile:

Rowland Pettit

Rowland Pettit (1927–1981) was born in Port Lincoln, Australia. He received two doctoral degrees, one from the University of Adelaide in 1952 and the second from the University of London in 1956, working with Michael Dewar. He then became professor of chemistry at the University of Texas, Austin (1957–1981).

- *Benzene* has six π electrons (4n + 2 = 6 when n = 1) and is aromatic:

Three double bonds; six π electrons

Benzene

- *Cyclooctatetraene* has eight π electrons and is not aromatic:

Four double bonds; eight π electrons

Cyclooctatetraene

Richard Willstätter

Richard Willstätter (1872–1942) was born in Karlsruhe, Germany, and obtained his Ph.D. from the Technische Hochschule, Munich (1895). He was professor of chemistry at the universities of Zurich, Berlin, and then Munich (1916–1924). In 1915, he won the Nobel Prize in chemistry for his work on elucidating the structure of chlorophyll. Nevertheless, as a Jew, he was subjected to anti-Semitic pressure that caused him to resign his position at Munich in 1924. He continued, however, to work privately.

Chemists in the early 1900s believed that the only requirement for aromaticity was the presence of a cyclic conjugated system. It was therefore expected that cyclooctatetraene, as a close analog of benzene, would also prove to be unusually stable. The facts, however, proved otherwise. When cyclooctatetraene was first prepared in 1911 by the German chemist Richard Willstätter, it was found not to be particularly stable but to resemble an open-chain polyene in its reactivity.

Cyclooctatetraene reacts readily with Br_2, $KMnO_4$, and HCl, just as other alkenes do. We now know, in fact, that cyclooctatetraene is not even conjugated. It is tub-shaped rather than planar and has no cyclic conjugation because neighboring *p* orbitals don't have the necessary parallel alignment for overlap (Figure 15.4). The π electrons are localized in four discrete C=C bonds rather than delocalized around the ring. X-ray studies show that the C–C single bonds are 147 pm long and the double bonds are 134 pm long. In addition, the ^{1}H NMR spectrum shows a single sharp resonance line at 5.7 δ, a value characteristic of an alkene rather than an aromatic molecule (Section 15.10).

FIGURE 15.4 ▼

Cyclooctatetraene is a tub-shaped molecule that has no cyclic conjugation because its *p* orbitals are not aligned properly for overlap.

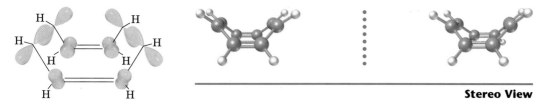

Stereo View

Problem 15.5 To be aromatic, a molecule must have 4*n* + 2 π electrons and must have cyclic conjugation. The cyclodecapentaene shown below in a stereo view fulfills one of these criteria but not the other, and has resisted all attempts at synthesis. Explain.

Cyclodecapentaene

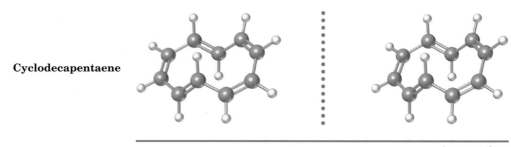

Stereo View

15.6 Aromatic Ions

According to the Hückel criteria for aromaticity described in the preceding section, a molecule must be cyclic, conjugated (that is, be nearly planar and have a p orbital on each carbon), and have $4n + 2$ π electrons. Nothing in this definition says that the numbers of p orbitals and π electrons must be the same. In fact, they can be different. The $4n + 2$ rule is broadly applicable to many kinds of molecules, not just to neutral hydrocarbons. For example, both the cyclopentadienyl *anion* and the cycloheptatrienyl *cation* are aromatic.

Cyclopentadienyl anion   **Cycloheptatrienyl cation**

Six π electrons; aromatic ions

Let's look first at the cyclopentadienyl anion. Cyclopentadiene itself is not aromatic because it is not fully conjugated. The $-CH_2-$ carbon in the ring is sp^3-hybridized, thus preventing complete cyclic conjugation. Imagine, though, that we remove one hydrogen from the saturated CH_2 group and let that carbon become sp^2-hybridized. The resultant species would have five p orbitals, one on each of the five carbons, and would be fully conjugated.

There are three ways we might imagine removing the hydrogen, as shown in Figure 15.5.

- We could remove the hydrogen atom and *both* electrons ($H:^-$) from the C–H bond, leaving a cyclopentadienyl cation.

- We could remove the hydrogen and *one* electron ($H\cdot$) from the C–H bond, leaving a cyclopentadienyl radical.

- We could remove a hydrogen ion with *no* electrons (H^+), leaving a cyclopentadienyl anion.

Although five equivalent resonance structures can be drawn for all three species, Hückel's rule predicts that *only the six-π-electron anion should be aromatic.* The four-π-electron cyclopentadienyl carbocation and the five-π-electron cyclopentadienyl radical are predicted to be unstable and antiaromatic.

In practice, both the cyclopentadienyl cation and the radical are highly reactive and difficult to prepare. Neither shows any sign of the stability expected for an aromatic system. The six-π-electron cyclopentadienyl anion, by contrast, is easily prepared and remarkably stable. In fact, cyclopenta-

FIGURE 15.5 ▼

Generating the
cyclopentadienyl cation,
radical, and anion by
removing a hydrogen
from cyclopentadiene.

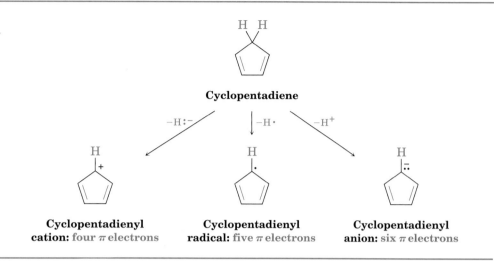

Cyclopentadiene

−H:⁻   −H·   −H⁺

**Cyclopentadienyl
cation:** four π electrons

**Cyclopentadienyl
radical:** five π electrons

**Cyclopentadienyl
anion:** six π electrons

diene is one of the most acidic hydrocarbons known. Although most hydrocarbons have a $pK_a > 45$, cyclopentadiene has $pK_a = 16$, a value comparable to that of water! Cyclopentadiene is acidic because the anion formed by
loss of H^+ is so stable (Figure 15.6).

FIGURE 15.6 ▼

An orbital view of the
aromatic cyclopentadienyl
anion, showing the cyclic
conjugation and six π
electrons in five *p* orbitals.
The electrostatic potential
map further indicates that
the ion is symmetrical and
that all five carbons are
electron-rich (red).

cyclopentadienyl anion

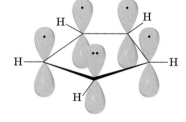

**Aromatic cyclopentadienyl anion
with six π electrons**

Similar arguments can be used to predict the relative stabilities of
the cycloheptatrienyl cation, radical, and anion. Removal of a hydrogen
from cycloheptatriene can generate the six-π-electron cation, the seven-π-
electron radical, or the eight-π-electron anion (Figure 15.7, p. 572). Once
again, all three species have numerous resonance forms, but Hückel's rule
predicts that only the six-π-electron cycloheptatrienyl cation should be
aromatic. The seven-π-electron cycloheptatrienyl radical and the eight-π-
electron anion are antiaromatic.

Generation of the
cycloheptatrienyl cation,
radical, and anion. Only
the six-π-electron cation
is aromatic.

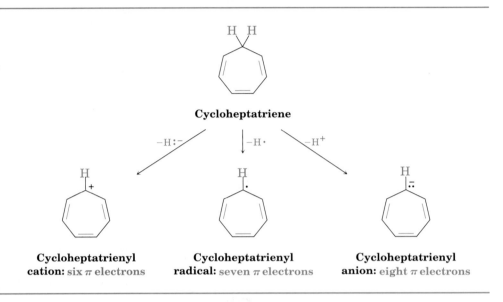

Both the cycloheptatrienyl radical and the anion are reactive and difficult to prepare. The six-π-electron cation, however, is extraordinarily stable. In fact, the cycloheptatrienyl cation was first prepared over a century
ago by reaction of Br_2 with cycloheptatriene (Figure 15.8), although its structure was not recognized at the time.

Reaction of cycloheptatriene with bromine yields cycloheptatrienylium bromide, an
ionic substance containing the cycloheptatrienyl cation. The electrostatic potential
map shows that all seven carbon atoms are equally charged.

cycloheptatrienyl cation

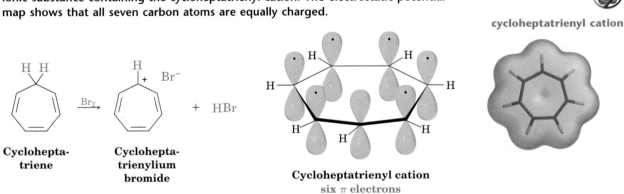

Problem 15.6   Draw the five resonance structures of the cyclopentadienyl anion. Are all carbon–
carbon bonds equivalent? How many absorption lines would you expect to see in the
^{1}H NMR and ^{13}C NMR spectra of the anion?

Problem 15.7   Cyclooctatetraene readily reacts with potassium metal to form the cyclooctatetraene dianion, $C_8H_8^{2-}$. Why do you suppose this reaction occurs so easily? What geometry do you expect for the cyclooctatetraene dianion?

15.7 Pyridine and Pyrrole: Two Aromatic Heterocycles

Biological Connection ▶

Look back once again at the definition of aromaticity in Section 15.6: . . . a cyclic, conjugated molecule containing $4n + 2\ \pi$ electrons. Nothing in this definition says that the atoms in the ring must be *carbon*. In fact, *heterocyclic* compounds can also be aromatic. A **heterocycle** is a cyclic compound that contains an atom or atoms other than carbon in its ring. The heteroatom is often nitrogen or oxygen, but sulfur, phosphorus, and other elements are also found. Pyridine, for example, is a six-membered heterocycle with a nitrogen atom in its ring.

Pyridine is much like benzene in its π electron structure. Each of the five sp^2-hybridized carbons has a p orbital perpendicular to the plane of the ring, and each p orbital contains one π electron. The nitrogen atom is also sp^2-hybridized and has one electron in a p orbital, bringing the total to six π electrons. The nitrogen lone-pair electrons are in an sp^2 orbital in the plane of the ring and are not involved with the aromatic π system (Figure 15.9).

FIGURE 15.9 ▼

Pyridine, an aromatic heterocycle, has a π electron arrangement much like that of benzene.

pyridine
(see MO's on CD-Rom)

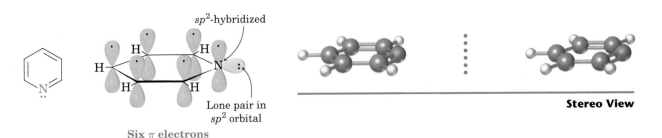

sp²-hybridized

Lone pair in *sp²* orbital

Six π electrons

Stereo View

Pyrrole (two *r*'s, one *l*), another example of an aromatic substance, is a five-membered heterocycle and has a π electron system similar to that of the cyclopentadienyl anion. Each of the four sp^2-hybridized carbons has a *p* orbital perpendicular to the ring, and each contributes one π electron. The nitrogen atom is also *sp²*-hybridized, and its lone pair of electrons occupies a *p* orbital. Thus, there are a total of six π electrons, making pyrrole an aromatic molecule. An orbital picture of pyrrole is shown in Figure 15.10.

FIGURE 15.10 ▼

Pyrrole, a five-membered aromatic heterocycle, has a π electron arrangement much like that of the cyclopentadienyl anion.

pyrrole
(see MO's on
CD-Rom)

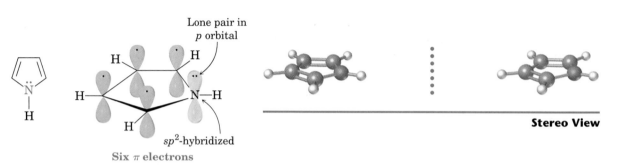

Stereo View

Note that the nitrogen atoms have different roles in pyridine and pyrrole even though both compounds are aromatic. The nitrogen atom in pyridine is in a double bond and therefore contributes only *one* π electron to the aromatic sextet, just as a carbon atom in benzene does. The nitrogen atom in pyrrole, however, is not in a double bond. Like one of the carbons in the cyclopentadienyl anion, the pyrrole nitrogen atom contributes *two* π electrons (the lone pair) to the aromatic sextet.

Pyridine, pyrrole, and other aromatic heterocycles are crucial to many biochemical processes. Their chemistry will be discussed in more detail in Chapter 28.

● ●

Practice Problem 15.1   Thiophene, a sulfur-containing heterocycle, undergoes typical aromatic substitution reactions rather than addition reactions. Explain why thiophene is aromatic.

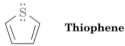
Thiophene

Strategy   Recall the requirements for aromaticity—a planar, cyclic, conjugated molecule with $4n + 2$ π electrons—and see how these requirements apply to thiophene.

Solution   Thiophene is the sulfur analog of pyrrole. The sulfur atom is *sp²*-hybridized and has a lone pair of electrons in a *p* orbital perpendicular to the plane of the ring, as shown at the top of the next page. (Sulfur also has a second lone pair of electrons in the ring plane.)

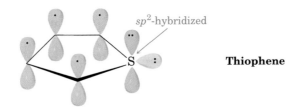

Thiophene

. . . .

Problem 15.8   The aromatic five-membered heterocycle imidazole is important in many biological processes. One of its nitrogen atoms is pyridine-like in that it contributes one π electron to the aromatic sextet, and the other nitrogen is pyrrole-like in that it contributes two π electrons. Draw an orbital picture of imidazole, and account for its aromaticity. Which atom is pyridine-like and which is pyrrole-like?

Imidazole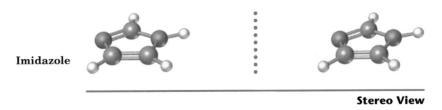

Stereo View

Problem 15.9   Draw an orbital picture of furan to show how the molecule is aromatic.

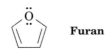

Furan

. . . .

15.8 Why 4*n* + 2?

What's so special about $4n + 2$ π electrons? Why do 2, 6, 10, 14, . . . π electrons lead to aromatic stability, while other numbers of electrons do not? The answer comes from molecular orbital theory.

When the energy levels of molecular orbitals for cyclic conjugated molecules are calculated, it turns out that there is always a *single* lowest-lying MO, above which the MO's come in degenerate *pairs*. Thus, when electrons fill the various molecular orbitals, it takes two electrons (one pair) to fill the lowest-lying orbital and four electrons (two pairs) to fill each of *n* succeeding energy levels—a total of $4n + 2$. Any other number would leave an energy level partially filled.

The six π molecular orbitals of benzene were shown previously in Figure 15.3, and their relative energies are shown again in Figure 15.11. The lowest-energy MO, ψ_1, occurs singly and contains two electrons. The next two lowest-energy orbitals, ψ_2 and ψ_3, are degenerate, and it therefore takes four electrons to fill them. The result is a stable six-π-electron aromatic molecule with filled bonding orbitals.

Energy levels of the six
benzene π molecular
orbitals. There is a single,
lowest-energy orbital,
above which the orbitals
come in degenerate pairs.

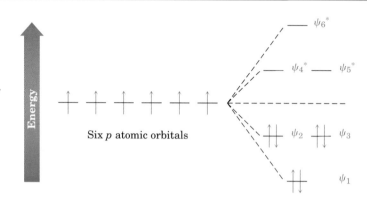

A similar line of reasoning carried out for the cyclopentadienyl cation,
radical, and anion is illustrated in Figure 15.12. The five atomic p orbitals
combine to give five π molecular orbitals, among which there is a single
lowest-energy orbital and higher-energy degenerate pairs of orbitals. In the
four-π-electron cation, there are two electrons in ψ_1 but only one electron
each in ψ_2 and ψ_3. Thus, the cation has two orbitals that are only partially
filled, and it is therefore antiaromatic. In the five-π-electron radical, ψ_1 and
ψ_2 are filled, but ψ_3 is still only half-full. Only in the six-π-electron cyclopentadienyl anion are all the bonding orbitals filled. Similar analyses can
be carried out for all other aromatic species.

Energy levels of the five cyclopentadienyl molecular orbitals. Only the six-π-electron
cyclopentadienyl anion has a filled-shell configuration leading to aromaticity.

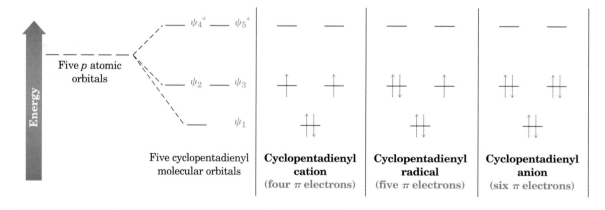

Problem 15.10   Show the relative energy levels of the seven π molecular orbitals of the cycloheptatrienyl system. Indicate which of the seven orbitals are filled in the cation,
radical, and anion, and account for the aromaticity of the cycloheptatrienyl cation.

15.9 Naphthalene: A Polycyclic Aromatic Compound

The Hückel rule is strictly applicable only to *monocyclic* compounds, but the general concept of aromaticity can be extended beyond simple monocyclic compounds to include **polycyclic aromatic compounds**. Naphthalene, with two benzene-like rings fused together, anthracene, 1,2-benzpyrene, and coronene are all well-known compounds. Benzo[a]pyrene is particularly interesting because it is one of the cancer-causing substances that has been isolated from tobacco smoke.

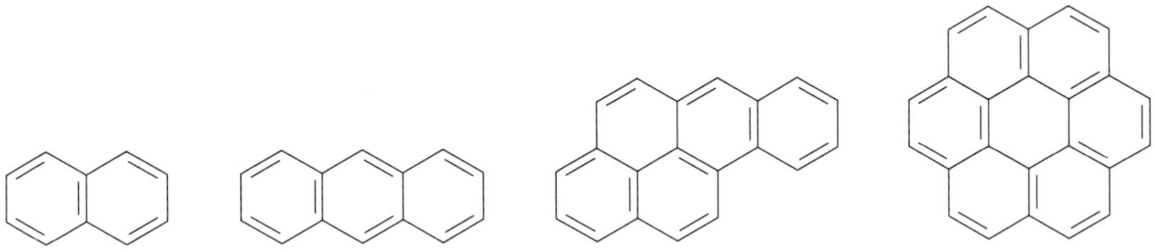

Naphthalene   **Anthracene**   **Benzo[a]pyrene**   **Coronene**

All polycyclic aromatic hydrocarbons can be represented by a number of different resonance forms. Naphthalene, for instance, has three:

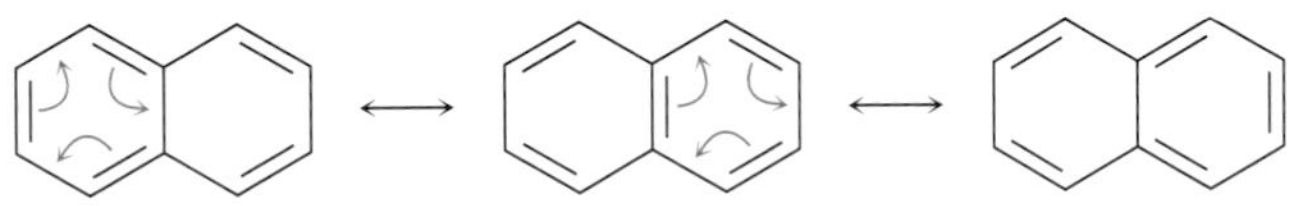

As was true for benzene with its two equivalent resonance forms, no individual structure is an accurate representation of naphthalene. The true structure of naphthalene is a hybrid of the three resonance forms.

Naphthalene and other polycyclic aromatic hydrocarbons show many of the chemical properties associated with aromaticity. Thus, heat of hydrogenation measurements show an aromatic stabilization energy of approximately 250 kJ/mol (60 kcal/mol). Furthermore, naphthalene reacts slowly with electrophiles such as Br_2 to give substitution products rather than double-bond addition products.

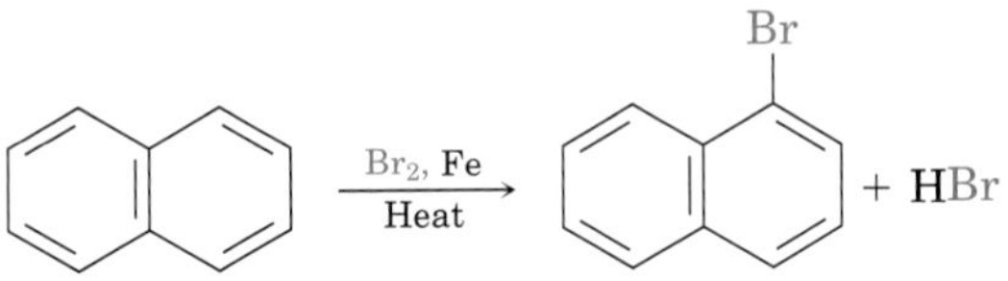

Naphthalene $\xrightarrow[\text{Heat}]{Br_2,\ Fe}$ **1-Bromonaphthalene (75%)** $+\ HBr$

The aromaticity of naphthalene is explained by the orbital picture in Figure 15.13 (p. 578). Naphthalene has a cyclic, conjugated π electron system, with p orbital overlap both around the ten-carbon periphery of the molecule and across the central bond. Since ten π electrons is a Hückel number, there is π electron delocalization and consequent aromaticity in naphthalene.

FIGURE 15.13 ▼

An orbital picture of naphthalene, showing that the ten π electrons are fully delocalized throughout both rings.

naphthalene
(see MO's on CD-Rom)

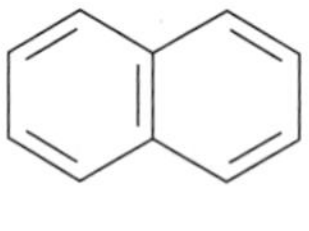

Naphthalene

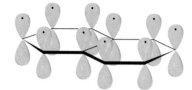

Problem 15.11   Azulene, a beautiful blue hydrocarbon, is an isomer of naphthalene. Is azulene aromatic? Draw a second resonance form of azulene in addition to that shown.

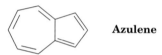   **Azulene**

Problem 15.12   Naphthalene is sometimes represented with circles in each ring to represent aromaticity:

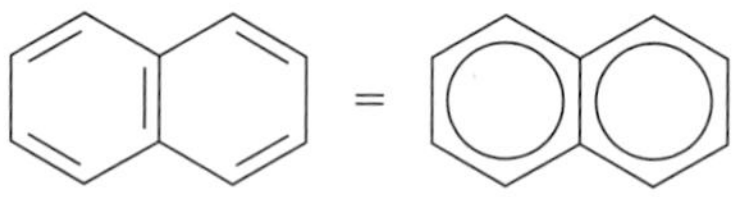

How many π electrons are in each circle?

15.10 Spectroscopy of Aromatic Compounds

Infrared Spectroscopy

Aromatic rings show a characteristic C–H stretching absorption at 3030 cm^{-1} and a characteristic series of peaks in the 1450–1600 cm^{-1} range of the infrared spectrum. The aromatic C–H band at 3030 cm^{-1} generally has low intensity and occurs just to the left of a typical saturated C–H band. As many as four absorptions are sometimes observed in the 1450–1600 cm^{-1} region because of complex molecular motions of the ring itself. Two bands, one at 1500 cm^{-1} and one at 1600 cm^{-1}, are usually the most intense. In addition, aromatic compounds show weak absorptions in the 1660–2000 cm^{-1} region and strong absorptions in the 690–900 cm^{-1} range due to C–H out-of-plane bending. The exact position of both sets of absorptions is diagnostic of the substitution pattern of the aromatic ring:

Monosubstituted:	690–710 cm^{-1}	m-Disubstituted:	690–710 cm^{-1}
	730–770 cm^{-1}		810–850 cm^{-1}
o-Disubstituted:	735–770 cm^{-1}	p-Disubstituted:	810–840 cm^{-1}

The IR spectrum of toluene in Figure 15.14 shows these characteristic absorptions.

FIGURE 15.14 ▼

The infrared spectrum of toluene.

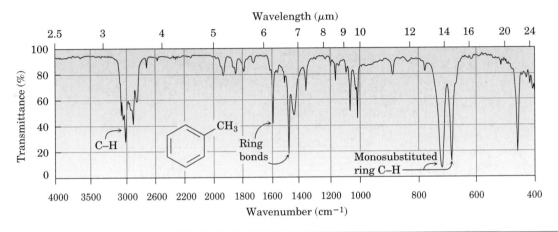

Ultraviolet Spectroscopy

Aromatic rings are detectable by ultraviolet spectroscopy because they contain a conjugated π electron system. In general, aromatic compounds show a series of bands, with a fairly intense absorption near 205 nm and a less intense absorption in the 255–275 nm range. The presence of these bands in the ultraviolet spectrum of a molecule of unknown structure is a sure indication of an aromatic ring.

Nuclear Magnetic Resonance Spectroscopy

Hydrogens directly bonded to an aromatic ring are easily identifiable in the ^{1}H NMR spectrum. Aromatic hydrogens are strongly deshielded by the ring and absorb between 6.5 and 8.0 δ. The spins of nonequivalent aromatic protons on substituted rings often couple with each other, giving rise to spin–spin splitting patterns that can give information about the substitution pattern of the ring.

Much of the difference in chemical shift between aromatic protons (6.5–8.0 δ) and vinylic protons (4.5–6.5 δ) is due to a property of aromatic rings called **ring current**. When an aromatic ring is oriented perpendicular to a strong magnetic field, the delocalized π electrons circulate around the ring, producing a small local magnetic field. This induced field *opposes* the applied field in the middle of the ring but *reinforces* the applied field outside the ring (Figure 15.15). Aromatic protons therefore experience an effective magnetic field greater than the applied field and come into resonance at a lower applied field.

FIGURE 15.15 ▼

The origin of aromatic ring current. Aromatic protons are deshielded by the induced magnetic field caused by delocalized π electrons circulating in the molecular orbitals of the aromatic ring.

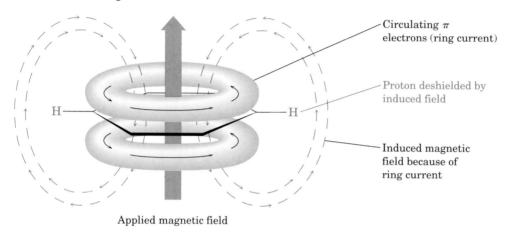

Circulating π electrons (ring current)

Proton deshielded by induced field

Induced magnetic field because of ring current

Applied magnetic field

Note that the aromatic ring current produces different effects inside and outside the ring. If a ring were large enough to have both "inside" and "outside" protons, those protons on the outside would be deshielded and absorb at a field lower than normal, but those protons on the inside would be *shielded* and absorb at a field higher than normal. This prediction has been strikingly verified by studies on [18]annulene, an 18-π-electron cyclic conjugated polyene that contains a Hückel number of electrons ($4n + 2 = 18$ when $n = 4$). The 6 inside protons of [18]annulene are strongly shielded by the aromatic ring current and absorb at $-3.0\,\delta$ (that is, 3.0 ppm *upfield* from TMS), while the 12 outside protons are strongly deshielded and absorb in the typical aromatic region at 9.3 ppm downfield from TMS.

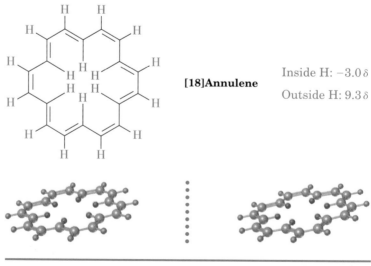

[18]Annulene

Inside H: $-3.0\,\delta$

Outside H: $9.3\,\delta$

Stereo View

The presence of a ring current is characteristic of all Hückel aromatic molecules and is a good test of aromaticity. For example, benzene, a six-π-electron aromatic molecule, absorbs at 7.37 δ, but cyclooctatetraene, an eight-π-electron nonaromatic molecule, absorbs at 5.78 δ.

Hydrogens on carbon next to aromatic rings also show distinctive absorptions in the NMR spectrum. Benzylic protons normally absorb downfield from other alkane protons in the region from 2.3 to 3.0 δ.

Aryl protons, 6.5–8.0 δ

Benzylic protons, 2.3–3.0 δ

The ^{1}H NMR spectrum of *p*-bromotoluene, shown in Figure 15.16, displays many of the features just discussed. The aromatic protons appear as two doublets at 7.02 and 7.45 δ, and the benzylic methyl protons absorb as a sharp singlet at 2.29 δ. Integration of the spectrum shows the expected 2:2:3 ratio of peak areas.

FIGURE 15.16 ▼

The ^{1}H NMR spectrum of *p*-bromotoluene.

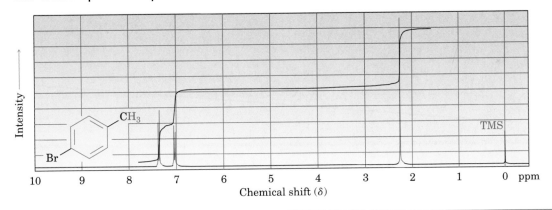

Carbon atoms of an aromatic ring absorb in the range 110–140 δ in the ^{13}C NMR spectrum, as indicated by the examples in Figure 15.17 (p. 582). These resonances are easily distinguished from those of alkane carbons but occur in the same range as alkene carbons. Thus, the presence of ^{13}C absorptions at 110–140 δ does not in itself establish the presence of an aromatic ring. Confirming evidence from infrared, ultraviolet, or ^{1}H NMR is needed.

FIGURE 15.17 ▼

Some ^{13}C NMR absorptions of aromatic compounds (δ units).

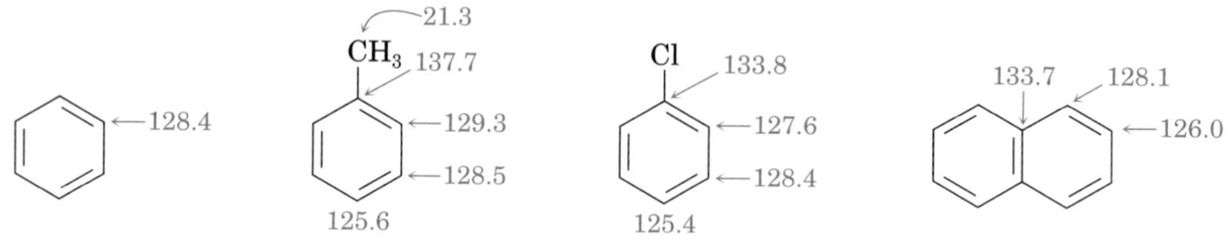

21.3
CH$_3$ 137.7   Cl   133.8   133.7   128.1
←128.4   ←129.3   ←127.6   ←126.0
←128.5   ←128.4
125.6   125.4

Benzene   Toluene   Chlorobenzene   Naphthalene

A summary of the kinds of information obtainable from different spectroscopic techniques is given in Table 15.2.

TABLE 15.2 Summary of Spectroscopic Information on Aromatic Compounds

Kind of spectroscopy	Absorption position	Interpretation
Infrared (cm^1)	3030	Aryl C–H stretch
	1500 and 1600	Two absorptions due to ring motions
	690–900	Intense C–H out-of-plane bending
Ultraviolet (nm)	205	Intense absorption
	255–275	Weak absorption
^{1}H NMR (δ)	2.3–3.0	Benzylic protons
	6.5–8.0	Aryl protons
^{13}C NMR (δ)	110–140	Aromatic ring carbons

CHEMISTRY @ WORK

Aspirin and Other Aromatic NSAID's

Whether from tennis elbow, a sprained ankle, or a wrenched knee, pain and inflammation seem to go together. They are, however, different in their origin, and powerful drugs are available for treating each separately. Codeine, for example, is a powerful *analgesic*, or pain reliever, used in the management of debilitating pain, while cortisone and related steroids are

(continued) ▶

potent *anti-inflammatory* agents, used for treating arthritis and other crippling inflammations. For minor pains and inflammation, both problems are often treated at the same time by using a common, over-the-counter medication called an *NSAID,* for *nonsteroidal anti-inflammatory drug.*

The most common NSAID is aspirin, or acetylsalicylic acid, whose use goes back to the late 1800s. It had been known from before the time of Hippocrates in 400 BC that fevers could be lowered by chewing the bark of willow trees. The active agent in willow bark was found in 1827 to be an aromatic compound called *salicin,* which could be converted by reaction with water (*hydrolysis*) to yield salicyl alcohol, and then oxidized to give salicyclic acid. Salicyclic acid turned out to be even more effective than salicin for reducing fevers and to have analgesic and anti-inflammatory action as well. Unfortunately, it also turned out to be too corrosive to the walls of the stomach for everyday use. Conversion of the phenol –OH group into an acetate ester, however, yielded acetylsalicylic acid, which proved just as potent as salicyclic acid but less corrosive to the stomach.

Salicyl alcohol → **Salicylic acid** → **Acetylsalicylic acid (Aspirin)**

Though extraordinary in its powers, aspirin is also more dangerous than commonly believed. Its toxicity is such that only about 15 g can be fatal to a small child, and it can cause stomach bleeding and allergic reactions in long-term users. Even more serious is a condition called *Reye's syndrome,* a potentially fatal reaction sometimes seen in children recovering from the flu. As a result of these problems, numerous other NSAID's have been developed in the last two decades, most notably ibuprofen and naproxen. Like aspirin, both ibuprofen and naproxen are relatively simple aromatic compounds containing a side-chain carboxylic acid group. Ibuprofen, sold under the names Advil, Motrin, Nuprin, and others, has roughly the same potency as aspirin but is much less prone to cause stomach upset. Naproxen, sold under the names Naprosyn and Aleve, also has about the same potency as aspirin but remains active in the body six times longer.

Ibuprofen (Advil, Motrin, Nuprin)   **Naproxen (Naprosyn, Aleve)**

Many athletes rely on NSAID's to help with pain and soreness.

Summary and Key Words

The term **aromatic** is used for historical reasons to refer to the class of compounds related structurally to benzene. Aromatic compounds are systematically named according to IUPAC rules, but many common names are also used. Disubstituted benzenes are named as **ortho** (1,2 disubstituted), **meta** (1,3 disubstituted), or **para** (1,4 disubstituted) derivatives. The C_6H_5- unit itself is referred to as a **phenyl** group, and the $C_6H_5CH_2-$ unit is a **benzyl** group.

Benzene is described by resonance theory as a resonance hybrid of two equivalent structures:

Benzene is described by molecular orbital theory as a planar, cyclic, conjugated molecule with six π electrons. According to the **Hückel rule**, a molecule must have **4n + 2 π electrons**, where $n = 0, 1, 2, 3$, and so on, to be aromatic. Planar, cyclic, conjugated molecules with other numbers of π electrons are **antiaromatic**.

Other kinds of molecules besides benzene-like compounds can also be aromatic. For example, the cyclopentadienyl anion and the cycloheptatrienyl cation are aromatic ions. Pyridine, a six-membered, nitrogen-containing **heterocycle**, is aromatic and resembles benzene electronically. Pyrrole, a five-membered heterocycle, resembles the cyclopentadienyl anion.

Aromatic compounds have the following characteristics:

• Aromatic compounds are cyclic, planar, and conjugated.

• Aromatic compounds are unusually stable. Benzene, for instance, has a heat of hydrogenation 150 kJ/mol less than we might expect for a cyclic triene.

• Aromatic compounds react with electrophiles to give substitution products, in which cyclic conjugation is retained, rather than addition products, in which conjugation is destroyed.

• Aromatic compounds have a Hückel number of π electrons, 4n + 2, which are delocalized over the entire ring.

Visualizing Chemistry

(Problems 15.1–15.12 appear within the chapter.)

15.13 Give IUPAC names for the following substances (red = O, blue = N):

(a) (b)

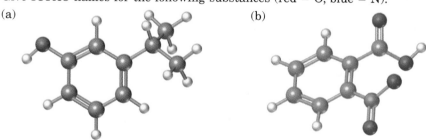

15.14 All-cis cyclodecapentaene is a stable molecule that shows a single absorption in its ^{1}H NMR spectrum at 5.67 δ. Tell whether it is aromatic, and explain its NMR spectrum.

All-cis cyclodecapentaene

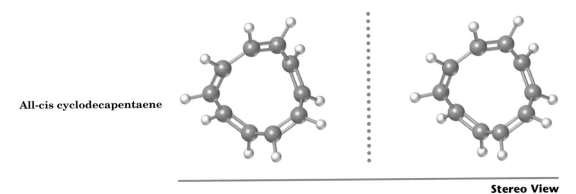

Stereo View

15.15 1,6-Methanonaphthalene has an interesting ^{1}H NMR spectrum in which the eight hydrogens around the perimeter absorb at 6.9–7.3 δ, while the two CH_2 protons absorb at –0.5 δ. Tell whether it is aromatic, and explain its NMR spectrum.

1,6-Methanonaphthalene

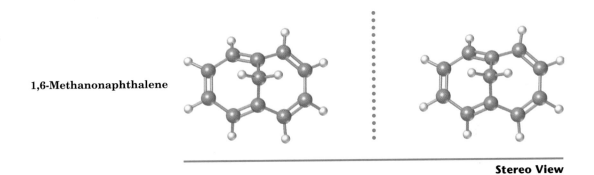

Stereo View

15.16 The following molecular model is that of a carbocation. Draw two resonance structures for the carbocation, indicating the positions of the double bonds.

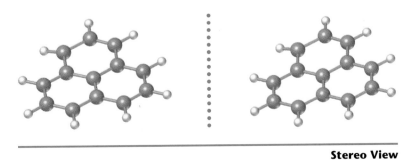

Stereo View

Additional Problems

• •

15.17 Give IUPAC names for the following compounds:

(a)

$$CH_3 \quad CH_3$$
$$CHCH_2CH_2CHCH_3$$

(b) CO_2H
  Br

(c) Br
  $H_3C \quad CH_3$

(d) Br
  $CH_2CH_2CH_3$

(e) F
  NO_2
  NO_2

(f) NH_2
  Cl

15.18 Draw structures corresponding to the following names:
(a) 3-Methyl-1,2-benzenediamine   (b) 1,3,5-Benzenetriol
(c) 3-Methyl-2-phenylhexane   (d) *o*-Aminobenzoic acid
(e) *m*-Bromophenol   (f) 2,4,6-Trinitrophenol (picric acid)
(g) *p*-Iodonitrobenzene

15.19 Draw and name all possible isomers of:
(a) Dinitrobenzene   (b) Bromodimethylbenzene   (c) Trinitrophenol

15.20 Draw and name all possible aromatic compounds with the formula C_7H_7Cl.

15.21 Draw and name all possible aromatic compounds with the formula C_8H_9Br. (There are 14.)

15.22 Propose structures for aromatic hydrocarbons that meet the following descriptions:
(a) C_9H_{12}; gives only one $C_9H_{11}Br$ product on substitution with bromine
(b) $C_{10}H_{14}$; gives only one $C_{10}H_{13}Cl$ product on substitution with chlorine
(c) C_8H_{10}; gives three C_8H_9Br products on substitution with bromine
(d) $C_{10}H_{14}$; gives two $C_{10}H_{13}Cl$ products on substitution with chlorine

15.23 There are four resonance structures for anthracene, one of which is shown. Draw the other three.

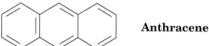

Anthracene

15.24 Look at the three resonance structures of naphthalene shown in Section 15.9, and account for the fact that not all carbon–carbon bonds have the same length. The C1–C2 bond is 136 pm long, whereas the C2–C3 bond is 139 pm long.

15.25 There are five resonance structures of phenanthrene, one of which is shown. Draw the other four.

Phenanthrene

15.26 Look at the five resonance structures for phenanthrene (Problem 15.25) and predict which of its carbon–carbon bonds is shortest.

15.27 Use the data in Figure 15.2 to calculate the heat of hydrogenation, $\delta H^{\circ}_{\text{hydrog}}$, for the partial hydrogenation of benzene to yield 1,3-cyclohexadiene. Is the reaction exothermic or endothermic?

15.28 In 1932, A. A. Levine and A. G. Cole studied the ozonolysis of *o*-xylene and isolated three products: glyoxal, 2,3-butanedione, and pyruvaldehyde:

In what ratio would you expect the three products to be formed if *o*-xylene is a resonance hybrid of two structures? The actual ratio found was 3 parts glyoxal, 1 part 2,3-butanedione, and 2 parts pyruvaldehyde. What conclusions can you draw about the structure of *o*-xylene?

15.29 3-Chlorocyclopropene, on treatment with $AgBF_4$, gives a precipitate of AgCl and a stable solution of a product that shows a single ^{1}H NMR absorption at 11.04 δ. What is a likely structure for the product, and what is its relation to Hückel's rule?

3-Chlorocyclopropene

15.30 Draw an energy diagram for the three molecular orbitals of the cyclopropenyl system (C_3H_3). How are these three molecular orbitals occupied in the cyclopropenyl anion, cation, and radical? Which of the three substances is aromatic according to Hückel's rule?

15.31 Cyclopropanone is highly reactive because of its large amount of angle strain. Methylcyclopropenone, although more strained than cyclopropanone, is nevertheless quite stable and can even be distilled. Explain, taking the polarity of the carbonyl group into account.

Cyclopropanone   **Methylcyclopropenone**

15.32 Cycloheptatrienone is stable, but cyclopentadienone is so reactive that it can't be isolated. Explain.

Cycloheptatrienone   **Cyclopentadienone**

15.33 Which would you expect to be most stable, cyclononatetraenyl radical, cation, or anion?

15.34 How might you convert 1,3,5,7-cyclononatetraene to an aromatic substance?

15.35 Compound A, C_8H_{10}, yields three substitution products, C_8H_9Br, on reaction with Br_2. Propose two possible structures for A. The ^1H NMR spectrum of A shows a complex four-proton multiplet at 7.0 δ and a six-proton singlet at 2.30 δ. What is the structure of A?

15.36 Azulene, an isomer of naphthalene, has a remarkably large dipole moment for a hydrocarbon (μ = 1.0 D). Explain, using resonance structures.

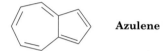

Azulene

15.37 Calicene, like azulene (Problem 15.36), has an unusually large dipole moment for a hydrocarbon. Explain, using resonance structures.

Calicene

15.38 Pentalene is a most elusive molecule and has never been isolated. The pentalene dianion, however, is well known and quite stable. Explain.

Pentalene   **Pentalene dianion**

15.39 Indole is an aromatic heterocycle that has a benzene ring fused to a pyrrole ring. Draw an orbital picture of indole.

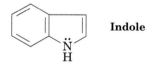

Indole

(a) How many π electrons does indole have?
(b) What is the electronic relationship of indole to naphthalene?

15.40 On reaction with acid, 4-pyrone is protonated on the carbonyl-group oxygen to give a stable cationic product. Explain why the protonated product is so stable.

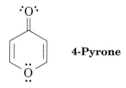

4-Pyrone

15.41 1-Phenyl-2-butene has an ultraviolet absorption at λ_{max} = 208 nm (ε = 8000). On treatment with a small amount of strong acid, isomerization occurs and a new substance with λ_{max} = 250 nm (ε = 15,800) is formed. Propose a structure for this isomer, and suggest a mechanism for its formation.

15.42 What is the structure of a hydrocarbon that has M⁺ = 120 in its mass spectrum and has the following ¹H NMR spectrum?

7.25 δ (5 H, broad singlet); 2.90 δ (1 H, septet, J = 7 Hz); 1.22 δ (6 H, doublet, J = 7 Hz)

15.43 Propose structures for compounds that fit the following descriptions:
(a) $C_{10}H_{14}$

¹H NMR: 7.18 δ (4 H, broad singlet); 2.70 δ (4 H, quartet, J = 7 Hz); 1.20 δ (6 H, triplet, J = 7 Hz)

IR: 745 cm⁻¹

(b) $C_{10}H_{14}$

¹H NMR: 7.0 δ (4 H, broad singlet); 2.85 δ (1 H, septet, J = 8 Hz); 2.28 δ (3 H, singlet); 1.20 δ (6 H, doublet, J = 8 Hz)

IR: 825 cm⁻¹

15.44 Propose structures for aromatic compounds that have the following ¹H NMR spectra:

(a) C_8H_9Br
IR: 820 cm⁻¹

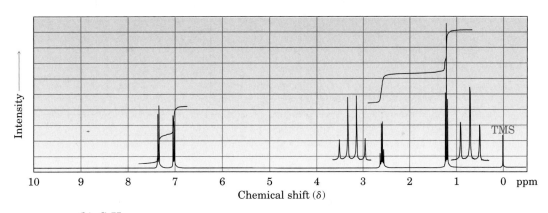

(b) C_9H_{12}
IR: 750 cm⁻¹

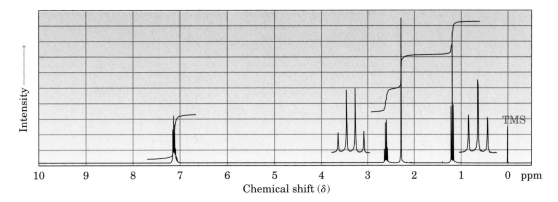

(c) $C_{11}H_{16}$
IR: 820 cm^{-1}

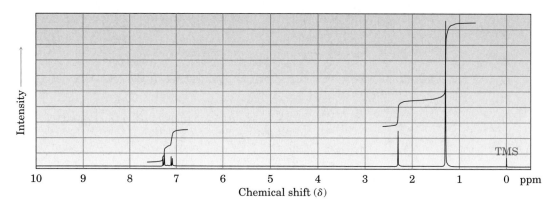

15.45   Propose a structure for a molecule $C_{14}H_{12}$ that has the following ^1H NMR spectrum and has IR absorptions at 700, 740, and 890 cm^{-1}:

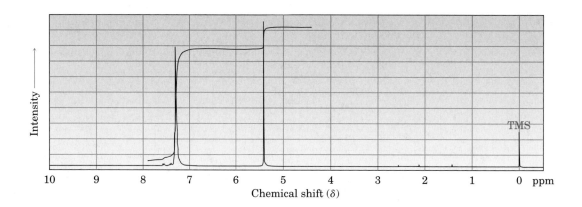

A Look Ahead

• •

15.46   Derivatives of the aromatic heterocycle purine are constituents of DNA and RNA. Why is purine aromatic? How many p electrons does each nitrogen donate to the aromatic π system? (See Section 28.7.)

Purine

15.47   Aromatic substitution reactions occur by addition of an electrophile such as Br$^+$ to the aromatic ring to yield an allylic carbocation intermediate, followed by loss of H$^+$. Show the structure of the intermediate formed by reaction of benzene with Br$^+$. (See Section 16.1.)

15.48 The substitution reaction of toluene with Br$_2$ can, in principle, lead to the formation of three isomeric bromotoluene products. In practice, however, only *o*- and *p*-bromotoluene are formed in substantial amounts. The meta isomer is not formed. Draw the structures of the three possible carbocation intermediates (Problem 15.47), and explain why ortho and para products predominate over meta. (See Sections 16.5 and 16.6.)

Molecular Modeling

● ●

15.49 Use SpartanView to examine electrostatic potential maps of the heterocycles tetrahydrothiophene, 2,3-dihydrothiophene, and thiophene. How does the electron distribution around sulfur change? Explain.

Tetrahydrothiophene   **2,3-Dihydrothiophene**   **Thiophene**

15.50 The ^{1}H NMR spectra of *cis*- and *trans*-1,2-diphenylbenzocyclobutene are similar except for the chemical shifts of the hydrogens on the four-membered ring (δ 4.4 in one isomer and δ 5.2 in the other.) Use SpartanBuild to build models of each isomer. How are cyclobutene hydrogens positioned with respect to the three benzene rings? Which isomer produces which spectrum?

     Ph

1,2-Diphenylbenzocyclobutene

Ph

15.51 Use SpartanView to compare electrostatic potential maps of 2-methylpropene, fulvene, and methylenecyclopropene. Assuming that 2-methylpropene contains a "normal" =CH$_2$ group, what effect does the conjugated ring have on the =CH$_2$? Explain.

2-Methylpropene   **Fulvene**   **Methylenecyclopropene**

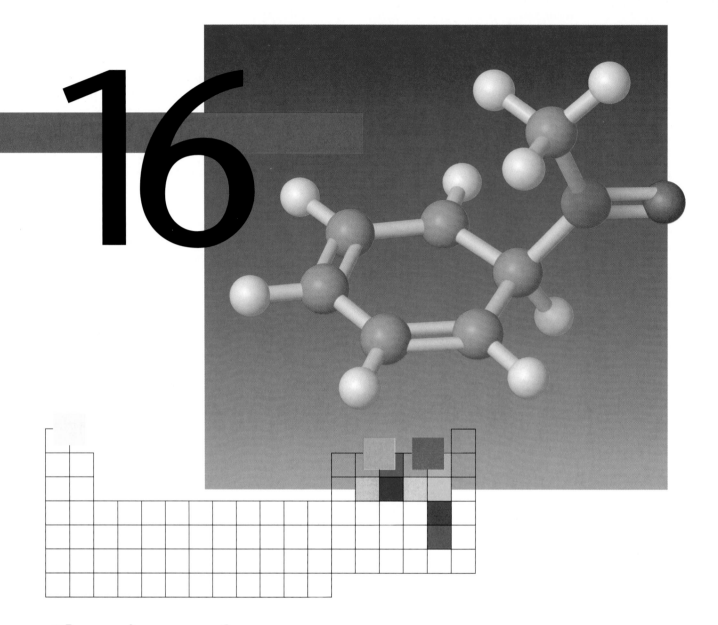

Chemistry of Benzene: Electrophilic Aromatic Substitution

The most common reaction of aromatic compounds is **electrophilic aromatic substitution**. That is, an electrophile (E^+) reacts with an aromatic ring and substitutes for one of the hydrogens:

Many different substituents can be introduced onto the aromatic ring by electrophilic substitution reactions. By choosing the proper reagents, it's possible to **halogenate** the aromatic ring (substitute a halogen: –F, –Cl, –Br, or –I), **nitrate** it (substitute a nitro group: $-NO_2$), **sulfonate** it (substitute a sulfonic acid group: $-SO_3H$), **alkylate** it (substitute an alkyl group: –R), or **acylate** it (substitute an acyl group: –COR). Starting from only a few simple materials, we can prepare many thousands of substituted aromatic compounds (Figure 16.1).

FIGURE 16.1 ▼

Some electrophilic aromatic substitution reactions.

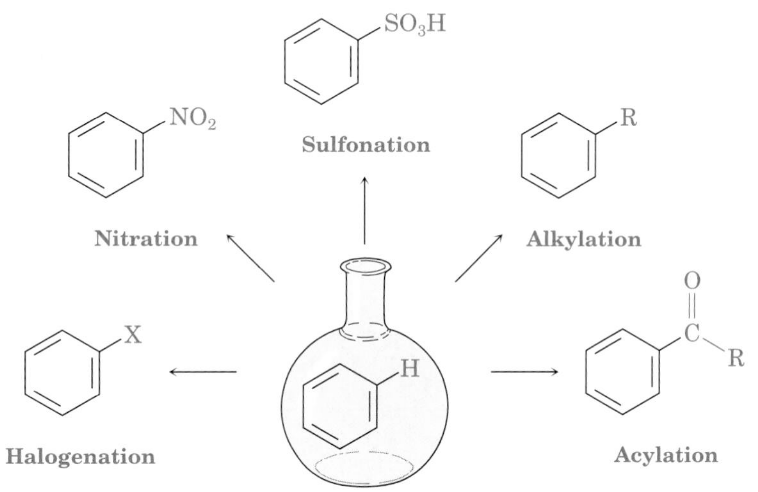

All these reactions—and many more as well—take place by a similar mechanism. Let's begin a study of the process by looking at one reaction in detail, the bromination of benzene.

16.1 Bromination of Aromatic Rings

A benzene ring, with its six π electrons in a cyclic conjugated system, is a site of electron density. Furthermore, the benzene π electrons are sterically accessible to attacking reagents because of their location above and below the plane of the ring. Thus, benzene acts as an electron donor (a Lewis base, or nucleophile) in most of its chemistry, and most of its reactions take place with electron acceptors (Lewis acids, or electrophiles). For example, benzene reacts with Br_2 in the presence of $FeBr_3$ as catalyst to yield the substitution product bromobenzene.

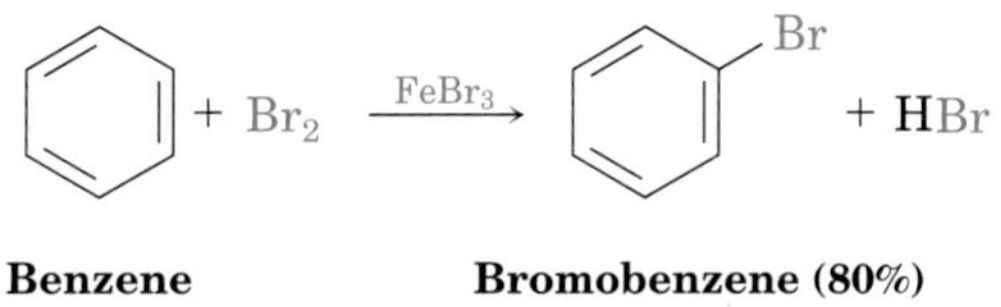

Benzene   **Bromobenzene (80%)**

Electrophilic substitution reactions are characteristic of all aromatic rings, not just of benzene and substituted benzenes. Indeed, the ability of a compound to undergo electrophilic substitution is a good test of aromaticity.

Before seeing how this electrophilic aromatic *substitution* occurs, let's briefly recall what was said in Chapter 6 about electrophilic alkene *additions*. When a reagent such as HCl adds to an alkene, the electrophilic H^+ approaches the *p* orbitals of the double bond and forms a bond to one carbon, leaving a positive charge at the other carbon. This carbocation intermediate is then attacked by the nucleophilic Cl^- ion to yield the addition product (Figure 16.2).

FIGURE 16.2 ▼

The mechanism of an alkene electrophilic addition reaction.

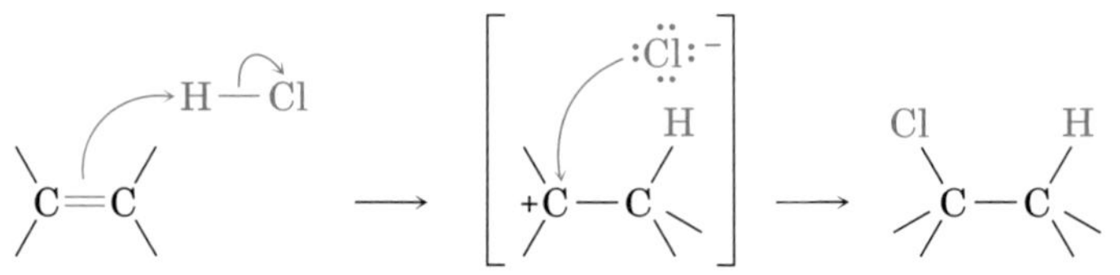

Alkene   **Carbocation intermediate**   **Addition product**

An electrophilic aromatic *substitution* reaction begins in a similar way, but there are a number of differences. One difference is that aromatic rings are less reactive toward electrophiles than alkenes are. For example, Br_2 in CH_2Cl_2 solution reacts instantly with most alkenes but does not react at room temperature with benzene. For bromination of benzene to take place, a catalyst such as $FeBr_3$ is needed. The catalyst makes the Br_2 molecule more electrophilic by polarizing it to give an $FeBr_4^-$ Br^+ species that reacts as if it were Br^+.

$$\overset{\delta-}{Br}\!-\!\overset{\delta+}{Br} \xrightarrow{FeBr_3} Br_3\overset{\delta-}{Fe}\cdots Br\cdots\overset{\delta+}{Br}$$

Bromine   **Polarized bromine**
(a weak electrophile)   (a strong electrophile)

The polarized Br_2 molecule is then attacked by the π electron system of the nucleophilic benzene ring in a slow, rate-limiting step to yield a nonaromatic carbocation intermediate. This carbocation is doubly allylic (recall the allyl cation, Section 11.9) and has three resonance forms:

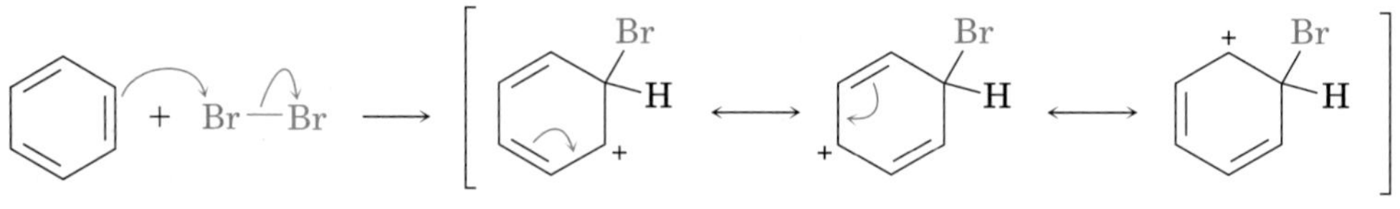

Although stable by comparison with typical alkyl carbocations, the intermediate in electrophilic aromatic substitution is nevertheless much less stable than the starting benzene ring itself with its 150 kJ/mol (36 kcal/mol) of aromatic stability. Thus, electrophilic attack on a benzene ring is endergonic, has a substantial activation energy, and is a rather slow reaction. Figure 16.3 gives reaction energy diagrams comparing the reaction of an electrophile with an alkene and with benzene. The benzene reaction is slower (higher $\Delta G^{\ddagger}$) because the starting material is more stable.

FIGURE 16.3 ▼

A comparison of the reactions of an electrophile (E^+) with an alkene and with benzene: $\Delta G^{\ddagger}_{alkene} < \Delta G^{\ddagger}_{benzene}$

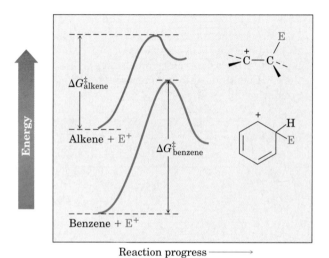

A second difference between alkene addition and aromatic substitution occurs after the carbocation intermediate has formed. Instead of adding Br^- to give an addition product, the carbocation intermediate loses H^+ from the bromine-bearing carbon to give a substitution product. Note that this loss of H^+ is similar to what occurs in the second step of an E1 reaction (Section 11.14). The net effect of reaction of Br_2 with benzene is the substitution of H^+ by Br^+; the overall mechanism is shown in Figure 16.4 (p. 596).

Why does the reaction of Br_2 with benzene take a different course than its reaction with an alkene? The answer is simple: If *addition* occurred, the 150 kJ/mol stabilization energy of the aromatic ring would be lost, and the overall reaction would be endergonic. When *substitution* occurs, though, the stability of the aromatic ring is retained and the reaction is exergonic. A reaction energy diagram for the overall process is shown in Figure 16.5.

There are many other kinds of electrophilic aromatic substitutions besides bromination, and all are thought to occur by the same general mechanism. We'll look at some of these other reactions briefly in the next section.

FIGURE 16.4 ▼

The mechanism of the electrophilic bromination of benzene. The reaction occurs in two steps and involves a resonance-stabilized carbocation intermediate.

refer to
**Mechanisms
& Movies**

An electron pair from the benzene ring attacks Br_2, forming a new C–Br bond and leaving a nonaromatic, carbocation intermediate.

The carbocation intermediate loses H^+, and the neutral substitution product forms as two electrons from the C–H bond move to regenerate the aromatic ring.

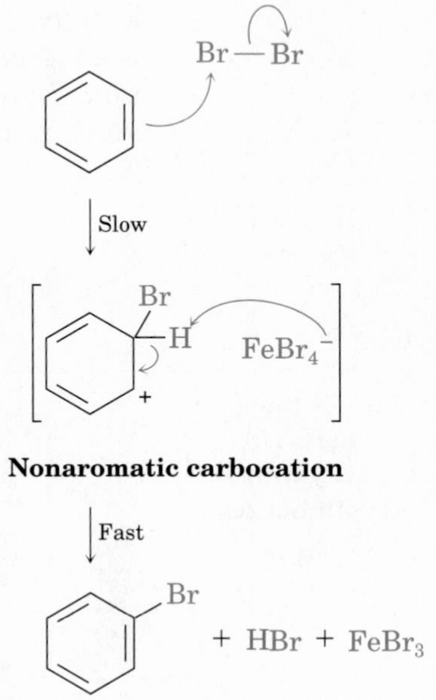

© 1999 JOHN MCMURRY

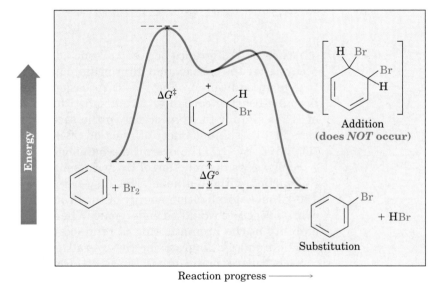

FIGURE 16.5 ▼

A reaction energy diagram for the electrophilic bromination of benzene. The overall process is exergonic.

Problem 16.1 Monobromination of toluene gives a mixture of three bromotoluene products. Draw and name them.

16.2 Other Aromatic Substitutions

Aromatic Chlorination and Iodination

Chlorine and iodine can be introduced into aromatic rings by electrophilic substitution reactions, but fluorine is too reactive, and only poor yields of monofluoroaromatic products are obtained by direct fluorination. Aromatic rings react with Cl_2 in the presence of $FeCl_3$ catalyst to yield chlorobenzenes. This kind of reaction is used in the synthesis of numerous pharmaceutical agents, including the tranquilizer diazepam (Valium).

Benzene   Chlorobenzene (86%)   Diazepam

Iodine itself is unreactive toward aromatic rings, and an oxidizing agent such as hydrogen peroxide or a copper salt such as $CuCl_2$ must be added to the reaction. These substances accelerate the iodination reaction by oxidizing I_2 to a more powerful electrophilic species that reacts as if it were I^+. The aromatic ring then attacks I^+ in the typical way, yielding a substitution product.

$$I_2 + 2\,Cu^{2+} \longrightarrow 2\,I^+ + 2\,Cu^+$$

Benzene   Iodobenzene (65%)

Aromatic Nitration

Aromatic rings can be nitrated by reaction with a mixture of concentrated nitric and sulfuric acids. The electrophile in this reaction is the nitronium ion, NO_2^+, which is generated from HNO_3 by protonation and loss of water. The nitronium ion reacts with benzene to yield a carbocation intermediate in much the same way as Br^+. Loss of H^+ from this intermediate gives the neutral substitution product, nitrobenzene (Figure 16.6).

FIGURE 16.6 ▼

The mechanism of electrophilic nitration of an aromatic ring. An electrostatic potential map of the reactive electrophile NO_2^+ shows that the nitrogen atom is most positive (blue).

NO2+

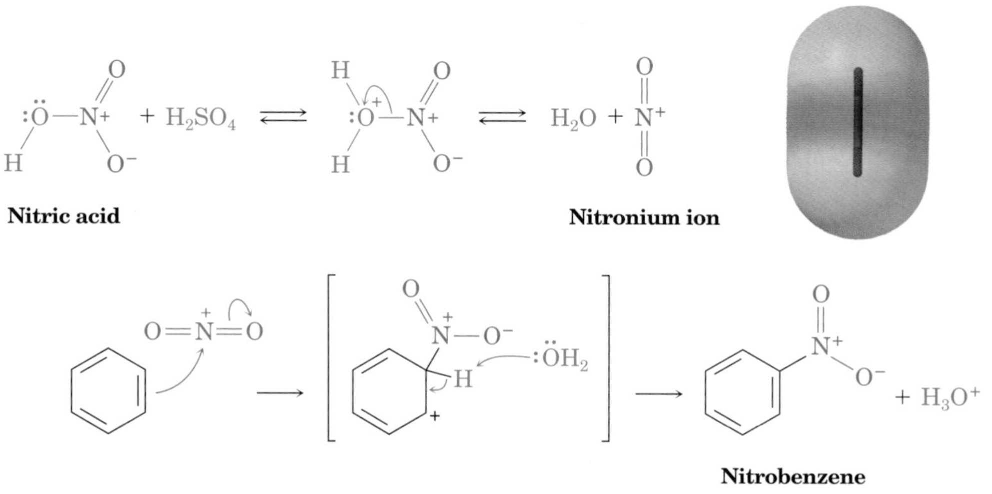

Nitration of an aromatic ring is a particularly important reaction because the nitro-substituted product can be reduced by reagents such as iron metal or $SnCl_2$ to yield an arylamine, $ArNH_2$. Attachment of a nitrogen to an aromatic ring by the two-step nitration/reduction sequence is a key part of the industrial synthesis of dyes and many pharmaceutical agents. We'll discuss this and other reactions of aromatic nitrogen compounds in Chapter 24.

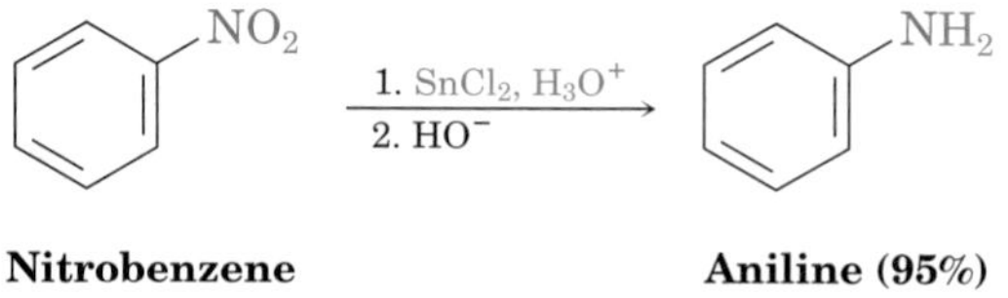

Aromatic Sulfonation

Aromatic rings can be sulfonated by reaction with fuming sulfuric acid, a mixture of H_2SO_4 and SO_3. The reactive electrophile is either HSO_3^+ or neutral SO_3, depending on reaction conditions. Substitution occurs by the same two-step mechanism seen previously for bromination and nitration (Figure 16.7). Note, however, that the sulfonation reaction is readily reversible; it can occur either forward or backward, depending on the reaction conditions.

FIGURE 16.7 ▼

The mechanism of electrophilic sulfonation of an aromatic ring. An electrostatic potential map of the reactive electrophile $HOSO_2^+$ shows that sulfur and hydrogen are the most positive atoms (blue).

HOSO2+

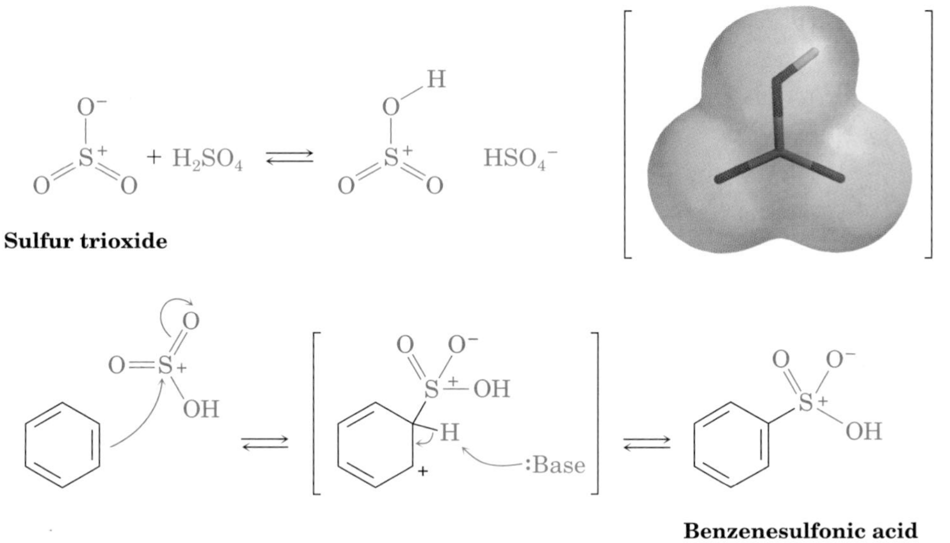

Sulfur trioxide

Benzenesulfonic acid

Sulfonation is favored in strong acid, but desulfonation is favored in hot, dilute aqueous acid.

Aromatic sulfonic acids are valuable intermediates in the preparation of dyes and pharmaceuticals. For example, the sulfa drugs, such as sulfanilamide, were among the first useful antibiotics. Although largely replaced today by more effective agents, sulfa drugs are still used in the treatment of meningitis and urinary-tract infections. These drugs are prepared commercially by a process that involves aromatic sulfonation as the key step.

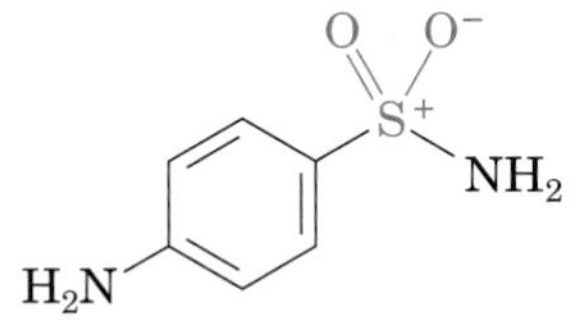

Sulfanilamide (an antibiotic)

Aromatic sulfonic acids are also useful because of the further chemistry they undergo. Heating an aromatic sulfonic acid with NaOH at 300°C in the absence of solvent effects a replacement of the –SO₃H group by –OH and gives a phenol. Yields in this so-called **alkali fusion** reaction are generally good, but the conditions are so vigorous that the reaction is not compatible with the presence of substituents other than alkyl groups on the aromatic ring.

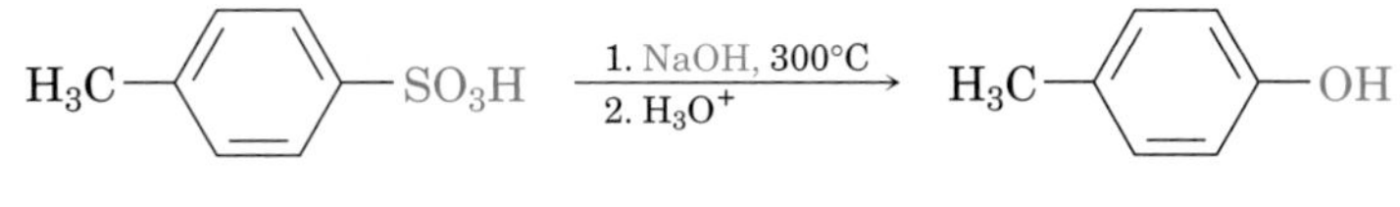

$$H_3C-\!\!\!\bigcirc\!\!\!-SO_3H \xrightarrow[\text{2. } H_3O^+]{\text{1. NaOH, 300°C}} H_3C-\!\!\!\bigcirc\!\!\!-OH$$

p-Toluenesulfonic acid   **p-Cresol (72%)**
                                   (a phenol)

• •

Problem 16.2   How many products might be formed on chlorination of *o*-xylene (dimethylbenzene), *m*-xylene, and *p*-xylene?

Problem 16.3   How can you account for the fact that deuterium slowly replaces all six hydrogens in the aromatic ring when benzene is treated with D_2SO_4?

• •

16.3 Alkylation of Aromatic Rings: The Friedel–Crafts Reaction

Charles Friedel

Charles Friedel (1832–1899) was born in Strasbourg, France, and studied at the Sorbonne in Paris. Trained as both a mineralogist and a chemist, he was among the first to attempt to manufacture synthetic diamonds. He was professor of mineralogy at the School of Mines before becoming professor of chemistry at the Sorbonne (1884–1899).

James Mason Crafts

James Mason Crafts (1839–1917) was born in Boston, Massachusetts, and graduated from Harvard in 1858. Although he did not receive a Ph.D., he studied with eminent chemists in Europe for several years and was appointed in 1868 as the first professor of chemistry at the newly founded Cornell University in Ithaca, New York. He soon moved to the Massachusetts Institute of Technology, however, where he served as president from 1897 to 1900.

One of the most useful of all electrophilic aromatic substitution reactions is *alkylation,* the attachment of an alkyl group to the benzene ring. Charles Friedel and James Crafts reported in 1877 that benzene rings can be alkylated by reaction with an alkyl chloride in the presence of aluminum chloride as catalyst. For example, benzene reacts with 2-chloropropane and $AlCl_3$ to yield isopropylbenzene, also called cumene.

Benzene   2-Chloropropane   Cumene (85%)
                                   (Isopropylbenzene)

The **Friedel–Crafts alkylation reaction** is an electrophilic aromatic substitution in which the electrophile is a carbocation, R^+. Aluminum chloride catalyzes the reaction by helping the alkyl halide to ionize in much the same way that $FeBr_3$ catalyzes aromatic brominations by polarizing Br_2 (Section 16.1). Loss of a proton then completes the reaction, as shown in Figure 16.8.

Though broadly useful for the synthesis of alkylbenzenes, the Friedel–Crafts alkylation nevertheless has strict limitations. One limitation is that only *alkyl* halides can be used. Alkyl fluorides, chlorides, bromides, and iodides all react well, but *aryl* halides and *vinylic* halides do not react. Aryl and vinylic carbocations are too high in energy to form under Friedel–Crafts conditions.

An aryl halide   **A vinylic halide**

NOT reactive

A second limitation is that the Friedel–Crafts reaction doesn't succeed on an aromatic ring that is substituted either by an amino group or by a strongly electron-withdrawing group. We'll see in Section 16.5 that the pres-

FIGURE 16.8 ▼

Mechanism of the Friedel–Crafts alkylation reaction. The electrophile is a carbocation, generated by AlCl₃-assisted ionization of an alkyl halide.

$$\underset{\overset{|}{\text{Cl}}}{\text{CH}_3\text{CHCH}_3} + \text{AlCl}_3 \longrightarrow \overset{+}{\text{CH}_3\text{CHCH}_3} \quad \text{AlCl}_4^-$$

An electron pair from the aromatic ring attacks the carbocation, forming a C–C bond and yielding a new carbocation intermediate.

$$\overset{+}{\text{CH}_3\text{CHCH}_3} \quad \text{AlCl}_4^-$$

Loss of a proton then gives the neutral alkylated substitution product.

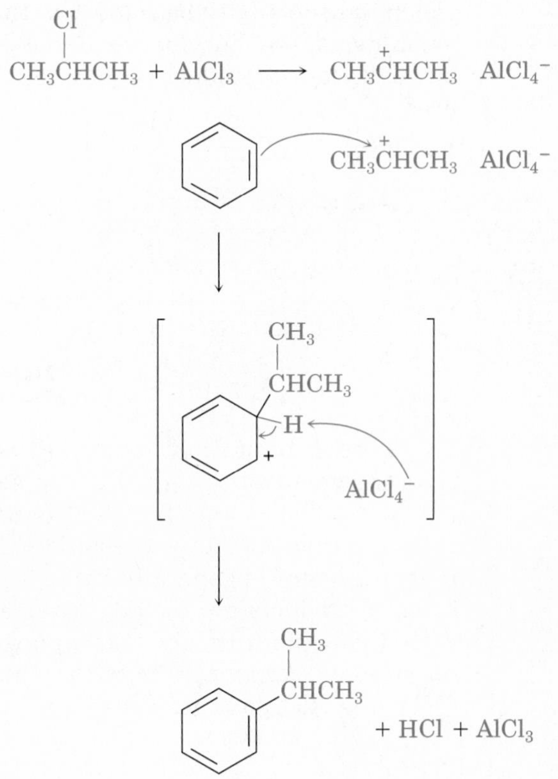

$$\left[\underset{\underset{+}{\overset{|}{\underset{\text{H}}{\overset{\text{CHCH}_3}{\overset{|}{\text{CH}_3}}}}}}{} \quad \text{AlCl}_4^- \right]$$

$$\underset{\overset{|}{\underset{\text{CHCH}_3}{\overset{|}{\text{CH}_3}}}}{} \quad + \text{HCl} + \text{AlCl}_3$$

© 1984 JOHN MCMURRY

ence of a substituent group already on a ring can have a dramatic effect on that ring's subsequent reactivity toward further electrophilic substitution. Rings that contain any of the substituents listed in Figure 16.9 are not reactive enough to undergo Friedel–Crafts alkylation.

FIGURE 16.9 ▼

Limitations on the aromatic substrate in Friedel–Crafts reactions. No reaction occurs if the substrate has either an electron-withdrawing substituent or an amino group, which reacts with the AlCl₃ catalyst in an acid–base reaction.

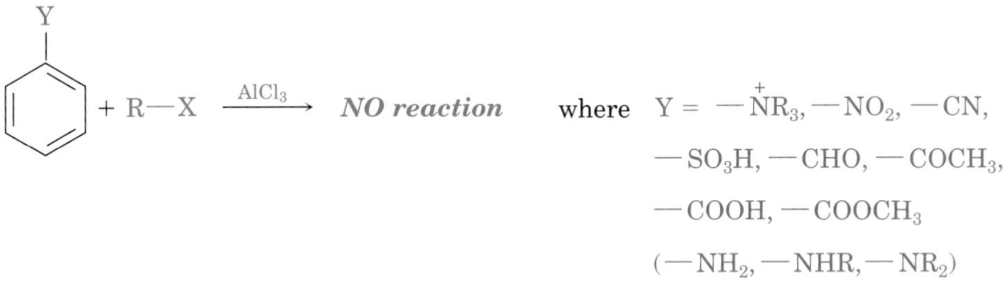

$$\underset{Y}{\bigcirc} + \text{R—X} \xrightarrow{\text{AlCl}_3} \textit{NO reaction} \qquad \text{where} \quad Y = -\overset{+}{\text{N}}\text{R}_3, -\text{NO}_2, -\text{CN},$$

$$-\text{SO}_3\text{H}, -\text{CHO}, -\text{COCH}_3,$$

$$-\text{COOH}, -\text{COOCH}_3$$

$$(-\text{NH}_2, -\text{NHR}, -\text{NR}_2)$$

Yet a third limitation of the Friedel–Crafts alkylation is that it's often difficult to stop the reaction after a single substitution. Once the first alkyl group is on the ring, a second substitution reaction is facilitated for reasons we'll discuss in the next section. Thus, we often observe *polyalkylation*. For example, reaction of benzene with 1 mol equiv of 2-chloro-2-methylpropane yields *p*-di-*tert*-butylbenzene as the major product, along with small amounts of *tert*-butylbenzene and unreacted benzene. A high yield of monoalkylation product is obtained only when a large excess of benzene is used.

| **Major product** | **Minor product** |

A final limitation to the Friedel–Crafts reaction is that skeletal rearrangement of the alkyl group sometimes occurs during reaction, particularly when a primary alkyl halide is used. The amount of rearrangement depends on catalyst, reaction temperature, and reaction solvent. Less rearrangement is generally found at lower reaction temperatures, but mixtures of products are usually obtained. For example, treatment of benzene with 1-chlorobutane gives an approximately 2:1 ratio of rearranged (*sec*-butyl) to unrearranged (butyl) products when the reaction is carried out at 0°C using $AlCl_3$ as catalyst.

Benzene   ***sec*-Butylbenzene** (65%)   **Butylbenzene** (35%)

These carbocation rearrangements are similar to those that occur during electrophilic additions to alkenes (Section 6.12). For example, the relatively unstable primary butyl carbocation produced by reaction of 1-chlorobutane with $AlCl_3$ rearranges to the more stable secondary butyl carbocation by shift of a hydrogen atom and its electron pair (a hydride ion, $H:^-$) from C2 to C1.

Primary butyl carbocation   **Secondary butyl carbocation**

Similarly, carbocation rearrangements can occur by *alkyl* shifts. For example, Friedel–Crafts alkylation of benzene with 1-chloro-2,2-dimethylpropane yields (1,1-dimethylpropyl)benzene as the sole product. The initially formed primary carbocation rearranges to a tertiary carbocation by shift of a methyl group and its electron pair from C2 to C1 (Figure 16.10).

FIGURE 16.10 ▼

The rearrangement of a primary to a tertiary carbocation during Friedel–Crafts reaction of benzene with 1-chloro-2,2-dimethylpropane occurs by shift of an alkyl group with its electron pair.

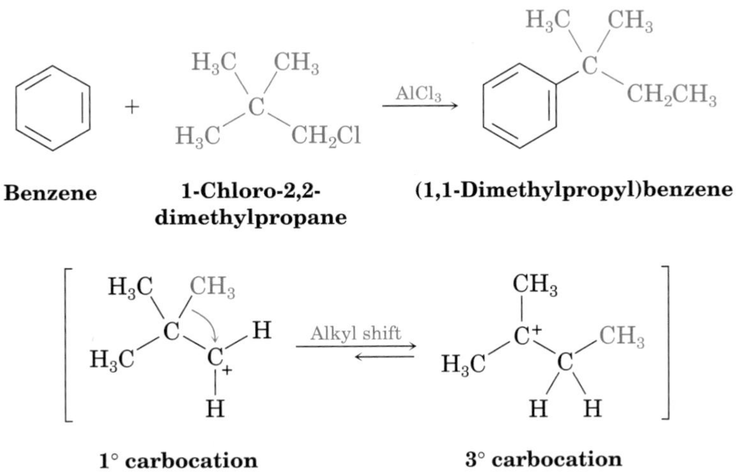

Benzene   1-Chloro-2,2-dimethylpropane   (1,1-Dimethylpropyl)benzene

1° carbocation   3° carbocation

• •

Practice Problem 16.1 The Friedel–Crafts reaction of benzene with 2-chloro-3-methylbutane in the presence of $AlCl_3$ occurs with a carbocation rearrangement. What is the structure of the product?

Strategy A Friedel–Crafts reaction involves initial formation of a carbocation, which can rearrange by either a hydride shift or an alkyl shift to give a more stable carbocation. Draw the initial carbocation, assess its stability, and see if the shift of a hydride ion or an alkyl group from a neighboring carbon will result in increased stability. In the present instance, the initial carbocation is a secondary one that can rearrange to a more stable tertiary one by a hydride shift:

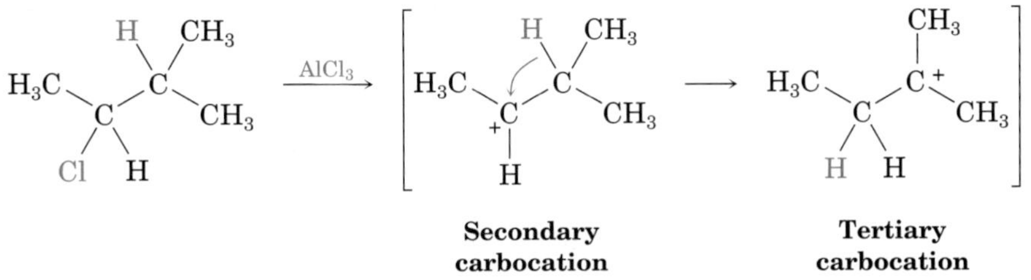

Secondary carbocation   Tertiary carbocation

Use this more stable tertiary carbocation to complete the Friedel–Crafts reaction.

Solution

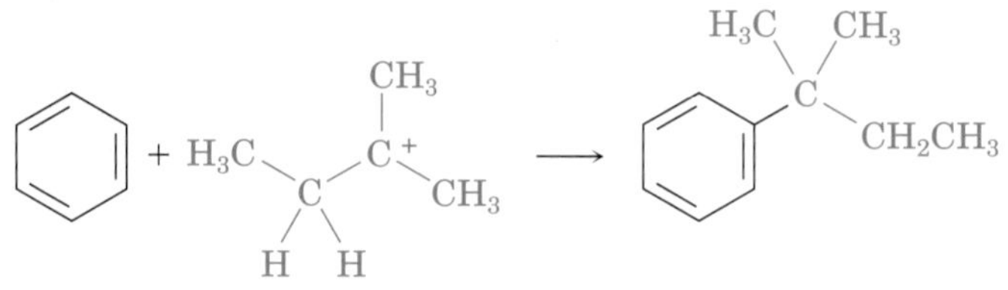

⋯⋯⋯⋯⋯⋯⋯⋯⋯⋯⋯⋯⋯⋯⋯⋯⋯

Problem 16.4 Which of the following alkyl halides would you expect to undergo Friedel–Crafts reaction *without* rearrangement? Explain.
(a) CH_3CH_2Cl
(b) $CH_3CH_2CH(Cl)CH_3$
(c) $CH_3CH_2CH_2Cl$
(d) $(CH_3)_3CCH_2Cl$
(e) Chlorocyclohexane

Problem 16.5 What is the major monosubstitution product from the Friedel–Crafts reaction of benzene with 1-chloro-2-methylpropane in the presence of $AlCl_3$?

⋯⋯⋯⋯⋯⋯⋯⋯⋯⋯⋯⋯⋯⋯⋯⋯⋯

16.4 Acylation of Aromatic Rings

An **acyl** group, –COR (pronounced **a**-sil), is introduced onto the ring when an aromatic compound reacts with a carboxylic acid chloride, RCOCl, in the presence of $AlCl_3$. For example, reaction of benzene with acetyl chloride yields the ketone, acetophenone.

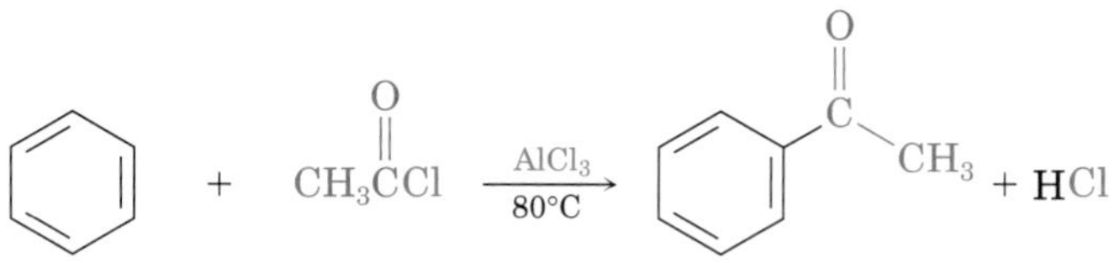

Benzene   **Acetyl chloride**   **Acetophenone (95%)**

The mechanism of **Friedel–Crafts acylation** is similar to that of Friedel–Crafts alkylation. The reactive electrophile is a resonance-stabilized acyl cation, generated by reaction between the acyl chloride and $AlCl_3$ (Figure 16.11). As the resonance structures in Figure 16.11 indicate, an acyl cation is stabilized by interaction of the vacant orbital on carbon with lone-pair electrons on the neighboring oxygen. Once formed, an acyl cation does not rearrange; rather, it is attacked by an aromatic ring to give unre-arranged substitution product.

FIGURE 16.11 ▼

Mechanism of the Friedel–Crafts acylation reaction. The electrophile is a resonance-stabilized acyl cation, whose electrostatic potential map indicates carbon as the most positive atom (blue).

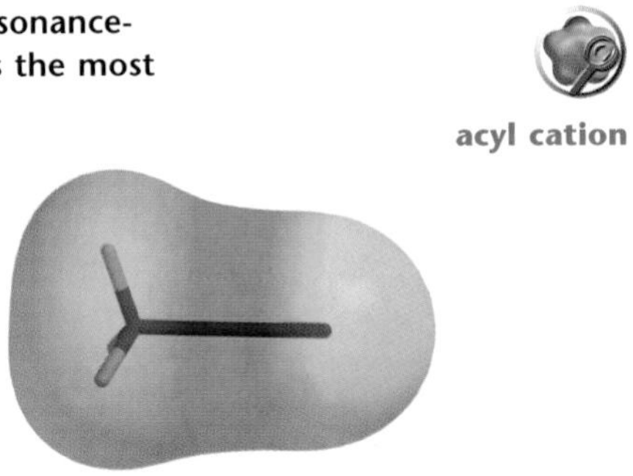

acyl cation

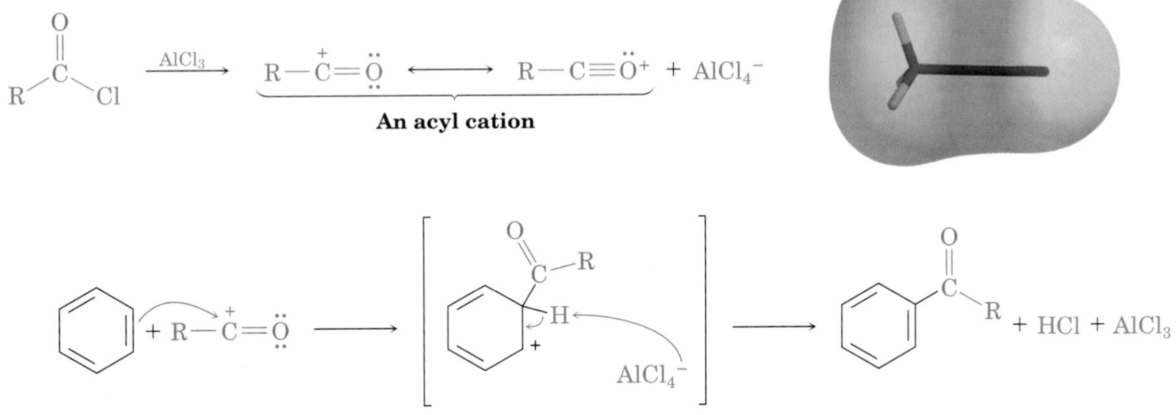

An acyl cation

Unlike the multiple substitutions that often occur in Friedel–Crafts alkylations, acylations never occur more than once on a ring because the product acylbenzene is always less reactive than the nonacylated starting material. We'll account for these reactivity differences in the next section.

Problem 16.6 Identify the carboxylic acid chloride that might be used in a Friedel–Crafts acylation reaction to prepare each of the following acylbenzenes:

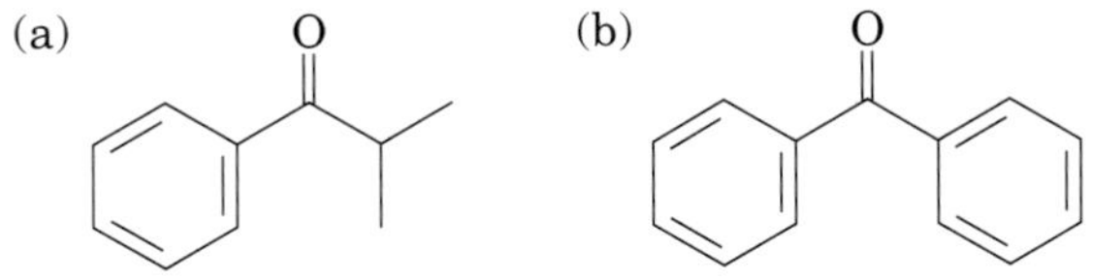

(a)   (b)

16.5 Substituent Effects in Substituted Aromatic Rings

Only one product can form when an electrophilic substitution occurs on benzene, but what would happen if we were to carry out a reaction on an aromatic ring that already has a substituent? A substituent already present on the ring has two effects:

1. A substituent affects the *reactivity* of the aromatic ring. Some substituents activate the ring, making it more reactive than benzene, and some deactivate the ring, making it less reactive than benzene. In aromatic nitration, for instance, an –OH substituent makes the ring 1000 times more reactive than benzene, while an –NO_2 substituent makes the ring more than 10 million times less reactive.

OH	H	Cl	NO_2

Relative rate of nitration   1000   1   0.033 6×10^{-8}

Reactivity

2. Substituents affect the *orientation* of the reaction. The three possible disubstituted products—ortho, meta, and para—are usually not formed in equal amounts. Instead, the nature of the substituent already present on the benzene ring determines the position of the second substitution. Table 16.1 lists experimental results for the nitration of some substituted benzenes and shows that some groups direct substitution

TABLE 16.1 Orientation of Nitration in Substituted Benzenes

$$\text{Y} \xrightarrow[\text{H}_2\text{SO}_4,\ 25°\text{C}]{\text{HNO}_3} \text{Y}—\text{NO}_2$$

	Product (%)				Product (%)		
	Ortho	**Meta**	**Para**		**Ortho**	**Meta**	**Para**
Meta-directing deactivators				**Ortho- and para-directing deactivators**			
—$\overset{+}{\text{N}}(CH_3)_3$	2	87	11	—F	13	1	86
—NO_2	7	91	2	—Cl	35	1	64
—COOH	22	76	2	—Br	43	1	56
—CN	17	81	2	—I	45	1	54
—$CO_2CH_2CH_3$	28	66	6	**Ortho- and para-directing activators**			
—$COCH_3$	26	72	2	—CH_3	63	3	34
—CHO	19	72	9	—ÖH	50	0	50
				—N̈HCOCH$_3$	19	2	79

primarily to the ortho and para positions, while other groups direct substitution primarily to the meta position.

Substituents can be classified into three groups: **ortho- and para-directing activators, ortho- and para-directing deactivators**, and **meta-directing deactivators**. There are no meta-directing activators. Figure 16.12 lists some groups in all three categories. Notice how the directing effects of the groups correlate with their reactivities. All meta-directing groups are strongly deactivating, and most ortho- and para-directing groups are activating. The halogens are unique in being ortho- and para-directing but weakly deactivating.

FIGURE 16.12 ▼

Classification of substituent effects in electrophilic aromatic substitution. All activating groups are ortho- and para-directing, and all deactivating groups other than halogen are meta-directing. The halogens are unique in being deactivating but ortho- and para-directing.

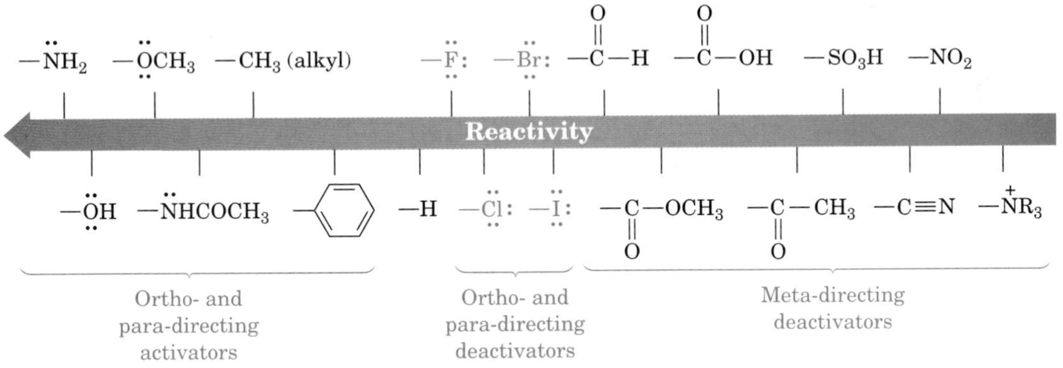

Reactivity and orientation in electrophilic aromatic substitutions are controlled by an interplay of *inductive effects* and *resonance effects*. As we saw in Sections 2.1 and 6.10, an **inductive effect** is the withdrawal or donation of electrons through a σ bond due to electronegativity and the polarity of bonds in functional groups. For example, halogens, carbonyl groups, cyano groups, and nitro groups inductively *withdraw* electrons through the σ bond linking the substituent to a benzene ring.

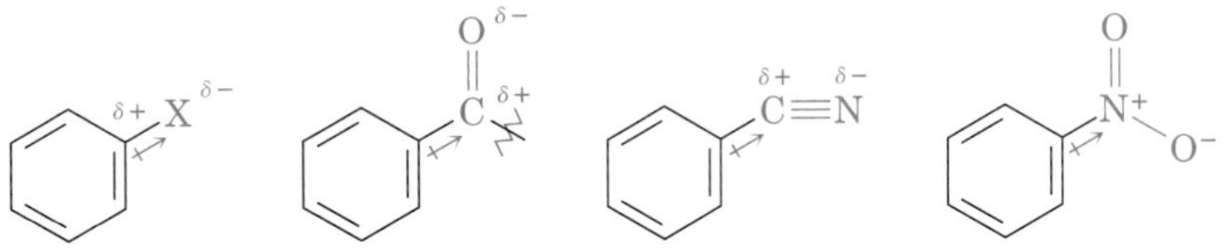

(X = F, Cl, Br, I)

The groups attached to the aromatic rings are inductively electron-withdrawing because of the polarity of their bonds.

Alkyl groups, on the other hand, inductively *donate* electrons. This is the same donating effect that causes alkyl substituents to stabilize alkenes (Section 6.7) and carbocations (Section 6.10).

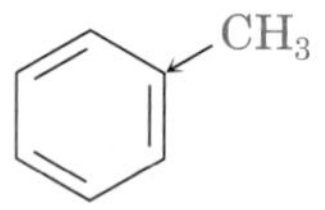

Alkyl group; inductively electron-donating

A **resonance effect** is the withdrawal or donation of electrons through a π bond due to the overlap of a p orbital on the substituent with a p orbital on the aromatic ring. Carbonyl, cyano, and nitro substituents, for example, *withdraw* electrons from the aromatic ring by resonance. Pi electrons flow from the rings to the substituents, leaving a positive charge in the ring. As shown by the following resonance structures for benzaldehyde, the effect is greatest at the ortho and para positions:

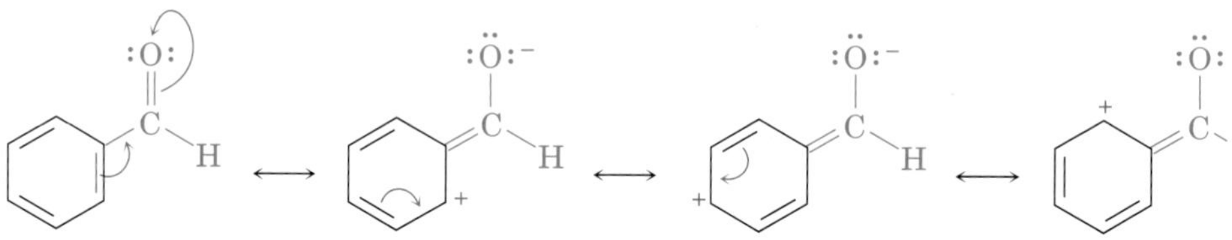

Benzaldehyde

Note that substituents with an electron-withdrawing resonance effect have the general structure –Y=Z, where the Z atom is more electronegative than Y:

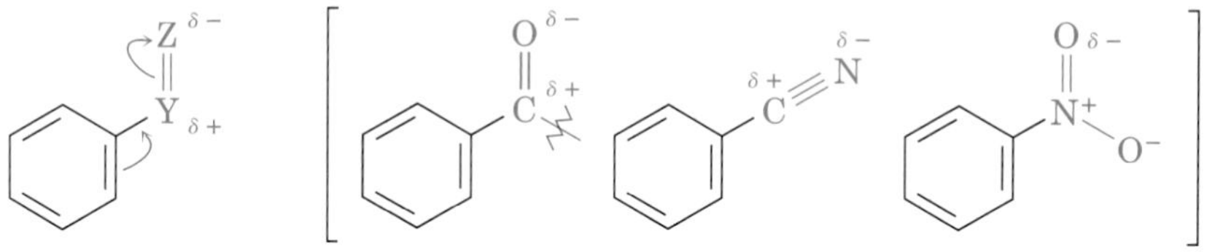

Rings substituted by a group with an electron-withdrawing resonance effect have this general structure.

Conversely, halogen, hydroxyl, alkoxyl (–OR), and amino substituents *donate* electrons to the aromatic ring by resonance. Pi electrons flow from the substituents to the ring, placing a negative charge in the ring, as shown by the following resonance structures for phenol. Again, the effect is greatest at the ortho and para positions.

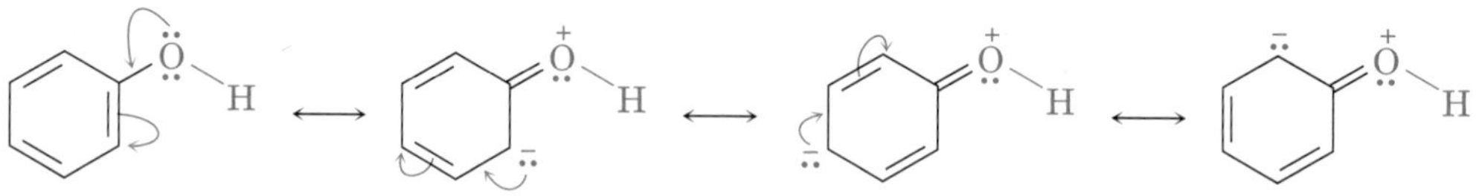

Phenol

Substituents with an electron-donating resonance effect have the general structure –$\ddot{Y}$, where the Y atom has a lone pair of electrons available for donation to the ring:

Rings substituted by a group with an electron-donating resonance effect have this general structure.

X = Halogen

One further point: *Inductive effects and resonance effects don't necessarily act in the same direction.* Halogen, hydroxyl, alkoxyl, and amino substituents, for example, have electron-*withdrawing* inductive effects because of the electronegativity of the –X, –O, or –N atom bonded to the aromatic ring but have electron-*donating* resonance effects because of the lone-pair electrons on those same –X, –O, or –N atoms.

Figure 16.13 compares electrostatic potential maps of benzene and several substituted benzenes. The ring becomes more negative when an electron-donating group such as –CH$_3$ or NH$_2$ is present, and more positive when an electron-withdrawing group such as –CN is present.

FIGURE 16.13 ▼

Electrostatic potential maps of benzene and several substituted benzenes show that an electon-donating group (–CH$_3$ or –NH$_2$) makes the ring more negative (red), while an electron-withdrawing group (–CN) makes the ring more positive (green).

benzene, toluene, aniline, benzonitrile

Benzene
(C$_6$H$_6$)

Toluene
(C$_6$H$_5$CH$_3$)

Aniline
(C$_6$H$_5$NH$_2$)

Benzonitrile
(C$_6$H$_5$C≡N)

● ●

Practice Problem 16.2   Predict the major product of the monosulfonation of toluene.

Strategy   Identify the substituent present on the ring, and decide whether it is ortho- and para-directing or meta-directing. According to Figure 16.12, an alkyl substituent is ortho- and para-directing. Monosulfonation of toluene will therefore give a mixture of *o*-toluenesulfonic acid and *p*-toluenesulfonic acid.

Solution

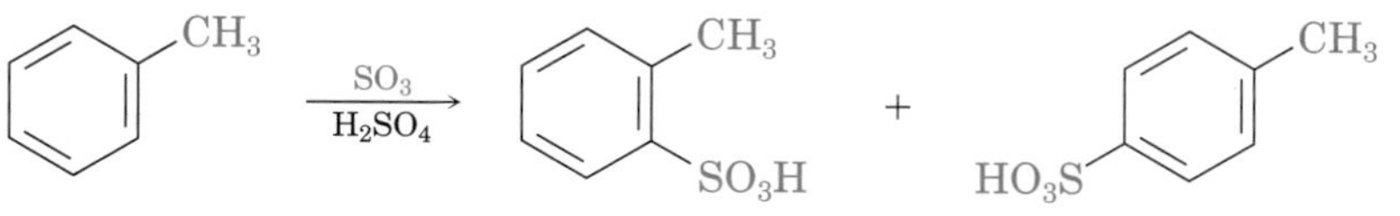

Toluene   *o*-**Toluenesulfonic acid**   *p*-**Toluenesulfonic acid**

● ●

Problem 16.7   Predict the major products of the following reactions:
(a) Mononitration of bromobenzene   (b) Monobromination of nitrobenzene
(c) Monochlorination of phenol   (d) Monobromination of aniline

Problem 16.8   Write resonance structures for nitrobenzene to show the electron-withdrawing resonance effect of the nitro group.

Problem 16.9   Write resonance structures for chlorobenzene to show the electron-donating resonance effect of the chloro group.

● ●

16.6   An Explanation of Substituent Effects

Activation and Deactivation of Aromatic Rings

How do inductive and resonance effects activate or deactivate an aromatic ring toward electrophilic substitution? *The common feature of all activating groups is that they donate electrons to the ring,* thereby stabilizing the carbocation intermediate from electrophilic addition and causing it to form faster. Hydroxyl, alkoxyl, and amino groups are activating because their stronger electron-donating resonance effect outweighs their weaker electron-withdrawing inductive effect. Alkyl groups are activating because of their electron-donating inductive effect.

The common feature of all deactivating groups is that they withdraw electrons from the ring, thereby destabilizing the carbocation intermediate and causing it to form more slowly. Carbonyl, cyano, and nitro groups are deactivating because of both electron-withdrawing resonance *and* inductive effects. Halogens are deactivating because their stronger electron-withdrawing inductive effect outweighs their weaker electron-donating resonance effect.

Y is an electron donor; carbocation intermediate is more stabilized, and ring is more reactive.

Y is an electron acceptor; carbocation intermediate is less stabilized, and ring is less reactive.

Reactivity

Problem 16.10   Rank the compounds in each group in order of their reactivity to electrophilic substitution:
(a) Nitrobenzene, phenol, toluene, benzene
(b) Phenol, benzene, chlorobenzene, benzoic acid
(c) Benzene, bromobenzene, benzaldehyde, aniline

Problem 16.11   Use Figure 16.12 to explain why Friedel–Crafts alkylations often give polysubstitution but Friedel–Crafts acylations do not.

Ortho- and Para-Directing Activators: Alkyl Groups

Inductive and resonance effects account for the directing ability of substituents as well as for their activating or deactivating ability. Take alkyl groups, for example, which have an electron-donating inductive effect and behave as ortho and para directors. The results of toluene nitration are shown in Figure 16.14 (p. 612).

Nitration of toluene might occur either ortho, meta, or para to the methyl group, giving the three carbocation intermediates shown in Figure 16.14. All three intermediates are resonance-stabilized, but *the ortho and para intermediates are the most stabilized.* For both ortho and para attack, but not for meta attack, a resonance form places the positive charge directly on the methyl-substituted carbon, where it is in a tertiary position and can best be stabilized by the electron-donating inductive effect of the methyl group. The ortho and para intermediates are thus lower in energy than the meta intermediate and therefore form faster.

FIGURE 16.14 ▼

Carbocation intermediates in the nitration of toluene. Ortho and para intermediates are more stable than the meta intermediate because the positive charge is on a tertiary carbon rather than a secondary carbon.

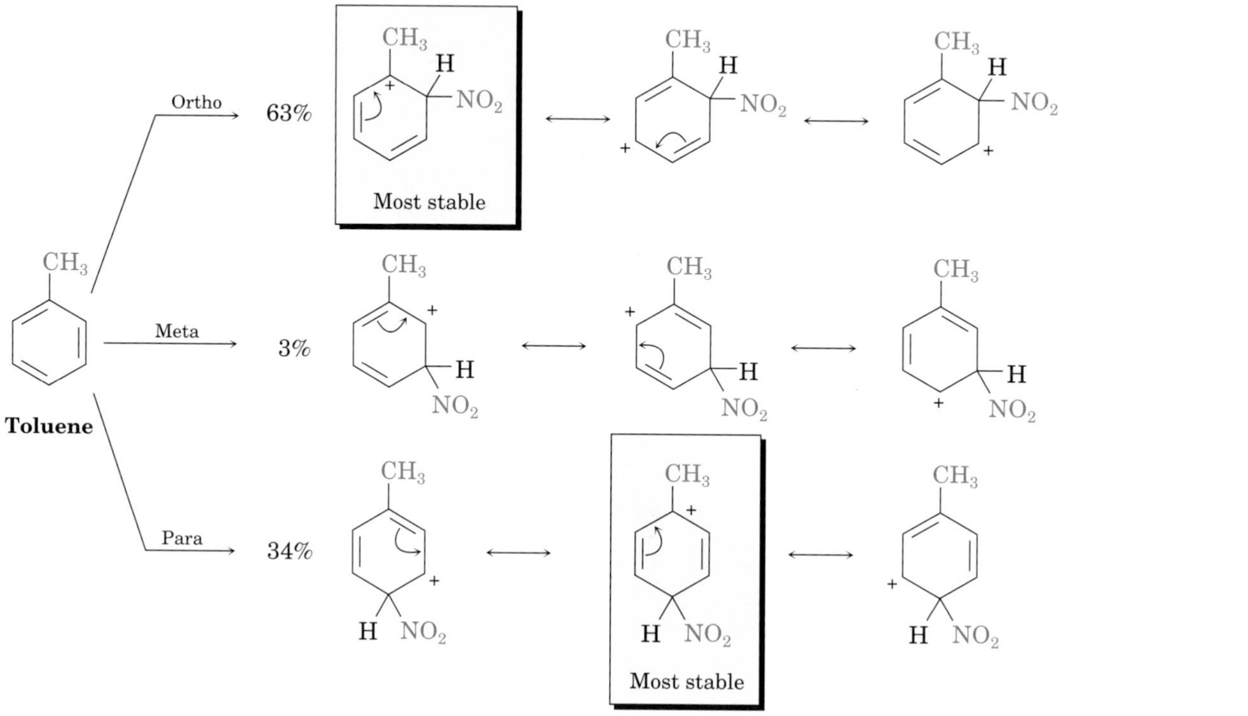

● ●

Problem 16.12   Which would you expect to be more reactive toward electrophilic substitution, toluene or (trifluoromethyl)benzene? Explain.

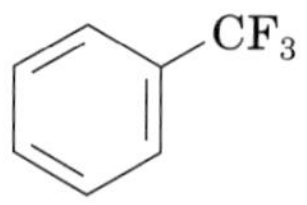

(Trifluoromethyl)benzene

● ●

Ortho- and Para-Directing Activators: OH and NH₂

Hydroxyl, alkoxyl, and amino groups are also ortho–para activators, but for a different reason than for alkyl groups. As mentioned in the previous section, hydroxyl, alkoxyl, and amino groups have a strong, electron-donating resonance effect that is most pronounced at the ortho and para positions and that outweighs a weaker electron-withdrawing inductive effect.

When phenol is nitrated, only ortho and para attack is observed, as shown in Figure 16.15. All three possible carbocation intermediates are stabilized by resonance, but the intermediates from ortho and para attack are stabilized most. Only in ortho and para attack are there resonance forms in which the positive charge is stabilized by donation of an electron pair from oxygen. The intermediate from meta attack has no such stabilization.

FIGURE 16.15 ▼

Carbocation intermediates in the nitration of phenol. The ortho and para intermediates are more stable than the meta intermediate because of resonance donation of electrons from oxygen.

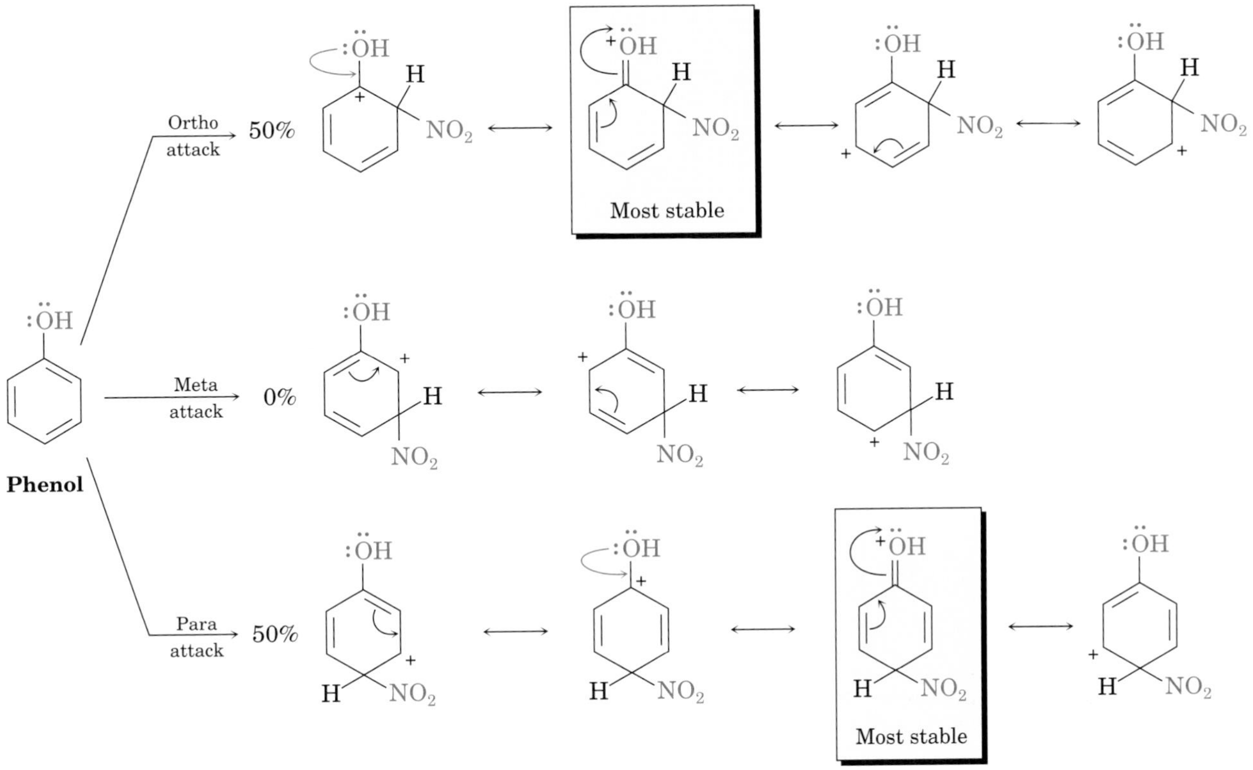

Problem 16.13   Acetanilide is less reactive than aniline toward electrophilic substitution. Explain.

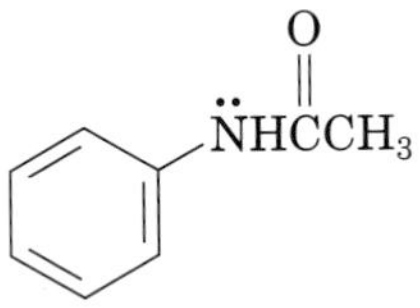

Acetanilide

Ortho- and Para-Directing Deactivators: Halogens

Halogens are deactivating because their stronger electron-withdrawing inductive effect outweighs their weaker electron-donating resonance effect. Though weak, that electron-donating resonance effect is felt only at the ortho and para positions (Figure 16.16). Thus, a halogen substituent can stabilize the positive charge of the carbocation intermediates from ortho and para attack in the same way that hydroxyl and amino substituents can. The meta intermediate, however, has no such stabilization and is therefore formed more slowly.

FIGURE 16.16 ▼

Carbocation intermediates in the nitration of chlorobenzene. The ortho and para intermediates are more stable than the meta intermediate because of electron donation of the halogen lone-pair electrons.

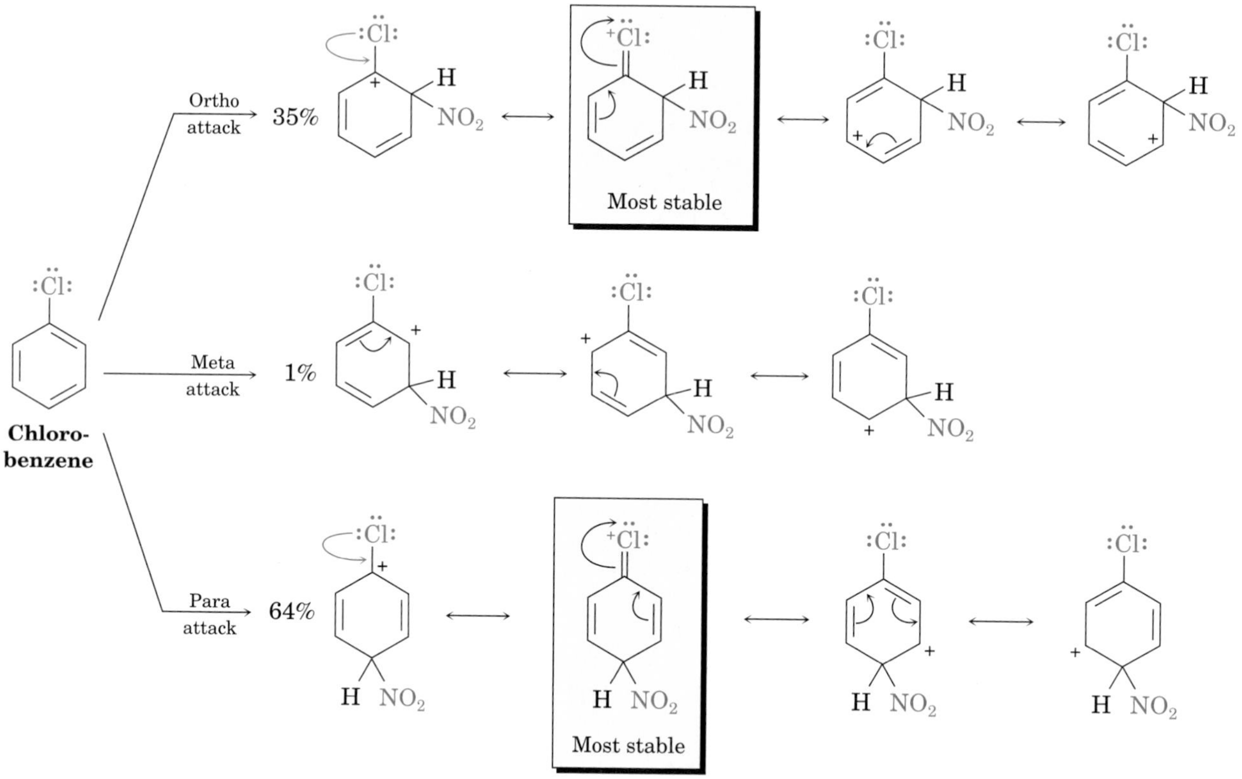

Note again that halogens, hydroxyl, alkoxyl, and amino groups all *withdraw* electrons inductively and *donate* electrons by resonance. Halogens have a stronger electron-withdrawing inductive effect but a weaker electron-donating resonance effect and are thus deactivators. Hydroxyl, alkoxyl, and amino groups have a weaker electron-withdrawing inductive effect but a stronger electron-donating resonance effect and are thus activators. All are ortho and para directors, however, because of the lone pair of electrons on the atom bonded to the aromatic ring.

Meta-Directing Deactivators

Meta-directing deactivators act through a combination of inductive and resonance effects that reinforce each other. Inductively, both ortho and para intermediates are destabilized because a resonance form places the positive charge of the carbocation intermediate directly on the ring carbon atom that bears the deactivating group (Figure 16.17). At the same time, resonance electron withdrawal is also felt at the ortho and para positions. Reaction with an electrophile therefore occurs at the meta position.

FIGURE 16.17 ▼

Carbocation intermediates in the chlorination of benzaldehyde. The meta intermediate is more stable than the ortho or para intermediate.

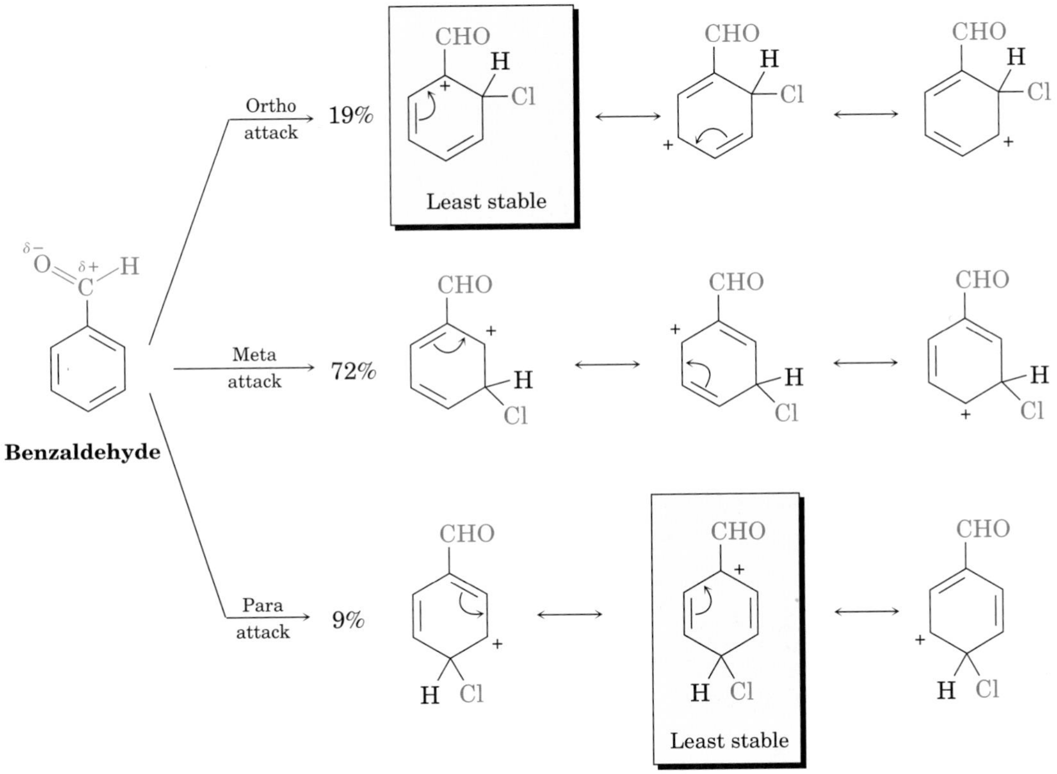

A Summary of Substituent Effects in Aromatic Substitution

A summary of the activating and directing effects of substituents in electrophilic aromatic substitution is shown in Table 16.2 (p. 616).

. . . .

Problem 16.14   Draw resonance structures for the intermediates from attack of an electrophile at the ortho, meta, and para positions of nitrobenzene. Which intermediates are most favored?

. . . .

TABLE 16.2   Substituent Effects in Electrophilic Aromatic Substitution

Substituent	Reactivity	Orientation	Inductive effect	Resonance effect
—CH$_3$	Activating	Ortho, para	Weak; electron-donating	None
–ÖH, –N̈H$_2$	Activating	Ortho, para	Weak; electron-withdrawing	Strong; electron-donating
–F̈:, –C̈l:, –B̈r:, –Ï:	Deactivating	Ortho, para	Strong; electron-withdrawing	Weak; electron-donating
–N̊(CH$_3$)$_3$	Deactivating	**Meta**	Strong; electron-withdrawing	None
—NO$_2$, —CN, —CHO, —CO$_2$CH$_3$, —COCH$_3$, —COOH	Deactivating	**Meta**	Strong; electron-withdrawing	Strong; electron-withdrawing

16.7   Trisubstituted Benzenes: Additivity of Effects

Further electrophilic substitution of a disubstituted benzene is governed by the same resonance and inductive effects just discussed. The only difference is that it's now necessary to consider the additive effects of two different groups. In practice, this isn't as difficult as it sounds; three rules are usually sufficient:

RULE 1   If the directing effects of the two groups reinforce each other, there is no problem. In *p*-nitrotoluene, for example, both the methyl and the nitro group direct further substitution to the same position (ortho to the methyl = meta to the nitro). A single product is thus formed by electrophilic substitution.

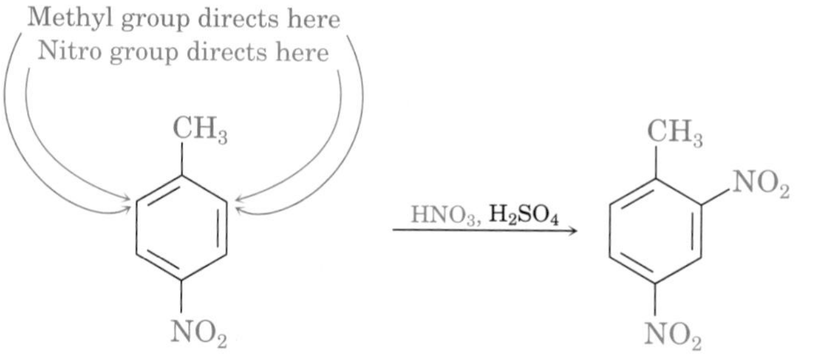

p-Nitrotoluene   2,4-Dinitrotoluene

RULE 2   If the directing effects of the two groups oppose each other, the more powerful activating group has the dominant influence, but mixtures of products often result. For example, bromination of *p*-methylphenol yields primarily 2-bromo-4-methylphenol because –OH is a more powerful activator than –CH$_3$.

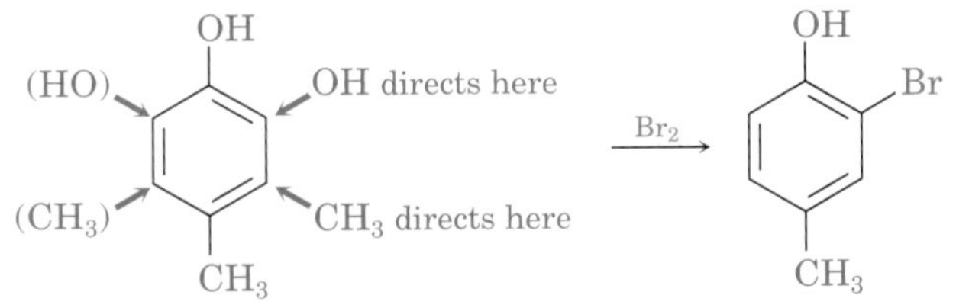

p-Methylphenol
(p-Cresol)

2-Bromo-4-methylphenol
(major product)

RULE 3   Further substitution rarely occurs between the two groups in a meta-disubstituted compound because this site is too hindered. Aromatic rings with three adjacent substituents must therefore be prepared by some other route, usually by substitution of an ortho-disubstituted compound.

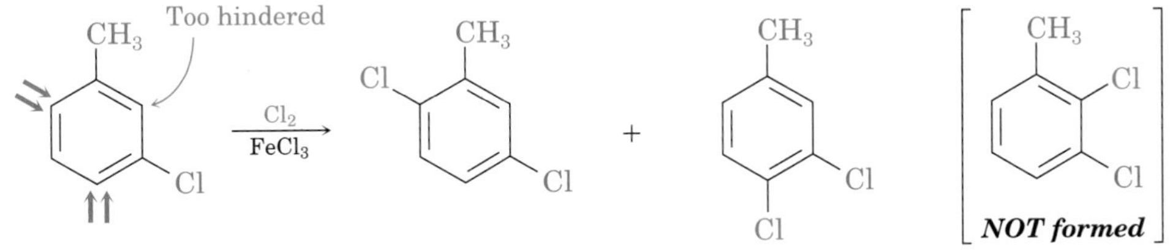

m-Chlorotoluene   **2,5-Dichlorotoluene**   **3,4-Dichlorotoluene**   *NOT formed*

But:

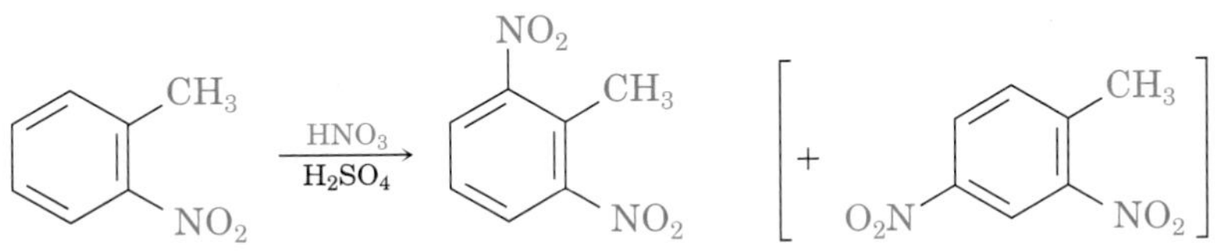

o-Nitrotoluene   **2,6-Dinitrotoluene**   **2,4-Dinitrotoluene**

· ·

Practice Problem 16.3   What product would you expect from bromination of *p*-methylbenzoic acid?

Strategy   Identify the two substituents present on the ring, decide the directing effect of each, and decide which substituent is the stronger activator. The carboxyl group (–COOH) is a meta director, and the methyl group is an ortho and para director. Both groups direct bromination to the position next to the methyl group, yielding 3-bromo-4-methylbenzoic acid.

Solution

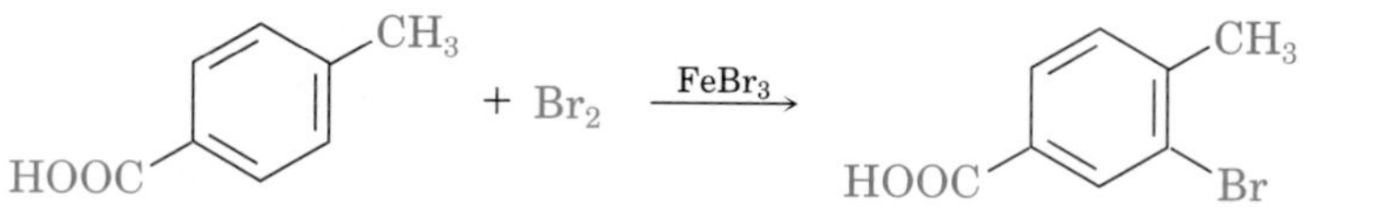

p-Methylbenzoic acid   3-Bromo-4-methylbenzoic acid

• •

Problem 16.15   Where would you expect electrophilic substitution to occur in the following substances?

(a) 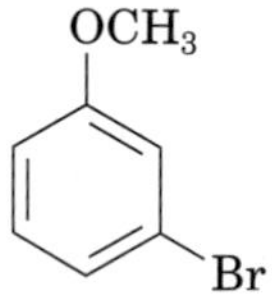   (b) 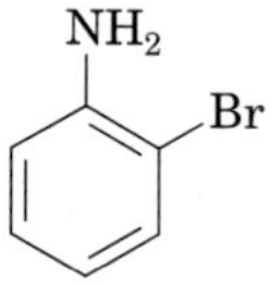   (c)

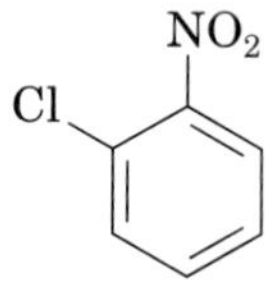

• •

16.8 Nucleophilic Aromatic Substitution

Aromatic substitution reactions usually occur by an *electrophilic* mechanism. Aryl halides that have electron-withdrawing substituents, however, can also undergo **nucleophilic aromatic substitution**. For example, 2,4,6-trinitrochlorobenzene reacts with aqueous NaOH at room temperature to give 2,4,6-trinitrophenol in 100% yield. The nucleophile OH⁻ has substituted for Cl⁻.

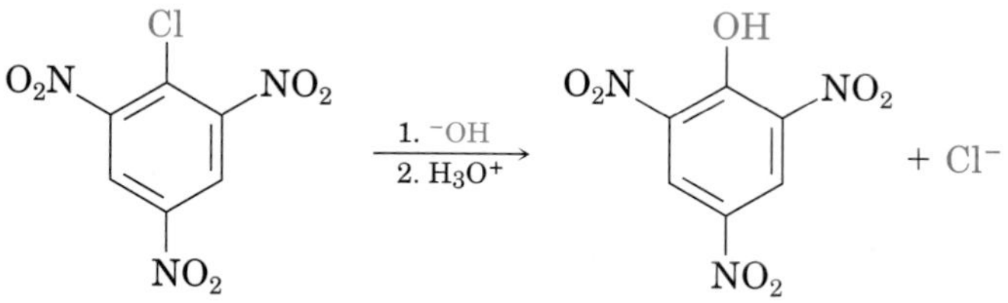

2,4,6-Trinitrochlorobenzene   **2,4,6-Trinitrophenol (100%)**

How does this reaction take place? Although it appears superficially similar to the S_N1 and S_N2 nucleophilic substitution reactions of alkyl halides discussed in Chapter 11, it must be different because aryl halides are inert to both S_N1 and S_N2 conditions. Aryl halides don't undergo S_N1 reactions because aryl cations are relatively unstable. The dissociation of an aryl halide is energetically unfavorable and does not occur easily.

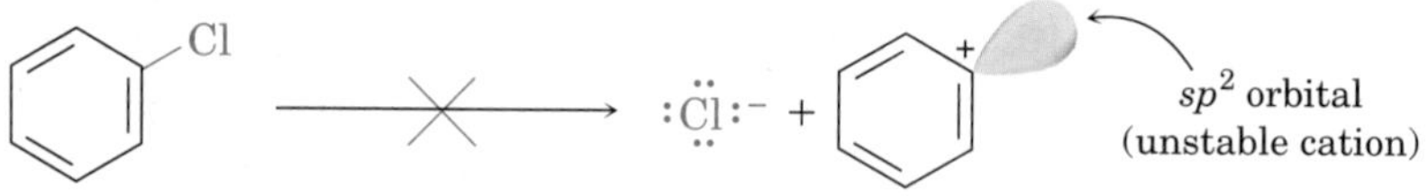

**Dissociation does *NOT* occur;
therefore, no S_N1 reaction**

sp^2 orbital
(unstable cation)

Aryl halides don't undergo S_N2 reactions because the halo-substituted carbon atom is sterically shielded from back-side attack by the aromatic ring. For a nucleophile to attack an aryl halide, it would have to approach directly through the aromatic ring and invert the stereochemistry of the aromatic ring—a geometric impossibility.

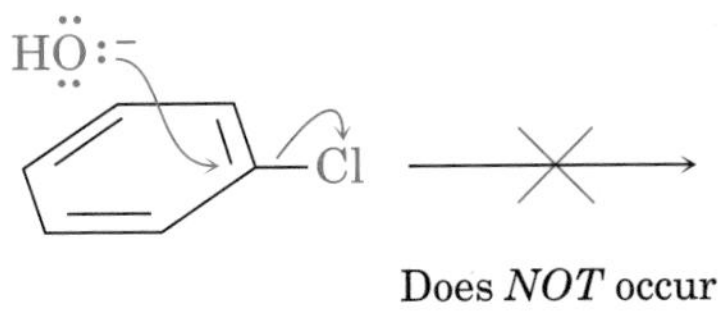

Does *NOT* occur

Nucleophilic substitutions on an aromatic ring proceed by the *addition/elimination* mechanism shown in Figure 16.18. The attacking nucleophile first adds to the electron-deficient aryl halide, forming a resonance-stabilized negatively charged intermediate called a *Meisenheimer complex*. Halide ion is then eliminated in the second step.

FIGURE 16.18 ▼

Mechanism of aromatic nucleophilic substitution. The reaction occurs in two steps and involves a resonance-stabilized carbanion intermediate.

 **refer to Mechanisms & Movies**

Nucleophilic addition of hydroxide ion to the electron-poor aromatic ring takes place yielding a stabilized carbanion intermediate.

The carbanion intermediate undergoes elimination of chloride ion in a second step to give the substitution product.

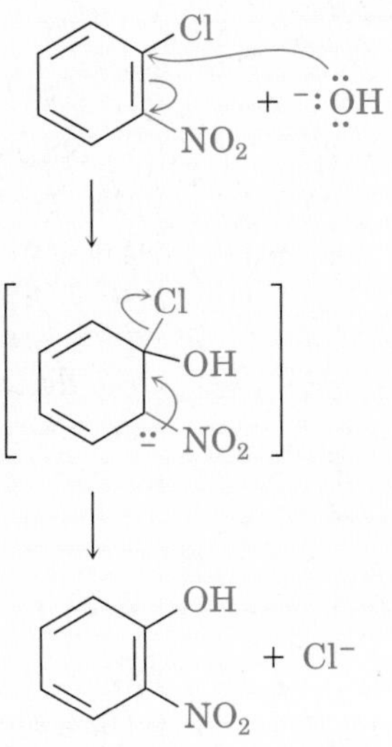

© 1984 JOHN MCMURRY

Jacob Meisenheimer

Jacob Meisenheimer (1876–1934) was born in Greisheim, Germany, and received his Ph.D. at Munich. He was professor of chemistry at the universities of Berlin and Tübingen.

Nucleophilic aromatic substitution occurs only if the aromatic ring has an electron-withdrawing substituent in a position ortho or para to the halogen. The more such substituents there are, the faster the reaction goes. As shown in Figure 16.19, only ortho and para electron-withdrawing substituents can stabilize the anion intermediate through resonance; a meta substituent offers no such resonance stabilization. Thus, *p*-chloronitrobenzene and *o*-chloronitrobenzene react with hydroxide ion at 130°C to yield substitution products, but *m*-chloronitrobenzene is inert to OH⁻.

FIGURE 16.19 ▼

Nucleophilic aromatic substitution on nitrochlorobenzenes. Only the ortho and para intermediate carbanions are resonance-stabilized, so only the ortho and para isomers undergo reaction.

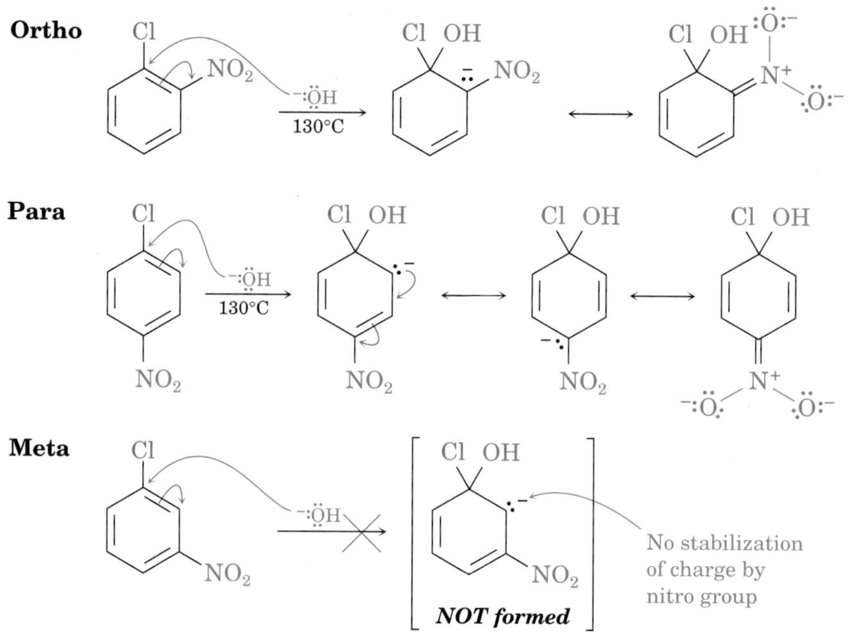

Note the differences between electrophilic and nucleophilic aromatic substitutions: *Electrophilic* substitutions are favored by electron-*donating* substituents, which stabilize the *carbocation* intermediate, while *nucleophilic* substitutions are favored by electron-*withdrawing* substituents, which stabilize a *carbanion* intermediate. The electron-withdrawing groups that deactivate rings for electrophilic substitution (nitro, carbonyl, cyano, and so on) activate them for nucleophilic substitution. What's more, these groups are meta directors in electrophilic substitution, but are ortho–para directors in nucleophilic substitution.

Problem 16.16   Propose a mechanism for the reaction of 1-chloroanthraquinone with methoxide ion to give the substitution product 1-methoxyanthraquinone. Use curved arrows to show the electron flow in each step.

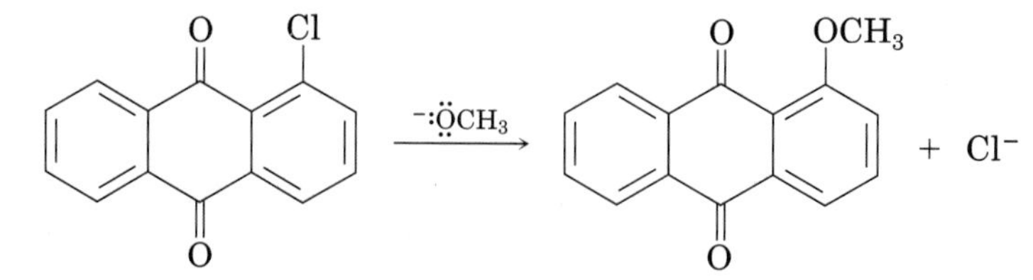

1-Chloroanthraquinone   **1-Methoxyanthraquinone**

16.9 Benzyne

Halobenzenes without electron-withdrawing substituents do not react with nucleophiles under most conditions. At high temperature and pressure, however, even chlorobenzene can be forced to react. Chemists at the Dow Chemical Company discovered in 1928 that phenol could be prepared on a large industrial scale by treatment of chlorobenzene with dilute aqueous NaOH at 340°C under 2500 psi (pounds per square inch) pressure.

Chlorobenzene → **Phenol**

1. NaOH, H_2O, 340°C, 2500 psi
2. H_3O^+

+ NaCl

This phenol synthesis is different from the nucleophilic aromatic substitutions discussed in the previous section because it takes place by an *elimination/addition* mechanism rather than an addition/elimination. Strong base first causes the elimination of HX from halobenzene in an E2 reaction, yielding a highly reactive **benzyne** intermediate, and a nucleophile then adds to benzyne in a second step to give the product. The two steps are similar to those in other nucleophilic aromatic substitutions, but their order is reversed: elimination before addition for the benzyne reaction rather than addition before elimination for the usual reaction.

Chlorobenzene   **Benzyne**   **Phenol**

−HCl
elimination

H_2O
addition

Evidence supporting the benzyne mechanism has been obtained by studying the reaction between bromobenzene and the strong base potassium amide (KNH_2) in liquid NH_3 solvent. When bromobenzene labeled with radioactive ^{14}C at the C1 position is used, the substitution product has the label scrambled between C1 and C2. The reaction must therefore proceed through a symmetrical intermediate in which C1 and C2 are equivalent— a requirement that only benzyne can meet.

Bromobenzene   **Benzyne**
                         (symmetrical)

$:\ddot{N}H_2^-$
NH_3
(−HBr)

NH_3

50%   50%

Aniline

Further evidence for a benzyne intermediate comes from trapping experiments. Although benzyne is too reactive to be isolated as a pure compound, it can be intercepted in a Diels–Alder reaction (Section 14.8) if a diene such as furan is present when benzyne is generated.

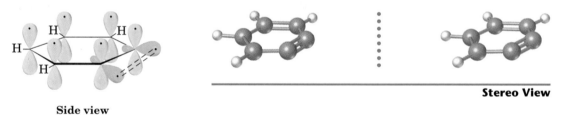

Benzyne   **Furan**   **Diels–Alder adduct**
(a dienophile) **(a diene)**

The electronic structure of benzyne, shown in Figure 16.20, is that of a highly distorted alkyne. Although a typical alkyne triple bond uses sp-hybridized carbon atoms, the benzyne triple bond uses sp^2-hybridized carbons. Furthermore, a typical alkyne triple bond has two mutually perpendicular π bonds formed by p–p overlap, but the benzyne triple bond has one π bond formed by p–p overlap and one π bond formed by sp^2–sp^2 overlap. The latter π bond is in the plane of the ring and is very weak.

FIGURE 16.20 ▼

An orbital picture of benzyne. The benzyne carbons are sp^2-hybridized, and the "third" bond results from weak overlap of two adjacent sp^2 orbitals.

H —
H   H

H

Side view

Stereo View

Problem 16.17 Treatment of *p*-bromotoluene with NaOH at 300°C yields a mixture of *two* products, but treatment of *m*-bromotoluene with NaOH yields a mixture of *three* products. Explain.

16.10 Oxidation of Aromatic Compounds

Oxidation of Alkylbenzene Side Chains

Despite its unsaturation, the benzene ring is inert to strong oxidizing agents such as $KMnO_4$ and $Na_2Cr_2O_7$, reagents that will cleave alkene carbon–carbon bonds (Section 7.8). It turns out, however, that the presence of the aromatic ring has a dramatic effect on alkyl-group side chains. Alkyl side chains are readily attacked by oxidizing agents and are converted into carboxyl groups, –COOH. The net effect is conversion of an alkylbenzene into a benzoic acid, Ar–R $\longrightarrow$ Ar–COOH. For example, *p*-nitrotoluene and butylbenzene are oxidized by aqueous $KMnO_4$ in high yield to give the corresponding benzoic acids.

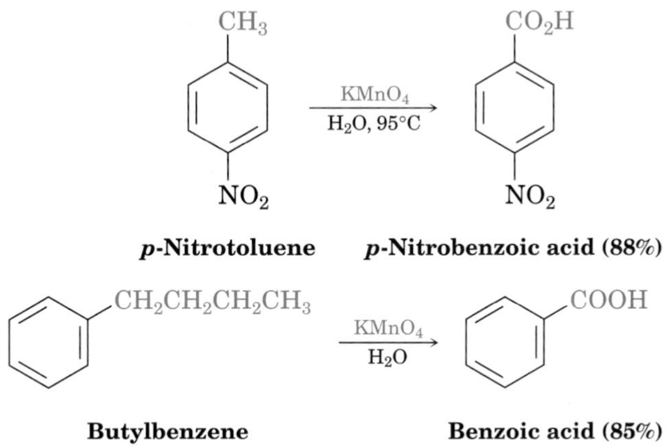

p-**Nitrotoluene**   *p*-**Nitrobenzoic acid (88%)**

Butylbenzene   **Benzoic acid (85%)**

A similar oxidation is employed industrially for the preparation of terephthalic acid, used in the production of polyester fibers (Section 21.10). Approximately 5 million tons per year of *p*-xylene are oxidized, using air as the oxidant and Co(III) salts as catalyst.

Industrial procedure

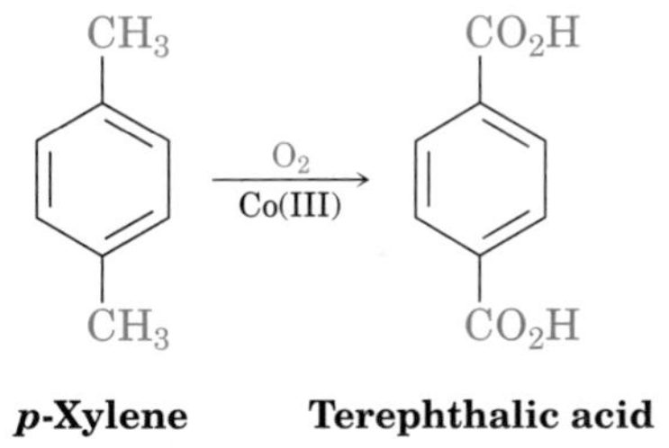

p-**Xylene**   **Terephthalic acid**

The mechanism of side-chain oxidation is complex and involves attack on C–H bonds at the position next to the aromatic ring to form intermediate benzylic radicals. *tert*-Butylbenzene has no benzylic hydrogens, however, and is therefore inert.

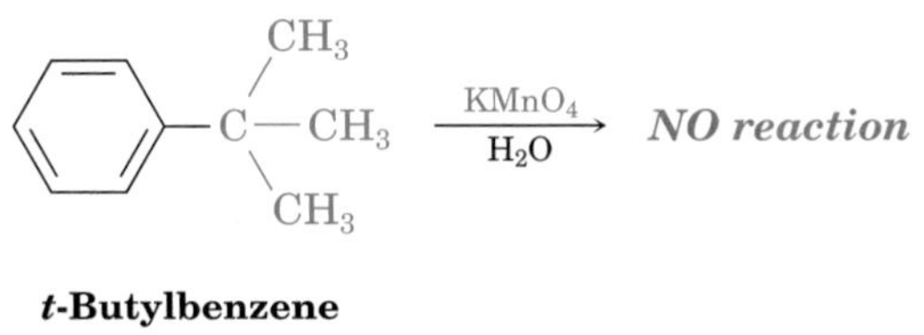

t-**Butylbenzene**

● ●

Problem 16.18   What aromatic products would you obtain from the $KMnO_4$ oxidation of the following substances?

(a) 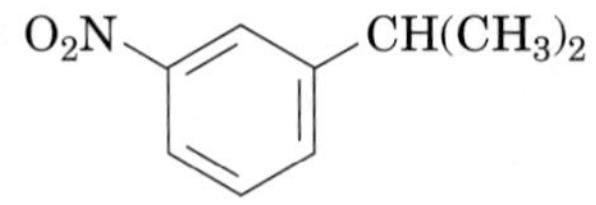 O_2N—⟨benzene ring⟩—$CH(CH_3)_2$ (b) 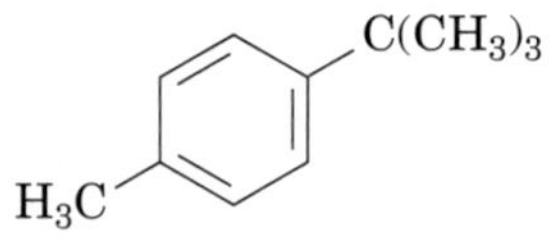 H_3C—⟨benzene ring⟩—$C(CH_3)_3$

Bromination of Alkylbenzene Side Chains

Side-chain bromination at the benzylic position occurs when an alkylbenzene is treated with *N*-bromosuccinimide (NBS). For example, propylbenzene gives (1-bromopropyl)benzene in 97% yield on reaction with NBS in the presence of benzoyl peroxide, $(PhCO_2)_2$, as a radical initiator. Bromination occurs exclusively in the benzylic position and does not give a mixture of products.

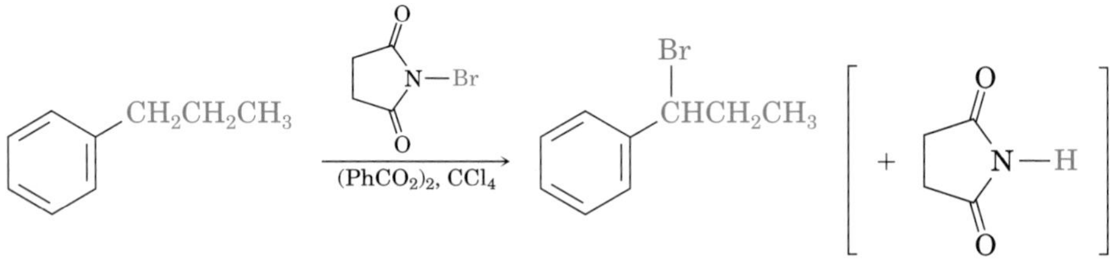

Propylbenzene   **(1-Bromopropyl)benzene**
                                                  (97%)

The mechanism of benzylic bromination is similar to that discussed in Section 10.5 for allylic bromination of alkenes. Abstraction of a benzylic hydrogen atom generates an intermediate benzylic radical, which reacts with Br_2 to yield product and a Br· radical that cycles back into the reaction to carry on the chain. The Br_2 necessary for reaction with the benzylic radical is produced by a concurrent reaction of HBr with NBS, as shown in Figure 16.21.

FIGURE 16.21 ▼

Mechanism of benzylic bromination with *N*-bromosuccinimide. The process is a
radical chain reaction and involves a benzylic radical as intermediate.

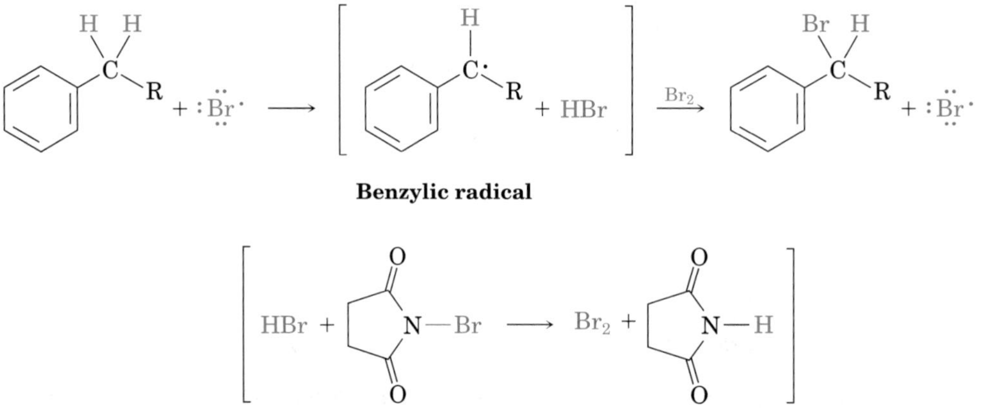

Benzylic radical

Reaction occurs exclusively at the benzylic position because the benzylic radical intermediate is highly stabilized by resonance. Figure 16.22 shows how the benzyl radical is stabilized by overlap of its *p* orbital with the ring π electron system.

FIGURE 16.22 ▼

A resonance-stabilized benzylic radical. The spin surface shows that the unpaired electron is shared by the ortho and para carbons in the ring.

benzylic radical

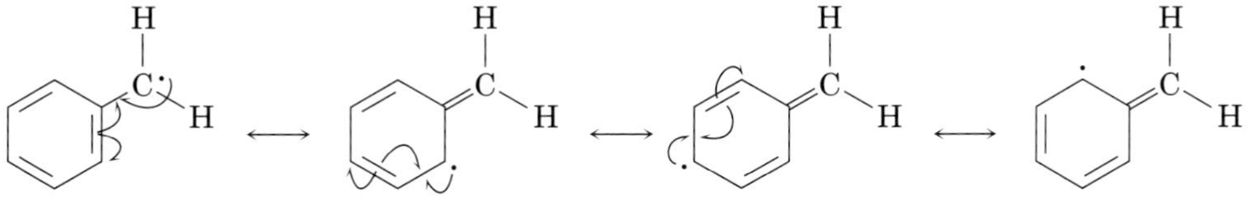

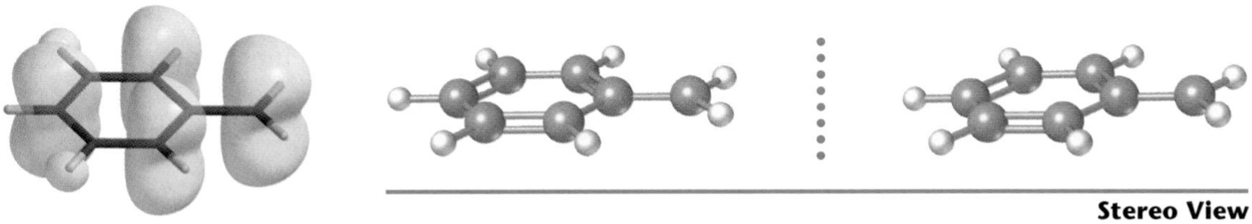

Stereo View

Problem 16.19   Styrene, the simplest alkenylbenzene, is prepared commercially for use in plastics manufacture by catalytic dehydrogenation of ethylbenzene. How might you prepare styrene from benzene using reactions you've studied?

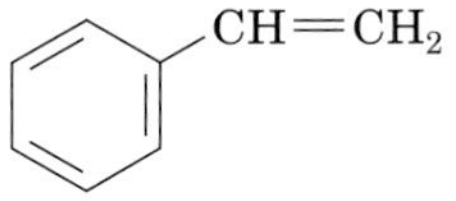

$$CH=CH_2$$

Styrene

Problem 16.20   Refer to Table 5.3 for a quantitative idea of the stability of a benzyl radical. How much more stable (in kJ/mol) is the benzyl radical than a primary alkyl radical? How does a benzyl radical compare in stability to an allyl radical?

16.11 Reduction of Aromatic Compounds

Catalytic Hydrogenation of Aromatic Rings

Just as aromatic rings are inert to oxidation under most conditions, they're also inert to catalytic hydrogenation under conditions that reduce typical alkene double bonds. As a result, it's possible to selectively reduce an alkene double bond in the presence of an aromatic ring. For example, 4-phenyl-3-buten-2-one is reduced to 4-phenyl-2-butanone when the reaction is carried out at room temperature and atmospheric pressure using a palladium catalyst. Neither the benzene ring nor the ketone carbonyl group is affected.

4-Phenyl-3-buten-2-one   **4-Phenyl-2-butanone (100%)**

To hydrogenate an aromatic ring, it's necessary either to use a platinum catalyst with hydrogen gas at several hundred atmospheres pressure or to use a more powerful catalyst such as rhodium on carbon. Under these conditions, aromatic rings are readily reduced to cyclohexanes. For example, *o*-xylene yields 1,2-dimethylcyclohexane, and 4-*tert*-butylphenol gives 4-*tert*-butylcyclohexanol.

o-**Xylene**   **1,2-Dimethylcyclohexane (100%)**

4-*tert*-Butylphenol   **4-*tert*-Butylcyclohexanol (100%)**

Reduction of Aryl Alkyl Ketones

Just as an aromatic ring activates a neighboring (benzylic) C–H position toward oxidation, it also activates a neighboring carbonyl group toward reduction. Thus, an aryl alkyl ketone prepared by Friedel–Crafts acylation of an aromatic ring can be converted into an alkylbenzene by catalytic hydrogenation over a palladium catalyst. For example, propiophenone is reduced to propylbenzene in 100% yield by catalytic hydrogenation. Since the net effect of Friedel–Crafts acylation followed by reduction is the preparation of a primary alkylbenzene, this two-step sequence of reactions makes it possible to circumvent the carbocation rearrangement problems associated with direct Friedel–Crafts alkylation using a primary alkyl halide (Section 16.3).

Propiophenone (95%)   **Propylbenzene (100%)**

Mixture of two products

Note that the conversion of a carbonyl group into a methylene group ($C{=}O \longrightarrow CH_2$) by catalytic hydrogenation is limited to *aryl* alkyl ketones; dialkyl ketones are not reduced under these conditions. Furthermore, the catalytic reduction of aryl alkyl ketones is not compatible with the presence of a nitro substituent on the aromatic ring, because a nitro group is reduced to an amino group under the reaction conditions. We'll see a more general method for reducing all ketone carbonyl groups to yield alkanes in Section 19.10.

***m*-Nitroacetophenone**   ***m*-Ethylaniline**

• •

Problem 16.21 How would you prepare diphenylmethane, $(Ph)_2CH_2$, from benzene and an appropriate acid chloride?

• •

16.12 Synthesis of Trisubstituted Benzenes

One of the surest ways to learn organic chemistry is to work synthesis problems. The ability to plan a successful multistep synthesis of a complex molecule requires a working knowledge of the uses and limitations of many hundreds of organic reactions. Not only must you know *which* reactions to use, you must also know *when* to use them. The order in which reactions are carried out is often critical to the success of the overall scheme.

The ability to plan a sequence of reactions in the right order is particularly valuable in the synthesis of substituted aromatic rings, where the introduction of a new substituent is strongly affected by the directing effects of other substituents. Planning syntheses of substituted aromatic compounds is therefore an excellent way to gain facility with the many reactions learned in the past few chapters.

During the previous discussion of strategies for working synthesis problems in Section 8.10, we said that it's usually best to work a problem *retrosynthetically*, or backward. Look at the target molecule and ask yourself, "What is an immediate precursor of this compound?" Choose a likely answer and continue working backward, one step at a time, until you arrive at a simple starting material. Let's try some examples.

● ●

Practice Problem 16.4   Synthesize *p*-bromobenzoic acid from benzene.

Strategy   As described in Section 8.10, synthesis problems are best solved by looking at the product, identifying the functional groups it contains, and then asking yourself how those functional groups can be prepared. Always work in a retrosynthetic sense, one step at a time.

p-Bromobenzoic acid

In this example, the two functional groups are a bromo group and a carboxylic acid. A bromo group can be introduced by bromination with $Br_2/FeBr_3$, and a carboxylic acid group can be introduced by Friedel–Crafts alkylation or acylation followed by oxidation.

Solution   Ask yourself, "What is an immediate precursor of *p*-bromobenzoic acid?" There are two substituents on the ring, a carboxyl group (COOH), which is meta-directing, and a bromine, which is ortho- and para-directing. We can't brominate benzoic acid, because the wrong isomer (*m*-bromobenzoic acid) would be produced. We know, however, that oxidation of an alkylbenzene yields a benzoic acid. Thus, an immediate precursor of our target molecule might be *p*-bromotoluene.

p-Bromotoluene   **p-Bromobenzoic acid**

Next ask yourself, "What is an immediate precursor of *p*-bromotoluene?" Perhaps toluene is an immediate precursor because the methyl group would direct bromination to the ortho and para positions, and the isomeric products could be separated. Alternatively, bromobenzene might be an immediate precursor because we could carry out a Friedel–Crafts methylation and obtain para product. Both answers are satisfactory, although, in view of the difficulties often observed with polyalkylation in Friedel–Crafts reactions, bromination of toluene may be the better route.

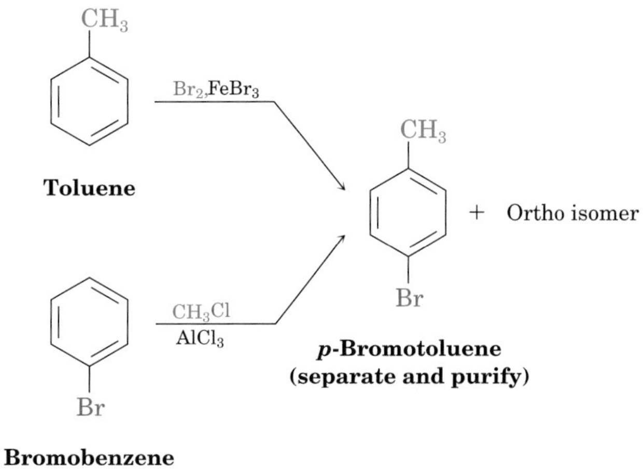

"What is an immediate precursor of toluene?" Benzene, which could be methylated in a Friedel–Crafts reaction.

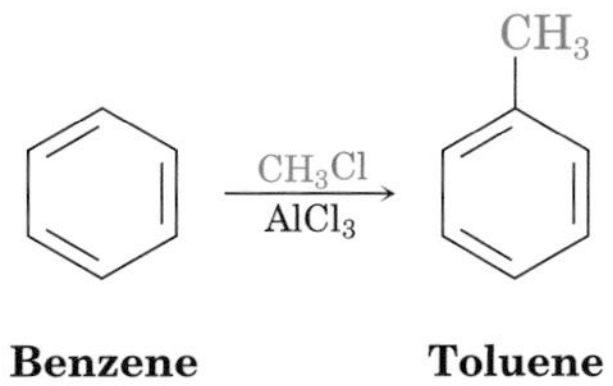

Alternatively, "What is an immediate precursor of bromobenzene?" Benzene, which could be brominated.

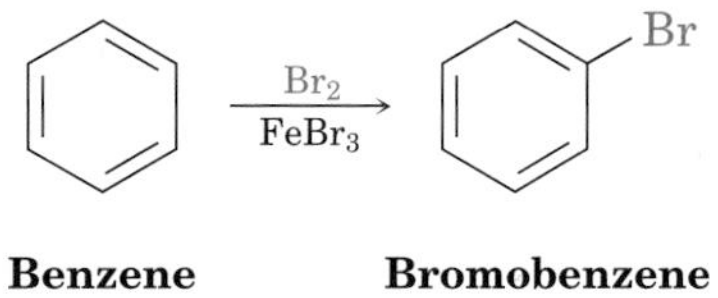

This retrosynthetic analysis has provided two valid routes from benzene to *p*-bromobenzoic acid (Figure 16.23).

FIGURE 16.23 ▼

Two routes for the synthesis of *p*-bromo-benzoic acid from benzene.

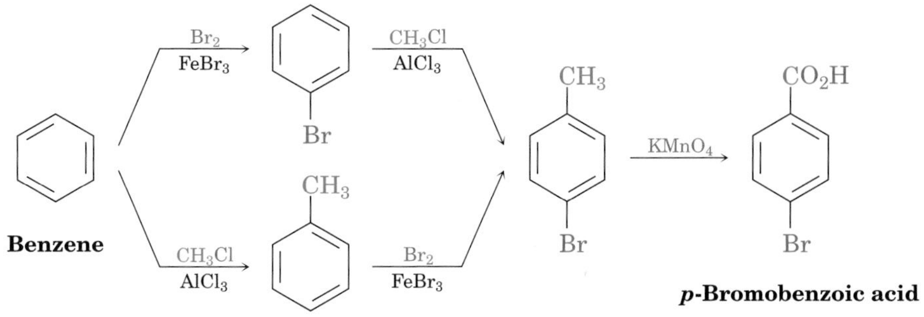

• •

Practice Problem 16.5   Propose a synthesis of 4-chloro-1-nitro-2-propylbenzene from benzene.

Strategy   Draw the target molecule and identify its substituents:

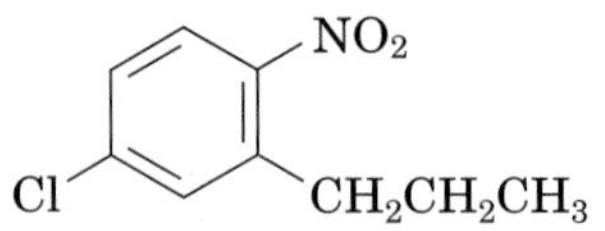

4-Chloro-1-nitro-2-propylbenzene

The three substituents on the ring are a chloro group, a nitro group, and a propyl group. A chloro group can be introduced by chlorination with $Cl_2/FeCl_3$; a nitro group can be introduced by nitration with HNO_3/H_2SO_4; and a propyl group can be introduced by Friedel–Crafts acylation followed by reduction.

Solution   "What is an immediate precursor of the target?" Because the final step will involve introduction of one of three groups—chloro, nitro, or propyl—we have to consider three possibilities. Of the three, we know that chlorination of o-nitropropylbenzene can't be used because the reaction would occur at the wrong position. Similarly, a Friedel–Crafts reaction can't be used as the final step because these reactions don't work on nitro-substituted (deactivated) benzenes. Thus, the immediate precursor of our desired product is probably m-chloropropylbenzene, which can be nitrated. This nitration gives a mixture of product isomers, which must then be separated (Figure 16.24).

FIGURE 16.24 ▼

| Possible routes for the synthesis of 4-chloro-1-nitro-2-propylbenzene. |

p-Chloronitrobenzene   **m-Chloropropylbenzene**   **o-Nitropropylbenzene**

(This deactivated ring will not undergo Friedel–Crafts reaction.) HNO_3 H_2SO_4 (This substance will not give the correct isomer on chlorination.)

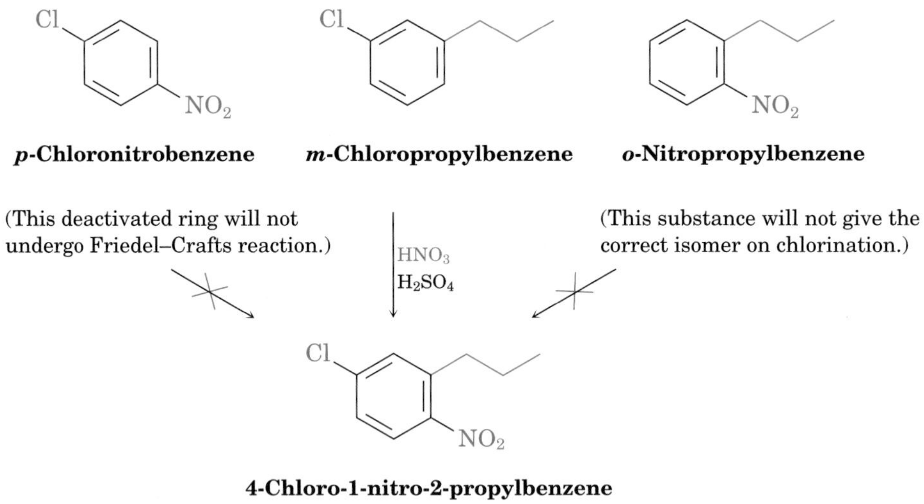

4-Chloro-1-nitro-2-propylbenzene

"What is an immediate precursor of m-chloropropylbenzene?" Because the two substituents have a meta relationship, the first substituent placed on the ring must be a meta director so that the second substitution will take place at the proper position. Furthermore, because primary alkyl groups such as propyl can't be introduced directly by Friedel–Crafts alkylation, the precursor of *m*-chloropropylbenzene is probably *m*-chloropropiophenone, which could be catalytically reduced.

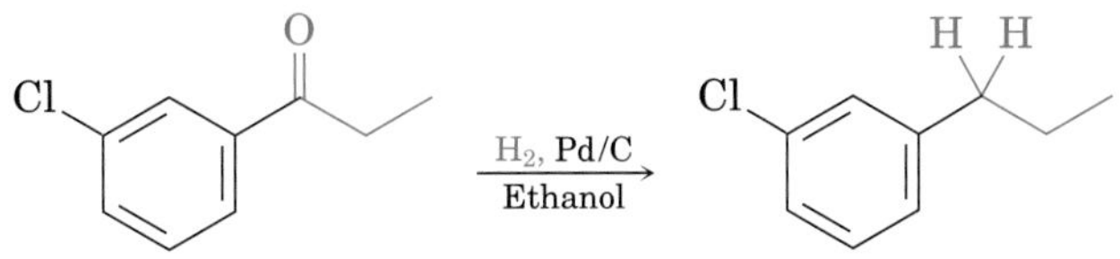

m-Chloropropiophenone   **m-Chloropropylbenzene**

"What is an immediate precursor of *m*-chloropropiophenone?" Propiophenone, which could be chlorinated.

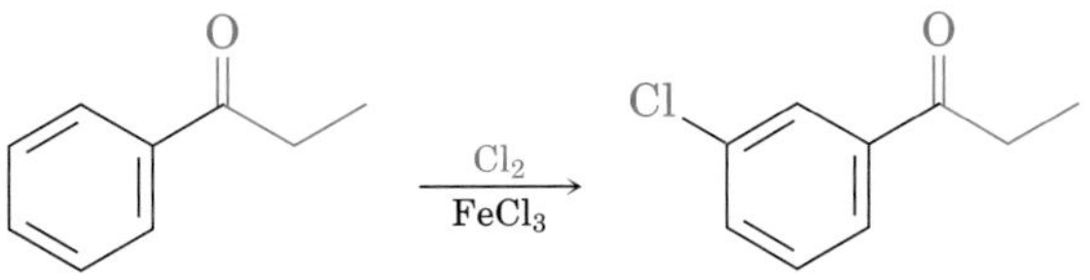

Propiophenone   **m-Chloropropiophenone**

"What is an immediate precursor of propiophenone?" Benzene, which could undergo Friedel–Crafts acylation with propanoyl chloride and AlCl₃.

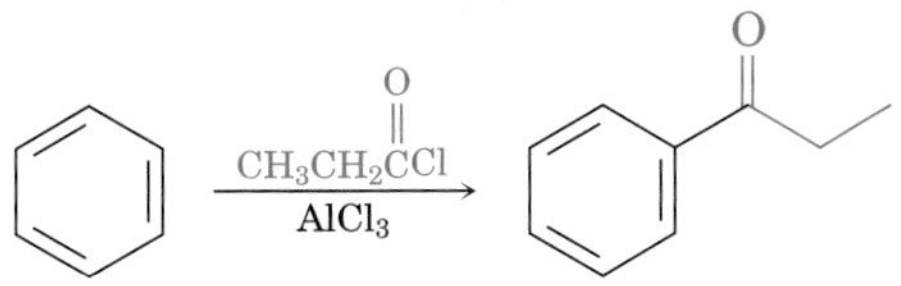

Benzene   **Propiophenone**

The final synthesis is a four-step route from benzene:

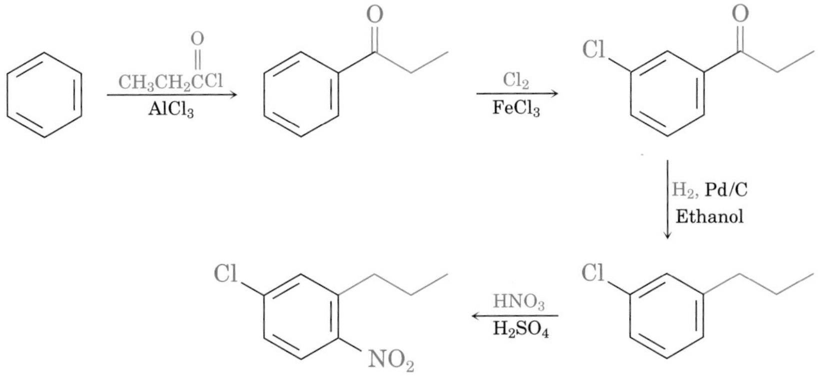

Planning organic syntheses has been compared to playing chess. There are no tricks; all that's required is a knowledge of the allowable moves (the organic reactions) and the discipline to evaluate carefully the consequences of each move. Practicing is not always easy, but there is no surer way to learn organic chemistry.

· ·

Problem 16.22   Propose syntheses of the following substances from benzene:
(a) *m*-Chloronitrobenzene   (b) *m*-Chloroethylbenzene
(c) *p*-Chloropropylbenzene

Problem 16.23   In planning a synthesis, it's as important to know what not to do as to know what to do. As written, the following reaction schemes have flaws in them. What is wrong with each?

(a)

$$\text{(benzonitrile)} \xrightarrow[\text{2. HNO}_3, \text{H}_2\text{SO}_4]{\text{1. CH}_3\text{CH}_2\text{CCl, AlCl}_3} \text{(product)}$$

(b)

$$\text{(chlorobenzene)} \xrightarrow[\text{2. Cl}_2, \text{FeCl}_3]{\text{1. CH}_3\text{CH}_2\text{CH}_2\text{Cl, AlCl}_3} \text{(product)}$$

· ·

CHEMISTRY @ WORK

Combinatorial Chemistry

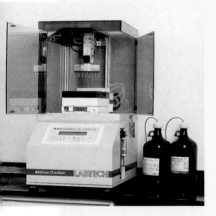

Traditionally, organic compounds have been synthesized one at a time. This works well for preparing large amounts of a few substances, but it doesn't work so well for preparing small amounts of a great many substances. This latter goal is particularly important in the pharmaceutical industry, where vast numbers of structurally similar compounds must be screened to find the optimum drug candidate.

To speed the process of drug discovery, *combinatorial chemistry* has been developed to prepare what are called *combinatorial libraries,* in which anywhere from a few dozen to several hundred thousand substances are prepared simultaneously. Among the early successes of combinatorial chemistry is the development of a benzodiazepine library, a class of aromatic compounds much used as anti-anxiety agents.

Organic chemistry by robot means no spilled flasks!

(continued) ▶

Benzodiazepine library (R₁–R₄ are various organic substituents)

There are two main approaches to combinatorial chemistry—*parallel synthesis* and *split synthesis*. In parallel synthesis, each compound is prepared independently. Typically, a reactant is first linked to the surface of polymer beads, which are then placed into small wells on a 96-well glass plate. Programmable robotic instruments add different sequences of building blocks to the different wells, thereby making 96 different products. When the reaction sequences are complete, the polymer beads are washed and their products are released.

In split synthesis, an initial reactant is again linked to the surface of polymer beads, which are then divided into several groups. A different building block is added to each group, the different groups are combined, and the reassembled mix is again split to form new groups. Another building block is added to each group, the groups are again combined and redivided, and the process continues. If, for example, the beads are divided into four groups at each step, the number of compounds increases in the progression $4 \rightarrow 16 \rightarrow 64 \rightarrow 256$. After 10 steps, more than 1 million compounds have been prepared.

Of course, with so many different products mixed together, the problem is to identify them. What structure is linked to what bead? Several approaches to this problem have been developed, all of which involve the attachment of encoding labels to each polymer bead to keep track of the chemistry each has undergone. Encoding labels thus far have included proteins, nucleic acids, halogenated aromatic compounds, and even computer chips.

The results of split combinatorial synthesis. Assuming that four different building blocks are used at each step, 64 compounds result after 3 steps, and more than 1,000,000 compounds result after 10 steps.

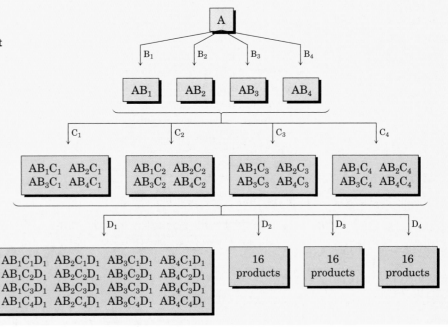

Summary and Key Words

An **electrophilic aromatic substitution reaction** takes place in two steps—initial reaction of an electrophile, E^+, with the aromatic ring, followed by loss of H^+ from the resonance-stabilized carbocation intermediate to regenerate the aromatic ring:

Many variations of the reaction can be carried out, including **halogenation**, **nitration**, and **sulfonation**. **Friedel–Crafts alkylation** and **acylation**, which involve reaction of an aromatic ring with carbocation electrophiles, are particularly useful. Both are limited, however, by the fact that the aromatic ring must be at least as reactive as a halobenzene. In addition, polyalkylation and carbocation rearrangements often occur in Friedel–Crafts alkylation.

Substituents on the benzene ring affect both the reactivity of the ring toward further substitution and the orientation of that substitution. Groups can be classified as **ortho- and para-directing activators, ortho- and para-directing deactivators**, or **meta-directing deactivators**. Substituents influence aromatic rings by a combination of resonance and inductive effects. **Resonance effects** are transmitted through π bonds; **inductive effects** are transmitted through σ bonds.

Halobenzenes undergo **nucleophilic aromatic substitution** through either of two mechanisms. If the halobenzene has a strongly electron-withdrawing substituent in the ortho or para position, substitution occurs by addition of a nucleophile to the ring followed by elimination of halide from the intermediate anion. If the halobenzene is not activated by an electron-withdrawing substituent, substitution can occur by elimination of HX, followed by addition of a nucleophile to the intermediate **benzyne**.

The benzylic position of alkylbenzenes can be brominated by reaction with *N*-bromosuccinimide, and the entire side chain can be degraded to a carboxyl group by oxidation with aqueous $KMnO_4$. Although aromatic rings are less reactive than isolated alkene double bonds, they can be reduced to cyclohexanes by hydrogenation over a platinum or rhodium catalyst. In addition, aryl alkyl ketones are reduced to alkylbenzenes by hydrogenation over a platinum catalyst.

Summary of Reactions

1. Electrophilic aromatic substitution
   (a) Bromination (Section 16.1)

(continued) ▶

(b) Chlorination (Section 16.2)

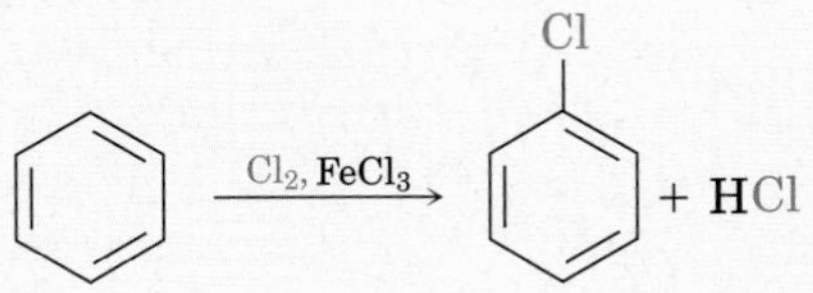
Cl
Cl_2, $FeCl_3$
+ HCl

(c) Iodination (Section 16.2)

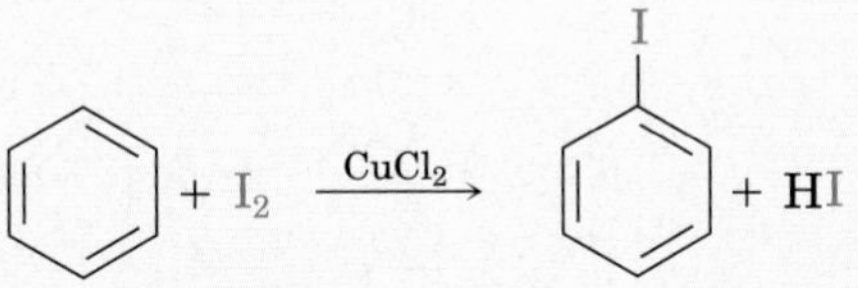
I
+ I_2
$CuCl_2$
+ HI

(d) Nitration (Section 16.2)

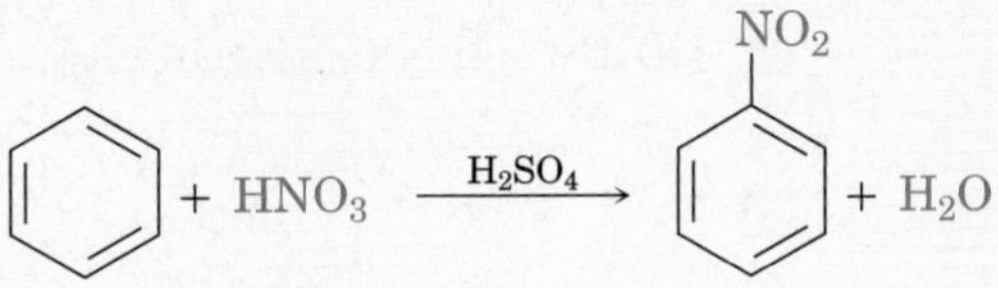
NO_2
+ HNO_3
H_2SO_4
+ H_2O

(e) Sulfonation (Section 16.2)

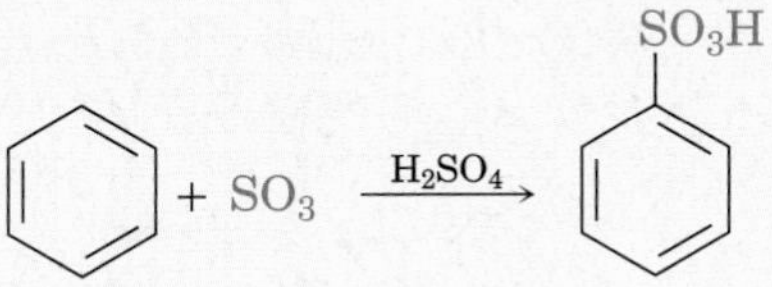
SO_3H
+ SO_3
H_2SO_4

(f) Friedel–Crafts alkylation (Section 16.3)

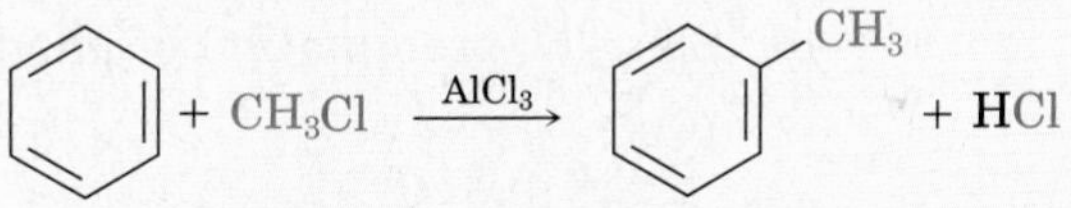
CH_3
+ CH_3Cl
$AlCl_3$
+ HCl

Aromatic ring: Must be at least as reactive as a halobenzene. Deactivated rings do not react.

Alkyl halide: Can be methyl, ethyl, 2°, or 3°; primary halides undergo carbocation rearrangement.

(g) Friedel–Crafts acylation (Section 16.4)

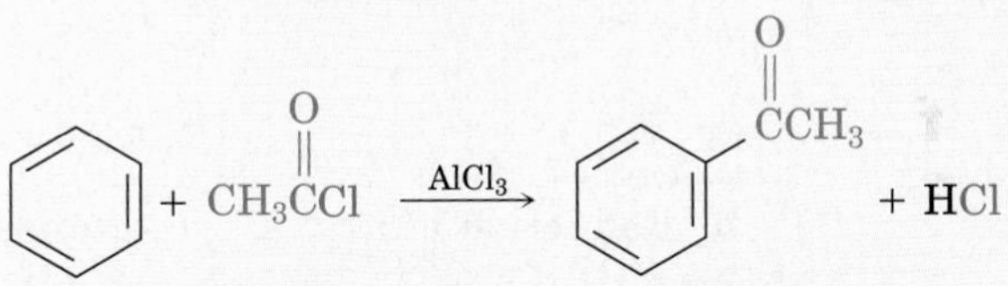
O
+ CH_3CCl
$AlCl_3$
O
CCH_3
+ HCl

2. Reduction of aromatic nitro groups (Section 16.2)

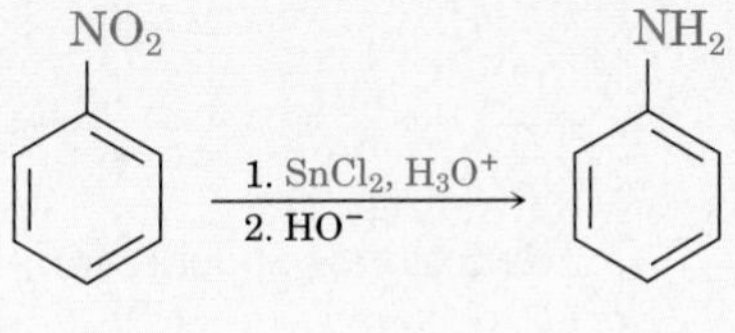
NO_2
1. $SnCl_2$, H_3O^+
2. HO^-
NH_2

(continued) ▶

3. Alkali fusion of aromatic sulfonates (Section 16.2)

4. Nucleophilic aromatic substitution
   (a) By addition/elimination to activated aryl halides (Section 16.8)

   (b) By benzyne intermediate for unactivated aryl halides (Section 16.9)

5. Oxidation of alkylbenzene side chains (Section 16.10)

Reaction occurs with 1° and 2°, but not 3°, alkyl side chains.

6. Benzylic bromination of alkylbenzenes (Section 16.10)

7. Catalytic hydrogenation of aromatic rings (Section 16.11)

8. Reduction of aryl alkyl ketones (Section 16.11)

Reaction is specific for alkyl aryl ketones; dialkyl ketones are not affected.

Visualizing Chemistry

• •

(Problems 16.1–16.23 appear within the chapter.)

16.24 Draw the product from reaction of each of the following substances with (i) Br_2, $FeBr_3$ and (ii) CH_3COCl, $AlCl_3$.

(a)   (b)

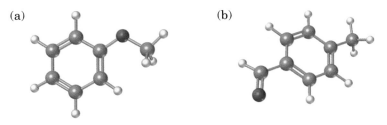

16.25 The following molecular model of acetophenone does *not* represent the lowest-energy, most stable conformation of the molecule. Explain.

Stereo View

16.26 The following molecular model of a dimethyl-substituted biphenyl represents the lowest-energy conformation of the molecule. Why are the two benzene rings not in the same plane so that their *p* orbitals can overlap? Why doesn't complete rotation around the single bond joining the two rings occur?

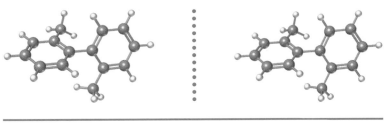

Stereo View

Additional Problems

• •

16.27 Predict the major product(s) of mononitration of the following substances. Which react faster than benzene, and which slower?
(a) Bromobenzene   (b) Benzonitrile   (c) Benzoic acid
(d) Nitrobenzene   (e) Benzenesulfonic acid (f) Methoxybenzene

16.28 Rank the compounds in each group according to their reactivity toward electrophilic substitution.
(a) Chlorobenzene, *o*-dichlorobenzene, benzene
(b) *p*-Bromonitrobenzene, nitrobenzene, phenol

(c) Fluorobenzene, benzaldehyde, *o*-xylene
(d) Benzonitrile, *p*-methylbenzonitrile, *p*-methoxybenzonitrile

16.29 Predict the major monoalkylation products you would expect to obtain from reaction of the following substances with chloromethane and $AlCl_3$:
(a) Bromobenzene   (b) *m*-Bromophenol
(c) *p*-Chloroaniline   (d) 2,4-Dichloronitrobenzene
(e) 2,4-Dichlorophenol   (f) Benzoic acid
(g) *p*-Methylbenzenesulfonic acid   (h) 2,5-Dibromotoluene

16.30 Name and draw the major product(s) of electrophilic monochlorination of the following substances:
(a) *m*-Nitrophenol   (b) *o*-Xylene
(c) *p*-Nitrobenzoic acid   (d) *p*-Bromobenzenesulfonic acid

16.31 Predict the major product(s) you would obtain from sulfonation of the following compounds:
(a) Fluorobenzene   (b) *m*-Bromophenol   (c) *m*-Dichlorobenzene
(d) 2,4-Dibromophenol

16.32 Rank the following aromatic compounds in the expected order of their reactivity toward Friedel–Crafts alkylation. Which compounds are unreactive?
(a) Bromobenzene   (b) Toluene   (c) Phenol
(d) Aniline   (e) Nitrobenzene   (f) *p*-Bromotoluene

16.33 What product(s) would you expect to obtain from the following reactions?

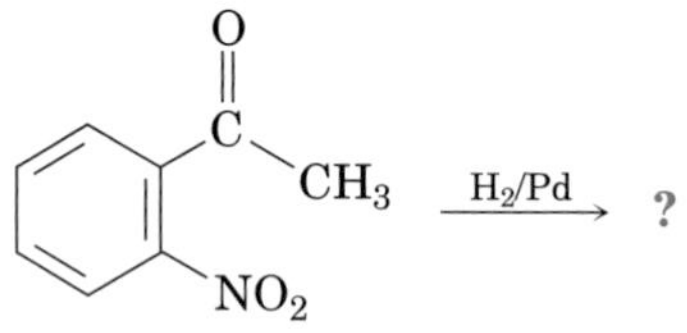

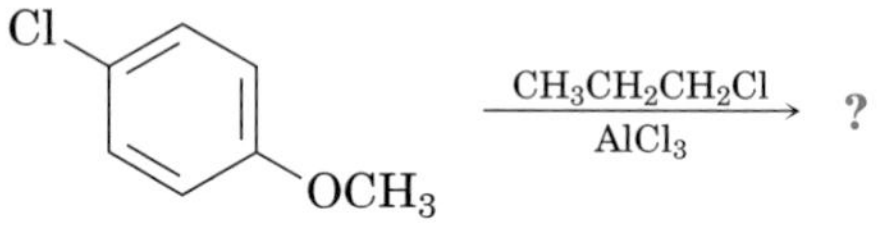

16.34 Aromatic iodination can be carried out with a number of reagents, including iodine monochloride, ICl. What is the direction of polarization of ICl? Propose a mechanism for the iodination of an aromatic ring with ICl.

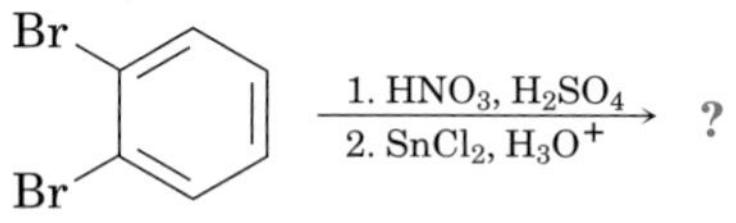

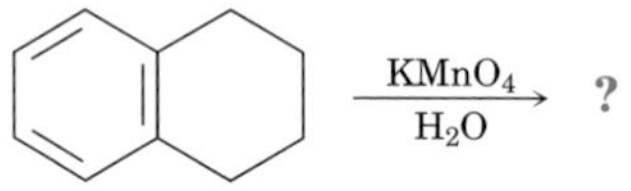

16.35 The sulfonation of an aromatic ring with SO_3 and H_2SO_4 is a reversible reaction. That is, heating benzenesulfonic acid with H_2SO_4 yields benzene. Show the mechanism of the desulfonation reaction. What is the electrophile?

16.36 The carbocation electrophile in a Friedel–Crafts reaction can be generated in ways other than by reaction of an alkyl chloride with $AlCl_3$. For example, reaction of benzene with 2-methylpropene in the presence of H_3PO_4 yields *tert*-butylbenzene. Propose a mechanism for this reaction.

16.37 The *N*,*N*,*N*-trimethylammonium group, $-\overset{+}{N}(CH_3)_3$, is one of the few groups that is a meta-directing deactivator yet has no electron-withdrawing resonance effect. Explain.

16.38 The nitroso group, $-N=O$, is one of the few nonhalogens that is an ortho- and para-directing deactivator. Explain by drawing resonance structures of the carbocation intermediates in ortho, meta, and para electrophilic attack on nitrosobenzene, $C_6H_5N=O$.

16.39 Using resonance structures of the intermediates, explain why bromination of biphenyl occurs at ortho and para positions rather than at meta.

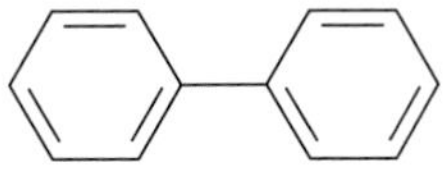

Biphenyl

16.40 At what position and on what ring do you expect nitration of 4-bromobiphenyl to occur? Explain, using resonance structures of the potential intermediates.

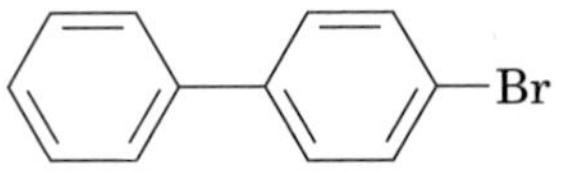

—Br   **4-Bromobiphenyl**

16.41 Electrophilic attack on 3-phenylpropanenitrile occurs at the ortho and para positions, but attack on 3-phenylpropenenitrile occurs at the meta position. Explain, using resonance structures of the intermediates.

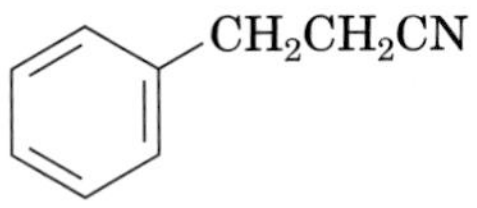

CH$_2$CH$_2$CN   CN

3-Phenylpropanenitrile 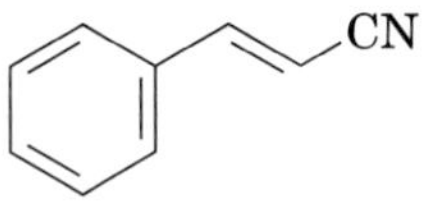   **3-Phenylpropenenitrile**

16.42 Addition of HBr to 1-phenylpropene yields only (1-bromopropyl)benzene. Propose a mechanism for the reaction, and explain why none of the other regioisomer is produced.

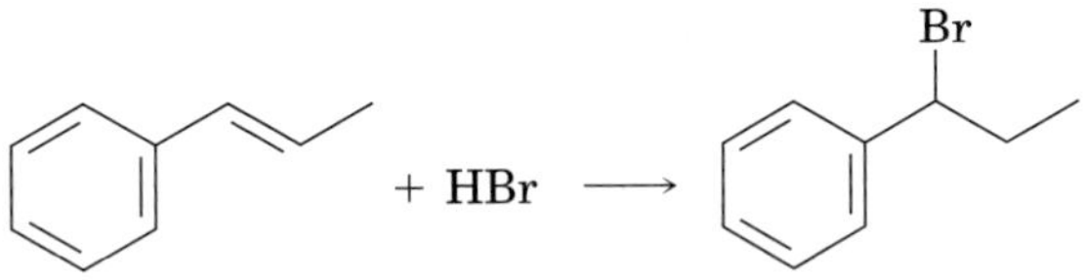

Br

+ HBr ⟶

16.43 Triphenylmethane can be prepared by reaction of benzene and chloroform in the presence of AlCl$_3$. Propose a mechanism for the reaction.

H

+ CHCl$_3$ $\xrightarrow{\text{AlCl}_3}$ C

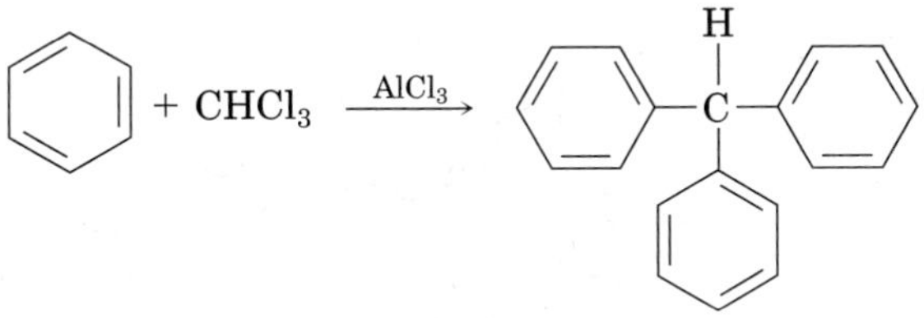

16.44 At what position, and on what ring, would you expect the following substances to undergo electrophilic substitution?

(a) 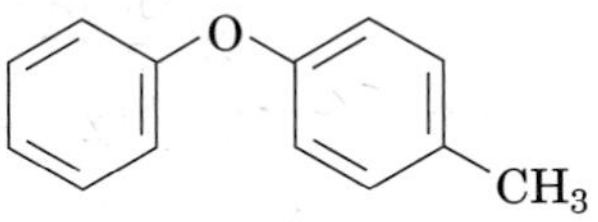 O

CH$_3$

(b) 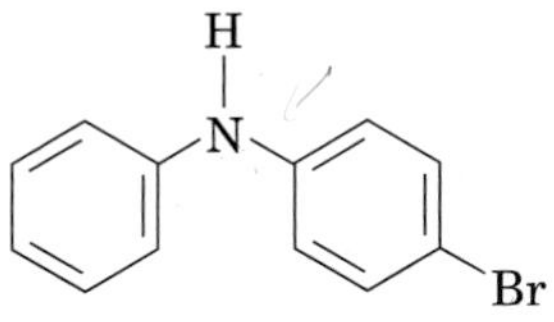 H
     N

Br

(c)

CH$_3$

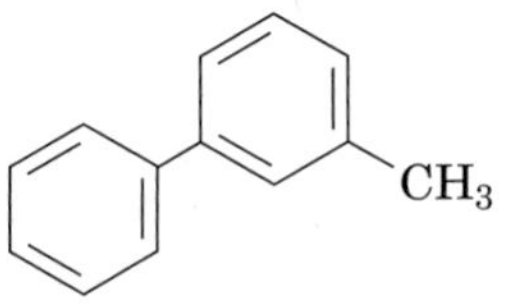

16.45 At what position, and on what ring, would you expect bromination of benzanilide to occur? Explain by drawing resonance structures of the intermediates.

Benzanilide

16.46 Would you expect the Friedel–Crafts reaction of benzene with (R)-2-chlorobutane to yield optically active or racemic product? Explain.

16.47 How would you synthesize the following substances starting from benzene? Assume that ortho- and para-substitution products can be separated.
  (a) o-Methylphenol   (b) 2,4,6-Trinitrophenol
  (c) 2,4,6-Trinitrobenzoic acid   (d) m-Bromoaniline

16.48 Starting with benzene as your only source of aromatic compounds, how would you synthesize the following substances? Assume that you can separate ortho and para isomers if necessary.
  (a) p-Chlorophenol   (b) m-Bromonitrobenzene
  (c) o-Bromobenzenesulfonic acid   (d) m-Chlorobenzenesulfonic acid

16.49 Starting with either benzene or toluene, how would you synthesize the following substances? Assume that ortho and para isomers can be separated.
  (a) 2-Bromo-4-nitrotoluene   (b) 1,3,5-Trinitrobenzene
  (c) 2,4,6-Tribromoaniline   (d) 2-Chloro-4-methylphenol

16.50 As written, the following syntheses have flaws. What is wrong with each?

(a) $\quad$ CH$_3$ $\xrightarrow[\text{2. KMnO}_4]{\text{1. Cl}_2, \text{FeCl}_3}$ CO$_2$H, Cl

(b) $\quad$ Cl $\xrightarrow[\begin{subarray}{l}\text{2. CH}_3\text{Cl, AlCl}_3\\ \text{3. SnCl}_2, \text{H}_3\text{O}^+\\ \text{4. NaOH, H}_2\text{O}\end{subarray}]{\text{1. HNO}_3, \text{H}_2\text{SO}_4}$ Cl, CH$_3$, NH$_2$

(c) $\quad$ CH$_3$ $\xrightarrow[\begin{subarray}{l}\text{2. HNO}_3, \text{H}_2\text{SO}_4\\ \text{3. H}_2/\text{Pd; ethanol}\end{subarray}]{\text{1. CH}_3\overset{\overset{\text{O}}{\|}}{\text{C}}\text{Cl, AlCl}_3}$ CH$_3$, NO$_2$, CH$_2$CH$_3$

16.51 How would you synthesize the following substances starting from benzene?

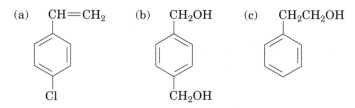

  (a) CH=CH$_2$ (b) CH$_2$OH   (c) CH$_2$CH$_2$OH

  (a) Cl   (b) CH$_2$OH

16.52 The compound MON-0585 is a nontoxic, biodegradable larvicide that is highly selective against mosquito larvae. Synthesize MON-0585 using only benzene as a source of the aromatic rings.

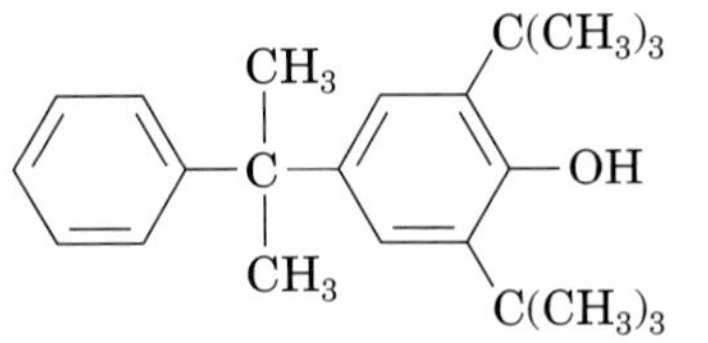

MON-0585

16.53 Hexachlorophene, a substance used in the manufacture of germicidal soaps, is prepared by reaction of 2,4,5-trichlorophenol with formaldehyde in the presence of concentrated sulfuric acid. Propose a mechanism for the reaction.

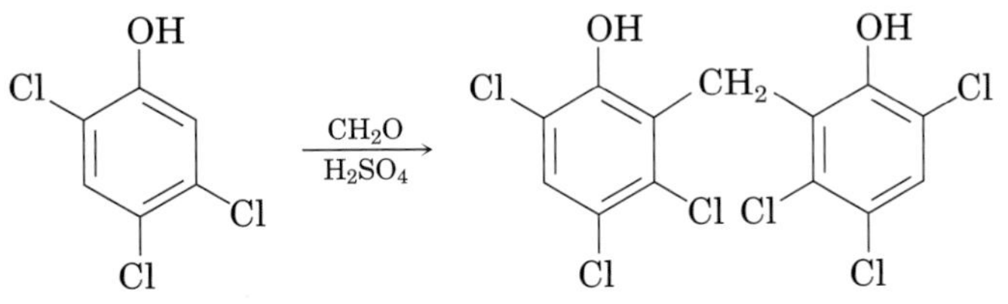

Hexachlorophene

16.54 Benzenediazonium carboxylate decomposes when heated to yield N_2, CO_2, and a reactive substance that can't be isolated. When benzenediazonium carboxylate is heated in the presence of furan, the following reaction is observed:

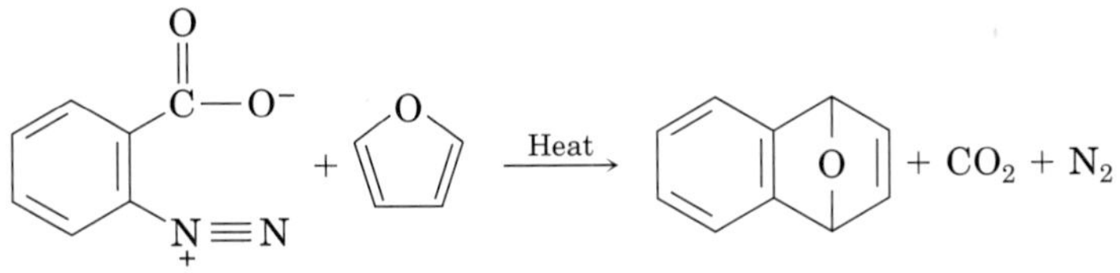

What intermediate is involved in this reaction? Propose a mechanism for its formation.

16.55 Phenylboronic acid, $C_6H_5B(OH)_2$, is nitrated to give 15% ortho-substitution product and 85% meta. Explain the meta-directing effect of the $-B(OH)_2$ group.

16.56 Draw resonance structures of the intermediate carbocations in the bromination of naphthalene, and account for the fact that naphthalene undergoes electrophilic attack at C1 rather than C2.

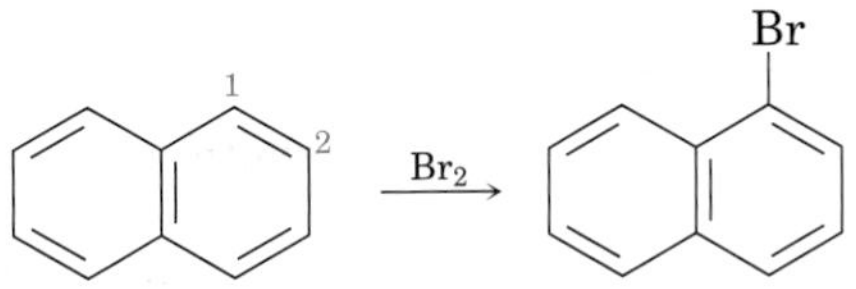

16.57 4-Chloropyridine undergoes reaction with dimethylamine to yield 4-dimethylamino-pyridine. Propose a mechanism for the reaction.

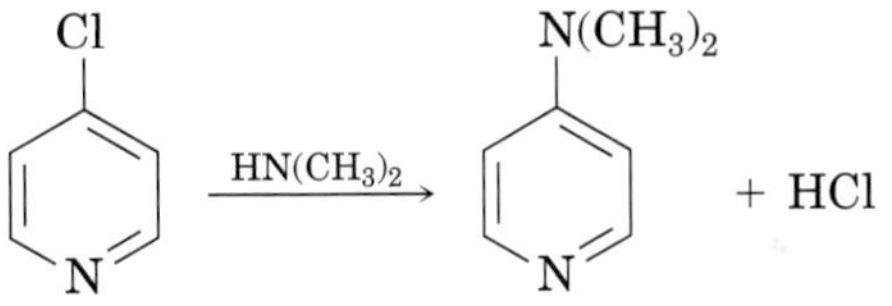

16.58   *p*-Bromotoluene reacts with potassium amide to give a mixture of *m*- and *p*-methyl-aniline. Explain.

16.59   Propose a synthesis of aspirin (acetylsalicylic acid) starting from benzene. You will need to use an *acetylation reaction* at some point in your scheme.

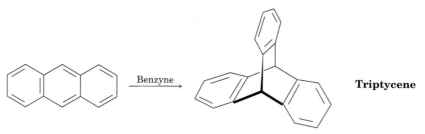

Aspirin

16.60   Propose a mechanism to account for the reaction of benzene with 2,2,5,5-tetra-methyltetrahydrofuran.

16.61   In the *Gatterman–Koch reaction,* a formyl group (–CHO) is introduced directly onto a benzene ring. For example, reaction of toluene with CO and HCl in the presence of mixed CuCl/AlCl₃ gives *p*-methylbenzaldehyde. Propose a mechanism.

16.62   Triptycene is an unusual molecule that has been prepared by reaction of benzyne with anthracene. What kind of reaction is involved? Show the mechanism.

Triptycene

16.63   Treatment of *p-tert*-butylphenol with a strong acid such as H₂SO₄ yields phenol and 2-methylpropene. Propose a mechanism.

16.64   Benzene and alkyl-substituted benzenes can be hydroxylated by reaction with H₂O₂ in the presence of an acidic catalyst. What is the structure of the reactive electrophile? Propose a mechanism for the reaction.

16.65 How would you synthesize the following compounds from benzene? Assume that ortho and para isomers can be separated.

(a)

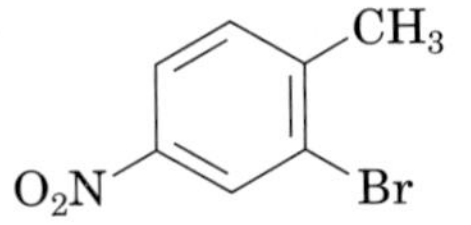

(b)

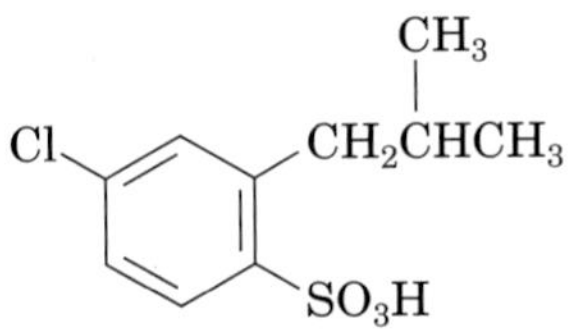

16.66 You know the mechanism of HBr addition to alkenes, and you know the effects of various substituent groups on aromatic substitution. Use this knowledge to predict which of the following two alkenes reacts faster with HBr. Explain your answer by drawing resonance structures of the carbocation intermediates.

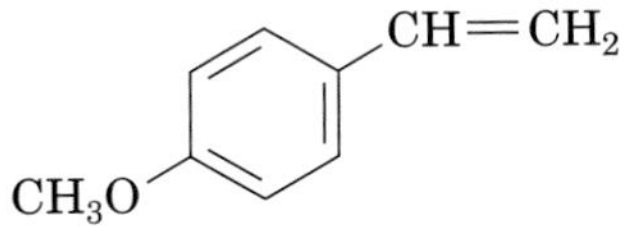

and

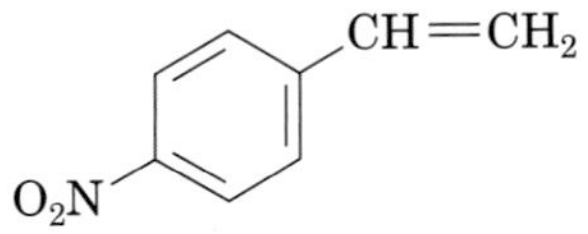

16.67 Draw a Fischer projection of (R)-2-phenylbutane, and predict the stereochemistry of its reaction with N-bromosuccinimide.

16.68 Benzyl bromide is converted into benzaldehyde by heating in dimethyl sulfoxide. Propose a structure for the intermediate, and show the mechanisms of the two steps in the reaction.

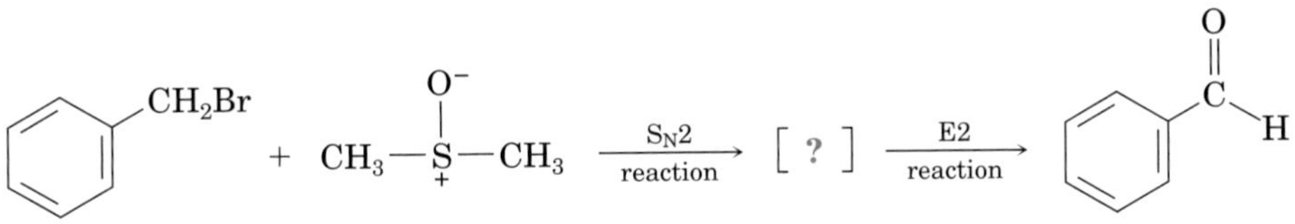

16.69 Use your knowledge of directing effects, along with the following data, to deduce the directions of the dipole moments in aniline and bromobenzene.

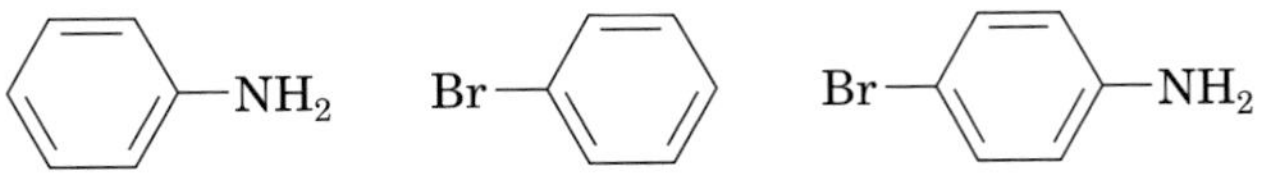

$\mu = 1.53\ D$ $\qquad$ $\mu = 1.52\ D$ $\qquad$ $\mu = 2.91\ D$

16.70 Identify the reagents represented by the letters a–e in the following scheme:

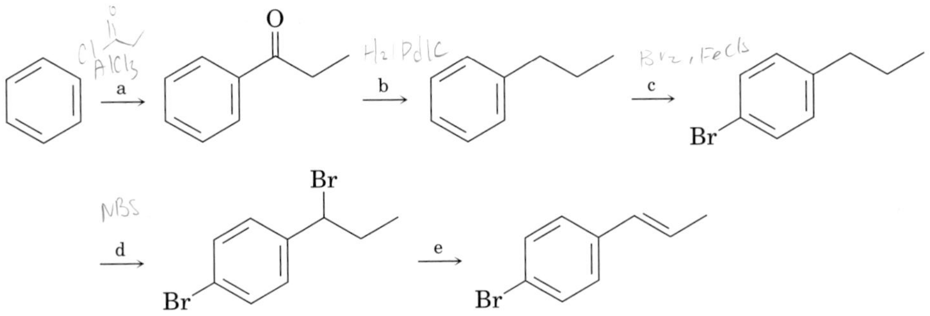

A Look Ahead

· ·

16.71   Phenols (ArOH) are relatively acidic, and the presence of a substituent group on the aromatic ring has a large effect. The pK_a of unsubstituted phenol, for example, is 9.89, while that of *p*-nitrophenol is 7.15. Draw resonance structures of the corresponding phenoxide anions and explain the data. (See Section 17.3.)

16.72   Would you expect *p*-methylphenol to be more acidic or less acidic than unsubstituted phenol? Explain. (See Problem 16.71.)

16.73   One method for determining the sequence of amino acids in a large protein molecule involves treatment of the protein with *Sanger's reagent*, 2,4-dinitrofluorobenzene. The reaction involves the –NH₂ group at the end of the protein chain. Predict the product, and tell what kind of reaction is taking place. (See Sections 26.8–26.10.)

2,4-Dinitrofluorobenzene   **A protein**

Molecular Modeling

· ·

16.74   Intramolecular Friedel–Crafts acylation can be used to make rings. Use Spartan-Build to build and minimize the ortho, meta, and para acylation products from the following reaction. Which product has the lowest strain energy? How do the higher-energy isomers reveal the presence of strain?

16.75   2,4,6-Trinitrotoluene (TNT) is made by multiple nitration of toluene. Use Spartan-View to compare electrostatic potential maps of toluene, 4-nitrotoluene, and 2,4-dinitrotoluene. Which substance has the most negative ring, and which the least? Are the second and third nitrations likely to be more or less difficult than the first one?

16.76   Bromination of *o*-methoxytoluene might give either of the two products shown. Use SpartanView to compare the energies of the carbocation intermediates in each reaction, and predict which product is likely to form faster.

o-Methoxytoluene

16.77   Use SpartanView to examine electrostatic potential maps of the intermediate Meisenheimer complexes in the reactions of *o*-fluoronitrobenzene and *m*-fluoronitrobenzene with CH₃O⁻. Which atoms in each complex are most negatively charged? Which reaction is more likely to occur?

A Brief Review of Organic Reactions

When you learn arithmetic, you have to memorize multiplication tables. When you learn a foreign language, you have to memorize vocabulary words. And when you learn organic chemistry, you have to know by memory a large number of reactions. The way to simplify the job, of course, is to organize the material. Just as the rules of grammar organize the words in a foreign language, *mechanisms* organize the reactions in organic chemistry. With our coverage of organic chemistry now half complete, this is a good time to review the reactions we've seen in the past several chapters and the common mechanisms that explain them. In upcoming chapters, particularly Chapters 19–23, which discuss the chemistry of carbonyl compounds, we'll see several more fundamental types of mechanisms.

I. A Summary of the Kinds of Organic Reactions

There are four main kinds of reactions: additions, eliminations, substitutions, and rearrangements. We've now seen examples of all four, as summarized in Review Tables 1–4.

Review Table 1: Some Addition Reactions

1. Additions to alkenes
   (a) Electrophilic addition of HX (X = Cl, Br, I; Sections 6.8 and 6.9)

$$CH_3CH{=}CH_2 \xrightarrow[\text{Ether}]{\text{HCl}} CH_3\overset{\overset{\displaystyle Cl}{|}}{C}HCH_3$$

   (b) Electrophilic addition of X_2 (X = Cl, Br; Section 7.2)

$$CH_3CH{=}CHCH_3 \xrightarrow[\text{CH}_2\text{Cl}_2]{\text{Br}_2} CH_3\overset{\overset{\displaystyle Br}{|}}{C}H{-}\overset{\overset{\displaystyle Br}{|}}{C}HCH_3$$

   (c) Electrophilic addition of HO–X (X = Cl, Br, I; Section 7.3)

$$CH_3CH{=}CH_2 \xrightarrow[\text{NaOH, H}_2\text{O}]{\text{Br}_2} CH_3\overset{\overset{\displaystyle OH}{|}}{C}HCH_2Br$$

   (d) Electrophilic addition of water by oxymercuration (Section 7.4)

$$CH_3CH{=}CH_2 \xrightarrow[\text{H}_2\text{O}]{\text{Hg(OAc)}_2} CH_3\overset{\overset{\displaystyle OH}{|}}{C}HCH_2HgOAc \xrightarrow{\text{NaBH}_4} CH_3\overset{\overset{\displaystyle OH}{|}}{C}HCH_3$$

(continued) ▶

Review Table 1 (*continued*)

(e) Addition of BH_3 (hydroboration; Section 7.5)

$$CH_3CH\!=\!CH_2 \xrightarrow{BH_3} CH_3\overset{\overset{\displaystyle H}{|}}{C}HCH_2BH_2 \xrightarrow[\text{NaOH}]{H_2O_2} CH_3CH_2CH_2OH$$

(f) Catalytic addition of H_2 (Section 7.7)

$$CH_3CH\!=\!CH_2 \xrightarrow[\text{Pd/C catalyst}]{H_2} CH_3\overset{\overset{\displaystyle H}{|}}{C}H\!-\!\overset{\overset{\displaystyle H}{|}}{C}H_2$$

(g) Hydroxylation with OsO_4 (Section 7.8)

$$CH_3CH\!=\!CH_2 \xrightarrow[\text{2. NaHSO}_3]{\text{1. OsO}_4} CH_3\overset{\overset{\displaystyle OH}{|}}{C}HCH_2OH$$

(h) Addition of carbenes; cyclopropane formation (Section 7.6)

$$CH_3CH\!=\!CH_2 \xrightarrow{CH_2I_2,\ Zn(Cu)} CH_3\overset{\overset{\displaystyle CH_2}{\diagup\ \ \diagdown}}{CH\!-\!CH_2}$$

(i) Cycloaddition: Diels–Alder reaction (Sections 14.8 and 14.9)

$$H_2C\!=\!CH\!-\!CH\!=\!CH_2 + H_2C\!=\!CHCOCH_3 \longrightarrow$$

(j) Radical addition (Section 7.10)

$$\diagdown\!\!\!\!-CH_2\dot{C}H_2 \xrightarrow{H_2C=CH_2} \diagdown\!\!\!\!-CH_2CH_2CH_2\dot{C}H_2$$

2. Additions to alkynes
   (a) Electrophilic additions of HX (X = Cl, Br, I; Section 8.4)

$$CH_3C\!\equiv\!CH \xrightarrow[\text{Ether}]{HBr} CH_3\overset{\overset{\displaystyle Br}{|}}{C}\!=\!CH_2$$

(b) Electrophilic addition of H_2O (Section 8.5)

$$CH_3C\!\equiv\!CH \xrightarrow[\text{HgSO}_4]{H_3O^+} CH_3\overset{\overset{\displaystyle OH}{|}}{C}\!=\!CH_2 \longrightarrow CH_3\overset{\overset{\displaystyle O}{\|}}{C}CH_3$$

(c) Addition of H_2 (Section 8.6)

$$CH_3C\!\equiv\!CCH_3 \xrightarrow[\text{Lindlar catalyst}]{H_2} CH_3\overset{\overset{\displaystyle H}{|}}{C}\!=\!\overset{\overset{\displaystyle H}{|}}{C}CH_3$$

Review Table 2: Some Elimination Reactions

1. Dehydrohalogenation of alkyl halides (Section 11.10)

$$\underset{\underset{\text{Br}}{|}}{CH_3CHCH_3} \xrightarrow{\text{KOH}} CH_3CH=CH_2 + KBr + H_2O$$

2. Dehydrohalogenation of vinylic halides (Section 8.3)

$$\underset{\underset{\text{Br}}{|}}{CH_3C}=CH_2 \xrightarrow{\text{NaNH}_2} CH_3C\equiv CH + NaBr + NH_3$$

3. Dehydrohalogenation of aryl halides: benzyne formation (Section 16.9)

Review Table 3: Some Substitution Reactions

1. S_N2 reactions of primary alkyl halides
   (a) General reaction (Sections 11.2–11.5)

$$CH_3CH_2CH_2X + :Nu^- \longrightarrow CH_3CH_2CH_2Nu + :X^-$$

   where   X = Cl, Br, I, OTos
   :Nu$^-$ = CH$_3$O$^-$, HO$^-$, CH$_3$S$^-$, HS$^-$, CN$^-$, CH$_3$COO$^-$,
   NH$_3$, (CH$_3$)$_3$N, etc.

   (b) Alkyne alkylation (Section 8.9)

$$CH_3C\equiv C:^- \; Na^+ \xrightarrow{\text{CH}_3\text{I}} CH_3C\equiv CCH_3 + Na^+ \, I^-$$

2. S_N1 reactions of tertiary alkyl halides
   (a) General reaction (Sections 11.6–11.9)

$$(CH_3)_3CX \xrightarrow{:NuH} (CH_3)_3CNu + HX$$

   (b) Preparation of alkyl halides from alcohols (Section 10.7)

$$(CH_3)_3COH \xrightarrow{\text{HBr}} (CH_3)_3CBr + H_2O$$

3. Electrophilic aromatic substitution (Sections 16.1–16.4)
   (a) Halogenation of aromatic compounds (Section 16.1)

(continued) ▶

Review Table 3 (continued)

(b) Nitration of aromatic compounds (Section 16.2)

$$\text{Ar}-H \xrightarrow[\text{H}_2\text{SO}_4]{\text{HNO}_3} \text{Ar}-NO_2 + H_2O$$

(c) Sulfonation of aromatic compounds (Section 16.2)

$$\text{Ar}-H \xrightarrow[\text{H}_2\text{SO}_4]{\text{SO}_3} \text{Ar}-SO_3H$$

(d) Alkylation of aromatic rings (Section 16.3)

$$\text{Ar}-H \xrightarrow[\text{AlCl}_3]{\text{CH}_3\text{Cl}} \text{Ar}-CH_3 + HCl$$

(e) Acylation of aromatic rings (Section 16.4)

$$\text{Ar}-H \xrightarrow[\text{AlCl}_3]{\text{CH}_3\text{COCl}} \text{Ar}-COCH_3 + HCl$$

4. Nucleophilic aromatic substitution (Section 16.8)

$$O_2N-\text{Ar}-Cl \xrightarrow[\text{H}_2\text{O}]{\text{NaOH}} O_2N-\text{Ar}-OH + NaCl$$

5. Radical substitution reactions
   (a) Chlorination of methane (Section 10.4)

$$CH_4 + Cl_2 \xrightarrow{\text{Light}} CH_3Cl + HCl$$

(b) NBS allylic bromination of alkenes (Section 10.5)

$$CH_3CH{=}CH_2 \xrightarrow[\text{Radicals}]{\text{NBS}} BrCH_2CH{=}CH_2$$

Review Table 4: Some Rearrangement Reactions

1. Carbocation rearrangement during electrophilic addition to alkenes (Section 6.12)

$$(CH_3)_2\overset{\text{H}}{\underset{|}{C}}CH{=}CH_2 + HCl \longrightarrow (CH_3)_2\overset{\text{Cl}}{\underset{|}{C}}CH_2CH_3$$

2. Carbocation rearrangement during Friedel–Crafts alkylation (Section 16.3)

$$\text{Ar} \xrightarrow[\text{AlCl}_3]{\text{CH}_3\text{CH}_2\text{CH}_2\text{Cl}} \text{Ar}-CH(CH_3)_2$$

II. A Summary of Organic Reaction Mechanisms

As we said in Chapter 5, there are three fundamental reaction types: *polar reactions, radical reactions,* and *pericyclic reactions.* Let's review each to see how the reactions we've covered fit the different categories.

A. Polar Reactions

Polar reactions take place between electron-rich reagents (nucleophiles/Lewis bases) and electron-poor reagents (electrophiles/Lewis acids). These reactions are heterolytic processes and involve species with an even number of electrons. Bonds are made when a nucleophile donates an electron pair to an electrophile; bonds are broken when one product leaves with an electron pair.

Heterogenic bond formation

$$A^+ \quad + \quad :B^- \longrightarrow A : B$$

Electrophile   Nucleophile

Heterolytic bond cleavage

$$A : B \longrightarrow A^+ + :B^-$$

The polar reactions we've studied can be grouped into five categories:

1. Electrophilic addition reactions
2. Elimination reactions
3. Nucleophilic alkyl substitution reactions
4. Electrophilic aromatic substitution reactions
5. Nucleophilic aromatic substitution reactions

1. Electrophilic Addition Reactions (Sections 6.8 and 6.9; Review Table 1, reactions 1a–1d, 2a–2b)

Alkenes react with electrophiles such as HBr to yield saturated addition products. The reaction occurs in two steps. The electrophile first adds to the alkene double bond to yield a carbocation intermediate, which reacts further to yield the addition product.

Alkene   Carbocation   Addition product

Many of the addition reactions listed in Review Table 1 take place by an electrophilic addition mechanism. The electrophile can be H^+, X^+, Hg^{2+}, but the basic process is the same. The remaining addition reactions in Review Table 1 occur by other mechanisms.

2. Elimination Reactions

(a) **E2 Reaction** (Sections 11.11–11.13; Review Table 2, reactions 1–3)
Alkyl halides undergo elimination of HX to yield alkenes on treatment
with base. When a strong base such as hydroxide ion (HO^-), alkoxide ion
(RO^-), or amide ion (NH_2^-) is used, alkyl halides react by the E2 mechanism. E2 reactions occur in a single step involving removal by base of
a neighboring hydrogen at the same time that the halide ion is leaving:

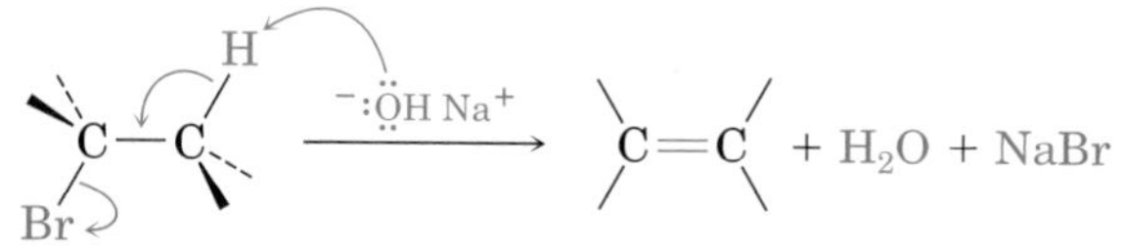

All the elimination reactions listed in Review Table 2 occur by the same
E2 mechanism. Though they appear different, the elimination of an
alkyl halide to yield an alkene (reaction 1), the elimination of a vinylic
halide to yield an alkyne (reaction 2), and the elimination of an aryl
halide to yield a benzyne (reaction 3) are all E2 reactions.

(b) **E1 Reaction** (Section 11.14) Tertiary alkyl halides undergo elimination by the E1 mechanism in competition with S_N1 substitution when
a nonbasic nucleophile is used in a hydroxylic solvent. The reaction
takes place in two steps: spontaneous dissociation of the alkyl halide,
followed by loss of H^+ from the carbocation intermediate:

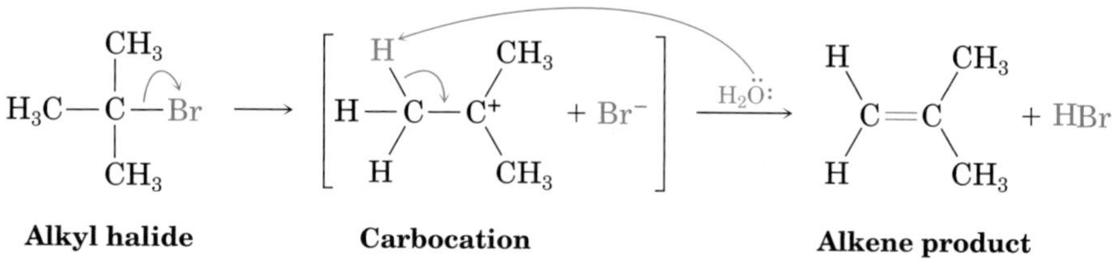

| Alkyl halide | Carbocation | Alkene product |

3. Nucleophilic Alkyl Substitution Reactions

(a) **S_N2 Reaction** (Sections 11.2–11.5; Review Table 3, reaction 1a) The
nucleophilic alkyl substitution reaction is one of the most common reactions encountered in organic chemistry. As illustrated in reaction 1a of
Review Table 3, most primary halides and tosylates, and some secondary ones, undergo substitution reactions with a variety of different
nucleophiles. Mechanistically, S_N2 reactions take place in a single step
involving attack of the incoming nucleophile from a direction 180° away
from the leaving group. This results in an umbrella-like inversion of
stereochemistry (Walden inversion).

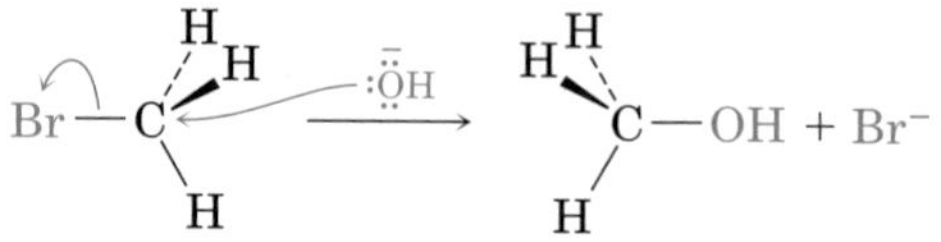

(b) **S$_N$1 Reaction**   (Sections 11.6–11.9; Review Table 3, reaction 2a)
Tertiary alkyl halides undergo nucleophilic substitution by the two-step S$_N$1 mechanism. Spontaneous dissociation of the alkyl halide to a carbocation intermediate takes place, followed by reaction of the carbocation with a nucleophile. The dissociation step is the slower of the two and is rate-limiting.

| Alkyl halide | Carbocation | Substitution product |

4. Electrophilic Aromatic Substitution Reactions   (Sections 16.1–16.4; Review Table 3, reaction 3)

All the electrophilic aromatic substitutions shown in reaction 3 of Review Table 3 occur by the same two-step mechanism. The first step is similar to the first step in electrophilic addition to alkenes: An electron-poor reagent reacts with the electron-rich aromatic ring. The second step is identical to what happens during E2 elimination: A base abstracts a hydrogen atom next to the positively charged carbon, and elimination of the proton occurs.

| Benzene | Carbocation | Bromobenzene |

5. Nucleophilic Aromatic Substitution Reactions   (Section 16.8; Review Table 3, reaction 4)

Nucleophilic aromatic substitution (reaction 4 in Review Table 3) occurs by addition of a nucleophile to an electrophilic aromatic ring, followed by elimination of the leaving group. The ring is made electrophilic, and hence reactive, only when substituted by strong electron-withdrawing groups such as nitro, cyano, and carbonyl.

B. Radical Reactions

Radical reactions are homolytic processes, which involve species with an odd number of electrons. Bonds are made when each reactant donates one electron, and bonds are broken when each product fragment leaves with one electron.

| **Homogenic bond formation** | $A \cdot + \cdot B \longrightarrow A : B$ |
| **Homolytic bond cleavage** | $A : B \longrightarrow A \cdot + \cdot B$ |

We've seen only a few examples of radical reactions because they're less common than polar reactions. Those we have studied can be classified as either radical addition reactions or radical substitution reactions. Radical additions, such as the benzoyl peroxide-catalyzed polymerization of alkene monomers (Review Table 1, reaction 1j), involve the addition of a radical to an unsaturated substrate. The reaction occurs through three kinds of steps, all of which involve odd-electron species: (1) initiation, (2) propagation, and (3) termination.

Initiation

Propagation

Termination

The reaction is initiated by homolytic cleavage of benzoyl peroxide to give two benzoyloxy radicals (BzO·). These radicals add to the alkene monomer, generating a new carbon radical and a C–O bond. The carbon radical then adds to another alkene monomer, which continues the chain.

Radical substitution reactions, such as the light-induced chlorination of methane and the allylic bromination of alkenes with *N*-bromosuccinimide (Review Table 3, reaction 5), are also common. The key step in all these reactions is that a radical abstracts an atom from a neutral molecule, leaving a new radical.

C. Pericyclic Reactions

Pericyclic reactions, such as the addition of a carbene to an alkene and the Diels–Alder cycloaddition (Review Table 1, reactions 1h and 1i), involve neither radicals nor nucleophile–electrophile interactions. Rather, these processes take place in a single step by a reorganization of bonding electrons through a cyclic transition state. We'll look at these reactions more closely in Chapter 30.

| 1,3-Butadiene | Methyl propenoate | Transition state | Methyl 3-cyclohexene-carboxylate |

Problems

1. We've seen in this brief review that reactions can be organized according to the mechanism by which they occur. Another way of organizing reactions is according to the product they form. List ways to synthesize the following kinds of products. Check your answers in the Appendix on Functional-Group Synthesis in the accompanying *Study Guide and Solutions Manual.*
   (a) Alkanes (3 ways)   (b) Alkenes (4 ways)
   (c) Alkynes (2 ways)   (d) Alkyl halides (8 ways)

2. Yet a third way to organize reactions is according to the functional group of the reactant. List reactions of the following functional groups. Check your answers in the Appendix on Functional-Group Reactions in the accompanying *Study Guide and Solutions Manual.*
   (a) Alkanes (2 reactions)   (b) Alkenes (10 reactions)
   (c) Alkynes (6 reactions)   (d) Aromatic compounds (10 reactions)
   (e) Alkyl halides (4 reactions)

3. List at least one use of each of the following reagents. Check your answers in the Appendix on Reagents in Organic Chemistry in the accompanying *Study Guide and Solutions Manual.*
   (a) Periodic acid, HIO_4 (b) Bromine, Br_2 (c) Chromium trioxide, CrO_3
   (d) Cuprous iodide, CuI   (e) Lithium, Li   (f) Diiodomethane, CH_2I_2
   (g) Ferric bromide, $FeBr_3$ (h) Ammonia, NH_3 (i) Thionyl chloride, $SOCl_2$
   (j) Magnesium, Mg   (k) Borane, BH_3 (l) Hydrogen peroxide, H_2O_2

4. The name of the discoverer is often associated with a major advance in organic chemistry. Give an example of each of the following reagents or reactions. Check your answers in the Appendix on Name Reactions in the accompanying *Study Guide and Solutions Manual.*
   (a) Grignard reagent   (b) Gilman reagent   (c) Diels–Alder reaction
   (d) Simmons–Smith reaction   (e) Walden inversion   (f) Friedel–Crafts reaction

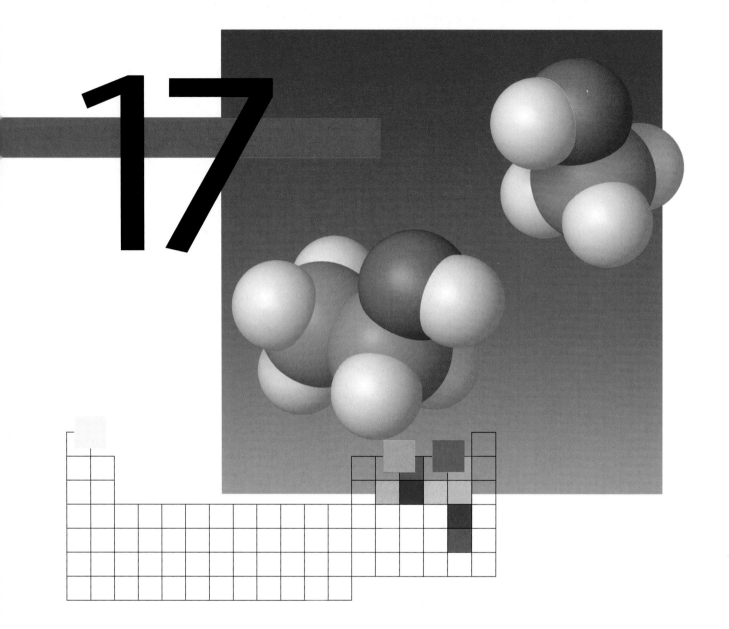

Alcohols and Phenols

Alcohols are compounds that have hydroxyl groups bonded to saturated, sp^3-hybridized carbon atoms, while **phenols** are compounds that have hydroxyl groups bonded to aromatic rings. Both can be thought of as organic derivatives of water in which one of the water hydrogens is replaced by an organic group: H–O–H versus R–O–H or Ar–O–H. Note that *enols,* compounds with an –OH group bonded to a vinylic carbon, are purposely excluded from discussion in this chapter because their chemistry is so different. We'll look at enols in Chapter 22.

An alcohol   **A phenol**   **An enol**

Alcohols occur widely in nature and have many industrial and pharmaceutical applications. Methanol and ethanol, for instance, are two of the most important of all industrial chemicals. Prior to the development of the modern chemical industry, methanol was prepared by heating wood in the absence of air and thus came to be called *wood alcohol*. Today, approximately 1.7 billion gallons of methanol are manufactured each year in the United States by catalytic reduction of carbon monoxide with hydrogen gas:

$$CO + 2\,H_2 \xrightarrow[\text{Zinc oxide/chromia}]{400°C} CH_3OH$$

Methanol is toxic to humans, causing blindness in low doses (15 mL) and death in larger amounts (100–250 mL). Industrially, it is used both as a solvent and as a starting material for production of formaldehyde (CH_2O), acetic acid (CH_3COOH), and the gasoline additive methyl *tert*-butyl ether [MTBE, $CH_3OC(CH_3)_3$].

Ethanol was one of the first organic chemicals to be prepared and purified. Its production by fermentation of grains and sugars has been carried out for millennia, and its purification by distillation goes back at least as far as the twelfth century. Ethanol for nonbeverage use is obtained by acid-catalyzed hydration of ethylene. Approximately 110 million gallons of ethanol a year are produced in the United States for use as a solvent or as a chemical intermediate in other industrial reactions.

$$H_2C=CH_2 + H_2O \xrightarrow[250°C]{H_3PO_4} CH_3CH_2OH$$

Phenols occur widely throughout nature and also serve as intermediates in the industrial synthesis of products as diverse as adhesives and antiseptics. Phenol itself is a general disinfectant found in coal tar; methyl salicylate is a flavoring agent found in oil of wintergreen; and the urushiols are the allergenic constituents of poison oak and poison ivy. Note that the word *phenol* is the name both of a specific compound and of a class of compounds.

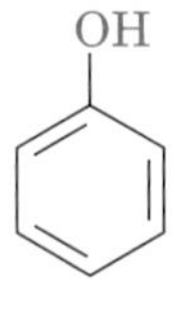

Phenol
(also known as
carbolic acid)

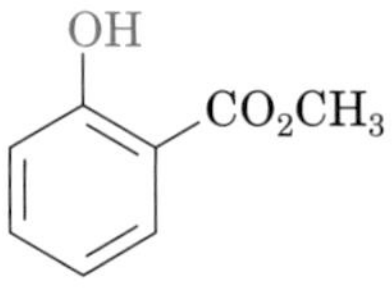

Methyl salicylate

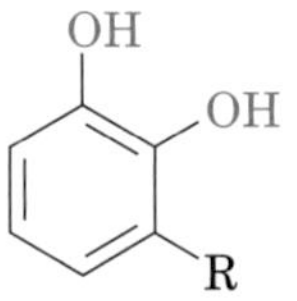

Urushiols
(R = different C_{15} alkyl
and alkenyl chains)

17.1 Naming Alcohols and Phenols

Naming Alcohols

Alcohols are classified as primary (1°), secondary (2°), or tertiary (3°), depending on the number of organic groups bonded to the hydroxyl-bearing carbon.

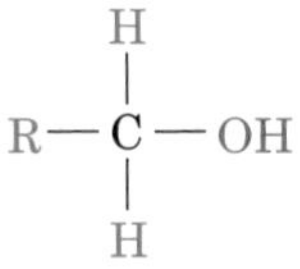

A primary alcohol (1°) 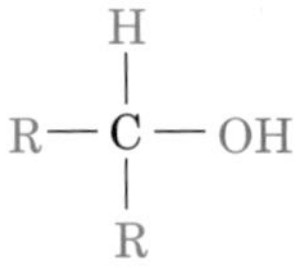   **A secondary alcohol (2°)** 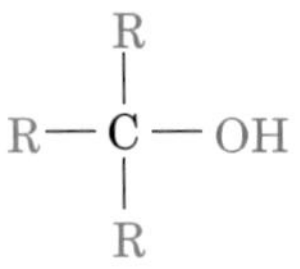   **A tertiary alcohol (3°)**

Simple alcohols are named by the IUPAC system as derivatives of the parent alkane, using the suffix *-ol*:

RULE 1   Select the longest carbon chain containing the hydroxyl group, and derive the parent name by replacing the *-e* ending of the corresponding alkane with *-ol*.

RULE 2   Number the alkane chain beginning at the end nearer the hydroxyl group.

RULE 3   Number the substituents according to their position on the chain, and write the name listing the substituents in alphabetical order.

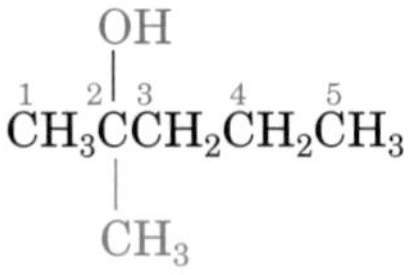

2-Methyl-2-pentanol 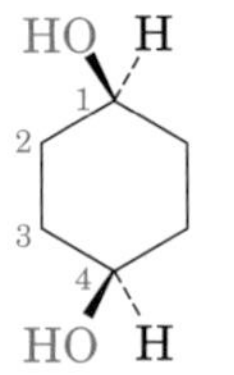   *cis*-1,4-**Cyclohexane**diol 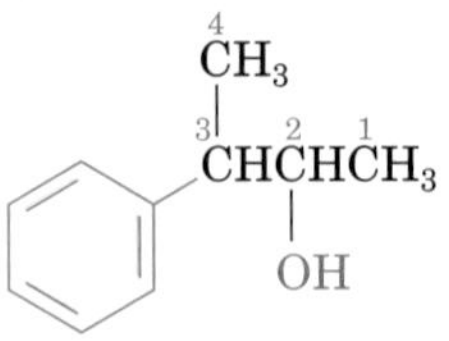   3-Phenyl-2-**butan**ol

Some simple and widely occurring alcohols have common names that are accepted by IUPAC. For example:

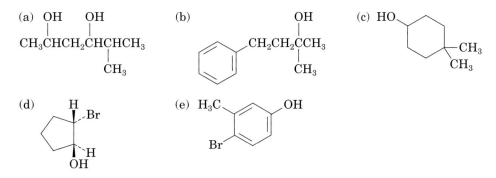

Benzyl alcohol (**Phenylmethanol**)	**Allyl alcohol** (**2-Propen-1-ol**)	*tert*-**Butyl alcohol** (**2-Methyl-2-propanol**)	**Ethylene glycol** (**1,2-Ethanediol**)	**Glycerol** (**1,2,3-Propanetriol**)

Naming Phenols

The word *phenol* is used both as the name of a specific substance (hydroxybenzene) and as the family name for hydroxy-substituted aromatic compounds, according to the rules discussed in Section 15.2. Note that -*phenol* is used as the parent name rather than -*benzene*.

_m_-Methylphenol
(**_m_-Cresol**)   **2,4-Dinitrophenol**

Problem 17.1 Give IUPAC names for the following compounds:

(a) OH   OH
     CH₃CHCH₂CHCHCH₃
                  |
                 CH₃

(b)   OH
          |
     CH₂CH₂CCH₃
          |
         CH₃

(c) HO   CH₃
               CH₃

(d)   H
       |—Br
       |
       `H
      OH

(e) H₃C   OH

     Br

Problem 17.2 Draw structures corresponding to the following IUPAC names:
(a) 2-Ethyl-2-buten-1-ol   (b) 3-Cyclohexen-1-ol
(c) *trans*-3-Chlorocycloheptanol   (d) 1,4-Pentanediol
(e) 2,6-Dimethylphenol   (f) *o*-(2-Hydroxyethyl)phenol

17.2 Properties of Alcohols and Phenols: Hydrogen Bonding

Alcohols and phenols have nearly the same geometry as water. The R–O–H bond angle has an approximately tetrahedral value (109° in methanol, for example), and the oxygen atom is sp^3-hybridized.

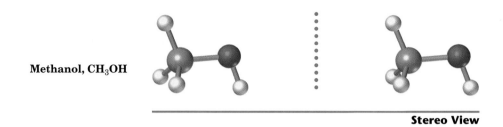

Methanol, CH_3OH

Alcohols and phenols are quite different from the hydrocarbons and alkyl halides we've studied thus far. Not only is their chemistry much richer, their physical properties are different as well. Figure 17.1, which provides a comparison of the boiling points of some simple alcohols, alkanes, and chloroalkanes, shows that alcohols have much higher boiling points. For example, 1-propanol (MW = 60), butane (MW = 58), and chloroethane (MW = 65) have similar molecular weights, yet 1-propanol boils at 97°C, compared to −0.5°C for the alkane and 12.5°C for the chloroalkane.

FIGURE 17.1 ▼

A comparison of boiling points for some alkanes, chloroalkanes, and alcohols. Alcohols generally have the higher boiling points.

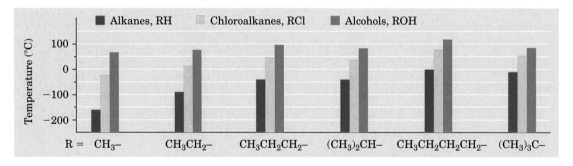

Phenols, too, have elevated boiling points relative to hydrocarbons. Phenol itself, for instance, boils at 181.7°C, while toluene boils at 110.6°C.

Phenol: bp = 181.7°C   **Toluene: bp = 110.6°C**

Alcohols and phenols have elevated boiling points because, like water, they form *hydrogen bonds* in the liquid state. A positively polarized –OH hydrogen atom from one molecule is attracted to a lone pair of electrons on a negatively polarized oxygen atom of another molecule, resulting in a weak force that holds the molecules together (Figure 17.2). These intermolecular attractions must be overcome for a molecule to break free from the liquid and enter the vapor state, so the boiling temperature is raised.

FIGURE 17.2 ▼

Hydrogen bonding in alcohols and phenols. A weak attraction between a positively polarized OH hydrogen and a negatively polarized oxygen holds molecules together. The electrostatic potential map of methanol clearly shows the positively polarized nature of the O–H hydrogen (blue).

methanol

• •

Problem 17.3   The following data for isomeric four-carbon alcohols show that there is a decrease in boiling point with increasing substitution. How might you account for this trend?

> 1-Butanol, bp 117.5°C
> 2-Butanol, bp 99.5°C
> 2-Methyl-2-propanol, bp 82.2°C

• •

17.3 Properties of Alcohols and Phenols: Acidity and Basicity

Like water, alcohols and phenols are both weakly basic and weakly acidic. As weak bases, they are reversibly protonated by strong acids to yield oxonium ions, $ROH_2{}^+$:

R–Ö:–H + H–X   ⇌   R–Ö⁺(H)(H)   :X⁻

An alcohol   **An oxonium ion**

$$\left[\text{or}\quad ArOH + HX \;\rightleftharpoons\; Ar\overset{+}{O}H_2\ X^- \right]$$

As weak acids, alcohols and phenols dissociate to a slight extent in dilute aqueous solution by donating a proton to water, generating H_3O^+ and an **alkoxide ion, RO^-,** or a **phenoxide ion, ArO^-:**

An alcohol   **An alkoxide ion**

A phenol   **A phenoxide ion**

Recall from our earlier discussion of acidity (Sections 2.7–2.11) that the strength of any acid HA in water can be expressed by an *acidity constant*, K_a:

$$K_a = \frac{[A^-][H_3O^+]}{[HA]} \qquad pK_a = -\log K_a$$

Compounds with a smaller K_a (or larger pK_a) are less acidic, whereas compounds with a larger K_a (or smaller pK_a) are more acidic. The data presented in Table 17.1 show that simple alcohols are about as acidic as water but that substituent groups can have a significant effect. For example, methanol and ethanol are similar to water in acidity, but *tert*-butyl alcohol is a weaker acid and 2,2,2-trifluoroethanol is stronger.

TABLE 17.1   Acidity Constants of Some Alcohols and Phenols

Alcohol or phenol	pK_a	
$(CH_3)_3COH$	18.00	Weaker acid
CH_3CH_2OH	16.00	
HOH (water)	(15.74)	
CH_3OH	15.54	
CF_3CH_2OH	12.43	
p-Aminophenol	10.46	
p-Methoxyphenol	10.21	
p-Methylphenol	10.17	
Phenol	9.89	
p-Chlorophenol	9.38	
p-Bromophenol	9.35	
p-Nitrophenol	7.15	
2,4,6-Trinitrophenol	0.60	Stronger acid

The effect of alkyl substitution on alcohol acidity is due primarily to solvation of the alkoxide ion that results from dissociation. The more easily the alkoxide ion is solvated by water, the more stable it is, the more its formation is energetically favored, and the greater the acidity of the parent alcohol. For example, the oxygen atom of an unhindered alkoxide ion, such as that from methanol, is sterically accessible and is easily solvated by water. The oxygen atom of a hindered alkoxide ion, however, such as that from *tert*-butyl alcohol, is less easily solvated and is therefore less stabilized.

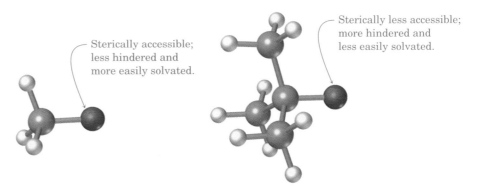

Sterically accessible; less hindered and more easily solvated.

Sterically less accessible; more hindered and less easily solvated.

Methoxide ion, CH$_3$O$^-$
(pK_a = 15.54)

***tert*-Butoxide ion, (CH$_3$)$_3$CO$^-$**
(pK_a = 18.00)

Inductive effects (Section 16.6) are also important in determining alcohol acidities. Electron-withdrawing halogen substituents, for example, stabilize an alkoxide ion by spreading the charge over a larger volume, thus making the alcohol more acidic. Compare, for example, the acidities of ethanol (pK_a = 16.00) and 2,2,2-trifluoroethanol (pK_a = 12.43), or of *tert*-butyl alcohol (pK_a = 18.0) and nonafluoro-*tert*-butyl alcohol (pK_a = 5.4).

Electron-withdrawing groups stabilize alkoxide and lower pK_a

$$CF_3 \leftarrow \overset{\uparrow CF_3}{\underset{\downarrow CF_3}{C}} - O^- \quad versus \quad CH_3 - \overset{CH_3}{\underset{CH_3}{C}} - O^-$$

pK_a = 5.4   **pK_a = 18**

Because alcohols are much less acidic than carboxylic acids or mineral acids, they don't react with weak bases such as amines or bicarbonate ion, and they react to only a limited extent with metal hydroxides such as NaOH. Alcohols do, however, react with alkali metals and with strong bases such as sodium hydride (NaH), sodium amide (NaNH$_2$), and Grignard reagents (RMgX). Alkoxides are themselves bases that are frequently used as reagents in organic chemistry.

$$2\ H_3C - \overset{CH_3}{\underset{CH_3}{C}} - OH + 2\ K \longrightarrow 2\ H_3C - \overset{CH_3}{\underset{CH_3}{C}} - O^-\ K^+ + H_2$$

***tert*-Butyl alcohol**   **Potassium *tert*-butoxide**

$$CH_3OH + NaH \longrightarrow CH_3O^- \ Na^+ + H_2$$

Methanol   **Sodium methoxide**

$$CH_3CH_2OH + NaNH_2 \longrightarrow CH_3CH_2O^- \ Na^+ + NH_3$$

Ethanol   **Sodium ethoxide**

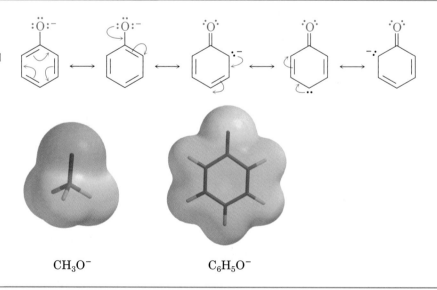

Cyclohexanol

Bromomagnesium cyclohexoxide

Phenols are about a million times more acidic than alcohols (Table 17.1). Indeed, some phenols, such as 2,4,6-trinitrophenol, even surpass the acidity of most carboxylic acids. One practical consequence of this acidity is that phenols are soluble in dilute aqueous NaOH. Thus, a phenolic component can often be separated from a mixture simply by basic extraction into aqueous solution, followed by reacidification.

Phenol   **Sodium phenoxide**

Phenols are more acidic than alcohols because the phenoxide anion is resonance-stabilized. Delocalization of the negative charge over the ortho and para positions of the aromatic ring results in increased stability of the phenoxide anion relative to undissociated phenol and in a consequently lower $\Delta G°$ for the dissociation reaction. Figure 17.3 compares electrostatic potential maps of an alkoxide ion (CH_3O^-) with phenoxide ion and shows how the negative charge in phenoxide ion is delocalized from oxygen to the ring.

FIGURE 17.3 ▼

The resonance-stabilized phenoxide ion is more stable than an alkoxide ion. Electrostatic potential maps show how the negative charge in the phenoxide ion is delocalized.

alkoxide ion,
phenoxide ion

CH_3O^- $C_6H_5O^-$

Substituted phenols can be either more acidic or less acidic than phenol itself. Phenols with an electron-withdrawing substituent are generally more acidic because these substituents stabilize the phenoxide ion by delocalizing the negative charge. Phenols with an electron-donating substituent are less acidic because these substituents destabilize the phenoxide ion by localizing the charge.

Electron-withdrawing groups (EWG)
stabilize phenoxide anion, resulting
in increased phenol acidity

Electron-donating groups (EDG)
destabilize phenoxide anion,
resulting in decreased phenol acidity

The acidifying effect of an electron-withdrawing substituent is particularly noticeable for phenols having a nitro group at the ortho or para position.

● ●

Practice Problem 17.1   Is *p*-cyanophenol more acidic or less acidic than phenol?

Strategy   Identify the substituent on the aromatic ring, and decide whether it is electron-donating or electron-withdrawing. Electron-withdrawing substituents make the phenol more acidic by stabilizing the phenoxide anion, and electron-donating substituents make the phenol less acidic.

Solution   We saw in Section 16.6 that a cyano group is electron-withdrawing. Thus, *p*-cyanophenol is more acidic ($pK_a = 7.97$) than phenol ($pK_a = 9.89$).

***p*-Cyanophenol**
($pK_a = 7.97$)

● ●

Problem 17.4   Rank the following substances in order of increasing acidity:
(a) $(CH_3)_2CHOH$, $HC{\equiv}CH$, $(CF_3)_2CHOH$, CH_3OH
(b) Phenol, *p*-methylphenol, *p*-(trifluoromethyl)phenol
(c) Benzyl alcohol, phenol, *p*-hydroxybenzoic acid

Problem 17.5 *p*-Nitrobenzyl alcohol is more acidic than benzyl alcohol but *p*-methoxybenzyl alcohol is less acidic. Explain.

17.4 Preparation of Alcohols: A Review

Alcohols occupy a central position in organic chemistry. They can be prepared from many other kinds of compounds (alkenes, alkyl halides, ketones, esters, and aldehydes, among others), and they can be transformed into an equally wide assortment of compounds (Figure 17.4).

FIGURE 17.4 ▼

The central position of alcohols in organic chemistry. Alcohols can be prepared from, and converted into, many other kinds of compounds.

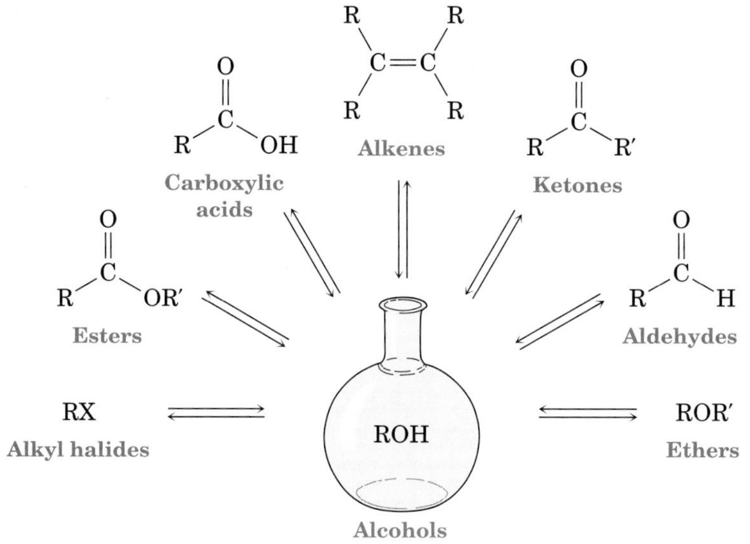

Let's review briefly some of the methods of alcohol preparation we've already seen:

• Alcohols can be prepared by hydration of alkenes. Because the direct hydration of alkenes with aqueous acid is generally a poor reaction in the laboratory, two indirect methods are commonly used. Hydroboration/oxidation yields the product of syn, non-Markovnikov hydration (Section 7.5), whereas oxymercuration/reduction yields the product of Markovnikov hydration (Section 7.4). Both reactions are generally applicable to most alkenes.

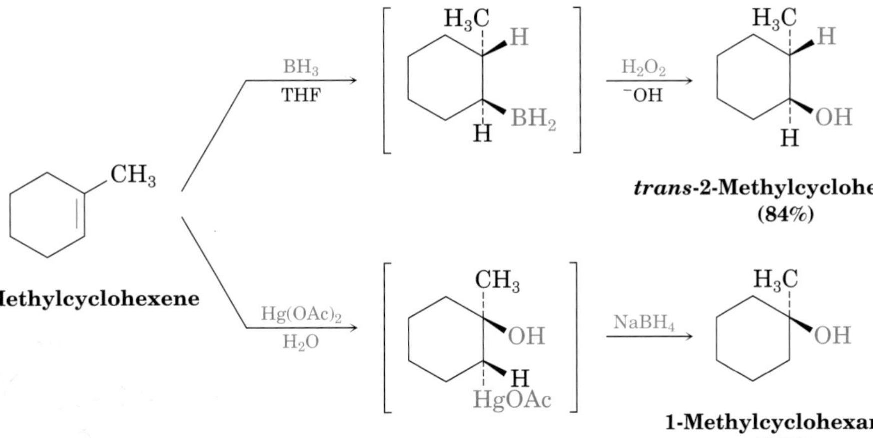

trans-2-Methylcyclohexanol
(84%)

1-Methylcyclohexene

1-Methylcyclohexanol
(90%)

• 1,2-Diols can be prepared by direct hydroxylation of an alkene with OsO$_4$ followed by reduction with NaHSO$_3$ (Section 7.8). The reaction takes place readily and occurs with syn stereochemistry. We'll see in the next chapter that 1,2-diols can also be prepared by acid-catalyzed hydrolysis of *epoxides*—compounds with a three-membered, oxygen-containing ring. Epoxide opening is complementary to direct hydroxylation because it occurs with anti stereochemistry.

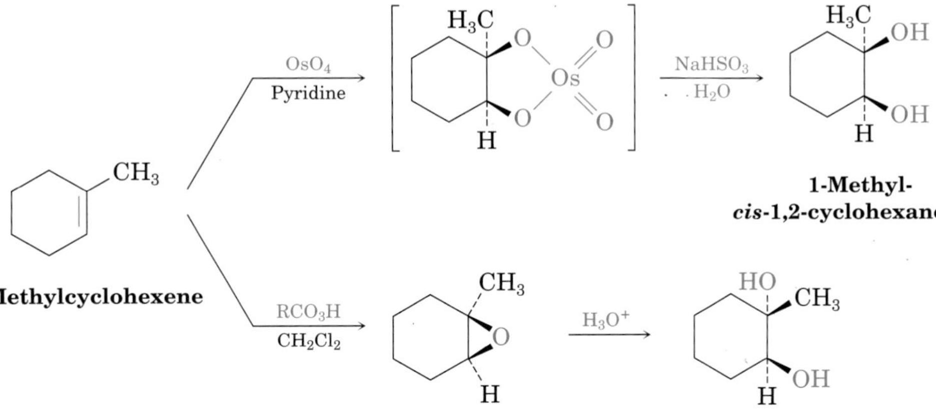

1-Methyl-
cis-1,2-cyclohexanediol

1-Methylcyclohexene

1-Methyl-1,2-epoxy-
cyclohexane

1-Methyl-
trans-1,2-cyclohexanediol

Problem 17.6 Predict the products of the following reactions:

(a)

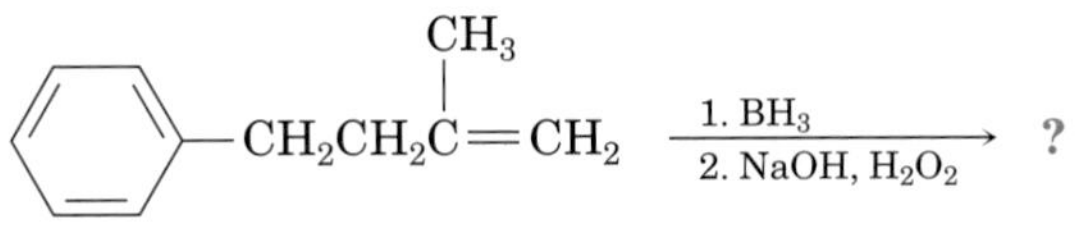

(b) $CH_3CH_2CH=C(CH_3)_2$ $\xrightarrow[\text{2. NaBH}_4]{\text{1. Hg(OAc)}_2, \text{H}_2\text{O}}$?

(c) Reaction of *cis*-5-decene with OsO$_4$, followed by NaHSO$_3$ reduction. Be sure to indicate the stereochemistry of the product.

17.5 Alcohols from Reduction of Carbonyl Compounds

One of the most general methods for preparing alcohols is by reduction of a carbonyl compound. As we saw in Section 10.10, an organic reduction is a reaction that adds hydrogen to a molecule:

where [H] is a generalized reducing agent

A carbonyl compound   **An alcohol**

All kinds of carbonyl compounds can be reduced, including aldehydes, ketones, carboxylic acids, and esters.

Reduction of Aldehydes and Ketones

Aldehydes and ketones are easily reduced to yield alcohols. Aldehydes are converted into primary alcohols, and ketones are converted into secondary alcohols.

An aldehyde   **A primary alcohol**   **A ketone**   **A secondary alcohol**

Many reagents are used to reduce ketones and aldehydes to alcohols, but sodium borohydride, $NaBH_4$, is usually chosen because of its safety and ease of handling. Sodium borohydride is a white, crystalline solid that can be weighed in the open atmosphere and used in either water or alcohol solution. High yields of products are usually obtained.

Aldehyde reduction

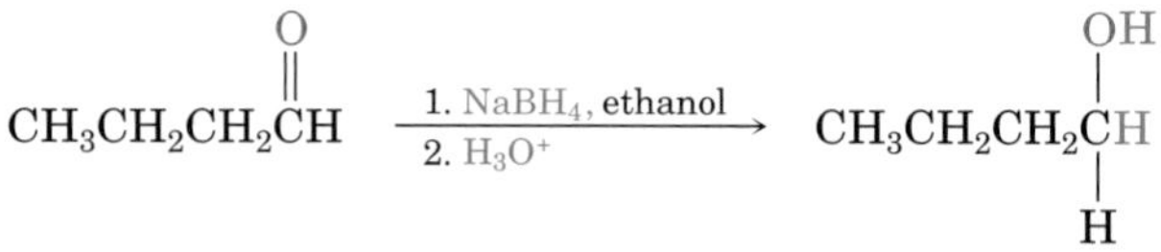

$$CH_3CH_2CH_2CH \xrightarrow[\text{2. } H_3O^+]{\text{1. } NaBH_4,\text{ ethanol}} CH_3CH_2CH_2CH$$

Butanal

1-Butanol (85%)
(a 1° alcohol)

Ketone reduction

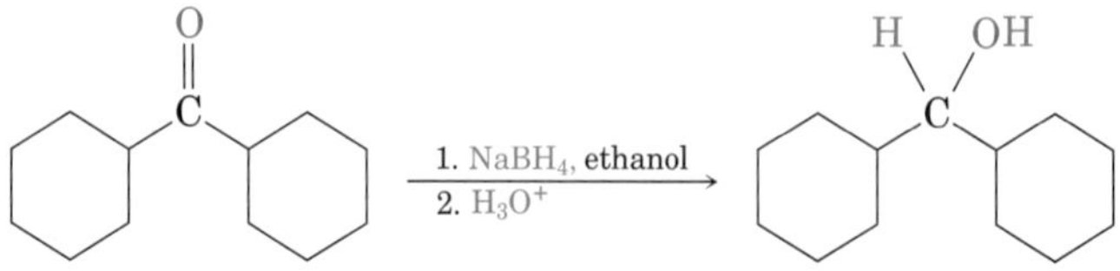

Dicyclohexyl ketone

Dicyclohexylmethanol (88%)
(a 2° alcohol)

Lithium aluminum hydride, $LiAlH_4$, is another reducing agent often used for reduction of ketones and aldehydes. A grayish powder soluble in ether and tetrahydrofuran, $LiAlH_4$ is much more reactive than $NaBH_4$ but also more dangerous. It reacts violently with water and decomposes explosively when heated above 120°C.

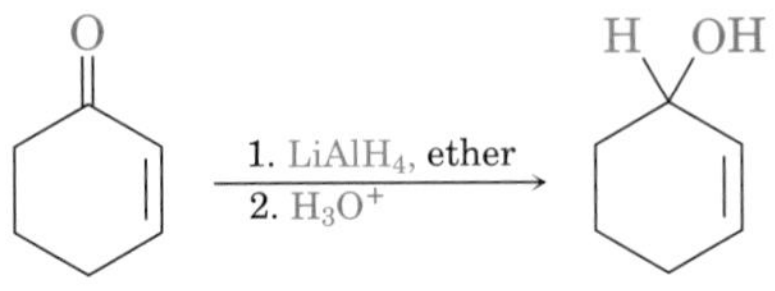

2-Cyclohexenone   **2-Cyclohexenol (94%)**

Reduction of Carboxylic Acids and Esters

Carboxylic acids and esters are reduced to give primary alcohols:

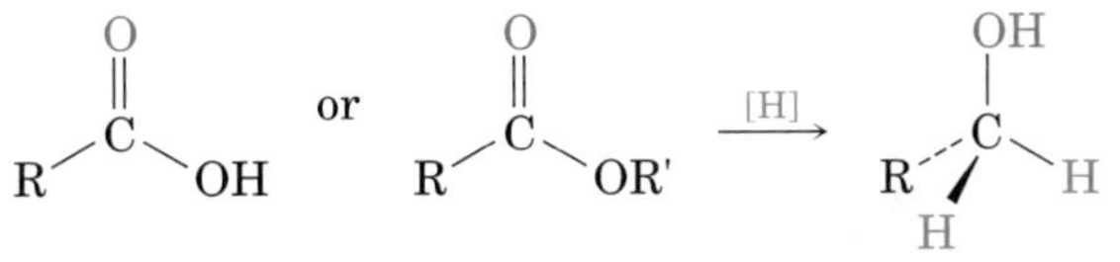

A carboxylic acid   **An ester**   **A primary alcohol**

These reactions are not as rapid as the analogous reductions of aldehydes and ketones: $NaBH_4$ reduces esters very slowly and does not reduce carboxylic acids at all. Carboxylic acid and ester reductions are therefore usually carried out with the more reactive reducing agent $LiAlH_4$. All carbonyl groups, including acids, esters, ketones, and aldehydes, are rapidly reduced by $LiAlH_4$. Note that one hydrogen atom is delivered to the carbonyl carbon atom during ketone and aldehyde reductions, but that two hydrogens become bonded to the former carbonyl carbon during carboxylic acid and ester reductions.

Carboxylic acid reduction

$$CH_3(CH_2)_7CH=CH(CH_2)_7\overset{\overset{O}{\|}}{C}OH \xrightarrow[\text{2. } H_3O^+]{\text{1. } LiAlH_4,\text{ ether}} CH_3(CH_2)_7CH=CH(CH_2)_7CH_2OH$$

9-Octadecenoic acid   **9-Octadecen-1-ol (87%)**
(Oleic acid)

Ester reduction

$$CH_3CH_2CH=CH\overset{\overset{O}{\|}}{C}OCH_3 \xrightarrow[\text{2. } H_3O^+]{\text{1. } LiAlH_4,\text{ ether}} CH_3CH_2CH=CHCH_2OH + CH_3OH$$

Methyl 2-pentenoate   **2-Penten-1-ol (91%)**

We'll defer until Chapter 19 a detailed discussion of the mechanisms by which carbonyl compounds are reduced to give alcohols. For the moment, we'll simply note that these reactions involve the addition of nucleophilic hydride ion ($:H^-$) to the positively polarized, electrophilic carbon atom of the carbonyl group. The initial product is an alkoxide ion, which is protonated by addition of H_3O^+ in a second step to yield the alcohol product.

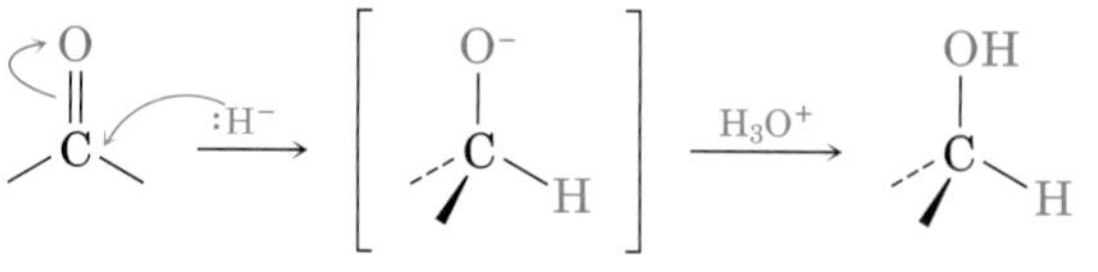

| A carbonyl compound | An alkoxide ion intermediate | An alcohol |

Practice Problem 17.2   What carbonyl compounds would you reduce to obtain the following alcohols?

(a) $CH_3CH_2CHCH_2CHCH_3$ with CH_3 and OH substituents   (b)

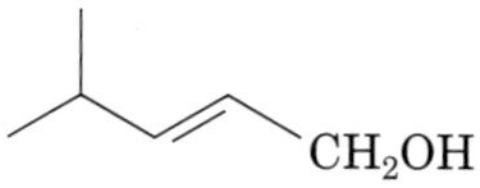

Strategy   Identify the alcohol as primary, secondary, or tertiary. A primary alcohol can be prepared by reduction of an aldehyde, an ester, or a carboxylic acid; a secondary alcohol can be prepared by reduction of a ketone; and a tertiary alcohol cannot be prepared by reduction.

Solution   (a) The target molecule is a secondary alcohol, which can only be prepared by reduction of a ketone. Either $NaBH_4$ or $LiAlH_4$ can be used.

$$CH_3CH_2CHCH_2CCH_3 \xrightarrow[\text{2. } H_3O^+]{\text{1. NaBH}_4 \text{ or LiAlH}_4} CH_3CH_2CHCH_2CHCH_3$$

(b) The target molecule is a primary alcohol, which can be prepared by reduction of an aldehyde, an ester, or a carboxylic acid. $LiAlH_4$ is needed for the ester and carboxylic acid reductions.

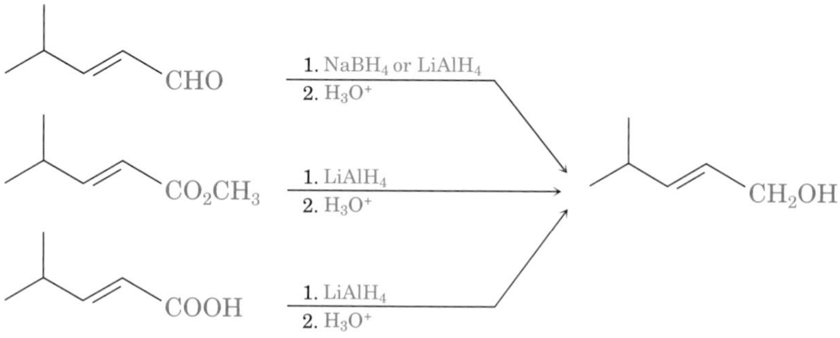

Problem 17.7   What reagent would you use to accomplish each of the following reactions?

(a) $CH_3CCH_2CH_2COCH_3 \xrightarrow{?} CH_3CHCH_2CH_2COCH_3$

(b) $CH_3\overset{O}{\overset{\|}{C}}CH_2CH_2\overset{O}{\overset{\|}{C}}OCH_3 \xrightarrow{?} CH_3\overset{OH}{\overset{|}{C}}HCH_2CH_2CH_2OH$

(c)

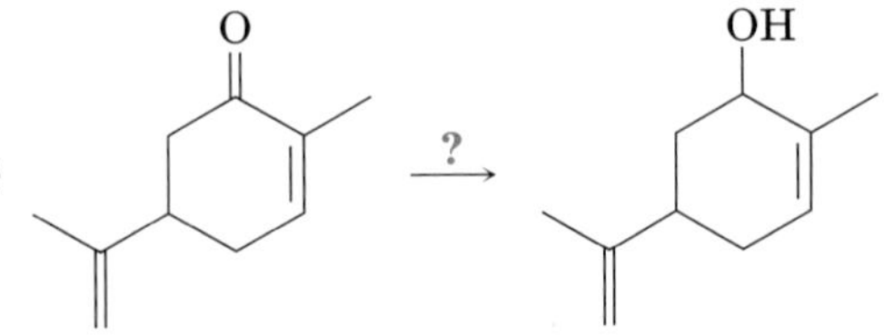

Problem 17.8 What carbonyl compounds give the following alcohols on reduction with $LiAlH_4$? Show all possibilities.

(a)

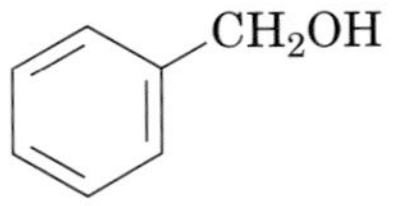

(b)

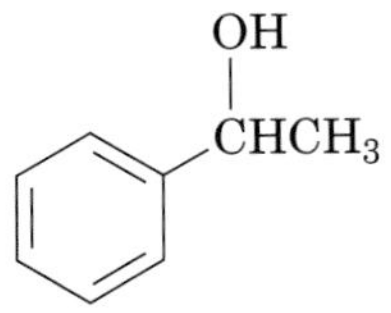

(c)

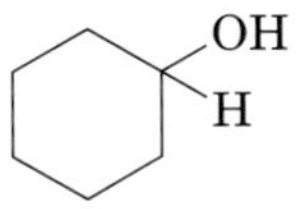

(d) $(CH_3)_2CHCH_2OH$

· ·

17.6 Alcohols from Reaction of Carbonyl Compounds with Grignard Reagents

We saw in Section 10.8 that alkyl, aryl, and vinylic halides react with magnesium in ether or tetrahydrofuran solution to generate Grignard reagents, RMgX. These Grignard reagents react with carbonyl compounds to yield alcohols in much the same way that hydride reducing agents do. The result is a useful and general method of alcohol synthesis.

Grignard formation

$$R—X + Mg \longrightarrow \overset{\delta-}{R}—\overset{\delta+}{MgX} \qquad R = 1°, 2°, \text{ or } 3° \text{ alkyl, aryl, or vinylic}$$
$$X = Cl, Br, \text{ or } I$$

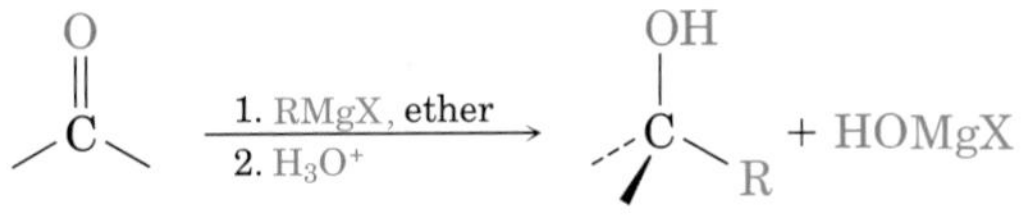

A Grignard reagent

A great many alcohols can be obtained from Grignard reactions, depending on the reactants. For example, Grignard reagents react with formaldehyde,

$H_2C{=}O$, to give primary alcohols, with aldehydes to give secondary alcohols, and with ketones to give tertiary alcohols:

Formaldehyde reaction

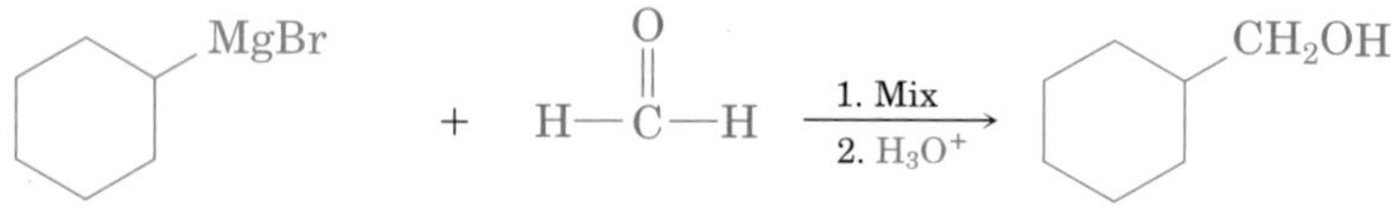

Cyclohexylmagnesium bromide **Formaldehyde** **Cyclohexylmethanol (65%) (a 1° alcohol)**

Aldehyde reaction

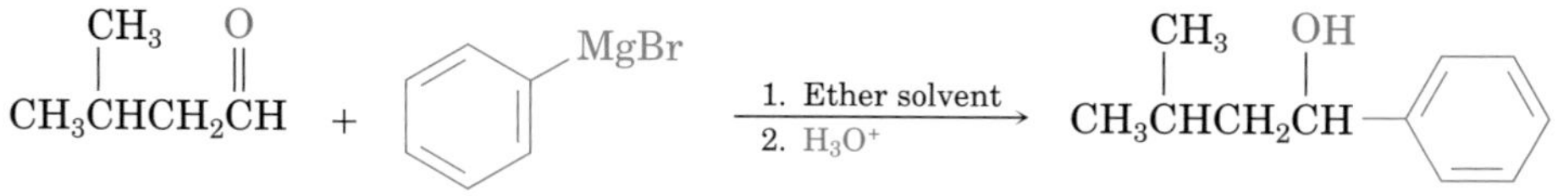

3-Methylbutanal **Phenylmagnesium bromide** **3-Methyl-1-phenyl-1-butanol (73%) (a 2° alcohol)**

Ketone reaction

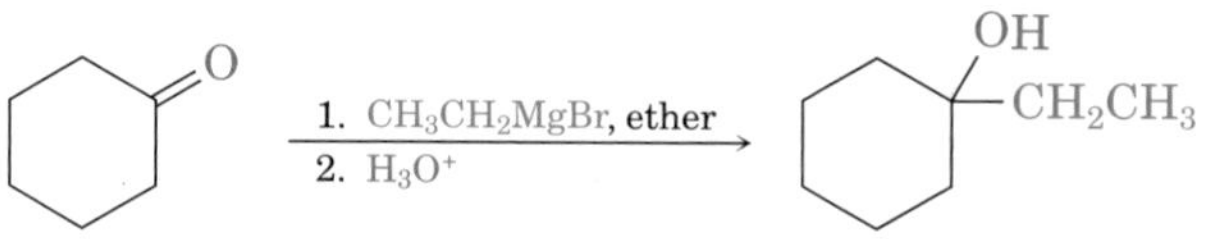

Cyclohexanone **1-Ethylcyclohexanol (89%) (a 3° alcohol)**

Esters react with Grignard reagents to yield tertiary alcohols in which *two* of the substituents bonded to the hydroxyl-bearing carbon have come from the Grignard reagent (just as $LiAlH_4$ reduction of an ester adds two hydrogens).

$$CH_3CH_2CH_2CH_2\overset{O}{\overset{\|}{C}}OCH_2CH_3 \xrightarrow[2.\ H_3O^+]{1.\ 2\ CH_3MgBr} CH_3CH_2CH_2CH_2C(OH)(CH_3)CH_3 + CH_3CH_2OH$$

Ethyl pentanoate **2-Methyl-2-hexanol (85%) (a 3° alcohol)**

Carboxylic acids don't give addition products with Grignard reagents because the acidic carboxyl hydrogen reacts with the basic Grignard reagent to yield a hydrocarbon and the magnesium salt of the acid. We saw this reaction in Section 10.8 as a means of reducing alkyl halides to alkanes.

$$RBr + Mg \longrightarrow RMgBr$$

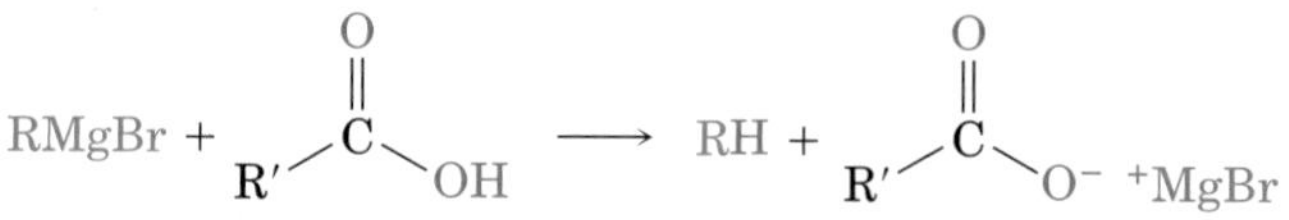

A carboxylic acid **A carboxylic acid salt**

The Grignard reaction, though useful, has several limitations. One major problem is that a Grignard reagent can't be prepared from an organohalide if there are other reactive functional groups in the same molecule. For example, a compound that is both an alkyl halide and a ketone can't form a Grignard reagent because it would react with itself. Similarly, a compound that is both an alkyl halide and a carboxylic acid, an alcohol, or an amine can't form a Grignard reagent because the acidic RCO_2H, ROH, or RNH_2 hydrogen present in the same molecule would react with the basic Grignard reagent as rapidly as it forms. In general, Grignard reagents can't be prepared from alkyl halides that contain the following functional groups (FG):

$$Br - \boxed{Molecule} - FG$$

where FG = —OH, —NH, —SH, —COOH } The Grignard reagent is protonated by these groups.

$$FG = \overset{O}{\underset{\|}{}}-CH, \overset{O}{\underset{\|}{}}-CR, \overset{O}{\underset{\|}{}}-CNR_2,$$
$$-C \equiv N, -NO_2, -SO_2R$$ } The Grignard reagent adds to these groups.

As with the reduction of carbonyl compounds discussed in the previous section, we'll defer a detailed treatment of the mechanism of Grignard reactions until Chapter 19. For the moment, it's sufficient to note that Grignard reagents act as nucleophilic carbon anions (carbanions, $:R^-$) and that the addition of a Grignard reagent to a carbonyl compound is analogous to the addition of hydride ion. The intermediate is an alkoxide ion, which is protonated by addition of H_3O^+ in a second step.

A carbonyl compound   An alkoxide ion intermediate   An alcohol

● ●

Practice Problem 17.3 How could you use the addition of a Grignard reagent to a ketone to synthesize 2-phenyl-2-propanol?

Strategy Draw the product, and identify the three groups bonded to the alcohol carbon atom. One of the three will have come from the Grignard reagent, and the remaining two will have come from the ketone.

Solution 2-Phenyl-2-propanol has two methyl groups (–CH₃) and one phenyl (–C₆H₅) attached to the alcohol carbon atom. Thus, the possibilities are addition of methylmagnesium bromide to acetophenone and addition of phenylmagnesium bromide to acetone:

Acetophenone

1. CH$_3$MgBr
2. H$_3$O$^+$

Acetone

1. C$_6$H$_5$MgBr
2. H$_3$O$^+$

2-Phenyl-2-propanol

· ·

Practice Problem 17.4   How could you use the reaction of a Grignard reagent with a carbonyl compound to synthesize 2-methyl-2-pentanol?

Strategy   Draw the product, and identify the three groups bonded to the alcohol carbon atom. If the three groups are all different, the starting carbonyl compound must be a ketone. If two of the three groups are identical, the starting carbonyl compound might be either a ketone or an ester.

Solution   In the present instance, the product is a tertiary alcohol with two methyl groups and one propyl group. Starting from a ketone, the possibilities are addition of methylmagnesium bromide to 2-pentanone and addition of propylmagnesium bromide to acetone:

$$CH_3CH_2CH_2\overset{\overset{O}{\|}}{C}CH_3$$

2-Pentanone

1. CH$_3$MgBr
2. H$_3$O$^+$

$$CH_3\overset{\overset{O}{\|}}{C}CH_3$$

Acetone

1. CH$_3$CH$_2$CH$_2$MgBr
2. H$_3$O$^+$

$$CH_3CH_2CH_2\overset{\overset{OH}{|}}{\underset{\underset{CH_3}{|}}{C}}CH_3$$

2-Methyl-2-pentanol

Starting from an ester, the only possibility is addition of methylmagnesium bromide to an ester of butanoic acid, such as methyl butanoate:

$$CH_3CH_2CH_2\overset{\overset{O}{\|}}{C}OCH_3 \quad \xrightarrow[\text{2. H}_3\text{O}^+]{\text{1. 2 CH}_3\text{MgBr}} \quad CH_3CH_2CH_2\overset{\overset{OH}{|}}{\underset{\underset{CH_3}{|}}{C}}CH_3 + CH_3OH$$

Methyl butanoate

2-Methyl-2-pentanol

Problem 17.9   Show the products obtained from addition of methylmagnesium bromide to the following compounds:
(a) Cyclopentanone   (b) Benzophenone (diphenyl ketone)   (c) 3-Hexanone

Problem 17.10   Use a Grignard reaction to prepare the following alcohols:
(a) 2-Methyl-2-propanol   (b) 1-Methylcyclohexanol   (c) 3-Methyl-3-pentanol
(d) 2-Phenyl-2-butanol   (e) Benzyl alcohol

Problem 17.11   Use the reaction of a Grignard reagent with a carbonyl compound to synthesize the following compound:

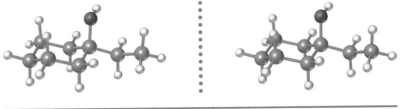

Stereo View

17.7   Some Reactions of Alcohols

Reactions of alcohols can be divided for convenience into two groups—those that occur at the C–O bond and those that occur at the O–H bond:

Let's begin looking at reactions of both types by reviewing some of the alcohol reactions seen in previous chapters.

Dehydration of Alcohols to Yield Alkenes

One of the most valuable C–O bond reactions of alcohols is dehydration to give alkenes. The C–O bond and a neighboring C–H are broken, and an alkene π bond is formed:

A dehydration reaction

Because of the usefulness of the reaction, a number of ways have been devised for carrying out dehydrations. One method that works particularly well for tertiary alcohols is the acid-catalyzed reaction discussed in Section 7.1. For example, treatment of 1-methylcyclohexanol with warm aqueous sulfuric acid in a solvent such as tetrahydrofuran results in loss of water and formation of 1-methylcyclohexene.

$$H_3O^+, \text{THF} \quad 50°C$$

1-Methylcyclohexanol   **1-Methylcyclohexene (91%)**

Acid-catalyzed dehydrations usually follow Zaitsev's rule (Section 11.10) and yield the more highly substituted alkene as the major product. Thus, 2-methyl-2-butanol gives primarily 2-methyl-2-butene (trisubstituted double bond) rather than 2-methyl-1-butene (disubstituted double bond).

$$H_3O^+, \text{THF} \quad 25°C$$

2-Methyl-2-butanol   **2-Methyl-2-butene (trisubstituted)**   **2-Methyl-1-butene (disubstituted)**

Major product   Minor product

Only tertiary alcohols are readily dehydrated with acid. Secondary alcohols can be made to react, but the conditions are severe (75% H_2SO_4, 100°C) and sensitive molecules don't survive. Primary alcohols are even less reactive than secondary ones, and very harsh conditions are necessary to cause dehydration (95% H_2SO_4, 150°C). Thus, the reactivity order for acid-catalyzed dehydrations is

Reactivity

The reasons for the observed reactivity order are best understood by looking at the mechanism of the reaction (Figure 17.5). Acid-catalyzed dehydrations are E1 reactions (Section 11.14), which occur by a three-step mechanism involving protonation of the alcohol oxygen, spontaneous loss of water to generate a carbocation intermediate, and final loss of a proton (H^+) from the neighboring carbon atom. Tertiary substrates *always* react fastest in E1 reactions because they lead to highly stabilized, tertiary carbocation intermediates.

FIGURE 17.5 ▼

Mechanism of the acid-catalyzed dehydration of an alcohol to yield an alkene. The process is an E1 reaction and involves a carbocation intermediate.

refer to
Mechanisms
& Movies

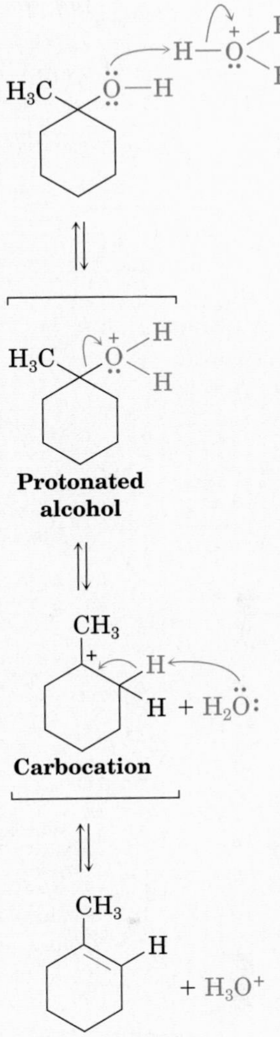

Two electrons from the oxygen atom bond to H$^+$, yielding a protonated alcohol intermediate.

Protonated alcohol

The carbon–oxygen bond breaks, and the two electrons from the bond stay with oxygen, leaving a carbocation intermediate.

Carbocation

Two electrons from a neighboring carbon–hydrogen bond form the alkene π bond, and H$^+$ (a proton) is eliminated.

© 1984 JOHN MCMURRY

To circumvent the need for strong acid and allow the dehydration of secondary alcohols in a gentler way, reagents have been developed that are effective under mild, basic conditions. One such reagent, phosphorus oxychloride (POCl$_3$) in the basic amine solvent pyridine, is often able to effect the dehydration of secondary and tertiary alcohols at 0°C.

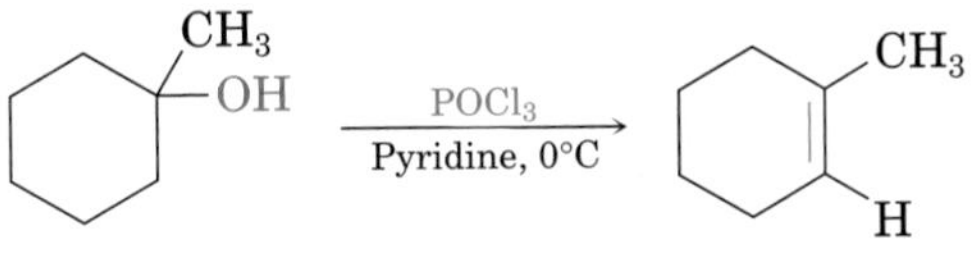

1-Methylcyclohexanol   **1-Methylcyclohexene (96%)**

Alcohol dehydrations carried out with $POCl_3$ in pyridine take place by the E2 mechanism shown in Figure 17.6. Because hydroxide ion is a poor leaving group (Section 11.5), direct E2 elimination of water from an alcohol does not occur. On reaction with $POCl_3$, however, the –OH group is converted into a dichlorophosphate ($-OPOCl_2$), which is an excellent leaving group and is readily eliminated to yield an alkene. Pyridine serves both as reaction solvent and as base to remove a neighboring proton in the E2 elimination step.

FIGURE 17.6 ▼

Mechanism of the dehydration of secondary and tertiary alcohols by reaction with $POCl_3$ in pyridine. The reaction is an E2 process.

refer to
Mechanisms
& Movies

The alcohol hydroxyl group reacts with $POCl_3$ to form a dichlorophosphate intermediate.

E2 elimination then occurs by the usual one-step mechanism as the amine base pyridine abstracts a proton from the neighboring carbon at the same time that the dichlorophosphate group is leaving.

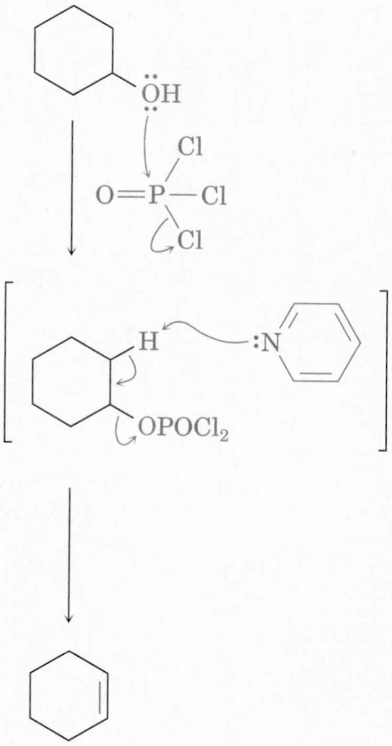

© 1984 JOHN MCMURRY

Problem 17.12   What product(s) would you expect from dehydration of the following alcohols with $POCl_3$ in pyridine? Indicate the major product in each case.

(a) $CH_3CH_2\overset{\displaystyle OH}{\underset{\displaystyle |}{C}}HCH(CH_3)_2$ (b) *trans*-2-Methylcyclohexanol

(c) *cis*-2-Methylcyclohexanol

Conversion of Alcohols into Alkyl Halides

Another C–O bond reaction of alcohols is their conversion into alkyl halides (Section 10.7). Tertiary alcohols are readily converted into alkyl halides by treatment with either HCl or HBr at 0°C. Primary and secondary alcohols are much more resistant to acid, however, and are best converted into halides by treatment with either $SOCl_2$ or PBr_3.

The reaction of a tertiary alcohol with HX takes place by an S_N1 mechanism. Acid protonates the hydroxyl oxygen atom, water is expelled to generate a carbocation, and the cation reacts with nucleophilic halide ion to give the alkyl halide product (Figure 17.7).

FIGURE 17.7 ▼

Mechanism of the reaction of a tertiary alcohol with HCl to give a tertiary alkyl chloride. The alcohol is first protonated, and the oxonium ion then undergoes an S_N1 reaction.

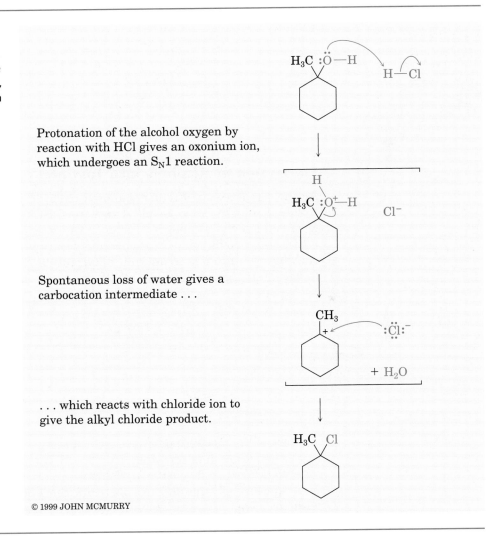

Protonation of the alcohol oxygen by reaction with HCl gives an oxonium ion, which undergoes an S_N1 reaction.

Spontaneous loss of water gives a carbocation intermediate . . .

. . . which reacts with chloride ion to give the alkyl chloride product.

© 1999 JOHN MCMURRY

The reactions of primary and secondary alcohols with $SOCl_2$ and PBr_3 take place by S_N2 mechanisms. Hydroxide ion itself is too poor a leaving group to be displaced by nucleophiles in S_N2 reactions, but reaction of an alcohol with $SOCl_2$ or PBr_3 converts the –OH into a much better leaving group—either –OSOCl or –OPBr$_2$—that is readily expelled by back-side nucleophilic attack (Figure 17.8).

FIGURE 17.8 ▼

Conversion of a primary alcohol into alkyl halides by S_N2 reactions with $SOCl_2$ and PBr_3.

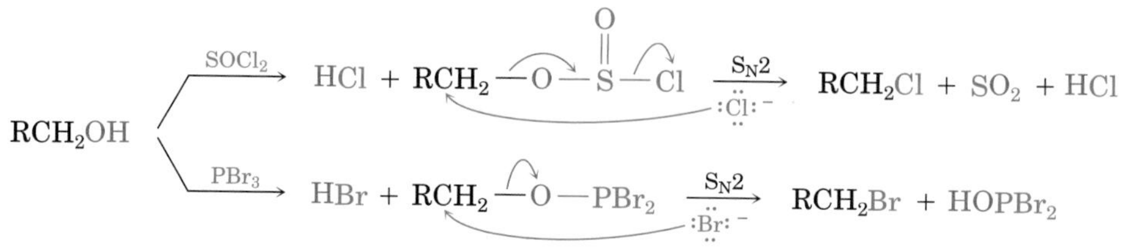

Conversion of Alcohols into Tosylates

Alcohols react with *p*-toluenesulfonyl chloride (tosyl chloride, *p*-TosCl) in pyridine solution to yield alkyl tosylates, ROTos (Section 11.2). Only the O–H bond of the alcohol is broken in this reaction; the C–O bond remains intact, and no change of configuration occurs if the oxygen is attached to a chirality center. The resultant alkyl tosylates behave much like alkyl halides, undergoing both S_N1 and S_N2 substitution reactions.

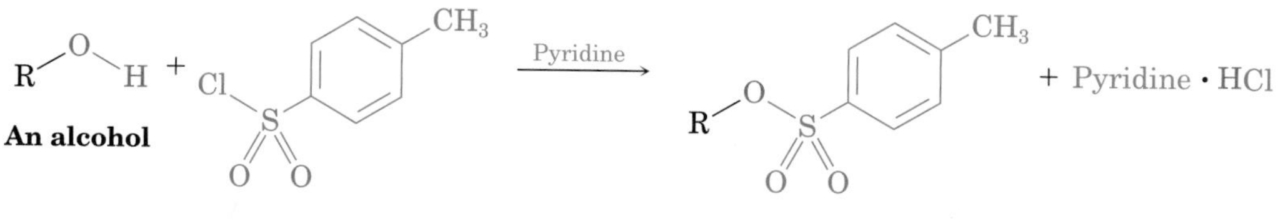

An alcohol

***p*-Toluenesulfonyl chloride**

A tosylate (ROTos)

One of the most important reasons for using tosylates instead of halides in S_N2 reactions is stereochemical. The S_N2 reaction of an alcohol via an alkyl halide proceeds with *two* Walden inversions—one to make the halide from the alcohol and one to substitute the halide—and yields a product with the same absolute stereochemistry as the starting alcohol. The S_N2 reaction of an alcohol via a tosylate, however, proceeds with only *one* Walden inversion and yields a product of opposite stereochemistry to the starting alcohol. Figure 17.9 shows a series of reactions on optically active 2-octanol that illustrates these stereochemical relationships.

FIGURE 17.9 ▼

Stereochemical consequences of S$_N$2 reactions on derivatives of (*R*)-2-octanol. Substitution via the halide gives a product with the same stereochemistry as the starting alcohol; substitution via the tosylate gives a product with opposite stereochemistry to the starting alcohol.

(*R*)-2-Octanol

PBr$_3$ / Ether → (*S*)-2-Bromooctane

CH$_3$CH$_2$O$^-$ Na$^+$ / S$_N$2 → Ethyl (*R*)-1-methylheptyl ether

p-TosCl / Pyridine → (*R*)-1-Methylheptyl tosylate

CH$_3$CH$_2$O$^-$ Na$^+$ / S$_N$2 → Ethyl (*S*)-1-methylheptyl ether

17.8 Oxidation of Alcohols

One of the most valuable reactions of alcohols is their oxidation to yield carbonyl compounds—the opposite of the reduction of a carbonyl compound to yield an alcohol:

An alcohol ⇌ (Oxidation / Reduction) ⇌ A carbonyl compound

Primary alcohols yield aldehydes or carboxylic acids, secondary alcohols yield ketones, but tertiary alcohols don't normally react with most oxidizing agents.

Primary alcohol

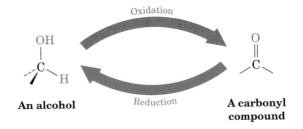

An aldehyde   A carboxylic acid

Secondary alcohol

A ketone

Tertiary alcohol

NO reaction

The oxidation of a primary or secondary alcohol can be accomplished by any of a large number of reagents, including $KMnO_4$, CrO_3, and $Na_2Cr_2O_7$. Which reagent is used in a specific case depends on such factors as cost, convenience, reaction yield, and alcohol sensitivity. For example, the large-scale oxidation of a simple, inexpensive alcohol such as cyclohexanol would best be done with a cheap oxidant such as $Na_2Cr_2O_7$. On the other hand, the small-scale oxidation of a delicate and expensive polyfunctional alcohol would best be done with a mild and high-yielding reagent, regardless of cost.

Primary alcohols are oxidized either to aldehydes or to carboxylic acids, depending on the reagents chosen and on the conditions used. Probably the best method for preparing an aldehyde from a primary alcohol on a laboratory scale (as opposed to an industrial scale) is by use of pyridinium chlorochromate (PCC, $C_5H_6NCrO_3Cl$) in dichloromethane solvent.

Citronellol (from rose oil)   **Citronellal (82%)**

Most other oxidizing agents, such as chromium trioxide (CrO_3) in aqueous acid, oxidize primary alcohols to carboxylic acids. An aldehyde is involved as an intermediate in this reaction but can't usually be isolated because it is further oxidized too rapidly.

$$CH_3(CH_2)_8CH_2OH \xrightarrow[\text{H}_3\text{O}^+,\ \text{acetone}]{\text{CrO}_3} CH_3(CH_2)_8\overset{\displaystyle O}{\overset{\displaystyle \|}{C}}OH$$

1-Decanol   **Decanoic acid (93%)**

Secondary alcohols are oxidized easily and in high yield to give ketones. For large-scale oxidations, an inexpensive reagent such as $Na_2Cr_2O_7$ in aqueous acetic acid is used.

4-*tert*-Butylcyclohexanol   **4-*tert*-Butylcyclohexanone (91%)**

For more sensitive alcohols, pyridinium chlorochromate is often used because the reaction is milder and occurs at lower temperatures.

Testosterone   **4-Androstene-3,17-dione (82%)**
(steroid; male sex hormone)

All these oxidations occur by a pathway that is closely related to the E2 reaction (Section 11.11). The first step involves reaction between the alcohol and a Cr(VI) reagent to form a *chromate* intermediate, which contains an O–Cr bond. Bimolecular elimination with expulsion of chromium as the leaving group then yields the carbonyl product.

An alcohol   **A chromate intermediate**   **Carbonyl product**

Although we usually think of the E2 reaction as a means of generating a carbon–*carbon* double bond by elimination of a halide leaving group, the reaction is also useful for generating a carbon–*oxygen* double bond by elimination of a metal as the leaving group. This is just one more example of how the same few fundamental mechanistic types keep reappearing in different variations.

Problem 17.13   What alcohols would give the following products on oxidation?

(a)   (b) CH_3CHCHO with CH_3 group   (c)

Problem 17.14 What products would you expect from oxidation of the following compounds with CrO_3 in aqueous acid? With pyridinium chlorochromate?
(a) 1-Hexanol (b) 2-Hexanol (c) Hexanal

17.9 Protection of Alcohols

It often happens, particularly during the synthesis of complex molecules, that one functional group in a molecule interferes with an intended reaction on a second functional group elsewhere in the same molecule. For example, we saw earlier in this chapter that a Grignard reagent can't be prepared from a halo alcohol because the C–Mg bond is not compatible with the presence of an acidic –OH group in the same molecule.

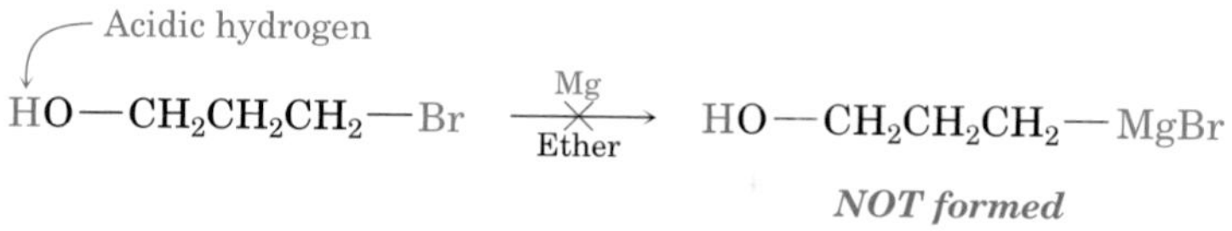

When this kind of incompatibility arises, it's sometimes possible to circumvent the problem by *protecting* the interfering functional group. Protection involves three steps: (1) introducing a **protecting group** to block the interfering function, (2) carrying out the desired reaction, and (3) removing the protecting group.

One of the most common methods of alcohol protection is reaction with chlorotrimethylsilane to yield a trimethylsilyl (TMS) ether. The reaction is carried out in the presence of a base (usually triethylamine) to help form the alkoxide anion from the alcohol and to remove the HCl by-product from the reaction.

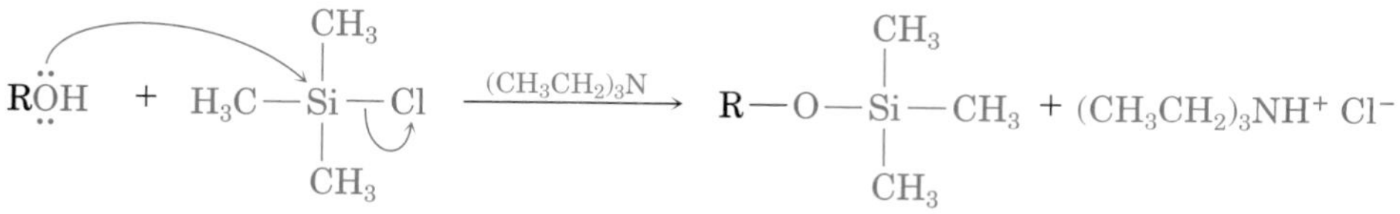

An alcohol **Chlorotrimethylsilane** **A trimethylsilyl (TMS) ether**

For example:

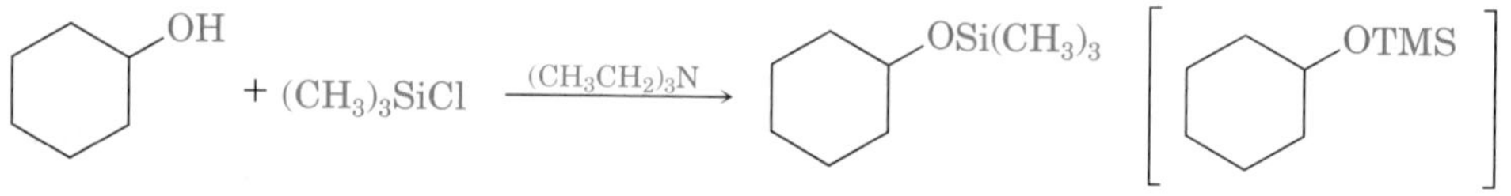

Cyclohexanol **Cyclohexyl trimethylsilyl ether (94%)**

The ether-forming step is an attack of the alkoxide ion on the silicon atom, with concurrent loss of a leaving chloride anion. Unlike most S_N2 reactions, though, this reaction takes place at a *tertiary* center—a trialkyl-substituted silicon atom. The reaction occurs because silicon, a third-row atom, is larger than carbon and forms longer bonds. The three methyl substituents attached to silicon thus offer less steric hindrance to attack than they do in the analogous *tert*-butyl chloride.

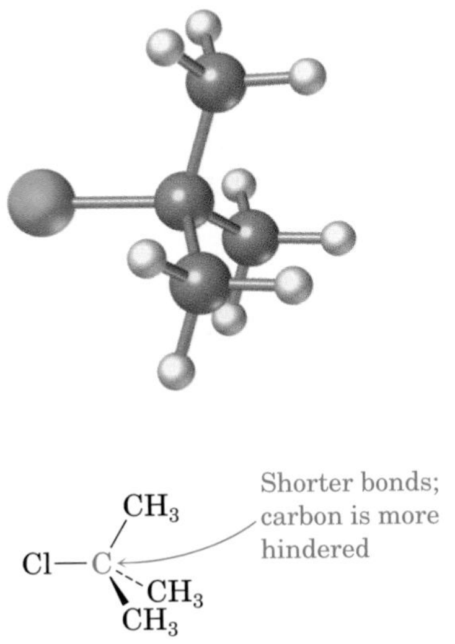

Shorter bonds; carbon is more hindered

Longer bonds; silicon is less hindered

C–C bond length: 154 pm

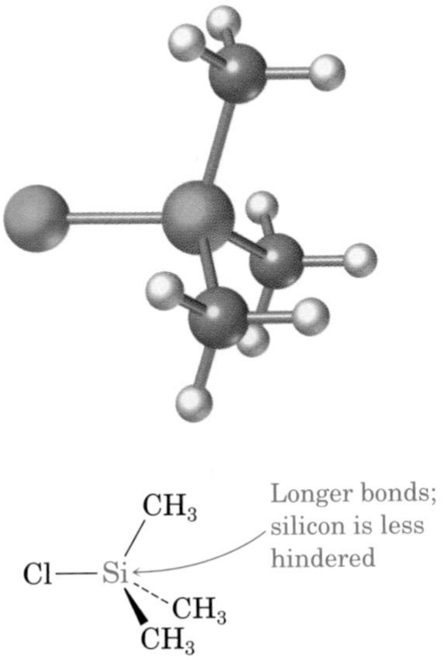

C–Si bond length: 195 pm

Like most other ethers that we'll study in the next chapter, TMS ethers are relatively unreactive. They have no acidic hydrogens and are therefore protected against reaction with oxidizing agents, reducing agents, and Grignard reagents. They do, however, react with aqueous acid or with fluoride ion to regenerate the alcohol.

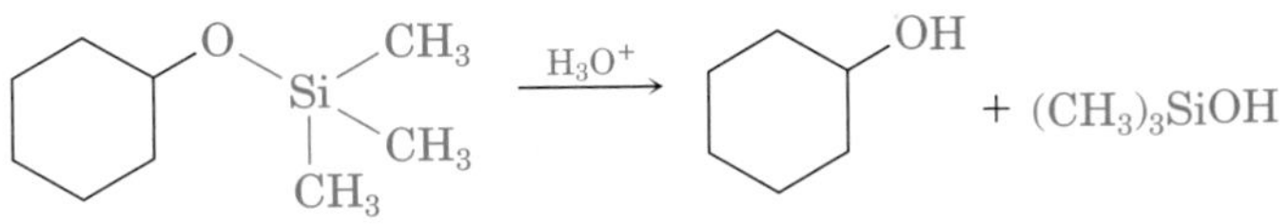

Cyclohexyl TMS ether

Cyclohexanol

To solve the problem posed at the beginning of this section, it's possible to use a halo alcohol in a Grignard reaction by employing a protection sequence. For example, we can add 3-bromo-1-propanol to acetaldehyde by the route shown in Figure 17.10.

FIGURE 17.10 ▼

Use of a TMS-protected alcohol during a Grignard reaction.

STEP 1   Protect alcohol:

$$HOCH_2CH_2CH_2Br + (CH_3)_3SiCl \xrightarrow{(CH_3CH_2)_3N} (CH_3)_3SiOCH_2CH_2CH_2Br$$

STEP 2a   Form Grignard reagent:

$$(CH_3)_3SiOCH_2CH_2CH_2Br \xrightarrow[\text{Ether}]{\text{Mg}} (CH_3)_3SiOCH_2CH_2CH_2MgBr$$

STEP 2b   Do Grignard reaction:

$$(CH_3)_3SiOCH_2CH_2CH_2MgBr \xrightarrow[\text{2. } H_3O^+]{\text{1. } CH_3\overset{\overset{\text{O}}{\|}}{C}H} (CH_3)_3SiOCH_2CH_2CH_2\overset{\overset{\text{OH}}{|}}{C}HCH_3$$

STEP 3   Remove protecting group:

$$(CH_3)_3SiOCH_2CH_2CH_2\overset{\overset{\text{OH}}{|}}{C}HCH_3 \xrightarrow{H_3O^+} HOCH_2CH_2CH_2\overset{\overset{\text{OH}}{|}}{C}HCH_3 + (CH_3)_3SiOH$$

· ·

Problem 17.15   TMS ethers can be removed by treatment with fluoride ion as well as by acid-catalyzed hydrolysis. Propose a mechanism for the reaction of cyclohexyl TMS ether with LiF. Fluorotrimethylsilane is a product.

· ·

17.10   Preparation and Uses of Phenols

The outbreak of World War I provided a stimulus for industrial preparation of large amounts of synthetic phenol, which was needed as a raw material to manufacture the explosive, picric acid (2,4,6-trinitrophenol). Today, more than 2 million tons of phenol are manufactured each year in the United States for use in such products as Bakelite resin and adhesives for binding plywood.

For many years, phenol was manufactured by the Dow process, in which chlorobenzene reacts with NaOH at high temperature and pressure (Section 16.9). Now, however, an alternative synthesis from isopropylbenzene (cumene) is used. Cumene reacts with air at high temperature by a radical mechanism to form cumene hydroperoxide, which is converted into phenol and acetone by treatment with acid. This is a particularly efficient process because two valuable chemicals are prepared at the same time.

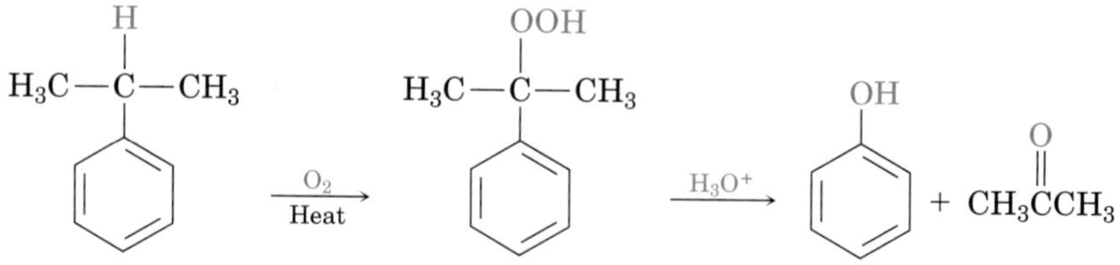

Cumene   Cumene hydroperoxide   Phenol   Acetone

The reaction occurs by protonation of oxygen, followed by rearrangement of the phenyl group from carbon to oxygen with simultaneous loss of water. Readdition of water then yields an intermediate called a *hemiacetal,* which breaks down to phenol and acetone (Figure 17.11). (A hemiacetal is

FIGURE 17.11 ▼

Mechanism of the
formation of phenol by
acid-catalyzed reaction of
cumene hydroperoxide.

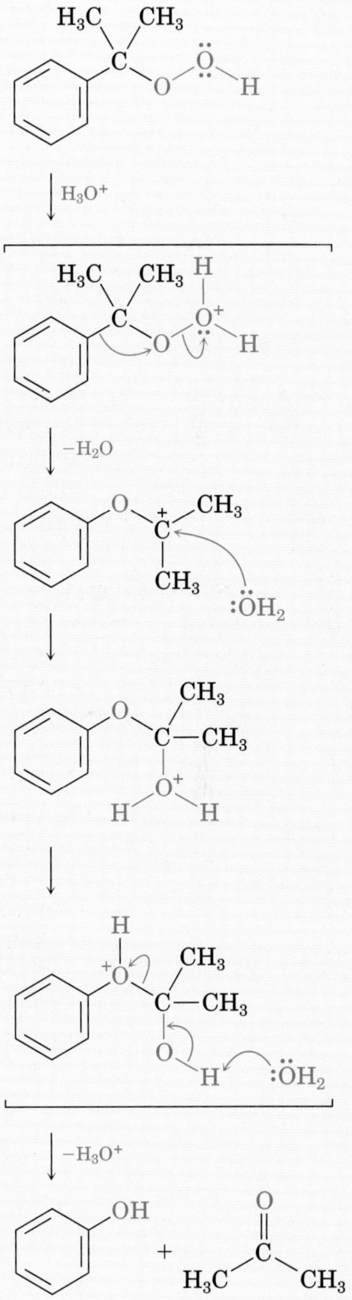

Protonation of the hydroperoxy group
on the terminal oxygen atom gives an
oxonium ion . . .

. . . which undergoes rearrangement
by migration of the phenyl group from
carbon to oxygen, yielding a carbocation
and expelling water as a leaving group.

Nucleophilic addition of water to the
carbocation yields another oxonium
ion . . .

. . . which rearranges by transfer of H⁺
from one oxygen to another, giving a
protonated hemiacetal.

Elimination of phenol from the
protonated hemiacetal gives acetone
as coproduct.

a compound that contains one –OR group and one –OH group bonded to the same carbon atom.)

In the laboratory, simple phenols can be prepared from aromatic sulfonic acids by melting with NaOH at high temperature (Section 16.2). Few functional groups can survive such harsh conditions, though, and the reaction is therefore limited to the preparation of alkyl-substituted phenols. We'll see a better method of phenol preparation from aromatic amines in Section 24.8.

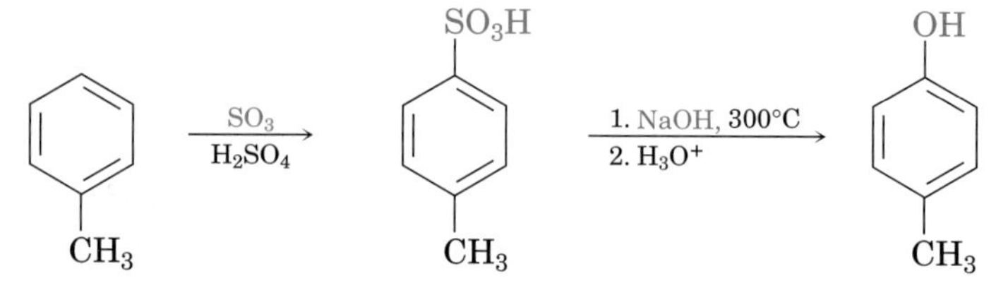

| Toluene | *p*-Toluenesulfonic acid | *p*-Methylphenol (72%) |

In addition to its use in resins and adhesives, phenol is also the starting material for the synthesis of chlorinated phenols and the food preservatives BHT (butylated hydroxytoluene) and BHA (butylated hydroxyanisole). Pentachlorophenol, a widely used wood preservative, is prepared by reaction of phenol with excess Cl_2. The herbicide 2,4-D (2,4-dichlorophenoxyacetic acid) is prepared from 2,4-dichlorophenol, and the hospital antiseptic agent hexachlorophene is prepared from 2,4,5-trichlorophenol.

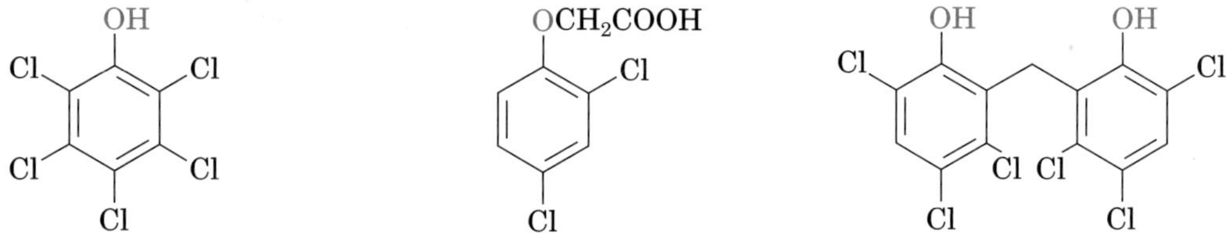

Pentachlorophenol
(wood preservative)

2,4-Dichlorophenoxyacetic acid,
2,4-D (herbicide)

Hexachlorophene
(antiseptic)

The food preservative BHT is prepared by Friedel–Crafts alkylation of *p*-methylphenol (*p*-cresol) with 2-methylpropene in the presence of acid; BHA is prepared similarly by alkylation of *p*-methoxyphenol.

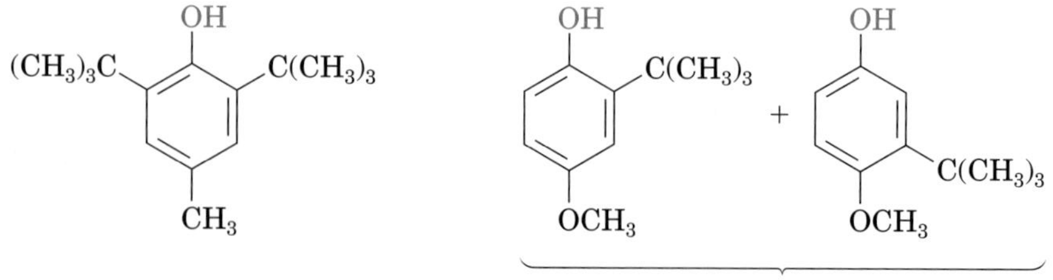

BHT

BHA

Problem 17.16   *p*-Cresol (*p*-methylphenol) is used both as an antiseptic and as a starting material to prepare the food additive BHT. How would you prepare *p*-cresol from benzene?

Problem 17.17   Show the mechanism of the reaction of *p*-methylphenol with 2-methylpropene and H_3PO_4 catalyst to yield the food additive BHT.

17.11   Reactions of Phenols

Electrophilic Aromatic Substitution Reactions

The hydroxyl group is a strongly activating, ortho- and para-directing substituent in electrophilic aromatic substitution reactions (Section 16.5). As a result, phenols are highly reactive substrates for electrophilic halogenation, nitration, sulfonation, and Friedel–Crafts reactions.

Oxidation of Phenols: Quinones

Phenols do not undergo oxidation in the same manner that alcohols do because they do not have a hydrogen atom on the hydroxyl-bearing carbon. Instead, reaction of a phenol with any of a number of strong oxidizing agents yields a 2,5-cyclohexadiene-1,4-dione, or **quinone**. Older procedures employed $Na_2Cr_2O_7$ as oxidant, but Fremy's salt [potassium nitrosodisulfonate, $(KSO_3)_2NO$] is now preferred. The reaction takes place under mild conditions through a radical mechanism, and good yields are normally obtained.

Phenol

$\xrightarrow[\text{H}_2\text{O}]{(KSO_3)_2NO}$

Benzoquinone (79%)

Quinones are an interesting and valuable class of compounds because of their oxidation–reduction (*redox*) properties. They can be easily reduced to **hydroquinones** (*p*-dihydroxybenzenes) by reagents such as $NaBH_4$ and $SnCl_2$, and hydroquinones can be easily reoxidized back to quinones by Fremy's salt.

$\underset{\text{Fremy's salt}}{\overset{\text{SnCl}_2,\ \text{H}_2\text{O}}{\rightleftharpoons}}$

Benzoquinone   **Hydroquinone**

**Biological
Connection** ▶

The redox properties of quinones are important to the functioning of living cells, where compounds called *ubiquinones* act as biochemical oxidizing agents to mediate the electron-transfer processes involved in energy production. Ubiquinones, also called *coenzymes Q*, are components of the cells of all aerobic organisms, from the simplest bacterium to humans. They are so named because of their ubiquitous occurrence in nature.

Ubiquinones (*n* = 1–10)

Ubiquinones function within the mitochondria of cells to mediate the respiration process in which electrons are transported from the biological reducing agent NADH to molecular oxygen. Although a complex series of steps is involved in the overall process, the ultimate result is a cycle whereby NADH is oxidized to NAD^+, O_2 is reduced to water, and energy is produced. Ubiquinone acts only as an intermediary and is itself unchanged.

STEP 1

$NADH + H^+ +$ **Reduced form** $\longrightarrow$ $+ \; NAD^+$ **Oxidized form**

STEP 2

$+ \frac{1}{2}O_2 \Longrightarrow$ $+ \; H_2O$

Net change: $NADH + \frac{1}{2}O_2 + H^+ \longrightarrow NAD^+ + H_2O$

17.12 Spectroscopy of Alcohols and Phenols

Infrared Spectroscopy

Alcohols have a characteristic O–H stretching absorption at 3300–3600 cm^{-1} in the infrared spectrum. The exact position of the absorption depends on the extent of hydrogen bonding in the sample. Unassociated alcohols show a fairly sharp absorption near 3600 cm^{-1}, whereas hydrogen-bonded alcohols show a broader absorption in the 3300–3400 cm^{-1} range. The hydrogen-bonded hydroxyl absorption appears at 3350 cm^{-1} in the infrared spectrum of cyclohexanol (Figure 17.12). Alcohols also show a strong C–O stretching absorption near 1050 cm^{-1}.

FIGURE 17.12 ▼

Infrared spectrum of cyclohexanol. Characteristic O–H and C–O stretching absorptions are indicated.

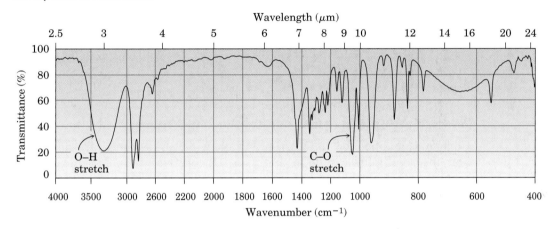

Phenols also show a characteristic broad IR absorption at 3500 cm^{-1} due to the –OH group, as well as the usual 1500 and 1600 cm^{-1} aromatic bands (Figure 17.13). In phenol itself, the monosubstituted aromatic-ring peaks at 690 and 760 cm^{-1} are visible.

FIGURE 17.13 ▼

Infrared spectrum of phenol.

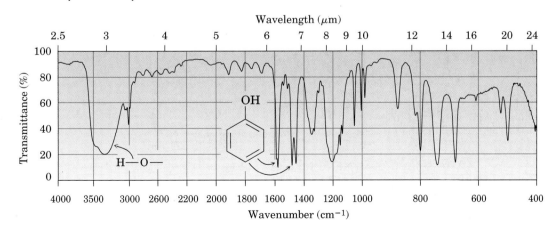

. . . .

Problem 17.18   Assume that you need to prepare 5-cholestene-3-one from cholesterol. How could you use infrared spectroscopy to tell if the reaction was successful? What differences would you look for in the infrared spectra of starting material and product?

$$\xrightarrow[\text{H}_3\text{O}^+]{\text{CrO}_3}$$

Cholesterol   **5-Cholestene-3-one**

. . . .

Nuclear Magnetic Resonance Spectroscopy

Carbon atoms bonded to electron-withdrawing –OH groups are deshielded and absorb at a lower field in the ^{13}C NMR spectrum than do typical alkane carbons. Most alcohol carbon absorptions fall in the range 50–80 δ, as the following data illustrate for cyclohexanol:

OH
69.50 δ
35.5 δ
24.4 δ
25.9 δ

Alcohols also show characteristic absorptions in the ^{1}H NMR spectrum. Hydrogens on the oxygen-bearing carbon atom are deshielded by the electron-withdrawing effect of the nearby oxygen, and their absorptions occur in the range 3.5–4.5 δ. Surprisingly, however, splitting is not usually observed between the O–H proton and the neighboring protons on carbon. Most samples contain small amounts of acidic impurities, which catalyze an exchange of the O–H proton on a time scale so rapid that the effect of spin–spin splitting is removed.

$$-\overset{|}{\underset{|}{\text{C}}}-\text{O}-\text{H} \xrightleftharpoons{\text{HA}} -\overset{|}{\underset{|}{\text{C}}}-\text{O}-\text{H} + \text{HA}$$

No NMR coupling observed

It's often possible to take advantage of this rapid proton exchange to identify the position of the O–H absorption. If a small amount of deuterated water, D_2O, is added to the NMR sample tube, the O–H proton is rapidly exchanged for deuterium, and the hydroxyl absorption disappears from the spectrum.

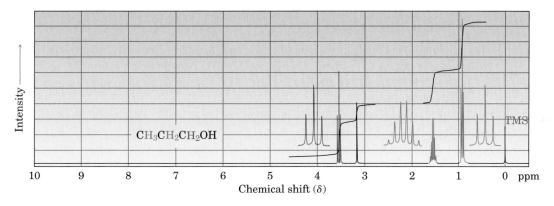

Spin–spin splitting *is* observed between protons on the oxygen-bearing carbon and other neighbors. For example, the signal of the two –CH₂O– protons in 1-propanol is split into a triplet by coupling with the neighboring –CH₂– protons (Figure 17.14).

FIGURE 17.14 ▼

¹H NMR spectrum of 1-propanol. The protons on the oxygen-bearing carbon are split into a triplet at 3.58 δ.

Phenols, like all aromatic compounds, show ¹H NMR absorptions near 7–8 δ, the expected position for aromatic-ring protons. In addition, phenol O–H protons absorb at 3–8 δ. In neither case are these absorptions uniquely diagnostic for phenols, since other kinds of protons absorb in the same range.

● ●

Problem 17.19   When the ¹H NMR spectrum of an alcohol is run in dimethyl sulfoxide (DMSO) solvent rather than in chloroform, exchange of the O–H proton is slow, and spin–spin splitting is seen between the O–H proton and C–H protons on the adjacent carbon. What spin multiplicities would you expect for the hydroxyl protons in the following alcohols?
(a) 2-Methyl-2-propanol (b) Cyclohexanol (c) Ethanol
(d) 2-Propanol (e) Cholesterol (f) 1-Methylcyclohexanol

● ●

Mass Spectrometry

Alcohols undergo fragmentation in the mass spectrometer by two characteristic pathways, *alpha cleavage* and *dehydration*. In the alpha-cleavage pathway, a C–C bond nearest the hydroxyl group is broken, yielding a neutral radical plus a charged oxygen-containing fragment:

Alpha cleavage

$$\left[\begin{array}{c} OH \\ | \\ R-\underset{\underset{R}{|}}{C_\alpha} \!\!\gtrless\!\! CH_2R \end{array} \right]^{+\cdot} \longrightarrow \left[\begin{array}{c} OH \\ | \\ R-\underset{\underset{R}{|}}{C} \end{array} \right]^{+} + \cdot CH_2R$$

In the dehydration pathway, water is eliminated, yielding an alkene radical cation:

Dehydration

$$\left[\begin{array}{c} H \quad\quad OH \\ \diagdown \quad\quad\quad\diagup \\ C-C \\ \diagup \quad\quad\quad\diagdown \end{array} \right]^{+\cdot} \longrightarrow H_2O + \left[\begin{array}{c} \diagdown \quad\quad\quad\diagup \\ C=C \\ \diagup \quad\quad\quad\diagdown \end{array} \right]^{+\cdot}$$

Both of these characteristic fragmentation modes are apparent in the mass spectrum of 1-butanol (Figure 17.15). The peak at $m/z = 56$ is due to loss of water from the molecular ion, and the peak at $m/z = 31$ is due to an alpha cleavage.

FIGURE 17.15 ▼

Mass spectrum of 1-butanol ($M^+ = 74$). Dehydration gives a peak at $m/z = 56$, and fragmentation by alpha cleavage gives a peak at $m/z = 31$.

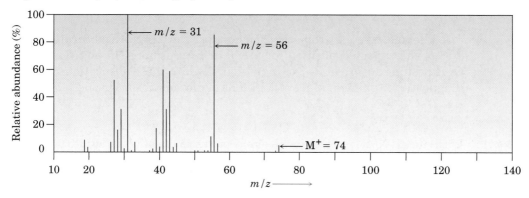

Ethanol: Chemical, Drug, and Poison

The production of ethanol by fermentation of grains and sugars is one of the oldest known organic reactions, going back at least 2500 years. Fermentation is carried out by adding yeast to an aqueous sugar solution, where enzymes break down carbohydrates into ethanol and CO_2:

$$C_6H_{12}O_6 \xrightarrow{\text{Yeast}} 2\ CH_3CH_2OH\ +\ 2\ CO_2$$

A carbohydrate

Approximately 110 million gallons of ethanol are produced each year in the United States, primarily for use as a solvent. Only about 5% of this industrial ethanol comes from fermentation, though; most is obtained by acid-catalyzed hydration of ethylene.

$$H_2C = CH_2\ +\ H_2O \xrightarrow[\text{catalyst}]{\text{Acid}} CH_3CH_2OH$$

Ethanol is classified for medical purposes as a central nervous system (CNS) depressant. Its effects (that is, being drunk) resemble the human response to anesthetics. There is an initial excitability and increase in sociable behavior, but this results from depression of inhibition rather than from stimulation. At a blood alcohol concentration of 0.1–0.3%, or 100–300 mg/dL, motor coordination is affected, accompanied by loss of balance, slurred speech, and amnesia. When blood alcohol concentration rises to 0.3–0.4%, nausea and loss of consciousness occur. Above 0.6%, spontaneous respiration and cardiovascular regulation are affected, ultimately leading to death. The LD_{50} of ethanol is 10.6 g/kg (see "Chemical Toxicity and Risk" at the end of Chapter 1).

The passage of ethanol through the body begins with its absorption in the stomach and small intestine, followed by rapid distribution to all body fluids and organs. In the pituitary gland, ethanol inhibits the production of a hormone that regulates urine flow, causing increased urine production and dehydration. In the stomach, ethanol stimulates production of acid. Throughout the body, ethanol causes blood vessels to dilate, resulting in flushing of the skin and a sensation of warmth as blood moves into capillaries beneath the surface. The result is not a warming of the body, but an increased loss of heat at the surface.

The metabolism of ethanol occurs mainly in the liver and proceeds by oxidation in two steps, first to acetaldehyde (CH_3CHO) and then to

(continued) ▶

acetic acid (CH₃COOH). Ethanol and acetaldehyde are toxic, leading to devastating physical and metabolic deterioration in chronic alcoholics. The liver usually suffers the worst damage since it is the major site of alcohol metabolism.

The quick and uniform distribution of ethanol in body fluids, the ease with which it crosses lung membranes, and its ready oxidizability provide the basis for simple tests for blood alcohol concentration. The *Breathalyzer test* measures alcohol concentration in expired air by the color change that occurs when the bright orange oxidizing agent potassium dichromate ($K_2Cr_2O_7$) is reduced to blue-green chromium(III). Alternatively, the *Intoxilyzer* test uses infrared spectroscopy to measure blood alcohol levels. In most states, driving with a blood alcohol level above 0.10% (100 mg/dL) is illegal, and some states have lowered the legal limit to 0.08%.

More than 2000 years ago, wine might have been stored in this amphora.

Summary and Key Words

KEY WORDS

alcohol, 654
alkoxide ion (RO⁻), 659
hydroquinone, 687
phenol, 654
phenoxide ion (ArO⁻), 659
protecting group, 682
quinone, 687

Alcohols are among the most versatile of all organic compounds. They occur widely in nature, are important industrially, and have an unusually rich chemistry. The most important methods of alcohol synthesis start with carbonyl compounds. Aldehydes, ketones, esters, and carboxylic acids are reduced by reaction with either $NaBH_4$ or $LiAlH_4$. Aldehydes, esters, and carboxylic acids yield primary alcohols (RCH_2OH) on reduction; ketones yield secondary alcohols (R_2CHOH).

The Grignard reaction with a carbonyl compound is another important method for preparing alcohols. Addition of a Grignard reagent to formaldehyde yields a primary alcohol, addition to an aldehyde yields a secondary alcohol, and addition to a ketone or an ester yields a tertiary alcohol. Carboxylic acids do not give Grignard addition products. The Grignard synthesis of alcohols is limited by the fact that Grignard reagents can't be prepared from alkyl halides that contain reactive functional groups in the same molecule. This problem can sometimes be avoided by **protecting** the interfering functional group. Alcohols are often protected by formation of trimethylsilyl (TMS) ethers.

Alcohols undergo a great many reactions. They can be dehydrated by treatment with $POCl_3$ and can be transformed into alkyl halides by treatment with PBr_3 or $SOCl_2$. Furthermore, alcohols are weakly acidic ($pK_a \approx 16$–18). They react with strong bases and with alkali metals to form **alkoxide anions**, which are used frequently in organic synthesis.

One of the most important reactions of alcohols is their oxidation to carbonyl compounds. Primary alcohols yield either aldehydes or carboxylic acids, secondary alcohols yield ketones, but tertiary alcohols are not normally oxidized. Pyridinium chlorochromate (PCC) in dichloromethane is often used for oxidizing primary alcohols to aldehydes and secondary alcohols to ketones. A solution of CrO_3 in aqueous acid is frequently used for oxidizing primary alcohols to carboxylic acids and secondary alcohols to ketones.

Phenols are aromatic counterparts of alcohols but are much more acidic ($pK_a \approx 10$) because phenoxide anions are stabilized by delocalization of the negative charge into the aromatic ring. Substitution of the aromatic ring by an electron-withdrawing group increases phenol acidity, and substitution by an electron-donating group decreases acidity. Phenols can be oxidized to **quinones** by reaction with Fremy's salt (potassium nitrosodisulfonate), and quinones can be reduced to **hydroquinones** by reaction with $NaBH_4$.

Summary of Reactions

1. Synthesis of alcohols
   (a) Reduction of carbonyl compounds (Section 17.5)
      (1) Aldehydes

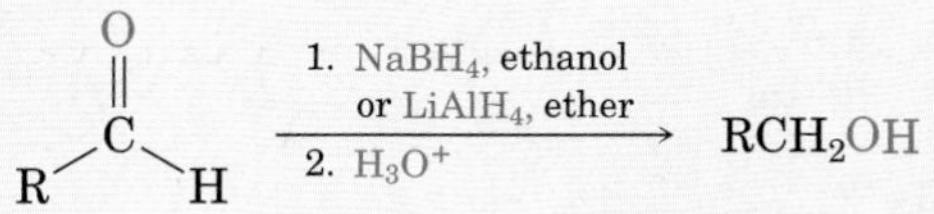

A primary alcohol

(2) Ketones

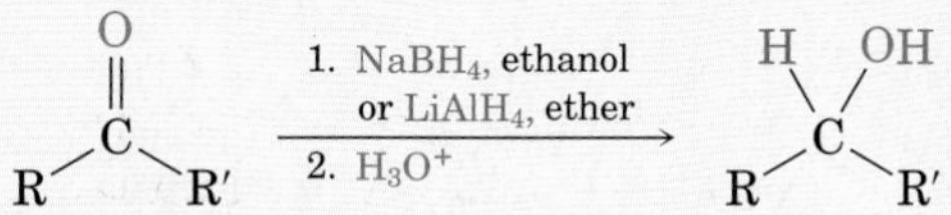

A secondary alcohol

(3) Esters

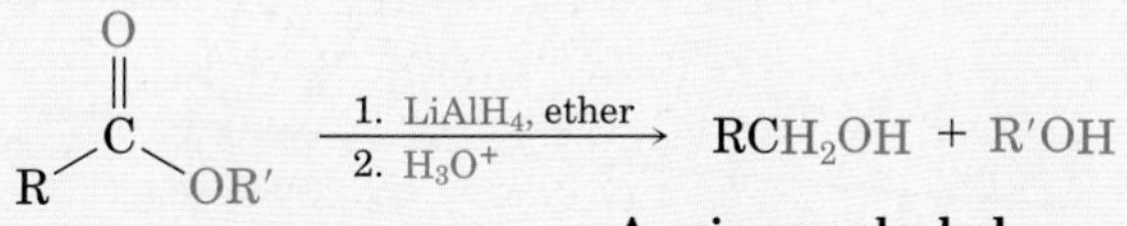

A primary alcohol

(4) Carboxylic acids

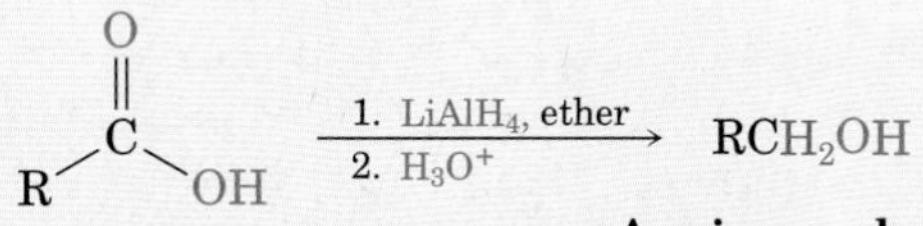

A primary alcohol

(b) Grignard addition to carbonyl compounds (Section 17.6)
   (1) Formaldehyde

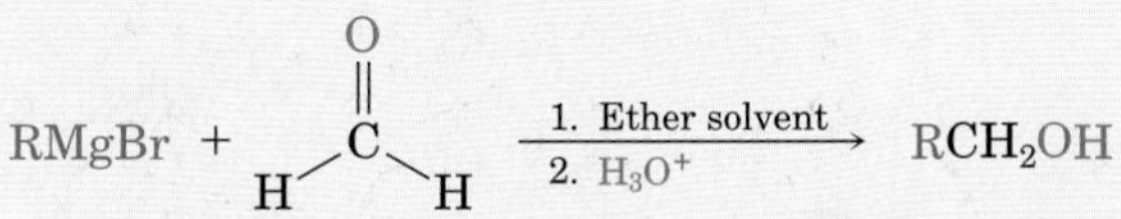

A primary alcohol

(continued) ▶

(2) Aldehydes

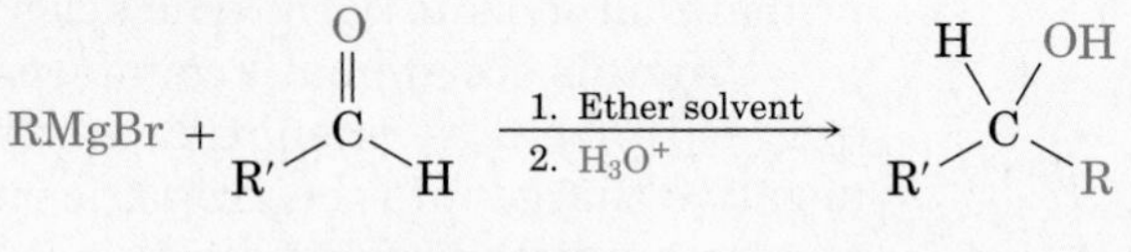

$$RMgBr + \underset{R'}{\overset{O}{\underset{\|}{C}}}H \xrightarrow[\text{2. } H_3O^+]{\text{1. Ether solvent}} \underset{R'}{\overset{H \quad OH}{\underset{R}{C}}}$$

A secondary alcohol

(3) Ketones

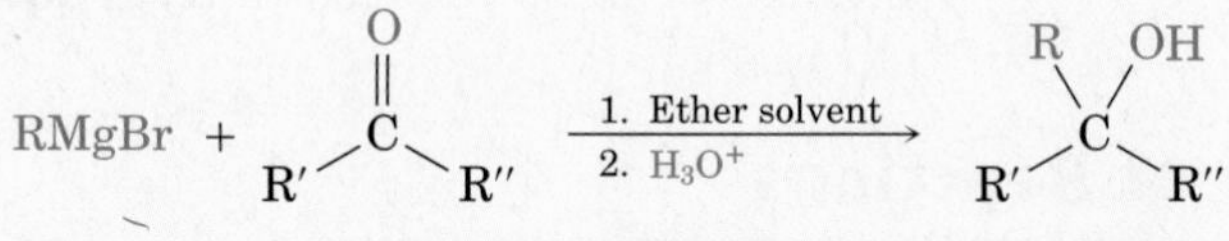

$$RMgBr + \underset{R'}{\overset{O}{\underset{\|}{C}}}R'' \xrightarrow[\text{2. } H_3O^+]{\text{1. Ether solvent}} \underset{R'}{\overset{R \quad OH}{\underset{R''}{C}}}$$

A tertiary alcohol

(4) Esters

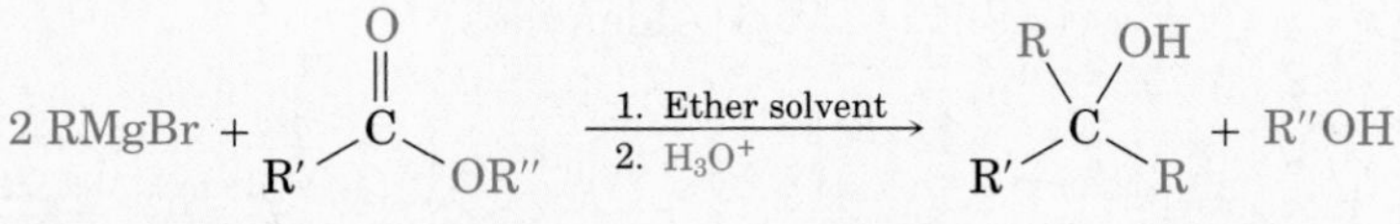

$$2\,RMgBr + \underset{R'}{\overset{O}{\underset{\|}{C}}}OR'' \xrightarrow[\text{2. } H_3O^+]{\text{1. Ether solvent}} \underset{R'}{\overset{R \quad OH}{\underset{R}{C}}} + R''OH$$

A tertiary alcohol

2. Reactions of alcohols
 (a) Acidity (Section 17.3)

$$ROH + NaH \longrightarrow RO^- Na^+ + H_2$$
$$2\,ROH + 2\,Na \longrightarrow 2\,RO^- Na^+ + H_2$$

(b) Dehydration (Section 17.7)
 (1) Tertiary alcohols

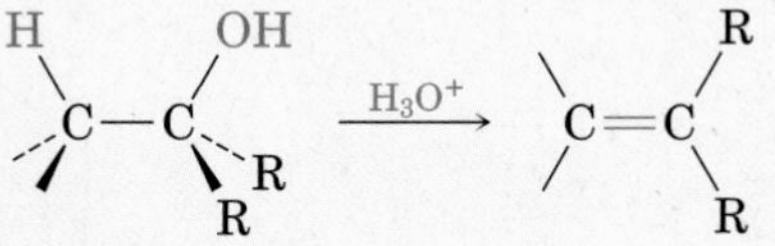

$$\underset{R}{\overset{H \quad OH}{\underset{R}{C-C}}} \xrightarrow{H_3O^+} \overset{R}{\underset{R}{C=C}}$$

(2) Secondary and tertiary alcohols

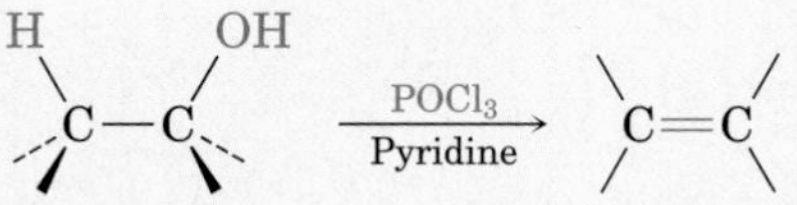

$$\overset{H \quad OH}{C-C} \xrightarrow[\text{Pyridine}]{POCl_3} C=C$$

(c) Oxidation (Section 17.8)
 (1) Primary alcohols

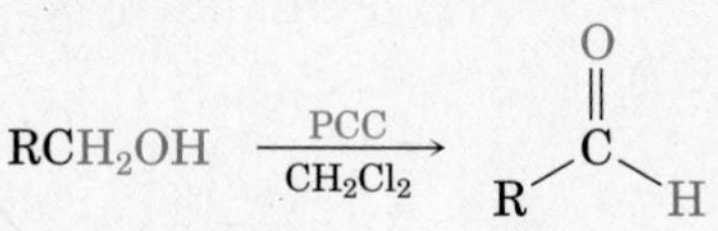

$$RCH_2OH \xrightarrow[CH_2Cl_2]{PCC} \underset{R}{\overset{O}{\underset{\|}{C}}}H$$

An aldehyde

(continued) ▶

$$RCH_2OH \xrightarrow[\text{H}_3\text{O}^+,\text{ acetone}]{\text{CrO}_3} \underset{\textbf{A carboxylic acid}}{R-\overset{\displaystyle O}{\overset{\|}{C}}-OH}$$

(2) Secondary alcohols

$$\underset{R}{\overset{H}{\underset{}{}}}\overset{OH}{\underset{R'}{C}} \xrightarrow[\text{CH}_2\text{Cl}_2]{\text{PCC}} \underset{\textbf{A ketone}}{R-\overset{\displaystyle O}{\overset{\|}{C}}-R'}$$

3. Preparation of phenols; alkali fusion of aryl sulfonates (Section 17.10)

$$\text{C}_6\text{H}_5\text{-SO}_3\text{H} + \text{NaOH} \xrightarrow{300°\text{C}} \text{C}_6\text{H}_5\text{-OH} + \text{NaHSO}_3$$

4. Reaction of phenols; oxidation of quinones (Section 17.11)

$$\text{C}_6\text{H}_5\text{-OH} \xrightarrow[\text{H}_2\text{O}]{(\text{KSO}_3)_2\text{NO}} \text{quinone}$$

Visualizing Chemistry

• •

(Problems 17.1–17.19 appear within the chapter.)

17.20 Give IUPAC names for the following compounds:

(a)

(b)

(c)

(d)

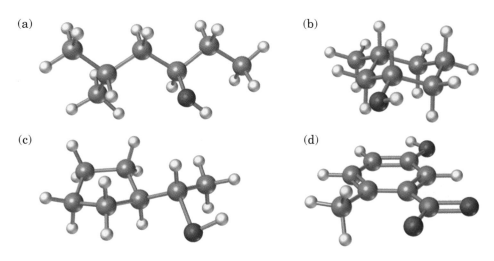

17.21 Predict the product from reaction of the following substance (reddish brown = Br) with:
(a) PBr_3 (b) Aqueous H_2SO_4 (c) $SOCl_2$ (d) PCC   (e) Br_2, $FeBr_3$

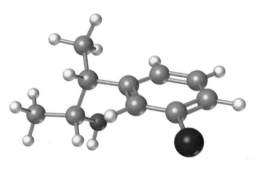

17.22 Predict the product from reaction of the following substance with:
(a) $NaBH_4$; then H_3O^+ (b) $LiAlH_4$; then H_3O^+ (c) CH_3CH_2MgBr; then H_3O^+

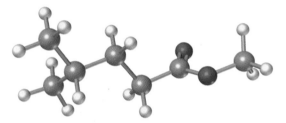

17.23 Name and assign R or S stereochemistry to the product(s) you would obtain by reaction of the following substance with ethylmagnesium bromide. Is the product chiral? Is it optically active? Explain.

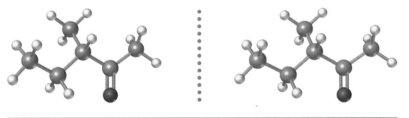

Stereo View

Additional Problems

17.24 Give IUPAC names for the following compounds:

(a) $HOCH_2CH_2\overset{\overset{\displaystyle CH_3}{|}}{C}HCH_2OH$ (b) $CH_3\overset{\overset{\displaystyle OH}{|}}{C}H\overset{\underset{\displaystyle CH_2CH_2CH_3}{|}}{C}HCH_2CH_3$ (c) ![cyclobutane with H, OH on one carbon and H, HO on adjacent carbon]

(d) (e) (f)

17.25 Draw and name the eight isomeric alcohols with formula $C_5H_{12}O$.

17.26 Which of the eight alcohols you identified in Problem 17.25 react with CrO_3 in aqueous acid? Show the products you would expect from each reaction.

17.27 How would you prepare the following compounds from 2-phenylethanol? More than one step may be required.
(a) Styrene ($PhCH=CH_2$) (b) Phenylacetaldehyde ($PhCH_2CHO$)
(c) Phenylacetic acid ($PhCH_2CO_2H$) (d) Benzoic acid
(e) Ethylbenzene (f) Benzaldehyde
(g) 1-Phenylethanol (h) 1-Bromo-2-phenylethane

17.28 How would you prepare the following compounds from 1-phenylethanol? More than one step may be required.
(a) Acetophenone ($PhCOCH_3$) (b) Benzyl alcohol
(c) *m*-Bromobenzoic acid (d) 2-Phenyl-2-propanol

17.29 What Grignard reagent and what carbonyl compound might you start with to prepare the following alcohols?

(a) $CH_3\overset{OH}{\underset{|}{C}}HCH_2CH_3$ (b) (c) $H_2C=\overset{CH_3}{\underset{|}{C}}-CH_2OH$

(d) $(Ph)_3COH$ (e) $CH_3\overset{OH}{\underset{|}{C}}HCH_2CH_2CH_2Br$

17.30 When 4-chloro-1-butanol is treated with a strong base such as sodium hydride, NaH, tetrahydrofuran is produced. Suggest a mechanism.

$$ClCH_2CH_2CH_2CH_2OH \xrightarrow[\text{Ether}]{\text{NaH}} \text{(structure)} + H_2 + NaCl$$

17.31 What carbonyl compounds would you reduce to prepare the following alcohols? List all possibilities.

(a) 2,2-Dimethyl-1-hexanol (b) 3,3-Dimethyl-2-butanol

(c) (d)

17.32 How would you carry out the following transformations?

(a)

(b)

a cinnamic acid with CO_2H group converts to cinnamyl alcohol with CH_2OH group

17.33 What carbonyl compounds might you start with to prepare the following compounds by Grignard reaction? List all possibilities.

(a) 2-Methyl-2-propanol

(b) 1-Ethylcyclohexanol

(c) 3-Phenyl-3-pentanol

(d) 2-Phenyl-2-pentanol

(e)

H_3C—benzene ring—CH_2CH_2OH

(f)

cyclopentane—$CH_2C(CH_3)_2$ with OH

17.34 What products would you obtain from reaction of 1-pentanol with the following reagents?
(a) PBr_3 (b) $SOCl_2$ (c) CrO_3, H_2O, H_2SO_4 (d) PCC

17.35 Evidence for the intermediate carbocations in the acid-catalyzed dehydration of alcohols comes from the observation that rearrangements sometimes occur. Propose a mechanism to account for the formation of 2,3-dimethyl-2-butene from 3,3-dimethyl-2-butanol. (Review Section 6.12 if necessary.)

$$CH_3-\underset{\underset{H_3C}{|}}{\overset{\overset{H_3C}{|}}{C}}-\overset{\overset{OH}{|}}{C}HCH_3 \xrightarrow{H_2SO_4} (CH_3)_2C=C(CH_3)_2 + H_2O$$

17.36 Acid-catalyzed dehydration of 2,2-dimethylcyclohexanol yields a mixture of 1,2-dimethylcyclohexene and isopropylidenecyclopentane. Propose a mechanism to account for the formation of both products.

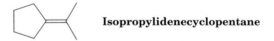 **Isopropylidenecyclopentane**

17.37 How would you prepare the following substances from cyclopentanol? More than one step may be required.
(a) Cyclopentanone   (b) Cyclopentene
(c) 1-Methylcyclopentanol   (d) *trans*−2-Methylcyclopentanol

17.38 What products would you expect to obtain from reaction of 1-methylcyclohexanol with the following reagents?
(a) HBr   (b) NaH   (c) H_2SO_4 (d) $Na_2Cr_2O_7$

17.39 Carvacrol (5-isopropyl-2-methylphenol) is a naturally occurring substance isolated from oregano, thyme, and marjoram. Propose a synthesis of carvacrol from benzene.

17.40 Benzoquinone is an excellent dienophile in the Diels–Alder reaction. What product would you expect from reaction of benzoquinone with 1 equivalent of butadiene? From reaction with 2 equivalents of butadiene?

17.41 Rank the following substituted phenols in order of increasing acidity, and explain your answer:

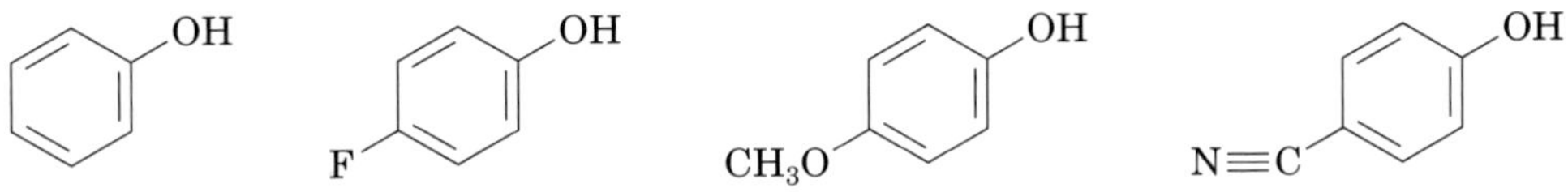

17.42 Benzyl chloride can be converted into benzaldehyde by treatment with nitromethane and base. The reaction involves initial conversion of nitromethane into its anion, followed by S$_N$2 reaction of the anion with benzyl chloride and subsequent E2 reaction. Write the mechanism in detail using curved arrows to indicate the electron flow in each step.

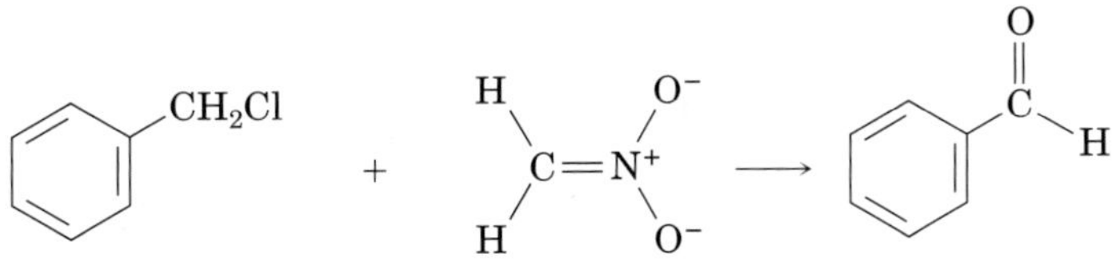

Benzyl chloride   **Nitromethane anion**   **Benzaldehyde**

17.43 Reduction of 2-butanone with NaBH$_4$ yields 2-butanol. Is the product chiral? Is it optically active? Explain.

17.44 Reaction of (S)-3-methyl-2-pentanone with methylmagnesium bromide followed by acidification yields 2,3-dimethyl-2-pentanol. What is the stereochemistry of the product? Is the product optically active?

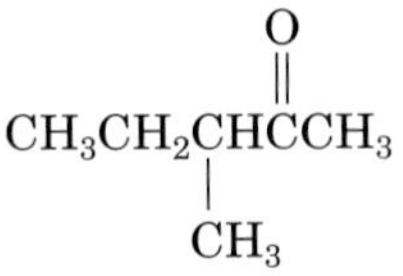

$$CH_3CH_2CHCCH_3$$

with O double bond on carbon and CH$_3$ substituent   **3-Methyl-2-pentanone**

17.45 Testosterone is one of the most important male steroid hormones. When testosterone is dehydrated by treatment with acid, rearrangement occurs to yield the product shown. Propose a mechanism to account for this reaction.

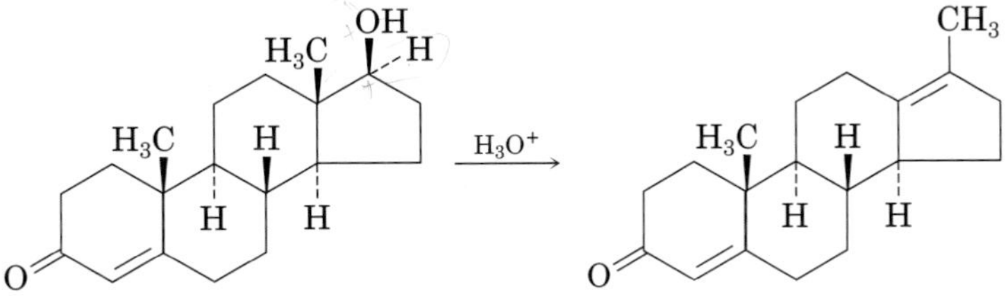

Testosterone

17.46 Starting from testosterone (Problem 17.45), how would you prepare the following substances?

(a)

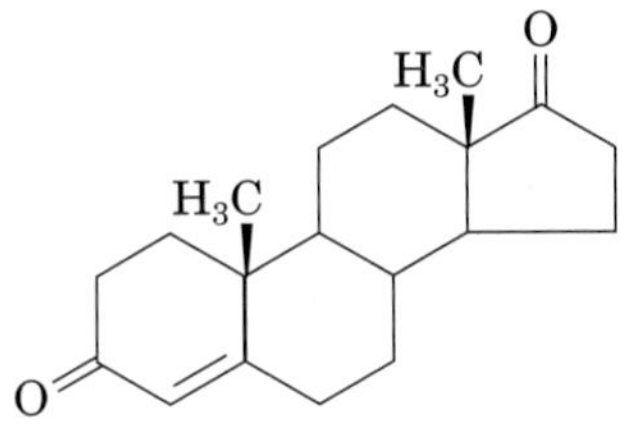

(b)

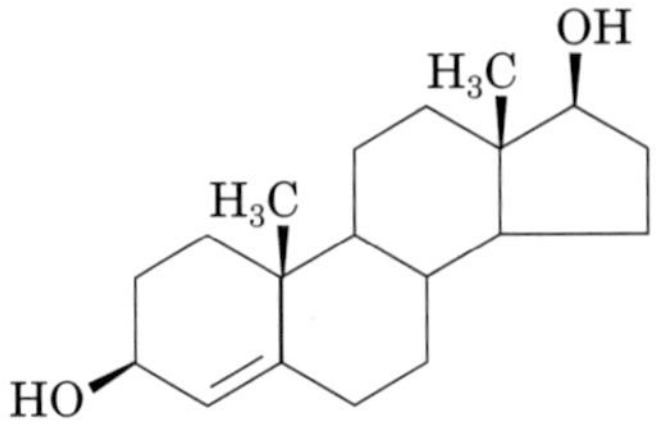

(c)

(d)

17.47   Compound A, $C_{10}H_{18}O$, undergoes reaction with dilute H_2SO_4 at 25°C to yield a mixture of two alkenes, $C_{10}H_{16}$. The major alkene product, B, gives only cyclopentanone after ozone treatment followed by reduction with zinc in acetic acid. Write the reactions involved, and identify A and B.

17.48   Dehydration of *trans*-2-methylcyclopentanol with $POCl_3$ in pyridine yields predominantly 3-methylcyclopentene. Is the stereochemistry of this dehydration syn or anti? Can you suggest a reason for formation of the observed product? (Make molecular models!)

17.49   How would you synthesize the following alcohols, starting with benzene and other alcohols of six or fewer carbons as your only organic reagents?

(a)

(b) $CH_3CH_2CH_2\overset{\overset{\displaystyle CH_3}{|}}{C}HCH_2CH_2OH$

(c)

(d) $CH_3\overset{\overset{\displaystyle CH_3}{|}}{C}HCH_2\overset{\overset{\displaystyle OH}{|}}{C}HCH_2CH_3$

17.50   2,3-Dimethyl-2,3-butanediol has the common name *pinacol*. On heating with aqueous acid, pinacol rearranges to *pinacolone*, 3,3-dimethyl-2-butanone. Suggest a mechanism for this reaction.

$$(CH_3)_2\overset{\overset{\displaystyle OH}{|}}{C}-\overset{\overset{\displaystyle OH}{|}}{C}(CH_3)_2 \quad \xrightarrow{H_3O^+} \quad CH_3\overset{\overset{\displaystyle O}{\|}}{C}C(CH_3)_3 + H_2O$$

Pinacol   **Pinacolone**

17.51   As a rule, axial alcohols oxidize somewhat faster than equatorial alcohols. Which would you expect to oxidize faster, *cis*-4-*tert*-butylcyclohexanol or *trans*-4-*tert*-butylcyclohexanol? Draw the more stable chair conformation of each molecule.

17.52   Propose a synthesis of bicyclohexylidene, starting from cyclohexanone as the only source of carbon.

Bicyclohexylidene

17.53   A problem often encountered in the oxidation of primary alcohols to acids is that esters are sometimes produced as by-products. For example, oxidation of ethanol yields acetic acid and ethyl acetate:

$$CH_3CH_2OH \quad \xrightarrow{CrO_3} \quad CH_3\overset{\overset{\displaystyle O}{\|}}{C}OH + CH_3\overset{\overset{\displaystyle O}{\|}}{C}OCH_2CH_3$$

Propose a mechanism to account for the formation of ethyl acetate. Take into account the reversible reaction between aldehydes and alcohols:

$$\underset{\overset{\|}{O}}{RCH} + R'OH \rightleftharpoons \underset{\overset{|}{OR'}}{\overset{\overset{OH}{|}}{RCH}}$$

17.54 Identify the reagents a–f in the following scheme:

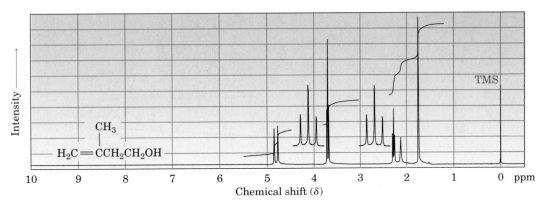

17.55 Propose a structure consistent with the following spectral data for a compound $C_8H_{18}O_2$:

IR: 3350 cm^{-1}

^{1}H NMR: 1.24 δ (12 H, singlet); 1.56 δ (4 H, singlet); 1.95 δ (2 H, singlet)

17.56 The ^{1}H NMR spectrum shown is that of 3-methyl-3-buten-1-ol. Assign all the observed resonance peaks to specific protons, and account for the splitting patterns.

CH$_3$
|
H$_2$C=CCH$_2$CH$_2$OH

TMS

Intensity →

Chemical shift (δ)

10   9   8   7   6   5   4   3   2   1   0   ppm

17.57 Compound A, $C_5H_{10}O$, is one of the basic building blocks of nature. All steroids and many other naturally occurring compounds are built from compound A. Spectroscopic analysis of A yields the following information:

IR: 3400 cm^{-1}; 1640 cm^{-1}

^{1}H NMR: 1.63 δ (3 H, singlet); 1.70 δ (3 H, singlet); 3.83 δ (1 H, broad singlet); 4.15 δ (2 H, doublet, $J = 7$ Hz); 5.70 δ (1 H, triplet, $J = 7$ Hz)

(a) How many double bonds and/or rings does A have?
(b) From the IR spectrum, what is the nature of the oxygen-containing functional group?
(c) What kinds of protons are responsible for the NMR absorptions listed?
(d) Propose a structure for A.

17.58   A compound of unknown structure gave the following spectroscopic data:

Mass spectrum: $M^+ = 88.1$

IR: 3600 cm^{-1}

^{1}H NMR: $1.4 \, \delta$ (2 H, quartet, $J = 7 \text{Hz}$); $1.2 \, \delta$ (6 H, singlet); $1.0 \, \delta$ (1 H, singlet); $0.9 \, \delta$ (3 H, triplet, $J = 7 \text{Hz}$)

^{13}C NMR: 74, 35, 27, 25 δ

(a) Assuming that the compound contains C and H, but may or may not contain O, give three possible molecular formulas.
(b) How many protons (H) does the compound contain?
(c) What functional group(s) does the compound contain?
(d) How many carbons does the compound contain?
(e) What is the molecular formula of the compound?
(f) What is the structure of the compound?
(g) Assign the peaks in the ^{1}H NMR spectrum of the molecule to specific protons.

17.59   The following ^{1}H NMR spectrum is that of an alcohol, $C_8H_{10}O$. Propose a structure.

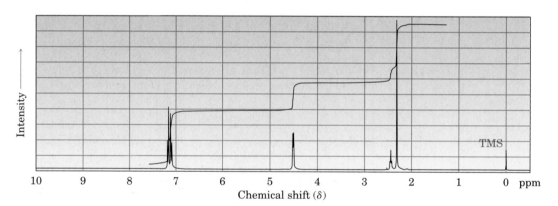

17.60   Propose structures for alcohols that have the following ^{1}H NMR spectra:
(a) $C_5H_{12}O$

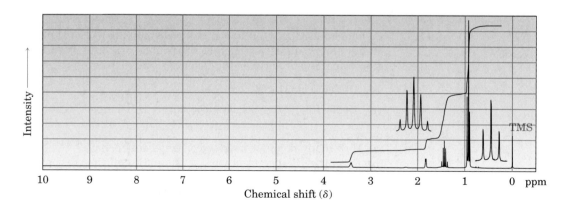

(b) $C_8H_{10}O$

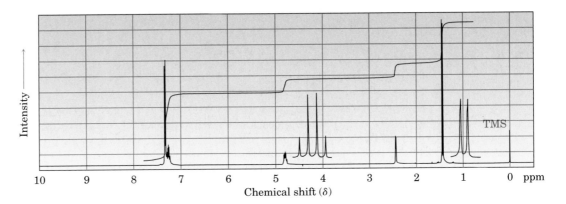

17.61 Propose structures for alcohols that have the following ^1H NMR spectra:

(a) $C_9H_{12}O$

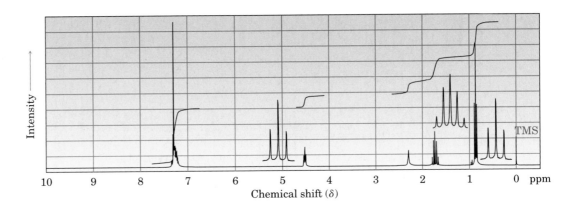

(b) $C_8H_{10}O_2$

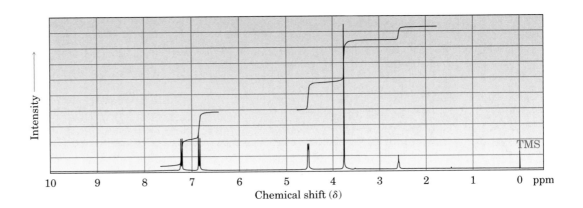

17.62   Compound A, $C_8H_{10}O$, has the IR and ^1H NMR spectra shown. Propose a structure consistent with the observed spectra, and assign each peak in the NMR spectrum. Note that the absorption at 5.5 δ disappears when D_2O is added.

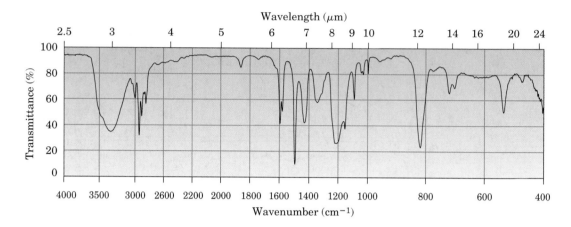

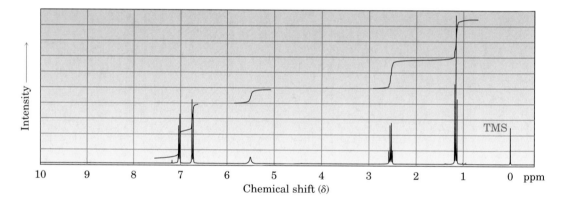

17.63   Propose a structure for a compound $C_{15}H_{24}O$ that has the following ^1H NMR spectrum. The peak marked by an asterisk disappears when D_2O is added to the sample.

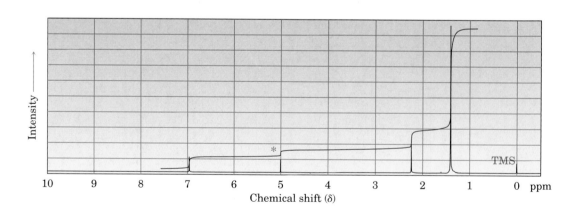

A Look Ahead

• •

17.64 The reduction of carbonyl compounds by reaction with hydride reagents (H :⁻) and the Grignard addition by reaction with organomagnesium halides (R :⁻ ⁺MgBr) are examples of *nucleophilic carbonyl addition reactions*. What analogous product do you think might result from reaction of cyanide ion with a ketone? (See Section 19.7.)

$$\underset{\text{C}}{\overset{\text{O}}{\parallel}} \xrightarrow[\text{H}_3\text{O}^+]{\text{CN}^-} \ ?$$

17.65 Ethers can be prepared by reaction of an alkoxide or phenoxide ion with a primary alkyl halide. Anisole, for instance, results from reaction of sodium phenoxide with iodomethane. What kind of reaction is occurring? Show the mechanism. (See Section 18.3.)

$$\text{O}^- \text{Na}^+ \quad + \text{ CH}_3\text{I} \longrightarrow \text{OCH}_3$$

Sodium phenoxide   **Anisole**

Molecular Modeling

• •

17.66 Use SpartanView to compare the energies of gauche and anti conformers of butane, 1,2-ethanediol, and 1,2-dimethoxyethane. Which molecules prefer the anti conformation about the C–C bond, and which prefer the gauche? Examine the electrostatic potential map of any molecule that prefers the gauche conformation, and explain why this conformation is preferred.

17.67 Use SpartanView to compare the electrostatic potential maps of the anions of phenol, 4-cyanophenol, and 4-cyanomethylphenol. Order them according to the amount of negative character on oxygen, and tell which phenol is most acidic.

17.68 Use SpartanView to measure the O–H bond distance and identify the O–H stretching vibration of *tert*-butyl alcohol. Next, measure the O–H bond distances and identify the two stretching vibrations in *tert*-butyl alcohol dimer. How does dimerization affect the hydrogen-bonded –OH group, and how does it affect the other –OH group?

17.69 Methanol reacts with thionyl chloride (SOCl₂) to give chloromethane. Use SpartanView to compare electrostatic potential maps and C–O bond distances for methanol and the intermediate formed from methanol and thionyl chloride. Why is the intermediate more reactive toward nucleophilic substitution than methanol itself? Is the –SOCl group electron-donating or electron-withdrawing? Explain.

$$\text{CH}_3-\text{O}-\text{H} \xrightarrow{\text{SOCl}_2} \left[\text{CH}_3-\text{O}-\underset{\parallel}{\overset{\text{O}}{\text{S}}}-\text{Cl} \right] \longrightarrow \text{CH}_3-\text{Cl}$$

Ethers and Epoxides;
Thiols and Sulfides

An **ether** is a substance that has two organic groups bonded to the same oxygen atom, R–O–R′. The organic groups may be alkyl, aryl, or vinylic, and the oxygen atom can be in either an open chain or a ring. Perhaps the most well-known ether is diethyl ether, a familiar substance that has been used medicinally as an anesthetic and is used industrially as a solvent. Other useful ethers include anisole, a pleasant-smelling aromatic ether used in perfumery, and tetrahydrofuran (THF), a cyclic ether often used as a solvent.

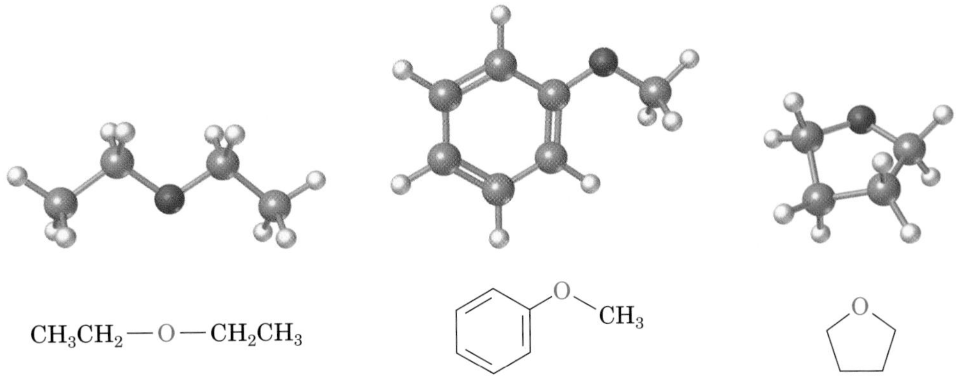
CH3CH2—O—CH2CH3
O
CH3
O

Diethyl ether | **Anisole (Methyl phenyl ether)** | **Tetrahydrofuran (a cyclic ether)**

Ethers are relatively stable and unreactive in many respects, but some ethers react slowly with air to give *peroxides*, compounds that contain an O–O bond. The peroxides from low-molecular-weight ethers such as diisopropyl ether and tetrahydrofuran are explosive and extremely dangerous, even in tiny amounts. Ethers are very useful as solvents in the laboratory, but they must always be treated with care.

Thiols (R–S–H) and *sulfides* (R–S–R′) are sulfur analogs of alcohols and ethers, respectively. Both functional groups are found in various biomolecules, though not as commonly as their oxygen-containing relatives. We'll take a brief look at both in this chapter.

18.1 Naming Ethers

Two systems for naming ethers are allowed by IUPAC rules. Simple ethers with no other functional groups are named by identifying the two organic substituents and adding the word *ether:*

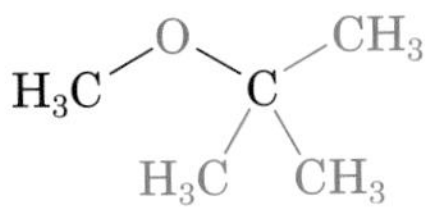
H3C
O
C
CH3
H3C
CH3

***tert*-Butyl methyl ether**

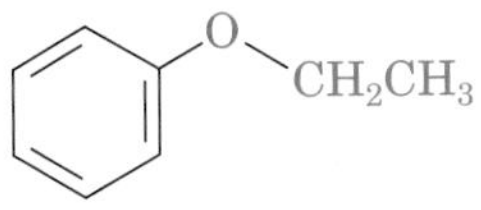
O
CH2CH3

Ethyl phenyl ether

If other functional groups are present, the ether part is considered an alkoxy substituent. For example:

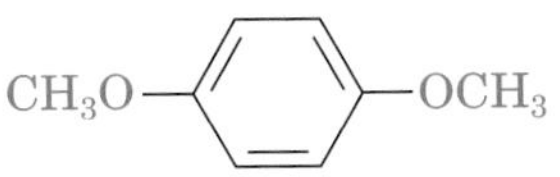
CH3O
OCH3

***p*-Dimethoxybenzene**

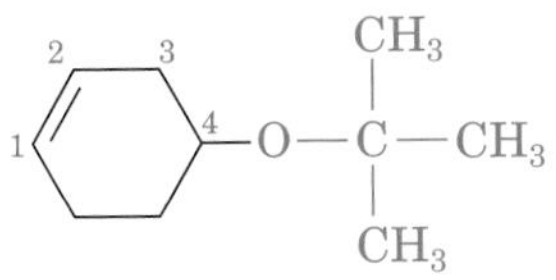
2
3
1
4
O
C
CH3
CH3
CH3

4-*tert*-Butoxy-1-cyclohexene

• •

Problem 18.1   Name the following ethers according to IUPAC rules:

(a) $CH_3\overset{\underset{|}{CH_3}}{C}HO\overset{\underset{|}{CH_3}}{C}HCH_3$

(b)   ⬠—$OCH_2CH_2CH_3$

(c)   [benzene ring with Br and OCH_3 substituents]

(d)   [cyclohexene ring with OCH_3]

(e) $CH_3\overset{\underset{|}{CH_3}}{C}HCH_2OCH_2CH_3$

(f) $H_2C\!=\!CHCH_2OCH\!=\!CH_2$

• •

18.2 Structure, Properties, and Sources of Ethers

Ethers can be thought of as organic derivatives of water in which the hydrogen atoms have been replaced by organic groups, H–O–H versus R–O–R. As such, ethers have nearly the same geometry as water. The R–O–R bonds have an approximately tetrahedral bond angle (112° in dimethyl ether), and the oxygen atom is sp^3-hybridized.

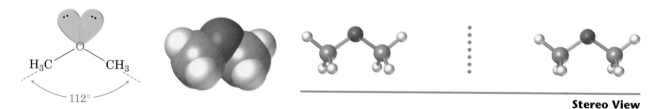

Stereo View

The electronegative oxygen atom gives ethers a slight dipole moment, and the boiling points of ethers are often somewhat higher than the boiling points of comparable alkanes. Table 18.1 compares the boiling points of some common ethers with the corresponding hydrocarbons in which the ether oxygen atom has been replaced by a CH_2 group.

Diethyl ether and other simple symmetrical ethers are prepared industrially by the sulfuric acid-catalyzed dehydration of alcohols:

$$2\ CH_3CH_2OH \xrightarrow{\ H_2SO_4\ } CH_3CH_2OCH_2CH_3\ +\ H_2O$$

Ethanol   **Diethyl ether**

The reaction occurs by S_N2 displacement of water from a protonated ethanol molecule by the oxygen atom of a second ethanol.

TABLE 18.1 Comparison of Boiling Points of Ethers and Hydrocarbons		

Ether	[Hydrocarbon]	Boiling point (°C)
CH_3OCH_3	$CH_3CH_2CH_3$	−25 [−45
$CH_3CH_2OCH_2CH_3$	$CH_3CH_2CH_2CH_2CH_3$	34.6	36
(O ring)	(pentagon)	65	49
(benzene)OCH₃	(benzene)CH₂CH₃	158	136]

$$CH_3CH_2-\overset{..}{\underset{..}{O}}: + CH_3CH_2-\overset{H}{\underset{H}{\overset{|}{O}}}{:}^+ \xrightarrow{S_N2} CH_3CH_2-\overset{H}{\underset{..}{\overset{|}{O}}}{}^+-CH_2CH_3 \longrightarrow CH_3CH_2-\overset{..}{\underset{..}{O}}-CH_2CH_3$$

This acid-catalyzed method is limited to the production of symmetrical ethers from primary alcohols because secondary and tertiary alcohols dehydrate to yield alkenes (Section 17.7). Thus, the method is of little practical value in the laboratory.

Problem 18.2 Why do you suppose only symmetrical ethers are prepared by the sulfuric acid-catalyzed dehydration procedure? What product(s) would you expect if ethanol and 1-propanol were allowed to react together? In what ratio would the products be formed if the two alcohols were of equal reactivity?

18.3 The Williamson Ether Synthesis

Metal alkoxides react with primary alkyl halides and tosylates by an S_N2 pathway to yield ethers, a process known as the **Williamson ether synthesis**. Discovered in 1850, the Williamson synthesis is still the best method for the preparation of ethers, both symmetrical and unsymmetrical.

The chemical reaction is shown:

Cyclopentoxide ion + CH₃—I $\xrightarrow[\text{solvent}]{\text{THF}}$ Cyclopentyl methyl ether + I⁻

Cyclopentoxide ion   **Cyclopentyl methyl ether**
                                                    (74%)

Alexander W. Williamson

Alexander W. Williamson (1824–1904) was born in London, England, and received his Ph.D. at the University of Giessen in 1846. His ability to work in the laboratory was hampered by a childhood injury that caused the loss of an arm. From 1849 until 1887, he was professor of chemistry at University College, London.

The alkoxides needed in the Williamson reaction are normally prepared by reaction of an alcohol with a strong base such as sodium hydride, NaH (Section 17.3). An acid–base reaction occurs between the alcohol and sodium hydride to generate the sodium salt of the alcohol.

$$\text{ROH} + \text{NaH} \longrightarrow \text{RO}^- \text{Na}^+ + \text{H}_2$$

A useful variation of the Williamson synthesis involves silver oxide, Ag_2O, as base rather than NaH. Under these conditions, the free alcohol reacts directly with alkyl halide, so there is no need to preform the metal alkoxide intermediate. For example, glucose reacts with iodomethane in the presence of Ag_2O to generate a *pentaether* in 85% yield.

α-**D-Glucose** $\xrightarrow[\text{Ag}_2\text{O}]{\text{CH}_3\text{I}}$ *α*-**D-Glucose pentamethyl ether** + AgI
                                                              (85%)

Mechanistically, the Williamson synthesis is simply an S_N2 displacement of halide ion by an alkoxide ion nucleophile. The Williamson synthesis is thus subject to all the usual constraints on S_N2 reactions discussed in Section 11.5. Primary halides and tosylates work best because competitive E2 elimination of HX can occur with more hindered substrates. Unsymmetrical ethers should therefore be synthesized by reaction between the more hindered alkoxide partner and less hindered halide partner rather than vice versa. For example, *tert*-butyl methyl ether, a substance used as an octane booster in gasoline, is best prepared by reaction of *tert*-butoxide ion with iodomethane rather than by reaction of methoxide ion with 2-chloro-2-methylpropane.

S_N2 **reaction**

CH₃—C(CH₃)(CH₃)—Ö⁻ + CH₃—I $\longrightarrow$ CH₃—C(CH₃)(CH₃)—O—CH₃ + I⁻

***tert*-Butoxide   Iodomethane   *tert*-Butyl methyl ether**
ion

E2 reaction

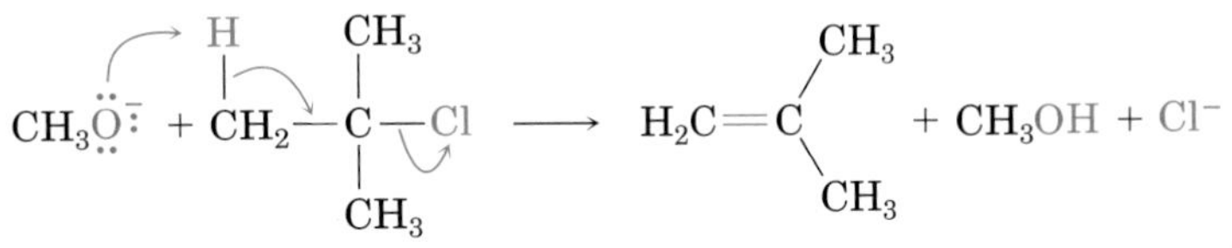

Methoxide ion   2-Chloro-2-methylpropane   2-Methylpropene

Problem 18.3 How would you prepare the following compounds using a Williamson synthesis?
(a) Methyl propyl ether   (b) Anisole (methyl phenyl ether)
(c) Benzyl isopropyl ether   (d) Ethyl 2,2-dimethylpropyl ether

Problem 18.4 Rank the following halides in order of their reactivity in the Williamson synthesis:
(a) Bromoethane, 2-bromopropane, bromobenzene
(b) Chloroethane, bromoethane, 1-iodopropene

18.4 Alkoxymercuration of Alkenes

We saw in Section 7.4 that alkenes react with water in the presence of mercuric acetate to yield a hydroxymercuration product. Subsequent treatment with $NaBH_4$ breaks the C–Hg bond and yields the alcohol. A similar **alkoxymercuration** reaction occurs when an alkene is treated with an *alcohol* in the presence of mercuric acetate. [Mercuric trifluoroacetate, $(CF_3CO_2)_2Hg$, works even better.] Demercuration by reaction with $NaBH_4$ then yields an ether. As indicated by the following examples, the net result is Markovnikov addition of the alcohol to the alkene.

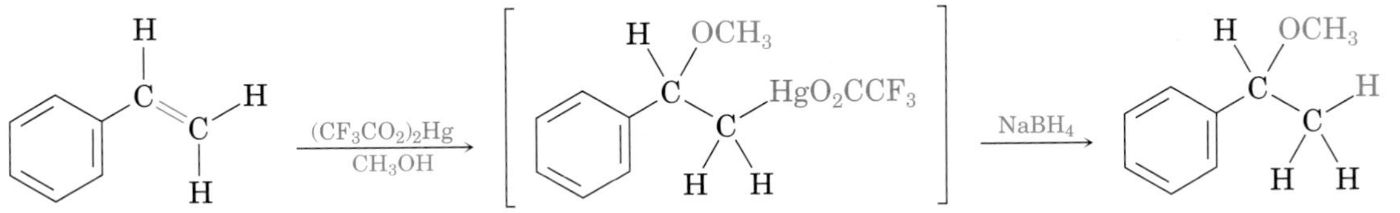

Styrene

1-Methoxy-1-phenylethane (97%)

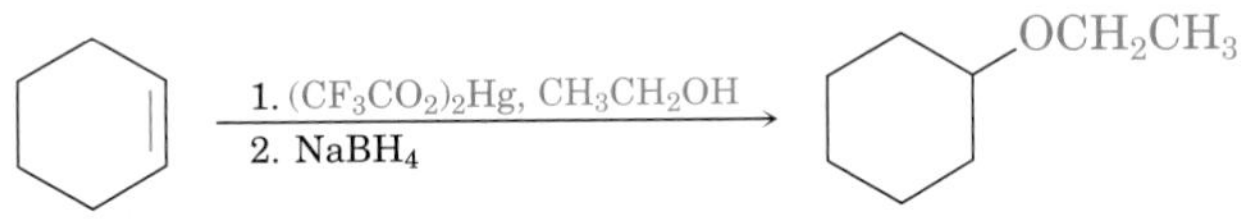

Cyclohexene

Cyclohexyl ethyl ether (100%)

The mechanism of the alkoxymercuration reaction is similar to that described in Section 7.4 for hydroxymercuration. The reaction is initiated by electrophilic addition of Hg^{2+} to the alkene, followed by reaction of the intermediate cation with alcohol. Reduction of the C–Hg bond by NaBH$_4$ completes the process.

A wide variety of alcohols and alkenes can be used in the alkoxymercuration reaction. Primary, secondary, and even tertiary alcohols react smoothly, but ditertiary ethers can't be prepared because of steric hindrance to reaction.

• •

Practice Problem 18.1   How would you prepare ethyl phenyl ether? Use whichever method you think is more appropriate, the Williamson synthesis or the alkoxymercuration reaction.

Strategy   Draw the target ether, identify the two groups attached to oxygen, and recall the limitations of the two methods for preparing ethers. The Williamson synthesis uses an S$_N$2 reaction and requires that one of the two groups attached to oxygen be either secondary or (preferably) primary. The alkoxymercuration reaction requires that one of the two groups come from an alkene precursor. Ethyl phenyl ether could be made by either method.

Solution

• •

Problem 18.5   Review the mechanism of oxymercuration shown in Figure 7.5, and then write the mechanism of the alkoxymercuration reaction of 1-methylcyclopentene with ethanol. Use curved arrows to show the electron flow in each step.

Problem 18.6   How would you prepare the following ethers? Use whichever method you think is more appropriate, the Williamson synthesis or the alkoxymercuration reaction.
(a) Butyl cyclohexyl ether   (b) Benzyl ethyl ether (C$_6$H$_5$CH$_2$OCH$_2$CH$_3$)
(c) *tert*-Butyl *sec*-butyl ether   (d) Tetrahydrofuran

• •

18.5 Reactions of Ethers: Acidic Cleavage

Ethers are unreactive to many reagents used in organic chemistry, a property that accounts for their wide use as reaction solvents. Halogens, dilute

Alexander M. Butlerov

Alexander M. Butlerov (1828–1886) was born in Tschistopol, Russia, and received his Ph.D. in 1854 from the University of Moscow. His mother died shortly after giving birth, and he was raised by his maternal grandfather. From 1854 to 1867, he was professor of chemistry at the University of Kazan, and from 1867 to 1880 he taught at the University of St. Petersburg. His many and varied interests ran from bee-keeping to a belief in spiritualism.

acids, bases, and nucleophiles have no effect on most ethers. In fact, ethers undergo only one reaction of general use—they are cleaved by strong acids.

The first example of acid-induced ether cleavage was observed in 1861 by Alexander Butlerov, who found that 2-ethoxypropanoic acid reacts with aqueous HI at 100°C to yield iodoethane and lactic acid:

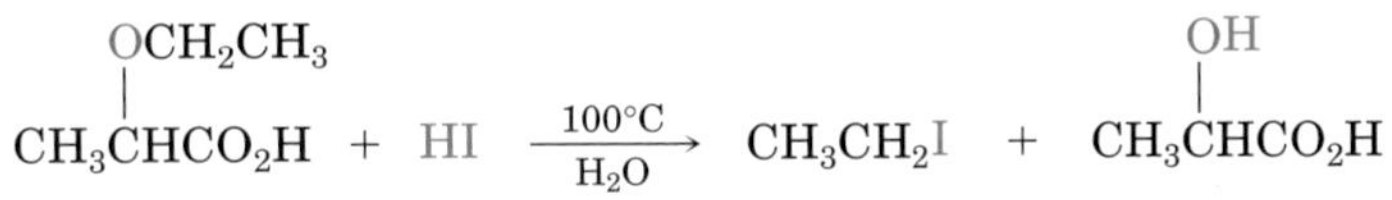

2-Ethoxypropanoic acid   **Iodoethane**   **Lactic acid**

In addition to HI, aqueous HBr also works well, but HCl does not cleave ethers.

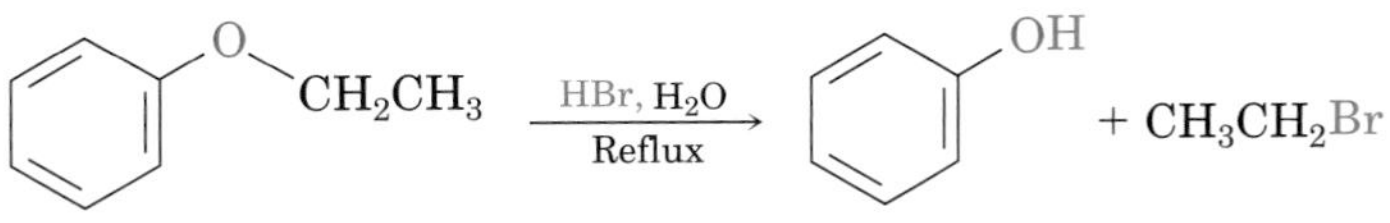

Ethyl phenyl ether   **Phenol**   **Bromoethane**

Acidic ether cleavages are typical nucleophilic substitution reactions, of the sort discussed in Chapter 11. Primary and secondary alkyl ethers react by an S_N2 mechanism, in which I^- or Br^- attacks the protonated ether at the less hindered site. This usually results in a selective cleavage into a single alcohol and a single alkyl halide. For example, ethyl isopropyl ether yields exclusively isopropyl alcohol and iodoethane on cleavage by HI, because nucleophilic attack by iodide ion occurs at the less hindered primary site rather than at the more hindered secondary site.

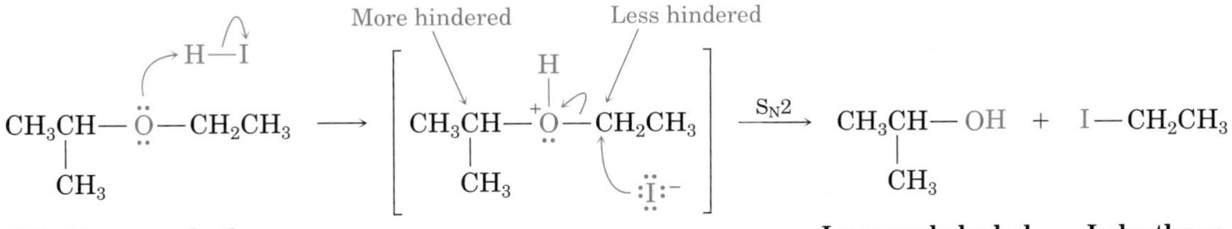

Ethyl isopropyl ether   **Isopropyl alcohol**   **Iodoethane**

Tertiary, benzylic, and allylic ethers cleave by an S_N1 or E1 mechanism because these substrates can produce stable intermediate carbocations. These reactions are often fast and take place at moderate temperatures. *tert*-Butyl ethers, for example, react by an E1 mechanism on treatment with trifluoroacetic acid at 0°C.

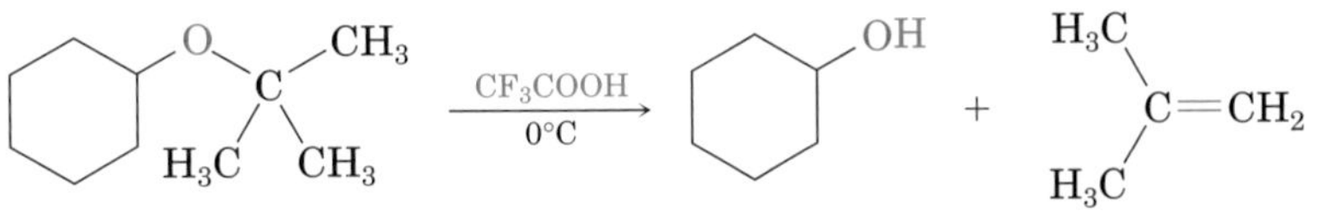

tert-**Butyl cyclohexyl ether**   **Cyclohexanol**   **2-Methylpropene**
                                              (90%)

· ·

Practice Problem 18.2   Predict the products of the following reaction:

$$CH_3C(CH_3)(CH_3)-O-CH_2CH_2CH_3 \xrightarrow{\ HBr\ } ?$$

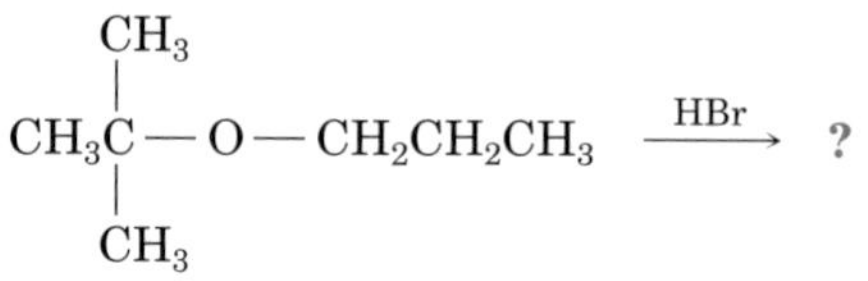

Strategy   Identify the substitution pattern of the two groups attached to oxygen—in this case a tertiary alkyl group and a primary alkyl group. Then recall the guidelines for ether cleavages. An ether with only primary and secondary alkyl groups usually undergoes cleavage by S_N2 attack of a nucleophile on the less hindered alkyl group, but an ether with a tertiary alkyl group usually undergoes cleavage by an S_N1 mechanism. In this case, an S_N1 cleavage of the tertiary C–O bond will occur, giving 1-propanol and a tertiary alkyl bromide.

Solution

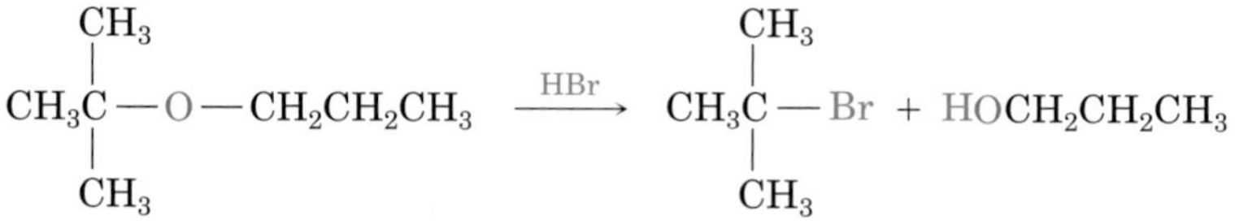

$$CH_3C(CH_3)(CH_3)-O-CH_2CH_2CH_3 \xrightarrow{\ HBr\ } CH_3C(CH_3)(CH_3)-Br\ +\ HOCH_2CH_2CH_3$$

tert-Butyl propyl ether   2-Bromo-2-   1-Propanol
                                                   methylpropane

· ·

Problem 18.7   Predict the products of each of the following reactions:

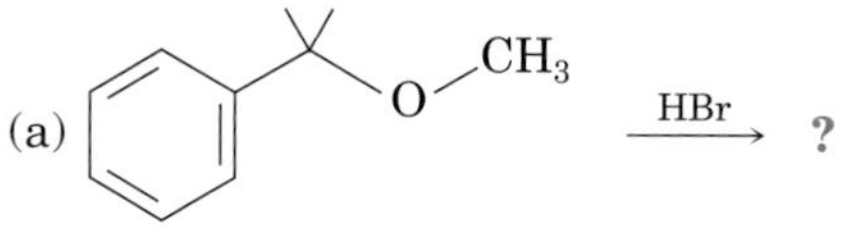

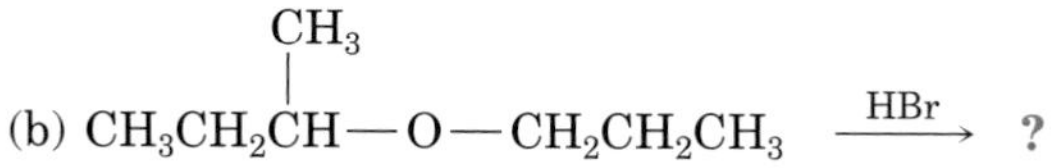

(a) [structure: 1-methyl-1-(methoxy)ethylbenzene] $\xrightarrow{\ HBr\ }$?

(b) $CH_3CH_2CH(CH_3)-O-CH_2CH_2CH_3 \xrightarrow{\ HBr\ }$?

Problem 18.8   Write the mechanism of the acid-catalyzed cleavage of *tert*-butyl cyclohexyl ether to yield cyclohexanol and 2-methylpropene.

Problem 18.9   Explain the observation that HI and HBr are more effective than HCl in cleaving ethers. (See Section 11.5.)

· ·

18.6 Reactions of Ethers: Claisen Rearrangement

Unlike the acid-catalyzed ether cleavage reaction discussed in the previous section, which is general to all ethers, the **Claisen rearrangement** is specific to allyl aryl ethers, Ar–O–CH$_2$CH=CH$_2$. Treatment of a phenoxide ion with 3-bromopropene (allyl bromide) results in a Williamson ether synthesis and production of an allyl aryl ether. Heating the allyl aryl ether to 200–250°C then effects Claisen rearrangement, leading to an *o*-allylphenol. The net result is alkylation of the phenol in an ortho position.

Phenol + NaH $\xrightarrow[\text{solution}]{\text{THF}}$ **Sodium phenoxide** $\xrightarrow{\text{BrCH}_2\text{CH}=\text{CH}_2}$ **Allyl phenyl ether**

Allyl phenyl ether $\xrightarrow[\text{250°C}]{\text{Claisen rearrangement}}$ *o*-**Allylphenol**

Like the Diels–Alder reaction (Section 14.8), the Claisen rearrangement reaction proceeds through a pericyclic mechanism in which a concerted reorganization of bonding electrons occurs by a six-membered, cyclic transition state. The 6-allyl-2,4-cyclohexadienone intermediate then isomerizes to *o*-allylphenol (Figure 18.1).

FIGURE 18.1 ▼

The mechanism of the Claisen rearrangement. The bond-density surface for the transition state shows that C–O bond-breaking and C–C bond-making occur simultaneously.

Claisen transition state

Allyl phenyl ether → **Transition state** → **Intermediate (6-Allyl-2,4-cyclohexadienone)** → *o*-**Allylphenol**

Evidence for this mechanism comes from the observation that the rearrangement takes place with an inversion of the allyl group. That is, allyl phenyl ether containing a ^{14}C label on the allyl *ether* carbon atom yields *o*-allylphenol in which the label is on the *terminal* carbon. It would be very difficult to explain this result by any mechanism other than a pericyclic one. We'll look at more details in Section 30.9.

● ●

Problem 18.10   What product would you expect from Claisen rearrangement of 2-butenyl phenyl ether?

$$\text{O}-\text{CH}_2\text{CH}=\text{CHCH}_3$$

$$\xrightarrow[250°\text{C}]{} \quad ?$$

2-Butenyl phenyl ether

● ●

18.7 Cyclic Ethers: Epoxides

For the most part, cyclic ethers behave like acyclic ethers. The chemistry of the ether functional group is the same, whether it's in an open chain or in a ring. Common cyclic ethers such as tetrahydrofuran and dioxane, for example, are often used as solvents because of their inertness, yet they can be cleaved by strong acids.

$$\begin{array}{c} \text{O} \\ \text{H}_2\text{C} \quad \text{CH}_2 \\ \text{H}_2\text{C} \quad \text{CH}_2 \\ \text{O} \end{array} \qquad\qquad \begin{array}{c} \text{O} \\ \text{H}_2\text{C} \quad \text{CH}_2 \\ \text{H}_2\text{C}-\text{CH}_2 \end{array}$$

1,4-Dioxane   **Tetrahydrofuran**

The only cyclic ethers that behave differently from open-chain ethers are the three-membered-ring compounds called **epoxides**, or **oxiranes**. The strain of the three-membered ring gives epoxides unique chemical reactivity.

Ethylene oxide, the simplest epoxide, is an intermediate in the manufacture of both ethylene glycol, used for automobile antifreeze, and polyester polymers. More than 4 million tons of ethylene oxide are produced each year in the United States by air oxidation of ethylene over a silver oxide catalyst at 300°C. This process is not useful for other epoxides, however, and is of little value in the laboratory. Note that the name *ethylene oxide* is not a systematic one because the *-ene* ending implies the presence of a double bond in the molecule. The name is frequently used, however, because ethylene oxide is derived *from* ethylene by addition of an oxygen atom. Other simple epoxides are named similarly. The systematic name for ethylene oxide is 1,2-epoxyethane.

$$\text{H}_2\text{C}=\text{CH}_2 \xrightarrow[\substack{\text{Ag}_2\text{O,} \\ 300°\text{C}}]{\text{O}_2} \overset{\text{O}}{\text{H}_2\text{C}-\text{CH}_2}$$

Ethylene   **Ethylene oxide**

Stereo View

In the laboratory, epoxides are prepared by treatment of an alkene with a **peroxyacid, RCO₃H**. Many different peroxyacids can be used to accomplish epoxidation, but *m*-chloroperoxybenzoic acid is the most common choice.

Cycloheptene   **1,2-Epoxycycloheptane**
                                              (78%)

Peroxyacids transfer oxygen to the alkene with syn stereochemistry through a one-step mechanism without intermediates. Studies have shown that the oxygen atom farthest from the carbonyl group is the one transferred.

Alkene   Peroxyacid   Epoxide   Acid

Another method for the synthesis of epoxides is through the use of halohydrins, prepared by electrophilic addition of HO–X to alkenes (Section 7.3). When halohydrins are treated with base, HX is eliminated, and an epoxide is produced.

Cyclohexene   ***trans*-2-Chloro-**   **1,2-Epoxycyclohexane**
                        cyclohexanol   **(73%)**

This formation of an epoxide by treatment of a halohydrin with base is just an *intramolecular* Williamson ether synthesis. The nucleophilic alkoxide ion and the electrophilic alkyl halide are in the same molecule.

A bromohydrin   **An epoxide**

● ●

Problem 18.11   What product would you expect from reaction of *cis*-2-butene with *m*-chloroperoxybenzoic acid? Show the stereochemistry.

Problem 18.12 Reaction of *trans*-2-butene with *m*-chloroperoxybenzoic acid yields an epoxide different from that obtained by reaction of the cis isomer (Problem 18.11). Explain.

18.8 Ring-Opening Reactions of Epoxides

Acid-Catalyzed Epoxide Opening

Epoxide rings are cleaved by treatment with acid just as other ethers are. The major difference is that epoxides react under much milder conditions because of ring strain. Dilute aqueous acid at room temperature is sufficient to cause the hydrolysis of epoxides to 1,2-diols, also called *vicinal glycols*. (The word *vicinal* means "adjacent," and a *glycol* is a diol.) More than 3 million tons of ethylene glycol, most of it used for automobile antifreeze, are produced each year in the United States by acid-catalyzed hydration of ethylene oxide. Note that the name *ethylene glycol* refers to the glycol derived from ethylene, just as ethylene oxide refers to the epoxide derived from ethylene.

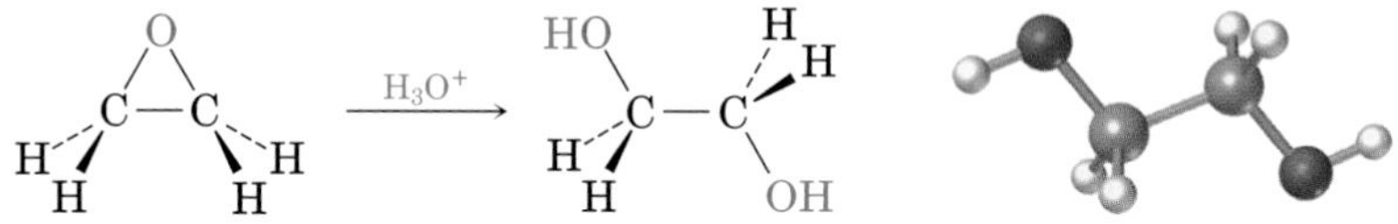

Ethylene oxide   **Ethylene glycol (1,2-Ethanediol)**

Acid-catalyzed epoxide cleavage takes place by back-side attack of a nucleophile on the protonated epoxide in a manner analogous to the final step of alkene bromination, in which a cyclic bromonium ion is opened by nucleophilic attack (Section 7.2). When an epoxycycloalkane is opened by aqueous acid, a *trans*-1,2-diol results, just as a *trans*-1,2-dibromide results from cycloalkene bromination.

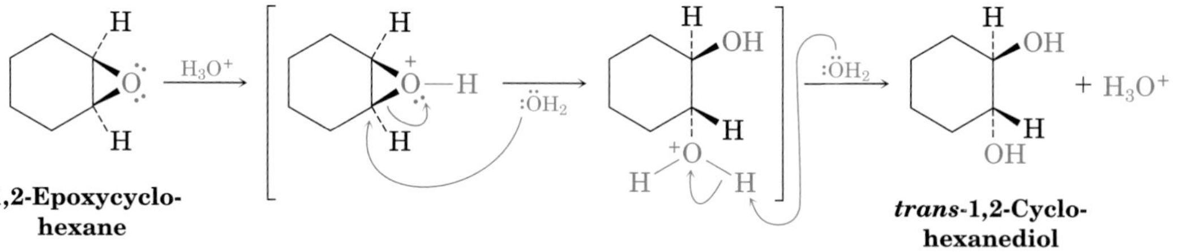

1,2-Epoxycyclohexane   **trans-1,2-Cyclohexanediol (86%)**

Recall the following:

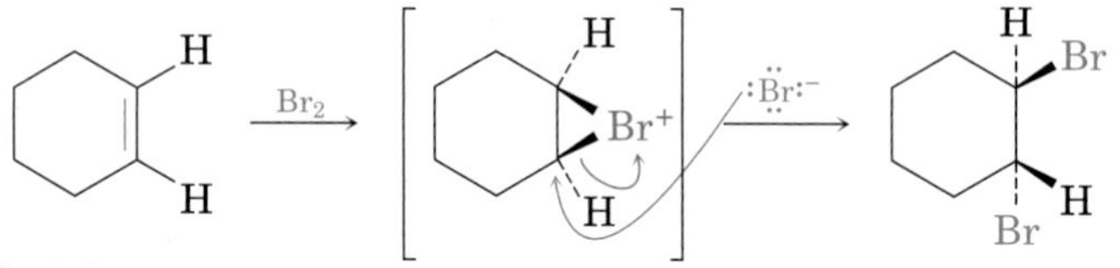

Cyclohexene   **trans-1,2-Dibromocyclohexane**

Epoxides can also be opened by reaction with acids other than H_3O^+. For example, if anhydrous HX is used, an epoxide is converted into a trans halohydrin:

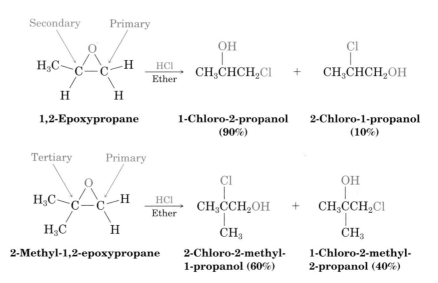

ANTI

A trans 2-halocyclohexanol

where   X = F, Br, Cl, or I

The regiochemistry of acid-catalyzed ring opening depends on the epoxide's structure, and a mixture of products is usually formed. When both epoxide carbon atoms are either primary or secondary, attack of the nucleophile occurs primarily at the *less* highly substituted site. When one of the epoxide carbon atoms is tertiary, however, nucleophilic attack occurs primarily at the *more* highly substituted site. Thus, 1,2-epoxypropane reacts with HCl to give primarily 1-chloro-2-propanol, but 2-methyl-1,2-epoxypropane gives 2-chloro-2-methyl-1-propanol as the major product.

Secondary   Primary

$$H_3C-\overset{H}{\underset{H}{C}}-\overset{H}{\underset{H}{C}}-H \quad \xrightarrow[\text{Ether}]{\text{HCl}} \quad CH_3\overset{OH}{C}HCH_2Cl \quad + \quad CH_3\overset{Cl}{C}HCH_2OH$$

1,2-Epoxypropane **1-Chloro-2-propanol** **2-Chloro-1-propanol**
(90%) (10%)

Tertiary   Primary

$$H_3C-\underset{H_3C}{\overset{}{C}}-\overset{}{\underset{H}{C}}-H \quad \xrightarrow[\text{Ether}]{\text{HCl}} \quad CH_3\overset{Cl}{\underset{CH_3}{C}}CH_2OH \quad + \quad CH_3\overset{OH}{\underset{CH_3}{C}}CH_2Cl$$

2-Methyl-1,2-epoxypropane **2-Chloro-2-methyl-1-propanol (60%)** **1-Chloro-2-methyl-2-propanol (40%)**

The mechanisms of these acid-catalyzed epoxide openings are interesting because they appear to be *midway* between typical S_N1 and S_N2 pathways and to have characteristics of both. Take the reaction of 1,2-epoxy-1-methylcyclohexane with HBr shown in Figure 18.2 (p. 722), for example. This reaction yields a single isomer of 2-bromo-2-methylcyclohexanol in which the –Br and –OH groups are trans. The fact that the product has the entering bromine and the leaving oxygen on opposite sides of the ring is an S_N2-like result (back-side displacement of the leaving group). But the fact that Br^- attacks the more hindered tertiary side of the epoxide rather than the less hindered secondary side is an S_N1-like result (more stable, tertiary carbocation involved).

FIGURE 18.2 ▼

Acid-induced ring opening of 1,2-epoxy-1-methylcyclohexane with HBr. There is a high degree of S_N1-like carbocation character in the transition state, which leads to back-side attack of the nucleophile at the tertiary center and to formation of the isomer of 2-bromo-2-methylcyclohexanol that has –Br and –OH groups trans.

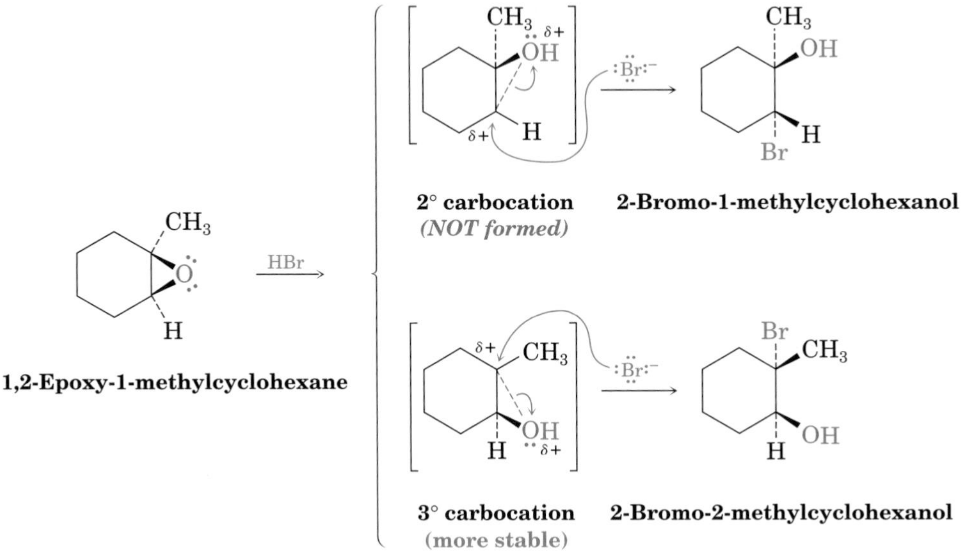

2° carbocation
(NOT formed)

2-Bromo-1-methylcyclohexanol

1,2-Epoxy-1-methylcyclohexane

3° carbocation
(more stable)

2-Bromo-2-methylcyclohexanol

Evidently, the transition state for acid-catalyzed epoxide opening has an S_N2-like geometry but also has a large amount of S_N1-like carbocationic character. Since the positive charge in the protonated epoxide is shared by the more highly substituted carbon atom, back-side attack of Br^- occurs at the more highly substituted site.

• •

Practice Problem 18.3   Predict the major product of the following reaction:

Strategy   Identify the substitution pattern of the two epoxide carbon atoms—in this case, one carbon is secondary and one is primary. Then recall the guidelines for epoxide cleavages. An epoxide with only primary and secondary carbons usually undergoes cleavage by S_N2 attack of a nucleophile on the less hindered carbon, but an epoxide with a tertiary carbon atom usually undergoes cleavage by an S_N1 mechanism. In this case, an S_N2 cleavage of the primary C–O epoxide bond will occur.

Solution

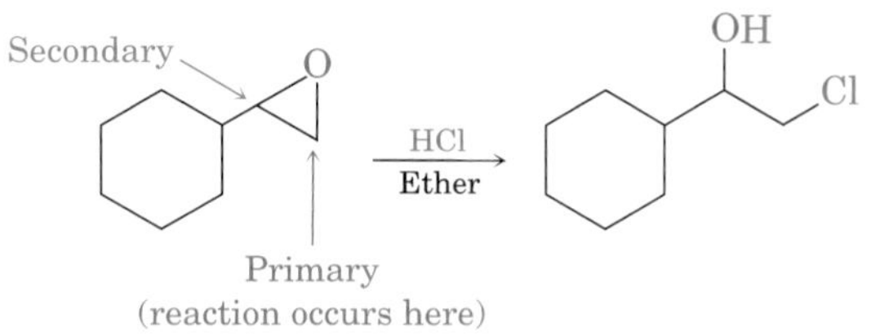
Secondary
O
HCl
Ether
Primary
(reaction occurs here)
OH
Cl

Problem 18.13 Predict the major product of the following reactions:

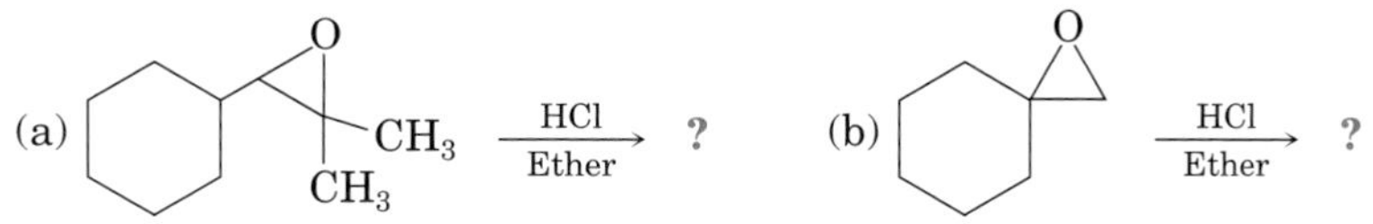
(a)
O
CH3
CH3
HCl
Ether
?
(b)
O
HCl
Ether
?

Problem 18.14 Write the mechanism of the hydrolysis of *cis*-5,6-epoxydecane by reaction with aqueous acid. What is the stereochemistry of the product, assuming normal back-side S_N2 attack?

Problem 18.15 What is the stereochemistry of the product from acid-catalyzed hydrolysis of *trans*-5,6-epoxydecane? How does the product differ from that formed in Problem 18.14?

Base-Catalyzed Epoxide Opening

Unlike other ethers, epoxide rings can be cleaved by base as well as by acid. Although an ether oxygen is normally a poor leaving group in an S_N2 reaction (Section 11.5), the reactivity of the three-membered ring is sufficient to allow epoxides to react with hydroxide ion at elevated temperatures.

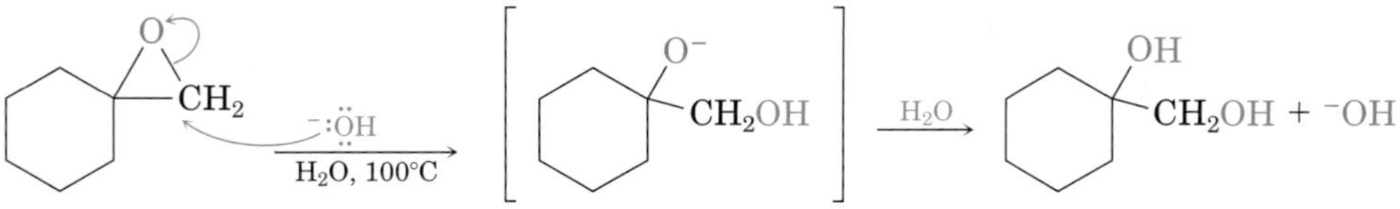
O
CH2
:ÖH
H2O, 100°C
O−
CH2OH
H2O
OH
CH2OH + −OH

Methylenecyclohexane oxide **1-Hydroxymethylcyclohexanol (70%)**

A similar nucleophilic ring opening occurs when epoxides are treated with Grignard reagents. Ethylene oxide is frequently used, thereby allowing the conversion of a Grignard reagent into a primary alcohol having two more carbons than the starting alkyl halide. 1-Bromobutane, for example, is converted into 1-hexanol by reaction of its Grignard reagent with ethylene oxide.

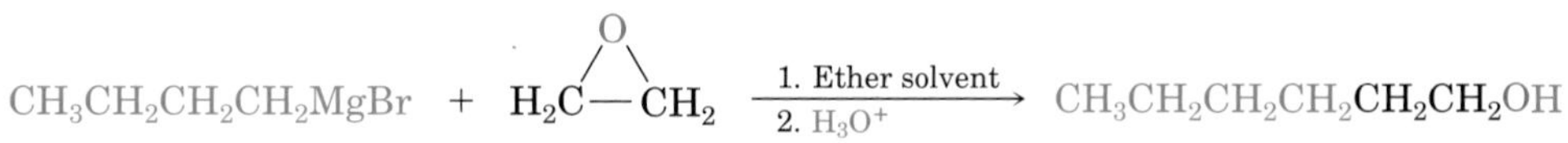
CH3CH2CH2CH2MgBr + H2C—CH2
O
1. Ether solvent
2. H3O+
CH3CH2CH2CH2CH2CH2OH

Butylmagnesium bromide **Ethylene oxide** **1-Hexanol (62%)**

Base-catalyzed epoxide opening is a typical S_N2 reaction in which attack of the nucleophile takes place at the less hindered epoxide carbon. For example, 1,2-epoxypropane reacts with ethoxide ion exclusively at the less highly substituted, primary carbon to give 1-ethoxy-2-propanol.

$$H_3C-\overset{\underset{|}{H}}{C}-\overset{\underset{|}{H}}{C}-H \xrightarrow[^-:\ddot{O}CH_2CH_3]{CH_3CH_2OH} CH_3\overset{\underset{|}{OH}}{C}HCH_2OCH_2CH_3$$

No attack here (2°)

1-Ethoxy-2-propanol (83%)

Problem 18.16   Predict the major product of the following reactions:

(a) $CH_3CH_2\overset{\underset{|}{CH_3}}{\underset{}{C}}\overset{O}{\overset{\diagup \diagdown}{-}}CH_2 \xrightarrow[H_2{}^{18}O]{NaOH}$ **?**   (b) $CH_3CH_2\overset{\underset{|}{CH_3}}{\underset{}{C}}\overset{O}{\overset{\diagup \diagdown}{-}}CH_2 \xrightarrow{H_3{}^{18}O^+}$ **?**

18.9 Crown Ethers

Discovered in the early 1960s by Charles Pedersen at the Du Pont Company, **crown ethers** are a relatively recent addition to the ether family. Crown ethers are named according to the general format *x*-crown-*y*, where *x* is the total number of atoms in the ring and *y* is the number of oxygen atoms. Thus, 18-crown-6 ether is an 18-membered ring containing 6 ether oxygen atoms. Note the size and negative (red) character of the crown ether cavity in the following electrostatic potential map.

18-crown-6 ether

18-Crown-6 ether

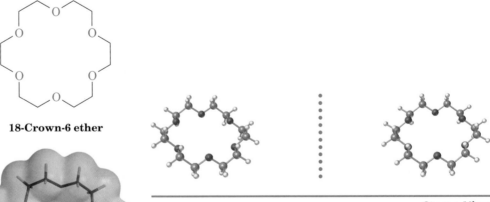

Stereo View

Charles John Pedersen

Charles John Pedersen (1904–1989) was born in Pusan, Korea, to a Korean mother and Norwegian father. A U.S. citizen, he moved to the United States in the early 1920s and received an M.Sc. at the Massachusetts Institute of Technology in 1927. He spent his entire scientific career at the Du Pont Company (1927–1969) and received the 1987 Nobel Prize in chemistry. He is among a very small handful of Nobel Prize-winning scientists who never received a formal doctorate.

The importance of crown ethers derives from their extraordinary ability to solvate metal cations by sequestering the metal in the center of the polyether cavity. For example, 18-crown-6 complexes strongly with potassium ion.

Complexes between crown ethers and ionic salts are soluble in nonpolar organic solvents, thus allowing many reactions to be carried out under aprotic conditions that would otherwise have to be carried out in aqueous solution. For example, the inorganic compound $KMnO_4$ actually dissolves in benzene in the presence of 18-crown-6. The resulting solution of "purple benzene" is a valuable reagent for oxidizing alkenes.

**$KMnO_4$ solvated by 18-crown-6
(this solvate is soluble in benzene)**

Many other inorganic salts, including KF, KCN, and NaN_3, can be dissolved in organic solvents with the help of crown ethers. The effect of using a crown ether to dissolve a salt in a hydrocarbon or ether solvent is similar to the effect of dissolving the salt in a polar aprotic solvent such as DMSO, DMF, or HMPA (Section 11.5). In both cases, the metal cation is strongly solvated, leaving the anion bare. Thus, the S_N2 reactivity of an anion is tremendously enhanced in the presence of a crown ether.

Problem 18.17   15-Crown-5 and 12-crown-4 ethers complex Na^+ and Li^+, respectively. Make models of these crown ethers, and compare the sizes of the cavities.

18.10 Spectroscopy of Ethers

Infrared Spectroscopy

Ethers are difficult to distinguish by IR spectroscopy. Although they show an absorption due to C–O single-bond stretching in the range 1050–1150 cm^{-1}, many other kinds of absorptions occur in the same range. Figure 18.3 shows the IR spectrum of diethyl ether and identifies the C–O stretch.

FIGURE 18.3 ▼

The infrared spectrum of diethyl ether, $CH_3CH_2OCH_2CH_3$.

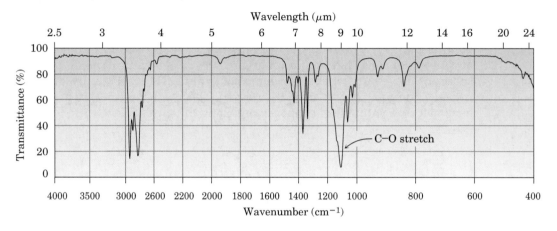

Nuclear Magnetic Resonance Spectroscopy

Hydrogens on carbon next to an ether oxygen are shifted downfield from the normal alkane resonance and show ^1H NMR absorptions in the region 3.4–4.5 δ. This downfield shift is clearly seen in the spectrum of dipropyl ether shown in Figure 18.4.

FIGURE 18.4 ▼

The ^1H NMR spectrum of dipropyl ether. Protons on carbon next to oxygen are shifted downfield to 3.4 δ.

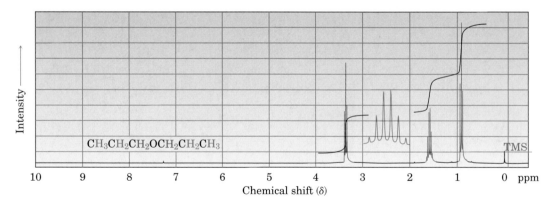

Epoxides absorb at a slightly higher field than other ethers and show characteristic resonances at 2.5–3.5 δ in their ^1H NMR spectra, as indicated for 1,2-epoxypropane in Figure 18.5.

FIGURE 18.5 ▼

The ¹H NMR spectrum of 1,2-epoxypropane.

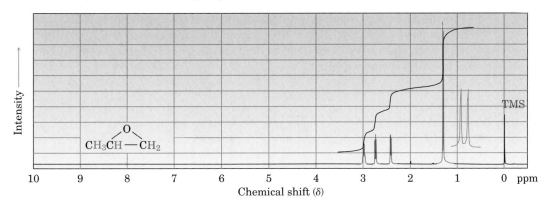

Ether carbon atoms also exhibit a downfield shift in the ¹³C NMR spectrum, where they usually absorb in the range 50–80 δ. For example, the carbon atoms next to oxygen in methyl propyl ether absorb at 58.5 and 74.8 δ. Similarly, the methyl carbon in anisole absorbs at 54.8 δ.

58.5 δ 74.8 δ 10.7 δ

$$CH_3 - O - CH_2 - CH_2 - CH_3$$

23.3 δ

129.5 δ 114.1 δ

120.7 δ 54.8 δ

$O - CH_3$

159.9 δ

● ●

Problem 18.18   The ¹H NMR spectrum shown is that of an ether with the formula C_4H_8O. Propose a structure.

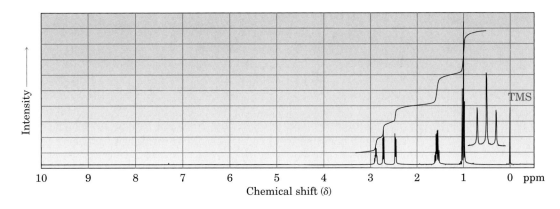

● ●

18.11 Thiols and Sulfides

Thiols, R–SH, are sulfur analogs of alcohols, and **sulfides, RSR′**, are sulfur analogs of ethers. Thiols are named by the same system used for alcohols, with the suffix *-thiol* used in place of *-ol*. The –SH group itself is referred to as a **mercapto group**.

CH₃CH₂SH

Ethanethiol   **Cyclohexanethiol**   *m*-**Mercaptobenzoic acid**

Sulfides are named by following the same rules used for ethers, with *sulfide* used in place of *ether* for simple compounds and *alkylthio* used in place of *alkoxy* for more complex substances.

CH₃—S—CH₃

Dimethyl sulfide   **Methyl phenyl sulfide**   **3-(Methylthio)cyclohexene**

Thiols

The most obvious characteristic of thiols is their appalling odor. Skunk scent, for example, is caused primarily by the simple thiols, 3-methyl-1-butanethiol and 2-butene-1-thiol. Volatile thiols are also added to natural gas to serve as an easily detectable warning in case of leaks.

Thiols are usually prepared from alkyl halides by S_N2 displacement with a sulfur nucleophile such as hydrosulfide anion, ⁻SH.

CH₃(CH₂)₆CH₂—Br + Na⁺ ⁻:S̈H ⟶ CH₃(CH₂)₆CH₂SH + NaBr

1-Bromooctane   **Sodium**   **1-Octanethiol**
                         hydrosulfide

Yields are often poor in this reaction unless an excess of the nucleophile is used, because the product thiol can undergo further S_N2 reaction with alkyl halide to give a symmetrical sulfide as a by-product. For this reason, thiourea, $(NH_2)_2C{=}S$, is often used as the nucleophile in the preparation of a thiol from an alkyl halide. The reaction occurs by displacement of the halide ion to yield an intermediate alkylisothiourea salt, which is hydrolyzed by subsequent reaction with aqueous base.

$$CH_3(CH_2)_6CH_2-Br \ + \ H_2N-\overset{\overset{\displaystyle :S:}{\|}}{C}-\ddot{N}H_2 \ \longrightarrow \ \left[CH_3(CH_2)_6CH_2-\overset{+}{S}=C-\overset{\overset{\displaystyle NH_2}{|}}{NH_2} \ Br^- \right]$$

1-Bromooctane   **Thiourea**   **Alkylisothiourea salt**

$$\Big\downarrow H_2O, \ NaOH$$

$$CH_3(CH_2)_6CH_2SH \ + \ H_2N-\overset{\overset{\displaystyle O}{\|}}{C}-NH_2$$

1-Octanethiol (83%)   **Urea**

Thiols can be oxidized by Br_2 or I_2 to yield **disulfides, RSSR**. The reaction is easily reversed, and a disulfide can be reduced back to a thiol by treatment with zinc and acid:

$$2 \ R-SH \ \underset{Zn, \ H^+}{\overset{I_2}{\rightleftarrows}} \ R-S-S-R \ + \ 2 \ HI$$

A thiol   **A disulfide**

Biological Connection ▶

We'll see later that the thiol–disulfide interconversion is extremely important in biochemistry, where disulfide "bridges" form the cross-links between protein chains that help stabilize the three-dimensional conformations of proteins.

Sulfides

Treatment of a thiol with a base, such as NaH, gives the corresponding **thiolate ion, RS⁻**, which undergoes reaction with a primary or secondary alkyl halide to give a sulfide. The reaction occurs by an S_N2 mechanism, analogous to the Williamson synthesis of ethers (Section 18.3). Thiolate anions are among the best nucleophiles known, and product yields are usually high in these S_N2 reactions.

Sodium benzenethiolate   **Methyl phenyl sulfide (96%)**

Because the valence electrons on sulfur are farther from the nucleus and are less tightly held than those on oxygen ($3p$ electrons versus $2p$ electrons), sulfur compounds are more nucleophilic than their oxygen analogs. Unlike dialkyl ethers, dialkyl sulfides are good nucleophiles that react rapidly with primary alkyl halides by an S_N2 mechanism to give **trialkylsulfonium salts (R_3S^+)**.

$$CH_3-\ddot{S}-CH_3 \ + \ CH_3-I \ \xrightarrow{THF} \ CH_3-\overset{\overset{\displaystyle CH_3}{|}}{\underset{..}{\overset{+}{S}}}-CH_3 \ I^-$$

Dimethyl sulfide   **Iodomethane**   **Trimethylsulfonium iodide**

Trialkylsulfonium salts are themselves useful alkylating agents because a nucleophile can attack one of the groups bonded to the positively charged sulfur, displacing a neutral sulfide as leaving group. Nature makes extensive use of the trialkylsulfonium salt *S*-adenosylmethionine as a biological methylating agent (see "Biological Substitution Reactions" at the end of Chapter 11).

S-Adenosylmethionine (a sulfonium salt)

Another difference between sulfides and ethers is that sulfides are easily oxidized. Treatment of a sulfide with hydrogen peroxide, H_2O_2, at room temperature yields the corresponding **sulfoxide (R_2SO)**, and further oxidation of the sulfoxide with a peroxyacid yields a **sulfone (R_2SO_2)**.

$$\text{Methyl phenyl sulfide} \xrightarrow[\text{H}_2\text{O, 25°C}]{\text{H}_2\text{O}_2} \text{Methyl phenyl sulfoxide} \xrightarrow{\text{CH}_3\text{CO}_3\text{H}} \text{Methyl phenyl sulfone}$$

Methyl phenyl sulfide   **Methyl phenyl sulfoxide**   **Methyl phenyl sulfone**

Dimethyl sulfoxide (DMSO) is a particularly well-known sulfoxide that is often used as a polar aprotic solvent. It must be handled with care, however, because it has a remarkable ability to penetrate the skin, carrying along whatever is dissolved in it.

**Dimethyl sulfoxide
(a polar aprotic solvent)**

Problem 18.19   Name the following compounds:

(a) $CH_3CH_2\overset{\underset{\displaystyle |}{CH_3}}{C}HSH$

(b) $CH_3\overset{\underset{\displaystyle |}{CH_3}}{\underset{\underset{\displaystyle |}{CH_3}}{C}}CH_2\overset{\underset{\displaystyle |}{SH}}{C}HCH_2\overset{\underset{\displaystyle |}{CH_3}}{C}HCH_3$

(c) cyclopentene with SH substituent

(d) $CH_3\overset{\underset{\displaystyle |}{CH_3}}{C}HSCH_2CH_3$

(e) benzene with two SCH₃ substituents (ortho)

Problem 18.20 2-Butene-1-thiol is one component of skunk spray. How would you synthesize this substance from methyl 2-butenoate? From 1,3-butadiene?

$$CH_3CH=CHCOCH_3 \longrightarrow CH_3CH=CHCH_2SH$$

Methyl 2-butenoate   **2-Butene-1-thiol**

Problem 18.21 How can you account for the fact that dimethyl sulfoxide has a boiling point of 189°C and is miscible with water, whereas dimethyl sulfide has a boiling point of 37°C and is immiscible with water?

• •

CHEMISTRY @ WORK

Epoxy Resins and Adhesives

Kayaks are often made of a high-strength polymer coated with epoxy resin.

Few people know what an epoxide is, but practically everyone has used an "epoxy glue" for household repairs or an epoxy resin for a protective coating. Epoxy resins and adhesives generally consist of two components that must be mixed prior to use. One component is a liquid "prepolymer," and the second is a "curing agent" that reacts with the prepolymer and causes it to solidify.

The most widely used epoxy resins and adhesives are based on a prepolymer made from bisphenol A and epichlorohydrin. On treatment with base under carefully controlled conditions, bisphenol A is converted into its anion, which acts as a nucleophile in an S_N2 reaction with epichlorohydrin. Each epichlorohydrin molecule can react with two molecules of bisphenol A, once by S_N2 displacement of chloride ion and once by opening of the epoxide ring. At the same time, each bisphenol A molecule can react with two epichlorohydrins, leading to a long polymer chain. Each end of a prepolymer chain has an unreacted epoxy group, and each chain has numerous secondary alcohol groups.

(continued) ▶

Bisphenol A   Epichlorohydrin

"Prepolymer"

When the epoxide is to be used, a basic curing agent such as an amine, R_3N, is added to cause the individual prepolymer chains to link together. This "cross-linking" of chains is simply a base-catalyzed epoxide ring-opening of an –OH group in the middle of one chain with an epoxide group on the end of another chain. The result of such cross-linking is formation of a vast, three-dimensional tangle that has enormous strength and chemical resistance.

Middle of
chain 1

End of
chain 2

"Cross-linked" chains

Summary and Key Words

Ethers are compounds that have two organic groups bonded to the same oxygen atom, ROR'. The organic groups can be alkyl, vinylic, or aryl, and the oxygen atom can be in a ring or in an open chain.

Ethers are prepared either by a Williamson synthesis or by an alkoxymercuration/demercuration sequence. The **Williamson ether synthesis** involves S_N2 attack of an alkoxide ion on a primary alkyl halide. The **alkoxymercuration** reaction involves the formation of an intermediate organomercury compound, followed by $NaBH_4$ reduction of the C–Hg bond. The net result is Markovnikov addition of an alcohol to an alkene.

Ethers are inert to most reagents but are attacked by strong acids to give cleavage products. Both HI and HBr are often used. The cleavage reaction takes place by an S$_N$2 mechanism if primary and secondary alkyl groups are bonded to the ether oxygen, but by an S$_N$1 or E1 mechanism if one of the alkyl groups bonded to oxygen is tertiary. Aryl allyl ethers undergo **Claisen rearrangement** to give *o*-allylphenols.

Epoxides are cyclic ethers with a three-membered, oxygen-containing ring. They differ from other ethers in their ease of cleavage. The high reactivity of the strained three-membered ether ring allows epoxide rings to be opened by nucleophilic attack of bases as well as acids. Base-catalyzed epoxide ring opening occurs by S$_N$2 attack of a nucleophile at the less hindered epoxide carbon, whereas acid-catalyzed epoxide ring opening occurs by S$_N$1-like attack at the more highly substituted epoxide carbon.

Thiols, RSH, the sulfur analogs of alcohols, are usually prepared by S$_N$2 reaction of an alkyl halide with thiourea. Mild oxidation of a thiol yields a **disulfide, RSSR**, and mild reduction of a disulfide gives back the thiol. **Sulfides, RSR′**, the sulfur analogs of ethers, are prepared by a Williamson-type S$_N$2 reaction between a thiolate anion and a primary or secondary alkyl halide. Sulfides are much more nucleophilic than ethers and can be oxidized to **sulfoxides (R$_2$SO)** and to **sulfones (R$_2$SO$_2$)**. Sulfides can also be alkylated by reaction with a primary alkyl halide to yield **sulfonium salts, R$_3$S$^+$**.

Summary of Reactions

1. Preparation of ethers
   (a) Williamson synthesis (Section 18.3)

$$RO^- + R'CH_2X \longrightarrow ROCH_2R' + X^-$$

   Alkyl halide should be primary.
   (b) Alkoxymercuration/demercuration (Section 18.4)

   Markovnikov orientation is observed.
   (c) Epoxidation of alkenes with peroxyacids (Section 18.7)

2. Reaction of ethers
   (a) Cleavage by HX (Section 18.5)

$$R-O-R' \xrightarrow[\text{H}_2\text{O}]{\text{HX}} RX + R'OH$$

   where HX = HBr or HI.

(continued) ▶

(b) Claisen rearrangement (Section 18.6)

(c) Acid-catalyzed hydrolysis of epoxides (Section 18.8)

Trans 1,2-diols are produced from cyclic epoxides.

(d) Acid-induced epoxide ring opening (Section 18.8)

(e) Base-catalyzed epoxide ring opening (Section 18.8)

Reaction occurs at least hindered site.

$$ RMgX + H_2C{-}CH_2 \xrightarrow[\text{2. } H_3O^+]{\text{1. Ether solvent}} RCH_2CH_2OH $$

3. Preparation of thiols (Section 18.11)

$$ RCH_2Br \xrightarrow[\text{2. } H_2O,\ NaOH]{\text{1. } (H_2N)_2C{=}S} RCH_2SH $$

4. Oxidation of thiols to disulfides (Section 18.11)

$$ 2\,RSH \xrightarrow{I_2,\ H_2O} RS{-}SR $$

5. Preparation of sulfides (Section 18.11)

$$ RS^- + R'CH_2Br \longrightarrow RSCH_2R' + Br^- $$

6. Oxidation of sulfides (Section 18.11)
   (a) Preparation of sulfoxides

(continued) ▶

(b) Preparation of sulfones

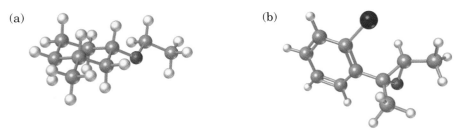

Visualizing Chemistry

● ●

(Problems 18.1–18.21 appear within the chapter.)

18.22   Give IUPAC names for the following compounds (reddish brown = Br):

(a)   (b)

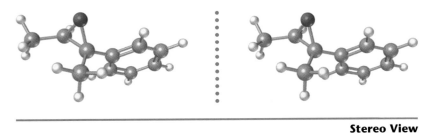

18.23   Show the product, including stereochemistry, that would result from reaction of the following epoxide with HBr:

Stereo View

18.24   Treatment of bornene with a peroxyacid yields a different epoxide from that obtained by reaction of bornene with aqueous Br₂ followed by base treatment. Propose structures for the two epoxides, and explain the result.

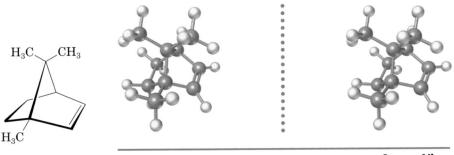

Bornene   **Stereo View**

Additional Problems

18.25 Draw structures corresponding to the following IUPAC names:
(a) Ethyl 1-ethylpropyl ether
(b) Di(*p*-chlorophenyl) ether
(c) 3,4-Dimethoxybenzoic acid
(d) Cyclopentyloxycyclohexane
(e) 4-Allyl-2-methoxyphenol (eugenol; from oil of cloves)

18.26 Give IUPAC names for the following structures:

(a) 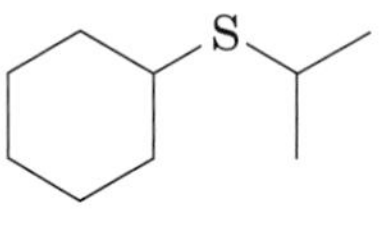

(b) 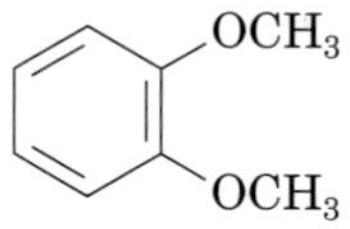

(c) 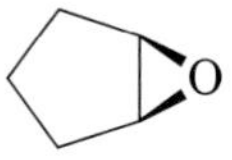

(d) 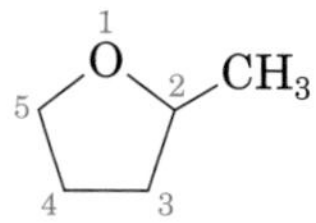

(e) 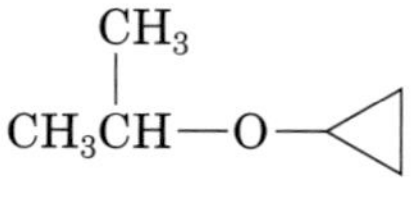

(f) 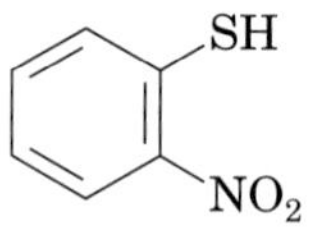

(g) 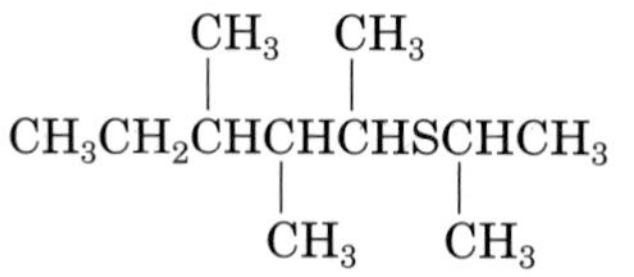

(h) 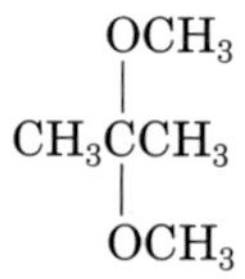

(i) 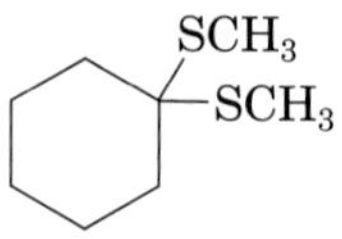

18.27 Predict the products of the following ether cleavage reactions:

(a) 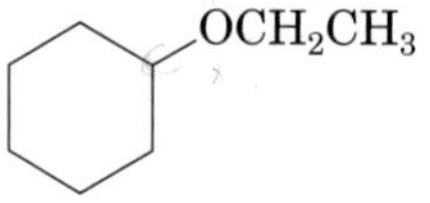 $\xrightarrow[H_2O]{HI}$?

(b) 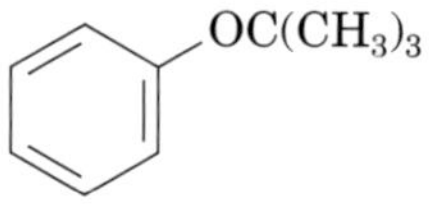 $\xrightarrow{CF_3CO_2H}$?

(c) $H_2C{=}CHOCH_2CH_3 \xrightarrow[H_2O]{HI}$?

(d) $(CH_3)_3CCH_2OCH_2CH_3 \xrightarrow[H_2O]{HI}$?

18.28 How would you prepare the following ethers?

(a) 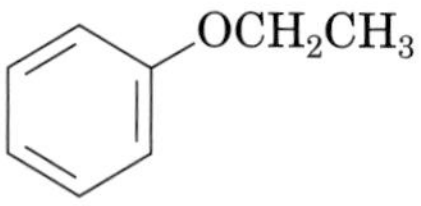

(b) 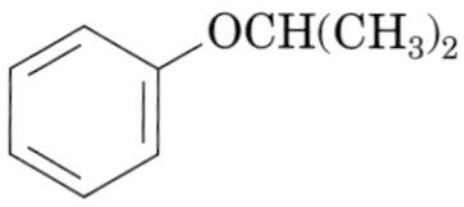

(c) 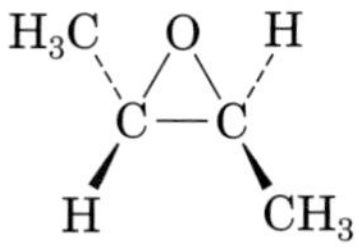

(d) 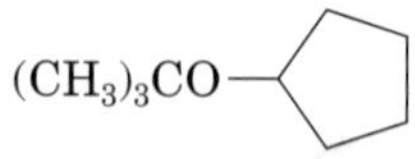

(e) 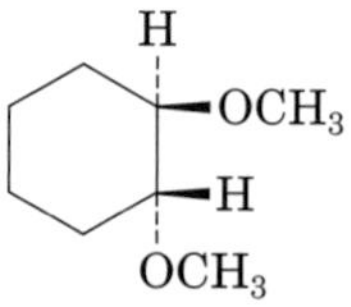

(f) 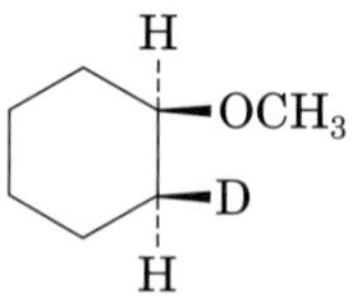

18.29 How would you prepare the following compounds from 1-phenylethanol?
(a) Methyl 1-phenylethyl ether
(b) Phenylepoxyethane
(c) *tert*-Butyl 1-phenylethyl ether
(d) 1-Phenylethanethiol

18.30 How would you carry out the following transformations? More than one step may be required.

(a) [cyclohexene structure] ⟶ [cyclohexane with OCH(CH₃)₂ structure]

$Hg(CF_3CO_2)_2, HOCH(CH_3)(CH_2)$
$NaBH_4$

(b) H₃C—[cyclohexane with OCH₃ and H substituents, H below] $\xrightarrow[H_2O]{Br_2}$ H₃C—[cyclohexane with H and Br substituents, H below]

✳ (c) [cyclohexene with C(CH₃)₃] ⟶ [cyclohexane with H, OH, H, OH and C(CH₃)₃]

(d) $CH_3CH_2CH_2CH_2C{\equiv}CH \longrightarrow CH_3CH_2CH_2CH_2CH_2CH_2OCH_3$

(e) $CH_3CH_2CH_2CH_2C{\equiv}CH \longrightarrow CH_3CH_2CH_2CH_2\overset{\displaystyle OCH_3}{\underset{|}{C}}HCH_3$

18.31 What product would you expect from cleavage of tetrahydrofuran with HI?

18.32 How could you prepare benzyl phenyl ether from benzene? More than one step is required.

18.33 When 2-methylpentane-2,5-diol is treated with sulfuric acid, dehydration occurs and 2,2-dimethyltetrahydrofuran is formed. Suggest a mechanism for this reaction. Which of the two oxygen atoms is most likely to be eliminated, and why?

[structure of 2,2-dimethyltetrahydrofuran with O, CH₃, CH₃] **2,2-Dimethyltetrahydrofuran**

18.34 Methyl aryl ethers, such as anisole, are cleaved to iodomethane and a phenoxide ion by treatment with LiI in hot DMF. Propose a mechanism for this reaction.

18.35 *tert*-Butyl ethers can be prepared by the reaction of an alcohol with 2-methylpropene in the presence of an acid catalyst. Propose a mechanism for this reaction.

18.36 *Meerwein's reagent*, triethyloxonium tetrafluoroborate, is a powerful ethylating agent that converts alcohols into ethyl ethers at neutral pH. Show the reaction of Meerwein's reagent with cyclohexanol, and account for the fact that trialkyloxonium salts are much more reactive alkylating agents than alkyl iodides.

$(CH_3CH_2)_3O^+\ BF_4^-$ **Meerwein's reagent**

18.37 Safrole, a substance isolated from oil of sassafras, is used as a perfumery agent. Propose a synthesis of safrole from catechol (1,2-benzenediol).

[structure of safrole: methylenedioxybenzene with CH₂CH=CH₂ substituent]

Safrole

18.38 Epoxides are reduced by treatment with lithium aluminum hydride to yield alcohols. Propose a mechanism for this reaction.

18.39 Show the structure and stereochemistry of the alcohol that would result if 1,2-epoxy-cyclohexane (Problem 18.38) were reduced with lithium aluminum deuteride, $LiAlD_4$.

18.40 Acid-catalyzed hydrolysis of a 1,2-epoxycyclohexane produces a trans-diaxial 1,2-diol. What product would you expect to obtain from acidic hydrolysis of *cis*-3-*tert*-butyl-1,2-epoxycyclohexane? (Recall that the bulky *tert*-butyl group locks the cyclohexane ring into a specific conformation.)

18.41 Grignard reagents react with oxetane, a four-membered cyclic ether, to yield primary alcohols, but the reaction is much slower than the corresponding reaction with ethylene oxide. Suggest a reason for the difference in reactivity between oxetane and ethylene oxide.

Oxetane

18.42 Treatment of *trans*-2-chlorocyclohexanol with NaOH yields 1,2-epoxycyclohexane, but reaction of the cis isomer under the same conditions yields cyclohexanone. Propose mechanisms for both reactions, and explain why the different results are obtained.

18.43 Ethers undergo an acid-catalyzed cleavage reaction when treated with the Lewis acid BBr_3 at room temperature. Propose a mechanism for the reaction.

18.44 The *Zeisel method* is an analytical procedure for determining the number of methoxyl groups in a compound. A weighed amount of the compound is heated with concentrated HI, ether cleavage occurs, and the iodomethane product is distilled off and passed into an alcohol solution of $AgNO_3$, where it reacts to form a precipitate of silver iodide. The AgI is then collected and weighed, and the percentage of methoxyl groups in the sample is thereby determined. For example, 1.06 g of vanillin, the material responsible for the characteristic odor of vanilla, yields 1.60 g of AgI. If vanillin has a molecular weight of 152, how many methoxyl groups does it contain?

18.45 Disparlure, $C_{19}H_{38}O$, is a sex attractant released by the female gypsy moth, *Lymantria dispar*. The ^1H NMR spectrum of disparlure shows a large absorption in the alkane region, 1–2 δ, and a triplet at 2.8 δ. Treatment of disparlure, first with aqueous acid and then with $KMnO_4$, yields two carboxylic acids identified as undecanoic acid and 6-methylheptanoic acid. ($KMnO_4$ cleaves 1,2-diols to yield carboxylic acids.) Neglecting stereochemistry, propose a structure for disparlure. The actual compound is a chiral molecule with 7R,8S stereochemistry. Draw disparlure, showing the correct stereochemistry.

18.46 How would you synthesize racemic disparlure (Problem 18.45) from compounds having ten or fewer carbons?

18.47 Treatment of 1,1-diphenyl-1,2-epoxyethane with aqueous acid yields diphenylacetaldehyde as the major product. Propose a mechanism for the reaction.

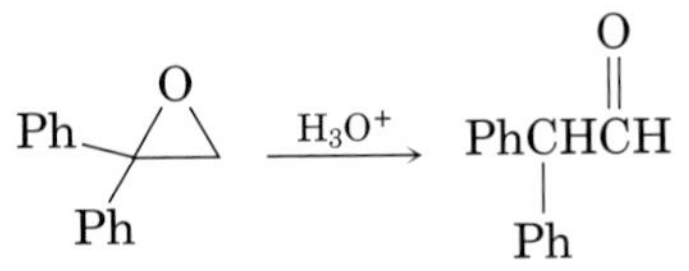

18.48 How would you prepare *o*-hydroxyphenylacetaldehyde from phenol? More than one step is required.

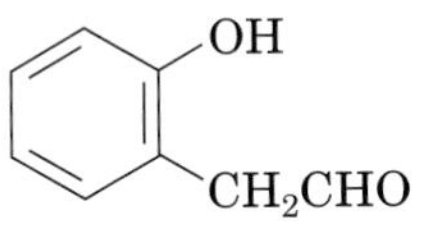

o-Hydroxyphenylacetaldehyde

18.49 Imagine that you have treated (2R,3R)-2,3-epoxy-3-methylpentane with aqueous acid to carry out a ring-opening reaction.

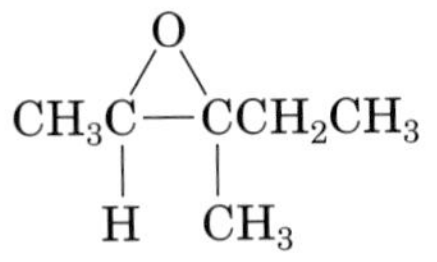

2,3-Epoxy-3-methylpentane
(no stereochemistry implied)

(a) Draw the epoxide, showing stereochemistry.
(b) Draw and name the product, showing stereochemistry.
(c) Is the product chiral? Explain.
(d) Is the product optically active? Explain.

18.50 Identify the reagents a–e in the following scheme:

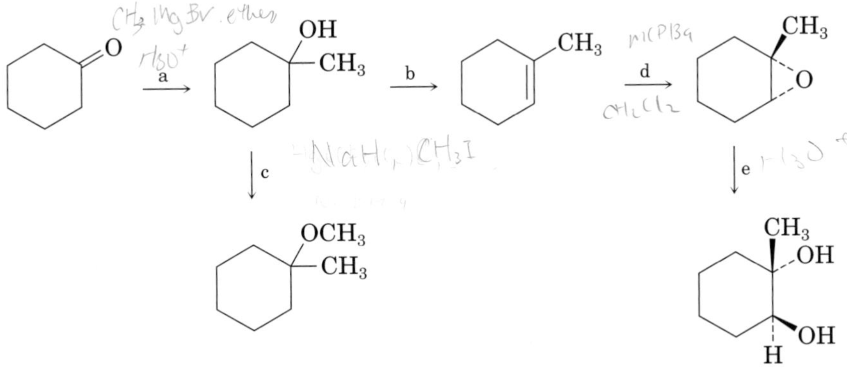

18.51 The red fox (*Vulpes vulpes*) uses a chemical communication system based on scent marks in urine. Recent work has shown one component of fox urine to be a sulfide. Mass spectral analysis of the pure scent-mark component shows $M^+ = 116$. IR spectroscopy shows an intense band at 890 cm^{-1}, and ^{1}H NMR spectroscopy reveals the following peaks:

1.74 δ (3 H, singlet); 2.11 δ (3 H, singlet); 2.27 δ (2 H, triplet, $J = 4.2$ Hz);
2.57 δ (2 H, triplet , $J = 4.2$ Hz); 4.73 δ (2 H, broad)

Propose a structure consistent with these data. [*Note:* (CH$_3$)$_2$S absorbs at 2.1 δ.]

18.52 Anethole, $C_{10}H_{12}O$, a major constituent of the oil of anise, has the ^1H NMR spectrum shown. On oxidation with $Na_2Cr_2O_7$, anethole yields p-methoxybenzoic acid. What is the structure of anethole? Assign all peaks in the NMR spectrum, and account for the observed splitting patterns.

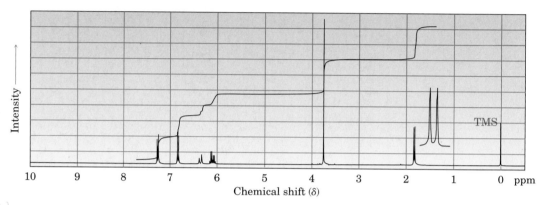

18.53 How would you synthesize anethole (Problem 18.52) from benzene?

aromatic
9.5

18.54 Propose structures for compounds that have the following ^1H NMR spectra:
(a) $C_5H_{12}S$

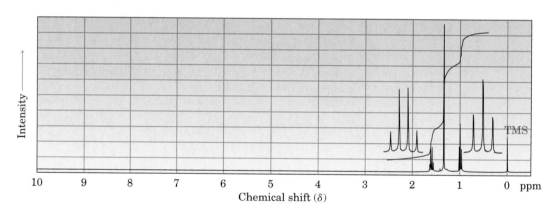

(b) $C_9H_{11}BrO$

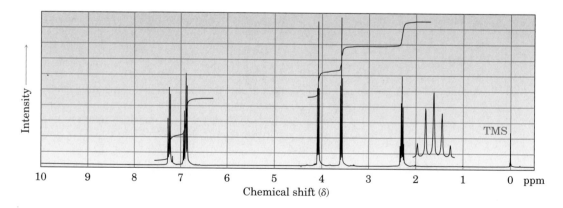

(c) $C_4H_{10}O_2$

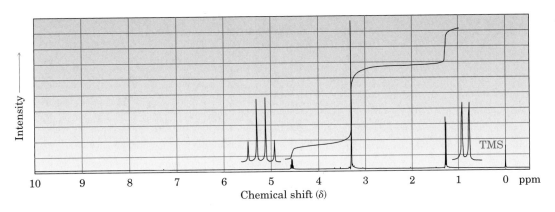

(d) $C_9H_{10}O$

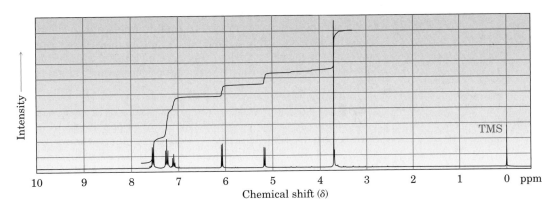

A Look Ahead

· ·

18.55 Aldehydes and ketones undergo acid-catalyzed reaction with alcohols to yield *hemiacetals*, compounds that have one alcohol-like oxygen and one ether-like oxygen bonded to the same carbon. Further reaction of a hemiacetal with alcohol then yields an *acetal*, a compound that has two ether-like oxygens bonded to the same carbon. (See Section 19.11.)

(a) Show the structures of the hemiacetal and acetal you would obtain by reaction of cyclohexanone with ethanol.

(b) Propose a mechanism for the conversion of a hemiacetal into an acetal.

18.56 We saw in Section 17.5 that ketones react with $NaBH_4$ to yield alcohols. We'll also see in Section 22.3 that ketones react with Br_2 to yield α-bromo ketones. Perhaps surprisingly, treatment with $NaBH_4$ of the α-bromo ketone from acetophenone yields an epoxide rather than a bromo alcohol. Show the structure of the epoxide, and explain its formation.

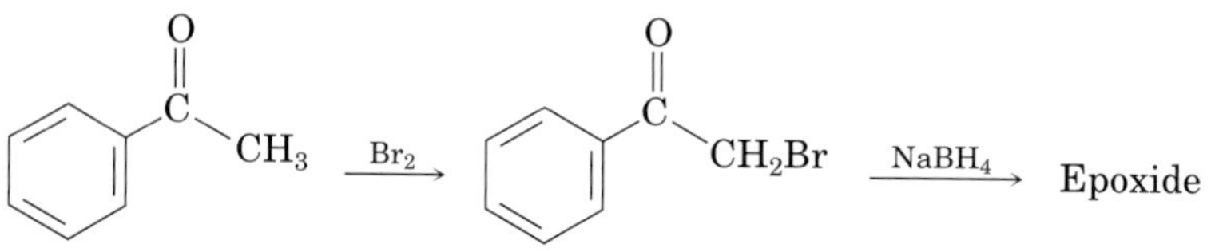

Acetophenone **An α-bromo ketone**

Molecular Modeling

18.57 Use SpartanView to examine two conformers A and B of 18-crown-6 ether. Which conformer adopts a shape that is useful for binding metal cations such as K^+? Which conformer is lower in energy? Compare electrostatic potential maps of the two conformers, and describe how electrostatic interactions might account for the energy difference.

18.58 Some diols can be converted into cyclic ethers by treatment with a strong acid catalyst. Use SpartanView to find the energies of the reactants and products of two such reactions: 1,2-ethanediol $\longrightarrow$ ethylene oxide + water, and 1,4-butanediol $\longrightarrow$ tetrahydrofuran + water. Which reaction has a more favorable $\Delta H°$, and which has a more favorable $\Delta S°$? Explain.

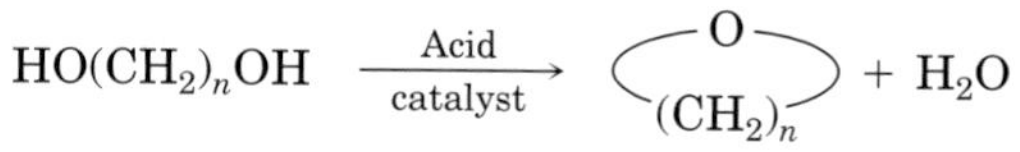

18.59 Many organic reactions must be carried out in solvents that are stable to strongly basic reagents. Use SpartanView to compare electrostatic potential maps of three possible solvents: ethanol, diethyl ether, and acetone (CH_3COCH_3). Which of the three is least likely to react with a strong base? Explain.

18.60 Crown ethers bind some metal cations more strongly than others. Binding is strongest when the cation is *slightly* larger than the ether's negative cavity, and binding is weakest when the cation is *much* larger than the cavity. Use Spartan-View to look at electrostatic potential maps of 12-crown-4 and 18-crown-6, and at space-filling models of Li^+, Na^+, and K^+ cations. Which cation will each crown ether bind most strongly?

A Preview of Carbonyl Compounds

In the next five chapters, we'll discuss the most important functional group in organic chemistry—the **carbonyl group, C=O** (pronounced car-bo-**neel**). Although there are many different kinds of carbonyl compounds and many different reactions, there are only a few fundamental principles that tie the entire field together. The purpose of this brief introduction is not to show details of specific reactions but rather to point out the principles and to provide a framework for learning carbonyl-group chemistry. Read through this overview now, and return to it regularly to remind yourself of the larger picture.

Carbonyl compounds are everywhere in nature. The majority of biologically important molecules contain carbonyl groups, as do most pharmaceutical agents and many of the synthetic chemicals that touch our everyday lives. Acetic acid, the chief component of vinegar; acetaminophen, the active ingredient in many over-the-counter headache remedies; and Dacron, the polyester material used in clothing, all contain different kinds of carbonyl groups.

Acetic acid
(a carboxylic acid)

Acetaminophen
(an amide)

Dacron
(a polyester)

I. Kinds of Carbonyl Compounds

There are many different kinds of carbonyl compounds, depending on what groups are bonded to the C=O unit. The chemistry of all carbonyl groups is similar, however, regardless of their exact structure.

Table 1 shows some of the many different kinds of carbonyl compounds. All contain an **acyl group, R—C—**, bonded to another residue. The R substituent of the acyl group may be alkyl, aryl, alkenyl, or alkynyl; the other substituent to which the acyl fragment is bonded may be a carbon, hydrogen, oxygen, halogen, nitrogen, sulfur, or other atom.

It's useful to classify carbonyl compounds into two general categories based on the kinds of chemistry they undergo. In one category are ketones and aldehydes; in the other are carboxylic acids and their derivatives. The acyl groups in ketones and aldehydes are bonded to atoms (C and H, respectively) that can't stabilize a negative charge and therefore can't act as leaving groups in substitution reactions. The acyl groups in carboxylic acids and

TABLE 1 Types of Carbonyl Compounds

Name	General formula	Name ending	Name	General formula	Name ending
Aldehyde	R–C(=O)–H	-al	Ester	R–C(=O)–O–R′	-oate
Ketone	R–C(=O)–R′	-one	Lactone (cyclic ester)	(cyclic ester structure)	None
Carboxylic acid	R–C(=O)–O–H	-oic acid	Amide	R–C(=O)–N	-amide
Acid halide	R–C(=O)–X (X = halogen)	-yl or -oyl halide			
Acid anhydride	R–C(=O)–O–C(=O)–R′	-oic anhydride	Lactam (cyclic amide)	(cyclic amide structure)	None

their derivatives are bonded to atoms (oxygen, halogen, nitrogen, and so forth) that *can* stabilize a negative charge and therefore *can* act as leaving groups in substitution reactions.

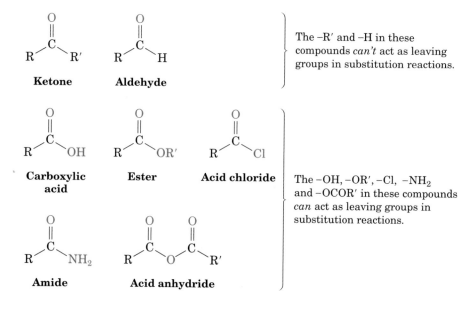

Ketone **Aldehyde**

The –R′ and –H in these compounds *can't* act as leaving groups in substitution reactions.

Carboxylic acid **Ester** **Acid chloride**

Amide **Acid anhydride**

The –OH, –OR′, –Cl, –NH$_2$ and –OCOR′ in these compounds *can* act as leaving groups in substitution reactions.

II. Nature of the Carbonyl Group

The carbon–oxygen double bond of carbonyl groups is similar in many respects to the carbon–carbon double bond of alkenes. The carbonyl carbon atom is sp^2-hybridized and forms three σ bonds. The fourth valence electron remains in a carbon p orbital and forms a π bond to oxygen by overlap with an oxygen p orbital. The oxygen atom also has two nonbonding pairs of electrons, which occupy its remaining two orbitals.

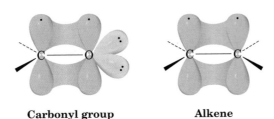

Carbonyl group   **Alkene**

Like alkenes, carbonyl compounds are planar about the double bond and have bond angles of approximately 120°. Figure 1 shows the structure of acetaldehyde and indicates the experimentally determined bond lengths and angles. As you might expect, the carbon–oxygen double bond is both shorter (122 pm versus 143 pm) and stronger [732 kJ/mol (175 kcal/mol) versus 385 kJ/mol (92 kcal/mol)] than a C–O single bond.

FIGURE 1 ▼

Structure of acetaldehyde.

Bond angle (°)		Bond length (pm)	
H—C—C	118	C=O	122
C—C=O	121	C—C	150
H—C=O	121	OC—H	109

Stereo View

Carbon–oxygen double bonds are polarized because of the high electronegativity of oxygen relative to carbon. Thus, all types of carbonyl compounds have substantial dipole moments, as listed in Table 2.

Carbonyl compound	Type of carbonyl compound	Observed dipole moment (D)
TABLE 2 Dipole Moments of Some Carbonyl Compounds, R_2CO		
CH_3CHO	Aldehyde	2.72
$(CH_3)_2CO$	Ketone	2.88
CH_3COOH	Carboxylic acid	1.74
CH_3COCl	Acid chloride	2.72
$CH_3CO_2CH_3$	Ester	1.72
CH_3CONH_2	Amide	3.76

The most important effect of carbonyl-group polarization is on the chemical reactivity of the C=O double bond. Because the carbonyl carbon carries a partial positive charge, it is an electrophilic (Lewis acidic) site and reacts with nucleophiles. Conversely, the carbonyl oxygen carries a partial negative charge, is a nucleophilic (Lewis basic) site, and reacts with electrophiles. The electrostatic potential map of acetone shown below clearly indicates this electron-rich (red) character of the carbonyl oxygen and electron-poor (blue) character of the carbonyl carbon. We'll see in the next five chapters that the majority of carbonyl-group reactions can be rationalized by simple bond-polarization arguments.

 acetone

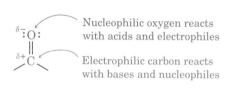

Nucleophilic oxygen reacts with acids and electrophiles

Electrophilic carbon reacts with bases and nucleophiles

III. General Reactions of Carbonyl Compounds

Most reactions of carbonyl groups occur by one of four general mechanisms: *nucleophilic addition, nucleophilic acyl substitution, alpha substitution,* and *carbonyl condensation.* These mechanisms have many variations, just as alkene electrophilic addition reactions and S_N2 reactions do, but the variations are much easier to learn when the fundamental features of the mechanisms are understood. Let's see what the four mechanisms are and what kinds of chemistry carbonyl groups undergo.

Nucleophilic Addition Reactions of Ketones and Aldehydes (Chapter 19)

The most common reaction of ketones and aldehydes is the **nucleophilic addition reaction,** in which a nucleophile, $:Nu^-$, adds to the electrophilic carbon of the carbonyl group. Since the nucleophile uses an electron pair to form a new bond to carbon, two electrons from the carbon–oxygen double bond must move toward the electronegative oxygen atom, where they can be stabilized on an alkoxide anion. The carbonyl carbon rehybridizes from sp^2 to sp^3 during the reaction, and the alkoxide ion product therefore has tetrahedral geometry.

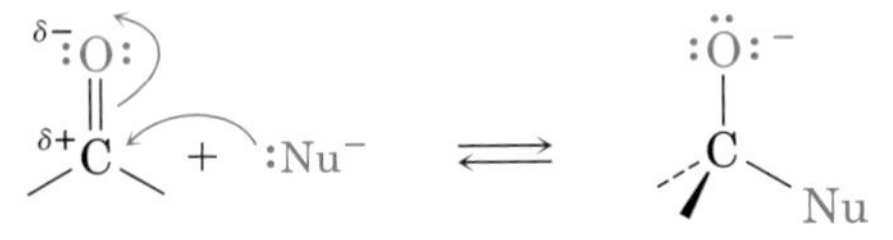

A carbonyl compound
(sp^2-hybridized carbon)

A tetrahedral intermediate
(sp^3-hybridized carbon)

Once formed, and depending on the nature of the nucleophile, the tetrahedral alkoxide intermediate can undergo either of the reactions shown in Figure 2. Often, the tetrahedral alkoxide intermediate is simply protonated by water or acid to form an alcohol product. Alternatively, the tetrahedral intermediate can expel the oxygen to form a new double bond between the carbonyl-group carbon and the nucleophile. We'll study both processes in detail in Chapter 19.

FIGURE 2 ▼

The addition reaction of a ketone or an aldehyde with a nucleophile. Depending on the nucleophile, either an alcohol or a compound with a C=Nu double bond is formed.

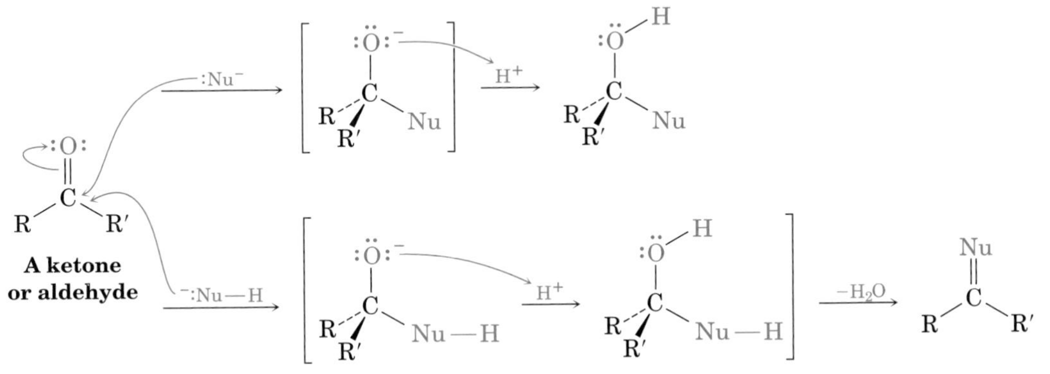

Formation of an Alcohol   The simplest reaction of a tetrahedral alkoxide intermediate is protonation to yield an alcohol. We've already seen two examples of this kind of process during reduction of ketones and aldehydes with hydride reagents such as $NaBH_4$ and $LiAlH_4$ (Section 17.5), and during Grignard reactions (Section 17.6). In the case of reduction, the nucleophile that adds to the carbonyl group is a hydride ion, $H:^-$, while in the case of Grignard reaction, the nucleophile is a carbanion, $R_3C:^-$.

Reduction

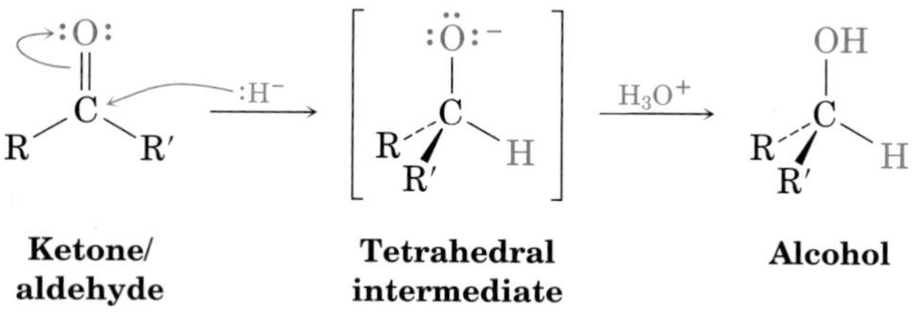

| Ketone/aldehyde | Tetrahedral intermediate | Alcohol |

Grignard reaction

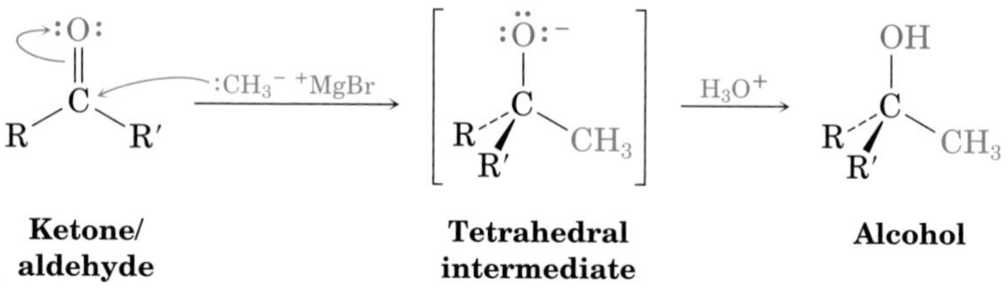

| Ketone/aldehyde | Tetrahedral intermediate | Alcohol |

Formation of C=Nu   The second mode of nucleophilic addition, which often occurs with amine nucleophiles, involves elimination of oxygen and formation of a C=Nu double bond. For example, ketones and aldehydes react with primary amines, RNH_2, to form *imines*, $R_2C=NR'$. These reactions proceed through exactly the same kind of tetrahedral intermediate as that formed during hydride reduction and Grignard reaction, but the initially formed alkoxide ion is not isolated. Instead, it loses water to form an imine, as shown in Figure 3.

FIGURE 3 ▼

Formation of an imine, $R_2C=NR'$, by reaction of an amine with a ketone or an aldehyde.

Addition to the ketone or aldehyde carbonyl group by the neutral amine nucleophile gives a dipolar tetrahedral intermediate.

Transfer of a proton from nitrogen to oxygen then yields a nonpolar amino alcohol intermediate.

Dehydration of the amino alcohol intermediate gives neutral imine plus water as final products.

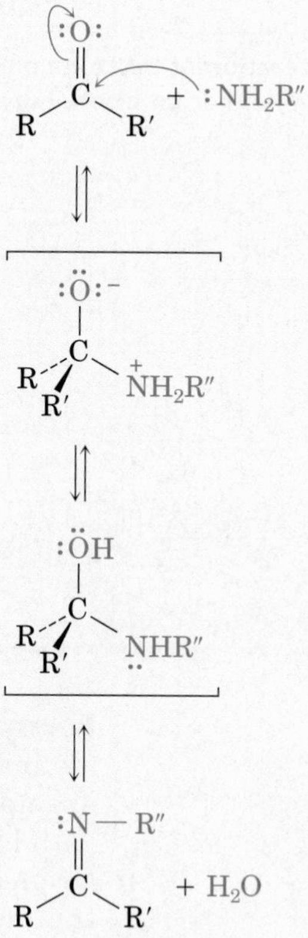

Nucleophilic Acyl Substitution Reactions of Carboxylic Acid Derivatives (Chapter 21)

A second fundamental reaction of carbonyl compounds, **nucleophilic acyl substitution**, is related to the nucleophilic addition reaction just discussed but occurs only with carboxylic acid derivatives rather than with ketones and aldehydes. When the carbonyl group of a carboxylic acid derivative reacts with a nucleophile, addition occurs in the usual way, but the initially formed tetrahedral alkoxide intermediate is not isolated. Because carboxylic acid derivatives have a leaving group bonded to the carbonyl-group carbon, the tetrahedral intermediate can react further by expelling the leaving group and forming a new carbonyl compound:

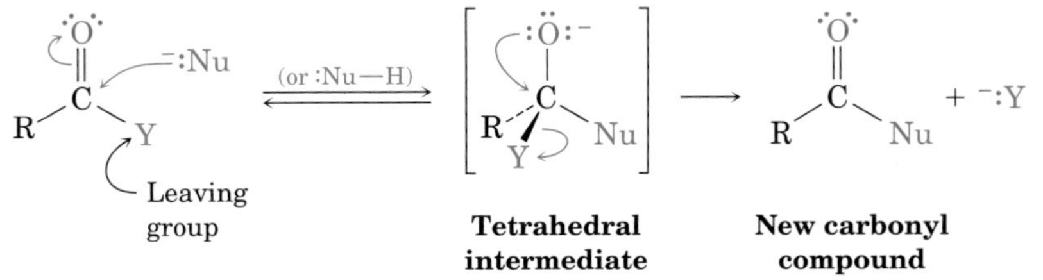

where   Y = −OR (ester), −Cl (acid chloride), −NH₂ (amide), or
         −OCOR′ (acid anhydride)

The net effect of nucleophilic acyl substitution is the replacement of the leaving group by the attacking nucleophile. We'll see in Chapter 21, for example, that acid chlorides are rapidly converted into esters by treatment with alkoxides (Figure 4).

FIGURE 4 ▼

The nucleophilic acyl substitution reaction of an acid chloride with an alkoxide ion yields an ester.

Nucleophilic addition of alkoxide ion to an acid chloride yields a tetrahedral intermediate.

An electron pair from oxygen expels chloride ion and yields the substitution product, an ester.

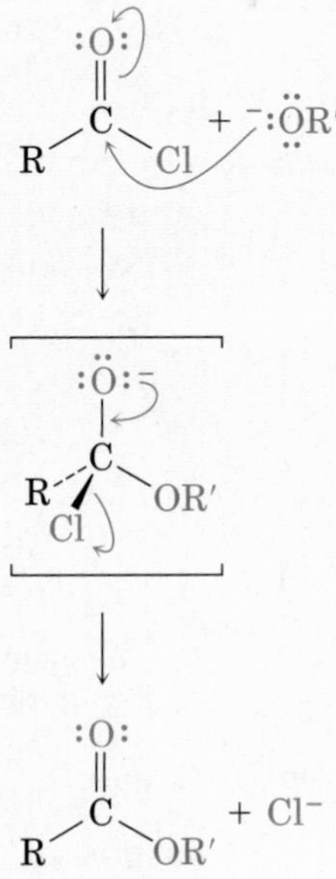

© 1984 JOHN MCMURRY

Alpha-Substitution Reactions (Chapter 22)

The third major reaction of carbonyl compounds, **alpha substitution**, occurs at the position *next to* the carbonyl group—the alpha (α) position. This reaction, which takes place with all carbonyl compounds regardless of structure, results in the substitution of an α hydrogen by an electrophile (E^+) and involves the formation of an intermediate *enol* or *enolate ion:*

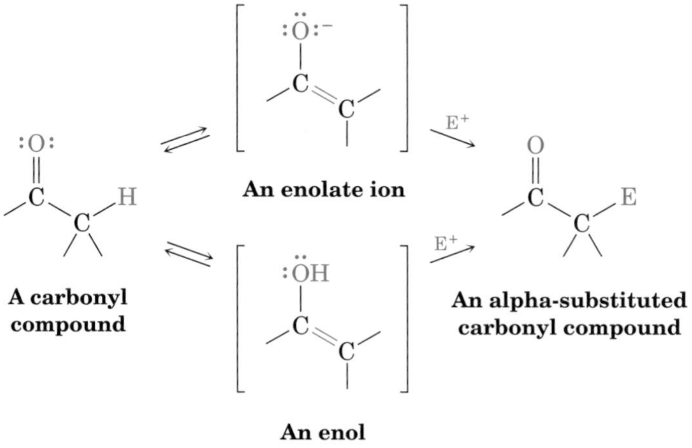

For reasons that we'll explore in Chapter 22, the presence of a carbonyl group renders the hydrogens on the α carbon acidic. Carbonyl compounds therefore react with strong base to yield enolate ions.

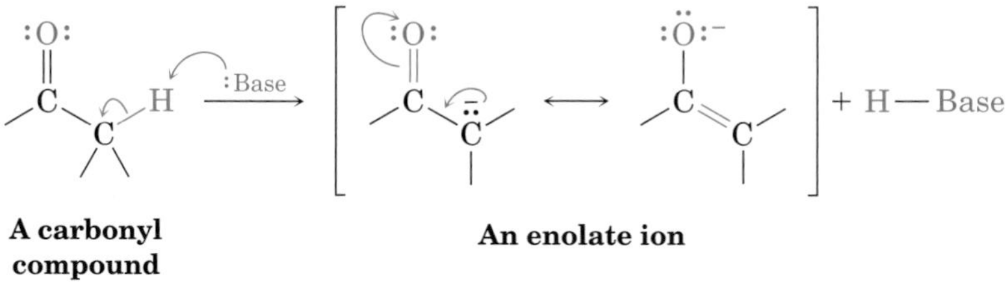

Since they are negatively charged, enolate ions behave as nucleophiles and undergo many of the reactions we've already studied. For example, enolates react with primary alkyl halides in the S_N2 reaction. The nucleophilic enolate ion displaces halide ion, and a new C–C bond forms:

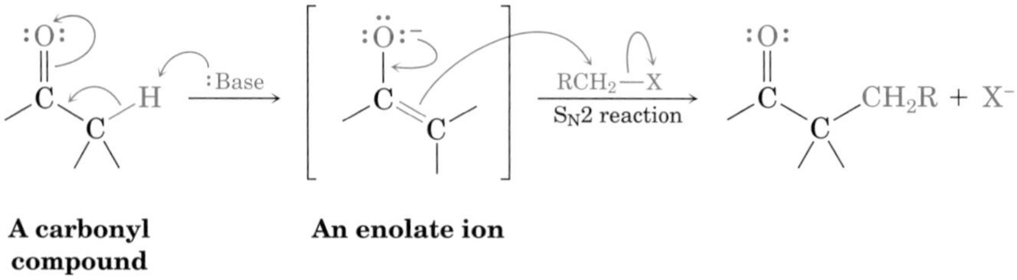

The S_N2 alkylation reaction between an enolate ion and an alkyl halide is one of the most powerful methods available for making C–C bonds, thereby building up larger molecules from smaller precursors. We'll study the alkylation of many kinds of carbonyl groups in Chapter 22.

Carbonyl Condensation Reactions (Chapter 23)

The fourth and last fundamental reaction of carbonyl groups, **carbonyl condensation**, takes place when two carbonyl compounds react with each other. For example, when acetaldehyde is treated with base, two molecules combine to yield the hydroxy aldehyde product known as *aldol* (*ald*ehyde + alco*hol*):

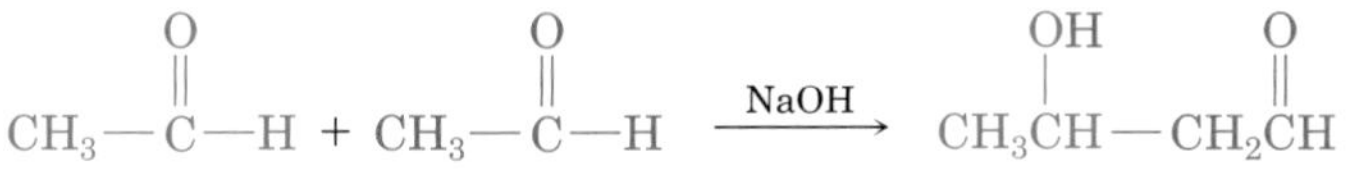

$$CH_3-\overset{\overset{\displaystyle O}{\|}}{C}-H \ + \ CH_3-\overset{\overset{\displaystyle O}{\|}}{C}-H \ \xrightarrow{\text{NaOH}} \ CH_3\overset{\overset{\displaystyle OH}{|}}{C}H-CH_2\overset{\overset{\displaystyle O}{\|}}{C}H$$

Two acetaldehydes   **Aldol**

    Although the carbonyl condensation reaction appears different from the three processes already discussed, it's actually quite similar. A carbonyl condensation reaction is simply a *combination* of a nucleophilic addition step and an α-substitution step. The initially formed enolate ion of one acetaldehyde acts as a nucleophile and adds to the carbonyl group of another acetaldehyde molecule. Reaction occurs by the pathway shown in Figure 5.

FIGURE 5 ▼

A carbonyl condensation reaction between two molecules of acetaldehyde yields a hydroxy aldehyde product.

Hydroxide ion abstracts an acidic alpha proton from one molecule of acetaldehyde, yielding an enolate ion.

The enolate ion adds as a nucleophile to the carbonyl group of a second molecule of acetaldehyde, producing a tetrahedral intermediate.

The intermediate is protonated by water solvent to yield the neutral aldol product and regenerate hydroxide ion.

© 1984 JOHN MCMURRY

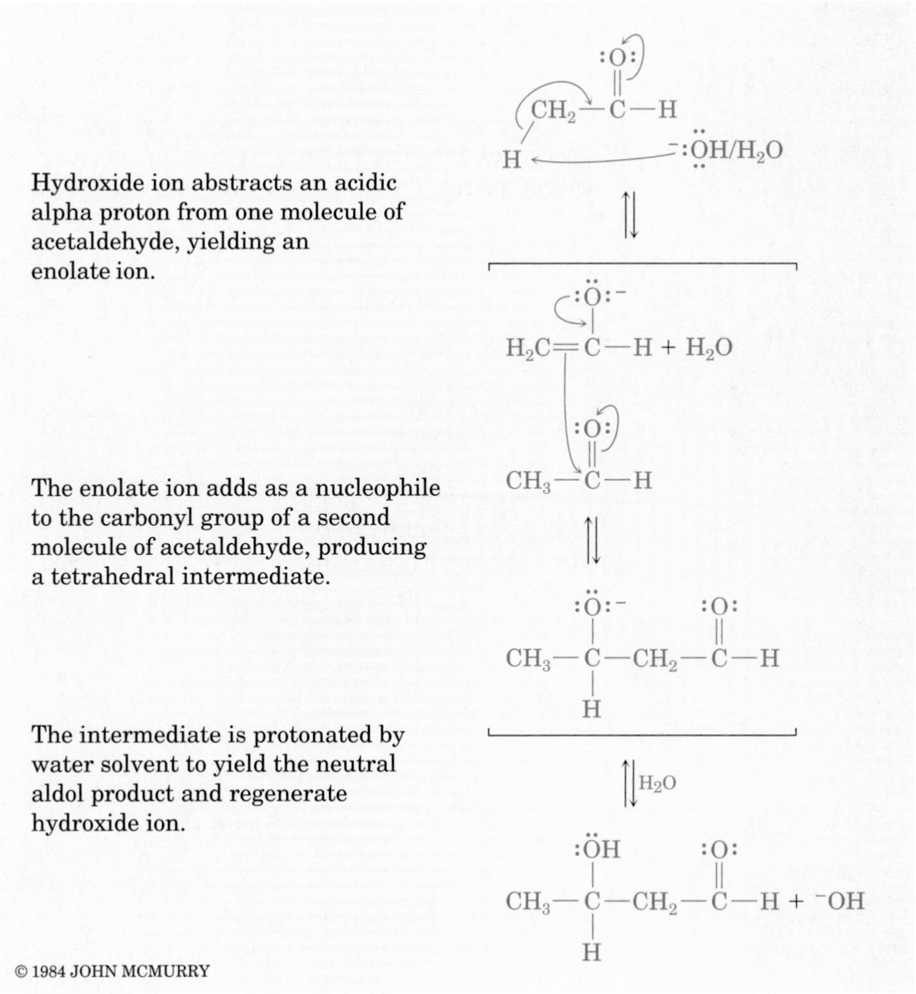

IV. Summary

The purpose of this short preview of carbonyl compounds is not to show details of specific reactions but rather to lay the groundwork for the next five chapters. All of the carbonyl-group reactions we'll be studying in Chapters 19–23 fall into one of the four fundamental categories discussed in this preview. Knowing where we'll be heading should help you to keep matters straight in understanding this most important of all functional groups.

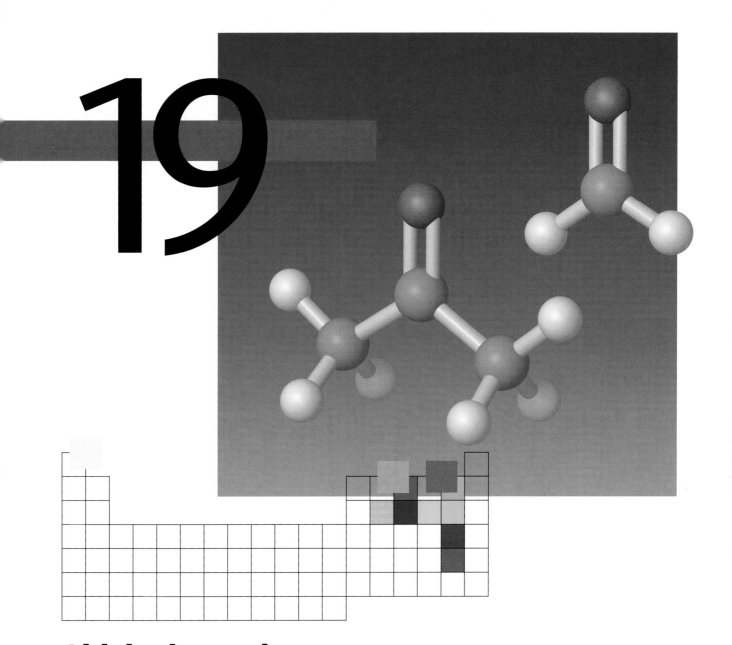

19

Aldehydes and Ketones: Nucleophilic Addition Reactions

Aldehydes and ketones are among the most widely occurring of all compounds, both in nature and in the chemical industry. In nature, many substances required by living organisms are aldehydes or ketones. In the chemical industry, simple aldehydes and ketones are produced in large quantities for use as solvents and as starting materials to prepare a host of other compounds. For example, more than 1.4 million tons per year of formaldehyde, $H_2C=O$, are produced in the United States for use in building insulation materials and in the adhesive resins that bind particle board and plywood. Acetone, $(CH_3)_2C{=}O$, is widely used as an industrial solvent; approximately 1.2 million tons per year are produced in the United States. Formaldehyde is synthesized industrially by catalytic oxidation of methanol, and one method of acetone preparation involves oxidation of 2-propanol.

Methanol   **Formaldehyde**   **Stereo View**

2-Propanol   **Acetone**   **Stereo View**

19.1 Naming Aldehydes and Ketones

Naming Aldehydes

Aldehydes are named by replacing the terminal -*e* of the corresponding alkane name with -*al*. The parent chain must contain the –CHO group, and the –CHO carbon is numbered as carbon 1. For example:

Ethanal   **Propanal**   2-Ethyl-4-methyl**pentanal**
(Acetaldehyde)   (Propionaldehyde)

Note that the longest chain in 2-ethyl-4-methylpentanal is a hexane, but this chain does not include the –CHO group and thus is not considered the parent.

For more complex aldehydes in which the –CHO group is attached to a ring, the suffix -*carbaldehyde* is used:

Cyclohexanecarbaldehyde   **2-Naphthalene**carbaldehyde

Certain simple and well-known aldehydes have common names that are recognized by IUPAC. Some of the more important common names are given in Table 19.1.

TABLE 19.1 Common Names of Some Simple Aldehydes

Formula	Common name	Systematic name
HCHO	Formaldehyde	Methanal
CH_3CHO	Acetaldehyde	Ethanal
CH_3CH_2CHO	Propionaldehyde	Propanal
$CH_3CH_2CH_2CHO$	Butyraldehyde	Butanal
$CH_3CH_2CH_2CH_2CHO$	Valeraldehyde	Pentanal
$H_2C{=}CHCHO$	Acrolein	Propenal
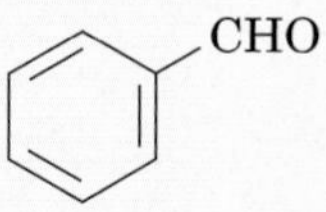 (benzene ring)–CHO	Benzaldehyde	Benzenecarbaldehyde

Naming Ketones

Ketones are named by replacing the terminal -e of the corresponding alkane name with -one. The parent chain is the longest one that contains the ketone group, and the numbering begins at the end nearer the carbonyl carbon. For example:

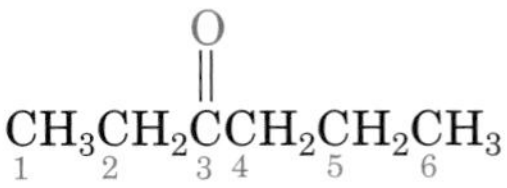

3-**Hexan**one   4-**Hex**en-2-one 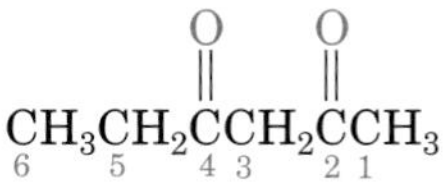   2,4-**Hexane**dione

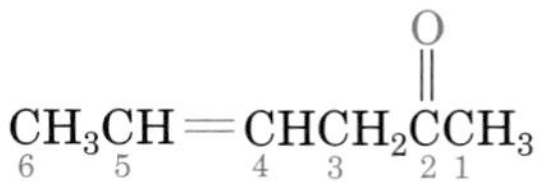

A few ketones are allowed by IUPAC to retain their common names:

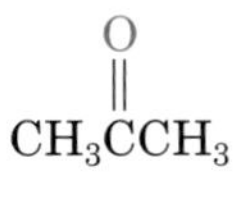

Acetone 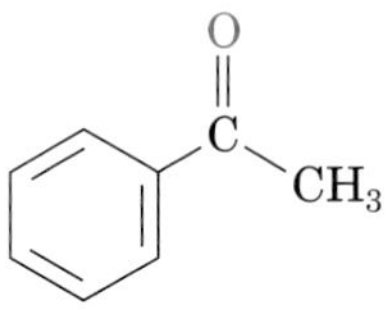   **Acetophenone** 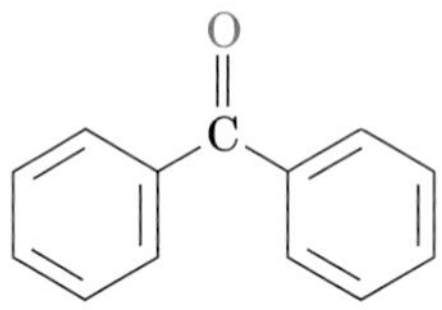   **Benzophenone**

When it's necessary to refer to the **RCO– group** as a substituent, the word **acyl** (**a**-sil) is used and the name ending -yl is attached. For example,

CH$_3$CO– is an **acetyl group**, –CHO is a **formyl group**, and C$_6$H$_5$CO– is a **benzoyl group**.

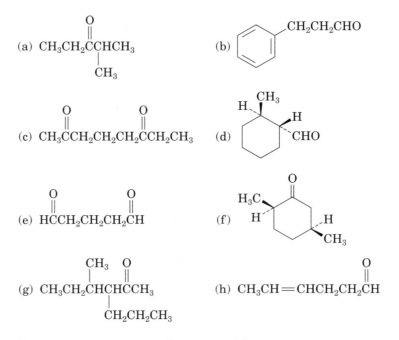

An acyl group   Acetyl   Formyl   Benzoyl

If other functional groups are present and the doubly bonded oxygen is considered a substituent, the prefix *oxo-* is used. For example:

$$\underset{6}{CH_3}\underset{5}{CH_2}\underset{4}{CH_2}\underset{3\ 2}{CCH_2}\underset{1}{COCH_3}$$

Methyl 3-oxohexanoate

• •

Problem 19.1   Name the following aldehydes and ketones according to IUPAC rules:

(a) CH$_3$CH$_2$CCHCH$_3$
          |
          CH$_3$

(b) CH$_2$CH$_2$CHO (on benzene ring)

(c) CH$_3$CCH$_2$CH$_2$CH$_2$CCH$_2$CH$_3$

(d) cyclohexane with H, CH$_3$ and H, CHO

(e) HCCH$_2$CH$_2$CH$_2$CH

(f) cyclohexanone with H$_3$C, H and H, CH$_3$

(g) CH$_3$CH$_2$CHCHCCH$_3$
          | |
          CH$_3$ CH$_2$CH$_2$CH$_3$

(h) CH$_3$CH=CHCH$_2$CH$_2$CH

Problem 19.2   Draw structures corresponding to the following names:
(a) 3-Methylbutanal   (b) 4-Chloro-2-pentanone
(c) Phenylacetaldehyde   (d) *cis*-3-*tert*-Butylcyclohexanecarbaldehyde
(e) 3-Methyl-3-butenal   (f) 2-(1-Chloroethyl)-5-methylheptanal

• •

19.2 Preparation of Aldehydes and Ketones

Preparing Aldehydes

We've already discussed two of the best methods of aldehyde synthesis: oxidation of primary alcohols and oxidative cleavage of alkenes. Let's review briefly.

- Primary alcohols can be oxidized to give aldehydes (Section 17.8). The reaction is often carried out using pyridinium chlorochromate (PCC) in dichloromethane solvent at room temperature:

Citronellol   →   **Citronellal (82%)**

$\xrightarrow[CH_2Cl_2]{PCC}$

- Alkenes with at least one vinylic hydrogen undergo oxidative cleavage when treated with ozone to yield aldehydes (Section 7.8). If the ozonolysis reaction is carried out on a cyclic alkene, a dicarbonyl compound results:

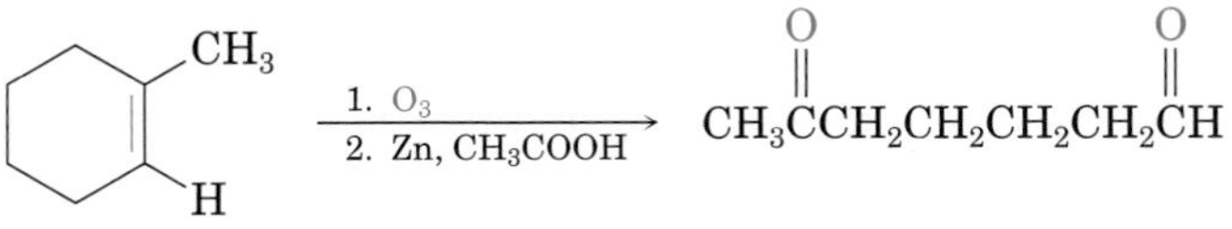

$$\xrightarrow[\text{2. Zn, CH}_3\text{COOH}]{\text{1. O}_3}$$

$$CH_3\overset{\displaystyle O}{\overset{\|}{C}}CH_2CH_2CH_2CH_2\overset{\displaystyle O}{\overset{\|}{C}}H$$

1-Methylcyclohexene   **6-Oxoheptanal (86%)**

A third method of aldehyde synthesis is one that we'll mention here just briefly and then return to for a more detailed explanation in Section 21.6. Certain carboxylic acid derivatives can be partially reduced to yield aldehydes:

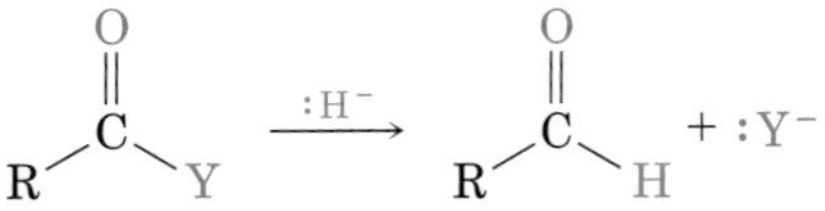

$$\underset{R}{\overset{\displaystyle O}{\overset{\|}{C}}}\underset{Y}{} \quad \xrightarrow{:H^-} \quad \underset{R}{\overset{\displaystyle O}{\overset{\|}{C}}}\underset{H}{} \quad + \; :Y^-$$

For example, the partial reduction of an ester by diisobutylaluminum hydride (DIBAH) is an important laboratory-scale method of aldehyde synthesis. The reaction is normally carried out at $-78°C$ (dry-ice temperature) in toluene solution.

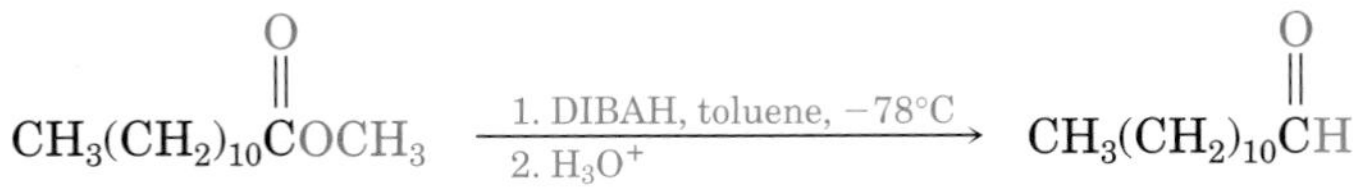

$$CH_3(CH_2)_{10}\overset{\displaystyle O}{\overset{\|}{C}}OCH_3 \quad \xrightarrow[\text{2. H}_3\text{O}^+]{\text{1. DIBAH, toluene, }-78°C} \quad CH_3(CH_2)_{10}\overset{\displaystyle O}{\overset{\|}{C}}H$$

Methyl dodecanoate   **Dodecanal (88%)**

$$\text{where DIBAH} = (CH_3)_2CHCH_2-\overset{\displaystyle H}{\overset{|}{Al}}-CH_2CH(CH_3)_2$$

• •

Problem 19.3 How would you prepare pentanal from the following starting materials?
(a) 1-Pentanol   (b) 1-Hexene   (c) $CH_3CH_2CH_2CH_2COOCH_3$

• •

Preparing Ketones

For the most part, methods of ketone synthesis are analogous to those for aldehydes:

- Secondary alcohols are oxidized by a variety of reagents to give ketones (Section 17.8). The choice of oxidant depends on such factors as reaction scale, cost, and acid or base sensitivity of the alcohol.

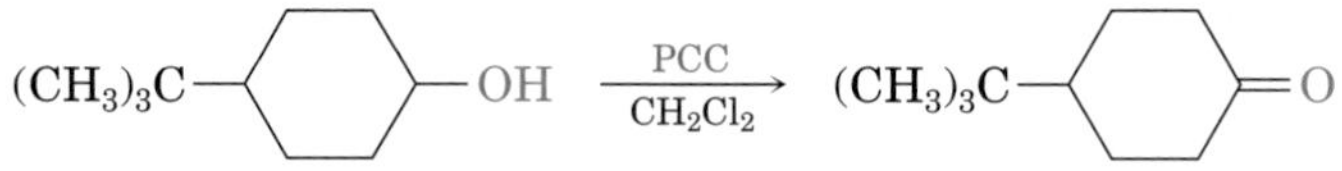

4-*tert*-Butylcyclohexanol   **4-*tert*-Butylcyclohexanone (90%)**

- Ozonolysis of alkenes yields ketones if one of the unsaturated carbon atoms is disubstituted (Section 7.8):

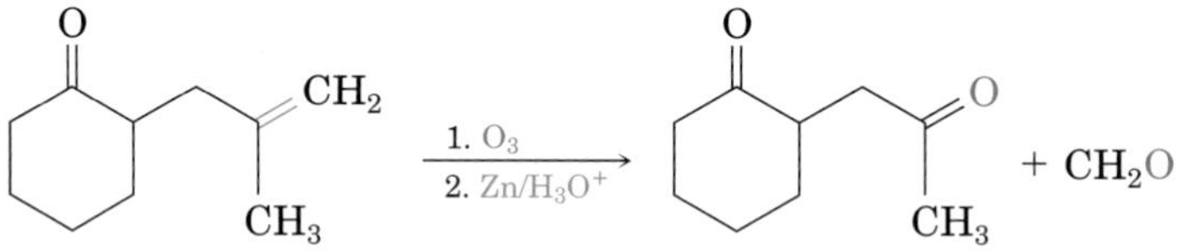

70%

- Aryl ketones are prepared by Friedel–Crafts acylation of an aromatic ring with an acid chloride in the presence of $AlCl_3$ catalyst (Section 16.4):

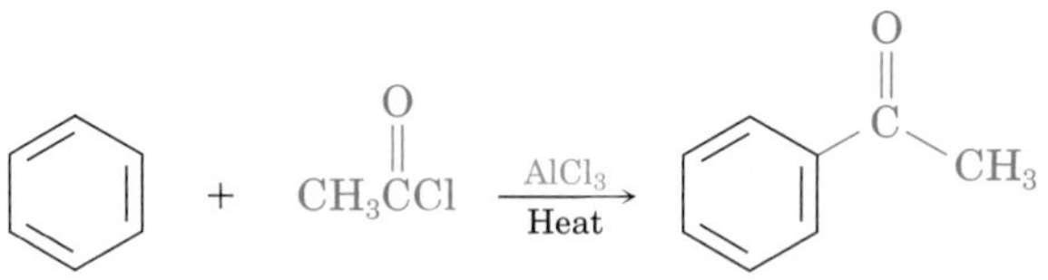

Benzene   **Acetyl chloride**   **Acetophenone (95%)**

- Methyl ketones are prepared by hydration of terminal alkynes in the presence of Hg^{2+} catalyst (Section 8.5):

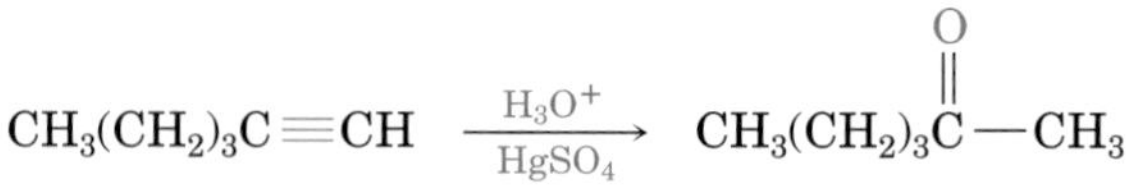

$$CH_3(CH_2)_3C\equiv CH \xrightarrow[HgSO_4]{H_3O^+} CH_3(CH_2)_3\overset{O}{\overset{\|}{C}}-CH_3$$

1-Hexyne   **2-Hexanone (78%)**

In addition to those methods already discussed, ketones can also be prepared from certain carboxylic acid derivatives, just as aldehydes can.

Among the most useful reactions of this type is that between an acid chloride and a diorganocopper reagent. We'll discuss this subject in more detail in Section 21.4.

$$CH_3(CH_2)_4\overset{\overset{\displaystyle O}{\displaystyle \|}}{C}Cl \quad + \quad (CH_3)_2\bar{C}u\,Li^+ \quad \longrightarrow \quad CH_3(CH_2)_4\overset{\overset{\displaystyle O}{\displaystyle \|}}{C}CH_3$$

Hexanoyl chloride   **Dimethylcopper lithium**   **2-Heptanone (81%)**

• •

Problem 19.4   How would you carry out the following reactions? More than one step may be required.
(a) 3-Hexyne $\longrightarrow$ 3-Hexanone
(b) Benzene $\longrightarrow$ *m*-Bromoacetophenone
(c) Bromobenzene $\longrightarrow$ Acetophenone
(d) 1-Methylcyclohexene $\longrightarrow$ 2-Methylcyclohexanone

• •

19.3   Oxidation of Aldehydes and Ketones

Bernhard Tollens

Bernhard Tollens (1841–1918) was born in Hamburg, Germany, received his Ph.D. at the University of Göttingen, and then became professor at the same institution.

Aldehydes are readily oxidized to yield carboxylic acids, but ketones are generally inert toward oxidation. The difference is a consequence of structure: Aldehydes have a –CHO proton that can be abstracted during oxidation, but ketones do not.

$$\underset{\textbf{An aldehyde}}{R-\overset{\overset{\displaystyle O}{\displaystyle \|}}{C}-H} \quad \xrightarrow{[O]} \quad R-\overset{\overset{\displaystyle O}{\displaystyle \|}}{C}-OH \qquad \left[\underset{\textbf{A ketone}}{R-\overset{\overset{\displaystyle O}{\displaystyle \|}}{C}-R'} \right]$$

Hydrogen here   No hydrogen here

Many oxidizing agents, including $KMnO_4$ and hot HNO_3, convert aldehydes into carboxylic acids, but CrO_3 in aqueous acid is a more common choice in the laboratory. The oxidation occurs rapidly at room temperature and results in good yields.

$$CH_3(CH_2)_4\overset{\overset{\displaystyle O}{\displaystyle \|}}{C}H \quad \xrightarrow[\text{Acetone, 0°C}]{CrO_3,\ H_3O^+} \quad CH_3(CH_2)_4\overset{\overset{\displaystyle O}{\displaystyle \|}}{C}OH$$

Hexanal   **Hexanoic acid (85%)**

One drawback to this CrO_3 oxidation is that it takes place under acidic conditions, and sensitive molecules sometimes undergo side reactions. In such cases, the laboratory oxidation of an aldehyde can be carried out using a solution of silver oxide, Ag_2O, in aqueous ammonia, the so-called **Tollens reagent**. Aldehydes are oxidized by the Tollens reagent in high yield without harming carbon–carbon double bonds or other functional groups in the molecule.

Benzaldehyde   **Benzoic acid**

Aldehyde oxidations occur through intermediate 1,1-diols, or *hydrates*, which are formed by a reversible nucleophilic addition of water to the carbonyl group. Even though formed to only a small extent at equilibrium, the hydrate reacts like any typical primary or secondary alcohol and is oxidized to a carbonyl compound (Section 17.8).

An aldehyde   **A hydrate**   **A carboxylic acid**

Ketones are inert to most oxidizing agents but undergo a slow cleavage reaction when treated with hot alkaline $KMnO_4$. The C–C bond next to the carbonyl group is broken, and carboxylic acids are produced. The reaction is useful primarily for symmetrical ketones such as cyclohexanone because product mixtures are formed from unsymmetrical ketones.

Cyclohexanone   **Hexanedioic acid (79%)**

19.4 Nucleophilic Addition Reactions of Aldehydes and Ketones

As we saw in Part III of "A Preview of Carbonyl Compounds," the most general reaction of aldehydes and ketones is the **nucleophilic addition reaction**. A nucleophile, :Nu⁻, attacks the electrophilic C=O carbon atom from a direction approximately 45° to the plane of the carbonyl group. At the same time, rehybridization of the carbonyl carbon from sp^2 to sp^3 occurs, an electron pair from the carbon–oxygen double bond moves toward the electronegative oxygen atom, and a tetrahedral alkoxide ion intermediate is produced (Figure 19.1).

FIGURE 19.1 ▼

A nucleophilic addition reaction to a ketone or aldehyde. The attacking nucleophile approaches the carbonyl group from a direction approximately 45° to the plane of the sp^2 orbitals, the carbonyl carbon rehybridizes from sp^2 to sp^3 and an alkoxide ion is formed. In an example of the reaction, electrostatic potential maps of acetone, cyanide ion nucleophile, and the alkoxide ion addition product show how electron density is transferred from the nucleophile to the oxygen atom.

acetone, cyanide ion, alkoxide ion product

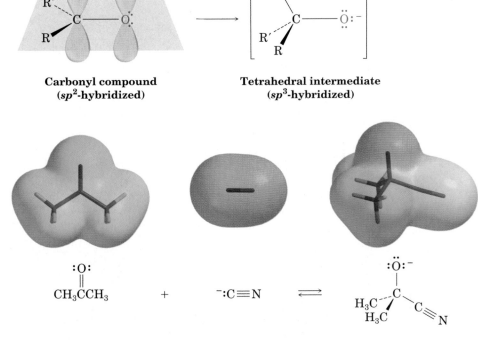

Carbonyl compound (sp^2-hybridized)　　　　　Tetrahedral intermediate (sp^3-hybridized)

$$CH_3\overset{\overset{\displaystyle :O:}{\|}}{C}CH_3 \quad + \quad {}^-:C\equiv N \quad \rightleftharpoons$$

The attacking nucleophile can be either negatively charged (:Nu⁻) or neutral (:Nu). If it's neutral, however, the nucleophile usually carries a hydrogen atom that can subsequently be eliminated, :Nu–H. For example:

Some negatively charged nucleophiles

- H$\ddot{\text{O}}$:⁻ (hydroxide ion)
- H:⁻ (hydride ion)
- R$_3$C:⁻ (a carbanion)
- R$\ddot{\text{O}}$:⁻ (an alkoxide ion)
- N≡C:⁻ (cyanide ion)

Some neutral nucleophiles

- H$\ddot{\text{O}}$H (water)
- R$\ddot{\text{O}}$H (an alcohol)
- H$_3$N: (ammonia)
- R$\ddot{\text{N}}$H$_2$ (an amine)

Nucleophilic additions to aldehydes and ketones have two general variations, as shown in Figure 19.2: (1) The tetrahedral intermediate can be protonated by water or acid to give an alcohol, or (2) the carbonyl oxygen atom can be eliminated as HO^- or H_2O to give a product with a C=Nu double bond.

FIGURE 19.2 ▼

Two general reaction pathways following addition of a nucleophile to a ketone or aldehyde. The top pathway leads to an alcohol product; the bottom pathway leads to a product with a C=Nu double bond.

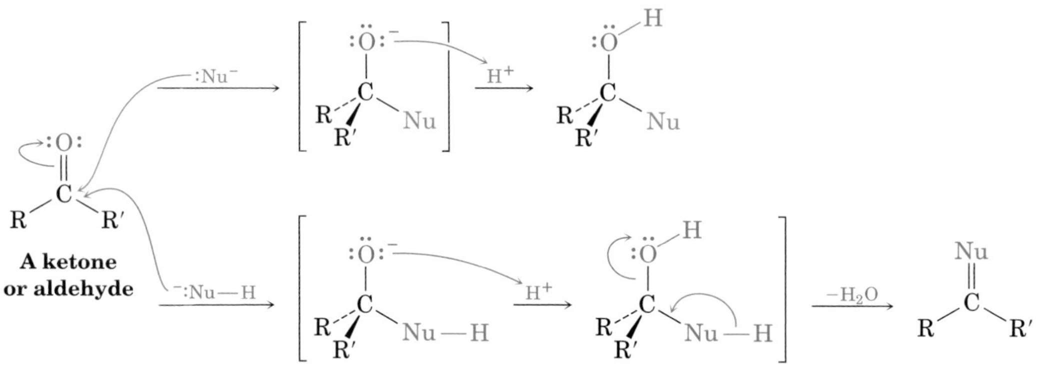

In the remainder of this chapter, we'll look at specific examples of nucleophilic addition reactions. In so doing, we'll be concerned both with the *reversibility* of a given reaction and with the acid or base *catalysis* of that reaction. Some nucleophilic addition reactions take place reversibly, and some do not. Some occur without catalysis, but many others require acid or base to proceed.

Problem 19.5   Treatment of a ketone or aldehyde with cyanide ion ($^-$:C≡N), followed by protonation of the tetrahedral alkoxide ion intermediate, gives a *cyanohydrin*. Show the structure of the cyanohydrin obtained from acetone.

19.5 Relative Reactivity of Aldehydes and Ketones

Aldehydes are generally more reactive than ketones in nucleophilic addition reactions for both steric and electronic reasons. Sterically, the presence of only one relatively large substituent bonded to the C=O carbon in an aldehyde versus two large substituents in a ketone means that an attacking nucleophile is able to approach an aldehyde more readily. Thus, the transition state leading to the tetrahedral intermediate is less crowded and lower in energy for an aldehyde than for a ketone (Figure 19.3).

FIGURE 19.3 ▼

(a) Nucleophilic attack on an aldehyde is sterically less hindered because only one relatively large substituent is attached to the carbonyl-group carbon. (b) A ketone, however, has two large substituents and is more hindered.

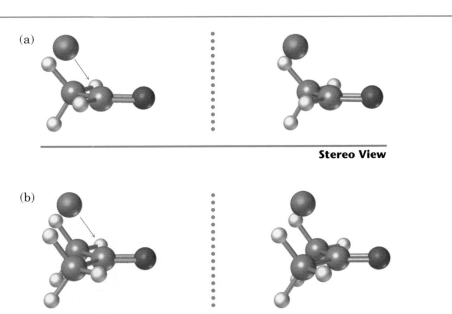

(a)

Stereo View

(b)

Stereo View

Electronically, aldehydes are more reactive than ketones because of the greater polarization of aldehyde carbonyl groups. To see this polarity difference, recall the stability order of carbocations (Section 6.10). A primary carbocation is less stable than a secondary carbocation because it has only one alkyl group inductively stabilizing the positive charge rather than two. In the same way, an aldehyde has only one alkyl group inductively stabilizing the partial positive charge on the carbonyl carbon rather than two. An aldehyde is thus more electrophilic and more reactive than a ketone.

1° carbocation
(less stable, more reactive)

2° carbocation
(more stable, less reactive)

Aldehyde
(less stabilization of $\delta+$, more reactive)

Ketone
(more stabilization of $\delta+$, less reactive)

One further comparison: Aromatic aldehydes, such as benzaldehyde, are less reactive in nucleophilic addition reactions than aliphatic aldehydes. The electron-donating resonance effect of the aromatic ring makes the carbonyl group less electrophilic than the carbonyl group of an aliphatic aldehyde. Comparing electrostatic potential maps of formaldehyde and benzaldehyde, for example, shows that the carbonyl carbon atom is less positive (less blue) in the aromatic aldehyde.

formaldehyde, benzaldehyde

$H_2C{=}O$

C_6H_5CHO

Problem 19.6   Which would you expect to be more reactive toward nucleophilic additions, *p*-methoxy-benzaldehyde or *p*-nitrobenzaldehyde? Explain.

19.6 Nucleophilic Addition of H₂O: Hydration

Biological Connection ▶

Aldehydes and ketones undergo reaction with water to yield 1,1-diols, or **geminal (gem) diols**. The hydration reaction is reversible, and a gem diol can eliminate water to regenerate a ketone or aldehyde.

Acetone (99.9%)   **Acetone hydrate (0.1%)**

The exact position of the equilibrium between a gem diol and a ketone or aldehyde depends on the structure of the carbonyl compound. Although the equilibrium generally favors the less crowded carbonyl compound for steric reasons, the gem diol is favored for a few simple aldehydes. For example, an aqueous solution of formaldehyde consists of 99.9% gem diol and 0.1% aldehyde, whereas an aqueous solution of acetone consists of only about 0.1% gem diol and 99.9% ketone.

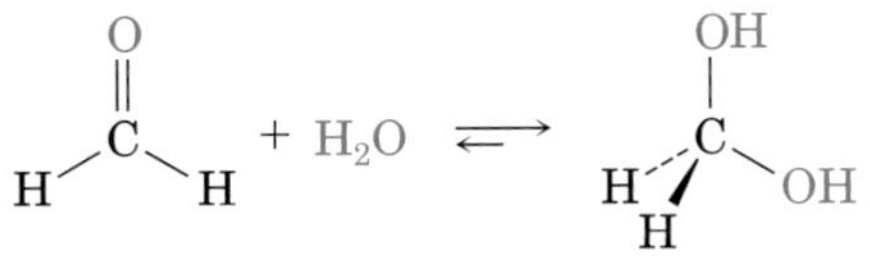

$$
\underset{\substack{\text{Formaldehyde (0.1\%)}}}{\overset{\displaystyle O \atop \displaystyle \underset{H}{\overset{\|}{\underset{\textstyle C}{}}} \; H}{}} + H_2O \;\rightleftharpoons\; \underset{\substack{\text{Formaldehyde hydrate (99.9\%)}}}{\overset{OH}{\underset{\underset{H}{\overset{|}{\underset{\textstyle C}{\,}}}}{H \diagup\!\!\diagdown OH}}}
$$

Formaldehyde (0.1%)   **Formaldehyde hydrate (99.9%)**

The nucleophilic addition of water to a ketone or aldehyde is slow in pure water but is catalyzed by both acid and base. Like all catalysts, acids and bases don't change the position of the equilibrium; they affect only the rate at which the hydration reaction occurs.

The base-catalyzed hydration reaction takes place as shown in Figure 19.4. The attacking nucleophile is the negatively charged hydroxide ion.

FIGURE 19.4 ▼

Mechanism of base-catalyzed hydration of a ketone or aldehyde. Hydroxide ion is a more reactive nucleophile than neutral water.

refer to
**Mechanisms
& Movies**

Hydroxide ion nucleophile adds to the ketone or aldehyde carbonyl group to yield an alkoxide ion intermediate.

The basic alkoxide ion intermediate abstracts a proton (H⁺) from water to yield gem diol product and regenerate hydroxide ion catalyst.

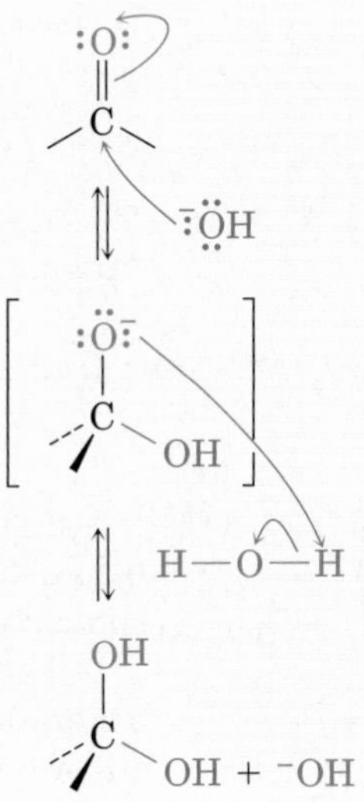

© 1984 JOHN MCMURRY

The acid-catalyzed hydration reaction begins with protonation of the carbonyl oxygen atom, which places a positive charge on oxygen and makes the carbonyl group more electrophilic. Subsequent nucleophilic addition of water to the protonated ketone or aldehyde then yields a protonated gem diol, which loses H⁺ to give the neutral product (Figure 19.5, p. 766).

Note the key difference between the base-catalyzed and the acid-catalyzed reactions. The base-catalyzed reaction takes place rapidly because water is converted into hydroxide ion, a much better nucleophilic electron *donor*. The

FIGURE 19.5 ▼

Mechanism of acid-catalyzed hydration of a ketone or aldehyde. Acid protonates the carbonyl group, thus making it more electrophilic and more reactive.

refer to
**Mechanisms
& Movies**

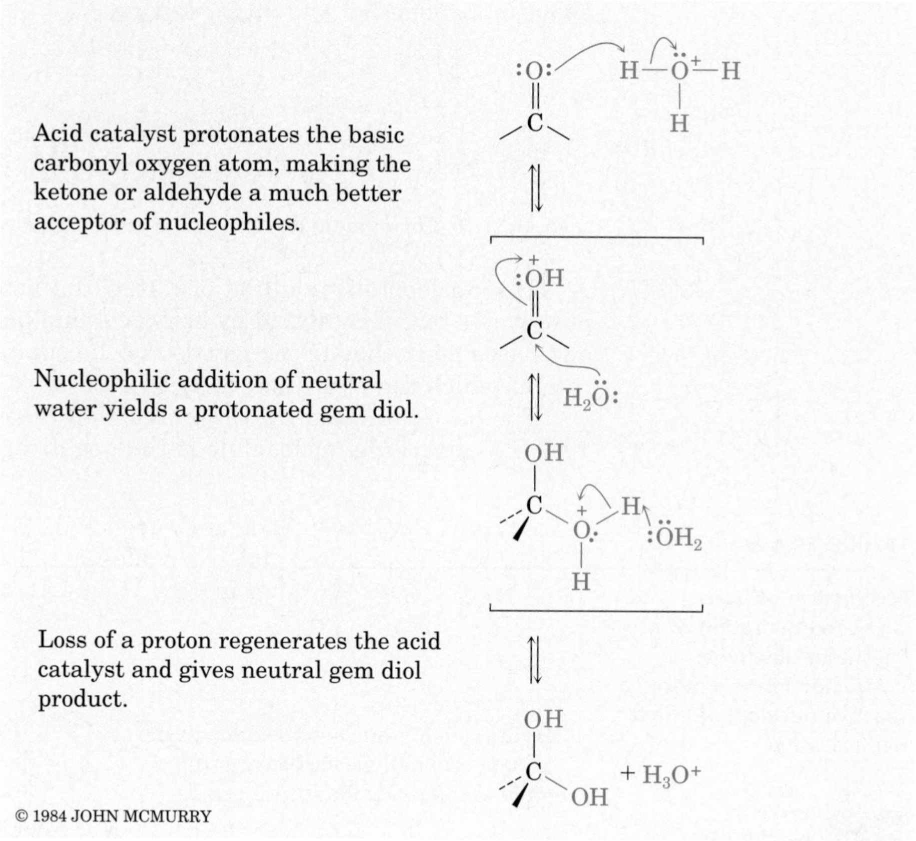

Acid catalyst protonates the basic carbonyl oxygen atom, making the ketone or aldehyde a much better acceptor of nucleophiles.

Nucleophilic addition of neutral water yields a protonated gem diol.

Loss of a proton regenerates the acid catalyst and gives neutral gem diol product.

© 1984 JOHN MCMURRY

acid-catalyzed reaction takes place rapidly because the carbonyl compound is converted by protonation into a much better electrophilic electron *acceptor*.

The hydration reaction just described is typical of what happens when a ketone or aldehyde is treated with a nucleophile of the type H–Y, where the Y atom is electronegative and can stabilize a negative charge (oxygen, halogen, or sulfur, for example). In such reactions, nucleophilic addition is reversible, with the equilibrium favoring the carbonyl reactant rather than the tetrahedral addition product. In other words, treatment of a ketone or aldehyde with CH_3OH, H_2O, HCl, HBr, or H_2SO_4 does not normally lead to an isolable addition product.

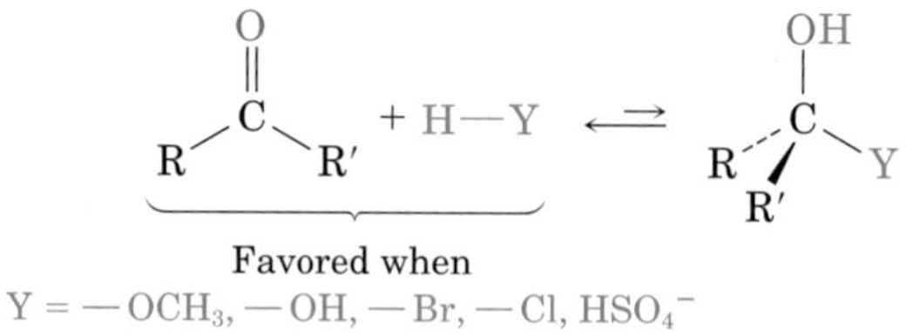

Favored when
$Y = -OCH_3, -OH, -Br, -Cl, HSO_4^-$

● ●

Problem 19.7   When dissolved in water, trichloroacetaldehyde (chloral, CCl_3CHO) exists primarily as chloral hydrate, $CCl_3CH(OH)_2$, better known by the non-IUPAC name "knockout drops." Show the structure of chloral hydrate.

Problem 19.8   The oxygen in water is primarily (99.8%) ^{16}O, but water enriched with the heavy isotope ^{18}O is also available. When a ketone or aldehyde is dissolved in ^{18}O-enriched water, the isotopic label becomes incorporated into the carbonyl group. Explain.

$$R_2C{=}O + H_2O^* \longrightarrow R_2C{=}O^* + H_2O \qquad \text{where} \quad O^* = {}^{18}O$$

● ●

19.7 Nucleophilic Addition of HCN: Cyanohydrin Formation

Aldehydes and unhindered ketones react with HCN to yield **cyanohydrins, RCH(OH)C≡N**. For example, benzaldehyde gives the cyanohydrin commonly called mandelonitrile in 88% yield on treatment with HCN:

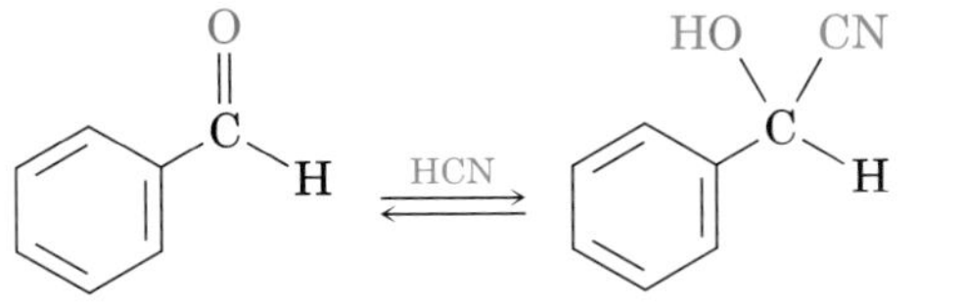

Benzaldehyde   **Mandelonitrile (88%)**
                          (a cyanohydrin)

Studies carried out in the early 1900s by Arthur Lapworth showed that cyanohydrin formation is reversible and base-catalyzed. Reaction occurs slowly when pure HCN is used but rapidly when a small amount of base is added to generate the nucleophilic cyanide ion, CN^-. Alternatively, a small amount of KCN can be added to HCN to catalyze the reaction.

Addition of CN^- to a ketone or aldehyde occurs by a typical nucleophilic addition pathway, yielding a tetrahedral intermediate that is protonated by HCN to give cyanohydrin product plus regenerated CN^-.

Arthur Lapworth

Arthur Lapworth (1872–1941) was born in Galashiels, Scotland, and received a D.Sc. at the City and Guilds Institute, London. He was professor of chemistry at the University of Manchester from 1909 until his retirement in 1937.

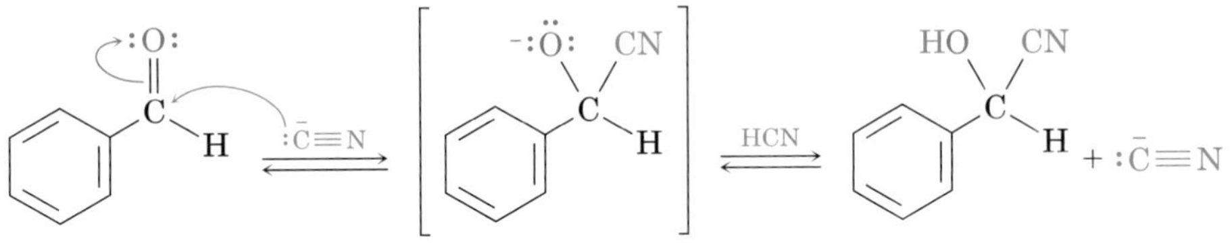

Benzaldehyde   **Tetrahedral**   **Mandelonitrile (88%)**
                          intermediate

Cyanohydrin formation is unusual because it is one of the few examples of the addition of a protic acid (H–Y) to a carbonyl group. As noted in the previous section, reagents such as H_2O, HBr, HCl, and H_2SO_4 don't normally form isolable carbonyl adducts because their equilibrium constant for reaction is unfavorable. With HCN, however, the equilibrium favors the cyanohydrin adduct.

Cyanohydrin formation is useful because of the further chemistry that can be carried out. For example, the nitrile group (–C≡N) can be reduced with $LiAlH_4$ to yield a primary amine (RCH_2NH_2) and can be hydrolyzed by hot aqueous acid to yield a carboxylic acid. Thus, cyanohydrin formation provides a method for transforming a ketone or aldehyde into a different functional group.

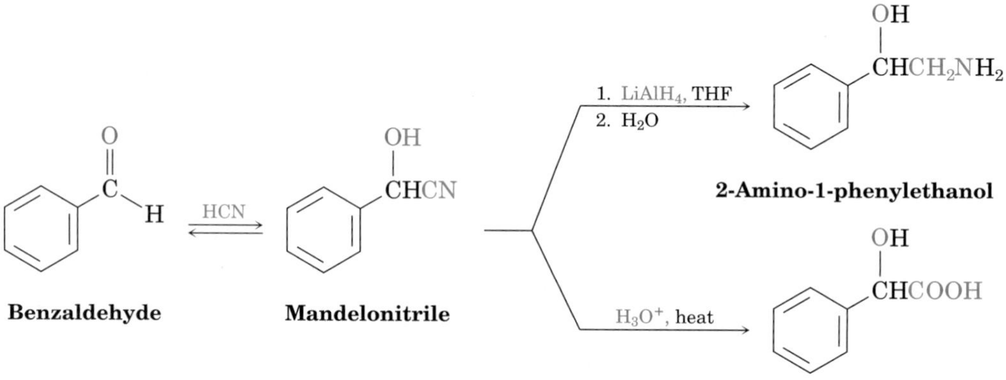

Benzaldehyde   **Mandelonitrile**

1. $LiAlH_4$, THF
2. H_2O

2-Amino-1-phenylethanol

H_3O^+, heat

Mandelic acid (90%)

· ·

Problem 19.9   Cyclohexanone forms a cyanohydrin in good yield but 2,2,6-trimethylcyclohexanone does not. Explain.

· ·

19.8 Nucleophilic Addition of Grignard Reagents and Hydride Reagents: Alcohol Formation

Treatment of a ketone or aldehyde with a Grignard reagent, RMgX, yields an alcohol by nucleophilic addition of a *carbon* anion, or *carbanion* (Section 17.6). The C–Mg bond in the Grignard reagent is so strongly polarized that a Grignard reagent acts for all practical purposes as R:⁻ ⁺MgX.

Acid–base complexation of Mg^{2+} with the carbonyl oxygen atom first serves to make the carbonyl group a better acceptor, and nucleophilic addition of R:⁻ then produces a tetrahedral magnesium alkoxide intermediate. Protonation by addition of water or dilute aqueous acid in a separate step yields the neutral alcohol (Figure 19.6). Unlike the nucleophilic additions of water and HCN, Grignard additions are irreversible because a carbanion is too poor a leaving group to be expelled in a reversal step.

Mechanism of the Grignard reaction. The nucleophilic addition of a carbanion to a ketone or aldehyde yields an alcohol.

The Lewis acid Mg^{2+} first forms an acid–base complex with the basic oxygen atom of the aldehyde or ketone, thereby making the carbonyl group a better acceptor.

Nucleophilic addition of an alkyl group $:R^-$ to the aldehyde or ketone produces a tetrahedral magnesium alkoxide intermediate . . .

. . . which undergoes hydrolysis when water is added in a separate step. The final product is a neutral alcohol.

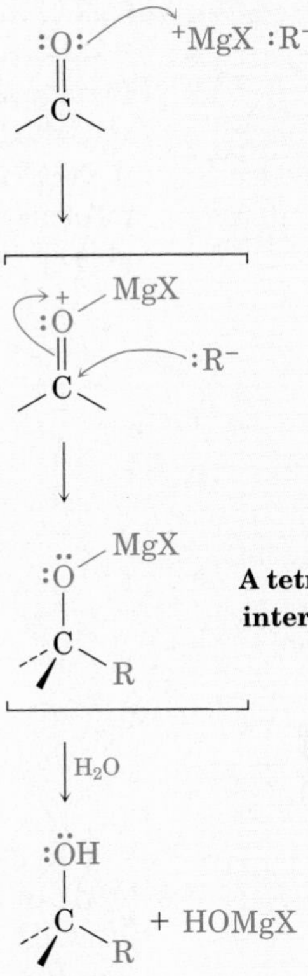

A tetrahedral intermediate

An alcohol

© 1999 JOHN MCMURRY

Treatment of a ketone or aldehyde with $LiAlH_4$ or $NaBH_4$ reduces the carbonyl group and yields an alcohol (Section 17.5). Although the exact details of carbonyl-group reduction are complex, $LiAlH_4$ and $NaBH_4$ act as if they were donors of hydride ion, $:H^-$, and the key step is a nucleophilic addition reaction (Figure 19.7). Addition of water or aqueous acid after the hydride addition step protonates the tetrahedral alkoxide intermediate and gives the alcohol product.

Mechanism of carbonyl-group reduction by nucleophilic addition of "hydride ion" from $NaBH_4$ or $LiAlH_4$.

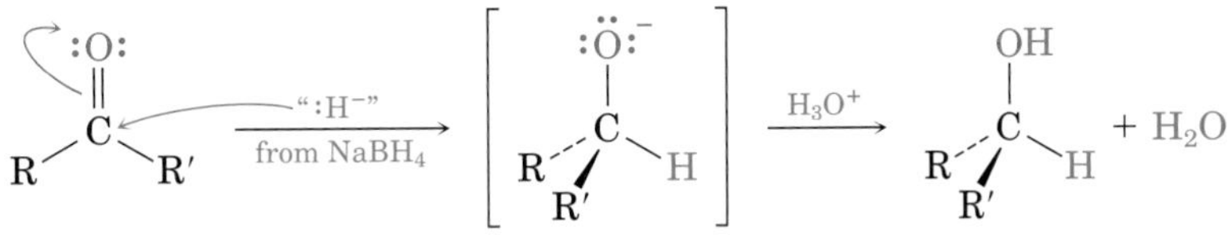

19.9 Nucleophilic Addition of Amines: Imine and Enamine Formation

Biological Connection ▶

Primary amines, RNH_2, add to aldehydes and ketones to yield **imines, $R_2C=NR$**. Secondary amines, R_2NH, add similarly to yield **enamines, $R_2N–CR=CR_2$** (*ene* + *amine* = unsaturated amine). Imines are important intermediates in many metabolic pathways, and we'll see frequent examples of their occurrence in Chapter 29.

A ketone or an aldehyde

An imine   **An enamine**

Imine formation and enamine formation appear different because one leads to a product with a C=N double bond and the other leads to a product with a C=C double bond. Actually, though, the reactions are quite similar. Both are typical examples of nucleophilic addition reactions in which water is eliminated from the initially formed tetrahedral intermediate and a new C=Nu double bond is formed.

Imines are formed in a reversible, acid-catalyzed process that begins with nucleophilic addition of the primary amine to the carbonyl group, followed by transfer of a proton from nitrogen to oxygen to yield a neutral amino alcohol, or **carbinolamine**. Protonation of the carbinolamine oxygen by an acid catalyst then converts the –OH into a better leaving group ($–OH_2^+$), and E1-like loss of water produces an iminium ion. Loss of a proton from nitrogen gives the final product and regenerates the acid catalyst (Figure 19.8).

Imine formation is slow at both high pH and low pH but reaches a maximum rate at a weakly acidic pH around 4–5. For example, the profile of pH versus rate obtained for the reaction between acetone and hydroxylamine, NH_2OH, shows that the maximum reaction rate is obtained at pH 4.5 (Figure 19.9, p. 772).

We can explain the observed pH dependence of imine formation by looking at each individual step in the mechanism. As indicated in Figure 19.8, an acid catalyst is required to protonate the intermediate carbinolamine, thereby converting the –OH into a better leaving group. Thus, reaction will be slow if there is not enough acid present (that is, at high pH). On the

FIGURE 19.8 ▼

Mechanism of imine
formation by reaction of a
ketone or aldehyde with a
primary amine. The key
step is nucleophilic
addition to yield a
carbinolamine
intermediate, which then
loses water to give the
imine.

refer to
Mechanisms
& Movies

Nucleophilic attack on the ketone or
aldehyde by the lone-pair electrons of
an amine leads to a dipolar tetrahedral intermediate.

A proton is then transferred from
nitrogen to oxygen, yielding a
neutral carbinolamine.

Acid catalyst protonates the hydroxyl
oxygen.

The nitrogen lone-pair electrons
expel water, giving an iminium ion.

Loss of H⁺ from nitrogen then gives
the neutral imine product.

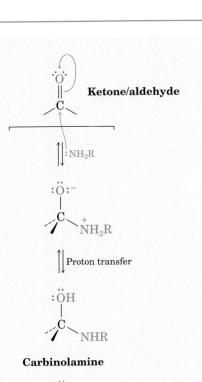

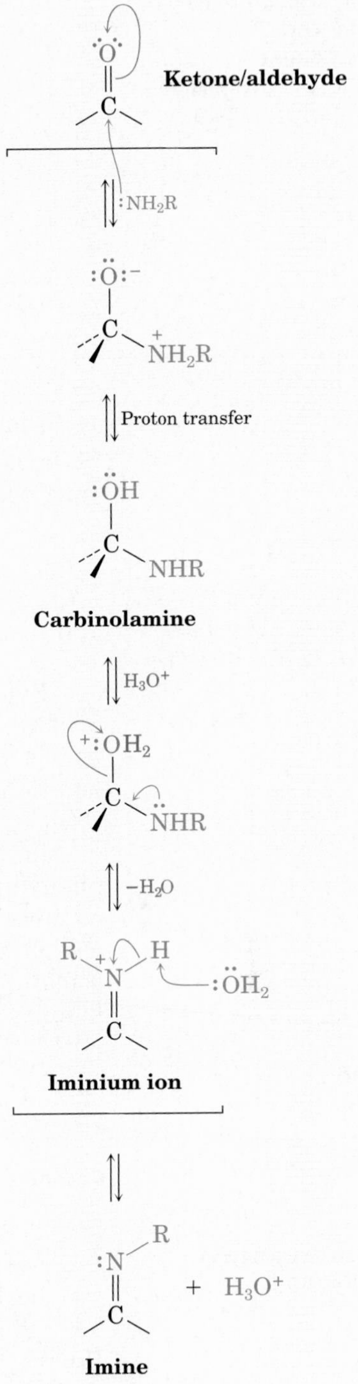

© 1984 JOHN MCMURRY

FIGURE 19.9 ▼

Dependence on pH of the rate of reaction between acetone and hydroxylamine: $(CH_3)_2C=O + NH_2OH \longrightarrow (CH_3)_2C=NOH + H_2O$.

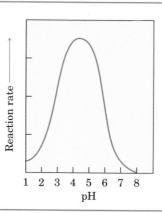

other hand, if *too much* acid is present (low pH), the attacking amine nucleophile is completely protonated so the initial nucleophilic addition step can't occur.

$$H_2\overset{..}{N}OH + H\overset{\frown}{-}A \longrightarrow H_3\overset{+}{N}OH + A^-$$

Base　　**Acid**　　　**Nonnucleophilic**

Evidently, pH 4.5 represents a compromise between the need for *some* acid to catalyze the rate-limiting dehydration step but *not too much* acid so as to avoid complete protonation of the amine. Each individual nucleophilic addition reaction has its own specific requirements, and reaction conditions must be optimized to obtain maximum reaction rates.

Imine formation from such reagents as hydroxylamine, and 2,4-dinitrophenylhydrazine are useful because the products of these reactions—**oximes** and **2,4-dinitrophenylhydrazones (2,4-DNP's)**, respectively—are often crystalline and easy to handle. Such crystalline derivatives are sometimes prepared as a means of purifying and characterizing liquid ketones or aldehydes.

Oxime

Cyclohexanone　　　**Hydroxylamine**　　　**Cyclohexanone oxime**
　　　　　　　　　　　　　　　　　　　　　　　　(mp 90°C)

2,4-Dinitrophenyl-hydrazone

Acetone　　　　**2,4-Dinitrophenyl-**　　　　**Acetone 2,4-dinitrophenyl-**
　　　　　　　　　　hydrazine　　　　　　　　**hydrazone (mp 126°C)**

Enamines are formed when a ketone or aldehyde reacts with a secondary amine, R_2NH. The process is identical to imine formation up to the iminium ion stage, but at this point there is no proton on nitrogen that can be lost to yield a neutral imine product. Instead, a proton is lost from the *neighboring* carbon (the α carbon), yielding an enamine (Figure 19.10).

FIGURE 19.10 ▼

Mechanism of enamine formation by reaction of a ketone or aldehyde with a secondary amine, R_2NH. The iminium ion intermediate has no hydrogen attached to N, and so must lose H^+ from the carbon two atoms away.

Nucleophilic addition of a secondary amine to the ketone or aldehyde, followed by proton transfer from nitrogen to oxygen, yields an intermediate carbinolamine in the normal way.

Protonation of the hydroxyl by acid catalyst converts it into a better leaving group.

Elimination of water by the lone-pair electrons on nitrogen then yields an intermediate iminium ion.

Loss of a proton from the alpha carbon atom yields the enamine product and regenerates the acid catalyst.

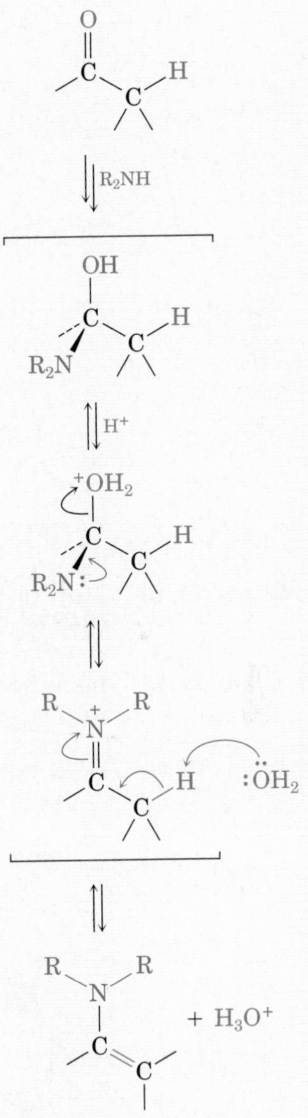

Enamine

© 1984 JOHN MCMURRY

Practice Problem 19.1   Show the products you would obtain by reaction of 3-pentanone with methyl-amine, CH_3NH_2, and with dimethylamine, $(CH_3)_2NH$.

Strategy   We've seen that a ketone or aldehyde reacts with a primary amine, RNH_2, to yield an imine, in which the carbonyl oxygen atom has been replaced by the =N–R group of the amine. Reaction of the same ketone or aldehyde with a secondary amine, R_2NH, yields an enamine, in which the oxygen atom has been replaced by the $-NR_2$ group of the amine and the double bond has moved to a position between the former carbonyl carbon and the neighboring carbon.

Solution

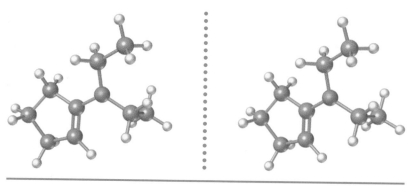

Problem 19.10   Show the products you would obtain by reaction of cyclohexanone with ethylamine, $CH_3CH_2NH_2$, and with diethylamine, $(CH_3CH_2)_2NH$.

Problem 19.11   Imine formation is reversible. Show all the steps involved in the reaction of an imine with water (hydrolysis) to yield a ketone or aldehyde plus primary amine.

Problem 19.12   Draw the following molecule as a standard line-bond structure, and show how it can be prepared from a ketone and an amine.

Stereo View

19.10 Nucleophilic Addition of Hydrazine: The Wolff–Kishner Reaction

A useful variant of the imine-forming reaction just discussed involves the treatment of a ketone or aldehyde with hydrazine, H_2NNH_2, in the presence of KOH. This reaction, discovered independently in 1911 by Ludwig Wolff in Germany and N. M. Kishner in Russia, is a valuable method for converting a ketone or aldehyde into an alkane, $R_2C=O \longrightarrow R_2CH_2$. The **Wolff–Kishner reaction** was originally carried out at temperatures as high as 240°C, but a modification in which dimethyl sulfoxide is used as solvent allows the process to take place near room temperature.

Propiophenone   Propylbenzene (82%)

Cyclopropanecarbaldehyde   Methylcyclo-propane (72%)

Ludwig Wolff

Ludwig Wolff (1857–1919) was born in Neustadt/Hardt, Germany, and received his Ph.D. from the University of Strasbourg working with Rudolf Fittig. He was professor of chemistry at the University of Jena.

N. M. Kishner

N. M. Kishner (1867–1935) was born in Moscow and received his Ph.D. at the University of Moscow working with Vladimir Markovnikov. He became professor, first at the University of Tomsk and then at the University of Moscow.

The Wolff–Kishner reaction involves formation of a hydrazone intermediate, $R_2C=NNH_2$, followed by base-catalyzed double-bond migration, loss of N_2 gas, and protonation to give the alkane product (Figure 19.11, p. 776). The double-bond migration takes place when base removes one of the weakly acidic NH protons to generate a hydrazone anion. Since the hydrazone anion has an allylic resonance structure that places the double bond between nitrogens and the negative charge on carbon, reprotonation can occur on carbon to generate the double-bond rearrangement product. The next step— loss of nitrogen and formation of an alkyl anion—is driven by the large thermodynamic stability of the N_2 molecule.

Note that the Wolff–Kishner reduction accomplishes the same overall transformation as the catalytic hydrogenation of an acylbenzene to yield an alkylbenzene (Section 16.11). The Wolff–Kishner reduction is more general and more useful than catalytic hydrogenation, however, because it works well with both alkyl and aryl ketones.

FIGURE 19.11 ▼

Mechanism of the Wolff–Kishner reduction of a ketone or aldehyde to yield an alkane.

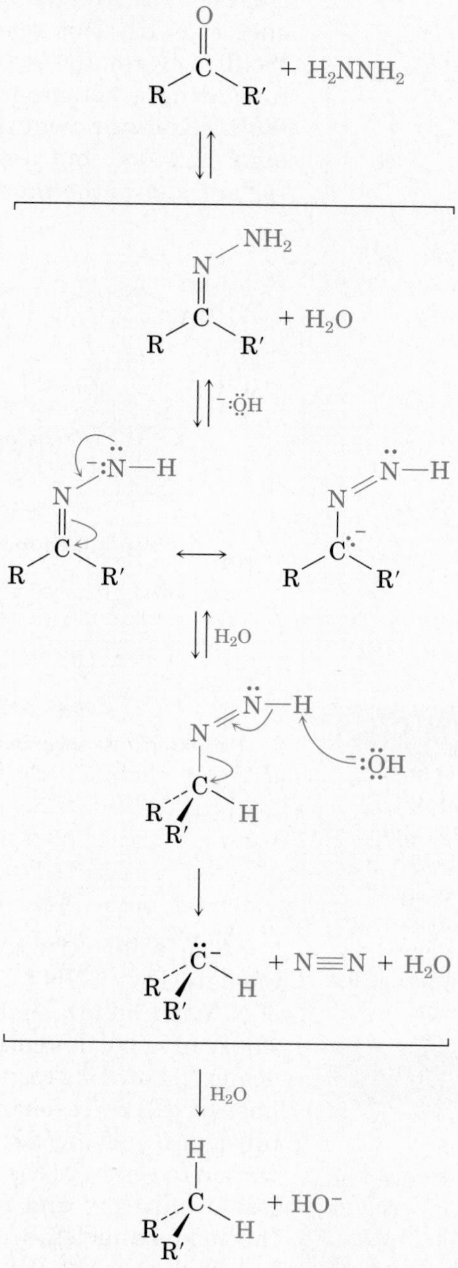

Reaction of the ketone or aldehyde with hydrazine yields a hydrazone in the normal way.

Base then abstracts one of the weakly acidic protons from $-NH_2$, yielding a hydrazone anion. This anion has an "allylic" resonance form that places the negative charge on carbon and the double bond between nitrogens.

Protonation of the hydrazone anion takes place on carbon to yield a neutral intermediate.

Base-induced loss of nitrogen then gives a carbanion . . .

. . . that is protonated to yield neutral alkane product.

© 1984 JOHN MCMURRY

19.11 Nucleophilic Addition of Alcohols: Acetal Formation

Biological Connection ▶

A ketone or aldehyde reacts reversibly with two equivalents of an alcohol in the presence of an acid catalyst to yield an **acetal, R$_2$C(OR')$_2$** (sometimes called a *ketal* if derived from a ketone).

$$\underset{\textbf{Ketone/aldehyde}}{\overset{\displaystyle O}{\underset{|}{\overset{\|}{C}}}} + 2\ R'OH \underset{\text{catalyst}}{\overset{\text{Acid}}{\rightleftharpoons}} \underset{\textbf{An acetal}}{\overset{\displaystyle OR'}{\underset{OR'}{C}}} + H_2O$$

Acetal formation is similar to the hydration reaction discussed in Section 19.6. Like water, alcohols are weak nucleophiles that add to aldehydes and ketones only slowly under neutral conditions. Under acidic conditions, however, the reactivity of the carbonyl group is increased by protonation so addition of an alcohol occurs rapidly.

Neutral carbonyl group (moderately electrophilic)

Protonated carbonyl group (strongly electrophilic and highly reactive toward nucleophiles)

Nucleophilic addition of an alcohol to the carbonyl group initially yields a hydroxy ether called a **hemiacetal**, analogous to the gem diol formed by addition of water (Section 19.6). Hemiacetals are formed reversibly, with the equilibrium normally favoring the carbonyl compound. In the presence of acid, however, a further reaction can occur. Protonation of the –OH group followed by an E1-like loss of water leads to an oxonium ion, R$_2$C=OR$^+$, which undergoes a second nucleophilic addition of alcohol to yield the acetal. For example, reaction of cyclohexanone with methanol yields the dimethyl acetal. The mechanism is shown in Figure 19.12 (p. 778).

Cyclohexanone   **A hemiacetal**   **Cyclohexanone dimethyl acetal**

Because all the steps in acetal formation are reversible, the reaction can be driven either forward (from carbonyl compound to acetal) or backward (from acetal to carbonyl compound), depending on the conditions. The forward reaction is favored by conditions that remove water from the medium and thus drive the equilibrium to the right. In practice, this is often

FIGURE 19.12 ▼

Mechanism of acid-catalyzed acetal formation by reaction of a ketone or aldehyde with an alcohol.

refer to
Mechanisms
& Movies

Protonation of the carbonyl oxygen strongly polarizes the carbonyl group and . . .

. . . activates the carbonyl group for nucleophilic attack by oxygen lone-pair electrons from alcohol.

Loss of a proton yields a neutral hemiacetal tetrahedral intermediate.

Hemiacetal

Protonation of the hemiacetal hydroxyl converts it into a good leaving group.

Dehydration yields an intermediate oxonium ion.

Addition of a second equivalent of alcohol gives protonated acetal.

Loss of a proton yields neutral acetal product.

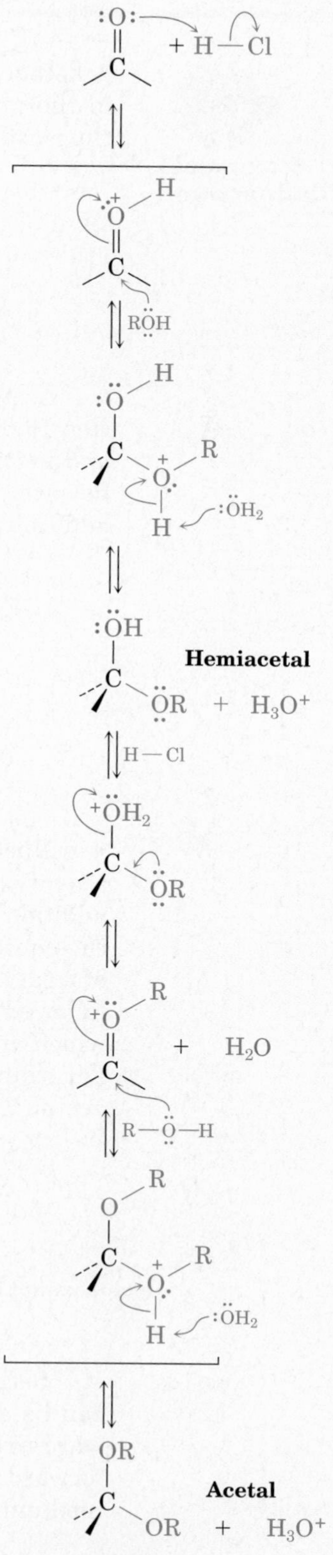

Acetal

© 1984 JOHN MCMURRY

done by distilling off water as it forms. The reverse reaction is favored by treating the acetal with a large excess of aqueous acid to drive the equilibrium to the left.

Acetals are useful because they can serve as protecting groups for aldehydes and ketones in the same way that trimethylsilyl ethers serve as protecting groups for alcohols (Section 17.9). As we saw previously, it sometimes happens that one functional group interferes with intended chemistry elsewhere in a complex molecule. For example, if we wanted to reduce only the ester group of ethyl 4-oxopentanoate, the ketone would interfere. Treatment of the starting keto ester with LiAlH$_4$ would reduce both the keto and the ester groups to give a diol product.

Ethyl 4-oxopentanoate   5-Hydroxy-2-pentanone

By protecting the keto group as an acetal, however, the problem can be circumvented. Like other ethers, acetals are unreactive to bases, hydride reducing agents, Grignard reagents, and catalytic reducing conditions. Thus, we can accomplish the selective reduction of the ester group in ethyl 4-oxopentanoate by first converting the keto group to an acetal, then reducing the ester with LiAlH$_4$, and then removing the acetal by treatment with aqueous acid.

In practice, it's convenient to use ethylene glycol as the alcohol and to form a *cyclic* acetal. The mechanism of cyclic acetal formation using 1 equivalent of ethylene glycol is exactly the same as that using 2 equivalents of methanol or other monoalcohol. The only difference is that both alcohol groups are now in the same molecule.

Practice Problem 19.2   Show the structure of the acetal you would obtain by acid-catalyzed reaction of 2-pentanone with propane-1,3-diol.

Strategy   Acid-catalyzed reaction of a ketone or aldehyde with 2 equivalents of a monoalcohol or 1 equivalent of a diol yields an acetal, in which the carbonyl oxygen atom is replaced by two –OR groups from the alcohol.

Solution

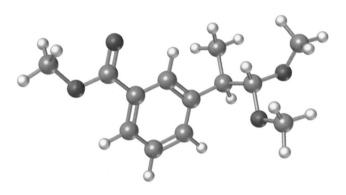

2-Pentanone

• •

Problem 19.13   Show all the steps in the acid-catalyzed formation of a cyclic acetal from ethylene glycol and a ketone or aldehyde.

Problem 19.14   Identify the carbonyl compound and the alcohol that were used to prepare the following acetal:

• •

19.12 Nucleophilic Addition of Phosphorus Ylides: The Wittig Reaction

A ketone or aldehyde is converted into an alkene by means of the **Wittig reaction**. In this process, a phosphorus **ylide**, $R_2\bar{C}-\overset{+}{P}(C_6H_5)_3$ (also called a **phosphorane**), adds to a ketone or aldehyde to yield a dipolar intermediate called a **betaine**. (An ylide—pronounced **ill**-id—is a neutral, dipolar compound with adjacent plus and minus charges. A betaine—pronounced **bay**-ta-een—is a neutral, dipolar compound with nonadjacent charges.)

The betaine intermediate is not isolated; rather, it spontaneously decomposes through a four-membered ring to yield alkene and triphenylphosphine oxide, $(Ph)_3P=O$. The net result is replacement of the carbonyl oxygen atom by the $R_2C=$ group originally bonded to phosphorus (Figure 19.13).

The phosphorus ylides necessary for Wittig reaction are easily prepared by S_N2 reaction of primary (and some secondary) alkyl halides with triphenylphosphine, followed by treatment with base. Triphenylphosphine,

FIGURE 19.13 ▼

The mechanism of the Wittig reaction between a phosphorus ylide and a ketone or aldehyde to yield an alkene.

The nucleophilic carbon atom of the phosphorus ylide adds to the carbonyl group of a ketone or aldehyde to give a betaine intermediate.

The betaine undergoes intramolecular O–P bond formation to produce a four-membered ring intermediate.

Spontaneous decomposition of the four-membered ring gives an alkene and triphenylphosphine oxide.

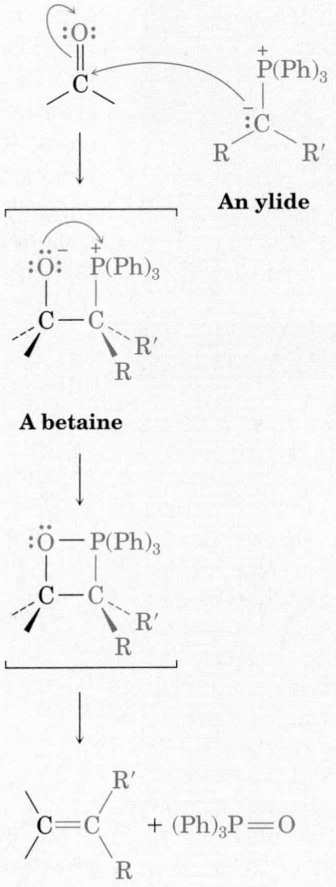

An ylide

A betaine

© 1999 JOHN MCMURRY

$(Ph)_3P$, is a good nucleophile in S_N2 reactions, and yields of the resultant alkyltriphenylphosphonium salts are high. The hydrogen on the carbon next to the positively charged phosphorus is weakly acidic and can be removed by a base such as butyllithium (BuLi) to generate the neutral ylide. For example:

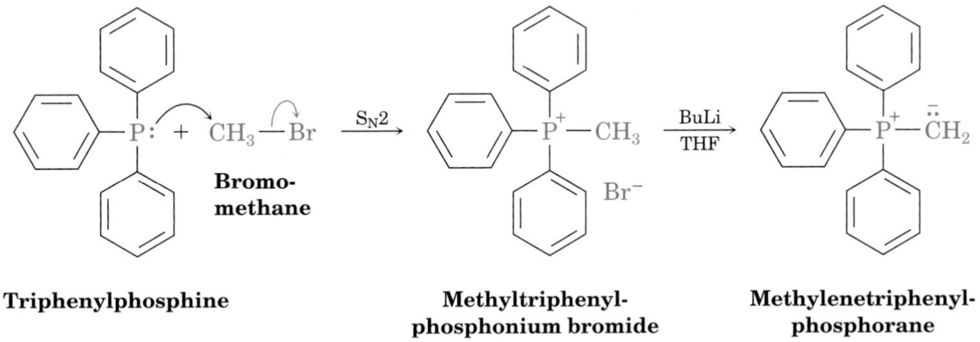

Triphenylphosphine   **Methyltriphenyl-**   **Methylenetriphenyl-**
                                      phosphonium bromide   **phosphorane**

Bromo-methane

Georg F. K. Wittig

Georg F. K. Wittig (1897–1987) was born in Berlin, Germany, and received his Ph.D. at the University of Marburg in 1926, working with von Auwers. He remained at Marburg for 6 years and then became professor of chemistry, first at the University of Braunschweig, and then in Freiburg, Tübingen, and Heidelberg. In 1979, he received the Nobel Prize in chemistry for his work on phosphorus-containing organic compounds.

The Wittig reaction is extremely general, and a great many mono-, di-, and trisubstituted alkenes can be prepared from the appropriate combination of phosphorane and ketone or aldehyde. Tetrasubstituted alkenes can't be prepared, however, because of steric hindrance during the reaction.

The real value of the Wittig reaction is that it yields a pure alkene of known structure. The C=C double bond is always exactly where the C=O group was in the precursor, and only a single product (not counting *E,Z* isomers) is formed. For example, Wittig reaction of cyclohexanone with methylenetriphenylphosphorane yields only the single alkene product, methylenecyclohexane. By contrast, addition of methylmagnesium bromide to cyclohexanone, followed by dehydration with POCl$_3$, yields a roughly 9:1 mixture of two alkenes:

Cyclohexanone

1. CH$_3$MgBr
2. POCl$_3$

1-Methylcyclohexene   **Methylenecyclohexane**

(9:1 ratio)

$(C_6H_5)_3\overset{+}{P}-\overset{-}{C}H_2$, THF solvent

Methylenecyclohexane
(84%)

$+$ $(C_6H_5)_3P=O$

Wittig reactions are used commercially in the synthesis of numerous pharmaceutical agents. For example, the Swiss chemical company Hoffmann-LaRoche prepares β-carotene, a yellow food-coloring agent and dietary source of vitamin A, by Wittig reaction between retinal and retinylidenetriphenylphosphorane.

CHO

$+$

$\overset{-}{C}HP(Ph)_3^{+}$

Retinal   **Retinylidenetriphenylphosphorane**

Wittig reaction

β-Carotene

• •

Practice Problem 19.3 What carbonyl compound and what phosphorus ylide might you use to prepare 3-ethyl-2-pentene?

Strategy A ketone or aldehyde reacts with a phosphorus ylide to yield an alkene in
which the oxygen atom of the carbonyl reactant is replaced by the $=CR_2$ of
the ylide. Preparation of the phosphorus ylide itself usually involves S_N2
reaction of a primary alkyl halide with triphenylphosphine, so the ylide is
typically primary, $RCHP(Ph)_3$. This means that the disubstituted alkene
carbon in the product comes from the carbonyl reactant, while the monosubstituted alkene carbon comes from the ylide.

Solution

Disubstituted; from ketone

Monosubstituted; from ylide

$$CH_3CH_2C=O \xrightarrow[\text{THF}]{(Ph)_3\overset{+}{P}-\overset{-}{C}HCH_3} CH_3CH_2C=CHCH_3$$
$$||$$
$$CH_2CH_3CH_2CH_3$$

3-Pentanone   **3-Ethyl-2-pentene**

• •

Problem 19.15 What carbonyl compound and what phosphorus ylide might you use to prepare each
of the following compounds?

(a) [structure with CH$_3$]   (b) [structure with CH$_2$]   (c) 2-Methyl-2-hexene

(d) $C_6H_5CH=C(CH_3)_2$ (e) 1,2-Diphenylethylene

Problem 19.16 Another route to β-carotene involves a *double* Wittig reaction between 2 equivalents
of β-ionylideneacetaldehyde and a *diylide*. Write the reaction, and show the structure of the diylide.

[structure with CHO]

β-Ionylideneacetaldehyde

• •

19.13 The Cannizzaro Reaction

Biological Connection ▶

We said in "A Preview of Carbonyl Compounds" that nucleophilic addition
reactions are characteristic of aldehydes and ketones but not of carboxylic
acid derivatives. The reason for the difference is structural. As shown in Figure 19.14, the tetrahedral intermediate produced by addition of a nucleophile

to a carboxylic acid derivative can eliminate a leaving group, leading to a net nucleophilic acyl substitution reaction. The tetrahedral intermediate produced by addition of a nucleophile to a ketone or aldehyde, however, has only alkyl or hydrogen substituents and thus can't usually expel a leaving group. The **Cannizzaro reaction**, discovered in 1853, is one exception to this rule.

FIGURE 19.14 ▼

Carboxylic acid derivatives have an electronegative substituent Y = −Br, −Cl, −OR, −NR$_2$ that can act as a leaving group and be expelled from the tetrahedral intermediate formed by nucleophilic addition. Aldehydes and ketones have no such leaving group, and thus do not usually undergo this reaction.

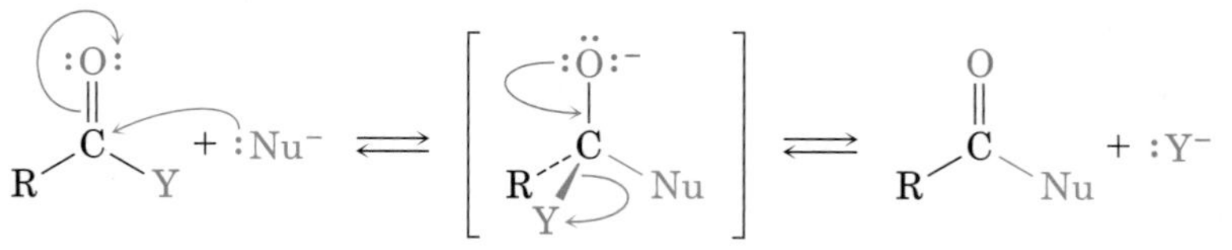

Reaction occurs when: Y = −Br, −Cl, −OR, −NR$_2$
Reaction *does NOT occur* when: Y = −H, −R

The Cannizzaro reaction takes place by nucleophilic addition of OH$^-$ to an aldehyde to give a tetrahedral intermediate, *which expels hydride ion as a leaving group*. A second aldehyde molecule accepts the hydride ion in another nucleophilic addition step, resulting in a simultaneous oxidation and reduction, or *disproportionation*. One molecule of aldehyde undergoes a substitution of H$^-$ by OH$^-$ and is thereby oxidized to an acid, while a second molecule of aldehyde undergoes an addition of H$^-$ and is thereby reduced to an alcohol. Benzaldehyde, for instance, yields a 1:1 mixture of benzoic acid and benzyl alcohol when heated with aqueous NaOH.

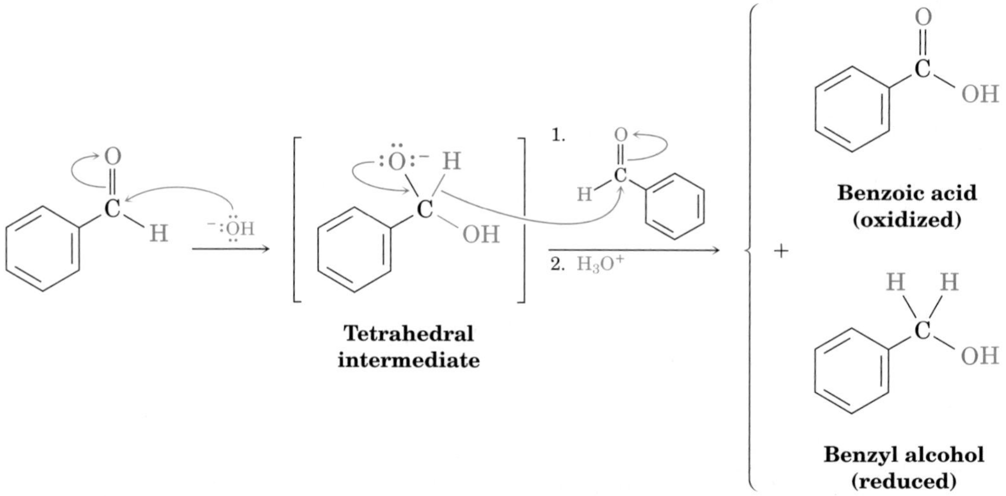

Tetrahedral intermediate

Benzoic acid (oxidized)

Benzyl alcohol (reduced)

Stanislao Cannizzaro

Stanislao Cannizzaro (1826–1910) was born in Palermo, Sicily, the son of the chief of police. He studied at the University of Pisa under Rafaelle Piria and also worked in Paris with Michel-Eugène Chevreul. As a youth, he took part in the Sicilian revolution of 1848 and was at one point condemned to death. He was professor of chemistry at the universities of Genoa, Palermo, and Rome, and is best known for being the first to clarify the distinction between atoms and molecules.

Biological Connection ▶

The Cannizzaro reaction has few practical applications and is limited to aldehydes such as formaldehyde and benzaldehyde, which have no hydrogen on the carbon next to the –CHO group. Nevertheless, the Cannizzaro reaction is interesting mechanistically because it serves as a simple laboratory analogy for an important biological pathway by which reductions occur in living organisms. In nature, one of the most important reducing agents is a substance called *reduced nicotinamide adenine dinucleotide,* abbreviated NADH.

Reduced nicotinamide adenine dinucleotide (NADH)

NADH donates H⁻ to aldehydes and ketones (and thereby reduces them) in much the same way that the tetrahedral intermediate in a Cannizzaro reaction does. The electron lone pair on a nitrogen atom of NADH pushes out H⁻, which adds to a carbonyl group in another molecule to cause a reduction. We'll see this reaction again in Chapter 29 when we look at the details of some metabolic pathways.

Problem 19.17   When *o*-phthalaldehyde is treated with base, *o*-(hydroxymethyl)benzoic acid is formed. Show the mechanism of this reaction.

o-Phthalaldehyde   *o*-(Hydroxymethyl)benzoic acid

19.14 Conjugate Nucleophilic Addition to α,β-Unsaturated Aldehydes and Ketones

The reactions we've been discussing have all involved the addition of a nucleophile *directly* to the carbonyl group. Closely related to this direct addition is the **conjugate addition** of a nucleophile to the C=C double bond of an α,β-unsaturated ketone or aldehyde. The two processes are often called *1,2 addition* and *1,4 addition,* respectively, as shown in Figure 19.15. (The carbon atom next to a carbonyl group is the *α carbon,* the next carbon is the *β carbon,* and so on. Thus, an α,β-*unsaturated* ketone or aldehyde is one that has a double bond conjugated with the carbonyl group.)

FIGURE 19.15 ▼

A comparison of direct (1,2) and conjugate (1,4) nucleophilic addition reactions.

Direct addition

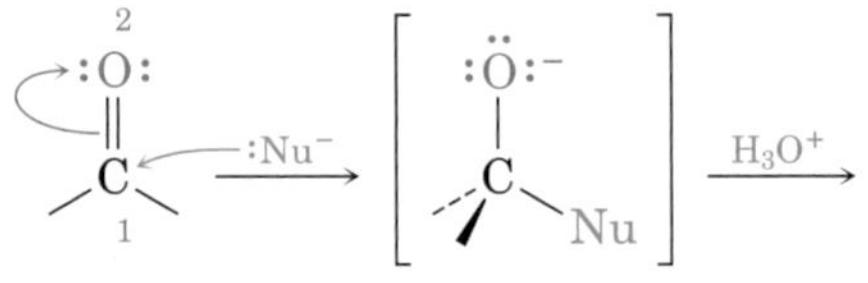

A direct addition
product

Conjugate addition

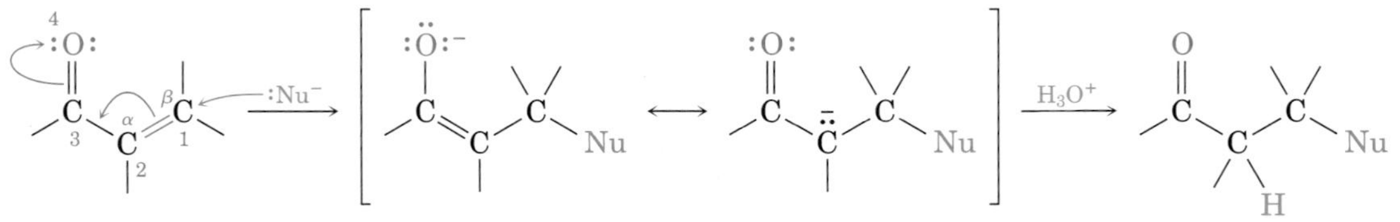

An α,β-unsaturated
carbonyl group

An enolate ion
intermediate

A conjugate
addition product

The conjugate addition of a nucleophile to an α,β-unsaturated ketone or aldehyde is due to the same electronic factors that are responsible for direct addition. We've seen that carbonyl groups are polarized so that the carbonyl carbon is positive, and we can even draw a dipolar resonance structure to underscore the point:

Carbonyl group

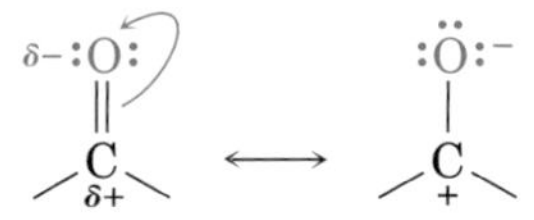

When we draw a similar resonance structure for an α,β-unsaturated carbonyl compound, however, the positive charge is allylic and can be shared
by the β carbon. In other words, the β carbon of an α,β-unsaturated carbonyl
compound is an electrophilic site and can react with nucleophiles. A com-
parison of electrostatic potential maps of ethylene with an α,β-unsaturated
ketone shows that the double-bond carbon atoms of the unsaturated ketone
are more positive (more green) than those of the isolated alkene ethylene.

**α,β-Unsaturated
carbonyl group**

ethylene,
unsaturated ketone

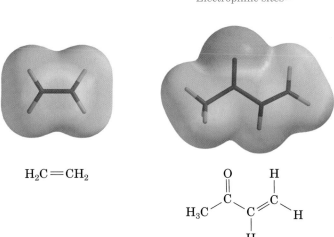

$$H_2C=CH_2$$

Conjugate addition of a nucleophile to the β carbon of an α,β-unsaturated
ketone or aldehyde leads to an enolate ion intermediate, which is protonated
on the α carbon to give the saturated product (Figure 19.15). The net effect
is addition of the nucleophile to the C=C double bond, with the carbonyl group
itself unchanged. In fact, of course, the carbonyl group is crucial to the success of the reaction. The C=C double bond would not be activated for addition, and no reaction would occur, without the carbonyl group.

Activated double bond

**Unactivated double
bond**

Conjugate Addition of Amines

Biological Connection

Primary and secondary amines add to α,β-unsaturated aldehydes and ketones to yield β-amino aldehydes and ketones. Reaction occurs rapidly under mild conditions, and yields are good. Note that the conjugate addition product is often obtained to the complete exclusion of the direct addition product.

$$\text{CH}_3\overset{\overset{\text{O}}{\|}}{\text{C}}\text{CH}{=}\text{CH}_2 \;+\; \text{H}\ddot{\text{N}}(\text{CH}_2\text{CH}_3)_2 \xrightarrow{\text{Ethanol}} \text{CH}_3\overset{\overset{\text{O}}{\|}}{\text{C}}\text{CH}_2\text{CH}_2\text{N}(\text{CH}_2\text{CH}_3)_2$$

3-Buten-2-one   **Diethylamine**   **4-*N*,*N*-Diethylamino-2-butanone**
                                                                                      (92%)

2-Cyclohexenone   **Methylamine**   **3-(*N*-Methylamino)cyclohexanone**

$$\text{+} \quad \text{CH}_3\ddot{\text{N}}\text{H}_2 \xrightarrow{\text{Ethanol}}$$

Conjugate Addition of Alkyl Groups: Organocopper Reactions

Conjugate addition of an alkyl group to an α,β-unsaturated ketone (but not aldehyde) is one of the most useful 1,4-addition reactions, just as direct addition of a Grignard reagent is one of the most useful 1,2 additions.

$$\xrightarrow[\text{2. H}_3\text{O}^+]{\text{1. ":R}^-\text{"}}$$

α,β-Unsaturated ketone

Conjugate addition of an alkyl group is carried out by treating the α,β-unsaturated ketone with a lithium diorganocopper reagent. As we saw in Section 10.9, diorganocopper reagents can be prepared by reaction between 1 equivalent of cuprous iodide and 2 equivalents of organolithium:

$$\text{RX} \xrightarrow[\text{Pentane}]{2\,\text{Li}} \text{RLi} + \text{Li}^+\,\text{X}^-$$

$$2\,\text{RLi} \xrightarrow[\text{Ether}]{\text{CuI}} \text{Li}^+(\text{R}\overset{-}{\text{Cu}}\text{R}) + \text{Li}^+\,\text{I}^-$$

A lithium diorganocopper (Gilman reagent)

Primary, secondary, and even tertiary alkyl groups undergo the addition reaction, as do aryl and alkenyl groups. Alkynyl groups, however, react poorly in the conjugate addition process.

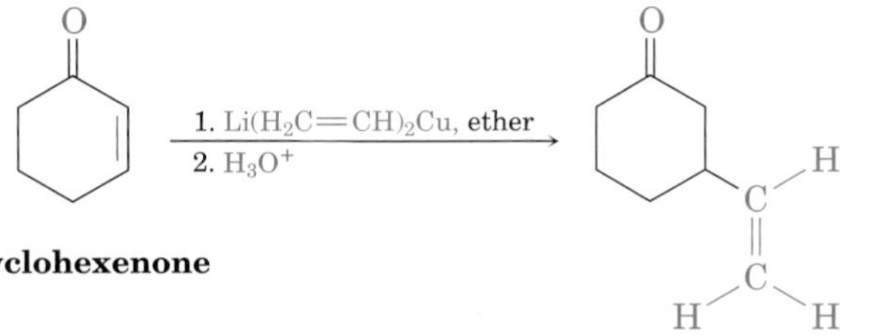

2-Cyclohexenone

1. Li(H₂C=CH)₂Cu, ether
2. H₃O⁺

3-Vinylcyclohexanone
(65%)

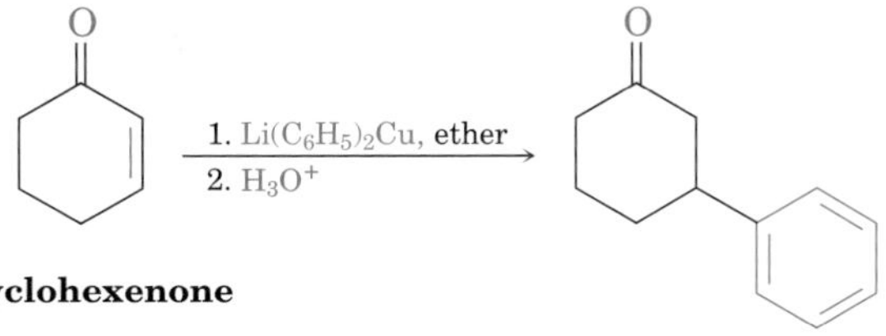

2-Cyclohexenone

1. Li(C₆H₅)₂Cu, ether
2. H₃O⁺

3-Phenylcyclohexanone
(70%)

Diorganocopper reagents are unique in their ability to give conjugate addition products. Other organometallic reagents, such as Grignard reagents and organolithiums, normally give direct carbonyl addition on reaction with α,β-unsaturated ketones.

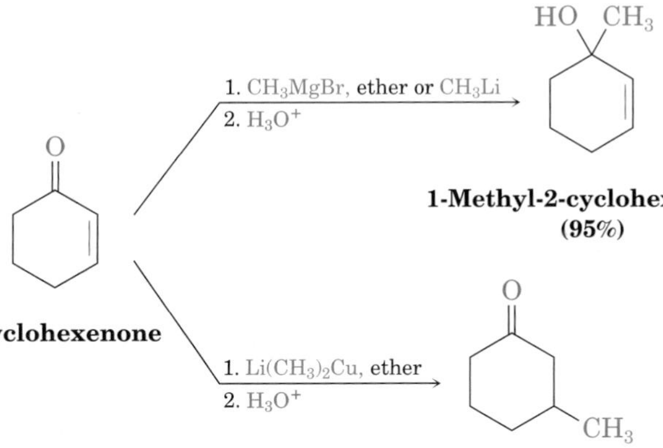

1. CH₃MgBr, ether or CH₃Li
2. H₃O⁺

1-Methyl-2-cyclohexen-1-ol
(95%)

2-Cyclohexenone

1. Li(CH₃)₂Cu, ether
2. H₃O⁺

3-Methylcyclohexanone
(97%)

The mechanism of the reaction is thought to involve conjugate nucleophilic addition of the diorganocopper anion, R_2Cu^-, to the enone to give a copper-containing intermediate. Transfer of an R group and elimination of a neutral organocopper species, RCu, gives the final product.

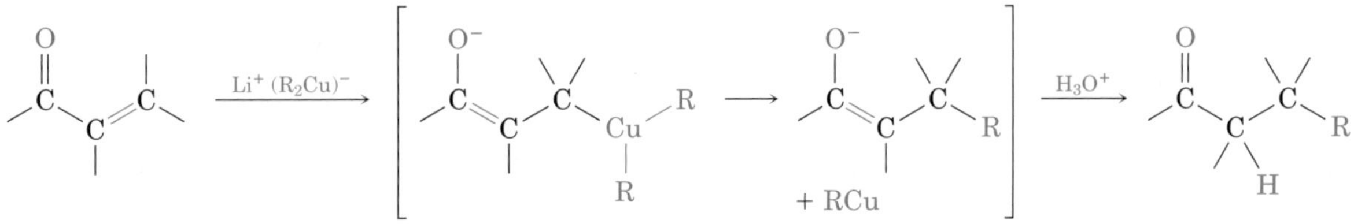

• •

Practice Problem 19.4   How might you use a conjugate addition reaction to prepare 2-methyl-3-propylcyclopentanone?

2-Methyl-3-propylcyclopentanone

Strategy   A ketone with a substituent group in its β position might be prepared by a conjugate addition of that group to an α,β-unsaturated ketone. In the present instance, the target molecule has a propyl substituent on the β carbon and might therefore be prepared from 2-methyl-2-cyclopentenone.

Solution

1. $Li(CH_3CH_2CH_2)_2Cu$, ether
2. H_3O^+

2-Methyl-2-cyclopentenone   **2-Methyl-3-propylcyclopentanone**

• •

Problem 19.18   How might conjugate addition reactions of lithium diorganocopper reagents be used to synthesize the following compounds?
(a) 2-Heptanone   (b) 3,3-Dimethylcyclohexanone
(c) 4-*tert*-Butyl-3-ethylcyclohexanone   (d)

• •

19.15 Some Biological Nucleophilic Addition Reactions

Biological Connection ▶

We'll see in Chapter 29 that living organisms use many of the same reactions that chemists use in the laboratory. This is particularly true of carbonyl-group reactions, where nucleophilic addition steps play a critical role in the biological synthesis of many vital molecules. For example, one of the pathways by which amino acids are made involves nucleophilic addition of an amine to α-keto acids. To choose a specific example, the bacterium *Bacillus subtilis* synthesizes the amino acid alanine from pyruvic acid.

The key step in this biological transformation is the nucleophilic addition of an amine to the ketone carbonyl group of pyruvic acid. The tetrahedral intermediate loses water to yield an imine, which is further reduced in a second nucleophilic addition step to yield alanine.

$$CH_3CCOOH + :NH_2R \longrightarrow \left[CH_3CCOOH \right] \xrightarrow[\text{enzyme}]{\text{Reducing}} CH_3CHCOOH$$

Pyruvic acid   **An imine**   **Alanine**

Another nucleophilic addition reaction—this time in reverse—is involved in the chemical defense mechanism by which the millipede *Apheloria corrugata* protects itself from predators. When attacked by ants, it secretes the cyanohydrin mandelonitrile and an enzyme that catalyzes the decomposition of mandelonitrile into benzaldehyde and HCN. The millipede actually protects itself by discharging poisonous HCN at its attackers.

Mandelonitrile
(from *Apheloria corrugata*)

19.16 Spectroscopy of Aldehydes and Ketones

Infrared Spectroscopy

Aldehydes and ketones show a strong C=O bond absorption in the infrared region from 1660 to 1770 cm^{-1}, as the spectra of benzaldehyde and cyclohexanone demonstrate (Figure 19.16, p. 792). In addition, aldehydes show two characteristic C–H absorptions in the range 2720–2820 cm^{-1}.

The exact position of the C=O absorption is highly diagnostic of the exact nature of the carbonyl group. As the data in Table 19.2 indicate,

TABLE 19.2 Infrared Absorptions of Some Aldehydes and Ketones

Carbonyl type	Example	Infrared absorption (cm^{-1})
Aliphatic aldehyde	Acetaldehyde	1730
Aromatic aldehyde	Benzaldehyde	1705
α,β-Unsaturated aldehyde	$H_2C{=}CH{-}CHO$	1705
Aliphatic ketone	Acetone	1715
Six-membered-ring ketone	Cyclohexanone	1715
Five-membered-ring ketone	Cyclopentanone	1750
Four-membered-ring ketone	Cyclobutanone	1785
Aromatic ketone		1690
α,β-Unsaturated ketone		1685

FIGURE 19.16 ▼

Infrared spectra of (a) benzaldehyde and (b) cyclohexanone.

(a)

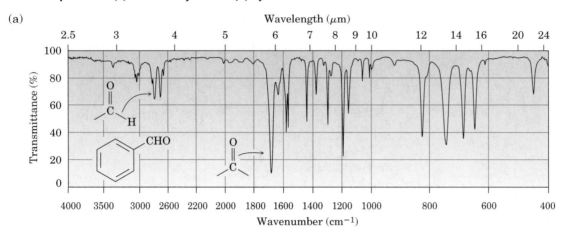

(b)

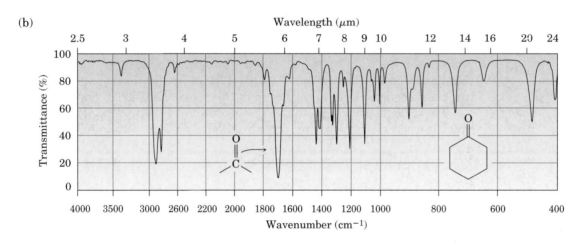

saturated aldehydes usually show carbonyl absorptions near 1730 cm^{-1} in the IR spectrum, but conjugation of the aldehyde to an aromatic ring or a double bond lowers the absorption by 25 cm^{-1} to near 1705 cm^{-1}. Saturated aliphatic ketones and cyclohexanones both absorb near 1715 cm^{-1}, and conjugation with a double bond or an aromatic ring again lowers the absorption by 30 cm^{-1} to 1685–1690 cm^{-1}. Angle strain in the carbonyl group caused by reducing the ring size of cyclic ketones to four or five raises the absorption position.

The values given in Table 19.2 are remarkably constant from one ketone or aldehyde to another. As a result, IR spectroscopy is a powerful tool for diagnosing the nature and chemical environment of a carbonyl group in a molecule of unknown structure. An unknown that shows an IR absorption at 1730 cm^{-1} is almost certainly an aldehyde rather than a ketone; an unknown that shows an IR absorption at 1750 cm^{-1} is almost certainly a cyclopentanone, and so on.

● ●

Problem 19.19   How might you use IR spectroscopy to determine whether reaction between 2-cyclohexenone and lithium dimethylcopper gives the direct addition product or the conjugate addition product?

Problem 19.20   Where would you expect each of the following compounds to absorb in the IR spectrum?
(a) 4-Penten-2-one
(b) 3-Penten-2-one
(c) 2,2-Dimethylcyclopentanone
(d) *m*-Chlorobenzaldehyde
(e) 3-Cyclohexenone
(f) 2-Hexenal

● ●

Nuclear Magnetic Resonance Spectroscopy

Aldehyde protons (RCHO) absorb near 10 δ in the ^{1}H NMR spectrum and are very distinctive, since no other absorptions occur in this region. The aldehyde proton shows spin–spin coupling with protons on the neighboring carbon, with coupling constant $J \approx 3\,\text{Hz}$. Acetaldehyde, for example, shows a quartet at 9.8 δ for the aldehyde proton, indicating that there are three protons neighboring the –CHO group (Figure 19.17).

FIGURE 19.17 ▼

^{1}H NMR spectrum of acetaldehyde. The absorption of the aldehyde proton appears at 9.8 δ and is split into a quartet.

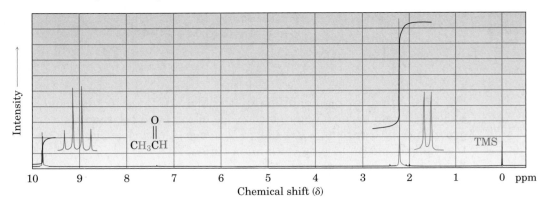

Hydrogens on the carbon next to a carbonyl group are slightly deshielded and normally absorb near 2.0–2.3 δ. (Note that the acetaldehyde methyl group in Figure 19.17 absorbs at 2.20 δ.) Methyl ketones are particularly distinctive because they always show a sharp three-proton singlet near 2.1 δ.

The carbonyl-group carbon atoms of aldehydes and ketones show characteristic ^{13}C NMR resonances in the range 190–215 δ. Since no other kinds of carbons absorb in this range, the presence of an NMR absorption near 200 δ is clear evidence for a carbonyl group. Saturated ketone or aldehyde carbons usually absorb in the region from 200 to 215 δ, while aromatic and α,β-unsaturated carbonyl carbons absorb in the 190–200 δ region.

Mass Spectrometry

Aliphatic aldehydes and ketones that have hydrogens on their gamma (γ) carbon atoms undergo a characteristic mass spectral cleavage called the **McLafferty rearrangement**. A hydrogen atom is transferred from the γ carbon to the carbonyl oxygen, the bond between the α and β carbons is broken, and a neutral alkene fragment is produced. The charge remains with the oxygen-containing fragment.

In addition to fragmentation by the McLafferty rearrangement, aldehydes and ketones also undergo cleavage of the bond between the carbonyl group and the α carbon, a so-called α *cleavage*. Alpha cleavage yields a neutral radical and an oxygen-containing cation.

Fred Warren McLafferty

Fred Warren McLafferty (1923–   ) was born in Evanston, Illinois, and received his Ph.D. in 1950 at Cornell University. He was a scientist at the Dow Chemical Company from 1950 to 1964 before becoming professor of chemistry at Purdue University. In 1968, he returned to Cornell University as professor.

Fragment ions from both α cleavage and McLafferty rearrangement are visible in the mass spectrum of 5-methyl-2-hexanone shown in Figure 19.18. Alpha cleavage occurs primarily at the more substituted side of the carbonyl group, leading to a $[CH_3CO]^+$ fragment with $m/z = 43$. McLafferty rearrangement and loss of 2-methylpropene yields a fragment with $m/z = 58$.

FIGURE 19.18 ▼

Mass spectrum of 5-methyl-2-hexanone. The abundant peak at $m/z = 43$ is due to α cleavage at the more highly substituted side of the carbonyl group. The peak at $m/z = 58$ is due to McLafferty rearrangement. Note that the peak due to the molecular ion is very small.

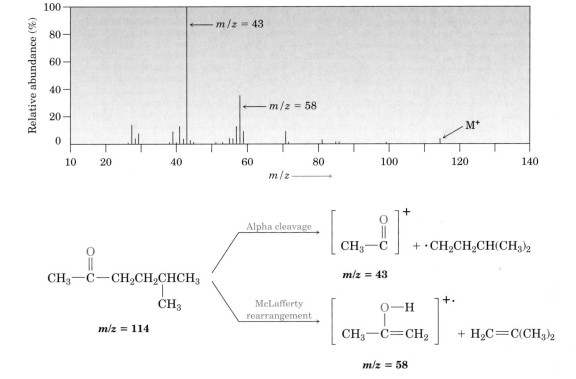

• •

Problem 19.21   How might you use mass spectrometry to distinguish between the following pairs of isomers?
(a) 3-Methyl-2-hexanone and 4-methyl-2-hexanone
(b) 3-Heptanone and 4-heptanone
(c) 2-Methylpentanal and 3-methylpentanal

• •

Enantioselective Synthesis

Whenever a chiral product is formed by reaction between achiral reagents, the product must be racemic. That is, both enantiomers of the product must be formed in equal amounts. The Grignard reaction of benzaldehyde with ethylmagnesium bromide, for instance, gives a racemic mixture of (R) and (S) alcohols, because both faces of the planar carbonyl group are equally accessible.

$$\underset{\text{O}}{\text{C}} \quad \text{H} \quad \xrightarrow[\text{2. H}_3\text{O}^+]{\text{1. CH}_3\text{CH}_2\text{MgBr}} \quad \underset{\text{HO} \quad \text{CH}_2\text{CH}_3}{\text{C}} \quad \text{H} \quad + \quad \underset{\text{CH}_3\text{CH}_2 \quad \text{OH}}{\text{C}} \quad \text{H}$$

50:50

Unfortunately, it's usually the case that only a *single* enantiomer of a given drug or other important substance has the desired biological properties. The other enantiomer might be inactive or even dangerous. Thus, much work is currently being done on developing *enantioselective* methods of synthesis, which yield only one of two possible enantiomers.

There are several approaches to enantioselective synthesis, but the most efficient are those that use chiral catalysts to temporarily hold a substrate molecule in an unsymmetrical environment. While in that unsymmetrical environment, the substrate may be more open to reaction on one side than on another, leading to an excess of one enantiomeric product over another. As an analogy, think about picking up a coffee mug in your right hand to take a drink. The mug by itself is achiral, but as soon as you pick it up by the handle, it becomes unsymmetrical. One side of the mug now faces toward you so you can drink from it, but the other side faces away. The two sides are different, with one side much more accessible to you than the other.

Among the thousands of enantioselective reactions now known, reaction of benzaldehyde with diethylzinc in the presence of a chiral titanium-containing catalyst gives 97% of the S addition product and only 3% of the R enantiomer. The catalyst is made from tartaric acid—the same substance Louis Pasteur isolated from wine more than 150 years

The tartaric acid found at the bottom of these wine vats catalyzes the enantioselective hydroxylation of alkenes.

(continued) ▶

ago (Section 9.5). We say that the major product is formed with an *enantiomeric excess* of 94%, meaning that 6% of the product is racemic (3% *R* and 3% *S*) and an extra 94% is *S*. The mechanistic details by which the chiral catalyst works are not fully understood, although it appears that a chiral dialkoxyethylzinc intermediate $[(RO)_2ZnCH_2CH_3]$ is involved.

S (97%) R (3%)

(Chiral catalyst)

Summary and Key Words

Aldehydes and ketones are among the most important of all compounds, both in biochemistry and in the chemical industry. Aldehydes are normally prepared in the laboratory by oxidative cleavage of alkenes, by oxidation of primary alcohols, or by partial reduction of esters. Ketones are similarly prepared by oxidative cleavage of alkenes, by oxidation of secondary alcohols, or by addition of diorganocopper reagents to acid chlorides.

The **nucleophilic addition reaction** is the most common reaction of aldehydes and ketones. As shown in Figure 19.19 (p. 798), many different kinds of products can be prepared by nucleophilic additions. Aldehydes and ketones are reduced by $NaBH_4$ or $LiAlH_4$ to yield secondary and primary alcohols, respectively. Addition of Grignard reagents to aldehydes and ketones also gives alcohols (tertiary and secondary, respectively), and addition of HCN yields **cyanohydrins**. Primary amines add to carbonyl compounds yielding **imines**, and secondary amines yield **enamines**. Reaction of a ketone or aldehyde with hydrazine and base yields an alkane (the **Wolff–Kishner reaction**). Alcohols add to carbonyl groups to yield **acetals**, which are valuable as protecting groups. Phosphoranes add to aldehydes and ketones to give alkenes (the **Wittig reaction**) in which the new C=C in the product is exactly where the C=O bond was in the starting material.

α,β-Unsaturated aldehydes and ketones often react with nucleophiles to give the product of **conjugate addition**, or *1,4 addition*. Particularly useful is the reaction with a diorganocopper reagent, which results in the addition of an alkyl, aryl, or alkenyl group.

Infrared spectroscopy is extremely useful for identifying aldehydes and ketones. Carbonyl groups absorb in the IR range 1660–1770 cm^{-1}, with the exact position highly diagnostic of the kind of carbonyl group

FIGURE 19.19 ▼

A summary of nucleophilic addition reactions of aldehydes and ketones.

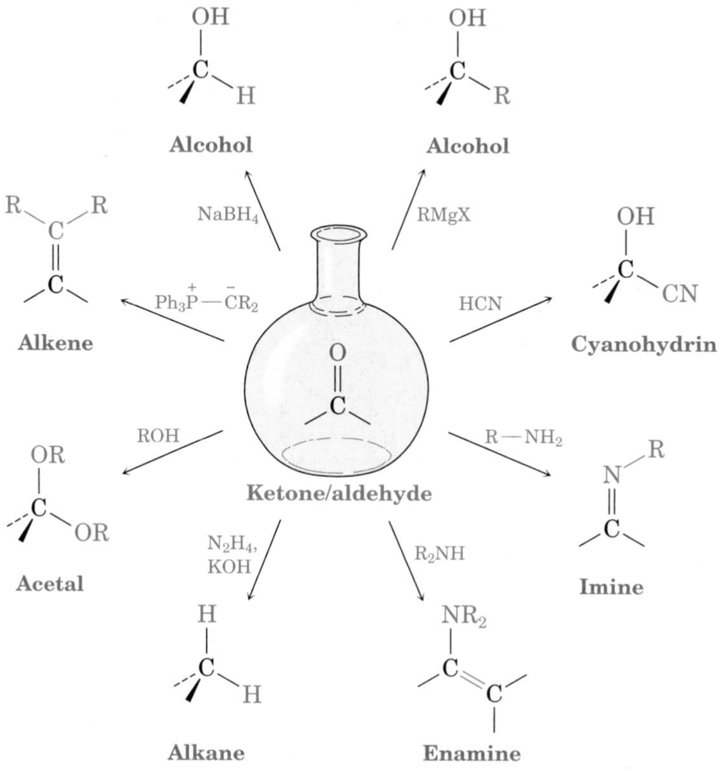

present in the molecule. ^{13}C NMR spectroscopy is also useful for aldehydes and ketones because their carbonyl carbons show resonances in the 190–215 δ range. ^{1}H NMR is useful largely for aldehyde –CHO protons, which absorb near 10 δ. Aldehydes and ketones undergo two characteristic kinds of fragmentation in the mass spectrometer: α cleavage and **McLafferty rearrangement**.

Summary of Reactions

1. Preparation of aldehydes (Section 19.2)
   (a) Oxidation of primary alcohols (Section 17.8)

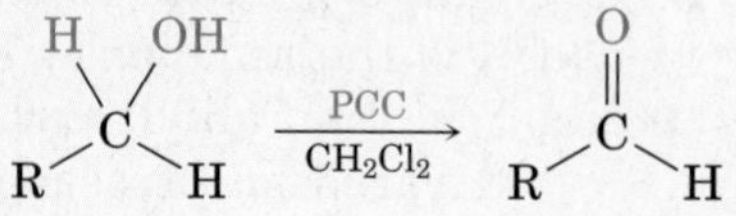

(continued) ▶

(b) Ozonolysis of alkenes (Section 7.8)

$$R\text{-C}=\text{C}\text{-H} \xrightarrow[\text{2. Zn, CH}_3\text{COOH}]{\text{1. O}_3} R\text{-C}=\text{O} + \text{O}=\text{C}\text{-H}$$

(c) Partial reduction of esters (Section 19.2)

$$R\text{-C(=O)-OR'} \xrightarrow[\text{2. H}_3\text{O}^+]{\text{1. DIBAH, toluene}} R\text{-C(=O)-H} + R'\text{OH}$$

2. Preparation of ketones (Section 19.2)
   (a) Oxidation of secondary alcohols (Section 17.8)

$$R\text{-C(H)(OH)-R'} \xrightarrow{\text{Cr(VI)}} R\text{-C(=O)-R'}$$

(b) Ozonolysis of alkenes (Section 7.8)

$$R_2\text{C}=\text{C}R_2 \xrightarrow[\text{2. Zn, CH}_3\text{COOH}]{\text{1. O}_3} R_2\text{C}=\text{O} + \text{O}=\text{C}R_2$$

(c) Friedel–Crafts acylation (Section 16.4)

$$\text{C}_6\text{H}_6 + R\text{-C(=O)-Cl} \xrightarrow{\text{AlCl}_3} \text{C}_6\text{H}_5\text{-C(=O)-R}$$

(d) Alkyne hydration (Section 8.5)

$$R\text{-C}\equiv\text{C-H} \xrightarrow[\text{H}_2\text{SO}_4,\ \text{H}_2\text{O}]{\text{Hg}^{2+}} R\text{-C(=O)-CH}_3$$

(e) Diorganocopper reaction with acid chlorides (Section 19.2)

$$R\text{-C(=O)-Cl} + R'_2\text{CuLi} \xrightarrow{\text{Ether}} R\text{-C(=O)-R'}$$

(continued) ▶

3. Reactions of aldehydes
   (a) Oxidation (Section 19.3)

$$
\underset{\substack{R}}{\overset{\substack{O \\ \|}}{C}}\diagdown H \xrightarrow[\text{or Ag}^+,\text{NH}_4\text{OH}]{\text{CrO}_3,\,\text{H}_3\text{O}^+} \underset{\substack{R}}{\overset{\substack{O \\ \|}}{C}}\diagdown \text{OH}
$$

   (b) Cannizzaro reaction (Section 19.13)

$$
2\ \underset{\substack{Ar}}{\overset{\substack{O \\ \|}}{C}}\diagdown H \xrightarrow[\text{2. H}_3\text{O}^+]{\text{1. NaOH, H}_2\text{O}} \underset{\substack{Ar}}{\overset{\substack{O \\ \|}}{C}}\diagdown \text{OH}\ +\ \underset{\substack{Ar}}{\overset{\substack{H\quad H}}{C}}\diagdown \text{OH}
$$

4. Nucleophilic addition reactions of aldehydes and ketones
   (a) Addition of hydride: reduction (Section 19.8)

$$
\underset{\substack{R}}{\overset{\substack{O \\ \|}}{C}}\diagdown R' \xrightarrow[\text{2. H}_3\text{O}^+]{\text{1. NaBH}_4,\text{ethanol}} \underset{\substack{R}}{\overset{\substack{H\quad OH}}{C}}\diagdown R'
$$

   (b) Addition of Grignard reagents (Section 19.8)

$$
\underset{\substack{R}}{\overset{\substack{O \\ \|}}{C}}\diagdown R' \xrightarrow[\text{2. H}_3\text{O}^+]{\text{1. R}''\text{MgX, ether}} \underset{\substack{R}}{\overset{\substack{R''\quad OH}}{C}}\diagdown R'
$$

   (c) Addition HCN: cyanohydrins (Section 19.7)

$$
\underset{\substack{R}}{\overset{\substack{O \\ \|}}{C}}\diagdown R' \underset{}{\overset{\text{HCN}}{\rightleftharpoons}} \underset{\substack{R}}{\overset{\substack{CN\quad OH}}{C}}\diagdown R'
$$

   (d) Addition of primary amines: imines (Section 19.9)

$$
\underset{\substack{R}}{\overset{\substack{O \\ \|}}{C}}\diagdown R' \underset{}{\overset{\text{R}''\text{NH}_2}{\rightleftharpoons}} \underset{\substack{R}}{\overset{\substack{NR'' \\ \|}}{C}}\diagdown R'\ +\ \text{H}_2\text{O}
$$

   For example:
   Oximes, $R_2C{=}N{-}OH$
   2,4-Dinitrophenylhydrazones, $R_2C{=}N{-}NH{-}C_6H_4(NO_2)_2$
   (e) Addition of secondary amines: enamines (Section 19.9)

$$
\underset{\substack{R}}{\overset{\substack{O \\ \|}}{C}}{-}\underset{}{\overset{}{C}}\diagdown H \underset{}{\overset{\text{HNR}'_2}{\rightleftharpoons}} \underset{\substack{R}}{\overset{\substack{NR'_2 \\ |}}{C}}{=}C\ +\ \text{H}_2\text{O}
$$

(continued) ▶

(f) Wolff–Kishner reaction (hydrazine addition) (Section 19.10)

$$\underset{R}{\overset{O}{\underset{\| }{C}}}\underset{R'}{}\quad\xrightarrow[\text{KOH}]{\text{H}_2\text{NNH}_2}\quad\underset{R}{\overset{H\quad H}{\underset{|\quad|}{C}}}\underset{R'}{}\quad+\ N_2\ +\ H_2O$$

(g) Addition of alcohols: acetals (Section 19.11)

$$\underset{R}{\overset{O}{\underset{\| }{C}}}\underset{R'}{}\quad+\ 2\ R''OH\quad\xrightarrow[\text{catalyst}]{\text{Acid}}\quad\underset{R}{\overset{R''O\quad OR''}{\underset{}{C}}}\underset{R'}{}\quad+\ H_2O$$

(h) Addition of phosphorus ylides: Wittig reaction (Section 19.12)

$$\underset{R}{\overset{O}{\underset{\| }{C}}}\underset{R'}{}\ +\ (C_6H_5)_3\overset{+}{P}-\overset{-}{C}HR''\ \xrightarrow{\text{THF}}\ \underset{R'}{\overset{R\quad\quad R''}{C=C}}\underset{H}{}\ +\ (C_6H_5)_3P=O$$

5. Conjugate additions to α,β-unsaturated ketones and aldehydes (Section 19.14)

   (a) Addition of amines

$$\underset{R}{\overset{O}{\underset{\| }{C}}}\overset{}{C}=\overset{}{C}\quad\underset{\xrightarrow{\quad}}{\overset{R'NH_2}{\rightleftharpoons}}\quad\underset{R}{\overset{O}{\underset{\| }{C}}}\underset{\underset{H}{|}}{C}\underset{NHR'}{C}$$

   (b) Addition of alkyl groups: diorganocopper reaction

$$\underset{R}{\overset{O}{\underset{\| }{C}}}\overset{}{C}=\overset{}{C}\quad\xrightarrow[\text{2. H}_3O^+]{\text{1. R}'_2\text{CuLi, ether}}\quad\underset{R}{\overset{O}{\underset{\| }{C}}}\underset{\underset{H}{|}}{C}\underset{R'}{C}$$

Visualizing Chemistry

• •

(Problems 19.1–19.21 appear within the chapter.)

19.22   Each of the following substances can be prepared by a nucleophilic addition reaction between a ketone or aldehyde and a nucleophile. Identify the reactants from which each was prepared. If the substance is an acetal, identify the carbonyl compound and the alcohol; if it is an imine, identify the carbonyl compound and the amine; and so forth.

(a)   (b)

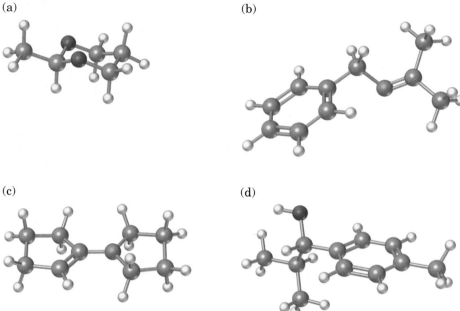

(c)   (d)

19.23   The following molecular model represents a tetrahedral intermediate resulting from addition of a nucleophile to a ketone or aldehyde. Identify the reactants, and write the structure of the final product when the nucleophilic addition reaction is complete.

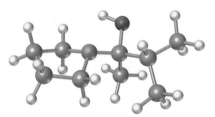

19.24   The enamine prepared from acetone and dimethylamine is shown below in its lowest-energy form.

(a) What is the geometry and hybridization of the nitrogen atom?

(b) What orbital on nitrogen holds the lone pair of electrons?

(c) What is the geometric relationship between the *p* orbitals of the double bond and the nitrogen orbital that holds the lone pair? Why do you think this geometry represents the minimum energy?

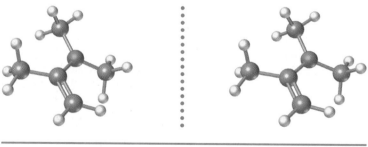

Stereo View

Additional Problems

• •

19.25 Draw structures corresponding to the following names:
(a) Bromoacetone
(b) 3,5-Dinitrobenzenecarbaldehyde
(c) 2-Methyl-3-heptanone
(d) 3,5-Dimethylcyclohexanone
(e) 2,2,4,4-Tetramethyl-3-pentanone
(f) 4-Methyl-3-penten-2-one
(g) Butanedial
(h) 3-Phenyl-2-propenal
(i) 6,6-Dimethyl-2,4-cyclohexadienone
(j) *p*-Nitroacetophenone
(k) (*S*)-2-Hydroxypropanal
(l) (2*S*,3*R*)-2,3,4-Trihydroxybutanal

19.26 Draw and name the seven aldehydes and ketones with the formula $C_5H_{10}O$.

19.27 Give IUPAC names for the following structures:

(a)

(b) CHO
H—C—OH
CH_2OH

(c)

(d) $CH_3CHCCH_2CH_3$
              |
            CH_3

(e) CH_3CHCH_2CH (with OH and O groups)

(f)

19.28 Give structures that fit the following descriptions:
(a) An α,β-unsaturated ketone, C_6H_8O
(b) An α-diketone
(c) An aromatic ketone, $C_9H_{10}O$
(d) A diene aldehyde, C_7H_8O

19.29 Predict the products of the reaction of phenylacetaldehyde with the following reagents:
(a) $NaBH_4$, then H_3O^+
(b) Tollens' reagent
(c) NH_2OH, HCl catalyst
(d) CH_3MgBr, then H_3O^+
(e) CH_3OH, HCl catalyst
(f) H_2NNH_2, KOH
(g) $(C_6H_5)_3P{=}CH_2$
(h) HCN, KCN

19.30 Answer Problem 19.29 for reaction with acetophenone.

19.31 How would you prepare the following substances from 2-cyclohexenone? More than one step may be required.

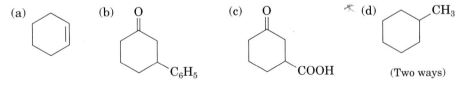

(a)   (b) O   (c) O   ✳ (d) CH_3
                    C_6H_5 COOH   (Two ways)

19.32 Show how the Wittig reaction might be used to prepare the following alkenes. Identify the alkyl halide and the carbonyl components that would be used.

(a) $C_6H_5CH{=}CH{-}CH{=}CHC_6H_5$ (b)

(c) CH$_2$

(d)

19.33 Why do you suppose tri*phenyl*phosphine rather than, say, tri*methyl*phosphine is used to prepare Wittig reagents? What problems might you run into if trimethylphosphine were used?

19.34 How would you use a Grignard reaction on a ketone or aldehyde to synthesize the following compounds?
(a) 2-Pentanol   (b) 1-Butanol
(c) 1-Phenylcyclohexanol (d) Diphenylmethanol

19.35 Aldehydes can be prepared by the Wittig reaction using (methoxymethylene)triphenylphosphorane as the Wittig reagent and then hydrolyzing the product with acid. For example,

O   + (C$_6$H$_5$)$_3$$\overset{+}{P}$—$\overset{-}{C}$HOCH$_3$ $\longrightarrow$ $\overset{H\diagdown \diagup OCH_3}{C}$ $\xrightarrow{\text{H}_3\text{O}^+}$ CHO

**(Methoxymethylene)-
triphenylphosphorane)**

(a) How would you prepare the required phosphorane?
(b) Propose a mechanism for the hydrolysis step.

19.36 When 4-hydroxybutanal is treated with methanol in the presence of an acid catalyst, 2-methoxytetrahydrofuran is formed. Explain.

HOCH$_2$CH$_2$CH$_2$CHO $\xrightarrow[\text{H}^+]{\text{CH}_3\text{OH}}$ $\overset{O}{\diagdown}$—OCH$_3$

19.37 How might you carry out the following selective transformations? One of the two schemes requires a protection step. (Recall from Section 19.5 that aldehydes are more reactive than ketones toward nucleophilic addition.)

(a) CH$_3$$\overset{\overset{\text{O}}{\|}}{C}CH_2CH_2CH_2$$\overset{\overset{\text{O}}{\|}}{C}$H $\longrightarrow$ CH$_3$$\overset{\overset{\text{O}}{\|}}{C}CH_2CH_2CH_2CH_2$OH

(b) CH$_3$$\overset{\overset{\text{O}}{\|}}{C}CH_2CH_2CH_2$$\overset{\overset{\text{O}}{\|}}{C}$H $\longrightarrow$ CH$_3$$\overset{\overset{\text{OH}}{|}}{C}HCH_2CH_2CH_2$$\overset{\overset{\text{O}}{\|}}{C}$H

19.38 How would you synthesize the following substances from benzaldehyde and any other reagents needed?

(a)

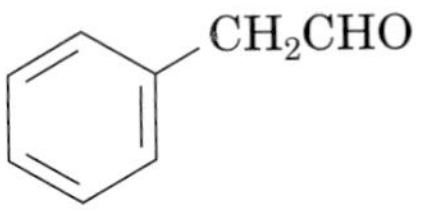

CH₂CHO

(b)

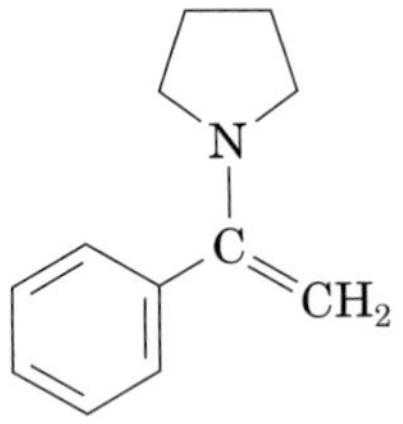

(c)

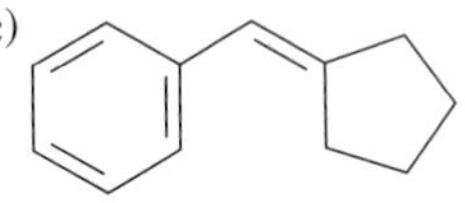

19.39 Carvone is the major constituent of spearmint oil. What products would you expect from reaction of carvone with the following reagents?

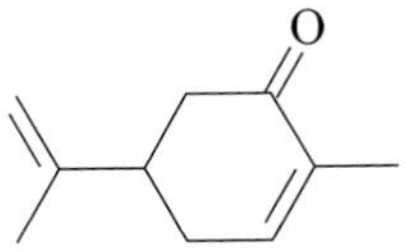

O

Carvone

(a) $(CH_3)_2Cu^- Li^+$, then H_3O^+ (b) $LiAlH_4$, then H_3O^+ (c) CH_3NH_2
(d) C_6H_5MgBr, then H_3O^+ (e) H_2/Pd (f) CrO_3, H_3O^+
(g) $(C_6H_5)_3\overset{+}{P}\overset{-}{C}HCH_3$ (h) $HOCH_2CH_2OH$, HCl

19.40 The S_N2 reaction of (dibromomethyl)benzene, $C_6H_5CHBr_2$, with NaOH yields benzaldehyde rather than (dihydroxymethyl)benzene, $C_6H_5CH(OH)_2$. Explain.

19.41 Give three methods for reducing a carbonyl group to a methylene group, $R_2C=O \longrightarrow R_2CH_2$. What are the advantages and disadvantages of each?

19.42 Reaction of 2-butanone with HCN yields a chiral product. What stereochemistry does the product have? Is it optically active?

19.43 How would you synthesize the following compounds from cyclohexanone?
(a) 1-Methylcyclohexene   (b) 2-Phenylcyclohexanone
(c) *cis*-1,2-Cyclohexanediol   (d) 1-Cyclohexylcyclohexanol

19.44 Each of the following reaction schemes contains one or more flaws. What is wrong in each case? How would you correct each scheme?

(a)

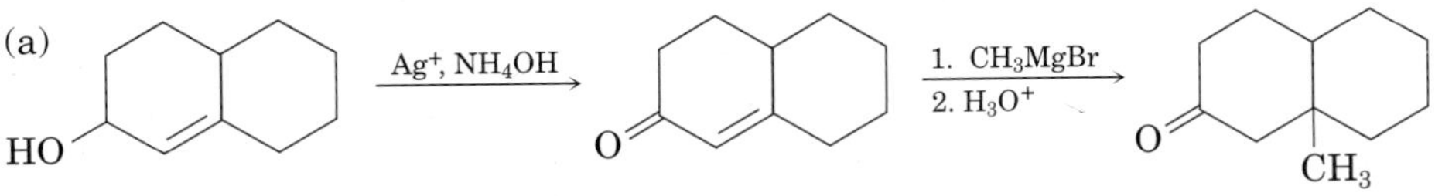

HO $\xrightarrow{Ag^+, NH_4OH}$ O $\xrightarrow[\text{2. } H_3O^+]{\text{1. } CH_3MgBr}$ O

CH₃

(b) $C_6H_5CH=CHCH_2OH \xrightarrow[H_3O^+]{CrO_3} C_6H_5CH=CHCHO$

$\downarrow H^+, CH_3OH$

$C_6H_5CH=CHCH(OCH_3)_2$

(c) $CH_3\overset{O}{\overset{||}{C}}CH_3 \xrightarrow[\text{Ethanol}]{\text{HCN, KCN}} CH_3\overset{OH}{\underset{CN}{\overset{|}{\underset{|}{C}}}}CH_3 \xrightarrow{H_3O^+} CH_3\overset{OH}{\underset{CH_2NH_2}{\overset{|}{\underset{|}{C}}}}CH_3$

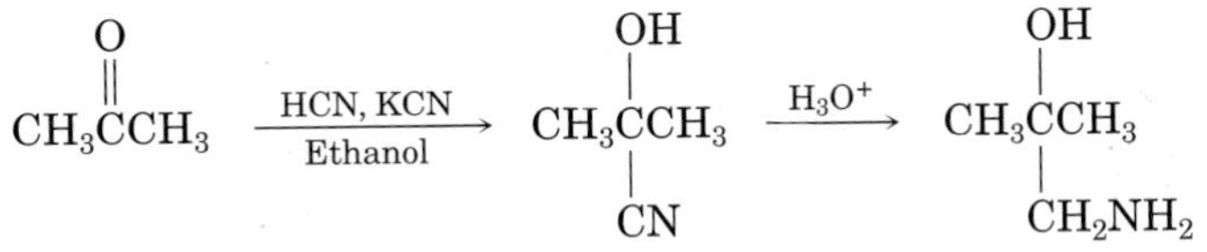

19.45   6-Methyl-5-hepten-2-one is a constituent of lemongrass oil. How could you synthesize this substance from methyl 4-oxopentanoate?

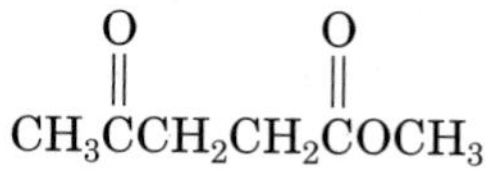

$$CH_3CCH_2CH_2COCH_3$$ **Methyl 4-oxopentanoate**

19.46   Aldehydes and ketones react with thiols to yield *thioacetals* just as they react with alcohols to yield acetals. Predict the product of the following reaction, and propose a mechanism:

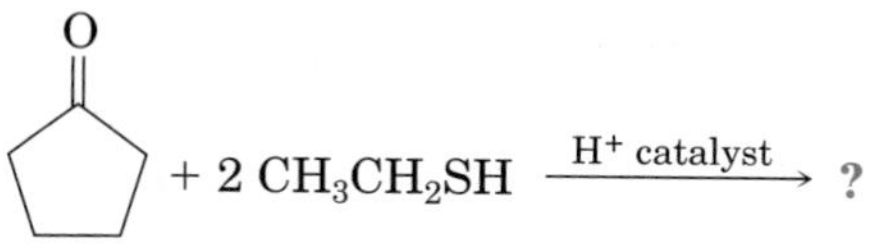

$$+ \ 2 \ CH_3CH_2SH \xrightarrow{\text{H}^+ \text{ catalyst}} ?$$

19.47   Ketones react with dimethylsulfonium methylide to yield epoxides. Suggest a mechanism for the reaction.

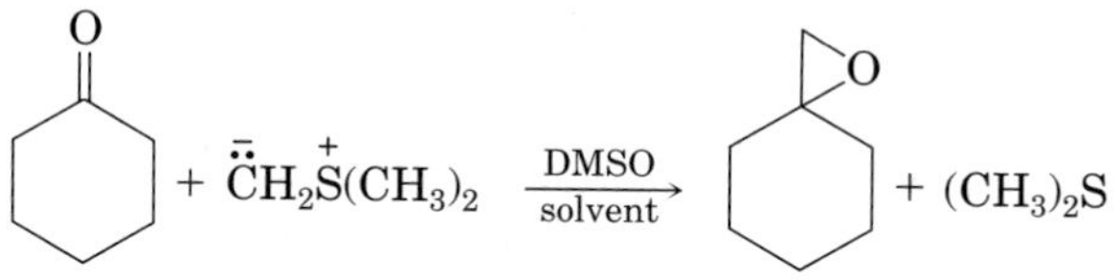

$$+ \ \ddot{\overset{-}{C}}H_2\overset{+}{S}(CH_3)_2 \xrightarrow[\text{solvent}]{\text{DMSO}} + (CH_3)_2S$$

**Dimethylsulfonium
methylide**

19.48   When cyclohexanone is heated in the presence of a large amount of acetone cyanohydrin and a small amount of base, cyclohexanone cyanohydrin and acetone are formed. Propose a mechanism.

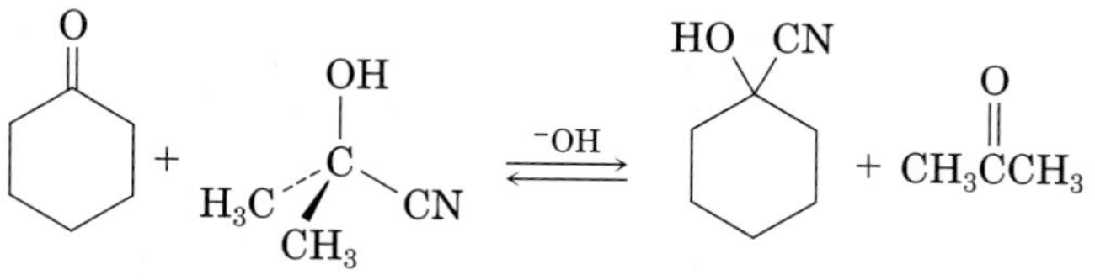

$$+ \quad H_3C{-}\overset{\overset{\displaystyle OH}{|}}{\underset{\underset{\displaystyle CH_3}{|}}{C}}{-}CN \ \rightleftharpoons \ + \ CH_3CCH_3$$

19.49   Treatment of an alcohol with dihydropyran yields an acetal called a tetrahydropyranyl ether, a reaction that can be used as a method of protecting alcohols (Section 17.9). Show the mechanism of the reaction.

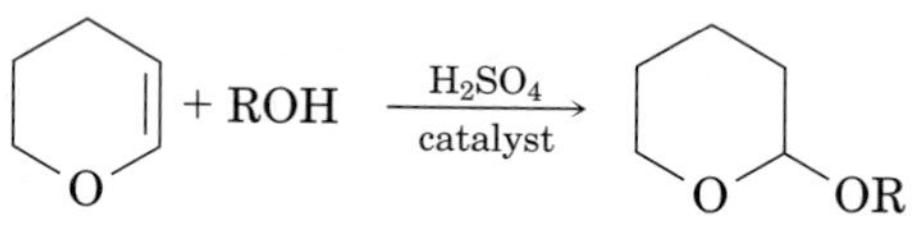

$$+ \ ROH \xrightarrow[\text{catalyst}]{\text{H}_2\text{SO}_4}$$

Dihydropyran   **A tetrahydropyranyl ether**

19.50   Tamoxifen is a drug used in the treatment of breast cancer. How would you prepare tamoxifen from benzene, the following ketone, and any other reagents needed?

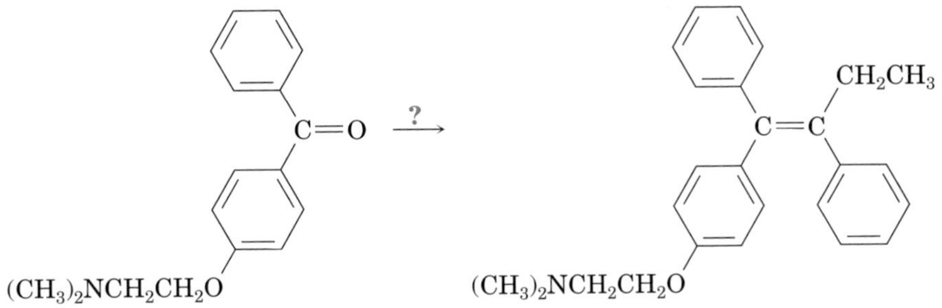

Tamoxifen

19.51 Paraldehyde, a sedative and hypnotic agent, is prepared by treatment of acetaldehyde with an acidic catalyst. Propose a mechanism for the reaction.

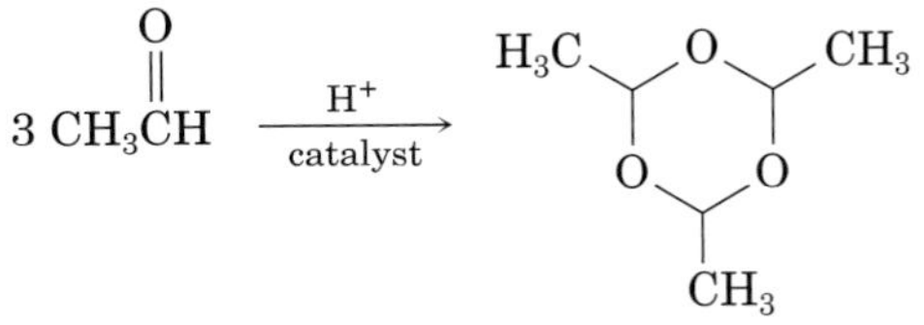

Paraldehyde

19.52 The Meerwein–Ponndorf–Verley reaction involves reduction of a ketone by treatment with an excess of aluminum triisopropoxide. The mechanism of the process is closely related to the Cannizzaro reaction in that a hydride ion acts as a leaving group. Propose a mechanism.

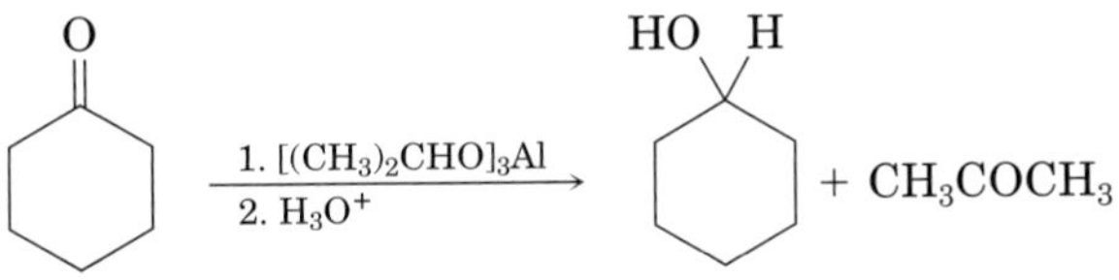

19.53 Propose a mechanism to account for the formation of 3,5-dimethylpyrazole from hydrazine and 2,4-pentanedione. Look carefully to see what has happened to each carbonyl carbon in going from starting material to product.

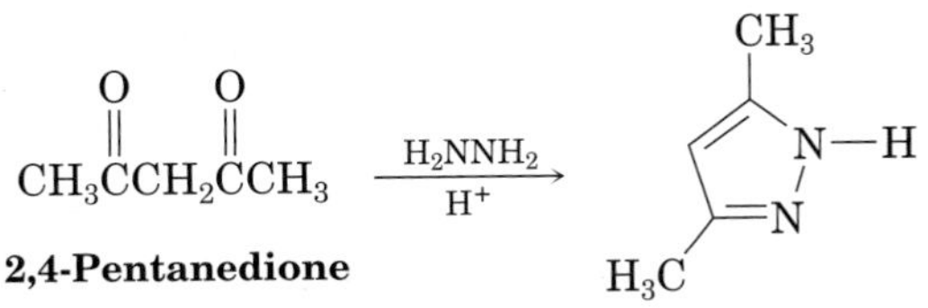

2,4-Pentanedione

3,5-Dimethylpyrazole

19.54 In light of your answer to Problem 19.53, propose a mechanism for the formation of 3,5-dimethylisoxazole from hydroxylamine and 2,4-pentanedione.

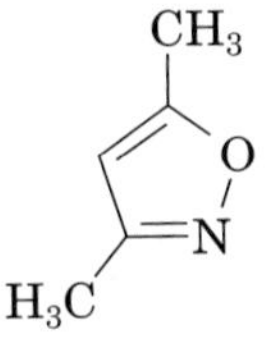

3,5-Dimethylisoxazole

19.55 Trans alkenes are converted into their cis isomers and vice versa on epoxidation followed by treatment of the epoxide with triphenylphosphine. Propose a mechanism for the epoxide ⟶ alkene reaction.

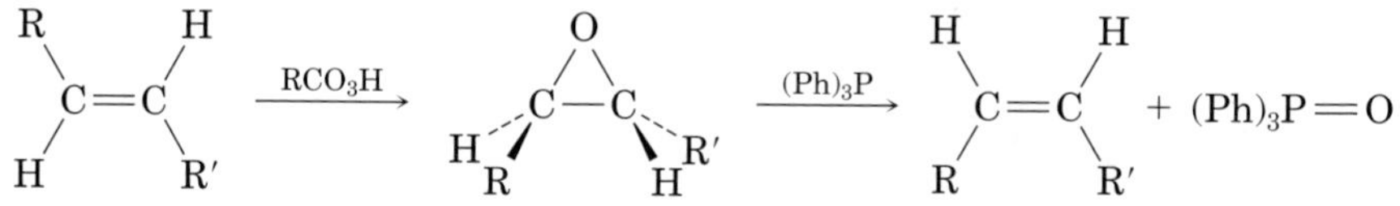

19.56 Treatment of an α,β-unsaturated ketone with basic aqueous hydrogen peroxide yields an epoxy ketone. The reaction is specific to unsaturated ketones; isolated alkene double bonds do not react. Propose a mechanism.

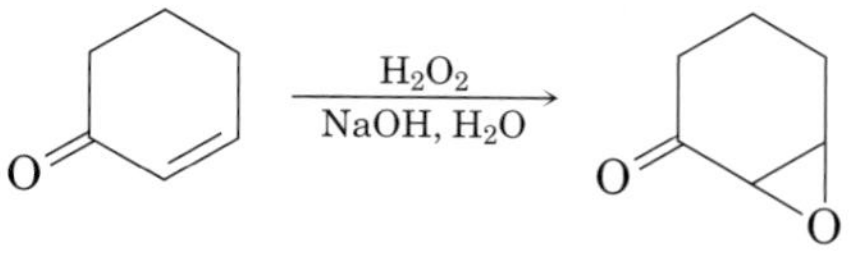

19.57 At what position would you expect to observe IR absorptions for the following molecules?

(a)

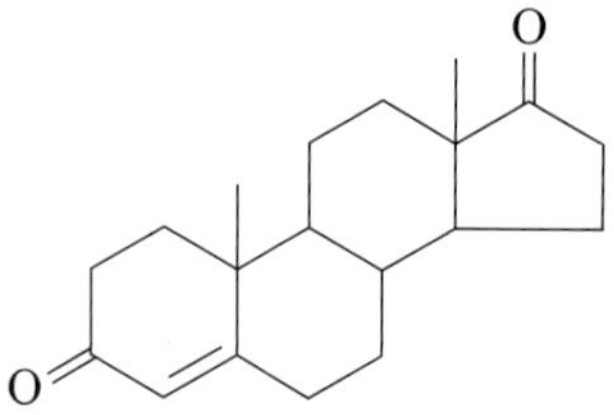

4-Androstene-3,17-dione

(b)

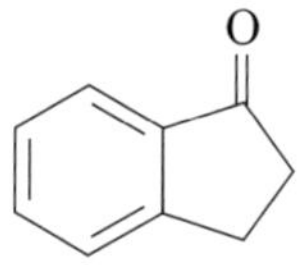

1-Indanone

(c)

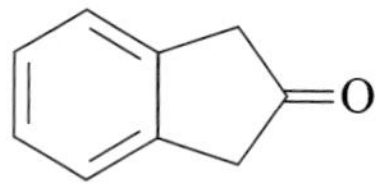

(d)

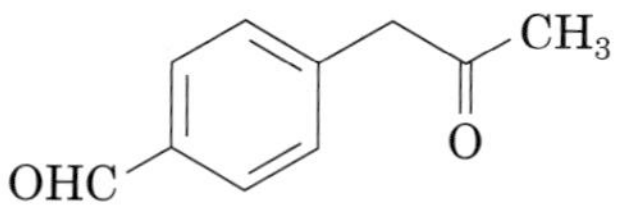

19.58 Acid-catalyzed dehydration of 3-hydroxy-3-phenylcyclohexanone leads to an unsaturated ketone. What possible structures are there for the product? At what position in the IR spectrum would you expect each to absorb? If the actual product has an absorption at 1670 cm^{-1}, what is its structure?

19.59 Compound A, MW = 86, shows an IR absorption at 1730 cm^{-1} and a very simple ^{1}H NMR spectrum with peaks at 9.7 δ (1 H, singlet) and 1.2 δ (9 H, singlet). Propose a structure for A.

19.60 Compound B is isomeric with A (Problem 19.59) and shows an IR peak at 1715 cm^{-1}. The ^{1}H NMR spectrum of B has peaks at 2.4 δ (1 H, septet, $J = 7$ Hz), 2.1 δ (3 H, singlet), and 1.2 δ (6 H, doublet, $J = 7$ Hz). What is the structure of B?

19.61 The ^1H NMR spectrum shown is that of a compound with formula $C_9H_{10}O$. How many double bonds and/or rings does this compound contain? If the unknown has an IR absorption at 1690 cm^{-1}, what is a likely structure?

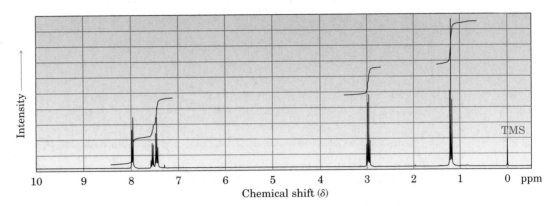

19.62 The ^1H NMR spectrum shown is that of a compound isomeric with the one in Problem 19.61. This isomer has an IR absorption at 1730 cm^{-1}. Propose a structure.

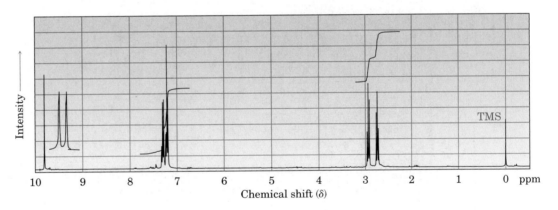

19.63 Propose structures for molecules that meet the following descriptions. Assume that the kinds of carbons (1°, 2°, 3°, or 4°) have been assigned by DEPT-NMR.
(a) $C_6H_{12}O$
IR: 1715 cm^{-1}
^{13}C NMR: 8.0 δ (1°), 18.5 δ (1°), 33.5 δ (2°), 40.6 δ (3°), 214.0 δ (4°)

(b) $C_5H_{10}O$
IR: 1730 cm^{-1}
^{13}C NMR: 22.6 δ (1°), 23.6 δ (3°), 52.8 δ (2°), 202.4 δ (3°)

(c) C_6H_8O
IR: 1680 cm^{-1}
^{13}C NMR: 22.9 δ (2°), 25.8 δ (2°), 38.2 δ (2°), 129.8 δ (3°), 150.6 δ (3°), 198.7 δ (4°)

19.64 Compound A, $C_8H_{10}O_2$, has an intense IR absorption at $1750 \, cm^{-1}$ and gives the ^{13}C NMR spectrum shown. Propose a structure for A.

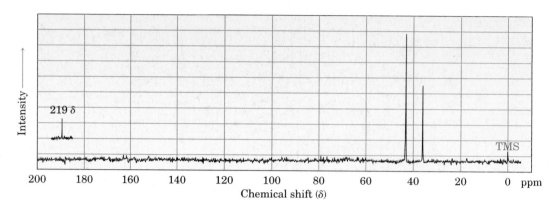

19.65 Propose structures for ketones or aldehydes that have the following ^1H NMR spectra:
   (a) C_4H_7ClO
       IR: $1715 \, cm^{-1}$

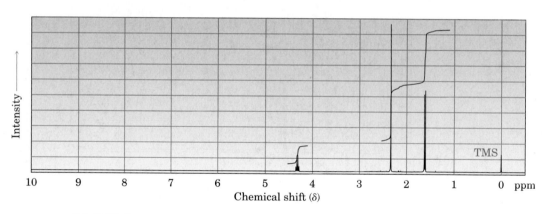

   (b) $C_7H_{14}O$
       IR: $1710 \, cm^{-1}$

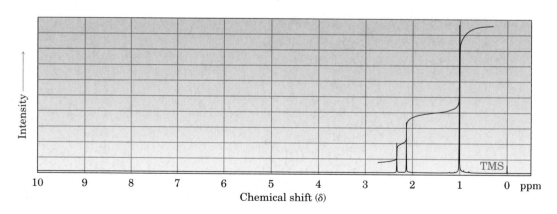

(c) $C_9H_{10}O_2$
IR: $1695\,cm^{-1}$

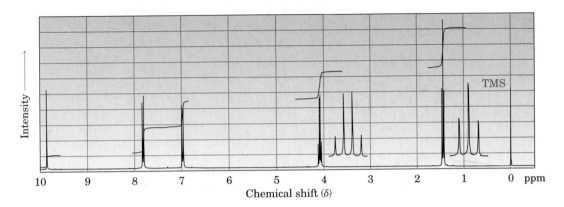

19.66 Propose structures for ketones or aldehydes that have the following ^1H NMR spectra.
(a) $C_{10}H_{12}O$
IR: $1710\,cm^{-1}$

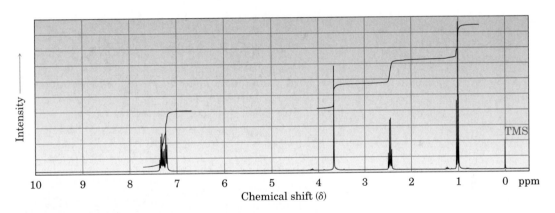

(b) $C_6H_{12}O_3$
IR: $1715\,cm^{-1}$

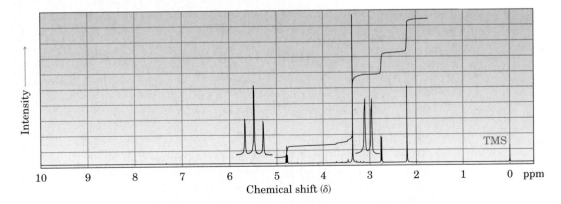

(c) C_4H_6O
   IR: $1690 \, cm^{-1}$

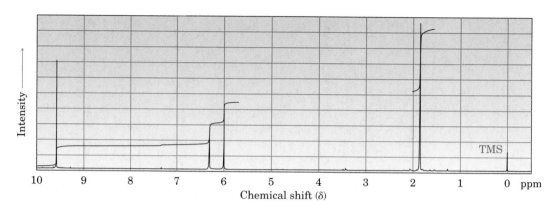

A Look Ahead

• •

19.67   Primary amines react with esters to yield amides:

$$RCO_2R' + R''NH_2 \longrightarrow RCONHR'' + R'OH$$

Propose a mechanism for the following reaction of an α,β-unsaturated ester. (See Section 21.6.)

19.68   When crystals of pure α-glucose are dissolved in water, isomerization slowly occurs to produce β-glucose. Propose a mechanism for the isomerization. (See Section 25.6.)

19.69   When glucose (Problem 19.68) is treated with $NaBH_4$, reaction occurs to yield *sorbitol*, a polyalcohol commonly used as a food additive. Show how this reduction occurs. (See Section 25.7.)

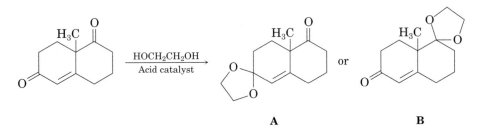

Glucose $\xrightarrow[\text{H}_2\text{O}]{\text{NaBH}_4}$ Sorbitol

Molecular Modeling

· ·

19.70 Use SpartanView to examine transition states for the nucleophilic addition of CN⁻ to formaldehyde, acetone, and benzophenone (PhCOPh). Assuming that the length of the developing C–C bond reflects steric repulsion between the nucleophile and the electrophile, which transition state is least strained and which is most strained?

19.71 The diketone shown below reacts selectively with 1 equivalent of 1,2-ethanediol to give a monoacetal. Since acetal formation is reversible, the reaction is under thermodynamic control. Use SpartanView to obtain the energies of monoacetals A and B, and predict which is favored.

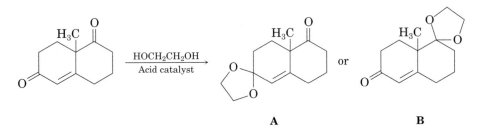

HOCH₂CH₂OH / Acid catalyst

A   **B**

19.72 Electron-withdrawing groups can affect nearby C=O infrared stretching frequencies. Use SpartanView to identify C=O stretching vibrations in cyclohexanone, axial 2-fluorocyclohexanone, equatorial 2-fluorocyclohexanone, equatorial 3-fluorocyclohexanone, and 2,2-difluorocyclohexanone. What effects do you observe?

19.73 The reaction of CH₃NH₂ with 2-methylpropanal can give three possible products: two isomeric imines and an enamine. Since the products form reversibly, the reaction is thermodynamically controlled. Use SpartanView to obtain the energies of imine A, imine B, and the enamine, and predict which is likely to be formed. What factors are responsible for the energy differences between the possible products?

19.74 Nucleophilic addition reactions occur by electron donation from the nucleophile to the π* antibonding orbital of the ketone. If the faces of the ketone are different, addition happens faster at the more accessible orbital lobe. Use SpartanView to display mesh electron-density surfaces of 2-norbornanone and camphor, and simultaneously display the π* antibonding orbital (LUMO) surface of each. Which face of each ketone is more reactive? What is the stereochemistry of the alcohol produced by reaction of each with NaBH₄?

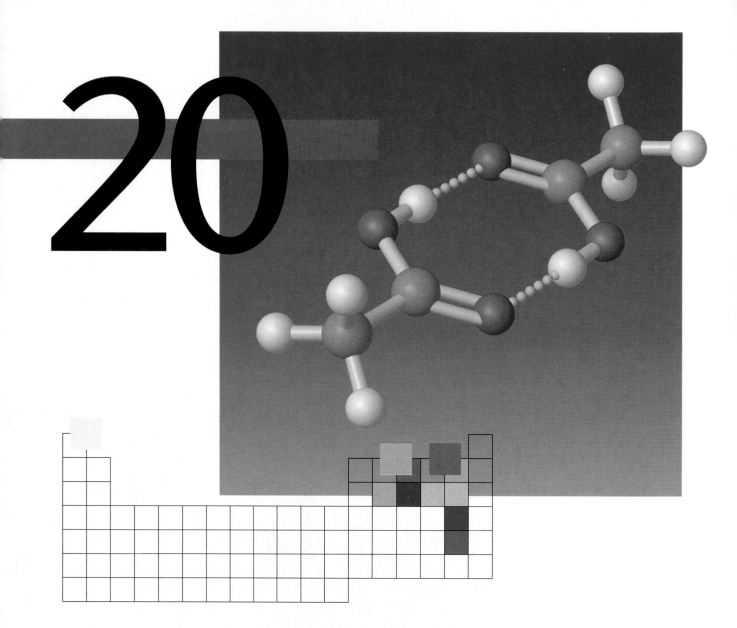

Carboxylic Acids

Carboxylic acids occupy a central place among carbonyl compounds. Not only are they important themselves, they also serve as starting materials for preparing numerous *acyl derivatives* such as esters, amides, and acid chlorides. A great many carboxylic acids are found in nature. For example, acetic acid, CH_3COOH, is the chief organic component of vinegar; butanoic acid, $CH_3CH_2CH_2COOH$, is responsible for the rancid odor of sour butter; and hexanoic acid (caproic acid), $CH_3(CH_2)_4COOH$, is responsible for the unmistakable aroma of goats and dirty gym socks (the name comes from the Latin *caper,* "goat"). Other examples are cholic acid, a major component of human bile, and long-chain aliphatic acids such as palmitic acid, $CH_3(CH_2)_{14}COOH$, a biological precursor of fats and other lipids.

Cholic acid

Approximately 2 million tons of acetic acid are produced each year in the United States for a variety of purposes, including preparation of the vinyl acetate polymer used in paints and adhesives. The industrial method of acetic acid synthesis involves a cobalt acetate-catalyzed air oxidation of acetaldehyde, but this method is not used in the laboratory.

$$\underset{CH_3\overset{\overset{\displaystyle O}{\|}}{C}H}{} + O_2 \xrightarrow[80°C]{\text{Cobalt acetate}} CH_3\overset{\overset{\displaystyle O}{\|}}{C}OH$$

The Monsanto Company has developed an even more efficient synthesis based on the rhodium-catalyzed reaction of methanol with carbon monoxide:

$$CH_3OH + CO \xrightarrow{\text{Rh catalyst}} CH_3\overset{\overset{\displaystyle O}{\|}}{C}OH$$

20.1 Naming Carboxylic Acids

IUPAC rules allow for two systems of nomenclature, depending on the complexity of the acid molecule. Carboxylic acids that are derived from open-chain alkanes are systematically named by replacing the terminal *-e* of the corresponding alkane name with *-oic acid*. The carboxyl carbon atom is numbered C1 in this system.

Propanoic acid   4-Methyl**pentan**oic acid   3-Ethyl-6-methyl**octane**dioic acid

Alternatively, compounds that have a –COOH group bonded to a ring are named using the suffix *-carboxylic acid*. The COOH carbon is attached to C1 and is not itself numbered in this system.

3-Bromo**cyclohexane**carboxylic acid   1-**Cyclopent**enecarboxylic acid

Because many carboxylic acids were among the first organic compounds to be isolated and purified, a large number of common names are recognized by IUPAC, some of which are given in Table 20.1. We'll use systematic names in this book, with a few exceptions, such as formic (methanoic) acid and acetic (ethanoic) acid, whose names are so well known that it makes little sense to refer to them any other way. Also listed in Table 20.1 are the common names used for acyl groups derived from the parent acids.

TABLE 20.1   Common Names of Some Carboxylic Acids and Acyl Groups

Carboxylic acid		Acyl group	
Structure	**Name**	**Name**	**Structure**
HCOOH	Formic	Formyl	HCO—
CH_3COOH	Acetic	Acetyl	CH_3CO-
CH_3CH_2COOH	Propionic	Propionyl	CH_3CH_2CO-
$CH_3CH_2CH_2COOH$	Butyric	Butyryl	$CH_3(CH_2)_2CO-$
$(CH_3)_3CCOOH$	Pivalic	Pivaloyl	$(CH_3)_3CCO-$
HOOCCOOH	Oxalic	Oxalyl	$-OCCO-$
$HOOCCH_2COOH$	Malonic	Malonyl	$-OCCH_2CO-$
$HOOCCH_2CH_2COOH$	Succinic	Succinyl	$-OC(CH_2)_2CO-$
$HOOCCH_2CH_2CH_2COOH$	Glutaric	Glutaryl	$-OC(CH_2)_3CO-$
$HOOCCH_2CH_2CH_2CH_2COOH$	Adipic	Adipoyl	$-OC(CH_2)_4CO-$
$H_2C=CHCOOH$	Acrylic	Acryloyl	$H_2C=CHCO-$
$H_2C=C(CH_3)COOH$	Methacrylic	Methacryloyl	$H_2C=C(CH_3)CO-$
HOOCCH=CHCOOH	{ *cis*-Maleic *trans*-Fumaric	Maleoyl } Fumaroyl }	$-OCCH=CHCO-$
	Benzoic	Benzoyl	
	Phthalic	Phthaloyl	

Problem 20.1 Give IUPAC names for the following compounds:

(a) $(CH_3)_2CHCH_2COOH$

(b) $CH_3CHCH_2CH_2COOH$
$\quad\quad\ \ |$
$\quad\quad\ \ Br$

(c) $CH_3CH{=}CHCH{=}CHCOOH$

(d) $\ \ \ \ \ \ \ \ \ \ COOH$
$\quad\quad\quad\quad\quad |$
$CH_3CH_2CHCH_2CH_2CH_3$

(e) HOOC   H

H

COOH

(f)   CH₃

COOH

Problem 20.2 Draw structures corresponding to the following IUPAC names:

(a) 2,3-Dimethylhexanoic acid

(b) 4-Methylpentanoic acid

(c) *trans*-1,2-Cyclobutanedicarboxylic acid

(d) *o*-Hydroxybenzoic acid

(e) (9Z,12Z)-9,12-Octadecadienoic acid

20.2 Structure and Physical Properties of Carboxylic Acids

Because the carboxylic acid functional group is structurally related to both ketones and alcohols, we might expect to see some familiar properties. Like ketones, the carboxyl carbon has sp^2 hybridization, and carboxylic acid groups are therefore planar with C–C=O and O=C–O bond angles of approximately 120°. The physical parameters of acetic acid are given in Table 20.2.

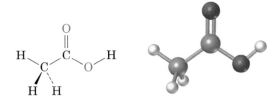

TABLE 20.2 Physical Parameters for Acetic Acid			
Bond angle (°)		**Bond length (pm)**	
C—C=O	119	C—C	152
C—C—OH	119	C=O	125
O=C—OH	122	C—OH	131

Like alcohols, carboxylic acids are strongly associated because of hydrogen bonding. Most carboxylic acids exist as cyclic dimers held together by two hydrogen bonds.

Acetic acid dimer

Stereo View

This strong hydrogen bonding has a noticeable effect on boiling points, making carboxylic acids much higher boiling than the corresponding alcohols. Table 20.3 lists the properties of some common acids.

TABLE 20.3   Physical Constants of Some Carboxylic Acids

Name	Structure	Melting point (°C)	Boiling point (°C)
Formic	$HCOOH$	8.4	100.7
Acetic	CH_3COOH	16.6	117.9
Propanoic	CH_3CH_2COOH	−20.8	141
Propenoic	$H_2C{=}CHCOOH$	13	141.6
Benzoic	C_6H_5COOH	122.1	249

20.3 Dissociation of Carboxylic Acids

As their name implies, carboxylic acids are *acidic*. They therefore react with bases such as $NaOH$ and $NaHCO_3$ to give metal carboxylate salts, RCO_2^- ^+M. Carboxylic acids with more than six carbons are only slightly soluble in water, but alkali metal salts of carboxylic acids are generally quite water-soluble because they are ionic. In fact, it's often possible to purify acids by extracting their salts into aqueous base, then reacidifying and extracting the pure acid back into an organic solvent.

A carboxylic acid
(water-insoluble)

A carboxylic acid salt
(water-soluble)

Like other Brønsted–Lowry acids discussed in Section 2.7, carboxylic acids dissociate slightly in dilute aqueous solution to give H_3O^+ and carboxylate anions, RCO_2^-. The exact extent of dissociation is given by an acidity constant, K_a:

$$R-C(=O)OH + H_2O \rightleftharpoons R-C(=O)O^- + H_3O^+$$

$$K_a = \frac{[\text{RCOO}^-][\text{H}_3\text{O}^+]}{[\text{RCOOH}]} \quad \text{and} \quad \text{p}K_a = -\log K_a$$

For most carboxylic acids, K_a is approximately 10^{-5}. Acetic acid, for example, has $K_a = 1.76 \times 10^{-5}$, which corresponds to a pK_a of 4.75. In practical terms, a K_a value near 10^{-5} means that only about 0.1% of the molecules in a 0.1 M solution are dissociated, as opposed to the 100% dissociation found with strong mineral acids such as HCl.

Although much weaker than mineral acids, carboxylic acids are nevertheless much stronger acids than alcohols. The K_a of ethanol, for example, is approximately 10^{-16}, making ethanol a weaker acid than acetic acid by a factor of 10^{11}.

$$\text{CH}_3\text{CH}_2\text{OH} \qquad \text{CH}_3\overset{\text{O}}{\overset{\|}{\text{C}}}\text{OH} \qquad \text{HCl}$$

$$\textbf{p}K_\textbf{a} = \textbf{16} \qquad \textbf{p}K_\textbf{a} = \textbf{4.75} \qquad \textbf{p}K_\textbf{a} = \textbf{-7}$$

Acidity →

Why are carboxylic acids so much more acidic than alcohols, even though both contain –OH groups? As noted in Section 2.10, an alcohol dissociates to give an alkoxide ion, in which the negative charge is localized on a single electronegative atom. A carboxylic acid, by contrast, gives a carboxylate ion, in which the negative charge is delocalized over *two* oxygen atoms. In resonance terms (Section 2.4), a carboxylate ion is a stabilized resonance hybrid of two equivalent Kekulé structures.

$$\text{CH}_3\text{CH}_2\overset{..}{\text{O}}\text{H} + \text{H}_2\text{O} \rightleftharpoons \text{CH}_3\text{CH}_2\overset{..}{\text{O}}:^- + \text{H}_3\text{O}^+$$

Alcohol **Unstabilized alkoxide ion**

$$\text{CH}_3-\text{C}\overset{:\text{O}:}{\underset{\overset{..}{\text{O}}-\text{H}}{}} + \text{H}_2\text{O} \rightleftharpoons \text{CH}_3-\text{C}\overset{:\text{O}:}{\underset{\overset{..}{\text{O}}:^-}{}} \longleftrightarrow \text{CH}_3-\text{C}\overset{:\overset{..}{\text{O}}:^-}{\underset{:\text{O}:}{}} + \text{H}_3\text{O}^+$$

Carboxylic acid   **Resonance-stabilized carboxylate ion (two equivalent resonance forms)**

Since a carboxylate ion is more stable than an alkoxide ion, it is lower in energy and more highly favored at equilibrium, as shown in the reaction energy diagram in Figure 20.1.

FIGURE 20.1 ▼

A reaction energy diagram for the dissociation of an alcohol (green curve) and a carboxylic acid (red curve). Resonance stabilization of the carboxylate anion lowers $\Delta G°$ for dissociation of the acid, leading to a more favorable K_a. (The starting energy levels of alcohol and acid are shown at the same point for ease of comparison.)

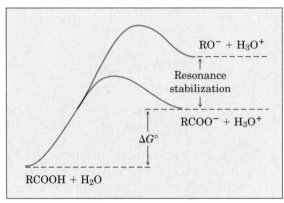

We can't really draw an accurate representation of the carboxylate resonance hybrid using Kekulé structures, but an orbital picture of acetate ion makes it clear that the carbon–oxygen bonds are equivalent and that each is intermediate between a single and a double bond (Figure 20.2). The p orbital on the carboxylate carbon atom overlaps equally well with p orbitals on both oxygens, and the four p electrons are delocalized throughout the three-atom π electron system.

FIGURE 20.2 ▼

An orbital picture and a stereo view of the acetate ion, showing the equivalence of the two oxygen atoms.

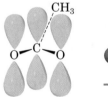  

Stereo View

Evidence for the equivalence of the two carboxylate oxygens comes from X-ray studies on sodium formate. Both carbon–oxygen bonds are 127 pm in length, midway between the C=O double bond (120 pm) and C–O single bond (134 pm) of formic acid. An electrostatic potential map of the formate ion also shows how the negative charge (red) is dispersed equally over both oxygens.

 formate ion

 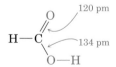

Sodium formate   **Formic acid**

● ●

Problem 20.3 Assume you have a mixture of naphthalene and benzoic acid that you want to separate. How might you take advantage of the acidity of one component in the mixture to effect a separation?

Problem 20.4 The K_a for dichloroacetic acid is 3.32×10^{-2}. Approximately what percentage of the acid is dissociated in a 0.10 M aqueous solution?

● ●

20.4 Substituent Effects on Acidity

A listing of pK_a values for different carboxylic acids indicates that there are substantial differences from one acid to another (Table 20.4). For example, trifluoroacetic acid ($K_a = 0.59$) is 33,000 times as strong as acetic acid ($K_a = 1.76 \times 10^{-5}$). How can we account for such differences?

TABLE 20.4 Acidity of Some Carboxylic Acids

Structure	K_a	pK_a	
F_3CCOOH	0.59	0.23	Stronger acid
FCH_2COOH	2.6×10^{-3}	2.59	
$ClCH_2COOH$	1.4×10^{-3}	2.85	
$BrCH_2COOH$	2.1×10^{-3}	2.68	
ICH_2COOH	7.5×10^{-4}	3.12	
$HCOOH$	1.77×10^{-4}	3.75	
$HOCH_2COOH$	1.5×10^{-4}	3.83	
C_6H_5COOH	6.46×10^{-5}	4.19	
$H_2C{=}CHCOOH$	5.6×10^{-5}	4.25	
CH_3COOH	1.76×10^{-5}	4.75	
CH_3CH_2COOH	1.34×10^{-5}	4.87	Weaker acid
CH_3CH_2OH (ethanol)[a]	(10^{-16})	(16)	

[a] Value for ethanol is shown for reference.

Since the dissociation of a carboxylic acid is an equilibrium process, any factor that stabilizes the carboxylate anion relative to undissociated carboxylic acid will drive the equilibrium toward increased dissociation and result in increased acidity. For example, an electron-*withdrawing* group attached to a carboxylate ion will delocalize the negative charge, thereby stabilizing the ion and increasing acidity. Conversely, any factor that destabilizes the carboxylate relative to undissociated acid will result in decreased acidity. An electron-*donating* group, for example, destabilizes the carboxylate anion and decreases acidity.

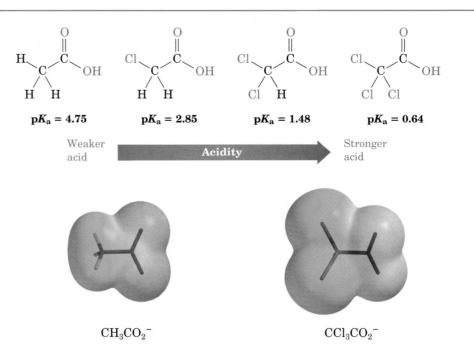

Electron-withdrawing group
stabilizes carboxylate
and strengthens acid

Electron-donating group
destabilizes carboxylate
and weakens acid

The data in Table 20.4 show exactly the expected effect. Electronegative substituents, such as the halogens, make the carboxylate anion more stable by inductively withdrawing electrons. Fluoroacetic, chloroacetic, bromoacetic, and iodoacetic acids are therefore stronger acids than acetic acid by factors of 50–150. Introduction of two electronegative substituents makes dichloroacetic acid some 3000 times as strong as acetic acid, and introduction of three chloro substituents makes trichloroacetic acid more than 12,000 times as strong (Figure 20.3).

FIGURE 20.3 ▼

Relative strengths of acetic acid and chlorosubstituted acetic acids. Electrostatic potential maps of $CH_3CO_2^-$ and $CCl_3CO_2^-$ show that the electron-withdrawing effect of the chlorine atoms makes the oxygen atoms in $CCl_3CO_2^-$ less negative and less basic than those in $CH_3CO_2^-$.

acetate ion,
trichloroacetate ion

$pK_a = 4.75$ $pK_a = 2.85$ $pK_a = 1.48$ $pK_a = 0.64$

Weaker acid → Acidity → Stronger acid

$CH_3CO_2^-$ $CCl_3CO_2^-$

Because inductive effects operate through σ bonds and are dependent on distance, the effect of halogen substitution decreases as the substituent moves farther from the carboxyl. For instance, 2-chlorobutanoic acid has $pK_a = 2.86$, 3-chlorobutanoic acid has $pK_a = 4.05$, and 4-chlorobutanoic acid has $pK_a = 4.52$, similar to that of butanoic acid itself (Table 20.5).

TABLE 20.5 Acidity of Chlorosubstituted Butanoic Acids

Structure	K_a	pK_a
Cl │ CH₃CH₂CHCOOH	1.39×10^{-3}	2.86
Cl │ CH₃CHCH₂COOH	8.9×10^{-5}	4.05
ClCH₂CH₂CH₂COOH	3.0×10^{-5}	4.52
CH₃CH₂CH₂COOH	1.5×10^{-5}	4.82

• •

Problem 20.5 Without looking at a table of pK_a values, rank the substances in each of the following groups in order of increasing acidity:
(a) CH_3CH_2COOH, $BrCH_2COOH$, FCH_2COOH
(b) CH_3CH_2OH, $CH_3CH_2NH_2$, CH_3CH_2COOH

Problem 20.6 Dicarboxylic acids have two dissociation constants, one for the initial dissociation into a monoanion and one for the second dissociation into a dianion. For oxalic acid, HOOC–COOH, the first ionization constant has $pK_1 = 1.2$ and the second ionization constant has $pK_2 = 4.2$. Why is the second carboxyl group so much less acidic than the first?

• •

20.5 Substituent Effects in Substituted Benzoic Acids

We saw during the discussion of electrophilic aromatic substitution in Section 16.5 that substituents on the aromatic ring dramatically affect reactivity. Aromatic rings with electron-donating groups are activated toward further electrophilic substitution, and aromatic rings with electron-withdrawing groups are deactivated. Exactly the same effects are noticed on the acidity of substituted benzoic acids (Table 20.6).

TABLE 20.6   Substituent Effects on Acidity of *p*-Substituted Benzoic Acids

$$Y-\!\!\!\!\bigcirc\!\!\!\!-COOH$$

	Y	K_a	pK_a	
Weaker acid	—OH	3.3×10^{-5}	4.48	Activating groups
	—OCH$_3$	3.5×10^{-5}	4.46	
	—CH$_3$	4.3×10^{-5}	4.34	
	—H	6.46×10^{-5}	4.19	
	—Cl	1.0×10^{-4}	4.0	Deactivating groups
	—Br	1.1×10^{-4}	3.96	
Stronger acid	—CHO	1.8×10^{-4}	3.75	
	—CN	2.8×10^{-4}	3.55	
	—NO$_2$	3.9×10^{-4}	3.41	

As Table 20.6 shows, an electron-withdrawing (deactivating) group such as nitro increases acidity by stabilizing the carboxylate anion, and an electron-donating (activating) group such as methoxy decreases acidity by destabilizing the carboxylate anion.

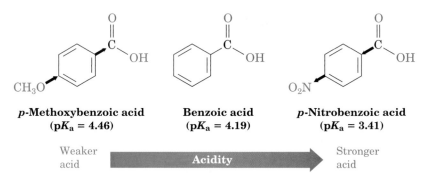

p-**Methoxybenzoic acid**   **Benzoic acid**   *p*-**Nitrobenzoic acid**
  (pK_a = 4.46)   (pK_a = 4.19)   (pK_a = 3.41)

Weaker acid   ——————→ **Acidity** ——————→   Stronger acid

Since it's much easier to measure the acidity of a substituted benzoic acid than to determine the relative reactivity of an aromatic ring toward electrophilic substitution, the correlation between the two effects is useful for predicting reactivity. If we want to know the effect of a certain substituent on electrophilic reactivity, we can simply find the acidity of the corresponding benzoic acid. Practice Problem 20.1 gives an example.

Finding the K_a of this acid . . .   . . . lets us predict the reactivity of this substituted benzene to electrophilic attack.

Practice Problem 20.1   The pK_a of p-(trifluoromethyl)benzoic acid is 3.6. Would you expect the tri-fluoromethyl substituent to be an activating or deactivating group in the Friedel–Crafts reaction?

Strategy   Decide whether p-(trifluoromethyl)benzoic acid is stronger or weaker than benzoic acid. A substituent that strengthens the acid is a deactivating group because it withdraws electrons, and a substituent that weakens the acid is an activating group because it donates electrons.

Solution   A pK_a of 3.6 means that p-(trifluoromethyl)benzoic acid is stronger than benzoic acid, whose pK_a is 4.19. Thus, the trifluoromethyl substituent favors dissociation by helping to stabilize the negative charge. Trifluoromethyl must therefore be an electron-withdrawing, deactivating group.

Problem 20.7   The pK_a of p-cyclopropylbenzoic acid is 4.45. Is cyclopropylbenzene likely to be more reactive or less reactive than benzene toward electrophilic bromination? Explain.

Problem 20.8   Rank the following compounds in order of increasing acidity. Don't look at a table of pK_a data to help with your answer.
(a) Benzoic acid, p-methylbenzoic acid, p-chlorobenzoic acid
(b) p-Nitrobenzoic acid, acetic acid, benzoic acid

20.6 Preparation of Carboxylic Acids

Let's review briefly some of the methods for preparing carboxylic acids that we've seen in past chapters:

- Oxidation of a substituted alkylbenzene with $KMnO_4$ or $Na_2Cr_2O_7$ gives a substituted benzoic acid (Section 16.10). Both primary and secondary alkyl groups can be oxidized, but tertiary groups are not affected.

p-Nitrotoluene   **p-Nitrobenzoic acid (88%)**

- Oxidative cleavage of an alkene with $KMnO_4$ gives a carboxylic acid if the alkene has at least one vinylic hydrogen (Section 7.8).

$$CH_3(CH_2)_7CH=CH(CH_2)_7COOH \xrightarrow[H_3O^+]{KMnO_4} CH_3(CH_2)_7COOH + HOOC(CH_2)_7COOH$$

Oleic acid   **Nonanoic acid   Nonanedioic acid**

- Oxidation of a primary alcohol or an aldehyde yields a carboxylic acid (Sections 17.8 and 19.3). Primary alcohols are often oxidized with CrO_3 in aqueous acid, and aldehydes are oxidized with either acidic CrO_3 or basic silver oxide (Tollens' reagent).

$$CH_3(CH_2)_8CH_2OH \xrightarrow[H_3O^+]{CrO_3} CH_3(CH_2)_8\overset{\displaystyle O}{\overset{\displaystyle \|}{C}}OH$$

1-Decanol   **Decanoic acid (93%)**

$$CH_3CH_2CH_2CH_2CH_2\overset{\displaystyle O}{\overset{\displaystyle \|}{C}}H \xrightarrow[NH_4OH]{Ag_2O} CH_3CH_2CH_2CH_2CH_2\overset{\displaystyle O}{\overset{\displaystyle \|}{C}}OH$$

Hexanal   **Hexanoic acid (85%)**

Hydrolysis of Nitriles

Nitriles, $R–C≡N$, can be hydrolyzed by strong, hot aqueous acid or base to yield carboxylic acids. Since nitriles themselves are usually prepared by S_N2 reaction of an alkyl halide with cyanide ion, the two-step sequence of cyanide displacement followed by nitrile hydrolysis is an excellent method for preparing a carboxylic acid from an alkyl halide ($RBr \longrightarrow RC≡N \longrightarrow RCOOH$). Note that the product acid has one more carbon than the starting alkyl halide.

$$RCH_2Br \xrightarrow[(S_N2)]{Na^+ \; ^-CN} RCH_2C≡N \xrightarrow{H_3O^+} RCH_2\overset{\displaystyle O}{\overset{\displaystyle \|}{C}}OH + NH_3$$

The method works best with primary halides because a competitive E2 elimination reaction can occur when a secondary or tertiary alkyl halide is used (Section 11.15). Nevertheless, some unhindered secondary halides react well. An example occurs in the commercial synthesis of fenoprofen, a nonsteroidal anti-inflammatory drug, or NSAID, marketed under the trade name Mylan. (See "Aspirin and Other Aromatic NSAID's" at the end of Chapter 15.)

Fenoprofen
(an antiarthritic agent)

Carboxylation of Grignard Reagents

An alternative method for preparing carboxylic acids is by reaction of a Grignard reagent with CO_2 to yield a metal carboxylate, followed by protonation to give the carboxylic acid. This **carboxylation** reaction is carried out either by pouring the Grignard reagent over dry ice (solid CO_2) or by bubbling a stream of dry CO_2 through a solution of the Grignard reagent. Grignard carboxylation generally gives good yields of acids from alkyl halides, but is of course limited to those alkyl halides that can form Grignard reagents in the first place (Section 17.6).

1-Bromo-2,4,6-trimethyl-benzene   **2,4,6-Trimethylbenzoic acid (87%)**

The mechanism of Grignard carboxylation is similar to that of other Grignard reactions (Section 19.8). The organomagnesium halide adds to a C=O bond of carbon dioxide in a typical nucleophilic addition reaction. Protonation of the carboxylate by addition of aqueous HCl in a separate step then gives the free carboxylic acid product.

• •

Practice Problem 20.2   How would you prepare phenylacetic acid ($PhCH_2COOH$) from benzyl bromide ($PhCH_2Br$)?

Strategy   We've seen two methods for preparing carboxylic acids from alkyl halides: (1) cyanide ion displacement followed by hydrolysis, and (2) formation of a Grignard reagent followed by carboxylation. The first method involves an S_N2 reaction and is therefore limited to use with primary alkyl halides. The second method involves formation of a Grignard reagent and is therefore limited to use with organic halides that have no acidic hydrogens or reactive functional groups. In the present instance, either method would work well.

Solution

. . . .

Problem 20.9   How would you prepare each of the following carboxylic acids?
(a) Benzoic acid from bromobenzene   (b) $(CH_3)_3CCOOH$ from $(CH_3)_3CCl$
(c) $CH_3CH_2CH_2COOH$ from $CH_3CH_2CH_2Br$

. . . .

20.7 Reactions of Carboxylic Acids: An Overview

We commented earlier in this chapter that carboxylic acids are similar in some respects to both alcohols and ketones. Like alcohols, carboxylic acids can be deprotonated to give anions, which are good nucleophiles in S_N2 reactions. Like ketones, carboxylic acids undergo attack by nucleophiles on the carbonyl group. In addition, carboxylic acids undergo other reactions characteristic neither of alcohols nor ketones. Figure 20.4 shows some of the general reactions of carboxylic acids.

FIGURE 20.4 ▼

Some general reactions of carboxylic acids.

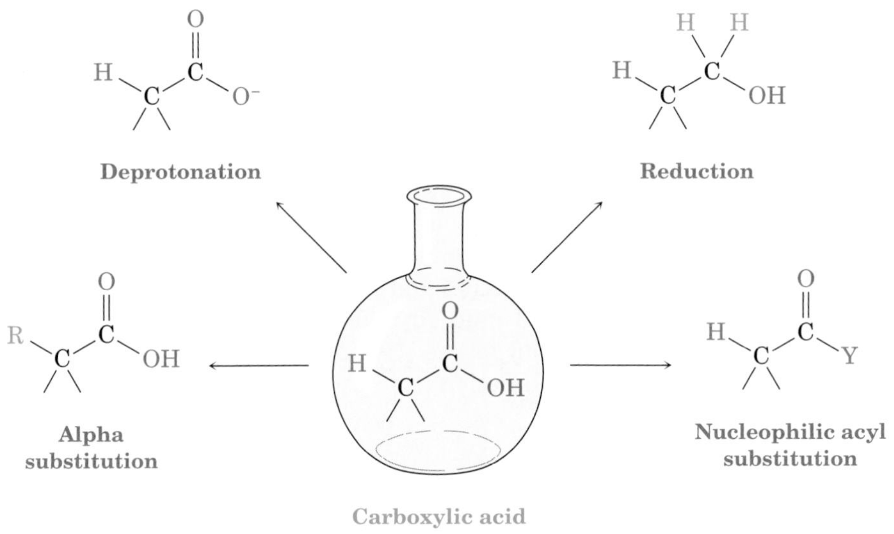

Reactions of carboxylic acids can be grouped into the four categories indicated in Figure 20.4. Of the four, we've already discussed the acidic behavior of carboxylic acids in Sections 20.3–20.5, and we'll discuss reduction in the next section. The remaining two categories are examples of fundamen-

tal carbonyl-group reaction mechanisms—nucleophilic acyl substitution and
α substitution—that will be discussed in detail in Chapters 21 and 22.

20.8 Reduction of Carboxylic Acids

Carboxylic acids are reduced by LiAlH$_4$ (but not by NaBH$_4$) to yield primary alcohols (Section 17.5). The reaction is difficult, however, and often
requires heating in tetrahydrofuran solvent to go to completion.

$$CH_3(CH_2)_7CH=CH(CH_2)_7\overset{\overset{\displaystyle O}{\|}}{C}OH \quad \xrightarrow[\text{2. H}_3\text{O}^+]{\text{1. LiAlH}_4,\ \text{THF}} \quad CH_3(CH_2)_7CH=CH(CH_2)_7CH_2OH$$

Oleic acid   ***cis*-9-Octadecen-1-ol (87%)**

Alternatively, borane in tetrahydrofuran (BH$_3$/THF) is a useful reagent
for reducing carboxylic acids to primary alcohols. Reaction of an acid with
BH$_3$/THF occurs rapidly at room temperature, and the procedure is often
preferred to reduction with LiAlH$_4$ because of its relative ease, safety, and
specificity. Borane reacts with carboxylic acids faster than with any other
functional group, thereby allowing selective transformations such as that
shown below on *p*-nitrophenylacetic acid. If the reduction of *p*-nitrophenyl-
acetic acid were done with LiAlH$_4$, both nitro and carboxyl groups would
be reduced.

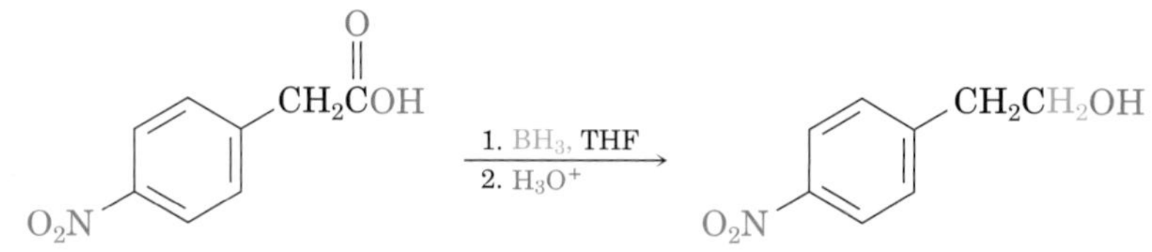

p-Nitrophenylacetic acid   **2-(p-Nitrophenyl)ethanol (94%)**

· ·

Problem 20.10   How might you prepare 2-phenylethanol from benzyl bromide? More than one step
is required.

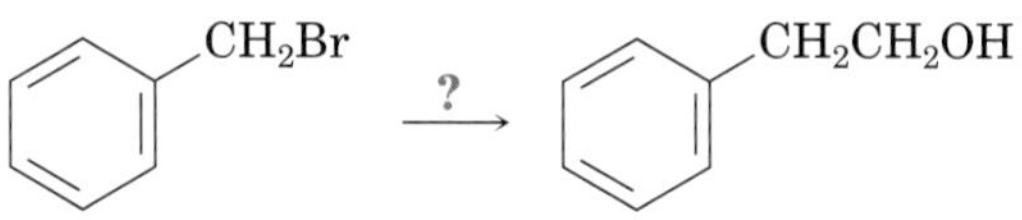

Problem 20.11   How might you carry out the following transformation? More than one step is
required.

· ·

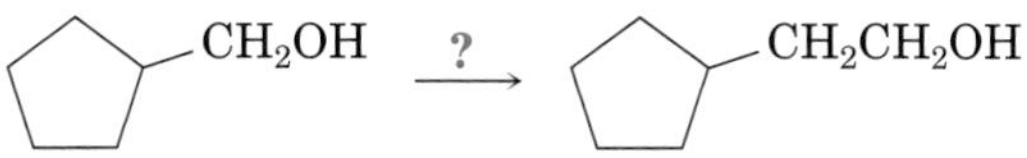

20.9 Spectroscopy of Carboxylic Acids

Infrared Spectroscopy

Carboxylic acids show two characteristic IR absorptions that make the –COOH group easily identifiable. The O–H bond of the carboxyl group gives rise to a very broad absorption over the range 2500–3300 cm^{-1}, and the C=O bond shows an absorption between 1710 cm^{-1} and 1760 cm^{-1}. The exact position of C=O absorption depends both on the structure of the molecule and on whether the acid is free (monomeric) or hydrogen-bonded (dimeric). Free carboxyl groups absorb at 1760 cm^{-1}, but the more commonly encountered dimeric carboxyl groups absorb in a broad band centered around 1710 cm^{-1}.

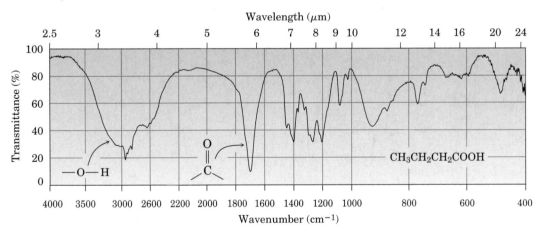

Both the broad O–H absorption and the C=O absorption at 1710 cm^{-1} (dimeric) are identified in the IR spectrum of butanoic acid shown in Figure 20.5.

FIGURE 20.5 ▼

Infrared spectrum of butanoic acid, CH$_3$CH$_2$CH$_2$COOH.

butanoic acid, butanoic acid dimer (see vibrations on CD-Rom)

....

Problem 20.12   Cyclopentanecarboxylic acid and 4-hydroxycyclohexanone have the same formula (C$_6$H$_{10}$O$_2$), and both contain an –OH and a C=O group. How could you distinguish between them by IR spectroscopy?

....

Nuclear Magnetic Resonance Spectroscopy

Carboxylic acid groups can be detected by both ^1H and ^{13}C NMR spectroscopy. Carboxyl carbon atoms absorb in the range 165–185 δ in the ^{13}C NMR spectrum, with aromatic and α,β-unsaturated acids near the upfield end of the range ($\sim$165 δ) and saturated aliphatic acids near the downfield end ($\sim$185 δ). The acidic –COO**H** proton normally absorbs as a singlet near 12 δ in the ^1H NMR spectrum. As with alcohols (Section 17.12), the –COOH proton can be replaced by deuterium when D_2O is added to the sample tube, causing the absorption to disappear from the NMR spectrum.

Figure 20.6 indicates the positions of the ^{13}C NMR absorptions for several carboxylic acids, and Figure 20.7 shows the ^1H NMR spectrum of phenylacetic acid. Note that the carboxyl proton absorption occurs at 12.0 δ.

FIGURE 20.6 ▼

Carbon-13 NMR absorptions for some carboxylic acids.

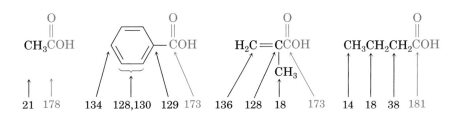

FIGURE 20.7 ▼

Proton NMR spectrum of phenylacetic acid, PhCH₂COOH.

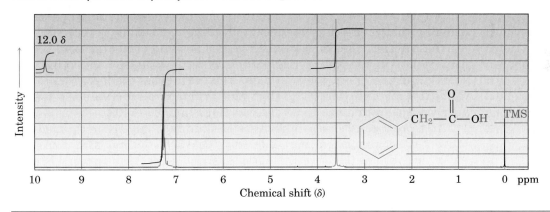

· ·

Problem 20.13 How could you distinguish between cyclopentanecarboxylic acid and 4-hydroxy-cyclohexanone by ^1H and ^{13}C NMR spectroscopy? (See Problem 20.12.)

· ·

CHEMISTRY @ WORK

Vitamin C

Vitamin C, or ascorbic acid, is surely the best known of all vitamins. It was the first vitamin to be discovered (1928), the first to be structurally characterized (1933), and the first to be synthesized in the laboratory (1933). Over 80 million pounds of vitamin C are now synthesized worldwide each year, more than the total amount of all other vitamins combined. In addition to its use as a vitamin supplement, vitamin C is used as a food preservative, a "flour improver" in bakeries, and an animal food additive.

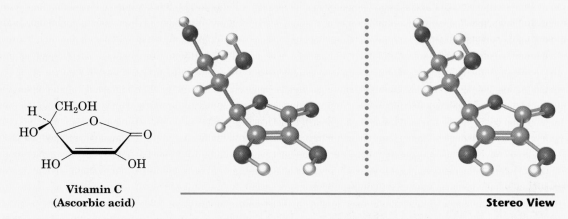

**Vitamin C
(Ascorbic acid)**

Stereo View

In addition to the hazards of weather, participants in early polar expeditions often suffered from scurvy, caused by a dietary vitamin C deficiency.

Vitamin C is perhaps most famous for its antiscorbutic properties, meaning that it prevents the onset of scurvy, a bleeding disease affecting those with a deficiency of fresh vegetables and citrus fruits in their diet. Sailors in the Age of Exploration were particularly susceptible to scurvy, and the death toll was high. The Portuguese explorer Vasco da Gama, for instance, lost more than half his crew to scurvy during his 2 year voyage around the Cape of Good Hope in 1497–1499.

In more recent times, large doses of vitamin C have been claimed to prevent the common cold, cure infertility, delay the onset of symptoms in AIDS, and inhibit the development of gastric and cervical cancers. Proof is still lacking for most of these claims, but a recent study in Europe did find statistical evidence for an inhibitory effect against gastric cancers. Although large daily doses of vitamin C are probably not warranted, the harmful side effects of vitamin C appear minimal, and many people have adopted a "better safe than sorry" approach.

(continued) ▶

The industrial preparation of vitamin C involves an unusual blend of biological and laboratory organic chemistry. The Hoffmann-LaRoche Company synthesizes ascorbic acid from glucose through the five-step route shown in Figure 20.8. Glucose, a pentahydroxy aldehyde, is first reduced to sorbitol, which is then oxidized by the microorganism *Acetobacter suboxydans*. No chemical reagent exists that is selective enough to oxidize only one of the six alcohol groups in sorbitol, so an enzymatic reaction is used. Treatment with acetone and an acid catalyst then protects four of the remaining hydroxyl groups in acetal linkages, and the unprotected hydroxyl group is chemically oxidized to the carboxylic acid by reaction with aqueous NaOCl (household bleach). Hydrolysis with acid then removes the two acetal groups and causes an internal ester-forming reaction to take place to give ascorbic acid. Each of the five steps takes place in better than 90% yield.

FIGURE 20.8 ▼

The industrial synthesis of ascorbic acid from glucose.

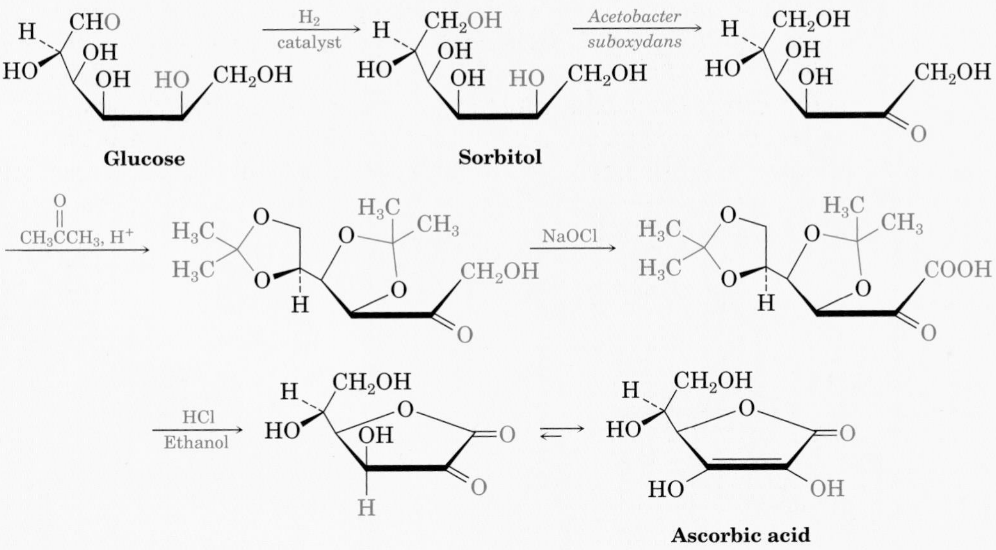

Summary and Key Words

OCOL

KEY WORDS

carboxylation, 826
carboxylic acid, 814

Carboxylic acids are among the most useful building blocks for synthesizing other molecules, both in nature and in the chemical laboratory. They are named systematically by replacing the terminal *-e* of the corresponding alkane name with *-oic acid*. Like aldehydes and ketones, the carbonyl carbon atom is sp^2-hybridized; like alcohols, carboxylic acids are associated through hydrogen bonding and therefore have high boiling points.

The distinguishing characteristic of carboxylic acids is their acidity. Although weaker than mineral acids such as HCl, carboxylic acids dissociate much more readily than alcohols because the resultant carboxylate ions are stabilized by resonance between two equivalent forms:

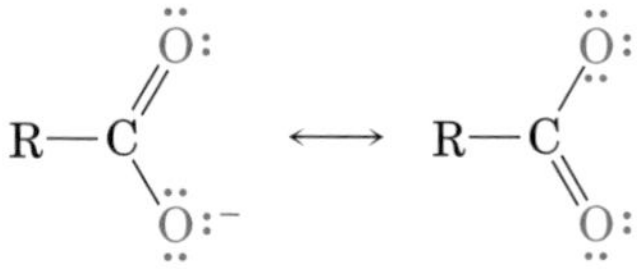

Most alkanoic acids have pK_a values near 5, but the exact pK_a of a given acid depends on structure. Carboxylic acids substituted by electron-withdrawing groups are more acidic (have a lower pK_a) because their carboxylate ions are stabilized. Carboxylic acids substituted by electron-donating groups are less acidic (have a higher pK_a) because their carboxylate ions are destabilized.

Methods of synthesis for carboxylic acids include: (1) oxidation of alkylbenzenes, (2) oxidative cleavage of alkenes, (3) oxidation of primary alcohols or aldehydes, (4) hydrolysis of nitriles, and (5) reaction of Grignard reagents with CO_2 (**carboxylation**). General reactions of carboxylic acids include: (1) loss of the acidic proton, (2) nucleophilic acyl substitution at the carbonyl group, (3) substitution on the α carbon, and (4) reduction.

Carboxylic acids are easily distinguished spectroscopically. They show characteristic IR absorptions at 2500–3300 cm^{-1} (due to the O–H) and at 1710–1760 cm^{-1} (due to the C=O). Acids also show ^{13}C NMR absorptions at 165–185 δ and ^{1}H NMR absorptions near 12 δ.

Summary of Reactions

1. Preparation of carboxylic acids (Section 20.6)
   (a) Oxidation of alkylbenzenes (Section 16.10)

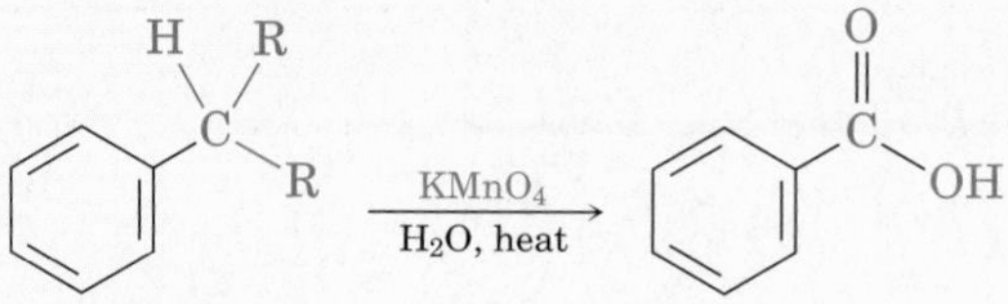

   (b) Oxidative cleavage of alkenes (Section 7.8)

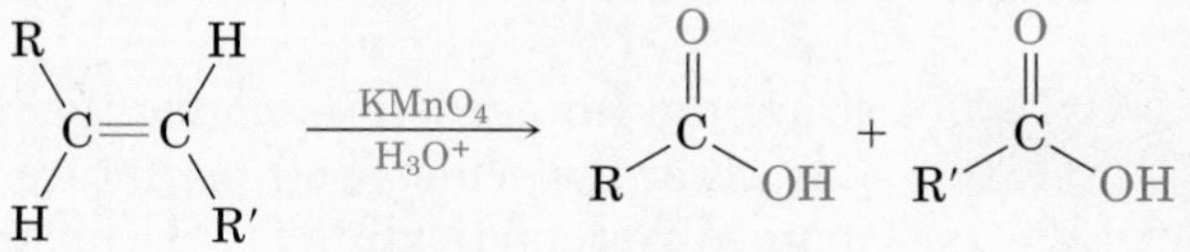

(continued) ▶

(c) Oxidation of primary alcohols (Section 17.8)

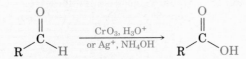

(d) Oxidation of aldehydes (Section 19.3)

(e) Hydrolysis of nitriles (Section 20.6)

(f) Carboxylation of Grignard reagents (Section 20.6)

2. Reactions of carboxylic acids
   (a) Deprotonation (Section 20.3)

   (b) Reduction to primary alcohols (Section 20.8)

Visualizing Chemistry

(Problems 20.1–20.13 appear within the chapter.)

20.14 Give IUPAC names for the following carboxylic acids (reddish brown = Br):

(a)   (b)

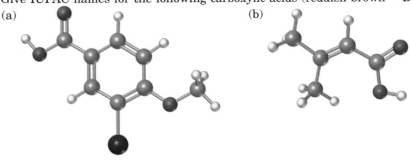

(c)

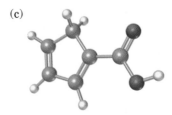

20.15 Would you expect the following carboxylic acids to be more acidic or less acidic than benzoic acid? Explain. (Reddish brown = Br.)

(a)   (b)

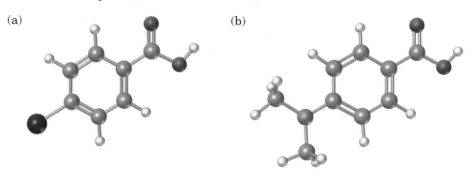

20.16 The following carboxylic acid can't be prepared from an alkyl halide by either the nitrile hydrolysis route or the Grignard carboxylation route. Explain.

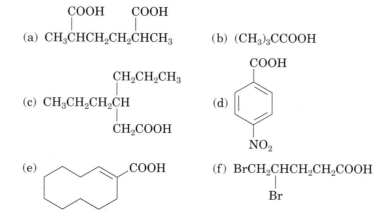

Stereo View

Additional Problems

20.17 Give IUPAC names for the following compounds:

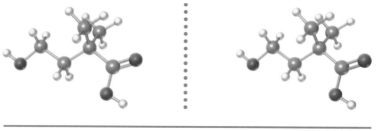

(a)
$$\underset{\underset{CH_3CHCH_2CH_2CHCH_3}{|}}{COOH} \quad \underset{\underset{}{|}}{COOH}$$

(b) $(CH_3)_3CCOOH$

(c)
$$\underset{\underset{CH_3CH_2CH_2CH}{|}}{CH_2CH_2CH_3} \\ \underset{}{|} \\ CH_2COOH$$

(d)
COOH

NO₂

(e) COOH

(f) $BrCH_2CHCH_2CH_2COOH$
        |
       Br

20.18 Draw structures corresponding to the following IUPAC names:
(a) *cis*-1,2-Cyclohexanedicarboxylic acid   (b) Heptanedioic acid
(c) 2-Hexen-4-ynoic acid   (d) 4-Ethyl-2-propyloctanoic acid
(e) 3-Chlorophthalic acid   (f) Triphenylacetic acid

20.19 Draw and name the eight carboxylic acid isomers with the formula $C_6H_{12}O_2$.

20.20 Order the compounds in each set with respect to increasing acidity:
(a) Acetic acid, oxalic acid, formic acid
(b) *p*-Bromobenzoic acid, *p*-nitrobenzoic acid, 2,4-dinitrobenzoic acid
(c) Fluoroacetic acid, 3-fluoropropanoic acid, iodoacetic acid

20.21 Arrange the compounds in each set in order of increasing basicity:
(a) Magnesium acetate, magnesium hydroxide, methylmagnesium bromide
(b) Sodium benzoate, sodium *p*-nitrobenzoate, sodium acetylide
(c) Lithium hydroxide, lithium ethoxide, lithium formate

20.22 How could you convert butanoic acid into the following compounds? Write each step showing the reagents needed.
(a) 1-Butanol   (b) 1-Bromobutane   (c) Pentanoic acid
(d) 1-Butene   (e) Octane

20.23 How could you convert each of the following compounds into butanoic acid? Write each step showing all reagents.
(a) 1-Butanol   (b) 1-Bromobutane   (c) 1-Butene
(d) 1-Bromopropane   (e) 4-Octene

20.24 How would you prepare the following compounds from benzene? More than one step is required in each case.
(a) *m*-Chlorobenzoic acid   (b) *p*-Bromobenzoic acid
(c) Phenylacetic acid, $C_6H_5CH_2COOH$

20.25 Calculate pK_a's for the following acids:
(a) Lactic acid, $K_a = 8.4 \times 10^{-4}$ (b) Acrylic acid, $K_a = 5.6 \times 10^{-6}$

20.26 Calculate K_a's for the following acids:
(a) Citric acid, p$K_a = 3.14$ (b) Tartaric acid, p$K_a = 2.98$

20.27 Use the equation $\Delta G° = -2.303\,RT \log K_a$ to calculate values of $\Delta G°$ for the dissociation of ethanol (p$K_a = 16.0$) and acetic acid (p$K_a = 4.75$) at 300 K (27°C). The gas constant R has the value 8.315 J/(K · mol).

20.28 Shown here are some pK_a data for simple dibasic acids. How can you account for the fact that the difference between the first and second ionization constants decreases with increasing distance between the carboxyl groups?

Name	Structure	pK_1	pK_2
Oxalic	HOOCCOOH	1.2	4.2
Succinic	HOOCCH$_2$CH$_2$COOH	4.2	5.6
Adipic	HOOC(CH$_2$)$_4$COOH	4.4	5.4

20.29 Predict the product of the reaction of *p*-methylbenzoic acid with each of the following:
(a) BH$_3$, then H$_3$O$^+$ (b) *N*-Bromosuccinimide in CCl$_4$
(c) CH$_3$MgBr in ether, then H$_3$O$^+$ (d) KMnO$_4$, H$_3$O$^+$
(e) LiAlH$_4$, then H$_3$O$^+$

20.30 Using $^{13}CO_2$ as your only source of labeled carbon, along with any other compounds needed, how would you synthesize the following compounds?
(a) $CH_3CH_2{}^{13}COOH$ (b) $CH_3{}^{13}CH_2COOH$

20.31 How would you carry out the following transformations?

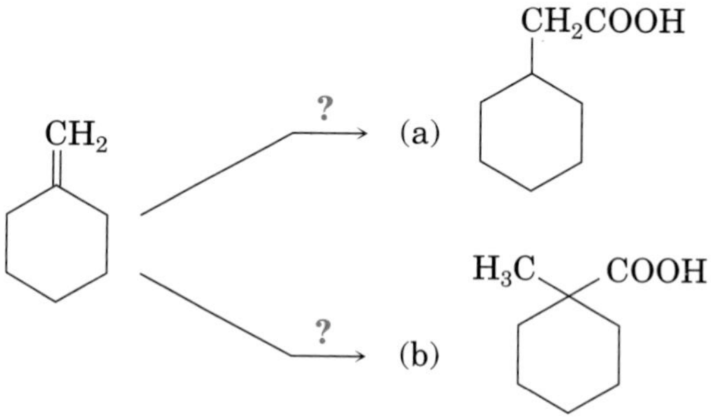

20.32 Which method—Grignard carboxylation or nitrile hydrolysis—would you use for each of the following reactions? Explain.

(a)

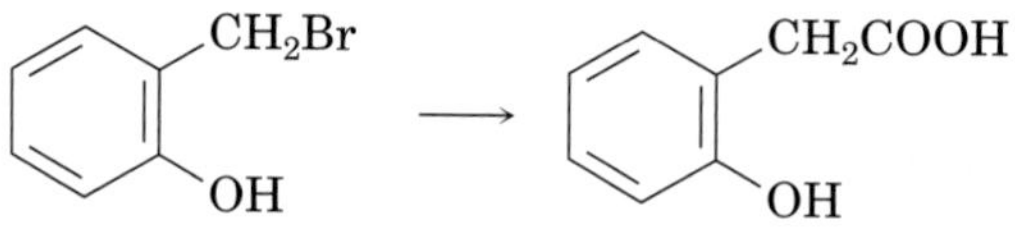

(b) $\underset{\underset{Br}{|}}{CH_3CH_2CHCH_3} \longrightarrow \underset{\underset{CH_3}{|}}{CH_3CH_2CHCOOH}$

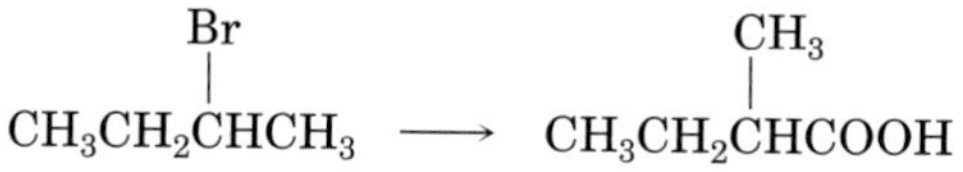

(c) $\underset{\overset{\parallel}{O}}{CH_3CCH_2CH_2CH_2I} \longrightarrow \underset{\overset{\parallel}{O}}{CH_3CCH_2CH_2CH_2COOH}$

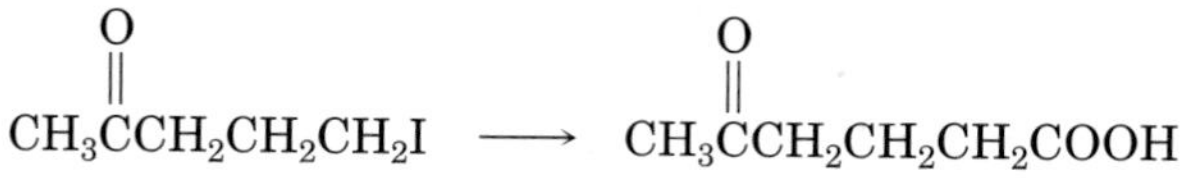

(d) $HOCH_2CH_2CH_2Br \longrightarrow HOCH_2CH_2CH_2COOH$

20.33 A chemist in need of 2,2-dimethylpentanoic acid decided to synthesize some by reaction of 2-chloro-2-methylpentane with NaCN, followed by hydrolysis of the product. After carrying out the reaction sequence, however, none of the desired product could be found. What do you suppose went wrong?

20.34 The following synthetic schemes all have at least one flaw in them. What is wrong with each?

(a)

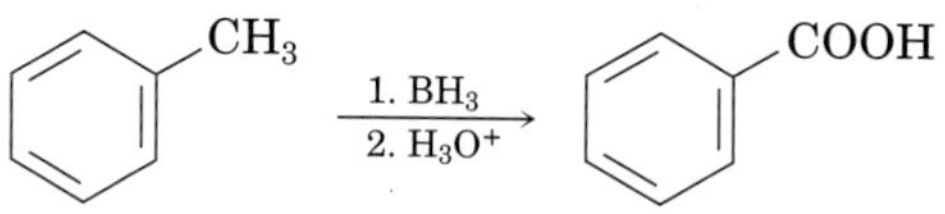

(b) $\underset{\overset{|}{Br}}{CH_3CH_2CHCH_2CH_3} \xrightarrow[\substack{2.\ NaCN \\ 3.\ H_3O^+}]{1.\ Mg} \underset{\overset{|}{COOH}}{CH_3CH_2CHCH_2CH_3}$

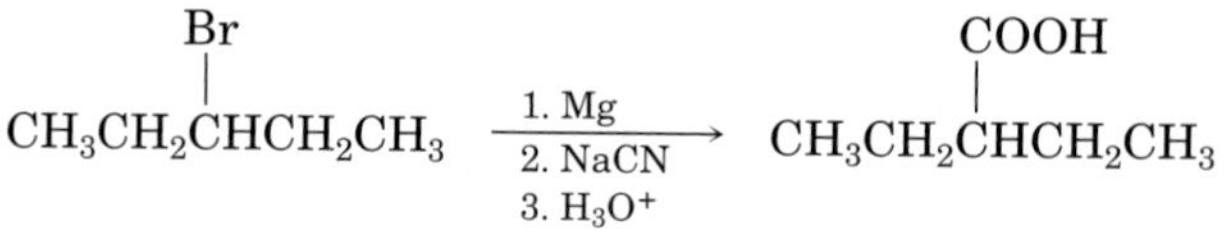

(c)

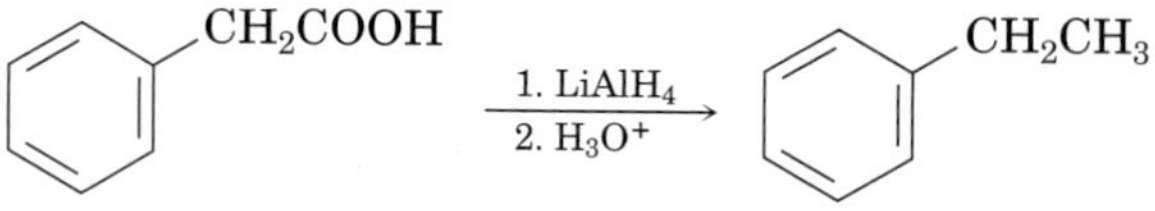

(d) $\underset{\overset{|}{CH_3}}{\underset{\overset{|}{OH}}{CH_3CCH_2CH_2Cl}} \xrightarrow[\substack{2.\ H_3O^+}]{1.\ NaCN} \underset{\overset{|}{CH_3}}{\underset{\overset{|}{OH}}{CH_3CCH_2CH_2\overset{\overset{\displaystyle O}{\parallel}}{C}OH}}$

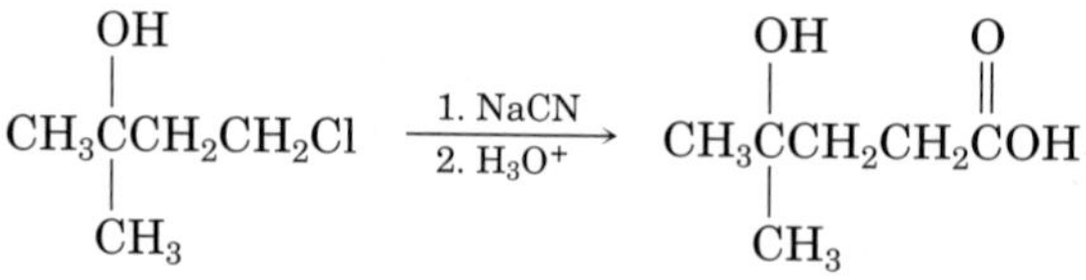

20.35 *p*-Aminobenzoic acid (PABA) is widely used as a sunscreen agent. Propose a synthesis of PABA starting from toluene.

20.36 Lithocholic acid is a steroid found in human bile:

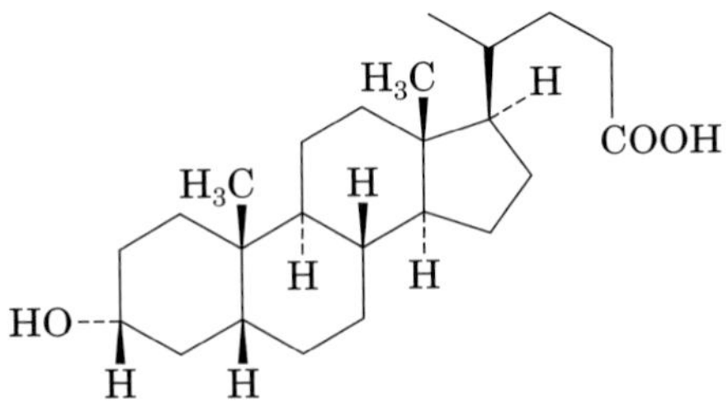

Lithocholic acid

Predict the product of reaction of lithocholic acid with each of the following reagents. Don't worry about the size of the molecule; just concentrate on the functional groups.
(a) CrO_3, H_3O^+ (b) Tollens' reagent
(c) BH_3, then H_3O^+ (d) $(CH_3)_3SiCl$, $(CH_3CH_2)_3N$
(e) CH_3MgBr, then H_3O^+ (f) $LiAlH_4$, then H_3O^+

20.37 Propose a synthesis of the anti-inflammatory drug Fenclorac from phenylcyclohexane.

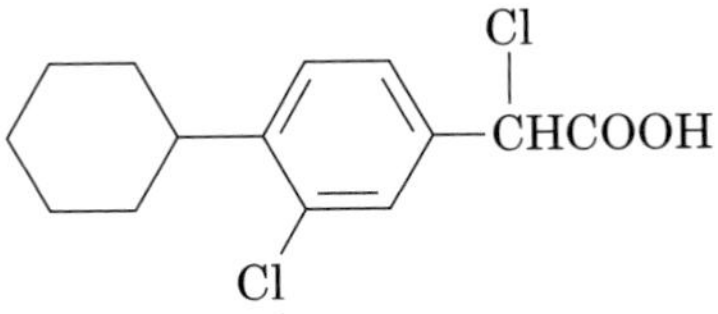

Fenclorac

20.38 The pK_a's of five *p*-substituted benzoic acids (YC_6H_4COOH) are given below. Rank the corresponding substituted benzenes (YC_6H_5) in order of their increasing reactivity toward electrophilic aromatic substitution. If benzoic acid has $pK_a = 4.19$, which of the substituents are activators and which are deactivators?

Substituent Y	pK_a of 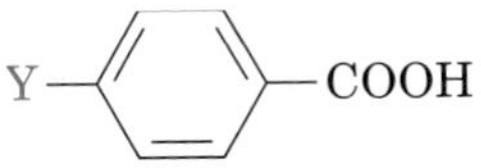
—Si(CH_3)_3	4.27
—CH=CHC≡N	4.03
—HgCH_3	4.10
—OSO_2CH_3	3.84
—PCl_2	3.59

20.39 How would you carry out the following transformations? More than one step is required in each case.

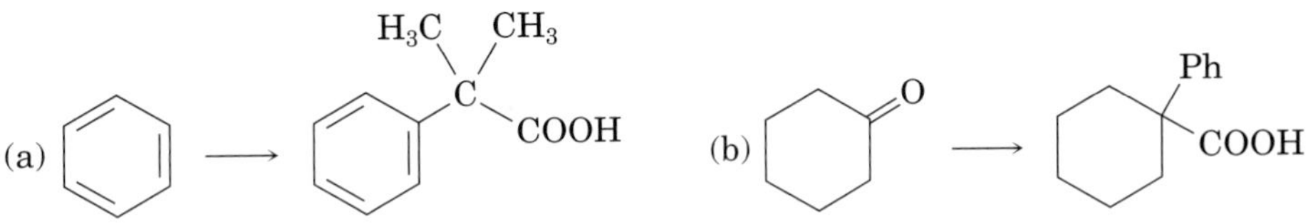

20.40 The following pK_a values have been measured. Explain why a hydroxyl group in the para position decreases the acidity while a hydroxyl group in the meta position increases the acidity.

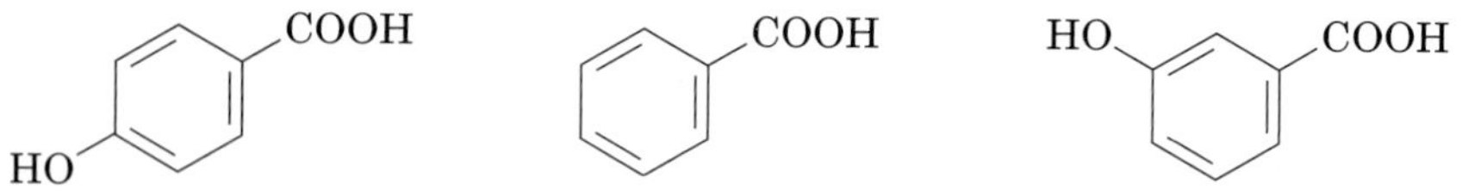

pK_a = 4.48   pK_a = 4.19   pK_a = 4.07

20.41 3-Methyl-2-hexenoic acid (mixture of *E* and *Z* isomers) has been identified as the substance responsible for the odor of human sweat. Synthesize the compound from starting materials having five or fewer carbons.

20.42 Identify the missing reagents a–f in the following scheme:

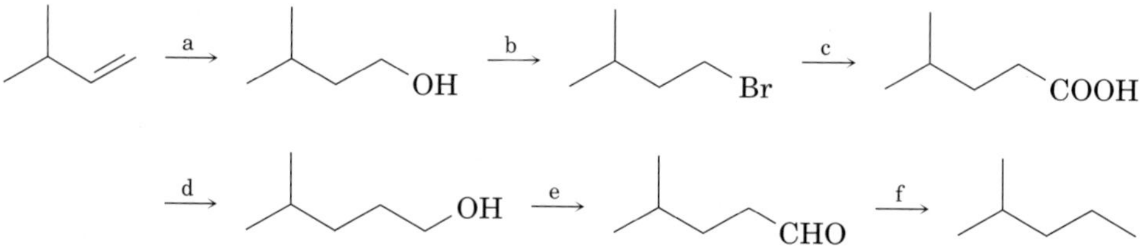

20.43 2-Bromo-6,6-dimethylcyclohexanone gives 2,2-dimethylcyclopentanecarboxylic acid on treatment with aqueous NaOH followed by acidification, a process called the *Favorskii reaction*. Propose a mechanism.

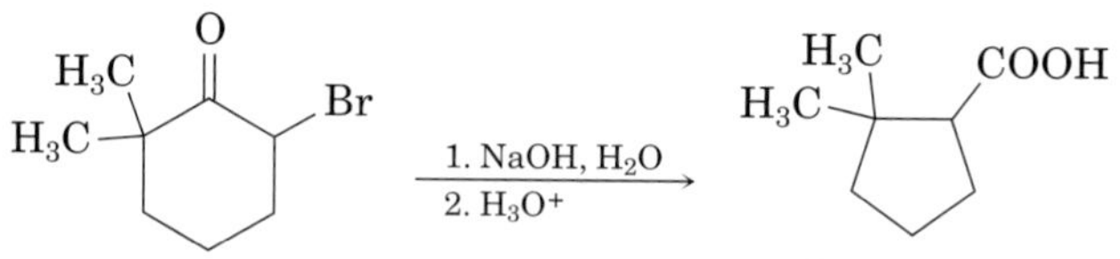

20.44 Propose a structure for an organic compound, $C_6H_{12}O_2$, that dissolves in dilute NaOH and shows the following ^1H NMR spectrum: 1.08 δ (9 H, singlet), 2.2 δ (2 H, singlet), and 11.2 δ (1 H, singlet).

20.45 What spectroscopic method could you use to distinguish among the following three isomeric acids? Tell what characteristic features you would expect for each acid.

$CH_3(CH_2)_3COOH$ $(CH_3)_2CHCH_2COOH$ $(CH_3)_3CCOOH$

Pentanoic acid   **3-Methylbutanoic acid**   **2,2-Dimethylpropanoic acid**

20.46 How would you use NMR (either ^{13}C or ^1H) to distinguish between the following isomeric pairs?
(a)

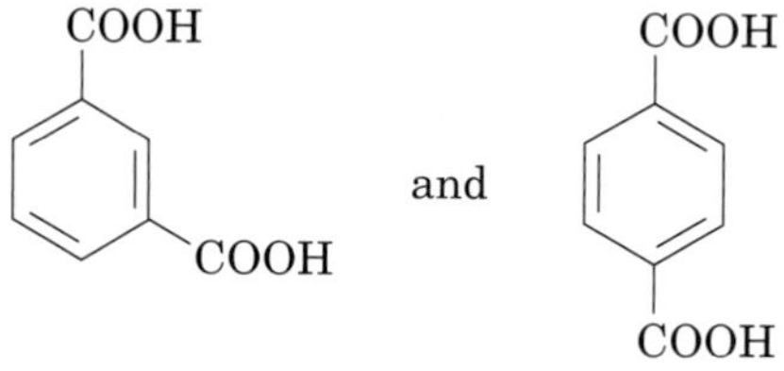

and

(b) $HOOCCH_2CH_2COOH$ and $CH_3CH(COOH)_2$
(c) $CH_3CH_2CH_2COOH$ and $HOCH_2CH_2CH_2CHO$

(d) $(CH_3)_2C{=}CHCH_2COOH$ and

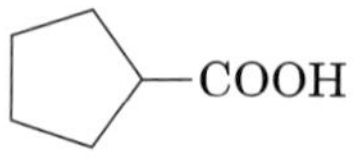

20.47 Compound A, $C_4H_8O_3$, has infrared absorptions at 1710 and 2500–3100 cm^{-1}, and has the ^{1}H NMR spectrum shown. Propose a structure for A.

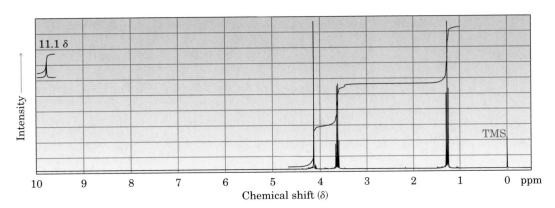

20.48 The two ^{1}H NMR spectra shown here belong to crotonic acid (*trans*-CH$_3$CH=CHCOOH) and methacrylic acid [H$_2$C=C(CH$_3$)COOH]. Which spectrum corresponds to which acid? Explain.

(a)

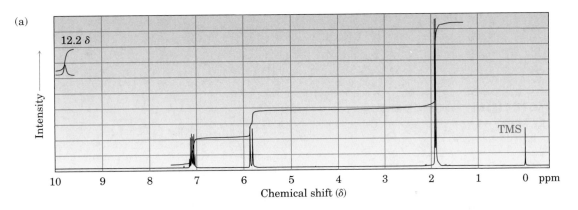

(b)

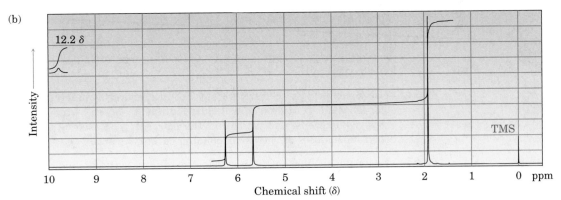

20.49 Propose structures for carboxylic acids that show the following peaks in their ^{13}C NMR spectra. Assume that the kinds of carbons (1°, 2°, 3°, or 4°) have been assigned by DEPT-NMR.
(a) $C_7H_{12}O_2$: 25.5 δ (2°), 25.9 δ (2°), 29.0 δ (2°), 43.1 δ (3°), 183.0 δ (4°)
(b) $C_8H_8O_2$: 21.4 δ (1°), 128.3 δ (4°), 129.0 δ (3°), 129.7 δ (3°), 143.1 δ (4°), 168.2 δ (4°)

A Look Ahead

• •

20.50 Carboxylic acids react with alcohols to yield esters:

$$RCO_2H + R'OH \longrightarrow RCO_2R'$$

Propose a mechanism for the following reaction. (See Section 21.3.)

20.51 Carboxylic acids that have a second carbonyl group two atoms away lose CO_2 (*decarboxylate*) through an intermediate enolate ion when treated with base. Write the mechanism of this decarboxylation reaction using curved arrows to show the electron flow in each step. (See Section 22.8.)

An enolate ion

Molecular Modeling

• •

20.52 Use SpartanView to compare electrostatic potential maps of acetic acid and thioacetic acid (CH_3COSH). Which contains a more negative carbonyl oxygen, and which contains a more positive hydrogen? Which molecule should have stronger intermolecular hydrogen bonds? Obtain the energies of acetic acid and thioacetic acid dimers, and calculate the energy required to break the hydrogen bonds in each. Are these energies consistent with the electrostatic potential maps?

20.53 Use SpartanView to compare geometries, bond-density surfaces, and electrostatic potential maps of pentanoic acid and sodium pentanoate. Why does H bond to only one oxygen in the acid while Na bonds to two oxygens in the salt?

20.54 Use SpartanView to compare electrostatic potential maps of butanoate anion, 2-chlorobutanoate anion, and 4-chlorobutanoate anion. Which anion has the most negative oxygens, and which has the least? Why? Based on these data, predict the relative acidities of the corresponding carboxylic acids.

20.55 Use SpartanView to compare electrostatic potential maps of benzoate anion and 4-nitrobenzoate anion. Is the nitro group electron-donating or electron-withdrawing? Next, compare electrostatic potential maps of phenoxide anion and 4-nitrophenoxide anion. Which is more strongly affected by the nitro group, benzoate anion or phenoxide anion? Explain, using resonance structures.

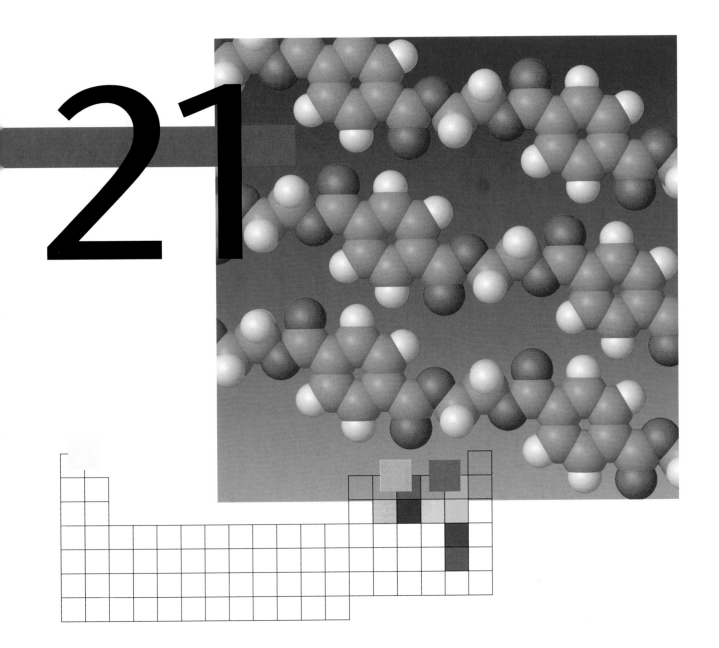

Carboxylic Acid Derivatives and Nucleophilic Acyl Substitution Reactions

Closely related to the carboxylic acids discussed in the previous chapter are **carboxylic acid derivatives**, compounds in which the acyl group is bonded to an electronegative atom or substituent –Y that can act as a leaving group in a substitution reaction. Many kinds of acid derivatives are known, but we'll be concerned only with four of the more common ones: *acid halides, acid anhydrides, esters,* and *amides.* Also in this chapter, we'll discuss *nitriles,* a class of compounds closely related to carboxylic acids.

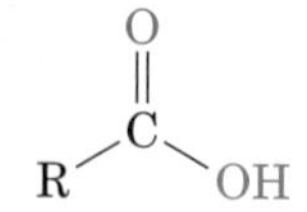

Carboxylic acid 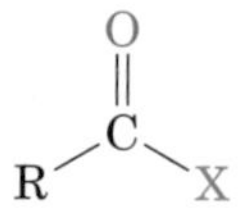   **Acid halide (X = F, Cl, Br, I)** 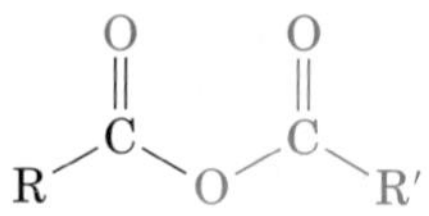   **Acid anhydride**

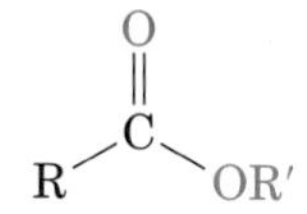

Ester 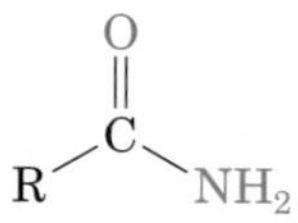   **Amide**   **Nitrile**

The chemistry of all acid derivatives is similar and is dominated by a single reaction—the **nucleophilic acyl substitution reaction** that we saw briefly in "A Preview of Carbonyl Compounds":

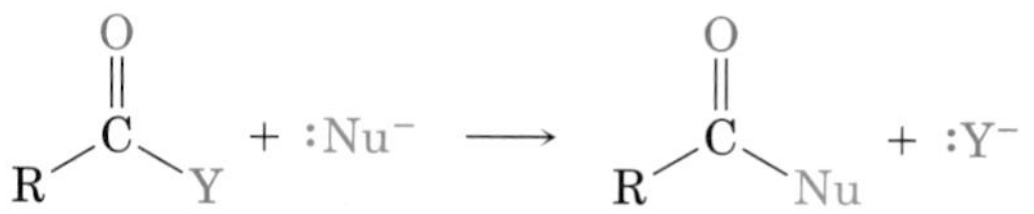

Let's first learn more about acid derivatives and then explore the chemistry of acyl substitution reactions.

21.1 Naming Carboxylic Acid Derivatives and Nitriles

Acid Halides: RCOX

Acid halides are named by identifying first the acyl group and then the halide. The acyl group name is derived from the carboxylic acid name by replacing the *-ic acid* ending with *-yl* or the *-carboxylic acid* ending with *-carbonyl.* For example:

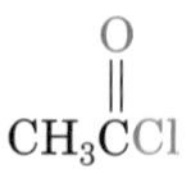

Acetyl chloride
(from acetic acid)

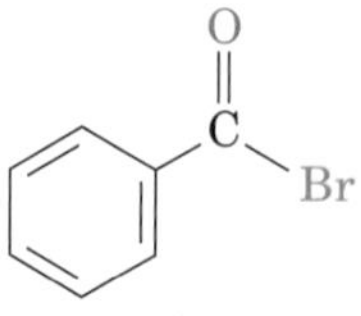

Benzoyl bromide
(from benzoic acid)

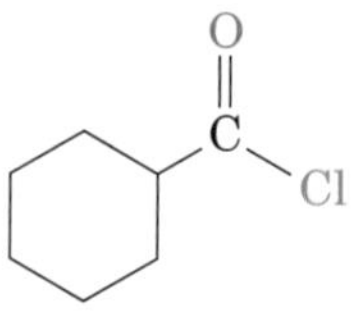

Cyclohexanecarbonyl chloride
(from cyclohexanecarboxylic acid)

Acid Anhydrides: RCO$_2$COR'

Symmetrical anhydrides of unsubstituted monocarboxylic acids and cyclic anhydrides of dicarboxylic acids are named by replacing the word *acid* with *anhydride*:

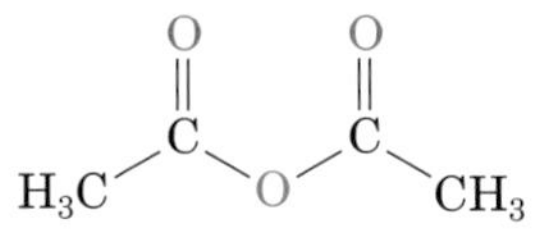

Acetic anhydride 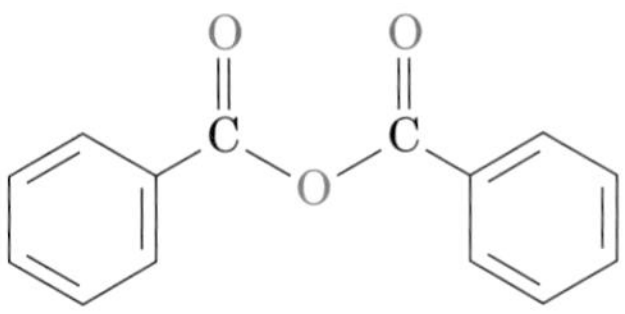   **Benzoic anhydride** 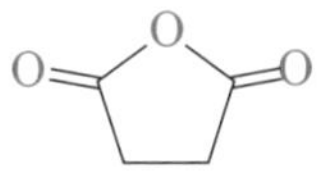   **Succinic anhydride**

Anhydrides derived from substituted monocarboxylic acids are named by adding the prefix *bis-* (meaning *two*) to the acid name:

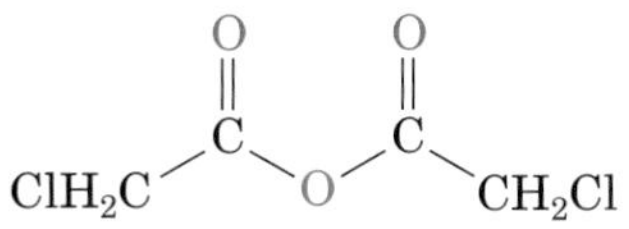

Bis(chloroacetic) anhydride

Unsymmetrical anhydrides—those prepared from two different carboxylic acids—are named by citing the two acids alphabetically:

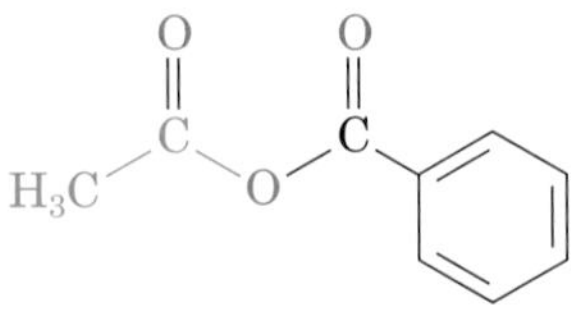

Acetic benzoic anhydride

Amides: RCONH$_2$

Amides with an unsubstituted –NH$_2$ group are named by replacing the *-oic acid* or *-ic acid* ending with *-amide,* or by replacing the *-carboxylic acid* ending with *-carboxamide.* For example:

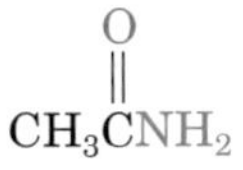

$$CH_3CNH_2$$ $$CH_3(CH_2)_4CNH_2$$

Acetamide   **Hexanamide**   **Cyclopentanecarboxamide**
(from acetic acid)   **(from hexanoic acid)**   **(from cyclopentanecarboxylic acid)**

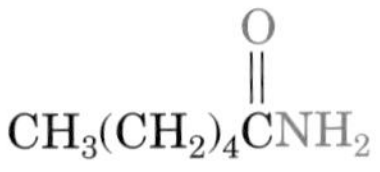

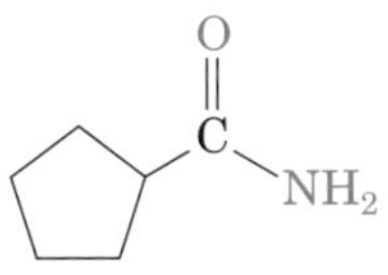

If the nitrogen atom is further substituted, the compound is named by first identifying the substituent groups and then the parent amide. The substituents are preceded by the letter *N* to identify them as being directly attached to nitrogen.

$$CH_3CH_2\overset{\overset{\displaystyle O}{\|}}{C}NHCH_3$$

N-Methylpropanamide

N,N-Diethylcyclohexanecarboxamide

Esters: RCO₂R′

Esters are named by first identifying the alkyl group attached to oxygen and then the carboxylic acid, with the *-ic acid* ending replaced by *-ate*:

$$CH_3\overset{\overset{\displaystyle O}{\|}}{C}OCH_2CH_3$$

**Ethyl acetate
(the ethyl ester of
acetic acid)**

$$CH_3O\overset{\overset{\displaystyle O}{\|}}{C}CH_2\overset{\overset{\displaystyle O}{\|}}{C}OCH_3$$

**Dimethyl malonate
(the dimethyl ester of
malonic acid)**

tert-**Butylcyclohexanecarboxylate
(the *tert*-butyl ester of
cyclohexanecarboxylic acid)**

Nitriles: RC≡N

Compounds containing the –C≡N functional group are called **nitriles**. Simple open-chain nitriles are named by adding *-nitrile* as a suffix to the alkane name, with the nitrile carbon numbered C1:

$$\underset{5\quad4\quad3\quad2\quad1}{CH_3\overset{\overset{\displaystyle CH_3}{|}}{C}HCH_2CH_2CN}\qquad \textbf{4-Methylpentanenitrile}$$

More complex nitriles are named as derivatives of carboxylic acids by replacing the *-ic acid* or *-oic acid* ending with *-onitrile*, or by replacing the *-carboxylic acid* ending with *-carbonitrile*. The nitrile carbon atom is attached to C1 but is not itself numbered in this system.

$$CH_3C\equiv N$$

**Acetonitrile
(from acetic acid)**

**Benzonitrile
(from benzoic acid)**

**2,2-Dimethylcyclohexanecarbonitrile
(from 2,2-dimethylcyclohexane-
carboxylic acid)**

A summary of nomenclature rules for carboxylic acid derivatives is given in Table 21.1.

TABLE 21.1 Nomenclature of Carboxylic Acid Derivatives and Nitriles

Functional group	Structure	Name ending
Carboxylic acid	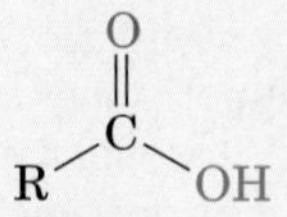 R—C(=O)—OH	-ic acid (-carboxylic acid)
Acid halide	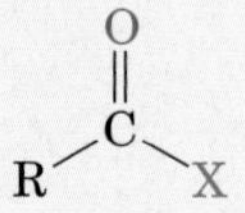 R—C(=O)—X	-yl halide (-carbonyl halide)
Acid anhydride	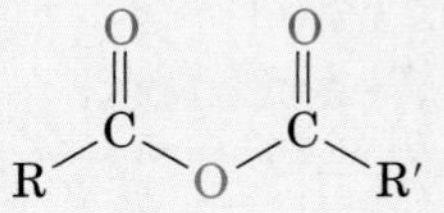 R—C(=O)—O—C(=O)—R′	anhydride
Amide	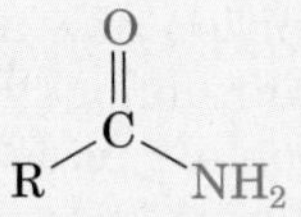 R—C(=O)—NH₂	-amide (-carboxamide)
Ester	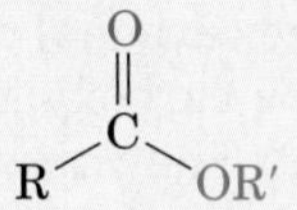 R—C(=O)—OR′	-ate (-carboxylate)
Nitrile	R—C≡N	-onitrile (-carbonitrile)

• •

Problem 21.1   Give IUPAC names for the following substances:

(a)

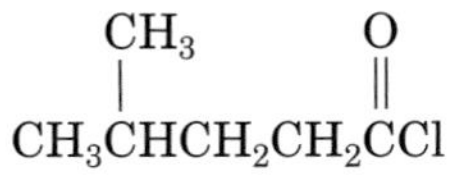

$$CH_3$$
$$|$$
$$CH_3CHCH_2CH_2CCl$$ (with C=O)

(b)

cyclohexyl—CH₂C(=O)NH₂

(c)

$$CH_3$$
$$|$$
$$CH_3CH_2CHCN$$

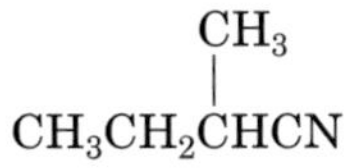

(d)

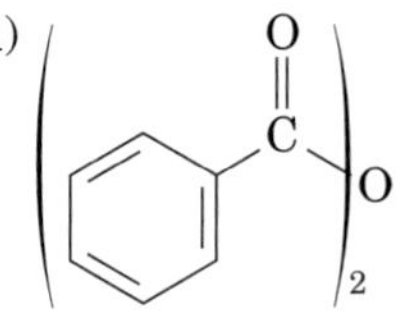

(phenyl—C(=O)—O)₂

(e)

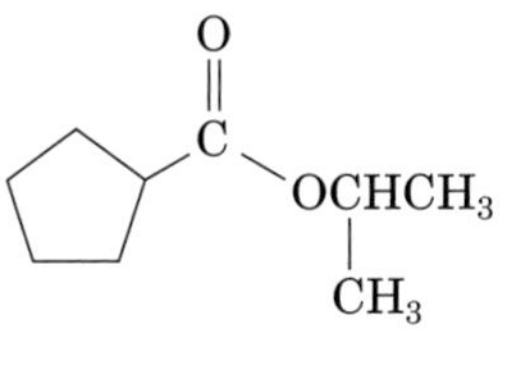

cyclopentyl—C(=O)—OCHCH₃
                              |
                            CH₃

(f)

cyclopentyl—O—C(=O)—CHCH₃
                              |
                            CH₃

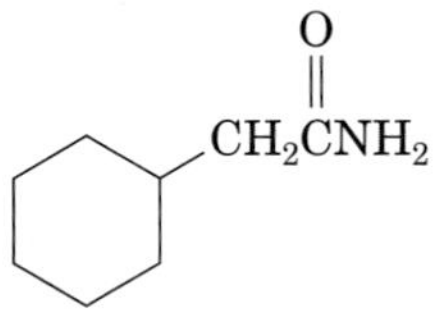

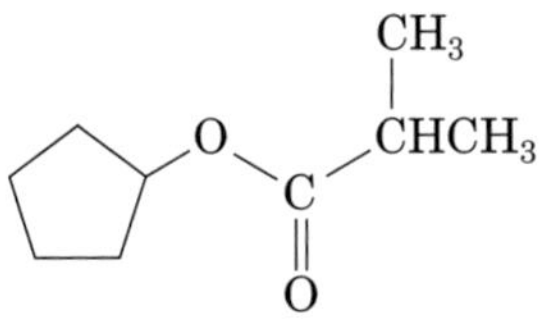

(g)

$$H_2C=CHCH_2CH_2\overset{\overset{\displaystyle O}{\|}}{C}NH_2$$

(h)

$$CH_3CH_2\overset{\overset{\displaystyle CN}{|}}{C}HCH_2CH_3$$

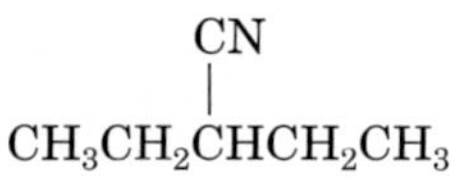

(i)

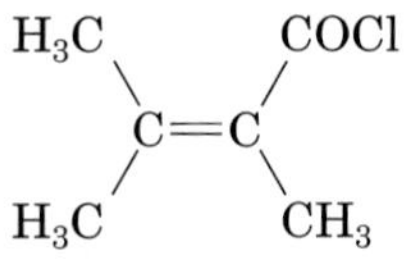

Problem 21.2 Draw structures corresponding to the following names:
(a) 2-Pentenenitrile
(b) *N*-Ethyl-*N*-methylbutanamide
(c) 2,4-Dimethylpentanoyl chloride
(d) Methyl 1-methylcyclohexanecarboxylate
(e) Ethyl 3-oxopentanoate
(f) Bis(*p*-bromobenzoic) anhydride
(g) Formic propanoic anhydride
(h) *cis*-2-Methylcyclopentanecarbonyl bromide

21.2 Nucleophilic Acyl Substitution Reactions

Biological Connection ▶

The addition of a nucleophile to a polar C=O bond is the key step in three of the four major carbonyl-group reactions. We saw in Chapter 19 that when a nucleophile adds to an aldehyde or ketone, the initially formed tetrahedral intermediate either can be protonated to yield an alcohol or can eliminate the carbonyl oxygen, leading to a new C=Nu bond. When a nucleophile adds to a carboxylic acid derivative, however, a different reaction course is followed. The initially formed tetrahedral intermediate eliminates one of the two substituents originally bonded to the carbonyl carbon, leading to a net nucleophilic acyl substitution (Figure 21.1).

FIGURE 21.1 ▼

General mechanism of a nucleophilic acyl substitution reaction.

refer to **Mechanisms & Movies**

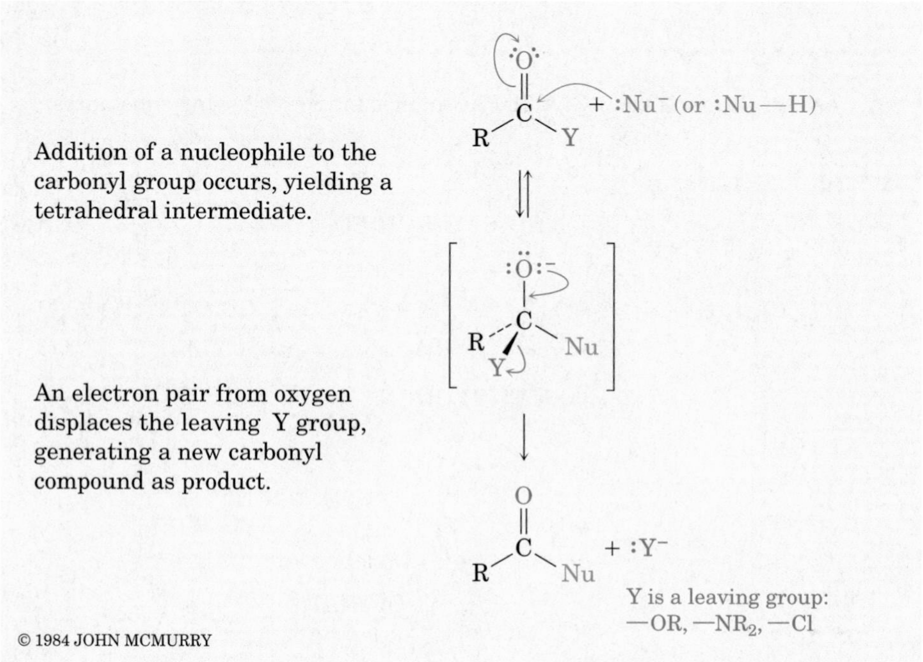

Addition of a nucleophile to the carbonyl group occurs, yielding a tetrahedral intermediate.

An electron pair from oxygen displaces the leaving Y group, generating a new carbonyl compound as product.

© 1984 JOHN MCMURRY

Y is a leaving group: —OR, —NR₂, —Cl

The difference in behavior between aldehydes/ketones and carboxylic acid derivatives is a consequence of structure. Carboxylic acid derivatives have an acyl carbon bonded to a potential leaving group –Y. As soon as the tetrahedral intermediate is formed, the leaving group is expelled to generate a new carbonyl compound. Aldehydes and ketones have no such leaving group, however, and therefore don't undergo substitution.

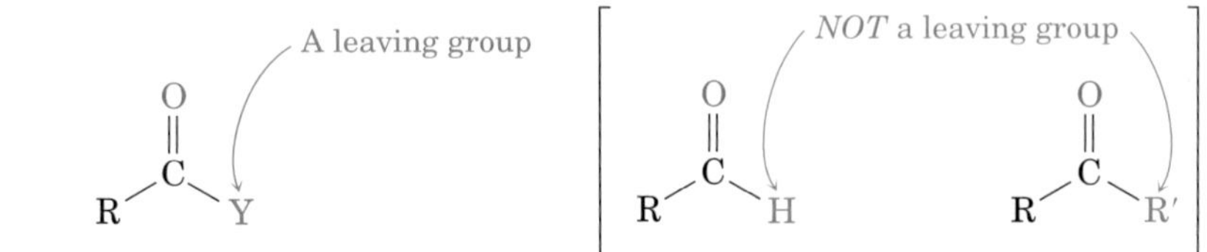

A carboxylic acid derivative   An aldehyde   A ketone

As shown in Figure 21.1, the net effect of the addition/elimination sequence is a substitution by the attacking nucleophile for the –Y group originally bonded to the acyl carbon. Thus, the overall reaction is superficially similar to the kind of nucleophilic substitution that occurs during an S_N2 reaction (Section 11.4), but the *mechanisms* of the two reactions are completely different. An S_N2 reaction occurs in a single step by back-side displacement of the leaving group; a nucleophilic acyl substitution takes place in two steps and involves a tetrahedral intermediate.

Problem 21.3 Propose a mechanism for the following nucleophilic acyl substitution reaction, using curved arrows to indicate the electron flow in each step:

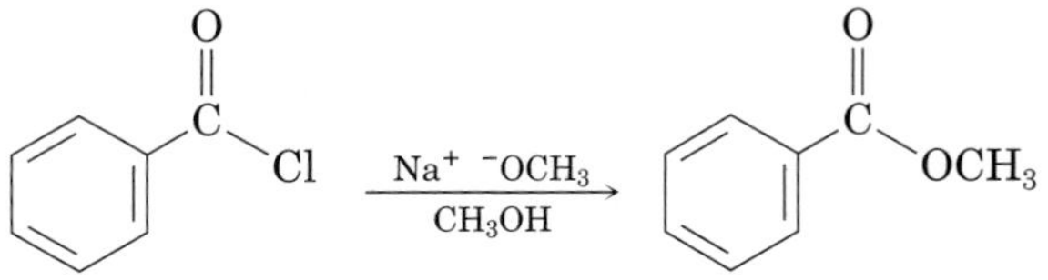

Relative Reactivity of Carboxylic Acid Derivatives

Both the addition step and the elimination step can affect the overall rate of a nucleophilic acyl substitution reaction, but the first step is generally rate-limiting. Thus, any factor that makes the carbonyl group more easily attacked by a nucleophile favors the reaction.

Steric and electronic factors are both important in determining reactivity. Sterically, we find within a series of similar acid derivatives that unhindered, accessible carbonyl groups react with nucleophiles more readily than do sterically hindered groups. For example, acetyl chloride, CH_3COCl, is much more reactive than 2,2-dimethylpropanoyl chloride, $(CH_3)_3CCOCl$. The reactivity order is:

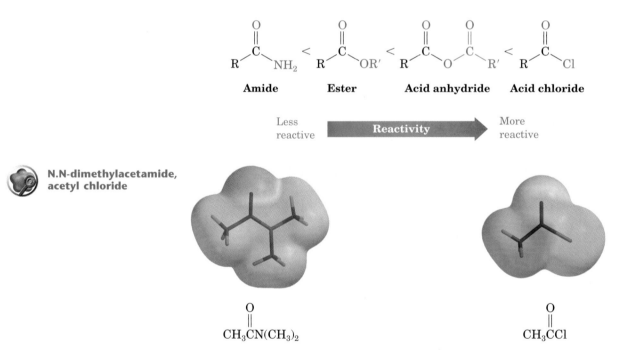

Electronically, we find that strongly polarized acid derivatives react more readily than less polar ones. Thus, acid chlorides are more reactive than esters, which are more reactive than amides, because the electronegative chlorine polarizes the carbonyl group more strongly than does an alkoxy group or an amino group. These polarity differences can be seen in electrostatic potential maps such as those of *N,N*-dimethylacetamide and acetyl chloride shown below. The carbonyl carbon is less positive and less reactive in the amide than in the acid chloride.

N.N-dimethylacetamide, acetyl chloride

The way in which various substituents affect the polarization of a carbonyl group is similar to the way they affect the reactivity of an aromatic ring toward electrophilic substitution (Section 16.6). A chlorine substituent, for example, inductively *withdraws* electrons from an acyl group in the same way that it withdraws electrons from an aromatic ring. Similarly, amino and methoxyl substituents *donate* electrons to acyl groups by resonance in the same way that they donate electrons to aromatic rings.

An important consequence of the observed reactivity order is that *it is usually possible to transform a more reactive acid derivative into a less reactive one*. As we'll see in the next few sections, acid chlorides can be directly converted into anhydrides, esters, and amides, but amides can't be directly converted into esters, anhydrides, or acid chlorides. Remembering the reactivity order is therefore a way to keep track of a large number of reactions. Figure 21.2 shows the kinds of transformations that can be carried out.

FIGURE 21.2 ▼

Interconversions of carboxylic acid derivatives. A more reactive acid derivative can be converted into a less reactive one, but not vice versa.

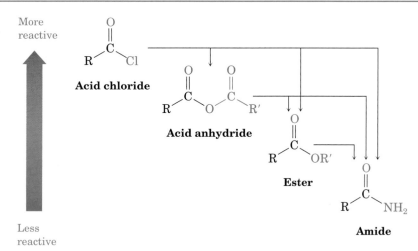

Another consequence of the reactivity differences among carboxylic acid derivatives is that only esters and amides are commonly found in nature. Acid halides and acid anhydrides undergo nucleophilic attack by water so rapidly that they can't exist in living organisms. Esters and amides, however, are stable enough to occur widely. All protein molecules, for example, contain amide functional groups.

Problem 21.4 Rank the compounds in each of the following sets in order of their expected reactivity toward nucleophilic acyl substitution:

(a) $CH_3\overset{O}{\overset{\|}{C}}Cl$, $CH_3\overset{O}{\overset{\|}{C}}OCH_3$, $CH_3\overset{O}{\overset{\|}{C}}NH_2$

(b) $CH_3\overset{O}{\overset{\|}{C}}OCH_3$, $CH_3\overset{O}{\overset{\|}{C}}OCH_2CCl_3$, $CH_3\overset{O}{\overset{\|}{C}}OCH(CF_3)_2$

Problem 21.5 Methyl trifluoroacetate, CF_3COOCH_3, is more reactive than methyl acetate, CH_3COOCH_3, in nucleophilic acyl substitution reactions. Explain.

Kinds of Nucleophilic Acyl Substitutions

In studying the chemistry of acid derivatives in the next few sections, we'll find that there are striking similarities among the various types of compounds. We'll be concerned largely with the reactions of just a few nucleophiles and will see that the same kinds of reactions keep occurring (Figure 21.3).

- **Hydrolysis:** Reaction with water to yield a carboxylic acid
- **Alcoholysis:** Reaction with an alcohol to yield an ester
- **Aminolysis:** Reaction with ammonia or an amine to yield an amide
- **Reduction:** Reaction with a hydride reducing agent to yield an aldehyde or an alcohol
- **Grignard reaction:** Reaction with an organometallic reagent to yield a ketone or an alcohol

FIGURE 21.3 ▼

Some general reactions of carboxylic acid derivatives.

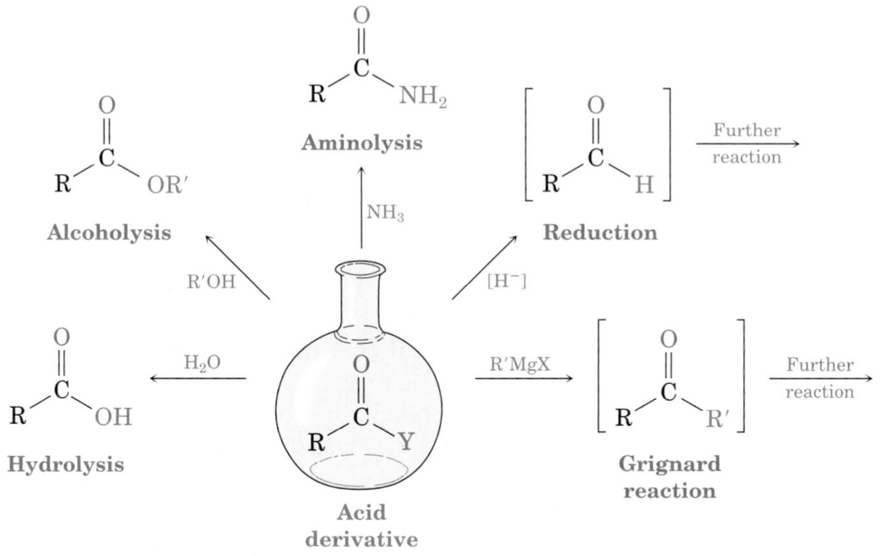

・ ・

Practice Problem 21.1   Predict the product of the following nucleophilic acyl substitution reaction of benzoyl chloride with 2-propanol:

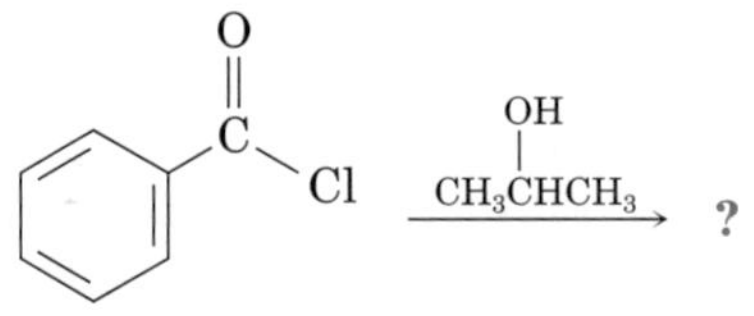

Benzoyl chloride

Strategy   A nucleophilic acyl substitution involves the substitution of a nucleophile for a leaving group in a carboxylic acid derivative. Identify the leaving group (Cl⁻ in the case of an acid chloride) and the nucleophile (an alcohol in this case), and replace one by the other. The product is isopropyl benzoate.

Solution

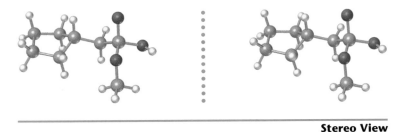

Benzoyl chloride   **Isopropyl benzoate**

● ●

Problem 21.6   Predict the product of each of the following nucleophilic acyl substitution reactions:

(a) $H_3C-\overset{\overset{\displaystyle O}{\|}}{C}-OCH_3 \xrightarrow[H_2O]{NaOH}$?

(b) $H_3C-\overset{\overset{\displaystyle O}{\|}}{C}-Cl \xrightarrow{NH_3}$?

(c) $H_3C-\overset{\overset{\displaystyle O}{\|}}{C}-O-\overset{\overset{\displaystyle O}{\|}}{C}-CH_3 \xrightarrow[CH_3OH]{Na^+ \ ^-OCH_3}$?

Problem 21.7   The following structure represents a tetrahedral alkoxide ion intermediate formed by addition of a nucleophile to a carboxylic acid derivative. Identify the nucleophile, the leaving group, the starting acid derivative, and the ultimate product.

Stereo View

● ●

21.3 Nucleophilic Acyl Substitution Reactions of Carboxylic Acids

Among the most important reactions of carboxylic acids are those that convert the carboxyl group into an acid derivative by a nucleophilic acyl substitution. Acid chlorides, anhydrides, esters, and amides can all be prepared from carboxylic acids (Figure 21.4).

FIGURE 21.4 ▼

Some nucleophilic acyl substitution reactions of carboxylic acids.

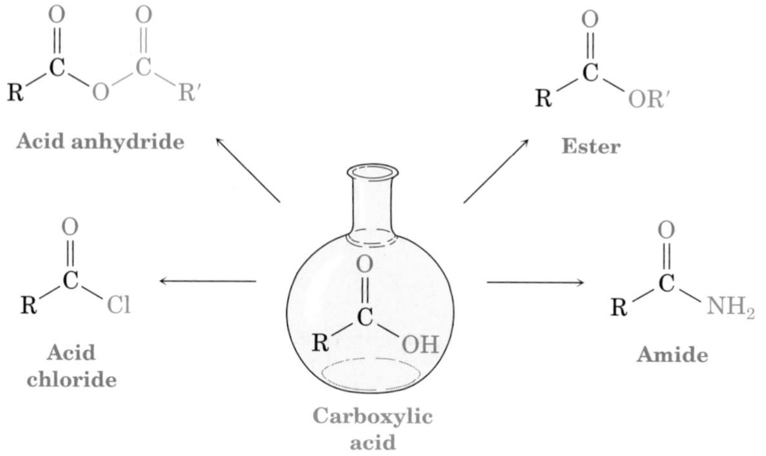

Conversion of Carboxylic Acids into Acid Chlorides (RCO₂H ⟶ RCOCl)

Carboxylic acids are converted into acid chlorides by treatment with thionyl chloride (SOCl₂):

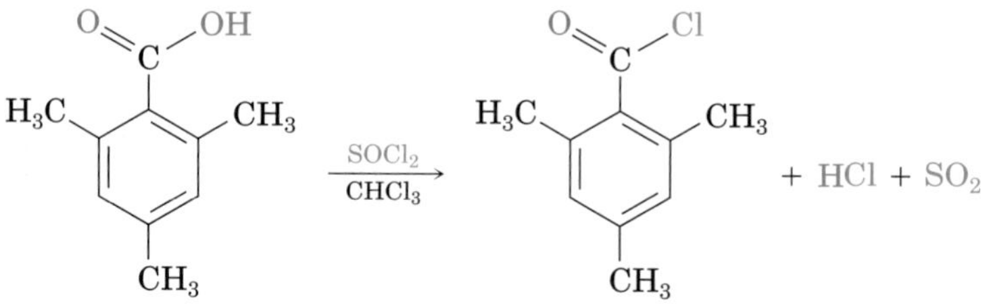

2,4,6-Trimethylbenzoic acid   2,4,6-Trimethylbenzoyl chloride (90%)

The reaction occurs by a nucleophilic acyl substitution pathway in which the carboxylic acid is first converted into a reactive *chlorosulfite* intermediate, which is then attacked by a nucleophilic chloride ion.

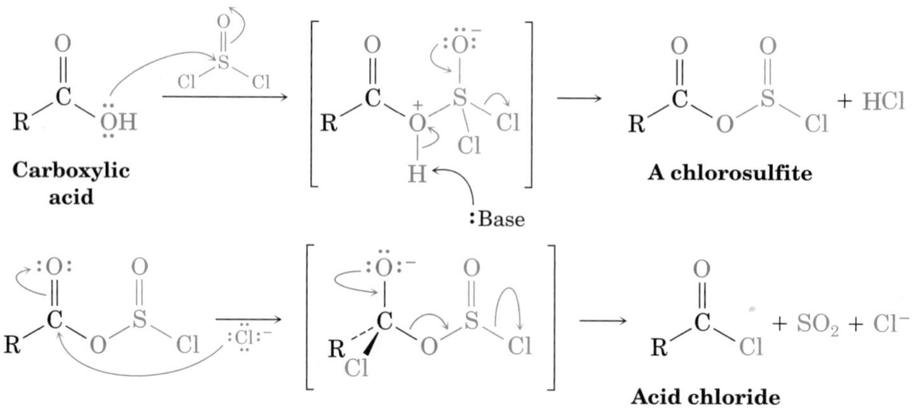

Conversion of Carboxylic Acids into Acid Anhydrides (RCO₂H ⟶ RCO₂COR′)

Acid anhydrides are derived from two molecules of carboxylic acid by heating to remove 1 equivalent of water. Acyclic anhydrides are difficult to prepare directly from the corresponding acids, however, and only acetic anhydride is commonly used.

Acetic anhydride

Cyclic anhydrides with five- or six-membered rings are obtained by high-temperature dehydration of the diacids.

Succinic acid   **Succinic anhydride**

Conversion of Carboxylic Acids into Esters (RCO₂H ⟶ RCO₂R′)

Perhaps the most useful reaction of carboxylic acids is their conversion into esters. There are many methods for accomplishing the transformation, including the S_N2 reaction of a carboxylate anion with a primary alkyl halide that we saw in Section 11.5.

$$\text{CH}_3\text{CH}_2\text{CH}_2\text{CO:}^- \text{ Na}^+ + \text{CH}_3-\text{I} \xrightarrow[\text{reaction}]{S_N2} \text{CH}_3\text{CH}_2\text{CH}_2\text{COCH}_3 + \text{NaI}$$

Sodium butanoate   **Methyl butanoate, an ester (97%)**

Biological Connection ▶

Esters can also be synthesized by a nucleophilic acyl substitution reaction of a carboxylic acid with an alcohol. Fischer and Speier discovered in 1895 that esters result simply from heating a carboxylic acid in alcohol solution containing a small amount of strong acid catalyst. Yields are good in this **Fischer esterification reaction**, but the need to use excess alcohol as solvent limits the method to the synthesis of methyl, ethyl, and propyl esters.

$$+ \text{ CH}_3\text{CH}_2\text{OH} \xrightarrow[\text{HCl}]{\text{Ethanol}} + \text{ H}_2\text{O}$$

Mandelic acid   **Ethyl mandelate (86%)**

The Fischer esterification reaction is a nucleophilic acyl substitution reaction carried out under acidic conditions, as shown in Figure 21.5. Carboxylic acids are not reactive enough to be attacked by neutral alcohols, but they can be made much more reactive in the presence of a strong acid such as HCl or H_2SO_4. The mineral acid protonates the carbonyl-group oxygen atom, thereby giving the carboxylic acid a positive charge and rendering it much more reactive toward nucleophilic attack by alcohol. Subsequent loss of water from the tetrahedral intermediate yields the ester product.

FIGURE 21.5 ▼

Mechanism of Fischer esterification. The reaction is an acid-catalyzed, nucleophilic acyl substitution of a carboxylic acid.

refer to
Mechanisms
& Movies

Protonation of the carbonyl oxygen activates the carboxylic acid . . .

. . . toward nucleophilic attack by alcohol, yielding a tetrahedral intermediate.

Transfer of a proton from one oxygen atom to another yields a second tetrahedral intermediate and converts the OH group into a good leaving group.

Loss of a proton and expulsion of H_2O regenerates the acid catalyst and gives the ester product.

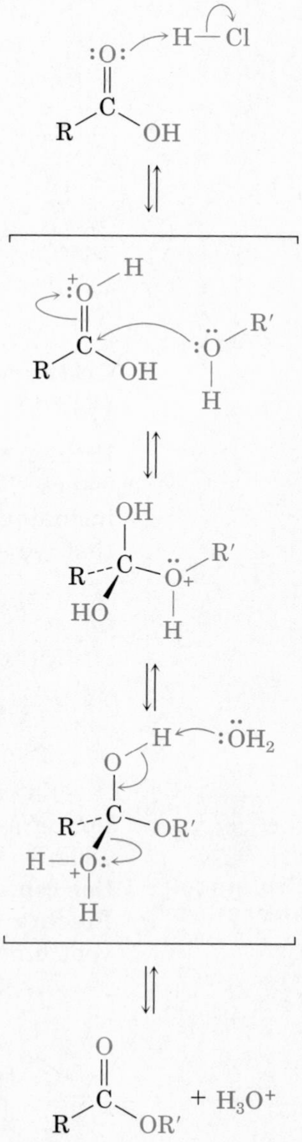

© 1984 JOHN MCMURRY

The net effect of Fischer esterification is substitution of an –OH group by OR'. All steps are reversible, and the reaction can be driven in either direction by choice of reaction conditions. Ester formation is favored when a large excess of alcohol is used as solvent, but carboxylic acid formation is favored when a large excess of water is present.

One of the best pieces of evidence in support of the mechanism shown in Figure 21.5 comes from isotope-labeling experiments. When ^{18}O-labeled methanol reacts with benzoic acid, the methyl benzoate produced is found to be ^{18}O-labeled, but the water produced is unlabeled. Thus, it is the C–OH bond of the carboxylic acid that is broken during the reaction rather than the CO–H bond, and the RO–H bond of the alcohol that is broken rather than the R–OH bond.

These bonds are broken

$$\text{(benzoic acid)–OH} + CH_3\overset{*}{O}\text{—H} \xrightarrow[\text{catalyst}]{HCl} \text{(benzoate)–}\overset{*}{O}CH_3 + HOH$$

Problem 21.8 Show how you might prepare the following esters:
(a) Butyl acetate   (b) Methyl butanoate

Problem 21.9 If 5-hydroxypentanoic acid is treated with acid catalyst, an intramolecular esterification reaction occurs. What is the structure of the product? (*Intramolecular* means within the same molecule.)

Conversion of Carboxylic Acids into Amides (RCO₂H ⟶ RCONH₂)

Biological Connection ▶

Amides are difficult to prepare by direct reaction of carboxylic acids with amines because amines are bases that convert acidic carboxyl groups into their carboxylate anions. Since the carboxylate anion has a negative charge, it is no longer likely to be attacked by a nucleophile. We'll see a better method for making amides from acids in Section 26.10 in connection with the synthesis of proteins from amino acids.

$$\underset{R}{\overset{O}{\parallel}}\!\!-\!\!C\!\!-\!\!OH + :NH_3 \rightleftharpoons \underset{R}{\overset{O}{\parallel}}\!\!-\!\!C\!\!-\!\!O^- \, NH_4^+$$

21.4 Chemistry of Acid Halides

Preparation of Acid Halides

Acid chlorides are prepared from carboxylic acids by reaction with thionyl chloride ($SOCl_2$), as we saw in the previous section. Reaction of a carboxylic acid with phosphorus tribromide (PBr_3) yields the acid bromide.

Reactions of Acid Halides

Acid halides are among the most reactive of carboxylic acid derivatives and can be converted into many other kinds of compounds. For example, we've already seen the value of acid chlorides in preparing aromatic alkyl ketones by the Friedel–Crafts acylation reaction (Section 16.4).

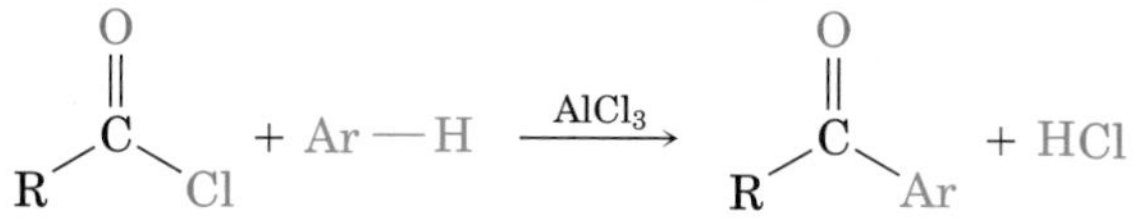

Most acid halide reactions occur by a nucleophilic acyl substitution mechanism. As shown in Figure 21.6, the halogen can be replaced by –OH to yield an acid, by –OR to yield an ester, or by –NH₂ to yield an amide. In addition, the reduction of an acid halide yields a primary alcohol, and reaction with a Grignard reagent yields a tertiary alcohol. Although the reactions we'll be discussing in this section are illustrated only for acid chlorides, they also occur with other acid halides.

FIGURE 21.6 ▼

Some nucleophilic acyl substitution reactions of acid chlorides.

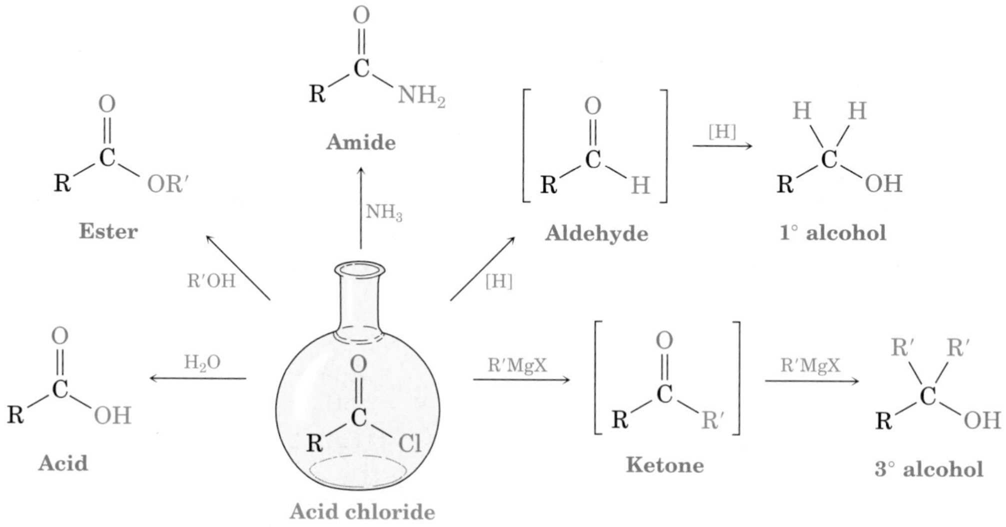

Hydrolysis: Conversion of Acid Halides into Acids (RCOX ⟶ RCO₂H) Acid chlorides react with water to yield carboxylic acids. This hydrolysis reaction is a typical nucleophilic acyl substitution process and is initiated by attack of water on the acid chloride carbonyl group. The tetra-hedral intermediate undergoes elimination of Cl⁻ and loss of H⁺ to give the product carboxylic acid plus HCl.

An acid chloride   **A carboxylic acid**

Since HCl is generated during the hydrolysis, the reaction is often carried out in the presence of a base such as pyridine or NaOH to remove the HCl and prevent it from causing side reactions.

Alcoholysis: Conversion of Acid Halides into Esters (RCOX ⟶ RCO₂R′)

Acid chlorides react with alcohols to yield esters in a process analogous to their reaction with water to yield acids. As with hydrolysis, alcoholysis reactions are usually carried out in the presence of pyridine or NaOH to react with the HCl formed.

Benzoyl chloride   **Cyclohexanol**   **Cyclohexyl benzoate (97%)**

The reaction of an alcohol with an acid chloride is strongly affected by steric hindrance. Bulky groups on either partner slow down the reaction considerably, resulting in a reactivity order among alcohols of primary > secondary > tertiary. As a result, it's often possible to esterify an unhindered alcohol selectively in the presence of a more hindered one. This can be important in complex syntheses where it is sometimes necessary to distinguish between similar functional groups. For example,

Primary alcohol (less hindered and more reactive)

Secondary alcohol (more hindered and less reactive)

• •

Problem 21.10   How might you prepare the following esters using a nucleophilic acyl substitution reaction of an acid chloride?
(a) $CH_3CH_2COOCH_3$ (b) $CH_3COOCH_2CH_3$ (c) Ethyl benzoate

Problem 21.11   Which method would you choose if you wanted to prepare cyclohexyl benzoate—Fischer esterification or reaction of an acid chloride with an alcohol? Explain.

• •

Aminolysis: Conversion of Acid Halides into Amides (RCOX ⟶ RCONH₂)

Acid chlorides react rapidly with ammonia and amines to give amides in good yield. Both mono- and disubstituted amines can be used, but not trisubstituted amines (R_3N).

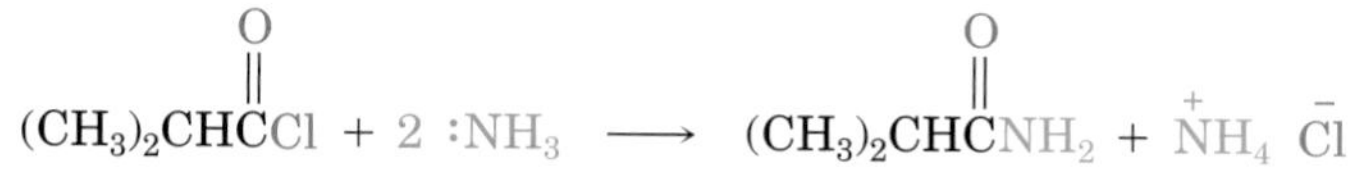

$$(CH_3)_2CHCCl + 2 :NH_3 \longrightarrow (CH_3)_2CHCNH_2 + \overset{+}{NH_4} \overset{-}{Cl}$$

2-Methylpropanoyl   **2-Methylpropanamide**
chloride   **(83%)**

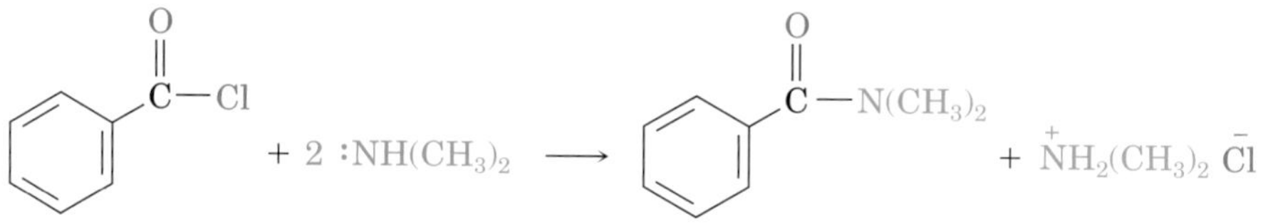

Benzoyl chloride   ***N,N*-Dimethylbenzamide**
                                          (92%)

Since HCl is formed during the reaction, 2 equivalents of the amine must be used. One equivalent reacts with the acid chloride, and 1 equivalent reacts with the HCl by-product to form an ammonium chloride salt. If, however, the amine component is valuable, amide synthesis is often carried out using 1 equivalent of the amine plus 1 equivalent of an inexpensive base such as NaOH. For example, the sedative trimetozine is prepared industrially by reaction of 3,4,5-trimethoxybenzoyl chloride with the amine morpholine in the presence of 1 equivalent of NaOH.

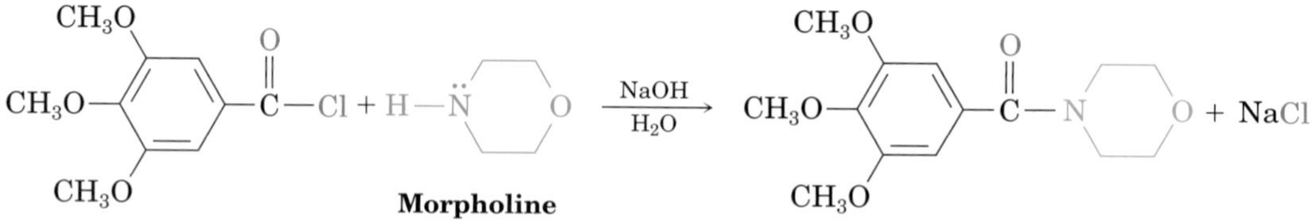

3,4,5-Trimethoxybenzoyl   **Morpholine**   **Trimetozine**
chloride   **(an amide)**

• •

Problem 21.12   Write the mechanism of the reaction shown above between 3,4,5-trimethoxybenzoyl chloride and morpholine to form trimetozine. Use curved arrows to show the electron flow in each step.

Problem 21.13 How could you prepare the following amides using an acid chloride and an amine or ammonia?
(a) $CH_3CH_2CONHCH_3$ (b) *N,N*-Diethylbenzamide
(c) Propanamide

. . . .

Reduction: Conversion of Acid Halides into Alcohols (RCOX ⟶ RCH$_2$OH)

Acid chlorides are reduced by LiAlH$_4$ to yield primary alcohols. The reaction is of little practical value, however, because the parent carboxylic acids are generally more readily available and can themselves be reduced by LiAlH$_4$ to yield alcohols.

Benzoyl chloride   Benzyl alcohol
(96%)

Reduction occurs via a typical nucleophilic acyl substitution mechanism in which a hydride ion (H:⁻) attacks the carbonyl group, yielding a tetrahedral intermediate that expels Cl⁻. The net effect is a substitution of –Cl by –H to yield an aldehyde, which is then immediately reduced by LiAlH$_4$ in a second step to yield the primary alcohol.

An acid   An aldehyde   A primary
chloride   (*NOT isolated*)   alcohol

The aldehyde intermediate can be isolated if a less powerful reducing agent such as lithium tri-*tert*-butoxyaluminum hydride is used in place of LiAlH$_4$. This reagent, which is obtained by reaction of LiAlH$_4$ with 3 equivalents of *tert*-butyl alcohol, is particularly effective for carrying out the partial reduction of acid chlorides to aldehydes (Section 19.2).

$$3\,(CH_3)_3COH + LiAlH_4 \longrightarrow Li^+\ {}^-AlH[OC(CH_3)_3]_3 + 3\,H_2$$

**Lithium tri-*tert*-butoxyaluminum
hydride**

p-Nitrobenzoyl chloride   *p*-Nitrobenzaldehyde
(81%)

Reaction of Acid Chlorides with Organometallic Reagents   Grignard reagents react with acid chlorides to yield tertiary alcohols in which two of the substituents are the same:

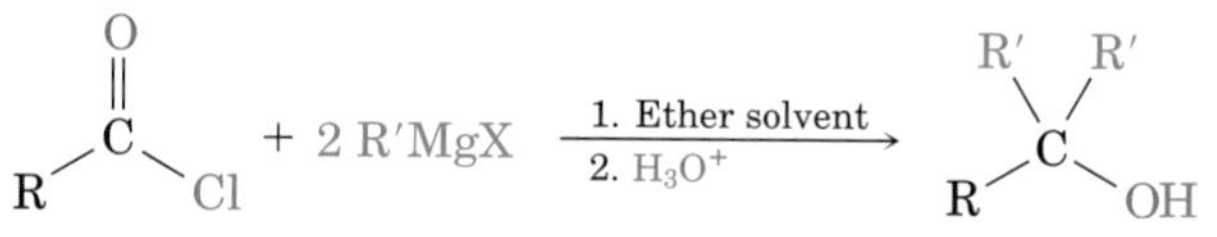

An acid chloride   **A 3° alcohol**

The mechanism of this Grignard reaction is similar to that of LiAlH$_4$ reduction. The first equivalent of Grignard reagent adds to the acid chloride, loss of Cl$^-$ from the tetrahedral intermediate yields a ketone, and a second equivalent of Grignard reagent immediately adds to the ketone to produce an alcohol.

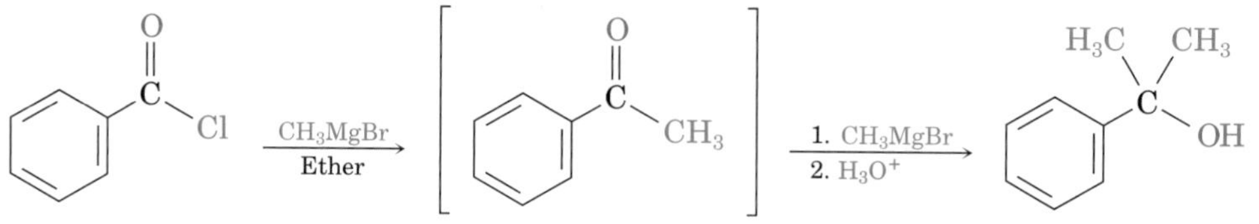

Benzoyl chloride   **Acetophenone**   **2-Phenyl-2-propanol**
                              (NOT isolated)   **(92%)**

The ketone intermediate formed during the Grignard reaction of an acid chloride can't usually be isolated because addition of the second equivalent of organomagnesium reagent occurs too rapidly. A ketone *can*, however, be isolated from the reaction of an acid chloride with a diorganocopper (Gilman) reagent (Section 19.2):

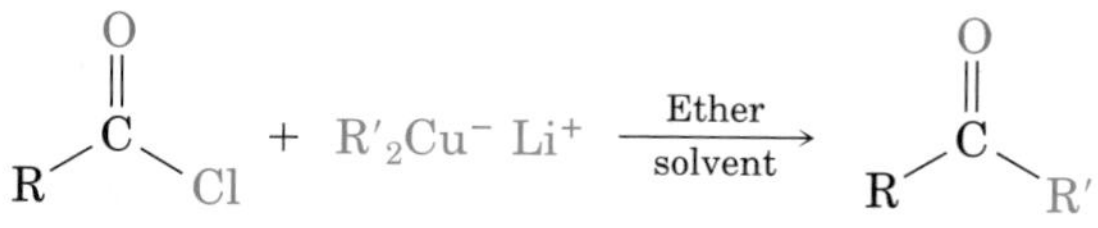

An acid   **A Gilman**   **A ketone**
chloride   **reagent**

The reaction occurs by initial nucleophilic acyl substitution on the acid chloride by the diorganocopper anion to yield an acyl diorganocopper intermediate, followed by loss of R'Cu and formation of the ketone: RCOCl + R'$_2$Cu$^-$ → RCOCuR'$_2$ → RCOR' + R'Cu. The reaction is generally carried out at −78°C in ether solution, and yields are often excellent. For example, manicone, a substance secreted by male ants to coordinate ant pairing and mating, has been synthesized by reaction of lithium diethylcopper with (*E*)-2,4-dimethyl-2-hexenoyl chloride:

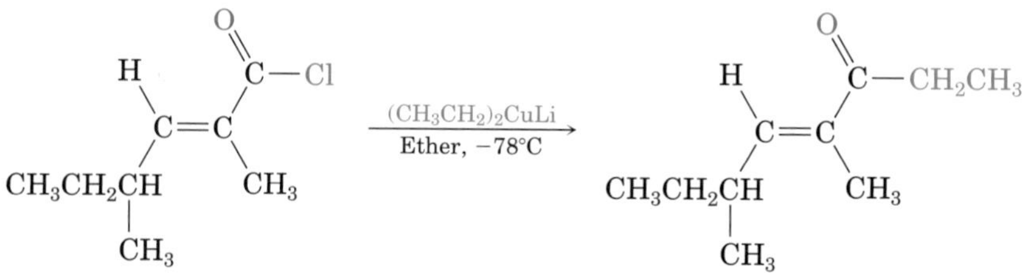

2,4-Dimethyl-2-hexenoyl   **Manicone (92%)**
chloride

Note that the diorganocopper reaction occurs only with acid chlorides. Carboxylic acids, esters, acid anhydrides, and amides do not react with diorganocopper reagents.

. . . .

Problem 21.14   How would you prepare the following ketones by reaction of an acid chloride with a lithium diorganocopper reagent?

(a) [structure: phenyl—C(=O)—CH(CH₃)₂]   (b) H_2C=$CHCCH_2CH_2CH_3$ with carbonyl O above the C

. . . .

21.5 Chemistry of Acid Anhydrides

Preparation of Acid Anhydrides

The most general method for preparing an acid anhydride is by nucleophilic acyl substitution reaction of an acid chloride with a carboxylate anion. Both symmetrical and unsymmetrical acid anhydrides can be prepared in this way.

Sodium formate   **Acetyl chloride**   **Acetic formic anhydride (64%)**

Reactions of Acid Anhydrides

The chemistry of acid anhydrides is similar to that of acid chlorides. Although anhydrides react more slowly than acid chlorides, the *kinds* of reactions the two groups undergo are the same. Thus, acid anhydrides react with water to form acids, with alcohols to form esters, with amines to form amides, and with LiAlH₄ to form primary alcohols (Figure 21.7, p. 864).

Acetic anhydride is often used to prepare acetate esters from alcohols and *N*-substituted acetamides from amines. For example, aspirin (acetylsalicylic acid) is prepared commercially by the acetylation of *o*-hydroxybenzoic acid (salicylic acid) with acetic anhydride. Acetaminophen, a drug used in over-the-counter analgesics such as Tylenol, is prepared by reaction of *p*-hydroxyaniline with acetic anhydride. Note that the more nucleophilic –NH₂ group reacts, rather than the less nucleophilic –OH group.

FIGURE 21.7 ▼

Some reactions of acid anhydrides.

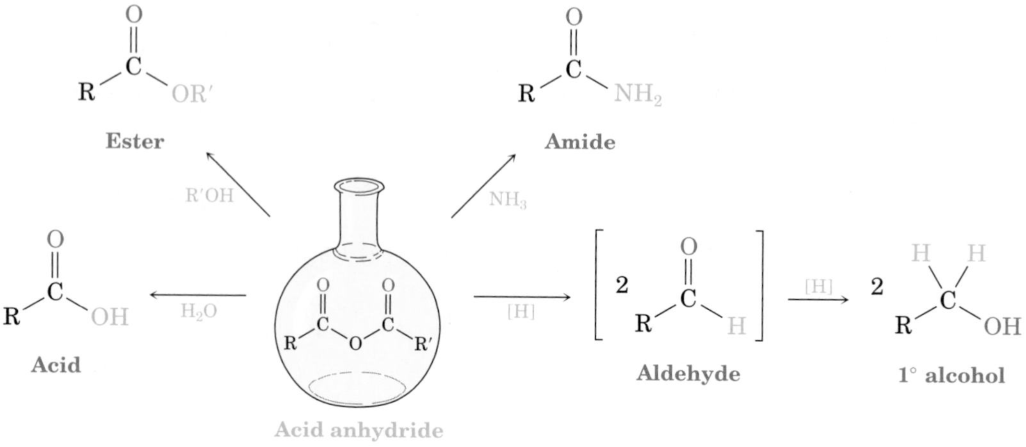

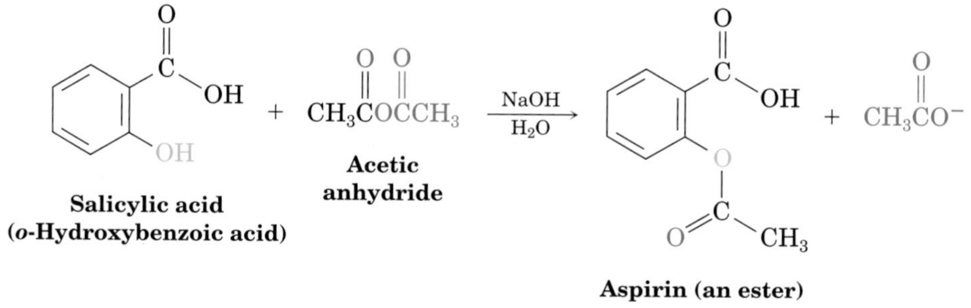

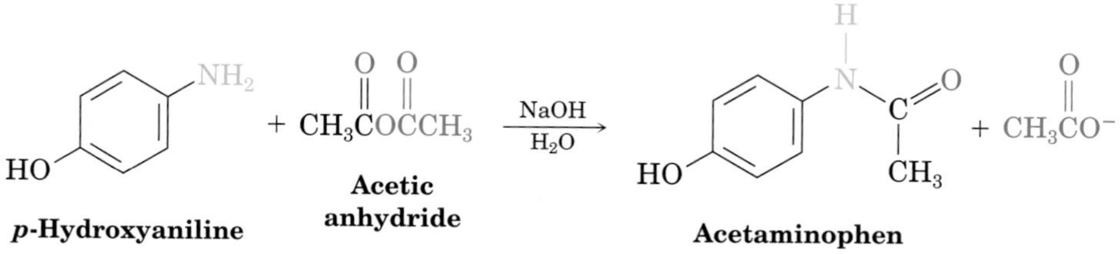

Notice in both these examples that only "half" of the anhydride molecule is used; the other half acts as the leaving group during the nucleophilic acyl substitution step and produces acetate ion as a by-product. Thus, anhydrides are inefficient to use, and acid chlorides are normally preferred for introducing acyl substituents other than acetyl groups.

· ·

Problem 21.15   What product would you expect from reaction of 1 equivalent of methanol with a cyclic anhydride, such as phthalic anhydride (1,2-benzenedicarboxylic anhydride)? What is the fate of the second "half" of the anhydride in this case?

Problem 21.16   Write the mechanism of the reaction shown above between *p*-hydroxyaniline and acetic anhydride to prepare acetaminophen.

Problem 21.17 Why is 1 equivalent of a base such as NaOH required for the reaction between an amine and an anhydride to go to completion? What would happen if no base were present?

21.6 Chemistry of Esters

Esters are among the most widespread of all naturally occurring compounds. Many simple esters are pleasant-smelling liquids that are responsible for the fragrant odors of fruits and flowers. For example, methyl butanoate is found in pineapple oil, and isopentyl acetate is a constituent of banana oil. The ester linkage is also present in animal fats and in many biologically important molecules.

$$CH_3CH_2CH_2\overset{\displaystyle O}{\overset{\|}{C}}OCH_3$$

Methyl butanoate
(from pineapples)

$$CH_3\overset{\displaystyle O}{\overset{\|}{C}}OCH_2CH_2\overset{\displaystyle CH_3}{\overset{|}{C}}HCH_3$$

Isopentyl acetate
(from bananas)

$$\begin{array}{c} CH_2O\overset{\displaystyle O}{\overset{\|}{C}}R \\ CHO\overset{\displaystyle O}{\overset{\|}{C}}R \\ CH_2O\overset{\displaystyle O}{\overset{\|}{C}}R \end{array}$$

A fat
(R = C$_{11-17}$ chains)

The chemical industry uses esters for a variety of purposes. Ethyl acetate, for example, is a common solvent found in nail-polish remover, and dialkyl phthalates are used as so-called *plasticizers* to keep polymers from becoming brittle.

$$\overset{\displaystyle O}{\overset{\|}{C}}OCH_2CH_2CH_2CH_3$$
$$\overset{\displaystyle }{\underset{\displaystyle O}{\overset{\|}{C}}}OCH_2CH_2CH_2CH_3$$

Dibutyl phthalate (a plasticizer)

Preparation of Esters

Esters are usually prepared from carboxylic acids by the methods already discussed. Thus, carboxylic acids are converted directly into esters by S$_N$2 reaction of a carboxylate ion with a primary alkyl halide or by Fischer esterification of a carboxylic acid with an alcohol in the presence of a mineral acid catalyst. In addition, acid chlorides are converted into esters by treatment with an alcohol in the presence of base (Section 21.4).

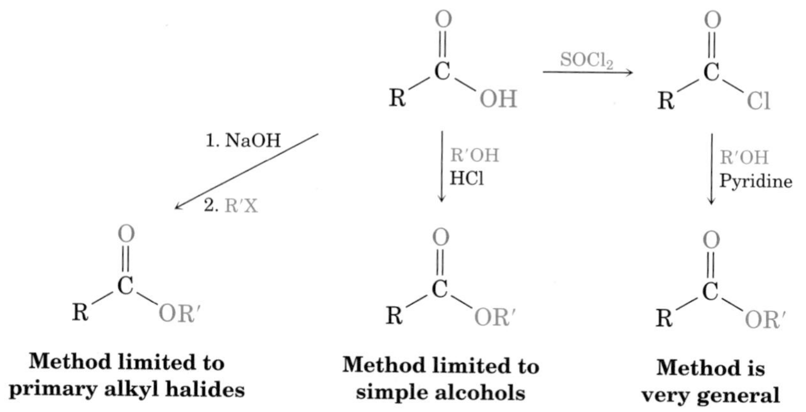

| Method limited to primary alkyl halides | Method limited to simple alcohols | Method is very general |

Reactions of Esters

Esters undergo the same kinds of reactions that we've seen for other carboxylic acid derivatives, but they are less reactive toward nucleophiles than either acid chlorides or anhydrides. Figure 21.8 shows some general reactions of esters, all of which are equally applicable to both acyclic and cyclic esters, called **lactones**.

FIGURE 21.8 ▼

Some reactions of esters.

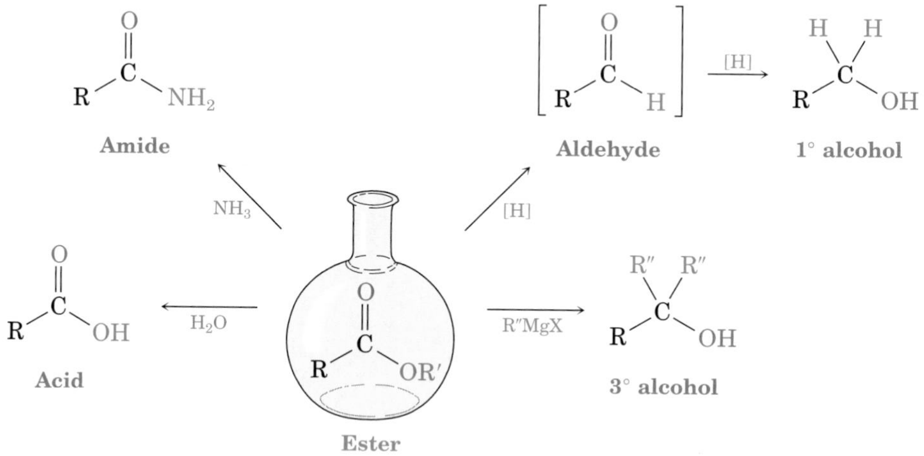

Hydrolysis: Conversion of Esters into Carboxylic Acids (RCO$_2$R′ ⟶ RCO$_2$H)   Esters are hydrolyzed, either by aqueous base or by aqueous acid, to yield carboxylic acids plus alcohols:

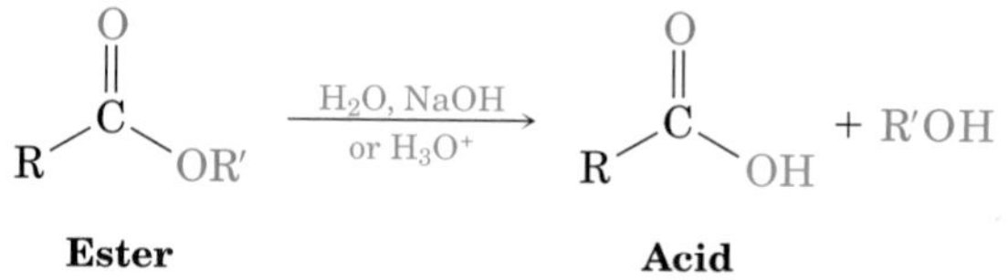

**Biological
Connection** ▶

Ester hydrolysis in basic solution is called **saponification**, after the Latin *sapo*, meaning "soap." As we'll see in Section 27.2, the boiling of animal fat with extract of wood ash to make soap is indeed a saponification, because wood ash contains base and fats have ester linkages.

Ester hydrolysis occurs through the nucleophilic acyl substitution pathway shown in Figure 21.9, in which hydroxide ion is the nucleophile that adds to the ester carbonyl group to give a tetrahedral intermediate. Loss of alkoxide ion then gives a carboxylic acid, which is deprotonated to give the carboxylate ion. Addition of aqueous HCl in a separate step after the saponification is complete then protonates the carboxylate ion and gives the carboxylic acid.

This mechanism is supported by isotope-labeling studies. When ethyl propanoate labeled with ^{18}O in the ether-like oxygen is hydrolyzed in aqueous NaOH, the ^{18}O label shows up exclusively in the ethanol product (p. 868). None of the label remains with the propanoic acid, indicating that saponification occurs by cleavage of the C–OR' bond rather than the CO–R' bond.

FIGURE 21.9 ▼

Mechanism of base-
induced ester hydrolysis
(saponification).

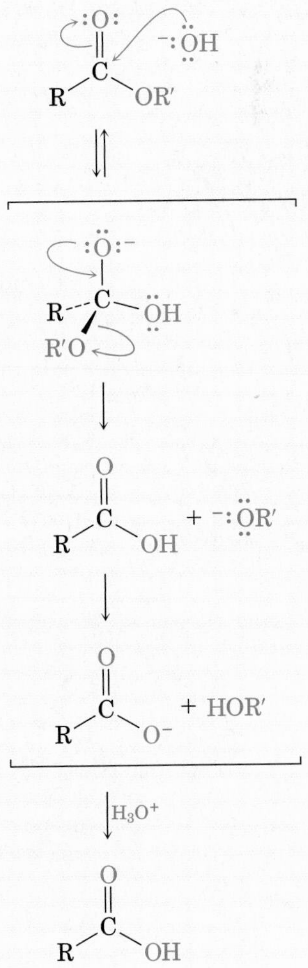

Nucleophilic addition of hydroxide
ion to the ester carbonyl group gives
the usual tetrahedral alkoxide
intermediate.

Elimination of alkoxide ion then
generates the carboxylic acid.

Alkoxide ion abstracts the acidic
proton from the carboxylic acid and
yields a carboxylate ion.

Protonation of the carboxylate ion
by addition of aqueous mineral acid
in a separate step then gives the free
carboxylic acid.

© 1984 JOHN MCMURRY

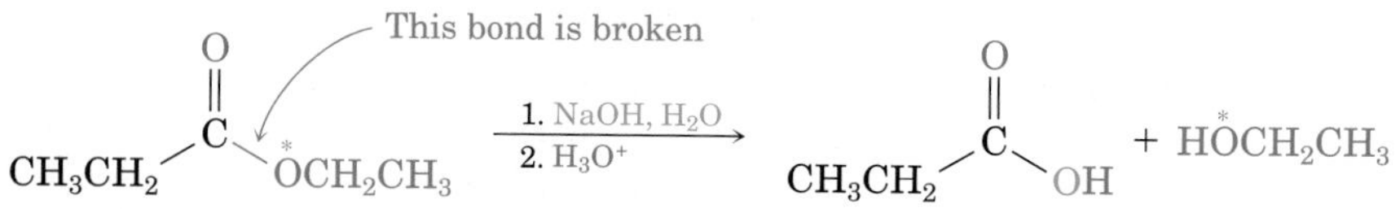

This bond is broken

$$CH_3CH_2 - \overset{\overset{\displaystyle O}{\|}}{C} - \overset{*}{O}CH_2CH_3 \xrightarrow[\text{2. } H_3O^+]{\text{1. NaOH, } H_2O} CH_3CH_2 - \overset{\overset{\displaystyle O}{\|}}{C} - OH + \overset{*}{H}OCH_2CH_3$$

Acid-catalyzed hydrolysis of esters can occur by more than one mechanism, depending on the structure of substrate. The usual pathway, however, is just the reverse of the Fischer esterification reaction (Section 21.3). The ester is first activated toward nucleophilic attack by protonation of the carboxyl oxygen atom, and nucleophilic attack by water then occurs. Transfer of a proton and elimination of alcohol yields the carboxylic acid (Figure 21.10). Since this hydrolysis reaction is the reverse of a Fischer esterification reaction, Figure 21.10 is the reverse of Figure 21.5.

FIGURE 21.10 ▼

Mechanism of acid-catalyzed ester hydrolysis. The forward reaction is a hydrolysis; the back-reaction is a Fischer esterification, and is thus the reverse of Figure 21.5.

 **refer to Mechanisms & Movies**

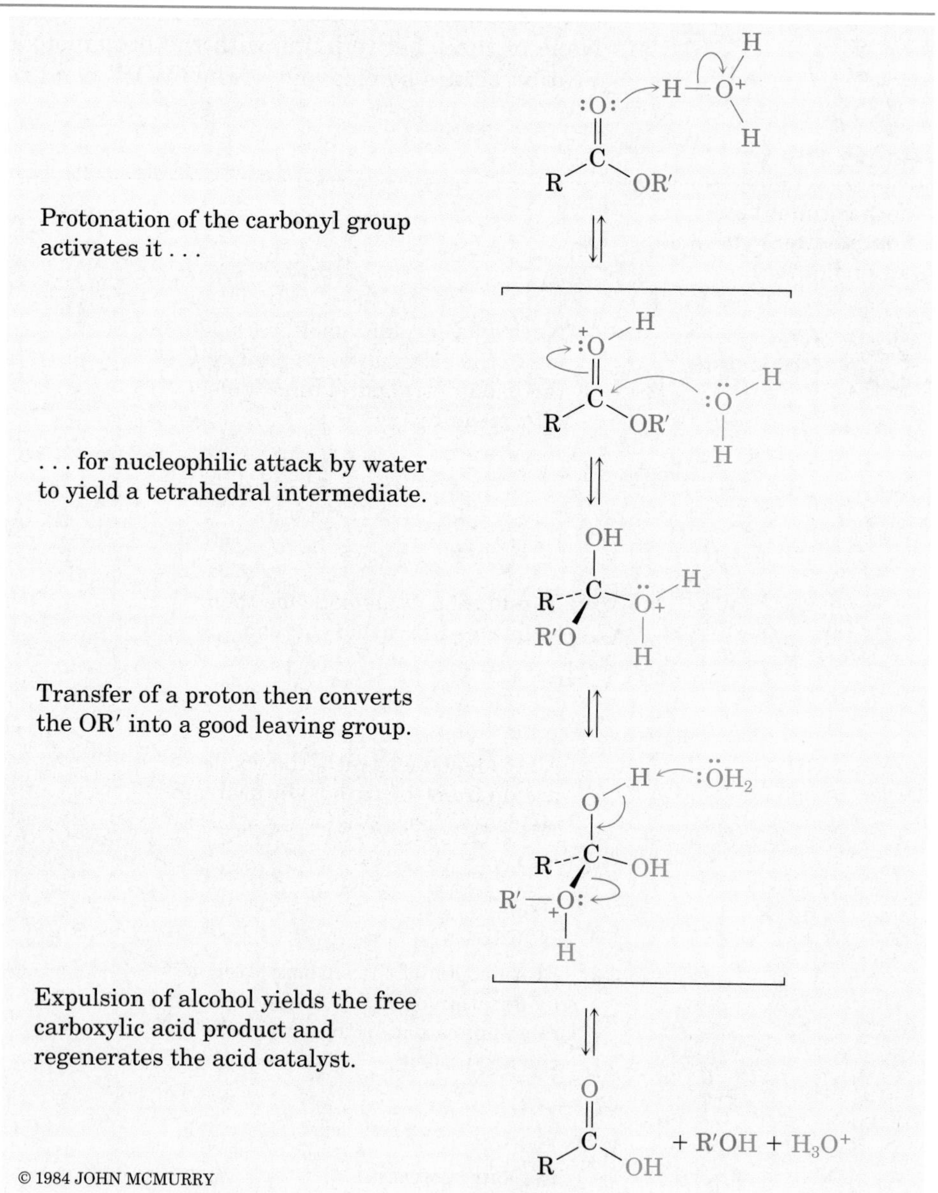

Protonation of the carbonyl group activates it . . .

. . . for nucleophilic attack by water to yield a tetrahedral intermediate.

Transfer of a proton then converts the OR′ into a good leaving group.

Expulsion of alcohol yields the free carboxylic acid product and regenerates the acid catalyst.

© 1984 JOHN MCMURRY

● ●

Problem 21.18   Why is the saponification of an ester irreversible? In other words, why doesn't treatment of a carboxylic acid with an alkoxide ion yield an ester?

● ●

Biological Connection ▶

Aminolysis: Conversion of Esters into Amides (RCO₂R' ⟶ RCONH₂)
Esters react with ammonia and amines to yield amides. The reaction is not often used, however, because it's usually easier to start with an acid chloride (Section 21.4).

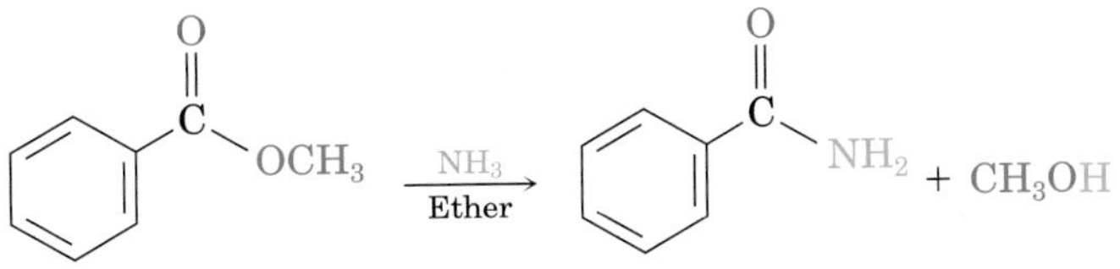

Methyl benzoate   Benzamide

Reduction: Conversion of Esters into Alcohols (RCO₂R' ⟶ RCH₂OH)
Esters are easily reduced by treatment with LiAlH₄ to yield primary alcohols (Section 17.5).

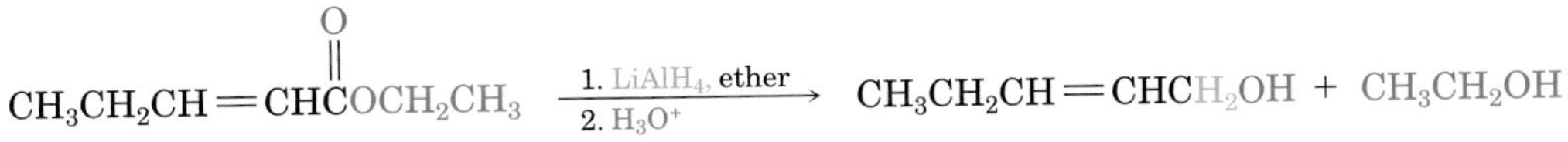

Ethyl 2-pentenoate   **2-Penten-1-ol (91%)**

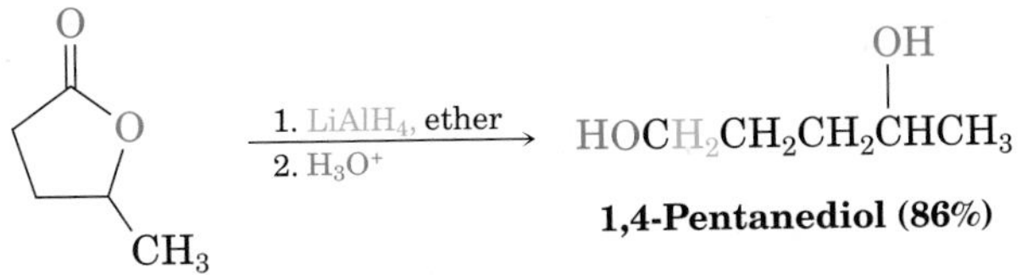

A lactone   **1,4-Pentanediol (86%)**

The mechanism of ester (and lactone) reduction is similar to that of acid chloride reduction. A hydride ion first adds to the carbonyl group, followed by elimination of alkoxide ion to yield an aldehyde. Further reduction of the aldehyde gives the primary alcohol.

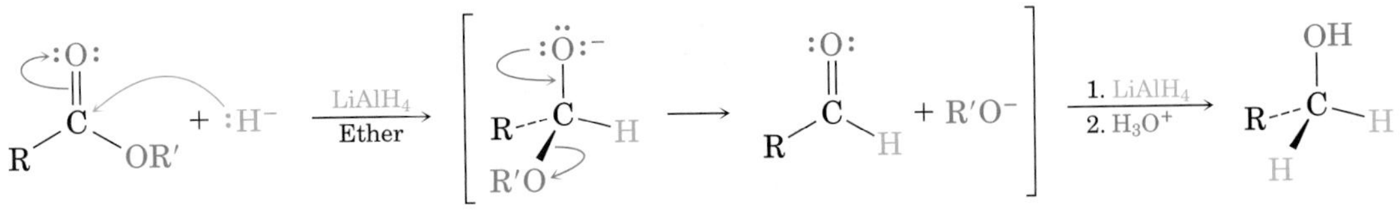

A primary alcohol

The aldehyde intermediate can be isolated if 1 equivalent of diisobutyl-aluminum hydride (DIBAH) is used as the reducing agent instead of LiAlH$_4$. The reaction has to be carried out at −78°C to avoid further reduction to the alcohol.

$$CH_3(CH_2)_{10}\overset{\displaystyle O}{\overset{\displaystyle \|}{C}}OCH_2CH_3 \quad \xrightarrow[\text{2. H}_3\text{O}^+]{\text{1. DIBAH in toluene}} \quad CH_3(CH_2)_{10}\overset{\displaystyle O}{\overset{\displaystyle \|}{C}}H + CH_3CH_2OH$$

Ethyl dodecanoate   **Dodecanal (88%)**

where   DIBAH = [(CH$_3$)$_2$CHCH$_2$]$_2$AlH

• •

Problem 21.19   What product would you expect from the reaction of butyrolactone with DIBAH at −78°C?

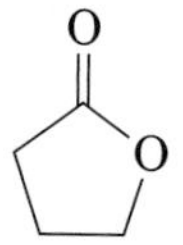

Butyrolactone

Problem 21.20   Show the products you would obtain by reduction of the following esters with LiAlH$_4$:

(a) CH$_3$CH$_2$CH$_2$$\overset{\displaystyle H_3C}{\overset{\displaystyle |}{C}}H\overset{\displaystyle O}{\overset{\displaystyle \|}{C}}OCH_3$

(b)

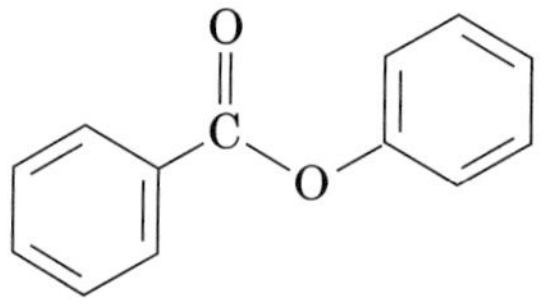

• •

Reaction of Esters with Grignard Reagents   Esters and lactones react with 2 equivalents of a Grignard reagent to yield a tertiary alcohol in which two of the substituents are identical (Section 17.6). The reaction occurs by the usual nucleophilic substitution mechanism to give an intermediate ketone, which reacts further with the Grignard reagent to yield a tertiary alcohol.

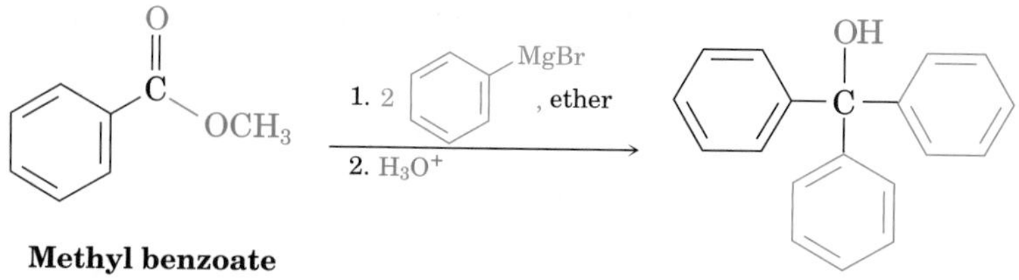

Methyl benzoate

Triphenylmethanol (96%)

O

$$\text{Valerolactone} \xrightarrow[\text{2. } H_3O^+]{\text{1. 2 } CH_3MgBr, \text{ ether}} CH_3\overset{\overset{\displaystyle OH}{|}}{\underset{\underset{\displaystyle CH_3}{|}}{C}}CH_2CH_2CH_2CH_2OH$$

Valerolactone

5-Methyl-1,5-hexanediol

Problem 21.21   What ester and what Grignard reagent might you start with to prepare the following alcohols?
(a) 2-Phenyl-2-propanol   (b) 1,1-Diphenylethanol   (c) 3-Ethyl-3-heptanol

21.7 Chemistry of Amides

Preparation of Amides

Amides are usually prepared by reaction of an acid chloride with an amine (Section 21.4). Ammonia, monosubstituted amines, and disubstituted amines all undergo the reaction.

$$\underset{R}{\overset{\overset{\displaystyle O}{\|}}{C}}Cl$$

NH_3 $R'NH_2$ R'_2NH

$$\underset{R}{\overset{\overset{\displaystyle O}{\|}}{C}}NH_2 \qquad \underset{R}{\overset{\overset{\displaystyle O}{\|}}{C}}NHR' \qquad \underset{R}{\overset{\overset{\displaystyle O}{\|}}{C}}NR'_2$$

Reactions of Amides

Hydrolysis: Conversion of Amides into Carboxylic Acids ($RCONH_2 \longrightarrow RCOOH$)

Biological Connection ▶

Amides are much less reactive than acid chlorides, acid anhydrides, or esters. We'll see in Chapter 26, for example, that the amide linkage is so stable that it serves as the basic unit from which proteins are made.

$$H_2N-\overset{\overset{\displaystyle R}{|}}{C}H-\overset{\overset{\displaystyle O}{\|}}{C}-OH \implies \xi NH-\overset{\overset{\displaystyle R}{|}}{C}H-\overset{\overset{\displaystyle O}{\|}}{C}-NH-\overset{\overset{\displaystyle R'}{|}}{C}H-\overset{\overset{\displaystyle O}{\|}}{C}-NH-\overset{\overset{\displaystyle R''}{|}}{C}H-\overset{\overset{\displaystyle O}{\|}}{C}\xi$$

Amino acids   **A protein (polyamide)**

Amides undergo hydrolysis to yield carboxylic acids plus amine on heating in either aqueous acid or aqueous base. The conditions required for amide hydrolysis are more severe than those required for the hydrolysis of acid chlorides or esters, but the mechanisms are similar. The acidic hydrolysis reaction occurs by nucleophilic addition of water to the protonated amide, followed by loss of ammonia. The basic hydrolysis occurs by nucleophilic addition of OH^- to the amide carbonyl group, followed by deprotonation of the –OH group and elimination of amide ion ($^-NH_2$).

Acidic hydrolysis

Basic hydrolysis

Reduction: Conversion of Amides into Amines ($RCONH_2 \longrightarrow RCH_2NH_2$)

Like other carboxylic acid derivatives, amides can be reduced by $LiAlH_4$. The product of the reduction, however, is an *amine* rather than an alcohol. The net effect of an amide reduction reaction is thus to convert the amide carbonyl group into a methylene group ($C=O \longrightarrow CH_2$). This kind of reaction is specific for amides and does not occur with other carboxylic acid derivatives.

$$CH_3(CH_2)_{10}\overset{\overset{\displaystyle O}{\displaystyle \|}}{C}NHCH_3 \quad \xrightarrow[\text{2. } H_2O]{\text{1. } LiAlH_4, \text{ ether}} \quad CH_3(CH_2)_{10}CH_2NHCH_3$$

N-Methyldodecanamide   **Dodecylmethylamine (95%)**

Amide reduction occurs by nucleophilic addition of hydride ion to the amide carbonyl group, followed by expulsion of the *oxygen* atom as an aluminate anion to give an iminium ion intermediate. The intermediate iminium ion is then further reduced by $LiAlH_4$ to yield the amine.

Amide **Iminium ion**

Lithium aluminum hydride reduction is equally effective with both acyclic and cyclic amides, or **lactams**. The reduction of a lactam is a good method for preparing a cyclic amine.

A lactam **A cyclic amine (80%)**

· ·

Problem 21.22 How would you convert *N*-ethylbenzamide to each of the following products?
(a) Benzoic acid
(b) Benzyl alcohol
(c) C₆H₅CH₂NHCH₂CH₃

Problem 21.23 How would you use the reaction of an amide with LiAlH₄ as the key step in going from bromocyclohexane to (*N,N*-dimethylaminomethyl)cyclohexane? Write all the steps in the reaction sequence.

(*N,N*-Dimethylaminomethyl)cyclohexane

· ·

21.8 Chemistry of Nitriles

Nitriles are analogous to carboxylic acids in that both have a carbon atom with three bonds to an electronegative atom, and both contain a π bond. Thus, the reactions of nitriles and carboxylic acid derivatives are similar.

A nitrile—three bonds to nitrogen **An acid—three bonds to two oxygens**

Preparation of Nitriles

The simplest method of nitrile preparation is S_N2 reaction of CN^- with a primary alkyl halide, as discussed in Section 20.6. This method is limited by the usual S_N2 steric constraints to the synthesis of α-unsubstituted nitriles, RCH_2CN.

$$RCH_2Br + Na^+ CN^- \xrightarrow[\text{reaction}]{S_N2} RCH_2CN + NaBr$$

Another method for preparing nitriles is by dehydration of a primary amide. Thionyl chloride is often used for the reaction, although other dehydrating agents such as $POCl_3$ and acetic anhydride also work.

$$CH_3CH_2CH_2CH_2\underset{\underset{CH_2CH_3}{|}}{CH}\overset{\overset{O}{\|}}{C}-NH_2 \xrightarrow[80°C]{SOCl_2, \text{ benzene}} CH_3CH_2CH_2CH_2\underset{\underset{CH_2CH_3}{|}}{CH}C\equiv N + SO_2 + 2\,HCl$$

2-Ethylhexanamide   **2-Ethylhexanenitrile (94%)**

The dehydration occurs by initial reaction of $SOCl_2$ on the amide oxygen atom, followed by an E2-like elimination reaction.

Both methods of nitrile synthesis—S_N2 displacement by CN^- on an alkyl halide and amide dehydration—are useful, but the synthesis from amides is more general because it is not limited by steric hindrance.

Reactions of Nitriles

The chemistry of nitriles is similar in many respects to the chemistry of carbonyl compounds. Like carbonyl groups, a nitrile group is strongly polarized, making the carbon atom electrophilic. Nitriles are therefore attacked by nucleophiles to yield sp^2-hybridized imine anions in a reaction analogous to the formation of an sp^3-hybridized alkoxide ion by nucleophilic addition to a carbonyl group.

Nitrile

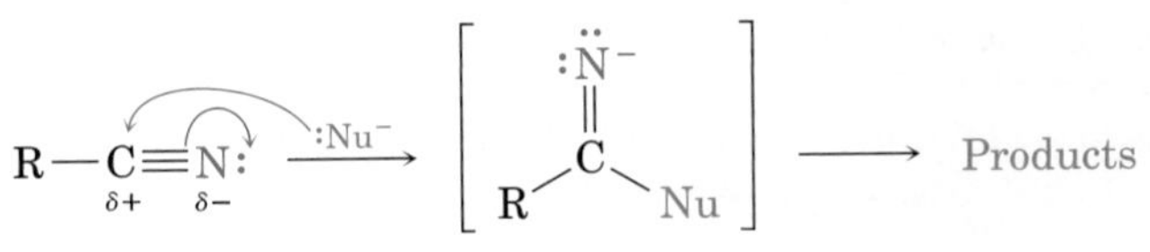

Among the important reactions of nitriles are hydrolysis, reduction, and Grignard reaction to yield ketones (Figure 21.11).

FIGURE 21.11 ▼

Some reactions of nitriles.

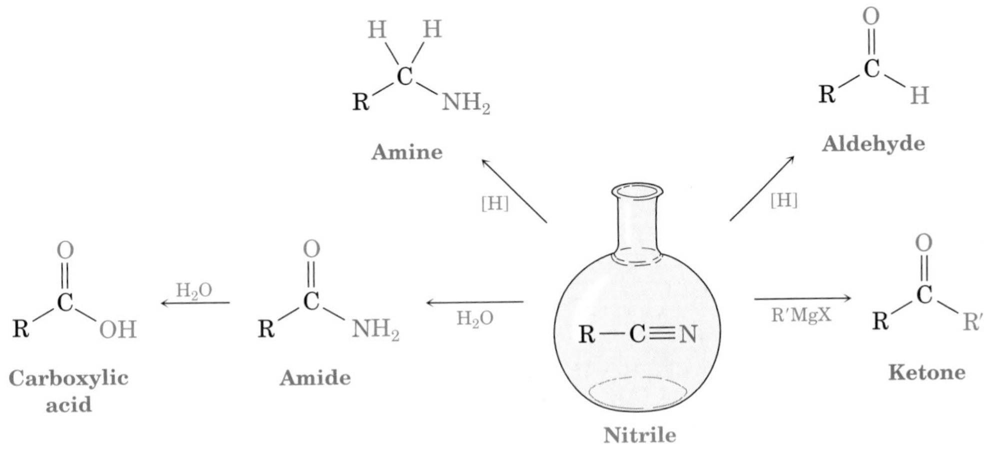

Hydrolysis: Conversion of Nitriles into Carboxylic Acids (RCN ⟶ RCO₂H)

Hydrolysis: Conversion of Nitriles into Carboxylic Acids (RCN ⟶ RCO$_2$H) Nitriles are hydrolyzed in either acidic or basic aqueous solution to yield carboxylic acids plus ammonia or an amine:

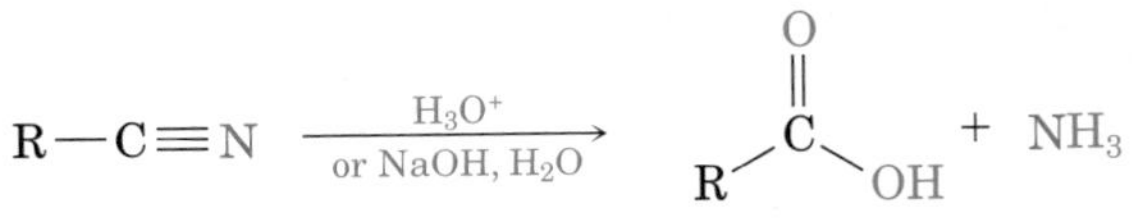

The mechanism of the basic hydrolysis involves nucleophilic addition of hydroxide ion to the polar C≡N bond in a manner analogous to that of nucleophilic addition to a polar carbonyl C=O bond. Next, the initially formed hydroxy imine isomerizes to an amide in a step similar to the isomerization of an enol to a ketone (Section 8.5). Further hydrolysis of the amide, as discussed in the previous section, then yields the carboxylic acid (Figure 21.12).

The conditions required for alkaline hydrolysis of a nitrile are severe (KOH, 200°C), so the amide intermediate can sometimes be isolated if milder conditions are used.

FIGURE 21.12 ▼

Mechanism of the basic hydrolysis of a nitrile to yield an amide, which is subsequently hydrolyzed further to a carboxylic acid.

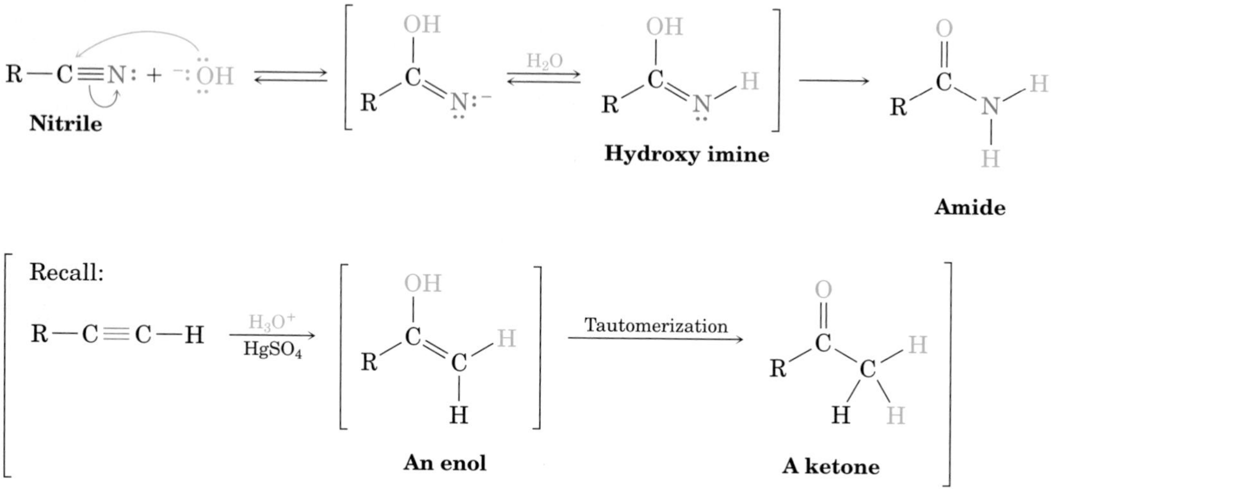

Nitrile   **Hydroxy imine**   **Amide**

Recall:

An enol   **A ketone**

• •

Problem 21.24   Acid-catalyzed nitrile hydrolysis is similar to amide hydrolysis and occurs by initial protonation of the nitrogen atom, followed by nucleophilic addition of water. Write all the steps involved in the acidic hydrolysis of a nitrile to yield a carboxylic acid, using curved arrows to represent electron flow in each step.

• •

Reduction: Conversion of Nitriles into Amines and Aldehydes

Reduction of a nitrile with LiAlH$_4$ gives a primary amine in high yield. For example:

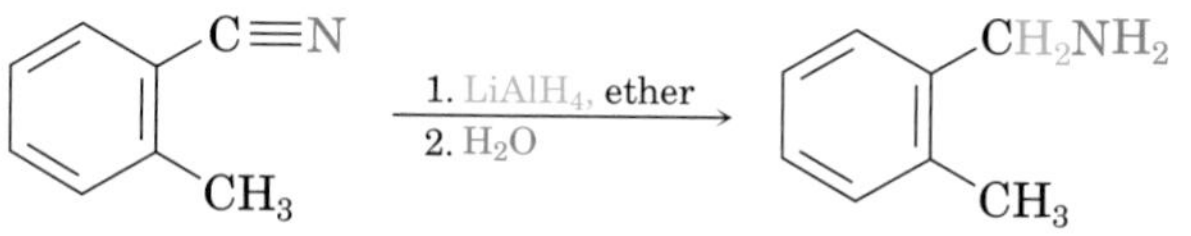

o-Methylbenzonitrile   *o*-Methylbenzylamine
(88%)

The reaction occurs by nucleophilic addition of hydride ion to the polar C≡N bond, yielding an imine anion that undergoes further addition of a second equivalent of hydride. If, however, a less powerful reducing agent such as DIBAH is used, the second addition of hydride does not occur, and the imine intermediate can be hydrolyzed by addition of water to yield an aldehyde.

The overall result—conversion of a nitrile to an aldehyde—is similar to the conversion of an ester to an aldehyde that we saw in Section 21.6.

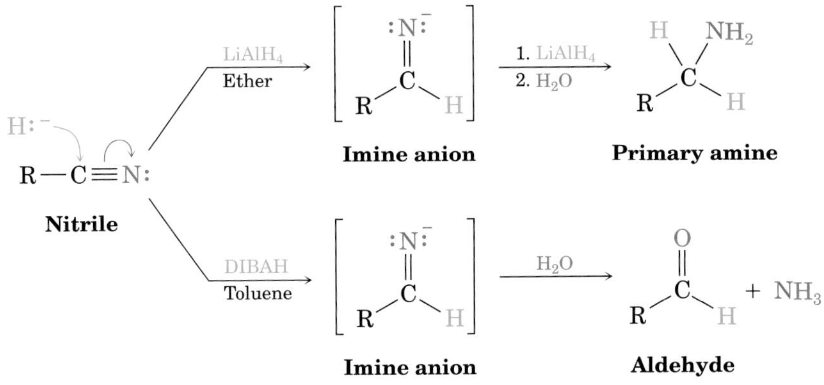

For example:

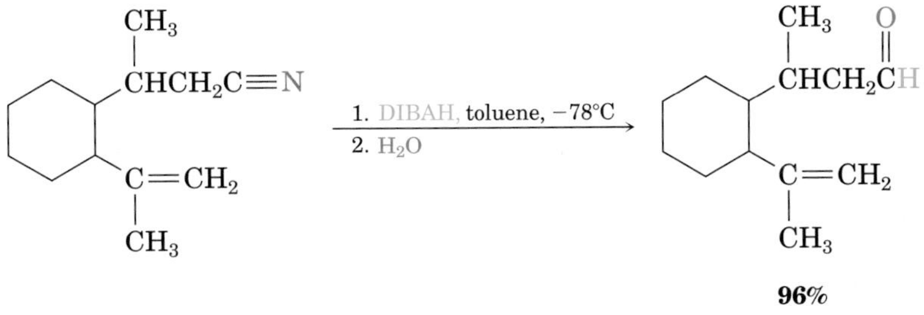

96%

Reaction of Nitriles with Organometallic Reagents
A Grignard reagent adds to a nitrile, giving an intermediate imine anion that can be hydrolyzed by addition of water to yield a ketone:

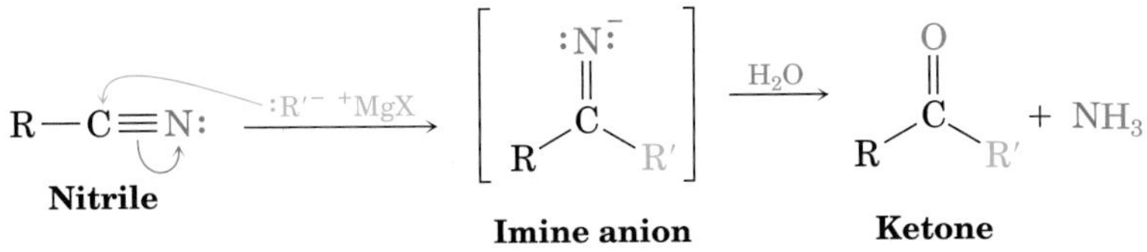

The reaction is similar to the DIBAH reduction of a nitrile to yield an aldehyde, except that the attacking nucleophile is a carbanion ($R:^-$) rather than a hydride ion. For example:

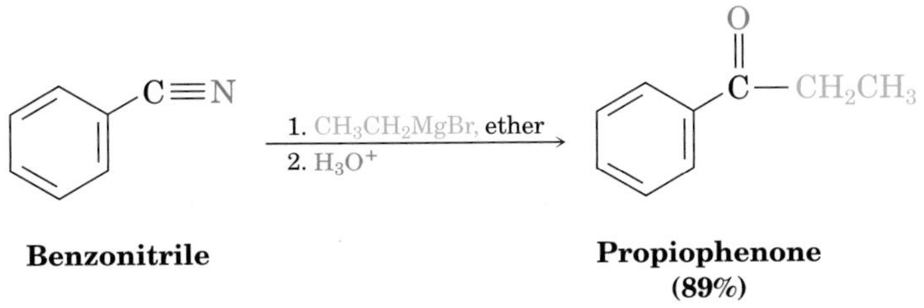

Benzonitrile   **Propiophenone**
                        (89%)

• •

Practice Problem 21.2   How would you prepare 2-methyl-3-pentanone from a nitrile?

$$\underset{\substack{| \\ CH_3}}{CH_3CH_2\overset{\displaystyle O}{\overset{\|}{C}}CHCH_3}$$ **2-Methyl-3-pentanone**

Strategy   A ketone results from the reaction between a Grignard reagent and a nitrile, with the C≡N carbon of the nitrile becoming the carbonyl carbon. Identify the two groups attached to the carbonyl carbon atom in the product. One will come from the Grignard reagent, and the other will come from the nitrile.

Solution   There are two possibilities:

$$\left.\begin{array}{c} CH_3CH_2C{\equiv}N \\ + \\ (CH_3)_2CHMgBr \end{array}\right\} \xrightarrow[\text{2. } H_3O^+]{\text{1. Grignard}} \underset{\substack{| \\ CH_3}}{CH_3CH_2\overset{\displaystyle O}{\overset{\|}{C}}CHCH_3} \xleftarrow[\text{2. } H_3O^+]{\text{1. Grignard}} \left\{\begin{array}{c} \overset{\displaystyle CH_3}{\underset{|}{}} \\ CH_3CHC{\equiv}N \\ + \\ CH_3CH_2MgBr \end{array}\right.$$

2-Methyl-3-pentanone

• •

Problem 21.25   How would you prepare each of the following carbonyl compounds from a nitrile?
(a) $CH_3CH_2COCH_2CH_3$ (b) $(CH_3)_2CHCHO$ (c) Acetophenone

Problem 21.26   How would you prepare 1-phenyl-2-butanone from benzyl bromide, $C_6H_5CH_2Br$? More than one step is required.

• •

21.9 Thiol Esters: Biological Carboxylic Acid Derivatives

Biological Connection ▶

Nucleophilic acyl substitution reactions take place in living organisms just as they take place in the chemical laboratory. The same principles apply in both cases. Nature, however, often uses a **thiol ester, RCOSR′**, as the acid derivative because it is intermediate in reactivity between an acid anhydride and an ester. Thiol esters aren't as reactive as anhydrides, yet they're more reactive than typical esters toward nucleophilic attack.

Acetyl coenzyme A (usually abbreviated **acetyl CoA**) is the most common thiol ester in nature. Acetyl CoA is a much more complex molecule than acetyl chloride or acetic anhydride, yet it serves exactly the same purpose as these simpler reagents. Nature uses acetyl CoA as a reactive acylating agent in nucleophilic acyl substitution reactions.

Acetyl CoA

As an example of how acetyl CoA is used in nature, *N*-acetylglucos-amine, an important constituent of surface membranes in mammalian cells, is synthesized by an aminolysis reaction between glucosamine and acetyl CoA. We'll look at some reactions of acetyl CoA in more detail in Chapter 29.

Glucosamine
(an amine)

N-Acetylglucosamine
(an amide)

21.10 Polyamides and Polyesters: Step-Growth Polymers

When an amine reacts with an acid chloride, an amide results (Section 21.4). What would happen, though, if a *diamine* and a *diacid chloride* were allowed to react? Each partner could form *two* amide bonds, linking more and more molecules together until a giant **polyamide** resulted. In the same way, reaction of a diol with a diacid would lead to a **polyester**.

$$H_2N(CH_2)_nNH_2 + ClC(CH_2)_mCCl \longrightarrow$$

A diamine   **A diacid chloride**   **A polyamide (nylon)**

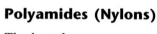

| **A diol** | **A diacid** | **A polyester** |

The alkene and diene polymers discussed in Sections 7.10 and 14.7 are called **chain-growth polymers** because they are produced by chain reactions. An initiator adds to a C=C bond to give a reactive intermediate, which adds to a second alkene molecule to produce a new intermediate, which adds to a third molecule, and so on. By contrast, polyamides and polyesters are said to be **step-growth polymers** because each bond in the polymer is formed independently of the others. A large number of different step-growth polymers have been made; some of the more important ones are shown in Table 21.2.

Polyamides (Nylons)

The best-known step-growth polymers are the polyamides, or **nylons**, first prepared by Wallace Carothers at the Du Pont Company by heating a diamine with a diacid. For example, nylon 66 is prepared by reaction of adipic acid with hexamethylenediamine at 280°C. The designation "66" tells the number of carbon atoms in the diamine (the first 6) and in the diacid (the second 6).

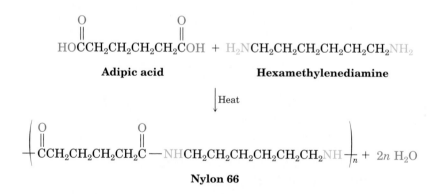

Nylon 66

Nylons are used both in engineering applications and in making fibers. A combination of high impact strength and abrasion resistance makes nylon an excellent metal substitute for bearings and gears. As fiber, nylon is used in a wide variety of applications, from clothing to tire cord to Perlon mountaineering ropes.

Polyesters

The most generally useful polyester is that made by reaction between dimethyl terephthalate and ethylene glycol. The product is used under the trade name Dacron to make clothing fiber and tire cord, and under the name

Wallace Hume Carothers

Wallace Hume Carothers (1896–1937) was born in Burlington, Iowa, and received his Ph.D. at the University of Illinois in 1924 with Roger Adams. He began his career with brief teaching positions at the University of South Dakota, the University of Illinois, and Harvard University, but moved to the Du Pont Company in 1928 to head their new chemistry research program in polymers. A prolonged struggle with depression led him to suicide after only 9 years at Du Pont.

TABLE 21.2 Some Common Step-Growth Polymers and Their Uses

Monomer name	Formula	Trade or common name of polymer	Uses
Hexamethylene-diamine	$H_2N(CH_2)_6NH_2$	Nylon 66	Fibers, clothing, tire cord, bearings
Adipic acid	$HOOC(CH_2)_4COOH$		
Ethylene glycol	$HOCH_2CH_2OH$	Dacron, Terylene, Mylar	Fibers, clothing, tire cord, film
Dimethyl terephthalate	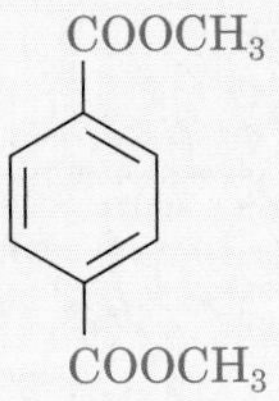		
Caprolactam	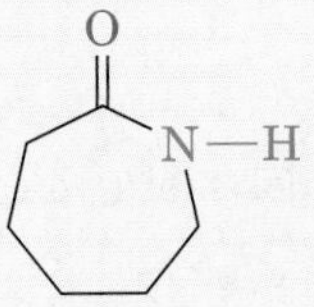	Nylon 6, Perlon	Fibers, large cast articles
Bisphenol A	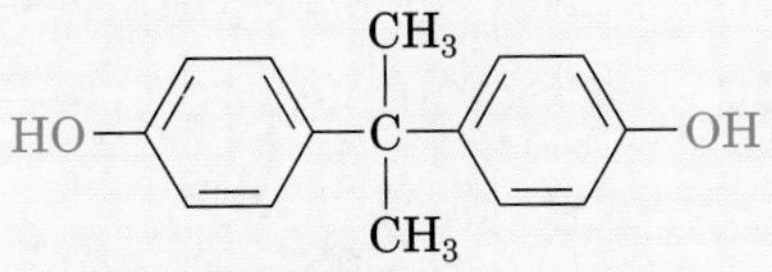	Lexan, polycarbonate	Molded articles, machine housings
Diphenyl carbonate	$C_6H_5OCOOC_6H_5$		
Poly(2-butene-1,4-diol)	$HO(CH_2CH=CHCH_2)_nOH$	Polyurethane, Spandex	Foams, fibers, coatings
Toluene-2,6-diisocyanate	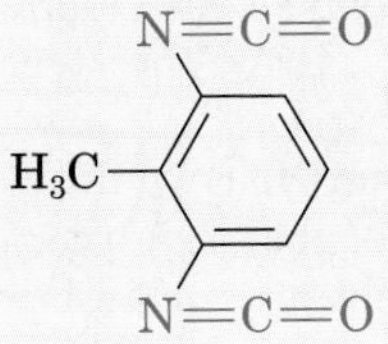		

Mylar to make recording tape. The tensile strength of poly(ethylene terephthalate) film is nearly equal to that of steel.

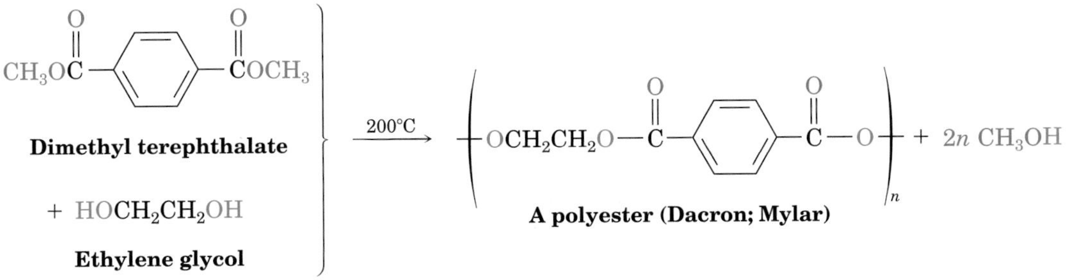

Dimethyl terephthalate

+ $HOCH_2CH_2OH$

Ethylene glycol

$\xrightarrow{200°C}$

A polyester (Dacron; Mylar) + $2n$ CH_3OH

Lexan, a polycarbonate prepared from diphenyl carbonate and bisphenol A, is another commercially valuable polyester. Lexan has an unusually high impact strength, making it valuable for use in machinery housings, telephones, and bicycle safety helmets.

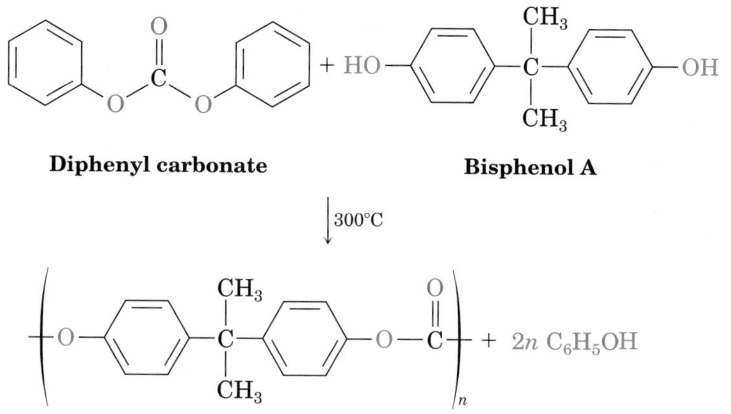

Diphenyl carbonate   **Bisphenol A**

300°C

Lexan

Problem 21.27   Draw structures of the step-growth polymers you would expect to obtain from the following reactions:

(a) $BrCH_2CH_2CH_2Br + HOCH_2CH_2CH_2OH \xrightarrow{\text{Base}}$?

(b) $HOCH_2CH_2OH + HOOC(CH_2)_6COOH \xrightarrow{\text{H}_2\text{SO}_4 \text{ catalyst}}$?

(c) $H_2N(CH_2)_6NH_2 + \overset{O}{\overset{\|}{Cl}C}(CH_2)_4\overset{O}{\overset{\|}{C}}Cl \longrightarrow$?

Problem 21.28   Kevlar, a nylon polymer prepared by reaction of 1,4-benzenedicarboxylic acid (terephthalic acid) with 1,4-diaminobenzene (*p*-phenylenediamine), is so strong that it's used to make bulletproof vests. Draw the structure of a segment of Kevlar.

Problem 21.29   Draw the structure of the polymer you would expect to obtain from reaction of dimethyl terephthalate with a triol such as glycerol. What structural feature would this new polymer have that was not present in Dacron? How do you think this new feature might affect the properties of the polymer?

21.11 Spectroscopy of Carboxylic Acid Derivatives and Nitriles

Infrared Spectroscopy

All carbonyl-containing compounds have intense IR absorptions in the range 1650–1850 cm^{-1}. As shown in Table 21.3, the exact position of the absorption provides information about the specific kind of carbonyl group. For comparison, the IR absorptions of aldehydes, ketones, and carboxylic acids are included in the table, along with values for carboxylic acid derivatives and nitriles.

TABLE 21.3 Infrared Absorptions of Some Carbonyl Compounds and Nitriles

Carbonyl type	Example	Infrared absorption of carbonyl (cm^{-1})
Aliphatic acid chloride	Acetyl chloride	1810
Aromatic acid chloride	Benzoyl chloride	1770
Aliphatic acid anhydride	Acetic anhydride	1820, 1760
Aliphatic ester	Ethyl acetate	1735
Aromatic ester	Ethyl benzoate	1720
Six-membered-ring lactone		1735
Aliphatic amide	Acetamide	1690
Aromatic amide	Benzamide	1675
N-Substituted amide	N-Methylacetamide	1680
N,N-Disubstituted amide	N,N-Dimethylacetamide	1650
Aliphatic nitrile	Acetonitrile	2250
Aromatic nitrile	Benzonitrile	2230
Aliphatic aldehyde	Acetaldehyde	1730
Aliphatic ketone	Acetone	1715
Aliphatic carboxylic acid	Acetic acid	1710

Acid chlorides are readily detected by their characteristic absorption near 1800 cm^{-1}. Acid anhydrides can be identified by the fact that they show two absorptions in the carbonyl region, one at 1820 cm^{-1} and another at 1760 cm^{-1}. Esters are detected by their absorption at 1735 cm^{-1}, a position somewhat higher than that for either aldehydes or ketones. Amides, by contrast, absorb near the low wavenumber end of the carbonyl region, with the degree of substitution on nitrogen affecting the exact position of the IR band. Nitriles are easily recognized by the presence of an intense absorption near 2250 cm^{-1}. Since few other functional groups absorb in this region, IR spectroscopy is highly diagnostic for nitriles.

• •

Problem 21.30 What kinds of functional groups might compounds have if they show the following IR absorptions?
(a) Absorption at 1735 cm^{-1} (b) Absorption at 1810 cm^{-1}
(c) Absorptions at 2500–3300 cm^{-1} and 1710 cm^{-1}
(d) Absorption at 2250 cm^{-1} (e) Absorption at 1715 cm^{-1}

Problem 21.31   Propose structures for compounds that have the following formulas and IR absorptions:

(a) C_3H_5N, 2250 cm^{-1} (b) $C_6H_{12}O_2$, 1735 cm^{-1}

(c) C_4H_9NO, 1650 cm^{-1} (d) C_4H_5ClO, 1780 cm^{-1}

· ·

Nuclear Magnetic Resonance Spectroscopy

Hydrogens on the carbon next to a carbonyl group are slightly deshielded and absorb near 2 δ in the ^{1}H NMR spectrum. The exact nature of the carbonyl group can't be determined by ^{1}H NMR, however, because all acid derivatives absorb in the same range. Figure 21.13 shows the ^{1}H NMR spectrum of ethyl acetate.

FIGURE 21.13 ▼

Proton NMR spectrum of ethyl acetate.

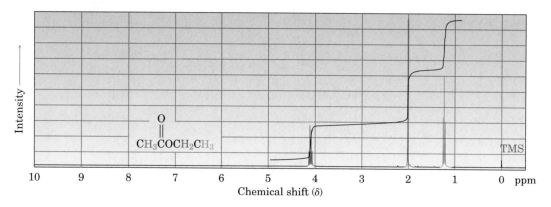

Although ^{13}C NMR is useful for determining the presence or absence of a carbonyl group in a molecule of unknown structure, precise information about the nature of the carbonyl group is difficult to obtain. Aldehydes and ketones absorb near 200 δ, while the carbonyl carbon atoms of various acid derivatives absorb in the range 160–180 δ (Table 21.4).

TABLE 21.4 ^{13}C NMR Absorptions in Some Carbonyl Compounds

Compound	Absorption (δ)	Compound	Absorption (δ)
Acetic acid	177.3	Acetic anhydride	166.9
Ethyl acetate	170.7	Acetonitrile	117.4
Acetyl chloride	170.3	Acetone	205.6
Acetamide	172.6	Acetaldehyde	201.0

β-Lactam Antibiotics

The value of hard work and logical thinking shouldn't be underestimated, but sheer good luck also plays a role in most real scientific breakthroughs. What has been called "the supreme example [of luck] in all scientific history" occurred in the late summer of 1928 when the Scottish bacteriologist Alexander Fleming went on vacation, leaving in his lab a culture plate recently inoculated with the bacterium *Staphylococcus aureus*.

While Fleming was away, an extraordinary chain of events occurred. First, a 9 day cold spell lowered the laboratory temperature to a point where the *Staphylococcus* on the plate could not grow. During this time, spores from a colony of the mold *Penicillium notatum* being grown on the floor below wafted up into Fleming's lab and landed in the culture plate. The temperature then rose, and both *Staphylococcus* and *Penicillium* began to grow. On returning from vacation, Fleming discarded the plate into a tray of antiseptic, intending to sterilize it. Evidently, though, the plate did not sink deeply enough into the antiseptic, because when Fleming happened to glance at it a few days later, what he saw changed the course of human history: He noticed that the growing *Penicillium* mold appeared to dissolve the colonies of staphylococci.

Fleming realized that the *Penicillium* mold must be producing a chemical that killed the *Staphylococcus* bacteria, and he spent several years trying to isolate the substance. Finally, in 1939, the Australian pathologist Howard Florey and the German refugee Ernst Chain managed to isolate the active substance, called *penicillin*. The dramatic ability of penicillin to cure infections in mice was soon demonstrated, and successful tests in humans followed shortly thereafter. By 1943, penicillin was being produced on a large scale for military use, and by 1944 it was being used on civilians. Fleming, Florey, and Chain shared the 1945 Nobel Prize in medicine.

Now called benzylpenicillin, or penicillin G, the substance first discovered by Fleming is but one member of a large class of so-called *β-lactam antibiotics*, compounds with a four-membered lactam (cyclic amide) ring. The four-membered lactam ring is fused to a five-membered, sulfur-containing ring, and the carbon atom next to the lactam carbonyl group is bonded to an acylamino substituent, RCONH–. This acylamino side chain can be varied in the laboratory to provide literally hundreds of penicillin analogs with different biological activity profiles. Ampicillin, for instance, has an α-aminophenylacetamido substituent [PhCH(NH$_2$)CONH–].

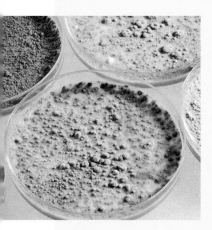

Penicillium mold growing in a petri dish.

(continued) ▶

Acylamino substituent

Benzylpenicillin
(Penicillin G)

β-Lactam ring

Closely related to the penicillins are the *cephalosporins*, a group of β-lactam antibiotics that contain an unsaturated six-membered, sulfur-containing ring. Cephalexin, marketed under the trade name Keflex, is an example. Cephalosporins generally have much greater antibacterial activity than penicillins, particularly against resistant strains of bacteria.

Cephalexin
(a cephalosporin)

The biological activity of penicillins and cephalosporins is due to the presence of the strained β-lactam ring, which reacts with and deactivates the *transpeptidase* enzyme needed to synthesize and repair bacterial cell walls. With the wall either incomplete or weakened, the bacterial cell ruptures and dies.

Summary and Key Words

KEY WORDS

acetyl coenzyme A
  (acetyl CoA), 878
acid anhydride
  (RCO$_2$COR′), 845
acid halide (RCOX),
  844
amide (RCONH$_2$),
  845
carboxylic acid
  derivative, 843
chain-growth
  polymer, 880
ester (RCO$_2$R′), 846
Fischer
  esterification, 855

Carboxylic acids can be transformed into a variety of **acid derivatives** in which the carboxyl –OH group has been replaced by another substituent. **Acid halides, acid anhydrides, esters**, and **amides** are the most common such derivatives.

The chemistry of carboxylic acid derivatives is dominated by the **nucleophilic acyl substitution reaction**. Mechanistically, these substitutions take place by addition of a nucleophile to the polar carbonyl group of the acid derivative, followed by expulsion of a leaving group from the tetrahedral intermediate.

where   Y = F, Cl, Br, I (acid halide); OR (ester); OCOR (anhydride);
        or NH$_2$ (amide)

The reactivity of an acid derivative toward substitution depends both on the steric environment near the carbonyl group and on the electronic nature of the substituent, Y. The reactivity order is:

Acid halide > Acid anhydride > Ester > Amide

The most common reactions of carboxylic acid derivatives are substitution by water (*hydrolysis*) to yield an acid, by an alcohol (*alcoholysis*) to yield an ester, by an amine (*aminolysis*) to yield an amide, by hydride ion to yield an alcohol (*reduction*), and by an organometallic reagent to yield an alcohol (*Grignard reaction*).

Nitriles undergo nucleophilic addition to the polar C≡N bond in the same way that carbonyl compounds do. The most important reactions of nitriles are their hydrolysis to carboxylic acids, reduction to primary amines, partial reduction to aldehydes, and reaction with organometallic reagents to yield ketones.

Nature employs nucleophilic acyl substitution reactions in the biosynthesis of many molecules, using **thiol esters** for the purpose. **Acetyl coenzyme A (acetyl CoA)** is a complex thiol ester that is employed in living systems to acetylate amines and alcohols.

Step-growth polymers, such as polyamides and polyesters, are prepared by reactions between difunctional molecules. **Polyamides (nylons)** are formed by step-growth polymerization between a diacid and a diamine; **polyesters** are formed from a diacid and a diol.

Infrared spectroscopy is a valuable tool for the structural analysis of acid derivatives. Acid chlorides, anhydrides, esters, amides, and nitriles all show characteristic infrared absorptions that can be used to identify these functional groups in unknowns.

Summary of Reactions

1. Reactions of carboxylic acids (Section 21.3)
    (a) Conversion into acid chlorides

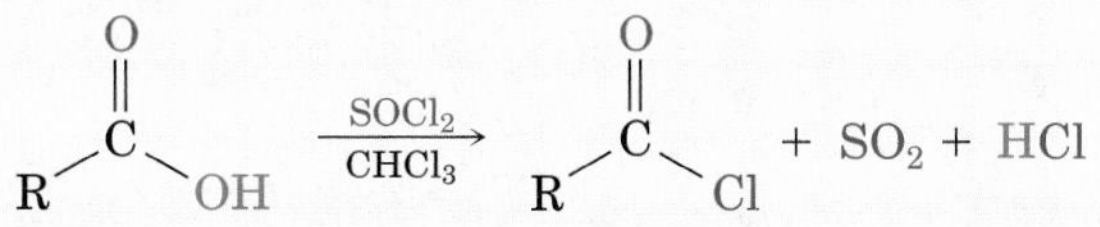

    (b) Conversion into cyclic acid anhydrides

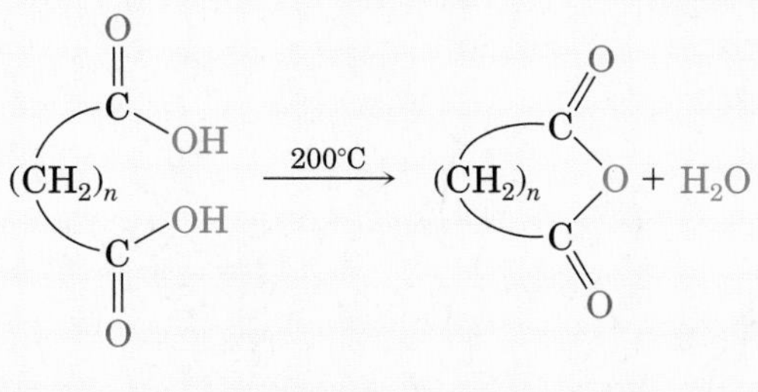

where n = 2 or 3

(continued) ▶

(c) Conversion into esters

$$\underset{R}{\overset{O}{\overset{\|}{C}}}O^- \; + \; R'X \quad \xrightarrow[S_N2 \text{ reaction}]{\text{Via}} \quad \underset{R}{\overset{O}{\overset{\|}{C}}}OR'$$

$$\underset{R}{\overset{O}{\overset{\|}{C}}}OH \; + \; R'OH \quad \xrightarrow[\text{catalyst}]{\text{Acid}} \quad \underset{R}{\overset{O}{\overset{\|}{C}}}OR' \; + \; H_2O$$

2. Reactions of acid chlorides (Section 21.4)
   (a) Hydrolysis to yield acids

$$\underset{R}{\overset{O}{\overset{\|}{C}}}Cl \; + \; H_2O \quad \longrightarrow \quad \underset{R}{\overset{O}{\overset{\|}{C}}}OH \; + \; HCl$$

(b) Alcoholysis to yield esters

$$\underset{R}{\overset{O}{\overset{\|}{C}}}Cl \; + \; R'OH \quad \xrightarrow{\text{Pyridine}} \quad \underset{R}{\overset{O}{\overset{\|}{C}}}OR' \; + \; HCl$$

(c) Aminolysis to yield amides

$$\underset{R}{\overset{O}{\overset{\|}{C}}}Cl \; + \; 2\,NH_3 \quad \longrightarrow \quad \underset{R}{\overset{O}{\overset{\|}{C}}}NH_2 \; + \; NH_4Cl$$

(d) Reduction to yield primary alcohols

$$\underset{R}{\overset{O}{\overset{\|}{C}}}Cl \quad \xrightarrow[\text{2. } H_3O^+]{\text{1. LiAlH}_4, \text{ ether}} \quad \underset{R}{\overset{H\;\;H}{\overset{|\;\;|}{C}}}OH$$

(e) Partial reduction to yield aldehydes

$$\underset{R}{\overset{O}{\overset{\|}{C}}}Cl \quad \xrightarrow[\text{2. } H_3O^+]{\text{1. LiAlH(O-}t\text{-Bu)}_3, \text{ ether}} \quad \underset{R}{\overset{O}{\overset{\|}{C}}}H$$

(f) Grignard reaction to yield tertiary alcohols

$$\underset{R}{\overset{O}{\overset{\|}{C}}}Cl \quad \xrightarrow[\text{2. } H_3O^+]{\text{1. 2 R'MgX, ether}} \quad \underset{R}{\overset{R'\;\;R'}{\overset{|\;\;|}{C}}}OH$$

(g) Diorganocopper reaction to yield ketones

$$\underset{R}{\overset{O}{\overset{\|}{C}}}Cl \quad \xrightarrow[\text{Ether}]{R'_2CuLi} \quad \underset{R}{\overset{O}{\overset{\|}{C}}}R'$$

(continued) ▶

3. Reactions of acid anhydrides (Section 21.5)
   (a) Hydrolysis to yield acids

$$R-\overset{O}{\underset{}{C}}-O-\overset{O}{\underset{}{C}}-R + H_2O \longrightarrow 2 \ R-\overset{O}{\underset{}{C}}-OH$$

   (b) Alcoholysis to yield esters

$$R-\overset{O}{\underset{}{C}}-O-\overset{O}{\underset{}{C}}-R + R'OH \longrightarrow R-\overset{O}{\underset{}{C}}-OR' + R-\overset{O}{\underset{}{C}}-OH$$

   (c) Aminolysis to yield amides

$$R-\overset{O}{\underset{}{C}}-O-\overset{O}{\underset{}{C}}-R + 2 \ NH_3 \longrightarrow R-\overset{O}{\underset{}{C}}-NH_2 + R-\overset{O}{\underset{}{C}}-O^- \ {}^+NH_4$$

   (d) Reduction to yield primary alcohols

$$R-\overset{O}{\underset{}{C}}-O-\overset{O}{\underset{}{C}}-R \xrightarrow[\text{2. H}_3O^+]{\text{1. LiAlH}_4, \text{ ether}} 2 \ R-\overset{H \ \ H}{\underset{}{C}}-OH$$

4. Reactions of esters and lactones (Section 21.6)
   (a) Hydrolysis to yield acids

$$R-\overset{O}{\underset{}{C}}-OR' \xrightarrow[\text{or NaOH, H}_2O]{\text{H}_3O^+} R-\overset{O}{\underset{}{C}}-OH + R'OH$$

   (b) Aminolysis to yield amides

$$R-\overset{O}{\underset{}{C}}-OR' + NH_3 \longrightarrow R-\overset{O}{\underset{}{C}}-NH_2 + R'OH$$

   (c) Reduction to yield primary alcohols

$$R-\overset{O}{\underset{}{C}}-OR' \xrightarrow[\text{2. H}_3O^+]{\text{1. LiAlH}_4, \text{ ether}} R-\overset{H \ \ H}{\underset{}{C}}-OH + R'OH$$

   (d) Partial reduction to yield aldehydes

$$R-\overset{O}{\underset{}{C}}-OR' \xrightarrow[\text{2. H}_3O^+]{\text{1. DIBAH, toluene}} R-\overset{O}{\underset{}{C}}-H + R'OH$$

(continued) ▶

(e) Grignard reaction to yield tertiary alcohols

$$
\underset{\substack{R \quad OR'}}{R-C(=O)-OR'} \xrightarrow[\text{2. H}_3\text{O}^+]{\text{1. 2 R''MgX, ether}} \underset{\substack{R \quad OH}}{R-C(R'')_2-OH} + R'OH
$$

5. Reactions of amides and lactams
   (a) Hydrolysis to yield acids (Section 21.7)

$$
\underset{\substack{R \quad NH_2}}{R-C(=O)-NH_2} \xrightarrow[\text{or NaOH, H}_2\text{O}]{\text{H}_3\text{O}^+} \underset{\substack{R \quad OH}}{R-C(=O)-OH} + NH_3
$$

   (b) Reduction to yield amines (Section 21.7)

$$
\underset{\substack{R \quad NH_2}}{R-C(=O)-NH_2} \xrightarrow[\text{2. H}_3\text{O}^+]{\text{1. LiAlH}_4\text{, ether}} \underset{\substack{R \quad NH_2}}{R-C(H)_2-NH_2}
$$

   (c) Dehydration of primary amides to yield nitriles (Section 21.8)

$$
\underset{\substack{R \quad NH_2}}{R-C(=O)-NH_2} \xrightarrow{\text{SOCl}_2} R-C\equiv N + SO_2 + HCl
$$

6. Reactions of nitriles (Section 21.8)
   (a) Hydrolysis to yield carboxylic acids

$$
R-C\equiv N + H_2O \xrightarrow[\text{or NaOH, H}_2\text{O}]{\text{H}_3\text{O}^+} \underset{\substack{R \quad OH}}{R-C(=O)-OH} + NH_3
$$

   (b) Partial hydrolysis to yield amides

$$
R-C\equiv N + H_2O \xrightarrow[\text{or NaOH, H}_2\text{O}]{\text{H}_3\text{O}^+} \underset{\substack{R \quad NH_2}}{R-C(=O)-NH_2}
$$

   (c) Reduction to yield primary amines

$$
R-C\equiv N \xrightarrow[\text{2. H}_3\text{O}^+]{\text{1. LiAlH}_4\text{, ether}} \underset{\substack{R \quad NH_2}}{R-C(H)_2-NH_2}
$$

   (d) Partial reduction to yield aldehydes

$$
R-C\equiv N \xrightarrow[\text{2. H}_3\text{O}^+]{\text{1. DIBAH, toluene}} \underset{\substack{R \quad H}}{R-C(=O)-H} + NH_3
$$

   (e) Reaction with Grignard reagents to yield ketones

$$
R-C\equiv N \xrightarrow[\text{2. H}_3\text{O}^+]{\text{1. R'MgX, ether}} \underset{\substack{R \quad R'}}{R-C(=O)-R'} + NH_3
$$

Visualizing Chemistry

• •

(Problems 21.1–21.31 appear within the chapter.)

21.32 Name the following compounds:
(a) (b)

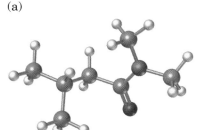
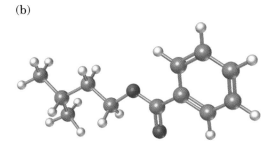

21.33 How would you prepare the following compounds starting with an appropriate carboxylic acid and any other reagents needed? (Reddish brown = Br.)
(a) (b)

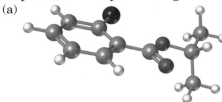
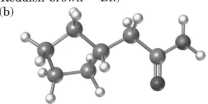

21.34 The following structure represents a tetrahedral alkoxide-ion intermediate formed by addition of a nucleophile to a carboxylic acid derivative. Identify the nucleophile, the leaving group, the starting acid derivative, and the ultimate product. (Yellow-green = Cl.)

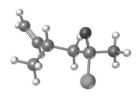

Stereo View

Additional Problems

• •

21.35 Give IUPAC names for the following compounds:

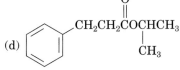

(a) [structure: 4-methylbenzamide — benzene ring with H₃C and C(=O)NH₂]

(b) $CH_3CH_2CHCH{=}CHCN$ with CH_2CH_3 substituent

(c) $CH_3OCCH_2CH_2COCH_3$ (with two C=O)

(d) [structure: benzene-$CH_2CH_2COCHCH_3$ with CH_3 and C=O]

(e) $\underset{\overset{|}{Br}}{CH_3CHCH_2}\overset{\overset{O}{\|}}{C}NHCH_3$ (f) [cyclopentene ring with —CN]

(g) [benzene ring]—$\overset{\overset{O}{\|}}{C}$—O—[benzene ring]   (h) [benzene ring]—$\overset{\overset{O}{\|}}{C}$—O—$\overset{\overset{O}{\|}}{C}$—$\underset{\overset{|}{CH_3}}{CHCH_3}$

21.36 Draw structures corresponding to the following names:
(a) *p*-Bromophenylacetamide   (b) *m*-Benzoylbenzonitrile
(c) 2,2-Dimethylhexanamide   (d) Cyclohexyl cyclohexanecarboxylate
(e) 2-Cyclobutenecarbonitrile   (f) 2-Propylbutanedioyl dichloride

21.37 Draw and name compounds that meet the following descriptions:
(a) Three acid chlorides having the formula C_6H_9ClO
(b) Three amides having the formula $C_7H_{11}NO$
(c) Three nitriles having the formula C_5H_7N

21.38 The following reactivity order has been found for the saponification of alkyl acetates by aqueous NaOH. Explain.

$$CH_3CO_2CH_3 > CH_3CO_2CH_2CH_3 > CH_3CO_2CH(CH_3)_2 > CH_3CO_2C(CH_3)_3$$

21.39 Explain the observation that attempted Fischer esterification of 2,4,6-trimethylbenzoic acid with methanol and HCl is unsuccessful. No ester is obtained, and the acid is recovered unchanged. What alternative method of esterification might be successful?

21.40 When a carboxylic acid is dissolved in isotopically labeled water, the label rapidly becomes incorporated into *both* oxygen atoms of the carboxylic acid. Explain.

$$\underset{R}{\overset{O}{\underset{}{\|}}}C\underset{OH}{} + H_2O^* \rightleftharpoons \underset{R}{\overset{O^*}{\underset{}{\|}}}C\underset{\overset{*}{OH}}{} + H_2O$$

21.41 Outline methods for the preparation of acetophenone (phenyl methyl ketone) starting from the following:
(a) Benzene   (b) Bromobenzene   (c) Methyl benzoate
(d) Benzonitrile   (e) Styrene

21.42 The following reactivity order has been found for the basic hydrolysis of *p*-substituted methyl benzoates:

$$Y = NO_2 > Br > H > CH_3 > OCH_3$$

How can you explain this reactivity order? Where would you expect $Y = C\equiv N$, $Y = CHO$, and $Y = NH_2$ to be in the reactivity list?

[benzene ring with CO_2CH_3 substituent, Y on other position] $\xrightarrow[\text{H}_2\text{O}]{^-\text{OH}}$ [benzene ring with COO^- substituent, Y] $+ CH_3OH$

COOH

21.43 How might you prepare the following compounds from butanoic acid?
(a) 1-Butanol (b) Butanal (c) 1-Bromobutane
(d) Pentanenitrile (e) 1-Butene (f) *N*-Methylpentanamide
(g) 2-Hexanone (h) Butylbenzene

21.44 What product would you expect to obtain from Grignard reaction of an excess of phenylmagnesium bromide with dimethyl carbonate, $CH_3OCOOCH_3$?

21.45 When *ethyl* benzoate is heated in methanol containing a small amount of HCl, *methyl* benzoate is formed. Propose a mechanism for the reaction.

21.46 *tert*-Butoxycarbonyl azide, a reagent used in protein synthesis, is prepared by treating *tert*-butoxycarbonyl chloride with sodium azide. Propose a mechanism for this reaction.

$$\begin{array}{ccc} H_3C & O & \\ | & \| & \\ CH_3COCCl + NaN_3 & \longrightarrow & CH_3COCN_3 + NaCl \\ | & & \\ H_3C & & \end{array}$$

$$\begin{array}{cc} H_3C & O \\ | & \| \\ & \\ | & \\ H_3C & \end{array}$$

21.47 Predict the product, if any, of reaction between propanoyl chloride and the following reagents:
(a) $(Ph)_2CuLi$ in ether (b) $LiAlH_4$, then H_3O^+ (c) CH_3MgBr, then H_3O^+
(d) $Li(O\text{-}t\text{-}Bu)_3AlH$ (e) H_3O^+ (f) Cyclohexanol
(g) Aniline (h) $CH_3COO^-\ {}^+Na$

21.48 Answer Problem 21.47 for reaction of the listed reagents with methyl propanoate.

21.49 Answer Problem 21.47 for reaction of the listed reagents with propanamide and with propanenitrile.

21.50 We said in Section 21.6 that mechanistic studies on ester hydrolysis have been carried out using ethyl propanoate labeled with ^{18}O in the ether-like oxygen. Assume that ^{18}O-labeled acetic acid is your only source of isotopic oxygen, and then propose a synthesis of the labeled ethyl propanoate.

21.51 Treatment of a carboxylic acid with trifluoroacetic anhydride leads to an unsymmetrical anhydride that rapidly reacts with alcohol to give an ester:

$$RCOOH + (CF_3CO)_2O \longrightarrow \overset{\displaystyle O \quad\quad O}{\underset{\displaystyle \|\quad\quad \|}{R-C-O-C-CF_3}} \xrightarrow{R'OH} \overset{\displaystyle O}{\underset{\displaystyle \|}{R-C-OR'}} + CF_3COOH$$

(a) Propose a mechanism for formation of the unsymmetrical anhydride.
(b) Why is the unsymmetrical anhydride unusually reactive?
(c) Why does the unsymmetrical anhydride react as indicated rather than giving a trifluoroacetate ester plus carboxylic acid?

21.52 How would you accomplish the conversion of pentanenitrile into the following substances? More than one step may be required.
(a) $CH_3CH_2CH_2CH_2CH_2NH_2$ (b) $CH_3CH_2CH_2CH_2CH_2N(CH_3)_2$
(c) $CH_3CH_2CH_2CH_2C(CH_3)_2OH$ (d) $CH_3CH_2CH_2CH_2CH(OH)CH_3$
(e) $CH_3CH_2CH_2CH_2CHO$

21.53 List as many ways as you can think of for transforming cyclohexanol into cyclohexanecarbaldehyde (try to get at least four).

21.54 One method for preparing the 1,6-hexanediamine needed for making nylon starts with 1,3-butadiene. How would you accomplish this synthesis?

$$H_2C=CHCH=CH_2 \xrightarrow{\;?\;} H_2NCH_2CH_2CH_2CH_2CH_2CH_2NH_2$$

21.55 Succinic anhydride yields succinimide when heated with ammonium chloride at 200°C. Propose a mechanism for this reaction. Why do you suppose such a high reaction temperature is required?

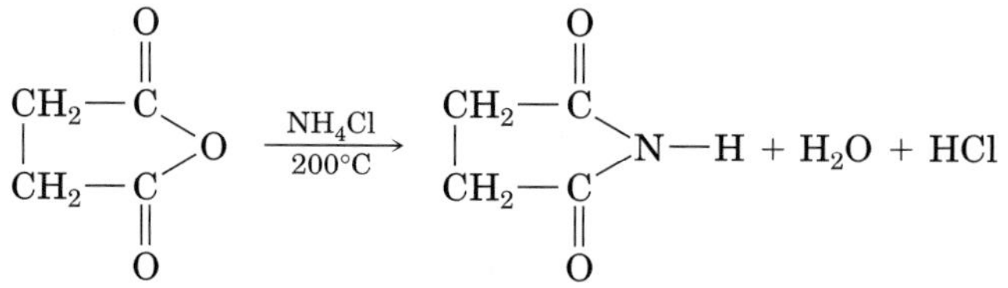

21.56 Butacetin is an analgesic (pain-killing) agent that is synthesized commercially from *p*-fluoronitrobenzene. Propose a synthesis.

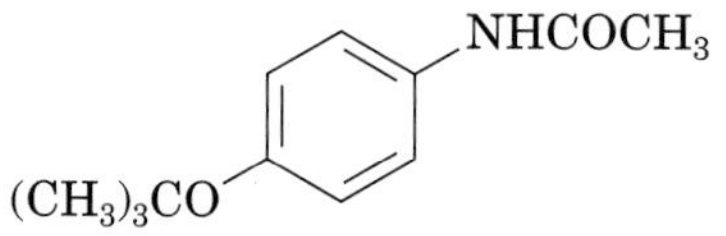

Butacetin

21.57 Phenyl 4-aminosalicylate is a drug used in the treatment of tuberculosis. Propose a synthesis of this compound starting from 4-nitrosalicylic acid.

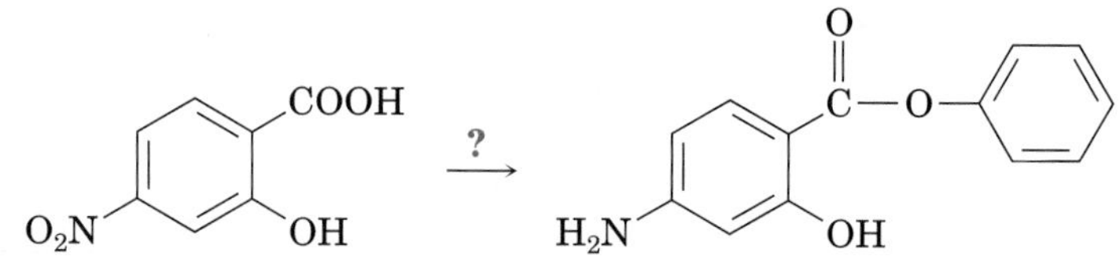

4-Nitrosalicylic acid   **Phenyl 4-aminosalicylate**

21.58 *N,N*-Diethyl-*m*-toluamide (DEET) is the active ingredient in many insect-repellent preparations. How might you synthesize this substance from *m*-bromotoluene?

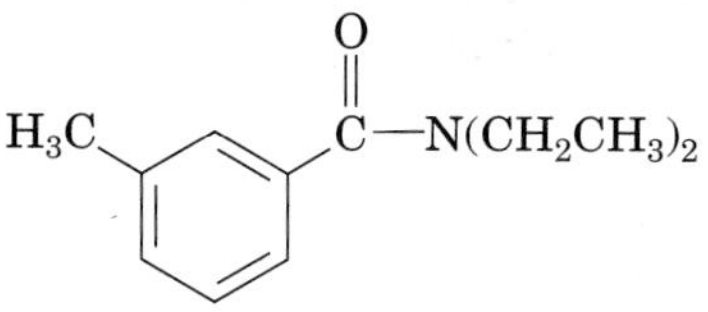

N,N-Diethyl-*m*-toluamide

21.59 Tranexamic acid, a drug useful against blood clotting, is prepared commercially from *p*-methylbenzonitrile. Formulate the steps likely to be used in the synthesis. (Don't worry about cis–trans isomers. Heating to 300°C interconverts the isomers.)

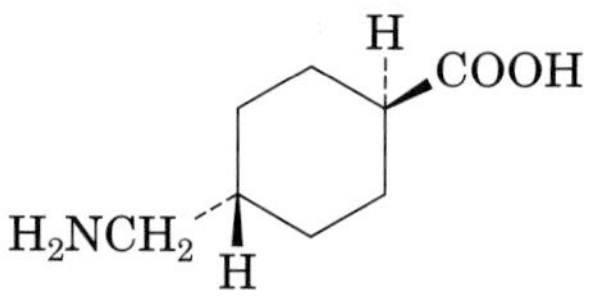

Tranexamic acid

21.60 One frequently used method for preparing methyl esters is by reaction of carboxylic acids with diazomethane, CH_2N_2:

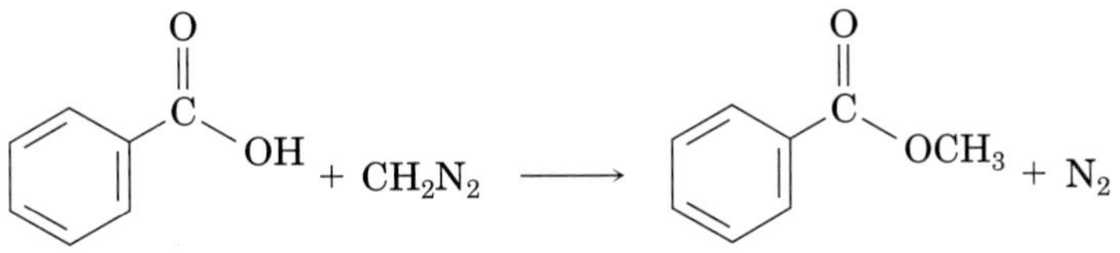

Benzoic acid   Diazomethane   **Methyl benzoate**
                                              (100%)

The reaction occurs in two steps: (1) protonation of diazomethane by the carboxylic acid to yield methyldiazonium ion, $CH_3N_2^+$, plus a carboxylate ion; and (2) reaction of the carboxylate ion with $CH_3N_2^+$.
(a) Draw two resonance structures of diazomethane and account for step 1.
(b) What kind of reaction occurs in step 2?

21.61 The following reaction, called the *benzilic acid rearrangement*, takes place by typical carbonyl-group reactions. Propose a mechanism (Ph = phenyl).

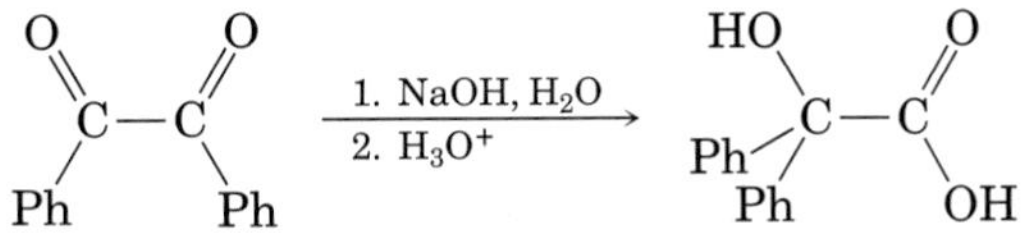

Benzil   **Benzylic acid**

21.62 In light of your answer to Problem 21.61, what product is likely to result from the following reaction?

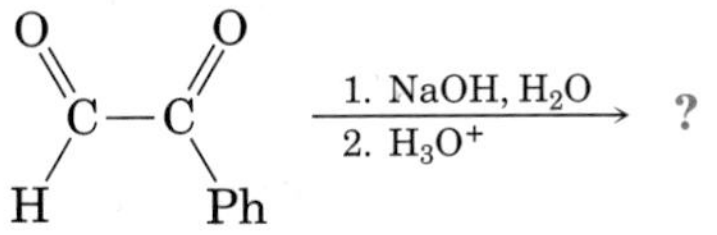

21.63 The step-growth polymer *nylon 6* is prepared from caprolactam. The reaction involves initial reaction of caprolactam with water to give an intermediate open-chain amino acid, followed by heating to form the polymer. Propose mechanisms for both steps, and show the structure of nylon 6.

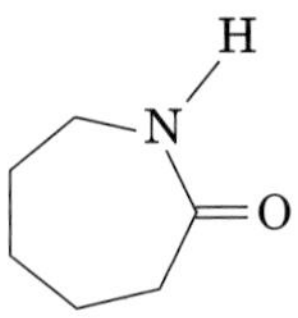

Caprolactam

21.64 *Qiana*, a polyamide fiber with a silky texture, has the following structure. What are the monomer units used in the synthesis of Qiana?

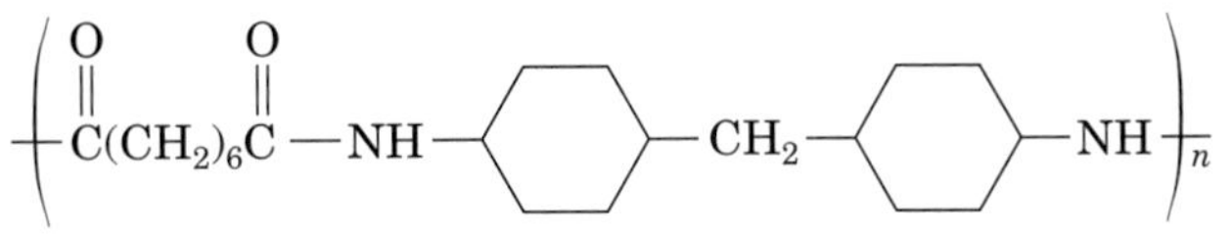

Qiana

21.65 What is the structure of the polymer produced by treatment of β-propiolactone with a small amount of hydroxide ion?

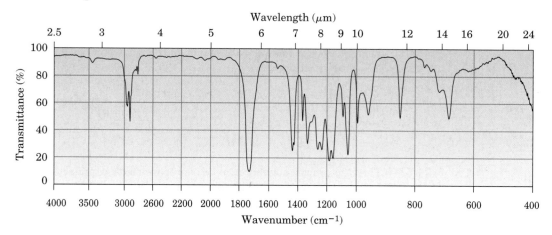

β-**Propiolactone**

21.66 Polyimides having the structure shown are used as coatings on glass and plastics to improve scratch resistance. How would you synthesize a polyimide? (See Problem 21.55.)

A polyimide

21.67 How would you distinguish spectroscopically between the following isomer pairs? Tell what differences you would expect to see.
(a) *N*-Methylpropanamide and *N,N*-dimethylacetamide
(b) 5-Hydroxypentanenitrile and cyclobutanecarboxamide
(c) 4-Chlorobutanoic acid and 3-methoxypropanoyl chloride
(d) Ethyl propanoate and propyl acetate

21.68 Propose a structure for a compound, $C_4H_7ClO_2$, that has the following IR and ^1H NMR spectra:

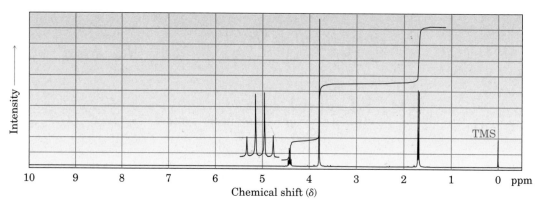

21.69 Propose a structure for a compound, C_4H_7N, that has the following IR and ^1H NMR spectra:

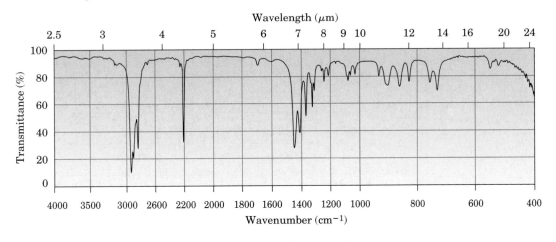

21.70 Assign structures to compounds with the following ^1H NMR spectra:
(a) C_4H_7ClO
IR: 1810 cm^{-1}

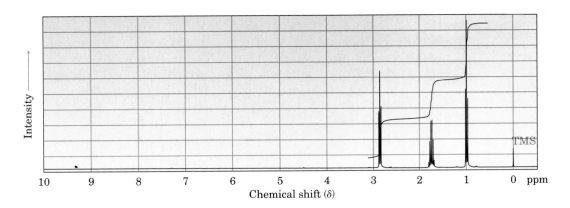

(b) $C_5H_7NO_2$
IR: 2250, 1735 cm^{-1}

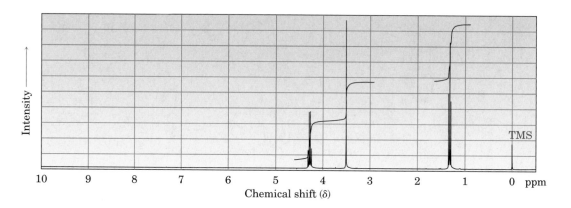

(c) $C_5H_{10}O_2$
IR: 1735 cm^{-1}

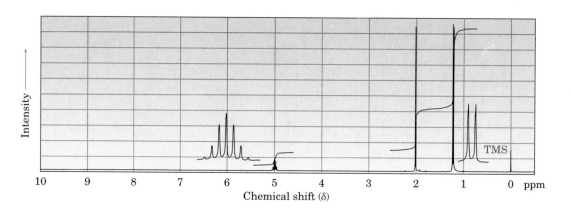

21.71   Propose structures for compounds with the following ^{1}H NMR spectra:
(a) $C_5H_9ClO_2$
IR: 1735 cm^{-1}

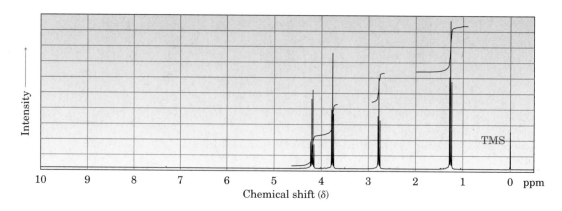

(b) $C_7H_{12}O_4$
IR: 1735 cm^{-1}

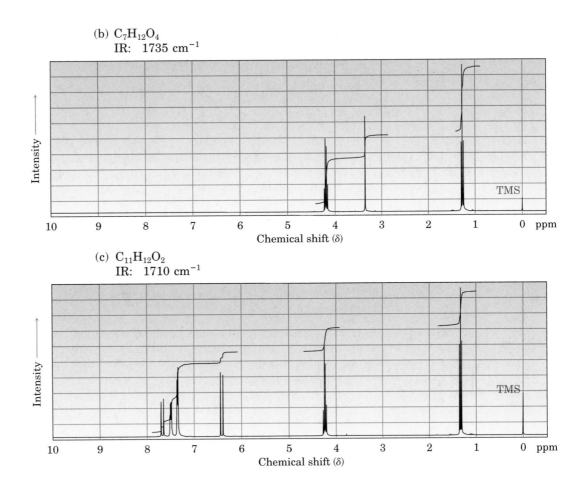

(c) $C_{11}H_{12}O_2$
IR: 1710 cm^{-1}

A Look Ahead

• •

21.72 Epoxy adhesives are prepared in two steps. S_N2 reaction of the disodium salt of bisphenol A with epichlorohydrin forms a "prepolymer," which is then "cured" by treatment with a triamine such as $H_2NCH_2CH_2NHCH_2CH_2NH_2$.

Bisphenol A   **Epichlorohydrin**

"Prepolymer"

Draw structures to show how addition of the triamine results in strengthening the polymer. Amines are good nucleophiles and can open epoxide rings in the same way other bases can. (See Sections 24.3, 24.4.)

21.73 In the *iodoform reaction*, a triiodomethyl ketone reacts with aqueous NaOH to yield a carboxylate ion and iodoform (triiodomethane). Propose a mechanism for this reaction. (See Section 22.7.)

$$
\underset{R}{\overset{O}{\underset{\|}{C}}}\text{–}CI_3 \xrightarrow[\text{H}_2\text{O}]{\text{OH}^-} \underset{R}{\overset{O}{\underset{\|}{C}}}\text{–}O^- + HCI_3
$$

Molecular Modeling

• •

21.74 Use SpartanView to examine electrostatic potential maps of ethyl acetate and *N*-methyl-2-pyrrolidinone. Identify the most basic atoms in each, and draw resonance structures of the protonated forms.

21.75 Use SpartanView to examine the energy profile for rotation about the C–OCH₃ bond in methyl acetate. Which geometries have the lowest energy? Next, examine two conformers (A and B) of the cyclic amido ester shown below, and account for the difference in conformer energy.

21.76 Nucleophiles add more rapidly to the amide group in the four-membered ring of penicillin than to typical alkylamides. Use SpartanView to compare the geometries and electrostatic potential maps of penicillin and *N,N*-dimethylacetamide, and explain the difference in reactivity.

A penicillin

21.77 Use SpartanView to identify the carbonyl stretching frequencies in acetone, ethyl acetate, and *N,N*-dimethylformamide. Is each compound easily distinguished by infrared spectroscopy?

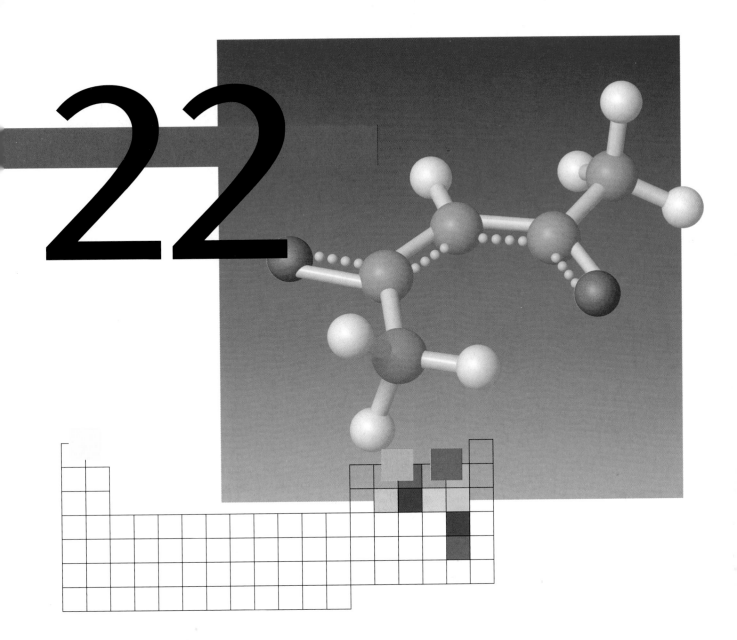

Carbonyl Alpha-Substitution Reactions

We said in "A Preview of Carbonyl Compounds" (p. 743) that much of the chemistry of carbonyl compounds can be explained by just four fundamental reaction types: nucleophilic additions, nucleophilic acyl substitutions, α substitutions, and carbonyl condensations. Having studied the characteristics of nucleophilic addition reactions and nucleophilic acyl substitution reactions in the past three chapters, let's now look in more detail at the third major carbonyl-group process—the **α-substitution reaction**.

Alpha-substitution reactions occur at the position *next to* the carbonyl group—the *α position*—and involve the substitution of an *α* hydrogen atom by an electrophile, E, through either an *enol* or *enolate ion* intermediate. Let's begin by learning more about these two species.

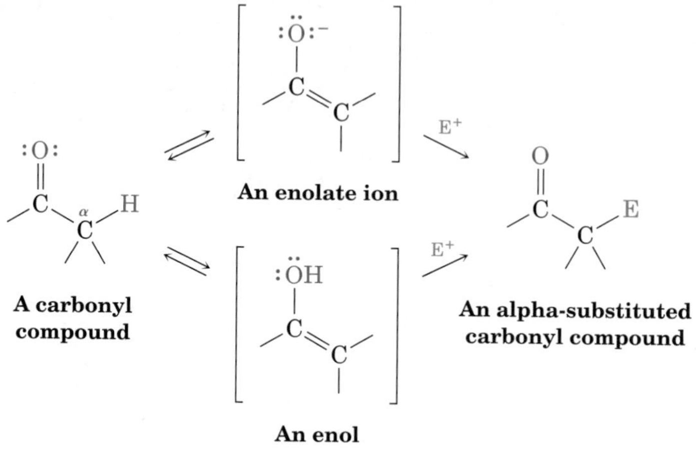

An enolate ion

A carbonyl compound

An enol

An alpha-substituted carbonyl compound

22.1 Keto–Enol Tautomerism

A carbonyl compound with a hydrogen atom on its *α* carbon rapidly equilibrates with its corresponding **enol** (Section 8.5). This rapid interconversion between two substances is a special kind of isomerism known as **tautomerism**, from the Greek *tauto,* "the same," and *meros,* "part." The individual isomers are called **tautomers**.

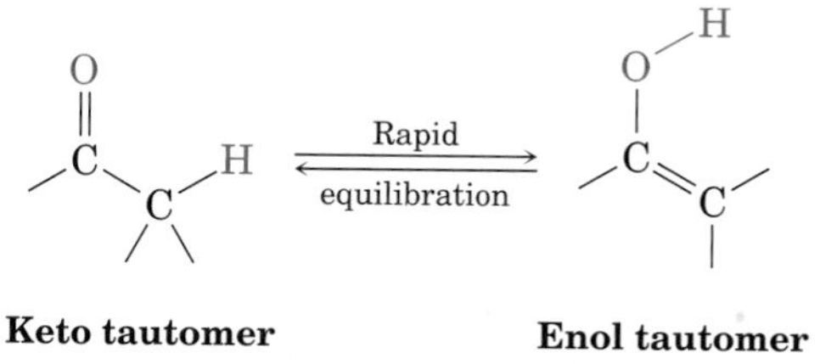

Keto tautomer   **Enol tautomer**

Note the difference between tautomers and resonance forms: Tautomers are different compounds (isomers) with different structures, while resonance forms are different representations of a single structure. Tautomers have their *atoms* arranged differently, while resonance forms differ only in the position of their *electrons.* Note also that tautomers are *rapidly* interconvertible. Thus, keto and enol isomers *are* tautomers, but alkene isomers such as 1-butene and 2-butene are not, because they don't interconvert rapidly under normal circumstances.

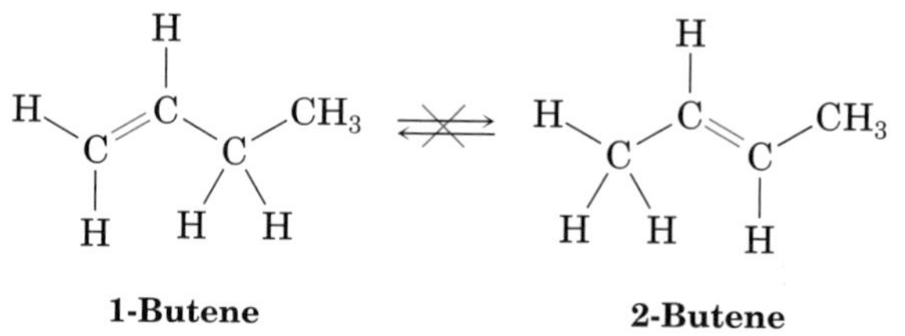

1-Butene   **2-Butene**

Most carbonyl compounds exist almost exclusively in the keto form at equilibrium, and it's usually difficult to isolate the pure enol. For example, cyclohexanone contains only about 0.0001% of its enol tautomer at room temperature, and acetone contains only about 0.000 000 1% enol. The percentage of enol tautomer is even less for carboxylic acids, esters, and amides. Even though enols are difficult to isolate and are present only to a small extent at equilibrium, they are nevertheless extremely important in much of the chemistry of carbonyl compounds because they are so reactive.

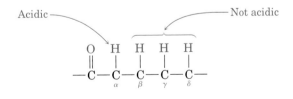

| 99.999 9% | 0.000 1% | 99.999 999 9% | 0.000 000 1% |

Cyclohexanone   **Acetone**

Keto–enol tautomerism of carbonyl compounds is catalyzed by both acids and bases. Acid catalysis occurs by protonation of the carbonyl oxygen atom (a Lewis base) to give an intermediate cation that can lose H⁺ from its α carbon to yield a neutral enol (Figure 22.1, p. 904). This proton loss from the cation intermediate is similar to what occurs during an E1 reaction when a carbocation loses H⁺ to form an alkene (Section 11.14).

Base-catalyzed enol formation occurs by an acid–base reaction between catalyst and carbonyl compound. The carbonyl compound acts as a weak protic acid and donates one of its α hydrogens to the base. The resultant anion—an **enolate ion**—is then reprotonated to yield a neutral compound. Since the enolate ion is a resonance hybrid of two forms, it can be protonated either on the α carbon to regenerate the keto tautomer or on oxygen to give the enol tautomer (Figure 22.2, p. 905).

Note that only the hydrogens on the α positions of carbonyl compounds are acidic. Hydrogens at β, γ, δ, and so on, are not acidic and can't be removed by base. We'll account for this unique behavior of α hydrogens shortly.

● ●

Problem 22.1 Draw structures for the enol tautomers of the following compounds:
(a) Cyclopentanone   (b) Acetyl chloride   (c) Ethyl acetate
(d) Propanal   (e) Acetic acid   (f) Phenylacetone
(g) Acetophenone (methyl phenyl ketone)

Problem 22.2 How many acidic hydrogens does each of the molecules listed in Problem 22.1 have? Identify them.

FIGURE 22.1 ▼

Mechanism of acid-catalyzed enol formation. The protonated intermediate can lose H⁺, either from the oxygen atom to regenerate keto tautomer or from the α carbon atom to yield an enol.

Protonation of the carbonyl oxygen atom by an acid catalyst HA yields a cation that can be represented by two resonance structures.

Loss of H⁺ from the α position by reaction with a base A⁻ then yields the enol tautomer and regenerates HA catalyst.

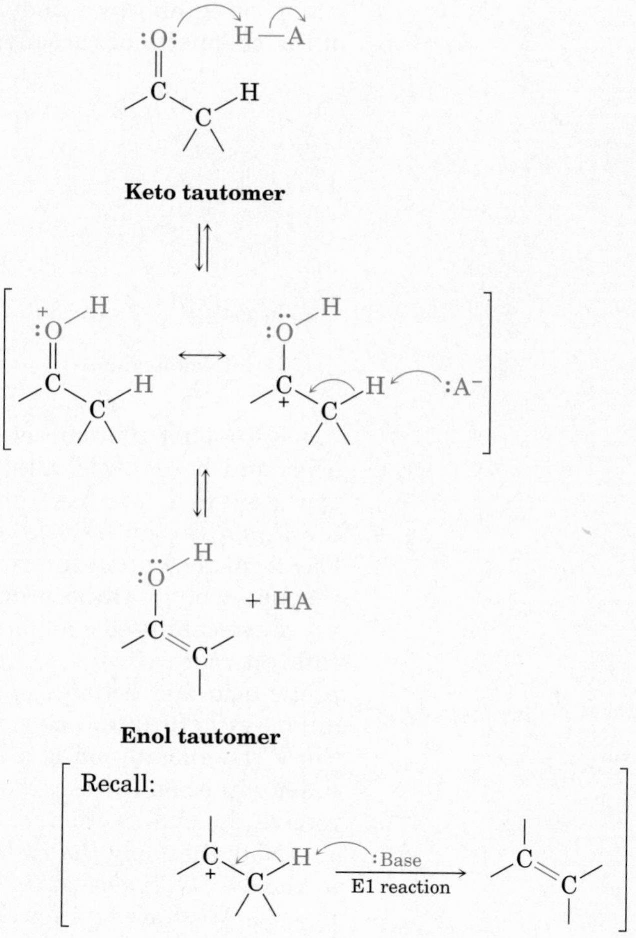

© 1999 JOHN MCMURRY

Problem 22.3 Draw structures for the monoenol forms of 1,3-cyclohexanedione. How many enol forms are possible? Which would you expect to be most stable? Explain.

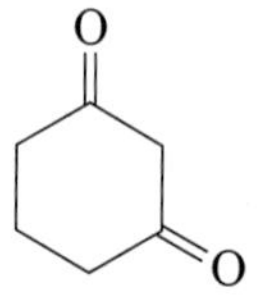

1,3-Cyclohexanedione

FIGURE 22.2 ▼

Mechanism of base-catalyzed enol formation. The intermediate enolate ion, a resonance hybrid of two forms, can be protonated either on carbon to regenerate the starting keto tautomer or on oxygen to give an enol.

Base removes an acidic hydrogen from the α position of the carbonyl compound, yielding an enolate anion that has two resonance structures.

Protonation of the enolate anion on the oxygen atom yields an enol and regenerates the base catalyst.

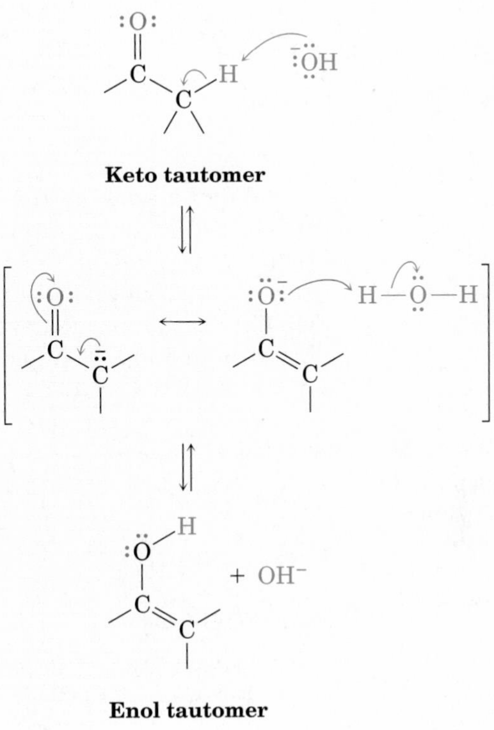

Keto tautomer

Enol tautomer

© 1999 JOHN MCMURRY

22.2 Reactivity of Enols: The Mechanism of Alpha-Substitution Reactions

What kind of chemistry do enols have? Since their double bonds are electron-rich, enols behave as nucleophiles and react with electrophiles in much the same way that alkenes do. But because of resonance electron donation of the lone-pair electrons on oxygen, enols are more electron-rich and correspondingly more reactive than alkenes. (This electron-donating effect of an –OH substituent in an enol is reminiscent of the activating, electron-donating effect of an –OH group in a phenol; Section 16.6.) Notice in the electrostatic potential map of ethenol ($H_2C=CHOH$) shown at the top of the next page that the electron density (red) is greater on the α carbon than on the hydroxyl-bearing carbon.

An enol

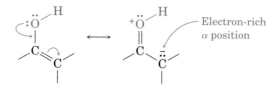

ethenol

Electron-rich
α position

A phenol

Electron-rich
ortho position

When an *alkene* reacts with an electrophile, such as Br_2, initial addition of Br^+ gives an intermediate cation, and subsequent reaction with Br^- then yields an addition product (Section 7.2). When an *enol* reacts with an electrophile, however, only the initial addition step is the same. The intermediate cation immediately loses the –OH proton to give a substituted carbonyl compound. The general mechanism is shown in Figure 22.3.

FIGURE 22.3 ▼

General mechanism of a
carbonyl α-substitution
reaction. The initially
formed cation loses H^+ to
regenerate a carbonyl
compound.

refer to
**Mechanisms
& Movies**

Acid-catalyzed enol formation occurs
by the usual mechanism.

Acid catalyst

An electron pair from the enol
oxygen attacks an electrophile (E^+),
forming a new bond and leaving a
cation intermediate that is stabilized
by resonance between two forms.

:Base

Loss of a proton from oxygen yields
the neutral alpha-substitution
product as a new C=O bond is formed.

© 1984 JOHN MCMURRY

22.3 Alpha Halogenation of Aldehydes and Ketones

Aldehydes and ketones can be halogenated at their α positions by reaction with Cl_2, Br_2, or I_2 in acidic solution. Bromine in acetic acid solvent is often used.

Acetophenone α-Bromoacetophenone (72%)

Cyclohexanone   2-Chlorocyclohexanone
                              (66%)

The α halogenation of an aldehyde or ketone is a typical α-substitution reaction that proceeds by acid-catalyzed formation of an enol intermediate, as shown in Figure 22.4 (p. 908).

A great deal of evidence has been obtained in support of the mechanism shown in Figure 22.4. For example, the rate of halogenation is independent of the halogen's identity. Chlorination, bromination, and iodination of a given aldehyde or ketone all occur at the same rate, indicating that the same rate-limiting step is involved and that the halogen has no part in that step.

Additional evidence is that acid-catalyzed halogenations show second-order kinetics and follow the rate law

Reaction rate $= k$ [Ketone][H^+]

In other words, the rate of halogenation depends only on the concentrations of ketone and acid, and is independent of halogen concentration. Halogen is not involved in the rate-limiting step.

A final piece of evidence comes from deuterium-exchange experiments. If an aldehyde or ketone is treated with D_3O^+ instead of H_3O^+, the acidic α hydrogens are replaced by deuterium. For a given ketone, the rate of deuterium exchange is identical to the rate of halogenation, indicating that the same intermediate is involved in both processes. That common intermediate can only be an enol (see the reaction mechanism at the top of p. 909).

FIGURE 22.4 ▼

Mechanism of the acid-catalyzed bromination of acetone.

The carbonyl oxygen atom is protonated by acid catalyst.

Loss of an acidic proton from the alpha carbon takes place in the normal way to yield an enol intermediate.

An electron pair from the enol attacks bromine, giving an intermediate cation that is stabilized by resonance between two forms.

Loss of the –OH proton then gives the alpha-halogenated product and generates more acid catalyst.

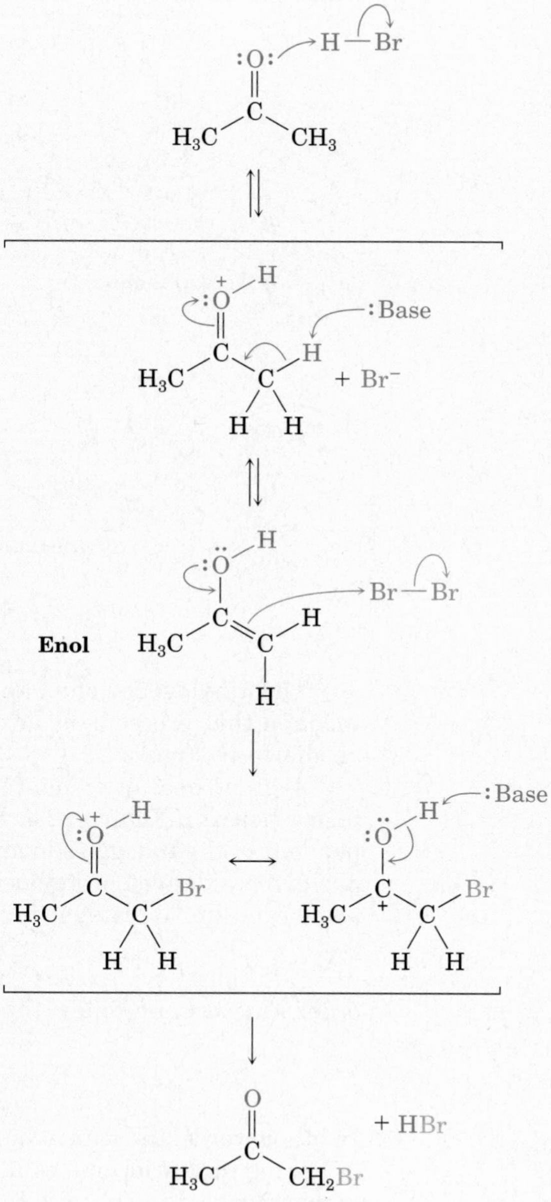

© 1984 JOHN MCMURRY

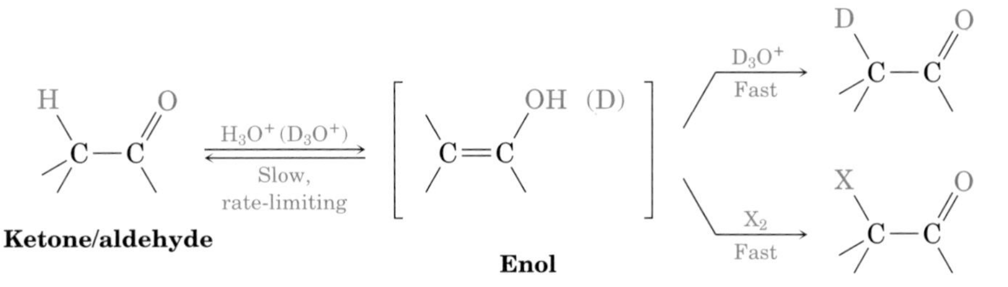

Ketone/aldehyde **Enol**

X = Cl, Br, or I

α-Bromo ketones are useful in organic synthesis because they can be dehydrobrominated by base treatment to yield α,β-unsaturated ketones. For example, 2-bromo-2-methylcyclohexanone gives 2-methyl-2-cyclohexenone in 62% yield when heated in pyridine. The reaction takes place by an E2 elimination pathway (Section 11.11) and is an excellent method for introducing C=C bonds into molecules.

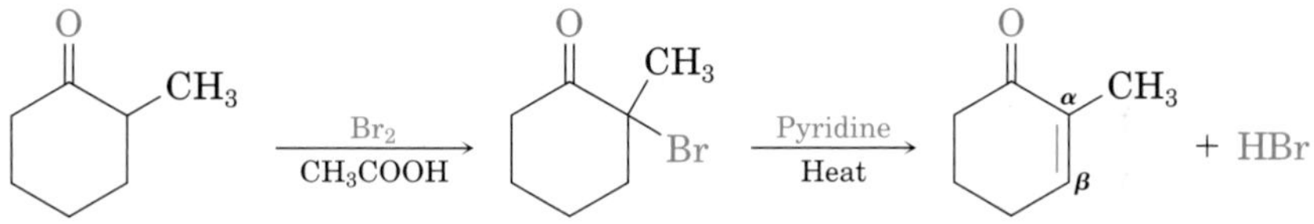

2-Methylcyclohexanone **2-Bromo-2-methylcyclohexanone** **2-Methyl-2-cyclohexenone (62%)**

Problem 22.4 Show the mechanism of the deuteration of acetone on treatment with D_3O^+.

$$CH_3\overset{\overset{\displaystyle O}{\|}}{C}CH_3 \xrightarrow{D_3O^+} CH_3\overset{\overset{\displaystyle O}{\|}}{C}CH_2D$$

Problem 22.5 How might you prepare 1-penten-3-one from 3-pentanone?

22.4 Alpha Bromination of Carboxylic Acids: The Hell–Volhard–Zelinskii Reaction

Direct α bromination of carbonyl compounds by Br_2 in acetic acid is limited to aldehydes and ketones because acids, esters, and amides don't enolize sufficiently for halogenation to take place. Carboxylic acids, however, can be α brominated by a mixture of Br_2 and PBr_3 in the **Hell–Volhard–Zelinskii (HVZ) reaction**.

$$CH_3CH_2CH_2CH_2CH_2CH_2\overset{\overset{\displaystyle O}{\|}}{C}OH \quad \xrightarrow[\text{2. } H_2O]{\text{1. } Br_2, PBr_3} \quad CH_3CH_2CH_2CH_2CH_2\underset{\underset{\displaystyle Br}{|}}{C}H\overset{\overset{\displaystyle O}{\|}}{C}OH$$

Heptanoic acid

2-Bromoheptanoic acid (90%)

The first step in the Hell–Volhard–Zelinskii reaction takes place between PBr_3 and a carboxylic acid to yield an intermediate acid bromide plus HBr (Section 21.4). The HBr thus formed catalyzes enolization of the acid bromide, and the resultant enol reacts rapidly with Br_2 in an α-substitution reaction. Addition of water results in hydrolysis of the α-bromo acid bromide (a nucleophilic acyl substitution reaction) and gives the α-bromo carboxylic acid product.

Acid bromide   **Acid bromide enol**

The overall result of the Hell–Volhard–Zelinskii reaction is the transformation of an acid into an α-bromo acid. Note, though, that the key step involves α substitution of an *acid bromide enol* rather than a carboxylic acid enol. The reaction is analogous in all respects to what occurs during ketone bromination.

Problem 22.6 If methanol rather than water is added at the end of a Hell–Volhard–Zelinskii reaction, an ester rather than an acid is produced. Show how you could prepare methyl 2-bromo-3-methylpentanoate from 3-methylpentanoic acid, and propose a mechanism for the ester-forming step.

22.5 Acidity of Alpha Hydrogen Atoms: Enolate Ion Formation

Biological Connection ▶

During the discussion of base-catalyzed enol formation in Section 22.1, we said that carbonyl compounds can act as weak protic acids. That is, a strong base can abstract an acidic α hydrogen atom from a carbonyl compound to yield an enolate ion.

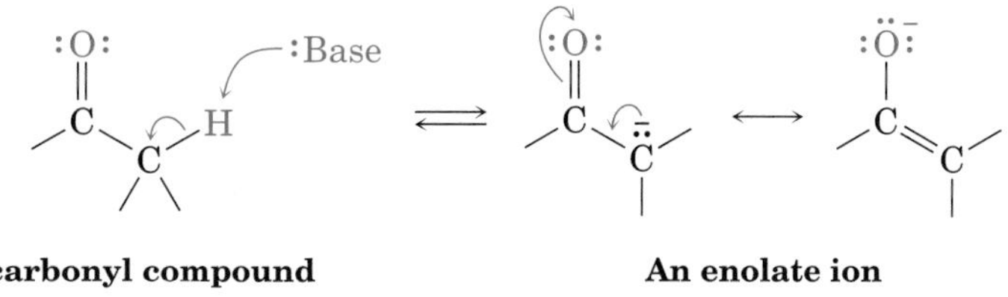

A carbonyl compound   **An enolate ion**

Why are carbonyl compounds somewhat acidic? If we compare acetone, ($pK_a = 19.3$) with ethane ($pK_a \approx 60$), we find that the presence of a neighboring carbonyl group increases the acidity of a ketone over an alkane by a factor of 10^{40}.

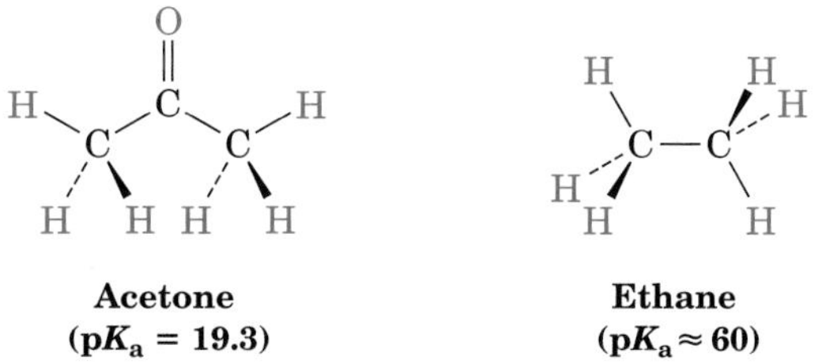

Acetone   **Ethane**
($pK_a = 19.3$)   **($pK_a \approx 60$)**

The reason for this increased acidity is best seen by looking at an orbital picture of the enolate ion (Figure 22.5). Proton abstraction from a carbonyl compound occurs when the α C–H bond is oriented roughly parallel to the

FIGURE 22.5 ▼

Mechanism of enolate ion formation by abstraction of an α proton from a carbonyl compound. The enolate ion is stabilized by resonance, and the negative charge (red) is shared by the oxygen and the α carbon atom, as indicated by the electrostatic potential map.

enolate ion

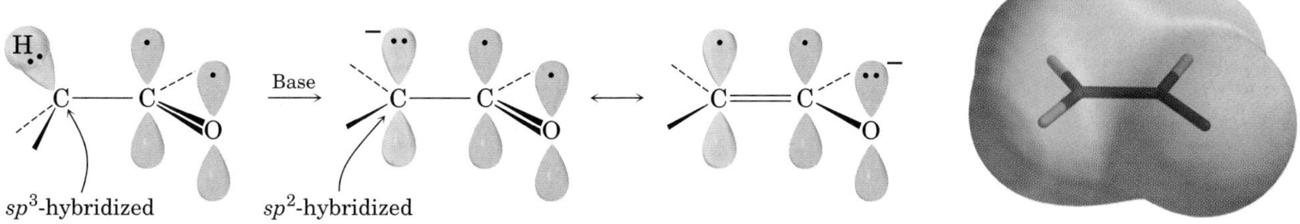

sp^3-hybridized sp^2-hybridized

carbonyl-group p orbitals. The α carbon atom of the enolate ion product is sp^2-hybridized and has a p orbital that overlaps the neighboring carbonyl-group p orbitals. Thus, the negative charge is shared by the electronegative oxygen atom, and the enolate ion is stabilized by resonance between two forms.

Carbonyl compounds are more acidic than alkanes for the same reason that carboxylic acids are more acidic than alcohols (Section 20.3). In both cases, the anions are stabilized by resonance. Enolate ions differ from carboxylate ions in that their two resonance forms are not equivalent—the form with the negative charge on oxygen is lower in energy than the form with the charge on carbon. Nevertheless, the principle behind resonance stabilization is the same in both cases.

CH$_3$CH$_3$ versus

Ethane
(p$K_a \approx$ 60)

Acetone
(pK_a = 19.3)

Nonequivalent resonance forms

CH$_3$OH   versus

Methanol
(pK_a = 15.5)

Acetic acid
(pK_a = 4.75)

Equivalent resonance forms

Because carbonyl compounds are only weakly acidic, strong bases are needed for enolate ion formation. If an alkoxide ion, such as sodium ethoxide, is used as base, deprotonation takes place only to the extent of about 0.1%, because acetone is a weaker acid than ethanol (pK_a = 16). If, however, a more powerful base such as sodium hydride (NaH) or lithium diisopropylamide [LiN(i-C$_3$H$_7$)$_2$] is used, a carbonyl compound can be completely converted into its enolate ion.

Cyclohexanone

Cyclohexanone
enolate ion (100%)

Lithium diisopropylamide (LDA) is easily prepared by reaction between butyllithium (BuLi) and diisopropylamine and is widely used as a base for preparing enolate ions from carbonyl compounds. LDA has nearly ideal properties:

• LDA is a very strong base because its conjugate acid, diisopropyl-amine, has p$K_a \approx$ 40.

- It is soluble in organic solvents such as THF.

- It is hindered so that it doesn't add to carbonyl compounds in competing nucleophilic addition reactions.

- It is effective even at $-78°C$.

$$C_4H_9Li \ + \ H-N\begin{matrix}CH(CH_3)_2 \\ \\ CH(CH_3)_2\end{matrix} \xrightarrow[\text{solvent}]{\text{THF}} Li^+ \ ^-:N\begin{matrix}CH(CH_3)_2 \\ \\ CH(CH_3)_2\end{matrix} \ + \ C_4H_{10}$$

Butyllithium   **Diisopropylamine**   **Lithium diisopropylamide (LDA)**

Many types of carbonyl compounds, including aldehydes, ketones, esters, acids, and amides, can be converted into enolate ions by reaction with LDA. Table 22.1 lists the approximate pK_a values of different types of carbonyl compounds and shows how these values compare to other acidic substances we've seen. Note that nitriles, too, are acidic and can be converted into "enolate-like" anions.

TABLE 22.1 Acidity Constants for Some Organic Compounds

Compound type	Compound	pK_a
Carboxylic acid	CH_3COOH	5
1,3-Diketone	$CH_2(COCH_3)_2$	9
1,3-Keto ester	$CH_3COCH_2CO_2C_2H_5$	11
1,3-Dinitrile	$CH_2(CN)_2$	11
1,3-Diester	$CH_2(CO_2C_2H_5)_2$	13
Alcohol	CH_3CH_2OH	16
Acid chloride	CH_3COCl	16
Aldehyde	CH_3CHO	17
Ketone	CH_3COCH_3	19
Ester	$CH_3CO_2C_2H_5$	25
Nitrile	CH_3CN	25
Dialkylamide	$CH_3CON(CH_3)_2$	30
Ammonia	NH_3	36
Dialkylamine	$HN(i\text{-}C_3H_7)_2$	40
Alkyne	$HC{\equiv}CH$	25
Alkene	$CH_2{=}CH_2$	44
Alkane	CH_3CH_3	60

When a hydrogen atom is flanked by two carbonyl groups, its acidity is enhanced even more. Table 22.1 thus shows that compounds such as 1,3-diketones (**β-diketones**), 3-oxo esters (**β-keto esters**), and 1,3-diesters are more acidic than water. This enhanced acidity of β-dicarbonyl compounds is due to the fact that the resultant enolate ions are stabilized by delocalization of the negative charge over *two* carbonyl groups. The enolate ion of 2,4-pentanedione, for example, has three resonance forms. Similar resonance forms can be drawn for other doubly stabilized enolate ions.

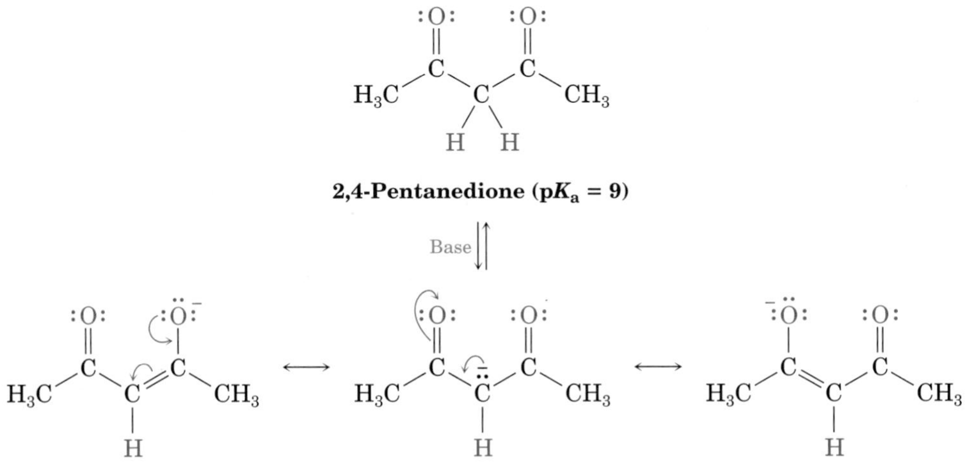

2,4-Pentanedione (pK_a = 9)

···

Practice Problem 22.1   Identify the most acidic hydrogens in each of the following compounds, and rank the compounds in order of increasing acidity:

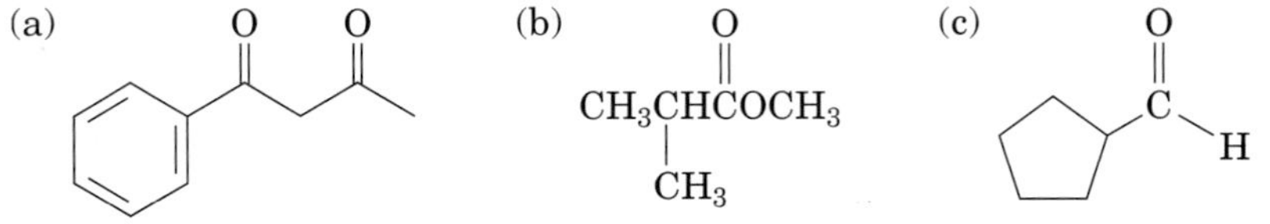

Strategy   Hydrogens on carbon next to a carbonyl group are acidic. In general, a β-dicarbonyl compound is most acidic, a ketone or aldehyde is next most acidic, and a carboxylic acid derivative is least acidic. Remember that alcohols, phenols, and carboxylic acids are also acidic because of their –OH hydrogens.

Solution   Acidic hydrogens are shown in red:

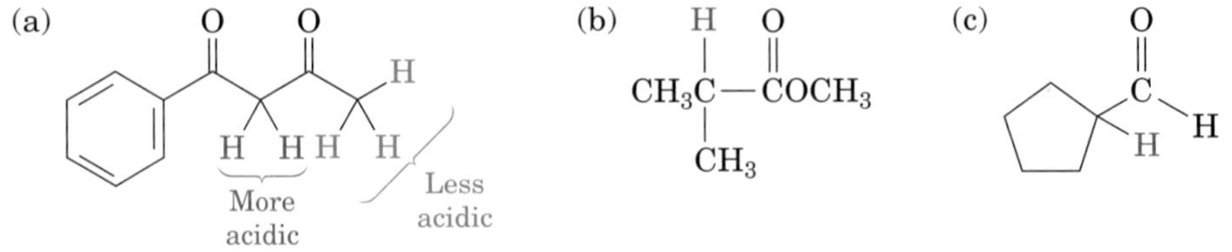

Acidity order: (a) > (c) > (b)

Problem 22.7 Identify the most acidic hydrogens in each of the following molecules:
(a) CH_3CH_2CHO (b) $(CH_3)_3CCOCH_3$ (c) CH_3COOH
(d) Benzamide (e) $CH_3CH_2CH_2CN$ (f) $CH_3CON(CH_3)_2$
(g) 1,3-Cyclohexanedione

Problem 22.8 Draw a resonance structure of the acetonitrile anion, $^-:CH_2C≡N$, and account for the acidity of nitriles.

22.6 Reactivity of Enolate Ions

Enolate ions are more useful than enols for two reasons. First, pure enols can't normally be isolated. They are usually generated only as short-lived intermediates in low concentration. By contrast, stable solutions of pure enolate ions are easily prepared from most carbonyl compounds by reaction with a strong base. Second and more important, enolate ions are much more reactive than enols and undergo many reactions that enols don't. Whereas enols are neutral, enolate ions are negatively charged, making them much better nucleophiles. Thus, the α carbon atom of an enolate ion is highly reactive toward electrophiles. An electrostatic potential map of acetone enolate ion, for instance, shows the electron-rich character (red) of the α carbon.

acetone enolate ion

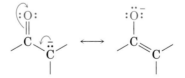

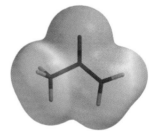

An enolate ion
(negatively charged; reactive;
easily prepared)

As resonance hybrids of two nonequivalent forms, enolate ions can be looked at either as α-keto carbanions ($^-C–C=O$) or as vinylic alkoxides ($C=C–O^-$). Thus, enolate ions can react with electrophiles either on carbon or on oxygen. Reaction on carbon yields an α-substituted carbonyl compound, while reaction on oxygen yields an enol derivative (Figure 22.6). Both kinds of reactivity are known, but reaction on carbon is more common.

FIGURE 22.6 ▼

Two modes of reaction of an enolate ion with an electrophile, E⁺. Reaction on carbon to yield an α-substituted carbonyl product is the more common path.

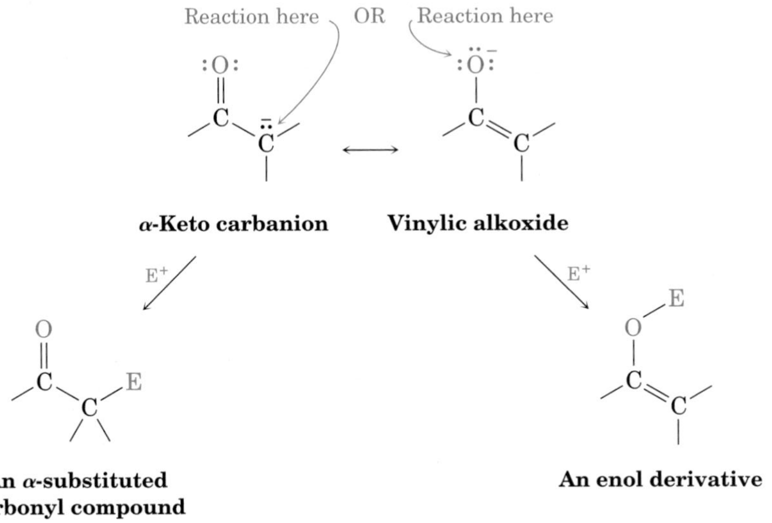

22.7 Halogenation of Enolate Ions: The Haloform Reaction

Halogenation of aldehydes and ketones occurs under both basic and acidic conditions. As you might expect, the base-promoted reaction occurs through an enolate ion intermediate. Even relatively weak bases such as hydroxide ion are effective for halogenation because it's not necessary to convert the ketone completely into its enolate ion. As soon as a small amount of enolate is generated, it reacts immediately with the halogen.

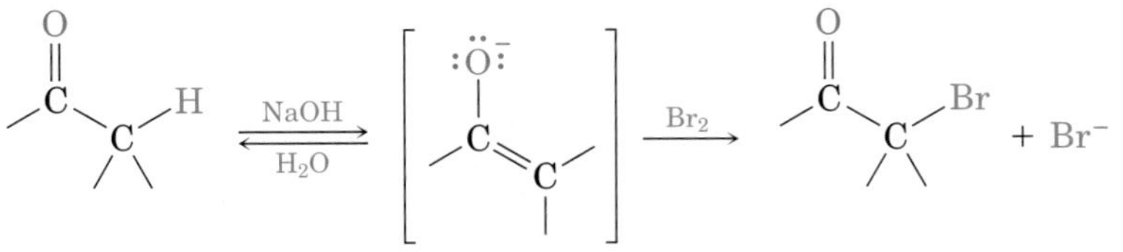

Base-promoted halogenation of aldehydes and ketones is little used in practice because it's difficult to stop the reaction at the monosubstituted product. An α-halogenated ketone is generally more acidic than the starting, unsubstituted ketone because of the electron-withdrawing inductive effect of the halogen atom. Thus, monohalogenated products are themselves rapidly turned into enolate ions and further halogenated.

If excess base and halogen are used, a methyl ketone is triply halogenated and then cleaved by base in the **haloform reaction**. The products are a carboxylic acid plus a so-called *haloform* (chloroform, $CHCl_3$; bromoform, $CHBr_3$; or iodoform, CHI_3). Note that the second step of the reaction is a nucleophilic acyl substitution of $^-CX_3$ by ^-OH. That is, a *carbanion* acts as a leaving group.

The reaction scheme showing halogenation of a methyl ketone:

$$\text{A methyl ketone} \xrightarrow[\text{NaOH}]{X_2} \left[\cdots \right] \longrightarrow$$

A methyl ketone

where   X = Cl, Br, I

- -

Problem 22.9   Why are ketone halogenations in acidic media referred to as being acid-*catalyzed*, whereas halogenations in basic media are base-*promoted*? In other words, why is a full equivalent of base required for halogenation?

Problem 22.10   As a rule, carbanions are poor leaving groups in nucleophilic substitution reactions. Why do you suppose the second step of the haloform reaction takes place so readily?

- -

22.8 Alkylation of Enolate Ions

Biological Connection ▶

Perhaps the single most important reaction of enolate ions is their *alkylation* by treatment with an alkyl halide or tosylate. The alkylation reaction is useful because it forms a new C–C bond, thereby joining two smaller pieces into one larger molecule. Alkylation occurs when the nucleophilic enolate ion reacts with the electrophilic alkyl halide in an S_N2 reaction and displaces the leaving group by back-side attack.

Enolate ion   **Alkyl halide** $\xrightarrow[\text{reaction}]{S_N2}$ ... $+ \ ^-:X$

Alkylation reactions are subject to the same constraints that affect all S_N2 reactions (Section 11.4). Thus, the leaving group X in the alkylating agent R–X can be chloride, bromide, iodide, or tosylate. The alkyl group R must be primary or methyl, and preferably should be allylic or benzylic. Secondary halides react poorly, and tertiary halides don't react at all because a competing E2 elimination of HX occurs instead. Vinylic and aryl halides are also unreactive, because back-side attack is sterically prevented.

$$R—X \begin{cases} —X\text{:} \quad \text{Tosylate} > —I > —Br > —Cl \\ R—\text{:} \quad \text{Allylic} \approx \text{Benzylic} > H_3C— > RCH_2— \end{cases}$$

The Malonic Ester Synthesis

One of the oldest and best-known carbonyl alkylation reactions is the **malonic ester synthesis**, an excellent method for preparing a carboxylic acid from an alkyl halide while lengthening the carbon chain by two atoms.

$$
\underset{\textbf{Alkyl halide}}{R-X} \quad \xrightarrow[\text{synthesis}]{\text{Malonic ester}} \quad \underset{\textbf{Carboxylic acid}}{R-CH_2\overset{\displaystyle O}{\overset{\displaystyle \|}{C}}OH}
$$

Diethyl propanedioate, commonly called diethyl malonate or *malonic ester*, is more acidic than monocarbonyl compounds ($pK_a = 13$) because its α hydrogens are flanked by two carbonyl groups. Thus, malonic ester is easily converted into its enolate ion by reaction with sodium ethoxide in ethanol. The enolate ion, in turn, is a good nucleophile that reacts rapidly with an alkyl halide to give an α-substituted malonic ester. Note in the following examples that the abbreviation "Et" is used for an ethyl group, $-CH_2CH_3$:

$$
\underset{\substack{\textbf{Diethyl propanedioate}\\ \textbf{(Malonic ester)}}}{H-\overset{\displaystyle CO_2Et}{\underset{\displaystyle H}{C}}-CO_2Et} \xrightarrow[\text{EtOH}]{\text{Na}^+ \ ^-\text{OEt}} \underset{\substack{\textbf{Sodio malonic ester}}}{\left[Na^+ \ ^-:\overset{\displaystyle CO_2Et}{\underset{\displaystyle H}{C}}-CO_2Et\right]} \xrightarrow{\text{RX}} \underset{\substack{\textbf{An alkylated}\\ \textbf{malonic ester}}}{R-\overset{\displaystyle CO_2Et}{\underset{\displaystyle H}{C}}-CO_2Et}
$$

The product of malonic ester alkylation has one acidic α hydrogen atom left, so the alkylation process can be repeated a second time to yield a dialkylated malonic ester:

$$
\underset{\substack{\textbf{An alkylated}\\ \textbf{malonic ester}}}{R-\overset{\displaystyle CO_2Et}{\underset{\displaystyle H}{C}}-CO_2Et} \xrightarrow[\text{2. R'X}]{\text{1. Na}^+ \ ^-\text{OEt}} \underset{\substack{\textbf{A dialkylated}\\ \textbf{malonic ester}}}{R-\overset{\displaystyle CO_2Et}{\underset{\displaystyle R'}{C}}-CO_2Et}
$$

On heating with aqueous hydrochloric acid, the alkylated (or dialkylated) malonic ester undergoes hydrolysis and *decarboxylation* (loss of CO_2) to yield a substituted monoacid:

$$
R-\overset{\displaystyle CO_2Et}{\underset{\displaystyle H}{C}}-CO_2Et \xrightarrow[\text{Heat}]{H_3O^+} R-\overset{\displaystyle H}{\underset{\displaystyle H}{C}}-COOH + CO_2 + 2\,EtOH
$$

This decarboxylation is not a general reaction of carboxylic acids. Rather, it is unique to compounds that have a *second* carbonyl group two atoms away from the –COOH. That is, only substituted malonic acids and β-keto acids undergo loss of CO_2 on heating. The decarboxylation reaction occurs by a cyclic mechanism and involves initial formation of an enol, thereby accounting for the need to have a second carbonyl group appropriately positioned.

A diacid → **An acid enol** → **A carboxylic acid**

A β-keto acid → **An enol** → **A ketone**

As noted previously, the overall effect of the malonic ester synthesis is to convert an alkyl halide into a carboxylic acid while lengthening the carbon chain by two atoms.

$CH_3CH_2CH_2CH_2Br$

1-Bromobutane
+
$CH_2(CO_2Et)_2$

Diethyl malonate

$\xrightarrow[\text{EtOH}]{Na^+ \; {}^-OEt}$ $CH_3CH_2CH_2CH_2CCO_2Et$ (with CO_2Et up, H down) $\xrightarrow[\text{Heat}]{H_3O^+}$ $CH_3CH_2CH_2CH_2CH_2COH$ + CO_2

Hexanoic acid (75%) + 2 EtOH

$CH_3CH_2CH_2CH_2CCO_2Et$ (with CO_2Et up, H down) $\xrightarrow[\text{2. } CH_3I]{\text{1. } Na^+ \; {}^-OEt}$ $CH_3CH_2CH_2CH_2CCO_2Et$ (with CO_2Et up, CH_3 down) $\xrightarrow[\text{Heat}]{H_3O^+}$ $CH_3CH_2CH_2CH_2CHCOH$ (with CH_3 down) + CO_2

2-Methylhexanoic acid (74%) + 2 EtOH

The malonic ester synthesis can also be used to prepare *cyclo*alkane-carboxylic acids. For example, when 1,4-dibromobutane is treated with diethyl malonate in the presence of 2 equivalents of sodium ethoxide base, the second alkylation step occurs *intramolecularly* to yield a cyclic product. Hydrolysis and decarboxylation then give cyclopentanecarboxylic acid. Three-, four-, five-, and six-membered rings can be prepared in this way, but yields decrease for larger ring sizes.

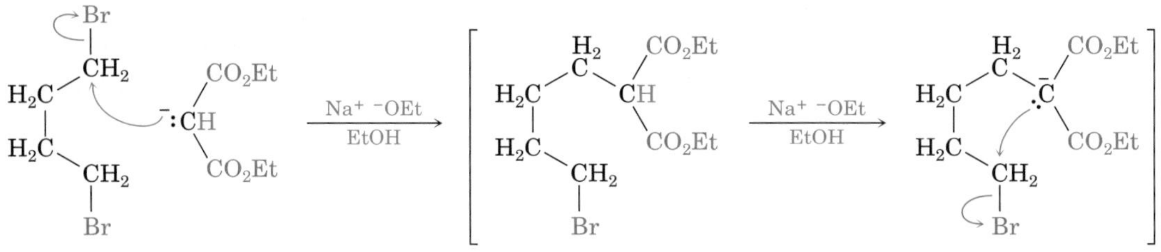

1,4-Dibromobutane

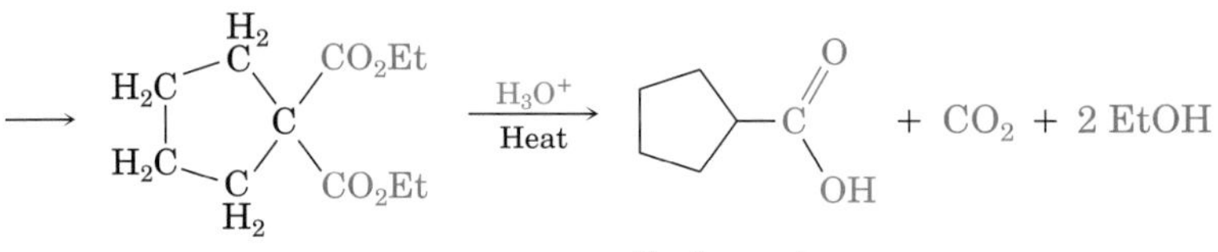

Cyclopentane-carboxylic acid

Practice Problem 22.2 How would you prepare heptanoic acid using a malonic ester synthesis?

Strategy The malonic ester synthesis converts an alkyl halide into a carboxylic acid having two more carbons. Thus, a *seven*-carbon acid chain must be derived from the *five*-carbon alkyl halide 1-bromopentane.

Solution

$$CH_3CH_2CH_2CH_2CH_2Br + CH_2(CO_2Et)_2 \xrightarrow[\text{2. } H_3O^+\text{, heat}]{\text{1. } Na^+ \ ^-OEt} CH_3CH_2CH_2CH_2CH_2\mathbf{CH_2}\overset{\overset{\displaystyle O}{\|}}{C}OH$$

Problem 22.11 How could you use a malonic ester synthesis to prepare the following compounds? Show all steps.

(a) 3-Phenylpropanoic acid (b) 2-Methylpentanoic acid
(c) 4-Methylpentanoic acid (d) Ethyl cyclobutanecarboxylate

Problem 22.12 Monoalkylated and dialkylated acetic acids can be prepared by the malonic ester synthesis, but trialkylated acetic acids (R_3CCOOH) can't be prepared. Explain.

The Acetoacetic Ester Synthesis

The **acetoacetic ester synthesis** is a method for converting an alkyl halide into a methyl ketone in the same way that the malonic ester synthesis is a method for converting an alkyl halide into a carboxylic acid.

$$R-X \xrightarrow[\text{synthesis}]{\text{Acetoacetic ester}} R-CH_2\overset{\overset{\displaystyle O}{\|}}{C}CH_3$$

Alkyl halide   **Methyl ketone**

Ethyl 3-oxobutanoate, commonly called ethyl acetoacetate or *acetoacetic ester*, is much like malonic ester in that its α hydrogens are flanked by two carbonyl groups. It is therefore readily converted into its enolate ion, which can be alkylated by reaction with an alkyl halide. A second alkylation can also be carried out if desired, since acetoacetic ester has two acidic α hydrogens.

Acetoacetic ester   **A monoalkylated
                                                          acetoacetic ester**

**A monoalkylated
acetoacetic ester**   **A dialkylated
                                                          acetoacetic ester**

On heating with aqueous HCl, the alkylated (or dialkylated) acetoacetic ester is hydrolyzed to a β-keto acid and then decarboxylated to yield the ketone product. The decarboxylation occurs in the same way as in the malonic ester synthesis and involves a ketone enol as initial product.

The three-step sequence of (1) enolate ion formation, (2) alkylation, and (3) hydrolysis/decarboxylation is applicable to all β-keto esters with acidic α hydrogens, not just to acetoacetic ester itself. For example, *cyclic β-keto* esters such as ethyl 2-oxocyclohexanecarboxylate can be alkylated and decarboxylated to give 2-substituted cyclohexanones.

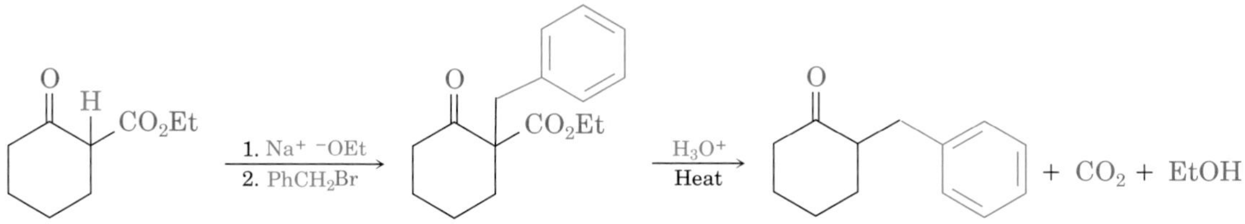

Ethyl 2-oxocyclohexane-carboxylate (a cyclic β-keto ester)

2-Benzylcyclohexanone (77%)

Practice Problem 22.3 How would you prepare 2-pentanone by an acetoacetic ester synthesis?

Strategy The acetoacetic ester synthesis yields a methyl ketone by adding three carbons to an alkyl halide:

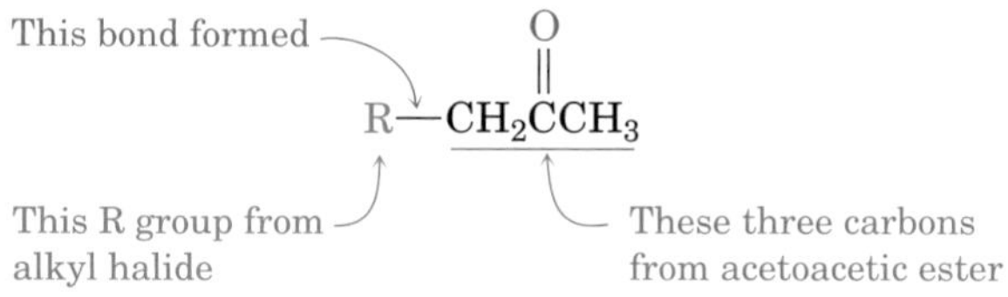

Thus, the acetoacetic ester synthesis of 2-pentanone must involve reaction of bromoethane.

Solution

$$CH_3CH_2Br + EtO\overset{O}{\overset{\|}{C}}CH_2\overset{O}{\overset{\|}{C}}CH_3 \xrightarrow[2.\ H_3O^+,\ \text{heat}]{1.\ Na^+\ {}^-OEt} CH_3CH_2CH_2\overset{O}{\overset{\|}{C}}CH_3$$

2-Pentanone

Problem 22.13 What alkyl halides would you use to prepare the following ketones by an acetoacetic ester synthesis?
(a) 5-Methyl-2-hexanone (b) 5-Phenyl-2-pentanone

Problem 22.14 How would you prepare methyl cyclopentyl ketone using an acetoacetic ester synthesis?

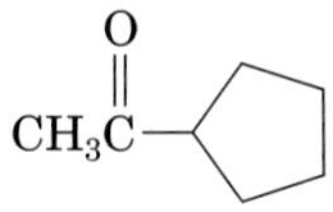

Methyl cyclopentyl ketone

Problem 22.15 Which of the following compounds can't be prepared by an acetoacetic ester synthesis? Explain.
(a) Phenylacetone   (b) Acetophenone   (c) 3,3-Dimethyl-2-butanone

• •

Direct Alkylation of Ketones, Esters, and Nitriles

Both the malonic ester synthesis and the acetoacetic ester synthesis are relatively easy to carry out because they involve unusually acidic carbonyl compounds. As a result, relatively mild bases like sodium ethoxide in an alcohol solvent can be used to prepare the necessary enolate ions. Alternatively, it's also possible in many cases to alkylate the α position of *mono*ketones, *mono*esters, and nitriles. A strong, sterically hindered base such as LDA is needed, so that complete conversion to the enolate ion takes place rather than a nucleophilic addition, and a nonprotic solvent must be used.

Ketones, esters, and nitriles can all be alkylated using LDA or related dialkylamide bases in THF. (Aldehydes rarely give high yields of pure products because their enolate ions undergo carbonyl condensation reactions instead of alkylation. We'll study this condensation reaction in the next chapter.) Some specific examples of alkylation reactions are shown below.

Lactone

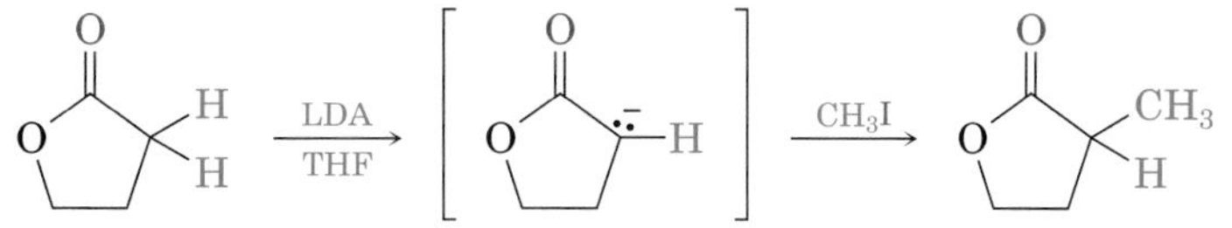

Butyrolactone   **2-Methylbutyrolactone (88%)**

Ester

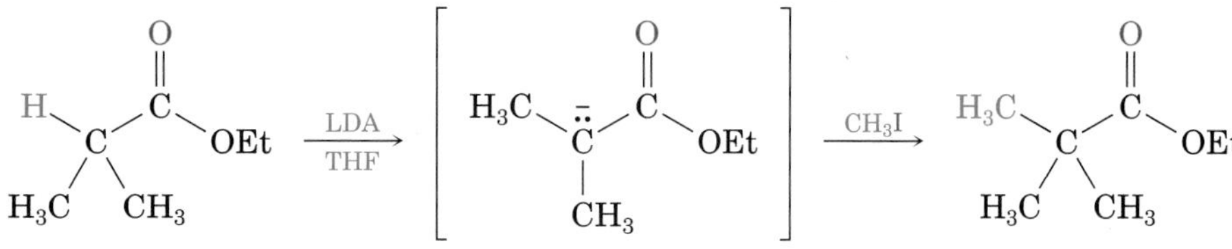

Ethyl 2-methylpropanoate   **Ethyl 2,2-dimethylpropanoate (87%)**

Ketone

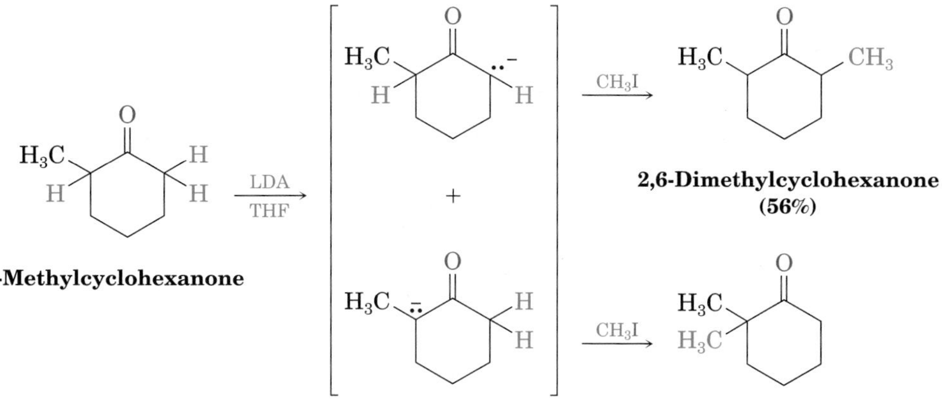

2-Methylcyclohexanone

2,6-Dimethylcyclohexanone (56%)

2,2-Dimethylcyclohexanone (6%)

Nitrile

Phenylacetonitrile   **2-Phenylpropanenitrile**
                                                                                 (71%)

   Note in one of the previous examples that alkylation of 2-methylcyclo-hexanone leads to a mixture of products because both possible enolate ions are formed. In general, the major product in such cases occurs by alkyla-tion at the less hindered, more accessible position. Thus, alkylation of 2-methylcyclohexanone occurs primarily at C6 (secondary) rather than at C2 (tertiary).

• •

Practice Problem 22.4   How might you use an alkylation reaction to prepare ethyl 1-methylcyclo-hexanecarboxylate?

Ethyl 1-methylcyclohexanecarboxylate

Strategy   An alkylation reaction is used to introduce a primary alkyl group onto the α position of a ketone, ester, or nitrile by S_N2 reaction of an enolate ion with an alkyl halide. Thus, we need to look at the target molecule and identify any primary alkyl groups attached to an α carbon. In the present instance, the target has an α methyl group, which might be introduced by alkylation of an ester enolate ion with iodomethane.

Solution

Ethyl cyclohexane-   **Ethyl 1-methylcyclo-**
carboxylate   **hexanecarboxylate**

• •

Problem 22.16   Show how you might prepare the following compounds using an alkylation reaction as the key step:
(a) 3-Phenyl-2-butanone   (b) 2-Ethylpentanenitrile
(c) 2-Allylcyclohexanone   (d) 2,2,6,6-Tetramethylcyclohexanone

• •

Unusual Elements in Organic Chemistry

Carbon, hydrogen, oxygen, nitrogen, sulfur, the halogens, and a few others—these are the elements usually thought of in connection with organic chemistry. In recent years, though, the list has grown to include such elements as boron, copper, and mercury among the more familiar elements, and also selenium, samarium, zirconium, titanium, thallium, cerium, and others among the less familiar elements. Modern organic chemistry, in fact, makes use of almost all the 90 naturally occurring elements except the noble gases (Figure 22.7).

FIGURE 22.7 ▼

The elements highlighted in green are all used in modern organic chemistry.

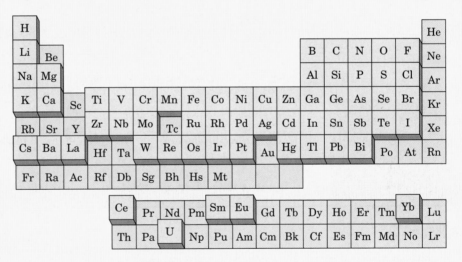

One of the more useful methods used in enolate ion chemistry is the finding that carbonyl compounds can be *selenenylated*. That is, a selenium atom can be introduced onto the α position of a carbonyl compound. Selenenylation is accomplished by allowing the carbonyl compound to react with LDA to generate an enolate ion, followed by addition of benzeneselenenyl bromide, C_6H_5SeBr. Immediate α-substitution reaction yields an α-phenylseleno-substituted product.

(continued) ▶

The value of the selenenylation reaction is that the product can be converted into an α,β-unsaturated carbonyl compound. On treatment with dilute H_2O_2 at room temperature, the selenium is oxidized, elimination occurs, and an α,β-unsaturated carbonyl compound is formed. The net result is introduction of a C=C bond into the α,β position of the carbonyl starting material. Yields are usually excellent, and the method is often superior to the alternative α-bromination/dehydrobromination route (Section 22.3). No added base is required (as in dehydrobromination), and the reaction occurs quickly at room temperature.

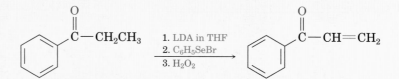

An α-phenylseleno ketone

For example:

Propiophenone   **Phenyl vinyl ketone**
                                                     (80%)

What is true of selenenylation is also true of many other newly discovered reactions. The elements involved may be unusual, but the chemistry is often mild, selective, and much superior to older, more classical methods.

Selenium, used in making photoelectric cells and photocopying machines, is also valuable in organic chemistry.

Summary and Key Words

The **α substitution** of a carbonyl compound through an **enol** or **enolate ion** intermediate is one of the four fundamental reaction types in carbonyl-group chemistry.

An enolate ion

**A carbonyl
compound**

An enol

**An α-substituted
carbonyl compound**

Carbonyl compounds rapidly equilibrate with their enols, a process called **tautomerism**. Although enol **tautomers** are normally present to only a small extent at equilibrium and can't usually be isolated in pure form, they nevertheless contain a highly nucleophilic double bond and react rapidly with electrophiles. For example, ketones and aldehydes are rapidly halogenated at the α position by reaction with Cl_2, Br_2, or I_2 in acetic acid solution. Alpha bromination of carboxylic acids can be similarly accomplished by the **Hell–Volhard–Zelinskii (HVZ) reaction**, in which an acid is treated with Br_2 and PBr_3. The α-halogenated products can then undergo base-induced E2 elimination to yield α,β-unsaturated carbonyl compounds.

Alpha hydrogen atoms of carbonyl compounds are acidic and can be removed by strong bases, such as lithium diisopropylamide (LDA), to yield nucleophilic enolate ions. The most important reaction of enolate ions is their S_N2 alkylation with alkyl halides. The **malonic ester synthesis** provides a method for converting an alkyl halide into a carboxylic acid with the addition of two carbon atoms. Similarly, the **acetoacetic ester synthesis** provides a method for converting an alkyl halide into a methyl ketone. In addition, many carbonyl compounds, including ketones, esters, and nitriles, can be directly alkylated by treatment with LDA and an alkyl halide.

Summary of Reactions

1. Ketone/aldehyde halogenation, where X = Cl, Br, or I (Section 22.3)

$$R-\overset{O}{\overset{\|}{C}}-\overset{H}{\underset{}{C}} + X_2 \xrightarrow{CH_3COOH} R-\overset{O}{\overset{\|}{C}}-\overset{X}{\underset{}{C}} + HX$$

(continued) ▶

2. Hell–Volhard–Zelinskii bromination of acids (Section 22.4)

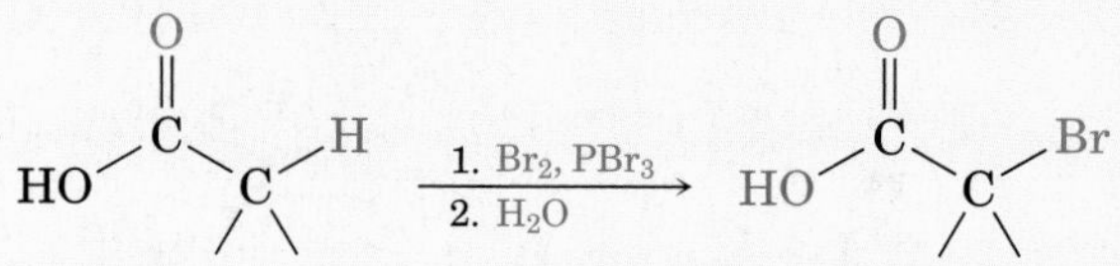

3. Dehydrobromination of α-bromo ketones (Section 22.3)

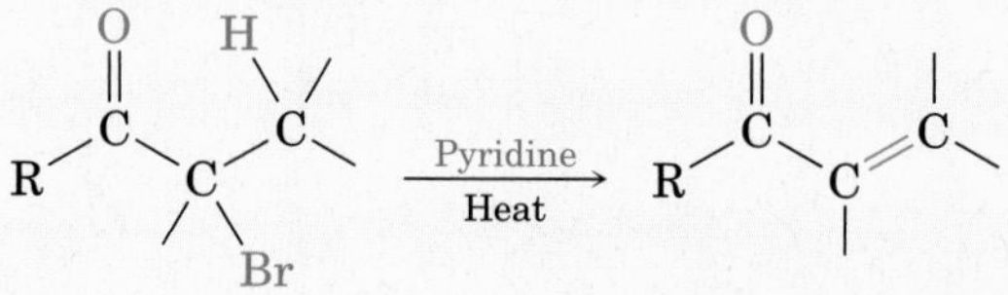

4. Haloform reaction, where X = Cl, Br, or I (Section 22.7)

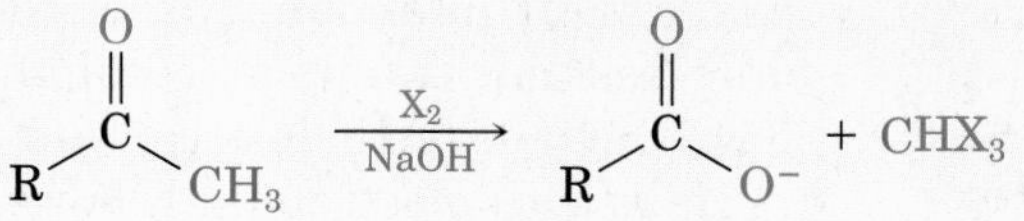

5. Alkylation of enolate ions (Section 22.8)
   (a) Malonic ester synthesis

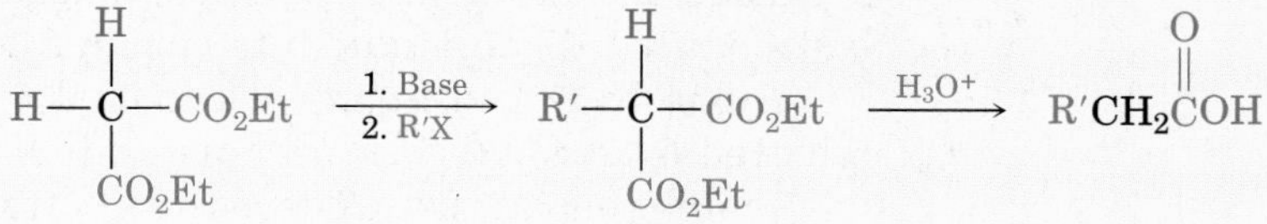

   (b) Acetoacetic ester synthesis

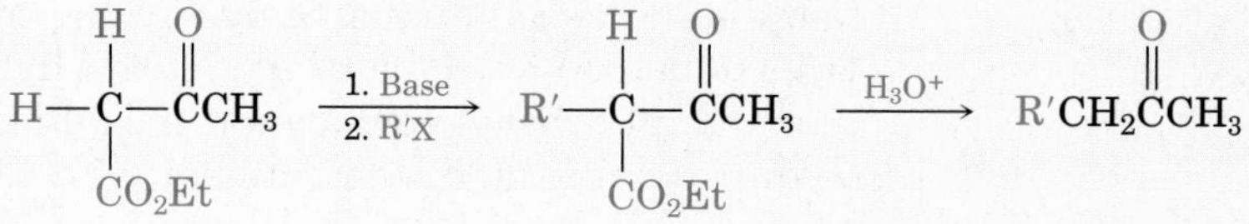

   (c) Alkylation of ketones

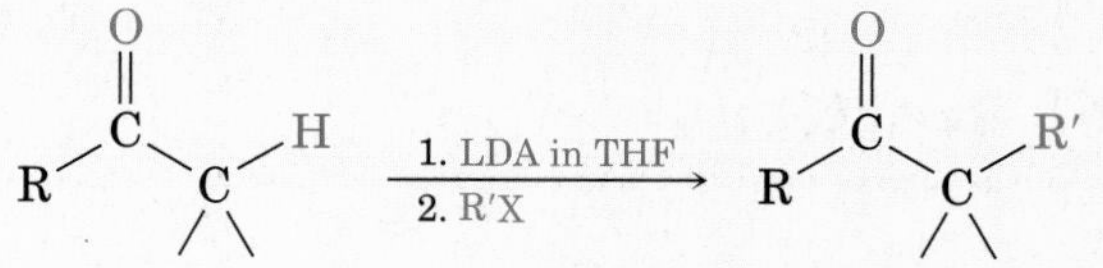

   (d) Alkylation of esters

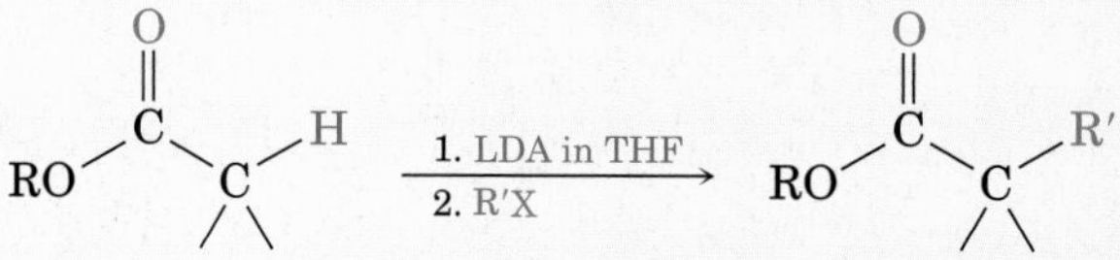

(continued) ▶

(e) Alkylation of nitriles

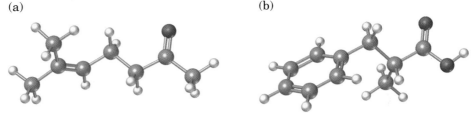

Visualizing Chemistry

● ●

(Problems 22.1–22.16 appear within the chapter.)

22.17 Show the steps in preparing each of the following substances, using either a malonic ester synthesis or an acetoacetic ester synthesis:

(a)   (b)

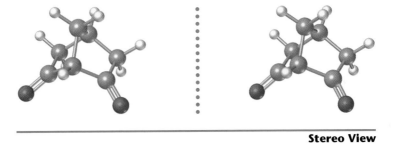

22.18 Unlike most β diketones, the following β diketone has no detectable enol content and is about as acidic as acetone. Explain.

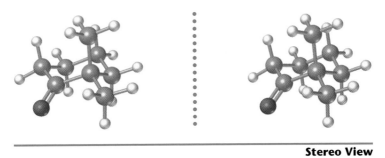

Stereo View

22.19 For a given α hydrogen atom to be acidic, the C–H bond must be parallel to the *p* orbitals of the C=O double bond (that is, perpendicular to the plane of the adjacent carbonyl group). Identify the most acidic hydrogen atom in the following structure. Is it axial or equatorial?

Stereo View

Additional Problems

· ·

22.20   Identify all the acidic hydrogens ($pK_a < 25$) in the following molecules:

(a)

$$CH_3CH_2\underset{\underset{CH_3}{|}}{C}HCCH_3$$ (with C=O)

(b) O=⬠=O

(c)

$$HOCH_2CH_2\overset{O}{\overset{||}{C}}C{\equiv}CCH_3$$

(d) benzene ring with CO_2CH_3 and CH_2CN substituents

(e) cyclopentane ring with $COCl$

(f)

$$CH_3CH_2\underset{\underset{CH_3}{|}}{\overset{\overset{O}{||}}{C}}C{=}CH_2$$

22.21   Rank the following compounds in order of increasing acidity:
(a) CH_3CH_2COOH (b) CH_3CH_2OH (c) $(CH_3CH_2)_2NH$

(d) CH_3COCH_3 (e) $CH_3\overset{O}{\overset{||}{C}}CH_2\overset{O}{\overset{||}{C}}CH_3$ (f) CCl_3COOH

22.22   Write resonance structures for the following anions:

(a) $CH_3\overset{O}{\overset{||}{C}}\overset{\cdot\cdot}{\overset{-}{C}}H\overset{O}{\overset{||}{C}}CH_3$ (b) $CH_3CH{=}CH\overset{\cdot\cdot}{\overset{-}{C}}H\overset{O}{\overset{||}{C}}CH_3$ (c) $N{\equiv}C\overset{\cdot\cdot}{\overset{-}{C}}H\overset{O}{\overset{||}{C}}OCH_3$

(d) benzene ring with $\overset{\cdot\cdot}{\overset{-}{C}}H\overset{O}{\overset{||}{C}}CH_3$

(e) indane-1,3-dione structure with $\overset{\cdot\cdot}{-}C$ bearing $\overset{O}{\overset{||}{}}$ and OCH_3

22.23   One way to determine the number of acidic hydrogens in a molecule is to treat the compound with NaOD in D_2O, isolate the product, and determine its molecular weight by mass spectrometry. For example, if cyclohexanone is treated with NaOD in D_2O, the product has MW = 102. Explain how this method works.

22.24   Base treatment of the following α,β-unsaturated carbonyl compound yields an anion by removal of H^+ from the γ carbon. Why are hydrogens on the γ carbon atom acidic?

benzene–$\overset{O}{\overset{||}{C}}$–CH=CHCH$_3$ $\xrightarrow{\text{LDA}}$ benzene–$\overset{O}{\overset{||}{C}}$–CH=CH$\overset{-}{C}H_2$

22.25   Treatment of 1-phenyl-2-propenone with a strong base such as LDA does *not* yield an anion, even though it contains a hydrogen on the carbon atom next to the carbonyl group. Explain.

benzene–$\overset{O}{\overset{||}{C}}$–C$\underset{H}{}$=CH$_2$ **1-Phenyl-2-propenone**

22.26 Predict the product(s) of the following reactions:

(a) 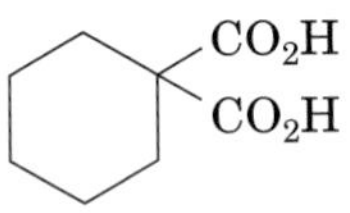 [cyclohexane with CO₂H, CO₂H substituents] $\xrightarrow{\text{Heat}}$?

(b) 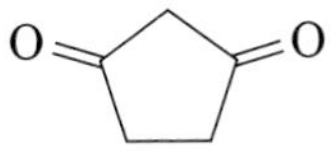 [cyclopentane-1,3-dione] $\xrightarrow[\text{2. CH}_3\text{I}]{\text{1. Na}^+\,{}^-\text{OEt}}$?

(c) $CH_3CH_2CH_2COOH$ $\xrightarrow{\text{Br}_2,\ \text{PBr}_3}$ **A?** $\xrightarrow{\text{H}_2\text{O}}$ **B?**

(d) 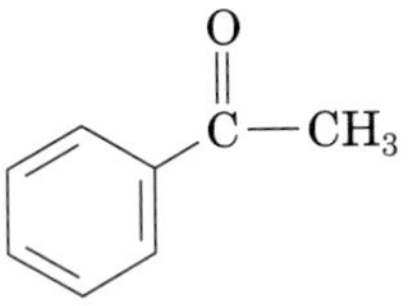 [phenyl ketone: C(=O)—CH₃ attached to benzene ring] $\xrightarrow[\text{I}_2]{{}^-\text{OH, H}_2\text{O}}$?

22.27 Base-promoted chlorination and bromination of a given ketone occur at the same rate. Explain.

22.28 Which of the following compounds can be prepared by a malonic ester synthesis? Show the alkyl halide you would use in each case.
(a) Ethyl pentanoate   (b) Ethyl 3-methylbutanoate
(c) Ethyl 2-methylbutanoate   (d) Ethyl 2,2-dimethylpropanoate

22.29 How would you prepare the following ketones using an acetoacetic ester synthesis?

(a) $CH_3CH_2\underset{\underset{\displaystyle CH_2CH_3}{|}}{CH}\overset{\displaystyle O}{\overset{\|}{C}}CH_3$

(b) $CH_3CH_2CH_2\underset{\underset{\displaystyle CH_3}{|}}{CH}\overset{\displaystyle O}{\overset{\|}{C}}CH_3$

22.30 How would you prepare the following compounds using either an acetoacetic ester synthesis or a malonic ester synthesis?

(a) $CH_3\underset{\underset{\displaystyle CO_2Et}{|}}{\overset{\overset{\displaystyle CH_3}{|}}{C}}CO_2Et$

(b) 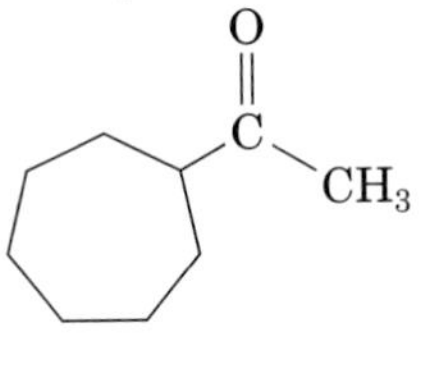 [cycloheptane with C(=O)CH₃ substituent]

(c) [cyclobutane with C(=O)OH substituent]

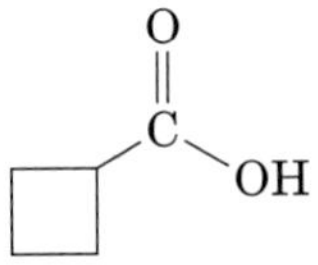

(d) $H_2C{=}CHCH_2CH_2\overset{\displaystyle O}{\overset{\|}{C}}CH_3$

22.31 Which of the following substances would give a positive haloform reaction?
(a) CH_3COCH_3 (b) Acetophenone   (c) CH_3CH_2CHO
(d) CH_3COOH (e) $CH_3C{\equiv}N$

22.32 When optically active (R)-2-methylcyclohexanone is treated with either aqueous base or acid, racemization occurs. Explain.

22.33 Would you expect optically active (S)-3-methylcyclohexanone to be racemized on acid or base treatment in the same way as 2-methylcyclohexanone (Problem 22.32)? Explain.

22.34 When an optically active carboxylic acid such as (R)-2-phenylpropanoic acid is brominated under Hell–Volhard–Zelinskii conditions, is the product optically active or racemic? Explain.

22.35 Fill in the reagents a–c that are missing from the following scheme:

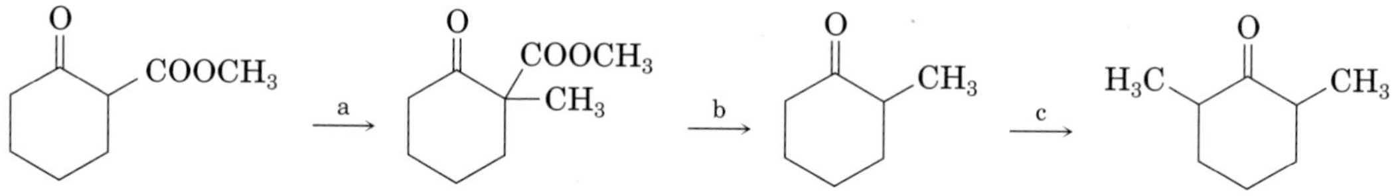

22.36 Nonconjugated β,γ-unsaturated ketones, such as 3-cyclohexenone, are in an acid-catalyzed equilibrium with their conjugated α,β-unsaturated isomers. Propose a mechanism for this isomerization.

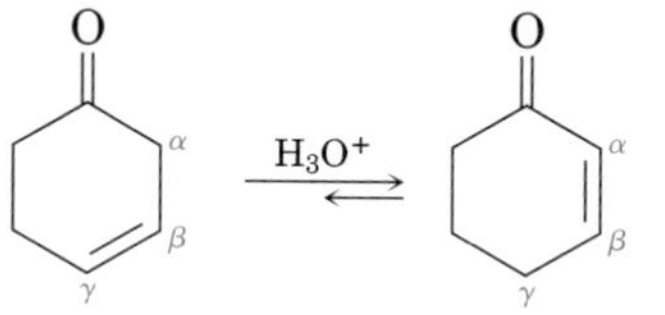

22.37 The interconversion of unsaturated ketones described in Problem 22.36 is also catalyzed by base. Explain.

22.38 An interesting consequence of the base-catalyzed isomerization of unsaturated ketones described in Problem 22.37 is that 2-substituted 2-cyclopentenones can be interconverted with 5-substituted 2-cyclopentenones. Propose a mechanism for this isomerization.

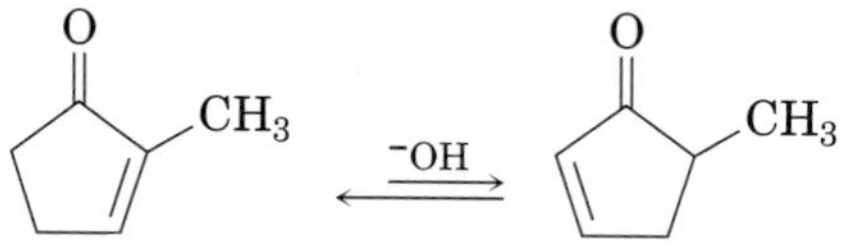

22.39 Although 2-substituted 2-cyclopentenones are in a base-catalyzed equilibrium with their 5-substituted 2-cyclopentenone isomers (Problem 22.38), the analogous isomerization is not observed for 2-substituted 2-cyclohexenones. Explain.

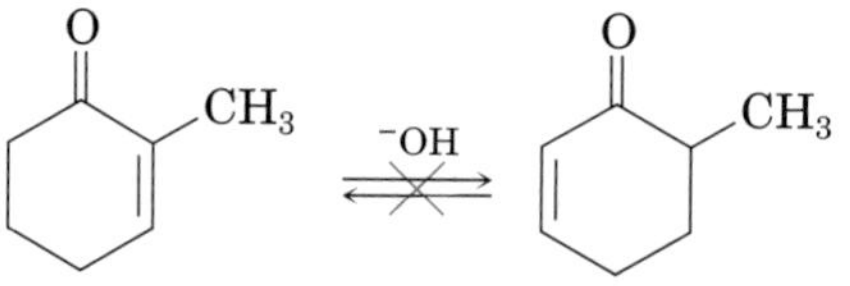

22.40 All attempts to isolate primary and secondary nitroso compounds result only in the formation of oximes. Tertiary nitroso compounds, however, are stable. Explain.

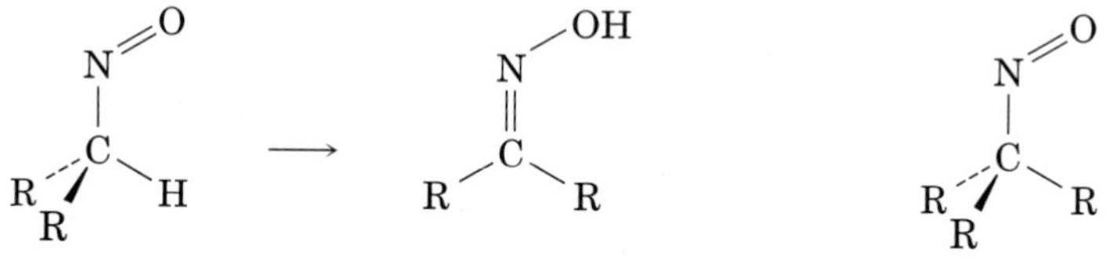

A 1° or 2° nitroso compound   An oxime   A 3° nitroso compound
(unstable)   (stable)

22.41 How might you convert geraniol into either ethyl geranylacetate or geranylacetone?

$\xrightarrow{?}$ $(CH_3)_2C{=}CHCH_2CH_2C(CH_3){=}CHCH_2CH_2CO_2Et$

Ethyl geranylacetate

$(CH_3)_2C{=}CHCH_2CH_2C(CH_3){=}CHCH_2OH$

Geraniol

$\xrightarrow{?}$ $(CH_3)_2C{=}CHCH_2CH_2C(CH_3){=}CHCH_2CH_2COCH_3$

Geranylacetone

22.42 How would you synthesize the following compounds from cyclohexanone? More than one step may be required.

(a) [cyclohexane ring with =CH$_2$]   (b) [cyclohexane ring with CH$_2$Br]   (c) [cyclohexanone ring with CH$_2$C$_6$H$_5$]

(d) [cyclohexane ring with CH$_2$CH$_2$COOH]   (e) [cyclohexene ring with COOH]   (f) [cyclohexenone ring]

(g) [cyclohexane ring with COOH]

22.43 The two isomers *cis*- and *trans*-4-*tert*-butyl-2-methylcyclohexanone are interconverted by base treatment. Which isomer do you think is more stable, and why?

22.44 The following synthetic routes are incorrect. What is wrong with each?

(a) $CH_3CH_2CH_2CH_2\overset{\overset{\textstyle O}{\|}}{C}OEt$ $\xrightarrow[\text{2. Pyridine, heat}]{\text{1. Br}_2\text{, CH}_3\text{CO}_2\text{H}}$ $CH_3CH_2CH{=}CH\overset{\overset{\textstyle O}{\|}}{C}OEt$

(b) $CH_3\overset{\overset{\textstyle CO_2Et}{|}}{C}HCO_2Et$ $\xrightarrow[\substack{\text{2. PhBr} \\ \text{3. H}_3\text{O}^+\text{, heat}}]{\text{1. Na}^+ {}^-\text{OEt}}$ [benzene ring]$-\overset{\overset{\textstyle CH_3}{|}}{C}HCOOH$

(c) $CH_3\overset{\overset{\textstyle O}{\|}}{C}CH_2\overset{\overset{\textstyle O}{\|}}{C}OEt$ $\xrightarrow[\substack{\text{2. H}_2\text{C}{=}\text{CHCH}_2\text{Br} \\ \text{3. H}_3\text{O}^+\text{, heat}}]{\text{1. Na}^+ {}^-\text{OEt}}$ $H_2C{=}CHCH_2CH_2\overset{\overset{\textstyle O}{\|}}{C}OH$

22.45 Attempted Grignard reaction of cyclohexanone with *tert*-butylmagnesium bromide gives only about 1% yield of the expected addition product along with 99% unreacted cyclohexanone. If D_3O^+ is added to the reaction mixture after a suitable period, however, the "unreacted" cyclohexanone is found to have one deuterium atom incorporated into it. Explain.

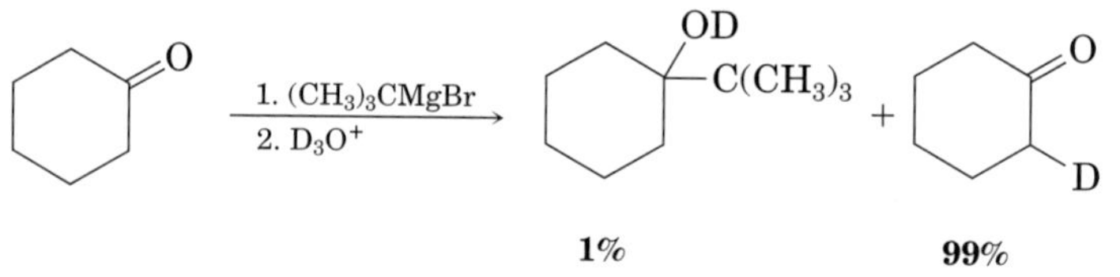

1% **99%**

22.46 The *Favorskii reaction* involves treatment of an α-bromo ketone with base to yield a ring-contracted product. For example, reaction of 2-bromocyclohexanone with aqueous NaOH yields cyclopentanecarboxylic acid. Propose a mechanism.

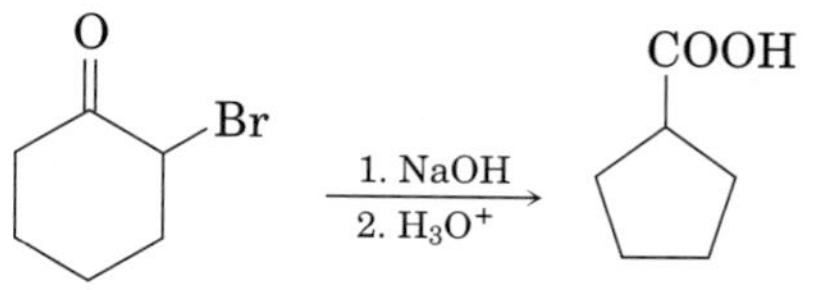

22.47 Treatment of a cyclic ketone with diazomethane is a method for accomplishing a *ring-expansion reaction*. For example, treatment of cyclohexanone with diazomethane yields cycloheptanone. Propose a mechanism.

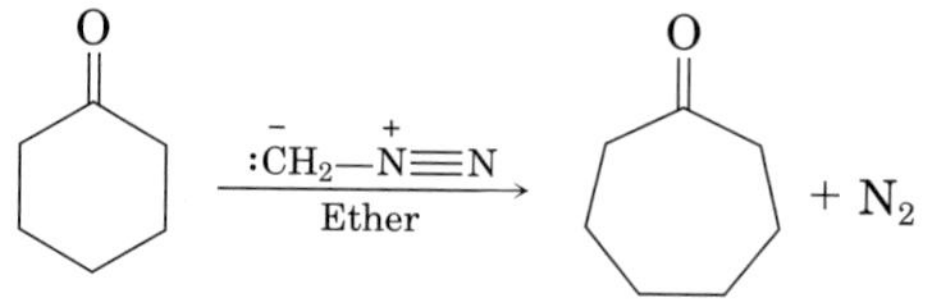

22.48 Ketones react slowly with benzeneselenenyl chloride in the presence of HCl to yield α-phenylseleno ketones. Propose a mechanism for this acid-catalyzed α-substitution reaction. (See "Unusual Elements in Organic Chemistry" at the end of this chapter.)

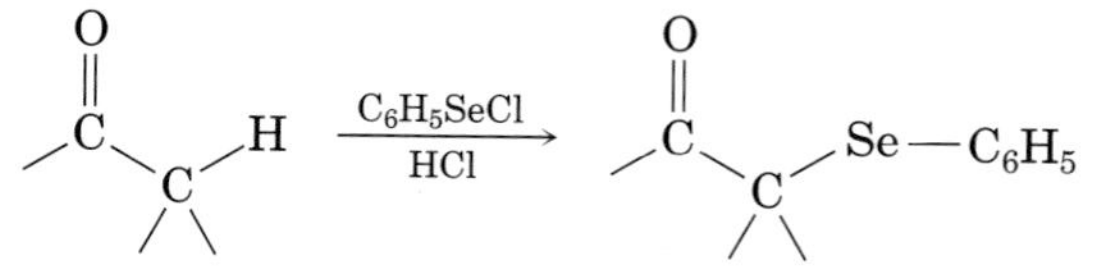

22.49 As far back as the sixteenth century, South American Incas chewed the leaves of the coca bush, *Erythroxylon coca*, to combat fatigue. Chemical studies of *Erythroxylon coca* by Friedrich Wöhler in 1862 resulted in the discovery of *cocaine*, $C_{17}H_{21}NO_4$, as the active component. Basic hydrolysis of cocaine leads to methanol, benzoic acid, and another compound called *ecgonine*, $C_9H_{15}NO_3$. Oxidation of ecgonine with CrO_3 yields a keto acid that readily loses CO_2 on heating, giving tropinone.

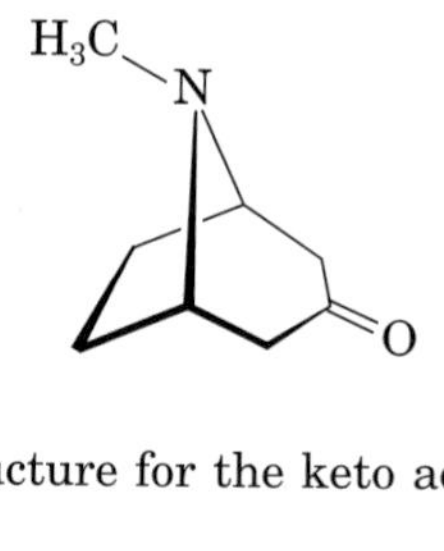

Tropinone

(a) What is a likely structure for the keto acid?

(b) What is a likely structure for ecgonine, neglecting stereochemistry?

(c) What is a likely structure for cocaine, neglecting stereochemistry?

22.50 The final step in an attempted synthesis of laurene, a hydrocarbon isolated from the marine alga *Laurencia glandulifera*, involved the Wittig reaction shown. The product obtained, however, was not laurene but an isomer. Propose a mechanism to account for these unexpected results.

Laurene
(*NOT formed*)

22.51 The key step in a reported laboratory synthesis of sativene, a hydrocarbon isolated from the mold *Helminthosporium sativum*, involves the following base treatment of a keto tosylate. What kind of reaction is occurring? How would you complete the synthesis?

A keto tosylate   **Sativene**

A Look Ahead

22.52 Amino acids can be prepared by reaction of alkyl halides with diethyl acetamido-malonate, followed by heating the initial alkylation product with aqueous HCl. Show how you would prepare alanine, $CH_3CH(NH_2)COOH$, one of the 20 amino acids found in proteins, and propose a mechanism for acid-catalyzed conversion of the initial alkylation product to the amino acid. (See Section 26.4.)

$$CH_3\overset{O}{\overset{\|}{C}}NHCH\overset{O}{\overset{\|}{C}}OEt \quad \text{\textbf{Diethyl acetamidomalonate}}$$
$$\underset{COOEt}{|}$$

22.53 Amino acids can also be prepared by a two-step sequence that involves Hell–Volhard–Zelinskii reaction of a carboxylic acid followed by treatment with ammonia. Show how you would prepare leucine, $(CH_3)_2CHCH_2CH(NH_2)COOH$, and identify the mechanism of the second step. (See Section 26.4.)

Molecular Modeling

22.54 Compare the energies of 2-butanone and its enol using SpartanView. Which tautomer is lower in energy, and by how much? Repeat the comparison for keto and enol tautomers of 2,4-pentanedione and 2,4-cyclohexadienone. Why does the keto–enol preference change?

22.55 Use SpartanView to compare electrostatic potential maps of acetone enol and lithium acetone enolate. Which of the two has a more negative carbon?

22.56 LDA reacts with 2-pentanone at −78°C to give mainly lithium enolate A rather than enolate B. Compare the energies of lithium enolate A and lithium enolate B using SpartanView, tell which is more stable, and explain the observed result.

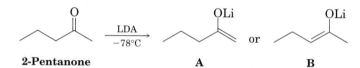

22.57 Treatment of the following enone with NaOD in D₂O results in the exchange of five hydrogens by deuterium. Use SpartanView to obtain the energies of all enolate anions that might be produced from the enone, draw resonance structures for them, and predict the structure of the deuterium-containing product.

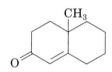

23

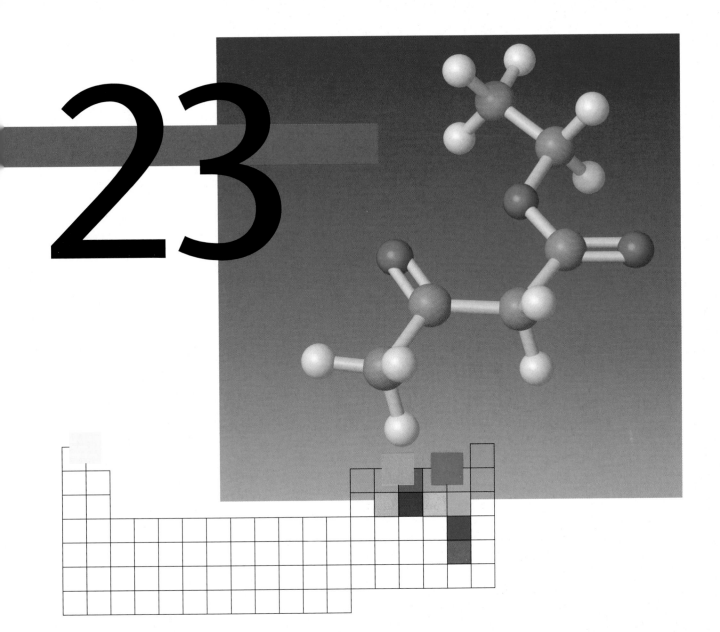

Carbonyl Condensation Reactions

At this point, we've seen three general kinds of carbonyl-group reactions and have studied two general kinds of behavior. In nucleophilic addition and nucleophilic acyl substitution reactions, a carbonyl compound behaves as an electrophile. In α-substitution reactions, a carbonyl compound behaves as a nucleophile when it is converted into its enol or enolate ion. In the **carbonyl condensation reactions** that we'll study in the present chapter, the carbonyl compound behaves *both* as an electrophile and as a nucleophile.

**Electrophilic carbonyl group
is attacked by nucleophiles**

**Nucleophilic enolate ion
attacks electrophiles**

We'll see later in this chapter and again in Chapter 29 that carbonyl condensation reactions occur frequently in metabolic pathways. Almost all classes of biomolecules—carbohydrates, lipids, proteins, nucleic acids, and many others—are biosynthesized through routes that involve carbonyl condensation reactions.

23.1 Mechanism of Carbonyl Condensation Reactions

Carbonyl condensation reactions take place between two carbonyl partners and involve a *combination* of nucleophilic addition and α-substitution steps. One partner (the nucleophilic donor) is converted into its enolate ion and undergoes an α-substitution reaction when it adds as a nucleophile to the second partner (the electrophilic acceptor). The general mechanism of a carbonyl condensation reaction is shown in Figure 23.1.

All kinds of carbonyl compounds, including aldehydes, ketones, esters, amides, acid anhydrides, and nitriles, enter into condensation reactions. Nature uses these same carbonyl condensation reactions in the biosynthesis of many naturally occurring compounds.

23.2 Condensations of Aldehydes and Ketones: The Aldol Reaction

**Biological
Connection** ▶

When acetaldehyde is treated with a base, such as sodium ethoxide or sodium hydroxide, a rapid and reversible condensation reaction occurs. The product is a β-hydroxy aldehyde, or *aldol* (*ald*ehyde + alcoh*ol*).

$$2 \; CH_3\overset{\overset{\textstyle O}{\|}}{C}H \quad \underset{CH_3CH_2OH}{\overset{NaOCH_2CH_3}{\rightleftharpoons}} \quad CH_3\underset{\beta}{\overset{\overset{\textstyle OH}{|}}{C}H} - CH_2\underset{\alpha}{\overset{\overset{\textstyle O}{\|}}{C}H}$$

Acetaldehyde

**Aldol
(a β-hydroxy aldehyde)**

Called the **aldol reaction**, base-catalyzed dimerization is a general reaction for all aldehydes and ketones with an α hydrogen atom. If the aldehyde or ketone does *not* have an α hydrogen atom, however, aldol conden-

FIGURE 23.1 ▼

The general mechanism of a carbonyl condensation reaction. One partner (the donor) acts as a nucleophile, while the other (the acceptor) acts as an electrophile.

refer to
Mechanisms
& Movies

One carbonyl partner with an alpha hydrogen atom is converted by base into its enolate ion.

This enolate ion acts as a nucleophilic donor and adds to the electrophilic carbonyl group of the acceptor partner.

Protonation of the tetrahedral alkoxide ion intermediate gives the neutral condensation product.

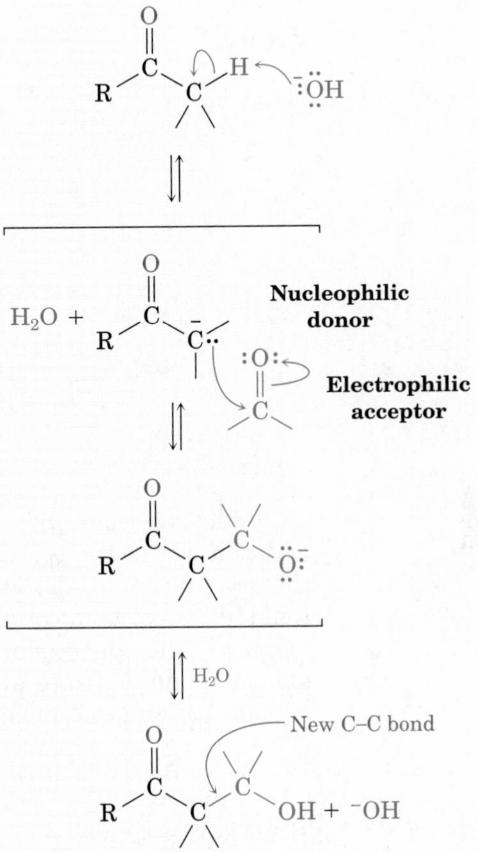

© 1984 JOHN MCMURRY

sation can't occur. As the following examples indicate, the aldol equilibrium generally favors condensation product in the case of aldehydes with no α substituent (RCH_2CHO), but favors starting material for more heavily substituted aldehydes and for most ketones. Steric factors are probably responsible for these trends, since increased substitution near the reaction site increases steric congestion in the aldol product.

Aldehydes

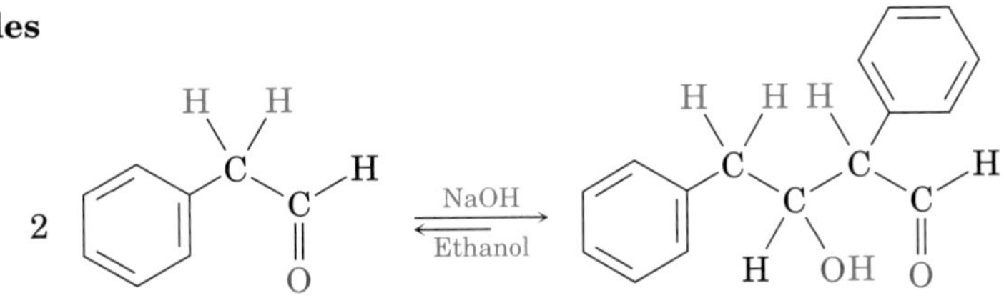

$$2 \quad \text{Phenylacetaldehyde} \quad \xrightleftharpoons[\text{Ethanol}]{\text{NaOH}} \quad \text{90\%}$$

Phenylacetaldehyde   90%

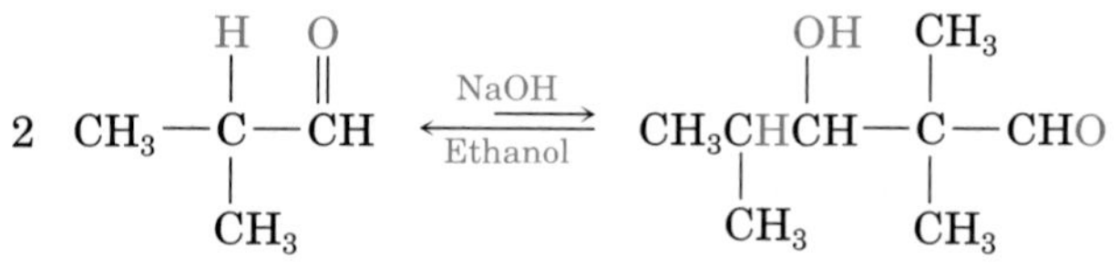

2-Methylpropanal   **Low yield**

Ketones

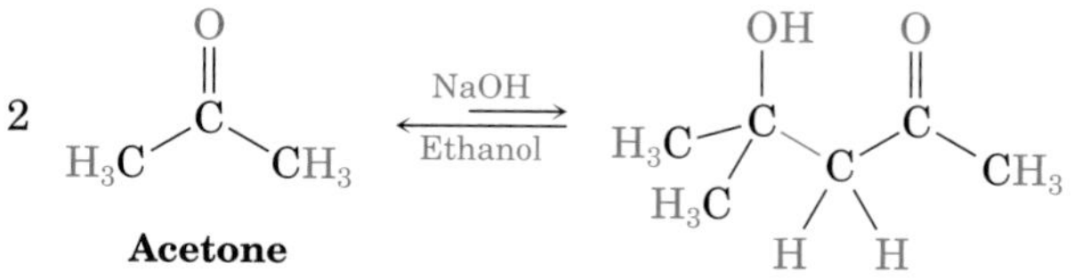

Acetone   **~5%**

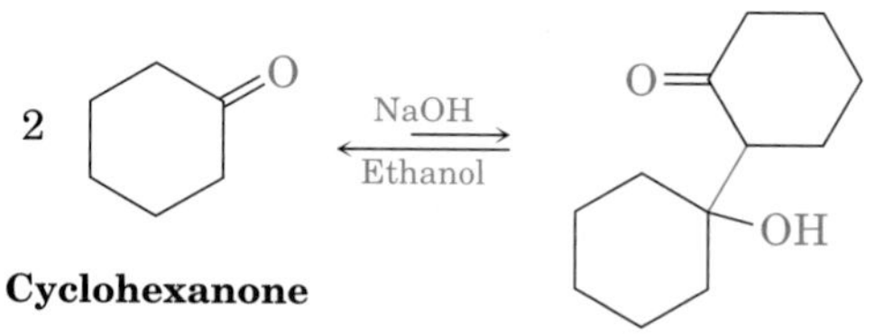

Cyclohexanone   **22%**

Aldol reactions, like other carbonyl condensations, occur by nucleophilic addition of the enolate ion of the donor molecule to the carbonyl group of the acceptor molecule. The resultant tetrahedral intermediate is then protonated to give an alcohol product (Figure 23.2). The reverse process occurs in exactly the opposite manner: Base abstracts the –OH hydrogen from the aldol to yield a β-keto alkoxide ion, which cleaves to give one molecule of enolate ion and one molecule of neutral carbonyl compound.

Practice Problem 23.1   What is the structure of the aldol product from propanal?

Strategy   An aldol reaction combines two molecules of reactant, forming a bond between the α carbon of one partner and the carbonyl carbon of the second partner.

Solution

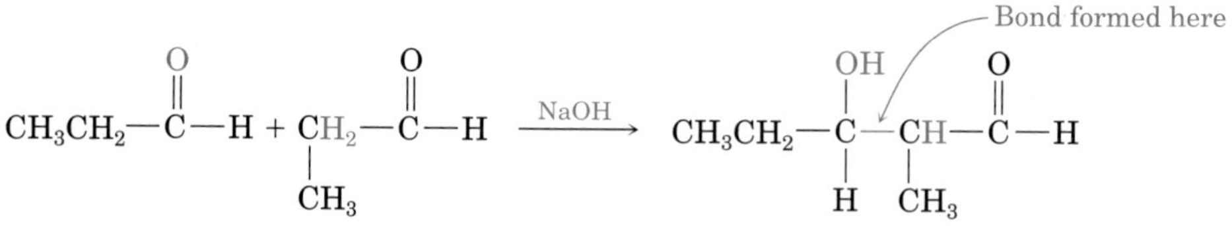

Problem 23.1   Predict the aldol reaction product of the following compounds:
(a) Butanal   (b) 2-Butanone   (c) Cyclopentanone

Problem 23.2   Using curved arrows to indicate the electron flow in each step, show how the base-catalyzed reverse aldol reaction of 4-hydroxy-4-methyl-2-pentanone takes place to yield 2 equivalents of acetone.

FIGURE 23.2 ▼

Mechanism of the aldol reaction, a typical carbonyl condensation.

refer to Mechanisms & Movies

Base removes an acidic alpha hydrogen from one aldehyde molecule, yielding a resonance-stabilized enolate ion.

The enolate ion attacks a second aldehyde molecule in a nucleophilic addition reaction to give a tetrahedral alkoxide ion intermediate.

Protonation of the alkoxide ion intermediate yields neutral aldol product and regenerates the base catalyst.

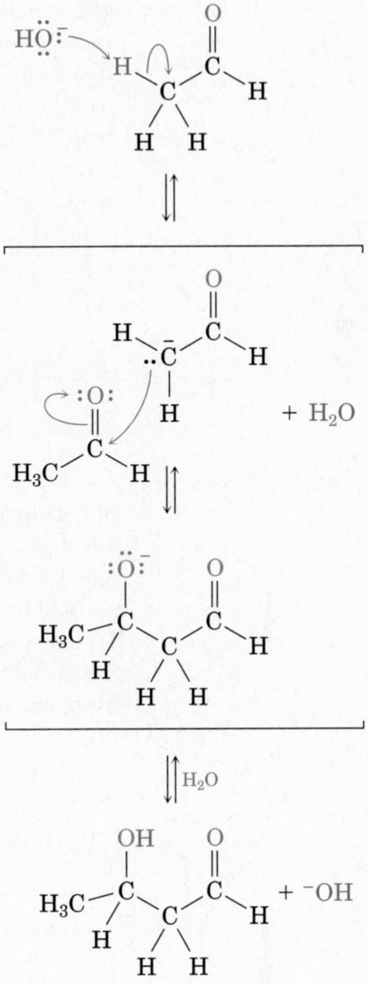

© 1984 JOHN MCMURRY

23.3 Carbonyl Condensation Reactions versus Alpha-Substitution Reactions

Two of the four general carbonyl-group reactions—carbonyl condensations and α substitutions—take place under basic conditions and involve enolate ion intermediates. Since the experimental conditions for the two reactions are so similar, how can we predict which will occur in a given case? When we generate an enolate ion with the intention of carrying out an α alkylation, how can we be sure that a carbonyl condensation reaction won't occur instead?

Although there is no simple answer to this question, the exact experimental conditions usually have much to do with the result. Alpha-substitution reactions require a full equivalent of strong base and are normally carried out so that the carbonyl compound is rapidly and completely converted into its enolate ion at a low temperature. An electrophile is then added rapidly to ensure that the reactive enolate ion is quenched quickly. In a ketone alkylation reaction, for instance, we might use 1 equivalent of lithium diisopropylamide (LDA) in tetrahydrofuran solution at −78°C. Rapid and complete generation of the ketone enolate ion would occur, and no unreacted ketone would be left so that no condensation reaction could take place. We would then immediately add an alkyl halide to complete the alkylation reaction.

On the other hand, we might want to carry out a carbonyl condensation reaction. Since we need to generate only a small amount of the enolate ion in the presence of unreacted carbonyl compound, the aldol reaction requires only a *catalytic* amount of a weaker base, rather than a full equivalent. Once a condensation has occurred, the basic catalyst is regenerated. To carry out an aldol reaction on propanal, for example, we might dissolve the aldehyde in methanol, add 0.05 equivalent of sodium methoxide, and then warm the mixture to give the aldol product.

23.4 Dehydration of Aldol Products: Synthesis of Enones

Biological Connection ▶

The β-hydroxy aldehydes and β-hydroxy ketones formed in aldol reactions can be easily dehydrated to yield conjugated enones. In fact, it's this loss of water that gives the aldol *condensation* its name, since water condenses out of the reaction when the enone product forms.

A β-hydroxy ketone   A conjugated
or aldehyde   enone

Most alcohols are resistant to dehydration by dilute acid or base (Section 17.7), but aldol products are special because of the carbonyl group. Under *basic* conditions, an acidic α hydrogen is removed, yielding an enolate ion that expels the ⁻OH leaving group in an E2-like reaction. Under *acidic* conditions, an enol is formed, the –OH group is protonated, and water is eliminated.

Base-catalyzed

Enolate ion

Acid-catalyzed

Enol

The conditions needed for aldol dehydration are often only a bit more vigorous (slightly higher temperature, for instance) than the conditions needed for the aldol formation itself. As a result, conjugated enones are often obtained directly from aldol reactions, and the intermediate β-hydroxy carbonyl compounds are not isolated.

Conjugated enones are more stable than nonconjugated enones for the same reason that conjugated dienes are more stable than nonconjugated dienes (Section 14.2). Interaction between the π electrons of the C=C bond and the π electrons of the C=O group leads to a molecular orbital description for a conjugated enone that shows a partial delocalization of the π electrons over all four atomic centers (Figure 23.3, p. 944).

The real value of aldol dehydration is that removal of water from the reaction mixture can be used to drive the aldol equilibrium toward product. Even though the initial aldol step itself may be unfavorable (as it usually is for ketones), the subsequent dehydration step nevertheless allows many aldol condensations to be carried out in good yield. Cyclohexanone,

FIGURE 23.3 ▼

The π bonding orbitals of conjugated alkenes (1,3-pentadiene) and conjugated enones (2-butenal) are similar in shape and are delocalized over the entire π system. In contrast, the π bonding orbitals of nonconjugated enones (3-butenal) are more localized and more closely resemble the orbitals of ethylene and formaldehyde.

1.3-pentadiene,
2-butenal,
3-butenal

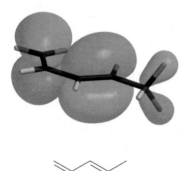 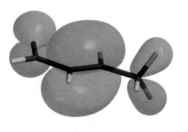 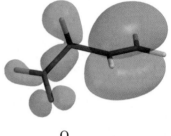

| 1,3-Pentadiene | 2-Butenal
(Conjugated; more stable) | 3-Butenal
(Nonconjugated; less stable) |

for example, gives cyclohexylidenecyclohexanone in 92% yield even though the initial equilibrium is unfavorable.

Cyclohexanone

Cyclohexylidenecyclohexanone
(92%)

• •

Practice Problem 23.2   What is the structure of the enone obtained from aldol condensation of acetaldehyde?

Strategy   In the aldol reaction, H₂O is eliminated and a double bond is formed by removing two hydrogens from the acidic α position of one partner and the carbonyl oxygen from the second partner.

Solution

2-Butenal

• •

Problem 23.3   What enone product would you expect from aldol condensation of each of the following compounds?
(a) Cyclopentanone   (b) Acetophenone   (c) 3-Methylbutanal

Problem 23.4   Aldol condensation of 3-methylcyclohexanone leads to a mixture of two enone products, not counting double-bond isomers. Draw them.

• •

23.5 Using Aldol Reactions in Synthesis

The aldol condensation reaction yields either a β-hydroxy aldehyde/ketone or an α,β-unsaturated aldehyde/ketone, depending on the reactant and on the experimental conditions. By learning how to think *backward*, it's possible to predict when the aldol reaction might be useful in synthesis. Any time the target molecule contains either a β-hydroxy aldehyde/ketone or a conjugated enone functional group, it might come from an aldol reaction.

Aldol products   **Aldol reactants**

We can extend this kind of reasoning even further by imagining that subsequent transformations might be carried out on the aldol products. For example, a saturated ketone might be prepared by catalytic hydrogenation of the enone product. A good example can be found in the industrial preparation of 2-ethyl-1-hexanol, an alcohol used in the synthesis of plasticizers for polymers. Although 2-ethyl-1-hexanol bears little resemblance to an aldol product at first glance, it is in fact prepared commercially from butanal by an aldol reaction. Working backward, we can reason that 2-ethyl-1-hexanol might come from 2-ethylhexanal by a reduction. 2-Ethylhexanal, in turn, might be prepared by catalytic reduction of 2-ethyl-2-hexenal, which is the aldol condensation product of butanal. The reactions that follow show the sequence in reverse order.

$$CH_3CH_2CH_2CH_2CHCH_2OH \xleftarrow[\text{(Industrially, H}_2\text{/Pt)}]{[H]} CH_3CH_2CH_2CH_2CHCH$$

with CH_2CH_3 substituent on left structure and O (double bond) and CH_2CH_3 on right structure

Target: 2-Ethyl-1-hexanol   **2-Ethylhexanal**

↑ H₂/Pt

$$CH_3CH_2CH_2CH \xrightarrow[\text{Ethanol}]{\text{KOH}} CH_3CH_2CH_2CH=CCH$$

with O (double bond) on butanal and O (double bond), CH_2CH_3 on product

Butanal   **2-Ethyl-2-hexenal**

• •

Problem 23.5   Which of the following compounds are aldol condensation products? What is the aldehyde or ketone precursor of each?
(a) 3-Hydroxy-2,2,3-trimethylbutanal   (b) 2-Hydroxy-2-Methylpentanal
(c) 5-Ethyl-4-methyl-4-hepten-3-one

Problem 23.6   1-Butanol is prepared commercially by a route that begins with an aldol reaction. Show the steps that are likely to be involved.

• •

23.6 Mixed Aldol Reactions

Biological Connection ▶

Until now, we've considered only *symmetrical* aldol reactions, in which the two carbonyl components have been the same. What would happen, though, if a *mixed* aldol reaction were carried out between two different carbonyl partners?

In general, a mixed aldol reaction between two similar aldehyde or ketone partners leads to a mixture of four possible products. For example, base treatment of a mixture of acetaldehyde and propanal gives a complex product mixture containing two "symmetrical" aldol products and two "mixed" aldol products. Clearly, such a reaction is of little practical value.

$$CH_3CHO + CH_3CH_2CHO \xrightarrow{\text{Base}}$$

Acetaldehyde   Propanal

OH
|
CH_3CHCH_2CHO +

OH
|
$CH_3CHCHCHO$
|
CH_3

+

+

OH
|
$CH_3CH_2CHCHCHO$ +
|
CH_3

OH
|
$CH_3CH_2CHCH_2CHO$

Mixed products

Symmetrical products

On the other hand, mixed aldol reactions *can* lead cleanly to a single product if either of two conditions is met:

- If one of the carbonyl partners contains no α hydrogens (and thus can't form an enolate ion to become a donor) but does contain a reactive carbonyl group that is a good acceptor of nucleophiles, then a mixed aldol reaction is likely to be successful. This is the case, for example, when either benzaldehyde or formaldehyde is used as one of the carbonyl partners.

  Neither benzaldehyde nor formaldehyde can form an enolate ion to condense with itself or with another partner, yet both compounds have an unhindered and reactive carbonyl group. The ketone 2-methylcyclohexanone, for instance, reacts preferentially with benzaldehyde to give the mixed aldol product.

2-Methylcyclohexanone
(donor)

Benzaldehyde
(acceptor)

78%

- If one of the carbonyl partners is much more acidic than the other and is easily transformed into its enolate ion, then a mixed aldol reaction is likely to be successful. Ethyl acetoacetate, for instance, is completely converted into its enolate ion in preference to enolate ion formation from other carbonyl partners. Thus, aldol condensations with ethyl acetoacetate occur preferentially to give the mixed product.

Cyclohexanone
(acceptor)

Ethyl acetoacetate
(donor)

80%

The situation can be summarized by saying that a mixed aldol reaction leads to a mixture of products unless one of the partners either has no α hydrogens but is a good electrophilic acceptor (such as benzaldehyde) or is an unusually good nucleophilic donor (such as ethyl acetoacetate).

Problem 23.7   Which of the following compounds can probably be prepared by a mixed aldol reaction? Show the reactants you would use in each case.

(a) $C_6H_5CH{=}CHCCH_3$ (b) $C_6H_5C{=}CHCCH_3$ (c)

23.7 Intramolecular Aldol Reactions

The aldol reactions we've seen up to this point have all been *inter*molecular. That is, they have taken place between two different molecules. When certain *di*carbonyl compounds are treated with base, however, an *intra*molecular aldol reaction can occur, leading to the formation of a cyclic product. For example, base treatment of a 1,4-diketone such as 2,5-hexanedione yields a cyclopentenone product, and base treatment of a 1,5-diketone such as 2,6-heptanedione yields a cyclohexenone.

2,5-Hexanedione   **3-Methyl-2-cyclopentenone**
(a 1,4-diketone)

2,6-Heptanedione   **3-Methyl-2-cyclohexenone**
(a 1,5-diketone)

The mechanism of these intramolecular aldol reactions is similar to that of intermolecular reactions. The only difference is that both the nucleophilic carbonyl anion donor and the electrophilic carbonyl acceptor are now in the same molecule.

In principle, many intramolecular aldol reactions can lead to a mixture of products, depending on which enolate ion is formed. For example, 2,5-hexanedione might yield either the five-membered-ring product 3-methyl-2-cyclopentenone or the three-membered-ring product (2-methylcyclopropenyl)ethanone (Figure 23.4). In practice, though, only the cyclopentenone is formed.

The selectivity observed in the intramolecular aldol reaction of 2,5-hexanedione is due to the fact that all steps in the mechanism are reversible and an equilibrium is reached. Thus, the relatively strain-free cyclopentenone product is considerably more stable than the highly strained cyclopropene alternative. For similar reasons, intramolecular aldol reactions of 1,5-diketones lead only to cyclohexenone products rather than to cyclobutenes.

• •

Problem 23.8   Why do you suppose that 1,3-diketones do not undergo intramolecular aldol condensation to yield cyclobutenones?

FIGURE 23.4 ▼

Intramolecular aldol reaction of 2,5-hexanedione yields 3-methyl-2-cyclopentenone rather than the alternative acetylcyclopropene.

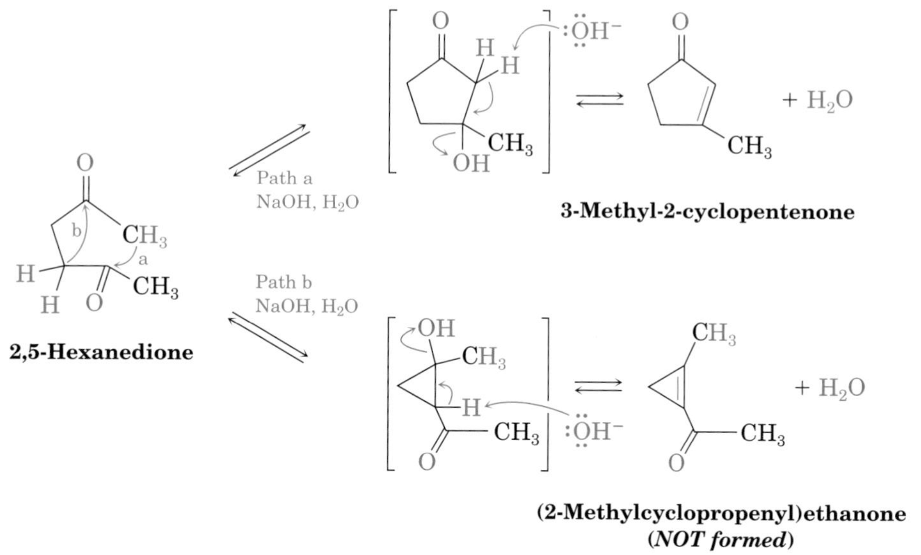

Path a
NaOH, H₂O

Path b
NaOH, H₂O

2,5-Hexanedione

+ H₂O

3-Methyl-2-cyclopentenone

+ H₂O

(2-Methylcyclopropenyl)ethanone
(*NOT formed*)

Problem 23.9   What product would you expect to obtain from base treatment of 1,6-cyclodecane-dione?

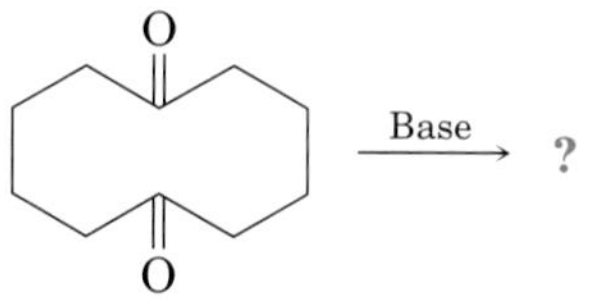

Base ⟶ ?

- -

23.8 The Claisen Condensation Reaction

Biological Connection ▶

Esters, like aldehydes and ketones, are weakly acidic. When an ester with an α hydrogen is treated with 1 equivalent of a base such as sodium ethoxide, a reversible condensation reaction occurs to yield a β-keto ester. For example, ethyl acetate yields ethyl acetoacetate on treatment with base. This reaction between two ester molecules is known as the **Claisen condensation reaction**. (We'll use ethyl esters for consistency, but other esters will also work.)

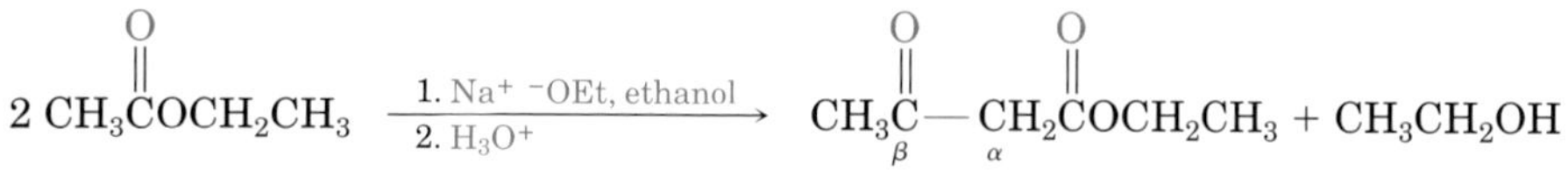

$$2\ CH_3COCH_2CH_3 \xrightarrow[\text{2. } H_3O^+]{\text{1. } Na^+\ {}^-OEt,\ \text{ethanol}} CH_3C \underset{\beta}{\overset{O}{\|}} CH_2 \underset{\alpha}{\overset{O}{\|}} COCH_2CH_3 + CH_3CH_2OH$$

Ethyl acetate

Ethyl acetoacetate, a β-keto ester (75%)

The mechanism of the Claisen condensation is similar to that of the aldol condensation. As shown in Figure 23.5, the Claisen condensation involves the nucleophilic acyl substitution of an ester enolate ion on the carbonyl group of a second ester molecule.

FIGURE 23.5 ▼

Mechanism of the Claisen condensation reaction.

refer to Mechanisms & Movies

Ethoxide base abstracts an acidic alpha hydrogen atom from an ester molecule, yielding an ester enolate ion.

$$CH_3\overset{O}{\overset{\|}{C}}OEt$$

$$\updownarrow \ ^-OEt$$

$$:\bar{C}H_2\overset{O}{\overset{\|}{C}}OEt + EtOH$$

Nucleophilic donor

In a nucleophilic addition, this ion adds to a second ester molecule, giving a tetrahedral intermediate.

$$CH_3\overset{:O:}{\overset{\|}{C}}OEt \quad \begin{array}{l} \text{Electrophilic} \\ \text{acceptor} \end{array}$$

$$CH_3\overset{:\overset{..}{O}:^-}{\underset{OEt}{\overset{|}{C}}}\!-CH_2\overset{O}{\overset{\|}{C}}OEt$$

The tetrahedral intermediate is not stable. It expels ethoxide ion to yield the new carbonyl compound, ethyl acetoacetate.

$$\updownarrow$$

But ethoxide ion is basic enough to convert the β-keto ester product into its enolate, thus shifting the equilibrium and driving the reaction to completion.

$$CH_3\overset{O}{\overset{\|}{C}}CH_2\overset{O}{\overset{\|}{C}}OEt + EtO^-$$

$$\updownarrow$$

$$CH_3\overset{O}{\overset{\|}{C}}\bar{C}H\overset{O}{\overset{\|}{C}}OEt + EtOH$$

Protonation by addition of acid in a separate step yields the final product.

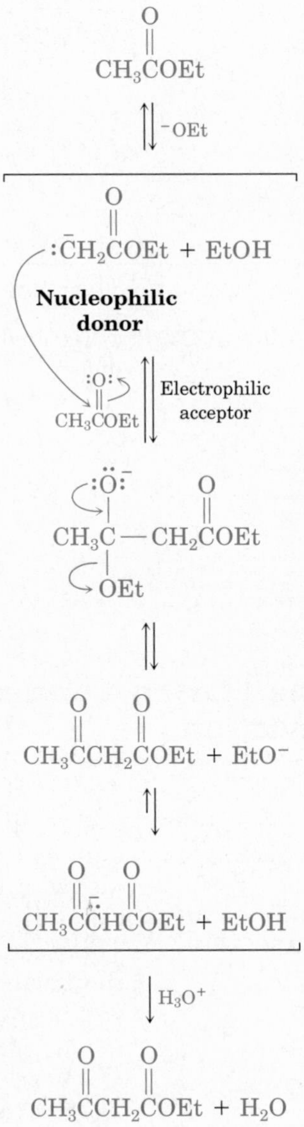

$$\downarrow H_3O^+$$

$$CH_3\overset{O}{\overset{\|}{C}}CH_2\overset{O}{\overset{\|}{C}}OEt + H_2O$$

© 1984 JOHN MCMURRY

The only difference between the aldol condensation of an aldehyde or ketone and the Claisen condensation of an ester involves the fate of the initially formed tetrahedral intermediate. The tetrahedral intermediate in the aldol reaction is protonated to give an alcohol product—exactly the behavior previously seen for aldehydes and ketones (Section 19.4). The tetrahedral intermediate in the Claisen reaction expels an alkoxide leaving group to yield an acyl substitution product—exactly the behavior previously seen for esters (Section 21.6).

If the starting ester has more than one acidic α hydrogen, the product β-keto ester has a highly acidic, doubly activated hydrogen atom that can be abstracted by base. This deprotonation of the product requires that a full equivalent of base rather than a catalytic amount be used in the reaction. Furthermore, the deprotonation serves to drive the Claisen equilibrium completely to the product side so that high yields are often obtained.

• •

Practice Problem 23.3 What product would you obtain from Claisen condensation of ethyl propanoate?

Strategy The Claisen condensation of an ester results in loss of one molecule of alcohol and formation of a product in which an acyl group of one reactant bonds to the α carbon of the second reactant.

Solution

$$CH_3CH_2\overset{\overset{\text{O}}{\|}}{C}\!-\!OEt + H\!-\!\underset{\underset{CH_3}{|}}{CH}COEt \xrightarrow[\text{2. H}_3O^+]{\text{1. Na}^+ \ ^-OEt} CH_3CH_2\overset{\overset{\text{O}}{\|}}{C}\!-\!\underset{\underset{CH_3}{|}}{CH}\overset{\overset{\text{O}}{\|}}{C}OEt + EtOH$$

2 Ethyl propanoate   **Ethyl 2-methyl-3-oxopentanoate**

• •

Problem 23.10 Show the products you would expect to obtain by Claisen condensation of the following esters:
(a) $(CH_3)_2CHCH_2CO_2Et$ (b) Ethyl phenylacetate   (c) Ethyl cyclohexylacetate

Problem 23.11 As shown in Figure 23.5, the Claisen reaction is reversible. That is, a β-keto ester can be cleaved by base into two fragments. Using curved arrows to indicate electron flow, show the mechanism by which this cleavage occurs.

• •

23.9 Mixed Claisen Condensations

Biological Connection ▶

The mixed Claisen condensation of two different esters is similar to the mixed aldol condensation of two different aldehydes or ketones (Section 23.6). Mixed Claisen reactions are successful only when one of the two ester components has no α hydrogens and thus can't form an enolate ion. For example, ethyl benzoate and ethyl formate can't form enolate ions and thus can't serve as donors. They can, however, act as the electrophilic acceptor components in reactions with other ester anions to give good yields of mixed β-keto ester products.

Ethyl benzoate
(acceptor)

Ethyl acetate
(donor)

Ethyl benzoylacetate

Mixed Claisen-like reactions can also be carried out between esters and ketones. The result is an excellent synthesis of β-diketones. The reaction works best when the ester component has no α hydrogens and thus can't act as the nucleophilic donor. For example, ethyl formate gives particularly high yields in mixed Claisen condensations with ketones.

2,2-Dimethylcyclohexanone
(donor)

Ethyl formate
(acceptor)

A β-keto aldehyde
(91%)

Practice Problem 23.4   Diethyl oxalate, (CO$_2$Et)$_2$, can give high yields in mixed Claisen reactions. What product would you expect to obtain from the reaction of ethyl acetate with diethyl oxalate?

Strategy   A mixed Claisen reaction is effective when only one of the two partners has an acidic α hydrogen atom. In the present instance, ethyl acetate can be converted into its enolate ion, but diethyl oxalate cannot. Thus, ethyl acetate acts as the donor and diethyl oxalate as the acceptor.

Solution

Diethyl oxalate

Ethyl acetate

• •

Problem 23.12   What product would you expect from a mixed Claisen-like reaction of 2,2-dimethyl-cyclohexanone with diethyl oxalate (Practice Problem 23.4)?

• •

23.10 Intramolecular Claisen Condensations: The Dieckmann Cyclization

Biological Connection ▶

Intramolecular Claisen condensations can be carried out with diesters, just as intramolecular aldol condensations can be carried out with diketones (Section 23.7). Called the **Dieckmann cyclization**, the reaction works best on 1,6-diesters and 1,7-diesters. Five-membered cyclic β-keto esters result from Dieckmann cyclization of 1,6-diesters, and six-membered cyclic β-keto esters result from cyclization of 1,7-diesters.

Diethyl hexanedioate
(a 1,6-diester)

1. Na⁺ ⁻OEt, ethanol
2. H₃O⁺

Ethyl 2-oxocyclopentanecarboxylate
(82%)

+ EtOH

Diethyl heptanedioate
(a 1,7-diester)

1. Na⁺ ⁻OEt, ethanol
2. H₃O⁺

Ethyl 2-oxocyclohexanecarboxylate

+ EtOH

The mechanism of the Dieckmann cyclization, shown in Figure 23.6 (p. 954), is analogous to that of the Claisen reaction. One of the two ester groups is converted into an enolate ion, which then carries out a nucleophilic acyl substitution on the second ester group at the other end of the molecule. A cyclic β-keto ester product results.

The product of a Dieckmann cyclization is a cyclic β-keto ester that can be further alkylated and decarboxylated by a series of reactions analogous to those in the acetoacetic ester synthesis (Section 22.8). For example, alkylation and subsequent decarboxylation of ethyl 2-oxocyclohexanecarboxylate yields a 2-alkylcyclohexanone. The overall sequence of (1) Dieckmann cyclization,

FIGURE 23.6 ▼

Mechanism of the
Dieckmann cyclization of
a 1,7-diester to yield a
cyclic β-keto ester
product.

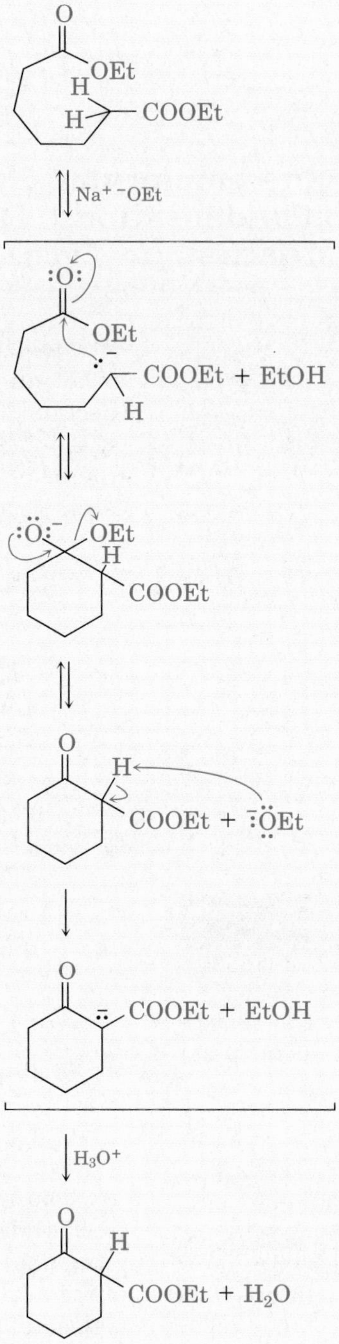

Base abstracts an acidic α proton
from the carbon atom next to one of
the ester groups, yielding an enolate
ion.

Intramolecular nucleophilic addition
of the ester enolate ion to the
carbonyl group of the second ester
group at the other end of the chain
then gives a cyclic tetrahedral
intermediate.

Loss of alkoxide ion from the
tetrahedral intermediate forms a
cyclic β-keto ester.

Deprotonation of the acidic β-keto
ester gives an enolate ion . . .

. . . which is protonated by addition
of aqueous acid at the end of the
reaction to generate the neutral
β-keto ester product.

© 1984 JOHN MCMURRY

(2) β-keto ester alkylation, and (3) decarboxylation is an excellent method for preparing 2-substituted cyclohexanones and cyclopentanones.

**Ethyl 2-oxocyclo-
hexanecarboxylate**

1. Na$^+$ $^-$OEt
2. H$_2$C=CHCH$_2$Br

H$_3$O$^+$
Heat

+ EtOH + CO$_2$

**2-Allylcyclohexanone
(83%)**

- -

Problem 23.13　What product would you expect on treatment of diethyl 4-methylheptanedioate with sodium ethoxide, followed by acidification?

Problem 23.14　Dieckmann cyclization of diethyl 3-methylheptanedioate gives a mixture of two β-keto ester products. What are their structures, and why is a mixture formed?

- -

23.11　The Michael Reaction

**Biological
Connection** ▶

We saw in Section 19.14 that nucleophiles can react with α,β-unsaturated aldehydes and ketones to give the conjugate addition product, rather than the direct addition product:

:Nu—H

**Conjugate addition
product**

Exactly the same kind of conjugate addition can occur when a nucleophilic enolate ion reacts with an α,β-unsaturated carbonyl compound—a process known as the **Michael reaction**.

The highest yielding Michael reactions are those that take place when a particularly stable enolate ion, such as that derived from a β-keto ester or malonic ester, adds to an unhindered α,β-unsaturated ketone. For example, ethyl acetoacetate reacts with 3-buten-2-one in the presence of sodium ethoxide catalyst to yield the conjugate addition product.

CH$_3$CCH$_2$　+　H$_2$C=CHCCH$_3$
　　|
　　CO$_2$Et

1. Na$^+$ $^-$OEt, ethanol
2. H$_3$O$^+$

CH$_3$CCHCH$_2$CH$_2$CCH$_3$
　　　|
　　　CO$_2$Et

**Ethyl
acetoacetate**　　　**3-Buten-2-one**　　　　　　　　　　**94%**

Michael reactions take place by addition of a nucleophilic enolate ion donor to the β carbon of an α,β-unsaturated carbonyl acceptor, according to the mechanism shown in Figure 23.7.

FIGURE 23.7 ▼

Mechanism of the Michael reaction between a β-keto ester and an α,β-unsaturated ketone.

refer to Mechanisms & Movies

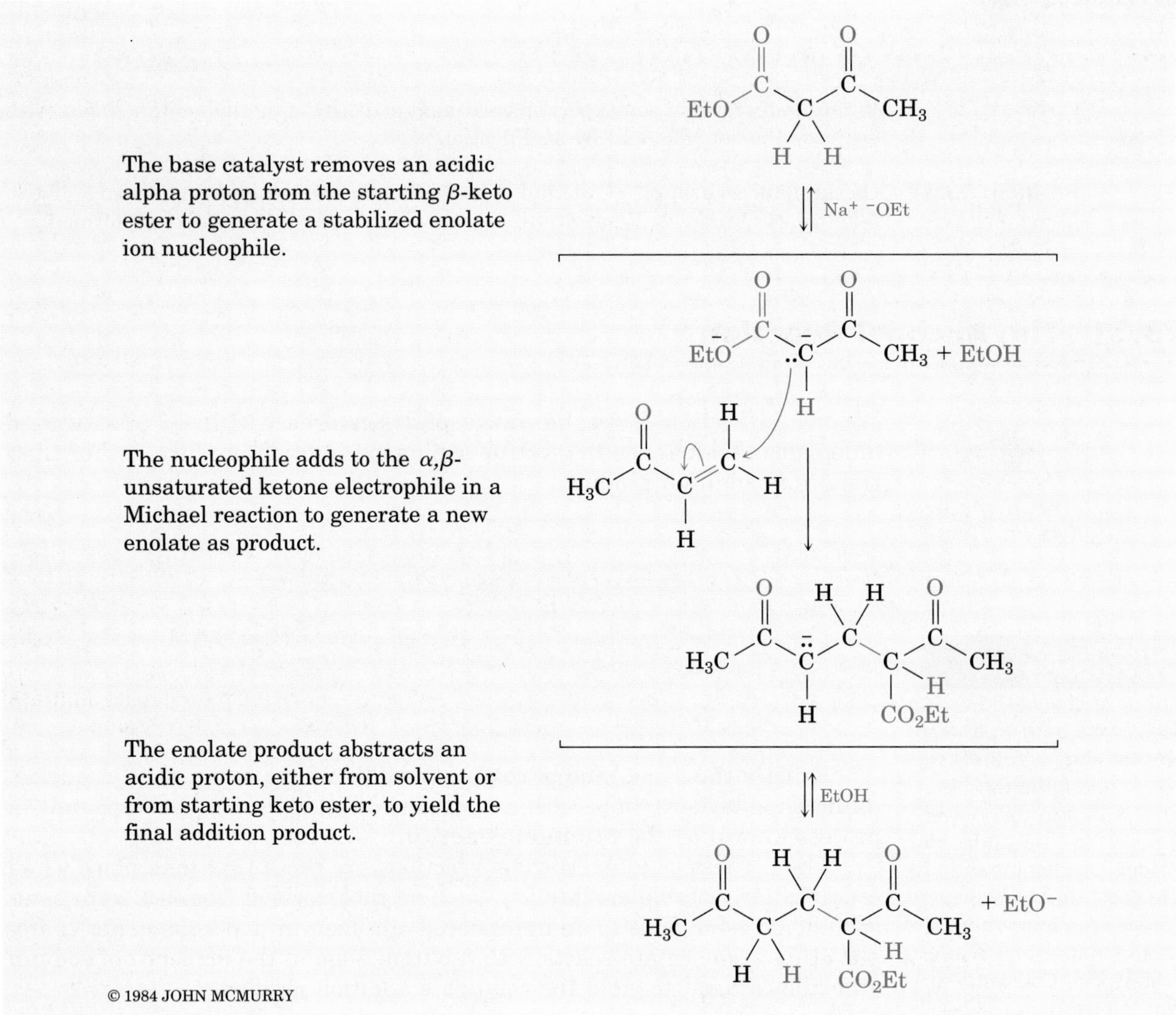

The base catalyst removes an acidic alpha proton from the starting β-keto ester to generate a stabilized enolate ion nucleophile.

The nucleophile adds to the α,β-unsaturated ketone electrophile in a Michael reaction to generate a new enolate as product.

The enolate product abstracts an acidic proton, either from solvent or from starting keto ester, to yield the final addition product.

© 1984 JOHN MCMURRY

The Michael reaction occurs with a wide variety of α,β-unsaturated carbonyl compounds, not just conjugated enones. Unsaturated aldehydes, esters, nitriles, amides, and nitro compounds can all act as the electrophilic acceptor component in Michael reactions (Table 23.1). Similarly, a variety

TABLE 23.1 Some Michael Acceptors and Michael Donors

Michael acceptors		Michael donors	
H₂C=CHCHO	Propenal	RCOCH₂COR′	β-Diketone
H₂C=CHCO₂Et	Ethyl propenoate	RCOCH₂CO₂Et	β-Keto ester
H₂C=CHC≡N	Propenenitrile	EtO₂CCH₂CO₂Et	Malonic ester
H₂C=CHCOCH₃	3-Buten-2-one	RCOCH₂C≡N	β-Keto nitrile
H₂C=CHNO₂	Nitroethylene	RCH₂NO₂	Nitro compound
H₂C=CHCONH₂	Propenamide		

of different donors can be used, including β-diketones, β-keto esters, malonic esters, β-keto nitriles, and nitro compounds.

Practice Problem 23.5 How might you obtain the following compound using a Michael reaction?

Strategy A Michael reaction involves the conjugate addition of a stable enolate ion donor to an α,β-unsaturated carbonyl acceptor, yielding a 1,5-dicarbonyl product. Usually, the stable enolate ion is derived from a β-diketone, β-keto ester, malonic ester, or similar compound. The C–C bond made in the conjugate addition step is the one between the α carbon of the acidic donor and the β carbon of the unsaturated acceptor. In the present instance:

This bond is formed in the Michael reaction.

Solution

Problem 23.15 What product would you obtain from a base-catalyzed Michael reaction of 2,4-pentanedione with each of the following α,β-unsaturated acceptors?
(a) 2-Cyclohexenone   (b) Propenenitrile   (c) Ethyl 2-butenoate

Problem 23.16　　What product would you obtain from a base-catalyzed Michael reaction of 3-buten-2-one with each of the following nucleophilic donors?

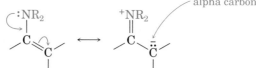

(a) EtOCCH₂COEt　　　(b)　　　　CO₂Et　　　(c) CH₃NO₂

. . . .

23.12　The Stork Enamine Reaction

Biological Connection ▶

In addition to enolate ions, other kinds of carbon nucleophiles add to α,β-unsaturated acceptors in the Michael reaction, greatly extending the usefulness and versatility of the process. Among the most important such nucleophiles are *enamines*. Recall from Section 19.9 that enamines are readily prepared by reaction between a ketone and a secondary amine:

$$R-C(=O)-C(CH_3)_2-H + HNR'_2 \rightleftharpoons R-C(NR'_2)=C + H_2O$$

For example:

Cyclohexanone　　**Pyrrolidine**　　**1-Pyrrolidinocyclohexene (87%)**
(an enamine)

As the following resonance structures indicate, enamines are electronically similar to enolate ions. Overlap of the nitrogen lone-pair orbital with the double-bond *p* orbitals leads to an increase in electron density on the α carbon atom, making that carbon strongly nucleophilic. An electrostatic potential map of *N,N*-dimethylaminoethylene shows that electron density (red) is shifted toward the α position.

An enolate ion

Nucleophilic alpha carbon

An enamine

N.N-dimethylaminoethylene

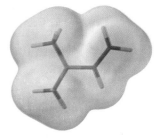

(CH₃)₂NCH=CH₂

Enamines behave in much the same way as enolate ions and enter into many of the same kinds of reactions. In the **Stork enamine reaction**, for example, an enamine adds to an α,β-unsaturated carbonyl acceptor in a Michael-type process. The initial product is then hydrolyzed by aqueous acid (Section 19.9) to yield a 1,5-dicarbonyl compound. The overall reaction is thus a three-step sequence:

STEP 1   Enamine formation from a ketone

STEP 2   Michael-type addition to an α,β-unsaturated carbonyl compound

STEP 3   Enamine hydrolysis back to a ketone

The net effect of the Stork enamine sequence is the Michael addition of a ketone to an α,β-unsaturated carbonyl compound. For example, cyclo-hexanone reacts with the cyclic amine pyrrolidine to yield an enamine; fur-ther reaction with an enone such as 3-buten-2-one yields a Michael-type adduct; and aqueous hydrolysis completes the sequence to provide a 1,5-diketone product (Figure 23.8).

FIGURE 23.8 ▼

The Stork enamine reaction between cyclohexanone and 3-buten-2-one. Cyclohexanone is first converted into an enamine; the enamine adds to the α,β-unsaturated ketone in a Michael reaction; and the conjugate addition product is hydrolyzed to yield a 1,5-diketone.

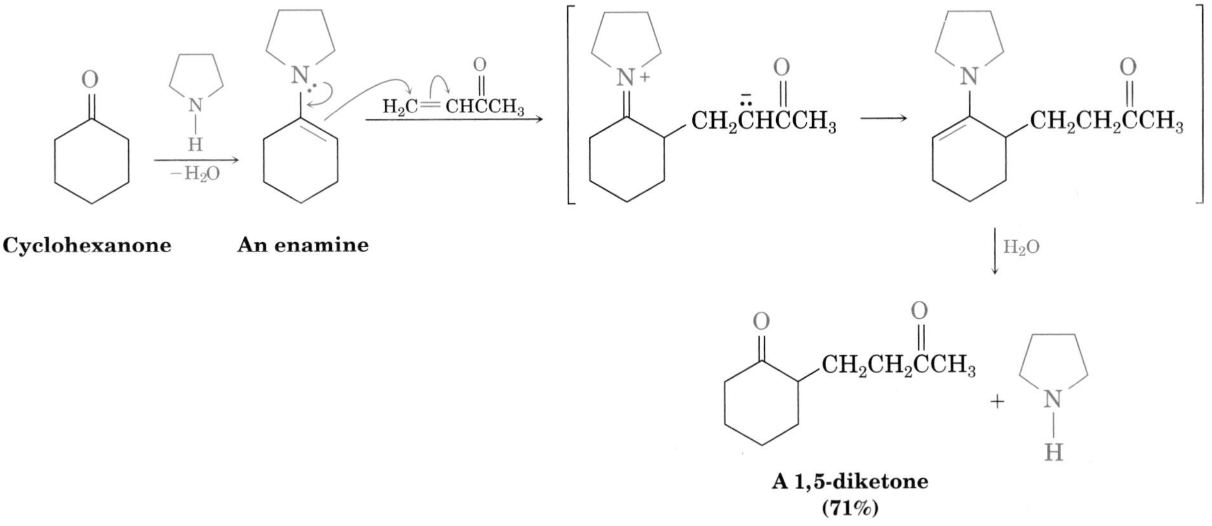

Cyclohexanone   **An enamine**

A 1,5-diketone
(71%)

• •

Practice Problem 23.6   How might you use an enamine reaction to prepare the following compound?

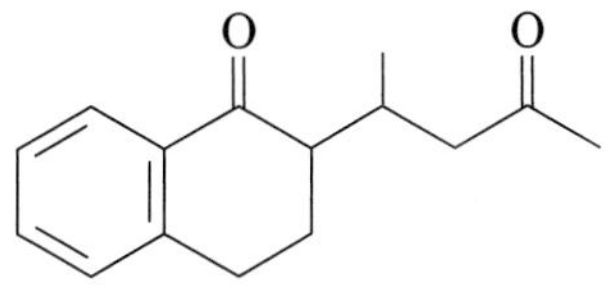

Strategy   The overall result of an enamine reaction is the Michael addition of a ketone as donor to an α,β-unsaturated carbonyl compound as acceptor, yielding a 1,5-dicarbonyl product. The C–C bond made in the Michael addition step is the one between the α carbon of the ketone donor and the β carbon of the unsaturated acceptor. In the present instance:

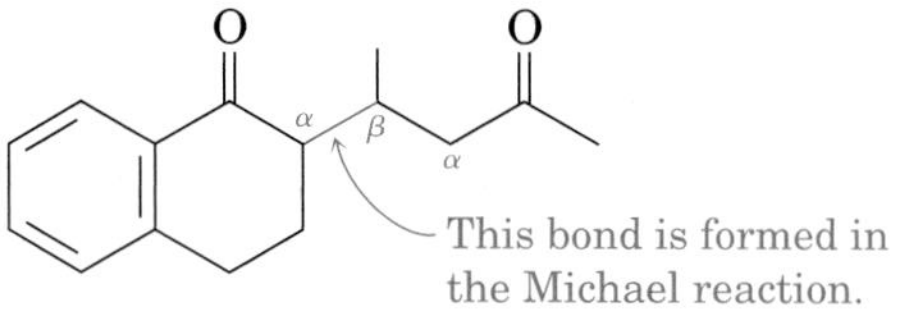

This bond is formed in the Michael reaction.

Solution

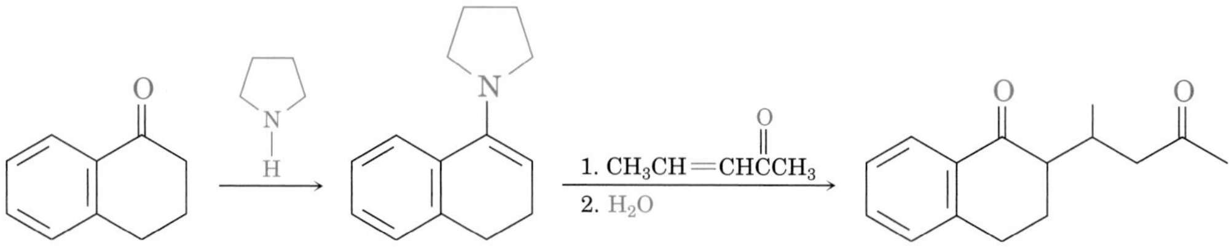

....

Problem 23.17   What products would result (after hydrolysis) from reaction of the enamine prepared from cyclopentanone and pyrrolidine with the following α,β-unsaturated acceptors?
(a) Ethyl propenoate   (b) Propenal (acrolein)

Problem 23.18   Show how you might use an enamine reaction to prepare each of the following compounds:

(a)

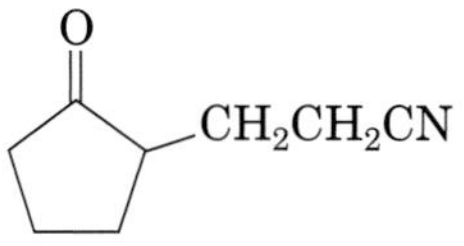

(b)

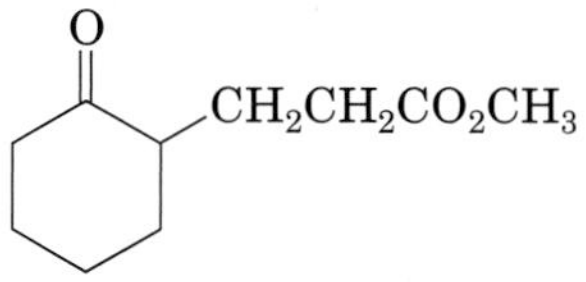

....

23.13 Carbonyl Condensation Reactions in Synthesis: The Robinson Annulation Reaction

Carbonyl condensation reactions are among the most valuable methods available for synthesizing complex molecules. By putting a few fundamental reactions together in the proper sequence, some remarkably useful transformations can be carried out. One such example is the **Robinson annulation reaction**, used for the synthesis of polycyclic molecules. (An *annulation* reaction, from the Latin *annulus*, meaning "ring," builds a new ring onto a molecule.)

The Robinson annulation is a two-step process that combines a Michael reaction with an intramolecular aldol reaction. It takes place between a nucleophilic donor, such as a β-keto ester, an enamine, or a β-diketone, and an α,β-unsaturated ketone acceptor, such as 3-buten-2-one. The product is a substituted 2-cyclohexenone.

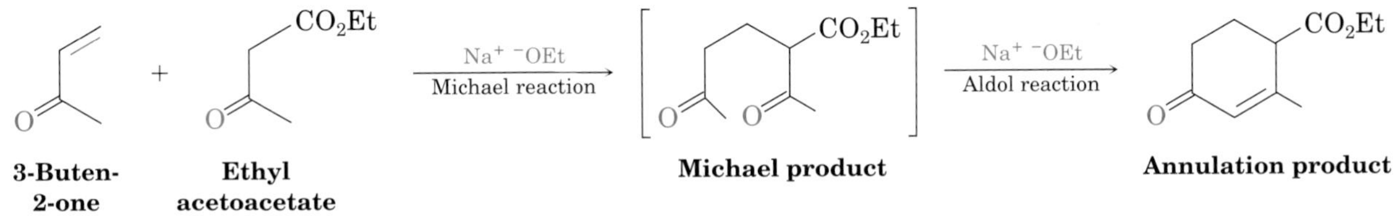

The first step of the Robinson annulation is simply a Michael reaction. An enamine or an enolate ion from the β-keto ester or β-diketone effects a conjugate addition to an α,β-unsaturated ketone, yielding a 1,5-diketone. But as we saw in Section 23.7, 1,5-diketones undergo intramolecular aldol condensation to yield cyclohexenones when treated with base. Thus, the final product contains a six-membered ring, and an annulation has been accomplished. An example occurs during the commercial synthesis of the steroid hormone estrone (Figure 23.9).

FIGURE 23.9 ▼

A Robinson annulation reaction used in the commercial synthesis of the steroid hormone estrone. The nucleophilic donor is a β-diketone.

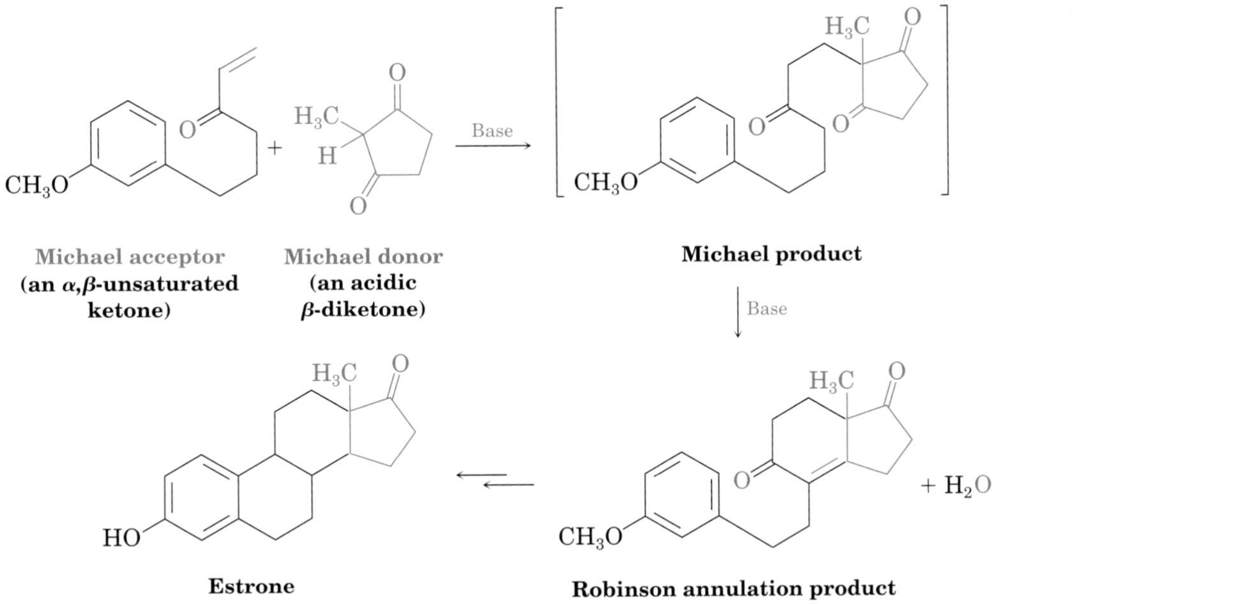

Sir Robert Robinson

Sir Robert Robinson (1886–1975) was born in Chesterfield, England, and received his D.Sc. from the University of Manchester with William Henry Perkin, Jr. After various appointments, he moved in 1930 to Oxford University, where he remained until his retirement in 1955. An accomplished mountain climber, Robinson was instrumental in developing the mechanistic descriptions of reactions that we use today. He received the 1947 Nobel Prize in chemistry.

In this example, 2-methyl-1,3-cyclopentanedione (a β-diketone) is used to generate the enolate ion required for Michael reaction, and an aryl-substituted α,β-unsaturated ketone is used as the acceptor. Base-catalyzed Michael reaction between the two partners yields an intermediate triketone, which then cyclizes in an intramolecular aldol condensation to give a Robinson annulation product. Several further transformations are required to complete the synthesis of estrone.

Problem 23.19   What product would you expect from a Robinson annulation reaction of 2-methyl-1,3-cyclopentanedione with 3-buten-2-one?

$$\text{—CH}_3 + \text{H}_2\text{C}=\text{CHCOCH}_3 \longrightarrow \quad ?$$

Problem 23.20   How would you prepare the following compound using a Robinson annulation reaction between a β-diketone and an α,β-unsaturated ketone? Draw the structures of both reactants and the intermediate Michael addition product.

23.14 Biological Carbonyl Condensation Reactions

Biological Connection ▶

Living organisms use carbonyl condensation reactions for the biological synthesis of a great many different molecules. Fats, amino acids, steroid hormones, and many other kinds of compounds are synthesized by plants and animals using carbonyl condensation reactions as the key step.

Nature uses the two-carbon acetate fragment of acetyl CoA as the major building block for synthesis. Acetyl CoA can act not only as an electrophilic acceptor, being attacked by nucleophiles at the carbonyl group, but also as a nucleophilic donor by loss of its acidic α hydrogen. Once formed, the enolate ion of acetyl CoA can add to another carbonyl group in a condensation reaction. For example, citric acid is biosynthesized by nucleophilic addition of acetyl CoA to the ketone carbonyl group of oxaloacetic acid (2-oxobutanedioic acid) in a kind of mixed aldol reaction.

$$CH_3\overset{O}{\underset{\|}{C}}SCoA \longrightarrow \left[\overset{O}{\underset{\|}{^-}CH_2\overset{}{C}SCoA} \right] + O = \overset{COOH}{\underset{\underset{COOH}{\overset{\|}{CH_2}}}{C}} \longrightarrow HO - \overset{COOH}{\underset{\underset{COOH}{\overset{\|}{CH_2}}}{C}} - CH_2COOH$$

Acetyl CoA
(a thiol ester)

Oxaloacetic acid   **Citric acid**

Acetyl CoA is also involved as the primary biological precursor in the biosynthesis of steroids, fats, and other lipids, where the key step is a Claisen-like condensation reaction. We'll go into more of the details of this process in Section 29.7.

$$\underset{H_3C}{\overset{:O:}{\overset{\|}{C}}} \underset{SCoA}{} + \underset{H_2C}{\overset{O}{\overset{\|}{C}}} \underset{SCoA}{} \longrightarrow \left[\underset{\underset{CoAS}{H_3C}}{\overset{:\overset{..}{O}:^-}{\overset{|}{C}}} \overset{O}{\underset{CH_2CSCoA}{}} \right] \longrightarrow CH_3\overset{O}{\overset{\|}{C}}CH_2\overset{O}{\overset{\|}{C}}SCoA + \ ^-SCoA$$

Acetyl CoA   **Acetyl CoA**
          anion

Acetoacetyl CoA

CHEMISTRY @ WORK

A Prologue to Metabolism

Biochemistry *is* carbonyl chemistry. Almost all metabolic processes used by living organisms involve one or more of the four fundamental carbonyl-group reactions. For example, the digestion and metabolic breakdown of all the major classes of food molecules—fats, carbohydrates, and proteins—take place by nucleophilic addition reactions, nucleophilic acyl substitutions, α substitutions, and carbonyl condensations. Similarly, hormones and other crucial biological molecules are built up from smaller precursors by these same carbonyl-group reactions.

Take *glycolysis*, for example, the metabolic pathway by which organisms convert glucose to pyruvate as the first step in extracting energy from carbohydrates:

$$\text{Glucose} \xrightarrow{\text{Glycolysis}} 2 \ CH_3 - \overset{O}{\overset{\|}{C}} - \overset{O}{\overset{\|}{C}} - O^-$$

Glucose   **Pyruvate**

(continued) ▶

Glycolysis is a ten-step process that begins with conversion of glucose from its cyclic hemiacetal form to its open-chain aldehyde form—a reverse nucleophilic addition reaction. The aldehyde then undergoes tautomerization to yield an enol, which undergoes yet another tautomerization to give the ketone fructose.

Glucose
(hemiacetal form)

Glucose
(aldehyde form)

Glucose enol

Fructose

Fructose, a β-hydroxy ketone, is then cleaved into two three-carbon molecules—one ketone and one aldehyde—by a reverse aldol reaction. Still further carbonyl-group reactions then occur until pyruvate results.

Fructose

The few examples just given are only an introduction; we'll look at several of the major metabolic pathways in much more detail in Chapter 29. You haven't seen the end of carbonyl-group chemistry, though. A good grasp of carbonyl-group reactions is crucial to an understanding of biochemistry.

You are what you eat. Food molecules are metabolized by pathways that involve the four major carbonyl-group reactions.

Summary and Key Words

A **carbonyl condensation reaction** takes place between two carbonyl partners and involves both nucleophilic addition and α-substitution steps. One carbonyl partner (the donor) is converted by base into a nucleophilic enolate ion, which adds to the electrophilic carbonyl group of the second partner (the acceptor). The donor molecule undergoes an α substitution, while the acceptor molecule undergoes a nucleophilic addition.

**Nucleophilic   Electrophilic
    donor   acceptor**

The **aldol reaction** is a carbonyl condensation that occurs between two aldehyde or ketone molecules. Aldol reactions are reversible, leading first to β-hydroxy aldehydes/ketones and then to α,β-unsaturated products. Mixed aldol condensations between two different aldehydes or ketones generally give a mixture of all four possible products. A mixed reaction *can* be successful, however, if one of the two partners is an unusually good donor (ethyl acetoacetate, for instance) or if it can act only as an acceptor (formaldehyde and benzaldehyde, for instance). Intramolecular aldol condensations of 1,4- and 1,5-diketones are also successful and provide a good way to make five- and six-membered rings.

The **Claisen reaction** is a carbonyl condensation that occurs between two ester components and gives a β-keto ester product. Mixed Claisen condensations between two different esters are successful only when one of the two partners has no acidic α hydrogens (ethyl benzoate and ethyl formate, for instance) and thus can function only as the acceptor partner. Intramolecular Claisen condensations, called **Dieckmann cyclization reactions**, provide excellent syntheses of five- and six-membered cyclic β-keto esters starting from 1,6- and 1,7-diesters.

The conjugate addition of a carbon nucleophile to an α,β-unsaturated acceptor is known as the **Michael reaction**. The best Michael reactions take place between unusually acidic donors (β-keto esters or β-diketones) and unhindered α,β-unsaturated acceptors. Enamines, prepared by reaction of a ketone with a disubstituted amine, are also good Michael donors.

Carbonyl condensation reactions are widely used in synthesis. One example of their versatility is the **Robinson annulation reaction**, which leads to the formation of substituted cyclohexenones. Treatment of a β-diketone or β-keto ester with an α,β-unsaturated ketone leads first to a Michael addition, which is followed by intramolecular aldol cyclization. Condensation reactions are also used widely in nature for the biosynthesis of such molecules as fats and steroids.

Summary of Reactions

1. Aldol reaction: a condensation between two ketones, two aldehydes, or one ketone and one aldehyde
   (a) Ketones (Section 23.2)

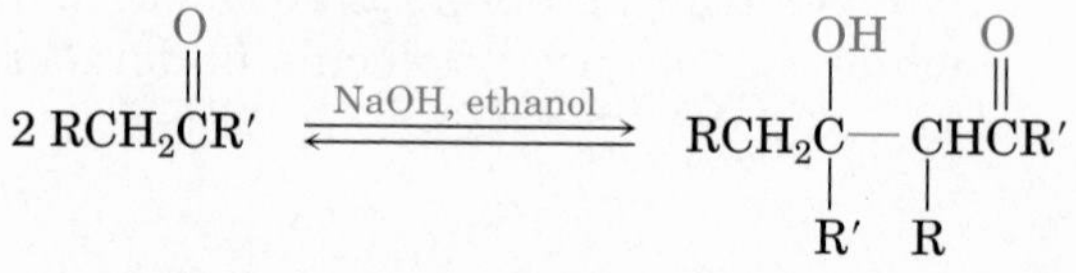

$$2\ RCH_2CR' \xrightleftharpoons{NaOH,\ ethanol} RCH_2\underset{R'}{\overset{OH}{C}}-\underset{R}{CH}CR'$$

   (b) Aldehydes (Section 23.2)

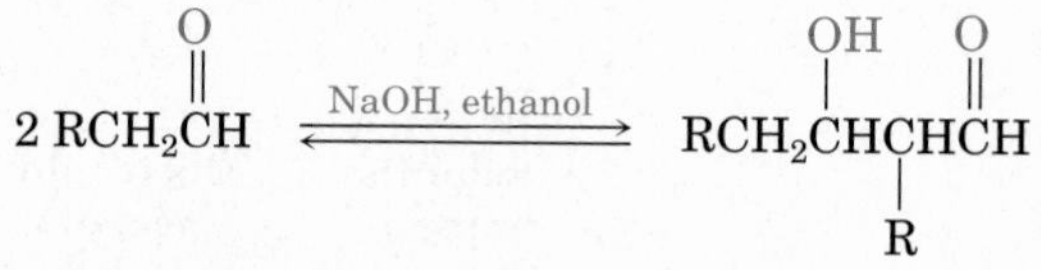

$$2\ RCH_2CH \xrightleftharpoons{NaOH,\ ethanol} RCH_2\overset{OH}{CH}\underset{R}{CH}CH$$

   (c) Mixed aldol reaction (Section 23.6)

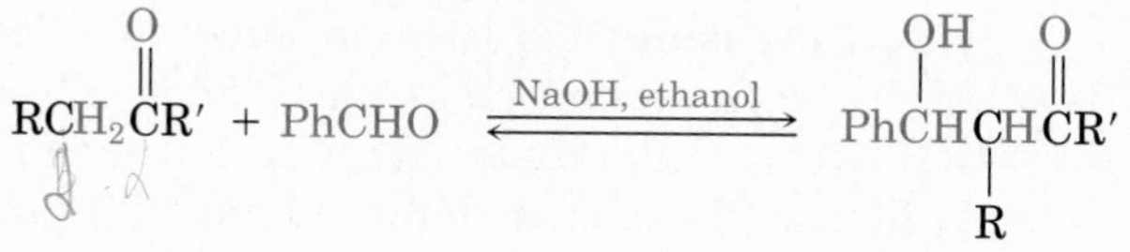

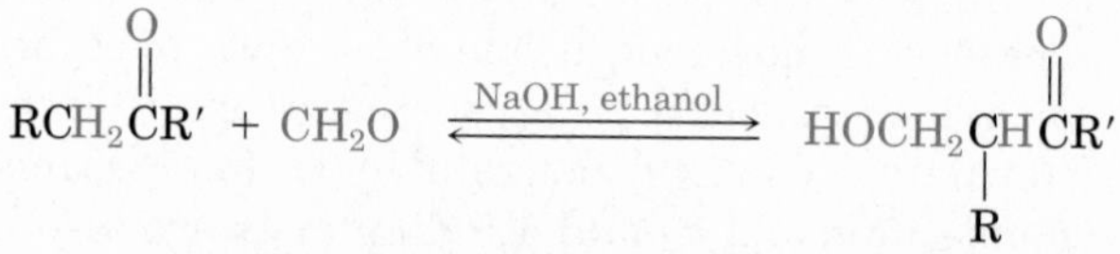

$$RCH_2CR' + PhCHO \xrightleftharpoons{NaOH,\ ethanol} PhCH\underset{R}{\overset{OH}{CH}}CR'$$

$$RCH_2CR' + CH_2O \xrightleftharpoons{NaOH,\ ethanol} HOCH_2\underset{R}{CH}CR'$$

   (d) Intramolecular aldol reaction (Section 23.7)

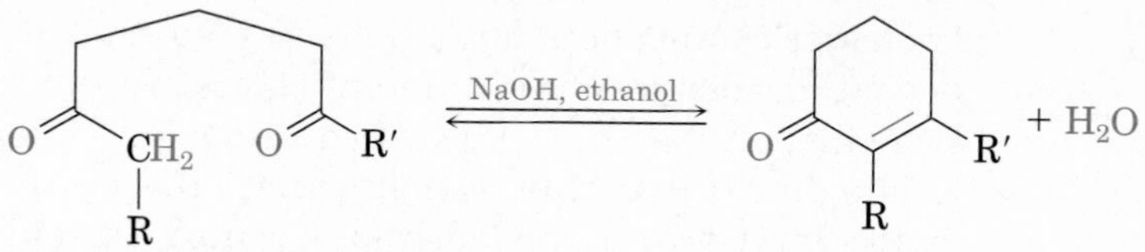

$$\xrightleftharpoons{NaOH,\ ethanol}$$ + H_2O

2. Dehydration of aldol products (Section 23.4)

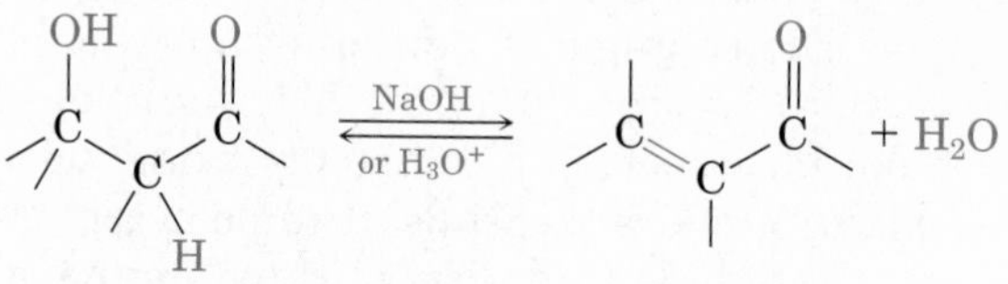

$$\xrightleftharpoons[or\ H_3O^+]{NaOH}$$ + H_2O

3. (a) Claisen reaction: the condensation between two esters or between one ester and one ketone (Section 23.8)

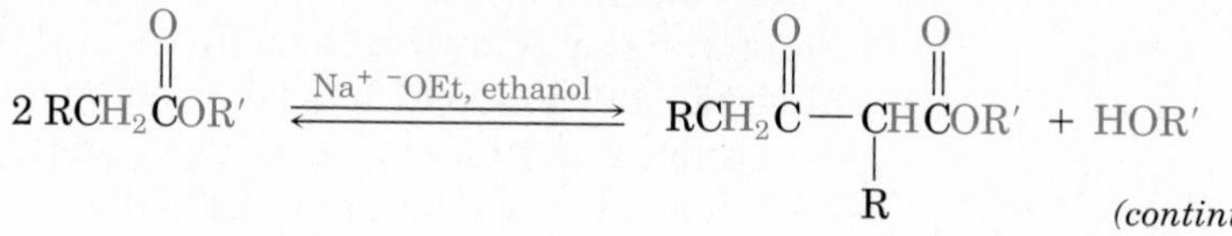

$$2\ RCH_2COR' \xrightleftharpoons{Na^+\ ^-OEt,\ ethanol} RCH_2C-\underset{R}{CH}COR' + HOR'$$

(continued) ▶

(b) Mixed Claisen reaction (Section 23.9)

$$\text{RCH}_2\overset{\overset{\displaystyle O}{\|}}{\text{C}}\text{OEt} + \text{H}\overset{\overset{\displaystyle O}{\|}}{\text{C}}\text{OEt} \underset{}{\overset{\text{Na}^+\ ^-\text{OEt, ethanol}}{\rightleftharpoons}} \text{H}\overset{\overset{\displaystyle O}{\|}}{\text{C}} - \underset{\underset{\displaystyle R}{|}}{\text{CH}}\overset{\overset{\displaystyle O}{\|}}{\text{C}}\text{OEt} + \text{HOEt}$$

4. Dieckmann cyclization; internal Claisen condensation (Section 23.10)

$$\text{EtO}\overset{\overset{\displaystyle O}{\|}}{\text{C}}(\text{CH}_2)_4\overset{\overset{\displaystyle O}{\|}}{\text{C}}\text{OEt} \rightleftharpoons^{\text{Na}^+\ ^-\text{OEt, ethanol}}$$ + HOEt

$$\text{EtO}\overset{\overset{\displaystyle O}{\|}}{\text{C}}(\text{CH}_2)_5\overset{\overset{\displaystyle O}{\|}}{\text{C}}\text{OEt} \rightleftharpoons^{\text{Na}^+\ ^-\text{OEt, ethanol}}$$ + HOEt

5. Michael reaction (Section 23.11)

6. Stork enamine reaction (Section 23.12)

7. Robinson annulation reaction (Section 23.13)

Visualizing Chemistry

• •

(Problems 23.1–23.20 appear within the chapter.)

23.21 What ketones or aldehydes might the following enones have been prepared from by aldol reaction?

(a)   (b)

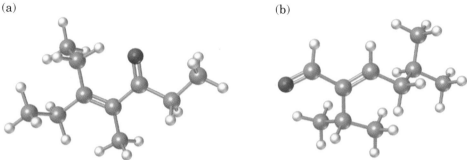

23.22 The following structure represents an intermediate formed by addition of an ester enolate ion to a second ester molecule. Identify the reactant, the leaving group, and the product.

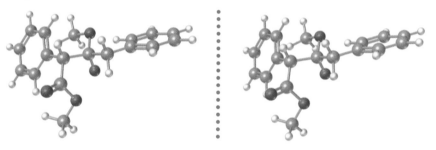

Stereo View

23.23 The following molecule was formed by an intramolecular aldol reaction. What dicarbonyl precursor was used for its preparation?

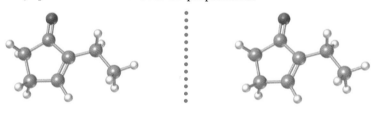

Stereo View

23.24 The following molecule was formed by a Robinson annulation reaction. What reactants were used?

Stereo View

Additional Problems

· ·

23.25 Which of the following compounds would you expect to undergo aldol self-conden-sation? Show the product of each successful reaction.
(a) Trimethylacetaldehyde   (b) Cyclobutanone
(c) Benzophenone (diphenyl ketone)   (d) 3-Pentanone
(e) Decanal   (f) 3-Phenyl-2-propenal

23.26 How might you synthesize each of the following compounds using an aldol reaction? In each case, show the structure of the starting aldehyde(s) or ketone(s) you would use.
(a) $C_6H_5CH=CHCOC_6H_5$ (b) 2-Cyclohexenone

(c)

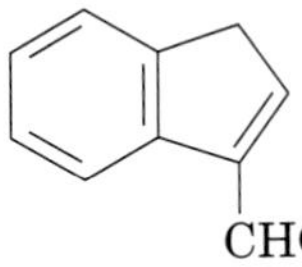

(d)

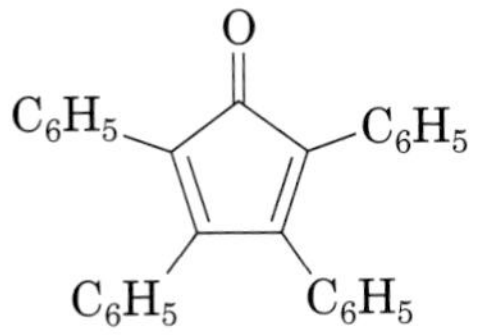

23.27 What product would you expect to obtain from aldol cyclization of hexanedial, $OHCCH_2CH_2CH_2CH_2CHO$?

23.28 Intramolecular aldol cyclization of 2,5-heptanedione with aqueous NaOH yields a mixture of two enone products in the approximate ratio 9:1. Write their structures, and show how each is formed.

23.29 The major product formed by intramolecular aldol cyclization of 2,5-heptanedione (Problem 23.28) has two singlet absorptions in the ^1H NMR spectrum at 1.65 δ and 1.90 δ, and has no absorptions in the range 3–10 δ. What is the structure?

23.30 Treatment of the minor product formed in the intramolecular aldol cyclization of 2,5-heptanedione (Problems 23.28 and 23.29) with aqueous NaOH converts it into the major product. Propose a mechanism to account for this base-catalyzed isomer-ization.

23.31 The aldol reaction is catalyzed by acid as well as by base. What is the reactive nucleo-phile in the acid-catalyzed aldol reaction? Propose a mechanism.

23.32 How can you account for the fact that 2,2,6-trimethylcyclohexanone yields no detectable aldol product even though it has an acidic α hydrogen?

23.33 Cinnamaldehyde, the aromatic constituent of cinnamon oil, can be synthesized by a mixed aldol condensation. Show the starting materials you would use, and write the reaction.

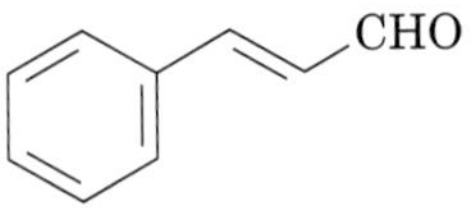

Cinnamaldehyde

23.34 The bicyclic ketone shown below does not undergo aldol self-condensation even though it has two α hydrogen atoms. Explain.

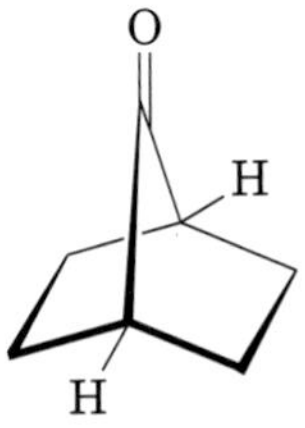

23.35 What condensation products would you expect to obtain by treatment of the following substances with sodium ethoxide in ethanol?
(a) Ethyl butanoate   (b) Cycloheptanone
(c) 3,7-Nonanedione   (d) 3-Phenylpropanal

23.36 In the mixed Claisen reaction of cyclopentanone with ethyl formate, a much higher yield of the desired product is obtained by first mixing the two carbonyl components and then adding base, rather than by first mixing base with cyclopentanone and then adding ethyl formate. Explain.

23.37 Give the structures of the possible Claisen condensation products from the following reactions. Tell which, if any, you would expect to predominate in each case.
(a) CH_3CO_2Et + $CH_3CH_2CO_2Et$ (b) $C_6H_5CO_2Et$ + $C_6H_5CH_2CO_2Et$
(c) $EtOCO_2Et$ + Cyclohexanone   (d) C_6H_5CHO + CH_3CO_2Et

23.38 Ethyl dimethylacetoacetate reacts instantly at room temperature when treated with ethoxide ion to yield two products, ethyl acetate and ethyl 2-methylpropanoate. Propose a mechanism for this cleavage reaction.

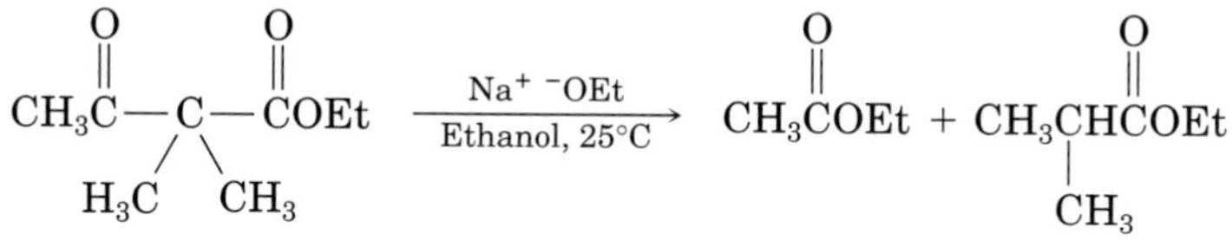

23.39 In contrast to the rapid reaction shown in Problem 23.38, ethyl acetoacetate requires a temperature over 150°C to undergo the same kind of cleavage reaction. How can you explain the difference in reactivity?

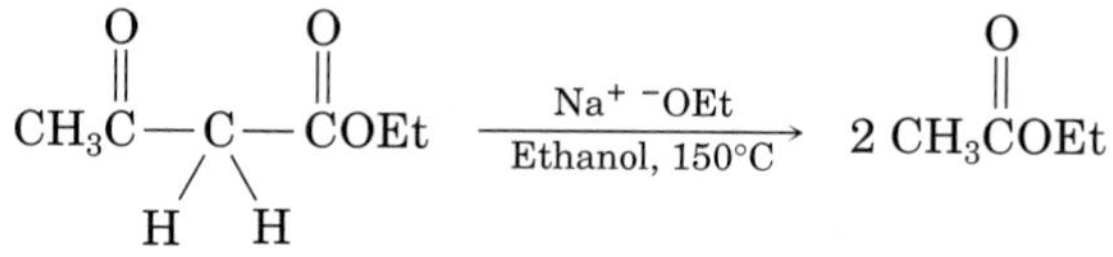

23.40 How might the following compounds be prepared using Michael reactions? Show the nucleophilic donor and the electrophilic acceptor in each case.

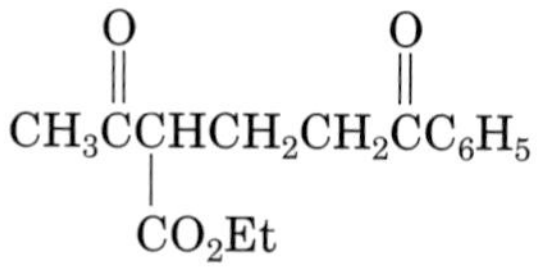

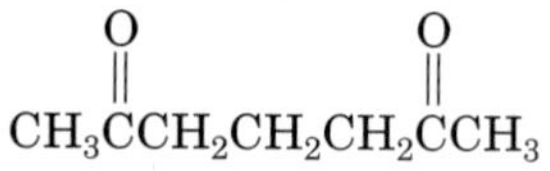

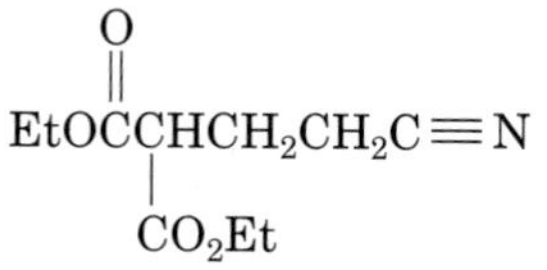

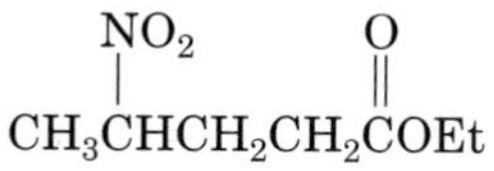

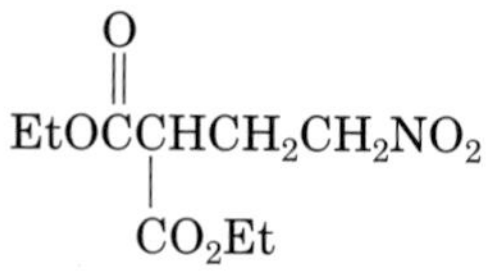

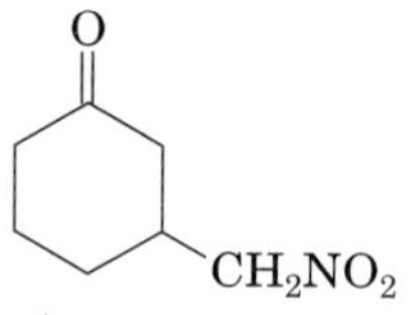

23.41 The so-called *Wieland–Miescher ketone* is a valuable starting material used in the synthesis of steroid hormones. How might you prepare it from 1,3-cyclohexane-dione?

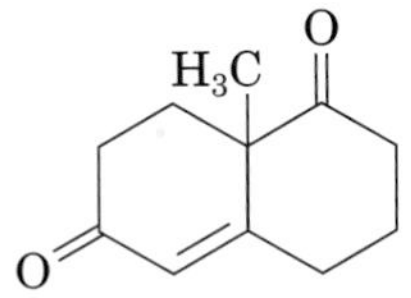

Wieland–Miescher ketone

23.42 The following reactions are unlikely to provide the indicated products in high yield. What is wrong with each?

(a) 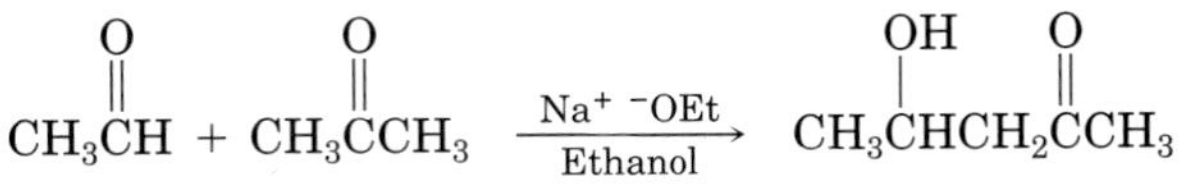 $CH_3\overset{O}{\overset{||}{C}}H + CH_3\overset{O}{\overset{||}{C}}CH_3 \xrightarrow[\text{Ethanol}]{Na^+ \ ^-OEt} CH_3\overset{OH}{\overset{|}{C}}HCH_2\overset{O}{\overset{||}{C}}CH_3$

(b)

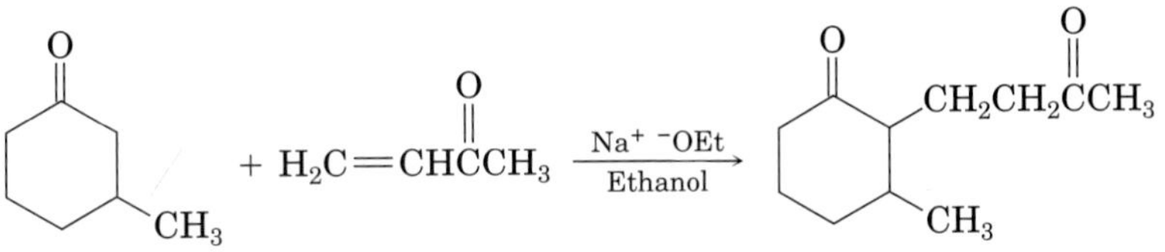

(c) 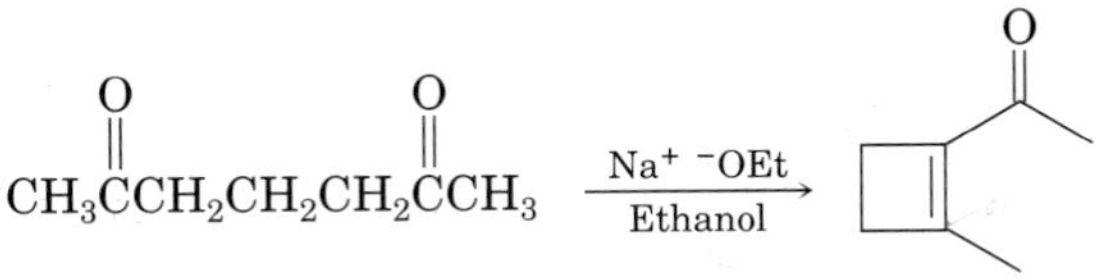 $CH_3\overset{O}{\overset{||}{C}}CH_2CH_2CH_2\overset{O}{\overset{||}{C}}CH_3 \xrightarrow[\text{Ethanol}]{Na^+ \ ^-OEt}$

23.43 Fill in the missing reagents a–h in the following scheme:

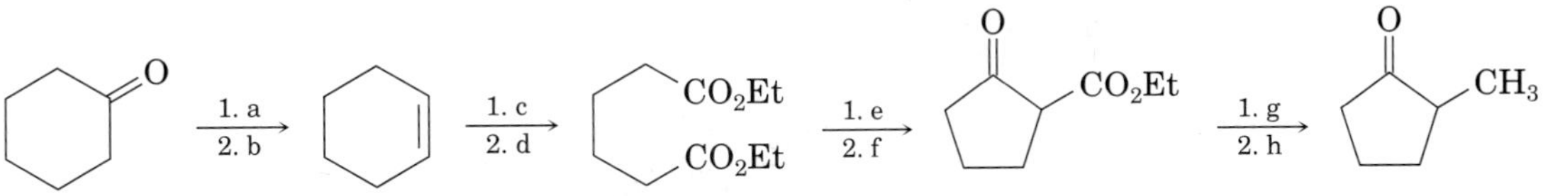

23.44 How would you prepare the following compounds from cyclohexanone?

(a)

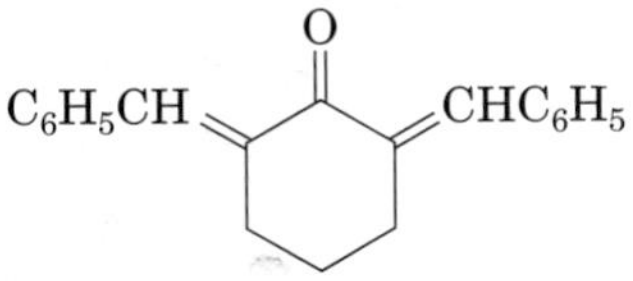

(b)

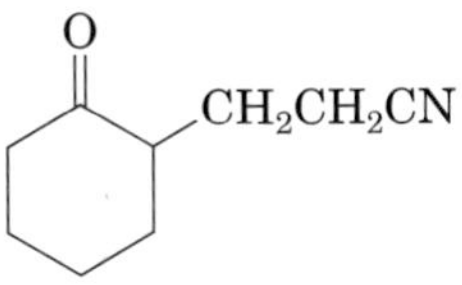

(c)

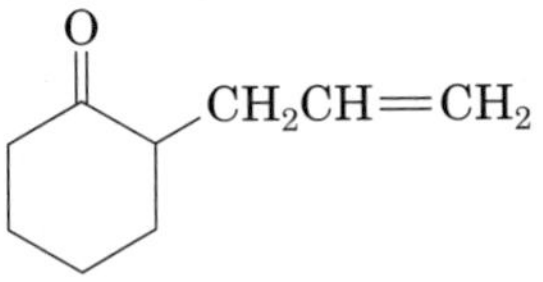

23.45 The compound known as *Hagemann's ester* is prepared by treatment of a mixture of formaldehyde and ethyl acetoacetate with base, followed by acid-catalyzed decarboxylation.

$$CH_3COCH_2CO_2Et + CH_2O \xrightarrow[\text{2. H}_3O^+]{\text{1. Na}^+ \ ^-OEt, \text{ ethanol}}$$

+ CO$_2$ + HOEt

Hagemann's ester

(a) The first step is an aldol-like condensation between ethyl acetoacetate and formaldehyde to yield an α,β-unsaturated product. Write the reaction, and show the structure of the product.
(b) The second step is a Michael reaction between ethyl acetoacetate and the unsaturated product of the first step. Show the structure of the product.

23.46 The third and fourth steps in the synthesis of Hagemann's ester from ethyl acetoacetate and formaldehyde (Problem 23.45) are an intramolecular aldol cyclization to yield a substituted cyclohexenone, and a decarboxylation reaction. Write both reactions, and show the products of each step.

23.47 When 2-methylcyclohexanone is converted into an enamine, only one product is formed despite the fact that the starting ketone is unsymmetrical. Build molecular models of the two possible products, and explain the fact that the sole product is the one with the double bond away from the methyl-substituted carbon.

23.48 The Stork enamine reaction and the intramolecular aldol reaction can be carried out in sequence to allow the synthesis of cyclohexenones. For example, reaction of the pyrrolidine enamine of cyclohexanone with 3-buten-2-one, followed by enamine hydrolysis and base treatment, yields the product indicated. Write each step, and show the mechanism of each.

23.49 How could you prepare the following cyclohexenones by combining a Stork enamine reaction with an intramolecular aldol condensation? (See Problem 23.48.)

(a) (b) (c)

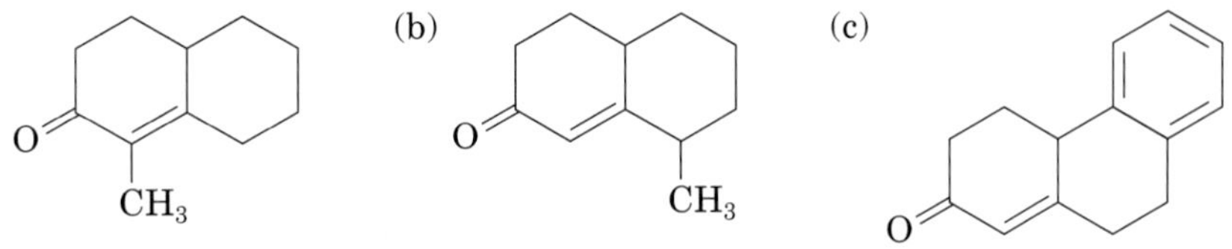

23.50 Griseofulvin, an antibiotic produced by the mold *Penicillium griseofulvum* (Dierckx), has been synthesized by a route that employs a twofold Michael reaction as the key step. Propose a mechanism for this transformation.

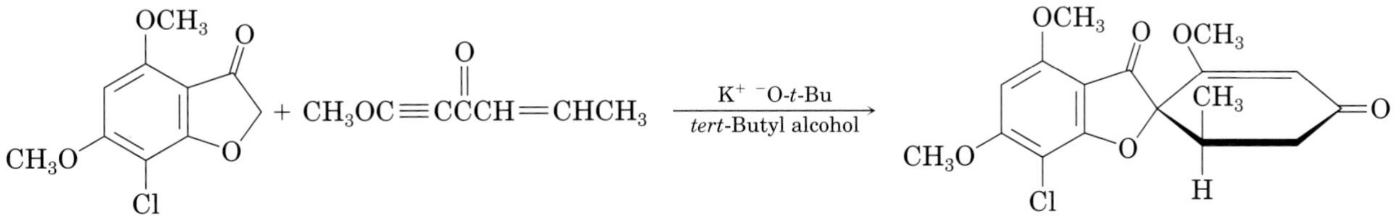

Griseofulvin

23.51 The *Knoevenagel reaction* is a carbonyl condensation reaction of an ester with an aldehyde or ketone to yield an α,β-unsaturated product. Show the mechanism of the Knoevenagel reaction of diethyl malonate with benzaldehyde.

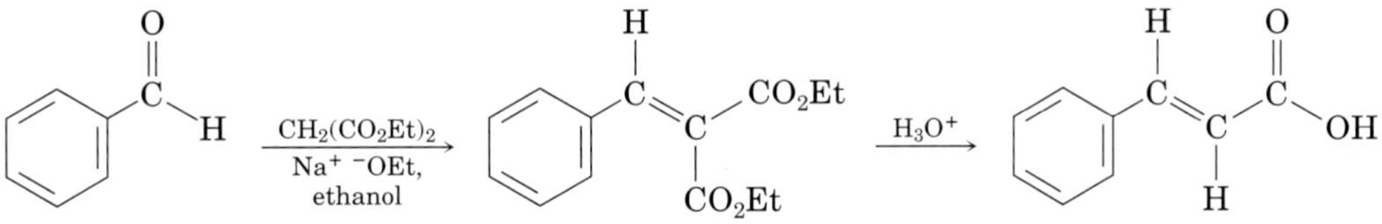

Benzaldehyde   **Cinnamic acid (91%)**

23.52 In the *Perkin reaction*, acetic anhydride condenses with an aromatic aldehyde to yield a cinnamic acid. The reaction takes place by a mixed carbonyl condensation of the anhydride with the aldehyde to yield an α,β-unsaturated intermediate that undergoes hydrolysis to yield the cinnamic acid. What is the structure of the unsaturated intermediate?

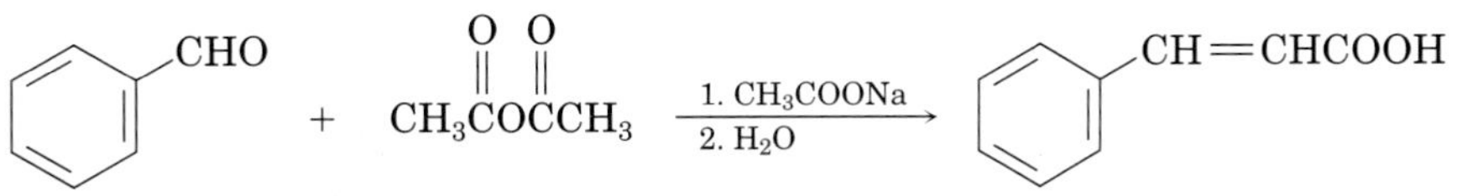

Benzaldehyde   Acetic anhydride   **Cinnamic acid (64%)**

23.53 The *Darzens reaction* involves a two-step, base-catalyzed condensation of ethyl chloroacetate with a ketone to yield an epoxy ester. The first step is a carbonyl condensation reaction, and the second step is an S_N2 reaction. Write both steps and show their mechanisms.

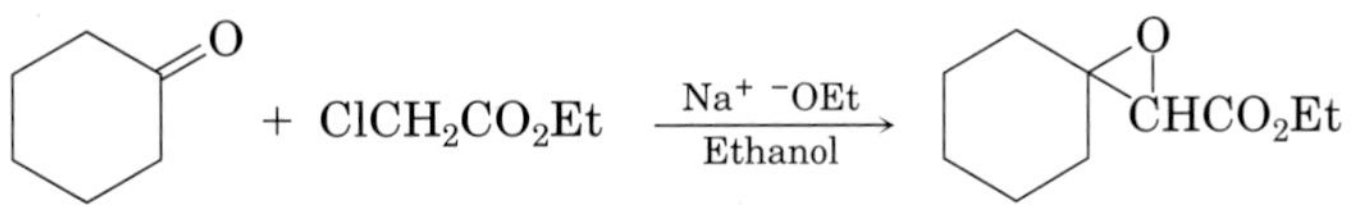

23.54 Propose a mechanism to account for the following reaction:

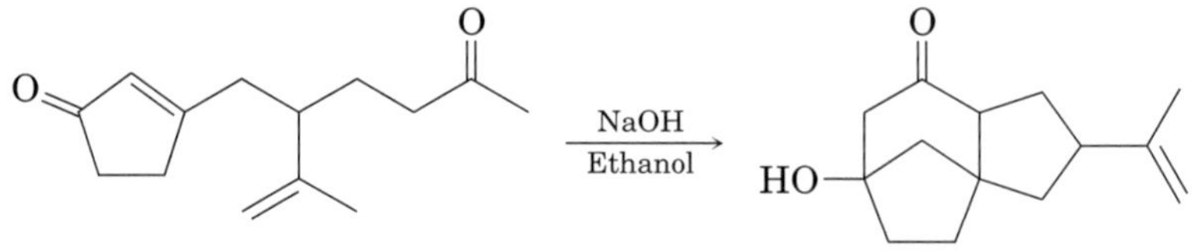

23.55 Propose a mechanism to account for the following reaction:

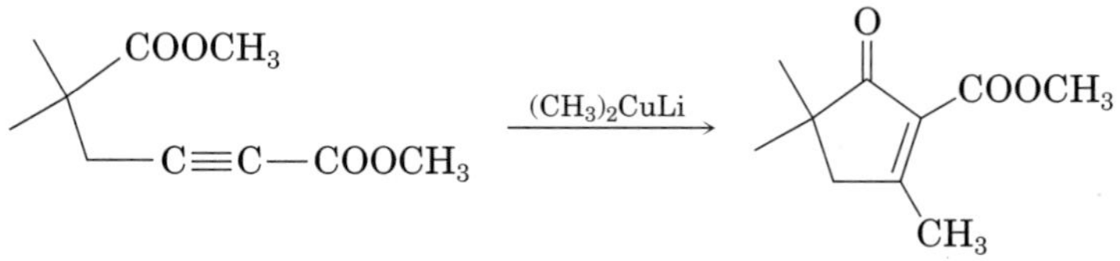

23.56 Propose a mechanism to account for the following reaction:

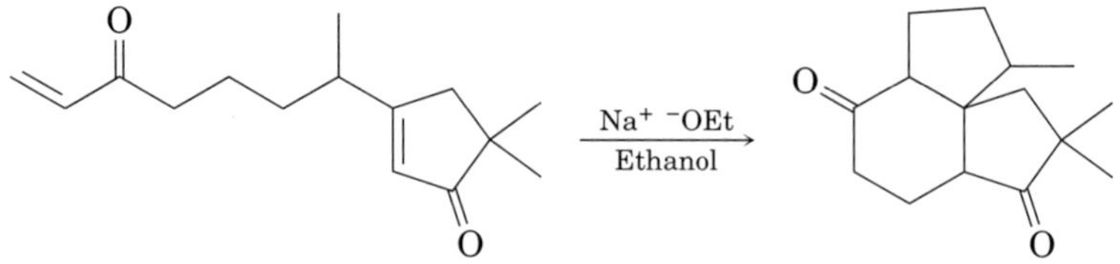

23.57 Propose a mechanism to account for the following reaction:

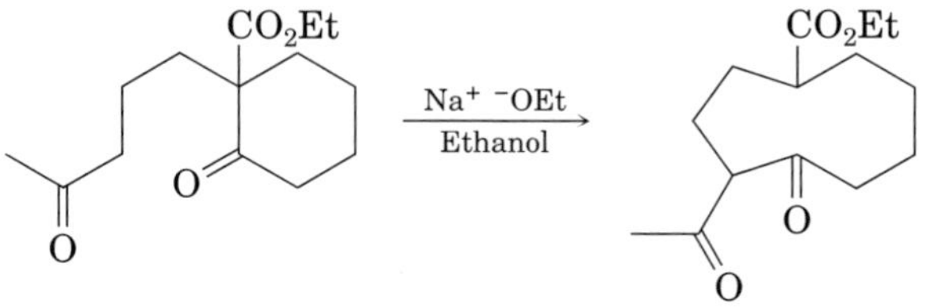

A Look Ahead

23.58 The *Mannich reaction* of a ketone, an amine, and an aldehyde is one of the few three-component reactions in organic chemistry. Cyclohexanone, for example, reacts with dimethylamine and acetaldehyde to yield an amino ketone:

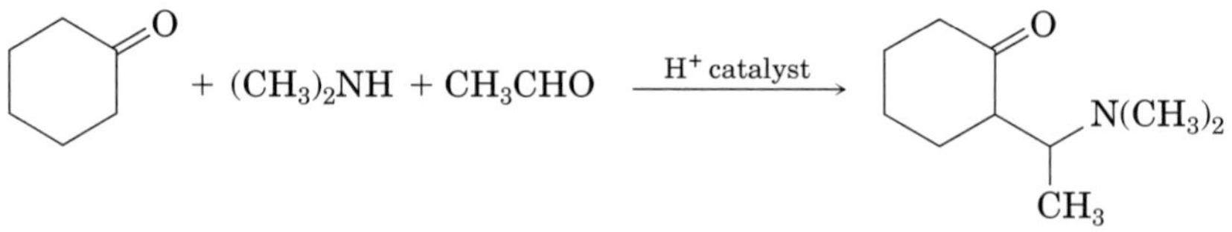

The reaction takes place in two steps, both of which are typical carbonyl-group reactions. The first step is the reaction between the aldehyde and the amine to yield an intermediate iminium ion ($R_2C=NR_2^+$) plus water. The second step is the reaction between the iminium ion intermediate and the ketone to yield the final product. Propose mechanisms for both steps, and show the structure of the intermediate iminium ion.

23.59 Cocaine has been prepared by a sequence beginning with a Mannich reaction (Problem 23.58) between dimethyl acetonedicarboxylate, an amine, and a dialdehyde. Show the structures of the amine and dialdehyde.

CH₃O₂C ⟍⟋ O ⟍⟋ CO₂CH₃

+ Amine + Dialdehyde ⟶

[cocaine synthesis structures]

Cocaine

Molecular Modeling

● ●

23.60 Intramolecular aldol condensation of 2,7-octanedione might give either of two cyclic products, A or B. Use SpartanView to compare the energies of transition state A and transition state B for the ring-forming step, and tell which product would be obtained if the reaction were kinetically controlled.

[structures]

Base ⟶

A or **B**

23.61 Dieckmann cyclization of diethyl adipate can be described by the two successive equilibria shown below. Use SpartanView to obtain the energies of diethyl adipate, the keto ester, the keto ester enolate ion, ethanol, and ethoxide anion, and calculate $\Delta H°$ for both steps. Which step is more favorable?

EtO₂C ⟍⟋⟍⟋ CO₂Et ⇌ [keto ester] + EtOH ⇌⁻OEt [enolate] + EtOH

Diethyl adipate

23.62 Imines can be deprotonated by LDA to give anions that react like enolate ions. Use SpartanView to compare electrostatic potential maps of an imine anion and an enamine. Which is the stronger nucleophile?

[structures]

LDA ⟶

An imine   **An imine anion**   **An enamine**

Amines

Amines are organic derivatives of ammonia, NH_3, in the same way that alcohols and ethers are organic derivatives of water, H_2O. Like ammonia, amines contain a nitrogen atom with a lone pair of electrons, making amines both basic and nucleophilic. We'll soon see that most of the chemistry of amines depends on the presence of this lone pair of electrons.

Amines occur widely throughout both plants and animals. Trimethylamine, for instance, occurs in animal tissues and is partially responsible for the distinctive odor of many fish; quinine is an important antimalarial drug isolated from the bark of the South American *Cinchona* tree; and codeine is an analgesic (painkiller) found in the opium poppy.

Quinine—an antimalarial

Codeine—an analgesic

24.1 Naming Amines

Amines are classified as **primary (RNH_2), secondary (R_2NH), or tertiary (R_3N)**, depending on the number of organic substituents attached to nitrogen. For example, methylamine (CH_3NH_2) is a primary amine, dimethylamine [$(CH_3)_2NH$] is a secondary amine, and trimethylamine [$(CH_3)_3N$] is a tertiary amine. Note that this usage of the terms *primary, secondary,* and *tertiary* is different from our previous usage. When we speak of a tertiary alcohol or alkyl halide, we refer to the degree of substitution at the alkyl carbon atom, but when we speak of a tertiary amine, we refer to the degree of substitution at the nitrogen atom.

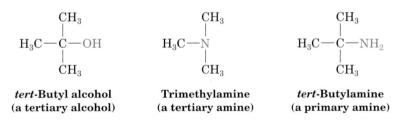

| ***tert*-Butyl alcohol** | **Trimethylamine** | ***tert*-Butylamine** |
| (a tertiary alcohol) | (a tertiary amine) | (a primary amine) |

Compounds containing a nitrogen atom with four attached groups also exist, but the nitrogen atom must carry a formal positive charge. Such compounds are called **quaternary ammonium salts**.

$$R-\overset{\overset{R}{|}}{\underset{\underset{R}{|}}{\overset{+}{N}}}-R \quad X^- \quad \textbf{A quaternary ammonium salt}$$

Primary amines are named in the IUPAC system in several ways, depending on their structure. For simple amines, the suffix *-amine* is added to the name of the alkyl substituent:

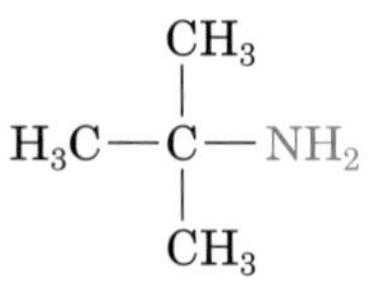

tert-Butylamine   **Cyclohexyl**amine   **1,4-Butane**diamine

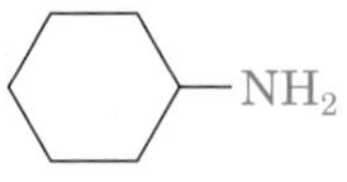

Alternatively, the suffix -*amine* can be used in place of the final -*e* in the name of the parent compound:

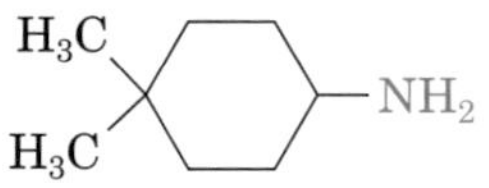

4,4-Dimethylcyclohexanamine

Amines with more than one functional group are named by considering the $-NH_2$ as an *amino* substituent on the parent molecule:

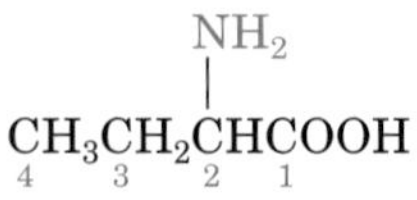

2-Aminobutanoic acid 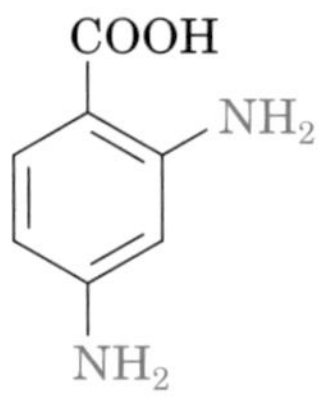   **2,4-Diamino**benzoic acid 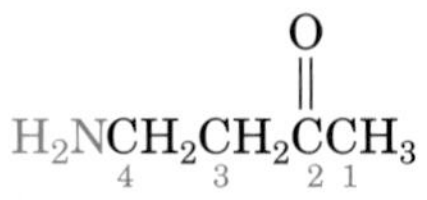   **4-Amino**-2-butanone

Symmetrical secondary and tertiary amines are named by adding the prefix *di-* or *tri-* to the alkyl group:

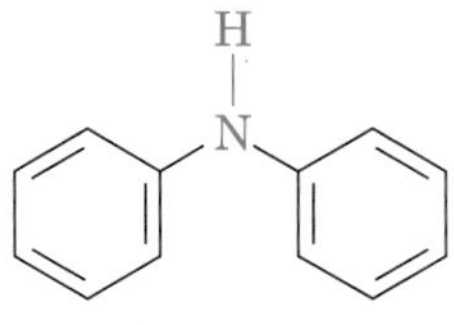

Diphenylamine 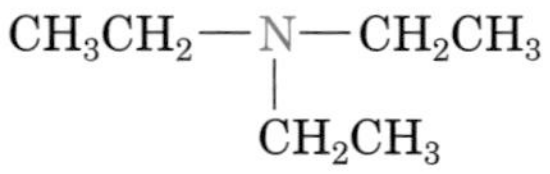   **Triethylamine**

Unsymmetrically substituted secondary and tertiary amines are named as *N*-substituted primary amines. The largest alkyl group is chosen as the parent name, and the other alkyl groups are considered *N*-substituents on the parent (*N* because they're attached to nitrogen).

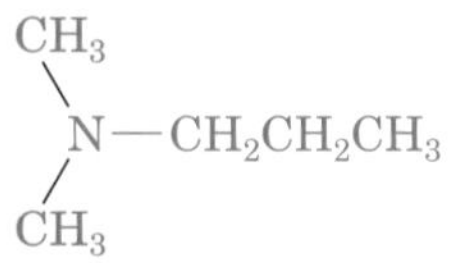

N,N-Dimethylpropylamine
(propylamine is the parent name; the two methyl groups are substituents on nitrogen)

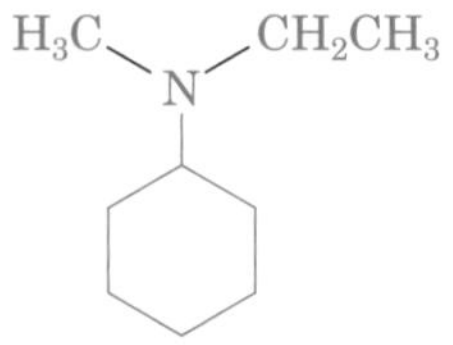

N-Ethyl-*N*-methylcyclohexylamine
(cyclohexylamine is the parent name; methyl and ethyl are *N*-substituents)

There are relatively few common names for alkylamines, but two of the simplest aromatic amines, or **arylamines**, are called *aniline* and *toluidine*.

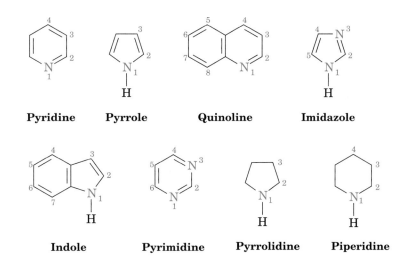

Aniline   ***o*-Toluidine**

Heterocyclic amines, compounds in which the nitrogen atom occurs as part of a ring, are also common, and each different heterocyclic ring system has its own parent name. The heterocyclic nitrogen atom is always numbered as position 1.

Pyridine   **Pyrrole**   **Quinoline**   **Imidazole**

Indole   **Pyrimidine**   **Pyrrolidine**   **Piperidine**

• •

Problem 24.1   Name the following compounds by IUPAC rules:

(a) CH₃NHCH₂CH₃

(b) [cyclohexyl-N]₃

(c) CH₃—N—CH₂CH₂CH₃ (with cyclohexyl substituent)

(d) pyrrolidine with N—CH₃

(e) [(CH₃)₂CH]₂NH

(f) H₂NCH₂CH₂CHNH₂ with CH₃

Problem 24.2   Draw structures corresponding to the following IUPAC names:
(a) Triethylamine   (b) Triallylamine
(c) *N*-Methylaniline   (d) *N*-Ethyl-*N*-methylcyclopentylamine
(e) *N*-Isopropylcyclohexylamine   (f) *N*-Ethylpyrrole

Problem 24.3   Draw structures for the following heterocyclic amines:

(a) 5-Methoxyindole (b) 1,3-Dimethylpyrrole
(c) 4-(*N*, *N*-Dimethylamino)pyridine (d) 5-Aminopyrimidine

• •

24.2 Structure and Bonding in Amines

The bonding in amines is similar to the bonding in ammonia. The nitrogen atom is sp^3-hybridized, with the three substituents occupying three corners of a tetrahedron and the lone pair of electrons occupying the fourth corner. As you might expect, the C–N–C bond angles are close to the 109° tetrahedral value. For trimethylamine, the C–N–C bond angle is 108°, and the C–N length is 147 pm. An electrostatic potential map of trimethylamine shows that the negative region (red) coincides with the lone-pair orbital on nitrogen.

trimethylamine

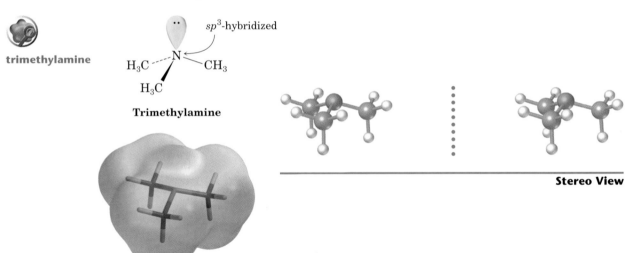

sp^3-hybridized

H_3C ⋯ N ← CH_3
H_3C

Trimethylamine

Stereo View

One consequence of tetrahedral geometry is that an amine with three different substituents on nitrogen is chiral. Such an amine has no plane of symmetry and therefore is not superimposable on its mirror image. If we consider the lone pair of electrons to be the fourth substituent on nitrogen, these chiral amines are analogous to chiral alkanes with four different substituents attached to carbon:

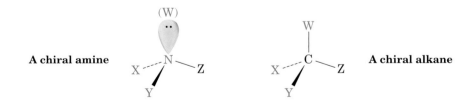

A chiral amine **A chiral alkane**

Unlike chiral carbon-based compounds, most chiral amines can't be resolved because the two enantiomeric forms rapidly interconvert by a *pyramidal inversion*, much as an alkyl halide inverts in an S_N2 reaction. Pyramidal inversion occurs by a momentary rehybridization of the nitrogen atom to planar, sp^2 geometry, followed by rehybridization of the planar intermediate to tetrahedral, sp^3 geometry (Figure 24.1).

FIGURE 24.1 ▼

Pyramidal inversion rapidly interconverts the two mirror-image (enantiomeric) forms of an amine.

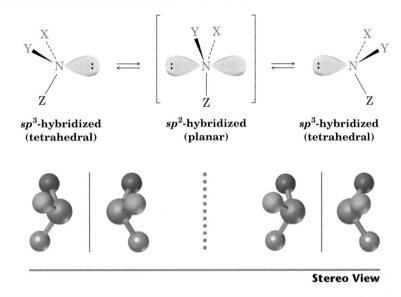

sp^3-hybridized
(tetrahedral)

sp^2-hybridized
(planar)

sp^3-hybridized
(tetrahedral)

Stereo View

Spectroscopic studies have shown that the barrier to nitrogen inversion is about 25 kJ/mol (6 kcal/mol), a figure only twice as large as the barrier to rotation about a C–C single bond. Pyramidal inversion is therefore so rapid at room temperature that the two enantiomeric forms can't normally be isolated.

24.3 Properties and Sources of Amines

Alkylamines have a variety of minor applications in the chemical industry as starting materials for the preparation of insecticides and pharmaceuticals.

For example, propranolol, a heart stimulant used in the control of cardiac arrhythmia, is prepared by S_N2 reaction of an epoxide with isopropylamine.

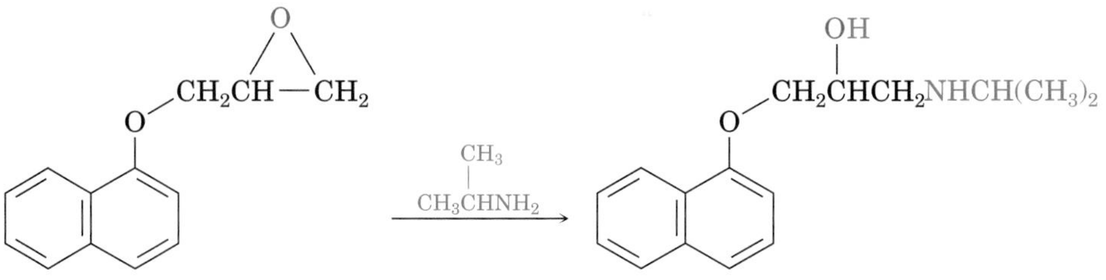

Propranolol (a heart stimulant)

Simple methylated amines are prepared by reaction of ammonia with methanol in the presence of an alumina catalyst. The reaction yields a mixture of mono-, di-, and trimethylated products but is nonetheless useful industrially because the separation of the three products by distillation is easy.

$$NH_3 + CH_3OH \xrightarrow[450°C]{Al_2O_3} CH_3NH_2 + CH_3\overset{H}{\underset{|}{N}}CH_3 + CH_3\overset{CH_3}{\underset{|}{N}}CH_3$$

Like alcohols, amines with fewer than five carbon atoms are generally water-soluble. Also like alcohols, primary and secondary amines form hydrogen bonds and are highly associated:

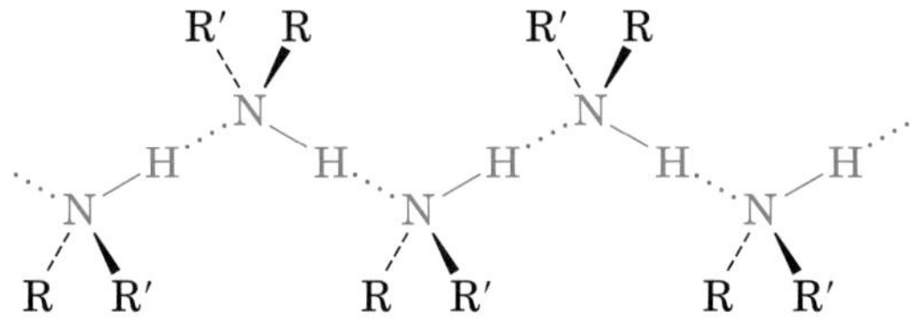

As a result, amines have higher boiling points than alkanes of similar molecular weight. Diethylamine, for example, boils at 56.3°C, while pentane boils at 36.1°C.

$$CH_3CH_2\overset{H}{\underset{|}{N}}CH_2CH_3 \qquad CH_3CH_2CH_2CH_2CH_3$$

Diethylamine, MW = 71.1 amu   **Pentane, MW = 72.1 amu**
bp = 56.3°C   **bp = 36.1°C**

Another characteristic of amines is their *odor*. Low-molecular-weight amines such as trimethylamine have a distinctive fishlike aroma, while diamines such as cadaverine (1,5-pentanediamine) have names that are self-explanatory.

24.4 Basicity of Amines

The chemistry of amines is dominated by the lone pair of electrons on nitrogen. Because of this lone pair, amines are both basic and nucleophilic. They react with acids to form acid–base salts, and they react with electrophiles in many of the polar reactions seen in past chapters.

An amine   **An acid**   **A salt**
(a Lewis base)

Amines are considerably more basic than alcohols, ethers, or water. When an amine is dissolved in water, an equilibrium is established in which water acts as an acid and transfers a proton to the amine. Just as the acid strength of a carboxylic acid can be measured by defining an acidity constant K_a (Section 2.8), the base strength of an amine can be measured by defining an analogous *basicity constant* K_b. The larger the value of K_b (and the smaller the value of pK_b), the more favorable the proton-transfer equilibrium and the stronger the base.

For the reaction:

$$RNH_2 + H_2O \ \rightleftharpoons\ RNH_3^+ + OH^-$$

$$K_b = \frac{[RNH_3^+][OH^-]}{[RNH_2]} \qquad pK_b = -\log K_b$$

In practice, K_b values (or pK_b values) are not often used. Instead, the most convenient way to measure the *basicity* of an amine (RNH_2) is to look at the *acidity* of the corresponding ammonium ion (RNH_3^+).

For the reaction:

$$RNH_3^+ + H_2O \ \rightleftharpoons\ RNH_2 + H_3O^+$$

$$K_a = \frac{[RNH_2][H_3O^+]}{[RNH_3^+]}$$

so:
$$K_a \cdot K_b = \left[\frac{[RNH_2][H_3O^+]}{[RNH_3^+]} \right]\left[\frac{[RNH_3^+][OH^-]}{[RNH_2]} \right]$$

$$= [H_3O^+][OH^-] = K_w = 1.00 \times 10^{-14}$$

Thus:
$$K_a = \frac{K_w}{K_b} \qquad \text{and} \qquad K_b = \frac{K_w}{K_a}$$

and:
$$pK_a + pK_b = 14$$

These equations say that the K_b of an amine multiplied by the K_a of the corresponding ammonium ion is equal to K_w, the ion-product constant for water (1.00×10^{-14}). Thus, if we know K_a for an ammonium ion, we also know K_b for the corresponding amine base because $K_b = K_w/K_a$. The more acidic the ammonium ion (larger K_a or smaller pK_a), the weaker the base. Thus, a weaker base has an ammonium ion with a smaller pK_a, and a stronger base has an ammonium ion with a larger pK_a.

> **Weaker base:**   Smaller pK_a for ammonium ion
> **Stronger base:**   Larger pK_a for ammonium ion

This relationship between the acidity of a conjugate acid (RNH_3^+) and the basicity of its conjugate base (RNH_2) is an example of the general relationship we saw in Section 2.9. A more strongly basic amine holds a proton more tightly, so its corresponding ammonium ion is less acidic. Conversely, a more weakly basic amine holds a proton less tightly, so its corresponding ammonium ion is more acidic.

If this ammonium salt has a smaller pK_a
(stronger acid), then this amine
is a weaker base.

$$R\!-\!NH_3^+ + H_2O \;\rightleftarrows\; R\!-\!\overset{..}{N}H_2 + H_3O^+$$

If this ammonium salt has a larger pK_a
(weaker acid), then this amine
is a stronger base.

Table 24.1 lists the pK_a's of some ammonium ions and indicates that there is a substantial range of amine basicities. Most simple alkylamines are similar in their base strength, with pK_a's for their ammonium ions in the narrow range 10–11. *Arylamines*, however, such as aniline, are considerably less basic than alkylamines, as are the heterocyclic amines pyridine and pyrrole.

The lower basicity of pyridine is due to the fact that the lone-pair electrons on nitrogen are in an sp^2 orbital, while those in an alkylamine are in an sp^3 orbital. Because s orbitals have their maximum electron density at the nucleus but p orbitals have a node at the nucleus (Section 1.2), electrons in an orbital with more s character are held more closely to the positively charged nucleus and are less available for bonding. As a result, the sp^2-hybridized nitrogen atom (33% s character) in pyridine is less basic than the sp^3-hybridized nitrogen in an alkylamine (25% s character).

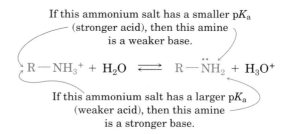

Pyridine

TABLE 24.1 Basicity of Some Common Amines

Name	Structure	pK_a of ammonium ion
Ammonia	NH_3	9.26
Primary alkylamine		
Ethylamine	$CH_3CH_2NH_2$	10.81
Methylamine	CH_3NH_2	10.66
Secondary alkylamine		
Pyrrolidine	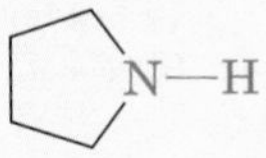 N—H	11.27
Dimethylamine	$(CH_3)_2NH$	10.73
Diethylamine	$(CH_3CH_2)_2NH$	10.49
Tertiary alkylamine		
Triethylamine	$(CH_3CH_2)_3N$	11.01
Trimethylamine	$(CH_3)_3N$	9.81
Arylamine		
Aniline	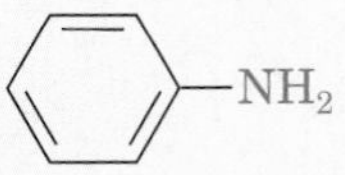 —NH_2	4.63
Heterocyclic amine		
Pyridine	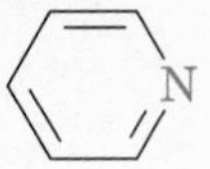 N	5.25
Pyrrole	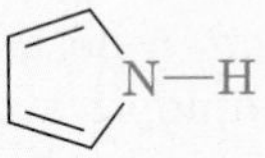 N—H	0.4

The almost complete lack of basicity in pyrrole is due to the fact that the lone-pair electrons on nitrogen are part of an aromatic sextet (Section 15.7). As a result, they are not available for bonding to an acid without disrupting the aromatic stability of the ring.

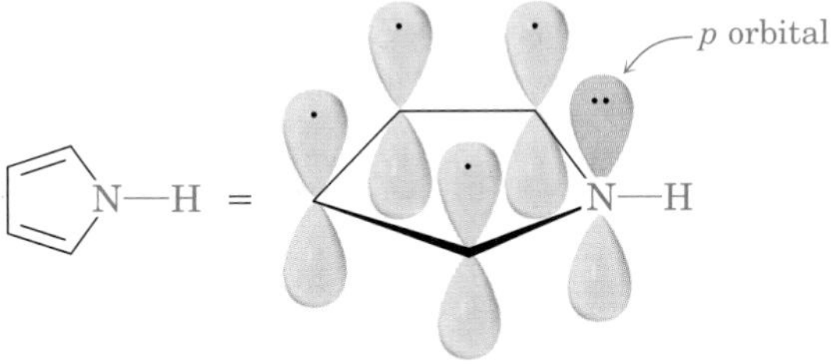

p orbital

Pyrrole

In contrast to amines, *amides* ($RCONH_2$) are nonbasic. Amides do not undergo protonation when treated with aqueous acids, and they are poor nucleophiles. The main reason for this difference in basicity between amines and amides is that an amide is stabilized by delocalization of the nitrogen lone-pair electrons through orbital overlap with the carbonyl group. Thus, electrostatic potential maps show that the amide nitrogen of formamide is much less negative than the nitrogen in methylamine.

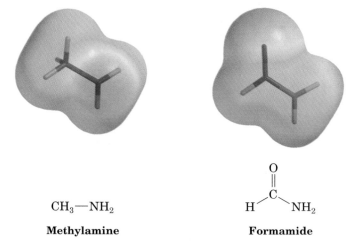

$$CH_3—NH_2$$

Methylamine

Formamide

In resonance language, there are two contributing forms. Since this amide resonance stabilization would be lost if the nitrogen atom were protonated, protonation is disfavored.

It's often possible to take advantage of their basicity to purify amines. For example, if a mixture of a basic amine and a neutral compound such as a ketone or alcohol is dissolved in an organic solvent and shaken with aqueous acid, the basic amine dissolves in the water layer as its protonated salt, while the neutral compound remains in the organic solvent layer. Separation, addition of base, and extraction of the aqueous layer with organic solvent then provides the pure amine (Figure 24.2).

FIGURE 24.2 ▼

Separation and purification of an amine component from a mixture.

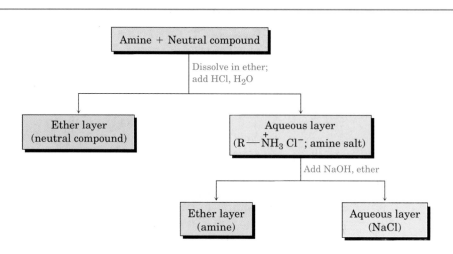

In addition to their behavior as bases, primary and secondary amines can also act as very weak *acids* because an N–H proton can be removed by a sufficiently strong base. We've already seen, for example, how diisopropylamine ($pK_a \approx 40$) reacts with butyllithium to yield lithium diisopropylamide (LDA; Section 22.5).

$$\text{C}_4\text{H}_9\text{Li} + \text{H}-\text{N} \xrightarrow[\text{solvent}]{\text{THF}} \text{Li}^+ \ ^-\text{:N} + \text{C}_4\text{H}_{10}$$

Butyllithium   **Diisopropylamine**   **Lithium diisopropylamide (LDA)**

Dialkylamine anions like LDA are extremely powerful bases that are much used in organic chemistry, particularly for the generation of enolate ions from carbonyl compounds (Section 22.8).

Problem 24.4 Which compound in each of the following pairs is more basic?
(a) $CH_3CH_2NH_2$ or $CH_3CH_2CONH_2$
(b) NaOH or CH_3NH_2
(c) CH_3NHCH_3 or pyridine

Problem 24.5 The benzylammonium ion ($C_6H_5CH_2NH_3^+$) has $pK_a = 9.33$, and the propylammonium ion has $pK_a = 10.71$. Which is the stronger base, benzylamine or propylamine? What are the pK_b's of benzylamine and propylamine?

24.5 Basicity of Substituted Arylamines

As indicated in Table 24.1, arylamines are generally less basic than alkylamines. Anilinium ion has $pK_a = 4.63$, for instance, whereas methylammonium ion has $pK_a = 10.66$. Arylamines are less basic than alkylamines because the nitrogen lone-pair electrons are delocalized by interaction with the aromatic ring π electron system and are less available for bonding to H$^+$. In resonance terms, arylamines are stabilized relative to alkylamines because of the five resonance structures:

Resonance stabilization is lost on protonation, though, because only two resonance structures are possible for the arylammonium ion:

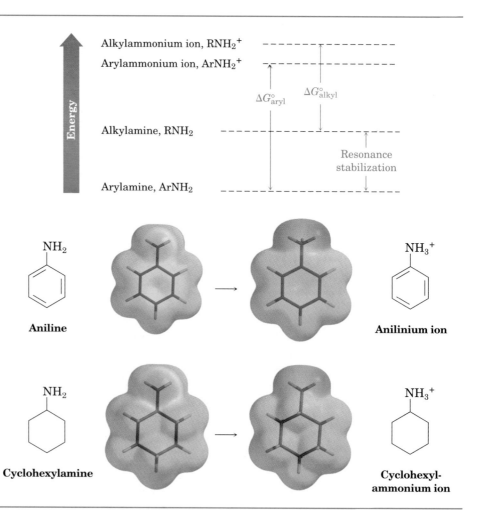

As a result, the energy difference $\Delta G°$ between protonated and nonprotonated forms is higher for arylamines than it is for alkylamines, and arylamines are therefore less basic. Figure 24.3 illustrates the difference in $\Delta G°$ for the two cases.

FIGURE 24.3 ▼

Arylamines have a larger positive $\Delta G°$ for protonation and are therefore less basic than alkylamines, primarily because of resonance stabilization of their ground state. Electrostatic potential maps show that lone-pair electron density is delocalized in aniline compared to cyclohexylamine but that the corresponding ammonium ions localize charge in the same way.

aniline,
anilinium ion,
cyclohexylamine,
cyclohexylammonium ion

Substituted arylamines can be either more basic or less basic than aniline, depending on the substituent. Electron-donating substituents, such as –CH$_3$, –NH$_2$, and –OCH$_3$, which increase the reactivity of an aromatic ring toward electrophilic substitution (Section 16.5), also increase the basicity of the corresponding arylamine. Electron-withdrawing substituents, such as –Cl, –NO$_2$, and –CN, which decrease ring reactivity toward electrophilic substitution, also decrease arylamine basicity. Table 24.2 considers only *p*-substituted anilines, but similar trends are observed for ortho and meta derivatives.

TABLE 24.2 Base Strength of Some *p*-Substituted Anilines

$$Y-\underset{}{\bigcirc}-\ddot{N}H_2 + H_2O \rightleftharpoons Y-\underset{}{\bigcirc}-\overset{+}{N}H_3 + {}^-OH$$

	Substituent, Y	pK$_a$	
Stronger base	—NH$_2$	6.15	Activating groups
	—OCH$_3$	5.34	
	—CH$_3$	5.08	
	—H	4.63	
	—Cl	3.98	Deactivating groups
	—Br	3.86	
	—C≡N	1.74	
Weaker base	—NO$_2$	1.00	

Problem 24.6 Rank the following compounds in order of ascending basicity. (Don't look at Table 24.2.)
(a) *p*-Nitroaniline, *p*-aminobenzaldehyde, *p*-bromoaniline
(b) *p*-Chloroaniline, *p*-aminoacetophenone, *p*-methylaniline
(c) *p*-(Trifluoromethyl)aniline, *p*-methylaniline, *p*-(fluoromethyl)aniline

24.6 Synthesis of Amines

Reduction of Nitriles, Amides, and Nitro Compounds

We've already seen in Sections 21.7 and 21.8 how amines can be prepared by reduction of amides and nitriles with LiAlH$_4$. The two-step sequence of S$_N$2 displacement with CN$^-$ followed by reduction is an excellent method for converting an alkyl halide into a primary alkylamine having one more carbon atom. Amide reduction provides an excellent method for converting carboxylic acids and their derivatives into amines with the same number of carbon atoms.

$$RX \xrightarrow{\text{NaCN}} RCN \xrightarrow[\text{2. H}_2\text{O}]{\text{1. LiAlH}_4,\text{ ether}} RCH_2NH_2$$

Alkyl halide   **1° amine**

$$\underset{\textbf{Carboxylic acid}}{R-\overset{\overset{\displaystyle O}{\|}}{C}-OH} \xrightarrow[\text{2. NH}_3]{\text{1. SOCl}_2} R-\overset{\overset{\displaystyle O}{\|}}{C}-NH_2 \xrightarrow[\text{2. H}_2\text{O}]{\text{1. LiAlH}_4,\text{ ether}} \underset{\textbf{1° amine}}{RCH_2NH_2}$$

Arylamines are usually prepared by nitration of an aromatic starting material, followed by reduction of the nitro group. The reduction step can be carried out in many different ways, depending on the circumstances. Catalytic hydrogenation over platinum is clean and gives high yields, but is often incompatible with the presence elsewhere in the molecule of other reducible groups, such as C=C bonds or carbonyl groups. Iron, zinc, tin, and stannous chloride ($SnCl_2$) are also effective when used in acidic aqueous solution. Stannous chloride is particularly mild and is often used when other reducible functional groups are present.

$$\text{H}_3\text{C}-\underset{\underset{\displaystyle \text{CH}_3}{|}}{\overset{\overset{\displaystyle \text{CH}_3}{|}}{\text{C}}}-\text{C}_6\text{H}_4-\text{NO}_2 \xrightarrow[\text{Ethanol}]{\text{H}_2,\text{ Pt catalyst}} \text{H}_3\text{C}-\underset{\underset{\displaystyle \text{CH}_3}{|}}{\overset{\overset{\displaystyle \text{CH}_3}{|}}{\text{C}}}-\text{C}_6\text{H}_4-\text{NH}_2$$

p-tert-**Butylnitrobenzene**   *p-tert*-**Butylaniline (100%)**

m-**Nitrobenzaldehyde**   *m*-**Aminobenzaldehyde**
                                                                    (90%)

● ●

Problem 24.7   Propose structures for either a nitrile or an amide that might be a precursor of each of the following amines:
(a) Propylamine   (b) Dipropylamine
(c) Benzylamine, $C_6H_5CH_2NH_2$ (d) *N*-Ethylaniline

● ●

S$_N$2 Reactions of Alkyl Halides

Ammonia and other amines are good nucleophiles in S$_N$2 reactions. As a result, the simplest method of alkylamine synthesis is by S$_N$2 alkylation of ammonia or an alkylamine with an alkyl halide. If ammonia is used, a primary amine results; if a primary amine is used, a secondary amine results; and so on. Even tertiary amines react rapidly with alkyl halides to yield quaternary ammonium salts, $R_4N^+ X^-$.

$$S_N2 \text{ reaction}$$

Ammonia $\quad \ddot{N}H_3 + R\!-\!X \longrightarrow RNH_3^+ X^- \xrightarrow{NaOH} RNH_2 \quad$ Primary

Primary $\quad R\ddot{N}H_2 + R\!-\!X \longrightarrow R_2NH_2^+ X^- \xrightarrow{NaOH} R_2NH \quad$ Secondary

Secondary $\quad R_2\ddot{N}H + R\!-\!X \longrightarrow R_3NH^+ X^- \xrightarrow{NaOH} R_3N \quad$ Tertiary

Tertiary $\quad R_3\ddot{N} + R\!-\!X \longrightarrow R_4N^+ X^- \quad$ Quaternary ammonium salt

Unfortunately, these reactions don't stop cleanly after a single alkylation has occurred. Because primary, secondary, and tertiary amines all have similar reactivity, the initially formed monoalkylated substance often undergoes further reaction to yield a mixture of products. For example, treatment of 1-bromooctane with a twofold excess of ammonia leads to a mixture containing only 45% of octylamine. A nearly equal amount of dioctylamine is produced by double alkylation, along with smaller amounts of trioctylamine and tetraoctylammonium bromide. Higher yields of monoalkylated product can sometimes be obtained by using a large excess of the starting amine, but even so the reaction is a poor one.

$$CH_3(CH_2)_6CH_2Br + :NH_3 \longrightarrow CH_3(CH_2)_6CH_2\ddot{N}H_2 + [CH_3(CH_2)_6CH_2]_2\ddot{N}H$$

1-Bromooctane $\qquad\qquad$ **Octylamine (45%)** $\qquad$ **Dioctylamine (43%)**

$$+ [CH_3(CH_2)_6CH_2]_3N: + [CH_3(CH_2)_6CH_2]_4\overset{+}{N}\ \overset{-}{Br}$$

$\qquad\qquad\qquad$ **Trace** $\qquad\qquad\qquad$ **Trace**

A better method for preparing primary amines is to use the **azide synthesis**, in which azide ion, N_3^-, is used for S_N2 displacement of a halide ion from a primary or secondary alkyl halide to give an alkyl azide, RN_3. Since alkyl azides are not nucleophilic, overalkylation can't occur. Reduction of the alkyl azide, either by catalytic hydrogenation over a palladium catalyst or by reaction with $LiAlH_4$, leads to the desired primary amine. Although the method works well, low-molecular-weight alkyl azides are explosive and must be handled carefully.

CH_2CH_2Br $\qquad\qquad\qquad$ $CH_2CH_2N\!=\!\overset{+}{N}\!=\!\overset{-}{N}$ $\qquad\qquad\qquad$ $CH_2CH_2NH_2$

$\qquad\qquad\qquad \xrightarrow[\text{Ethanol}]{NaN_3} \qquad\qquad\qquad \xrightarrow[\text{2. } H_2O]{\text{1. } LiAlH_4, \text{ ether}}$

1-Bromo-2-phenylethane $\qquad$ **2-Phenylethyl azide** $\qquad$ **2-Phenylethylamine (89%)**

An alternative to the azide synthesis is the **Gabriel amine synthesis**, which uses a *phthalimide* alkylation for preparing a primary amine from an alkyl halide. **Imides (–CONHCO–)** are similar to ethyl acetoacetate in that

Siegmund Gabriel

Siegmund Gabriel (1851–1924) was born in Berlin, Germany, and received his Ph.D. in 1874 at the University of Berlin, working with August von Hofmann. After further work with Robert Bunsen, he became Professor of Chemistry at the University of Berlin.

the N–H hydrogen is flanked by two carbonyl groups. Thus, imides are deprotonated by such bases as KOH, and the resultant anions are readily alkylated in a reaction similar to the acetoacetic ester synthesis (Section 22.8). Basic hydrolysis of the N-alkylated imide then yields a primary amine product. Note that the imide hydrolysis step is closely analogous to the hydrolysis of an amide (Section 21.7).

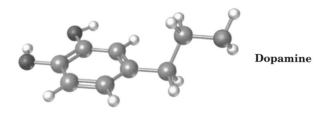

Phthalimide   **Potassium phthalimide**

For example,

Benzyl bromide   **Benzylamine (81%)**

Problem 24.8   Write the mechanism of the last step in the Gabriel amine synthesis, the base-promoted hydrolysis of a phthalimide to yield an amine plus phthalate ion.

Problem 24.9   Show two methods for the synthesis of dopamine, a neurotransmitter involved in regulation of the central nervous system. Use any alkyl halide needed.

Dopamine

Biological Connection ▶

Reductive Amination of Aldehydes and Ketones

Amines can be synthesized in a single step by treatment of an aldehyde or ketone with ammonia or an amine in the presence of a reducing agent, a process called **reductive amination**. For example, amphetamine, a central nervous system stimulant, is prepared commercially by reductive amination of phenyl-2-propanone with ammonia, using hydrogen gas over a nickel catalyst as the reducing agent.

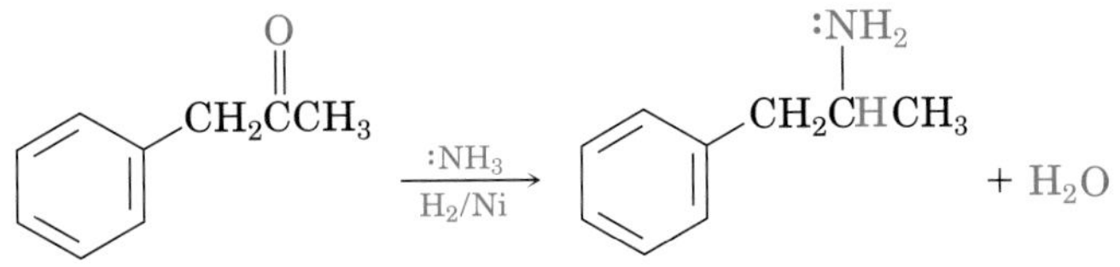

Phenyl-2-propanone   **Amphetamine**

Reductive amination takes place by the pathway shown in Figure 24.4. An imine intermediate is first formed by a nucleophilic addition reaction (Section 19.9), and the imine is then reduced.

FIGURE 24.4 ▼

Mechanism of reductive amination of a ketone to yield an amine. (Details of the imine-forming step are shown in Figure 19.8.)

Ammonia attacks the carbonyl group in a nucleophilic addition reaction to yield an intermediate carbinolamine.

The intermediate loses water to give an imine.

The imine is reduced catalytically over nickel to yield the amine product.

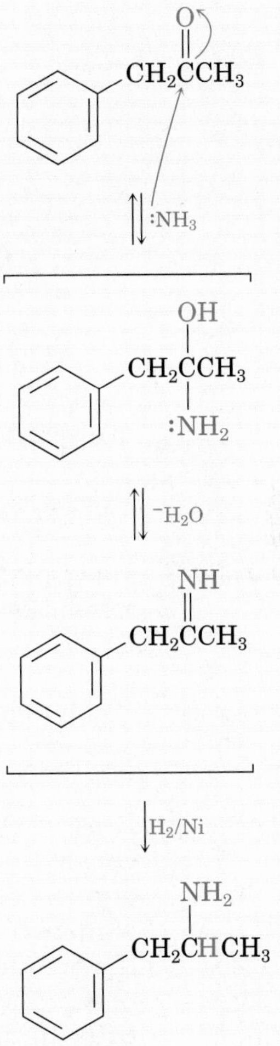

Ammonia, primary amines, and secondary amines can all be used in the reductive amination reaction, yielding primary, secondary, and tertiary amines, respectively.

Primary amine   **Secondary amine**   **Tertiary amine**

Many different reducing agents are effective, but the most common choice in the laboratory is sodium cyanoborohydride, $NaBH_3CN$, a relative of $NaBH_4$.

Cyclohexanone   **N,N-Dimethylcyclohexylamine**
                                                              (85%)

We'll see in Section 29.6 that a process closely related to reductive amination occurs frequently in the biological pathways by which amino acids are synthesized.

Practice Problem 24.1   How might you prepare *N*-methyl-2-phenylethylamine using a reductive amination reaction?

N-Methyl-2-phenylethylamine

Strategy   Look at the target molecule, and identify the groups attached to nitrogen. One of the groups must be derived from the aldehyde or ketone component, and the other must be derived from the amine component. In the case of *N*-methyl-2-phenylethylamine, there are two combinations that can lead to the product: phenylacetaldehyde plus methylamine or formaldehyde plus 2-phenylethylamine. In general, it's usually better to choose the combination with the simpler amine component—methylamine in this case—and to use an excess of that amine as reactant.

Solution

Benzaldehyde (PhCH$_2$CHO) $+$ CH$_3$NH$_2$ $\xrightarrow{\text{NaBH}_3\text{CN}}$ PhCH$_2$CH$_2$NHCH$_3$ $\xleftarrow{\text{NaBH}_3\text{CN}}$ PhCH$_2$CH$_2$NH$_2$ $+$ CH$_2$O

· ·

Problem 24.10   How might the following amines be prepared using reductive amination reactions? Show all precursors if more than one is possible.
(a) CH$_3$CH$_2$NHCH(CH$_3$)$_2$ (b) *N*-Ethylaniline   (c) *N*-Methylcyclopentylamine

· ·

Hofmann and Curtius Rearrangements

Carboxylic acid derivatives can be converted into primary amines with loss of one carbon atom by both the **Hofmann rearrangement** and the **Curtius rearrangement**. Although the Hofmann rearrangement involves a primary amide and the Curtius rearrangement involves an acyl azide, both proceed through similar mechanisms.

Hofmann rearrangement

An amide $\xrightarrow[\text{H}_2\text{O}]{\text{NaOH, Br}_2}$ RNH$_2$ + CO$_2$

Curtius rearrangement

An acyl azide $\xrightarrow[\text{Heat}]{\text{H}_2\text{O}}$ RNH$_2$ + CO$_2$ + N$_2$

Hofmann rearrangement occurs when a primary amide, RCONH$_2$, is treated with Br$_2$ and base (Figure 24.5, p. 996). The overall mechanism is lengthy, but most of the individual steps have been encountered before. Thus, the bromination of an amide in steps 1 and 2 is analogous to the base-promoted bromination of a ketone enolate ion (Section 22.7), and the rearrangement of the bromoamide anion in step 4 is analogous to a carbocation rearrangement (Section 6.12). The main difference between the migration step in a Hofmann rearrangement and that in a carbocation rearrangement is that the –R group begins its migration to the neighboring atom *at the same time* the bromide ion is leaving, rather than after it has left. Nucleophilic addition of water to the isocyanate carbonyl group in step 5 is a typical carbonyl-group process (Section 19.6), as is the final decarboxylation step (Section 22.8).

FIGURE 24.5 ▼

Mechanism of the Hofmann rearrangement of an amide to an amine. Each step is analogous to a reaction studied previously.

refer to
**Mechanisms
& Movies**

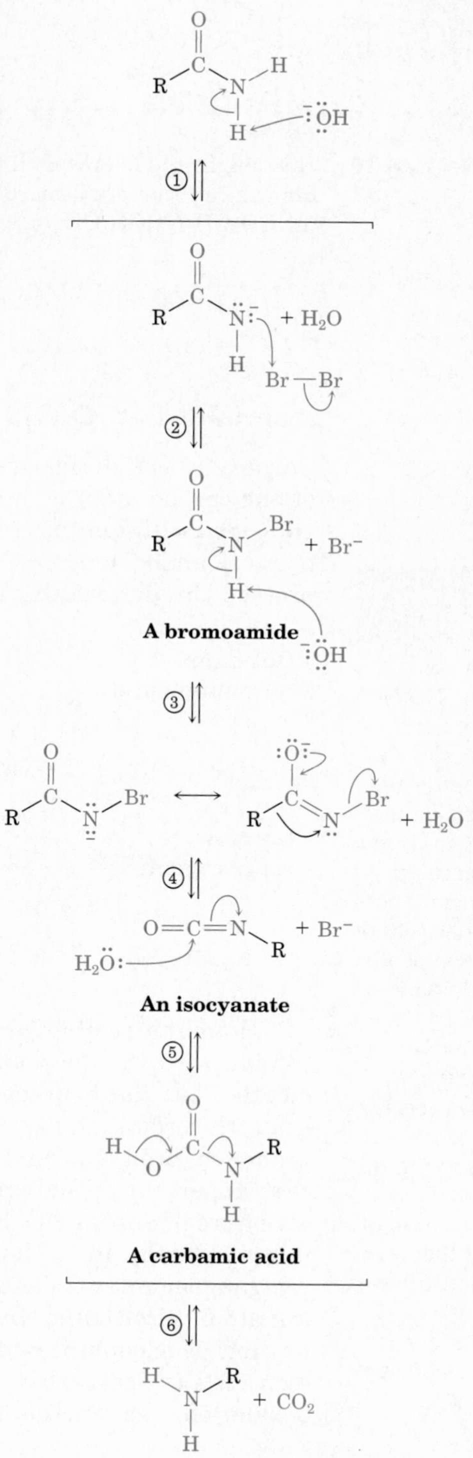

Base abstracts an acidic N–H proton, yielding an anion.

The anion reacts with bromine in an alpha-substitution reaction to give an *N*-bromoamide.

A bromoamide

Base abstraction of the remaining amide proton gives a bromoamide anion.

The bromoamide anion rearranges as the R group attached to the carbonyl carbon migrates to nitrogen at the same time the bromide ion leaves, giving an isocyanate.

An isocyanate

The isocyanate adds water in a nucleophilic addition step to yield a carbamic acid.

A carbamic acid

The carbamic acid spontaneously loses CO_2, yielding the amine product.

© 1984 JOHN MCMURRY

Theodor Curtius

Theodor Curtius (1857–1928) was born in Duisberg, Germany, and received his doctorate at the University of Leipzig working with Herman Kolbe. He was professor at the universities of Kiel, Bonn, and Heidelberg (1898–1926).

Despite its mechanistic complexity, the Hofmann rearrangement often gives high yields of both aryl- and alkylamines. For example, the appetite-suppressant drug phentermine is prepared commercially by Hofmann rearrangement of a primary amide. Commonly known by the name *fen–phen*, the combination of phentermine with another appetite-suppressant, fenfluramine, is suspected of causing heart damage.

2,2-Dimethyl-3-phenylpropanamide   **Phentermine**

The Curtius rearrangement, like the Hofmann rearrangement, involves migration of an –R group from the C=O carbon atom to the neighboring nitrogen with simultaneous loss of a leaving group. The reaction takes place on heating an acyl azide that is itself prepared by nucleophilic acyl substitution of an acid chloride.

Acid chloride   **Acyl azide**   **Isocyanate**   **Amine**

Like the Hofmann rearrangement, the Curtius rearrangement is often used commercially. For example, the antidepressant drug tranylcypromine is made by Curtius rearrangement of 2-phenylcyclopropanecarbonyl chloride.

trans-2-Phenylcyclopropanecarbonyl chloride   **Tranylcypromine**

· ·

Practice Problem 24.2   How would you prepare *o*-methylbenzylamine from a carboxylic acid, using both Hofmann and Curtius rearrangements?

Strategy   Both Hofmann and Curtius rearrangements convert a carboxylic acid derivative—either an amide (Hofmann) or an acid chloride (Curtius)—into a primary amine with loss of one carbon, $RCOY \longrightarrow RNH_2$. Both reactions begin with the same carboxylic acid, which can be identified by replacing the $-NH_2$ group of the amine product by a $-COOH$ group. In the present instance, *o*-methylphenylacetic acid is needed.

Solution

o-Methylphenylacetic acid → (SOCl₂) → acid chloride → o-Methylbenzylamine

$$\text{(with CH}_2\text{CO}_2\text{H and CH}_3\text{ on benzene ring)} \xrightarrow{\text{SOCl}_2} \text{(CH}_2\text{COCl and CH}_3\text{)}$$

Reagents over arrows:
- 1. NH₃
- 2. Br₂, NaOH, H₂O

and

- 1. NaN₃
- 2. H₂O, heat

to give **o-Methylbenzylamine** (CH₂NH₂ and CH₃ on benzene ring)

o-Methylphenylacetic acid

o-Methylbenzylamine

• •

Problem 24.11   Show the mechanism of the Curtius rearrangement of an acyl azide to an isocyanate, using curved arrows to indicate electron flow. Also show the mechanism of the addition of water to an isocyanate to yield a carbamic acid.

Problem 24.12   How would you prepare the following amines, using both Hofmann and Curtius rearrangements on a carboxylic acid derivative?

(a) $(CH_3)_3CCH_2CH_2NH_2$ (b) $H_3C\text{—}\langle\text{benzene ring}\rangle\text{—}NH_2$

• •

24.7   Reactions of Amines

Biological Connection ▶

Alkylation and Acylation

We've already studied the two most general reactions of amines—alkylation and acylation. As we saw earlier in this chapter, primary, secondary, and tertiary amines can be alkylated by reaction with a primary alkyl halide. Alkylations of primary and secondary amines are difficult to control and often give mixtures of products, but tertiary amines are cleanly alkylated to give quaternary ammonium salts. Primary and secondary (but not tertiary) amines can also be acylated by reaction with acid chlorides or acid anhydrides to yield amides (Sections 21.4 and 21.5).

$$\underset{R}{\overset{O}{\underset{\|}{C}}}\text{—Cl} + NH_3 \xrightarrow[\text{solvent}]{\text{Pyridine}} \underset{R}{\overset{O}{\underset{\|}{C}}}\text{—}\underset{\underset{H}{|}}{N}\text{—H} + HCl$$

$$\underset{R}{\overset{O}{\underset{\|}{C}}}\text{—Cl} + R'NH_2 \xrightarrow[\text{solvent}]{\text{Pyridine}} \underset{R}{\overset{O}{\underset{\|}{C}}}\text{—}\underset{\underset{H}{|}}{N}\text{—R}' + HCl$$

$$\underset{R}{\overset{O}{\underset{\|}{C}}}\text{—Cl} + R'_2NH \xrightarrow[\text{solvent}]{\text{Pyridine}} \underset{R}{\overset{O}{\underset{\|}{C}}}\text{—}\underset{\underset{R'}{|}}{N}\text{—R}' + HCl$$

Hofmann Elimination

Like alcohols, amines can be converted into alkenes by an elimination reaction. Because an amide ion, NH_2^-, is such a poor leaving group, however, it must first be converted into a better leaving group. In the **Hofmann elimination reaction**, an amine is methylated by reaction with excess iodomethane to produce a quaternary ammonium salt, which then undergoes elimination to give an alkene on heating with silver oxide, Ag_2O, as base. For example, hexylamine is converted into 1-hexene in 60% yield.

$$CH_3CH_2CH_2CH_2CH_2CH_2NH_2 \xrightarrow[\text{(excess)}]{CH_3I} CH_3CH_2CH_2CH_2CH_2CH_2\overset{+}{N}(CH_3)_3I^-$$

Hexylamine   **Hexyltrimethylammonium iodide**

$$\downarrow \begin{array}{c} Ag_2O \\ H_2O, \text{heat} \end{array}$$

$$CH_3CH_2CH_2CH_2CH=CH_2 + N(CH_3)_3$$

1-Hexene (60%)

Silver oxide functions by exchanging hydroxide ion for iodide ion in the quaternary salt, thus providing the base necessary to cause elimination. The actual elimination step is an E2 reaction (Section 11.11) in which hydroxide ion removes a proton at the same time that the positively charged nitrogen atom acts as the leaving group.

**Quaternary ammonium
salt**

Alkene

An interesting feature of the Hofmann elimination is that it gives products different from those of most other E2 reactions. Whereas the *more* highly substituted alkene product generally predominates in the E2 reaction of an alkyl halide (Zaitsev's rule; Section 11.10), the *less* highly substituted alkene predominates in the Hofmann elimination of a quaternary ammonium salt. For example, (1-methylbutyl)trimethylammonium hydroxide yields 1-pentene and 2-pentene in a 94:6 ratio. The reason for this selectivity is probably steric. Because of the large size of the trialkylamine leaving group, the attacking base must abstract a hydrogen from the most sterically accessible, least hindered position. (See the stereo view on p. 1000.)

$$CH_3CH_2CH_2CH=CH_2 + CH_3CH_2CH=CHCH_3$$

1-Pentene   **2-Pentene**

94:6 ratio

More hindered;
less accessible

Less hindered;
more accessible

**(1-Methylbutyl)trimethylammonium
hydroxide**

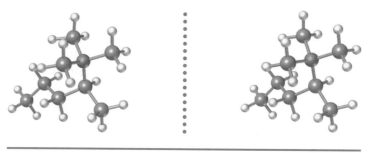

Stereo View

The Hofmann elimination reaction is important primarily because of its historical use as a degradative tool in the structure determination of many complex naturally occurring amines. The reaction is not often used today because the product alkenes can be made more easily in other ways.

Practice Problem 24.3   What product would you expect from Hofmann elimination of the following amine?

$$H\diagdown \overset{\displaystyle CH_2CH_3}{\underset{\displaystyle N}{\diagup}}$$

Strategy   The Hofmann elimination is an E2 reaction that converts an amine into an alkene. It occurs with non-Zaitsev regiochemistry to form the least highly substituted double bond. To predict the product, look at the reactant and identify the positions from which elimination might occur (the positions two carbons removed from nitrogen). Then carry out an elimination using the most accessible hydrogen. In the present instance, there are three possible positions from which elimination might occur—one primary, one secondary, and one tertiary. The primary position is the most accessible and leads to the least highly substituted alkene, ethylene.

Solution

1. Excess CH_3I
2. Ag_2O, H_2O, heat

$N(CH_3)_2$ $+ \ H_2C{=}CH_2$

Problem 24.13   What products would you expect to obtain from Hofmann elimination of the following amines? If more than one product is formed, indicate which is major.
(a) $CH_3CH_2CH_2CH(NH_2)CH_2CH_2CH_2CH_3$ (b) Cyclohexylamine
(c) $CH_3CH_2CH_2CH(NH_2)CH_2CH_2CH_3$ (d) *N*-Ethylcyclohexylamine

Problem 24.14 What product would you expect from Hofmann elimination of a heterocyclic amine such as piperidine? Write all the steps.

$$
\text{piperidine (NH)} \quad \xrightarrow[\text{2. Ag}_2\text{O, H}_2\text{O, heat}]{\text{1. CH}_3\text{I (excess)}} \quad ?
$$

24.8 Reactions of Arylamines

Electrophilic Aromatic Substitution

Amino substituents are strongly activating, ortho- and para-directing groups in electrophilic aromatic substitution reactions (Section 16.5). The high reactivity of amino-substituted benzenes can be a drawback at times because it's sometimes difficult to prevent polysubstitution. For example, reaction of aniline with Br_2 takes place rapidly and yields the 2,4,6-tribrominated product. The amino group is so strongly activating that it's not possible to stop at the monobromo stage.

$$
\text{Aniline} \quad \xrightarrow[\text{H}_2\text{O}]{\text{Br}_2} \quad \text{2,4,6-Tribromoaniline (100\%)}
$$

Another drawback to the use of amino-substituted benzenes in electrophilic aromatic substitution reactions is that Friedel–Crafts reactions are not successful (Section 16.3). The amino group forms an acid–base complex with the $AlCl_3$ catalyst, which prevents further reaction from occurring. Both drawbacks—high reactivity and amine basicity—can be overcome by carrying out electrophilic aromatic substitution reactions on the corresponding amide rather than on the free amine.

As we saw in Section 21.5, treatment of an amine with acetic anhydride yields an N-acetylated product. Though still activating and ortho-, para-directing, *amido* substituents (–NHCOR) are less strongly activating and less basic than amino groups because their nitrogen lone-pair electrons are delocalized by the neighboring carbonyl group. As a result, bromination of an N-arylamide occurs cleanly to give a monobromo product, and hydrolysis with aqueous base then gives the free amine. For example, p-toluidine (4-methylaniline) can be acetylated, brominated, and hydrolyzed to yield

2-bromo-4-methylaniline in 79% yield. None of the 2,6-dibrominated product is obtained.

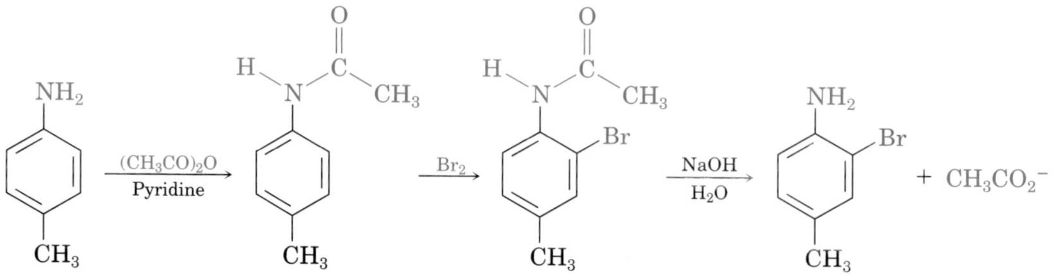

p-Toluidine

2-Bromo-4-methyl-aniline (79%)

Friedel–Crafts alkylations and acylations of *N*-arylamides also proceed normally. For example, benzoylation of acetanilide (*N*-acetylaniline) under Friedel–Crafts conditions gives 4-aminobenzophenone in 80% yield after hydrolysis:

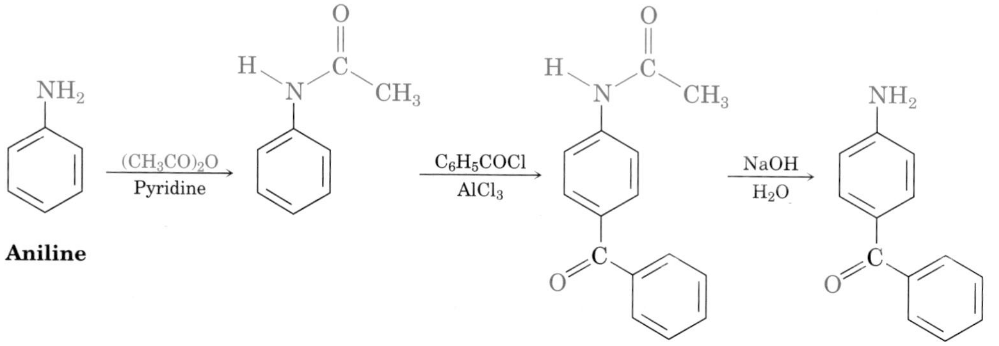

Aniline

4-Aminobenzophenone (80%)

Modulating the reactivity of an amino-substituted benzene by forming an amide is a useful trick that allows many kinds of electrophilic aromatic substitutions to be carried out that would otherwise be impossible. A good example is the preparation of the so-called *sulfa drugs*.

Sulfa drugs, such as sulfanilamide, were among the first pharmaceutical agents to be used clinically against infection. Although they have largely been replaced by safer and more powerful antibiotics, sulfa drugs were widely used in the 1940s and were credited with saving the lives of thousands of wounded during World War II. They are prepared by chlorosulfonation of acetanilide, followed by reaction of *p*-(*N*-acetylamino)-benzenesulfonyl chloride with ammonia or some other amine to give a sulfonamide. Hydrolysis of the amide then yields the sulfa drug. Note that this amide hydrolysis can be carried out in the presence of the sulfonamide group because sulfonamides hydrolyze very slowly.

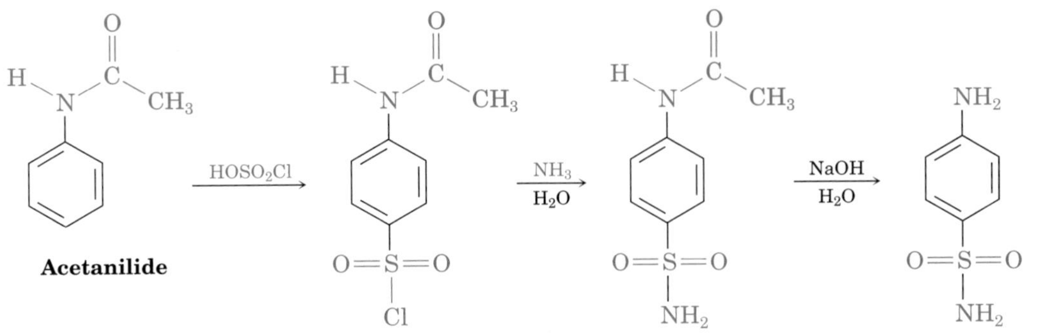

Acetanilide

$\xrightarrow{\text{HOSO}_2\text{Cl}}$

$\xrightarrow[\text{H}_2\text{O}]{\text{NH}_3}$

$\xrightarrow[\text{H}_2\text{O}]{\text{NaOH}}$

Sulfanilamide
(a sulfa drug)

Problem 24.15 Propose a synthesis of sulfathiazole from benzene and any necessary amine.

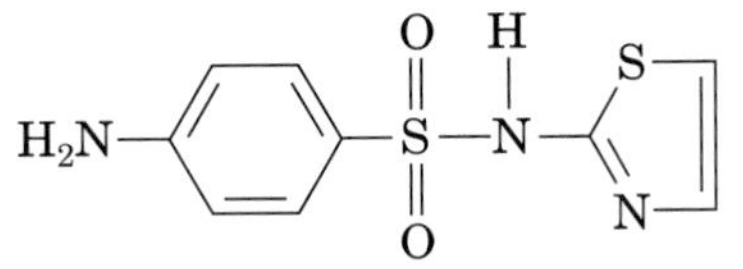

Sulfathiazole

Problem 24.16 Propose syntheses of the following compounds from benzene:
(a) *N,N*-Dimethylaniline   (b) *p*-Chloroaniline
(c) *m*-Chloroaniline   (d) 2,4-Dimethylaniline

Diazonium Salts: The Sandmeyer Reaction

Primary arylamines react with nitrous acid, HNO_2, to yield stable **arene-diazonium salts, Ar–N≡N X⁻**. This *diazotization* reaction is compatible with the presence of a wide variety of substituents on the aromatic ring.

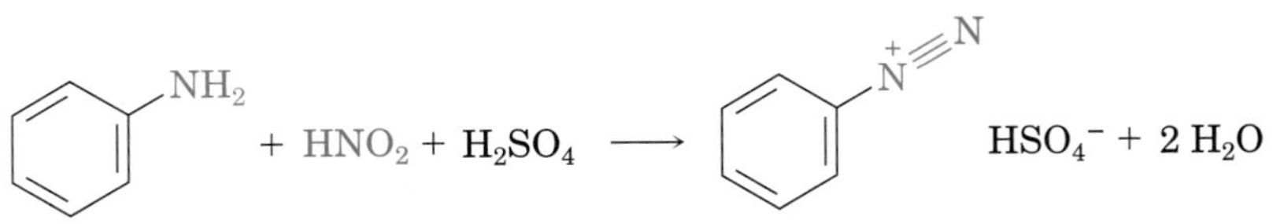

$+ \text{HNO}_2 + \text{H}_2\text{SO}_4 \longrightarrow$

$\text{HSO}_4^- + 2 \text{H}_2\text{O}$

*Alkyl*amines also react with nitrous acid, but the alkanediazonium products of these reactions are so reactive they can't be isolated. Instead, they lose nitrogen instantly to yield carbocations. The analogous loss of N_2 from an arenediazonium ion to yield an aryl cation is disfavored by the instability of the cation.

Arenediazonium salts are extremely useful because the diazonio group (N_2) can be replaced by a nucleophile in a radical substitution reaction:

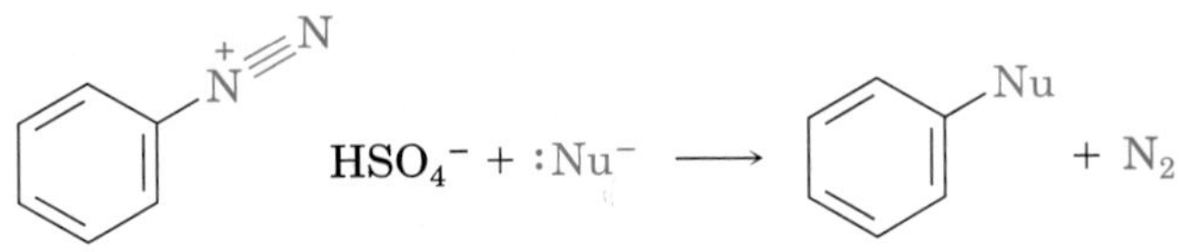

$\text{HSO}_4^- + :\text{Nu}^- \longrightarrow$

$+ N_2$

Many different nucleophiles react with arenediazonium salts, yielding many different kinds of substituted benzenes. The overall sequence of (1) nitration, (2) reduction, (3) diazotization, and (4) nucleophilic substitution is probably the single most versatile method of aromatic substitution (Figure 24.6).

FIGURE 24.6

Preparation of substituted aromatic compounds by diazonio replacement reactions.

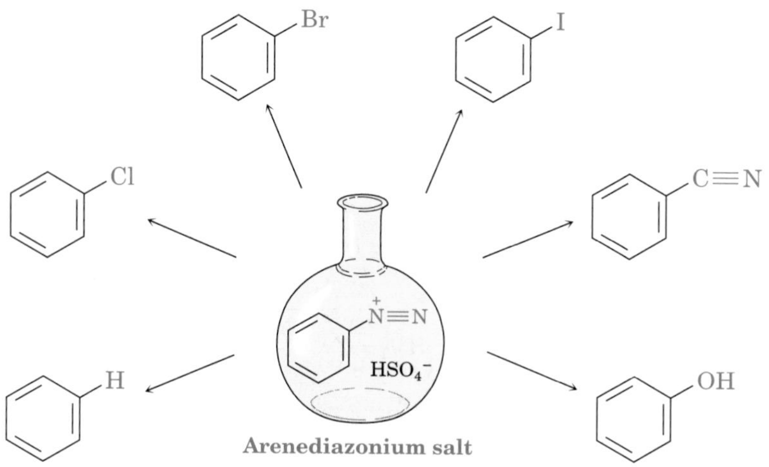

Aryl chlorides and bromides are prepared by reaction of an arenediazonium salt with the corresponding cuprous halide, CuX, a process called the **Sandmeyer reaction**. Aryl iodides can be prepared by direct reaction with NaI without using a cuprous salt. Yields generally fall in the range 60–80%.

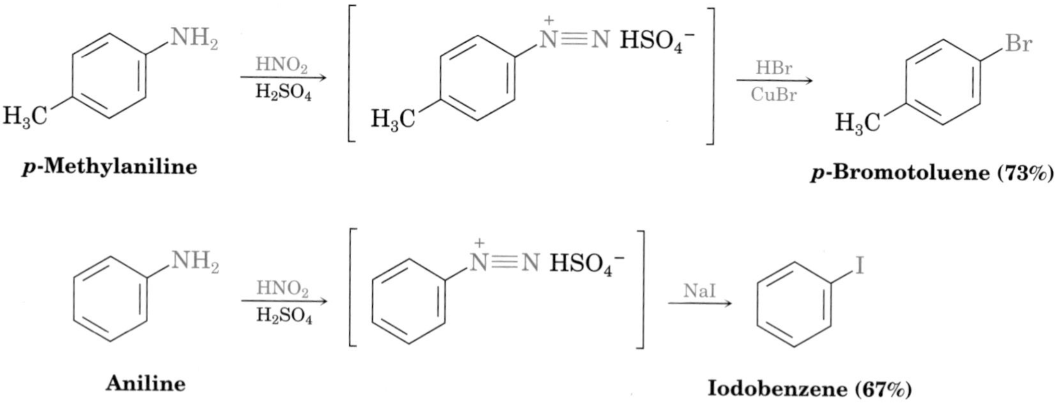

Similar treatment of an arenediazonium salt with CuCN yields the nitrile, ArCN. The nitrile can then be further converted into other functional groups such as carboxyl. For example, Sandmeyer reaction of *o*-methylbenzenediazonium bisulfate with CuCN yields *o*-methylbenzonitrile, which can be hydrolyzed to give *o*-methylbenzoic acid. This product can't be prepared from *o*-xylene by the usual side-chain oxidation route because both methyl groups would be oxidized.

o-Methylaniline → *o*-Methylbenzene-diazonium bisulfate → *o*-Methylbenzonitrile → *o*-Methylbenzoic acid

HNO₂/H₂SO₄ ; KCN/CuCN ; H₃O⁺

o-Methylaniline — **o-Methylbenzene-diazonium bisulfate** — **o-Methylbenzonitrile** — **o-Methylbenzoic acid**

Traugott Sandmeyer

Traugott Sandmeyer (1854–1922) was born in Wettingen, Switzerland, and received his Ph.D. at the University of Heidelberg. He spent his professional career doing pharmaceutical research at the Geigy Company in Basel, Switzerland.

The diazonio group can also be replaced by –OH to yield a phenol and by –H to yield an arene. A phenol is prepared by reaction of the arenediazonium salt with copper(I) oxide in an aqueous solution of copper(II) nitrate, a reaction that is especially useful because few other general methods exist for introducing an –OH group onto an aromatic ring.

p-Methylaniline (p-Toluidine) — **p-Cresol (93%)**

HNO₂/H₂SO₄ ; Cu₂O/Cu(NO₃)₂, H₂O

Reduction of a diazonium salt to give an arene occurs on treatment with hypophosphorous acid, H_3PO_2. This reaction is used primarily when there is a need for temporarily introducing an amino substituent onto a ring to take advantage of its directing effect. Suppose, for example, that you needed to make 3,5-dibromotoluene. The product can't be made by direct bromination of toluene because reaction would occur at positions 2 and 4. Starting with *p*-methylaniline (*p*-toluidine), however, dibromination occurs ortho to the strongly directing amino substituent, and diazotization followed by treatment with H_3PO_2 yields the desired product.

p-Methylaniline — **3,5-Dibromotoluene**

2 Br₂ ; HNO₂/H₂SO₄ ; H₃PO₂

Toluene — **2,4-Dibromotoluene**

2 Br₂/FeBr₃

● ●

Practice Problem 24.4   How would you prepare *m*-hydroxyacetophenone from benzene, using a diazonium replacement reaction in your scheme?

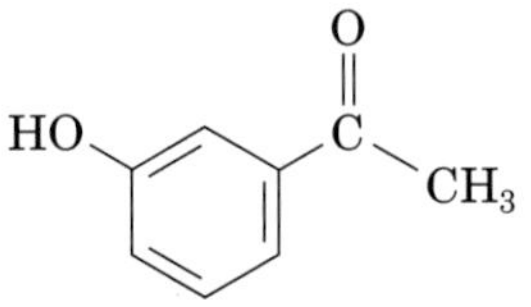

***m*-Hydroxyacetophenone**

Strategy   As always, organic syntheses are planned by working backward from the final product, one step at a time. First, identify the functional groups in the product and recall how those groups can be synthesized. *m*-Hydroxyacetophenone has an –OH group and a –COCH$_3$ group in a meta relationship on a benzene ring. A hydroxyl group is generally introduced onto an aromatic ring by a four-step sequence of nitration, reduction, diazotization, and diazonio replacement. An acetyl group is introduced by a Friedel–Crafts acylation reaction.

Next, ask yourself what an immediate precursor of the product might be. Since an acetyl group is a meta director while a hydroxyl group is an ortho and para director, acetophenone might be a precursor of *m*-hydroxyacetophenone. Benzene, in turn, is a precursor of acetophenone.

Solution

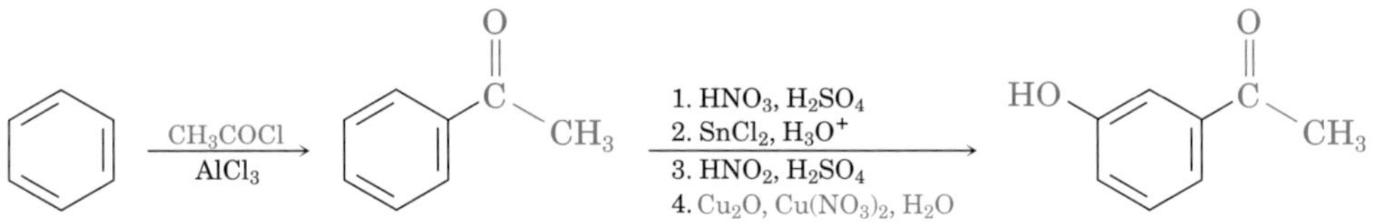

Benzene   **Acetophenone**   ***m*-Hydroxyacetophenone**

● ●

Problem 24.17   How would you prepare the following compounds from benzene, using a diazonium replacement reaction in your scheme?
(a) *p*-Bromobenzoic acid   (b) *m*-Bromobenzoic acid
(c) *m*-Bromochlorobenzene   (d) *p*-Methylbenzoic acid
(e) 1,2,4-Tribromobenzene

● ●

Diazonium Coupling Reactions

Arenediazonium salts undergo a coupling reaction with activated aromatic rings to yield brightly colored **azo compounds, Ar–N=N–Ar′**:

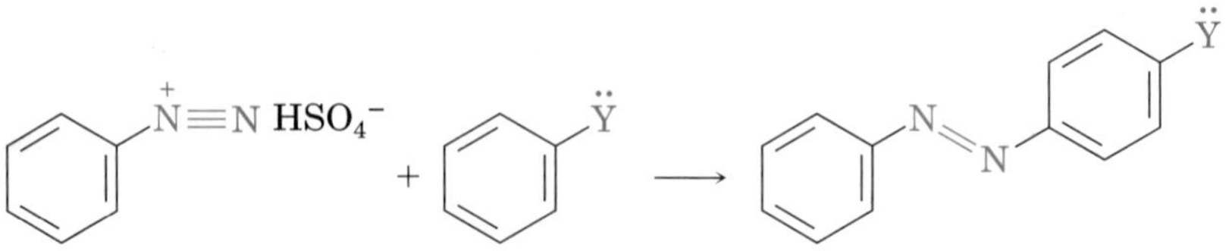

An azo compound

where   Y = –OH or –NR$_2$

Diazonium coupling reactions are typical electrophilic aromatic substitutions in which the positively charged diazonium ion is the electrophile that reacts with the electron-rich ring of a phenol or arylamine. Reaction usually occurs at the para position, although ortho attack can take place if the para position is blocked.

Benzenediazonium bisulfate   **Phenol**

p-**Hydroxyazobenzene**
(orange crystals, mp 152°C)

Azo-coupled products are widely used as dyes because their extended conjugated π electron system causes them to absorb in the visible region of the electromagnetic spectrum (Section 14.13). *p*-(Dimethylamino)azobenzene, for instance, is a bright yellow compound that was at one time used as a coloring agent in margarine.

Benzenediazonium bisulfate   ***N,N*-Dimethylaniline**   *p*-**(Dimethylamino)azobenzene**
(yellow crystals, mp 127°C)

Problem 24.18 Propose a synthesis of *p*-(dimethylamino)azobenzene from benzene as your only organic starting material.

24.9 Tetraalkylammonium Salts as Phase-Transfer Catalysts

Tetraalkylammonium salts, $R_4N^+ X^-$, are used as catalysts for many different kinds of organic reactions. As an example, imagine an experiment in which cyclohexene is dissolved in chloroform and treated with aqueous

NaOH. Since the organic layer and the water layer are immiscible, the base in the aqueous phase does not come into contact with chloroform in the organic phase, and there is no reaction. If, however, a small amount of benzyltri-ethylammonium chloride is added, an immediate reaction occurs. The chloroform reacts with NaOH to generate dichlorocarbene, which adds to the cyclohexene double bond to give a dichlorocyclopropane in 77% yield (Section 7.6).

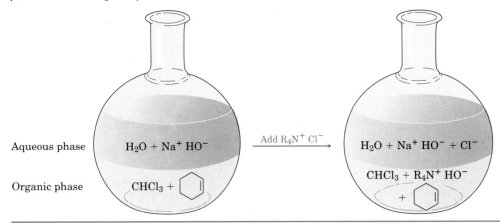

77%

How does the tetraalkylammonium salt catalyze the cyclopropanation reaction? Benzyltriethylammonium ion, even though charged, is soluble in organic solvents because of the four hydrocarbon substituents on nitrogen. But when the *positively* charged tetraalkylammonium ion goes into the organic layer, a *negatively* charged ion must follow to preserve charge neutrality. Hydroxide ion, present in far greater amount than chloride ion, is thus transferred from the aqueous phase into the organic phase where reaction with chloroform immediately occurs (Figure 24.7).

FIGURE 24.7 ▼

Phase-transfer catalysis. Addition of a small amount of a tetraalkylammonium salt to a two-phase mixture allows an inorganic anion to be transferred from the aqueous phase into the organic phase, where a reaction can occur.

Aqueous phase $H_2O + Na^+ HO^-$ $\xrightarrow{\text{Add } R_4N^+ Cl^-}$ $H_2O + Na^+ HO^- + Cl^-$

Organic phase $CHCl_3 +$ $CHCl_3 + R_4N^+ HO^-$

$+$

The transfer of an inorganic ion such as OH^- from one phase to another is called **phase transfer**, and the tetraalkylammonium salt is referred to as a *phase-transfer catalyst*. Many different kinds of organic reactions, including oxidations, reductions, carbonyl-group alkylations, and S_N2 reactions, are subject to phase-transfer catalysis, often with considerable improvements in yield. S_N2 reactions are particularly good candidates for phase-transfer catalysis because inorganic nucleophiles can be transferred from an aqueous (protic) phase to an organic (aprotic) phase, where they are much more reactive. For example:

$$CH_3(CH_2)_6CH_2Br + NaCN \xrightarrow[\underset{+}{C_6H_5CH_2N(CH_2CH_3)_3\ Cl^-}]{H_2O,\ benzene} CH_3(CH_2)_6CH_2CN + NaBr$$

1-Bromooctane   **Nonanenitrile (92%)**

24.10 Spectroscopy of Amines

Infrared Spectroscopy

Primary and secondary amines can be identified by characteristic N–H stretching absorptions in the 3300–3500 cm^{-1} range of the IR spectrum. Alcohols also absorb in this range (Section 17.12), but amine absorption bands are generally sharper and less intense than hydroxyl bands. Primary amines show a pair of bands at about 3350 and 3450 cm^{-1}, and secondary amines show a single band at 3350 cm^{-1}. Tertiary amines show no absorption in this region because they have no N–H bonds. Representative IR spectra of both primary and secondary amines are shown in Figure 24.8.

FIGURE 24.8 ▼

Infrared spectra of (a) cyclohexylamine and (b) diisopropylamine.

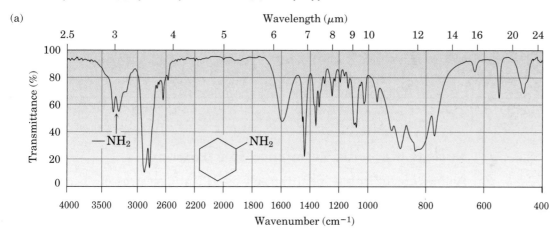

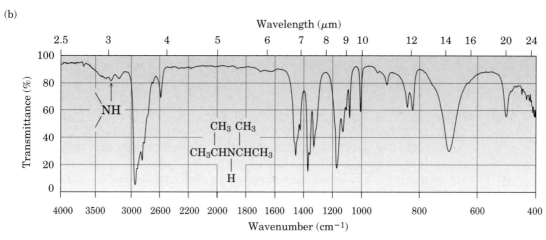

In addition to looking for characteristic N–H absorptions, there's also a simple trick for telling whether a compound is an amine. Addition of a small amount of HCl produces a broad and strong ammonium band in the 2200–3000 cm^{-1} range if the sample contains an amino group. All protonated amines show this readily observable absorption caused by the ammonium R_3N-H^+ bond. Figure 24.9 gives an example.

FIGURE 24.9 ▼

Infrared spectrum of trimethylammonium chloride.

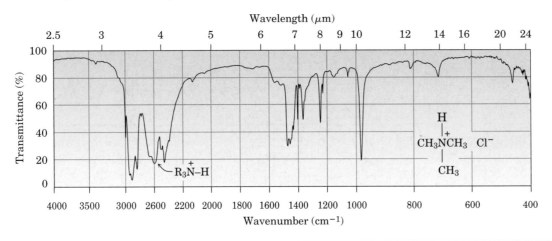

Nuclear Magnetic Resonance Spectroscopy

Amines are difficult to identify solely by ^{1}H NMR spectroscopy because N–H hydrogens tend to appear as broad signals without clear-cut coupling to neighboring C–H hydrogens. As with O–H absorptions (Section 17.12), amine N–H absorptions can appear over a wide range and are best identified by adding a small amount of D_2O to the sample tube. Exchange of N–D for N–H occurs, and the N–H signal disappears from the NMR spectrum.

$$\underset{/}{\overset{\backslash}{N}}-H \quad \underset{}{\overset{D_2O}{\rightleftharpoons}} \quad \underset{/}{\overset{\backslash}{N}}-D + HDO$$

Hydrogens on the carbon next to nitrogen are somewhat deshielded because of the electron-withdrawing effect of the nitrogen, and they therefore absorb at lower field than alkane hydrogens. *N*-Methyl groups are particularly distinctive because they absorb as a sharp three-proton singlet at 2.2–2.6 δ. This *N*-methyl resonance at 2.42 δ is easily seen in the ^{1}H NMR spectrum of *N*-methylcyclohexylamine (Figure 24.10).

FIGURE 24.10 ▼

Proton NMR spectrum of *N*-methylcyclohexylamine.

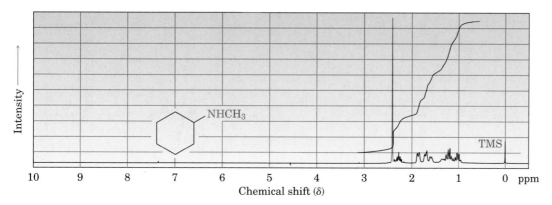

Carbons next to amine nitrogens are slightly deshielded in the ^{13}C NMR spectrum and absorb about 20 ppm downfield from where they would absorb in an alkane of similar structure. In *N*-methylcyclohexylamine, for example, the ring carbon to which nitrogen is attached absorbs at a position 24 ppm lower than that of any other ring carbon.

Problem 24.19 Compound A, $C_6H_{12}O$, has an IR absorption at 1715 cm^{-1} and gives compound B, $C_6H_{15}N$, when treated with ammonia and NaBH$_3$CN. The IR and ^{1}H NMR spectra of B are shown. What are the structures of A and B?

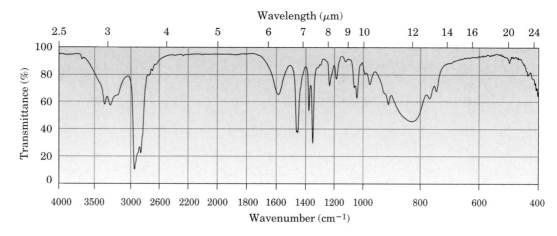

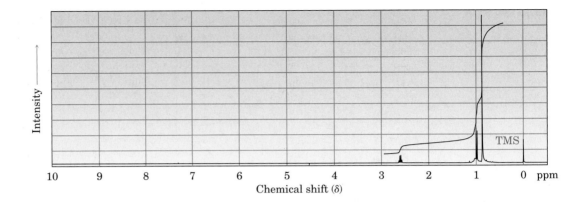

Mass Spectrometry

The **nitrogen rule** of mass spectrometry says that a compound with an odd number of nitrogen atoms has an odd-numbered molecular weight. Thus, the presence of nitrogen in a molecule is detected simply by observing its mass spectrum. An odd-numbered molecular ion usually means that the unknown compound has one or three nitrogen atoms, and an even-numbered molecular ion usually means that a compound has either zero or two nitrogen atoms. The logic behind the rule derives from the fact that nitrogen is trivalent, thus requiring an odd number of hydrogen atoms in a molecule. For example, methylamine has the formula CH_5N and a molecular weight of 31 amu; morphine has the formula $C_{17}H_{19}NO_3$ and a molecular weight of 285 amu.

Aliphatic amines undergo a characteristic α cleavage in the mass spectrometer, similar to the cleavage observed for aliphatic alcohols (Section 17.12). A C–C bond nearest the nitrogen atom is broken, yielding an alkyl radical and a nitrogen-containing cation:

$$\left[RCH_2 \overset{}{\underset{\alpha}{\gtrless}} CH_2 - N \begin{matrix} R' \\ \\ R' \end{matrix} \right]^{+\cdot} \xrightarrow{\text{Alpha cleavage}} RCH_2\cdot + \left[CH_2 = N \begin{matrix} R' \\ \\ R' \end{matrix} \right]^{+}$$

As an example, the mass spectrum of *N*-ethylpropylamine shown in Figure 24.11 has peaks at *m/z* = 58 and *m/z* = 72, corresponding to the two possible modes of α cleavage.

FIGURE 24.11 ▼

Mass spectrum of *N*-ethylpropylamine. The two possible modes of α cleavage lead to the observed fragment ions at $m/z = 58$ and $m/z = 72$.

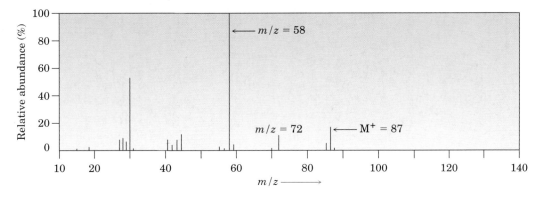

CHEMISTRY @ WORK

Morphine Alkaloids

The medical uses of morphine alkaloids have been known at least since the seventeenth century, when crude extracts of the opium poppy, *Papaver somniferum,* were used for the relief of pain. Morphine was the first pure alkaloid to be isolated from the poppy, but its close relative, codeine, also occurs naturally. Codeine, which is simply the methyl ether of morphine and is converted to morphine in the body, is used in prescription cough medicines and as an analgesic. Heroin, another close relative of morphine, does not occur naturally but is synthesized by diacetylation of morphine.

(continued) ▶

Morphine   **Codeine**   **Heroin**

Chemical investigations into the structure of morphine occupied some of the finest chemical minds of the nineteenth and early twentieth centuries, but it was not until 1924 that the puzzle was finally solved by Robert Robinson. The key reaction used to establish structure was the Hofmann elimination.

Morphine and its relatives are extremely useful pharmaceutical agents, yet they also pose an enormous social problem because of their addictive properties. Much effort has therefore gone into understanding how morphine works and into developing modified morphine analogs that retain the analgesic activity but don't cause physical dependence. Our present understanding is that morphine binds to opiate receptor sites in the brain. It doesn't interfere with the transmission of a pain signal to the brain but rather changes the brain's reception of the signal.

Hundreds of morphine-like molecules have been synthesized and tested for their analgesic properties. Research has shown that not all the complex framework of morphine is necessary for biological activity. According to the "morphine rule," biological activity requires: (1) an aromatic ring attached to (2) a quaternary carbon atom and (3) a tertiary amine situated (4) two carbon atoms farther away. Meperidine (Demerol), a widely used analgesic, and methadone, a substance used in the treatment of heroin addiction, are two compounds that fit the morphine rule.

The alkaloid morphine is isolated from the opium poppy, *Papaver somniferum*.

The morphine rule: an aromatic ring, attached to a quaternary carbon, attached to two more carbons, attached to a tertiary amine

Methadone   **Meperidine**

Summary and Key Words

Amines are organic derivatives of ammonia. They are named in the IUPAC system either by adding the suffix *-amine* to the names of the alkyl substituents or by considering the amino group as a substituent on a more complex parent molecule.

The bonding in amines is similar to that in ammonia. The nitrogen atom is sp^3-hybridized, the three substituents are directed to three corners of a tetrahedron, and the lone pair of nonbonding electrons occupies the fourth corner of the tetrahedron. An interesting feature of this tetrahedral structure is that amines undergo a rapid pyramidal inversion, which interconverts mirror-image structures.

The chemistry of amines is dominated by the lone-pair electrons on nitrogen, which makes amines both basic and nucleophilic. The base strength of **arylamines** is generally lower than that of aliphatic amines because the nitrogen lone-pair electrons are delocalized by interaction with the aromatic π system. Electron-withdrawing substituents on the aromatic ring further weaken the basicity of a substituted aniline, while electron-donating substituents increase basicity.

Arylamines are prepared by nitration of an aromatic ring followed by reduction. Alkylamines are prepared by S_N2 reaction of ammonia or an amine with an alkyl halide. This method often gives poor yields, however, and an alternative such as the **Gabriel amine synthesis** is preferred. Amines can also be prepared by a number of reductive methods, including $LiAlH_4$ reduction of amides, nitriles, and azides. Even more important is the **reductive amination** reaction in which a ketone or an aldehyde is treated with an amine in the presence of a reducing agent such as $NaBH_3CN$. In addition, amines result from the **Hofmann** and **Curtius rearrangements** of carboxylic acid derivatives. Both methods involve migration of the –R group bonded to the carbonyl carbon and yield a product that has one less carbon atom than the starting material.

Many of the reactions of amines are familiar from past chapters. Thus, amines react with alkyl halides in S_N2 reactions and with acid chlorides in nucleophilic acyl substitution reactions. Amines also undergo E2 elimination to yield alkenes if they are first quaternized by treatment with iodomethane and then heated with silver oxide (the **Hofmann elimination**).

The most useful reaction of arylamines is conversion by diazotization with nitrous acid into **arenediazonium salts**, $ArN_2^+ X^-$. The diazonio group can then be replaced by many other substituents in the **Sandmeyer reaction** to give a wide variety of substituted aromatic compounds. Aryl chlorides, bromides, iodides, and nitriles can be prepared from arenediazonium salts, as can arenes and phenols. In addition to their reactivity toward substitution reactions, diazonium salts undergo coupling with phenols and arylamines to give brightly colored azo dyes.

Summary of Reactions

1. Preparation of amines (Section 24.6)
   (a) Reduction of nitriles

$$RCH_2X + Na^+ {}^-CN \xrightarrow[\text{solvent}]{\text{DMF}} RCH_2C \equiv N \xrightarrow[\text{2. H}_2\text{O}]{\text{1. LiAlH}_4,\text{ ether}} RCH_2CH_2NH_2$$

   (b) Reduction of amides

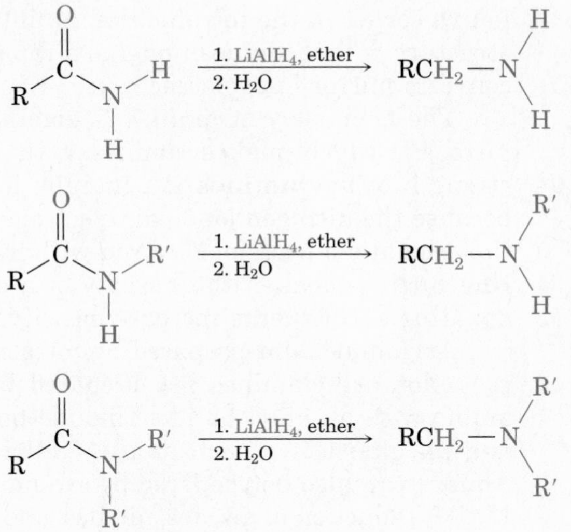

Primary

Secondary

Tertiary

   (c) Reduction of nitrobenzenes

$$ArNO_2 + H_2 \xrightarrow[\text{Ethanol}]{\text{Pt}} ArNH_2$$

$$ArNO_2 + Fe \xrightarrow[\text{2. HO}^-]{\text{1. H}_3\text{O}^+} ArNH_2$$

$$ArNO_2 + SnCl_2 \xrightarrow[\text{2. HO}^-]{\text{1. H}_3\text{O}^+} ArNH_2$$

   (d) The S_N2 alkylation of alkyl halides

Ammonia $:NH_3 + RX \longrightarrow RNH_3^+ X^- \xrightarrow{\text{NaOH}} RNH_2$ Primary

Primary $:NH_2R + RX \longrightarrow R_2NH_2^+ X^- \xrightarrow{\text{NaOH}} R_2NH$ Secondary

Secondary $:NHR_2 + RX \longrightarrow R_3NH^+ X^- \xrightarrow{\text{NaOH}} R_3N$ Tertiary

Tertiary $:NR_3 + RX \longrightarrow R_4N^+ X^-$ Quaternary ammonium salt

   (e) Gabriel amine synthesis

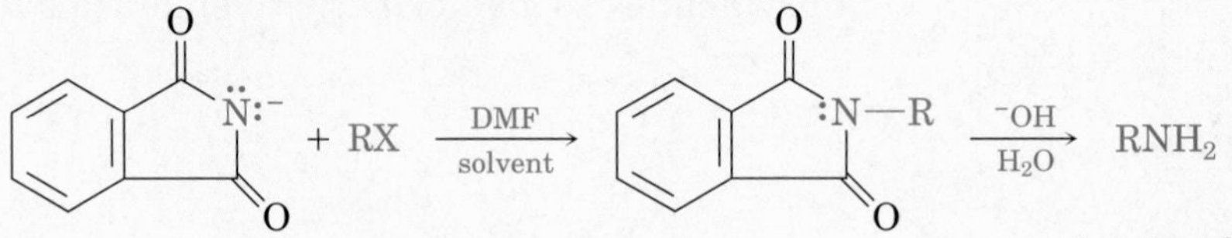

(continued) ▶

(f) Reduction of azides

$$RCH_2X + Na^+ \ ^-N_3 \xrightarrow[\text{solvent}]{\text{Ethanol}} RCH_2-N=\overset{+}{N}=\overset{-}{N} \xrightarrow[\text{2. }H_2O]{\text{1. LiAlH}_4,\text{ ether}} RCH_2NH_2$$

(g) Reductive amination of ketones/aldehydes

$$\underset{R}{\overset{O}{\underset{\|}{R-C}}} + NH_3 \xrightarrow{\text{NaBH}_3\text{CN}} RCHNH_2 \qquad \text{Primary}$$

$$\underset{R}{\overset{O}{\underset{\|}{R-C}}} + R'NH_2 \xrightarrow{\text{NaBH}_3\text{CN}} RCHNHR' \qquad \text{Secondary}$$

$$\underset{R}{\overset{O}{\underset{\|}{R-C}}} + R'_2NH \xrightarrow{\text{NaBH}_3\text{CN}} RCHNR'_2 \qquad \text{Tertiary}$$

(h) Hofmann rearrangement of amides

$$\underset{R \quad NH_2}{\overset{O}{\underset{\|}{C}}} \xrightarrow[\text{H}_2\text{O}]{\text{Br}_2,\text{ NaOH}} RNH_2 + CO_2$$

(i) Curtius rearrangement of acyl azides

$$\underset{R \quad Cl}{\overset{O}{\underset{\|}{C}}} + Na^+ \ ^-N_3 \xrightarrow[\text{solvent}]{\text{Ethanol}} \underset{R \quad N=\overset{+}{N}=\overset{-}{N}}{\overset{O}{\underset{\|}{C}}} \xrightarrow[\text{Heat}]{\text{H}_2\text{O}} RNH_2 + CO_2 + N_2$$

2. **Reactions of amines**
   (a) Alkylation of alkyl halides; see reaction 1(d) (Section 24.7)
   (b) Nucleophilic acyl substitution (Sections 21.4 and 24.7)

Ammonia
$$\underset{R \quad Cl}{\overset{O}{\underset{\|}{C}}} + NH_3 \xrightarrow[\text{solvent}]{\text{Pyridine}} \underset{R \quad \underset{H}{\overset{|}{N}}}{\overset{O}{\underset{\|}{C}}}H + HCl$$

Primary
$$\underset{R \quad Cl}{\overset{O}{\underset{\|}{C}}} + R'NH_2 \xrightarrow[\text{solvent}]{\text{Pyridine}} \underset{R \quad \underset{H}{\overset{|}{N}}}{\overset{O}{\underset{\|}{C}}}R' + HCl$$

Secondary
$$\underset{R \quad Cl}{\overset{O}{\underset{\|}{C}}} + R'_2NH \xrightarrow[\text{solvent}]{\text{Pyridine}} \underset{R \quad \underset{R'}{\overset{|}{N}}}{\overset{O}{\underset{\|}{C}}}R' + HCl$$

(continued) ▶

(c) Hofmann elimination (Section 24.7)

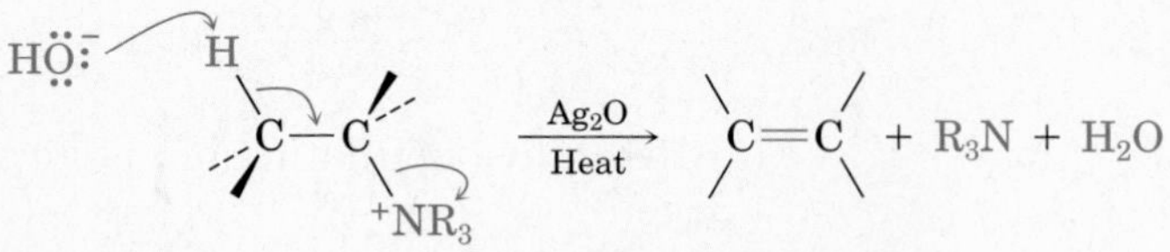

The less highly substituted alkene product is favored.

(d) Electrophilic aromatic substitution (Sections 16.5 and 24.8)

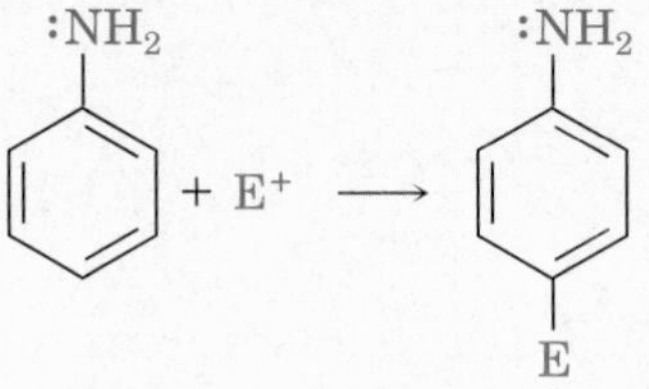

Ortho- and para-directing

(e) Formation of arenediazonium salts (Section 24.8)

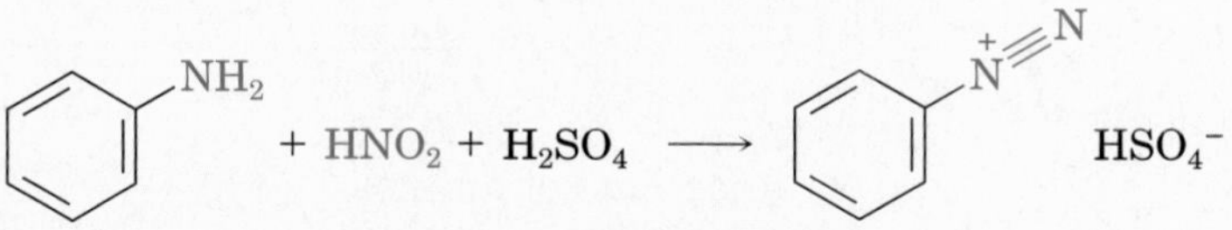

(f) Reactions of arenediazonium salts (Section 24.8)
  (1) Sandmeyer-type reactions

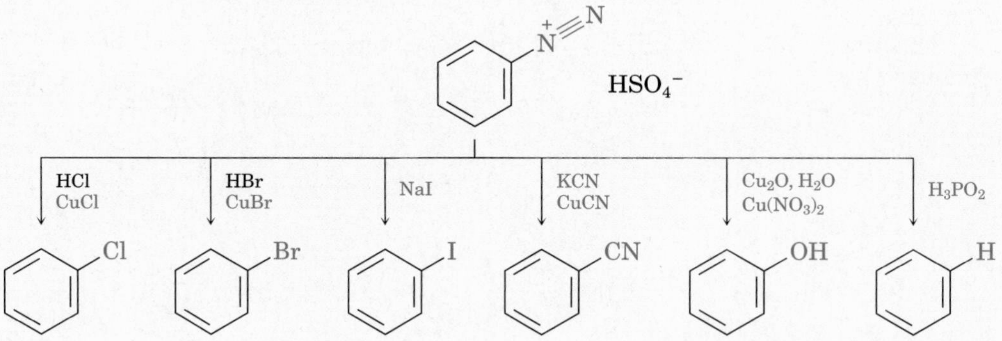

(continued) ▶

(2) Diazonium coupling

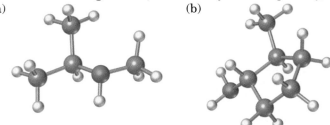

Visualizing Chemistry

• •

(Problems 24.1–24.19 appear within the chapter.)

24.20 Name the following amines, and identify each as primary, secondary, or tertiary:

(a)   (b)

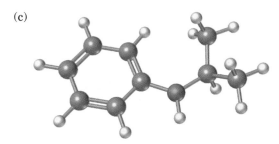

(c)

24.21 The following compound contains three nitrogen atoms. Rank them in order of increasing basicity.

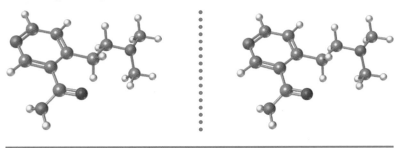

Stereo View

24.22 Name the following amine, including *R,S* stereochemistry, and draw the product of its reaction with excess iodomethane followed by heating with Ag$_2$O (Hofmann elimination). Is the stereochemistry of the alkene product *Z* or *E*? Explain.

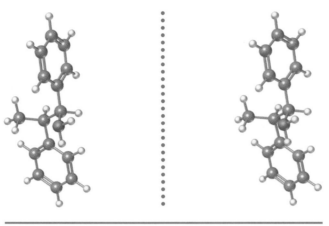

Stereo View

Additional Problems

24.23 Classify each of the amine nitrogen atoms in the following substances as primary, secondary, or tertiary:

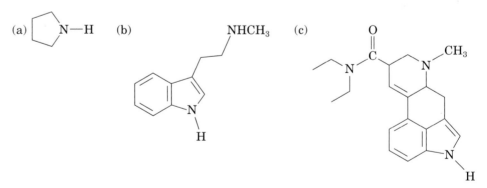

(a) N—H   (b)   NHCH$_3$ (c)

Lysergic acid diethylamide

24.24 Draw structures corresponding to the following IUPAC names:
(a) *N,N*-Dimethylaniline
(b) (Cyclohexylmethyl)amine
(c) *N*-Methylcyclohexylamine
(d) (2-Methylcyclohexyl)amine
(e) 3-(*N,N*-Dimethylamino)propanoic acid
(f) *N*-Isopropyl-*N*-methylcyclohexylamine

24.25 Name the following compounds:

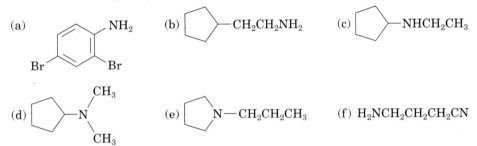

24.26 Propose structures for substances that fit the following descriptions:
(a) A chiral quaternary ammonium salt
(b) A five-membered heterocyclic amine
(c) A secondary amine, $C_6H_{11}N$

24.27 Give the structures of the major organic products you would expect from reaction of *m*-toluidine (*m*-methylaniline) with the following reagents:
(a) Br_2 (1 equiv)   (b) CH_3I (excess)   (c) $CH_3Cl + AlCl_3$
(d) CH_3COCl in pyridine   (e) The product of (d), then HSO_3Cl

24.28 Show the products from reaction of *p*-bromoaniline with the following reagents:
(a) Excess CH_3I (b) HCl   (c) HNO_2, H_2SO_4
(d) CH_3COCl (e) CH_3MgBr (f) CH_3CH_2Cl, $AlCl_3$
(g) Product of (c) with CuCl   (h) Product of (d) with CH_3CH_2Cl, $AlCl_3$

24.29 How would you prepare the following substances from 1-butanol?
(a) Butylamine   (b) Dibutylamine   (c) Propylamine
(d) Pentylamine   (e) *N*,*N*-Dimethylbutylamine   (f) Propene

24.30 How would you prepare the following substances from pentanoic acid?
(a) Pentanamide   (b) Butylamine   (c) Pentylamine
(d) 2-Bromopentanoic acid   (e) Hexanenitrile   (f) Hexylamine

24.31 How would you prepare aniline from the following starting materials?
(a) Benzene   (b) Benzamide   * (c) Toluene

24.32 How would you convert aniline into each of the products listed in Problem 24.31?

24.33 How might you prepare pentylamine from the following starting materials?
(a) Pentanamide   (b) Pentanenitrile   (c) 1-Butene
(d) Hexanamide   (e) 1-Butanol   (f) 5-Decene
(g) Pentanoic acid

24.34 What are the major products you would expect from Hofmann elimination of the following amines?
(a) *N*-Methylcyclopentylamine
(b) *N*-Phenyl-*N*-(1-methyl)pentylamine

$$\overset{\displaystyle CH_3}{\underset{\displaystyle NH_2}{\text{(c) } CH_3CHCHCH_2CH_2CH_3}}$$

24.35 Predict the product(s) of the following reactions. If more than one product is formed, tell which is major.

(a)

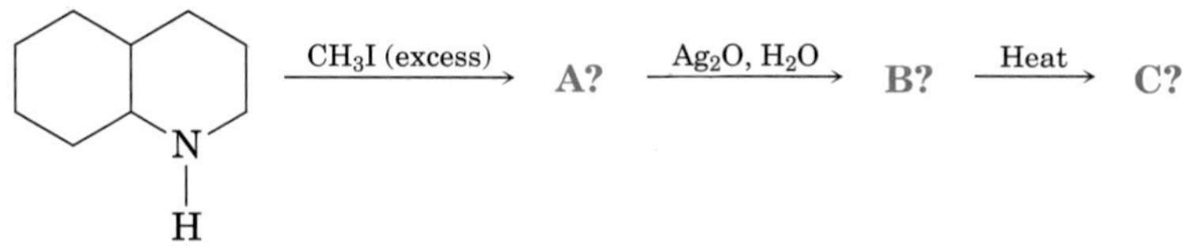

$\xrightarrow{\text{CH}_3\text{I (excess)}}$ A? $\xrightarrow{\text{Ag}_2\text{O, H}_2\text{O}}$ B? $\xrightarrow{\text{Heat}}$ C?

(b)

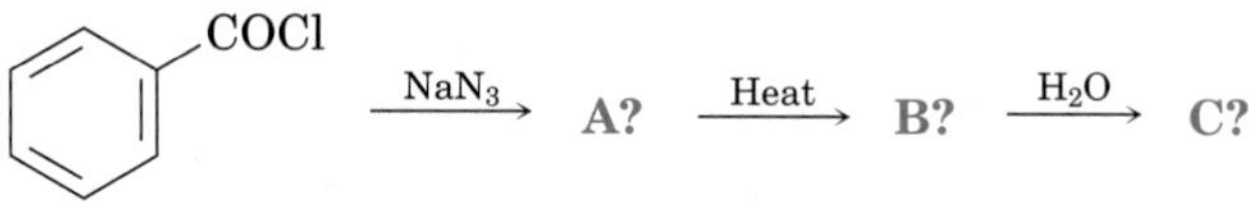

$\xrightarrow{\text{NaN}_3}$ A? $\xrightarrow{\text{Heat}}$ B? $\xrightarrow{\text{H}_2\text{O}}$ C?

(c)

$\xrightarrow{\text{KOH}}$ A? $\xrightarrow{\text{C}_6\text{H}_5\text{CH}_2\text{Br}}$ B? $\xrightarrow[\text{H}_2\text{O}]{\text{KOH}}$ C?

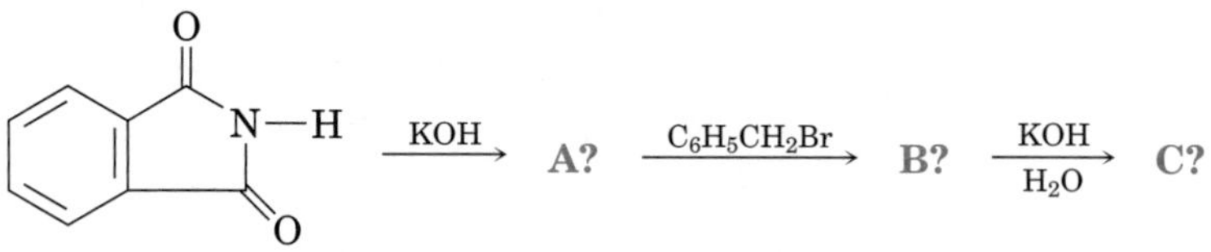

(d) $BrCH_2CH_2CH_2CH_2Br$ + 1 equiv CH_3NH_2 $\xrightarrow[\text{H}_2\text{O}]{\text{NaOH}}$?

24.36 Fill in the missing reagents a–e in the following scheme:

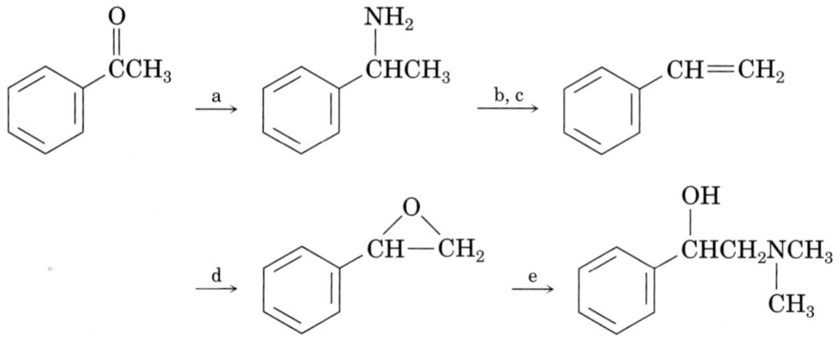

24.37 The following syntheses are incorrect. What is wrong with each?

(a) $CH_3CH_2CONH_2$ $\xrightarrow{\text{Br}_2, \text{NaOH}, \text{H}_2\text{O}}$ $CH_3CH_2CH_2NH_2$

(b)

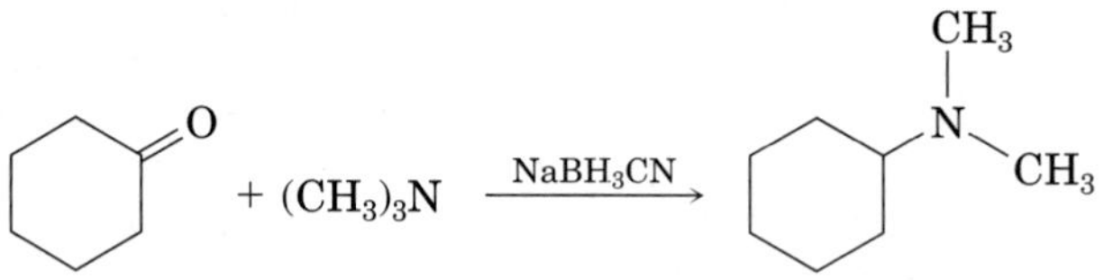

+ $(CH_3)_3N$ $\xrightarrow{\text{NaBH}_3\text{CN}}$

(c) $(CH_3)_3C-Br$ + NH_3 $\longrightarrow$ $(CH_3)_3C-NH_2$

(d)

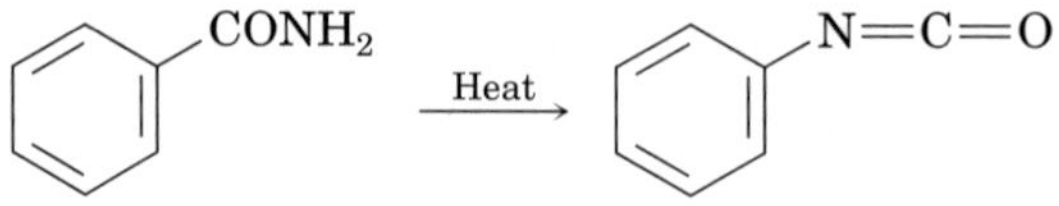

$\xrightarrow{\text{Heat}}$

(e) $CH_3CH_2CH_2\overset{\overset{\displaystyle NH_2}{|}}{C}HCH_3$ $\xrightarrow[\substack{\text{2. Ag}_2\text{O} \\ \text{3. Heat}}]{\text{1. CH}_3\text{I (excess)}}$ $CH_3CH_2CH=CHCH_3$

24.38 Protonation of an amide occurs on oxygen rather than on nitrogen. Suggest a reason for this behavior, taking resonance into account.

$$\underset{\text{R}-\overset{\displaystyle :\text{O}:}{\underset{\displaystyle \|}{\text{C}}}-\ddot{\text{N}}\text{H}_2}{} \;\overset{\text{H}_2\text{SO}_4}{\underset{\longleftarrow}{\rightleftharpoons}}\; \underset{\text{R}-\overset{\displaystyle {}^+\ddot{\text{O}}-\text{H}}{\underset{\displaystyle \|}{\text{C}}}-\ddot{\text{N}}\text{H}_2 + \text{HSO}_4{}^-}{}$$

24.39 How can you account for the fact that diphenylamine does not dissolve in dilute aqueous HCl and appears to be nonbasic?

24.40 Account for the fact that *p*-nitroaniline (pK_a = 1.0) is less basic than *m*-nitroaniline (pK_a = 2.5) by a factor of 30. Draw resonance structures to support your argument. (The pK_a values refer to the corresponding ammonium ions.)

24.41 Most chiral trisubstituted amines can't be resolved into enantiomers because nitrogen pyramidal inversion occurs too rapidly, but the substance known as *Tröger's base* is an exception. Make a molecular model of Tröger's base, and then explain why it is resolvable into enantiomers.

Tröger's base

24.42 Show the mechanism of reductive amination of cyclohexanone and dimethylamine with NaBH$_3$CN.

24.43 How might a reductive amination be used to synthesize ephedrine, an amino alcohol that is widely used for the treatment of bronchial asthma?

Ephedrine

24.44 One problem with reductive amination as a method of amine synthesis is that by-products are sometimes obtained. For example, reductive amination of benzaldehyde with methylamine leads to a mixture of *N*-methylbenzylamine and *N*-methyldibenzylamine. How do you suppose the tertiary amine by-product is formed? Propose a mechanism.

24.45 Propose a route for the synthesis of 1-bromo-2,4-dimethylbenzene from benzene.

24.46 Prontosil is an antibacterial azo dye that was once used for urinary tract infections. How would you prepare prontosil from benzene?

Prontosil

24.47 Cyclopentamine is an amphetamine-like central nervous system stimulant. Propose a synthesis of cyclopentamine from materials of five carbons or less.

CH₃ structure — **Cyclopentamine**

24.48 Tetracaine is a substance used medicinally as a spinal anesthetic during lumbar punctures (spinal taps).

$CH_3CH_2CH_2CH_2$—N— ... —$COCH_2CH_2N(CH_3)_2$ **Tetracaine**

(a) How would you prepare tetracaine from the corresponding aniline derivative, $ArNH_2$?
(b) How would you prepare tetracaine from *p*-nitrobenzoic acid?
(c) How would you prepare tetracaine from benzene?

24.49 Atropine, $C_{17}H_{23}NO_3$, is a poisonous alkaloid isolated from the leaves and roots of *Atropa belladonna*, the deadly nightshade. In low doses, atropine acts as a muscle relaxant; 0.5 ng (nanogram, 10^{-9} g) is sufficient to cause pupil dilation. On basic hydrolysis, atropine yields tropic acid, $C_6H_5CH(CH_2OH)COOH$, and tropine, $C_8H_{15}NO$. Tropine is an optically inactive alcohol that yields tropidene on dehydration with H_2SO_4. Propose a structure for atropine.

Tropidene

24.50 Tropidene (Problem 24.49) can be converted by a series of steps into tropilidene (1,3,5-cycloheptatriene). How would you accomplish this conversion?

24.51 Propose a structure for the product with formula $C_9H_{17}N$ that results when 2-(2-cyanoethyl)cyclohexanone is reduced catalytically.

$\xrightarrow{H_2/Pt}$ $C_9H_{17}N$

24.52 Coniine, $C_8H_{17}N$, is the toxic principle of the poison hemlock drunk by Socrates. When subjected to Hofmann elimination, coniine yields 5-(*N,N*-dimethylamino)-1-octene. If coniine is a secondary amine, what is its structure?

24.53 How would you synthesize coniine (Problem 24.52) from acrylonitrile ($H_2C=CHCN$) and ethyl 3-oxohexanoate ($CH_3CH_2CH_2COCH_2CO_2Et$)? (*Hint*: See Problem 24.51.)

24.54 How would you synthesize the heart stimulant propranolol starting from 1-hydroxy-naphthalene and any other reagents needed?

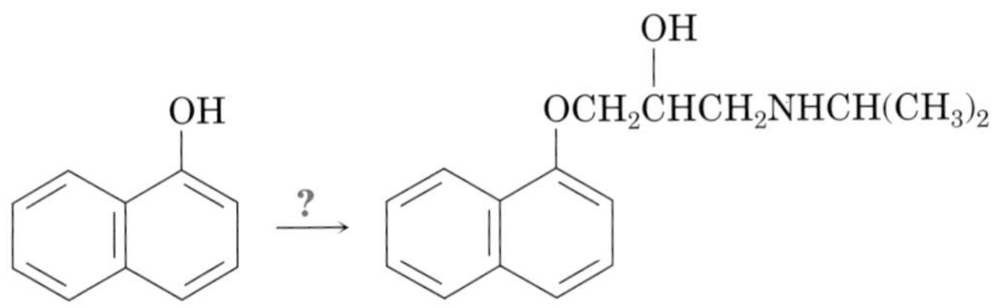

1-Hydroxynaphthalene   **Propranolol**

24.55 Tyramine is an alkaloid found, among other places, in mistletoe and ripe cheese. How would you synthesize tyramine from benzene? From toluene?

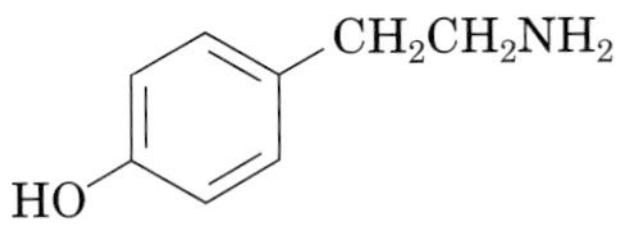

Tyramine

24.56 How would you prepare the following compounds from toluene? A diazonio replacement reaction is needed in some instances.

(a) 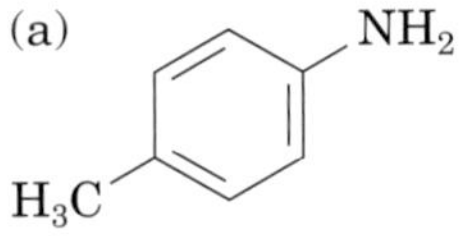   (b) 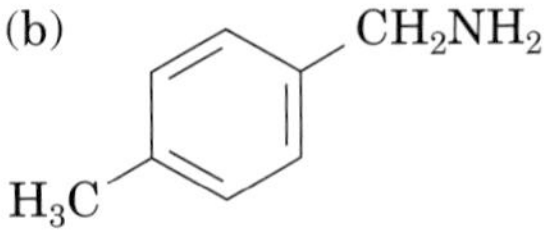   (c)

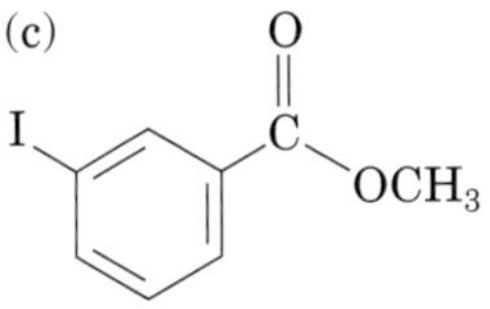

24.57 Mephenesin is a drug used as a muscle relaxant and sedative. Propose a synthesis of mephenesin from benzene and any other reagents needed.

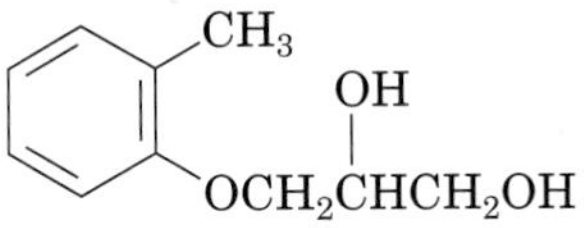

Mephenesin

24.58 Reaction of anthranilic acid (*o*-aminobenzoic acid) with HNO_2 and H_2SO_4 yields a diazonium salt that can be treated with base to yield a neutral diazonium carboxylate.

(a) What is the structure of the neutral diazonium carboxylate?

(b) Heating the diazonium carboxylate results in the formation of CO_2, N_2, and an intermediate that reacts with 1,3-cyclopentadiene to yield the following product:

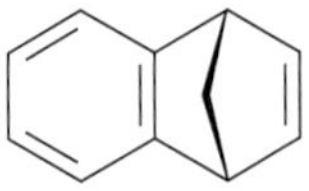

What is the structure of the intermediate, and what kind of reaction does it undergo with cyclopentadiene?

24.59 Cyclooctatetraene was first synthesized in 1911 by a route that involved the following transformation:

How might you use the Hofmann elimination to accomplish this reaction? How would you finish the synthesis by converting cyclooctatriene into cyclooctatetraene?

24.60 When an α-hydroxy amide is treated with Br_2 in aqueous NaOH under Hofmann rearrangement conditions, loss of CO_2 occurs and a chain-shortened aldehyde is formed. Propose a mechanism.

24.61 Propose a mechanism for the following reaction:

24.62 Propose a mechanism for the following reaction:

24.63 Phenacetin, a substance formerly used in over-the-counter headache remedies, has the formula $C_{10}H_{13}NO_2$. Phenacetin is neutral and does not dissolve in either acid or base. When warmed with aqueous NaOH, phenacetin yields an amine, $C_8H_{11}NO$, whose ^1H NMR spectrum is shown. When heated with HI, the amine is cleaved to an aminophenol, C_6H_7NO. What is the structure of phenacetin, and what are the structures of the amine and the aminophenol?

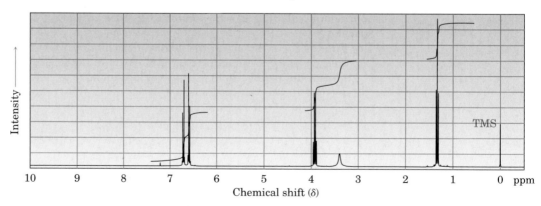

24.64 Propose structures for amines with the following ^{1}H NMR spectra:

(a) C_3H_9NO

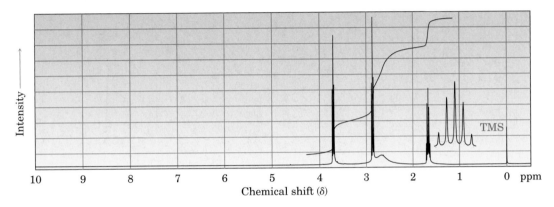

(b) $C_4H_{11}NO_2$

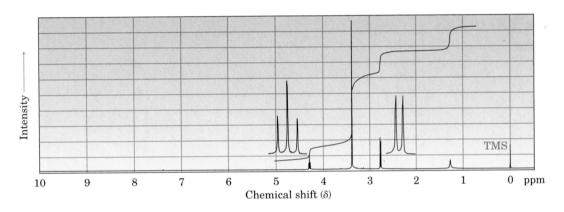

24.65 Propose structures for compounds that show the following ^{1}H NMR spectra.

(a) $C_9H_{13}N$

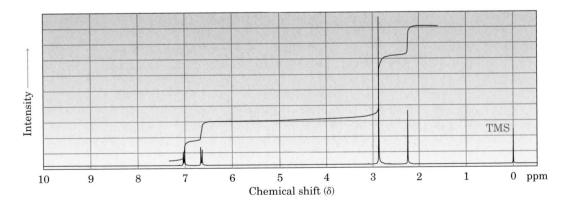

(b) $C_{15}H_{17}N$

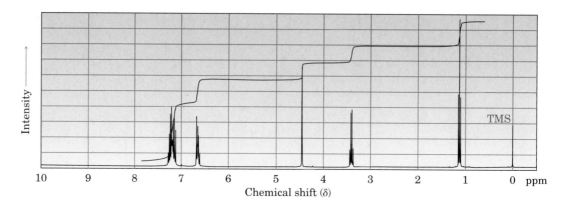

A Look Ahead

• •

24.66 One of the reactions used in determining the sequence of nucleotides in a strand of DNA is reaction with hydrazine. Propose a mechanism for the following reaction. (See Section 28.15.)

24.67 α-Amino acids can be prepared by the *Strecker synthesis*, a two-step process in which an aldehyde is treated with ammonium cyanide followed by hydrolysis of the amino nitrile intermediate with aqueous acid. Propose a mechanism for the reaction. (See Section 26.3.)

An α-amino acid

Molecular Modeling

24.68 Use SpartanView to obtain the dipole moments of aniline, nitrobenzene, and 4-nitro-aniline. Is the dipole moment of 4-nitroaniline the sum of the dipole moments of aniline and nitrobenzene? Compare electrostatic potential maps and geometries for all three structures, and draw resonance structures for 4-nitroaniline that account for your observations.

24.69 Many medicinal compounds contain a basic nitrogen atom. Use SpartanView to examine electrostatic potential maps of ergotamine and mitomycin C, and identify the most basic nitrogen in each.

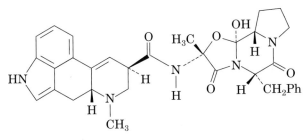

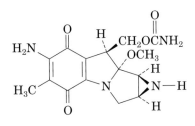

Ergotamine   **Mitomycin C**

24.70 A good phase-transfer catalyst must be soluble in both aqueous and nonpolar organic solvents. Use SpartanView to compare electrostatic potential maps of benzyltrimethyl-ammonium ion, tetraethylammonium ion, and tetrabenzylammonium ion. Which ion is likely to be the most water-soluble, and which the least water-soluble?

24.71 Use SpartanView to compare electrostatic potential maps of the reactants and the transition state for the S_N2 reaction of ammonia with iodomethane. Which is more polar? How would the reaction rate change if a polar aprotic solvent such as DMSO were replaced with a polar protic solvent such as water?

25

Biomolecules: Carbohydrates

Carbohydrates occur in every living organism. The sugar and starch in food, and the cellulose in wood, paper, and cotton, are nearly pure carbohydrates. Modified carbohydrates form part of the coating around living cells, other carbohydrates are part of the nucleic acids that carry our genetic information, and still others are used as medicines.

The word *carbohydrate* derives historically from the fact that glucose, the first simple carbohydrate to be obtained pure, has the molecular formula $C_6H_{12}O_6$ and was originally thought to be a "hydrate of carbon, $C_6(H_2O)_6$."

This view was soon abandoned, but the name persisted. Today, the term **carbohydrate** is used to refer loosely to the broad class of polyhydroxylated aldehydes and ketones commonly called *sugars*.

$$
\begin{array}{c}
\text{H}\diagdown\underset{|}{\text{C}}{\diagup}^{\text{O}} \\
\text{H}-\underset{|}{\text{C}}-\text{OH} \\
\text{HO}-\underset{|}{\text{C}}-\text{H} \\
\text{H}-\underset{|}{\text{C}}-\text{OH} \\
\text{H}-\underset{|}{\text{C}}-\text{OH} \\
\text{CH}_2\text{OH}
\end{array}
$$

Glucose (also called dextrose), a pentahydroxyhexanal

Carbohydrates are synthesized by green plants during photosynthesis, a complex process in which sunlight provides the energy to convert carbon dioxide and water into glucose plus oxygen. Many molecules of glucose are then chemically linked for storage by the plant in the form of either cellulose or starch. It has been estimated that more than 50% of the dry weight of the earth's biomass—all plants and animals—consists of glucose polymers. When eaten and metabolized, carbohydrates provide the major source of energy required by organisms. Thus, carbohydrates act as the chemical intermediaries by which solar energy is stored and used to support life.

$$6\,CO_2 \;+\; 6\,H_2O \;\xrightarrow{\text{Sunlight}}\; 6\,O_2 \;+\; \underset{\textbf{Glucose}}{C_6H_{12}O_6} \;\longrightarrow\; \text{Cellulose, starch}$$

Because humans and most other mammals lack the enzymes needed for digestion of cellulose, they require starch as their dietary source of carbohydrates. Grazing animals such as cows, however, have in their first stomach microorganisms that are able to digest cellulose. The energy stored in cellulose is thus moved up the biological food chain when these ruminant animals eat grass and are then used for food.

25.1 Classification of Carbohydrates

Carbohydrates are generally classed into two groups, *simple* and *complex*. **Simple sugars**, or **monosaccharides**, are carbohydrates like glucose and fructose that can't be converted into smaller sugars by hydrolysis. **Complex carbohydrates** are made of two or more simple sugars linked together. Sucrose (table sugar), for example, is a **disaccharide** made up of one glucose linked to one fructose. Similarly, cellulose is a **polysaccharide** made up of several thousand glucose units linked together. Hydrolysis of a polysaccharide breaks it down into its constituent monosaccharides.

$$1 \text{ Sucrose} \xrightarrow{\text{H}_3\text{O}^+} 1 \text{ Glucose} + 1 \text{ Fructose}$$

$$\text{Cellulose} \xrightarrow{\text{H}_3\text{O}^+} \sim 3000 \text{ Glucose}$$

Monosaccharides are further classified as either **aldoses** or **ketoses**. The *-ose* suffix designates a carbohydrate, and the *aldo-* and *keto-* prefixes identify the nature of the carbonyl group. The number of carbon atoms in the monosaccharide is indicated by using *tri-*, *tetr-*, *pent-*, *hex-*, and so forth in the name. For example, glucose is an *aldohexose* (a six-carbon aldehydic sugar); fructose is a *ketohexose* (a six-carbon ketonic sugar); and ribose is an *aldopentose* (a five-carbon aldehydic sugar). Most of the commonly occurring sugars are either aldopentoses or aldohexoses.

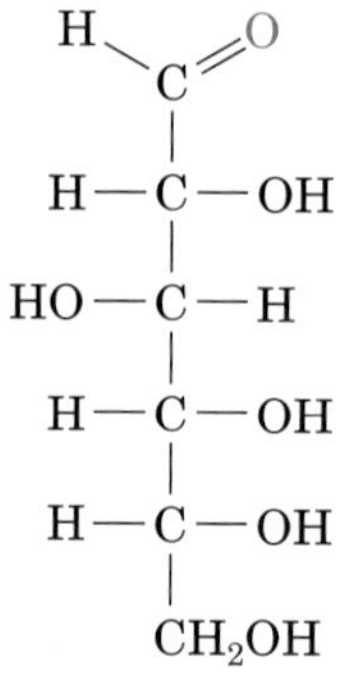

Glucose
(an aldohexose)

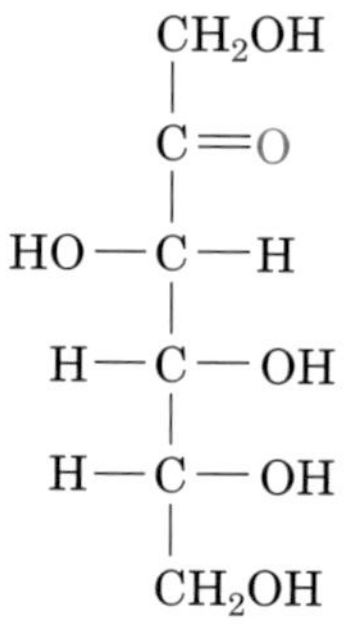

Fructose
(a ketohexose)

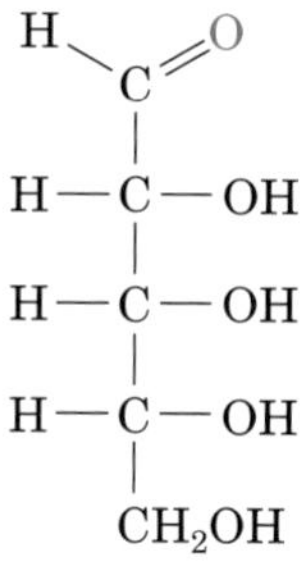

Ribose
(an aldopentose)

Problem 25.1   Classify each of the following monosaccharides:

(a) **Threose**

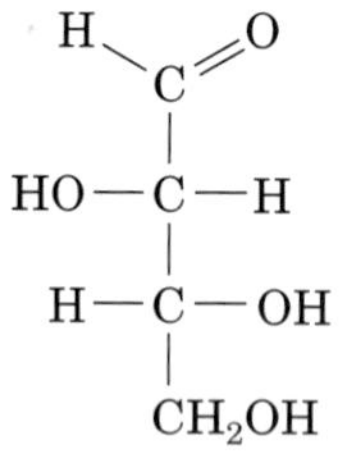

(b) **Ribulose**

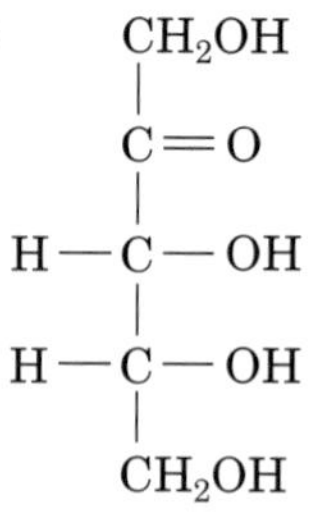

(c) **Tagatose**

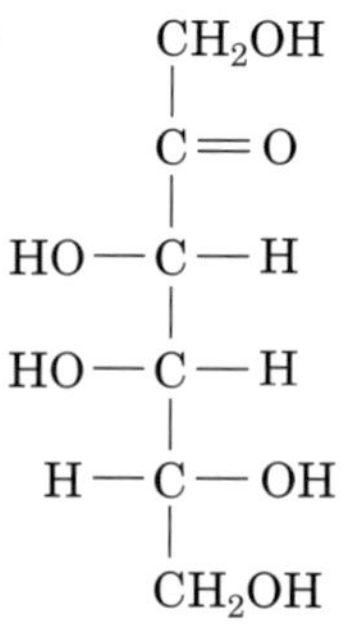

(d) **2-Deoxyribose**

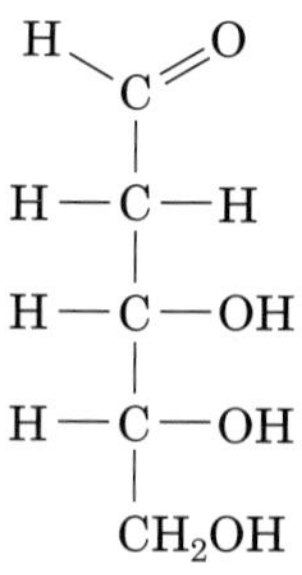

25.2 Configurations of Monosaccharides: Fischer Projections

Since all carbohydrates have chiral carbon atoms, it was recognized long ago that a standard method of representation is needed to describe carbohydrate stereochemistry. The method most commonly used employs Fischer projections for depicting chirality centers on a flat page.

Recall from Section 9.13 that a tetrahedral carbon atom is represented in a Fischer projection by two crossed lines. The horizontal lines represent bonds coming out of the page, and the vertical lines represent bonds going into the page. By convention, the carbonyl carbon is placed at or near the top in Fischer projections. Thus, (*R*)-glyceraldehyde, the simplest monosaccharide, is drawn as shown in Figure 25.1.

FIGURE 25.1 ▼

A Fischer projection of (*R*)-glyceraldehyde.

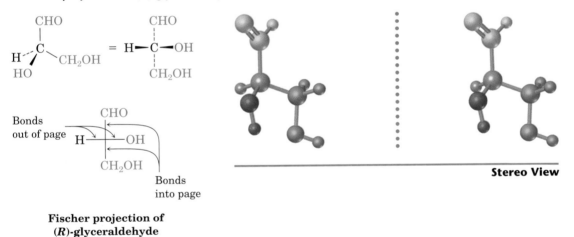

Fischer projection of
(*R*)-glyceraldehyde

Stereo View

Recall also that Fischer projections can be rotated on the page by 180° without changing their meaning, but not by 90° or 270°.

180° (CHO — H——OH — CH₂OH)   same as   HO——H with CH₂OH top and CHO bottom

(*R*)-Glyceraldehyde

Carbohydrates with more than one chirality center are shown by stacking the centers on top of one another, with the carbonyl carbon again placed either at or near the top. Glucose, for example, has four chirality centers stacked on top of one another in a Fischer projection. Such representations don't, of course, give an accurate picture of the true conformation of a molecule, which actually is curled around on itself like a bracelet.

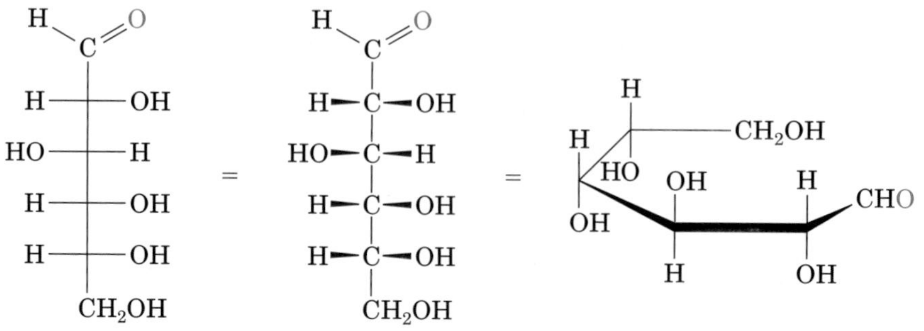

Glucose
(carbonyl group at top)

Problem 25.2 Which of the following Fischer projections of glyceraldehyde represent the same enantiomer?

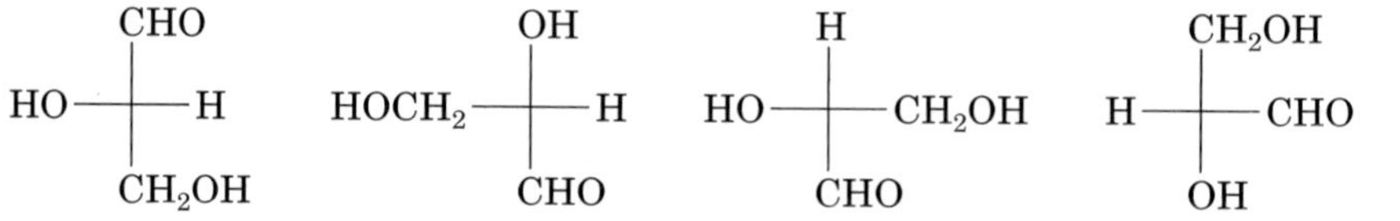

A   **B**   **C**   **D**

Problem 25.3 Convert the following Fischer projections into tetrahedral representations, and assign R or S stereochemistry to each:

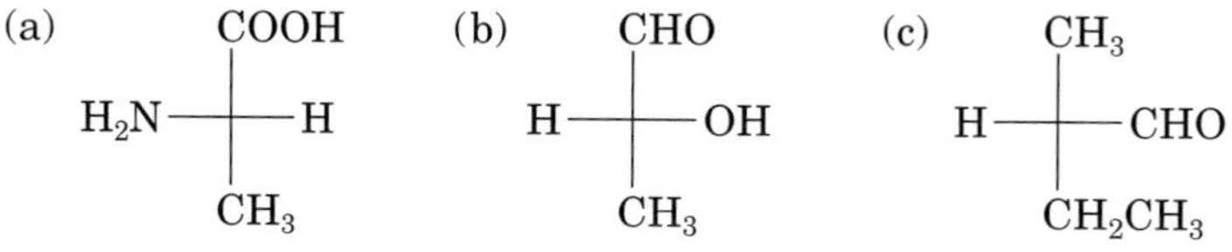

25.3 D,L Sugars

Glyceraldehyde, the simplest aldose, has only one chirality center and thus has two enantiomeric (mirror-image) forms. Only the dextrorotatory enantiomer occurs naturally, however. That is, a sample of naturally occurring glyceraldehyde placed in a polarimeter rotates plane-polarized light in a clockwise direction, denoted (+).

Since (+)-glyceraldehyde has been shown to have an R configuration at C2, it can be represented in a Fischer projection as shown in Figure 25.1. For historical reasons dating back long before the adoption of the R,S system, (R)-(+)-glyceraldehyde is also referred to as D-glyceraldehyde (D for dextrorotatory). The other enantiomer, (S)-(−)-glyceraldehyde, is known as L-glyceraldehyde (L for levorotatory).

Because of the way monosaccharides are biosynthesized in nature, glucose, fructose, and almost all other naturally occurring monosaccharides have the same R stereochemical configuration as D-glyceraldehyde at the chirality center farthest from the carbonyl group. In Fischer projections, therefore, most naturally occurring sugars have the hydroxyl group at the lowest chirality center pointing to the right (Figure 25.2). All such compounds are referred to as **D sugars.**

FIGURE 25.2 ▼

Some naturally occurring D sugars. The hydroxyl group at the chirality center farthest from the carbonyl group has the same R configuration as that in (+)-glyceraldehyde. When the molecule is drawn in Fischer projection with the carbonyl group at or near the top, the –OH group at the lowest chirality center points toward the right.

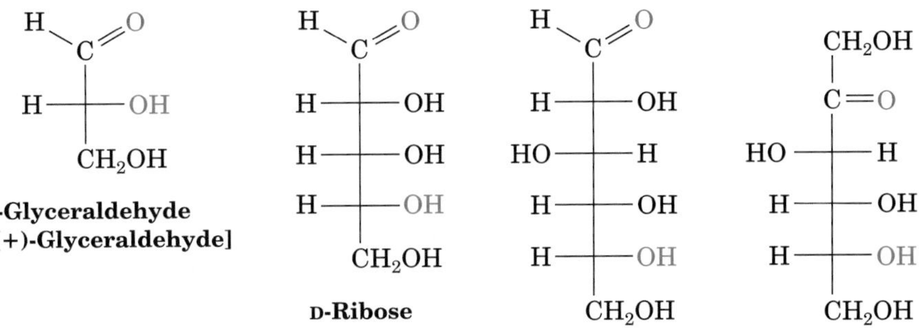

D-Glyceraldehyde
[(R)-(+)-Glyceraldehyde]

D-Ribose

D-Glucose

D-Fructose

In contrast to D sugars, **L sugars** have an S configuration at the lowest chirality center, with the –OH group pointing to the *left* in Fischer projections. Thus, an L sugar is the mirror image (enantiomer) of the corresponding D sugar and has the opposite configuration from the D sugar at all chirality centers. Note that the D and L notations have no relation to the direction in which a given sugar rotates plane-polarized light; a D sugar can be either dextrorotatory or levorotatory. The prefix D indicates only that the –OH group at the lowest chirality center is to the right when the molecule is drawn in a Fischer projection with the carbonyl group at or near the top.

Mirror

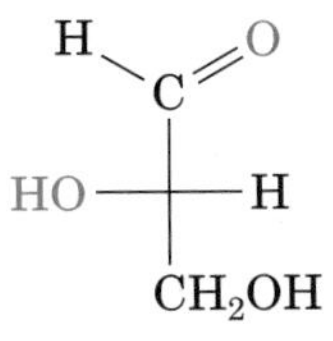

L-Glyceraldehyde
[(S)-(–)-Glyceraldehyde]

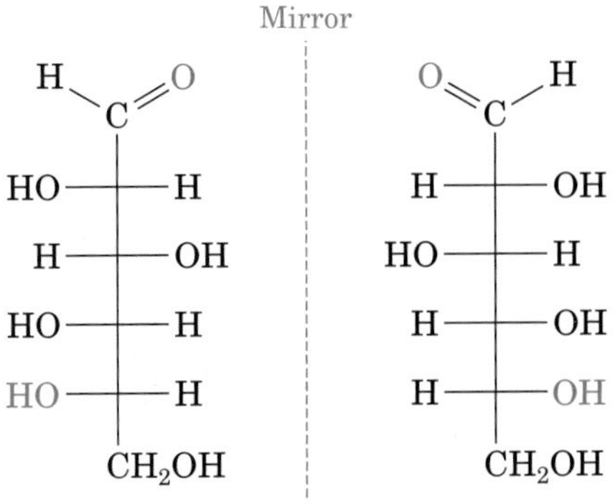

L-Glucose
(not naturally occurring)

D-Glucose

Note also that the D,L system of carbohydrate nomenclature describes the configuration at only *one* chirality center and says nothing about the configuration of other chirality centers that may be present. The advantage of the system, though, is that it allows us to relate one sugar to another rapidly and visually.

• •

Problem 25.4   Assign R or S configuration to each chirality center in the following sugars, and tell whether each is a D sugar or an L sugar:

(a)
```
        CHO
         |
HO ——————— H
         |
HO ——————— H
         |
       CH₂OH
```

(b)
```
        CHO
         |
 H ——————— OH
         |
HO ——————— H
         |
 H ——————— OH
         |
       CH₂OH
```

(c)
```
       CH₂OH
         |
        C=O
         |
HO ——————— H
         |
 H ——————— OH
         |
       CH₂OH
```

Problem 25.5   (+)-Arabinose, an aldopentose that is widely distributed in plants, is systematically named (2R,3S,4S)-2,3,4,5-tetrahydroxypentanal. Draw a Fischer projection of (+)-arabinose, and identify it as a D sugar or an L sugar.

• •

25.4 Configurations of the Aldoses

Aldotetroses are four-carbon sugars with two chirality centers. There are $2^2 = 4$ possible stereoisomeric aldotetroses, or two D,L pairs of enantiomers, called *erythrose* and *threose*.

Aldopentoses have three chirality centers and a total of $2^3 = 8$ possible stereoisomers, or four D,L pairs of enantiomers. These four pairs are called *ribose, arabinose, xylose,* and *lyxose.* All except lyxose occur widely. D-Ribose is an important constituent of RNA (ribonucleic acid), L-arabinose is found in many plants, and D-xylose is found in wood.

Aldohexoses have four chirality centers and a total of $2^4 = 16$ possible stereoisomers, or eight D,L pairs of enantiomers. The names of the eight are *allose, altrose, glucose, mannose, gulose, idose, galactose,* and *talose.* Of the eight, only D-glucose (from starch and cellulose) and D-galactose (from gums and fruit pectins) are found widely in nature. D-Mannose and D-talose also occur naturally, but in lesser abundance.

Fischer projections of the four-, five-, and six-carbon D aldoses are shown in Figure 25.3. Starting from D-glyceraldehyde, we can imagine constructing the two D aldotetroses by inserting a new chirality center just below the aldehyde carbon. Each of the two D aldotetroses leads to two D aldopentoses (four total), and each of the four D aldopentoses leads to two D aldohexoses (eight total). Each of the D aldoses in Figure 25.3 has an L enantiomer, which is not shown.

FIGURE 25.3 ▼

Configurations of D aldoses. The structures are arranged in order from left to right so that the –OH groups on C2 alternate right/left (R/L) in going across a series. Similarly, the –OH groups at C3 alternate two right/two left (2R/2L); the –OH groups at C4 alternate 4R/4L; and the –OH groups at C5 are to the right in all eight (8R). Each D aldose has a corresponding L enantiomer, which is not shown.

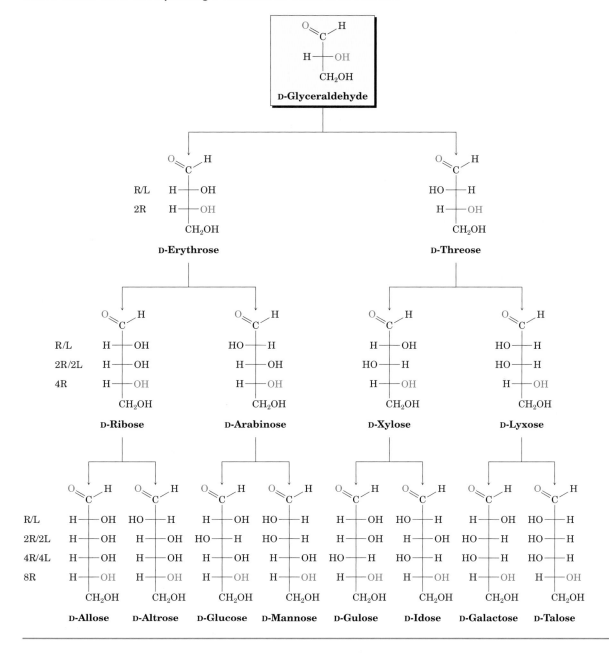

Louis F. Fieser

Louis F. Fieser (1899–1977) was born in Columbus, Ohio, and received his Ph.D. at Harvard University in 1924 with James B. Conant. He was Professor of Chemistry at Bryn Mawr College and then at Harvard University. While at Bryn Mawr, he met his future wife, Mary. The two Fiesers wrote numerous chemistry texts and monographs. Among his scientific contributions, Fieser was known for his work in steroid chemistry and in carrying out the first synthesis of vitamin K. He was also the inventor of jellied gasoline, or napalm, which was developed at Harvard during World War II.

Louis Fieser of Harvard University suggested the following procedure for remembering the names and structures of the eight D aldohexoses (Figure 25.3):

STEP 1 Set up eight Fischer projections with the –CHO group on top and the –CH₂OH group at the bottom.

STEP 2 Indicate stereochemistry at C5 by placing all eight –OH groups to the right (D series).

STEP 3 Indicate stereochemistry at C4 by alternating four –OH groups to the right and four to the left.

STEP 4 Indicate stereochemistry at C3 by alternating two –OH groups to the right, two to the left, and so on.

STEP 5 Indicate stereochemistry at C2 by alternating –OH groups right, left, right, left, and so on.

STEP 6 Name the eight isomers using the mnemonic "All altruists gladly make gum in gallon tanks."

The structures of the four D aldopentoses can be generated in a similar way and can be named by the mnemonic suggested by a Cornell undergraduate: "Ribs are extra lean."

Practice Problem 25.1 Draw a Fischer projection of L-fructose.

Strategy Since L-fructose is the enantiomer of D-fructose, simply look at the structure of D-fructose and reverse the configuration at each chirality center.

Solution

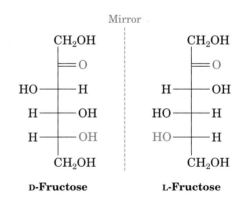

Problem 25.6 Only the D sugars are shown in Figure 25.3. Draw Fischer projections for the following L sugars:
(a) L-Xylose   (b) L-Galactose   (c) L-Allose

Problem 25.7   How many aldoheptoses are there? How many are D sugars, and how many are L sugars?

Problem 25.8   The following model is that of an aldopentose. Draw a Fischer projection of the sugar, and identify it. Is it a D sugar or an L sugar?

Stereo View

25.5 Cyclic Structures of Monosaccharides: Hemiacetal Formation

We said in Section 19.11 that aldehydes and ketones undergo a rapid and reversible nucleophilic addition reaction with alcohols to form hemiacetals:

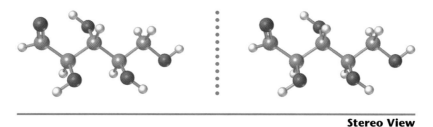

An aldehyde   A hemiacetal

If the carbonyl and the hydroxyl groups are in the same molecule, an *intramolecular* nucleophilic addition can take place, leading to the formation of a *cyclic* hemiacetal. Five- and six-membered cyclic hemiacetals are particularly stable, and many carbohydrates therefore exist in an equilibrium between open-chain and cyclic forms. Glucose, for example, exists in aqueous solution primarily as the six-membered, or **pyranose**, form resulting from intramolecular nucleophilic addition of the –OH group at C5 to the C1 carbonyl group. Fructose, on the other hand, exists to the extent of about 80% in the pyranose form and about 20% as the five-membered, or **furanose**, form resulting from addition of the –OH group at C5 to the C2 carbonyl. The words *pyranose* for a six-membered ring and *furanose* for a five-membered ring are derived from the names of the simple cyclic ethers pyran and furan. The cyclic forms of glucose and fructose are shown in Figure 25.4.

Like cyclohexane rings (Section 4.9), pyranose rings have a chair-like geometry with axial and equatorial substituents. By convention, the rings are usually drawn by placing the hemiacetal oxygen atom at the right rear, as shown in Figure 25.4. Note that an –OH group on the *right* in a Fischer projection is on the *bottom* face of the pyranose ring, and an –OH group on the *left* in a Fischer projection is on the *top* face of the ring. For D sugars,

FIGURE 25.4 ▼

Glucose and fructose in their cyclic pyranose and furanose forms.

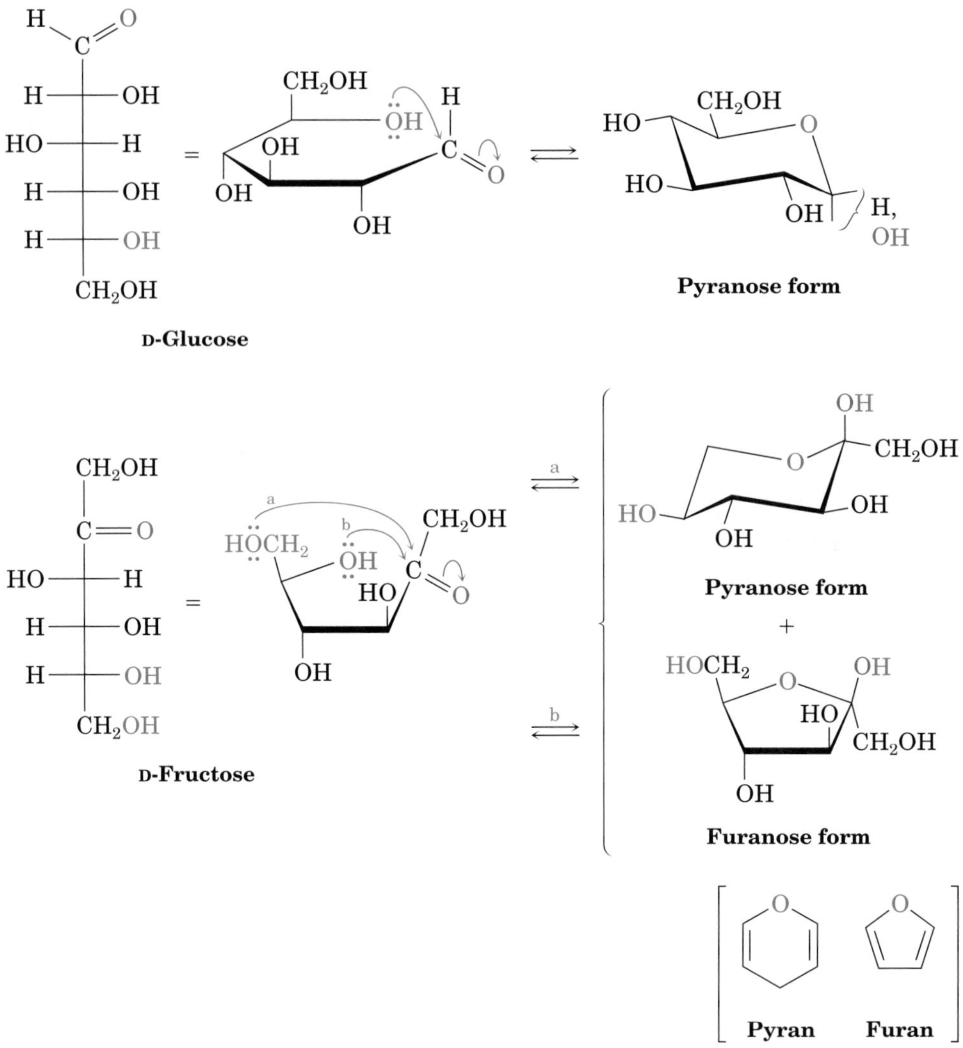

D-Glucose

Pyranose form

D-Fructose

Pyranose form

+

Furanose form

Pyran   Furan

the terminal –CH_2OH group is on the top of the ring, whereas for L sugars, the –CH_2OH group is on the bottom.

• •

Practice Problem 25.2   D-Mannose differs from D-glucose in its stereochemistry at C2. Draw D-mannose in its chair-like pyranose form.

Strategy   First draw a Fischer projection of D-mannose. Then lay it on its side, and curl it around so that the –CHO group (C1) is on the right front and the –CH_2OH group (C6) is toward the left rear. Now, connect the –OH at C5 to the C1 carbonyl group to form the pyranose ring. In drawing the chair form, raise the leftmost carbon (C4) up and drop the rightmost carbon (C1) down.

Solution

D-**Mannose**   **Pyranose form**

- -

Problem 25.9   D-Allose differs from D-glucose in its stereochemistry at C3. Draw D-allose in its pyranose form.

Problem 25.10   Draw D-ribose in its furanose form.

- -

25.6 Monosaccharide Anomers: Mutarotation

When an open-chain monosaccharide cyclizes to a pyranose or furanose form, a new chirality center is generated at the former carbonyl carbon. The two diastereomers produced are called **anomers**, and the hemiacetal carbon atom is referred to as the **anomeric center**. For example, glucose cyclizes reversibly in aqueous solution to a 36:64 mixture of two anomers. The minor anomer, which has the –OH group at C1 trans to the –CH₂OH substituent at C5, is called the **α anomer**; its full name is α-D-glucopyranose. The major anomer, which has the –OH group at C1 cis to the –CH₂OH substituent at C5, is called the **β anomer**; its full name is β-D-glucopyranose (Figure 25.5, p. 1042). Note that in β-D-glucopyranose, all the substituents on the ring are equatorial. Thus, β-D-glucopyranose is the least sterically crowded and most stable of the eight D aldohexoses.

Both anomers of D-glucopyranose can be crystallized and purified. Pure α-D-glucopyranose has a melting point of 146°C and a specific rotation, $[\alpha]_D$, of +112.2°; pure β-D-glucopyranose has a melting point of 148–155°C and a specific rotation of +18.7°. When a sample of either pure anomer is dissolved in water, however, its optical rotation slowly changes and ultimately reaches a constant value of +52.6°. The specific rotation of the α-anomer solution decreases from +112.2° to +52.6°, and the specific rotation of the β-anomer solution increases from +18.7° to +52.6°. Called **mutarotation**, this change in optical rotation is due to the slow conversion of the pure anomers into a 36:64 equilibrium mixture.

FIGURE 25.5 ▼

Alpha and beta anomers of glucose.

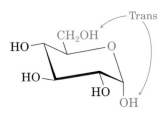

α-D-Glucopyranose (36%)
(α anomer: OH and
CH₂OH are trans)

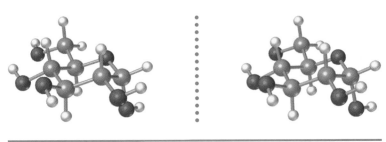

Stereo View

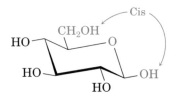

β-D-Glucopyranose (64%)
(β anomer: OH and
CH₂OH are cis)

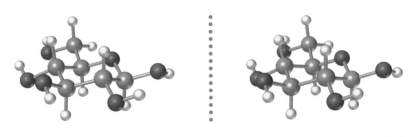

Stereo View

Mutarotation occurs by a reversible ring-opening of each anomer to the open-chain aldehyde, followed by reclosure. Although equilibration is slow at neutral pH, it is catalyzed by both acid and base.

α-D-Glucopyranose (36%)
$[\alpha]_D = +112.2°$

D-Glucose

β-D-Glucopyranose (64%)
$[\alpha]_D = +18.7°$

Practice Problem 25.3   Draw β-L-glucopyranose in its more stable chair conformation.

Strategy   It's probably easiest to begin by drawing the chair conformation of β-D-glucopyranose. Then draw its mirror image by changing the stereochemistry at every position on the ring, and carry out a ring-flip to give the more stable chair conformation. Note that the –CH₂OH group is on the bottom face of the ring in the L enantiomer.

Solution

β-D-Glucopyranose

β-L-Glucopyranose

• •

Problem 25.11 Draw both anomers of D-fructose in their furanose forms.

Problem 25.12 Draw β-D-galactopyranose and β-D-mannopyranose in their more stable chair conformations. Label each ring substituent as either axial or equatorial. Which would you expect to be more stable, galactose or mannose?

Problem 25.13 Draw β-L-galactopyranose in its more stable chair conformation, and label the substituents as either axial or equatorial.

• •

25.7 Reactions of Monosaccharides

Since monosaccharides contain only two kinds of functional groups, carbonyls and hydroxyls, most of the chemistry of monosaccharides is the now familiar chemistry of these two groups.

Ester and Ether Formation

Monosaccharides behave as simple alcohols in much of their chemistry. For example, carbohydrate –OH groups can be converted into esters and ethers, which are often easier to work with than the free sugars. Because of their many hydroxyl groups, monosaccharides are usually soluble in water but insoluble in organic solvents such as ether. They are also difficult to purify and have a tendency to form syrups rather than crystals when water is removed. Ester and ether derivatives, however, are soluble in organic solvents and are easily purified and crystallized.

Esterification is normally carried out by treating the carbohydrate with an acid chloride or acid anhydride in the presence of a base (Sections 21.4 and 21.5). All the hydroxyl groups react, including the anomeric one. For example, β-D-glucopyranose is converted into its pentaacetate by treatment with acetic anhydride in pyridine solution.

β-D-Glucopyranose

Penta-O-acetyl-β-D-glucopyranose
(91%)

Carbohydrates are converted into ethers by treatment with an alkyl halide in the presence of base—the Williamson ether synthesis (Section 18.3). Standard Williamson conditions using a strong base tend to degrade sensitive sugar molecules, but silver oxide works well and gives high yields of ethers. For example, α-D-glucopyranose is converted into its pentamethyl ether in 85% yield on reaction with iodomethane and Ag_2O.

α-D-Glucopyranose　　　　**α-D-Glucopyranose pentamethyl ether (85%)**

- -

Problem 25.14　Draw the products you would obtain by reaction of β-D-ribofuranose with:
(a) CH_3I, Ag_2O　　(b) $(CH_3CO)_2O$, pyridine

β-D-Ribofuranose

- -

Glycoside Formation

We saw in Section 19.11 that treatment of a hemiacetal with an alcohol and an acid catalyst yields an acetal:

$$OH \qquad\qquad OR$$

C—OR + ROH $\underset{}{\overset{HCl}{\rightleftarrows}}$ C—OR + H_2O

Biological Connection ▶

In the same way, treatment of a monosaccharide hemiacetal with an alcohol and an acid catalyst yields an acetal in which the anomeric –OH has been replaced by an –OR group. For example, reaction of β-D-glucopyranose with methanol gives a mixture of α and β methyl D-glucopyranosides:

β-D-Glucopyranose (a cyclic hemiacetal)　　　　**Methyl α-D-glucopyranoside (66%)**　　+　　**Methyl β-D-glucopyranoside (33%)**

Called **glycosides**, carbohydrate acetals are named by first citing the alkyl group and replacing the *-ose* ending of the sugar with *-oside*. Like all acetals, glycosides are stable to neutral water. They aren't in equilibrium with an open-chain form, and they don't show mutarotation. They can, however, be converted back to the free monosaccharide by hydrolysis with aqueous acid.

Glycosides are widespread in nature, and many biologically important molecules contain glycosidic linkages. For example, digitoxin, the active component of the digitalis preparations used for treatment of heart disease, is a glycoside consisting of a complex steroid alcohol linked to a trisaccharide. Note also that the three sugars are linked to each other by glycosidic bonds.

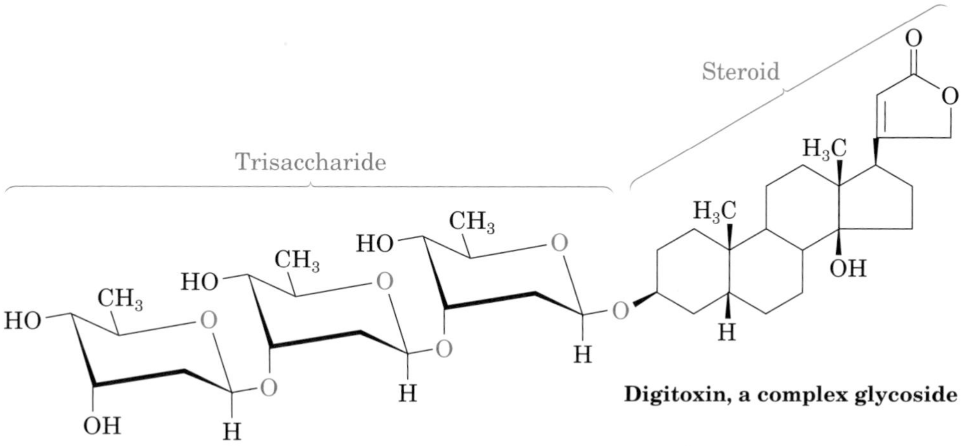

Digitoxin, a complex glycoside

The laboratory synthesis of glycosides is often difficult, but one method that is particularly suitable for preparing glucose β-glycosides involves treatment of glucose pentaacetate with HBr, followed by addition of the appropriate alcohol in the presence of silver oxide. Called the **Koenigs–Knorr reaction**, the sequence involves formation of a pyranosyl bromide, followed by nucleophilic substitution. For example, methylarbutin, a glycoside found in pears, has been prepared by reaction of tetraacetyl-α-D-glucopyranosyl bromide with *p*-methoxyphenol:

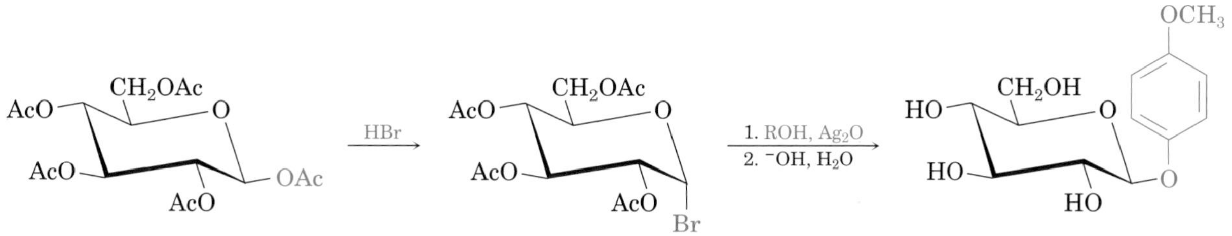

Pentaacetyl-β-D-glucopyranose   **Tetraacetyl-α-D-glucopyranosyl bromide**   **Methylarbutin**

Although the Koenigs–Knorr reaction appears to involve a simple backside S$_N$2 displacement of bromide ion by alkoxide ion, the situation is actually more complex. Both α and β anomers of tetraacetyl-D-glucopyranosyl bromide give the *same* β-glycoside product, implying that both anomers react by a common pathway.

The results can be understood by assuming that tetraacetyl-D-glucopyranosyl bromide (either α or β anomer) undergoes a spontaneous loss of Br⁻, followed by internal reaction with the ester group at C2 to form an oxonium ion. Since the acetate at C2 is on the bottom of the glucose ring, the new carbon–oxygen bond also forms from the bottom. An S_N2 displacement of the oxonium ion by back-side attack at C1 then occurs with the usual inversion of configuration, yielding a β-glycoside and regenerating the acetate at C2 (Figure 25.6).

FIGURE 25.6 ▼

Mechanism of the Koenigs–Knorr reaction, showing the neighboring-group effect of a nearby acetate.

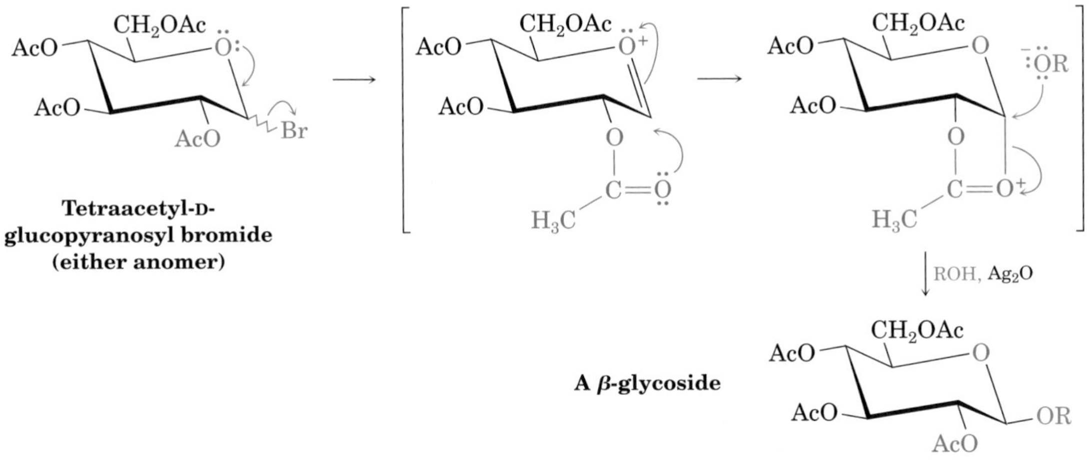

Tetraacetyl-D-glucopyranosyl bromide (either anomer)

ROH, Ag_2O

A β-glycoside

The participation shown by the nearby acetate in the Koenigs–Knorr reaction is referred to as a *neighboring-group effect* and is a common occurrence in organic chemistry. Neighboring-group effects are usually noticeable only because they affect the rate or stereochemistry of a reaction; the nearby group itself does not undergo any evident change during the reaction.

Reduction of Monosaccharides

Biological Connection ▶

Treatment of a monosaccharide with $NaBH_4$ reduces it to a polyalcohol called an **alditol**. The reduction occurs by interception of the open-chain form present in the aldehyde/ketone–hemiacetal equilibrium. Although only a small amount of the open-chain form is present at any given time, that small amount is reduced; then more is produced by opening of the pyranose form and that additional amount is reduced; and so on, until the entire sample has undergone reaction.

β-D-Glucopyranose   D-Glucose   D-Glucitol (D-Sorbitol), an alditol

D-Glucitol, the alditol produced by reduction of D-glucose, is itself a naturally occurring substance present in many fruits and berries. It is used under its alternative name D-sorbitol as an artificial sweetener and sugar substitute in foods.

Problem 25.15 How can you account for the fact that reduction of D-galactose with $NaBH_4$ yields an alditol that is optically inactive?

Problem 25.16 Reduction of L-gulose with $NaBH_4$ leads to the same alditol (D-glucitol) as reduction of D-glucose. Explain.

Oxidation of Monosaccharides

Like other aldehydes, aldoses are easily oxidized to yield the corresponding monocarboxylic acids, called **aldonic acids**. Aldoses react with Tollens' reagent (Ag^+ in aqueous NH_3), Fehling's reagent (Cu^{2+} in aqueous sodium tartrate), or Benedict's reagent (Cu^{2+} in aqueous sodium citrate) to yield the oxidized sugar and a reduced metallic species. All three reactions serve as simple chemical tests for **reducing sugars**—*reducing* because the sugar reduces the oxidizing reagent.

If Tollens' reagent is used, metallic silver is produced as a shiny mirror on the walls of the reaction flask or test tube. If Fehling's or Benedict's reagent is used, a reddish precipitate of Cu_2O signals a positive result. Some diabetes self-test kits sold for home use still employ the Benedict test, although more modern methods have largely replaced the chemical test. As little as 0.1% glucose in urine gives a positive test.

All aldoses are reducing sugars because they contain an aldehyde carbonyl group, but some ketoses are reducing sugars as well. Fructose reduces Tollens' reagent, for example, even though it contains no aldehyde group. Reduction occurs because fructose is readily isomerized to an aldose in basic solution by a series of keto–enol tautomeric shifts (Figure 25.7). Glycosides, however, are nonreducing. They don't react with Tollens' reagent because the acetal group is not hydrolyzed to an aldehyde under basic conditions.

FIGURE 25.7 ▼

Fructose is a reducing sugar because it undergoes two base-catalyzed keto–enol tautomerizations that result in conversion to an aldohexose. (The wavy bonds indicate unknown stereochemistry.)

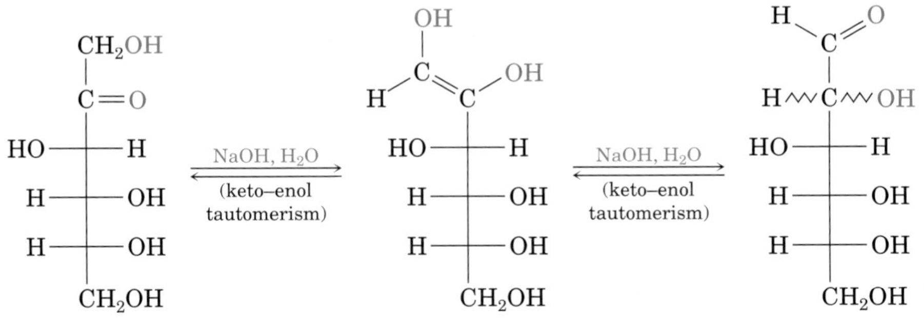

D-Fructose **An enediol** **An aldohexose**

Although the Tollens and Fehling reactions serve as useful tests for reducing sugars, they don't give good yields of aldonic acid products because the alkaline conditions cause decomposition of the carbohydrate. For preparative purposes, a buffered solution of aqueous Br_2 is a better oxidant. The reaction is specific for aldoses; ketoses are not oxidized by aqueous Br_2.

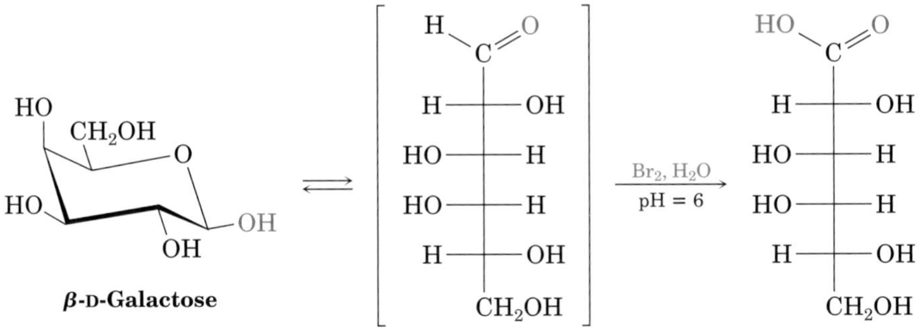

β-D-Galactose

D-Galactonic acid (an aldonic acid)

If a more powerful oxidizing agent such as warm dilute HNO_3 is used, aldoses are oxidized to dicarboxylic acids, called **aldaric acids**. Both the –CHO group at C1 and the terminal $–CH_2OH$ group are oxidized in this reaction.

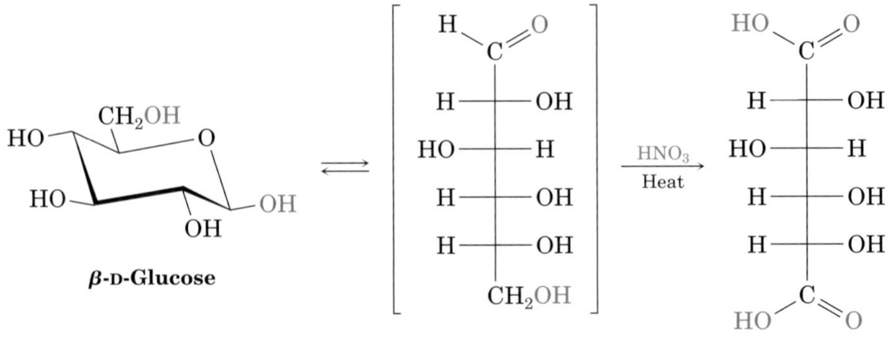

β-D-Glucose

D-Glucaric acid (an aldaric acid)

Problem 25.17 D-Glucose yields an optically active aldaric acid on treatment with HNO_3, but D-allose yields an optically inactive aldaric acid. Explain.

Problem 25.18 Which of the other six D aldohexoses yield optically active aldaric acids on oxidation, and which yield meso aldaric acids? (See Problem 25.17.)

Chain Lengthening: The Kiliani–Fischer Synthesis

Much early activity in carbohydrate chemistry was devoted to unraveling the stereochemical relationships among monosaccharides. One of the most important methods used was the **Kiliani–Fischer synthesis**, which results in the lengthening of an aldose chain by one carbon atom. The C1 aldehyde group of the starting sugar becomes C2 of the chain-lengthened sugar, and a new C1 carbon is added. For example, an aldo*pent*ose is converted by the Kiliani–Fischer synthesis into an aldo*hex*ose.

Discovery of the chain-lengthening sequence was initiated by the observation of Heinrich Kiliani in 1886 that aldoses react with HCN to form cyanohydrins (Section 19.7). Emil Fischer immediately realized the importance of Kiliani's discovery and devised a method for converting the cyanohydrin nitrile group into an aldehyde.

An aldose   A cyanohydrin   A chain-lengthened aldose

Fischer's original method for conversion of the nitrile into an aldehyde involved hydrolysis to a carboxylic acid, ring closure to a cyclic ester (lactone), and subsequent reduction. A modern improvement is to reduce the nitrile over a palladium catalyst, yielding an imine intermediate that is hydrolyzed. Note that the cyanohydrin is formed as a mixture of stereoisomers at the new chirality center. Thus, *two* new aldoses, differing only in their stereochemistry at C2, result from Kiliani–Fischer synthesis. Chain extension of D-arabinose, for example, yields a mixture of D-glucose and D-mannose (Figure 25.8, p. 1050).

Problem 25.19 What product(s) would you expect from Kiliani–Fischer reaction of D-ribose?

Problem 25.20 What aldopentose would give a mixture of L-gulose and L-idose on Kiliani–Fischer chain extension?

FIGURE 25.8 ▼

Kiliani–Fischer chain lengthening of D-arabinose leads to a mixture of D-glucose and D-mannose.

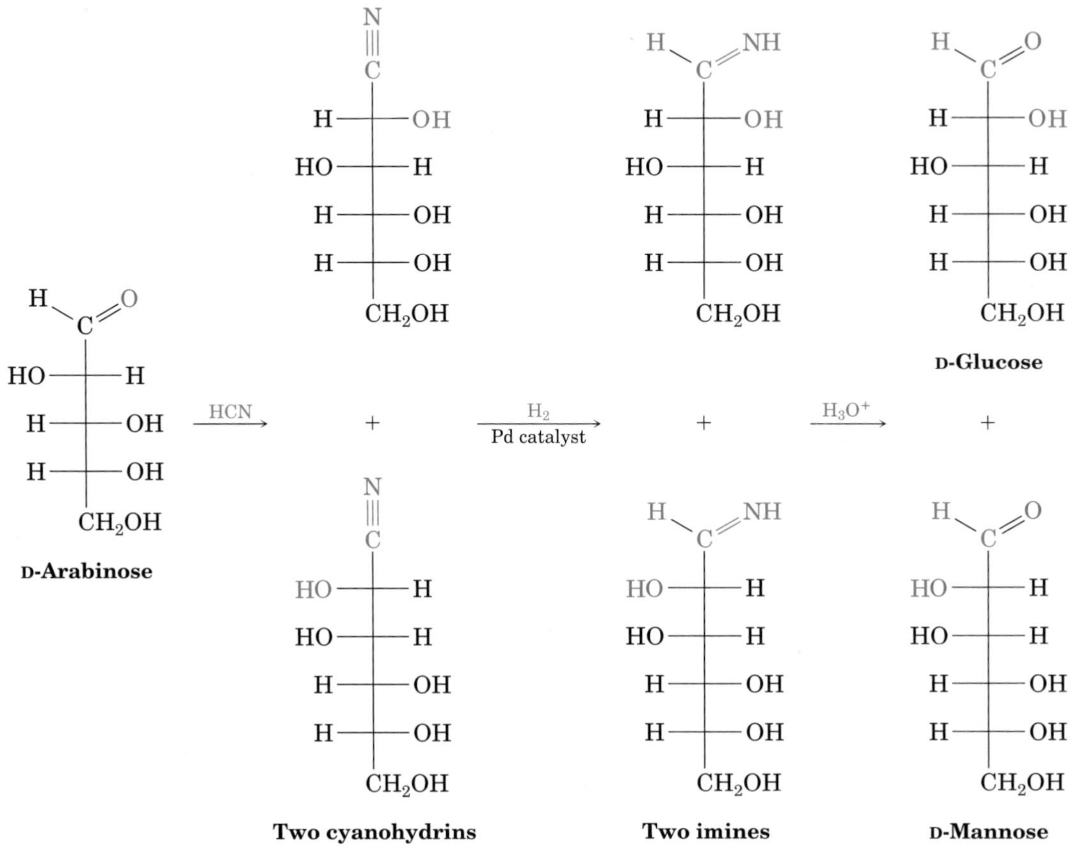

Chain Shortening: The Wohl Degradation

Alfred Wohl

Alfred Wohl (1863–1933) was born in Graudenz, West Prussia, now part of Poland. He received his Ph.D. at the University of Berlin in 1886 with August von Hofmann and became Professor of Chemistry at the Technical University of Danzig.

Just as the Kiliani–Fischer synthesis lengthens an aldose chain by one carbon, the **Wohl degradation** shortens an aldose chain by one carbon. The Wohl degradation is almost exactly the opposite of the Kiliani–Fischer sequence: The aldose aldehyde carbonyl group is first converted into a nitrile, and the resulting cyanohydrin loses HCN under basic conditions—the reverse of a nucleophilic addition reaction.

Conversion of the aldehyde into a nitrile is accomplished by treatment of an aldose with hydroxylamine to give an oxime (Section 19.9), followed by dehydration of the oxime with acetic anhydride. The Wohl degradation does not give particularly high yields of chain-shortened aldoses, but the reaction is general for all aldopentoses and aldohexoses. For example, D-galactose is converted by Wohl degradation into D-lyxose:

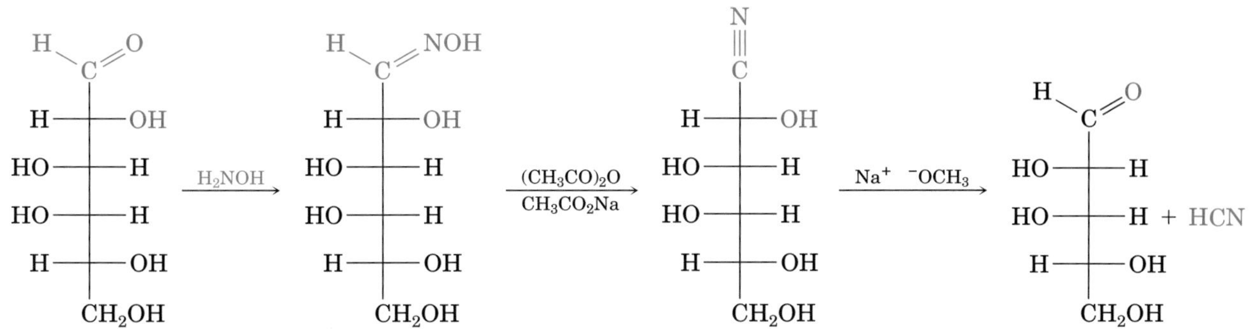

The reaction scheme (left to right):

D-Galactose

H—C=O
H—OH
HO—H
HO—H
H—OH
CH₂OH

$\xrightarrow{H_2NOH}$

D-Galactose oxime

H—C=NOH
H—OH
HO—H
HO—H
H—OH
CH₂OH

$\xrightarrow{(CH_3CO)_2O \atop CH_3CO_2Na}$

A cyanohydrin

N
|||
C
H—OH
HO—H
HO—H
H—OH
CH₂OH

$\xrightarrow{Na^+ \ ^-OCH_3}$

D-Lyxose (37%)

H—C=O
HO—H
HO—H
H—OH
CH₂OH + HCN

● ●

Problem 25.21   What two D aldopentoses yield D-threose on Wohl degradation?

● ●

25.8   Stereochemistry of Glucose: The Fischer Proof

In the late 1800s, the stereochemical theories of van't Hoff and Le Bel on the tetrahedral geometry of carbon were barely a decade old, modern methods of product purification were unknown, and modern spectroscopic techniques of structure determination were undreamed of. Despite these obstacles, Emil Fischer published in 1891 what remains today perhaps the finest use of chemical logic ever recorded—a structure proof of the stereochemistry of naturally occurring (+)-glucose. Let's follow Fischer's logic and see how he arrived at his conclusions.

1. **(+)-Glucose is an aldohexose.** (+)-Glucose has four chirality centers and can therefore be any one of $2^4 = 16$ possible stereoisomers. Since no method was available at the time for determining the absolute three-dimensional stereochemistry of a molecule, Fischer decided to simplify matters by considering only the eight enantiomers having the C5 hydroxyl group on the right in Fischer projections—what we now call D sugars. Fischer knew that this arbitrary choice of D-series stereochemistry had only a 50:50 chance of being right, but it was finally shown in 1953 by X-ray spectroscopy that the choice was indeed correct.

   The four D aldopentoses and the eight D aldohexoses derived from them by Kiliani–Fischer synthesis are shown in Figure 25.9 (p. 1052). One of the eight aldohexoses is glucose, but which one?

2. **Arabinose, an aldopentose, is converted by Kiliani–Fischer chain extension into a mixture of glucose and mannose.** This means that glucose and mannose have the same stereochemistry at C3, C4, and C5, and differ only at C2. Glucose and mannose are therefore represented by one of the pairs of structures 1 and 2, 3 and 4, 5 and 6, or 7 and 8 in Figure 25.9.

FIGURE 25.9 ▼

The four D aldopentoses and the eight D aldohexoses derived from them by
Kiliani–Fischer chain extension.

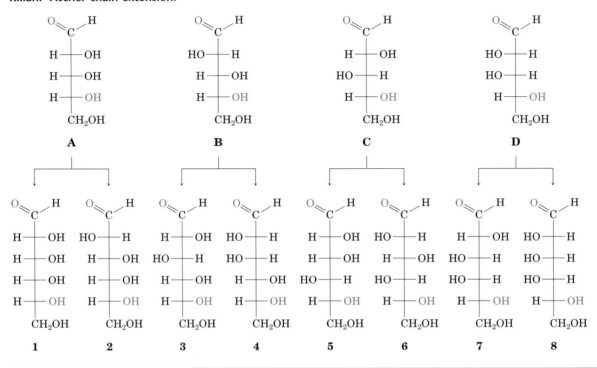

3. **Arabinose is oxidized by warm HNO₃ to an optically active aldaric acid.** Of the four aldopentoses (A, B, C, and D in Figure 25.9), A and C give optically inactive meso aldaric acids when oxidized, whereas B and D give optically active products. Thus, arabinose must be either B or D, and mannose and glucose must be either 3 and 4 or 7 and 8 (Figure 25.10).

4. **Both glucose and mannose are oxidized by warm HNO₃ to optically active aldaric acids.** Of the possibilities left at this point, the pair represented by structures 3 and 4 would both be oxidized to optically active aldaric acids, but the pair represented by 7 and 8 would not *both* give optically active products. Compound 7 would give an optically inactive, meso aldaric acid (Figure 25.11). Thus, glucose and mannose must be 3 and 4, though we can't yet tell which is which.

5. **One of the other 15 aldohexose stereoisomers gives the same aldaric acid as that derived from glucose on oxidation.** How can two aldohexoses give the same aldaric acid? Since aldaric acids have –COOH groups at both ends of the carbon chain, there is no way to tell which was originally the –CHO end and which was the –CH₂OH end. Thus, a given aldaric acid has two different precursors. The aldaric acid from compound 3, for example, might also come from oxidation of a

FIGURE 25.10 ▼

Oxidation of aldopentoses to aldaric acids. Only structures B and D lead to optically active products.

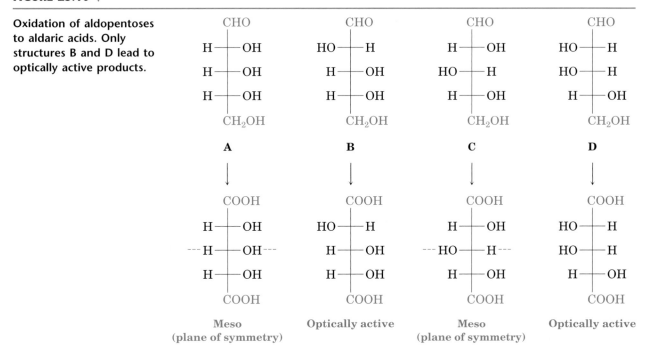

FIGURE 25.11 ▼

Oxidation of aldohexoses to aldaric acids. Only the pair of structures 3 and 4 both give optically active products.

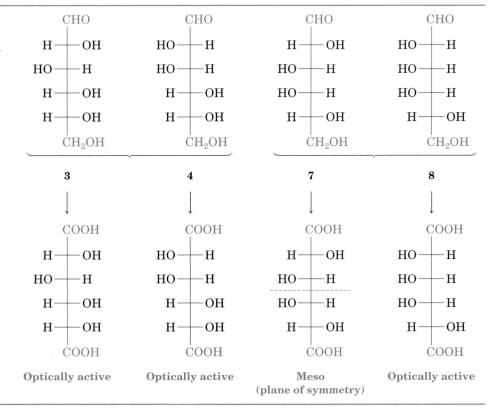

second aldohexose, and the aldaric acid from compound 4 might come from oxidation of a second aldohexose (Figure 25.12).

If we look carefully at the aldaric acids derived from compounds 3 and 4, we find that the aldaric acid derived from 3 could also come from oxidation of another aldohexose (L-gulose), but that the aldaric acid derived from 4 could not. The "other" aldohexose that could produce the same aldaric acid as that from compound 4 is in fact identical to 4. Thus, glucose must have structure 3 and mannose must have structure 4 (Figure 25.12).

FIGURE 25.12 ▼

There is another aldohexose (L-gulose) that can produce the same aldaric acid as that from compound 3, but there is no other aldohexose that can produce the same aldaric acid as that from compound 4. Thus, glucose has structure 3 and mannose has structure 4.

Rotate 180°

```
    CHO                      COOH                   CH2OH                  CHO
 H──┼──OH                 H──┼──OH               H──┼──OH           HO──┼──H
HO──┼──H      HNO3       HO──┼──H      HNO3     HO──┼──H           HO──┼──H
 H──┼──OH     ───→        H──┼──OH     ←───      H──┼──OH     =     H──┼──OH
 H──┼──OH                 H──┼──OH               H──┼──OH          HO──┼──H
   CH2OH                    COOH                    CHO                CH2OH

 3 (D-Glucose)          Glucaric acid                  (L-Gulose)
```

Rotate 180°

```
    CHO                      COOH                   CH2OH                  CHO
HO──┼──H                 HO──┼──H               HO──┼──H           HO──┼──H
HO──┼──H      HNO3       HO──┼──H      HNO3     HO──┼──H           HO──┼──H
 H──┼──OH     ───→        H──┼──OH     ←───      H──┼──OH     =     H──┼──OH
 H──┼──OH                 H──┼──OH               H──┼──OH           H──┼──OH
   CH2OH                    COOH                    CHO                CH2OH

 4 (D-Mannose)         Mannaric acid                 (Also D-mannose)
```

Further reasoning allowed Fischer to determine the stereochemistry of 12 of the 16 aldohexoses. For this remarkable achievement, he was awarded the 1902 Nobel Prize in chemistry.

Problem 25.22 The structures of the four aldopentoses, A, B, C, and D, are shown in Figure 25.9. In light of point 2 presented by Fischer, what is the structure of arabinose? In light of point 3, what is the structure of lyxose, another aldopentose that yields an optically active aldaric acid?

Problem 25.23 The aldotetrose D-erythrose yields a mixture of D-ribose and D-arabinose on Kiliani–
Fischer chain extension.
(a) What is the structure of D-ribose?
(b) What is the structure of D-xylose, the fourth possible aldopentose?
(c) What is the structure of D-erythrose?
(d) What is the structure of D-threose, the other possible aldotetrose?

25.9 Disaccharides

We saw in Section 25.7 that reaction of a monosaccharide with an alcohol
yields a glycoside in which the anomeric –OH group is replaced by an –OR
substituent. If the alcohol is itself a sugar, the glycosidic product is a disaccharide.

Cellobiose and Maltose

Disaccharides contain a glycosidic acetal bond between the anomeric carbon (the carbonyl carbon) of one sugar and an –OH group at *any* position
on the other sugar. A glycosidic bond between C1 of the first sugar and the
–OH at C4 of the second sugar is particularly common. Such a bond is called
a **1,4′ link** (read as "one, four-prime"). The prime superscript indicates that
the 4′ position is on a different sugar than the 1 position.

A glycosidic bond to the anomeric carbon can be either α or β. Maltose,
the disaccharide obtained by enzyme-catalyzed hydrolysis of starch, consists of two D-glucopyranose units joined by a 1,4′-α-glycoside bond. Cellobiose, the disaccharide obtained by partial hydrolysis of cellulose, consists
of two D-glucopyranose units joined by a 1,4′-β-glycoside bond.

Stereo View

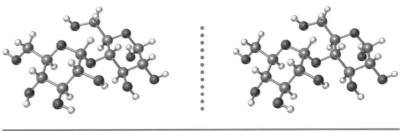

Maltose, a 1,4′-α-glycoside
[4-*O*-(α-D-Glucopyranosyl)-α-D-glucopyranose]

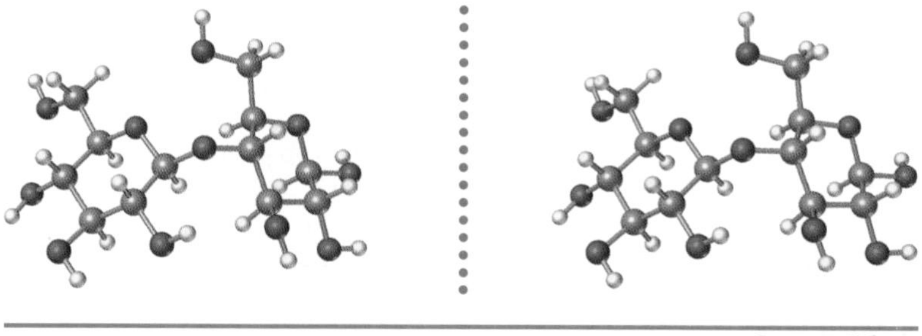

Stereo View

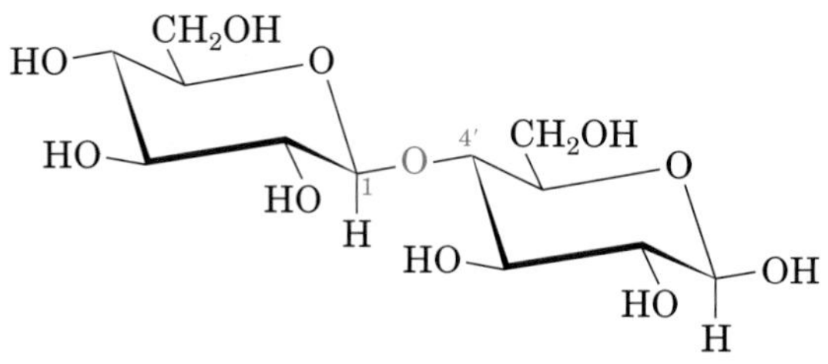

Cellobiose, a 1,4'-β-glycoside
[4-*O*-(β-D-Glucopyranosyl)-β-D-glucopyranose]

Maltose and cellobiose are both reducing sugars because the anomeric carbons on the right-hand glucopyranose units have hemiacetal groups. Both are therefore in equilibrium with aldehyde forms, which can reduce Tollens' or Fehling's reagent. For a similar reason, both maltose and cellobiose exhibit mutarotation of α and β anomers of the glucopyranose unit on the right (Figure 25.13).

FIGURE 25.13 ▼

Mutarotation of maltose and cellobiose.

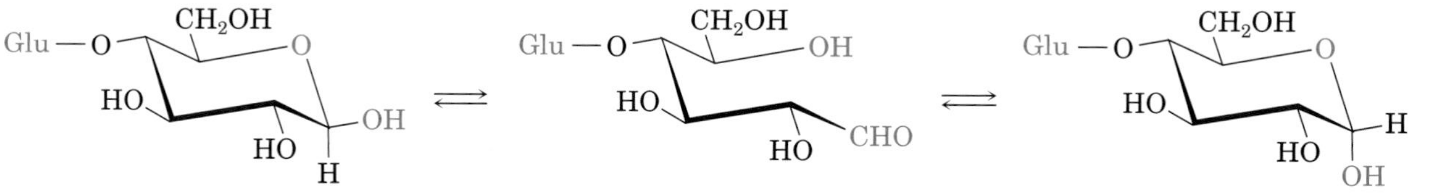

Maltose or cellobiose
(β anomers)

Maltose or cellobiose
(aldehydes)

Maltose or cellobiose
(α anomers)

Despite the similarities of their structures, cellobiose and maltose have dramatically different biological properties. Cellobiose can't be digested by humans and can't be fermented by yeast. Maltose, however, is digested without difficulty and is fermented readily.

Problem 25.24   Show the product you would obtain from the reaction of cellobiose with the following reagents:
(a) NaBH₄   (b) Br₂, H₂O   (c) CH₃COCl, pyridine

Lactose

Lactose is a disaccharide that occurs naturally in both human and cow's milk. It is widely used in baking and in commercial milk formulas for infants. Like cellobiose and maltose, lactose is a reducing sugar. It exhibits mutarotation and is a 1,4'-β-linked glycoside. Unlike cellobiose and maltose, however, lactose contains two *different* monosaccharide units—D-glucose and D-galactose—joined by a β-glycosidic bond between C1 of galactose and C4 of glucose.

Stereo View

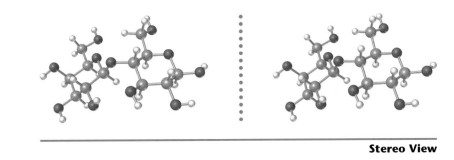

Lactose, a 1,4'-β-glycoside
[4-*O*-(β-D-Galactopyranosyl)-β-D-glucopyranose]

Sucrose

Sucrose, or ordinary table sugar, is among the most abundant pure organic chemicals in the world and is the one most widely known to nonchemists. Whether from sugar cane (20% by weight) or sugar beets (15% by weight), and whether raw or refined, all table sugar is sucrose.

Sucrose is a disaccharide that yields 1 equivalent of glucose and 1 equivalent of fructose on hydrolysis. This 1:1 mixture of glucose and fructose is often referred to as *invert sugar* because the sign of optical rotation changes (inverts) during the hydrolysis from sucrose ($[\alpha]_D = +66.5°$) to a glucose/fructose mixture ($[\alpha]_D \approx -22.0°$). Insects such as honeybees have enzymes called *invertases* that catalyze the hydrolysis of sucrose to a glucose + fructose mixture. Honey, in fact, is primarily a mixture of glucose, fructose, and sucrose.

Unlike most other disaccharides, sucrose is not a reducing sugar and does not exhibit mutarotation. These observations imply that sucrose is not a hemiacetal and suggest that glucose and fructose must *both* be glycosides. This can happen only if the two sugars are joined by a glycoside link between the anomeric carbons of both sugars—C1 of glucose and C2 of fructose.

Stereo View

Sucrose, a 1,2′-glycoside
[2-_O_-(α-D-Glucopyranosyl)-β-D-fructofuranoside]

25.10 Polysaccharides and Their Synthesis

Polysaccharides are carbohydrates in which tens, hundreds, or even thousands of simple sugars are linked together through glycoside bonds. Since they have only the one free anomeric –OH group at the end of a very long chain, polysaccharides aren't reducing sugars and don't show noticeable mutarotation. Cellulose and starch are the two most widely occurring polysaccharides.

Cellulose

Cellulose consists of several thousand D-glucose units linked by 1,4′-β-glycoside bonds like those in cellobiose. Different cellulose molecules then interact to form a large aggregate structure held together by hydrogen bonds.

Cellulose, a 1,4′-_O_-(β-D-glucopyranoside) polymer

Nature uses cellulose primarily as a structural material to impart strength and rigidity to plants. Leaves, grasses, and cotton are primarily cellulose. Cellulose also serves as raw material for the manufacture of cellulose acetate, known commercially as acetate rayon, and cellulose nitrate, known as guncotton. Guncotton is the major ingredient in smokeless powder, the explosive propellant used in artillery shells and in ammunition for firearms.

Starch and Glycogen

Potatoes, corn, and cereal grains contain large amounts of *starch*, a polymer of glucose in which the monosaccharide units are linked by 1,4'-α-glycoside bonds like those in maltose. Starch can be separated into two fractions: *amylose*, which is insoluble in cold water, and *amylopectin*, which *is* soluble in cold water. Amylose accounts for about 20% by weight of starch and consists of several hundred glucose molecules linked together by 1,4'-α-glycoside bonds.

Amylose, a 1,4'-*O*-(α-D-glucopyranoside) polymer

Amylopectin accounts for the remaining 80% of starch and is more complex in structure than amylose. Unlike cellulose and amylose, which are linear polymers, amylopectin contains 1,6'-α-glycoside *branches* approximately every 25 glucose units.

Amylopectin

Starch is digested in the mouth and stomach by *glycosidase* enzymes, which catalyze the hydrolysis of glycoside bonds and release individual molecules of glucose. Like most enzymes, glycosidases are highly selective in their action. They hydrolyze only the α-glycoside links in starch and leave the β-glycoside links in cellulose untouched. Thus, humans can eat potatoes and grains but not grass and leaves.

Glycogen is a polysaccharide that serves the same energy storage function in animals that starch serves in plants. Dietary carbohydrates not needed for immediate energy are converted by the body to glycogen for long-term storage. Like the amylopectin found in starch, glycogen contains a complex branching structure with both 1,4' and 1,6' links (Figure 25.14). Glycogen molecules are larger than those of amylopectin—up to 100,000 glucose units—and contain even more branches.

FIGURE 25.14 ▼

A representation of the structure of glycogen. The hexagons represent glucose units linked by 1,4' and 1,6' acetal bonds.

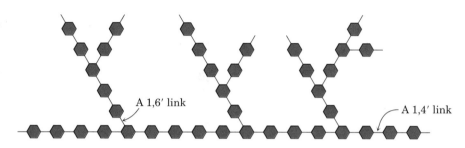

Polysaccharide Synthesis

With numerous –OH groups of similar reactivity, polysaccharides are so structurally complex that their laboratory synthesis has been a particularly difficult problem. Several methods are now under development, however, that appear poised to revolutionize the field. Among the most promising of these new approaches is the *glycal assembly method*.

Easily prepared from the appropriate monosaccharide, a *glycal* is an unsaturated sugar with a C1–C2 double bond. To ready it for use in polysaccharide synthesis, the glycal is first protected at its primary –OH group by formation of a silyl ether (Section 17.9) and at its two adjacent secondary –OH groups by formation of a cyclic carbonate. Then, the protected glycal is epoxidized.

A glycal   A protected glycal   An epoxide

Treatment of the glycal epoxide in the presence of $ZnCl_2$ with a *second* glycal having a free –OH group causes acid-catalyzed opening of the epoxide ring by back-side attack (Section 18.8) and yields a disaccharide. The disaccharide is itself a glycal, so it can be epoxidized and coupled again to yield a trisaccharide, and so on. Using the appropriate sugars at each step, a great variety of polysaccharides can, in principle, be prepared.

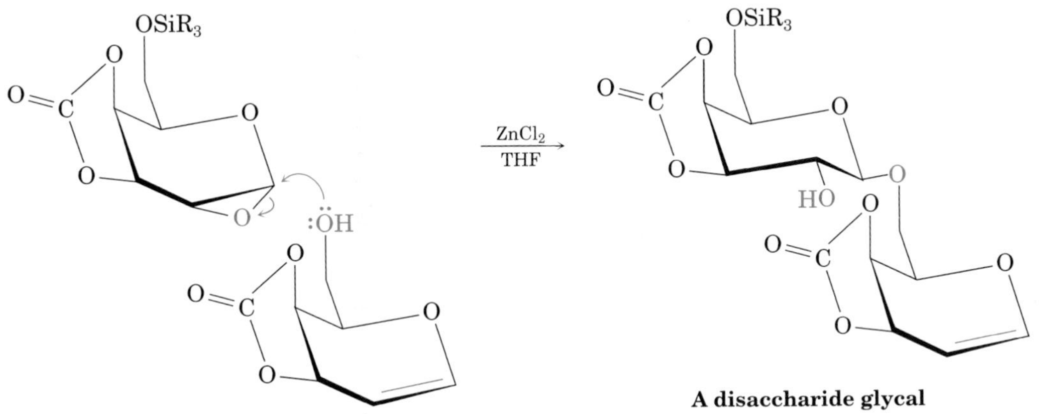

A disaccharide glycal

25.11 Other Important Carbohydrates

In addition to the common carbohydrates mentioned in previous sections, there are a variety of important carbohydrate-derived materials. Their structural resemblance to sugars is clear, but they aren't simple aldoses or ketoses.

Deoxy sugars have an oxygen atom "missing." That is, an –OH group is replaced by an –H. The most common deoxy sugar is 2-deoxyribose, a sugar found in DNA (deoxyribonucleic acid). Note that 2-deoxyribose adopts a furanose (five-membered) form.

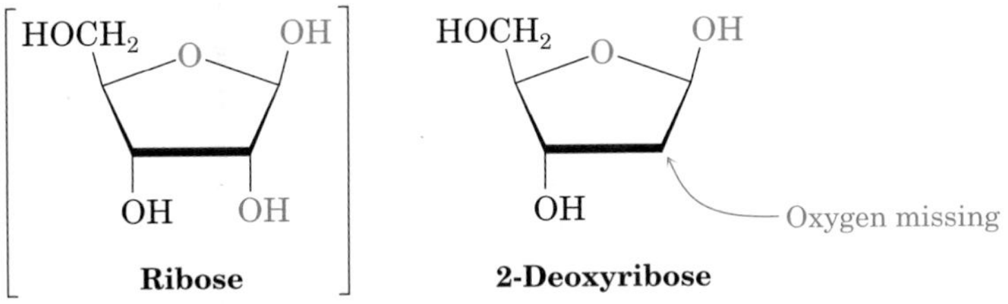

Ribose   **2-Deoxyribose**   Oxygen missing

Amino sugars, such as D-glucosamine, have an –OH group replaced by an –NH$_2$. The *N*-acetyl amide derived from D-glucosamine is the monosaccharide unit from which *chitin*, the hard crust that protects insects and shellfish, is built. Still other amino sugars are found in antibiotics such as streptomycin and gentamicin.

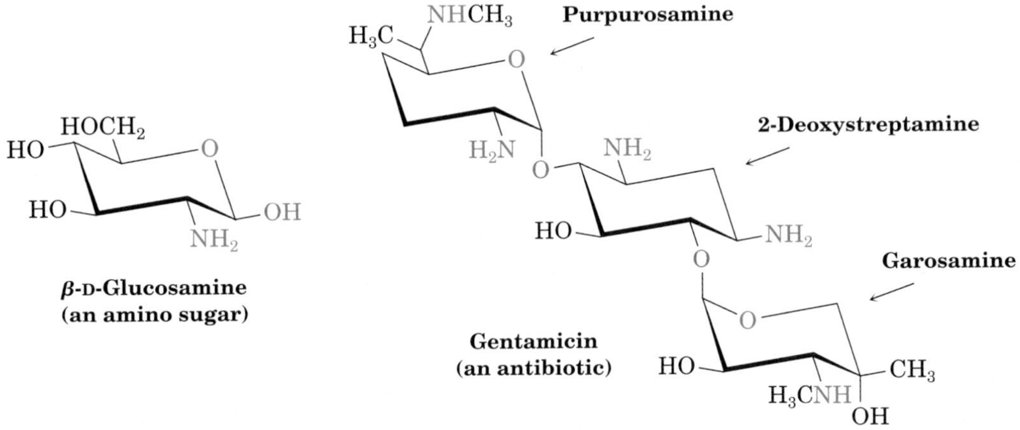

β-D-Glucosamine
(an amino sugar)

Gentamicin
(an antibiotic)

Purpurosamine

2-Deoxystreptamine

Garosamine

25.12 Cell-Surface Carbohydrates and Carbohydrate Vaccines

It was once thought that the only biological roles of carbohydrates were as structural materials and energy sources. Although carbohydrates do indeed serve these two purposes, they also have many other important biochemical functions. For example, polysaccharides are centrally involved in cell recognition, the critical process by which one type of cell distinguishes another. Small polysaccharide chains, covalently bound by glycosidic links to hydroxyl groups on proteins (*glycoproteins*), act as biochemical markers on cell surfaces, as illustrated by the human blood-group antigens.

It has been known for over a century that human blood can be classified into four blood-group types (A, B, AB, and O), and that blood from a donor of one type can't be transfused into a recipient with another type unless the two types are compatible (Table 25.1). Should an incompatible mix be made, the red blood cells clump together, or *agglutinate*.

TABLE 25.1 Human Blood-Group Compatibilities

Donor blood type	Acceptor blood type			
	A	**B**	**AB**	**O**
A	o	x	o	x
B	x	o	o	x
AB	x	x	o	x
O	o	o	o	o

o = Compatible; x = Incompatible.

The agglutination of incompatible red blood cells, which indicates that the body's immune system has recognized the presence of foreign cells in the body and has formed antibodies against them, results from the presence of polysaccharide markers on the surface of the cells. Types A, B, and O red blood cells each have characteristic markers, called *antigenic determinants*; type AB cells have both type A and type B markers. The structures of all three blood-group determinants are shown in Figure 25.15.

Note that some unusual carbohydrates are involved. All three blood-group antigenic determinants contain *N*-acetylamino sugars as well as the unusual monosaccharide L-fucose.

β-D-*N*-Acetylglucosamine
(D-2-Acetamino-2-deoxyglucose)

β-D-*N*-Acetylgalactosamine
(D-2-Acetamino-2-deoxygalactose)

α-L-Fucose
(L-6-Deoxygalactose)

FIGURE 25.15 ▼

Structures of the A, B, and O blood-group antigenic determinants.

Blood group A

| L-Fucose | 1,2′ link | D-Galactose | 1,4′ link | N-Acetyl-D-glucosamine | — Protein |

1,3′ link

N-Acetyl-D-galactosamine

Blood group B

| L-Fucose | 1,2′ link | D-Galactose | 1,4′ link | N-Acetyl-D-glucosamine | — Protein |

1,3′ link

D-Galactose

Blood group O

| L-Fucose | 1,2′ link | D-Galactose | 1,4′ link | N-Acetyl-D-glucosamine | — Protein |

Elucidation of the role of carbohydrates in cell recognition is a vigorous area of current research that offers hope of breakthroughs in the understanding of a wide range of diseases from bacterial infections to cancer. Particularly exciting is the possibility of developing useful anticancer vaccines to help mobilize the body's immune system against tumor cells. Recent advances along these lines have included a laboratory synthesis of the so-called *globo H antigen*, found on the surface of human breast, prostate, colon, and pancreatic cancer cells. Mice treated with the synthetic globo H hexasaccharide linked to a carrier protein developed large amounts of antibodies, which then recognized tumor cells.

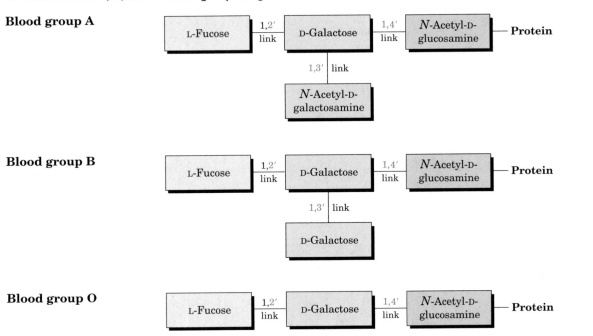

Globo H antigen

CHEMISTRY @ WORK

Sweetness

Say the word *sugar* and most people immediately think of sweet-tasting candies, desserts, and such. In fact, most simple carbohydrates *do* taste sweet, but the degree of sweetness varies greatly from one sugar to another. With sucrose (table sugar) as a reference point, fructose is nearly twice as sweet, but lactose is only about one-sixth as sweet. Comparisons are difficult, though, because perceived sweetness varies depending on the concentration of the solution being tasted. Nevertheless, the ordering in Table 25.2 is generally accepted.

TABLE 25.2   Sweetness of Some Sugars and Sugar Substitutes

Name	Type	Sweetness
Lactose	Disaccharide	0.16
Glucose	Monosaccharide	0.75
Sucrose	Disaccharide	1.00
Fructose	Monosaccharide	1.75
Aspartame	Synthetic	180
Acesulfame-K	Synthetic	200
Saccharin	Synthetic	350

The desire of many people to cut their caloric intake has led to the development of synthetic sweeteners such as saccharin, aspartame, and acesulfame. All are far sweeter than natural sugars, so the choice of one or another depends on personal taste, government regulations, and (for baked goods) heat stability. Saccharin, the oldest synthetic sweetener, has been used for over a century, although it has a somewhat metallic aftertaste. Doubts about its safety and potential carcinogenicity were raised in the early 1970s, but it has now been cleared of suspicion. Acesulfame potassium, the most recently approved sweetener, is proving to be extremely popular in soft drinks because it has little aftertaste. None of the three synthetic sweeteners listed in Table 25.2 has any structural resemblance to a carbohydrate.

The real thing comes from cane fields like this one.

(continued) ▶

Saccharin   **Aspartame**   **Acesulfame potassium**

Summary and Key Words

Carbohydrates are polyhydroxy aldehydes and ketones. They are classified according to the number of carbon atoms and the kind of carbonyl group they contain. Glucose, for example, is an aldohexose, a six-carbon aldehydic sugar. **Monosaccharides** are further classified as either D or L **sugars**, depending on the stereochemistry of the chirality center farthest from the carbonyl group.

Monosaccharides normally exist as cyclic hemiacetals rather than as open-chain aldehydes or ketones. The hemiacetal linkage results from reaction of the carbonyl group with an –OH group three or four carbon atoms away. A five-membered cyclic hemiacetal is called a **furanose**, and a six-membered cyclic hemiacetal is called a **pyranose**. Cyclization leads to the formation of a new chirality center and production of two diastereomeric hemiacetals, called α and β **anomers**.

Much of the chemistry of monosaccharides is the familiar chemistry of alcohols and aldehydes/ketones. Thus, the hydroxyl groups of carbohydrates form esters and ethers. The carbonyl group of a monosaccharide can be reduced with $NaBH_4$ to form an **alditol**, oxidized with aqueous Br_2 to form an **aldonic acid**, oxidized with HNO_3 to form an **aldaric acid**, or treated with an alcohol in the presence of acid to form a **glycoside**. Monosaccharides can also be chain-lengthened by the multistep **Kiliani–Fischer synthesis** and can be chain-shortened by the **Wohl degradation**.

Disaccharides are complex carbohydrates in which two simple sugars are linked by a glycoside bond between the **anomeric** carbon of one unit and a hydroxyl of the second unit. The two sugars can be the same, as in maltose and cellobiose, or different, as in lactose and sucrose. The glycosidic bond can be either α (maltose) or β (cellobiose, lactose) and can involve any hydroxyl of the second sugar. A **1,4′ link** is most common (cellobiose, maltose), but others such as 1,2′ (sucrose) are also known.

Summary of Reactions

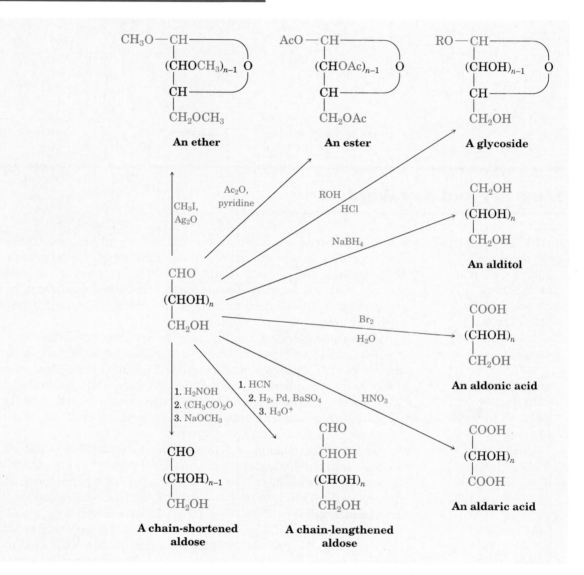

An ether

An ester

A glycoside

An alditol

An aldonic acid

An aldaric acid

A chain-shortened aldose

A chain-lengthened aldose

Visualizing Chemistry

(Problems 25.1–25.24 appear within the chapter.)

25.25 Identify the following aldoses, and tell whether each is a D or L sugar.

(a)   (b)

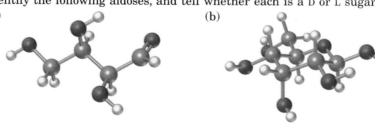

25.26 Draw Fischer projections of the following molecules, placing the carbonyl group at the top in the usual way. Identify each as a D or L sugar.

(a)   (b)

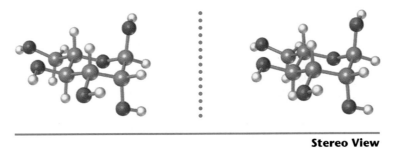

25.27 The following structure is that of an L aldohexose in its pyranose form. Identify it, and tell whether it is an α or β anomer.

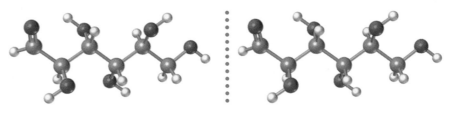

Stereo View

25.28 The following model is that of an aldohexose:

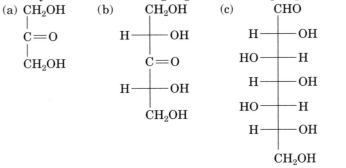

Stereo View

(a) Draw Fischer projections of the sugar, its enantiomer, and a diastereomer.
(b) Is this a D sugar or an L sugar? Explain.
(c) Draw the β anomer of the sugar in its furanose form.

Additional Problems

25.29 Classify each of the following sugars. (For example, glucose is an aldohexose.)

(a) CH_2OH
   |
   $C=O$
   |
   CH_2OH

(b) CH_2OH
H———OH
   $C=O$
H———OH
   CH_2OH

(c) CHO
H———OH
HO———H
H———OH
HO———H
H———OH
   CH_2OH

25.30 Write open-chain structures for the following:
   (a) A ketotetrose   (b) A ketopentose
   (c) A deoxyaldohexose   (d) A five-carbon amino sugar
   (e) An α anomer   (f) A β-1,4′-linked disaccharide

25.31 Does ascorbic acid (vitamin C) have a D or L configuration?

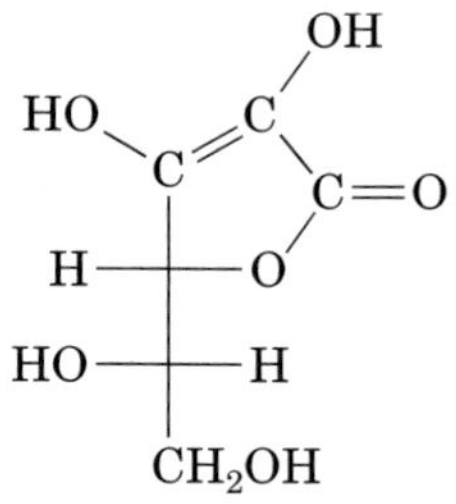

Ascorbic acid

25.32 Draw the three-dimensional furanose form of ascorbic acid (Problem 25.31), and assign *R* or *S* stereochemistry to each chirality center.

25.33 Draw Fischer projections for the two D aldoheptoses whose stereochemistry at C3, C4, C5, and C6 is the same as that of D-glucose at C2, C3, C4, and C5.

25.34 The following cyclic structure is that of allose. Is this a furanose or pyranose form? Is it an α or β anomer? Is it a D or L sugar?

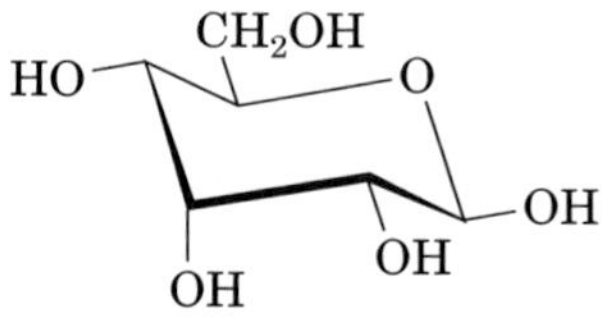

Allose

25.35 What is the complete name of the following sugar?

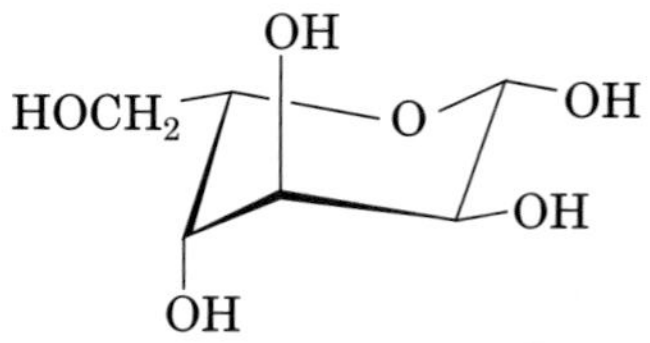

25.36 Write the following sugars in their open-chain forms:

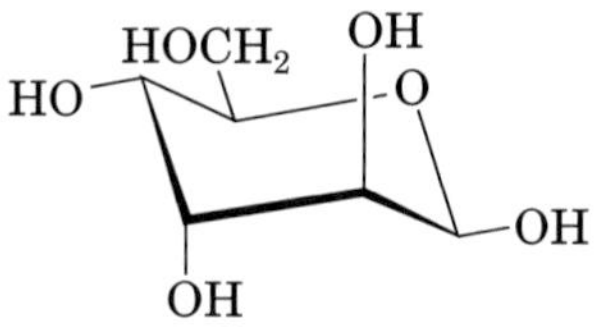

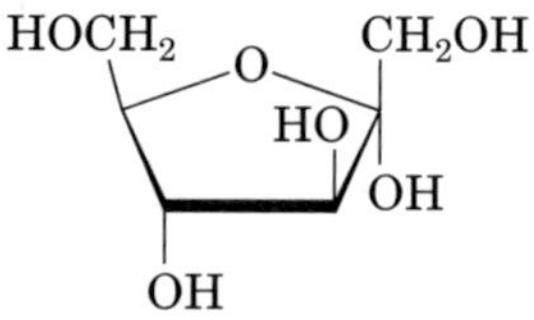

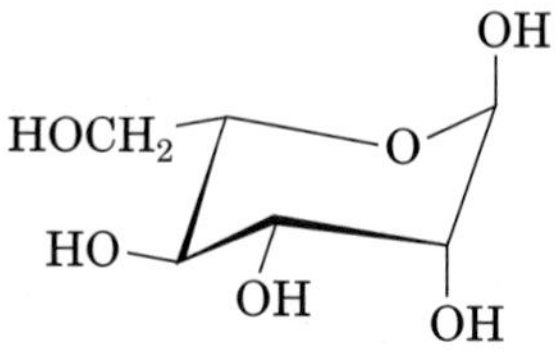

25.37 Draw D-ribulose in its five-membered cyclic β-hemiacetal form.

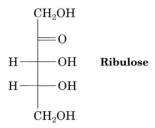

Ribulose

25.38 Look up the structure of D-talose in Figure 25.3, and draw the β anomer in its pyranose form. Identify the ring substituents as axial or equatorial.

25.39 Draw structures for the products you would expect to obtain from reaction of β-D-talopyranose with each of the following reagents:
(a) NaBH₄ in H₂O   (b) Warm dilute HNO₃   (c) Br₂, H₂O
(d) CH₃CH₂OH, HCl   (e) CH₃I, Ag₂O   (f) (CH₃CO)₂O, pyridine

25.40 Many other sugars besides glucose exhibit mutarotation. For example, α-D-galactopyranose has $[\alpha]_D$ = +150.7°, and β-D-galactopyranose has $[\alpha]_D$ = +52.8°. If either anomer is dissolved in water and allowed to reach equilibrium, the specific rotation of the solution is +80.2°. What are the percentages of each anomer at equilibrium? Draw the pyranose forms of both anomers.

25.41 How many D-2-ketohexoses are possible? Draw them.

25.42 One of the D-2-ketohexoses is called *sorbose*. On treatment with NaBH₄, sorbose yields a mixture of gulitol and iditol. What is the structure of sorbose? (See Problem 25.41.)

25.43 Another D-2-ketohexose, *psicose*, yields a mixture of allitol and altritol when reduced with NaBH₄. What is the structure of psicose? (See Problem 25.41.)

25.44 Fischer prepared the L-gulose needed for his structure proof of glucose in the following way: D-Glucose was oxidized to D-glucaric acid, which can form two six-membered-ring lactones. These were separated and reduced with sodium amalgam to give D-glucose and L-gulose. What are the structures of the two lactones, and which one is reduced to L-gulose?

25.45 What other D aldohexose gives the same alditol as D-talose?

25.46 Which of the eight D aldohexoses give the same aldaric acids as their L enantiomers?

25.47 Which of the other three D aldopentoses gives the same aldaric acid as D-lyxose?

25.48 Draw the structure of L-galactose, and then answer the following questions:
(a) Which other aldohexose gives the same aldaric acid as L-galactose on oxidation with warm HNO₃?
(b) Is this other aldohexose a D sugar or an L sugar?
(c) Draw this other aldohexose in its most stable pyranose conformation.

25.49 Gentiobiose, a rare disaccharide found in saffron and gentian, is a reducing sugar and forms only D-glucose on hydrolysis with aqueous acid. Reaction of gentiobiose with iodomethane and Ag₂O yields an octamethyl derivative, which can be hydrolyzed with aqueous acid to give 1 equivalent of 2,3,4,6-tetra-*O*-methyl-D-glucopyranose and 1 equivalent of 2,3,4-tri-*O*-methyl-D-glucopyranose. If gentiobiose contains a β-glycoside link, what is its structure?

25.50   Amygdalin, or laetrile, is a glycoside isolated in 1830 from almond and apricot seeds. It is known as a *cyanogenic glycoside* because acidic hydrolysis liberates HCN, along with benzaldehyde and 2 equivalents of D-glucose. Structural studies have shown amygdalin to be a β-glycoside of benzaldehyde cyanohydrin with gentiobiose (Problem 25.49). Draw the structure of amygdalin.

25.51   Trehalose is a nonreducing disaccharide that is hydrolyzed by aqueous acid to yield 2 equivalents of D-glucose. Methylation followed by hydrolysis yields 2 equivalents of 2,3,4,6-tetra-*O*-methylglucose. How many structures are possible for trehalose?

25.52   Trehalose (Problem 25.51) is cleaved by enzymes that hydrolyze α-glycosides but not by enzymes that hydrolyze β-glycosides. What is the structure and systematic name of trehalose?

25.53   Isotrehalose and neotrehalose are chemically similar to trehalose (Problems 25.51 and 25.52) except that neotrehalose is hydrolyzed only by β-glycosidase enzymes, whereas isotrehalose is hydrolyzed by both α- and β-glycosidase enzymes. What are the structures of isotrehalose and neotrehalose?

25.54   D-Glucose reacts with acetone in the presence of acid to yield the nonreducing 1,2:5,6-diisopropylidene-D-glucofuranose. Propose a mechanism.

1,2:5,6-Diisopropylidene-D-glucofuranose

25.55   D-Mannose reacts with acetone to give a diisopropylidene derivative (see Problem 25.54) that is still reducing toward Tollens' reagent. Propose a likely structure for this derivative.

25.56   Propose a mechanism to account for the fact that D-gluconic acid and D-mannonic acid are interconverted when either is heated in pyridine solvent.

25.57   The *cyclitols* are a group of carbocyclic sugar derivatives having the general formulation 1,2,3,4,5,6-cyclohexanehexol. How many stereoisomeric cyclitols are possible? Draw them in their chair forms.

25.58   Compound A is a D aldopentose that can be oxidized to an optically inactive aldaric acid B. On Kiliani–Fischer chain extension, A is converted into C and D; C can be oxidized to an optically active aldaric acid E, but D is oxidized to an optically inactive aldaric acid F. What are the structures of A–F?

25.59 Simple sugars undergo reaction with phenylhydrazine, $PhNHNH_2$, to yield crystalline derivatives called *osazones*. The reaction is a bit complex, however, as shown by the fact that glucose and fructose yield the same osazone.

CHO
H——OH
HO——H   3 equiv
H——OH   PhNHNH₂ →
H——OH
CH₂OH
D-Glucose

H⟍   N—NHPh
    ⟋
    ══N—NHPh
HO——H
H——OH   ← 3 equiv
H——OH   PhNHNH₂
CH₂OH
+ NH₃ + PhNH₂
+ 2 H₂O

CH₂OH
══O
HO——H
H——OH
H——OH
CH₂OH
D-Fructose

(a) Draw the structure of a third sugar that yields the same osazone as glucose and fructose.

(b) Using glucose as the example, the first step in osazone formation is reaction of the sugar with phenylhydrazine to yield an imine called a *phenylhydrazone*. Draw the structure of the product.

(c) The second and third steps in osazone formation are tautomerization of the phenylhydrazone to give an enol, followed by elimination of aniline to give a keto imine. Draw the structures of both the enol tautomer and the keto imine.

(d) The final step is reaction of the keto imine with 2 equivalents of phenylhydrazine to yield the osazone plus ammonia. Propose a mechanism for this step.

25.60 When heated to 100°C, D-idose undergoes a reversible loss of water and exists primarily as 1,6-anhydro-D-idopyranose.

CHO
HO——H
H——OH   100°C
HO——H   ⇌
H——OH
CH₂OH
D-Idose

⎛ CH ⎞
HO——H
H——OH   + H₂O
HO——H
H——O
⎝ OCH₂ ⎠
1,6-Anhydro-D-idopyranose

(a) Draw D-idose in its pyranose form, showing the more stable chair conformation of the ring.

(b) Which is more stable, α-D-idopyranose or β-D-idopyranose? Explain.

(c) Draw 1,6-anhydro-D-idopyranose in its most stable conformation.

(d) When heated to 100°C under the same conditions as those used for D-idose, D-glucose does not lose water and does not exist in a 1,6-anhydro form. Explain.

A Look Ahead

. . . .

25.61   Acetyl coenzyme A (acetyl CoA) is the key intermediate in food metabolism. What sugar is present in acetyl CoA? (See Section 29.5.)

Acetyl coenzyme A

Molecular Modeling

. . . .

25.62   Use SpartanView to examine an electrostatic potential map of ascorbic acid. Identify the most acidic hydrogen, and then examine the geometry and electrostatic potential map of ascorbate anion. Draw resonance structures for this ion.

Ascorbic acid

25.63   Use SpartanView to compare the energies of the six-membered-ring (pyranose) and seven-membered-ring cyclization products of glucose. Which of the two products is favored thermodynamically?

Glucose

Biomolecules: Amino Acids, Peptides, and Proteins

Proteins are large biomolecules that occur in every living organism. They are of many different types and have many different biological functions. The keratin of skin and fingernails, the fibroin of silk and spider webs, and most enzymes that catalyze the thousands of biological reactions within cells are all proteins. Regardless of their function, all proteins are made up of many *amino acid* units linked together into a long chain.

Amino acids, as their name implies, are difunctional. They contain both a basic amino group and an acidic carboxyl group:

H₃C   H
H_2N—C—C=O
OH

Alanine, an amino acid

Stereo View

Their value as biological building blocks stems from the fact that amino acids can join together into long chains by forming amide bonds between the $-NH_2$ of one amino acid and the $-COOH$ of another. For classification purposes, chains with fewer than 50 amino acids are often called **peptides**, while the term **protein** is reserved for larger chains.

$$\text{Many } H_2NCHCOH \longrightarrow \text{NHCHC—NHCHC—NHCHC}$$

A peptide (many amide bonds)

26.1 Structures of Amino Acids

Since amino acids contain both an acidic and a basic group, they undergo an intramolecular acid–base reaction and exist primarily in the form of a dipolar ion, or **zwitterion** (German *zwitter*, "hybrid"):

zwitterion

$$H-\ddot{N}-CH-C-O-H \rightleftharpoons H-\overset{+}{N}-CH-C-O^-$$

A zwitterion

$H_3\overset{+}{N}CH_2CO_2^-$

Amino acid zwitterions are internal salts and therefore have many of the physical properties associated with salts. They have large dipole moments, are soluble in water but insoluble in hydrocarbons, and are crystalline substances with high melting points. In addition, amino acids are *amphoteric:* They can react either as acids or as bases, depending on the circumstances. In aqueous acid solution, an amino acid zwitterion is a base that *accepts* a proton to yield a cation; in aqueous base solution, the zwitterion is an acid that *loses* a proton to form an anion.

In acid solution

$$H-\overset{+}{N}\text{—}\underset{R}{\overset{H}{\underset{|}{\overset{|}{C}}}}\text{—}\overset{O}{\overset{\|}{C}}\text{—}O^- + H_3O^+ \rightleftharpoons H-\overset{+}{N}\text{—}\underset{R}{\overset{H}{\underset{|}{\overset{|}{C}}}}\text{—}\overset{O}{\overset{\|}{C}}\text{—}OH + H_2O$$

In base solution

$$H-\overset{+}{N}\text{—}\underset{R}{\overset{H}{\underset{|}{\overset{|}{C}}}}\text{—}\overset{O}{\overset{\|}{C}}\text{—}O^- + \ ^-OH \rightleftharpoons H-N\text{—}\underset{R}{\overset{H}{\underset{|}{\overset{|}{C}}}}\text{—}\overset{O}{\overset{\|}{C}}\text{—}O^- + H_2O$$

Note that it is the carboxylate, –COO⁻, rather than the amino group that acts as the basic site and accepts a proton in acid solution. Similarly, it is the ammonium cation, –NH₃⁺, rather than the carboxyl group that acts as the acidic site and donates a proton in base solution.

The structures of the 20 amino acids commonly found in proteins are shown in Table 26.1 (p. 1076) in the form that predominates within cells at pH 7.3. All are **α-amino acids**, meaning that the amino group in each is a substituent on the α carbon atom—the one next to the carbonyl group. Note that 19 of the 20 amino acids are primary amines, RNH_2, and differ only in the nature of the substituent attached to the α carbon, called the **side chain**. Proline, however, is a secondary amine whose nitrogen and α carbon atoms are part of a five-membered pyrrolidine ring.

A primary α-amino acid
(R = a side chain)

Proline, a secondary
α-amino acid

Note also that each of the amino acids in Table 26.1 is referred to by a three-letter shorthand code: Ala for alanine, Gly for glycine, and so on. In addition, a one-letter code is also used, as shown in parentheses in the table.

With the exception of glycine, H_2NCH_2COOH, the α carbons of the amino acids are centers of chirality. Two enantiomeric forms are therefore possible, but nature uses only a single enantiomer to build proteins. In Fischer projections, naturally occurring amino acids are represented by placing the –COOH group at the top and the side chain down, as if drawing a carbohydrate (Section 25.2), and then placing the –NH₂ group on the left.

TABLE 26.1 The 20 Common Amino Acids in Proteins
(The forms shown are those that predominate at pH = 7.3.)

Name	Abbreviations		MW	Structure	pK_{a1} α-COOH	pK_{a2} α-NH_3^+	pK_a side chain	Isoelectric point
Neutral amino acids								
Alanine	Ala	(A)	89		2.34	9.69	—	6.01
Asparagine	Asn	(N)	132		2.02	8.80	—	5.41
Cysteine	Cys	(C)	121		1.96	10.28	8.18	5.07
Glutamine	Gln	(Q)	146		2.17	9.13	—	5.65
Glycine	Gly	(G)	75		2.34	9.60	—	5.97
Isoleucine	Ile	(I)	131		2.36	9.60	—	6.02
Leucine	Leu	(L)	131		2.36	9.60	—	5.98
Methionine	Met	(M)	149		2.28	9.21	—	5.74
Phenylalanine	Phe	(F)	165		1.83	9.13	—	5.48
Proline	Pro	(P)	115		1.99	10.60	—	6.30
Serine	Ser	(S)	105		2.21	9.15	—	5.68

Name	Abbreviations		MW	Structure	pK_{a1} α-COOH	pK_{a2} α-NH$_3^+$	pK_a side chain	Isoelectric point
Threonine	Thr	(T)	119		2.09	9.10	—	5.60
Tryptophan	Trp	(W)	204	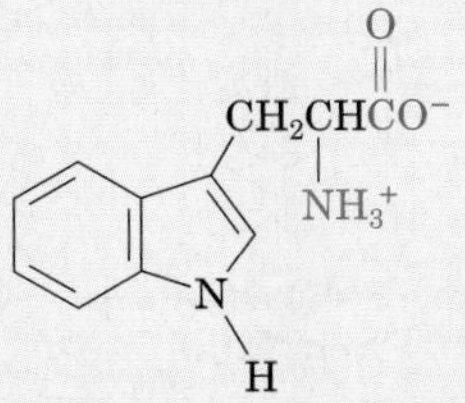	2.83	9.39	—	5.89
Tyrosine	Tyr	(Y)	181	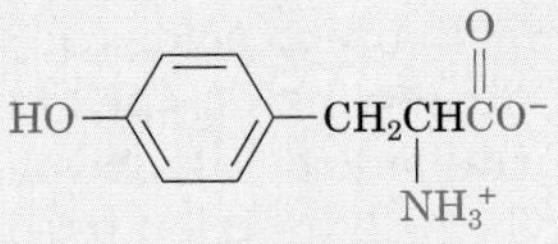	2.20	9.11	10.07	5.66
Valine	Val	(V)	117		2.32	9.62	—	5.96

Acidic amino acids

Name	Abbreviations		MW	Structure	pK_{a1} α-COOH	pK_{a2} α-NH$_3^+$	pK_a side chain	Isoelectric point
Aspartic acid	Asp	(D)	133		1.88	9.60	3.65	2.77
Glutamic acid	Glu	(E)	147		2.19	9.67	4.25	3.22

Basic amino acids

Name	Abbreviations		MW	Structure	pK_{a1} α-COOH	pK_{a2} α-NH$_3^+$	pK_a side chain	Isoelectric point
Arginine	Arg	(R)	174		2.17	9.04	12.48	10.76
Histidine	His	(H)	155	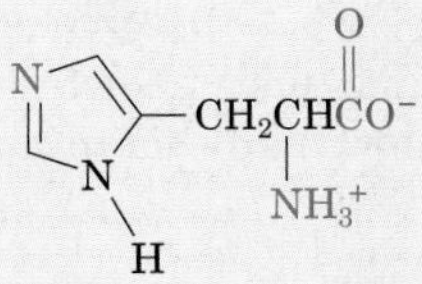	1.82	9.17	6.00	7.59
Lysine	Lys	(K)	146		2.18	8.95	10.53	9.74

Because of their stereochemical similarity to L sugars (Section 25.3), the naturally occurring α-amino acids are often referred to as L amino acids.

(S)-Alanine
(L-Alanine)

(S)-Serine
(L-Serine)

**Stereochemically
similar to
L-glyceraldehyde**

(S)-Phenylalanine
(L-Phenylalanine)

The 20 common amino acids can be further classified as either neutral, acidic, or basic, depending on the structure of their side chains. Fifteen of the 20 have neutral side chains, two (aspartic acid and glutamic acid) have an extra carboxylic acid function in their side chains, and three (lysine, arginine, and histidine) have basic amino groups in their side chains. Note, however, that both cysteine and tyrosine, though classified as neutral amino acids, nevertheless have weakly acidic side chains and can be deprotonated in strongly basic solution.

At the pH of 7.3 found within cells, the side-chain carboxylic groups of aspartic acid and glutamic acid are dissociated and exist as carboxylate ions, $-CO_2^-$. Similarly, the basic side-chain nitrogens of lysine and arginine are protonated at pH = 7.3 and exist as ammonium ions, $-NH_3^+$. Histidine, however, which contains a heterocyclic imidazole ring in its side chain, is not quite basic enough to be protonated at pH 7.3. Note that only the pyridine-like, doubly bonded nitrogen in histidine is basic. The pyrrole-like singly bonded nitrogen is nonbasic because its lone pair of electrons is part of the six-π-electron aromatic imidazole ring (Section 24.4).

Basic; pyridine-like

Nonbasic; pyrrole-like

Histidine

Imidazole ring

Problem 26.1   How many of the α-amino acids shown in Table 26.1 contain aromatic rings? How many contain sulfur? How many contain alcohols? How many contain hydrocarbon side chains?

Problem 26.2   Eighteen of the 19 L amino acids have the S configuration at the α carbon. Cysteine is the only L amino acid that has an R configuration. Explain.

Problem 26.3   The amino acid threonine, (2S,3R)-2-amino-3-hydroxybutanoic acid, has two chirality centers. Draw a Fischer projection of threonine.

Problem 26.4 Draw the Fischer projection of a threonine diastereomer, and label its chirality centers as R or S (Problem 26.3).

• •

26.2 Isoelectric Points

In acid solution, an amino acid is protonated and exists primarily as a cation. In basic solution, an amino acid is deprotonated and exists primarily as an anion. Thus, there must be some intermediate pH at which the amino acid is exactly balanced between anionic and cationic forms and exists primarily as the neutral, dipolar zwitterion. This pH is called the amino acid's **isoelectric point, pI**.

The isoelectric point of an amino acid depends on its structure, with values for the 20 common amino acids given in Table 26.1. The 15 amino acids with neutral or weakly acidic side chains have isoelectric points near neutrality, in the pH range 5.0–6.5. The two amino acids with more strongly acidic side chains have isoelectric points at lower pH, so that dissociation of the extra –COOH in the side chain is suppressed; and the three amino acids with basic side chains have isoelectric points at higher pH, so that protonation of the extra amino group is suppressed.

Notice that the isoelectric points of the 13 amino acids without an acidic or basic side chain are simply the average of the two dissociation constants, pK_{a1} and pK_{a2}. Alanine, for instance, has pK_{a1} = 2.34 and pK_{a2} = 9.69, so the pI of alanine is (2.34 + 9.69)/2, or 6.01. For the four amino acids with either a strongly or weakly acidic side chain, the pI is the average of the two *lowest* pK_a values, and for the three amino acids with a basic side chain, the pI is the average of the two *highest* pK_a values.

Acidic

pK_a = 3.65   pK_a = 1.88

$$HOCCH_2CHCOH$$

$$NH_3^+ \quad pK_a = 9.60$$

Aspartic acid

$$pI = \frac{1.88 + 3.65}{2} = 2.77$$

Basic

pK_a = 10.53   pK_a = 2.18

$$^+H_3NCH_2CH_2CH_2CH_2CHCOH$$

$$NH_3^+ \quad pK_a = 8.95$$

Lysine

$$pI = \frac{8.95 + 10.53}{2} = 9.74$$

We can take advantage of the differences in isoelectric points to separate a mixture of amino acids (or a mixture of proteins) into its pure constituents. Using a technique known as **electrophoresis**, a solution of different amino acids or proteins is placed near the center of a strip of paper or gel. The paper or gel is moistened with an aqueous buffer of a given pH, and electrodes are connected to the ends of the strip. When an electric potential is applied, those amino acids with negative charges (those that are deprotonated because the pH of the buffer is above their isoelectric points) migrate slowly toward the positive electrode. At the same time, those amino acids with positive charges (those that are protonated because the pH of the buffer is below their isoelectric points) migrate toward the negative electrode.

Different amino acids migrate at different rates, depending on their isoelectric point and on the pH of the aqueous buffer. Thus, the different amino acids can be separated. Figure 26.1 illustrates this separation for a mixture of lysine (basic), glycine (neutral), and aspartic acid (acidic).

FIGURE 26.1 ▼

Separation of an amino acid mixture by electrophoresis. At pH = 5.97, glycine molecules are neutral and don't migrate; lysine molecules are protonated and migrate toward the negative electrode; and aspartic acid molecules are deprotonated and migrate toward the positive electrode.

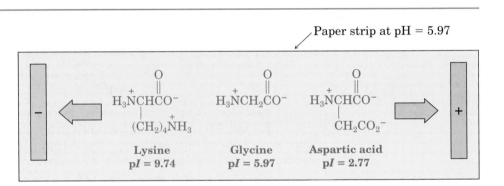

If exact pK_a values for the acidic sites of an amino acid are known (Table 26.1), the percentages of protonated, neutral, and deprotonated forms in a solution of a given pH can be calculated.

For any acid HA, we have

$$pK_a = -\log\frac{[H_3O^+][A^-]}{[HA]}$$

$$= -\log[H_3O^+] - \log\frac{[A^-]}{[HA]}$$

$$= pH - \log\frac{[A^-]}{[HA]}$$

Rearranging gives the **Henderson–Hasselbalch equation**:

$$pH = pK_a + \log\frac{[A^-]}{[HA]} \qquad \text{or} \qquad \log\frac{[A^-]}{[HA]} = pH - pK_a$$

According to the Henderson–Hasselbalch equation, the logarithm of the conjugate base concentration [A⁻] divided by the acid concentration [HA] is equal to the pH of the solution minus the pK_a of the acid. Thus, if we know both the pH of the solution and the pK_a of the acid, we can calculate the ratio of [A⁻] to [HA] in the solution. Furthermore, when pH = pK_a, the two forms HA and A⁻ are present in equal amounts.

To see how to use the Henderson–Hasselbalch equation, let's find out what species are present in a 1.00 M solution of alanine at pH = 9.00. According to Table 26.1, protonated alanine [⁺H₃NCH(CH₃)COOH] has pK_{a1} = 2.34, and neutral, zwitterionic alanine [⁺H₃NCH(CH₃)CO₂⁻] has pK_{a2} = 9.69.

$$
\underset{\substack{|\\CH_3}}{\overset{\overset{O}{\parallel}}{H_3\overset{+}{N}CHCOH}} + H_2O \;\rightleftharpoons\; \underset{\substack{|\\CH_3}}{\overset{\overset{O}{\parallel}}{H_3\overset{+}{N}CHCO^-}} + H_3O^+ \qquad pK_{a1} = 2.34
$$

$$
\underset{\substack{|\\CH_3}}{\overset{\overset{O}{\parallel}}{H_3\overset{+}{N}CHCO^-}} + H_2O \;\rightleftharpoons\; \underset{\substack{|\\CH_3}}{\overset{\overset{O}{\parallel}}{H_2NCHCO^-}} + H_3O^+ \qquad pK_{a2} = 9.69
$$

Since the pH of our solution is much closer to pK_{a2} than to pK_{a1}, we need to use pK_{a2} for our calculation. From the Henderson–Hasselbalch equation, we have:

$$
\log \frac{[A^-]}{[HA]} = pH - pK_a = 9.00 - 9.69 = -0.69
$$

so

$$
\frac{[A^-]}{[HA]} = \text{antilog}(-0.69) = 0.20 \qquad \text{and} \qquad [A^-] = (0.20)[HA]
$$

In addition, we know that

$$
[A^-] + [HA] = 1.00\,M
$$

so we have two simultaneous equations, which can be solved to give [HA] = 0.83 and [A⁻] = 0.17. In other words, at pH = 9.00, 83% of alanine molecules in a 1.00 M solution are neutral (zwitterionic) and 17% are deprotonated. Similar calculations can be done at any other pH, leading to the *titration curve* shown in Figure 26.2.

Each leg of the titration curve is calculated separately. The first leg, from pH 1 to 6, corresponds to the dissociation of protonated alanine, H₂A⁺; the second leg, from pH 6 to 11, corresponds to the dissociation of zwitterionic alanine, HA. Exactly halfway between the two legs is the isoelectric point at 6.01. In essence, it's as if we started with H₂A⁺ at low pH and then titrated with NaOH. When 0.5 equiv of NaOH is added, the deprotonation of H₂A⁺ is 50% done; when 1.0 equiv of NaOH is added, the deprotonation of H₂A⁺ is complete and HA predominates (the isoelectric point); when

FIGURE 26.2 ▼

A titration curve for alanine, plotted using the Henderson–Hasselbalch equation.
Each of the two legs is plotted separately. At pH < 1, alanine is entirely protonated;
at pH = 2.34, alanine is a 50:50 mix of protonated and neutral forms; at pH = 6.01,
alanine is entirely neutral; at pH = 9.69, alanine is a 50:50 mix of neutral and
deprotonated forms; at pH > 11.5, alanine is entirely deprotonated.

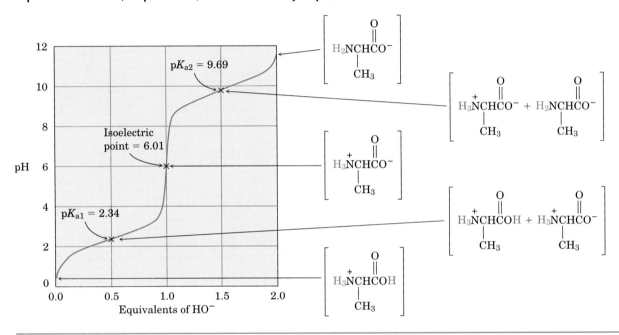

1.5 equiv of NaOH is added, the deprotonation of HA is 50% done; and when
2.0 equiv of NaOH is added, the deprotonation of HA is complete.

Problem 26.5   For the following mixtures of amino acids, predict the direction and relative rate of
migration of each component:
(a) Valine, glutamic acid, and histidine at pH = 7.6
(b) Glycine, phenylalanine, and serine at pH = 5.7
(c) Glycine, phenylalanine, and serine at pH = 5.5
(d) Glycine, phenylalanine, and serine at pH = 6.0

Problem 26.6   Threonine has $pK_{a1} = 2.09$ and $pK_{a2} = 9.10$. Use the Henderson–Hasselbalch equation to calculate the ratio of protonated and neutral forms at pH = 1.50. Calculate
the ratio of neutral and deprotonated forms at pH = 10.00.

26.3 Synthesis of α-Amino Acids

α-Amino acids can be synthesized using some of the reactions discussed in
previous chapters. One of the oldest methods of α-amino acid synthesis

begins with α bromination of a carboxylic acid by treatment with Br_2 and PBr_3 (the Hell–Volhard–Zelinskii reaction, Section 22.4). S_N2 substitution of the α-bromo acid with ammonia then yields an α-amino acid.

4-Methylpentanoic acid   **2-Bromo-4-methyl-**   **(R,S)-Leucine (45%)**
                                    pentanoic acid

Alternatively, higher product yields are obtained when the bromide displacement reaction is carried out by the Gabriel phthalimide method (Section 24.6) rather than by the ammonia method.

• •

Problem 26.7 Show how you could prepare the following α-amino acids from the appropriate carboxylic acids:
(a) Phenylalanine   (b) Valine

• •

Adolph Friedrich Ludwig Strecker

Adolph Friedrich Ludwig Strecker (1822–1871) was born in Darmstadt, Germany, and received his Ph.D. in 1842 at the University of Giessen with Justus von Liebig. Following a period as assistant to Liebig, Strecker became Professor of Chemistry at the University of Christiania and then at the University of Tübingen.

The Strecker Synthesis

Another method for preparing racemic α-amino acids is the **Strecker synthesis**, developed in 1850. This two-step process involves treatment of an aldehyde with KCN and aqueous ammonia to yield an intermediate α-amino nitrile, which is hydrolyzed to give an α-amino acid.

The Strecker synthesis occurs by initial reaction of the aldehyde with ammonia to give an imine intermediate (Section 19.9), which then adds HCN in a nucleophilic addition step similar to what occurs in cyanohydrin formation (Section 19.7). The α-amino nitrile that results undergoes hydrolysis in the usual way (Section 21.8).

Phenylacetaldehyde   **An imine**   **An α-amino nitrile**

(R,S)-Phenylalanine (53%)

● ●

Problem 26.8   Show how you might synthesize leucine using the Strecker synthesis.

● ●

The Amidomalonate Synthesis

The most general method of preparation for α-amino acids is the **amido-malonate synthesis**, a straightforward extension of the malonic ester synthesis (Section 22.8). The reaction begins with conversion of diethyl acetamidomalonate into an enolate ion by treatment with base, followed by S_N2 alkylation with a primary alkyl halide. Hydrolysis of both the amide protecting group and the esters occurs when the alkylated product is warmed with aqueous acid, and decarboxylation then takes place to yield an α-amino acid. For example, aspartic acid can be prepared from ethyl bromoacetate:

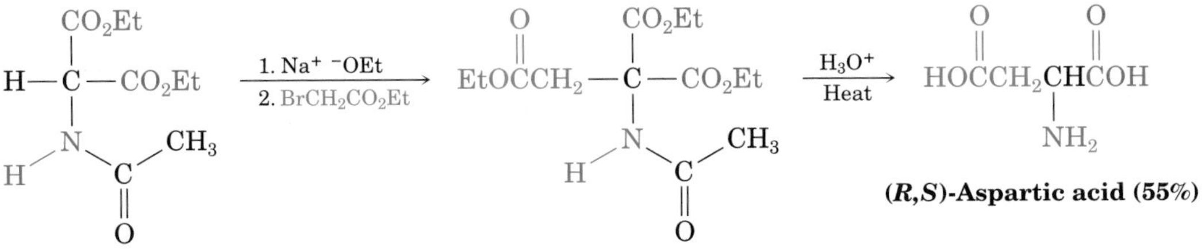

**Diethyl
acetamidomalonate**

(R,S)-Aspartic acid (55%)

● ●

Problem 26.9   What alkyl halides would you use to prepare the following α-amino acids by the amidomalonate method?
(a) Leucine   (b) Histidine   (c) Tryptophan   (d) Methionine

● ●

Reductive Amination of α-Keto Acids: Biosynthesis

Yet a fourth method for the synthesis of α-amino acids is by reductive amination of an α-keto acid with ammonia and a reducing agent (Section 24.6):

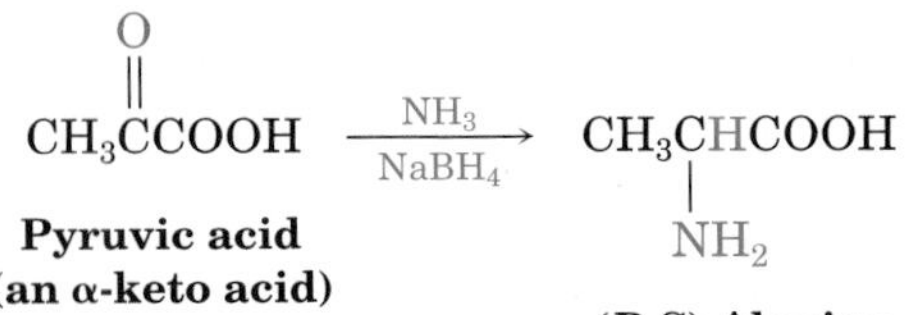

**Pyruvic acid
(an α-keto acid)**

(R,S)-Alanine

This method is particularly interesting because it is a close laboratory analogy of a pathway by which some amino acids are biosynthesized in nature. For example, the major route for glutamic acid synthesis in most organisms is by reductive amination of α-ketoglutaric acid. The biological reducing agent is a complex molecule called *nicotinamide adenine dinu-cleotide* (NADH), and the reaction is catalyzed by an enzyme, L-glutamate dehydrogenase. We'll look at a related process in Section 29.6.

$$\underset{\textbf{α-Ketoglutaric acid}}{HOOCCH_2CH_2\overset{\overset{\displaystyle O}{\|}}{C}COOH} + \text{"}NH_3\text{"} \quad \xrightarrow[\substack{\text{L-Glutamate} \\ \text{dehydrogenase}}]{NADH} \quad \underset{\textbf{(S)-Glutamic acid}}{HOOCCH_2CH_2\underset{\underset{\displaystyle NH_2}{|}}{C}HCOOH}$$

26.4 Resolution of R,S Amino Acids

The synthesis of a chiral amino acid from an achiral precursor by any of the methods described in the previous section yields a racemic mixture—an equal mixture of *S* and *R* products. To use these synthetic amino acids for the laboratory synthesis of naturally occurring proteins, however, the racemic mixture must first be resolved into pure enantiomers. Sometimes this resolution can be done by allowing the racemic amino acid to undergo a reaction that forms two diastereomers, which are then separated and con-verted back to the amino acid (Section 9.10).

Alternatively, biological methods of resolution are often used. A num-ber of enzymes are available that selectively catalyze the hydrolysis of an amide formed from an *S* amino acid, while leaving the related amide from an *R* amino acid untouched. We can therefore resolve an *R,S* mixture of an amino acid by forming an *N*-acetyl derivative, carrying out an enzyme-catalyzed hydrolysis, and separating the *S* amino acid from unreacted *R* amide.

An *R,S* mixture   An *R,S* mixture   An *S* amino acid   An *R* amido acid

26.5 Peptides and Proteins

Proteins and peptides are amino acid polymers in which the individual amino acid units, called **residues**, are linked together by amide bonds, or *peptide bonds*. An amino group from one residue forms an amide bond with the carboxyl of a second residue; the amino group of the second forms an amide bond with the carboxyl of a third, and so on. For example, alanylserine is the *dipeptide* that results when an amide bond is formed between the alanine carboxyl and the serine amino group:

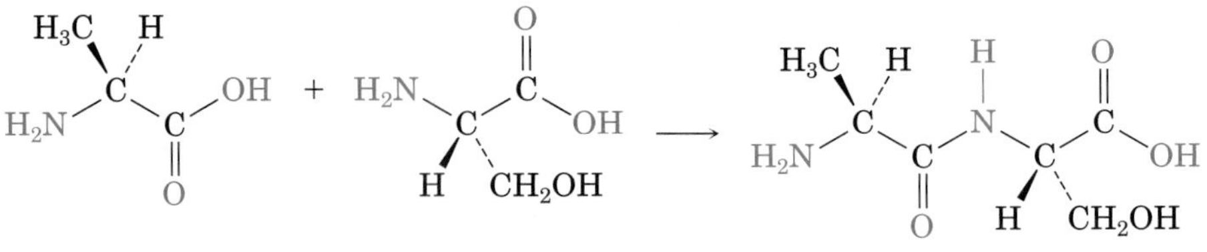

Alanine (Ala)   **Serine (Ser)**   **Alanylserine (Ala-Ser)**

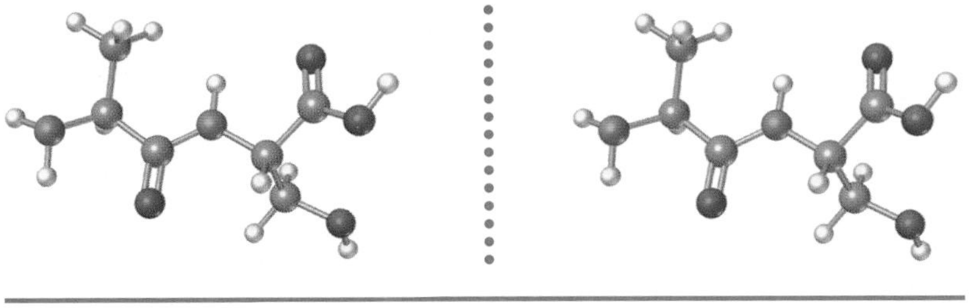

Stereo View

Note that two dipeptides can result from reaction between alanine and serine, depending on which carboxyl group reacts with which amino group. If the alanine amino group reacts with the serine carboxyl, serylalanine results:

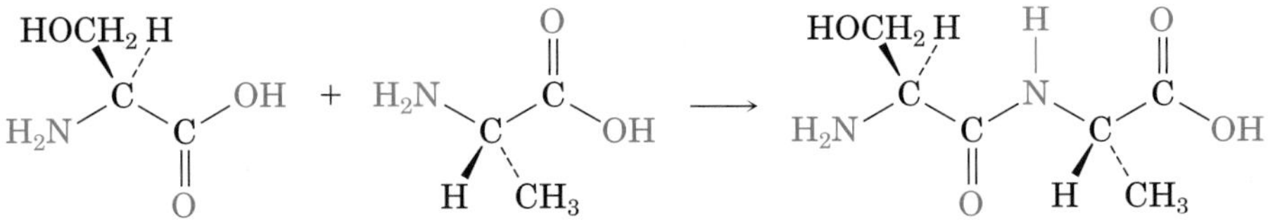

Serine (Ser)   **Alanine (Ala)**   **Serylalanine (Ser-Ala)**

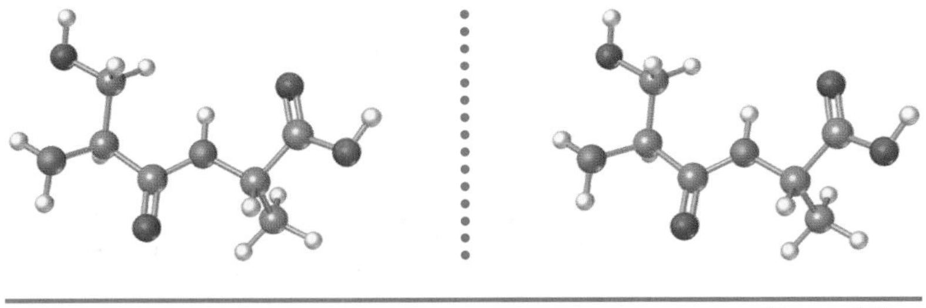

Stereo View

The long, repetitive sequence of –N–CH–CO– atoms that make up a continuous chain is called the protein's **backbone**. By convention, peptides are always written with the **N-terminal amino acid** (the one with the free –NH$_2$ group) on the left and the **C-terminal amino acid** (the one with the free –COOH group) on the right. The name of the peptide is indicated by using the abbreviations listed in Table 26.1 for each amino acid. Thus, alanylserine is abbreviated Ala-Ser (or A-S) and serylalanine is abbreviated Ser-Ala (or S-A).

Problem 26.10 There are six possible isomeric tripeptides that contain valine, tyrosine, and glycine. Name them using both three-letter and one-letter abbreviations.

Problem 26.11 Draw the full structure of Met-Pro-Val-Gly, and indicate the amide bonds.

26.6 Covalent Bonding in Peptides

The amide bond that links different amino acids together in peptides is no different from any other amide bond (Section 24.4). Amide nitrogens are nonbasic because their unshared electron pair is delocalized by interaction with the carbonyl group. This overlap of the nitrogen p orbital with the p orbitals of the carbonyl group imparts a certain amount of double-bond character to the C–N bond and restricts rotation around it. As indicated by the stereo views of alanylserine and serylalanine shown in the previous section, the amide bond is planar and the N–H is oriented 180° to the C=O.

Restricted rotation

Planar

A second kind of covalent bonding in peptides occurs when a disulfide linkage, RS–SR, is formed between two cysteine residues. As we saw in Section 18.11, disulfide bonds are easily formed by mild oxidation of thiols, RSH, and are easily cleaved by mild reduction.

Two cysteines (thiols)

A disulfide

A disulfide bond between cysteines in two different peptide chains links the otherwise separate chains together. Alternatively, a disulfide bond between two cysteines within the same chain causes a loop in the chain.

Such is the case with vasopressin, an antidiuretic hormone found in the pituitary gland. Note that the C-terminal end of vasopressin occurs as the primary amide, –CONH$_2$, rather than as the free acid.

Cys-Tyr-Phe-Glu-Asn-Cys-Pro-Arg-Gly-NH$_2$

Vasopressin

26.7 Structure Determination of Peptides: Amino Acid Analysis

William Howard Stein

William Howard Stein (1911–1980) was born in New York City and received his Ph.D. in 1938 from the Columbia College of Physicians and Surgeons. He immediately joined the faculty of the Rockefeller Institute, where he remained until his death. In 1972, he shared the Nobel Prize in chemistry for his work with Stanford Moore on developing methods of amino acid analysis and for determining the structure of ribonuclease.

Determining the structure of a peptide requires answering three questions: What amino acids are present? How much of each is present? In what sequence do the amino acids occur in the peptide chain? The answers to the first two questions are provided by an instrument called an *amino acid analyzer*.

An amino acid analyzer is an automated instrument based on analytical techniques worked out in the 1950s by William Stein and Stanford Moore at the Rockefeller Institute (now the Rockefeller University). In preparation for analysis, the peptide is broken into its constituent amino acids by reducing all disulfide bonds, capping the –SH groups of cysteine residues by S$_N$2 reaction with iodoacetic acid, and hydrolyzing the amide bonds by heating with aqueous HCl. The resultant amino acid mixture is then analyzed by placing it at the top of a glass column (a *chromatography* column) filled with a special adsorbent material. When a series of aqueous buffers is pumped through the column, the various amino acids migrate down the column at different rates depending on their structures and isoelectric points, thereby separating them.

As each amino acid exits (*elutes*) from the end of the chromatography column, it reacts with a solution of *ninhydrin*, giving an intense purple color. The color is detected by a spectrometer, and a plot of elution time versus spectrometer absorbance is obtained.

Ninhydrin + H$_2$NCHCOH (An α-amino acid) $\xrightarrow[\text{H}_2\text{O}]{^-\text{OH}}$ (Purple color) + RCH + CO$_2$

Stanford Moore

Stanford Moore (1913–1982) was born in Chicago, Illinois, and received his Ph.D. from the University of Wisconsin in 1938. He was a professor at the Rockefeller Institute.

Because the amount of time required for a given amino acid to elute from the chromatography column is reproducible, the identity of the amino acids in a peptide of unknown composition can be determined simply by noting the various elution times. The amount of each amino acid in the sample is determined by measuring the intensity of the purple color resulting from its reaction with ninhydrin. Figure 26.3 shows the results of amino acid analysis of a standard equimolar mixture of 17 α-amino acids. Typically, amino acid analysis requires about 150 picomoles (4–5 μg) of sample for a protein containing about 200 residues.

FIGURE 26.3 ▼

Amino acid analysis of an equimolar mixture of 17 amino acids.

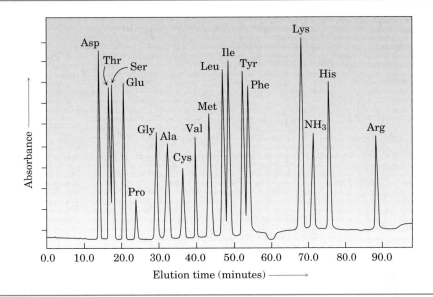

. . . .

Problem 26.12   Show the structure of the product you would expect to obtain by S_N2 reaction of a cysteine residue with iodoacetic acid.

Problem 26.13   Show the structures of the products obtained on reaction of valine with ninhydrin.

. . . .

26.8 Sequencing of Peptides: The Edman Degradation

With the identity and amount of the amino acids known, the next task of structure determination is to *sequence* the peptide—that is, to find out in what order the amino acids are linked together. The general idea of peptide sequencing is to cleave one amino acid at a time from the end of the peptide chain (at either the N terminus or the C terminus). That terminal amino acid is then separated and identified, and the cleavage reactions are repeated on the chain-shortened peptide until the entire peptide sequence is determined.

Most peptide sequencing is now done by **Edman degradation**, an efficient method of N-terminal analysis. Automated Edman protein sequencers are available that allow as many as 50 repetitive sequencing cycles to be carried out before a buildup of unwanted by-products interferes with the

Pehr Victor Edman

Pehr Victor Edman (1916–1977) was born in Stockholm, Sweden, and received an M.D. in 1946 at the Karolinska Institute. After a year in the United States at the Rockefeller Institute, he returned to Sweden as professor at the University of Lund. In 1957, he moved to St. Vincent's School of Medical Research in Melbourne, Australia, where he developed and automated the method of peptide sequencing that now bears his name. A reclusive man, he never received the prizes or recognition merited by the importance of his work.

results. So efficient are these instruments that sequence information can be obtained from as little as 1–5 picomoles of sample—less than $0.1\,\mu g$.

Edman degradation involves treatment of a peptide with phenyl isothiocyanate (PITC), $C_6H_5-N=C=S$, followed by mild acid hydrolysis, as shown in Figure 26.4. The first step attaches the PITC to the $-NH_2$ group of the N-terminal amino acid, and the second step splits the N-terminal residue from the peptide chain, yielding an anilinothiazolinone (ATZ) derivative plus the chain-shortened peptide. Further acid-catalyzed rearrangement of the ATZ derivative converts it into a *phenylthiohydantoin* (PTH), which is identified chromatographically by comparison of its elution time with the known elution times of PTH derivatives of all 20 common amino acids. The chain-shortened peptide is then automatically resubmitted to another round of Edman degradation.

Complete sequencing of large peptides and proteins by Edman degradation is impractical because the buildup of unwanted by-products limits the method to a maximum of 50 cycles. Instead, a large peptide chain is first cleaved by partial hydrolysis into a number of smaller fragments, the sequence of each fragment is determined, and the individual fragments are fitted together by matching the overlapping ends. In this way, protein chains with more than 400 amino acids have been sequenced.

Partial hydrolysis of a peptide can be carried out either chemically with aqueous acid, or enzymatically. Acidic hydrolysis is unselective and leads to a more or less random mixture of small fragments. Enzymic hydrolysis, however, is quite specific. The enzyme *trypsin*, for instance, catalyzes hydrolysis of peptides only at the carboxyl side of the basic amino acids arginine and lysine; *chymotrypsin* cleaves only at the carboxyl side of the aryl-substituted amino acids phenylalanine, tyrosine, and tryptophan.

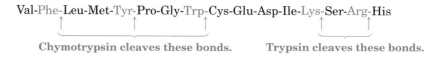

Val-Phe-Leu-Met-Tyr-Pro-Gly-Trp-Cys-Glu-Asp-Ile-Lys-Ser-Arg-His

Chymotrypsin cleaves these bonds.   **Trypsin cleaves these bonds.**

Practice Problem 26.1   Amino acid analysis of the peptide angiotensin II shows the presence of eight different amino acids in equimolar amounts: Arg, Asp, His, Ile, Phe, Pro, Tyr, and Val. Partial hydrolysis of angiotensin II with dilute hydrochloric acid yields the following fragments:

(1) Asp-Arg-Val-Tyr   (2) Ile-His-Pro   (3) Pro-Phe   (4) Val-Tyr-Ile-His

What is the sequence of angiotensin II?

Strategy   Line up the fragments to identify the overlapping regions, and then write the sequence.

Asp-Arg-Val-Tyr

Val-Tyr-Ile-His

Ile-His-Pro

Pro-Phe

Solution   Asp-Arg-Val-Tyr-Ile-His-Pro-Phe

FIGURE 26.4 ▼

Mechanism of the Edman degradation for N-terminal analysis of peptides.

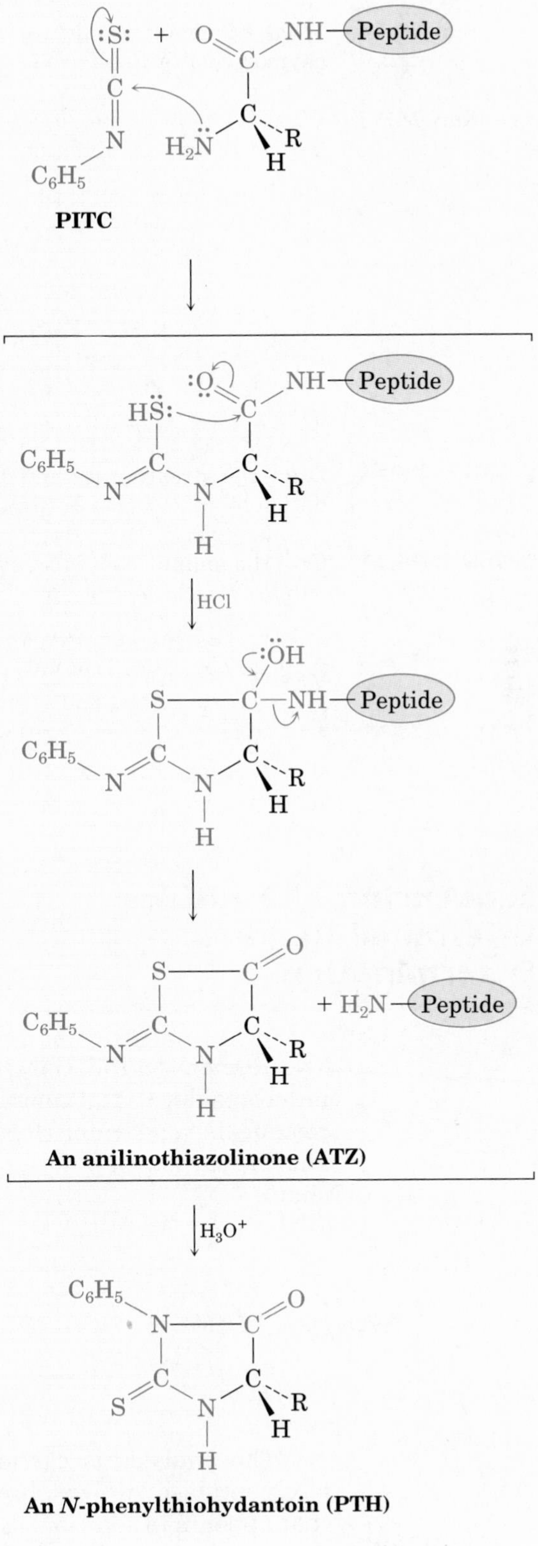

Nucleophilic addition of the peptide terminal amino group to phenyl isothiocyanate (PITC) yields an *N*-phenylthiourea derivative.

Acid-catalyzed cyclization then yields a tetrahedral intermediate . . .

. . . which expels the chain-shortened peptide and forms an anilinothiazolinone (ATZ).

An anilinothiazolinone (ATZ)

The ATZ rearranges in the presence of aqueous acid to yield the final *N*-phenylthiohydantoin (PTH) derivative.

An *N*-phenylthiohydantoin (PTH)

● ●

Problem 26.14   What fragments would result if angiotensin II were cleaved with trypsin? With chymotrypsin? (See Practice Problem 26.1.)

Problem 26.15   What is the N-terminal residue on a peptide that gives the following PTH derivative on Edman degradation?

$$C_6H_5 \qquad O$$

N—C

S=C   C‑‑CH$_2$CH$_2$SCH$_3$

N   H

H

Problem 26.16   Draw the structure of the PTH derivative that would be formed on Edman degradation of angiotensin II (Practice Problem 26.1).

Problem 26.17   Give the amino acid sequence of hexapeptides that produce the following fragments on partial acid hydrolysis:

(a) Arg, Gly, Ile, Leu, Pro, Val gives Pro-Leu-Gly, Arg-Pro, Gly-Ile-Val
(b) Asp, Leu, Met, Trp, Val$_2$ gives Val-Leu, Val-Met-Trp, Trp-Asp-Val

● ●

26.9 Sequencing of Peptides: C-Terminal Residue Determination

The Edman degradation is an excellent method of analysis for the N-terminal residue, but a complementary method of analysis for the C-terminal residue is also valuable. The best method currently available uses the enzyme carboxypeptidase to cleave the C-terminal amide bond in a peptide chain.

$$\text{Peptide}-\text{NHCHC}-\text{NHCHCOH} \xrightarrow[\text{Carboxy-peptidase}]{H_2O} \text{Peptide}-\text{NHCHCOH} + H_2\text{NCHCOH}$$

The analysis is carried out by incubating the polypeptide with carboxypeptidase and watching for the appearance of the first free amino acid that appears in solution. (Some further degradation also occurs, since a new C-terminus is produced when the first amino acid is cleaved off.)

Problem 26.18   A hexapeptide with the composition Arg, Gly, Leu, Pro$_3$ is found to have proline at both C-terminal and N-terminal positions. Partial hydrolysis gives the following fragments:

Gly-Pro-Arg,   Arg-Pro,   Pro-Leu-Gly

What is the structure of the hexapeptide?

Problem 26.19   Propose two structures for a tripeptide that gives Leu, Ala, and Phe on hydrolysis but does not react with carboxypeptidase and does not react with phenyl isothiocyanate.

26.10 Synthesis of Peptides

With the structure known, synthesis of a peptide can then be undertaken—perhaps to obtain larger amounts for biological evaluation. Although simple amides are usually formed by reaction between amines and acid chlorides (Section 21.7), peptide synthesis is more difficult because many different amide bonds must be formed in a specific order rather than at random.

The solution to the specificity problem is *protection* (Section 17.9). For example, if we wanted to couple alanine with leucine to synthesize Ala-Leu, we could protect the –NH$_2$ group of alanine and the –COOH group of leucine to render them unreactive, then form the desired amide bond, and finally remove the protecting groups.

Many different amino- and carboxyl-protecting groups have been devised, but only a few are widely used. Carboxyls are often protected simply by converting them into methyl or benzyl esters. Both groups are easily introduced by standard methods of ester formation (Section 21.6) and are easily removed by mild hydrolysis with aqueous NaOH. Benzyl esters can also be cleaved by catalytic *hydrogenolysis* of the weak benzylic C–O bond ($RCOO–CH_2Ph + H_2 \longrightarrow RCOOH + PhCH_3$).

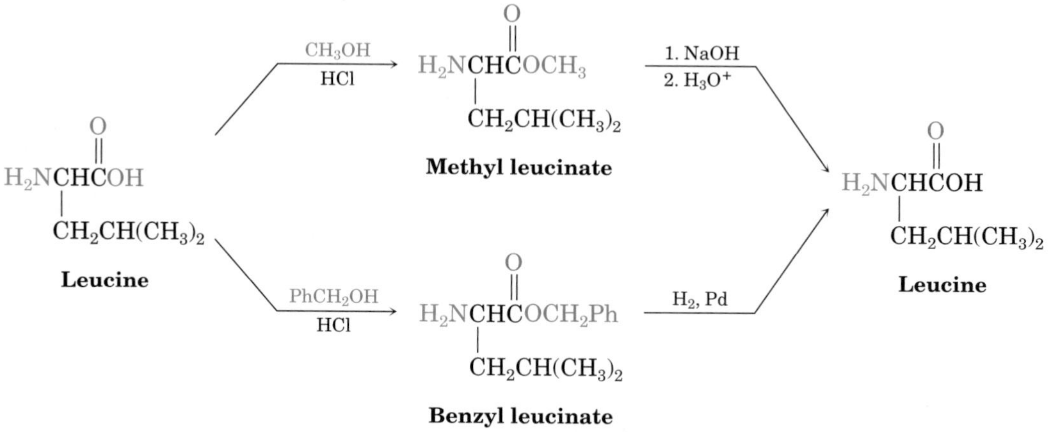

$$
\begin{array}{ccc}
& \xrightarrow[\text{HCl}]{CH_3OH} & \underset{\substack{|\\CH_2CH(CH_3)_2}}{H_2N\overset{O}{\overset{\|}{C}H\overset{}{C}OCH_3}} \xrightarrow[\text{2. } H_3O^+]{\text{1. NaOH}} \\
\underset{\substack{|\\CH_2CH(CH_3)_2}}{H_2N\overset{O}{\overset{\|}{C}H\overset{}{C}OH}} & & \textbf{Methyl leucinate} \\
\textbf{Leucine} & & \\
& \xrightarrow[\text{HCl}]{PhCH_2OH} & \underset{\substack{|\\CH_2CH(CH_3)_2}}{H_2N\overset{O}{\overset{\|}{C}H\overset{}{C}OCH_2Ph}} \xrightarrow{H_2,\ Pd} \\
& & \textbf{Benzyl leucinate}
\end{array}
$$

(right product)

$$\underset{\substack{|\\CH_2CH(CH_3)_2}}{H_2N\overset{O}{\overset{\|}{C}H\overset{}{C}OH}}$$

Leucine

Amino groups are often protected as their *tert*-butoxycarbonyl amide (BOC) derivatives. The BOC protecting group is introduced by reaction of the amino acid with di-*tert*-butyl dicarbonate in a nucleophilic acyl substitution reaction (Section 21.5) and is removed by brief treatment with a strong organic acid such as trifluoroacetic acid, CF_3COOH.

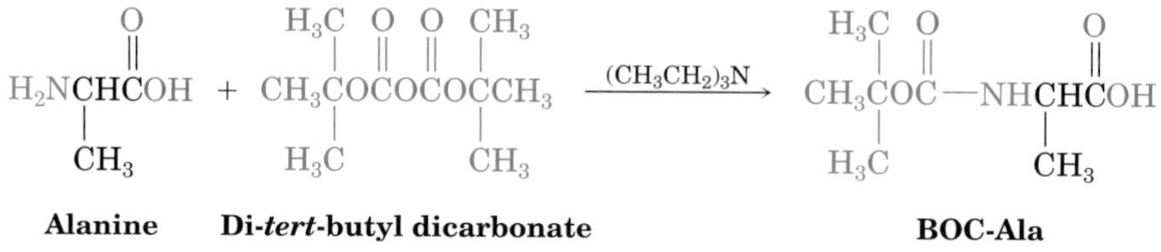

$$
\underset{\substack{|\\CH_3}}{H_2N\overset{O}{\overset{\|}{C}H\overset{}{C}OH}} + \underset{\substack{|\\H_3C}}{CH_3\overset{CH_3}{\overset{|}{C}}\overset{O}{\overset{\|}{O}}\overset{O}{\overset{\|}{C}}O\overset{O}{\overset{\|}{C}}\overset{CH_3}{\overset{|}{O}}CCH_3\overset{CH_3}{\underset{|}{}}} \xrightarrow{(CH_3CH_2)_3N} \underset{\substack{|\\H_3C}}{CH_3\overset{CH_3}{\overset{|}{C}}\overset{O}{\overset{\|}{O}}C-NH\underset{\substack{|\\CH_3}}{\overset{O}{\overset{\|}{C}H\overset{}{C}OH}}}
$$

Alanine   **Di-*tert*-butyl dicarbonate**   **BOC-Ala**

The peptide bond is usually formed by treating a mixture of protected acid and amine with dicyclohexylcarbodiimide (DCC). As shown in Figure 26.5, DCC functions by converting the carboxylic acid group into a reactive acylating agent, which then undergoes a further nucleophilic acyl substitution with the amine.

To summarize, five steps are needed to synthesize a dipeptide such as Ala-Leu:

STEPS 1–2 The amino group of alanine is protected as the BOC derivative, and the carboxyl group of leucine is protected as the methyl ester.

$$
\begin{array}{cc}
Ala + (t\text{-BuO}\overset{O}{\overset{\|}{C}})_2O & Leu + CH_3OH \\
\downarrow & \downarrow \text{ } {\substack{H^+\\ \text{catalyst}}} \\
BOC\text{–}Ala & Leu\text{–}OCH_3
\end{array}
$$

STEP 3 The two protected amino acids are coupled using DCC.

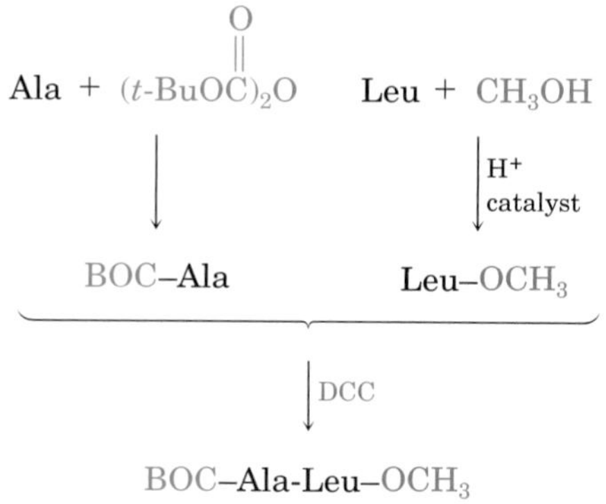

$$\underbrace{\phantom{BOC\text{–}Ala \qquad Leu\text{–}OCH_3}}$$

$$\downarrow DCC$$

$$BOC\text{–}Ala\text{-}Leu\text{–}OCH_3$$

FIGURE 26.5 ▼

The mechanism of amide formation by reaction of a carboxylic acid and an amine with DCC (dicyclohexylcarbodiimide).

Dicyclohexylcarbodiimide (DCC)

The carboxylic acid first adds to the carbodiimide reagent to yield a reactive acylating agent.

Nucleophilic attack of the amine on the acylating agent gives a tetrahedral intermediate.

$H_2\ddot{N}-R'$

The intermediate loses dicyclohexylurea and yields the desired amide.

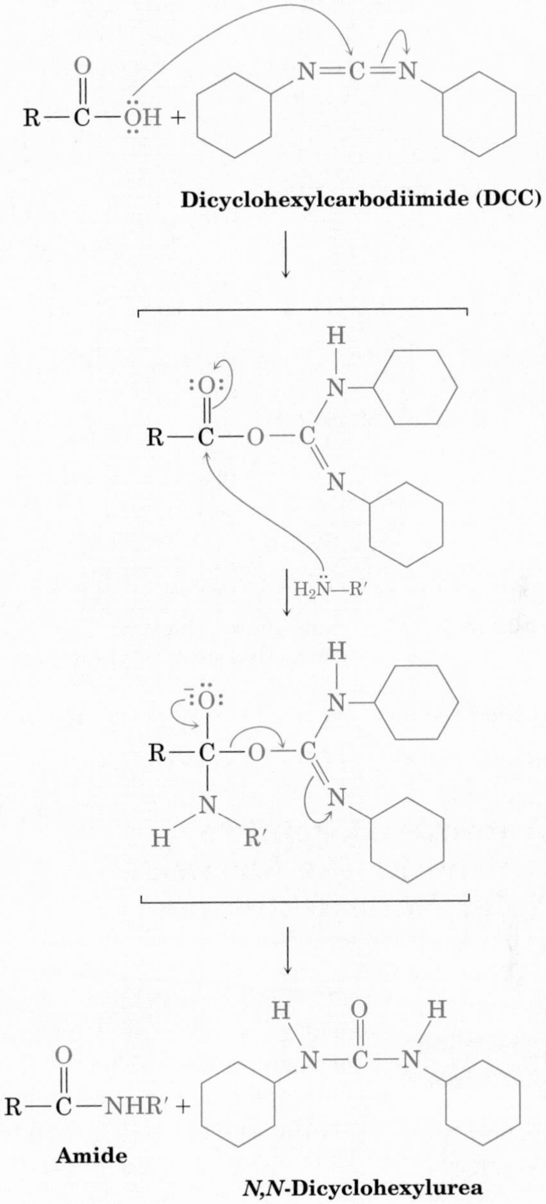

Amide

N,N-Dicyclohexylurea

© 1984 JOHN MCMURRY

STEP 4 The BOC protecting group is removed by acid treatment.

CF_3COOH

Ala-Leu–OCH_3

STEP 5 The methyl ester is removed by basic hydrolysis.

NaOH
H_2O

Ala-Leu

These steps can be repeated to add one amino acid at a time to the growing chain or to link two peptide chains together. Many remarkable achievements in peptide synthesis have been reported, including a complete synthesis of human insulin. Insulin is composed of two chains totaling 51 amino acids linked by two disulfide bridges. Its structure was determined by Frederick Sanger, who received the 1958 Nobel Prize in chemistry for his work.

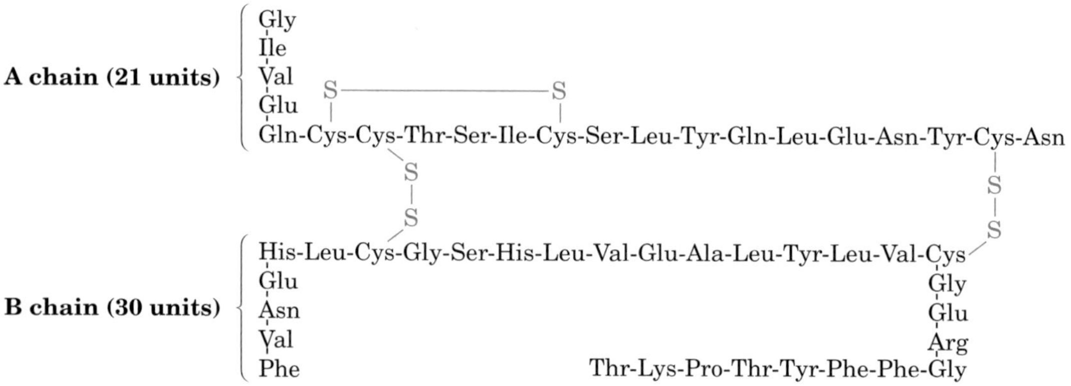

Insulin

Problem 26.20   Show the mechanism for formation of a BOC derivative by reaction of an amino acid with di-*tert*-butyl dicarbonate.

Problem 26.21   Write all five steps required for the synthesis of Leu-Ala from alanine and leucine.

26.11 Automated Peptide Synthesis: The Merrifield Solid-Phase Technique

Robert Bruce Merrifield

Robert Bruce Merrifield (1921–) was born in Fort Worth, Texas, and received his Ph.D. at the University of California, Los Angeles, in 1949. He then joined the faculty at the Rockefeller Institute. In 1984, he was awarded the Nobel Prize in chemistry for his development of methods for automated peptide synthesis.

The synthesis of large peptide chains by sequential addition of one amino acid at a time is a long and arduous task. An immense simplification is possible, however, using the *solid-phase* method introduced by R. Bruce Merrifield at the Rockefeller Institute. In the Merrifield method, peptide synthesis is carried out on solid polymer beads of polystyrene, prepared so that one of every 100 or so benzene rings bears a chloromethyl (–CH_2Cl) group:

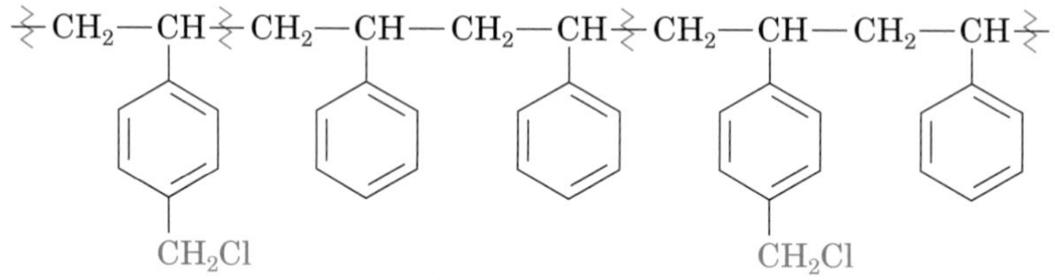

Chloromethylated polystyrene

In the standard solution-phase method discussed in the previous section, a methyl ester was used to protect the carboxyl group during formation of the amide bond. In the solid-phase method, however, a solid *polymer* particle is the ester protecting group. Four steps are required in solid-phase peptide synthesis:

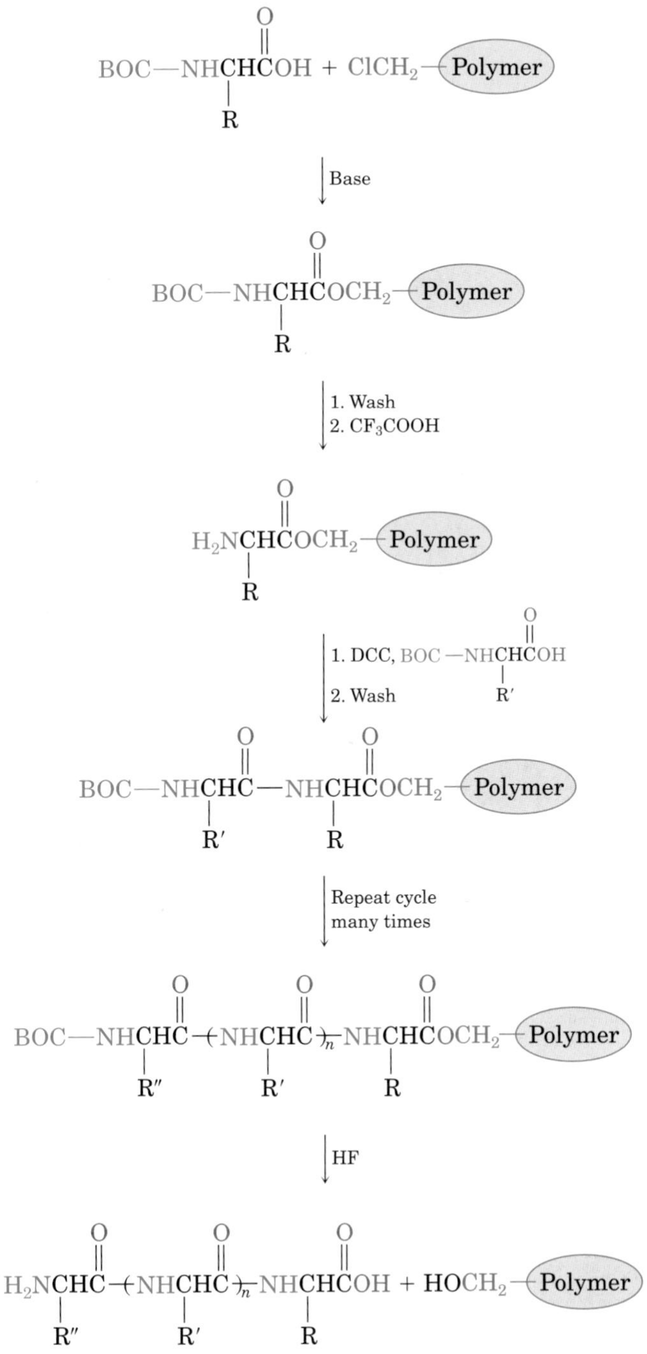
BOC—NHCHCOH + ClCH2—Polymer
R
Base
BOC—NHCHCOCH2—Polymer
R
1. Wash
2. CF3COOH
H2NCHCOCH2—Polymer
R
1. DCC, BOC—NHCHCOH
R′
2. Wash
BOC—NHCHC—NHCHCOCH2—Polymer
R′
R
Repeat cycle
many times
BOC—NHCHC—(NHCHC)n—NHCHCOCH2—Polymer
R″
R′
R
HF
H2NCHC—(NHCHC)n—NHCHCOH + HOCH2—Polymer
R″
R′
R

STEP 1 A BOC-protected amino acid is covalently linked to the polystyrene polymer by formation of an ester bond (S_N2 reaction).

STEP 2 The polymer-bonded amino acid is washed free of excess reagent and then treated with trifluoroacetic acid to remove the BOC group.

STEP 3 A second BOC-protected amino acid is coupled to the first by reaction with DCC. Excess reagents are removed by washing them from the insoluble polymer.

The cycle of deprotection, coupling, and washing is repeated as many times as desired to add amino acid units to the growing chain.

STEP 4 After the desired peptide has been made, treatment with anhydrous HF removes the final BOC group and cleaves the ester bond to the polymer, yielding the free peptide.

The solid-phase technique has now been automated, and computer-controlled peptide synthesizers are available for automatically repeating the coupling and deprotection steps with different amino acids. Each step occurs in high yield, and mechanical losses are minimized because the peptide intermediates are never removed from the insoluble polymer until the final step. Among the many remarkable achievements recorded by Merrifield is the synthesis of bovine pancreatic ribonuclease, a protein containing 124 amino acid units. The entire synthesis required only 6 weeks and took place in 17% overall yield.

26.12 Classification of Proteins

Proteins are classified into two major types according to their composition. **Simple proteins**, such as blood serum albumin, are those that yield only amino acids on hydrolysis. **Conjugated proteins**, which are much more common than simple proteins, yield other compounds such as carbohydrates, fats, or nucleic acids in addition to amino acids on hydrolysis.

Another way to classify proteins is as either *fibrous* or *globular*, according to their three-dimensional shape. **Fibrous proteins**, such as collagen and keratin, consist of polypeptide chains arranged side by side in long filaments. Because these proteins are tough and insoluble in water, they are used in nature for structural materials such as tendons, hooves, horns, and muscles. **Globular proteins**, by contrast, are usually coiled into compact, roughly spherical shapes. These proteins are generally soluble in water and are mobile within cells. Most of the several thousand known enzymes are globular proteins. Table 26.2 lists some common examples of both kinds.

TABLE 26.2 Some Common Fibrous and Globular Proteins

Name	Occurrence and use
Fibrous proteins (insoluble)	
Collagens	Animal hide, tendons, connective tissues
Elastins	Blood vessels, ligaments
Fibrinogen	Necessary for blood clotting
Keratins	Skin, wool, feathers, hooves, silk, fingernails
Myosins	Muscle tissue
Globular proteins (soluble)	
Hemoglobin	Involved in oxygen transport
Immunoglobulins	Involved in immune response
Insulin	Hormone for controlling glucose metabolism
Ribonuclease	Enzyme for controlling RNA synthesis

26.13 Protein Structure

Proteins are so large that the word *structure* takes on a broader meaning than it does with most other organic compounds. In fact, chemists speak of four different levels of structure when describing proteins:

- The **primary structure** of a protein is simply the amino acid sequence.

- The **secondary structure** of a protein describes how *segments* of the peptide backbone orient into a regular pattern.

- The **tertiary structure** describes how the *entire* protein molecule coils into an overall three-dimensional shape.

- The **quaternary structure** describes how individual protein molecules come together to yield large aggregate structures.

Let's look at three examples—α-keratin (fibrous), fibroin (fibrous), and myoglobin (globular)—to see how higher structure affects protein properties.

α-Keratin

α-Keratin is the fibrous structural protein found in wool, hair, nails, and feathers. X-ray studies have shown that segments of the α-keratin chain are coiled into a right-handed helical secondary structure like that of a telephone cord. Illustrated in Figure 26.6, this so-called **α-helix** is stabilized

FIGURE 26.6 ▼

The helical secondary structure present in α-keratin.

alpha-helix

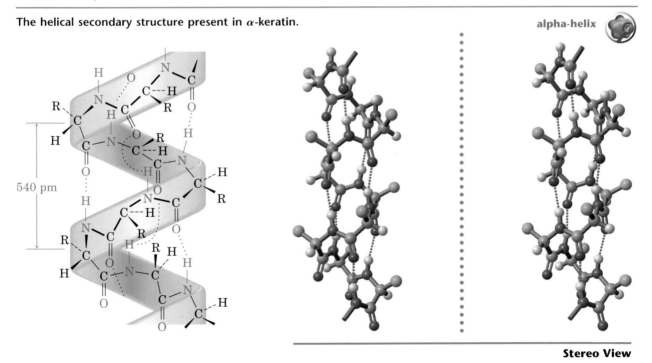

540 pm

Stereo View

by hydrogen bonding between amide N–H groups and C=O groups four residues away. Each coil of the helix contains 3.6 amino acids, and the distance between coils (the *repeat distance*) is 540 pm, or 5.40 Å. Almost all globular proteins contain α-helical segments in their chains.

Fibroin

Fibroin, the fibrous protein found in silk, has a secondary structure called a **β-pleated sheet** in which polypeptide chains line up in a parallel arrangement held together by hydrogen bonds between chains (Figure 26.7). Although not as common as the α-helix, small β-pleated-sheet regions are often found in proteins where sections of peptide chains double back on themselves.

FIGURE 26.7 ▼

The β-pleated-sheet structure present in silk fibroin.

beta-pleated sheet

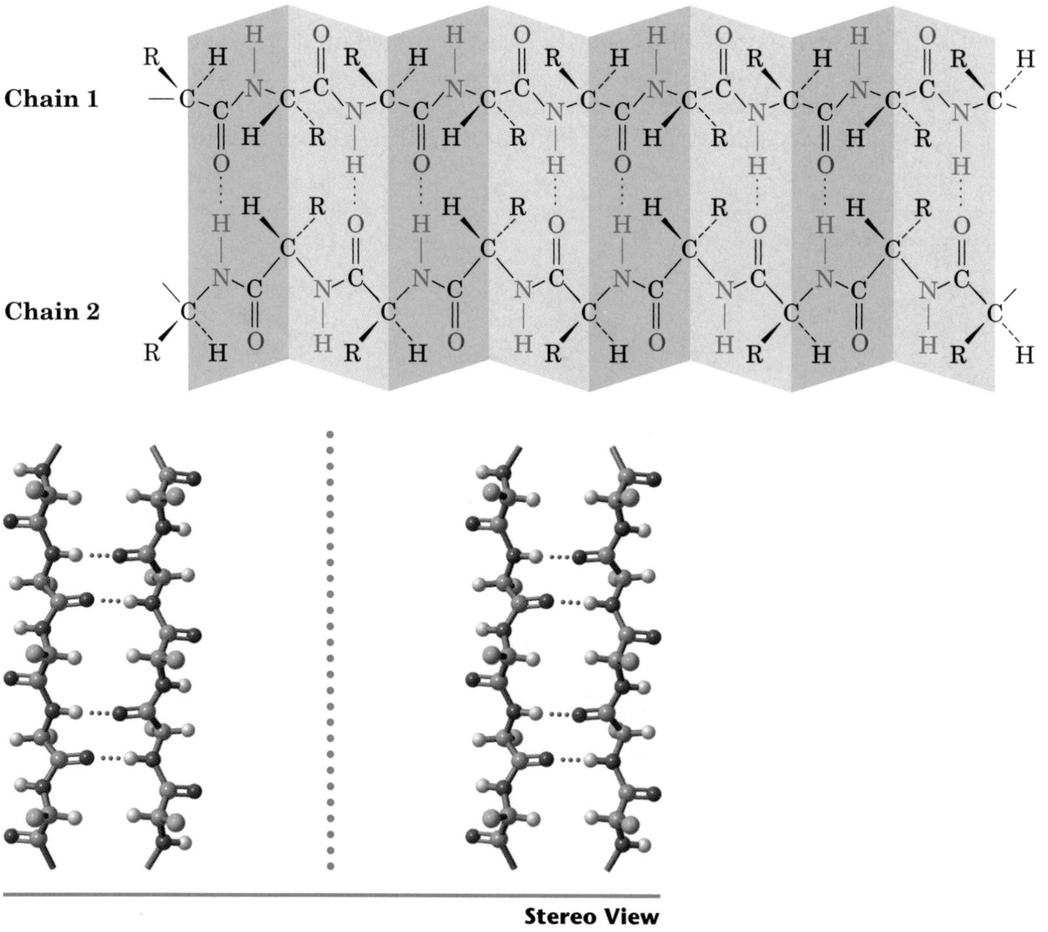

Stereo View

Myoglobin

Myoglobin is a small globular protein containing 153 amino acids in a single chain. A relative of hemoglobin, myoglobin is found in the skeletal muscles of sea mammals, where it stores oxygen needed to sustain the animals during long dives. X-ray evidence obtained by Sir John Kendrew and Max Perutz has shown that myoglobin consists of eight helical segments connected by bends to form a compact, nearly spherical, tertiary structure (Figure 26.8).

Why does myoglobin adopt the shape it does? The forces that determine the tertiary structure of myoglobin and other globular proteins are the same simple forces that act on all molecules, regardless of size, to provide maximum stability. Particularly important are the hydrophobic (water-repelling) interactions of hydrocarbon side chains on neutral amino acids. Those amino acids with neutral, nonpolar side chains have a strong tendency to congregate on the hydrocarbon-like interior of a protein molecule, away from the aqueous medium. Those acidic or basic amino acids with charged side chains, by contrast, tend to congregate on the exterior of the protein where they can be solvated by water.

Sir John Cowdery Kendrew

Sir John Cowdery Kendrew (1917–1997) was born in Oxford, England, and received a Ph.D. in physics in 1949 working with Sir Laurence Bragg at the University of Cambridge. He then served on the faculty at Cambridge before joining the Medical Research Council. In 1962, he received the Nobel Prize for his work on determining the structure of myoglobin.

FIGURE 26.8 ▼

Secondary and tertiary structures of myoglobin, a globular protein.

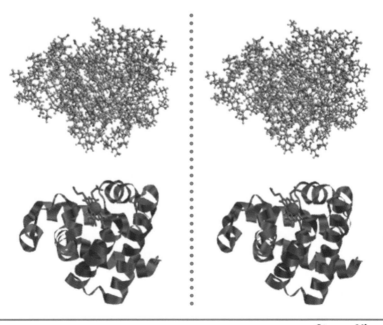

Stereo View

Also important for stabilizing a protein's tertiary structure are the formation of disulfide bridges between cysteine residues, the formation of hydrogen bonds between nearby amino acid residues, and the development of ionic attractions, called *salt bridges*, between positively and negatively charged sites on various amino acid side chains within the protein.

Note that myoglobin is a conjugated protein that contains a covalently bound organic group (a **prosthetic group**) called *heme*. A great many proteins contain such prosthetic groups, which are crucial to their mechanism of action.

Heme

26.14 Enzymes

An **enzyme** is a substance—usually a protein—that acts as a catalyst for a biological reaction. Like all catalysts, enzymes don't affect the equilibrium constant of a reaction and can't bring about a chemical change that is otherwise unfavorable. Enzymes act only to lower the activation energy for a reaction, thereby making the reaction take place more rapidly.

Unlike many of the catalysts that chemists use in the laboratory, enzymes are usually specific in their action. Often, in fact, an enzyme will catalyze only a single reaction of a single compound, called the enzyme's *substrate*. For example, the enzyme amylase found in the human digestive tract catalyzes only the hydrolysis of starch to yield glucose; cellulose and other polysaccharides are untouched by amylase.

Different enzymes have different specificities. Some, such as amylase, are specific for a single substrate, but others operate on a range of substrates. Papain, for instance, a globular protein of 212 amino acids isolated from papaya fruit, catalyzes the hydrolysis of many kinds of peptide bonds. In fact, it's this ability to hydrolyze peptide bonds that makes papain useful as a meat tenderizer and a cleaner for contact lenses.

Most of the more than 2000 known enzymes are globular proteins. In addition to the protein part, most enzymes also have a small nonprotein part called a **cofactor**. The protein part in such an enzyme is called an **apoenzyme**, and the combination of apoenzyme plus cofactor is called a **holoenzyme**. Only holoenzymes have biological activity; neither cofactor nor apoenzyme can catalyze reactions by themselves.

A cofactor can be either an inorganic ion, such as Zn^{2+}, or a small organic molecule, called a **coenzyme**. The requirement of many enzymes for inorganic cofactors is the main reason for our dietary need of trace minerals. Iron, zinc, copper, manganese, and numerous other metal ions are all essential minerals that act as enzyme cofactors, though the exact biological role is not known in all cases.

A variety of organic molecules act as coenzymes. Many, though not all, coenzymes are **vitamins**, small organic molecules that must be obtained in the diet and are required in trace amounts for proper growth. Table 26.3 lists the 13 known vitamins required in the human diet and their enzyme functions.

TABLE 26.3 Vitamins and Their Enzyme Functions

Vitamin	Enzyme function	Deficiency symptoms
Water-soluble vitamins		
Ascorbic acid (vitamin C)	Hydrolases	Bleeding gums, bruising
Thiamin (vitamin B_1)	Reductases	Fatigue, depression
Riboflavin (vitamin B_2)	Reductases	Cracked lips, scaly skin
Pyridoxine (vitamin B_6)	Transaminases	Anemia, irritability
Niacin	Reductases	Dermatitis, dementia
Folic acid (vitamin M)	Methyltransferases	Megaloblastic anemia
Vitamin B_{12}	Isomerases	Megaloblastic anemia, neurodegeneration
Pantothenic acid	Acyltransferases	Weight loss, irritability
Biotin (vitamin H)	Carboxylases	Dermatitis, anorexia, depression
Fat-soluble vitamins		
Vitamin A	Visual system	Night blindness, dry skin
Vitamin D	Calcium metabolism	Rickets, osteomalacia
Vitamin E	Antioxidant	Hemolysis of red blood cells
Vitamin K	Blood clotting	Hemorrhage, delayed blood clotting

Enzymes are grouped into six classes according to the kind of reaction they catalyze (Table 26.4). *Hydrolases* catalyze hydrolysis reactions; *isomerases* catalyze isomerizations; *ligases* catalyze the bonding together of two molecules; *lyases* catalyze the breaking away of a small molecule such as H_2O from a substrate; *oxidoreductases* catalyze oxidations and reductions; and *transferases* catalyze the transfer of a group from one substrate to another.

TABLE 26.4   Classification of Enzymes

Main class	Some subclasses	Type of reaction catalyzed
Hydrolases	Lipases	Hydrolysis of an ester group
	Nucleases	Hydrolysis of a phosphate group
	Proteases	Hydrolysis of an amide group
Isomerases	Epimerases	Isomerization of a chirality center
Ligases	Carboxylases	Addition of CO_2
	Synthetases	Formation of new bond
Lyases	Decarboxylases	Loss of CO_2
	Dehydrases	Loss of H_2O
Oxidoreductases	Dehydrogenases	Introduction of double bond by removal of H_2
	Oxidases	Oxidation
	Reductases	Reduction
Transferases	Kinases	Transfer of a phosphate group
	Transaminases	Transfer of an amino group

Although some enzymes, like papain and trypsin, have uninformative common names, the systematic name of an enzyme has two parts, ending with -*ase*. The first part identifies the enzyme's substrate, and the second part identifies its class. For example, *hexose kinase* is an enzyme that catalyzes the transfer of a phosphate group from adenosine triphosphate (ATP) to glucose.

. . . .

Problem 26.22   To what classes do the following enzymes belong?
(a) Pyruvate decarboxylase   (b) Chymotrypsin   (c) Alcohol dehydrogenase

. . . .

26.15   How Do Enzymes Work? Citrate Synthase

Enzymes exert their catalytic activity by bringing reactant molecules together, holding them in the orientation necessary for reaction, and providing any necessary acidic or basic sites to catalyze specific steps. Let's look, for example, at *citrate synthase*, an enzyme that catalyzes the aldol-like addition of acetyl CoA to oxaloacetate to give citrate (Section 23.14). This reaction is the first step in the so-called *citric acid cycle*, in which acetyl groups produced by degradation of food molecules are metabolically "burned" to yield CO_2 and H_2O. We'll look at the details of the citric acid cycle in Section 29.5.

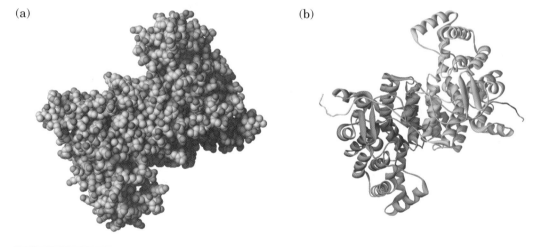

$$^-O_2CCH_2\overset{\overset{\displaystyle O}{\|}}{C}CO_2{}^- + CH_3\overset{\overset{\displaystyle O}{\|}}{C}SCoA \xrightarrow[\text{synthase}]{\text{Citrate}} {}^-O_2CCH_2\underset{\underset{\displaystyle CO_2{}^-}{|}}{\overset{\overset{\displaystyle OH}{|}}{C}}CH_2CO_2{}^- + HSCoA$$

Oxaloacetate   **Acetyl CoA**   **Citrate**

Citrate synthase is a globular protein with a deep cleft lined by an array of functional groups that can bind to oxaloacetate. Upon binding oxaloacetate, the original cleft closes and another opens up to bind acetyl CoA. This second cleft is also lined by appropriate functional groups, including a histidine at position 274 and an aspartic acid at position 375. The two reactants are now held by the enzyme in close proximity and with a suitable orientation for reaction. Figure 26.9 shows the structure of citrate synthase as determined by X-ray crystallography.

FIGURE 26.9 ▼

Computer-generated models of citrate synthase. Part (a) is a space-filling model, which shows the deep clefts in the enzyme. Part (b) is a ribbon model, which emphasizes the α-helical segments of the protein chain and indicates that the enzyme is *dimeric*; that is, it consists of two identical chains held together by hydrogen bonds and other intermolecular attractions.

(a)   (b)

The first step in the aldol reaction is generation of the enol of acetyl CoA. The side-chain carboxyl of Asp 375 acts as base to abstract an acidic α proton, while at the same time the side-chain imidazole ring of His 274 donates H^+ to the carbonyl oxygen. The enol thus produced then does a nucleophilic addition to the ketone carbonyl group of oxaloacetate. The His 274 acts as a base to remove the –OH hydrogen from the enol, while another histidine residue at position 320 simultaneously donates a proton to the oxaloacetate carbonyl group, giving citryl CoA. Water then hydrolyzes the thiol ester group in citryl CoA, releasing citrate and coenzyme A as the final products. The mechanism is shown in Figure 26.10.

FIGURE 26.10 ▼

Mechanism of action of the enzyme citrate synthase.

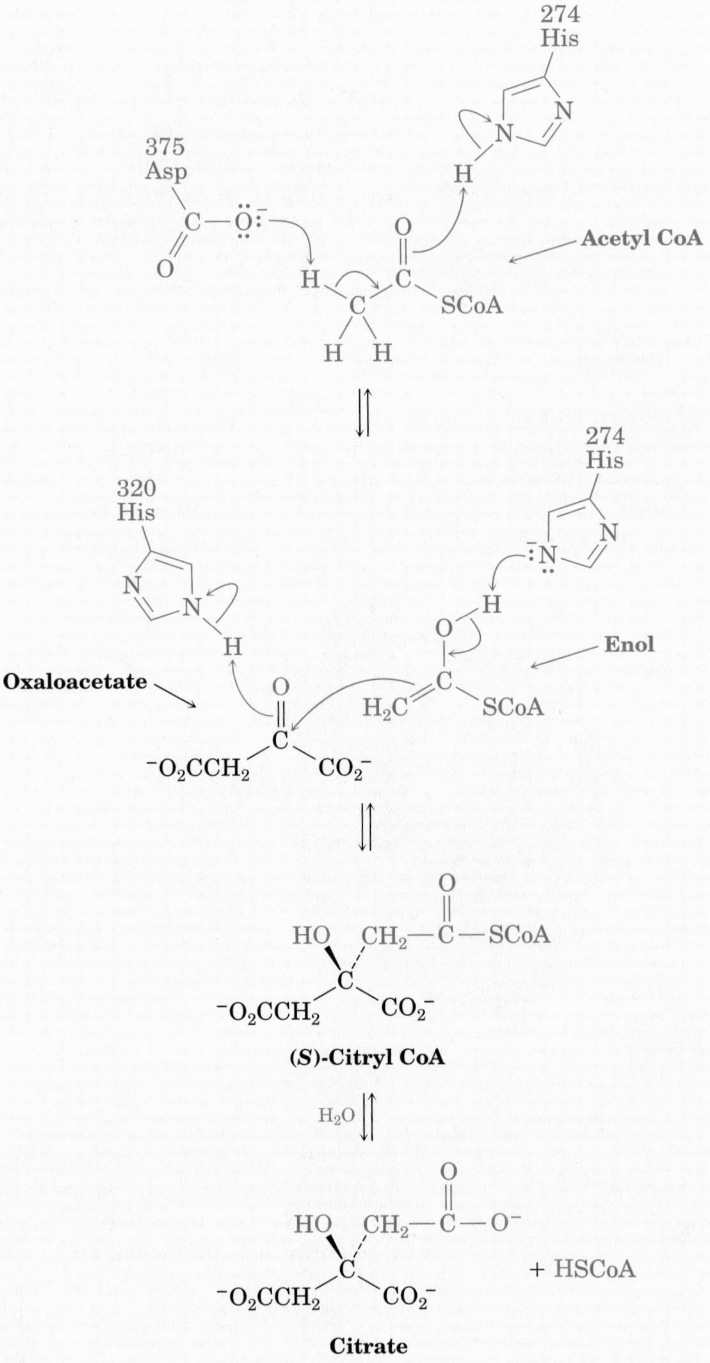

Acetyl CoA is held in the cleft of the citrate synthase enzyme with His 274 and Asp 375 nearby. The side-chain carboxylate group of Asp 375 acts as a base and removes an acidic α proton, while an N–H group on the side chain of His 274 acts as an acid and donates a proton to the carbonyl oxygen. The net result is formation of an enol.

A nitrogen atom on the His 274 side chain acts as a base to deprotonate the acetyl CoA enol, which adds to the ketone carbonyl group of oxaloacetate in an aldol-like reaction. Simultaneously, an acidic N–H proton on the side chain of His 320 protonates the carbonyl oxygen, producing citryl CoA.

The thiol ester group of citryl CoA is hydrolyzed in a typical nucleophilic acyl substitution step, breaking the C–S bond and producing citrate plus coenzyme A.

26.16 Protein Denaturation

The tertiary structure of a globular protein is delicately held together by weak intramolecular attractions. Often, a modest change in temperature or pH will disrupt the tertiary structure and cause the protein to become **denatured**. Denaturation occurs under such mild conditions that the primary structure remains intact but the tertiary structure unfolds from a specific globular shape to a randomly looped chain (Figure 26.11).

FIGURE 26.11 ▼

Schematic representation of protein denaturation. A globular protein loses its specific three-dimensional shape and becomes randomly looped.

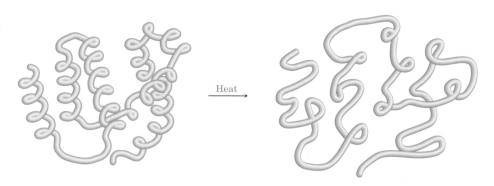

Heat →

Denaturation is accompanied by changes in both physical and biological properties. Solubility is drastically decreased, as occurs when egg white is cooked and the albumins unfold and coagulate. Most enzymes also lose all catalytic activity when denatured, since a precisely defined tertiary structure is required for their action. Although most denaturation is irreversible, some cases have now been found where spontaneous *renaturation* of an unfolded protein to its stable tertiary structure occurs. Renaturation is accompanied by a full recovery of biological activity.

 **CHEMISTRY @ WORK**

Protein and Nutrition

Everyone—from infants to weightlifters—needs protein. Children need large amounts of protein for proper growth, and adults need protein to replace what is lost each day by the body's normal biochemical reactions. Dietary protein is necessary because our bodies can synthesize only 10 of the 20 common amino acids from simple precursor molecules; the other

(continued) ▶

10 amino acids (called *essential amino acids*) must be obtained from food by digestion of edible proteins. Table 26.5 shows the estimated essential amino acid requirements of an infant and an adult.

TABLE 26.5 Estimated Essential Amino Acid Requirements		
	Daily requirement (mg/kg body weight)	
Amino acid	**Infant**	**Adult**
Arginine	?	None
Histidine	33	?
Isoleucine	83	12
Leucine	35	16
Lysine	99	12
Methionine	49	10
Phenylalanine	141	16
Threonine	68	8
Tryptophan	21	3
Valine	92	14

Not all foods provide sufficient amounts of the 10 essential amino acids to meet our minimum daily needs. Most meat and dairy products are satisfactory, but many vegetable sources, such as wheat and corn, are *incomplete*; that is, many vegetable proteins contain too little of one or more essential amino acids to sustain the growth of laboratory animals. Wheat is low in lysine, for example, and corn is low in both lysine and tryptophan.

Using an incomplete food as the sole source of protein can cause nutritional deficiencies, particularly in children. Vegetarians must therefore be careful to adopt a varied diet that provides proteins from several sources. Legumes and nuts, for example, are useful for overcoming the deficiencies of wheat and grains. Some of the limiting amino acids found in various foods are listed in Table 26.6.

TABLE 26.6 Limiting Amino Acids in Some Foods	
Food	**Limiting amino acid**
Wheat, grains	Lysine, threonine
Peas, beans, legumes	Methionine, tryptophan
Nuts, seeds	Lysine
Leafy green vegetables	Methionine

Hard work and dietary protein are both necessary to build muscle mass.

Summary and Key Words

Proteins are large biomolecules made up of **α-amino acid residues** linked together by amide, or *peptide*, bonds. Chains with fewer than 50 amino acids are often called **peptides**, while the term **protein** is reserved for larger chains. Twenty amino acids are commonly found in proteins; all are α-amino acids, and all except glycine have S stereochemistry similar to that of L sugars. In neutral solution, amino acids exist as dipolar **zwitterions**.

Amino acids can be synthesized by several methods, including ammonolysis of an α-bromo acid, **Strecker reaction** of an aldehyde with KCN/NH_3 followed by hydrolysis, alkylation of diethyl acetamidomalonate, and reductive amination of an α-keto acid. Resolution of the synthetic racemate is necessary to obtain an optically active amino acid.

Determining the structure of a peptide or protein is carried out in several steps. The identity and amount of each amino acid present in a peptide is determined by **amino acid analysis**. The peptide is hydrolyzed to its constituent α-amino acids, which are then separated and identified. Next, the peptide is sequenced. **Edman degradation** by treatment with phenyl isothiocyanate (PITC) cleaves one residue from the N terminus of the peptide and forms an easily identifiable phenylthiohydantoin (PTH) derivative of the **N-terminal amino acid**. A series of sequential Edman degradations allows the sequencing of a peptide chain up to 50 residues in length.

Peptide synthesis is made possible by the use of selective protecting groups. An N-protected amino acid with a free carboxyl group is coupled to an O-protected amino acid with a free amino group in the presence of dicyclohexylcarbodiimide (DCC). Amide formation occurs, the protecting groups are removed, and the sequence is repeated. Amines are usually protected as their *tert*-butoxycarbonyl (BOC) derivatives, and acids are protected as esters. This synthetic sequence is often carried out by the **Merrifield solid-phase technique**, in which the peptide is esterified to an insoluble polymeric support.

Proteins have four levels of structure. **Primary structure** describes a protein's amino acid sequence; **secondary structure** describes how segments of the protein chain orient into regular patterns—either **α-helix** or **β-pleated sheet**; **tertiary structure** describes how the *entire* protein molecule coils into an overall three-dimensional shape; and **quaternary structure** describes how individual protein molecules aggregate into larger structures.

Proteins are classified as either globular or fibrous. **Fibrous proteins** such as α-keratin are tough, rigid, and water-insoluble; **globular proteins** such as myoglobin are water-soluble and roughly spherical in shape. Many globular proteins are **enzymes**—substances that act as catalysts for biological reactions. Enzymes are grouped into six classes according to the kind of reaction they catalyze. They exert their catalytic activity by bringing reactant molecules together, holding them in the orientation necessary for reaction, and providing any necessary acidic or basic sites to catalyze specific steps.

Summary of Reactions

1. Amino acid synthesis (Section 26.3)
   (a) From α-bromo acids

$$\underset{\text{O}}{\overset{\text{O}}{\underset{\|}{\text{RCH}_2\text{COH}}}} \xrightarrow[\text{PBr}_3]{\text{Br}_2} \underset{\text{Br}}{\overset{\text{Br}}{\underset{|}{\text{RCHCOOH}}}} \xrightarrow{\text{NH}_3} \underset{\text{NH}_2}{\overset{\text{NH}_2}{\underset{|}{\text{RCHCOOH}}}}$$

   (b) Strecker synthesis

$$\underset{\text{O}}{\overset{\text{O}}{\underset{\|}{\text{RCH}}}} \xrightarrow[\text{NH}_3]{\text{KCN}} \underset{\text{NH}_2}{\overset{\text{NH}_2}{\underset{|}{\text{RCHCN}}}} \xrightarrow{\text{H}_3\text{O}^+} \underset{\text{NH}_2}{\overset{\text{NH}_2}{\underset{|}{\text{RCHCOOH}}}}$$

   (c) Diethyl acetamidomalonate synthesis

$$\underset{\text{O}}{\overset{\text{O}}{\underset{\|}{\text{CH}_3\text{CNHCH(CO}_2\text{Et)}_2}}} \xrightarrow[\substack{\text{2. RX} \\ \text{3. H}_3\text{O}^+}]{\text{1. Na}^+\ ^-\text{OEt}} \underset{\text{R}}{\overset{\text{R}}{\underset{|}{\text{H}_2\text{NCHCOOH}}}}$$

   (d) Reductive amination

$$\underset{\text{O}}{\overset{\text{O}}{\underset{\|}{\text{RCCOOH}}}} \xrightarrow[\text{NaBH}_4]{\text{NH}_3} \underset{\text{NH}_2}{\overset{\text{NH}_2}{\underset{|}{\text{RCHCOOH}}}}$$

2. Peptide sequencing: Edman degradation (Section 26.8)

3. Peptide synthesis (Section 26.10)
   (a) Nitrogen protection

$$\underset{\text{R O}}{\overset{\text{R O}}{\underset{|\ \|}{\text{H}_2\text{NCHCOH}}}} + [(\text{CH}_3)_3\text{COC}]_2\text{O} \longrightarrow (\text{CH}_3)_3\text{COC}-\underset{\text{R O}}{\overset{\text{R O}}{\underset{|\ \|}{\text{NHCHCOH}}}}$$

   BOC-protected amino acid

The BOC protecting group can be removed by acid treatment:

$$(\text{CH}_3)_3\text{COC}-\underset{\text{R O}}{\overset{\text{R O}}{\underset{|\ \|}{\text{NHCHCOH}}}} \xrightarrow{\text{CF}_3\text{COOH}} \underset{\text{R O}}{\overset{\text{R O}}{\underset{|\ \|}{\text{H}_2\text{NCHCOH}}}} + \text{CO}_2 + (\text{CH}_3)_2\text{C}=\text{CH}_2$$

(continued) ▶

(b) Oxygen protection

$$\underset{\text{H}_2\text{NCHCOH}}{\overset{\text{R} \quad \text{O}}{|\quad||}} + \text{CH}_3\text{OH} \xrightarrow{\text{HCl}} \underset{\text{H}_2\text{NCHCOCH}_3}{\overset{\text{R} \quad \text{O}}{|\quad||}}$$

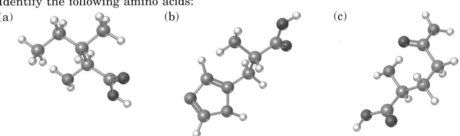

$$\underset{\text{H}_2\text{NCHCOH}}{\overset{\text{R} \quad \text{O}}{|\quad||}} + \text{CH}_2\text{OH} \xrightarrow{\text{HCl}} \underset{\text{H}_2\text{NCHCOCH}_2\text{C}_6\text{H}_5}{\overset{\text{R} \quad \text{O}}{|\quad||}}$$

The ester protecting group can be removed by base hydrolysis:

$$\underset{\text{H}_2\text{NCHCOCH}_3}{\overset{\text{R} \quad \text{O}}{|\quad||}} \xrightarrow[\text{H}_2\text{O}]{\text{-OH}} \underset{\text{H}_2\text{NCHCO}^-}{\overset{\text{R} \quad \text{O}}{|\quad||}} + \text{CH}_3\text{OH}$$

(c) Amide bond formation

$$\underset{\text{BOC}-\text{NHCHCOH}}{\overset{\text{R} \quad \text{O}}{|\quad||}} + \underset{\text{H}_2\text{NCHCOCH}_3}{\overset{\text{R}' \quad \text{O}}{|\quad||}} \xrightarrow{\text{DCC}} \underset{\text{BOC}-\text{NHCHC}}{\overset{\text{R} \quad \text{O}}{|\quad||}} - \underset{\text{NHCHCOCH}_3}{\overset{\text{R}' \text{O}}{|\quad||}}$$

Visualizing Chemistry

• •

(Problems 26.1–26.22 appear within the chapter.)

26.23 Identify the following amino acids:

(a)   (b)   (c)

26.24 Give the sequence of the following tetrapeptide (yellow = S)

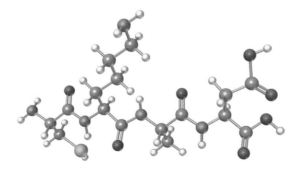

26.25   Isoleucine and threonine (Problem 26.3) are the only two amino acids with two chirality centers. Assign R or S configuration to the methyl-bearing carbon atom of isoleucine:

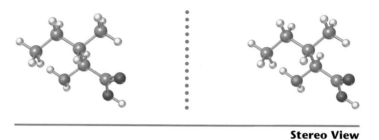

Stereo View

26.26   Is the following structure a D amino acid or an L amino acid? Identify it.

Additional Problems

26.27   Except for cysteine, only S amino acids occur in proteins. Several R amino acids are also found in nature, however. (R)-Serine is found in earthworms, and (R)-alanine is found in insect larvae. Draw Fischer projections of (R)-serine and (R)-alanine. Are these D or L amino acids?

26.28   Cysteine is the only amino acid that has L stereochemistry but an R configuration. Design another L amino acid that also has an R configuration.

26.29   Draw a Fischer projection of (S)-proline.

26.30   Show the structures of the following amino acids in their zwitterionic forms:
(a) Trp   (b) Ile   (c) Cys   (d) His

26.31   Explain the observation that amino acids exist as dipolar zwitterions in aqueous solution but exist largely as nonpolar amino carboxylic acids in chloroform solution.

26.32   At what pH would you carry out an electrophoresis experiment if you wanted to separate a mixture of histidine, serine, and glutamic acid? Explain.

26.33   Proline has $pK_{a1} = 1.99$ and $pK_{a2} = 10.60$. Use the Henderson–Hasselbalch equation to calculate the ratio of protonated and neutral forms at pH = 2.50. Calculate the ratio of neutral and deprotonated forms at pH = 9.70.

26.34   Using the three-letter code names for amino acids, write the structures of all possible peptides containing the following amino acids:
(a) Val, Ser, Leu   (b) Ser, Leu$_2$, Pro

26.35   Predict the product of the reaction of valine with the following reagents:
(a) CH_3CH_2OH, acid   (b) Di-*tert*-butyl dicarbonate
(c) KOH, H_2O (d) CH_3COCl, pyridine; then H_2O

26.36 Show how you could use the Strecker synthesis to prepare the following amino acids:
(a) Glycine   (b) Valine

26.37 Show how you could use the acetamidomalonate method to prepare the following amino acids:
(a) Leucine   (b) Tryptophan

26.38 Show how you could prepare the following amino acids using a reductive amination:
(a) Methionine   (b) Isoleucine

26.39 Serine can be synthesized by a simple variation of the amidomalonate method using formaldehyde rather than an alkyl halide. How might this be done?

26.40 Write full structures for the following peptides:
(a) Val-Phe-Cys-Ala   (b) Glu-Pro-Ile-Leu

26.41 Show the steps involved in a synthesis of Phe-Ala-Val using the Merrifield procedure.

26.42 Draw the structure of the PTH derivative product you would obtain by Edman degradation of the following peptides:
(a) Ile-Leu-Pro-Phe   (b) Asp-Thr-Ser-Gly-Ala

26.43 The following drawing shows a titration curve for an amino acid.
(a) What are the approximate values of pK_{a1} and pK_{a2}, and what is the isoelectric point for this amino acid?
(b) Is this a neutral, acidic, or basic amino acid?

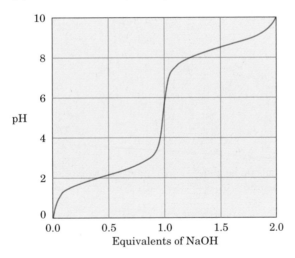

26.44 The α-helical parts of myoglobin and other proteins stop whenever a proline residue is encountered in the chain. Why is proline never present in a protein α-helix?

26.45 Which amide bonds in the following polypeptide are cleaved by trypsin? By chymotrypsin?

Phe-Leu-Met-Lys-Tyr-Asp-Gly-Gly-Arg-Val-Ile-Pro-Tyr

26.46 What kinds of reactions do the following classes of enzymes catalyze?
(a) Hydrolases   (b) Lyases   (c) Transferases

26.47   Which of the following amino acids are more likely to be found on the outside of a globular protein, and which on the inside? Explain.
(a) Valine   (b) Aspartic acid   (c) Phenylalanine   (d) Lysine

26.48   The chloromethylated polystyrene resin used for Merrifield solid-phase peptide synthesis is prepared by treatment of polystyrene with chloromethyl methyl ether and a Lewis acid catalyst. Propose a mechanism for the reaction.

26.49   The *Sanger end-group determination* is sometimes used as an alternative to the Edman degradation. In the Sanger method, a peptide is allowed to react with 2,4-dinitrofluorobenzene, the peptide is hydrolyzed, and the N-terminal amino acid is identified by separation as its *N*-2,4-dinitrophenyl derivative. Propose a mechanism to account for the initial reaction between peptide and dinitrofluorobenzene.

26.50   Proteins can be cleaved specifically at the amide bond on the carboxyl side of methionine residues by reaction with cyanogen bromide, $BrC{\equiv}N$.

The reaction occurs in several steps:
(a) The first step is a nucleophilic substitution reaction of the sulfur on the methionine side chain with BrCN to give a cyanosulfonium ion, R_2SCN^+. Show the structure of the product, and propose a mechanism for the reaction.

(b) The second step is an internal S_N2 reaction, with the carbonyl oxygen of the methionine residue displacing the positively charged sulfur leaving group and forming a five-membered ring product. Show the structure of the product and the mechanism of its formation.

(c) The third step is a hydrolysis reaction to split the peptide chain. The carboxyl group of the former methionine residue is now part of a lactone (cyclic ester) ring. Show the structure of the lactone product and the mechanism of its formation.

(d) The final step is a hydrolysis of the lactone to give the product shown. Show the mechanism of the reaction.

26.51 When an α-amino acid is treated with dicyclohexylcarbodiimide (DCC), a 2,5-diketopiperazine results. Propose a mechanism.

A 2,5-diketopiperazine

26.52 Arginine, the most basic of the 20 common amino acids, contains a *guanidino* functional group in its side chain. Explain, using resonance structures to show how the protonated guanidino group is stabilized.

Arginine

26.53 Cytochrome *c* is an enzyme found in the cells of all aerobic organisms. Elemental analysis of cytochrome *c* shows that it contains 0.43% iron. What is the minimum molecular weight of this enzyme?

26.54 Evidence for restricted rotation around amide CO–N bonds comes from NMR studies. At room temperature, the ^1H NMR spectrum of *N,N*-dimethylformamide shows three peaks: 2.9 δ (singlet, 3 H), 3.0 δ (singlet, 3 H), 8.0 δ (singlet, 1 H). As the temperature is raised, however, the two singlets at 2.9 δ and 3.0 δ slowly merge. At 180°C, the ^1H NMR spectrum shows only two peaks: 2.95 δ (singlet, 6 H) and 8.0 δ (singlet, 1 H). Explain this temperature-dependent behavior.

N,N-**Dimethylformamide**

26.55 Propose a structure for an octapeptide that shows the composition Asp, Gly$_2$, Leu, Phe, Pro$_2$, Val on amino acid analysis. Edman analysis shows a glycine N-terminal group, and carboxypeptidase cleavage produces leucine as the first amino acid to appear. Acidic hydrolysis gives the following fragments:

Val-Pro-Leu, Gly, Gly-Asp-Phe-Pro, Phe-Pro-Val

26.56 The reaction of ninhydrin with an α-amino acid occurs in several steps:
(a) The first step is formation of an imine by reaction of the amino acid with ninhydrin. Show its structure and the mechanism of its formation.
(b) The second step is a decarboxylation. Show the structure of the product and the mechanism of the decarboxylation reaction.
(c) The third step is hydrolysis of an imine to yield an amine and an aldehyde. Show the structures of both products and the mechanism of the hydrolysis reaction.
(d) The final step is formation of the purple anion. Show the mechanism of the reaction.

26.57 Draw resonance forms for the purple anion obtained by reaction of ninhydrin with an α-amino acid (Problem 26.56).

26.58 Look up the structure of human insulin (Section 26.10), and indicate where in each chain the molecule is cleaved by trypsin and chymotrypsin.

26.59 What is the structure of a nonapeptide that gives the following fragments when cleaved?

Trypsin cleavage: Val-Val-Pro-Tyr-Leu-Arg, Ser-Ile-Arg
Chymotrypsin cleavage: Leu-Arg, Ser-Ile-Arg-Val-Val-Pro-Tyr

26.60 Oxytocin, a nonapeptide hormone secreted by the pituitary gland, functions by stimulating uterine contraction and lactation during childbirth. Its sequence was determined from the following evidence:

1. Oxytocin is a cyclic compound containing a disulfide bridge between two cysteine residues.
2. When the disulfide bridge is reduced, oxytocin has the constitution Asn, Cys_2, Gln, Gly, Ile, Leu, Pro, Tyr.
3. Partial hydrolysis of reduced oxytocin yields seven fragments:

Asp-Cys,   Ile-Glu,   Cys-Tyr,   Leu-Gly,   Tyr-Ile-Glu,   Glu-Asp-Cys,   Cys-Pro-Leu

4. Gly is the C-terminal group.
5. Both Glu and Asp are present as their side-chain amides (Gln and Asn) rather than as free side-chain acids.

What is the amino acid sequence of reduced oxytocin? What is the structure of oxytocin itself?

26.61 *Aspartame*, a nonnutritive sweetener marketed under the trade name NutraSweet, is the methyl ester of a simple dipeptide, Asp-Phe-OCH_3.
(a) Draw the structure of aspartame.

(b) The isoelectric point of aspartame is 5.9. Draw the principal structure present in aqueous solution at this pH.

(c) Draw the principal form of aspartame present at physiological pH = 7.3.

26.62 Refer to Figure 26.4 and propose a mechanism for the final step in the Edman degradation—the acid-catalyzed rearrangement of the ATZ derivative to the PTH derivative.

A Look Ahead

· ·

26.63 Amino acids are metabolized by a transamination reaction in which the –NH₂ group of the amino acid changes places with the keto group of an α-keto acid. The products are a new amino acid and a new α-keto acid. Show the product from transamination of isoleucine. (See Section 29.6.)

Molecular Modeling

· ·

26.64 Glycine is a zwitterion in aqueous solution, but what about the gas phase? Compare the gas-phase energies of zwitterionic and neutral glycine using SpartanView, and then tell which is more stable. Use electrostatic potential maps to account for your observations.

26.65 Use SpartanView to examine two peptides, one an alpha helix and the other a beta sheet. What is the amino acid sequence in each peptide? Display the helix as a tube and as a space-filling model. Is there any empty space in the middle of the helix?

26.66 The imidazole ring in the histidine side chain contains two nitrogen atoms. Use SpartanView to compare energies and electrostatic potential maps of imidazolium ion A and imidazolium ion B. Which nitrogen of imidazole is more basic, and why?

A   **B**

Biomolecules: Lipids

Lipids are the naturally occurring organic molecules isolated from cells and tissues by extraction with nonpolar organic solvents. Note that this definition differs from the sort used for carbohydrates and proteins in that lipids are defined by a physical property (solubility) rather than by structure.

Lipids are classified into two general types: those like fats and waxes, which contain ester linkages and can be hydrolyzed, and those like cholesterol and other steroids, which don't have ester linkages and can't be hydrolyzed.

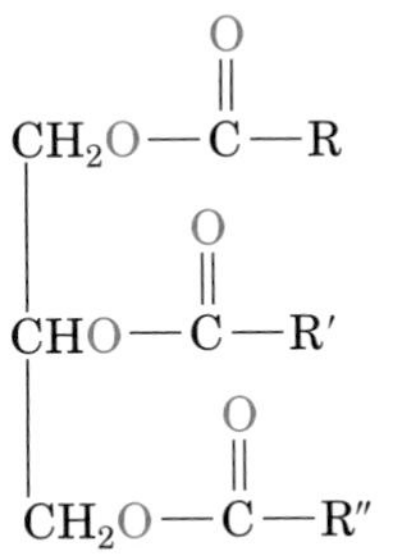

Animal fat—an ester
(R, R′, R″ = C$_{11}$–C$_{19}$ chains)

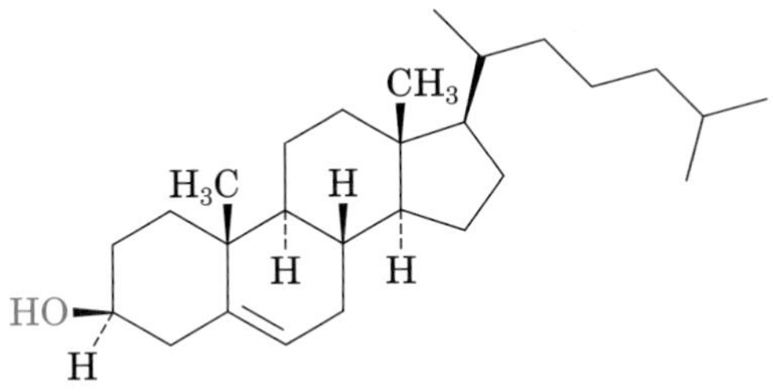

Cholesterol

27.1 Waxes, Fats, and Oils

Waxes are mixtures of esters of long-chain carboxylic acids with long-chain alcohols. The carboxylic acid usually has an even number of carbons from 16 through 36, while the alcohol has an even number of carbons from 24 through 36. One of the major components of beeswax, for instance, is triacontyl hexadecanoate, the ester of the C$_{30}$ alcohol triacontanol and the C$_{16}$ acid hexadecanoic acid. The waxy protective coatings on most fruits, berries, leaves, and animal furs have similar structures.

$$CH_3(CH_2)_{14}\overset{\displaystyle O}{\overset{\displaystyle \|}{C}}O(CH_2)_{29}CH_3$$

Triacontyl hexadecanoate (from beeswax)

Animal fats and vegetable oils are the most widely occurring lipids. Although they appear different—animal fats like butter and lard are solids, whereas vegetable oils like corn and peanut oil are liquid—their structures are closely related. **Fats** and **oils** are **triacylglycerols** (**TAG's**, also called *triglycerides*), triesters of glycerol with three long-chain carboxylic acids. Animals use fats for long-term energy storage because they are much less highly oxidized than carbohydrates and provide about six times as much energy as an equal weight of stored, hydrated glycogen.

Hydrolysis of a fat or oil with aqueous NaOH yields glycerol and three **fatty acids**:

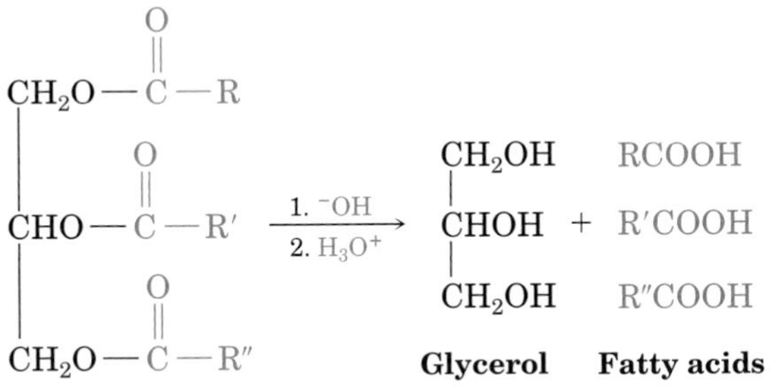

A fat

The fatty acids are generally unbranched and contain an even number of carbon atoms between 12 and 20. If double bonds are present, they usually have Z (cis) geometry. The three fatty acids of a specific triacylglycerol molecule need not be the same, and the fat or oil from a given source is likely to be a complex mixture of many different triacylglycerols. Table 27.1 lists some of the commonly occurring fatty acids, and Table 27.2 lists the approximate composition of fats and oils from different sources.

TABLE 27.1   Structures of Some Common Fatty Acids

Name	Number of carbons	Structure	Melting point (°C)
Saturated			
Lauric	12	$CH_3(CH_2)_{10}COOH$	44
Myristic	14	$CH_3(CH_2)_{12}COOH$	58
Palmitic	16	$CH_3(CH_2)_{14}COOH$	63
Stearic	18	$CH_3(CH_2)_{16}COOH$	70
Arachidic	20	$CH_3(CH_2)_{18}COOH$	75
Unsaturated			
Palmitoleic	16	$CH_3(CH_2)_5CH{=}CH(CH_2)_7COOH$ (cis)	32
Oleic	18	$CH_3(CH_2)_7CH{=}CH(CH_2)_7COOH$ (cis)	16
Ricinoleic	18	$CH_3(CH_2)_5CH(OH)CH_2CH{=}CH(CH_2)_7COOH$ (cis)	5
Linoleic	18	$CH_3(CH_2)_4CH{=}CHCH_2CH{=}CH(CH_2)_7COOH$ (cis, cis)	−5
Arachidonic	20	$CH_3(CH_2)_4(CH{=}CHCH_2)_4CH_2CH_2COOH$ (all cis)	−50

TABLE 27.2   Approximate Composition of Some Fats and Oils

Source	Saturated fatty acids (%) C_{12} Lauric	C_{14} Myristic	C_{16} Palmitic	C_{18} Stearic	Unsaturated fatty acids (%) C_{18} Oleic	C_{18} Ricinoleic	C_{18} Linoleic
Animal fat							
Lard	—	1	25	15	50	—	6
Butter	2	10	25	10	25	—	5
Human fat	1	3	25	8	46	—	10
Whale blubber	—	8	12	3	35	—	10
Vegetable oil							
Coconut	50	18	8	2	6	—	1
Corn	—	1	10	4	35	—	45
Olive	—	1	5	5	80	—	7
Peanut	—	—	7	5	60	—	20
Linseed	—	—	5	3	20	—	20
Castor bean	—	—	—	1	8	85	4

More than 100 different fatty acids have been identified; about 40 of them occur widely. Palmitic acid (C_{16}) and stearic acid (C_{18}) are the most abundant saturated fatty acids; oleic and linoleic acids (C_{18}) are the most abundant unsaturated ones. Oleic acid is monounsaturated since it has only one double bond, whereas linoleic, linolenic, and arachidonic acids are **polyunsaturated fatty acids**, or **PUFA's**, because they have more than one double bond. Linoleic and linolenic acids occur in cream and are essential in the human diet; infants grow poorly and develop skin lesions if fed a diet of nonfat milk for prolonged periods.

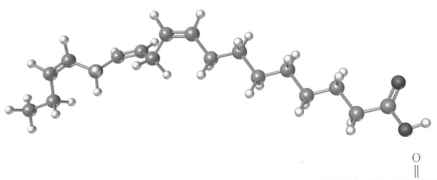

$$CH_3CH_2CH_2CH_2CH_2CH_2CH_2CH_2CH_2CH_2CH_2CH_2CH_2CH_2CH_2CH_2CH_2\overset{\textstyle O}{\overset{\textstyle \|}{C}}OH$$

Stearic acid

$$CH_3CH_2CH{=}CHCH_2CH{=}CHCH_2CH{=}CHCH_2CH_2CH_2CH_2CH_2CH_2\overset{\textstyle O}{\overset{\textstyle \|}{C}}OH$$

Linolenic acid, a polyunsaturated fatty acid (PUFA)

The data in Table 27.1 show that unsaturated fatty acids generally have lower melting points than their saturated counterparts, a trend that also holds true for triacylglycerols. Since vegetable oils generally have a higher proportion of unsaturated to saturated fatty acids than animal fats (Table 27.2), they have lower melting points. The difference is a consequence of structure. Saturated fats have a uniform shape that allows them to pack together efficiently in a crystal lattice. In unsaturated vegetable oils, however, the C=C bonds introduce bends and kinks into the hydrocarbon chains, making crystal formation more difficult. The more double bonds there are, the harder it is for the molecules to crystallize and the lower the melting point of the oil.

The C=C bonds in vegetable oils can be reduced by catalytic hydrogenation (Section 7.7) to produce saturated solid or semisolid fats. Margarine and solid cooking fats, such as Crisco, are produced by hydrogenating soybean, peanut, or cottonseed oil until the proper consistency is obtained.

• •

Problem 27.1   Carnauba wax, used in floor and furniture polishes, contains an ester of a C_{32} straight-chain alcohol with a C_{20} straight-chain carboxylic acid. Draw its structure.

Problem 27.2   Draw structures of glyceryl tripalmitate and glyceryl trioleate. Which would you expect to have a higher melting point?

• •

27.2 Soap

Soap has been known since at least 600 BC, when the Phoenicians prepared a curdy material by boiling goat fat with extracts of wood ash. The cleansing properties of soap weren't generally recognized, however, and the use of soap did not become widespread until the eighteenth century. Chemically, soap is a mixture of the sodium or potassium salts of the long-chain fatty acids produced by hydrolysis (*saponification*) of animal fat with alkali. Wood ash was used as a source of alkali until the mid-1800s, when the development of the LeBlanc process made NaOH commercially available.

$$
\begin{array}{c}
\text{CH}_2\text{OCR} \\
\overset{\displaystyle O}{\overset{\|}{\text{CHOCR}}} \xrightarrow[\text{H}_2\text{O}]{\text{NaOH}} 3\ \text{RCO}^- \ \text{Na}^+ + \overset{\text{CH}_2\text{OH}}{\underset{\text{CH}_2\text{OH}}{\text{CHOH}}} \\
\overset{\displaystyle O}{\overset{\|}{\text{CH}_2\text{OCR}}}
\end{array}
$$

A fat
(R = C_{11}–C_{19} aliphatic chains)

Soap   **Glycerol**

Crude soap curds contain glycerol and excess alkali as well as soap but can be purified by boiling with water and adding NaCl or KCl to precipitate the pure carboxylate salts. The smooth soap that precipitates is dried, perfumed, and pressed into bars for household use. Dyes are added to make colored soaps, antiseptics are added for medicated soaps, pumice is added for scouring soaps, and air is blown in for soaps that float. Regardless of these extra treatments and regardless of price, though, all soaps are basically the same.

Soaps act as cleansers because the two ends of a soap molecule are so different. The anionic carboxylate end of the long-chain molecule is ionic and therefore *hydrophilic* (water-loving). As a result, it tries to dissolve in water. The long hydrocarbon portion of the molecule, however, is nonpolar and *hydrophobic* (water-fearing). It therefore tries to avoid water and is attracted to grease. The net effect of these two opposing tendencies is that soaps are attracted to both grease and water and are therefore valuable as cleansers.

When soaps are dispersed in water, the long hydrocarbon tails cluster together on the inside of a tangled, hydrophobic ball, while the ionic heads on the surface of the cluster stick out into the water layer. These spherical clusters, called **micelles**, are shown schematically in Figure 27.1. Grease and oil droplets are solubilized in water when they are coated by the non-polar tails of soap molecules in the center of micelles. Once solubilized, the grease and dirt can be rinsed away.

FIGURE 27.1 ▼

A soap micelle solubilizing a grease particle in water. An electrostatic potential map of a fatty acid carboxylate shows how the negative charge is located in the head group.

fatty acid carboxylate

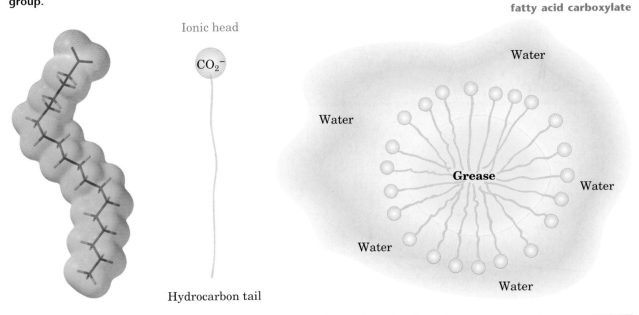

Ionic head

CO_2^-

Water

Water

Water

Grease

Water

Water

Water

Hydrocarbon tail

Soaps make life more pleasant than it might otherwise be, but they also have drawbacks. In hard water, which contains metal ions, soluble sodium carboxylates are converted into insoluble magnesium and calcium salts, leaving the familiar ring of scum around bathtubs and the gray tinge on white clothes. Chemists have circumvented these problems by synthe-sizing a class of synthetic detergents based on salts of long-chain alkyl-benzenesulfonic acids. The principle of synthetic detergents is the same as that of soaps: The alkylbenzene end of the molecule is attracted to grease, while the anionic sulfonate end is attracted to water. Unlike soaps, though, sulfonate detergents don't form insoluble metal salts in hard water and don't leave an unpleasant scum. Note that the electrostatic potential map of a detergent is similar to that of a soap molecule, with the negative charge located in the ionic head group.

detergent

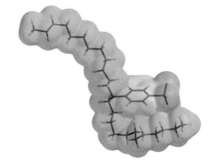

A synthetic detergent
(R = a mixture of C$_{12}$ aliphatic chains)

· ·

Problem 27.3   Draw the structure of magnesium oleate, a component of bathtub scum.

· ·

27.3 Phospholipids

Just as waxes, fats, and oils are esters of carboxylic acids, **phospholipids** are diesters of phosphoric acid, H_3PO_4:

**A phosphoric
acid monoester**

**A phosphoric
acid diester**

**A phosphoric
acid triester**

**A carboxylic
acid ester**

There are two main kinds of phospholipids: *phosphoglycerides* and *sphingolipids*. **Phosphoglycerides** are closely related to fats and oils in that they contain a glycerol backbone linked by ester bonds to two fatty acids and one phosphoric acid. Although the fatty acid residues can be any of the C_{12}–C_{20} units typically present in fats, the acyl group at C1 is usually saturated and the one at C2 is usually unsaturated. The phosphate group at C3 is also bound by a separate ester link to an amino alcohol such as choline, $[(CH_3)_3NCH_2CH_2OH]^+$, or ethanolamine, $H_2NCH_2CH_2OH$. The most important phosphoglycerides are the *lecithins* and the *cephalins*. Note that these compounds are chiral and that they have an L, or R, configuration at C2.

L configuration

$$CH_2O-\overset{\overset{\displaystyle O}{\|}}{C}-R$$

$$R'-\overset{\overset{\displaystyle O}{\|}}{C}-O-\overset{|}{\underset{|}{C}}-H$$

$$CH_2O-\overset{\overset{\displaystyle O}{\|}}{\underset{\underset{\displaystyle O^-}{|}}{P}}-O-CH_2CH_2\overset{+}{N}(CH_3)_3$$

Phosphatidylcholine, a lecithin

$$CH_2O-\overset{\overset{\displaystyle O}{\|}}{C}-R$$

$$R'-\overset{\overset{\displaystyle O}{\|}}{C}-O-\overset{|}{\underset{|}{C}}-H$$

$$CH_2O-\overset{\overset{\displaystyle O}{\|}}{\underset{\underset{\displaystyle O^-}{|}}{P}}-O-CH_2CH_2\overset{+}{N}H_3$$

Phosphatidylethanolamine, a cephalin

where R is saturated and R′ is unsaturated

Found widely in both plant and animal tissues, phosphoglycerides comprise the major lipid component of cell membranes (approximately 40%). Like soaps, phosphoglycerides have a long, nonpolar hydrocarbon tail bound to a polar ionic head (the phosphate group). Cell membranes are composed mostly of phosphoglycerides organized into a **lipid bilayer** about 5.0 nm, or 50 Å, thick. As shown in Figure 27.2, the hydrophobic tails aggregate in the center of the bilayer in much the same way that soap tails aggregate in the center of a micelle. This bilayer serves as an effective barrier to the passage of water, ions, and other components into and out of cells.

FIGURE 27.2 ▼

Aggregation of phosphoglycerides into the lipid bilayer that composes cell membranes.

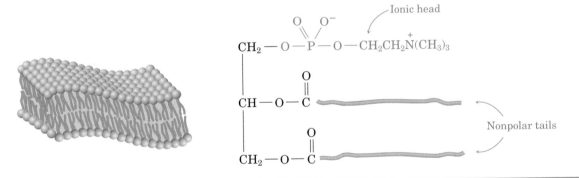

Sphingolipids are the second major group of phospholipids. These compounds, which have sphingosine or a related dihydroxyamine as their backbone, are constituents of plant and animal cell membranes. They are particularly abundant in brain and nerve tissue, where compounds called *sphingomyelins* are a major constituent of the coating around nerve fibers.

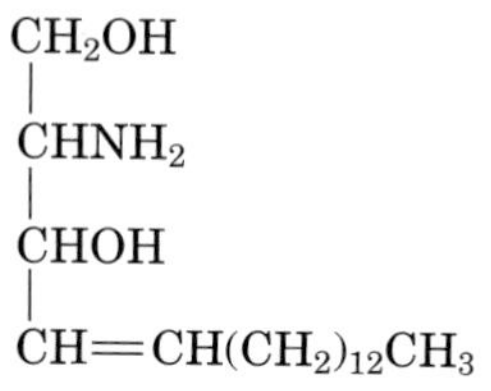

$$CH_2OH$$
$$|$$
$$CHNH_2$$
$$|$$
$$CHOH$$
$$|$$
$$CH{=}CH(CH_2)_{12}CH_3$$

Sphingosine

$$\begin{array}{c} O \\ \| \end{array}$$
$$CH_2O{-}P{-}OCH_2CH_2\overset{+}{N}(CH_3)_3$$
$$|$$
$$O^-$$
$$CHNHCO(CH_2)_{16-24}CH_3$$
$$|$$
$$CHOH$$
$$|$$
$$CH{=}CH(CH_2)_{12}CH_3$$

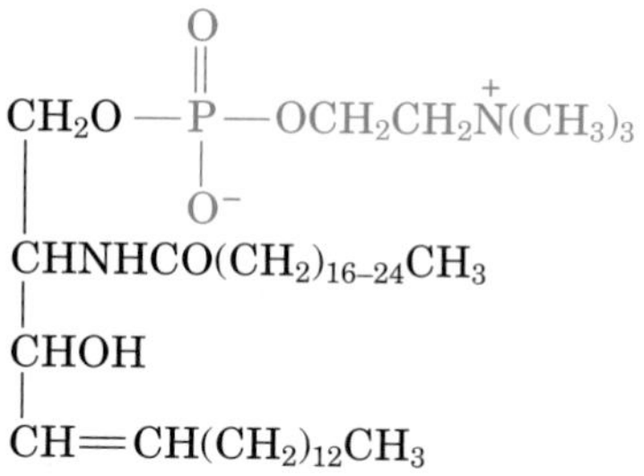

Sphingomyelin, a sphingolipid

27.4   Prostaglandins

Ulf Svante von Euler

Ulf Svante von Euler (1905–1983) was born in Stockholm, Sweden, to a distinguished academic family. His father, Hans von Euler-Chelpin, received the 1929 Nobel Prize in chemistry; his godfather, Svante Arrhenius, received the 1903 Nobel Prize in chemistry; and his mother had a Ph.D. in botany. Von Euler received an M.D. from the Karolinska Institute in 1930, and then remained there his entire career (1930–1971). He received the 1970 Nobel Prize in medicine for his work on the chemical transmission of nerve impulses.

The **prostaglandins** are a group of C_{20} carboxylic acids that contain a five-membered ring with two long side chains. First isolated in the 1930s by Ulf von Euler at the Karolinska Institute in Sweden, much of the structural and chemical work on the prostaglandins was carried out by Sune Bergström and Bengt Samuelsson at the Karolinska Institute. The name *prostaglandin* derives from the fact that the compounds were first isolated from sheep prostate glands, but they have subsequently been shown to be present in small amounts in all body tissues and fluids. Prostaglandin E_1 (PGE_1) and prostaglandin $F_{2\alpha}$ ($PGF_{2\alpha}$) are representative structures:

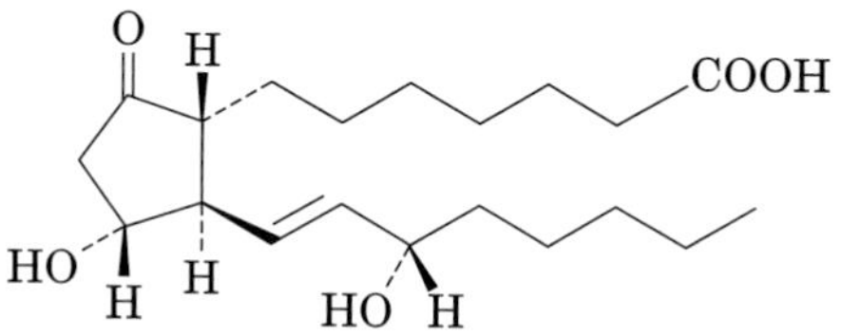

Prostaglandin E_1 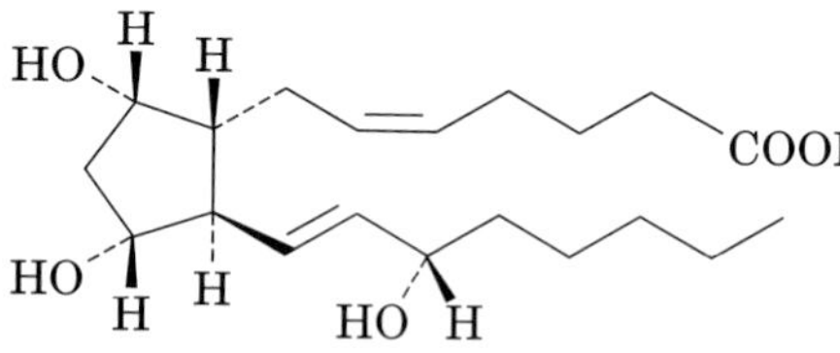   **Prostaglandin $F_{2\alpha}$**

The several dozen known prostaglandins have an extraordinarily wide range of biological effects. They can lower blood pressure, affect blood-platelet aggregation during clotting, lower gastric secretions, control inflammation, affect kidney function, affect reproductive systems, and stimulate uterine contractions during childbirth. In addition, compounds that are closely related to the prostaglandins have still other effects. Interest has centered particularly on the thromboxanes, on prostacyclin, and on the leukotrienes, compounds whose release in the body appears to trigger the asthmatic response.

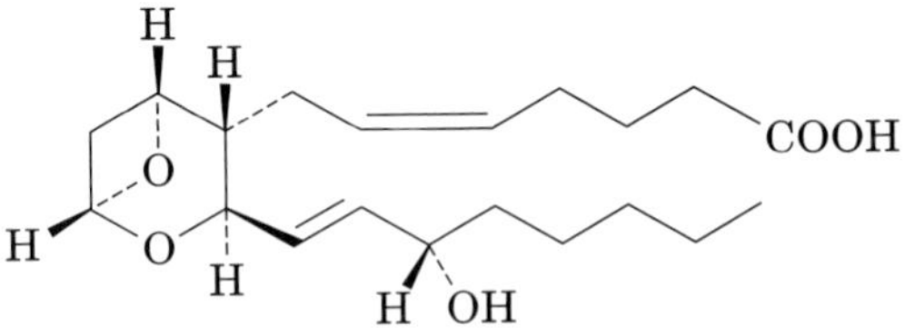

Bengt Samuelsson

Bengt Samuelsson (1934–) was born in Halmstad, Sweden, and received both Ph.D. (1960) and M.D. (1961) degrees from the Karolinska Institute, where he worked with Sune Bergström. He remained at the Karolinska Institute as professor, and shared the 1982 Nobel Prize in medicine with Bergström and John R. Vane.

Sune K. Bergström

Sune K. Bergström (1916–) was born in Stockholm, Sweden, and received an M.D. from the Karolinska Institute in 1944. He was professor at the University of Lund (1947–1958) before moving back to the Karolinska Institute in 1958. He shared the 1982 Nobel Prize in medicine for his work on identifying and studying the prostaglandins.

Thromboxane A$_2$

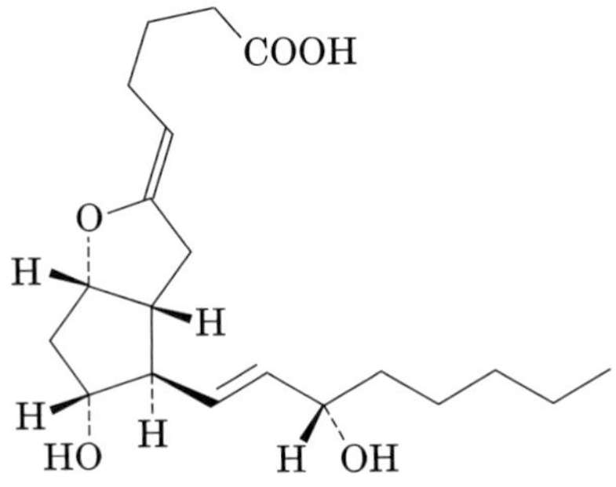

Prostacyclin

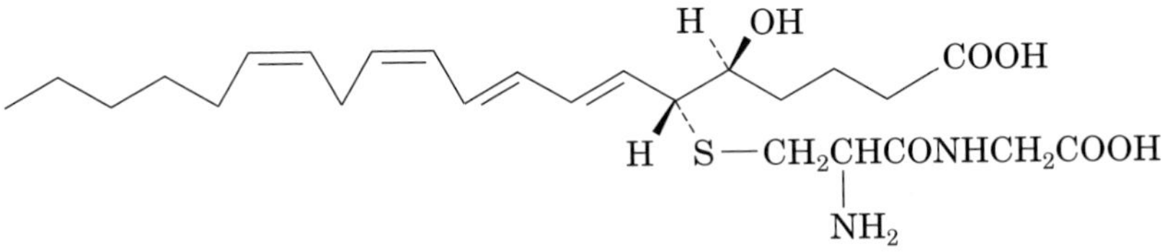

Leukotriene D$_4$

The prostaglandins are biosynthesized in nature from the C$_{20}$ unsaturated fatty acid, arachidonic acid, which is itself synthesized from linoleic acid. The transformation of arachidonic acid to prostaglandins is catalyzed by the *cyclooxygenase* (COX) enzyme, also called PGH synthase, and involves two steps. First, an enzyme-catalyzed cyclooxygenation of arachidonic acid by reaction with molecular oxygen gives prostaglandin G$_2$ (PGG$_2$), and reduction with the same enzyme then gives PGH$_2$. Further transformations by other enzymes yield a variety of other prostaglandins.

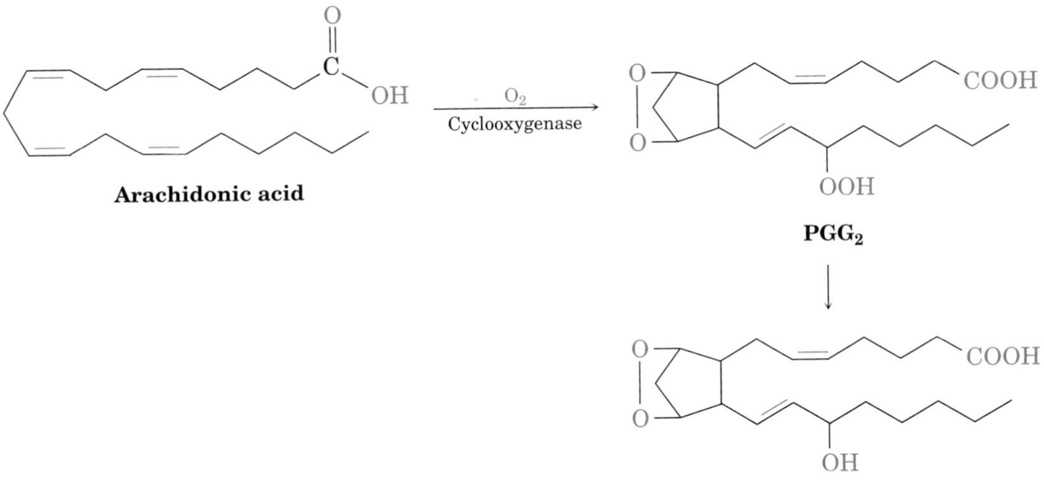

Arachidonic acid

O_2
Cyclooxygenase

PGG$_2$

PGH$_2$

There are two forms of the cyclooxygenase enzyme. Cyclooxygenase-1 (COX-1) carries out the normal physiological production of prostaglandins, and cyclooxygenase-2 (COX-2) produces additional prostaglandin in response to arthritis or other inflammatory conditions. Aspirin and other nonsteroidal anti-inflammatory drugs (NSAID's; see "Aspirin and Other Aromatic NSAID's" at the end of Chapter 15) function by blocking the COX enzymes, thereby decreasing the body's response to inflammation. Unfortunately, *both* COX-1 and COX-2 enzymes are blocked by aspirin, ibuprofen, and other NSAID's, thereby shutting down not only the response to inflammation but also various protective functions, including the control mechanism for production of acid in the stomach. As a result, aspirin causes stomach acidity to rise and can cause stomach lesions.

Recently, medicinal chemists have devised a number of drugs that act as selective inhibitors of the COX-2 enzyme. Inflammation is thereby controlled without harmful side effects. Celecoxib, introduced by Monsanto under the name Celebrex, and rofecoxib, introduced by Merck under the name Vioxx, appear poised to revolutionize the medical treatment of arthritis and other inflammatory conditions.

Celecoxib
(Celebrex)

Rofecoxib
(Vioxx)

Problem 27.4   Assign *R* or *S* configuration to each chirality center in prostaglandin E_2, the most common and biologically potent of mammalian prostaglandins.

Prostaglandin E_2

27.5 Terpenes

It has been known for centuries that codistillation of many plant materials with steam, a technique called *steam distillation*, produces a fragrant mixture of liquids called plant **essential oils**. For thousands of years, such plant extracts have been used as medicines, spices, and perfumes. The investigation of essential oils also played a major role in the emergence of organic chemistry as a science during the nineteenth century.

Chemically, plant essential oils consist largely of mixtures of lipids called *terpenes*. **Terpenes** are small organic molecules with an immense diversity of structure. Thousands of different terpenes are known. Some are hydrocarbons, and others contain oxygen; some are open-chain molecules, and others contain rings. Figure 27.3 gives some examples.

FIGURE 27.3 ▼

The structures of some terpenes isolated from plant essential oils.

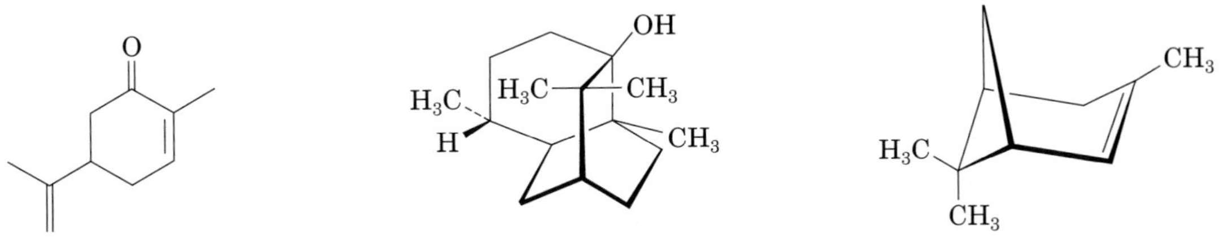

| Carvone (oil of spearmint) | Patchouli alcohol (patchouli oil) | α-Pinene (turpentine) |

All terpenes are related, regardless of their apparent structural differences. According to the *isoprene rule* proposed by Leopold Ruzicka, terpenes can be thought of as arising from head-to-tail joining of five-carbon isoprene (2-methyl-1,3-butadiene) units. Carbon 1 is called the head and carbon 4 is the tail. For example, myrcene contains two isoprene units joined head to tail, forming an eight-carbon chain with two one-carbon branches. α-Pinene similarly contains two isoprene units assembled into a more complex cyclic structure.

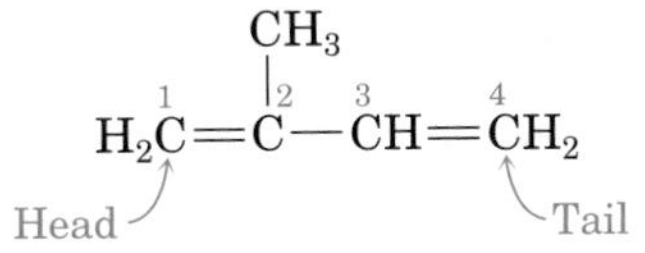

Isoprene
(2-Methyl-1,3-butadiene)

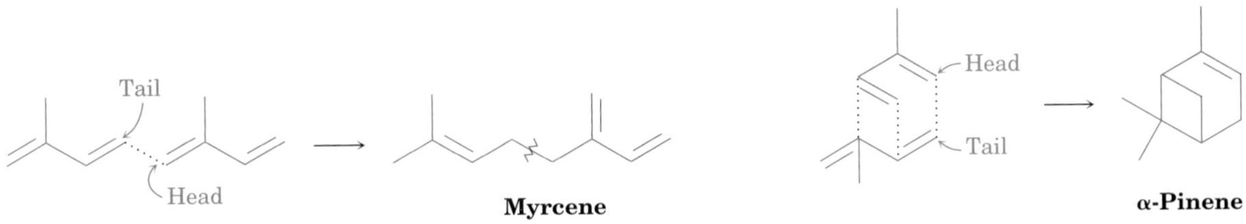

Myrcene

α-Pinene

Terpenes are classified according to the number of isoprene units they contain. Thus, **monoterpenes** are 10-carbon substances biosynthesized from two isoprene units, **sesquiterpenes** are 15-carbon molecules from three isoprene units, and so on (Table 27.3).

Leopold Stephen Ruzicka

Leopold Stephen Ruzicka (1887–1976) was born in Vukovar, Croatia. Though few in his family had much formal schooling, he nonetheless decided to study chemistry and ultimately received his Ph.D. in 1910 at the University of Karlsruhe with Hermann Staudinger. He followed Staudinger to the Swiss Federal Institute (E.T.H.) in Zurich and later became professor there (1929–1957). He was the first to show that rings of more than eight carbons are possible, and he opened up the entire field of terpene chemistry. With Adolf Butanandt of Germany, he received the 1939 Nobel Prize in chemistry.

TABLE 27.3 Classification of Terpenes		
Carbon atoms	**Isoprene units**	**Classification**
10	2	Monoterpene
15	3	Sesquiterpene
20	4	Diterpene
25	5	Sesterterpene
30	6	Triterpene
40	8	Tetraterpene

   Mono- and sesquiterpenes are found primarily in plants, but the higher terpenes occur in both plants and animals, and many have important biological roles. The triterpene lanosterol, for example, is the precursor from which all steroid hormones are made; the tetraterpene β-carotene is a major dietary source of vitamin A.

Lanosterol, a triterpene (C_{30})

β-Carotene, a tetraterpene (C_{40})

Problem 27.5 Show the positions of the isoprene units in the following terpenes:

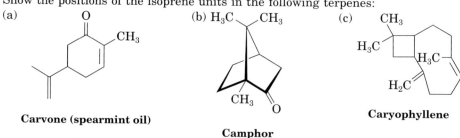

(a) **Carvone (spearmint oil)**

(b) **Camphor**

(c) **Caryophyllene**

27.6 Biosynthesis of Terpenes

The isoprene rule is a convenient formalism, but isoprene itself is not the biological precursor of terpenes. Nature instead uses two "isoprene equivalents"—isopentenyl pyrophosphate and dimethylallyl pyrophosphate. These five-carbon molecules are themselves made from condensation of three acetyl CoA units (Section 21.9).

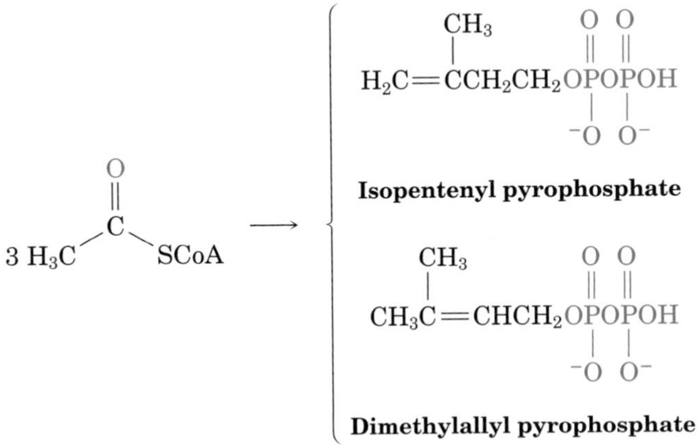

Isopentenyl pyrophosphate

Dimethylallyl pyrophosphate

Dimethylallyl pyrophosphate is an effective alkylating agent in S_N2-like reactions because the primary, allylic pyrophosphate (abbreviated OPP) can be displaced as a leaving group. Thus, displacement of the pyrophosphate group by the nucleophilic C=C bond of isopentenyl pyrophosphate, followed by loss of a proton from the carbocation reaction intermediate, leads to the head-to-tail coupled 10-carbon unit, geranyl pyrophosphate. The corresponding alcohol, geraniol, is itself a fragrant terpene that occurs in rose oil.

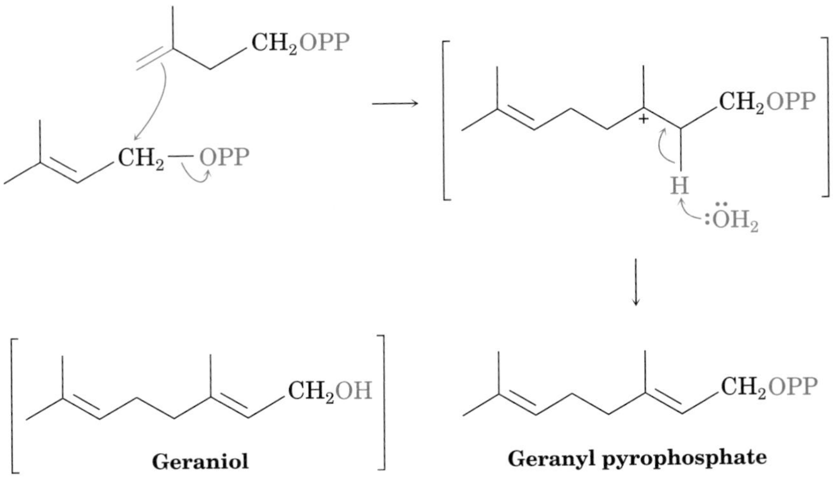

Geraniol

Geranyl pyrophosphate

Geranyl pyrophosphate is the precursor of all monoterpenes. Limonene, for instance, a monoterpene found in many citrus oils, arises from geranyl pyrophosphate by a cis-to-trans double-bond isomerization to give neryl pyrophosphate, followed by internal nucleophilic displacement of the pyrophosphate group and subsequent loss of a proton.

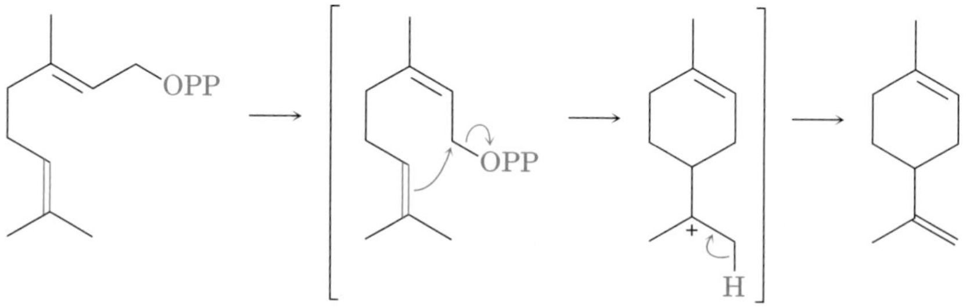

Geranyl pyrophosphate → **Neryl pyrophosphate** → **Limonene**

Reaction of geranyl pyrophosphate with isopentenyl pyrophosphate yields the 15-carbon farnesyl pyrophosphate, the precursor of all sesquiterpenes. Farnesol, the corresponding alcohol, is found in citronella oil and lemon oil.

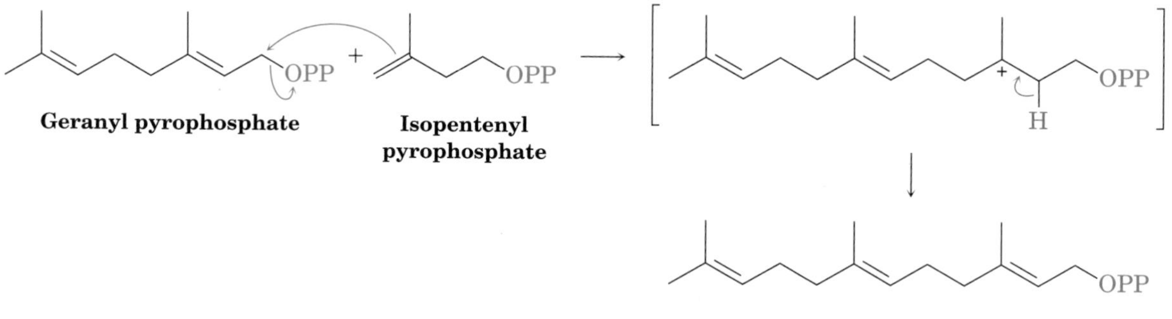

Geranyl pyrophosphate   **Isopentenyl pyrophosphate**

Farnesyl pyrophosphate

Further reactions of farnesyl pyrophosphate with yet other isopentenyl pyrophosphate molecules give the 20-carbon and 25-carbon units that serve as precursors of diterpenes and sesterterpenes, respectively. Triterpenes, however, arise not by further reaction with isopentenyl pyrophosphate but by a reductive tail-to-tail coupling of two 15-carbon farnesyl pyrophosphates to give squalene, a 30-carbon hexaene. Squalene, a major constituent of shark oil, is the precursor from which all triterpenes and steroids arise.

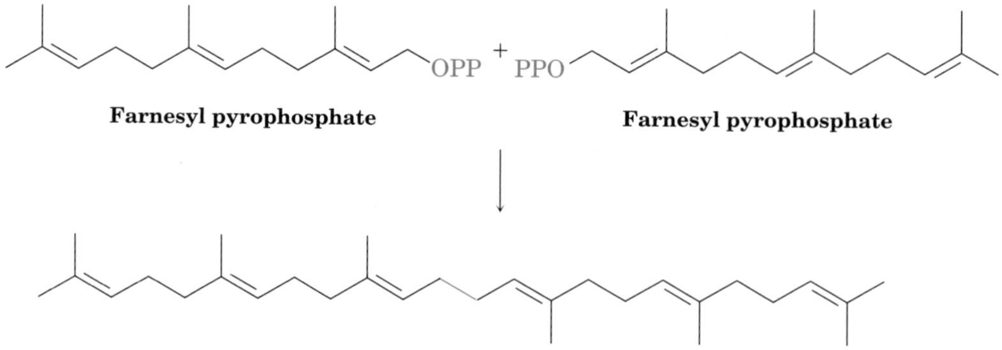

Farnesyl pyrophosphate   **Farnesyl pyrophosphate**

Squalene

....

Practice Problem 27.1 Propose a mechanistic pathway for the biosynthesis of α-terpineol.

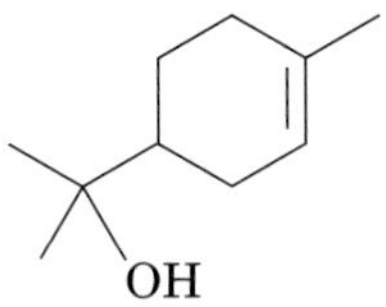

α-Terpineol

Strategy α-Terpineol, a monoterpene, must be derived biologically from geranyl pyrophosphate through its cis–trans isomer neryl pyrophosphate. Draw the pyrophosphate precursor in a conformation that approximates the structure of the target molecule, and then carry out a cationic cyclization, using a double bond to displace the pyrophosphate leaving group. Since the target is an alcohol, the carbocation resulting from cyclization must react with water.

Solution

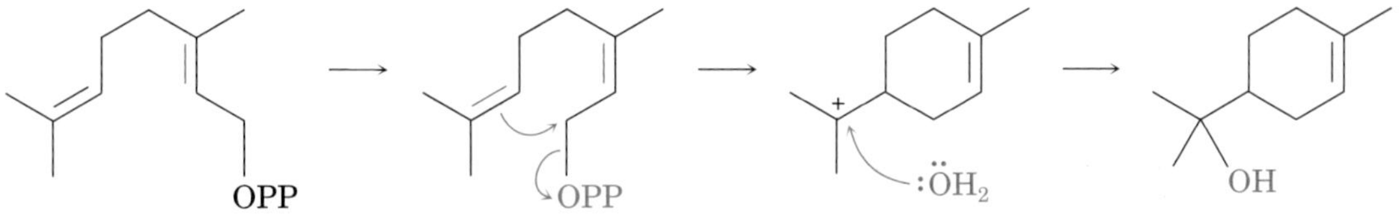

Geranyl pyrophosphate **Neryl pyrophosphate** *α*-Terpineol

....

Problem 27.6 Propose mechanistic pathways for the biosynthetic formation of the following terpenes:

(a)   (b)

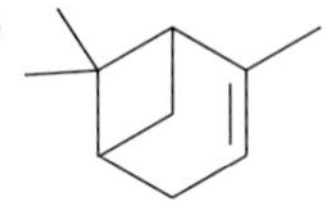

α-Pinene

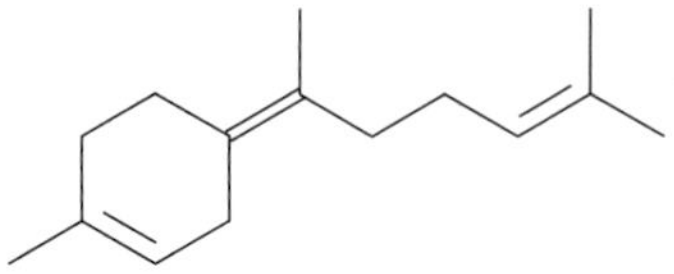

γ-Bisabolene

....

27.7 Steroids

In addition to fats, phospholipids, and terpenes, the lipid extracts of plants and animals also contain **steroids**, molecules whose structures are based on the tetracyclic ring system shown at the top of the next page. The four rings are designated A, B, C, and D, beginning at the lower left, and the carbon atoms are numbered beginning in the A ring. The three six-membered rings (A, B, and C) adopt chair conformations but are prevented by their rigid geometry from undergoing the usual cyclohexane ring-flips (Section 4.11).

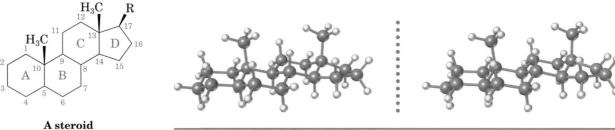

A steroid
(R = various side chains)

Stereo View

In humans, most steroids function as **hormones**, chemical messengers that are secreted by endocrine glands and carried through the bloodstream to target tissues. There are two main classes of steroid hormones: the *sex hormones*, which control maturation, tissue growth, and reproduction; and the *adrenocortical hormones*, which regulate a variety of metabolic processes.

Sex Hormones

Testosterone and androsterone are the two most important male sex hormones, or **androgens**. Androgens are responsible for the development of male secondary sex characteristics during puberty and for promoting tissue and muscle growth. Both are synthesized in the testes from cholesterol. Androstenedione is another minor hormone that has received particular attention because of its use by prominent athletes.

Testosterone

Androsterone

Androstenedione

(Androgens)

Estrone and estradiol are the two most important female sex hormones, or **estrogens**. Synthesized in the ovaries from testosterone, estrogenic hormones are responsible for the development of female secondary sex characteristics and for regulation of the menstrual cycle. Note that both have a benzene-like aromatic A ring. In addition, another kind of sex hormone called a *progestin* is essential for preparing the uterus for implantation of a fertilized ovum during pregnancy. Progesterone is the most important progestin.

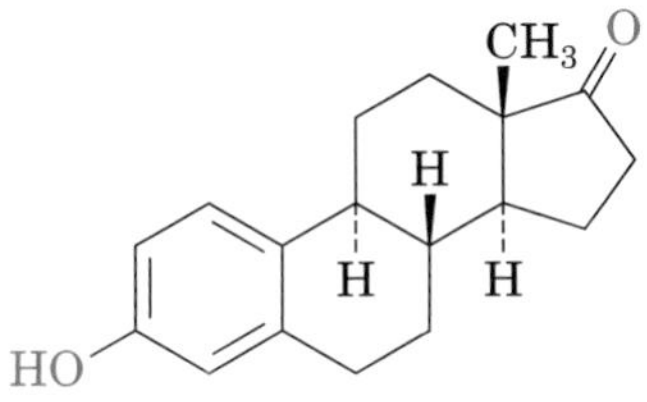

Estrone

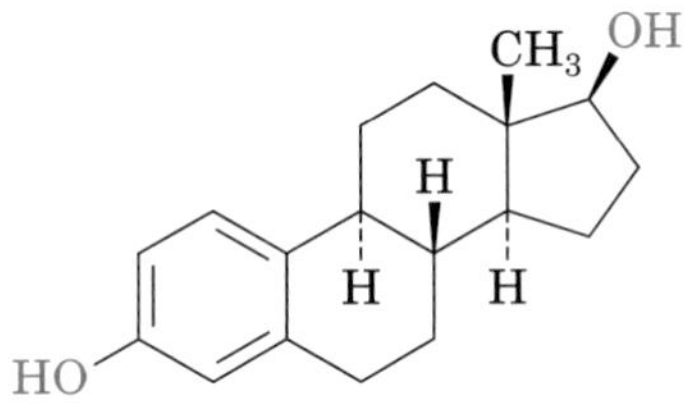

Estradiol

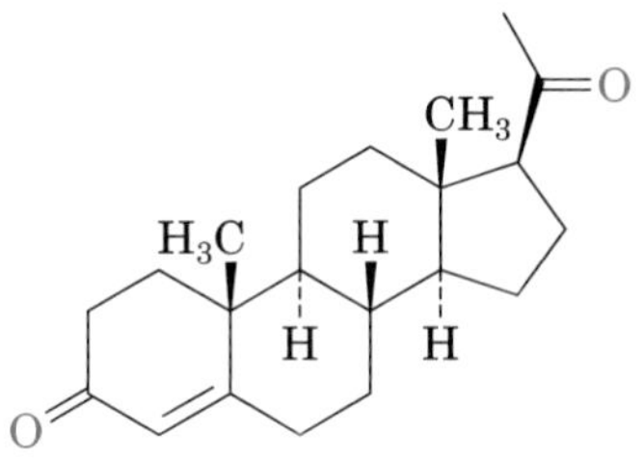

Progesterone
(a progestin)

(Estrogens)

Adrenocortical Hormones

Adrenocortical steroids are secreted by the adrenal glands, small organs located near the upper end of each kidney. There are two types of adrenocortical steroids, called *mineralocorticoids* and *glucocorticoids*. Mineralocorticoids, such as aldosterone, control tissue swelling by regulating cellular salt balance between Na^+ and K^+. Glucocorticoids, such as hydrocortisone, are involved in the regulation of glucose metabolism and in the control of inflammation. Glucocorticoid ointments are widely used to bring down the swelling from exposure to poison oak or poison ivy.

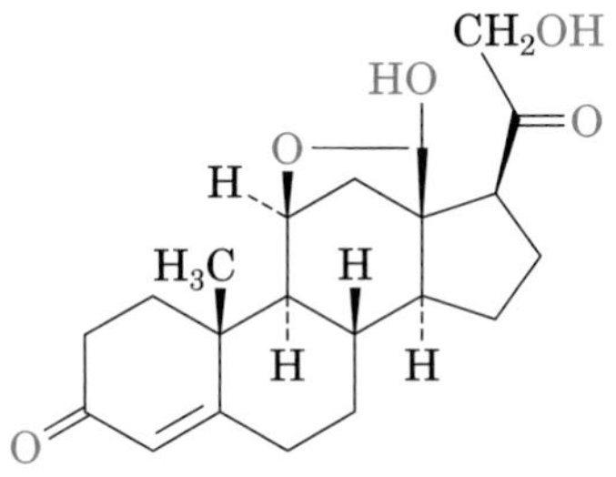

Aldosterone
(a mineralocorticoid)

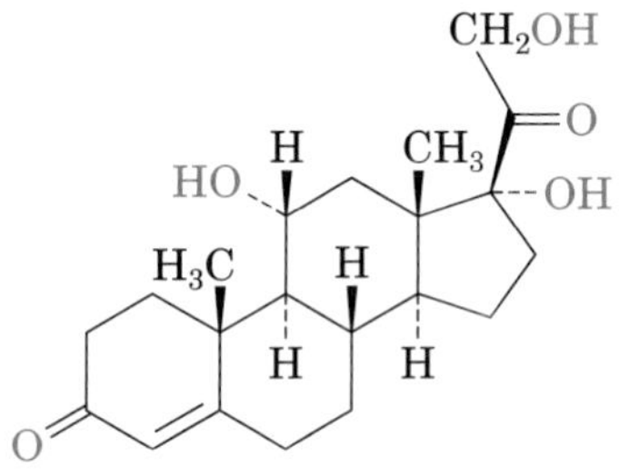

Hydrocortisone
(a glucocorticoid)

Synthetic Steroids

In addition to the many hundreds of steroids isolated from plants and animals, thousands more have been synthesized in pharmaceutical laboratories in a search for new drugs. Among the best-known synthetic steroids are the oral contraceptives and anabolic agents. Most birth-control pills are a mixture of two compounds, a synthetic estrogen, such as ethynylestradiol, and a synthetic progestin, such as norethindrone. Anabolic steroids, such as stanozolol and methandrostenolone (Dianabol), are synthetic androgens that mimic the tissue-building effects of natural testosterone.

Ethynylestradiol
(a synthetic estrogen)

Norethindrone
(a synthetic progestin)

Stanozolol
(an anabolic agent)

Methandrostenolone
(Dianabol)

27.8 Stereochemistry of Steroids

Two cyclohexane rings can be joined in either a cis or a trans manner. In *cis*-decalin, both groups at the ring-junction positions (the *angular* groups) are on the same side of the two rings. In *trans*-decalin, the groups at the ring junctions are on opposite sides. These spatial relationships are best grasped by building molecular models.

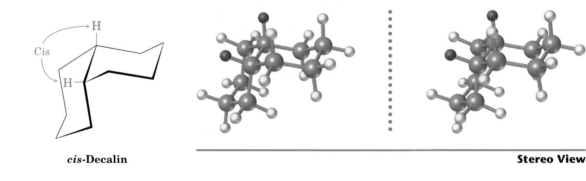

cis-Decalin

Stereo View

trans-Decalin

Stereo View

As shown in Figure 27.4, steroids can have either a cis or a trans fusion of the A and B rings, but the other ring fusions (B–C and C–D) are usually trans. An A–B trans steroid has the C19 angular methyl group "up" (denoted β) and the hydrogen atom at C5 "down" (denoted α) on opposite sides of the molecule. An A–B cis steroid, by contrast, has both the C19 angular methyl group and the C5 hydrogen atom on the same side (β) of the molecule. Both kinds of steroids are relatively long, flat molecules that have their two methyl groups (C18 and C19) protruding axially above the ring system. The A–B trans steroids are by far the more common, though A–B cis steroids are found in liver bile.

FIGURE 27.4 ▼

Steroid conformations. The three six-membered rings have chair conformations but are unable to undergo ring-flips. The A and B rings can be either cis-fused or trans-fused.

An A–B trans steroid

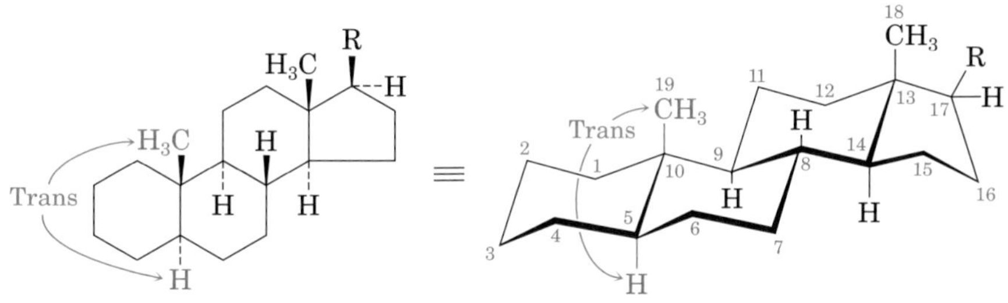

An A–B cis steroid

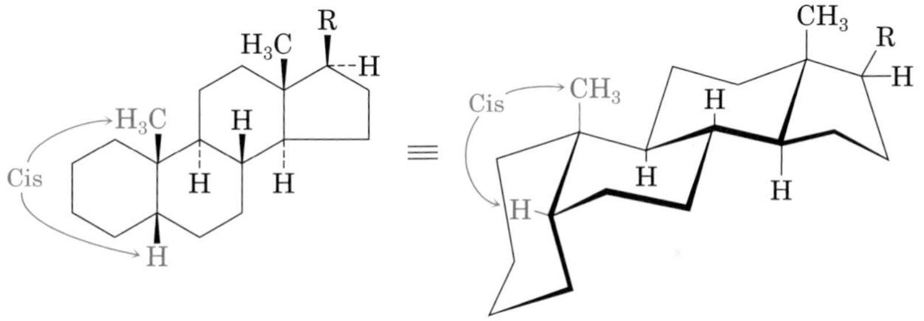

Substituent groups on the steroid ring system can be either axial or equatorial. As with simple cyclohexanes (Section 4.12), equatorial substitution is generally more favorable than axial substitution for steric reasons. The hydroxyl group at C3 of cholesterol, for example, has the more stable equatorial orientation (Figure 27.5, p. 1138).

● ●

Problem 27.7 Draw the following molecules in chair conformations, and tell whether the ring substituents are axial or equatorial:

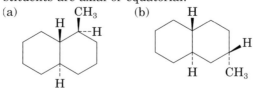

FIGURE 27.5

The stereochemistry of cholesterol. The –OH group at C3 is equatorial.

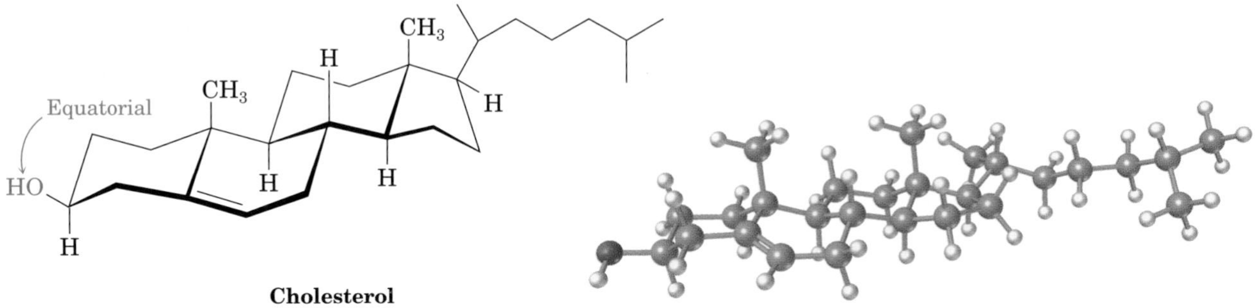

Cholesterol

Problem 27.8   Lithocholic acid is an A–B cis steroid found in human bile. Draw lithocholic acid showing chair conformations as in Figure 27.4, and tell whether the hydroxyl group at C3 is axial or equatorial.

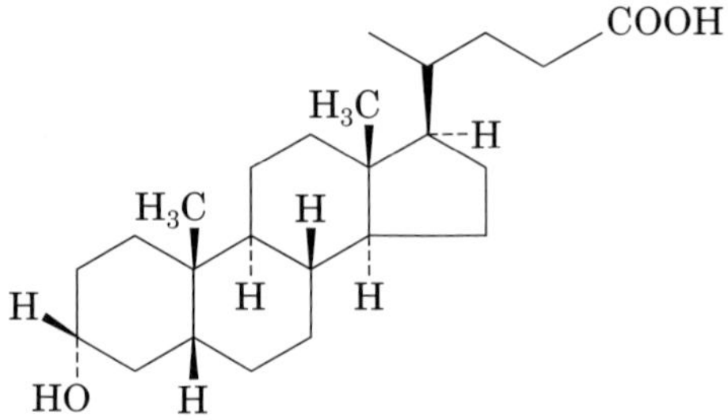

Lithocholic acid

27.9 Steroid Biosynthesis

Steroids are heavily modified triterpenes that are biosynthesized in living organisms from the acyclic hydrocarbon squalene (Section 27.6). The exact pathway by which this remarkable transformation is accomplished is lengthy and complex, but the key steps have now been worked out, with notable contributions made by Konrad Bloch and John Cornforth, who received Nobel Prizes for their efforts.

Steroid biosynthesis occurs by enzyme-catalyzed epoxidation of squalene to yield squalene oxide, followed by acid-catalyzed cyclization and an extraordinary cascade of nine sequential carbocation reactions to yield lanosterol (Figure 27.6). Lanosterol is then degraded by other enzymes to

Konrad Emil Bloch

Konrad Emil Bloch (1912–) was born in Neisse, Germany, and began his study at the Technische Hochschule in Munich. He then immigrated to the United States in 1936 and obtained his Ph.D. from Columbia University College of Physicians and Surgeons in 1938. After first serving as professor at the University of Chicago, he moved to Harvard University in 1954. He is best known for his work on cholesterol biosynthesis, for which he shared the 1964 Nobel Prize in medicine.

FIGURE 27.6

Biosynthesis of lanosterol from squalene.

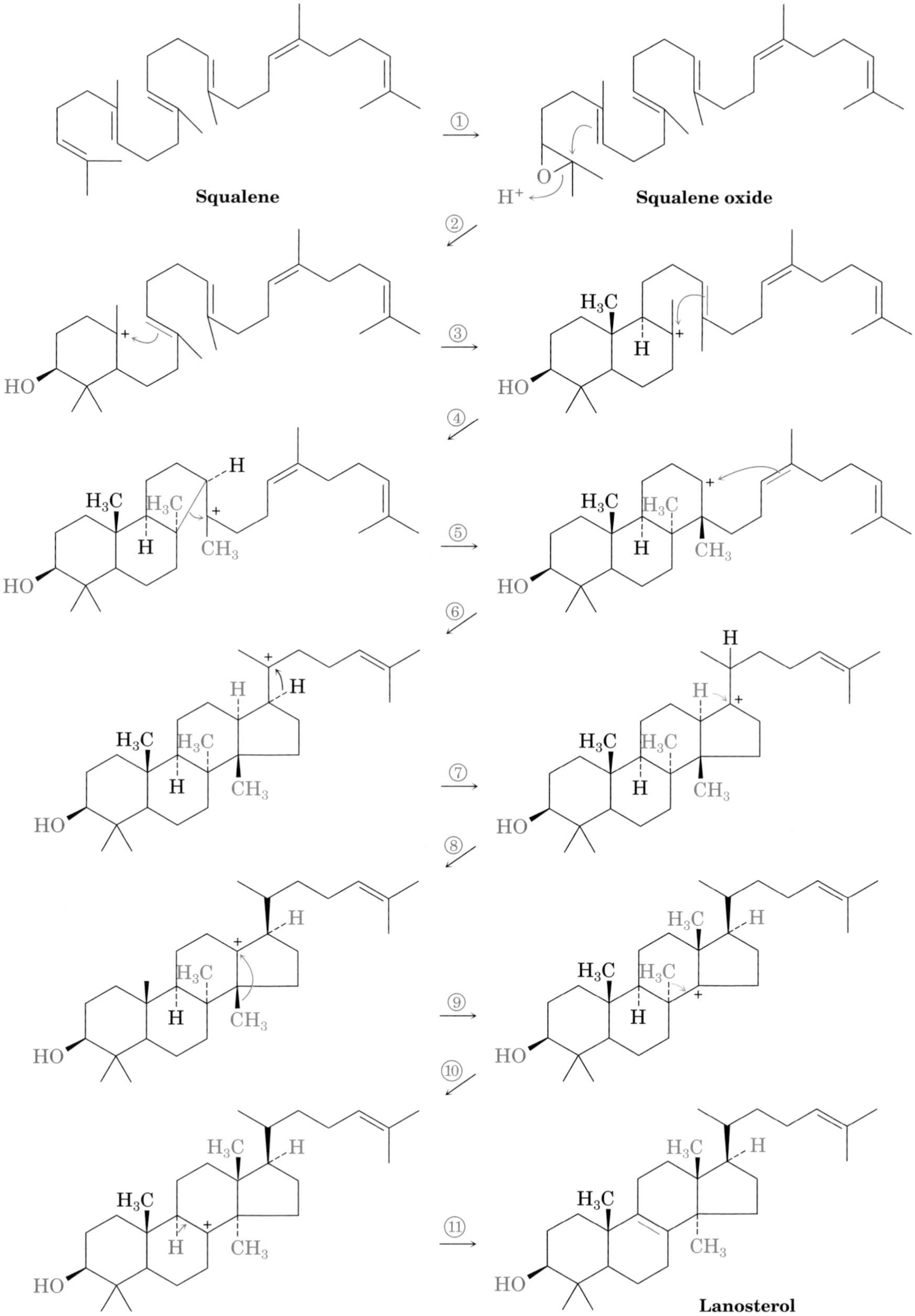

Sir John Warcup Cornforth

Sir John Warcup Cornforth (1917–) was born in Sydney, Australia, and earned his Ph.D. from Oxford University in 1941 working with Sir Robert Robinson. He was on the staff of the National Institute for Medical Research in London from 1946 to 1962, at Shell Research Ltd. (1962–1975), and ultimately at Sussex University (1975–1982). Completely deaf for most of his life, he worked in constant collaboration with his wife, Rita Harradence. He received the 1975 Nobel Prize in chemistry.

produce cholesterol, which is itself converted by various enzymes to produce a host of different steroids.

The series of processes involved in the biosynthetic conversion of squalene to lanosterol in Figure 27.6 is written in a stepwise format for convenience, but the cyclization sequence in steps 2–4 appears to take place at one time without intermediates. Similarly, the carbocation rearrangements (Section 6.12) in steps 7–10 take place at essentially the same time without intermediates.

STEP 1   The enzyme *squalene oxidase* selectively epoxidizes a terminal double bond of squalene to yield squalene oxide.

STEP 2   Squalene oxide is protonated on oxygen, and the epoxide ring is opened by nucleophilic attack of the double bond six carbons away, to yield a six-membered, cyclic carbocation intermediate. This step is similar to the acid-catalyzed epoxide openings we saw in Section 18.8.

STEP 3   The tertiary carbocation intermediate produced in step 2 undergoes further cyclization by nucleophilic attack by another double bond six carbons away from the positively charged carbon to yield a second carbocation intermediate.

STEP 4   A third cyclization occurs by attack of an appropriately positioned double bond on the positively charged carbon, forming a *five*-membered ring, and yielding another tertiary carbocation.

STEP 5   The tertiary carbocation produced in step 4 rearranges with expansion of the five-membered ring to a six-membered ring and formation of a secondary carbocation (Section 6.12).

STEP 6   A fourth and last cyclization takes place, this one giving another five-membered ring.

STEP 7   A carbocation rearrangement occurs by a hydride shift (Section 6.12).

STEP 8   A second hydride shift within the five-membered ring gives yet another carbocation.

STEP 9   A third carbocation rearrangement occurs by shift of a methyl group.

STEP 10   A second methyl-group shift gives a final carbocation intermediate.

STEP 11   Loss of a proton (E1 reaction) from the carbon next to the cationic center gives lanosterol.

● ●

Problem 27.9   We saw in Section 6.12 that carbocation rearrangements normally involve the conversion of a less stable cation into a more stable one. One of the steps in lanosterol biosynthesis, however, converts a more stable ion into a less stable one. Which step is the unusual one?

Problem 27.10   Compare the structures of lanosterol and cholesterol, and catalog the changes that have occurred in the transformation.

● ●

CHEMISTRY @ WORK

Cholesterol and Heart Disease

We read a lot about the relationship between cholesterol and heart disease. What are the facts? It's well established that a diet rich in saturated animal fats often leads to an increase in blood serum cholesterol, at least in sedentary, overweight people. Conversely, a diet lower in saturated fats and higher in polyunsaturated fats (PUFA's) leads to a lower serum cholesterol level. Studies have shown that a serum cholesterol level greater than 240 mg/dL (a normal value is 120–200 mg/dL) is weakly correlated with an increased incidence of *atherosclerosis*, a form of heart disease in which cholesterol deposits build up on the inner walls of coronary arteries, blocking the flow of blood to the heart muscles.

A better indication of a person's risk of heart disease comes from a measurement of blood lipoprotein levels. *Lipoproteins* are complex molecules with both lipid and protein parts that transport lipids through the body. They can be divided into four types according to density, as shown in Table 27.4. People with a high serum level of high-density lipoproteins (HDL's) seem to have a decreased risk of heart disease. As a rule of thumb, a person's risk drops about 25% for each increase of 5 mg/dL in HDL concentration. Normal values are about 45 mg/dL for men and 55 mg/dL for women, perhaps explaining why women are generally less susceptible than men to heart disease.

TABLE 27.4 Serum Lipoproteins

Name	Density (g/mL)	% Lipid	% Protein
Chylomicrons	< 0.94	98	2
VLDL's (very-low-density lipoproteins)	0.940–1.006	90	10
LDL's (low-density lipoproteins)	1.006–1.063	75	25
HDL's (high-density lipoproteins)	1.063–1.210	60	40

Chylomicrons and very-low-density lipoproteins (VLDL's) act primarily as carriers of triglycerides from the intestines to peripheral tissues, whereas LDL's and HDL's act as carriers of cholesterol to and from the liver. Present evidence suggests that LDL's transport cholesterol as its

(continued) ▶

fatty acid ester *to* peripheral tissues, whereas HDL's remove cholesterol as its stearate ester *from* dying cells and transport it back to the liver.

If LDL's deliver more cholesterol than is needed, and if insufficient HDL's are present to remove it, the excess is deposited in arteries. The higher the HDL level, the less the likelihood of deposits and the lower the risk of heart disease. In addition, HDL contains an enzyme that has antioxidant properties, offering further protection against heart disease.

Not surprisingly, the most important factor in gaining high HDL levels is a generally healthy lifestyle. Obesity, smoking, and lack of exercise lead to low HDL levels, whereas regular exercise and a sensible diet lead to high HDL levels. Distance runners and other endurance athletes have HDL levels nearly 50% higher than the general population.

It's hard to resist, but a high intake of saturated animal fat doesn't do much for your cholesterol level.

Summary and Key Words

Lipids are the naturally occurring materials isolated from plants and animals by extraction with organic solvents. **Animal fats** and **vegetable oils** are the most widely occurring lipids. Both are **triacylglycerols**—triesters of glycerol with long-chain **fatty acids**. Animal fats are usually saturated, whereas vegetable oils usually have unsaturated fatty acid residues.

Phosphoglycerides such as *lecithin* and *cephalin* are closely related to fats. The glycerol backbone in these molecules is esterified to two fatty acids (one saturated and one unsaturated) and to one phosphate ester. **Sphingolipids**, another major class of **phospholipids**, have an amino alcohol such as sphingosine for their backbone. These compounds are important constituents of cell membranes.

Prostaglandins and **terpenes** are still other classes of lipids. Prostaglandins, which are found in all body tissues, have a wide range of physiological actions. Terpenes are often isolated from the **essential oils** of plants. They have an immense diversity of structure and are produced biosynthetically by head-to-tail coupling of two five-carbon "isoprene equivalents"—isopentenyl pyrophosphate and dimethylallyl pyrophosphate.

Steroids are plant and animal lipids with a characteristic tetracyclic carbon skeleton. Like the prostaglandins, steroids occur widely in body tissues and have a large variety of physiological activities. Steroids are closely related to terpenes and arise biosynthetically from the triterpene precursor lanosterol. Lanosterol, in turn, arises from cationic cyclization of the acyclic hydrocarbon squalene.

Visualizing Chemistry

• •

(Problems 27.1–27.10 appear within the chapter.)

27.11 The following model is that of cholic acid, a constituent of human bile. Locate the three hydroxyl groups, and identify each as axial or equatorial. Is cholic acid an A–B trans steroid or an A–B cis steroid?

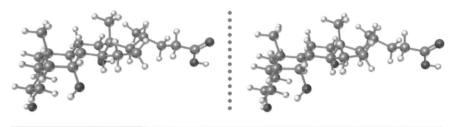

Stereo View

27.12 Propose a biosynthetic pathway for the sesquiterpene helminthogermacrene from farnesyl pyrophosphate:

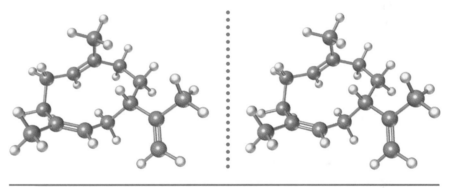

Stereo View

Additional Problems

• •

27.13 Fats can be either optically active or optically inactive, depending on their structure. Draw the structure of an optically active fat that yields 2 equivalents of stearic acid and 1 equivalent of oleic acid on hydrolysis. Draw the structure of an optically inactive fat that yields the same products.

27.14 Spermaceti, a fragrant substance from sperm whales, was much used in cosmetics until it was banned in 1976 to protect the whales from extinction. Chemically, spermaceti is cetyl palmitate, the ester of cetyl alcohol (n-$C_{16}H_{33}OH$) with palmitic acid. Draw its structure.

27.15   The *plasmalogens* are a group of lipids found in nerve and muscle cells. How do plasmalogens differ from fats, lecithins, and cephalins?

$$CH_2OCH{=}CHR$$
$$CHOCR'$$ **A plasmalogen**
$$CH_2OCR''$$

27.16   What product would you obtain from hydrolysis of a plasmalogen (Problem 27.15) with aqueous NaOH? With H_3O^+?

27.17   *Cardiolipins* are a group of lipids found in heart muscles. What products would be formed if all ester bonds were saponified by treatment with aqueous NaOH?

$$RCOCH_2 \qquad CH_2OCOR''$$
$$R'COCH \quad O \qquad O \; CHOCR'''$$ **A cardiolipin**
$$CH_2OPOCH_2CHCH_2OPOCH_2$$
$$O^- \qquad OH \qquad O^-$$

27.18   Stearolic acid, $C_{18}H_{32}O_2$, yields stearic acid on catalytic hydrogenation and undergoes oxidative cleavage with ozone to yield nonanoic acid and nonanedioic acid. What is the structure of stearolic acid?

27.19   How would you synthesize stearolic acid (Problem 27.18) from 1-decyne and 1-chloro-7-iodoheptane?

27.20   Show the products you would expect to obtain from reaction of glyceryl trioleate with the following reagents:
(a) Excess Br_2 in CH_2Cl_2 (b) H_2/Pd (c) NaOH/H_2O
(d) O_3, then Zn/CH_3COOH (e) LiAlH$_4$, then H_3O^+ (f) CH_3MgBr, then H_3O^+

27.21   How would you convert oleic acid into the following substances?
(a) Methyl oleate   (b) Methyl stearate
(c) Nonanal   (d) Nonanedioic acid
(e) 9-Octadecynoic acid (stearolic acid)   (f) 2-Bromostearic acid
(g) 18-Pentatriacontanone, $CH_3(CH_2)_{16}CO(CH_2)_{16}CH_3$

27.22   Show the location of the isoprene units in the following terpenes:

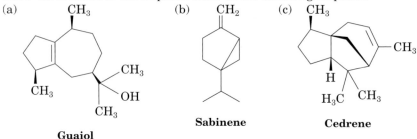

(a) Guaiol   (b) Sabinene   (c) Cedrene

27.23 Indicate by asterisks the chirality centers present in each of the terpenes shown in Problem 27.22. What is the maximum possible number of stereoisomers for each?

27.24 Assume that the three terpenes in Problem 27.22 are derived biosynthetically from isopentenyl pyrophosphate and dimethylallyl pyrophosphate, each of which was isotopically labeled at the pyrophosphate-bearing carbon atom (C1). At what positions would the terpenes be isotopically labeled?

27.25 Propose a mechanistic pathway for the biosynthesis of caryophyllene, a substance found in clove oil.

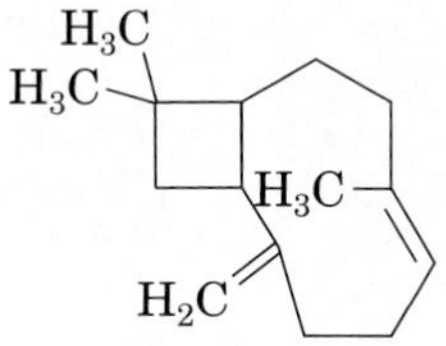

Caryophyllene

27.26 Flexibilene, a compound isolated from marine coral, is the only known terpene to contain a 15-membered ring. What is the structure of the acyclic biosynthetic precursor of flexibilene? Show the mechanistic pathway for the biosynthesis.

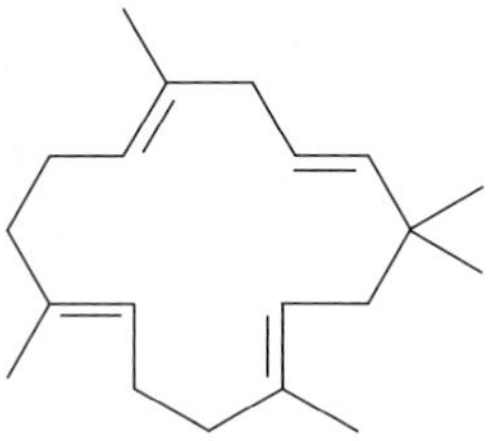

Flexibilene

27.27 Suggest a mechanism by which ψ-ionone is transformed into β-ionone on treatment with acid.

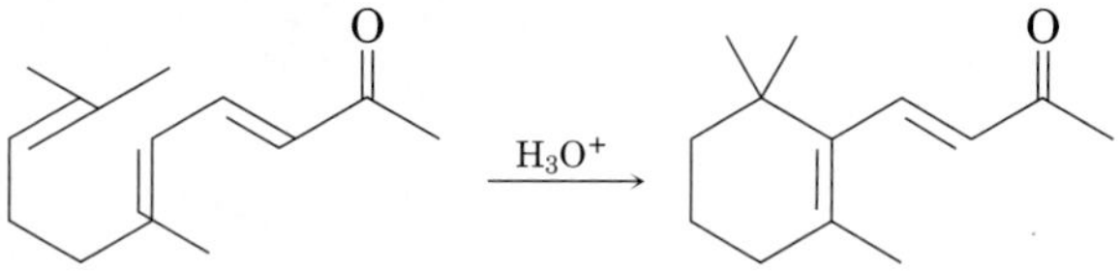

ψ-**Ionone**   β-**Ionone**

27.28 Draw the most stable chair conformation of dihydrocarvone.

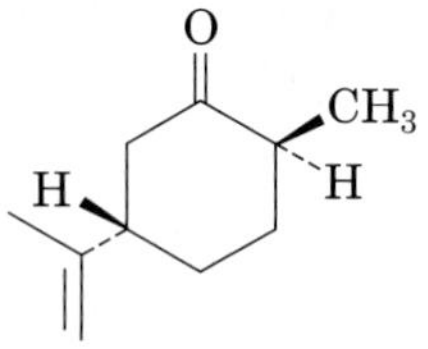

Dihydrocarvone

27.29 Draw the most stable chair conformation of menthol, and label each substituent as axial or equatorial.

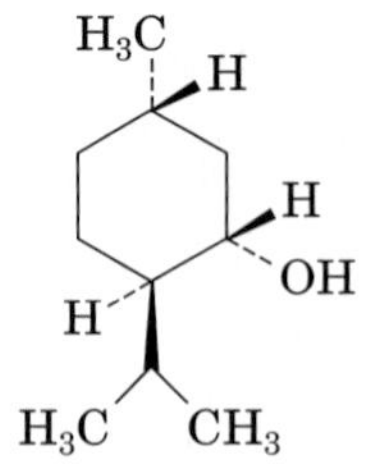

Menthol (from peppermint oil)

27.30 As a general rule, equatorial alcohols are esterified more readily than axial alcohols. What product would you expect to obtain from reaction of the following two compounds with 1 equivalent of acetic anhydride?

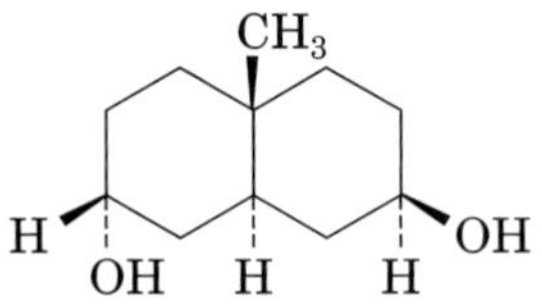

27.31 Propose a mechanistic pathway for the biosynthesis of isoborneol. A carbocation rearrangement is needed at one point in the scheme.

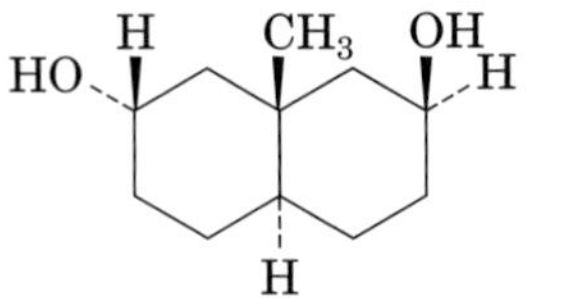

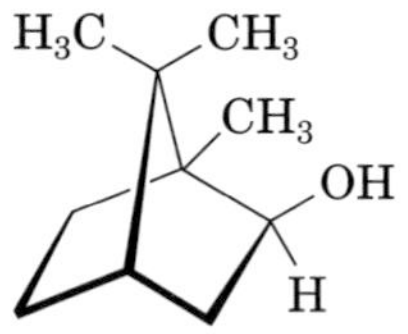

Isoborneol

27.32 Isoborneol (Problem 27.31) is converted into camphene on treatment with dilute sulfuric acid. Propose a mechanism for the reaction, which involves a carbocation rearrangement.

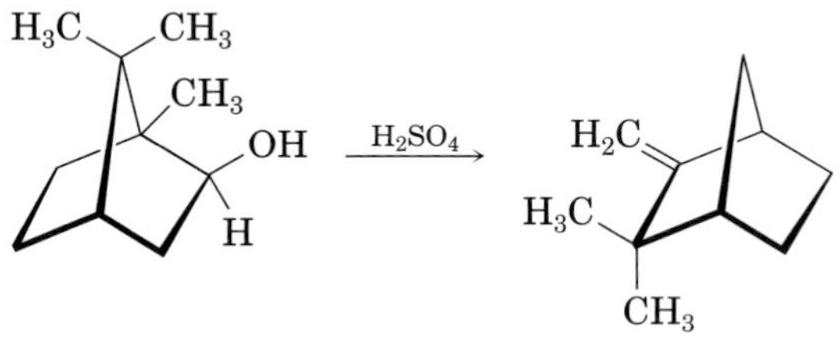

Isoborneol   **Camphene**

27.33 Digitoxigenin is a heart stimulant obtained from the purple foxglove *Digitalis purpurea* and used in the treatment of heart disease. Draw the three-dimensional conformation of digitoxigenin, and identify the two –OH groups as axial or equatorial.

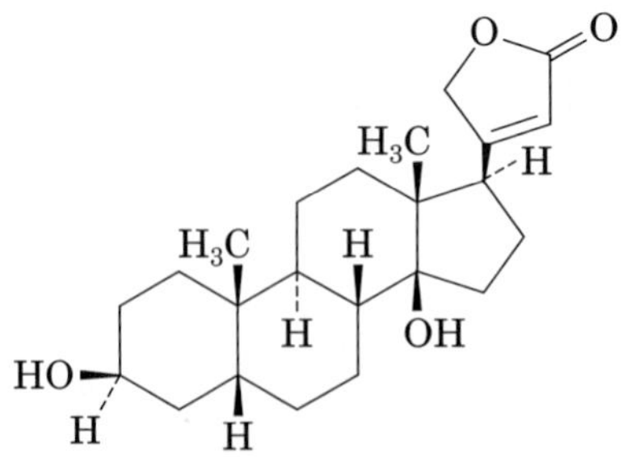

Digitoxigenin

27.34 What product would you obtain by reduction of digitoxigenin (Problem 27.33) with LiAlH₄? By oxidation with pyridinium chlorochromate?

27.35 Vaccenic acid, $C_{18}H_{34}O_2$, is a rare fatty acid that gives heptanal and 11-oxoundecanoic acid [OHC(CH₂)₉COOH] on ozonolysis followed by zinc treatment. When allowed to react with CH₂I₂/Zn(Cu), vaccenic acid is converted into lactobacillic acid. What are the structures of vaccenic and lactobacillic acids?

27.36 Eleostearic acid, $C_{18}H_{30}O_2$, is a rare fatty acid found in the tung oil used for finishing furniture. On ozonolysis followed by treatment with zinc, eleostearic acid furnishes one part pentanal, two parts glyoxal (OHC–CHO), and one part 9-oxononanoic acid [OHC(CH₂)₇COOH]. What is the structure of eleostearic acid?

27.37 Diethylstilbestrol (DES) has estrogenic activity even though it is structurally unrelated to steroids. Once used as an additive in animal feed, DES has been implicated as a causative agent in several types of cancer. Look up the structure of estradiol (Section 27.7), and show how DES can be drawn so that it is sterically similar to estradiol.

$$CH_3CH_2$$

Diethylstilbestrol

27.38 Propose a synthesis of diethylstilbestrol (Problem 27.37) from phenol and any other organic compound required.

27.39 What products would you expect from reaction of estradiol (Section 27.7) with the following reagents?
(a) NaH, then CH₃I   (b) CH₃COCl, pyridine
(c) Br₂, FeBr₃   (d) Pyridinium chlorochromate in CH₂Cl₂

27.40 Cembrene, $C_{20}H_{32}$, is a diterpene hydrocarbon isolated from pine resin. Cembrene has a UV absorption at 245 nm, but dihydrocembrene ($C_{20}H_{34}$), the product of hydrogenation with 1 equiv H₂, has no UV absorption. On exhaustive hydrogenation, 4 equiv H₂ react, and octahydrocembrene, $C_{20}H_{40}$, is produced. On ozonolysis of cembrene, followed by treatment of the ozonide with zinc, four carbonyl-containing products are obtained:

$$CH_3CCH_2CH_2CH + CH_3CCHO + HCCH_2CH + CH_3CCH_2CH_2CHCHCH_3$$
$$CH_3$$

Propose a structure for cembrene that is consistent with the isoprene rule.

27.41 α-Fenchone is a pleasant-smelling terpene isolated from oil of lavender. Propose a pathway for the formation of α-fenchone from geranyl pyrophosphate. (*Hint*: A carbocation rearrangement is required.)

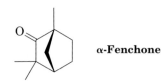

α-Fenchone

A Look Ahead

• •

27.42 Fatty acids are synthesized by a multistep route that starts with acetate. The first step is a reaction between protein-bound acetyl and malonyl units to give a 3-keto-butyryl unit. Show the mechanism, and tell what kind of reaction is occurring. (See Section 29.7).

$$^-OCCH_2\overset{\text{O}}{\underset{\parallel}{C}}-S-Protein \ + \ CH_3\overset{\text{O}}{\underset{\parallel}{C}}-S-Protein \ \longrightarrow \ \left[CH_3\overset{\text{O}}{\underset{\parallel}{C}}CHC\overset{\text{O}}{\underset{\parallel}{}}-S-Protein \right]$$

Malonyl   **Acetyl** CO_2^-

$$CH_3\overset{\text{O}}{\underset{\parallel}{C}}CH_2\overset{\text{O}}{\underset{\parallel}{C}}-S-Protein \ + \ CO_2$$

3-Ketobutyryl

27.43 Isopentenyl pyrophosphate arises biosynthetically by loss of CO_2 from *mevalonate*, which itself results from a reaction of acetate with 3-ketobutyrate. Show the mechanism, and tell what kind of reaction is occurring in the formation of mevalonate. (See Section 29.5.)

$$CH_3\overset{\text{O}}{\underset{\parallel}{C}}CH_2\overset{\text{O}}{\underset{\parallel}{C}}O^- \ + \ CH_3\overset{\text{O}}{\underset{\parallel}{C}}O^- \ \longrightarrow \ ^-OCCH_2\underset{\underset{CH_3}{|}}{\overset{OH}{\underset{|}{C}}}CH_2\overset{\text{O}}{\underset{\parallel}{C}}O^-$$

3-Ketobutyrate   **Acetate**

Mevalonate

Molecular Modeling

 •

27.44 Cholic acid (Problem 27.11) has distinct hydrophobic and hydrophilic regions, and acts like a "soap" inside the body to solubilize fats. Use SpartanBuild to build cholic acid, and minimize its energy. Then, examine the positions of polar groups, and identify the hydrophobic and hydrophilic regions.

Cholic acid

27.45 Use SpartanView to examine the structures of lipids A–D. Identify the class of lipid in each case (triacylglycerol, phospholipid, prostaglandin, or steroid), and identify the fatty acids in the triacylglycerol. Examine the electrostatic potential map of each lipid, and decide if it contains a polar head group.

27.46 Vitamins are classifed as either water-soluble or fat-soluble. Use SpartanView to examine electrostatic potential maps of vitamin A, vitamin B_6, vitamin C, and vitamin E, and determine which class each vitamin belongs to.

28

Biomolecules: Heterocycles and Nucleic Acids

Cyclic organic compounds are classified as **carbocycles** or as **heterocycles**. Carbocyclic rings contain only carbon atoms, but heterocyclic rings contain one or more different atoms in addition to carbon. Nitrogen, oxygen, and sulfur are the most common heteroatoms.

Heterocyclic compounds are common in organic chemistry, and many have important biological properties. For example, the antibiotic penicillin, the sedative phenobarbital, and the nonnutritive sweetener saccharin all have heterocyclic rings.

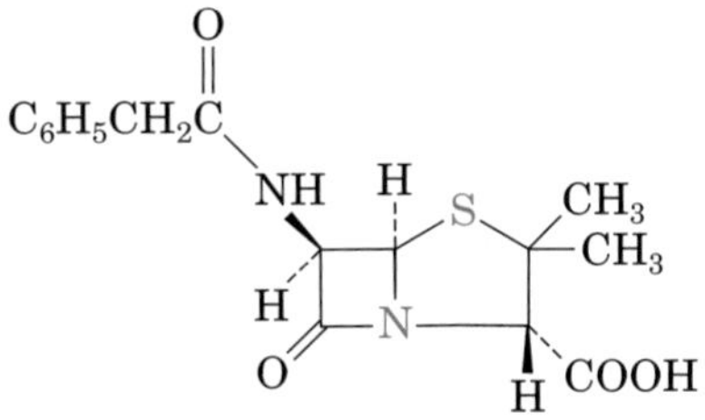

Penicillin G 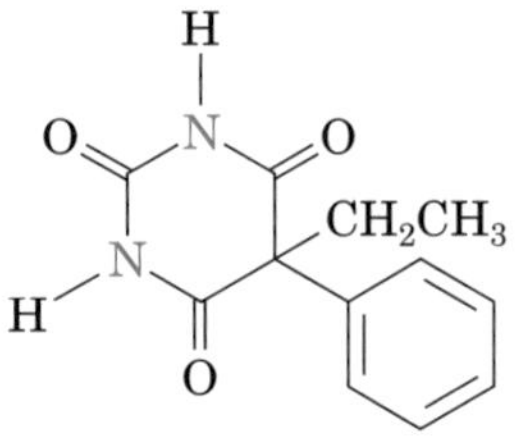   **Phenobarbital** 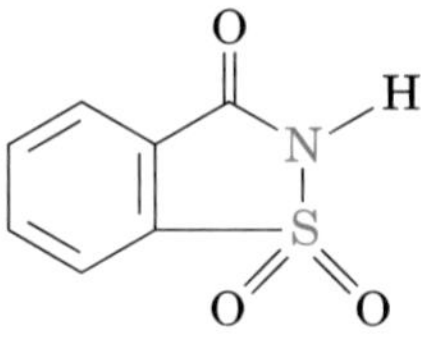   **Saccharin**

Heterocycles aren't new at this point; we've encountered them many times in previous chapters, usually without comment. Thus, epoxides (three-membered cyclic ethers), lactones (cyclic esters), and lactams (cyclic amides) are heterocycles, as are the solvents tetrahydrofuran (a cyclic ether) and pyridine (an aromatic cyclic amine). In addition, carbohydrates exist as heterocyclic hemiacetals (Section 25.5).

Most heterocycles have the same chemistry as their open-chain counterparts. Lactones and acyclic esters behave similarly, lactams and acyclic amides behave similarly, and cyclic and acyclic ethers behave similarly. In certain cases, however, particularly when the ring is unsaturated, heterocycles have unique and interesting properties. Let's look first at the five-membered unsaturated heterocycles.

28.1 Five-Membered Unsaturated Heterocycles

Pyrrole, furan, and thiophene are the most common five-membered unsaturated heterocycles. Each has two double bonds and one heteroatom (N, O, or S).

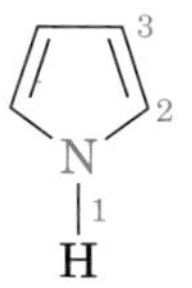

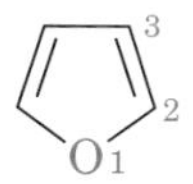

Furan 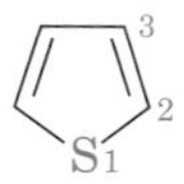   **Thiophene**

Pyrrole

Pyrrole is obtained commercially either directly from coal tar or by treatment of furan with ammonia over an alumina catalyst at 400°C.

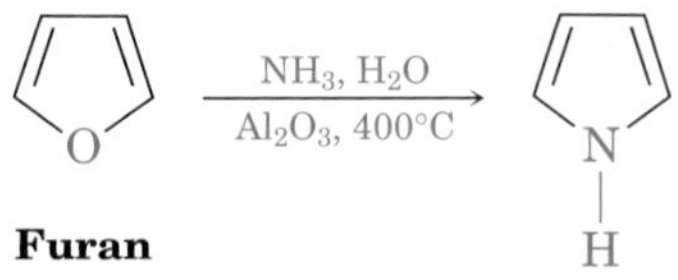

Furan

Pyrrole

Furan is synthesized by loss of carbon monoxide (*decarbonylation*) from furfural, which is itself prepared by acidic dehydration of the pentose sugars found in oat hulls and corncobs.

$$C_5H_{10}O_5 \xrightarrow{H_3O^+} \underset{\text{Furfural}}{\text{(O)}\!-\!\text{CHO}} \xrightarrow[280°C]{\text{Ni catalyst}} \underset{\text{Furan}}{\text{(O)}} + CO$$

Pentose mixture   **Furfural**   **Furan**

Thiophene is found in small amounts in coal tar and is synthesized industrially by cyclization of butane or butadiene with sulfur at 600°C.

$$\underset{\text{1,3-Butadiene}}{\overset{\text{CH}\!-\!\text{CH}}{\underset{\text{CH}_2\qquad\text{CH}_2}{}}} \xrightarrow[600°C]{S} \underset{\text{Thiophene}}{\text{(S)}} + H_2S$$

1,3-Butadiene   **Thiophene**

The chemistry of all three heterocyclic ring systems contains some surprises. Pyrrole, for example, is both an amine and a conjugated diene, yet its chemical properties are not consistent with either of these structural features. Unlike most other amines, pyrrole is not basic (Section 24.4); unlike most other conjugated dienes, pyrrole undergoes electrophilic substitution rather than addition reactions. The same is true of furan and thiophene: Both react with electrophiles to give substitution products.

28.2 Structures of Pyrrole, Furan, and Thiophene

Pyrrole, furan, and thiophene give electrophilic substitution products because they're *aromatic* (Section 15.7). Each has six π electrons in a cyclic conjugated system of overlapping p orbitals. Taking pyrrole as an example, each of the four carbon atoms of pyrrole contributes one π electron, and the sp^2-hybridized nitrogen atom contributes two (its lone pair). The six π electrons occupy p orbitals, with lobes above and below the plane of the ring, as shown in Figure 28.1. Overlap of the five p orbitals forms aromatic molecular orbitals just as in benzene.

Note that the pyrrole nitrogen atom uses all five valence electrons in bonding. Three electrons are used in forming three σ bonds (two to carbon and one to hydrogen), and the two lone-pair electrons are involved in aromatic π bonding. Because the nitrogen lone pair is a part of the aromatic sextet, protonation on nitrogen would destroy the aromaticity of the ring. The nitrogen atom in pyrrole is therefore less electron-rich, less basic, and less nucleophilic than the nitrogen in an aliphatic amine (pK_a of pyrrolinium ion = 0.4). By the same token, the *carbon* atoms of pyrrole are *more* electron-rich and more nucleophilic than typical double-bond carbons. The pyrrole ring is therefore reactive toward electrophiles in the same way that activated benzene rings are reactive.

FIGURE 28.1 ▼

Pyrrole, a six-π-electron aromatic heterocycle.

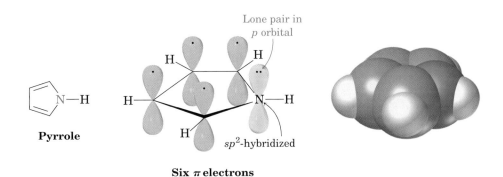

Six π electrons

pyrrole, pyrrolidine, 1,3-cyclopentadiene

Electrostatic potential maps show both trends, indicating that the pyrrole nitrogen is electron-poor (less red) compared to the nitrogen in its saturated counterpart pyrrolidine, while the pyrrole carbons are electron-rich (more red) compared to the carbons in 1,3-cyclopentadiene.

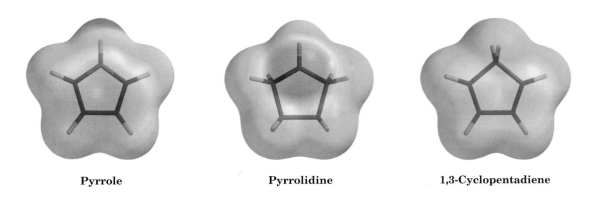

Pyrrole   Pyrrolidine   1,3-Cyclopentadiene

Problem 28.1 Draw an orbital picture of furan. Assume that the oxygen atom is sp^2-hybridized, and show the orbitals that the two oxygen lone pairs occupy.

28.3 Electrophilic Substitution Reactions of Pyrrole, Furan, and Thiophene

The chemistry of pyrrole, furan, and thiophene is similar to that of activated benzene rings. In general, however, the heterocycles are more reactive toward electrophiles than benzene rings are, and low temperatures are often necessary to control the reactions. Halogenation, nitration, sulfonation, and

Friedel–Crafts acylation can all be accomplished if the proper reaction conditions are chosen. The usual reactivity order is furan > pyrrole > thiophene.

Bromination

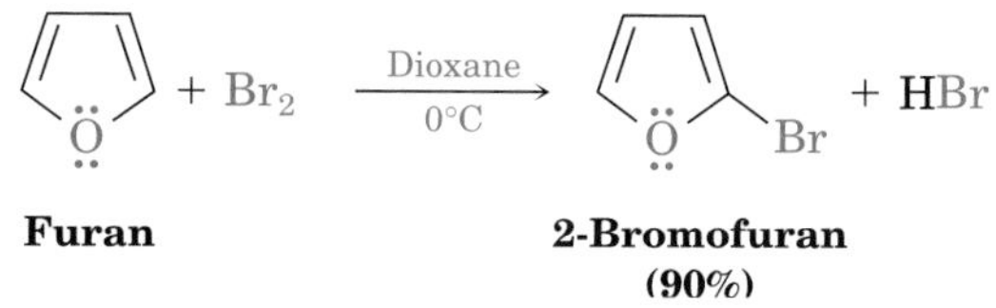

Furan   2-Bromofuran
(90%)

Nitration

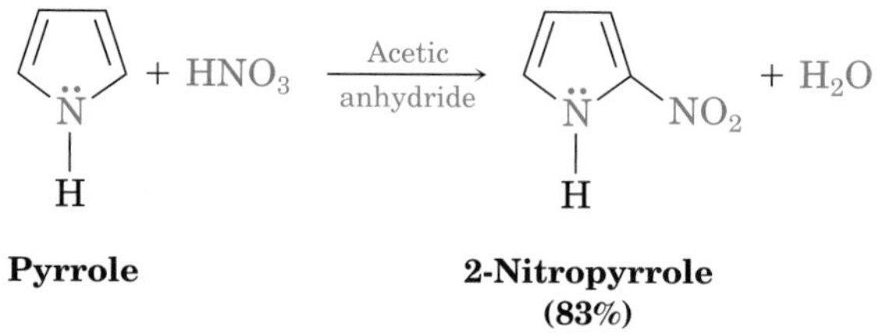

Pyrrole   2-Nitropyrrole
(83%)

Friedel–Crafts acylation

Thiophene

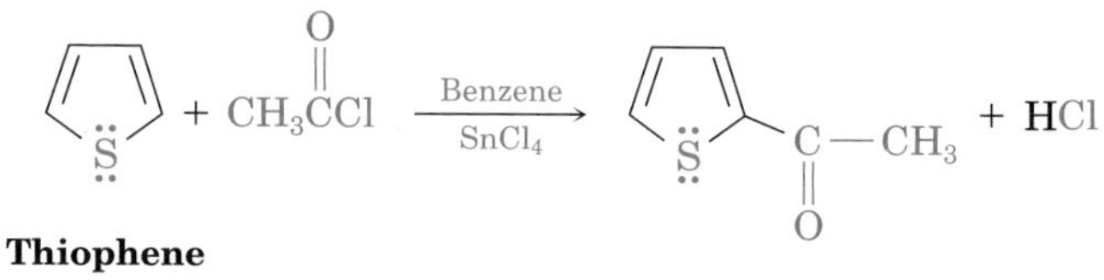

Electrophilic substitution normally occurs at C2, the position next to the heteroatom, because reaction at this position leads to a more stable intermediate cation having three resonance forms, while attack at C3 gives a less stable cation with only two resonance forms (Figure 28.2).

FIGURE 28.2 ▼

Electrophilic nitration of pyrrole. The intermediate produced by reaction at C2 is more stable than that produced by reaction at C3.

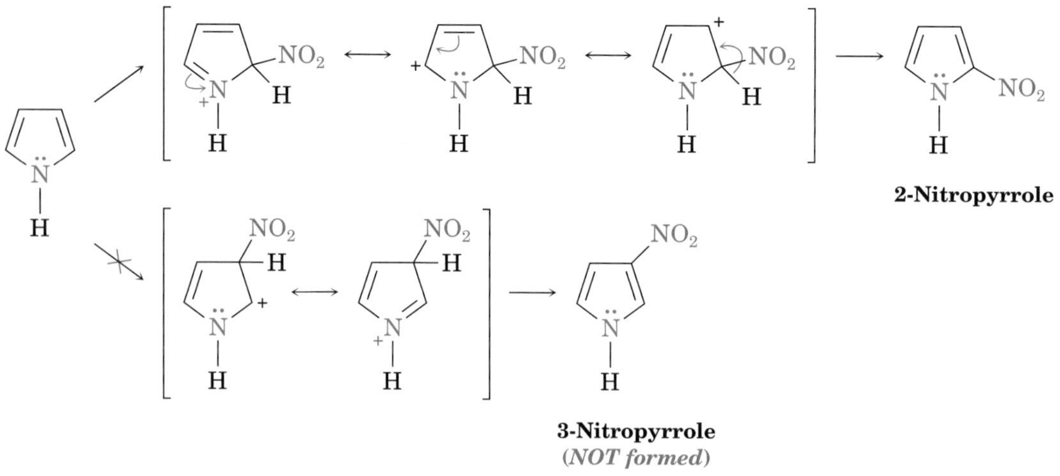

2-Nitropyrrole

3-Nitropyrrole
(*NOT formed*)

• •

Problem 28.2   Treatment of pyrrole with deuteriosulfuric acid, D_2SO_4, leads to formation of 2-deuteriopyrrole. Propose a mechanism.

• •

28.4 Pyridine, a Six-Membered Heterocycle

Pyridine, obtained commercially by distillation of coal tar, is the nitrogen-containing heterocyclic analog of benzene. Like benzene, pyridine is a flat, aromatic molecule with bond angles of 120° and C–C bond lengths of 139 pm, intermediate between typical single and double bonds. The five carbon atoms and the sp^2-hybridized nitrogen atom each contribute one π electron to the aromatic sextet. Unlike the situation in pyrrole, the lone pair of electrons on the pyridine nitrogen atom occupies an sp^2 orbital in the plane of the ring and is not involved in bonding (Figure 28.3).

FIGURE 28.3 ▼

Electronic structure of pyridine, a six-π-electron, nitrogen-containing analog of benzene. The electrostatic potential map shows that the nitrogen is the most negative atom (red).

pyridine

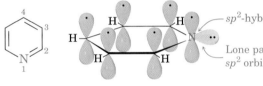 

Six π electrons

As noted in Section 24.4, pyridine is a stronger base than pyrrole but a weaker base than alkylamines. The sp^2-hybridized nitrogen atom in pyridine, with 33% s character, holds the lone-pair electrons more tightly than the sp^3-hybridized nitrogen in an alkylamine (25% s character).

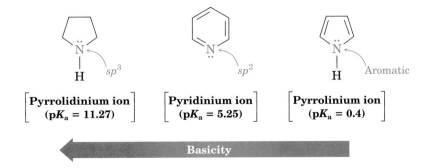

• •

Problem 28.3   Imidazolium ion has pK_a = 6.95. Draw an orbital picture of imidazole, and tell which nitrogen is more basic. (See Section 26.1.)

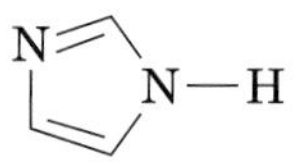

Imidazole

• •

28.5 Electrophilic Substitution of Pyridine

The pyridine ring undergoes electrophilic aromatic substitution reactions with great difficulty. Halogenation and sulfonation can be carried out under drastic conditions, but nitration occurs in very low yield, and Friedel–Crafts reactions are not successful. Reactions usually give the 3-substituted product.

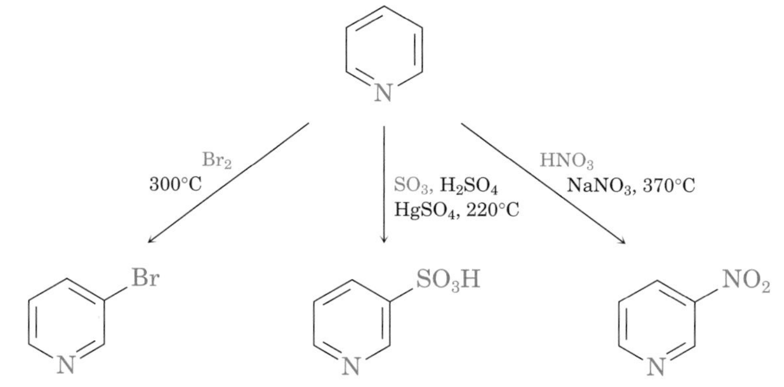

3-Bromopyridine
(30%)

3-Pyridinesulfonic acid (70%)

3-Nitropyridine
(5%)

The low reactivity of pyridine toward electrophilic aromatic substitution is due to a combination of factors. Most important is that the electron density of the ring is decreased by the electron-withdrawing inductive effect of the electronegative nitrogen atom. Thus, pyridine has a substantial dipole moment (μ = 2.26 D), with the ring carbons acting as the positive end of the dipole. Electrophilic attack on the positively polarized carbon atoms is therefore difficult.

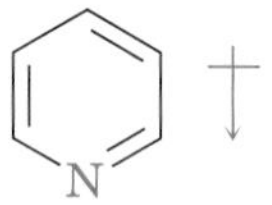

μ = **2.26 D**

A second factor that decreases the reactivity of the pyridine ring toward electrophilic attack is that acid–base complexation between the basic ring nitrogen atom and the attacking electrophile places a positive charge on the ring, further deactivating it.

• •

Problem 28.4   Electrophilic aromatic substitution reactions of pyridine normally occur at C3. Draw the carbocation intermediates resulting from electrophilic attack at C1, C2, and C3, and explain the observed result.

• •

28.6 Nucleophilic Substitution of Pyridine

In contrast to their lack of reactivity toward *electrophilic* substitution, 2- and 4-substituted (but not 3-substituted) halopyridines undergo *nucleophilic* aromatic substitution easily.

4-Chloropyridine   **4-Ethoxypyridine**
                                        (75%)

2-Bromopyridine   **2-Aminopyridine**
                                        (67%)

These reactions are typical nucleophilic aromatic substitutions, similar to those we saw earlier for halobenzenes (Section 16.8). Reaction occurs by addition of the nucleophile to the C=N bond, followed by loss of halide ion from the anion intermediate.

This nucleophilic aromatic substitution is in some ways analogous to the nucleophilic acyl substitution of acid chlorides (Section 21.4). In both cases, the initial addition step is favored by the ability of the electronegative atom (nitrogen or oxygen) to stabilize the anion intermediate. The intermediate then expels chloride ion to yield the substitution product.

2-Chloropyridine Stabilized anion 2-Aminopyridine

Acid chloride Stabilized anion Amide

Problem 28.5 Draw the anion intermediates from nucleophilic attack at C4 of a 4-halopyridine and at C3 of a 3-halopyridine. Why does substitution of the 4-halopyridine occur so much more easily?

28.7 Fused-Ring Heterocycles

Quinoline, isoquinoline, and indole are **fused-ring heterocycles** containing both a benzene ring and a heterocyclic aromatic ring. All three ring systems occur commonly in nature, and many compounds with these rings have pronounced physiological activity. The quinoline alkaloid quinine, for instance, is widely used as an antimalarial drug, and the indole alkaloid *N,N*-dimethyltryptamine is a powerful hallucinogen.

Quinoline Isoquinoline Indole

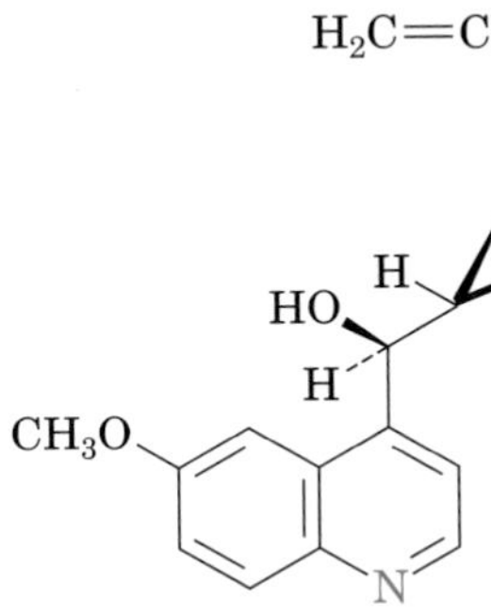

**Quinine, an antimalarial drug
(a quinoline alkaloid)**

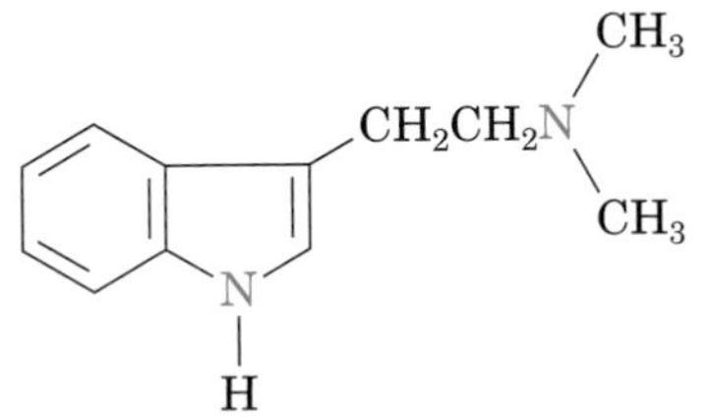

**N,N-Dimethyltryptamine, a hallucinogen
(an indole alkaloid)**

The chemistry of these fused-ring heterocycles is just what you might expect from a knowledge of the simpler heterocycles pyridine and pyrrole. Quinoline and isoquinoline both have basic, pyridine-like nitrogen atoms, and both undergo electrophilic substitutions, although less easily than benzene. Reaction occurs on the benzene ring rather than on the pyridine ring, and a mixture of substitution products is obtained.

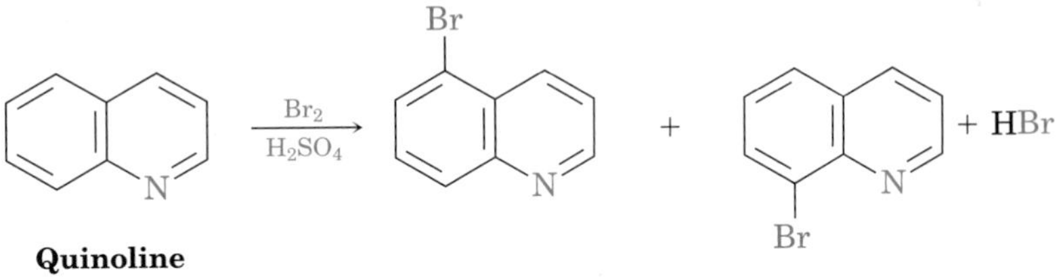

Quinoline

5-Bromoquinoline   **8-Bromoquinoline**

A 51:49 ratio

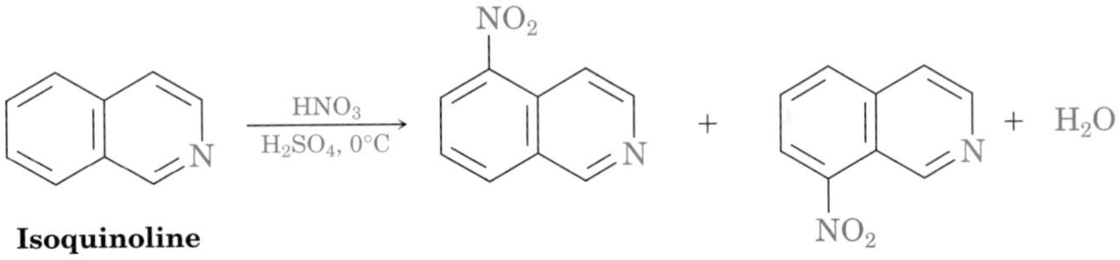

Isoquinoline

5-Nitroisoquinoline   **8-Nitroisoquinoline**

A 90:10 ratio

Indole has a nonbasic, pyrrole-like nitrogen and undergoes electrophilic substitution more easily than benzene. Substitution occurs at C3 of the electron-rich pyrrole ring, rather than on the benzene ring.

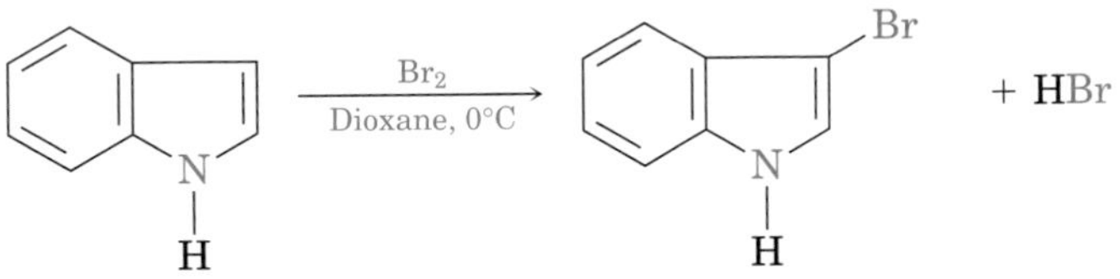

Indole   **3-Bromoindole**

Perhaps the most important heterocyclic ring systems from a biological viewpoint are *pyrimidine* and *purine*. **Pyrimidine** contains two pyridine-like nitrogens in a six-membered aromatic ring, while **purine** has four nitrogens in a fused-ring structure. Three of the purine nitrogens are basic and pyridine-like in having their lone-pair electrons in sp^2 orbitals in the plane of the ring. The remaining purine nitrogen is nonbasic and pyrrole-like in having its lone-pair electrons as part of the aromatic π electron system. Both heterocycles are essential components of the last major class of biomolecules we'll consider—the nucleic acids.

Pyrimidine

Purine

. . . .

Problem 28.6   Which nitrogen atom in *N,N*-dimethyltryptamine is more basic? Explain.

Problem 28.7   Indole reacts with electrophiles at C3 rather than at C2. Draw resonance forms of the intermediate cations resulting from attack at C2 and C3, and explain the observed results.

. . . .

28.8 Nucleic Acids and Nucleotides

The nucleic acids, **deoxyribonucleic acid (DNA)** and **ribonucleic acid (RNA)**, are the chemical carriers of a cell's genetic information. Coded in a cell's DNA is all the information that determines the nature of the cell, controls cell growth and division, and directs biosynthesis of the enzymes and other proteins required for all cellular functions.

Just as proteins are biopolymers made of amino acid units, nucleic acids are biopolymers made of **nucleotides** joined together to form a long chain. Each nucleotide is composed of a **nucleoside** bonded to a phosphate group, and each nucleoside is composed of an aldopentose sugar linked to a heterocyclic purine or pyrimidine base.

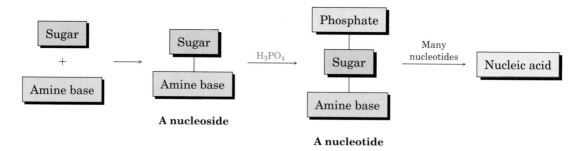

The sugar component in RNA is ribose, and the sugar in DNA is 2′-deoxyribose. (The prefix *2′-deoxy* indicates that oxygen is missing from the 2′ position of ribose. Numbers with a prime superscript refer to positions on the sugar of a nucleotide, and numbers without a prime refer to positions on the heterocyclic amine base.)

Ribose　　　　**2′-Deoxyribose**

There are four different heterocyclic amine bases in deoxyribonucleotides. Two are substituted purines (**adenine** and **guanine**), and two are substituted pyrimidines (**cytosine** and **thymine**). Adenine, guanine, and cytosine also occur in RNA, but thymine is replaced in RNA by a different pyrimidine base called **uracil**.

Adenine (A)　　　　**Guanine (G)**　　　　**Cytosine (C)**
DNA　　　　**DNA**　　　　**DNA**
RNA　　　　**RNA**　　　　**RNA**

Thymine (T)　　　　**Uracil (U)**
DNA　　　　**RNA**

In both DNA and RNA, the heterocyclic amine base is bonded to C1′ of the sugar, and the phosphoric acid is bonded by a phosphate ester linkage to the C5′ sugar position. The names and structures of all four deoxyribonucleotides and all four ribonucleotides are shown in Figure 28.4 (p. 1162).

Though chemically similar, DNA and RNA differ in size and have different roles within the cell. Molecules of DNA are enormous. They have molecular weights of up to 150 *billion* and lengths of up to 12 cm when stretched out, and they are found mostly in the nucleus of cells. Molecules of RNA, by contrast, are much smaller (as low as 35,000 in molecular weight) and are found mostly outside the cell nucleus. We'll consider the two kinds of nucleic acids separately, beginning with DNA.

FIGURE 28.4 ▼

Names and structures of the four deoxyribonucleotides and the four ribonucleotides.

Deoxyribonucleotides

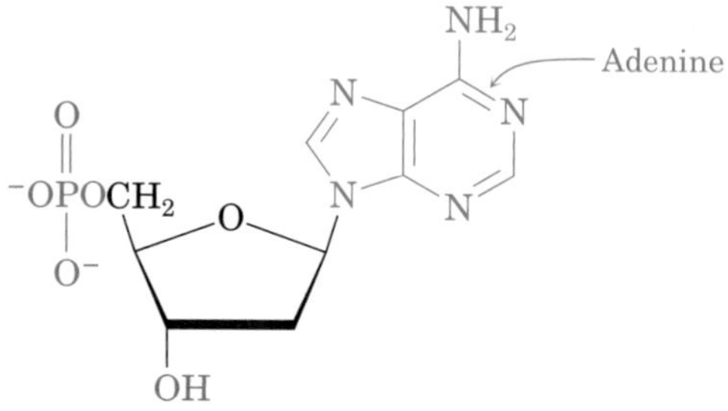

2′-Deoxyadenosine 5′-phosphate

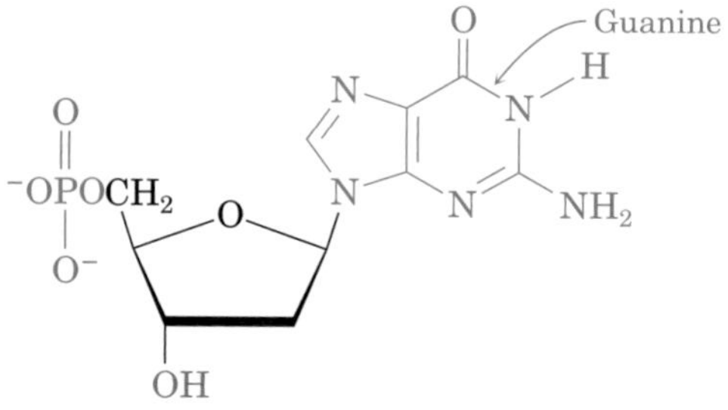

2′-Deoxyguanosine 5′-phosphate

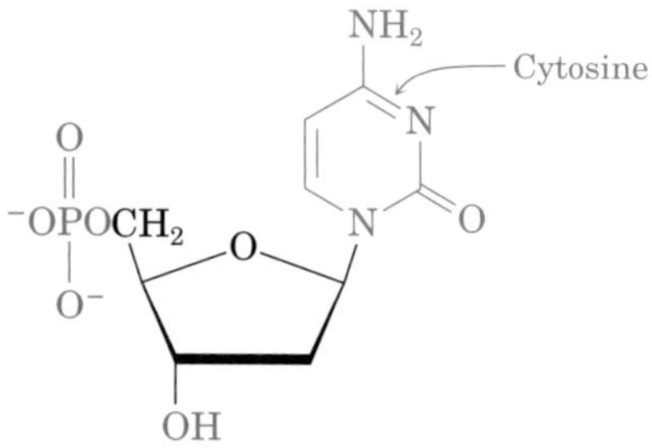

2′-Deoxycytidine 5′-phosphate

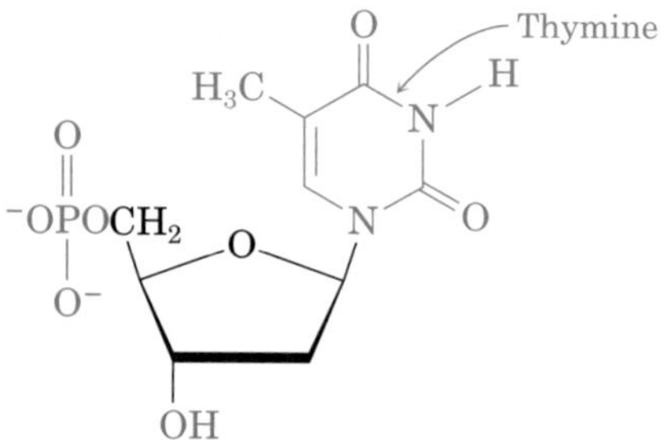

2′-Deoxythymidine 5′-phosphate

Ribonucleotides

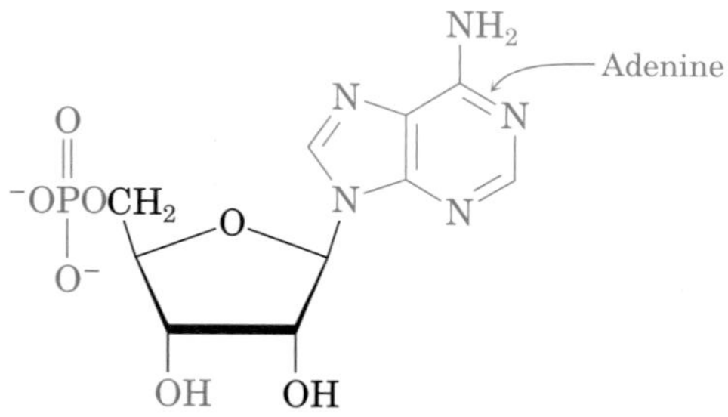

Adenosine 5′-phosphate

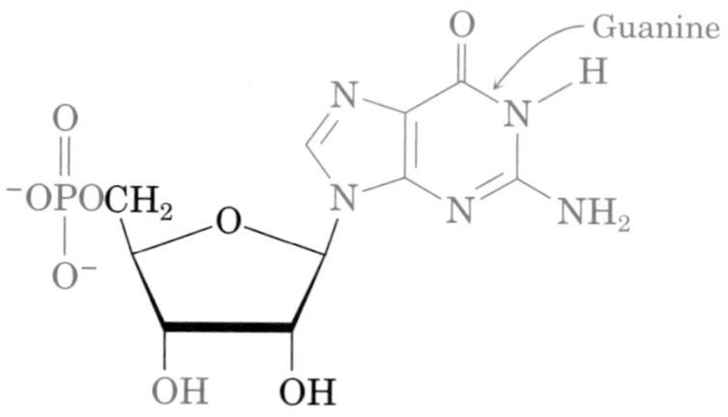

Guanosine 5′-phosphate

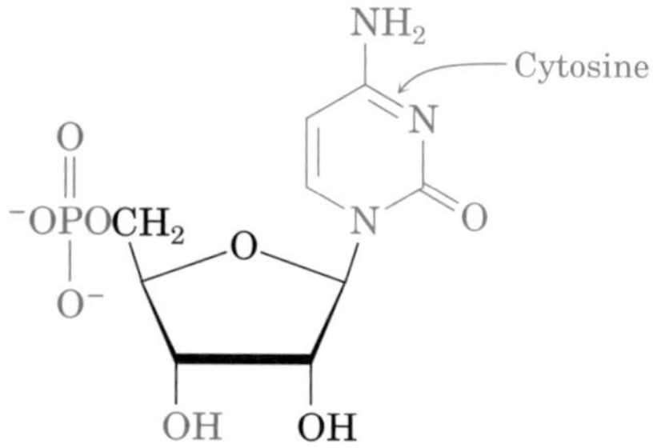

Cytidine 5′-phosphate

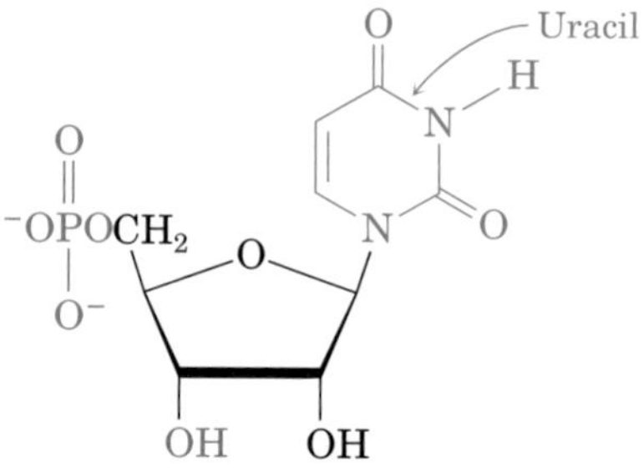

Uridine 5′-phosphate

28.9 Structure of Nucleic Acids

Nucleotides join together in DNA and RNA by forming a phosphate ester bond between the 5′-phosphate group on one nucleotide and the 3′-hydroxyl group on the sugar of another nucleotide (Figure 28.5). One end of the nucleic acid polymer has a free hydroxyl at C3′ (the **3′ end**), and the other end has a phosphate at C5′ (the **5′ end**).

FIGURE 28.5 ▼

Generalized structure of DNA.

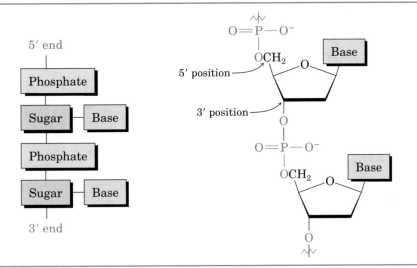

Just as the structure of a protein depends on the sequence in which individual amino acids are connected, the structure of a nucleic acid depends on the sequence of individual nucleotides. To carry the analogy further, just as a protein has a polyamide backbone with different side chains attached to it, a nucleic acid has an alternating sugar–phosphate backbone with different amine bases attached.

A protein

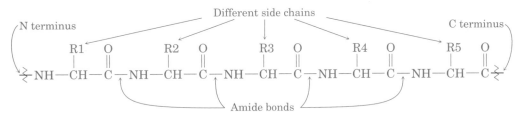

A nucleic acid

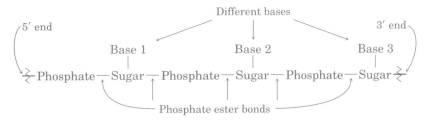

The sequence of nucleotides in a chain is described by starting at the 5′ end and identifying the bases in order of occurrence. Rather than write the full name of each nucleotide, though, it's more convenient to use abbreviations: A for adenosine, T for thymine, G for guanosine, and C for cytidine. Thus, a typical DNA sequence might be written as TAGGCT.

Problem 28.8   Draw the full structure of the DNA dinucleotide AG.

Problem 28.9   Draw the full structure of the RNA dinucleotide UA.

28.10   Base Pairing in DNA: The Watson–Crick Model

Samples of DNA isolated from different tissues of the same species have the same proportions of heterocyclic bases, but samples from different species often have different proportions of bases. Human DNA, for example, contains about 30% each of adenine and thymine, and about 20% each of guanine and cytosine. The bacterium *Clostridium perfringens*, however, contains about 37% each of adenine and thymine, and only 13% each of guanine and cytosine. Note that in both examples the bases occur in pairs. Adenine and thymine are usually present in equal amounts, as are cytosine and guanine. Why should this be?

In 1953, James Watson and Francis Crick made their now classic proposal for the secondary structure of DNA. According to the Watson–Crick model, DNA consists of two polynucleotide strands coiled around each other in a **double helix** like the handrails on a spiral staircase. The two strands run in opposite directions and are held together by hydrogen bonds between specific pairs of bases. Adenine (A) and thymine (T) form strong hydrogen bonds to each other but not to C or G, while guanine (G) and cytosine (C) form strong hydrogen bonds to each other but not to A or T (Figure 28.6).

James Dewey Watson and Francis Harry Compton Crick

James Dewey Watson (1928–) was born in Chicago, Illinois, and enrolled in the University of Chicago at age 15. He received his Ph.D. in 1950 at the University of Indiana and then worked at Cambridge University in England from 1951 to 1953, where he and Francis Crick deduced the structure of DNA. After more than 20 years as professor at Harvard University, he moved in 1976 to the Laboratory of Quantitative Biology at Cold Spring Harbor, Long Island, New York. He shared the 1960 Nobel Prize in medicine for his work on nucleic acids.

Francis Harry Compton Crick (1916–) was born in Northampton, England, and began his scientific career as a physicist. Following an interruption in his studies caused by World War II, he switched to biology and received his Ph.D. in 1954 at Cambridge University. He then remained at Cambridge University as professor. He shared the 1960 Nobel Prize in medicine.

FIGURE 28.6 ▼

Hydrogen bonding between base pairs in the DNA double helix. Electrostatic potential maps show that the faces of the bases are relatively neutral, while the edges contain positive (blue) and negative (red) regions. Pairing G with C and A with T brings together oppositely charged regions.

adenine, cytosine, guanine, thymine

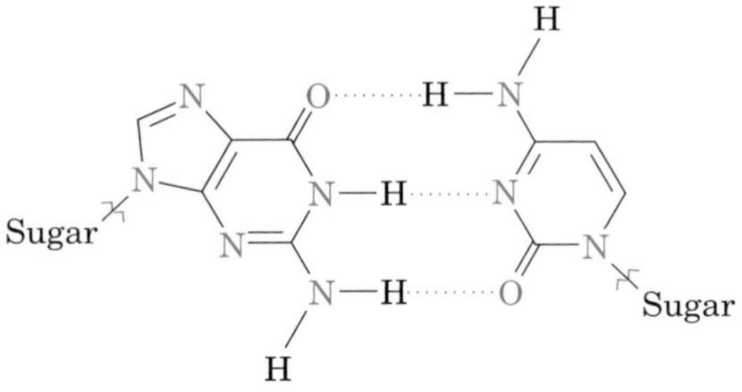

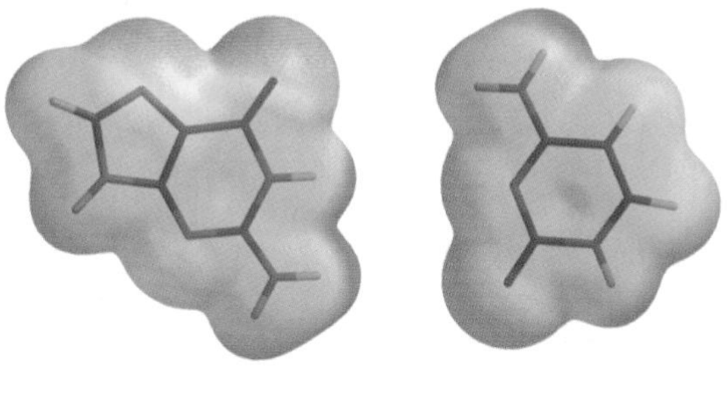

(Guanine) G ⋮⋮⋮⋮⋮⋮ C (Cytosine)

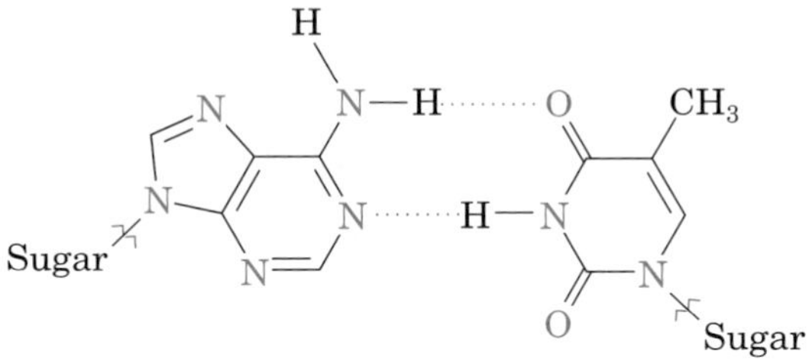

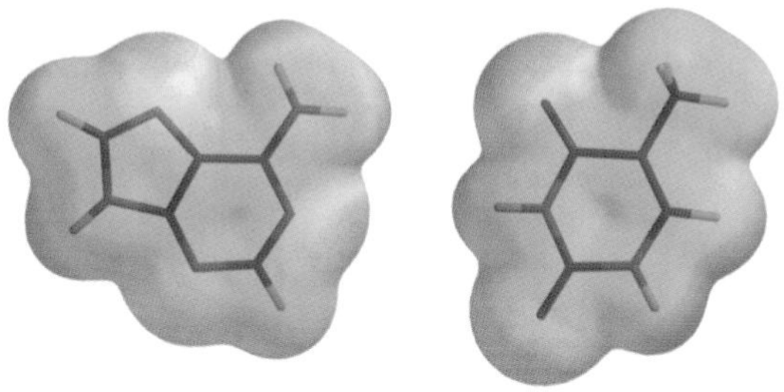

(Adenine) A ⋮⋮⋮⋮⋮ T (Thymine)

The two strands of the DNA double helix aren't identical; rather, they're *complementary*. Whenever a C base occurs in one strand, a G base occurs opposite it in the other strand. When an A base occurs in one strand, a T appears opposite it in the other strand. This complementary pairing of bases explains why A and T are always found in equal amounts, as are C and G. Figure 28.7 (p. 1166) illustrates this base pairing, showing how the two complementary strands are coiled into the double helix. X-ray measurements show that the DNA double helix is 2.0 nm (20 Å) wide, that there are 10 base pairs in each full turn, and that each turn is 3.4 nm in height.

A helpful mnemonic device to remember the nature of the hydrogen bonding between the four DNA bases is the simple phrase "Pure silver taxi."

Pure	Silver	Taxi
Pur	Ag	TC
The purine bases,	A and G,	hydrogen bond to T and C.

Notice in Figure 28.7 that the two strands of the double helix coil in such a way that two kinds of "grooves" result, a *major groove* 1.2 nm wide and a *minor groove* 600 pm wide. Interestingly, a variety of flat, polycyclic aromatic molecules are able to fit sideways into the groove between the

FIGURE 28.7 ▼

Complementarity in base pairing in the DNA double helix. The sugar–phosphate backbone runs along the outside of the helix, while the amine bases hydrogen-bond to one another on the inside. Both major and minor grooves are visible.

DNA double helix

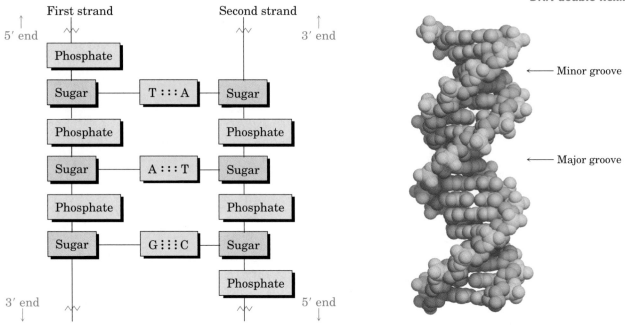

strands and *intercalate,* or insert themselves, between the stacked base pairs. Many cancer-causing and cancer-preventing agents function by intercalating with DNA in this way.

● ●

Problem 28.10   What sequence of bases on one strand of DNA is complementary to the following sequence on another strand? (Remember, the 5′ end is on the left and the 3′ end is on the right.)

GGCTAATCCGT

● ●

28.11 Nucleic Acids and Heredity

The genetic information of an organism is stored as a sequence of deoxyribonucleotides strung together in the DNA chain. For this information to be preserved and passed on to future generations, a mechanism must exist for copying DNA. For the information to be used, a mechanism must exist for decoding the DNA message and for implementing the instructions it contains.

Crick's "central dogma of molecular genetics" says that the function of DNA is to store information and pass it on to RNA at the proper time. The function of RNA, in turn, is to read, decode, and use the information to make proteins. By decoding the right bit of DNA at the right time in the right place, an organism can use genetic information to synthesize the many thousands of proteins necessary for carrying out its biochemical reactions.

Three fundamental processes take place in the transfer of genetic information:

- **Replication** is the process by which identical copies of DNA are made so that information can be preserved and handed down to off-spring.

- **Transcription** is the process by which the genetic messages contained in DNA are read and carried out of the nucleus to parts of the cell called *ribosomes* where protein synthesis occurs.

- **Translation** is the process by which the genetic messages are decoded and used to build proteins.

28.12 Replication of DNA

Replication of DNA is an enzyme-catalyzed process that begins by a partial unwinding of the double helix. As the strands separate and bases are exposed, new nucleotides line up on each strand in a complementary manner, A to T and C to G, and two new strands begin to grow. Each new strand is complementary to its old template strand, and two new identical DNA double helices are produced (Figure 28.8, p. 1168). Since each of the new DNA molecules contains one strand of old DNA and one strand of new DNA, the process is described as **semiconservative replication**. Crick described the process best when he used the analogy of the two DNA strands fitting together like a hand in a glove. The hand and glove separate, a new hand forms inside the glove, and a new glove forms around the hand. Two identical copies now exist where only one existed before.

The process by which the individual nucleotides are joined to create new DNA strands involves many steps and enzymes. Addition of new nucleotide units to the growing chain takes place in the 5' $\longrightarrow$ 3' direction and is catalyzed by the enzyme DNA polymerase. The key step is the addition of a 5'-mononucleoside triphosphate to the free 3'-hydroxyl group of the growing chain as the 3'-hydroxyl attacks the triphosphate and expels a diphosphate leaving group.

FIGURE 28.8 ▼

Schematic representation of DNA replication. The original double-stranded DNA partially unwinds, bases are exposed, nucleotides line up on each strand in a complementary manner, and two new strands begin to grow.

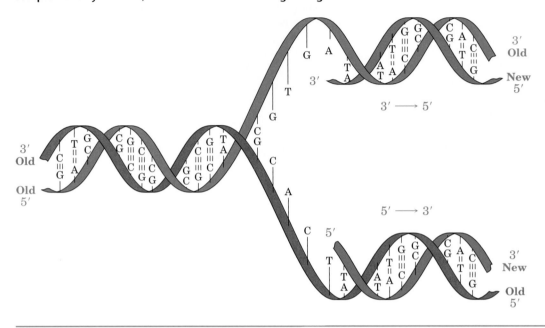

Both new DNA strands are synthesized in the same $5' \longrightarrow 3'$ direction, which implies that they can't be made in exactly the same way. Since the two complementary DNA strands are lined up in opposite directions, one strand must have its 3' end near the point of unraveling (the **replication fork**) while the other strand has its 5' end near the replication fork. What happens is that *both* strands are synthesized in the 5' ⟶ 3' direction, but the complement of the original 5' ⟶ 3' strand is synthesized continuously in a single piece while the complement of the original 5' ⟶ 3' strand is synthesized discontinuously in small pieces. The pieces are then linked by DNA ligase enzymes.

The magnitude of the replication process is staggering. The nucleus of every human cell contains 46 chromosomes (23 pairs), each of which consists of one very large DNA molecule. Each chromosome, in turn, is made up of several thousand DNA segments called *genes*, and the sum of all genes in a human cell (the human *genome*) is estimated to be 3 billion base pairs. A single DNA chain might have a length of over 12 cm and contain up to 250 million pairs of bases. Despite the size of these enormous molecules, their base sequence is faithfully copied during replication. The copying process takes only minutes, and an error occurs only about once each 10–100 billion bases.

28.13 Structure and Synthesis of RNA: Transcription

As noted previously, RNA is structurally similar to DNA but contains ribose rather than deoxyribose and uracil rather than thymine. There are three major kinds of RNA, each of which serves a specific function. All three are much smaller molecules than DNA, and all remain single-stranded rather than double-stranded.

- **Messenger RNA (mRNA)** carries genetic messages from DNA to ribosomes, small granular particles in the cytoplasm of a cell where protein synthesis takes place.

- **Ribosomal RNA (rRNA)** complexed with protein provides the physical makeup of the ribosomes.

- **Transfer RNA (tRNA)** transports amino acids to the ribosomes where they are joined together to make proteins.

The conversion of the information in DNA into proteins begins in the nucleus of cells with the synthesis of mRNA by **transcription** of DNA. Several turns of the DNA double helix unwind, forming a "bubble" and exposing the bases of the two strands. Ribonucleotides line up in the proper order by hydrogen bonding to their complementary bases on DNA, bond formation occurs in the 5' ⟶ 3' direction, and the growing RNA molecule unwinds from DNA (Figure 28.9, p. 1170).

Unlike what happens in DNA replication, where both strands are copied, only one of the two DNA strands is transcribed into mRNA. The strand that contains the gene is called the **coding strand**, or **sense strand**, and the strand that gets transcribed is called the **template strand**, or **antisense strand**. Since the template strand and the coding strand are

FIGURE 28.9 ▼

Biosynthesis of RNA using a DNA segment as template.

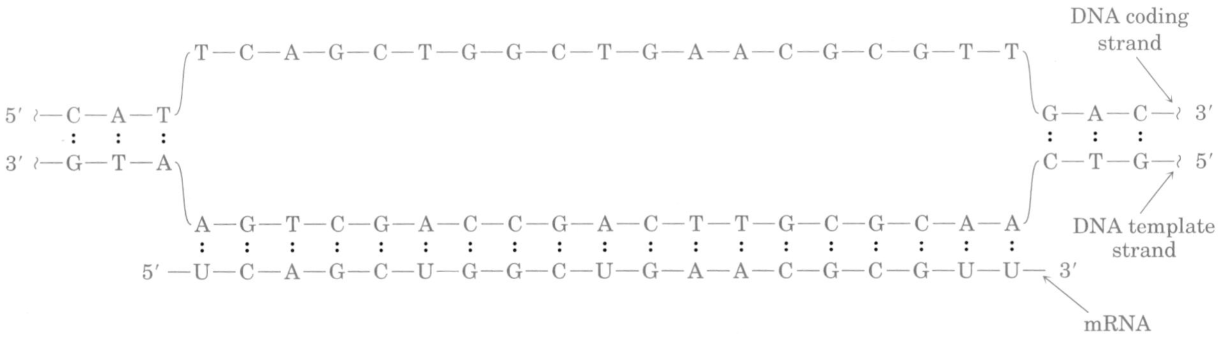

```
                    ⌈T—C—A—G—C—T—G—G—C—T—G—A—A—C—G—C—G—T—T⌉         DNA coding
                    /                                          \         strand
5′ ⸾—C—A—T/                                                     \G—A—C—⸾ 3′
      :   :   :                                                    :   :   :
3′ ⸾—G—T—A                                                         C—T—G—⸾ 5′
          \                                                      /
           ⌊A—G—T—C—G—A—C—C—G—A—C—T—T—G—C—G—C—A—A⌋         DNA template
            :   :   :   :   :   :   :   :   :   :   :   :   :   :   :   :   :   :   :         strand
        5′ —U—C—A—G—C—U—G—G—C—U—G—A—A—C—G—C—G—U—U— 3′
                                                              mRNA
```

complementary, and since the template strand and the RNA molecule are also complementary, *the RNA molecule produced during transcription is a copy of the coding strand.* The only difference is that the RNA molecule has a U everywhere the DNA coding strand has a T.

Transcription of DNA by the process just discussed raises many questions. How does the DNA know where to unwind? Where along the chain does one gene stop and the next one start? How do the ribonucleotides know the right place along the template strand to begin lining up and the right place to stop? A DNA chain contains specific base sequences called *promoter sites* that lie at positions 10 base pairs and 35 base pairs upstream from the beginning of the coding region and signal the beginning of a gene. Similarly, there are other base sequences near the end of the gene that signal a stop.

Another part of the picture is that genes are not necessarily continuous segments of the DNA chain. Often a gene will begin in one small section of DNA called an **exon**, then be interrupted by a seemingly nonsensical section called an **intron**, and then take up again farther down the chain in another exon. The final mRNA molecule results only after the nonsense sections are cut out and the remaining pieces are spliced together. Current evidence is that up to 98% of human DNA is made up of introns and only about 2% of DNA actually contains genetic instructions.

• •

Problem 28.11   Show how uracil can form strong hydrogen bonds to adenine.

Problem 28.12   What RNA base sequence is complementary to the following DNA base sequence?

GATTACCGTA

• •

28.14   RNA and Protein Biosynthesis: Translation

The primary cellular function of RNA is to direct biosynthesis of the thousands of diverse peptides and proteins required by an organism—at least 100,000 in a human. The mechanics of protein biosynthesis appear to be catalyzed by mRNA rather than by protein-based enzymes and take place on ribosomes, small granular particles in the cytoplasm of a cell that consist of about 60% ribosomal RNA and 40% protein. On the ribosome, mRNA serves as a template to pass on the genetic information it has transcribed from DNA.

The specific ribonucleotide sequence in mRNA forms a message that determines the order in which different amino acid residues are to be joined. Each "word," or **codon**, along the mRNA chain consists of a sequence of three ribonucleotides that is specific for a given amino acid. For example, the series U-U-C on mRNA is a codon directing incorporation of the amino acid phenylalanine into the growing protein. Of the $4^3 = 64$ possible triplets of the four bases in RNA, 61 code for specific amino acids, and 3 code for chain termination. Table 28.1 shows the meaning of each codon.

TABLE 28.1   Codon Assignments of Base Triplets

First base (5' end)	Second base	Third base (3' end)			
		U	C	A	G
U	U	Phe	Phe	Leu	Leu
	C	Ser	Ser	Ser	Ser
	A	Tyr	Tyr	Stop	Stop
	G	Cys	Cys	Stop	Trp
C	U	Leu	Leu	Leu	Leu
	C	Pro	Pro	Pro	Pro
	A	His	His	Gln	Gln
	G	Arg	Arg	Arg	Arg
A	U	Ile	Ile	Ile	Met
	C	Thr	Thr	Thr	Thr
	A	Asn	Asn	Lys	Lys
	G	Ser	Ser	Arg	Arg
G	U	Val	Val	Val	Val
	C	Ala	Ala	Ala	Ala
	A	Asp	Asp	Glu	Glu
	G	Gly	Gly	Gly	Gly

The message carried by mRNA is read by transfer RNA (tRNA) in a process called **translation**. There are 61 different tRNA's, one for each of the 61 codons that specifies an amino acid. A typical tRNA is roughly the shape of a cloverleaf, as shown in Figure 28.10. It consists of about 70–100 ribonucleotides and is bonded to a specific amino acid by an ester linkage through the 3′ hydroxyl on ribose at the 3′ end of the tRNA. Each tRNA also contains on its middle leaf a segment called an **anticodon**, a sequence of three ribonucleotides complementary to the codon sequence. For example, the codon sequence UUC present on mRNA is read by a phenylalanine-bearing tRNA having the complementary anticodon base sequence GAA. (Remember that nucleotide sequences are written in the 5′ ⟶ 3′ direction, so the sequence in an anticodon must be reversed. That is, the complement to (5′)-UUC-(3′) is (3′)-AAG-(5′), which is written as (5′)-GAA-(3′).

FIGURE 28.10 ▼

Structure of a tRNA molecule. The tRNA is a roughly cloverleaf-shaped molecule containing an anticodon triplet on one "leaf" and an amino acid attached covalently at its 3′ end. The example shown is a yeast tRNA that codes for phenylalanine. The nucleotides not specifically identified are chemically modified analogs of the four standard nucleotides.

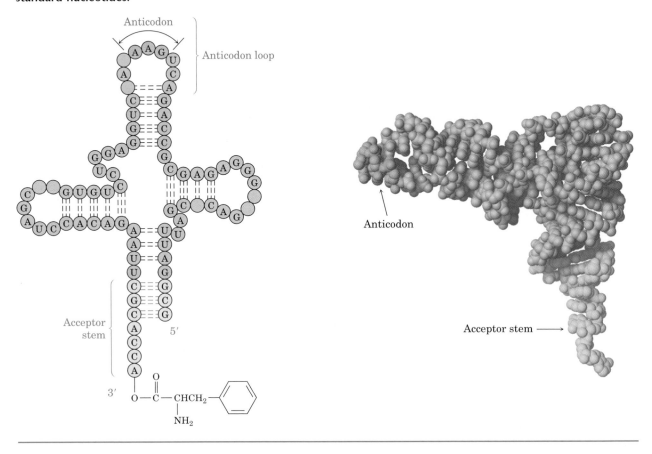

As each successive codon on mRNA is read, different tRNA's bring the correct amino acids into position for enzyme-mediated transfer to the growing peptide. When synthesis of the proper protein is completed, a "stop" codon signals the end, and the protein is released from the ribosome. The process is illustrated schematically in Figure 28.11.

FIGURE 28.11 ▼

A schematic representation of protein biosynthesis. The codon base sequences on mRNA are read by tRNA's containing complementary anticodon base sequences. Transfer RNA's assemble the proper amino acids into position for incorporation into the growing peptide.

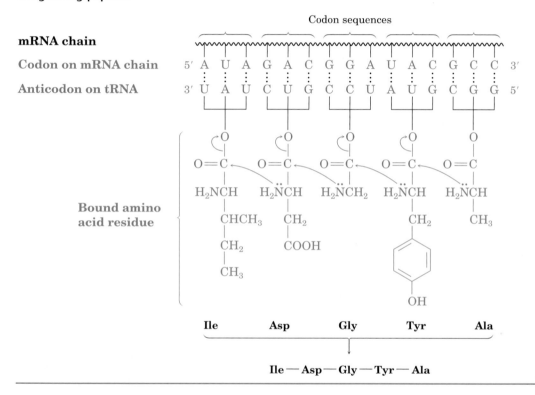

• •

Practice Problem 28.1 What amino acid sequence is coded by the following segment of a DNA coding strand?

CTA-ACT-AGC-GGG-TCG-CCG

Strategy The mRNA produced during translation is a copy of the DNA coding strand, with each T replaced by U. Thus, the mRNA has the sequence

CUA-ACU-AGC-GGG-UCG-CCG

Each set of three bases forms a codon, whose meaning can be found in Table 29.1.

Solution Leu-Thr-Ser-Gly-Ser-Pro

Problem 28.13   List codon sequences for the following amino acids:
(a) Ala   (b) Phe   (c) Leu   (d) Tyr

Problem 28.14   List anticodon sequences on the tRNA's carrying the amino acids shown in Problem 28.13.

Problem 28.15   What amino acid sequence is coded by the following mRNA base sequence?

CUU-AUG-GCU-UGG-CCC-UAA

Problem 28.16   What is the base sequence in the original DNA strand on which the mRNA sequence in Problem 28.15 was made?

28.15 DNA Sequencing

Walter Gilbert

Walter Gilbert (1932–) was born in Boston, Massachusetts, and received a Ph.D. in mathematics at Cambridge University in 1957. He then joined the faculty at Harvard University as professor of physics, but his research interests soon shifted to biochemistry. While still carrying out his academic work at Harvard, he founded Biogen, one of the earliest commercial biotechnology companies. He received the 1980 Nobel Prize in chemistry for his development of DNA sequencing methods.

When we work out the structure of DNA molecules, we examine the fundamental level that underlies all processes in living cells. DNA is the information store that ultimately dictates the structure of every gene product, delineates every part of the organism. The order of the bases along DNA contains the complete set of instructions that make up the genetic inheritance. *(Walter Gilbert, Nobel Prize Lecture, 1980)*

One of the greatest scientific revolutions in history is now underway in molecular biology, as scientists are learning how to manipulate and harness the genetic machinery of organisms. None of the extraordinary advances of the past two decades would have been possible, however, were it not for the discovery in 1977 of methods for sequencing immense DNA chains to find the messages they contain.

Two methods of DNA sequencing are in general use. Both operate along similar lines, but the *Maxam–Gilbert method* uses chemical techniques, while the *Sanger dideoxy method* uses enzymatic reactions. The Maxam–Gilbert method is preferred for some specialized uses, but the Sanger method is preferred for large-scale sequencing. Let's look at both.

Maxam–Gilbert DNA Sequencing

There are five steps to the **Maxam–Gilbert method** of DNA sequencing:

STEP 1   The first problem in DNA sequencing is to cleave the enormous DNA chain at predictable points to produce smaller, more manageable pieces, a task accomplished by the use of enzymes called **restriction endonucleases**. Each different restriction enzyme, of which more than 200 are available, cleaves a DNA molecule at well-defined points in the chain wherever a specific base sequence occurs. For example, the restriction enzyme *Alu*I cleaves between G and C in the four-base sequence AG-CT (Figure 28.12). Note that the sequence is a *palindrome*, meaning that it reads the same from left to right and right to left; that is, the *sequence* (5')-AG-CT-(3') is identical to its *complement*, (3')-TC-GA-(5'). The same is true for other restriction endonucleases.

If the original DNA molecule is cut with another restriction enzyme having a different specificity for cleavage, still other segments are produced whose sequences partially overlap those produced by the first enzyme. Sequencing of all the segments, followed by identification of the overlapping sequences, then allows complete DNA sequencing.

FIGURE 28.12 ▼

Cleavage of a double-stranded DNA molecule with the restriction enzyme *Alu*I cleaves at the sequence AG-CT. After cleavage, the fragments are isolated and each is radioactively labeled at its 5' end by enzyme-catalyzed formation of a ^{32}P-containing phosphate ester. The strands are then separated.

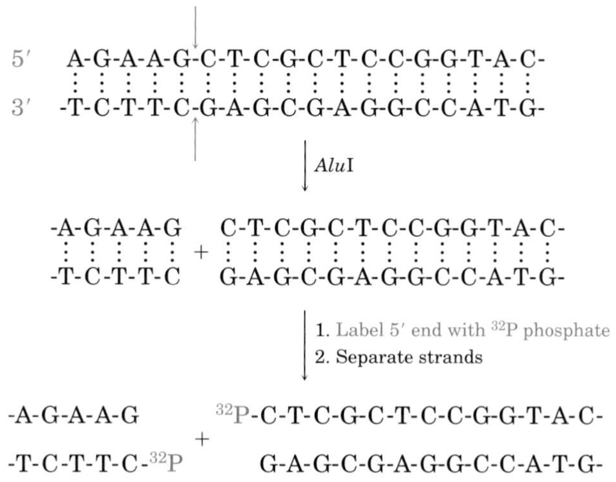

STEP 2 After cleavage of the DNA into smaller pieces, called *restriction fragments*, each piece is radioactively tagged by enzymatically incorporating a ^{32}P-labeled phosphate group onto the 5'-hydroxyl group of the terminal nucleotide. The double-stranded restriction fragments are then separated into single strands by heating, and the strands are isolated. Imagine, for example, that we now have a single-stranded DNA fragment with the following partial structure:

$$5' \quad ^{32}\text{P-GATCAGCGAT----} \quad 3'$$

STEP 3 The labeled DNA sample is divided into four subsamples and subjected to four parallel sets of chemical reactions under conditions that cause:

(a) Splitting of the DNA chain next to A

(b) Splitting of the DNA chain next to G

(c) Splitting of the DNA chain next to C

(d) Splitting of the DNA chain next to *both* T and C

Mild reaction conditions are used so that *only a few of the many possible splittings occur*. Literally hundreds of different product fragments result from a cleavage reaction, but *only those fragments that retain a ^{32}P label are important for sequencing*. In our example, the labeled pieces shown in Table 28.2 might be produced.

TABLE 28.2 Splitting of a DNA Fragment Under Four Conditions

Cleavage conditions	Labeled DNA pieces produced
Original DNA fragment	^{32}P-G-A-T-C-A-G-C-G-A-T-
Next to A	^{32}P-G ^{32}P-G-A-T-C ^{32}P-G-A-T-C-A-G-C-G + Larger pieces
Next to G	^{32}P-G-A-T-C-A ^{32}P-G-A-T-C-A-G-C + Larger pieces
Next to C	^{32}P-G-A-T ^{32}P-G-A-T-C-A-G + Larger pieces
Next to C + T	^{32}P-G-A ^{32}P-G-A-T ^{32}P-G-A-T-C-A-G ^{32}P-G-A-T-C-A-G-C-G-A + Larger pieces

Cleavages next to A and G are accomplished by treating a restriction fragment with dimethyl sulfate [$(CH_3O)_2SO_2$]. Deoxyadenosine (A) is methylated at N3 (S_N2 reaction), and deoxyguanosine (G) is methylated at N7, but T and C aren't affected.

Deoxyadenosine

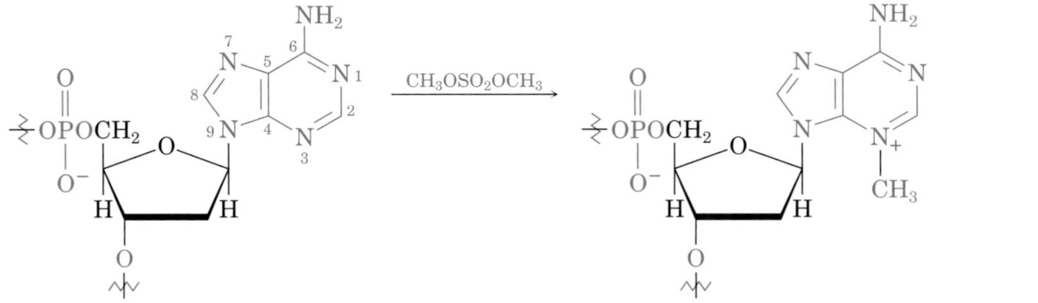

Deoxyguanosine

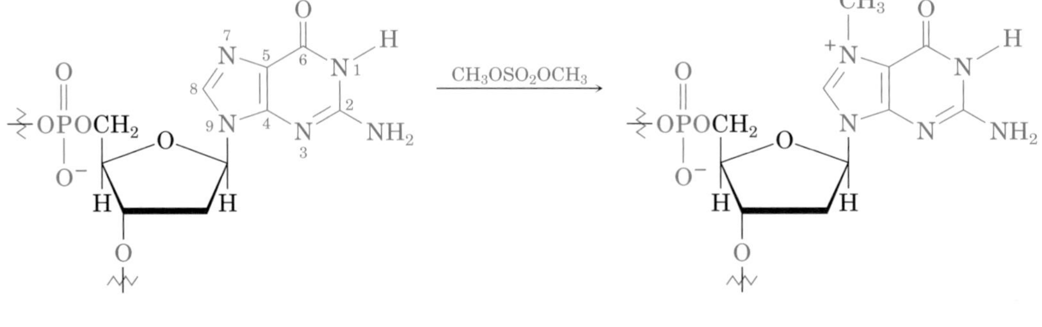

Treatment of methylated DNA with an aqueous solution of the secondary amine piperidine then brings about destruction of the methylated nucleotides and opening of the DNA chain at both the 3′ and 5′ positions next to the methylated bases. The mechanism of the cleavage process is shown in Figure 28.13 for deoxyguanosine.

1. A lone-pair of electrons on the sugar oxygen atom eliminates the methylated base to give an oxonium ion intermediate.

2. Addition of water to the sugar oxonium ion then opens the sugar and gives an aldehyde.

3. Formation of an enamine between piperidine and the 2-deoxyribose aldehyde group occurs (Section 19.9).

4. Lone-pair electrons of the enamine nitrogen atom break open the DNA chain by expelling the 3′ oxygen as leaving group.

5. A second, E2-like elimination of the 5′ oxygen occurs, destroying the deoxyribose sugar and further breaking the DNA chain.

FIGURE 28.13 ▼

Mechanism of DNA cleavage at deoxyguanosine (G).

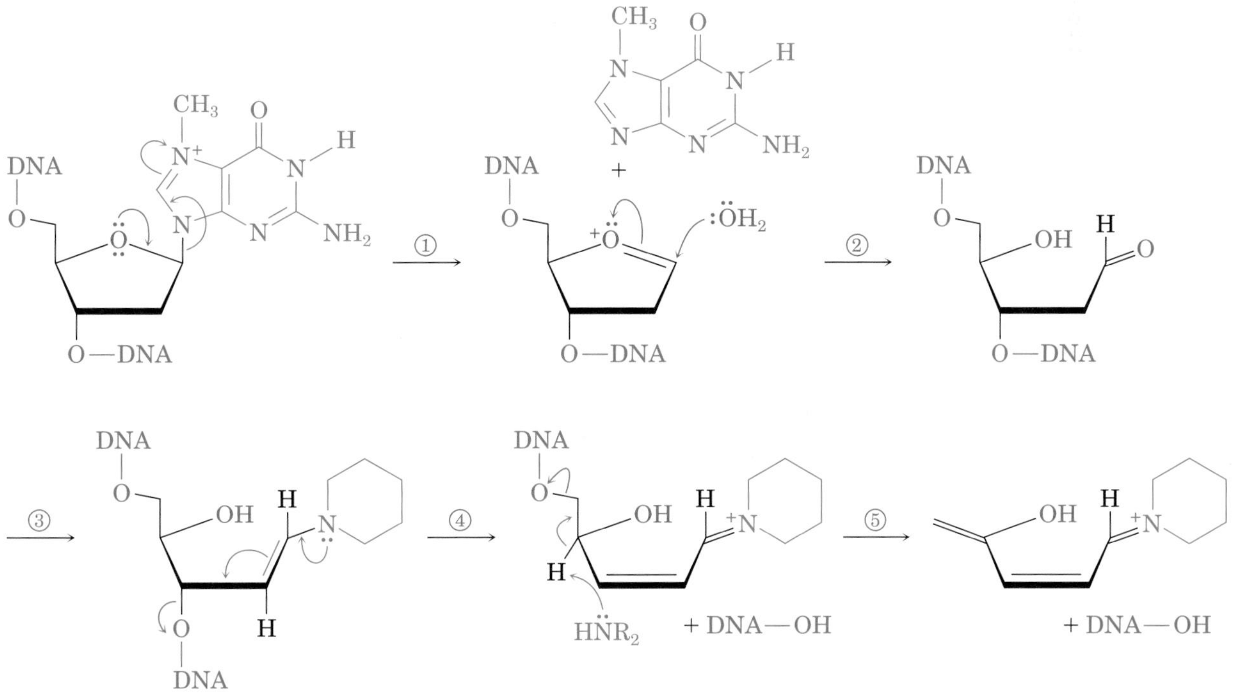

By working carefully, Maxam and Gilbert were able to find reaction conditions that are selective for cleavage at either A or G. (They found that G methylates five times as rapidly as A, but the hydrolytic breakdown of methylated A occurs more rapidly than the corresponding breakdown of methylated G if the product is first heated with dilute acid prior to base treatment.)

Breaking the DNA chain next to the pyrimidine nucleotides C and T is accomplished by treatment of DNA with hydrazine, H_2NNH_2, followed by heating with aqueous piperidine. Although no conditions have been found that are selective for cleavage next to T, a selective cleavage next to C can be accomplished by carrying out the hydrazine reaction in 5 M NaCl solution.

STEP 4 Each of the product mixtures from the four cleavage reactions is separated by electrophoresis (Section 26.2). When a mixture of DNA cleavage products is placed at one end of a strip of buffered polyacrylamide gel and a voltage is applied, electrically charged pieces move along the gel. Each piece moves at a rate that depends on the number of negatively charged phosphate groups (the number of nucleotides) it contains. Smaller pieces move rapidly, and larger pieces move more slowly. The technique is so sensitive that up to 600 DNA pieces, differing in size by only one nucleotide, can be separated.

Once separated, the locations of the DNA cleavage products are detected by exposing the gel to a photographic plate, a process called **autoradiography**. Each radioactive end piece containing a ^{32}P label appears as a dark band on the photographic plate, but nonradioactive pieces from the middle of the chain aren't seen. The gel electrophoresis pattern shown in Figure 28.14 would be obtained in our hypothetical example.

FIGURE 28.14 ▼

Representation of a gel electrophoresis pattern. The products of the four cleavage experiments are placed at the top of the gel, and a voltage is applied between top and bottom. Smaller products migrate along the gel at a faster rate and thus appear at the bottom. The DNA sequence can be read from the positions of the radioactive spots.

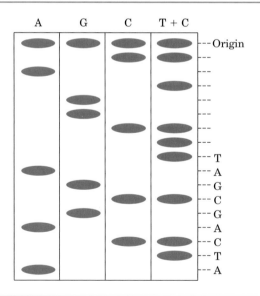

STEP 5 The DNA sequence is read directly from the gel. The band that appears farthest from the origin is the 5′ terminal mononucleotide (the smallest piece) and can't be identified. Because the terminal mononucleotide appears in the A column, though, *it must have been produced by splitting next to an A.* Thus, the *second* nucleotide in the sequence is an A.

The second farthest band from the origin is a dinucleotide that appears only in the T + C column. It is produced by splitting next to the third nucleotide, which must therefore be a T or a C. But because this piece doesn't appear in the C column, the third nucleotide must be a T and not a C. The third farthest band appears in both C and T + C columns, meaning that the fourth nucleotide is a C. Continuing in this manner, the entire sequence of the DNA can be read from the gel simply by noting in what columns the successively larger labeled polynucleotide pieces appear. Once read, the entire sequence can be checked by determining the sequence of the complementary strand.

Problem 28.17 Show the labeled cleavage products you would expect to obtain if the following DNA segment were subjected to each of the four cleavage reactions:

$$^{32}\text{P-AACATGGCGCTTATGACGA}$$

Problem 28.18 Sketch what you would expect the gel electrophoresis pattern to look like if the DNA segment in Problem 28.17 were sequenced.

Problem 28.19 Finish assigning the sequence to the gel electrophoresis pattern shown in Figure 28.14.

Sanger Dideoxy DNA Sequencing

All large-scale DNA sequencing is now done by the **Sanger dideoxy method**, which has several different variants. One particularly important variant used in commercial sequencing instruments begins with a mixture of the following:

- The restriction fragment to be sequenced

- A small piece of DNA called a *primer*, whose sequence is complementary to that on the 3′ end of the restriction fragment

- The four 2′-deoxyribonucleoside triphosphates (dNTP's)

- Very small amounts of the four 2′,3′-dideoxyribonucleoside triphosphates (ddNTP's), each of which is labeled with a fluorescent dye of a different color (A 2′,3′-*dideoxy*ribonucleoside triphosphate is one in which both 2′ and 3′ –OH groups are missing from ribose.)

**A 2'-deoxyribonucleoside
triphosphate (dNTP)**

**A labeled 2',3'-dideoxyribonucleoside
triphosphate (ddNTP)**

DNA polymerase enzyme is then added to this mix, and a strand of DNA complementary to the restriction fragment begins to grow from the end of the primer. Most of the time, only normal deoxyribonucleotides are incorporated into the growing chain, but every so often, a *dideoxy*ribonucleotide is incorporated. When that happens, DNA synthesis stops because the chain end no longer has a 3' hydroxyl group for adding further nucleotides.

When reaction is complete, the product consists of a mixture of DNA fragments of all possible lengths, each terminated by one of the four dye-labeled dideoxyribonucleotides. After separation by electrophoresis, the identity of the terminal dideoxyribonucleotide in each piece—and thus the sequence of the restriction fragment—can be identified simply by noting the color with which it fluoresces. Figure 28.15 shows a typical result.

So efficient is the automated dideoxy method that sequences up to 1000 nucleotides in length can be rapidly sequenced with 98% accuracy. The entire genome of the nematode worm *Caenorhabditis elegans* containing 19,000 genes and 97 million base pairs has now been sequenced, and substantial progress has been made on sequencing the human genome with approximately 140,000 genes and 3 billion base pairs. Completion of the work is scheduled for 2003 at the latest.

FIGURE 28.15 ▼

The sequence of a restriction fragment determined by the Sanger dideoxy method can be read simply by noting the colors of the dye attached to each of the various terminal nucleotides. (Courtesy of PE Biosystems.)

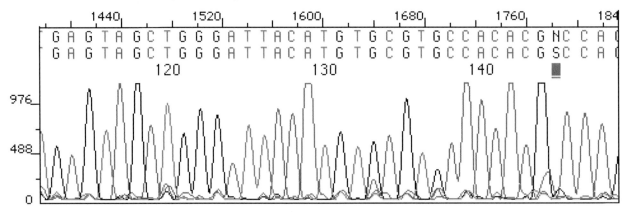

28.16 DNA Synthesis

The recent revolution in molecular biology has brought with it an increased demand for the efficient chemical synthesis of short DNA segments, called *oligonucleotides*. The problems of DNA synthesis are similar to those of protein synthesis (Section 26.10) but are more difficult because of the complexity of the nucleotide monomers. Each nucleotide has multiple reactive sites that must be selectively protected and deprotected at the proper times, and coupling of the four nucleotides must be carried out in the proper sequence. Automated DNA synthesizers are now available, however, that allow the fast and reliable synthesis of DNA segments up to 200 nucleotides in length.

DNA synthesizers operate on a principle similar to that of the Merrifield solid-phase peptide synthesizer (Section 26.11). In essence, a protected nucleotide is covalently bound to a solid support, and one nucleotide at a time is added to the chain by the use of a coupling reagent. After the final nucleotide has been added, the protecting groups are removed, and the synthetic DNA is cleaved from the solid support. Five steps are needed:

STEP 1 The first step in DNA synthesis involves attachment of a protected deoxynucleoside to a silica (SiO_2) support by an ester linkage to the 3′ –OH group of the deoxynucleoside. Both the 5′ –OH group on the sugar and free –NH_2 groups on the heterocyclic bases must be protected. Adenine and cytosine bases are protected by benzoyl groups, guanine is protected by an isobutyryl group, and thymine requires no protection. The deoxyribose 5′ –OH is protected as its *p*-dimethoxytrityl (DMT) ether.

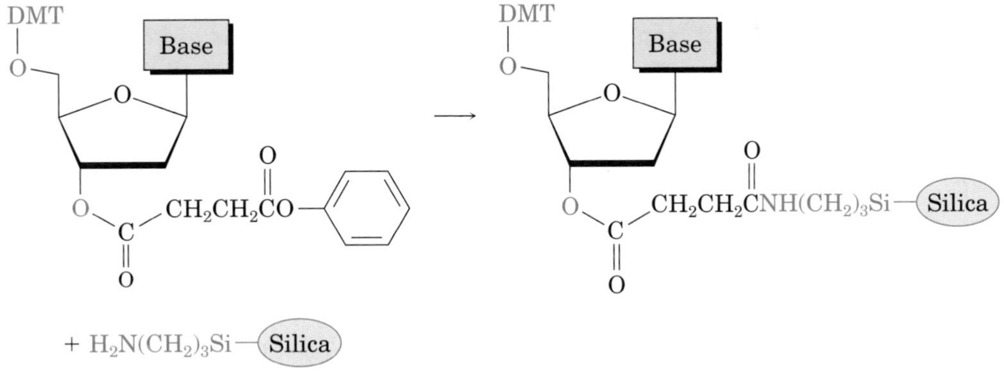

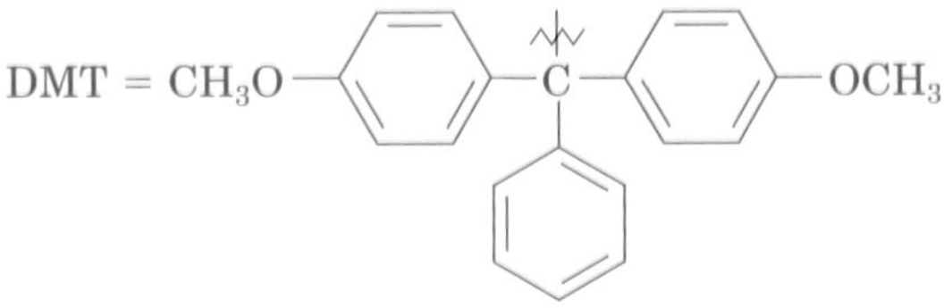

Base =

**N-protected
adenine**   **N-protected
guanine**   **N-protected
cytosine**   **Thymine**

STEP 2   The second step involves removal of the DMT protecting group by treatment with dichloroacetic acid in CH_2Cl_2. The reaction occurs by an S_N1 mechanism and proceeds rapidly because of the stability of the tertiary, benzylic dimethoxytrityl cation.

STEP 3   The third step involves coupling of the polymer-bonded deoxynucleoside with a protected deoxynucleoside containing a *phosphoramidite* group at its 3′ position. [A phosphoramidite has the structure $R_2NP(OR)_2$.] The coupling reaction takes place in the polar aprotic solvent acetonitrile, requires catalysis by the heterocyclic amine tetrazole, and yields a *phosphite*, $P(OR)_3$, as product. Note that one of the phosphorus oxygen atoms is protected by a β-cyanoethyl group, $-OCH_2CH_2C\equiv N$. The coupling step takes place in better than 99% yield.

STEP 4 With the coupling accomplished, the phosphite product is oxidized to a phosphate by treatment with iodine. The reaction is carried out in aqueous tetrahydrofuran in the presence of 2,6-dimethylpyridine. The cycle (1) deprotection, (2) coupling, and (3) oxidation is then repeated until an oligonucleotide chain of the desired sequence has been built.

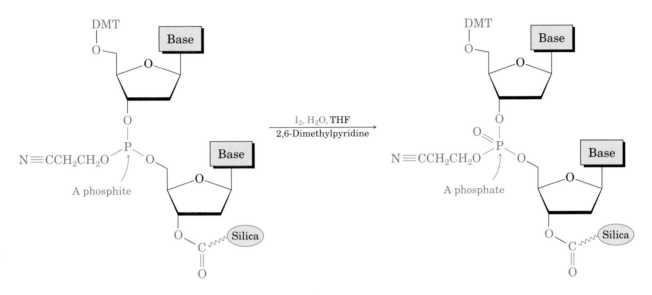

STEP 5 The final step is removal of all protecting groups and cleavage of the ester bond holding the DNA to the silica. All these reactions are done at the same time by treatment with aqueous NH$_3$. Purification by electrophoresis then yields the synthetic DNA.

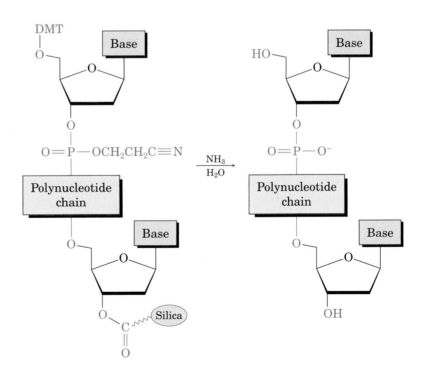

● ●

Problem 28.20   *p*-Dimethoxytrityl (DMT) ethers are easily cleaved by mild acid treatment. Show the mechanism of the cleavage reaction in detail.

Problem 28.21   Propose a mechanism to account for cleavage of the *β*-cyanoethyl protecting group from the phosphate groups on treatment with aqueous ammonia. (Acrylonitrile, $H_2C=CHCN$, is a by-product.) What kind of reaction is occurring?

● ●

28.17 The Polymerase Chain Reaction

Kary Banks Mullis

Kary Banks Mullis (1944–) was born in rural Lenoir, North Carolina, did undergraduate work at Georgia Tech., and received his Ph.D. at the University of California, Berkeley, in 1973. From 1979 to 1986 he worked at Cetus Corp., where his work on developing PCR was carried out. Since 1988, he has followed his own drummer as self-employed consultant and writer. He received the 1993 Nobel Prize in chemistry.

The invention of the **polymerase chain reaction (PCR)** by Kary Mullis in 1986 has been described as being to genes what Gutenberg's invention of the printing press was to the written word. Just as the printing press produces multiple copies of a book, PCR produces multiple copies of a given DNA sequence. Starting from less than 1 *picogram* of DNA with a chain length of 10,000 nucleotides (1 pg = 10^{-12} g; about 100,000 molecules), PCR makes it possible to obtain several micrograms (1 μg = 10^{-6} g; about 10^{11} molecules) in just a few hours.

The key to the polymerase chain reaction is *Taq* DNA polymerase, a heat-stable enzyme isolated from the thermophilic bacterium *Thermus aquaticus* found in a hot spring in Yellowstone National Park. *Taq* polymerase is able to take a single strand of DNA that has a short, primer segment of complementary chain at one end and then finish constructing the entire complementary strand. The overall process takes three steps, as shown schematically in Figure 28.16. (More recently, improved heat-stable DNA polymerase enzymes have become available, including Vent polymerase and *Pfu* polymerase, both isolated from bacteria growing near geothermal vents in the ocean floor. The error rate of both enzymes is substantially less than that of *Taq*.)

STEP 1   The double-stranded DNA to be amplified is first heated to 95°C in the presence of *Taq* polymerase, Mg^{2+} ion, the four deoxynucleotide triphosphate monomers (dNTP's), and a large excess of two short oligonucleotide primers of about 20 bases each. Each primer is complementary to the sequence at the end of one of the target DNA segments. At a temperature of 95°C, double-stranded DNA denatures, spontaneously breaking apart into two single strands.

STEP 2   The temperature is lowered to between 37°C and 50°C, allowing the primers, because of their relatively high concentration, to anneal by hydrogen bonding to their complementary sequence at the end of each target strand.

STEP 3   The temperature is then raised to 72°C, and *Taq* polymerase catalyzes the addition of further nucleotides to the two primed DNA strands. When replication of each strand is finished, *two* copies of the original DNA now exist. Repeating the denature–anneal–synthesize cycle a second time yields four DNA copies, repeating a third time yields eight copies, and so on, in an exponential series.

PCR has been automated, and 30 or so cycles can be carried out in an hour, resulting in a theoretical amplification factor of 2^{30} ($\sim 10^9$). In practice, however, the efficiency of each cycle is less than 100%, and an experimental amplification of about 10^6–10^8 is routinely achieved for 30 cycles.

FIGURE 28.16 ▼

The polymerase chain reaction. Double-stranded DNA is heated to 95°C in the presence of two short oligonucleotide primer sequences, each of which is complementary to the end of one of the strands. After the DNA denatures, the temperature is lowered and the primer sequences anneal to the strand ends. Raising the temperature in the presence of *Taq* polymerase, Mg²⁺, and a mixture of the four deoxynucleotide triphosphates (dNTP's) effects strand replication, producing two DNA copies. Each further repetition of the sequence again doubles the number of copies.

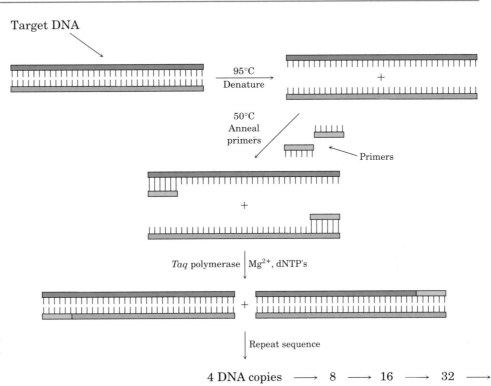

CHEMISTRY @ WORK

DNA Fingerprinting

The technique of DNA fingerprinting arose from the discovery in 1984 that human genes contain short, repeating sequences of noncoding DNA, called *short tandem repeat* (STR) loci. Furthermore, the STR loci are slightly different for every individual (except identical twins). By sequencing these loci, a pattern unique to each person can be obtained.

Forensic laboratories in the United States have agreed on 13 core STR loci that are most accurate for identification of an individual. Based

(continued) ▶

on these 13 loci, a Combined DNA Index System (CODIS) has been established to serve as a registry of convicted offenders. When a DNA sample is obtained from a crime scene—from blood, hair, skin, or semen, for example—the sample is subjected to cleavage with restriction endonucleases to cut out fragments containing the STR loci, the fragments are amplified using the polymerase chain reaction, and the sequences of the fragments are determined. If the profile of sequences from a known individual and the profile from DNA obtained at a crime scene match, the probability is approximately 82 billion to 1 that the DNA is from the same individual. In paternity cases, where the DNA of father and offspring are related but not fully identical, the identity of the father can be established with a probability of 100,000 to 1.

Historians have wondered for many years whether Thomas Jefferson fathered a child by Sally Hemmings. DNA fingerprinting evidence obtained in 1998 suggests that he did.

Summary and Key Words

A **heterocycle** is a compound with a ring that has more than one kind of atom. Nitrogen, oxygen, and sulfur are often found along with carbon in heterocyclic rings. Saturated heterocyclic amines, ethers, and sulfides usually have the same chemistry as their open-chain analogs, but unsaturated heterocycles such as pyrrole, furan, and thiophene are aromatic. All three are unusually stable, and all three undergo aromatic substitution on reaction with electrophiles.

Pyridine is the six-membered-ring, nitrogen-containing heterocyclic analog of benzene. The pyridine ring is electron-poor and undergoes electrophilic aromatic substitution reactions with difficulty. Nucleophilic aromatic substitutions of 2- or 4-halopyridines take place readily, however.

The nucleic acids **DNA (deoxyribonucleic acid)** and **RNA (ribonucleic acid)** are biological polymers that act as chemical carriers of an organism's genetic information. Enzyme-catalyzed hydrolysis of nucleic acids yields **nucleotides**, the monomer units from which RNA and DNA are constructed. Each nucleotide consists of a **purine** or **pyrimidine base** linked to C1′ of an aldopentose sugar (ribose in RNA and 2′-deoxyribose in DNA), with the sugar in turn linked through its C5′-hydroxyl to a phosphate group. The nucleotides are joined by phosphate links between the phosphate of one nucleotide and the 3′-hydroxyl on the sugar of another nucleotide.

Molecules of DNA consist of two polynucleotide strands held together by hydrogen bonds between heterocyclic bases on the different strands and coiled into a **double helix**. **Adenine** and **thymine** form hydrogen bonds to each other, as do **cytosine** and **guanine**. The two strands of DNA are not identical but are complementary.

Three processes take place in deciphering the genetic information of DNA:

- **Replication** of DNA is the process by which identical DNA copies are made and genetic information is preserved. This occurs when the DNA double helix unwinds, complementary deoxyribonucleotides line up in order, and two new DNA molecules are produced.

- **Transcription** is the process by which RNA is produced in order to carry genetic information from the nucleus to the ribosomes. This occurs when a short segment of the DNA double helix unwinds and complementary ribonucleotides line up to produce **messenger RNA (mRNA)**.

- **Translation** is the process by which mRNA directs protein synthesis. Each mRNA is divided into **codons**, ribonucleotide triplets that are recognized by small amino acid-carrying molecules of **transfer RNA (tRNA)**, which deliver the appropriate amino acids needed for protein synthesis.

Small DNA segments can be synthesized in the laboratory, and commercial instruments are available for automating the work. Sequencing of DNA can be carried out either by the **Maxam-Gilbert method**, which uses chemical techniques, or by the **Sanger dideoxy method**, which uses enzymatic techniques. Small amounts of DNA can be amplified by factors of 10^6 using the **polymerase chain reaction (PCR)**.

Summary of Reactions

1. Electrophilic aromatic substitution (Section 28.3)
   (a) Bromination

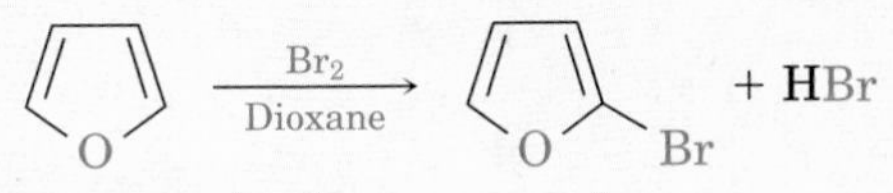

   Furan

   (b) Nitration

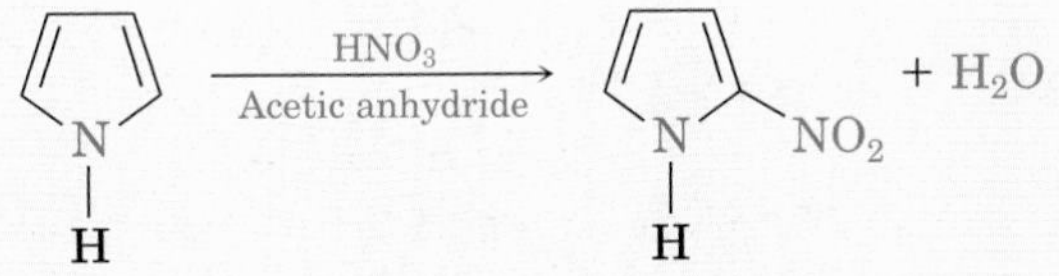

   Pyrrole

(continued) ▶

(c) Friedel–Crafts acylation

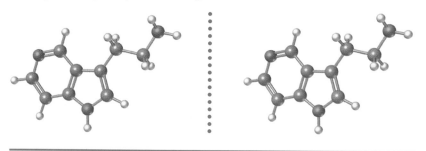

Thiophene

2. Nucleophilic aromatic substitution of halopyridines (Section 28.6)

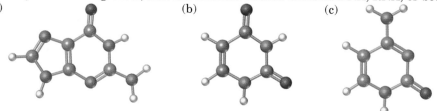

Visualizing Chemistry

•••••••••••••••••••••••••••••••••••••

(Problems 28.1–28.21 appear within the chapter.)

28.22   The following molecule has three nitrogen atoms. List them in order of increasing basicity, and explain your ordering.

Stereo View

28.23   Identify the following bases, and tell whether each is found in DNA, RNA, or both:

(a)   (b)   (c)

28.24   Identify the following nucleotide, and tell how it is used:

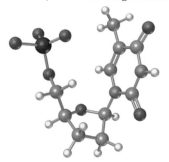

Additional Problems

● ●

28.25 Although pyrrole is a much weaker base than most other amines, it is a much stronger acid ($pK_a \approx 15$ for pyrrole versus 35 for diethylamine). The N–H proton is readily abstracted by base to yield the pyrrole anion, $C_4H_4N^-$. Explain.

28.26 Oxazole is a five-membered aromatic heterocycle. Draw an orbital picture of oxazole, showing all p orbitals and all lone-pair orbitals. Would you expect oxazole to be more basic or less basic than pyrrole? Explain.

Oxazole

28.27 Write the products of the reaction of furan with each of the following reagents:
(a) Br_2, dioxane, 0°C   (b) HNO_3, acetic anhydride   (c) CH_3COCl, $SnCl_4$
(d) H_2/Pd   (e) SO_3, pyridine

28.28 Pyrrole has a dipole moment $\mu = 1.8\,D$, with the *nitrogen* atom at the positive end of the dipole. Explain.

28.29 If 3-bromopyridine is heated with $NaNH_2$, a mixture of 3- and 4-aminopyridine is obtained. Explain.

28.30 Nitrofuroxime is a pharmaceutical agent used in the treatment of urinary tract infections. Propose a synthesis of nitrofuroxime from furfural.

Furfural   **Nitrofuroxime**

28.31 Substituted pyrroles are often prepared by treatment of a 1,4-diketone with ammonia. Suggest a mechanism.

28.32 3,5-Dimethylisoxazole is prepared by reaction of 2,4-pentanedione with hydroxylamine. Propose a mechanism.

28.33 Isoquinolines are often synthesized by the *Bischler–Napieralski cyclization* of an *N*-acyl-2-phenylethyl amine with strong acid and P_2O_5, followed by oxidation of the initially formed dihydroisoquinoline. Suggest a mechanism for the cyclization.

A dihydroisoquinoline   **1-Methylisoquinoline**

28.34 Quinolines are often prepared by the *Skraup synthesis*, in which an aniline reacts with an α,β-unsaturated aldehyde and the dihydroquinoline product is oxidized. Suggest a mechanism.

1,2-Dihydroquinoline   **Quinoline**

28.35 Human and horse insulin both have two polypeptide chains, with one chain containing 21 amino acids and the other containing 30 amino acids. How many nitrogen bases are present in the DNA that codes for each chain?

28.36 Human and horse insulin (Problem 28.35) differ in primary structure at two places. At position 9 in one chain, human insulin has Ser and horse insulin has Gly; at position 30 in the other chain, human insulin has Thr and horse insulin has Ala. How must the DNA for the two insulins differ?

28.37 The DNA of sea urchins contains about 32% A. What percentages of the other three bases would you expect in sea urchin DNA? Explain.

28.38 The codon UAA stops protein synthesis. Why does the sequence UAA in the following stretch of mRNA not cause any problems?

-GCA-UUC-GAG-GUA-ACG-CCC-

28.39 Which of the following base sequences would most likely be recognized by a restriction endonuclease? Explain.
(a) GAATTC   (b) GATTACA   (c) CTCGAG

28.40 For what amino acids do the following ribonucleotide triplets code?
(a) AAU   (b) GAG   (c) UCC   (d) CAU

28.41 From what DNA sequences were each of the mRNA codons in Problem 28.40 transcribed?

28.42 What anticodon sequences of tRNA's are coded for by the codons in Problem 28.40?

28.43 Draw the complete structure of the ribonucleotide codon UAC. For what amino acid does this sequence code?

28.44 Draw the complete structure of the deoxyribonucleotide sequence from which the mRNA codon in Problem 28.43 was transcribed.

28.45 Give an mRNA sequence that will code for synthesis of metenkephalin:

Tyr-Gly-Gly-Phe-Met

28.46 Give an mRNA sequence that will code for the synthesis of angiotensin II:

Asp-Arg-Val-Tyr-Ile-His-Pro-Phe

28.47 What amino acid sequence is coded for by the following DNA coding strand?

CTT-CGA-CCA-GAC-AGC-TTT

28.48 What amino acid sequence is coded for by the following mRNA base sequence?

CUA-GAC-CGU-UCC-AAG-UGA

28.49 If the DNA gene sequence -TAA-CCG-GAT- were miscopied during replication and became -TGA-CCG-GAT-, what effect would there be on the sequence of the protein produced?

28.50 Show the steps involved in a laboratory synthesis of the DNA fragment with the sequence CTAG.

28.51 Sodium nitrite, a food preservative used in meats, causes the mutation of cytosine into uracil under acidic conditions. Propose a mechanism (see Section 24.8).

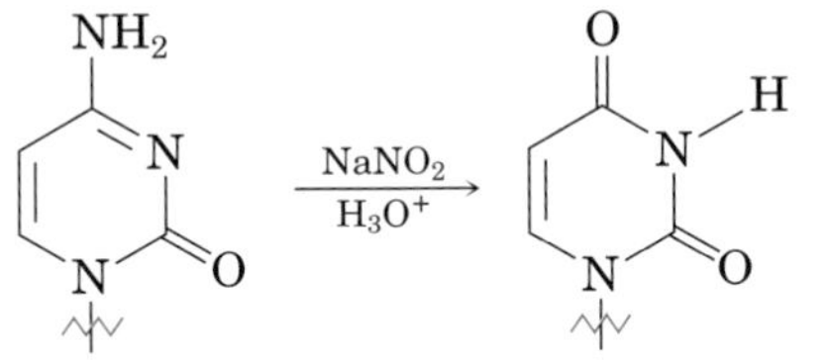

28.52 The final step in DNA synthesis is deprotection by treatment with aqueous ammonia. Show the mechanisms by which deprotection occurs at the points indicated in the following structure:

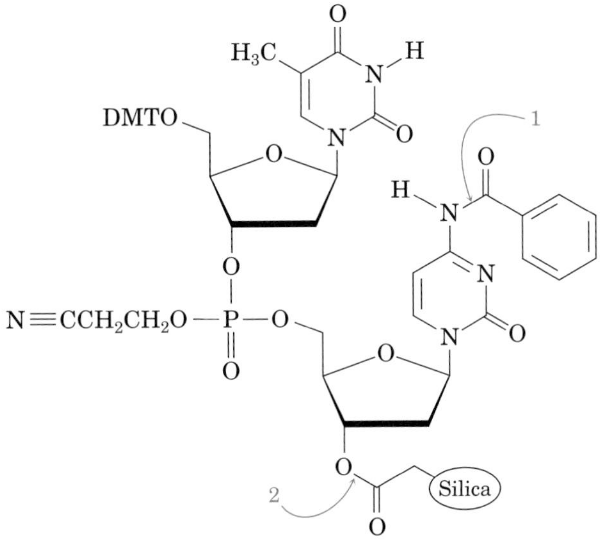

28.53 Review the mechanism shown in Figure 28.13 for the cleavage of deoxyguanosine residues, and propose a mechanism to account for the similar cleavage of deoxyadenosine residues in a DNA chain. Recall that deoxyadenosine is first methylated at N3 prior to hydrolysis.

A Look Ahead

∙∙∙

28.54 Draw the structure of cyclic adenosine monophosphate (cAMP), a messenger involved in the regulation of glucose production in the body. Cyclic AMP has a phosphate ring connecting the 3'- and 5'-hydroxyl groups on adenosine. (See Section 29.3.)

Molecular Modeling

∙∙∙∙∙∙∙∙∙∙∙∙∙∙∙∙∙∙∙∙∙∙∙∙∙∙∙∙∙∙∙∙∙∙∙∙∙∙∙

28.55 Use SpartanView to compare electrostatic potential maps and N–CHO bond distances in *N*-formylpyrrole and *N*-formylpyrrolidine. Account for any differences using resonance structures.

N-Formylpyrrole   **N-Formylpyrrolidine**

28.56 Maxam–Gilbert DNA sequencing relies on methylation of guanosine and adenosine nitrogens. Use SpartanView to examine electrostatic potential maps of 9-methylguanine and 9-methyladenine, and compare the electrostatic potentials at positions 7 on guanine and 3 on adenine. Which molecule is the better nucleophile, and why?

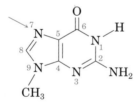

9-Methylguanine   **9-Methyladenine**

28.57 Mistakes in DNA replication lead to base-pair mismatches. Use SpartanView to obtain the energies of adenine, thymine, guanine, cytosine, an A–T pair, a G–C pair, an A–G pair, and a T–C pair. Calculate the binding energy for each base pair, and account for any differences. Which base pairs are flat, and which are twisted? Why?

28.58 Use SpartanView to examine a typical double helix in a DNA model. Tell how many base pairs there are in the model, the base sequence in each strand, and whether the strands are complementary.

29

The Organic Chemistry of Metabolic Pathways

The organic chemical reactions that take place in even the smallest and simplest living organism are more complex than those carried out in any laboratory. Yet the reactions in living organisms, regardless of their complexity, follow the same rules of reactivity and proceed by the same mechanisms that we've seen in the preceding chapters.

In this chapter, we'll look at some of the pathways by which organisms carry out their chemistry, focusing primarily on how they metabolize fats and carbohydrates. Our emphasis will not be on describing the details and subtleties of the pathways or on the enzymes that catalyze them. Instead, we'll be interested primarily in recognizing the similarities between the mechanisms of biological reactions and the mechanisms of analogous laboratory reactions.

29.1 An Overview of Metabolism and Biochemical Energy

The many reactions that go on in the cells of living organisms are collectively called **metabolism**. The pathways that break down larger molecules into smaller ones are called **catabolism**, while the pathways that synthesize larger biomolecules from smaller ones are known as **anabolism**. Catabolic reaction pathways usually release energy, while anabolic reaction pathways often absorb energy. Catabolism can be divided into the four stages shown in Figure 29.1.

In the first catabolic stage, **digestion**, food is broken down in the mouth, stomach, and small intestine by hydrolysis of ester, glycoside (acetal), and peptide (amide) bonds to yield primarily fatty acids, simple sugars, and amino acids. These smaller molecules are further degraded in the cytoplasm of cells to yield acetyl groups attached by a thiol ester bond (Section 21.9) to the large carrier molecule *coenzyme A*. The resultant compound, *acetyl coenzyme A (acetyl CoA)*, is an intermediate in the breakdown of all main classes of food molecules.

Acetyl coenzyme A

Acetyl groups are oxidized inside cellular mitochondria in the third stage of catabolism, the **citric acid cycle**, to yield CO_2. This stage also releases a large amount of energy that is used in the fourth stage, the **electron-transport chain**, to produce molecules of the nucleotide *adenosine triphosphate, ATP*. The final result of food catabolism, ATP has been called the "energy currency" of the cell. Catabolic reactions "pay off" in ATP by synthesizing it from *adenosine diphosphate, ADP*, plus hydrogen phosphate ion, HPO_4^{2-} (abbreviated P_i). Anabolic reactions "spend" ATP by transferring a phosphate group to another molecule, thereby regenerating ADP. Energy production and use in living organisms thus revolves around the ATP $\rightleftharpoons$ ADP interconversion (see p. 1196).

FIGURE 29.1 ▼

An overview of catabolic pathways for the degradation of food and the production of biochemical energy. The ultimate products of food catabolism are CO_2, H_2O, and adenosine triphosphate (ATP).

STAGE 1 Bulk food is digested in the stomach and small intestine to yield small molecules.

STAGE 2 Small sugar, fatty acid, and amino acid molecules are degraded in cells to yield acetyl CoA.

STAGE 3 Acetyl CoA is oxidized in the citric acid cycle to yield CO_2 and energy.

STAGE 4 The energy produced in stage 3 is used by the electron-transport chain to make ATP.

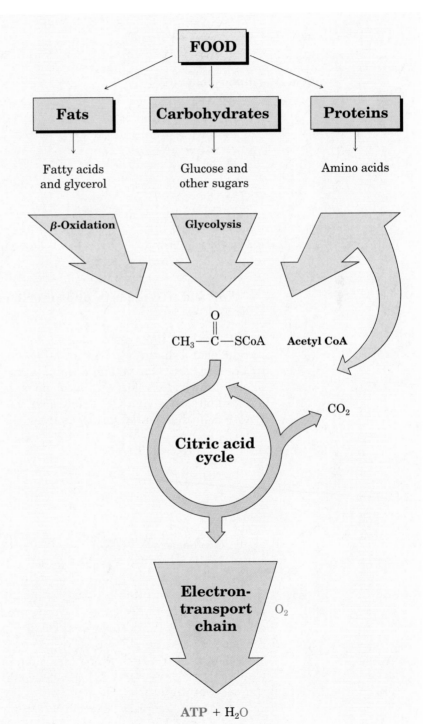

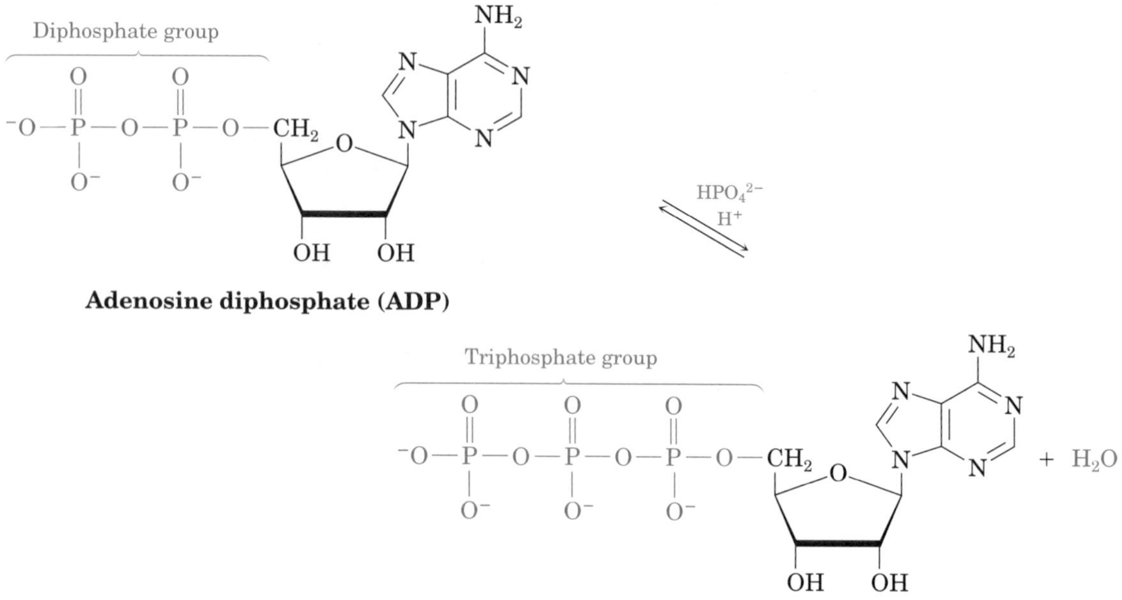

Adenosine diphosphate (ADP)

Adenosine triphosphate (ATP)

ADP and ATP are both **phosphoric acid anhydrides**, which contain

$$-\overset{\overset{\displaystyle O}{\|}}{P}-O-\overset{\overset{\displaystyle O}{\|}}{P}-$$ linkages analogous to the $$-\overset{\overset{\displaystyle O}{\|}}{C}-O-\overset{\overset{\displaystyle O}{\|}}{C}-$$ linkage in carboxylic acid anhydrides. Just as carboxylic anhydrides react with alcohols by breaking a C–O bond and forming a carboxylic ester (Section 21.5), phosphoric anhydrides react with alcohols by breaking a P–O bond and forming a phosphate ester, $ROPO_3{}^{2-}$. Note that the reactants and products are written in their dissociated forms, as they exist at physiological pH.

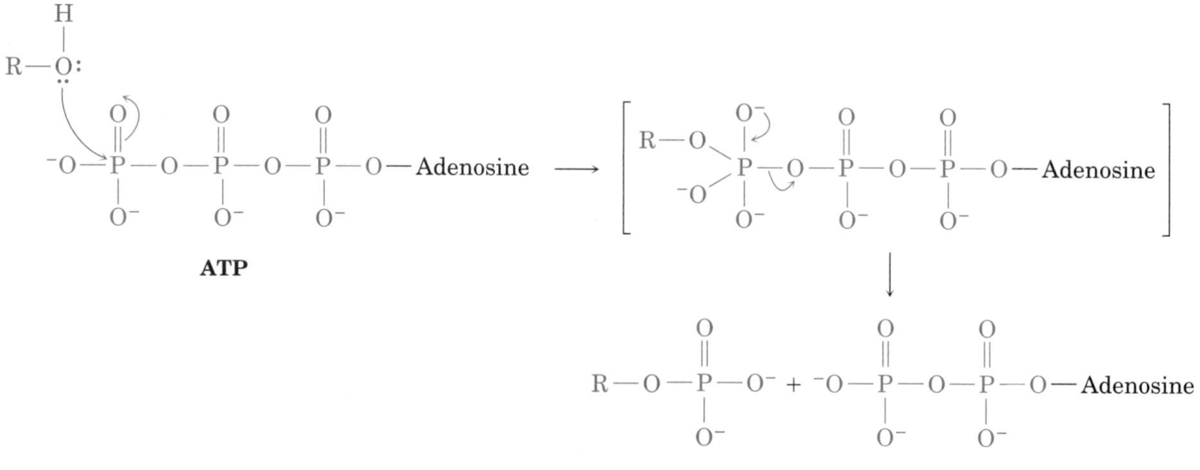

ATP

A phosphate ester   ADP

How does the body use ATP? Recall from Section 5.7 that the free-energy change ΔG must be negative, and energy must be released, for a reaction to occur spontaneously. If ΔG is positive, then the reaction is unfavorable and the process can't occur spontaneously.

What normally happens for an energetically unfavorable reaction to occur is that it is "coupled" to an energetically favorable reaction so that the *overall* free-energy change for the two reactions together is favorable. Take the phosphorylation reaction of glucose to yield glucose 6-phosphate plus water, an important step in the breakdown of dietary carbohydrates. The reaction of glucose with HPO_4^{2-} does not occur spontaneously because it is energetically unfavorable, with $\Delta G^{\circ\prime} = +13.8$ kJ/mol (3.3 kcal/mol). (The standard free-energy change for a biological reaction, denoted $\Delta G^{\circ\prime}$, refers to a process in which reactants and products have a concentration of 1.0 M in a solution with pH = 7.)

$$\underset{\textbf{Glucose}}{\underset{\substack{|\quad\ |\quad\ |\\ HO\ \ OH\ \ \ OH}}{HOCH_2CHCHCHCHCH}}\overset{\substack{OH\ \ \ O\\ |\quad\ \ \parallel}}{} \ \xrightleftharpoons{HPO_4^{2-}} \ \underset{\textbf{Glucose 6-phosphate}}{\underset{\substack{|\quad\ \ \ \ |\quad\ |\quad\ |\\ O^-\ \ \ HO\ \ OH\ \ \ OH}}{^-OPOCH_2CHCHCHCHCH}}\overset{\substack{O\quad\quad\quad OH\ \ \ O\\ \parallel\quad\quad\quad |\quad\ \ \parallel}}{} + H_2O \qquad \Delta G^{\circ\prime} = +13.8 \text{ kJ}$$

With ATP, however, glucose undergoes an energetically favorable reaction to yield glucose 6-phosphate plus ADP. The overall effect is the same as if HPO_4^{2-} reacted with glucose and ATP then reacted with the water byproduct, making the *coupled* process favorable by about 16.7 kJ/mol (4.0 kcal/mol). We therefore say that ATP "drives" the phosphorylation reaction of glucose:

$$\begin{array}{ll} \text{Glucose} + HPO_4^{2-} \longrightarrow \text{Glucose 6-phosphate} + H_2O & \Delta G^{\circ\prime} = +13.8 \text{ kJ/mol} \\ ATP + H_2O \longrightarrow ADP + HPO_4^{2-} + H^+ & \Delta G^{\circ\prime} = -30.5 \text{ kJ/mol} \\ \hline \text{Net:}\quad \text{Glucose} + ATP \longrightarrow \text{Glucose 6-phosphate} + ADP + H^+ & \Delta G^{\circ\prime} = -16.7 \text{ kJ/mol} \end{array}$$

It's this ability to drive otherwise unfavorable phosphorylation reactions that makes ATP so useful. The resultant phosphates are much more reactive molecules than the corresponding compounds they are derived from and therefore more likely to do chemically useful things.

• •

Problem 29.1   One of the steps in fat metabolism is the reaction of glycerol (1,2,3-propanetriol) with ATP to yield glycerol 1-phosphate. Write the reaction, and draw the structure of glycerol 1-phosphate.

• •

29.2 Catabolism of Fats: β-Oxidation

The metabolic breakdown of triacylglycerols begins with their hydrolysis in the stomach and small intestine to yield glycerol plus fatty acids. Glycerol is first phosphorylated by reaction with ATP and is then oxidized to yield glyceraldehyde 3-phosphate, which enters the carbohydrate catabolic pathway. (We'll discuss this in detail in Section 29.3.)

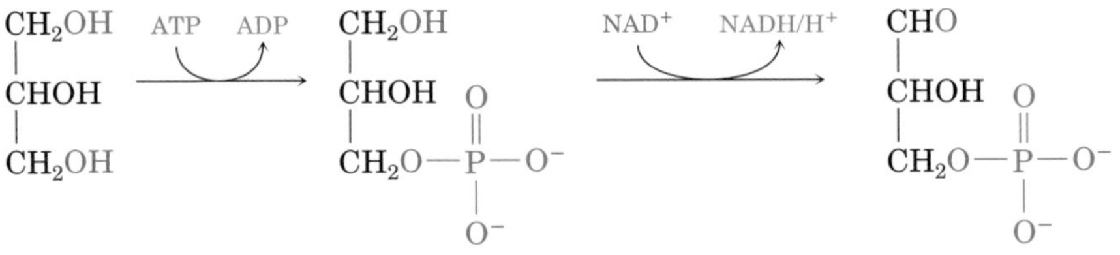

Glycerol   **Glycerol 1-phosphate**   **Glyceraldehyde 3-phosphate**

Note how the above reactions are written. It's common when writing biochemical transformations to show only the structures of the reactant and product, while abbreviating the structures of coenzymes and other reactants. The curved arrow intersecting the usual straight reaction arrow in the first step shows that ATP is also a reactant and that ADP is a product. The coenzyme *nicotinamide adenine dinucleotide* (*NAD⁺*) is required in the second step, and *reduced nicotinamide adenine dinucleotide* (*NADH*) plus a proton are products. We'll see shortly that NAD⁺ is often involved as a biochemical oxidizing agent for converting alcohols to ketones or aldehydes.

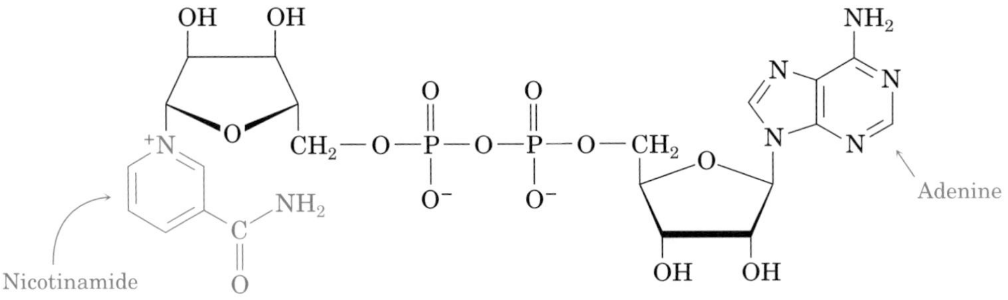

Nicotinamide adenine dinucleotide (NAD⁺)

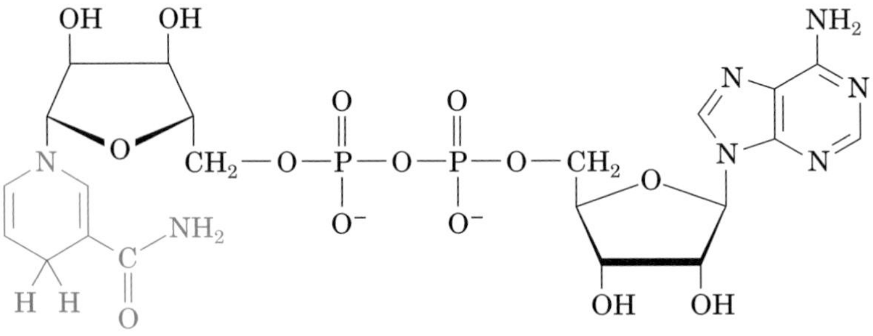

Reduced nicotinamide adenine dinucleotide (NADH)

Fatty acids are catabolized by a repetitive four-step sequence of enzyme-catalyzed reactions called the *fatty acid spiral*, or **β-oxidation pathway**, shown in Figure 29.2. Each passage along the pathway results in the cleavage of an acetyl group from the end of the fatty acid chain, until the entire molecule is ultimately degraded. As each acetyl group is produced, it enters the citric acid cycle and is further degraded, as we'll see in Section 29.5.

FIGURE 29.2 ▼

The four steps of the β-oxidation pathway, resulting in the cleavage of an acetyl group from the end of the fatty acid chain. The key chain-shortening step is a retro-Claisen reaction of a β-keto thiol ester.

$$O$$
$$\parallel$$
$$RCH_2CH_2CH_2CH_2CSCoA$$

Fatty acyl CoA

STEP 1 A double bond is introduced by enzyme-catalyzed removal of hydrogens from C2 and C3.

FAD

FADH₂

$$O$$
$$\parallel$$
$$RCH_2CH_2CH=CHCSCoA$$

Unsaturated acyl CoA

STEP 2 Water adds to the double bond in a conjugate addition reaction to yield an alcohol.

H₂O

$$OH \quad\quad O$$
$$| \quad\quad\quad \parallel$$
$$RCH_2CH_2CHCH_2CSCoA$$

β-Hydroxy acyl CoA

STEP 3 The alcohol is oxidized by NAD⁺ to give a β-keto thiol ester.

NAD⁺

NADH/H⁺

$$O \quad\quad O$$
$$\parallel \quad\quad \parallel$$
$$RCH_2CH_2CCH_2CSCoA$$

β-Keto acyl CoA

STEP 4 The bond between C2 and C3 is broken by nucleophilic attack of coenzyme A on the C3 carbonyl group in a retro-Claisen reaction to yield acetyl CoA and a chain-shortened fatty acid.

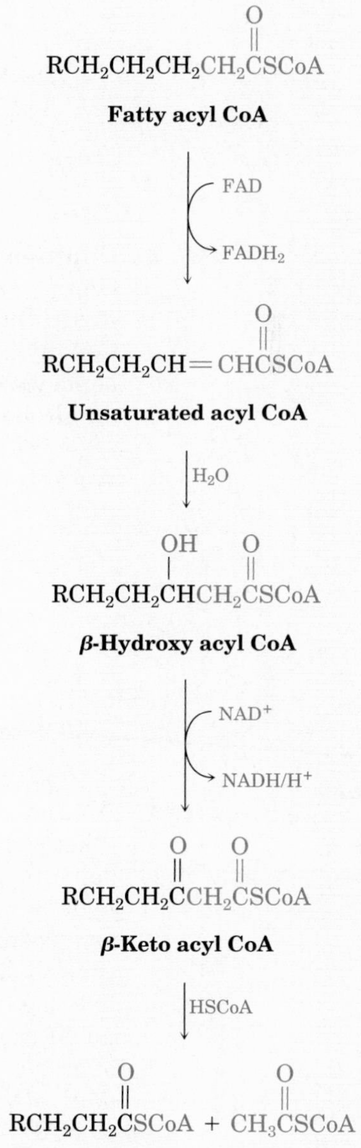

HSCoA

$$O \quad\quad\quad\quad\quad\quad O$$
$$\parallel \quad\quad\quad\quad\quad\quad \parallel$$
$$RCH_2CH_2CSCoA + CH_3CSCoA$$

© 1995 JOHN MCMURRY

STEP 1 **Introduction of a double bond.** The β-oxidation pathway begins when a fatty acid forms a thiol ester with coenzyme A to give a fatty acyl CoA. Two hydrogen atoms are then removed from carbons 2 and 3 by an acyl CoA dehydrogenase enzyme to yield an α,β-unsaturated acyl CoA. This kind of oxidation—the introduction of a conjugated double bond into a carbonyl compound—occurs frequently in biochemical pathways and is usually carried out by the coenzyme *flavin adenine dinucleotide (FAD)*. Reduced FADH$_2$ is the by-product.

FAD **FADH$_2$**

STEP 2 **Conjugate addition of water.** The α,β-unsaturated acyl CoA produced in step 1 reacts with water by a conjugate addition pathway (Section 19.14) to yield a β-hydroxy acyl CoA in a process catalyzed by the enzyme enoyl CoA hydratase. Water as nucleophile adds to the β carbon of the double bond, yielding an enolate ion intermediate, which is then protonated to yield an alcohol.

α,β-Unsaturated carbonyl *β*-Hydroxy carbonyl

STEP 3 **Alcohol oxidation.** The β-hydroxy acyl CoA from step 2 is oxidized to a β-keto acyl CoA in a reaction catalyzed by the enzyme L-3-hydroxyacyl CoA dehydrogenase. As in the oxidation of glycerol 1-phosphate to glyceraldehyde 3-phosphate mentioned earlier, this alcohol oxidation requires NAD$^+$ as a coenzyme and yields reduced NADH/H$^+$ as by-product.

It's often useful when thinking about enzyme-catalyzed redox reactions to recognize that a hydrogen *atom* is equivalent to a hydrogen *ion*, H$^+$, plus an electron, e$^-$. Thus, for the two hydrogen atoms removed in the oxidation of an alcohol, 2 H atoms = 2 H$^+$ + 2 e$^-$. When NAD$^+$ is involved as the oxidant, both electrons accompany one H$^+$, in effect adding a hydride ion, H:$^-$, to NAD$^+$ to give NADH. The second hydrogen removed from the oxidized substrate enters the solution as H$^+$.

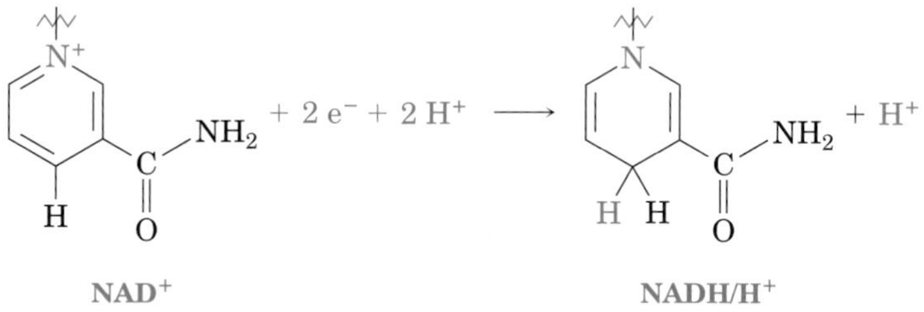

$$\text{NAD}^+ \quad + 2\,e^- + 2\,H^+ \longrightarrow \quad \text{NADH/H}^+ \quad + H^+$$

The mechanism of alcohol oxidation with NAD^+ has several analogies in laboratory chemistry. A base removes the O–H proton from the alcohol and generates an alkoxide ion, which expels a hydride ion leaving group as in the Cannizzaro reaction (Section 19.13). The nucleophilic hydride ion then adds to the C=C–C=N$^+$ part of NAD^+ in a conjugate addition reaction, much the same as water adds to the C=C–C=O part of the α,β-unsaturated acyl CoA in step 2.

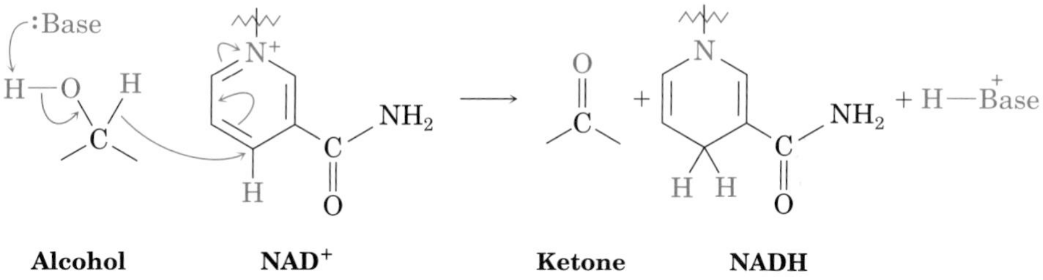

Alcohol   **NAD$^+$**   **Ketone**   **NADH**

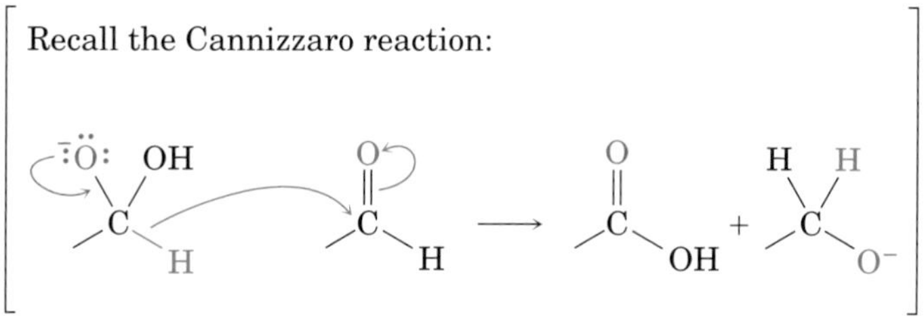

Recall the Cannizzaro reaction:

STEP 4   **Chain cleavage.** Acetyl CoA is split off from the acyl chain in the final step of β-oxidation, leaving behind an acyl CoA that is two carbon atoms shorter than the original. The reaction is catalyzed by the enzyme β-keto thiolase and is mechanistically the reverse of a Claisen condensation reaction (Section 23.8). In the *forward* direction, a Claisen condensation joins two esters together to form a β-keto ester product. In the *reverse* direction, a retro-Claisen reaction splits a β-keto ester (or β-keto thiol ester) apart to form two esters (or two thiol esters).

The reaction occurs by nucleophilic addition of coenzyme A to the keto group of the β-keto acyl CoA to yield an alkoxide ion intermediate, followed by cleavage of the C2–C3 bond with expulsion of an acetyl CoA enolate ion. Protonation of the enolate ion gives acetyl CoA, and the chain-shortened acyl CoA enters another round of the β-oxidation pathway for further degradation.

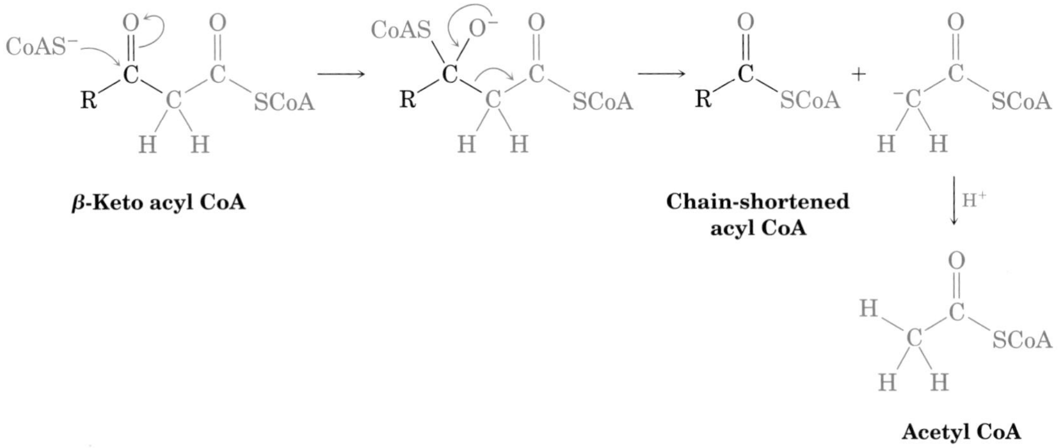

β-Keto acyl CoA   **Chain-shortened acyl CoA**

Acetyl CoA

Look at the catabolism of myristic acid shown in Figure 29.3 to see the overall results of the β-oxidation pathway. The first passage along the pathway converts the 14-carbon myristyl CoA into the 12-carbon lauryl CoA plus acetyl CoA; the second passage converts lauryl CoA into the 10-carbon capryl CoA plus acetyl CoA; the third passage converts capryl CoA into the 8-carbon caprylyl CoA; and so on. Note that the final passage produces *two* molecules of acetyl CoA because the precursor has four carbons.

FIGURE 29.3 ▼

Catabolism of the 14-carbon myristic acid by the β-oxidation pathway yields seven molecules of acetyl CoA after six passages.

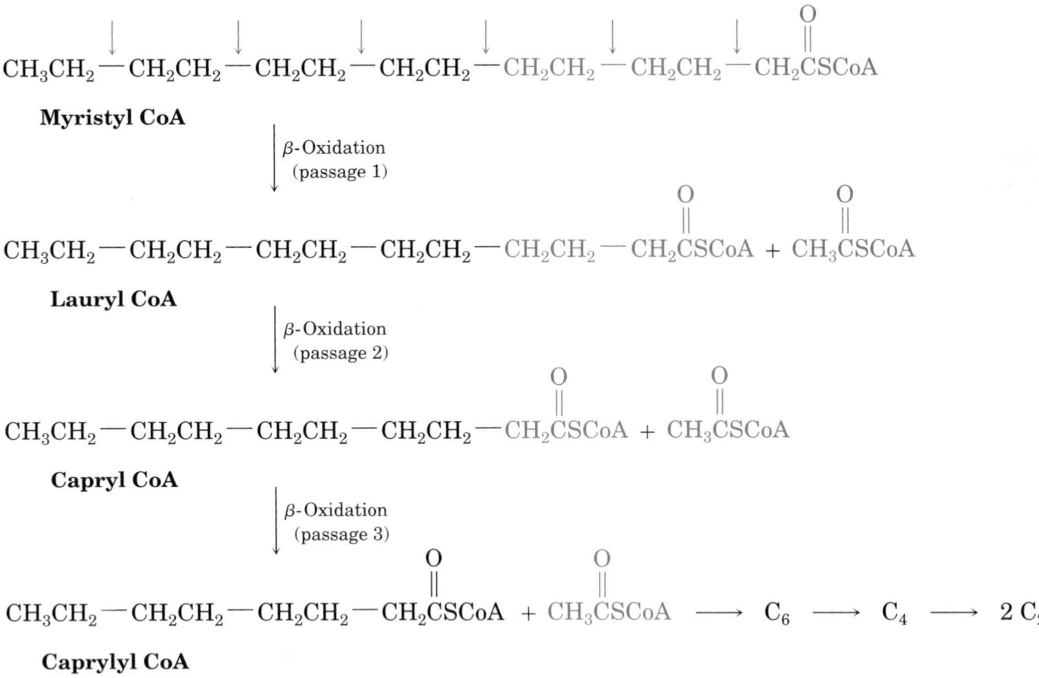

$CH_3CH_2—CH_2CH_2—CH_2CH_2—CH_2CH_2—CH_2CH_2—CH_2CH_2—CH_2CSCoA$

   Myristyl CoA

               β-Oxidation
               (passage 1)

$CH_3CH_2—CH_2CH_2—CH_2CH_2—CH_2CH_2—CH_2CH_2—CH_2CSCoA + CH_3CSCoA$

   Lauryl CoA

               β-Oxidation
               (passage 2)

$CH_3CH_2—CH_2CH_2—CH_2CH_2—CH_2CH_2—CH_2CSCoA + CH_3CSCoA$

   Capryl CoA

               β-Oxidation
               (passage 3)

$CH_3CH_2—CH_2CH_2—CH_2CH_2—CH_2CSCoA + CH_3CSCoA \longrightarrow C_6 \longrightarrow C_4 \longrightarrow 2 C_2$

   Caprylyl CoA

You can predict how many molecules of acetyl CoA will be obtained from a given fatty acid simply by counting the number of carbon atoms and dividing by two. For example, the 14-carbon myristic acid yields seven molecules of acetyl CoA after six passages through the β-oxidation pathway. The number of passages is always one less than the number of acetyl CoA molecules produced because the last passage cleaves a four-carbon chain into two acetyl CoA's.

Most fatty acids have an even number of carbon atoms, so that none are left over after β-oxidation. Those fatty acids with an odd number of carbon atoms or with double bonds require additional steps for degradation, but all carbon atoms are ultimately released for further oxidation in the citric acid cycle.

Problem 29.2 Write the equations for the remaining passages of the β-oxidation pathway following those shown in Figure 29.3.

Problem 29.3 How many molecules of acetyl CoA are produced by catabolism of the following fatty acids, and how many passages of the β-oxidation pathway are needed?
(a) Palmitic acid, $CH_3(CH_2)_{14}COOH$ (b) Arachidic acid, $CH_3(CH_2)_{18}COOH$

29.3 Catabolism of Carbohydrates: Glycolysis

Glycolysis is a series of ten enzyme-catalyzed reactions that break down glucose into 2 equivalents of pyruvate, $CH_3COCO_2^-$. The steps of glycolysis, also called the *Embden–Meyerhoff pathway* after its discoverers, are summarized in Figure 29.4 (p. 1204).

STEPS 1–3 **Phosphorylation and isomerization.** Glucose, produced by the digestion of dietary carbohydrates, is first phosphorylated at the hydroxyl group on C6 by reaction with ATP in a process catalyzed by the enzyme hexokinase. The glucose 6-phosphate that results is isomerized by glucose 6-phosphate isomerase to fructose 6-phosphate. As the open-chain structures in Figure 29.4 show, this isomerization reaction takes place by keto–enol tautomerism (Section 22.1), since both glucose and fructose share a common enol:

| Glucose | Glucose/fructose enol | Fructose |

Fructose 6-phosphate is then converted to fructose 1,6-bisphosphate by phosphofructokinase-catalyzed reaction with ATP (the prefix "bis" means two). The result is a molecule ready to be split into the two three-carbon intermediates that will ultimately become two molecules of pyruvate.

FIGURE 29.4 ▼

The ten-step glycolysis pathway for catabolizing glucose to pyruvate. The individual steps are described in more detail in the text.

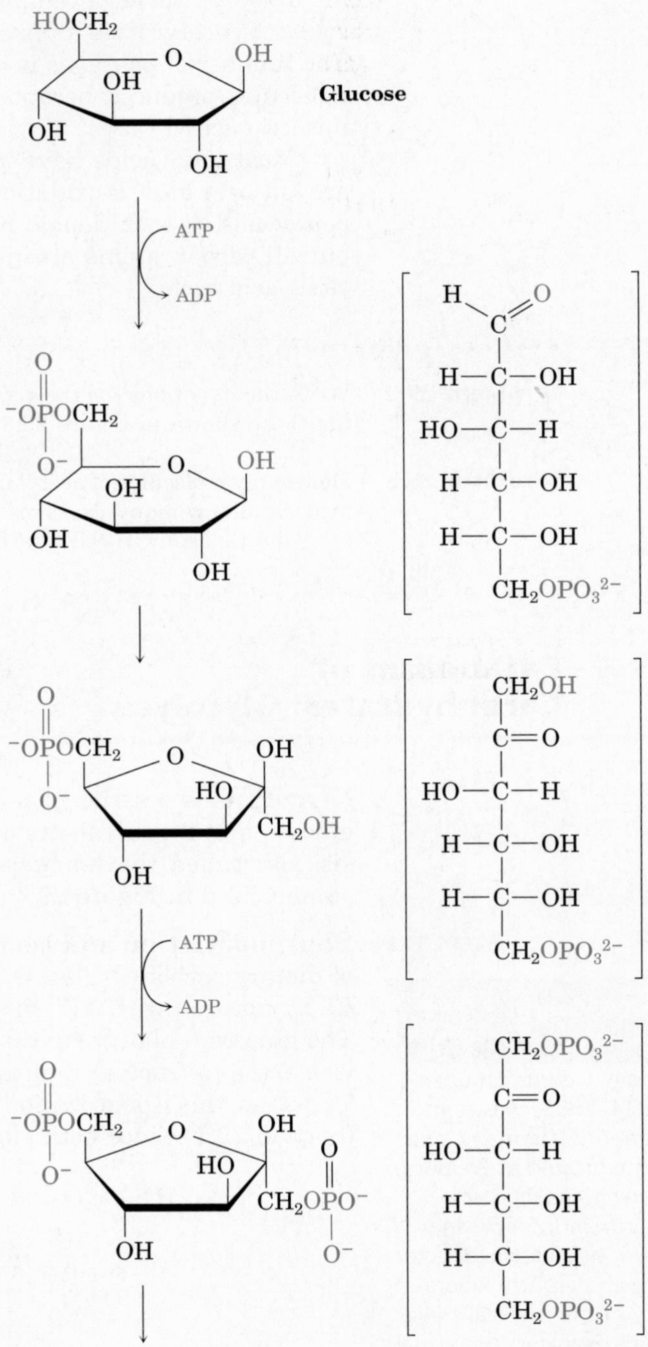

STEP 1 Glucose is phosphorylated by reaction with ATP to yield glucose 6-phosphate.

STEP 2 Glucose 6-phosphate is isomerized to fructose 6-phosphate.

STEP 3 Fructose 6-phosphate is phosphorylated by reaction with ATP to yield fructose 1,6-bisphosphate.

STEP 4 Fructose 1,6-bisphosphate is cleaved into two three-carbon pieces by the enzyme aldolase.

STEP 5 Dihydroxyacetone phosphate, one of the products of step 4, is isomerized to glyceraldehyde 3-phosphate, the other product of step 4.

$$^{2-}O_3POCH_2\overset{\underset{|}{HO}}{C}H\overset{\underset{\|}{O}}{C}H \rightleftharpoons {}^{2-}O_3POCH_2\overset{\underset{\|}{O}}{C}CH_2OH$$

Glyceraldehyde 3-phosphate　　　　**Dihydroxyacetone phosphate**

STEP 6 Glyceraldehyde 3-phosphate is oxidized and phosphorylated to yield 3-phosphoglyceroyl phosphate.

NAD^+, HPO_4^{2-}

$NADH/H^+$

3-Phosphoglyceroyl phosphate　　$^{2-}O_3POCH_2\overset{\underset{|}{HO}}{C}H\overset{\underset{\|}{O}}{C}OPO_3^{2-}$

$$\begin{bmatrix} O=\overset{\displaystyle OPO_3^{2-}}{\underset{\displaystyle |}{C}} \\ H-\overset{|}{C}-OH \\ CH_2OPO_3^{2-} \end{bmatrix}$$

STEP 7 A phosphate is transferred from the carboxyl group to ADP, resulting in synthesis of an ATP and yielding 3-phosphoglycerate.

ADP

ATP

3-Phosphoglycerate　　$^{2-}O_3POCH_2\overset{\underset{|}{HO}}{C}H\overset{\underset{\|}{O}}{C}O^-$

$$\begin{bmatrix} O=\overset{\displaystyle O^-}{\underset{\displaystyle |}{C}} \\ H-\overset{|}{C}-OH \\ CH_2OPO_3^{2-} \end{bmatrix}$$

STEP 8 Isomerization of 3-phosphoglycerate gives 2-phosphoglycerate.

2-Phosphoglycerate　　$HOCH_2\overset{\underset{|}{^{2-}O_3PO}}{C}H\overset{\underset{\|}{O}}{C}O^-$

$$\begin{bmatrix} O=\overset{\displaystyle O^-}{\underset{\displaystyle |}{C}} \\ H-\overset{|}{C}-OPO_3^{2-} \\ CH_2OH \end{bmatrix}$$

STEP 9 Dehydration occurs to yield phosphoenolpyruvate (PEP).

H_2O

Phosphoenolpyruvate　　$H_2C=\overset{\underset{|}{^{2-}O_3PO}}{C}-\overset{\underset{\|}{O}}{C}O^-$

$$\begin{bmatrix} O=\overset{\displaystyle O^-}{\underset{\displaystyle |}{C}} \\ \overset{|}{C}-OPO_3^{2-} \\ \| \\ CH_2 \end{bmatrix}$$

STEP 10 A phosphate is transferred from PEP to ADP, yielding pyruvate and ATP.

ADP

ATP

Pyruvate　　$CH_3\overset{\underset{\|}{O}}{C}-\overset{\underset{\|}{O}}{C}O^-$

$$\begin{bmatrix} O=\overset{\displaystyle O^-}{\underset{\displaystyle |}{C}} \\ C=O \\ CH_3 \end{bmatrix}$$

1205

STEPS 4–5

Cleavage and isomerization. Fructose 1,6-bisphosphate is cleaved in step 4 into two, three-carbon monophosphates, one an aldose and one a ketose. The bond between carbons 3 and 4 of fructose 1,6-bisphosphate breaks, and a C–O group is formed. Mechanistically, the cleavage is the reverse of an aldol reaction (Section 23.2) and is carried out by an aldolase enzyme. (A *forward* aldol reaction joins two aldehydes or ketones to give a β-hydroxy carbonyl compound; a *retro* aldol reaction cleaves a β-hydroxy carbonyl compound into two aldehydes or ketones.)

Otto Fritz Meyerhof

Otto Fritz Meyerhof (1884–1951) was born in Hanover, Germany, and received an M.D. from the University of Heidelberg. After holding several posts in Germany, he fled to the United States in 1940 and became Research Professor at the University of Pennsylvania. He received the 1922 Nobel Prize for medicine for his work on the relationship between oxygen uptake and lactic acid metabolism in muscles.

Fructose 1,6-bisphosphate (a β-hydroxy ketone)

Glyceraldehyde 3-phosphate

Dihydroxyacetone phosphate

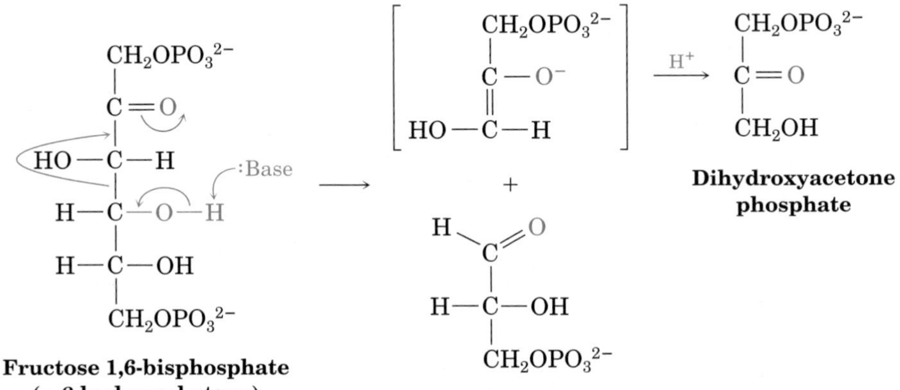

Actually, the reaction is a bit more complex than shown above because it does not take place on the free ketone. Instead, fructose 1,6-bisphosphate undergoes reaction with the side-chain –NH$_2$ group of a lysine residue on the aldolase enzyme to yield an imine (Section 19.9), sometimes called a *Schiff base*. Protonation of the imine makes it more reactive; a retro aldol-like reaction ensues, giving glyceraldehyde 3-phosphate and the imine of dihydroxyacetone phosphate; and the imine is then hydrolyzed to yield dihydroxyacetone phosphate.

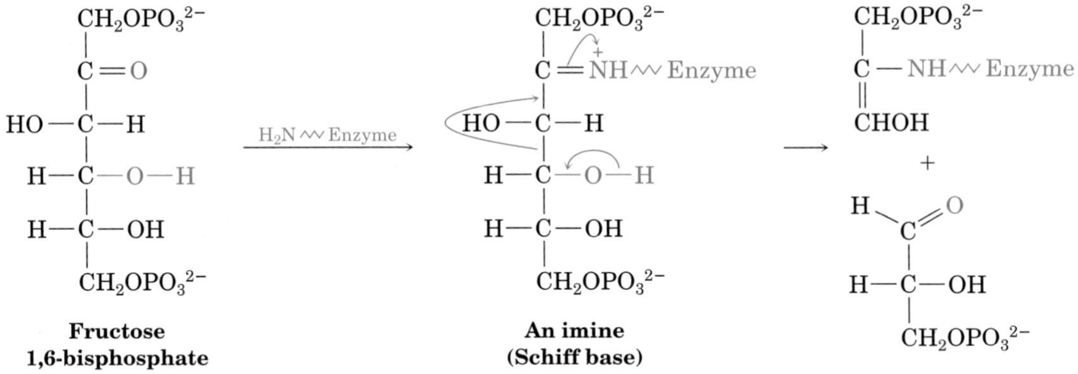

Fructose 1,6-bisphosphate

An imine (Schiff base)

Glyceraldehyde 3-phosphate continues on in the glycolysis pathway, but dihydroxyacetone phosphate is first isomerized by the enzyme triose phosphate isomerase. As in the glucose-to-fructose conversion of step 2, the isomerization of dihydroxyacetone phosphate to glyceraldehyde 3-phosphate takes place by keto–enol tautomerization through a common enol.

| Dihydroxyacetone phosphate | Enol | Glyceraldehyde 3-phosphate |

The net result of steps 4 and 5 is the production of *two* glyceraldehyde 3-phosphate molecules, both of which pass down the rest of the pathway. Thus, each of the remaining five steps of glycolysis takes place twice for every glucose molecule that enters at step 1.

STEPS 6–8 **Oxidation and phosphorylation.** Glyceraldehyde 3-phosphate is oxidized and phosphorylated by the coenzyme NAD^+ in the presence of the enzyme glyceraldehyde 3-phosphate dehydrogenase and hydrogen phosphate ion, HPO_4^{2-}. The reaction occurs when a thiol group (–SH) on the enzyme adds to the aldehyde carbonyl group in a nucleophilic addition step to yield a *hemithioacetal*, the sulfur analog of a hemiacetal (Section 19.11). Oxidation of the hemithioacetal –OH group by NAD^+ then yields a thiol ester intermediate. The reaction is thus similar mechanistically to the laboratory oxidation of an aldehyde to a carboxylic acid (Section 19.3).

Glyceraldehyde 3-phosphate   **A hemithioacetal**   **A thiol ester**

The thiol ester intermediate resulting from oxidation of glyceraldehyde 3-phosphate next reacts with phosphate ion in a nucleophilic acyl substitution step (Section 21.2) to yield 3-phosphoglyceroyl phosphate, a mixed anhydride between a carboxylic acid and phosphoric acid.

3-Phosphoglyceroyl phosphate

Like all anhydrides, the mixed carboxylic–phosphoric anhydride is a reactive substrate in nucleophilic acyl substitution reactions (Section 21.5). Reaction of 3-phosphoglyceroyl phosphate with ADP occurs with nucleophilic attack on phosphorus and results in transfer of a phosphate group to yield ATP and 3-phosphoglycerate. The process is catalyzed by the enzyme phosphoglycerate kinase. Note that the carboxylic acid group is written in its dissociated form to reflect the state in which it exists at physiological pH.

3-Phosphoglyceroyl phosphate   **3-Phosphoglycerate**

Isomerization of 3-phosphoglycerate then gives 2-phosphoglycerate in a step catalyzed by the enzyme phosphoglycerate mutase. The process involves formation of a 2,3-bisphosphoglycerate intermediate, followed by selective transfer of the 3-phosphate.

3-Phosphoglycerate   **2,3-Bisphosphoglycerate**   **2-Phosphoglycerate**

STEPS 9–10 **Dehydration and dephosphorylation.** Like the β-hydroxy carbonyl compounds produced in aldol reactions, 2-phosphoglycerate undergoes a ready dehydration (Section 23.4). The process is catalyzed by enolase, and the product is phosphoenolpyruvate, abbreviated PEP.

2-Phosphoglycerate   **Phosphoenolpyruvate (PEP)**
(a β-hydroxy carbonyl compound)

Transfer of the phosphate group to ADP then generates ATP and gives pyruvate, a reaction catalyzed by pyruvate kinase.

Phosphoenolpyruvate   **Pyruvate**

The net result of glycolysis can be summarized by the following equation:

$$C_6H_{12}O_6 \;+\; 2\,NAD^+ \;+\; 2\,HPO_4{}^{2-} \;+\; 2\,ADP \;\longrightarrow\; 2\,\underset{\text{Pyruvate}}{CH_3\overset{O}{\underset{}{C}}-CO^-} \;+\; 2\,NADH \;+\; 2\,ATP \;+\; 2\,H_2O \;+\; 2\,H^+$$

Glucose

- -

Problem 29.4   Identify the two steps in glycolysis in which ATP is produced.

Problem 29.5   Look at the entire glycolysis pathway and make a list of the kinds of organic reactions that take place—nucleophilic acyl substitutions, aldol reactions, E2 reactions, and so forth.

- -

29.4   The Conversion of Pyruvate to Acetyl CoA

Pyruvate, produced in the catabolism of glucose, can undergo several further transformations depending on the conditions and on the organism. In the absence of oxygen, pyruvate is reduced to lactate [$CH_3CH(OH)CO_2{}^-$]. In yeast, pyruvate is fermented to give ethanol. Most commonly, however, pyruvate is converted to acetyl CoA plus CO_2 through a multistep sequence of reactions catalyzed by a complex of enzymes and cofactors called the *pyruvate dehydrogenase complex* (Figure 29.5, p. 1210). All the steps have simple laboratory analogies.

STEP 1   **Addition of thiamine.** The conversion of pyruvate to acetyl CoA begins by reaction of pyruvate with thiamine pyrophosphate, a derivative of vitamin B_1. The hydrogen on the heterocyclic (thiazole) ring of thiamine pyrophosphate is weakly acidic and can be removed by reaction with base to yield a nucleophilic ylide much like the phosphorus ylides used in Wittig reactions (Section 19.12). This nucleophilic ylide adds to the ketone carbonyl group of pyruvate to yield a tetrahedral intermediate.

FIGURE 29.5 ▼

Mechanism of the conversion of pyruvate to acetyl CoA through a multistep sequence of reactions that requires three different enzymes and four different coenzymes. The individual steps are explained in more detail in the text.

STEP 1 Nucleophilic addition of thiamine pyrophosphate to the ketone carbonyl group of pyruvate yields an intermediate addition product.

STEP 2 Decarboxylation occurs, analogous to the loss of CO_2 from a β-keto acid, yielding an enamine intermediate.

STEP 3 The nucleophilic enamine double bond attacks a sulfur atom of lipoamide and does an S_N2-like displacement of the second sulfur atom.

STEP 4 Elimination of thiamine pyrophosphate from the tetrahedral intermediate then yields acetyl dihydrolipoamide.

STEP 5 Reaction with coenzyme A exchanges one thiol ester for another, giving acetyl CoA and dihydrolipoamide.

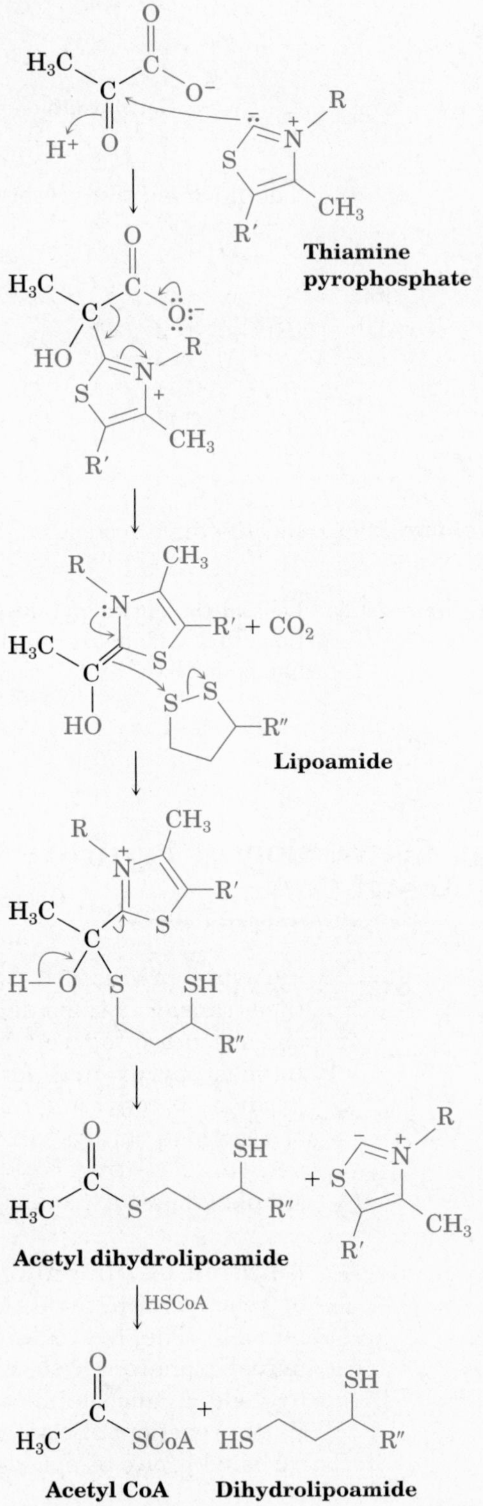

Thiamine pyrophosphate

Lipoamide

Acetyl dihydrolipoamide

HSCoA

Acetyl CoA   **Dihydrolipoamide**

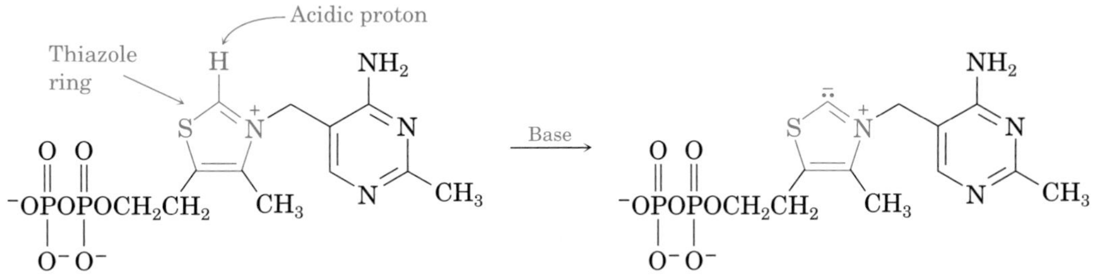

Thiamine pyrophosphate

STEP 2 **Decarboxylation.** Decarboxylation of the pyruvate/thiamine addition product occurs in much the same way that decarboxylation of a β-keto acid intermediate occurs in the acetoacetic ester synthesis (Section 22.8). The $C=N^+$ double bond of the pyruvate addition product acts like the $C=O$ double bond of a β-keto acid to accept electrons as CO_2 leaves.

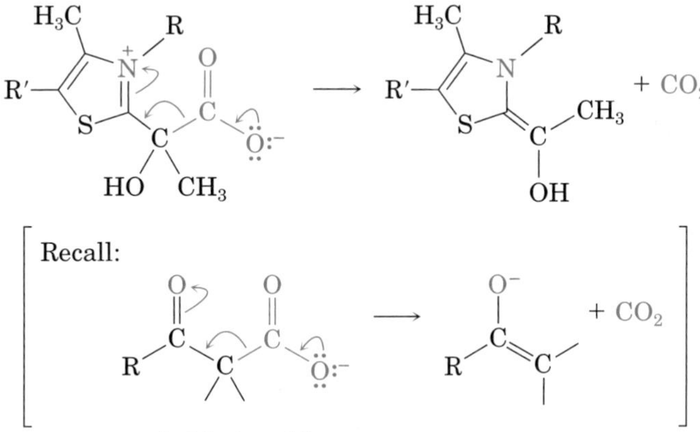

A β-keto acid

STEP 3 **Reaction with lipoamide.** The decarboxylation product is an enamine ($R_2N-C=C$), which, like all enamines, is strongly nucleophilic (Section 23.12). The enamine undergoes reaction with the cyclic disulfide lipoamide by nucleophilic attack on a sulfur atom, displacing the second sulfur in an S_N2-like process.

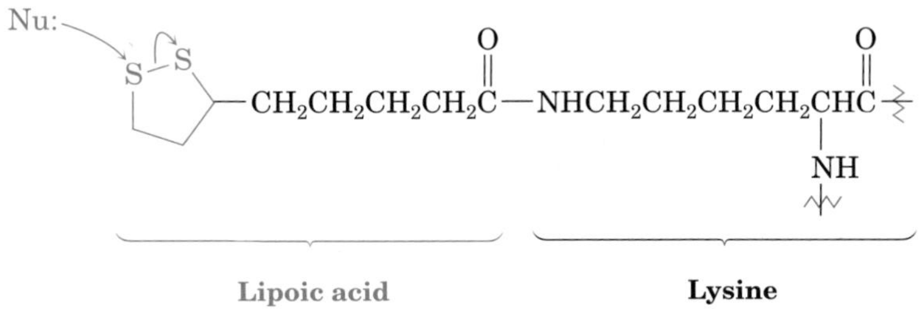

Lipoic acid   Lysine

Lipoamide: Lipoic acid is linked through an amide bond to the side-chain NH_2 group of a lysine residue in dihydrolipoyl transacetylase

STEP 4 **Elimination of thiamine.** The product of the enamine reaction with lipoamide is itself a tetrahedral carbonyl addition product, which can eliminate thiamine pyrophosphate. This elimination, the exact reverse of step 1, generates the carbonyl compound acetyl dihydrolipoamide.

STEP 5 **Acyl transfer.** Acetyl dihydrolipoamide, a thiol ester, undergoes a nucleophilic acyl substitution reaction with coenzyme A to yield acetyl CoA plus dihydrolipoamide. The dihydrolipoamide is then oxidized back to lipoamide by FAD, and the $FADH_2$ that results is in turn oxidized back to FAD by NAD^+.

Dihydrolipoamide   **Lipoamide**

Problem 29.6 Write a mechanism for the reaction of acetyl dihydrolipoamide with coenzyme A to yield acetyl CoA in step 5 of Figure 29.5.

Problem 29.7 Which carbon atoms in glucose end up as $-CH_3$ carbons in acetyl CoA? Which carbons end up as CO_2?

29.5 The Citric Acid Cycle

The first two stages of catabolism result in the conversion of fats and carbohydrates into acetyl groups that are bonded through a thiol ester link to coenzyme A. These acetyl groups now enter the third stage of catabolism—the **citric acid cycle**, also called the *tricarboxylic acid (TCA) cycle*, or *Krebs cycle*. The steps of the citric acid cycle are given in Figure 29.6.

As its name implies, the citric acid *cycle* is a closed loop of reactions in which the product of the final step is a reactant in the first step. The intermediates are constantly regenerated and flow continuously through the cycle, which operates as long as the oxidizing coenzymes NAD^+ and FAD are available. To meet this condition, the reduced coenzymes NADH and $FADH_2$ must be reoxidized via the electron-transport chain, which in turn relies on oxygen as the ultimate electron acceptor. Thus, the cycle is dependent on the availability of oxygen and on the operation of the electron-transport chain.

STEPS 1–2 **Addition to oxaloacetate.** Acetyl CoA enters the citric acid cycle in step 1 by nucleophilic addition to the ketone carbonyl group of oxaloacetate to give citryl CoA (Section 26.15). The addition is an aldol reaction of an enolate ion from acetyl CoA, and is catalyzed by the enzyme citrate synthase, as discussed in Section 26.15. Citryl CoA is then hydrolyzed to citrate.

FIGURE 29.6 ▼

The citric acid cycle is an eight-step series of reactions that results in the conversion of an acetyl group into two molecules of CO_2 plus reduced coenzymes. Individual steps are explained in more detail in the text.

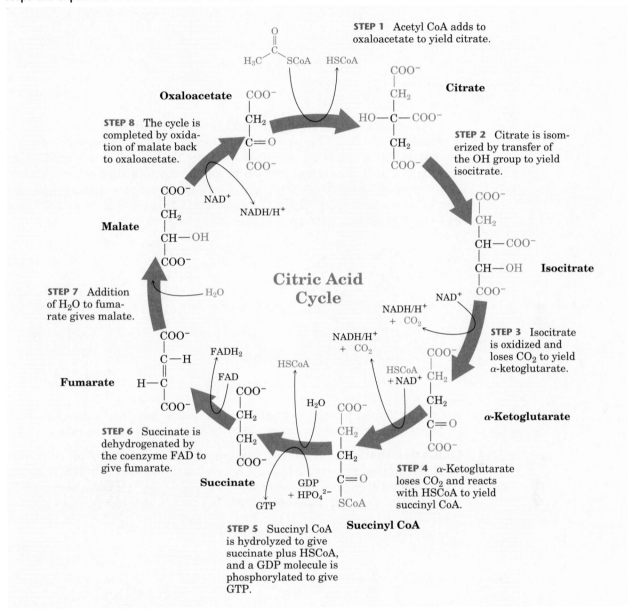

STEP 1 Acetyl CoA adds to oxaloacetate to yield citrate.

Citrate

STEP 2 Citrate is isomerized by transfer of the OH group to yield isocitrate.

STEP 3 Isocitrate is oxidized and loses CO_2 to yield α-ketoglutarate.

Isocitrate

α-Ketoglutarate

STEP 4 α-Ketoglutarate loses CO_2 and reacts with HSCoA to yield succinyl CoA.

Succinyl CoA

STEP 5 Succinyl CoA is hydrolyzed to give succinate plus HSCoA, and a GDP molecule is phosphorylated to give GTP.

Succinate

STEP 6 Succinate is dehydrogenated by the coenzyme FAD to give fumarate.

Fumarate

STEP 7 Addition of H_2O to fumarate gives malate.

Malate

STEP 8 The cycle is completed by oxidation of malate back to oxaloacetate.

Oxaloacetate

Citric Acid Cycle

Sir Hans Adolf Krebs

Sir Hans Adolf Krebs (1900–1981) was born in Hildesheim, Germany, and received an M.D. in 1925 from the University of Hamburg. In 1933 he moved to England, first at the University of Cambridge, then at the University of Sheffield (1935–1954), and finally at the University of Oxford (1954–1967). He received the 1953 Nobel Prize in medicine for his work on elucidating pathways in intermediary metabolism.

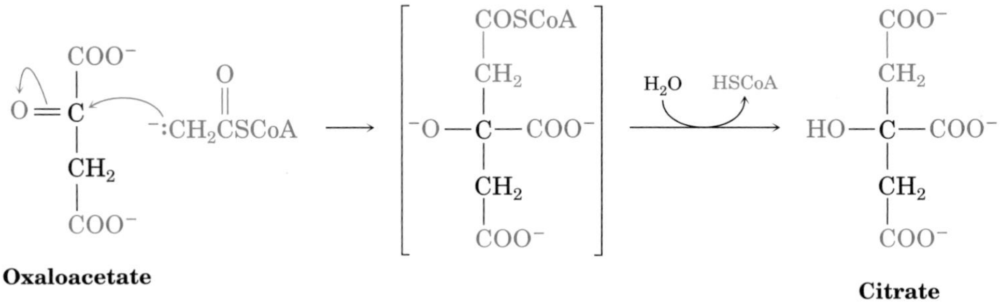

Oxaloacetate

Citrate

Citrate, a tertiary alcohol, is next converted into its isomer, isocitrate, a secondary alcohol. The isomerization occurs in two steps, both of which are catalyzed by the same aconitase enzyme. The initial step is an E2 dehydration of a β-hydroxy acid, the same sort of reaction that occurs in step 9 of glycolysis (Figure 29.4). The second step is a conjugate nucleophilic addition of water to the C=C bond, the same sort of reaction that occurs in step 2 of the β-oxidation pathway (Figure 29.2).

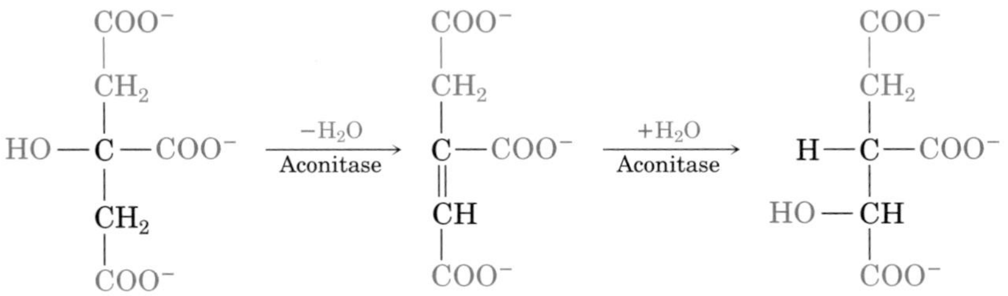

Citrate   **Aconitate**   **Isocitrate**

Note that the dehydration of citrate takes place specifically *away* from the carbon atoms of the acetyl group that added to oxaloacetate in step 1.

STEPS 3–4   **Oxidative decarboxylations.** Isocitrate, a secondary alcohol, is oxidized by NAD⁺ in step 3 to give a ketone, which loses CO_2 to give α-ketoglutarate. Catalyzed by the enzyme isocitrate dehydrogenase, the decarboxylation is a typical reaction of a β-keto acid, just like that in the acetoacetic ester synthesis (Section 22.8).

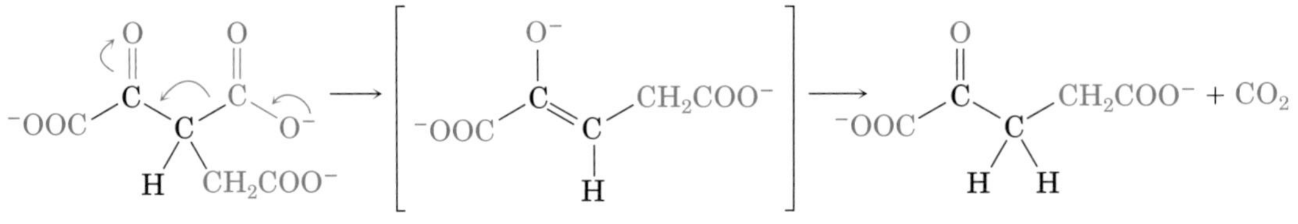

α-Ketoglutarate

The transformation of α-ketoglutarate to succinyl CoA in step 4 is a multistep process catalyzed by an enzyme complex, analogous to the transformation of pyruvate to acetyl CoA that we saw in the previous section. In both cases, an α-keto acid loses CO_2 in a step catalyzed by thiamine pyrophosphate.

STEPS 5–6   **Hydrolysis and dehydrogenation of succinyl CoA.** Succinyl CoA is hydrolyzed to succinate in step 5. The reaction is catalyzed by succinyl CoA synthetase and is coupled with phosphorylation of guanosine diphosphate (GDP) to give guanosine triphosphate (GTP). The overall transformation is similar to that of step 8 in glycolysis (Figure 29.4), in which a thiol ester is converted into an acyl phosphate and a phosphate group is then transferred to ADP.

Succinyl CoA   **An acyl phosphate**   **Succinate**

Succinate is next dehydrogenated by FAD and the enzyme succinate dehydrogenase to give fumarate, a process analogous to that of step 1 in the fatty acid β-oxidation pathway.

STEPS 7–8   **Regeneration of oxaloacetate.** Catalyzed by the enzyme fumarase, conjugate nucleophilic addition of water to fumarate yields L-malate in a reaction similar to that of step 2 in the fatty acid β-oxidation pathway. Oxidation with NAD$^+$ then gives oxaloacetate in a step catalyzed by malate dehydrogenase, and the citric acid cycle has returned to its starting point, ready to revolve again.

The net result of the cycle can be summarized as:

Acetyl CoA + 3 NAD$^+$ + FAD + ADP + HPO$_4^{2-}$ + 2 H$_2$O $\longrightarrow$

HSCoA + 3 NADH + 3 H$^+$ + FADH$_2$ + ATP + 2 CO$_2$

· ·

Problem 29.8   Which of the substances in the citric acid cycle are tricarboxylic acids, thus giving the cycle its alternative name?

Problem 29.9   Write mechanisms for step 2 of the citric acid cycle, the dehydration of citrate and the addition of water to aconitate.

Problem 29.10   Write a mechanism for the conversion of succinyl CoA to succinate in step 5 of the citric acid cycle.

· ·

29.6 Catabolism of Proteins: Transamination

The catabolism of proteins is more complex than that of fats and carbohydrates because each of the 20 amino acids is degraded through its own unique pathway. The general idea, however, is that the amino nitrogen atom is removed and the substance that remains is converted into a compound that enters the citric acid cycle.

Most amino acids lose their nitrogen atom by a **transamination** reaction in which the $-NH_2$ group of the amino acid changes places with the keto group of α-ketoglutarate. The products are a new α-keto acid and glutamate:

$$\underset{\textbf{An amino acid}}{\overset{\overset{\displaystyle NH_3^+}{|}}{RCHCOO^-}} + \underset{\textbf{α-Ketoglutarate}}{\overset{\overset{\displaystyle O}{\|}}{^-OOCCH_2CH_2CCOO^-}} \rightleftharpoons \underset{\textbf{An α-keto acid}}{\overset{\overset{\displaystyle O}{\|}}{RCCOO^-}} + \underset{\textbf{Glutamate}}{\overset{\overset{\displaystyle NH_3^+}{|}}{^-OOCCH_2CH_2CHCOO^-}}$$

Transaminations use pyridoxal phosphate, a derivative of vitamin B_6, as cofactor. As shown in Figure 29.7 for the reaction of alanine, the key step in transamination is nucleophilic addition of $-NH_2$ group to the pyridoxal aldehyde group to yield an imine (Section 19.9). Loss of a proton from the α carbon then results in a bond rearrangement to give a new imine, which is hydrolyzed (the exact reverse of imine formation) to yield pyruvate and a nitrogen-containing derivative of pyridoxal phosphate. Pyruvate is converted into acetyl CoA (Section 29.4), which enters the citric acid cycle for further catabolism. The pyridoxal phosphate derivative transfers its nitrogen atom to α-ketoglutarate by the reverse of the steps in Figure 29.7, thereby forming glutamate and regenerating pyridoxal phosphate for further use.

Glutamate, which now contains the nitrogen atom of the former amino acid, next undergoes an oxidative deamination to yield ammonium ion and regenerated α-ketoglutarate. The oxidation of the amine to an imine is mechanistically similar to the oxidation of a secondary alcohol to a ketone and is carried out by NAD^+. The imine is then hydrolyzed in the usual way.

$$\underset{\textbf{Glutamate}}{\overset{\overset{\displaystyle NH_3^+}{|}}{^-O_2CCH_2CH_2CHCO_2^-}} \xrightarrow[\text{NAD}^+ \quad \text{NADH/H}^+]{} \left[\overset{\overset{\displaystyle NH}{\|}}{^-O_2CCH_2CH_2CCO_2^-} \right] \xrightarrow{\text{H}_2\text{O, H}^+} \underset{\textbf{α-Ketoglutarate}}{\overset{\overset{\displaystyle O}{\|}}{^-O_2CCH_2CH_2CCO_2^-}} + NH_4^+$$

• •

Problem 29.11   What α-keto acid is formed on transamination of leucine?

• •

FIGURE 29.7 ▼

Oxidative deamination of alanine requires the cofactor pyridoxal phosphate and yields pyruvate as product.

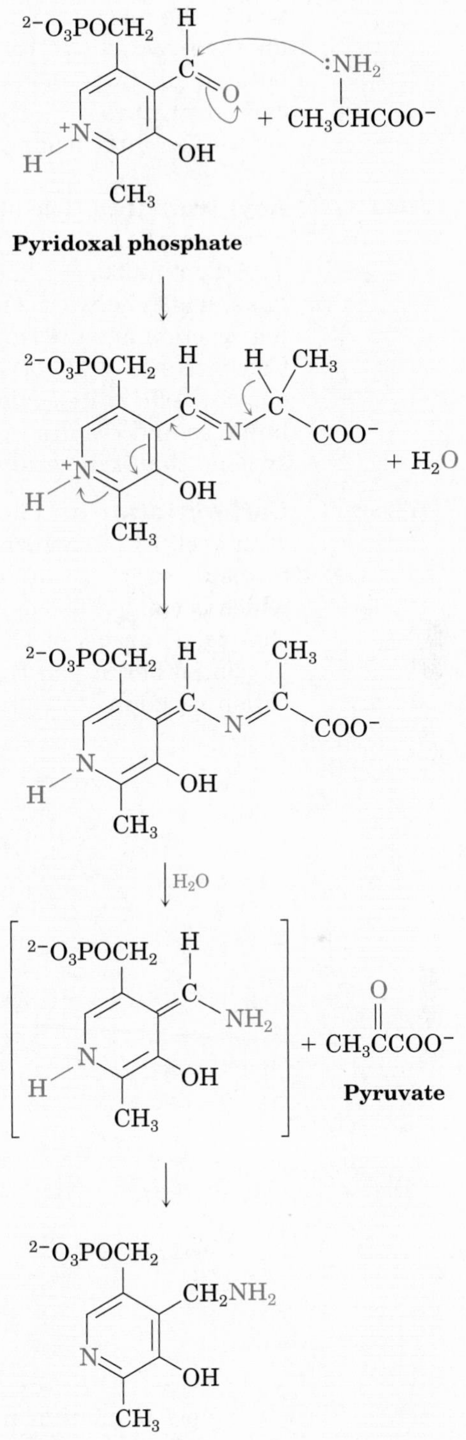

Nucleophilic attack of the amino acid on the pyridoxal phosphate carbonyl group gives an imine.

Loss of a proton moves the double bonds and gives a second imine intermediate.

Hydrolysis of the imine then yields an α-keto acid along with a nitrogen-containing pyridoxal phosphate derivative.

Bond tautomerization regenerates an aromatic pyridine ring.

29.7 Anabolism of Fatty Acids

One of the most striking features of the common fatty acids is that all have an even number of carbon atoms (Table 27.1). This even number results because all fatty acids are derived biosynthetically from the simple two-carbon precursor, acetyl CoA. The anabolic pathway by which organisms synthesize fatty acids is shown in Figure 29.8.

STEPS 1–2   **Acyl transfers.** The starting material for fatty acid synthesis is the thiol ester acetyl CoA, which is prepared in nature by decarboxylation of pyruvate (Section 29.4). The synthetic pathway begins with several *priming reactions*, which convert acetyl CoA into more reactive species. The first priming reaction is a nucleophilic acyl substitution reaction that converts acetyl CoA into acetyl ACP (acyl carrier protein). The reaction is catalyzed by the enzyme ACP transacylase. Step 2 involves a further exchange of thiol ester linkages and results in covalent bonding of the acetyl group to a synthase enzyme that will catalyze the upcoming condensation step.

STEPS 3–4   **Carboxylation and acyl transfer.** The third priming reaction again starts with acetyl CoA, which is carboxylated by reaction with HCO_3^- and ATP to yield malonyl CoA plus ADP. This step involves the coenzyme biotin, which is bonded to the lysine residue of acetyl CoA carboxylase enzyme and acts as a carrier of CO_2. The enolate ion of acetyl CoA reacts with carboxylated biotin and transfers the CO_2 group in a nucleophilic acyl substitution reaction.

Step 4 is another nucleophilic acyl substitution reaction that converts malonyl CoA into the more reactive malonyl ACP.

FIGURE 29.8 ▼

Biological pathway for fatty acid synthesis from the two-carbon precursor, acetyl CoA. Individual steps are explained in more detail in the text.

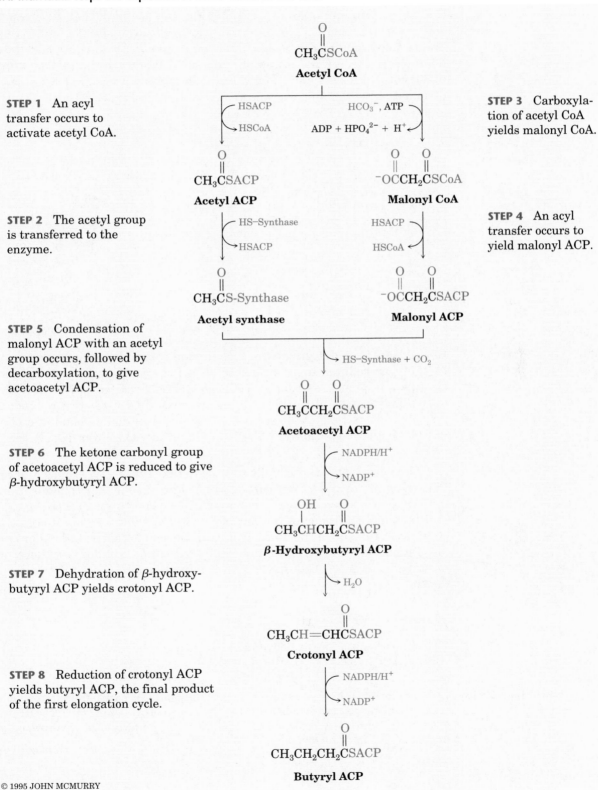

STEP 1 An acyl transfer occurs to activate acetyl CoA.

STEP 2 The acetyl group is transferred to the enzyme.

STEP 3 Carboxylation of acetyl CoA yields malonyl CoA.

STEP 4 An acyl transfer occurs to yield malonyl ACP.

STEP 5 Condensation of malonyl ACP with an acetyl group occurs, followed by decarboxylation, to give acetoacetyl ACP.

STEP 6 The ketone carbonyl group of acetoacetyl ACP is reduced to give β-hydroxybutyryl ACP.

STEP 7 Dehydration of β-hydroxybutyryl ACP yields crotonyl ACP.

STEP 8 Reduction of crotonyl ACP yields butyryl ACP, the final product of the first elongation cycle.

© 1995 JOHN MCMURRY

STEP 5   **Condensation.** The key carbon–carbon bond-forming reaction that builds the fatty acid chain occurs in step 5. This step is simply a Claisen condensation (Section 23.8) between acetyl synthase as the electrophilic acceptor and malonyl ACP as the nucleophilic donor. An enolate ion derived from the doubly activated –CH$_2$– group of malonyl ACP adds to the carbonyl group of acetyl synthase, yielding an intermediate β-keto acid that loses carbon dioxide to give the four-carbon product acetoacetyl ACP.

Acetoacetyl ACP

STEPS 6–8   **Reduction and dehydration.** The ketone carbonyl group in acetoacetyl ACP is next reduced to an alcohol by NADPH (nicotinamide adenine dinucleotide phosphate), a reducing coenzyme closely related to NADH. Subsequent dehydration of the resulting β-hydroxy thiol ester (E2 reaction) in step 7 yields crotonyl ACP, and the carbon–carbon double bond of crotonyl ACP is further reduced by NADPH in step 8 to yield butyryl ACP.

   The net effect of these eight steps is to take two acetyl groups and combine them into a single four-carbon butyryl group. Further condensation of butyryl synthase with another malonyl ACP yields a six-carbon unit, and still further repetitions of the pathway add two more carbon atoms to the chain each time until the 16-carbon palmitic acid is reached. Further chain elongation of palmitic acid occurs by reactions similar to those just described, but acetyl CoA itself rather than malonyl ACP is the two-carbon donor.

● ●

Problem 29.12   Write a mechanism for the dehydration reaction of β-hydroxybutyryl ACP to yield crotonyl ACP in step 7 of fatty acid synthesis.

Problem 29.13   Evidence for the role of acetate in fatty acid biosynthesis comes from isotope-labeling experiments. If acetate labeled with ^{13}C in the methyl group (^{13}CH$_3$COOH) were incorporated into fatty acids, at what positions in the fatty acid chain would you expect the ^{13}C label to appear?

● ●

29.8 Anabolism of Carbohydrates: Gluconeogenesis

As a rule, the anabolic pathway by which an organism makes a substance is not the reverse of the catabolic pathway by which the organism degrades the same substance. For example, the β-oxidation pathway for fatty acid degradation (Figure 29.2) and the cycle for fatty acid synthesis (Figure 29.8) are clearly related, but one is not the exact reverse of the other. Fatty acid synthesis involves carboxylation and decarboxylation reactions, for example, but β-oxidation does not.

The differences between catabolic and anabolic pathways are due to differences in energy. As noted previously, the overall free-energy change ΔG for any reaction sequence must be negative for the sequence to proceed spontaneously. But if ΔG for a sequence is *negative* in one direction, it must be *positive* in the reverse direction, implying that the reverse sequence can't proceed spontaneously. To both catabolize and anabolize a substance, an organism must use different reaction sequences, both of which have favorable free-energy changes.

Just as fatty acids are catabolized and anabolized by different pathways, so too are carbohydrates. **Gluconeogenesis**, the anabolic pathway by which organisms make glucose from pyruvate, is related to glycolysis but is not its exact reverse. The gluconeogenesis pathway is shown in Figure 29.9 (p. 1222).

STEP 1 **Carboxylation.** Gluconeogenesis begins with the carboxylation of pyruvate to yield oxaloacetate. As in the third step of fatty acid synthesis (Figure 29.8), the reaction requires ATP and the coenzyme biotin, acting as a carrier of CO_2.

STEP 2 **Decarboxylation and phosphorylation.** Decarboxylation of oxaloacetate, a β-keto acid, occurs by a mechanism similar to that of step 4 in the citric acid cycle (Figure 29.6), but phosphorylation of the resultant pyruvate enolate ion occurs concurrently to give phosphoenolpyruvate.

STEPS 3–4 **Hydration and isomerization.** Conjugate addition of water to the double bond of phosphoenolpyruvate takes place in a process similar to that of step 2 in the β-oxidation pathway (Figure 29.2). Isomerization then occurs by transfer of a phosphate group from C2 to C3, yielding 3-phosphoglycerate.

STEPS 5–7 **Phosphorylation, reduction, and tautomerization.** Reaction of 3-phosphoglycerate with ATP generates the corresponding acyl phosphate, which is reduced by NADH/H$^+$ to an aldehyde. Keto–enol tautomerization of the aldehyde gives dihydroxyacetone phosphate, the same reaction as step 5 of glycolysis (Figure 29.4).

FIGURE 29.9 ▼

The gluconeogenesis pathway for biosynthesis of glucose from pyruvate. The individual steps are explained in more detail in the text.

$$\underset{\displaystyle CH_3\overset{\displaystyle O}{\overset{\|}{C}}\!-\!\overset{\displaystyle O}{\overset{\|}{C}}O^-}{}$$

STEP 1   Pyruvate undergoes biotin-dependent carboxylation to give oxaloacetate.

HCO$_3^-$, ATP

ADP, HPO$_4^{2-}$, H$^+$

$$^-O\overset{\displaystyle O}{\overset{\|}{C}}CH_2\overset{\displaystyle O}{\overset{\|}{C}}\!-\!\overset{\displaystyle O}{\overset{\|}{C}}O^- \quad \textbf{Oxaloacetate}$$

STEP 2   Phosphorylation and decarboxylation then produce phosphoenolpyruvate.

GTP

GDP, CO$_2$

$$\underset{\displaystyle H_2C\!=\!\overset{\displaystyle ^{2-}O_3PO}{\overset{|}{C}}\!-\!\overset{\displaystyle O}{\overset{\|}{C}}O^-}{} \quad \textbf{Phosphoenolpyruvate}$$

STEP 3   Conjugate addition of water to the double bond of phosphoenol-pyruvate gives 2-phosphoglycerate.

H$_2$O

$$\underset{\displaystyle HOCH_2\overset{\displaystyle ^{2-}O_3PO}{\overset{|}{C}}H\overset{\displaystyle O}{\overset{\|}{C}}O^-}{} \quad \textbf{2-Phosphoglycerate}$$

STEP 4   Isomerization by transfer of a phosphate group yields 3-phospho-glycerate.

$$\underset{\displaystyle ^{2-}O_3POCH_2\overset{\displaystyle HO}{\overset{|}{C}}H\overset{\displaystyle O}{\overset{\|}{C}}O^-}{} \quad \textbf{3-Phosphoglycerate}$$

STEP 5   Phosphorylation with ATP gives 3-phosphoglyceroyl phosphate.

ATP

ADP

$$\underset{\displaystyle ^{2-}O_3POCH_2\overset{\displaystyle HO}{\overset{|}{C}}H\overset{\displaystyle O}{\overset{\|}{C}}OPO_3^{2-}}{} \quad \textbf{3-Phosphoglyceroyl phosphate}$$

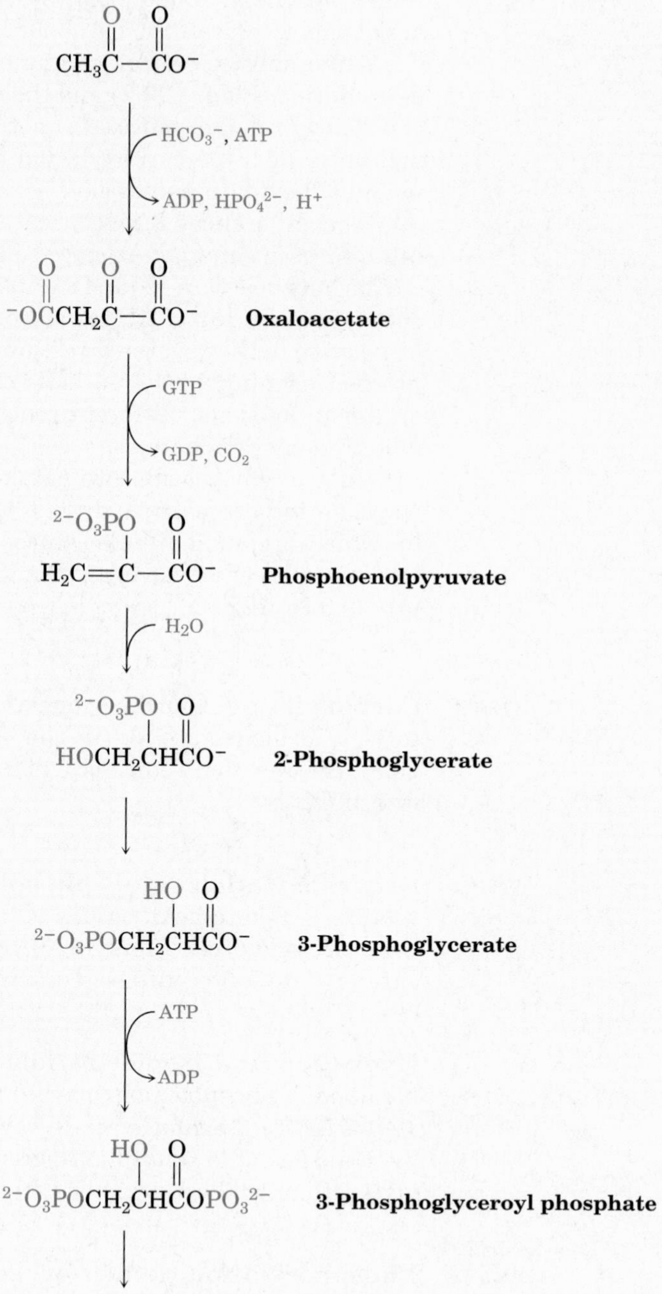

STEPS 6–7 Reduction yields glyceraldehyde 3-phosphate, which undergoes keto–enol tautomerization to give dihydroxyacetone phosphate.

$$- NADH/H^+$$

$$\searrow NAD^+, HPO_4^{2-}$$

$$\underset{\textbf{Glyceraldehyde}}{\underset{\textbf{3-phosphate}}{{}^{2-}O_3POCH_2\overset{HO}{\underset{|}{C}}H\overset{O}{\overset{||}{C}}H}} \longrightarrow \underset{\textbf{Dihydroxyacetone}}{\underset{\textbf{phosphate}}{{}^{2-}O_3POCH_2\overset{O}{\overset{||}{C}}CH_2OH}}$$

STEP 8 Two three-carbon units join in an aldol reaction to yield fructose 1,6-bisphosphate.

$$\underset{\underset{HO\ \ OH}{|\ \ \ |}}{{}^{2-}O_3POCH_2\overset{HO}{\underset{|}{C}}HCH\overset{O}{\overset{||}{C}}H CCH_2OPO_3{}^{2-}} \qquad \textbf{Fructose 1,6-bisphosphate}$$

STEP 9 Hydrolysis of the phosphate group at C1 occurs, giving fructose 6-phosphate.

$$\leftarrow H_2O$$

$$\underset{\underset{HO\ \ OH}{|\ \ \ |}}{{}^{2-}O_3POCH_2\overset{HO}{\underset{|}{C}}HCH\overset{O}{\overset{||}{C}}CH_2OH} \qquad \textbf{Fructose 6-phosphate}$$

STEP 10 Keto–enol tautomerization shifts the carbonyl group from C2 to C1, yielding glucose 6-phosphate.

$$\underset{\underset{HO\ \ OH\ \ \ OH}{|\ \ \ |\ \ \ \ |}}{{}^{2-}O_3POCH_2\overset{OH}{\underset{|}{C}}HCHCHCH\overset{O}{\overset{||}{C}}H}$$

Glucose 6-phosphate

STEP 8 **Aldol condensation.** Dihydroxyacetone phosphate and glyceraldehyde 3-phosphate, the two three-carbon units produced in step 7, join in step 8 to give fructose 1,6-bisphosphate. The reaction looks like an aldol condensation (Section 23.2) between the enolate ion from dihydroxyacetone phosphate and the carbonyl group of glyceraldehyde 3-phosphate.

$$\underset{\underset{\textbf{Glyceraldehyde 3-phosphate}}{HO}}{{}^{2-}O_3POCH_2\overset{O}{\overset{||}{C}}HC H} + \underset{\underset{\textbf{Dihydroxyacetone phosphate}}{OH}}{:\overset{-}{C}HC\overset{O}{\overset{||}{C}}CH_2OPO_3{}^{2-}} \longrightarrow \underset{\underset{\textbf{Fructose 1,6-bisphosphate}}{HO\ \ OH}}{{}^{2-}O_3POCH_2\overset{HO}{C}HCH\overset{O}{\overset{||}{C}}CCH_2OPO_3{}^{2-}}$$

As was true in step 4 of glycolysis (Figure 29.4), this "aldol" reaction actually takes place not on the free ketone but on an imine (Schiff base) formed by reaction of dihydroxyacetone phosphate with a side-chain –NH₂ group on the enzyme. Loss of a proton from the neighboring carbon then generates an enamine (Section 19.9), an aldol-like reaction ensues, and the product is hydrolyzed.

$$
\begin{array}{ccc}
\text{CH}_2\text{OPO}_3{}^{2-} & & \text{CH}_2\text{OPO}_3{}^{2-} \\
| & \xrightarrow{\text{H}_2\text{N}\sim\text{Enzyme}} & | \\
\text{C}=\text{O} & & \text{C}=\text{N}\sim\text{Enzyme} \\
| & & | \\
\text{CH}_2\text{OH} & & \text{CH}_2\text{OH}
\end{array}
$$

Dihydroxyacetone phosphate   **An imine (Schiff base)**

$$
\begin{array}{ccc}
\text{CH}_2\text{OPO}_3{}^{2-} & & \text{CH}_2\text{OPO}_3{}^{2-} \\
| & & | \\
\text{C}=\overset{+}{\text{N}}\text{H}\sim\text{Enzyme} & \xrightarrow{\text{H}_2\text{O}} & \text{C}=\text{O} \qquad + \text{H}_2\text{N}\sim\text{Enzyme}\\
| & & | \\
\text{HO}-\text{C}-\text{H} & & \text{HO}-\text{C}-\text{H} \\
| & & | \\
\text{H}-\text{C}-\text{OH} & & \text{H}-\text{C}-\text{OH} \\
| & & | \\
\text{H}-\text{C}-\text{OH} & & \text{H}-\text{C}-\text{OH} \\
| & & | \\
\text{CH}_2\text{OPO}_3{}^{2-} & & \text{CH}_2\text{OPO}_3{}^{2-}
\end{array}
$$

Fructose 1,6-bisphosphate

STEPS 9–10   **Hydrolysis and isomerization.** Hydrolysis of the phosphate group at C1 of fructose 1,6-bisphosphate, followed by keto–enol isomerization of the carbonyl group from C2 to C1, then completes gluconeogenesis. The isomerization is the exact reverse of step 2 in glycolysis (Figure 29.4).

• •

Problem 29.14   Write a mechanism for step 6 of gluconeogenesis, the reduction of 3-phosphoglyceroyl phosphate with NADH/H⁺ to yield glyceraldehyde 3-phosphate.

Problem 29.15   In which carbons of glucose do the –CH₃ carbons of pyruvate end up?

• •

29.9 Some Conclusions About Biological Chemistry

After examining metabolic pathways, perhaps the main conclusion about biological chemistry is the near identity between the mechanisms of biological reactions and the mechanisms of laboratory reactions. In all the pathways described in this chapter, terms like "imine formation," "aldol reaction," "nucleophilic acyl substitution reaction," "E2 reaction," and "Claisen reaction" appear constantly. Biological reactions are not mysterious—the vitalistic theory described on page 2 of this text died long ago. There are clear, understandable reasons for the reactions carried out within living organisms.

A further interesting conclusion from looking at the metabolic pathways discussed in this chapter is that all the reaction sequences are linear except one. Only the citric acid cycle is different, consisting of a closed loop of reactions that starts and ends with the same substance. Why does nature use a different strategy for the citric acid cycle? Although there's probably no one "right" answer, part of the reason may again be that chemistry in living organisms must follow the same rules of reactivity as chemistry in the laboratory. When a relatively large, multifunctional molecule like glucose is degraded, the reaction choices are numerous, and an efficient linear pathway is possible. When a small, monofunctional molecule like acetyl CoA is degraded, however, the mechanistic options are limited. There are relatively few reaction choices available for degrading acetyl CoA, so a linear pathway may not be energetically feasible. By employing a cyclic pathway with multifunctional *intermediates*, however, the range of available reaction choices becomes much larger.

CHEMISTRY @ WORK

Basal Metabolism

The minimum amount of energy per unit time an organism must expend to stay alive is called the organism's *basal metabolic rate (BMR)*. This rate is measured by monitoring respiration and finding the rate of oxygen consumption, which is proportional to the amount of energy used. Assuming an average dietary mix of fats, carbohydrates, and proteins, approximately 4.82 kcal are required for each liter of oxygen consumed.

(continued) ▶

The average basal metabolic rate for humans is about 65 kcal/h, or 1600 kcal/day. Obviously, the rate varies for different people depending on sex, age, weight, and physical condition. As a rule, the BMR is lower for older people than for younger people, is lower for females than for males, and is lower for people in good physical condition than for those who are out of shape and overweight. A BMR substantially above the expected value indicates an unusually rapid metabolism, perhaps caused by a fever or some biochemical abnormality.

The total number of calories a person needs each day is the sum of the basal requirement plus the energy used for physical activities, as shown in Table 29.1. A relatively inactive person needs about 30% above basal requirements per day, a lightly active person needs about 50% above basal, and a very active person such as an athlete or construction worker may need 100% above basal requirements. Some endurance athletes in ultradistance events can use as many as 10,000 kcal/day above the basal level. Each day that your caloric intake is above what you use, fat is stored in your body and your weight rises. Each day that your caloric intake is below what you use, fat in your body is metabolized and your weight drops.

TABLE 29.1   Energy Cost of Various Activities[a]

Activity	Energy cost (kcal/min)
Sleeping	1.2
Sitting, reading	1.6
Standing still	1.8
Walking	3–6
Tennis	7–9
Basketball	9–10
Walking up stairs	10–18
Running	9–22

[a] For a 70 kg man.

Tarahumara Indians living in remote villages of the Barranca del Cobre (Copper Canyon), Mexico, are among the world's finest endurance runners, using up to 10,000 kcal in runs of over 100 miles.

Summary and Key Words

Metabolism is the sum of all chemical reactions in the body. Reactions that break down large molecules into smaller fragments are called **catabolism**; reactions that build up large molecules from small pieces are called **anabolism**. Although the details of specific biochemical pathways are sometimes complex, all the reactions that occur follow the normal rules of organic chemical reactivity.

The catabolism of fats begins with **digestion**, in which ester bonds are hydrolyzed to give glycerol and fatty acids. The fatty acids are degraded in the four-step **β-oxidation pathway** by removal of two carbons at a time, yielding acetyl CoA. Catabolism of carbohydrates begins with the hydrolysis of glycoside bonds to give glucose, which is degraded in the ten-step **glycolysis** pathway. Pyruvate, the initial product of glycolysis, is then converted into acetyl CoA. The acetyl groups produced by degradation of fats and carbohydrates next enter the eight-step **citric acid cycle**, where they are further degraded into CO_2.

Catabolism of proteins is more complex than that of fats or carbohydrates because each of the 20 different amino acids is degraded by its own unique pathway. In general, though, the amino nitrogen atoms are removed and the substances that remain are converted into compounds that enter the citric acid cycle. Most amino acids lose their nitrogen atom by **transamination**, a reaction in which the –NH_2 group of the amino acid changes places with the keto group of an α-keto acid such as α-ketoglutarate. The products are a new α-keto acid and glutamate. Oxidation and hydrolysis of glutamate then yield ammonium ion.

The energy released in catabolic pathways is used in the **electron-transport chain** to make molecules of adenosine triphosphate (ATP). ATP, the final result of food catabolism, couples to and drives many otherwise unfavorable reactions.

Biomolecules can be synthesized as well as degraded, although the pathways for anabolism and catabolism are not the exact reverse of one another. Fatty acids are biosynthesized from acetic acid by an eight-step pathway, and carbohydrates are made from pyruvate by the ten-step **gluconeogenesis** pathway.

Visualizing Chemistry

• •

(Problems 29.1–29.15 appear within the chapter.)

29.16 Identify the amino acid that is a catabolic precursor of each of the following α-keto acids (yellow = S):

(a)   (b)

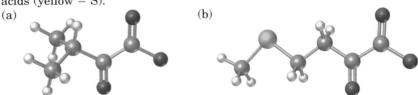

29.17 Identify the following intermediate in the citric acid cycle, and tell whether it has R or S stereochemistry:

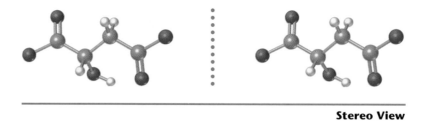

Stereo View

Additional Problems

29.18 What chemical events occur during the digestion of food?

29.19 What is the difference between digestion and metabolism?

29.20 What is the difference between anabolism and catabolism?

29.21 Draw the structure of adenosine *mono*phosphate (AMP), an intermediate in some biochemical pathways.

29.22 Cyclic adenosine monophosphate (cyclic AMP), a modulator of hormone action, is related to AMP (Problem 29.21) but has its phosphate group linked to *two* hydroxyl groups at C3′ and C5′ of the sugar. Draw the structure of cyclic AMP.

29.23 What general kind of reaction does ATP carry out?

29.24 What general kind of reaction does NAD^+ carry out?

29.25 What general kind of reaction does FAD carry out?

29.26 Why aren't the glycolysis and gluconeogenesis pathways the exact reverse of one another?

29.27 Lactate, a product of glucose catabolism in oxygen-starved muscles, can be converted into pyruvate by oxidation. What coenzyme do you think is needed? Write the equation in the normal biochemical format using a curved arrow.

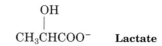

$$\underset{\textbf{Lactate}}{CH_3\overset{\displaystyle OH}{\overset{|}{C}}HCOO^-}$$

29.28 How many moles of acetyl CoA are produced by catabolism of the following substances?
(a) 1.0 mol glucose   (b) 1.0 mol palmitic acid   (c) 1.0 mol maltose

29.29 How many grams of acetyl CoA (MW = 809.6 amu) are produced by catabolism of the following substances?
(a) 100.0 g glucose   (b) 100.0 g palmitic acid   (c) 100.0 g maltose

29.30 Which of the substances listed in Problem 29.29 is the most efficient precursor of acetyl CoA on a weight basis?

29.31 List the sequence of intermediates involved in the catabolism of glycerol from hydrolyzed fats to yield acetyl CoA.

29.32 Write the equation for the final step in the β-oxidation pathway of any fatty acid with an even number of carbon atoms.

29.33 Show the products of each of the following reactions:

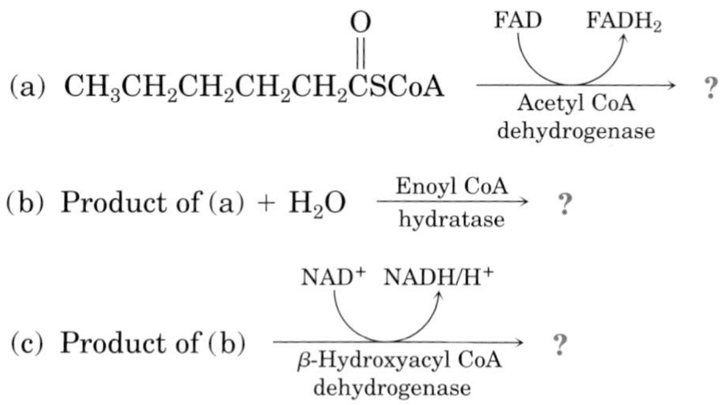

(a) $\underset{\displaystyle \text{CH}_3\text{CH}_2\text{CH}_2\text{CH}_2\text{CH}_2\overset{\displaystyle \overset{O}{\|}}{\text{C}}\text{SCoA}}{}$ $\xrightarrow[\substack{\text{Acetyl CoA} \\ \text{dehydrogenase}}]{\text{FAD} \quad \text{FADH}_2}$?

(b) Product of (a) + H_2O $\xrightarrow[\text{hydratase}]{\text{Enoyl CoA}}$?

(c) Product of (b) $\xrightarrow[\substack{\beta\text{-Hydroxyacyl CoA} \\ \text{dehydrogenase}}]{\text{NAD}^+ \quad \text{NADH/H}^+}$?

29.34 What is the structure of the α-keto acid formed by transamination of each of the following amino acids?
(a) Threonine   (b) Phenylalanine   (c) Asparagine

29.35 What enzyme cofactor is associated with each of the following kinds of reactions?
(a) Transamination   (b) Carboxylation of a ketone
(c) Decarboxylation of an α-keto acid

29.36 The glycolysis pathway shown in Figure 29.4 has a number of intermediates that contain phosphate groups. Why can 3-phosphoglyceroyl phosphate and phospho-enolpyruvate transfer a phosphate group to ADP while glucose 6-phosphate cannot?

29.37 In the *pentose phosphate* pathway for degrading sugars, ribulose 5-phosphate is converted to ribose 5-phosphate. Propose a mechanism for the isomerization.

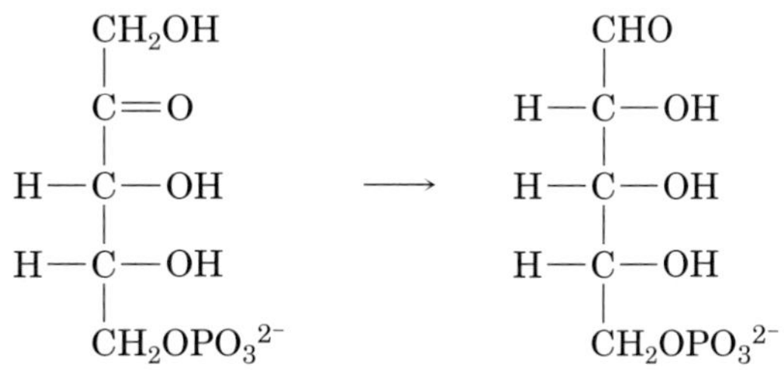

CH_2OH	CHO
$\text{C}{=}\text{O}$	$\text{H}{-}\text{C}{-}\text{OH}$
$\text{H}{-}\text{C}{-}\text{OH}$	$\text{H}{-}\text{C}{-}\text{OH}$
$\text{H}{-}\text{C}{-}\text{OH}$	$\text{H}{-}\text{C}{-}\text{OH}$
$\text{CH}_2\text{OPO}_3{}^{2-}$	$\text{CH}_2\text{OPO}_3{}^{2-}$
Ribulose 5-phosphate	**Ribose 5-phosphate**

29.38 Another step in the pentose phosphate pathway for degrading sugars (see Problem 29.37) is the conversion of ribose 5-phosphate to glyceraldehyde 3-phosphate. What kind of organic process is occurring? Propose a mechanism for the conversion.

Ribose 5-phosphate → Glyceraldehyde 3-phosphate

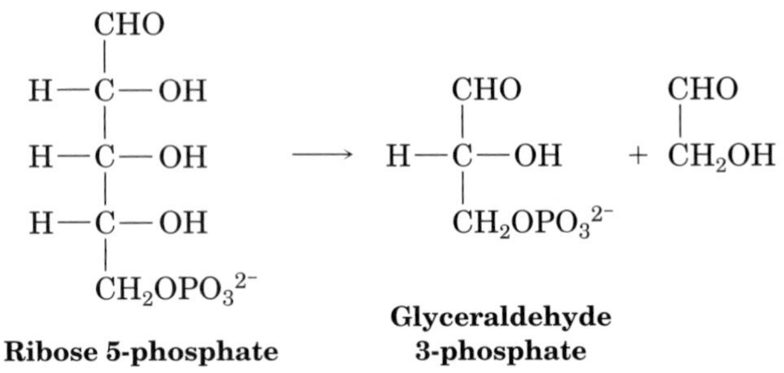

29.39 Write a mechanism for the conversion of α-ketoglutarate to succinyl CoA in step 4 of the citric acid cycle (Figure 29.6).

29.40 The primary fate of acetyl CoA under normal metabolic conditions is degradation in the citric acid cycle to yield CO_2. When the body is stressed by prolonged starvation, however, acetyl CoA is converted into compounds called *ketone bodies*, which can be used by the brain as a temporary fuel. Fill in the missing information indicated by the four question marks in the following biochemical pathway for the synthesis of ketone bodies from acetyl CoA:

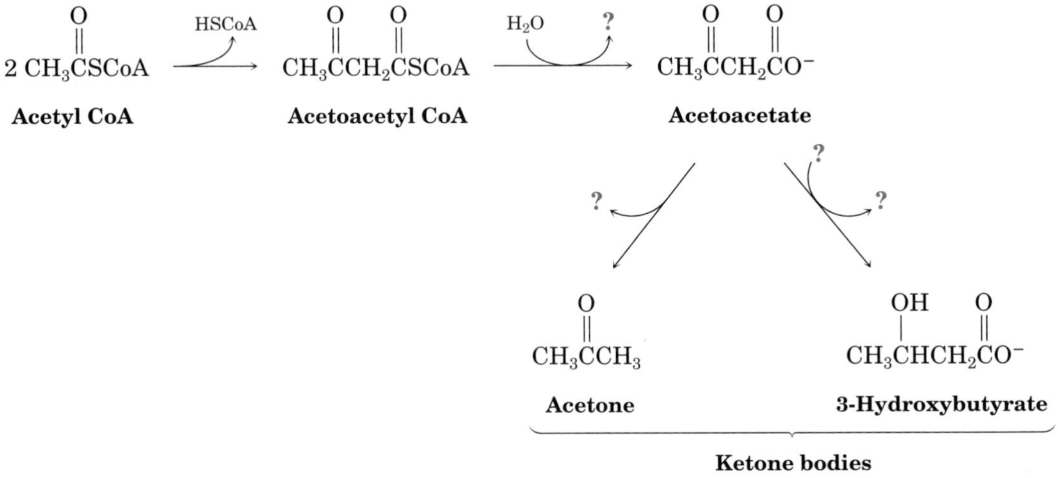

29.41 The initial reaction in Problem 29.40, conversion of two molecules of acetyl CoA to one molecule of acetoacetyl CoA, is a Claisen reaction. Assuming that there is a base present, show the mechanism of the reaction.

29.42 Sedoheptulose 7-phosphate reacts with glyceraldehyde 3-phosphate in the presence of a transaldolase enzyme to yield erythrose 4-phosphate and fructose 6-phosphate. Propose a mechanism.

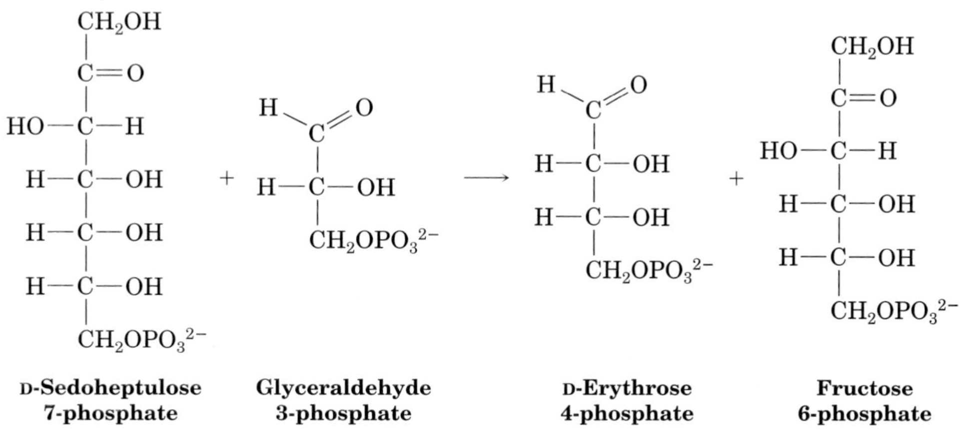

| **D-Sedoheptulose 7-phosphate** | **Glyceraldehyde 3-phosphate** | **D-Erythrose 4-phosphate** | **Fructose 6-phosphate** |

29.43 The amino acid tyrosine is biologically degraded by a series of steps that include the following transformations:

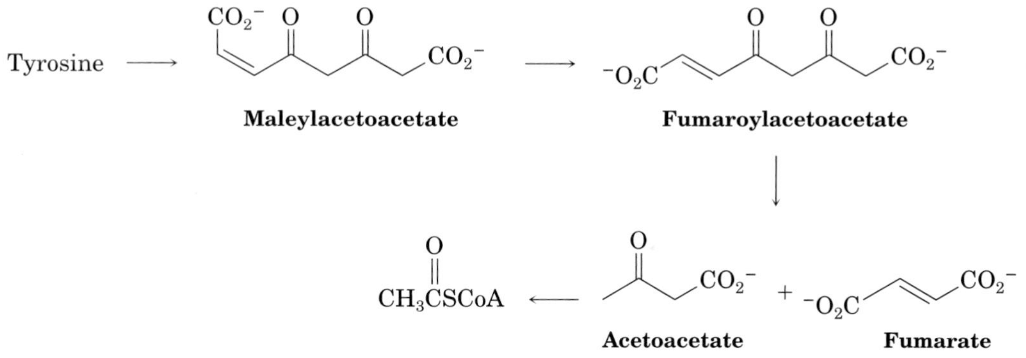

The double-bond isomerization of maleylacetoacetate to fumaroyl acetoacetate is catalyzed by practically any nucleophile, :Nu⁻. Propose a mechanism.

29.44 Propose a chemically reasonable mechanism for the biological conversion of fumaroyl-acetoacetate to fumarate plus acetoacetate (Problem 29.43).

29.45 Propose a chemically reasonable mechanism for the biological conversion of aceto-acetate to acetyl CoA (Problem 29.43).

29.46 We saw in Section 29.6 that the first step in the metabolism of most amino acids is transamination, a process in which the amino group of the amino acid exchanges with the keto group of α-ketoglutarate. Reaction of the amino acid with pyridoxal phosphate gives pyridoxamine phosphate, which then reacts with α-ketoglutarate to give glutamate and regenerated pyridoxal phosphate.

Pyridoxamine phosphate   **α-Ketoglutarate**   **Pyridoxal phosphate**   **Glutamate**

(a) The first step in the reaction of pyridoxamine phosphate with α-ketoglutarate is formation of an imine. Draw its structure.

(b) The imine then rearranges to a different imine, which undergoes hydrolysis to give pyridoxal phosphate and glutamate. Draw the structure of the second imine, and propose a mechanism for its formation.

29.47   Design your own degradative pathway. You know the rules (organic mechanisms) and you've seen the kinds of reactions that occur in the biological degradation of fats and carbohydrates into acetyl CoA. If you were Mother Nature, what series of steps would you use to degrade the amino acid serine into acetyl CoA?

$$\underset{\textbf{Serine}}{HOCH_2\overset{\overset{\displaystyle H_2N}{|}}{C}H\overset{\overset{\displaystyle O}{\|}}{C}OH} \xrightarrow{\ ?\ } \underset{\textbf{Acetyl CoA}}{CH_3\overset{\overset{\displaystyle O}{\|}}{C}SCoA}$$

29.48   The amino acid serine is biosynthesized by a route that involves reaction of 3-phosphohydroxypyruvate with glutamate. Propose a mechanism.

$$\underset{\textbf{3-Phosphohydroxypyruvate}}{\overset{\displaystyle COO^-}{\underset{\displaystyle CH_2OPO_3^{2-}}{\overset{|}{\underset{|}{C=O}}}}} \xrightarrow[\text{Glutamate}\quad \alpha\text{-Ketoglutarate}]{} \underset{\textbf{3-Phosphoserine}}{\overset{\displaystyle COO^-}{\underset{\displaystyle CH_2OPO_3^{2-}}{\overset{|}{\underset{|}{H_3\overset{+}{N}-C-H}}}}}$$

29.49   The amino acid leucine is biosynthesized from α-ketoisocaproate, which is itself prepared from α-ketoisovalerate by a multistep route that involves: (1) reaction with acetyl CoA, (2) hydrolysis, (3) dehydration, (4) hydration, (5) oxidation, and (6) decarboxylation. Show the steps in the transformation, and propose a mechanism for each.

α-Ketoisovalerate   **α-Ketoisocaproate**

Acetyl CoA, H$_2$O, NAD$^+$; HSCoA, CO$_2$, NADH/H$^+$

29.50 The amino acid cysteine, $C_3H_7NO_2S$, is biosynthesized from a substance called cystathionine by a multistep pathway:

$$\underset{\underset{NH_3^+}{|}}{\overset{\overset{O}{\|}}{^-OCCHCH_2}}CH_2SCH_2\underset{\underset{NH_3^+}{|}}{\overset{\overset{O}{\|}}{CHCO^-}} \longrightarrow NH_4^+ + ? + Cysteine$$

Cystathionine

(a) The first step is a transamination. What is the product?

(b) The second step is an E2 reaction. Show the products and the mechanism of the reaction.

(c) The final step is a double-bond reduction. What organic cofactor is required for this reaction, and what is the product represented by the question mark in the equation?

Molecular Modeling

• •

29.51 The complete oxidation of a hydrocarbon to CO_2 and H_2O is highly exothermic, but intermediate stages in the oxidation may be endothermic. Use SpartanView to obtain the appropriate energies, and calculate $\Delta H°$ for the two oxidation reactions shown below. Is either of them endothermic?

(a) $CH_3CH_2CH_3 + O_2 \longrightarrow CH_3CH_2CHO + H_2O$

(b) $CH_3CH_2CHO + O_2 \longrightarrow \underset{\underset{\overset{|}{OH}}{}}{HOCH_2CHCHO}$

29.52 There are several points in the glycolysis pathway where an α-hydroxy ketone equilibrates with an α-hydroxy aldehyde through its enol tautomer. Use SpartanView to obtain energies for 1-hydroxypropanone and its enol. Compare these values to the energies for acetone and its enol. For which ketone is enolization more favored?

$$\underset{\textbf{1-Hydroxypropanone}}{\overset{\overset{O}{\|}}{CH_3CCH_2OH}} \rightleftharpoons \overset{\overset{OH}{|}}{CH_3C}=CHOH \rightleftharpoons \underset{}{\overset{\overset{HO\ \ O}{|\ \ \ \|}}{CH_3CHCH}}$$

29.53 Acetyl CoA is synthesized by a multistep sequence that begins with nucleophilic addition of thiamine to pyruvate. Use SpartanView to examine electrostatic potential maps of a thiamine model and pyruvate. Is the negative charge in thiamine located primarily in the σ or π system? Is it delocalized? Which carbonyl carbon in pyruvate is more positive?

$$H_3C \underset{\underset{CH_3}{|}}{\overset{S}{\underset{N^+}{\diagdown}}}\!\!:^- \quad + \quad CH_3-\overset{\overset{\displaystyle O}{\|}}{C}-CO_2^- \quad \longrightarrow \quad ?$$

Pyruvate

Thiamine model

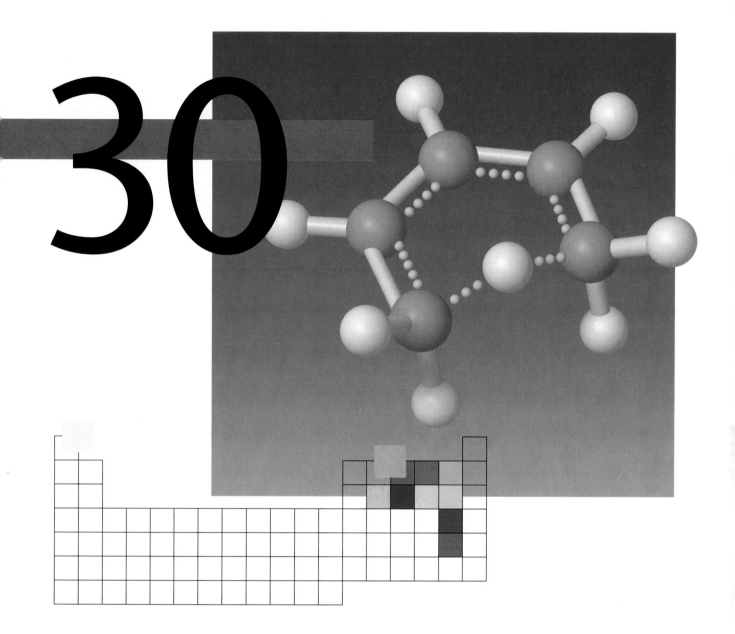

30

Orbitals and Organic Chemistry: Pericyclic Reactions

Most organic reactions take place by polar mechanisms in which a nucleophile donates two electrons to an electrophile in forming a new bond. Other reactions take place by radical mechanisms in which each of two reactants donates one electron in forming a new bond. Although there is always more to learn, both classes have been studied for many years and are relatively well understood.

By contrast, the fundamental principles of *pericyclic reactions*, the third major class of organic reaction mechanisms, have been worked out more recently. A **pericyclic reaction** is one that occurs by a concerted process through a cyclic transition state. The word *concerted* means that all bonding changes occur at the same time and in a single step; no intermediates are involved. Rather than try to expand this definition now, let's begin by briefly reviewing some of the ideas of molecular orbital theory introduced in Sections 1.9 and 14.3. We'll then look individually at the three main classes of pericyclic reactions: *electrocyclic reactions*, *cycloadditions*, and *sigmatropic rearrangements*.

30.1 Molecular Orbitals of Conjugated π Systems

A *conjugated* diene or polyene, as we saw in Chapter 14, is one with alternating double and single bonds. According to molecular orbital (MO) theory (Section 14.3), the *p* orbitals on the sp^2-hybridized carbons of a conjugated polyene interact to form a set of π molecular orbitals whose energies depend on the number of *nodes* they have between nuclei. Those molecular orbitals with fewer nodes are lower in energy than the isolated *p* atomic orbitals and are *bonding MO's;* those molecular orbitals with more nodes are higher in energy than the isolated *p* orbitals and are *antibonding MO's*. The π molecular orbitals of ethylene and 1,3-butadiene were shown in Figures 14.1 and 14.2.

A similar sort of molecular orbital description can be derived for any conjugated π electron system. 1,3,5-Hexatriene, for example, has three double bonds and six π MO's, which are shown in Figure 30.1. In the *ground state*, only the three bonding orbitals, ψ_1, ψ_2, and ψ_3, are filled. On irradiation with ultraviolet light, however, an electron is promoted from the highest-energy filled orbital (ψ_3) to the lowest-energy unfilled orbital (ψ_4^*) to give an *excited state* (Section 14.11). In this excited state, ψ_3 and ψ_4^* are each half-filled. (An asterisk denotes an antibonding orbital.)

30.2 Molecular Orbitals and Pericyclic Reactions

What do molecular orbitals and their nodes have to do with pericyclic reactions? The answer is, *everything*. According to a series of rules formulated in the mid-1960s by R. B. Woodward and Roald Hoffmann, a pericyclic reaction can take place only if the symmetries of the reactant MO's are the same as the symmetries of the product MO's. In other words, *the lobes of reactant MO's must be of the correct algebraic sign for bonding to occur in the transition state leading to product.*

If the symmetries of both reactant and product orbitals match up, or "correlate," the reaction is **symmetry-allowed**. If the symmetries don't correlate, the reaction is **symmetry-disallowed**. Symmetry-allowed reactions often occur under relatively mild conditions, but symmetry-disallowed reactions can't occur by concerted paths. Either they take place by nonconcerted pathways, or they don't take place at all.

The Woodward–Hoffmann rules for pericyclic reactions require an analysis of all reactant and product molecular orbitals, but Kenichi Fukui

FIGURE 30.1 ▼

The six π molecular orbitals of 1,3,5-hexatriene. In the ground state, only the three bonding MO's are occupied. In the excited state, ψ_3 and ψ_4^* each have one electron.

1.3.5-hexatriene
(see MO's on CD-Rom)

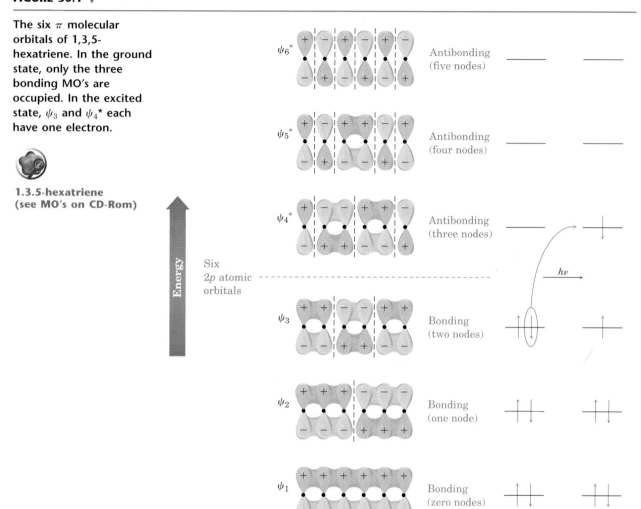

at Kyoto Imperial University in Japan introduced a simplified version. According to Fukui, we need to consider only *two* molecular orbitals, called the **frontier orbitals**. These frontier orbitals are the **highest occupied molecular orbital (HOMO)** and the **lowest unoccupied molecular orbital (LUMO)**. In ground-state 1,3,5-hexatriene, for example, ψ_3 is the HOMO and ψ_4^* is the LUMO (Figure 30.1). In *excited-state* 1,3,5-hexatriene, however, ψ_4^* is the HOMO and ψ_5^* is the LUMO.

Robert Burns Woodward

Robert Burns Woodward (1917–1979) was born in Boston, Massachusetts. He entered MIT at age 16, was expelled, reentered, obtained a B.S. degree at age 19, and received a Ph.D. at age 20. He then moved to Harvard University, where he joined the faculty at age 23. His vast scientific contributions included determining the structure of penicillin and turning the field of synthetic organic chemistry into an art form. He received the 1965 Nobel Prize for his work in organic synthesis.

• •

Problem 30.1   Look back at Figures 14.1 and 14.2, and tell which molecular orbital is the HOMO and which is the LUMO for both ground and excited states of ethylene and 1,3-butadiene.

• •

30.3 Electrocyclic Reactions

The best way to understand how orbital symmetry affects pericyclic reactions is to look at some examples. Let's look first at a group of polyene rearrangements called *electrocyclic reactions*. An **electrocyclic reaction** is a pericyclic process that involves the cyclization of a conjugated polyene. One π bond is broken, the other π bonds change position, a new σ bond is formed, and a cyclic compound results. For example, a conjugated triene can be converted into a cyclohexadiene, and a conjugated diene can be converted into a cyclobutene.

Both reactions are reversible, and the position of the equilibrium depends on the specific case. In general, the triene $\rightleftarrows$ cyclohexadiene equilibrium favors the ring-closed product, whereas the diene $\rightleftarrows$ cyclobutene equilibrium favors the unstrained ring-opened product.

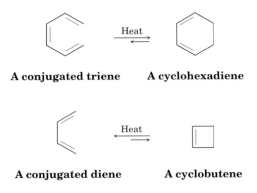

A conjugated triene   A cyclohexadiene

A conjugated diene   A cyclobutene

The most striking feature of electrocyclic reactions is their stereochemistry. For example, (2E,4Z,6E)-octatriene yields only *cis*-5,6-dimethyl-1,3-cyclohexadiene when heated, and (2E,4Z,6Z)-octatriene yields only *trans*-5,6-dimethyl-1,3-cyclohexadiene. Remarkably, however, the stereochemical results change completely when the reactions are carried out under what are called *photochemical*, rather than thermal, conditions. Irradiation of (2E,4Z,6E)-octatriene with ultraviolet light yields *trans*-5,6-dimethyl-1,3-cyclohexadiene (Figure 30.2).

A similar result is obtained for the thermal electrocyclic ring opening of 3,4-dimethylcyclobutene. The trans isomer yields only (2E,4E)-hexadiene when heated, and the cis isomer yields only (2E,4Z)-hexadiene. On UV irradiation, however, the results are opposite: Cyclization of the 2E,4E isomer under photochemical conditions yields cis product (Figure 30.3).

To account for these results, we need to look at the two outermost lobes of the polyene MO's—the lobes that interact when bonding occurs. There are two possibilities: The lobes of like sign can be either on the same side or on opposite sides of the molecule.

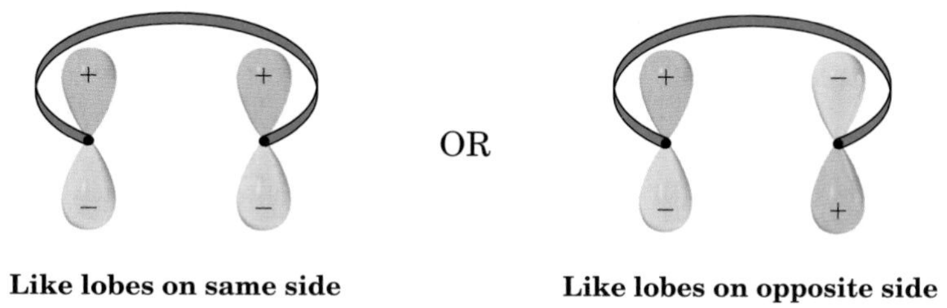

Like lobes on same side   OR   Like lobes on opposite side

FIGURE 30.2 ▼

Electrocyclic
interconversions of
2,4,6-octatriene isomers
and 5,6-dimethyl-1,3-
cyclohexadiene isomers.

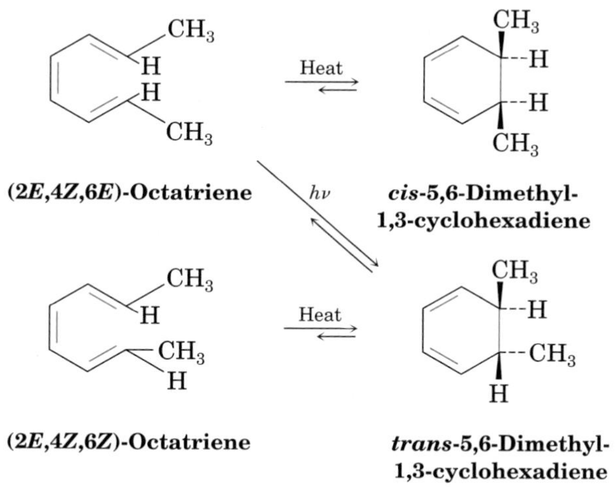

(2E,4Z,6E)-Octatriene → Heat ⇌ → **cis-5,6-Dimethyl-1,3-cyclohexadiene**

hν

(2E,4Z,6Z)-Octatriene → Heat ⇌ → **trans-5,6-Dimethyl-1,3-cyclohexadiene**

FIGURE 30.3 ▼

Electrocyclic
interconversions of 2,4-
hexadiene isomers and
3,4-dimethylcyclobutene
isomers.

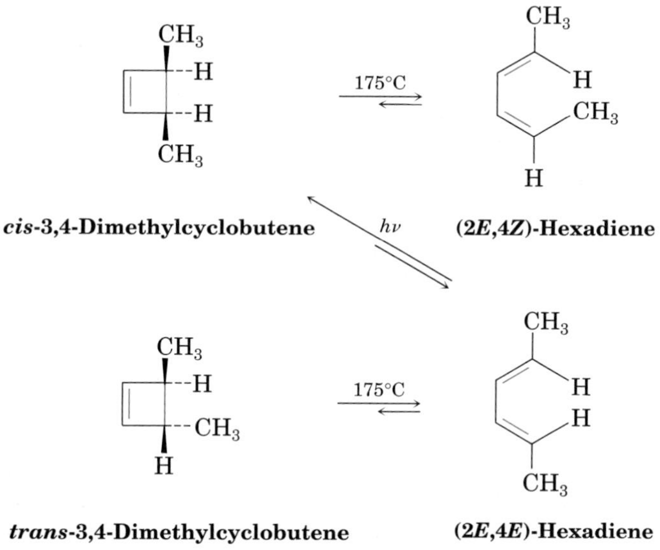

cis-3,4-Dimethylcyclobutene → 175°C ⇌ → **(2E,4Z)-Hexadiene**

hν

trans-3,4-Dimethylcyclobutene → 175°C ⇌ → **(2E,4E)-Hexadiene**

For a bond to form, the outermost π lobes must rotate so that favorable bonding interaction is achieved—a positive lobe with a positive lobe or a negative lobe with a negative lobe. If two lobes of like sign are on the *same* side of the molecule, the two orbitals must rotate in *opposite* directions—one clockwise, and one counterclockwise. This kind of motion is referred to as **disrotatory**.

Clockwise Counterclockwise

Conversely, if lobes of like sign are on *opposite* sides of the molecule, both orbitals must rotate in the *same* direction, either both clockwise or both counterclockwise. This kind of motion is called **conrotatory**.

Clockwise   Clockwise

30.4 Stereochemistry of Thermal Electrocyclic Reactions

How can we predict whether conrotatory or disrotatory motion will occur in a given case? According to frontier orbital theory, *the stereochemistry of an electrocyclic reaction is determined by the symmetry of the polyene HOMO.* The electrons in the HOMO are the highest-energy, most loosely held electrons, and are therefore most easily moved during reaction. For thermal reactions, the ground-state electronic configuration is used to identify the HOMO; for photochemical reactions, the excited-state electronic configuration is used.

Let's look again at the thermal ring closure of conjugated trienes. According to Figure 30.1, the HOMO of a conjugated triene in its ground state has lobes of like sign on the same side of the molecule, a symmetry that predicts disrotatory ring closure. This disrotatory cyclization is exactly what is observed in the thermal cyclization of 2,4,6-octatriene. The 2E,4Z,6E isomer yields cis product; the 2E,4Z,6Z isomer yields trans product (Figure 30.4).

In the same way, the ground-state HOMO of a conjugated diene (Figure 14.2) has a symmetry that predicts conrotatory ring closure. In practice, however, the conjugated diene reaction can only be observed in the reverse direction (cyclobutene → diene) because of the position of the

FIGURE 30.4 ▼

Thermal cyclizations of 2,4,6-octatrienes occur by disrotatory ring closures.

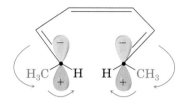

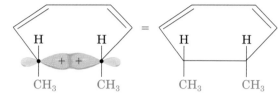

(2E,4Z,6E)-Octatriene   *cis*-5,6-Dimethyl-1,3-cyclohexadiene

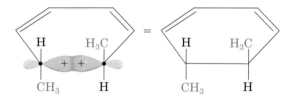

(2E,4Z,6Z)-Octatriene   *trans*-5,6-Dimethyl-1,3-cyclohexadiene

equilibrium. We thus predict that the 3,4-dimethylcyclobutene ring will *open* in a conrotatory fashion, which is exactly what is observed. *cis*-3,4-Dimethylcyclobutene yields (2E,4Z)-hexadiene, and *trans*-3,4-dimethylcyclobutene yields (2E,4E)-hexadiene by conrotatory opening (Figure 30.5).

FIGURE 30.5 ▼

Thermal ring openings of *cis*- and *trans*-dimethylcyclobutene occur by conrotatory paths.

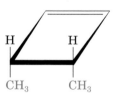

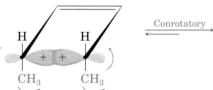

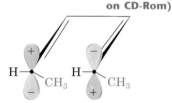

cis-3,4-Dimethylcyclobutene   **(2E,4Z)-Hexadiene**

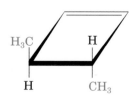

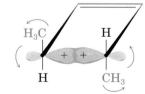

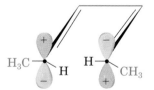

trans-3,4-Dimethylcyclobutene   **(2E,4E)-Hexadiene**

Note that a conjugated diene and a conjugated triene react in opposite stereochemical senses. The diene opens and closes by a conrotatory path, whereas the triene opens and closes by a disrotatory path. This difference, of course, is due to the different symmetries of the diene and triene HOMO's:

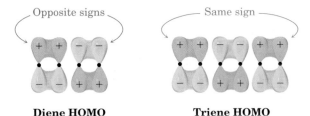

Diene HOMO   **Triene HOMO**

It turns out that there is an alternating relationship between the number of electron pairs (double bonds) undergoing bond reorganization and the stereochemistry of ring opening or closure. Polyenes with an even number of electron pairs undergo thermal electrocyclic reactions in a conrotatory sense, whereas polyenes with an odd number of electron pairs undergo the same reactions in a disrotatory sense.

- -

Problem 30.2   Draw the products you would expect from conrotatory and disrotatory cyclizations of (2Z,4Z,6Z)-octatriene. Which of the two paths would you expect the thermal reaction to follow?

Problem 30.3   *trans*-3,4-Dimethylcyclobutene can open by two conrotatory paths to give either (2E,4E)-hexadiene or (2Z,4Z)-hexadiene. Explain why both products are symmetry-allowed, and then account for the fact that only the 2E,4E isomer is obtained in practice.

- -

30.5 Photochemical Electrocyclic Reactions

We noted previously that photochemical electrocyclic reactions take a different stereochemical course than their thermal counterparts, and we can now explain this difference. Ultraviolet irradiation of a polyene causes an excitation of one electron from the ground-state HOMO to the ground-state LUMO. For example, irradiation of a conjugated diene excites an electron from ψ_2 to ψ_3^*, and irradiation of a conjugated triene excites an electron from ψ_3 to ψ_4^* (Figure 30.6).

Since electronic excitation changes the symmetries of HOMO and LUMO, it also changes the reaction stereochemistry. (2E,4E)-Hexadiene, for example, undergoes photochemical cyclization by a disrotatory path, whereas the thermal reaction is conrotatory. Similarly, (2E,4Z,6E)-octatriene undergoes photochemical cyclization by a conrotatory path, whereas the thermal reaction is disrotatory.

FIGURE 30.6 ▼

Ground-state and excited-state electronic configurations of conjugated dienes and trienes.

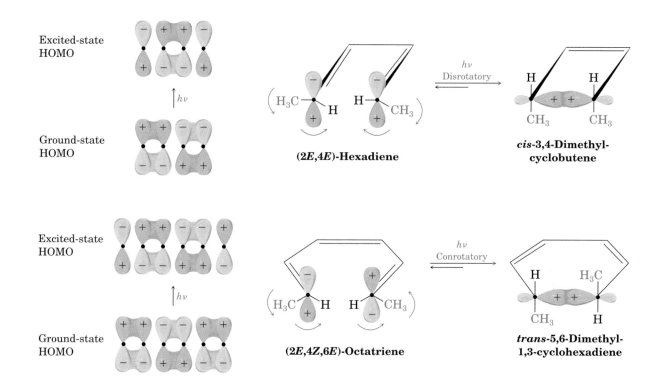

Thermal and photochemical electrocyclic reactions *always* take place with opposite stereochemistry because the symmetries of the frontier orbitals are always different. Table 30.1 gives some simple rules that make it possible to predict the stereochemistry of electrocyclic reactions.

TABLE 30.1   Stereochemical Rules for Electrocyclic Reactions

Electron pairs (double bonds)	Thermal reaction	Photochemical reaction
Even number	Conrotatory	Disrotatory
Odd number	Disrotatory	Conrotatory

Problem 30.4   What product would you expect to obtain from the photochemical cyclization of (2*E*,4*Z*,6*E*)-octatriene? Of (2*E*,4*Z*,6*Z*)-octatriene?

30.6 Cycloaddition Reactions

A **cycloaddition reaction** is one in which two unsaturated molecules add to one another, yielding a cyclic product. As with electrocyclic reactions, cycloadditions are controlled by the orbital symmetry of the reactants. Symmetry-allowed processes often take place readily, but symmetry-disallowed processes take place with great difficulty, if at all, and then only by nonconcerted pathways. Let's look at two examples to see how they differ.

The Diels–Alder cycloaddition reaction (Section 14.8) is a pericyclic process that takes place between a diene (four π electrons) and a dienophile (two π electrons) to yield a cyclohexene product. Thousands of examples of Diels–Alder reactions are known. They often take place easily at room temperature or slightly above, and they are stereospecific with respect to substituents. For example, room-temperature reaction between 1,3-butadiene and diethyl maleate (cis) yields exclusively the cis-disubstituted cyclohexene product. Similar reaction between 1,3-butadiene and diethyl fumarate (trans) yields exclusively the trans-disubstituted product (Figure 30.7).

FIGURE 30.7 ▼

Diels–Alder cycloaddition reactions of diethyl maleate (cis) and diethyl fumarate (trans). The reactions are stereospecific.

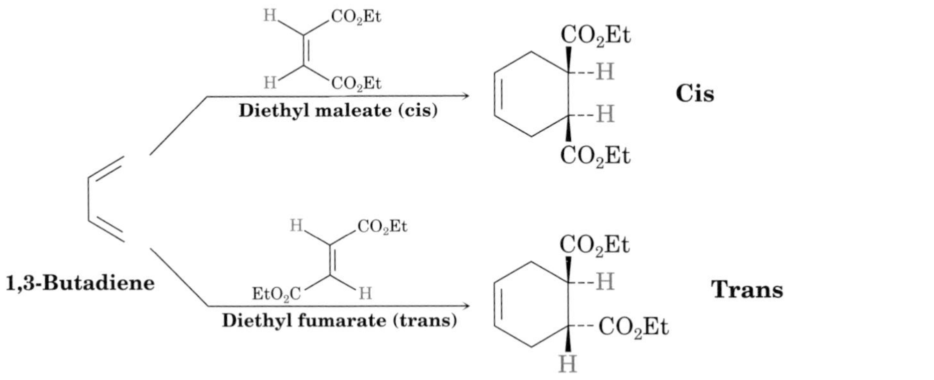

In contrast to the [4 + 2]-π-electron Diels–Alder reaction, [2 + 2] thermal cycloaddition between two alkenes does not occur. Only the *photochemical* [2 + 2] cycloaddition takes place to yield cyclobutane products.

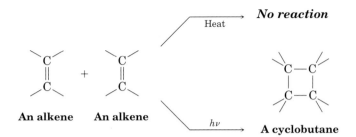

For a successful cycloaddition to take place, the terminal π lobes of the two reactants must have the correct symmetry for bonding to occur. This can happen in either of two ways, denoted *suprafacial* and *antarafacial*. **Suprafacial** cycloadditions take place when a bonding interaction occurs between lobes on the *same* face of one reactant and lobes on the *same* face of the other reactant (Figure 30.8).

FIGURE 30.8 ▼

Suprafacial cycloaddition occurs when there is bonding between lobes on the same face of one reactant and lobes on the same face of the second reactant.

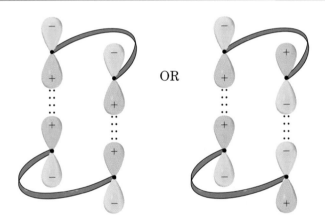

Antarafacial cycloadditions take place when a bonding interaction occurs between lobes on the *same* face of one reactant and lobes on *opposite* faces of the other reactant (Figure 30.9).

FIGURE 30.9 ▼

Antarafacial cycloaddition occurs when there is bonding between lobes on the same face of one reactant and lobes on opposite faces of the second reactant.

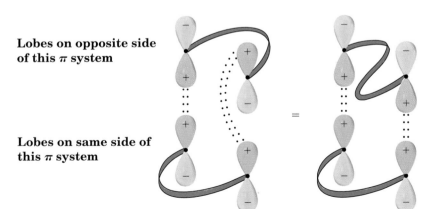

Note that both suprafacial and antarafacial cycloadditions are symmetry-allowed. Geometric constraints often make antarafacial reactions difficult, however, because there must be a twisting of the π orbital system. Thus, suprafacial cycloadditions are most common for small π systems.

30.7 Stereochemistry of Cycloadditions

How can we predict whether a given cycloaddition reaction will occur with suprafacial or with antarafacial geometry? According to frontier orbital theory, a cycloaddition reaction takes place when a bonding interaction occurs between the HOMO of one reactant and the LUMO of the other. An intuitive explanation of this rule is to imagine that one reactant donates electrons to the other. As with electrocyclic reactions, it's the electrons in the HOMO of the first reactant that are least tightly held and most likely to be donated. But when the second reactant accepts those electrons, they must go into a *vacant* orbital—the LUMO.

For a [4 + 2]-π-electron cycloaddition (Diels–Alder reaction), let's arbitrarily select the diene LUMO and the alkene HOMO. (We could equally well use the diene HOMO and the alkene LUMO.) The symmetries of the two ground-state orbitals are such that bonding of the terminal lobes can occur with suprafacial geometry (Figure 30.10). The Diels–Alder reaction therefore takes place readily under thermal conditions. Note that, as with electrocyclic reactions, we need be concerned only with the *terminal* lobes. For purposes of prediction, interactions among the interior lobes need not be considered.

FIGURE 30.10 ▼

Interaction of diene LUMO and alkene HOMO in a suprafacial [4 + 2] cycloaddition reaction (Diels–Alder reaction).

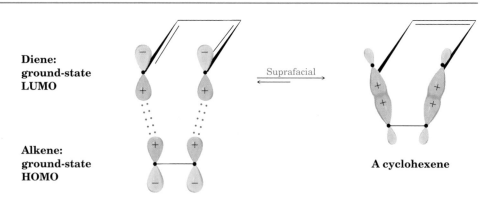

Diene: ground-state LUMO

Alkene: ground-state HOMO

Suprafacial

A cyclohexene

In contrast to the thermal [4 + 2] Diels–Alder reaction, the [2 + 2] cycloaddition of two alkenes to yield a cyclobutane can only be observed photochemically. The explanation follows from orbital-symmetry arguments.

Looking at the ground-state HOMO of one alkene and the LUMO of the second alkene, it's apparent that a thermal [2 + 2] cycloaddition must take place by an antarafacial pathway (Figure 30.11). Geometric constraints make the antarafacial transition state difficult, however, and so concerted thermal [2 + 2] cycloadditions are not observed.

FIGURE 30.11 ▼

Interaction of HOMO and LUMO in a potential thermal [2 + 2] cycloaddition. The reaction does not occur because antarafacial geometry is too strained.

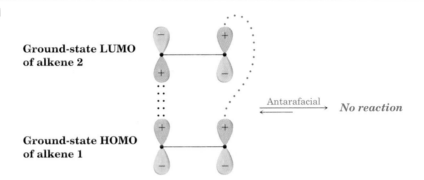

Ground-state LUMO of alkene 2

Ground-state HOMO of alkene 1

Antarafacial · · · · · · *No reaction*

In contrast to the thermal process, *photochemical* [2 + 2] cycloadditions *are* observed. Irradiation of an alkene with UV light excites an electron from ψ_1, the ground-state HOMO, to $\psi_2{}^*$, which becomes the excited-state HOMO. Interaction between the excited-state HOMO of one alkene and the LUMO of the second alkene indicates that a photochemical [2 + 2] cycloaddition reaction can occur by a suprafacial pathway (Figure 30.12).

FIGURE 30.12 ▼

Interaction of excited-state HOMO and LUMO in photochemical [2 + 2] cycloaddition reactions. The reaction occurs with suprafacial geometry.

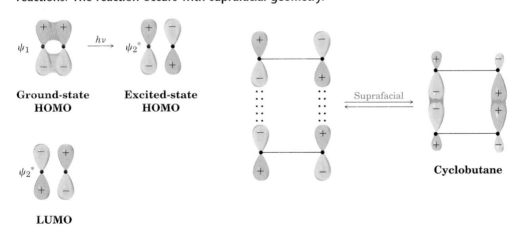

ψ_1 $\xrightarrow{h\nu}$ $\psi_2{}^*$

Ground-state HOMO   Excited-state HOMO

$\psi_2{}^*$

LUMO

Suprafacial

Cyclobutane

The photochemical [2 + 2] cycloaddition reaction occurs smoothly and represents one of the best methods known for synthesizing cyclobutane rings. For example:

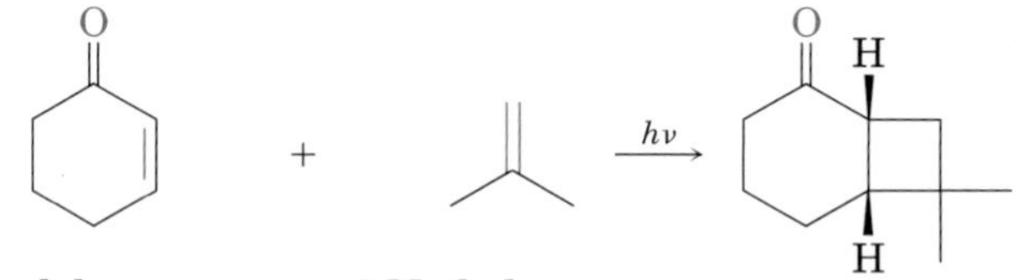

2-Cyclohexenone   **2-Methylpropene**

40%

Thermal and photochemical cycloaddition reactions *always* take place by opposite stereochemical pathways. As with electrocyclic reactions, we can categorize cycloadditions according to the total number of electron pairs (double bonds) involved in the rearrangement. Thus, a thermal Diels–Alder [4 + 2] reaction between a diene and a dienophile involves an odd number (three) of electron pairs and takes place by a suprafacial pathway. A thermal [2 + 2] reaction between two alkenes involves an even number (two) of electron pairs and must take place by an antarafacial pathway. For photochemical cyclizations, these selectivities are reversed. The general rules are given in Table 30.2.

TABLE 30.2 Stereochemical Rules for Cycloaddition Reactions

Electron pairs (double bonds)	Thermal reaction	Photochemical reaction
Even number	Antarafacial	Suprafacial
Odd number	Suprafacial	Antarafacial

Problem 30.5 What stereochemistry would you expect for the product of the Diels–Alder reaction between (2*E*,4*E*)-hexadiene and ethylene? What stereochemistry would you expect if (2*E*,4*Z*)-hexadiene were used instead?

Problem 30.6 Cyclopentadiene reacts with cycloheptatrienone to give the product shown. Tell what kind of reaction is involved, and explain the observed result. Is the reaction suprafacial or antarafacial?

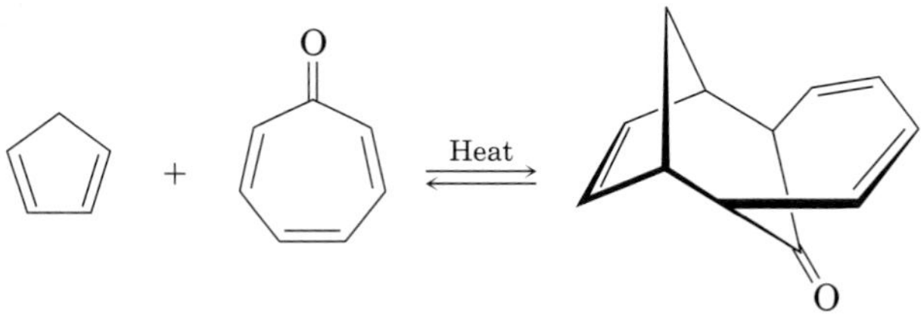

30.8 Sigmatropic Rearrangements

A **sigmatropic rearrangement**, the third general kind of pericyclic reaction, is a process in which a σ-bonded substituent atom or group migrates across a π electron system from one position to another. A σ bond is broken in the reactant, the π bonds move, and a new σ bond is formed in the product. The σ-bonded group can be either at the end or in the middle of the π system, as the following [1,5] and [3,3] rearrangements illustrate:

**1.5 rearrangement,
3.3 rearrangement
(see transition states on
CD-Rom)**

A [1,5] rearrangement

σ bond broken

A 1,5-diene   **Cyclic transition state**   **A 1,5-diene**

A [3,3] rearrangement

σ bond broken

An allyl vinyl ether   **Cyclic transition state**   **An unsaturated ketone**

The notations [1,5] and [3,3] describe the kind of rearrangement that is occurring. The numbers refer to the two groups connected by the σ bond and designate the positions in those groups *to which migration occurs*. For example, in the [1,5] sigmatropic rearrangement of a diene, the two groups connected by the σ bond are a hydrogen atom and a pentadienyl group. Migration occurs to position 1 of the H group (the only possibility) and to position 5 of the pentadienyl group. In the [3,3] Claisen rearrangement (Section 18.6), the two groups connected by the σ bond are an allyl group and a vinylic ether group. Migration occurs to position 3 of the allyl group and also to position 3 of the vinylic ether.

Sigmatropic rearrangements, like electrocyclic reactions and cyclo-additions, are controlled by orbital-symmetry considerations. There are two possible modes of reaction: Migration of a group across the same face of the π system is called a *suprafacial* rearrangement, and migration of a group from one face of the π system to the other face is called an *antarafacial* rearrangement (Figure 30.13).

FIGURE 30.13 ▼

Suprafacial and
antarafacial sigmatropic
rearrangements.

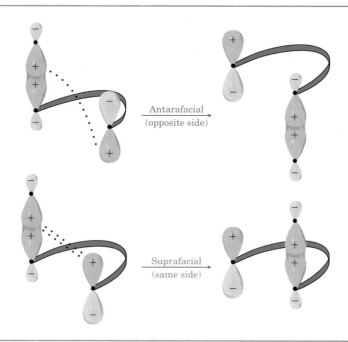

Both suprafacial and antarafacial sigmatropic rearrangements are symmetry-allowed, but suprafacial rearrangements are often easier for geometric reasons. The rules for sigmatropic rearrangements are identical to those for cycloaddition reactions (Table 30.3).

TABLE 30.3 Sterochemical Rules for Sigmatropic Rearrangements		
Electron pairs	**Thermal reaction**	**Photochemical reaction**
Even number	Antarafacial	Suprafacial
Odd number	Suprafacial	Antarafacial

Problem 30.7   Classify the following sigmatropic reaction by order [x,y], and tell whether it will proceed with suprafacial or antarafacial stereochemistry:

$$CH_2-H \rightleftharpoons CH_2-H$$

30.9 Some Examples of Sigmatropic Rearrangements

Since a [1,5] sigmatropic rearrangement involves three electron pairs (two π bonds and one σ bond), the orbital-symmetry rules in Table 30.3 predict a suprafacial reaction. In fact, the [1,5] suprafacial shift of a hydrogen atom across two double bonds of a π system is one of the most commonly observed of all sigmatropic rearrangements. For example, 5-methylcyclopentadiene rapidly rearranges at room temperature to yield a mixture of 1-methyl-, 2-methyl-, and 5-methyl-substituted products.

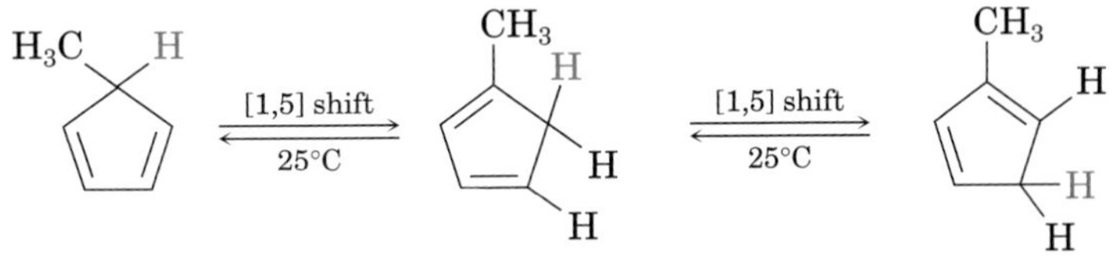

As another example, heating 5,5,5-trideuterio-(1,3Z)-pentadiene causes scrambling of deuterium between positions 1 and 5.

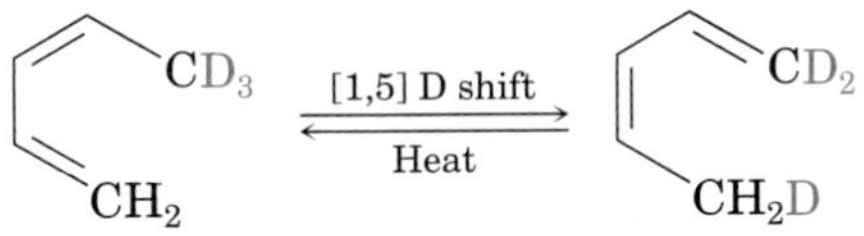

Both of these [1,5] hydrogen shifts occur by a symmetry-allowed suprafacial rearrangement, as illustrated in Figure 30.14.

FIGURE 30.14 ▼

An orbital view of a suprafacial [1,5] hydrogen shift.

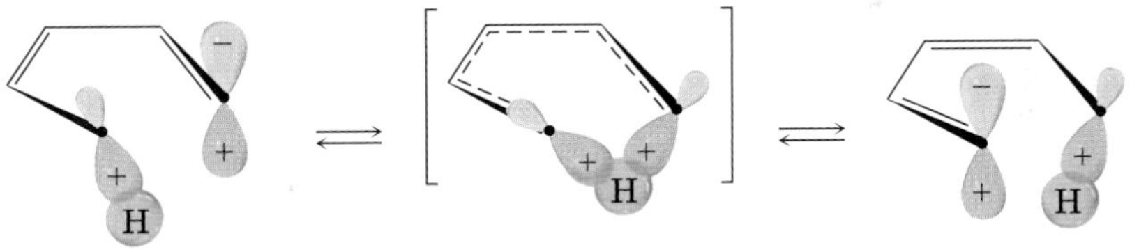

Transition state

In contrast to thermal [1,5] sigmatropic hydrogen shifts, thermal [1,3] hydrogen shifts are unknown. Were they to occur, they would have to proceed by a strained antarafacial reaction pathway.

Two other important sigmatropic reactions are the **Cope rearrangement** of a 1,5-hexadiene and the **Claisen rearrangement** of an allyl aryl ether (Section 18.6). These two, along with the Diels–Alder reaction, are the most useful pericyclic reactions for organic synthesis; many thousands of examples of all three are known. Note that the Claisen rearrangement works well with both allyl *aryl* ethers and with allyl *vinylic* ethers.

Claisen rearrangement

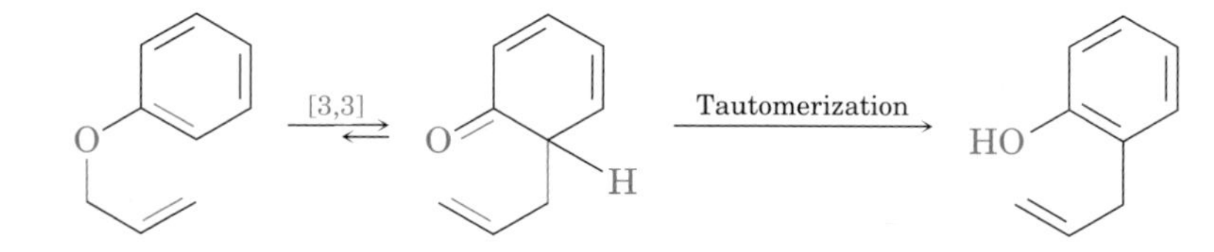

An allyl aryl ether   An o-allylphenol

Claisen rearrangement

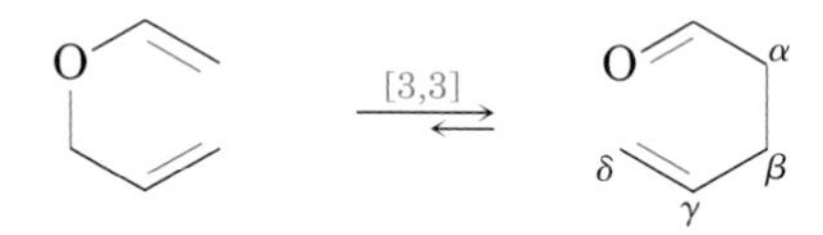

An allyl vinylic ether   A γ,δ-unsaturated
                              carbonyl compound

Cope rearrangement

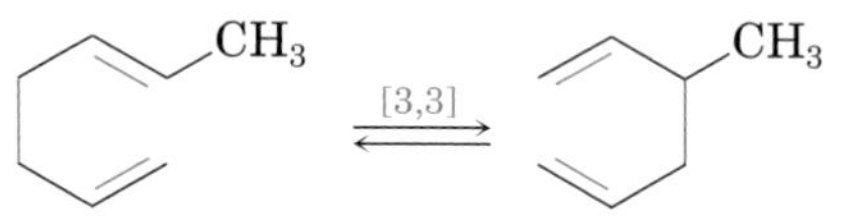

A 1,5-diene   A new 1,5-diene

Both Cope and Claisen rearrangements involve reorganization of an odd number of electron pairs (two π bonds and one σ bond), and both react by suprafacial pathways (Figure 30.15).

FIGURE 30.15 ▼

Suprafacial [3,3] Cope and Claisen rearrangements.

(a)

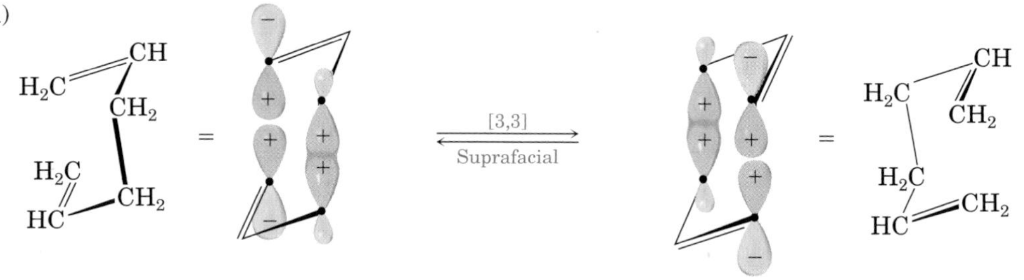

Cope rearrangement of a 1,5-hexadiene

(b)

Claisen rearrangement of an allyl vinyl ether

• •

Problem 30.8 Propose a mechanism to account for the fact that heating 1-deuterioindene scrambles the isotope label to all three positions on the five-membered ring.

Problem 30.9 When a 2,6-disubstituted allyl phenyl ether is heated in an attempted Claisen rearrangement, migration occurs to give the *p*-allyl product as the result of two sequential pericyclic reactions. Explain.

• •

30.10 A Summary of Rules for Pericyclic Reactions

How can you keep straight all the rules about pericyclic reactions? The summary information in Tables 30.1–30.3 can be distilled into one mnemonic phrase that provides an easy way to predict the stereochemical outcome of any pericyclic reaction:

The Electrons Circle Around (TECA)

Thermal reactions with an *Even* number of electron pairs are either *Conrotatory* or *Antarafacial.*

A change either from thermal to photochemical or from an even to an odd number of electron pairs changes the outcome from conrotatory/antarafacial to disrotatory/suprafacial. A change from both thermal and even to photochemical and odd causes no change, since two negatives make a positive.

These selection rules are summarized in Table 30.4, thereby giving you the ability to predict the stereochemistry of literally thousands of pericyclic reactions.

TABLE 30.4 Stereochemical Rules for Pericyclic Reactions		
Electron state	**Electron pairs**	**Stereochemistry**
Ground state (thermal)	Even number	Antara-con
	Odd number	Supra-dis
Excited state (photochemical)	Even number	Supra-dis
	Odd number	Antara-con

● ●

Problem 30.10   Predict the stereochemistry of the following pericyclic reactions:
(a) The thermal cyclization of a conjugated tetraene
(b) The photochemical cyclization of a conjugated tetraene
(c) A photochemical [4 + 4] cycloaddition
(d) A thermal [2 + 6] cycloaddition
(e) A photochemical [3,5] sigmatropic rearrangement

● ●

CHEMISTRY @ WORK

Vitamin D, the Sunshine Vitamin

Vitamin D, discovered in 1918, is a general name for two related compounds, *cholecalciferol* (vitamin D_3) and *ergocalciferol* (vitamin D_2). Both are steroids (Section 27.7) and differ only in the nature of the hydrocarbon side chain attached to the five-membered ring. Cholecalciferol comes from dairy products and fish; ergocalciferol comes from some vegetables. Their function in the body is to control the calcification of bones by increasing intestinal absorption of calcium. When sufficient vitamin D is present, approximately 30% of ingested calcium is absorbed, but in the absence of vitamin D, calcium absorption falls to about 10%. A deficiency of vitamin D thus leads to poor bone growth and to the childhood disease known as *rickets*.

Actually, neither vitamin D_2 nor D_3 is present in foods. Rather, foods contain the precursor molecules 7-dehydrocholesterol and ergosterol. In the presence of sunlight, however, both precursors are converted under the skin to the active vitamins, hence the nickname "sunshine vitamin."

(continued) ▶

hv → [1,7] H shift →

7-Dehydrocholesterol
Ergosterol

R = $CH(CH_3)CH_2CH_2CH_2CH(CH_3)_2$
R = $CH(CH_3)CH$=$CHCH(CH_3)CH(CH_3)_2$

Cholecalciferol
Ergocalciferol

Pericyclic reactions are unusual in living organisms, and the photo-chemical synthesis of vitamin D is one of the few well-studied examples. The reaction takes place in two steps: an electrocyclic ring opening of a cyclohexadiene to yield a hexatriene, followed by a sigmatropic [1,7] H shift to yield an isomeric hexatriene. Further metabolic processing in the liver and the kidney introduces several –OH groups to give the active form of the vitamin.

Synthesizing vitamin D takes dedication and hard work.

Summary and Key Words

A **pericyclic reaction** is one that takes place in a single step through a cyclic transition state without intermediates. There are three major classes of pericyclic processes: *electrocyclic reactions, cycloaddition reactions,* and *sigmatropic rearrangements.* The stereochemistry of these reactions is controlled by the symmetry of the orbitals involved in bond reorganization.

Electrocyclic reactions involve the cyclization of conjugated polyenes. For example, 1,3,5-hexatriene cyclizes to 1,3-cyclohexadiene on heating. Electrocyclic reactions can occur by either **conrotatory** or **disrotatory** paths, depending on the symmetry of the terminal lobes of the π system. Conrotatory cyclization requires that both lobes rotate in the same direction, whereas disrotatory cyclization requires that the lobes rotate in opposite directions. The reaction course in a specific case can be found by looking at the symmetry of the **highest occupied molecular orbital (HOMO)**.

Cycloaddition reactions are those in which two unsaturated molecules add together to yield a cyclic product. For example, Diels–Alder reaction between a diene (four π electrons) and a dienophile (two π electrons) yields a cyclohexene. Cycloadditions can take place either by **suprafacial** or **antarafacial** pathways. Suprafacial cycloaddition involves interaction between lobes on the same face of one component and on the same face of the second component. Antarafacial cycloaddition involves interaction between lobes on the same face of one component and on opposite faces of the other component. The reaction course in a specific case can be found by looking at the symmetry of the HOMO of one component and the **lowest unoccupied molecular orbital (LUMO)** of the other component.

Sigmatropic rearrangements involve the migration of a σ-bonded group across a π electron system. For example, **Claisen rearrangement** of an allyl vinylic ether yields an unsaturated carbonyl compound, and **Cope rearrangement** of a 1,5-hexadiene yields a new 1,5-hexadiene. Sigmatropic rearrangements can occur with either suprafacial or antarafacial stereochemistry; the selection rules for a given case are the same as those for cycloaddition reactions.

The stereochemistry of any pericyclic reaction can be predicted by counting the total number of electron pairs (bonds) involved in bond reorganization and then applying the mnemonic "The Electrons Circle Around." That is, **thermal** (ground-state) reactions involving an even number of electron pairs occur with either conrotatory or antarafacial stereochemistry. Exactly the opposite rules apply to **photochemical** (excited-state) reactions.

Visualizing Chemistry

● ●

(Problems 30.1–30.10 appear within the chapter.)

30.11   Predict the product obtained when the following substance is heated:

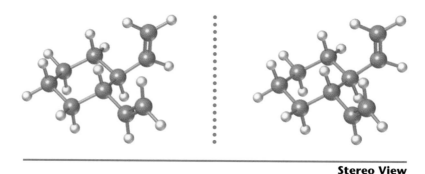

Stereo View

30.12 The ^{13}C NMR spectrum of homotropilidene taken at room temperature shows only three peaks. Explain.

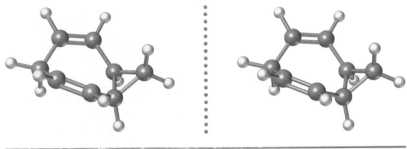

Stereo View

Additional Problems

30.13 Have the following reactions taken place in a conrotatory or disrotatory manner? Under what conditions, thermal or photochemical, would you carry out each reaction?

(a)

(b)

30.14 What stereochemistry—antarafacial or suprafacial—would you expect to observe in the following reactions?
(a) A photochemical [1,5] sigmatropic rearrangement
(b) A thermal [4 + 6] cycloaddition
(c) A thermal [1,7] sigmatropic rearrangement
(d) A photochemical [2 + 6] cycloaddition

30.15 The following thermal isomerization occurs under relatively mild conditions. Identify the pericyclic reactions involved, and show how the rearrangement occurs.

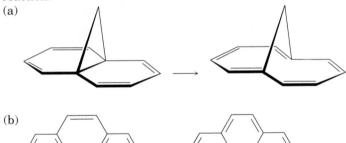

30.16 Would you expect the following reaction to proceed in a conrotatory or disrotatory manner? Show the stereochemistry of the cyclobutene product, and explain your answer.

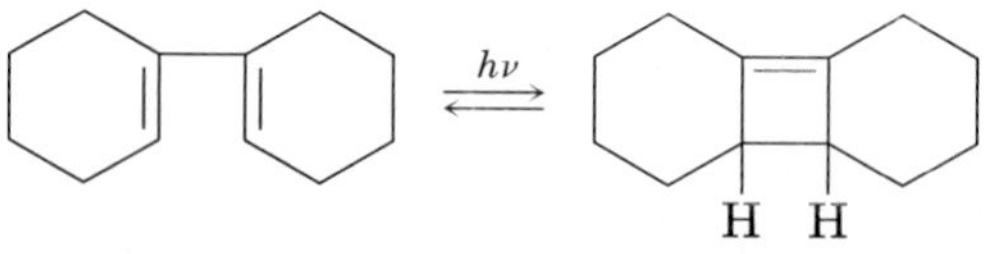

30.17 Heating (1Z,3Z,5Z)-cyclononatriene to 100°C causes cyclization and formation of a bicyclic product. Is the reaction conrotatory or disrotatory? What is the stereochemical relationship of the two hydrogens at the ring junctions, cis or trans?

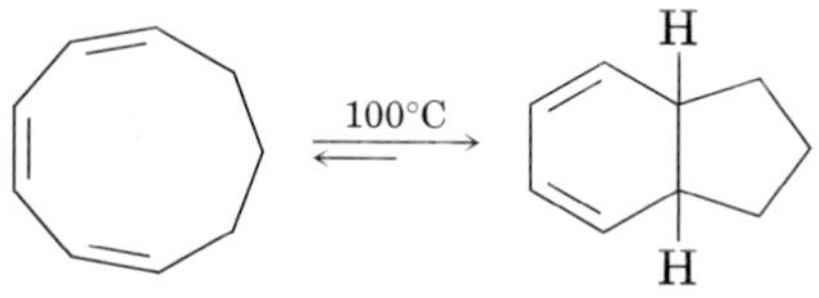

30.18 (2E,4Z,6Z,8E)-Decatetraene has been cyclized to give 7,8-dimethyl-1,3,5-cyclooctatriene. Predict the manner of ring closure—conrotatory or disrotatory—for both thermal and photochemical reactions, and predict the stereochemistry of the product in each case.

30.19 Answer Problem 30.18 for the thermal and photochemical cyclizations of the compound (2E,4Z,6Z,8Z)-decatetraene.

30.20 The cyclohexadecaoctaene shown isomerizes to two different isomers, depending on reaction conditions. Explain the observed results, and indicate whether each reaction is conrotatory or disrotatory.

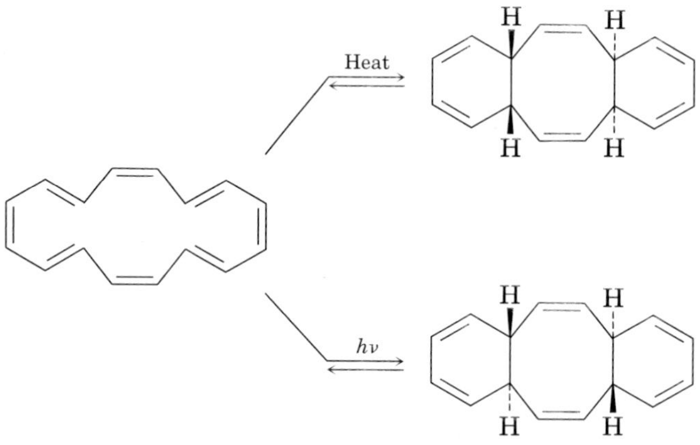

30.21 Which of the following reactions is more likely to occur? Explain.

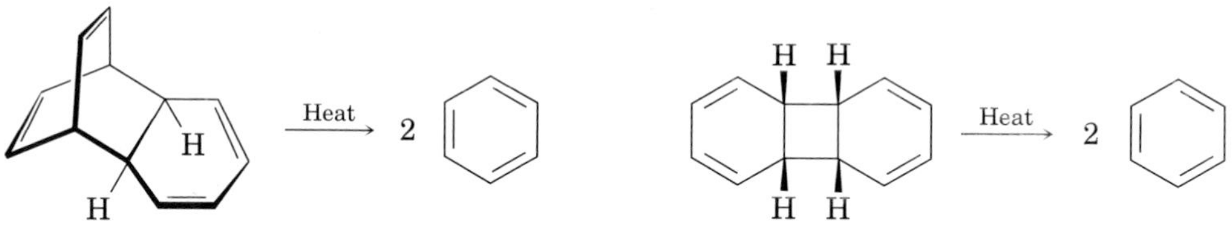

30.22 Bicyclohexadiene, also known as *Dewar benzene,* is extremely stable in spite of the fact that its rearrangement to benzene is energetically favored. Explain why the rearrangement is so slow.

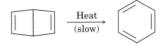

Dewar benzene   **Benzene**

30.23 The following thermal rearrangement involves two pericyclic reactions in sequence. Identify them, and propose a mechanism to account for the observed result.

30.24 Predict the product of the following pericyclic reaction. Is this [5,5] shift a supra-facial or an antarafacial process?

30.25 Ring opening of the *trans*-cyclobutene isomer shown takes place at much lower temperature than a similar ring opening of the *cis*-cyclobutene isomer. Explain the temperature effect, and identify the stereochemistry of each reaction as either conrotatory or disrotatory.

30.26 Photolysis of the *cis*-cyclobutene isomer in Problem 30.25 yields *cis*-cyclododecaen-7-yne, but photolysis of the trans isomer yields *trans*-cyclododecaen-7-yne. Explain these results, and identify the type and stereochemistry of the pericyclic reaction.

30.27 Vinyl-substituted cyclopropanes undergo thermal rearrangement to yield cyclopentenes. Propose a mechanism for the reaction, and identify the pericyclic process involved.

Vinylcyclopropane   **Cyclopentene**

30.28 The following reaction takes place in two steps, one of which is a cycloaddition and the other of which is a *reverse* cycloaddition. Identify the two pericyclic reactions, and show how they occur.

30.29 Two sequential pericyclic reactions are involved in the following furan synthesis. Identify them, and propose a mechanism for the transformation.

30.30 The following synthesis of dienones occurs readily. Propose a mechanism to account for the results, and identify the kind of pericyclic reaction involved.

30.31 Karahanaenone, a terpene isolated from oil of hops, has been synthesized by the thermal reaction shown. Identify the kind of pericyclic reaction, and explain how karahanaenone is formed.

Karahanaenone

30.32 The ^{1}H NMR spectrum of bullvalene at 100°C consists only of a single peak at 4.22 δ. Explain.

Bullvalene

30.33 The following rearrangement was devised and carried out to prove the stereo-chemistry of [1,5] sigmatropic hydrogen shifts. Explain how the observed result confirms the predictions of orbital symmetry.

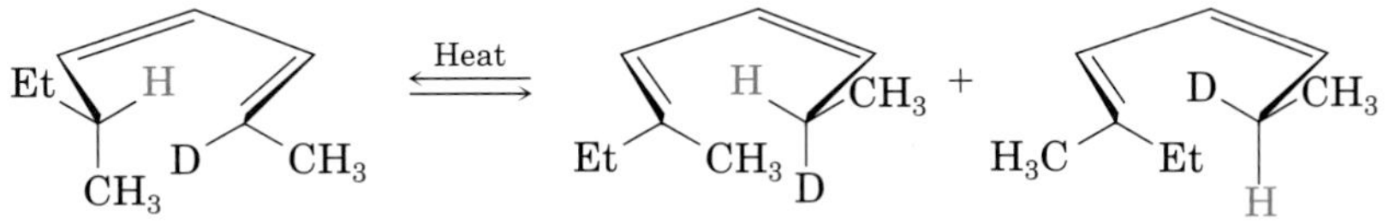

30.34 The following reaction is an example of a [2,3] sigmatropic rearrangement. Would you expect the reaction to be suprafacial or antarafacial? Explain.

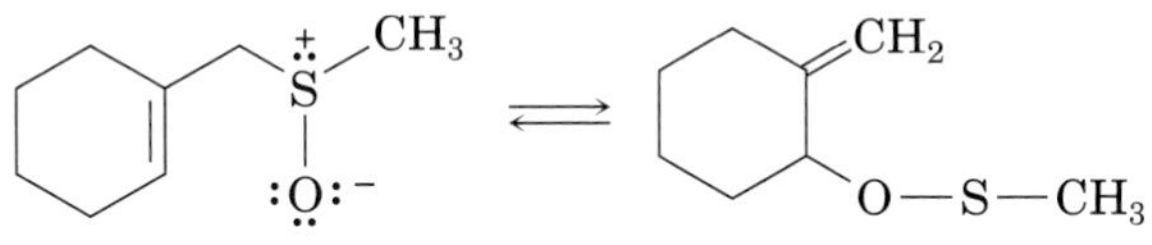

30.35 When the compound having a cyclobutene fused to a five-membered ring is heated, (1Z,3Z)-cycloheptadiene is formed. When the related compound having a cyclobutene fused to an eight-membered ring is heated, however, (1E,3Z)-cyclodecadiene is formed. Explain these results, and suggest a reason why opening of the eight-membered ring occurs at a lower temperature.

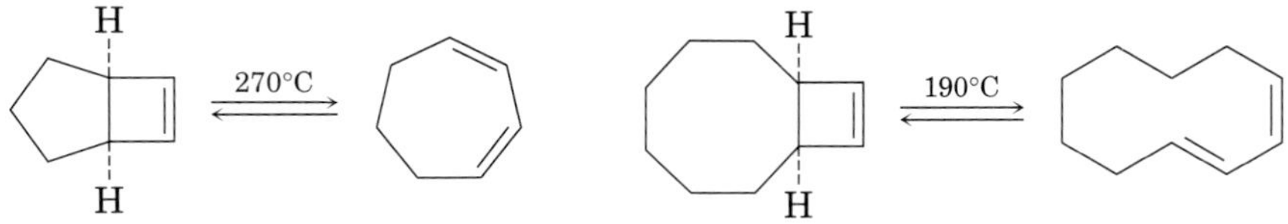

30.36 In light of your answer to Problem 30.35, explain why a mixture of products occurs in the following reaction:

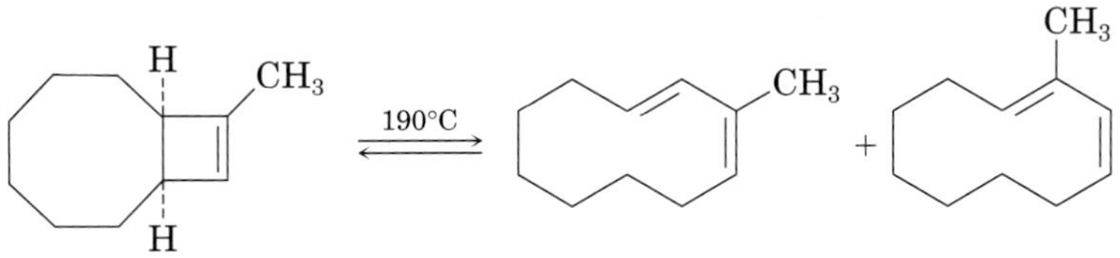

30.37 The sex hormone estrone has been synthesized by a route that involves the following step. Identify the pericyclic reactions involved, and propose a mechanism.

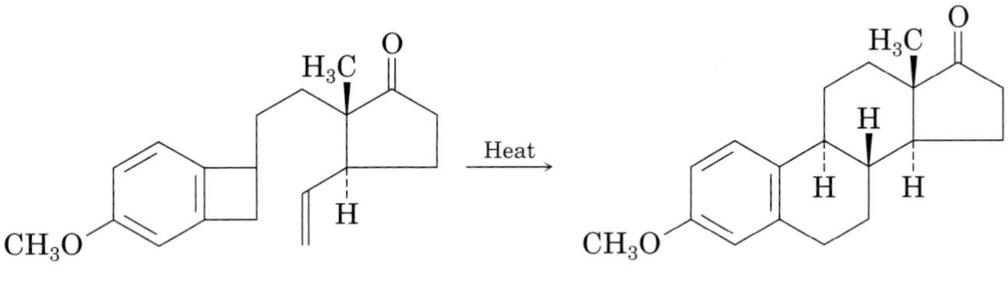

Estrone methyl ether

30.38 Coronafacic acid, a bacterial toxin, was synthesized using a key step that involves three sequential pericyclic reactions. Identify them, and propose a mechanism for the overall transformation. How would you complete the synthesis?

185°C

92%   **Coronafacic acid**

30.39 The following rearrangement of *N*-allyl-*N,N*-dimethylanilinium ion has been observed. Propose a mechanism for the reaction.

Heat

N-Allyl-*N,N*-dimethylanilinium ion   *o*-Allyl-*N,N*-dimethylanilinium ion

Molecular Modeling

30.40 Cyclobutenes undergo conrotatory ring opening to give 1,3-butadienes, but different products can be obtained depending on the direction of rotation. Use Spartan-View to compare the energies of transition states A and B for the following reaction when R = CH₃ and when R = CHO. Assuming kinetic control of the reaction, what is the major product in each case?

Heat

R   or   **A**   **B**

30.41 The transition state in a Diels–Alder reaction involves transfer of electrons from one molecule to the other. Using SpartanView to compare electrostatic potential maps of cyclopentadiene, tetracyanoethylene, and their Diels–Alder transition state, describe the direction of electron transfer.

30.42 The Diels–Alder reaction of 2-methoxy-1,3-butadiene and acrylonitrile can give either the 1,3- or the 1,4-disubstituted product. Use SpartanView to examine the HOMO surface of 2-methoxy-1,3-butadiene. Which terminal diene carbon has the larger HOMO lobe and is thus a better electron donor? Next, simultaneously display the density surface and LUMO surface of acrylonitrile, and look at how the LUMO extends beyond the density surface. Which dienophile carbon is a better electron acceptor? Which of the two products should form most rapidly?

30.43 Semibullvalene undergoes a degenerate Cope rearrangement to give a product that is identical to the reactant. Use SpartanView to obtain the energies of semibullvalene and its Cope transition state, and calculate the energy barrier for the reaction. How many different signals would you expect to see in the ^{13}C NMR spectrum of semibullvalene at room temperature? (A single signal is observed if the barrier for two carbons to exchange environments is less than 60 kJ/mol.)

Semibullvalene

31

Synthetic Polymers

A *polymer*, as we've seen, is a large molecule built up by repetitive bonding together of many smaller units, or *monomers*. Polyethylene, for instance, is a saturated polymer made from ethylene units (Section 7.10), rubber is an unsaturated polymer made from isoprene units (Section 14.7), and Dacron is a polyester made from acid and alcohol units (Section 21.10).

Polyethylene   **Rubber**   **Dacron**

Note that polymers are drawn by indicating their repeating unit in parentheses. The repeat unit in polystyrene, for example, comes from the monomer styrene.

Polystyrene   comes from   **Styrene**

Our treatment of polymers has thus far been dispersed over several chapters, but it's now time to take a more comprehensive view. In the present chapter, we'll look further at how polymers are made, and we'll see how polymer structure correlates with physical properties.

31.1 Chain-Growth Polymers

Synthetic polymers are classified by their method of synthesis as either *chain-growth* or *step-growth*. The categories are somewhat imprecise but nevertheless provide a useful distinction. **Chain-growth polymers** are produced by chain-reaction polymerization in which an initiator adds to a carbon–carbon double bond of an unsaturated substrate (a *vinyl monomer*) to yield a reactive intermediate. This intermediate reacts with a second molecule of monomer to yield a new intermediate, which reacts with a third monomer unit, and so on.

The initiator can be a radical, as in ethylene polymerization (Section 7.10), an acid, as in isobutylene polymerization (Section 7.10), or an anion. Radical polymerization is the most common and can be carried out with practically any vinyl monomer. Acid-catalyzed (cationic) polymerization, however, is effective only with vinyl monomers that contain an electron-donating group (EDG) capable of stabilizing the chain-carrying carbocation intermediate. Thus, isobutylene polymerizes rapidly under cationic conditions, but ethylene, vinyl chloride, and acrylonitrile do not. Isobutylene polymerization is carried out commercially at $-80°C$, using BF_3 and a small amount of water to generate $BF_3OH^-\ H^+$ catalyst.

where   BzO· = Benzoyloxy, $PhCO_2$·

where   EDG = an electron-donating group

Vinyl monomers with electron-withdrawing substituents (EWG) can be polymerized by anionic catalysts. The key chain-carrying step is nucleophilic addition of an anion to the unsaturated monomer by a Michael reaction (Section 23.11).

$$Nu\!:^- + H_2C=\!CH \longrightarrow \left[Nu-CH_2-CH\!:^- \xrightarrow{H_2C=CH} NuCH_2CH-CH_2CH\!:^- \right] \xrightarrow{Repeat} \left(CH_2-CH \right)_n$$

where   EWG = an electron-withdrawing group

Acrylonitrile ($H_2C=CHCN$), methyl methacrylate [$H_2C=C(CH_3)CO_2CH_3$], and styrene ($H_2C=CHC_6H_5$) can all be polymerized anionically. The polystyrene used in foam coffee cups, for example, is prepared by anionic polymerization of styrene using butyllithium as catalyst.

$$Bu\!:^- Li^+ \quad H_2C=CH \longrightarrow \left[Bu-CH_2-CH\!:^- \xrightarrow{H_2C=CH} \right] \xrightarrow{Repeat} \left(CH_2-CH \right)_n$$

Styrene   **Polystyrene**

One interesting example of anionic polymerization accounts for the remarkable properties of "super glue," one drop of which can support up to 2000 lb. Super glue is simply a solution of pure methyl α-cyanoacrylate. Since the carbon–carbon double bond has two electron-withdrawing groups, anionic addition is particularly easy. Trace amounts of water or bases on the surface of an object are sufficient to initiate polymerization of the cyanoacrylate and bind articles together. Skin is a good source of the necessary basic initiators, and many people have found their fingers stuck together after inadvertently touching super glue.

$$HO\!:^- + H_2C=C \overset{C\equiv N}{\underset{C-OCH_3}{\big|}} \longrightarrow \left[HO-CH_2-\underset{CO_2CH_3}{\overset{CN}{\underset{|}{\overset{|}{C}}}}\!:^- \right] \xrightarrow{Repeat} \left(CH_2-\underset{CO_2CH_3}{\overset{CN}{\underset{|}{\overset{|}{C}}}} \right)_n$$

Methyl α-cyanoacrylate
(Super glue)

• •

Problem 31.1   Order the following monomers with respect to their expected reactivity toward cationic polymerization, and explain your answer:

$$H_2C=CHCH_3, \quad H_2C=CHCl, \quad H_2C=CH–C_6H_5, \quad H_2C=CHCO_2CH_3$$

Problem 31.2   Order the following monomers with respect to their expected reactivity toward anionic polymerization, and explain your answer:

$$H_2C=CHCH_3, \quad H_2C=CHC\equiv N, \quad H_2C=CHC_6H_5$$

Problem 31.3   Polystyrene is produced commercially by reaction of styrene with butyllithium as an anionic initiator. Explain how the chain-carrying intermediate is stabilized.

•••••••••••••••••••••••••••••••••••••••

31.2 Stereochemistry of Polymerization: Ziegler–Natta Catalysts

The polymerization of a substituted vinyl monomer can lead to a polymer with numerous chirality centers on its chain. For example, propylene might polymerize with any of the three stereochemical outcomes shown in Figure 31.1. The product having all methyl groups on the same side of the zigzag backbone is called **isotactic**; that in which the methyl groups regularly alternate on opposite sides of the backbone is called **syndiotactic**; and the polymer having the methyl groups randomly oriented is called **atactic**.

FIGURE 31.1 ▼

Isotactic, syndiotactic, and atactic forms of polypropylene.

isotactic polypropylene, syndiotactic polypropylene, atactic polypropylene

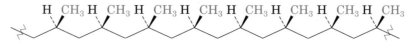

Isotactic (same side)

Syndiotactic (alternating sides)

Atactic (random)

The three different stereochemical forms of polypropylene all have somewhat different properties, and all three can be made by the proper choice of polymerization catalyst. Propylene polymerization using radical initiators

Giulio Natta

Giulio Natta (1903–1979) was born in Imperia, near Genoa, Italy, and received his Ph.D. in chemical engineering at Milan Polytechnic in 1924. After holding positions at the universities of Pavia, Rome, and Turin, he returned to Milan in 1938 as Professor of Industrial Chemistry. For his work on developing methods of polymer synthesis, he shared the 1963 Nobel Prize in chemistry with Karl Ziegler.

does not work well, but polymerization using *Ziegler–Natta catalysts* allows preparation of isotatic, syndiotactic, and atactic polypropylene.

Ziegler–Natta catalysts, which are of many different formulations, are organometallic transition-metal complexes prepared by treatment of an alkylaluminum with a titanium compound. Triethylaluminum and titanium tetrachloride form a typical preparation.

$$(CH_3CH_2)_3Al + TiCl_4 \rightarrow \text{A Ziegler–Natta catalyst}$$

Introduced in 1953, Ziegler–Natta catalysts immediately revolutionized the field of polymer chemistry, largely because of two advantages:

- Ziegler–Natta polymers are linear and have practically no chain branching.

- Ziegler–Natta polymers are stereochemically controllable. Isotactic, syndiotactic, and atactic forms can all be produced, depending on the catalyst system used.

The active form of a Ziegler–Natta catalyst is an alkyltitanium intermediate with a vacant coordination site on the metal. Coordination of alkene monomer to the titanium occurs, and the coordinated alkene then inserts into the carbon–titanium bond to extend the alkyl chain. Since a new coordination site opens up during the insertion step, the process repeats indefinitely.

The linear polyethylene produced by the Ziegler–Natta process (called *high-density polyethylene*) is a highly crystalline polymer with 4000–7000 ethylene units per chain and molecular weights in the range 100,000–200,000 amu. High-density polyethylene has greater strength and heat resistance than the branched product of radical-induced polymerization (*low-density polyethylene*) and is used to produce plastic squeeze bottles and molded housewares.

Polyethylenes of even higher molecular weights are produced for specialty applications using Ziegler–Natta catalysis. So-called high-molecular-weight (HMW) polyethylene contains 10,000–18,000 monomer units per chain (MW = 300,000–500,000 amu) and is used for pipes and large containers. Ultrahigh-molecular-weight (UHMW) polyethylene contains over 100,000 monomer units per chain and has molecular weights in the range 3,000,000–6,000,000 amu. It is used in bearings, conveyor belts, and other applications requiring unusual wear resistance.

• •

Problem 31.4   Vinylidene chloride, $H_2C=CCl_2$, does not polymerize in isotactic, syndiotactic, and atactic forms. Explain.

Problem 31.5 Polymers such as polypropylene contain a large number of chirality centers. Would you therefore expect samples of isotactic, syndiotactic, or atactic polypropylene to rotate plane-polarized light? Explain.

31.3 Copolymers

Up to this point we've discussed only **homopolymers**—polymers that are made up of *identical* repeating units. In practice, however, *copolymers* are more important commercially. **Copolymers** are obtained when two or more different monomers are allowed to polymerize together. For example, copolymerization of vinyl chloride with vinylidene chloride (1,1-dichloroethylene) in a 1:4 ratio leads to the well-known polymer Saran.

$$H_2C=\overset{\overset{\displaystyle Cl}{|}}{CH} + H_2C=CCl_2 \longrightarrow \left(CH_2\overset{\overset{\displaystyle Cl}{|}}{CH}\right)_m\left(CH_2\overset{\overset{\displaystyle Cl}{|}}{\underset{\underset{\displaystyle Cl}{|}}{C}}\right)_n$$

Vinyl chloride **Vinylidene chloride**   **Saran**

Copolymerization of monomer mixtures often leads to materials with properties quite different from those of either corresponding homopolymer, giving the polymer chemist a vast amount of flexibility for devising new materials. Table 31.1 lists some common copolymers and their commercial applications.

TABLE 31.1 Some Common Copolymers and Their Uses

Monomers	Structures	Trade name	Uses
Vinyl chloride (20%) Vinylidene chloride (80%)	$H_2C=CHCl$ $H_2C=CCl_2$	Saran	Food wrappings, fibers
Styrene (25%) 1,3-Butadiene (75%)	$H_2C=CHC_6H_5$ $H_2C=CHCH=CH_2$	SBR (styrene–butadiene rubber)	Tires, rubber articles
Hexafluoropropene Vinylidene fluoride	$F_2C=CFCF_3$ $H_2C=CF_2$	Viton	Gaskets, seals
Acrylonitrile 1,3-Butadiene	$H_2C=CHCN$ $H_2C=CHCH=CH_2$	Nitrile rubber	Adhesives, gasoline hoses
Isobutylene Isoprene	$H_2C=C(CH_3)_2$ $H_2C=C(CH_3)CH=CH_2$	Butyl rubber	Inner tubes
Acrylonitrile 1,3-Butadiene Styrene	$H_2C=CHCN$ $H_2C=CHCH=CH_2$ $H_2C=CHC_6H_5$	ABS (initials of monomers)	Pipes, high-impact applications

Several different types of copolymers can be defined, depending on the distribution of monomer units in the chain. If monomer A is copolymerized with monomer B, for instance, the resultant product might have a random distribution of the two units throughout the chain, or it might have an alternating distribution:

$$+A—A—A—B—A—B—B—A—B—A—A—A—B—B—B+$$

Random copolymer

$$+A—B—A—B—A—B—A—B—A—B—A—B—A—B—A+$$

Alternating copolymer

The exact distribution of monomer units depends on the initial proportions of the two reactant monomers and their relative reactivities. In practice, neither perfectly random nor perfectly alternating copolymers are usually found. Most copolymers have many random imperfections.

Two other forms of copolymers that can be prepared under certain conditions are called *block copolymers* and *graft copolymers*. **Block copolymers** are those in which different blocks of identical monomer units alternate with each other; **graft copolymers** are those in which homopolymer branches of one monomer unit are "grafted" onto a homopolymer chain of another monomer unit.

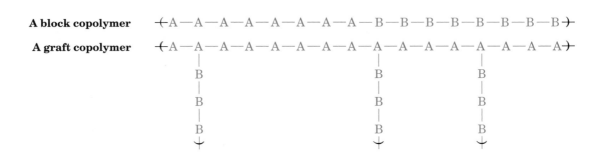

Block copolymers are prepared by initiating the polymerization of one monomer as if growing a homopolymer chain, and then adding an excess of the second monomer to the still-active reaction mix. Graft copolymers are made by gamma irradiation of a completed homopolymer chain in the presence of the second monomer. The high-energy irradiation knocks hydrogen atoms off the homopolymer chain at random points, thus generating radical sites that can initiate polymerization of the added monomer.

• •

Problem 31.6   Draw the structure of an alternating segment of butyl rubber, a copolymer of isoprene and isobutylene prepared using a cationic initiator.

Problem 31.7   Irradiation of poly(1,3-butadiene), followed by addition of styrene, yields a graft
copolymer that is used to make rubber soles for shoes. Draw the structure of a representative segment of this styrene–butadiene graft copolymer.

• •

31.4 Step-Growth Polymers

Step-growth polymers are produced by reactions in which each bond in
the polymer is formed independently of the others. The nylons (polyamides)
and polyesters that we saw in Section 21.10 are examples.

Most step-growth polymers are produced by reaction between two
difunctional reactants. Nylon 66, for instance, is made by reaction between
the 6-carbon adipic acid and the 6-carbon hexamethylenediamine. Alternatively, a single reactant with two different functional groups can polymerize. Nylon 6, a close relative of nylon 66, is made by polymerization of caprolactam. The reaction is initiated by addition of a small amount of water,
which hydrolyzes some caprolactam to 6-aminohexanoic acid. Nucleophilic
attack of the amino group on caprolactam then propagates the polymerization.

$$\underset{\substack{\text{Adipic acid} \\ \text{(Hexanedioic acid)}}}{\text{HOCCH}_2\text{CH}_2\text{CH}_2\text{CH}_2\text{COH}} + \underset{\substack{\text{Hexamethylenediamine} \\ \text{(1,6-Hexanediamine)}}}{\text{H}_2\text{NCH}_2\text{CH}_2\text{CH}_2\text{CH}_2\text{CH}_2\text{CH}_2\text{NH}_2}$$

Heat

$$\left(\text{CCH}_2\text{CH}_2\text{CH}_2\text{CH}_2\text{C}-\text{NHCH}_2\text{CH}_2\text{CH}_2\text{CH}_2\text{CH}_2\text{CH}_2\text{NH}\right)_n$$

Nylon 66

Caprolactam $\xrightarrow[\text{Heat}]{\text{H}_2\text{O}}$ $\left[\text{HOCCH}_2\text{CH}_2\text{CH}_2\text{CH}_2\text{CH}_2\ddot{\text{N}}\text{H}_2\right]$ $\xrightarrow[\text{Heat}]{}$ $\left(\text{CCH}_2\text{CH}_2\text{CH}_2\text{CH}_2\text{CH}_2\text{NH}\right)_n$

Caprolactam   **6-Aminohexanoic acid**   **Nylon 6**

Polycarbonates

Polycarbonates are like polyesters, but their carbonyl group is linked to two –OR groups, [O=C(OR)$_2$]. Lexan, for instance, is a polycarbonate prepared from diphenyl carbonate and a diphenol called *bisphenol A*. Lexan has an unusually high impact strength, making it valuable for use in machinery housings, telephones, bicycle safety helmets, and "bullet-proof" glass.

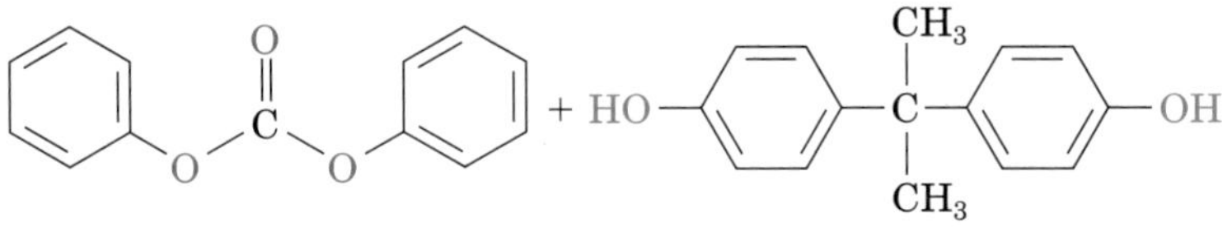

Diphenyl carbonate   **Bisphenol A**

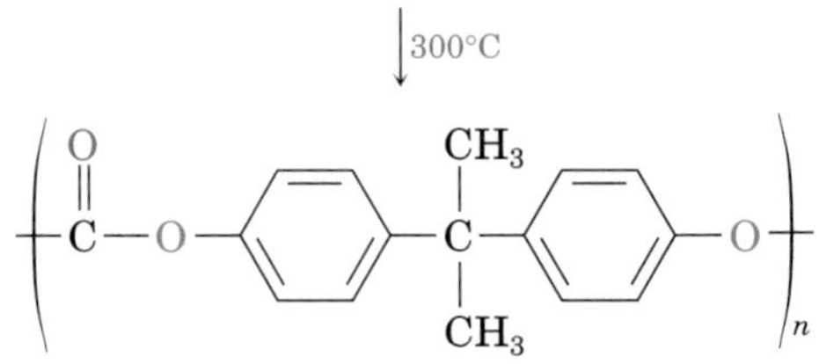

Lexan

Polyurethanes

A *urethane* is a carbonyl-containing functional group in which the carbonyl carbon is bonded to both an –OR group and an –NR$_2$ group. As such, a urethane is halfway between a carbonate and a urea:

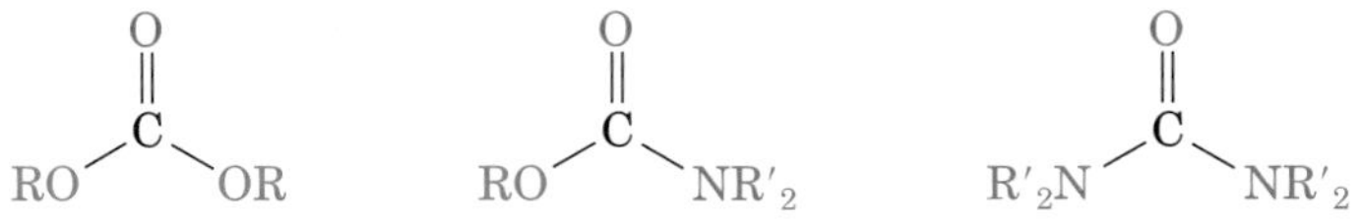

A carbonate   **A urethane**   **A urea**

Urethanes are typically prepared by nucleophilic addition reaction between an alcohol and an isocyanate (R–N=C=O), so a *polyurethane* is prepared by reaction between a diol and a diisocyanate. The diol is usually a low-molecular-weight polymer (MW ≈ 1000 amu) with hydroxyl end groups; the diisocyanate is often toluene-2,4-diisocyanate.

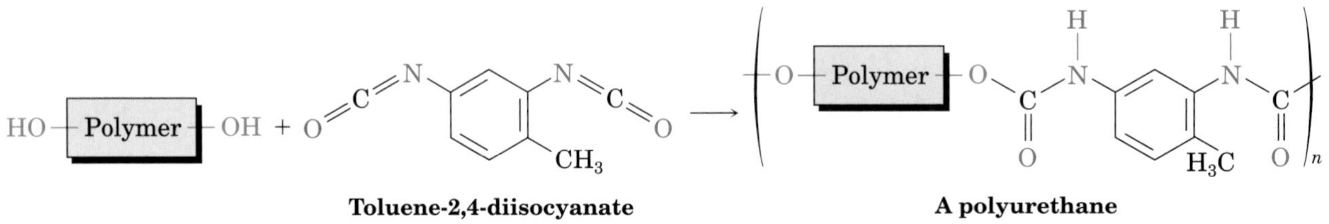

Toluene-2,4-diisocyanate   **A polyurethane**

Several different kinds of polyurethanes are produced, depending on the nature of the polymeric alcohol used. One major use of polyurethane is in the stretchable spandex fibers used for bathing suits and leotards. These polyurethanes have a fairly low degree of cross-linking so that the resultant polymer is soft and elastic. A second major use of polyurethanes is in the foams used for insulation. Foaming occurs when a small amount of water is added during polymerization, giving a carbamic acid intermediate that spontaneously loses bubbles of CO_2.

A carbamic acid

Polyurethane foams are generally made using a *poly*alcohol rather than a diol as the monomer so that the polymer has a high amount of three-dimensional cross-linking. The result is a rigid but very light foam suitable for use as thermal insulation in building construction and in portable ice chests.

· ·

Problem 31.8   Poly(ethylene terephthalate), or PET, is a polyester used to make soft-drink bottles. It is prepared by reaction of ethylene glycol with 1,4-benzenedicarboxylic acid (terephthalic acid). Draw the structure of PET.

Problem 31.9   Show the mechanism of the nucleophilic addition reaction of an alcohol with an isocyanate to yield a urethane.

· ·

31.5 Polymer Structure and Physical Properties

Polymers aren't really that different from other organic molecules. They're much larger, of course, but their chemistry is similar to that of analogous small molecules. Thus, the alkane chains of polyethylene undergo radical-initiated halogenation; the aromatic rings of polystyrene undergo typical electrophilic aromatic substitution reactions; and the amide linkages of a nylon are hydrolyzed by base.

The major difference between small and large organic molecules is in their physical properties. For instance, their large size means that polymers experience substantially larger van der Waals forces than do small molecules (Section 3.5). But because van der Waals forces operate only at close distances, they are strongest in polymers like high-density polyethylene in which chains can pack together closely in a regular way. Many polymers, in fact, have regions that are essentially crystalline. These regions, called **crystallites**, consist of highly ordered portions in which the zigzag polymer chains are held together by van der Waals forces (Figure 31.2).

FIGURE 31.2 ▼

Crystallites in linear polyethylene. The long polymer chains are arranged in parallel lines in the crystallite regions.

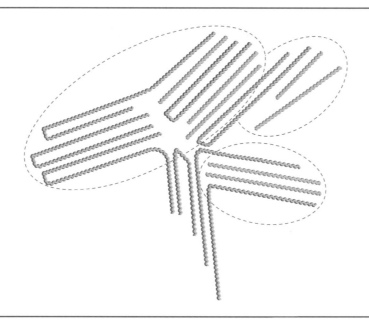

As you might expect, polymer crystallinity is strongly affected by the steric requirements of substituent groups on the chains. Linear polyethylene is highly crystalline, but poly(methyl methacrylate) is noncrystalline because the chains can't pack closely together in a regular way. Polymers with a high degree of crystallinity are generally hard and durable. When heated, the crystalline regions melt at the *melt transition temperature, T_m,* to give an amorphous material.

Noncrystalline, amorphous polymers like poly(methyl methacrylate), sold under the trade name Plexiglas, have little or no long-range ordering among chains, but can nevertheless be very hard at room temperature. When heated, the hard amorphous polymer becomes soft and flexible at a point called the *glass transition temperature, T_g.* Much of the art in polymer synthesis is in finding methods for controlling the degree of crystallinity and the glass transition temperature, thereby imparting useful properties to the polymer.

In general, polymers can be divided into four major categories, depending on their physical behavior: *thermoplastics, fibers, elastomers,* and *thermosetting resins.* **Thermoplastics** are the polymers most people think of when the word *plastic* is mentioned. These polymers have a high T_g and are therefore hard at room temperature, but they become soft and viscous when heated. As a result, they can be molded into toys, beads, telephone housings, or into any of a thousand other items. Because thermoplastics have little or no cross-linking, the individual chains can slip past one another in the melt. Some thermoplastic polymers, such as polystyrene and poly(methyl methacrylate), are amorphous and noncrystalline; others, such as polyethylene and nylon, are partially crystalline. Among the better-known thermoplastics is poly(ethylene terephthalate), or PET, used for making plastic soft-drink bottles.

Poly(ethylene terephthalate)

Plasticizers—small organic molecules that act as lubricants between chains—are usually added to thermoplastics to keep them from becoming brittle at room temperature. A good example is poly(vinyl chloride), which is brittle when pure but becomes supple and pliable when a plasticizer is added. In fact, the drip bags used in hospitals to deliver intravenous solutions are made of poly(vinyl chloride). Dialkyl phthalates such as di(2-ethylhexyl) phthalate are commonly used for this purpose, although questions about their safety have recently been raised.

Di(2-ethylhexyl) phthalate, a plasticizer

Fibers are thin threads produced by extruding a molten polymer through small holes in a die, or *spinneret*. The fibers are then cooled and drawn out, which orients the crystallite regions along the axis of the fiber and adds considerable tensile strength (Figure 31.3). Nylon, Dacron, and polyethylene all have the semicrystalline structure necessary for drawing into oriented fibers.

FIGURE 31.3 ▼

Oriented crystallite regions in a polymer fiber.

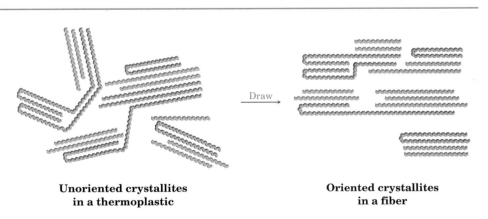

Draw

Unoriented crystallites in a thermoplastic

Oriented crystallites in a fiber

Elastomers are amorphous polymers that have the ability to stretch out and spring back to their original shapes. These polymers must have low T_g values and a small amount of cross-linking to prevent the chains from slipping over one another. In addition, the chains must have an irregular shape to prevent crystallite formation. When stretched, the randomly coiled chains straighten out and orient along the direction of the pull. Van der Waals forces are too weak and too few to maintain this orientation, however, and the elastomer therefore reverts to its random coiled state when the stretching force is released (Figure 31.4).

FIGURE 31.4 ▼

Unstretched and stretched forms of an elastomer.

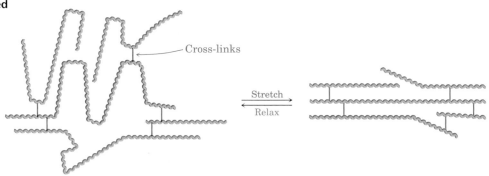

Natural rubber (Section 14.7) is the most common example of an elastomer. Rubber has the long chains and occasional cross-links needed for elasticity, but its irregular geometry prevents close packing of the chains into crystallites. Gutta-percha, by contrast, is highly crystalline and is not an elastomer (Figure 31.5).

FIGURE 31.5 ▼

(a) Natural rubber is elastic and noncrystalline because of its cis double-bond geometry, but (b) gutta-percha is nonelastic and crystalline because its geometry allows for better packing together of chains.

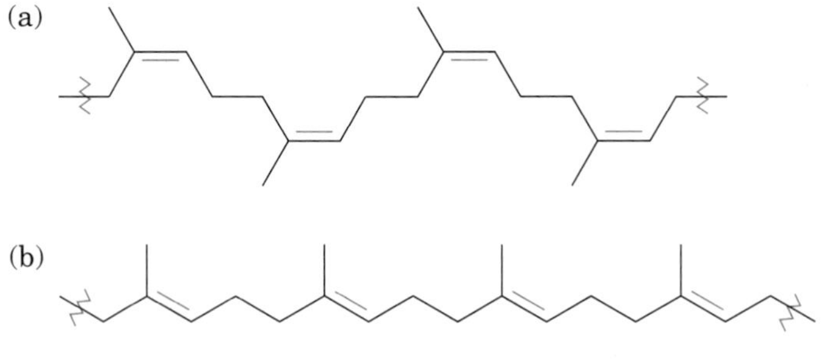

(a)

(b)

Thermosetting resins are polymers that become highly cross-linked and solidify into a hard, insoluble mass when heated. *Bakelite*, a thermosetting resin first produced in 1907, has been in commercial use longer than any other synthetic polymer. It is widely used for molded parts, adhesives, coatings, and even high-temperature applications such as missile nose cones.

Chemically, Bakelite is a *phenolic resin*, produced by reaction of phenol and formaldehyde. On heating, water is eliminated, many cross-links form, and the polymer sets into a rock-like mass. The cross-linking in Bakelite and other thermosetting resins is three-dimensional and is so extensive that we can't really speak of polymer "chains." A piece of Bakelite is essentially one large molecule.

Bakelite

∙ ∙

Problem 31.10   What product would you expect to obtain from catalytic hydrogenation of natural rubber? Would the product be syndiotactic, atactic, or isotactic?

Problem 31.11   Propose a mechanism to account for the formation of Bakelite from acid-catalyzed polymerization of phenol and formaldehyde.

∙ ∙

CHEMISTRY @ WORK

Biodegradable Polymers

The high chemical stability of many polymers is both a blessing and a curse. Heat resistance, wear resistance, and long life are valuable characteristics of clothing fibers, bicycle helmets, underground pipes, food wrappers, and many other items. Yet when those items outlive their usefulness, disposal can become a problem.

(continued) ▶

Recycling of unwanted polymers is the best solution, and six types of plastics in common use are frequently stamped with identifying codes assigned by the Society of the Plastics Industry (Table 31.2). After sorting by type, the items to be recycled are shredded into small chips, washed, dried, and melted for reuse. Soft-drink bottles, for instance, are made from recycled poly(ethylene terephthalate), trash bags are made from recycled low-density polyethylene, and garden furniture is made from recycled polypropylene and mixed plastics.

TABLE 31.2 Recyclable Plastics

Polymer	Recycling code	Use
Poly(ethylene terephthalate)	1–PET	Soft-drink bottles
High-density polyethylene	2–HDPE	Bottles
Poly(vinyl chloride)	3–V	Floor mats
Low-density polyethylene	4–DPE	Grocery bags
Polypropylene	5–PP	Furniture
Polystyrene	6–PS	Molded articles
Mixed plastics	7	Benches, plastic lumber

Frequently, however, plastics are simply thrown away rather than recycled, and much work has therefore been carried out on developing *biodegradable* polymers, which can be broken down rapidly by soil microorganisms. Among the most common biodegradable polymers are polyglycolic acid (PGA), polylactic acid (PLA), and polyhydroxybutyrate (PHB). All are polyesters and are therefore susceptible toward hydrolysis of their ester links. Copolymers of PGA with PLA have found a particularly wide range of uses. A 90/10 copolymer of polyglycolic acid with polylactic acid is used to make absorbable sutures, for instance. The sutures are entirely degraded and absorbed by the body within 90 days after surgery.

What happens to the plastics that end up here?

(continued) ▶

$$HOCH_2COH$$ (O double bond on C)

$$HOCHCOH$$ with CH_3

$$HOCHCH_2COH$$ with CH_3

Heat → Heat → Heat

$$\left(OCH_2C \right)_n$$

$$\left(OCHC \right)_n$$ with CH_3

$$\left(OCHCH_2C \right)_n$$ with CH_3

Polyglycolic acid   **Polylactic acid**   **Polyhydroxybutyrate**

In Europe, interest has centered particularly on polyhydroxybutyrate, which can be made into films for packaging as well as into molded items. The polymer degrades within 4 weeks in landfills, both by ester hydrolysis and by an elimination reaction of the oxygen atom β to the carbonyl group. This elimination is similar to what occurs during the dehydration step in an aldol reaction (Section 23.4). The use of polyhydroxybutyrate is limited at present, however, by its cost—about four times as much as polypropylene.

Summary and Key Words

KEY WORDS

Synthetic polymers can be classified as either chain-growth polymers or step-growth polymers. **Chain-growth polymers** are prepared by chain-reaction polymerization of *vinyl monomers* in the presence of a radical, an anion, or a cation initiator. Radical polymerization is the most commonly used method, but alkenes such as 2-methylpropene that have electron-donating substituents on the double bond polymerize easily by a cationic route. Similarly, monomers such as methyl α-cyanoacrylate that have electron-withdrawing substituents on the double bond polymerize by an anionic (Michael reaction) pathway.

Copolymerization of two monomers gives a product with properties different from those of either homopolymer. **Graft copolymers** and **block copolymers** are two examples.

Alkene polymerization can be carried out in a controlled manner using a **Ziegler–Natta catalyst**. Ziegler–Natta polymerization minimizes the amount of chain branching in the polymer and leads to stereoregular chains—either **isotactic** (substituents on the same side of the chain) or **syndiotactic** (substituents on alternate sides of the chain), rather than **atactic** (substituents randomly disposed).

Step-growth polymers, the second major class of polymers, are prepared by reactions between difunctional molecules; the individual bonds in the polymer are formed independently of one another. *Polycarbonates* are formed from a diester and a diol, and *polyurethanes* are formed from a diisocyanate and a diol.

The chemistry of synthetic polymers is similar to the chemistry of small molecules with the same functional groups, but the physical properties of polymers are greatly affected by size. Polymers can be classified by physical property into four groups: **thermoplastics**, **fibers**, **elastomers**, and **thermosetting resins**. The properties of each group can be accounted for by the structure, the degree of crystallinity, and the amount of crosslinking they contain.

Visualizing Chemistry

(Problems 31.1–31.11 appear within the chapter.)

31.12   Identify the structural class to which the following polymer belongs, and show the structure of the monomer units used to make it:

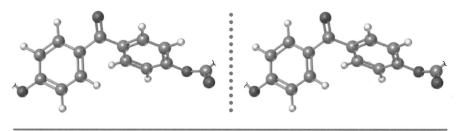

Stereo View

31.13   Show the structures of the polymers that could be made from the following monomers (yellow-green = Cl):

(a)   (b)

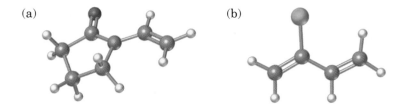

Additional Problems

• •

31.14 Identify the monomer units from which each of the following polymers is made, and tell whether each is a chain-growth or a step-growth polymer.

(a) $+CH_2-O\frac{}{n}$ (b) $+CF_2-CFCl\frac{}{n}$ (c) $\left(NHCH_2CH_2CH_2\overset{O}{\underset{\|}{C}}\right)_n$

(d) $\left(O-\bigcirc-\overset{O}{\underset{\|}{C}}\right)_n$ (e) $\left(O-\bigcirc-O-\overset{O}{\underset{\|}{C}}\right)_n$

31.15 Draw a three-dimensional representation of segments of the following polymers: (a) Syndiotactic polyacrylonitrile (b) Atactic poly(methyl methacrylate) (c) Isotactic poly(vinyl chloride)

31.16 Draw the structure of *Kodel*, a polyester prepared by heating dimethyl 1,4-benzene-dicarboxylate with 1,4-bis(hydroxymethyl)cyclohexane.

$HOCH_2-\bigcirc-CH_2OH$ **1,4-Bis(hydroxymethyl)cyclohexane**

31.17 Show the structure of the polymer that results from heating the following diepoxide and diamine:

$+ H_2N-\bigcirc-NH_2 \xrightarrow{\text{Heat}} ?$

31.18 *Nomex,* a polyamide used for such applications as high-performance tires, is prepared by reaction of 1,3-benzenediamine with 1,3-benzenedicarbonyl chloride. Show the structure of Nomex.

31.19 Nylon 10,10 is an extremely tough, strong polymer used to make reinforcing rods for concrete. Draw a segment of nylon 10,10, and show its monomer units.

31.20 Cyclopentadiene undergoes thermal polymerization to yield a polymer that has no double bonds in the chain. On strong heating, the polymer breaks down to regenerate cyclopentadiene. Propose a structure for the polymer.

31.21 When styrene, $C_6H_5CH=CH_2$, is copolymerized in the presence of a few percent *p*-divinylbenzene, a hard, insoluble, cross-linked polymer is obtained. Show how this cross-linking of polystyrene chains occurs.

31.22 Poly(ethylene glycol), or Carbowax, is made by anionic polymerization of ethylene oxide using NaOH as catalyst. Propose a mechanism for the reaction.

$+O-CH_2CH_2\frac{}{n}$ **Poly(ethylene glycol)**

31.23 Nitroethylene, $H_2C=CHNO_2$, is a sensitive compound that must be prepared with great care. Attempted purification of nitroethylene by distillation often results in low recovery of product and a white coating on the inner walls of the distillation apparatus. Explain.

31.24 Poly(vinyl butyral) is used as the plastic laminate in the preparation of automobile windshield safety glass. How would you synthesize this polymer?

Poly(vinyl butyral)

31.25 What is the structure of the polymer produced by anionic polymerization of β-propiolactone using NaOH as catalyst?

β-Propiolactone

31.26 *Glyptal* is a highly cross-linked thermosetting resin produced by heating glycerol and phthalic anhydride (1,2-benzenedicarboxylic acid anhydride). Show the structure of a representative segment of glyptal.

31.27 *Melmac*, a thermosetting resin often used to make plastic dishes, is prepared by heating melamine with formaldehyde. Look at the structure of Bakelite shown in Section 31.5, and then propose a structure for Melmac.

Melamine

31.28 Epoxy adhesives are cross-linked resins prepared in two steps. The first step involves S_N2 reaction of the disodium salt of bisphenol A with epichlorohydrin to form a low-molecular-weight prepolymer. This prepolymer is then "cured" into a cross-linked resin by treatment with a triamine such as $H_2NCH_2CH_2NHCH_2CH_2NH_2$.

Bisphenol A

Epichlorohydrin

(a) What is the structure of the prepolymer?
(b) How does addition of the triamine to the prepolymer result in cross-linking?

31.29 The polyurethane foam used for home insulation uses methanediphenyldiisocyanate (MDI) as monomer. The MDI is prepared by acid-catalyzed reaction of aniline with formaldehyde, followed by treatment with phosgene, $COCl_2$. Propose mechanisms for both steps.

MDI

31.30 Write the structure of a representative segment of polyurethane prepared by reaction of ethylene glycol with MDI (Problem 31.29).

31.31 The smoking salons of the Hindenburg and other hydrogen-filled dirigibles of the 1930s were insulated with urea–formaldehyde polymer foams. The structure of this polymer is highly cross-linked, like that of Bakelite (Section 31.5). Propose a structure.

31.32 The polymeric resin used for Merrifield solid-phase peptide synthesis (Section 26.11) is prepared by treating polystyrene with *N*-(hydroxymethyl)phthalimide and trifluoromethanesulfonic acid, followed by reaction with hydrazine. Propose a mechanism for both steps.

31.33 2-Ethyl-1-hexanol, used in the synthesis of di(2-ethylhexyl) phthalate plasticizer, is made commercially from butanal. Show the likely synthesis route.

Molecular Modeling

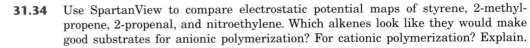

31.34   Use SpartanView to compare electrostatic potential maps of styrene, 2-methyl-propene, 2-propenal, and nitroethylene. Which alkenes look like they would make good substrates for anionic polymerization? For cationic polymerization? Explain.

31.35   Use SpartanView to compare electrostatic potential maps of styrene + hydride anion, 2-vinylpyridine + hydride anion, and 3-vinylfuran + hydride anion. Are either of the two heterocycles as effective as styrene at delocalizing the developing negative charge during anionic polymerization? Next, compare electrostatic potential maps of neutral styrene, 2-vinylpyridine, and 3-vinylfuran. Why don't the heterocyclic alkenes lend themselves to cationic polymerization?

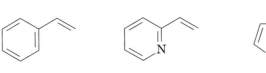

Styrene   **2-Vinylpyridine**   **3-Vinylfuran**

31.36   Use SpartanView to compare chain-growth polymers A–E. Identify the monomers found in each, and tell whether each is a homopolymer or a copolymer. If a copolymer, identify it as either random, alternating, or block.

APPENDIX A
Nomenclature of Polyfunctional Organic Compounds

Judging from the number of incorrect names that appear in the chemical literature, it's probably safe to say that relatively few practicing organic chemists are fully conversant with the rules of organic nomenclature. Simple hydrocarbons and monofunctional compounds present few difficulties because the basic rules for naming such compounds are logical and easy to understand. Problems, however, are often encountered with polyfunctional compounds. Whereas most chemists could correctly identify hydrocarbon **1** as 3-ethyl-2,5-dimethylheptane, rather few could correctly identify polyfunctional compound **2**. Should we consider **2** as an ether? As an ethyl ester? As a ketone? As an alkene? It is, of course, all four, but it has only one correct name: ethyl 3-(4-methoxy-2-oxo-3-cyclohexenyl)propanoate.

CH₃CHCH₃
CH₃CH₂CHCH₂CHCH₂CH₃
CH₃

1. 3-Ethyl-2,5-dimethylheptane

2. Ethyl 3-(4-methoxy-2-oxo-3-cyclohexenyl)propanoate

Naming polyfunctional organic compounds isn't really much harder than naming monofunctional ones. All that's required is a knowledge of nomenclature for monofunctional compounds and a set of additional rules. In the following discussion, it's assumed that you have a good command of the rules of nomenclature for monofunctional compounds that were given throughout the text as each new functional group was introduced. A list of where these rules can be found is shown in Table A.1.

TABLE A.1 Where to Find Nomenclature Rules for Simple Functional Groups

Functional group	Text section	Functional group	Text section
Acid anhydrides	21.1	Amines	24.1
Acid halides	21.1	Aromatic compounds	15.2
Alcohols	17.1	Carboxylic acids	20.1
Aldehydes	19.1	Cycloalkanes	3.7
Alkanes	3.4	Esters	21.1
Alkenes	6.3	Ethers	18.1
Alkyl halides	10.1	Ketones	19.1
Alkynes	8.2	Nitriles	21.1
Amides	21.1	Phenols	17.1

The name of a polyfunctional organic molecule has four parts:

1. **Suffix**—the part that identifies the principal functional-group class to which the molecule belongs.
2. **Parent**—the part that identifies the size of the main chain or ring.
3. **Substituent prefixes**—parts that identify what substituents are located on the main chain or ring.
4. **Locants**—numbers that tell where substituents are located on the main chain or ring.

To arrive at the correct name for a complex molecule, you must identify the four name parts and then express them in the proper order and format. Let's look at the four parts.

The Suffix—Functional-Group Precedence

A polyfunctional organic molecule can contain many different kinds of functional groups, but for nomenclature purposes, we must choose just one suffix. It's not correct to use two suffixes. Thus, keto ester **3** must be named either as a ketone with an -*one* suffix or as an ester with an -*oate* suffix but can't be named as an -*onoate*. Similarly, amino alcohol **4** must be named either as an alcohol (-*ol*) or as an amine (-*amine*) but can't properly be named as an -*olamine*. The only exception to this rule is in naming compounds that have double or triple bonds. For example, the unsaturated acid $H_2C=CHCH_2COOH$ is 3-butenoic acid, and the acetylenic alcohol $HC\equiv CCH_2CH_2CH_2OH$ is 5-hexyn-1-ol.

$$\underset{CH_3\overset{O}{\overset{\|}{C}}CH_2CH_2\overset{O}{\overset{\|}{C}}OCH_3}{}$$

3. Named as an ester with a keto (oxo) substituent: methyl 4-oxopentanoate

$$\underset{CH_3\overset{OH}{\overset{|}{C}}HCH_2CH_2CH_2NH_2}{}$$

4. Named as an alcohol with an amino substituent: 5-amino-2-pentanol

How do we choose which suffix to use? Functional groups are divided into two classes, **principal groups** and **subordinate groups**, as shown in Table A.2. Principal groups are those that may be cited either as prefixes or as suffixes, whereas subordinate groups are those that may be cited only as prefixes. Within the principal groups, an order of precedence has been established. The proper suffix for a given compound is determined by identifying all of the functional groups present and then choosing the principal group of highest priority. For example, Table A.2 indicates that keto ester **3** must be named as an ester rather than as a ketone, since an ester functional group is higher in priority than a ketone is. Similarly, amino alcohol **4** must be named as an alcohol rather than as an amine. The correct name of **3** is methyl 4-oxopentanoate, and the correct name of **4** is 5-amino-2-pentanol. Further examples are shown below and at the top of p. A-4.

5. Named as a cyclohexanecarboxylic acid with an oxo substituent: 4-oxocyclohexanecarboxylic acid

TABLE A.2 Classification of Functional Groups for Purposes of Nomenclature[a]

Functional group	Name as suffix	Name as prefix
Principal groups		
Carboxylic acids	-oic acid -carboxylic acid	carboxy
Acid anhydrides	-oic anhydride -carboxylic anhydride	
Esters	-oate -carboxylate	alkoxycarbonyl
Acid halides	-oyl halide -carbonyl halide	halocarbonyl
Amides	-amide -carboxamide	amido
Nitriles	-nitrile -carbonitrile	cyano
Aldehydes	-al -carbaldehyde	oxo
Ketones	-one	oxo
Alcohols	-ol	hydroxy
Phenols	-ol	hydroxy
Thiols	-thiol	mercapto
Amines	-amine	amino
Imines	-imine	imino
Alkenes	-ene	alkenyl
Alkynes	-yne	alkynyl
Alkanes	-ane	alkyl
Subordinate groups		
Azides		azido
Diazo		diazo
Ethers		alkoxy
Halides		halo
Nitro		nitro
Sulfides		alkylthio

[a]Principal functional groups are listed in order of decreasing priority; subordinate functional groups have no established priority order.

$$\underset{\overset{|}{CH_3}}{\overset{\overset{O}{\parallel}}{HOC}} - \underset{\overset{|}{CH_3}}{\overset{CH_3}{\underset{|}{C}}} - CH_2CH_2CH_2\overset{\overset{O}{\parallel}}{C}Cl$$

6. Named as a carboxylic acid with a chlorocarbonyl substituent:
5-chlorocarbonyl-2,2-dimethylpentanoic acid

$$CH_3\overset{\overset{CHO}{|}}{C}HCH_2CH_2CH_2\overset{\overset{O}{\parallel}}{C}OCH_3$$

7. Named as an ester with an oxo substituent:
methyl **5-methyl-6-oxohexanoate**

The Parent—Selecting the Main Chain or Ring

The parent or base name of a polyfunctional organic compound is usually easy to identify. If the group of highest priority is *part of* an open chain, we simply select the longest chain that contains the largest number of principal functional groups. If the highest-priority group is *attached to* a ring, we use the name of that ring system as the parent. For example, compounds **8** and **9** are isomeric aldehydo acids, and both must be named as acids rather than as aldehydes according to Table A.2. The longest chain in compound **8** has seven carbons, and the substance is therefore named 6-methyl-7-oxoheptanoic acid. Compound **9** also has a chain of seven carbons, but the longest chain that contains both of the principal functional groups has only three carbons. The correct name of **9** is 3-oxo-2-pentylpropanoic acid.

8. Named as a substituted heptanoic acid:
6-methyl-7-oxoheptanoic acid

9. Named as a substituted propanoic acid:
3-oxo-2-pentylpropanoic acid

Similar rules apply for compounds **10–13**, which contain rings. Compounds **10** and **11** are isomeric keto nitriles, and both must be named as nitriles according to Table A.2. Substance **10** is named as a benzonitrile since the –CN functional group is a substituent on the aromatic ring, but substance **11** is named as an acetonitrile since the –CN functional group is part of an open chain. The correct names are 2-acetyl-4-methylbenzonitrile (**10**) and (2-acetyl-4-methylphenyl)acetonitrile (**11**). Compounds **12** and **13** are both keto acids and must be named as acids. The correct names are 3-(2-oxocyclohexyl)propanoic acid (**12**) and 2-(3-oxopropyl)cyclohexane-carboxylic acid (**13**).

10. Named as a substituted benzonitrile:
2-acetyl-4-methylbenzonitrile

11. Named as a substituted acetonitrile:
(2-acetyl-4-methylphenyl)acetonitrile

12. Named as a carboxylic acid:
3-(2-oxocyclohexyl)propanoic acid

13. Named as a carboxylic acid:
2-(3-oxopropyl)cyclohexanecarboxylic acid

The Prefixes and Locants

With the suffix and parent name established, the next step is to identify and number all substituents on the parent chain or ring. These substituents include all alkyl groups and all functional groups other than the one cited in the suffix. For example, compound **14** contains three different functional groups (carboxyl, keto, and double bond). Because the carboxyl group is highest in priority, and because the longest chain containing the functional groups is seven carbons long, **14** is a heptenoic acid. In addition, the main chain has an oxo (keto) substituent and three methyl groups. Numbering from the end nearer the highest-priority functional group, we find that **14** is 2,5,5-trimethyl-4-oxo-2-heptenoic acid. Note that the final *-e* of heptene is deleted in the word *heptenoic*. This deletion occurs only when the name would have two adjacent vowels (thus, *heptenoic* has the final *e* deleted, but *heptenenitrile* retains the *-e*). Look back at some of the other compounds we've named to see other examples of how prefixes and locants are assigned.

14. Named as a heptenoic acid:
2,5,5-trimethyl-4-oxo-2-heptenoic acid

Writing the Name

Once the name parts have been established, the entire name is written out. Several additional rules apply:

RULE 1   **Order of prefixes.** When the substituents have been identified, the main chain has been numbered, and the proper multipliers such as *di-* and *tri-* have been assigned, the name is written with the substituents listed in alphabetical, rather than numerical, order. Multipliers such as *di-* and *tri-* are not used for alphabetization purposes, but the prefix *iso-* is used.

$$CH_3$$
$$|$$
$$H_2NCH_2CH_2CHCHCH_3$$
$$|$$
$$OH$$

15. 5-Amino-3-methyl-2-pentanol
(*NOT* 3-methyl-5-amino-2-pentanol)

RULE 2   **Use of hyphens; single- and multiple-word names.** The general rule in such cases is to determine whether the principal functional group is itself an element or compound. If it is, then the name is written as a single word; if it isn't, then the name is written as multiple words. For example, methylbenzene (one word) is correct because the parent—benzene—is itself a compound. Diethyl ether, however, is written as two words because the parent—ether—is a class name rather than a compound name. Some further examples are shown below.

$$H_3C — Mg — CH_3$$

16. Dimethylmagnesium
(one word, since magnesium is an element)

$$O$$
$$||$$
$$CH_3CHCOH$$
$$|$$
$$Br$$

17. 2-Bromopropanoic acid
(two words, since "acid" is not a compound)

18. 4-(Dimethylamino)pyridine
(one word, since pyridine is a compound)

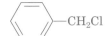

19. Methyl cyclopentanecarboxylate

RULE 3   **Parentheses.** Parentheses are used to denote complex substituents when ambiguity would otherwise arise. For example, chloromethylbenzene has two substituents on a benzene ring, but (chloromethyl)benzene has only one complex substituent. Note that the expression in parentheses is not set off by hyphens from the rest of the name.

20. *p*-Chloromethylbenzene
(two substituents)

21. (Chloromethyl)benzene
(one complex substituent)

$$
\overset{\text{O}}{\overset{\|}{\text{HOCCHCH}_2\text{CH}_2\text{COH}}} \quad \overset{\text{O}}{\overset{\|}{}}
$$

$$
\underset{\text{CH}_3\text{CHCH}_2\text{CH}_3}{|}
$$

22. 2-(1-Methylpropyl)pentanedioic acid
(The 1-methylpropyl group is a complex
substituent on C2 of the main chain.)

Additional Reading

Further explanations of the rules of organic nomenclature can be found in the following references:

1. "A Guide to IUPAC Nomenclature of Organic Compounds," CRC Press, Boca Raton, FL, 1993.

2. "Nomenclature of Organic Chemistry, Sections A, B, C, D, E, F, and H," International Union of Pure and Applied Chemistry, Pergamon Press, Oxford, 1979.

APPENDIX B
Acidity Constants for Some Organic Compounds

Compound	pK_a	Compound	pK_a	Compound	pK_a
CH_3SO_3H	−1.8	CH_2ICOOH	3.2		4.5
$CH(NO_2)_3$	0.1	$CHOCOOH$	3.2		
	0.3		3.4		
			3.5	$H_2C=C(CH_3)COOH$	4.7
				CH_3COOH	4.8
CCl_3COOH	0.5	$HSCH_2COOH$	3.5; 10.2	CH_3CH_2COOH	4.8
CF_3COOH	0.5	$CH_2(NO_2)_2$	3.6	$(CH_3)_3CCOOH$	5.0
CBr_3COOH	0.7	CH_3OCH_2COOH	3.6	$CH_3COCH_2NO_2$	5.1
$HOOCC≡CCOOH$	1.2; 2.5	CH_3COCH_2COOH	3.6		
$HOOCCOOH$	1.2; 3.7	$HOCH_2COOH$	3.7		5.3
$CHCl_2COOH$	1.3	$HCOOH$	3.7		
$CH_2(NO_2)COOH$	1.3				
$HC≡CCOOH$	1.9		3.8	$O_2NCH_2COOCH_3$	5.8
Z HOOCCH=CHCOOH	1.9; 6.3				
	2.4		4.0		5.8
$CH_3COCOOH$	2.4				
$NCCH_2COOH$	2.5	CH_2BrCH_2COOH	4.0		6.2
$CH_3C≡CCOOH$	2.6				
CH_2FCOOH	2.7				
$CH_2ClCOOH$	2.8		4.1		
$HOOCCH_2COOH$	2.8; 5.6				
$CH_2BrCOOH$	2.9		4.2		6.6
	3.0			HCO_3H	7.1
	3.0	$H_2C=CHCOOH$	4.2		7.2
		$HOOCCH_2CH_2COOH$	4.2; 5.7		
		$HOOCCH_2CH_2CH_2COOH$	4.3; 5.4		

Compound	pKa	Compound	pKa	Compound	pKa
$(CH_3)_2CHNO_2$	7.7	![phenol]OH	9.9	CH_3COCH_2Br	16.1
![2,4-dichlorophenol] Cl—OH, Cl	7.8	$CH_3COCH_2SOCH_3$	10.0	![cyclohexanone] =O	16.7
CH_3CO_3H	8.2	![o-cresol] OH, CH_3	10.3	CH_3CHO	17
![o-chlorophenol] OH, Cl	8.5	CH_3NO_2	10.3	$(CH_3)_2CHCHO$	17
		CH_3SH	10.3	$(CH_3)_2CHOH$	17.1
		$CH_3COCH_2COOCH_3$	10.6	$(CH_3)_3COH$	18.0
$CH_3CH_2NO_2$	8.5	CH_3COCHO	11.0	CH_3COCH_3	19.3
![p-CF3 phenol] F_3C—OH	8.7	$CH_2(CN)_2$	11.2	![fluorene]	23
		CCl_3CH_2OH	12.2		
		Glucose	12.3	$CH_3COOCH_2CH_3$	25
$CH_3COCH_2COCH_3$	9.0	$(CH_3)_2C=NOH$	12.4	$HC≡CH$	25
![resorcinol] HO—OH	9.3; 11.1	$CH_2(COOCH_3)_2$	12.9	CH_3CN	25
		$CHCl_2CH_2OH$	12.9	$CH_3SO_2CH_3$	28
		$CH_2(OH)_2$	13.3	$(C_6H_5)_3CH$	32
![catechol] OH, OH	9.3; 12.6	$HOCH_2CH(OH)CH_2OH$	14.1	$(C_6H_5)_2CH_2$	34
		CH_2ClCH_2OH	14.3	CH_3SOCH_3	35
![benzyl mercaptan] CH_2SH	9.4	![cyclopentadiene]	15.0	NH_3	36
				$CH_3CH_2NH_2$	36
				$(CH_3CH_2)_2NH$	40
		![benzyl alcohol] —CH_2OH	15.4	![toluene] CH_3	41
![hydroquinone] OH, HO	9.9; 11.5	CH_3OH	15.5	![benzene]	43
		$H_2C=CHCH_2OH$	15.5		
		CH_3CH_2OH	16.0	$H_2C=CH_2$	44
		$CH_3CH_2CH_2OH$	16.1	CH_4	~60

An acidity list covering more than 5000 organic compounds has been published: E.P. Serjeant and B. Dempsey (eds.), "Ionization Constants of Organic Acids in Aqueous Solution," IUPAC Chemical Data Series No. 23, Pergamon Press, Oxford, 1979.

APPENDIX C
How to Use
SpartanBuild

SpartanBuild, a computer program for building molecular models on Windows and Power Macintosh computers, can be found on the CD that is included with this book. It is intended to be used in conjunction with the Molecular Modeling problems found at the end of each chapter, as indicated by the icon shown in the margin.

SpartanBuild shares many features with SpartanView, and the following instructions assume that you already know how to use SpartanView (see "How to Use SpartanView and Interpret Molecular Modeling Data"). For example, you install, start, and quit both programs in exactly the same way. Likewise, both programs are CD-protected. (The copy and sale of both programs are prohibited by their license agreements.)

Build a Model Using Atoms

SpartanBuild models can be assembled in several ways. The simplest method is to build a model by starting with one atom and then adding additional atoms, one at a time, as needed. For example, 2-chloroacetaldehyde can be built from Cl, sp^3 C, sp^2 C, and sp^2 O in the following way:

Cl → Add sp^3 C → Cl–C → Add sp^2 C → Cl–C–C → Add sp^2 O →

$$\underset{\text{2-Chloroacetaldehyde}}{Cl-\overset{H\;H}{\underset{}{C}}-\overset{O}{\underset{}{C}}-H}$$

Note that each atom is introduced with a particular assortment of "dangling bonds" or unfilled valences. These are used to build new chemical bonds. For example, in the above scheme, chlorine is introduced as an atom with one unfilled valence. When sp^3 carbon, an atom with four unfilled valences, is added to the molecule, the new chlorine–carbon bond uses up one unfilled valence on each atom. Thus, the new structure has no unfilled valences on chlorine and only three unfilled valences on carbon. Unused valences are automatically converted into hydrogen atoms.

Computer Instruction	Comments
Start SpartanBuild	Starting the program opens a large SpartanBuild window (blank initially), a model kit, and a tool bar.

Start building 2-chloroacetaldehyde

Step 1. Click on –Cl in the model kit. | The –Cl button becomes highlighted.
Step 2. Click anywhere in the SpartanBuild window. | A chlorine atom with one unfilled valence (white) appears in the window as a Ball and Wire model.

If you make a mistake at any point, you can either undo the last operation by selecting Undo from the Edit menu, or you can start over by selecting Clear from the Edit menu.

In order to finish building 2-chloroacetaldehyde, you need to add two carbons and one oxygen. Look at the model kit and identify all the carbon and oxygen atoms that are available. There are five different carbon atoms and three different oxygen atoms. Each atom is defined by a unique combination of unfilled valences and a particular ideal geometry for these valences (see the following table).

Atom Button	$\backslash$C— $/$	$\backslash$C= $/$	—C≡	$\backslash$C— $/$	$\backslash$C— $/$
Atom Label	sp^3 C	sp^2 C	sp C	Delocalized C	Trigonal C
Unfilled Valences	4 single	2 single 1 double	1 single 1 triple	1 single 2 partial double	3 single
Ideal Bond Angles	109.5°	120°	180°	120°	120°

Computer Instruction | **Comments**

Finish building 2-chloroacetaldehyde

Step 1. Click on sp^3 C in the model kit. | This selects the carbon atom with four single unfilled valences.

Step 2. Click on the tip of chlorine's unfilled valence. | This makes a chlorine–carbon single bond (the new bond appears as a dashed line).

Step 3. Click on sp^2 C in the model kit. | This selects the carbon atom with two single and one double unfilled valences.

Step 4. Click on the tip of any one of carbon's unfilled valences. | This makes a carbon–carbon single bond. Bonds can only be made between unfilled valences of the same type (single + single, double + double, etc., are allowed, but single + double is not).

Step 5. Click on sp^2 O in the model kit. | This selects the oxygen atom with one double unfilled valence.

Step 6. Click on the tip of carbon's double unfilled valence. | This makes a carbon–oxygen double bond and completes the model. *Note:* If you cannot see which of carbon's unfilled valences is the double-bond valence, then rotate the molecule with the mouse first.

Change Model Display, Move and Scale Model, Measure Geometry

SpartanBuild adopts many of the same procedures used by SpartanView. For example, commands for changing the model display are located on the Model menu, commands for measuring geometry are located on the Geometry menu, and models are moved and scaled using the same combination of mouse and keyboard commands used in SpartanView.

SpartanBuild also provides a second method for measuring model geometry. Clicking on the appropriate tool bar button performs the same function as items located on the Geometry menu (see table).

Geometry Menu	PC	Mac
Distance		
Angle		
Dihedral		

Build a Model Using Groups

Although it is always possible to build models using atoms, this process can become tedious for large models. SpartanBuild simplifies the construction of large models by providing Groups of atoms that can be combined with atoms and with each other. As an example, 2-nitroacetic acid contains seven atoms (not counting hydrogens), but it can be constructed in just three steps using the Nitro and Carboxylic Acid groups.

2-Nitroacetic acid

Computer Instruction	Comments

Select Clear from the Edit menu

This removes the existing model from the SpartanBuild window.

Build 2-Nitroacetic Acid

Step 1. Click on sp^3 C in the model kit; then click in the SpartanBuild window.

Step 2. Click on the Groups button in the model kit.

Step 3. Select Nitro from the Groups menu. — The Nitro group appears in the model kit.

Step 4. Click on the tip of any one of carbon's unfilled valences. — This adds the entire Nitro group to the model

Step 5. Select Carboxylic Acid from the Groups menu. — This group will appear in the model kit.

Step 6. Examine the unfilled valences of the Carboxylic Acid group, and find the one marked by a small circle. If necessary, click on the group to make this circle move to the unfilled valence on carbon. — This group has two structurally distinct unfilled valences that can be used to connect it to the model. The "active" unfilled valence is indicated by a small circle, and this site can be changed by clicking anywhere on the group's structure in the model kit.

Step 7. Click on the tip of any one of carbon's unfilled valences. — A new carbon–carbon bond forms and the entire Carboxylic Acid group is added to the model.

Build a Model Using Rings

Models that contain one or more rings, such as phenylcyclohexane, are most easily built using SpartanBuild's Rings.

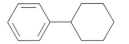 **Phenylcyclohexane**

Computer Instruction	Comments

Select Clear from the Edit menu

This removes the existing model from the SpartanBuild window.

Build Phenylcyclohexane

Step 1. Click on the Rings button.

Step 2. Select Benzene from the Rings menu. — The Benzene ring appears in the model kit.

Step 3. Click anywhere in the SpartanBuild window. — This places an entire benzene ring in the window.

Step 4. Select Cyclohexane from the Rings menu. — This ring will appear in the model kit.

Step 5. Examine the label that appears with the Cyclohexane ring. If necessary, click on the ring to make this label "eq." — This ring has two structurally distinct unfilled valences that can be used to connect the ring to the model. The "active" unfilled valence is indicated by a small label: "eq" (equatorial) or "ax" (axial). The label can be changed by clicking anywhere on the formula.

Step 6. Click on the tip of any one of the benzene ring's unfilled valences. — A new carbon–carbon bond forms, and an entire cyclohexane ring is added to the model.

Additional Tools

Some models require special techniques (or are more easily built) using some of the SpartanBuild tools described below.

Tool	PC	MAC	Use	Example
Make Bond			Click on two unfilled valences. Unfilled valences are replaced by a bond.	
Break Bond			Click on bond. Bond is replaced by two unfilled valences.	
Delete			Click on atom or unfilled valence. Deleting an atom removes all unfilled valences associated with that atom.	
Internal Rotation			Click on bond to select it for rotation. Simultaneously hold down Alt key (PC) or space bar (Mac) and the left mouse button while moving the mouse. One part of the model will rotate about the selected bond relative to the other part.	
Atom Replacement			Select atom from model kit; then double-click on the atom in the model. The valences of the atom in the model kit and the bonds in the model must match, or replacement will not occur.	

Minimize: Generating Realistic Structures and Strain Energy

SpartanBuild's building procedures typically lead to distorted and unrealistic structures. This is mainly because of the stepwise process by which a model takes shape. For example, using Make Bond to change ethylbenzene into benzocyclobutene gives a four-membered ring with an unrealistic C–C bond distance (see below). This absurd structure can be improved by clicking on Minimize in the model kit. (Minimize can also be performed by clicking on the bottom tool bar button.) Minimize starts a structure–energy, or molecular mechanics, calculation that finds the least strained geometry for any model.

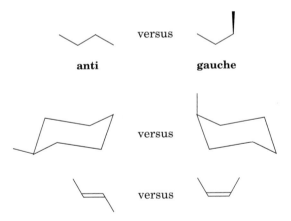

The strain energy values that SpartanBuild calculates have another use besides structure refinement. They also can be used to compare the energies of models that share the same molecular formula—that is, stereoisomers or conformational isomers. Allowed comparisons are shown below. Strain energy differences between these pairs of molecules correspond closely to differences in heat of formation and to differences in free energy. SpartanBuild reports strain energies in kcal/mol (1 kcal/mol = 4.184 kJ/mol) in the lower left-hand corner of the SpartanBuild window.

anti   versus   gauche

versus

versus

Since it is usually difficult to anticipate which models are distorted, we recommend that Minimize be applied routinely to all completed models.

APPENDIX D
Glossary

Absolute configuration (Section 9.6): The exact three-dimensional structure of a chiral molecule. Absolute configurations are specified verbally by the Cahn–Ingold–Prelog R,S convention and are represented on paper by Fischer projections.

Absorbance (Section 14.11): In optical spectroscopy, the logarithm of the intensity of the incident light divided by the intensity of the light transmitted through a sample; $A = \log I_0/I$.

Absorption spectrum (Section 12.5): A plot of wavelength of incident light versus amount of light absorbed. Organic molecules show absorption spectra in both the infrared and the ultraviolet regions of the electromagnetic spectrum.

Acetal (Section 19.11): A functional group consisting of two –OR groups bonded to the same carbon. Acetals are often used as protecting groups for ketones and aldehydes.

Acetoacetic ester synthesis (Section 22.8): The synthesis of a methyl ketone by alkylation of an alkyl halide, followed by hydrolysis and decarboxylation.

Acetyl group (Section 19.1): The CH_3CO- group.

Acetylide anion (Section 8.8): The anion formed by removal of a proton from a terminal alkyne.

Achiral (Section 9.2): Having a lack of handedness. A molecule is achiral if it has a plane of symmetry and is thus superimposable on its mirror image.

Acidity constant, K_a (Section 2.8): A measure of acid strength. For any acid HA, the acidity constant is given by the expression

$$K_a = K_{eq}[H_2O] = \frac{[H_3O^+][A^-]}{[HA]}$$

Activating group (Section 16.5): An electron-donating group such as hydroxyl (–OH) or amino (–NH$_2$) that increases the reactivity of an aromatic ring toward electrophilic aromatic substitution.

Activation energy (Section 5.9): The difference in energy between ground state and transition state in a reaction. The amount of activation energy determines the rate at which the reaction proceeds. Most organic reactions have activation energies of 40–100 kJ/mol.

Acyl group (Sections 16.4, 19.1): A –COR group.

Acylation (Section 16.4): The introduction of an acyl group, –COR, onto a molecule. For example, acylation of an alcohol yields an ester, acylation of an amine yields an amide, and acylation of an aromatic ring yields an alkyl aryl ketone.

Acylium ion (Section 16.4): A resonance-stabilized carbocation in which the positive charge is located at a carbonyl-group carbon, $R-C^+=O \longleftrightarrow R-C\equiv O^+$. Acylium ions are strongly electrophilic and are involved as intermediates in Friedel–Crafts acylation reactions.

Adams catalyst (Section 7.7): The PtO_2 catalyst used for hydrogenations.

1,2-Addition (Section 14.5): The addition of a reactant to the two ends of a double bond.

1,4-Addition (Sections 14.5, 19.14): Addition of a reactant to the ends of a conjugated π system. Conjugated dienes yield 1,4 adducts when treated with electrophiles such as HCl. Conjugated enones yield 1,4 adducts when treated with nucleophiles such as cyanide ion.

Addition reaction (Section 5.1): The reaction that occurs when two reactants add together to form a single new product with no atoms "left over."

Adrenocortical hormone (Section 27.7): A steroid hormone secreted by the adrenal glands. There are two types of adrenocortical hormones: mineralocorticoids and glucocorticoids.

Alcohol (Chapter 17 introduction): A compound with an –OH group bonded to a saturated, alkane-like carbon.

Aldaric acid (Section 25.8): The dicarboxylic acid resulting from oxidation of an aldose.

Aldehyde (Section 19.1): A compound containing the –CHO functional group.

Alditol (Section 25.8): The polyalcohol resulting from reduction of the carbonyl group of a sugar.

Aldol reaction (Section 23.2): The carbonyl condensation reaction of an aldehyde or ketone to give a β-hydroxy carbonyl compound.

Aldonic acid (Section 25.8): The monocarboxylic acid resulting from mild oxidation of an aldose.

Aldose (Section 25.1): A carbohydrate with an aldehyde functional group.

Alicyclic (Section 3.6): An aliphatic cyclic hydrocarbon such as a cycloalkane or cycloalkene.

Aliphatic (Section 3.2): A nonaromatic hydrocarbon such as a simple alkane, alkene, or alkyne.

Alkali fusion (Section 16.2): A process for converting an aryl halide into a phenol by melting with NaOH.

Alkaloid (Chapter 2 Chemistry @ Work): A naturally occurring organic base, such as morphine.

Alkane (Section 3.2): A compound of carbon and hydrogen that contains only single bonds.

Alkene (Chapter 6 introduction): A hydrocarbon that contains a carbon–carbon double bond.

Alkoxide ion (Section 17.3): The anion RO^- formed by deprotonation of an alcohol.

Alkoxymercuration reaction (Section 18.4): A method for synthesizing ethers by addition of an alcohol to an alkene.

Alkyl group (Section 3.3): The partial structure that remains when a hydrogen atom is removed from an alkane.

Alkylation (Sections 8.9, 16.3, 18.3, 22.8): Introduction of an alkyl group onto a molecule. For example, aromatic rings can be alkylated to yield arenes, and enolate anions can be alkylated to yield α-substituted carbonyl compounds.

Alkyne (Chapter 8 introduction): A hydrocarbon that contains a carbon–carbon triple bond.

Allyl group (Section 6.3): A $H_2C=CHCH_2-$ substituent.

Allylic (Section 10.5): The position next to a double bond. For example, $H_2C=CHCH_2Br$ is an allylic bromide.

α-helix (Section 26.13): The coiled secondary structure of a protein.

Alpha (α) position (Chapter 22 introduction): The position next to a carbonyl group.

Alpha-substitution reaction (Section 22.2): The substitution of the α hydrogen atom of a carbonyl compound by reaction with an electrophile.

Amide (Chapter 21 introduction): A compound containing the $-CONR_2$ functional group.

Amidomalonate synthesis (Section 26.3): A method for preparing α-amino acids by alkylation of diethyl amidomalonate with an alkyl halide.

Amine (Chapter 24 introduction): A compound containing one or more organic substituents bonded to a nitrogen atom, RNH_2, R_2NH, or R_3N.

α-Amino acid (Section 26.1): A difunctional compound, $RCH(NH_2)COOH$, with an $-NH_2$ group as a substituent on the carbon atom next to a $-COOH$ group.

Amino sugar (Section 25.12): A sugar with one of its $-OH$ groups replaced by $-NH_2$.

Amphoteric (Section 26.1): Capable of acting either as an acid or as a base. Amino acids are amphoteric.

Amplitude (Section 12.5): The height of a wave measured from the midpoint to the maximum. The intensity of radiant energy is proportional to the square of the wave's amplitude.

Anabolism (Section 29.1): The group of metabolic pathways that build up larger molecules from smaller ones.

Androgen (Section 27.7): A male steroid sex hormone.

Angle strain (Section 4.4): The strain introduced into a molecule when a bond angle is deformed from its ideal value. Angle strain is particularly important in small-ring cycloalkanes, where it results from compression of bond angles to less than their ideal tetrahedral values.

Annulation (Section 23.10): The building of a new ring onto an existing molecule.

Anomers (Section 25.6): Cyclic stereoisomers of sugars that differ only in their configuration at the hemiacetal (anomeric) carbon.

Antarafacial (Section 30.6): A pericyclic reaction that takes place on opposite faces of the two ends of a π electron system.

Anti conformation (Section 4.3): The geometric arrangement around a carbon–carbon single bond in which the two largest substituents are 180° apart as viewed in a Newman projection.

Anti stereochemistry (Section 7.2): The opposite of syn. An anti addition reaction is one in which the two ends of the double bond are attacked from different sides. An anti elimination reaction is one in which the two groups leave from opposite sides of the molecule.

Antiaromatic (Section 15.5): Referring to a planar, conjugated molecule with $4n$ π electrons. Delocalization of the π electrons leads to an increase in energy.

Antibonding orbital (Section 1.6): A molecular orbital that is higher in energy than the atomic orbitals from which it is formed.

Anticodon (Section 28.14): A sequence of three bases on tRNA that reads the codons on mRNA and brings the correct amino acids into position for protein synthesis.

Antisense strand (Section 28.13): The strand of double-helical DNA that does not contain the gene.

Apoenzyme (Section 26.14): The protein part of an enzyme that also contains a cofactor.

Arene (Section 15.2): An alkyl-substituted benzene.

Aromaticity (Chapter 15): The special characteristics of cyclic conjugated π electron molecules. These characteristics include unusual stability, the presence of a ring current in the ^{1}H NMR spectrum, and a tendency to undergo substitution reactions rather than addition reactions on treatment with electrophiles. Aromatic molecules are planar, cyclic, conjugated species that have $4n + 2$ π electrons.

Arylamine (Section 24.1): An amino-substituted aromatic compound, $ArNH_2$.

Atactic (Section 31.2): A chain-growth polymer in which the substituents are randomly oriented along the backbone.

Atomic number, Z (Section 1.1): The number of protons in the nucleus of an atom.

Atomic weight (Section 1.1): The average mass number of the atoms of an element.

ATZ derivative (Section 26.8): An anilinothiazolinone, formed from an amino acid.

Aufbau principle (Section 1.3): The rules for determining the electron configuration of an atom.

Autoradiography (Section 28.15): A method for visualizing radioactive compounds that have been separated by gel electrophoresis.

Axial bond (Section 4.10): A bond to chair cyclohexane that lies along the ring axis perpendicular to the rough plane of the ring.

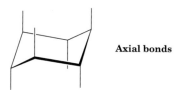

Axial bonds

Azide synthesis (Section 24.6): A method for preparing amines by S_N2 reaction of an alkyl halide with azide ion, followed by reduction.

Azo compound (Section 24.8): A compound with the general structure R–N=N–R′.

Backbone (Section 26.5): The continuous chain of atoms running the length of a polymer.

Basal metabolic rate (Chapter 29 Chemistry @ Work): The minimum amount of energy per unit time an organism expends to stay alive.

Base peak (Section 12.1): The most intense peak in a mass spectrum.

Bent bonds (Section 4.7): The bonds in small rings such as cyclopropane that bend away from the internuclear line and overlap at a slight angle, rather than head-on. Bent bonds are highly strained and highly reactive.

Benzoyl group (Section 19.1): The C_6H_5CO- group.

Benzylic (Sections 11.9, 16.10): The position next to an aromatic ring.

Benzyne (Section 16.9): An unstable compound having a triple bond in a benzene ring.

β-pleated sheet (Section 26.13): A type of secondary structure of a protein.

Betaine (Section 19.12): A neutral dipolar molecule with nonadjacent positive and negative charges. For example, the adduct of a Wittig reagent with a carbonyl compound is a betaine.

$$O^- \quad \overset{+}{P}Ph_3$$
$$\diagdown C - C \diagup$$

A Wittig betaine

Bicycloalkane (Section 4.15): A cycloalkane that contains two rings.

Bimolecular reaction (Section 11.4): A reaction whose rate-limiting step occurs between two reactants.

Block copolymer (Section 31.3): A polymer in which different blocks of identical monomer units alternate with each other.

Boat cyclohexane (Section 4.14): A conformation of cyclohexane that bears a slight resemblance to a boat. Boat cyclohexane has no angle strain, but has a large number of eclipsing interactions that make it less stable than chair cyclohexane.

Boat cyclohexane

BOC derivative (Section 26.10): A butyloxycarbonyl amide protected amino acid.

Bond angle (Section 1.7): The angle formed between two adjacent bonds.

Bond dissociation energy, D (Section 5.8): The amount of energy needed to break a bond homolytically and produce two radical fragments.

Bond length (Section 1.6): The equilibrium distance between the nuclei of two atoms that are bonded to each other.

Bond strength (Section 1.6): An alternative name for bond dissociation energy.

Bonding orbital (Section 1.6): A molecular orbital that is lower in energy than the atomic orbitals from which it is formed.

Branched-chain alkane (Section 3.2): An alkane that contains a branching connection of carbons as opposed to a straight-chain alkane.

Bridgehead atoms (Section 4.15): Atoms that are shared by more than one ring in a polycyclic molecule.

Bromohydrin (Section 7.3): A 1,2-disubstituted bromoalcohol; obtained by addition of HOBr to an alkene.

Bromonium ion (Section 7.2): A species with a divalent, positively charged bromine, R_2Br^+.

Brønsted–Lowry acid (Section 2.7): A substance that donates a hydrogen ion (proton; H^+) to a base.

Brønsted–Lowry base (Section 2.7): A substance that accepts H^+ from an acid.

C-terminal amino acid (Section 26.5): The amino acid with a free –COOH group at the end of a protein chain.

Cahn–Ingold–Prelog sequence rules (Sections 6.6, 9.6): A series of rules for assigning relative priorities to substituent groups on a double-bond carbon atom or on a chirality center.

Cannizzaro reaction (Section 19.13): The disproportionation reaction of an aldehyde to yield an alcohol and a carboxylic acid on treatment with base.

Carbanion (Section 10.8): A carbon anion, or substance that contains a trivalent, negatively charged carbon atom ($R_3C:^-$). Carbanions are sp^3-hybridized and have eight electrons in the outer shell of the negatively charged carbon.

Carbene (Section 7.6): A neutral substance that contains a divalent carbon atom having only six electrons in its outer shell ($R_2C:$).

Carbinolamine (Section 19.9): A molecule that contains the $R_2C(OH)NH_2$ functional group. Carbinolamines are produced as intermediates during the nucleophilic addition of amines to carbonyl compounds.

Carbocation (Sections 5.6, 6.10): A carbon cation, or substance that contains a trivalent, positively charged carbon atom having six electrons in its outer shell (R_3C^+).

Carbocycle (Section 15.9): A cyclic molecule that has only carbon atoms in its ring.

Carbohydrate (Section 25.1): A polyhydroxy aldehyde or ketone. Carbohydrates can be either simple sugars, such as glucose, or complex sugars, such as cellulose.

Carbonyl group (Section 2.1): The C=O functional group.

Carboxylation (Section 20.6): The addition of CO_2 to a molecule.

Carboxylic acid (Chapter 20 introduction): A compound containing the –COOH functional group.

Catabolism (Section 29.1): The group of metabolic pathways that break down larger molecules into smaller ones.

Chain-growth polymer (Sections 21.10, 31.1): A polymer whose bonds are produced by chain reactions. Polyethylene and other alkene polymers are examples.

Chain reaction (Section 5.3): A reaction that, once initiated, sustains itself in an endlessly repeating cycle of propagation steps. The radical chlorination of alkanes is an example of a chain reaction that is initiated by irradiation with light and then continues in a series of propagation steps.

Chair cyclohexane (Section 4.9): A three-dimensional conformation of cyclohexane that resembles the rough shape of a chair. The chair form of cyclohexane is the lowest-energy conformation of the molecule.

Chair cyclohexane

Chemical shift (Section 13.3): The position on the NMR chart where a nucleus absorbs. By convention, the chemical shift of tetramethylsilane (TMS) is arbitrarily set at zero, and all other absorptions usually occur downfield (to the left on the chart). Chemical shifts are expressed in delta units, δ, where 1 δ equals 1 ppm of the spectrometer operating frequency.

Chiral (Section 9.2): Having handedness. Chiral molecules are those that do not have a plane of symmetry and are therefore not superimposable on their mirror image. A chiral molecule thus exists in two forms, one right-handed and one left-handed. The most common cause of chirality in a molecule is the presence of a carbon atom that is bonded to four different substituents.

Chirality center (Section 9.2): An atom (usually carbon) that is bonded to four different groups.

Chlorohydrin (Section 7.3): A 1,2-disubstituted chloroalcohol; obtained by addition of HOCl to an alkene.

Chromatography (Chapter 12 Chemistry @ Work, Section 26.7): A technique for separating a mixture of compounds into pure components. Different compounds adsorb to a stationary support phase and are then carried along it at different rates by a mobile phase.

Cis–trans isomers (Sections 3.8, 6.5): Stereoisomers that differ in their stereochemistry about a double bond or ring.

Citric acid cycle (Section 29.5): The metabolic pathway by which acetyl CoA is degraded to CO_2.

Claisen condensation reaction (Section 23.8): The carbonyl condensation reaction of an ester to give a β-keto ester product.

Claisen rearrangement reaction (Sections 18.6, 30.8) The pericyclic conversion of an allyl phenyl ether to an *o*-allylphenol by heating.

Coding strand (Section 28.13): The strand of double-helical DNA that contains the gene.

Codon (Section 28.14): A three-base sequence on a messenger RNA chain that encodes the genetic information necessary to cause a specific amino acid to be incorporated into a protein. Codons on mRNA are read by complementary anticodons on tRNA.

Coenzyme (Section 26.14): A small organic molecule that acts as a cofactor.

Cofactor (Section 26.14): A small nonprotein part of an enzyme that is necessary for biological activity.

Complex carbohydrate (Section 25.1): A carbohydrate that is made of two or more simple sugars linked together.

Concerted (Section 30.1): A reaction that takes place in a single step without intermediates. For example, the Diels–Alder cycloaddition reaction is a concerted process.

Condensed structure (Section 2.12): A shorthand way of writing structures in which carbon–hydrogen and carbon–carbon bonds are understood rather than shown explicitly. Propane, for example, has the condensed structure $CH_3CH_2CH_3$.

Configuration (Section 9.6): The three-dimensional arrangement of atoms bonded to a chirality center.

Conformation (Section 4.1): The three-dimensional shape of a molecule at any given instant, assuming that rotation around single bonds is frozen.

Conformational analysis (Section 4.13): A means of assessing the energy of a substituted cycloalkane by totaling the steric interactions present in the molecule.

Conformer (Section 4.1): A conformational isomer.

Conjugate acid (Section 2.7): The product that results from protonation of a Brønsted–Lowry base.

Conjugate addition (Section 19.14): Addition of a nucleophile to the β carbon atom of an α,β-unsaturated carbonyl compound.

Conjugate base (Section 2.7): The anion that results from deprotonation of a Brønsted–Lowry acid.

Conjugated protein (Section 26.12): A protein that yields other compounds, such as carbohydrates, fats, or nucleic acids, in addition to amino acids on hydrolysis.

Conjugation (Chapter 14 introduction): A series of overlapping p orbitals, usually in alternating single and multiple bonds. For example, 1,3-butadiene is a conjugated diene, 3-buten-2-one is a conjugated enone, and benzene is a cyclic conjugated triene.

Conrotatory (Section 30.3): A term used to indicate that p orbitals must rotate in the same direction during electrocyclic ring opening or ring closure.

Constitutional isomers (Sections 3.2, 9.12): Isomers that have their atoms connected in a different order. For example, butane and 2-methylpropane are constitutional isomers.

Cope rearrangement (Section 30.8): The sigmatropic rearrangement of a 1,5-hexadiene.

Copolymer (Section 31.3): A polymer obtained when two or more different monomers are allowed to polymerize together.

Coupling constant, J (Section 13.11): The magnitude (expressed in hertz) of the interaction between nuclei whose spins are coupled.

Covalent bond (Section 1.5): A bond formed by sharing electrons between atoms.

Cracking (Chapter 3 Chemistry @ Work): A process used in petroleum refining in which large alkanes are thermally cracked into smaller fragments.

Crown ether (Section 18.9): A large-ring polyether; used as a phase-transfer catalyst.

Curtius rearrangement (Section 24.6): The conversion of an acid chloride into an amine by reaction with azide ion, followed by heating with water.

Cycloaddition (Sections 14.8, 30.6): A pericyclic reaction in which two reactants add together in a single step to yield a cyclic product. The Diels–Alder reaction between a diene and a dienophile to give a cyclohexene is an example.

Cycloalkane (Section 3.6): An alkane that contains a ring of carbons.

D sugar (Section 25.3): A sugar whose hydroxyl group at the chirality center farthest from the car-

bonyl group points to the right when drawn in Fischer projection.

d,l form (Section 9.10): The racemic modification of a compound.

Deactivating group (Section 16.5): An electron-withdrawing substituent that decreases the reactivity of an aromatic ring toward electrophilic aromatic substitution.

Debye, D (Section 2.2): A unit for measuring dipole moments; $1\ D = 3.336 \times 10^{-30}$ coulomb meter (C · m).

Decarbonylation (Section 28.1): The loss of carbon monoxide from a molecule.

Decarboxylation (Section 22.8): The loss of carbon dioxide from a molecule. β-Keto acids decarboxylate readily on heating.

Degenerate orbitals (Section 15.4): Two or more orbitals that have the same energy level.

Degree of unsaturation (Section 6.2): The number of rings and/or multiple bonds in a molecule.

Dehydration (Sections 7.1, 17.7): The loss of water from an alcohol. Alcohols can be dehydrated to yield alkenes.

Dehydrohalogenation (Sections 7.1, 11.10): The loss of HX from an alkyl halide. Alkyl halides undergo dehydrohalogenation to yield alkenes on treatment with strong base.

Delocalization (Section 10.6): A spreading out of electron density over a conjugated π electron system. For example, allylic cations and allylic anions are delocalized because their charges are spread out over the entire π electron system.

Delta scale (Section 13.3): An arbitrary scale used to calibrate NMR charts. One delta unit (δ) is equal to 1 part per million (ppm) of the spectrometer operating frequency.

Denaturation (Section 26.16): The physical changes that occur in a protein when secondary and tertiary structures are disrupted.

Deoxy sugar (Section 25.12): A sugar with one of its –OH groups replaced by an –H.

DEPT-NMR (Section 13.6): An NMR method for distinguishing among signals due to CH_3, CH_2, CH, and quaternary carbons. That is, the number of hydrogens attached to each carbon can be determined.

Deshielding (Section 13.2): An effect observed in NMR that causes a nucleus to absorb downfield (to the left) of tetramethylsilane (TMS) standard. Deshielding is caused by a withdrawal of electron density from the nucleus.

Deuterium isotope effect (Section 11.13): A tool used in mechanistic investigations to establish whether a C–H bond is broken in the rate-limiting step of a reaction.

Dextrorotatory (Section 9.3): A word used to describe an optically active substance that rotates the plane of polarization of plane-polarized light in a right-handed (clockwise) direction.

Diastereomer (Section 9.7): A term that indicates the relationship between non-mirror-image stereoisomers. Diastereomers are stereoisomers that have the same configuration at one or more chirality centers but differ at other chirality centers.

1,3-Diaxial interaction (Section 4.12): The strain energy caused by a steric interaction between axial groups three carbon atoms apart in chair cyclohexane.

Diazonium salt (Section 24.8): A compound with the general structure $RN_2^+ X^-$.

Diazotization (Section 24.8): The conversion of a primary amine, RNH_2, into a diazonium ion, RN_2^+, by treatment with nitrous acid.

Dideoxy DNA sequencing (Section 28.15): A biochemical method for sequencing DNA strands.

Dieckmann cyclization reaction (Section 23.10): An intramolecular Claisen condensation reaction to give a cyclic β-keto ester.

Dielectric polarization (Section 11.9): A measure of the ability of a solvent to act as an insulator of electric charge.

Diels–Alder reaction (Sections 14.8, 30.6): The cycloaddition reaction of a diene with a dienophile to yield a cyclohexene.

Dienophile (Section 14.9): A compound containing a double bond that can take part in the Diels–Alder cycloaddition reaction. The most reactive dienophiles are those that have electron-withdrawing groups on the double bond.

Digestion (Section 29.1): The first stage of catabolism, in which food is broken down by hydrolysis of ester, glycoside (acetal), and peptide (amide) bonds to yield fatty acids, simple sugars, and amino acids.

Dipolar molecule (Section 2.3): A molecule that is neutral overall but has plus and minus charges on individual atoms.

Dipole moment, μ (Section 2.2): A measure of the net polarity of a molecule. A dipole moment arises when the centers of mass of positive and negative charges within a molecule do not coincide.

Disrotatory (Section 30.3): A term used to indicate that p orbitals rotate in opposite directions during electrocyclic ring opening or ring closing.

Disulfide (Section 18.11): A compound of the general structure RSSR′.

DNA (Section 28.9): Deoxyribonucleic acid; the biopolymer consisting of deoxyribonucleotide units linked together through phosphate–sugar bonds. Found in the nucleus of cells, DNA contains an organism's genetic information.

Double helix (Section 28.10): The structure of DNA in which two polynucleotide strands coil around one another.

Doublet (Section 13.7): A two-line NMR absorption caused by spin–spin splitting when the spin of the nucleus under observation couples with the spin of a neighboring magnetic nucleus.

Downfield (Section 13.3): Referring to the left-hand portion of the NMR chart.

E1 reaction (Section 11.14): A unimolecular elimination reaction.

E2 reaction (Section 11.11): A bimolecular elimination reaction.

Eclipsed conformation (Section 4.1): The geometric arrangement around a carbon–carbon single bond in which the bonds to substituents on one carbon are parallel to the bonds to substituents on the neighboring carbon as viewed in a Newman projection.

Eclipsed conformation

Eclipsing strain (Section 4.1): The strain energy in a molecule caused by electron repulsions between eclipsed bonds. Eclipsing strain is also called torsional strain.

Edman degradation (Section 26.8): A method for N-terminal sequencing of peptide chains.

Elastomer (Section 31.5): An amorphous polymer that has the ability to stretch out and spring back to its original shape.

Electrocyclic reaction (Section 30.3): A unimolecular pericyclic reaction in which a ring is formed or broken by a concerted reorganization of electrons through a cyclic transition state. For example, the cyclization of 1,3,5-hexatriene to yield 1,3-cyclohexadiene is an electrocyclic reaction.

Electromagnetic spectrum (Section 12.5): The range of electromagnetic energy, including infrared, ultraviolet, and visible radiation.

Electron configuration (Section 1.3): A list of the orbitals occupied by electrons in an atom.

Electron-dot structure (Section 1.5): A representation of a molecule showing valence electrons as dots.

Electronegativity (Section 2.1): The ability of an atom to attract electrons in a covalent bond. Electronegativity increases across the periodic table from right to left and from bottom to top.

Electrophile (Section 5.4): An "electron-lover," or substance that accepts an electron pair from a nucleophile in a polar bond-forming reaction.

Electrophilic addition reaction (Section 6.8): The addition of an electrophile to an alkene to yield a saturated product.

Electrophilic aromatic substitution (Chapter 16 introduction): A reaction in which an electrophile (E^+) reacts with an aromatic ring and substitutes for one of the ring hydrogens.

Electrophoresis (Section 26.2): A technique used for separating charged organic molecules, particularly proteins and amino acids. The mixture to be separated is placed on a buffered gel or paper, and an electric potential is applied across the ends of the apparatus. Negatively charged molecules migrate toward the positive electrode, and positively charged molecules migrate toward the negative electrode.

Electrostatic potential map (Section 2.2): A molecular representation that uses color to indicate the charge distribution in the molecule as derived by quantum-mechanical calculations.

Elimination reaction (Section 5.1): What occurs when a single reactant splits into two products.

Elution (Chapter 12 Chemistry @ Work): The removal of a substance from a chromatography column.

Embden–Meyerhof pathway (Section 29.3): An alternative name for glycolysis.

Enamine (Section 19.9): A compound with the $R_2N–CR=CR_2$ functional group.

Enantiomers (Section 9.1): Stereoisomers of a chiral substance that have a mirror-image relationship. Enantiomers must have opposite configurations at all chirality centers.

Endergonic (Section 5.7): A reaction that has a positive free-energy change and is therefore nonspontaneous. In a reaction energy diagram, the product of an endergonic reaction has a higher energy level than the reactants.

Endo (Section 14.9): A term indicating the stereochemistry of a substituent in a bridged bicycloalkane. An endo substituent is syn to the larger of the two bridges.

Endothermic (Section 5.7): A reaction that absorbs heat and therefore has a positive enthalpy change.

Enol (Sections 8.5, 22.1): A vinylic alcohol that is in equilibrium with a carbonyl compound.

Enolate ion (Section 22.1): The anion of an enol.

Entgegen, *E* (Section 6.6): A term used to describe the stereochemistry of a carbon–carbon double bond. The two groups on each carbon are assigned priorities according to the Cahn–Ingold–Prelog sequence rules, and the two carbons are compared. If the high-priority groups on each carbon are on opposite sides of the double bond, the bond has *E* geometry.

Enthalpy change, ΔH (Section 5.7): The heat of reaction. The enthalpy change that occurs during a reaction is a measure of the difference in total bond energy between reactants and products.

Entropy change, ΔS (Section 5.7): The change in amount of disorder. The entropy change that occurs during a reaction is a measure of the difference in disorder between reactants and products.

Enzyme (Section 26.14): A biological catalyst. Enzymes are large proteins that catalyze specific biochemical reactions.

Epoxide (Section 18.7): A three-membered-ring ether functional group.

Equatorial bond (Section 4.10):   A bond to cyclo-hexane that lies along the rough equator of the ring.

Equatorial bonds

Equilibrium constant, K_{eq} (Section 5.7):   A measure of the equilibrium position for a reaction. The equilibrium constant for the reaction $aA + bB \rightarrow cC + dD$ is given by the expression

$$K_{eq} = \frac{[\text{Products}]}{[\text{Reactants}]} = \frac{[C]^c[D]^d}{[A]^a[B]^b}$$

Essential oil (Section 27.5):   The volatile oil obtained by steam distillation of a plant extract.

Ester (Chapter 21 introduction):   A compound containing the –COOR functional group.

Estrogen (Section 27.7):   A female steroid sex hormone.

Ether (Chapter 18 introduction):   A compound that has two organic substituents bonded to the same oxygen atom, ROR′.

Exergonic (Section 5.7):   A reaction that has a negative free-energy change and is therefore spontaneous. On a reaction energy diagram, the product of an exergonic reaction has a lower energy level than that of the reactants.

Exo (Section 14.9):   A term indicating the stereochemistry of a substituent in a bridged bicycloalkane. An exo substituent is anti to the larger of the two bridges.

Exon (Section 28.13):   A section of DNA that contains genetic information.

Exothermic (Section 5.7):   A reaction that releases heat and therefore has a negative enthalpy change.

Fat (Section 27.1):   A solid triacylglycerol derived from animal sources.

Fatty acid (Section 27.1):   A long, straight-chain carboxylic acid found in fats and oils.

Fiber (Section 31.5):   A thin thread produced by extruding a molten polymer through small holes in a die.

Fibrous protein (Section 26.12):   A protein that consists of polypeptide chains arranged side by side in long threads. Such proteins are tough, insoluble in water, and used in nature for structural materials such as hair, hooves, and fingernails.

Fingerprint region (Section 12.7):   The complex region of the infrared spectrum from 1500 cm^{-1} to 400 cm^{-1}.

First-order reaction (Section 11.7):   A reaction whose rate-limiting step is unimolecular and whose kinetics therefore depend on the concentration of only one reactant.

Fischer esterification reaction (Section 21.3):   The acid-catalyzed reaction of an alcohol with a carboxylic acid to yield an ester.

Fischer projection (Sections 9.13, 25.2):   A means of depicting the absolute configuration of a chiral molecule on a flat page. A Fischer projection uses a cross to represent the chirality center. The horizontal arms of the cross represent bonds coming out of the plane of the page, and the vertical arms of the cross represent bonds going back into the plane of the page.

$$\underset{\displaystyle D}{\overset{\displaystyle A}{E \!-\!\!\!+\!\!\!- B}} \quad = \quad \underset{\displaystyle D}{\overset{\displaystyle A}{E \!\blacktriangleleft\! C \!\blacktriangleright\! B}}$$

Fischer projection

Formal charge (Section 2.3):   The difference in the number of electrons owned by an atom in a molecule and by the same atom in its elemental state. The formal charge on an atom is given by the formula:

Formal charge =
$$\begin{bmatrix} \text{Number of} \\ \text{outer-shell electrons} \\ \text{in free atom} \end{bmatrix} - \begin{bmatrix} \text{Number of} \\ \text{outer-shell electrons} \\ \text{in bonded atom} \end{bmatrix}$$

Formyl group (Section 19.1):   A –CHO group.

Frequency (Section 12.5):   The number of electromagnetic wave cycles that travel past a fixed point in a given unit of time. Frequencies are expressed in units of cycles per second, or hertz.

Friedel–Crafts reaction (Section 16.3):   An electrophilic aromatic substitution reaction to alkylate or acylate an aromatic ring.

Frontier orbitals (Section 30.1):   The highest occupied (HOMO) and lowest unoccupied (LUMO) molecular orbitals.

FT-NMR (Section 13.4): Fourier-transform NMR; a rapid technique for recording NMR spectra in which all magnetic nuclei absorb at the same time.

Functional group (Section 3.1): An atom or group of atoms that is part of a larger molecule and that has a characteristic chemical reactivity.

Furanose (Section 25.5): The five-membered-ring form of a simple sugar.

Gabriel synthesis (Section 24.6): A method for preparing an amine by S_N2 reaction of an alkyl halide with potassium phthalimide, followed by hydrolysis.

Gauche conformation (Section 4.3): The conformation of butane in which the two methyl groups lie 60° apart as viewed in a Newman projection. This conformation has 3.8 kJ/mol steric strain.

Gauche conformation

Geminal (Section 19.6): Referring to two groups attached to the same carbon atom. For example, 1,1-dibromopropane is a geminal dibromide.

Geometric isomers (Sections 3.8, 6.5): An old term for cis–trans isomers.

Gibbs free-energy change, ΔG (Section 5.7): The free-energy change that occurs during a reaction, given by the equation $\Delta G = \Delta H - T\Delta S$. A reaction with a negative free-energy change is spontaneous, and a reaction with a positive free-energy change is nonspontaneous.

Gilman reagent (Section 10.9): A diorganocopper reagent, R_2CuLi.

Glass transition temperature, T_g (Section 31.5): The temperature at which a hard, amorphous polymer becomes soft and flexible.

Globular protein (Section 26.12): A protein that is coiled into a compact, nearly spherical shape. These proteins, which are generally water-soluble and mobile within the cell, are the structural class to which enzymes belong.

Gluconeogenesis (Section 29.8): The anabolic pathway by which organisms make glucose from simple precursors.

Glycal assembly method (Section 25.10): A method for linking monosaccharides together to synthesize polysaccharides.

Glycol (Section 7.8): A diol, such as ethylene glycol, $HOCH_2CH_2OH$.

Glycolysis (Section 29.3): A series of ten enzyme-catalyzed reactions that break down glucose into 2 equivalents of pyruvate, $CH_3COCO_2^-$.

Glycoside (Section 25.8): A cyclic acetal formed by reaction of a sugar with another alcohol.

Graft copolymer (Section 31.3): A copolymer in which homopolymer branches of one monomer unit are "grafted" onto a homopolymer chain of another monomer unit.

Grignard reagent (Section 10.8): An organomagnesium halide, RMgX.

Ground state (Section 1.3): The most stable, lowest-energy electron configuration of a molecule or atom.

Haloform (Section 22.7): A trihalomethane, such as $CHCl_3$, $CHBr_3$, or CHI_3.

Halohydrin (Section 7.3): A 1,2-disubstituted haloalcohol, such as that obtained on addition of HOBr to an alkene.

Halonium ion (Section 7.2): A species containing a positively charged, divalent halogen. Three-membered-ring bromonium ions are implicated as intermediates in the electrophilic addition of Br_2 to alkenes.

Hammond postulate (Section 6.11): A postulate stating that we can get a picture of what a given transition state looks like by looking at the structure of the nearest stable species. Exergonic reactions have transition states that resemble reactant; endergonic reactions have transition states that resemble product.

Heat of combustion (Section 4.5): The amount of heat released when a compound is burned in a calorimeter.

Heat of hydrogenation (Section 6.7): The amount of heat released when a carbon–carbon double bond is hydrogenated.

Heat of reaction (Section 5.7): An alternative name for the enthalpy change in a reaction, ΔH.

Hell–Volhard–Zelinskii (HVZ) reaction (Section 22.4): The reaction of a carboxylic acid with Br_2 and phosphorus to give an α-bromo carboxylic acid.

Hemiacetal (Section 19.11): A functional group consisting of one –OR and one –OH group bonded to the same carbon.

Henderson–Hasselbalch equation (Section 26.2): An equation for determining the extent of deprotonation of a weak acid at various pH values.

Heterocycle (Sections 15.7, 28.1): A cyclic molecule whose ring contains more than one kind of atom. For example, pyridine is a heterocycle that contains five carbon atoms and one nitrogen atom in its ring.

Heterogenic bond formation (Section 5.2): What occurs when one reaction partner donates both electrons in forming a new bond. Polar reactions always involve heterogenic bond formation: $A^+ + B{:}^- \rightarrow A{:}B$

Heterolytic bond breakage (Section 5.2): The kind of bond breaking that occurs in polar reactions when one fragment leaves with both of the bonding electrons, as in the equation $A{:}B \rightarrow A^+ + B{:}^-$.

Hofmann elimination (Section 24.7): The elimination reaction of an amine to yield an alkene by reaction with iodomethane, followed by heating with Ag_2O.

Hofmann rearrangement (Section 24.6): The conversion of an amide into an amine by reaction with Br_2 and base.

Holoenzyme (Section 26.14): The combination of apoenzyme plus cofactor.

HOMO (Sections 14.11, 30.2): An acronym for highest occupied molecular orbital. The symmetries of the HOMO and LUMO are important in pericyclic reactions.

Homogenic bond formation (Section 5.2): What occurs in radical reactions when each reactant donates one electron to the new bond: $A{\cdot} + B{\cdot} \rightarrow A{:}B$

Homolytic bond breakage (Section 5.2): The kind of bond breaking that occurs in radical reactions when each fragment leaves with one bonding electron: $A{:}B \rightarrow A{\cdot} + B{\cdot}$

Homopolymer (Section 31.3): A polymer made up of identical repeating units.

Hormone (Section 27.7): A chemical messenger that is secreted by an endocrine gland and carried through the bloodstream to a target tissue.

Hückel's rule (Section 15.5): A rule stating that monocyclic conjugated molecules having $4n + 2\,\pi$ electrons (n = an integer) are aromatic.

Hund's rule (Section 1.3): If two or more empty orbitals of equal energy are available, one electron occupies each, with their spins parallel, until all are half-full.

Hybrid orbital (Section 1.7): An orbital derived from a combination of atomic orbitals. Hybrid orbitals, such as the sp^3, sp^2, and sp hybrids of carbon, are strongly directed and form stronger bonds than atomic orbitals do.

Hydration (Section 7.4): Addition of water to a molecule, such as occurs when alkenes are treated with aqueous sulfuric acid to give alcohols.

Hydride shift (Section 6.12): The shift of a hydrogen atom and its electron pair to a nearby cationic center.

Hydroboration (Section 7.5): Addition of borane (BH_3) or an alkylborane to an alkene. The resultant trialkylborane products are useful synthetic intermediates that can be oxidized to yield alcohols.

Hydrocarbon (Section 3.2): A compound that contains only carbon and hydrogen.

Hydrogen bond (Section 17.2): A weak attraction between a hydrogen atom bonded to an electronegative atom and an electron lone pair on another electronegative atom.

Hydrogenation (Section 7.7): Addition of hydrogen to a double or triple bond to yield a saturated product.

Hydrogenolysis (Section 26.10): Cleavage of a bond by reaction with hydrogen. Benzylic ethers and esters, for instance, are cleaved by hydrogenolysis.

Hydrophilic (Section 27.2): Water-loving; attracted to water.

Hydrophobic (Section 27.2): Water-fearing; repelled by water.

Hydroquinone (Section 17.11): A 1,4-dihydroxybenzene.

Hydroxylation (Section 7.8): Addition of two –OH groups to a double bond.

Hyperconjugation (Section 6.7): An interaction that results from overlap of a vacant p orbital on one atom with a neighboring C–H σ bond. Hyperconjugation is important in stabilizing carbocations and in stabilizing substituted alkenes.

Imide (Section 24.6): A compound with the –CONHCO– functional group.

Imine (Section 19.9): A compound with the $R_2C=NR$ functional group.

Inductive effect (Sections 2.1, 6.10, 16.6): The electron-attracting or electron-withdrawing effect transmitted through σ bonds. Electronegative elements have an electron-withdrawing inductive effect.

Infrared (IR) spectroscopy (Section 12.5): A kind of optical spectroscopy that uses infrared energy. IR spectroscopy is particularly useful in organic chemistry for determining the kinds of functional groups present in molecules.

Initiator (Section 5.3): A substance with an easily broken bond that is used to initiate a radical chain reaction. For example, radical chlorination of alkanes is initiated when light energy breaks the weak Cl–Cl bond to form Cl· radicals.

Integration (Section 13.10): A technique for measuring the area under an NMR peak to determine the relative number of each kind of proton in a molecule. Integrated peak areas are superimposed over the spectrum as a "stair-step" line, with the height of each step proportional to the area underneath the peak.

Intermediate (Section 5.10): A species that is formed during the course of a multistep reaction but is not the final product. Intermediates are more stable than transition states, but may or may not be stable enough to isolate.

Intramolecular, intermolecular (Section 23.7): A reaction that occurs within the same molecule is intramolecular; a reaction that occurs between two molecules is intermolecular.

Intron (Section 28.13): A section of DNA that does not contain genetic information.

Ion pair (Section 11.8): A loose complex between two ions in solution. Ion pairs are implicated as intermediates in S_N1 reactions to account for the partial retention of stereochemistry that is often observed.

Isoelectric point, p*I* (Section 26.2): The pH at which the number of positive charges and the number of negative charges on a protein or an amino acid are equal.

Isomers (Section 3.2): Compounds that have the same molecular formula but different structures.

Isoprene rule (Section 27.6): An observation to the effect that terpenoids appear to be made up of isoprene (2-methyl-1,3-butadiene) units connected head-to-tail.

Isotactic (Section 31.2): A chain-growth polymer in which the substituents are regularly oriented on the same side of the backbone.

Isotopes (Section 1.1): Atoms of the same element that have different mass numbers.

IUPAC nomenclature (Section 3.4): Rules for naming compounds, devised by the International Union of Pure and Applied Chemistry.

Kekulé structure (Section 1.5): A method of representing molecules in which a line between atoms indicates a bond.

Keto–enol tautomerism (Sections 8.5, 22.1): The rapid equilibration between a carbonyl form and vinylic alcohol form of a molecule.

Ketone (Section 19.1): A compound with two organic substituents bonded to a carbonyl group, $R_2C=O$.

Ketose (Section 25.1): A carbohydrate with a ketone functional group.

Kiliani–Fischer synthesis (Section 25.8): A method for lengthening the chain of an aldose sugar.

Kinetic control (Section 14.6): A reaction that follows the lowest activation energy pathway is said to be kinetically controlled. The product is the most rapidly formed, but is not necessarily the most stable.

Kinetics (Section 11.3): Referring to reaction rates. Kinetic measurements are useful for helping to determine reaction mechanisms.

Koenigs–Knorr reaction (Section 25.7): A method for the synthesis of glycosides by reaction of an alcohol with a pyranosyl bromide.

Krebs cycle (Section 29.5): An alternative name for the citric acid cycle, by which acetyl CoA is degraded to CO_2.

L sugar (Section 25.3): A sugar whose hydroxyl group at the chirality center farthest from the carbonyl group points to the left when drawn in Fischer projection.

Lactam (Section 21.7): A cyclic amide.

Lactone (Section 21.6): A cyclic ester.

Leaving group (Section 11.5): The group that is replaced in a substitution reaction.

Levorotatory (Section 9.3): An optically active substance that rotates the plane of polarization of plane-polarized light in a left-handed (counterclockwise) direction.

Lewis acid (Section 2.11):   A substance with a vacant low-energy orbital that can accept an electron pair from a base. All electrophiles are Lewis acids.

Lewis base (Section 2.11):   A substance that donates an electron lone pair to an acid. All nucleophiles are Lewis bases.

Lewis structure (Section 1.5):   A representation of a molecule showing valence electrons as dots.

Lindlar catalyst (Section 8.6):   A hydrogenation catalyst used to convert alkynes to cis alkenes.

Line-bond structure (Section 1.5):   A representation of a molecule showing covalent bonds as lines between atoms.

Lipid (Section 27.1):   A naturally occurring substance isolated from cells and tissues by extraction with a nonpolar solvent. Lipids belong to many different structural classes, including fats, terpenes, prostaglandins, and steroids.

Lipid bilayer (Section 27.3):   The ordered lipid structure that forms a cell membrane.

Lipoprotein (Chapter 27 Chemistry @ Work):   A complex molecule with both lipid and protein parts that transports lipids through the body.

Lone-pair electrons (Section 1.5):   Nonbonding valence-shell electron pairs. Lone-pair electrons are used by nucleophiles in their reactions with electrophiles.

LUMO (Sections 14.11, 30.2):   An acronym for lowest unoccupied molecular orbital. The symmetries of the LUMO and the HOMO are important in determining the stereochemistry of pericyclic reactions.

Magnetic resonance imaging, MRI (Chapter 13 Chemistry @ Work):   A medical diagnostic technique based on nuclear magnetic resonance.

Malonic ester synthesis (Section 22.8):   The synthesis of a carboxylic acid by alkylation of an alkyl halide, followed by hydrolysis and decarboxylation.

Markovnikov's rule (Section 6.9):   A guide for determining the regiochemistry (orientation) of electrophilic addition reactions. In the addition of HX to an alkene, the hydrogen atom bonds to the alkene carbon that has fewer alkyl substituents.

Mass number, A (Section 1.1):   The total of protons plus neutrons in an atom.

Mass spectrometry (Section 12.1):   A technique for measuring the mass, and therefore the molecular weight (MW), of ions.

Maxam–Gilbert DNA sequencing (Section 28.15):   A chemical method for sequencing DNA strands.

McLafferty rearrangement (Section 12.4):   A mass-spectral fragmentation pathway for carbonyl compounds.

Mechanism (Section 5.2):   A complete description of how a reaction occurs. A mechanism must account for all starting materials and all products, and must describe the details of each individual step in the overall reaction process.

Meisenheimer complex (Section 16.8):   An intermediate formed by addition of a nucleophile to a halo-substituted aromatic ring.

Melt transition temperature, T_m (Section 31.5):   The temperature at which crystalline regions of a polymer melt to give an amorphous material.

Mercapto group (Section 18.11):   An alternative name for the thiol group, –SH.

Meso compound (Section 9.8):   A compound that contains chirality centers but is nevertheless achiral by virtue of a symmetry plane.

Metabolism (Section 29.1):   A collective name for the many reactions that go on in the cells of living organisms.

Methylene group (Section 6.3):   A $-CH_2-$ or $=CH_2$ group.

Micelle (Section 27.2):   A spherical cluster of soap-like molecules that aggregate in aqueous solution. The ionic heads of the molecules lie on the outside where they are solvated by water, and the organic tails bunch together on the inside of the micelle.

Michael reaction (Section 23.11):   The conjugate addition reaction of an enolate ion to an unsaturated carbonyl compound.

Molar absorptivity (Section 14.11):   A quantitative measure of the amount of UV light absorbed by a sample.

Molecular ion (Section 12.1):   The cation produced in the mass spectrometer by loss of an electron from the parent molecule. The mass of the molecular ion corresponds to the molecular weight of the sample.

Molecular mechanics (Chapter 4 Chemistry @ Work): A computer-based method for calculating the minimum-energy conformation of a molecule.

Molecular orbital, MO (Section 1.6): An orbital that is the property of the entire molecule rather than an individual atom. Molecular orbitals result from interaction of two or more atomic orbitals when bonds are formed.

Molecular orbital (MO) theory (Section 1.6): A description of covalent bond formation as resulting from a mathematical combination of atomic orbitals (wave functions) to form molecular orbitals.

Molecule (Section 1.5): A neutral collection of atoms held together by covalent bonds.

Molozonide (Section 7.8): The initial addition product of ozone with an alkene.

Monomer (Section 7.10, Chapter 31 introduction): The simple starting unit from which a polymer is made.

Multiplet (Section 13.7): A pattern of peaks in an NMR spectrum that arises by spin–spin splitting of a single absorption because of coupling between neighboring magnetic nuclei.

Mutarotation (Section 25.6): The change in optical rotation observed when a pure anomer of a sugar is dissolved in water. Mutarotation is caused by the reversible opening and closing of the acetal linkage, which yields an equilibrium mixture of anomers.

n + 1 rule (Section 13.11): A hydrogen with n other hydrogens on neighboring carbons shows n + 1 peaks in its ^{1}H NMR spectrum.

N-terminal amino acid (Section 26.5): The amino acid with a free $-NH_2$ group at the end of a protein chain.

Newman projection (Section 4.1): A means of indicating stereochemical relationships between substituent groups on neighboring carbons. The carbon–carbon bond is viewed end-on, and the carbons are indicated by a circle. Bonds radiating from the center of the circle are attached to the front carbon, and bonds radiating from the edge of the circle are attached to the rear carbon.

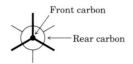

Front carbon
← Rear carbon

Nitrogen rule (Section 24.10): A compound with an odd number of nitrogen atoms has an odd-numbered molecular weight.

Node (Section 1.2): A surface of zero electron density within an orbital. For example, a p orbital has a nodal plane passing through the center of the nucleus, perpendicular to the axis of the orbital.

Nonbonding electrons (Section 1.5): Valence electrons that are not used in forming covalent bonds.

Normal alkane (Section 3.2): A straight-chain alkane, as opposed to a branched alkane. Normal alkanes are denoted by the suffix n, as in n-C_4H_{10} (n-butane).

NSAID (Chapter 15 Chemistry @ Work): A nonsteroidal anti-inflammatory drug, such as aspirin or ibuprofen.

Nuclear magnetic resonance, NMR (Chapter 13): A spectroscopic technique that provides information about the carbon–hydrogen framework of a molecule. NMR works by detecting the energy absorption accompanying the transition between nuclear spin states that occurs when a molecule is placed in a strong magnetic field and irradiated with radiofrequency waves.

Nucleophile (Section 5.4): A "nucleus-lover," or species that donates an electron pair to an electrophile in a polar bond-forming reaction. Nucleophiles are also Lewis bases.

Nucleophilic acyl substitution reaction (Section 21.2): A reaction in which a nucleophile attacks a carbonyl compound and substitutes for a leaving group bonded to the carbonyl carbon.

Nucleophilic addition reaction (Section 19.4): A reaction in which a nucleophile adds to the electrophilic carbonyl group of a ketone or aldehyde to give an alcohol.

Nucleophilic aromatic substitution reaction (Section 16.8): The substitution reaction of an aryl halide by a nucleophile.

Nucleophilic substitution reaction (Section 11.1): A reaction in which one nucleophile replaces another attached to a saturated carbon atom.

Nucleophilicity (Section 11.5): The ability of a substance to act as a nucleophile in an S_N2 reaction.

Nucleoside (Section 28.8): A nucleic acid constituent, consisting of a sugar residue bonded to a heterocyclic purine or pyrimidine base.

Nucleotide (Section 28.8): A nucleic acid constituent, consisting of a sugar residue bonded both to a heterocyclic purine or pyrimidine base and to a phosphoric acid. Nucleotides are the monomer units from which DNA and RNA are constructed.

Nylon (Section 21.10): A synthetic polyamide step-growth polymer.

Olefin (Chapter 6 introduction): An alternative name for an alkene.

Optical isomers (Section 9.5): An alternative name for enantiomers. Optical isomers are isomers that have a mirror-image relationship.

Optically active (Section 9.3): A substance that rotates the plane of polarization of plane-polarized light.

Orbital (Section 1.2): A wave function, which describes the volume of space around a nucleus in which an electron is most likely to be found.

Organic chemistry: The study of carbon compounds.

Oxidation (Section 10.10): A reaction that causes a decrease in electron ownership by carbon, either by bond formation between carbon and a more electronegative atom (usually oxygen, nitrogen, or a halogen) or by bond breaking between carbon and a less electronegative atom (usually hydrogen).

β-Oxidation pathway (Section 29.2): The repetitive four-step sequence of enzyme-catalyzed reactions for catabolism of fatty acids.

Oxime (Section 19.9): A compound with the $R_2C=NOH$ functional group.

Oxirane (Section 18.7): An alternative name for an epoxide.

Oxymercuration (Section 7.4): A method for double-bond hydration using aqueous mercuric acetate as the reagent.

Ozonide (Section 7.8): The product formed by addition of ozone to a carbon–carbon double bond. Ozonides are usually treated with a reducing agent, such as zinc in acetic acid, to produce carbonyl compounds.

Paraffin (Section 3.5): A common name for alkanes.

Parent peak (Section 12.1): The peak in a mass spectrum corresponding to the molecular ion. The mass of the parent peak therefore represents the molecular weight of the compound.

Pauli exclusion principle (Section 1.3): No more than two electrons can occupy the same orbital, and those two must have spins of opposite sign.

Peptide (Section 26.5): A short amino acid polymer in which the individual amino acid residues are linked by amide bonds.

Peptide bond (Section 26.5): An amide bond in a peptide chain.

Pericyclic reaction (Chapter 30): A reaction that occurs by a concerted reorganization of bonding electrons in a cyclic transition state.

Periplanar (Section 11.11): A conformation in which bonds to neighboring atoms have a parallel arrangement. In an eclipsed conformation, the neighboring bonds are syn periplanar; in a staggered conformation, the bonds are anti periplanar.

Anti periplanar   Syn periplanar

Peroxide (Section 18.2): A molecule containing an oxygen–oxygen bond functional group, ROOR′ or ROOH.

Peroxyacid (Section 18.7): A compound with the $-CO_3H$ functional group.

Phase-transfer catalysts (Section 24.9): Substances that cause the transfer of ions between water and organic phases, thus catalyzing reactions. Tetraalkylammonium salts, R_4N^+ X^-, are often used.

Phenol (Chapter 17 introduction): A compound with an −OH group directly bonded to an aromatic ring, ArOH.

Phenyl (Section 15.2): The name for the $-C_6H_5$ unit when the benzene ring is considered as a substituent. A phenyl group is abbreviated as −Ph.

Phospholipid (Section 27.3): A lipid that contains a phosphate residue. For example, phosphoglycerides contain a glycerol backbone linked to two fatty acids and a phosphoric acid.

Photochemical reaction (Section 30.3): A reaction carried out by irradiating the reactants with light.

Pi (π) bond (Section 1.9): The covalent bond formed by sideways overlap of atomic orbitals. For example, carbon–carbon double bonds contain a π bond formed by sideways overlap of two *p* orbitals.

PITC (Section 26.8): Phenylisothiocyanate, used in the Edman degradation.

Plane of symmetry (Section 9.2): A plane that bisects a molecule such that one half of the molecule is the mirror image of the other half. Molecules containing a plane of symmetry are achiral.

Plane-polarized light (Section 9.3): Ordinary light that has electromagnetic waves oscillating in a single plane rather than in random planes. The plane of polarization is rotated when the light is passed through a solution of a chiral substance.

Polar aprotic solvent (Section 11.5): A polar solvent that can't function as a hydrogen ion donor. Polar aprotic solvents such as dimethyl sulfoxide (DMSO), hexamethylphosphoramide (HMPA), and dimethylformamide (DMF) are particularly useful in S_N2 reactions because of their ability to solvate cations.

Polar covalent bond (Section 2.1): A covalent bond in which the electron distribution between atoms is unsymmetrical.

Polar reaction (Section 5.2): A reaction in which bonds are made when a nucleophile donates two electrons to an electrophile, and in which bonds are broken when one fragment leaves with both electrons from the bond.

Polarity (Section 2.1): The unsymmetrical distribution of electrons in a molecule that results when one atom attracts electrons more strongly than another.

Polarizability (Section 5.4): The measure of the change in a molecule's electron distribution in response to changing electric interactions with solvents or ionic reagents.

Polycyclic (Section 4.15): A compound that contains more than one ring.

Polycyclic aromatic compound (Section 15.9): A compound with two or more benzene-like aromatic rings fused together.

Polymer (Section 7.10, Chapter 31): A large molecule made up of repeating smaller units. For example, polyethylene is a synthetic polymer made from repeating ethylene units, and DNA is a biopolymer made of repeating deoxyribonucleotide units.

Polymerase chain reaction, PCR (Section 28.17): A method for amplifying small amounts of DNA to produce larger amounts.

Polysaccharide (Section 25.1): A carbohydrate that is made of many simple sugars linked together.

Polyunsaturated fatty acid, PUFA (Section 27.1): A fatty acid containing two or more double bonds.

Primary, secondary, tertiary, quaternary (Section 3.3): Terms used to describe the substitution pattern at a specific site. A primary site has one organic substituent attached to it, a secondary site has two organic substituents, a tertiary site has three, and a quaternary site has four.

	Primary	Secondary	Tertiary	Quaternary
Carbon	RCH_3	R_2CH_2	R_3CH	R_4C
Carbocation	RCH_2^+	R_2CH^+	R_3C^+	
Hydrogen	RCH_3	R_2CH_2	R_3CH	
Alcohol	RCH_2OH	R_2CHOH	R_3COH	
Amine	RNH_2	R_2NH	R_3N	

Primary structure (Section 26.13): The amino acid sequence in a protein.

Propagation step (Section 5.3): The step or series of steps in a radical chain reaction that carry on the chain. The propagation steps must yield both product and a reactive intermediate.

Prostaglandin (Section 27.4): A lipid with the general carbon skeleton

Prostaglandins are present in nearly all body tissues and fluids, where they serve many important hormonal functions.

Prosthetic group (Section 26.13): A covalently bound organic group attached to a protein.

Protecting group (Sections 17.9, 26.10): A group that is introduced to protect a sensitive functional group toward reaction elsewhere in the molecule. After serving its protective function, the group is removed.

Protein (Section 26.5): A large peptide containing 50 or more amino acid residues. Proteins serve both as structural materials and as enzymes that control an organism's chemistry.

Protic solvent (Section 11.9): A solvent such as water or alcohol that can act as a proton donor.

Pyranose (Section 25.5): The six-membered-ring form of a simple sugar.

Quartet (Section 13.7): A set of four peaks in an NMR spectrum, caused by spin–spin splitting of a signal by three adjacent nuclear spins.

Quaternary (*see* Primary)

Quaternary structure (Section 26.13): The highest level of protein structure, involving a specific aggregation of individual proteins into a larger cluster.

Quinone (Section 17.11): A 2,5-cyclohexadiene-1,4-dione.

R group (Section 3.3): A generalized abbreviation for an organic partial structure.

R,S convention (Section 9.6): A method for defining the absolute configuration at chirality centers using the Cahn–Ingold–Prelog sequence rules.

Racemic mixture (Section 9.10): A mixture consisting of equal parts (+) and (−) enantiomers of a chiral substance.

Radical (Section 5.2): A species that has an odd number of electrons, such as the chlorine radical, Cl·.

Radical reaction (Section 5.2): A reaction in which bonds are made by donation of one electron from each of two reactants and in which bonds are broken when each fragment leaves with one electron.

Rate constant (Section 11.3): The constant k in a rate equation.

Rate equation (Section 11.3): An equation that expresses the dependence of a reaction's rate on the concentration of reactants.

Rate-limiting step (Section 11.7): The slowest step in a multistep reaction sequence. The rate-limiting step acts as a kind of bottleneck in multistep reactions.

Reaction energy diagram (Section 5.9): A representation of the course of a reaction, in which free energy is plotted as a function of reaction progress. Reactants, transition states, intermediates, and products are represented, and their appropriate energy levels are indicated.

Rearrangement reaction (Section 5.1): What occurs when a single reactant undergoes a reorganization of bonds and atoms to yield an isomeric product.

Reducing sugar (Section 25.8): A sugar that reduces silver ion in the Tollens test or cupric ion in the Fehling or Benedict tests.

Reduction (Section 10.10): A reaction that causes an increase of electron ownership by carbon, either by bond breaking between carbon and a more electronegative atom or by bond formation between carbon and a less electronegative atom.

Reductive amination (Sections 24.6, 26.3): A method for preparing an amine by reaction of an aldehyde or ketone with ammonia and a reducing agent.

Refining (Chapter 3 Chemistry @ Work): The process by which petroleum is converted into gasoline and other useful products.

Regiochemistry (Section 6.9): A term describing the orientation of a reaction that occurs on an unsymmetrical substrate.

Regiospecific (Section 6.9): A term describing a reaction that occurs with a specific regiochemistry to give a single product rather than a mixture of products.

Replication (Section 28.12): The process by which double-stranded DNA uncoils and is replicated to produce two new copies.

Replication fork (Section 28.12): The point of unraveling in a DNA chain where replication occurs.

Residue (Section 26.5): An amino acid in a protein chain.

Resolution (Section 9.10): The process by which a racemic mixture is separated into its two pure enantiomers.

Resonance effect (Section 16.5): The donation or withdrawal of electrons through orbital overlap with neighboring π bonds. For example, an oxygen or nitrogen substituent donates electrons to an aromatic ring by overlap of the O or N orbital with the aromatic ring p orbitals.

Resonance form (Section 2.4): An individual Lewis structure of a resonance hybrid.

Resonance hybrid (Section 2.4): A molecule, such as benzene, that can't be represented adequately by a single Kekulé structure but must instead be considered as an average of two or more resonance structures. The resonance structures themselves differ only in the positions of their electrons, not their nuclei.

Restriction endonuclease (Section 28.15): An enzyme that is able to cleave a DNA molecule at points in the chain where a specific base sequence occurs.

Retrosynthetic (Sections 8.10, 16.12): A technique for planning organic syntheses by working backward from the final product to the starting material.

Ring current (Section 15.10): The circulation of π electrons induced in aromatic rings by an external magnetic field. This effect accounts for the downfield shift of aromatic ring protons in the ^{1}H NMR spectrum.

Ring-flip (Section 4.11): A molecular motion that converts one chair conformation of cyclohexane into another chair conformation. The effect of a ring-flip is to convert an axial substituent into an equatorial substituent.

RNA (Section 28.8): Ribonucleic acid; the biopolymer found in cells that serves to transcribe the genetic information found in DNA and uses that information to direct the synthesis of proteins.

Robinson annulation reaction (Section 23.13): A synthesis of cyclohexenones by sequential Michael reaction and intramolecular aldol reaction.

s-cis conformation (Section 14.9): The conformation of a conjugated diene that is cis-like around the single bond.

Saccharide (Section 25.1): A sugar.

Salt bridge (Section 26.13): The ionic attraction between two oppositely charged groups in a protein chain.

Sandmeyer reaction (Section 24.7): The nucleophilic substitution reaction of an arenediazonium salt with a cuprous halide to yield an aryl halide.

Saponification (Section 21.6): An old term for the base-induced hydrolysis of an ester to yield a carboxylic acid salt.

Saturated (Section 3.2): A molecule that has only single bonds and thus can't undergo addition reactions. Alkanes are saturated, but alkenes are unsaturated.

Sawhorse structure (Section 4.1): A manner of representing stereochemistry that uses a stick drawing and gives a perspective view of the conformation around a single bond.

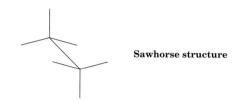

Sawhorse structure

Schiff base (Section 29.3): An alternative name for an imine, $R_2C=NR'$, used primarily in biochemistry.

Second-order reaction (Section 11.3): A reaction whose rate-limiting step is bimolecular and whose kinetics are therefore dependent on the concentration of two reactants.

Secondary (*see* Primary)

Secondary structure (Section 26.13): The level of protein substructure that involves organization of chain sections into ordered arrangements such as β-pleated sheets or α-helices.

Semiconservative replication (Section 28.12): The process by which DNA molecules are made, containing one strand of old DNA and one strand of new DNA.

Sense strand (Section 28.13): The strand of double-helical DNA that contains the gene.

Sequence rules (Sections 6.6, 9.6): A series of rules for assigning relative priorities to substituent groups on a double-bond carbon atom or on a chirality center.

Shell (electron) (Section 1.2): A group of an atom's electrons with the same principal quantum number.

Shielding (Section 13.2): An effect observed in NMR that causes a nucleus to absorb toward the right (upfield) side of the chart. Shielding is caused by donation of electron density to the nucleus.

Sigma (σ) bond (Section 1.6): A covalent bond formed by head-on overlap of atomic orbitals.

Sigmatropic reaction (Section 30.8): A pericyclic reaction that involves the migration of a group from one end of a π electron system to the other.

Simple protein (Section 26.12): A protein that yields only amino acids on hydrolysis.

Skeletal structure (Section 2.12): A shorthand way of writing structures in which carbon atoms are assumed to be at each intersection of two lines (bonds) and at the end of each line.

S$_N$1 reaction (Section 11.7): A unimolecular nucleophilic substitution reaction.

S$_N$2 reaction (Section 11.4): A bimolecular nucleophilic substitution reaction.

Solid-phase synthesis (Section 26.11): A technique of synthesis whereby the starting material is covalently bound to a solid polymer bead and reactions are carried out on the bound substrate. After the desired transformations have been effected, the product is cleaved from the polymer.

Solvation (Sections 5.8, 11.5): The clustering of solvent molecules around a solute particle to stabilize it.

sp **orbital** (Section 1.10): A hybrid orbital derived from the combination of an *s* and a *p* atomic orbital. The two *sp* orbitals that result from hybridization are oriented at an angle of 180° to each other.

sp^2 **orbital** (Section 1.9): A hybrid orbital derived by combination of an *s* atomic orbital with two *p* atomic orbitals. The three *sp*^{2} hybrid orbitals that result lie in a plane at angles of 120° to each other.

sp^3 **orbital** (Section 1.7): A hybrid orbital derived by combination of an *s* atomic orbital with three *p* atomic orbitals. The four *sp*^{3} hybrid orbitals that result are directed toward the corners of a regular tetrahedron at angles of 109° to each other.

Specific rotation, [α]$_D$ (Section 9.4): The specific rotation of a chiral compound is a physical constant that is defined by the equation

$$[\alpha]_D = \frac{\text{Observed rotation}}{\text{Pathlength} \times \text{Concentration}} = \frac{\alpha}{l \times C}$$

where the pathlength *l* of the sample solution is expressed in decimeters and the concentration *C* of the sample solution is expressed in grams per milliliter.

Sphingolipid (Section 27.3): A phospholipid that has sphingosine or a related dihydroxyamine as its backbone.

Spin–spin splitting (Section 13.11): The splitting of an NMR signal into a multiplet because of an interaction between nearby magnetic nuclei whose spins are coupled. The magnitude of spin–spin splitting is given by the coupling constant, *J*.

Staggered conformation (Section 4.1): The three-dimensional arrangement of atoms around a carbon–carbon single bond in which the bonds on one carbon bisect the bond angles on the second carbon as viewed end-on.

Staggered conformation

Step-growth polymer (Sections 21.10, 31.4): A polymer in which each bond is formed independently of the others. Polyesters and polyamides (nylons) are examples.

Stereochemistry (Chapters 4, 9): The branch of chemistry concerned with the three-dimensional arrangement of atoms in molecules.

Stereoisomers (Section 3.8): Isomers that have their atoms connected in the same order but have different three-dimensional arrangements. The term *stereoisomer* includes both enantiomers and diastereomers.

Stereospecific (Section 7.6): A term indicating that only a single stereoisomer is produced in a given reaction rather than a mixture.

Steric strain (Sections 4.3, 4.12): The strain imposed on a molecule when two groups are too close together and try to occupy the same space. Steric strain is responsible both for the greater stability of trans versus cis alkenes and for the greater stability of equatorially substituted versus axially substituted cyclohexanes.

Steroid (Section 27.7): A lipid whose structure is based on the tetracyclic carbon skeleton:

Steroids occur in both plants and animals and have a variety of important hormonal functions.

Straight-chain alkane (Section 3.2): An alkane whose carbon atoms are connected without branching.

Strecker synthesis (Section 26.3): A method for preparing an α-amino acid by treatment of an α-keto acid with NH$_3$ and KCN, followed by hydrolysis.

Substitution reaction (Section 5.1): What occurs when two reactants exchange parts to give two new products. S_N1 and S_N2 reactions are examples.

Sulfide (Section 18.11): A compound that has two organic substituents bonded to the same sulfur atom, RSR′.

Sulfone (Section 18.11): A compound of the general structure $RSO_2R′$.

Sulfoxide (Section 18.11): A compound of the general structure RSOR′.

Suprafacial (Section 30.6): A word used to describe the geometry of pericyclic reactions. Suprafacial reactions take place on the same side of the two ends of a π electron system.

Symmetry-allowed, symmetry-disallowed (Section 30.2): A symmetry-allowed reaction is a pericyclic process that has a favorable orbital symmetry for reaction through a concerted pathway. A symmetry-disallowed reaction is one that does not have favorable orbital symmetry for reaction through a concerted pathway.

Symmetry plane (Section 9.2): A plane that bisects a molecule such that one half of the molecule is the mirror image of the other half. Molecules containing a plane of symmetry are achiral.

Syn stereochemistry (Section 7.5): The opposite of anti. A syn addition reaction is one in which the two ends of the double bond are attacked from the same side. A syn elimination is one in which the two groups leave from the same side of the molecule.

Syndiotactic (Section 31.2): A chain-growth polymer in which the substituents regularly alternate on opposite sides of the backbone.

Tautomers (Sections 8.5, 22.1): Isomers that are rapidly interconverted.

Template strand (Section 28.13): The strand of double-helical DNA that does not contain the gene.

Terpene (Section 27.5): A lipid that is formally derived by head-to-tail polymerization of isoprene units.

Tertiary (*see* Primary)

Tertiary structure (Section 26.13): The level of protein structure that involves the manner in which the entire protein chain is folded into a specific three-dimensional arrangement.

Thermodynamic control (Section 14.6): An equilibrium reaction that yields the lowest-energy, most stable product is said to be thermodynamically controlled.

Thermoplastic (Section 31.5): A polymer that has a high T_g and is therefore hard at room temperature, but becomes soft and viscous when heated.

Thermosetting resin (Section 31.5): A polymer that becomes highly cross-linked and solidifies into a hard, insoluble mass when heated.

Thiol (Section 18.11): A compound containing the −SH functional group.

Thiol ester (Section 21.9): A compound with the RCOSR′ functional group.

Thiolate ion (Section 18.11): The anion of a thiol, RS^-.

TMS (Section 13.3): Tetramethylsilane, used as an NMR calibration standard.

Tollens' reagent (Section 19.3): A solution of Ag_2O in aqueous ammonia; used to oxidize aldehydes to carboxylic acids.

Torsional strain (Section 4.1): The strain in a molecule caused by electron repulsion between eclipsed bonds. Torsional strain is also called eclipsing strain.

Tosylate (Section 11.2): A *p*-toluenesulfonate ester.

Transamination (Section 29.6): The exchange of an amino group and a keto group between reactants.

Transcription (Section 28.13): The process by which the genetic information encoded in DNA is read and used to synthesize RNA in the nucleus of the cell. A small portion of double-stranded DNA uncoils, and complementary ribonucleotides line up in the correct sequence for RNA synthesis.

Transition state (Section 5.9): An activated complex between reactants, representing the highest energy point on a reaction curve. Transition states are unstable complexes that can't be isolated.

Translation (Section 28.14): The process by which the genetic information transcribed from DNA onto mRNA is read by tRNA and used to direct protein synthesis.

Tree diagram (Section 13.12): A diagram used in NMR to sort out the complicated splitting patterns that can arise from multiple couplings.

Triacylglycerol (Section 27.1):   A lipid, such as found in animal fat and vegetable oil, that is a tri-ester of glycerol with long-chain fatty acids.

Tricarboxylic acid cycle (Section 29.5):   An alternative name for the citric acid cycle by which acetyl CoA is degraded to CO_2.

Triplet (Section 13.7):   A symmetrical three-line splitting pattern observed in the ^{1}H NMR spectrum when a proton has two equivalent neighbor protons.

Twist-boat conformation (Section 4.14):   A conformation of cyclohexane that is somewhat more stable than a pure boat conformation.

Ultraviolet (UV) spectroscopy (Section 14.10): An optical spectroscopy employing ultraviolet irradiation. UV spectroscopy provides structural information about the extent of π electron conjugation in organic molecules.

Unimolecular reaction (Section 11.7):   A reaction that occurs by spontaneous transformation of the starting material without the intervention of other reactants. For example, the dissociation of a tertiary alkyl halide in the S_N1 reaction is a unimolecular process.

Unsaturated (Section 6.2):   A molecule that has one or more multiple bonds.

Upfield (Section 13.3):   The right-hand portion of the NMR chart.

Urethane (Section 31.4):   A functional group in which a carbonyl group is bonded to both an −OR group and an −NR$_2$ group.

Valence bond theory (Section 1.6):   A bonding theory that describes a covalent bond as resulting from the overlap of two atomic orbitals.

Valence shell (Section 1.5):   The outermost electron shell of an atom.

Van der Waals forces (Section 3.5):   Intermolecular forces that are responsible for holding molecules together in the liquid and solid states.

Vicinal (Section 8.3):   A term used to refer to a 1,2-disubstitution pattern. For example, 1,2-dibromoethane is a vicinal dibromide.

Vinyl group (Section 6.3):   A $H_2C=CH-$ substituent.

Vinyl monomer (Section 31.1):   A substituted alkene monomer used to make chain-growth polymers.

Vinylic (Section 8.4):   A term that refers to a substituent at a double-bond carbon atom. For example, chloroethylene is a vinylic chloride, and enols are vinylic alcohols.

Vitamin (Section 26.14):   A small organic molecule that must be obtained in the diet and is required in trace amounts for proper growth and functioning.

Vulcanization (Section 14.7):   A technique for cross-linking and hardening a diene polymer by heating with a few percent by weight of sulfur.

Walden inversion (Section 11.1):   The inversion of configuration at a chirality center that accompanies an S_N2 reaction.

Wave equation (Section 1.2):   A mathematical expression that defines the behavior of an electron in an atom.

Wave function (Section 1.2):   A solution to the wave equation for defining the behavior of an electron in an atom. The square of the wave function defines the shape of an orbital.

Wavelength (Section 12.5):   The length of a wave from peak to peak. The wavelength of electromagnetic radiation is inversely proportional to frequency and inversely proportional to energy.

Wavenumber (Section 12.6):   The reciprocal of the wavelength in centimeters.

Wax (Section 27.1):   A mixture of esters of long-chain carboxylic acids with long-chain alcohols.

Williamson ether synthesis (Section 18.3):   A method for synthesizing ethers by S_N2 reaction of an alkyl halide with an alkoxide ion.

Wittig reaction (Section 19.12):   The reaction of a phosphorus ylide with a ketone or aldehyde to yield an alkene.

Wohl degradation (Section 25.8):   A method for shortening the chain of an aldose sugar.

Wolff–Kishner reaction (Section 19.10):   The conversion of an aldehyde or ketone into an alkane by reaction with hydrazine and base.

Ylide (Section 19.12):   A neutral dipolar molecule with adjacent positive and negative charges. The phosphoranes used in Wittig reactions are ylides.

Zaitsev's rule (Section 11.10):   A rule stating that E2 elimination reactions normally yield the more highly substituted alkene as major product.

Ziegler–Natta catalyst (Section 31.2): A catalyst of an alkylaluminum and a titanium compound used for preparing alkene polymers.

Zusammen, Z (Section 6.6): A term used to describe the stereochemistry of a carbon–carbon double bond. The two groups on each carbon are assigned priorities according to the Cahn–Ingold–Prelog sequence rules, and the two carbons are compared. If the high-priority groups on each carbon are on the same side of the double bond, the bond has Z geometry.

Zwitterion (Section 26.1): A neutral dipolar molecule in which the positive and negative charges are not adjacent. For example, amino acids exist as zwitterions, H_3N^+–CHR–COO$^-$. Zwitterions are also called *betaines*.

APPENDIX E
Answers to Selected In-Text Problems

The following answers are meant only as a quick check while you study. Full answers for all problems are provided in the accompanying *Study Guide and Solutions Manual.*

CHAPTER 1

1.1 (a) $1s^2\ 2s^2\ 2p^1$ (b) $1s^2\ 2s^2\ 2p^6\ 3s^2\ 3p^3$
(c) $1s^2\ 2s^2\ 2p^4$ (d) $1s^2\ 2s^2\ 2p^6\ 3s^2\ 3p^5$

1.2 (a) 1 (b) 3 (c) 8

1.3

1.4

1.5 (a) $GeCl_4$ (b) AlH_3 (c) CH_2Cl_2 (d) SiF_4
(e) CH_3NH_2

1.6 (a)

(b)

(c)

(d) Na:H   Na—H   (e)

1.7 C_2H_7 has too many hydrogens for a compound with 2 carbons.

1.8 All bond angles are near 109°.

1.9 The CH_3 carbon is sp^3; the double-bond carbons are sp^2; the C=C–C and C=C–H bond angles are approximately 120°; other bond angles are near 109°.

1.10 All carbons are sp^2; all bond angles are near 120°.

1.11

1.12 All carbons except CH_3 are sp^2.

1.13 The CH_3 carbon is sp^3; the triple-bond carbons are sp; the C≡C–C and H–C≡C bond angles are approximately 180°.

1.14 The nitrogen atom is sp^2.

1.15 All are sp^3-hybridized and have roughly tetrahedral geometry.

CHAPTER 2

2.1 (a) H (b) Br (c) Cl (d) C

2.2 (a) $\overset{\delta+}{C}-\overset{\delta-}{Br}$ (b) $\overset{\delta+}{C}-\overset{\delta-}{N}$ (c) $\overset{\delta-}{C}-\overset{\delta+}{Li}$ (d) $\overset{\delta-}{N}-\overset{\delta+}{H}$
(e) $\overset{\delta+}{C}-\overset{\delta-}{O}$ (f) $\overset{\delta-}{C}-\overset{\delta+}{Mg}$ (g) $\overset{\delta+}{C}-\overset{\delta-}{F}$

2.3 H_3C–OH < H_3C–MgBr < H_3C–Li = H_3C–F < H_3C–K

2.4

2.5 The two C=O dipoles cancel because of the 180° O=C=O bond angle.

A-38

2.6 (a)

No dipole moment

(b)

(c)

(d)

2.7 For sulfur: $FC = 6 - \dfrac{6}{2} - 2 = +1$;

for oxygen: $FC = 6 - \dfrac{2}{2} - 6 = -1$

2.8 (a) For carbon: $FC = 4 - \dfrac{8}{2} - 0 = 0$;

for the middle nitrogen:

$FC = 5 - \dfrac{8}{2} - 0 = +1$;

for the end nitrogen: $FC = 5 - \dfrac{4}{2} - 4 = -1$

(b) For nitrogen: $FC = 5 - \dfrac{8}{2} - 0 = +1$;

for oxygen: $FC = 6 - \dfrac{2}{2} - 6 = -1$

(c) For nitrogen: $FC = 5 - \dfrac{8}{2} - 0 = +1$;

for the end carbon: $FC = 4 - \dfrac{6}{2} - 2 = -1$

2.9 (a)

(b) $H_2C=CH-CH_2^+ \longleftrightarrow H_2\overset{+}{C}-CH=CH_2$

(c) $:N\equiv\overset{+}{N}-\overset{..}{\underset{..}{N}}^--H \longleftrightarrow {}^-\overset{..}{:}N=\overset{+}{N}=\overset{..}{N}-H$

(d)

2.10 $HNO_3 + NH_3 \rightarrow NH_4^+ + NO_3^-$

Acid Base Conjugate Conjugate
            acid   base

2.11 Picric acid is stronger.

2.12 Water is a stronger acid.

2.13 Neither reaction will take place.
2.14 Reaction will take place.
2.15 $K_a = 4.9 \times 10^{-10}$

2.16 (a) $CH_3CH_2\overset{..}{\underset{..}{O}}H + H-Cl \longrightarrow CH_3CH_2\overset{H}{\overset{|+}{\underset{..}{O}}}H + Cl^-$

or $(CH_3)_2\overset{..}{N}H$ or $(CH_3)_3P:$

(b) $H_3C^+ + {}^-\overset{..}{:}OH \longrightarrow H_3C-OH$

or $(CH_3)_3B$ or Br_2Mg

2.17 (a) For boron: $FC = 3 - \dfrac{8}{2} - 0 = -1$;

for oxygen: $FC = 6 - \dfrac{6}{2} - 2 = +1$

(b) For aluminum: $FC = 3 - \dfrac{8}{2} - 0 = -1$;

for nitrogen: $FC = 5 - \dfrac{8}{2} - 0 = +1$

2.18 (a) $C_9H_{13}NO_3$ (b) $C_{18}H_{22}O_2$

2.19 (a)

$\overset{\overset{\displaystyle CH_3}{|}}{CH_3CHCH_2CH_3}$

(b) $CH_3CH_2NH_2$

(c)

(d)

$\overset{\overset{\displaystyle Cl}{|}}{CH_3CHCH_2CH_3}$

2.20

CHAPTER 3

3.1 (a) Amide, double bond
(b) Amine, carboxylic acid
(c) Double bond, ketone, ester
(d) Aromatic ring, double bond, alcohol

3.2 (a) CH_3OH (b)

(c) $CH_3\overset{\overset{\displaystyle O}{\|}}{C}OH$

(d) CH_3NH_2 (e) $CH_3\overset{\overset{\displaystyle O}{\|}}{C}CH_2CH_2NH_2$

(f) $H_2C=CHCH=CH_2$

3.3

$C_8H_{13}NO_2$

3.4

CH$_3$CH$_2$CH$_2$CH$_2$CH$_2$CH$_3$ CH$_3$CHCH$_2$CH$_2$CH$_3$
with CH$_3$ above

CH$_3$CH$_2$CHCH$_2$CH$_3$ CH$_3$CCH$_2$CH$_3$
with CH$_3$ above (left); CH$_3$ above and CH$_3$ below (right)

CH$_3$CHCHCH$_3$
with CH$_3$ above and CH$_3$ below

3.5   CH$_3$CH$_2$CH$_2$CH$_2$OH   CH$_3$CHCH$_2$CH$_3$ (with OH above)

CH$_3$CHCH$_2$OH (with CH$_3$ above)   CH$_3$CCH$_3$ (with OH above, CH$_3$ below)   CH$_3$OCH$_2$CH$_2$CH$_3$

CH$_3$CH$_2$OCH$_2$CH$_3$ CH$_3$CHCH$_3$ (with OCH$_3$ above)

3.6   Part (a) has nine possible answers.

(a) CH$_3$CH$_2$CH$_2$COCH$_3$ (with O double bond above the C)   CH$_3$CH$_2$COCH$_2$CH$_3$ (with O above)

(b) CH$_3$CH$_2$CH$_2$C≡N   CH$_3$CHC≡N (with CH$_3$ above)

3.7   (a) Two   (b) Four

3.8   CH$_3$CH$_2$CH$_2$CH$_2$CH$_2$—   CH$_3$CH$_2$CH$_2$CH— (with CH$_3$ below)

CH$_3$CH$_2$CH— (with CH$_2$CH$_3$ below)   CH$_3$CHCH$_2$CH$_2$— (with CH$_3$ above)

CH$_3$CH$_2$CHCH$_2$— (with CH$_3$ above)   CH$_3$CH$_2$C— (with CH$_3$ above, CH$_3$ below)   CH$_3$CHCH— (with CH$_3$ above, CH$_3$ below)

CH$_3$CCH$_2$— (with CH$_3$ above, CH$_3$ below)

3.9   (a) 3 primary, 2 secondary, 1 tertiary
(b) 4 primary, 2 secondary, 2 tertiary
(c) 5 primary, 1 secondary, 1 tertiary,
    1 quaternary

3.10   Primary carbons have primary hydrogens,
secondary carbons have secondary
hydrogens, and tertiary carbons have tertiary
hydrogens.

3.11   (a) CH$_3$CHCHCH$_3$ (with CH$_3$ below)   (b) CH$_3$CH$_2$CHCH$_2$CH$_3$ (with CH$_3$CHCH$_3$ above)

(c) CH$_3$CCH$_2$CH$_3$ (with CH$_3$ above and CH$_3$ below)

3.12   (a) Pentane, 2-methylbutane, 2,2-
dimethylpropane   (b) 3,4-Dimethylhexane
(c) 2,4-Dimethylpentane
(d) 2,2,5-Trimethylheptane

3.13   (a) CH$_3$CH$_2$CHCHCH$_2$CH$_2$CH$_2$CH$_3$ (with CH$_3$ above and CH$_3$ below)

(b) CH$_3$CH$_2$CH$_2$C—CHCH$_2$CH$_3$ (with H$_3$C and CH$_2$CH$_3$ above, H$_3$C below)

(c) CH$_3$CCH$_2$CHCH$_2$CH$_2$CH$_2$CH$_3$ (with CH$_3$ and CH$_2$CH$_2$CH$_3$ above, CH$_3$ below)

(d) CH$_3$CCH$_2$CHCH$_3$ (with CH$_3$ and CH$_3$ above, CH$_3$ below)

3.14   Pentyl, 1-methylbutyl, 1-ethylpropyl,
3-methylbutyl, 2-methylbutyl,
1,1-dimethylpropyl, 1,2-dimethylpropyl,
2,2-dimethylpropyl

3.15

CH$_3$CH$_2$C—CHCHCH$_2$CH$_3$ (with H$_3$C and CH$_3$ above, H$_3$C and CH$_3$ below)   **3,3,4,5-Tetramethyl-heptane**

3.16   (a) 1,4-Dimethylcyclohexane
(b) 1-Methyl-3-propylcyclopentane
(c) 3-Cyclobutylpentane
(d) 1-Bromo-4-ethylcyclodecane
(e) 1-Isopropyl-2-methylcyclohexane
(f) 4-Bromo-1-*tert*-butyl-2-methylcycloheptane

3.17   (a)   (b)   (c)   (d)

3.18 (a) *trans*-1-Chloro-4-methylcyclohexane
(b) *cis*-1-Ethyl-3-methylcycloheptane

3.19 (a) [structure] (b) [structure]

(c) [structure]

CHAPTER 4

4.3 (a) [structure] (b) [structure]

4.4 [structure]

4.5 Strain = 3 × 3.8 kJ/mol = 11.4 kJ/mol
[structure]

4.6 Cyclopropane

4.7 Six interactions; 21% of strain

4.8 The cis isomer is less stable because the methyl groups eclipse each other.

4.9 Ten eclipsing interactions; 40 kJ/mol; 35% is relieved.

4.10 [structures]

Most stable   **Least stable**

4.11 [structures]

4.12–4.13 Axial and equatorial positions alternate around the ring on each side.

4.14 [structures]

4.15 With a methyl, an ethyl, and an isopropyl group, a hydrogen points in toward the ring. With a *tert*-butyl group, a methyl points in.

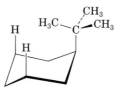

4.16 Cyano group points straight up.

4.17 Equatorial = 70%; axial = 30%

4.18 (a) 2.0 kJ/mol   (b) 11.4 kJ/mol
(c) 2.0 kJ/mol   (d) 8.0 kJ/mol

4.19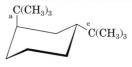

**1-Chloro-2,4-dimethyl-
cyclohexane
(less stable chair form)**

4.20 Both *tert*-butyl groups can be equatorial in the twist-boat conformation.

[structure] **Chair**

[structure] **Twist-boat**

4.21 *trans*-Decalin is more stable because it has no 1,3-diaxial interactions.

CHAPTER 5

5.1 (a) Substitution   (b) Elimination
(c) Addition

5.2 1-Chloro-2-methylpentane, 2-chloro-2-methylpentane, 3-chloro-2-methylpentane, 2-chloro-4-methylpentane, 1-chloro-4-methylpentane

5.3 Pentane has three types of hydrogens; neopentane has only one.

5.4 Electrophiles: HCl, CH_3CHO, CH_3SH; nucleophiles: CH_3NH_2, CH_3SH

5.5 Bromocyclohexane; chlorocyclohexane

5.6 (a) $\overset{\frown}{Cl}-\overset{\frown}{Cl} + :NH_3 \longrightarrow ClNH_3^+ + Cl^-$

(b) $CH_3\ddot{\overset{..}{O}}:^- + H_3C\overset{\frown}{-}Br \longrightarrow CH_3OCH_3 + Br^-$

(c)

5.7 $H_2O + H_2C=CH_2 + Br^-$

5.8 Negative $\Delta G°$ more favored

5.9 Larger K_{eq} more exergonic

5.10 $\Delta G° = -17.1$ kJ/mol, 0 kJ/mol, +17.1 kJ/mol; $K_{eq} = 1.0 \times 10^7$, 1, 1.0×10^{-7}

5.11 $\Delta H° = -28$ kJ/mol; less exothermic

5.12 (a) $\Delta H° = -33$ kJ/mol
(b) $\Delta H° = +33$ kJ/mol

5.13 Lower $\Delta G^{\ddagger}$ is faster. Can't predict K_{eq}.

CHAPTER 6

6.1 (a) 2 (b) 3 (c) 3 (d) 5 (e) 13
6.2 (a) 1 (b) 2 (c) 2
6.3 (a) 5 (b) 5 (c) 3 (d) 1 (e) 6 (f) 5
6.4 (a) 3,4,4-Trimethyl-1-pentene
(b) 3-Methyl-3-hexene
(c) 4,7-Dimethyl-2,5-octadiene

6.5 (a) $\overset{\quad\;\; CH_3}{H_2C{=}CCH_2CH_2CH{=}CH_2}$

(b) $\overset{\qquad\qquad\quad CH_2CH_3}{CH_3CH_2CH_2CH{=}CC(CH_3)_3}$

(c) $\underset{\quad\;\; H_3C}{\overset{\quad\;\; H_3C\quad CH_3}{CH_3CH{=}CHCH{=}CHC-C{=}CH_2}}$

(d) $\underset{(CH_3)_2CH\qquad CH(CH_3)_2}{\overset{(CH_3)_2CH\qquad CH(CH_3)_2}{C{=}C}}$

(e) $\underset{}{\overset{CH_3\quad C(CH_3)_3}{CH_3CHCH_2CHCH_2CH_2CH_3}}$

6.6 (a) 1,2-Dimethylcyclohexene
(b) 4,4-Dimethylcycloheptene
(c) 3-Isopropylcyclopentene

6.7 Compounds (c), (e), and (f) have cis–trans isomers.

6.8 *trans*-Cyclohexene is too strained.

6.9 (a) –Br (b) –Br (c) –CH$_2$CH$_3$ (d) –OH
(e) –CH$_2$OH (f) –CH=O

6.10 (a) –Cl, –OH, –CH$_3$, –H
(b) –CH$_2$OH, –CH=CH$_2$, –CH$_2$CH$_3$, –CH$_3$
(c) –COOH, –CH$_2$OH, –C≡N, –CH$_2$NH$_2$
(d) –CH$_2$OCH$_3$, –C≡N, –C≡CH, –CH$_2$CH$_3$

6.11 (a) Z (b) E (c) Z (d) E

6.12 $\underset{(CH_3)_2CH\qquad CH_2OH}{\overset{H_2C{=}CH\qquad CO_2CH_3}{C{=}C}}$ z

6.13 (a) 2-Methylpropene (b) E isomer
(c) 1-Methylcyclohexene

6.14 (a) Chlorocyclohexane
(b) 2-Bromo-2-methylpentane
(c) 2-Iodopentane
(d) 1-Bromo-1-methylcyclohexane

6.15 (a) Cyclopentene (b) 1-Ethylcyclohexene or ethylidenecyclohexane (c) 3-Hexene
(d) Cyclohexylethylene

6.16 (a) $\underset{}{\overset{CH_3\quad CH_3}{CH_3CH_2\overset{+}{C}CH_2CHCH_3}}$ (b)

6.17 In the conformation shown, only the methyl-group C–H that is parallel to the carbocation p orbital can show hyperconjugation.

6.18 The second step is exergonic; the transition state resembles the carbocation.

6.19

CHAPTER 7

7.1 2-Methyl-2-butene and 2-methyl-1-butene
7.2 Five
7.3 *trans*-1,2-Dichloro-1,2-dimethylcyclohexane
7.4

7.5 *trans*-2-Bromocyclopentanol
7.6 Markovnikov
7.7 (a) 2-Pentanol (b) 2-Methyl-2-pentanol
7.8 (a) 2-Methyl-1-hexene or 2-methyl-2-hexene
(b) Cyclohexylethylene

7.9

7.10 (a) 3-Methyl-1-butene
(b) 2-Methyl-2-butene
(c) Methylenecyclohexane

7.11

7.12 (a) [structure: spirocyclohexane cyclopropane with two Cl] (b) CH_3 CH_2 / $CH_3CHCH_2CH—CHCH_3$

7.13 (a) 2-Methylpentane
(b) 1,1-Dimethylcyclopentane

7.14 (a) 1-Methylcyclohexene
(b) 2-Methyl-2-pentene   (c) 1,3-Butadiene

7.15 (a) $CH_3COCH_2CH_2CH_2CH_2COOH$
(b) $CH_3COCH_2CH_2CH_2CH_2CHO$

7.16 (a) 2-Methylpropene   (b) 3-Hexene

7.17 (a) $H_2C=CHOCH_3$ (b) $ClCH=CHCl$

7.18

$\cdot CH_2\dot CH_2 + \cdot CH—CH_2 \longrightarrow$

$\cdot CH_2CH_3 + \cdot CH=CH_2$

7.19

$\left(—CH_2CH— \right)_n$ with $OC(CH_3)_3$

CHAPTER 8

8.1 (a) 2,5-Dimethyl-3-hexyne
(b) 3,3-Dimethyl-1-butyne
(c) 2,4-Octadiene-6-yne
(d) 3,3-Dimethyl-4-octyne
(e) 2,5,5-Trimethyl-3-heptyne
(f) 6-Isopropylcyclodecyne

8.2 1-Hexyne, 2-hexyne, 3-hexyne, 3-methyl-1-pentyne, 4-methyl-1-pentyne, 4-methyl-2-pentyne, 3,3-dimethyl-1-butyne

8.3 (a) 1,1,2,2-Tetrachloropentane
(b) 1-Bromo-1-cyclopentylethylene
(c) 2-Bromo-2-heptene and 3-bromo-2-heptene

8.4 4-Octanone; 2-methyl-4-octanone and 7-methyl-4-octanone

8.5 (a) 1-Pentyne   (b) 2-Pentyne

8.6 (a) $C_6H_5C\equiv CH$ (b) 2,5-Dimethyl-3-hexyne

8.7 (a) Reduce 2-octyne with Li/NH_3
(b) Reduce 3-heptyne with H_2/Lindlar catalyst   (c) Reduce 3-methyl-1-pentyne

8.8 (a) $C_6H_5C\equiv CH$
(b) $CH_3(CH_2)_7C\equiv C(CH_2)_7C\equiv C(CH_2)_7CH_3$

8.9 No: (a), (c), (d); yes: (b)

8.10 (a) 1-Pentyne + CH_3I, or propyne + $CH_3CH_2CH_2I$
(b) 3-Methyl-1-butyne + CH_3CH_2I
(c) Cyclohexylacetylene + CH_3I
(d) 4-Methyl-1-pentyne + CH_3I, or propyne + $(CH_3)_2CHCH_2I$
(e) 3,3-Dimethyl-1-butyne + CH_3CH_2I

8.11 $CH_3C\equiv CH \xrightarrow[\text{2. } CH_3I]{\text{1. } NaNH_2} CH_3C\equiv CCH_3$

$\xrightarrow[\substack{\text{Lindlar} \\ \text{cat.}}]{H_2} cis\text{-}CH_3CH=CHCH_3$

8.12 (a) $KMnO_4$, H_3O^+ (b) H_2/Lindlar catalyst
(c) 1. H_2/Lindlar catalyst; 2. HBr
(d) 1. H_2/Lindlar catalyst; 2. BH_3; 3. NaOH, H_2O_2
(e) 1. H_2/Lindlar catalyst; 2. Cl_2

8.13 (a) 1. $HC\equiv CH$ + $NaNH_2$; 2. $CH_3(CH_2)_7Cl$; 3. $2 H_2$/Pd
(b) 1. $HC\equiv CH$ + $NaNH_2$; 2. $(CH_3)_3CCH_2CH_2I$; 3. $2 H_2$/Pd
(c) 1. $HC\equiv CH$ + $NaNH_2$; 2. $CH_3CH_2CH_2CH_2I$; 3. BH_3; 4. H_2O_2
(d) 1. $HC\equiv CH$ + $NaNH_2$; 2. $CH_3CH_2CH_2CH_2CH_2I$; 3. $HgSO_4$, H_3O^+

CHAPTER 9

9.1 Chiral: screw, beanstalk, shoe

9.2 Chiral: (b)

9.3 (a) [structure: cyclohexane with HO, CH_3, isopropyl groups, marked with asterisks] (b) [bicyclic camphor-like structure with =O]

(c) [steroid-like structure with CH_3O, NCH_3, H]

9.4 [two structures:] $COOH$ / C / H, CH_3, H_2N and $COOH$ / C / H_3C, H, NH_2

9.5 $+16.1°$

9.6 (a) $-Br$, $-CH_2CH_2OH$, $-CH_2CH_3$, $-H$
(b) $-OH$, $-CO_2CH_3$, $-CO_2H$, $-CH_2OH$
(c) $-NH_2$, $-CN$, $-CH_2NHCH_3$, $-CH_2NH_2$
(d) $-Br$, $-Cl$, $-CH_2Br$, $-CH_2Cl$

9.7 (a) S (b) R (c) S

9.8 (a) S (b) S (c) R

9.9 [structure: H / C / HO, H_3C, $CH_2CH_2CH_3$]

9.10 S

9.11 (a) R,R (b) S,R (c) R,S (d) S,S
Compounds (a) and (d) are enantiomers and are diastereomeric to (b) and (c).

9.12 R,R　**9.13** S,S

9.14 (a), (d)　**9.15** (a), (c)

9.16

H_3C　CH_3

Meso

OH

9.17 Five chirality centers; 32 stereoisomers

9.18 Two diastereomeric salts: (R)-lactic acid plus (S)-1-phenylethylamine and (S)-lactic acid plus (S)-1-phenylethylamine

9.19 (a) Constitutional isomers　(b) Diastereomers

9.20 A and B are identical; C and D are identical

9.21 (a) Enantiomers　(b) Enantiomers

9.22

H

H_3C ── CH_2CH_3

Cl

9.23 (a) S　(b) S　(c) R

9.24

CH_2CH_3

H ── Cl　　R

CH_3

9.25 Non-50:50 mixture of a racemic pair

9.26 Non-50:50 mixture of a racemic pair

9.27 Non-50:50 mixture of two racemic pairs; optically inactive

9.28 Non-50:50 mixture of two racemic pairs

CHAPTER 10

10.1 (a) 1-Iodobutane

(b) 1-Chloro-3-methylbutane

(c) 1,5-Dibromo-2,2-dimethylpentane

(d) 1,3-Dichloro-3-methylbutane

(e) 1-Chloro-3-ethyl-4-iodopentane

(f) 2-Bromo-5-chlorohexane

10.2 (a) $CH_3CH_2CH_2C(CH_3)_2CH(Cl)CH_3$

(b) $CH_3CH_2CH_2C(Cl)_2CH(CH_3)_2$

(c) $CH_3CH_2C(Br)(CH_2CH_3)_2$

(d)

Br

Br

(e)

$CH_3CHCH_2CH_3$

$CH_3CH_2CH_2CH_2CH_2CHCH_2CHCH_3$

Cl

(f)

Br

Br

10.3 Chiral: 1-chloro-2-methylpentane, 3-chloro-2-methylpentane, 2-chloro-4-methylpentane
Achiral: 2-chloro-2-methylpentane, 1-chloro-4-methylpentane

10.4 1-Chloro-2-methylbutane (29%), 1-chloro-3-methylbutane (14%), 2-chloro-2-methylbutane (24%), 2-chloro-3-methylbutane (33%)

10.5 For Cl·, $\Delta H° = -31\,kJ/mol$; for Br·, $\Delta H° = +35\,kJ/mol$. Bromination is more selective.

10.6

10.7 The intermediate allylic radical reacts at the more accessible site and gives the more highly substituted double bond.

10.8 (a) 3-Bromo-5-methylcycloheptene and 3-bromo-6-methylcycloheptene

(b) Four products

10.9 (a) 2-Methyl-2-propanol + HCl

(b) 4-Methyl-2-pentanol + PBr_3

(c) 5-Methyl-1-pentanol + PBr_3

(d) 2,4-Dimethyl-2-hexanol + HCl

10.10 Both reactions occur.

10.11 React Grignard reagent with D_2O.

10.12 (a) 1. NBS; 2. $(CH_3)_2CuLi$　(b) 1. Li; 2. CuI; 3. $CH_3CH_2CH_2CH_2Br$　(c) 1. BH_3; 2. H_2O_2, NaOH; 3. PBr_3; 4. Li, then CuI; 5. $CH_3(CH_2)_4Br$

10.13 (a)

(b) $CH_3CH_2NH_2$ < $H_2NCH_2CH_2NH_2$ < $CH_3C≡N$

10.14 (a) Reduction　(b) Neither

CHAPTER 11

11.1 (R)-1-Methylpentyl acetate, $CH_3CO_2CH(CH_3)CH_2CH_2CH_2CH_3$

11.2 (S)-2-Butanol

11.3 (S)-2-Bromo-4-methylpentane ⟶

CH_3　SCH_3

(R) $CH_3CHCH_2CHCH_3$

11.4 Back-side attack is too hindered.

11.5 (a) 1-Iodobutane　(b) 1-Butanol
(c) 1-Hexyne　(d) Butylammonium bromide

11.6 (a) $(CH_3)_2N^-$　(b) $(CH_3)_3N$　(c) H_2S

11.7 CH_3OTos > CH_3Br > $(CH_3)_2CHCl$ > $(CH_3)_3CCl$

11.8 Similar to protic solvents

11.9 Racemic 1-ethyl-1-methylhexyl acetate

11.10

(S)-Bromide ⟶ **Racemic**

11.11 90.1% racemization; 9.9% inversion

11.12 $H_2C=CHCH(Br)CH_3 > CH_3CH(Br)CH_3 > CH_3CH_2Br > H_2C=CHBr$

11.13 The same allylic carbocation intermediate is formed.

11.14 The rate-limiting step of this S_N1 reaction does not involve the nucleophile.

11.15 (a) S_N1 (b) S_N2

11.16 (a) 2-Methyl-2-pentene
(b) 2,3,5-Trimethyl-2-hexene
(c) Ethylidenecyclohexane

11.17 (Z)-1-Bromo-1,2-diphenylethylene

11.18 (Z)-3-Methyl-2-pentene

11.19 Cis isomer reacts faster because the bromine is axial.

11.20 (a) S_N2 (b) E2 (c) S_N1

CHAPTER 12

12.1 (a) C_6H_{14}, $C_5H_{10}O$, $C_4H_6O_2$, $C_3H_2O_3$
(b) C_9H_{20}, C_9H_4O, $C_{10}H_8$, $C_8H_{16}O$, $C_7H_{12}O_2$, $C_6H_8O_3$, $C_5H_4O_4$
(c) $C_{11}H_{24}$, $C_{12}H_{12}$, $C_{11}H_8O$, $C_{10}H_{20}O$, $C_{10}H_4O_2$, $C_9H_{16}O_2$, $C_8H_{12}O_3$, $C_7H_8O_4$, $C_6H_4O_5$

12.2 $C_{15}H_{22}O$, $C_{14}H_{18}O_2$, $C_{13}H_{14}O_3$, $C_{12}H_{10}O_4$, $C_{11}H_6O_5$, $C_{16}H_{10}$

12.3 (a) 2-Methyl-2-pentene (b) 2-Hexene

12.4 (a) 43, 71 (b) 82 (c) 58 (d) 86

12.5 X-ray energy is higher.

12.6 $\lambda = 9.0 \times 10^{-6}$ m is higher in energy.

12.7 (a) 2.4×10^6 kJ/mol (b) 4.0×10^4 kJ/mol
(c) 2.4×10^3 kJ/mol (d) 2.8×10^2 kJ/mol
(e) 6.0 kJ/mol (f) 4.0×10^{-2} kJ/mol

12.8 (a) $3225\,cm^{-1}$ (b) $1710\,cm^{-1}$ (c) $4.44\,\mu m$
(d) $10.3\,\mu m$

12.9 (a) Ketone or aldehyde
(b) Nitro compound
(c) Carboxylic acid

12.10 (a) CH_3CH_2OH has an –OH absorption.
(b) 1-Hexene has a double-bond absorption.
(c) CH_3CH_2COOH has a very broad –OH absorption.

12.11 $1450–1600\,cm^{-1}$: aromatic ring;
$2100\,cm^{-1}$: C≡C; $3300\,cm^{-1}$: C≡C–H

12.12 (a) $1715\,cm^{-1}$ (b) 1730, 2100, $3300\,cm^{-1}$
(c) 1720, $2500–3100\,cm^{-1}$, $3400–3650\,cm^{-1}$

12.13 1690, 1650, $2230\,cm^{-1}$

CHAPTER 13

13.1 2.2×10^{-5} kJ/mol for ^{19}F; 2.4×10^{-5} kJ/mol for ^1H

13.2 4.0×10^{-5} kJ/mol

13.3 The vinylic C–H protons are nonequivalent.

13.4 (a) 126 Hz (b) 2.1 δ (c) 210 Hz

13.5 (a) 7.27 δ (b) 3.05 δ (c) 3.47 δ
(d) 5.30 δ

13.6 $-CH_3$, 9.3 δ; $-CH_2-$, 27.6 δ; C=O, 174.6 δ; $-OCH_3$, 51.4 δ

13.7 (a) 4 (b) 7 (c) 4 (d) 5

13.8 (a) 1,3-Dimethylcyclopentene
(b) 2-Methylpentane
(c) 1-Chloro-2-methylpropane

13.9

13.10

13.11

13.12 A DEPT-90 spectrum would show two absorptions for the non-Markovnikov product (RCH=CHBr) but no absorptions for the Markovnikov product ($RBrC=CH_2$).

13.13 (a) 2 (b) 4 (c) 3 (d) 4 (e) 5 (f) 3

13.14 5

13.15 (a) 1.43 δ (b) 2.17 δ (c) 7.37 δ (d) 9.70 δ
(e) 5.30 δ (f) 2.12 δ

13.16 Seven kinds of protons

13.17 Two peaks; 3:2 ratio

13.18 (a) $-CHBr_2$, quartet; $-CH_3$, doublet
(b) CH_3O-, singlet; $-OCH_2-$, triplet; $-CH_2Br$, triplet
(c) $ClCH_2-$, triplet; $-CH_2-$, quintet
(d) CH_3-, triplet; $-CH_2-$, quartet; $-CH-$, septet; $(CH_3)_2$, doublet
(e) CH_3-, triplet; $-CH_2-$, quartet; $-CH-$, septet; $(CH_3)_2$, doublet
(f) =CH, triplet, $-CH_2-$, doublet, aromatic C–H, doublet

13.19 (a) CH_3OCH_3 (b) $CH_3CH(Cl)CH_3$
(c) $ClCH_2CH_2OCH_2CH_2Cl$
(d) $CH_3CH_2CO_2CH_3$ or $CH_3CO_2CH_2CH_3$

13.20 $CH_3CH_2OCH_2CH_3$
13.21 $J_{1-2} = 16$ Hz; $J_{2-3} = 8$ Hz
13.22 1-Chloro-1-methylcyclohexane has a singlet methyl absorption.

CHAPTER 14

14.1 Conjugated: (b), (c), (d), (f)
14.2 Expected ΔH_{hydrog} for allene is -252 kJ/mol. Allene is less stable than either a conjugated or a nonconjugated diene.
14.3 1-Chloro-2-pentene, 3-chloro-1-pentene, 4-chloro-2-pentene
14.4 4-Chloro-2-pentene predominates in both.
14.5 Interconversion occurs by S_N1 dissociation to a common intermediate cation.
14.6 The double bond is more highly substituted.
14.7

14.8

14.9

14.10 Good dienophiles: (a), (d)
14.11 Compound (a) is *s*-cis. Compound (c) can rotate to *s*-cis.
14.12

14.13 300–600 kJ/mol
14.14 UV energy is greater than IR or NMR energy.
14.15 1.46×10^{-5} M
14.16 All except (a) have UV absorptions.

CHAPTER 15

15.1 (a) Meta   (b) Para   (c) Ortho
15.2 (a) *m*-Bromochlorobenzene
        (b) (3-Methylbutyl)benzene

(c) *p*-Bromoaniline
(d) 2,5-Dichlorotoluene
(e) 1-Ethyl-2,4-dinitrobenzene
(f) 1,2,3,5-Tetramethylbenzene

15.3 (a)   (b)

15.4 Pyridine has an aromatic sextet of electrons.
15.5 Cyclodecapentaene is not flat because of steric interactions.
15.6 All C–C bonds are equivalent; one resonance line in both ^1H and ^{13}C NMR spectra.
15.7 The cyclooctatetraenyl dianion is aromatic (ten π electrons) and flat.
15.8 The singly bonded nitrogen is pyrrole-like, and the doubly bonded nitrogen is pyridine-like.
15.9

15.10

15.11

15.12 5

CHAPTER 16

16.1 *o*-, *m*-, and *p*-Bromotoluene
16.2 *o*-Xylene: 2; *p*-xylene: 1; *m*-xylene: 3
16.3 D^+ does electrophilic substitutions on the ring.
16.4 No rearrangement: (a), (b), (e)
16.5 *tert*-Butylbenzene
16.6 (a) $(CH_3)_2CHCOCl$ (b) PhCOCl

16.7 (a) *o*- and *p*-Bromonitrobenzene
(b) *m*-Bromonitrobenzene
(c) *o*- and *p*-Chlorophenol
(d) *o*- and *p*-Bromoaniline

16.8

and others

16.9

and others

16.10 (a) Phenol > Toluene > Benzene > Nitrobenzene
(b) Phenol > Benzene > Chlorobenzene > Benzoic acid
(c) Aniline > Benzene > Bromobenzene > Benzaldehyde

16.11 Alkylbenzenes are more reactive than benzene itself, but acylbenzenes are less reactive.

16.12 Toluene is more reactive; the trifluoromethyl group is electron-withdrawing.

16.13 The nitrogen electrons are donated to the nearby carbonyl group and are less available to the ring.

16.14 The meta intermediate is most favored.

16.15 (a) Ortho and para to $-OCH_3$
(b) Ortho and para to $-NH_2$
(c) Ortho and para to $-Cl$

16.16 Addition of $^-OCH_3$, followed by elimination of Cl^-.

16.17 Only one benzyne intermediate can form from *p*-bromotoluene; two different benzyne intermediates can form from *m*-bromotoluene.

16.18 (a) *m*-Nitrobenzoic acid
(b) *p-tert*-Butylbenzoic acid

16.19 1. CH_3CH_2Cl, $AlCl_3$; 2. NBS; 3. KOH, ethanol

16.20 A benzyl radical is more stable than a primary alkyl radical by 52 kJ/mol and is similar in stability to an allyl radical.

16.21 1. PhCOCl, $AlCl_3$; 2. H_2/Pd

16.22 (a) 1. HNO_3, H_2SO_4; 2. Cl_2, $FeCl_3$
(b) 1. CH_3COCl, $AlCl_3$; 2. Cl_2, $FeCl_3$; 3. H_2/Pd
(c) 1. Cl_2, $FeCl_3$; 2. CH_3CH_2COCl, $AlCl_3$; 3. H_2/Pd

16.23 (a) Friedel–Crafts acylation does not occur on a deactivated ring.
(b) Rearrangement occurs during Friedel–Crafts alkylation with primary halides; chlorination occurs ortho to the alkyl group.

CHAPTER 17

17.1 (a) 5-Methyl-2,4-hexanediol
(b) 2-Methyl-4-phenyl-2-butanol
(c) 4,4-Dimethylcyclohexanol
(d) *trans*-2-Bromocyclopentanol
(e) 4-Bromo-3-methylphenol

17.2 (a)

(b)

(c)

(d)

(e)

(f)

17.3 Hydrogen bonding is more difficult in hindered alcohols.

17.4 (a) $HC{\equiv}CH < (CH_3)_2CHOH < CH_3OH < (CF_3)_2CHOH$
(b) *p*-Methylphenol < Phenol < *p*-(Trifluoromethyl)phenol
(c) Benzyl alcohol < Phenol < *p*-Hydroxybenzoic acid

17.5 The electron-withdrawing nitro group stabilizes an alkoxide ion, but the electron-donating methoxyl group destabilizes the anion.

17.6 (a) 2-Methyl-4-phenyl-1-butanol
(b) 2-Methyl-2-pentanol
(c) *meso*-5,6-Decanediol

17.7 (a) $NaBH_4$ (b) $LiAlH_4$ (c) $LiAlH_4$

17.8 (a) Benzaldehyde or benzoic acid (or ester)
(b) Acetophenone (c) Cyclohexanone
(d) 2-Methylpropanal or 2-methylpropanoic acid (or ester)

17.9 (a) 1-Methylcyclopentanol
(b) 1,1-Diphenylethanol
(c) 3-Methyl-3-hexanol

17.10 (a) Acetone + CH_3MgBr, or ethyl acetate + 2 CH_3MgBr
(b) Cyclohexanone + CH_3MgBr
(c) 3-Pentanone + CH_3MgBr, or 2-butanone + CH_3CH_2MgBr, or ethyl acetate + 2 CH_3CH_2MgBr
(d) 2-Butanone + PhMgBr, or ethyl phenyl ketone + CH_3MgBr, or acetophenone + CH_3CH_2MgBr
(e) Formaldehyde + PhMgBr

17.11 Cyclohexanone + CH_3CH_2MgBr

17.12 (a) 2-Methyl-2-pentene
(b) 3-Methylcyclohexene
(c) 1-Methylcyclohexene

17.13 (a) 1-Phenylethanol (b) 2-Methyl-1-propanol (c) Cyclopentanol

17.14 (a) Hexanoic acid, hexanal
(b) 2-Hexanone
(c) Hexanoic acid, no reaction

17.15 S_N2 reaction of F^- on silicon with displacement of alkoxide ion

17.16 1. CH_3Cl, $AlCl_3$; 2. SO_3, H_2SO_4; 3. NaOH, 200°C

17.17 Protonation of 2-methylpropene gives the *tert*-butyl cation, which carries out an electrophilic aromatic substitution reaction.

17.18 Disappearance of –OH absorption; appearance of C=O

17.19 (a) Singlet (b) Doublet (c) Triplet
(d) Doublet (e) Doublet (f) Singlet

CHAPTER 18

18.1 (a) Diisopropyl ether (b) Cyclopentyl propyl ether (c) *p*-Bromoanisole or 4-bromo-1-methoxybenzene
(d) 1-Methoxycyclohexene (e) Ethyl isobutyl ether (f) Allyl vinyl ether

18.2 A mixture of diethyl ether, dipropyl ether, and ethyl propyl ether is formed in a 1:1:2 ratio.

18.3 (a) $CH_3CH_2CH_2O^- + CH_3Br$
(b) $PhO^- + CH_3Br$
(c) $(CH_3)_2CHO^- + PhCH_2Br$
(d) $(CH_3)_3CCH_2O^- + CH_3CH_2Br$

18.4 (a) Bromoethane > 2-Bromopropane > Bromobenzene
(b) Bromoethane > Chloroethane > 1-Iodopropene

18.6 (a) Either method (b) Williamson
(c) Alkoxymercuration (d) Williamson

18.7 (a)

(b) $CH_3CH_2CHOH + CH_3CH_2CH_2Br$ (with CH_3 substituent)

18.8 Protonation of the oxygen atom, followed by E1 reaction

18.9 Br^- and I^- are better nucleophiles than Cl^-.

18.10 *o*-(1-Methylallyl)phenol

18.11 *cis*-2,3-Epoxybutane

18.12 *trans*-2,3-Epoxybutane

18.13 (a) (b)

18.14 Racemic 5,6-decanediol is formed.

18.15 *meso*-5,6-Decanediol is formed.

18.16 (a) $CH_3CH_2\overset{OH}{\underset{CH_3}{C}}-CH_2$ (b) $CH_3CH_2\overset{OH}{\underset{CH_3}{C}}-CH_2$

18.18 1,2-Epoxybutane

18.19 (a) 2-Butanethiol
(b) 2,2,6-Trimethyl-4-heptanethiol
(c) 2-Cyclopentene-1-thiol
(d) Ethyl isopropyl sulfide
(e) *o*-Di(methylthio)benzene

18.20 (a) 1. $LiAlH_4$; 2. PBr_3; 3. $(H_2N)_2C=S$; 4. H_2O, NaOH
(b) 1. HBr; 2. $(H_2N)_2C=S$; 3. H_2O, NaOH

18.21 Dimethyl sulfoxide is highly polar.

CHAPTER 19

19.1 (a) 2-Methyl-3-pentanone
(b) 3-Phenylpropanal (c) 2,6-Octanedione
(d) *trans*-2-Methylcyclohexanecarbaldehyde
(e) Pentanedial
(f) *cis*-2,5-Dimethylcyclohexanone
(g) 4-Methyl-3-propyl-2-hexanone
(h) 4-Hexenal

19.2 (a) $(CH_3)_2CHCH_2CHO$

(b) $CH_3CH(Cl)CH_2COCH_3$ (c) $PhCH_2CHO$

(d) (e) $H_2C=CCH_2CHO$ (with CH_3 substituent)

(f) $CH_3CH_2\overset{CH_3}{\underset{}{C}}HCH_2CH_2\overset{CH(Cl)CH_3}{\underset{}{C}}HCHO$

19.3 (a) PCC (b) 1. O_3; 2. Zn (c) DIBAH

19.4 (a) $Hg(OAc)_2$, H_3O^+
(b) 1. CH_3COCl, $AlCl_3$; 2. Br_2, $FeBr_3$
(c) 1. Mg; 2. CH_3CHO; 3. H_3O^+; 4. PCC
(d) 1. BH_3; 2. H_2O_2, NaOH; 3. PCC

19.5 $(CH_3)_2C(OH)C\equiv N$

19.6 *p*-Nitrobenzaldehyde, because its carbonyl group is more polarized.

19.7 $CCl_3CH(OH)_2$

19.8 Labeled water adds reversibly to the carbonyl group.

19.9 The equilibrium is unfavorable for sterically hindered ketones.

19.10

and

19.11 The steps are the exact reverse of the forward reaction.

19.12

19.14

+ CH$_3$OH

19.15 (a) Cyclohexanone + (Ph)$_3$P=CHCH$_3$
(b) 2-Cyclohexenone + (Ph)$_3$P=CH$_2$
(c) Acetone + (Ph)$_3$P=CHCH$_2$CH$_2$CH$_3$
(d) Acetone + (Ph)$_3$P=CHPh
(e) Benzaldehyde + (Ph)$_3$P=CHPh

19.16 (Ph)$_3$P=CHC(CH$_3$)=CHCH=CHCH= C(CH$_3$)CH=P(Ph)$_3$

19.17 Intramolecular Cannizzaro reaction

19.18 (a) 3-Buten-2-one + (CH$_3$CH$_2$CH$_2$)$_2$CuLi
(b) 3-Methyl-2-cyclohexenone + (CH$_3$)$_2$CuLi
(c) 4-*tert*-Butyl-2-cyclohexenone + (CH$_3$CH$_2$)$_2$CuLi
(d) Unsaturated ketone + (H$_2$C=CH)$_2$CuLi

19.19 Look for appearance of either an alcohol or a saturated ketone in the product.

19.20 (a) 1715 cm^{-1} (b) 1685 cm^{-1}
(c) 1750 cm^{-1} (d) 1705 cm^{-1}
(e) 1715 cm^{-1} (f) 1705 cm^{-1}

19.21 (a) Different peaks due to McLafferty rearrangement
(b) Different peaks due to α cleavage and McLafferty rearrangement
(c) Different peaks due to McLafferty rearrangement

CHAPTER 20

20.1 (a) 3-Methylbutanoic acid
(b) 4-Bromopentanoic acid
(c) 2,4-Hexadienoic acid
(d) 2-Ethylpentanoic acid
(e) *cis*-1,3-Cyclopentanedicarboxylic acid
(f) 2-Phenylpropanoic acid

20.2 (a)

(b)

(c)

(d)

(e)

20.3 Dissolve the mixture in ether, extract with aqueous NaOH, separate and acidify the aqueous layer, and extract with ether.

20.4 43%

20.5 (a) CH$_3$CH$_2$COOH < BrCH$_2$COOH < FCH$_2$COOH
(b) CH$_3$CH$_2$NH$_2$ < CH$_3$CH$_2$OH < CH$_3$CH$_2$COOH

20.6 The dianion is destabilized by repulsion between charges.

20.7 More reactive

20.8 (a) *p*-Methylbenzoic acid < Benzoic acid < *p*-Chlorobenzoic acid
(b) Acetic acid < Benzoic acid < *p*-Nitrobenzoic acid

20.9 (a) 1. Mg; 2. CO$_2$; 3. H$_3$O$^+$
(b) 1. Mg; 2. CO$_2$; 3. H$_3$O$^+$
(c) 1. Mg; 2. CO$_2$; 3. H$_3$O$^+$ or 1. NaCN; 2. H$_3$O$^+$

20.10 1. NaCN; 2. H$_3$O$^+$; 3. LiAlH$_4$

20.11 1. PBr$_3$; 2. NaCN; 3. H$_3$O$^+$; 4. LiAlH$_4$

20.12 A carboxylic acid has a very broad –OH absorption at 2500–3300 cm^{-1}.

20.13 4-Hydroxycyclohexanone: **H**–C–O absorption near 4 δ in ^{1}H spectrum and **C**=O absorption near 210 δ in ^{13}C spectrum. Cyclopentanecarboxylic acid: –COO**H** absorption near 12 δ in ^{1}H spectrum and –**C**OOH absorption near 170 δ in ^{13}C spectrum.

CHAPTER 21

21.1 (a) 4-Methylpentanoyl chloride
(b) Cyclohexylacetamide
(c) 2-Methylbutanenitrile
(d) Benzoic anhydride
(e) Isopropyl cyclopentanecarboxylate
(f) Cyclopentyl 2-methylpropanoate
(g) 4-Pentenamide
(h) 2-Ethylbutanenitrile

(i) 2,3-Dimethyl-2-butenoyl chloride

21.2 (a) $CH_3CH_2CH=CHCN$

(b) $CH_3CH_2CH_2CON(CH_3)CH_2CH_3$

(c) $(CH_3)_2CHCH_2CH(CH_3)COCl$

(d)

(e) $CH_3CH_2COCH_2CO_2CH_2CH_3$

(f)

(g) $HCO_2COCH_2CH_3$

(h)

21.3

21.4 (a) Acetyl chloride > Methyl acetate > Acetamide

(b) Hexafluoroisopropyl acetate > 2,2,2-Trichloroethyl acetate > Methyl acetate

21.5 The electron-withdrawing trifluoromethyl group polarizes the carbonyl carbon.

21.6 (a) $CH_3COO^- Na^+$ (b) CH_3CONH_2

(c) $CH_3CO_2CH_3 + CH_3COO^- Na^+$

21.7

21.8 (a) Acetic acid + 1-butanol

(b) Butanoic acid + methanol

21.9

21.10 (a) Propanoyl chloride + methanol

(b) Acetyl cloride + ethanol

(c) Benzoyl chloride + ethanol

21.11 Benzoyl chloride + cyclohexanol

21.13 (a) Propanoyl chloride + methylamine

(b) Benzoyl chloride + diethylamine

(c) Propanoyl chloride + ammonia

21.14 (a) Benzoyl chloride + $[(CH_3)_2CH]_2CuLi$, or 2-methylpropanoyl chloride + Ph_2CuLi

(b) Propenoyl chloride + $(CH_3CH_2CH_2)_2CuLi$, or butanoyl chloride + $(H_2C=CH)_2CuLi$

21.15 Monomethyl ester of benzene-1,2-dicarboxylic acid

21.17 If no added base were present, half of the reactant amine would form a salt.

21.18 Reaction of a carboxylic acid with an alkoxide ion gives the carboxylate ion.

21.19 $HOCH_2CH_2CH_2CHO$

21.20 (a) $CH_3CH_2CH_2CH(CH_3)CH_2OH$

(b) $PhOH + PhCH_2OH$

21.21 (a) Ethyl benzoate + 2 CH_3MgBr

(b) Ethyl acetate + 2 $PhMgBr$

(c) Ethyl pentanoate + 2 CH_3CH_2MgBr

21.22 (a) H_2O, NaOH   (b) Benzoic acid + BH_3

(c) $LiAlH_4$

21.23 1. Mg; 2. CO_2, then H_3O^+; 3. $SOCl_2$; 4. $(CH_3)_2NH$; 5. $LiAlH_4$

21.25 (a) $CH_3CH_2CN + CH_3CH_2MgBr$

(b) $(CH_3)_2CHCN + DIBAH$

(c) $PhCN + CH_3MgBr$ or $CH_3CN + PhMgBr$

21.26 1. NaCN; 2. CH_3CH_2MgBr; 3. H_2O

21.27 (a)

(b)

(c)

21.28

21.29 The product has a large amount of cross-linking.

21.30 (a) Ester   (b) Acid chloride

(c) Carboxylic acid   (d) Nitrile

(e) Aliphatic ketone or cyclohexanone

21.31 (a) CH_3CH_2CN

(b) $CH_3CH_2CH_2CO_2CH_2CH_3$ and other possibilities

(c) $CH_3CON(CH_3)_2$

(d) $CH_3CH=CHCOCl$ or $H_2C=C(CH_3)COCl$

CHAPTER 22

22.1 (a) (b) $H_2C=CCl$ with OH

(c) $H_2C=COCH_2CH_3$ with OH   (d) $CH_3CH=CHOH$

(e) $CH_3CH=COH$ with OH

(f) $PhCH=CCH_3$ with OH or $PhCH_2C=CH_2$ with OH

(g) $PhC=CH_2$ with OH

22.2 (a) 4 (b) 3 (c) 3 (d) 2 (e) 4 (f) 5
(g) 3

22.3 **More stable**

22.5 1. Br_2; 2. Pyridine, heat
22.6 The intermediate α-bromo acid bromide undergoes a nucleophilic acyl substitution reaction to give an α-bromo ester.
22.7 (a) CH_3CH_2CHO (b) $(CH_3)_3CCOCH_3$
(c) CH_3COOH (d) $PhCONH_2$
(e) $CH_3CH_2CH_2CN$ (f) $CH_3CON(CH_3)_2$
(g) $-COCH_2CO-$
22.8 $^-\!:CH_2C\equiv N: \longleftrightarrow H_2C=C=\ddot{N}\!:^-$
22.9 Acid is regenerated, but base is used stoichiometrically.
22.10 The CCl_3^- ion is stabilized by the electron-withdrawing chlorine atoms.
22.11 (a) 1. $Na^+\ ^-OEt$; 2. $PhCH_2Br$; 3. H_3O^+
(b) 1. $Na^+\ ^-OEt$; 2. $CH_3CH_2CH_2Br$;
3. $Na^+\ ^-OEt$; 4. CH_3Br; 5. H_3O^+
(c) 1. $Na^+\ ^-OEt$; 2. $(CH_3)_2CHCH_2Br$; 3. H_3O^+
(d) 1. 2 $Na^+\ ^-OEt$; 2. $BrCH_2CH_2CH_2Br$;
3. H_3O^+
22.12 Malonic ester has only two acidic hydrogens.
22.13 (a) $(CH_3)_2CHCH_2Br$ (b) $PhCH_2CH_2Br$
22.14 1. 2 $Na^+\ ^-OEt$; 2. $BrCH_2CH_2CH_2CH_2Br$;
3. H_3O^+
22.15 None can be prepared.
22.16 (a) Alkylate phenylacetone with CH_3I.
(b) Alkylate pentanenitrile with CH_3CH_2I.
(c) Alkylate cyclohexanone with $H_2C=CHCH_2Br$.
(d) Alkylate cyclohexanone with excess CH_3I.

CHAPTER 23

23.1 (a) $CH_3CH_2CH_2CHCHCH$ with OH, O, and CH_2CH_3

(b) $CH_3CH_2CCH_2CCH_2CH_3$ (with OH, O, CH_3) $+ CH_3CH_2C-CHCCH_3$ (with OH, O, H_3C, CH_3)

(c)

23.2 The reverse reaction is the exact opposite of the forward reaction.
23.3 (a) (b) $PhCCH=CCH_3$ with O and Ph

(c) $CH_3CHCH_2CH=CCH$ with CH_3, O, and $CH(CH_3)_2$

23.4 and

23.5 (c) 3-Pentanone
23.6 1. NaOH; 2. $LiAlH_4$; 3. H_2/Pd
23.7 (a)
23.8 The CH_2 position between the two carbonyl groups is most acidic.
23.9

23.10 (a) $CH_3CHCH_2CCHCOEt$ with CH_3, O, O, and $CH(CH_3)_2$

(b) $PhCH_2CCHCOEt$ with O, O, and Ph (c) $C_6H_{11}CH_2CCHCOEt$ with O, O, and C_6H_{11}

23.11 The cleavage reaction is the exact reverse of the forward reaction.

23.12

23.13 Ethyl 5-methyl-2-oxocyclohexanecarboxylate
23.14 Ethyl 4-methyl-2-oxocyclohexanecarboxylate and ethyl 6-methyl-2-oxocyclohexane-carboxylate
23.15 (a) O=⟨⟩—CH(COCH$_3$)$_2$

(b) (CH$_3$CO)$_2$CHCH$_2$CH$_2$CN

(c) (CH$_3$CO)$_2$CHCHCH$_2$COEt
          |
          CH$_3$

23.16 (a) (EtO$_2$C)$_2$CHCH$_2$CH$_2$CCH$_3$

(b)

(c) O$_2$NCH$_2$CH$_2$CH$_2$CCH$_3$

23.17 (a)

(b)

23.18 (a) Cyclopentanone enamine + propenenitrile
(b) Cyclohexanone enamine + methyl propenoate

23.19

23.20 2,5,5-Trimethyl-1,3-cyclohexanedione + 1-penten-3-one

CHAPTER 24

24.1 (a) N-Methylethylamine
(b) Tricyclohexylamine
(c) N-Methyl-N-propylcyclohexylamine

(d) N-Methylpyrrolidine
(e) Diisopropylamine
(f) 1,3-Butanediamine

24.2 (a) (CH$_3$CH$_2$)$_3$N

(b) (H$_2$C=CHCH$_2$)$_3$N

(c)

(d)

(e)

(f)

24.3 (a)

(b)

(c)

(d)

24.4 (a) CH$_3$CH$_2$NH$_2$ (b) NaOH
(c) CH$_3$NHCH$_3$
24.5 Propylamine is stronger; benzylamine pK_b = 4.67; propylamine pK_b = 3.29
24.6 (a) p-Nitroaniline < p-Aminobenzaldehyde < p-Bromoaniline
(b) p-Aminoacetophenone < p-Chloroaniline < p-Methylaniline
(c) p-(Trifluoromethyl)aniline < p-(Fluoromethyl)aniline < p-Methylaniline
24.7 (a) Propanenitrile or propanamide
(b) N-Propylpropanamide
(c) Benzonitrile or benzamide
(d) N-Phenylacetamide
24.9

24.10 (a) Ethylamine + acetone, or isopropylamine + acetaldehyde
(b) Aniline + acetaldehyde
(c) Cyclopentylamine + formaldehyde, or methylamine + cyclopentanone
24.12 (a) 4,4-Dimethylpentanamide or 4,4-dimethylpentanoyl azide
(b) p-Methylbenzamide or p-methylbenzoyl azide
24.13 (a) 3-Octene and 4-octene   (b) Cyclohexene
(c) 3-Heptene
(d) Ethylene and cyclohexene
24.14 H$_2$C=CHCH$_2$CH$_2$CH$_2$N(CH$_3$)$_2$

24.15 1. HNO_3, H_2SO_4; 2. H_2/PtO_2; 3. $(CH_3CO)_2O$;
4. $HOSO_2Cl$; 5. aminothiazole; 6. H_2O, NaOH

24.16 (a) 1. HNO_3, H_2SO_4; 2. H_2/PtO_2; 3. 2 CH_3Br
(b) 1. HNO_3, H_2SO_4; 2. H_2/PtO_2;
    3. $(CH_3CO)_2O$; 4. Cl_2; 5. H_2O, NaOH
(c) 1. HNO_3, H_2SO_4; 2. Cl_2, $FeCl_3$; 3. $SnCl_2$
(d) 1. HNO_3, H_2SO_4; 2. H_2/PtO_2;
    3. $(CH_3CO)_2O$; 4. 2 CH_3Cl, $AlCl_3$;
    5. H_2O, NaOH

24.17 (a) 1. CH_3Cl, $AlCl_3$; 2. HNO_3, H_2SO_4;
    3. $SnCl_2$; 4. $NaNO_2$, H_2SO_4; 5. CuBr;
    6. $KMnO_4$; H_2O
(b) 1. HNO_3, H_2SO_4; 2. Br_2, $FeBr_3$; 3. $SnCl_2$,
    H_3O^+; 4. $NaNO_2$, H_2SO_4; 5. CuCN; 6. H_3O^+
(c) 1. HNO_3, H_2SO_4; 2. Cl_2, $FeCl_3$; 3. $SnCl_2$;
    4. $NaNO_2$, H_2SO_4; 5. CuBr
(d) 1. CH_3Cl, $AlCl_3$; 2. HNO_3, H_2SO_4; 3. $SnCl_2$;
    4. $NaNO_2$, H_2SO_4; 5. CuCN; 6. H_3O^+
(e) 1. HNO_3, H_2SO_4; 2. H_2/PtO_2;
    3. $(CH_3CO)_2O$; 4. 2 Br_2; 5. H_2O, NaOH;
    6. $NaNO_2$, H_2SO_4; 7. CuBr

24.19 $(CH_3)_3CCOCH_3 \rightarrow (CH_3)_3CCH(NH_2)CH_3$

CHAPTER 25

25.1 (a) Aldotetrose (b) Ketopentose
    (c) Ketohexose (d) Aldopentose
25.2 A, B, and C are the same.
25.3 (a) S (b) R (c) S
25.4 (a) L-Erythrose; $2S,3S$
    (b) D-Xylose; $2R,3S,4R$
    (c) D-Xylulose; $3S,4R$
25.5

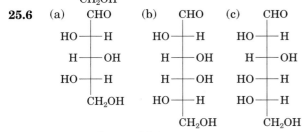

25.6 (a), (b), (c)

25.7 16 D and 16 L aldoheptoses
25.8

25.9

25.10

25.11

α anomer   β anomer

25.12

β-D-Galactopyranose   β-D-Mannopyranose

25.13

25.14

25.15 D-Galactitol has a plane of symmetry.
25.16 The −CHO end of L-gulose corresponds to the −CH_2OH end of D-glucose after reduction.
25.17 D-Allaric acid has a symmetry plane but D-glucaric acid does not.
25.18 D-Allose and D-galactose yield meso aldaric acids; the other six D-hexoses yield optically active aldaric acids.
25.19 D-Allose + D-altrose   **25.20** L-Xylose
25.21 D-Xylose and D-lyxose
25.22 See Figure 25.3.   **25.23** See Figure 25.3.

CHAPTER 26

26.1 Aromatic: Phe, Tyr, Trp, His; sulfur-containing: Cys, Met; alcohols: Ser, Thr; hydrocarbon side chains: Ala, Ile, Leu, Val, Phe

26.2 The sulfur atom in the $-CH_2SH$ group of cysteine makes the side chain higher in priority than the $-COOH$ group.

26.3

$$
\begin{array}{c}
COOH \\
H_2N \!-\!\!\!-\! H \\
H \!-\!\!\!-\! OH \\
CH_3
\end{array}
$$

26.4

$$
\begin{array}{c}
COOH \\
H \overset{S}{-\!\!\!-} NH_2 \\
H \overset{S}{-\!\!\!-} OH \\
CH_3
\end{array}
\quad \text{and} \quad
\begin{array}{c}
COOH \\
H_2N \overset{R}{-\!\!\!-} H \\
HO \overset{R}{-\!\!\!-} H \\
CH_3
\end{array}
$$

26.5 (a) Toward +: Glu > Val; toward −: none
(b) Toward +: Phe; toward −: Gly
(c) Toward +: none; toward −: Gly > Ser
(d) Toward +: Phe > Ser; toward −: none

26.6 At pH 1.5: 20% neutral and 80% protonated; at pH 10.0: 11% neutral and 89% deprotonated

26.7 (a) Start with 3-phenylpropanoic acid: 1. Br_2, PBr_3; 2. NH_3
(b) Start with 3-methylbutanoic acid: 1. Br_2, PBr_3; 2. NH_3

26.8 Start with 3-methylbutanal: 1. NH_3, KCN; 2. H_3O^+

26.9 (a) $(CH_3)_2CHCH_2Br$

(b)

(c)

(d) $CH_3SCH_2CH_2Br$

26.10 Val-Tyr-Gly (VTG), Tyr-Gly-Val (TGV), Gly-Val-Tyr (GVT), Val-Gly-Tyr (VGT), Tyr-Val-Gly (TVG), Gly-Tyr-Val (GTV)

26.11

26.12

26.13

26.14 Trypsin: Asp-Arg + Val-Tyr-Ile-His-Pro-Phe
Chymotrypsin: Asp-Arg-Val-Tyr + Ile-His-Pro-Phe

26.15 Methionine

26.16

26.17 (a) Arg-Pro-Leu-Gly-Ile-Val
(b) Val-Met-Trp-Asp-Val-Leu

26.18 Pro-Leu-Gly-Pro-Arg-Pro

26.19

26.22 (a) Lyase (b) Hydrolase
(c) Oxidoreductase

CHAPTER 27

27.1 $CH_3(CH_2)_{18}CO_2CH_2(CH_2)_{30}CH_3$

27.2 Glyceryl tripalmitate is higher melting.

27.3 $[CH_3(CH_2)_7CH{=}CH(CH_2)_7CO_2^-]_2\ Mg^{2+}$

27.4

27.5 (a)

(b)

(c)

27.7 (a)

(b)

27.8

27.9 Step 5

CHAPTER 28

28.1 One lone pair is in an sp^2 orbital in the plane of the ring; the other lone pair is in a p orbital perpendicular to the plane of the ring.

28.2 Electrophilic aromatic substitution mechanism by D^+.

28.3 The pyridine-like, doubly bonded nitrogen is more basic because its lone pair of electrons is in an sp^2 orbital in the plane of the ring.

28.4 The intermediate from attack at C3 is best because it does not have the positive charge on the electronegative nitrogen atom.

28.5 The intermediate from attack at C4 has the negative charge on nitrogen.

28.6 The side-chain nitrogen atom is more basic because the lone pair of electrons is in an sp^3 orbital.

28.7 The intermediate from attack at C3 is stabilized by resonance involving the nitrogen atom. The intermediate from attack at C2 is not as stabilized.

C2 attack   **C3 attack**

28.10 (5') ACGGATTAGCC (3')

28.11

28.12 (5') UACGGUAAUC (3')

28.13 (a) GCU, GCC, GCA, GCG (b) UUU, UUC
(c) UUA, UUG, CUU, CUC, CUA, CUG
(d) UAU, UAC

28.14 (a) AGC, GGC, UGC, CGC (b) AAA, GAA
(c) UAA, CAA, GAA, GAG, UAG, CAG
(d) AUA, GUA

28.15 Leu-Met-Ala-Trp-Pro-Stop

28.16 (5') TTA-GGG-CCA-AGC-CAT-AAG (3')

28.17 A cleavage: ^{32}P-A, ^{32}P-AAC,
^{32}P-AACATGGCGCTT,
^{32}P-AACATGGCGCTTATG,
^{32}P-AACATGGCGCTTATGACG
G cleavage: ^{32}P-AACAT, ^{32}P-AACATG,
^{32}P-AACATGGC ^{32}P-AACATGGCGCTTAT,
^{32}P-AACATGGCGCTTATGAC
C cleavage: ^{32}P-AA, ^{32}P-AACATGG,
^{32}P-AACATGGCG,
^{32}P-AACATGGCGCTTATGA
C + T cleavage: ^{32}P-AA, ^{32}P-AACA,
^{32}P-AACATGG, ^{32}P-AACATGGCG,
^{32}P-AACATGGCGC, ^{32}P-AACATGGCGCT,
^{32}P-AACATGGCGCTTA,
^{32}P-AACATGGCGCTTATGA

28.19 TCGGTAC

28.20 The cleavage is an S_N1 reaction.

28.21 E2 reaction

CHAPTER 29

29.1 $HOCH_2CH(OH)CH_2OH + ATP \rightarrow$
$HOCH_2CH(OH)CH_2OPO_3^{2-} + ADP$

29.2 Caprylyl CoA $\rightarrow$ Hexanoyl CoA $\rightarrow$
Butyryl CoA $\rightarrow$ 2 Acetyl CoA

29.3 (a) 8 acetyl CoA; 7 passages
(b) 10 acetyl CoA; 9 passages

29.4 Steps 7 and 10

29.5 Steps 1, 3: phosphate transfers; steps 2, 5, 8: isomerizations; step 4: retro-aldol reaction; step 5: oxidation and nucleophilic acyl substitution; steps 7, 10: phosphate transfers; step 9: E2 dehydration

29.6 Nucleophilic acyl substitution of acetyl dihydrolipoamide by coenzyme A

29.7 C1 and C6 of glucose become $-CH_3$ groups; C3 and C4 become CO_2.

29.8 Citrate and isocitrate

29.9 E2 elimination of water, followed by conjugate addition

29.11 $(CH_3)_2CHCH_2COCO_2^-$

29.12 E2 reaction

29.13 At C2, C4, C6, C8, and so forth

29.14 Nucleophilic acyl substitution of phosphate ion by hydride ion donated from NADH

29.15 C1 and C6

CHAPTER 30

30.1 Ethylene: ψ_1 is the HOMO and ψ_2^* is the LUMO in the ground state; ψ_2^* is the HOMO and there is no LUMO in the excited state. 1,3-Butadiene: ψ_2 is the HOMO and ψ_3^* is the LUMO in the ground state; ψ_3^* is the HOMO and ψ_4^* is the LUMO in the excited state.

30.2 Disrotatory: *cis*-5,6-dimethyl-1,3-cyclohexadiene; conrotatory: *trans*-5,6-dimethyl-1,3-cyclohexadiene. Disrotatory closure occurs.

30.3 The more stable of two allowed products is formed.

30.4 *trans*-5,6-Dimethyl-1,3-cyclohexadiene; *cis*-5,6-dimethyl-1,3-cyclohexadiene

30.5 *cis*-3,6-Dimethylcyclohexene; *trans*-3,6-dimethylcyclohexene

30.6 A [6 + 4] suprafacial cycloaddition

30.7 An antarafacial [1,7] sigmatropic rearrangement

30.8 A series of [1,5] hydrogen shifts occur.

30.9 Claisen rearrangement is followed by a Cope rearrangement.

30.10 (a) Conrotatory (b) Disrotatory
(c) Suprafacial (d) Antarafacial
(e) Suprafacial

CHAPTER 31

31.1 $H_2C=CHCO_2CH_3 < H_2C=CHCl <$
$H_2C=CHCH_3 < H_2C=CH-C_6H_5$

31.2 $H_2C=CHCH_3 < H_2C=CHC_6H_5 <$
$H_2C=CHC≡N$

31.3 The intermediate is a resonance-stabilized benzylic carbanion, Ph–$\ddot{C}$HR.

31.4 The polymer has no chirality centers.

31.5 No, the polymers are racemic.

31.6

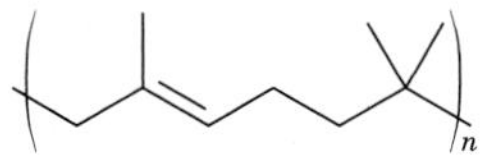

31.7

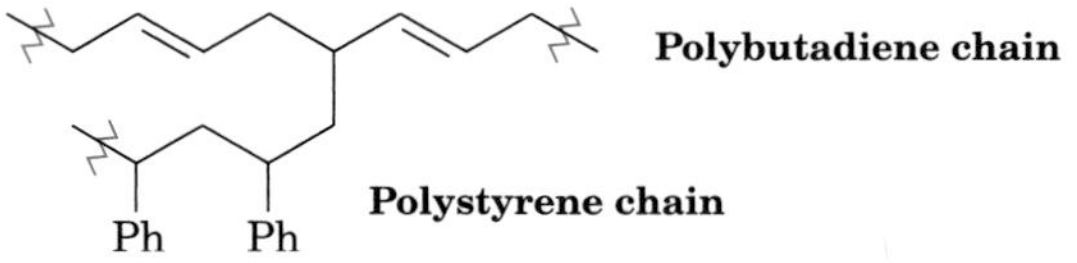

Polybutadiene chain

Polystyrene chain

31.8

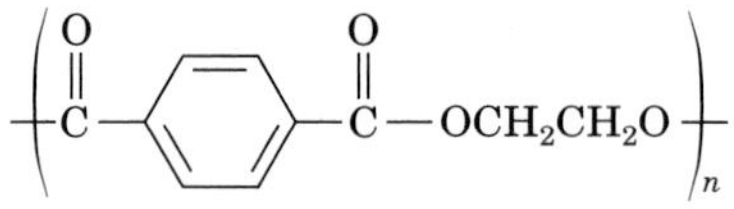

31.9

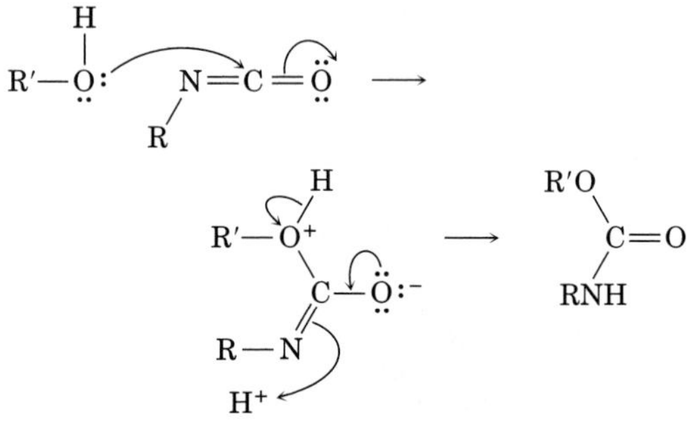

31.10

Atactic

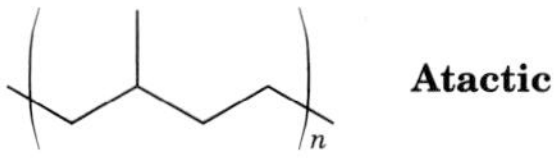

31.11

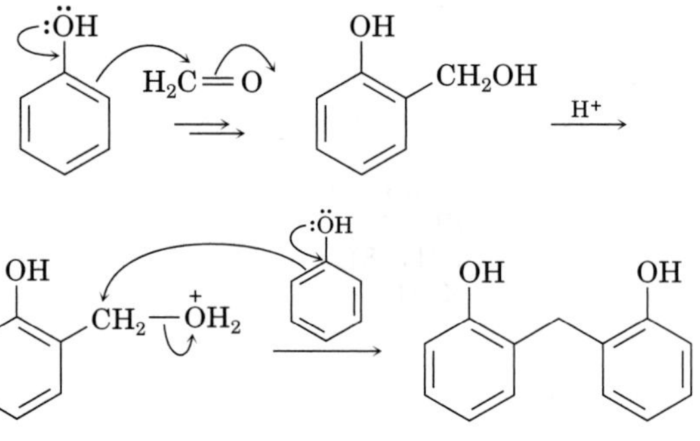

INDEX

The references in boldface type refer either to pages where terms are defined or to biographies.

PHOTO CREDITS

Periodic Table of the Elements

Key (example):
- Atomic number — 22
- Name — Titanium
- Symbol — **Ti**
- Atomic weight — 47.90

Atomic weights are based on carbon-12. Atomic weights in parentheses indicate the most stable or best-known isotope.

[a] Mass number of most stable or best-known isotope
[b] Mass of the isotope of longest half-life

Transition elements (3B–2B), 8B spanning groups 8, 9, 10.

Period	1A	2A	3B	4B	5B	6B	7B	8B	8B	8B	1B	2B	3A	4A	5A	6A	7A	8A
1	1 Hydrogen **H** 1.0079																	2 Helium **He** 4.00260
2	3 Lithium **Li** 6.941	4 Beryllium **Be** 9.01218											5 Boron **B** 10.81	6 Carbon **C** 12.011	7 Nitrogen **N** 14.0067	8 Oxygen **O** 15.9994	9 Fluorine **F** 18.99840	10 Neon **Ne** 20.179
3	11 Sodium **Na** 22.98977	12 Magnesium **Mg** 24.305											13 Aluminum **Al** 26.98154	14 Silicon **Si** 28.086	15 Phosphorus **P** 30.97376	16 Sulfur **S** 32.06	17 Chlorine **Cl** 35.453	18 Argon **Ar** 39.948
4	19 Potassium **K** 39.098	20 Calcium **Ca** 40.08	21 Scandium **Sc** 44.9559	22 Titanium **Ti** 47.90	23 Vanadium **V** 50.9414	24 Chromium **Cr** 51.996	25 Manganese **Mn** 54.9380	26 Iron **Fe** 55.847	27 Cobalt **Co** 58.9332	28 Nickel **Ni** 58.69	29 Copper **Cu** 63.546	30 Zinc **Zn** 65.38	31 Gallium **Ga** 69.72	32 Germanium **Ge** 72.61	33 Arsenic **As** 74.9216	34 Selenium **Se** 78.96	35 Bromine **Br** 79.904	36 Krypton **Kr** 83.80
5	37 Rubidium **Rb** 85.4678	38 Strontium **Sr** 87.62	39 Yttrium **Y** 88.9059	40 Zirconium **Zr** 91.22	41 Niobium **Nb** 92.9064	42 Molybdenum **Mo** 95.94	43 Technetium **Tc** 98.9062[b]	44 Ruthenium **Ru** 101.07	45 Rhodium **Rh** 102.9055	46 Palladium **Pd** 106.4	47 Silver **Ag** 107.868	48 Cadmium **Cd** 112.40	49 Indium **In** 114.82	50 Tin **Sn** 118.71	51 Antimony **Sb** 121.75	52 Tellurium **Te** 127.60	53 Iodine **I** 126.9045	54 Xenon **Xe** 131.30
6	55 Cesium **Cs** 132.9054	56 Barium **Ba** 137.34	*57 Lanthanum **La** 138.9055	72 Hafnium **Hf** 178.49	73 Tantalum **Ta** 180.9479	74 Wolfram (Tungsten) **W** 183.85	75 Rhenium **Re** 186.2	76 Osmium **Os** 190.2	77 Iridium **Ir** 192.22	78 Platinum **Pt** 195.09	79 Gold **Au** 196.9665	80 Mercury **Hg** 200.59	81 Thallium **Tl** 204.37	82 Lead **Pb** 207.2	83 Bismuth **Bi** 208.9804	84 Polonium **Po** (210)[a]	85 Astatine **At** (210)[a]	86 Radon **Rn** (222)[a]
7	87 Francium **Fr** (223)[a]	88 Radium **Ra** 226.0254[b]	**89 Actinium **Ac** (227)[a]	104 Rutherfordium **Rf** (261)[a]	105 Dubnium **Db** (262)[a]	106 Seaborgium **Sg** (263)[a]	107 Bohrium **Bh** (262)	108 Hassium **Hs** (265)	109 Meitnerium **Mt** (268)	110 — (269)	111 — (272)	112 — (277)						

Inner transition elements

Lanthanide series (6) *

58 Cerium **Ce** 140.12	59 Praseodymium **Pr** 140.9077	60 Neodymium **Nd** 144.24	61 Promethium **Pm** (145)[a]	62 Samarium **Sm** 150.4	63 Europium **Eu** 151.96	64 Gadolinium **Gd** 157.25	65 Terbium **Tb** 158.9254	66 Dysprosium **Dy** 162.50	67 Holmium **Ho** 164.9304	68 Erbium **Er** 167.26	69 Thulium **Tm** 168.9342	70 Ytterbium **Yb** 173.04	71 Lutetium **Lu** 174.97

Actinide series (7) **

90 Thorium **Th** 232.0381[b]	91 Protactinium **Pa** 231.0359[b]	92 Uranium **U** 238.029	93 Neptunium **Np** 237.0482	94 Plutonium **Pu** (242)[a]	95 Americium **Am** (243)[a]	96 Curium **Cm** (247)[a]	97 Berkelium **Bk** (249)[a]	98 Californium **Cf** (251)[a]	99 Einsteinium **Es** (254)[a]	100 Fermium **Fm** (253)[a]	101 Mendelevium **Md** (256)[a]	102 Nobelium **No** (254)[a]	103 Lawrencium **Lr** (257)[a]